Organic Reactions

Organic Reactions

VOLUME 106

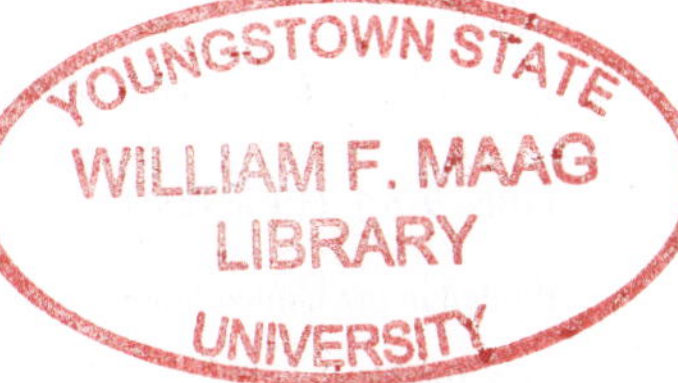

WILEY

Published by John Wiley & Sons, Inc., Hoboken, New Jersey
Published simultaneously in Canada

Library of Congress Cataloging-in-Publication Data:

ISBN: 978-1-119-77123-4

Printed in the United States of America

SKY10032213_122821

INTRODUCTION TO THE SERIES BY ROGER ADAMS, 1942

In the course of nearly every program of research in organic chemistry, the investigator finds it necessary to use several of the better-known synthetic reactions. To discover the optimum conditions for the application of even the most familiar one to a compound not previously subjected to the reaction often requires an extensive search of the literature; even then a series of experiments may be necessary. When the results of the investigation are published, the synthesis, which may have required months of work, is usually described without comment. The background of knowledge and experience gained in the literature search and experimentation is thus lost to those who subsequently have occasion to apply the general method. The student of preparative organic chemistry faces similar difficulties. The textbooks and laboratory manuals furnish numerous examples of the application of various syntheses, but only rarely do they convey an accurate conception of the scope and usefulness of the processes.

For many years American organic chemists have discussed these problems. The plan of compiling critical discussions of the more important reactions thus was evolved. The volumes of *Organic Reactions* are collections of chapters each devoted to a single reaction, or a definite phase of a reaction, of wide applicability. The authors have had experience with the processes surveyed. The subjects are presented from the preparative viewpoint, and particular attention is given to limitations, interfering influences, effects of structure, and the selection of experimental techniques. Each chapter includes several detailed procedures illustrating the significant modifications of the method. Most of these procedures have been found satisfactory by the author or one of the editors, but unlike those in *Organic Syntheses*, they have not been subjected to careful testing in two or more laboratories. Each chapter contains tables that include all the examples of the reaction under consideration that the author has been able to find. It is inevitable, however, that in the search of the literature some examples will be missed, especially when the reaction is used as one step in an extended synthesis. Nevertheless, the investigator will be able to use the tables and their accompanying bibliographies in place of most or all of the literature search so often required. Because of the systematic arrangement of the material in the chapters and the entries in the tables, users of the books will be able to find information desired by reference to the table of contents of the appropriate chapter. In the interest of economy, the entries in the indices have been kept to a minimum, and, in particular, the compounds listed in the tables are not repeated in the indices.

The success of this publication, which will appear periodically, depends upon the cooperation of organic chemists and their willingness to devote time and effort to the preparation of the chapters. They have manifested their interest already by the almost unanimous acceptance of invitations to contribute to the work. The editors will welcome their continued interest and their suggestions for improvements in *Organic Reactions*.

INTRODUCTION TO THE SERIES BY SCOTT E. DENMARK, 2008

In the intervening years since "The Chief" wrote this introduction to the second of his publishing creations, much in the world of chemistry has changed. In particular, the last decade has witnessed a revolution in the generation, dissemination, and availability of the chemical literature with the advent of electronic publication and abstracting services. Although the exponential growth in the chemical literature was one of the motivations for the creation of *Organic Reactions*, Adams could never have anticipated the impact of electronic access to the literature. Yet, as often happens with visionary advances, the value of this critical resource is now even greater than at its inception.

From 1942 to the 1980's the challenge that *Organic Reactions* successfully addressed was the difficulty in compiling an authoritative summary of a preparatively useful organic reaction from the primary literature. Practitioners interested in executing such a reaction (or simply learning about the features, advantages, and limitations of this process) would have a valuable resource to guide their experimentation. As abstracting services, in particular *Chemical Abstracts* and later *Beilstein*, entered the electronic age, the challenge for the practitioner was no longer to locate all of the literature on the subject. However, *Organic Reactions* chapters are much more than a surfeit of primary references; they constitute a distillation of this avalanche of information into the knowledge needed to correctly implement a reaction. It is in this capacity, namely to provide focused, scholarly, and comprehensive overviews of a given transformation, that *Organic Reactions* takes on even greater significance for the practice of chemical experimentation in the 21st century.

Adams' description of the content of the intended chapters is still remarkably relevant today. The development of new chemical reactions over the past decades has greatly accelerated and has embraced more sophisticated reagents derived from elements representing all reaches of the Periodic Table. Accordingly, the successful implementation of these transformations requires more stringent adherence to important experimental details and conditions. The suitability of a given reaction for an unknown application is best judged from the informed vantage point provided by precedent and guidelines offered by a knowledgeable author.

As Adams clearly understood, the ultimate success of the enterprise depends on the willingness of organic chemists to devote their time and efforts to the preparation of chapters. The fact that, at the dawn of the 21st century, the series continues to thrive is fitting testimony to those chemists whose contributions serve as the foundation of this edifice. Chemists who are considering the preparation of a manuscript for submission to *Organic Reactions* are urged to contact the Editor-in-Chief.

PREFACE TO VOLUME 106

Something old, something new,
Something borrowed, something blue,
and a sixpence in her shoe.

Anon (1871)
St James' Magazine, London

The two chapters in this volume of *Organic Reactions* epitomize the sentiment of the traditional old English rhyme, in that the combination of *something old* with *something new* often provides a different perspective, and thus, important new possibilities. The first two lines of the verse were originally published in a magazine article on *Marriage Superstitions* in 1871. The *something old* is meant to ward off evil spirits, whereas *something new* offers optimism for the future. The *something borrowed* is thought to bring good luck, and the *something blue* represents purity and fidelity. The final line, *and a sixpence in her shoe*, was introduced during the Victorian era to represent prosperity. Although there were originally minor regional variations, this version has now been endorsed internationally, which exemplifies the impact of a simple piece of prose. The development of a synthetic transformation has many of the attributes delineated within this simple rhyme, wherein the combination of *something old* with *something new* can significantly extend the reaction scope and synthetic utility. The *something borrowed* and *something blue* could represent the serendipitous discoveries that often result in a very fertile area of investigation that advances a process in the context of efficiency and selectivity. In some cases, these findings can lead to commercial applications that can be lucrative to the inventor, which ties into the rhyme's last line. The two chapters in this volume focus on transformations connected by the preparation and reactions of alkenes. Notably, the cross-metathesis reaction involves the catalytic construction of acyclic alkenes, whereas the Stetter reaction entails the catalytic enantioselective conjugate addition of aldehydes to alkenes.

The first chapter by Karol Grela, Anna Kajetanowicz, Anna Szadkowska, and Justyna Czaban-Jóźwiak provides an extensive review on alkene cross-metathesis, which completes the trilogy of chapters dealing with metathesis reactions. The two earlier chapters were on olefin ring-closing metathesis by Larry Yet (Volume 89) and alkyne metathesis by Daesung Lee, Ivan Volchkov, and Sang Young Yun (Volume 102). The term "olefin metathesis" was coined by Calderon in 1967, wherein the word metathesis means "transposition" and is the combination of two Greek words–*change* (meta) and *position* (thesis). The reaction was serendipitously discovered in the 1950s and is often referred to as the "child of industry" because of the strong industrial connection. Chemists at Du Pont, Standard Oil, and Phillips

Petroleum independently reported the metathesis of propene and other variants. In the late 1980s and early 1990s, Grubbs and Schrock, who received the 2005 Nobel Prize in Chemistry with Yves Chauvin for their work on olefin metathesis, developed well-behaved ruthenium, molybdenum and tungsten catalysts that were compatible with functionalized alkenes required for fine chemical synthesis and the field exploded exponentially. This chapter delineates the historical development of the cross-metathesis reaction from an academic/industrial curiosity to a sophisticated modern transformation for the catalytic construction of acyclic alkenes in a highly selective manner. The Mechanism and Stereochemistry section provides an insightful account of reaction initiation, catalyst regeneration, and stereocontrol. For example, the difference between the three initiation processes, the so-called "boomerang" mechanism, and the various aspects of diastereo- and enantioselective cross-metathesis reactions are discussed. Importantly, the latter part outlines the catalyst requirements for controlling (*E*)- and (*Z*)-geometry, the desymmetrization of prochiral *meso*-compounds through asymmetric ring-opening cross-metathesis (AROCM), and asymmetric cross-metathesis (ACM). The Scope and Limitations component delineates the classification of alkenes (Types I-IV according to Grubbs), the consequences of their ability to undergo cross-metathesis, and the impact of the classification on the mechanism in terms of homodimerization. This section is organized by various alkene substituents, including boron, nitrogen, oxygen, halogens, silicon, tin, phosphorus, sulfur, etc. There is also a substantial section on (*Z*)-selective processes that have proven particularly challenging, and, as such, makes the addition of this section very timely. The Applications to Synthesis section describes several impressive natural product syntheses that use this reaction, and the Comparison with Other Methods section provides a comprehensive assessment of more classical methods that are commonly deployed to construct alkenes. The Tabular Survey incorporates reactions reported up to March 2020. The tables are organized by the type of olefin metathesis (according to Grubbs) and then further subdivided by functional groups, which makes identifying a particular combination effortless. Overall, this is a superb chapter on a fundamentally important process that will be an invaluable resource to anyone wishing to construct an olefin using a cross-metathesis reaction.

The second chapter by Darrin M. Flanigan, Kerem E. Ozboya, Alberto Muñoz, Tomislav Rovis, Subhash D. Tanpure, and Paul R. Blakemore chronicles the development of the catalytic enantioselective Stetter reaction, which represents an update to the original chapter by Stetter and Kuhlmann in Volume 40. This process is closely related to the venerable acyloin condensation, which is catalyzed by cyanide and vitamin B_1 (thiamine) in Nature. While plants tend to employ cyanide derived from cyanoglucosides, other living organisms use non-toxic thiamine and the reaction proceeds via the so-called "Breslow Intermediate" from the addition of the deprotonated quaternary thiazolium salt to pyruvate. The recognition that an aldehyde's natural electrophilicity can be reversed to afford the requisite acyl anion equivalent prompted Hermann Stetter to examine the cyanide-catalyzed addition of aromatic aldehydes to α,β-unsaturated esters, ketones, and nitriles in the early 1970s. He later demonstrated that thiazolylidenes also catalyze this process to permit the coupling of two formally electrophilic species for construction of 1,4-dicarbonyls that constitute useful intermediates for target-directed synthesis. Although the catalyst requirement mitigates any background reaction, the first enantioselective process was not described until the mid-1990s by Dieter Enders using a chiral triazolium catalyst. Consequently, this important and seminal report set the wheels in motion for others to develop more selective and versatile catalysts that significantly broaden the substrate scope. This chapter catalogs the development of the enantioselective variant that has emerged as a powerful synthetic tool for target-directed synthesis. The Mechanism and Stereochemistry section outlines the various methods for generating the NHC catalysts from the corresponding salts and the subsequent formation and reactivity of Breslow intermediates, including associated mechanistic experiments, namely, deuterium isotope and competition studies. The section on stereochemistry presents a series of models that rationalize the origin of stereocontrol, including computational studies on the preferred enaminol geometry and the transition state for the key C-C bond-forming event. The Scope and Limitations component documents the preparation of triazolium salts, and the remainder of the discussion is organized by the different intra- and intermolecular variants. The authors subdivide this section by the type of aldehyde donor (aryl, aliphatic, α,β-unsaturated, etc.). A section on the recent aza-Stetter reaction with imine donors may also be of interest to the reader. The Applications to Synthesis section provides some unique applications to the synthesis of natural products, and the Comparison with Other Methods section provides a detailed comparison with several alternative methods. The organization of the Tabular Survey mirrors the Scope and Limitations section, thereby making it easy for the reader to identify a specific transformation. Overall, this is an outstanding chapter on a particularly important and useful process that will be a valuable resource to the synthetic community.

I would be remiss if I did not acknowledge the entire *Organic Reactions* Editorial Board for their collective efforts in steering this volume through the various stages of the editorial process. I want to thank Steven M. Weinreb (Chapter 1 and 2), the Responsible Editor for the two chapters, albeit I had some early input into Chapter 2 through the early phases of development. I am also deeply indebted to Dr. Danielle Soenen for her heroic efforts as the Editorial Coordinator; her knowledge of *Organic Reactions* is critical to maintaining consistency in the series. Dr. Dena Lindsay (Secretary to the Editorial Board) is thanked for coordinating the author's, editor's, and publisher's contributions. In addition, the *Organic Reactions* enterprise could not maintain the quality of production without the efforts of Dr. Steven M. Weinreb (Executive Editor), Dr. Engelbert Ciganek (Editorial Advisor), Dr. Landy Blasdel (Processing Editor), and Dr. Debra Dolliver (Processing Editor). I would also like to acknowledge Dr. Barry R. Snider (Secretary) for keeping everyone on task and Dr. Jeffery Press (Treasurer) for making sure that we are fiscally solvent!

I am also indebted to past and present members of the Board of Editors and Directors for ensuring the enduring quality of *Organic Reactions*. The unique format of the chapters, in conjunction with the collated tables of examples, make this series of reviews both unique and exceptionally valuable to the practicing synthetic organic chemist.

P. Andrew Evans
Kingston
Ontario, Canada

CONTENTS

CHAPTER 1

ALKENE CROSS-METATHESIS REACTIONS

KAROL GRELA, ANNA KAJETANOWICZ, AND ANNA SZADKOWSKA

Biological and Chemical Research Centre, Faculty of Chemistry, University of Warsaw, Żwirki i Wigury 101, 02-089 Warsaw, Poland

ANNA KAJETANOWICZ AND JUSTYNA CZABAN-JÓŹWIAK

Institute of Organic Chemistry, Polish Academy of Sciences, Kasprzaka 44/52, 01-224 Warsaw, Poland

Edited by STEVEN M. WEINREB

CONTENTS

---

karol.grela@gmail.com
How to cite: Grela, K.; Kajetanowicz, A.; Szadkowska, A.; Czaban-Jóźwiak, J. Alkene Cross-Metathesis Reactions. *Org. React.* **2021**, *106*, 1–1190.

Doi:10.1002/0471264180.or106.01

ACKNOWLEDGMENTS

A. K. thanks the Foundation for Polish Science for a Homing Plus grant (HOMING PLUS/2013-7/6). K. G. thanks the Foundation for Polish Science for a "Wyjazdowe Stypendium Naukowe" (WSN-2016) Professorship. J. C.-J. thanks the Foundation for Polish Science for a Ventures grant (Ventures/2011–7/3).

INTRODUCTION

The term metathesis means "transposition" and comes from the Greek words μετα ("meta", meaning "change") and θεσις ("thesis", meaning "position"). The term "olefin metathesis" was coined in 1967[1] and refers to the metal-catalyzed redistribution of carbon–carbon double bonds. Thus, in olefin metathesis, carbons at the termini of two alkenes are exchanged to afford two new alkenes (Scheme 1).

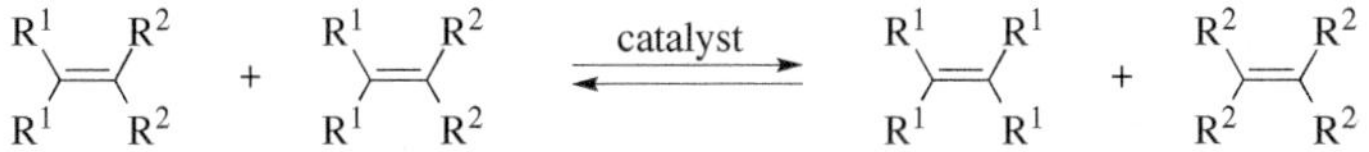

$$R^1R^1C{=}CR^2R^2 + R^1R^1C{=}CR^2R^2 \underset{}{\overset{\text{catalyst}}{\rightleftharpoons}} R^1R^1C{=}CR^1R^1 + R^2R^2C{=}CR^2R^2$$

Scheme 1

Transition-metal-catalyzed olefin metathesis is one of the most powerful tools in organic chemistry for the formation of carbon–carbon double bonds.[2–4] This method has revolutionized the synthesis of a wide variety of useful organic molecules, with far-reaching applications in natural product synthesis,[5–8] medicinal chemistry,[9] macromolecular architectures,[10,11] solid-phase chemistry,[12,13] polymerization,[14–17] fine chemicals,[18] and target-oriented synthesis.[19] The spectacular success of this reaction culminated in the 2005 Nobel Prize in Chemistry being awarded to three primary contributors to this field: Grubbs, Schrock, and Chauvin.[20]

Types of Olefin Metathesis

The main advantage of this method is that several types of olefin metathesis transformations can be performed with the same catalysts, depending on the reaction conditions and on the structural features of the substrates, as illustrated in Scheme 2.[3,7,19,22] Since the metathesis process is energetically neutral and reversible, a mixture of both starting substrates and products is obtained at equilibrium. Nevertheless, this intrinsic problem can be circumvented by the judicious choice of substrates and/or reaction conditions, so that one product type can be prepared selectively.

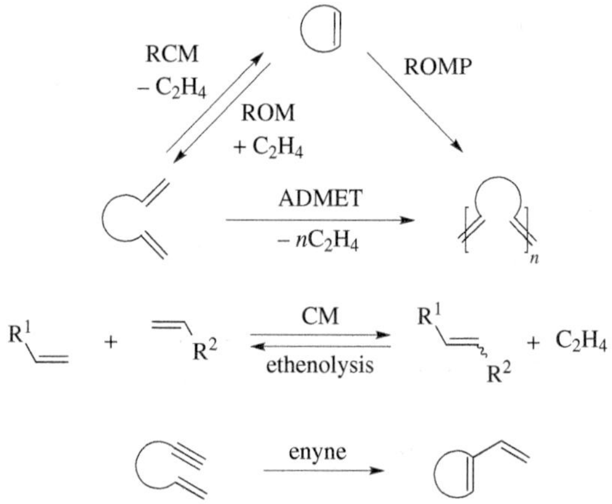

RCM
$- C_2H_4$
ROM
$+ C_2H_4$
ROMP
ADMET
$- nC_2H_4$
CM
ethenolysis
$+ C_2H_4$
enyne

Scheme 2

To date, the most widely used olefin metathesis reaction among synthetic organic chemists is the ring-closing metathesis (RCM) of dienes.[10,23–26] The formation of a ring is usually accompanied by the production of an equivalent of a volatile alkene (e.g., ethylene), which can be easily removed, thereby shifting the equilibrium toward the desired product. RCM facilitates the formation of many ring sizes, from five-membered rings to macrocycles of greater than 20 atoms.

The energy gained upon release of strain with cyclic alkenes (e.g., norbornene derivatives) is the driving force behind ring-opening metathesis polymerization (ROMP).[14,16,27] In many cases, the ROMP of strained cyclic alkenes initiated by metal alkylidene complexes exhibits the characteristic features of a living polymerization, and therefore, block copolymers can be prepared by sequential addition of different alkene monomers.

Acyclic diene metathesis polymerization (ADMET) is a second metathetical method for making polymers.[28–31] Dienes are polymerized with concomitant release of a low-molecular-weight, volatile alkene, usually ethylene. The polymer chain can grow further, usually in a living manner, by reaction with the double bond of a second diene molecule. ROMP and ADMET have been used to construct polymeric materials with a wide range of properties, including those with biological activity.

Intermolecular mutual exchange of alkylidene (or carbene) fragments between two alkene partners is known as cross-metathesis (CM).[32–36] The biggest challenge in cross-metathesis is the chemo- and stereoselective formation of the desired compound from several potential reaction products. There are a few variations on the theme of cross-metathesis (Scheme 3): (1) cross-metathesis, (2) ring-opening cross-metathesis (ROCM),[37,38] and (3) enyne metathesis.

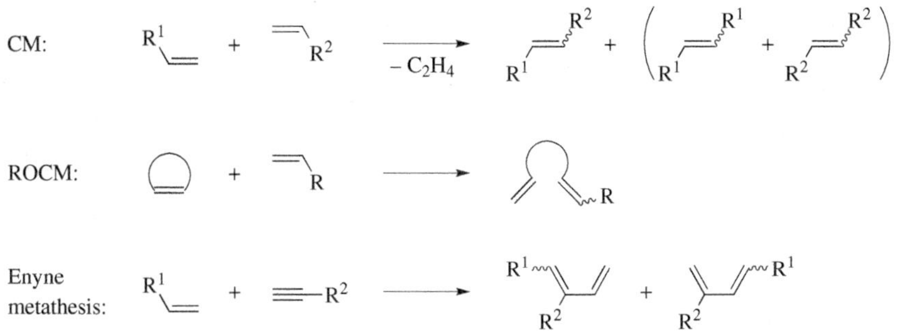

CM: $-C_2H_4$

ROCM:

Enyne metathesis:

Scheme 3

Enyne metatheses are unique and interesting transformations that involve the reaction of an alkene and an alkyne. Two main types of enyne metathesis exist, namely, intermolecular and intramolecular variations. The products from these processes are synthetically useful butadiene derivatives, which lend themselves to structural elaboration, for example by Diels–Alder reactions and other cycloaddition processes. The intramolecular variant (sometimes called metathetical cycloisomerization of enynes) leads to cyclic products, including four-membered rings.[39,40]

Compounds containing several double and triple carbon–carbon bonds can undergo sequential or domino metathesis.[41–44]

Due to the importance of cross-metathesis reactions, several reviews have appeared on this topic.[45–50] Reviews by Blechert,[45] Astruc,[2,51] Fürstner,[19,52] and Grela[47] discuss olefin metathesis in general, whereas reviews by Grubbs,[46] Blechert,[34] Grela,[48,49] and O'Leary[50] address only cross-metathesis. This chapter focuses on state of the art methods in cross-metathesis, and the Tables provide examples through March 2020.

Metathesis (Pre)-Catalysts

Olefin metathesis catalysts are based on tungsten, molybdenum, and ruthenium. In general, ruthenium complexes are more stable towards air and moisture, while tungsten and molybdenum catalysts are more active and more resistant towards ethylene (the ruthenium methylidene complexes, $[Ru]{=}CH_2$, are relatively short-lived).[53–56] The most commonly used metathesis catalysts are presented in Figure 1 and are described in detail in "Metathesis Catalysts"

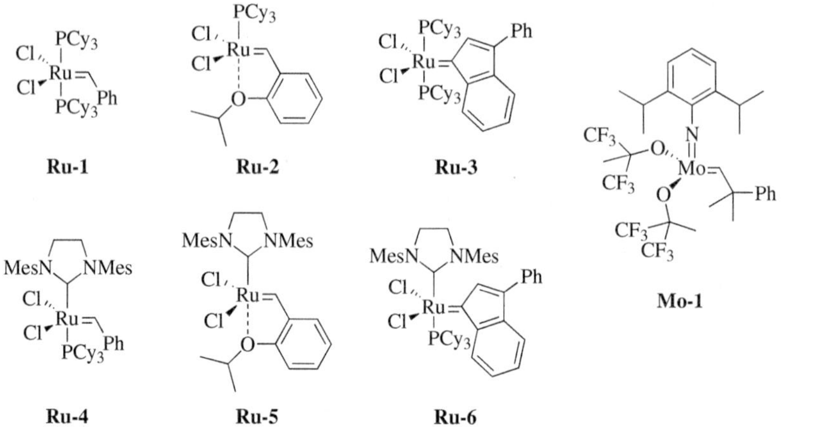

Figure 1. Grubbs (**Ru-1**, **Ru-4**), Hoveyda–Grubbs (**Ru-2**, **Ru-5**), indenylidene (**Ru-3**, **Ru-6**) and Schrock (**Mo-1**) complexes.

MECHANISM AND STEREOCHEMISTRY

General Mechanism for Olefin Metathesis

Several mechanistic hypotheses were proposed during early olefin metathesis exploration.[1,57,58] The commonly accepted mechanism involves first an alkylidene exchange between two reacting alkenes mediated by a transition-metal complex through formation of metallacyclobutanes as the pivotal intermediates (Scheme 4). The CM reaction is initiated when a metal alkylidene (methylidene) of type **4**—the

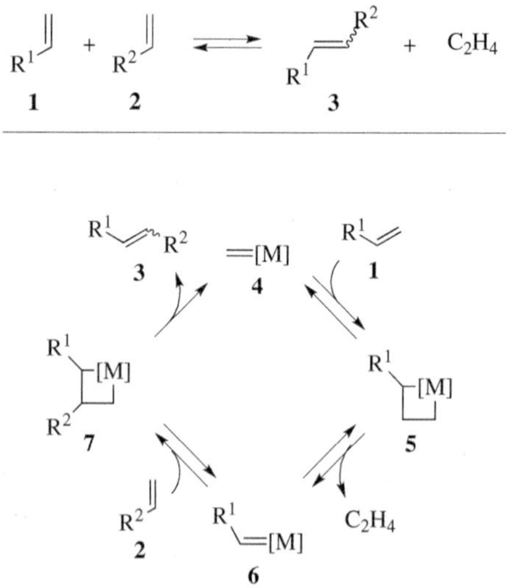

Scheme 4

active catalyst—reacts with the alkene substrate (**1**). Metallacyclobutane **5** is formed as an intermediate, in turn producing either ethylene and a new metal alkylidene **6**, or reverting back to the initiating species. Intermediate **6** reacts with the second alkene (**2**) to give, via metallacyclobutane **7**, the final product **3** and the complex **4**, which restarts the cycle.

Mechanism for Initiation of Metathesis Catalysts

Understanding the mechanisms for metathesis catalyst initiation is crucial for explaining the differences between Schrock (e.g., **Mo-1**) and Grubbs-type (e.g., **Ru-1** and **Ru-4**) complexes and can thus lead to the design of more active catalysts. Schrock-type complexes undergo the [2 + 2] cycloaddition with the alkene substrate directly to form the metallacyclobutane intermediate, which can be observed by NMR spectroscopy and, in some cases, can be isolated. An important feature of the Schrock $M(=NR^1)(=CHR^2)(OR^3)_2$ (e.g., **Mo-1**) species is the possibility of forming a *syn*-alkylidene, in which the alkylidene substituent R^2 is directed toward the imido ligand, or an *anti*-alkylidene, in which the alkylidene substituent points away from the imido nitrogen.[59] The fact that these alkylidenes can interconvert during a metathesis reaction has important consequences for developing (*Z*)-selective metathesis reactions and is discussed below (see Schemes 7 and 8).

The mechanism can be even more complicated with ruthenium carbene complexes that bear neutral ligands capable of dissociation. Three mechanisms of ruthenium catalyst initiation have been proposed, based on experimental and theoretical studies: associative, dissociative, and interchange.[60] In the associative mechanism, the alkene binds ruthenium prior to the dissociation of a neutral ligand or its fragment; in contrast, the dissociative mechanism involves dissociation of the ligand first, and the resulting coordinatively unsaturated complex binds the alkene. In the interchange mechanism, the binding of the alkene and the loss of a ligand occur simultaneously (Scheme 5).

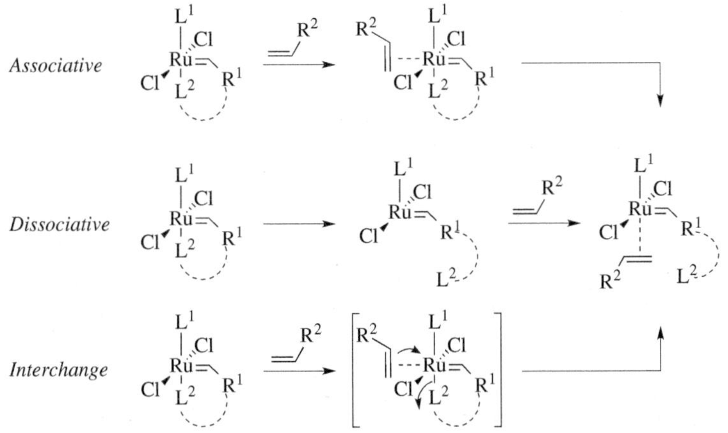

Associative

Dissociative

Interchange

Scheme 5

Associative Mechanism. Experimental[60–62] and computational[63,64] studies indicate that the associative mechanism is generally disfavored owing to large energy barriers and steric repulsion between the ligands at an overcrowded six-coordinated center.

Dissociative Mechanism. According to the generally accepted mechanism, phosphine-containing catalysts (Grubbs-type, e.g., **Ru-1** and **Ru-4**, and indenylidene-type complexes, e.g., **Ru-3** and **Ru-6**) initiate by the dissociation of a neutral ligand to form 14-electron complex **8** (Scheme 6).[61] In the case of Hoveyda–Grubbs-type complexes (e.g., **Ru-2** or **Ru-5**), initiation according to the dissociative mechanism requires breakage of the oxygen–ruthenium coordination as a first step to form complex **9**.[65] Importantly, Grubbs-, indenylidene-, and Hoveyda–Grubbs-type catalysts should generate identical propagating species (**10a**) after a single turnover.

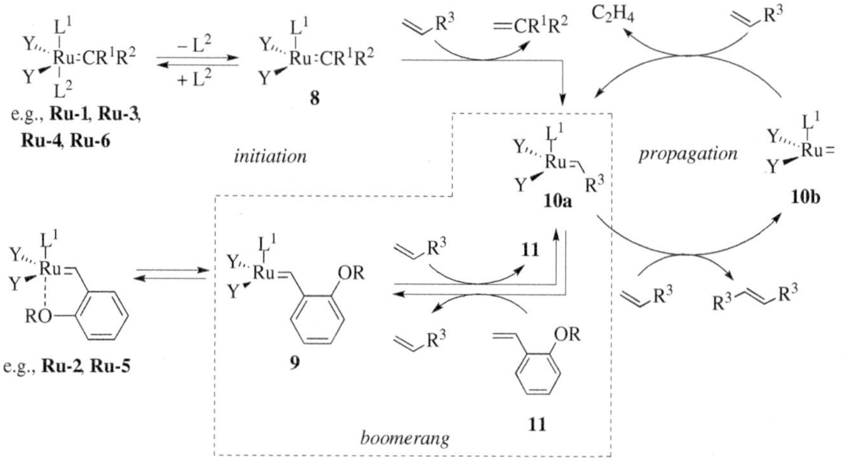

Scheme 6

The initiation of pyridine-bearing complexes[66] has been studied using DFT calculations, which conclude that the dissociative initiation mechanism is favored, although the associative mechanism is plausible with small substrates.[67]

Recent results challenge the belief that initiation of all ruthenium alkylidene complexes is identical and always begins with dissociation of a neutral ligand. For example, the simple dissociative mechanism has been questioned for Hoveyda–Grubbs-type catalysts,[68] and a process other than a dissociative mechanism was proposed for at least one of the phosphine-containing catalysts based on experimental and DFT studies.[69]

Interchange Mechanism. As mentioned above, the dissociative mechanism for Hoveyda–Grubbs-type catalysts has been challenged by experimental and in silico studies.[60] Computational analysis suggests that the interchange mechanism is the most favorable from an energetic point of view.[70] Nevertheless, while most studies rule out the associative mechanism, in some cases the interchange and dissociative mechanisms cannot be clearly distinguished.[71] Initiation of Hoveyda–Grubbs-type catalysts is generally assumed to follow either the dissociative or the interchange pathway, depending on the nature and concentration of the alkene substrate.[60] Sterically small alkenes are thought to follow only the interchange mechanism, whereas larger ones typically follow the dissociative mechanism.[72]

"Boomerang" Mechanism of Regeneration of Hoveyda–Grubbs-Type Complexes. To account for the high stability and the recoverability of Hoveyda–Grubbs-type complexes (e.g., **Ru-2**, **Ru-5**), a plausible "release–return" mechanism has been proposed.[73] The initiation step involves the release of alkoxystyrene **11** (Scheme 6); after the metathesis reaction is complete, the released alkoxystyrene **11** reacts with propagating species **10a** and **10b** to re-form the initial (pre)catalyst **9**. This release–return or "boomerang" mechanism has led to the design of a number of immobilized catalysts.[73] A study using deuterium labeling provides evidencc that this mechanism is indeed operative under high catalyst-loading conditions (5 mol %).[74] However, the "boomerang" mechanism has been challenged, since there is no evidence for a release–return pathway when the olefin metathesis is carried out with a fluorophore-tagged Hoveyda–Grubbs catalyst.[75] A recent study involving crossover studies with a ^{13}C-labeled Hoveyda–Grubbs ligand demonstrates that the release–return mechanism is possible, at least under some conditions.[76] Consequently, there is currently no consensus on the validity of the release–return or "boomerang" mechanism, but regardless, this class of catalysts has been frequently, and with success, used in a number of challenging CM reactions.

Stereochemistry

Diastereoselectivity. A longstanding problem in olefin metathesis is lack of stereocontrol in the alkene product. The reversible character of olefin metathesis usually results in the formation of (*E*)- and (*Z*)-product mixtures. A higher proportion of the (*E*)-product is generally obtained using the ruthenium second-generation pre-catalysts, owing to their higher activity[77] and ability to equilibrate (*E*)/(*Z*) mixtures of products in the CM reaction to the thermodynamically more stable (*E*)-isomer. Similarly, α,β-unsaturated carbonyl compounds and alkenes with large substituents at the carbon–carbon double bond (such as vinyl sulfones, vinyl phosphine oxides, or alkenes with a *tert*-butyl group) give predominantly or exclusively (*E*)-products in CM, in which selective access to the corresponding (*Z*)-alkenes is limited in most cases.

The (*E*)/(*Z*) selectivity in CM reactions can be linked to several factors. The geometry-determining step in the catalytic cycle is the formation of the metallacyclobutane. When substituents are placed in a *cis*-arrangement in the metallacyclobutane (Scheme 7), the (*Z*)-alkene forms upon cycloreversion of this intermediate. The *anti*-metallacycle produces the (*E*)-product, accordingly. Large ligands placed in the proximity of the forming metallacyclobutane can force all the substituents into a *cis*-arrangement, leading to high (*Z*)-selectivity. However, the (*Z*)-alkene can re-enter the catalytic cycle and, in a so-called post-metathetic isomerization, can be transformed to the (*E*)-configured product, thereby eroding the stereoselectivity of the final product. The (*E*)-alkene, in contrast, is reluctant to coordinate with metal due to the distribution of substituents, leading to a kinetically controlled process.[78]

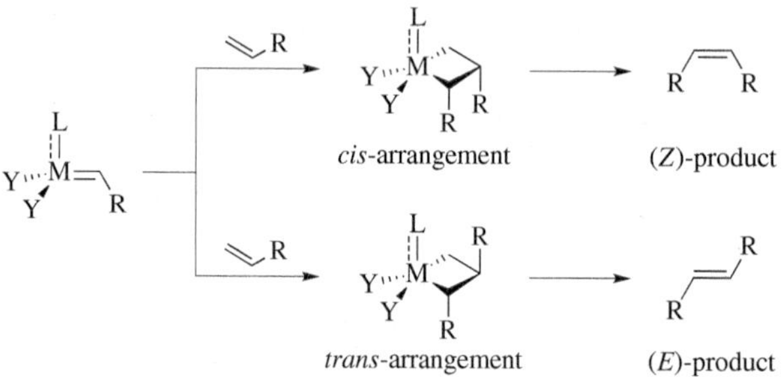

cis-arrangement (*Z*)-product

trans-arrangement (*E*)-product

Scheme 7

Different families of catalysts, such as molybdenum-, tungsten-, and ruthenium-based complexes have been developed for (*Z*)-selective olefin metathesis reactions. The first (*Z*)-selective molybdenum-based olefin metathesis catalyst was described in 2008;[79] since then, significant advances in (*Z*)-selective catalysts have been described.[78] Typical molybdenum- and tungsten-based chiral complexes contain aryloxide ligands (usually biphenol or BINOL derivatives with different substitution patterns) and a pyrrolide moiety, and are referred to as monoalkoxide pyrrolide (MAP) complexes (Figure 2).

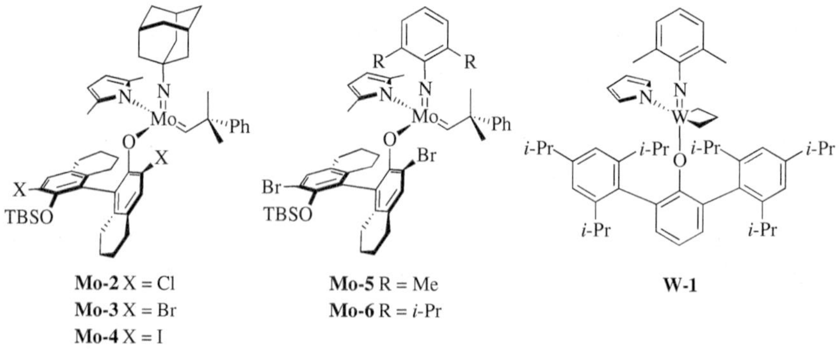

Mo-2 X = Cl
Mo-3 X = Br
Mo-4 X = I

Mo-5 R = Me
Mo-6 R = *i*-Pr

W-1

Figure 2. Representative monoalkoxide pyrrolide (MAP) molybdenum and tungsten complexes.

(Z)-Selectivity in the molybdenum-based catalyst family is the result of the interplay between the small imido group and the large and flexible monodentate aryloxide ligand. Rotation of the aryloxide ligand forces the substituents to orient toward the imido ligand, favoring formation of (Z)-alkenes (Scheme 8).[80] In homodimerizations of terminal alkenes, tungsten-based complexes usually lead to higher (Z)-selectivities than the analogous molybdenum catalysts.[81]

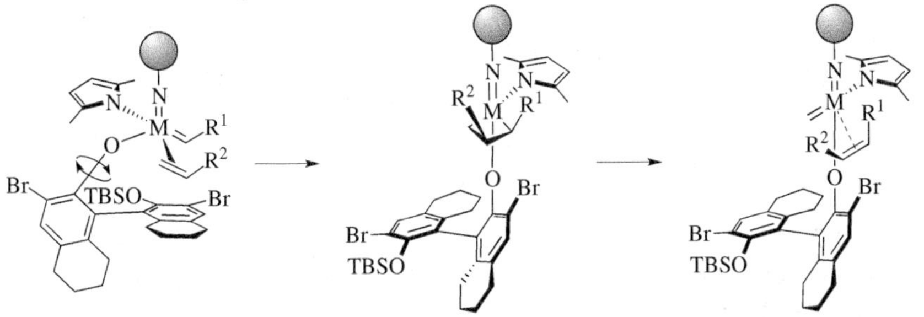

Scheme 8

Some carefully designed ruthenium complexes also exhibit enhanced (Z)-selectivity (Figure 3).[82–85]

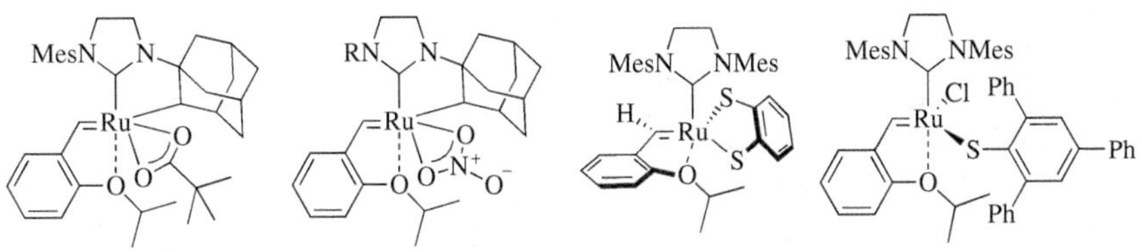

Ru-7

Ru-8, R = Mes
Ru-9, R = Mipp
Ru-10, R = Dipp

Ru-11

Ru-12

Figure 3. (Z)-Selective, ruthenium-based olefin metathesis catalysts.

Computational and experimental studies provide explanations for the (Z)-selectivity imparted by these catalysts.[86,87] The structure of these complexes favors formation of the *cis*-metallacycle, which leads to the (Z)-product. Formation of the (*E*)-product requires generation of a *trans*-substituted ruthenacycle, which in these cases is sterically unfavorable because the *N*-aryl group blocks the top of the ruthenacycle (Scheme 9).

In the case of complex **Ru-12**, which possesses a 2,4,6-triphenylbenzenethiolate ligand, the rationale for the observed diastereoselectivity is similar.[85] A very large

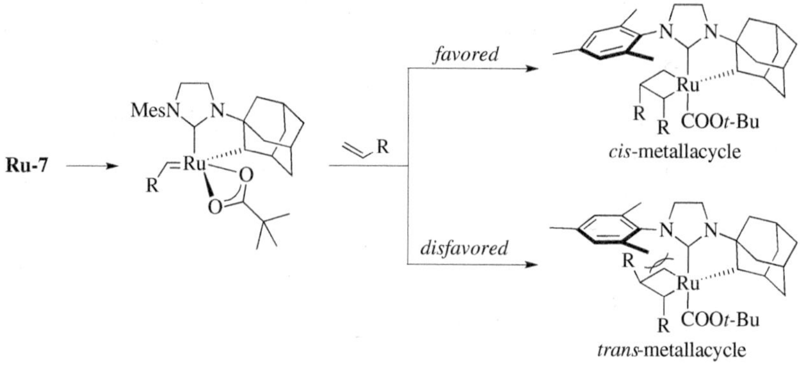

Ru-7 → MesN N Ru R O O ; R ; favored → cis-metallacycle (R R COOt-Bu) ; disfavored → trans-metallacycle (R R COOt-Bu)

Scheme 9

Y-type ligand blocks one side of the catalyst, forcing all of the substituents on the metallacycle to reside in a *cis* arrangement. However, so-called stereoretentive complexes like **Ru-11** exhibit preferential (*Z*)-selectivity with (*Z*)-configured substrates and (*E*)-selectivity when (*E*)-configured alkenes are used.[88,89]

Enantioselectivity. A chiral catalyst can, in principle, metathesize the two enantiomers of an alkene at different rates, thereby leading to a kinetic resolution; some examples exist in the context of RCM.[90] Stereocenters can also be created by CM-based desymmetrization of prochiral *meso*-compounds, as in the asymmetric ring-opening cross-metathesis (AROCM) and asymmetric cross-metathesis (ACM) reactions shown in Scheme 10. Figure 4 depicts chiral molybdenum-based catalysts that are used in enantioselective metathetical processes; tungsten catalysts may also be employed.[91]

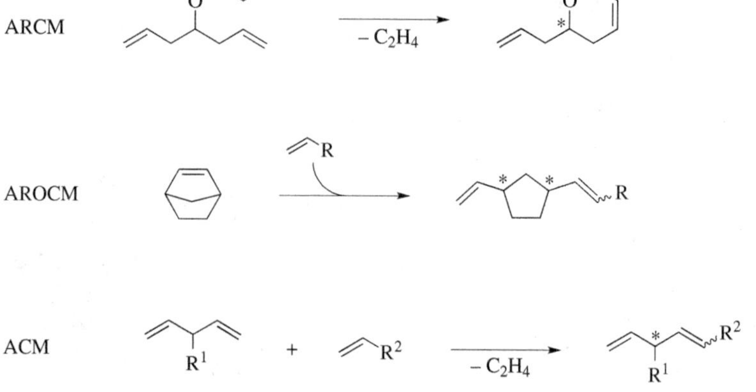

ARCM — $-\ C_2H_4$ →

AROCM — R → R

ACM — R^1 + R^2 — $-\ C_2H_4$ → R^2 R^1

Scheme 10

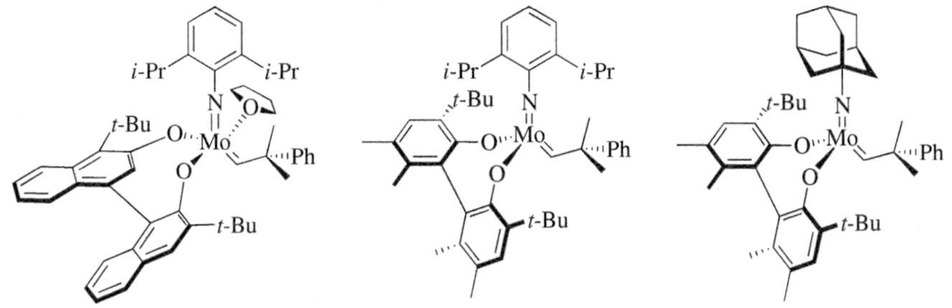

Mo-7 **Mo-8** **Mo-9**

Figure 4. Chiral molybdenum-based complexes.

In ruthenium-based complexes, C_1- and C_2-symmetrical *N*-heterocyclic carbene ligands are employed to prepare chiral catalysts. The first chiral ruthenium-based metathesis catalyst was reported in 2001,[92] and a number of catalysts bearing C_2-symmetric NHC ligands have been developed. Bulky *ortho*-substituents on the *N*-aryl groups (Figure 5) exert beneficial influences on enantioselectivity in metathesis reactions.

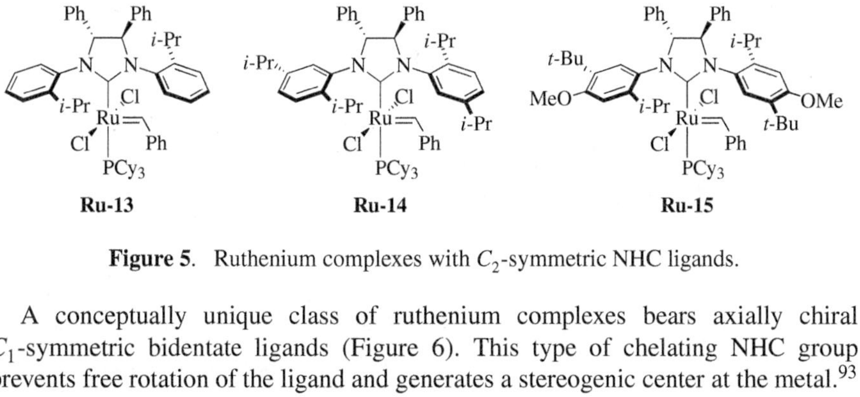

Ru-13 **Ru-14** **Ru-15**

Figure 5. Ruthenium complexes with C_2-symmetric NHC ligands.

A conceptually unique class of ruthenium complexes bears axially chiral C_1-symmetric bidentate ligands (Figure 6). This type of chelating NHC group prevents free rotation of the ligand and generates a stereogenic center at the metal.[93] Complexes with a C_1-symmetric ligand provide higher enantioselectivity in model reactions compared to those with a C_2-symmetric ligand.[94–96]

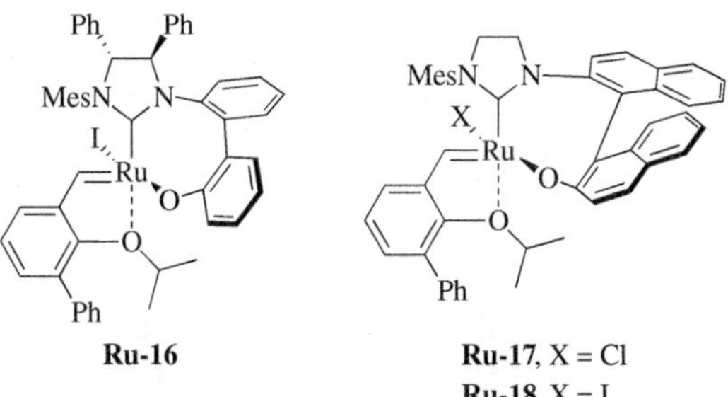

Figure 6. Ruthenium complexes with C_1-symmetric NHC ligands.

Asymmetric Cross-Metathesis (ACM). ACM processes are the most challenging enantioselective reactions because secondary metathesis events, which would generate symmetrically disubstituted products, must be avoided. Only terminal dienes can be used in such transformations. For example, the ACM of diene **12** can be accomplished with C_1- and C_2-symmetric catalysts (Scheme 11).[97,98] In the case of **Ru-8**, the product of ACM is formed with exclusive (*E*)-selectivity and is obtained in 35% yield and with 75:25 er.

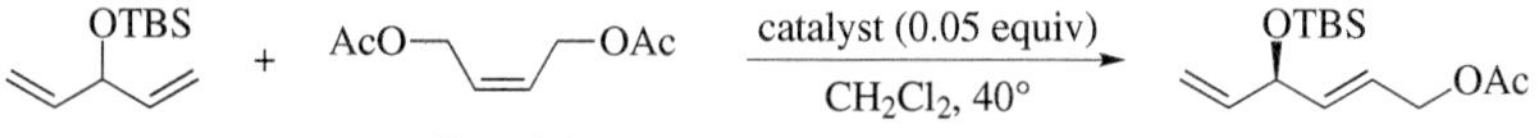

OTBS **12** + AcO–CH=CH–OAc (5 equiv) → catalyst (0.05 equiv), CH_2Cl_2, 40° → OTBS ... OAc

Cat.	Yield (%)	er
Ru-8	35	75:25
Ru-13	15	61:39
Ru-14	28	72:28
Ru-15	12	72:28

Scheme 11

Asymmetric Ring-Opening Cross-Metathesis (AROCM). AROCM reactions can be classified as a variant of ACM reactions, in which one of the substrates is a prochiral cycloalkene. A representative example is illustrated in Scheme 12.[99] The AROCM process is conducted using an achiral norbornene derivative in the presence of a chiral molybdenum complex, and the product is formed with excellent geometrical selectivity and enantioselectivity.

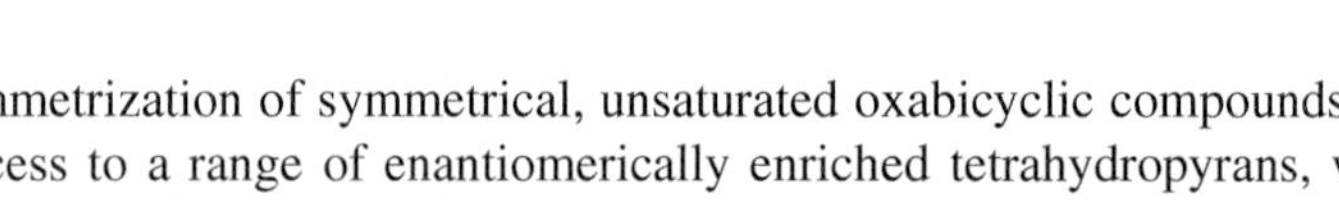

OTMS + Ph → **Mo-8** (0.05 equiv), benzene, 22° → OTMS ... Ph

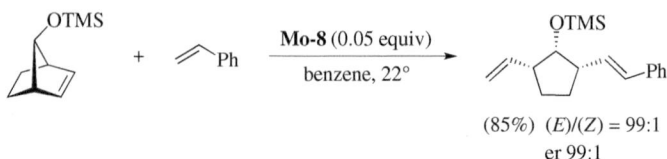

(85%) (*E*)/(*Z*) = 99:1
er 99:1

Scheme 12

Desymmetrization of symmetrical, unsaturated oxabicyclic compounds provides direct access to a range of enantiomerically enriched tetrahydropyrans, which are versatile building blocks for target-oriented synthesis. Scheme 13 illustrates the efficiency and selectivity of various chiral metathesis catalysts.[91,100,101]

AROCM of azabicyclic compounds provides access to chiral *cis*-2,6-disubstituted piperidines, which are particularly important intermediates (Scheme 14).[102] The ruthenium-based catalyst **Ru-17** leads to higher enantioselectivity than **Mo-9** for the *N*-benzyl protected substrate, whereas the opposite is true for the *N*-methyl derivative. The Cbz-protected azabicycle reacts only in the presence of the molybdenum complex, and no product is obtained using **Ru-17**.

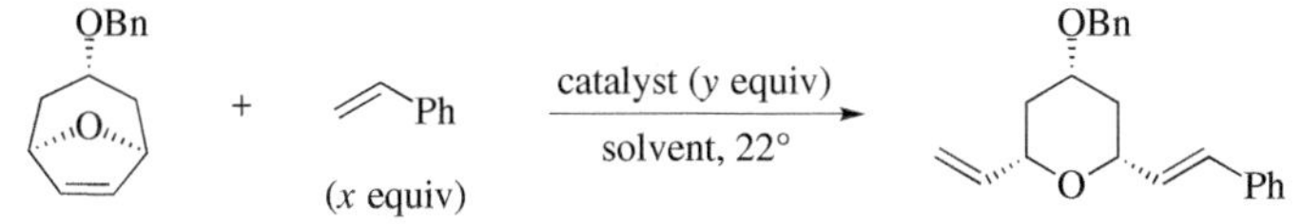

OBn + Ph (x equiv) → catalyst (y equiv), solvent, 22° → OBn, O, Ph

x	Cat.	y	Solvent	Time (h)	Yield (%)	er
5	**Mo-9**	0.05	benzene	1	0	—
5	**Ru-16**	0.02	neat	1	90	94:6
5	**Ru-17**	0.02	neat	1	70	98:2
2	**Ru-17**	0.02	neat	6	87	90:10
2	**Ru-18**	0.05	neat	36	85	95:5

Scheme 13

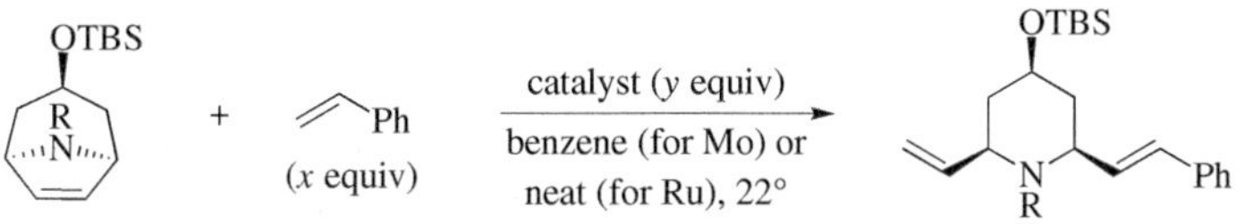

OTBS, R, N + Ph (x equiv) → catalyst (y equiv), benzene (for Mo) or neat (for Ru), 22° → OTBS, N, R, Ph

R	x	Cat.	y	Time (h)	Yield (%)	er
Me	10	**Mo-9**	0.05	1	95	97:3
Me	20	**Ru-17**	0.05	36	30	66.5:33.5
Bn	10	**Mo-9**	0.05	1	85	60:40
Bn	20	**Ru-17**	0.05	24	80	>99:1
Cbz	10	**Mo-9**	0.1	12	93	95:5
Cbz	10	**Ru-17**	0.1	12	0	—

Scheme 14

SCOPE AND LIMITATIONS

Introduction

Formation of a new carbon–carbon double bond by CM facilitates access to complex organic compounds from simple molecules.[103] Nevertheless, in addition to the desired reaction between cross partners, one or both of the starting alkenes can undergo unwanted homodimerization, thereby leading to complex mixtures of products. Studies by many academic and industrial research groups have led to the introduction of methods and reaction conditions for selective cross-metathesis. One of the key factors was the development of rules for attaining selectivity in CM reactions. If both of the CM partners exhibit similar reactivities with a given metathesis catalyst, the simplest way to achieve an acceptable yield of the cross-product is to employ the cheaper or more accessible alkene in excess (sometimes >10 equivalents) and then separate the resulting mixture of the homodimers and CM products. However, not all alkenes exhibit the same reactivity in metathesis and, for example, acrylonitrile and some trisubstituted alkenes do not readily undergo homodimerization. Therefore, a judicious choice of alkene partners can lead to

Table A Alkene Types for Selective Cross-Metathesis (as defined in 2003 by Grubbs et al.[36] Please note that classifications may differ for some catalysts).

Type	Properties	Examples
I	fast homodimerization homodimers consumed	α-alkenes, terminal alkenes with functional groups at distant positions, allylic silanes, allylic halides, primary allylic alcohols, allyl boronates, allylic phosphonates, protected allylic amines, styrenes (with no large *ortho* substituent), allylic sulfides, allyl ethers
II	slow homodimerization homodimers sparingly consumed	styrenes with a large *ortho* substituent, allylic stannanes, acrylates, acrylamides, acrylic acid, acrolein, vinyl ketones, vinyl epoxides, vinyl boronates, vinyl dioxolanes, alkenes bearing perfluoroalkyl groups, secondary allylic alcohols, unprotected tertiary allylic alcohols
III	no homodimerization	tertiary allylic amines, vinyl sulfones, vinyl sulfonamides,* 1,1-disubstituted alkenes, acrylonitrile, vinyl siloxanes, vinyl phosphonates, protected tertiary allylic alcohols, quaternary allylic carbons (all alkyl substituents), vinyl halides,* allylic nitro compounds*
IV	spectators to cross-metathesis	1,1-disubstituted alkenes, vinyl nitroalkenes, quaternary allylic carbon-containing alkenes, perfluorinated alkyl alkenes, protected tertiary allylic amines, *O*-protected trisubstituted allylic alcohols

(*denotes newer additions to this classification table)

enhancement of the CM reaction selectivity. A commonly accepted classification system of CM partners that can be used to predict selectivity was established by Grubbs et al. in 2003.[36] CM partners are divided into four categories depending on their rate of homodimerization with three prototypical catalysts: Schrock molybdenum alkylidene (**Mo-1**) and Grubbs first-generation (**Ru-1**) and second-generation (**Ru-4**) complexes. The four categories include CM partners that (I) undergo rapid homodimerization, (II) undergo slow homodimerization, (III) undergo no homodimerization, or (IV) do not participate in CM (Table A).

Using this table, the reactivity of each class of CM partners in a reaction with another class can be predicted. For example, alkenes of Type III should undergo selective CM with alkenes of Types I and II (Table B). In case of Type IV alkenes, CM reactions do not proceed but also do not cause deactivation of the metathesis catalyst

Table B Rules for Selectivity in CM.

	Type I	Type II	Type III	Type IV
Type IV	no reaction	no reaction	no reaction	no reaction
Type III	selective	selective	non-selective	no reaction
Type II	selective	non-selective	selective	no reaction
Type I	non-selective	selective	selective	no reaction

(those alkenes that do were omitted from the Grubbs classification). However, new complexes have been developed since 2003, and reactions of some alkenes initially classified as Type IV are now possible. Nonetheless, this model serves as a useful predictor of which alkenes will undergo selective cross-metathesis and, similarly, which combinations of substrates should be avoided.

The list of CM partners is updated regularly (e.g., substrates added recently are marked in Table A with an asterisk). In addition, some alkenes previously classified as inert in cross-metathesis (e.g., vinyl chlorides) can now successfully be employed using more recently developed molybdenum and ruthenium catalysts.[104–106]

In the following discussion, the most common CM partners belonging to Types I–III are arranged by the functional groups that they contain. With only a few exceptions, a functional group impacts reactivity only if it is attached directly to the double bond or in an allylic position; substituents situated in homoallylic or more distal positions usually have negligible effects on the metathesis reaction. Some Lewis basic groups present in the alkene can, however, form stable five- and six-membered ring chelates with the propagating species (Scheme 15), or can simply complex the organometallic catalyst, thereby partially or completely deactivating catalytic turnover.

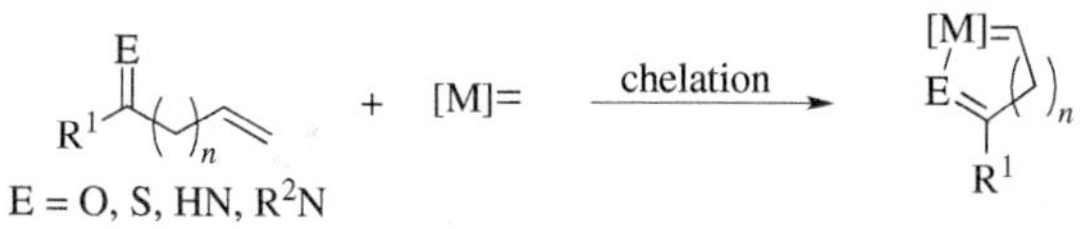

E R¹ n + [M]= chelation → [M]= E R¹ n

E = O, S, HN, R^2N

Scheme 15

Metathesis Catalysts

Generally, olefin metathesis catalysts are based on tungsten, molybdenum, or ruthenium. Ruthenium complexes are generally less active, but they are more stable toward air and moisture than tungsten and molybdenum complexes, and as a result, the use of ruthenium catalysts is more common than that of the other metal catalysts. The development of active and robust ruthenium complexes has permitted the synthesis of many important building blocks. The most common ruthenium catalysts employed in organic synthesis are presented in Figure 1.

More recent studies have led to the development of ruthenium complexes with high activity, including the sterically activated complex **Ru-19** (Figure 7), the electronically altered complex **Ru-20**, and complexes with modified ethereal fragments (**Ru-21, Ru-22, Ru-23**). These catalysts have been successfully applied in target-oriented synthesis and in the pharmaceutical industry. Other modifications of the Hoveyda–Grubbs catalyst (**Ru-24, Ru-25, Ru-26**) have been made to increase solubility of the ruthenium catalyst in aqueous or protic solvents.

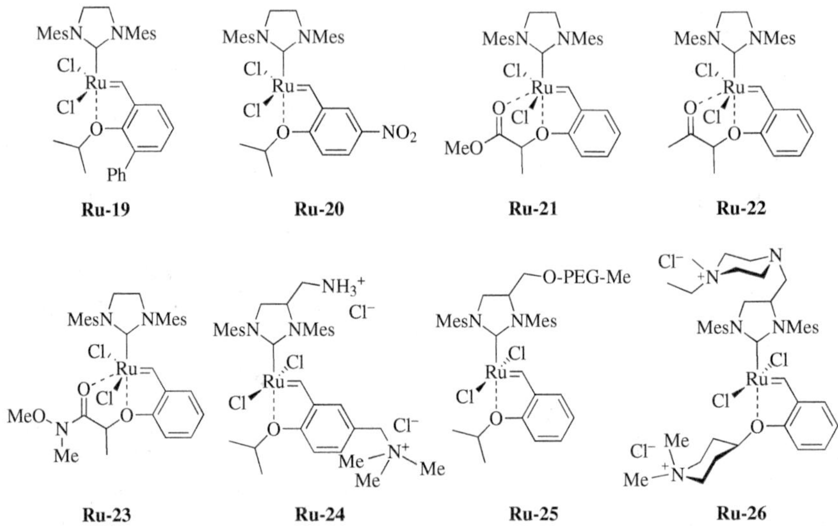

Figure 7. Analogues of the Hoveyda–Grubbs catalyst.

Molybdenum-based catalysts (Figure 8)[59,107] are generally more reactive than ruthenium-based catalysts, but they are sensitive to air and moisture. Nonetheless, modern molybdenum complexes have been widely used in enantioselective and diastereoselective metathesis,[2] and new, more user-friendly formulations of molybdenum catalysts have been introduced in recent years.[108,109]

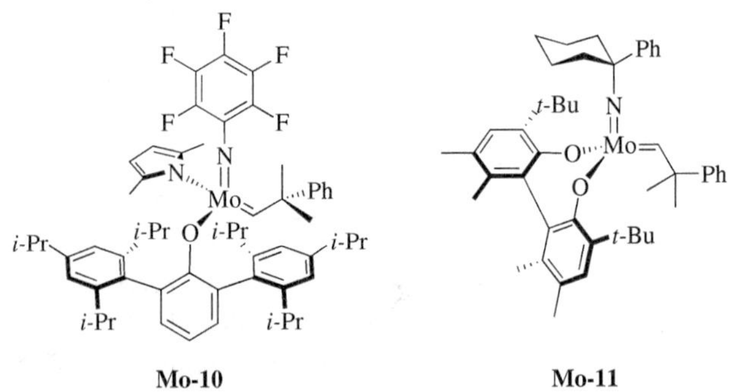

Figure 8. Selected molybdenum-based complexes used in metathesis.

Influence of Functional Groups on Alkene Classification and Selectivity

Hydrogen as a Functional Group (Ethenolysis). A synthetically valuable type of cross-metathesis involves the ethenolysis reaction, which uses ethylene as a cross-partner. Although not originally included in Table 1, ethylene can be classified as a terminal alkene with hydrogen as a functional group and can be formally recognized as a Type I partner, even though it cannot undergo homodimerization.[36] Ethenolysis of an internal alkene facilitates the selective production of a pair of terminal alkenes and has been successfully implemented in a number of syntheses.[110,111] One important industrial application of this process is the ethenolysis of unsaturated fatty acid esters, resulting in products (α-alkenes and unsaturated esters) that are valuable intermediates in the production of polymers and specialized materials and chemicals (Figure 9).[112–115]

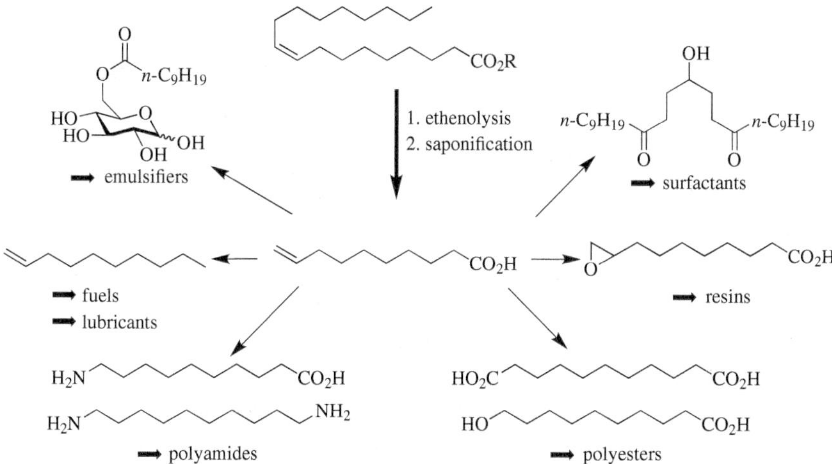

Figure 9. Valuable intermediates that can be derived from ethenolysis products.

Ethylene can also be used in other types of CM reactions.[116–118] For example, several β-lactams **13** are prepared in a ROM/CM process with ethylene (300 psi) at room temperature (Scheme 16).[119] The most effective catalyst for this particular transformation is Grubbs' first-generation complex **Ru-1**.

excess ethylene, **Ru-1** (0.1 equiv), CH_2Cl_2, rt, 4 h → **13**

R	Yield (%)
H	50
TBS	89
TBDPS	60

Scheme 16

Boron-Bearing Alkenes. Alkenyl and allyl boronate esters are compounds of importance in organic synthesis, because they can be easily transformed into many valuable products.[120–122] In addition, boron compounds are very useful intermediates in Suzuki-type cross-coupling reactions.[123] According to the general model of selectivity, allylic boronates are Type I alkenes, and alkenylboronates are Type II (e.g. **14**) or Type III (e.g. **17**) (they exhibit slow or no homodimerization).[36] Because of difficulties in preparing highly functionalized alkenylboron compounds, the cross-metathesis reaction represents a promising strategy for the formation of substituted alkenylboronates. Thus, reactions of pinacol boronic esters **14** and **17** with Type I terminal alkenes **15** and **18** in CH_2Cl_2 under reflux conditions lead to substituted alkenes **16** and **19**, respectively, in moderate yields (67% and 58%), favoring the less-hindered (*E*)-alkenes (Scheme 17).[124,125]

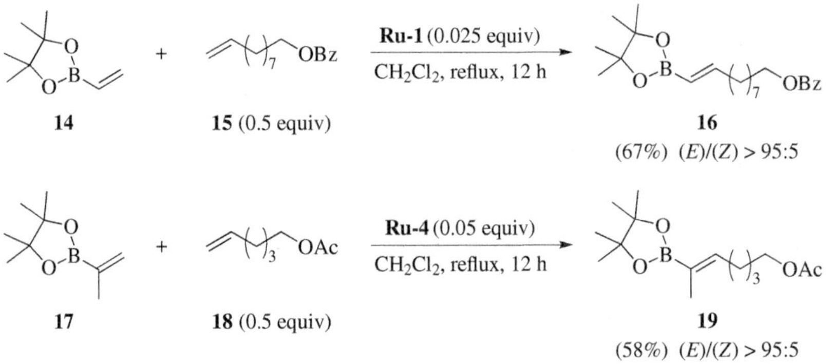

14 + **15** (0.5 equiv) —**Ru-1** (0.025 equiv), CH_2Cl_2, reflux, 12 h→ **16** (67%) (*E*)/(*Z*) > 95:5

17 + **18** (0.5 equiv) —**Ru-4** (0.05 equiv), CH_2Cl_2, reflux, 12 h→ **19** (58%) (*E*)/(*Z*) > 95:5

Scheme 17

Metathetical functionalization of allylboronates (Type I alkenes) proceeds more easily than CM of alkenylboronates, usually in higher yields and with high selectivities. In CM protocols, a pinacol allylboronate is usually treated with an excess of the alkene in the presence of the highly reactive Grubbs' second-generation catalyst (e.g., **Ru-4**).

This approach has been employed in the one-pot formation of functionalized homoallylic alcohols that are synthetically useful intermediates. Thus, the reaction between pinacol allylboronate (**20**) and 2-vinyl-1,3-dioxolane (**21**) in the presence of Grubbs' second-generation catalyst **Ru-4** affords CM-product **22**, which is converted directly into homoallylic alcohol **23** in 69% yield and with excellent diastereoselectivity (Scheme 18).[126] Notably, reducing the loading of the catalyst from 5 mol % to 2 mol % does not impact the yield for this process.

Alkenes with Carbon-Based Functional Groups.

Alkenes Containing Linear Alkyl Groups. Terminal alkenes are classified as Type I substrates, and a typical procedure for CM is illustrated in Scheme 19 for the formation of lipidic amino acids.[127] In the CM of *N*-Fmoc-protected allyl glycine **24** and 1-decene, the best yields are obtained in either dichloromethane or ethyl acetate under reflux conditions with Hoveyda–Grubbs catalyst **Ru-5**. Optimization studies

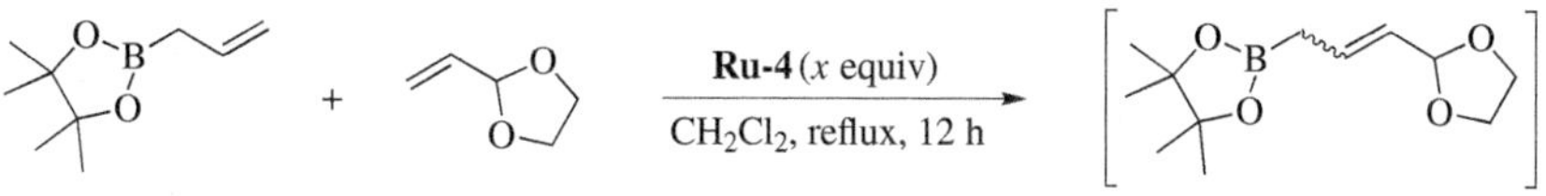

20 + **21** (3 equiv) → **Ru-4** (x equiv), CH_2Cl_2, reflux, 12 h → [**22**]

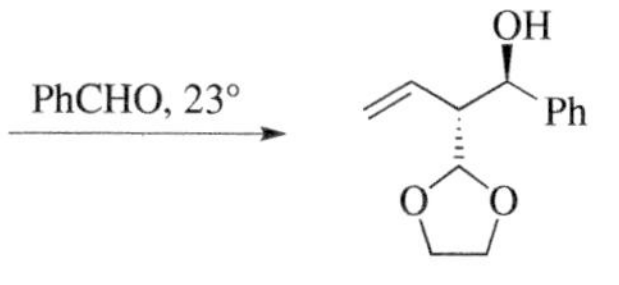

→ PhCHO, 23° → **23**

x	Yield (%)	*anti*/*syn*
0.02	69	>95:5
0.05	68	>95:5

Scheme 18

indicate that five equivalents of the n-alkene are required to minimize homodimerization of alkene **24** and thus afford the product **25** with high cross-selectivity.

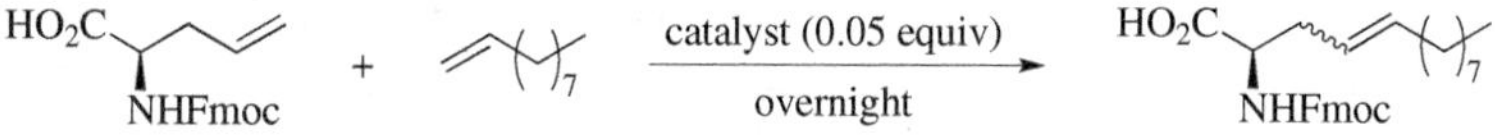

24 + (x equiv) → catalyst (0.05 equiv), overnight → **25**

x	Cat.	Solvent	Temp	Conv (%)
1	**Ru-5**	CH_2Cl_2	rt	32
2	**Ru-5**	CH_2Cl_2	rt	65
5	**Ru-5**	CH_2Cl_2	reflux	93
5	**Ru-5**	EtOAc	reflux	90
5	**Ru-5**	CH_2Cl_2	rt	66
5	**Ru-4**	EtOAc	rt	52
5	**Ru-5**	EtOAc	rt	62
5	**Ru-5**	CH_2Cl_2	rt	52[a]
5	**Ru-5**	CH_2Cl_2	rt	87[b]
10	**Ru-5**	CH_2Cl_2	rt	70

[a] Molecular sieves were added to the reaction.
[b] The reaction was performed with an N_2 bleed.

Scheme 19

Another approach for selective CM with terminal alkenes that undergo fast homodimerization is illustrated in an elegant synthesis of an isocarbacyclin analogue (Scheme 20).[128] When bicyclic alkene **26** is treated with terminal alkene **27**, the expected product **29** is obtained in only 54% yield and with an (*E*)/(*Z*) selectivity of 88:12. However, changing the CM partner from terminal alkene **27** to symmetrical alkene **28**, which contains an internal double bond (formally a product of homodimerization of **27**), affords **29** in higher yield (69%) and with improved selectivity ((*E*)/(*Z*) = 90:10).

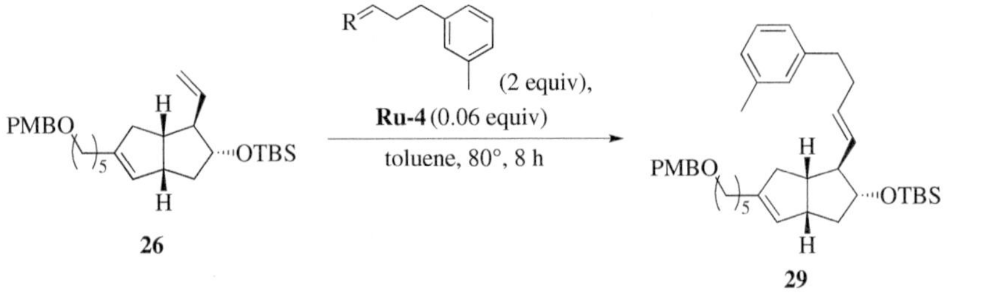

	R	Yield (%)	(*E*)/(*Z*)
27	CH_2	54	88:12
28	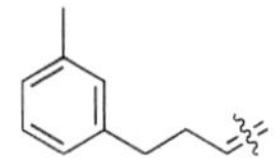	69	90:10

Scheme 20

Sterically Hindered Alkenes. Carbon–carbon double bonds bearing bulky substituents are often unreactive in CM reactions. Sterically hindered di- or trisubstituted alkenes (Type III) do not undergo homodimerization but, due to the development of ruthenium complexes bearing NHC ligands, are effective in CM reactions with Type I and Type II alkenes leading to 1,1-disubstituted alkenes. Hence, the prenylation of terminal alkene **18** with isobutylene leads to **30** in high yield (Scheme 21);[129] isobutylene also reacts with other terminal carbon–carbon double bonds, as exemplified by the reaction with the prochiral homoallylic acetate **31** to form **32**. 2-Methyl-2-butene is a very useful CM partner that can be used in place of isobutylene, as illustrated

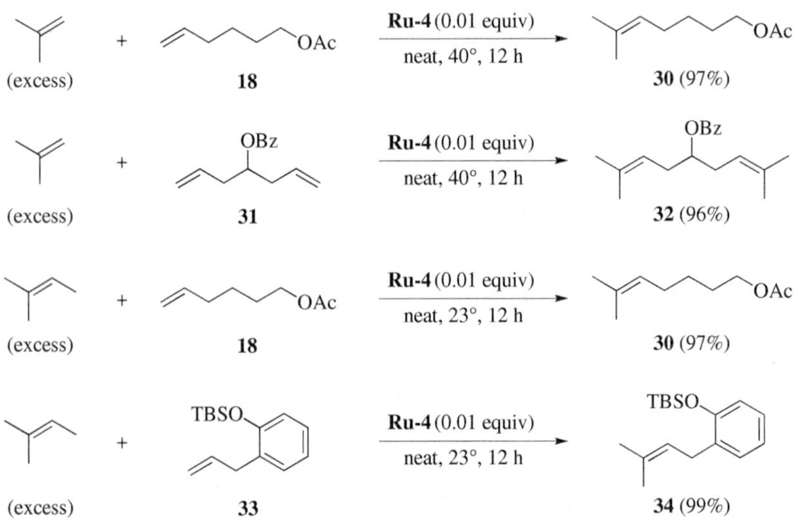

Scheme 21

in the reaction with alkene **18**. Similarly, 2-methyl-2-butene undergoes CM with the 2-substituted allylbenzene **33** to generate **34** in nearly quantitative yield (99%). In all of these cases, the Grubbs catalyst **Ru-4** provides the optimal results. Isobutylene and 2-methyl-2-butene are usually used in excess or the CM reactions are run neat, and very good product yields are obtained even with 1 mol % of the catalyst. Furthermore, because isobutylene and 2-methyl-2-butene have such low boiling points (–6.9° and 38.5°, respectively), they can be recovered from the crude reaction mixture.

Alkenes containing sterically hindered allylic carbon centers (Type III alkenes) react smoothly with Type I and Type II alkenes using Grubbs' second-generation **Ru-4** and related catalysts. Steric hindrance at the allylic position inhibits homodimerization of alkene **35**, which can be used in excess, and the corresponding reaction with Type I substrate **36** provides the expected product **37** in high yield as only the (*E*)-isomer (Scheme 22).[36] Selective cross-metathesis between hindered alkenes (Type III) with Type II or Type III partners usually results in diminished conversions,[130] but good yields can be achieved with careful selection of reaction conditions. For example, in the CM with acrylic ester **38** (Type II), an excess of alkene **35** is required to afford a 73% yield of product **39**.[36]

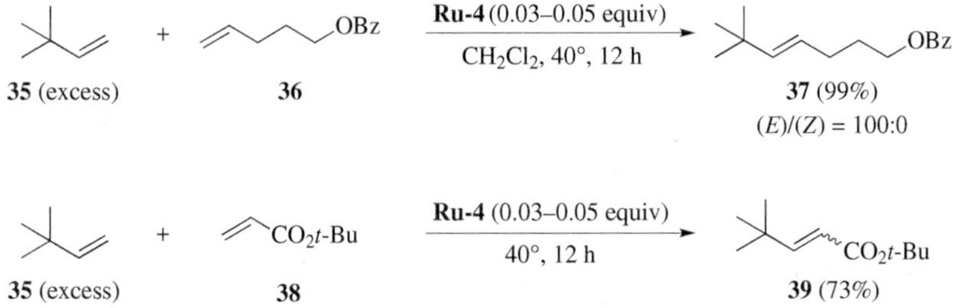

OBz; **35** (excess) + **36** → **Ru-4** (0.03–0.05 equiv), CH_2Cl_2, 40°, 12 h → OBz; **37** (99%), (*E*)/(*Z*) = 100:0

CO_2t-Bu; **35** (excess) + **38** → **Ru-4** (0.03–0.05 equiv), 40°, 12 h → CO_2t-Bu; **39** (73%)

Scheme 22

Alkenes Bearing Additional Carbon–Carbon Double Bonds. Cross-metathesis of conjugated dienes is possible, but in order for the reaction to occur with selectivity, one of the double bonds in the substrate must be significantly more reactive. For example, good selectivity is observed in reactions using a diene substituted with either an electron-withdrawing group or a bulky substituent. In general, the less-hindered double bond participates in the CM reaction, although sometimes a complex mixture of products is obtained. For example, methyl sorbate reacts with allyl bromide (a Type I alkene) to selectively afford the new diene **40**, albeit in modest yield (48%) (Scheme 23).[131] The cross-metathesis of ethyl sorbate (**41**) with another Type I partner, 5-hexenyl acetate (**18**), requires refluxing temperatures with a higher catalyst loading (10 mol %); the resulting nonselective metathesis of both carbon–carbon double bonds gives a mixture of products **42** and **43**.[132] If the two double bonds have sufficiently different reactivities, it is usually possible to obtain a single product in good yield. For example, the reaction of 2-substituted 1,3-butadiene **44** with terminal alkene **18** provides product **45** in 75% yield and with excellent (*E*)-selectivity.[132]

In these cases, the second-generation catalysts **Ru-4** and **Ru-5** exhibit better activities in the CM reactions, whereas the first-generation Grubbs catalyst either promotes homodimerization of the Type I partners (hex-5-en-1-yl acetate (**18**) or allyl bromide) or is deactivated by alkenes bearing electron-withdrawing groups.

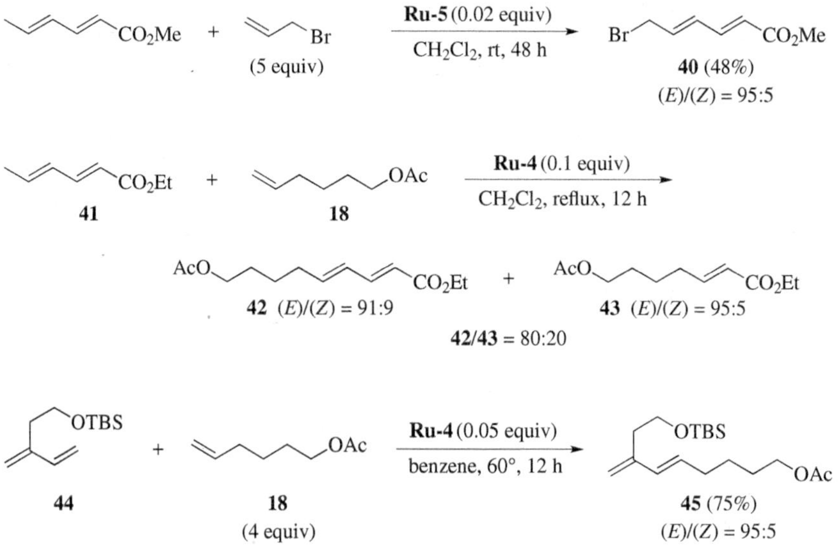

CO_2Me + Br (5 equiv) → **Ru-5** (0.02 equiv), CH_2Cl_2, rt, 48 h → Br CO_2Me **40** (48%) (*E*)/(*Z*) = 95:5

CO_2Et **41** + OAc **18** → **Ru-4** (0.1 equiv), CH_2Cl_2, reflux, 12 h →

AcO CO_2Et **42** (*E*)/(*Z*) = 91:9 + AcO CO_2Et **43** (*E*)/(*Z*) = 95:5

42/**43** = 80:20

OTBS **44** + OAc **18** (4 equiv) → **Ru-4** (0.05 equiv), benzene, 60°, 12 h → OTBS OAc **45** (75%) (*E*)/(*Z*) = 95:5

Scheme 23

Alkenes Containing Perfluoroalkyl Groups. A number of examples that demonstrate the reactivity of perfluoroalkyl-substituted alkenes have been reported using both Hoveyda–Grubbs (**Ru-5**) and Grubbs (**Ru-4**) catalysts.[133–139] The CM reaction of Type I alkene **18** with **46** (Type II) results in the formation of the 1,2-disubstituted alkene **47** in 34% yield with moderate (*E*)/(*Z*)-selectivity (Scheme 24).[139] In the reaction between trifluoropropene (**48**, Type II) and Type I alkene **49**, improved yields of products **50** and **51** are obtained because a fluorinated aromatic hydrocarbon (FAH) is used as the reaction medium.[137] FAH solvents promote the initiation of the catalyst and also stabilize the propagating species, and therefore, can significantly improve the outcome of challenging cross-metathesis reactions.[140,141] Good yields can be also obtained when the reaction is conducted at high temperature in a sealed tube, as in the reaction between 4-(trifluoromethyl)styrene with the allylic fluoride **52**.[142] In some cases, the alkene partner that undergoes the faster homodimerization is used in excess.

Alkenes Bearing Aryl and Heteroaryl Substituents. Alkenes possessing an aromatic moiety at various distances from the reacting carbon–carbon double bond are common partners in olefin cross-metathesis. Such alkenes are typically classified as Type I and undergo facile CM and homodimerization reactions. Exceptions to this rule are styrenes bearing bulky substituents in the 2-position, which are considered to be Type II alkenes.

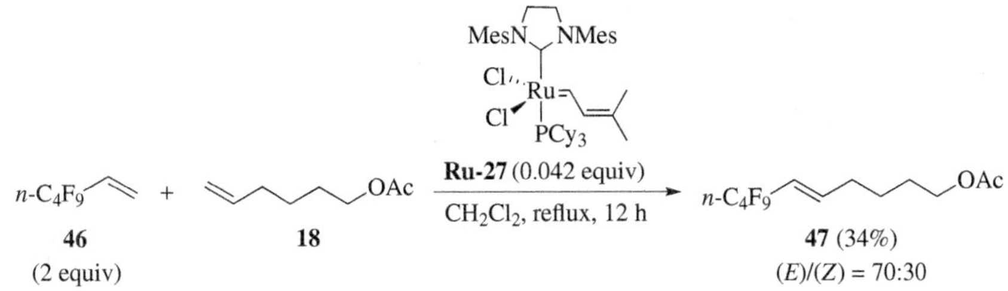

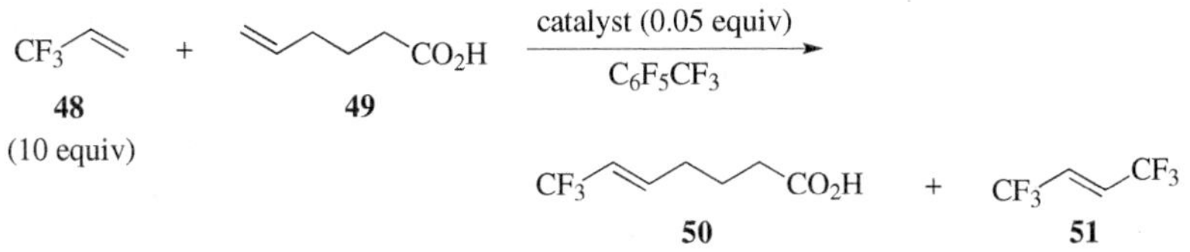

MesN NMes
Cl Ru Cl PCy$_3$

n-C_4F_9 (**46**, 2 equiv) + **18** (OAc) → **Ru-27** (0.042 equiv), CH_2Cl_2, reflux, 12 h → n-C_4F_9 ... OAc **47** (34%) (E)/(Z) = 70:30

CF_3 (**48**, 10 equiv) + **49** (CO_2H) → catalyst (0.05 equiv), $C_6F_5CF_3$ → CF_3 ... CO_2H **50** + CF_3 ... CF_3 **51**

Cat.	Temp (°)	Time (h)	**50** Yield (%)	(E)/(Z)	**51** Yield (%)
Ru-4	60	4	58	>95:5	23
Ru-5	45	3	70	>95:5	19

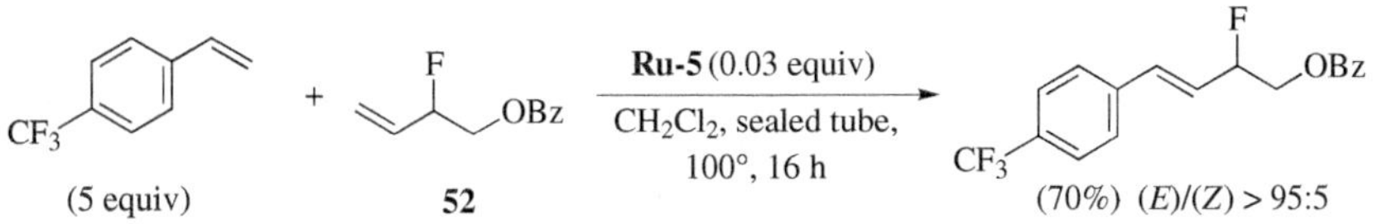

CF_3-styrene (5 equiv) + **52** (F, OBz) → **Ru-5** (0.03 equiv), CH_2Cl_2, sealed tube, 100°, 16 h → CF_3 ... F ... OBz (70%) (E)/(Z) > 95:5

Scheme 24

Coupling of Type I styrenes catalyzed by **Ru-4** has been applied in the synthesis of stilbenes, albeit they can sometimes suffer from statistical product distributions.[143] For example, Scheme 25 depicts the CM reaction of two similar styrenes (Type I) that results in a mixture of the desired product **53** and the homodimers **54** and **55**, which demonstrates a crucial limitation with specific types of CM reactions.

Type I terminal alkenes readily react with styrenes, as do α,β-unsaturated ketones **56**, terminal nitro alkenes **57**, α,β-unsaturated esters **58,** and vinylphosphonates **59** (i.e., alkenes of Types II and III) (Scheme 26).[144–146] These reactions are typically catalyzed by **Ru-4** or related second-generation catalysts, and provide the expected CM products in good-to-high yields and with stereoselectivities in favor of the (E)-isomer. Typically, the cheaper or more available CM partner is used in excess to diminish the homodimerization process of the other (more valuable) partner, thereby maximizing the yield for the desired CM product. Because many of the popular CM partners (acrylates, methyl vinyl ketone, acrolein, 1,4-dichloro-2-butene) are volatile, they can be readily removed from the reaction mixture, even when they are employed in excess.

For CM reactions between 2-substituted styrenes and simple Type I alkenes, the stoichiometry can dramatically impact the yield.[36] 2-Fluorostyrene (**60**) reacts

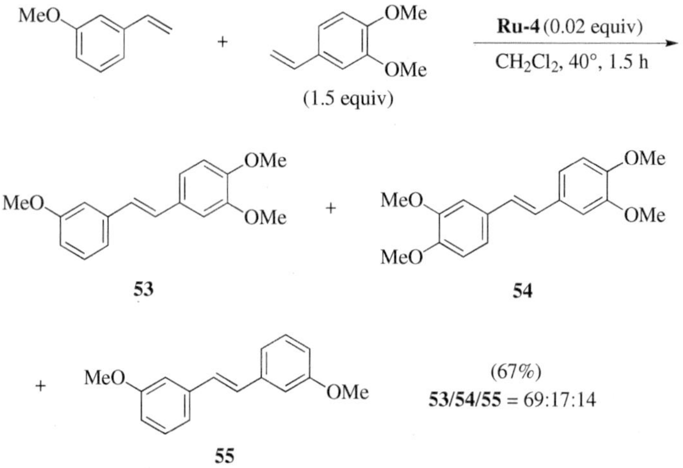

MeO + OMe OMe (1.5 equiv) Ru-4 (0.02 equiv) CH_2Cl_2, 40°, 1.5 h

OMe MeO OMe 53 + OMe MeO MeO OMe 54

+ MeO OMe 55 (67%) 53/54/55 = 69:17:14

Scheme 25

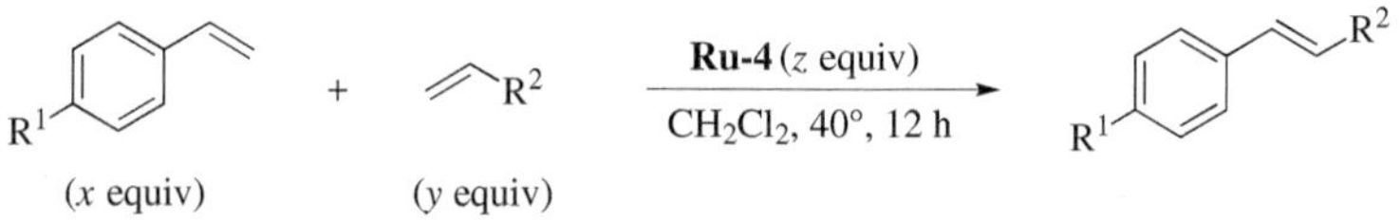

R^1 (x equiv) + R^2 (y equiv) Ru-4 (z equiv) CH_2Cl_2, 40°, 12 h R^1 R^2

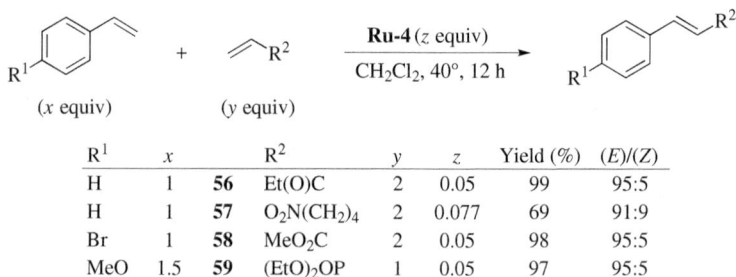

R^1	x		R^2	y	z	Yield (%)	$(E)/(Z)$
H	1	**56**	Et(O)C	2	0.05	99	95:5
H	1	**57**	$O_2N(CH_2)_4$	2	0.077	69	91:9
Br	1	**58**	MeO_2C	2	0.05	98	95:5
MeO	1.5	**59**	$(EtO)_2OP$	1	0.05	97	95:5

Scheme 26

with two equivalents of 1,4-diacetoxy-2-butene (**62**) (a cheap and easily removed CM partner) to furnish the expected product **63** in 98% yield and with exclusive (*E*)-selectivity (Scheme 27).[36] A nearly quantitative yield (98%) of CM product **64** is obtained in the reaction between 2-bromostyrene (**61**) and hex-5-en-1-yl acetate (**18**) when the former is used in excess. However, the yield decreases to 80% when a 1:1 ratio of **61** and **18** is employed.

Similar to styrenes, vinyl-substituted heterocycles, such as **65**, are often used in cross-metathesis reactions (Scheme 28).[147] These molecules are classified as Type I or II partners and readily participate in CM transformations, usually proceeding in good yields and (*E*)/(*Z*) selectivities. Substrates containing heteroatoms that can coordinate to the metal catalysts (pyridine, thiophene, etc.) can be problematic, but such issues can often be circumvented with either an additive (e.g., a Lewis acid) or by modifying the reaction conditions (e.g., concentration, solvent, or the use of microwave irradiation).[148]

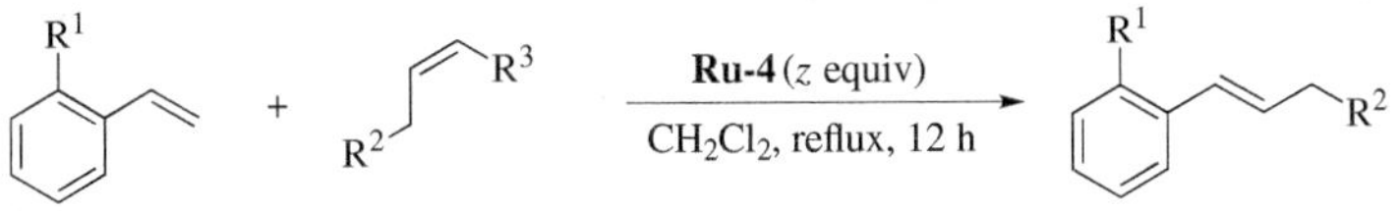

(x equiv) + (y equiv) —Ru-4 (z equiv), CH_2Cl_2, reflux, 12 h→

	R^1	x		R^2	R^3	y	z		Yield (%)	(E)/(Z)
60	F	1	**62**	AcO	$AcOCH_2$	2	0.05	**63**	98	>99:1
61	Br	3	**18**	$AcO(CH_2)_3$	H	1	0.045	**64**	98	>99:1
61	Br	1	**18**	$AcO(CH_2)_3$	H	1	0.045	**64**	80	>99:1

Scheme 27

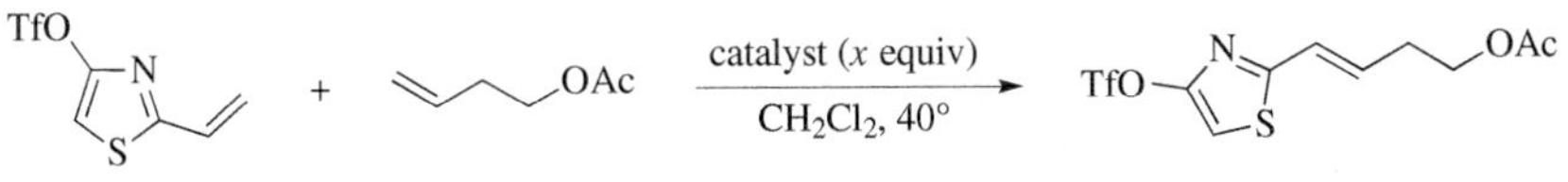

65 + (1.5 equiv) —catalyst (x equiv), CH_2Cl_2, 40°→

Cat.	x	Time (h)	Yield (%)	(E)/(Z)
Ru-4	0.1	18	63	89:11
Ru-5	0.05	24	58	87:13

Scheme 28

Unsaturated Nitriles. Alkenes bearing a cyano group are particularly useful CM partners because this substituent can be readily transformed into other functional groups, such as carboxylic acids, amides, and amines. The most common unsaturated nitrile used in CM reactions is acrylonitrile (Scheme 29).[149–151] Because acrylonitrile possesses a very electron-deficient double bond, it does not homodimerize easily and is classified as a Type III partner, which permits good chemoselectivity with most cross-coupling partners of Types I and II. Nevertheless, the use of acrylonitrile in a large excess is not advisable, as it can poison the more commonly employed ruthenium catalysts.[152] Interestingly, in contrast to the reactions with acrylic esters, the (*Z*)-product typically predominates in the CM reactions with acrylonitrile, albeit the stereoselectivity is rather poor ((*E*)/(*Z*) = 1:3 to 1:1).

Examples of CM with acrylonitrile reveal that the best results are achieved using Schrock's catalyst **Mo-1**.[149] Grubbs' second-generation catalyst usually affords lower yields compared to the **Ru-5** complex,[153] although the inclusion of copper(I) chloride can improve the product yields.[150] Surprisingly, a reasonable yield (61%) is observed for the CM reaction between acrylonitrile and methyl methacrylate (Type III) in the presence of the Grubbs' second-generation catalyst **Ru-28**, albeit the addition of the Lewis acid titanium(IV) isopropoxide (20 mol %) is necessary.[151] Cyclic alkyl amino carbene catalysts have also proven successful in this regard.[152]

The ability to employ a higher nitrile homologue, allyl cyanide, in CM has proven particularly challenging (Scheme 30).[154] The ligation of the CN group to the ruthenium is thought to decrease the efficiency of unsaturated nitriles as cross-metathesis partners.

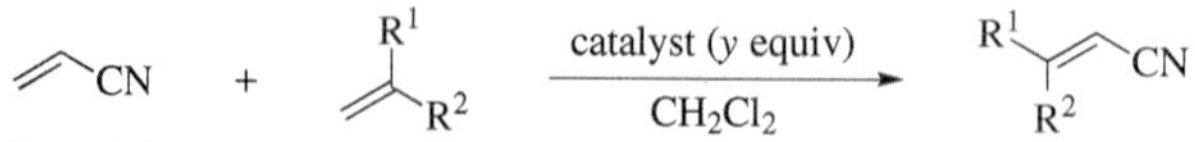

R^1	R^2	x	Cat.	y	Additive	Temp (°)	Time (h)	Yield (%)	(E)/(Z)
H	$TMSCH_2$	2	**Mo-1**	0.05	—	rt	3	76	25:75
H	$Br(CH_2)_3$	2	**Mo-1**	0.05	—	rt	3	45	12:88
H	$HO(CH_2)_3$	1	**Ru-4**	0.01	—	40	overnight	29	—
H	$HO(CH_2)_3$	1	**Ru-4**	0.01	CuCl	40	overnight	53	—
Me	MeO_2C	1	**Ru-28**	0.02	—	45	12	36	50:50
Me	MeO_2C	1	**Ru-28**	0.02	$Ti(Oi\text{-}Pr)_4$	45	12	61	50:50

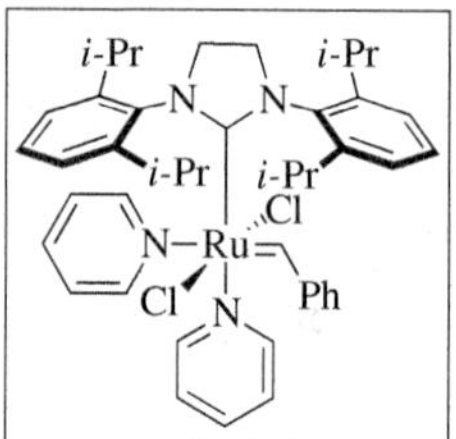

Ru-28

Scheme 29

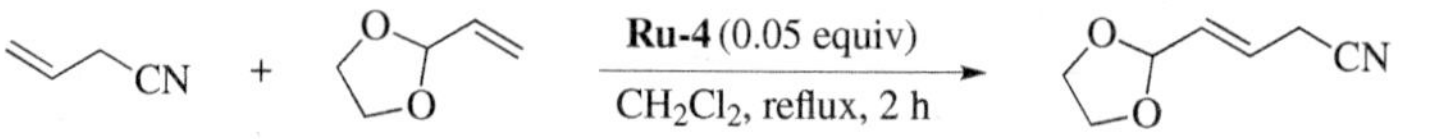

(2 equiv) **21** (26%) (E)/(Z) = 71:29

Ru-4 (0.05 equiv), CH_2Cl_2, reflux, 2 h

Scheme 30

Unsaturated Aldehydes. Acrolein, the simplest unsaturated aldehyde (Type II), is difficult to work with because of its unpleasant odor and ease of polymerization, but it can be readily replaced by crotonaldehyde.[155] Although CM reactions with acrolein are catalyzed by **Ru-1**,[156] the second-generation Grubbs and Hoveyda–Grubbs catalysts provide improved results in most reactions; one such CM reaction is illustrated in Scheme 31.[157]

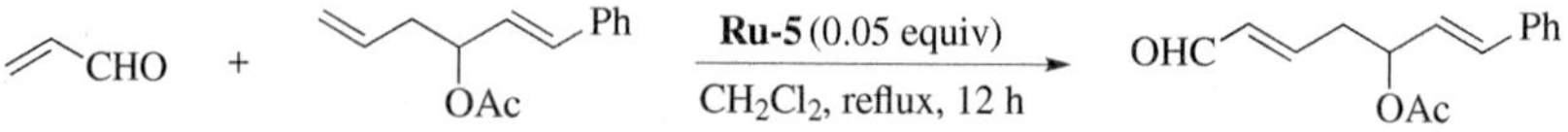

Ru-5 (0.05 equiv), CH_2Cl_2, reflux, 12 h

(3 equiv) (71%) (E)/(Z) = 97:3

Scheme 31

Methacrolein can be employed in CM reactions with Grubbs second-generation complex **Ru-4** as the catalyst.[36] This transformation is characterized by high stereoselectivity in favor of the (*E*)-isomer (Scheme 32). Notably, methacrolein is a Type III alkene and is thus very slow to homodimerize.

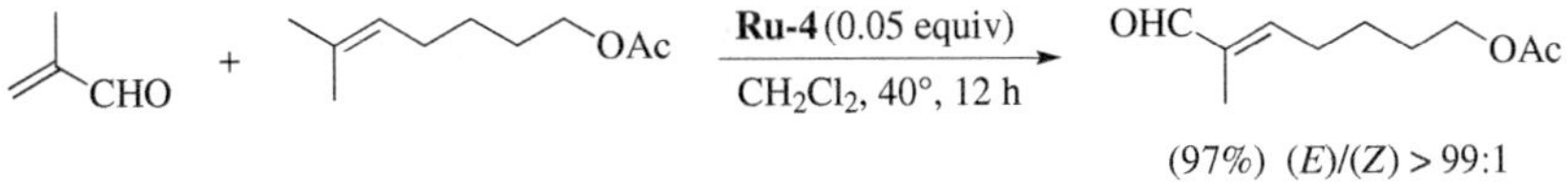

CHO + OAc —**Ru-4** (0.05 equiv), CH_2Cl_2, 40°, 12 h→ OHC OAc (97%) (*E*)/(*Z*) > 99:1

Scheme 32

Alkenes containing a formyl group in an allylic or more distant position have reactivity similar to terminal alkenes (Type I) and can be used in CM without difficulty.[158,159]

Unsaturated Ketones. α,β-Unsaturated ketones are excellent Type II partners in cross-metathesis. Methyl vinyl ketone reacts in good yield and with high stereoselectivity ((*E*)/(*Z*) = 95:5) with alkenes of Type I (e.g., **66**, Scheme 33) and Type II. Hoveyda–Grubbs-type complexes usually provide the best results.[160] However, reactions of α,β-unsaturated ketones with less reactive cross-partners are more problematic, as is exemplified in the CM of ketones **56** and **67** with **68**, which afford poor geometrical control.[161]

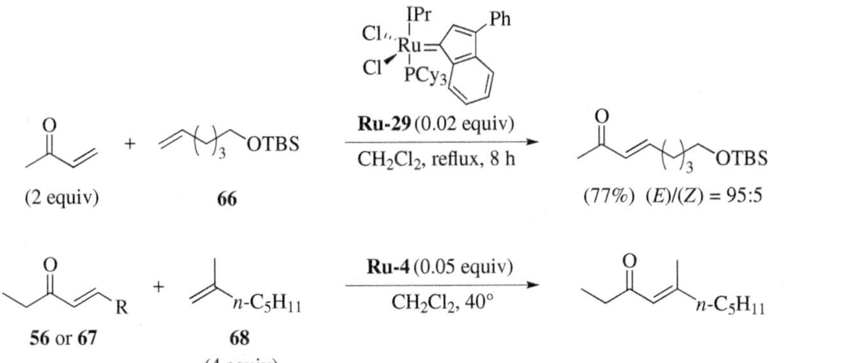

IPr, Cl, Cl, Ru, PCy_3, Ph

(2 equiv) + OTBS **66** —**Ru-29** (0.02 equiv), CH_2Cl_2, reflux, 8 h→ OTBS (77%) (*E*)/(*Z*) = 95:5

56 or **67** (R) + n-C_5H_{11} **68** (4 equiv) —**Ru-4** (0.05 equiv), CH_2Cl_2, 40°→ n-C_5H_{11}

	R	Time (h)	Yield (%)	(*E*)/(*Z*)
56	H	12	26	67:33
67	Me	3	68	67:33

Scheme 33

Terminal alkenes containing a ketone moiety in a distal position are classified as Type I alkenes and react smoothly with cross partners of Types I–III.[162–164]

Unsaturated Carboxylic Acids and Their Derivatives. Applications of acrylic acid derivatives in cross-metathesis are widespread; these compounds, along with α,β-unsaturated acids, acid chlorides, esters, and amides, are Type II alkenes.

CM reactions between acrylic acid derivatives and terminal alkenes provide products in good-to-excellent yields and with very high (*E*)/(*Z*) selectivities.

A study on the reactivity of α,β-unsaturated amides with other types of alkenes is presented in Scheme 34.[165,166] Unsaturated amides are more demanding partners than esters and thus require more forcing conditions, such as reflux temperatures or higher catalyst loadings.[166] Unsaturated amides such as **69** and **70** react efficiently with terminal alkenes using Grubbs (**Ru-4**) and Hoveyda–Grubbs (**Ru-5**) complexes to afford the products with excellent (*E*)-selectivities.

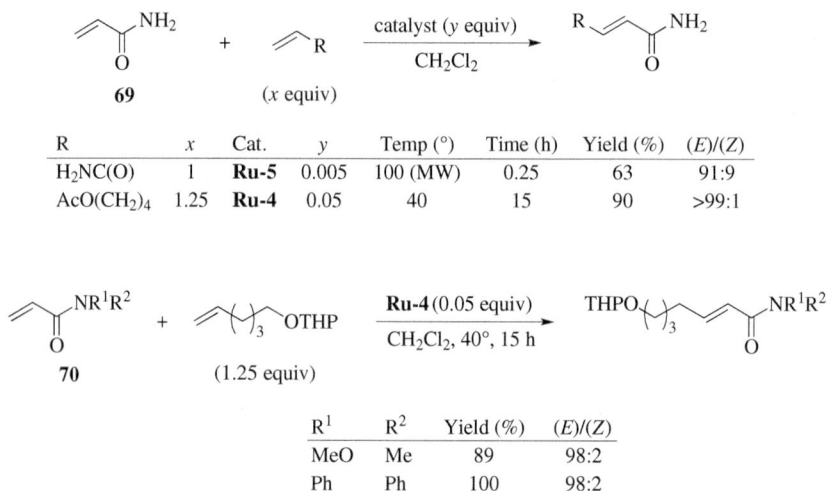

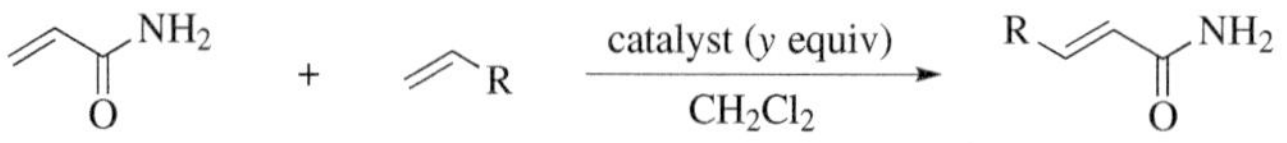

69 + R-alkene (*x* equiv) → catalyst (*y* equiv), CH_2Cl_2

R	*x*	Cat.	*y*	Temp (°)	Time (h)	Yield (%)	(*E*)/(*Z*)
$H_2NC(O)$	1	**Ru-5**	0.005	100 (MW)	0.25	63	91:9
$AcO(CH_2)_4$	1.25	**Ru-4**	0.05	40	15	90	>99:1

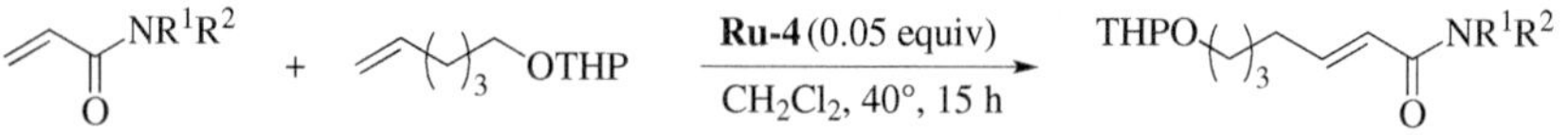

70 + OTHP alkene (1.25 equiv) → **Ru-4** (0.05 equiv), CH_2Cl_2, 40°, 15 h

R^1	R^2	Yield (%)	(*E*)/(*Z*)
MeO	Me	89	98:2
Ph	Ph	100	98:2

Scheme 34

Acrylic acid is a Type II alkene and is well tolerated by most ruthenium catalysts. However, because of the electron-deficient character of the double bond, some transformations with less reactive alkenes are challenging. The reactions of acrylic acid with allylsilanes **71** and **72**, as well as with the allylbenzene derivative **73**, proceed in very good yields and with high (*E*)-selectivities (≥95:5) with second-generation ruthenium catalysts (**Ru-4** and **Ru-5**) (Scheme 35).[36,167,168] In the case of methacrylic acid (a Type III alkene), the CM yield decreases because of the increased steric bulk on this alkene component.[36]

Compounds bearing a carboxyl functionality at an allylic position (as in **74**) or in a more distant position typically exhibit the same reactivity as simple terminal alkenes (Scheme 36).[169]

Cross-metathesis reactions of α,β-unsaturated esters and lactones provide important intermediates that are frequently used in target-oriented syntheses.[34] In this class of molecules, the most common substrates are acrylate esters **38** and **58**, which are classified as Type II alkenes (Scheme 37).[36,161,170,171] These substrates react with various alkenes using a range of catalysts, including Grubbs- (**Ru-4**), indenylidene- (**Ru-29**), and Hoveyda–Grubbs-type (**Ru-20**) complexes. Since the methyl and

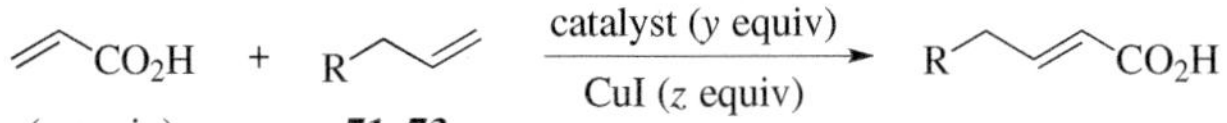

(*x* equiv) **71–73**

	R	*x*	Cat.	*y*	*z*	Solvent	Temp	Time (h)	Yield (%)	(*E*)/(*Z*)
71	TMS	1	**Ru-5**	0.05	—	CH_2Cl_2	rt	12	60	97:3
72	Ph_3Si	1	**Ru-5**	0.05	—	CH_2Cl_2	rt	12	70	97:3
73	2-$TBSOC_6H_4$	3	**Ru-4**	0.02	0.03	Et_2O	35°	3	82	>95:5

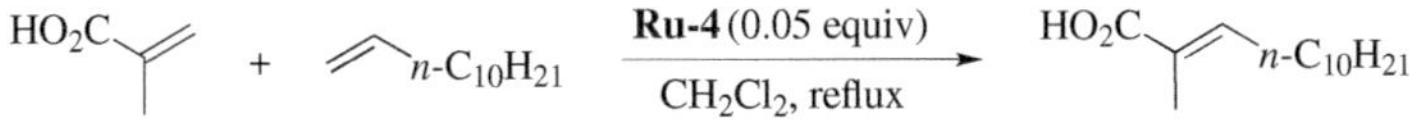

(1.1 equiv) (23%) (*E*)/(*Z*) = 80:20

Scheme 35

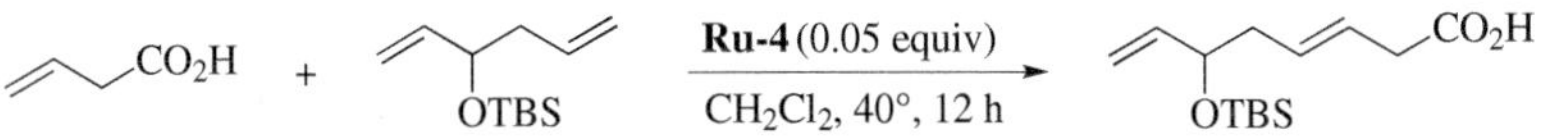

74 (3 equiv) (77%) (*E*)/(*Z*) = 75:25

Scheme 36

tert-butyl acrylic acid esters are volatile, they can be used in excess and then removed under vacuum after the reaction is complete. In general, α,β-unsaturated esters react with alkene partners to afford (*E*)-products with excellent stereoselectivities ((*E*)/(*Z*) = 95:5 to 99:1).

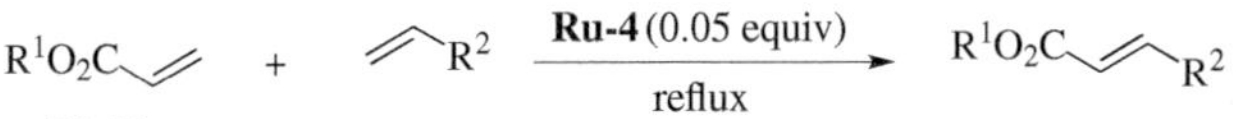

58, 38
(*x* equiv)

	R^1	*x*	R^2	Solvent	Time (h)	Yield (%)	(*E*)/(*Z*)
58	Me	2	Ac	CH_2Cl_2	3	41	91:9
38	*t*-Bu	excess	*t*-Bu	—	12	73	—

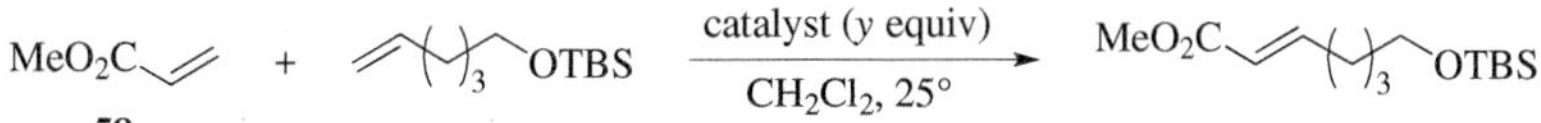

58
(*x* equiv)

x	Cat.	*y*	Time (h)	Yield (%)	(*E*)/(*Z*)
0.5	**Ru-20**	0.01	0.5	95	95:5
2	**Ru-29**	0.02	3	88	>95:5

Scheme 37

Methylenelactones **75**[172] and **77**[173] (Type III alkenes) seem to be rather problematic partners in CM reactions, but transformations with less demanding substrates such as **71**, **76**, or **78** have provided good results (Scheme 38). In the case of lactone

75, 5 mol % of Grubbs' second-generation catalyst was needed to obtain the desired product in high yield (93%). Compound **77** was more problematic,[173] and in the metathesis reaction with 4-methylpentene, the catalysts **Ru-1** and **Ru-5** did not yield any of the desired product. The use of **Ru-4** was more successful, providing the desired product in up to 55% yield, along with unreacted substrate **77** and its isomer, 3-methyl-2(5*H*)-furanone. In an effort to further improve the yield, the influence of different additives was evaluated, and the Lewis acid *B*-chlorocatecholborane (**79**) was found to provide the best results. In the presence of this additive and **Ru-4**, the metathesis reactions of **71** and **78** with **77** generate the desired products in 81% and 59% yields, respectively (Scheme 38).

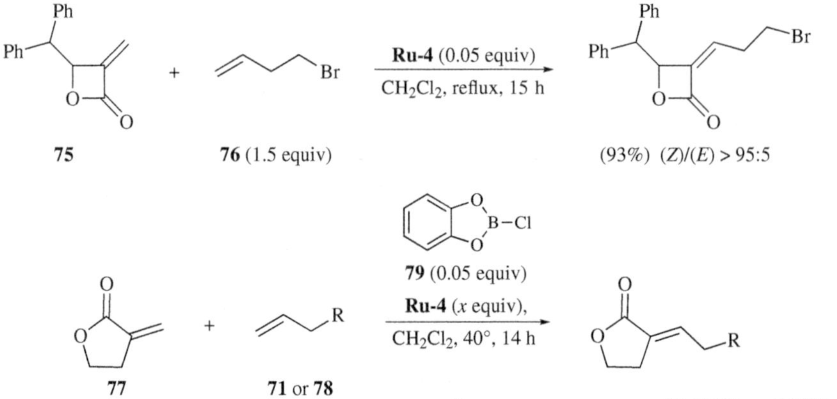

75 + **76** (1.5 equiv) → **Ru-4** (0.05 equiv), CH_2Cl_2, reflux, 15 h → (93%) (*Z*)/(*E*) > 95:5

77 + **71** or **78** (1.5 equiv) → **79** (0.05 equiv), **Ru-4** (*x* equiv), CH_2Cl_2, 40°, 14 h

	R	*x*	Yield (%)	(*E*)/(*Z*)
71	TMS	2 × 0.025	81	83:17
78	$(EtO)_2OP$	2 × 0.025	59	83:17

Scheme 38

Interestingly, the cross-metathesis of highly reactive acryloyl chloride (**80**) is efficiently catalyzed by the Hoveyda–Grubbs second-generation catalyst **Ru-5**, providing a product that can be easily functionalized further. To this end, the freshly distilled acryloyl chloride (**80**) reacts with various terminal alkenes (such as **71, 81–83**) to provide an efficient one-pot preparation of α,β-unsaturated amides (Scheme 39).[174]

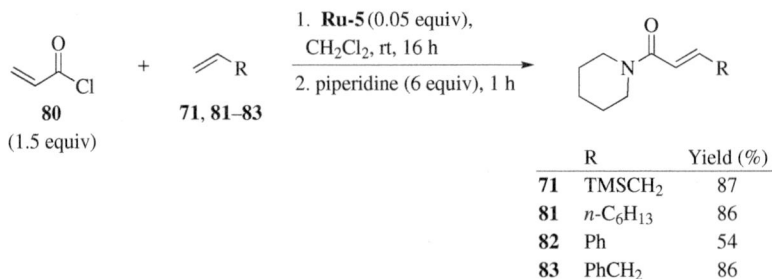

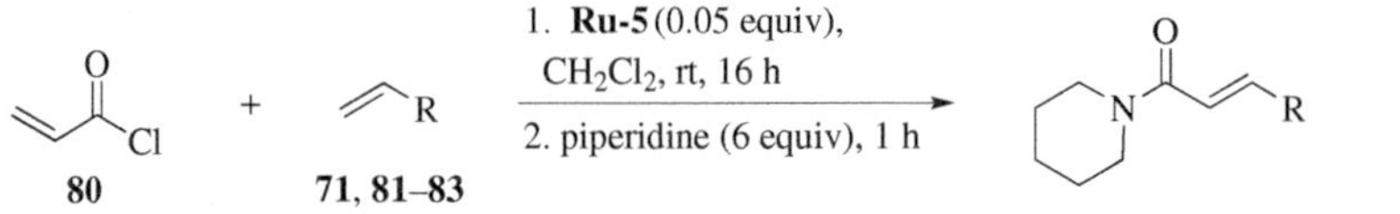

80 (1.5 equiv) + **71, 81–83** → 1. **Ru-5** (0.05 equiv), CH_2Cl_2, rt, 16 h; 2. piperidine (6 equiv), 1 h

	R	Yield (%)
71	$TMSCH_2$	87
81	n-C_6H_{13}	86
82	Ph	54
83	$PhCH_2$	86

Scheme 39

Alkenes That Have Nitrogen-Based Substituents. Alkenes containing a basic nitrogen atom (e.g., an unsaturated amine) are problematic substrates in cross-metathesis reactions because they form stable complexes with ruthenium-based catalysts. To circumvent this problem, the basic nitrogen is usually protected with Ts, Boc, Cbz, or other electron-withdrawing groups, or it is converted into an ammonium salt.

Allylic and Homoallylic Amines. *N*-Protected homoallyl- and allylamines (e.g., **84** and **86**, respectively) are classified as Type I alkenes. As illustrated in Scheme 40, Hoveyda–Grubbs catalyst **Ru-5** promotes CM reactions between the *N*-protected homoallylamine **84** with alkenes of Type II (e.g., **85**) in good yields (87% for R = *n*-Bu and 89% for R = Et).[175] Similarly, the *N*-protected allylamine **86** participates in CM with dec-9-en-1-yl benzoate (**15**) with Grubbs' first-generation complex **Ru-1** to furnishes **87** in 71% yield, albeit with moderate stereoselectivity ((*E*)/(*Z*) = 75:25) (Scheme 40).[176] Allylamines with additional substitution at the allylic position are classified as Type III alkenes and are challenging partners for this process. For example, *N*-benzoyl amine **88** reacts with Type I alkene **89** in

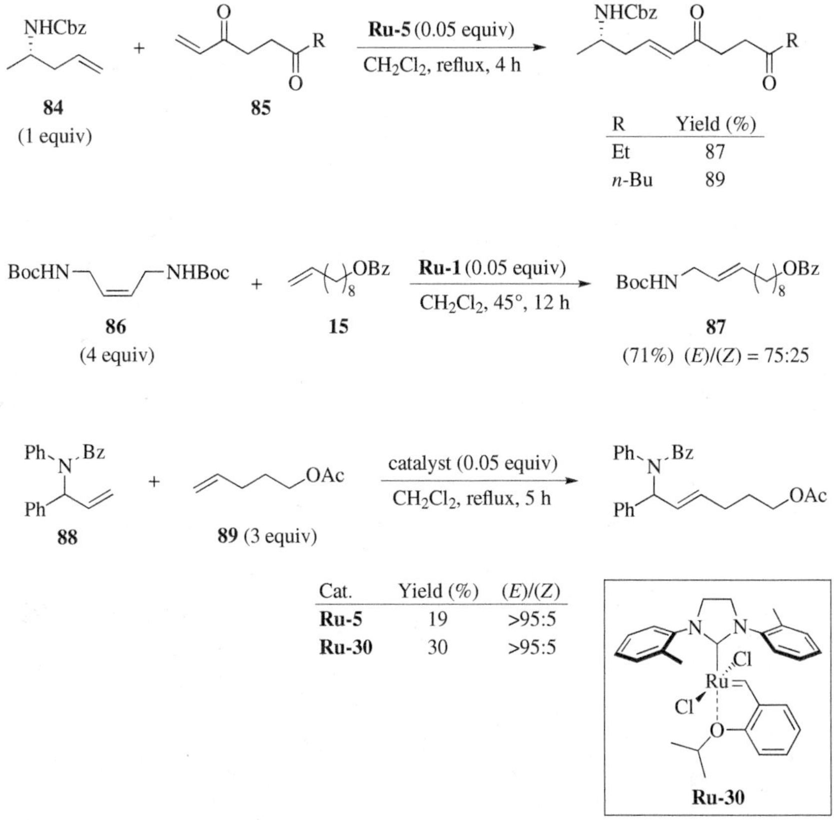

NHCbz + 85 → **Ru-5** (0.05 equiv), CH_2Cl_2, reflux, 4 h

84 (1 equiv) **85**

R	Yield (%)
Et	87
n-Bu	89

BocHN–NHBoc + OBz → **Ru-1** (0.05 equiv), CH_2Cl_2, 45°, 12 h → BocHN OBz

86 (4 equiv) **15** **87** (71%) (*E*)/(*Z*) = 75:25

88 + OAc **89** (3 equiv) → catalyst (0.05 equiv), CH_2Cl_2, reflux, 5 h → Ph, N, Bz, Ph, OAc

Cat.	Yield (%)	(*E*)/(*Z*)
Ru-5	19	>95:5
Ru-30	30	>95:5

Ru-30

Scheme 40

the presence of either Hoveyda–Grubbs second-generation complex **Ru-5** or its analogue, **Ru-30**, to produce the CM product in low-to-moderate yield, but with good (*E*)/(*Z*) selectivity.[177]

Unsaturated Azides. Unsaturated azides are uncommon as CM substrates. In one example, the bisazide **90** is subjected to CM using third-generation Grubbs catalyst **Ru-31** to furnish the product in good yield (Scheme 41).[178]

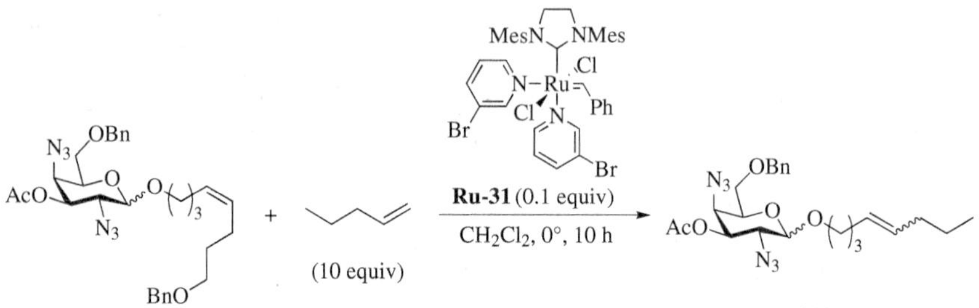

90 + (10 equiv) → **Ru-31** (0.1 equiv), CH_2Cl_2, 0°, 10 h → (84%)

Scheme 41

Unsaturated Nitro Compounds. A nitro group can provide access to many other valuable functional groups. Although nitro-2-propene (Scheme 42) and higher homologues (**92**; see Scheme 43 below),[146] have been employed successfully in CM reactions, the use of nitroethylene has not been reported. According to the general model of reactivity, nitroethylene should be classified as a Type IV alkene (spectator), nitro-2-propene as a Type III alkene, and higher aliphatic nitro compounds as Type I alkenes. Nitro-2-propene undergoes CM with Type I substrates such as terminal alkene **66** to give the desired product in low-to-moderate yield (Scheme 42).[179] The most useful catalyst for this transformation is the nitro-analogue of Hoveyda–Grubbs second-generation catalyst, **Ru-20**; in this case, a Lewis acid (0.1 equiv) is required to prevent the propagating catalytic species from forming the stable five-membered chelate **91** (Figure 10), which inhibits the CM reaction.

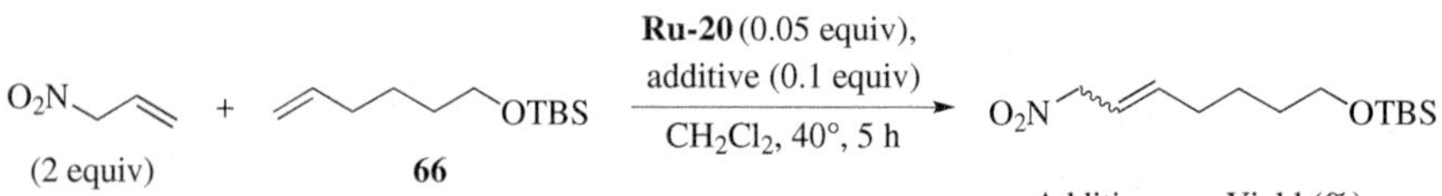

O_2N (2 equiv) + **66** (OTBS) → **Ru-20** (0.05 equiv), additive (0.1 equiv), CH_2Cl_2, 40°, 5 h → O_2N … OTBS

Additive	Yield (%)
$Ti(Oi\text{-}Pr)_4$	16
Ph_2SnCl_2	58
$BF_3{\cdot}Et_2O$	58
$B(OMe)_3$	26
$B(OPh)_3$	68

Scheme 42

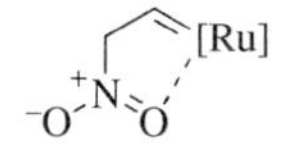

Figure 10. Structure of the chelate that forms in the absence of a Lewis acid.

Grubbs second-generation catalyst **Ru-4** promotes CM of the higher homologues of 1-nitro-2-propene to afford the expected products in 51–85% yields (Scheme 43).[146,179]

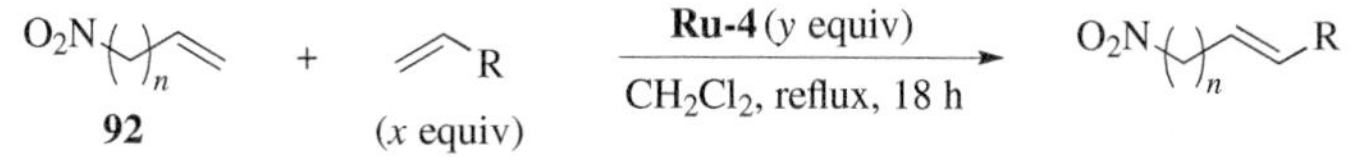

n	R	*x*	*y*	Yield (%)	(*E*)/(*Z*)
4	OHC	2.5	0.065	51	>99:1
4	*t*-BuO_2C	4	0.052	85	>99:1
2	Ph	4	0.048	68	>99:1
3	Ph	3.6	0.052	63	>99:1
4	Ph	3.5	0.077	69	>99:1

Scheme 43

Alkenes Containing an Oxygen Moiety. Free and protected alcohols, phenols, diols, ethers, acetals, epoxides, and similar oxygen-based functional groups are present in numerous CM partners.

Unsaturated Alcohols. In general, the free hydroxyl group of alcohols and phenols is compatible with all ruthenium catalysts. 2-Buten-1,4-diol (**93**) is often used as a replacement for 3-propen-1-ol,[180] and it is also used for CM in aqueous media.[181] Homoallylic alcohols and higher unsaturated alcohol analogues (Type I alkenes) undergo CM with most of the established catalysts. Secondary and tertiary allylic alcohols (Type II alkenes) usually engage in CM with reactive alkenes, namely, Type I, but the more challenging CM partners (e.g., 3-butenenitrile) afford lower yields (Scheme 44).[36,182] Unless (*Z*)-selective catalysts are used, (*E*)-products are formed with moderate-to-good selectivities.

In contrast, the (*Z*)-selective ruthenium-, tungsten-, and molybdenum-catalysts facilitate stereoselective CM reactions that favor the (*Z*)-configured products with very good selectivity (Scheme 45).[183]

(*Z*)-Selective molybdenum- and tungsten-complexes can be used in the synthesis of substituted allylic alcohols with excellent (*Z*)/(*E*)-selectivity (Scheme 46).[184] However, in these cases, the free hydroxyl group in the substrate must be protected prior to the CM step. Molybdenum-catalysts (e.g., **Mo-3**) afford very high (*Z*)-selectivities even in CM reactions involving sterically crowded secondary allylic silyl ethers (Type II alkenes), such as **94**, with terminal alkenes.

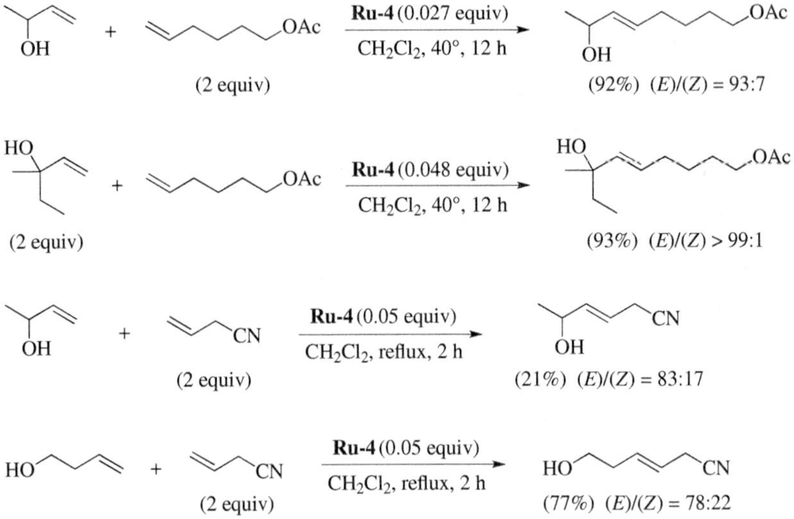

OH + OAc (2 equiv) — **Ru-4** (0.027 equiv), CH_2Cl_2, 40°, 12 h → OH, OAc (92%) (*E*)/(*Z*) = 93:7

HO (2 equiv) + OAc — **Ru-4** (0.048 equiv), CH_2Cl_2, 40°, 12 h → HO, OAc (93%) (*E*)/(*Z*) > 99:1

OH + CN (2 equiv) — **Ru-4** (0.05 equiv), CH_2Cl_2, reflux, 2 h → OH, CN (21%) (*E*)/(*Z*) = 83:17

HO + CN (2 equiv) — **Ru-4** (0.05 equiv), CH_2Cl_2, reflux, 2 h → HO, CN (77%) (*E*)/(*Z*) = 78:22

Scheme 44

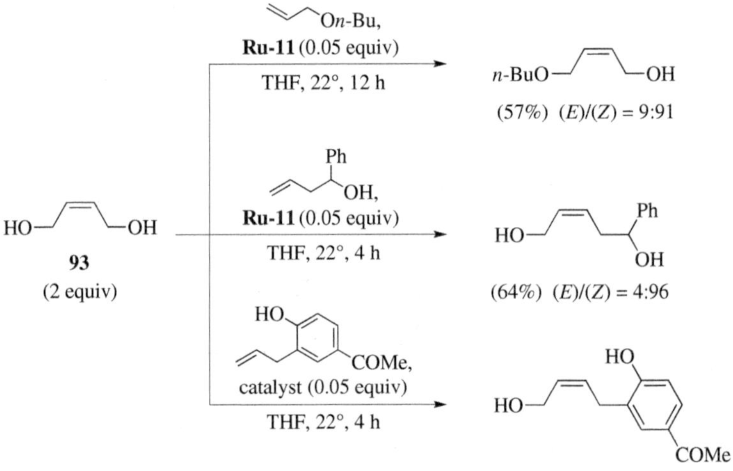

HO–, –OH **93** (2 equiv)

On-Bu, **Ru-11** (0.05 equiv), THF, 22°, 12 h → *n*-BuO–, –OH (57%) (*E*)/(*Z*) = 9:91

Ph, OH, **Ru-11** (0.05 equiv), THF, 22°, 4 h → HO–, Ph, OH (64%) (*E*)/(*Z*) = 4:96

HO, COMe, catalyst (0.05 equiv), THF, 22°, 4 h → HO–, HO, COMe

Cat.	Yield (%)	(*E*)/(*Z*)
Ru-8	68	2:98
Ru-11	50	18:82

Scheme 45

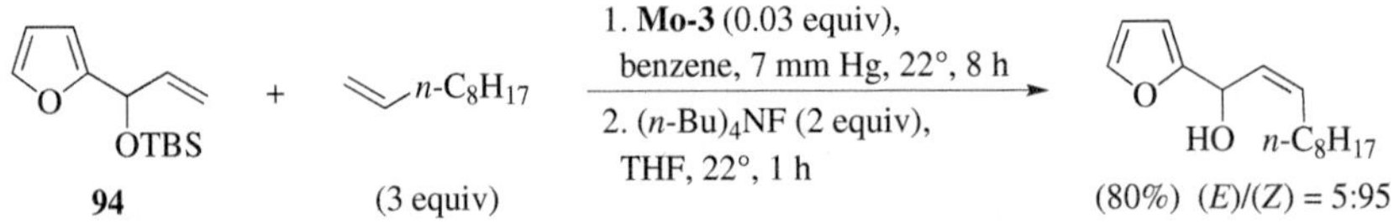

O, OTBS **94** + *n*-C_8H_{17} (3 equiv) — 1. **Mo-3** (0.03 equiv), benzene, 7 mm Hg, 22°, 8 h; 2. $(n\text{-}Bu)_4NF$ (2 equiv), THF, 22°, 1 h → O, HO *n*-C_8H_{17} (80%) (*E*)/(*Z*) = 5:95

Scheme 46

Unsaturated Ethers, Epoxides, and Peroxides. Vinyl ethers are problematic CM substrates for the current ruthenium-based catalysts, although there are some examples of successful RCM reactions with substrates that contain a vinyl ether fragment.[185] Using the standard Schrock catalysts, the reactions proceed in good yield, albeit with low (*E*)/(*Z*)-selectivity. However, modern (*Z*)-selective molybdenum catalysts make stereoselective CM involving vinyl ethers possible (Scheme 47).[5]

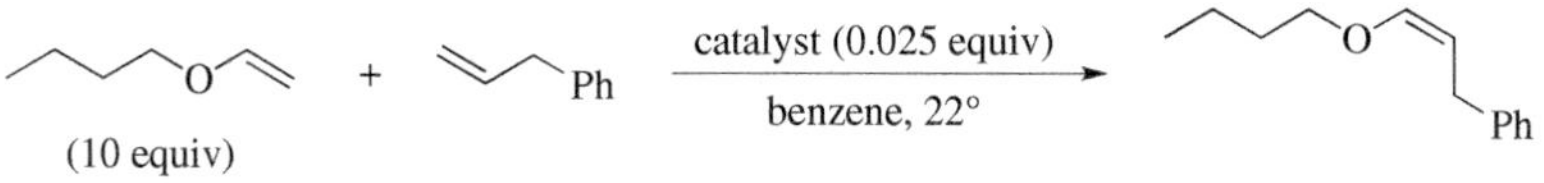

(10 equiv) + Ph — catalyst (0.025 equiv), benzene, 22° → Ph

Cat.	Yield (%)	(*E*)/(*Z*)
Ru-5	<2	—
Mo-1	80	52:48
Mo-6	73	2:98

Scheme 47

Allylic and homoallylic ethers, as well as higher protected unsaturated alcohols (Type I or II alkenes), are generally not problematic CM substrates and have been employed in numerous examples. In the case of CM reactions with Type II and III partners, bulky *O*-protected allylic alcohols react more slowly. This observation can be used to selectively functionalize the more reactive carbon–carbon double bond in a Type I alkene in the presence of an allylic alkene (Scheme 48).[169] Notably, trisubstituted allylic alcohols protected with bulky groups are unreactive in CM (i.e., they are Type IV alkenes).[36]

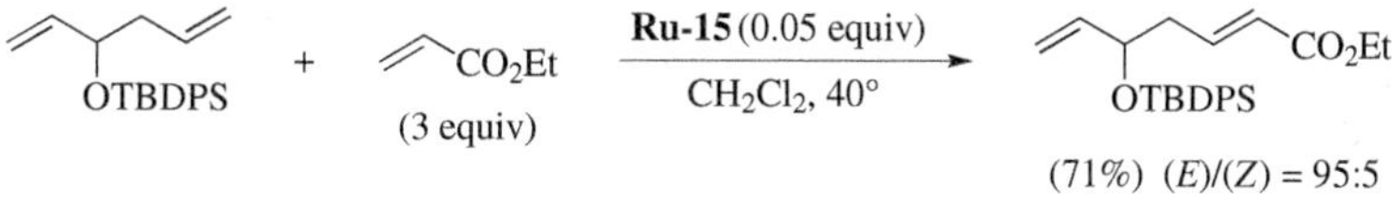

OTBDPS + CO_2Et (3 equiv) — **Ru-15** (0.05 equiv), CH_2Cl_2, 40° → OTBDPS, CO_2Et (71%) (*E*)/(*Z*) = 95:5

Scheme 48

Homodimerization of allylic ethers present in molecules of pharmaceutical interest (including some antibiotics, amino acids, peptides, carbohydrates, and steroids) is employed to prepare analogues with potentially increased biological activity (Scheme 49).[8,186–189]

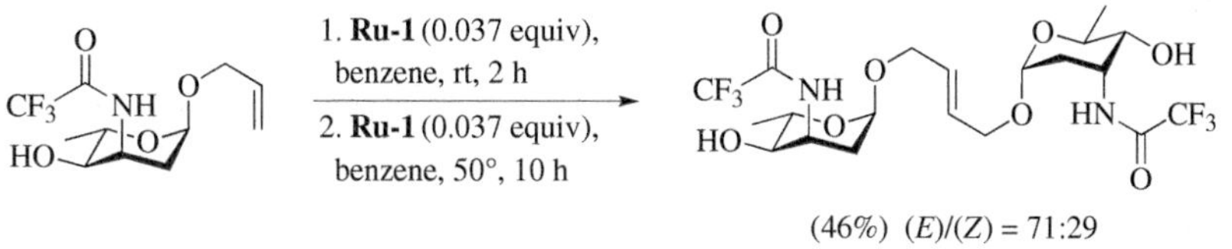

1. **Ru-1** (0.037 equiv), benzene, rt, 2 h
2. **Ru-1** (0.037 equiv), benzene, 50°, 10 h

(46%) (*E*)/(*Z*) = 71:29

Scheme 49

Epoxides are generally well tolerated by modern olefin metathesis catalysts; the epoxide may be present at various distances relative to the reacting carbon–carbon double bond. Nevertheless, in the case of vinyloxiranes, the epoxide functionality deactivates the carbon–carbon double bond toward CM. For example, the CM reaction between vinyloxirane **95** and the crowded alkene **96** affords only a modest yield of the desired product (Scheme 50).[190]

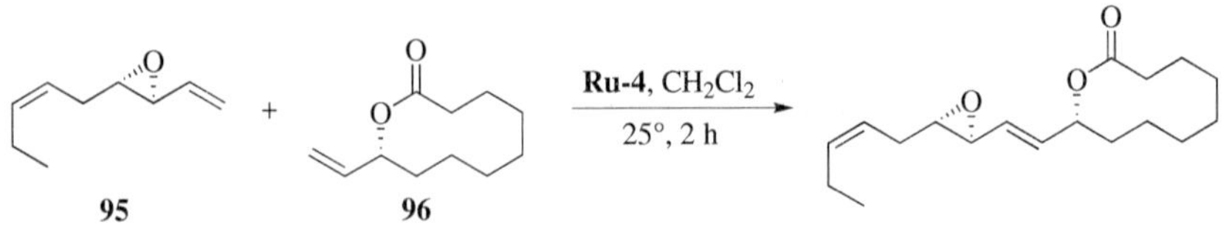

95 + **96** → **Ru-4**, CH_2Cl_2, 25°, 2 h → (40%) (*E*)/(*Z*) > 99:1

Scheme 50

Cross-metathesis of vinyl oxiranes with (*Z*)-selective catalysts (e.g., **Ru-8**) provides access to synthetically valuable (*Z*)-vinyloxiranes in moderate-to-good yields and with excellent stereoselectivities (Scheme 51).[191]

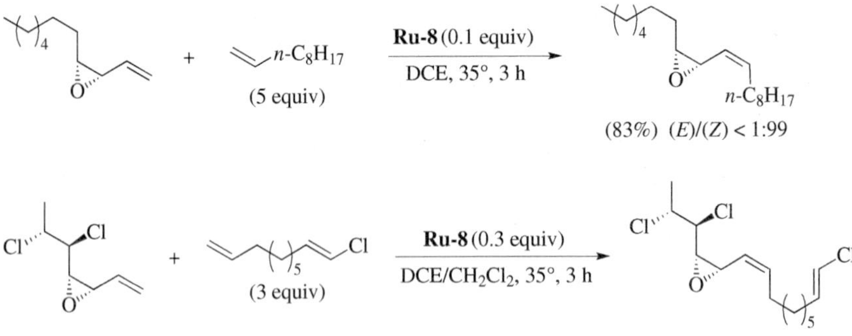

+ n-C_8H_{17} (5 equiv) → **Ru-8** (0.1 equiv), DCE, 35°, 3 h → n-C_8H_{17} (83%) (*E*)/(*Z*) < 1:99

+ Cl (3 equiv) → **Ru-8** (0.3 equiv), DCE/CH_2Cl_2, 35°, 3 h → (32%) (*E*)/(*Z*) = 5:95

Scheme 51

Many metathesis catalysts are destroyed by peroxides,[192–194] but some are compatible with medicinally important ones, including an artemisinin derivative[195] and tetraoxane fragment **97** (present in potential antimalarial drugs) (Scheme 52).[193]

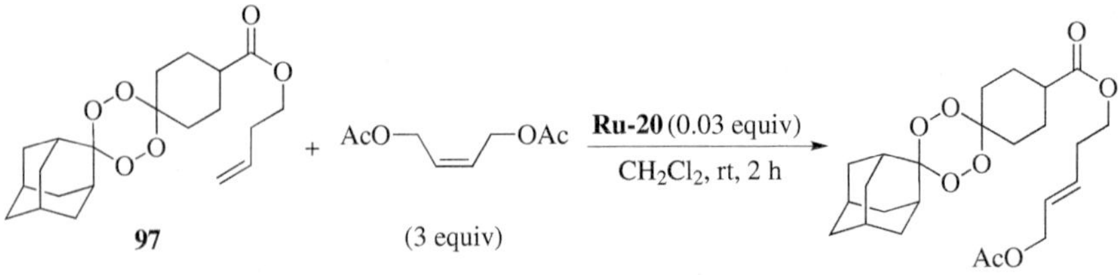

97 + AcO–CH=CH–OAc (3 equiv) → **Ru-20** (0.03 equiv), CH_2Cl_2, rt, 2 h → (87%) (*E*)/(*Z*) = 89:11

Scheme 52

The ability of the CM catalysts to tolerate the presence of a peroxide is significant because CM can be employed at a late stage in the synthesis of these biologically important compounds to readily enable analogue preparation.

Haloalkenes. The ability to employ vinyl halides and 1,2-dihaloethylenes (Type III alkenes) in CM has not been widely explored, but a handful of examples have been reported. Molybdenum or phosphine-free ruthenium complexes must be employed as catalysts, because Grubbs-type complexes are deactivated by haloalkenes.[196] Cross-metathesis reactions with (*E*)-1,2-dichloroethene (**98**) are performed neat in the presence of Hoveyda-type catalysts using an excess of **98** that can be recovered after the reaction.[104,105] The CM with **98** is not stereoselective with terminal alkenes such as **99**, but electron-rich alkenes provide the corresponding products with a high degree of (*Z*)-selectivity, as exemplified in the reaction with phenyl vinyl sulfide to generate **100** (Scheme 53).[105] Relatively high catalyst loading is required for the ruthenium-catalyzed CM with vinyl chlorides, and unfortunately, vinyl bromides are even less reactive in the analogous CM reactions.

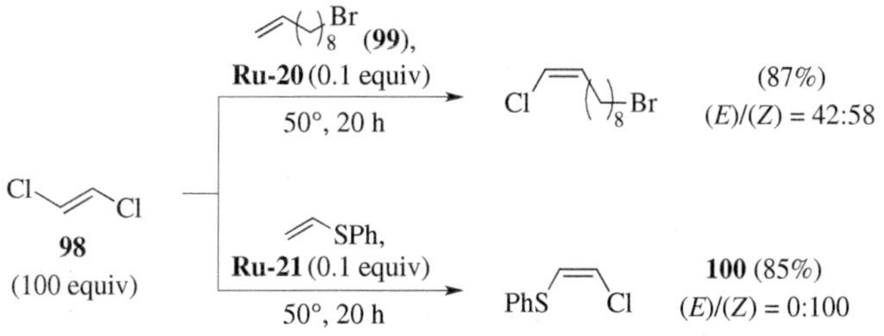

98 (100 equiv)
(**99**), **Ru-20** (0.1 equiv), 50°, 20 h → (87%) (*E*)/(*Z*) = 42:58
SPh, **Ru-21** (0.1 equiv), 50°, 20 h → **100** (85%) (*E*)/(*Z*) = 0:100

Scheme 53

The best yields and selectivities are obtained with the (*Z*)-selective molybdenum MAP catalysts and (*Z*)-1,2-dichloroethene (**101**). (*Z*)-Vinyl fluorides and even (*Z*)-vinyl bromides can be obtained with good-to-high stereocontrol (Scheme 54).[106] The reactivity of vinyl halides decreases in the order F > Cl > Br, and, in general, vinyl halides are challenging CM partners for ruthenium complexes, so the more active molybdenum catalysts are recommended in these cases.

Allyl halides (Type I) are very useful CM partners, and the commercially available (*Z*)-1,4-dichloro-2-butene serves as a convenient alternative to allyl chloride.[197–199] Halogen atoms at more distant positions do not impact the reactivity of an alkene double bond in CM.

Unsaturated Compounds Containing Silicon and Tin.

Vinylsilanes. In the presence of a metathesis catalyst, vinylsilanes undergo CM with various alkenes, leading to functionalized alkenylsilicon compounds,[200] although with some ruthenium hydride complexes, the same products can be obtained by an alternative mechanism that is based on a silylative coupling process.[201]

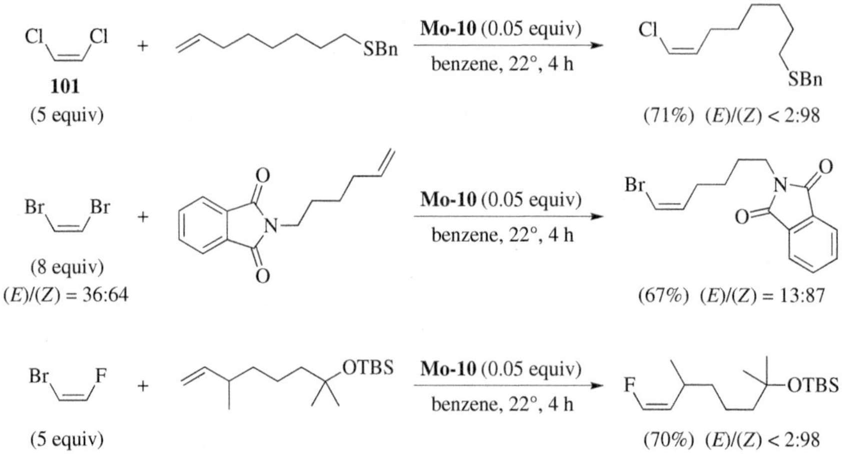

Cl/Cl (Z)-alkene **101** (5 equiv) + SBn-terminated alkene → **Mo-10** (0.05 equiv), benzene, 22°, 4 h → Cl-(Z)-alkene SBn product (71%) (*E*)/(*Z*) < 2:98

Br/Br alkene (8 equiv) (*E*)/(*Z*) = 36:64 + N-alkenylphthalimide → **Mo-10** (0.05 equiv), benzene, 22°, 4 h → Br-(Z)-alkene phthalimide product (67%) (*E*)/(*Z*) = 13:87

Br/F alkene (5 equiv) + OTBS alkene → **Mo-10** (0.05 equiv), benzene, 22°, 4 h → F-(Z)-alkene OTBS product (70%) (*E*)/(*Z*) < 2:98

Scheme 54

Vinylsiloxanes (Scheme 55),[202] vinylsilanes, polymeric oligosiloxanes,[202] divinylsilanes,[203] and vinyl-functionalized cubic silsesquioxanes (i.e., POSS)[204–207] also undergo CM reactions in the presence of Grubbs and Schrock catalysts.

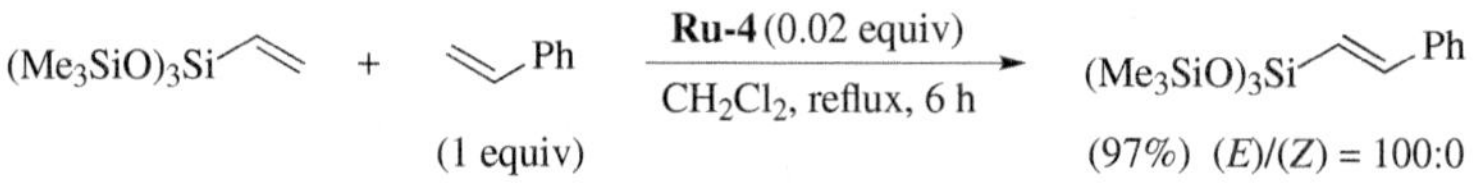

$(Me_3SiO)_3Si$-vinyl + styrene (1 equiv) → **Ru-4** (0.02 equiv), CH_2Cl_2, reflux, 6 h → $(Me_3SiO)_3Si$-CH=CH-Ph (97%) (*E*)/(*Z*) = 100:0

Scheme 55

Allylsilanes. Allylsilanes, such as allyltrimethylsilane[208–210] and allyltriphenylsilane,[168] are very useful CM partners. These substrates can be classified as Type I partners, because they undergo efficient homodimerization in the presence of both ruthenium and molybdenum catalysts.[108] Cross-metathesis affords functionalized allylsilane products with good (*E*)-selectivities, which represent useful synthetic intermediates for target-orientated synthesis (Scheme 56).[211]

Allyltriphenylstannane and allyltributylstannane successfully participate in CM reactions with a variety of alkenes using molybdenum catalyst **Mo-1**.[212] Moderate-to-good product yields are obtained in CM with a number of highly functionalized alkenes containing ester, cyano, acetal, and other functional groups. Notably, no significant quantities of allylstannane homodimerization products are formed in these experiments.[212]

Unsaturated Phosphorus Compounds.

Alkenes Containing a Phosphine–Borane Moiety. Free phosphines bind strongly to transition metals, which necessitates the protection of the Lewis basic phosphorus atom with a borane prior to a CM reaction. In the presence of the

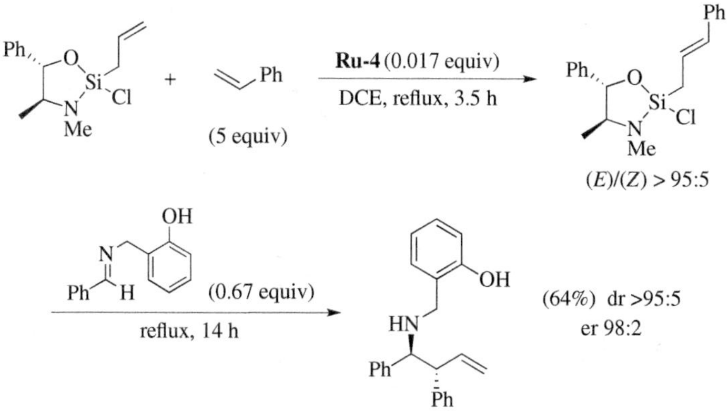

Scheme 56

catalyst **Ru-4**, vinyl- and allyl-substituted phosphine–boranes undergo smooth CM with various alkene partners, providing products with exclusive (*E*)-selectivity (Scheme 57).[213] Interestingly, no homodimerization of vinylphosphine–boranes (Type III substrates) is observed, whereas the allylic homologues (Type I) undergo facile dimerization.

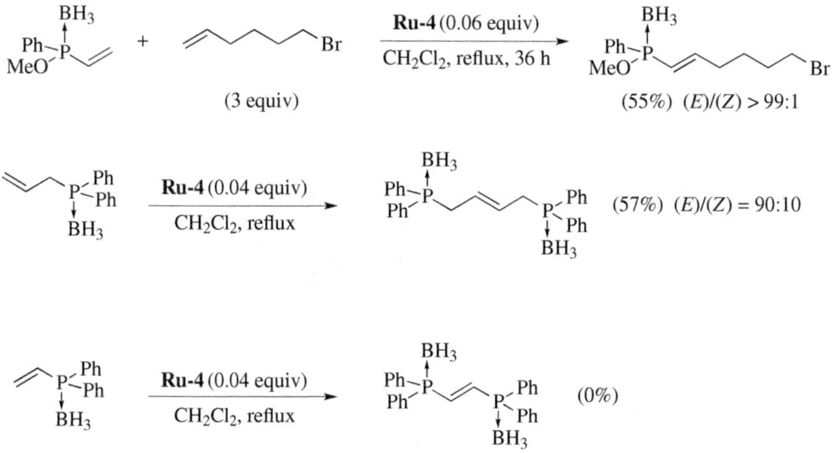

Scheme 57

Alkenes Containing a Phosphine Oxide Moiety. Cross-metathesis reactions of substituted vinyl and allylic phosphine oxides catalyzed by ruthenium complexes provide a number of valuable products, including some *P*-ligand precursors, with exclusive (*E*)-stereochemistry,[164,214–216] one example of which is shown in Scheme 58.[215]

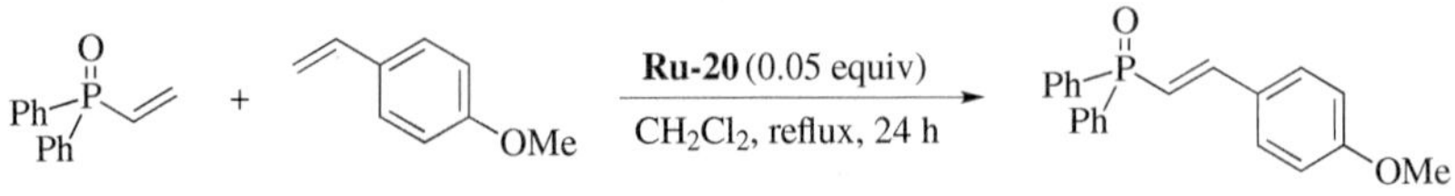

102 (3 equiv) Ru-20 (0.05 equiv), CH_2Cl_2, reflux, 24 h (81%) (*E*)/(*Z*) > 99:1

Scheme 58

Cross-metathesis functionalization of di- and tri(vinyl) phosphine oxides with various alkene partners is known as well.[217] This reaction permits direct access to novel racemic *P*-stereogenic products bearing two or three different alkenyl groups.

The usefulness of bisphosphines as bidentate ligands in catalysis has prompted the investigation of the homodimerization of unsaturated phosphine oxides. Despite many attempts, no dimerization of vinylphosphine oxide **102** has been achieved, thereby making it a Type III alkene partner. In contrast, the dimerization of the nonconjugated phosphine oxides **103** and **104** is promoted by **Ru-4** to afford the homodimeric products in high yields and with good (*E*)-selectivities (Scheme 59).[214]

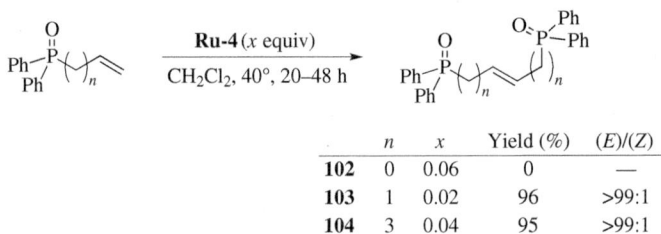

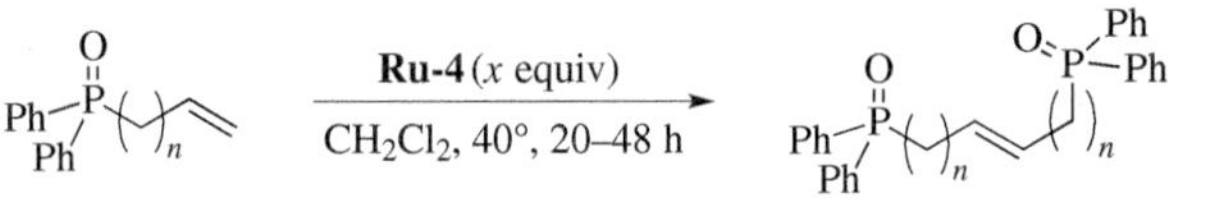

Ru-4 (*x* equiv), CH_2Cl_2, 40°, 20–48 h

	n	*x*	Yield (%)	(*E*)/(*Z*)
102	0	0.06	0	—
103	1	0.02	96	>99:1
104	3	0.04	95	>99:1

Scheme 59

Vinyl phosphine oxides that are less bulky than the diphenyl-substituted phosphine oxide **102** are successfully dimerized in the presence of the nitro analogue of the Hoveyda–Grubbs catalysts, which is exemplified in the dimerization of **105** (Scheme 60).[215] Hence, these substrates are between Types II and III in the Grubbs alkene classification for CM reactions.[215,216]

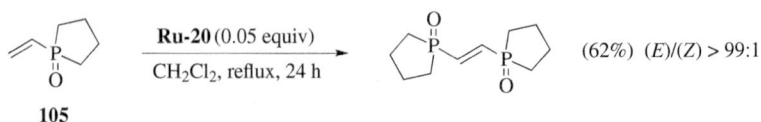

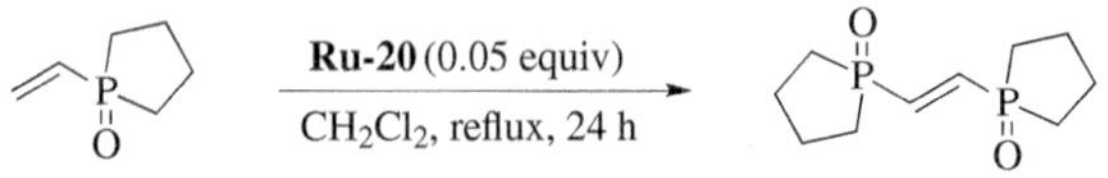

105 Ru-20 (0.05 equiv), CH_2Cl_2, reflux, 24 h (62%) (*E*)/(*Z*) > 99:1

Scheme 60

P-Stereogenic vinylphosphine oxides undergo CM and homodimerization reactions to furnish products with exclusive (*E*)-selectivities and without epimerization at the phosphorus center.[164] This finding provides a potentially useful method for preparing chiral, nonracemic, functionalized bisphosphine ligands, which are omnipresent in asymmetric synthesis.

Unsaturated Phosphonates. Examples of CM involving vinyl- and allylphosphonates and related compounds have been reported,[144,218–220] and allylphosphonates undergo homodimerization.[221] For example, the reaction between phosphonate **106** and an excess of 1-heptene (seven equivalents) proceeds smoothly in the presence of Grubbs catalyst **Ru-4** and copper(I) iodide, which acts as a phosphine scavenger (Scheme 61).[220] (Copper iodide is sometimes used as an additive to enhance the cross-metathesis reaction, acting as an activator for Grubbs-type phosphine catalysts[167]).

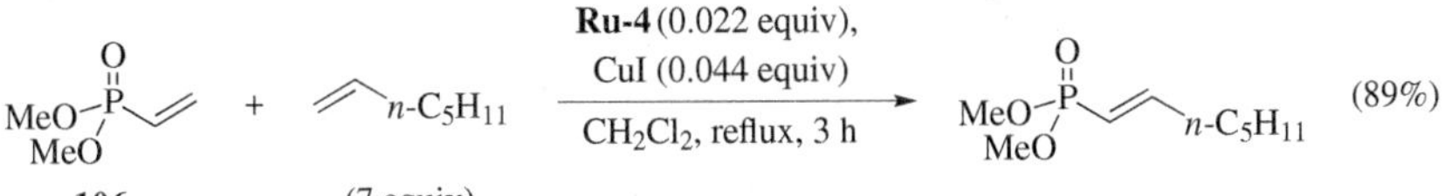

106 + n-C_5H_{11} (7 equiv) → **Ru-4** (0.022 equiv), CuI (0.044 equiv), CH_2Cl_2, reflux, 3 h → (89%)

Scheme 61

Alkenes Containing Sulfur and Selenium Moieties.

Unsaturated Thiols and Sulfides. Vinyl sulfides and selenides, with very few exceptions,[105,222,223] do not participate in olefin cross-metathesis reactions that are catalyzed by ruthenium complexes. In contrast, allyl sulfides—and even allyl mercaptan—undergo homodimerization,[224] and these compounds, along with their selenium analogues, are efficient partners in CM with various alkenes. Notably, these reactions can be conducted in aqueous media (Scheme 62),[225] which make them particularly useful for protein modification.[226] Interestingly, the reactivity of homoallyl sulfides is noticeably diminished as compared to allyl sulfides, likely owing to the formation of stable cyclic ruthenium chelates that attenuate the catalytic activity.[227]

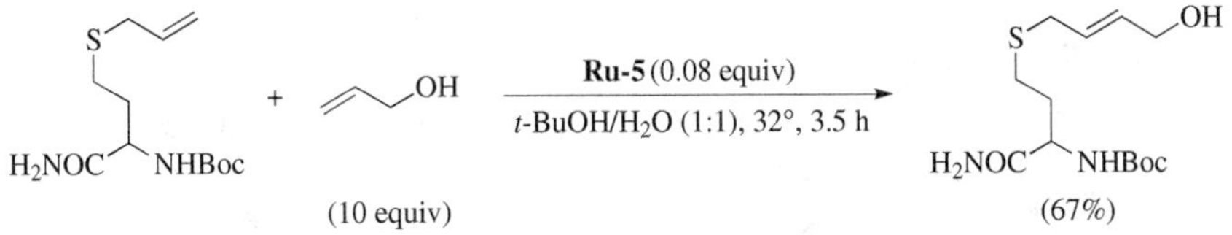

H_2NOC NHBoc + OH (10 equiv) → **Ru-5** (0.08 equiv), t-BuOH/H_2O (1:1), 32°, 3.5 h → H_2NOC NHBoc (67%)

Scheme 62

Unsaturated Sulfoxides. Both vinyl and allyl sulfoxides are unreactive in CM with terminal alkenes in the presence of ruthenium catalysts.[228,229] While the origin of the observed impotence is not clear, it has been suggested that the formation of ruthenium chelates may render the catalyst unreactive. However, higher homologues of unsaturated sulfoxides undergo cross-metathesis and homodimerization relatively easily (Scheme 63).[230]

Unsaturated Sulfones. The CM reactions of phenyl vinyl sulfone (Scheme 64)[231] or divinyl sulfone (both Type III alkenes) with a variety of terminal alkenes

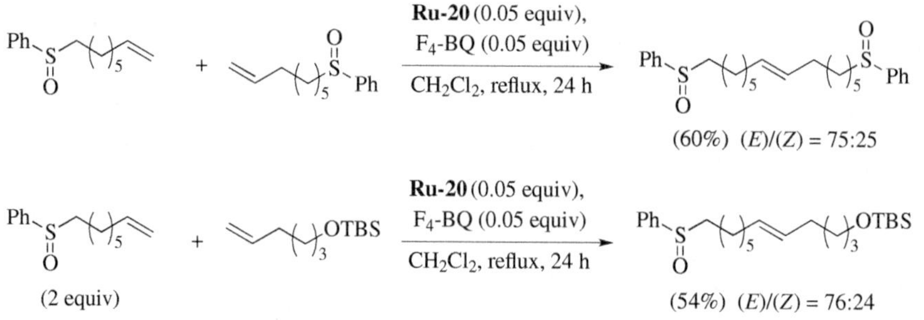

Ru-20 (0.05 equiv), F_4-BQ (0.05 equiv), CH_2Cl_2, reflux, 24 h

(60%) (*E*)/(*Z*) = 75:25

(2 equiv)

Ru-20 (0.05 equiv), F_4-BQ (0.05 equiv), CH_2Cl_2, reflux, 24 h

(54%) (*E*)/(*Z*) = 76:24

Scheme 63

PhO_2S (6 equiv)

Ru-5 (0.025 equiv), CH_2Cl_2, 50°, 96 h

SO_2Ph SO_2Ph

(73%) (*E*)/(*Z*) > 99:1

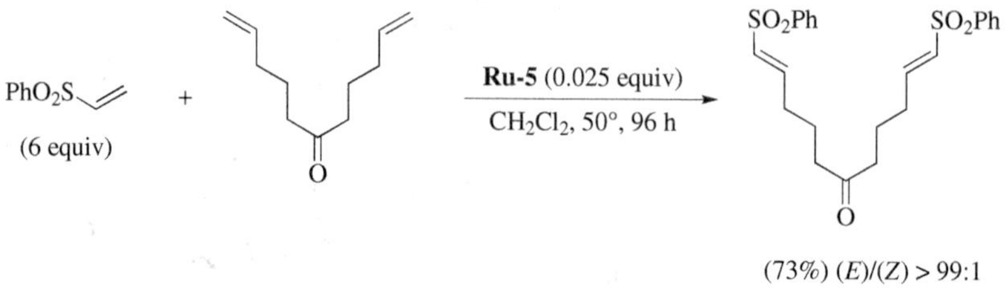

Scheme 64

proceeds in moderate-to-high yields and with exclusive (*E*)-selectivity.[228,232,233] Cross-metathesis reactions of vinyl sulfones are particularly useful and have been featured in several target-oriented syntheses.[231,234–237] The introduction of a vinyl sulfone pharmacophore (sometimes called a warhead)[238] is interesting from a medicinal chemistry point of view[239,240] because it can act as a Michael acceptor, thereby allowing, for example, the addition of a thiol group of an active-site cysteine residue. Moreover, compounds bearing this functional group can be readily prepared by CM using vinyl sulfones.[238,239] Cross-metathesis reactions of phenyl allyl sulfone,[241] other allyl sulfones, and alkenes bearing a sulfone group in more distal position proceed uneventfully, which identifies such substrates as Type I alkenes.[229]

Unsaturated Sulfonamides. Vinyl sulfonamides (Type III partners) undergo CM with a number of terminal alkenes in the presence of second-generation Hoveyda–Grubbs complexes (Scheme 65).[154,230] Cross-metathesis reactions of allyl sulfonamides have also been demonstrated both in solution and on solid phase in the presence of up to 20 mol % of **Ru-4** or **Ru-5**. The presence of a Lewis acid (e.g., Cy_2BCl) increases the CM yields in some cases.[154]

(Z)-Selective Metathesis. The first successful (Z)-selective metathesis was the enantio- and diastereoselective ROCM reaction of cyclic ether **107** with styrene catalyzed by **Mo-3** (Scheme 66).[242]

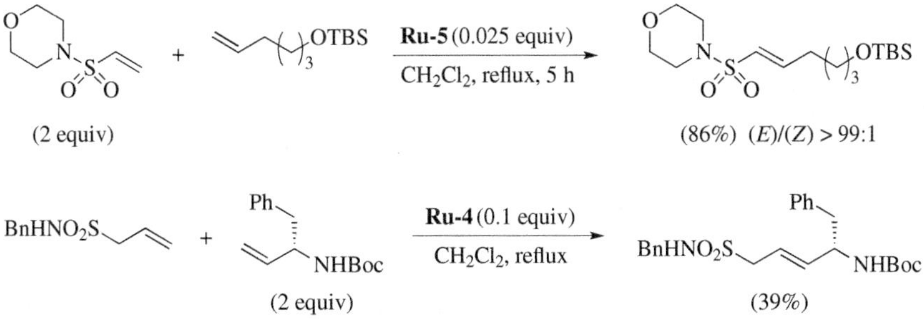

(2 equiv)

Ru-5 (0.025 equiv)
CH_2Cl_2, reflux, 5 h

(86%) (*E*)/(*Z*) > 99:1

$BnHNO_2S$ + (2 equiv) NHBoc

Ru-4 (0.1 equiv)
CH_2Cl_2, reflux

(39%)

Scheme 65

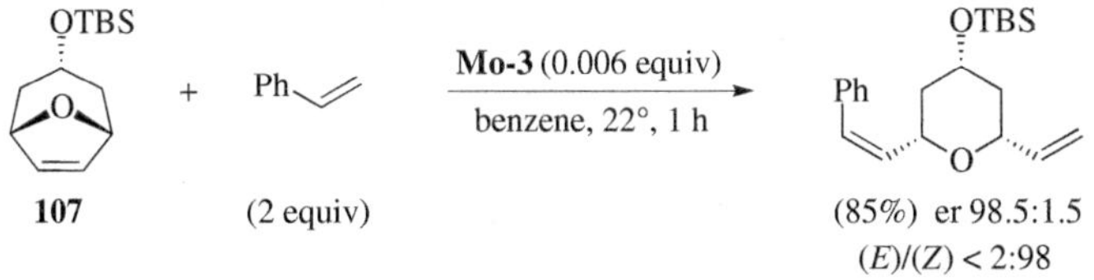

107 + (2 equiv)

Mo-3 (0.006 equiv)
benzene, 22°, 1 h

(85%) er 98.5:1.5
(*E*)/(*Z*) < 2:98

Scheme 66

Low pressure conditions can have a dramatic effect on the yield and selectivity of a CM reaction. For instance, when oxabicycle **107** and (allyloxy)(*tert*-butyl)dimethylsilane (**108**) are combined in the presence of chiral complex **Mo-3** under atmospheric pressure, the desired product **109** is obtained in 45% yield, with good (*Z*)-selectivity ((*Z*)/(*E*) = 95:5) and modest enantioselectivity (er 84:16) (Scheme 67).[242] In surprising contrast, when the analogous process is carried out at lower pressure (7 mm Hg), the results are significantly improved (75% yield, (*Z*)/(*E*) = 98:2, er 93.5:6.5), even with a much shorter reaction time (a recent review explains the role of ethylene in various metathesis processes[243]).

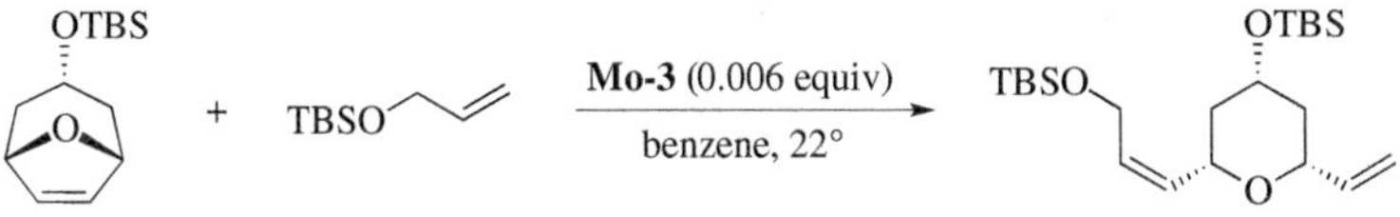

107 + **108** (10 equiv)

Mo-3 (0.006 equiv)
benzene, 22°

109

Pressure (mm Hg)	Time (min)	Yield (%)	(*E*)/(*Z*)	er
ambient	60	45	5:95	84:16
7	10	75	2:98	93.5:6.5

Scheme 67

The (*Z*)-selective CM of two different alkenes is difficult to achieve, because up to six different products can be generated (Scheme 68), which contrasts with homocoupling ($R^1 = R^2$) where only two stereoisomeric alkenes can be formed.[5]

Notably, molybdenum complexes promote (*Z*)-selective CM of enol ethers with Type 1 alkenes (Scheme 69).[5] Consequently, while the Hoveyda–Grubbs ruthenium catalyst is ineffective with these substrates, the **Mo-5** is effective in this process. With larger cross-partners, the 2,6-dimethylphenylimido complex **Mo-5** is more suitable than other molybdenum complexes that were examined.

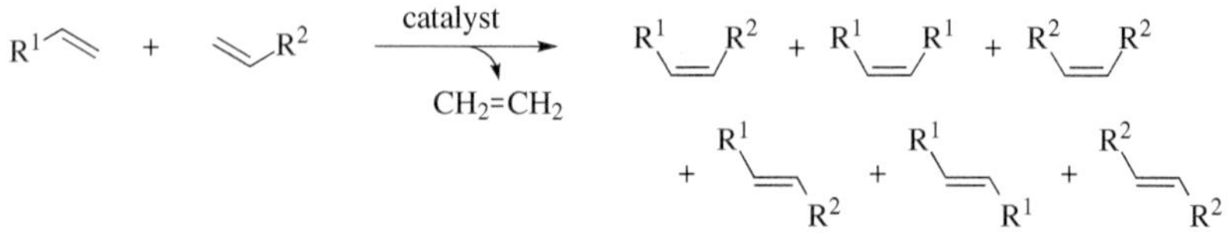

R^1 + R^2 —catalyst ($CH_2{=}CH_2$)→ R^1 R^2 + R^1 R^1 + R^2 R^2 + R^1 R^2 + R^1 R^1 + R^2 R^2

Scheme 68

n-BuO (10 equiv) + Bn —catalyst (0.025 equiv), benzene, 22°→ *n*-BuO Bn

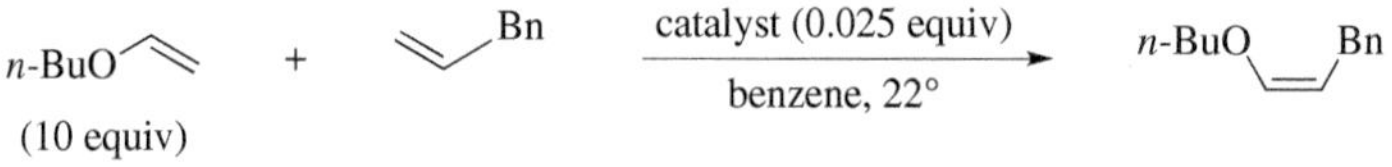

Cat.	Time	Conv (%)	Yield (%)	(*E*)/(*Z*)
Mo-1	10 min	80	—	52.5:47.5
Mo-3	2 h	37	—	<2:98
Mo-5	2 h	85	73	2:98
Mo-6	2 h	47	—	<2:98
Ru-5	24 h	<2	—	—

Scheme 69

Tungsten MAP complexes, such as **W-1** (Figure 2), catalyze the homocoupling of alkenes with high (*Z*)-selectivities (Scheme 70).[244] Such catalysts are active at room temperature, and the reaction is usually performed neat in order to increase the rate and conversion.[244]

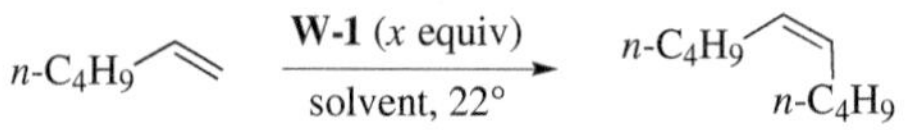

n-C_4H_9 —**W-1** (*x* equiv), solvent, 22°→ *n*-C_4H_9 *n*-C_4H_9

x	Conc (M)	Solvent	Time (h)	Conv (%)	(*E*)/(*Z*)
0.04	0.66	C_6D_6	16	32	<1:99
0.02	neat	—	4	74	<1:99

Scheme 70

Similar conversions and selectivities can be obtained with ruthenium-based, (*Z*)-selective catalysts (Figure 3). In contrast to the molybdenum-based complexes, ruthenium catalysts are able to dimerize more challenging substrates, such as unprotected alcohols (Scheme 71).[82,245]

A (*Z*)-selective metathesis reaction was used in the preparation of an allylic alcohol in the synthesis of (+)-neopeltolide (Scheme 72).[246] A tungsten MAP complex

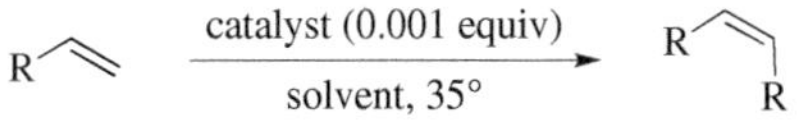

R-CH=CH2 → catalyst (0.001 equiv), solvent, 35° → RCH=CHR

R	Cat.	Solvent	Time (h)	Conv (%)	(*E*)/(*Z*)
$HO(CH_2)_3$	**Ru-7**	THF	12	81	57:43
$HO(CH_2)_3$	**Ru-8**	THF	12	67	19:81
$HO(CH_2)_3$	**Ru-9**	DCE	2	7	<5:95
$HO(CH_2)_3$	**Ru-10**	THF	2	77	<5:95
$HO(CH_2)_3$	**Ru-10**	DCE	2	79	8:92
$MeO_2C(CH_2)_8$	**Ru-7**	THF	12	16	10:90
$MeO_2C(CH_2)_8$	**Ru-8**	THF	12	85	9:91
$MeO_2C(CH_2)_8$	**Ru-9**	DCE	6	65	<5:95
$MeO_2C(CH_2)_8$	**Ru-10**	THF	6	>95	<5:95

Scheme 71

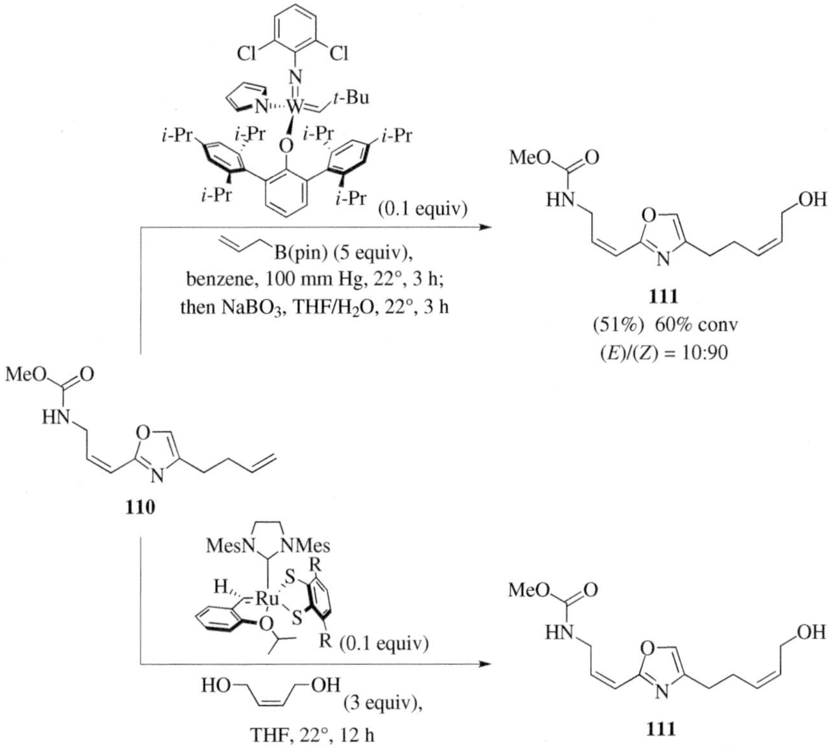

110 → W catalyst (0.1 equiv), B(pin) (5 equiv), benzene, 100 mm Hg, 22°, 3 h; then $NaBO_3$, THF/H_2O, 22°, 3 h → **111** (51%) 60% conv, (*E*)/(*Z*) = 10:90

110 → Ru catalyst (0.1 equiv), HO–CH=CH–OH (3 equiv), THF, 22°, 12 h → **111**

R	Conv (%)	Yield (%)	(*E*)/(*Z*)
H	61	55	3:97
Cl	85	70	2:98

Scheme 72

promotes the CM reaction of alkene **110** with allyl pinacol boronate, leading to 60% conversion after three hours (a longer reaction time did not result in higher conversion). After oxidation, the expected product **111** is obtained in 51% overall yield with a 10:90 (*E*)/(*Z*) ratio. In contrast, the direct CM with 1,4-but-2-endiol using a ruthenium catechothiolate catalyst furnishes the same product in 70% yield and with excellent stereoselectivity ((*E*)/(*Z*) = 2:98).

APPLICATIONS TO SYNTHESIS

Cross-Metathesis in Natural Products Synthesis

Cross-metathesis reactions are widely utilized in the total synthesis of complex natural products, not only because the reactions can be carried out under mild conditions, but because of their high tolerance to a variety of functional groups present in the molecules.[7,247]

A synthesis of epicoccamides A and D, which are potential anticancer agents, utilizes an olefin cross-metathesis as a key step in linking two terminal alkenes that do not contain any proximal functional groups (Scheme 73).[248] Hence, the reaction of alkenes **112a/b** and **113** in the presence of 1 mol % of Grubbs second-generation catalyst **Ru-4** gives the desired products **114a/b**, which were subjected to acetal hydrolysis with TFA and subsequent hydrogenolysis of the benzyl ether. Hydrogenation of the alkene with Pearlman's catalyst delivers the target natural products.

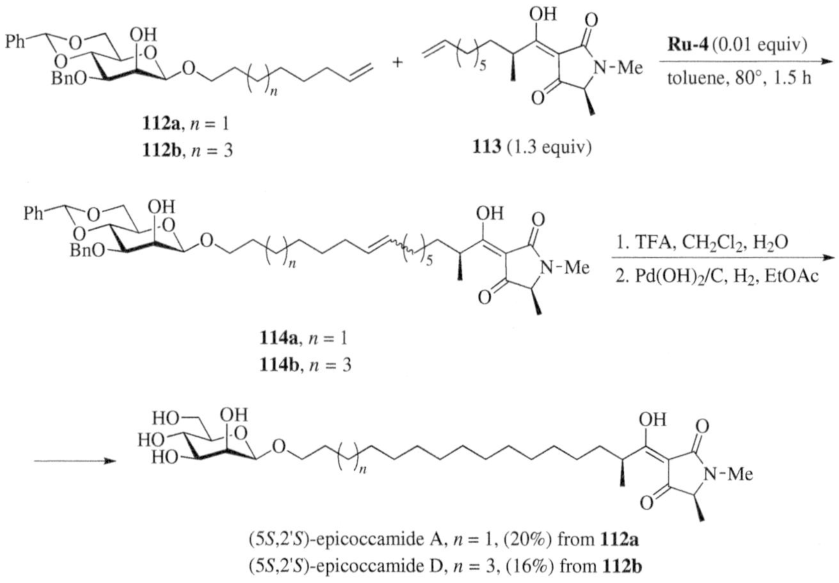

112a, *n* = 1
112b, *n* = 3

113 (1.3 equiv)

Ru-4 (0.01 equiv), toluene, 80°, 1.5 h

114a, *n* = 1
114b, *n* = 3

1. TFA, CH_2Cl_2, H_2O
2. $Pd(OH)_2/C$, H_2, EtOAc

(5*S*,2′*S*)-epicoccamide A, *n* = 1, (20%) from **112a**
(5*S*,2′*S*)-epicoccamide D, *n* = 3, (16%) from **112b**

Scheme 73

The same strategy is used in an enantioselective synthesis of some paraconic acids, which are biologically active compounds that exhibit interesting antitumor and antibiotic properties.[249] For instance, in the synthesis of (+)-nephrosteranic acid, the lactone **115** undergoes CM with 1-decene in the presence of 5 mol % of Grubbs second-generation catalyst **Ru-4** (Scheme 74). Using 1.5 equivalents of 1-decene, the CM product **116** is obtained in only 45% yield along with 30% of recovered starting material. This yield can be improved by using four equivalents of 1-decene, providing **116** in 53% yield along with 27% of recovered **115**. Three additional steps (alkene hydrogenation, aldehyde oxidation, and lactone α-methylation) furnish (+)-nephrosteranic acid in 58% yield from **116**.

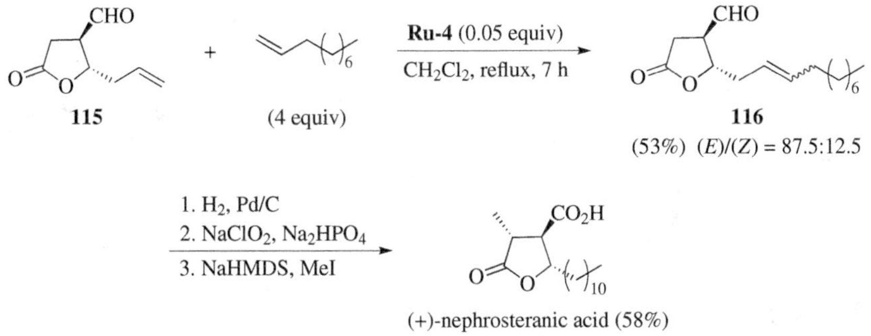

115 + (4 equiv) → **Ru-4** (0.05 equiv), CH_2Cl_2, reflux, 7 h → **116** (53%) (*E*)/(*Z*) = 87.5:12.5

1. H_2, Pd/C
2. $NaClO_2$, Na_2HPO_4
3. NaHMDS, MeI

(+)-nephrosteranic acid (58%)

Scheme 74

The synthesis of (+)-roccellaric acid also begins from lactone **115**. Protection of the formyl group generates acetal **117**, which enables a CM reaction using the less reactive Grubbs first-generation catalyst and 1.5 equivalents of 1-dodecene to afford alkene **118** in 57% yield and as a 78:22 mixture of (*E*)- and (*Z*)-isomers (Scheme 75).[249] Notably, catalyst **Ru-4** gives higher (*E*)-selectivity in the cross-coupling than **Ru-1**. The synthesis of (+)-roccellaric acid is then completed using a similar sequence as for (+)-nephrosteranic acid.

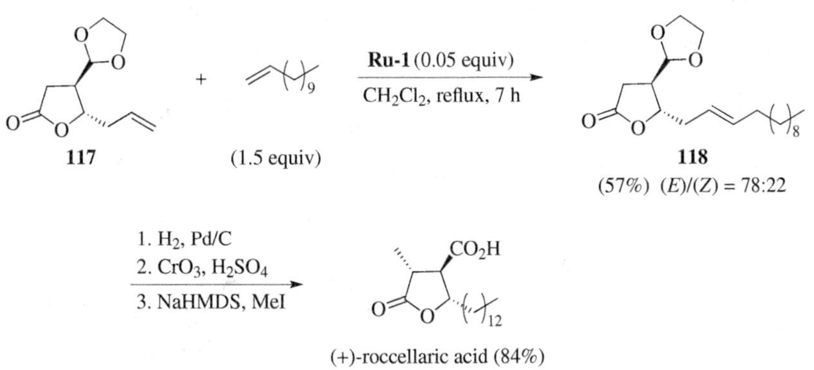

117 + (1.5 equiv) → **Ru-1** (0.05 equiv), CH_2Cl_2, reflux, 7 h → **118** (57%) (*E*)/(*Z*) = 78:22

1. H_2, Pd/C
2. CrO_3, H_2SO_4
3. NaHMDS, MeI

(+)-roccellaric acid (84%)

Scheme 75

The cladospolide family of natural products belongs to a class of 12-membered lactones that exhibit interesting plant-growth-regulatory activity as well as being plant-growth promoters. In the synthesis of (+)-cladospolide A,[250] alcohol **119** is subjected to olefin cross-metathesis with (*S*)-hept-6-en-2-yl acetate (**120**) to furnish the coupled product **121** in 76% yield (Scheme 76).[250] Several additional steps convert the metathesis product to (+)-cladospolide A in 9% overall yield over the 11-linear-step sequence from D-ribose, a popular building block.[251]

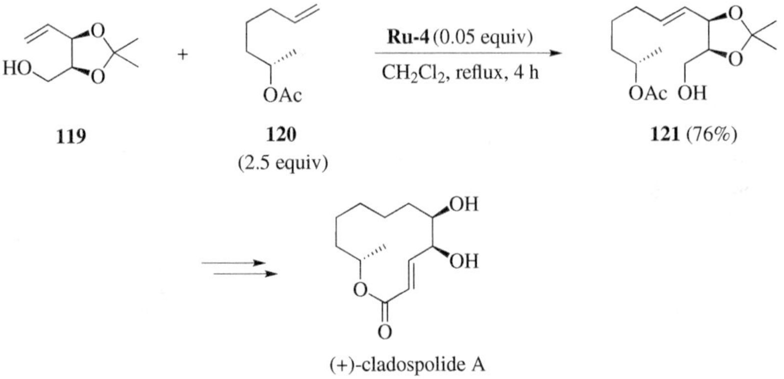

Ru-4 (0.05 equiv), CH_2Cl_2, reflux, 4 h

119 + **120** (2.5 equiv) → **121** (76%) ⇉ (+)-cladospolide A

Scheme 76

Prostaglandins are naturally occurring compounds that provide a promising platform for the development of new therapeutic agents for a number of diseases.[252] Among the strategies developed for the asymmetric synthesis of prostaglandin analogues, an olefin cross-metathesis-based strategy is attractive for the installation of the ω-chain.[128,253,254] In a synthesis of 8-aza prostaglandin E_1, olefin cross-metathesis is used to combine the 5-vinyl-γ-lactam subunit **122a** with the pre-ω-chain **123** (Scheme 77). The reaction is performed in the presence of 19 mol %

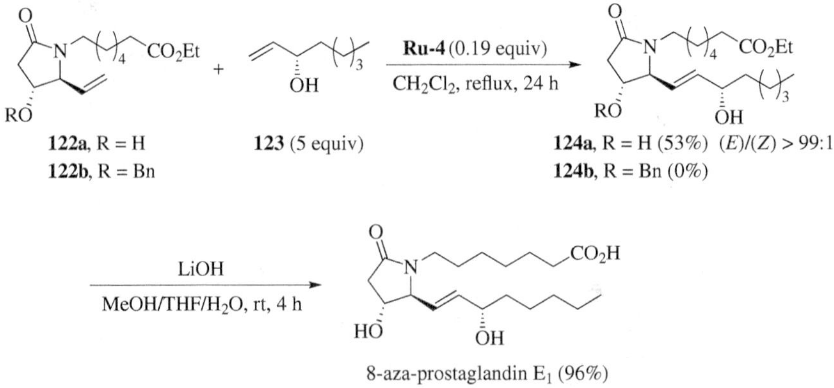

Ru-4 (0.19 equiv), CH_2Cl_2, reflux, 24 h

122a, R = H; **122b**, R = Bn + **123** (5 equiv) → **124a**, R = H (53%) (*E*)/(*Z*) > 99:1; **124b**, R = Bn (0%)

LiOH, MeOH/THF/H_2O, rt, 4 h → 8-aza-prostaglandin E_1 (96%)

Scheme 77

of Grubbs second-generation catalyst **Ru-4** to give the desired alkene **124a** in 53% yield and with >99:1 (*E*)/(*Z*) selectivity.[255] In contrast, the metathesis reaction with the benzyl-protected derivative **122b** is unsuccessful; this result may be attributed to increased steric hindrance of this substrate. The synthesis of 8-aza prostaglandin E_1 is completed by the hydrolysis of the ethyl ester to afford the desired acid in excellent yield.

(−)-Mucocin belongs to the anonnaceous acetogenin family, which exhibits antimitotic and cytotoxic properties. The acetogenins are comprised of polycyclic ethers that contain either adjacent or non-adjacent tetrahydrofuran or tetrahydropyran rings, which are biosynthetically derived from an unsaturated fatty acid.[256] In a pertinent application of cross-metathesis to the synthesis of (−)-mucocin, the reaction between alkene **125**, which has an unprotected hydroxyl group, and an excess of the easily accessible, less reactive alkene **126**, which bears an acetate-protected hydroxyl group, was conceived. The formation of the cross-metathesis product **127** (together with unreacted **126**) was anticipated to be favored over the homodimeric products on the basis of statistical and reactivity considerations.[32,36] For instance, the reaction of **125** with three equivalents of **126** leads to heterodimer **127** as the major metathesis product in 51% yield; the homodimeric products are formed in less than 5% yield (Scheme 78).[257] Alkene **127** is obtained essentially as a single diastereoisomer, presumably the (*E*)-isomer, which is subjected to hydrogenation and several functional group transformations to complete the synthesis of (−)-mucocin.

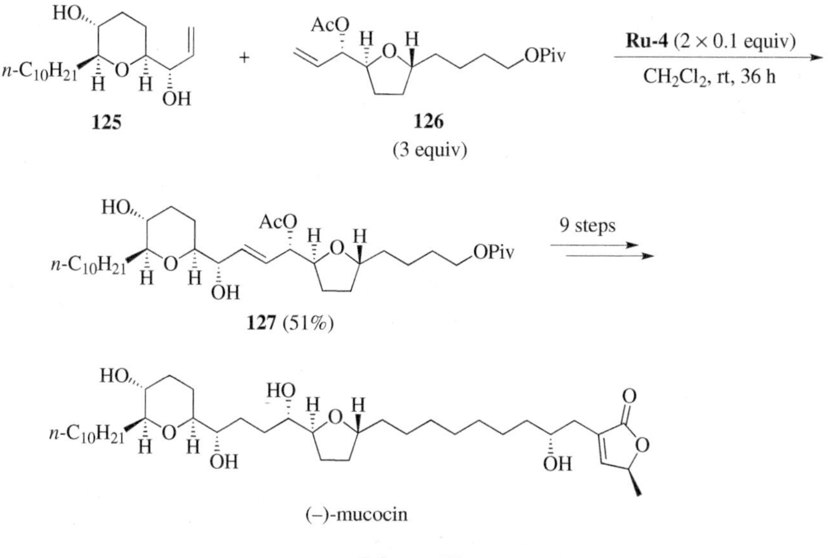

Scheme 78

A synthesis of apoptolidinone, an analogue of the potent anticancer agent apoptolidin A,[258] illustrates an example of a selective cross-metathesis in the presence of differently substituted double bonds (Scheme 79).[259] To this end, the trisubstituted, conjugated alkene moieties of tetraene **128** and the trisubstituted alkene of diene **129** were expected to be unreactive under the cross-metathesis conditions. Thus, the cross-metathesis reaction between the terminal alkenes in each of these compounds was anticipated to furnish the desired intermediate **130**.[36] In the event, exposure of alkenes **128** and **129** to the Grubbs second-generation catalyst **Ru-32** provides a 60% yield of the desired (*E*)-isomer **130**, which was further elaborated to the target natural product.

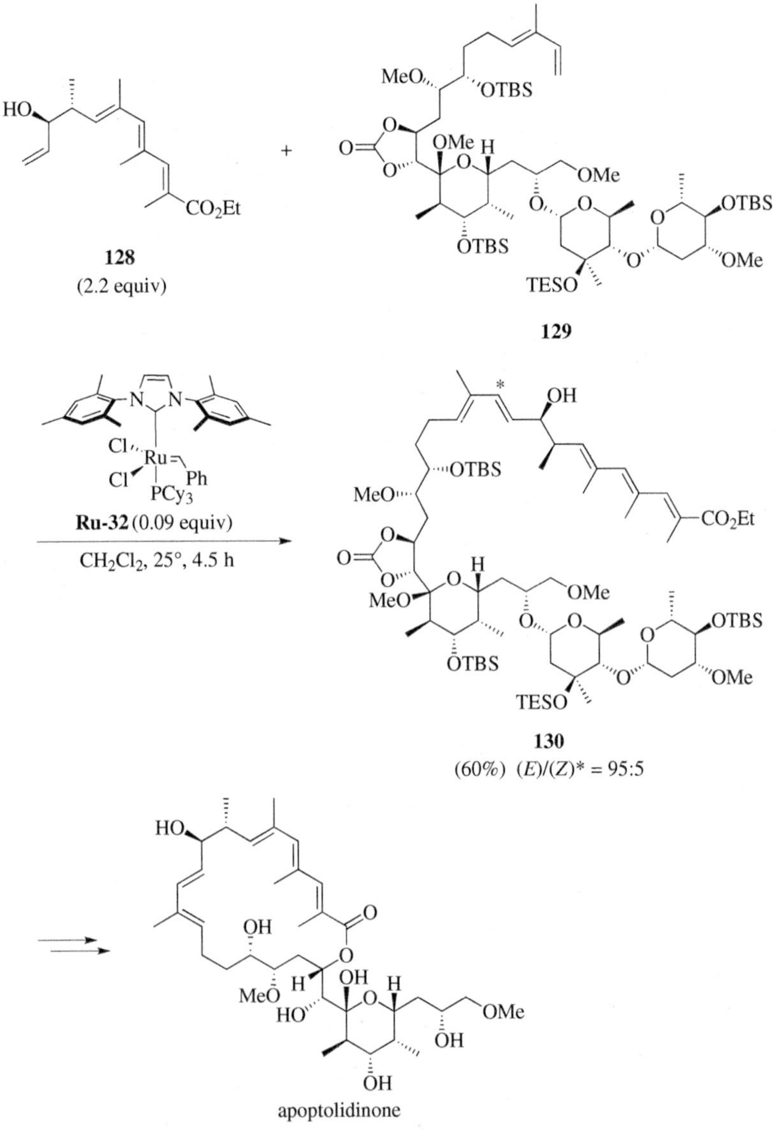

HO CO_2Et **128** (2.2 equiv) + MeO OTBS O O O OMe H O OMe O O OTBS O OTBS OTBS O OMe TESO **129**

Cl Ru Ph Cl PCy_3 N N **Ru-32** (0.09 equiv) CH_2Cl_2, 25°, 4.5 h

* OH OTBS MeO CO_2Et O O O MeO O H OMe O O OTBS OTBS O OMe TESO **130** (60%) (*E*)/(*Z*)* = 95:5

HO O OH O MeO H OH H HO O OMe OH OH apoptolidinone

Scheme 79

Cross-metathesis of α,β-unsaturated ketone **131** is employed as a key step in the synthesis of curvulone B, an antibacterial and antifungal agent (Scheme 80).[260] The metathesis reaction between acrylate **131** and the TES-protected alcohol **132** is performed in the presence of Grubbs second-generation catalyst **Ru-4** to deliver the desired product **133** in good yield. Subsequent oxy-Michael cyclization, isomerization, and functional group manipulation afford curvulone B.

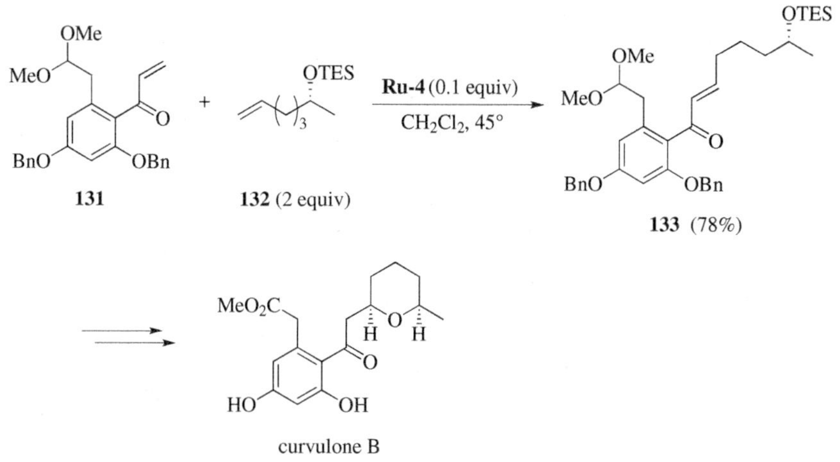

Scheme 80

Multiple cross-metathesis reactions are employed in a synthesis of the C1–C14 fragment of the mis-assigned structure of amphidinol 3 (Scheme 81).[157] Aldehyde **135** is obtained stereoselectively ((*E*)/(*Z*) = 98:2) in 79% yield from the coupling of the secondary allylic alcohol **134** with three equivalents of acrolein in the presence of Hoveyda–Grubbs second-generation catalyst **Ru-5** (2.5 mol %). Conversion into bis-acetate **137** is accomplished with the allyltitanium complex (*S,S*)-**136**,[261] followed by acetylation. Compound **137** is then subjected to a second cross-metathesis with acrolein under the same conditions, leading to aldehyde **138** in 63% yield ((*E*)/(*Z*) = 98:2).[157] The formation of homoallylic acetate **139** and a third cross-metathesis reaction, this time with ethyl acrylate, affords the unsaturated C1–C14 fragment of amphidinol 3 in an overall yield of 17%.

Cross-metathesis reactions are well suited to cascade and tandem reactions in natural products synthesis.[262] A cross-metathesis/ring-closing metathesis sequence is used to assemble the core structure of (−)-cylindrocyclophanes, which are potent anticancer agents.[263] Consequently, the dimerization of diene **140** with Schrock's catalyst **Mo-1** and subsequent ring-closing metathesis of the resulting dimer provides a single (*E,E*)-isomer of the [7,7]-paracyclophane **141** in 77% yield (Scheme 82).[157] The alternative "head-to-head" dimerization products are not detected in these experiments, suggesting that the [7,7]-paracyclophane skeleton is thermodynamically favored. This hypothesis has been proven by independently preparing the "head-to-head" and "head-to-tail" dimers of an analogue of **140** (with H in place of

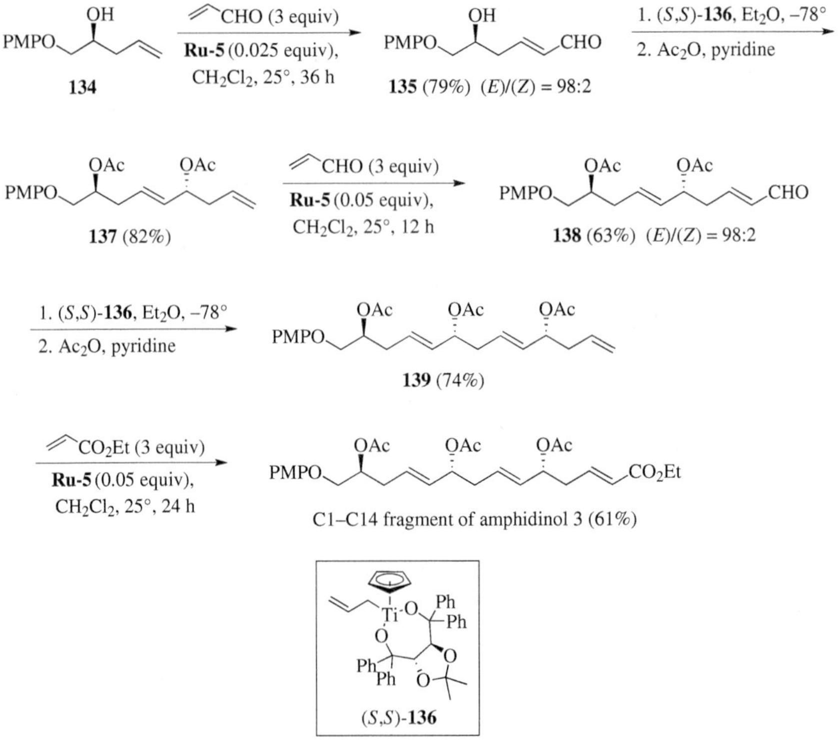

134 → (CHO (3 equiv), **Ru-5** (0.025 equiv), CH_2Cl_2, 25°, 36 h) → **135** (79%) (*E*)/(*Z*) = 98:2

135 → (1. (*S*,*S*)-**136**, Et_2O, –78°; 2. Ac_2O, pyridine) → **137** (82%)

137 → (CHO (3 equiv), **Ru-5** (0.05 equiv), CH_2Cl_2, 25°, 12 h) → **138** (63%) (*E*)/(*Z*) = 98:2

138 → (1. (*S*,*S*)-**136**, Et_2O, –78°; 2. Ac_2O, pyridine) → **139** (74%)

139 → (CO_2Et (3 equiv), **Ru-5** (0.05 equiv), CH_2Cl_2, 25°, 24 h) → C1–C14 fragment of amphidinol 3 (61%)

(*S*,*S*)-**136**

Scheme 81

140 → (**Mo-1** (0.34 equiv), benzene, rt, 1 h) → **141** (77%)

141 → (1. TBAF, THF; 2. H_2, PtO_2, EtOH; 3. PhSH, K_2CO_3, NMP, 215°) → (–)-cylindrocyclophane A (60%)

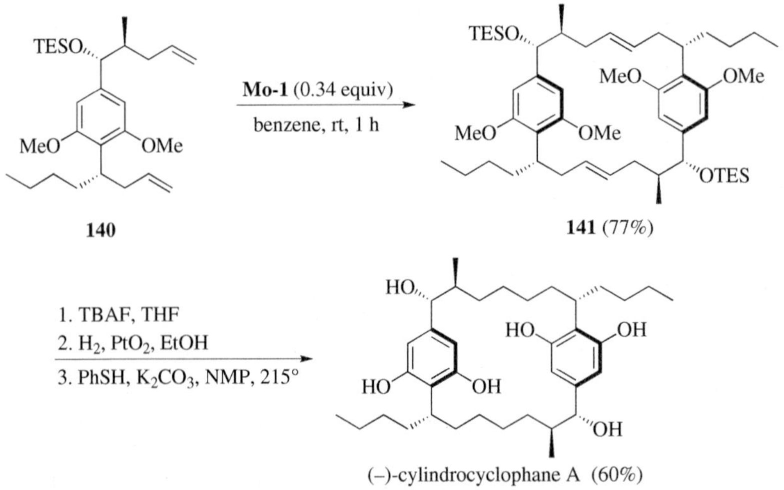

Scheme 82

the TESO group), and then subjecting them to RCM in the presence of either Grubbs **Ru-1** and **Ru-4** or Schrock **Mo-1** catalysts to converge on the [7,7]-paracyclophane skeleton.[263] The synthesis of cylindrocyclophane A is accomplished in three steps from intermediate **141**.

Bistramide A exhibits significant neuro- and cytotoxic properties in addition to profound effects on cell cycle regulation.[264] The key spiroketal domain in the natural product is prepared by a sequence involving ROM followed by two CM reactions (Scheme 83).[265] In the first step, a mixture of terminal alkene **142** and cyclopropene acetal **143** undergo a ring-opening/cross-metathesis cascade in the presence of Grubbs second-generation catalyst **Ru-4** to afford, after acid-mediated removal of the ketal protecting group, divinyl ketone **144** as an inconsequential mixture of geometric isomers favoring the (*Z*)-isomer ((*E*)/(*Z*) = 40:60). Treatment of α,β-unsaturated compound **144** with the metathesis partner **145**, again using Grubbs second-generation catalyst **Ru-4**, affords the desired cross-metathesis product **146** in 68% yield. Hydrogenation of dienone **146** with concomitant cleavage of the three benzyl ethers induces the spiroketalization, which was followed by a Dess–Martin oxidation to afford the spiroketal **147** as a single diastereomer. This intermediate is converted to bistramide A in several additional steps.

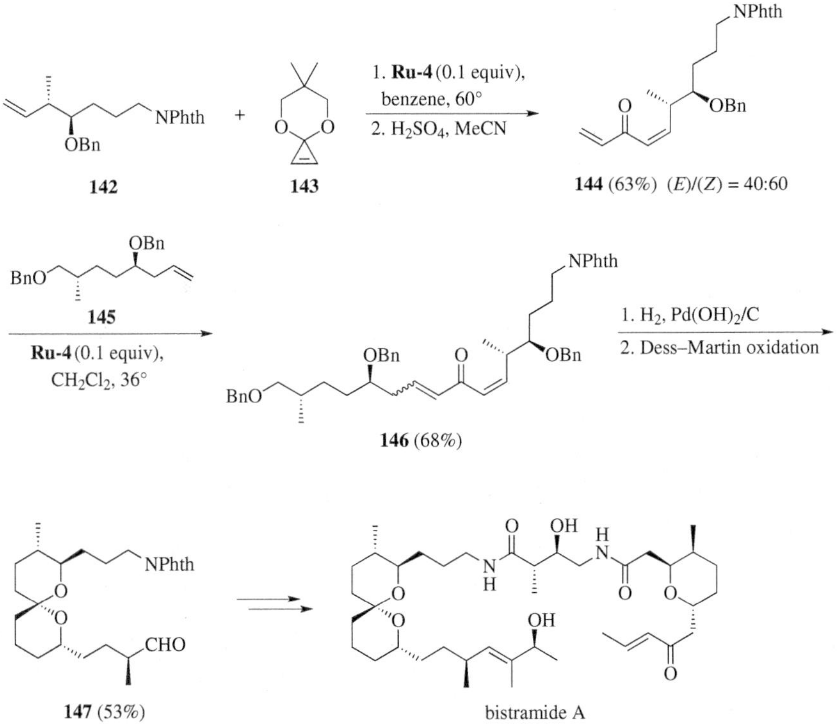

NPhth OBn NPhth + O O 1. **Ru-4** (0.1 equiv), benzene, 60° 2. H_2SO_4, MeCN O OBn

142 **143** **144** (63%) (*E*)/(*Z*) = 40:60

OBn BnO **145** **Ru-4** (0.1 equiv), CH_2Cl_2, 36° NPhth OBn O OBn BnO 1. H_2, $Pd(OH)_2/C$ 2. Dess–Martin oxidation

146 (68%)

NPhth O O CHO O OH H N N H O O O O OH O

147 (53%) bistramide A

Scheme 83

An elegant one-pot reaction sequence is used in a total synthesis of puraquinonic acid, a potent anticancer agent.[266] An enyne/cross-metathesis reaction of precursor **148** with protected 3-buten-1-ol derivative **149** in the presence of Grubbs first-generation catalyst **Ru-1** delivers diene **150** in 89% yield favoring only the (*E*)-isomer (Scheme 84). A Diels–Alder reaction with dimethyl acetylenedicarboxylate followed by oxidation by DDQ furnishes the desired indane skeleton **151** in 87% yield. Alternatively, **151** can be obtained in 83% yield via a one-pot sequence from **148** by simply removing the dichloromethane after the initial metathesis sequence and proceeding with the Diels–Alder and oxidation reactions. Compound **151** is then converted to puraquinonic acid in a few additional steps.

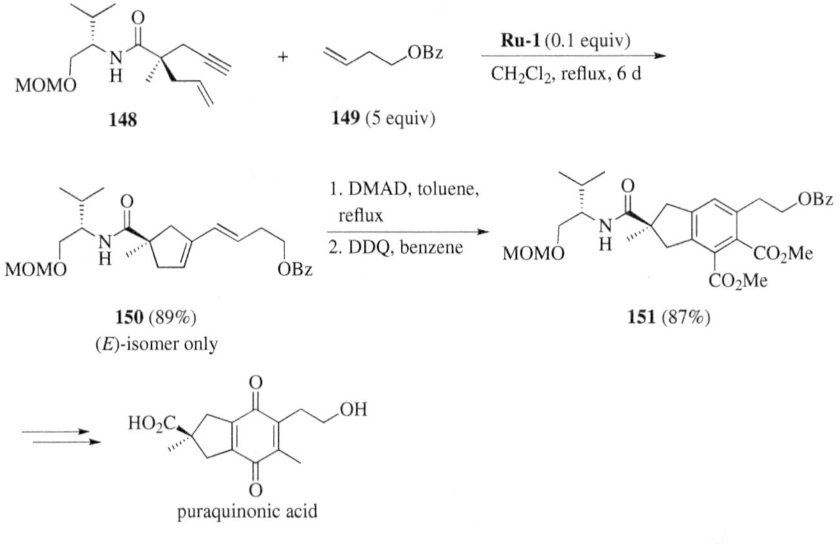

148 + **149** (5 equiv) → **Ru-1** (0.1 equiv), CH_2Cl_2, reflux, 6 d → **150** (89%) (*E*)-isomer only

150 → 1. DMAD, toluene, reflux; 2. DDQ, benzene → **151** (87%) ⟶⟶ puraquinonic acid

Scheme 84

A synthesis of (+)-mycoepoxydiene, which exhibits antimicrobial and anticancer activities, has been accomplished by a sequential ROM/CM/RCM process to construct the 9-oxabicyclo[4.2.1]nona-2,4-diene core (Scheme 85).[267] In initial investigations, the reactions were performed in a stepwise manner: ring-opening metathesis of **152** in the presence of 1,3-butadiene furnishes a mixture of trienes **153a** and **153b** in 75% yield; however, because only the (*Z*)-isomer can undergo subsequent ring closure, this reaction was optimized for the formation of this product using the Grubbs first-generation catalyst **Ru-1** in benzene at room temperature. In contrast, Grubbs second-generation catalyst **Ru-4** preferentially produces the undesired (*E*)-isomers, whereas the formation of cyclooctadiene **154** requires the more active Grubbs second-generation catalyst **Ru-4**. In a one-pot procedure, this complex is added to the reaction mixture after consumption of substrate **152** and removal of 1,3-butadiene to furnish the cyclooctadiene **154**, which is further elaborated to the target natural product.

A sequential ROM/RCM/CM sequence is a key feature in a total synthesis of (+)-cylindramide A, a potent anticancer agent (Scheme 86).[268] A mixture of the

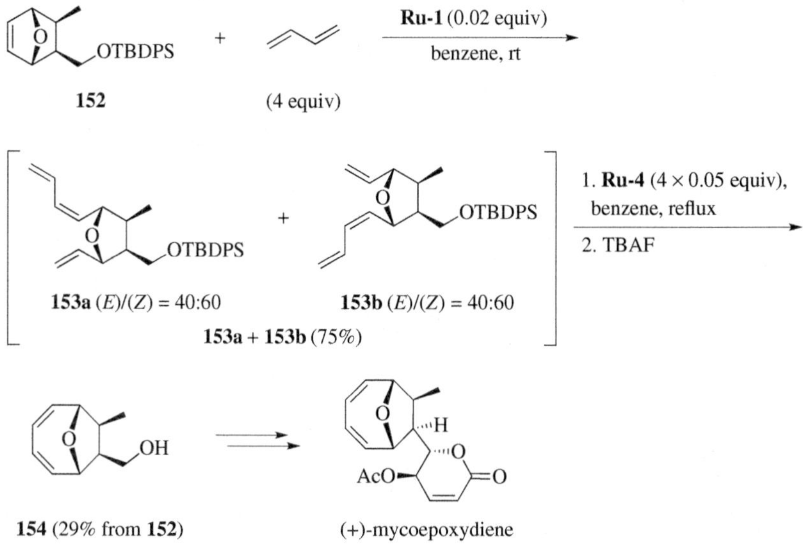

OTBDPS + (4 equiv) **Ru-1** (0.02 equiv) benzene, rt

152

OTBDPS OTBDPS

153a (*E*)/(*Z*) = 40:60 **153b** (*E*)/(*Z*) = 40:60

153a + **153b** (75%)

1. **Ru-4** (4 × 0.05 equiv), benzene, reflux
2. TBAF

OH H O AcO O

154 (29% from **152**) (+)-mycoepoxydiene

Scheme 85

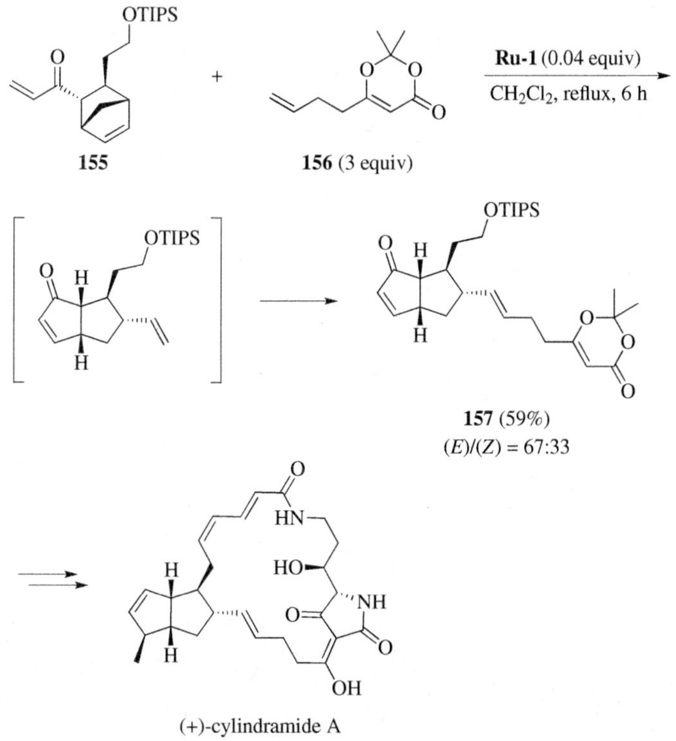

OTIPS O O O + O **Ru-1** (0.04 equiv) CH_2Cl_2, reflux, 6 h

155 **156** (3 equiv)

OTIPS O H H OTIPS O H H O O O

157 (59%)
(*E*)/(*Z*) = 67:33

O HN H HO NH O O H OH

(+)-cylindramide A

Scheme 86

norbornene derivative **155** and the terminal alkene **156** is treated with 4 mol % of Grubbs first-generation catalyst **Ru-1** to promote ROM/RCM followed by cross-metathesis to afford **157** in 59% yield as a 67:33 mixture of separable (*E*)/(*Z*)-isomers. Compound **157** is converted into (+)-cylindramide A in several steps.

A synthesis of (+)-asteriscanolide involves a sequential ROM/CM/Cope rearrangement.[269] Consequently, treatment of cyclobutene **158** with 5 mol % of Grubbs second-generation catalyst **Ru-4** in benzene under an atmosphere of ethylene promotes ring-opening metathesis followed by cross-metathesis with ethylene to furnish the Cope-rearrangement precursor **159** (Scheme 87). This intermediate is not isolated, and under the reaction conditions, undergoes a Cope rearrangement to generate cyclooctadiene **160** in 74% yield. The synthesis of (+)-asteriscanolide is then completed from the tricyclic core by a series of functional group transformations.

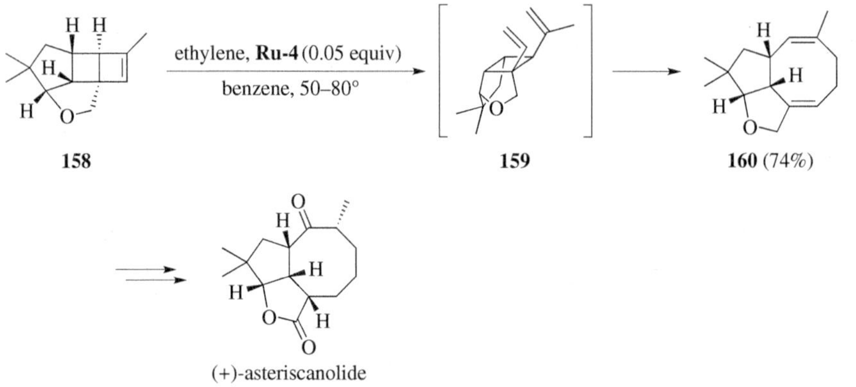

ethylene, **Ru-4** (0.05 equiv)
benzene, 50–80°

158 **159** **160** (74%)

(+)-asteriscanolide

Scheme 87

COMPARISON WITH OTHER METHODS

The formation of carbon–carbon double bonds is a key type of transformation on which much synthetic chemistry is based, and as such, several methods have been developed during the last century. The following is a brief discussion of three of the most important strategies for the construction of alkenes: (1) elimination reactions, (2) alkenation ("olefination") of carbonyl compounds, and (3) transition-metal-catalyzed cross-coupling reactions.[270–274]

Elimination Reactions

A number of molecules containing good leaving groups can undergo elimination reactions in the presence of bases,[275] at elevated temperatures,[276] or under the influence of agents such as cerium chloride[276] or samarium iodide.[277] This method usually exhibits low regio- and stereoselectivity, and in most cases a mixture of products is obtained. When the elimination process is base-dependent, the position of the double bond is determined by Zaitsev's rule,[278] which will lead predominantly to the more highly substituted alkene (where the base is sterically unhindered). However, if steric nature of the base prohibits removal of a proton from the most substituted carbon, the less substituted alkene is preferred. Methods of converting 1,2-diols into alkenes have been comprehensively reviewed.[279] One important method is the elimination of cyclic thiocarbonates promoted by trialkyl phosphites (Corey–Winter reaction)[280,281] or metal carbonyls.[282] Diols are converted into alkenes stereospecifically by heating with trialkyl orthoformates and a catalytic amount of a carboxylic acid.[283] Elimination (extrusion) of sulfur dioxide in the Ramberg–Bäcklund reaction is another powerful method for the synthesis of alkenes that have a broad range of substitution patterns, including tetrasubstituted alkenes, cyclobutenes, and polyenes.[284]

Alkenation of Carbonyl Compounds

Carbonyl olefination reactions are site-selective and enable the formation of the double bond with no ambiguity as to the position of the resulting alkene. One drawback of this process is a low level of atom economy: commonly used reagents lead to phosphorus-, silicon-, sulfur-, or titanium-containing byproducts in equimolar quantities.

The Wittig reaction[285] couples aldehydes and ketones (often containing other functional groups such as alcohols, ethers, arenes, nitro, and ester groups) with phosphine ylides to provide mono-, di-, and trisubstituted alkenes, typically in good yields. Synthesis of tetrasubstituted alkenes is best performed by means other than the Wittig reaction (for example by McMurry reaction between two carbonyl compounds and low-valent titanium).[286,287] Because simple phosphoranes are very reactive and sensitive to moisture and oxygen, they are usually prepared in a dry solvent (typically THF) under an atmosphere of either nitrogen or argon, and the carbonyl compound is added immediately after the phosphorane has been formed. Phosphoranes containing groups that can stabilize a negative charge are more stable and much easier to handle. Simple, nonstabilized phosphonium ylides (e.g., **161**, **163**) mainly provide the thermodynamically less stable (*Z*)-isomers (Scheme 88).[288] (*E*)-Alkenes can be obtained either by using a stabilized ylide (e.g., **162**) (Scheme 89)[289] or by employing the Schlosser modification of the Wittig reaction with a nonstabilized ylide (Scheme 90).[290]

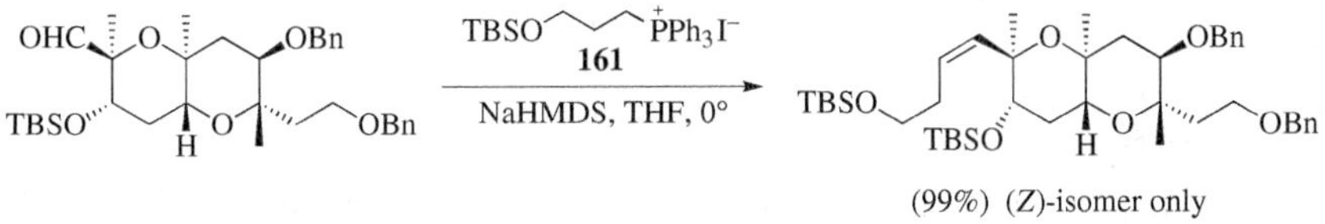

OHC, O, OBn, TBSO, H, O, OBn
TBSO ~ $\overset{+}{P}Ph_3I^-$ **161**
NaHMDS, THF, 0°
TBSO, O, OBn, TBSO, H, O, OBn
(99%) (*Z*)-isomer only

Scheme 88

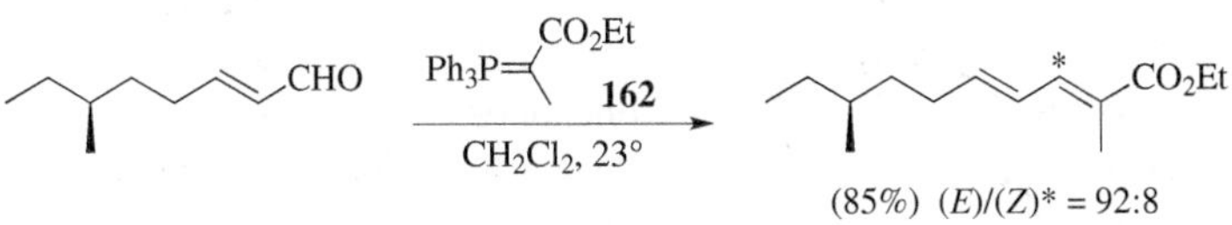

CHO
$Ph_3P=C(CO_2Et)$ **162**
CH_2Cl_2, 23°
CO_2Et
(85%) (*E*)/(*Z*)* = 92:8

Scheme 89

1. PhLi, THF, 0°
2. CHO ... SiMe$_3$, −78°
3. PhLi, Et_2O, −78° to 0°

163

PPh_3I

$SiMe_3$

(71%) (*E*)/(*Z*)* = 96:4

Scheme 90

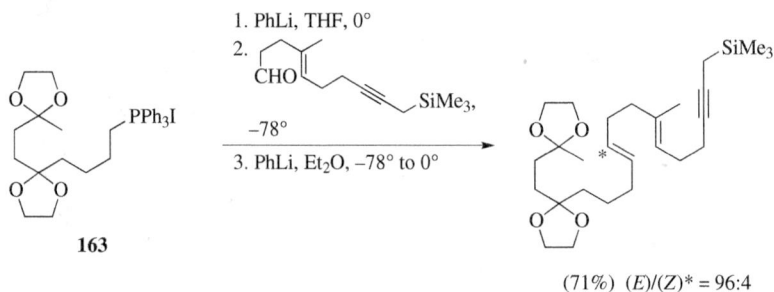

The Horner–Wadsworth–Emmons (HWE)[291] reaction is a modification of the Wittig reaction that employs phosphonate-stabilized carbanions. A particular advantage of this reaction is the broad substrate scope. For example, α,β-unsaturated ketones and sterically hindered ketones can be employed as substrates. Furthermore, purifications of HWE reactions are easier than those of Wittig reactions because the dialkylphosphate byproduct is readily soluble in water. As in the Wittig reaction, careful selection of the olefination conditions and reagents allows the synthesis of (*E*)- and (*Z*)-alkenes with a high degree of stereoselectivity.[292] (*Z*)-Isomers of alkenes are favored under conditions of kinetic control and in the presence of phosphonates with electron-withdrawing groups (e.g., trifluoroethyl phosphonates),[293] whereas more stable (*E*)-isomers of alkenes are favored under thermodynamic conditions (Scheme 91).[294]

CHO + MeO_2C P(O)(OMe)OMe

NaH, THF, 0° → CO_2Me (80%)

n-BuLi, THF, −78° → MeO_2C (39%)

Scheme 91

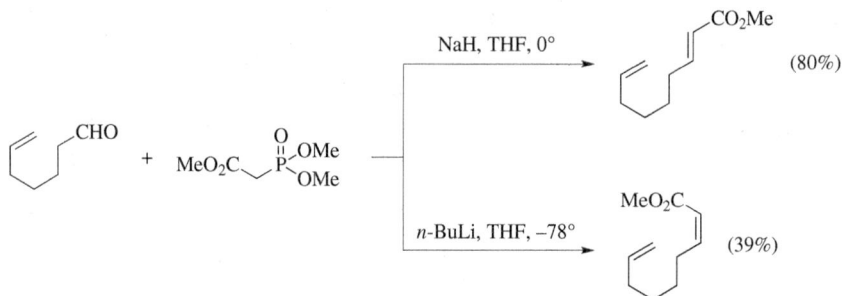

The Peterson reaction[295–297] employs α-silylcarbanions, usually generated in situ, for the synthesis of alkenes from aldehydes and ketones via the intermediate β-hydroxysilanes (Scheme 92). The Peterson reaction is characterized by simpler workup and purification procedures than the Wittig reaction. In addition, because the reagents are both more basic and more nucleophilic, as well as less sterically hindered, they are more reactive than phosphorus ylides. However, the Peterson process does have some drawbacks: α-silylcarbanion formation can be difficult, and a mixture of (*E*)- and (*Z*)-alkene isomers is often obtained. Thus, the addition of α-silylcarbanions **164** to carbonyl compounds **165** usually leads to a diastereomeric mixture of β-hydroxysilanes **166a/b** which, depending on the substituents, may be difficult to separate. Each diastereoisomer **166a/b** undergoes stereospecific decomposition to afford either the (*E*)- or (*Z*)-isomer (alkene **167a** and **167b**), depending on the elimination conditions.

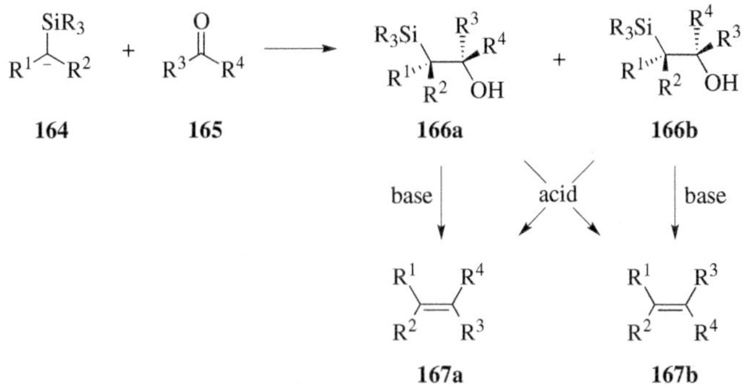

Scheme 92

Another method for alkenation of carbonyl compounds is the Julia reaction,[298] which involves the reaction of an aldehyde or ketone with a phenyl sulfone and yields an alkene after reductive elimination with sodium amalgam (Scheme 93). The reaction consists of multiple steps: (1) generation of a sulfonyl carbanion, (2) formation of a β-hydroxy sulfone, (3) transformation of the hydroxyl group into a good leaving group (e.g., acetate, benzoate, 4-toluenesulfonate), and (4) reduction of the hydroxy sulfone derivative with sodium amalgam. This last step limits the scope of the process to substrates containing functional groups resistant to reduction. As an alternative to sodium amalgam, reducing agents such as samarium diiodide or magnesium diiodide can be utilized.[299] The use of heteroaryl sulfones like benzothiazol-2-yl sulfones,[300] or 1-phenyl-1*H*-tetrazol-5-yl sulfones[301] (Julia–Kocienski olefination) enables the synthesis of alkenes in a one-step procedure, usually with good (*E*)-selectivity, and also circumvents the need for a reductive elimination step (Scheme 94).

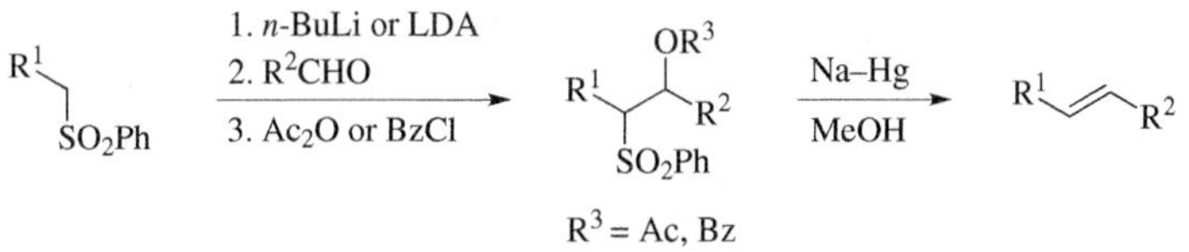

Scheme 93

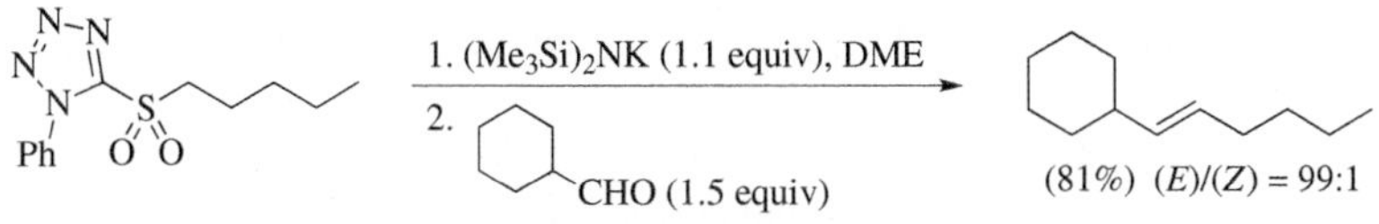

Scheme 94

Tebbe and Petasis olefinations use titanium alkylidene complexes to methylenate a variety of carbonyl compounds (Scheme 95).[302–304] The method is useful not only for aldehydes and ketones but also for carboxylic acid derivatives, such as esters and amides. Thus, vinyl ethers[305] and enamines[306] can be obtained by this process. The Tebbe olefination can be applied to the synthesis of 1,1-disubstituted alkenes containing bulky substituents, but the reaction can be problematic in systems with high steric hindrance. Whereas this approach is applicable primarily for methylenations, some analogues of the Tebbe and Petasis reagents bearing alkyl chains have been developed to produce trisubstituted alkenes,[307] albeit low stereoselectivity can be a problem. Because the titanium carbenoids require rigorously air-free conditions, an alternative approach relies on generating the titanium carbenoids in situ with a CH_2X_2–TiX_4–Mg reagent.[308]

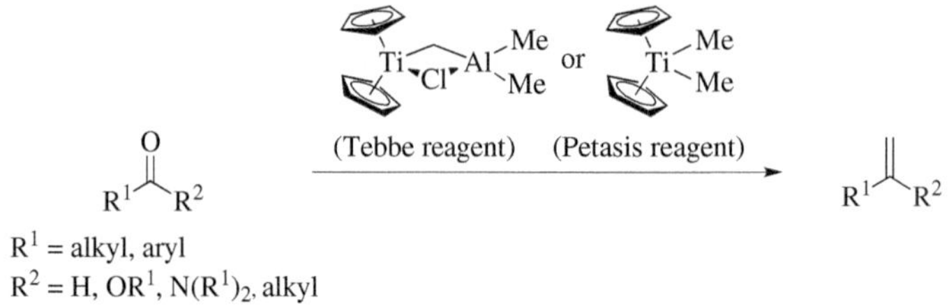

O
R^1 R^2
R^1 = alkyl, aryl
R^2 = H, OR^1, $N(R^1)_2$, alkyl

Ti Cl Al Me Me or Ti Me Me
(Tebbe reagent) (Petasis reagent)

R^1 R^2

Scheme 95

Transition-Metal-Catalyzed Cross-Coupling Reactions

Transition-metal-promoted couplings arguably constitute the most powerful means to generate carbon–carbon, carbon–nitrogen, carbon–oxygen, and carbon–sulfur bonds. In contrast to the previously described methods for creating a double bond, the transition-metal-catalyzed reactions described in this section involve bond formation to a double bond already present in one of the reacting partners. Although the carbon–carbon double bond is introduced indirectly, the enormous power of these transformations in organic synthesis warrants their inclusion.

The Heck reaction[309,310] involves insertion of an alkene into a carbon–palladium bond, which is formed from an aryl halide, vinyl halide, or related derivative (Scheme 96). The Heck reaction has a broad functional-group tolerance and, like cross-metathesis, usually proceeds with high (*E*)-selectivity. The regioselectivity of the process is affected by the electronic properties of the alkenes used, the nature of the group into which palladium inserts, the ligands,[311,312] and the additives used. A significant limitation to this approach is that it is not applicable to halide substrates containing a β-hydrogen atom, although some exceptions have been reported.[313,314]

The coupling of an organometaloid or organometallic compound with a halide or pseudohalide is one of the most general and reliable methods for the formation of carbon–carbon bonds (Scheme 97). The main limitation of this methodology is

R^1 + R^2Y —Pd^0→ R^2–CH=CH–R^1

R^1 = aryl, alkyl
R^2 = aryl, vinyl
Y = halide, OTf

Scheme 96

+ → + MX

M = MgX (Kumada coupling)
M = SnR_3 (Stille coupling)
M = BR_2 or $B(OR)_2$ (Suzuki–Miyaura coupling)
M = SiR_3 (Hiyama–Denmark coupling)
M = ZnX (Negishi coupling)

Scheme 97

the production of an equimolar amount of MX byproducts that are often difficult to remove and, in some cases, may be toxic.

The Kumada coupling[315,316] employs Grignard reagents that are often commercially available or readily prepared from the corresponding halides. The reaction can be performed at room temperature or below in the presence of relatively inexpensive nickel complexes. However, the applicability of this reaction is limited owing to the reactivity of Grignard reagents with a wide range of functional groups. Nevertheless, the Kumada reaction can be conducted both on small and large scales, including industrial scales, as exemplified in the preparation of the hypertension drug aliskiren (Scheme 98).[317]

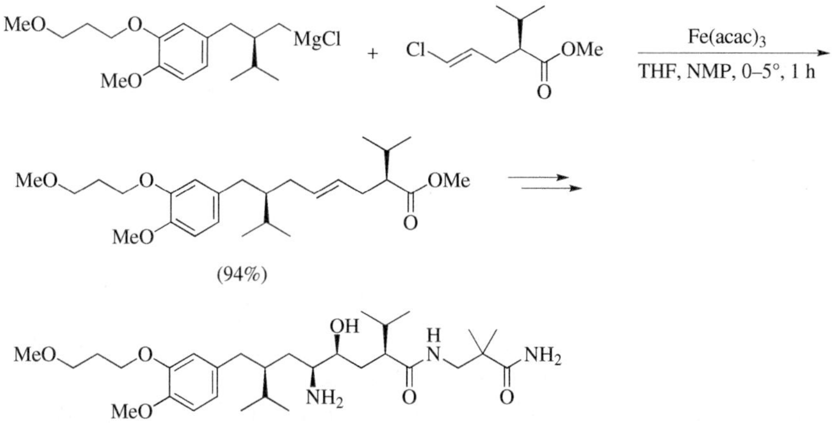

MeO, O, MeO, MgCl + Cl, OMe, O —$Fe(acac)_3$, THF, NMP, 0–5°, 1 h→

MeO, O, MeO, OMe, O (94%) ⇉

MeO, O, MeO, OH, NH_2, H, N, O, O, NH_2

aliskiren

Scheme 98

The Stille reaction[318] couples an organostannane with a variety of organic electrophiles. The air and moisture stability of organotin compounds, combined with their ready availability, make the Stille reaction one of the most extensively used coupling reactions. Moreover, because of the low reactivity of organotin reagents and their good functional group compatibility, this method is ideal for the synthesis of complex molecules. The configuration of the alkene is typically retained in the process, except when forcing reaction conditions are employed. One major drawback of the organotin-based reagents is their high toxicity. The synthesis of ripostatin A, a bacterial RNA-polymerase inhibitor, utilizes a double Stille cross-coupling sequence of diiodide **169** for the construction of a macrolactone ring precursor **170** (Scheme 99).[319]

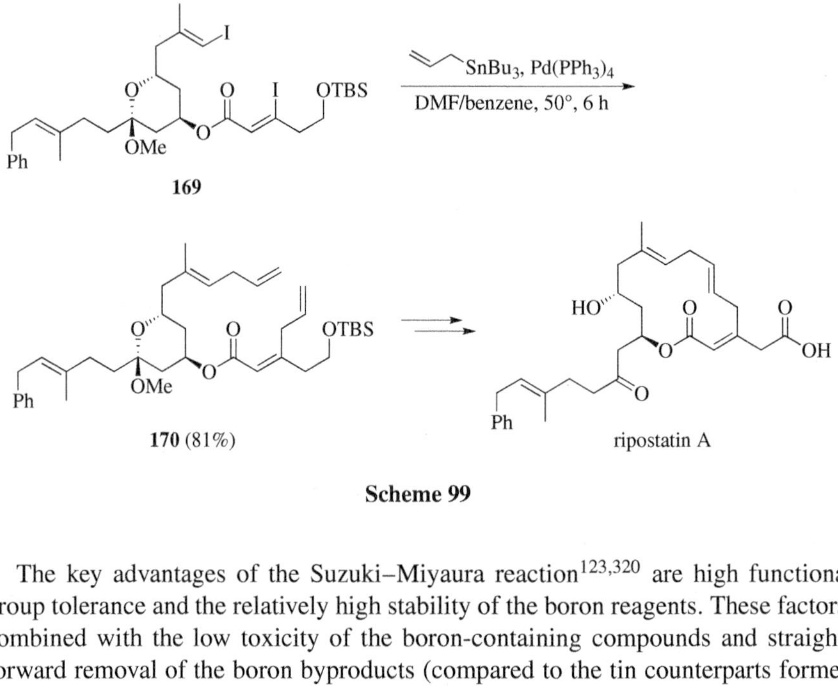

$SnBu_3$, $Pd(PPh_3)_4$
DMF/benzene, 50°, 6 h

169

170 (81%)

ripostatin A

Scheme 99

The key advantages of the Suzuki–Miyaura reaction[123,320] are high functional group tolerance and the relatively high stability of the boron reagents. These factors, combined with the low toxicity of the boron-containing compounds and straightforward removal of the boron byproducts (compared to the tin counterparts formed in the Stille coupling), often make this the reaction of choice. Since neutral boron compounds do not participate in coupling reactions, the use of Lewis bases is essential to generate an active boronate species. The Suzuki–Miyaura coupling also occurs with retention of the electrophile double-bond geometry. Thus, the stereospecific cross-coupling of (*E*)- and (*Z*)-tosylates **172** and **174** with aryl boronic acids **171** give the corresponding trisubstituted (*E*)- and (*Z*)-α,β-unsaturated esters **173** and **175**, respectively, in moderate-to-high yields (Scheme 100).[321]

Scheme 101 depicts a synthesis of synechoxanthin that involves a sequence of four Suzuki–Miyaura couplings, beginning with the reaction of aryl iodide **176** and functionalized building block **177**, in which the *N*-methyliminodiacetic acid (MIDA) boronate motif serves as a masked electrophile.[322] MIDA boronate **178** is

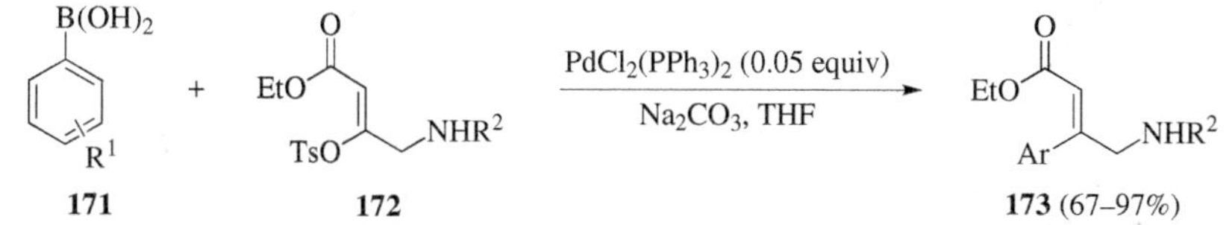

$B(OH)_2$ + EtO, O, TsO, NHR^2 → $PdCl_2(PPh_3)_2$ (0.05 equiv), Na_2CO_3, THF → EtO, O, Ar, NHR^2

171 R^1 **172** **173** (67–97%)

R^1 = H, F, MeO_2C, Me, HCO, NC–, MeO
R^2 = Cbz, Boc

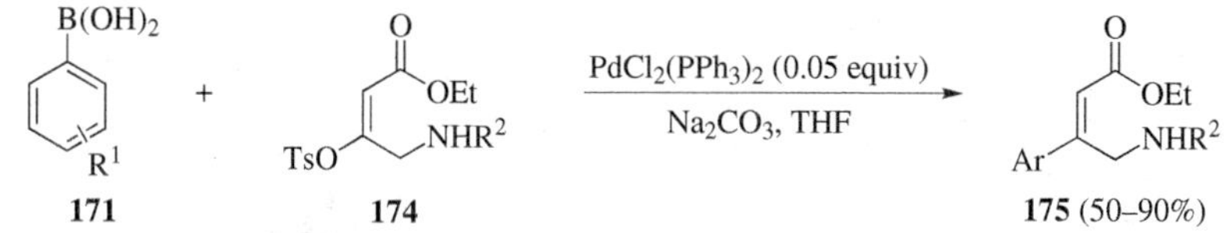

$B(OH)_2$ + O, OEt, TsO, NHR^2 → $PdCl_2(PPh_3)_2$ (0.05 equiv), Na_2CO_3, THF → O, OEt, Ar, NHR^2

171 R^1 **174** **175** (50–90%)

R^1 = H, F, MeO_2C, Me, MeS, MeO
R^2 = Cbz, Boc

Scheme 100

then halodeborylated and subjected to a second cross-coupling reaction with another molecule of **177**. To complete the polyene framework of synechoxanthin, an in situ MIDA boronate hydrolysis/two-directional double cross-coupling sequence of **179** and **180** yields synechoxanthin bismethyl ester; hydrolysis of the methyl esters affords the natural product.

The Hiyama–Denmark coupling involves the palladium-mediated reaction of organosilanes with organic halides.[323] Organosilanes are often inexpensive, and the silicon-containing byproducts are almost completely nontoxic. Interestingly, the Hiyama–Denmark reaction is less frequently employed than other coupling reactions because of the rather harsh conditions originally required. Inclusion of either a Lewis base (often a fluoride source due to its high affinity for silicon) or the use of an organosilane with an electronegative atom (such as fluoride or oxygen) is required because neutral silanes exhibit low reactivity towards transmetalation with palladium. A fluoride-free variant of the Hiyama–Denmark coupling has been introduced, and the reaction is now compatible with substrates bearing silyl protecting groups.[324,325]

In the Negishi reaction,[326] organozinc compounds are coupled with organic halides in the presence of palladium or nickel complexes to give alkenes. Organozinc compounds are moisture- and air-sensitive, and therefore these reactions must be performed in an oxygen- and water-free environment. However, the high functional-group tolerance and increased reactivity relative to other cross-coupling reactions make the Negishi coupling ideal for combining complex intermediates in the synthesis of natural products. This method has been employed for a gram-scale synthesis of the clinically relevant microtubule-stabilizing agent (+)-discodermolide (Scheme 102).[327,328] In this approach, in situ prepared organozinc **181** is coupled with the vinyl iodide **182** to forge the C14–C15 bond of the target product.

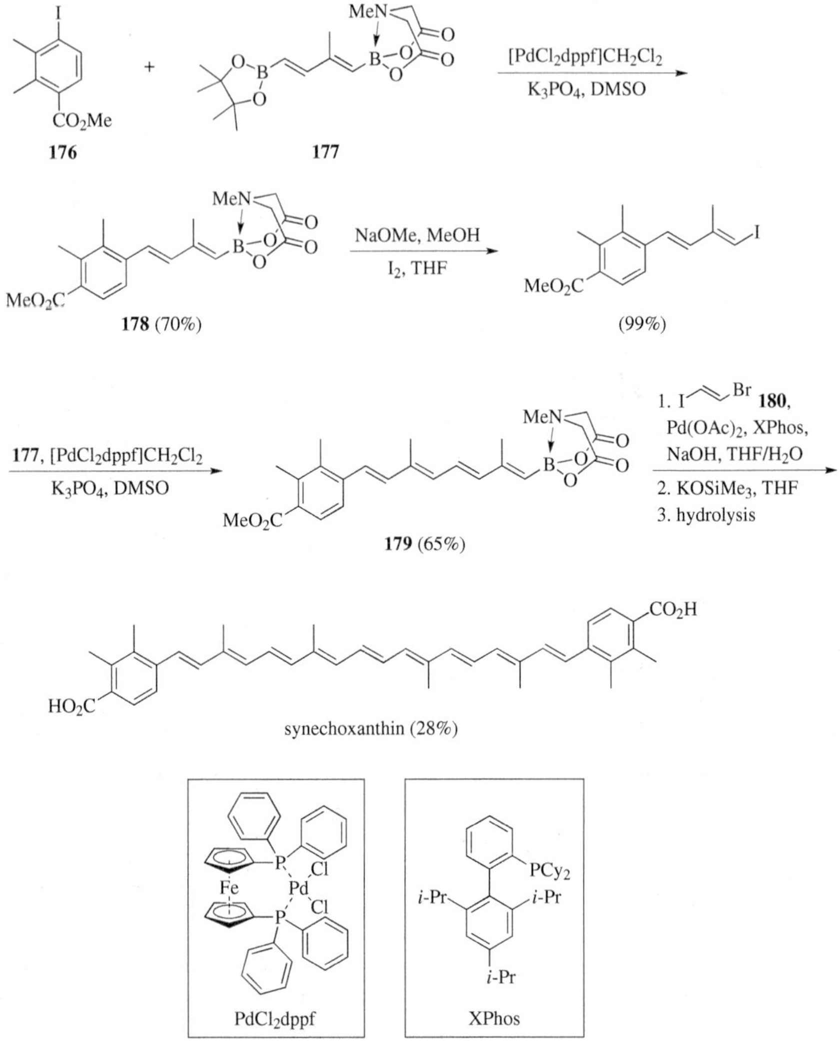

Scheme 101

EXPERIMENTAL CONDITIONS

Cross-metathesis reactions are typically performed under experimentally simple conditions in various solvents (including dichloromethane, 1,2-dichloroethane, benzene, toluene, ethyl acetate, alkyl carbonates, 2-methyltetrahydrofuran, and neat) and at a wide range of temperatures depending on the reactivity of the reactants and catalysts. Most of the reactions proceed best in oven-dried glassware, in moisture- and oxygen-free solvents, and under an inert atmosphere of either N_2 or Ar. Because of the increased activity and stability of metathesis complexes, a growing number of examples of metathesis processes are carried out in technical-grade solvents, as

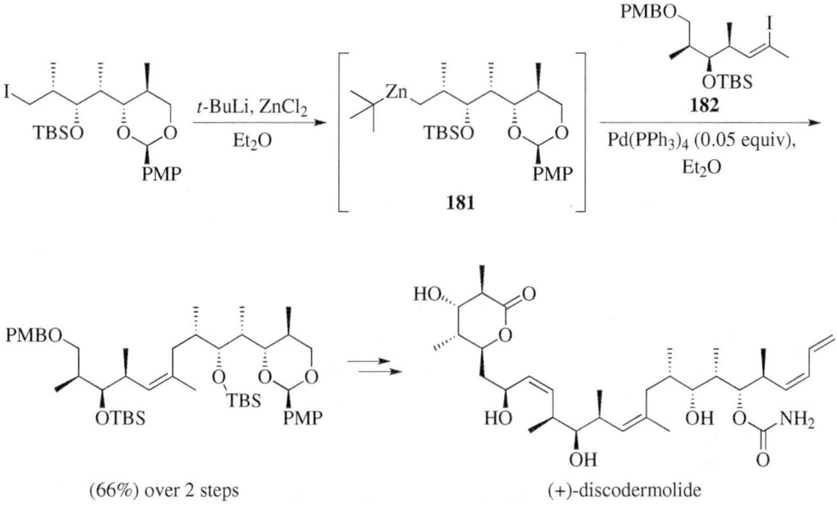

t-BuLi, $ZnCl_2$
Et_2O
181
182
$Pd(PPh_3)_4$ (0.05 equiv), Et_2O
(66%) over 2 steps
(+)-discodermolide

Scheme 102

well as in aqueous media. The olefin metathesis catalysts are usually stable in solid form, but their storage at reduced temperature in a refrigerator and/or in an inert gas atmosphere is strongly recommended. Owing to the large variety of metathesis catalysts used, it is difficult to provide general conditions for their handling, and instructions given by suppliers should be followed. Additives, such as Lewis acids (e.g., $Ti(Oi\text{-}Pr)_4$, $B(OPh)_3$, CuCl), phenols, or quinones can sometimes be used to increase the yield in the case of problematic CM reactions.

CAUTION: If a cross-metathesis reaction that leads to a volatile alkene byproduct (e.g., ethylene) is done on a large scale, it is important that appropriate precautions are taken to deal with the large volume of the flammable gas that is evolved.

The work-up for CM reactions usually involves either distillation or simple filtration through a short silica gel pad. In most cases, the reaction mixture is treated with ethyl vinyl ether or other scavengers before purification to deactivate the catalyst residues; self-scavenging complexes that are easily removable by filtration have gained in popularity.[329]

EXPERIMENTAL PROCEDURES

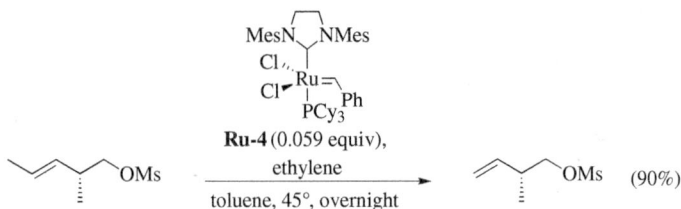

MesN NMes Cl Ru Ph Cl PCy_3
Ru-4 (0.059 equiv), ethylene
toluene, 45°, overnight
OMs → OMs (90%)

(*R*)-2-Methylbut-3-enyl Methanesulfonate [Ethenolysis of a Mesylate (Type I Alkene)].[330] A solution of (*R*)-2-methylpent-3-enyl methanesulfonate (40 mg, 0.22 mmol) in toluene (3.5 mL) was saturated with ethylene by bubbling ethylene gas through the solution for 10 min. The system was then heated to 45°, and a continuous ethylene flow was maintained. Into the solution was added a solution of catalyst **Ru-4** (11.5 mg, 0.013 mmol, 0.059 equiv) in toluene (1.5 mL) in three portions over an interval of 2 h. The mixture was stirred at 45° overnight while maintaining the ethylene flow. Upon completion of the reaction, the solvent was removed under reduced pressure, and the residue was purified by column chromatography on silica gel (hexane/EtOAc, 9:1) to give (*R*)-2-methylbut-3-enyl methanesulfonate as a colorless oil (33 mg, 90%): $[\alpha]^{22}{}_D$ + 5.0 (*c* 0.6, $CHCl_3$); ^1H NMR (400 MHz, $CDCl_3$) δ 5.74 (ddd, *J* = 17.4, 10.4, 7.0 Hz, 1H), 5.15 (dd, *J* = 17.3, 1.3 Hz, 2H), 4.12 (dd, *J* = 9.7, 6.6 Hz, 1H), 4.05 (dd, *J* = 9.6, 6.6 Hz, 1H), 2.99 (s, 3H), 2.64–2.57 (m, 1H), 1.09 (d, *J* = 6.84, 3H); ^{13}C NMR (100 MHz, $CDCl_3$) δ 138.3, 116.2, 73.3, 37.3, 37.2, 16.0; IR 3085, 1350, 1171, 957 cm^{-1}; LRMS (*m/z*): $[M + Na]^+$ 187, $[M + H]^+$ 165, 117, 102; HRMS (*m/z*): $[M + H]^+$ calcd for $C_6H_{13}O_3S$, 165.0585; found, 165.0581.

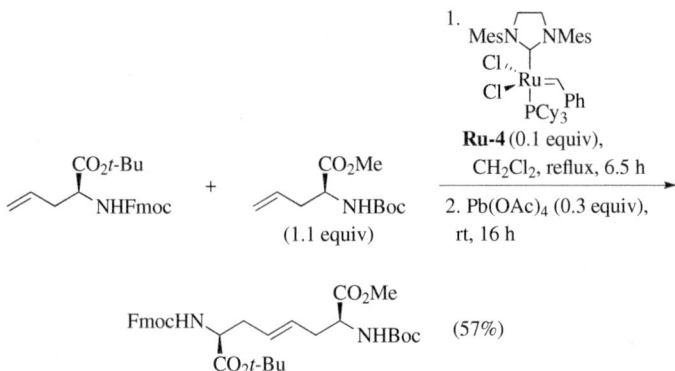

CO_2t-Bu, NHFmoc + CO_2Me, NHBoc (1.1 equiv)
1. MesN NMes Cl Ru Ph Cl PCy_3 **Ru-4** (0.1 equiv), CH_2Cl_2, reflux, 6.5 h
2. $Pb(OAc)_4$ (0.3 equiv), rt, 16 h
→ FmocHN, CO_2t-Bu, CO_2Me, NHBoc (57%)

(2*S*,7*S*,*E*)-*tert*-Butyl 7-(*tert*-Butoxycarbonylamino)-2-(9*H*-fluoren-9-ylmethoxycarbonylamino)-7-(methoxycarbonyl)hept-4-enoate [Cross-Metathesis of Two Type I Alkenes].[331] *tert*-Butyl (*S*)-2-((((9*H*-fluoren-9-yl)methoxy)carbonyl)amino)pent-4-enoate (335 mg, 0.852 mmol) and methyl (*S*)-2-((*tert*-butoxycarbonyl) amino)pent-4-enoate (216 mg, 0.942 mmol, 1.1 equiv) were added to an oven-dried, round-bottomed flask under Ar. Dichloromethane (8.5 mL) and catalyst **Ru-4** (73 mg, 0.085 mmol, 0.1 equiv) were added, and the reaction mixture was heated at reflux for 6.5 h under Ar. The reaction mixture was cooled to rt, and $Pb(OAc)_4$ (112 mg, 0.252 mmol, 0.3 equiv) was added. The mixture was stirred for 16 h at rt. The mixture was diluted with CH_2Cl_2 (ca. 15 mL) and filtered through a pad of Celite/silica gel (1:1), washing with CH_2Cl_2 (ca. 100 mL). The filtrate was evaporated under vacuum, and the crude product was purified via column chromatography on silica gel (toluene/diethyl ether, 9:1) to yield the title product (287 mg, 57%) as a colorless oil: $[\alpha]_D$ + 1.2 (*c* 0.5, MeOH); ^{1}H NMR (300 MHz, $CDCl_3$) δ 7.76 (d, J = 7.4 Hz, 2H), 7.63 (d, J = 7.3 Hz, 2H), 7.40 (app t, J = 7.4 Hz, 2H), 7.31 (app td, J = 7.3, 1.1 Hz, 2H), 5.52–5.35 (m, 3H), 5.18 (d, J = 9.5 Hz, 1H), 4.50–4.18 (m, 5H), 3.72 (s, 3H), 2.57–2.37 (m, 4H), 1.47 (s, 9H), 1.44 (s, 9H); ^{13}C NMR (75 MHz, $CDCl_3$) δ 172.7, 170.9, 155.9, 155.4, 144.1, 141.5, 128.8, 128.6, 127.9, 127.3, 125.4, 120.2, 82.5, 80.2, 67.2, 54.0, 53.3, 52.5, 47.4, 35.83, 35.78, 28.5, 28.3; IR (KBr) 3354 (br), 3065, 3041, 2978, 2934, 1716, 1508, 1451, 1367, 1352, 1249, 1221, 1161, 1081, 972, 847, 760, 740, 703 cm^{-1}; LRMS–EI (*m/z*): M$^+$ 594 (20), 550 (35), 494 (40), 465 (40), 393 (100), 351 (45). Anal. Calcd for $C_{33}H_{42}N_2O_8$•1.2 H_2O: C, 64.31; H, 7.26; N, 4.55. Found: C, 64.13; H, 7.33; N, 4.38.

2,3,4,5-Tetra-*O*-acetyl-α-pentadec-2-enyl D-Galactoside [Cross-Metathesis of an Allylic Ether (Type I Alkene)].[332] 2,3,4,5-Tetra-*O*-acetyl-α-allyl D-galactoside (388.7 mg, 1 mmol) was dissolved in 5 mL of dry CH_2Cl_2 in an oven-dried flask. A solution of 1-tetradecene (785.5 mg, 4 mmol, 4 equiv) and catalyst **Ru-1** (57.6 mg, 0.07 mmol, 0.07 equiv) in dry CH_2Cl_2 (5 mL) was added via syringe, and the reaction mixture was heated at reflux for about 16 h. Conversion of the starting material was monitored by TLC or by ^{1}H NMR spectroscopic analysis of an aliquot of the reaction mixture (e.g., using APT spectra). When complete consumption of the starting material had occurred, the solution was concentrated under vacuum and then directly chromatographed on a silica gel column (hexane, then hexane/EtOAc) to provide the title product (406.4 mg, 73%, (*E*)/(*Z*) = 88:12): ^{1}H NMR (CD_3OD, 400 MHz) δ 5.61 (dt, J = 15.0, 7.0 Hz, 1H), 5.38 (m, 1H), 5.35 (dd, J = 3.3, 1.0 Hz, 1H), 5.26 (m, 1H), 5.05–5.00 (m, 2H), 4.15 (t, J = 6.6 Hz, 1H), 4.07–3.94 (m, 3H), 3.85 (dd, J = 12.2, 6.9 Hz, 1H), 2.03 (s, 3H), 1.99–1.92 (m, 2H),

1.97 (s, 3H), 1.94 (s, 3H), 1.87 (s, 3H), 1.26 (m, 2H), 1.21–1.11 (m, 18H), 0.77 (t, J = 6.5 Hz, 3H); ^{13}C NMR ($CDCl_3$, 125 MHz) δ 170.0, 170.0, 169.9, 169.6, 135.9, 124.5, 94.7, 68.4, 67.9, 67.9, 67.5, 66.1, 61.5, 32.1, 31.7, 29.4, 29.4, 29.4, 29.3, 29.1, 30.0, 28.8, 22.5, 20.5, 20.4, 20.3, 13.9; MALDI–FTMS (m/z): $[M + Na]^+$ calcd for $C_{29}H_{48}O_{10}Na$, 579.3139; found, 579.3143.

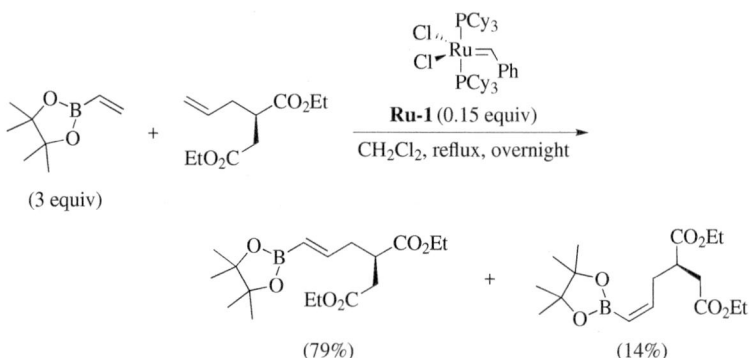

Ru-1 (0.15 equiv)
CH_2Cl_2, reflux, overnight
(3 equiv)
(79%) (14%)

Diethyl 2-(*R*)-(3-(4,4,5,5-Tetramethyl-1,3,2-dioxaborolan-2-yl)allyl)succinate [Cross-Metathesis of a Vinyl Boronic Ester (Type II Alkene)].[333] To a solution of diethyl (*R*)-2-allyl succinate (500 mg, 2.34 mmol) in CH_2Cl_2 (23 mL) under Ar was added pinacol vinylboronate (1.17 mL, 6.9 mmol, 3 equiv) and catalyst **Ru-1** (287 mg, 0.35 mmol, 0.15 equiv). The mixture was stirred at reflux overnight and then was cooled to ambient temperature, diluted with CH_2Cl_2, and filtered through a Celite pad. Concentration of the filtrate and purification of the residue by column chromatography on silica gel (heptane/EtOAc, 9:1 to 8:2) provided pure (*E*)-isomer (633 mg, 79%) and (*Z*)-isomer (111 mg, 14%) as white amorphous solids:

(*E*)-Isomer: ^{1}H NMR (300 MHz, $CDCl_3$) δ 6.51 (dt, J = 18.0, 13.0 Hz, 1H), 5.50 (d, J = 18.0 Hz, 1H), 4.15 (m, 4H), 2.97 (m, 1H), 2.69 (dd, J = 16.0, 9.5 Hz, 1H), 2.51 (m, 1H), 2.47 (dd, J = 16.0, 6.0 Hz, 1H), 2.38 (m, 1H), 1.26 (m, 18H); ^{13}C NMR (75 MHz, $CDCl_3$) δ 174.3, 172.1, 149.9, 83.4, 83.1, 60.6, 60.5, 41.0, 35.1, 34.1, 24.9, 24.8, 24.7, 14.2, 14.1; LRMS–ESI (m/z): $[M + Na]^+$ 363.

(*Z*)-Isomer: ^{1}H NMR (300 MHz, $CDCl_3$) δ 6.36 (dt, J = 14.5, 13.5 Hz, 1H), 5.48 (d, J = 13.5 Hz, 1H), 4.15 (m, 4H), 2.94 (m, 1H), 2.73 (m, 3H), 2.45 (dd, J = 16.0, 4.0 Hz, 1H), 1.27 (m, 18H); ^{13}C NMR (75 MHz, $CDCl_3$) δ 174.5, 172.3, 150.1, 83.6, 83.3, 60.8, 60.7, 41.2, 35.3, 34.3, 25.0, 24.9, 24.8, 14.4, 4.3; LRMS–ESI (m/z): $[M + Na]^+$ 363.

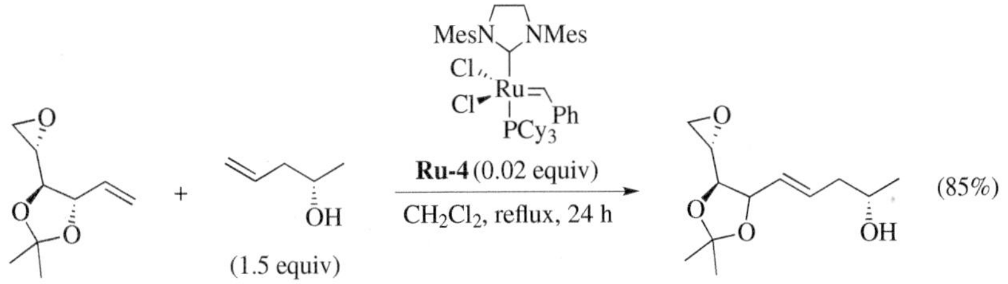

(S,E)-5-((4S,5R)-2,2-Dimethyl-5-((S)-oxiran-2-yl)-1,3-dioxolan-4-yl)pent-4-en-2-ol [Cross-Metathesis of an Acetal (Type II Alkene)].[334] To a stirred solution of (4*R*,5*S*)-2,2-dimethyl-4-((*S*)-oxiran-2-yl)-5-vinyl-1,3-dioxolane (1.10 g, 6.9 mmol) and (*S*)-pent-4-en-2-ol (885 mg, 10.3 mmol, 1.5 equiv) in dry CH_2Cl_2 (2 mL) was added catalyst **Ru-4** (117.16 mg, 0.138 mmol, 0.02 equiv) under an N_2 atmosphere. The reaction mixture was heated to reflux and stirred at that temperature for 24 h. The reaction mixture was concentrated under reduced pressure, and the residue was purified by column chromatography on silica gel (petroleum ether/EtOAc, 6:1) to give the title product as a colorless oil (1.33 g, 85%): $[\alpha]_D$ + 64.7 (*c* 1.3, $CHCl_3$); ^1H NMR (400 MHz, $CDCl_3$) δ 5.93–5.85 (m, 1H), 5.62–5.56 (m, 1H), 4.37 (t, *J* = 8.0 Hz, 1H), 3.91–3.83 (m, 1H), 3.51 (dd, *J* = 8.4, 5.6 Hz, 1H), 3.06 (ddd, *J* = 6.4, 3.6, 2.4 Hz, 1H), 2.82 (dd, *J* = 4.8, 4.0 Hz, 1H), 2.70 (dd, *J* = 4.8, 2.4 Hz, 1H), 2.25–2.23 (m, 2H), 1.67 (br s, 1H), 1.44 (s, 3H), 1.43 (s, 3H), 1.21 (d, *J* = 6.0 Hz, 3H); ^{13}C NMR (100 MHz, $CDCl_3$) δ 132.8, 129.8, 109.8, 80.3, 80.2, 67.2, 51.3, 44.9, 42.2, 27.1, 26.8, 23.0; IR 3433, 2985, 2921, 1597, 1454, 1376, 1219, 1160, 1112, 1058 cm^{-1}; HRMS–ESI (*m/z*): $[M + Na]^+$ calcd for $C_{12}H_{20}O_4Na$, 251.1253; found 251.1264.

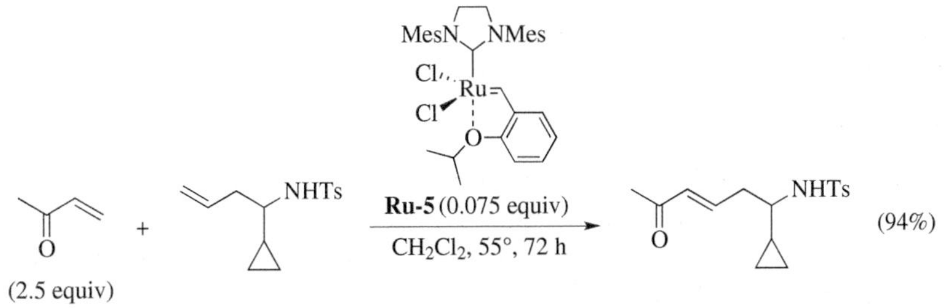

(E)-N-(1-Cyclopropyl-5-oxohex-3-en-1-yl)-4-methylbenzenesulfonamide [Cross-Metathesis of an α,β-Unsaturated Ketone (Type II Alkene)].[335] A resealable reaction tube was fitted with a magnetic stir bar and then charged with catalyst **Ru-5** (70.9 mg, 0.113 mmol, 0.075 equiv). The tube was then sealed with a rubber septum and purged with Ar. Argon-sparged CH_2Cl_2 (0.25 M with respect to the homoallylic sulfonamide employed), *N*-(1-cyclopropylbut-3-en-1-yl)-4-methylbenzenesulfonamide (400 mg, 1.51 mmol), and methyl vinyl ketone (0.31 mL, 3.77 mmol, 2.5 equiv) were then added sequentially via syringe. The rubber septum was replaced with a screw cap, and the tube was heated at 55° (oil-bath temperature) for 72 h. After cooling to rt, the reaction mixture was concentrated in vacuo, and the crude material was purified directly by column chromatography on silica gel (petroleum ether/EtOAc, 4:1) to afford the title product as a brown oil (436 mg, 94%): ^{1}H NMR ($CDCl_3$, 300 MHz) δ 7.62 (d, J = 8.0 Hz, 2H), 7.17 (d, J = 8.0 Hz, 2H), 6.61 (q, J = 7.5 Hz, 1H), 5.88 (d, J = 16.0 Hz, 1H), 5.18–5.05 (m, 1H), 2.58–2.26 (m, 6H), 2.05 (s, 3H), 0.74–0.60 (m, 1H), 0.43–0.31 (m, 1H), 0.26–0.18 (m, 1H), 0.23–0.14 (m, 1H), 0.07–0.04 (m, 1H); ^{13}C NMR ($CDCl_3$, 75 MHz) δ 198.4, 143.4, 143.2, 137.8, 133.8, 129.6, 127.0, 58.2, 38.8, 26.7, 21.5, 16.1, 4.1, 3.6; IR 1670, 1361, 1326, 1156, 1093, 981, 816, 665 cm^{-1}; HRMS–ESI (*m/z*): [M + Na]$^+$ calcd for $C_{16}H_{21}NNaO_3SNa$, 330.1134; found, 330.1141.

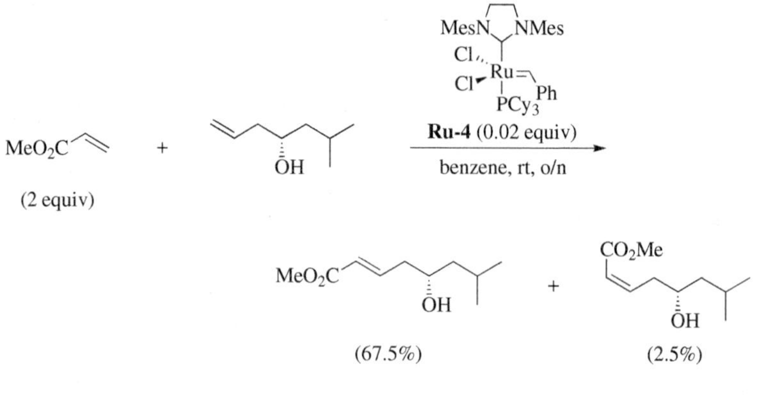

(5*R*)-5-Hydroxy-7-methyl-oct-2-enoic Acid Methyl Ester [Cross-Metathesis of Methyl Acrylate (Type II Alkene)].[336] 6-Methylhept-1-en-4-ol (200 mg, 1.56 mmol) was dissolved in benzene (7.8 mL) under N_2, and methyl acrylate (0.28 mL, 3.1 mmol, 2 equiv) was added, followed by catalyst **Ru-4** (26.5 mg, 31.2 μmol, 0.02 equiv). The reaction was stirred at rt overnight, and then was quenched by exposure to air. The solvent was removed under reduced pressure, and the crude product was purified by column chromatography on silica gel (hexane/EtOAc, 9:1 to 8:2) to give the (*E*)-product (196 mg) and the (*Z*)-product (7 mg) (203.4 mg, total yield 70%): R_f (hexane/EtOAc, 8:2) 0.36; ^{1}H NMR (500 MHz, $CDCl_3$) δ 6.94 (dddd, J = 16.0, 7.0, 7.0, 1.5 Hz, 1H), 5.85 (dd, J = 16.0, 1.5 Hz, 1H), 3.80–3.75 (m, 1H), 3.67 (s, 3H), 2.36–2.23 (m, 2H), 2.07 (br s, 1H), 1.76–1.67 (m, 1H), 1.37 (dddd, J = 15.5, 7.5, 5.5, 1.5 Hz, 1H), 1.19 (dddd, J = 15.0, 7.5, 4.5, 1.5 Hz, 1H), 0.87 (dd, J = 7.0, 1.5 Hz, 3H), 0.85 (dd, J = 7.0, 1.5 Hz, 3H); ^{13}C NMR (125 MHz, $CDCl_3$) δ 166.9, 145.8, 123.3, 68.5, 51.5, 46.2, 40.7, 24.5, 23.3, 21.9; IR (thin film) 3465, 2954, 2870, 1726, 1657, 1436, 1323, 1273, 1210, 1168, 1040, 986 cm^{-1}; HRMS–CI

(m/z): $[M + H]^+$ calcd for $C_{10}H_{19}O_3$, 187.1334; found, 187.332; $[M + NH_4]^+$ calcd for $C_{10}H_{18}O_3$, 204.1600; found, 204.1595.

MesN NMes
Cl
Ru
Cl
O
Ph
(30 equiv)
NHBz
CO_2Me
Ru-5 (0.05 equiv)
benzene, partial vacuum,
100°, 24 h
Ph
NHBz
CO_2Me
(97%)

Methyl (*S*)-2-Benzamido-5-phenylhex-4-enoate [Cross-Metathesis of α-Methylstyrene (Type III Alkene)].[337] Methyl (*S*)-2-benzamido-5-methylhex-4-enoate (50.0 mg, 0.191 mmol), α-methylstyrene (677 mg, 5.74 mmol, 30 equiv), and benzene (1 mL) were added to an oven-dried Schlenk vessel equipped with a magnetic stir bar. The vessel was sealed and subjected to a freeze–pump–thaw cycle to remove traces of oxygen. Catalyst **Ru-5** (6.0 mg, 0.0096 mmol, 0.05 equiv) was then added under a flow of N_2. The reaction vessel was placed under a partial vacuum until the solvent began to bubble. The tube was then sealed and heated to 100° for 24 h. The reaction mixture was cooled to rt, and then was distilled (0.2 mm Hg) with a water or liquid N_2 condenser to recover excess α-methylstyrene. The residue was purified by column chromatography on silica gel (hexane/EtOAc, 5:1 to 3:1) to afford the title product as a colorless oil (59.9 mg, 97%): ^1H NMR (400 MHz, $CDCl_3$, mixture of isomers) δ 7.80–7.78 (m, 2H), 7.51 (tt, J = 7.2, 1.6 Hz, 1H), 7.46–7.41 (m, 2H), 7.36–7.22 (m, 5H), 6.80 (d, J = 7.6 Hz, 1H), 5.70 (tq, J = 7.6, 1.2 Hz, 1H), 4.99 (dt, J = 7.6, 5.8 Hz, 1H), 3.80 (s, 3H), 2.99–2.77 (m, 2H), 2.04 (s, 3H); ^{13}C NMR (100 MHz, $CDCl_3$, mixture of isomers) δ 172.6, 167.1, 143.4, 139.4, 134.1, 131.9, 128.7, 128.4, 127.2, 127.1, 125.9, 121.2, 60.5, 52.7, 52.6, 31.7, 16.3; IR (neat) 3345, 1739, 1646, 1602, 1533, 1489, 1436, 1266, 1217, 1179 cm^{-1}; HRMS–ESI (m/z): $[M + Na]^+$ calcd for $C_{20}H_{21}NO_3Na$, 346.1414; found, 346.1417.

79 (0.05 equiv), Ru-4 (0.025 equiv)

CH_2Cl_2, 40°, 14 h

(1.5 equiv)

(86%) (*E*)/(*Z*) > 95:5

MesN NMes Cl Ru Cl PCy3 Ph

Ru-4

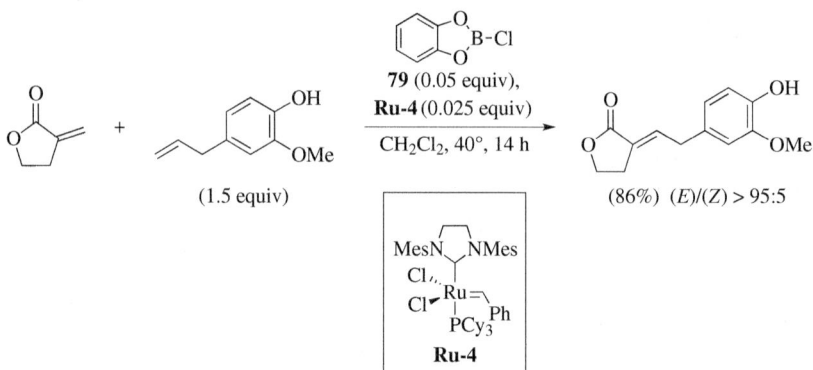

Dihydro-3-(2-(4-hydroxy-3-methoxyphenyl)ethylidene)furan-2(3*H*)-one [Cross-Metathesis of a Lactone with an Exocyclic Double Bond (Type III Alkene)].[173] To a stirred solution of α-methylene-γ-butyrolactone (0.5 mmol, 44 μL) and 4-allyl-2-methoxyphenol (0.75 mmol, 123 mg, 1.5 equiv) in CH_2Cl_2 (1.2 mL) was added B-chlorocatecholborane (**79**) (3.86 mg, 0.025 mmol, 0.05 equiv) and catalyst **Ru-4** (10.19 mg, 0.012 mmol, 0.025 equiv). The reaction mixture was heated at 40° under N_2 for 14 h. The solvent was then removed under reduced pressure, and the residue was purified by column chromatography on silica gel (hexane/EtOAc, 20:1, changing to EtOAc/MeOH, 4:1) to provide the title product as a colorless oil (100.7 mg, 86%, (*E*)/(*Z*) > 95:5): ^{1}H NMR (400 MHz, $CDCl_3$) δ 6.82 (m, 1H), 6.76 (d, J = 8.5 Hz, 1H), 6.62–6.57 (m, 2H), 5.66 (br s, 1H, OH), 4.31 (app t, J = 7.5 Hz, 2H), 3.78 (s, 3H), 3.38 (d, J = 7.5 Hz, 2H), 2.84–2.80 (m, 2H); ^{13}C NMR (400 MHz, $CDCl_3$) δ 171.3 (s), 146.8 (s), 144.5 (s), 138.9 (d), 129.4 (s), 125.8 (s), 121.1 (d), 114.5 (d), 111.0 (d), 65.5 (t), 56.0 (q), 35.9 (t), 25.1 (t); IR 3396, 2920, 1742, 1676, 1601, 1513, 1430, 1379, 1268, 1205, 1177, 1150, 1122, 1027 cm^{-1}; LCMS–EI (*m/z*): M$^+$ 234 (100), 189 (60), 175 (14), 157 (20), 129 (21), 115 (18); HRMS–EI (*m/z*): [M + H]$^+$ calcd for $C_{13}H_{15}O_4$, 235.09649; found, 235.09674.

MesN NMes
Cl Ru Cl O NO2

Ru-20 (0.05 equiv), $B(OPh)_3$ (0.25 equiv)

O_2N (2.1 equiv) + OTBS → CH_2Cl_2, 40°, 5 h → O_2N OTBS (70%) (*E*)/(*Z*) = 88:12

***tert*-Butyldimethyl((7-nitrohept-5-en-1-yl)oxy)silane [Cross-Metathesis of an Allylic Nitro Compound (Type III Alkene)].**[179] To a solution of 5-hexenyl *tert*-butyldimethylsilyl ether (236.1 mg, 1.10 mmol) and 3-nitropropene (199.4 mg, 2.29 mmol, 2.1 equiv) in CH_2Cl_2 (4.5 mL) was added triphenyl borate (80.2 mg, 0.28 mmol, 0.25 equiv). Catalyst **Ru-20** was then added in one portion (37.3 mg, 0.06 mmol, 0.05 equiv). The resulting mixture was heated at 40° for 5 h under an Ar atmosphere. After cooling the reaction mixture to rt, the solvent was removed under reduced pressure. The crude product was purified by preparative TLC (cyclohexane/EtOAc, 20:1), and the title product was obtained as a yellow oil (210.8 mg, 70%, (*E*)/(*Z*) = 88:12) that solidified in a refrigerator: ^1H NMR ($CDCl_3$, 400 MHz, (*E*)-isomer) δ 5.92 (dtt, $J = 15.2, 6.7, 1.0$ Hz, 1H), 5.75 (dtt, $J = 15.3, 7.2, 1.4$ Hz, 1H), 4.88 (dd, $J = 7.2, 0.9$ Hz, 2H), 3.61 (t, $J = 6.1$ Hz, 2H), 2.18–2.13 (m, 2H), 1.58–1.40 (m, 4H), 0.89 (s, 9H), 0.47 (s, 6H); ^{13}C NMR ($CDCl_3$, 100 MHz, (*E*)-isomer) δ 141.6, 118.7, 77.6, 62.8, 32.2, 32.0, 25.9, 24.8, 18.4, 5.3; IR (film) 2952, 2931, 2897, 2858, 1608, 1557, 1519, 1483, 1472, 1463, 1429, 1375, 1361, 1341, 1271, 1256, 1137, 1104, 1006, 972, 952, 911, 836, 813, 799, 776, 742, 661, 653 cm^{-1}; LRMS–ESI (*m/z*): $[M + Na]^+$ 296; HRMS–ESI (*m/z*): $[M + Na]^+$ calcd for $C_{13}H_{27}NO_3SiNa$, 296.1652; found, 296.1651.

MesN NMes
Cl
Ru
O Cl O
MeO

Cl–CH=CH–Cl (100 equiv) + 4-methoxystyrene →[Ru-21 (0.15 equiv); neat, reflux, 6 h] 1-(2-chloroethenyl)-4-methoxybenzene (91%) (*E*)/(*Z*) = 17:83

1-[(*E*,*Z*)-2-Chloroethenyl]-4-methoxybenzene [Cross-Metathesis of a Vinyl Halide (Type III Alkene)].[105] To a solution of 4-methoxystyrene (80.5 mg, 0.60 mmol) in (*E*)-1,2-dichloroethene (5.8 g, 4.6 mL, 60 mmol, 100 equiv) was added catalyst **Ru-21** (60.3 mg, 0.09 mmol, 0.15 equiv). The resulting mixture was stirred at reflux for 6 h, cooled to rt, and (*E*)-1,2-dichloroethene was removed under reduced pressure. The crude product was purified by preparative TLC (cyclohexane/EtOAc, 20:1) to give the title product as a colorless oil (92 mg, 91%, (*E*)/(*Z*) = 17:83): ^{1}H NMR (400 MHz, $CDCl_3$) δ 7.68–7.64 (m, 2H), 7.25–7.21 (m, 2H), 6.93–6.89 (m, 2H), 6.88–6.84 (m, 2H), 6.78 (d, J = 13.7 Hz, 1H), 6.57 (d, J = 8.1 Hz, 1H), 6.51 (d, J = 13.7 Hz, 1H), 6.16 (d, J = 8.1 Hz, 1H), 3.83 (s, 3H), 3.81 (s, 3H); ^{13}C NMR (100 MHz, $CDCl_3$) δ 159.5, 159.3, 132.7, 130.7, 128.6, 127.6, 127.3, 126.9, 116.4, 115.4, 114.2, 113.6, 55.2; IR (film) 3079, 3005, 2957, 2837, 1606, 1574, 1510, 1462, 1345, 1304, 1257, 1174, 1033, 833, 787, 691, 536 cm^{-1}; LRMS–EI (*m*/*z*): M$^+$ 168 (100), 153 (56), 125 (40), 89 (33), 63 (21); HRMS–EI (*m*/*z*): M$^+$ calcd for C_9H_9ClO, 168.0342; found, 168.0347. Anal. Calcd for C_9H_9ClO: C, 64.11; H, 5.38; Cl, 21.02. Found: C, 64.25; H, 5.39; Cl, 20.84.

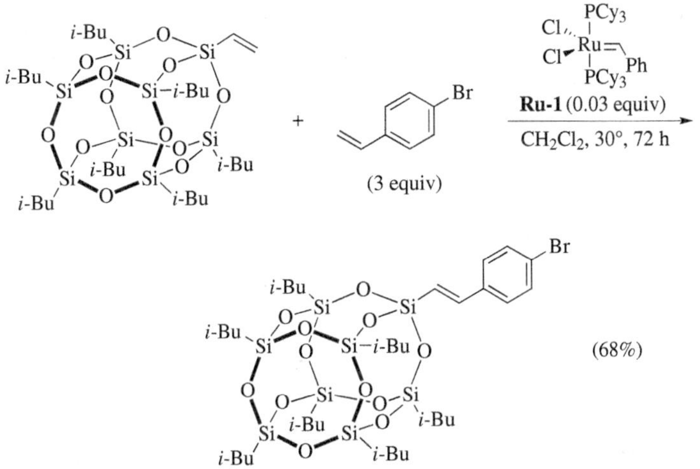

(*E*)-2-(4-Bromophenyl)vinylheptaisobutyl-T8-silsesquioxane [Cross-Metathesis of a Vinyl Siloxane (Type III Alkene)].[338] Vinylhepta-isobutyl-T8-silsesquioxane (3.5 g, 4.15 mmol) and catalyst **Ru-1** (0.068 g, 0.083 mmol, 0.02 equiv) were placed in a Schlenk flask at rt under N_2. Dry CH_2Cl_2 (4.9 mL) and 4-bromostyrene (2.27 g, 1.63 mL, 12.4 mmol, 3 equiv) were added to the flask by syringe, and the mixture was magnetically stirred at 30°. After 48 h, an additional portion of catalyst **Ru-1** (0.034 g, 0.041 mmol, 0.01 equiv) was added, and the mixture was further stirred at 30° for 24 h. After cooling to rt, the reaction solution was poured into MeOH (100 mL), and the crude product was collected by filtration, washed with MeOH, and purified by column chromatography on silica gel (*n*-hexane) to provide the title product (68%). The product was determined to be the (*E*)-isomer by ^{1}H NMR spectroscopic analysis: ^{1}H NMR ($CDCl_3$, 400 MHz) δ 7.47 (d, J = 8.4 Hz, 2H), 7.30 (d, J = 8.4 Hz, 2H), 7.10 (d, J = 19.1 Hz, 1H), 6.12 (d, J = 19.1 Hz, 1H), 1.88 (m, 7H), 0.97 (d, J = 6.6 Hz, 18H), 0.96 (d, J = 6.6 Hz, 24H), 0.65 (d, J = 7.0 Hz, 6H), 0.62 (d, J = 7.1 Hz, 8H); ^{13}C NMR ($CDCl_3$, 100 MHz) δ 146.66, 136.56, 131.69, 128.27, 122.61, 119.67, 25.71, 23.88, 23.86, 22.50, 22.43; ^{29}Si NMR ($CDCl_3$, 80 MHz) δ −67.39, −67.83, −80.17; FTIR (KBr) 2954, 2926, 2906, 2870, 1610, 1486, 1464, 1400, 1383, 1367, 1332, 1230, 1105, 1039, 1010, 989, 952, 839, 782, 742 cm^{-1}; HRMS–FAB (*m/z*): $[M + H]^+$ calcd for $C_{36}H_{70}BrO_{12}Si_8$, 997.2205; found, 997.2209.

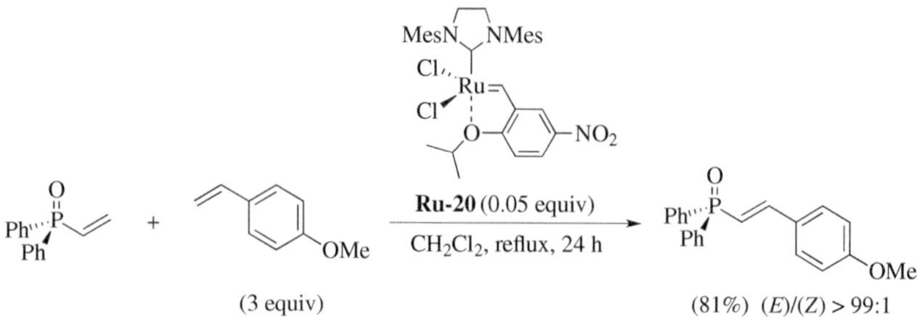

(*E*)-[2-(4-Methoxyphenyl)vinyl]diphenylphosphine Oxide [Cross-Metathesis of a Vinyl Phosphine Oxide (Type III Alkene)].[215] To a mixture of the vinylphosphine oxide (62 mg, 0.27 mmol) and catalyst **Ru-20** (9.1 mg, 0.014 mmol, 0.05 equiv) in CH_2Cl_2 (13 mL) was added 4-methoxystyrene (108.7 mg, 0.11 mL, 0.81 mmol, 3 equiv) via syringe. The resulting mixture was stirred at reflux for 24 h, and then the solvent was removed under reduced pressure. The crude product was purified by column chromatography on silica gel (EtOAc, then EtOAc/methanol, 10:1 to 5:1) to afford the title product as a colorless solid (73 mg, 81%, (*E*)/(*Z*) > 99:1): ^{1}H NMR (400 MHz, $CDCl_3$) δ 7.79–7.72 (m, 4H), 7.57–7.51 (m, 2H), 7.51–7.45 (m, 6H), 7.45–7.38 (m, 1H), 6.94–6.86 (m, 2H), 6.67 (dd, J = 22.3, 17.3 Hz, 1H), 3.84 (s, 3H); ^{13}C NMR (125 MHz, $CDCl_3$) δ 161.2 (s), 147.2 (d, J = 4.2 Hz), 133.1 (d, J = 106.0 Hz), 131.8 (d, J = 2.6 Hz), 131.4 (d, J = 10.3 Hz), 129.4 (s), 128.5 (d, J = 12.1 Hz), 127.9 (d, J = 17.7 Hz), 116.0 (d, J = 106.0 Hz), 114.2 (s), 55.4 (s); ^{31}P NMR (162 MHz, $CDCl_3$) δ 24.9; LRMS–EI (*m/z*): M^+ 334 (100), 319 (11.81), 257 (9.83), 202 (50.93), 198 (13), 183 (11.37), 155 (20.90), 77 (10.11), 47 (11.35); HRMS–EI (*m/z*): M^+ calcd for $C_{21}H_{19}O_2P$, 334.1122; found, 334.1116. Anal. Calcd for $C_{21}H_{19}O_2P$: C, 75.45; H, 5.68. Found: C, 75.51; H, 5.69.

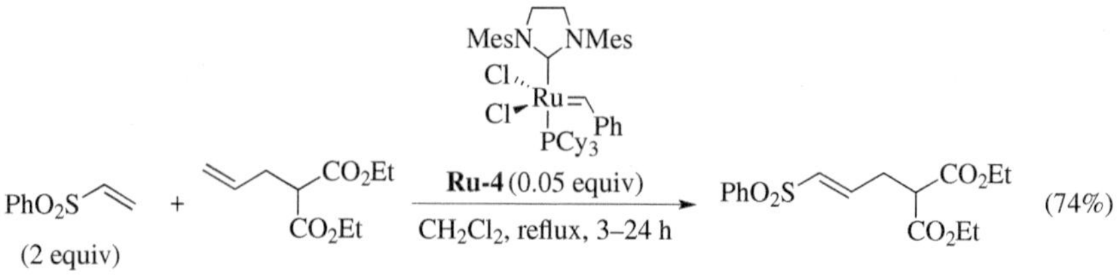

Diethyl 2-[(*E*)-3-(Phenylsulfonyl)-2-propenyl]malonate [Cross-Metathesis of a Vinyl Sulfone (Type III Alkene)].[228] To a mixture of diethyl 2-allylmalonate (50.1 mg, 0.25 mmol) and vinyl sulfone (84.1 mg, 0.50 mmol, 2 equiv) in CH_2Cl_2 (15 mL) was added a solution of Grubbs second-generation catalyst **Ru-4** (10.6 mg, 0.0125 mmol, 0.05 equiv) in CH_2Cl_2 (5 mL). The resulting mixture was stirred at reflux for 3–24 h, and the solvent was removed under reduced pressure. The crude product was purified by column chromatography on silica gel (cyclohexane/EtOAc) to give the title compound as a colorless oil (63 mg, 74%): ^{1}H NMR (200 MHz, $CDCl_3$) δ 7.78–7.14 (m, 5H), 6.48 (dt, J = 15.8, 1.2 Hz, 1H), 6.15 (dt, J = 15.8, 7.0 Hz, 1H), 4.21 (q, J = 7.1 Hz, 4H), 3.48 (t, J = 7.4 Hz, 1H), 2.80 (td, J = 7.4, 1.2 Hz, 2H), 1.23 (t, J = 7.1 Hz, 6H); ^{13}C NMR (50 MHz, $CDCl_3$) δ 168.9, 137.1, 132.8, 128.5, 127.4, 126.2, 125.6, 77.5, 77.3, 76.7, 61.5, 52.0, 32.2, 14.1; IR (film) 2984, 2938, 1731, 1447, 1321, 1308, 1148, 1087, 821, 596 cm^{-1}; LRMS–EI (m/z): 276 (23), 202 (36), 157 (16), 130 (16), 129 (100), 128 (31), 117 (34), 115 (20), 91 (11); HRMS–LSI (m/z): $[M + H]^+$ calcd for $C_{16}H_{21}O_6S$, 341.1059; found, 341.1040.

n-BuO (10 equiv) + TBSO OTBS → Mo-3 (0.01 equiv), benzene-d_6, 22°, 30 min → TBSO On-Bu OTBS (90%) er 86:14 (E)/(Z) = 4:96

Mo-3: N, N, Mo, Ph, O, Br, Br, TBSO

(5*R*,6*S*)-5-((*Z*)-2-Butoxyvinyl)-2,2,3,3,8,8,9,9-octamethyl-6-vinyl-4,7-dioxa-3,8-disiladecane [AROCM of a Vinyl Ether (Type III Alkene)].[339] In a N_2-filled glovebox, an oven-dried 4-mL vial equipped with a magnetic stir bar was charged with the catalyst **Mo-3** (16 μL, 0.02 M in benzene, 0.32 μmol) and *n*-butyl vinyl ether (32.0 mg, 0.320 mmol, 10 equiv). The resulting mixture was stirred for 5 min and then was added by syringe to a solution of 3,4-bis(*tert*-butyldimethylsilyloxy)-*cis*-cyclobutene (10.0 mg, 31.8 μmol) in benzene-d_6 in a 4-mL vial, and the resulting solution was stirred for 30 min at 22°. When the starting material had been consumed, the reaction was quenched by exposure to air or by addition of wet diethyl ether, and then the mixture was concentrated under vacuum (conversion was determined by ^{1}H NMR spectroscopic analysis). The resulting brown oil was purified by column chromatography on silica gel (hexane/Et_2O, 9:1) to afford the desired product as a colorless oil (11.1 mg, 90% yield, (*E*)/(*Z*) = 4:96, er 86:14): $[\alpha]_D^{20}$ – 19.2 (*c* 0.500, $CHCl_3$); ^{1}H NMR (400 MHz, $CDCl_3$) δ 5.90 (dd, J = 6.3, 1.0 Hz, 1H), 5.83 (ddd, J = 17.2, 10.4, 6.1 Hz, 1H), 5.14 (ddd, J = 17.4, 3.5, 2.2 Hz, 1H), 5.10 (ddd, J = 10.5, 3.0, 2.1 Hz, 1H), 4.52 (ddd, J = 8.9, 4.1, 1.1 Hz, 1H), 4.33 (dd, J = 9.3, 6.4 Hz, 1H), 4.00 (ddd, J = 7.4, 6.1, 2.8 Hz, 1H), 3.70 (ddd, J = 13.2, 7.0, 4.0 Hz, 2H), 1.60–1.52 (m, 2H), 1.40 (ddd, J = 22.1, 14.5, 7.5 Hz, 2H), 0.90 (t, J = 7.2 Hz, 3H), 0.87 (s, 9H),

0.85 (s, 9H), 0.034 (s, 3H), 0.03 (s, 3H), 0.01 (s, 3H), 0.005 (s, 3H); ^{13}C NMR (100 MHz, $CDCl_3$) δ 146.0, 139.1, 115.2, 108.0, 78.2, 72.4, 71.0, 32.1, 26.3, 26.2, 19.3, 18.7, 18.5, 14.0, –4.2, –4.3, –4.6; IR (neat) 2956, 2929, 2857, 1663, 1472, 1463, 1405, 1377, 1251, 1217, 1112, 1083, 1005, 940, 919, 834, 776, 671 cm^{-1}; HRMS–ESI (*m/z*): $[M + Na]^+$ calcd for $C_{22}H_{46}O_3Si_2$, 437.2899; found, 437.2907. The enantiomeric ratio was determined by HPLC analysis of the corresponding alcohol (after removal of the silyl group) by comparison with racemic material (t_R (major) 17.34 min, t_R (minor) = 16.43 min; OJ column, hexane/*i*-PrOH, 98/2, 1.0 mL/min, 220 nm).

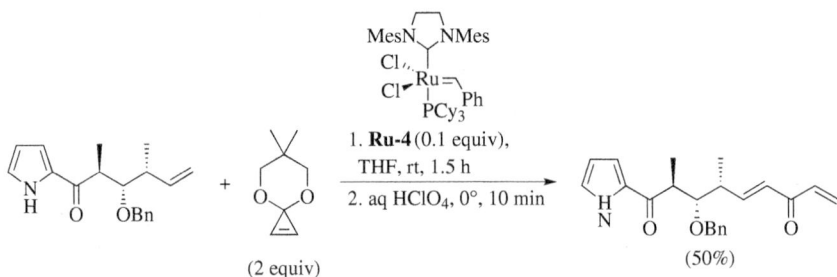

MesN NMes Cl Ru Ph Cl PCy$_3$

1. **Ru-4** (0.1 equiv), THF, rt, 1.5 h

2. aq $HClO_4$, 0°, 10 min

(2 equiv) (50%)

3-Benzyloxy-2,4-dimethyl-1-(1*H*-pyrrol-2-yl)nona-5,8-diene-1,7-dione [Tandem ROCM of a Cyclopropenone Ketal (Type II Alkene)].[340] Catalyst **Ru-4** (25 mg, 0.030 mmol, 0.1 equiv) was dissolved in anhydrous THF (2 mL) and treated dropwise over 30 min with a solution of 3-benzyloxy-2,4-dimethyl-1-(1*H*-pyrrol-2-yl) -hex-5-en-1-one (90 mg, 0.30 mmol) and the cyclopropenone ketal (84 mg, 0.60 mmol, 2 equiv) in THF (3 mL) at rt. After 1.5 h, the reaction mixture was cooled to 0°, treated with $HClO_4$ (0.050 mL, 30% aqueous solution), and stirred at 0° for 10 min, at which point TLC analysis indicated complete consumption of the starting material. The reaction mixture was quenched with saturated $NaHCO_3$ solution and then diluted with EtOAc/hexane, and the biphasic mixture was washed with brine. The organic layer was dried over $MgSO_4$, filtered, and concentrated. The residue was purified by column chromatography on silica gel (hexane/EtOAc, 5:1 to 3:1) to provide the title compound as a pale-brown oil (53 mg, 50%): $[\alpha]^{26}_D$ – 31.2 (*c* 1.0, $CHCl_3$); ^{1}H NMR (500 MHz, $CDCl_3$) δ 9.33 (br s, 1H), 7.22–7.16 (m, 3H), 7.08 (dd, *J* = 16.0, 7.0 Hz, 1H), 7.08–7.04 (m, 1H), 7.02–6.97 (m, 3H), 6.58 (dd, *J* = 17.5, 11.0 Hz, 1H), 6.43 (dd, *J* = 16.0, 1.0 Hz, 1H), 6.31 (dd, *J* = 4.5, 3.0 Hz, 1H), 6.29 (dd, *J* = 17.5, 1.0 Hz, 1H), 5.83 (dd, *J* = 11.0, 1.0 Hz, 1H), 4.38 (d, *J* = 11.0 Hz, 1H), 4.34 (d, *J* = 11.0 Hz, 1H), 3.95 (dd, *J* = 9.5, 3.0 Hz, 1H), 3.50 (dq, *J* = 9.5, 7.0 Hz, 1H), 2.76–2.68 (m, 1H), 1.22 (d, *J* = 7.0 Hz, 3H), 1.18 (d, *J* = 7.0 Hz, 3H); ^{13}C NMR (125 MHz, $CDCl_3$) δ 192.9 (s), 189.6 (s), 151.7 (d), 138.0 (s), 134.7 (d), 132.1 (s), 128.4 (t), 128.0 (d), 127.5 (d), 127.3 (d), 127.2 (d), 125.5 (d), 117.2 (d), 110.6 (d), 83.9 (d), 74.6 (t), 44.2 (d), 38.6 (d), 14.8 (q), 11.9 (q); LRMS–APCI (*m/z*): $[M + H]^+$ 352.

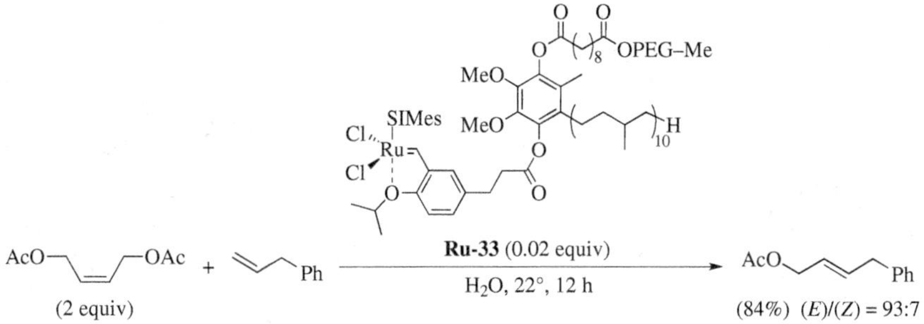

4-Phenyl-2-butenyl Acetate [Cross-Metathesis in Water].[32,341] Allylbenzene (31 mg, 0.26 mmol), (*Z*)-but-2-ene-1,4-diyl diacetate (92 mg, 0.53 mmol, 2 equiv), and the catalyst **Ru-33** (19 mg, 0.005 mmol, 0.02 equiv) were sequentially added to a Biotage 2–5 mL microwave reactor vial containing a Teflon-coated stir bar at rt. The vial was sealed with a septum, and H_2O (0.5 mL) was added via syringe. The resulting solution was stirred at 22° for 12 h. The homogeneous reaction mixture was then diluted with EtOAc (5 mL), filtered through a pad of silica gel layered over a pad of Celite, and the pad was washed with EtOAc (3 × 10 mL). The resulting solution was concentrated in vacuo, and the crude product was purified by column chromatography on silica gel (hexane/EtOAc, 96:4) to afford the title product as a colorless oil (42 mg, 84%, (*E*)/(*Z*) = 93:7): ^{1}H NMR (300 MHz, $CDCl_3$) δ 7.34–7.17 (m, 5H), 5.92 (m, 1H), 5.65 (m, 1H), 4.54 (m, 2H), 3.41 (d, J = 3.3 Hz, 2H), 2.06 (br s, 3H); ^{13}C NMR (75 MHz, $CDCl_3$) δ 171.4, 135.1, 134.0, 129.2, 129.1, 126.8, 125.8, 65.5, 60.8, 39.2, 21.6; HRMS–EI (*m/z*): $[M - H]^-$ calcd for $C_{12}H_{13}O_2$, 189.0916; found, 189.0916.

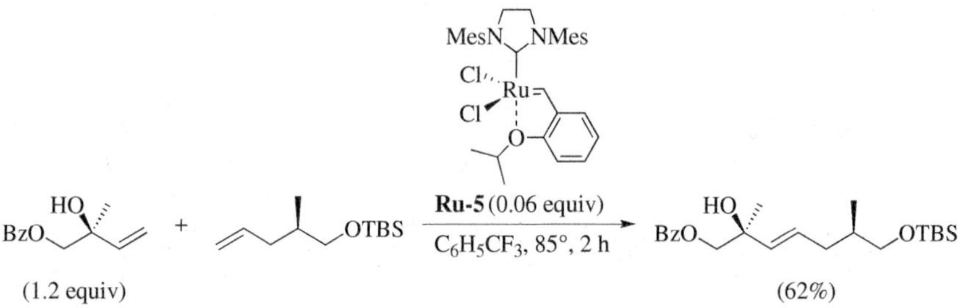

(2*S*,6*R*,*E*)-7-(*tert*-Butyldimethylsilyloxy)-2-hydroxy-2,6-dimethylhept-3-enyl Benzoate [Cross-Metathesis in a Fluorinated Solvent].[342] Catalyst **Ru-5** (37.6 mg, 0.06 mmol, 0.06 equiv) was added to a mixture of (*R*)-*tert*-butyldimethyl(2-methylpent-4-enyloxy)silane (0.21 g, 1.0 mmol) and (*S*)-2-hydroxy-2-methylbut-3-enyl benzoate (0.25 g, 1.2 mmol, 1.2 equiv) in $C_6H_5CF_3$ (10 mL) under an N_2 atmosphere at rt. The mixture was heated to 85° and stirred at that temperature for 2 h. The reaction mixture was allowed to cool to rt and then was concentrated under vacuum. The residual crude product was purified by column chromatography on silica gel (hexane/EtOAc, 10:1) to afford the title product as colorless oil (0.24 g, 62%): $[\alpha]_D$ + 4.4 (*c* 1.10, $CHCl_3$); ^{1}H NMR (300 MHz, $CDCl_3$) δ 8.11–7.99 (m, 2H), 7.63–7.36 (m, 3H), 5.90–5.70 (m, 1H), 5.60 (d, J = 15.55 Hz, 1H), 4.25 (q, J = 11.12 Hz, 2H), 3.39 (d, J = 6.06 Hz, 1H), 2.20 (m, 1H), 1.96–1.58 (m, 2H), 1.39 (s, 3H), 0.88 (s, 9H), 0.82 (d, J = 6.70 Hz, 3H), 0.02 (s, 6H); ^{13}C NMR (75 MHz, $CDCl_3$) δ 166.5, 134.5, 133.1, 129.9, 129.6, 129.0, 128.4, 72.1, 71.5, 67.6, 36.0, 35.9, 25.9, 25.1, 18.3, 16.3, −5.4; IR (neat) 3447, 2956, 2857, 1723, 1603, 1452, 1274, 1114, 836, 711 cm^{-1}; HRMS–ESI (*m*/*z*): $[M + Na]^+$ calcd for $C_{22}H_{36}O_4SiNa$, 415.2281; found, 415.2267.

MesN NMes
Cl Ru Ph
Cl
Cy_3P

EtO$_2$C + AcO OAc (2 equiv)

Ru-6 (4 × 0.05 equiv)
$C_6F_5CF_3$, 120°,
MW (4 × 10 min)

EtO$_2$C OAc
(51%) (*E*)/(*Z*) = 82:18

(*E*/*Z*)-Ethyl 3-(3-Acetoxyprop-1-enyl)-2,2-dimethylcyclopropanecarboxylate [Cross-Metathesis Assisted by Microwave Irradiation].[40] To a microwave vial charged with *cis*-1,4-diacetoxy-2-butene (172.2 mg, 1 mmol, 2 equiv), ethyl chrysanthemate (98 mg, 0.5 mmol), and $C_6H_5CF_3$ (2 mL) was added the catalyst **Ru-6** (23.7 mg, 0.025 mmol, 0.05 equiv). The vial was flushed with Ar, placed in a microwave reactor, and heated at 120° for 10 min. Progress of the reaction was monitored by TLC and ^{1}H NMR spectroscopic analysis. Three additional portions of **Ru-6** (23.7 mg, 0.025 mmol, 0.05 equiv) were added; each addition was followed by heating in the microwave reactor at 120° for 10 min. The vial contents were transferred to a round-bottomed flask, and the solvent was removed under vacuum. The crude residue was purified by column chromatography on silica gel (cyclohexane/EtOAc, 20:1) to provide the title compound as a brownish oil (61.3 mg, 51%, (*E*)/(*Z*) = 82:18): ^{1}H NMR (500 MHz, $CDCl_3$) δ 5.75 (td, J = 15.3, 6.5 Hz, 1H), 5.53 (dd, J = 15.3, 8.7 Hz, 1H), 4.53 (d, J = 6.5 Hz, 2H), 4.18–4.07 (m, 2H), 2.06 (s, 3H), 1.57 (d, J = 5.3 Hz, 1H), 1.29–1.24 (m, 7H), 1.17 (s, 3H); ^{13}C NMR (125 MHz, $CDCl_3$) δ 171.7, 170.8, 133.1, 125.9, 64.8, 60.4, 35.3, 34.2, 28.7, 22.1, 21.0, 20.2, 14.3; IR (film) 2981, 2953, 2876, 1741, 1726, 1668, 1459, 1448, 1430, 1380, 1341, 1277, 1231, 1177, 1148, 1115, 1096, 1075, 1057, 1026, 967, 844, 607 cm^{-1}; HRMS–ESI (*m*/*z*): $[M + Na]^+$ calcd for $C_{13}H_{20}O_4Na$, 263.1254; found, 263.1264. Anal. Calcd for $C_{13}H_{20}O_4$: C, 64.98; H, 8.39. Found: C, 64.99; H, 8.24.

MesN NMes
Cl Ru
Cl O

MeO_2C + CO_2Me

Ru-5 (2 × 0.005 equiv),
NaCl (3.9 equiv)
milling, 2 × 45 min

CO_2Me
MeO_2C
(93%)

Dimethyl 4,4′-Stilbenedicarboxylate [Mechanochemically Induced Cross-Metathesis].[343] Methyl 4-vinylbenzoate (324.4 mg, 2 mmol) was added to a 10-mL Teflon jar charged with a single stainless steel ball bearing. Sodium chloride (450 mg, 150 wt %, 3.9 equiv) and catalyst **Ru-5** (6.0 mg, 0.01 mmol, 0.005 equiv) were added to the jar. The reaction mixture was then milled in a Retsch MM400 mill for 45 min. Another portion of catalyst **Ru-5** (6.0 mg, 0.01 mmol, 0.005 equiv) was added to the reaction mixture, which was then milled for an additional 45 min. The catalyst and salt were then removed from the purple mixture by adding diethylene glycol vinyl ether (a catalyst quenching agent) (500 ppm) and deionized water (3–4 mL) to the crude reaction mixture. After milling at 30 Hz for an additional 15–60 minutes (depending on the quenching agent used), the solid was collected by filtration and washed with deionized water. Any remaining purple residues and salt were removed by washing with cold EtOAc and water, and the resulting white solid was dried in air to afford the product as a white powder (275.4 mg, 93%): ^{1}H NMR (300 MHz, $CDCl_3$) δ 8.05 (d, $J = 7.95$ Hz, 4H), 7.59 (d, $J = 7.62$ Hz, 4H), 7.23 (s, 2H), 3.93 (s, 6H); ^{13}C NMR (75 MHz, $CDCl_3$) δ 166.8, 141.2, 130.1, 130.0, 129.5, 126.6, 52.2; HRMS (*m*/*z*): $[M]^+$ calcd for $C_{18}H_{17}O_4$, 297.1121; found, 297.1124.

TABULAR SURVEY

Efforts were made to include all cross-metathesis (CM) reactions that appeared in the literature through the beginning of 2020. Tables are organized according to the classification presented in the Scope and Limitations section.

It is difficult to systematize examples of the cross metathesis since in each reaction two alkene substrates are used. To facilitate the positioning of a given reaction, the tables are arranged according to the CM alkene type. The alkenes are assigned to one of three types according to the classification of Grubbs (Type I: alkenes that undergo rapid homodimerization; Type II: alkenes that undergo slow homodimerization; Type III: alkenes that do not undergo homodimerization). As Type IV alkenes do not undergo CM or homodimerization, they are not included. The first table contains CM reactions of ethylene and of alkenes of Type I that do not possess functional groups in an allylic or vinylic position. It is assumed, in most cases, that functional groups that are more remote from the reacting C–C double bond do not affect the metathesis reactions. The second table shows CM of allylic boronates. The third table shows CM of protected and unprotected allylic amines, etc. Table 33 contains examples of CM with vinylic organosulfur compounds—challenging partners of Type III in Grubbs' classification. The titles of the individual tables are listed in the Table of Contents.

To find a given CM reaction, locate each of the two CM partners using the table of contents, and then examine the table of the higher number. For example, if the first of the CM partners of a given reaction is a primary allylic alcohol (Type I, Table 4) and the second CM partner is a 4-substituted styrene (Type II, Table 10), this CM reaction is placed in Table 10. Similarly, a CM reaction between ethyl acrylate (Type II, Table 20A) and an allyl boronic ester (Type I, Table 2) is placed in Table 20A. Therefore, any given CM reaction is set in the Tabular Survey according to its more senior CM partner.

A consequence of the classification system used is that the same alkene may be found in several different tables. The only exception from the rules given above are ethenolysis reactions (i.e., CM with ethylene), which are all included in Table 1 for simplicity.

Within each table the individual entries are arranged by increasing carbon count of the alkene which appears in the first column. For alkenes with the same core structure, the cross partners are arranged by increasing carbon count. The carbon count does not include protecting groups, nor heteroatom substituents when their omission allows a better comparison of similar core structures. Compounds within the same carbon count are grouped based on structural similarities. The yields for the reactions are given in parentheses, followed by the ratio of (*E*)/(*Z*) isomers or enantiomers where applicable. Unspecified yields are denoted by (—).

The Charts preceding the Tables contain the structures of the bold-numbered catalysts; the numbers in the Tables do not correspond to those in the text. In addition to the abbreviations listed in "*The Journal of Organic Chemistry* Standard Abbreviations and Acronyms" the following are used in the Tables:

ADMET	acyclic diene metathesis polymerization
Ad	adamantyl
BHT	butylated hydroxytoluene, dibutylhydroxytoluene
BMI	bismaleimide
BMIM	1-butyl-3-methylimidazolium ion
BOM	benzyloxymethyl
B(pin)	boronic acid pinacol ester
BQ	1,4-benzoquinone
Brij30	polyoxyethylene(4)lauryl ether (surfactant)
CAM	carbamoylmethyl
ClBcat	B-chlorocatecholborane
2-Cl-BQ	2-chloro-1,4-benzoquinone
2,6-Cl_2-BQ	2,6-dichloro-1,4-benzoquinone
CM	cross-metathesis
CSA	camphorsulfonic acid
DB24C8	dibenzo-24-crown-8
DCB	dichlorobenzene
Dipp	2,6-diisopropylphenyl
DMC	dimethylcarbonate
DMP	Dess–Martin periodinane
2,2-DMP	2,2-dimethoxypropane
DVB	divinylbenzene
FAH	fluorinated aromatic hydrocarbon
F_4-BQ	2,3,5,6-tetrafluoro-1,4-benzoquinone
Fc	ferrocene
IC	incomplete conversion
IMes	1,3-bis(2,4,6-trimethylphenyl)imidazolium
IPr	1,3-bis-(2,6-diisopropylphenyl)imidazole-2-ylidene
MAP	monoalkoxidepyrrolide
MIDA	*N*-methyliminodiacetic acid
Mipp	2-methyl-6-isopropylphenyl
MS	molecular sieves
MW	microwave irradiation
Naph	naphthyl
NB	2-nitrobenzyl
o-NBSH	2-nitrobenzenesulfonyl hydrazine
NMM	*N*-methylmorpholine
p-NOS	4-nitrobenzensulfonyl (nosyl)
o-Ns	2-nitrobenzensulfonyl
PAC	phenacyl

PBB	polybrominated biphenyl
Pbf	pentamethyl-2,3-dihydrobenzofuran-5-sulfonyl
PC	propylene carbonate
Phoban	9-phosphabicyclo-[3.3.1]-nonane
Phth	phthaloyl
PMP	4-methoxyphenyl
PNB	4-nitrobenzoate
n-Pn	*n*-pentyl
PNP	4-nitrophenyl
PPTS	pyridinium *p*-toluenesulfonate
PSS	polyoxyethanyl β-sitosterol sebacate (surfactant)
PTS	polyoxyethanyl α-tocopheryl sebacate (surfactant)
RCM	ring-closing cross-metathesis
ROMP	ring-opening metathesis polymerization
rr	regioisomer ratio
SAMP	(*S*)-*N*-amino-2-(methoxymethyl)pyrrolidyl
SDS	sodium dodecyl sulfate (surfactant)
SES	2-(trimethylsilyl)ethylsulfonyl
SIMes	1,3-bis(2,4,6-trimethylphenyl)-4,5-dihydroimidazol-2-ylidene
TBDPS	*tert*-butyldiphenylsilyl
TDS	(dimethyl)thexylsilyl
TIS	triisopropylsilane
TPGS	*D*-α-tocopherol polyethylene glycol succinate (surfactant)
TPS	tripropylsilyl
Trip	2,4,6-triisopropylphenyl
Troc	trichloroethyloxycarbonyl
Trt	trityl
TxDMS	thexyldimethylsilyl

CHART 1. RUTHENIUM CATALYSTS USED IN THE TABLES

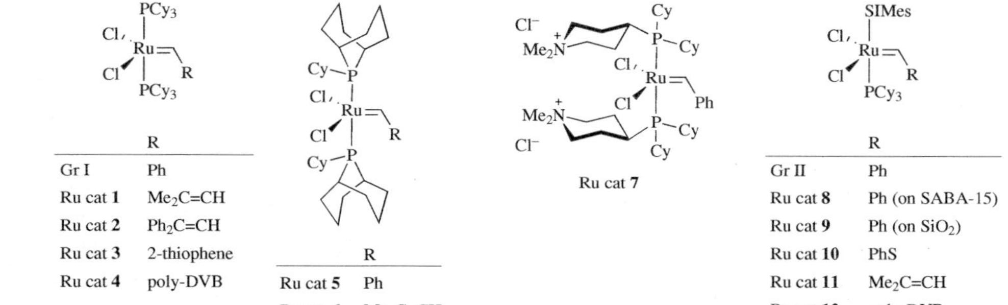

	R
Gr I	Ph
Ru cat **1**	Me_2C=CH
Ru cat **2**	Ph_2C=CH
Ru cat **3**	2-thiophene
Ru cat **4**	poly-DVB

	R
Ru cat **5**	Ph
Ru cat **6**	Me_2C=CH

Ru cat **7**

	R
Gr II	Ph
Ru cat **8**	Ph (on SABA-15)
Ru cat **9**	Ph (on SiO_2)
Ru cat **10**	PhS
Ru cat **11**	Me_2C=CH
Ru cat **12**	poly-DVB

	R^1	R^2
Ru cat **13**	Me	H
Ru cat **14**	F	F
Ru cat **15**	*i*-Pr	*i*-Pr

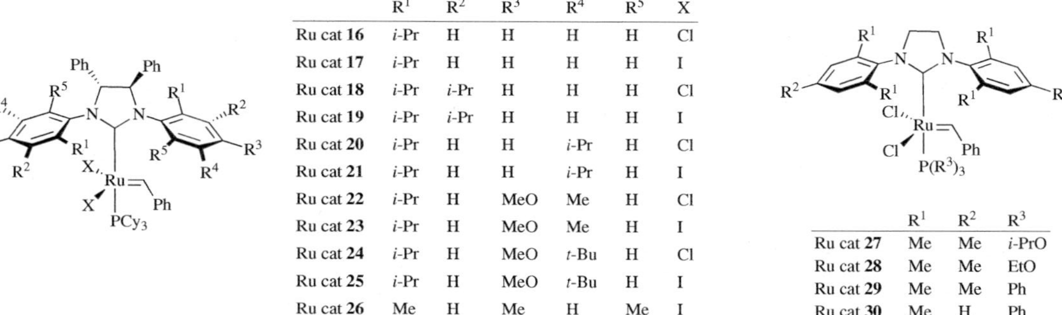

	R^1	R^2	R^3	R^4	R^5	X
Ru cat **16**	*i*-Pr	H	H	H	H	Cl
Ru cat **17**	*i*-Pr	H	H	H	H	I
Ru cat **18**	*i*-Pr	*i*-Pr	H	H	H	Cl
Ru cat **19**	*i*-Pr	*i*-Pr	H	H	H	I
Ru cat **20**	*i*-Pr	H	H	*i*-Pr	H	Cl
Ru cat **21**	*i*-Pr	H	H	*i*-Pr	H	I
Ru cat **22**	*i*-Pr	H	MeO	Me	H	Cl
Ru cat **23**	*i*-Pr	H	MeO	Me	H	I
Ru cat **24**	*i*-Pr	H	MeO	*t*-Bu	H	Cl
Ru cat **25**	*i*-Pr	H	MeO	*t*-Bu	H	I
Ru cat **26**	Me	H	Me	H	Me	I

	R^1	R^2	R^3
Ru cat **27**	Me	Me	*i*-PrO
Ru cat **28**	Me	Me	EtO
Ru cat **29**	Me	Me	Ph
Ru cat **30**	Me	H	Ph

CHART 1. RUTHENIUM CATALYSTS USED IN THE TABLES (*Continued*)

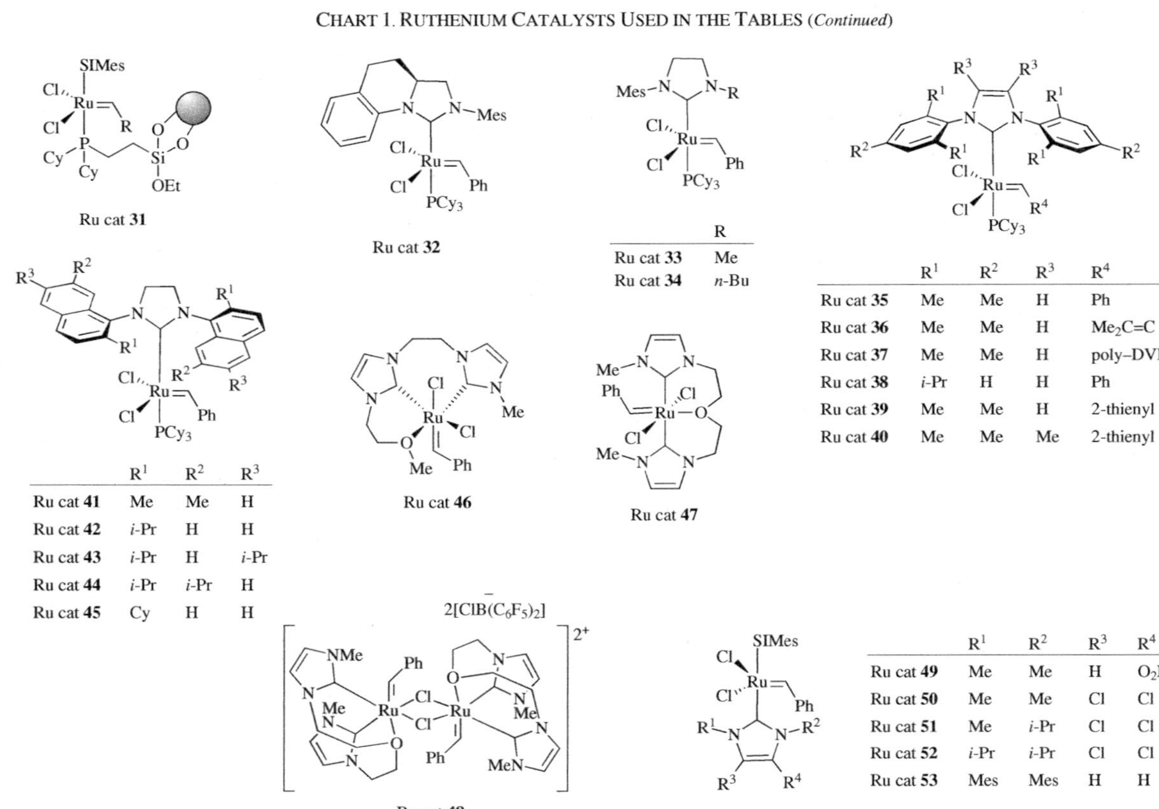

	R
Ru cat **33**	Me
Ru cat **34**	*n*-Bu

	R^1	R^2	R^3	R^4
Ru cat **35**	Me	Me	H	Ph
Ru cat **36**	Me	Me	H	Me_2C=C
Ru cat **37**	Me	Me	H	poly–DVB
Ru cat **38**	*i*-Pr	H	H	Ph
Ru cat **39**	Me	Me	H	2-thienyl
Ru cat **40**	Me	Me	Me	2-thienyl

	R^1	R^2	R^3
Ru cat **41**	Me	Me	H
Ru cat **42**	*i*-Pr	H	H
Ru cat **43**	*i*-Pr	H	*i*-Pr
Ru cat **44**	*i*-Pr	*i*-Pr	H
Ru cat **45**	Cy	H	H

	R^1	R^2	R^3	R^4
Ru cat **49**	Me	Me	H	O_2N
Ru cat **50**	Me	Me	Cl	Cl
Ru cat **51**	Me	*i*-Pr	Cl	Cl
Ru cat **52**	*i*-Pr	*i*-Pr	Cl	Cl
Ru cat **53**	Mes	Mes	H	H

	R^1	R^2	R^3
Ru cat **54**	H	H	H
Ru cat **55**	H	Br	H
Ru cat **56**	Me	H	H
Ru cat **57**	Me	H	Me

	R
Ru cat **58**	IMes
Ru cat **59**	SIMes

Ru cat **60**

	L	R^1	R^2
Ru cat **61**	PCy_3	Ph	Me
Ru cat **62**	SIMes	Ph	Me
Ru cat **63**	SIMes	Ph	Ph
Ru cat **64**	SIMes	PhS	Me

	R^1	R^2	R^3	R^4	R^5
Ru cat **65**	Me	H	H	H	H
Ru cat **66**	Me	Me	H	H	H
Ru cat **67**	Me	Me	Br	Br	H
Ru cat **68**	Me	Me	Br	H	polyvinyl
Ru cat **69**	*i*-Pr	H	H	H	H

Ru cat **70**

Ru cat **71**

Ru cat **72**

Ru cat **73**

Ru cat **74**

	L
Ru cat **75**	Cy_3P
Ru cat **76**	IPr

CHART 1. RUTHENIUM CATALYSTS USED IN THE TABLES (*Continued*)

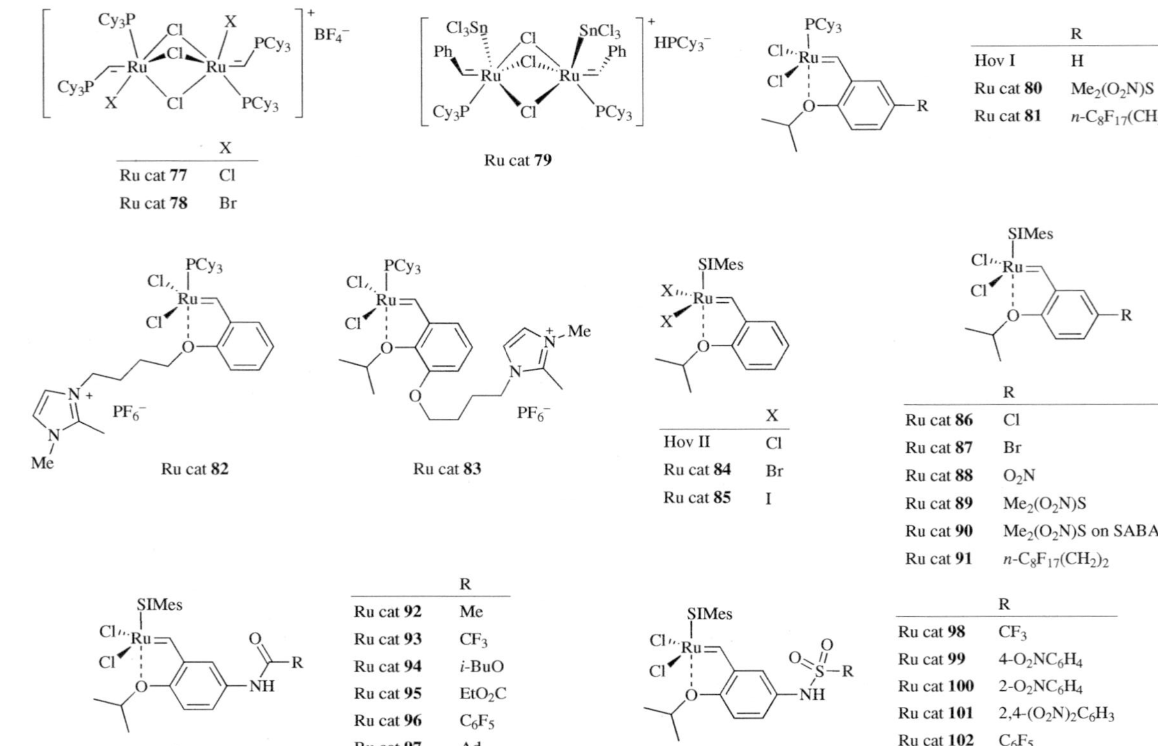

	X
Ru cat **77**	Cl
Ru cat **78**	Br

	R
Hov I	H
Ru cat **80**	$Me_2(O_2N)S$
Ru cat **81**	n-$C_8F_{17}(CH_2)_3$

	X
Hov II	Cl
Ru cat **84**	Br
Ru cat **85**	I

	R
Ru cat **86**	Cl
Ru cat **87**	Br
Ru cat **88**	O_2N
Ru cat **89**	$Me_2(O_2N)S$
Ru cat **90**	$Me_2(O_2N)S$ on SABA-15
Ru cat **91**	n-$C_8F_{17}(CH_2)_2$

	R
Ru cat **92**	Me
Ru cat **93**	CF_3
Ru cat **94**	*i*-BuO
Ru cat **95**	EtO_2C
Ru cat **96**	C_6F_5
Ru cat **97**	Ad

	R
Ru cat **98**	CF_3
Ru cat **99**	4-$O_2NC_6H_4$
Ru cat **100**	2-$O_2NC_6H_4$
Ru cat **101**	2,4-$(O_2N)_2C_6H_3$
Ru cat **102**	C_6F_5

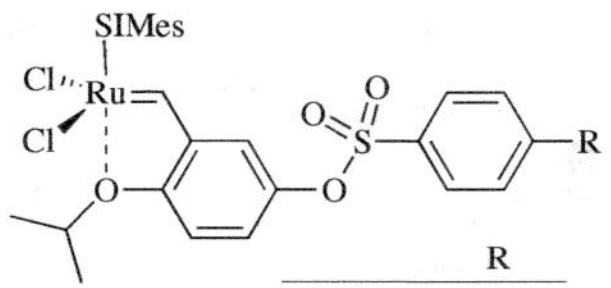

	R
Ru cat **103**	H
Ru cat **104**	MeO
Ru cat **105**	O_2N

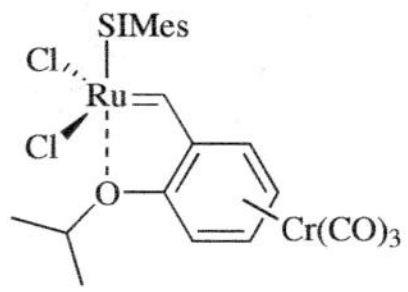

Ru cat **106**

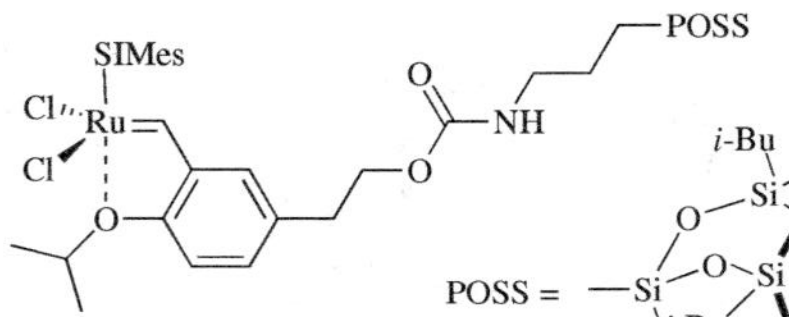

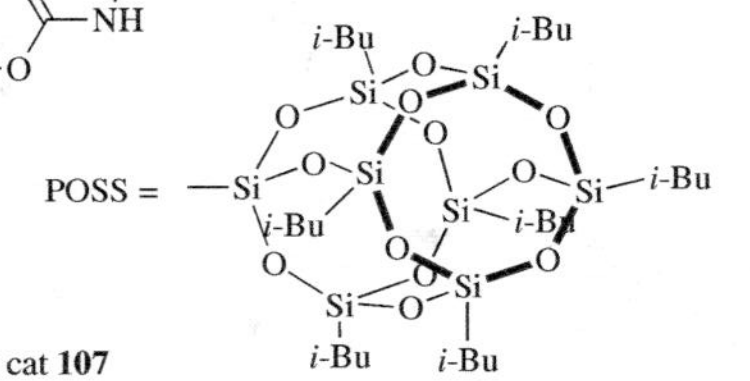

POSS =

Ru cat **107**

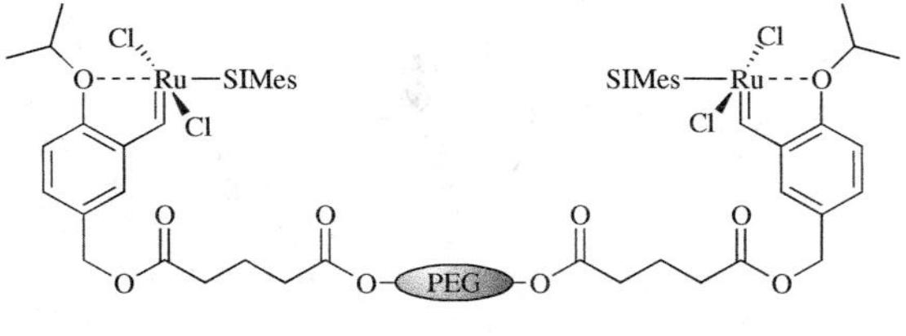

Ru cat **108**

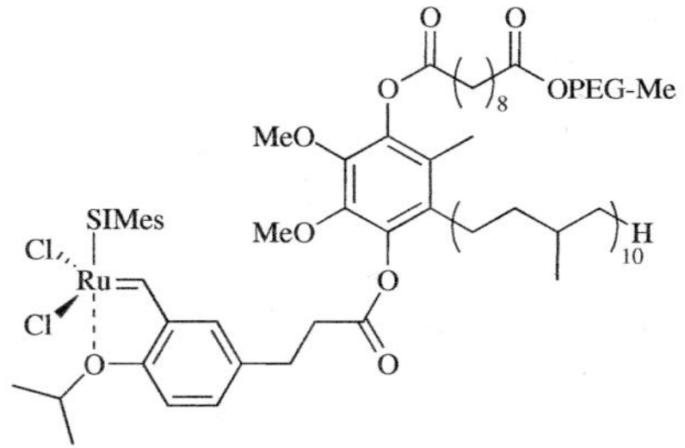

Ru cat **109**

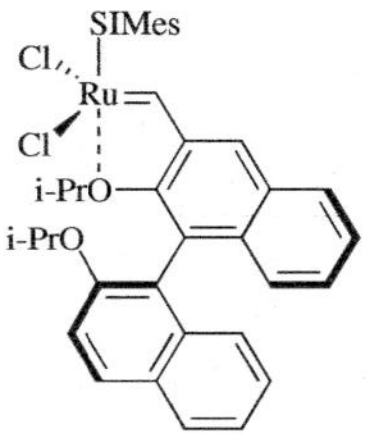

Ru cat **110**

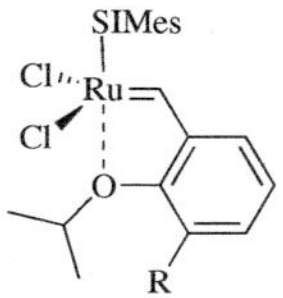

	R
Ru cat **111**	MeO
Ru cat **112**	Ph

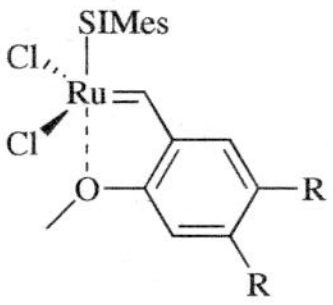

	R
Ru cat **113**	H
Ru cat **114**	MeO

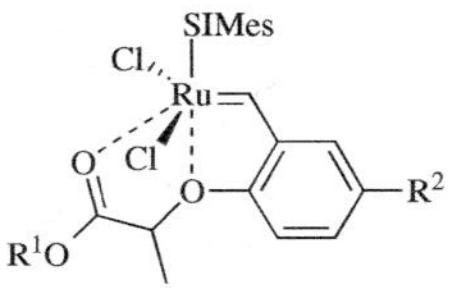

	R^1	R^2
Ru cat **115**	Me	H
Ru cat **116**	Me	O_2N
Ru cat **117**	Me	Ac
Ru cat **118**	Et	H
Ru cat **119**	Et	Me
Ru cat **120**	Et	MeO
Ru cat **121**	Et	O_2N

CHART 1. RUTHENIUM CATALYSTS USED IN THE TABLES (*Continued*)

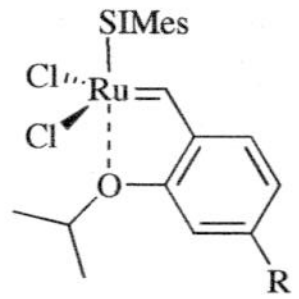

	R
Ru cat **122**	O_2N
Ru cat **123**	$Et_2HN^+TsO^-$
Ru cat **124**	$Et_2MeN^+Cl^-$
Ru cat **125**	Et_2MeN^+poly-O_3S^-
Ru cat **126**	$Et_2MeN^+I^-$
Ru cat **127**	$Et_2MeN^+PF_6^-$

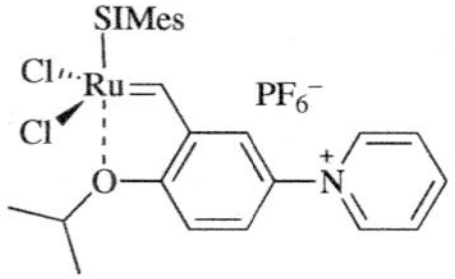

Ru cat **128**

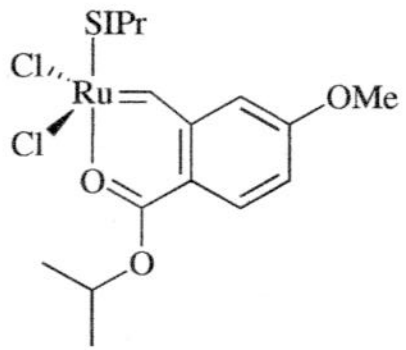

Ru cat **129**

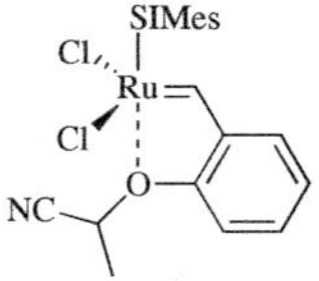

Ru cat **130**

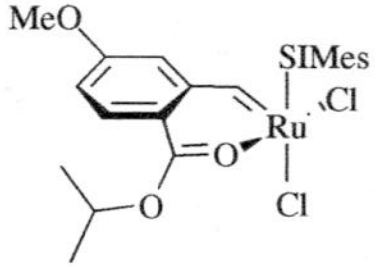

Ru cat **131**

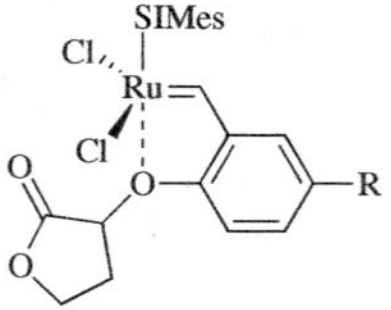

	R
Ru cat **132**	H
Ru cat **133**	MeO

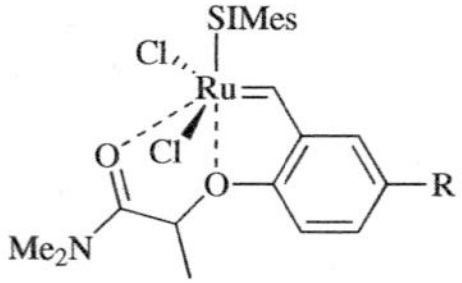

	R
Ru cat **134**	H
Ru cat **135**	MeO

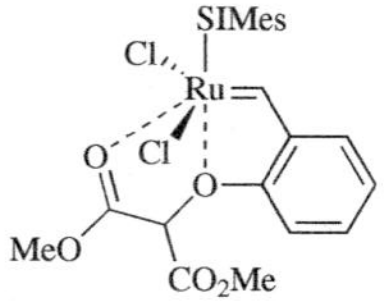

Ru cat **136**

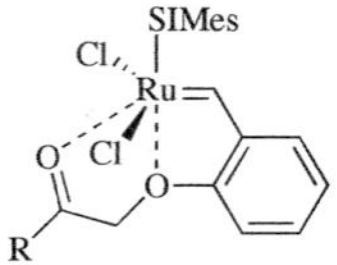

	R
Ru cat **137**	Me
Ru cat **138**	Et

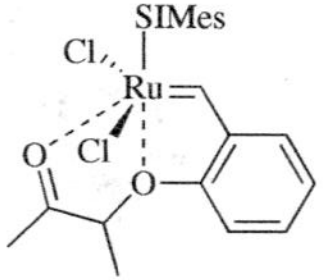

Ru cat **139**

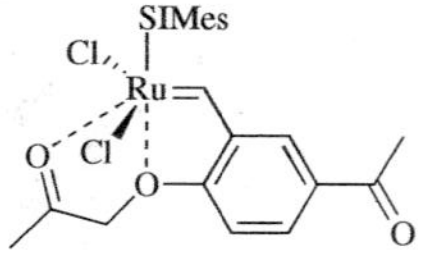

Ru cat **140**

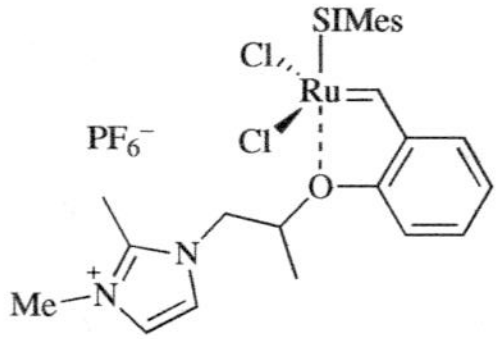

Ru cat **141**

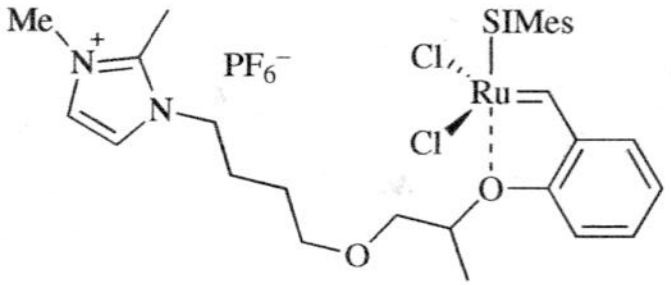

Ru cat **142**

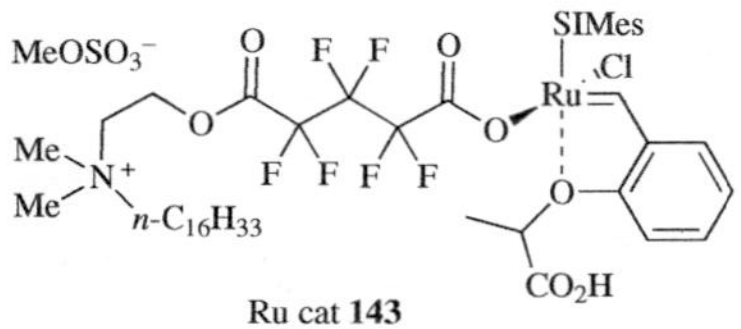

Ru cat **143**

	R
Ru cat **144**	Cl_3Sn
Ru cat **145**	$MesSO_3$
Ru cat **146**	$NapSO_3$
Ru cat **147**	$(PhO)_2PO_2$

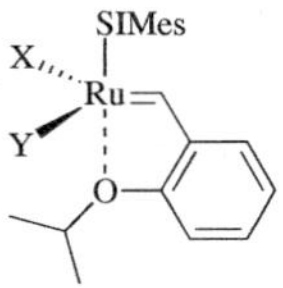

Ru cat **148**, X = Y =

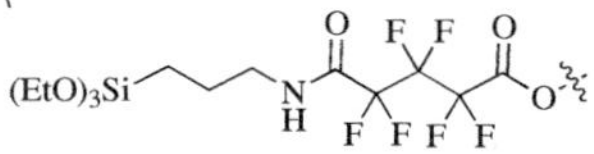

Ru cat **149**, X = Cl, Y =

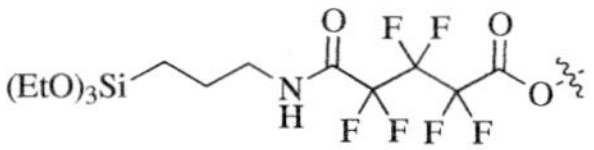

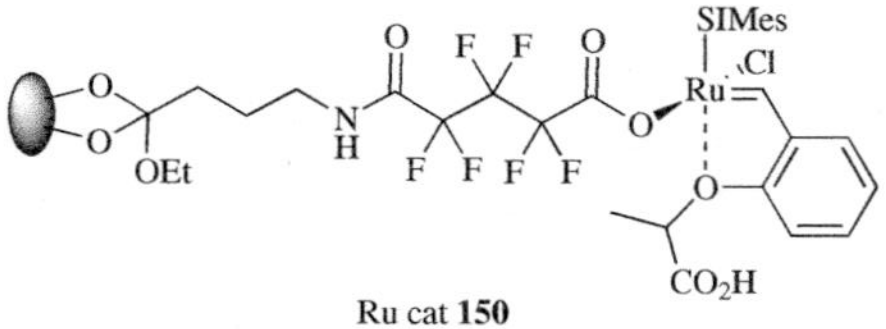

Ru cat **150**

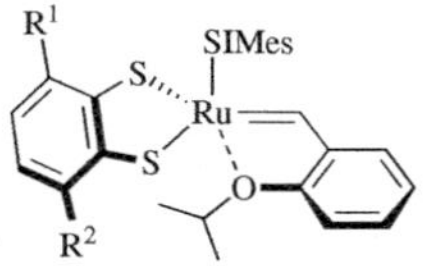

	R^1	R^2
Ru cat **151**	H	H
Ru cat **152**	Cl	Cl
Ru cat **153**	Br	Cl

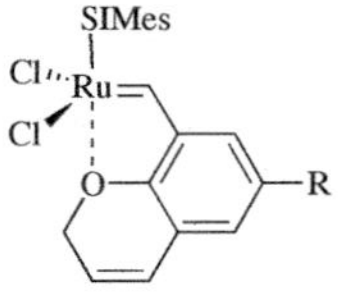

	R
Ru cat **154**	Me
Ru cat **155**	O_2N

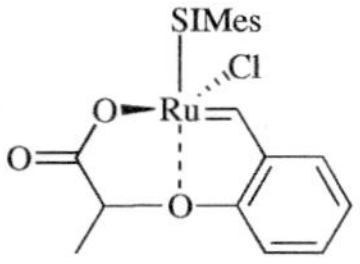

Ru cat **156**

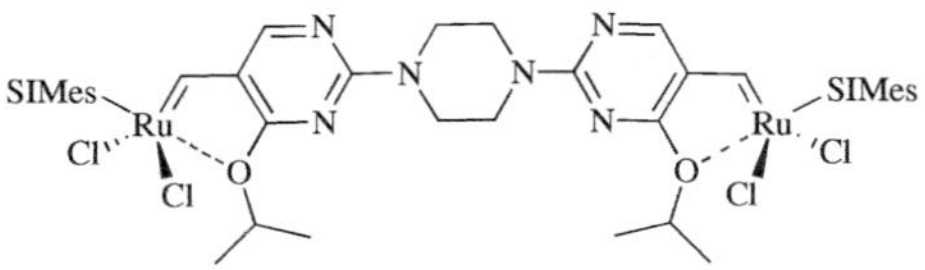

Ru cat **157**

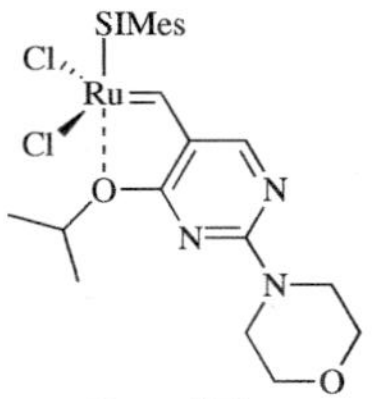

Ru cat **158**

Chart 1. Ruthenium Catalysts Used in the Tables (*Continued*)

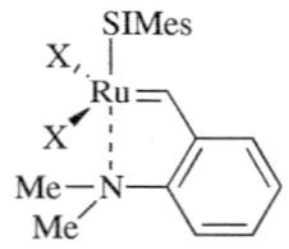

	X
Ru cat **159**	Cl
Ru cat **160**	Br
Ru cat **160**	I

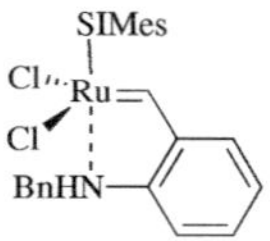

Ru cat **162**

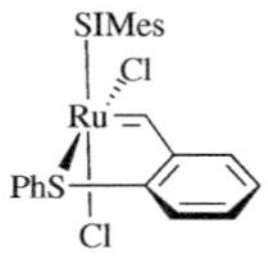

Ru cat **163**

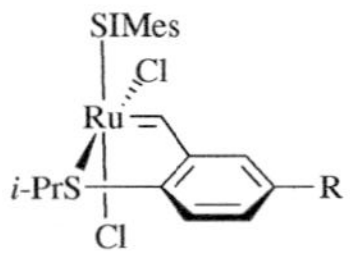

	R
Ru cat **164**	H
Ru cat **165**	O_2N

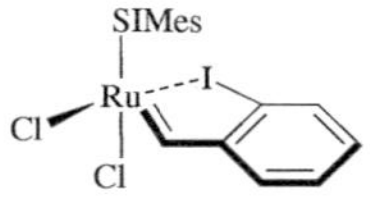

Ru cat **166**

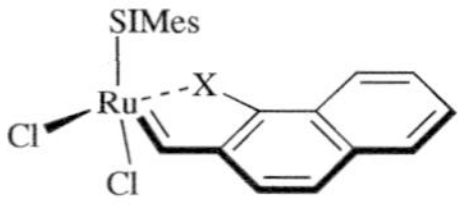

	X
Ru cat **167**	Br
Ru cat **168**	I

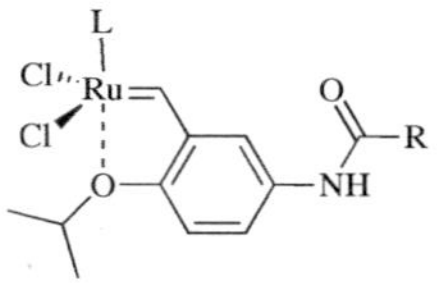

	L	R
Ru cat **169**	IMes	CF_3
Ru cat **170**	SIPr	CF_3
Ru cat **171**	IPr	CF_3
Ru cat **172**	SIPr	*i*-BuO
Ru cat **173**	SIPr	EtO_2C

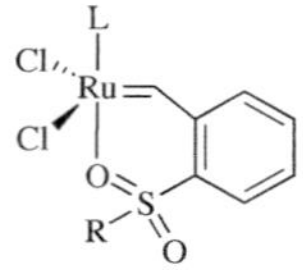

	L	R
Ru cat **174**	SIMes	*t*-Bu
Ru cat **175**	SIPr	*t*-Bu
Ru cat **176**	SIPr	Ph
Ru cat **177**	SIPr	Me_2N
Ru cat **178**	IMes	*t*-Bu

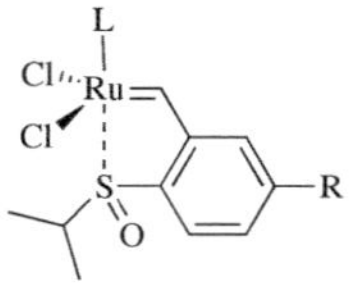

	L	R
Ru cat **179**	SIMes	H
Ru cat **180**	SIMes	O_2N
Ru cat **181**	IMes	H

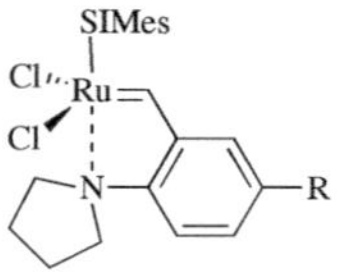

	R
Ru cat **182**	H
Ru cat **183**	MeO
Ru cat **184**	O_2N

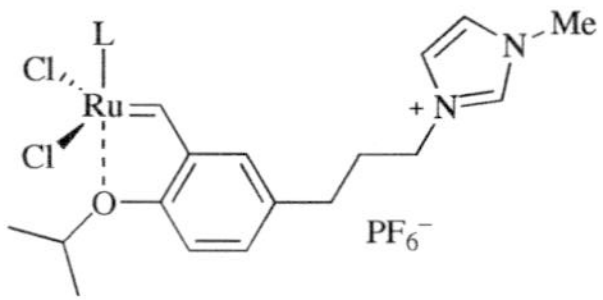

PF_6^-

	L
Ru cat **185**	SIMes
Ru cat **186**	IMes

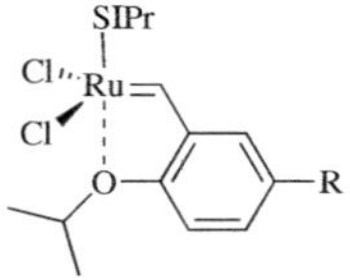

	R
Ru cat **187**	H
Ru cat **188**	O_2N

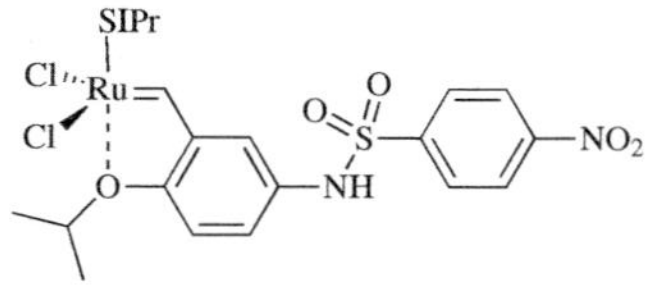

Ru cat **189**

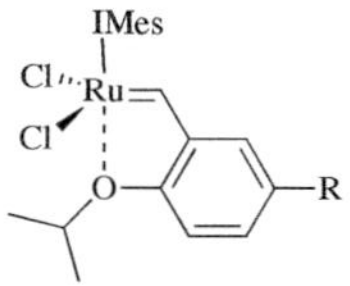

	R
Ru cat **190**	H
Ru cat **191**	poly-CH_2O

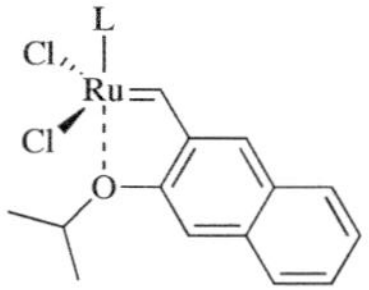

	L
Ru cat **192**	IMes
Ru cat **193**	Me_2IMes

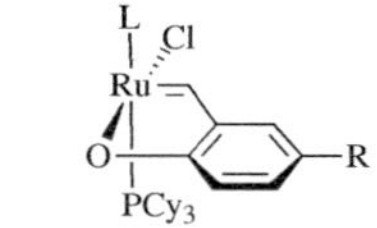

	L	R
Ru cat **194**	SIMes	H
Ru cat **195**	SIMes	O_2N
Ru cat **196**	Me_2IMes	H

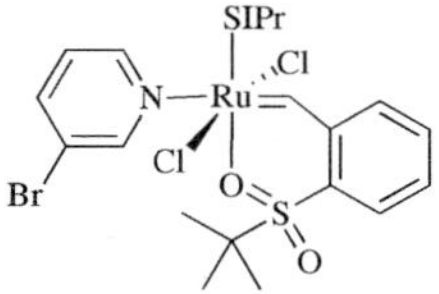

Ru cat **197**

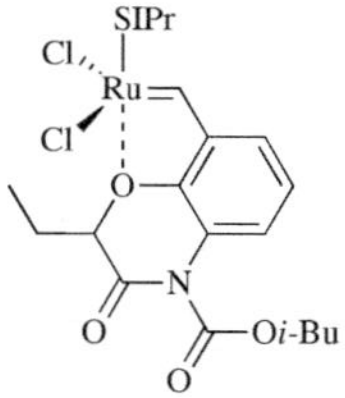

Ru cat **198**

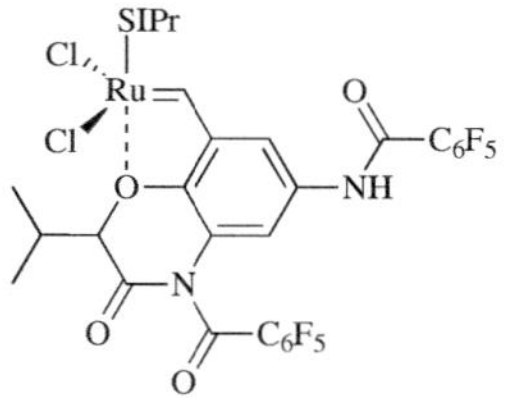

Ru cat **199**

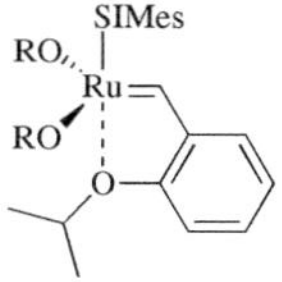

	R
Ru cat **200**	TBS
Ru cat **201**	TMS
Ru cat **202**	TIPS

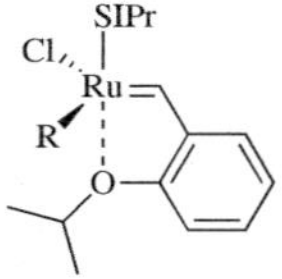

	R
Ru cat **203**	$MesSO_3$
Ru cat **204**	$TripSO_3$
Ru cat **205**	*p*-$TolSO_3$
Ru cat **206**	$CamSO_3$
Ru cat **207**	$BINAPPO_4$

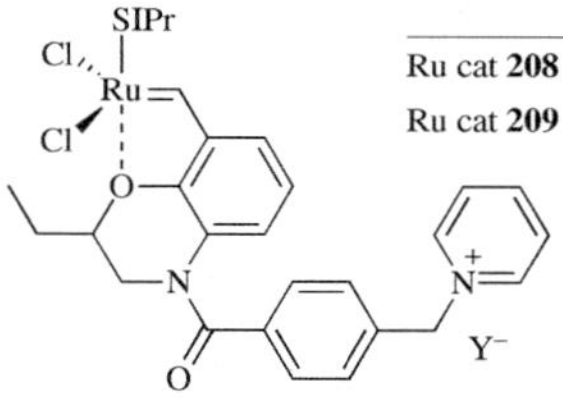

	Y
Ru cat **208**	PF_6
Ru cat **209**	[Si]-$CH_2C_6H_4SO_3$

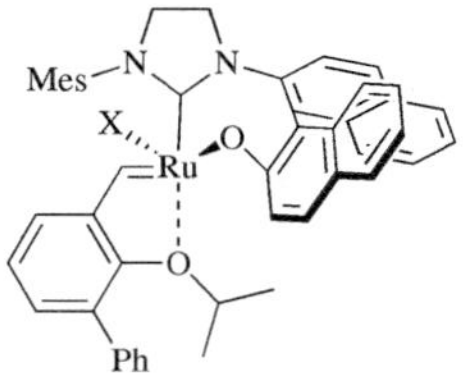

	X
Ru cat **210**	Cl
Ru cat **211**	I

CHART 1. RUTHENIUM CATALYSTS USED IN THE TABLES (*Continued*)

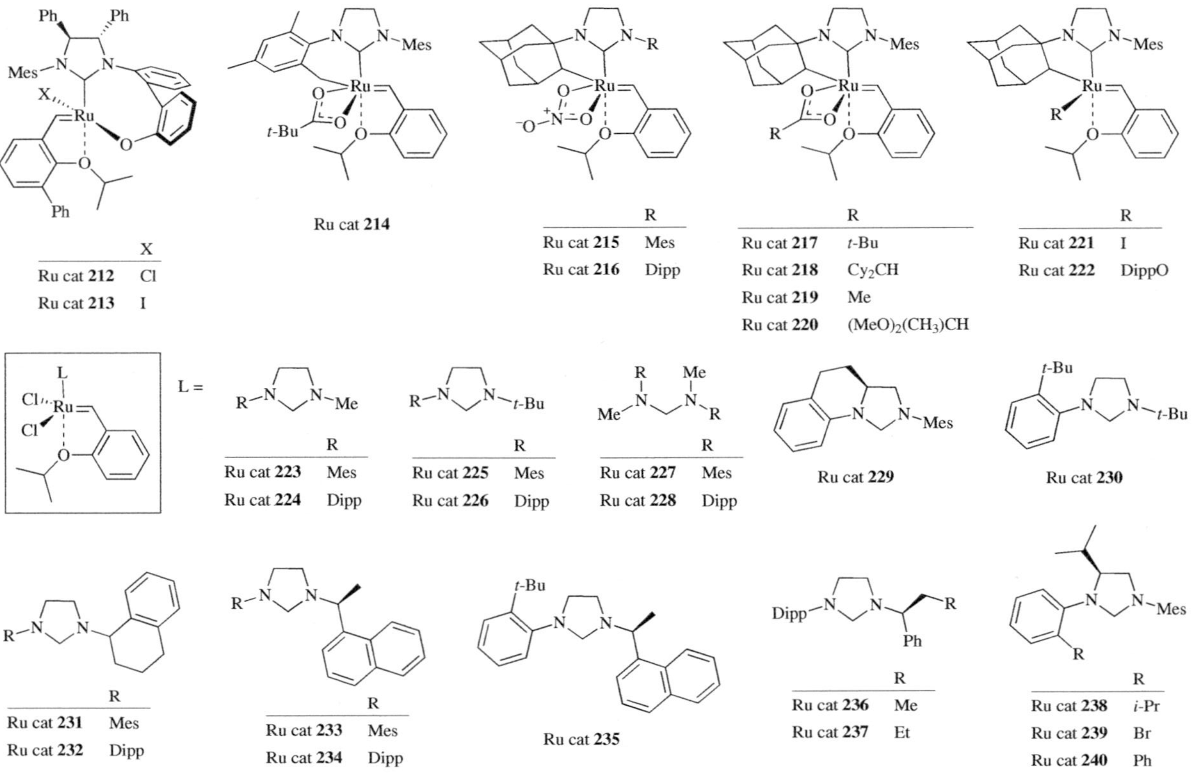

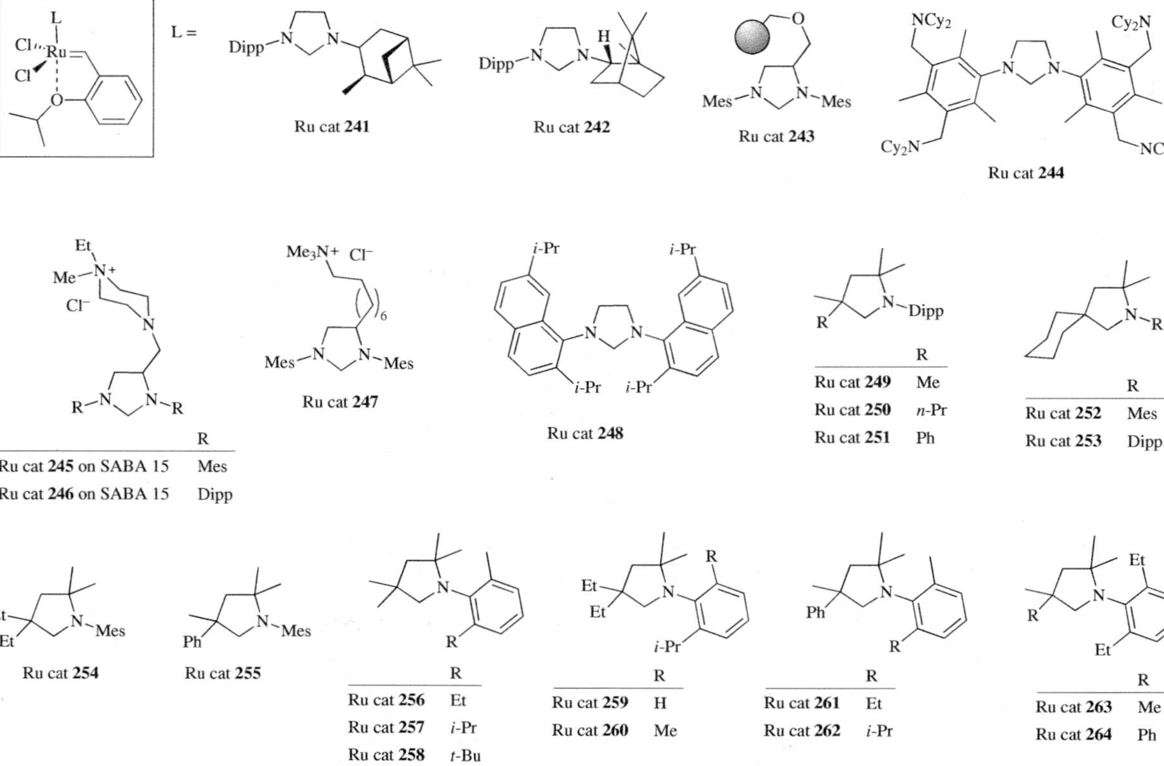
L =
Ru cat 241
Ru cat 242
Ru cat 243
Ru cat 244
R
Ru cat 245 on SABA 15 Mes
Ru cat 246 on SABA 15 Dipp
Ru cat 247
Ru cat 248
R
Ru cat 249 Me
Ru cat 250 n-Pr
Ru cat 251 Ph
R
Ru cat 252 Mes
Ru cat 253 Dipp
Ru cat 254
Ru cat 255
R
Ru cat 256 Et
Ru cat 257 i-Pr
Ru cat 258 t-Bu
R
Ru cat 259 H
Ru cat 260 Me
R
Ru cat 261 Et
Ru cat 262 i-Pr
R
Ru cat 263 Me
Ru cat 264 Ph

Chart 1. Ruthenium Catalysts Used in the Tables (*Continued*)

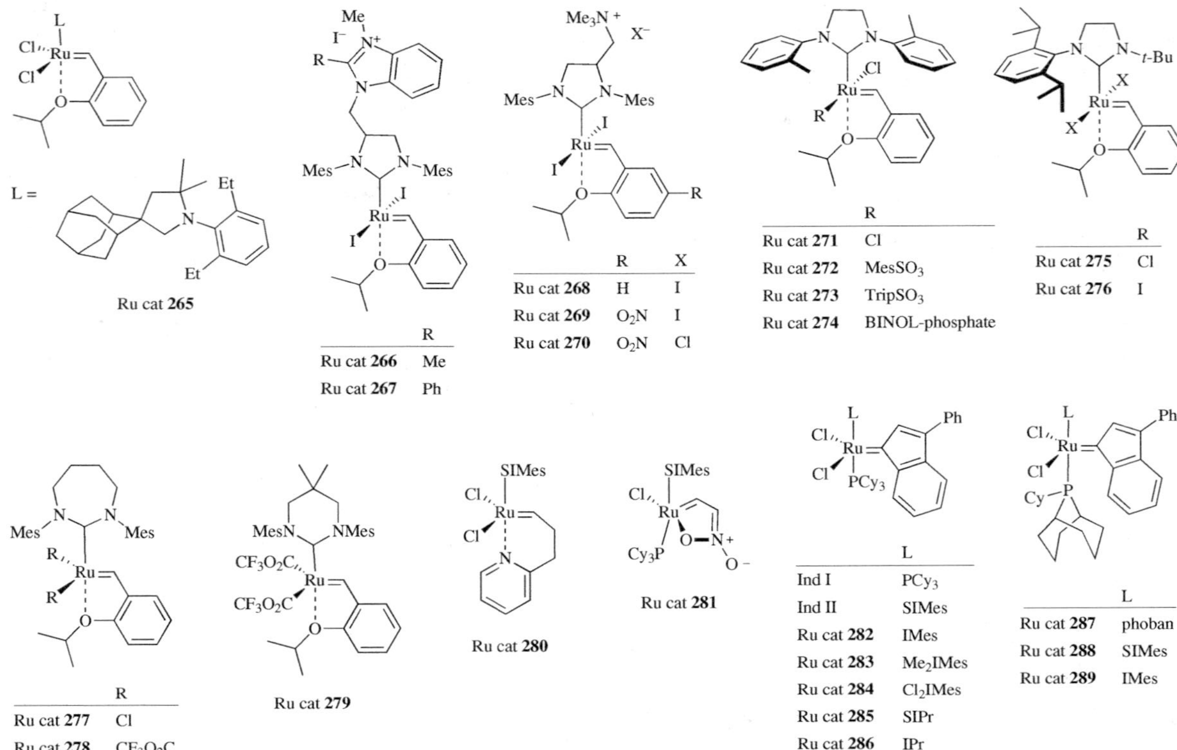

	R
Ru cat **266**	Me
Ru cat **267**	Ph

	R	X
Ru cat **268**	H	I
Ru cat **269**	O_2N	I
Ru cat **270**	O_2N	Cl

	R
Ru cat **271**	Cl
Ru cat **272**	$MesSO_3$
Ru cat **273**	$TripSO_3$
Ru cat **274**	BINOL-phosphate

	R
Ru cat **275**	Cl
Ru cat **276**	I

	R
Ru cat **277**	Cl
Ru cat **278**	CF_3O_2C

	L
Ind I	PCy_3
Ind II	SIMes
Ru cat **282**	IMes
Ru cat **283**	Me_2IMes
Ru cat **284**	Cl_2IMes
Ru cat **285**	SIPr
Ru cat **286**	IPr

	L
Ru cat **287**	phoban
Ru cat **288**	SIMes
Ru cat **289**	IMes

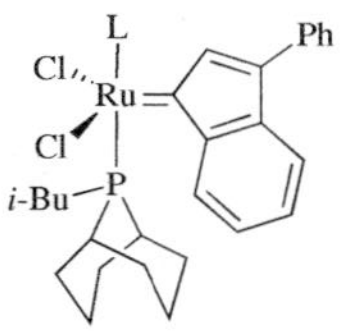

	L
Ru cat **290**	phoban
Ru cat **291**	SIMes
Ru cat **292**	IMes

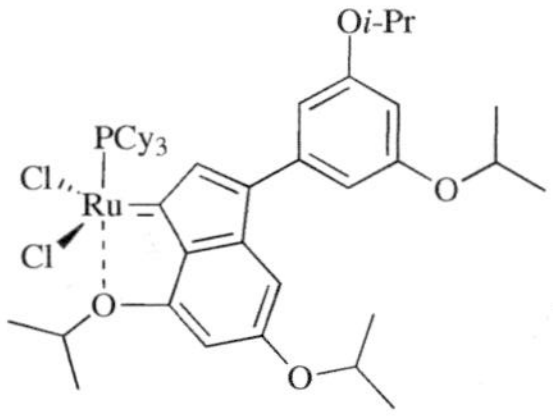

Ru cat **293**

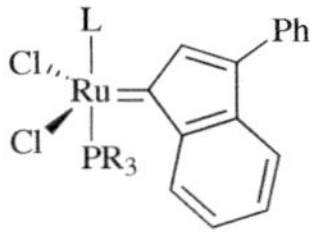

	L	R
Ru cat **294**	SIMes	Ph
Ru cat **295**	SIMes	4-$MeOC_6H_4$
Ru cat **296**	SIMes	4-FC_6H_4
Ru cat **297**	SIPr	Ph
Ru cat **298**	IPr	Ph

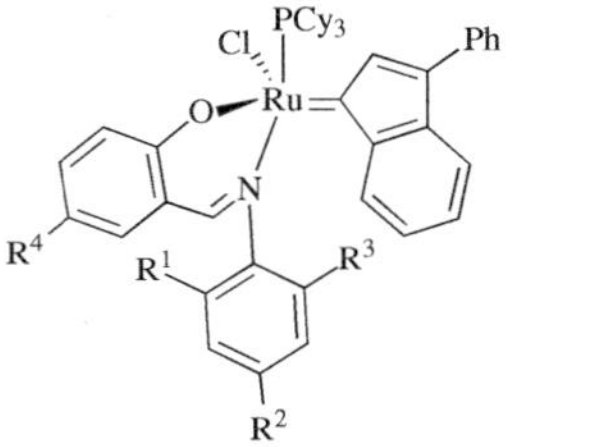

	R^1	R^2	R^3	R^4
Ru cat **299**	Me	H	H	H
Ru cat **300**	Me	Me	Me	H
Ru cat **301**	Me	H	Me	O_2N
Ru cat **302**	Cl	H	H	O_2N
Ru cat **303**	Me	Me	Me	O_2N
Ru cat **304**	*i*-Pr	H	*i*-Pr	O_2N
Ru cat **305**	H	Me	H	O_2N
Ru cat **306**	H	Me	H	H
Ru cat **307**	*i*-Pr	H	*i*-Pr	H

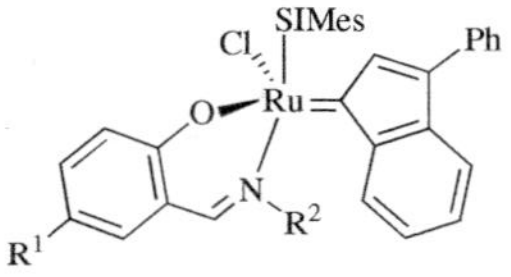

	R^1	R^2
Ru cat **308**	H	Ph
Ru cat **309**	H	2-MeC_6H_4
Ru cat **310**	H	4-MeC_6H_4
Ru cat **311**	O_2N	2-MeC_6H_4
Ru cat **312**	O_2N	4-MeC_6H_4
Ru cat **313**	O_2N	Cy

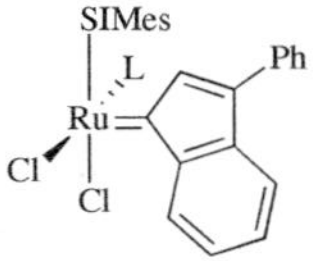

	L
Ru cat **314**	$(MeO)_3P$
Ru cat **315**	$(EtO)_3P$
Ru cat **316**	$(i\text{-}PrO)_3P$
Ru cat **317**	$(PhO)_3P$

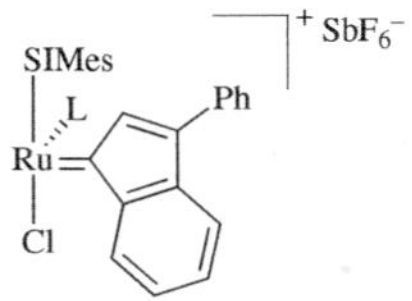

	L
Ru cat **318**	$(i\text{-}PrO)_3P$

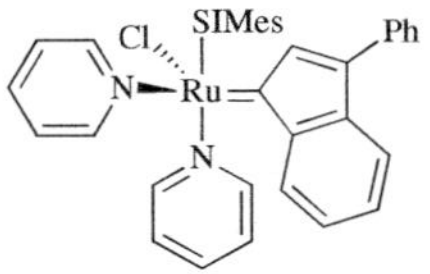

Ru cat **319**

CHART 1. RUTHENIUM CATALYSTS USED IN THE TABLES (*Continued*)

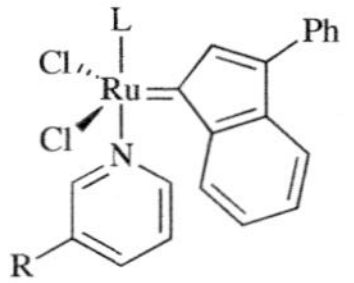

	L	R
Ru cat **320**	SIMes	H
Ru cat **321**	SIPr	H
Ru cat **322**	SIPr	Br
Ru cat **323**	IPr	H

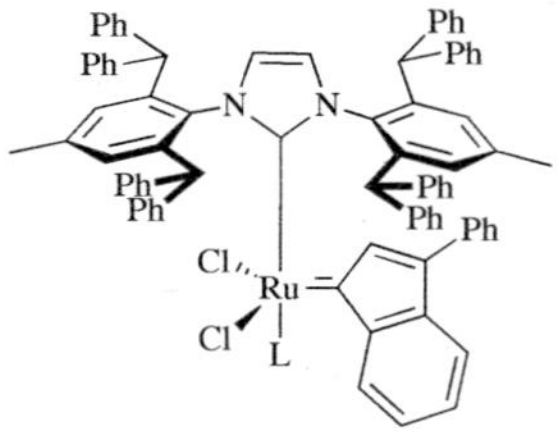

	L
Ru cat **324**	PPh_3
Ru cat **325**	Py

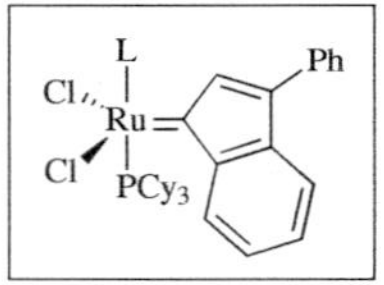

L =

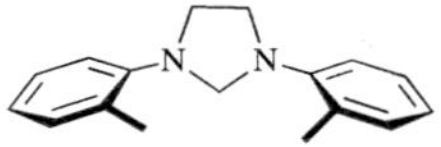

Ru cat **326**

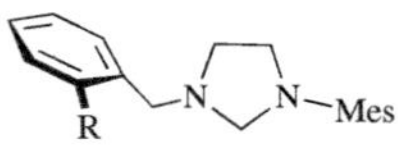

	R
Ru cat **327**	MeO
Ru cat **328**	Me_2N

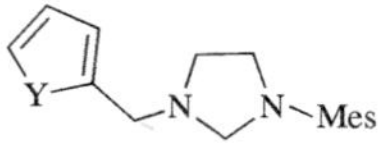

	Y
Ru cat **329**	O
Ru cat **330**	S

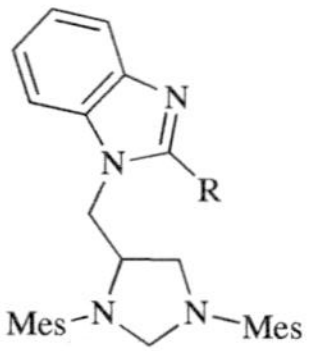

	R
Ru cat **331**	Me
Ru cat **332**	Ph

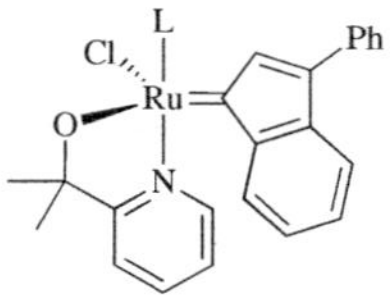

	L
Ru cat **333**	SIMes
Ru cat **334**	IMes

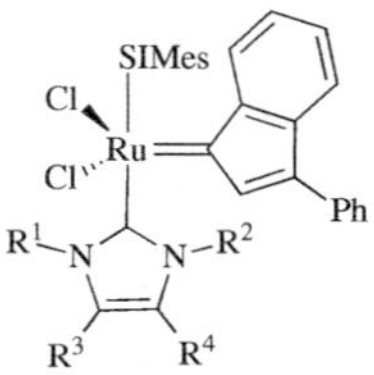

	R^1	R^2	R^3	R^4
Ru cat **335**	Me	Me	H	H
Ru cat **336**	Me	Me	H	O_2N
Ru cat **337**	Me	Me	Cl	Cl
Ru cat **338**	Me	Me	NC–	NC–
Ru cat **339**	Me	Me	Me	Me
Ru cat **340**	Me	*i*-Pr	Cl	Cl
Ru cat **341**	Et	Et	Cl	Cl
Ru cat **342**	*i*-Pr	*i*-Pr	Me	Me
Ru cat **343**	Cy	Cy	H	H

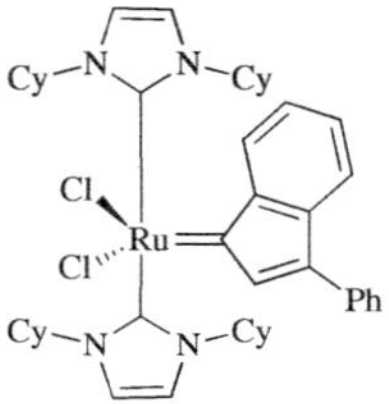

Ru cat **344**

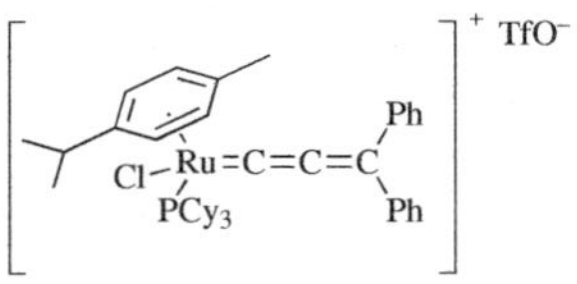

Ru cat **345**

$[RuCl_2(p\text{-cymene})]_2 + IMesH_2Cl + CsCO_2$

Ru cat **346**

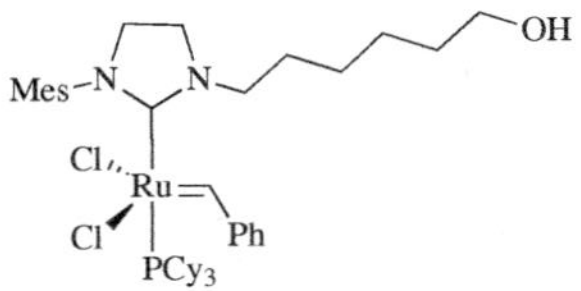

Ru cat **347**

Ru cat **348** = Ru cat **347** immobilized

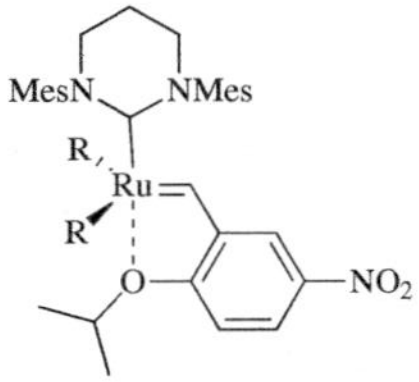

	R
Ru cat **349**	Cl
Ru cat **350**	CF_3O_2C

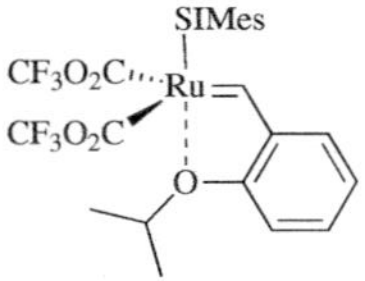

Ru cat **351**

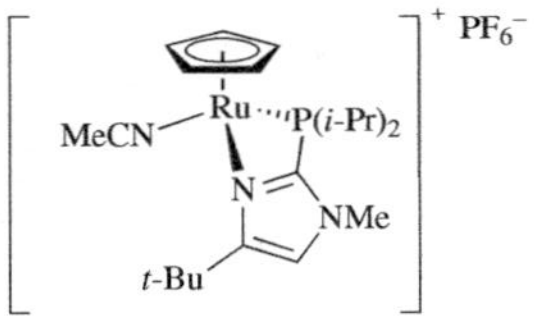

Ru cat **352**

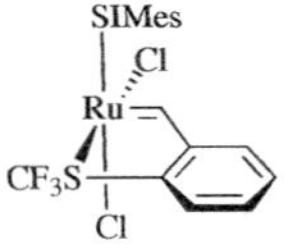

Ru cat **353**

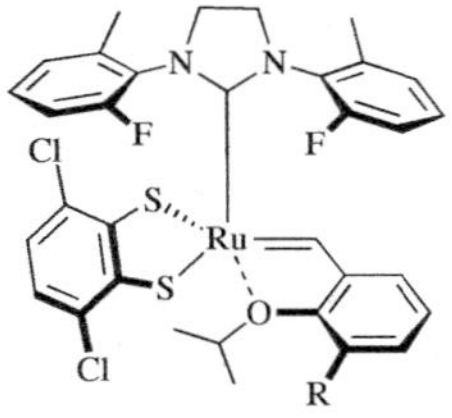

	R
Ru cat **354**	H
Ru cat **355**	Ph

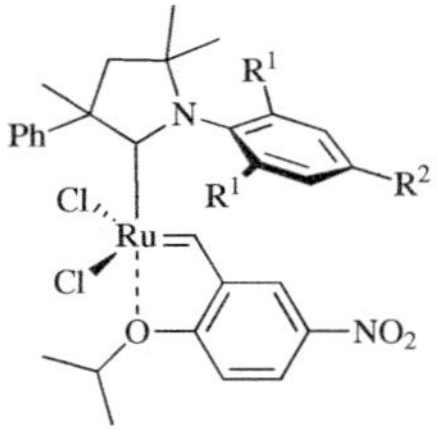

	R^1	R^2
Ru cat **356**	Me	Me
Ru cat **357**	Et	H

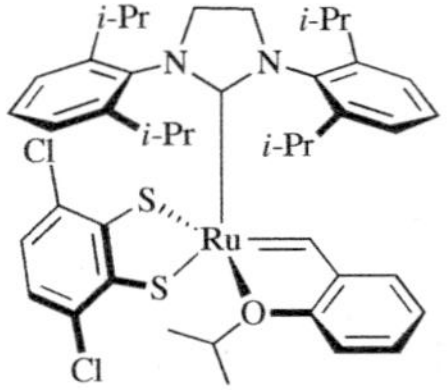

Ru cat **358**

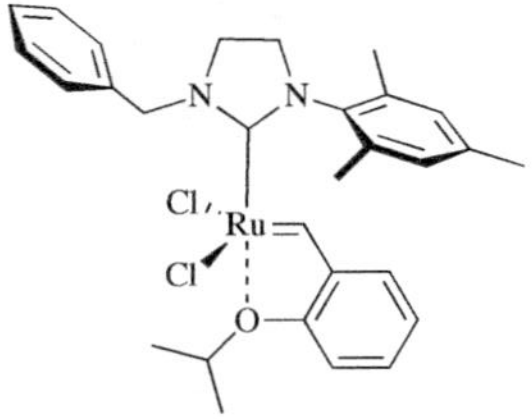

Ru cat **359**

CHART 1. RUTHENIUM CATALYSTS USED IN THE TABLES (*Continued*)

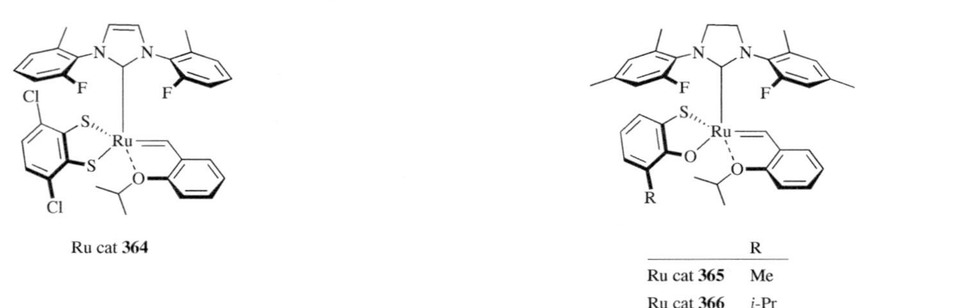

Ru cat **360**

Ru cat **361**

Ru cat **362**

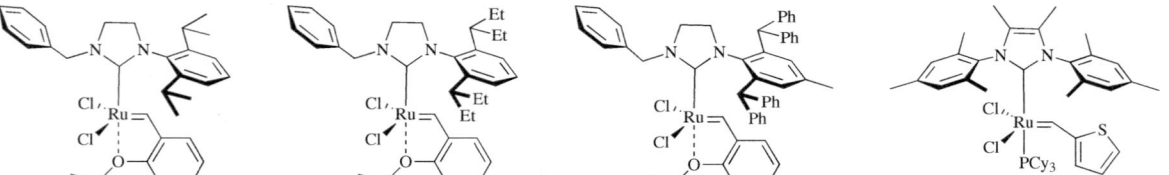

Ru cat **363**

Ru cat **364**

	R
Ru cat **365**	Me
Ru cat **366**	*i*-Pr
Ru cat **367**	*t*-Bu

CHART 2. MOLYBDENUM CATALYSTS USED IN THE TABLES

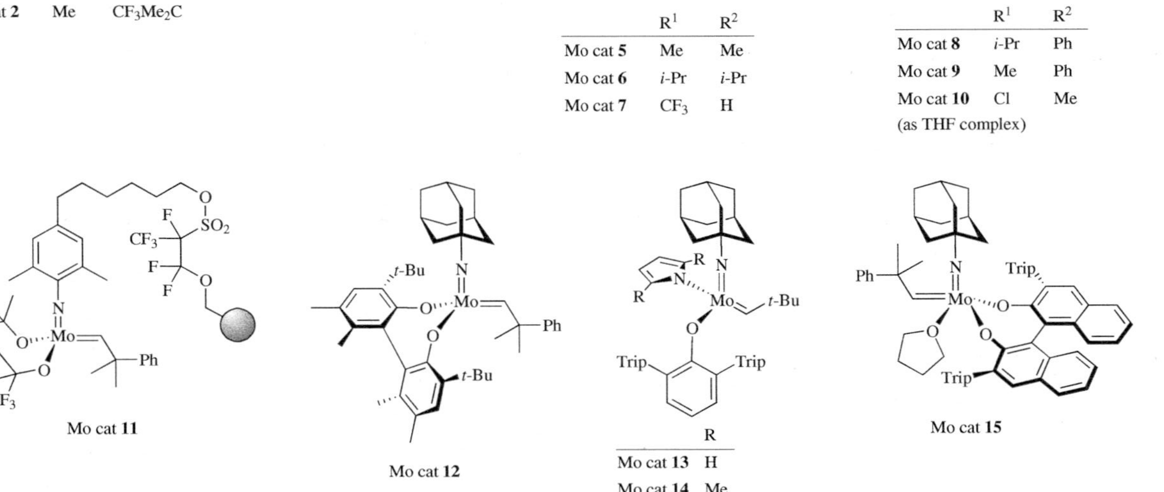

	R^1	R^2
Mo cat **1**	*i*-Pr	$(CF_3)_2MeC$
Mo cat **2**	Me	CF_3Me_2C

	R
Mo cat **3**	$(CF_3)_2MeC$
Mo cat **4**	Ph_3Si

	R^1	R^2
Mo cat **5**	Me	Me
Mo cat **6**	*i*-Pr	*i*-Pr
Mo cat **7**	CF_3	H

	R^1	R^2
Mo cat **8**	*i*-Pr	Ph
Mo cat **9**	Me	Ph
Mo cat **10** (as THF complex)	Cl	Me

Mo cat **11**

Mo cat **12**

	R
Mo cat **13**	H
Mo cat **14**	Me

Mo cat **15**

CHART 2. MOLYBDENUM CATALYSTS USED IN THE TABLES (*Continued*)

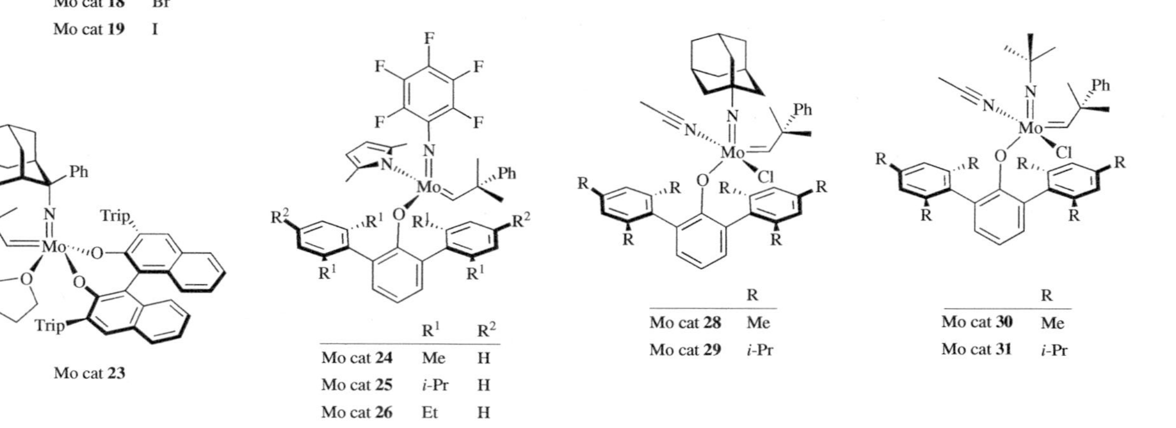

	X
Mo cat **16**	F
Mo cat **17**	Cl
Mo cat **18**	Br
Mo cat **19**	I

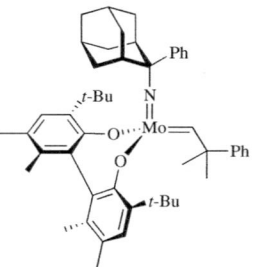

Mo cat **20**

Mo cat **21**

Mo cat **22**

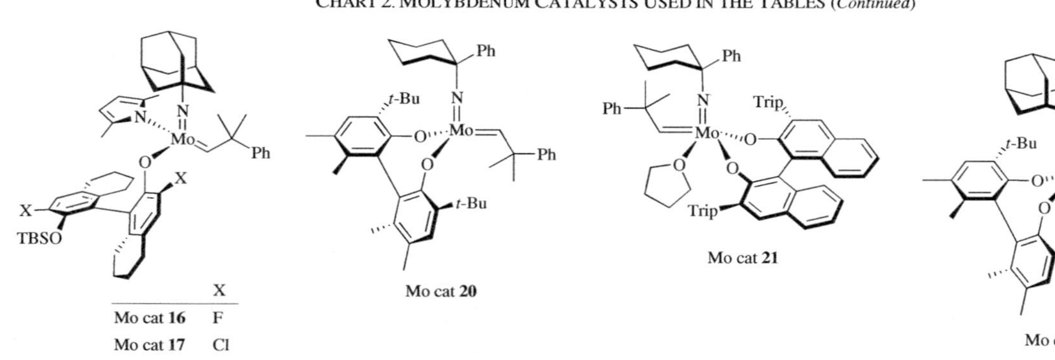

Mo cat **23**

	R^1	R^2
Mo cat **24**	Me	H
Mo cat **25**	*i*-Pr	H
Mo cat **26**	Et	H
Mo cat **27**	Et	Et

	R
Mo cat **28**	Me
Mo cat **29**	*i*-Pr

	R
Mo cat **30**	Me
Mo cat **31**	*i*-Pr

CHART 3. TUNGSTEN CATALYSTS USED IN THE TABLES

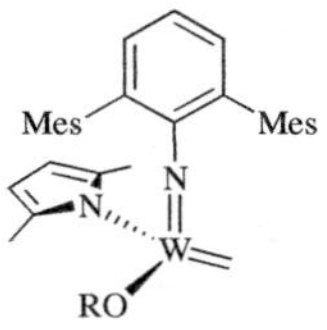

	R
W cat **1**	Ph_3Si
W cat **2**	2,6-$Me_2C_6H_3$

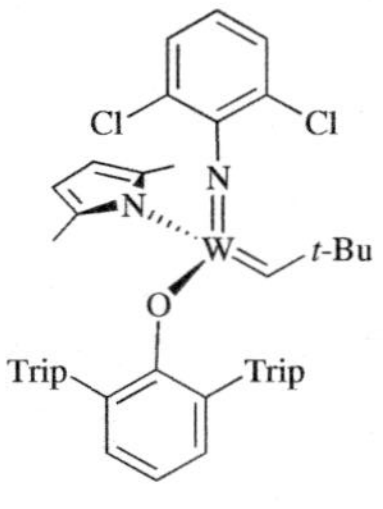

W cat **3**

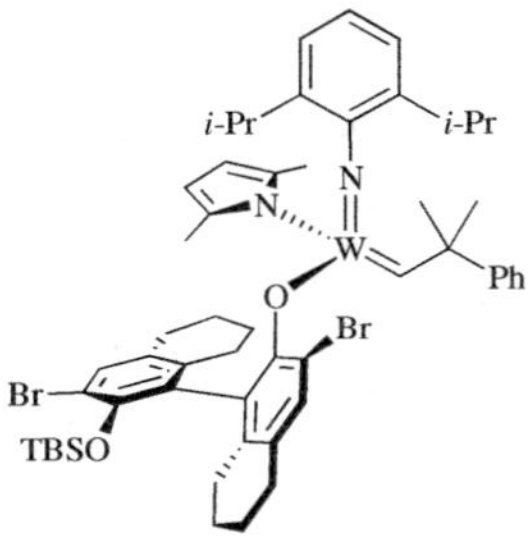

W cat **4**

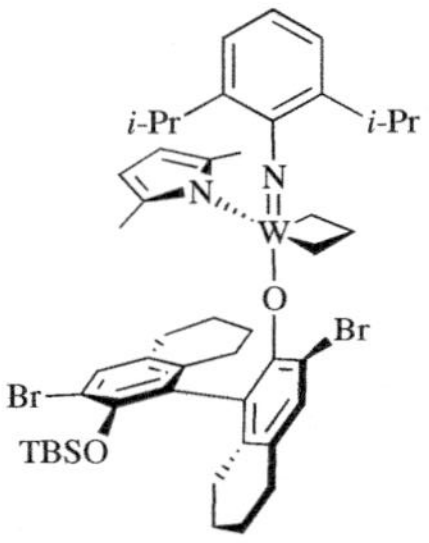

W cat **5**

TABLE 1. CROSS-METATHESIS OF UNSATURATED COMPOUNDS WITH FUNCTIONAL GROUPS IN HOMOALLYLIC AND MORE DISTANT POSITIONS

A. ETHYLENE AND COMPOUNDS WITH A TERMINAL DOUBLE BOND

Terminal Alkene	Cross Partner	Conditions	Product(s) and Yield(s) (%)	Refs.
Please refer to the charts preceding the tables for the catalyst structures.				
C_2				
excess	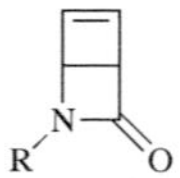	Gr I (0.1 eq), DCM, rt	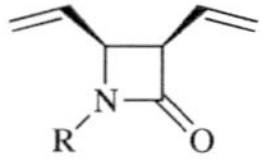 R: H (50); TMS (—); TBDPS (60); TBS (89)	119
excess	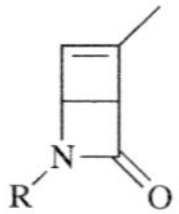	Gr I (0.1 eq), DCM, rt	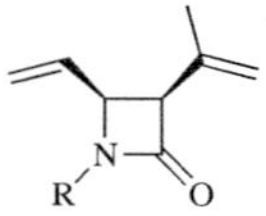 R: H (28); TMS (—); TBDPS (53); TBS (90)	119
1 bar	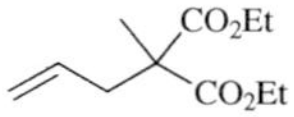 CO_2Et, CO_2Et	Hov II (0.1 eq), $Pd_2Br_2[P(t\text{-}Bu)_3]_2$ (0.005 eq), toluene, 100°, 12 h	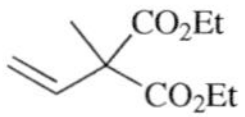 CO_2Et, CO_2Et (77)	344
	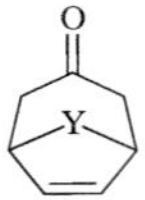	Gr I (0.01 eq), DCM, rt, 26 h	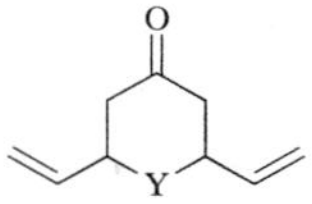 Y: O (43); CH_2 (50)	345

x bar

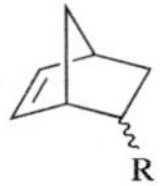

Catalyst (y eq), toluene, rt

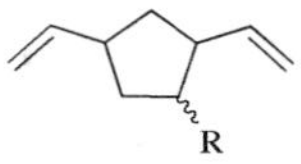

R

R	x	Catalyst	y	Time (h)		Refs.
$(EtO)_3Si$	4	Ru cat **73**	0.00025	24	(99)	346
Me_3SiOMe_2Si	4	Ru cat **73**	0.001	24	(99)	346
$(EtO)_3Si$	3	Gr I	0.00025	24	(97)	347
$CH_2{=}CHCH_2$	2	Gr I	0.0014	24	(88)	347
Ac	2	Gr I	0.001	48	(92)	347

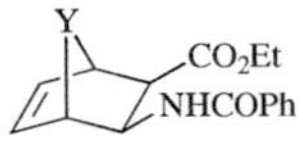

Y, CO_2Et, NHCOPh

Catalyst (0.05 eq), DCM, 20°, 2 h

Y, CO_2Et, NHCOPh

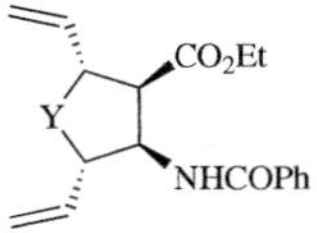

Y	Catalyst		
O	Hov I	(79)	
CH_2	Gr I	(37)	348
CH_2	Gr II	(33)	
CH_2	Hov I	(38)	
CH_2	Hov II	(41)	

Y, NHCOPh, CO_2Et

Catalyst (0.05 eq), DCM, 20°, 2 h

Y, CO_2Et, NHCOPh

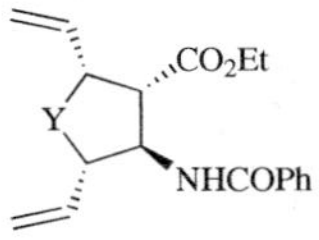

Y	Catalyst		
O	Hov I	(68)	
CH_2	Gr I	(46)	348
CH_2	Gr II	(20)	
CH_2	Hov I	(80)	
CH_2	Hov II	(28)	

TABLE 1. CROSS-METATHESIS OF UNSATURATED COMPOUNDS WITH FUNCTIONAL GROUPS IN HOMOALLYLIC AND MORE DISTANT POSITIONS *(Continued)*

A. ETHYLENE AND COMPOUNDS WITH A TERMINAL DOUBLE BOND *(Continued)*

Terminal Alkene	Cross Partner	Conditions	Product(s) and Yield(s) (%)	Refs.

Please refer to the charts preceding the tables for the catalyst structures.

C_2

Terminal Alkene	Cross Partner	Conditions	Product(s) and Yield(s) (%)	Refs.
25 atm	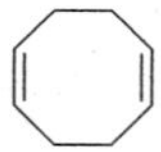	1. $MoCl_5/SiO_2$ (0.06 eq), 24 h 2. $SnMe_4$ (8 eq), 24 h	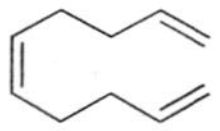 (68) $(Z)/(E) = 99:1$	349
x bar	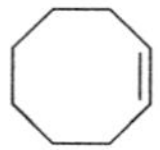	Mo cat **6** (*y* eq), neat, rt	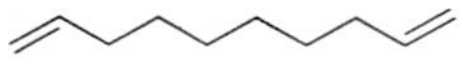	350
x atm	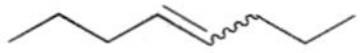 (*E*)/(*Z*) mixture	Mo cat **6** (*y* eq)	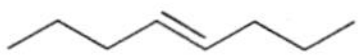	351

Ref. 350:

x	*y*	Time (h)	Conv. (%)[a]	
10	0.0002	16	98	(90)
10	0.0001	20	98	(80)
20	0.0001	20	93	(93)
20	0.00005	16	88	(88)
20	0.000033	20	75	(75)

Ref. 351:

x	*y*	Time (h)		(*E*)/(*Z*)
4	0.004	0.25	(79)	>98:2
4	0.001	1	(76)	91:9
20	0.001	1	(77)	97:3
20	0.002	1	(71)	89:11
20	0.0005	4	(62)	>98:2
20	0.0002	18	(77)	>98:2

20 atm	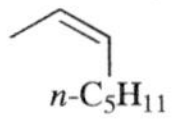 n-C_5H_{11} (Z)/(E) = 80:20	Mo cat **6** (0.0002 eq), 18 h	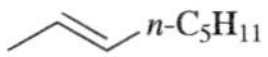 n-C_5H_{11} (20) (E)/(Z) = 98:2	351
x atm	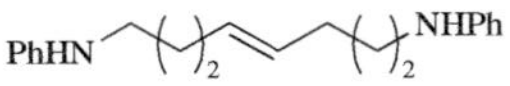 PhHN–$(\,)_2$–CH=CH–$(\,)_2$–NHPh **I**	Ru cat **215** (y eq), THF, 35°	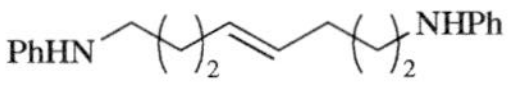 PhHN–$(\,)_2$–CH=CH–$(\,)_2$–NHPh **II** (—)	352

x	(E)/(Z) **I**	y	Time (h)	(E)/(Z) **II**
1	80:20	0.005	4	92:8
5	80:20	0.01	4	>95:5
5	60:40	0.01	6	86:14

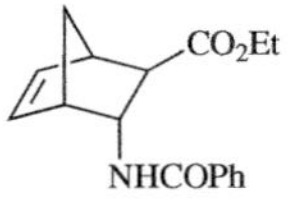

ethylene	CO_2Et, NHCOPh	Catalyst (0.05 eq), DCM, 20°, 2 h	CO_2Et, NHCOPh	348
	CO_2Et, NHCOPh	Catalyst (0.05 eq), DCM, 20°, 2 h	CO_2Et, NHCOPh	348
	EtO_2C, PhOCHN	Hov I (0.05 eq), DCM, 20°, 2 h	EtO_2C, PhOCHN (39)	348

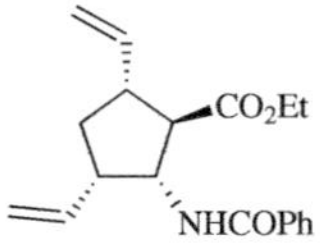

Catalyst	
Gr I	(68)
Gr II	(29)
Hov I	(45)
Hov II	(16)

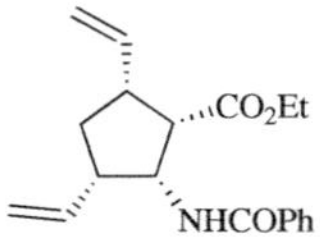

Catalyst	
Gr I	(6)
Gr II	(26)
Hov I	(29)
Hov II	(31)

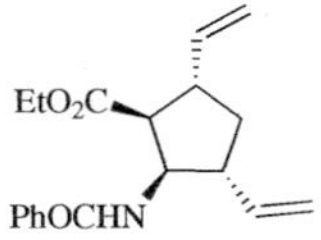

TABLE 1. CROSS-METATHESIS OF UNSATURATED COMPOUNDS WITH FUNCTIONAL GROUPS IN HOMOALLYLIC AND MORE DISTANT POSITIONS (*Continued*)

A. ETHYLENE AND COMPOUNDS WITH A TERMINAL DOUBLE BOND (*Continued*)

Terminal Alkene	Cross Partner	Conditions	Product(s) and Yield(s) (%)	Refs.
Please refer to the charts preceding the tables for the catalyst structures.				
C_2				
	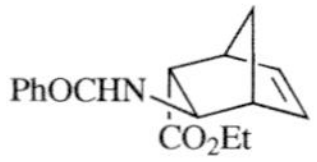 PhOCHN, CO_2Et	Hov I (0.05 eq), DCM, 20°, 2 h	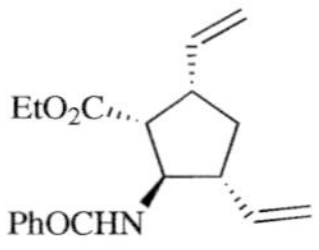 EtO_2C, PhOCHN (64)	348
x bar	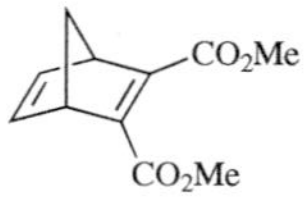 CO_2Me, CO_2Me	Catalyst (0.01 eq), benzene	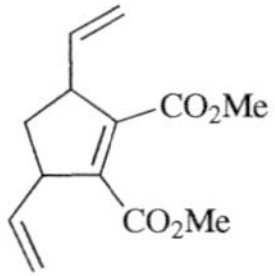 CO_2Me, CO_2Me	353

x	Catalyst	Temp (°)	Time (h)	
1	W cat **1**	20	16	(22)
3.7	W cat **1**	20	16	(24)
20	W cat **1**	20	16	(3)
1	W cat **2**	20	16	(25)
3.7	W cat **2**	20	16	(41)
3.7	W cat **2**	20	22	(47)
3.7	W cat **2**	20	48	(53)
20	W cat **2**	20	16	(24)
1	W cat **2**	20	22	(45)
1	W cat **2**	20	24	(63)
1	W cat **2**	60	21	(85)
1	W cat **2**	60	48	(70)
1	W cat **2**	60	93	(>98)
1	W cat **2**	80	21	(52)

1 atm

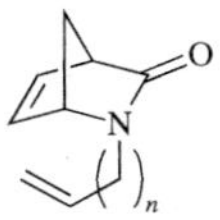

Catalyst (x eq),
DCM, rt, 24 h

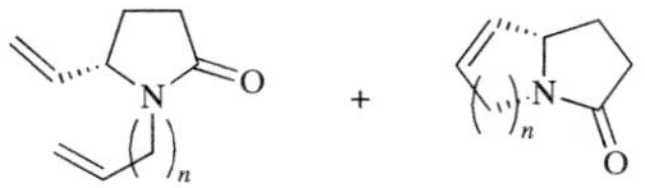

I + **II**

354

n	Catalyst	x	Conv. (%)[b]	**I**	**II**
1	Gr I	0.05	20	(10)	(—)
1	Gr I	0.1	50	(25)	(—)
1	Ru cat **35**	0.05	20	(5)	(—)
2	Gr I	0.05	70	(7)	(60)
2	Gr I	0.1	70	(5)	(65)
2	Ru cat **35**	0.05	80	(35)	(55)
3	Gr I	0.1	80	(10)	(60)
4	Gr I	0.05	80	(30)	(—)
4	Gr I	0.1	80	(60)	(—)

1 bar

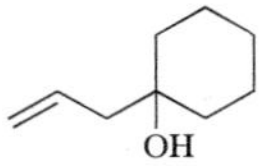

OH

Hov II (0.03 eq),
$Pd_2Br_2[P(t\text{-}Bu)_3]_2$
(0.005 eq),
toluene, 100°, 12 h

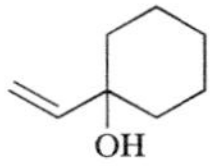

OH (86)

344

TABLE 1. CROSS-METATHESIS OF UNSATURATED COMPOUNDS WITH FUNCTIONAL GROUPS IN HOMOALLYLIC AND MORE DISTANT POSITIONS (*Continued*)

A. ETHYLENE AND COMPOUNDS WITH A TERMINAL DOUBLE BOND (*Continued*)

Terminal Alkene	Cross Partner	Conditions	Product(s) and Yield(s) (%)	Refs.

Please refer to the charts preceding the tables for the catalyst structures.

C_2

Terminal Alkene	Cross Partner	Conditions	Product(s) and Yield(s) (%)	Refs.
x eq	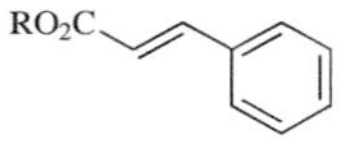 RO_2C	Hov II (*y* eq), DCM, 40°	 RO_2C +	356
10 bar	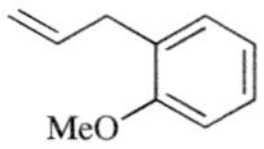 MeO	Hov II (0.03 eq), $Pd_2Br_2[P(t\text{-}Bu)_3]_2$ (0.005 eq), THF, 60°, 16 h	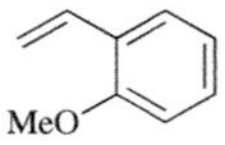 MeO (87)	344

x	R	*y*	Time (h)	Conv. (%)[a]
0.5	H	0.125	48	29
1	H	0.125	24	23
1	Et	0.05	24	16
1	Bu	0.125	24	38
5	Bu	0.05	24	29

x bar	OMe, OH	Hov II (y eq), $Pd_2Br_2[P(t\text{-}Bu)_3]_2$ (z eq), THF	344

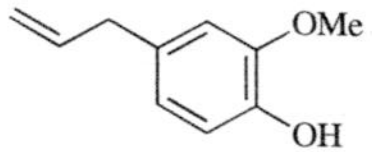

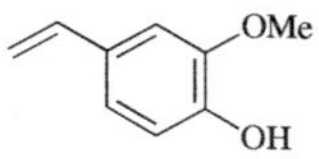

x	y	z	Time (h)	Temp (°)	Conv. (%)[a]	
10	0.03	0.005	16	60	>99	(97)
10	0.03	0.001	16	60	>99	(92)
10	0.01	0.005	16	60	>99	(91)
10	0.03	0.005	4	60	>99	(87)
5	0.03	0.005	16	60	>99	(83)
10	0.03	0.005	0.25	60 (MW)	>99	(90)
10	0.03	—	16	60	44	(33)

10 bar	R^1, R^2	Hov II (0.03 eq), $Pd_2Br_2[P(t\text{-}Bu)_3]_2$ (0.005 eq), THF, 60°, 16 h	344

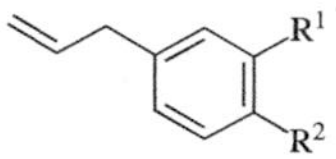

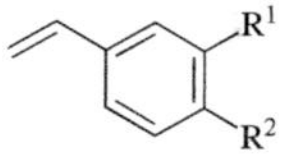

R^1	R^2	
H	MeO	(89)
H	Cl	(92)
H	F	(95)
H	Me	(94)
MeO	HO	(85)
MeO	MeO	(90)
CF_3	H	(95)
n-PrO_2C	MeO	(93)

TABLE 1. CROSS-METATHESIS OF UNSATURATED COMPOUNDS WITH FUNCTIONAL GROUPS IN HOMOALLYLIC AND MORE DISTANT POSITIONS (*Continued*)

A. ETHYLENE AND COMPOUNDS WITH A TERMINAL DOUBLE BOND (*Continued*)

Terminal Alkene	Cross Partner	Conditions	Product(s) and Yield(s) (%)	Refs.
Please refer to the charts preceding the tables for the catalyst structures.				
C_2				
10 bar		Catalyst (0.03 eq), solvent, 60°, 16 h	OMe, OH	344
10 bar		Hov II (0.03 eq), $Pd_2Br_2[P(t\text{-}Bu)_3]_2$ (0.005 eq), THF, 60°, 16 h	O, O (92)	344

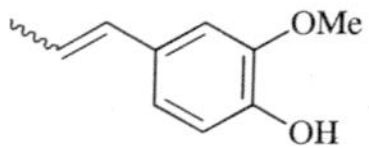

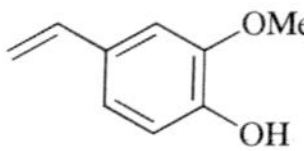

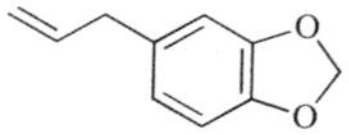

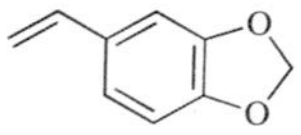

Catalyst	Solvent	Conv. (%)[c]	
Gr I	toluene	46	(39)
Hov I	toluene	51	(35)
Gr II	toluene	>99	(44)
Hov II	toluene	>99	(55)
Ru cat **139**	toluene	>99	(47)
Ind I	toluene	30	(26)
Ru cat **320**	toluene	57	(31)
Ru cat **312**	toluene	41	(35)
Hov II	MeOH	82	(—)
Hov II	NMP	62	(62)
Hov II	acetone	82	(80)
Hov II	diglyme	>99	(93)
Hov II	THF	>99	(96)

x atm

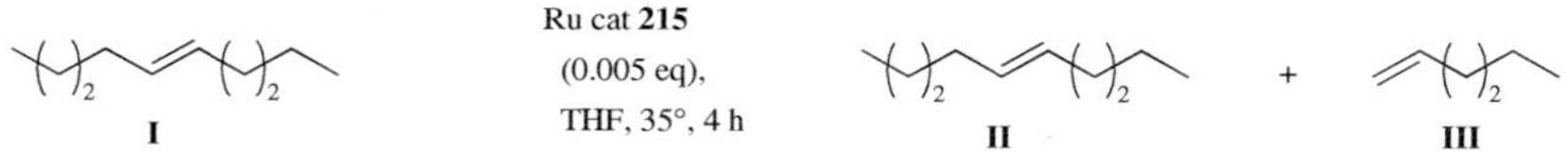

I

Ru cat **215** (0.005 eq), THF, 35°, 4 h

II + **III**

352

x	(*E*)/(Z) **I**	(*E*)/(Z) **II**
1	79:21	90:10
5	79:21	>95:5
5	52:48	90:10

x atm

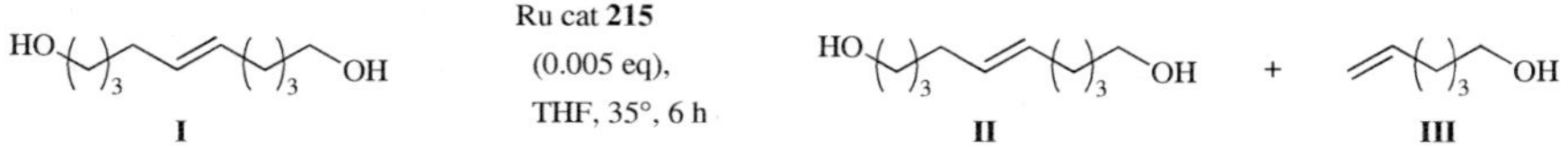

HO, OH, **I**

Ru cat **215** (0.005 eq), THF, 35°, 6 h

HO, OH **II** + OH **III**

352

x	(*E*)/(Z) **I**	(*E*)/(Z) **II**
1	82:18	92:8
5	82:18	>95:5
5	68:32	90:10

x atm

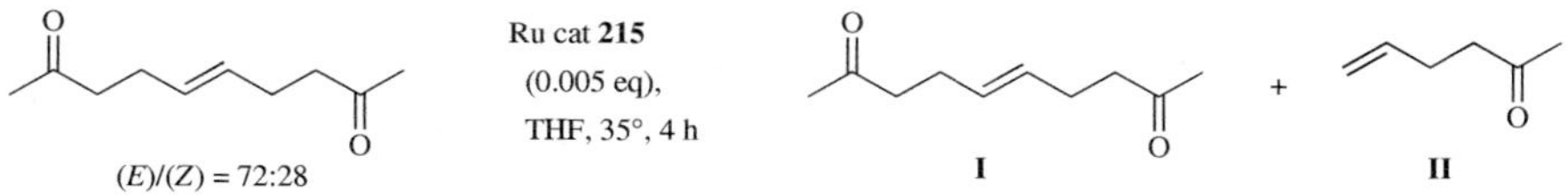

O, O

(*E*)/(Z) = 72:28

Ru cat **215** (0.005 eq), THF, 35°, 4 h

O, O **I** + O **II**

352

x	**I + II**	(*E*)/(Z) **I**
1	(85)	90:10
5	(85)	>95:5

TABLE 1. CROSS-METATHESIS OF UNSATURATED COMPOUNDS WITH FUNCTIONAL GROUPS IN HOMOALLYLIC AND MORE DISTANT POSITIONS (*Continued*)

A. ETHYLENE AND COMPOUNDS WITH A TERMINAL DOUBLE BOND (*Continued*)

Terminal Alkene	Cross Partner	Conditions	Product(s) and Yield(s) (%)	Refs.
Please refer to the charts preceding the tables for the catalyst structures.				
C_2				
	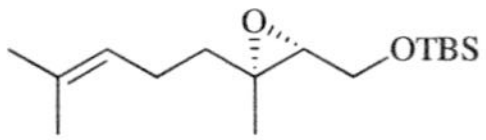 OTBS	1. Gr II (0.03 eq), DCM, 40°, 24 h 2. *m*-CPBA (2 eq), DCM, 0.5 h	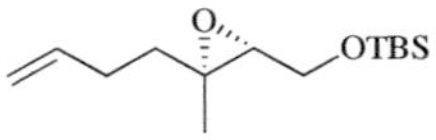 OTBS (54)	355
1 atm	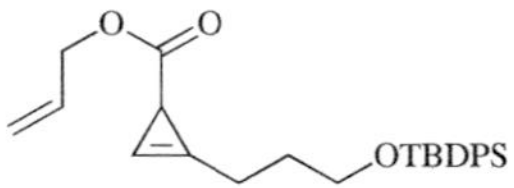 O, O, OTBDPS	Gr II (0.025 eq), toluene, reflux	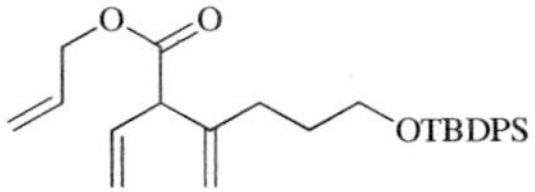 O, O, OTBDPS (43)	357
1 atm	 O, OTBDPS	Gr II (0.025 eq), DCM, reflux	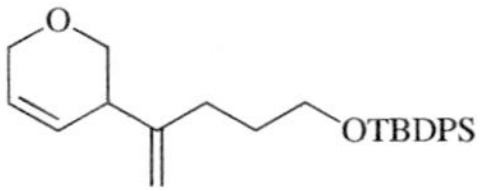 O, OTBDPS (57)	357
1 atm	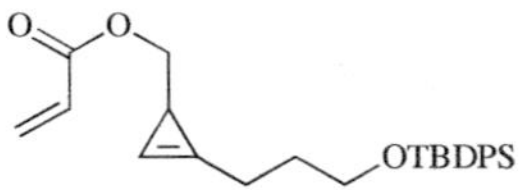 O, O, OTBDPS	Gr II (0.025 eq), benzene, reflux	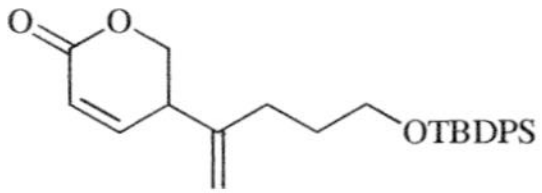 O, O, OTBDPS (64)	357
1 atm	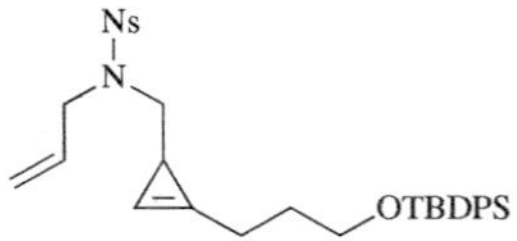 Ns–N, OTBDPS	Gr II (0.025 eq), benzene, reflux	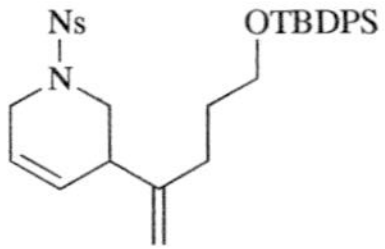 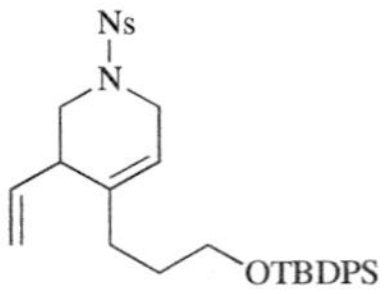 Ns–N, OTBDPS **I** + Ns–N, OTBDPS **II** (68) **I/II** = 87:13	357

1 atm	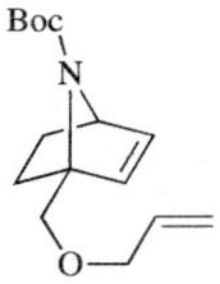 Boc, N	Gr II (0.05 eq), toluene, 80°, 8 h	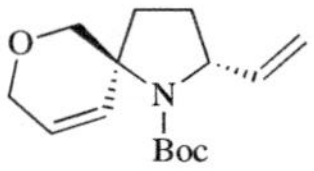 O, N, Boc (91)	358
1 atm	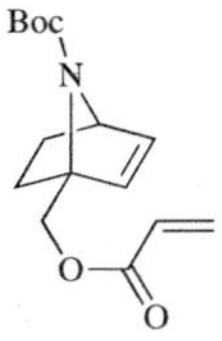 Boc, N, O, O	Gr II (0.05 eq), toluene, 80°, 14 h	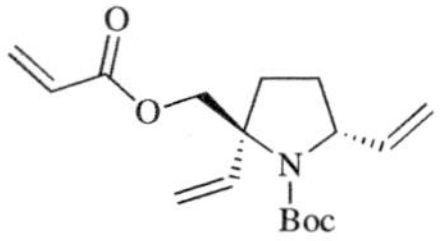 O, O, N, Boc (79)	358
1 atm	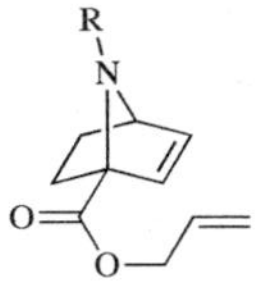 R, N, O, O	Gr II (0.05 eq), toluene, 80°, 14 h	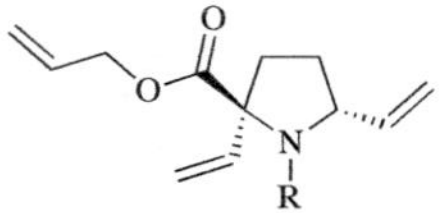 O, O, N, R; R: Boc (88), Ac (73)	358
1 atm	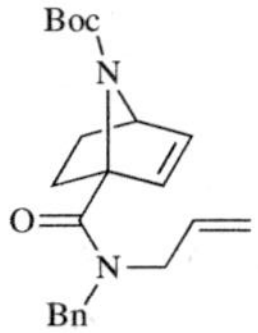 Boc, N, O, N, Bn	Gr II (0.05 eq), toluene, 80°, 14 h	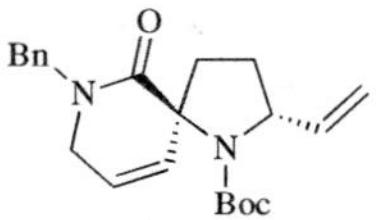 Bn, N, O, N, Boc (91)	358

TABLE 1. CROSS-METATHESIS OF UNSATURATED COMPOUNDS WITH FUNCTIONAL GROUPS IN HOMOALLYLIC AND MORE DISTANT POSITIONS (*Continued*)

A. ETHYLENE AND COMPOUNDS WITH A TERMINAL DOUBLE BOND (*Continued*)

Terminal Alkene	Cross Partner	Conditions	Product(s) and Yield(s) (%)	Refs.
Please refer to the charts preceding the tables for the catalyst structures.				
C_2				
1 atm	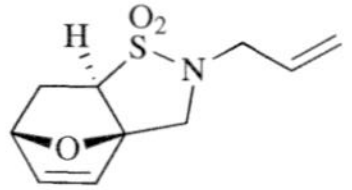 H, O_2, S, N; (±)	Gr II (0.05 eq), DCM, 45°, 3h	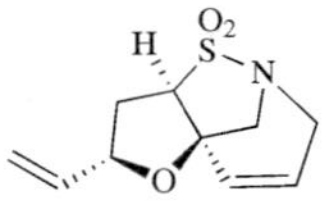 H, O_2, S, N, O; (±) (74)	359
1 atm	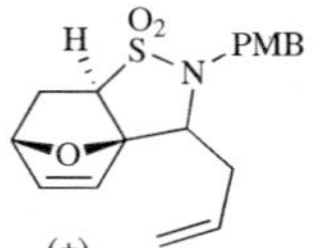 H, O_2, S, N–PMB, O; (±)	Gr II (0.1 eq), DCM, 45°, 3h	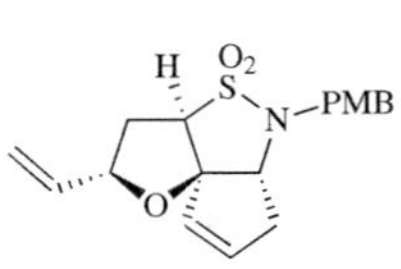 H, O_2, S, N–PMB, O; (±) (54) 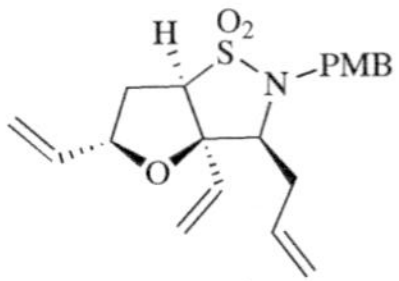 + H, O_2, S, N–PMB, O; (±) (16)	359
4 atm	 $(\ )_2$, $(\ )_2$, OAc; $(E)/(Z)$ = 83:17	Mo cat **6** (0.001 eq), 0.25 h	 $(\ )_2$, $(\ )_2$, OAc (75) $(E)/(Z)$ > 98:2	351
1 atm	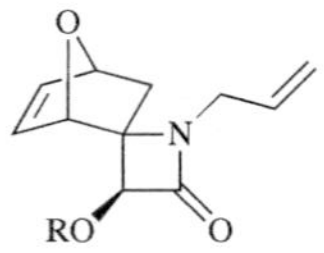 O, N, RO, O	Gr I (0.1 eq), DCM, rt, 12 h	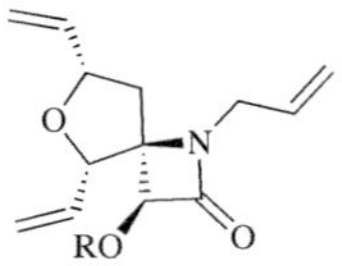 O, N, RO, O R: Ph (85); Bn (85)	360
1 atm	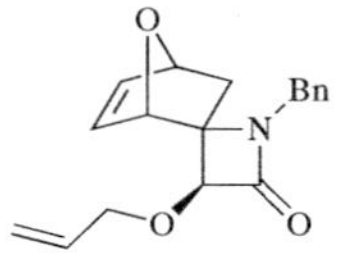 O, N, Bn, O, O	Gr I (0.1 eq), DCM, rt, 12 h	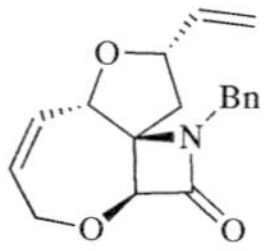 O, Bn, N, O, O (98)	360

1 atm	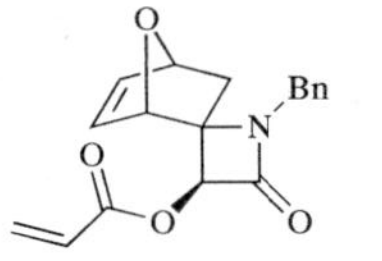	Gr II (0.1 eq), DCM, rt, 12 h		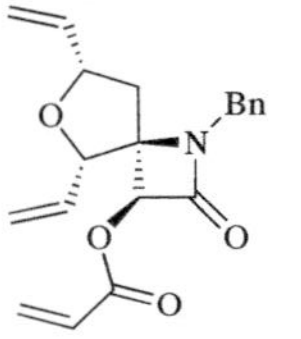 (70)	360
1 atm	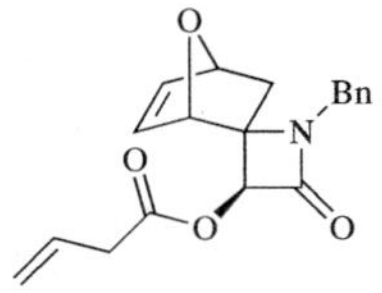	Gr II (0.1 eq), DCM, rt, 12 h		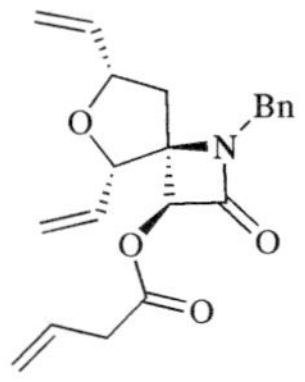 (70)	360
1 atm	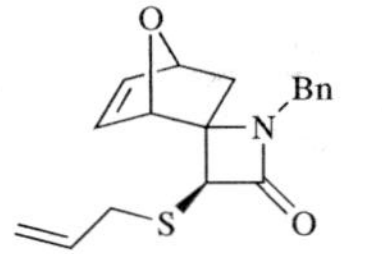	Gr II (0.1 eq), DCM, rt, 12 h		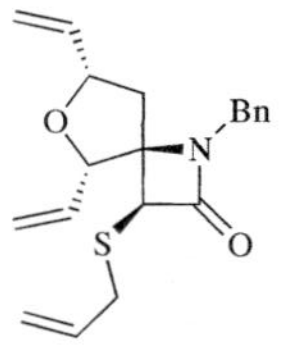 (50)	360
1 atm	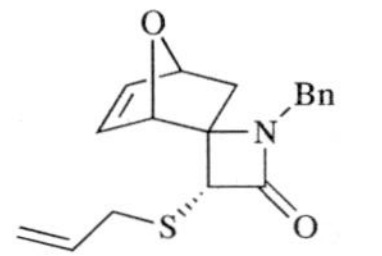	Gr II (0.1 eq), DCM, rt, 12 h		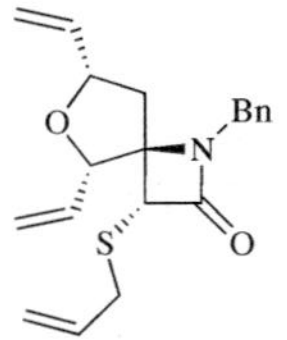 (70)	360

TABLE 1. CROSS-METATHESIS OF UNSATURATED COMPOUNDS WITH FUNCTIONAL GROUPS IN HOMOALLYLIC AND MORE DISTANT POSITIONS (*Continued*)

A. ETHYLENE AND COMPOUNDS WITH A TERMINAL DOUBLE BOND (*Continued*)

Terminal Alkene	Cross Partner	Conditions	Product(s) and Yield(s) (%)	Refs.
Please refer to the charts preceding the tables for the catalyst structures.				
C_2				
	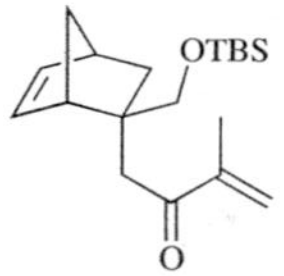 OTBS, O	Gr I (0.025 eq), DCM, rt, 6 h	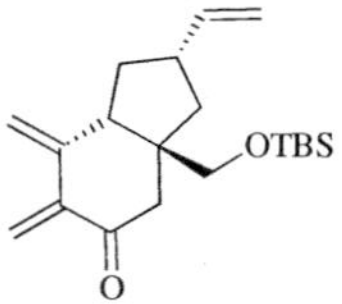 OTBS, O (99)	361
1 atm	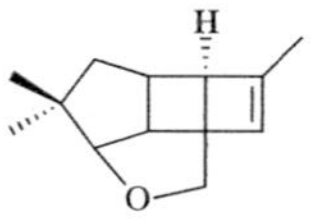 H, O	Gr II (0.05 eq), benzene, 50–80°, 6 + 10 h	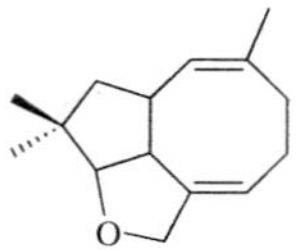 O (74)	269
	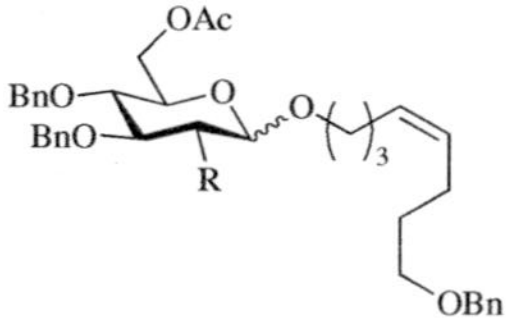 OAc, BnO, BnO, O, O, R, 3, OBn	Catalyst (0.1 eq), solvent, 48 h	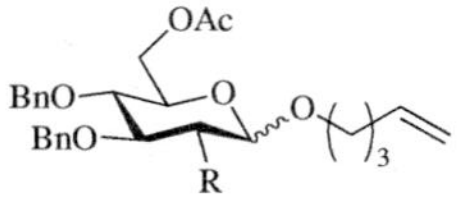 OAc, BnO, BnO, O, O, R, 3	178

R	Catalyst	Solvent	Temp (°)	
BnO	Gr I	DCM	rt	(90)
BnO	Gr II	DCM	40	(50)
BnO	Ru cat **66**	DCE	80	(tr)
N_3	Gr II	DCM	rt	(0)
N_3	Ru cat **66**	DCE	80	(tr)
AcHN	Gr I	DCM	rt	(92)

1 atm

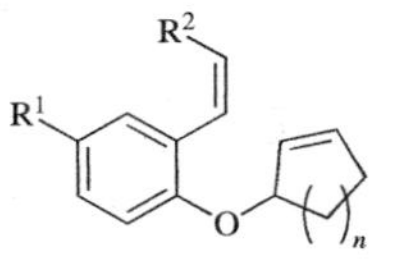

Catalyst (x eq),
DCM, 22°

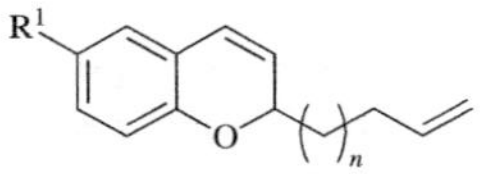

R^1	R^2	n	Catalyst	x	Time (h)		
H	H	1	Gr I	0.05	16	(0)	363
H	H	2	Gr I	0.05	16	(35)	363
H	H	3	Gr I	0.05	10–14	(92)	363
H	H	4	Gr I	0.05	10–14	(90)	363
MeO	H	3	Gr I	0.05	10–14	(100)	363
F	H	3	Gr I	0.05	10–14	(86)	363
O_2N	H	4	Gr I	0.05	10–14	(84)	363
H	H	1	Gr I	0.1	—	(<5)	364
H	Me	1	Gr I	0.1	50	(78)	364
MeO	H	3	Ru cat **2**	0.05	14	(99)	364
MeO	H	4	Ru cat **2**	0.05	14	(90)	364
O_2N	H	3	Ru cat **2**	0.05	14	(91)	364

TABLE 1. CROSS-METATHESIS OF UNSATURATED COMPOUNDS WITH FUNCTIONAL GROUPS IN HOMOALLYLIC AND MORE DISTANT POSITIONS *(Continued)*

A. ETHYLENE AND COMPOUNDS WITH A TERMINAL DOUBLE BOND *(Continued)*

Terminal Alkene	Cross Partner	Conditions	Product(s) and Yield(s) (%)	Refs.

Please refer to the charts preceding the tables for the catalyst structures.

C_2

Terminal Alkene	Cross Partner	Conditions	Product(s) and Yield(s) (%)	Refs.
1 atm	OR, *n*	Catalyst (*x* eq), solvent, 22°	OR, *n*	364

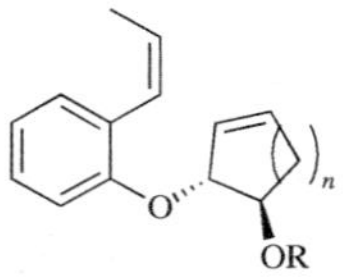

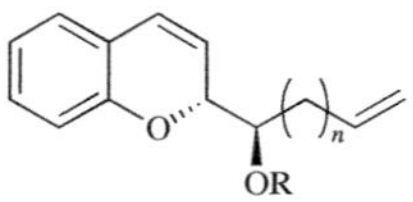

R	*n*	Catalyst	*x*	Solvent	Time (h)	
Bn	1	Gr I	0.09	DCM	36	(35)
TBS	1	Gr I	0.09	DCM	36	(12)
Bn	2	Gr I	0.09	DCM	36	(36)
TBS	2	Gr I	0.09	DCM	36	(32)
H	3	Gr I	0.09	DCM	36	(54)
Bn	3	Gr I	0.09	DCM	36	(82)
TBS	3	Gr I	0.09	DCM	36	(88)
Bn	1	Mo cat **1**	0.1	benzene	24	(91)
TBS	1	Mo cat **1**	0.1	benzene	24	(93)
Bn	2	Mo cat **1**	0.1	benzene	24	(79)
TBS	2	Mo cat **1**	0.1	benzene	24	(90)
H	3	Mo cat **1**	0.1	benzene	24	(<5)
Bn	3	Mo cat **1**	0.1	benzene	24	(75)
TBS	3	Mo cat **1**	0.1	benzene	24	(97)

Olefin	Substrate	Conditions	Product(s) and Yield(s) (%)	Refs.
3 bar		Gr I (0.0003 eq), toluene, rt, 24 h	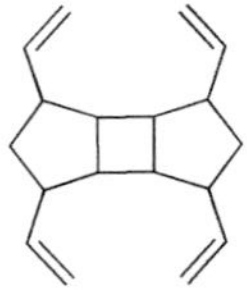 (99)	347
4 bar	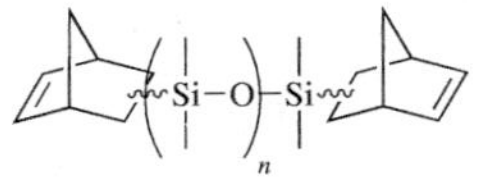 ~Si–O–Si~ (n)	Gr I (x eq), toluene, rt, 24 h	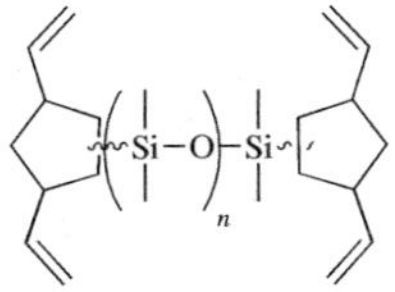 ~Si–O–Si~ (n); n = 1, x = 0.0005 (99); n = 12, x = 0.0007 (99); n = 13, x = 0.0007 (99)	347
=	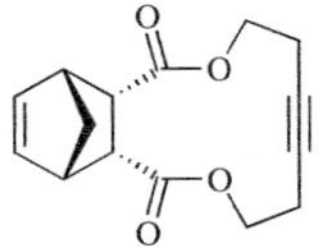 O, O, O, O	Gr I (0.05 eq), rt, 20 h	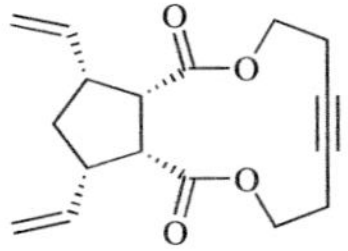 O, O, O, O (76)	385
	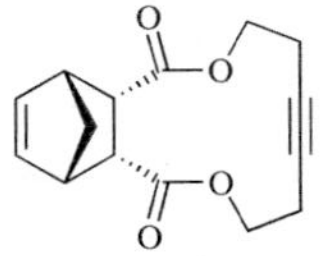 O, O, O, O	Gr II (0.05 eq), 60°, 20 h	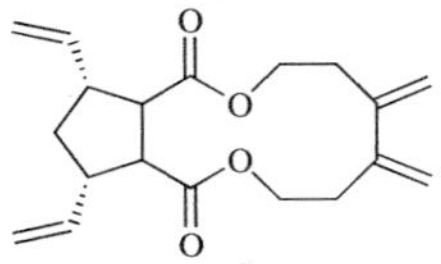 O, O, O, O (100)	385
x atm	AcO–$(CH_2)_6$–CH=CH–$(CH_2)_6$–OAc, $(E)/(Z)$ = 78:22	Ru cat **215** (0.005 eq), THF, 35°, 4 h	AcO–$(CH_2)_6$–CH=CH–$(CH_2)_6$–OAc **I** + CH_2=CH–$(CH_2)_6$–OAc **II**	352

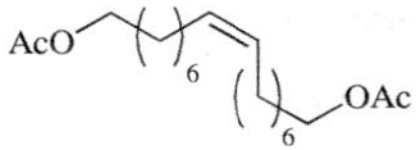

x	$(E)/(Z)$ **I**
1	93:7
5	>95:5

TABLE 1. CROSS-METATHESIS OF UNSATURATED COMPOUNDS WITH FUNCTIONAL GROUPS IN HOMOALLYLIC AND MORE DISTANT POSITIONS (*Continued*)

A. ETHYLENE AND COMPOUNDS WITH A TERMINAL DOUBLE BOND (*Continued*)

Terminal Alkene	Cross Partner	Conditions	Product(s) and Yield(s) (%)	Refs.

Please refer to the charts preceding the tables for the catalyst structures.

C_2

Terminal Alkene	Cross Partner	Conditions	Product(s) and Yield(s) (%)	Refs.
x atm	$MeO_2C(CH_2)_5CH=CH(CH_2)_5CO_2Me$ (*E*)/(*Z*) = 80:20	Ru cat **215** (0.005 eq), THF, 35°, 6 h	$MeO_2C(CH_2)_5CH=CH(CH_2)_5CO_2Me$ **I** + $CH_2=CH(CH_2)_5CO_2Me$ **II**	352
x atm	$AcO(CH_2)_6CH=CH(CH_2)_6OAc$ (*Z*)/(*E*) = 75:25 + *n*-Bu(*Z*)-CH=CH-*n*-Bu	Catalyst (*y* eq), THF, 35°	*n*-BuCH=CH$(CH_2)_6$OAc **I** + $CH_2=CH(CH_2)_6OAc$ **II**	352
x atm	$AcO(CH_2)_6CH=CH(CH_2)_6OAc$ (*Z*)/(*E*) = 75:25 + *n*-Bu(*E*)-CH=CH-*n*-Bu	Catalyst (*y* eq), THF, 35°	*n*-BuCH=CH$(CH_2)_6$OAc **I** + $CH_2=CH(CH_2)_6OAc$ **II**	352

First entry:

x	(*E*)/(*Z*) **I**
1	88:12
5	>95:5

Second entry:

x	Catalyst	*y*	Time (h)	**I**	(*E*)/(*Z*) **I**	**II**
0	Ru cat **215**	0.01	24	(<1)	—	(<1)
1	Ru cat **215**	0.01	4.5	(21)	5:95	(8)
0	Ru cat **217**	0.025	2	(37)	69:31	(<1)

Third entry:

x	Catalyst	*y*	Time (h)	**I**	(*E*)/(*Z*) **I**	**II**
0	Ru cat **215**	0.01	24	(<1)	—	(<1)
1	Ru cat **215**	0.01	24	(<1)	—	(2)
0	Ru cat **217**	0.025	2	(30)	69:31	(<1)

8 bar	$RO_2C(CH_2)_7CH{=}CHCH_2$-n-C_5H_{11}	Ind II (0.0005 eq), DCM, 20°, 6 h	$RO_2C(CH_2)_7CH{=}CH_2$	365
1 atm	OR	Catalyst (x eq), solvent, 22°	O, OR	364

R	Conv. (%)[c]
Me	47
4-HOC_6H_4	7

R	Catalyst	x	Solvent	Time (h)	
H	Gr I	0.09	DCM	36	(10)
Bn	Gr I	0.09	DCM	36	(65)
TBS	Gr I	0.09	DCM	36	(55)
H	Mo cat **1**	0.1	benzene	24	(<5)
Bn	Mo cat **1**	0.1	benzene	24	(98)
TBS	Mo cat **1**	0.1	benzene	24	(95)

TABLE 1. CROSS-METATHESIS OF UNSATURATED COMPOUNDS WITH FUNCTIONAL GROUPS IN HOMOALLYLIC AND MORE DISTANT POSITIONS (*Continued*)

A. ETHYLENE AND COMPOUNDS WITH A TERMINAL DOUBLE BOND (*Continued*)

Terminal Alkene	Cross Partner	Conditions	Product(s) and Yield(s) (%)	Refs.

Please refer to the charts preceding the tables for the catalyst structures.

C_2

1 atm

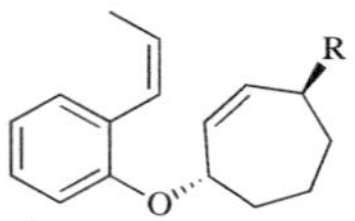

Catalyst (*x* eq), solvent, 22°

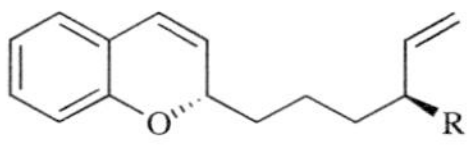

364

R	Catalyst	*x*	Solvent	Time (h)	
H	Ru cat **2**	0.1	DCM	24	(81)
BnO	Gr I	0.09	DCM	36	(25)
TBSO	Gr I	0.09	DCM	36	(29)
BnO	Mo cat **1**	0.1	benzene	24	(88)
TBSO	Mo cat **1**	0.1	benzene	24	(92)

x atm

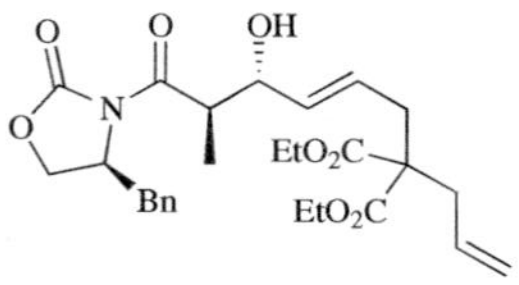

Catalyst (0.05 eq), solvent

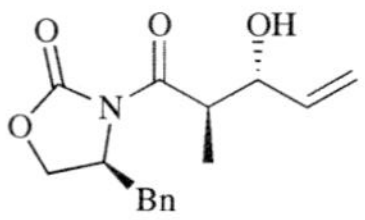

366

x	Catalyst	Solvent	Temp (°)	Time (h)	
1	Gr II	DCE	40	12	(56)
—	Gr II	DCE	40	12	(61)
—	Gr II	DCE	60	12	(10)
1	Gr II	DCE	60	12	(7)
1	Hov II	DCE	40	12	(69)
1	Hov II	DCE	60	12	(62)
—	Hov II	DCE	60	12	(12)
1	Hov II	Toluene	60	12	(83)
—	Hov II	Toluene	60	12	(46)
1	Hov II	Toluene	60	3	(77)
1	Hov II	Toluene	60	6	(74)

x bar n-C_7H_{15} (internal alkene) → Catalyst (y eq), solvent → n-C_8H_{17} (terminal alkene) **I** + n-C_7H_{15} (terminal alkene) **II** 367

x	Catalyst	y	Solvent	Temp (°)	Time (h)	Conv. (%)[d]	**I + II**
1	Gr I	0.049	DCM	40	20	94	(41)
1	Hov I	0.049	DCM	40	20	96	(59)
1	Hov II	0.034	DCM	40	20	97	(61)
10	Hov II	0.034	DCM	40	20	98	(73)
5	Hov II	0.037	DCM	40	20	99	(80)
5	Hov II	0.035	MeOH	40	20	94	(0)
5	Hov II	0.035	toluene	40	20	99	(71)
5	Hov II	0.035	hexane	40	20	99	(91)
5	Hov II	0.035	acetic acid	40	20	95	(32)
5	Hov II	0.034	ethyl acetate	40	20	95	(10)
5	Hov II	0.035	dimethyl carbonate	40	20	98	(21)
5	Hov II	0.034	hexane	40	1	99	(81)
5	Hov II	0.034	hexane	40	3	99	(84)
5	Hov II	0.01	hexane	40	1	98	(66)
5	Hov II	0.01	hexane	40	20	99	(87)
5	Hov II	0.001	hexane	40	1	96	(30)
5	Hov II	0.036	hexane	RT	1	93	(39)
5	Hov II	0.034	hexane	RT	3	95	(32)
10	Hov II	0.034	hexane	RT	20	98	(78)
5	Hov II	0.034	hexane	RT	20	100	(96)
1	Hov II	0.034	hexane	RT	20	99	(79)
1	Hov II	0.034	hexane	40	20	98	(69)
5	Hov II	0.001	—	40	20	94	(25)

TABLE 1. CROSS-METATHESIS OF UNSATURATED COMPOUNDS WITH FUNCTIONAL GROUPS IN HOMOALLYLIC AND MORE DISTANT POSITIONS (*Continued*)

A. ETHYLENE AND COMPOUNDS WITH A TERMINAL DOUBLE BOND (*Continued*)

Terminal Alkene	Cross Partner	Conditions	Product(s) and Yield(s) (%)	Refs.
Please refer to the charts preceding the tables for the catalyst structures.				
C_2				
(ethylene)	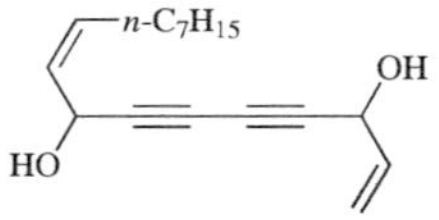 HO, OH, n-C_7H_{15}	Gr II (0.1 eq), DCM, rt, 16 h	 HO, OH (81)	117
	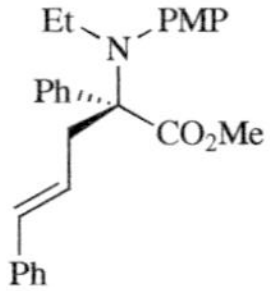 Et, N, PMP, Ph, CO_2Me, Ph	Hov II (0.2 eq), toluene, 90° (MW), 0.5 h	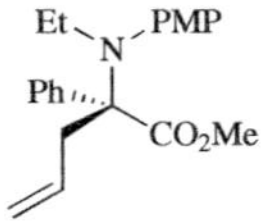 Et, N, PMP, Ph, CO_2Me (24)	368
1 atm	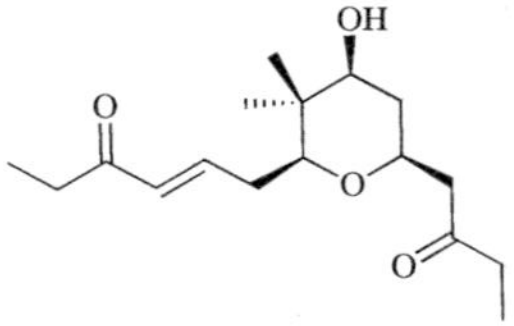 O, OH, O, O	Hov II (0.1 eq), toluene, 60°	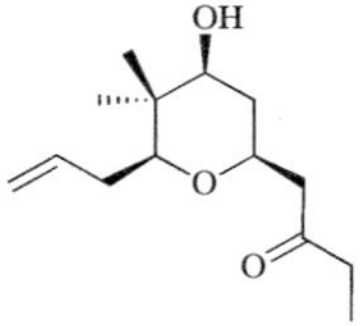 OH, O, O (70)	369

x bar NC(CH2)7CH=CH-n-C_8H_{17} → Catalyst (y eq), solvent → NC(CH2)7CH=CH2 (**I**) + CH2=CH-n-C_8H_{17} (**II**) 370

x	Catalyst	y	Solvent	Temp (°)	Time (h)	Conv. (%)[c]	**I**	**II**
1	Gr I	0.025	toluene	rt	3	45	(—)	(—)
1	Hov I	0.025	toluene	rt	3	80	(—)	(—)
1	Ind I	0.025	toluene	rt	3	45	(—)	(—)
1	Ru cat **293**	0.025	toluene	rt	3	82	(—)	(—)
1	Gr II	0.025	toluene	rt	3	82	(—)	(—)
1	Hov II	0.025	toluene	rt	3	91	(—)	(—)
1	Ru cat **89**	0.025	toluene	rt	3	91	(—)	(—)
1	Ind II	0.025	toluene	rt	3	83	(—)	(—)
1	Ru cat **139**	0.025	toluene	rt	3	85	(—)	(—)
1	Ru cat **137**	0.025	toluene	rt	3	40	(—)	(—)
1	Hov I	0.01	toluene	rt	3	70	(64)	(66)
1	Ru cat **293**	0.01	toluene	rt	3	73	(66)	(70)
1	Hov II	0.01	toluene	rt	3	82	(69)	(77)
1	Hov II	0.01	pentane	rt	3	43	(—)	(—)
1	Hov II	0.01	cyclohexane	rt	3	36	(—)	(—)
1	Hov II	0.01	dibuthyl ether	rt	3	52	(—)	(—)
1	Hov II	0.01	CPME	rt	3	63	(—)	(—)
1	Hov II	0.01	DMC	rt	3	71	(—)	(—)
1	Ru cat **89**	0.01	toluene	rt	3	83	(70)	(76)
1	Ru cat **139**	0.01	toluene	rt	3	83	(79)	(75)
1	Hov II	0.005	toluene	rt	3	79	(61)	(68)
1	Ru cat **89**	0.005	toluene	rt	3	82	(67)	(75)
1	Ru cat **139**	0.005	toluene	rt	3	79	(63)	(70)
20	Gr I	0.0025	toluene	120	16	98	(—)	(—)
20	Gr I	0.00125	toluene	120	16	86	(—)	(—)

TABLE 1. CROSS-METATHESIS OF UNSATURATED COMPOUNDS WITH FUNCTIONAL GROUPS IN HOMOALLYLIC AND MORE DISTANT POSITIONS (*Continued*)

A. ETHYLENE AND COMPOUNDS WITH A TERMINAL DOUBLE BOND (*Continued*)

Terminal Alkene	Cross Partner	Conditions	Product(s) and Yield(s) (%)	Refs.

Please refer to the charts preceding the tables for the catalyst structures.

C_2

Ethylene (x bar) + $MeO_2C(CH_2)_7CH=CH$-n-C_8H_{17} → Catalyst (y eq), conditions → $MeO_2C(CH_2)_7CH=CH_2$ (I) + $CH_2=CH$-n-C_8H_{17} (II)

x	Catalyst	y	Solvent	Temp (°)	Time (h)	Conv. (%)[d]	I	II	I + II	Refs.
1	Gr I	0.025	toluene	70	3.5	45	(40)	(42)	(—)	371
1	Hov I	0.025	toluene	70	3.5	91	(89)	(88)	(—)	371
1	Ru cat **345**	0.025	toluene	70	3.5	25	(19)	(20)	(—)	371
1	Ru cat **346**	0.025	toluene	70	3.5	97	(17)	(54)	(—)	371
10	Hov I	0.025	toluene	70	3.5	97	(97)	(97)	(—)	371
10	Hov I	0.025	toluene	70	3.5	97	(97)	(85)	(—)	371
1	Hov I	0.025	toluene	70	3.5	93	(93)	(93)	(—)	371
10	Gr I	0.065	[bmim][OTf]	20	2	19	(—)	(—)	(—)	371
10	Gr I	0.065	[bmim][NTf_2]	20	2	79	(—)	(—)	(—)	371
10	Gr I	0.065	[bdmim][NTf_2]	20	2	83	(—)	(—)	(—)	371
10	Ru cat **82**	0.05	[bdmim][NTf_2]	20	2	38	(—)	(—)	(—)	371
10	Ru cat **83**	0.05	[bdmim][NTf_2]	20	2	89	(—)	(—)	(—)	371
10	Gr I	0.0001	—	40	2	58	(—)	(—)	(54)	372
10	Gr I	0.0001	—	60	0.5	54	(—)	(—)	(48)	372
10	Hov I	0.0001	—	40	0.5	51	(—)	(—)	(48)	372
10	Gr II	0.0001	—	40	2	64	(—)	(—)	(28)	372
10	Gr II	0.0001	—	60	<0.25	64	(—)	(—)	(28)	372
10	Hov II	0.0001	—	40	0.5	60	(—)	(—)	(20)	372
10	Hov II	0.0001	—	60	<0.25	68	(—)	(—)	(32)	372
10	Ru cat **15**	0.0001	—	40	1	69	(—)	(—)	(38)	372
10	Ru cat **187**	0.0001	—	60	0.25	79	(—)	(—)	(56)	372
55	Ru cat **187**	0.0001	—	25	6	87	(—)	(—)	(70)	372
10	Ru cat **69**	0.0001	—	40	0.25	62	(—)	(—)	(21)	372
10	Ru cat **66**	0.0001	—	40	0.25	50	(—)	(—)	(3)	372

12	Ru cat **159**	0.0001	—	60	6.5	79	(—)	(—)	(57)	373
12	Ru cat **159**	0.0001	—	80	2	81	(—)	(—)	(54)	373
10	Ru cat **160**	0.0001	—	60	3	78	(—)	(—)	(57)	373
10	Ru cat **161**	0.0001	—	60	4	59	(—)	(—)	(53)	373
10	Ru cat **279**	0.0001	—	60	0.5	35	(—)	(—)	(30)	373
10	Ru cat **71**	0.0001	—	60	4	77	(—)	(—)	(51)	373
10	Ru cat **249**	0.00005	—	40	20	61	(—)	(—)	(57)	373
10	Ru cat **253**	0.0001	—	40	6	46	(—)	(—)	(43)	373
10	Ru cat **263**	0.000035	—	40	1	75	(—)	(—)	(56)	373
10	Ru cat **72**	0.0001	—	40	22	60	(—)	(—)	(54)	373
10	Ru cat **256**	0.0001	—	40	3	37	(—)	(—)	(32)	374
10	Ru cat **263**	0.0001	—	40	3	42	(—)	(—)	(37)	374
10	Ru cat **257**	0.000003	—	40	3	59	(—)	(—)	(54)	374
10	Ru cat **258**	0.000003	—	40	3	18	(—)	(—)	(17)	374
10	Ru cat **249**	0.000003	—	40	3	19	(—)	(—)	(18)	374
10	Ru cat **259**	0.000003	—	40	3	22	(—)	(—)	(14)	374
10	Ru cat **254**	0.000003	—	40	3	26	(—)	(—)	(22)	374
10	Ru cat **260**	0.000003	—	40	3	42	(—)	(—)	(39)	374
10	Ru cat **252**	0.000003	—	40	3	19	(—)	(—)	(14)	374
10	Ru cat **253**	0.000003	—	40	3	13	(—)	(—)	(13)	374
10	Ru cat **250**	0.000003	—	40	3	16	(—)	(—)	(15)	374
10	Ru cat **255**	0.000003	—	40	3	41	(—)	(—)	(34)	374
10	Ru cat **261**	0.000003	—	40	3	46	(—)	(—)	(39)	374
10	Ru cat **264**	0.000003	—	40	3	48	(—)	(—)	(43)	374
10	Ru cat **262**	0.000003	—	40	3	57	(—)	(—)	(54)	374
10	Ru cat **251**	0.000003	—	40	3	47	(—)	(—)	(46)	374
10	Ru cat **249**	0.00005	—	40	20	61	(—)	(—)	(57)	375
10	Ru cat **253**	0.0001	—	40	6	46	(—)	(—)	(43)	375
10	Ru cat **263**	0.000035	—	40	1	75	(—)	(—)	(56)	375

TABLE 1. CROSS-METATHESIS OF UNSATURATED COMPOUNDS WITH FUNCTIONAL GROUPS IN HOMOALLYLIC AND MORE DISTANT POSITIONS *(Continued)*

A. ETHYLENE AND COMPOUNDS WITH A TERMINAL DOUBLE BOND *(Continued)*

Terminal Alkene	Cross Partner	Conditions	Product(s) and Yield(s) (%)	Refs.

Please refer to the charts preceding the tables for the catalyst structures.

C_2

Ethylene, *x* bar; Cross partner: $MeO_2C(CH_2)_7CH=CH$-*n*-C_8H_{17}; Conditions: Catalyst (*y* eq), ; Products: $MeO_2C(CH_2)_7CH=CH_2$ **I** + $CH_2=CH$-*n*-C_8H_{17} **II**

x	Catalyst	*y*	Solvent	Temp (°)	Time (h)	Conv. (%)[d]	I	II	I + II	Refs.
4	Gr I	0.0003	—	rt	2	65	(—)	(—)	(63)	113
4	Hov I	0.0003	—	rt	2	69	(—)	(—)	(65)	113
4	Ru cat **187**	0.00005	—	rt	1	54	(—)	(—)	(28)	113
4	Ru cat **15**	0.00005	—	rt	0.67	61	(—)	(—)	(35)	113
4	Ru cat **263**	0.00005	—	rt	1	80	(—)	(—)	(70)	376
10	Mo cat **6**	0.002	—	rt	15	95	(—)	(—)	(95)	376
4	Mo cat **18**	0.0002	—	rt	18	96	(—)	(—)	(94)	376
4	W cat **5**	0.002	—	rt	17	48	(—)	(—)	(48)	376
10	Ru cat **5**	0.00003	—	60	2	43	(—)	(—)	(—)	377
10	Ru cat **287**	0.0002	—	50	2	60	(—)	(—)	(—)	376
10	Ru cat **287**	0.0001	—	50	2	70	(—)	(—)	(—)	376
10	Ru cat **287**	0.00005	—	50	2	64	(—)	(—)	(—)	376
4	Gr I	0.0002	—	30	18	75	(—)	(—)	(—)	378
1	Ru cat **75**	0.01	—	80	12	99	(96)	(97)	(—)	379
1	Ru cat **76**	0.01	—	80	12	99	(92)	(92)	(—)	379
10	Ru cat **11**	0.0001	—	60	2	—	(24)	(—)	(—)	380
9	Ru cat **3**	0.00017	—	13	4	61	(—)	(—)	(—)	381
50	Ru cat **3**	0.00002	—	90	4	20	(—)	(—)	(—)	381
9	Ru cat **192**	0.00017	—	12	4	18	(—)	(—)	(—)	381

9	Ru cat **193**	0.00017	—	13	4	18	(—)	(—)	(—)	381
50	Ru cat **193**	0.00016	—	25	4	23	(—)	(—)	(—)	381
7	Ru cat **40**	0.005	—	60	0.3	25	(—)	(—)	(—)	382
7	Ru cat **40**	0.005	toluene	60	0.3	32	(—)	(—)	(—)	382
7	Ru cat **40**	0.005	DCM	60	1	46	(—)	(—)	(—)	382
10	Ru cat **234**	0.0001	—	40	6	54	(—)	(—)	(46)	383
10	Ru cat **233**	0.0001	—	40	6	11	(—)	(—)	(9)	383
10	Ru cat **236**	0.0001	—	40	6	52	(—)	(—)	(45)	383
10	Ru cat **237**	0.0001	—	40	6	42	(—)	(—)	(36)	383
10	Ru cat **232**	0.0001	—	40	6	59	(—)	(—)	(51)	383
10	Ru cat **241**	0.0001	—	40	6	52	(—)	(—)	(46)	383
10	Ru cat **242**	0.0001	—	40	6	15	(—)	(—)	(15)	383
10	Ru cat **231**	0.0001	—	40	6	17	(—)	(—)	(11)	383
10	Ru cat **225**	0.0001	—	40	6	40	(—)	(—)	(31)	383
10	Ru cat **242**	0.0005	—	40	6	48	(—)	(—)	(46)	383
10	Ru cat **232**	0.0005	—	40	6	89	(—)	(—)	(78)	383
10	Ru cat **237**	0.0005	—	40	2	83	(—)	(—)	(72)	383
10	Ru cat **225**	0.0005	—	40	4	65	(—)	(—)	(53)	383
10	Ru cat **235**	0.0005	—	40	6	70	(—)	(—)	(41)	383
10	Ru cat **241**	0.0005	—	40	6	86	(—)	(—)	(75)	383
1	Hov I	0.025	DCM	rt	3	88	(—)	(—)	(—)	384
10.3	Ru cat **215**	0.001	DCM	40	1	—	(—)	(—)	(80)	352
10.3	Ru cat **215**	0.0001	DCM	40	1	—	(—)	(—)	(12)	352

TABLE 1. CROSS-METATHESIS OF UNSATURATED COMPOUNDS WITH FUNCTIONAL GROUPS IN HOMOALLYLIC AND MORE DISTANT POSITIONS *(Continued)*

A. ETHYLENE AND COMPOUNDS WITH A TERMINAL DOUBLE BOND *(Continued)*

Terminal Alkene	Cross Partner	Conditions	Product(s) and Yield(s) (%)	Refs.

Please refer to the charts preceding the tables for the catalyst structures.

C_2

Terminal Alkene	Cross Partner	Conditions	Product(s) and Yield(s) (%)	Refs.
1 atm	MeO_2C, n-C_8H_{17} + MeO_2C, n-Bu + MeO_2C	Catalyst (0.01 eq), toluene, 4 h	MeO_2C + n-C_8H_{17} + n-Bu + MeO_2C + n-Bu + + + MeO_2C	379

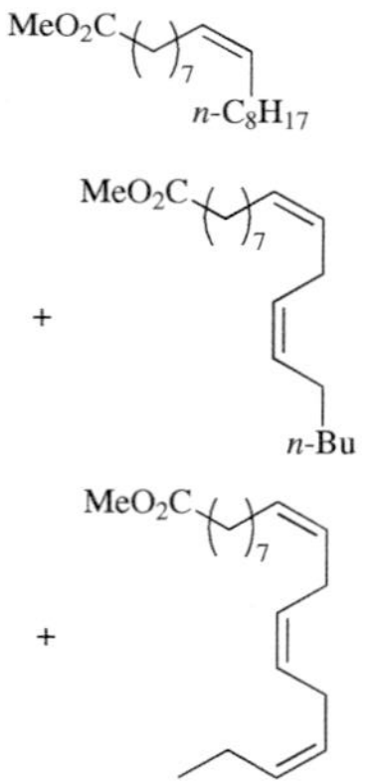

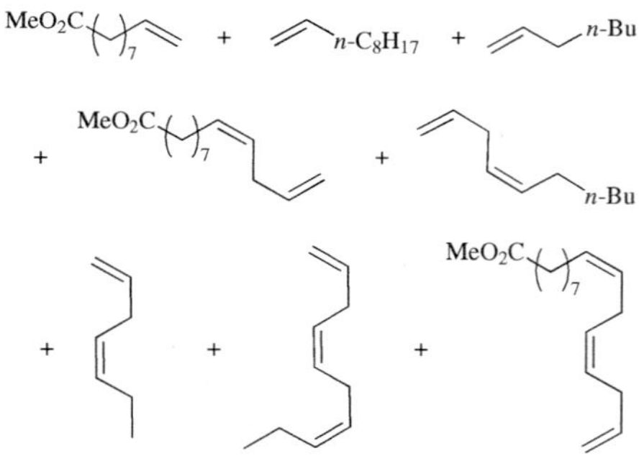

Catalyst	Temp (°)	Conv. (%)[c]	Select (%)[c]
Gr I	25	82	90
Gr I	50	95	92
Gr I	70	99	93
Ru cat **38**	25	90	60
Ru cat **38**	50	99	55
Ru cat **38**	70	99	50
Ru cat **75**	25	99	96
Ru cat **75**	50	99	90
Ru cat **75**	70	99	80
Ru cat **76**	25	99	90
Ru cat **76**	50	99	88
Ru cat **76**	70	99	74

20 bar	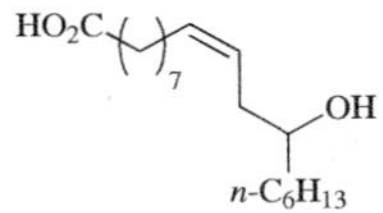 $HO_2C(CH_2)_7$ … OH, n-C_6H_{13}	Ind II (0.001 eq), toluene, 80°, 3 h	 HO_2C (7) (74) 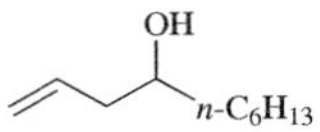 + OH, n-C_6H_{13} (87)	112
20 bar	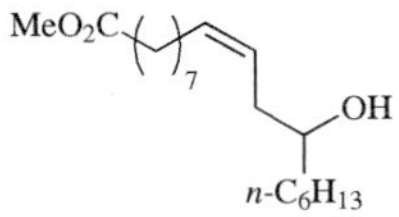 $MeO_2C(CH_2)_7$ … OH, n-C_6H_{13}	Catalyst (0.01 eq), toluene, 50°, 8 h	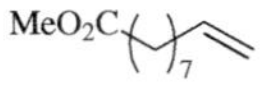 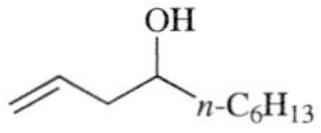 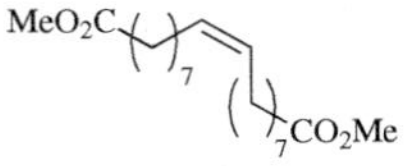 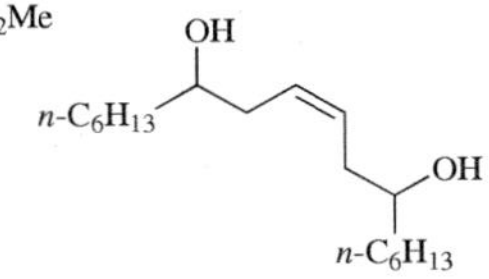 MeO_2C (7) **I** + OH, n-C_6H_{13} **II** + MeO_2C (7) … (7) CO_2Me **III** + OH, n-C_6H_{13}, OH, n-C_6H_{13} **IV**	112

Catalyst	Conv (%)[c]	**I**	**II**	**III**	**IV**
Gr I	26	(16)	(19)	(—)	(1)
Gr II	98	(54)	(76)	(8)	(2)
Ind I	26	(19)	(22)	(—)	(—)
Ind II	98	(63)	(78)	(4)	(1)
Ru cat **319**	98	(33)	(79)	(20)	(2)
Ru cat **290**	42	(30)	(34)	(—)	(—)
Ru cat **320**	99	(36)	(81)	(18)	(1)
Ru cat **292**	84	(52)	(48)	(6)	(10)
Ru cat **39**	93	(66)	(66)	(3)	(7)
Ru cat **40**	91	(62)	(58)	(3)	(8)
Ru cat **139**	89	(36)	(81)	(18)	(1)

TABLE 1. CROSS-METATHESIS OF UNSATURATED COMPOUNDS WITH FUNCTIONAL GROUPS IN HOMOALLYLIC AND MORE DISTANT POSITIONS *(Continued)*

A. ETHYLENE AND COMPOUNDS WITH A TERMINAL DOUBLE BOND *(Continued)*

Terminal Alkene	Cross Partner	Conditions	Product(s) and Yield(s) (%)	Refs.
Please refer to the charts preceding the tables for the catalyst structures.				
C_2				
1 atm	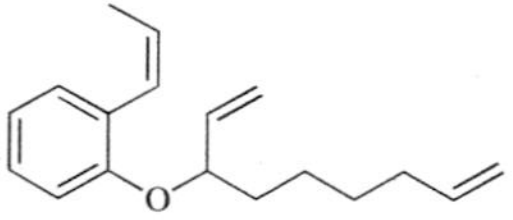	Gr I (0.05 eq), DCM, 22°	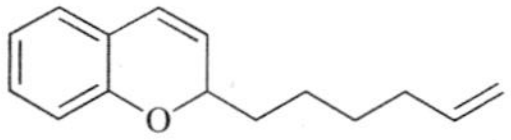 O (83)	364
300 kPa	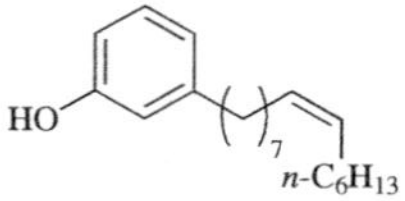 HO, 7, n-C_6H_{13}	Catalyst (x eq), toluene, 60°, 2 h	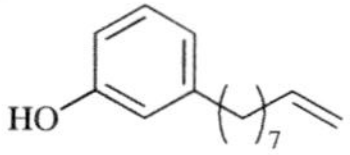 HO, 7	387

Catalyst	x	Conv. (%)[e]
Gr II	0.02	90
Ru cat **8**	0.02	71
Ru cat **8**	0.01	60
Ru cat **8**	0.004	56
Ru cat **8**	0.002	46
Ru cat **8**	0.001	26
Ru cat **89**	0.02	92
Ru cat **90**	0.02	89
Ru cat **90**	0.01	86
Ru cat **90**	0.004	81
Ru cat **90**	0.002	73
Ru cat **90**	0.001	56

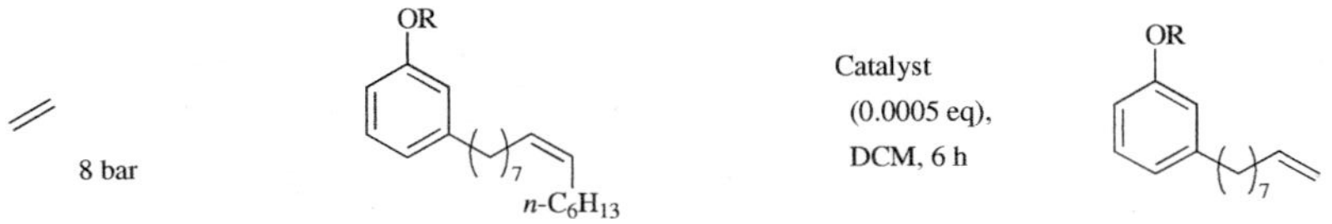

8 bar

OR

n-C_6H_{13}

Catalyst (0.0005 eq), DCM, 6 h

OR

365

R	Catalyst	Temp (°)	Conv. (%)[e]	
H	Ru cat **316**	70	28	(13)
H	Ru cat **139**	70	72	(9)
Me	Ind I	rt	53	(50)
Me	Ru cat **316**	rt	71	(6)
Me	Ru cat **139**	rt	77	(2)

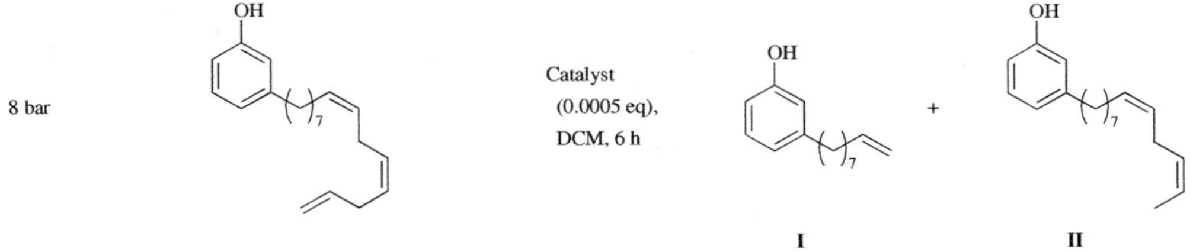

8 bar

OH

Catalyst (0.0005 eq), DCM, 6 h

OH

I

+

OH

II

365

Catalyst	Temp (°)	Conv. (%)[e]	**I**	**II**
Ind I	20	95	(94)	(2)
Gr II	20	90	(91)	(5)
Ind II	70	84	(68)	(12)
Ru cat **294**	40	83	(54)	(20)
Ru cat **320**	70	87	(59)	(22)
Ru cat **316**	90	92	(75)	(12)
Ru cat **139**	20	82	(50)	(12)
Ru cat **137**	40	95	(24)	(10)

TABLE 1. CROSS-METATHESIS OF UNSATURATED COMPOUNDS WITH FUNCTIONAL GROUPS IN HOMOALLYLIC AND MORE DISTANT POSITIONS *(Continued)*

A. ETHYLENE AND COMPOUNDS WITH A TERMINAL DOUBLE BOND *(Continued)*

Terminal Alkene	Cross Partner	Conditions	Product(s) and Yield(s) (%)	Refs.

Please refer to the charts preceding the tables for the catalyst structures.

C_2

Terminal Alkene	Cross Partner	Conditions	Product(s) and Yield(s) (%)	Refs.
8 bar	OMe, CO_2Me, n-C_6H_{13}	Catalyst (0.01 eq), solvent, 16 h	OMe, CO_2Me	365
		Hov II (5 × 0.05 eq), TsOH (1 eq), DCM, reflux, 60 h	(81)	116

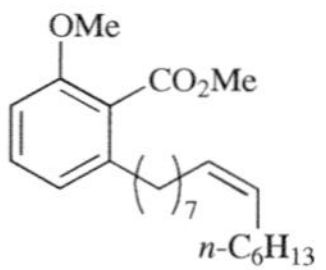

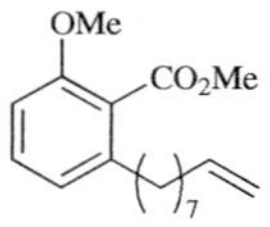

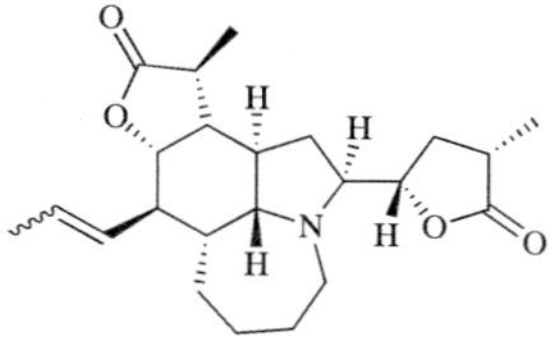

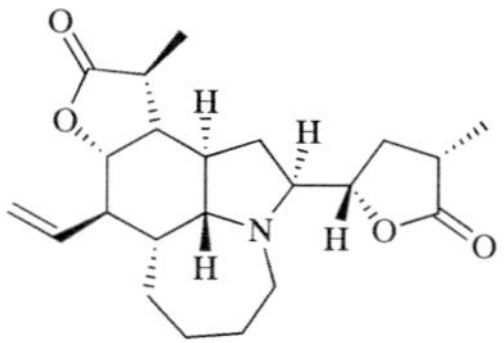

Catalyst	Solvent	Temp (°)	Conv. (%)[e]
Hov I	DCM	25	93
Ru cat **320**	DCM	25	17
Ru cat **312**	DCM	25	30
Ru cat **139**	DCM	25	96
Ru cat **95**	DCM	25	90
Hov I	THF	60	74
Ru cat **320**	THF	60	88
Ru cat **312**	THF	60	56
Ru cat **139**	THF	60	44
Ru cat **95**	THF	60	93

8 bar	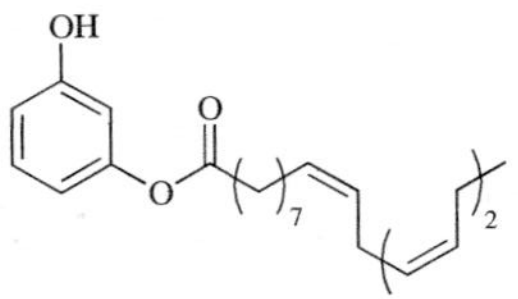 OH, O, O, 7, 2	Ind II (0.0005 eq), DCM, 20°, 6 h	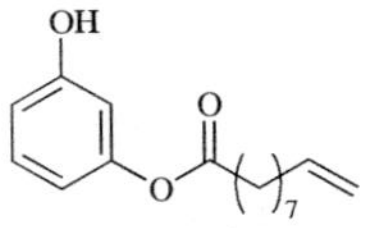 OH, O, O, 7	36[c]	365
excess	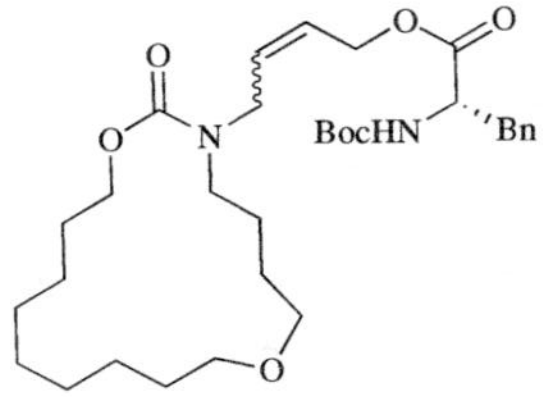 O, O, N, O, O, BocHN, Bn, O	Gr II (0.02 eq), DCM, reflux, 2.5 h	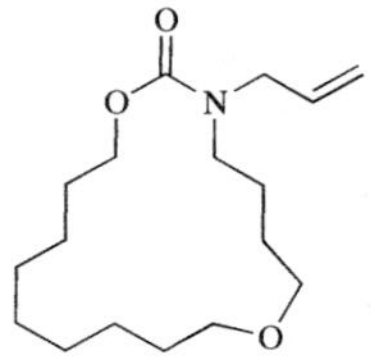 O, O, N, O (36) 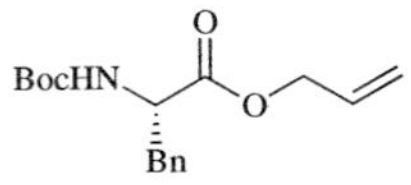 + BocHN, O, O, Bn (37) 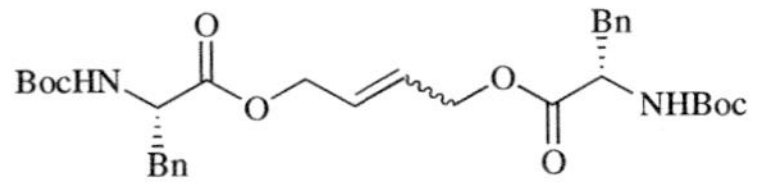 + BocHN, O, O, Bn, Bn, O, O, NHBoc (4)		386

TABLE 1. CROSS-METATHESIS OF UNSATURATED COMPOUNDS WITH FUNCTIONAL GROUPS IN HOMOALLYLIC AND MORE DISTANT POSITIONS (*Continued*)

A. ETHYLENE AND COMPOUNDS WITH A TERMINAL DOUBLE BOND (*Continued*)

Terminal Alkene	Cross Partner	Conditions	Product(s) and Yield(s) (%)	Refs.

Please refer to the charts preceding the tables for the catalyst structures.

C_2

7 bar

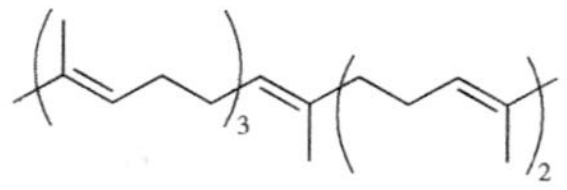

Catalyst (x eq), solvent

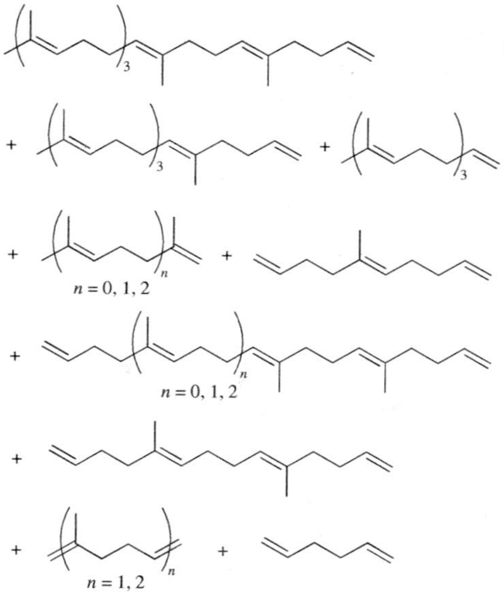

$n = 0, 1, 2$

$n = 0, 1, 2$

$n = 1, 2$

388

Catalyst	x	Solvent	Temp (°)	Time (h)	Conv. (%)[c]
Ru cat **341**	0.001	C_6D_6	100	6	99
Ru cat **341**	0.0001	C_6D_6	120	3	65
Ru cat **341**	0.0001	toluene	120	3	65
Ru cat **340**	0.0001	toluene	120	3	33
Ru cat **336**	0.0001	toluene	120	3	75
Ru cat **335**	0.0001	toluene	120	3	48
Ru cat **51**	0.0001	toluene	120	3	33
Ru cat **49**	0.000005	toluene	120	3	6
Ru cat **50**	0.0001	toluene	120	3	55
Ru cat **53**	0.000005	toluene	120	3	3
Gr II	0.0001	toluene	120	3	35
Hov II	0.0001	toluene	120	3	85

	Olefin	Substrate	Conditions	Product(s) and Yield(s) (%)	Refs.
	20 bar	(triglyceride; n-C_6H_{13}, HO, O, 7)	Ind II (0.001eq), toluene, 80°, 3 h	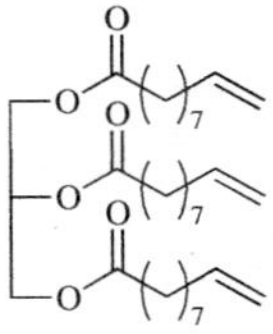 (triester, O, 7) (0) 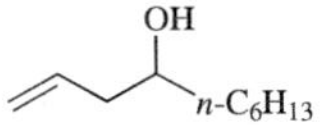 + (OH, n-C_6H_{13}) (87)	112
C_{3-4}					
	(n) 5 eq	(NHR, CO_2H)	1. Hov II (0.05 eq), DCM, rt 2. H_2 (60 psi), MeOH	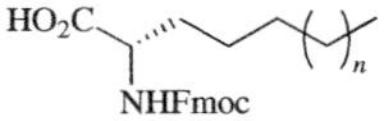 (HO_2C, NHFmoc, n)	127
				n / R: 0 / Fmoc (93); 0 / Boc (78); 1 / Fmoc (86)	
C_4					
	 excess	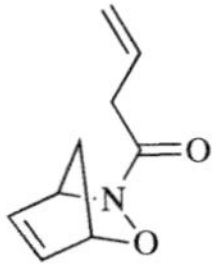 (N, O, O)	Gr II (2 × 0.05 eq), toluene, 80°, 2 × 3 h	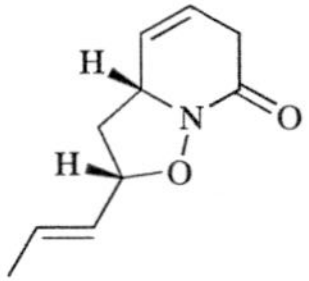 (H, H, N, O, O) (75)	389

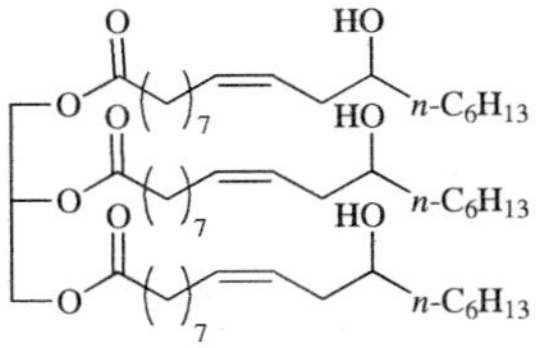

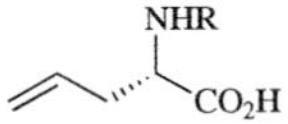

TABLE 1. CROSS-METATHESIS OF UNSATURATED COMPOUNDS WITH FUNCTIONAL GROUPS IN HOMOALLYLIC AND MORE DISTANT POSITIONS (*Continued*)

A. ETHYLENE AND COMPOUNDS WITH A TERMINAL DOUBLE BOND (*Continued*)

Terminal Alkene	Cross Partner	Conditions	Product(s) and Yield(s) (%)	Refs.

Please refer to the charts preceding the tables for the catalyst structures.

C_{4-5}

Terminal Alkene	Cross Partner	Conditions	Product(s) and Yield(s) (%)	Refs.
CN-terminated alkene (n), 2 eq	OR-terminated alkene (m)	Catalyst (0.05 eq), DCM, reflux, 2 h	RO–(m)–CH=CH–(n)–CN	182

n	R	m	Catalyst	Conc (M)		$(E)/(Z)$
1	H	2	Gr I	1.2	(2)	78:22
1	H	2	Gr II	0.05	(38)	86:14
1	H	2	Gr II	0.5	(81)	86:14
1	H	2	Hov II	0.05	(72)	67:33
1	H	2	Hov II	0.5	(65)	86:14
1	H	1	Gr II	0.5	(77)	78:22
1	H	7	Gr II	0.5	(16)	71:29
1	Ph_3C	2	Gr II	0.5	(23)	71:29
2	H	2	Gr II	0.5	(72)	83:17
2	H	2	Hov II	0.05	(48)	80:20

C_4

Terminal Alkene	Cross Partner	Conditions	Product(s) and Yield(s) (%)	Refs.
NHBoc-butene, 2 eq	CO_2Me / NHCbz allyl glycine	Gr I (0.1 eq), DCM, reflux, 16 h	MeO_2C (NHCbz) ... NHBoc (69)	390
CO_2H-butene	n-C_6H_{13} alkene	Gr II (2 × 0.025 eq), DCM, 40°, 24 h	HO_2C ... n-C_6H_{13} (98) $(E)/(Z)$ = 80:20	391
CO_2H alkene (n)	OMe, n-C_7H_{15} alkene, 5 eq	Gr II (0.05 eq), DCM, reflux	HO_2C–(n)– ... OMe, n-C_7H_{15}	416

n		$(E)/(Z)$
0	(<5)	95:5
1	(84)	95:5

OEt OEt | OEt OEt | Gr I (0.05 eq), DCM, 45°, 12 h | EtO, OEt, OEt, OEt | (37) $(E)/(Z) = 52{:}48$ | 124

CO_2Me 2 eq | R, N-PMP, Cl, F F | Gr II (0.1 eq), solvent, reflux | MeO_2C, R, N-PMP, Cl, F F | | 392

R	Solvent	Time (h)		$(E)/(Z)$
H	DCM	4	(60)	91:9
H	DCM	15	(70)	87.5:12.5
H	toluene	15	(—)	—
Me	DCM	15	(75)	86:14

CO_2Bn *x* eq | R | Catalyst (*y* eq), DCM, reflux | BnO_2C, R

x	R	Catalyst	*y*	Time (h)		$(E)/(Z)$	
2	PMP-N=C(Cl)-CF_2	Gr II	0.1	4	(60)	91:9	392
5	MeO_2C, OH	Ru cat **35**	0.02	4–12	(69)	85:15	393

TABLE 1. CROSS-METATHESIS OF UNSATURATED COMPOUNDS WITH FUNCTIONAL GROUPS IN HOMOALLYLIC AND MORE DISTANT POSITIONS (*Continued*)

A. ETHYLENE AND COMPOUNDS WITH A TERMINAL DOUBLE BOND (*Continued*)

Terminal Alkene	Cross Partner	Conditions	Product(s) and Yield(s) (%)	Refs.
Please refer to the charts preceding the tables for the catalyst structures.				
C_4				
CO_2Bn 3 eq		Gr II (0.025 eq), DCM, 40°, 16 h	CO_2Bn (60)	394
C_{4-11}				
R^1	R^2 3 eq	1. Ru cat **152** (0.01 eq), (*Z*)-butene (20 eq), THF, 22°, time 2. Ru cat **152** (0.04 eq), THF, 100 torr, 22°, 8 h	R^1, R^2	395
C_4				
OH, CO_2Me	OTBDPS 5 eq	Ru cat **35** (0.02 eq), DCM, reflux, 4–12 h	TBDPSO, OH, CO_2Me (61) (*E*)/(*Z*) = 60:40	393

R^1	R^2	Time (h)		(*Z*)/(*E*)
$HO(CH_2)_2$	$BnO_2C(CH_2)_2$	16	(56)	95:5
$BnO_2C(CH_2)_2$	$HO_2C(CH_2)_2$	1	(74)	97:3
$HO(CH_2)_8$	$HO_2C(CH_2)_2$	1	(58)	>98:2
$OHC(CH_2)_8$	$HO_2C(CH_2)_2$	1	(66)	>98:2

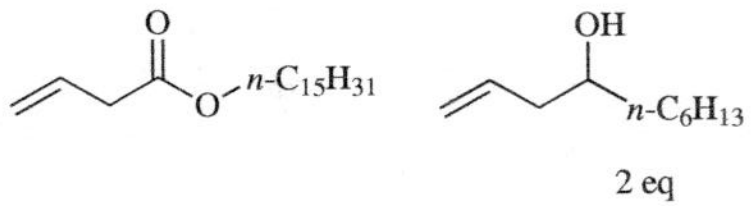

n-$C_{15}H_{31}$ OH n-C_6H_{13} 2 eq

Gr II (0.05 eq), DCM, reflux, 15 h

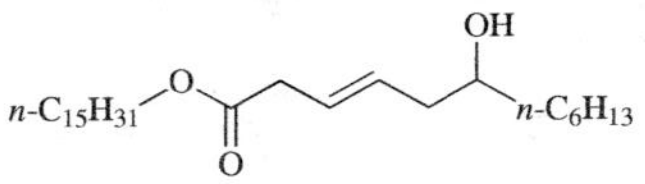

OH n-$C_{15}H_{31}$ n-C_6H_{13}

(49) (*E*) only

396

OTIPS   1 eq

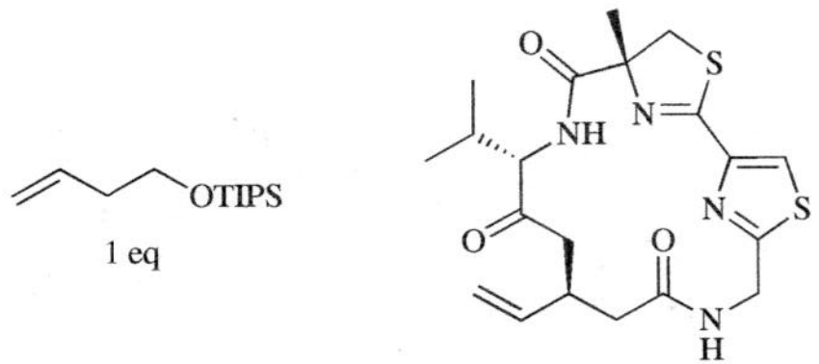

Gr II (0.3 eq), toluene, reflux, 3 h

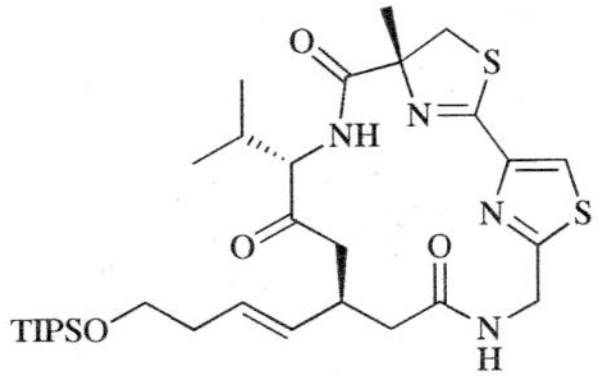

TIPSO

(50) (*E*)/(*Z*) = 90:10

397

n-$C_{16}H_{33}$ 2 eq   R   OMe

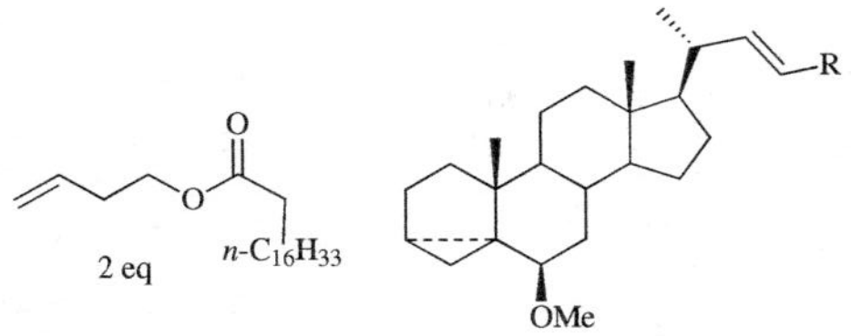

Hov II (0.2 eq), toluene, 80°, 48 h

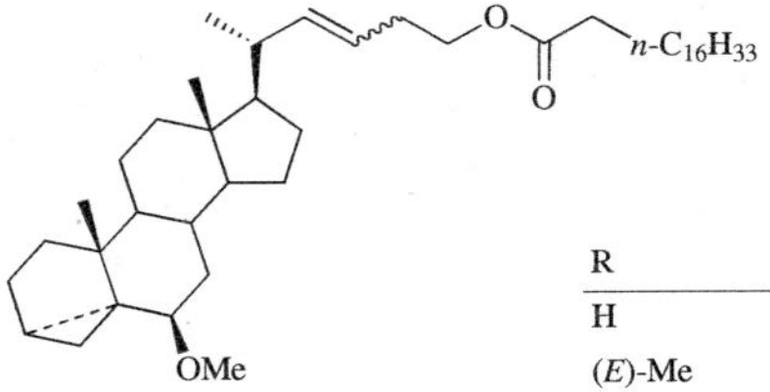

n-$C_{16}H_{33}$ OMe

R	
H	(43)
(*E*)-Me	(44)
(*E*)-*i*-Bu	(12)
(*Z*)-*i*-Bu	(13)

398

TABLE 1. CROSS-METATHESIS OF UNSATURATED COMPOUNDS WITH FUNCTIONAL GROUPS IN HOMOALLYLIC AND MORE DISTANT POSITIONS (*Continued*)

A. ETHYLENE AND COMPOUNDS WITH A TERMINAL DOUBLE BOND (*Continued*)

Terminal Alkene	Cross Partner	Conditions	Product(s) and Yield(s) (%)	Refs.
Please refer to the charts preceding the tables for the catalyst structures.				
C_{4-5}				
 R (4 eq)	 N-Ph, N-Ac	Gr II (0.02 eq), 2,6-Cl_2-BQ (0.1 eq), DCM, reflux, 7 d	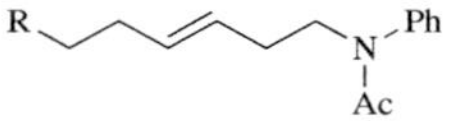 R, N-Ph, N-Ac R / (E)/(Z) TBSO (66) 85:15 BnO (59) 85:15 MeO (60) 85:15 $BnOCH_2$ (45) 80:20 AcPhN (60) 85:15 HO (71) 75:25	399
C_4				
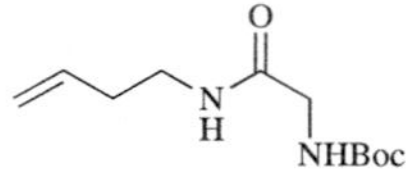 O, N-H, NHBoc	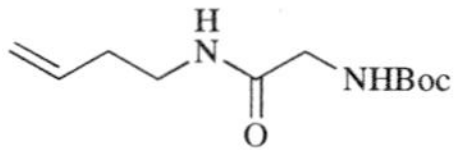 H-N, NHBoc, O	1. Ru cat **152** (0.01 eq), (*Z*)-butene (20 eq), THF, 22°, 4 h 2. Ru cat **152** (0.04 eq), THF, 100 torr, 22°, 1 h	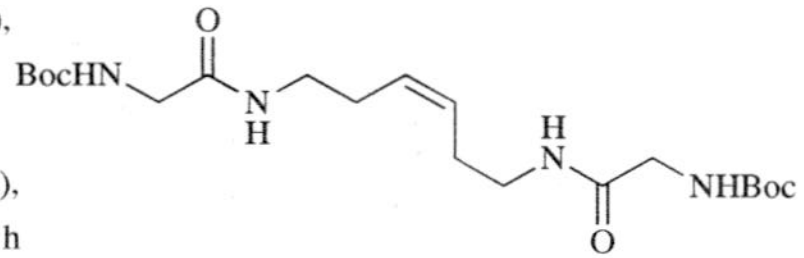 BocHN, O, N-H, H-N, NHBoc, O (88) (*Z*)/(*E*) > 98:2	395
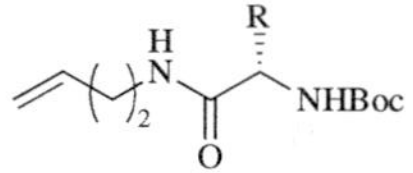 H-N, R, NHBoc, O, 2	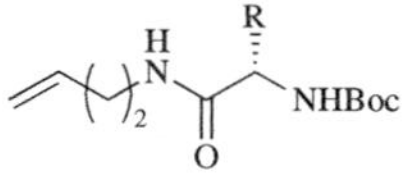 H-N, R, NHBoc, O, 2	Gr I (0.1 eq), DCM, rt, 12 h	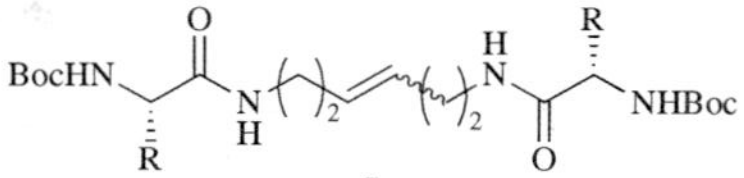 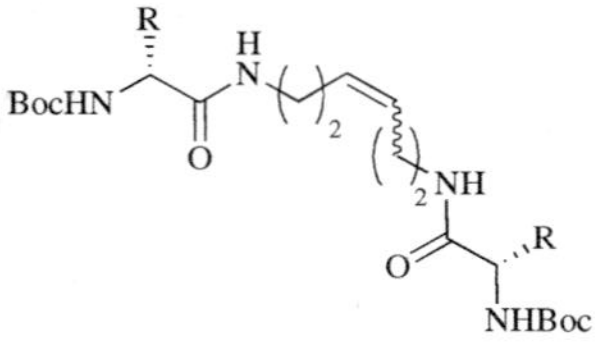 BocHN, O, N-H, R, 2, 2, H-N, R, NHBoc, O **I** + R, BocHN, H-N, O, 2, 2, NH, R, O, NHBoc **II**	400

R	I	II
H	(0)	(0)
Bn	(19)	(58)
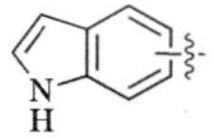	(22)	(67)
	(0)	(0)
i-PrCH$_2$	(0)	(0)
2-napthalen-CH$_2$	(35)	(56)
$C_6F_5CH_2$	(0)	(0)
i-Pr	(0)	(0)

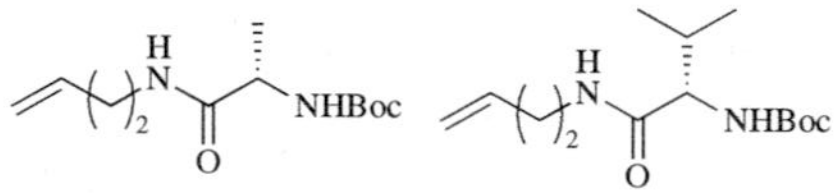

Catalyst (0.075 eq),
THF, 40°, 4 h

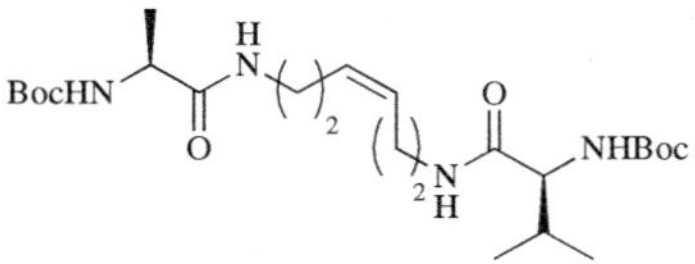

x	y	Catalyst		(Z)/(E)
1	1	Ru cat **215**	(41)	90:10
1	2	Ru cat **215**	(44)	86:14
1	4	Ru cat **215**	(48)	91:9
1	6	Ru cat **215**	(58)	88:12
2	1	Ru cat **215**	(44)	90:10
4	1	Ru cat **215**	(52)	91:9
6	1	Ru cat **215**	(51)	87:13
1	1	Ru cat **216**	(47)	93:7
1	2	Ru cat **216**	(48)	91:9
1	4	Ru cat **216**	(41)	90:10
1	6	Ru cat **216**	(60)	90:10
2	1	Ru cat **216**	(58)	93:10
4	1	Ru cat **216**	(57)	91:9
6	1	Ru cat **216**	(60)	90:10

401

TABLE 1. CROSS-METATHESIS OF UNSATURATED COMPOUNDS WITH FUNCTIONAL GROUPS IN HOMOALLYLIC AND MORE DISTANT POSITIONS (*Continued*)

A. ETHYLENE AND COMPOUNDS WITH A TERMINAL DOUBLE BOND (*Continued*)

Terminal Alkene	Cross Partner	Conditions	Product(s) and Yield(s) (%)	Refs.

Please refer to the charts preceding the tables for the catalyst structures.

C_4

Terminal Alkene	Cross Partner	Conditions	Product(s) and Yield(s) (%)	Refs.
NHBoc *x* eq	NHPbf *y* eq	Catalyst (0.075 eq), THF, 40°, 4 h	BocHN ... NHPbf	401

x	*y*	Catalyst		(*Z*)/(*E*)
1	1	Ru cat **215**	(46)	90:10
1	2	Ru cat **215**	(43)	84:16
1	4	Ru cat **215**	(38)	84:16
1	6	Ru cat **215**	(34)	72:28
2	1	Ru cat **215**	(47)	88:12
4	1	Ru cat **215**	(58)	90:10
6	1	Ru cat **215**	(62)	87:13
1	1	Ru cat **216**	(47)	93:7
1	2	Ru cat **216**	(48)	90:10
1	4	Ru cat **216**	(41)	91:9
1	6	Ru cat **216**	(38)	88:12
2	1	Ru cat **216**	(58)	91:9
4	1	Ru cat **216**	(60)	92:8
6	1	Ru cat **216**	(66)	90:10

C_{4-5}

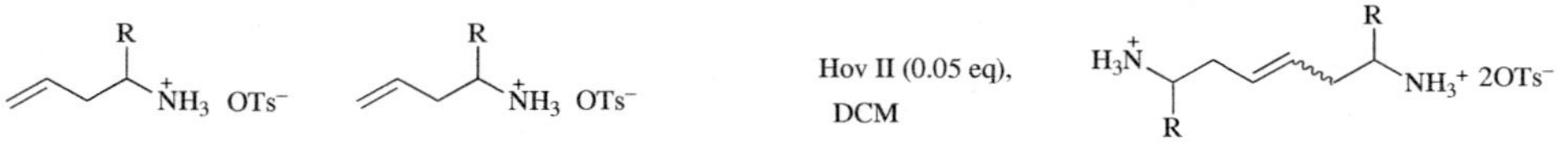

R–CH(NH_3^+) OTs⁻ + R–CH(NH_3^+) OTs⁻ → Hov II (0.05 eq), DCM → H_3N^+ … NH_3^+ 2OTs⁻ 402

R	Temp (°)	Time (h)	Conv. (%)[a]
H	40	24	92
H	100 (MW)	2	95
HO_2C	40	24	0
MeO_2C	40	24	92
HO_2C	100 (MW)	2	0
MeO_2C	100 (MW)	2	90

C_4

SCN + SCN → Catalyst (x eq), DCM → NCS … SCN 403

Catalyst	x	Temp	Time	
Gr II	0.025 + 0.025	reflux	16 h	(5)
Gr II	0.04 + 0.01	rt (MW)	20 + 10 min	(22)
Hov II	0.025 + 0.025	reflux	16 h	(40)
Hov II	0.04 + 0.01	rt (MW)	20 + 10 min	(81)

TABLE 1. CROSS-METATHESIS OF UNSATURATED COMPOUNDS WITH FUNCTIONAL GROUPS IN HOMOALLYLIC AND MORE DISTANT POSITIONS *(Continued)*

A. ETHYLENE AND COMPOUNDS WITH A TERMINAL DOUBLE BOND *(Continued)*

Terminal Alkene	Cross Partner	Conditions	Product(s) and Yield(s) (%)	Refs.

Please refer to the charts preceding the tables for the catalyst structures.

C_4

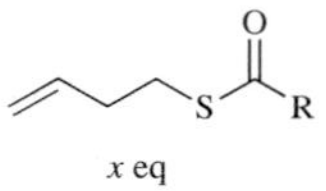

Terminal Alkene	Cross Partner	Conditions	Product(s) and Yield(s) (%)	Refs.
x eq	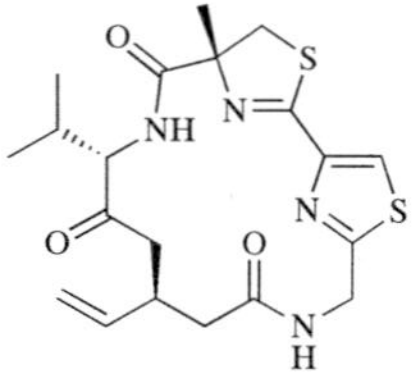	Gr II (*y* eq), toluene		

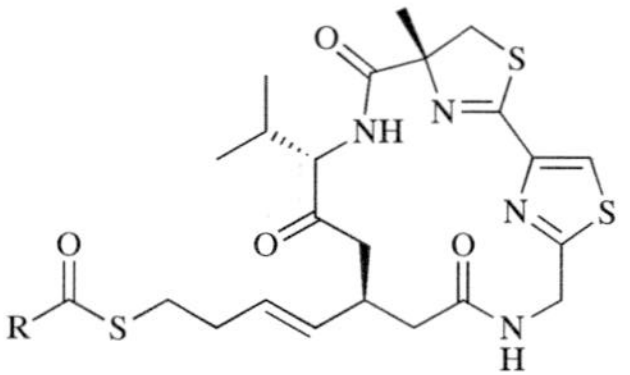

x	R	*y*	Temp	Time (h)		(*E*)/(*Z*)	Refs.
2	Me	0.5	reflux	4	(54)	(*E*) only	397
2	$CH_3(CH_2)_6$	0.5	reflux	4	(41)	(*E*) only	397
1	$CH_3(CH_2)_6$	0.2	60°	24	(34)	—	404

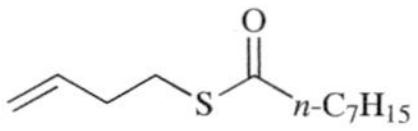

Terminal Alkene	Cross Partner	Conditions	Product(s) and Yield(s) (%)	Refs.
4 eq	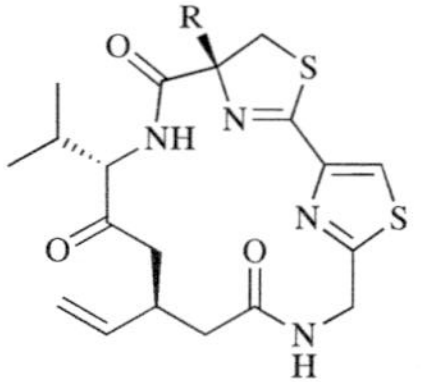	Hov II (*x* eq), DCE, 90°, overnight, sealed tube		405

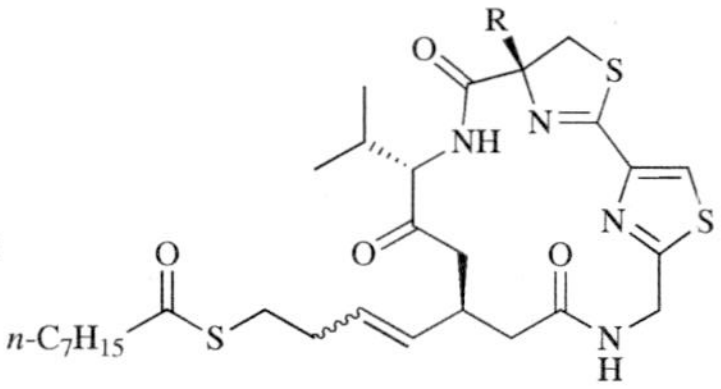

R	*x*	
H	0.14	(35)
H	0.18	(20)
Me	0.14	(27)
Et	0.18	(23)
Bn	0.18	(21)

C_{4-12}

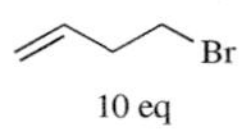

10 eq

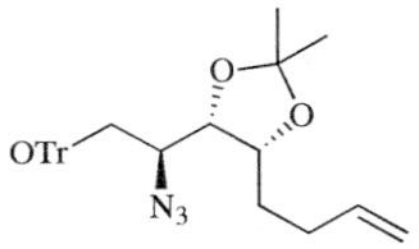

Gr II (0.1 eq), DCM, rt, 18 h

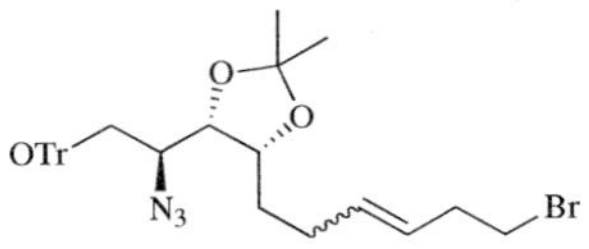

(56) 406

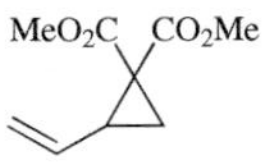

x eq

Catalyst (y eq), CuI (z eq), solvent

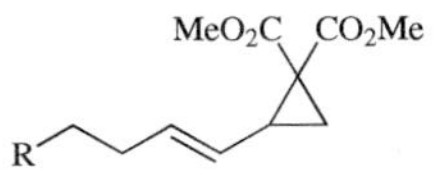

407

R	x	Catalyst	y	z	Solvent	Temp	Time (h)		$(E)/(Z)$
Br	1.4	Gr II	0.01	—	DCM	reflux	—	(78)	80:20
Et	1.4	Gr II	0.05	—	DCM	reflux	2	(70)	86:14
Et	1.4	Gr II	0.01	—	DCM	reflux	3	(67)	86:14
Et	1.4	Gr I	0.01	—	DCM	reflux	24	IC	—
Et	1.4	Gr I	0.1	—	DCM	reflux	24	IC	—
Et	1.4	Hov II	0.01	—	DCM	reflux	24	IC	—
Et	1.2	Gr II	0.01	—	DCM	reflux	2	(53)	86:14
Et	2.0	Gr II	0.01	—	DCM	reflux	2	(67)	86:14
Et	1.4	Gr II	0.01	—	DCM	rt	24	(50)	66:34
Et	1.4	Gr II	0.01	—	DCM	0°	72	IC	—
Et	1.4	Gr II	0.01	0.03	Et_2O	reflux	40	IC	—
n-Pr	1.4	Gr II	0.01	—	DCM	reflux	—	(74)	86:14
n-C_8H_{17}	1.4	Gr II	0.01	—	DCM	reflux	—	(69)	75:25
$TBSO(CH_2)_2$	1.4	Gr II	0.01	—	DCM	reflux	—	(73)	75:25

TABLE 1. CROSS-METATHESIS OF UNSATURATED COMPOUNDS WITH FUNCTIONAL GROUPS IN HOMOALLYLIC AND MORE DISTANT POSITIONS (*Continued*)

A. ETHYLENE AND COMPOUNDS WITH A TERMINAL DOUBLE BOND (*Continued*)

Terminal Alkene	Cross Partner	Conditions	Product(s) and Yield(s) (%)	Refs.

Please refer to the charts preceding the tables for the catalyst structures.

C_{5-6}

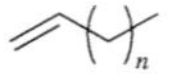

Terminal Alkene	Cross Partner	Conditions	Product(s) and Yield(s) (%)	Refs.
5 eq		Gr II (0.025 eq), DCM, 40°, 16 h		408

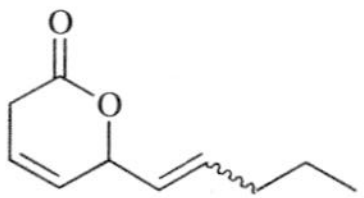

n	
2	(60)
3	(61)

C_{5-12}

Terminal Alkene	Cross Partner	Conditions	Product(s) and Yield(s) (%)	Refs.
 *x* eq	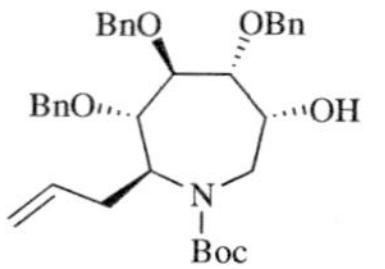 BnO, OBn, BnO, OH, N–Boc	Catalyst (*y* eq), DCM, 40°, 24 h	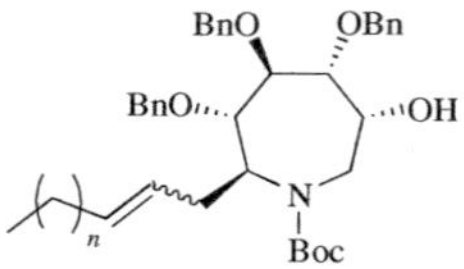 BnO, OBn, BnO, OH, N–Boc	409

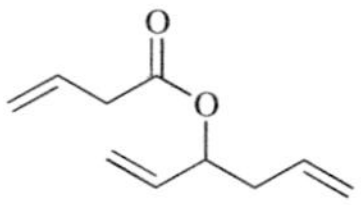

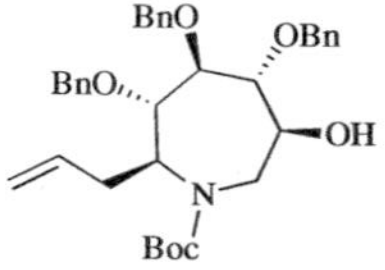

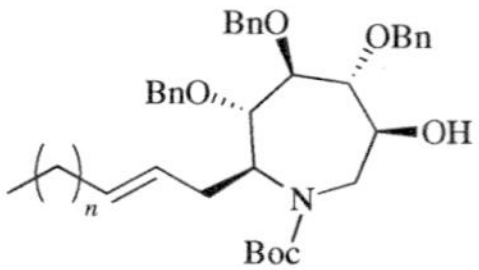

n	*x*	Catalyst	*y*	
2	13	Hov I	0.06	(42)
5	4	Hov II	0.035	(46)
9	4	Hov II	0.03	(67)

Terminal Alkene	Cross Partner	Conditions	Product(s) and Yield(s) (%)	Refs.
x eq	BnO, OBn, BnO, OH, N–Boc	Catalyst (*y* eq), DCM, 40°, 24 h	BnO, OBn, BnO, OH, N–Boc	409

n	*x*	Catalyst	*y*	
2	13	Hov I	0.12	(60)
5	4	Hov II	0.4	(69)
9	4	Hov II	0.3	(83)

C_5

N_3 OBn AcO O N_3 O ( )$_3$ BnO

10 eq

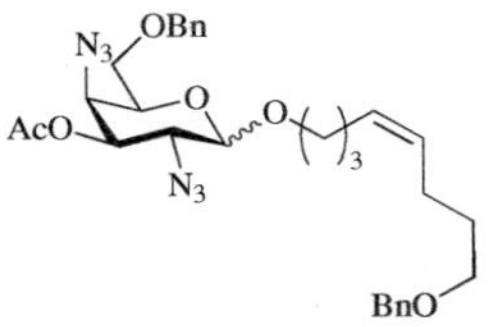

Ru cat **66** (0.1 eq), DCM, 0°, 10 h

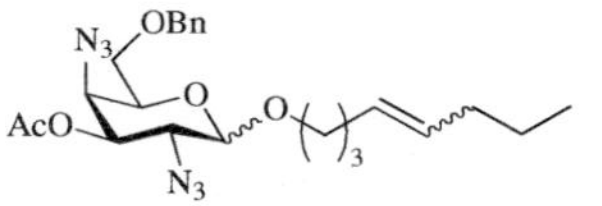

N_3 OBn AcO O N_3 O ( )$_3$

(84) 178

OAc BnO BnO O R O ( )$_3$ BnO

10 eq

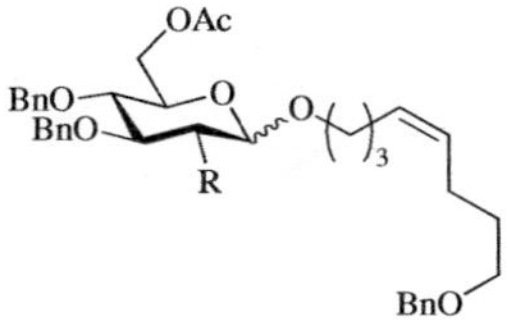

Catalyst (0.1 eq), solvent

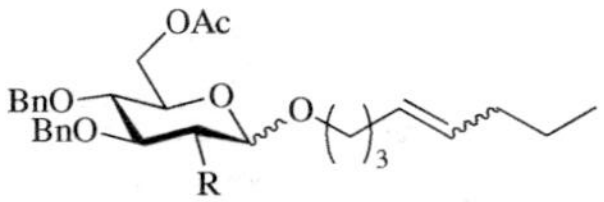

OAc BnO BnO O R O ( )$_3$

178

R	Catalyst	Solvent	Temp (°)	Time (h)	
BnO	Ru cat **66**	DCE	80	12	(73)
N_3	Gr I	DCE	rt	24	(22)
N_3	Gr II	DCE	rt	24	(47)
N_3	Ru cat **66**	DCE	rt	16	(62)
N_3	Ru cat **66**	DCM	0	10	(79)

A. ETHYLENE AND COMPOUNDS WITH A TERMINAL DOUBLE BOND (*Continued*)

Terminal Alkene	Cross Partner	Conditions	Product(s) and Yield(s) (%)	Refs.

Please refer to the charts preceding the tables for the catalyst structures.

C_{5-8}

Terminal Alkene	Cross Partner	Conditions	Product(s) and Yield(s) (%)	Refs.
	CO₂Me, OAc, OAc, OAc, AcO	Gr II (0.1 eq), toluene, 100°	CO₂Me, OAc, OAc, OAc, AcO	410

n		(E)/(Z)
2	(97)	95:5
3	(92)	95:5
5	(90)	95:5

C_5

Terminal Alkene	Cross Partner	Conditions	Product(s) and Yield(s) (%)	Refs.
OH (x eq)	OR	Catalyst (y eq), solvent	HO ... OR	

x	R	Catalyst	y	Solvent	Temp (°)	Time (h)		(E)/(Z)	Refs.
—	H	Ru cat **217**	0.02	THF	35	1	(72)	28:72	84
2	Ph_3C	Gr II	0.05	DCM	reflux	2	(60)	71:29	182

Terminal Alkene	Cross Partner	Conditions	Product(s) and Yield(s) (%)	Refs.
	Ph, Ph, Si, Ot-Bu, OMe	Gr II (0.05 eq), DCM, 40°, 1 h	HO, Ph, Ph, Si, Ot-Bu, OMe (42) (E)/(Z) = 75:25	411

OAc
3 eq

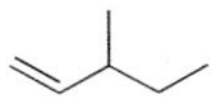

Catalyst (0.05 eq), DCM, reflux, 6 h

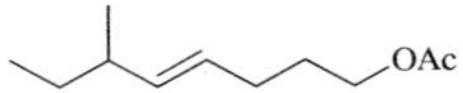

OAc

(E)/(Z) = 95:5

Catalyst	
Hov II	(85)
Ru cat **271**	(98)

177

OBn
10 eq

$n\text{-}C_{10}H_{21}$

Gr I (0.05 eq), DCM, reflux, 4 h

BnO $n\text{-}C_{10}H_{21}$

(85) (E)/(Z) = 75:25

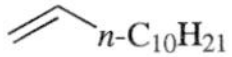

412

OAc
OAc
2 eq

OH
PMP

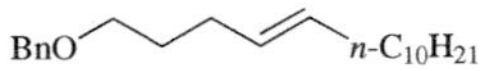

Hov II (0.05 eq), DCM, rt, 24 h

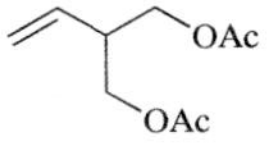

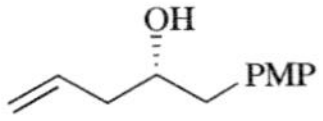

AcO
AcO
OH
PMP

(67)

413

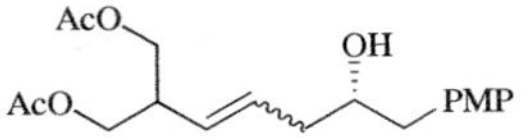

OH

OPiv
x eq

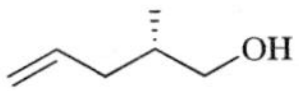

Gr II (y eq), DCM

HO
OPiv

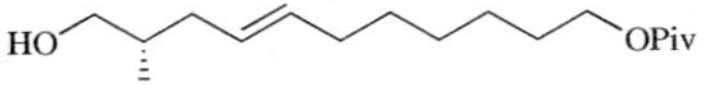

414

x	y	Temp (°)	Time (h)		(E)/(Z)
3	0.05	25	12	(60)	82:18
1	0.05	25	12	(33)	83:17
3	0.05	40	12	(55)	50:50
3	0.01	25	36	(40)	93:7
5	3 × 0.005	25	168	(68)	94:6

TABLE 1. CROSS-METATHESIS OF UNSATURATED COMPOUNDS WITH FUNCTIONAL GROUPS IN HOMOALLYLIC AND MORE DISTANT POSITIONS *(Continued)*

A. ETHYLENE AND COMPOUNDS WITH A TERMINAL DOUBLE BOND *(Continued)*

Terminal Alkene	Cross Partner	Conditions	Product(s) and Yield(s) (%)	Refs.
Please refer to the charts preceding the tables for the catalyst structures.				
C_5				
 CO_2H	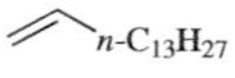 $n\text{-}C_{13}H_{27}$ 5 eq	Gr II (0.05 eq), DCM, 40°, 12 h	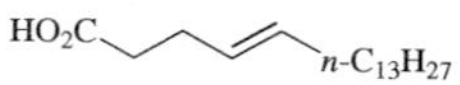 HO_2C, $n\text{-}C_{13}H_{27}$ (95) $(E)/(Z) = 95:5$	415
 CO_2Et 2 eq	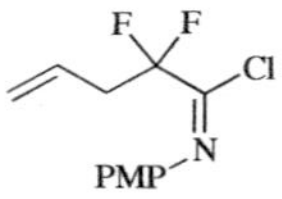 F, F, Cl, N, PMP	Gr II (0.1 eq), DCM, reflux, 3 h	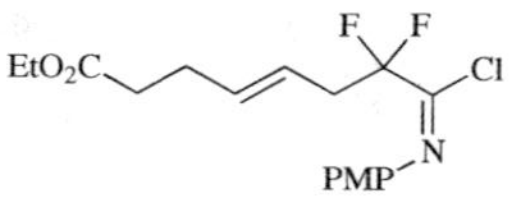 EtO_2C, F, F, Cl, N, PMP (70) $(E)/(Z) = 75:25$	392
7 eq	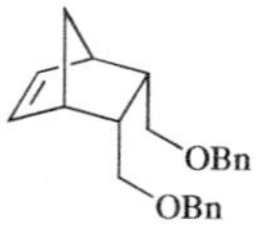 OBn, OBn	Ru cat **215** (0.01 eq), THF, 23°, 1 h	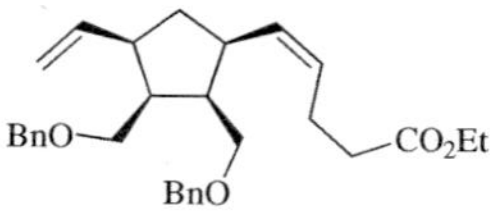 BnO, BnO, CO_2Et (65) er 95.5:4.5, $(Z)/(E) = 95:5$	417
 CO_2Bn 3 eq	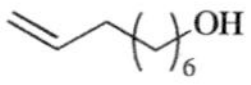 OH, 6	1. Ru cat **355** (0.075 eq), (*E*)-butene (75 eq), THF, 22°, 1 h 2. Ru cat **355** (0.05 eq), THF, 100 torr, 22°, 4 h	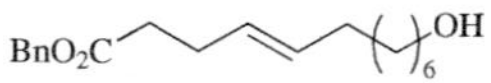 BnO_2C, OH, 6 (66) $(E)/(Z) = 95:5$	395

Reactant	Reactant	Conditions	Product(s) and Yield(s) (%)	Refs.
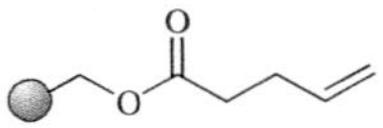 O, O	 Ph 5 eq	1. Gr II (0.05 eq), DCM, reflux, 20 h 2. TFA (0.1 eq), DCM, 1 h 3. CH_2N_2, DCM, 0°, 0.5 h	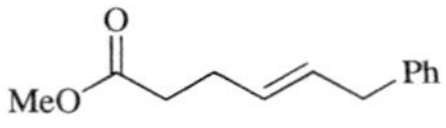 O, MeO, Ph (18) (*E*) only	418
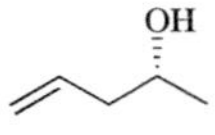 OH 1.5 eq	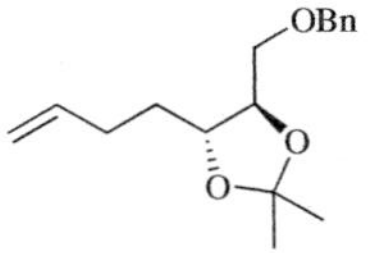 OBn, O, O	Gr II (0.05 eq), DCM, 40°, 22 h 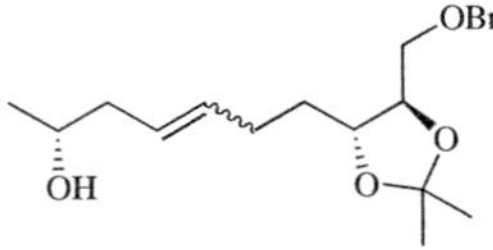	OH, OBn, O, O (64)	419
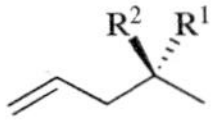 R^2, R^1 1.5 eq	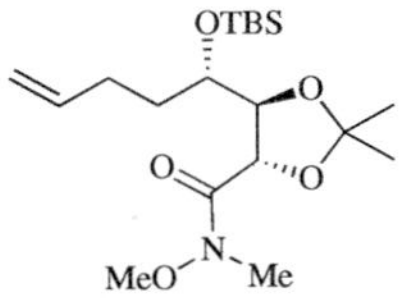 OTBS, O, O, O, MeO, N, Me	Gr II (0.055 eq), DCM, 40°, 7 h	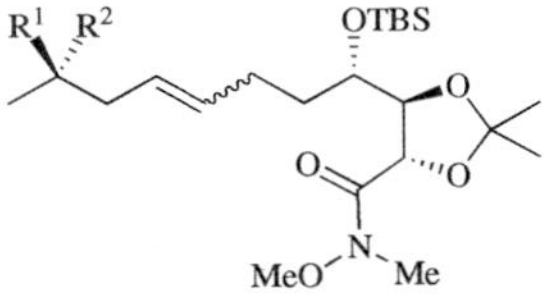 R^1, R^2, OTBS, O, O, O, MeO, N, Me R^1 / R^2: H / HO (69); HO / H (64)	420
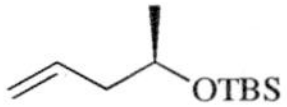 OTBS	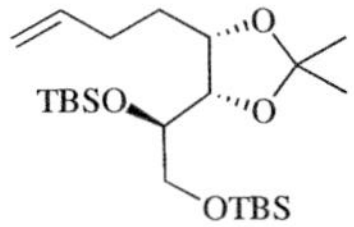 O, O, TBSO, OTBS	1. Gr II (0.05 eq), toluene, 110°, 6 h 2. Pd/C (0.05 eq), H_2, EtOH, rt, 2 h 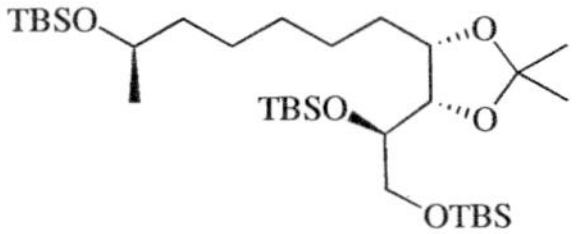	TBSO, O, O, TBSO, OTBS (42)	421

TABLE 1. CROSS-METATHESIS OF UNSATURATED COMPOUNDS WITH FUNCTIONAL GROUPS IN HOMOALLYLIC AND MORE DISTANT POSITIONS (*Continued*)

A. ETHYLENE AND COMPOUNDS WITH A TERMINAL DOUBLE BOND (*Continued*)

Terminal Alkene	Cross Partner	Conditions	Product(s) and Yield(s) (%)	Refs.
Please refer to the charts preceding the tables for the catalyst structures.				
C_5				
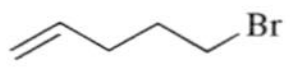 Br; 5 eq	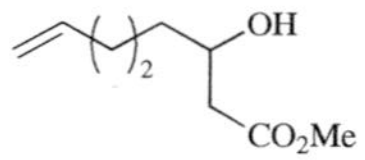 OH; CO_2Me; 2	Ru cat **35** (0.02 eq), DCM, reflux, 4–12 h	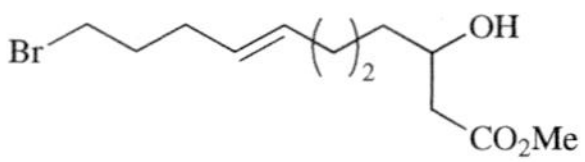 Br; OH; CO_2Me; 2 (83) $(E)/(Z)$ = 90:10	393
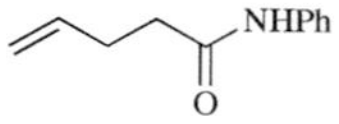 NHPh; O; *x* eq	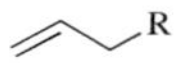 R	Gr II (0.05 eq), DCM, reflux	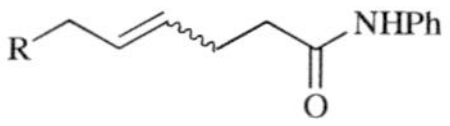 R; NHPh; O	422

x	R	Time (h)	
2.5	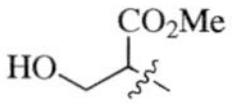 HO; CO_2Me	20	(47)
2.15	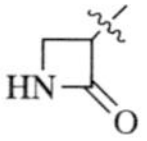 HN; O	12	(58)
2.5	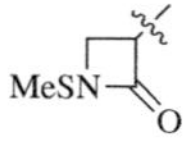 MeSN; O	4	(37)
2.5	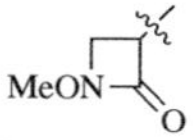 MeON; O	20	(44)

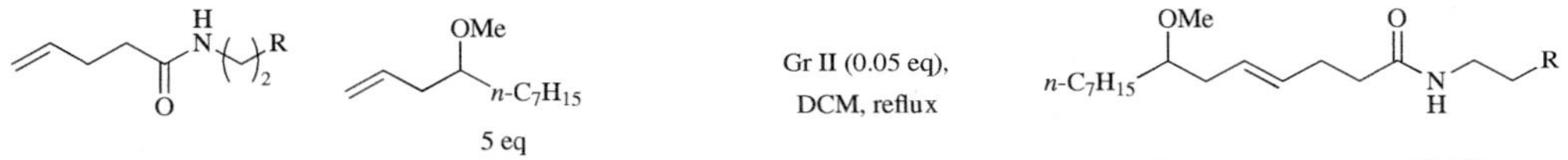

5 eq

Gr II (0.05 eq), DCM, reflux

OMe O n-C_7H_{15} N H R

R		$(E)/(Z)$
Ph	(54)	95:5
3-indolyl	(44)	95:5

416

NHFmoc CO_2H n-C_8H_{17}

x eq

Catalyst (0.05 eq), solvent, overnight

HO_2C n-C_8H_{17} NHFmoc

x	Catalyst	Solvent	Temp	Conv. (%)[a]
5	Hov II	DCM	reflux	93
5	Hov II	EtOAc	reflux	90
5	Hov II	DCM	rt	66
5	Gr II	EtOAc	rt	52
5	Hov II	EtOAc	rt	62
5	Hov II	DCM	rt	52
1	Hov II	DCM	rt	32
2	Hov II	DCM	rt	65
10	Hov II	DCM	rt	70
5	Hov II	DCM	rt	87

127

TABLE 1. CROSS-METATHESIS OF UNSATURATED COMPOUNDS WITH FUNCTIONAL GROUPS IN HOMOALLYLIC AND MORE DISTANT POSITIONS (Continued)

A. ETHYLENE AND COMPOUNDS WITH A TERMINAL DOUBLE BOND (Continued)

Terminal Alkene	Cross Partner	Conditions	Product(s) and Yield(s) (%)	Refs.

Please refer to the charts preceding the tables for the catalyst structures.

C_5

Terminal Alkene	Cross Partner	Conditions	Product(s) and Yield(s) (%)	Refs.
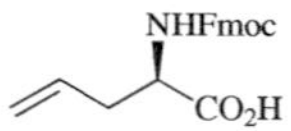 NHFmoc, CO_2H	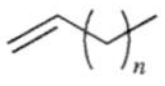 5 eq	1. Hov II (0.05 eq), DCM, rt 2. H_2 (60 psi), MeOH	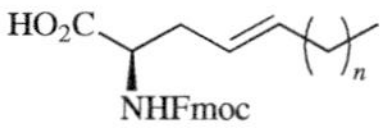 HO_2C, NHFmoc, n n = 2 (65); 3 (81); 4 (73); 5 (76); 6 (80); 7 (75)	127
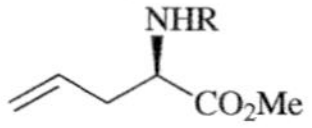 NHR, CO_2Me	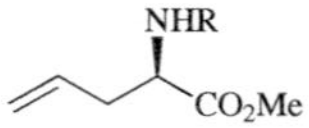 NHR, CO_2Me	Gr II (0.1 eq), DCM, rt, 2.5 h	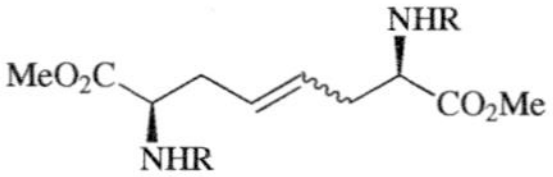 MeO_2C, NHR, NHR, CO_2Me R = Boc (76); Fmoc (86)	423
CO_2t-Bu, NHFmoc	CO_2Me, NHBoc 1.1 eq	Gr II (x eq), DCM	FmocHN, CO_2t-Bu, CO_2Me, NHBoc	331

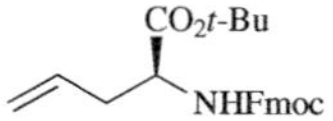

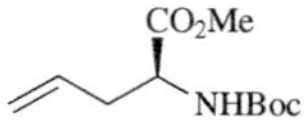

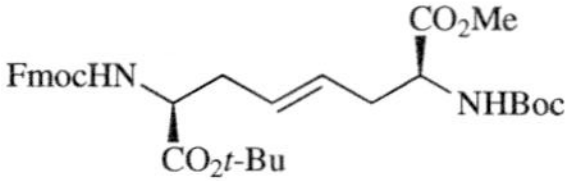

x	Temp	Time (h)	
0.1	reflux	6.5	(57)
0.1	MW	—	(36)
0.05	reflux	6.5	(46)

Boc–N(R) / OTHP	Boc–N(R) / OTHP	Gr I (0.05 eq) DCM, 40°, 48 h	THPO / R–N–Boc / Boc–N–R / OTHP	424
ONPhth / OTHP	ONPhth / OTHP	Gr I (0.05 eq) DCM, 40°, 48 h	THPO / ONPhth / ONPhth / OTHP (80) $(E)/(Z) = 60:40$	424
NHBoc / CO_2Bn 3 eq	OTBDPS / O	Gr II (0.05 eq), DCM, rt, 7 d	OTBDPS / O / CO_2Bn / NHBoc (55)	425
NHBoc / CO_2Me x eq	OBn / OH / BnO / BnO / R	Gr II (y eq), DCM, reflux	OBn / OH / BnO / BnO / R / BocHN / CO_2Me	426

R		$(E)/(Z)$
BocHN	(80)	55:45
(o-NB)O	(94)	55:45

x	R	y	Time (h)	
0.5	HO	0.15	15	(48)
1	HO	0.15	15	(45)
2	HO	0.15	15	(54)
4	HO	0.15	15	(70)
2	AcHN	0.2	16	(16)

TABLE 1. CROSS-METATHESIS OF UNSATURATED COMPOUNDS WITH FUNCTIONAL GROUPS IN HOMOALLYLIC AND MORE DISTANT POSITIONS *(Continued)*

A. ETHYLENE AND COMPOUNDS WITH A TERMINAL DOUBLE BOND *(Continued)*

Terminal Alkene	Cross Partner	Conditions	Product(s) and Yield(s) (%)	Refs.

Please refer to the charts preceding the tables for the catalyst structures.

C_5

Terminal Alkene	Cross Partner	Conditions	Product(s) and Yield(s) (%)	Refs.
NHR^1, CO_2Me; *x* eq	OR^2, R^2O, R^2O, O, AcHN	Catalyst (*y* eq), DCM, reflux	OR^2, R^2O, R^2O, O, AcHN, NHR^1, CO_2Me	427

R^1	*x*	R^2	Catalyst	*y*	Time (h)	
Fmoc	2	Ac	Gr I	0.2	24	(0)
Fmoc	2	Ac	Gr II	0.2	48	(65)
Fmoc	3	Ac	Gr II	0.2	48	(70)
Boc	2	Ac	Gr II	0.2	48	(60)
Fmoc	2	Bn	Gr I	0.2	22	(41)
Boc	2	Bn	Gr I	0.2	12	(50)
Fmoc	2	Bn	Gr II	0.1	16	(74)
Boc	2	Bn	Gr II	0.1	16	(78)
Fmoc	1	Bn	Gr II	0.1	16	(49)

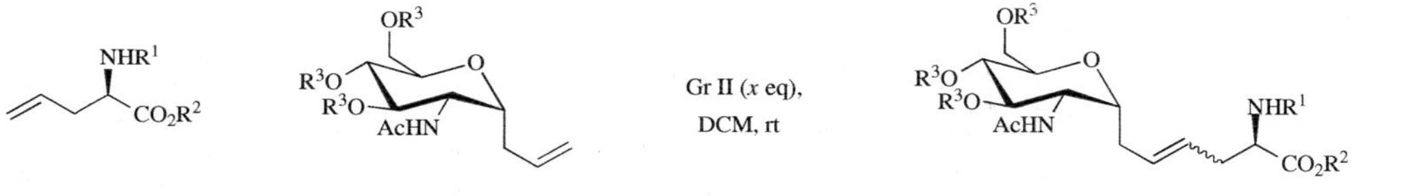

Gr II (x eq), DCM, rt

423

R^1	R^2	R^3	x	Time (h)	
Fmoc	Me	Bn	0.1	24	(74)
Boc	Me	Bn	0.1	24	(78)
Fmoc	Me	Ac	0.2	48	(70)
Boc	Me	Ac	0.2	48	(60)
Cbz	Bn	Ac	0.2	48	(70)

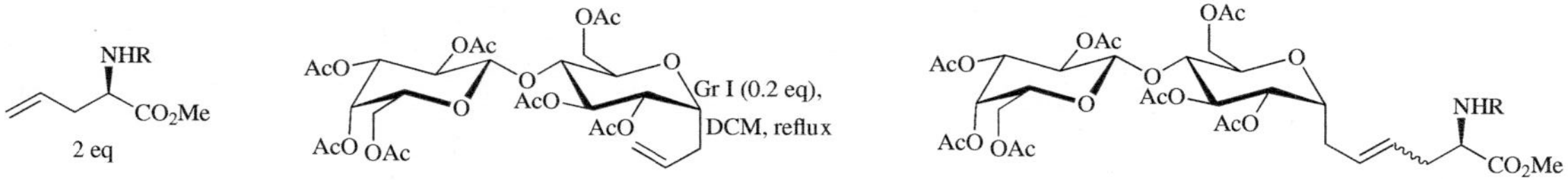

2 eq

Gr I (0.2 eq), DCM, reflux

427

R	Time (h)	
Fmoc	48	(57)
Boc	24	(73)

TABLE 1. CROSS-METATHESIS OF UNSATURATED COMPOUNDS WITH FUNCTIONAL GROUPS IN HOMOALLYLIC AND MORE DISTANT POSITIONS (*Continued*)

A. ETHYLENE AND COMPOUNDS WITH A TERMINAL DOUBLE BOND (*Continued*)

Terminal Alkene	Cross Partner	Conditions	Product(s) and Yield(s) (%)	Refs.
Please refer to the charts preceding the tables for the catalyst structures.				
C_5				
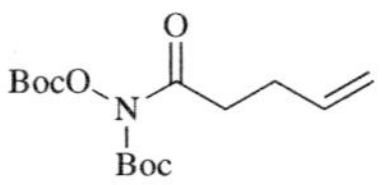 BocO, N, Boc, O; 2 eq	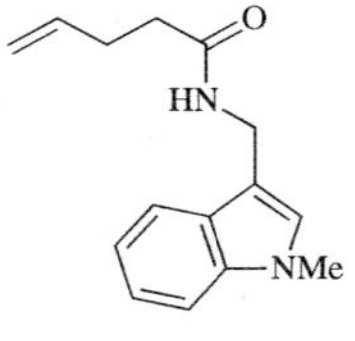 HN, O, NMe	Gr I (0.15 eq), DCM, reflux, 7 h	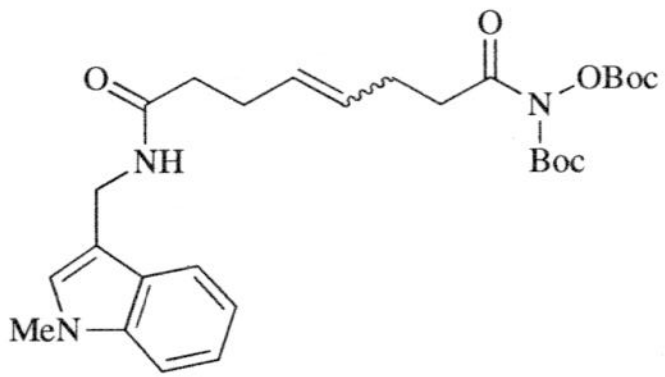 O, NH, MeN, O, N, OBoc, Boc (43)	428
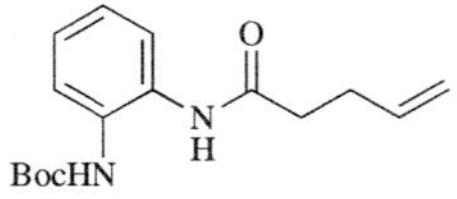 BocHN, N, H, O; 1 eq	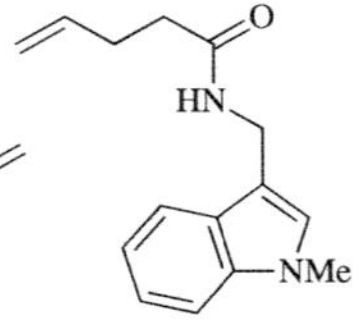 HN, O, NMe	Gr II (0.15 eq), DCM, reflux, 7 h	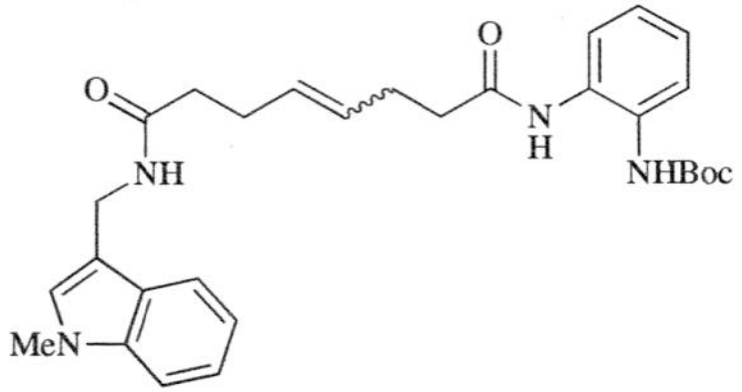 O, NH, MeN, O, N, H, NHBoc (27)	428
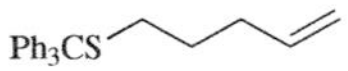 Ph_3CS; 2 eq	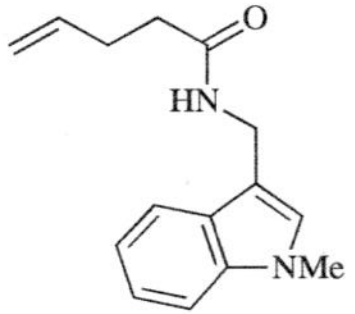 HN, O, NMe	Gr I (0.36 eq), DCM, reflux, 7 h	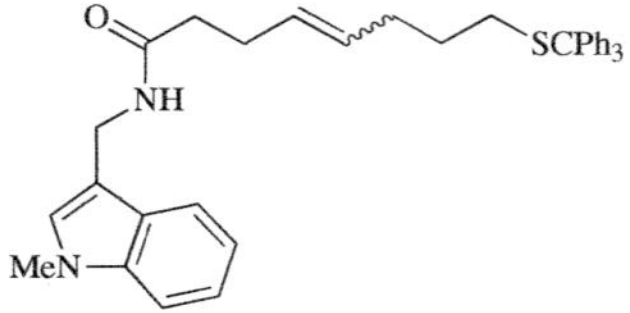 O, NH, MeN, $SCPh_3$ (32)	428
2 eq	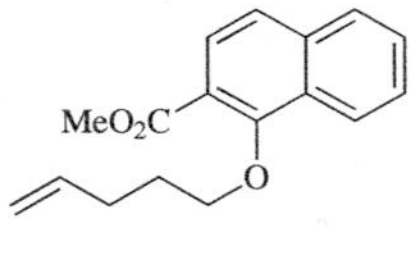 MeO_2C, O	Gr I (0.36 eq), DCM, reflux, 7 h	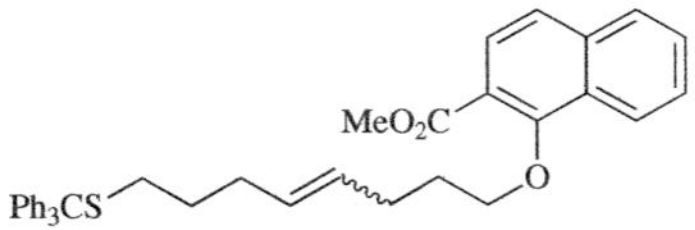 Ph_3CS, MeO_2C, O (48)	428

Olefin 1	Olefin 2	Conditions	Product (%)	Ref.
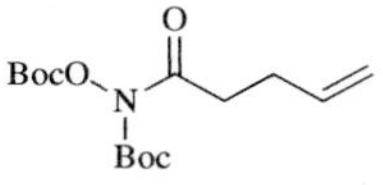 BocO–N(Boc)–C(O)–; 2 eq	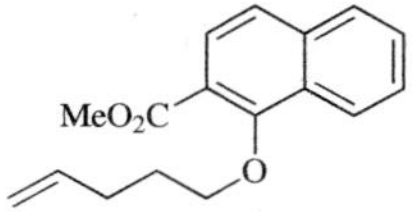 MeO_2C, O	Gr I (0.15 eq), DCM, reflux, 7 h 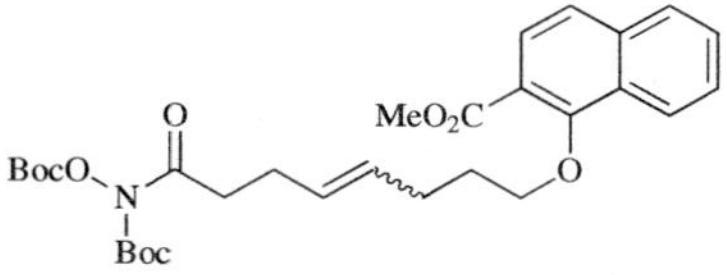	BocO–N(Boc)–C(O)–, MeO_2C, O (50)	428
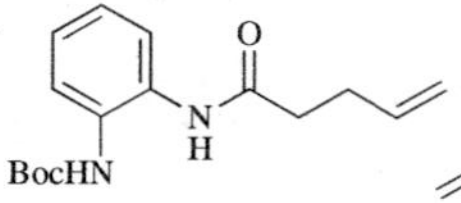 BocHN, N–H, O; 1 eq	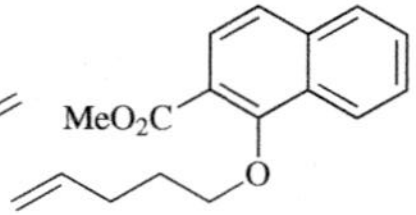 MeO_2C, O	Gr II (0.17 eq), DCM, reflux, 7 h 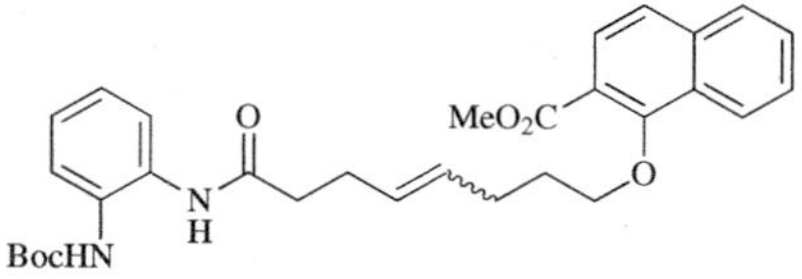	BocHN, N–H, O, MeO_2C, O (36)	428
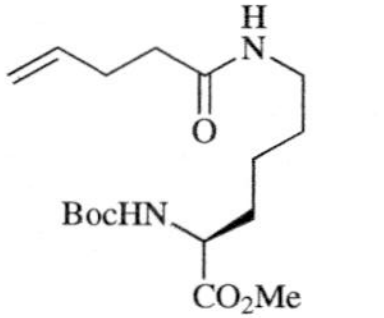 H–N, O, BocHN, CO_2Me	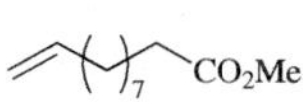 $(\ )_7$ CO_2Me; 3 eq	Gr II (0.2 eq), DCM, 40°, 6 h 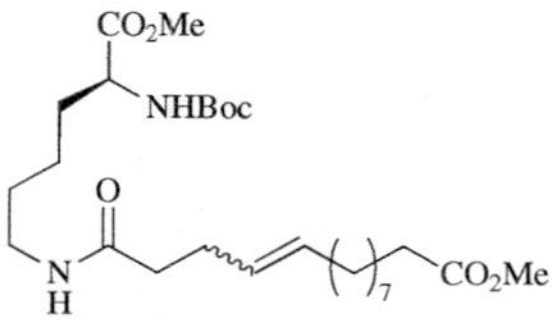	CO_2Me, NHBoc, O, N–H, $(\ )_7$ CO_2Me (45)	429

TABLE 1. CROSS-METATHESIS OF UNSATURATED COMPOUNDS WITH FUNCTIONAL GROUPS IN HOMOALLYLIC AND MORE DISTANT POSITIONS *(Continued)*

A. ETHYLENE AND COMPOUNDS WITH A TERMINAL DOUBLE BOND *(Continued)*

Terminal Alkene	Cross Partner	Conditions	Product(s) and Yield(s) (%)	Refs.
Please refer to the charts preceding the tables for the catalyst structures.				
C_5				
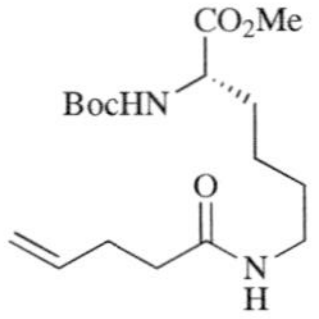 BocHN, CO_2Me, O, N H	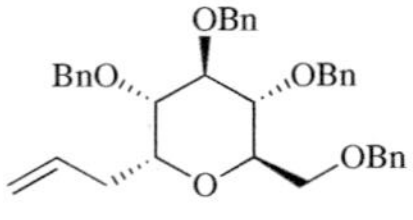 BnO, OBn, OBn, OBn, O; 3 eq	Gr II (0.2 eq), DCM, 40°, 6 h	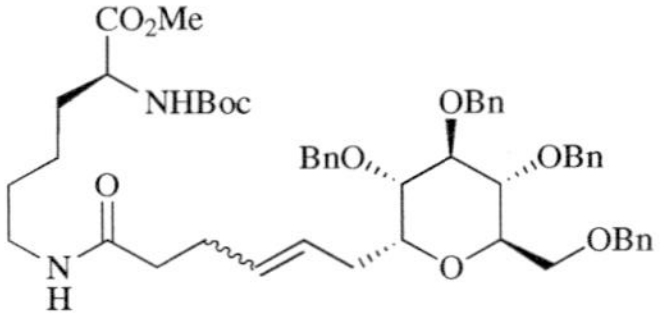 CO_2Me, NHBoc, O, N H, BnO, OBn, OBn, OBn, O (33)	429
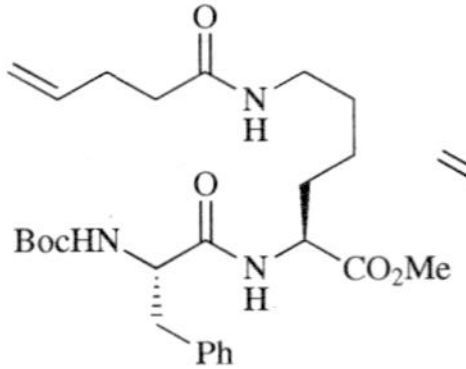 O, N H, O, BocHN, N H, CO_2Me, Ph	 CO_2Me, 7; 3 eq	Gr II (0.2 eq), DCM, 40°, 6 h	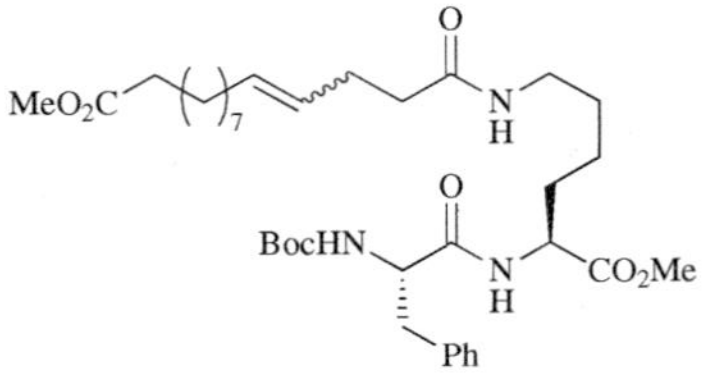 MeO_2C, 7, O, N H, O, BocHN, N H, CO_2Me, Ph (39)	429
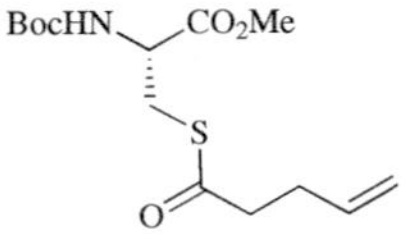 BocHN, CO_2Me, S, O	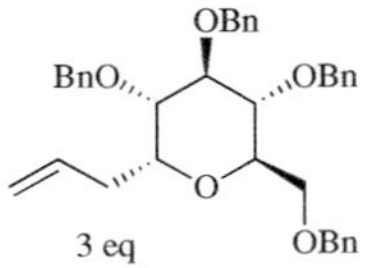 BnO, OBn, OBn, O, OBn; 3 eq	Gr II (0.2 eq), DCM, 40°, 6 h	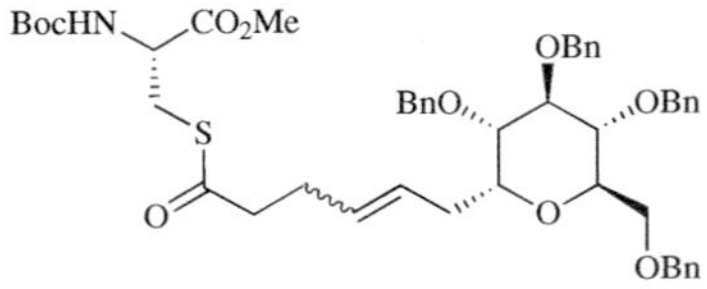 BocHN, CO_2Me, S, O, BnO, OBn, OBn, O, OBn (43)	429

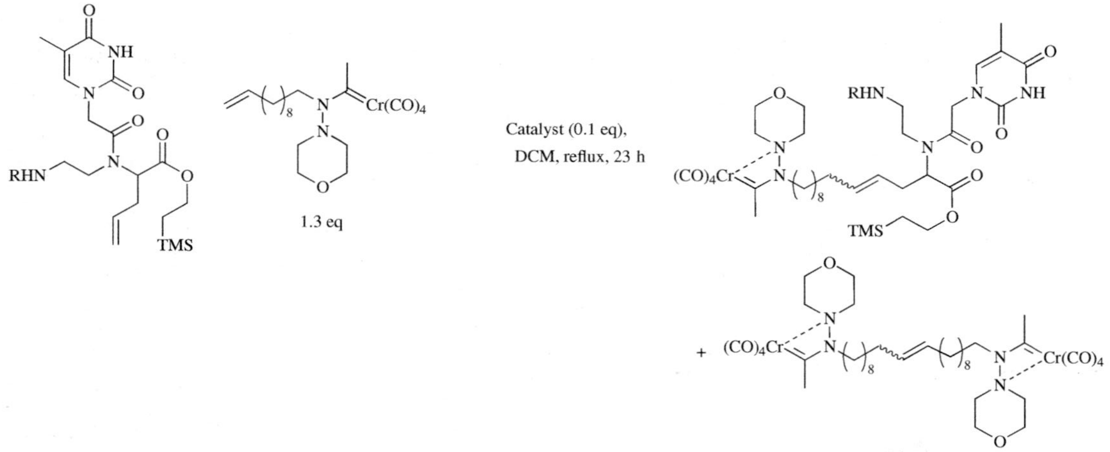

1.3 eq

Catalyst (0.1 eq), DCM, reflux, 23 h

(26–77)

R	Catalyst	
Fmoc	Gr I	(34)
Fmoc	Gr II	(40)
Phth	Gr I	(25)
Phth	Gr II	(35)

430

TABLE 1. CROSS-METATHESIS OF UNSATURATED COMPOUNDS WITH FUNCTIONAL GROUPS IN HOMOALLYLIC AND MORE DISTANT POSITIONS (*Continued*)

A. ETHYLENE AND COMPOUNDS WITH A TERMINAL DOUBLE BOND (*Continued*)

Terminal Alkene	Cross Partner	Conditions	Product(s) and Yield(s) (%)	Refs.
Please refer to the charts preceding the tables for the catalyst structures.				
C_5				
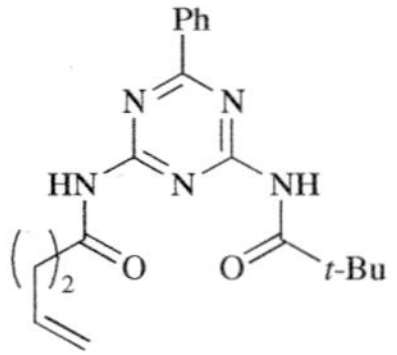	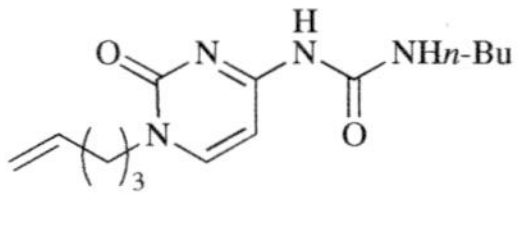	Gr II, DCM, reflux	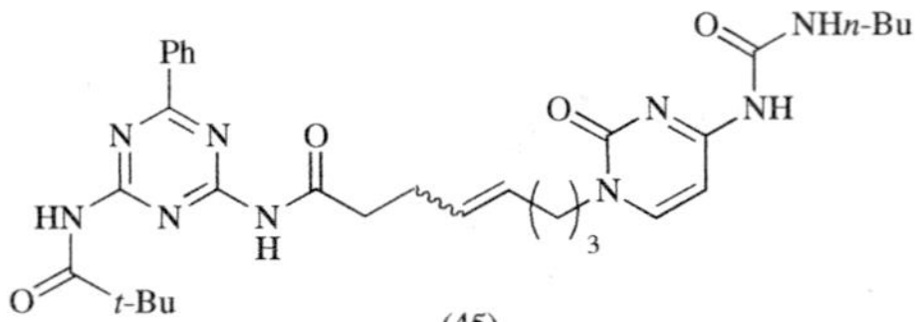 (45)	431
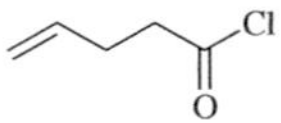 20 eq	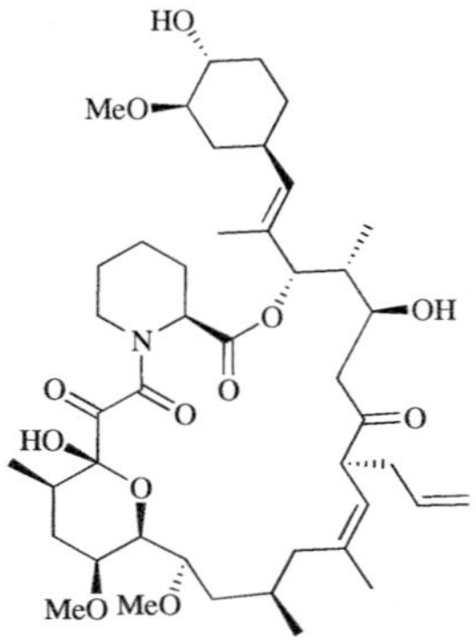	Gr II (0.1 eq), DCB, MW	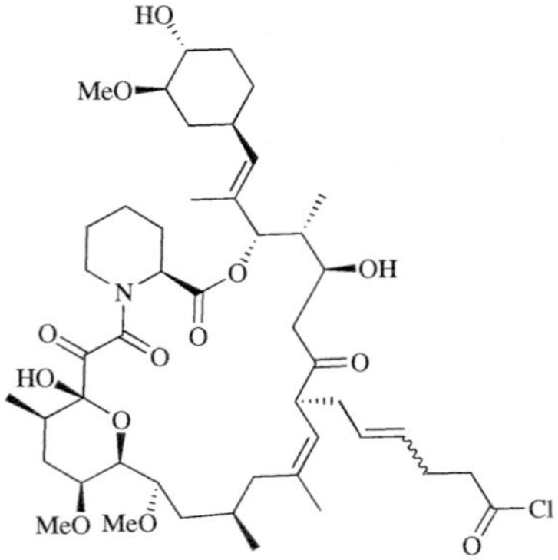 (25)	432

C_6

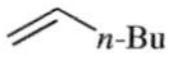

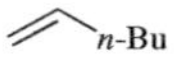

n-Bu | n-Bu | Catalyst (0.02 eq), benzene, 24 h | n-Bu | 433

Catalyst	Temp (°)	Conv. (%)[a]
Ru cat **230**	25	<5
Ru cat **230**	40	5
Ru cat **230**	50	11
Ru cat **230**	60	35
Ru cat **230**	85	90
Ru cat **226**	85	41

$C_{6–8}$

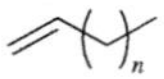

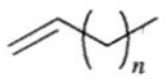

Ru cat **217** (0.02 eq), THF, 35°, sealed tube — 84

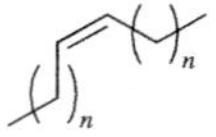

n	Time (h)		(*Z*)/(*E*)
3	3	(21)	69:31
5	4	(79)	83:17

C_6

n-Bu (2 eq) | CO_2Me, NHBoc | Gr I (0.05 eq), DCE, rt, 30 h | BocHN, CO_2Me, n-Bu (55) (*E*)/(*Z*) = 80:20 | 434

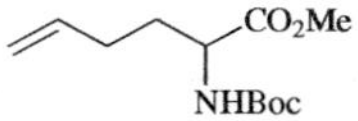

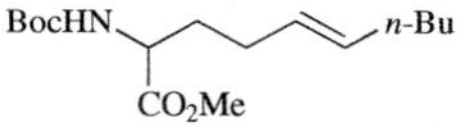

+ BocHN, CO_2Me, NHBoc, CO_2Me (43)

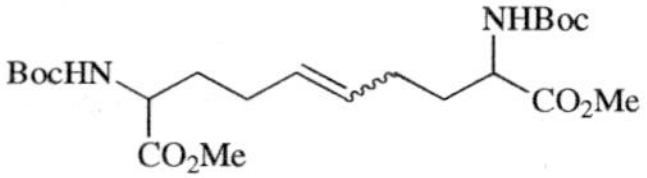

TABLE 1. CROSS-METATHESIS OF UNSATURATED COMPOUNDS WITH FUNCTIONAL GROUPS IN HOMOALLYLIC AND MORE DISTANT POSITIONS (*Continued*)

A. ETHYLENE AND COMPOUNDS WITH A TERMINAL DOUBLE BOND (*Continued*)

Terminal Alkene	Cross Partner	Conditions	Product(s) and Yield(s) (%)	Refs.
Please refer to the charts preceding the tables for the catalyst structures.				
C_6				
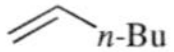 n-Bu 5 eq	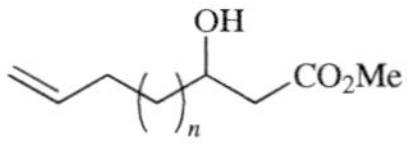 OH, CO_2Me, n	Ru cat **35** (0.02 eq), DCM, reflux, 4–12 h	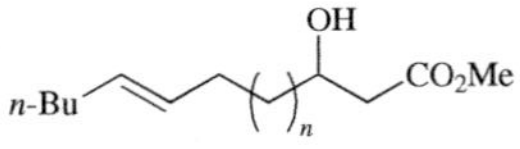 n-Bu, OH, CO_2Me, n *n* / (*E*)/(*Z*) 1 (70) 90:10 2 (82) 90:10	393
5 eq	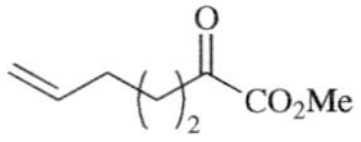 O, CO_2Me, 2	Ru cat **35** (0.02 eq), DCM, reflux, 4–12 h	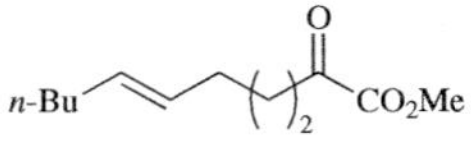 n-Bu, O, CO_2Me, 2 (78) (*E*)/(*Z*) = 95:5	393
5 eq	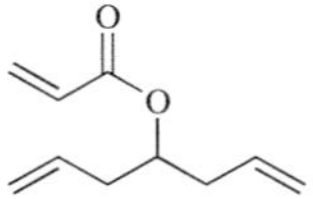 O, O	Gr II (0.025 eq), DCM, 40°, 16 h	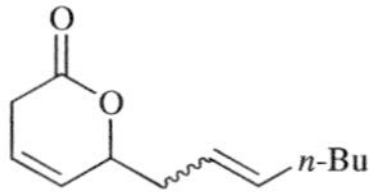 O, O, n-Bu (21)	394
5 eq	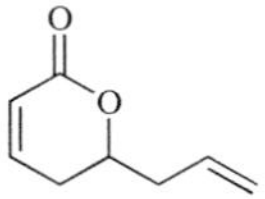 O, O	Gr II (0.025 eq), DCM, 40°, 16 h	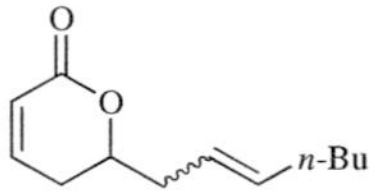 O, O, n-Bu (61)	408

		Conditions	Product(s) and Yield(s) (%)	Refs.
2 eq	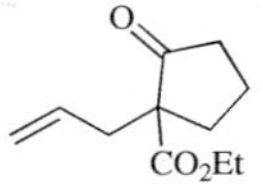 CO_2Et	Gr II (0.02 eq), neat	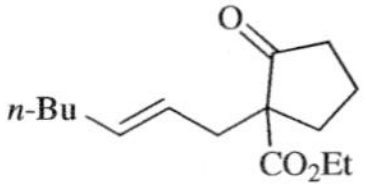 *n*-Bu, CO_2Et	435
7 eq	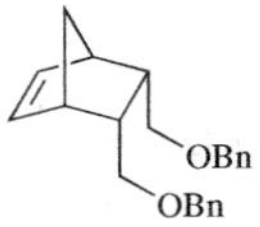 OBn, OBn	Ru cat **215** (0.01 eq), THF, 23°, 1 h	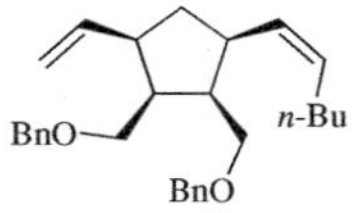 BnO, BnO, *n*-Bu (62) er 94.5:5.5, (*Z*)/(*E*) = 95:5	417
13 eq	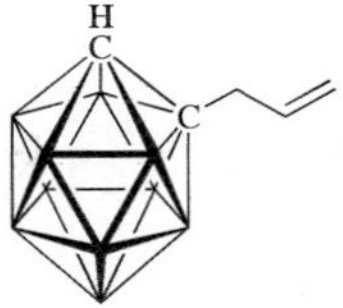 H C, C	Hov II (0.005 eq), DCM, reflux, 14 h	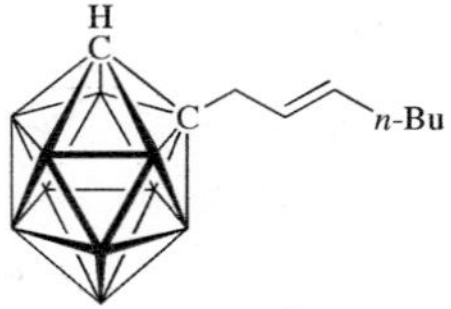 H C, C, *n*-Bu (88) (*E*)/(*Z*) = 88:12	436

Temp (°)	Time (min)		(*E*)/(*Z*)
90	960	(50)	80:20
60 (MW)	1.25	(68)	80:20

TABLE 1. CROSS-METATHESIS OF UNSATURATED COMPOUNDS WITH FUNCTIONAL GROUPS IN HOMOALLYLIC AND MORE DISTANT POSITIONS (*Continued*)

A. ETHYLENE AND COMPOUNDS WITH A TERMINAL DOUBLE BOND (*Continued*)

Terminal Alkene	Cross Partner	Conditions	Product(s) and Yield(s) (%)	Refs.

Please refer to the charts preceding the tables for the catalyst structures.

C_6

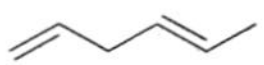

Terminal Alkene	Cross Partner	Conditions	Product(s) and Yield(s) (%)	Refs.
9 eq	R, n	Catalyst (x eq), THF, rt, 5 h	R, n	437

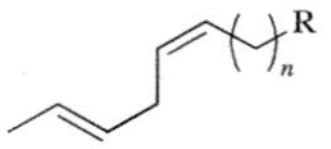

R	n	Catalyst	x		$(Z)/(E)$
Ph	1	Ru cat **216**	0.01	(63)	>95:5
Ac	2	Ru cat **216**	0.01	(49)	>95:5
HO	7	Ru cat **215**	0.01	(87)	89:11
HO	7	Ru cat **216**	0.01	(80)	>95:5
HO	7	Ru cat **216**	0.005	(77)	>95:5
HO	7	Ru cat **216**	0.001	(50)	>95:5
HO	7	Hov II	0.01	(<5)	—
HO	7	Hov I	0.01	(28)	29:71
MeO_2C	7	Ru cat **216**	0.01	(82)	>95:5
OHC	8	Ru cat **216**	0.01	(70)	>95:5

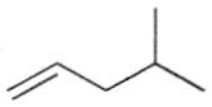

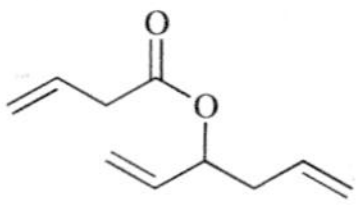

Terminal Alkene	Cross Partner	Conditions	Product(s) and Yield(s) (%)	Refs.
	O, O	Gr II (0.025 eq), DCM, 40°, 16 h	O, O (61)	394

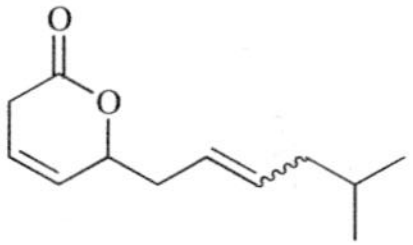

5 eq	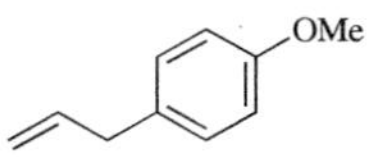 OMe	1. Gr II (0.025 eq), 2,6-Cl_2BQ (0.1 eq), DCM, 40°, 3 h 2. BBr_3, DCM, 0°, 0.5 h 3. BnBr (1.1 eq), K_2CO_3 (4 eq), acetone, reflux, 6 h	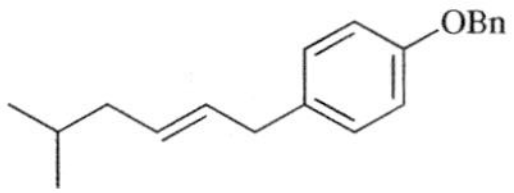 OBn	(64) (*E*)/(Z) = 88:12	438
3 eq	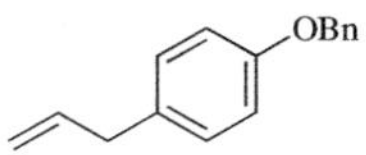 OBn	1. Ru cat **215** (0.025 eq), THF, reflux, 6 h 2. OsO_4, NMO (2 eq), THF/H_2O (99:1), 5 h	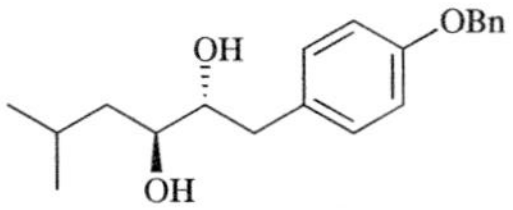 OH, OH, OBn	(6) dr > 98:2	438
5 eq	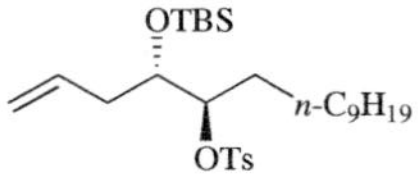 OTBS, OTs, *n*-C_9H_{19}	Gr II (0.05 eq), DCM, reflux, 5 h	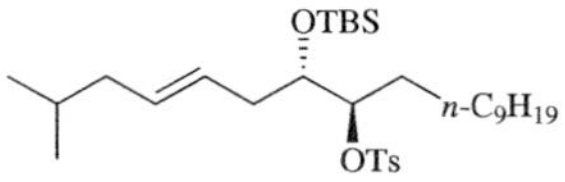 OTBS, OTs, *n*-C_9H_{19}	(84) (*E*)/(Z) = 83:17	439
5 eq	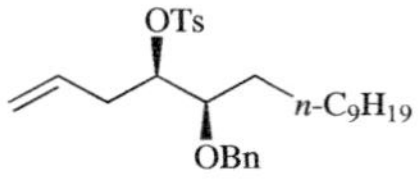 OTs, OBn, *n*-C_9H_{19}	Gr II (0.05 eq), DCM, reflux, 8 h	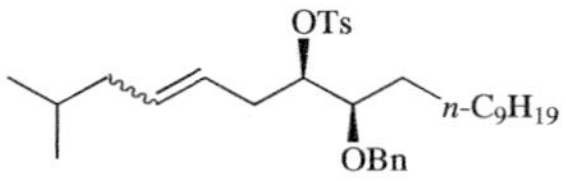 OTs, OBn, *n*-C_9H_{19}	(92)	440

TABLE 1. CROSS-METATHESIS OF UNSATURATED COMPOUNDS WITH FUNCTIONAL GROUPS IN HOMOALLYLIC AND MORE DISTANT POSITIONS (*Continued*)

A. ETHYLENE AND COMPOUNDS WITH A TERMINAL DOUBLE BOND (*Continued*)

Terminal Alkene	Cross Partner	Conditions	Product(s) and Yield(s) (%)	Refs.
Please refer to the charts preceding the tables for the catalyst structures.				
C_6				
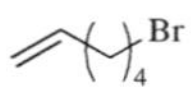 Br, 4; 2 eq	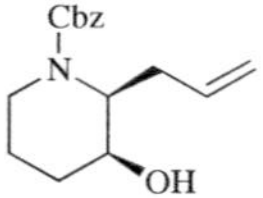 Cbz, N, OH	Gr II (0.1 eq) DCM, reflux, overnight	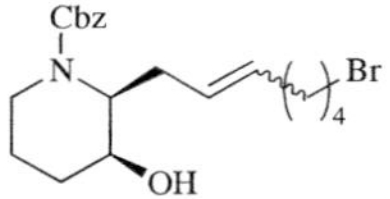 Cbz, N, Br, 4, OH (53)	441
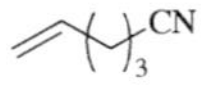 CN, 3; 17.5 eq	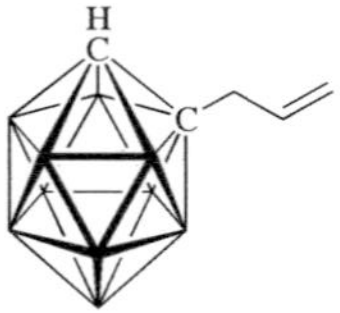 H, C, C	Hov II (0.025 eq), DCM, reflux, 14 h	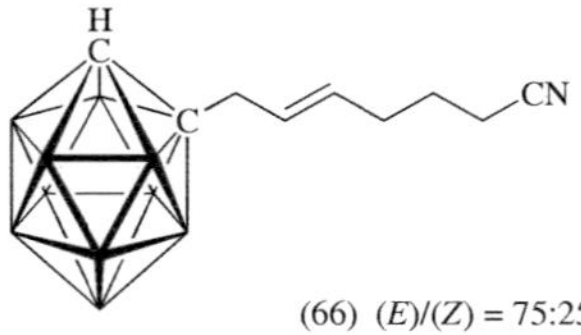 H, C, C, CN (66) (*E*)/(*Z*) = 75:25	436
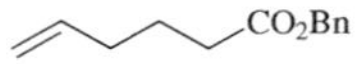 CO_2Bn; 3 eq	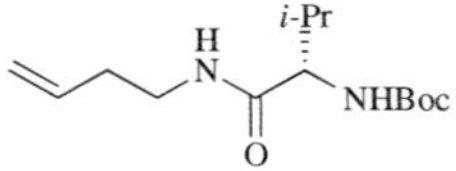 H, N, *i*-Pr, NHBoc, O	1. Ru cat **152** (0.01 eq), (*Z*)-butene (20 eq), THF, 22°, 16 h 2. Ru cat **152** (0.04 eq), THF, 100 torr, 22°, 8 h	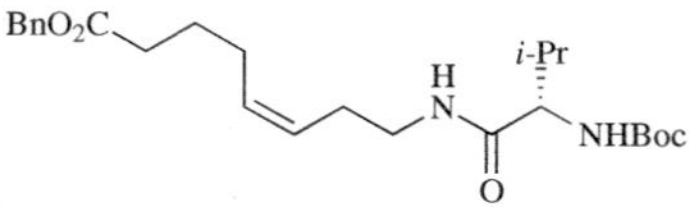 BnO_2C, H, N, *i*-Pr, NHBoc, O (47) (*Z*)/(*E*) = 91:9	395

		Conditions	Product(s) and Yield(s) (%)		Refs.
3 eq	NHBoc, H N, O, SMe	1. Ru cat **152** (0.01 eq), (*Z*)-butene (20 eq), THF, 22°, 16 h 2. Ru cat **152** (0.04 eq), THF, 100 torr, 22°, 8 h	BnO_2C, H N, NHBoc, O, SMe (51) (*Z*)/(*E*) > 98:2		395
CO_2R^1 3 eq	R^2, R^3, R^4	1. Ru cat **152** (0.01 eq), (*Z*)-butene (20 eq), THF, 22°, 1 h 2. Ru cat **152** (0.04 eq), THF, 100 torr, 22°, 8 h	R^1O_2C, R^2, R^3, R^4		395
CO_2Bn	OH, OBn, *n*-Bu 5 eq	Gr II (0.15 eq), DCM, rt, 16 h	BnO_2C, OH, OBn, *n*-Bu	(70)	442

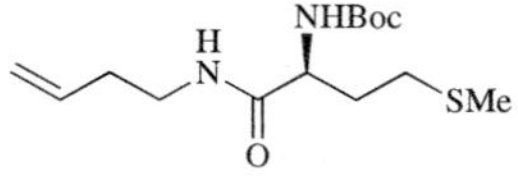

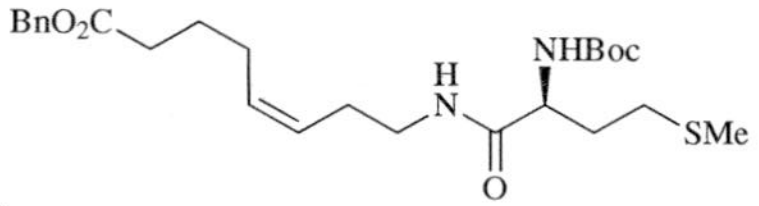

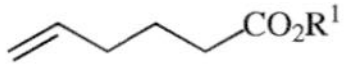

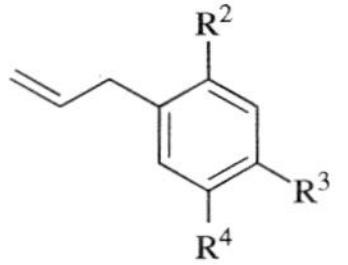

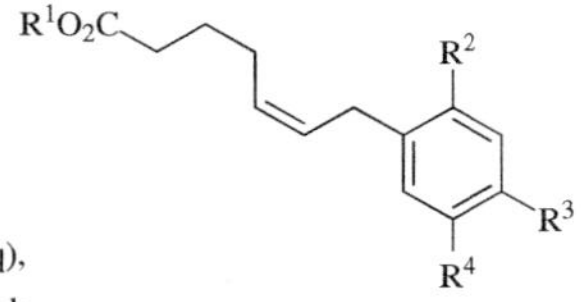

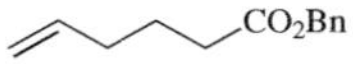

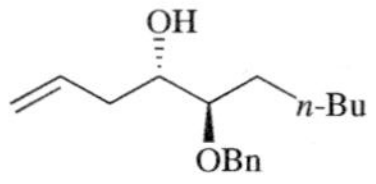

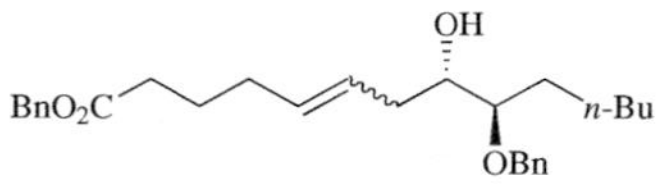

R^1	R^2	R^3	R^4		(*Z*)/(*E*)
H	H	MeO	MeO	(64)	98:2
Bn	HO	H	Ac	(63)	97:3

TABLE 1. CROSS-METATHESIS OF UNSATURATED COMPOUNDS WITH FUNCTIONAL GROUPS IN HOMOALLYLIC AND MORE DISTANT POSITIONS (*Continued*)

A. ETHYLENE AND COMPOUNDS WITH A TERMINAL DOUBLE BOND (*Continued*)

Terminal Alkene	Cross Partner	Conditions	Product(s) and Yield(s) (%)	Refs.
Please refer to the charts preceding the tables for the catalyst structures.				
C_6				
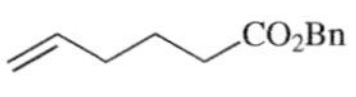 CO_2Bn **I**	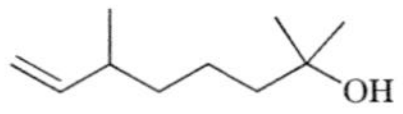 OH	1. Ru cat **152** (0.01 eq), (*Z*)-butene (20 eq), THF, 22°, 1 h 2. Ru cat **152** (0.04 eq), **I** (3 eq), THF, 100 torr, 22°, 8 h	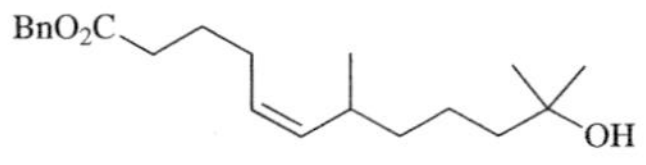 BnO_2C OH (58) (*Z*)/(*E*) = 98:2	395
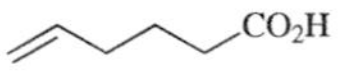 CO_2H 3 eq	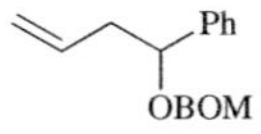 Ph OBOM	1. Ru cat **152** (0.01 eq), (*Z*)-butene (20 eq), THF, 22°, 1 h 2. Ru cat **152** (0.04 eq), THF, 100 torr, 22°, 8 h	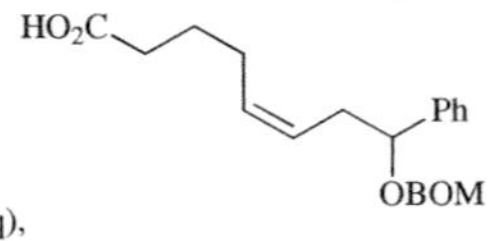 HO_2C Ph OBOM (64) (*Z*)/(*E*) > 98:2	395
3 eq	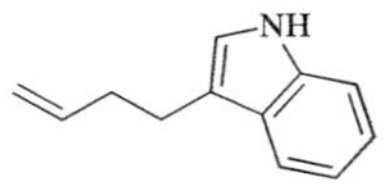 NH	1. Ru cat **152** (0.01 eq), (*Z*)-butene (20 eq), THF, 22°, 1 h 2. Ru cat **152** (0.04 eq), THF, 100 torr, 22°, 8 h	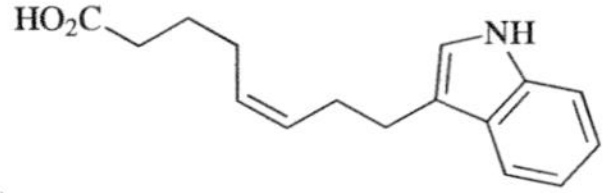 HO_2C NH (58) (*Z*)/(*E*) > 98:2	395

I	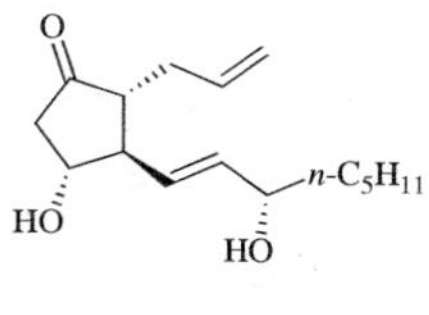 1'. Ru cat **152** (0.02 eq), (*Z*)-butene (20 eq), THF, 22°, 16 h 1''. Ru cat **152** (0.01 eq), **I** (3 eq), (*Z*)-butene (20 eq), THF, 22°, 1 h 2. Ru cat **152** (0.15 eq), THF, 100 torr, 22°, 1 h 3. 400 torr, 7 h	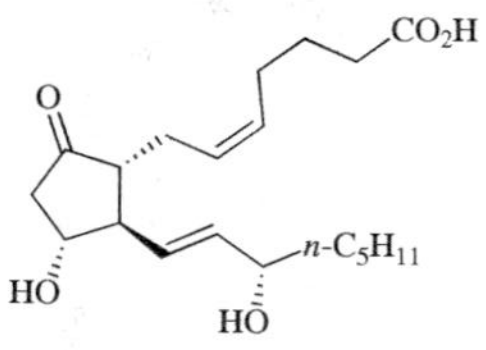	(51) (*Z*)/(*E*) > 98:2	395
I	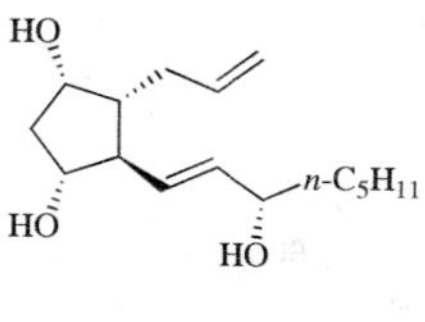 1'. Ru cat **152** (0.02 eq), (*Z*)-butene (20 eq), THF, 22°, 16 h 1''. Ru cat **152** (0.01 eq), **I** (3 eq), (*Z*)-butene (20 eq), THF, 22°, 1 h 2. Ru cat **152** (0.15 eq), THF, 100 torr, 22°,1 h 3. 400 torr, 7 h	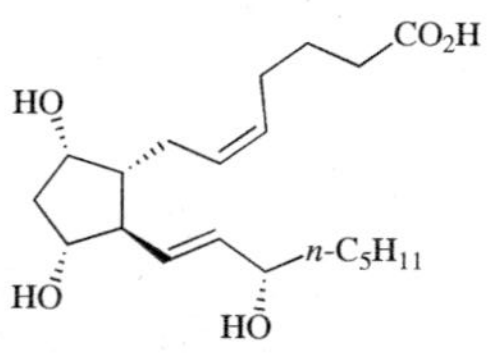	(59) (*Z*)/(*E*) > 98:2	395

TABLE 1. CROSS-METATHESIS OF UNSATURATED COMPOUNDS WITH FUNCTIONAL GROUPS IN HOMOALLYLIC AND MORE DISTANT POSITIONS (*Continued*)

A. ETHYLENE AND COMPOUNDS WITH A TERMINAL DOUBLE BOND (*Continued*)

Terminal Alkene	Cross Partner	Conditions	Product(s) and Yield(s) (%)	Refs.
Please refer to the charts preceding the tables for the catalyst structures.				
C_6				
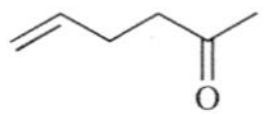 3 eq	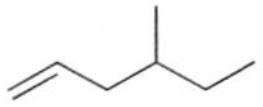	Gr I (0.01 eq), DCM, reflux, 5 h	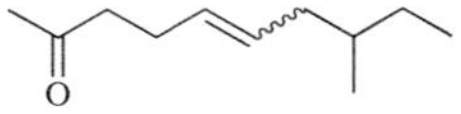 O (24)	443
6 eq	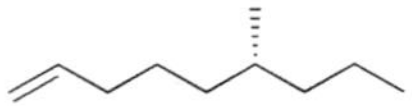	Gr I (0.1 eq), DCM, reflux, 9.5 h	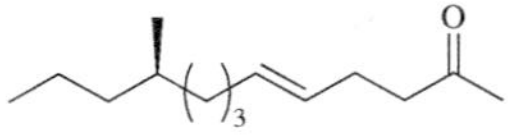 O 3 (73) $(E)/(Z) = 66:34$	444
8.3 eq	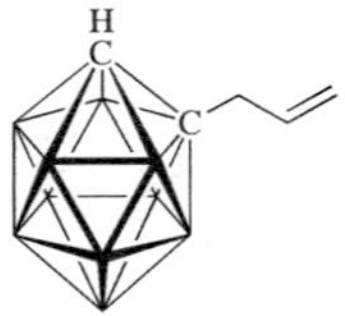 H C C	Hov II (0.025 eq), DCM, reflux, 18 h	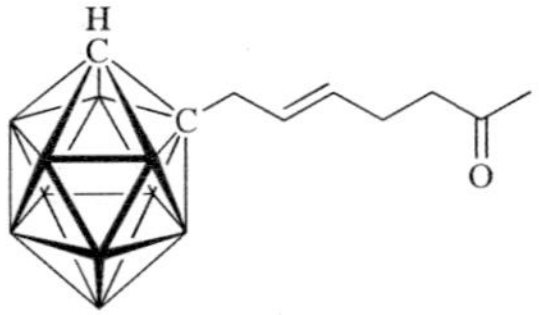 H C C O (69) $(E)/(Z) = 95:5$	436
2 eq	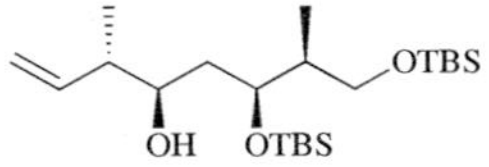 OTBS OH OTBS	1. Hov II (0.2 eq), DCM, reflux, 20 h 2. H_2, Pd/C, AcOEt, rt	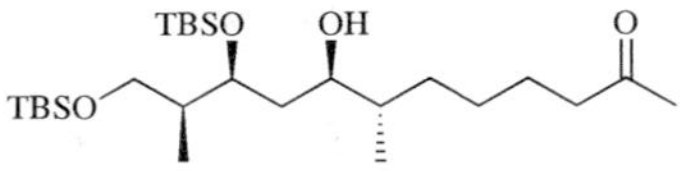 TBSO OH TBSO O (64)	162

10 eq

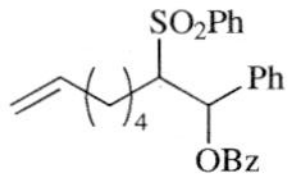

SO_2Ph, Ph, OBz, $(\ )_4$

1. Gr II (0.1 eq), toluene, 60°, 12 h
2. SmI_2 (3 eq), THF/DMPU (4:1), –78°, 1 h

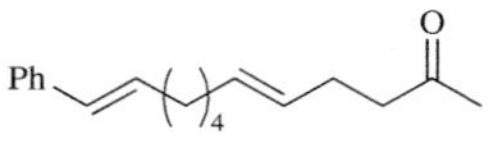

Ph, $(\ )_4$, O

(70) $(E,E)/(Z,E)$ = 88:12 445

OH

O, NMe, OH, O, $(\ )_6$

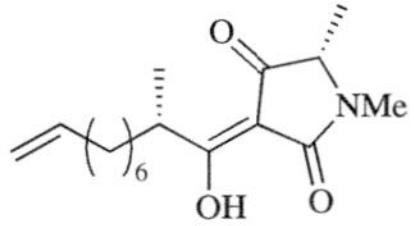

1. Gr II (0.015 eq), DCM, reflux, 6 h
2. H_2, Pd/C, MeOH

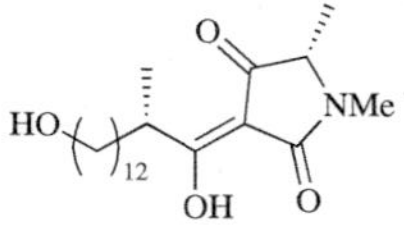

HO, $(\ )_{12}$, O, NMe, OH, O

(28) 248

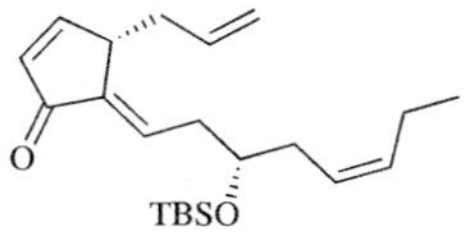

8 eq

O, TBSO

Ru cat **216** (0.2 eq), THF, 40°, 4 h

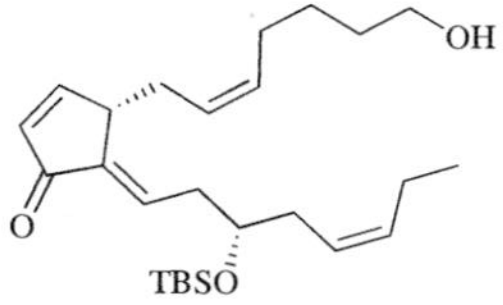

OH, O, TBSO

(52) $(Z)/(E)$ > 99:1 446

C_{6-13}

R, $(\ )_n$

n-C_6H_{17}

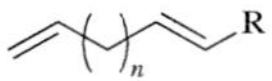

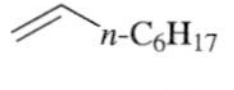

9 eq

Ru cat **216** (0.01 eq), THF, rt, 5 h

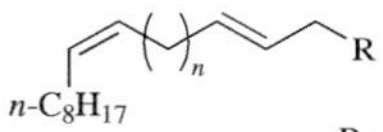

n-C_8H_{17}, $(\ )_n$, R

437

R	n		$(Z)/(E)$
$HOCH_2$	1	(<5)	—
$BzOCH_2$	1	(<5)	—
$HOCH_2$	2	(51)	>95:5
MeO_2C	8	(80)	>95:5
$HOCH_2$	8	(60)	>95:5

TABLE 1. CROSS-METATHESIS OF UNSATURATED COMPOUNDS WITH FUNCTIONAL GROUPS IN HOMOALLYLIC AND MORE DISTANT POSITIONS (*Continued*)

A. ETHYLENE AND COMPOUNDS WITH A TERMINAL DOUBLE BOND (*Continued*)

Terminal Alkene	Cross Partner	Conditions	Product(s) and Yield(s) (%)	Refs.
Please refer to the charts preceding the tables for the catalyst structures.				
C_6				
OBz	OBz	Gr I (0.2 eq), DCM, reflux, 12 h	BzO … OBz (64) (*E*)/(*Z*) = 92:8	447
OAc (4)	OAc (4)	Catalyst (*x* eq), solvent	AcO … OAc	
OTMS (4)	Ph	Ru cat **162** (0.02 eq), toluene, 80°, 1 h	TMSO … Ph (79)	449
O–n-$C_{12}H_{25}$ (4), 1.21 eq	EtO_2C (tetrahydrofuranyl)	Gr I (0.24 eq), DCE, reflux, 62 h	n-$C_{12}H_{25}$–O (3) … EtO_2C (46) (*E*)/(*Z*) = 75:25 + n-$C_{12}H_{25}$–O (3) … (4) O–n-$C_{12}H_{25}$ (42) (*E*)/(*Z*) = 75:25	450

Catalyst	*x*	Solvent	Temp (°)	Time (h)	Conv. (%)[a]	Refs.
Ru cat **230**	0.02	benzene	85	22	62	433
Ru cat **150**	0.0007	toluene	30	0.5	28	448
Ru cat **150**	0.0007	toluene	30	1	57	448
Ru cat **150**	0.0007	toluene	30	5	69	448

C_{6-18}

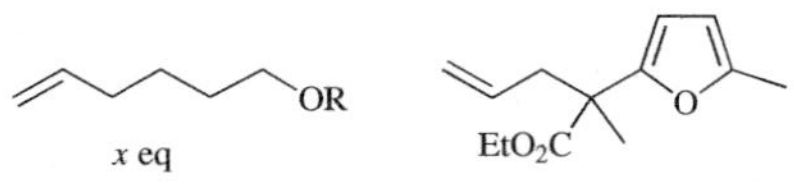

x eq

Gr I (0.1 eq),
DCM, reflux, 20 h

EtO_2C

I

+

II

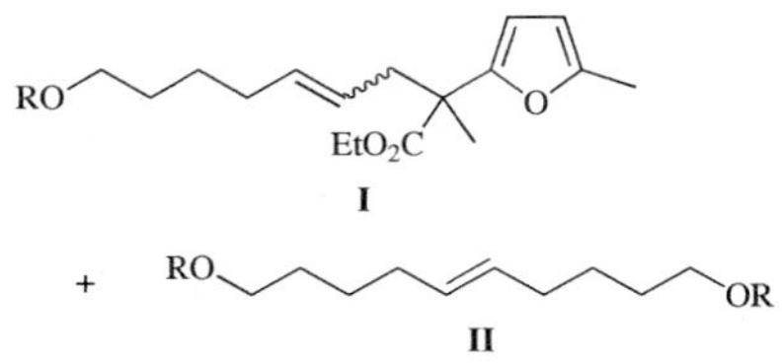

450

R	x	**I**	**II**	$(E)/(Z)$ **I**
H	1.6	(81)	(—)	65:35
n-$C_{12}H_{25}$	1.2	(54)	(21)	65:35
n-$C_{12}H_{25}$	1	(30)	(3)	65:35

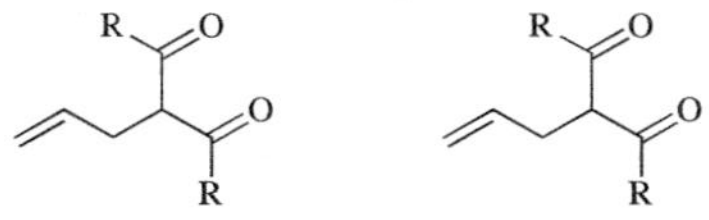

Catalyst (0.1 eq),
solvent, 24 h

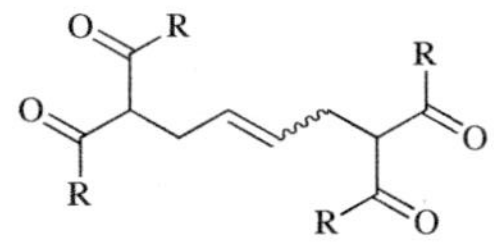

492

R	Catalyst	Solvent	Temp (°)	Conv. (%)	
EtO	Gr I	DCM	40	69	(—)
EtO	Gr II	DCM	40	95	(—)
EtO	Ru cat **67**	DCM	40	99	(—)
EtO	Hov II	toluene	110	99	(—)
Me	Gr I	DCM	40	37	(—)
Me	Gr II	DCM	40	99	(83)
Me	Gr III	DCM	40	99	(—)
Me	Hov II	benzene	80	85	(—)
Et	Gr II	DCM	40	99	(71)
Ph	Gr II	DCM	40	74	(61)

TABLE 1. CROSS-METATHESIS OF UNSATURATED COMPOUNDS WITH FUNCTIONAL GROUPS IN HOMOALLYLIC AND MORE DISTANT POSITIONS (*Continued*)

A. ETHYLENE AND COMPOUNDS WITH A TERMINAL DOUBLE BOND (*Continued*)

Terminal Alkene	Cross Partner	Conditions	Product(s) and Yield(s) (%)	Refs.

Please refer to the charts preceding the tables for the catalyst structures.

C_6

Terminal Alkene	Cross Partner	Conditions	Product(s) and Yield(s) (%)	Refs.
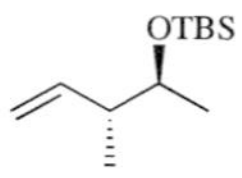 OTBS 6.7 eq	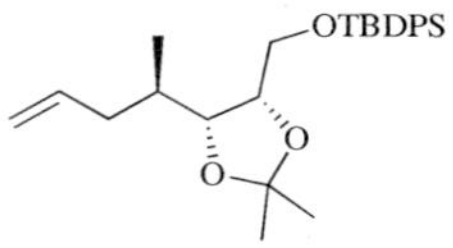 OTBDPS	Gr II (x eq), solvent, reflux, 24 h	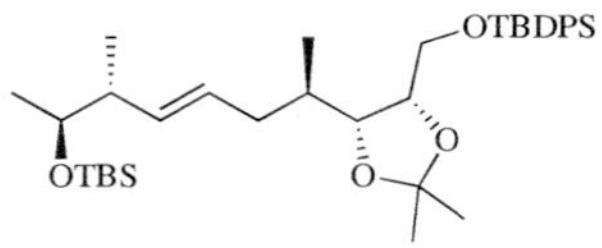 OTBS, OTBDPS	452
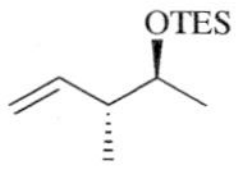 OTES 2 eq	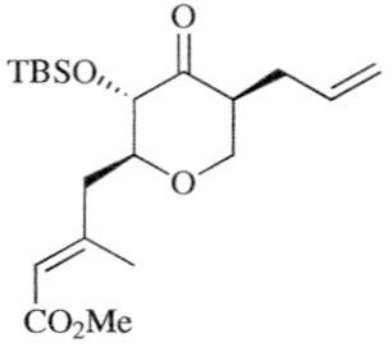 TBSO, O, O, CO_2Me	Gr II (0.05 eq), toluene, 50°, 5 h	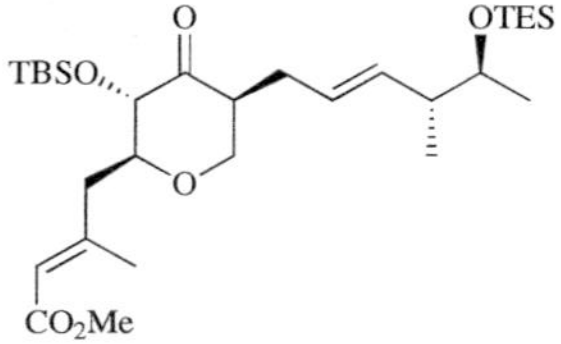 TBSO, O, OTES, O, CO_2Me (56) (E) only	453

x	Solvent		(E)/(Z)
0.05	DCM	(—)	—
0.15	DCM	(—)	—
0.05	benzene	(71)	—
0.15	benzene	(74)	—
0.05	toluene	(82)	—
0.1	toluene	(90	95:5
0.15	toluene	(90)	95:5

Reactant	Reactant	Conditions	Product(s) and Yield(s) (%)		Refs.
	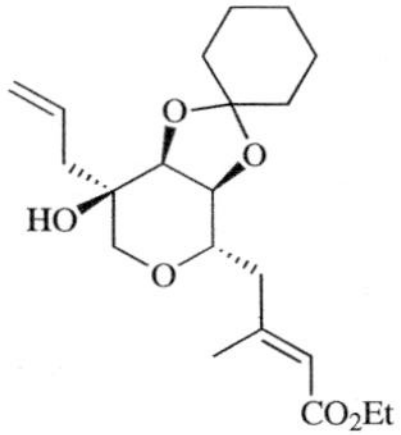 HO, CO_2Et	Gr II, toluene, 55°	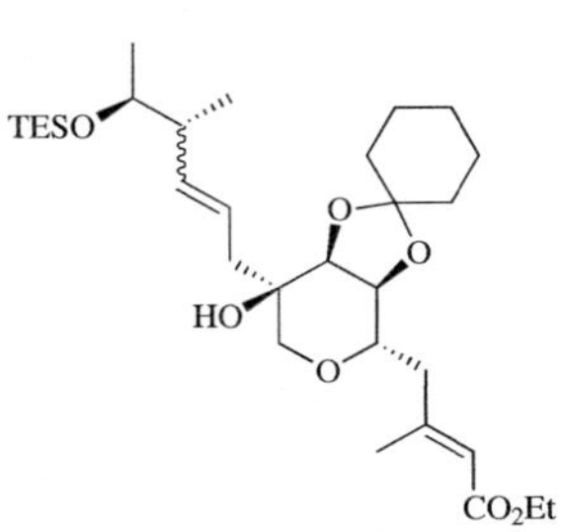 TESO, HO, CO_2Et	(66)	454
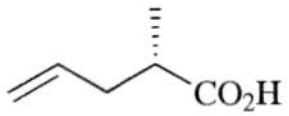 CO_2H 4 eq	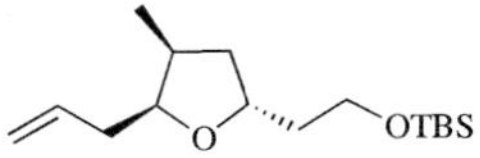 OTBS	1. Gr II, DCM, 40°, 5 h 2. Pd/C (0.1 eq), H_2, EtOAc, rt, 1 h	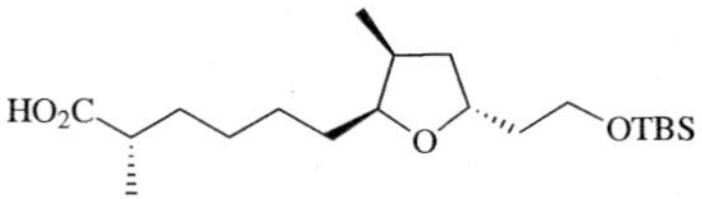 HO_2C, OTBS	(61)	456

TABLE 1. CROSS-METATHESIS OF UNSATURATED COMPOUNDS WITH FUNCTIONAL GROUPS IN HOMOALLYLIC AND MORE DISTANT POSITIONS (*Continued*)

A. ETHYLENE AND COMPOUNDS WITH A TERMINAL DOUBLE BOND (*Continued*)

Terminal Alkene	Cross Partner	Conditions	Product(s) and Yield(s) (%)	Refs.

Please refer to the charts preceding the tables for the catalyst structures.

C_6

Terminal Alkene	Cross Partner	Conditions	Product(s) and Yield(s) (%)	Refs.
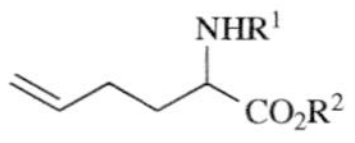 NHR[1], CO_2R^2	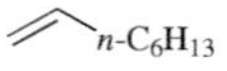 n-C_6H_{13}, 2 eq	Gr I (0.05 eq), DCE, rt, 30 h	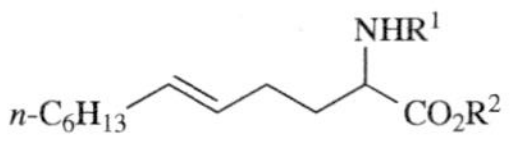 n-C_6H_{13} … NHR[1], CO_2R^2 **I** (*E*)/(*Z*) = 80:20 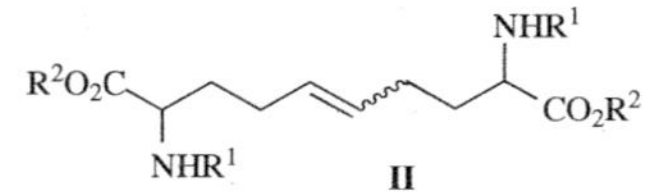 + R^2O_2C … NHR[1] … NHR[1], CO_2R^2 **II**	434
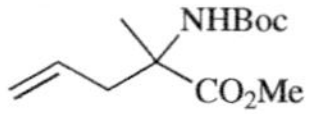 NHBoc, CO_2Me	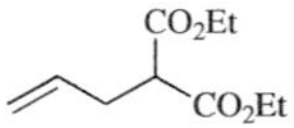 CO_2Et, CO_2Et, 4 eq	Gr II (0.05 eq), toluene, rt, 17 h	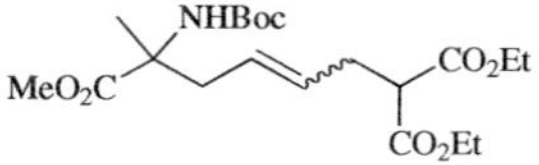 MeO_2C, NHBoc … CO_2Et, CO_2Et (55)	457
 NHCHO, CO_2Me	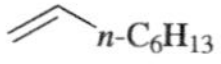 n-C_6H_{13}, 4 eq	Gr II (0.05 eq), toluene, rt, 17 h	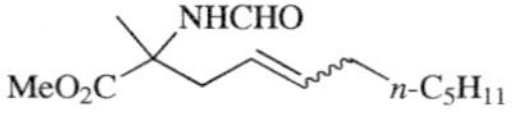 MeO_2C, NHCHO … n-C_5H_{11} (55)	457

R[1]	R[2]	I	II
Fmoc	H	(58)	(18)
Fmoc	Me	(58)	(25)
Boc	t-Bu	(66)	(28)

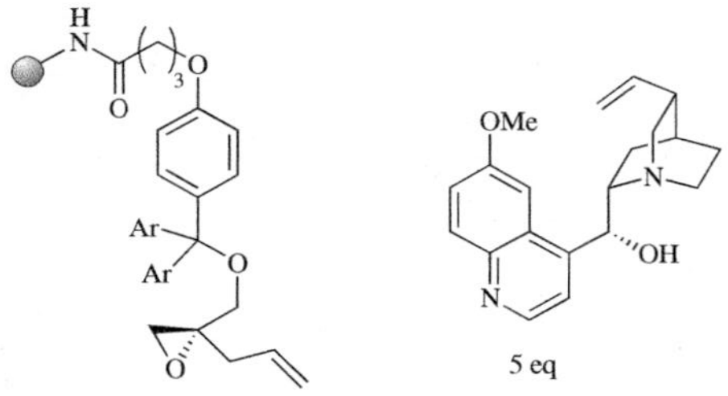

Ar = 4-ClC_6H_4

5 eq

1. Hov II (0.05 eq), $Ti(Oi\text{-}Pr)_4$, DCE, rt, 24 h
2. TFA

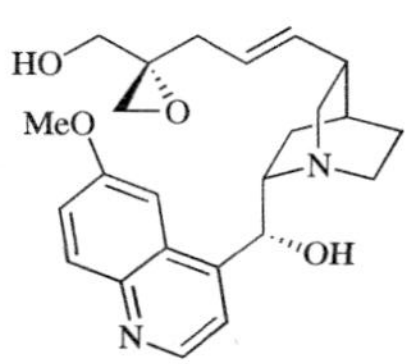

(60) 458

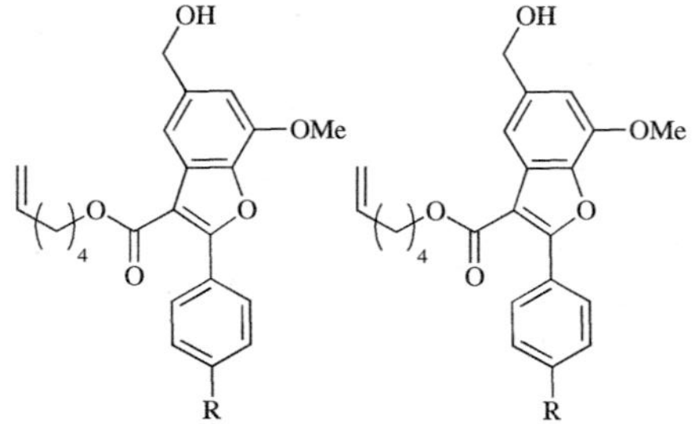

Gr I (0.1 eq), DCM, 40°, 24 h

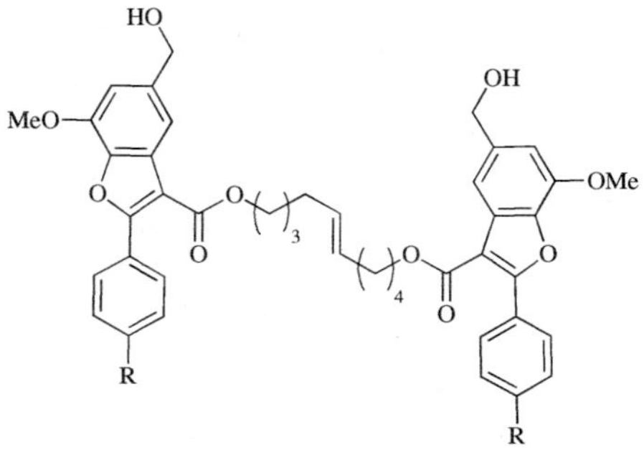

459

R		(*E*)/(*Z*)
H	(75)	66:34
MeO	(80)	91:9
Me	(80)	91:9
n-Bu	(80)	91:9
t-Bu	(80)	60:40

TABLE 1. CROSS-METATHESIS OF UNSATURATED COMPOUNDS WITH FUNCTIONAL GROUPS IN HOMOALLYLIC AND MORE DISTANT POSITIONS (*Continued*)

A. ETHYLENE AND COMPOUNDS WITH A TERMINAL DOUBLE BOND (*Continued*)

Terminal Alkene	Cross Partner	Conditions	Product(s) and Yield(s) (%)	Refs.
Please refer to the charts preceding the tables for the catalyst structures.				
C_7				
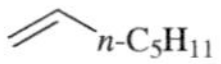 n-C_5H_{11} 2 eq	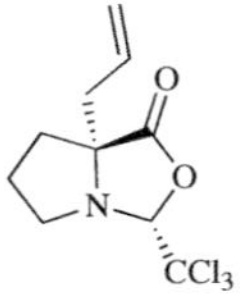 CCl_3	Gr II (0.05 eq), DCM, 80° (MW), 20 min	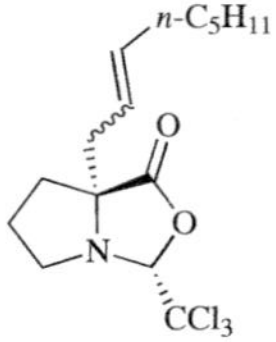 n-C_5H_{11}, CCl_3 (74) $(E)/(Z)$ = 74:26	460
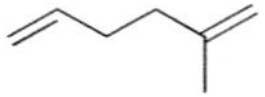 x eq	 OH, 7 y eq	Ru cat **216** (0.01 eq), THF, rt, 5 h	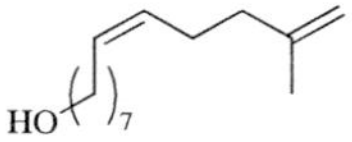 HO, 7 x | y | | $(Z)/(E)$ 1 | 5 | (45) | >95:5 7 | 1 | (66) | >95:5	437
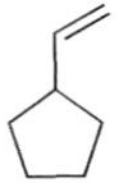	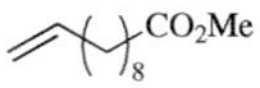 CO_2Me, 8 4 eq	Ru cat **216** (0.02 eq), THF, 20°, 5 h	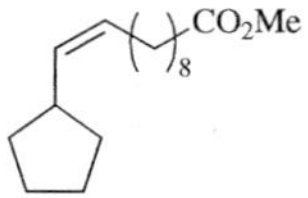 CO_2Me, 8 (44) $(Z)/(E)$ = 94:6	461

 20 eq	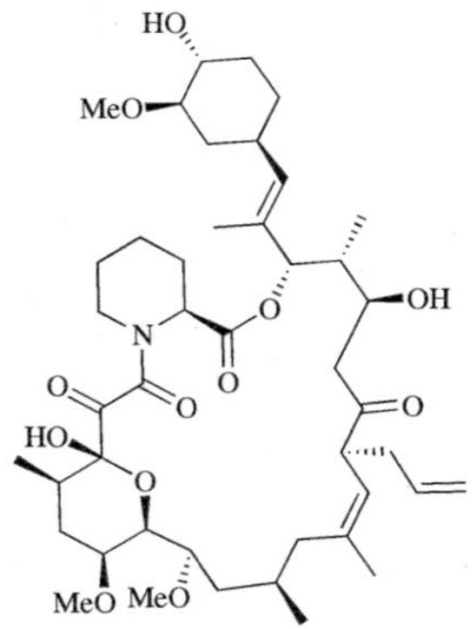	Gr II (0.1 eq), DCB, MW	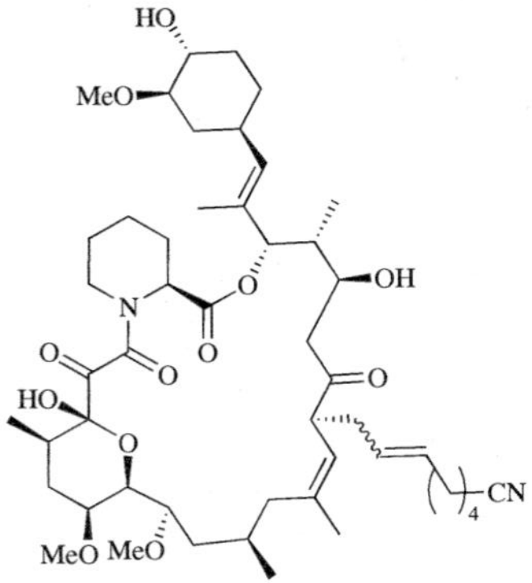 (36)	432
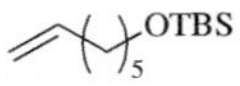 3 eq	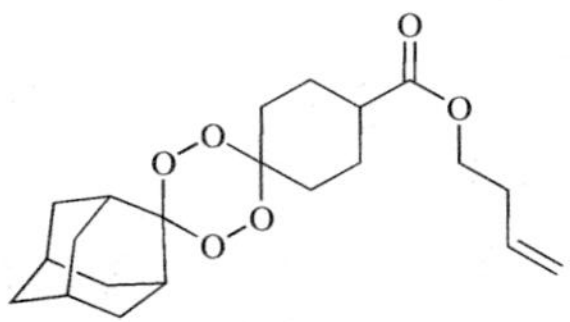	Ru cat **88** (0.1 eq), DCM, 45°, 2 h	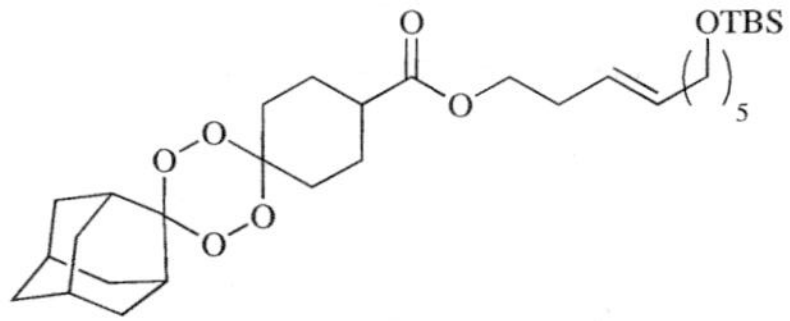 (68) $(E)/(Z) = 89:11$	193

TABLE 1. CROSS-METATHESIS OF UNSATURATED COMPOUNDS WITH FUNCTIONAL GROUPS IN HOMOALLYLIC AND MORE DISTANT POSITIONS (*Continued*)

A. ETHYLENE AND COMPOUNDS WITH A TERMINAL DOUBLE BOND (*Continued*)

Terminal Alkene	Cross Partner	Conditions	Product(s) and Yield(s) (%)	Refs.
Please refer to the charts preceding the tables for the catalyst structures.				
C_7				
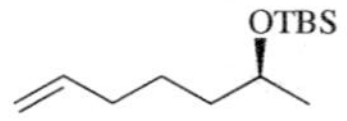 OTBS	 O, O, R H	Gr I, DCM, reflux	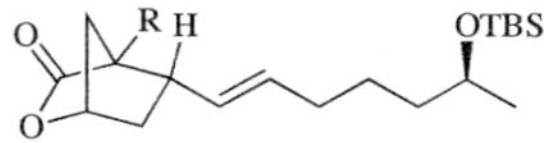 O, O, R H, OTBS R / Time (h) / (*E*)/(*Z*) H / 24 / (79) / 88:12 PhO_2S / 48 / (15) / 83:17	462
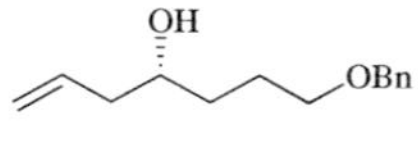 OH, OBn 2 eq	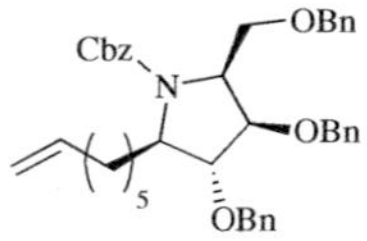 Cbz, N, OBn, OBn, OBn, 5	Gr II (0.01 eq), DCM, reflux, 4 h	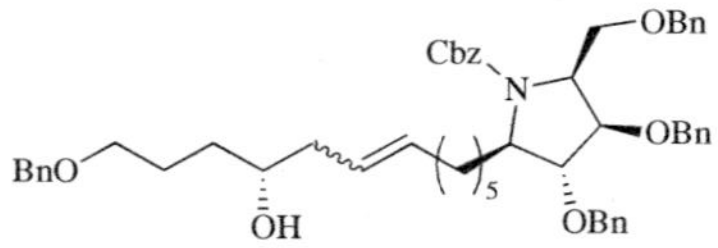 BnO, OH, Cbz, N, OBn, OBn, OBn, 5 (43)	463
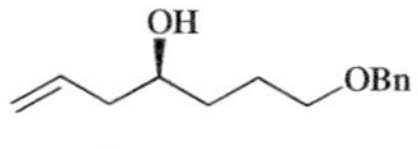 OH, OBn 2 eq	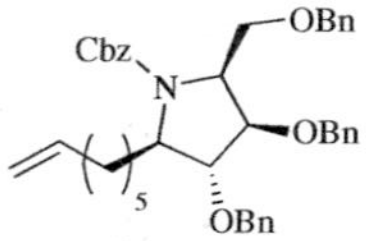 Cbz, N, OBn, OBn, OBn, 5	Gr II (0.01 eq), DCM, reflux, 4 h	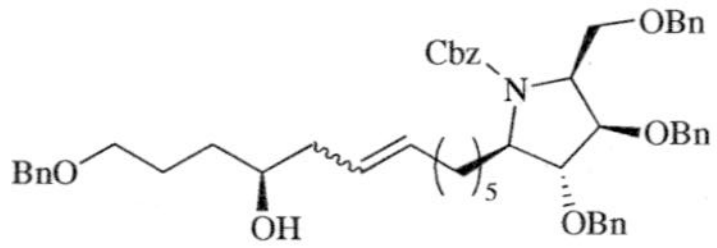 BnO, OH, Cbz, N, OBn, OBn, OBn, 5 (43)	463

OH, OBn, 2 eq	Cbz, N, OBn, OBn, OBn, 5	Gr II (0.01 eq), DCM, reflux, 4 h	BnO, OH, Cbz, N, OBn, OBn, OBn, 5	(41)	463
OH, OBn, 2 eq	Cbz, N, OBn, OBn, OBn, 5	Gr II (0.01 eq), DCM, reflux, 4 h	BnO, OH, Cbz, N, OBn, OBn, OBn, 5	(45)	463
OH, OBn, 2 eq	Cbz, N, OBn, OBn, OBn, 5	Gr II (0.01 eq), DCM, reflux, 4 h	BnO, OH, Cbz, N, OBn, OBn, OBn, 5	(40)	463

TABLE 1. CROSS-METATHESIS OF UNSATURATED COMPOUNDS WITH FUNCTIONAL GROUPS IN HOMOALLYLIC AND MORE DISTANT POSITIONS (*Continued*)

A. ETHYLENE AND COMPOUNDS WITH A TERMINAL DOUBLE BOND (*Continued*)

Terminal Alkene	Cross Partner	Conditions	Product(s) and Yield(s) (%)		Refs.
Please refer to the charts preceding the tables for the catalyst structures.					
C_7					
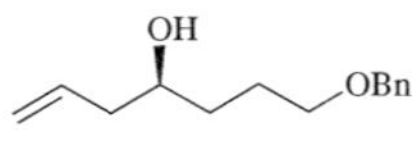 OH, OBn; 2 eq	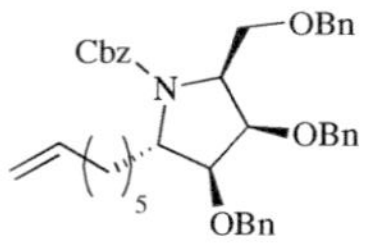 Cbz, N, OBn, OBn, OBn, 5	Gr II (0.01 eq), DCM, reflux, 4 h	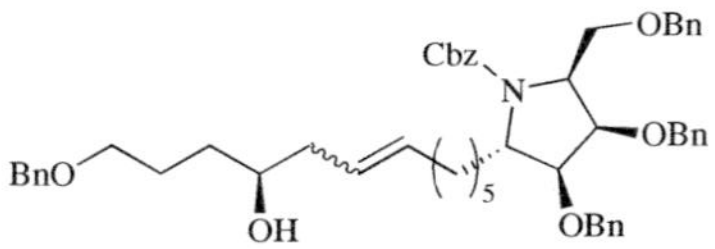 BnO, OH, 5, Cbz, N, OBn, OBn, OBn	(41)	463
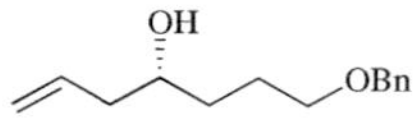 OH, OBn; 2 eq	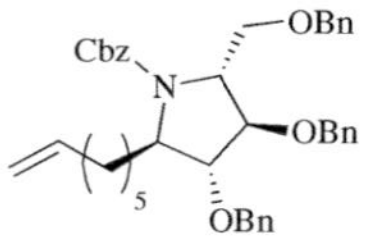 Cbz, N, OBn, OBn, OBn, 5	Gr II (0.01 eq), DCM, reflux, 4 h	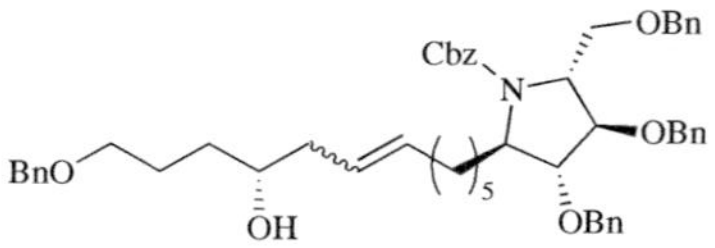 BnO, OH, 5, Cbz, N, OBn, OBn, OBn	(43)	463
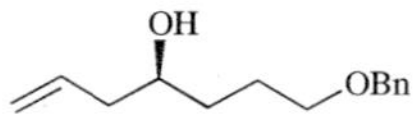 OH, OBn; 2 eq	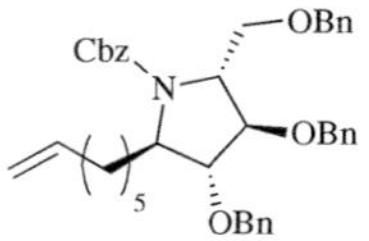 Cbz, N, OBn, OBn, OBn, 5	Gr II (0.01 eq), DCM, reflux, 4 h	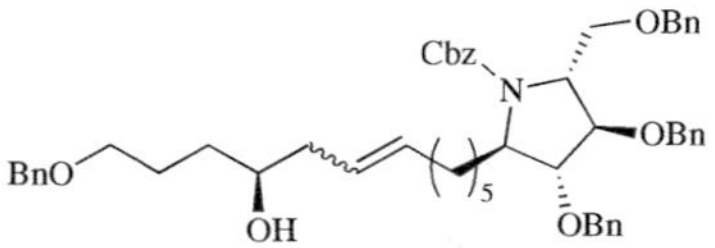 BnO, OH, 5, Cbz, N, OBn, OBn, OBn	(43)	463

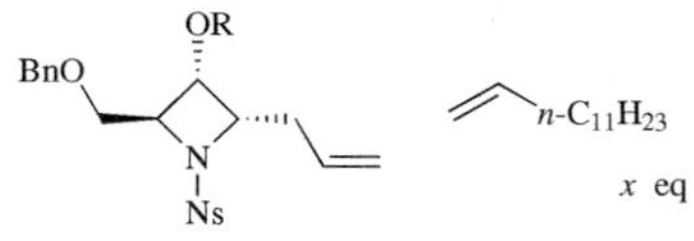

BnO OR N Ns n-$C_{11}H_{23}$ x eq

Gr II (y eq), DCM, reflux

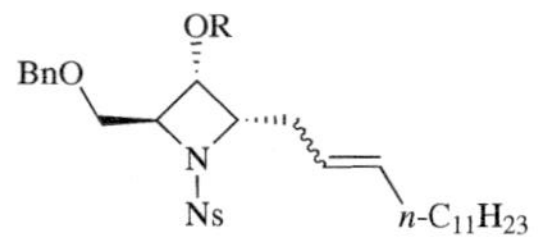

BnO OR N Ns n-$C_{11}H_{23}$ 464

R	x	y	Time (h)	
H	4	0.05	2	(63)
PMB	12	0.2	48	(<30)

BnO OH N Ns OBn 6.15 eq

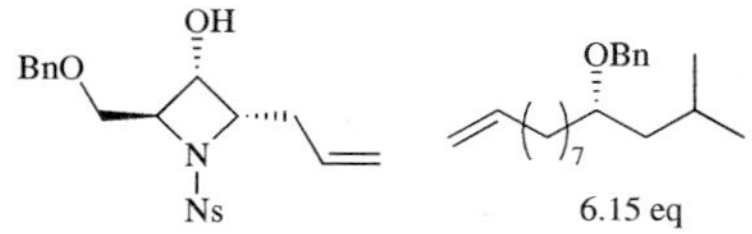

Gr II (0.05 eq), DCM, reflux, 3.5 h

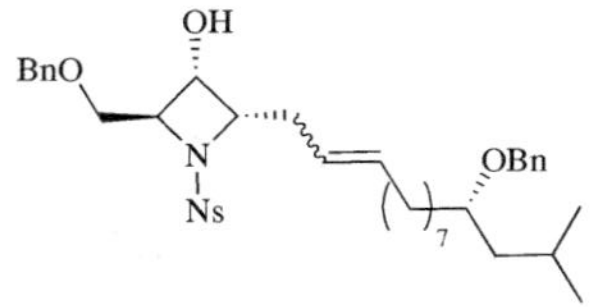

BnO OH N Ns OBn (74) 464

OTBDPS OTHP R

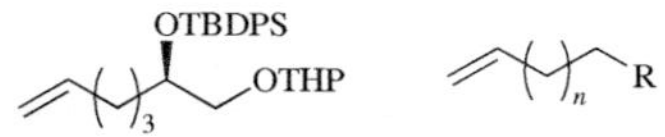

Gr I (0.05 eq), BQ (0.2 eq), DCM, 40°

THPO OTBDPS R 465

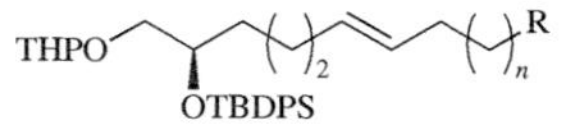

n	R	Time (h)		(E)/(Z)
7	Bn-Gly-C(O)	12	(80)	80:20
8	TBSO	20	(65)	64:36

TABLE 1. CROSS-METATHESIS OF UNSATURATED COMPOUNDS WITH FUNCTIONAL GROUPS IN HOMOALLYLIC AND MORE DISTANT POSITIONS (*Continued*)

A. ETHYLENE AND COMPOUNDS WITH A TERMINAL DOUBLE BOND (*Continued*)

Terminal Alkene	Cross Partner	Conditions	Product(s) and Yield(s) (%)	Refs.
Please refer to the charts preceding the tables for the catalyst structures.				
C_7				
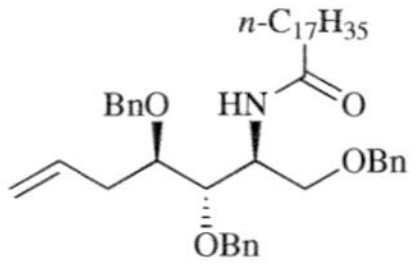 BnO, HN, n-$C_{17}H_{35}$, O, OBn, OBn	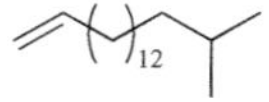 x eq	Gr I (0.05 eq), DCM, reflux, 6 h	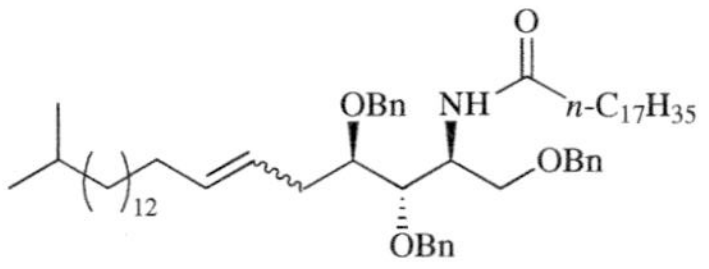 OBn, NH, O, n-$C_{17}H_{35}$, OBn, OBn x — 1 (62); 2 (70); 3 (79); 4 (81)	466
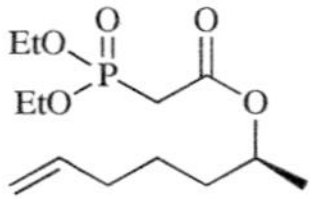 EtO, EtO, P, O, O, O 3 eq	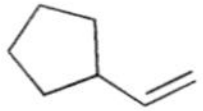	Catalyst (0.025 eq), DCM, reflux, 22 h	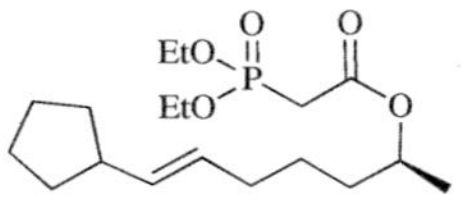 EtO, EtO, P, O, O, O $(E)/(Z)$ = 80:20 Catalyst — Gr I (74); Gr II (44)	467
x eq	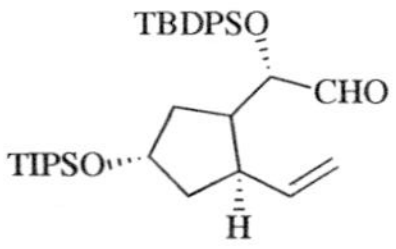 TBDPSO, CHO, TIPSO, H	Catalyst (0.025 eq), DCM, reflux, 22 h	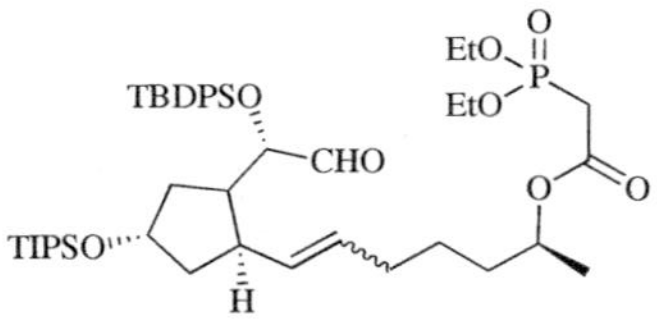 TBDPSO, CHO, TIPSO, H, EtO, EtO, P, O, O, O	467

x	Catalyst	
1	Gr I	(17)
0.25	Gr I	(30)
3	Gr I	(20)
4	Gr I	(50)
2	Gr II	(84)
2	Gr II	(100)

 $Fe(CO)_3$	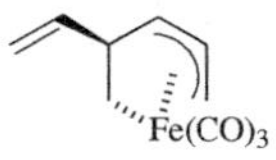 $Fe(CO)_3$	Gr I (0.05 + 0.05 eq), DCM, 40°, 20 h	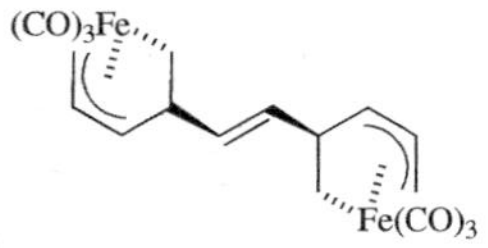 $(CO)_3Fe$ $Fe(CO)_3$	(88)	468
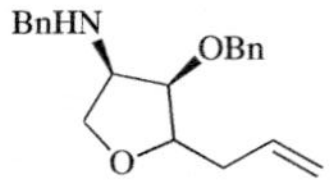 BnHN OBn O	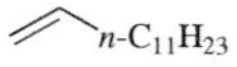 n-$C_{11}H_{23}$ 5 eq	Gr I (0.3 eq), DCM, reflux, 12 h	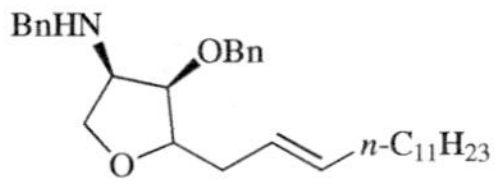 BnHN OBn O n-$C_{11}H_{23}$	(78)	469
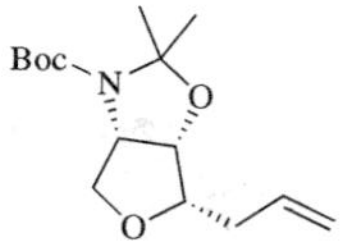 Boc N O O	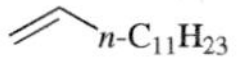 n-$C_{11}H_{23}$ 2.5 eq	Gr II (0.05 + 0.03 eq), DCM, 45°, 8 h	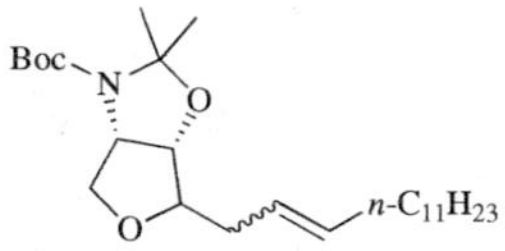 Boc N O O n-$C_{11}H_{23}$	(82)	470
C_8					
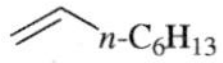 n-C_6H_{13}	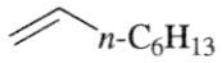 n-C_6H_{13}	Ru cat **185** (x eq), BMIM•NTf_2, 50°	 n-C_5H_{11} n-C_6H_{13}		471

x	Flow rate/cm^{-1}	Time (h)	Conv. (%)[c]
0.022	0.2	6	0
0.024	0.1	5	40

TABLE 1. CROSS-METATHESIS OF UNSATURATED COMPOUNDS WITH FUNCTIONAL GROUPS IN HOMOALLYLIC AND MORE DISTANT POSITIONS (*Continued*)

A. ETHYLENE AND COMPOUNDS WITH A TERMINAL DOUBLE BOND (*Continued*)

Terminal Alkene	Cross Partner	Conditions	Product(s) and Yield(s) (%)	Refs.
Please refer to the charts preceding the tables for the catalyst structures.				
C_8				
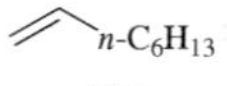 n-C_6H_{13} 10 eq	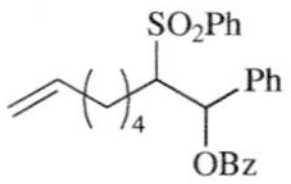 SO_2Ph, Ph, OBz	1. Gr II (0.1 eq), toluene, 60°, 12 h 2. SmI_2 (3 eq), THF/DMPU (4:1), –78°, 1 h	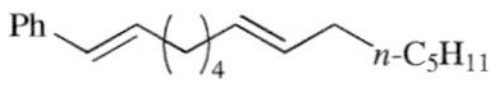 Ph, n-C_5H_{11} (80) $(E,E)/(Z,E)$ = 91:9	445
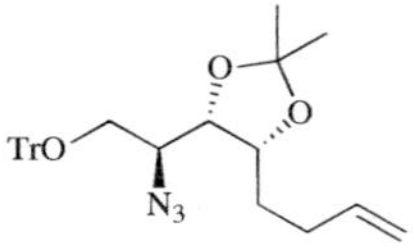 TrO, N_3	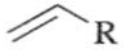 R 10 eq	Gr II (0.1 eq), DCM, rt, 18 h	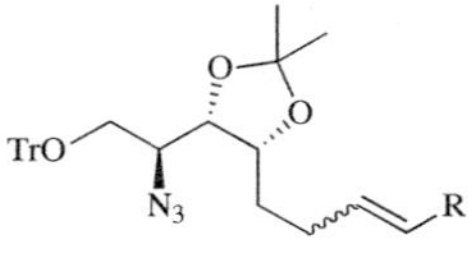 TrO, N_3, R R: n-C_7H_{15} (57); $Ph(CH_2)_2$ (57); n-$C_{10}H_{21}$ (60); n-$C_{15}H_{31}$ (59)	406
	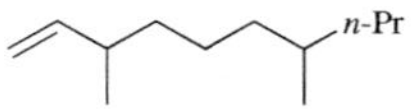 n-Pr 10 eq	Gr II (0.1 eq), DCM, rt, 18 h	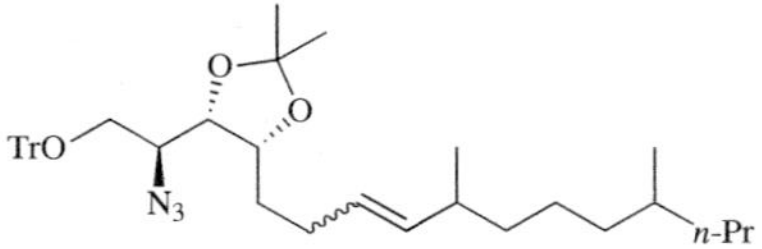 TrO, N_3, n-Pr (25)	406

C_{8-15}

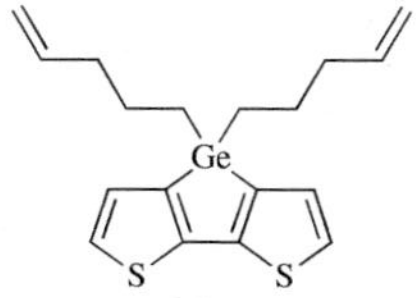

Gr II (0.06 eq), DCM, reflux, 24 h

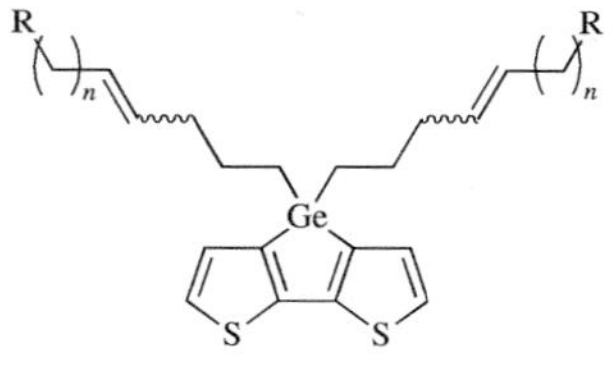

16 eq

R	n		
Br	6	(90)	472
H	9	(90)	
H	13	(92)	

C_8

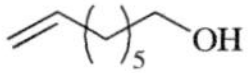

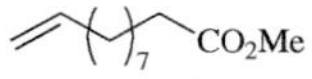

3 eq

Gr II (0.07 eq), DCM, rt, 24 h

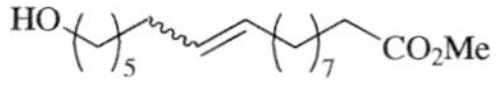

CO_2Me (51) 473

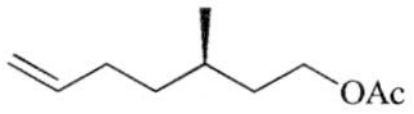

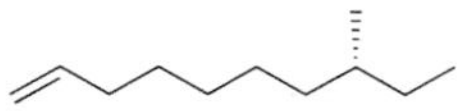

10 eq

Gr I (0.05 eq), DCM, reflux, 6 h

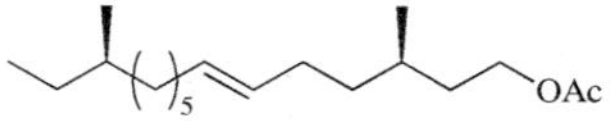

(88) (*E*)/(*Z*) = 83:17 474

TABLE 1. CROSS-METATHESIS OF UNSATURATED COMPOUNDS WITH FUNCTIONAL GROUPS IN HOMOALLYLIC AND MORE DISTANT POSITIONS (*Continued*)

A. ETHYLENE AND COMPOUNDS WITH A TERMINAL DOUBLE BOND (*Continued*)

Terminal Alkene	Cross Partner	Conditions	Product(s) and Yield(s) (%)	Refs.
Please refer to the charts preceding the tables for the catalyst structures.				
C_8				
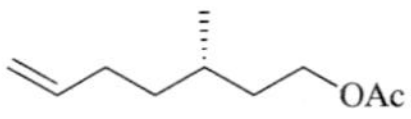 OAc	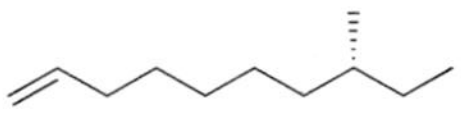 10 eq	Gr I (0.05 eq), DCM, reflux, 6 h	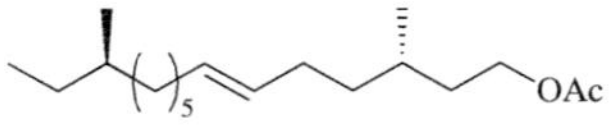 OAc, $(\ )_5$ (86) $(E)/(Z)$ = 83:11	474
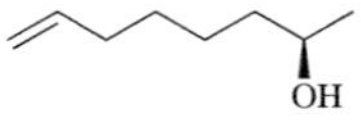 OH	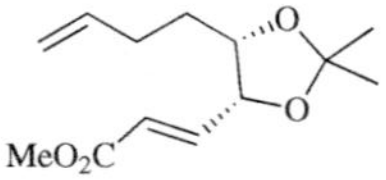 MeO_2C, O, O	Gr I (0.1 eq), DCM, reflux, 16 h	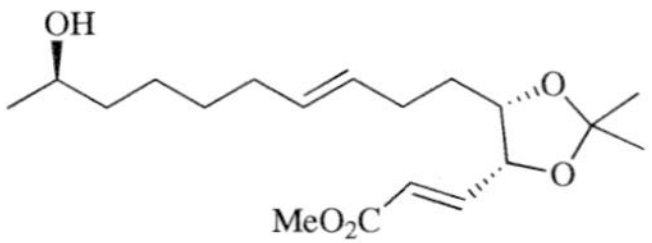 OH, O, O, MeO_2C (75) $(E)/(Z)$ = 75:25	475
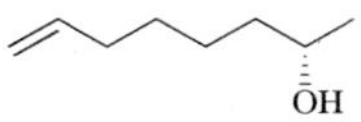 OH, 2 eq	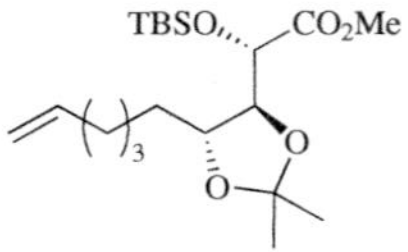 TBSO, CO_2Me, $(\ )_3$, O, O	Gr II (0.1 eq), DCM, reflux, 15 h	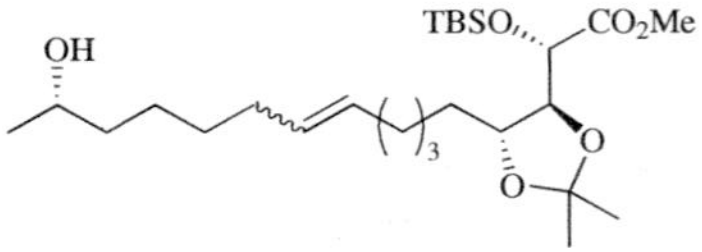 OH, TBSO, CO_2Me, $(\ )_3$, O, O (71)	476

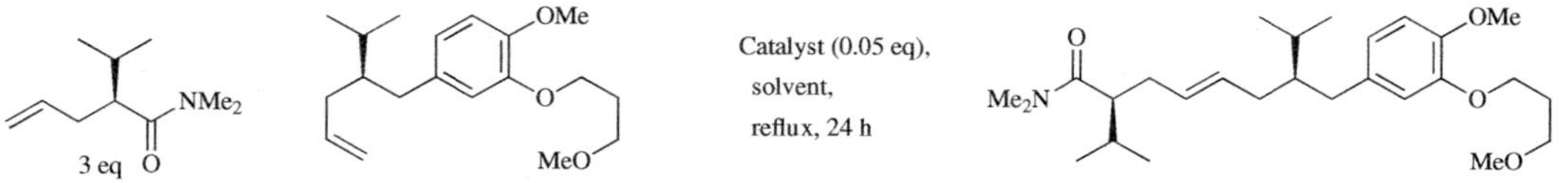

3 eq

Catalyst (0.05 eq), solvent, reflux, 24 h

477

Catalyst	Solvent		$(E)/(Z)$
Gr I	DCM	(18)	—
Gr II	DCM	(76)	76:24
Gr II	*n*-hexane	(66)	84:16
Hov II	DCM	(65)	77:23
Hov II	*n*-hexane	(56)	77:23
Gr II	PE	(66)	85:15
Gr II	toluene	(53)	81:19
Gr II	DCE	(30)	—
Gr II	cyclohexane	(28)	—

4 eq

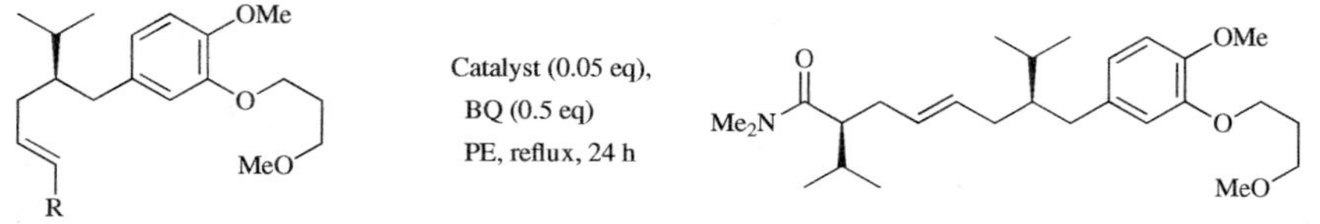

Catalyst (0.05 eq), BQ (0.5 eq) PE, reflux, 24 h

477

R	Catalyst		$(E)/(Z)$
H	Hov II	(88)	79:21
H	Gr II	(72)	81:19
Me	Hov II	(72)	82:18
Me	Gr II	(78)	81:19

TABLE 1. CROSS-METATHESIS OF UNSATURATED COMPOUNDS WITH FUNCTIONAL GROUPS IN HOMOALLYLIC AND MORE DISTANT POSITIONS (*Continued*)

A. ETHYLENE AND COMPOUNDS WITH A TERMINAL DOUBLE BOND (*Continued*)

Terminal Alkene	Cross Partner	Conditions	Product(s) and Yield(s) (%)	Refs.
Please refer to the charts preceding the tables for the catalyst structures.				
C_8				
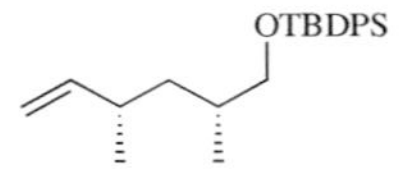 OTBDPS 2 eq	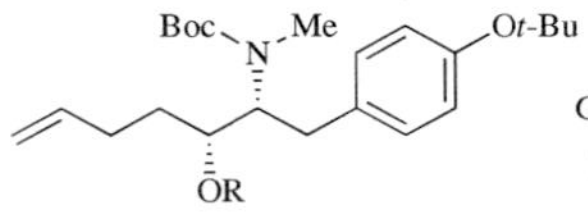 Boc, N, Me, O*t*-Bu, OR	Gr II (0.05 eq), DCM, 40°, 19 h	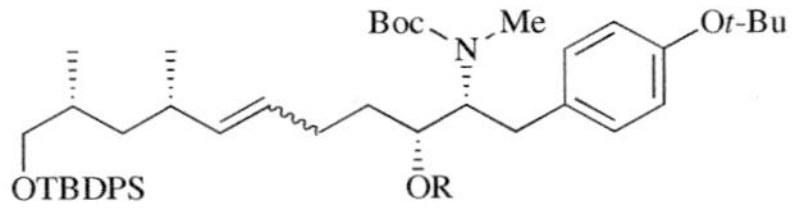 Boc, N, Me, O*t*-Bu, OTBDPS, OR R: H (<25); Ac (89)	478
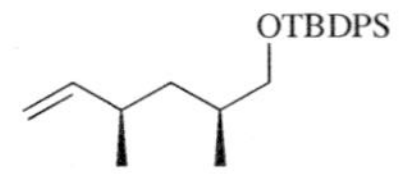 OTBDPS 2 eq	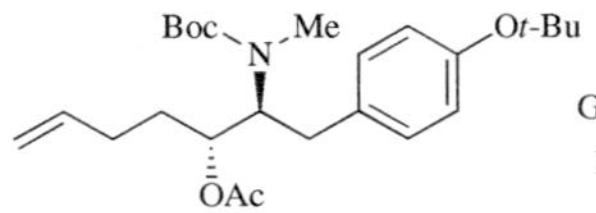 Boc, N, Me, O*t*-Bu, OAc	Gr II (0.05 eq), DCM, 40°, 19 h	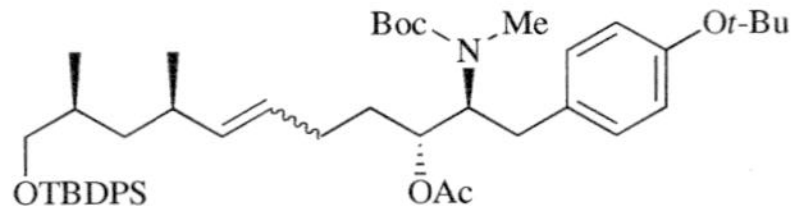 Boc, N, Me, O*t*-Bu, OTBDPS, OAc (88)	478
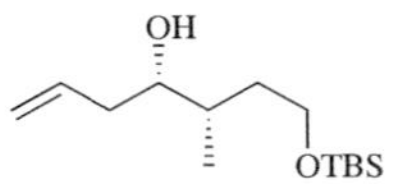 OH, OTBS 1.3 eq	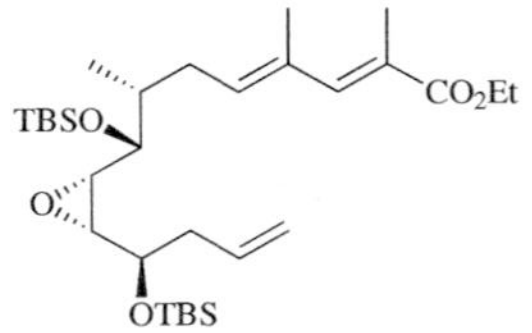 TBSO, CO_2Et, O, OTBS	Gr II (0.019 eq), DCM, 40°, 8 h	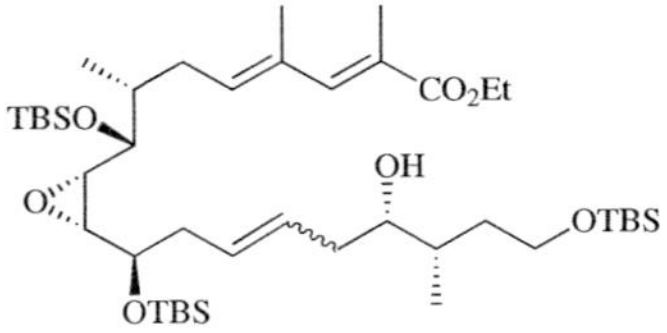 TBSO, CO_2Et, O, OH, OTBS, OTBS (65)	479
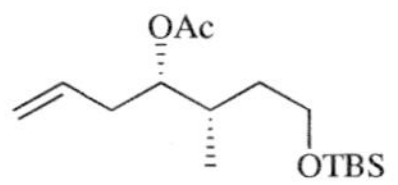 OAc, OTBS 3 eq	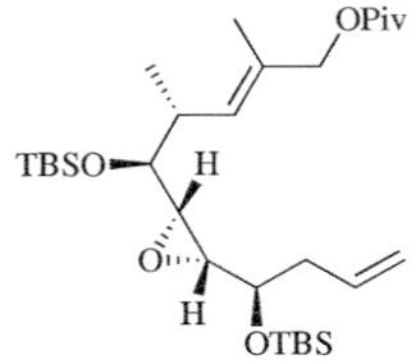 OPiv, TBSO, H, O, H, OTBS	Ru cat **35** (0.05 eq), DCM, 40°	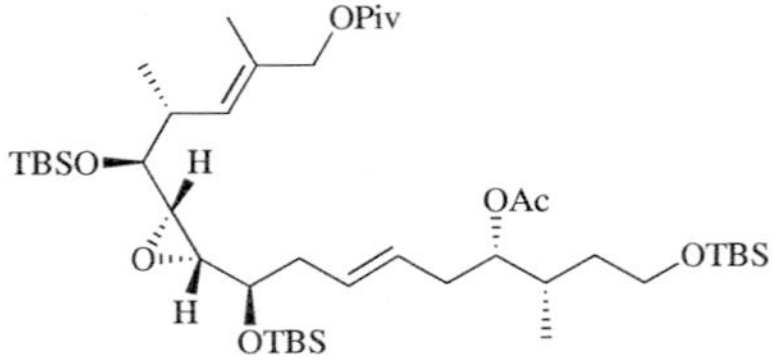 OPiv, TBSO, H, O, H, OTBS, OAc, OTBS (78) (*E*)/(*Z*) = 87:13	480

3 eq

Catalyst (0.2 eq), DCM, reflux

n-C_6H_{13}

Catalyst	Time (h)	
Gr I	24	(62)
Gr II	18	(46)
Hov I	24	(54)
Hov II	18	(46)

508

2 eq

Gr II (0.05 eq), BQ (0.2 eq), DCM, reflux, 5 h

R^1	R^2	
H	Bn	(85)
H	Bz	(85)
Me	Bn	(86)
Me	Bz	(84)

481

2 eq

Gr II (0.05 eq), BQ (0.2 eq), DCM, reflux, 5 h

R	
H	(87)
Me	(85)

481

TABLE 1. CROSS-METATHESIS OF UNSATURATED COMPOUNDS WITH FUNCTIONAL GROUPS IN HOMOALLYLIC AND MORE DISTANT POSITIONS (*Continued*)

A. ETHYLENE AND COMPOUNDS WITH A TERMINAL DOUBLE BOND (*Continued*)

Terminal Alkene	Cross Partner	Conditions	Product(s) and Yield(s) (%)	Refs.
Please refer to the charts preceding the tables for the catalyst structures.				
C_8				
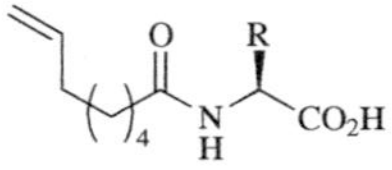 O, R, N, H, CO_2H, 4; 2.08 eq	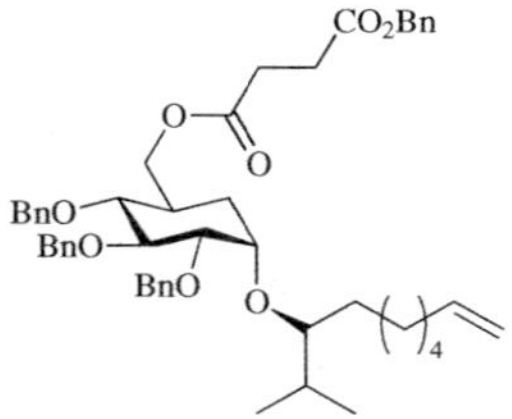 CO_2Bn, O, O, BnO, BnO, BnO, O, 4	Gr II (0.05 eq), BQ (0.02 eq) DCM, reflux, 5 h	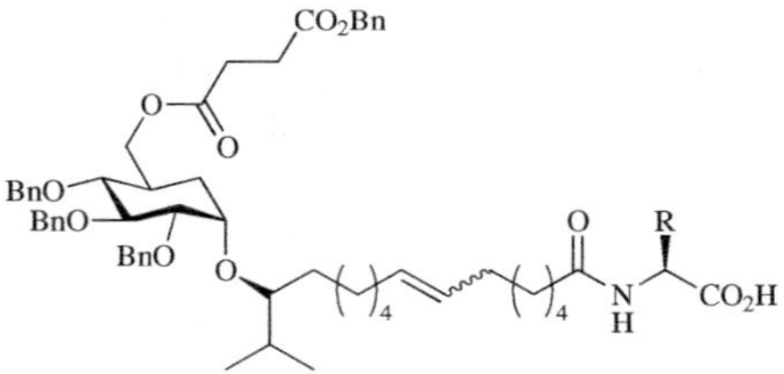 CO_2Bn, O, O, BnO, BnO, BnO, O, 4, 4, O, R, N, H, CO_2H; R: H (84), Me (82)	482
2.08 eq	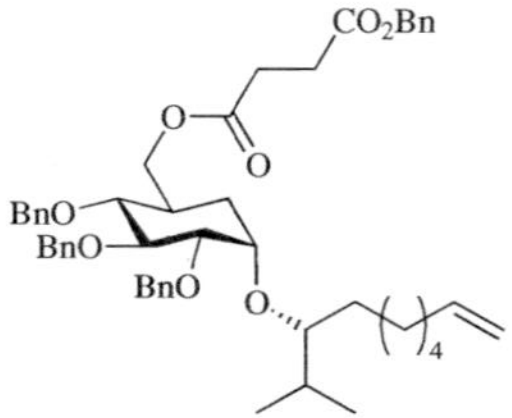 CO_2Bn, O, O, BnO, BnO, BnO, O, 4	Gr II (0.05 eq), BQ (0.02 eq) DCM, reflux, 5 h	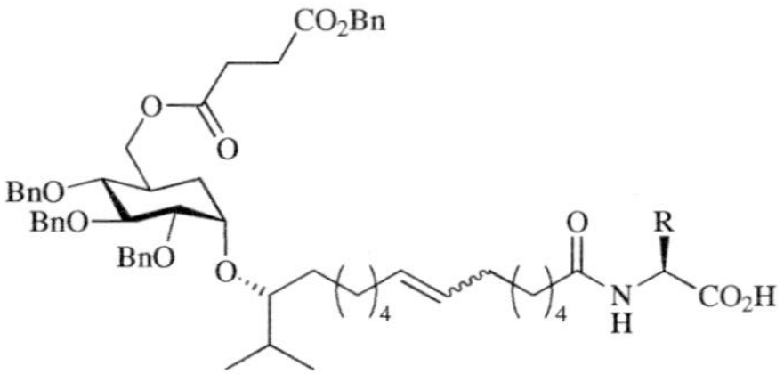 CO_2Bn, O, O, BnO, BnO, BnO, O, 4, 4, O, R, N, H, CO_2H; R: H (85), Me (84)	482

Substrate	Olefin	Conditions	Product(s) and Yield(s) (%)		Refs.
MeO_2C, OH, CO_2Me, CO_2Me	O, 6, n-C_7H_{15}	Gr II (0.2 eq), toluene, 80°	O, n-C_7H_{15}, 5, MeO_2C, OH, CO_2Me, CO_2Me (62) (E) only, $syn/anti$ = 7:1		483
O, O, O, t-Bu, HO_2C, CO_2t-Bu 2 eq	O, O, 6, 5, 6, 6	Ru cat **88** (0.2 eq), DCM, 55° (MW), 1 h	O, 6, 5, O, O, O, t-Bu, HO_2C, CO_2t-Bu	(57)	483
O, O, O, t-Bu, CO_2t-Bu, O, NH, R, t-BuO_2C 2 eq	O, O, 6, 5, 6, 6	Ru cat **88** (0.2 eq), DCM, 65° (MW), 4 h	O, 6, 5, O, O, O, t-Bu, CO_2t-Bu, O, NH, R, t-BuO_2C	(E) only	483

R	
4-t-BuC_6H_4	(87)
Ph	(96)
3-indolyl	(92)

TABLE 1. CROSS-METATHESIS OF UNSATURATED COMPOUNDS WITH FUNCTIONAL GROUPS IN HOMOALLYLIC AND MORE DISTANT POSITIONS (*Continued*)

A. ETHYLENE AND COMPOUNDS WITH A TERMINAL DOUBLE BOND (*Continued*)

Terminal Alkene	Cross Partner	Conditions	Product(s) and Yield(s) (%)	Refs.
Please refer to the charts preceding the tables for the catalyst structures.				
C_8				
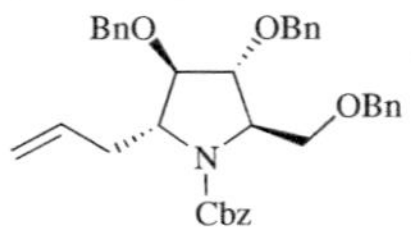 BnO, OBn, OBn, N, Cbz	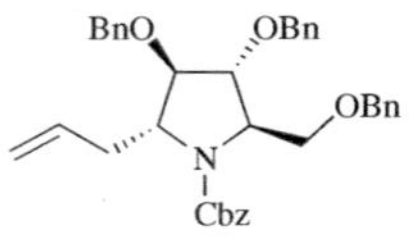 BnO, OBn, OBn, N, Cbz	Hov II (0.05 eq), DCM, rt, overnight	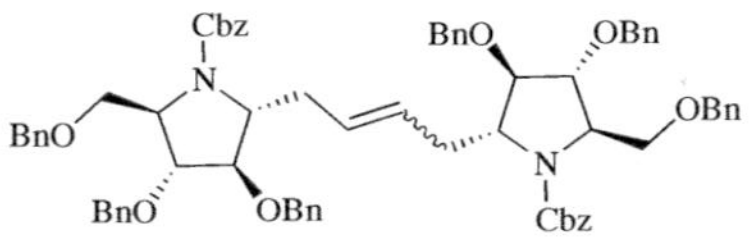 Cbz, N, BnO, BnO, OBn, BnO, OBn, OBn, N, Cbz (54)	484
	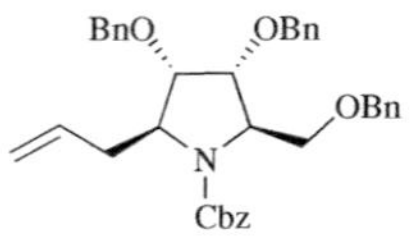 BnO, OBn, OBn, N, Cbz	Hov II (0.05 eq), DCM, rt, overnight	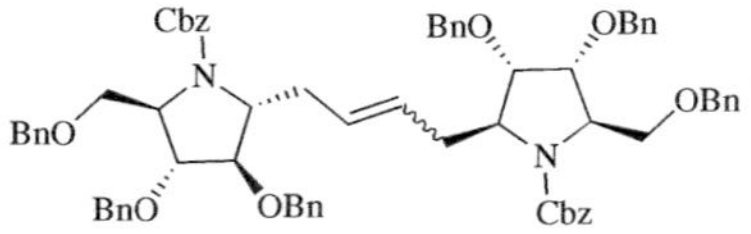 Cbz, N, BnO, BnO, OBn, BnO, OBn, OBn, N, Cbz (12) 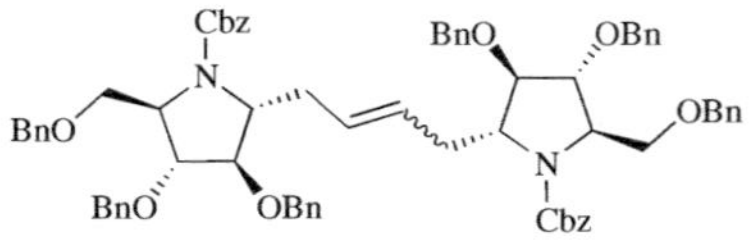 + Cbz, N, BnO, BnO, OBn, BnO, OBn, OBn, N, Cbz (20) 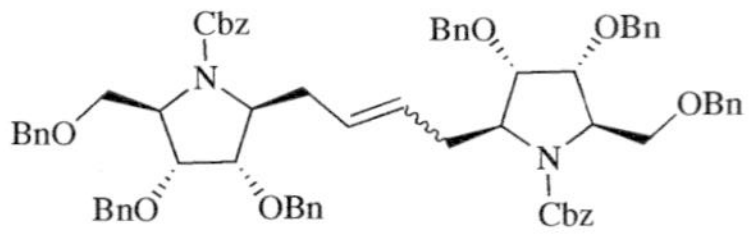 + Cbz, N, BnO, BnO, OBn, BnO, OBn, OBn, N, Cbz (20)	484
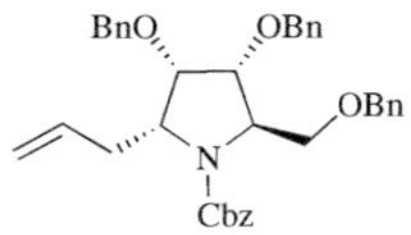 BnO, OBn, OBn, N, Cbz	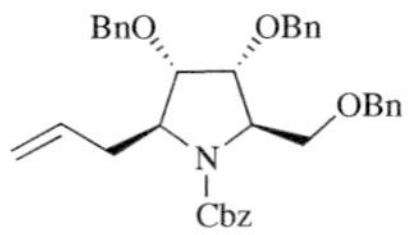 BnO, OBn, OBn, N, Cbz	Hov II (0.05 eq), DCM, rt, overnight	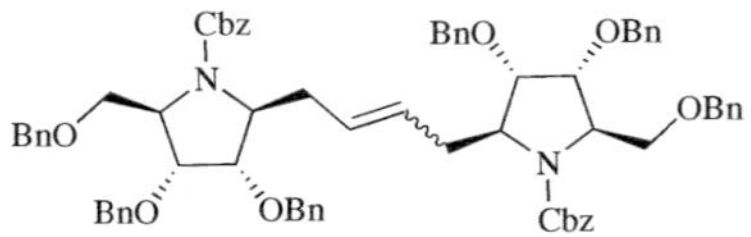 Cbz, N, BnO, BnO, OBn, BnO, OBn, OBn, N, Cbz (42)	484

C$_9$

Substrate 1	Substrate 2	Conditions	Product(s) and Yield(s) (%)	Refs.
OHC, R, x eq		Gr II (y eq), DCM, reflux	OHC, R, 6; $(E)/(Z)$ = 88:12, dr 95:5	249
	n-$C_{10}H_{21}$, 1.5 eq	Gr I (0.05 eq), DCM, reflux, 18 h	n-$C_{10}H_{21}$ (57) $(E)/(Z)$ = 78:22, dr 95:5	249
t-Bu	t-Bu	Ru cat **230** (0.02 eq), benzene, 22 h	t-Bu, t-Bu	433
OAc, 7	1.4 eq	Gr I (0.0.12 eq), DCM, reflux, 6 h	AcO, 6 (56) $(E)/(Z)$ = 80:20	485

R	x	y	Time (h)	
Me	4	0.05	30	(53)
$MeO_2C(CH_2)_4$	1.3	0.027	24	(38)

Temp (°)	Conv. (%)[a]
25	0
85	3

TABLE 1. CROSS-METATHESIS OF UNSATURATED COMPOUNDS WITH FUNCTIONAL GROUPS IN HOMOALLYLIC AND MORE DISTANT POSITIONS (*Continued*)

A. ETHYLENE AND COMPOUNDS WITH A TERMINAL DOUBLE BOND (*Continued*)

Terminal Alkene	Cross Partner	Conditions	Product(s) and Yield(s) (%)	Refs.
Please refer to the charts preceding the tables for the catalyst structures.				
C_9				
OAc (7)	1.4 eq	Gr II (0.012 eq), DCM, reflux, 7 h	AcO (6) (47) $(E)/(Z)$ = 80:20	485
	BnO, BnO, OBn, OBn; 2 eq	Gr I (0.05 eq), DCM, 45°, 4 h	BnO, BnO, OBn, OBn, OAc (7) (73) $(E)/(Z)$ = 74:26	124
OTs (7)	OTs (7)	Gr II (0.075 eq), DCM, 20°, 16 h	TsO (6), OTs (7) (79)	403
OTBS; 3.2 eq	OPMB, OMe, H, OMe, N, O, N, O	Gr II, DCM, 60°	OPMB, OMe, H, OMe, N, O, N, O, OTBS (59)	486
OBn, MPMO; 2 eq	OTBS, OTBS, OTBS	Gr II (0.1 eq), DCM, rt, 36 h	BnO, OMPM, OTBS, OTBS, OTBS (50)	487

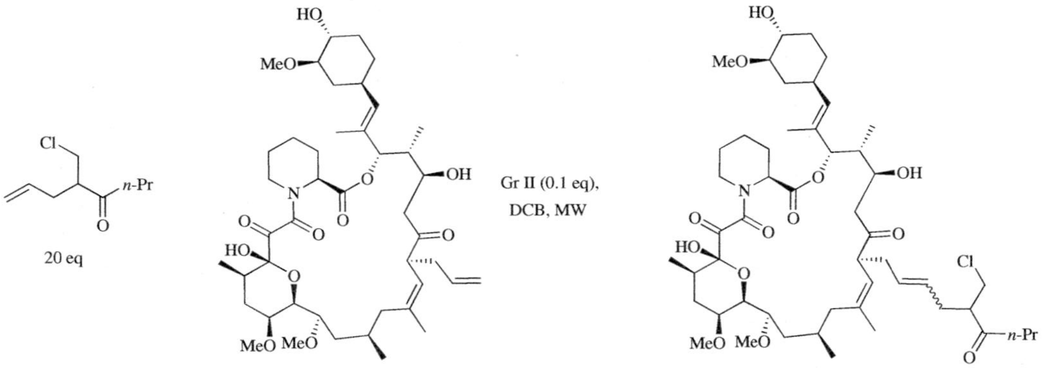

Cl, n-Pr, O

20 eq

HO, MeO, OH, N, O, HO, MeO MeO

Gr II (0.1 eq), DCB, MW

HO, MeO, OH, N, O, HO, MeO MeO, Cl, n-Pr

(44) 432

Ph

Ph

Catalyst (x eq), solvent

Ph, Ph

Catalyst	x	Solvent	Temp (°)	Time (h)	Conv. (%)[a]		$(Z)/(E)$	
Ru cat **217**	0.02	THF	70	6	>95	(—)	83:17	84
Ru cat **217**	0.02	MeCN	70	2.5	12	(—)	95:5	84
Ru cat **217**	0.02	THF	35	1	>95	(81)	92:8	84
Ru cat **217**	0.001	THF	35	3	—	(73)	86:14	488
Ru cat **215**	0.001	THF	35	3	—	(89)	92:8	488
Gr I	0.05	DCM	40	12	—	(82)	20:80	156

TABLE 1. CROSS-METATHESIS OF UNSATURATED COMPOUNDS WITH FUNCTIONAL GROUPS IN HOMOALLYLIC AND MORE DISTANT POSITIONS (*Continued*)

A. ETHYLENE AND COMPOUNDS WITH A TERMINAL DOUBLE BOND (*Continued*)

Terminal Alkene	Cross Partner	Conditions	Product(s) and Yield(s) (%)	Refs.

Please refer to the charts preceding the tables for the catalyst structures.

C_9

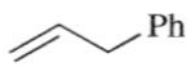

Ph
2 eq

OBz
8

Catalyst (*x* eq),
DCM

Ph OBz 8

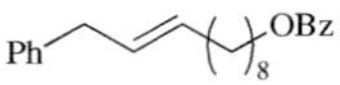

Catalyst	*x*	Temp	Time (h)	Conv (%)[a]		(*E*)/(*Z*)	Refs.
Ind II	0.01	rt	2.5	14	(—)	60:40	489
Ru cat **282**	0.01	rt	2.5	8	(—)	57:43	489
Ru cat **284**	0.01	rt	20	0	(—)	—	489
Ru cat **283**	0.01	rt	20	0	(—)	—	489
Ru cat **285**	0.01	rt	2.5	30	(—)	73:27	489
Ru cat **286**	0.01	rt	2.5	12	(—)	78:22	489
Gr I	0.05	45°	4	69	(68)	79:21	124

Cy
3 eq

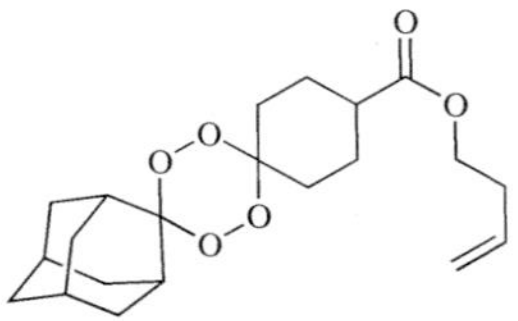

Ru cat **88** (0.05 eq)
DCM, 45°, 2 h

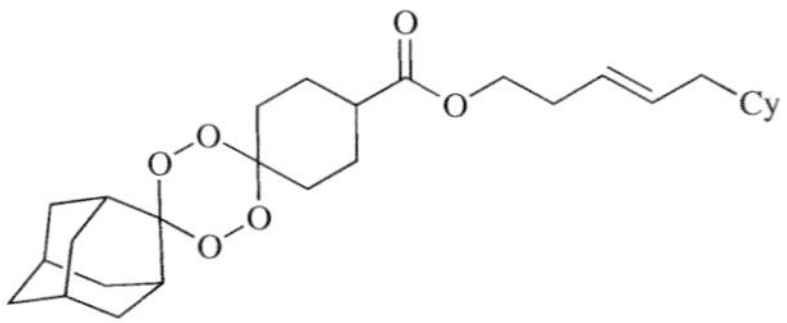

(87) (*E*)/(*Z*) = 80:20

193

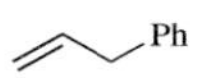 Ph; 3 eq	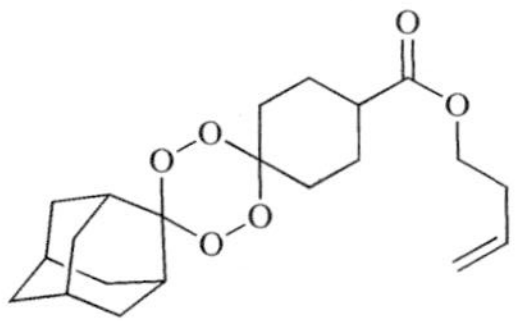 O; O; O; O; O; O	Ru cat **88** (0.05 eq) DCM, 45°, 2 h	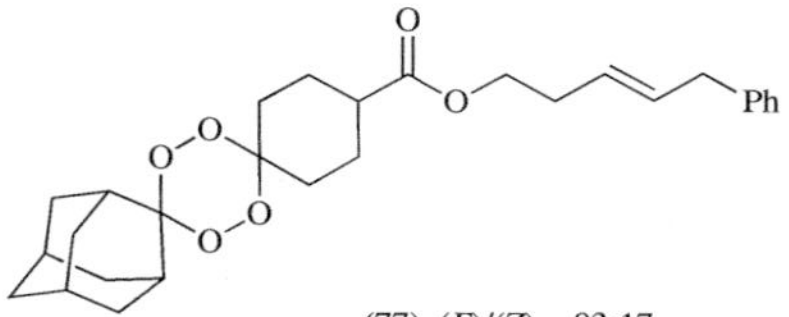 O; O; O; O; O; O; Ph; (77) (*E*)/(*Z*) = 83:17	193
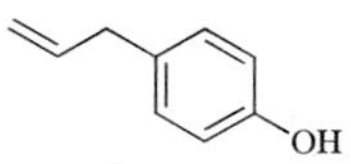 OH; 5 eq	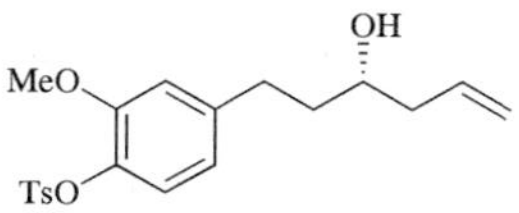 OH; MeO; TsO	Gr II (0.05 eq), DCM, 40°, 2 h	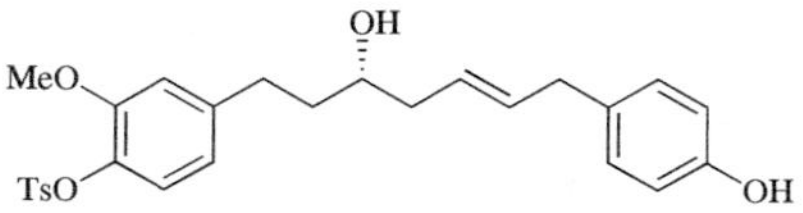 OH; MeO; TsO; OH; (81) (*E*)/(*Z*) = 90:10	490
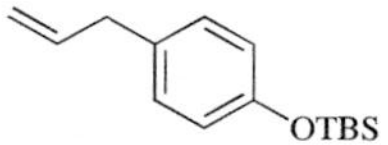 OTBS; 2 eq	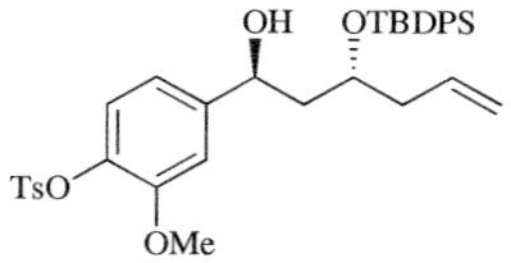 OH; OTBDPS; TsO; OMe	Gr II (0.05 eq), DCM, rt, 6 h	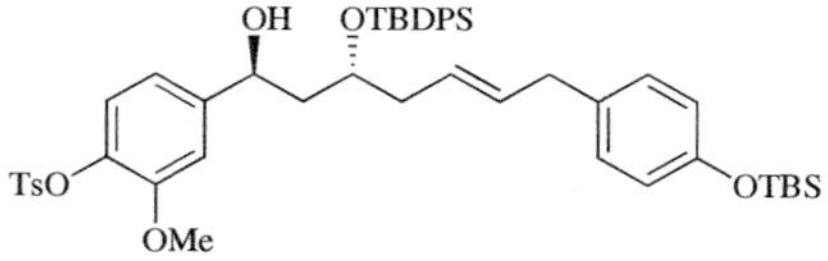 OH; OTBDPS; TsO; OMe; OTBS; (65) (*E*)/(*Z*) = 86:14	451
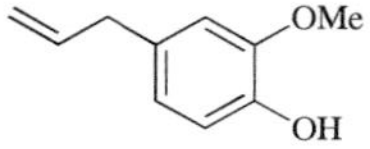 OMe; OH; 3 eq	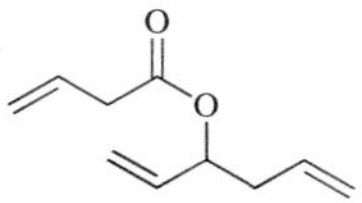 O; O	Gr II (0.025 eq), DCM, 40°, 12 h	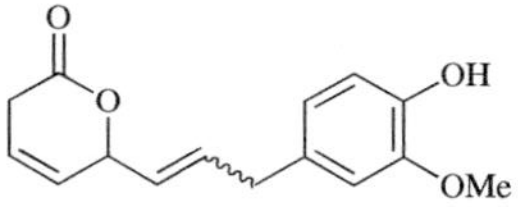 O; O; OH; OMe; (70)	394

TABLE 1. CROSS-METATHESIS OF UNSATURATED COMPOUNDS WITH FUNCTIONAL GROUPS IN HOMOALLYLIC AND MORE DISTANT POSITIONS *(Continued)*

A. ETHYLENE AND COMPOUNDS WITH A TERMINAL DOUBLE BOND *(Continued)*

Terminal Alkene	Cross Partner	Conditions	Product(s) and Yield(s) (%)	Refs.

Please refer to the charts preceding the tables for the catalyst structures.

C_9

Terminal Alkene	Cross Partner	Conditions	Product(s) and Yield(s) (%)	Refs.
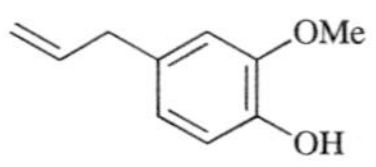 OMe, OH	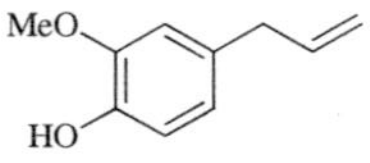 MeO, HO	Ru cat **40** (*x* eq), solvent	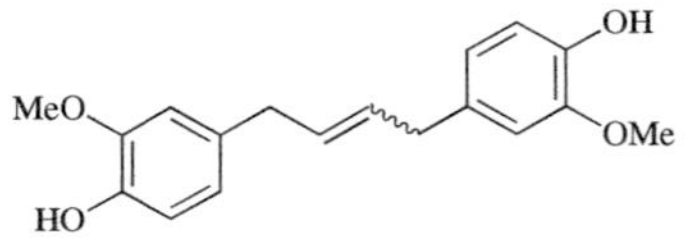 MeO, HO, OH, OMe	382
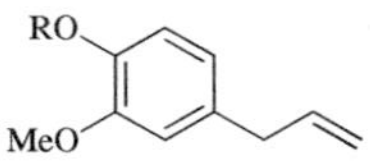 RO, MeO	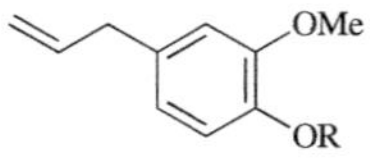 OMe, OR	Ru cat **360** (0.005 eq), toluene, 40°, 15 h	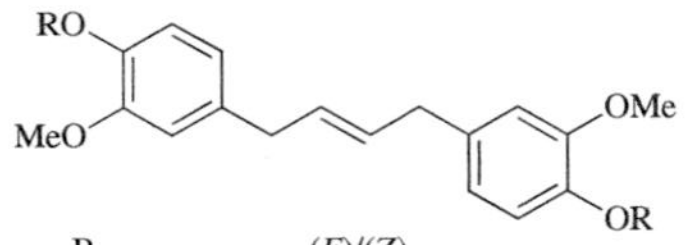 RO, MeO, OMe, OR	491

x	Solvent	Temp (°)	Time (h)	Conv. (%)[a]
0.05	toluene	rt	1	48
0.005	toluene	80	1	60
2 x 0.005	toluene	80	1 + 1	93
0.005	DCM	80	1	0

R		(*E*)/(*Z*)
Ac	(85)	83:17
Me	(96)	84:16

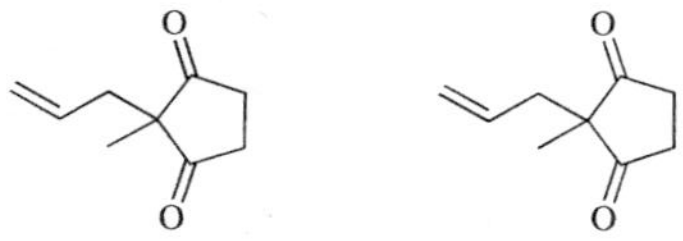

Reactant	Reactant	Conditions	Product(s) and Yield(s) (%)	Refs.
		Gr II (0.05 eq), toluene, 110°, 24 h	(79) (*E*)/(*Z*) = 95:5	492
		Gr II (0.5 eq), DCM, reflux, 12 h	(55)	493
		Gr II (0.5 eq), DCM, reflux, 12 h	(60)	493
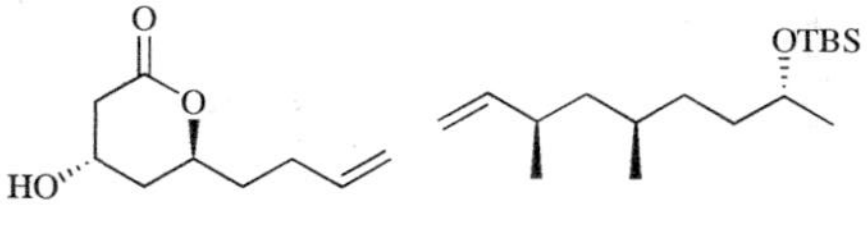 1.03 eq		Gr II (0.5 eq), DCM, reflux, 12 h	(60)	493

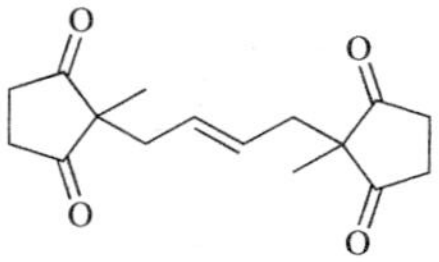

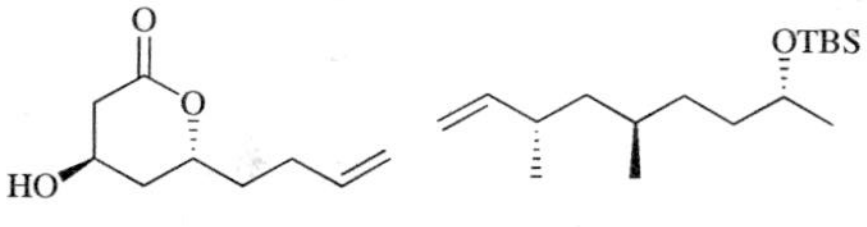

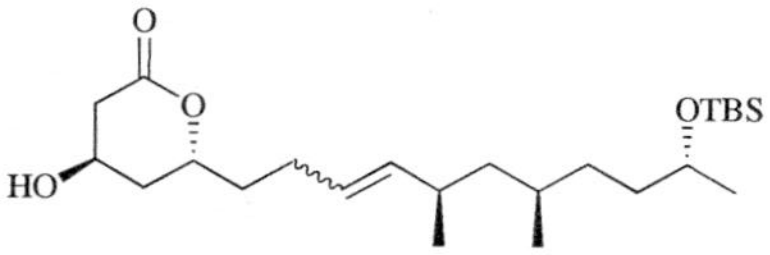

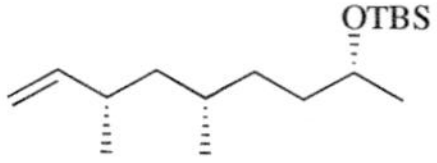

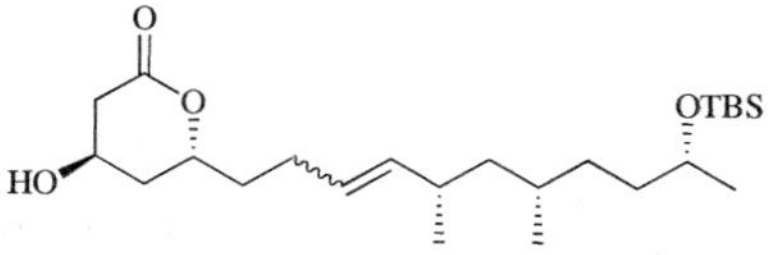

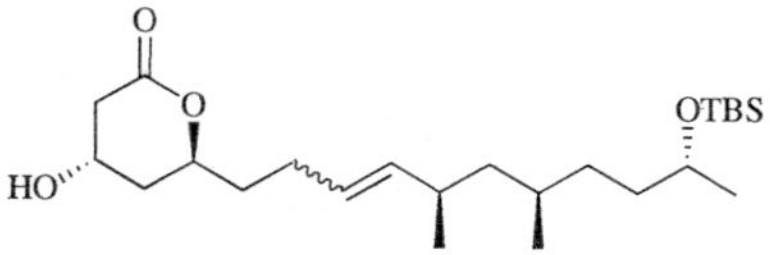

TABLE 1. CROSS-METATHESIS OF UNSATURATED COMPOUNDS WITH FUNCTIONAL GROUPS IN HOMOALLYLIC AND MORE DISTANT POSITIONS (*Continued*)

A. ETHYLENE AND COMPOUNDS WITH A TERMINAL DOUBLE BOND (*Continued*)

Terminal Alkene	Cross Partner	Conditions	Product(s) and Yield(s) (%)	Refs.
Please refer to the charts preceding the tables for the catalyst structures.				
C_9				
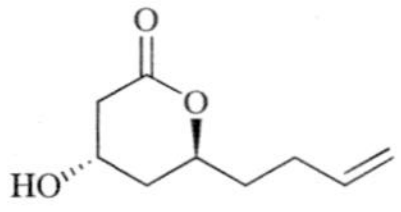 O O HO 1 eq	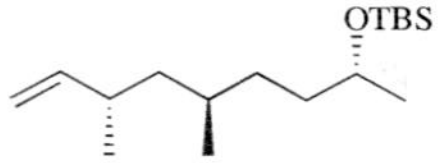 OTBS	Gr II (0.5 eq), DCM, reflux, 12 h	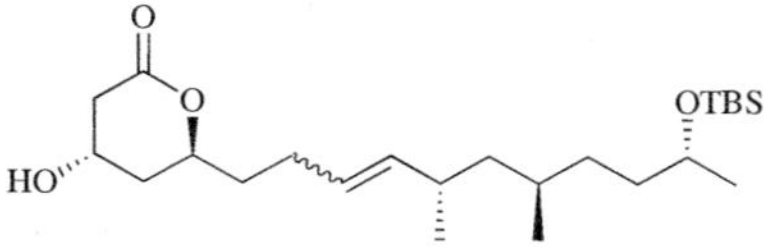 O O HO OTBS (55)	493
1 eq	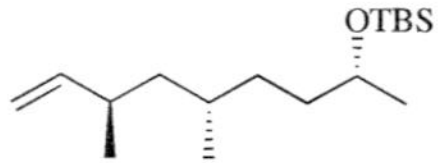 OTBS	Gr II (0.5 eq), DCM, reflux, 12 h	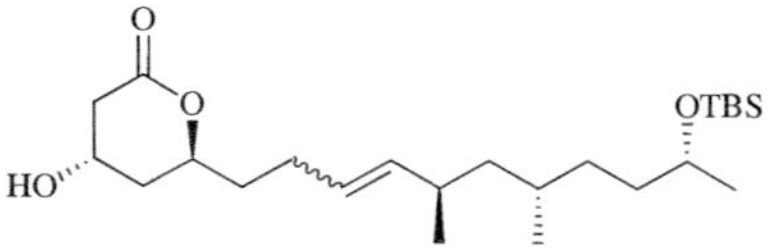 O O HO OTBS (54)	493
1 eq	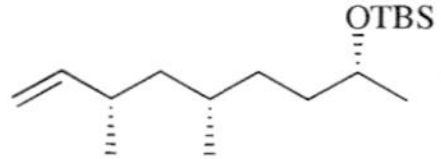 OTBS	Gr II (0.5 eq), DCM, reflux, 12 h	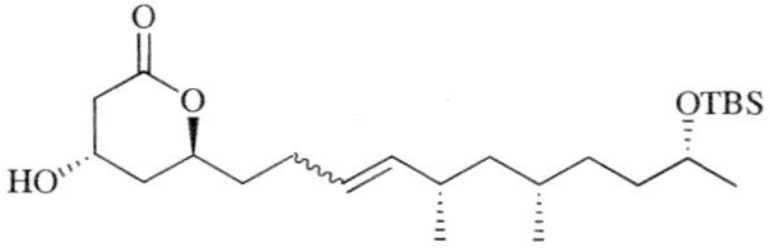 O O HO OTBS (58)	493
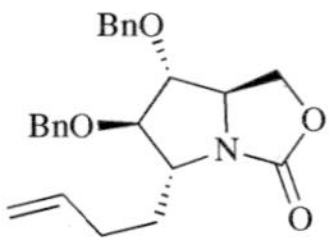 BnO BnO N O O	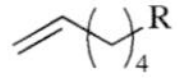 R 4 2 eq	Gr II (0.1 eq), DCM, 110° (MW), 12 h	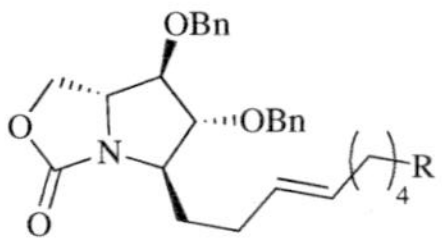 OBn OBn O N O R 4 (E)/(Z) = 80:20	494
			R: 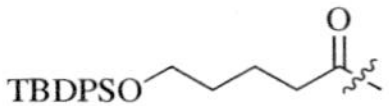 TBDPSO O (62)	
			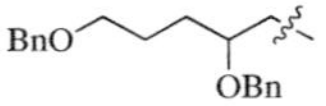 BnO OBn (60)	

Reactant(s)	Conditions	Product(s) and Yield(s) (%)	Refs.
2 eq	Gr II (0.05 eq), DCM, reflux, 4 h	(58)	495, 496
	Hov II (0.05 eq), DCM, rt, overnight	(51)	484
	Gr I (0.05 eq), DCM, 45°, 4 h	(93) $(E)/(Z) = 80{:}20$	124

A. ETHYLENE AND COMPOUNDS WITH A TERMINAL DOUBLE BOND (Continued)

Terminal Alkene	Cross Partner	Conditions	Product(s) and Yield(s) (%)	Refs.
Please refer to the charts preceding the tables for the catalyst structures.				
C_9				
3 eq	BocHN, CO_2Me	Gr II (0.2 eq), DCM, reflux, 6 h	(64)	429
		Gr II (x eq), DCM, rt	R / x / Time (h): Bn, 0.1, 24, (78); Ac, 0.2, 48, (81)	423
		Gr I (0.1 eq), DCM, reflux, 8 h	(83) $(E)/(Z) = 50:50$	497

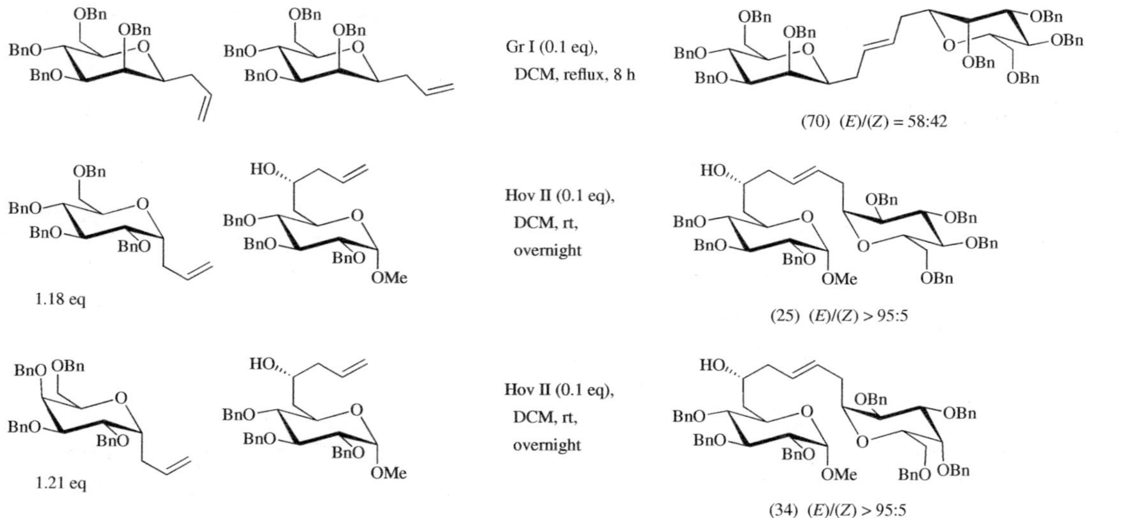
Gr I (0.1 eq),
DCM, reflux, 8 h
(70) (E)/(Z) = 58:42
1.18 eq
Hov II (0.1 eq),
DCM, rt,
overnight
(25) (E)/(Z) > 95:5
1.21 eq
Hov II (0.1 eq),
DCM, rt,
overnight
(34) (E)/(Z) > 95:5

497

498

498

TABLE 1. CROSS-METATHESIS OF UNSATURATED COMPOUNDS WITH FUNCTIONAL GROUPS IN HOMOALLYLIC AND MORE DISTANT POSITIONS (*Continued*)

A. ETHYLENE AND COMPOUNDS WITH A TERMINAL DOUBLE BOND (*Continued*)

Terminal Alkene	Cross Partner	Conditions	Product(s) and Yield(s) (%)	Refs.
Please refer to the charts preceding the tables for the catalyst structures.				
C_9				
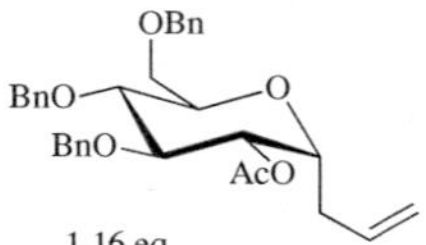 OBn, BnO, BnO, AcO; 1.16 eq	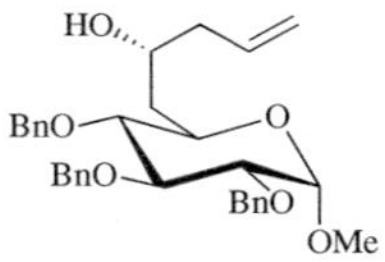 HO, BnO, BnO, BnO, OMe	Hov II (0.1 eq), DCM, rt, overnight	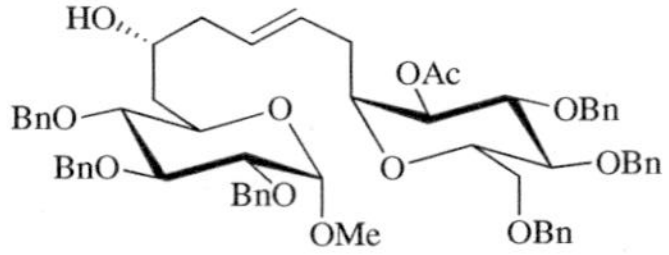 HO, BnO, BnO, BnO, OMe, OAc, OBn, OBn, OBn; (40) $(E)/(Z) > 95:5$	498
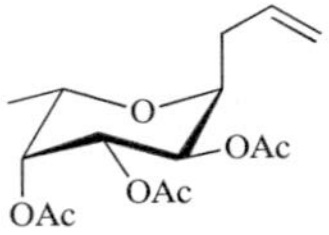 OAc, OAc, OAc	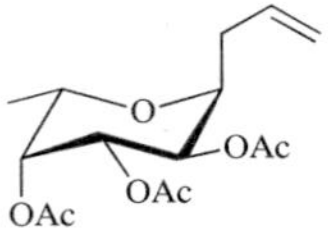 OAc, OAc, OAc	Gr I (0.1 eq), DCM, reflux, 8 h	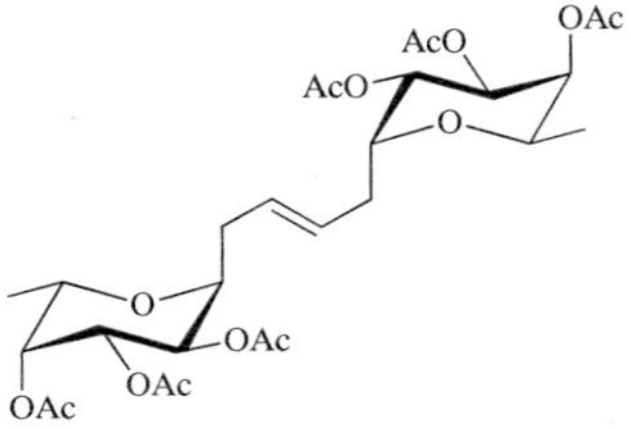 AcO, AcO, OAc, OAc, OAc, OAc; (95) $(E)/(Z) = 75:25$	497
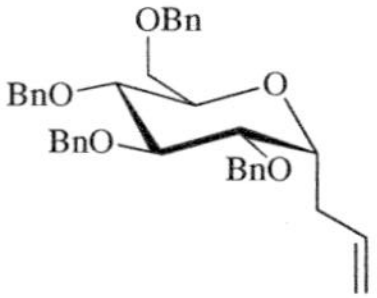 OBn, BnO, BnO, BnO	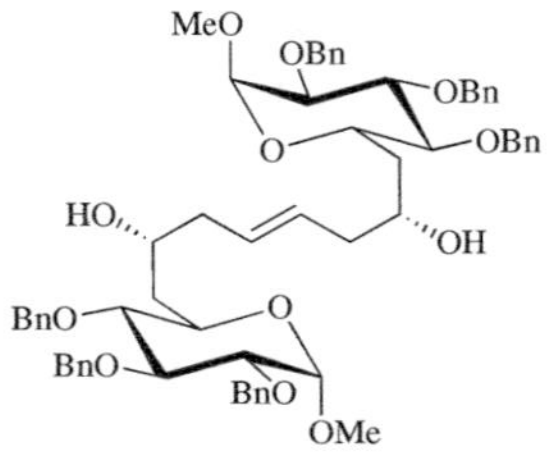 MeO, OBn, OBn, OBn, OH, HO, BnO, BnO, BnO, OMe; 2 eq	Hov II (0.1 eq), DCM, rt	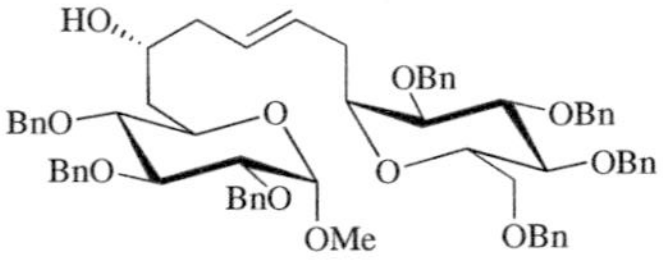 HO, BnO, BnO, BnO, OMe, OBn, OBn, OBn, OBn; (55) $(E)/(Z) = >95:5$	498

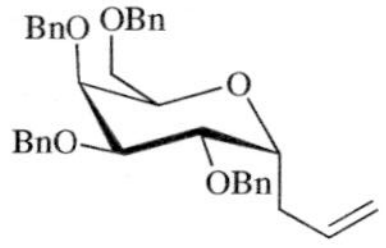	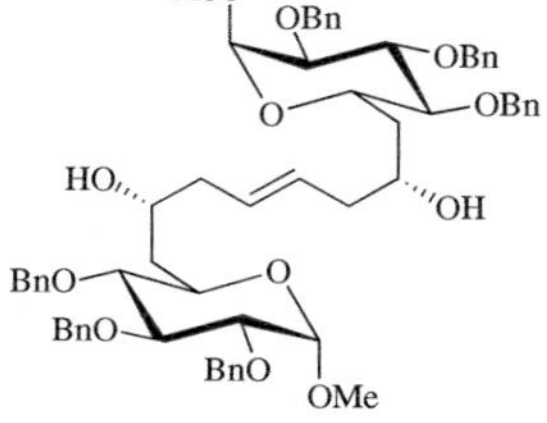 2 eq	Hov II (0.1 eq), DCM, rt	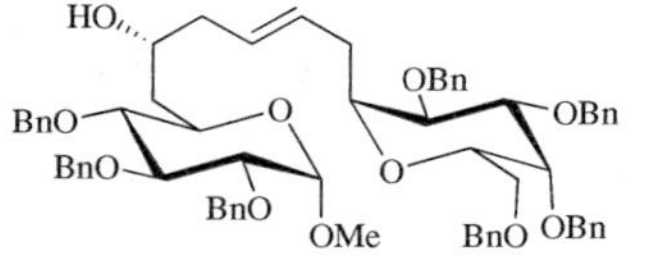 (64) $(E)/(Z) > 95:5$	498
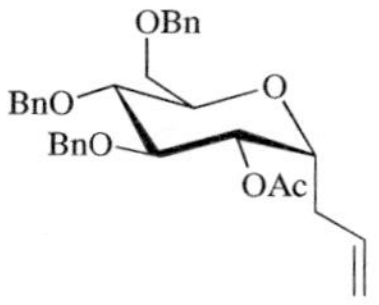	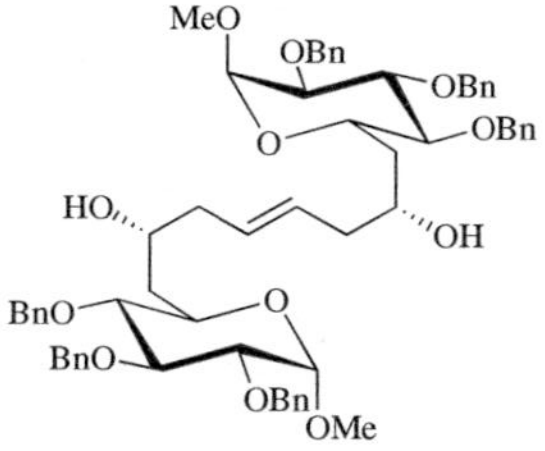 2 eq	Hov II (0.1 eq), DCM, rt	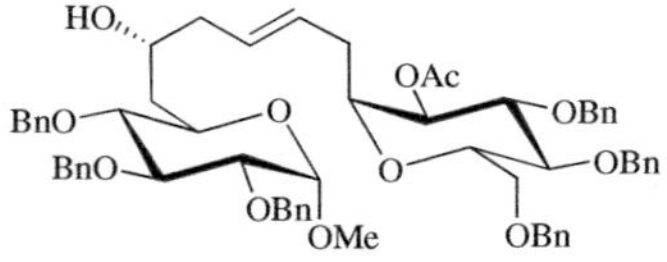 (68) $(E)/(Z) > 95:5$	498

TABLE 1. CROSS-METATHESIS OF UNSATURATED COMPOUNDS WITH FUNCTIONAL GROUPS IN HOMOALLYLIC AND MORE DISTANT POSITIONS (*Continued*)

A. ETHYLENE AND COMPOUNDS WITH A TERMINAL DOUBLE BOND (*Continued*)

Terminal Alkene	Cross Partner	Conditions	Product(s) and Yield(s) (%)	Refs.
Please refer to the charts preceding the tables for the catalyst structures.				
C_9				
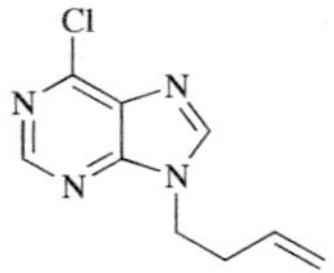 3 eq	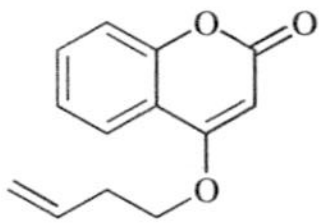	Gr II (0.1 + 0.04 eq), DCM, 100° (MW), 2 + 1 h	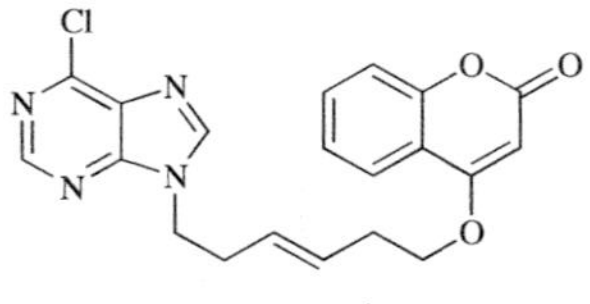 (76) (*E*)/(*Z*) = 76:24 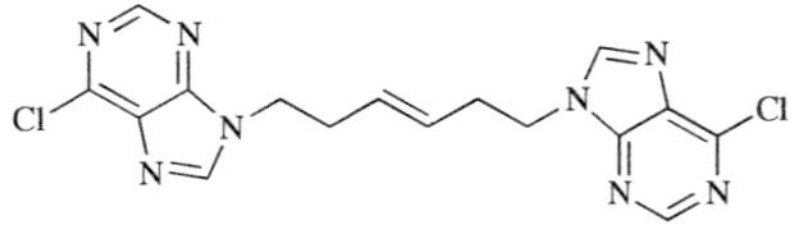 + (51) (*E*)/(*Z*) = 75:25	499
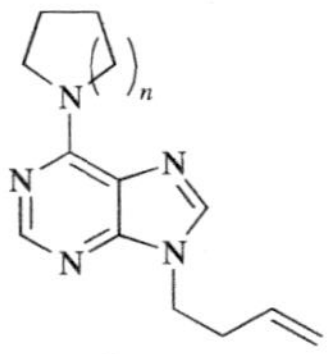 3 eq	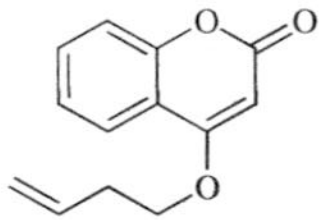	Gr II (0.1 + 0.04 eq), DCM, 100° (MW), 2 + 1 h	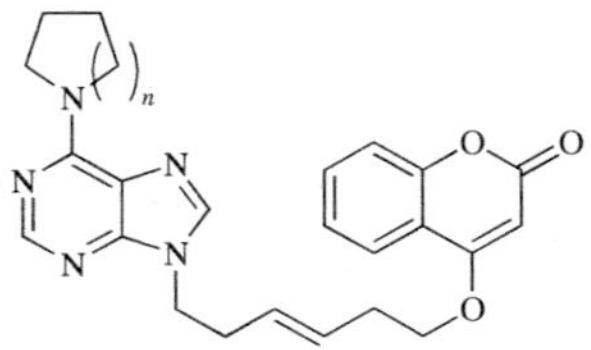 *n* = 1 (89) (*E*)/(*Z*) 75:25; *n* = 2 (86) (*E*)/(*Z*) 76:24	499

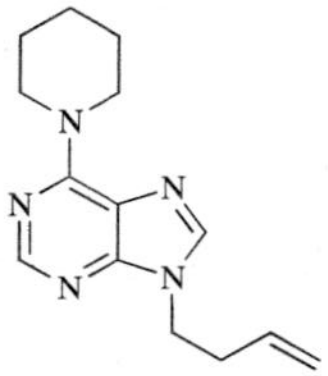

3 eq

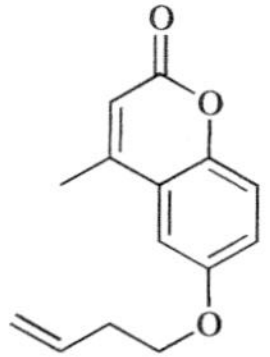

Gr II (0.1 + 0.04 eq), DCM, 100° (MW), 2 + 1 h

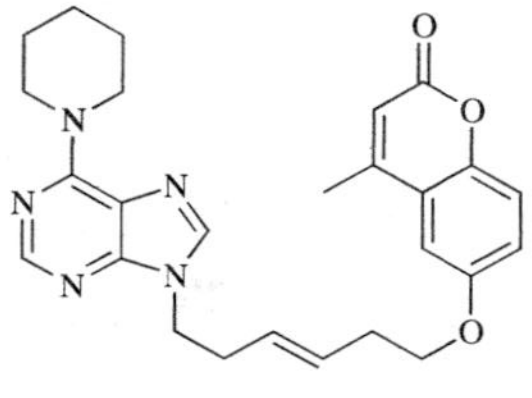

(76) (*E*)/(*Z*) = 76:24 | 499

3 eq

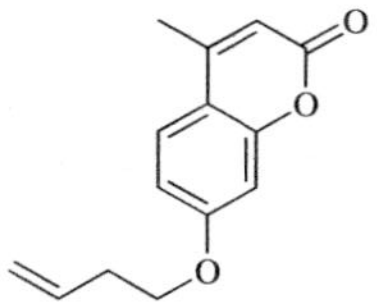

Gr II (0.1 + 0.04 eq), DCM, 100° (MW), 2 + 1 h

(74) (*E*)/(*Z*) = 75:25 | 499

C_{10}

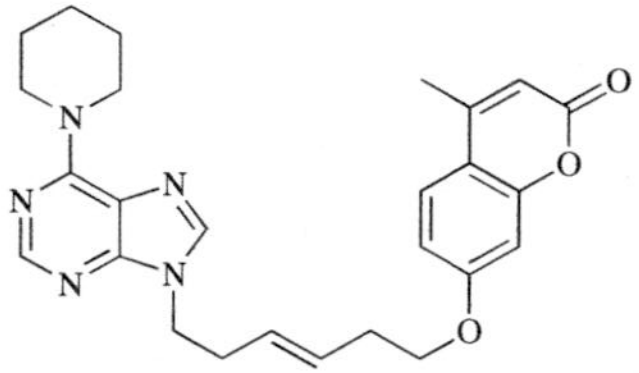

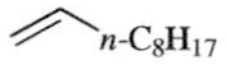

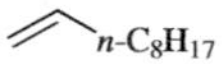

n-C_8H_{17} | n-C_8H_{17}

Catalyst (0.0001 eq), 60°, 4 h

n-C_7H_{15} n-C_8H_{17}

Catalyst	Conv. (%)[c]
Gr I	8
Gr II	52
Ru cat **5**	77

377

TABLE 1. CROSS-METATHESIS OF UNSATURATED COMPOUNDS WITH FUNCTIONAL GROUPS IN HOMOALLYLIC AND MORE DISTANT POSITIONS (*Continued*)

A. ETHYLENE AND COMPOUNDS WITH A TERMINAL DOUBLE BOND (*Continued*)

Terminal Alkene	Cross Partner	Conditions	Product(s) and Yield(s) (%)	Refs.
Please refer to the charts preceding the tables for the catalyst structures.				
C_{10}				
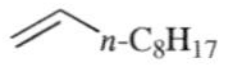 n-C_8H_{17} 6.25 eq	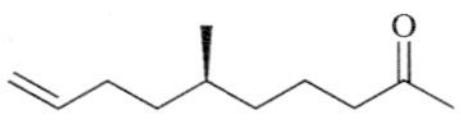	Gr I (0.03 eq), DCM, reflux, 4 h	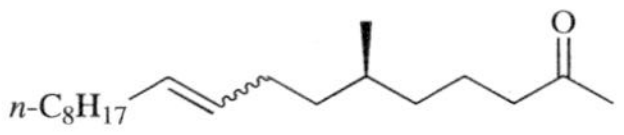 n-C_8H_{17} (83)	500
6.25 eq	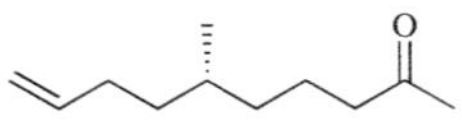	Gr I (0.03 eq), DCM, reflux, 4 h	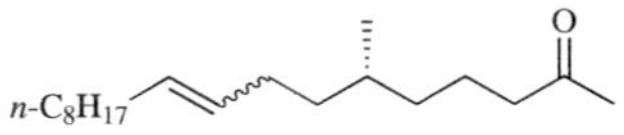 n-C_8H_{17} (80)	500
4 eq	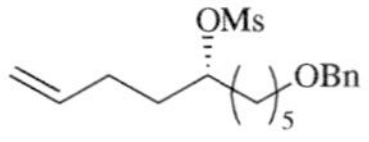 OMs, OBn, 5	Gr II (0.03 eq), DCM, reflux, 10 h	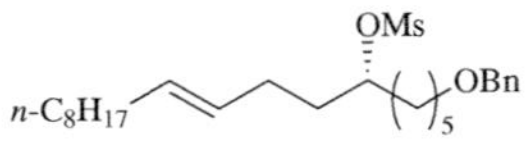 n-C_8H_{17}, OMs, OBn, 5 (75) $(E)/(Z)$ = 94:6	501
5 eq	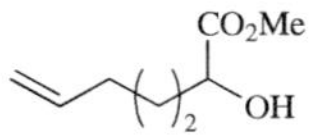 CO_2Me, OH, 2	Ru cat **35** (0.02 eq), DCM, reflux, 4–12 h	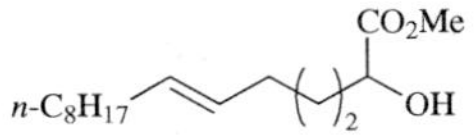 n-C_8H_{17}, CO_2Me, OH, 2 (76) $(E)/(Z)$ = 52:48	393
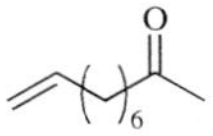 O, 6	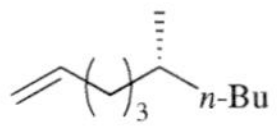 3, n-Bu 6.2 eq	Gr I (0.067 eq), DCM, reflux, 5 h	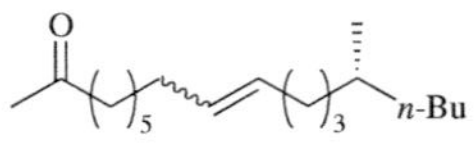 O, 5, 3, n-Bu (78)	500
	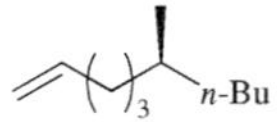 3, n-Bu 4.4 eq	Gr I (0.063 eq), DCM, reflux, 5 h	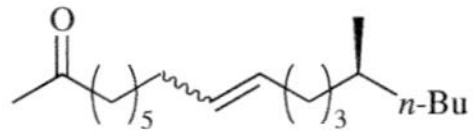 O, 5, 3, n-Bu (81)	500

Reactant 1	Reactant 2	Conditions	Product(s) and Yield(s) (%)	Refs.
OR (C8 chain), terminal alkene	OR (C8 chain), terminal alkene	Gr I (0.05 eq), DCM, 45°, 12 h	RO–(C7)–CH=CH–(C8)–OR; R = Ac (95) $(E)/(Z)$ 82:18; R = Bz (94) $(E)/(Z)$ 79:21	124
OAc (C8 chain), terminal alkene, x eq	NHBoc glycinate ester (C8 chain), terminal alkene	Gr I (0.05 eq), DCM, 45°, 12 h	AcO–(C7)–CH=CH–(C8)–O–C(=O)–CH2–NHBoc; x = 1 (45); x = 2 (69); x = 4 (80)	124
OBz (C8 chain), terminal alkene	Ph–CH2–CH=CH–CH2–Ph, 2 eq	Catalyst (0.01 eq), DCM, rt	Ph–CH2–CH=CH–(C8)–OBz	489

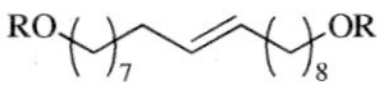

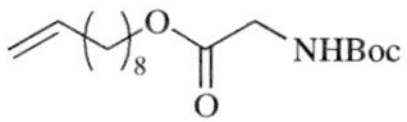

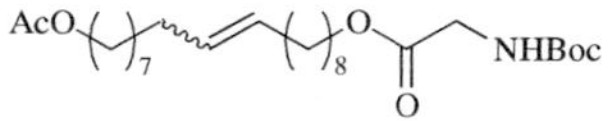

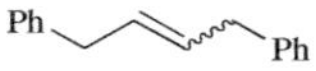

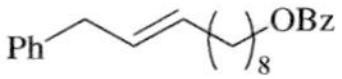

Catalyst	Time (h)	Conv (%)[a]	$(E)/(Z)$
Ind II	2.5	92	52:48
Ru cat **282**	2.5	59	38:63
Ru cat **284**	20	88	44:56
Ru cat **283**	20	63	52:48
Ru cat **285**	2.5	97	71:29
Ru cat **286**	2.5	92	58:42

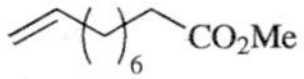

Reactant 1	Reactant 2	Conditions	Product(s) and Yield(s) (%)	Refs.
CO_2Me (C6 chain), terminal alkene, 3 eq	tetraoxane (cycloheptane spiro) cyclohexanecarboxylate but-3-enyl ester	Ru cat **88** (0.03 eq), DCM, 45°, 2 h	tetraoxane ester cross-metathesis product, –(C7)–CO_2Me; (55) (E)/(Z) = 93:7	193

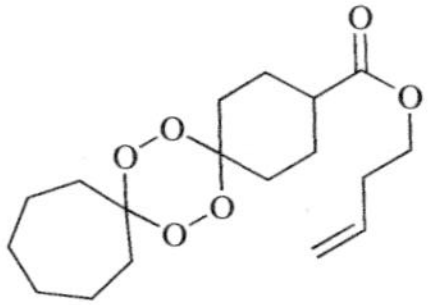

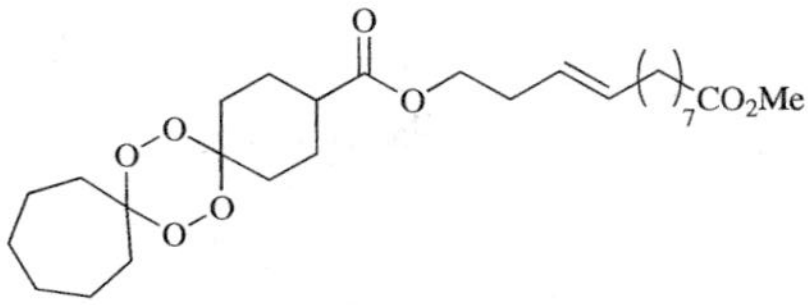

TABLE 1. CROSS-METATHESIS OF UNSATURATED COMPOUNDS WITH FUNCTIONAL GROUPS IN HOMOALLYLIC AND MORE DISTANT POSITIONS (*Continued*)

A. ETHYLENE AND COMPOUNDS WITH A TERMINAL DOUBLE BOND (*Continued*)

Terminal Alkene	Cross Partner	Conditions	Product(s) and Yield(s) (%)	Refs.
Please refer to the charts preceding the tables for the catalyst structures.				
C_{10}				
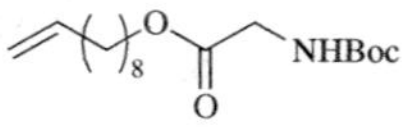 NHBoc	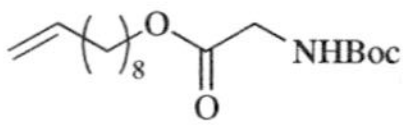 NHBoc	Gr I (0.003 eq), DCM, rt, 12 h	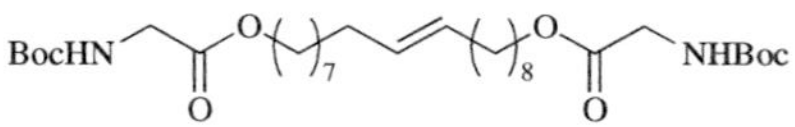 BocHN … NHBoc (93) (E)/(Z) = 79:21	124
	 AcO … OAc, *x* eq	Gr I (0.05 eq), DCM, 45°, 12 h	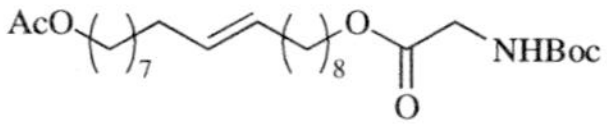 AcO … NHBoc *x* 0.5 (28) (E)/(Z) — *x* 1 (47) (E)/(Z) — *x* 2 (72) (E)/(Z) 77:23	124
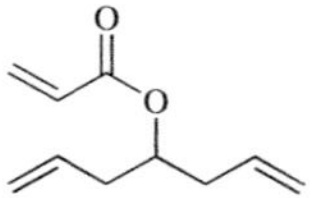	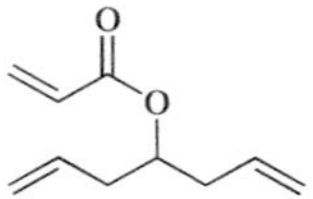	Gr II (0.025 eq), DCM, reflux, 16 h	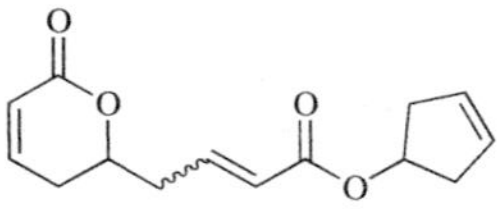 (53)	394

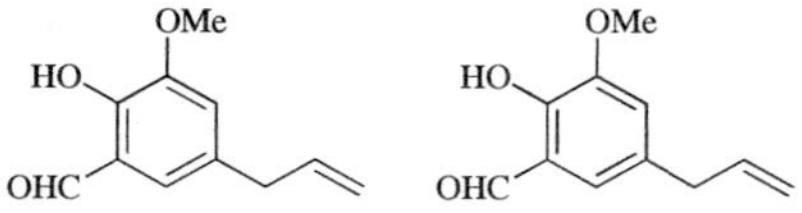

OMe HO OHC OMe HO OHC

Catalyst (0.01 eq), toluene, 80°, 20 h

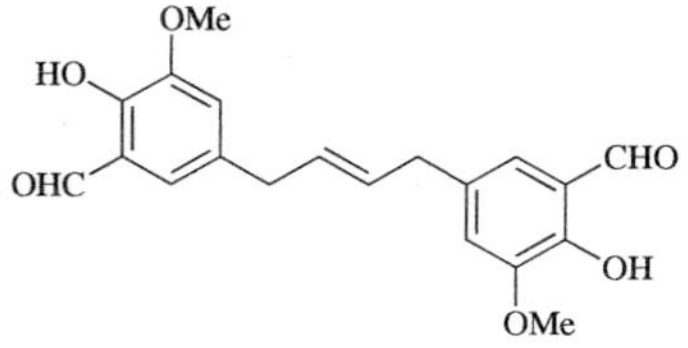

OMe HO OHC CHO OH OMe

Catalyst		(E)/(Z)
Ru cat **359**	(55)	87:13
Ru cat **360**	(66)	88:12
Ru cat **361**	(58)	90:10
Ru cat **362**	(64)	84:16

491

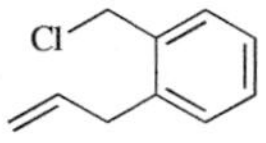

Cl

2 eq

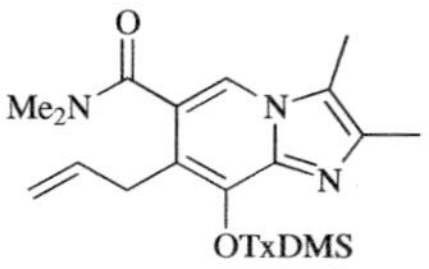

O Me_2N N N OTxDMS

Hov II (0.08 eq), DCM, 45°, 20 h

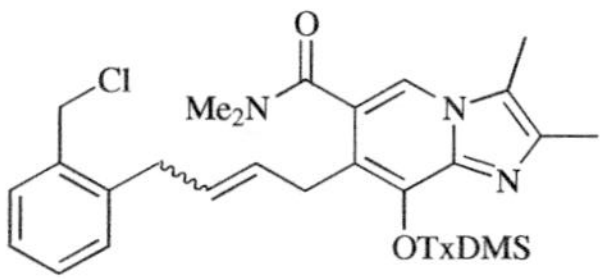

Cl O Me_2N N N OTxDMS (72)

502

TABLE 1. CROSS-METATHESIS OF UNSATURATED COMPOUNDS WITH FUNCTIONAL GROUPS IN HOMOALLYLIC AND MORE DISTANT POSITIONS *(Continued)*

A. ETHYLENE AND COMPOUNDS WITH A TERMINAL DOUBLE BOND *(Continued)*

Terminal Alkene	Cross Partner	Conditions	Product(s) and Yield(s) (%)	Refs.

Please refer to the charts preceding the tables for the catalyst structures.

C_{10}

CCl$_3$ | R, n, x eq | Gr II (y eq), solvent | R, n, O, N, CCl$_3$ | 460

R	n	x	y	Solvent	Conc (M)	Temp (°)	Time (min)	
Me	9	7	0.05	DCM	0.1	80 (MW)	40	(86)
MeO_2C	8	2	0.05	DCM	0.1	80 (MW)	40	(63)
MeO_2C	8	1	0.05	DCM	0.1	80 (MW)	20	(57)
Et	8	1	0.05	DCM	0.1	40	150	(58)
HO	9	2	0.05	DCM	0.1	80 (MW)	40	(40)
HO	9	5	0.1	THF	0.5	40 (MW)	40	(56)
TBDPSO	9	2	0.1	DCM	0.5	80 (MW)	40	(40)
GlcNHC(O)	8	2	0.1	THF	0.3	80 (MW)	40	(48)

BnO, OBn, N, O | OTPS, 5, 3, 2 eq | Gr II (0.1 eq), DCM | BnO, OBn, N, O, OTPS, 5, 3; (E)/(Z) = 80:20 | 503

Temp (°)	Time (h)	
reflux	24	(61)
100 (MW)	1	(60)

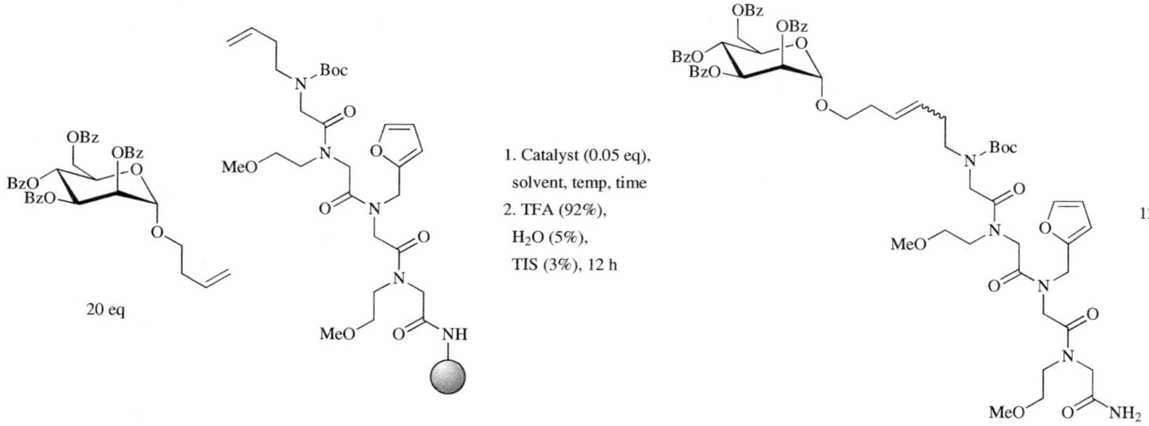

20 eq

1. Catalyst (0.05 eq), solvent, temp, time
2. TFA (92%), H_2O (5%), TIS (3%), 12 h

12

Catalyst	Solvent	Temp (°)	Time	Conv (%)[e]
Gr I	DCM	40	8 h	18
Gr II	DCM	40	8 h	22
Hov II	DCM	40	8 h	47
Gr I	DCB	40 (MW)	3 min	<5
Gr II	DCB	40 (MW)	3 min	<5
Hov II	DCB	40 (MW)	3 min	25

TABLE 1. CROSS-METATHESIS OF UNSATURATED COMPOUNDS WITH FUNCTIONAL GROUPS IN HOMOALLYLIC AND MORE DISTANT POSITIONS *(Continued)*

A. ETHYLENE AND COMPOUNDS WITH A TERMINAL DOUBLE BOND *(Continued)*

Terminal Alkene	Cross Partner	Conditions	Product(s) and Yield(s) (%)	Refs.
Please refer to the charts preceding the tables for the catalyst structures.				
C_{10-11}				
20 eq		1. Hov II (0.05 eq), DCM, 40°, 8 h 2. TFA (92%), H_2O (5%), TIS (3%), 1–2 h		12

n	R	Conv. (%)[e]
1	Bz	47
1	Ac	15
2	Bz	39
2	Ac	30

20 eq

1. Hov II (0.05 eq), DCM, 40°, 8 h
2. TFA (92%), H_2O (5%), TIS (3%), 1–2 h

12

n	Conv. (%)[e]
1	40
2	46

TABLE 1. CROSS-METATHESIS OF UNSATURATED COMPOUNDS WITH FUNCTIONAL GROUPS IN HOMOALLYLIC AND MORE DISTANT POSITIONS (*Continued*)

A. ETHYLENE AND COMPOUNDS WITH A TERMINAL DOUBLE BOND (*Continued*)

Terminal Alkene	Cross Partner	Conditions	Product(s) and Yield(s) (%)	Refs.

Please refer to the charts preceding the tables for the catalyst structures.

C_{10-11}

Terminal Alkene	Cross Partner	Conditions	Product(s) and Yield(s) (%)	Refs.
BzO, BzO, OBz, OBz, O, n *x* eq	Boc, N, OMe, O, N, O, N, O, N, O, MeO, N, O, O, HN	1. Hov II (0.05 eq), DCM, 40°, 8 h 2. TFA (92%), H_2O (5%), TIS (3%), 1–2 h	BzO, BzO, OBz, OBz, O, n, Boc, N, OMe, O, N, O, N, O, N, O, MeO, N, O, O, NH_2	12

n	*x*	Resin	Conv. (%)[c]
1	20	Rink Amide (0.4 mmol/g)	46
1	20	Rink Amide (0.2 mmol/g)	37
1	40	Rink Amide (0.2 mmol/g)	50
1	56	Rink Amide (0.2 mmol/g)	49
1	80	Rink Amide (0.2 mmol/g)	51
2	20	Rink Amide (0.4 mmol/g)	44
2	20	Rink Amide (0.2 mmol/g)	40

20 eq

1. Hov II (0.05 eq), DCM, 40°, 8 h
2. TFA (92%), H_2O (5%), TIS (3%), 1–2 h

12

n	Conv. (%)[e]
1	49
2	32

TABLE 1. CROSS-METATHESIS OF UNSATURATED COMPOUNDS WITH FUNCTIONAL GROUPS IN HOMOALLYLIC AND MORE DISTANT POSITIONS (*Continued*)

A. ETHYLENE AND COMPOUNDS WITH A TERMINAL DOUBLE BOND (*Continued*)

Terminal Alkene	Cross Partner	Conditions	Product(s) and Yield(s) (%)	Refs.
Please refer to the charts preceding the tables for the catalyst structures.				
C_{10-11}				
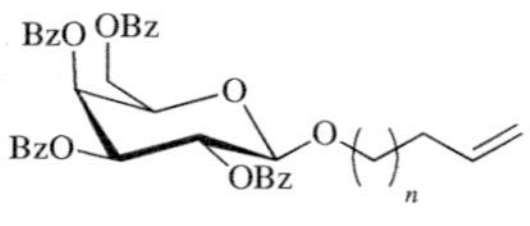 BzO, OBz, BzO, OBz, O, O, *n* 20 eq	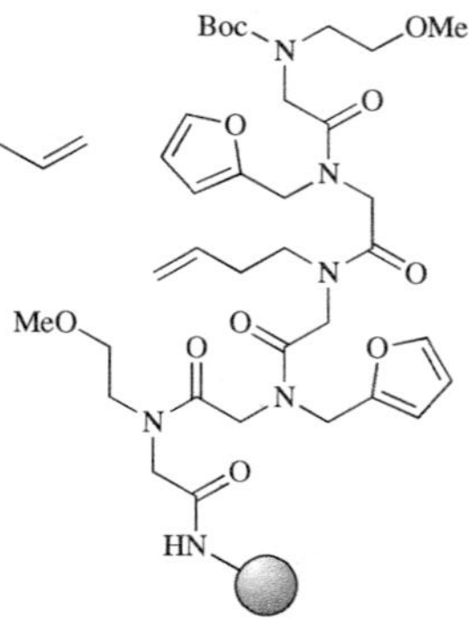 Boc, N, OMe, O, O, N, O, N, O, MeO, O, O, N, N, O, O, HN	1. Hov II (0.05 eq), DCM, 40°, 8 h 2. TFA (92%), H_2O (5%), TIS (3%), 1–2 h	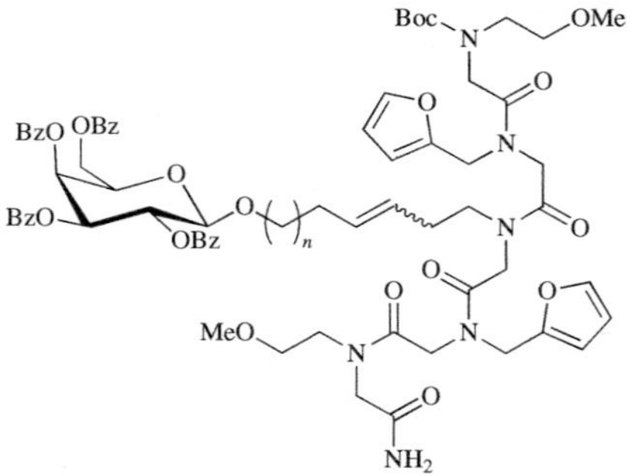 BzO, OBz, BzO, OBz, O, O, *n*, Boc, N, OMe, O, O, N, N, O, O, O, MeO, N, N, O, O, NH_2 *n* / Conv (%)[e] 1 / 47 2 / 49	12

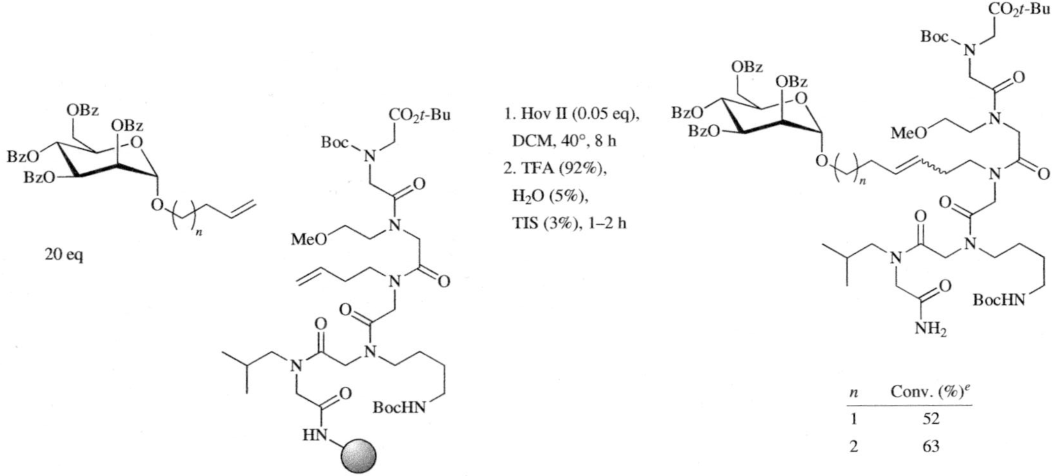

OBz OBz BzO BzO O O n

20 eq

CO2t-Bu Boc N O MeO N N O O HN O N O N BocHN

1. Hov II (0.05 eq), DCM, 40°, 8 h
2. TFA (92%), H_2O (5%), TIS (3%), 1–2 h

OBz OBz BzO BzO O O n CO2t-Bu Boc N O MeO N N O O O N N O BocHN NH2

n	Conv. (%)[e]
1	52
2	63

12

TABLE 1. CROSS-METATHESIS OF UNSATURATED COMPOUNDS WITH FUNCTIONAL GROUPS IN HOMOALLYLIC AND MORE DISTANT POSITIONS (*Continued*)

A. ETHYLENE AND COMPOUNDS WITH A TERMINAL DOUBLE BOND (*Continued*)

Terminal Alkene	Cross Partner	Conditions	Product(s) and Yield(s) (%)	Refs.
Please refer to the charts preceding the tables for the catalyst structures.				
C_{10-11}				
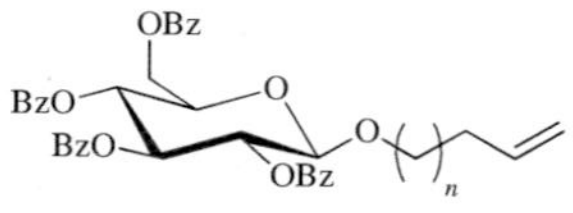 BzO, BzO, OBz, OBz, O, n 20 eq	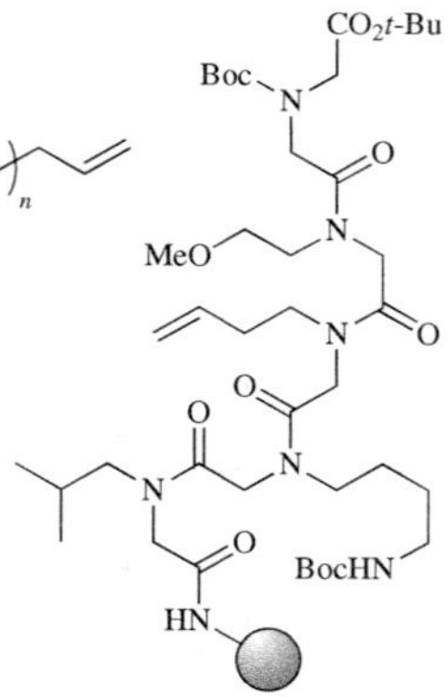 Boc, N, CO_2t-Bu, O, N, MeO, N, O, O, N, O, N, BocHN, O, HN	1. Hov II (0.05 eq), DCM, 40°, 8 h 2. TFA (92%), H_2O (5%), TIS (3%), 1–2 h	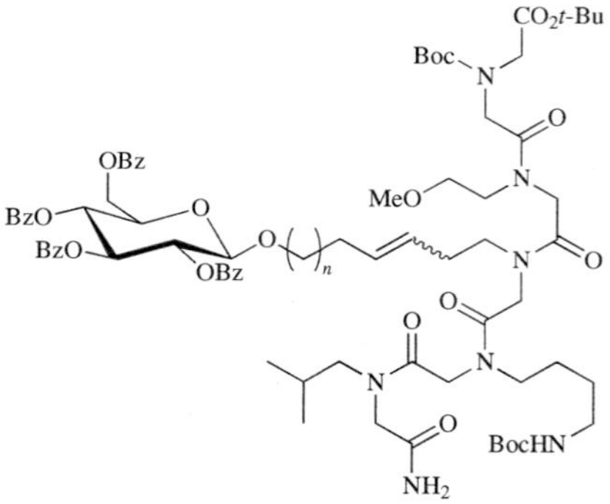 OBz, BzO, BzO, OBz, O, n, MeO, Boc, N, CO_2t-Bu, N, O, N, O, O, N, O, N, BocHN, O, NH_2 n — Conv. (%)[e] 1 — 46 2 — 53	12

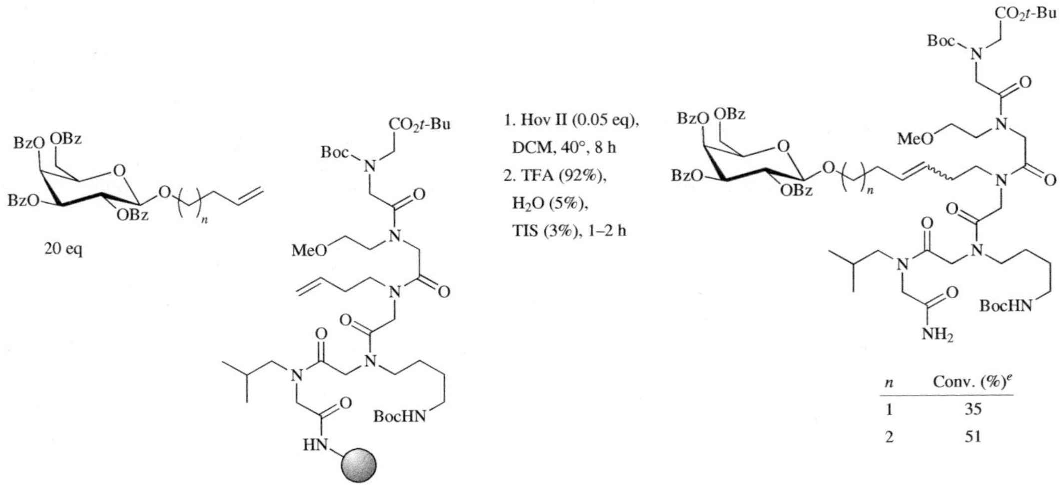

20 eq

1. Hov II (0.05 eq), DCM, 40°, 8 h
2. TFA (92%), H_2O (5%), TIS (3%), 1–2 h

12

n	Conv. (%)[e]
1	35
2	51

TABLE 1. CROSS-METATHESIS OF UNSATURATED COMPOUNDS WITH FUNCTIONAL GROUPS IN HOMOALLYLIC AND MORE DISTANT POSITIONS (*Continued*)

A. ETHYLENE AND COMPOUNDS WITH A TERMINAL DOUBLE BOND (*Continued*)

Terminal Alkene	Cross Partner	Conditions	Product(s) and Yield(s) (%)	Refs.
Please refer to the charts preceding the tables for the catalyst structures.				
C_{11}				
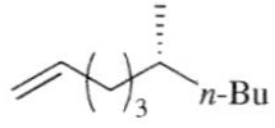 *n*-Bu, 3; 6.25 eq	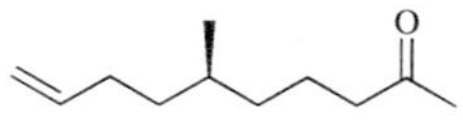 O	Gr I (0.03 eq), DCM, reflux, 4 h	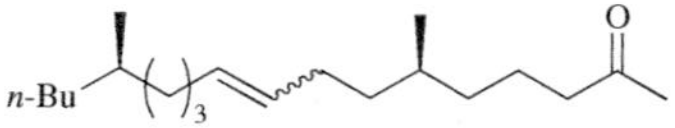 *n*-Bu, 3, O (86)	500
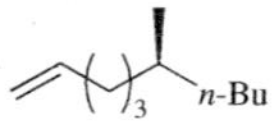 *n*-Bu, 3; 6.25 eq	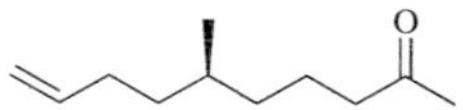 O	Gr I (0.03 eq), DCM, reflux, 4 h	 *n*-Bu, 3, O (92)	500
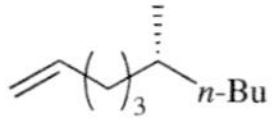 *n*-Bu, 3; 6.25 eq	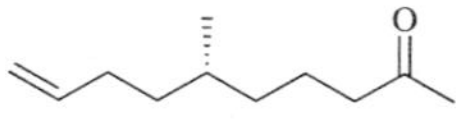 O	Gr I (0.03 eq), DCM, reflux, 4 h	 *n*-Bu, 3, O (76)	500
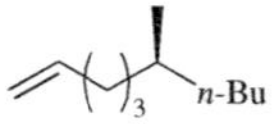 *n*-Bu, 3; 6.25 eq	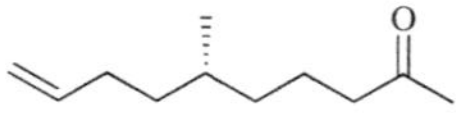 O	Gr I (0.03 eq), DCM, reflux, 4 h	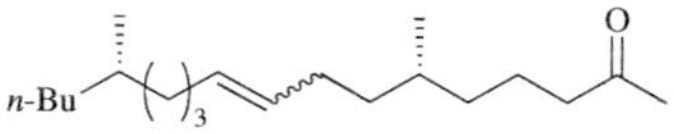 *n*-Bu, 3, O (73)	500
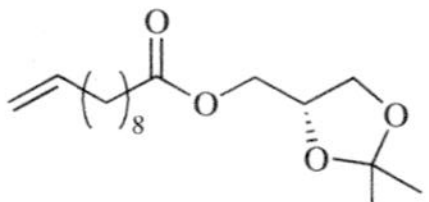 O, 8, O, O, O	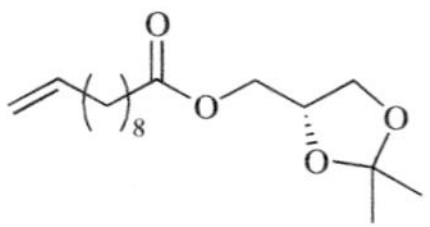 O, 8, O, O, O	Hov II (0.05 eq), DCM, rt, 24 h	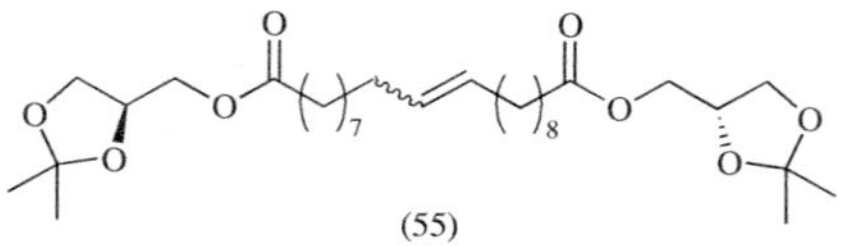 O, O, O, 7, 8, O, O, O (55)	473

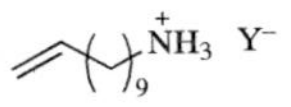

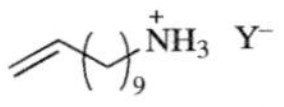

Y^-	Y^-	Hov II (0.05 eq), DCM	$2Y^-$	402

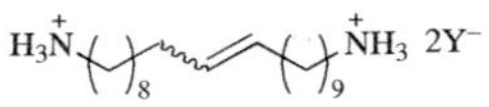

Y	Temp (°)	Time (h)	Conv. (%)[a]
TsO	40	24	92
Cl	40	24	81
TsO	100 (MW)	2	88
Cl	100 (MW)	2	82

		Gr I (0.05 eq), DCM, reflux	(62) $(E)/(Z) = 86:14$	447
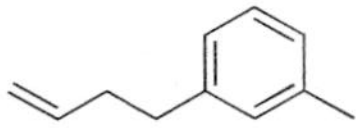	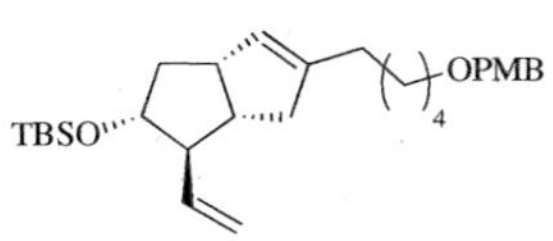	Gr II (0.06 eq), toluene, 80°, 8 h	(54) $(E)/(Z) = 88:12$	128
	2 eq	Catalyst (0.025 eq), DCM, 45°, 2 h		504

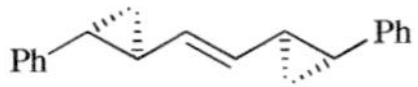

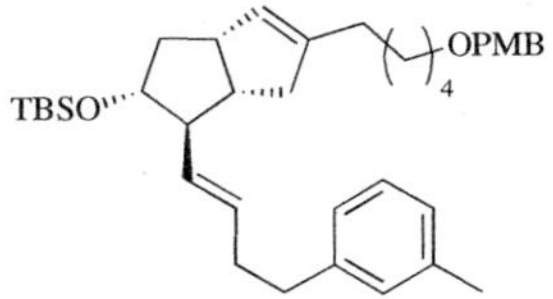

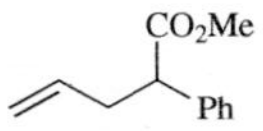

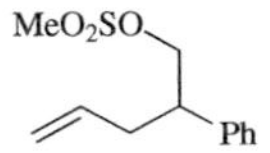

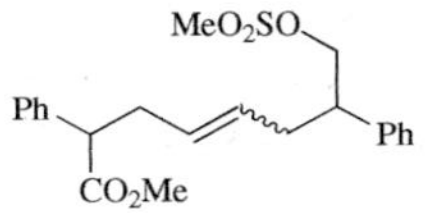

Catalyst	Conv. (%)[a]
Gr I	9
Gr II	44
Hov II	72
Ind II	28

TABLE 1. CROSS-METATHESIS OF UNSATURATED COMPOUNDS WITH FUNCTIONAL GROUPS IN HOMOALLYLIC AND MORE DISTANT POSITIONS (*Continued*)

A. ETHYLENE AND COMPOUNDS WITH A TERMINAL DOUBLE BOND (*Continued*)

Terminal Alkene	Cross Partner	Conditions	Product(s) and Yield(s) (%)	Refs.

Please refer to the charts preceding the tables for the catalyst structures.

C_{11}

Terminal Alkene	Cross Partner	Conditions	Product(s) and Yield(s) (%)	Refs.
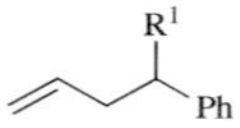 R[1], Ph	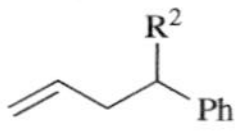 R[2], Ph, *x* eq	Hov II (0.025 eq), DCM, 45°, 2 h	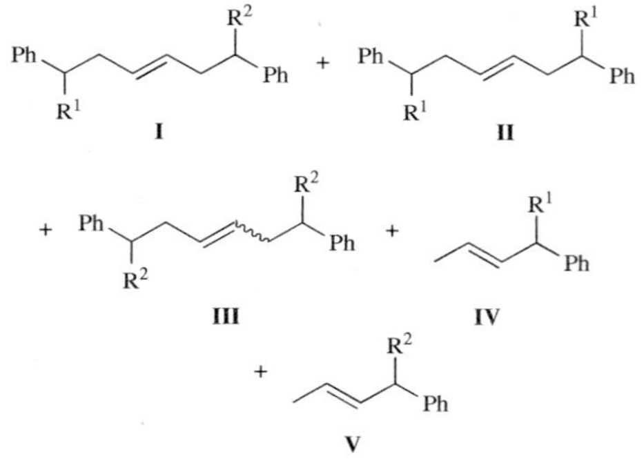 I + II + III + IV + V	504

R^1	R^2	*x*	I	(*E*)/(*Z*) I	II	(*E*)/(*Z*) II	III	IV	(*E*)/(*Z*) IV	V	(*E*)/(*Z*) V
MeO_2C	$MsOCH_2$	0.5	(28)	80:20	(35)	82:18	(—)	(—)	—	(—)	—
BzO	$HOCH_2$	1.2	(36)	80:20	(—)	—	(—)	(12)	68:32	(20)	71:29
MeO_2C	$BrCH_2$	2	(40)	—	(—)	—	(25)	(—)	—	(—)	—
$MsOCH_2$	$HOCH_2$	2	(—)	—	(—)	—	(—)	(30)	67:33	(21)	89:11
$HOCH_2$	Bn_2NOC	0.5	(16)	—	(3)	—	(—)	(24)	86:14	(41)	76:24
MeO_2C	Bn_2NOC	0.5	(17)	60:40	(10)	68:32	(—)	(—)	—	(10)	67:33
$MsOCH_2$	Bn_2NOC	0.5	(3)	—	(—)	—	(—)	(—)	—	(—)	—
$MsOCH_2$	Me(MeO)NOC	2	(—)	—	(—)	—	(—)	(—)	—	(—)	—

Terminal Alkene	Cross Partner	Conditions	Product(s) and Yield(s) (%)	Refs.
 CHO (n = 8)	 CHO (n = 8)	Catalyst (*x* eq), toluene, 40°	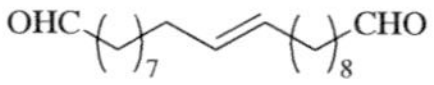 OHC–(7)–CH=CH–(8)–CHO (*E*) only	1088

Catalyst	*x*	Time (h)	Conv. (%)[c]
Gr I	0.01	48	81
Gr I	0.005	48	74
Ru cat **79**	0.0165	16	72
Ru cat **79**	0.0066	4	61

Ru cat **79**	0.0033	48	70
Hov II	0.025	23	84
Ru cat **144**	0.0125	13	72
Ru cat **144**	0.025	22	85

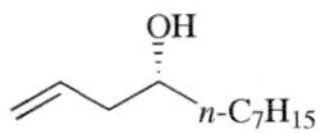

OH / n-C$_7$H$_{15}$ OH / n-C$_7$H$_{15}$ 1. Ac_2O, pyridine 2. Ru cat **35** (0.007 eq), DCM, 50° (MW), 1 h n-C$_7$H$_{15}$ OAc OAc n-C$_7$H$_{15}$ (98) $(E)/(Z)$ = 77:23   505

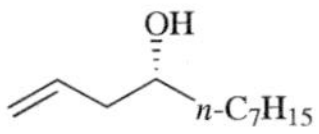

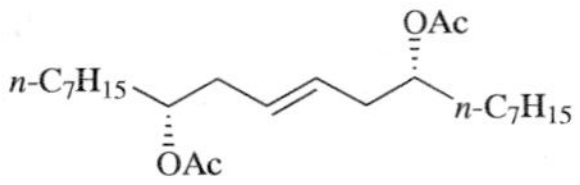

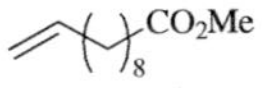

CO_2Me (8) CO_2Me (8)   Catalyst (x eq), solvent MeO_2C (8) CO_2Me (7)

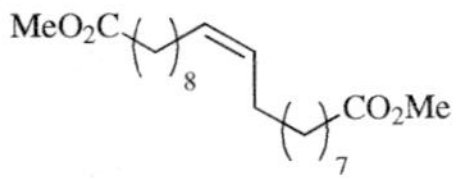

Catalyst	x	Solvent	Temp (°)	Time (h)	Conv. (%)[a]		$(Z)/(E)$	
Ru cat **217**	0.001	THF	35	12	16	(—)	90:10	488
Ru cat **218**	0.02	THF	35	6	67	(—)	81:19	488
Ru cat **219**	0.001	THF	35	12	3	(—)	>95:5	488
Ru cat **220**	0.001	THF	35	12	8.4	(—)	>95:5	488
Ru cat **217**	0.001	THF	35	12	—	(13)	90:10	488
Ru cat **215**	0.001	THF	35	12	—	(85)	91:9	488
Ru cat **216**	0.001	THF	35	12	—	(94)	92:8	488
Ru cat **217**	0.02	THF	35	5.5	95	(>95)	73:27	84
Ru cat **217**	0.04	THF	70	4	78	(—)	87:13	84
Ru cat **217**	0.02	MeCN	70	2.5	7	(—)	95:5	84
Ru cat **217**	0.02	MeCN	25	3	19	(—)	94:6	84
Ru cat **217**	0.02	MeOH	25	3	49	(—)	88:12	84
Ru cat **217**	0.02	EtOH	25	3	50	(—)	89:11	84
Ru cat **217**	0.02	benzene	25	3	13	(—)	>95:5	84
Ru cat **217**	0.02	Et_2O	25	3	50	(—)	93:7	84
Ru cat **217**	0.02	DMF	25	3	44	(—)	92:8	84
Ru cat **217**	0.02	DCM	25	3	35	(—)	93:7	84
Ru cat **217**	0.02	diglyme	25	3	31	(—)	95:5	84
Gr I	0.003	DCM	25	12	90	(—)	83:17	124

TABLE 1. CROSS-METATHESIS OF UNSATURATED COMPOUNDS WITH FUNCTIONAL GROUPS IN HOMOALLYLIC AND MORE DISTANT POSITIONS (*Continued*)

A. ETHYLENE AND COMPOUNDS WITH A TERMINAL DOUBLE BOND (*Continued*)

Terminal Alkene	Cross Partner	Conditions	Product(s) and Yield(s) (%)	Refs.
Please refer to the charts preceding the tables for the catalyst structures.				
C_{11}				
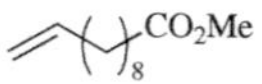 CO_2Me (3 eq)	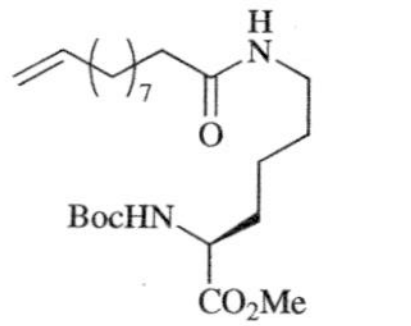 BocHN, H N, O, CO_2Me	Gr II (0.2 eq), DCM, 40°, 6 h	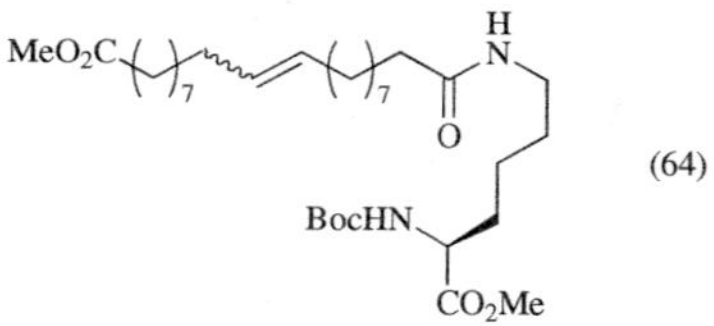 MeO_2C, H N, O, BocHN, CO_2Me (64)	429
3 eq	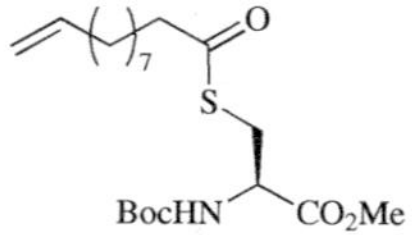 O, S, BocHN, CO_2Me	Gr II (0.2 eq), DCM, 40°, 6 h	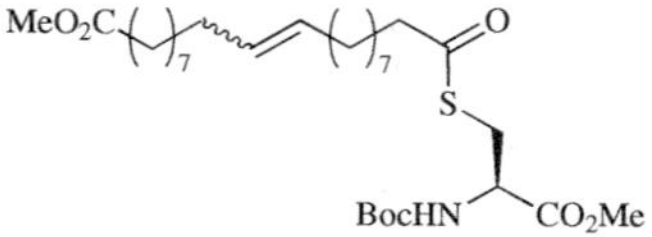 MeO_2C, O, S, BocHN, CO_2Me (61)	429
3 eq	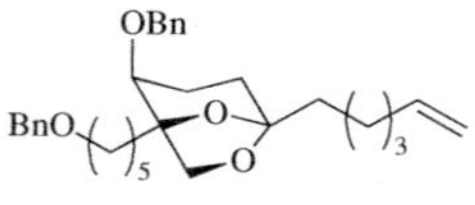 OBn, BnO, O, O	Gr II (0.15 eq), DCM, reflux, 24 h	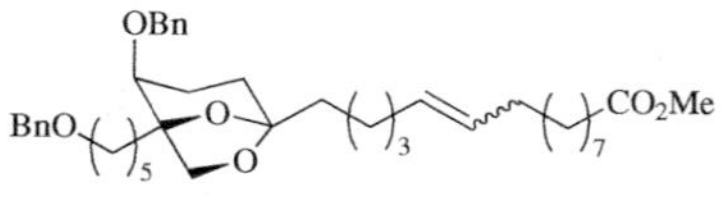 OBn, BnO, O, O, CO_2Me (80)	506

OTIPS

3 eq

OTIPS

n-$C_{12}H_{25}$

Ph_2Si

Gr II (0.2 eq),
DCM, 45°, 9 h

OTIPS

n-$C_{12}H_{25}$

Ph_2Si

OTIPS

(63) 507

TABLE 1. CROSS-METATHESIS OF UNSATURATED COMPOUNDS WITH FUNCTIONAL GROUPS IN HOMOALLYLIC AND MORE DISTANT POSITIONS (*Continued*)

A. ETHYLENE AND COMPOUNDS WITH A TERMINAL DOUBLE BOND (*Continued*)

Terminal Alkene	Cross Partner	Conditions	Product(s) and Yield(s) (%)	Refs.

Please refer to the charts preceding the tables for the catalyst structures.

C_{11}

Terminal Alkene	Cross Partner	Conditions	Product(s) and Yield(s) (%)	Refs.
BnO, BnO, BnO, N_3, O, O, HO	BnO, BnO, BnO, N_3, O, O, HO	Hov II (0.05 eq), DCM, 40°, 24 h	BnO, BnO, BnO, N_3, O, O, OH, OH, O, O, OBn, OBn, N_3, OBn (77) (*E*)/(*Z*) = 75:25	509
OH, OH, HO, HO, O, O, 3	OH, OH, HO, HO, O, O, 3	Gr I (0.1 eq), DCM/MeOH	OH, OH, HO, HO, O, O, 3, 3, O, OH, OH, O, OH, OH	497

Temp (°)	Time (h)		(*E*)/(*Z*)
rt	8	(67)	57:43
40	16	(42)	52:48

C_{12}					
	$n\text{-}C_{10}H_{21}$	$n\text{-}C_{10}H_{21}$	Ru cat **156** (0.00002 eq), 2-Cl-BQ (0.00004 eq), 60°, 6 h	$n\text{-}C_{10}H_{21}$ $n\text{-}C_{10}H_{21}$ (91)	510
	CO_2Bn 9 10 eq	$n\text{-}C_{10}H_{21}$ OTBS TBSO OTBS	Gr I (0.1 eq), DCM, 40°, 4 h	$n\text{-}C_{10}H_{21}$ OTBS TBSO BnO_2C 9 OTBS (73)	412
	O 3 OMe	O 3 OMe	Gr II (0.1 eq), paraffin, DCM, 40°, 12 h	O 2 O 3 OMe OMe (46) $(E)/(Z)$ = 95:5	511
C_{13}					
	$n\text{-}C_{11}H_{23}$ 10 eq	OH OH OMe	Gr II (0.05 eq), DCM, reflux, 12 h	$n\text{-}C_{11}H_{23}$ OH OH OMe (85)	512

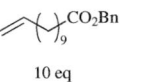

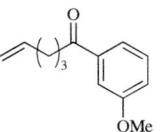

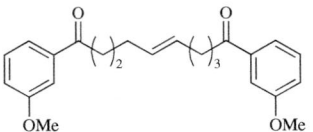

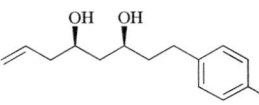

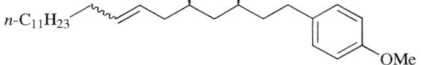

TABLE 1. CROSS-METATHESIS OF UNSATURATED COMPOUNDS WITH FUNCTIONAL GROUPS IN HOMOALLYLIC AND MORE DISTANT POSITIONS (*Continued*)

A. ETHYLENE AND COMPOUNDS WITH A TERMINAL DOUBLE BOND (*Continued*)

Terminal Alkene	Cross Partner	Conditions	Product(s) and Yield(s) (%)	Refs.
Please refer to the charts preceding the tables for the catalyst structures.				
C_{13}				
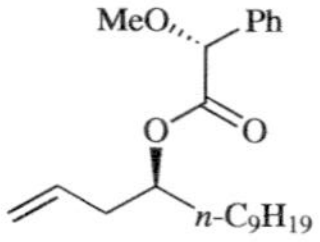 MeO, Ph, O, O, n-C_9H_{19}	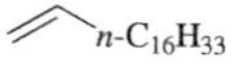 n-$C_{16}H_{33}$ 2 eq	Gr II (0.17 eq), DCM, reflux, 4 h	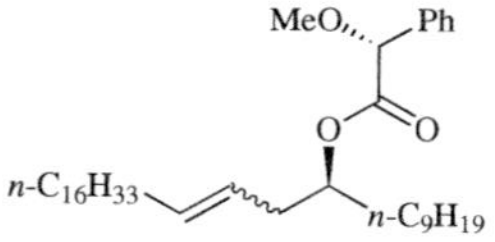 MeO, Ph, O, O, n-$C_{16}H_{33}$, n-C_9H_{19} (58)	513
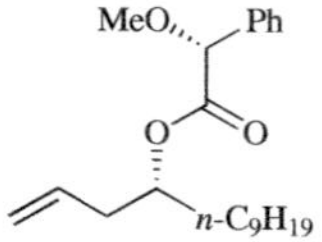 MeO, Ph, O, O, n-C_9H_{19}	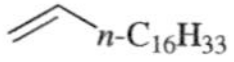 n-$C_{16}H_{33}$ 2 eq	Gr II (0.17 eq), DCM, reflux, 4 h	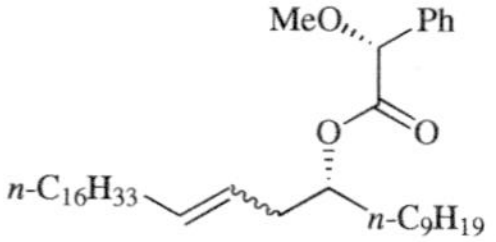 MeO, Ph, O, O, n-$C_{16}H_{33}$, n-C_9H_{19} (87)	513
C_{14}				
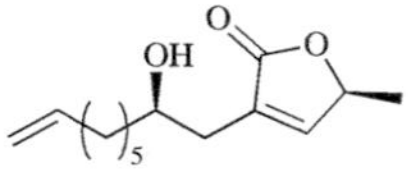 OH, O, O, 5 4 eq	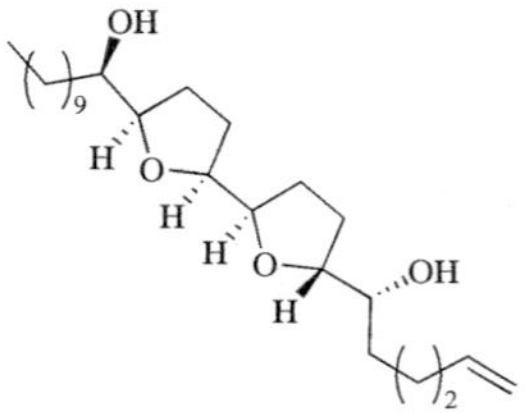 OH, 9, H, O, H, H, O, H, OH, 2	1. Gr I (0.1 eq), DCM, reflux, 6 h 2. DMSO, rt, 15 h	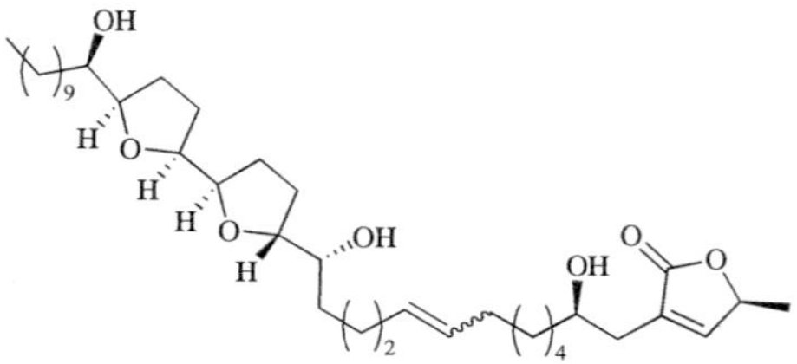 OH, 9, H, O, H, H, O, H, OH, 2, 4, OH, O, O (73)	514

Reactant	Conditions	Product(s) and Yield(s) (%)		Refs.
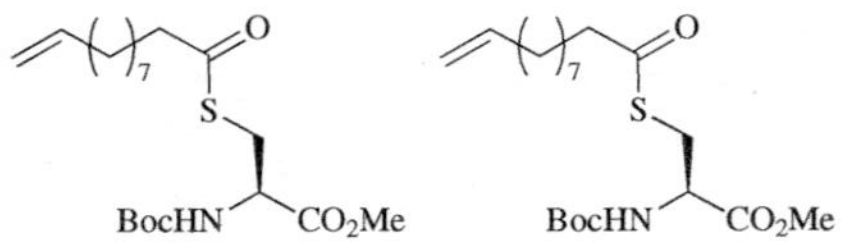	Gr II (0.2 eq), DCM, reflux	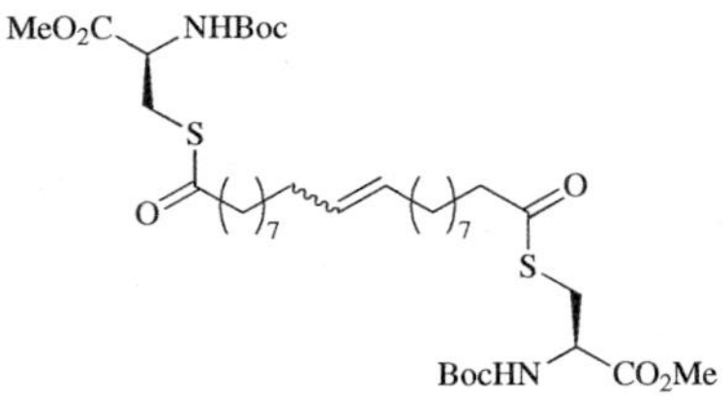	(36)	429
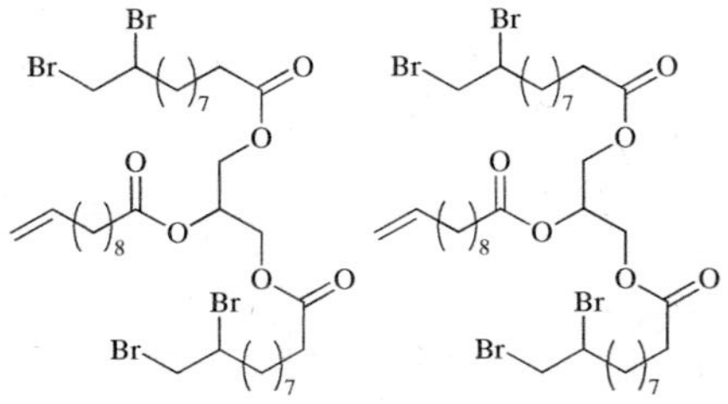	Gr II (0.05 eq), DCM, 40°, 8 h	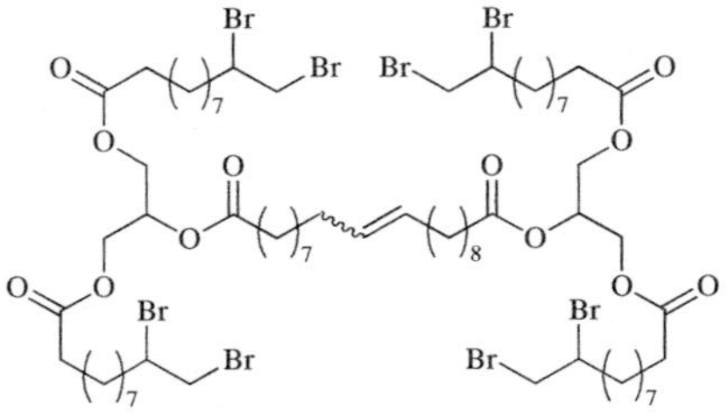	(67)	515
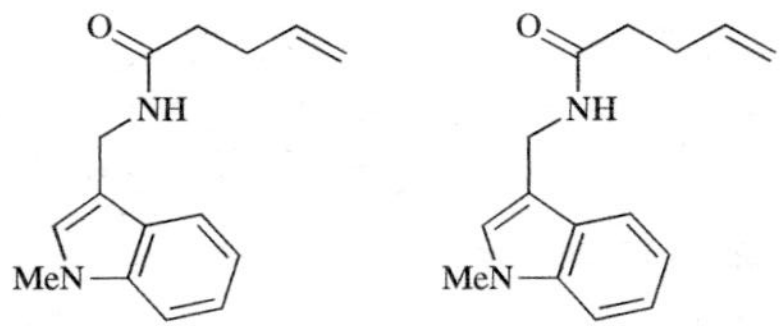	Gr II (0.08 eq), DCM, reflux, 7 h	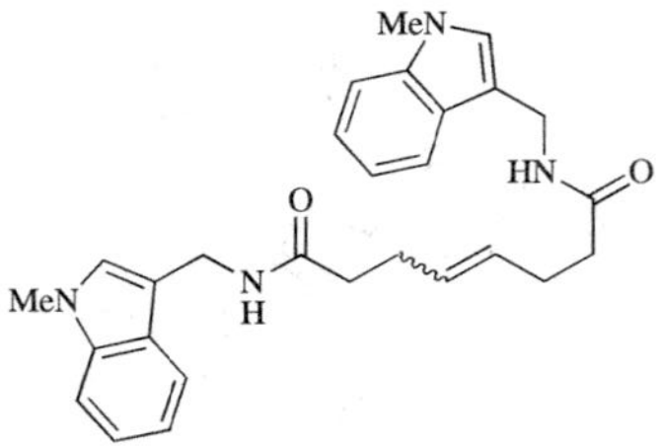	(20)	428

TABLE 1. CROSS-METATHESIS OF UNSATURATED COMPOUNDS WITH FUNCTIONAL GROUPS IN HOMOALLYLIC AND MORE DISTANT POSITIONS (*Continued*)

A. ETHYLENE AND COMPOUNDS WITH A TERMINAL DOUBLE BOND (*Continued*)

Terminal Alkene	Cross Partner	Conditions	Product(s) and Yield(s) (%)	Refs.
Please refer to the charts preceding the tables for the catalyst structures.				
C_{15}				
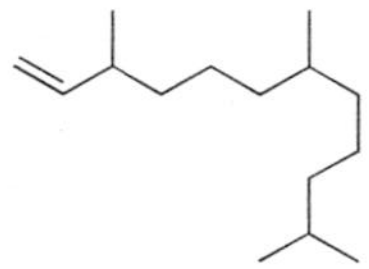 2.9 eq	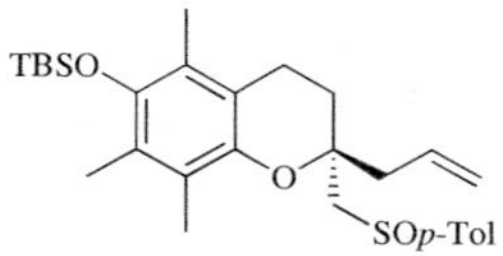 TBSO, SO*p*-Tol	Ru cat **271** (0.2 eq), DCM, 65° (MW), 1 h	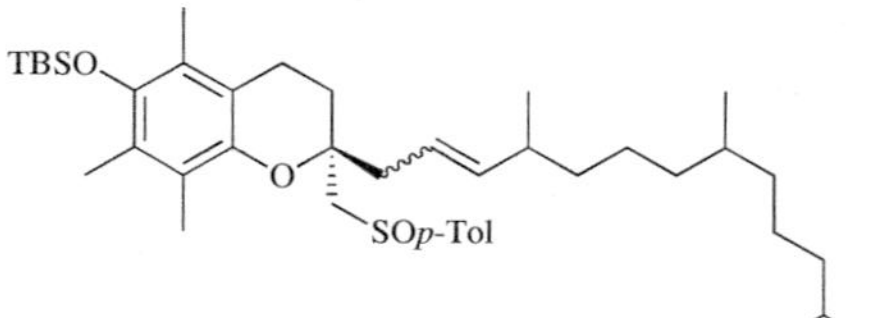 TBSO, SO*p*-Tol (63)	516
C_{16}				
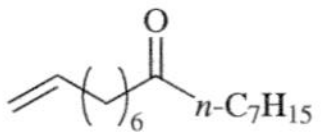 O, *n*-C_7H_{15}, 6	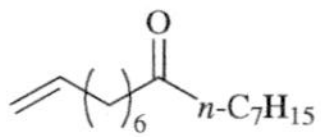 O, *n*-C_7H_{15}, 6	Gr II (0.1 eq), DCM, 25°, 4 h	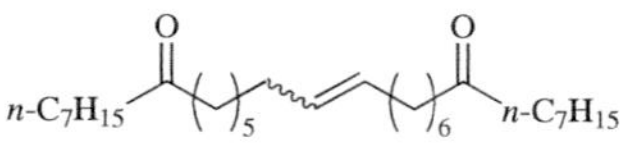 O, O, *n*-C_7H_{15}, 5, 6, *n*-C_7H_{15} (92)	483
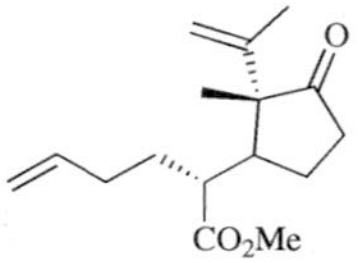 O, CO_2Me	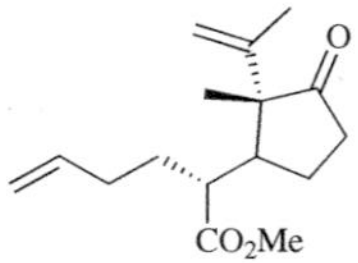 O, CO_2Me	Ru cat **126** (0.5 eq), DCM, rt, 12 h	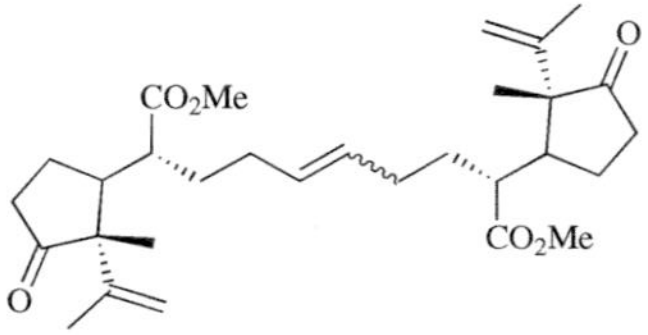 CO_2Me, O, O, CO_2Me (55)	517

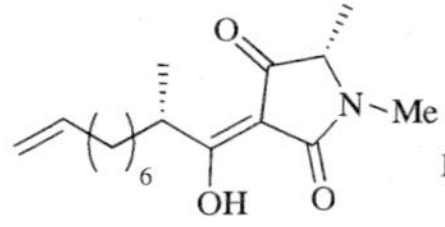 1.3 eq	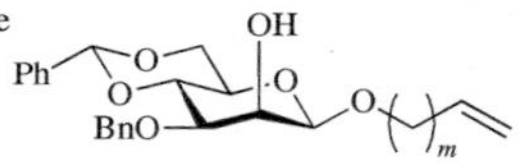	1. Gr II (0.01 eq), toluene, 80°, 1.5 h 2. TFA, DCM, H_2O, rt, 1.5 h 3. H_2, $Pd(OH)_2/C$, EtOAc, rt, 20 min	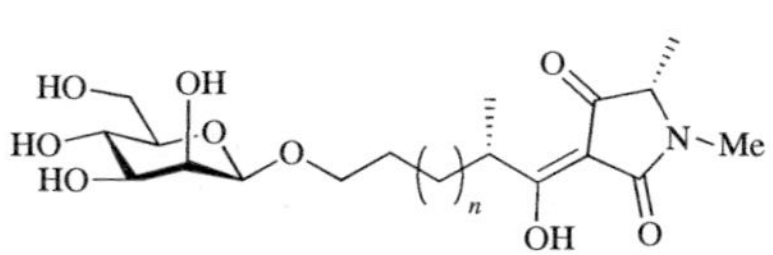 *m* *n* 6 12 (20) 8 14 (16)	248
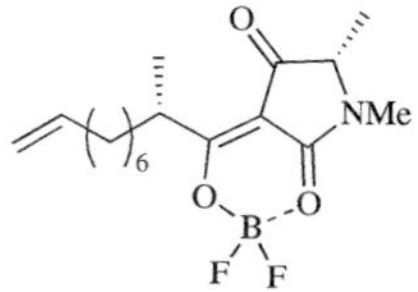	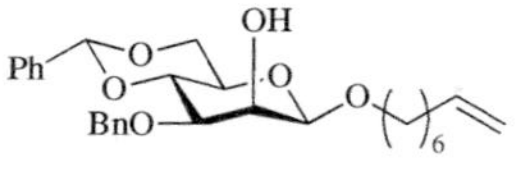	Gr II (0.015 eq), toluene, reflux	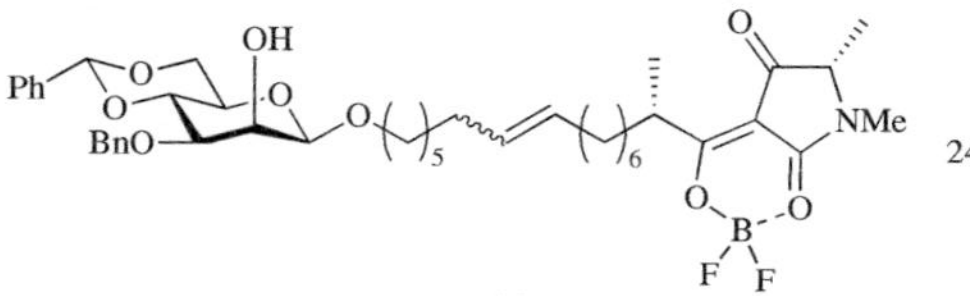 (47)	248
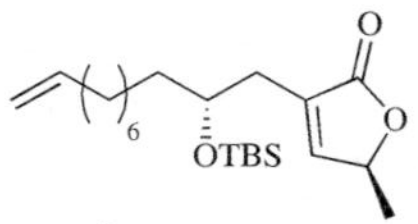 4 eq	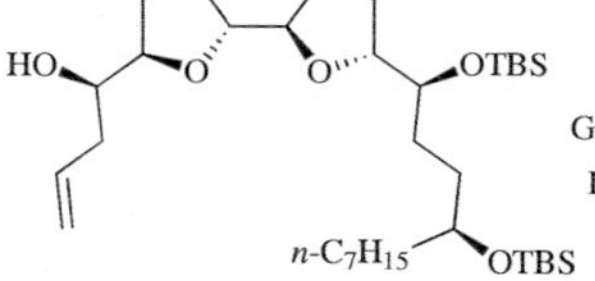	Gr II (0.2 eq), DCM, reflux, 12 h	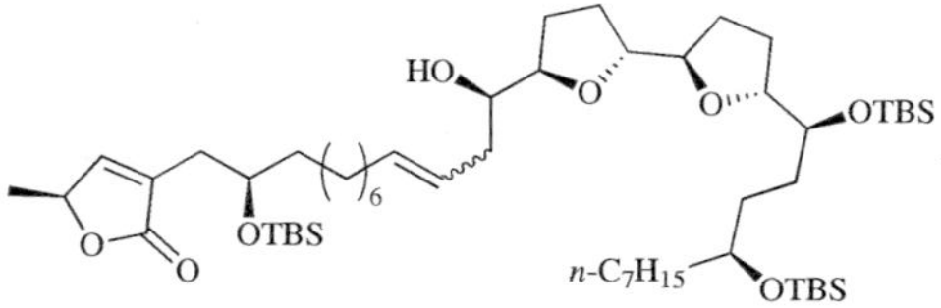 (66)	518

TABLE 1. CROSS-METATHESIS OF UNSATURATED COMPOUNDS WITH FUNCTIONAL GROUPS IN HOMOALLYLIC AND MORE DISTANT POSITIONS (*Continued*)

A. ETHYLENE AND COMPOUNDS WITH A TERMINAL DOUBLE BOND (*Continued*)

Terminal Alkene	Cross Partner	Conditions	Product(s) and Yield(s) (%)	Refs.
Please refer to the charts preceding the tables for the catalyst structures.				
C_{16}				
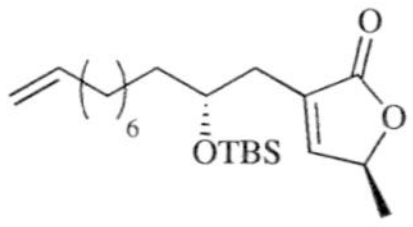 OTBS; 4 eq	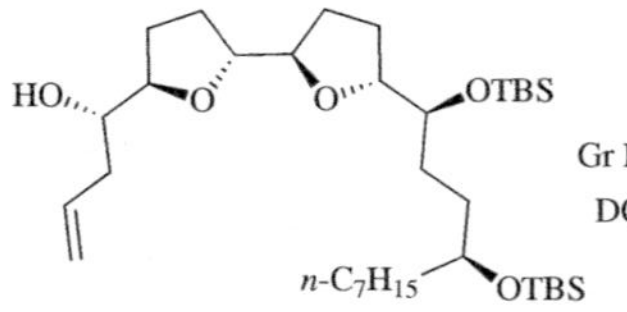 HO, O, O, OTBS, n-C_7H_{15}, OTBS	Gr II (0.2 eq), DCM, reflux, 12 h	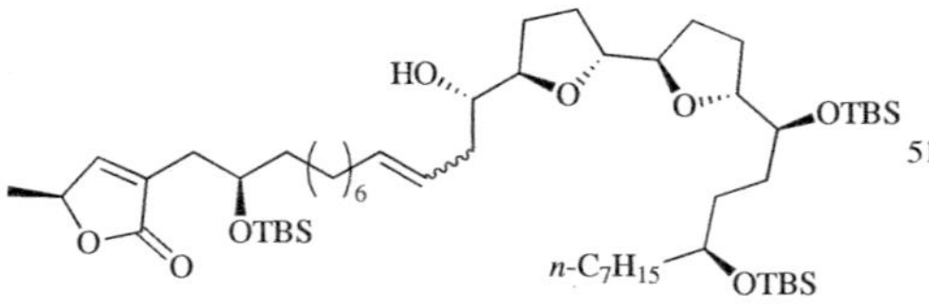 O, OTBS, HO, O, O, OTBS, n-C_7H_{15}, OTBS (58)	518
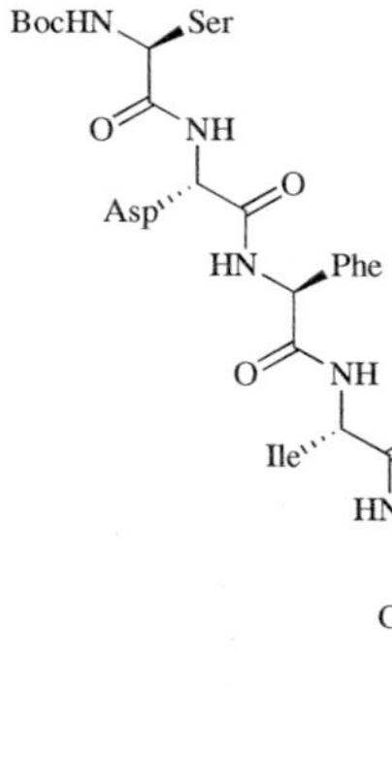 BocHN, Ser, NH, Asp, HN, Phe, NH, Ile, HN, Gln, NH, Val, HN	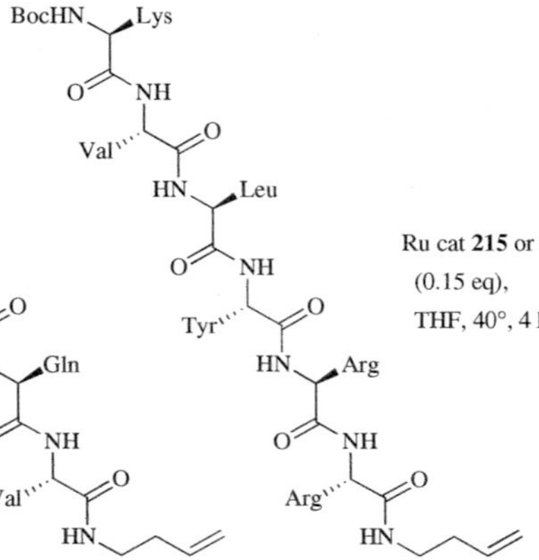 BocHN, Lys, NH, Val, HN, Leu, NH, Tyr, HN, Arg, NH, Arg, HN	Ru cat **215** or **216** (0.15 eq), THF, 40°, 4 h	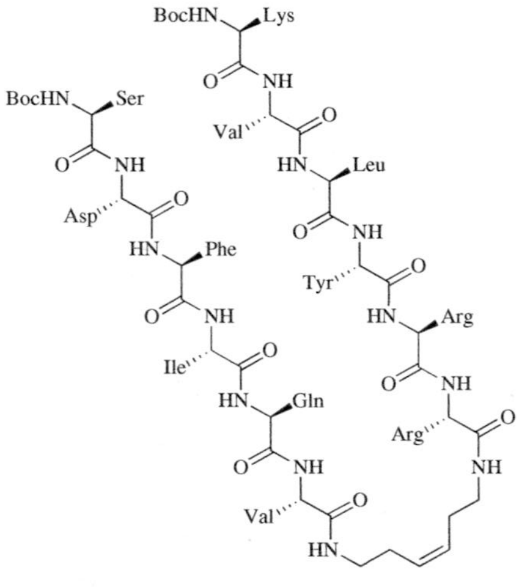 BocHN, Ser, NH, Asp, HN, Phe, NH, Ile, HN, Gln, NH, Val, HN, BocHN, Lys, NH, Val, HN, Leu, NH, Tyr, HN, Arg, NH, Arg, NH; 60% conv.[a], $(Z)/(E)$ = 60:40	401

C_{17}

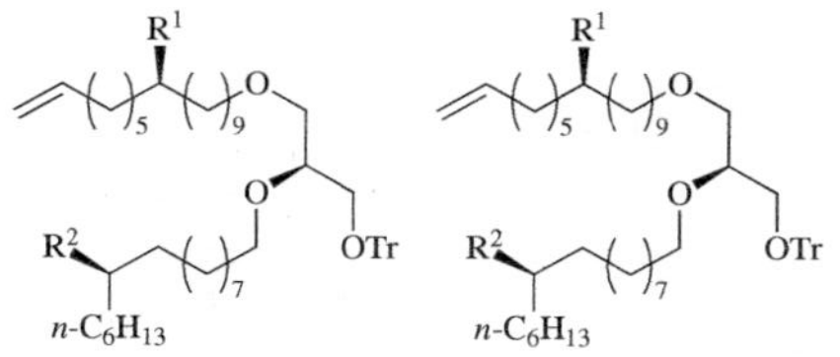

Gr I (0.29 eq), DCM, reflux, 24 h

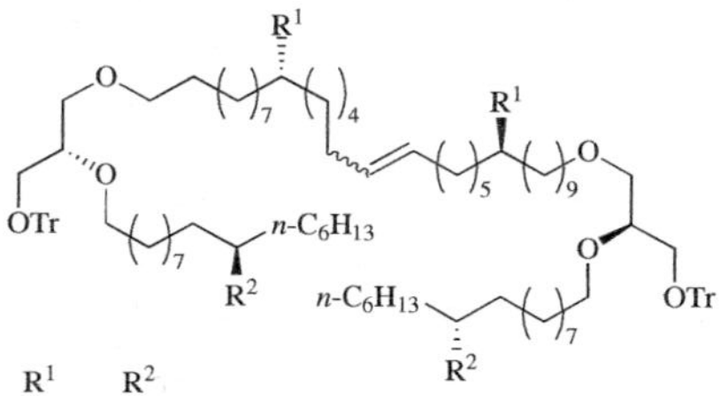

519

R^1	R^2	
H	Me	(31)
Me	H	(27)
Me	Me	(35)

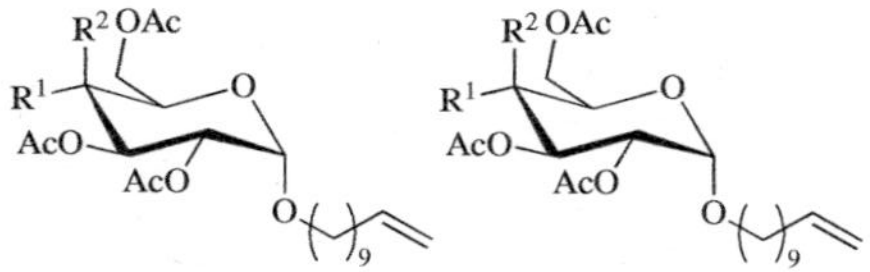

Gr I (x eq), DCM, rt

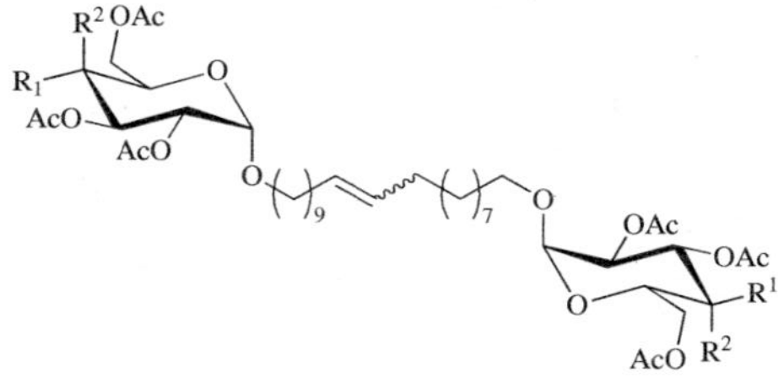

520

R^1	R^2	x	Time (h)	
H	AcO	0.115	20.5	(83)
AcO	H	0.078	15	(85)
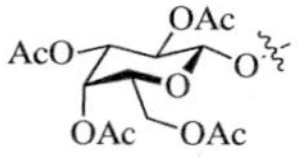 AcO, OAc, OAc, OAc (glycosyloxy structure)	H	0.176	27	(61)

TABLE 1. CROSS-METATHESIS OF UNSATURATED COMPOUNDS WITH FUNCTIONAL GROUPS IN HOMOALLYLIC AND MORE DISTANT POSITIONS (*Continued*)

A. ETHYLENE AND COMPOUNDS WITH A TERMINAL DOUBLE BOND (*Continued*)

Terminal Alkene	Cross Partner	Conditions	Product(s) and Yield(s) (%)	Refs.

Please refer to the charts preceding the tables for the catalyst structures.

C_{17}

Terminal Alkene	Cross Partner	Conditions	Product(s) and Yield(s) (%)	Refs.
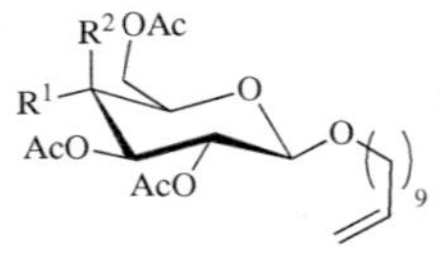 R^2 OAc, R^1, AcO, AcO, O, O, 9	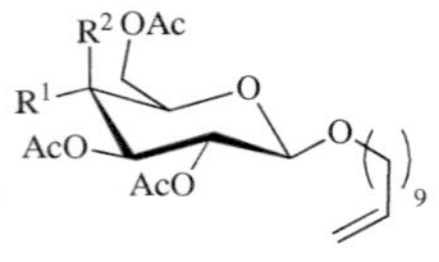 R^2 OAc, R^1, AcO, AcO, O, O, 9	Gr I (x eq), DCM, rt	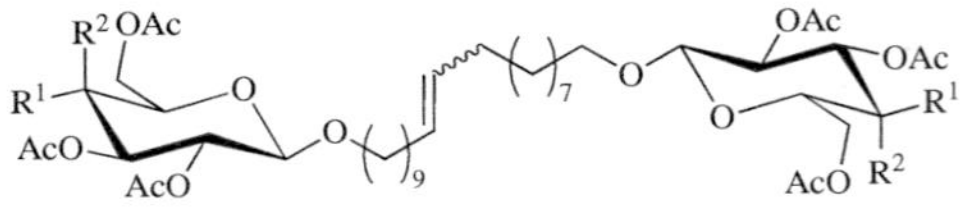 R^2 OAc, R^1, AcO, AcO, O, O, 9, 7, O, OAc, OAc, O, R^1, AcO R^2	520

R^1	R^2	x	Time (h)	
H	AcO	0.126	18	(62)
AcO	H	0.086	15	(95)
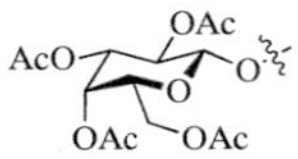 AcO, OAc, O, O, OAc, OAc (glycosyl)	H	0.152	21	(50)

C_{18}

Terminal Alkene	Cross Partner	Conditions	Product(s) and Yield(s) (%)	Refs.
n-$C_{12}H_{25}$, OMs, OTBS	OTBS, 10, O, O 4 eq	Gr I (0.4 eq), benzene, 50°, 13 h	n-$C_{12}H_{25}$, OMs, OTBS, OTBS, 10, O, O (62) (E) only	521

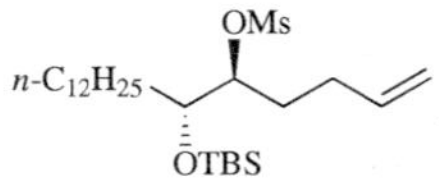

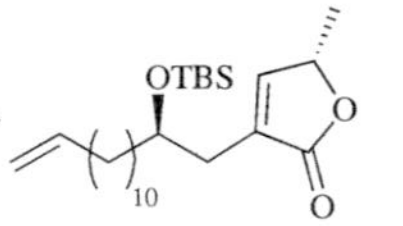

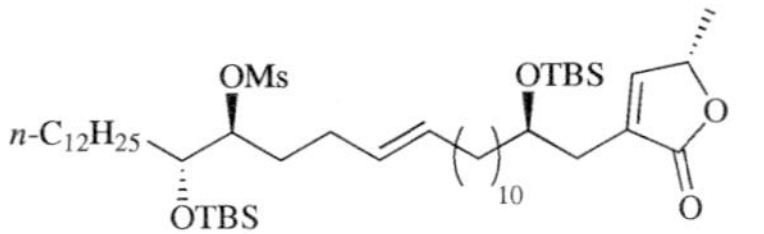

Reactant(s)	Conditions	Product(s) and Yield(s) (%)	Refs.
1 eq	Hov II (0.1 eq), DCM, rt, 14 h	(16)	522
	Gr I (0.1 eq), DCM, reflux, 2 h	(78) $(E)/(Z) = 63:37$	497

TABLE 1. CROSS-METATHESIS OF UNSATURATED COMPOUNDS WITH FUNCTIONAL GROUPS IN HOMOALLYLIC AND MORE DISTANT POSITIONS *(Continued)*

A. ETHYLENE AND COMPOUNDS WITH A TERMINAL DOUBLE BOND *(Continued)*

Terminal Alkene	Cross Partner	Conditions	Product(s) and Yield(s) (%)	Refs.

Please refer to the charts preceding the tables for the catalyst structures.

Terminal Alkene	Cross Partner	Conditions	Product(s) and Yield(s) (%)	Refs.
C_{18}				
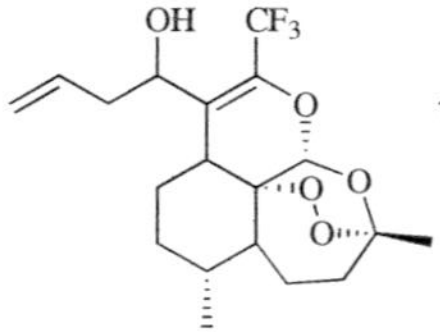 OH, CF_3	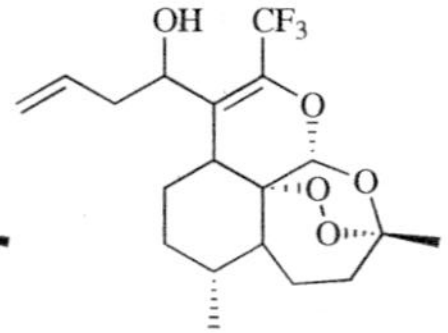 OH, CF_3	Ru cat **35** (0.1 eq), DCM	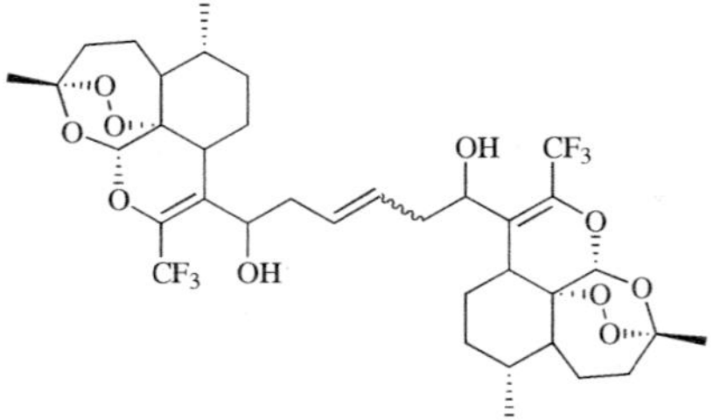 OH, CF_3, OH, CF_3 (50)	195
C_{20}				
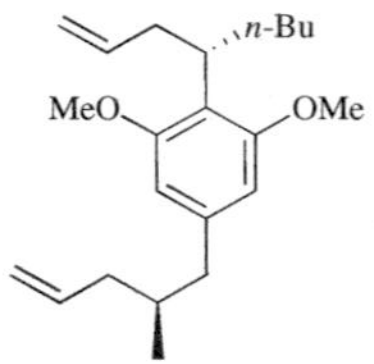 *n*-Bu, MeO, OMe	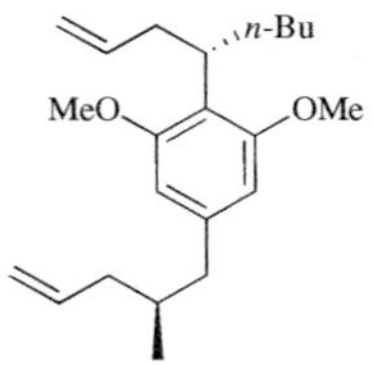 *n*-Bu, MeO, OMe	Catalyst (*x* eq), solvent	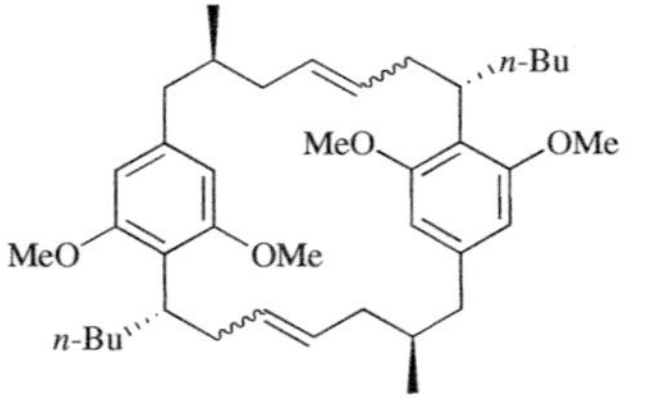 *n*-Bu, MeO, OMe, MeO, OMe, *n*-Bu	263

Catalyst	*x*	Solvent	Temp (°)	Time (h)	
Gr I	0.15	DCM	20	25	(55)
Gr I	0.2	DCM	20	75	(61)
Gr II	0.15	DCM	40	4	(48)
Gr II	0.15	benzene	40	27	(58)
Mo cat **1**	0.3	benzene	20	2	(72)

C_{21}

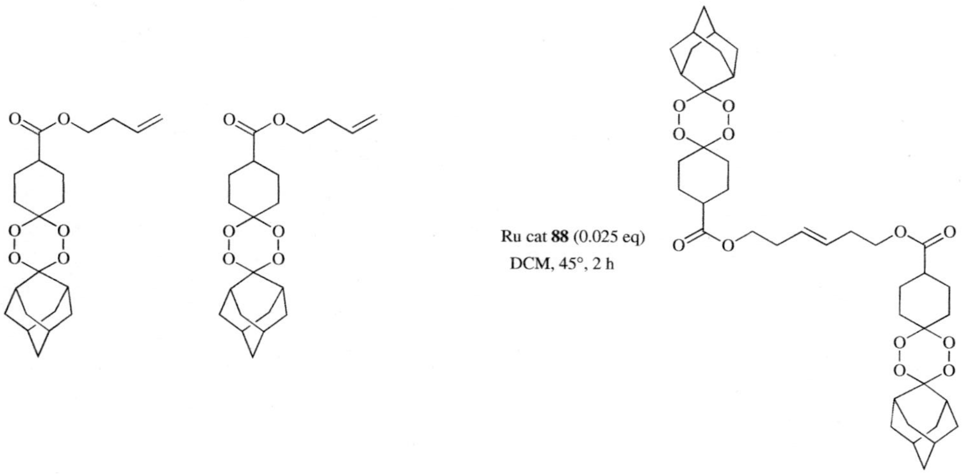

O O O O O O O O O O O O

Ru cat **88** (0.025 eq)
DCM, 45°, 2 h

O O O O O O O O O O O O

(87) (*E*)/(*Z*) = 67:33 193

TABLE 1. CROSS-METATHESIS OF UNSATURATED COMPOUNDS WITH FUNCTIONAL GROUPS IN HOMOALLYLIC AND MORE DISTANT POSITIONS (*Continued*)

A. ETHYLENE AND COMPOUNDS WITH A TERMINAL DOUBLE BOND (*Continued*)

Terminal Alkene	Cross Partner	Conditions	Product(s) and Yield(s) (%)	Refs.
Please refer to the charts preceding the tables for the catalyst structures.				
C_{21}				
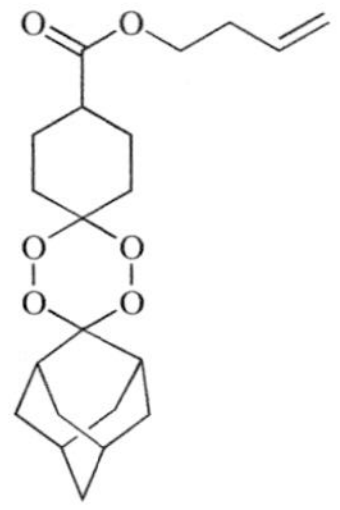 3 eq	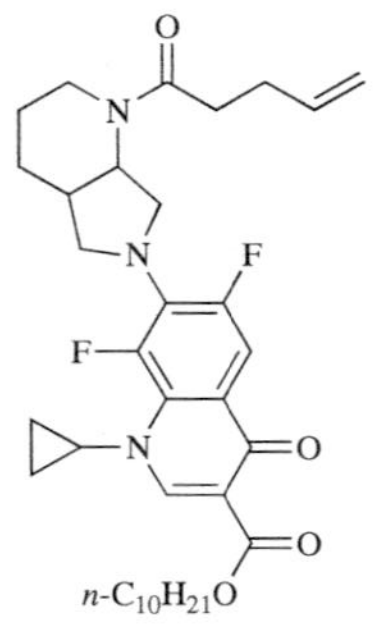 n-$C_{10}H_{21}O$	Ru cat **88** (0.1 eq) DCM, 45°, 2 h	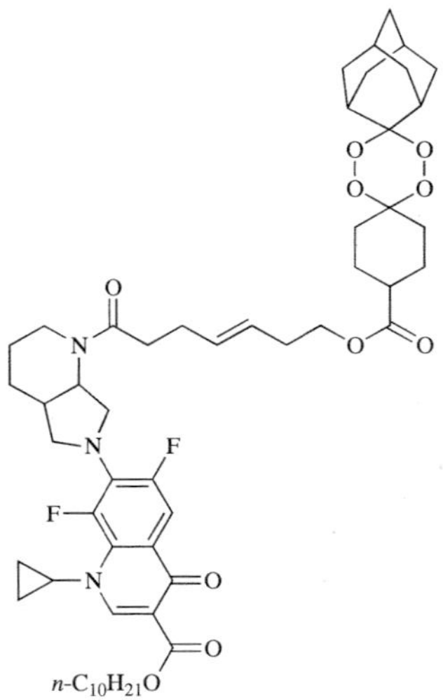 n-$C_{10}H_{21}O$ (40) $(E)/(Z)$ = 80:20	193

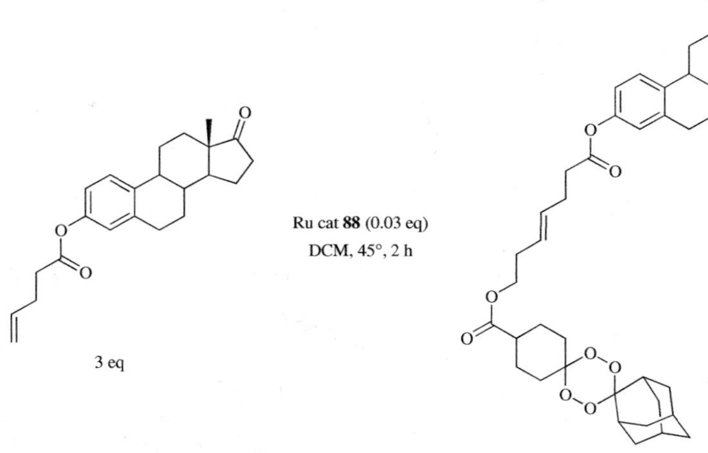

3 eq

Ru cat **88** (0.03 eq)
DCM, 45°, 2 h

(67) $(E)/(Z) = 67{:}33$ 193

TABLE 1. CROSS-METATHESIS OF UNSATURATED COMPOUNDS WITH FUNCTIONAL GROUPS IN HOMOALLYLIC AND MORE DISTANT POSITIONS (*Continued*)

A. ETHYLENE AND COMPOUNDS WITH A TERMINAL DOUBLE BOND (*Continued*)

Terminal Alkene	Cross Partner	Conditions	Product(s) and Yield(s) (%)	Refs.
Please refer to the charts preceding the tables for the catalyst structures.				
C_{21}				
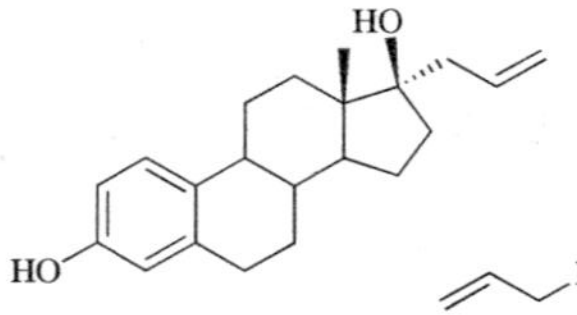 HO, HO; 1 eq	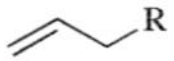 R	Hov II (0.1 eq), DCM, reflux, 12 h	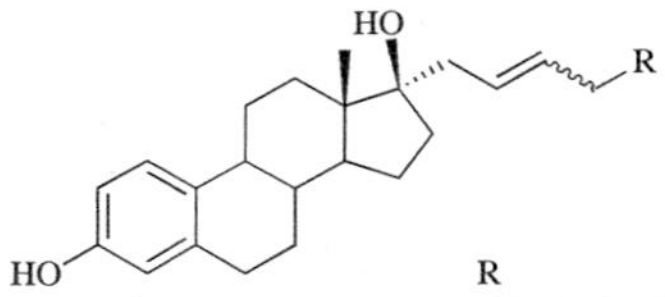   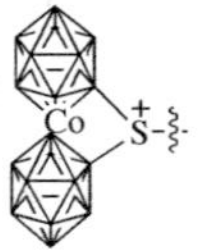 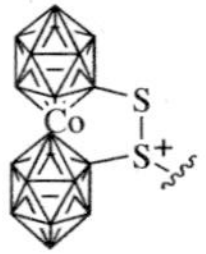 HO, HO, R; R: H C C (57); H C C (55); Co S (44); Co S S (52)	523

C_{22}

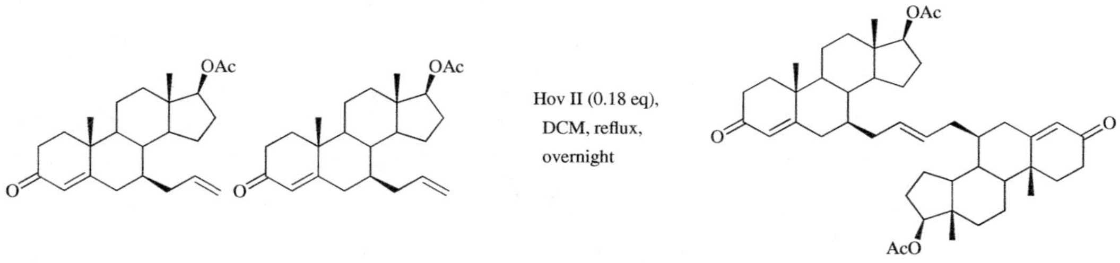

OAc OAc

O O

Hov II (0.18 eq), DCM, reflux, overnight

OAc

O O

AcO

(75) $(E)/(Z) = 67:33$

524

TABLE 1. CROSS-METATHESIS OF UNSATURATED COMPOUNDS WITH FUNCTIONAL GROUPS IN HOMOALLYLIC AND MORE DISTANT POSITIONS (*Continued*)

A. ETHYLENE AND COMPOUNDS WITH A TERMINAL DOUBLE BOND (*Continued*)

Terminal Alkene	Cross Partner	Conditions	Product(s) and Yield(s) (%)	Refs.
Please refer to the charts preceding the tables for the catalyst structures.				
C_{41}				
HO, MeO, HO, MeO MeO, N, O, OH	HO, MeO, HO, MeO MeO, N, O, OH	Gr II (0.1 eq), DCB, MW	(4)	432

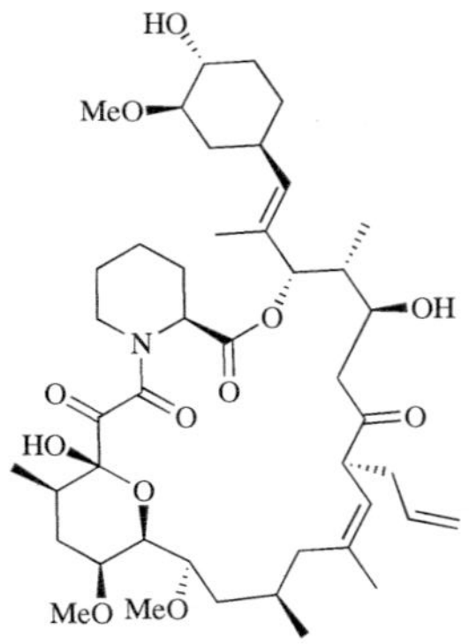

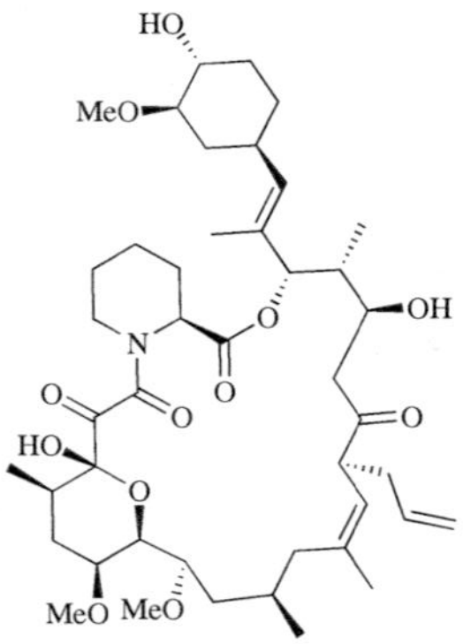

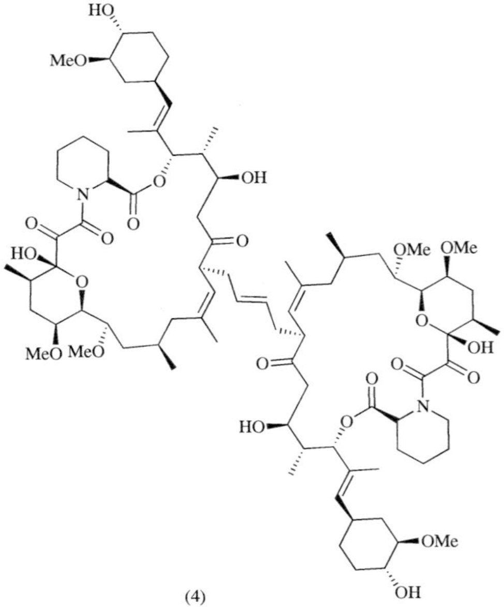

[a] The conversion was determined by ^{1}H NMR analysis.

[b] The conversion was based on recovered starting material.

[c] The conversion was determined by GC analysis.

[d] The conversion was determined by GC-MS analysis.

[e] The conversion was determined by HPLC analysis.

TABLE 1. CROSS-METATHESIS OF UNSATURATED COMPOUNDS WITH FUNCTIONAL GROUPS IN HOMOALLYLIC AND MORE DISTANT POSITIONS

(Continued)

B. COMPOUNDS WITH AN INTERNAL DOUBLE BOND

Internal Alkene	Cross Partner	Conditions	Product(s) and Yield(s) (%)	Refs.
Please refer to the charts preceding the tables for the catalyst structures.				
C_4				
 excess	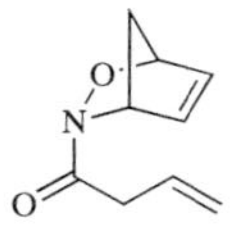 *rac*	Gr II (2 × 0.03 eq), toluene, 50–60°, 1 + 2 + 2 h	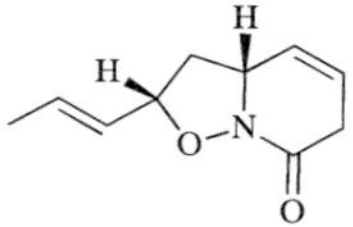 *rac* (75) (*E*)/(*Z*) = 90:10	525
excess	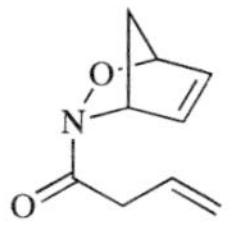 er 96.0:4.0	Gr II (2 x 0.03 eq), toluene, 50–60°, 1 + 2 + 2 h	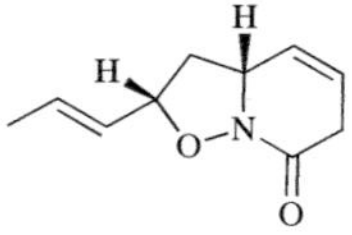 (55) (*E*)/(*Z*) = 90:10, er 96.0:4.0	525
	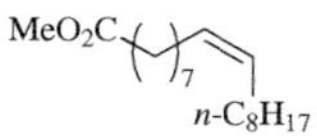 MeO_2C ( )$_7$ n-C_8H_{17}	Catalyst (0.02 eq), TMSCl (*x* eq) toluene, 80°, 18 h	 MeO_2C ( )$_7$ + n-C_8H_{17}	526

Catalyst	*x*	Conv. (%)[a]
Hov II	—	95
Ru cat **194**	0.2	31
Ru cat **195**	0.2	91
Ru cat **196**	0.2	13

TABLE 1. CROSS-METATHESIS OF UNSATURATED COMPOUNDS WITH FUNCTIONAL GROUPS IN HOMOALLYLIC AND MORE DISTANT POSITIONS (*Continued*)

B. COMPOUNDS WITH AN INTERNAL DOUBLE BOND (*Continued*)

Internal Alkene	Cross Partner	Conditions	Product(s) and Yield(s) (%)	Refs.
Please refer to the charts preceding the tables for the catalyst structures.				
C_6				
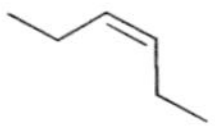 4 eq	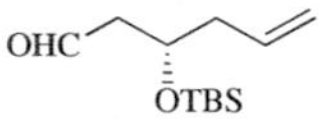 OHC, OTBS	Ru cat **358** (0.04 eq), THF, 23°, 1 h	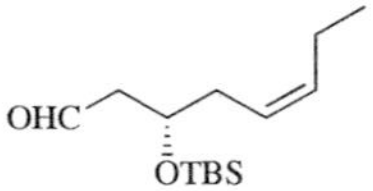 OHC, OTBS (88) (Z)/(E) > 99:1	446
4 eq	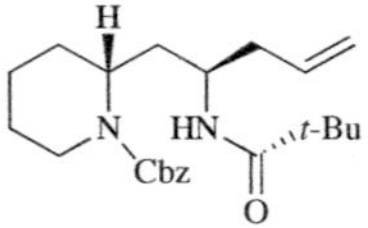 H, N, Cbz, HN, t-Bu, O	Gr II (0.031 eq), Ti(i-PrO)$_4$ (0.31 eq), DCM, 40°, 6 h	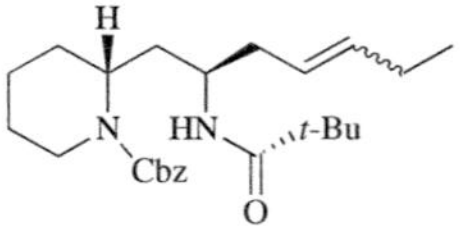 H, N, Cbz, HN, t-Bu, O (79)	527
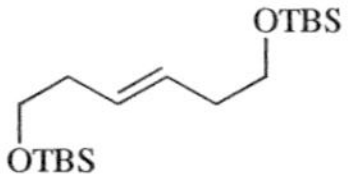 OTBS, OTBS 2 eq	 OBz, 8	Gr I (0.05 eq), DCM, 45°, 12 h	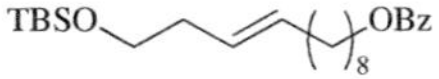 TBSO, OBz, 8 (49) (E)/(Z) = 74:26	124
C_8				
OH 8 eq	O, n-C_5H_{11}	Ru cat **358** (0.05 eq), THF, 40°, 16 h, static vacuum	OH, O, n-C_5H_{11} (47) (Z)/(E) > 99:1	528
			+ O, n-C_5H_{11} (53) (Z)/(E) > 99:1	

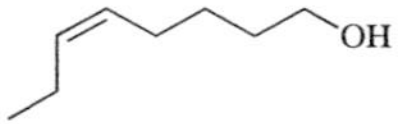

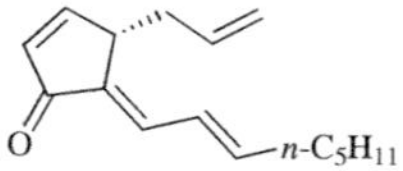

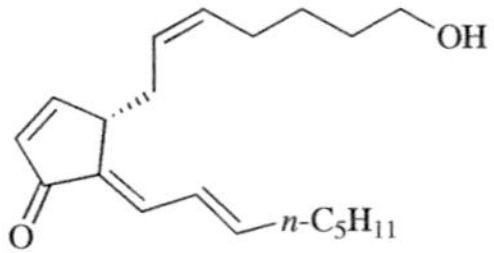

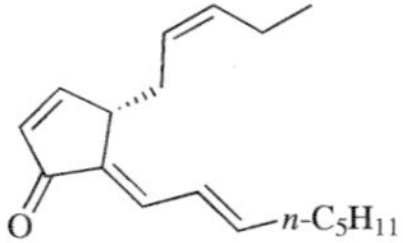

C_9

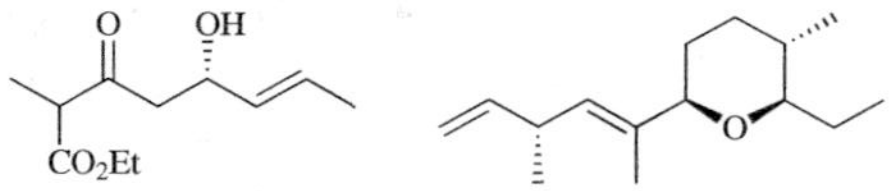

O OH

CO_2Et

4.67 eq

Ru cat **88**
(0.06 + 0.09 eq),
$C_6F_5CF_3$, 70°, 4 h

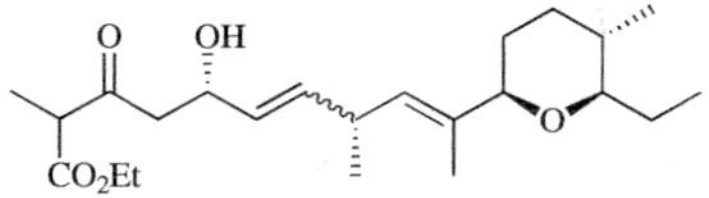

O OH

CO_2Et

(75) 529

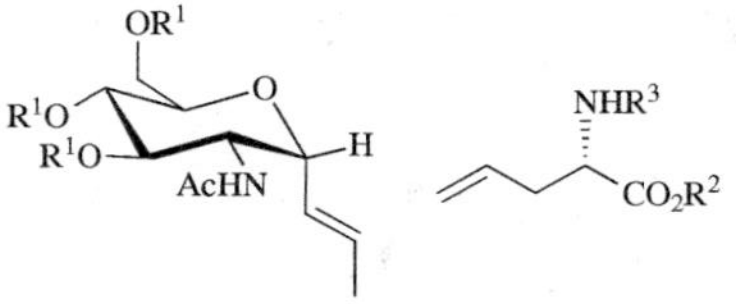

OR[1]
R[1]O
R[1]O
AcHN
H

NHR[3]
CO_2R^2

Gr II (*x* eq),
DCM, rt

OR[1]
R[1]O
R[1]O
AcHN
H
NHR[3]
CO_2R^2

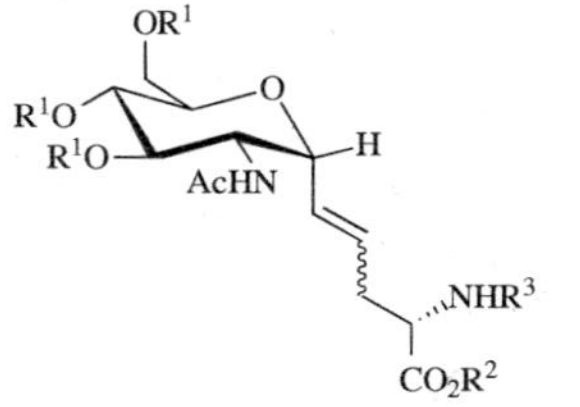

423

R[1]	R[2]	R[3]	*x*	Time (h)	
Bn	Me	Fmoc	0.1	24	(82)
Bn	Me	Boc	0.1	24	(77)
Ac	Me	Fmoc	0.2	48	(69)
Ac	Me	Boc	0.2	48	(77)
Ac	Bn	Cbz	0.2	48	(62)

TABLE 1. CROSS-METATHESIS OF UNSATURATED COMPOUNDS WITH FUNCTIONAL GROUPS IN HOMOALLYLIC AND MORE DISTANT POSITIONS

(Continued)

B. COMPOUNDS WITH AN INTERNAL DOUBLE BOND *(Continued)*

Internal Alkene	Cross Partner	Conditions	Product(s) and Yield(s) (%)	Refs.

Please refer to the charts preceding the tables for the catalyst structures.

C_{10}

Internal Alkene	Cross Partner	Conditions	Product(s) and Yield(s) (%)	Refs.
n-Pr, n-Bu; 5 eq	OAc ($_4$)	Catalyst (0.01 eq), DCM, 25°, 8 h, 50 mbar	n-Pr ... OAc ($_4$)	530
2 eq	OAc ($_7$)	Catalyst (x eq), THF, 35°	n-Pr ... OAc ($_7$) **I** + AcO ($_6$) ... OAc ($_7$) **II**	352
n-Pr, n-Bu; 2 eq	OAc ($_7$)	Catalyst (x eq), THF, 35°	n-Pr ... OAc ($_7$) **I** + AcO ($_6$) ... OAc ($_7$) **II**	352

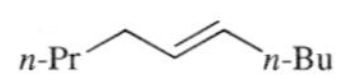

Product table for the first entry (Ref. 530):

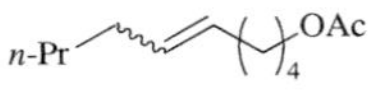

Catalyst	Conv. (%)[a]
Gr I	10
Gr II	85
Ind II	78
Ru cat **282**	35

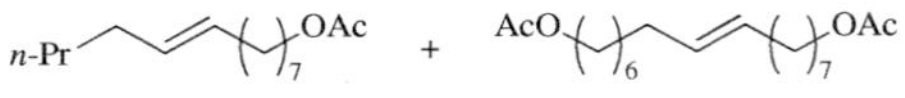

Product table for the second entry (Ref. 352, first):

Catalyst	x	Time (h)	I	(E)/(Z) I	II	(E)/(Z) II
Ru cat **215**	0.002	24	(<1)	—	(19)	13:87
Ru cat **214**	0.025	2	(53)	75:25	(17)	77:23

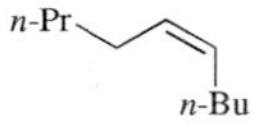

Product table for the third entry (Ref. 352, second):

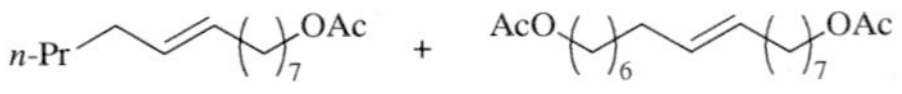

Catalyst	x	Time (h)	I	(E)/(Z) I	II	(E)/(Z) II
Ru cat **215**	0.002	6	(57)	9:91	(21)	17:83
Ru cat **214**	0.025	2	(69)	77:23	(14)	78:22

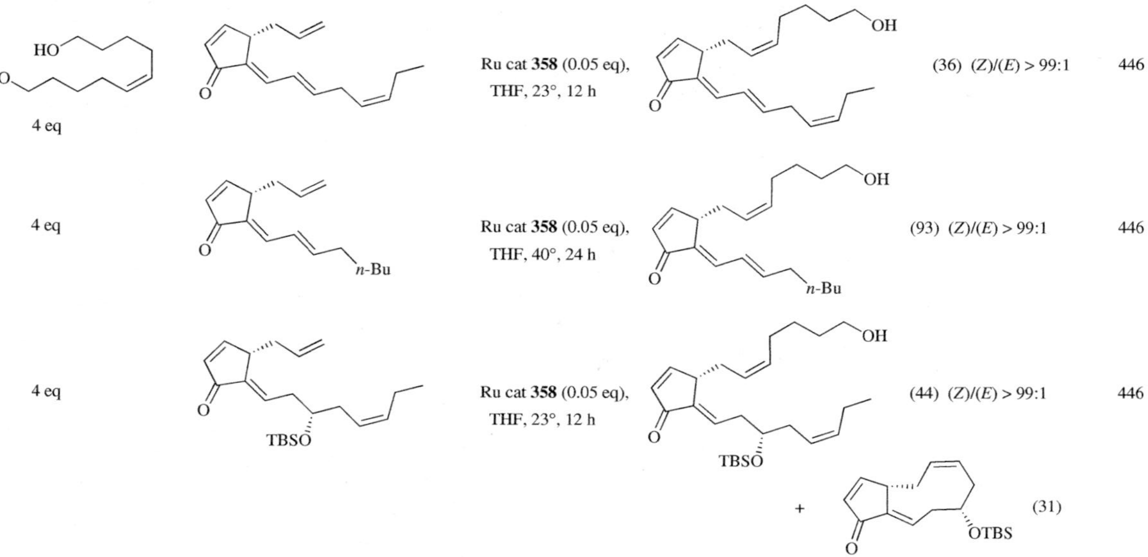

HO
HO
4 eq
O
Ru cat 358 (0.05 eq),
THF, 23°, 12 h
OH
O
(36) (Z)/(E) > 99:1
446
4 eq
O
n-Bu
Ru cat 358 (0.05 eq),
THF, 40°, 24 h
OH
O
n-Bu
(93) (Z)/(E) > 99:1
446
4 eq
O
TBSO
Ru cat 358 (0.05 eq),
THF, 23°, 12 h
OH
O
TBSO
(44) (Z)/(E) > 99:1
446
+
O
OTBS
(31)

TABLE 1. CROSS-METATHESIS OF UNSATURATED COMPOUNDS WITH FUNCTIONAL GROUPS IN HOMOALLYLIC AND MORE DISTANT POSITIONS (*Continued*)

B. COMPOUNDS WITH AN INTERNAL DOUBLE BOND (*Continued*)

Internal Alkene	Cross Partner	Conditions	Product(s) and Yield(s) (%)	Refs.
Please refer to the charts preceding the tables for the catalyst structures.				
C_{10}				
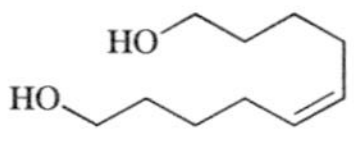 HO, HO; 4 eq	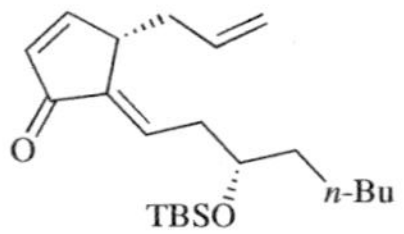 O, TBSO, n-Bu	Ru cat **358** (0.05 eq), THF, 40°, 24 h	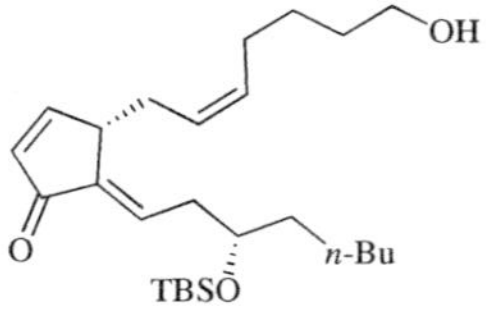 OH, O, TBSO, n-Bu (95) (Z)/(E) > 99:1	446
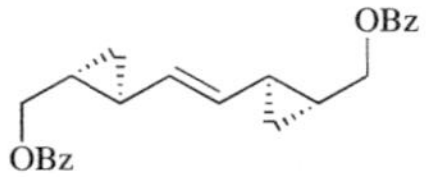 OBz, OBz; 2 eq	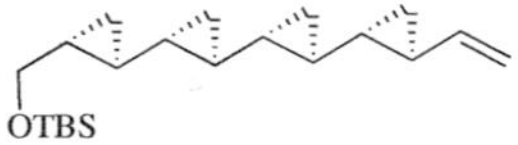 OTBS	Gr I (0.05 eq), DCM, reflux, 12 h	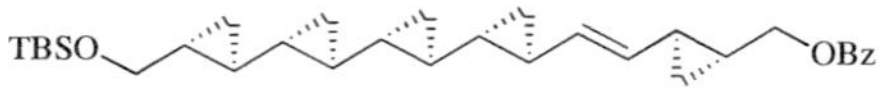 TBSO, OBz (82) (E)/(Z) = 83:17	447
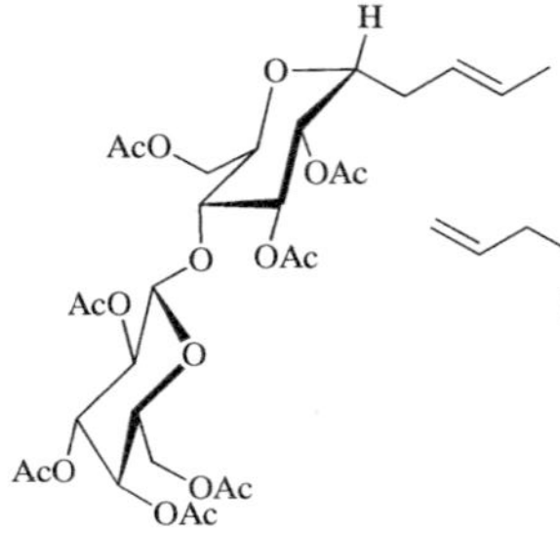 H, O, AcO, OAc, OAc, O, AcO, O, AcO, OAc, OAc	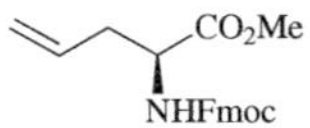 CO_2Me, NHFmoc; 2 eq	Gr I (0.2 eq), DCM, reflux, 48 h	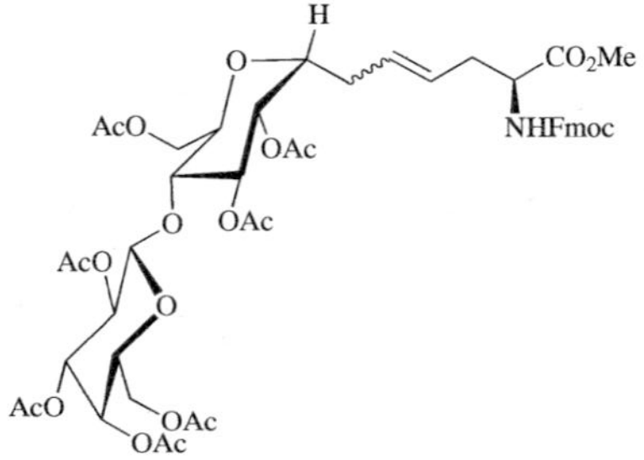 H, O, CO_2Me, NHFmoc, AcO, OAc, OAc, O, AcO, O, AcO, OAc, OAc (60)	427

C_{11}

MeO_2C–$(CH_2)_7$–CH₂CH₂CH₂CH=CH₂ + CH₂=CHCH₂CN (x eq) → Catalyst (y eq), additive (z eq), solvent → MeO_2C–$(CH_2)_7$–CH₂CH=CHCH₂CN

536

x	Catalyst	y	Additive	z	Solvent	Temp (°)	Time (h)	Conv. (%)[a]
5	Hov II	0.01	—	—	toluene	80	17	7
2	Hov II	0.01	—	—	toluene	95	21	95
5	Hov II	0.01	—	—	toluene	95	8	92
5	Hov II	0.01	BQ	0.1	toluene	95	5	79
5	Hov II	0.02	BQ	0.1	toluene	95	4	92
5	Hov II	0.02	acetic acid	0.1	toluene	95	6	48
5	Hov II	0.02	BQ	0.1	toluene	110	6	84
5	Hov II	0.02	BQ	0.1	toluene	95	6	43
5	Hov II	0.02	BQ	0.1	toluene	95	6	49
5	Gr II	0.02	BQ	0.1	toluene	110	4	94
5	Hov II	0.02	BQ	0.1	toluene	95	6	65
10	Hov II	0.02	BQ	0.1	toluene	110	6	38
5	Hov II	0.02	BQ	0.5	toluene	110	4	85
5	Hov II	0.02	BQ	0.5	toluene	110	4	70
5	Hov II	0.02	BQ	0.1	toluene	110	4	79
5	Hov II	0.02	BQ	0.1	DCE	80	8	41
5	Hov II	0.02	BQ	0.1	C_6F_5Cl	110	2	95
5	Hov II	0.02	BQ	0.1	C_6H_5Cl	110	2	90
5	Hov II	0.02	BQ	0.1	$C_6F_5CF_3$	110	5	77
5	Hov II	0.045	BQ	0.1	C_6H_5Cl	110	3	94

TABLE 1. CROSS-METATHESIS OF UNSATURATED COMPOUNDS WITH FUNCTIONAL GROUPS IN HOMOALLYLIC AND MORE DISTANT POSITIONS (*Continued*)

B. COMPOUNDS WITH AN INTERNAL DOUBLE BOND (*Continued*)

Internal Alkene	Cross Partner	Conditions	Product(s) and Yield(s) (%)	Refs.

Please refer to the charts preceding the tables for the catalyst structures.

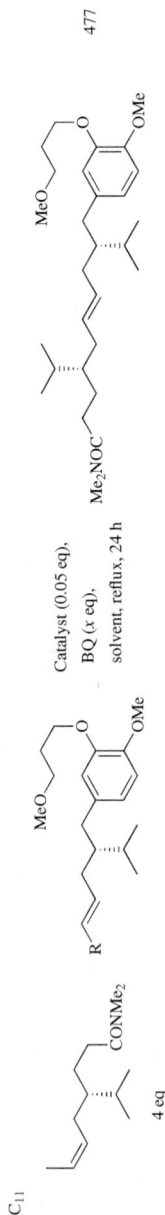

C_{11}

Internal Alkene	Cross Partner	Conditions	Product(s) and Yield(s) (%)	Refs.
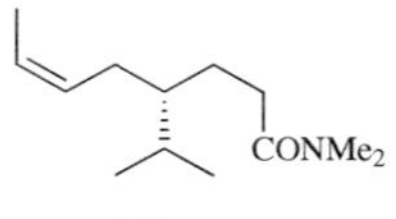 $CONMe_2$ 4 eq	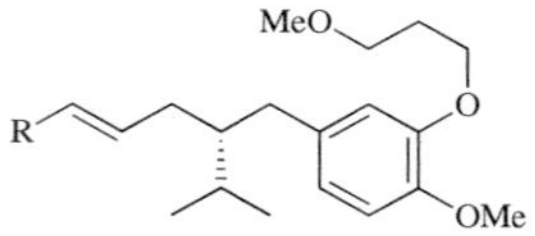 R, MeO, O, OMe	Catalyst (0.05 eq), BQ (*x* eq), solvent, reflux, 24 h	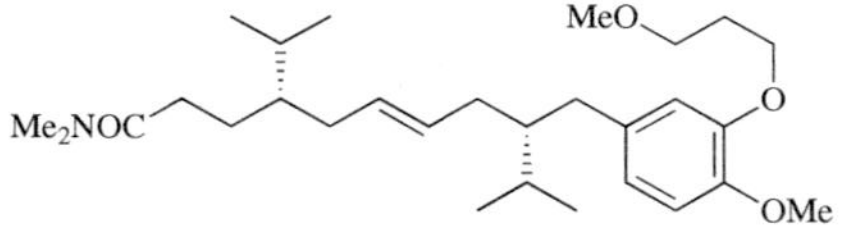 Me_2NOC, MeO, O, OMe	477

R	Catalyst	Solvent	*x*		(*E*)/(*Z*)
H	Hov II	PE	0.5	(82)	75:25
H	Gr II	PE	0.5	(80)	81:19
Me	Hov II	PE	0.5	(72)	80:20
Me	Gr II	PE	0.5	(80)	81:19
Me	Hov II	DCM	0.5	(68)	85:15
Me	Gr II	DCM	0.5	(62)	87:13
Me	Gr II	DCM	0.6	(62)	88:12
Me	Gr II	DCM	0.7	(67)	88:12
Me	Gr II	DCM	0.8	(66)	87:13

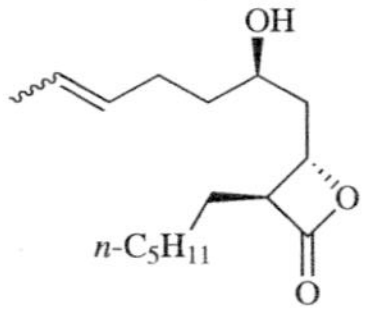

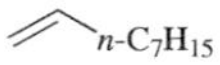

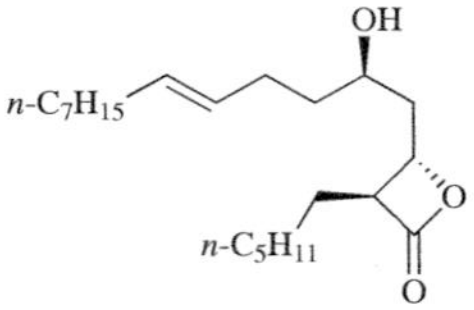

C_{16}

Internal Alkene	Cross Partner	Conditions	Product(s) and Yield(s) (%)	Refs.
OH, O, n-C_5H_{11}, O	n-C_7H_{15} 5 eq	Gr I (0.15 eq), DCM, 40°	n-C_7H_{15}, OH, O, n-C_5H_{11}, O (20) (*E*)/(*Z*) = 80:20	531

Reactants	Conditions	Product(s) and Yield(s) (%)	Refs.
Ph–CH2CH=CHCH2–Ph (4 eq); allyl-carborane (H–C cage–C–CH2CH=CH2)	Gr II (0.0056 eq), DCM, reflux, 24 h	H–C cage–C–CH2CH=CHCH2Ph (62) (E)/(Z) = 94:6	436

C_{18}

Reactants	Conditions	Product(s) and Yield(s) (%)	Refs.
n-C_8H_{17}CH=CH$(CH_2)_7CO_2Me$; CH_2=CH–n-Bu (x eq)	Ru cat **8** (0.003 eq), cyclohexane, 30°	n-Pr~CH=CH$(CH_2)_7CO_2Me$ **I** + n-Pr~CH=CH–n-C_8H_{17} **II** + CH_2=CH$(CH_2)_7CO_2Me$ **III** + CH_2=CH–n-C_8H_{17} **IV** + n-Pr~CH=CH–n-Bu **V** + $MeO_2C(CH_2)_6$~CH=CH$(CH_2)_7CO_2Me$ **VI** + n-C_7H_{15}~CH=CH–n-C_8H_{17} **VII**	532

x	**I + II + III + IV**[b]	**V**[b]	**VI + VII**[b]
1	(46)	(23)	(21)
2	(55)	(22)	(20)
3	(69)	(20)	(15)
4	(78)	(22)	(10)
5	(83)	(19)	(8)
6	(86)	(17)	(6)
7	(87)	(15)	(6)

TABLE 1. CROSS-METATHESIS OF UNSATURATED COMPOUNDS WITH FUNCTIONAL GROUPS IN HOMOALLYLIC AND MORE DISTANT POSITIONS (*Continued*)

B. COMPOUNDS WITH AN INTERNAL DOUBLE BOND (*Continued*)

Internal Alkene	Cross Partner	Conditions	Product(s) and Yield(s) (%)	Refs.
Please refer to the charts preceding the tables for the catalyst structures.				
C_{18}				
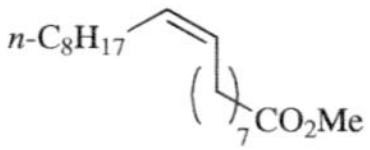 n-C_8H_{17} (Z)-alkene, $(CH_2)_7CO_2Me$	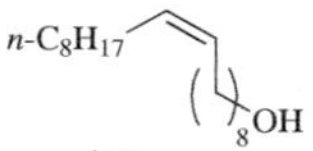 n-C_8H_{17} (Z)-alkene, $(CH_2)_8OH$ (1 eq)	Gr II (0.01 eq), 45°, 24.5 h	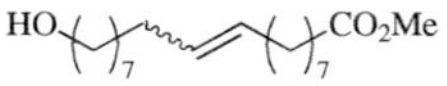 $HO(CH_2)_7$CH=CH$(CH_2)_7CO_2Me$ (40)	533
	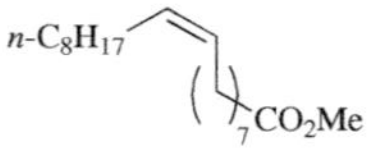 n-C_8H_{17} (Z)-alkene, $(CH_2)_7CO_2Me$	Catalyst (0.01 eq), dodecane (0.6 eq), toluene, 50°	  $MeO_2C(CH_2)_6$CH=CH$(CH_2)_7CO_2Me$ **I** + n-C_7H_{15}CH=CH-n-C_8H_{17} **II**	534

Catalyst	Conv. (%)^c (1 h)	Selectivity **I**/**II**^{c} (1 h)	**I**^{c} (1 h)	**II**^{c} (1 h)	Conv. (%)^c (5 h)	Selectivity **I**/**II**^{c} (5 h)	**I**^{c} (5 h)	**II**^{c} (5 h)
Ind II	52	87:66	(45)	(34)	53	88:67	(47)	(36)
Ru cat **283**	54	88:73	(47)	(39)	55	89:75	(49)	(41)
Ru cat **332**	53	89:69	(47)	(36)	54	89:75	(49)	(38)
Ru cat **297**	50	89:58	(44)	(29)	53	89:58	(48)	(32)
Ru cat **331**	53	88:69	(47)	(37)	53	88:69	(47)	(32)
Ru cat **320**	51	84:56	(43)	(29)	53	87:60	(46)	(32)
Ru cat **282**	5	100:100	(5)	(5)	52	91:57	(47)	(30)
Ru cat **327**	50	88:60	(44)	(30)	52	88:60	(47)	(31)
Ru cat **326**	48	68:90	(33)	(43)	50	67:92	(34)	(46)
Ru cat **328**	51	81:47	(38)	(22)	55	88:53	(42)	(26)
Ru cat **330**	43	88:63	(37)	(27)	43	89:72	(38)	(31)
Ru cat **294**	0	—	(0)	(0)	30	100:95	(30)	(29)
Ru cat **329**	26	89:69	(23)	(18)	28	93:72	(26)	(20)
Ru cat **316**	5	100:100	(5)	(5)	5	91:71	(5)	(4)
Gr II	49	65:69	(32)	(33)	50	58:68	(29)	(34)
Hov II	52	87:65	(34)	(45)	54	88:66	(48)	(36)

n-C_8H_{17} ... $(\ )_7CO_2Me$ + n-C_8H_{17} ... $(\ )_7CO_2Me$ → Catalyst (x eq), solvent → $MeO_2C(\ )_6$~~$(\ )_7CO_2Me$ **I** + n-C_7H_{15}~~n-C_8H_{17} **II**

Catalyst	x	Solvent	Temp (°)	Time (h)	Conv. (%)[a]	I	II	
Gr I	0.000025	—	60	4	7	(—)	(—)	377
Gr II	0.000025	—	60	4	51	(—)	(—)	377
Ru cat **5**	0.000025	—	60	4	47	(—)	(—)	376
Ru cat **287**	0.000025	—	50	3	49	(—)	(—)	376
Ru cat **287**	0.000006	—	50	3	37	(—)	(—)	376
Ru cat **75**	0.01	—	80	12	99	(96)	(96)	379
Ru cat **76**	0.01	—	80	12	99	(85)	(80)	379
Hov II	0.01	[bmim][PF_6]	40	4	78	(—)	(—)	535
Gr II	0.01	[bmim][PF_6]	60	4	75	(—)	(—)	535
Hov II	0.01	[bmim][PF_6]	60	4	87	(—)	(—)	535
Hov II	0.01	[bmim][BF_4]	40	4	79	(—)	(—)	535
Gr II	0.01	[bmim][BF_4]	60	4	78	(—)	(—)	535
Hov II	0.01	[bmim][BF_4]	60	4	87	(—)	(—)	535
Hov II	0.01	[bmim][NTf_2]	40	4	74	(—)	(—)	535
Gr II	0.01	[bmim][NTf_2]	60	4	73	(—)	(—)	535
Hov II	0.01	[bmim][NTf_2]	60	4	85	(—)	(—)	535
Hov II	0.01	[bdmim][PF_6]	40	4	81	(—)	(—)	535
Hov II	0.01	[bdmim][PF_6]	60	4	90	(—)	(—)	535
Hov II	0.01	[bdmim][BF_4]	40	4	85	(—)	(—)	535
Hov II	0.01	[bdmim][BF_4]	40	4	92	(—)	(—)	535

TABLE 1. CROSS-METATHESIS OF UNSATURATED COMPOUNDS WITH FUNCTIONAL GROUPS IN HOMOALLYLIC AND MORE DISTANT POSITIONS (*Continued*)

B. COMPOUNDS WITH AN INTERNAL DOUBLE BOND (*Continued*)

Internal Alkene	Cross Partner	Conditions	Product(s) and Yield(s) (%)	Refs.
Please refer to the charts preceding the tables for the catalyst structures.				
C_{18}				
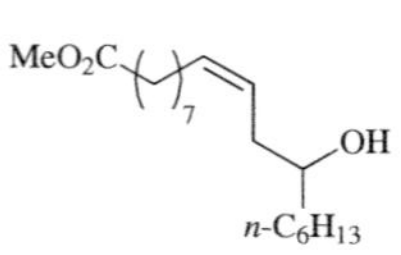 MeO_2C, 7, OH, n-C_6H_{13}	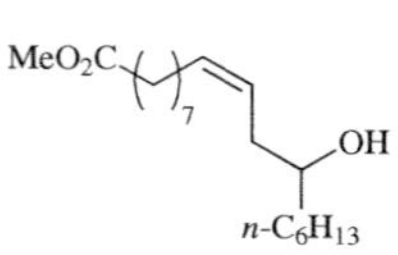 MeO_2C, 7, OH, n-C_6H_{13}	Catalyst (0.01eq), solvent, 60°, 4 h	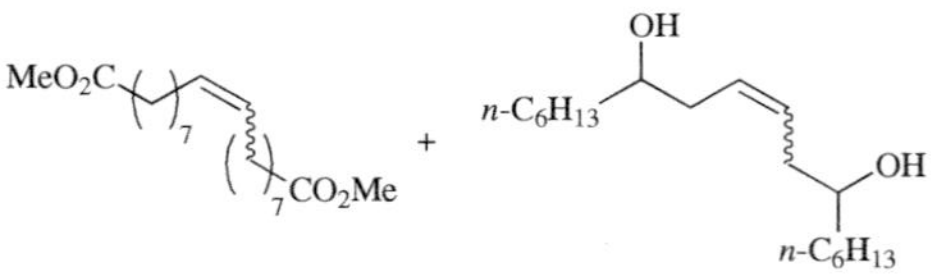 MeO_2C, 7, 7, CO_2Me + n-C_6H_{13}, OH, OH, n-C_6H_{13}	535

Catalyst	Solvent	Conv. (%)[a]
Gr I	[bmim][PF_6]	45
Gr I	[bmim][BF_4]	41
Gr I	[bmim][NTf_2]	38
Gr II	[bmim][PF_6]	58
Gr II	[bmim][BF_4]	56
Gr II	[bmim][NTf_2]	55

MeO_2C–$(\ )_7$ ... n-C_7H_{15} + CN (x eq)

Hov II (0.02 eq), BQ (0.5 eq), PhCl, 110°

MeO_2C–$(\ )_7$ ... CN

536

x	Time (h)	Conv. (%)[a]
0.15	6	69
0.25	6	79
0.25	8	74
0.25	4	89
0.3	6	78
0.5	6	23

C_{20}

TBSO ... $(\ )_4$OPMB

Gr II (0.06 eq), toluene, 80°, 8 h

TBSO ... $(\ )_4$OPMB

(69) $(E)/(Z)$ = 90:10

128

TABLE 1. CROSS-METATHESIS OF UNSATURATED COMPOUNDS WITH FUNCTIONAL GROUPS IN HOMOALLYLIC AND MORE DISTANT POSITIONS (*Continued*)

B. COMPOUNDS WITH AN INTERNAL DOUBLE BOND (*Continued*)

Internal Alkene	Cross Partner	Conditions	Product(s) and Yield(s) (%)	Refs.
Please refer to the charts preceding the tables for the catalyst structures.				
C_{20}				
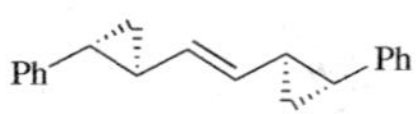 Ph, Ph; 2 eq	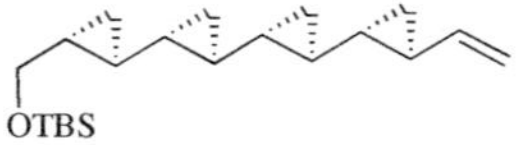 OTBS	Gr I (0.05 eq), DCM, reflux, 12 h	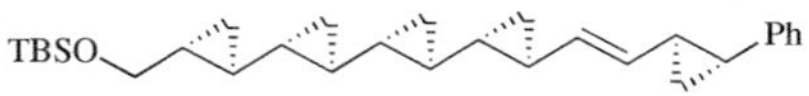 TBSO, Ph; (94) (*E*)/(*Z*) = 78:22	447
C_{22-32}				
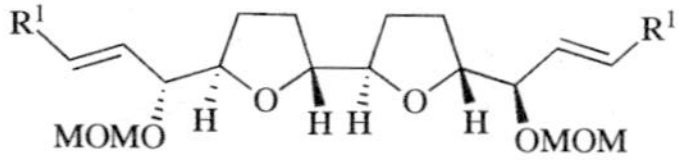 R[1], MOMO, H, O, H H, O, H, OMOM, R[1]; 1 eq	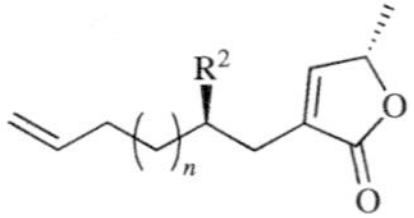 R[2], *n*, O, O	Gr II (0.1 eq), BQ (0.2 eq), DCM	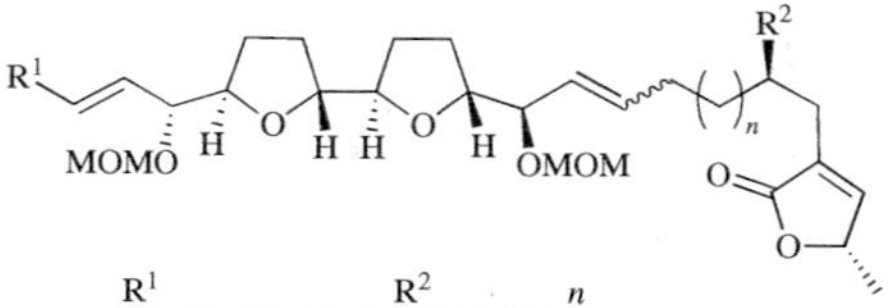 R[1], MOMO, H, O, H H, O, H, OMOM, *n*, R[2], O, O	537

R[1]	R[2]	*n*	
$TBSO(CH_2)_4$	TBSO	7	(41)
$TBSO(CH_2)_9$	TBSO	7	(40)
$TBSO(CH_2)_9$	H	6	(41)

[a] The conversion was determined by GC analysis.

[b] The yield was determined by HPLC analysis.

[c] The conversion, selectivity, and yield were determined by GC analysis.

TABLE 1. CROSS-METATHESIS OF UNSATURATED COMPOUNDS WITH FUNCTIONAL GROUPS IN HOMOALLYLIC AND MORE DISTANT POSITIONS (*Continued*)

C. COMPOUNDS WITH CONJUGATED DOUBLE BONDS

Alkene	Cross Partner	Conditions	Product(s) and Yield(s) (%)	Refs.
Please refer to the charts preceding the tables for the catalyst structures.				
C_{4-10}				
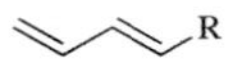 R (10 eq)	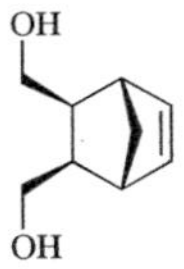 OH, OH	Ru cat **151** (x eq), DCM, 22°, 2 h	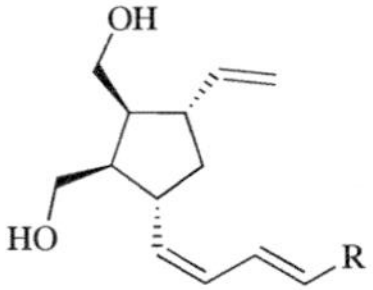 OH, HO, R R / x / / (Z)/(E) MeO / 0.02 / (84) / 91:9 n-C_6H_{13} / 0.05 / (80) / 98:2	83
C_6				
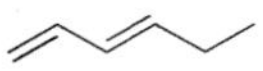 1.5 eq	 OAc, 8	Ru cat **216** (0.04 eq) DCE, 50°, 24 h	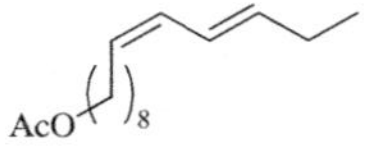 AcO, 8 (56) (Z)/(E) > 95:5	538
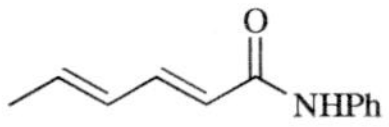 NHPh, O (4 eq)	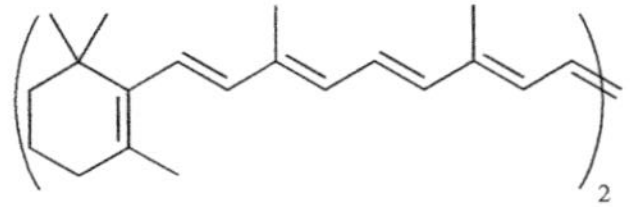 2	Hov II (0.15 eq), toluene/DCM, rt, 72 h	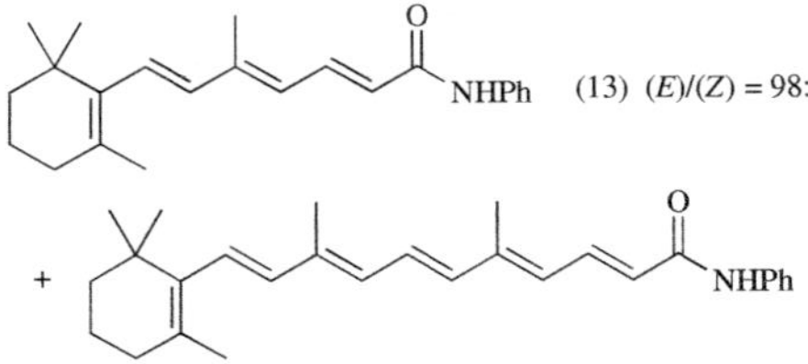 O, NHPh (13) (E)/(Z) = 98:2 + O, NHPh (22) (E)/(Z) =91:9	539

TABLE 1. CROSS-METATHESIS OF UNSATURATED COMPOUNDS WITH FUNCTIONAL GROUPS IN HOMOALLYLIC AND MORE DISTANT POSITIONS (Continued)

C. COMPOUNDS WITH CONJUGATED DOUBLE BONDS (Continued)

Alkene	Cross Partner	Conditions	Product(s) and Yield(s) (%)	Refs.
Please refer to the charts preceding the tables for the catalyst structures.				
C_6				
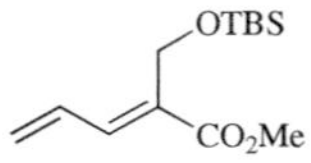 OTBS, CO_2Me; 2 eq	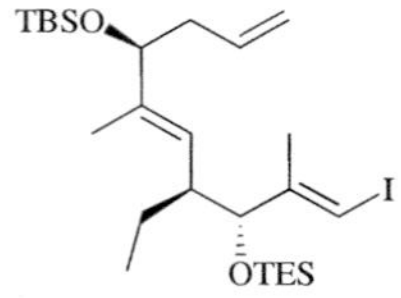 TBSO, OTES, I	Hov II (0.15 eq), EtOAc, rt, 3.5 h	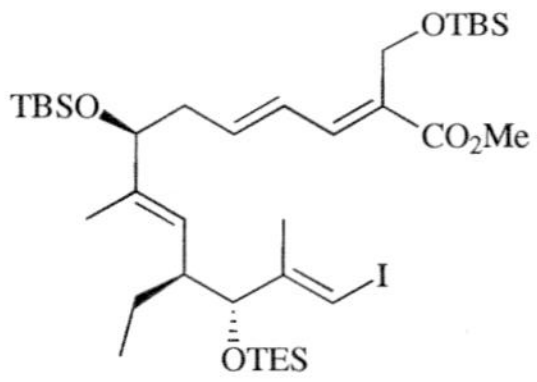 TBSO, OTBS, CO_2Me, OTES, I (56) $(E)/(Z)$ = 87:13	540
C_7				
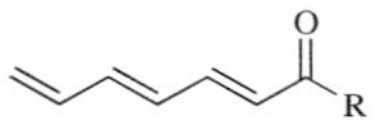 O, R; *x* eq	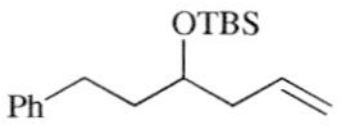 Ph, OTBS	Gr I (0.01 eq), DCM, reflux, 5 h	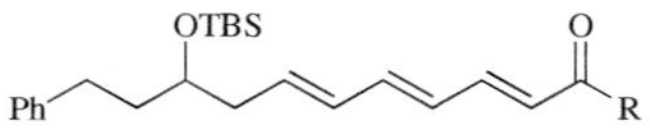 Ph, OTBS, O, R	541
			x = 2, R = H (72) $(E)/(Z)$ = 80:20	
			x = 1.5, R = EtO (84) $(E)/(Z)$ = 83:17	
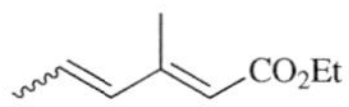 CO_2Et; 3 eq	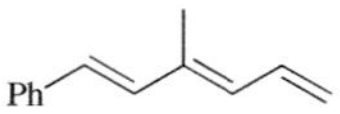 Ph	Hov II (0.1 eq), toluene, rt	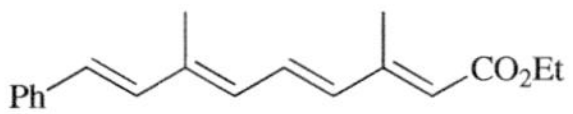 Ph, CO_2Et (90) $(E)/(Z)$ = 99:1	542

3 eq	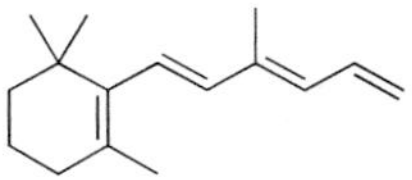	Hov II (0.1 eq), toluene, rt	CO_2Et (88) (*E*)/(*Z*) = 97:3	542
3 eq	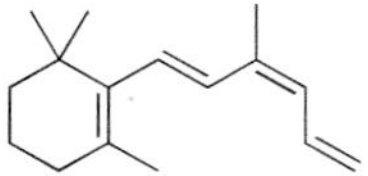	Hov II (0.1 eq), toluene, rt	CO_2Et (37)	542
CO_2R 4 eq	OAc	Catalyst (0.15 eq), toluene, rt, 96 h	CO_2R	539

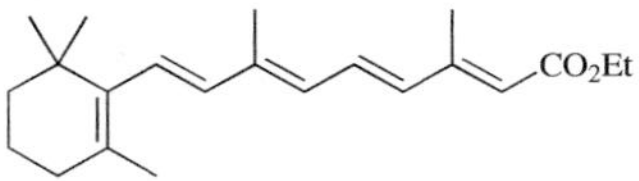

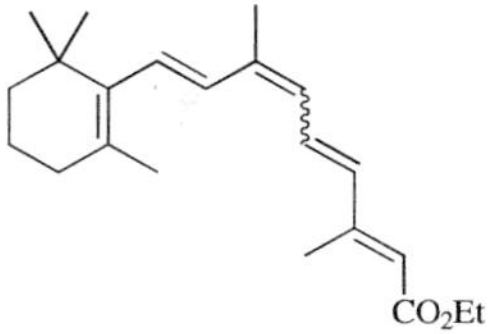

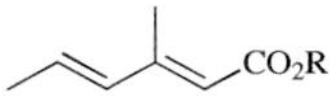

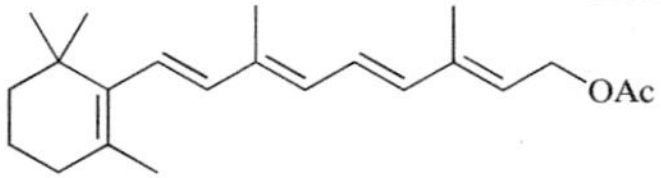

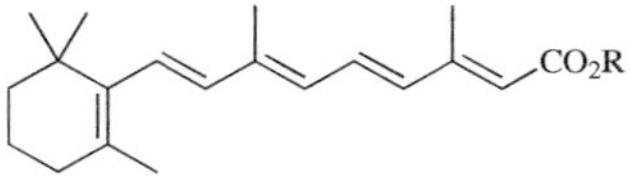

R	Catalyst		(*E*)/(*Z*)
Et	Hov II	(55)	97:3
Et	Gr II	(35)	99:1
Ph	Hov II	(40)	98:2

TABLE 1. CROSS-METATHESIS OF UNSATURATED COMPOUNDS WITH FUNCTIONAL GROUPS IN HOMOALLYLIC AND MORE DISTANT POSITIONS (*Continued*)

C. COMPOUNDS WITH CONJUGATED DOUBLE BONDS (*Continued*)

Alkene	Cross Partner	Conditions	Product(s) and Yield(s) (%)	Refs.

Please refer to the charts preceding the tables for the catalyst structures.

C_7

Alkene	Cross Partner	Conditions	Product(s) and Yield(s) (%)	Refs.
CO_2Et, 4 eq	( )$_2$	Hov II (0.15 eq), toluene, rt	CO_2Et (13) (*E*)/(*Z*) = 94:6 + CO_2Et (21) (*E*)/(*Z*) =88:17	539
CO_2R, 4 eq	( )$_2$	Catalyst (0.15 eq), toluene, rt	CO_2R **I** + CO_2R **II**	539

R	Catalyst	Time (h)	**I**	(*E*)/(*Z*) **I**	**II**	(*E*)/(*Z*) **II**
Et	Hov II	24	(9)	97:3	(38)	96:4
Et	Hov II	96	(31)	90:10	(15)	94:6
Ph	Hov II	24	(6)	95:5	(50)	99:1
Ph	Hov II	96	(33)	97:3	(57)	98:2
Ph	Gr II	96	(4)	97:3	(23)	99:1

Reactant 1	Reactant 2	Conditions	Product (yield %)	Ref.
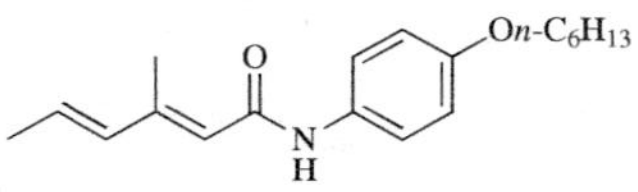 1.2 eq		Hov II (0.15 eq), toluene/DCM, rt, 96 h	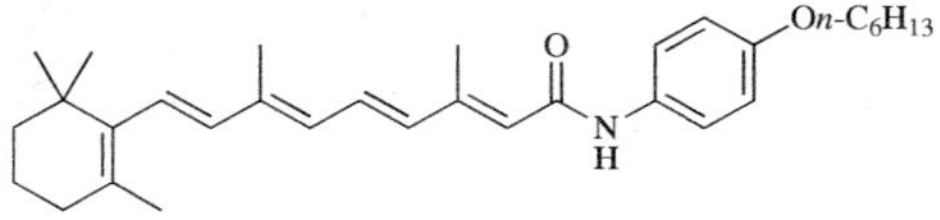 (26) (*E*)/(*Z*) = 99:1	539
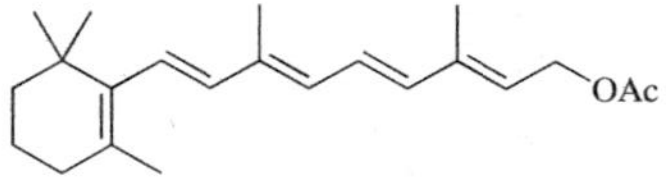 2 eq		Hov II (0.15 eq), DCM, rt, 96 h	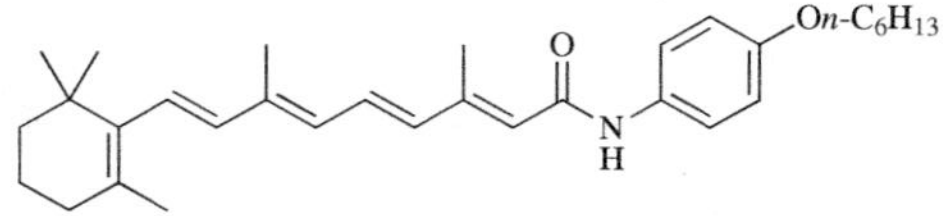 (22) (*E*)/(*Z*) = 99:1	539
			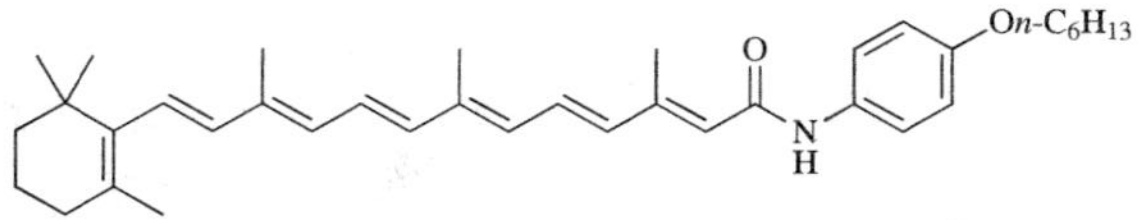 (32) (*E*)/(*Z*) = 95:5	
C_8				
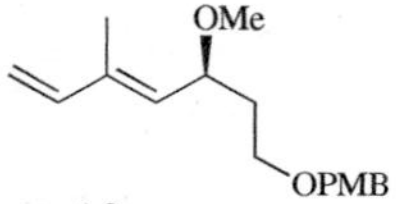 1.2 eq 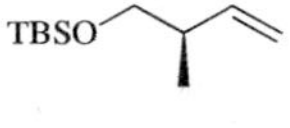		Gr II (0.01 eq), toluene, reflux, 3 h	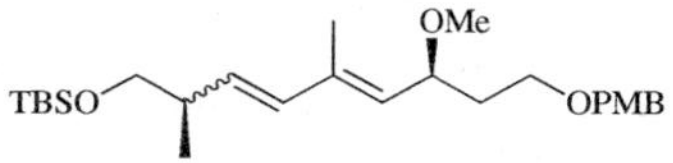 (65)	543

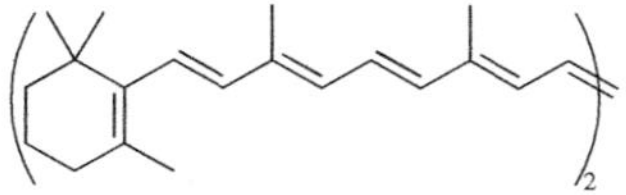

Structure labels: On-C_6H_{13}; N H; O; OAc; ( )$_2$ +; OMe; OPMB; TBSO

TABLE 1. CROSS-METATHESIS OF UNSATURATED COMPOUNDS WITH FUNCTIONAL GROUPS IN HOMOALLYLIC AND MORE DISTANT POSITIONS (*Continued*)

C. COMPOUNDS WITH CONJUGATED DOUBLE BONDS (*Continued*)

Alkene	Cross Partner	Conditions	Product(s) and Yield(s) (%)	Refs.
Please refer to the charts preceding the tables for the catalyst structures.				
C_{8-10}				
R² (3 eq)	CO₂R¹	1. Ru cat **152** (0.01 eq), (*Z*)-butene (20 eq), THF, 22°, 1 h 2. Ru cat **152** (0.04 eq), THF, 100 torr, 22°, 8 h	R¹O₂C … R² R¹ / R² / (*Z*)/(*E*) H / HO(CH₂)₄ / (59) / 95:5 Bn / Ph / (80) / 97:3 Bn / 4-MeOC₆H₄ / (56) / 97:3	395
C_8				
OR (n = 4)	OR (n = 4)	Ru cat **216** (0.027 eq), 50°, 24 h	RO … OR R / (*Z*)/(*E*) H / (57) / >95:5 Ac / (74) / >95:5	538
OAc (n = 4) (1.5 eq)	*n*-Bu	Ru cat **216** (0.04 eq), DCE, 50°, 24 h	*n*-Bu … OAc (30) (*Z*)/(*E*) > 95:5	538
C_{8-10}				
R (1.5 eq)	OH (n = 7)	Ru cat **216** (0.04 eq), 50°, 24 h	HO … R R / (*Z*)/(*E*) AcO(CH₂)₄ / (75) / >95:5 Ph / (15) / >95:5 4-MeOC₆H₄ / (25) / >95:5	538

C_9

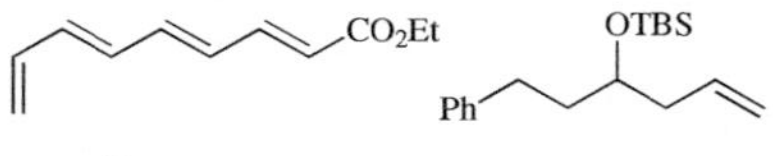

CO_2Et 2 eq; Ph, OTBS

Gr I (0.01 eq), DCM, reflux, 5 h

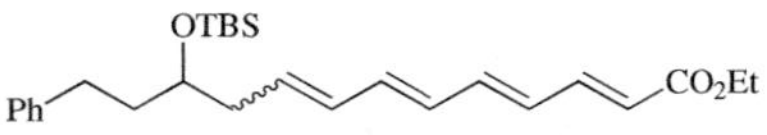

Ph, OTBS, CO_2Et

62% conv.[a]

541

OMe, OTBS, CO_2Me; TBSO, N, O, OMe

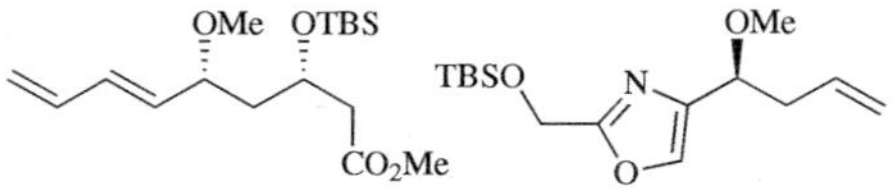

3 eq

Catalyst (0.1 eq), solvent

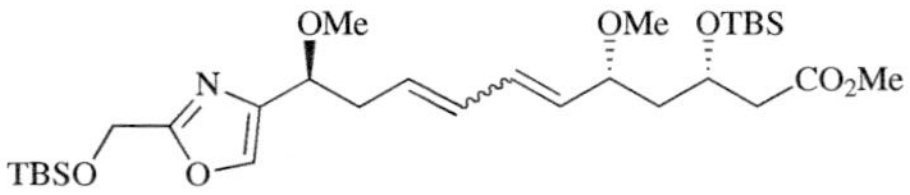

TBSO, N, O, OMe, OMe, OTBS, CO_2Me

544

Catalyst	Solvent	Temp (°)	
Gr I	toluene	80	(—)
Gr II	DCM	40	(—)
Hov II	DCM	40	(—)
Gr II	toluene	60	(24)
Hov II	toluene	60	(40)
Hov II	toluene	80	(51)

C_{11}

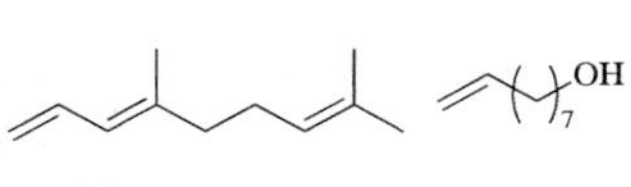

1.5 eq; $(\ )_7$ OH

Ru cat **216** (0.02 eq) 50°, 24 h

HO $(\ )_7$

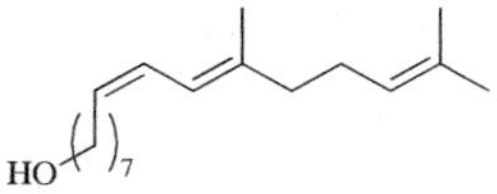

(>95) (*Z*)/(*E*) > 95:5

538

TABLE 1. CROSS-METATHESIS OF UNSATURATED COMPOUNDS WITH FUNCTIONAL GROUPS IN HOMOALLYLIC AND MORE DISTANT POSITIONS (*Continued*)

C. COMPOUNDS WITH CONJUGATED DOUBLE BONDS (*Continued*)

Alkene	Cross Partner	Conditions	Product(s) and Yield(s) (%)	Refs.
Please refer to the charts preceding the tables for the catalyst structures.				
C_{13}				
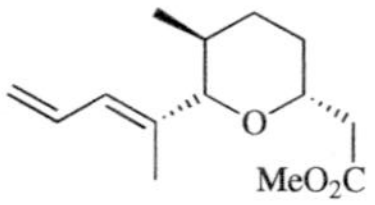 MeO$_2$C; 3 eq	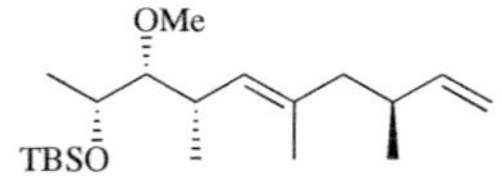 OMe, TBSO	Hov II (0.05 eq), DCM, rt, 48 h	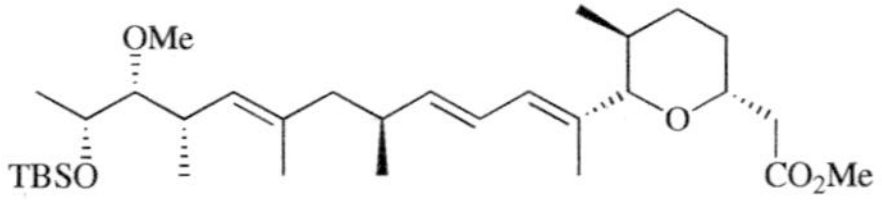 OMe, TBSO, CO$_2$Me; (76) (*E*)/(*Z*) = 93:7	545
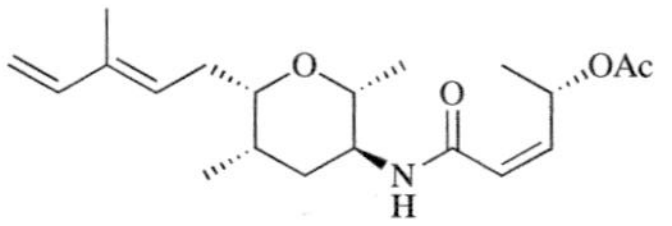 O, OAc, N H	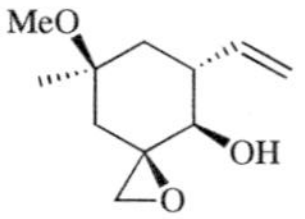 MeO, OH, O; 1.2 eq	Gr II (0.08 eq), DCM, reflux, 5 h	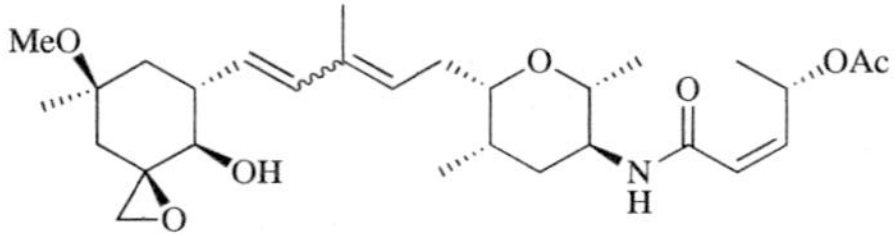 MeO, OH, O, OAc, N H; (52)	546
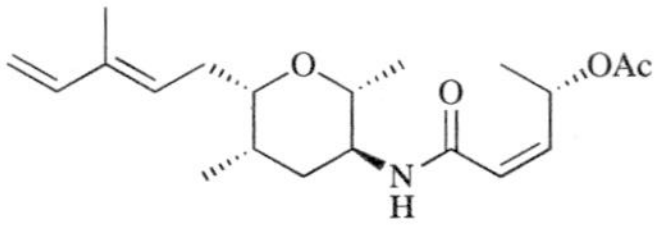 O, OAc, N H	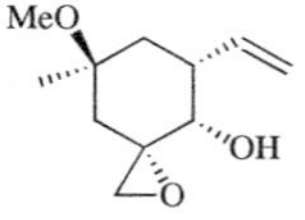 MeO, OH, O; 1.2 eq	Gr II (0.08 eq), DCM, reflux, 5 h	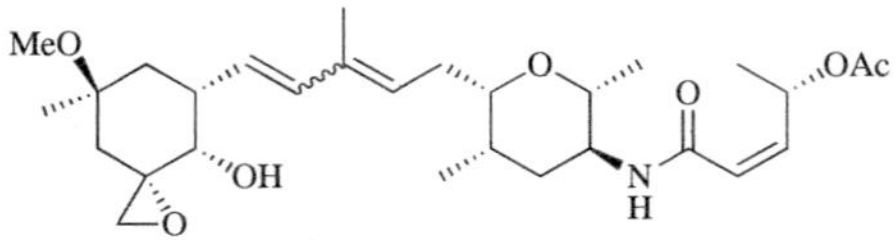 MeO, OH, O, OAc, N H; (55)	546

C_{13}	OAc, O, N H, MeO, OH, O 1.2 eq	Gr II (0.08 eq), DCM, reflux, 5 h	MeO, OH, O, O, O, N H, OAc (57)	546
	OAc, O, N H, MeO, OH, O 1.2 eq	Gr II (0.08 eq), DCM, reflux, 5 h	MeO, OH, O, O, O, N H, OAc (53)	546

TABLE 1. CROSS-METATHESIS OF UNSATURATED COMPOUNDS WITH FUNCTIONAL GROUPS IN HOMOALLYLIC AND MORE DISTANT POSITIONS (*Continued*)

C. COMPOUNDS WITH CONJUGATED DOUBLE BONDS (*Continued*)

Alkene	Cross Partner	Conditions	Product(s) and Yield(s) (%)	Refs.
Please refer to the charts preceding the tables for the catalyst structures.				
C_{14}				
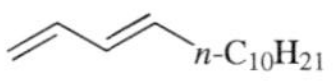 n-$C_{10}H_{21}$ (diene) 1.5 eq	 R (alkene)	Ru cat **216** (*x* eq) 50°, 24 h	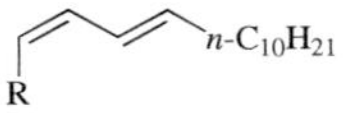 R, n-$C_{10}H_{21}$ (diene)	538
			R | *x* | | (*Z*)/(*E*)	
			$HO(CH_2)_7$ | 0.02 | (84) | >95:5	
			$OHC(CH_2)_8$ | 0.04 | (59) | >95:5	
			$AcO(CH_2)_4$ | 0.04 | (69) | >95:5	
			$Ac(CH_2)_2$ | 0.04 | (41) | >95:5	
			$Br(CH_2)_6$ | 0.04 | (34) | >95:5	
			$NC(CH_2)_2$ | 0.04 | (12) | >95:5	
	n-$C_{10}H_{21}$ (diene)	Ru cat **216** (0.027 eq) DCE, 50°, 24 h	n-$C_{10}H_{21}$, n-$C_{10}H_{21}$ (triene) (62) (*Z*)/(*E*) > 95:5	538
C_{20}				
OAc (retinyl acetate)	CO_2Et (alkene) 2 eq	Hov II (0.15 eq), toluene, rt, 96 h	CO_2Et (product) (10) (*E*)/(*Z*) = 83:17	539

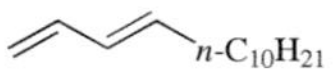

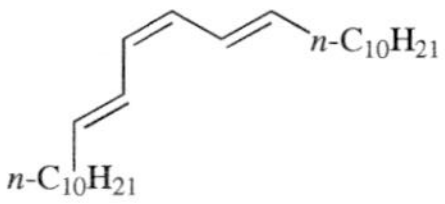

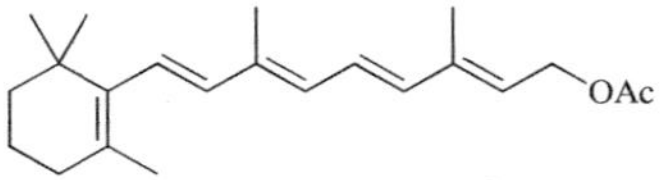

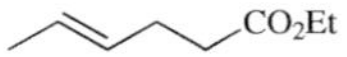

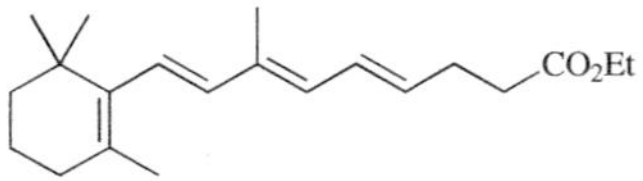

C_{40}

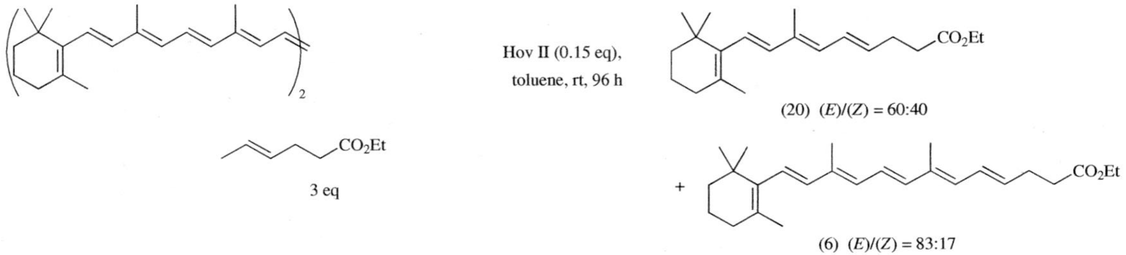

CO_2Et

3 eq

Hov II (0.15 eq), toluene, rt, 96 h

CO_2Et

(20) (*E*)/(*Z*) = 60:40

+

CO_2Et

(6) (*E*)/(*Z*) = 83:17

539

[a] The conversion was determined by ^{1}H NMR analysis.

TABLE 2. CROSS-METATHESIS OF ALLYLIC BORONATES

Allylic Boronate	Cross Partner	Conditions	Product(s) and Yield(s) (%)	Refs.
Please refer to the charts preceding the tables for the catalyst structures.				
C_3				
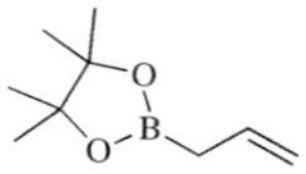 O-B-O	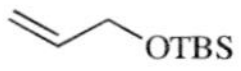 OTBS 5 eq	1. W cat **3** (0.05 eq), benzene, 100 torr, 22°, 2 h 2. PhCHO (2 eq), THF, rt, 2 h	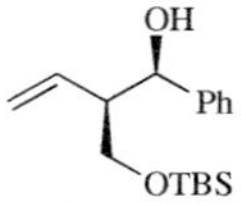 OH, Ph, OTBS (79) dr 91:9	547
	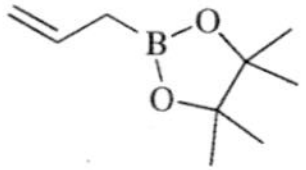 B-O-O	Catalyst (0.02 eq), THF, 35°	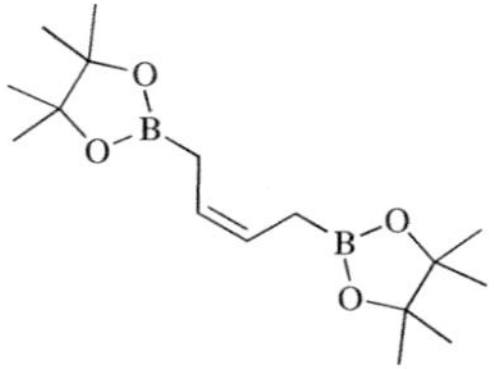 O-B-O, B-O-O	
			Catalyst: Ru cat **217**; Time (h): 4; (74); $(Z)/(E)$: 95:5	84
			Catalyst: Ru cat **215**; Time (h): 3; (36); $(Z)/(E)$: 95:5	488
	 CO_2Ph 5 eq	1. W cat **3** (0.05 eq), benzene, 100 torr, 22°, 2 h 2. NaOH, H_2O	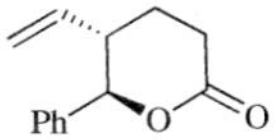 Ph, O, O (68) dr > 98:2	547
	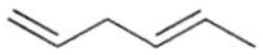 9 eq	Ru cat **216** (0.01 eq), THF, rt, 5 h	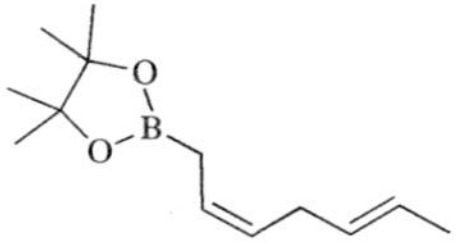 O-B-O (65) $(Z)/(E)$ > 95:5	437

R, n; 5 eq

1. W cat **3** (0.05 eq), benzene, 100 torr, 22°, 2 h
2. PhCHO (2 eq), THF, rt, 2 h

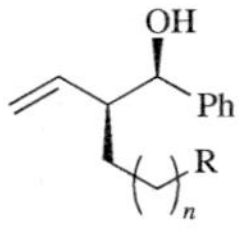

OH, Ph, R, n

R	n		dr
Br	5	(82)	95:5
PMBO	5	(84)	96:4
Me	6	(72)	92:8

547

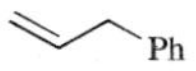

Ph; 5 eq

1. W cat **3** (0.05 eq), benzene, 100 torr, 22°, 2 h
2. PhCHO (2 eq), THF, rt, 2 h

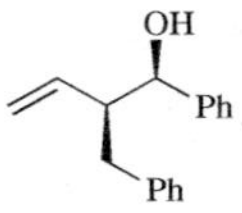

OH, Ph, Ph

(91) dr 95:5

547

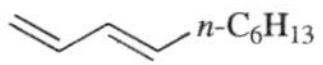

n-C_6H_{13}; 5 eq

1. W cat **3** (0.05 eq), benzene, 100 torr, 22°, 2 h
2. PhCHO (2 eq), THF, rt, 2 h

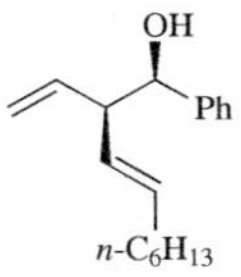

OH, Ph, n-C_6H_{13}

(72) dr 95:5

547

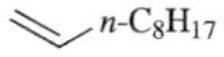

n-C_8H_{17}; 5 eq

1. Catalyst (0.05 eq), benzene, 100 torr, 22°, 2 h
2. H_2O_2, NaOH, 22°, 2 h

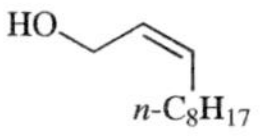

HO, n-C_8H_{17}

Catalyst		(*E*)/(*Z*)
Mo cat **1**	(54)	74:26
Hov II	(31)	85:15
Gr II	(16)	80:20
W cat **3**	(65)	5:95

547

TABLE 2. CROSS-METATHESIS OF ALLYLIC BORONATES (*Continued*)

Allylic Boronate	Cross Partner	Conditions	Product(s) and Yield(s) (%)	Refs.
Please refer to the charts preceding the tables for the catalyst structures.				
C_3				
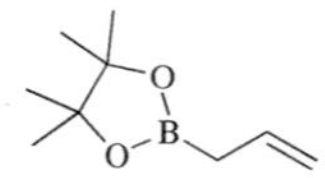 O, B, O	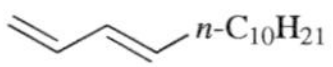 n-$C_{10}H_{21}$ 1.5 eq	Ru cat **216** (0.04 eq) 50°, 24 h	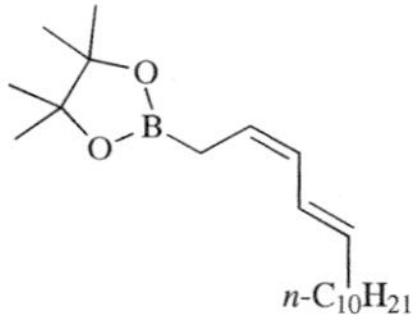 O, B, O, n-$C_{10}H_{21}$ (60) (*Z*)/(*E*) > 95:5	538
3.5 eq	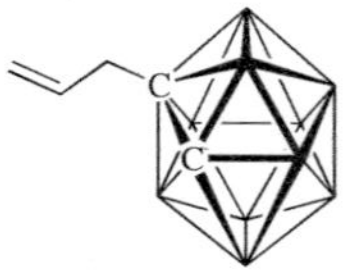 C, C	Gr II (0.013 eq), DCM, reflux, 14 h	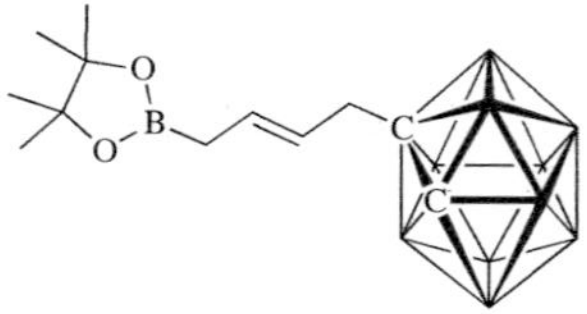 O, B, O, C, C (55) (*E*)/(*Z*) = 89:11	436

TABLE 3. CROSS-METATHESIS OF ALLYLIC AMINES (NOT INCLUDING AMINES WITH A SUBSTITUTED CARBON IN THE α-POSITION)

A. PRIMARY AMINES

Primary Amine	Cross Partner	Conditions	Product(s) and Yield(s) (%)	Refs.
Please refer to the charts preceding the tables for the catalyst structures.				
C_{3-11}				
 R–CH=CH–$(CH_2)_n$–NH_3^+ Y^-	 R–CH=CH–$(CH_2)_n$–NH_3^+ Y^-	1. Hov II (0.05 eq) EtOAc, reflux, 16 h 2. H_2 (60 psi), MeOH rt, 16 h	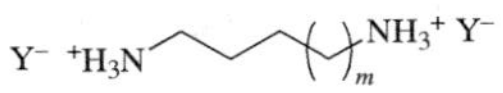 Y^- ^+H_3N–…–$(CH_2)_m$–NH_3^+ Y^- R / Y / n / m H / TfO / 1 / 1 (91) H / BF_4 / 1 / 1 (>99) H / TsO / 2 / 3 (90) Me / TsO / 8 / 15 (>99)	548
C_3				
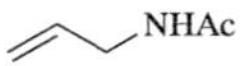 NHAc (2 eq)	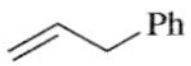 Ph	Gr I (0.05 eq), DCM, reflux, 16 h	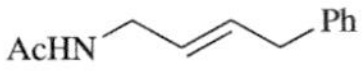 AcHN … Ph (49) (*E*)/(*Z*) = 90:10	156
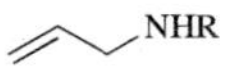 NHR	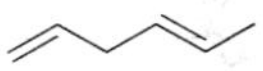 (9 eq)	Ru cat **216** (0.01 eq), THF, rt, 5 h	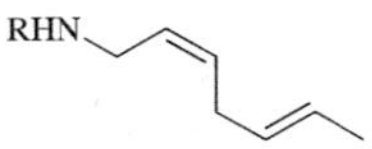 RHN … R / (*Z*)/(*E*) Ph (68) >95:5 Boc (54) >95:5	437
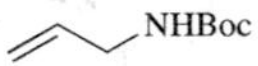 NHBoc	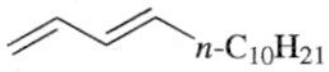 n-$C_{10}H_{21}$ (1.5 eq)	Ru cat **216** (0.04 eq), 50°, 24 h	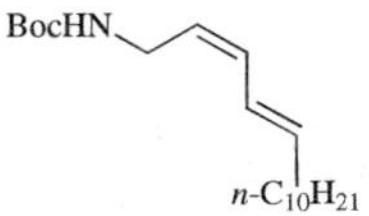 BocHN … n-$C_{10}H_{21}$ (60) (*Z*)/(*E*) > 95:5	538

TABLE 3. CROSS-METATHESIS OF ALLYLIC AMINES (NOT INCLUDING AMINES WITH A SUBSTITUTED CARBON IN THE α-POSITION) (*Continued*)
A. PRIMARY AMINES (*Continued*)

Primary Amine	Cross Partner	Conditions	Product(s) and Yield(s) (%)	Refs.
Please refer to the charts preceding the tables for the catalyst structures.				
C_3				
 NHBoc 3 eq	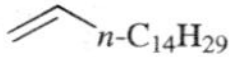 n-$C_{14}H_{29}$	Mo cat **18** (0.03 eq), benzene, 7.0 torr, 5 h, 22°	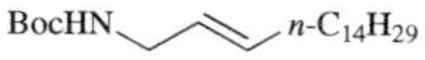 BocHN n-$C_{14}H_{29}$ (75) $(Z)/(E)$ = 81:19	5
20 eq	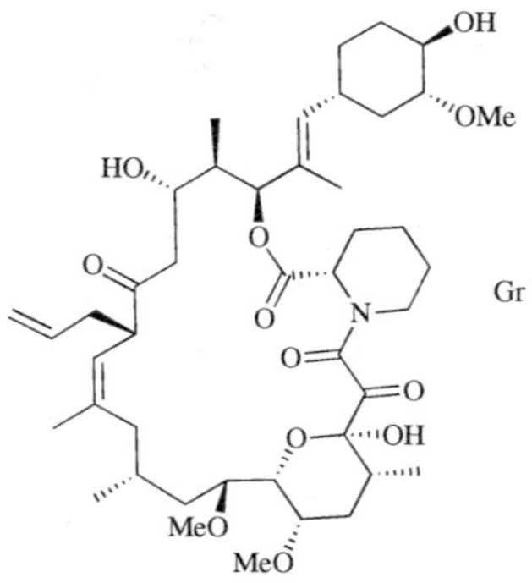 HO, OH, OMe, O, N, O, O, OH, MeO, MeO	Gr II (0.1 eq), DCB	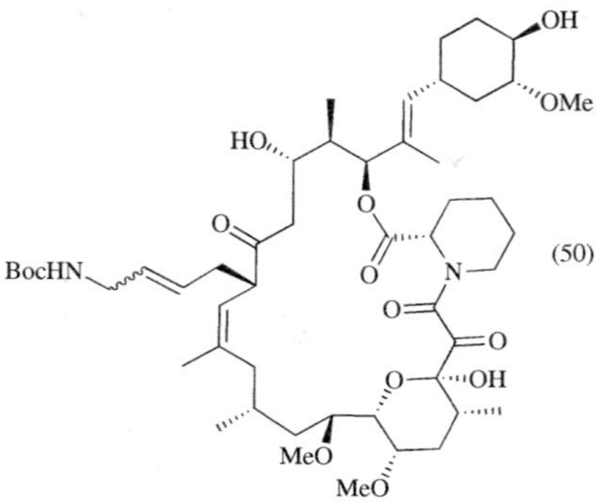 BocHN, HO, OH, OMe, O, N, O, O, OH, MeO, MeO (50)	432
5.5 eq	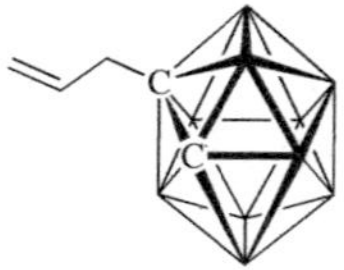 C, C	Gr II (0.036 eq), DCM, reflux, 48 h	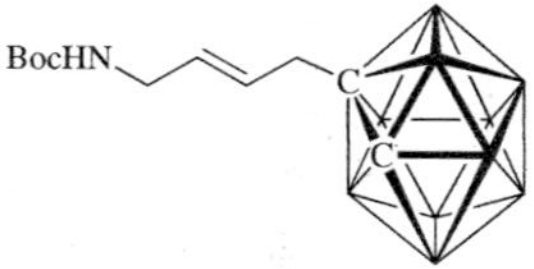 BocHN, C, C (70) $(E)/(Z)$ > 95:5	436

NHTs 2 eq | OBz | Catalyst (x eq), DCM, 40° | TsHN~OBz

Catalyst	x	Time (h)		(E)/(Z)	Refs.
Ru cat **286**	0.05	8	(39)	>95:5	161
Ru cat **93**	0.02	24	(60)	>95:5	550

5 eq | CO_2Me | Hov II (0.05 eq), DCM, rt, 48 h | TsHN~CO_2Me (36) (E)/(Z) = 95:5 + TsHN~CO_2Me (9) | 549

2 eq | OTBS | Ru cat **286** (x eq), DCM, 40°, 8 h | TsHN~OTBS | 161

x		(E)/(Z)
0.02	(18)	>95:5
0.05	(45)	>95:5

O, N, O | n 5 eq | Catalyst (x eq), solvent | O, N, O, n | 551

n	Catalyst	x	Solvent	Temp (°)		(E)/(Z)
3	Hov II	0.05	DCM	40	(70)	88:12
7	Ru cat **216**	0.02	THF	rt	(62)	>5:95

TABLE 3. CROSS-METATHESIS OF ALLYLIC AMINES (NOT INCLUDING AMINES WITH A SUBSTITUTED CARBON IN THE α-POSITION) (*Continued*)
A. PRIMARY AMINES (*Continued*)

Primary Amine	Cross Partner	Conditions	Product(s) and Yield(s) (%)	Refs.
Please refer to the charts preceding the tables for the catalyst structures.				
C_3				
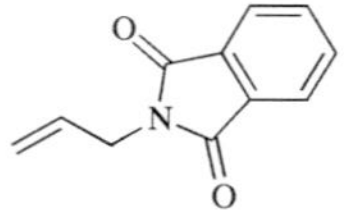 O, N, O	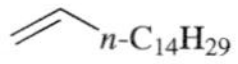 n-$C_{14}H_{29}$ 3 eq	Mo cat **18** (0.03 eq), benzene, 7.0 torr, 5 h, 22°	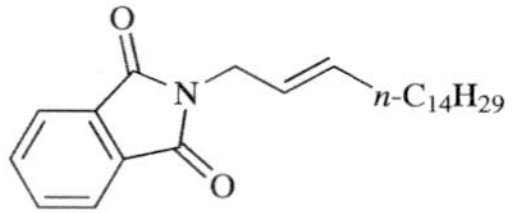 O, N, O, n-$C_{14}H_{29}$ (87) $(Z)/(E)$ = 85:15	5
4.4 eq	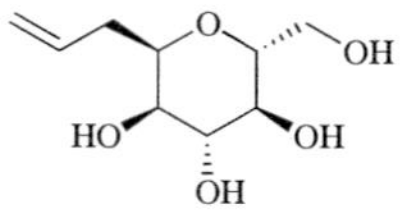 O, OH, HO, OH, OH	Hov II (0.1 eq), MeOH, $CHCl_3$, 40°, 36 h	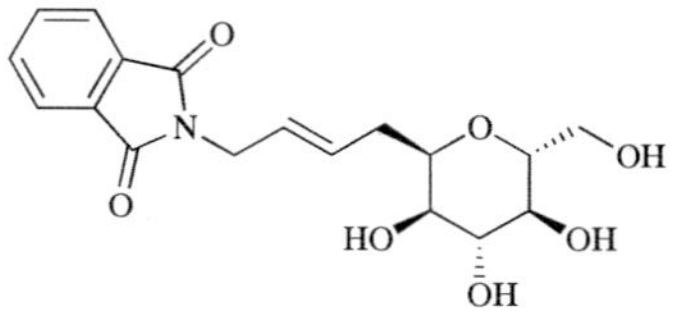 O, N, O, O, OH, HO, OH, OH (57) $(E)/(Z)$ = 80:20	552
C_4				
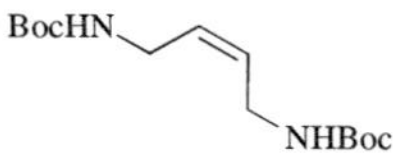 BocHN, NHBoc 4 eq	 OBz, 8	Gr II (0.05 eq), DCM, 45°	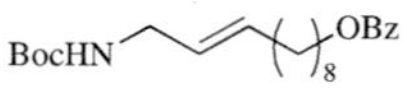 BocHN, OBz, 8 (71) $(E)/(Z)$ = 75:25	176

TABLE 3. CROSS-METATHESIS OF ALLYLIC AMINES (NOT INCLUDING AMINES WITH A SUBSTITUTED CARBON IN THE α-POSITION) (*Continued*)

B. SECONDARY AMINES

Secondary Amine	Cross Partner	Conditions	Product(s) and Yield(s) (%)	Refs.

Please refer to the charts preceding the tables for the catalyst structures.

C_3

NHPh | NHPh | Catalyst (0.02 eq), THF, 35° | PhHN NHPh

Catalyst	Time (h)		(Z)/(E)	Refs.
Ru cat **217**	2	(67)	71:29	84
Ru cat **215**	12	(12)	90:10	488

Bz N CO_2Me R | Bz N CO_2Me R | Gr II (*x* eq), DCE | R Bz N CO_2Me / CO_2Me N Bz R | 553

R	*x*	Temp (°)	Time		(E)/(Z)
i-Pr	0.15	24	18 h	(0)	—
i-Pr	0.15	80	72 h	(27)	81:19
i-Pr	0.20	100 (MW)	10 min	(40)	83:17
i-Bu	0.15	24	18 h	(0)	—
i-Bu	0.15	80	72 h	(32)	75:25
i-Bu	0.20	100 (MW)	10 min	(54)	68:32
Bn	0.075	24	18 h	(0)	—
Bn	0.075	24	72 h	(0)	—
Bn	0.075	80	18 h	(29)	70:30
Bn	0.075	80	72 h	(34)	70:30
Bn	0.05	100 (MW)	10 min	(48)	73:27

TABLE 3. CROSS-METATHESIS OF ALLYLIC AMINES (NOT INCLUDING AMINES WITH A SUBSTITUTED CARBON IN THE α-POSITION) (*Continued*)
B. SECONDARY AMINES (*Continued*)

Secondary Amine	Cross Partner	Conditions	Product(s) and Yield(s) (%)	Refs.

Please refer to the charts preceding the tables for the catalyst structures.

C_{11}

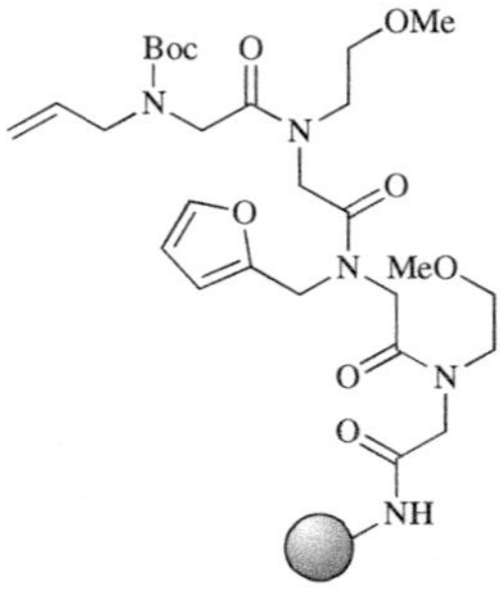

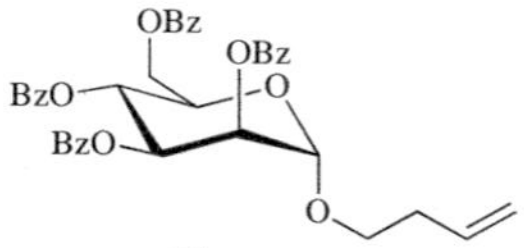

Boc O OMe N N O O N MeO O N O NH

OBz OBz BzO O BzO O

20 eq

1. Catalyst (0.05 eq), solvent, time
2. $TFA/H_2O/TIS$ (92:5:3), 1–2 h

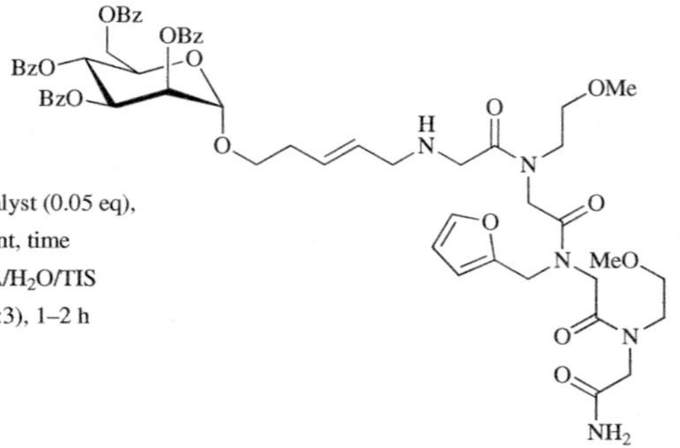

OBz OBz BzO O BzO O H N O OMe N O O N MeO O N O NH_2

12

Catalyst	Solvent	Temp (°)	Time	Conv. (%)[a]
Gr I	DCM	40	8 h	<5
Gr II	DCM	40	8 h	10
Hov II	DCM	40	8 h	29
Gr I	DCB	MW	3 min	<5
Gr II	DCB	MW	3 min	<5
Hov II	DCB	MW	3 min	20

[a] The conversion was determined by HPLC analysis.

TABLE 3. CROSS-METATHESIS OF ALLYLIC AMINES (NOT INCLUDING AMINES WITH A SUBSTITUTED CARBON IN THE α-POSITION) (*Continued*)

C. TERTIARY AMINES

Tertiary Amine	Cross Partner	Conditions	Product(s) and Yield(s) (%)	Refs.

Please refer to the charts preceding the tables for the catalyst structures.

C_{10}

Tertiary Amine	Cross Partner	Conditions	Product(s) and Yield(s) (%)	Refs.
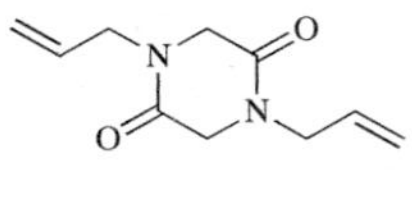	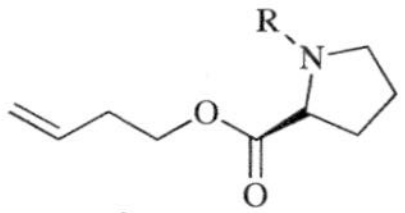 3 eq	Gr II (0.10 eq), $CHCl_3$, reflux, 10 h	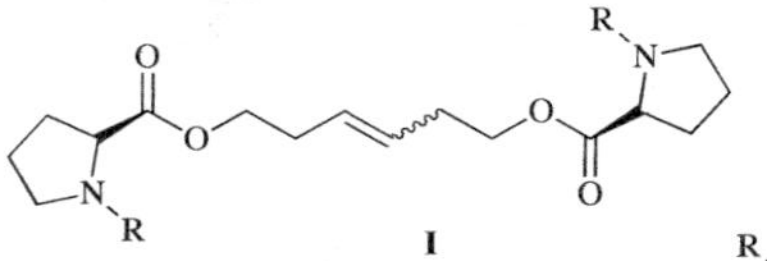 I 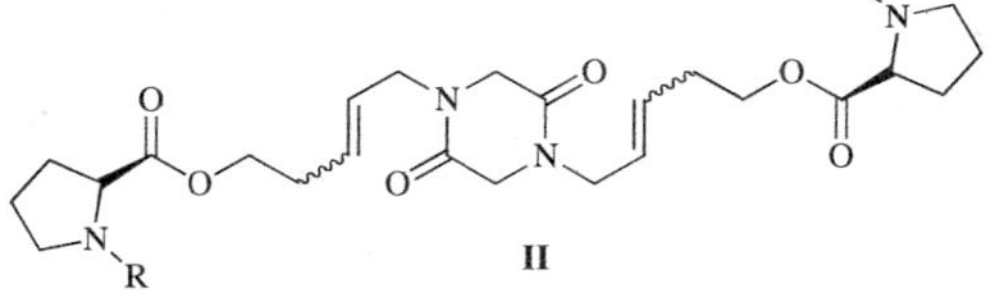 + II	554
3 eq	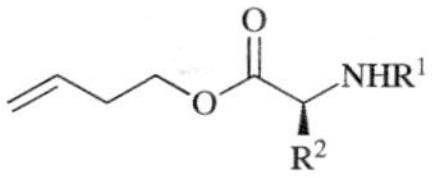	Gr II (0.10 eq), $CHCl_3$, reflux, 10 h	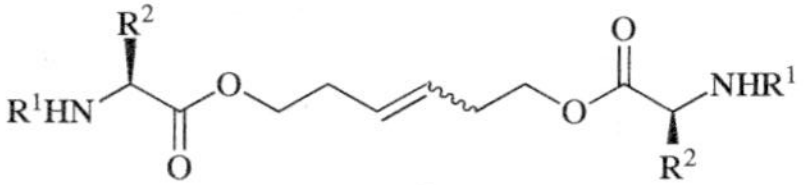 I 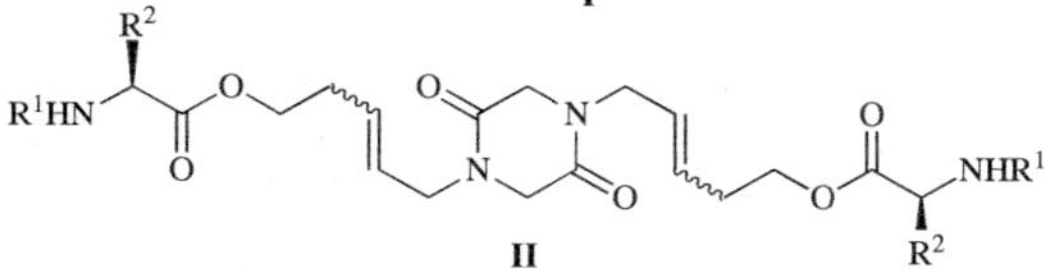 + II	554

R	I	II
Boc	(55)	(35)
Fmoc	(52)	(46)

R^1	R^2	I	II
Boc	$PhCH_2$	(50)	(44)
Boc	Me	(66)	(39)
Boc	$CH_3S(CH_2)_2$	(14)	(0)
Boc	i-$Pr(CH_2)_2$	(59)	(30)
Fmoc	$PhCH_2$	(48)	(42)
Fmoc	H	(44)	(30)

TABLE 3. CROSS-METATHESIS OF ALLYLIC AMINES (NOT INCLUDING AMINES WITH A SUBSTITUTED CARBON IN THE α-POSITION) (*Continued*)

C. TERTIARY AMINES (*Continued*)

Tertiary Amine	Cross Partner	Conditions	Product(s) and Yield(s) (%)	Refs.
Please refer to the charts preceding the tables for the catalyst structures.				
C_{10}				
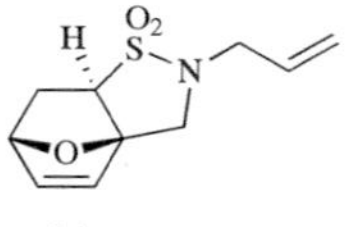 (±)	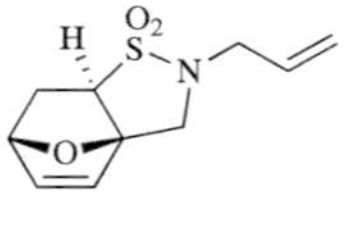 (±)	Gr II (0.1 eq), DCM, 50°, 3 h	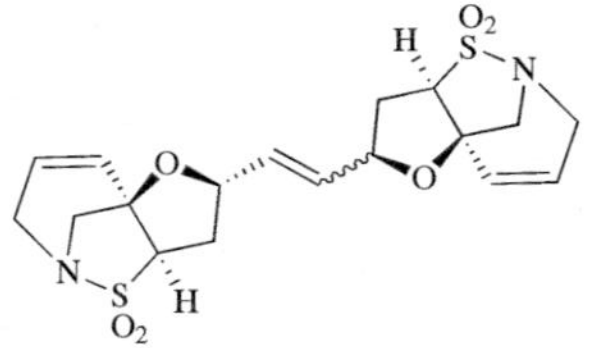 (84)	359
C_{12}				
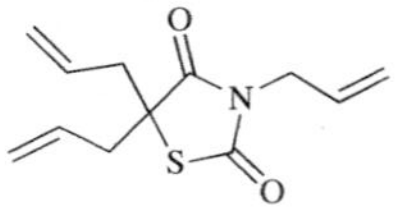	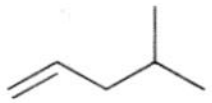 5 eq	Hov II (0.02 eq), DCM, rt, 8 h	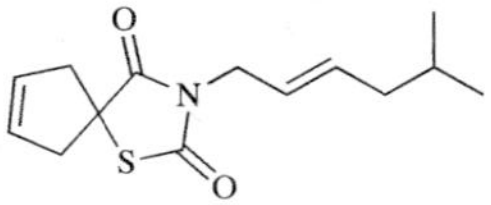 (57) $(E)/(Z) = 87:13$ 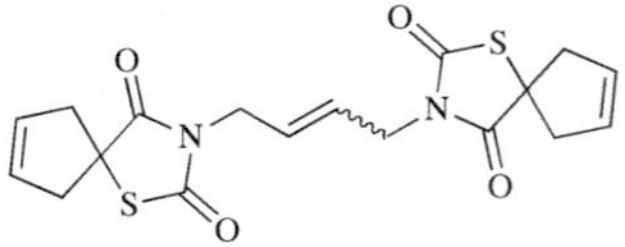 + (14)	555
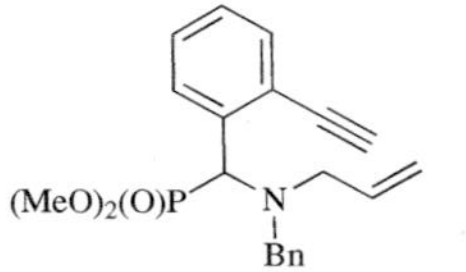 $(MeO)_2(O)P$, Bn	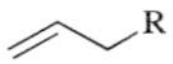 R, 5 eq	Gr II (0.1 eq), DCM, reflux, 2 + 4 h	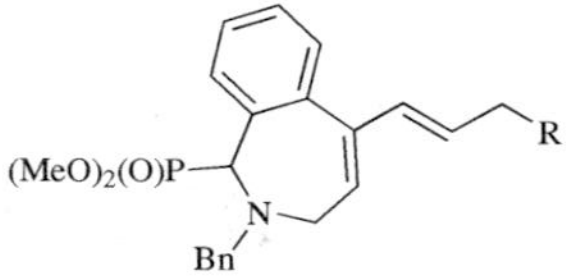 $(MeO)_2(O)P$, Bn, R	556

R		$(E)/(Z)$
TMS	(74)	66:34
$AcCH_2$	(68)	67:33

C$_{13}$

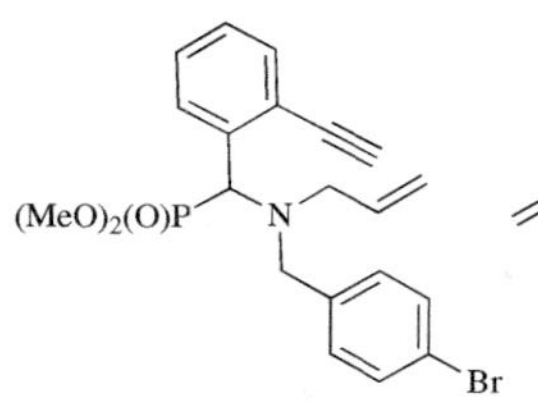

Substrate	Olefin	Conditions	Product(s) and Yield(s) (%)	Refs.
	5 eq	Gr II (0.1 eq), DCM, reflux, 2 + 4 h	(*E*) only R: Br (50); Et (70)	556
	10 eq	Gr II (0.1 eq), DCM, rt, 3–12 h	*n* = 2 (81), (*E*)/(*Z*) 60:40 *n* = 3 (82), (*E*)/(*Z*) 67:33	557
	10 eq	Gr II (0.1 eq), DCM, rt, 3–12 h	*n* = 2 (30), (*E*)/(*Z*) 50:50 *n* = 3 (41), (*E*)/(*Z*) 55:45	557

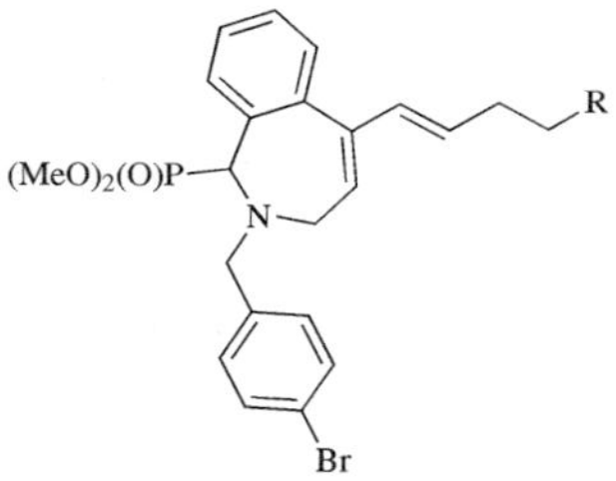

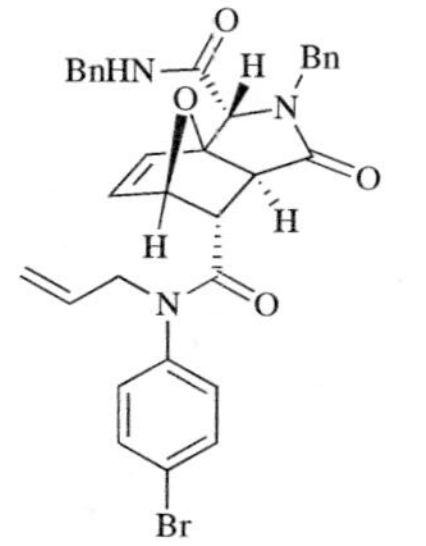

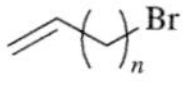

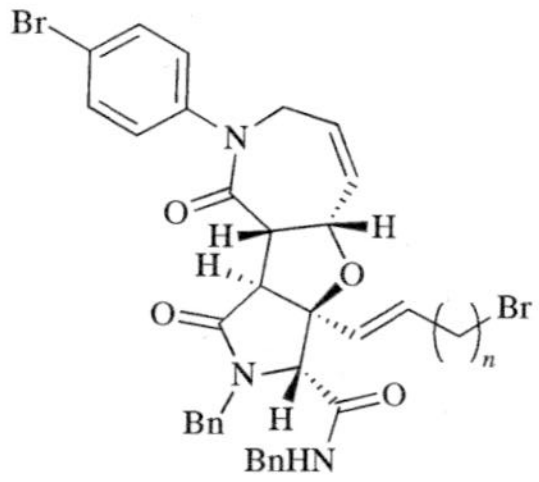

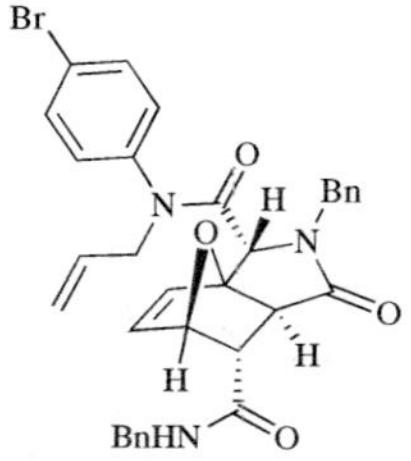

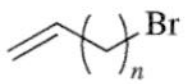

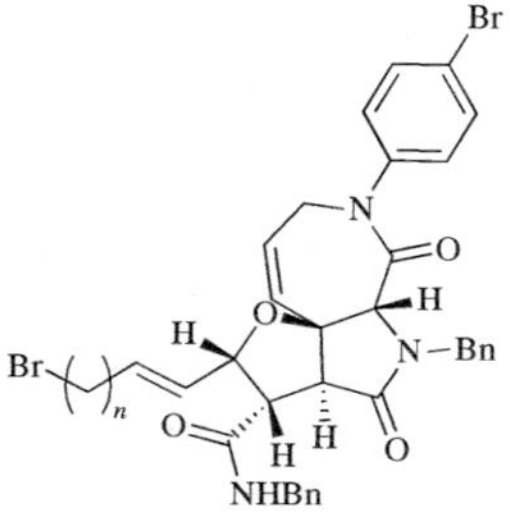

TABLE 3. CROSS-METATHESIS OF ALLYLIC AMINES (NOT INCLUDING AMINES WITH A SUBSTITUTED CARBON IN THE α-POSITION) (*Continued*)

C. TERTIARY AMINES (*Continued*)

Tertiary Amine	Cross Partner	Conditions	Product(s) and Yield(s) (%)	Refs.
Please refer to the charts preceding the tables for the catalyst structures.				
C_{36}				
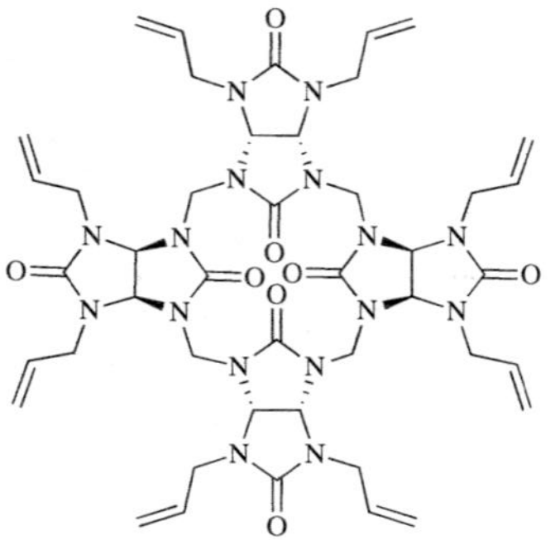 [structure: tetraallyl glycoluril macrocycle]	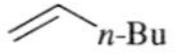 n-Bu-substituted alkene, 2.5 eq	Gr II (0.1 eq), DCM	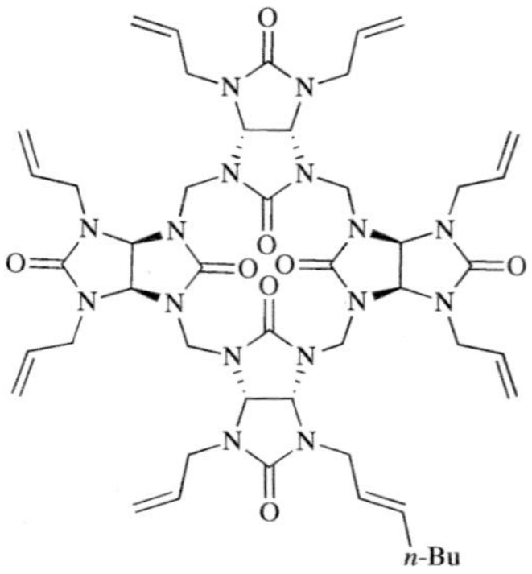 [structure: cross-metathesis product with n-Bu] (20)	558

TABLE 4. CROSS-METATHESIS OF PRIMARY ALLYLIC ALCOHOLS AND DERIVATIVES

A. ALLYLIC ALCOHOLS

Allylic alcohol	Cross Partner	Conditions	Product(s) and Yield(s) (%)	Refs.
Please refer to the charts preceding the tables for the catalyst structures.				
C_3				
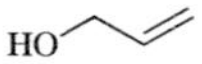 HO, 10 eq	 SBn, n	Hov II (0.06 eq), t-BuOH/H_2O (1:1), 32°, 3.5 h	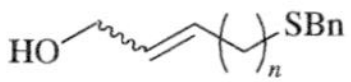 HO SBn, n: 0 (0); 1 (52); 2 (19); 3 (8)	559
10 eq	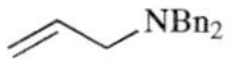 NBn_2	Hov II (0.06 eq), t-BuOH/H_2O (1:1), 32°, 3.5 h	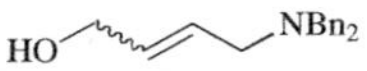 HO NBn_2 (11)	559
2 eq	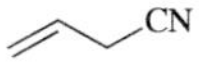 CN	Gr II (0.05 eq), DCM, reflux, 2 h	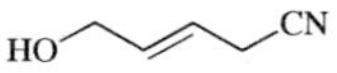 HO CN (38) $(E)/(Z) = 94:6$	182
2 eq	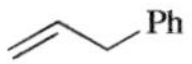 Ph	Gr I (0.05 eq), DCM, reflux, 16 h	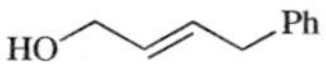 HO Ph (72) $(E)/(Z) = 75:25$	156
C_4				
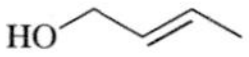 HO	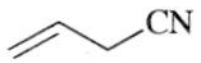 CN, 2 eq	Gr II (0.05 eq), DCM, reflux, 2 h	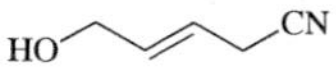 HO CN (46) $(E)/(Z) = 94:6$	182

TABLE 4. CROSS-METATHESIS OF PRIMARY ALLYLIC ALCOHOLS AND DERIVATIVES (*Continued*)

A. ALLYLIC ALCOHOLS (*Continued*)

Allylic alcohol	Cross Partner	Conditions	Product(s) and Yield(s) (%)	Refs.
Please refer to the charts preceding the tables for the catalyst structures.				
C_4				
HO OH (2 eq)	OBz (n)	Catalyst (0.05 eq), $AlCl_3$ (x eq), THF, rt, 6 h	HO * OBz (n)	560
2 eq	Br	Hov II (0.01 eq), DCM, 23°, 4 h	HO Br (97)	180
3 eq	OPMB	Hov II (0.01 eq), DCM, rt, 4 h	HO OPMB (85)	561
2 eq	OR	Ru cat **152** (0.05 eq), THF, 22°, 4 h	HO OR	183

n	Catalyst	x		*$(E)/(Z)$
0	Ru cat **366**	0	(19)	80:20
0	Ru cat **366**	0.05	(84)	4:96
3	Ru cat **365**	0	(27)	88:12
3	Ru cat **366**	0	(35)	86:14
3	Ru cat **366**	0.05	(77)	10:90
3	Ru cat **367**	0	(30)	84:16
8	Ru cat **366**	0	(38)	89:11
8	Ru cat **366**	0.05	(63)	10:90

R		$(Z)/(E)$
TBS	(65)	93:7
4-$O_2NC_6H_4$	(74)	96:4

2 eq	R	Catalyst (0.05 eq), THF, 22°		183

R	Catalyst	Time (h)		(Z)(E)
MeO	Ru cat **152**	8	(63)	92:8
t-BuO_2C	Ru cat **152**	8	(56)	96:4
n-C_5H_{11}	Ru cat **152**	8	(54)	87:13
n-C_6H_{13}	Ru cat **152**	8	(66)	95:5
n-C_6H_{13}	Ru cat **215**	12	(53)	93:7

1 eq	OBz	Catalyst (0.025 eq), DCM, rt, 5 h	HO, OBz (E)/(Z) > 95:5	562

Catalyst	
Ru cat **41**	(75)
Ru cat **44**	(85)
Ru cat **45**	(84)

2 eq	CO_2Bn, NHFmoc	Ru cat **152** (0.05 eq), THF, 22°, 4 h	HO, CO_2Bn, NHFmoc (73) (Z)/(E) = 98:2	183
2 eq	R	Catalyst (0.05 eq), THF, 22°, 4 h	HO, R	183

R	Catalyst		(Z)/(E)
BnO_2C	Ru cat **152**	(80)	98:2
HO_2CCH_2	Ru cat **152**	(70)	96:4
HO_2CCH_2	Ru cat **215**	(<2)	—
$OHC(CH_2)_6$	Ru cat **152**	(80)	94:6
$OHC(CH_2)_6$	Ru cat **215**	(30)	87:13
n-C_8H_{17}	Ru cat **152**	(72)	96:4

TABLE 4. CROSS-METATHESIS OF PRIMARY ALLYLIC ALCOHOLS AND DERIVATIVES (*Continued*)

A. ALLYLIC ALCOHOLS (*Continued*)

Allylic alcohol	Cross Partner	Conditions	Product(s) and Yield(s) (%)	Refs.
Please refer to the charts preceding the tables for the catalyst structures.				
C_4				
HO, OH; 2 eq	OTBS	Ru cat **286** (0.02 eq), DCM, 40°, 8 h	 HO, OTBS (39) $(E)/(Z)$ = 91:9	161
2 eq	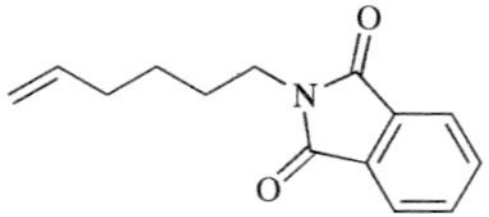 N, O, O	Ru cat **152** (0.05 eq), THF, 22°, 4 h	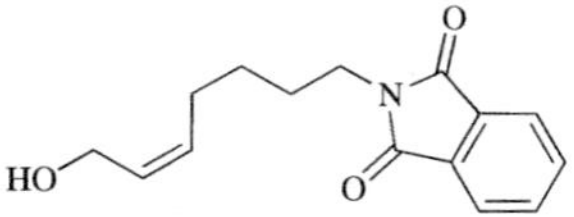 HO, N, O, O (64) $(Z)/(E)$ = 98:2	183
2 eq	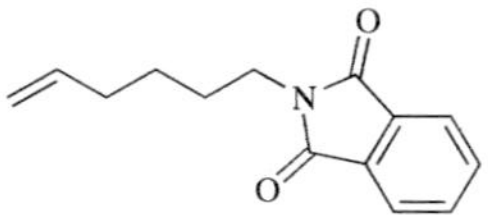 N, O, O	Catalyst (0.05 eq), $AlCl_3$ (*x* eq), THF, rt, 6 h	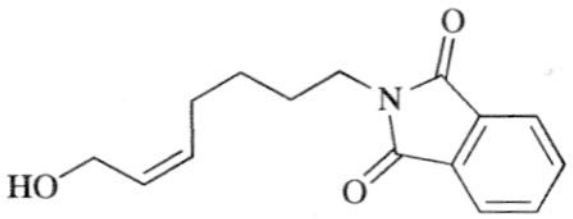 HO, N, O, O	560

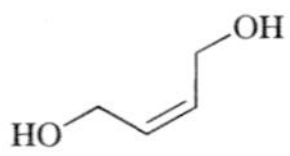

Catalyst	*x*		$(Z)/(E)$
Ru cat **365**	0	(13)	86:14
Ru cat **366**	0	(22)	83:17
Ru cat **366**	0.05	(87)	9:91
Ru cat **367**	0	(18)	80:20

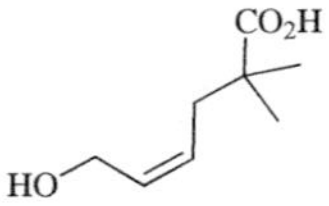

Allylic alcohol	Cross Partner	Conditions	Product(s) and Yield(s) (%)	Refs.
2 eq	CO_2H	Ru cat **152** (0.05 eq), THF, 22°, 4 h	HO, CO_2H (61) $(Z)/(E)$ = 98:2	183

2 eq

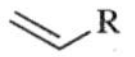

Ru cat **152** (0.05 eq), THF, 22°, 4 h

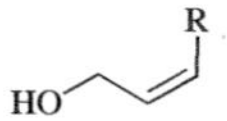

$(Z)/(E) = 98:2$

R	
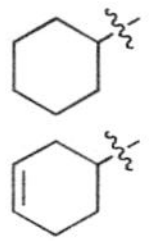 cyclohexyl	(59)
cyclohexenyl	(63)

183

2 eq

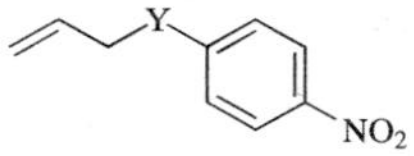

Catalyst (0.05 eq), $AlCl_3$ (x eq), THF, rt, 6 h

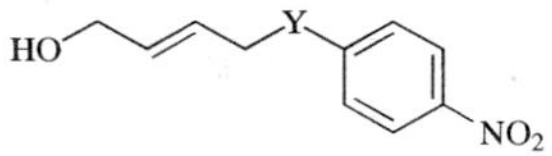

560

Y	Catalyst	x		$(E)/(Z)$
HN	Ru cat **366**	0.05	(71)	95:5
O	Ru cat **365**	0	(19)	23:77
O	Ru cat **366**	0	(28)	21:79
O	Ru cat **366**	0.05	(88)	89:11
O	Ru cat **367**	0	(23)	30:70

2 eq

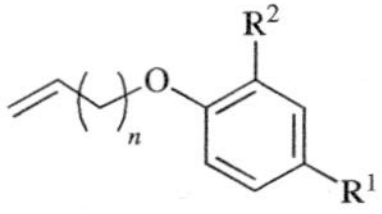

Ru cat **366** (0.05 eq), $AlCl_3$ (0.05 eq), THF, rt, 6 h

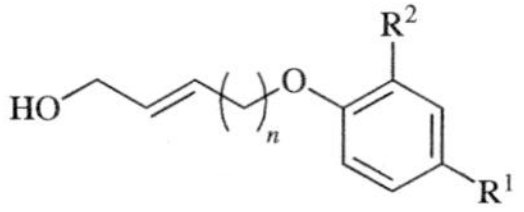

560

n	R^1	R^2		$(E)/(Z)$
2	NC–	H	(51)	79:21
5	H	Br	(75)	92:8

TABLE 4. CROSS-METATHESIS OF PRIMARY ALLYLIC ALCOHOLS AND DERIVATIVES (*Continued*)

A. ALLYLIC ALCOHOLS (*Continued*)

Allylic alcohol	Cross Partner	Conditions	Product(s) and Yield(s) (%)	Refs.
Please refer to the charts preceding the tables for the catalyst structures.				
C_4				
HO, OH; 6.3 eq	OH, OMe	Gr II (0.02 eq), PE, rt, 23 h	HO, OH, OMe (86)	511
2 eq	OH, Ph	Ru cat **152** (0.05 eq), THF, 22°, 4 h	OH, Ph, HO (64) (*Z*)/(*E*) = 96:4	183
2 eq	O, N, $NHCO_2Me$	Catalyst (0.05 eq), THF, 22°, 12 h	HO, O, N, $NHCO_2Me$	183

Catalyst		(*Z*)/(*E*)
Ru cat **215**	(—)	61:39
Ru cat **151**	(55)	97:3
Ru cat **152**	(70)	98:2

Reactant	Structure	Conditions	Product(s) and Yield(s) (%)	Refs.
2 eq	HO, Ac	Catalyst (0.05 eq), THF, 22°, 4 h	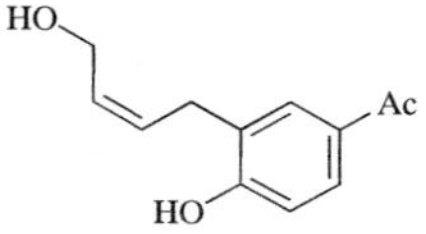 HO, Ac, HO	183
			Catalyst / (Z)/(E): Ru cat **215** (50) 82:18; Ru car **152** (68) 98:2	
2 eq	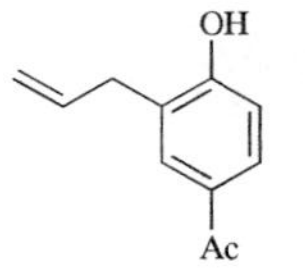 OH, Ac	Ru cat **366** (0.05 eq), $AlCl_3$ (0.05 eq), THF, rt, 6 h	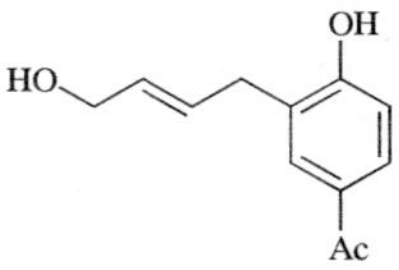 HO, OH, Ac (51) (*E*)/(*Z*) = 93:7	560
2 eq	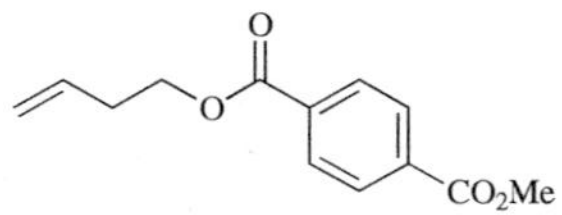 O, O, CO_2Me	Ru cat **366** (0.05 eq), $AlCl_3$ (0.05 eq), THF, rt, 6 h	HO, O, O, CO_2Me (66) (*E*)/(*Z*) = 94:6	560

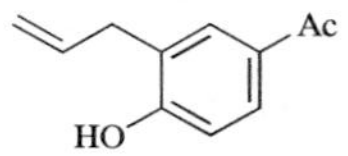

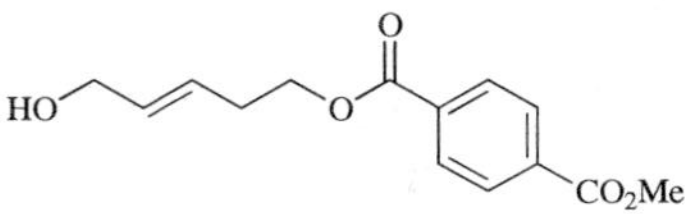

TABLE 4. CROSS-METATHESIS OF PRIMARY ALLYLIC ALCOHOLS AND DERIVATIVES (*Continued*)

A. ALLYLIC ALCOHOLS (*Continued*)

Allylic alcohol	Cross Partner	Conditions	Product(s) and Yield(s) (%)	Refs.
Please refer to the charts preceding the tables for the catalyst structures.				
C_4				
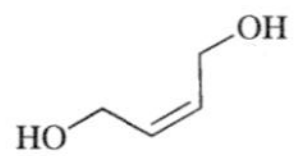 HO, OH; 2 eq	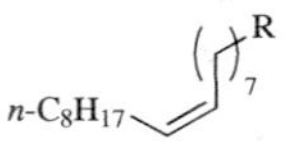 n-C_8H_{17}, R, 7	Ru cat **152** (0.05 eq), THF, 22°, 6 h	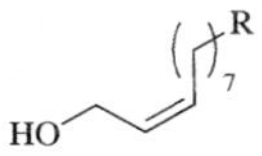 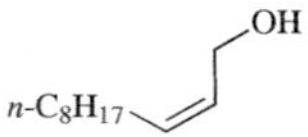 HO, R, 7 **I** + n-C_8H_{17}, OH **II** R / I / (Z)/(E) I / II / (Z)/(E) II HO / (62) / 96:4 / (59) / 94:6 MeO_2C / (65) / 94:6 / (60) / 94:6	183
6.3 eq	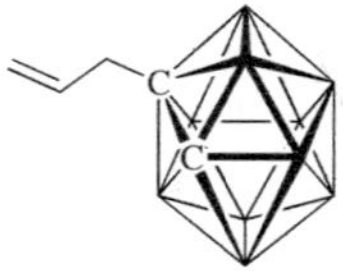 C, C	Gr II (0.014 eq), DCM, THF, reflux, 16 h	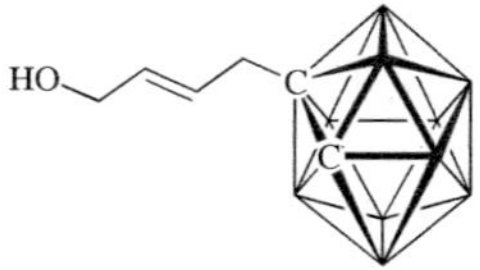 HO, C, C (76) (*E*)/(*Z*) = 93:7	436
C_5				
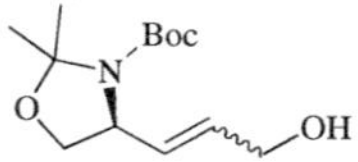 O, N, Boc, OH; 5 eq	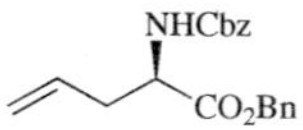 NHCbz, CO_2Bn	Gr II (0.05 eq), toluene, reflux, 3 h	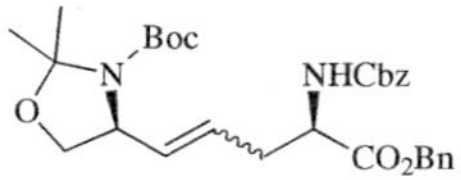 O, N, Boc, NHCbz, CO_2Bn (9)	563

TABLE 4. CROSS-METATHESIS OF PRIMARY ALLYLIC ALCOHOLS AND DERIVATIVES (*Continued*)

B. ALLYLIC ETHERS

Allylic Ether	Cross Partner	Conditions	Product(s) and Yield(s) (%)	Refs.
Please refer to the charts preceding the tables for the catalyst structures.				
C_3				
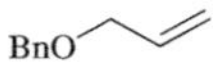 BnO (allyl)	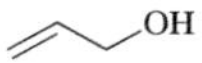 allyl OH, 10 eq	Hov II (0.06 eq), t-BuOH/H_2O (1:1), 32°, 3.5 h	 BnO~CH=CH~OH (31)	559
 RO (allyl)	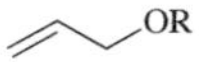 allyl OR	Gr I (0.05 eq), benzene, 40°	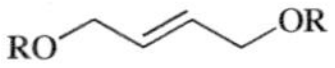 RO~CH=CH~OR	564
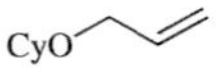 CyO (allyl)	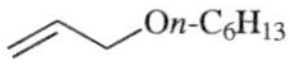 allyl On-C_6H_{13}	Catalyst (0.025 eq), DCM, 3 h	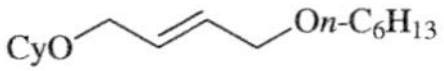 CyO~CH=CH~On-C_6H_{13}	565

R	Time (min)		(E)/(Z)
Bn	5	(68)	81:19
Ph_3C	5	(23)	66:34
TBS	5	(35)	89:11
Bn	30	(66)	86:14
Ph_3C	30	(41)	75:25
TBS	30	(46)	91:9
Bn	240	(69)	88:12
Ph_3C	240	(66)	86:14
TBS	240	(59)	94:6
Bn	1440	(62)	88:12
Ph_3C	1440	(61)	89:11
TBS	1440	(51)	95:5

Catalyst	Temp (°)		(E)/(Z)
Gr II	40	(21)	86:14
Hov II	40	(18)	89:11
Ru cat **154**	40	(35)	89:11
Ru cat **155**	20	(10)	93:7

TABLE 4. CROSS-METATHESIS OF PRIMARY ALLYLIC ALCOHOLS AND DERIVATIVES (*Continued*)

B. ALLYLIC ETHERS (*Continued*)

Allylic Ether	Cross Partner	Conditions	Product(s) and Yield(s) (%)	Refs.
Please refer to the charts preceding the tables for the catalyst structures.				
C_3				
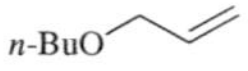 n-BuO	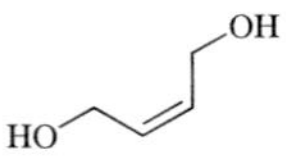 HO, OH; 2 eq	Ru cat **152** (0.05 eq), THF, 22°, 12 h	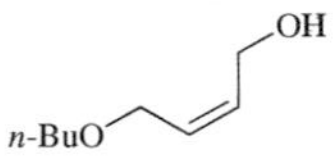 n-BuO, OH (57) (*Z*)/(*E*) = 91:9	183
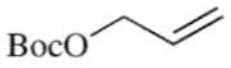 BocO; 1.4 eq	 MeO_2C, CO_2Me	Gr II (0.01 eq), DCM, reflux	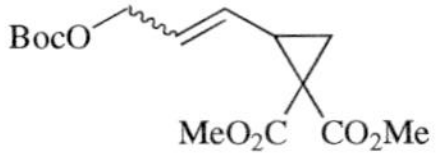 BocO, MeO_2C, CO_2Me (6)	407
 PhO	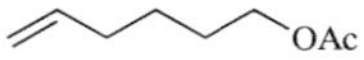 OAc	Catalyst (0.025 eq), DCM, 20°, 3 h	 PhO, OAc Catalyst / (*E*)/(*Z*) Gr II (47) 88:12 Hov II (53) 83:17 Ru cat **155** (43) 90:10	565
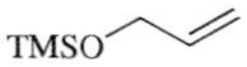 TMSO; 2 eq	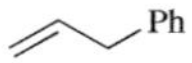 Ph	Gr I (0.05 eq), DCM, reflux, 16 h	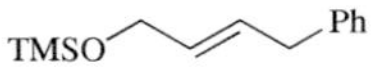 TMSO, Ph (79) (*E*)/(*Z*) = 67:33	156

C_4

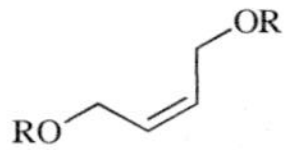

RO–CH=CH–OR (cis)
3 eq

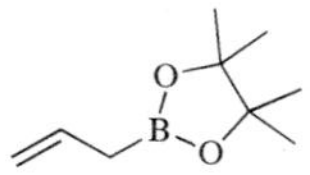

1. Gr II (x eq), DCM, reflux, 2–12 h
2. PhCHO (1.5 eq), 23°, 12–48 h

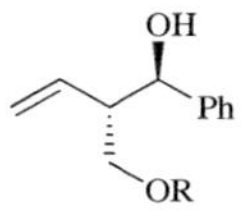

R	x		*anti/syn*[a]
TBS	0.05	(44-67)	80:20
Bn	0.05	(60)	82:18
Bn	0.02	(63)	76:24

126

2 eq

Gr I (0.05 eq), DCM, 45°

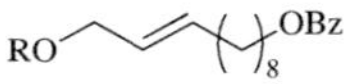

R		$(E)/(Z)$
TBS	(77)	91:9
t-Bu	(90)	88:12
Bn	(71)	90:10

176

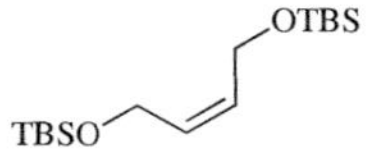

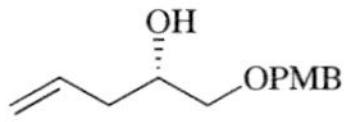

Gr II (0.1 eq), DCM, reflux, , 1 h

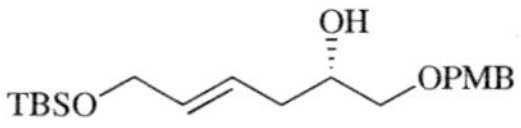

(50) $(E)/(Z) = 71{:}29$ 566

C_5

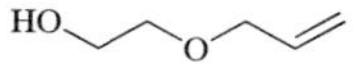

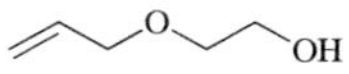

Catalyst (x eq), THF, 35°

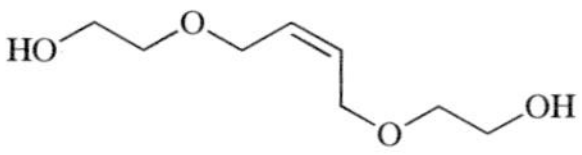

$(Z)/(E) = 67{:}33$

Catalyst	x	Time (h)		
Ru cat **217**	0.02	1	(73)	84
Ru cat **215**	0.001	12	(30)	488

TABLE 4. CROSS-METATHESIS OF PRIMARY ALLYLIC ALCOHOLS AND DERIVATIVES (*Continued*)

B. ALLYLIC ETHERS (*Continued*)

Allylic Ether	Cross Partner	Conditions	Product(s) and Yield(s) (%)	Refs.
Please refer to the charts preceding the tables for the catalyst structures.				
C_6				
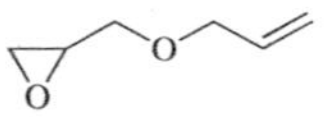 O, O	 OBz, 8	Gr I (0.05 eq), DCM, 45°, 12 h	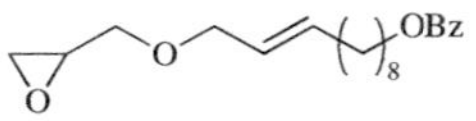 O, O, OBz, 8 (61) (*E*)/(*Z*) = 75:25	124
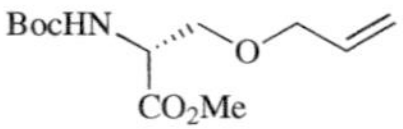 BocHN, O, CO_2Me	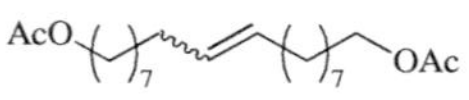 AcO, 7, 7, OAc 2 eq	Gr I (0.05 eq), DCM, 45°	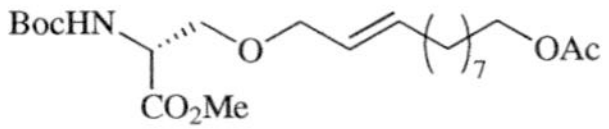 BocHN, O, 7, OAc, CO_2Me (86) (*E*)/(*Z*) = 86:14	176
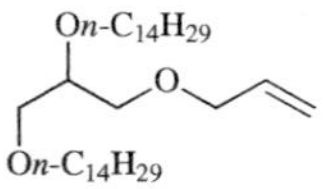 O*n*-$C_{14}H_{29}$, O, O*n*-$C_{14}H_{29}$ 2 eq	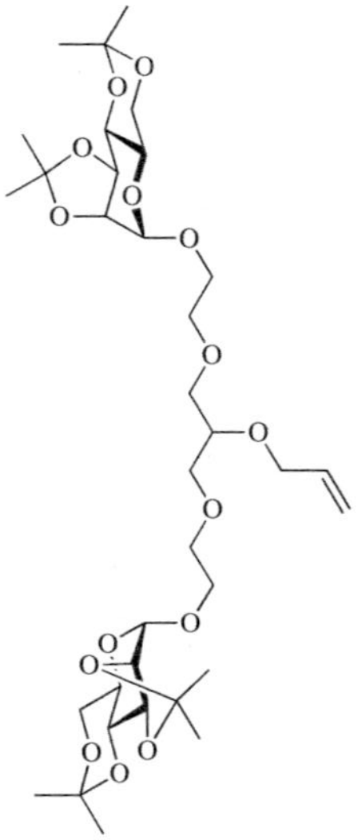	Gr II (0.1 eq), DCM, reflux, 14 h	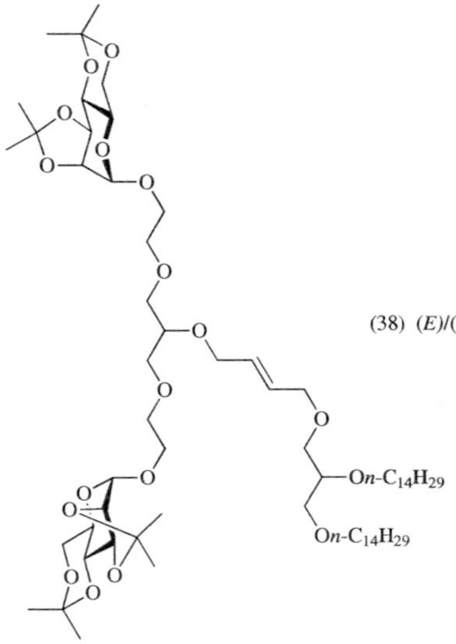 O*n*-$C_{14}H_{29}$, O*n*-$C_{14}H_{29}$ (38) (*E*)/(*Z*) = 73:27	567

$On\text{-}C_{14}H_{29}$

$On\text{-}C_{14}H_{29}$

2 eq

Gr II (0.1 eq),
DCM,
reflux, 22 h

567

$On\text{-}C_{14}H_{29}$

$On\text{-}C_{14}H_{29}$

(38) (*E*)/(*Z*) = 81:19

TABLE 4. CROSS-METATHESIS OF PRIMARY ALLYLIC ALCOHOLS AND DERIVATIVES (*Continued*)

B. ALLYLIC ETHERS (*Continued*)

Allylic Ether	Cross Partner	Conditions	Product(s) and Yield(s) (%)	Refs.

Please refer to the charts preceding the tables for the catalyst structures.

C_8

Allylic Ether	Cross Partner	Conditions	Product(s) and Yield(s) (%)	Refs.
O, CO_2Et, N_2, O	Cy; 10 eq	Gr II (0.05 eq), DCM, reflux, 18 h	O, CO_2Et, N_2, O, Cy; (89) (*E*)/(*Z*) = 90:10	569
TIPS, O; 5 × 1.5 eq	H, X, H, O, H, H, Br, H, O, H	1. Hov II (5 × 0.07 eq), benzene, 70°, 6 h 2. DMSO, rt, 12 h	H, X, H, O, H, H, Br, H, O, H, TIPS	570

X	*x*		(*Z*)/(*E*)
Cl	0.07	(88)	79:21
Br	0.09	(75)	80:20

C_9

Allylic Ether	Cross Partner	Conditions	Product(s) and Yield(s) (%)	Refs.
NO_2, O; 4 eq	R	Gr I (0.05 eq), DCM, 45°, 14 h	NO_2, O, R	568

R		(*E*)/(*Z*)
TBSOCH$_2$	(47)	92:8
Br(CH$_2$)$_3$	(47)	77:23
Cy	(37)	(*E*) only
Ph(CH$_2$)$_2$	(79)	77:23

Allylic Ether	Cross Partner	Conditions	Product(s) and Yield(s) (%)	Refs.
4 eq	R, O	Gr I (0.05 eq), DCM, 45°, 14 h	NO_2, O, R, O	568

R		(*E*)/(*Z*)
Me	(51)	76:14
MeO	(57)	52:48

4 eq

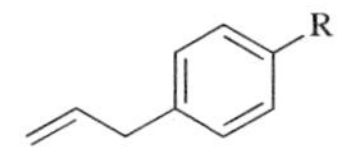

R

Gr I (0.05 eq),
DCM, 45°, 14 h

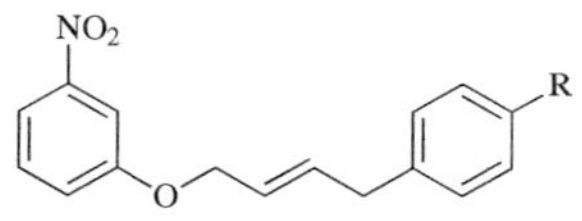

NO_2 O R

R		(E)/(Z)
H	(31)	82:18
MeO	(49)	85:15

568

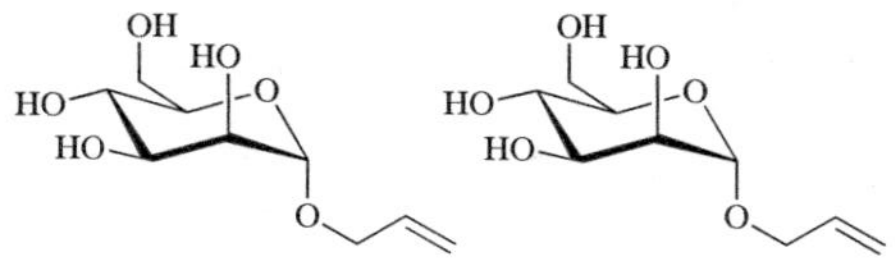

OH HO HO O HO O

OH HO HO O HO O

Gr I (0.1 eq),
DCM/MeOH (3:1)

OH HO HO O HO O O OH OH O OH OH

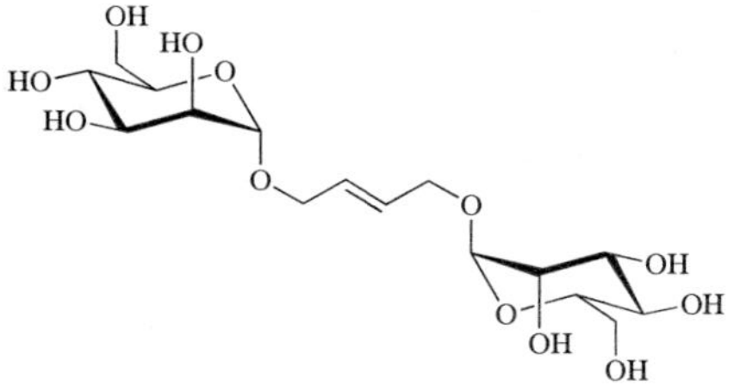

Temp (°)	Time (h)		(E)/(Z)
rt	8	(60)	63:37
40	18	(34)	58:42

497

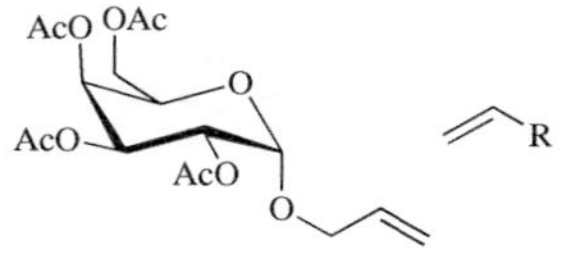

AcO OAc O AcO AcO O

R

4 eq

Gr I (0.07 eq),
DCM, reflux, 16 h

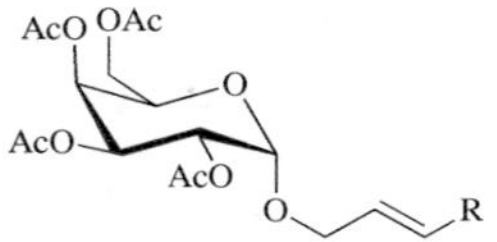

AcO OAc O AcO AcO O R

R		(E)/(Z)
n-C_6H_{13}	(74)	91:9
n-C_8H_{17}	(73)	87:13
n-$C_{10}H_{21}$	(77)	88:12
n-$C_{12}H_{25}$	(73)	89:11
n-$C_{14}H_{29}$	(77)	88:12
n-$C_{16}H_{33}$	(73)	86:14

332

TABLE 4. CROSS-METATHESIS OF PRIMARY ALLYLIC ALCOHOLS AND DERIVATIVES (*Continued*)

B. ALLYLIC ETHERS (*Continued*)

Allylic Ether	Cross Partner	Conditions	Product(s) and Yield(s) (%)	Refs.

Please refer to the charts preceding the tables for the catalyst structures.

C_9

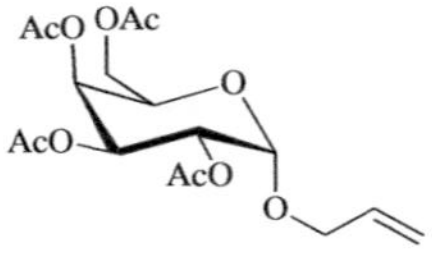

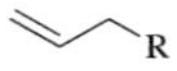

AcO OAc AcO AcO O O — R, *x* eq — Gr I (*y* eq), DCM — AcO OAc AcO AcO O O R

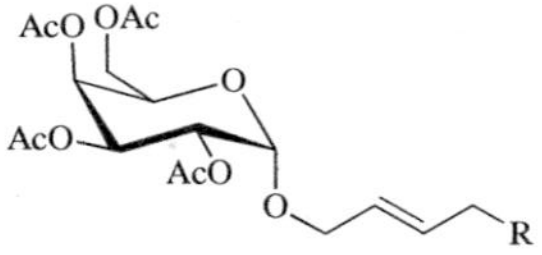

R	*x*	*y*	Temp	Time (h)		(*E*)/(*Z*)	Refs.
BocHN	2	0.2	reflux	6	(30)	80:20	571
BocHN	2	0.2	rt	15	(57)	80:20	571
CbzHN	2	0.2	reflux	6	(39)	80:20	571
CbzHN	2	0.2	rt	15	(65)	80:20	571
TBSO	2	0.2	reflux	6	(69)	95:5	571
BnO	4	0.07	reflux	16	(69)	87:13	332
PhO	4	0.07	reflux	16	(77)	80:20	332
Ph	4	0.07	reflux	16	(69)	84:16	332

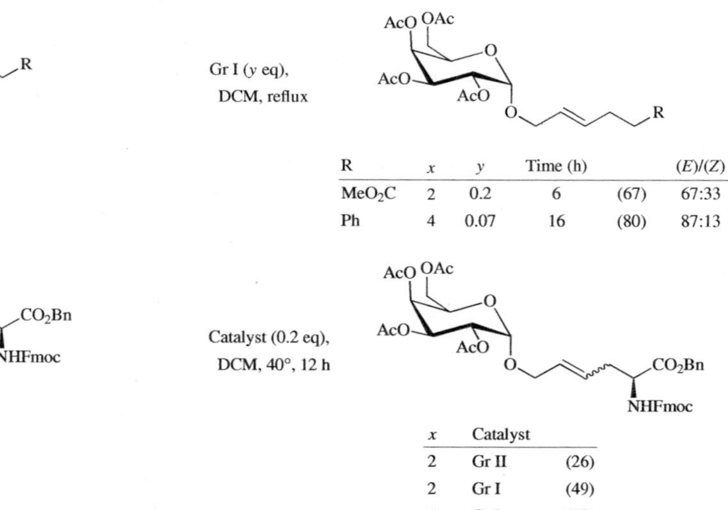

R	*x*	*y*	Time (h)		(*E*)/(*Z*)	
MeO_2C	2	0.2	6	(67)	67:33	571
Ph	4	0.07	16	(80)	87:13	332

x	Catalyst		
2	Gr II	(26)	572
2	Gr I	(49)	
5	Gr I	(70)	

TABLE 4. CROSS-METATHESIS OF PRIMARY ALLYLIC ALCOHOLS AND DERIVATIVES (*Continued*)

B. ALLYLIC ETHERS (*Continued*)

Allylic Ether	Cross Partner	Conditions	Product(s) and Yield(s) (%)	Refs.
Please refer to the charts preceding the tables for the catalyst structures.				
C_9				
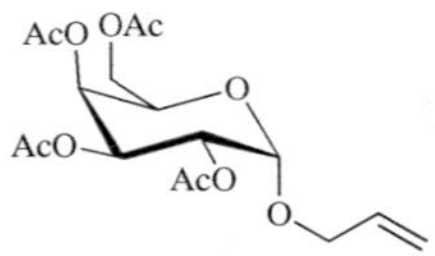 AcO OAc, AcO, AcO, O	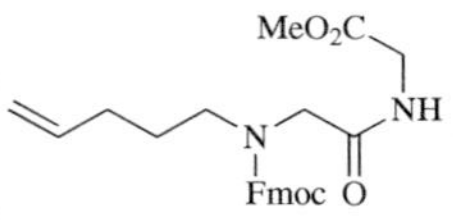 MeO_2C, NH, N, Fmoc O	Gr I (0.1 eq), DCM, rt, 24 h	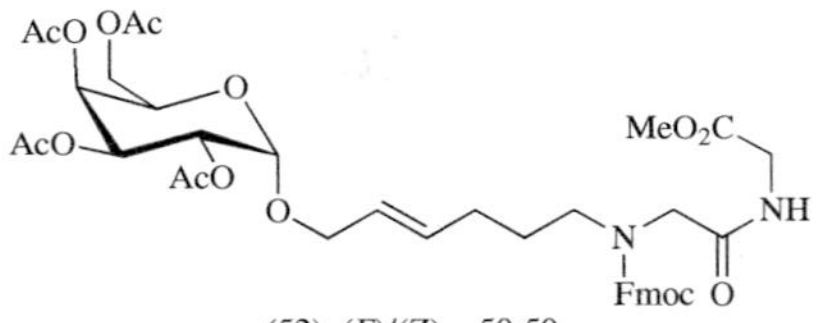 AcO OAc, AcO, AcO, O, MeO_2C, NH, N, Fmoc O (52) $(E)/(Z)$ = 50:50 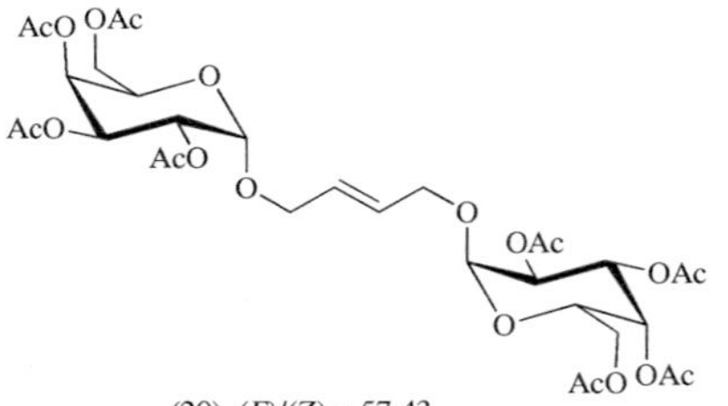 + AcO OAc, AcO, AcO, O, O, OAc, OAc, AcO OAc (20) $(E)/(Z)$ = 57:43 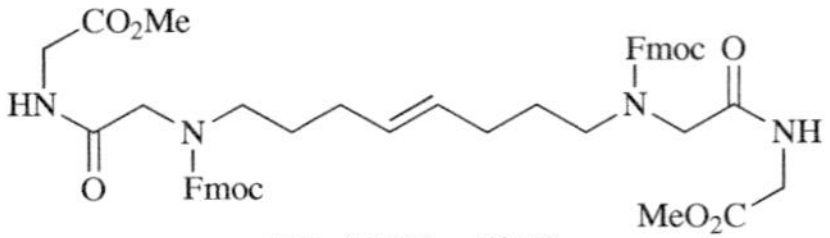 + CO_2Me, HN, O, N, Fmoc, Fmoc O, N, NH, MeO_2C (18) $(E)/(Z)$ = 50:50	573
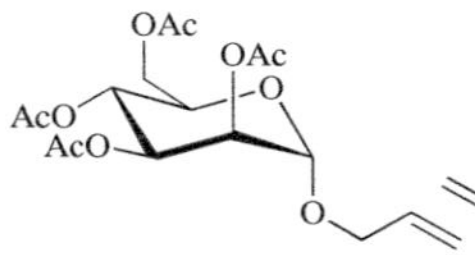 OAc, OAc, AcO, AcO, O	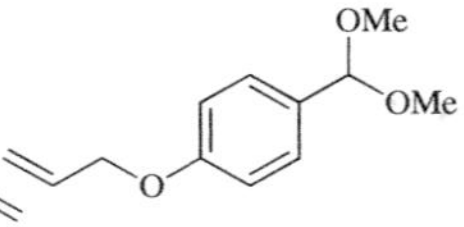 OMe, OMe, O 5 eq	Gr II (0.1 eq), DCM, reflux, 6 h	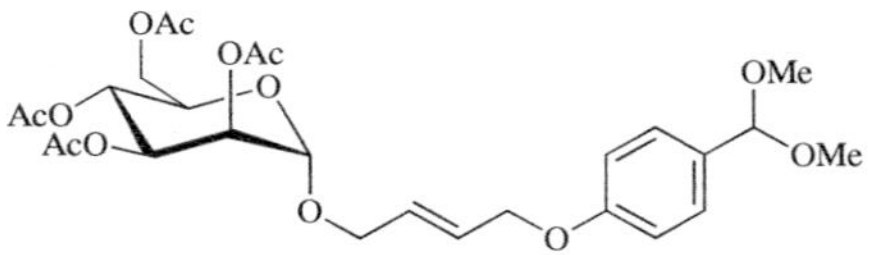 OAc, OAc, AcO, AcO, O, O, OMe, OMe (87) (E) only	574

Reactant	Conditions	Product(s) and Yield(s) (%)	Refs.
20 eq	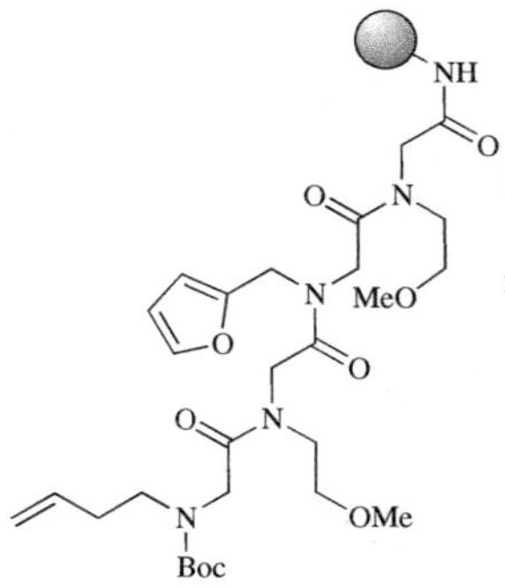 1. Hov II (0.05 eq), DCM, 40°, 8 h 2. TFA (92%), H_2O (5%), TIS (3%), 1–2 h	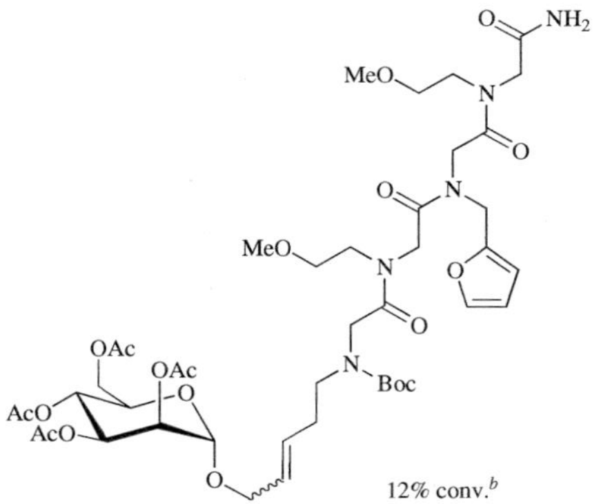 12% conv.[b]	12
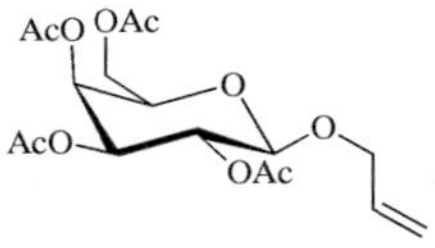 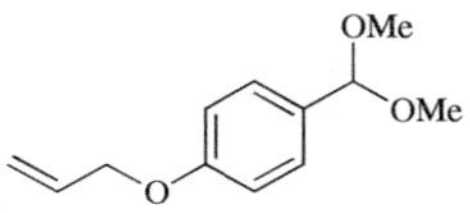 5.54 eq	Gr II (0.1 eq), DCM, reflux, 6 h	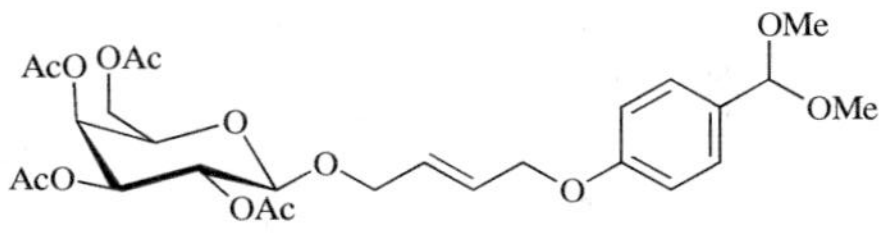 (75) (*E*) only	574

Allylic Ether	Cross Partner	Conditions	Product(s) and Yield(s) (%)	Refs.

Please refer to the charts preceding the tables for the catalyst structures.

C_9

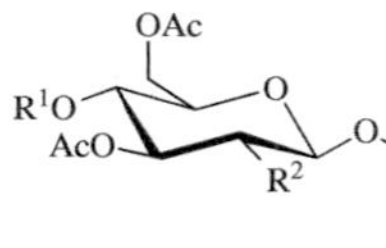

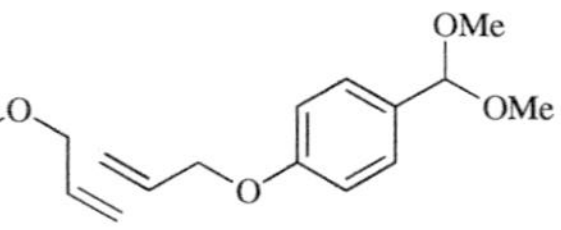

OAc, R¹O, AcO, O, R², O | OMe, OMe, O, *x* eq | Catalyst (0.1 eq), DCM | OAc, R¹O, AcO, O, R², O, O, OMe, OMe; (*E*) only | 574

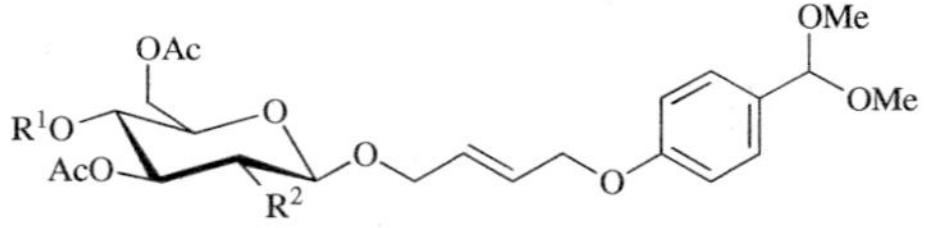

x	R¹	R²	Catalyst	Time (h)	Temp (°)	
5	Ac	AcO	Hov II	3	25	(36)
5	Ac	AcO	Hov II	3	40	(55)
5	Ac	AcO	Hov II	6	40	(92)
5	Ac	AcO	Hov II	12	40	(89)
5	Ac	AcO	Gr II	6	40	(54)
5	Ac	AcNH	Gr II	6	40	(81)
7.48	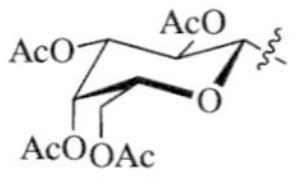 AcO, AcO, O, AcO, OAc	AcNH	Gr II	6	40	(12)
8	AcO, AcO, O, AcO, OAc	AcO	Gr II	6	40	(52)
8	AcO, AcO, AcO, O, OAc	AcO	Gr II	6	40	(46)

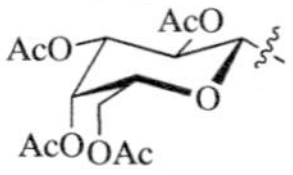

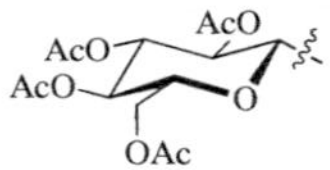

AcO AcO OR[1] O NHAc O

N Fmoc O R[2]

Gr I (0.1 eq), DCM, rt, 24 h

I

\+ II

\+ III (*E*)/(*Z*) = 67:33

573

R^1	R^2	**I**	(*E*)/(*Z*) **I**	**II**	(*E*)/(*Z*) **II**	**III**
Ac	HO	(41)	75:25	(10)	50:50	(10)
Tr	PfpO	(46)	50:50	(16)	50:50	(12)
Ac	MeO_2CCH_2HN	(49)	50:50	(27)	55:45	(11)

TABLE 4. CROSS-METATHESIS OF PRIMARY ALLYLIC ALCOHOLS AND DERIVATIVES (*Continued*)

B. ALLYLIC ETHERS (*Continued*)

Allylic Ether	Cross Partner	Conditions	Product(s) and Yield(s) (%)	Refs.
Please refer to the charts preceding the tables for the catalyst structures.				
C_9				
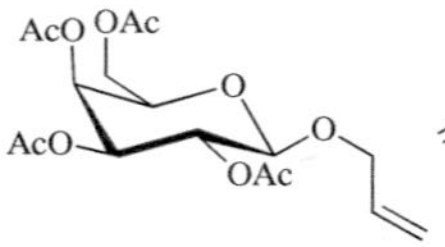 AcO OAc, AcO, OAc, O, O	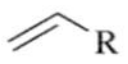 R; 4 eq	Gr I (0.07 eq), DCM, reflux, 16 h	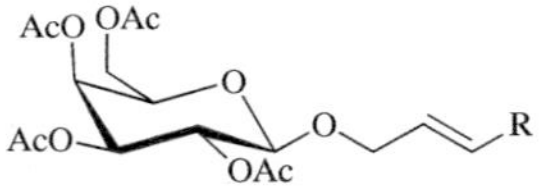 AcO OAc, AcO, OAc, O, O, R R / (*E*)/(*Z*) *n*-$C_{12}H_{25}$ (62) 87:13 *n*-$C_{14}H_{29}$ (64) 88:12	332
	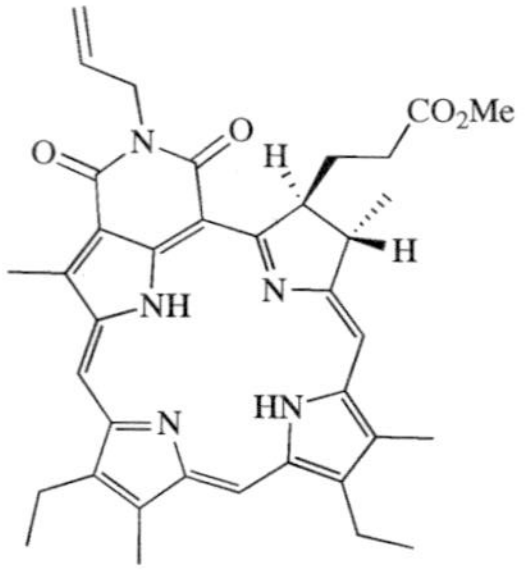 O, N, O, H, H, CO_2Me, NH, N, N, HN; 20 eq	Gr II (1 eq), DCM, reflux, 4 h	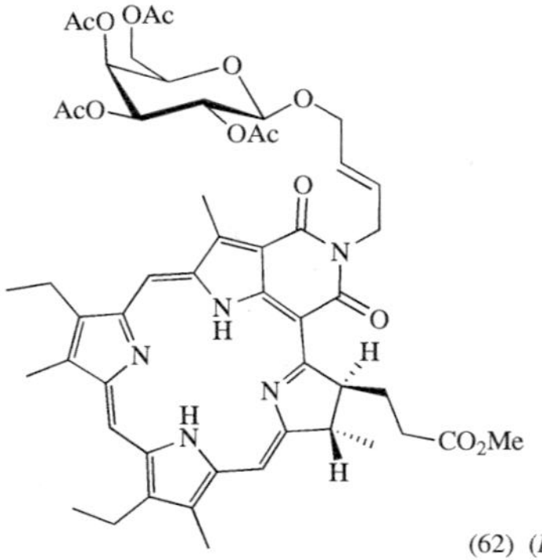 AcO OAc, AcO, OAc, O, O, O, N, O, N H, N, N, H N, H, H, CO_2Me (62) (*E*)/(*Z*) = 86:14	575

5 eq	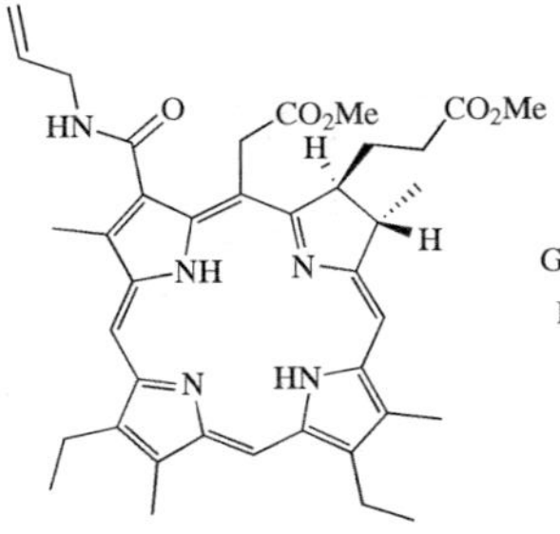	Gr II (0.05 eq), DCM, reflux, 4 h	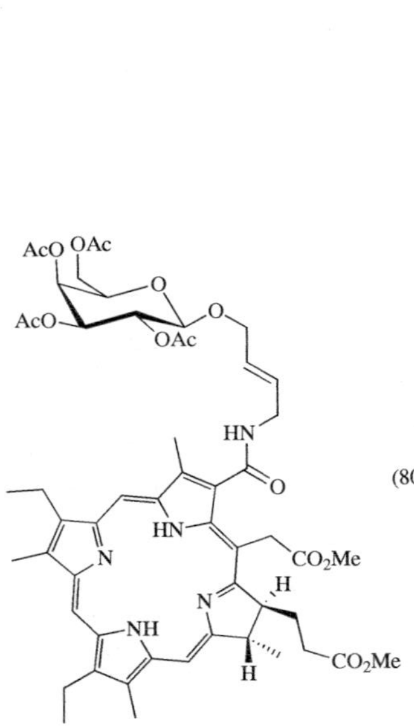	(80) (*E*)/(*Z*) = 83:17	575
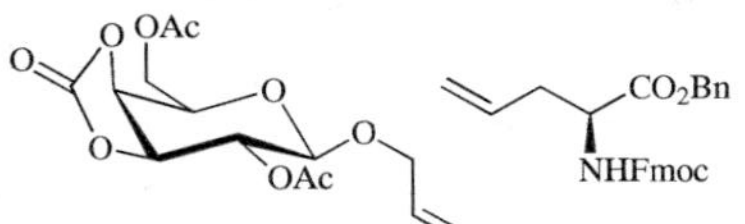 5 eq		Gr I (0.2 eq), DCM, 40°, 12 h	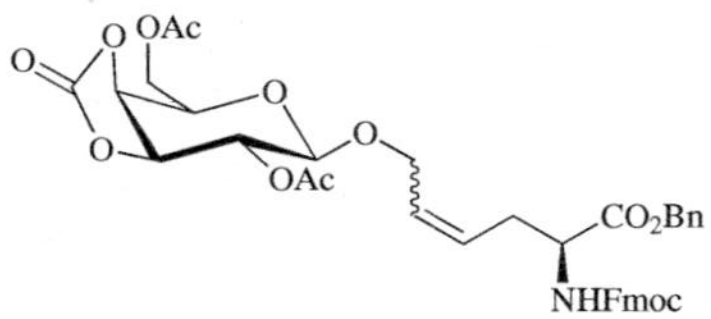	(66)	572

TABLE 4. CROSS-METATHESIS OF PRIMARY ALLYLIC ALCOHOLS AND DERIVATIVES (*Continued*)

B. ALLYLIC ETHERS (*Continued*)

Allylic Ether	Cross Partner	Conditions	Product(s) and Yield(s) (%)	Refs.
Please refer to the charts preceding the tables for the catalyst structures.				
C_9				
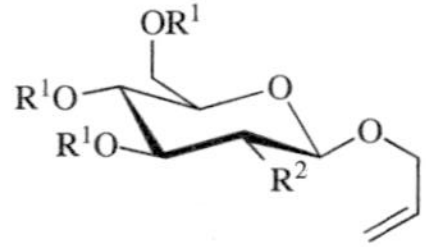 OR[1], R[1]O, R[1]O, R[2]; 1.2 eq	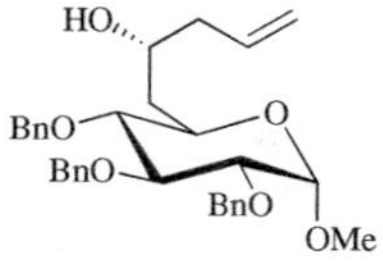 HO, BnO, BnO, BnO, OMe	Hov II (0.1 eq), DCM, rt	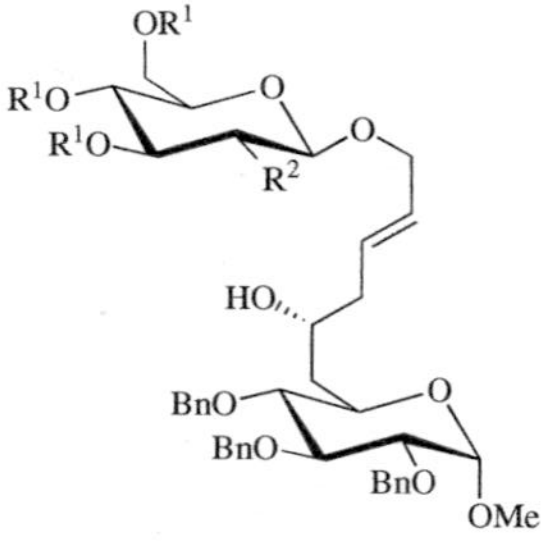 OR[1], R[1]O, R[1]O, R[2], HO, BnO, BnO, BnO, OMe; $(E)/(Z) > 95:5$; R[1] / R[2]: Ac / N_3 (62); Bn / BnO (54)	498
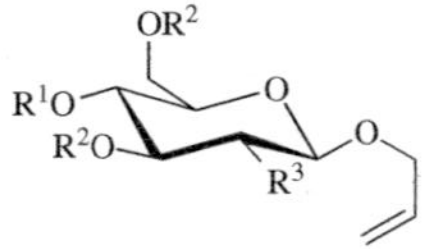 OR[2], R[1]O, R[2]O, R[3]	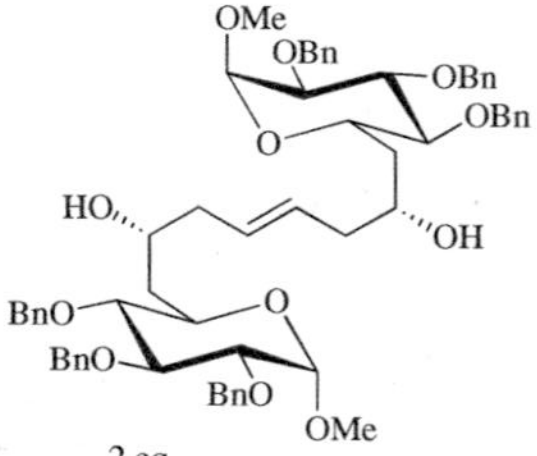 OMe, OBn, OBn, OBn, HO, OH, BnO, BnO, BnO, OMe; 2 eq	Hov II (0.1 eq), DCM, rt	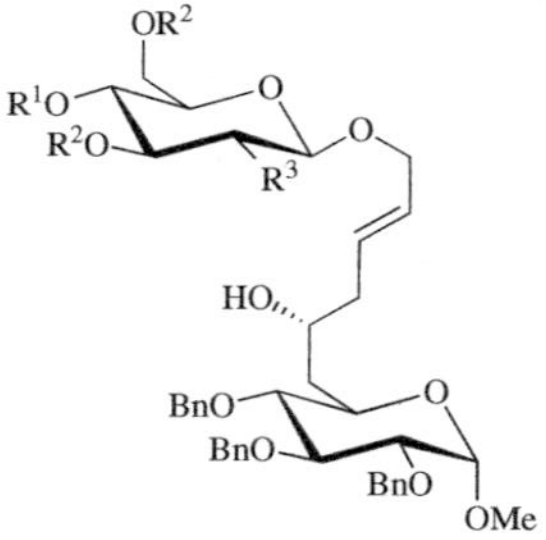 OR[2], R[1]O, R[2]O, R[3], HO, BnO, BnO, BnO, OMe; $(E)/(Z) > 95:5$	498

R^1	R^2	R^3	
Ac	Ac	N_3	(45)
Bn	Bn	BnO	(42)
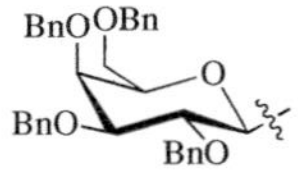 BnO OBn / BnO / BnO (sugar residue)	Bn	BnO	(40)

Reactants	Conditions	Product (yield %)	Ref.
Ph, O, O, O, O, CbzHN, MeO; 1.13 eq; HO, BnO, BnO, O, BnO, OMe	Hov II (0.1 eq), DCM, rt	Ph, O, O, O, O, HO, CbzHN, OMe, BnO, BnO, O, BnO, OMe; (41) $(E)/(Z) > 95{:}5$	498
OAc, O, O, AcO, OAc, OAc; $(\ )_9$ OPh, 4 eq	Gr I (0.07 eq), DCM, reflux, 14 h	OAc, O, O, $(\ )_9$ OPh, AcO, OAc, OAc (76)	576

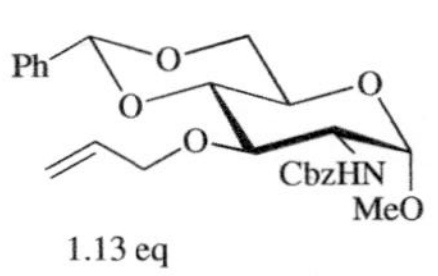

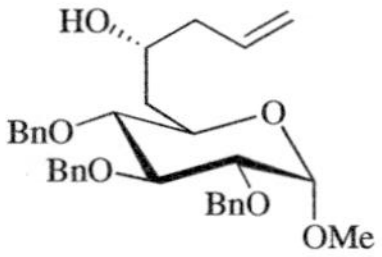

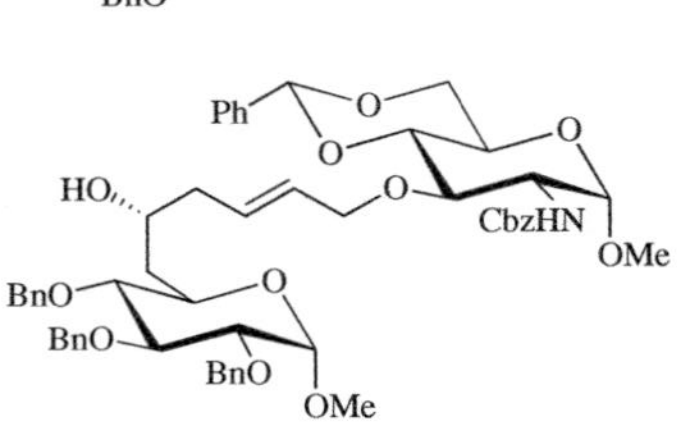

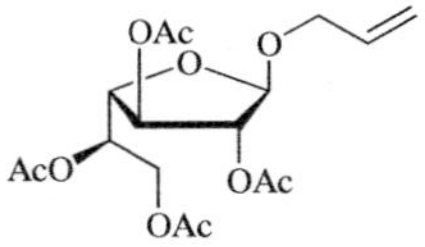

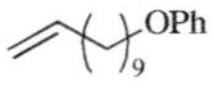

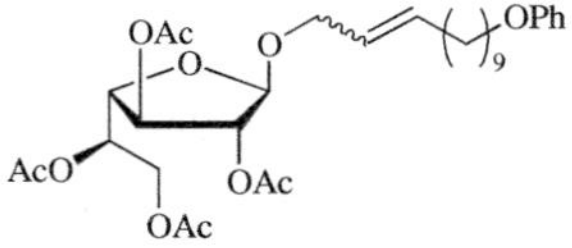

TABLE 4. CROSS-METATHESIS OF PRIMARY ALLYLIC ALCOHOLS AND DERIVATIVES *(Continued)*

B. ALLYLIC ETHERS *(Continued)*

Allylic Ether	Cross Partner	Conditions	Product(s) and Yield(s) (%)	Refs.
Please refer to the charts preceding the tables for the catalyst structures.				
C_9				
OH, BnO, BnO, BnO, O, O	OH, BnO, BnO, BnO, O, O	Gr I, DCM, reflux	OH, BnO, BnO, BnO, O, O, O, OBn, OBn, OBn, O, OH (84) $(E)/(Z) = 80:20$	577
BnO, OBn, BnO, BnO, O, O 1.23 eq	HO, BnO, BnO, BnO, O, OMe	Hov II (0.1 eq), DCM, rt	BnO, OBn, BnO, BnO, O, O, HO, BnO, BnO, BnO, O, OMe (45) $(E)/(Z) > 95:5$	498

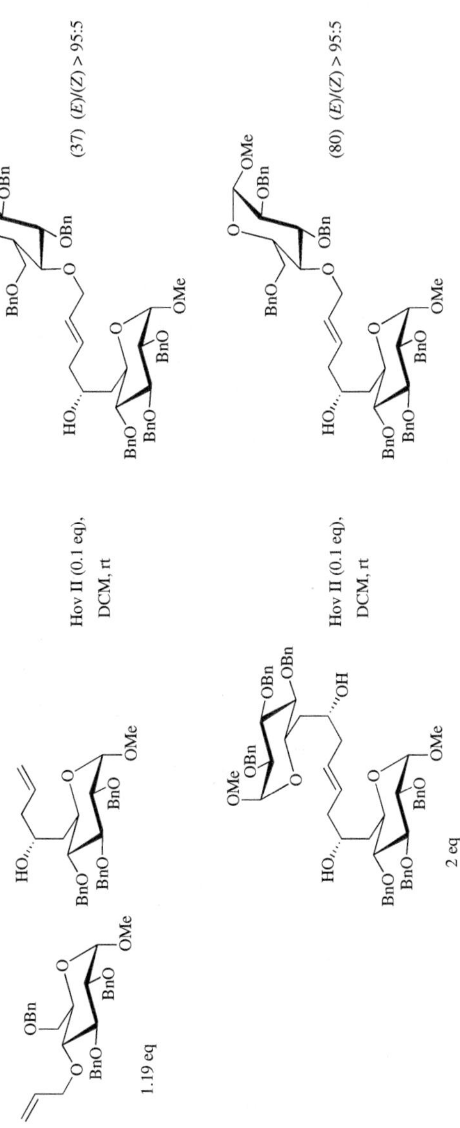
1.19 eq
Hov II (0.1 eq),
DCM, rt
(37) (E)/(Z) > 95:5
498
2 eq
Hov II (0.1 eq),
DCM, rt
(80) (E)/(Z) > 95:5
498

TABLE 4. CROSS-METATHESIS OF PRIMARY ALLYLIC ALCOHOLS AND DERIVATIVES (*Continued*)

B. ALLYLIC ETHERS (*Continued*)

Allylic Ether	Cross Partner	Conditions	Product(s) and Yield(s) (%)	Refs.
Please refer to the charts preceding the tables for the catalyst structures.				
C_9				
BnO OBn, BnO, BnO, O, O	MeO, OBn, OBn, OBn, HO, OH, O, O, BnO, BnO, BnO, OMe 2 eq	Hov II (0.1 eq), DCM, rt	OBn, OBn, OBn, BnO, O, O, HO, BnO, BnO, BnO, O, OMe (40) (*E*)/(*Z*) > 95:5	498
Ph, O, O, O, O, CbzHN, MeO	MeO, OBn, OBn, OBn, HO, OH, O, O, BnO, BnO, BnO, OMe 2 eq	Hov II (0.1 eq), DCM, rt	Ph, O, O, O, O, HO, CbzHN, OMe, BnO, BnO, BnO, O, OMe (34) (*E*)/(*Z*) > 95:5	498

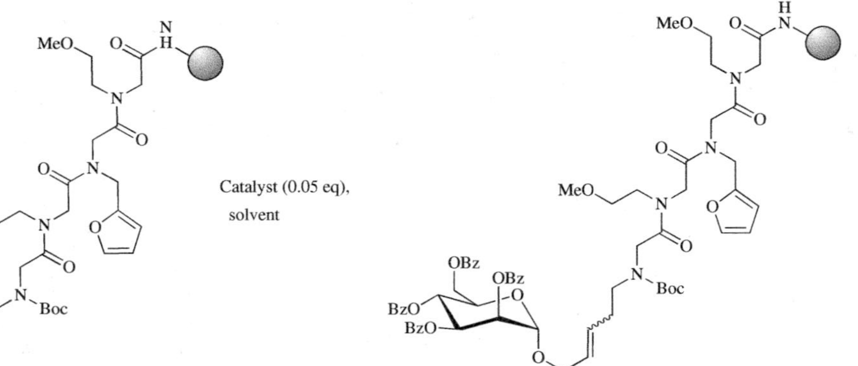

Catalyst	Solvent	Temp (°)	Time	Conv. (%)[b]
Gr I	DCM	40	8 h	10
Gr II	DCM	40	8 h	<5
Hov II	DCM	40	8 h	27
Gr I	DCB	MW	3 min	<5
Gr II	DCB	MW	3 min	<5
Hov II	DCB	MW	3 min	<5

TABLE 4. CROSS-METATHESIS OF PRIMARY ALLYLIC ALCOHOLS AND DERIVATIVES (*Continued*)
B. ALLYLIC ETHERS (*Continued*)

Allylic Ether	Cross Partner	Conditions	Product(s) and Yield(s) (%)	Refs.

Please refer to the charts preceding the tables for the catalyst structures.

C_9

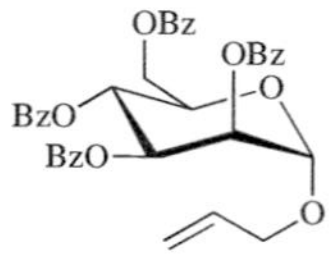

Allylic Ether	Cross Partner	Conditions	Product(s) and Yield(s) (%)	Refs.
BzO, BzO, OBz, OBz, O, O 20 eq	MeO, O, NH, N, O, O, N, MeO, N, O, O, Boc, N	Catalyst (0.05 eq), solvent	OBz, OBz, O, BzO, BzO, O, Boc, N, OMe, N, O, O, N, O, O, N, MeO, O, NH	12

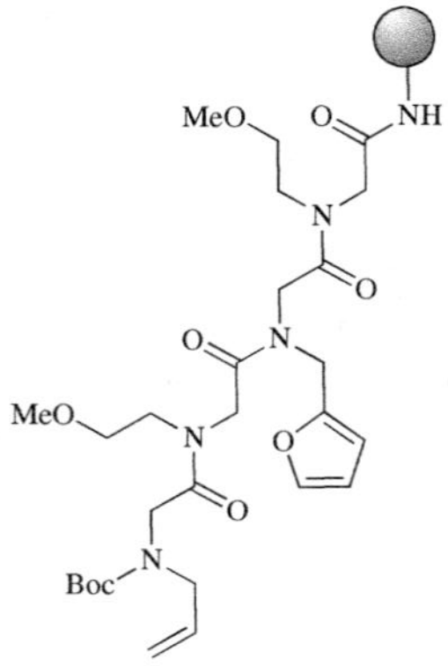

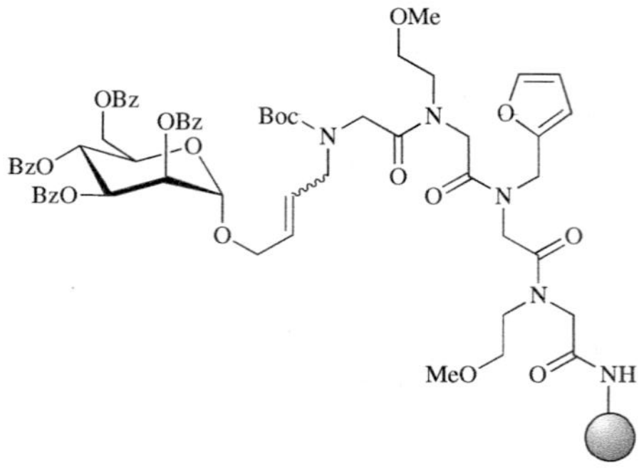

Catalyst	Solvent	Temp (°)	Time	Conv. (%)[b]
Gr I	DCM	40	8 h	<5
Gr II	DCM	40	8 h	<5
Hov II	DCM	40	8 h	13
Gr I	DCB	MW	3 min	<5
Gr II	DCB	MW	3 min	<5
Hov II	DCB	MW	3 min	<5

20 eq

1. Hov II (0.05 eq),
DCM, 40°, 8 h
2. TFA (92%),
H_2O (5%),
TIS (3%), 1–2 h

(29) 12

20 eq

1. Hov II (0.05 eq),
DCM, 40°, 8 h
2. TFA (92%),
H_2O (5%),
TIS (3%), 1–2 h

(28)

12

TABLE 4. CROSS-METATHESIS OF PRIMARY ALLYLIC ALCOHOLS AND DERIVATIVES (*Continued*)

B. ALLYLIC ETHERS (*Continued*)

Allylic Ether	Cross Partner	Conditions	Product(s) and Yield(s) (%)	Refs.
Please refer to the charts preceding the tables for the catalyst structures.				
C_9				
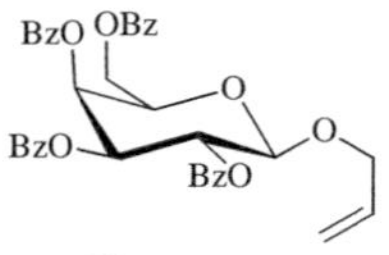 BzO OBz, BzO, BzO, O, O; 20 eq	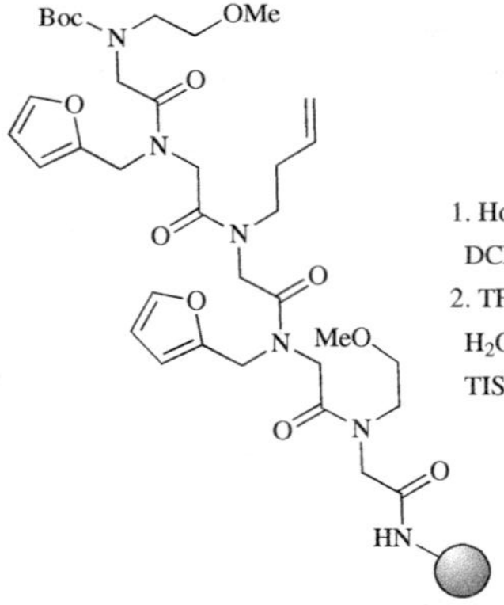 Boc, N, OMe, O, O, N, O, N, O, O, N, MeO, O, N, O, HN	1. Hov II (0.05 eq), DCM, 40°, 8 h 2. TFA (92%), H_2O (5%), TIS (3%), 1–2 h	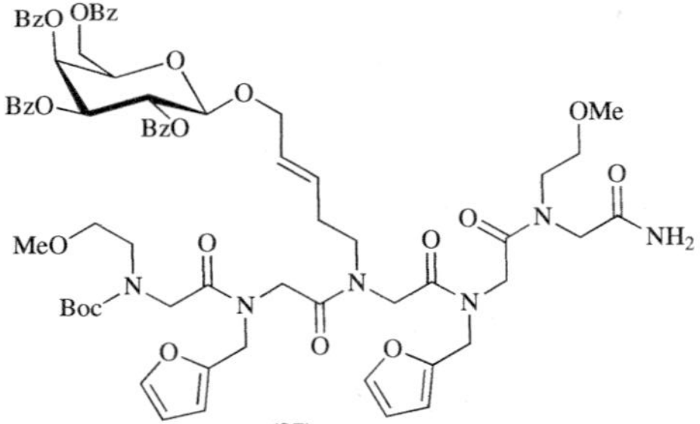 BzO OBz, BzO, BzO, O, O, OMe, O, NH_2, MeO, Boc, N, O, N, O, N, O, N, O, O, N, O; (27)	12
C_{12}				
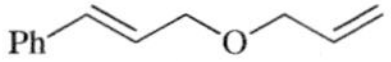 Ph, O	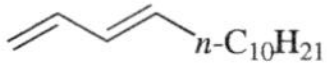 n-$C_{10}H_{21}$; 1.5 eq	Ru cat **216** (0.03 eq) 50°, 24 h	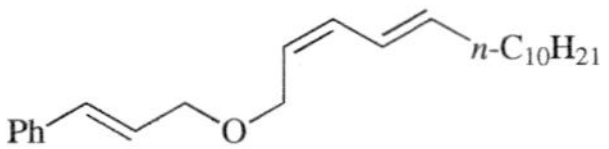 Ph, O, n-$C_{10}H_{21}$; (82) $(Z)/(E) > 95:5$	538
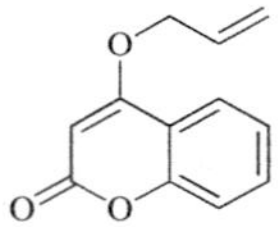 O, O, O	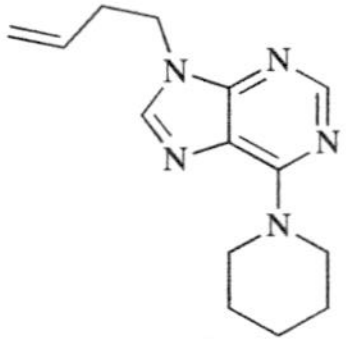 N, N, N, N, N; 3 eq	Gr II (0.1 + 0.04 eq), DCM, 100° (MW), 2 + 1 h	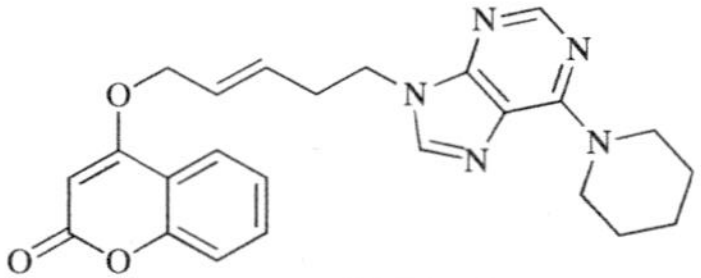 O, N, N, N, N, N, O, O; (63) (E) only	499

3 eq

RO

OH

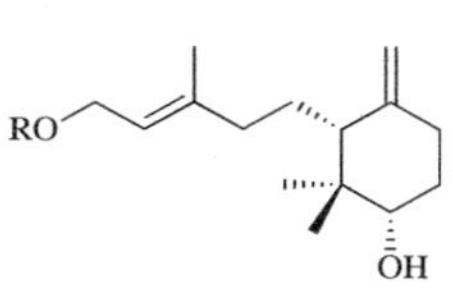

Catalyst (0.1 eq),
DCM, reflux, 48 h

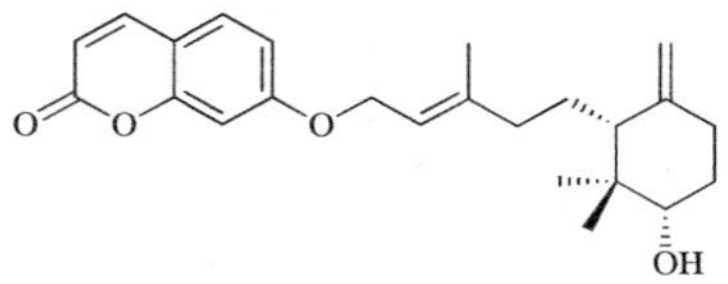

OH

578

R	Catalyst		$(E)/(Z)$
H	Gr I	(0)	—
H	Gr II	(0)	—
Ac	Gr II	(19)	60:40

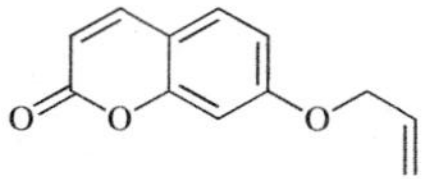

3 eq

AcO

OH

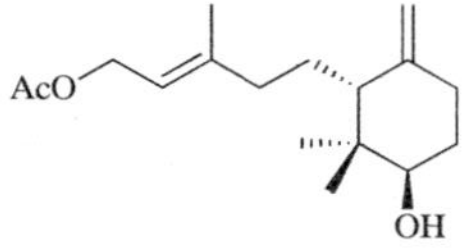

Gr II (0.1 eq),
DCM, reflux, 48 h

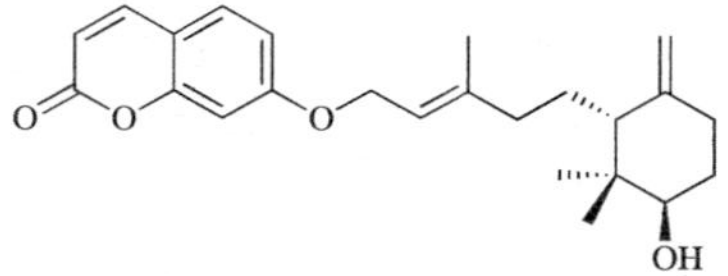

OH

(19) $(E)/(Z)$ = 60:40

578

BnO_2C CO_2Bn BnO BnO BnO

BnO_2C CO_2Bn BnO BnO BnO

Gr I, DCM, reflux

BnO_2C CO_2Bn BnO BnO BnO OBn OBn OBn BnO_2C CO_2Bn

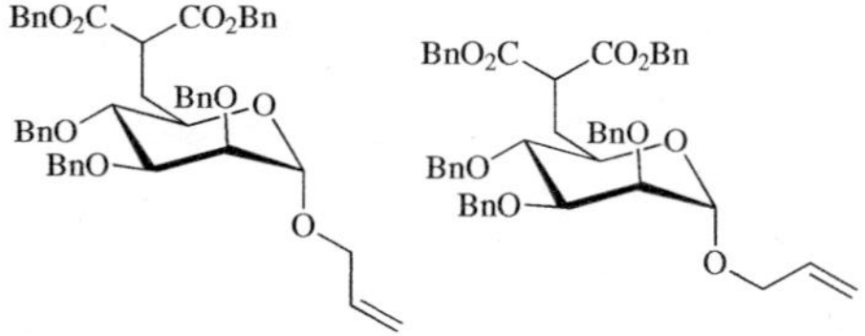

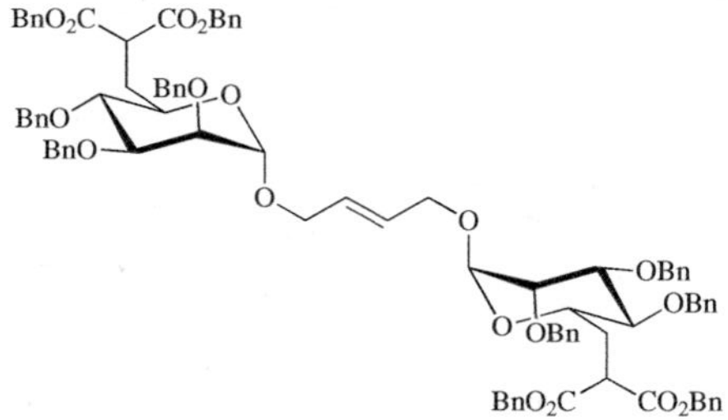

(67) $(E)/(Z)$ = 80:20

577

TABLE 4. CROSS-METATHESIS OF PRIMARY ALLYLIC ALCOHOLS AND DERIVATIVES (*Continued*)

B. ALLYLIC ETHERS (*Continued*)

Allylic Ether	Cross Partner	Conditions	Product(s) and Yield(s) (%)	Refs.
Please refer to the charts preceding the tables for the catalyst structures.				
C_{13}				
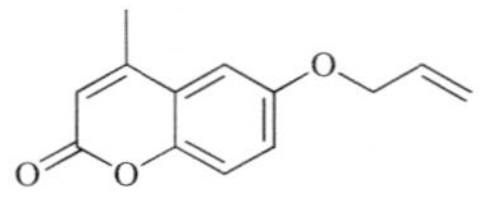	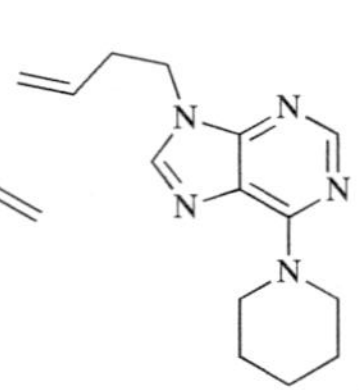 3 eq	Gr II (0.1 + 0.04 eq), DCM, 100° (MW), 2 + 1 h	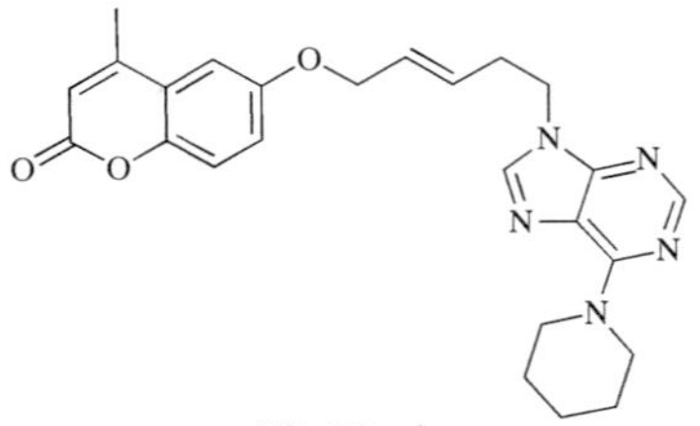 (58) (*E*) only	499
C_{14}				
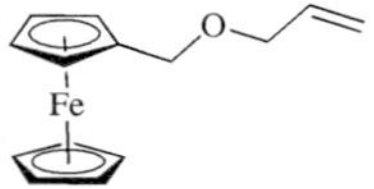	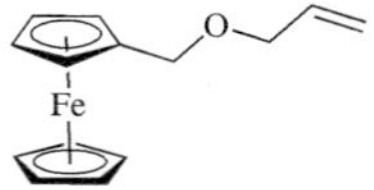	Gr I (0.05 eq), DCM, 45°, 12 h	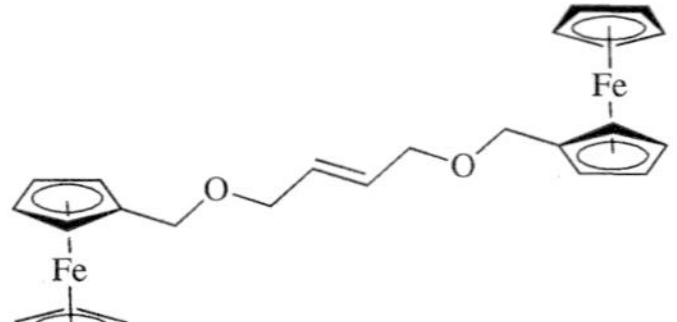 (77) (*E*)/(*Z*) = 89:11	124
C_{18}				
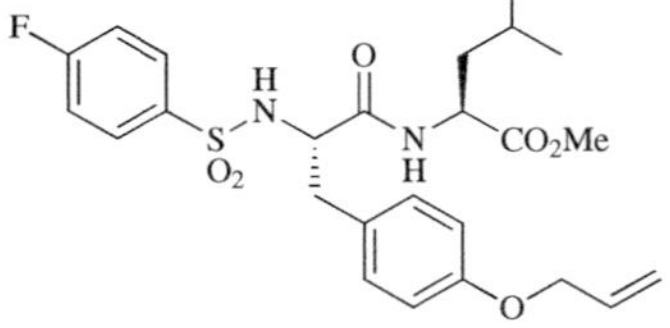	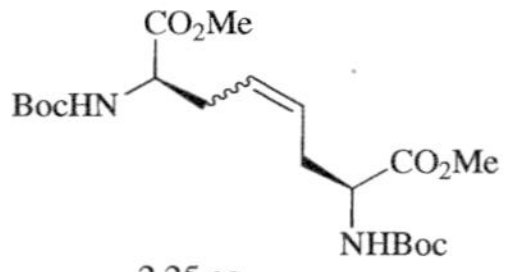 2.25 eq	Gr II (1.25 eq), DCM, reflux, 18 h	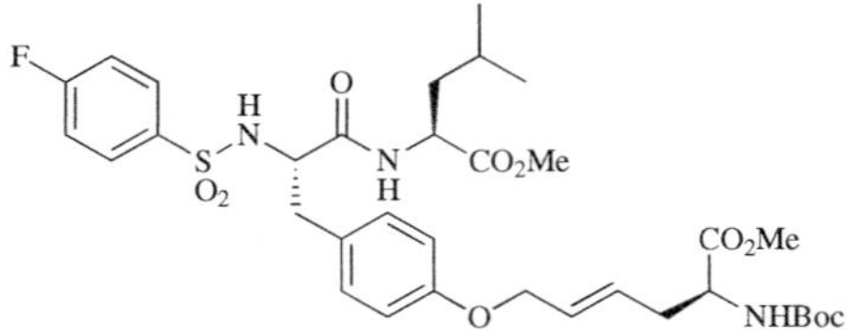 (33) (*E*)/(*Z*) = 92:8	579

	Gr II (4 x 0.04 eq), BCl_2Cy_2 (0.15 eq), DCM, MW, 4 × 2 min	(87) (*E*)/(*Z*) = 81:19	579
	Gr I (0.1 eq), DCM, rt, 24 h	(83) (*E*)/(*Z*) > 90:10	195

TABLE 4. CROSS-METATHESIS OF PRIMARY ALLYLIC ALCOHOLS AND DERIVATIVES (*Continued*)

B. ALLYLIC ETHERS (*Continued*)

Allylic Ether	Cross Partner	Conditions	Product(s) and Yield(s) (%)	Refs.
Please refer to the charts preceding the tables for the catalyst structures.				
C_{19}				
		Catalyst (x eq), DCM	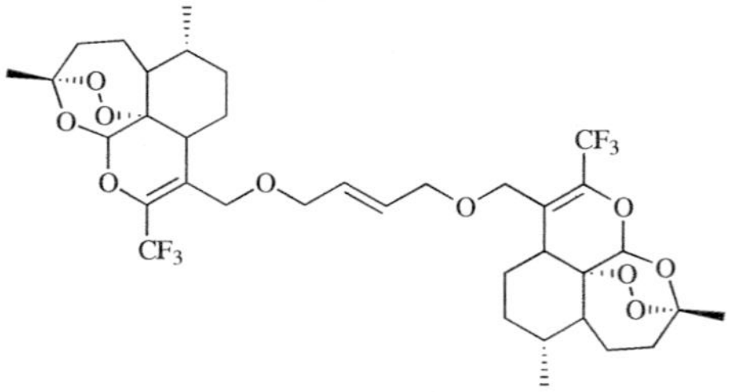	195
	NHBoc, 5 eq	Gr II (0.1 eq), DCM, reflux, 0.5 h	NHBoc (52)	580

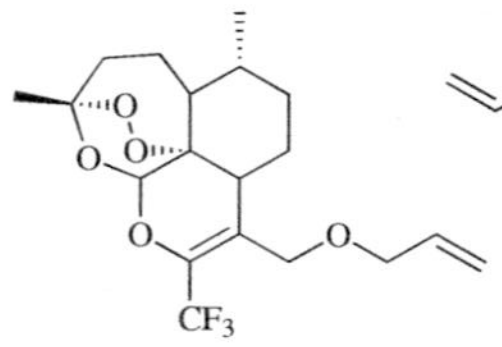

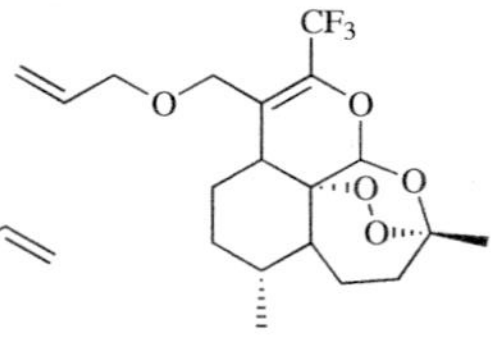

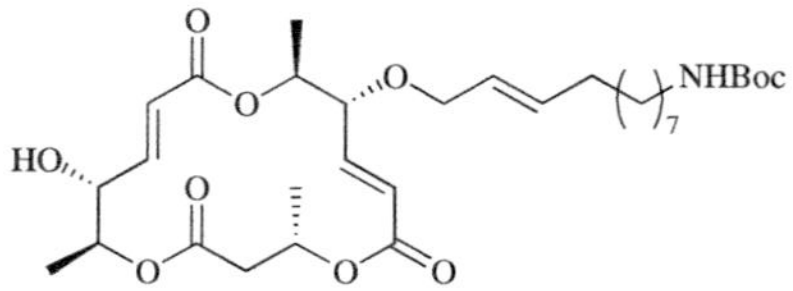

Catalyst	x	Temp	Time (h)	Conv. (%)[c]		(E)/(Z)
Gr I	0.05	rt	6	80	(—)	90:10
Gr I	0.1	rt	4	80	(—)	90:10
Gr I	0.05	reflux	1.5	100	(71)	90:10
Ru cat **35**	0.1	rt	24	90	(69)	(*E*) only

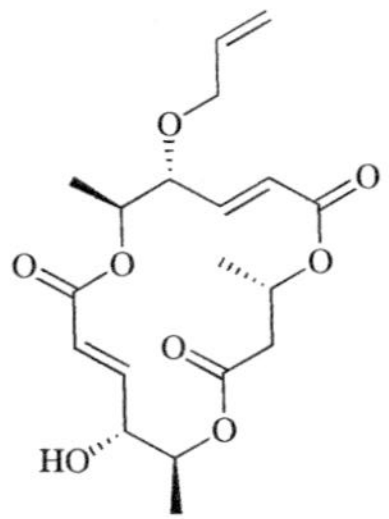

Entry	Reactants	Conditions	Product	Ref.
C_{20}	2 eq	Gr I (0.25 eq), DCM, 20°, 72 h	(60) (*E*) only	581
C_{21}	2 eq	Gr I (0.05 eq), DCM, reflux, 18–40 h	(65) (*E*)/(*Z*) = 75:25	582
	5 eq	Gr I (0.2 eq), DCM, 40°, 12 h	(70)	572

TABLE 4. CROSS-METATHESIS OF PRIMARY ALLYLIC ALCOHOLS AND DERIVATIVES (*Continued*)

B. ALLYLIC ETHERS (*Continued*)

Allylic Ether	Cross Partner	Conditions	Product(s) and Yield(s) (%)	Refs.
Please refer to the charts preceding the tables for the catalyst structures.				
C_{21}				
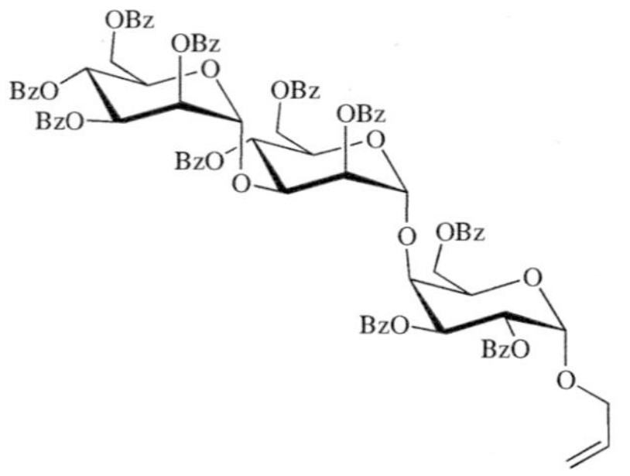 BzO, BzO, OBz, OBz, BzO, O, OBz, OBz, O, OBz, BzO, BzO, O	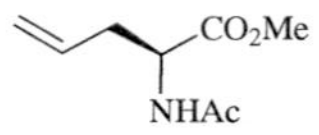 CO_2Me, NHAc 5 eq	1. Gr I (0.1 eq), DCM, reflux, 8 h 2. 10% Pd/C, MeOH, 60°	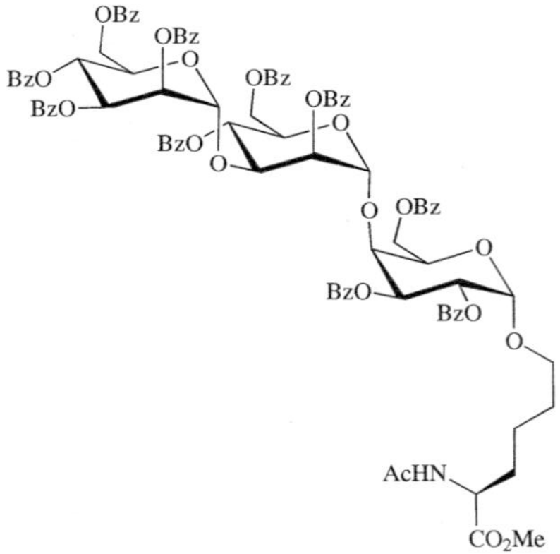 BzO, BzO, OBz, OBz, BzO, O, OBz, OBz, O, OBz, BzO, BzO, O, AcHN, CO_2Me (60)	583

C_{22}

4 eq

Gr II (x eq),
DCM, reflux

567

n	x	Time (h)	
7	0.05	16	(81)
12	0.1	4	(64)

TABLE 4. CROSS-METATHESIS OF PRIMARY ALLYLIC ALCOHOLS AND DERIVATIVES (*Continued*)

B. ALLYLIC ETHERS (*Continued*)

Allylic Ether	Cross Partner	Conditions	Product(s) and Yield(s) (%)	Refs.
Please refer to the charts preceding the tables for the catalyst structures.				
C_{24}				
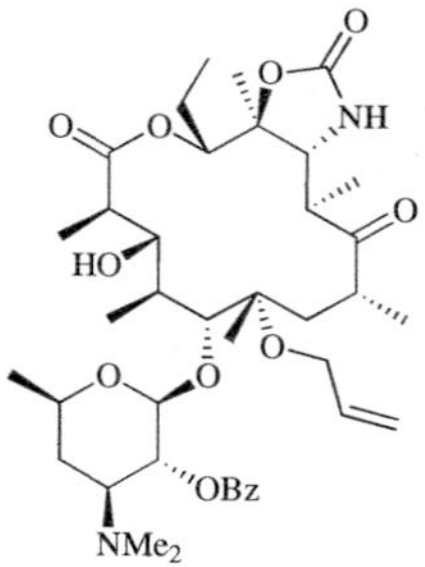 HO, O, O, NH, O, O, O, O, OBz, NMe$_2$	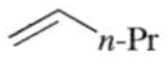 *n*-Pr 10 eq	Gr I (0.1 eq), DCM, 40°, 20 h	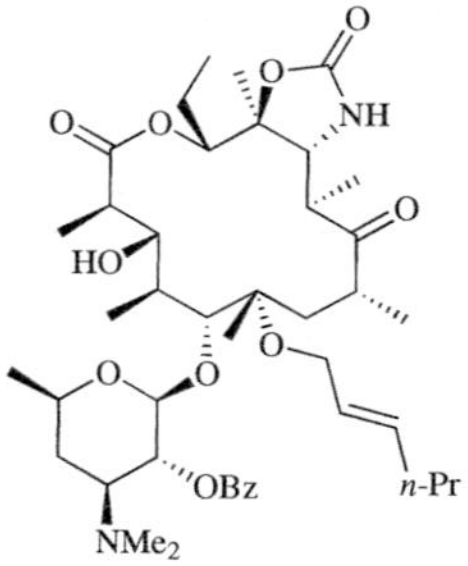 HO, O, O, NH, O, O, O, O, OBz, NMe$_2$, *n*-Pr (84) (*E*)/(*Z*) = 64:36	584
C_{34}				
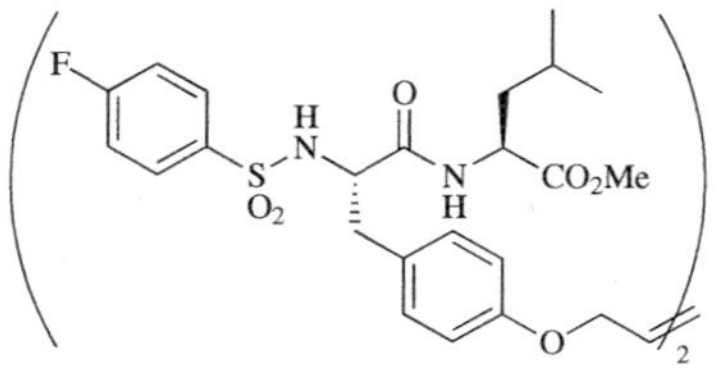 F, H, N, S, O_2, O, N, H, CO_2Me, O, 2	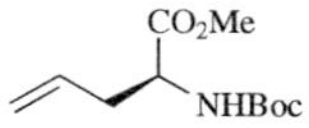 CO_2Me, NHBoc 5.7 eq	Gr II (0.09 eq), DCM, reflux, 23 h	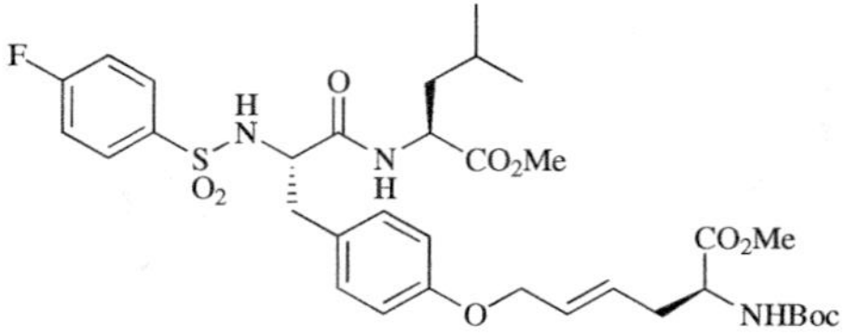 F, H, N, S, O_2, O, N, H, CO_2Me, O, CO_2Me, NHBoc (51) (*E*)/(*Z*) = 87:13	579

C_{39}

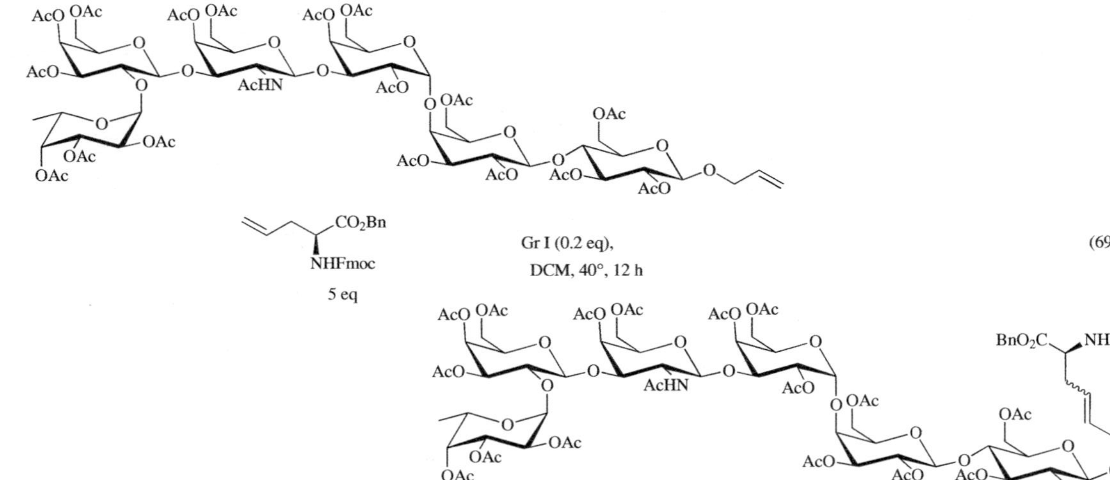

(69) 572

TABLE 4. CROSS-METATHESIS OF PRIMARY ALLYLIC ALCOHOLS AND DERIVATIVES (*Continued*)

B. ALLYLIC ETHERS (*Continued*)

Allylic Ether	Cross Partner	Conditions	Product(s) and Yield(s) (%)	Refs.
Please refer to the charts preceding the tables for the catalyst structures.				
C_{44}				
	x eq	Gr II (y eq), DCM, reflux		567

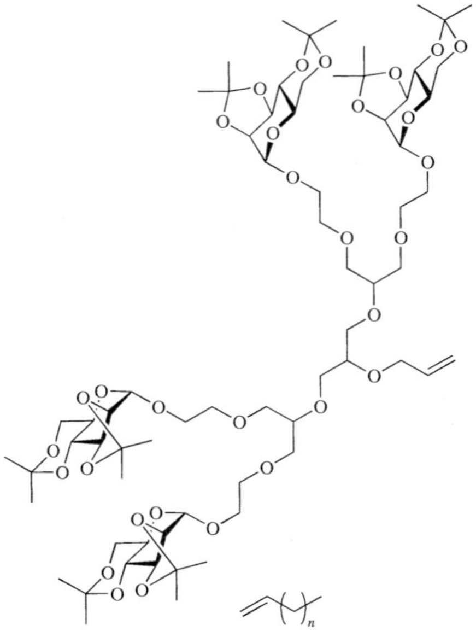

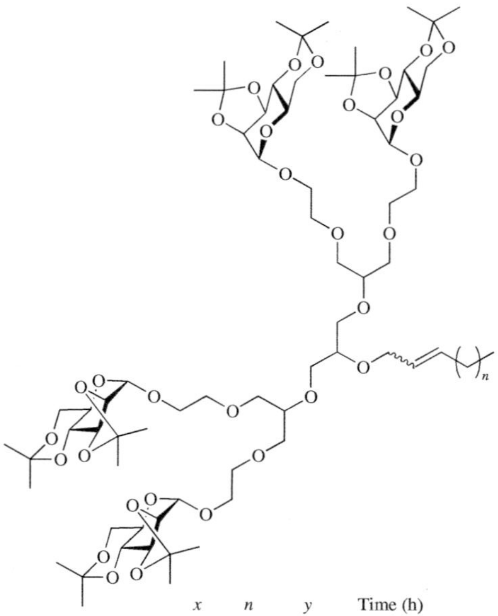

x	n	y	Time (h)	
2	7	0.1	16	(77)
4	12	0.05	5	(43)

C_{45}

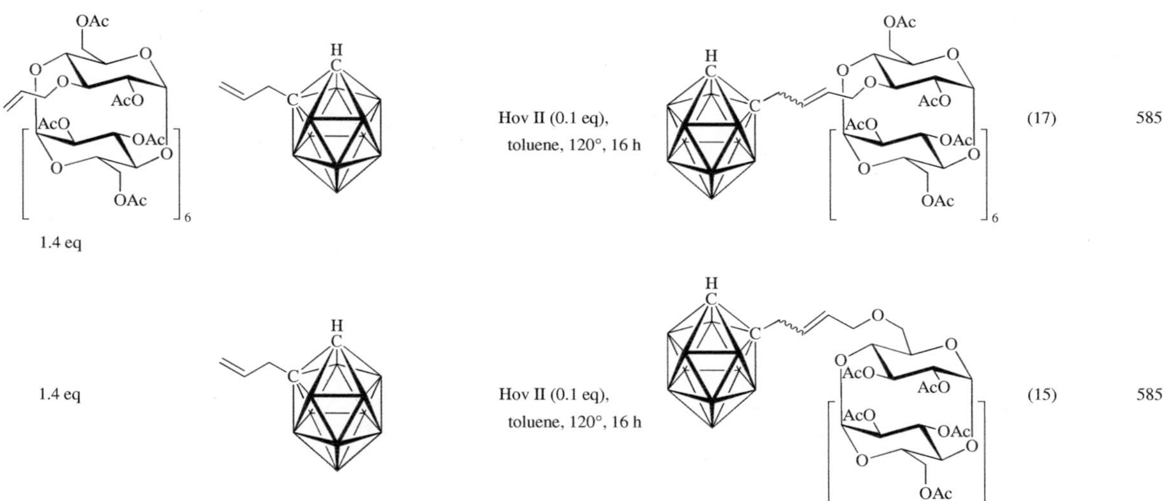

OAc
AcO
AcO
OAc
OAc
1.4 eq
H
C
C
Hov II (0.1 eq),
toluene, 120°, 16 h
(17)
1.4 eq
(15)

585

585

TABLE 4. CROSS-METATHESIS OF PRIMARY ALLYLIC ALCOHOLS AND DERIVATIVES (*Continued*)

B. ALLYLIC ETHERS (*Continued*)

Allylic Ether	Cross Partner	Conditions	Product(s) and Yield(s) (%)		Refs.
Please refer to the charts preceding the tables for the catalyst structures.					
C_{45}					
OAc, O, AcO, O, O, OAc, OAc, O, O, OAc, 6; 1.4 eq	H, C, C	Hov II (0.1 eq), toluene, 120°, 16 h	OAc, O, AcO, O, O, OAc, OAc, O, O, OAc, 6, H, C, C	(24)	585

[a] The ratios were determined by ^{1}H NMR spectroscopy.

[b] The conversion was determined by HPLC analysis.

[c] The conversion was determined by ^{1}H NMR analysis.

TABLE 4. CROSS-METATHESIS OF PRIMARY ALLYLIC ALCOHOLS AND DERIVATIVES (*Continued*)

C. ALLYLIC ESTERS

Allylic ester	Cross Partner	Conditions	Product(s) and Yield(s) (%)	Refs.

Please refer to the charts preceding the tables for the catalyst structures.

C_3

Allylic ester: allyl acetate (OAc); Cross partner: allyl acetate (OAc); Conditions: Catalyst (*x* eq), solvent; Product: AcO–CH₂CH=CHCH₂–OAc

Catalyst	*x*	Solvent	Temp (°)	Time	Conv. (%)[a]	Yield	(*E*)/(*Z*)	Refs.
Ru cat **217**	0.02	THF	70	3 h	53	(—)	11:89	84
Ru cat **217**	0.02	THF	70	6 h	60	(—)	17:83	84
Ru cat **217**	0.02	THF	35	4 h	>95	(62)	11:89	84
Ru cat **215**	0.001	THF	35	12 h	8	(—)	5:95	488
Gr I	0.05	benzene	40	5 min	—	(16)	81:19	564
Gr I	0.05	benzene	40	0.5 h	—	(27)	82:18	564
Gr I	0.05	benzene	40	6 h	—	(34)	84:16	564
Gr I	0.05	benzene	40	24 h	—	(58)	85:15	564

Allylic ester: 5 eq; Cross partner: alkene–$(CH_2)_n$–CH₂CH(NHBoc)CO₂Me; Conditions: Catalyst (0.075 eq), additive (2 mM), H_2O/*t*-BuOH (*x*/*y*) THF, 40°, 4 h; Product: AcO–CH₂CH=CH–$(CH_2)_n$–CH₂CH(NHBoc)CO₂Me; Refs. 401

n	Catalyst	Additive	*x*/*y*	Yield	(*Z*)/(*E*)
0	Ru cat **215**	—	—	(40)	88:12
0	Ru cat **215**	—	1:1	(38)	90:10
0	Ru cat **215**	LiCl	1:1	(31)	76:24
0	Ru cat **215**	$MgCl_2$	1:1	(30)	72:28
0	Ru cat **216**	—	—	(42)	90:10
0	Ru cat **216**	—	1:1	(36)	92:8
0	Ru cat **216**	LiCl	1:1	(31)	84:16
0	Ru cat **216**	$MgCl_2$	1:1	(34)	83:17
1	Ru cat **215**	—	—	(56)	90:10
1	Ru cat **215**	—	1:1	(53)	90:10
1	Ru cat **215**	LiCl	1:1	(48)	84:16
1	Ru cat **215**	$MgCl_2$	1:1	(44)	79:21
1	Ru cat **216**	—	—	(55)	95:5
1	Ru cat **216**	—	1:1	(51)	93:7
1	Ru cat **216**	LiCl	1:1	(50)	87:13
1	Ru cat **216**	$MgCl_2$	1:1	(46)	84:16

TABLE 4. CROSS-METATHESIS OF PRIMARY ALLYLIC ALCOHOLS AND DERIVATIVES (*Continued*)

C. ALLYLIC ESTERS (*Continued*)

Allylic ester	Cross Partner	Conditions	Product(s) and Yield(s) (%)	Refs.

Please refer to the charts preceding the tables for the catalyst structures.

C_3

Allylic ester	Cross Partner	Conditions	Product(s) and Yield(s) (%)	Refs.
OAc, 5 eq	O, CO_2Me, NHBoc	Catalyst (0.075 eq), additive (2 mM), H_2O/*t*-BuOH (*x*/*y*), THF, 40°, 4 h	AcO, O, CO_2Me, NHBoc	401

Catalyst	Additive	*x*/*y*		(*Z*)/(*E*)
Ru cat **215**	—	—	(63)	88:12
Ru cat **215**	—	1:1	(64)	87:13
Ru cat **215**	LiCl	1:1	(54)	67:33
Ru cat **215**	$MgCl_2$	1:1	(56)	63:37
Ru cat **216**	—	—	(66)	92:8
Ru cat **216**	—	1:1	(67)	93:7
Ru cat **216**	LiCl	1:1	(61)	79:21
Ru cat **216**	$MgCl_2$	1:1	(61)	84:16

Allylic ester	Cross Partner	Conditions	Product(s) and Yield(s) (%)	Refs.
5 eq	O, O, CO_2Me	Gr II (0.05 eq), DCM, reflux, 22 h	AcO, O, O, CO_2Me (56) (*E*)/(*Z*) = 90:10	586
7 eq	OR, OR	Ru cat **215** (0.01 eq), THF, 23°, 1 h	RO, RO, OAc	417

R		(*Z*)/(*E*)	er
Ac	(45)	97:3	91.0:9.0
Bn	(64)	95:5	93.0:7.0

7 eq		Ru cat **215** (0.01 eq), THF, 23°, 1 h		(63) $(Z)/(E)$ = 94:6, er 60.0:40.0	417
7 eq		Ru cat **215** (0.01 eq), THF, 23°, 1 h		R / $(Z)/(E)$ / er H (58) 98:2 75.0:25.0 Me (65) 96:4 95.0:5.0	417
7 eq		Ru cat **215** (0.01 eq), THF, 23°, 1 h		(40) $(Z)/(E)$ = 70:30 er (Z) 95.0:5.0 er (E) 95.0:5.0	417
2 eq		Gr I (0.19 eq), DCM, reflux, 17–24 h	*n* / $(E)/(Z)$ 1 (70) 75.0:25.0 2 (63) 75.0:25.0		587

TABLE 4. CROSS-METATHESIS OF PRIMARY ALLYLIC ALCOHOLS AND DERIVATIVES *(Continued)*

C. ALLYLIC ESTERS *(Continued)*

Allylic ester	Cross Partner	Conditions	Product(s) and Yield(s) (%)	Refs.
Please refer to the charts preceding the tables for the catalyst structures.				
C_3				
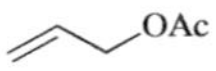 OAc 2 eq	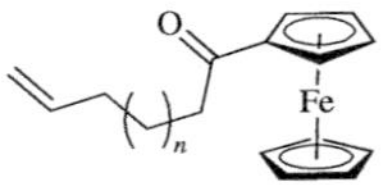 O, Fe, n	Gr I (0.19 eq), DCM, reflux, 17–24 h	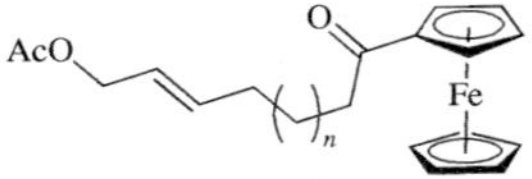 AcO, O, Fe, n n = 1: (58), $(E)/(Z)$ 70.0:30.0 n = 2: (68), $(E)/(Z)$ 75.0:25.0	587
C_{3-4}				
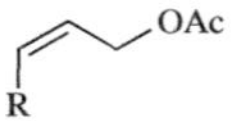 R, OAc x eq	 OBz, 6	Gr I (0.05 eq), solvent, 45°, 12 h	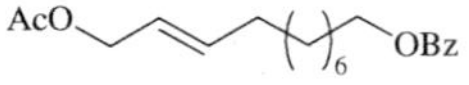 AcO, OBz, 6 R = H, x = 4, DCM: (81), $(E)/(Z)$ 75:25 R = H, x = 2, DCM: (80), $(E)/(Z)$ 80:20 R = H, x = 1, DCM: (59), $(E)/(Z)$ 85:15 R = $AcOCH_2$, x = 55, —: (91), $(E)/(Z)$ 75:25 R = $AcOCH_2$, x = 2, DCM: (89), $(E)/(Z)$ 82:18 R = $AcOCH_2$, x = 1, DCM: (77), $(E)/(Z)$ 83:17	176
 OBz	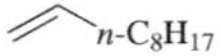 n-C_8H_{17} 5 eq	Ru cat **216** (0.02 eq), THF, 35°	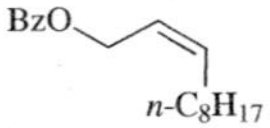 BzO, n-C_8H_{17} (71) $(Z)/(E)$ > 95:5	551

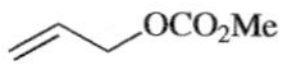OCO$_2$Me	 9 eq	Ru cat **216** (0.01 eq), THF, rt, 5 h	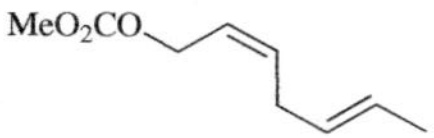MeO$_2$CO	(79) (*Z*)/(*E*) > 95:5	437
	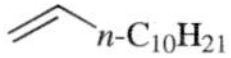*n*-C$_{10}$H$_{21}$ 5 eq	Hov II (0.05 eq), DCM, 40°	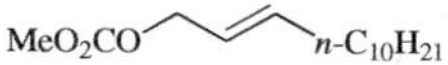MeO$_2$CO *n*-C$_{10}$H$_{21}$	(68) (*E*)/(*Z*) = 91:9	551
	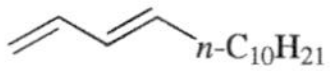*n*-C$_{10}$H$_{21}$ 1.5 eq	Ru cat **216** (0.04 eq) 50°, 24 h	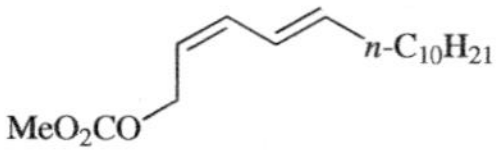*n*-C$_{10}$H$_{21}$ MeO$_2$CO	(42) (*Z*)/(*E*) > 95:5	538
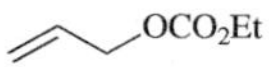OCO$_2$Et 2 eq	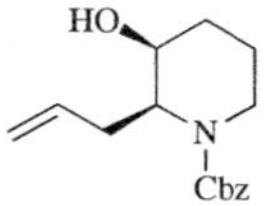HO N Cbz	Gr II (0.1 eq), DCM, reflux, 24 h	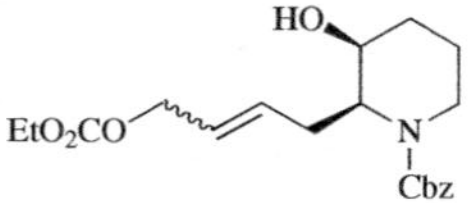HO EtO$_2$CO N Cbz	(72)	441
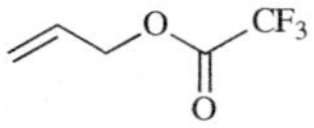O CF$_3$ O 6 eq	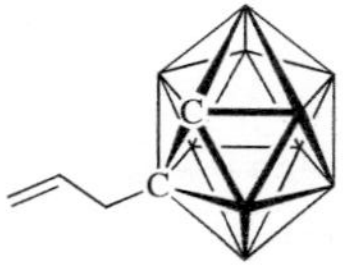C C	Gr II (0.02 eq), DCM, reflux, 5 h	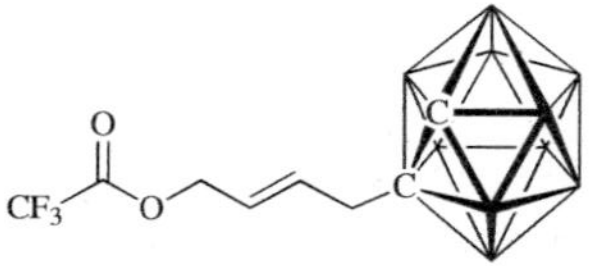O CF$_3$ O C C (77) (*E*)/(*Z*) > 95:5		436

TABLE 4. CROSS-METATHESIS OF PRIMARY ALLYLIC ALCOHOLS AND DERIVATIVES (*Continued*)

C. ALLYLIC ESTERS (*Continued*)

Allylic ester	Cross Partner	Conditions	Product(s) and Yield(s) (%)	Refs.
Please refer to the charts preceding the tables for the catalyst structures.				
C_4				
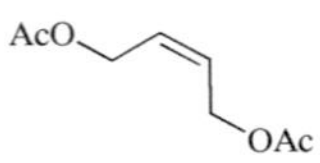 AcO–CH₂CH=CHCH₂–OAc (*Z*), *x* eq	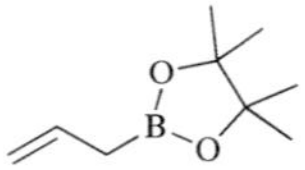 allyl pinacol boronate	1. Gr II (*y* eq), DCM, 40°, 2–12 h 2. PhCHO (*z* eq), 23°, 12–48 h	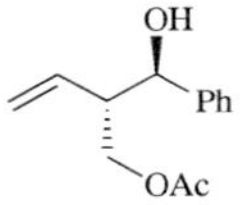 [structure: OH, Ph, OAc]	126
2 eq	 CH₂=CHCH₂OPh	Catalyst (0.025 eq), DCM, 20°, 3 h	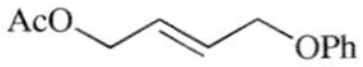 AcO–CH₂CH=CHCH₂–OPh	565
2 eq	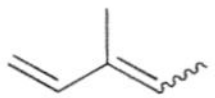 [diene]	Gr II (0.05 eq), DCM, reflux, 12 h	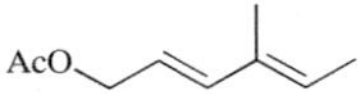 AcO–[diene product] (82) (*E*) only	132
2 eq	 CH₂=CH(CH₂)₄OH	Ru cat **143** (0.05 eq), H_2O, 30°, 12 h	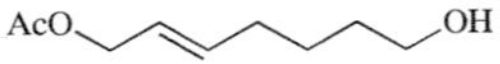 AcO–CH₂CH=CH(CH₂)₄–OH (76) (*E*)/(*Z*) = 90:10	588

x	*y*	*z*		*anti/syn*[b]
3	1.5	0.05	(32)	64:36
3	1.5	0.05	(75)	82:18
3	1.5	0.02	(75)	82:18
0.5	1.5	0.05	(57)	82:18
3	0.75	0.05	(75)	82:18

Catalyst		(*E*)/(*Z*)
Gr II	(76)	92:8
Hov II	(75)	92:8
Ru cat **155**	(67)	92:8

— eq	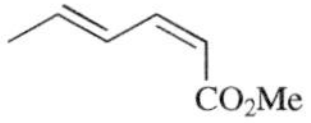 CO_2Me	Gr II (0.075 eq), DCM, 40°, 8 h	AcO … CO_2Me	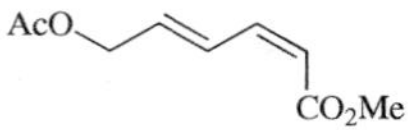 (68) (*E*) only	589
5 eq	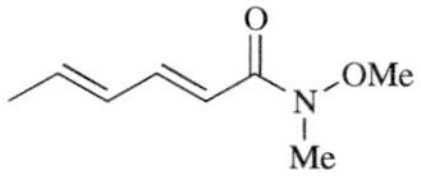 O, N, OMe, Me	Hov II (0.1 eq), DCM, rt, 24 h	AcO … O, N, OMe, Me	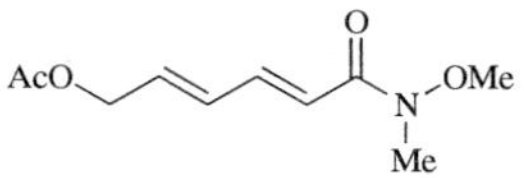 (42) (*E*)/(*Z*) = 83:17	549
2 eq	 OTBS	Catalyst (*x* eq), solvent	AcO … OTBS		

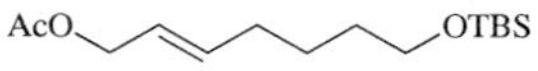

Catalyst	*x*	Solvent	Temp (°)	Time (h)		(*E*)/(*Z*)	
Ru cat **158**	0.01	DCM	25	2	(90)	93:7	590
Ru cat **157**	0.005	DCM	25	2	(93)	93:7	590
Ru cat **316**	0.02	toluene	80	3.5	(59)	86:14	591
Ru cat **286**	0.02	DCM	40	8	(71)	91:9	161
Ru cat **88**	0.025	DCM	25	5	(67)	89:11	171

3 eq	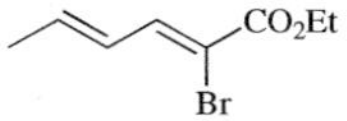 CO_2Et, Br	Gr II (0.1 eq), DCM, reflux, 12 h	AcO … CO_2Et, Br	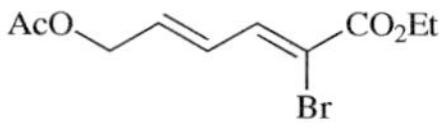 (70) (*E*)/(*Z*) = 88:12	132
5 eq	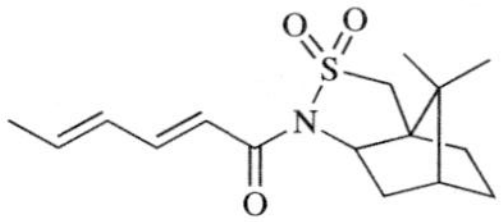 O, O, S, N, O	Hov II (0.1 eq), DCM, rt, 24 h	AcO … O, O, S, N, O	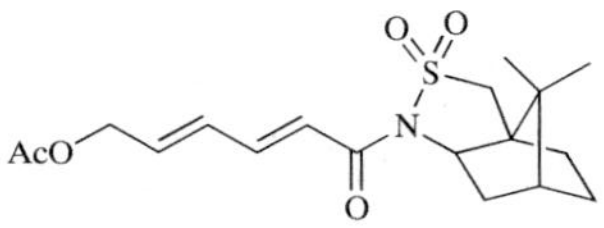 (40) (*E*)/(*Z*) = 83:17	549

TABLE 4. CROSS-METATHESIS OF PRIMARY ALLYLIC ALCOHOLS AND DERIVATIVES (*Continued*)

C. ALLYLIC ESTERS (*Continued*)

Allylic ester	Cross Partner	Conditions	Product(s) and Yield(s) (%)	Refs.

Please refer to the charts preceding the tables for the catalyst structures.

C_4

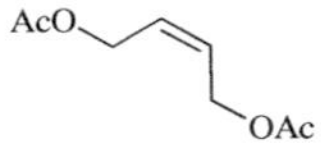

Allylic ester	Cross Partner	Conditions	Product(s) and Yield(s) (%)	Refs.
AcO, OAc; 2 eq	O	Gr I (0.1 eq), DCM, 40°, 12 h	AcO, O; (78) (*E*)/(*Z*) = 82:18	36
2 eq	Ph	Catalyst (0.01 eq), DCM, N_2, rt	AcO, Ph	

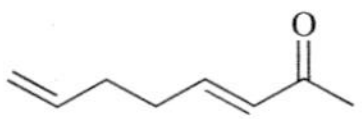

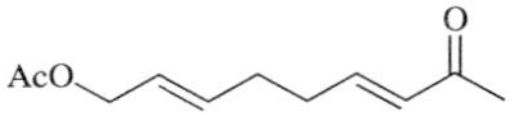

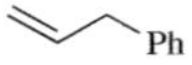

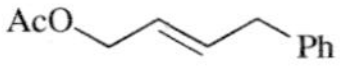

Catalyst	Time (h)	Conv. (%)[a]	(*E*)/(*Z*)	Refs.
Ind II	1	50	52:48	77
Ru cat **294**	1	96	88:12	77
Ru cat **295**	1	97	90:10	77
Ru cat **296**	1	97	91:9	77
Ru cat **285**	1	75	89:11	77
Ru cat **297**	1	97	90.5:9.5	77
Ru cat **343**	20	>99	64:36	77
Ru cat **339**	20	67	76:24	77
Ru cat **342**	20	55	65:35	77
Ru cat **344**	5	89	56.5:43.5	77
Ind II	2.5	98	75:25	489
Ru cat **282**	20	97	71:29	489
Ru cat **284**	20	90	64:36	489
Ru cat **283**	20	61	69:31	489
Ru cat **285**	2.5	95	89:11	489
Ru cat **286**	2.5	97	84:16	489

2 eq Ph (allylbenzene) — Catalyst (0.01 eq), toluene, N_2, 80° → AcO–CH$_2$CH=CHCH$_2$–Ph

Catalyst	Time (h)	Conv. (%)[a]	(*E*)/(*Z*)	
Ru cat **343**	1	98	85:15	77
Ru cat **343**	20	99	87.5:12.5	77
Ru cat **339**	1	98	74:26	77
Ru cat **339**	20	99	84:16	77
Ru cat **342**	1	>99	77:23	77
Ru cat **342**	20	99	84.5:15.5	77
Ru cat **344**	5	>99	69:31	77
Ru cat **344**	20	>99	70:30	77
Ru cat **283**	1	100	80:20	489
Ru cat **283**	20	100	87.5:12.5	489

2 eq Ph (allylbenzene) — Catalyst (*x* eq), THF → AcO–CH$_2$CH=CHCH$_2$–Ph (**I**) + Ph–CH$_2$CH=CHCH$_2$–Ph (**II**) 488

Catalyst	*x*	Temp (°)	Time (h)	Conv. **I** (%)[c]	(*Z*)/(*E*) **I**	Conv. **II** (%)[c]	(*Z*)/(*E*) **II**
Ru cat **217**	0.05	70	9	37	89:11	26	96:4
Ru cat **217**	0.05	35	6	50	86:14	19	>95:5
Ru cat **218**	0.05	70	6	48	82:18	33	91:9
Ru cat **218**	0.05	35	9	45	87:13	23	>95:5
Ru cat **220**	0.05	70	3	57	75:25	42	94:6
Ru cat **220**	0.05	35	6	64	79:21	22	>95:5
Ru cat **219**	0.05	35	7	54	83:17	17	>95:5
Ru cat **215**	0.01	35	9	58	91:9	28	>95:5

TABLE 4. CROSS-METATHESIS OF PRIMARY ALLYLIC ALCOHOLS AND DERIVATIVES (*Continued*)

C. ALLYLIC ESTERS (*Continued*)

Allylic ester	Cross Partner	Conditions	Product(s) and Yield(s) (%)	Refs.

Please refer to the charts preceding the tables for the catalyst structures.

C_4

AcO–CH₂CH=CHCH₂–OAc (2 eq) + CH₂=CHCH₂Ph; Catalyst (*x* eq), DCM → AcOCH₂CH=CHCH₂Ph

Catalyst	*x*	Temp (°)	Time (h)	Conv. (%)		(*E*)/(*Z*)	Refs.
Ru cat **327**	0.025	30	20	—	(80)	89:11	592, 593
Ru cat **328**	0.025	30	20	—	(74)	90:10	592, 593
Ind II	0.025	30	20	—	(74)	92:8	592, 593
Ru cat **14**	0.025	40	16	—	(47)	86:14	594
Gr II	0.03	40	12	—	(80)	88:12	36
Gr I	0.03	40	12	—	(80)	76:24	36
Ru cat **288**	0.025	25	24	—	(79)	87:13	595
Ru cat **289**	0.025	25	23	—	(76)	86:14	595
Ru cat **107**	0.01	25	5	—	(63)	85:15	596
Ru cat **247**	0.01	40	1	61	(—)	80:20	597
Ru cat **247**	0.01	40	5	62	(—)	80:20	

(2 eq) + CH₂=CHCH₂Ph; Catalyst (0.02 eq), 2-MeTHF, 70°, 0.5 h → AcOCH₂CH=CHCH₂Ph

Catalyst		(*E*)/(*Z*)	Refs.
Ru cat **103**	(72)	86:14	598
Ru cat **104**	(70)	85:15	
Ru cat **105**	(70)	88:12	
Hov II	(73)	87:13	

2 eq Ph Catalyst (0.01 eq), benzene, 23° AcO Ph 599

Catalyst	Time (min)	Conv. (%)[c]	(*E*)/(*Z*)
Ru cat **271**	5	87	90.5:9.5
Hov II	5	87	91:9
Ru cat **187**	5	97	86:14
Ru cat **272**	15	74	77:23
Ru cat **145**	25	70	76:24
Ru cat **203**	240	78	73:27
Ru cat **273**	30	78	78:22
Ru cat **204**	75	80	75:25
Ru cat **205**	60	75	76:24
Ru cat **146**	15	72	86:14
Ru cat **206**	45	73	75:25
Ru cat **147**	10	83	85:15
Ru cat **274**	120	74	80:20
Ru cat **207**	90	71	75:25

2.4 eq Ph Catalyst (0.025 eq), DCM, ethylene, 12 h AcO Ph 600

Catalyst	Temp (°)	Conv. (%)[c]	(*E*)/(*Z*)
Ru cat **228**	60	75	50:50
Ru cat **227**	23	80	86:14
Ru cat **223**	23	80	86:14
Ru cat **224**	23	75	71:29

TABLE 4. CROSS-METATHESIS OF PRIMARY ALLYLIC ALCOHOLS AND DERIVATIVES (*Continued*)

C. ALLYLIC ESTERS (*Continued*)

Allylic ester	Cross Partner	Conditions	Product(s) and Yield(s) (%)	Refs.

Please refer to the charts preceding the tables for the catalyst structures.

C_4

Allylic ester	Cross Partner	Conditions	Product(s) and Yield(s) (%)	Refs.
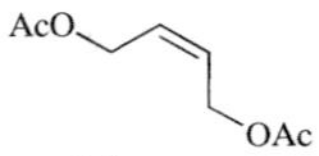 AcO, OAc (2.4 eq)	 Ph	Gr II (x eq), solvent, 25°, 3 h	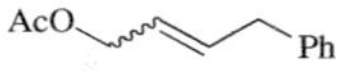 AcO, Ph	601

x	Solvent	Conv. (%)[c]
0.005	DCM	64
0.005	methyl decanoate	79
0.005	neat	82
0.025	DCM	87
0.025	methyl decanoate	95
0.025	neat	86
0.001	neat	55

Allylic ester	Cross Partner	Conditions	Product(s) and Yield(s) (%)	Refs.
2 eq	 Ph	Hov II (x eq), solvent, 25°, 3 h	 AcO, Ph	601

x	Solvent	Conv. (%)[c]
0.005	DCM	88
0.025	DCM	91
0.025	methyl decanoate	91
0.025	neat	91
0.001	DCM	15
0.005	methyl decanoate	72
0.005	neat	88
0.001	neat	85

2 eq Ph (allylbenzene) — Ru cat **89** (x eq), solvent, 25°, 3 h — AcO~Ph 601

x	Solvent	Conv. (%)[c]
0.005	DCM	72
0.005	methyl decanoate	79
0.001	methyl decanoate	36
0.005	none	89
0.001	none	82
0.025	DCM	92
0.025	methyl decanoate	90
0.025	none	90

2 eq Ph (allylbenzene) — Catalyst (x eq), additive (y eq), solvent — AcO~Ph

Catalyst	x	Additive	y	Solvent	Temp (°)	Time (h)	Conv. (%)[c]		$(E)/(Z)$	
Ru cat **281**	0.02	C_2Cl_6	0.08	toluene	80	24	—	(65)	80:20	602
Hov II	0.01	—	—	DCM	40	2	63	(—)	90:10	526
Ru cat **194**	0.01	TMSCl	0.1	DCM	40	2	55	(—)	90:10	526
Ru cat **195**	0.01	TMSCl	0.1	DCM	40	2	81	(—)	90:10	526
Ru cat **196**	0.01	TMSCl	0.1	DCM	40	2	52	(—)	83:17	526
Ru cat **150**	0.005	—	—	toluene	70	0.25	—	(84)	90:10	448
Ru cat **150**	0.005	—	—	toluene	70	0.5	—	(87)	90:10	448

TABLE 4. CROSS-METATHESIS OF PRIMARY ALLYLIC ALCOHOLS AND DERIVATIVES (*Continued*)

C. ALLYLIC ESTERS (*Continued*)

Allylic ester	Cross Partner	Conditions	Product(s) and Yield(s) (%)	Refs.
Please refer to the charts preceding the tables for the catalyst structures.				
C_4				
AcO, OAc; 2 eq	Ph	Ru cat **109** (0.02 eq), H_2O, 22°, 12 h	AcO, Ph (84) $(E)/(Z) = 93:7$	341
	OMe, OH	Catalyst (0.01 eq), DCM, rt, 5 h	AcO, OMe, OH; $(E)/(Z) = 91:9$; Catalyst: Ru cat **41** (80), Ru cat **44** (78), Ru cat **45** (86)	562
3 eq	OR, Cl, N	Hov II (0.1 eq), DCM, 50°, 24 h	AcO, OR, Cl, N; R: H (79), Ac (80), TBS (71)	603
2 eq	AcO, OAc, O, AcO, AcO, O	Gr I (0.2 eq), DCM, reflux, 6 h	AcO, OAc, O, AcO, AcO, O, OAc (70) $(E)/(Z) = 83:17$	571

3 eq

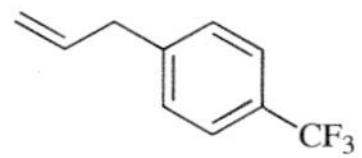

Catalyst (0.01 eq), toluene, 80°, 20 h

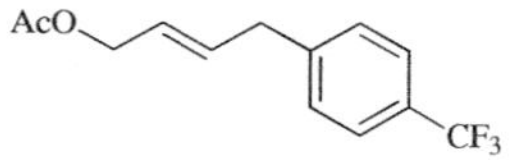

Catalyst		(*E*)/(*Z*)
Hov II	(82)	85:15
Ru cat **359**	(79)	77:23
Ru cat **360**	(84)	86:14
Ru cat **361**	(83)	86:14
Ru cat **362**	(77)	75:25

491

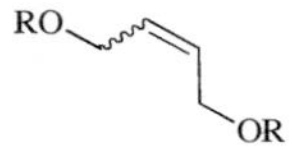

2 eq

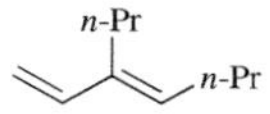

Gr II (0.05 eq), solvent, 12 h

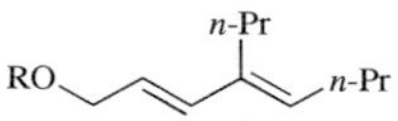

R	Solvent	Temp (°)		(*E*)/(*Z*)
Ac	DCM	40	(51)	—
Ac	benzene	60	(72)	—
Ac	benzene	80	(63)	—
Bz	DCM	reflux	(79)	(*E*) only

132

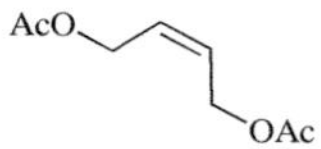

5 eq

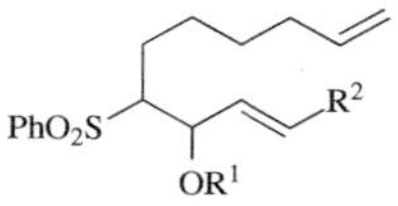

1. Gr II (0.05 eq), solvent, temp, 6 h
2. DMPU (5.0 eq), SmI_2 (3.0 eq), –78°, 1 h

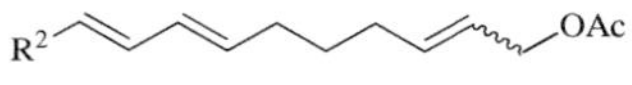

I

II

R^1	R^2	Solvent	Temp (°)	**I + II**	**I/II**
Ac	H	DCM	rt	(65)	>91:9
Bz	H	toluene	80	(67)	83:17
Bz	Me	toluene	80	(70)	>91:9
Bz	n-C_5H_{11}	toluene	80	(64)	>91:9
Bz	Ph	toluene	80	(68)	>91:9

604

TABLE 4. CROSS-METATHESIS OF PRIMARY ALLYLIC ALCOHOLS AND DERIVATIVES (*Continued*)
C. ALLYLIC ESTERS (*Continued*)

Allylic ester	Cross Partner	Conditions	Product(s) and Yield(s) (%)	Refs.
Please refer to the charts preceding the tables for the catalyst structures.				
C_4				
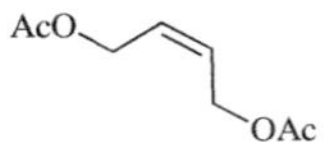 AcO, OAc 10 eq	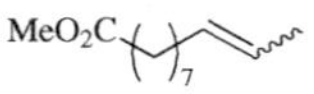 MeO_2C, 7	Hov II (0.01 eq), DCM, rt, 4 h	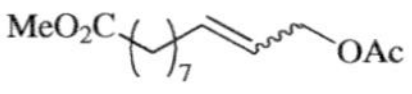 MeO_2C, 7, OAc (90)	605
4 eq	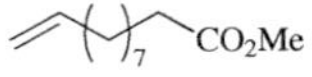 7, CO_2Me	Ru cat **246** (0.00025 eq), toluene, 60°, 10 min	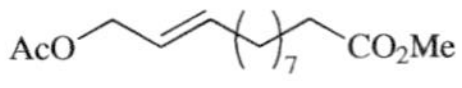 AcO, 7, CO_2Me 68% conv.[a] (*E*)/(*Z*)= 70:30	606
3 eq	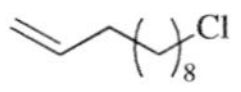 8, Cl	Catalyst (0.01 eq), toluene, 80°, 20 h	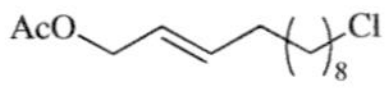 AcO, 8, Cl	491

Catalyst		(*E*)/(*Z*)
Hov II	(90)	85:15
Ru cat **359**	(85)	87:13
Ru cat **360**	(85)	84:16
Ru cat **361**	(82)	77:23
Ru cat **362**	(73)	74:16

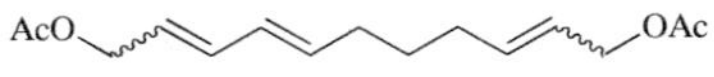

Allylic ester	Cross Partner	Conditions	Product(s) and Yield(s) (%)	Refs.
5 eq	PhO_2S, R^2, R^1, O	1. Gr II (0.05 eq), toluene, 80°, 6 h 2. 5% Na in Hg (5.5 eq), NaH_2PO_4 (6.4 eq), MeOH, THF, rt, 1 h	R^2, OAc **I** + AcO, OAc **II**	604

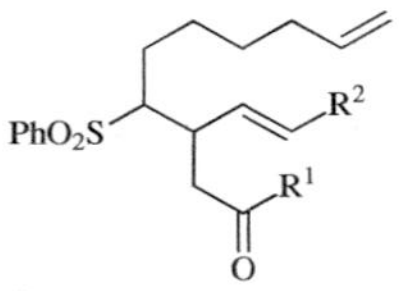

R^1	R^2	**I + II**	**I/II**
Me	H	(74)	65:35
Me	Me	(80)	>91:9
Me	n-C_5H_{11}	(83)	>91:9
Me	Ph	(72)	>91:9
Ph	Ph	(74)	>91:9

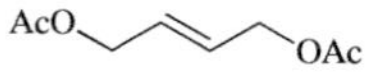

AcO OAc
10 eq

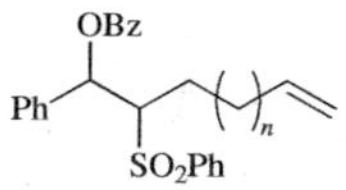

OBz
Ph
SO_2Ph

1. Gr II (0.1 eq),
toluene, 60°, 12 h
2. SmI_2 (x eq),
solvent, –78°

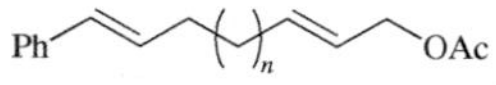

Ph OAc

445

n	x	Solvent	Time (h)		$(E,E)/(Z,E)$
2	2	HMPA (6 eq), THF	0.5	(72)	88:12
3	2	HMPA (6 eq), THF	0.5	(68)	90:10
3	3	THF/DMPU (4:1)	1	(61)	90:10

10 eq

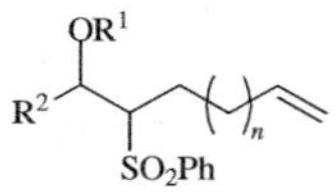

OR^1
R^2
SO_2Ph

1. Gr II (0.1 eq),
toluene, 60°, 12 h
2. Na/Hg (5%),
NaH_2PO_4,
THF, MeOH, rt

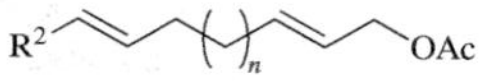

R^2 OAc

445

n	R^1	R^2		$(E,E)/(Z,E)$
2	Ac	n-C_6H_{13}	(85)	88:12
2	Bz	n-C_6H_{13}	(83)	90:10
3	Bz	Ph	(70)	88:12

TABLE 4. CROSS-METATHESIS OF PRIMARY ALLYLIC ALCOHOLS AND DERIVATIVES (*Continued*)

C. ALLYLIC ESTERS (*Continued*)

Allylic ester	Cross Partner	Conditions	Product(s) and Yield(s) (%)	Refs.
Please refer to the charts preceding the tables for the catalyst structures.				
C_4				
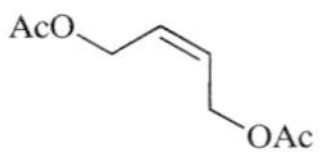 AcO, OAc 2 eq	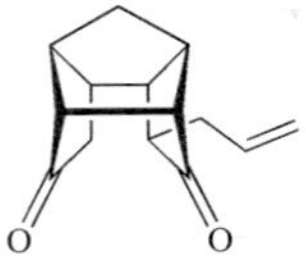	Ru cat **271** (0.05–0.07 eq), toluene, reflux, 12 h	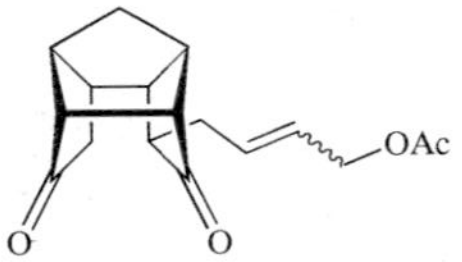 OAc (62)	607
2 eq	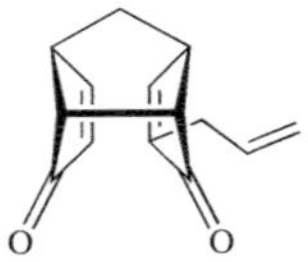	Ru cat **271** (0.05–0.07 eq), toluene, reflux, 12 h	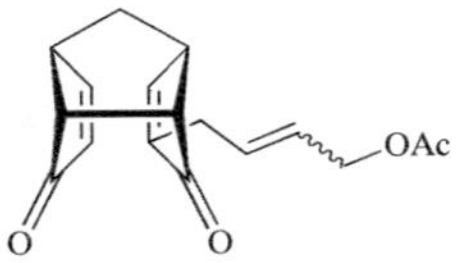 OAc (63)	607
2 eq	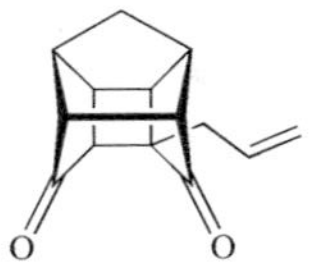	Ru cat **271** (0.05–0.07 eq), toluene, reflux, 12 h	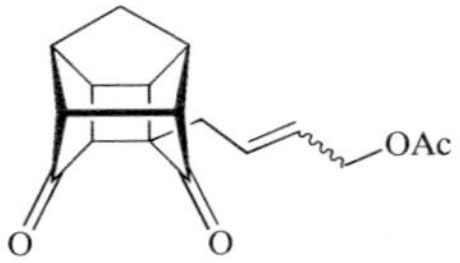 OAc (64)	607
10 eq	O, H, N, H, R^2, R^1, N, O	Gr II (0.15 eq), DCM, reflux, 2 h	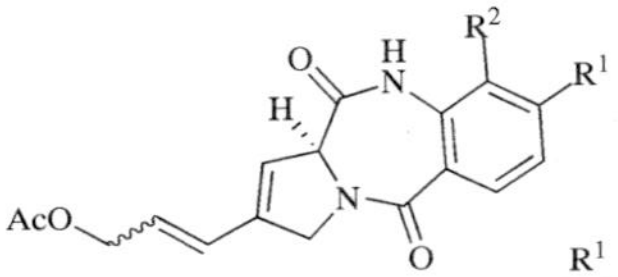 AcO, O, H, N, H, R^2, R^1, N, O R^1 | R^2 | H | H | (89) Me | BnO | (91)	608

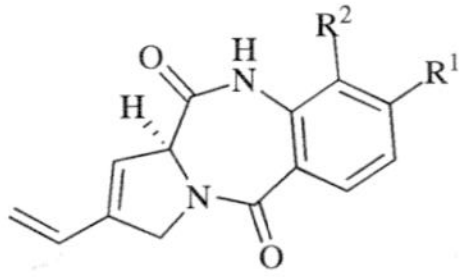

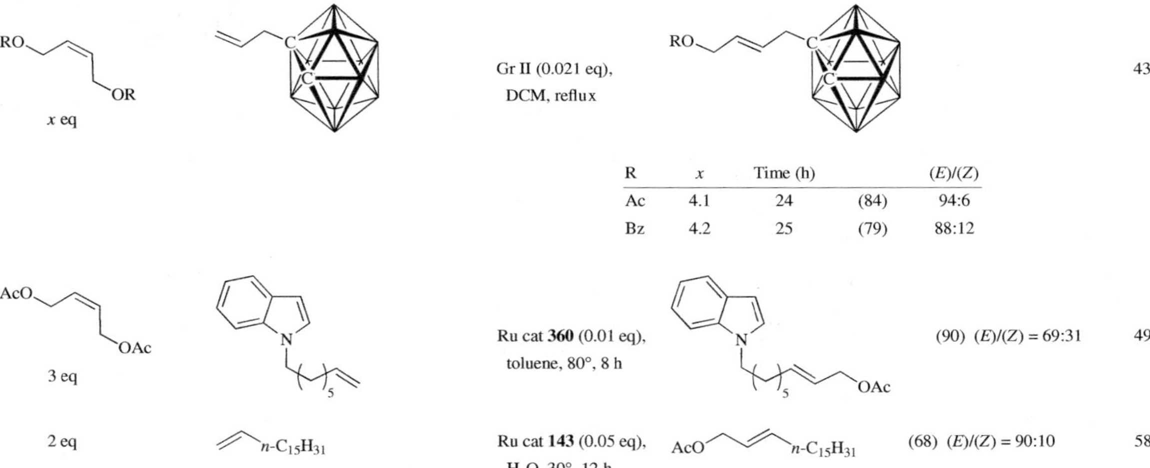

R	x	Time (h)		$(E)/(Z)$
Ac	4.1	24	(84)	94:6
Bz	4.2	25	(79)	88:12

TABLE 4. CROSS-METATHESIS OF PRIMARY ALLYLIC ALCOHOLS AND DERIVATIVES (*Continued*)

C. ALLYLIC ESTERS (*Continued*)

Allylic ester	Cross Partner	Conditions	Product(s) and Yield(s) (%)	Refs.

Please refer to the charts preceding the tables for the catalyst structures.

C_4

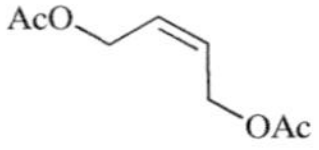

AcO, OAc; *x* eq 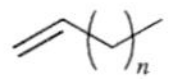 — Catalyst (*y* eq), DCM — AcO

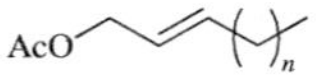

x	*n*	Catalyst	*y*	Temp (°)	Time (h)		(*E*)/(*Z*)	Refs.
2	14	Ru cat **88**	0.03	28	3	(85)	88:12	609
2	14	Ru cat **88**	0.01	25	2	(55)	88:12	609
2	14	Ru cat **130**	0.01	25	3	(38)	88:12	609
2	14	Ru cat **116**	0.01	25	2	(97)	88:12	609
2	14	Ru cat **116**	0.003	25	0.5	(73)	88:12	609
2	14	Ru cat **117**	0.01	25	2	(64)	88:12	609
—	15	Hov II	0.01	24	6	(37)	—	610
—	15	Ru cat **88**	0.01	24	6	(89)	—	610
2	14	Ru cat **175**	0.02	22	0.5	(73)	91:9	611
2	14	Ru cat **174**	0.02	22	0.5	(69)	91:9	611
2	14	Ru cat **177**	0.02	22	0.5	(64)	91:9	611
2	14	Ru cat **176**	0.02	22	0.5	(55)	91:9	611
2	14	Ru cat **178**	0.02	22	0.5	(36)	91:9	611
2	14	Ru cat **197**	0.02	22	0.5	(70)	91:9	611

Allylic ester	Cross Partner	Conditions	Product(s) and Yield(s) (%)	Refs.
2 eq	O, O	Ru cat **271** (0.05–0.07 eq), toluene, reflux, 12 h	AcO, OAc, O, O (55)	607

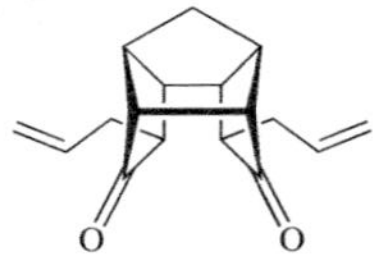

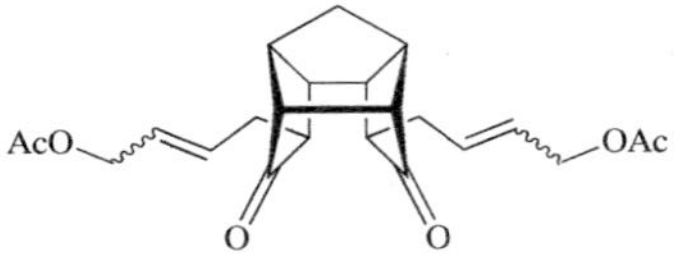

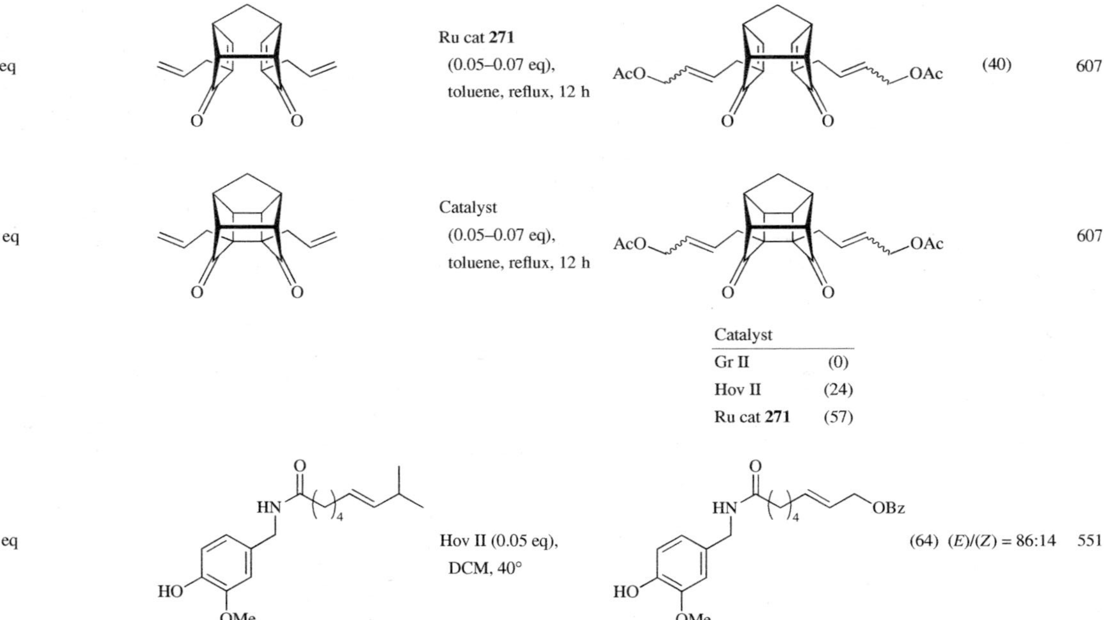

2 eq		Ru cat **271** (0.05–0.07 eq), toluene, reflux, 12 h		(40)	607
2 eq		Catalyst (0.05–0.07 eq), toluene, reflux, 12 h			607
			Catalyst: Gr II (0); Hov II (24); Ru cat **271** (57)		
5 eq		Hov II (0.05 eq), DCM, 40°		(64) $(E)/(Z) = 86:14$	551

TABLE 4. CROSS-METATHESIS OF PRIMARY ALLYLIC ALCOHOLS AND DERIVATIVES *(Continued)*

C. ALLYLIC ESTERS *(Continued)*

Allylic ester	Cross Partner	Conditions	Product(s) and Yield(s) (%)	Refs.

Please refer to the charts preceding the tables for the catalyst structures.

C_4

Allylic ester	Cross Partner	Conditions	Product(s) and Yield(s) (%)	Refs.
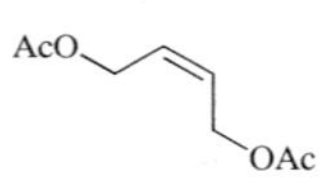 AcO, OAc; 3 eq	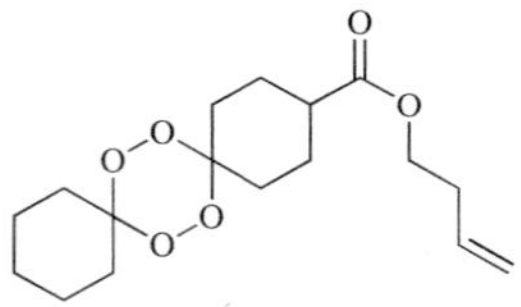 O, O, O, O, O, O	Ru cat **88** (0.03 eq), DCM, 45°, 2 h	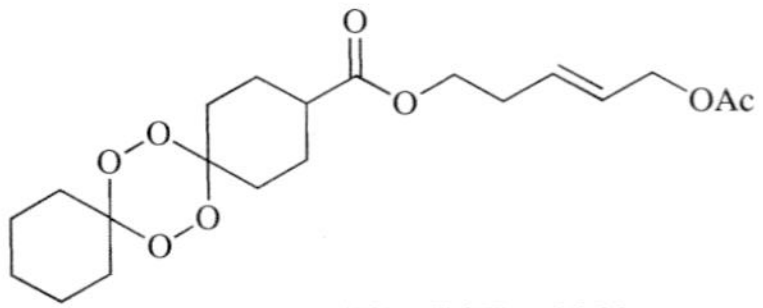 O, O, O, O, O, O, OAc; (89) $(E)/(Z)$ = 80:20	193
	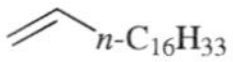 n-$C_{16}H_{33}$	Catalyst (0.01 eq), toluene	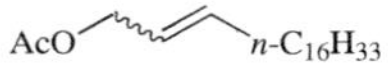 AcO, n-$C_{16}H_{33}$	610

Catalyst	Temp (°)	Time (h)	
Ru cat **182**	55	5	(56)
Ru cat **182**	55	24	(78)
Ru cat **184**	24	5	(63)
Ru cat **184**	24	24	(85)
Ru cat **183**	55	5	(35)
Ru cat **183**	55	24	(52)
Ru cat **164**	90	47	(64)
Ru cat **165**	90	42	(57)
Ru cat **179**	24	24	(19)
Ru cat **180**	24	24	(53)

BzO, OBz (5 eq) | HO$(CH_2)_8$, n-C_8H_{17} | Hov II (0.05 eq), DCM, 40° | BzO, n-C_8H_{17} (62) $(E)/(Z)$ = 88:12 + HO$(CH_2)_8$, OBz (63) $(E)/(Z)$ = 89:11 | 551

AcO, OAc (10 eq) | $MeO_2C(CH_2)_7$, n-C_7H_{15} | Hov II (0.01 eq), DCM, rt, 4 h | $MeO_2C(CH_2)_7$, OAc (80) | 1214

x eq | $MeO_2C(CH_2)_7$, n-C_7H_{15} | Ind II (0.01 eq), tetradecane (0.6 eq), solvent, 5 h | n-C_8H_{17}, OAc **I** + $MeO_2C(CH_2)_7$, OAc **II** | 534

x	Solvent	Temp (°)	Conv. (%)[e]	Selectivity (%) **I/II**[e]	**I**[e]	**II**[e]
2	toluene	25	7	100:100	(7)	(7)
2	toluene	50	72	86:83	(62)	(58)
2	toluene	80	80	80:77	(64)	(61)
2	DCE	50	67	80:78	(54)	(52)
3	—	50	67	80:78	(54)	(52)

TABLE 4. CROSS-METATHESIS OF PRIMARY ALLYLIC ALCOHOLS AND DERIVATIVES (*Continued*)

C. ALLYLIC ESTERS (*Continued*)

Allylic ester	Cross Partner	Conditions	Product(s) and Yield(s) (%)	Refs.

Please refer to the charts preceding the tables for the catalyst structures.

C_4

Allylic ester	Cross Partner	Conditions	Product(s) and Yield(s) (%)	Refs.
AcO–CH_2CH=CHCH_2–OAc (2 eq)	$MeO_2C(CH_2)_7$CH=CHCH_2–n-C_7H_{15}	Catalyst (0.01 eq), tetradecane (0.6 eq), toluene, 50°	n-C_8H_{17}CH=CHCH_2OAc (I) + $MeO_2C(CH_2)_7$CH=CHCH_2OAc (II)	534

Catalyst	Conv. (%)[e] (1 h)	Selectivity I/II[e] (1 h)	I[e] (1 h)	II[e] (1 h)	Conv. (%)[e] (5 h)	Selectivity I/II[e] (5 h)	I[e] (1 h)	II[e] (1 h)
Ind II	43	85:75	(36)	(32)	72	86:83	(62)	(58)
Ru cat **283**	35	86;91	(30)	(32)	58	86:82	(50)	(48)
Ru cat **331**	58	80:76	(46)	(44)	77	82:79	(63)	(61)
Ru cat **297**	56	83:80	(46)	(45)	75	87:82	(65)	(62)
Ru cat **330**	50	84:82	(42)	(41)	70	84:82	(59)	(57)
Ru cat **319**	30	81:77	(24)	(23)	41	83:78	(34)	(32)
Ru cat **282**	8	100:100	(8)	(8)	34	78:77	(27)	(26)
Ru cat **326**	66	88:89	(58)	(59)	70	90:83	(63)	(58)
Ru cat **325**	71	81:75	(56)	(53)	73	82:76	(60)	(55)
Ru cat **327**	51	81:83	(41)	(42)	55	89:84	(49)	(46)
Ru cat **329**	40	91:92	(36)	(37)	44	90:87	(40)	(35)
Ru cat **294**	0	—	(0)	(0)	26	83:81	(22)	(21)
Ru cat **328**	16	89:87	(14)	(15)	16	100:100	(16)	(16)
Ru cat **315**	5	100:100	(5)	(5)	13	85:83	(11)	(10)
Gr II	63	83:86	(52)	(54)	65	80:84	(52)	(55)
Hov II	47	85:87	(40)	(41)	49	83:87	(41)	(42)

x eq $MeO_2C(CH_2)_7CH{=}CHCH_2$-*n*-$C_7H_{15}$ → Catalyst (*y* eq), $PhSiCl_3$ (*z* eq), toluene → $MeO_2C(CH_2)_7CH{=}CHOAc$ (**I**) + *n*-$C_8H_{17}CH{=}CHOAc$ (**II**) + *n*-$C_8H_{17}CH{=}CH$-*n*-C_7H_{15} (**III**) + $MeO_2C(CH_2)_7CH{=}CH(CH_2)_6CO_2Me$ (**IV**) 1214

x	Catalyst	*y*	*z*	Temp (°)	Time (h)	Conv. (%)[c]	**I**	**II**	**III**	**IV**
5	Gr I	0.01	—	50	5	14	(3)	(4)	(6)	(5)
5	Gr II	0.01	—	50	5	48	(29)	(28)	(10)	(10)
5	Ind I	0.01	—	50	5	15	(4)	(4)	(6)	(6)
5	Ind II	0.01	—	50	5	42	(24)	(26)	(9)	(10)
5	Ru cat **304**	0.01	100	50	5	84	(53)	(52)	(16)	(16)
5	Ru cat **307**	0.01	100	50	5	90	(59)	(58)	(16)	(15)
5	Ru cat **307**	0.001	100	50	5	15	(1)	(2)	(7)	(7)
5	Ru cat **307**	0.005	100	50	5	32	(6)	(6)	(12)	(13)
5	Ru cat **307**	0.015	100	50	5	94	(66)	(65)	(15)	(14)
5	Ru cat **307**	0.02	100	50	5	96	(64)	(63)	(15)	(15)
2	Ru cat **307**	0.015	100	50	5	96	(45)	(45)	(25)	(26)
4	Ru cat **307**	0.015	100	50	5	92	(54)	(54)	(19)	(18)
8	Ru cat **307**	0.015	100	50	5	96	(76)	(78)	(10)	(9)
10	Ru cat **307**	0.015	100	50	5	96	(77)	(79)	(10)	(10)
8	Ru cat **307**	0.015	100	50	1	94	(58)	(61)	(18)	(17)
8	Ru cat **307**	0.015	100	50	3	94	(67)	(71)	(13)	(14)
8	Ru cat **307**	0.015	100	50	7	93	(73)	(78)	(10)	(10)
8	Ru cat **307**	0.015	100	50	9	93	(79)	(71)	(11)	(10)

TABLE 4. CROSS-METATHESIS OF PRIMARY ALLYLIC ALCOHOLS AND DERIVATIVES (*Continued*)

C. ALLYLIC ESTERS (*Continued*)

Allylic ester	Cross Partner	Conditions	Product(s) and Yield(s) (%)	Refs.
Please refer to the charts preceding the tables for the catalyst structures.				
C_4				
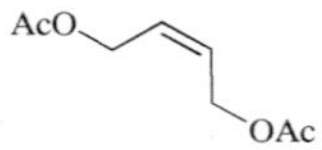 AcO, OAc 3 eq	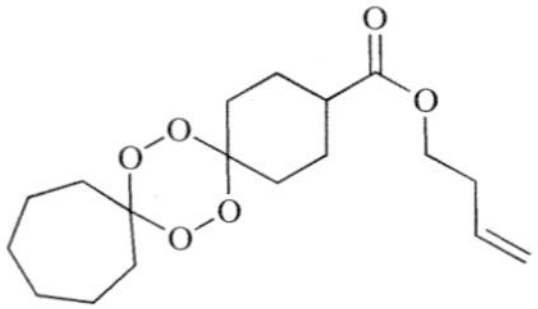 O, O, O–O, O–O	Ru cat **88** (0.03 eq) DCM, 45°, 2 h	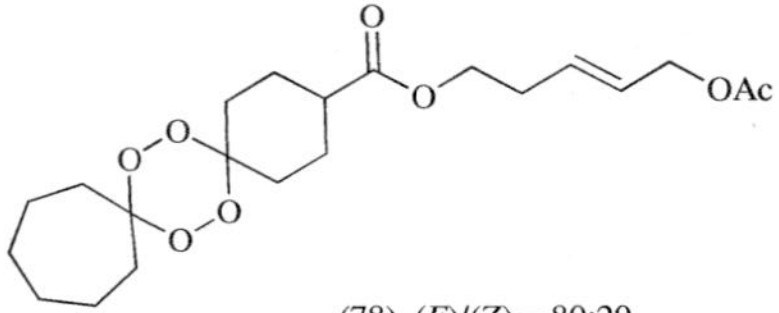 O, O, O–O, O–O, OAc (78) (*E*)/(*Z*) = 80:20	193
5 eq	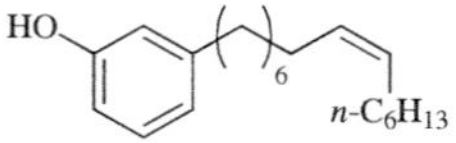 HO, 6, n-C_6H_{13}	Catalyst (x eq), toluene, 60°	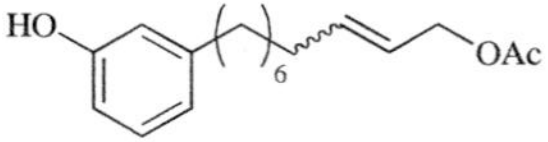 HO, 6, OAc	387

Catalyst	x	Time (h)	Conv. (%)[d]
Gr II	0.02	1	90
Gr II	0.02	2	91
Gr II	0.02	5	91
Ru cat **8**	0.02	1	74
Ru cat **8**	0.02	2	75
Ru cat **8**	0.02	5	81
Ru cat **8**	0.002	1	42
Ru cat **8**	0.002	2	47
Ru cat **8**	0.002	5	54
Ru cat **89**	0.02	1	98
Ru cat **89**	0.02	2	99
Ru cat **89**	0.02	5	99
Ru cat **90**	0.02	1	95
Ru cat **90**	0.02	2	96
Ru cat **90**	0.02	5	96
Ru cat **90**	0.002	1	60
Ru cat **90**	0.002	2	63
Ru cat **90**	0.002	5	66

3 eq	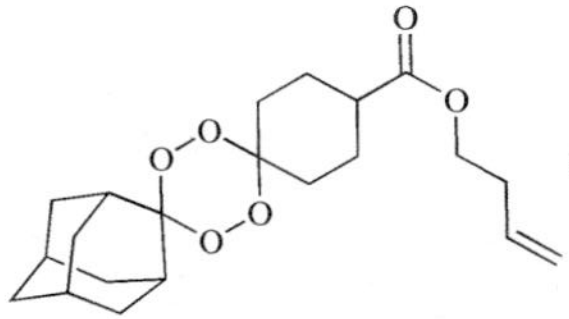	Ru cat **88** (0.03 eq) DCM, 45°, 2 h	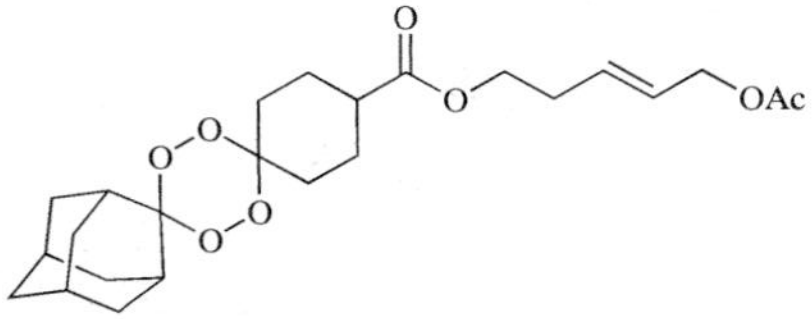	(90%) $(E)/(Z) = 89:11$	193
3 eq	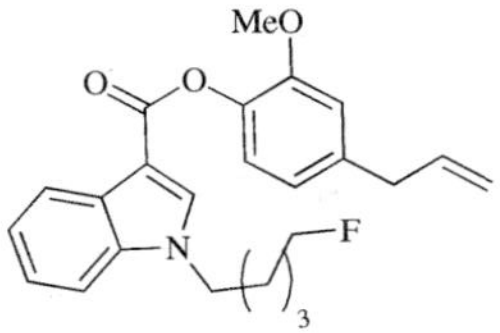	Ru cat **360** (0.01 eq), toluene, 80°, 8 h	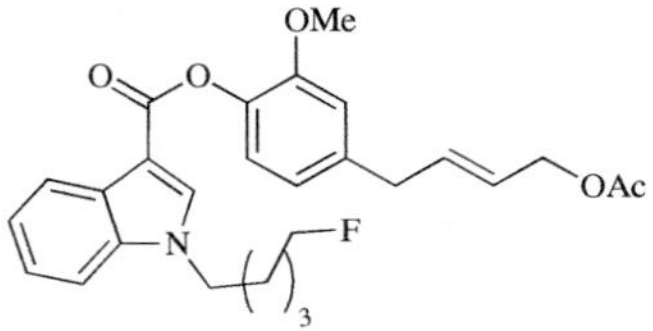	(92) $(E)/(Z) = 77:23$	491
10 eq	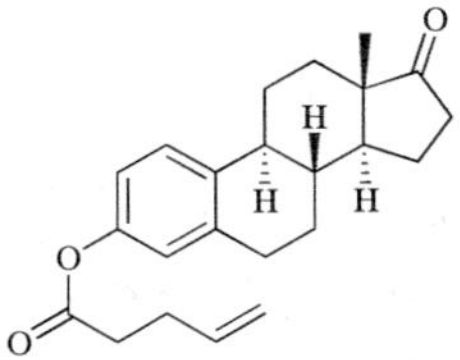	Ru cat **143** (0.05 eq), H_2O, 30°, 12 h	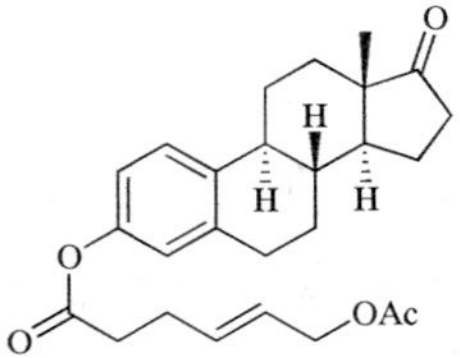	(75) $(E)/(Z) = 90:10$	588
3 eq	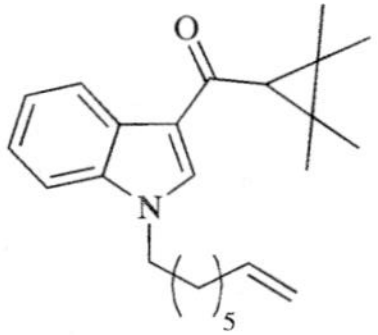	Ru cat **360** (0.01 eq), toluene, 80°, 8 h	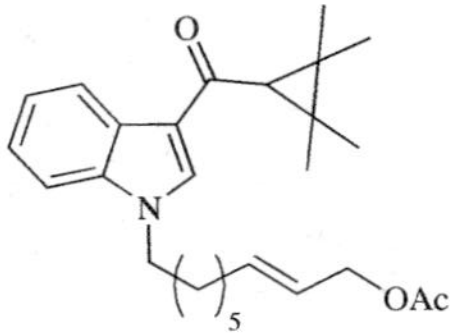	(85) $(E)/(Z) = 80:20$	491

TABLE 4. CROSS-METATHESIS OF PRIMARY ALLYLIC ALCOHOLS AND DERIVATIVES (*Continued*)

C. ALLYLIC ESTERS (*Continued*)

Allylic ester	Cross Partner	Conditions	Product(s) and Yield(s) (%)	Refs.
Please refer to the charts preceding the tables for the catalyst structures.				
C_4				
AcO, OAc; 3 eq	O, O, N, 5	Ru cat **360** (0.01 eq), toluene, 80°, 8 h	O, O, N, 5, OAc; (90) (*E*)/(*Z*) = 67:33	491
4 eq	O, *n*, n-C_8F_{17}, n-C_8F_{17}, n-C_8F_{17}	Hov II (0.02 eq), Et_2O, reflux	AcO, O, *n*, n-C_8F_{17}, n-C_8F_{17}, n-C_8F_{17}; *n* / Time (h): 1 / 18 (76); 3 / 24 (87)	613
10 eq	N, OH, 50, O, 8	Hov II (0.05 eq), DCM, 40°, 2 h	N, OH, 50, O, 8, OAc; (88)	614

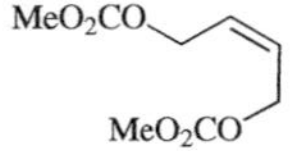

MeO_2CO MeO_2CO

2 eq

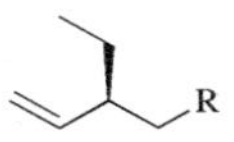

R

Gr II (0.03 eq), DCM, reflux, 20 h

MeO_2CO R

615

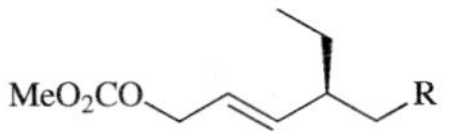

R		(E)/(Z)
BnO	(65)	90:10
$BnOCH_2$	(67)	90:10
$Ph(CH_2)_2$	(61)	87:13

2 eq

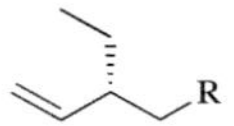

R

Gr II (0.03 eq), DCM, reflux, 20 h

MeO_2CO R

615

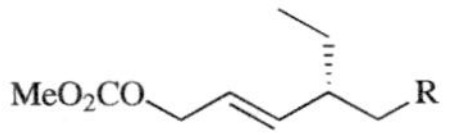

R		(E)/(Z)
BnO	(65)	90:10
$BnOCH_2$	(69)	90:10
$Ph(CH_2)_2$	(63)	87:13

2 eq

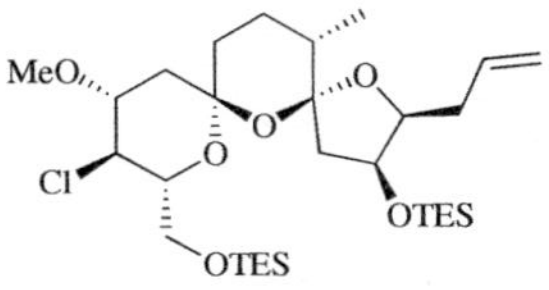

MeO Cl O O O OTES OTES

Gr II (0.05 eq), DCM, 40°, 3 h

MeO Cl O O O OTES OTES OCO_2Me

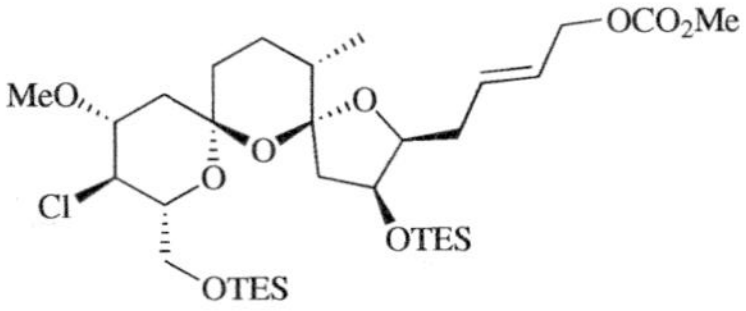

(95) (E)/(Z) = 87.5:12.5

616

TABLE 4. CROSS-METATHESIS OF PRIMARY ALLYLIC ALCOHOLS AND DERIVATIVES *(Continued)*

C. ALLYLIC ESTERS *(Continued)*

Allylic ester	Cross Partner	Conditions	Product(s) and Yield(s) (%)	Refs.
Please refer to the charts preceding the tables for the catalyst structures.				
C_4				
MeO_2CO / MeO_2CO (Z-alkene) 15 eq	R^1–R^2 = (acetonide)	Gr II (0.2 eq), toluene, 80°, 16 h	(58) $(E)/(Z)$ = 86:14	616
C_{6-12}				
BocHN, R; 3 eq		Gr II (0.10 eq), $CHCl_3$, reflux, 10 h	I + II	554

R	I	II
Me	(40)	(42)
$MeS(CH_2)_2$	(0)	(0)
Bn	(37)	(38)

C_8

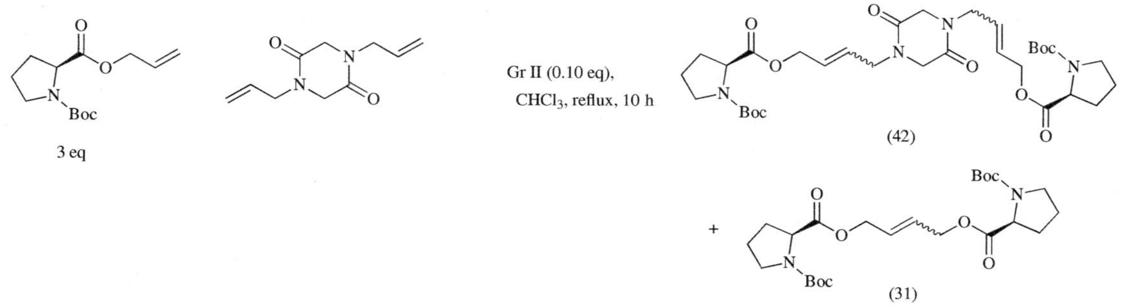

Boc, 3 eq

Gr II (0.10 eq), $CHCl_3$, reflux, 10 h

(42)

+ (31)

554

C_9

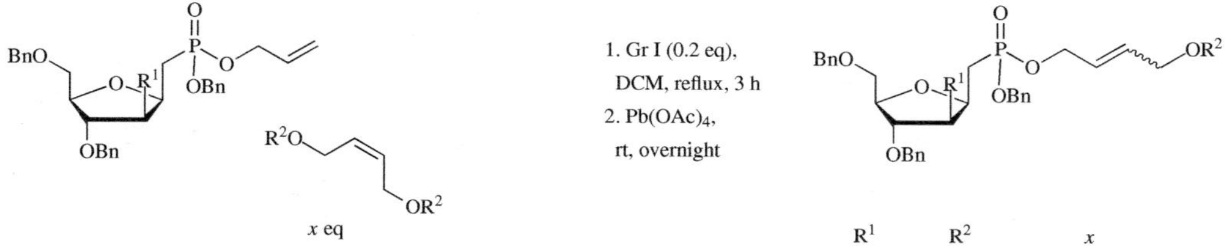

BnO, OBn, R^1, R^2O, OR^2, *x* eq

1. Gr I (0.2 eq), DCM, reflux, 3 h
2. $Pb(OAc)_4$, rt, overnight

617

R^1	R^2	*x*	
BnO	n-C_7H_{15}	3	(55)
BnO	n-C_8H_{17}	3	(63)
BnO	n-C_9H_{19}	4	(57)
BnO	n-$C_{10}H_{21}$	4	(51)
BnO	n-$C_{12}H_{25}$	4	(66)
BnO	n-$C_{16}H_{33}$	4	(62)

TABLE 4. CROSS-METATHESIS OF PRIMARY ALLYLIC ALCOHOLS AND DERIVATIVES (*Continued*)

C. ALLYLIC ESTERS (*Continued*)

Allylic ester	Cross Partner	Conditions	Product(s) and Yield(s) (%)	Refs.
Please refer to the charts preceding the tables for the catalyst structures.				
C_{11}				
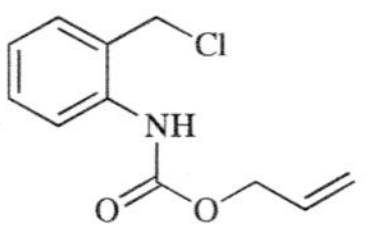	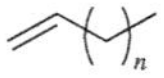	Hov II (0.05 + 0.02 eq), DCM, 40°, 16 h	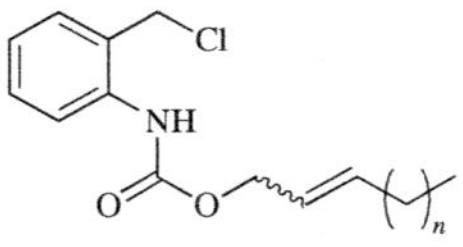 *n* 3 (49); 5 (62)	618
C_{12}				
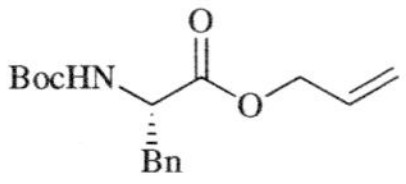 2 eq	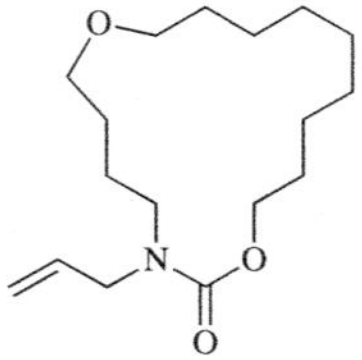	Gr II (0.05 eq), $CHCl_3$, reflux, 21 h	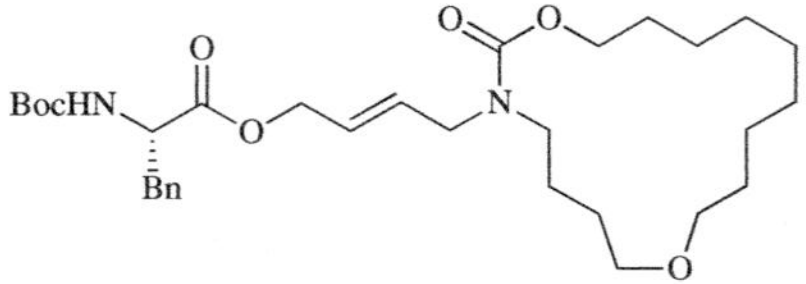 (48) (*E*)/(*Z*) = 55:45 + 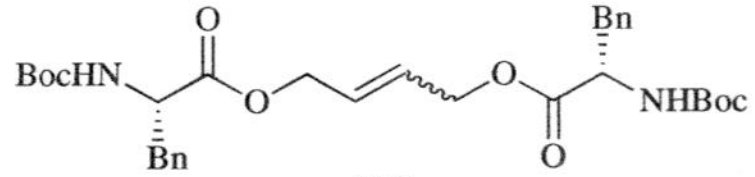 (45)	386

[a] The conversion was determined by ^{1}H NMR analysis.

[b] The ratios were determined by ^{1}H NMR spectroscopy.

[c] The conversion was determined by GC analysis.

[d] The conversion was determined by HPLC analysis.

[e] The conversion, selectivity and yield were determined by GC analysis.

TABLE 5. CROSS-METATHESIS OF PRIMARY ALLYLIC HALIDES AND 1,4-DIHALOBUT-2-ENES

Allyl Halide	Cross Partner	Conditions	Product(s) and Yield(s) (%)	Refs.
Please refer to the charts preceding the tables for the catalyst structures.				
C_3 X 2 eq	OR	Hov II (0.02 eq), DCM, rt, 3–8 h	X OR	197
2 eq	O	Hov II (0.02 eq), DCM, rt, 3–8 h	X O	197
Cl 3.5 eq	CO_2Me NHBoc	Gr II (0.1 eq), DCM, 40°, 14 h	Cl CO_2Me NHBoc (67) $(E)/(Z)$ = 86:14	619
6 eq	CO_2Me 8	Ru cat **139** (0.02 eq), DCM, reflux, 5 h	Cl CO_2Me 8 (80) $(E)/(Z)$ = 89:11	620
X 3 eq	Ph H OH	1. Gr II (0.1 eq), DCM, reflux, time 1 2. Toluene, reflux, time 2	Ph H O	621

First entry products:

X	R		$(E)/(Z)$
Cl	TBS	(39)	95:5
Br	TBS	(35)	76:24
Cl	Bn	(65)	89:11
Br	Bn	(58)	86:14

Second entry products:

X		$(E)/(Z)$
Cl	(49)	93:7
Br	(46)	80:20

Last entry products:

X	Time 1 (h)	Time 2 (h)		dr (*cis:trans*)
Cl	16	—	(66)	67:33
Cl	3	12	(79)	67:33
Br	2	10	(78)	67:33

TABLE 5. CROSS-METATHESIS OF PRIMARY ALLYLIC HALIDES AND 1,4-DIHALOBUT-2-ENES (*Continued*)

Allyl Halide	Cross Partner	Conditions	Product(s) and Yield(s) (%)	Refs.
Please refer to the charts preceding the tables for the catalyst structures.				
C_3				
Cl (3 + 3 eq)	HO, H, H, O, O	Gr II (4 x 0.05 eq), DCM, reflux, 30 h	O, H, H, O, O (53)	622
Br (2 eq)	OR	Hov II (0.02 eq), DCM, rt, 3–8 h	Br, OR; R = H (37) (*E*)/(*Z*) 83:17; R = Bz (49) (*E*)/(*Z*) 78:22	197
2 eq	Br	Hov II (0.02 eq), DCM, rt, 3–8 h	Br, Br (51) (*E*)/(*Z*) = 85:15	197
5 eq	CO_2Me	Hov II (0.02 eq), DCM, rt, 48 h	Br, CO_2Me (48) (*E*)/(*Z*) = 95:5	549, 131, 623
5 eq	Me, N, OMe, O	Hov II (0.05 eq), DCM, rt, 14 h	Br, Me, N, OMe, O (39) (*E*)/(*Z*) = 95:5	549
2 eq	CN	Hov II (0.02 eq), DCM, rt, 3–8 h	Br, CN (30) (*E*)/(*Z*) = 83:17	197

10 eq

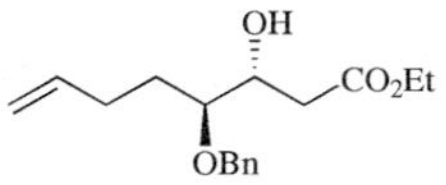

1. Gr II (0.1 eq), DCM, 40°, 5 h
2. Toluene, 100°, 10 h

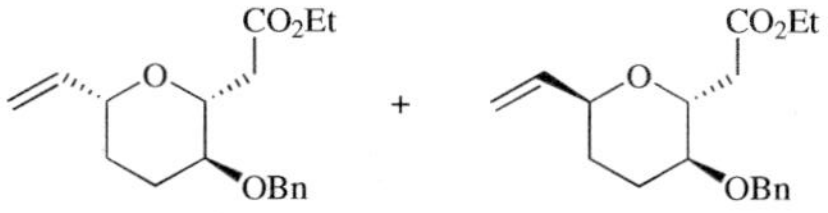

I **I** + **II** (71), **I**/**II** = 75:25 **II** 624

3 eq

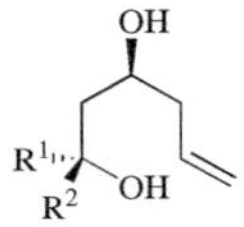

1. Gr II (0.1 eq), DCM, reflux, 2 h
2. Toluene, reflux

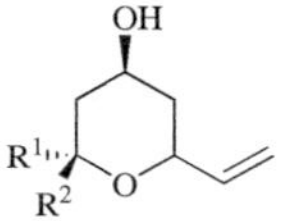

621

R^1	R^2	Time (h)		dr (*cis*:*trans*)
$BnO(CH_2)_2$	H	3–10	(95)	80:20
i-Pr	H	3–10	(80)	75:25
$BnOCH_2C(Me)_2$	H	3–10	(85)	80:20
Ph	H	3	(83)	83:17
Ph	Ph	12	(29)	72:28

2 eq

Gr I (0.05 eq), DCM, reflux, 16 h

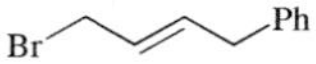

(65) (*E*)/(*Z*) = 77:23 156

2 eq

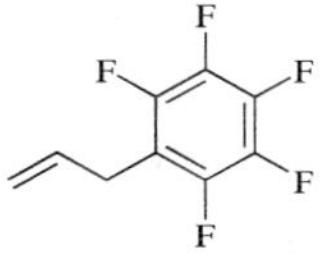

Hov II (0.02 eq), DCM, rt, 3–8 h

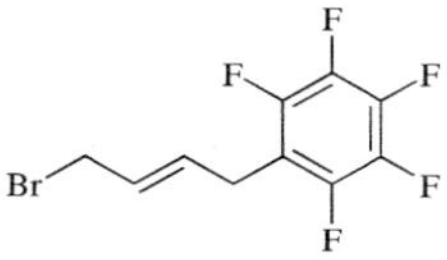

(39) (*E*)/(*Z*) = 75:25 197

TABLE 5. CROSS-METATHESIS OF PRIMARY ALLYLIC HALIDES AND 1,4-DIHALOBUT-2-ENES (*Continued*)

Allyl Halide	Cross Partner	Conditions	Product(s) and Yield(s) (%)	Refs.
Please refer to the charts preceding the tables for the catalyst structures.				
C_3				
Br 6 eq		Ru cat **190** (2 x 0.025 eq), DCM, rt, 192 h	Br, O, Br (35)	231
3 eq	OH, Ph, H, OH	1. Gr II (0.1 eq), DCM, reflux, 3 h 2. Toluene, reflux, 10 h	OH, Ph, H, O (62) dr (*cis:trans*) = 67:33	621
20 eq	MeO, OMe, OH, O, N, HO, OMe, OH	Gr II (0.1 eq), DCB, 150° (MW), 5 min	Br, MeO, OMe, OH, O, N, HO, OMe, OH (11)	432
C_4				
Cl, Cl 3 eq	O, B, O	1.Gr II (*x* eq), DCM, reflux, 2–12 h 2. PhCHO (1.5 eq), 23°, 12–48 h	OH, Ph, Cl *x* / (yield) / *anti/syn*[a] 0.05 (78) 83:17 0.02 (79) 82:18	126

2 eq	SO_2Ph	Catalyst (x eq), DCM, 25°	Cl, SO_2Ph (E)/(Z) = 90:10	609
	CO_2Me	Gr II (0.05 eq), DCM, 40°, 8 h	Cl, CO_2Me (63) (E) only	589
2 eq	Br, CO_2Et	Gr II (0.05 eq), DCM, reflux, 12 h	Cl, Br, CO_2Et (48) (E)/(Z) = 86:14	132
5 eq	Me, N, OMe, O	Hov II (x eq), DCM, rt	Cl, Me, N, OMe, O	549

Catalyst	x	Time (h)	
Hov II	0.01	1	(100)
Hov II	0.002	3	(67)
Ru cat **115**	0.002	3	(82)
Ru cat **115**	0.003	4	(89)

x	Time (h)		(E)/(Z)
0.1	24	(70)	95:5
0.05	0.5	(59)	99:1

TABLE 5. CROSS-METATHESIS OF PRIMARY ALLYLIC HALIDES AND 1,4-DIHALOBUT-2-ENES (*Continued*)

Allyl Halide	Cross Partner	Conditions	Product(s) and Yield(s) (%)	Refs.
Please refer to the charts preceding the tables for the catalyst structures.				
C_4				
Cl, Cl; 5 eq		Hov II (0.1 eq), DCM, rt, 24 h	Cl (57) $(E)/(Z) = 94:6$	549
3 eq	OH, Cl, N	Hov II (0.1 eq), DCM, 50°, 24 h	Cl, OH, Cl, N (38)	603
3 eq		Ru cat **88** (0.03 eq) DCM, 45°, 2 h	Cl (89) $(E)/(Z) = 86:14$	193
3 eq		Ru cat **88** (0.03 eq) DCM, 45°, 2 h	Cl (87) $(E)/(Z) = 86:14$	193

Reactant	Reactant	Conditions	Product	Yield (%)	Refs.
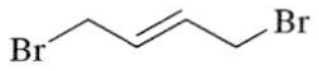 Br, Br (3 eq)	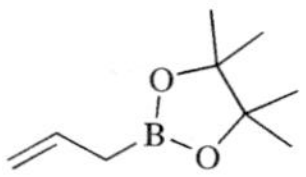	1.Gr II (x eq), DCM, reflux, 2–12 h 2. PhCHO (1.5 eq), 23°, 12–48 h	OH, Ph, Br	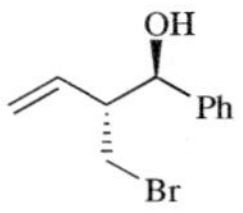 *x* / *anti/syn*[a] 0.05 (73) 79:21 0.02 (72) 78:22	126
2 eq	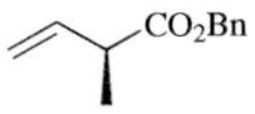 CO_2Bn	Hov II (0.05 eq), 2,6-Cl_2-BQ (0.1 eq), DCM, 40°, 16 h	Br, CO_2Bn	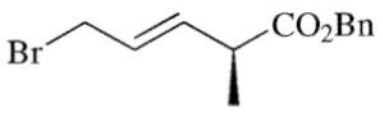 (73) (*E*) only	555

[a] The ratios were determined by ^{1}H NMR spectroscopy.

TABLE 6. CROSS-METATHESIS OF ALLYLSILANES

Allylsilane	Cross Partner	Conditions	Product(s) and Yield(s) (%)	Refs.

Please refer to the charts preceding the tables for the catalyst structures.

C_3

TMS-allyl + allyl-TMS; Catalyst (0.02 eq), THF, 35°; product: TMS…TMS, $(Z)/(E) = 95:5$

Catalyst	Time (h)		Refs.
Ru cat **217**	3	(54)	84
Ru cat **215**	9	(14)	488

2 eq; cross partner: alkene-$(CH_2)_n$-R; Mo cat **1** (0.02 eq), DME, rt, 4 h; products: TMS…R (**I**) + R…R (**II**); Ref. 208

R	n	I	$(E)/(Z)$ I	II
NC–	0	(34)	80:20	(0)
NC–	1	(37)	81:19	(15)
NC–	2	(62)	83:17	(14)
BnO_2C	1	(61)	75:25	(30)
Br	1	(45)	74:26	(12)
Br	2	(60)	76:24	(19)

R^1-allyl, 2 eq; cross partner: alkene-$(CH_2)_n$-OR^2; Mo cat **1** (0.02 eq), DME, rt, 4 h; products: R^1…OR^2 (**I**) + R^2O…OR^2 (**II**); Ref. 208

R^1	R^2	n	I	$(E)/(Z)$ I	II
TMS	TBS	1	(62)	76:24	(23)
TMS	TBS	2	(54)	79:21	(10)
TMS	TBS	3	(56)	74:26	(19)
TMS	Bn	1	(67)	79:21	(8)
TMS	Bn	2	(60)	82:18	(26)
TMS	Bn	3	(66)	72:28	(6)
TMS	Ph	2	(72)	72:28	(5)
TMS	Ph	3	(72)	72:28	(—)
TIPS	Ph	3	(77)	88:12	(—)

Reactant 1	Reactant 2	Conditions	Product(s)	Yield (%), ratio	Ref.
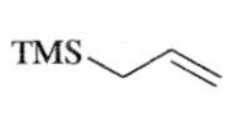 TMS 0.9 eq	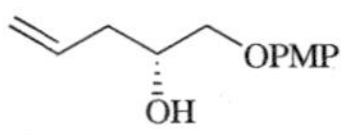 OPMP, OH	Hov **II** (0.025 eq), DCM, rt, 36 h	TMS, OPMP, OH	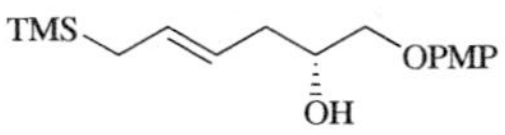 (76) $(E)/(Z) = 75:25$	626
2 eq	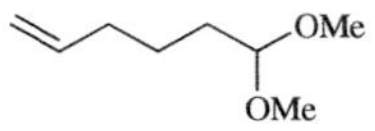 OMe, OMe	Mo cat **1** (0.02 eq), DME, rt, 4 h	TMS, OMe, OMe	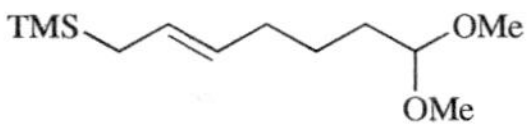 (55) $(E)/(Z) = 73:27$ 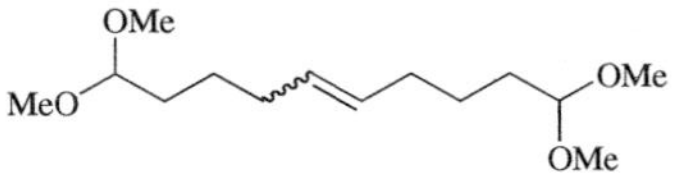	208
			+ MeO, OMe, OMe, OMe	(5)	
5 eq	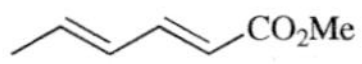 CO_2Me	Hov **II** (0.05 eq), DCM, rt, 48 h	TMS, CO_2Me	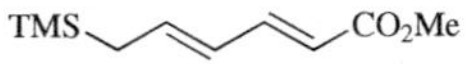 (61) $(E)/(Z) = 83:17$ 	549
			+ TMS, CO_2Me	(10)	
5 eq	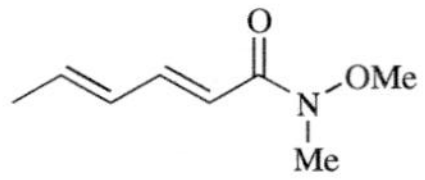 O, N, OMe, Me	Hov **II** (0.1 eq), DCM, rt, 24 h	TMS, O, N, OMe, Me	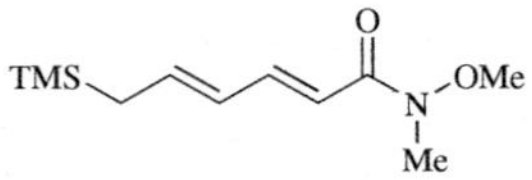 (45) $(E)/(Z) = 80:20$	549

TABLE 6. CROSS-METATHESIS OF ALLYLSILANES (*Continued*)

Allylsilane	Cross Partner	Conditions	Product(s) and Yield(s) (%)	Refs.
Please refer to the charts preceding the tables for the catalyst structures.				
C_3				
TMS 4 eq	CO_2Me NHBoc	Gr II (0.05 eq), toluene, rt, 17 h	TMS CO_2Me NHBoc (70)	457
1.4 eq	MeO_2C CO_2Me	Gr II (0.01 eq), DCM, reflux	TMS MeO_2C CO_2Me (7)	407
2 eq	HO N Cbz	Gr II (0.1 eq), DCM, reflux, 24 h	TMS HO N Cbz (70)	441
2 eq	MeO	Mo cat **1** (0.02 eq), DME, rt, 4 h	TMS MeO (73) (*E*)/(*Z*) = 76:24 + OMe MeO (7)	208

2 eq

N
R

Catalyst (x eq),
solvent

TMS

N
R

96

R	Catalyst	x	Solvent	Temp (°)	Time (h)	Conv. (%)[a]	(*E*)/(*Z*)	er
Tos	Ru cat **32**	0.03	DCM	rt	12	>99	40:60	87.5:12.5
Tos	Ru cat **32**	0.03	MTBE	rt	12	>99	67:33	79.0:21.0
Tos	Ru cat **32**	0.03	THF	rt	12	>99	50:50	81.5:18.5
Tos	Ru cat **32**	0.03	benzene	rt	12	>99	67:33	76.5:23.5
Tos	Ru cat **32**	0.03	toluene	rt	12	>99	67:33	78.5:21.5
Tos	Ru cat **32**	0.03	$C_6H_5CF_3$	rt	12	92	60:40	84.5:15.5
Tos	Ru cat **32**	0.025	DCM	40	1	>99	40:60	87.5:12.5
Tos	Ru cat **32**	0.025	DCM	rt	2	>99	40:60	87.5:12.5
Tos	Ru cat **32**	0.025	DCM	0	2	>99	33:67	89.5:10.5
Tos	Ru cat **32**	0.025	DCM	–10	48	>99	33:67	89.5:10.5
Tos	Ru cat **229**	0.025	DCM	rt	18	>99	50:50	86.5:13.5
Tos	Ru cat **24**	0.025	DCM	rt	2	>99	75:24	67.0:33.0
CF_3CO	Ru cat **32**	0.025	DCM	rt	2	>99	67:33	90.0:10.0
CF_3CO	Ru cat **32**	0.025	DCM	0	6	>99	67:33	91.0:9.0

TABLE 6. CROSS-METATHESIS OF ALLYLSILANES (*Continued*)

Allylsilane	Cross Partner	Conditions	Product(s) and Yield(s) (%)	Refs.
Please refer to the charts preceding the tables for the catalyst structures.				
C_3				
TMS (2 eq)	AcO OAc, AcO, AcO, O	Gr I (0.2 eq), DCM, reflux, 6 h	AcO OAc, AcO, AcO, O, TMS (94) $(E)/(Z) = 80:20$	571
4 eq	OH, Ph	Gr II (0.05 eq), DCM, reflux, 4 h	TMS, OH, Ph (84) $(E)/(Z) = 70:30$	627
4 eq	R^1, OR^2, Ph	Gr II (0.05 eq), DCM, reflux	TMS, R^1, OR^2, Ph	627

R^1	R^2	Time (h)	Conv. (%)[a]		$(E)/(Z)$
Me	H	4	—	(81)	80:20
EtO_2C	H	4	—	(74)	80:20
EtO_2C	TMS	24	33–45	(—)	92:8

	Conditions	Product(s) and Yield(s) (%)	Refs.
4 eq	Gr II (0.05 eq), DCM, reflux		627
4 eq	Gr II (0.05 eq), DCM, reflux, 4 h	(75) $(E)/(Z) > 95:5$	627
4 eq	Gr II (0.05 eq), DCM, reflux, 18 h	(—) $(E)/(Z) = 54:46$	627

R^1	R^2	Time (h)	Conv. (%)[a]		$(E)/(Z)$
Me	H	4	—	(86)	92:8
EtO_2C	H	4	—	(67)	92:8
EtO_2C	TMS	24	33–45	(—)	89:11

TABLE 6. CROSS-METATHESIS OF ALLYLSILANES (*Continued*)

Allylsilane	Cross Partner	Conditions	Product(s) and Yield(s) (%)	Refs.
Please refer to the charts preceding the tables for the catalyst structures.				
C_3				
TMS 2 eq	OH Fe *n*	Gr I (0.19 eq), DCM, reflux, 17–24 h	OH Fe *n* TMS *n* / (E)/(Z) 1 (59) 62:38 2 (45) 63:37	587
2 eq	O Fe *n*	Gr I (0.19 eq), DCM, reflux, 17–24 h	O Fe *n* TMS *n* / (E)/(Z) 1 (62) 60:40 2 (58) 58:42	587
Ph O Si Cl N Me 1.5 eq	Cy 7.5 eq + OH N Ph H **I**	1. Gr II (0.026 eq), DCM, reflux, 7 h 2. **I**, reflux, 14 h	OH HN Ph Cy (58) dr > 95:5, er = 99.5:0.5	211

[a] The conversion was determined by ^{1}H NMR analysis.

TABLE 7. CROSS-METATHESIS OF ALLYLIC ORGANOPHOSPHORUS COMPOUNDS

A. ALLYLPHOSPHONATES

Allylphosphonate	Cross Partner	Conditions	Product(s) and Yield(s) (%)	Refs.
Please refer to the charts preceding the tables for the catalyst structures.				
C_3				
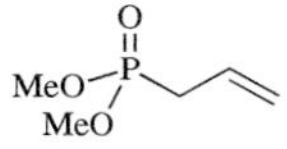 MeO–P(=O)(MeO) 5 eq	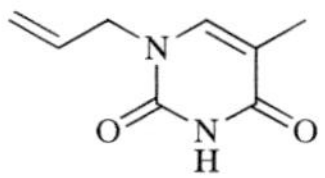	Ru cat **35** (0.1 eq), DCM, 40°, 5 h	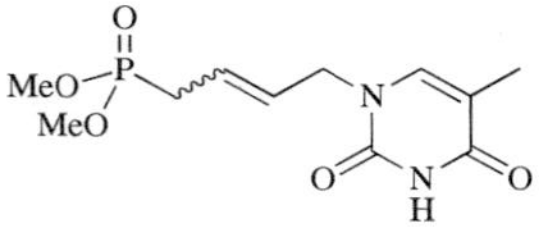 MeO–P(=O)(MeO) (0)	628
4 eq	R, Bz	Gr II (0.05 eq), DCM, reflux, 16 h	MeO–P(=O)(MeO) R, Bz	629
6 eq	Boc, R	Ru cat **35** (0.05 eq), DCM, 60°, 12 h	MeO–P(=O)(MeO) Boc, R	630

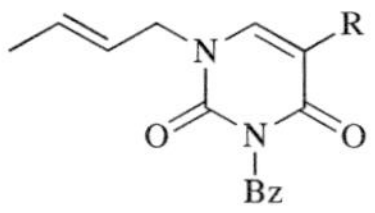

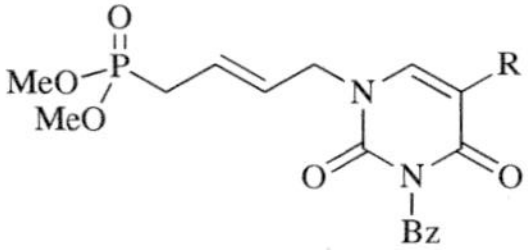

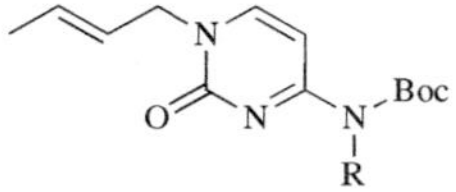

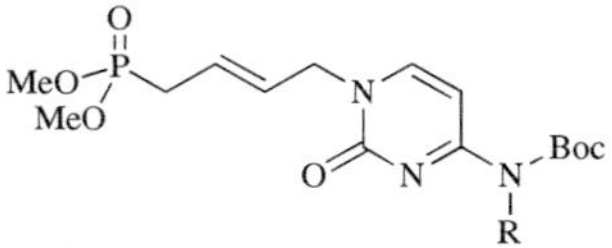

Ref. 629 products:

R		(E)/(Z)
H	(87)	80:20
Cl	(87)	83:17
Br	(74)	83:17
PhS	(74)	83:17

Ref. 630 products:

R		(E)/(Z)
n-C_4H_9	(59)	95:5
n-C_5H_{11}	(53)	95:5
Cy	(66)	95:5
n-C_9H_{19}	(46)	95:5

TABLE 7. CROSS-METATHESIS OF ALLYLIC ORGANOPHOSPHORUS COMPOUNDS (*Continued*)

A. ALLYLPHOSPHONATES (*Continued*)

Allylphosphonate	Cross Partner	Conditions	Product(s) and Yield(s) (%)	Refs.
Please refer to the charts preceding the tables for the catalyst structures.				
C_3				
MeO–P(=O)(MeO) 5 eq		Ru cat **35** (0.06 eq), DCM, reflux, 7 h	(90) (*E*) only	631
4 eq		Gr II (0.05 eq), DCM, reflux, 16 h		629
			R: H (94) (*E*)/(*Z*) 83:17; F (81) 83:17; MeO (92) 83:17	
6 eq		Ru cat **35** (0.05 eq), DCM, 60°, 12 h		630
			R: H (52) (*E*)/(*Z*) 95:5; Cl (67) 95:5; F (67) 95:5	

		Conditions	Product(s) and Yield(s) (%)		Refs.
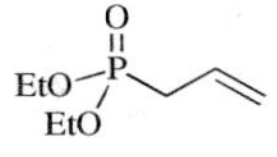 5 eq	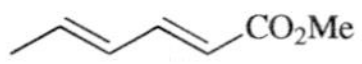	Hov II (0.05 eq), DCM, rt, 48 h	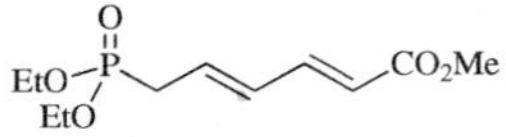 **I** $(E,E)/(E,Z) = 85{:}15$ 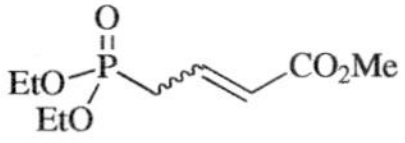 + **II**; **I** + **II** (56), **I**/**II** = 80:20		549
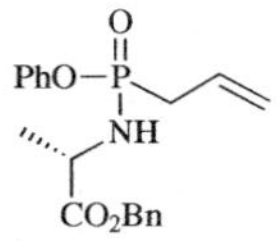	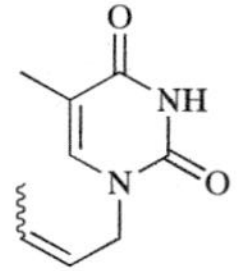 2 eq	Gr II (3 x 0.06 eq), H_2O, PTS (x eq), ultrasound, 55°, 20 h		x = 0 (41); x = 0.025 (40)	632
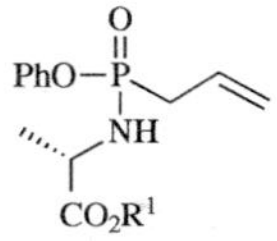	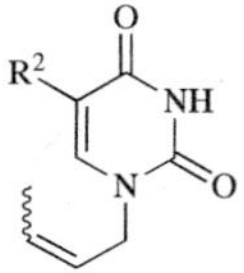 2 eq	Gr II (3 x 0.06 eq), H_2O, PTS (0.025 eq), ultrasound, 55°, 20 h		R^1 = Me, R^2 = Me (41); R^1 = *i*-Pr, R^2 = Me (44); R^1 = Bn, R^2 = F (36); R^1 = Bn, R^2 = Cl (35)	632

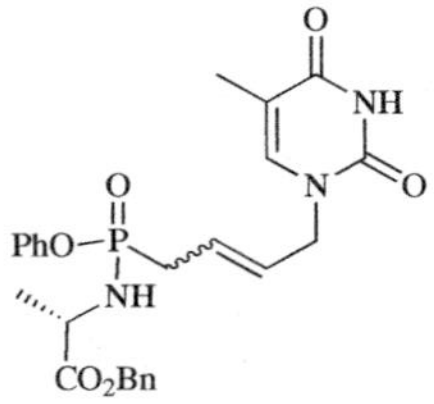

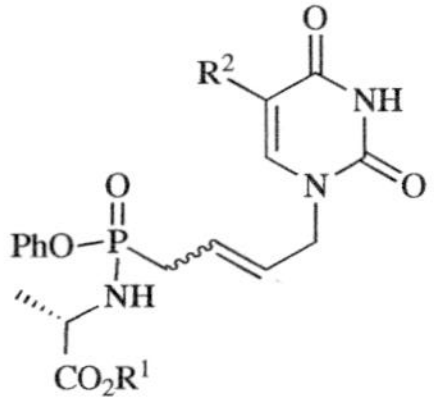

TABLE 7. CROSS-METATHESIS OF ALLYLIC ORGANOPHOSPHORUS COMPOUNDS (*Continued*)

A. ALLYLPHOSPHONATES (*Continued*)

Allylphosphonate	Cross Partner	Conditions	Product(s) and Yield(s) (%)	Refs.

Please refer to the charts preceding the tables for the catalyst structures.

C_3

Allylphosphonate	Cross Partner	Conditions	Product(s) and Yield(s) (%)	Refs.
O O Br O NH O O=P O N O O O O 1.3 eq		Catalyst (0.05 eq), DCM	O O Br NH O O N O O=P O O O O O	219

Catalyst	Temp (°)	Time (h)	
Ru cat **35**	rt	24	(tr)
Ru cat **282**	rt	24	(tr)
Ind II	rt	24	(tr)
Ru cat **38**	rt	0.5	(44)
Ru cat **286**	rt	0.5	(46)
Ru cat **285**	rt	0.5	(46)
Ru cat **35**	40	18	(36)
Ru cat **282**	40	18	(59)
Ind II	40	18	(29)
Ru cat **285**	40	18	(23)

1.3 eq

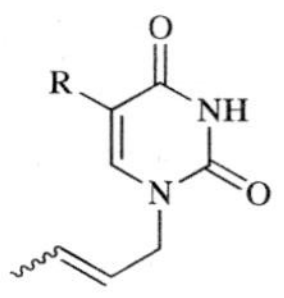

Ru cat **285** (0.05 eq), DCM, rt

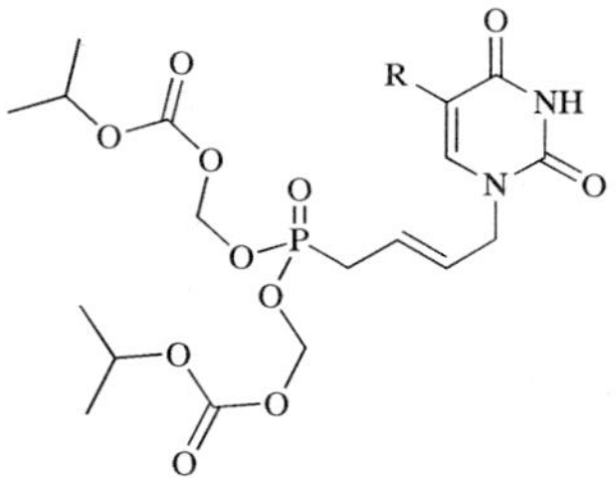

219

R	Time (min)		(E)/(Z)
H	45	(46)	85:15
F	60	(52)	85:15
Cl	40	(53)	85:15
Br	30	(56)	85:15
CH_3	30	(47)	85:15

1.3 eq

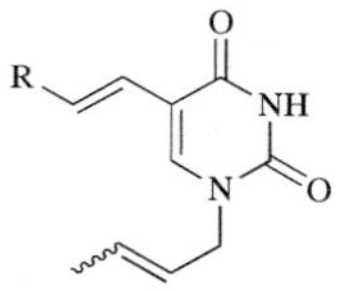

Ru cat **285** (0.05 eq), DCM, rt, 3.5 h

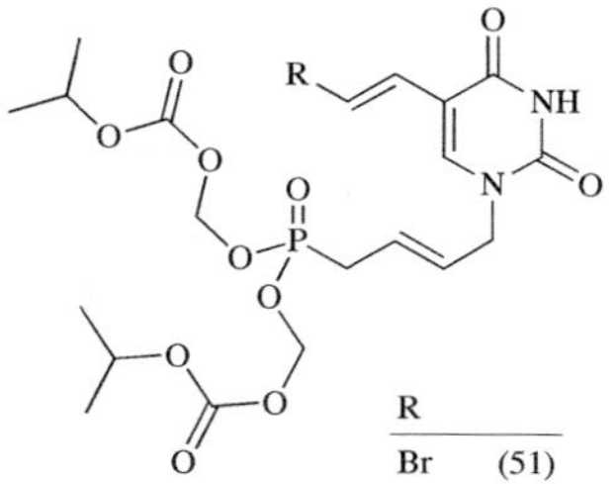

R	
Br	(51)
I	(47)

(*E*) only

633

TABLE 7. CROSS-METATHESIS OF ALLYLIC ORGANOPHOSPHORUS COMPOUNDS (*Continued*)

A. ALLYLPHOSPHONATES (*Continued*)

Allylphosphonate	Cross Partner	Conditions	Product(s) and Yield(s) (%)	Refs.
Please refer to the charts preceding the tables for the catalyst structures.				
C_3				
1.3 eq		Ru cat **285** (0.05 eq), DCM, rt, 3.5 h	(*E*) only	633
			R^1 / R^2: I / I (51); I / Br (40); I / Cl (45); Br / Br (48)	
1.3 eq		Ru cat **285** (0.05 eq), DCM, reflux, 16 h	(*E*) only	633
			R: n-C_3H_7 (51); n-C_5H_{11} (49); n-C_8H_{17} (51)	

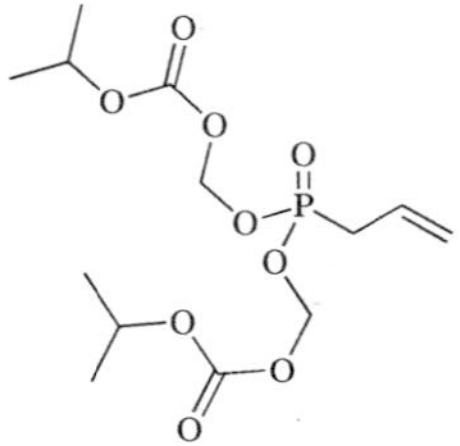

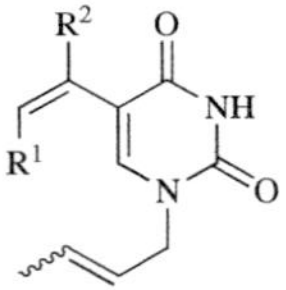

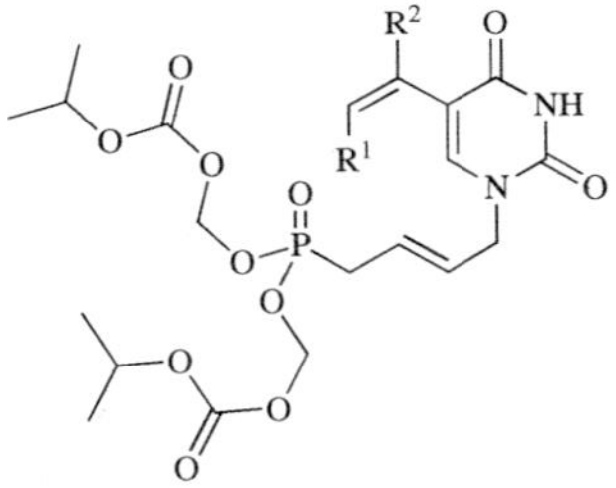

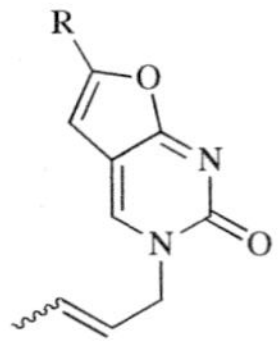

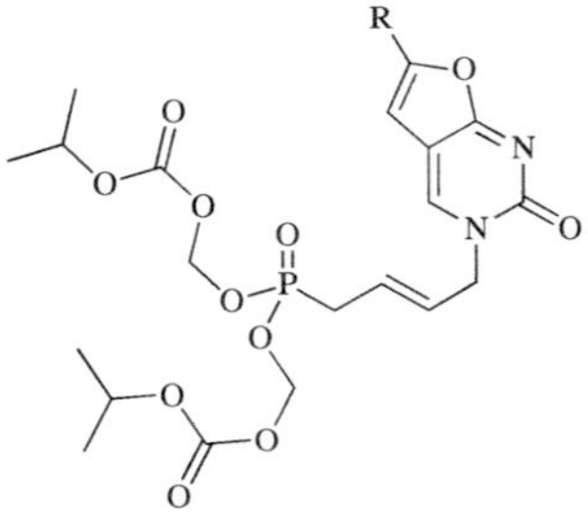

R^1	R^2	
I	I	(51)
I	Br	(40)
I	Cl	(45)
Br	Br	(48)

R	
n-C_3H_7	(51)
n-C_5H_{11}	(49)
n-C_8H_{17}	(51)

1.3 eq

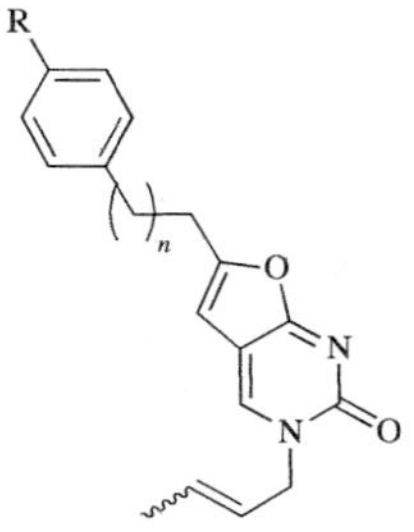

Ru cat **285** (0.05 eq), DCM, reflux, 16 h

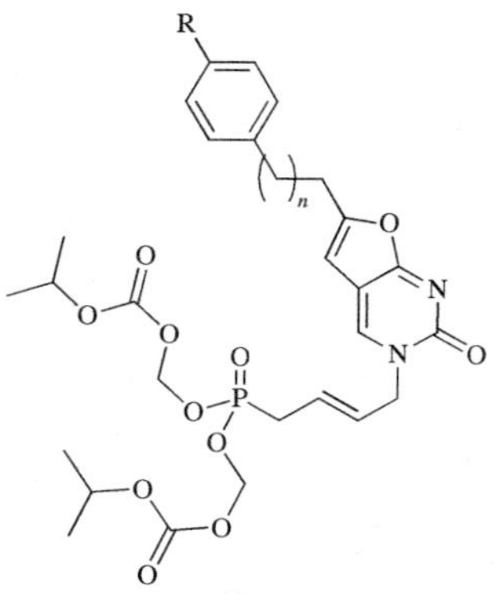

R	n	
H	1	(58)
Me	1	(52)
Me	3	(55)

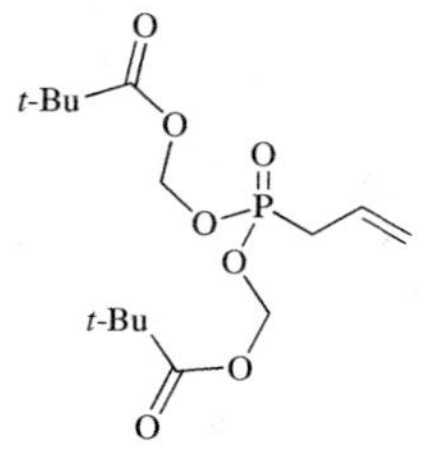

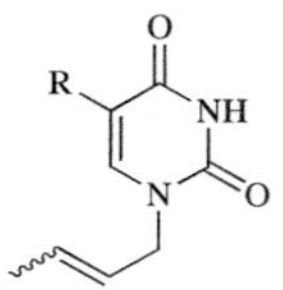

(*E*) only 633

1.3 eq

Ru cat **285** (0.05 eq), DCM, 40°

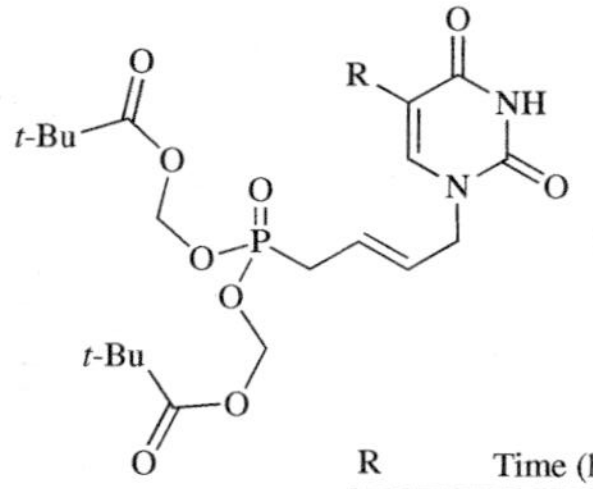

R	Time (h)	
H	1	(56)
F	0.75	(60)
Cl	1	(53)
Me	0.5	(63)

(*E*)/(*Z*) = 85:15 219

TABLE 7. CROSS-METATHESIS OF ALLYLIC ORGANOPHOSPHORUS COMPOUNDS (*Continued*)

A. ALLYLPHOSPHONATES (*Continued*)

Allylphosphonate	Cross Partner	Conditions	Product(s) and Yield(s) (%)	Refs.
Please refer to the charts preceding the tables for the catalyst structures.				
C_3				
1.3 eq	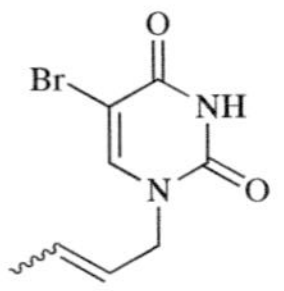	Catalyst (0.05 eq), DCM	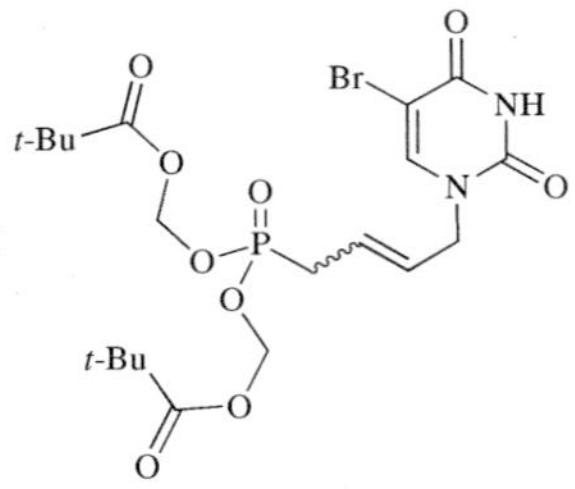	219

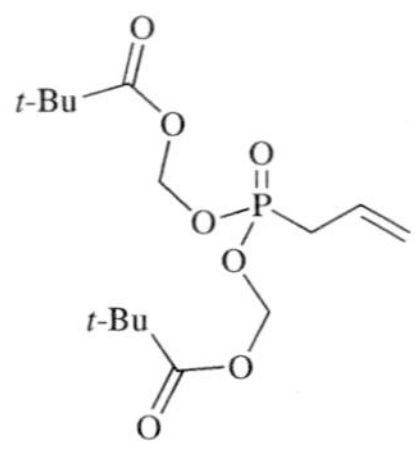

Catalyst	Temp (°)	Time (h)		(*E*)/(*Z*)
Gr I	rt	24	(0)	—
Ind I	rt	24	(0)	—
Ru cat **35**	rt	24	(tr)	—
Ru cat **282**	rt	24	(tr)	—
Ind II	rt	24	(tr)	—
Ru cat **38**	rt	1	(50)	—
Ru cat **286**	rt	1.5	(54)	—
Ru cat **285**	rt	1	(47)	—
Gr I	40	24	(0)	—
Ind I	40	24	(0)	—
Ru cat **35**	40	0.3	(69)	—
Ru cat **282**	40	4	(74)	—
Ind II	40	3	(71)	—
Ru cat **285**	40	1	(36)	—
Ru cat **285**	40	0.5	(69)	85:15

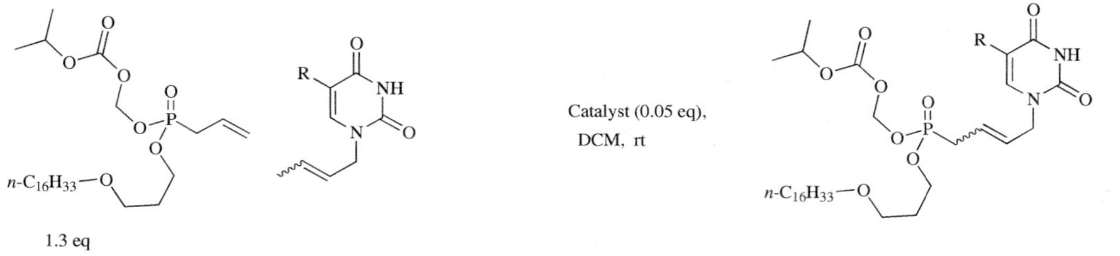

O O O O P O O n-$C_{16}H_{33}$

1.3 eq

R O NH O N

Catalyst (0.05 eq),
DCM, rt

O O O O P O O R O NH O N n-$C_{16}H_{33}$

R	Catalyst	Time (min)	
H	Ru cat **285**	35	(41)
F	Ru cat **285**	35	(45)
Cl	Ru cat **35**	—	[a]
Cl	Ru cat **282**	—	(tr)
Cl	Ru cat **38**	30	(47)
Cl	Ru cat **286**	45	(50)
Cl	Ru cat **285**	30	(48)
Cl	Ru cat **285**	35	(47)
Br	Ru cat **285**	35	(48)
Me	Ru cat **285**	35	(44)

219

[a] The reaction resulted in low conversion.

TABLE 7. CROSS-METATHESIS OF ALLYLIC ORGANOPHOSPHORUS COMPOUNDS (*Continued*)

B. ALLYL PHOSPHINE OXIDES

Allyl Phosphine Oxide	Cross Partner	Conditions	Product(s) and Yield(s) (%)	Refs.
Please refer to the charts preceding the tables for the catalyst structures.				
C_3				
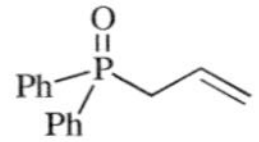 Ph, Ph, P, O	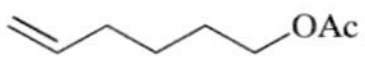 OAc 3 eq	Gr II (0.02–0.06 eq), DCM, 40°, 20–48 h	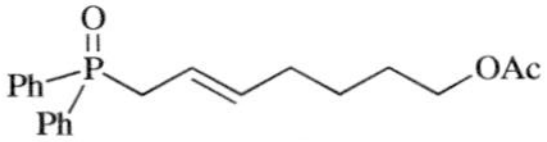 Ph, Ph, P, O, OAc (70) (*E*)/(*Z*) = 80:20	214
— eq	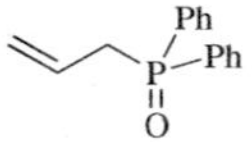 Ph, Ph, P, O	Gr II (0.02 eq), DCM, 40°, 20–48 h	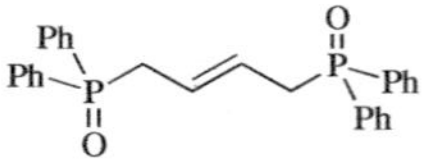 Ph, Ph, P, O, Ph, Ph, P, O (96) (*E*) only	214
3 eq	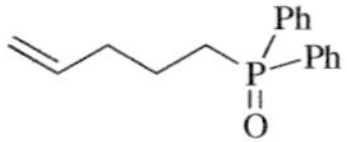 Ph, Ph, P, O	Gr II (0.04 eq), DCM, 40°, 20–48 h	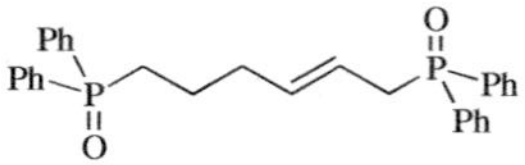 Ph, Ph, P, O, Ph, Ph, P, O (94) (*E*) only	214

TABLE 7. CROSS-METATHESIS OF ALLYLIC ORGANOPHOSPHORUS COMPOUNDS (*Continued*)

C. ALLYL PHOSPHINE BORANES

Allyl Phosphine Borane	Cross Partner	Conditions	Product(s) and Yield(s) (%)	Refs.
Please refer to the charts preceding the tables for the catalyst structures.				
C_3				
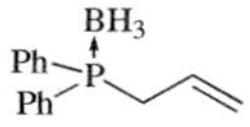 BH_3 Ph–P(Ph) allyl	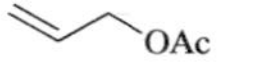 OAc, 3 eq	Gr II (0.08 eq), DCM, reflux, 96 h	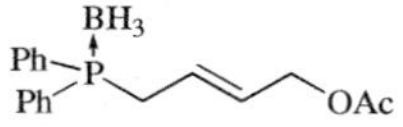 BH_3 Ph–P(Ph) OAc (0)	213
	 TMS, 3 eq	Gr II (0.04 eq), DCM, reflux, 24 h	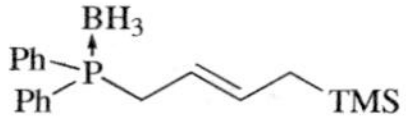 BH_3 Ph–P(Ph) TMS (85) (*E*)/(*Z*) = 83:17	213
	 Br, 3 eq	Gr II (0.08 eq), DCM, reflux, 48 h	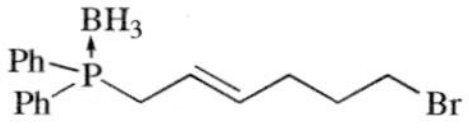 BH_3 Ph–P(Ph) Br (41) (*E*) only	213
	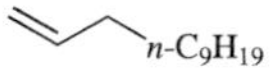 *n*-C_9H_{19}, 3 eq	Gr II (0.04 eq), DCM, reflux, 24 h	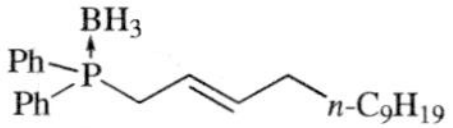 BH_3 Ph–P(Ph) *n*-C_9H_{19} (60) (*E*)/(*Z*) = 90:10	213

TABLE 8. CROSS-METATHESIS OF PRIMARY ALLYLIC ORGANOSULFUR AND ORGANOSELENIUM COMPOUNDS

A. ALLYLIC SULFIDES AND SELENIDES

Sulfide or Selenide	Cross Partner	Conditions	Product(s) and Yield(s) (%)	Refs.
Please refer to the charts preceding the tables for the catalyst structures.				
C_3				
 RS	 OH, 10 eq	Hov II (0.06 eq), t-BuOH/H_2O (1:1), 32°, 3.5 h	 RS … OH; R: Ph (28), Bn (52)	559
 PhS, 3 eq	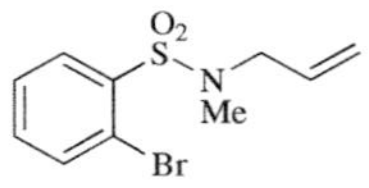 O_2 S N Me Br	Gr II (0.05 eq), DCM, 40°, 3 h	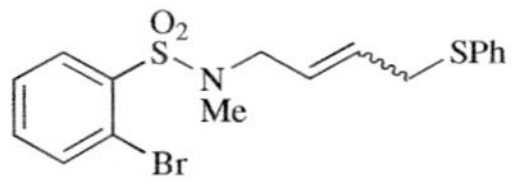 O_2 S N Me Br SPh (72)	634
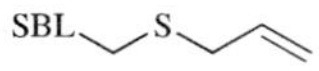 SBL S	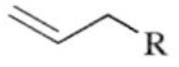 R, x eq	Hov II (y eq), additive (z eq), 30% t-BuOH/H_2O, pH 8.0	 SBL S R	

R	x	y	Additive	z	Time (h)	Temp (°)	Conv. (%)[a]	Refs.
HO	9250	187.5	—	—	5	rt	0	559
HO	9250	187.5	$MgCl_2$•$6H_2O$	8750	5	rt	>90	559
HO	9259	187.5	NaCl	8750	5	rt	0	559
HO	10,000	~200	$MgCl_2$•$6H_2O$	10,000	2	rt	>95	226
AcHN	5009	~210	$MgCl_2$•$6H_2O$	10,600	1	37	0	226
$Cl^-Me_3N^+$	5009	~210	$MgCl_2$•$6H_2O$	10,600	1	37	0	226
$Cl^-HMeN^+(CH_2)_2O$	5009	~210	$MgCl_2$•$6H_2O$	10,600	1	37	0	226

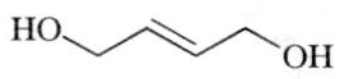

Substrate	Conditions	Product	Conv.	Ref.
HO–CH2CH=CHCH2–OH, 10,000 eq	Hov II (~200 eq), $MgCl_2 \cdot 6H_2O$ (10,000 eq), 30% *t*-BuOH/H_2O, pH 8.0, rt, 2 h	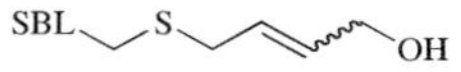 SBL–CH2–S–CH2CH=CH–CH2OH	28% conv.[a]	226
allyl–O–(CH2CH2O)$_n$–CH2CH2CH2–OR, *x* eq	Hov II (~210 eq), $MgCl_2 \cdot 6H_2O$ (10,600 eq), 30% *t*-BuOH/H_2O, pH 8.0, rt + 37°	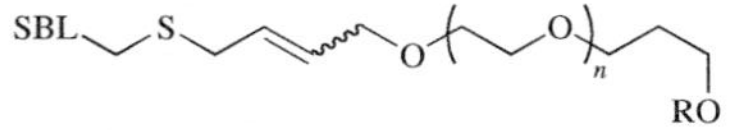 SBL–CH2–S–CH2CH=CH–CH2–O–(CH2CH2O)$_n$–CH2CH2CH2–OR		

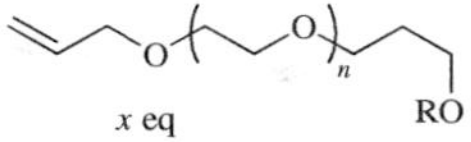

n	R	*x*	Time (h)	Conv. (%)[a]	Ref.
2	Me	5400	2 + 1	55	559
3	H	5400	2 + 1	60	559
3	H	5009	2 + 0.5	65	226

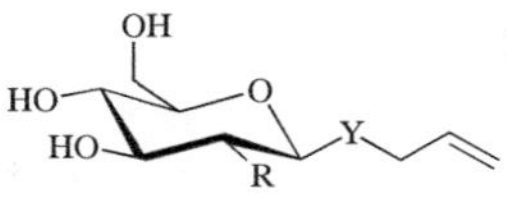

Substrate	Conditions	Product
HO, HO, OH-substituted pyranose (2-R) with Y–CH2CH=CH2, *x* eq	Hov II (*y* eq), $MgCl_2 \cdot 6H_2O$ (*z* eq), 30% *t*-BuOH/H_2O, pH 8.0	pyranose–Y–CH2CH=CH–CH2–S–CH2–SBL

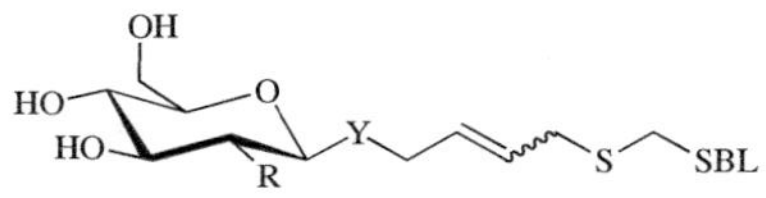

R	Y	*x*	*y*	*z*	Temp (°)	Time (h)	Conv. (%)[a]	Ref.
HO	O	2750	~183	7500	rt + 37	1 + 4	50	559
HO	O	4971	~200	10,000	37	1	30	226
AcHN	O	5009	~210	10,800	rt + 37	2	0	226
AcHN	S	5009	~210	10,800	rt + 37	1	0	226

TABLE 8. CROSS-METATHESIS OF PRIMARY ALLYLIC ORGANOSULFUR AND ORGANOSELENIUM COMPOUNDS (*Continued*)

A. ALLYLIC SULFIDES AND SELENIDES (*Continued*)

Sulfide or Selenide	Cross Partner	Conditions	Product(s) and Yield(s) (%)	Refs.

Please refer to the charts preceding the tables for the catalyst structures.

C_3

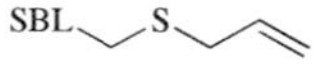

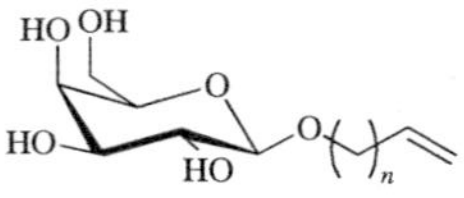

Sulfide or Selenide	Cross Partner	Conditions	Product(s) and Yield(s) (%)	Refs.
SBL–S–allyl	HO, OH, HO, HO glycoside–O–(CH₂)$_n$–CH=CH₂; 5009 eq	Hov II (~200 eq), $MgCl_2$•$6H_2O$ (10,130 eq), 30% *t*-BuOH/H_2O, pH 8.0, 37°, 1 h	glycoside–O–(CH₂)$_n$–CH=CH–CH₂–S–SBL	226

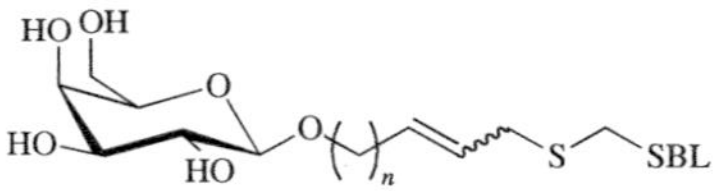

n	Conv. (%)[a]
1	30
4	>95

Sulfide or Selenide	Cross Partner	Conditions	Product(s) and Yield(s) (%)	Refs.
	OH, OH, HO, HO glycoside–O–allyl; *x* eq	Hov II (~210 eq), $MgCl_2$•$6H_2O$ (10,600 eq), 30% *t*-BuOH/H_2O, pH 8.0	glycoside–O–CH₂–CH=CH–CH₂–S–SBL	

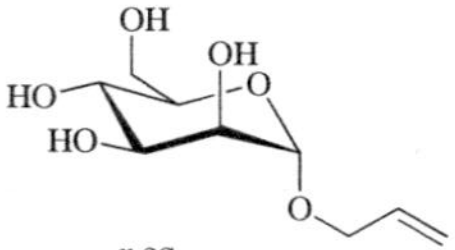

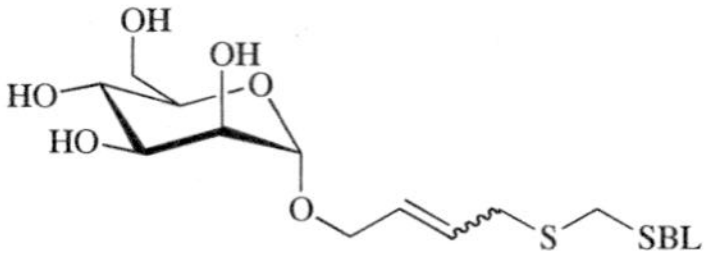

x	Temp (°)	Time (h)	Conv. (%)[a]	Refs.
10,600	rt + 37	1 + 4	60	559
5009	37	1	30	226

C_6

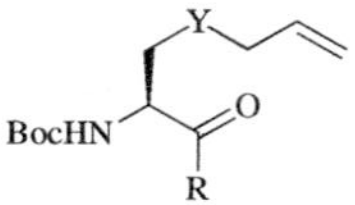

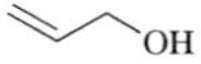

Sulfide or Selenide	Cross Partner	Conditions	Product(s) and Yield(s) (%)	Refs.
BocHN–CH(COR)–CH₂–Y–allyl	allyl alcohol (OH); 10 eq	Hov II (*x* eq), *t*-BuOH/H_2O (*y*/*z*), 32°	BocHN–CH(COR)–CH₂–Y–CH₂–CH=CH–CH₂OH	

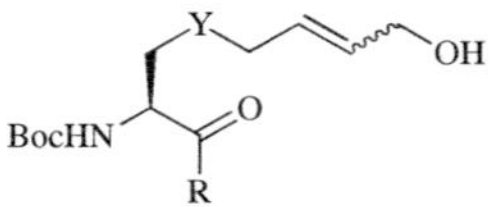

R	Y	*x*	*y*/*z*	Time (h)	Conv. (%)[a]	Refs.
H_2N	S	0.08	1:1	3.5	56	559
H_2N	S	0.08	3:7	3.5	68	559
H_2N	S	0.06	1:1	2.5	56	226
H_2N	Se	0.06	1:1	2.5	72	226
MeO	S	2 x 0.1	1:1	2.5	72	227

BocHN CO$_2$Me (S-allyl) + OAc (allyl acetate), 5 eq

Catalyst (0.075 eq), additive (2 mM), H_2O/*t*-BuOH (*x*/*y*) THF, 40°, 4 h

→ BocHN CO$_2$Me S OAc (401)

Catalyst	Additive	*x*/*y*		(*Z*)/(*E*)
Ru cat **215**	—	—	(62)	88:12
Ru cat **215**	—	1:1	(63)	86:14
Ru cat **215**	LiCl	1:1	(55)	74:26
Ru cat **215**	$MgCl_2$	1:1	(58)	76:24
Ru cat **216**	—	—	(61)	90:10
Ru cat **216**	—	1:1	(67)	92:8
Ru cat **216**	LiCl	1:1	(60)	88:12
Ru cat **216**	$MgCl_2$	1:1	(60)	82:18

BocHN CO$_2$Me Y + AcO, AcO, OAc, OAc, O (allyl glycoside)

Hov II (0.06 eq), *t*-BuOH/H_2O (1:1), $MgCl_2$, 32°, 2.5 h

→ MeO$_2$C NHBoc Y; AcO, AcO, OAc, OAc, O — **I** + MeO$_2$C NHBoc Y Y BocHN CO$_2$Me — **II** (226)

Y	**I**	**II**
S	(45)	(0)
Se	(53)	(20)

TABLE 8. CROSS-METATHESIS OF PRIMARY ALLYLIC ORGANOSULFUR AND ORGANOSELENIUM COMPOUNDS (*Continued*)

A. ALLYLIC SULFIDES AND SELENIDES (*Continued*)

Sulfide or Selenide	Cross Partner	Conditions	Product(s) and Yield(s) (%)	Refs.
Please refer to the charts preceding the tables for the catalyst structures.				
C_7				
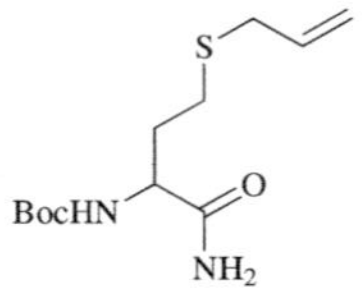 BocHN, S, O, NH_2	 OH 10 eq	Hov II (0.08 eq), *t*-BuOH/H_2O (1:1), 32°, 3.5 h	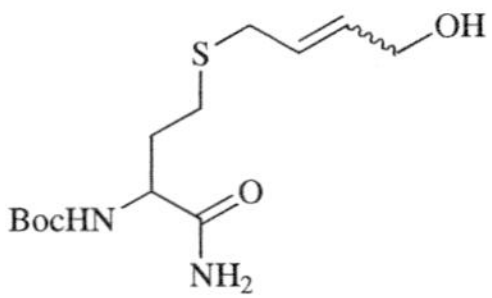 S, OH, BocHN, O, NH_2 (67)	559
C_9				
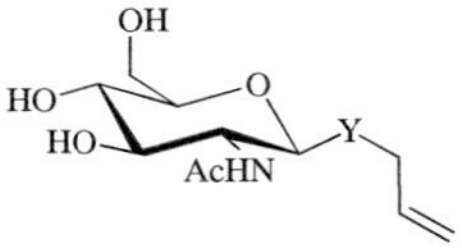 OH, HO, O, HO, Y, AcHN	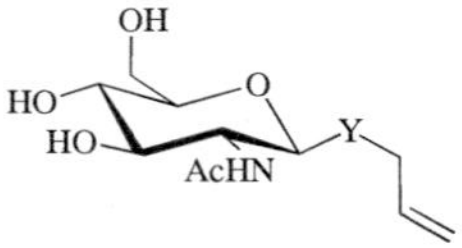 OH, HO, O, HO, Y, AcHN	Hov II (0.06 eq), *t*-BuOH/H_2O (1:1), 32°, 2.5 h	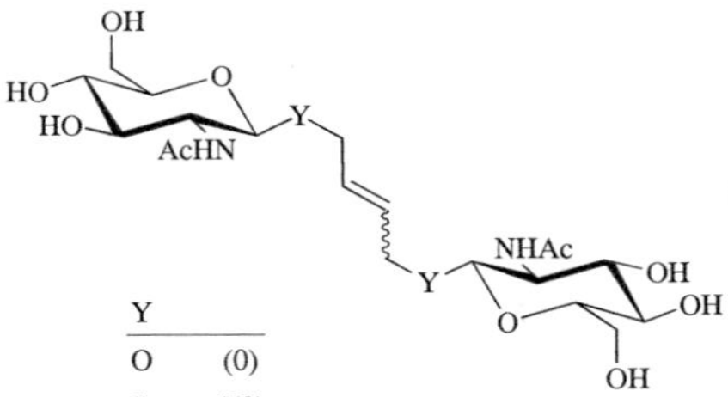 OH, HO, O, HO, Y, AcHN, NHAc, Y, OH, OH, O, OH Y: O (0); S (59)	226

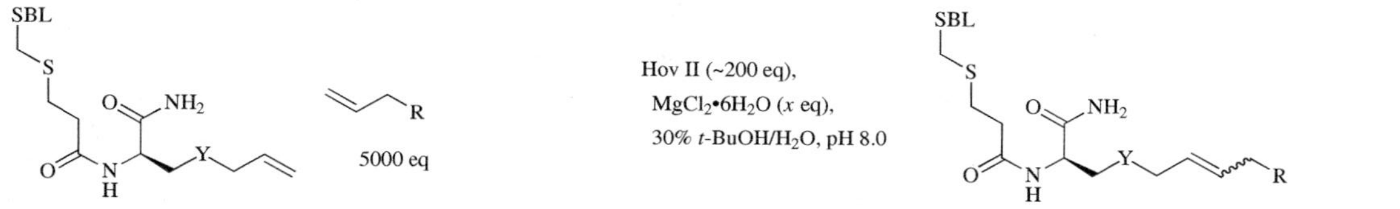

5000 eq

Hov II (~200 eq),
$MgCl_2 \cdot 6H_2O$ (x eq),
30% t-BuOH/H_2O, pH 8.0

226

Y	R	x	Time (h)	Temp (°)	Conv. (%)[a]
S	AcHN	9794	0.5	37	0
S	$Cl^-Me_3N^+$	9794	1	37	0
Se	AcHN	9794	0.5	37	90
Se	$Cl^-Me_3N^+$	9794	1	37	0
S	HO	9770	0.5	rt	>95
Se	HO	9770	0.25	rt	>95
S	$Cl^-HMeN^+(CH_2)_2O$	9770	0.5	37	29
Se	$Cl^-HMeN^+(CH_2)_2O$	9770	0.5	37	>95

TABLE 8. CROSS-METATHESIS OF PRIMARY ALLYLIC ORGANOSULFUR AND ORGANOSELENIUM COMPOUNDS (*Continued*)

A. ALLYLIC SULFIDES AND SELENIDES (*Continued*)

Sulfide or Selenide	Cross Partner	Conditions	Product(s) and Yield(s) (%)	Refs.

Please refer to the charts preceding the tables for the catalyst structures.

C_9

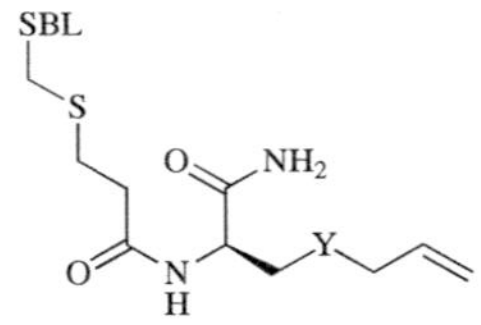

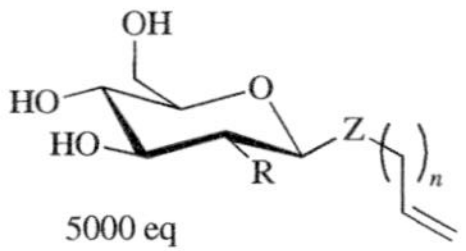

Sulfide or Selenide	Cross Partner	Conditions	Product(s) and Yield(s) (%)	Refs.
SBL, S, O, NH₂, N, H, Y	OH, HO, HO, O, R, Z, n; 5000 eq	Hov II (~200 eq), $MgCl_2$•$6H_2O$ (9770 eq), 30% t-BuOH/H_2O, pH 8.0	OH, HO, HO, O, R, Z, n, Y, H, N, SBL, S, H_2N, O	226
	OH, OH, HO, HO, O, O; 5000 eq	Hov II (~200 eq), $MgCl_2$•$6H_2O$ (9770 eq), 30% t-BuOH/H_2O, pH 8.0, rt, 1 h	OH, OH, HO, HO, O, O, Y, O, NH_2, HN, O, S, SBL	226

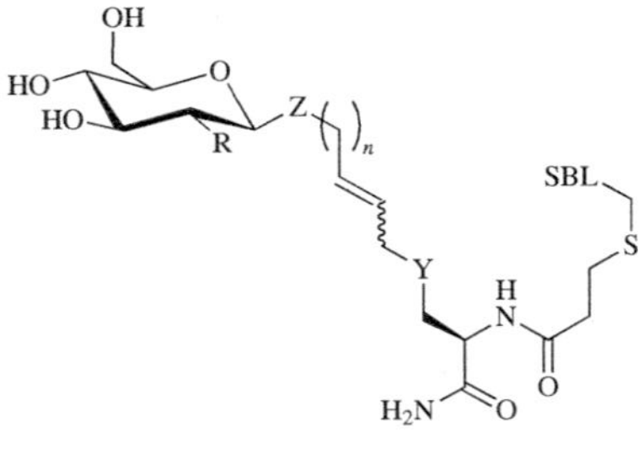

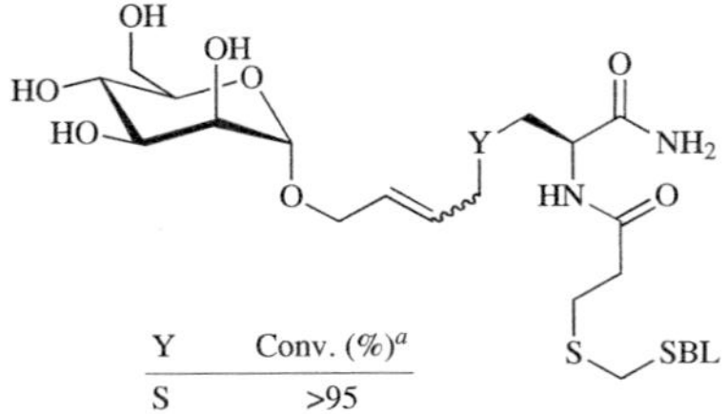

Y	R	Z	n	Temp (°)	Time (h)	Conv. (%)[a]
S	HO	O	1	37	1	>95
Se	HO	O	1	37	1	>95
S	HO	O	4	rt	0.5	>95
Se	HO	O	4	rt	0.5	>95
S	AcHN	O	1	37	2	53
S	AcHN	S	1	37	1	0
Se	AcHN	O	1	37	1	>95
Se	AcHN	S	1	rt	1	0

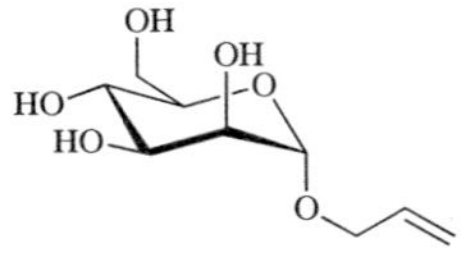

Y	Conv. (%)[a]
S	>95
Se	>95

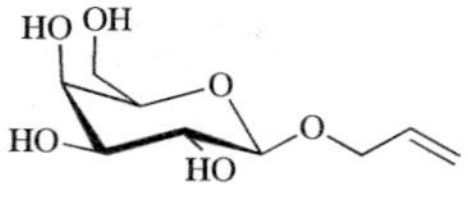

5000 eq

Hov II (~200 eq), $MgCl_2 \cdot 6H_2O$ (9770 eq), 30% *t*-BuOH/H_2O, pH 8.0, 37°, 1 h

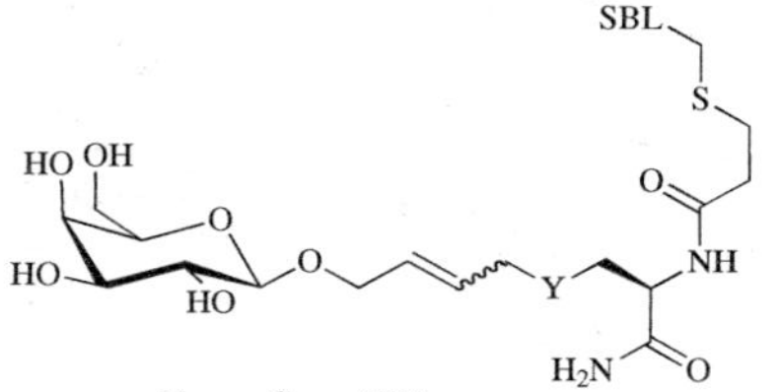

Y	Conv. (%)[a]
S	>95
Se	>95

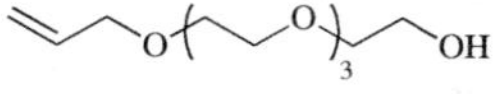

226

5000 eq

Hov II (~200 eq), $MgCl_2 \cdot 6H_2O$ (9770 eq), 30% *t*-BuOH/H_2O, pH 8.0, rt

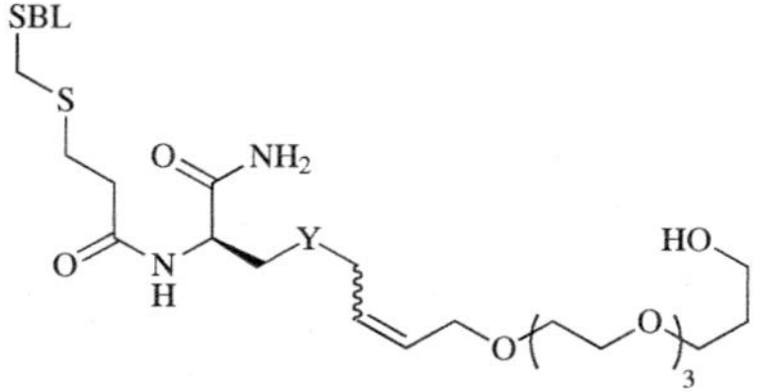

Y	Time (h)	Conv. (%)[a]
S	2	>95
Se	1	>95

226

C_{12}

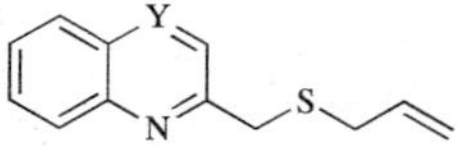

20 eq

Hov II (2 × 0.1 eq), *t*-BuOH/H_2O (1:1), 32°, 2.5 h

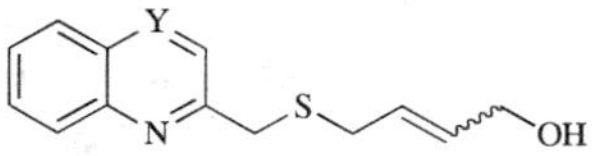

Y	
N	(31)
CH	(53)

227

TABLE 8. CROSS-METATHESIS OF PRIMARY ALLYLIC ORGANOSULFUR AND ORGANOSELENIUM COMPOUNDS (*Continued*)

A. ALLYLIC SULFIDES AND SELENIDES (*Continued*)

Sulfide or Selenide	Cross Partner	Conditions	Product(s) and Yield(s) (%)	Refs.
Please refer to the charts preceding the tables for the catalyst structures.				
C_{17}				
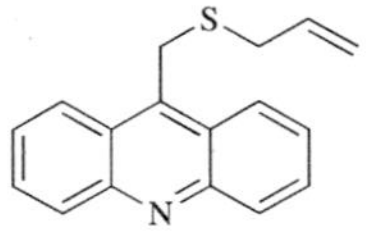 N; S	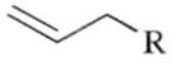 R; *x* eq	Hov II (2 x 0.1 eq), *t*-BuOH/H_2O (1:1), 32°, 2.5 h	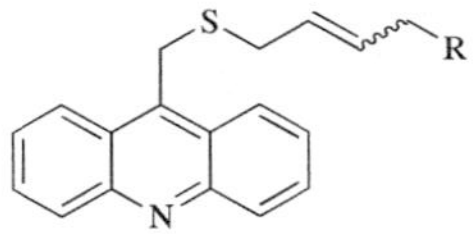 N; S; R R, *x*: AcHN 10 (30) HO 20 (65)	227
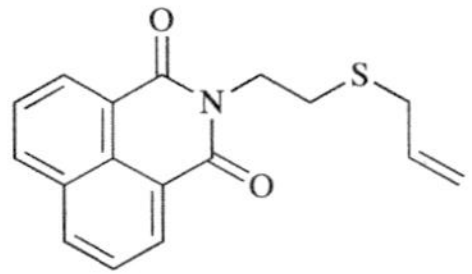 O; N; S; O	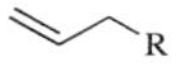 R; 10 eq	Hov II (2 x 0.1 eq), *t*-BuOH/H_2O (1:1), 32°, 2.5 h	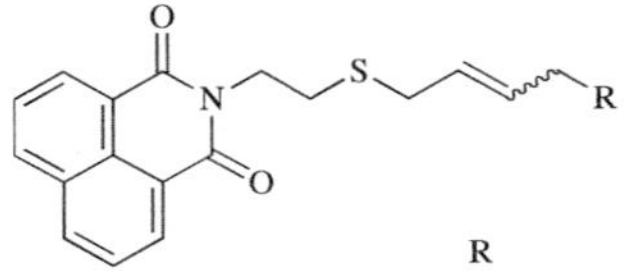 O; N; S; R; O R: AcHN (96) HO (94)	227
	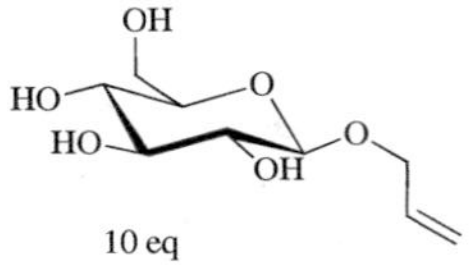 OH; HO; O; HO; OH; O; 10 eq	Hov II (2 x 0.1 eq), *t*-BuOH/H_2O (1:1), 32°, 2.5 h	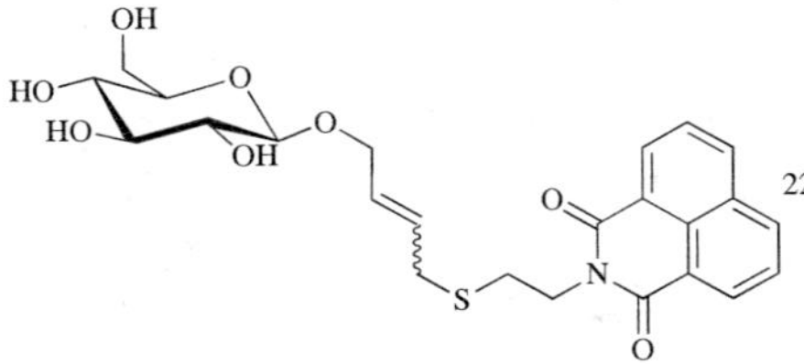 OH; HO; O; HO; OH; O; O; S; N; O (90)	227

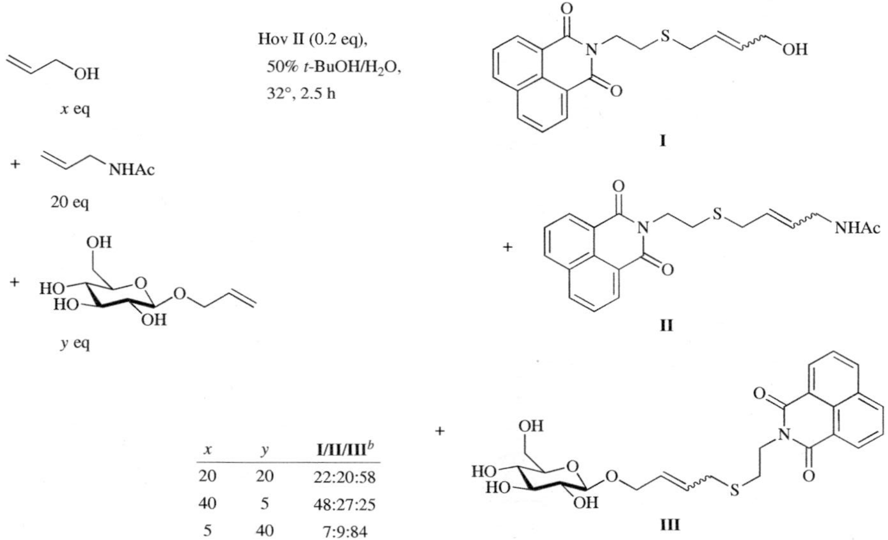

OH

x eq

\+ NHAc

20 eq

\+ OH HO HO O O OH

y eq

Hov II (0.2 eq),
50% *t*-BuOH/H_2O,
32°, 2.5 h

O N S OH O

I

\+ O N S NHAc O

II

\+ OH HO HO O O OH S N O O

III

227

x	*y*	**I/II/III**[b]
20	20	22:20:58
40	5	48:27:25
5	40	7:9:84

[a] The conversion was determined by LC-MS analysis.

[b] The ratio was determined by LC-MS analysis.

TABLE 8. CROSS-METATHESIS OF PRIMARY ALLYLIC ORGANOSULFUR AND ORGANOSELENIUM COMPOUNDS (*Continued*)

B. SULFONIUM SALTS

Sulfonium Salt	Cross Partner	Conditions	Product(s) and Yield(s) (%)	Refs.
Please refer to the charts preceding the tables for the catalyst structures.				
C_{25}				
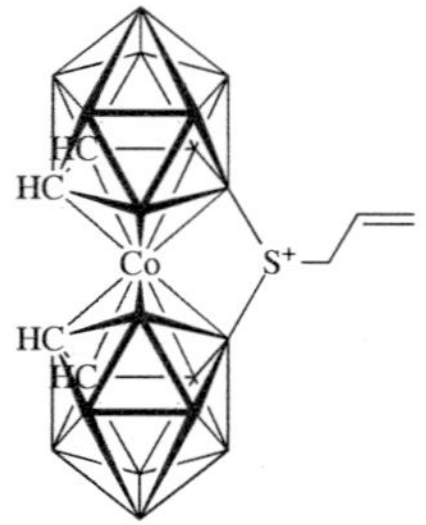 HC, HC, Co, HC, HC, S⁺	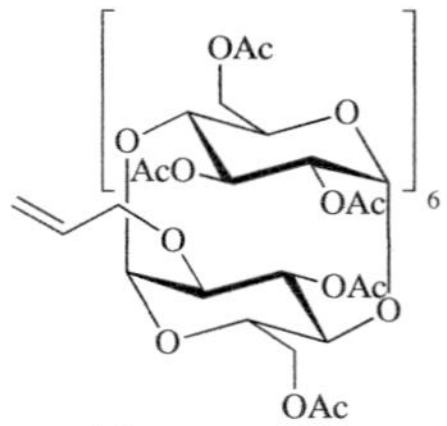 OAc, O, AcO, OAc, O, OAc, O, O, OAc, 6; 1.4 eq	Hov II (0.1 eq), toluene, 120°, 16 h	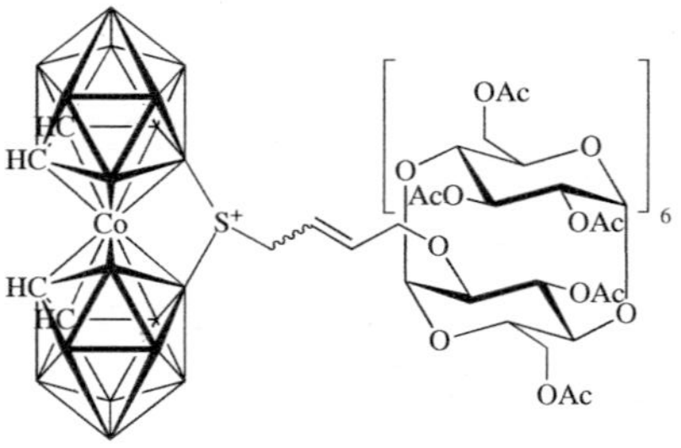 HC, HC, Co, HC, HC, S⁺, OAc, O, AcO, OAc, O, OAc, O, O, OAc, 6 (20)	585
	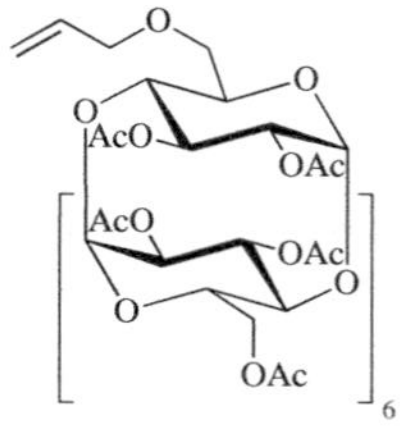 O, O, AcO, OAc, AcO, OAc, O, O, OAc, 6; 1.4 eq	Hov II (0.1 eq), toluene, 120°, 16 h	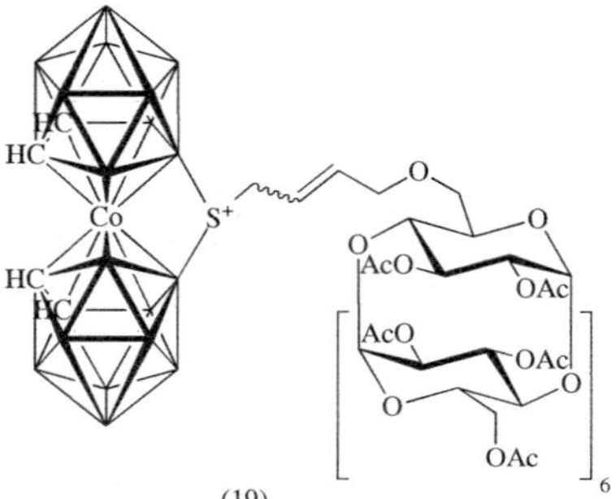 HC, HC, Co, HC, HC, S⁺, O, O, O, AcO, OAc, AcO, OAc, O, O, OAc, 6 (19)	585

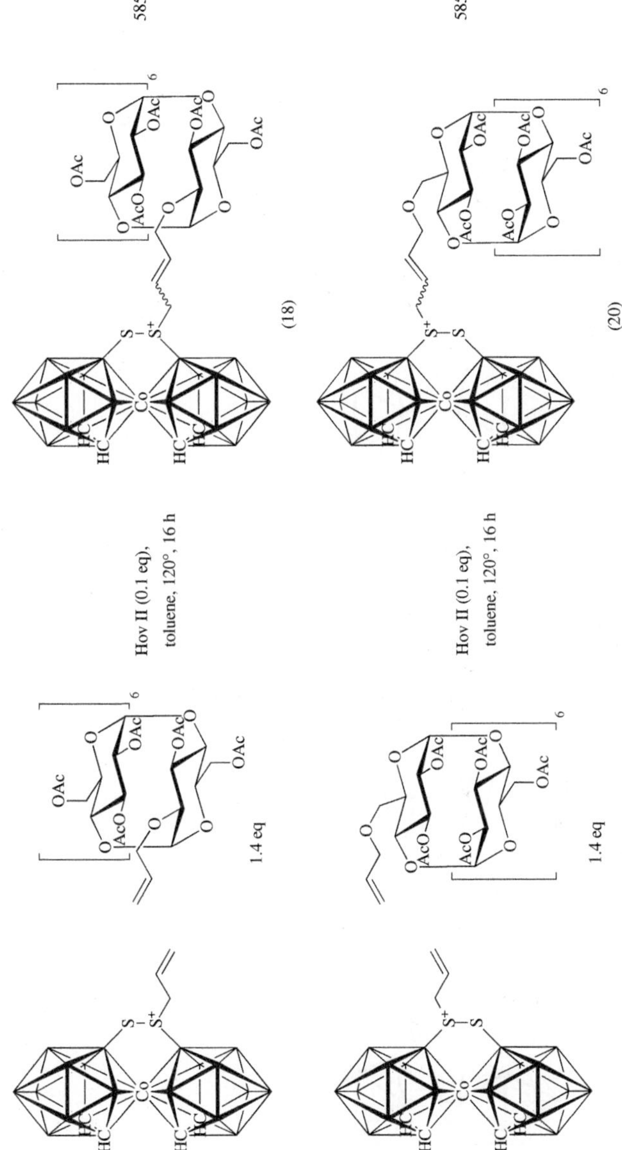
1.4 eq
Hov II (0.1 eq),
toluene, 120°, 16 h
(18)
585
1.4 eq
Hov II (0.1 eq),
toluene, 120°, 16 h
(20)
585

TABLE 8. CROSS-METATHESIS OF PRIMARY ALLYLIC ORGANOSULFUR AND ORGANOSELENIUM COMPOUNDS (*Continued*)

B. SULFONIUM SALTS (*Continued*)

Sulfonium Salt	Cross Partner	Conditions	Product(s) and Yield(s) (%)	Refs.
Please refer to the charts preceding the tables for the catalyst structures.				
C_{25}				
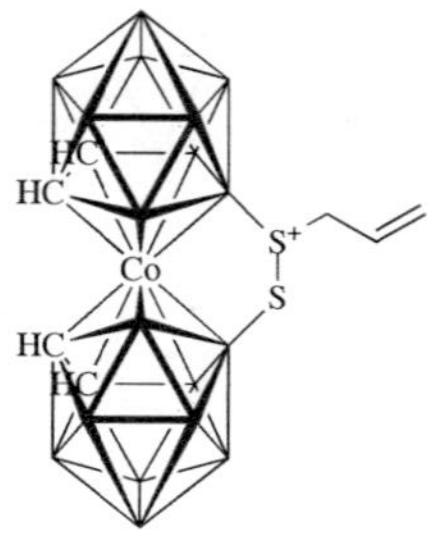 HC HC Co HC HC S+ S	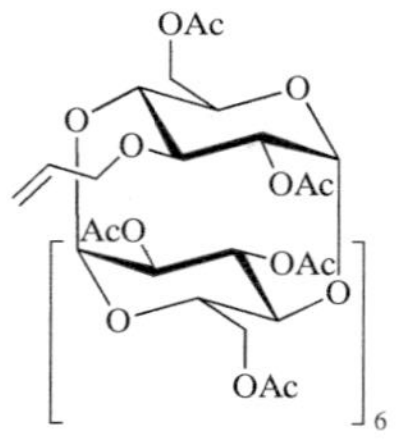 OAc O O O OAc AcO OAc O O OAc 6 1.4 eq	Hov II (0.1 eq), toluene, 120°, 16 h	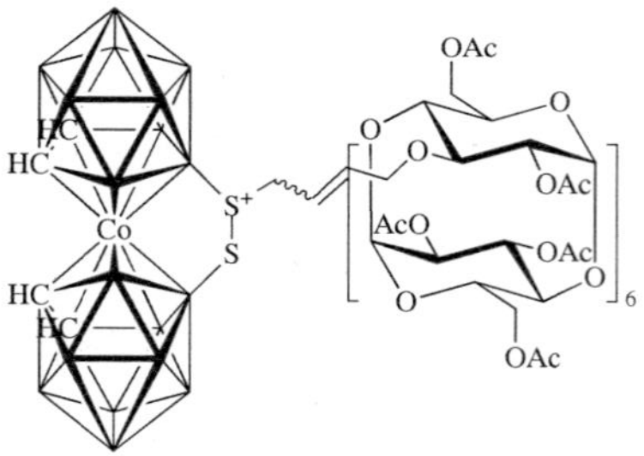 HC HC Co HC HC S+ S O O OAc AcO OAc O O OAc OAc 6 (19)	585

TABLE 8. CROSS-METATHESIS OF PRIMARY ALLYLIC ORGANOSULFUR AND ORGANOSELENIUM COMPOUNDS (*Continued*)

C. ALLYLIC SULFOXIDES, SULFONES AND SULFONAMIDES

Sulfur Compound	Cross Partner	Conditions	Product(s) and Yield(s) (%)	Refs.
Please refer to the charts preceding the tables for the catalyst structures.				
C_3				
Bn–S(=O)–allyl	OTBS-terminated alkene (n = 4), 1.5 eq	Gr II (0.1 eq), $(PhO)_3B$ (*x* eq), solvent, 70°, 6 h	Bn–S(=O) cross product, OTBS (n = 4); (*E*) only *x* / Solvent: — toluene (9); 0.1 toluene (12); — C_6F_6 (31)	229
PhO_2S–allyl	allyl–SO_2Ph	Gr I (0.05 eq), DCM, 45°	PhO_2S cross product, SO_2Ph (37) (*E*)/(*Z*) = 89:11	124
2 eq	OBz-terminated alkene (n = 8)	Gr I (0.025 eq), DCM, 45°, 4 h	PhO_2S cross product, OBz (n = 8) (90) (*E*)/(*Z*) = 89:11	124
resin–NH–SO_2–allyl	allylic NHFmoc (R), *x* eq	Catalyst (*y* eq), additive, solvent	resin–NH–SO_2 cross product, NHFmoc (R)	154

R	*x*	Catalyst	*y*	Additive	Solvent	Temp (°)	Time (h)	Loading (%)
H	5	Gr II	0.5	—	DCM	40	16	17
H	10	Gr II	1	—	DCM	40	16	11
H	5	Gr II	0.5	—	DCM/MeOH	40	16	0
H	5	Hov II	0.5	—	DCM/MeOH	40	16	11
H	5	Gr II	0.2	—	DCM	150 (MW)	0.5	16
H	5	Gr II	0.2	—	toluene	150 (MW)	0.5	32
H	5	Gr II	0.2	Cy_2BCl	toluene	150 (MW)	0.5	21
H	5	Gr II	0.2	Cy_2BCl	toluene	80	16	19
H	5	Hov II	0.2	Cy_2BCl	toluene	80	16	35
Bn	5	Hov II	0.2	Cy_2BCl	toluene	80	16	13
Bn	5	Hov II	0.5	Cy_2BCl	toluene	80	16	26

TABLE 9. CROSS-METATHESIS OF VINYL-SUBSTITUTED HETEROAROMATIC COMPOUNDS

Heteroaromatic Compound	Cross Partner	Conditions	Product(s) and Yield(s) (%)	Refs.

Please refer to the charts preceding the tables for the catalyst structures.

C_5

Cross partner: x eq. Conditions: Catalyst (y eq), DCM, 40°. Ref. 147

R	n	x	Catalyst	y	Time (h)		$(E)/(Z)$
TMS	1	3	Gr II	0.1	2	(89)	83:17
TMS	1	3	Hov II	0.05	12	(83)	83:17
$(EtO)_2P(O)$	1	1.5	Gr II	0.1	24	(43)	83:17
$(EtO)_2P(O)$	1	1.5	Hov II	0.05	24	(37)	87:13
AcO	2	1.5	Gr II	0.1	18	(63)	89:11
AcO	2	1.5	Hov II	0.05	24	(58)	87:13
Me	2	3	Gr II	0.1	4	(82)	>95:5
Me	2	3	Hov II	0.05	8	(79)	>95:5

Cross partner: 1.5 eq. Conditions: Catalyst (x eq), DCM, 40°. Ref. 147

R^1	R^2	Catalyst	x	Time (h)		$(E)/(Z)$
MeO_2C	MeO_2C	Gr II	0.1	24	(46)	80:20
MeO_2C	MeO_2C	Hov II	0.05	24	(54)	75:25
TBSO	$TBSOCH_2$	Gr II	0.1	18	(57)	>95:5
TBSO	$TBSOCH_2$	Hov II	0.05	8	(51)	>95:5
HO	Ph	Gr II	0.1	24	(22)	80:20
HO	Ph	Hov II	0.05	24	(19)	75:25
TBSO	Ph	Gr II	0.1	12	(65)	>95:5
TBSO	Ph	Hov II	0.05	24	(83)	>95:5

C$_6$

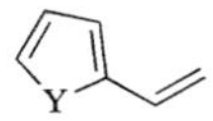

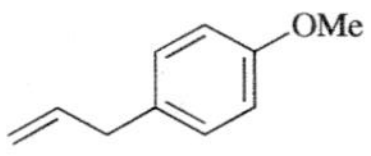

1.5 eq

Catalyst (x eq), DCM, 40°

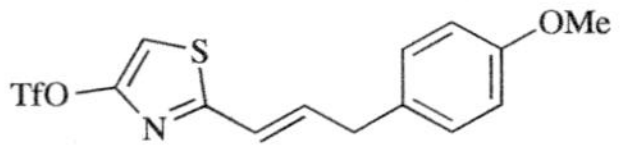

Catalyst	x	Time (h)		$(E)/(Z)$
Gr II	0.1	12	(89)	>95:5
Hov II	0.05	24	(86)	>95:5

147

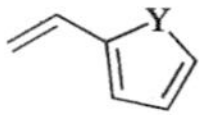

Catalyst (0.05 eq), toluene, rt, 17 h

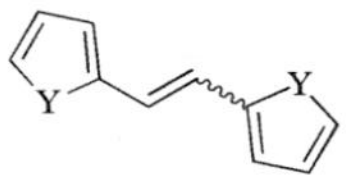

Y	Catalyst	Gas	Conv. (%)
O	Gr I	O_2	0
O	Gr I	N_2	0
O	Mo cat **1**	N_2	2
S	Gr I	O_2	0
S	Gr I	N_2	0
S	Mo cat **1**	N_2	4

636

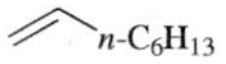

n-C_6H_{13}

6 eq

Catalyst (0.05 eq), toluene, rt, 17 h

n-C_6H_{13}

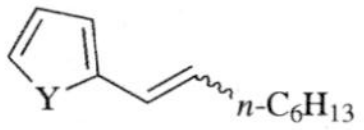

Y	Catalyst	Gas	
O	Gr I	N_2	(50)
O	Mo cat **1**	N_2	(78)
S	Gr I	O_2	(50)
S	Gr I	N_2	(51)
S	Mo cat **1**	N_2	(98)

636

TABLE 9. CROSS-METATHESIS OF VINYL-SUBSTITUTED HETEROAROMATIC COMPOUNDS (*Continued*)

Heteroaromatic Compound	Cross Partner	Conditions	Product(s) and Yield(s) (%)	Refs.
Please refer to the charts preceding the tables for the catalyst structures.				
C_6				
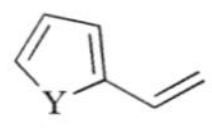 1 eq	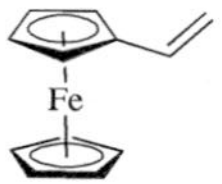	Mo cat **1** (0.017–0.01 eq), rt, 15 h	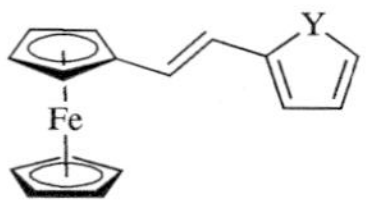 **I** + 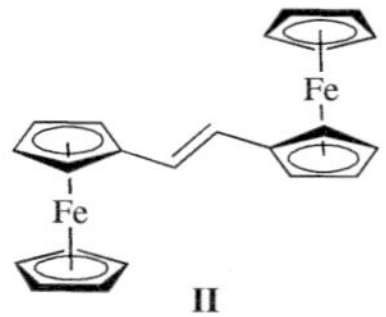 **II** + 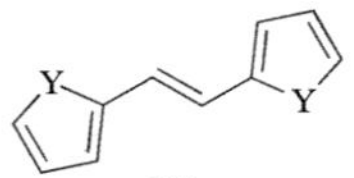 **III**	637
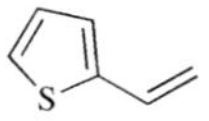 7.5 eq	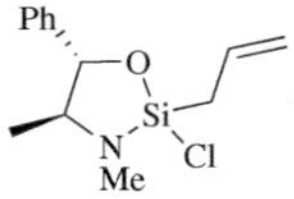 1.5 eq + 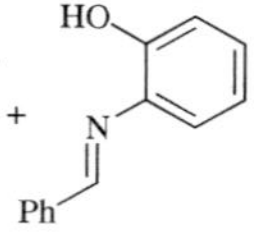 1 eq	1. Gr II (0.026 eq), DCE, reflux, 3.5 h 2. Imine, reflux, 14 h	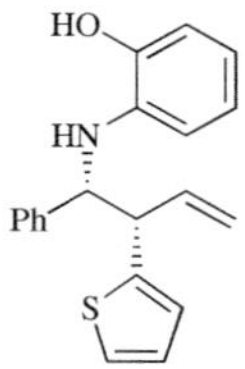 (54) dr 83:17, er 97.0:3.0	211

Y	I	I/II/III	(*E*)/(*Z*)
S	(51)	91:0:9	(*E*) only
O	(65)	79:7:14	(*E*) only

7.5 eq	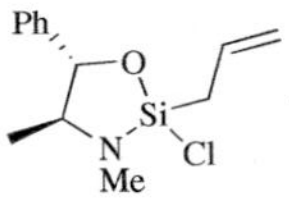 Ph, O, Si, Cl, N, Me 1.5 eq 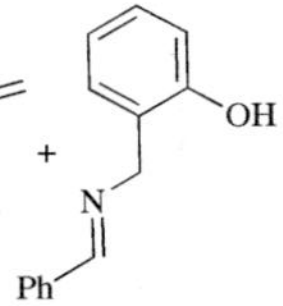 + OH, N, Ph 1 eq	1. Gr II (0.026 eq), $CHCl_3$, reflux, 5 h 2. Imine, reflux, 14 h	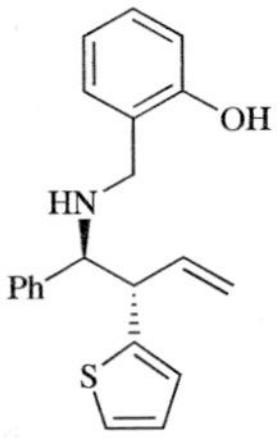 OH, HN, Ph, S	(65) dr > 95:5, er 98.5:1.5	211
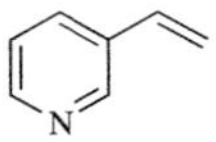 N	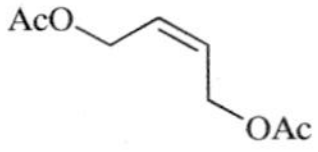 AcO, OAc 10 eq	Gr II (0.02 eq), DCM, 45°, 3 h	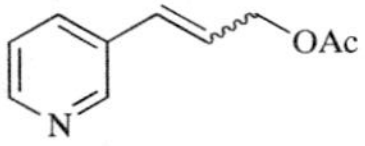 N, OAc	(85)	638
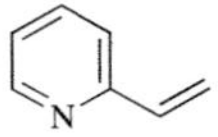 N	 OAc, 4 2 eq	Gr II (0.05 eq), DCM, reflux, 12 h	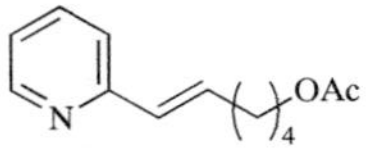 N, OAc, 4	(45) (*E*)/(*Z*) > 95:5	145
C_7					
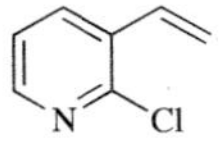 N, Cl	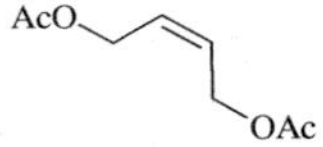 AcO, OAc 3 eq	Hov II (0.1 eq), DCM, 50°, 24 h	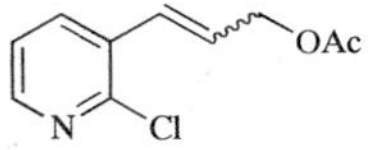 N, Cl, OAc	(56)	603

TABLE 9. CROSS-METATHESIS OF VINYL-SUBSTITUTED HETEROAROMATIC COMPOUNDS (*Continued*)

Heteroaromatic Compound	Cross Partner	Conditions	Product(s) and Yield(s) (%)	Refs.
Please refer to the charts preceding the tables for the catalyst structures.				
C_7				
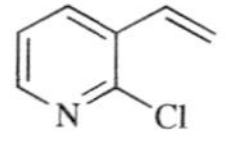 7.5 eq	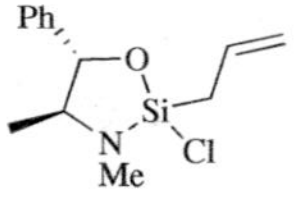 1.5 eq + 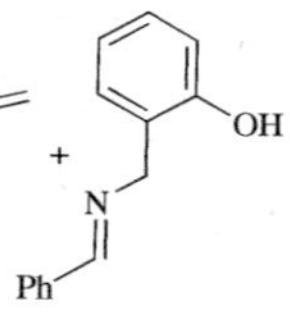 1 eq	1. Gr II (0.026 eq), $CHCl_3$, reflux, 5 h 2. Imine, reflux, 14 h	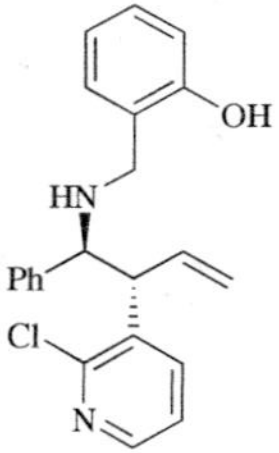 (54) dr > 95:5, er 97.5:2.5	211
7.5 eq	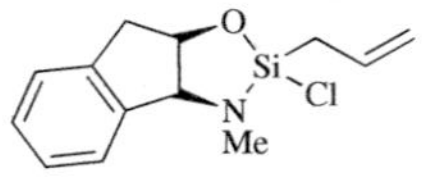 1.5 eq + 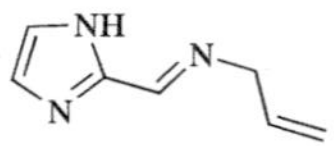 1 eq	1. Hov II (0.036 eq), benzene, reflux, 2 h 2. Imine, rt, 14 h	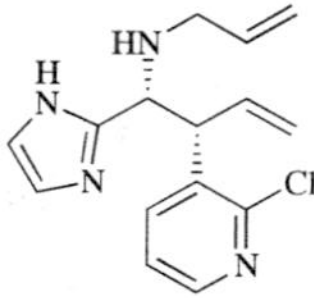 (58) dr 92:8, er 93.5:6.5	211
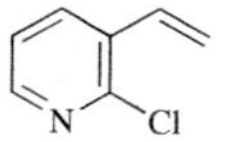 3 eq	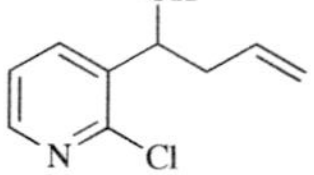	Hov II (0.1 eq), DCM, 50°, 24 h	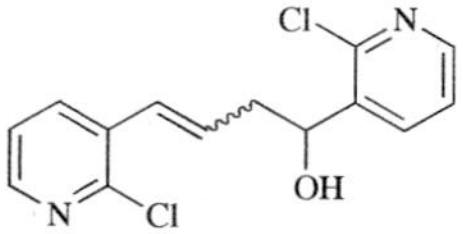 (55)	603
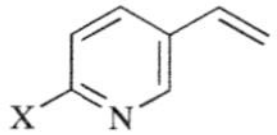	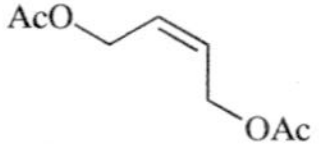 3 eq	Hov II (0.1 eq), DCM, 50°, 24 h	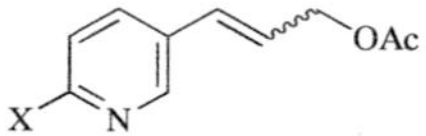 X: F (22); Cl (61); Br (65)	603

Entry	Reactant 1	Reactant 2	Conditions	Product	Yield (%)	Ref.
C_9	CO_2Me, O, N, N, O	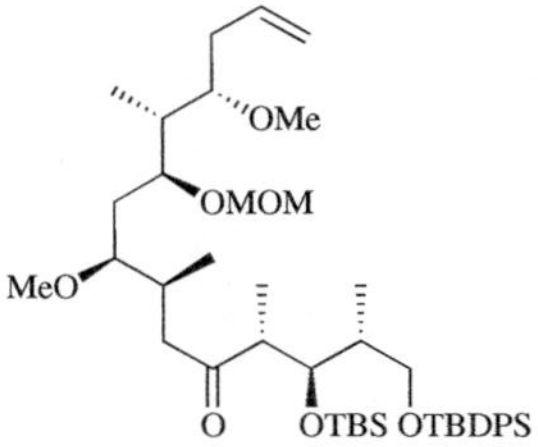 MeO, OMe, OMOM, O, OTBS, OTBDPS 2 eq	Gr II (0.15 eq), DCM, 40°, 16 h sealed tube	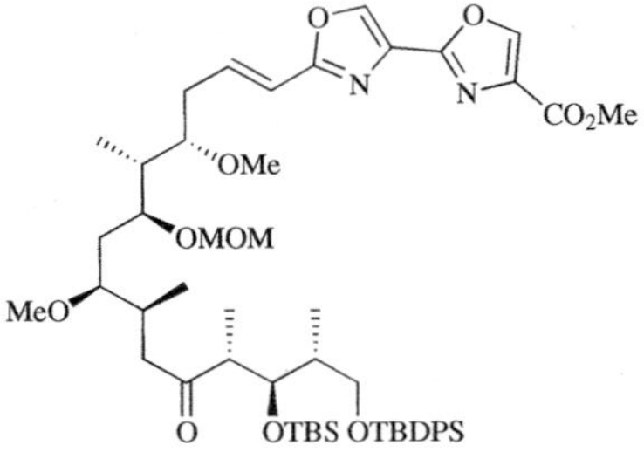 O, N, O, N, CO_2Me, OMe, OMOM, MeO, O, OTBS, OTBDPS	(57) (*E*)/(*Z*) = 95:5	639
C_{10}	N, Ts 1.4 eq	MeO_2C, CO_2Me	Gr II (0.01 eq), DCM, reflux	MeO_2C, CO_2Me, N, Ts	(5)	407
C_{11}	N	*n*-Pr 10 eq	Gr I (0.1 eq), DCM, 40°, 20 h	*n*-Pr, N	(73) (*E*)/(*Z*) = 95:5	584

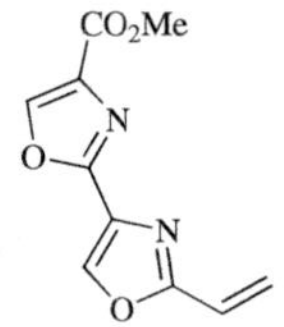

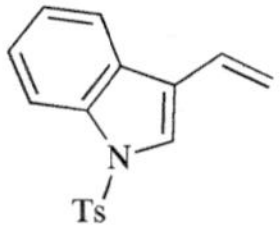

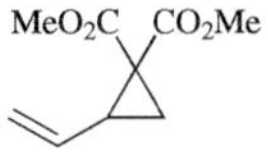

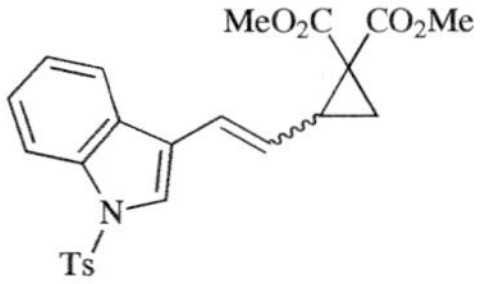

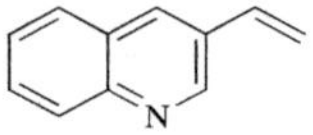

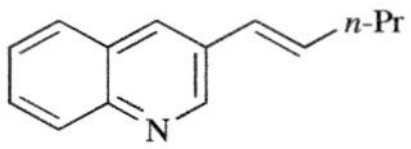

TABLE 9. CROSS-METATHESIS OF VINYL-SUBSTITUTED HETEROAROMATIC COMPOUNDS (*Continued*)

Heteroaromatic Compound	Cross Partner	Conditions	Product(s) and Yield(s) (%)	Refs.
Please refer to the charts preceding the tables for the catalyst structures.				
C_{11}				
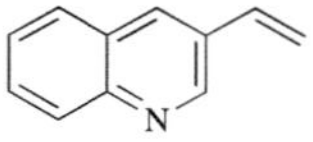 N 3 eq	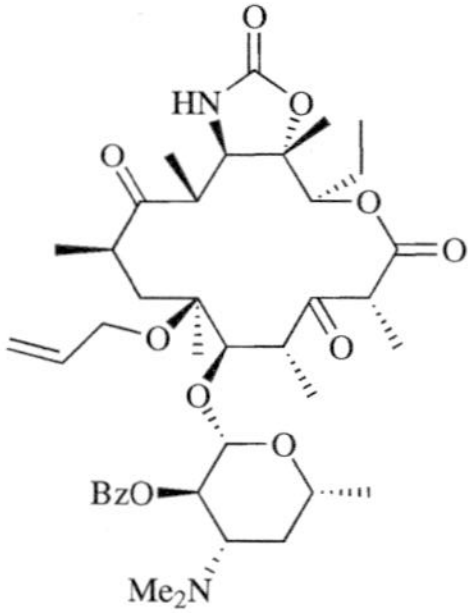 HN O O O O O BzO O Me$_2$N	Gr I (0.1 eq), DCM, 40°, 20 h	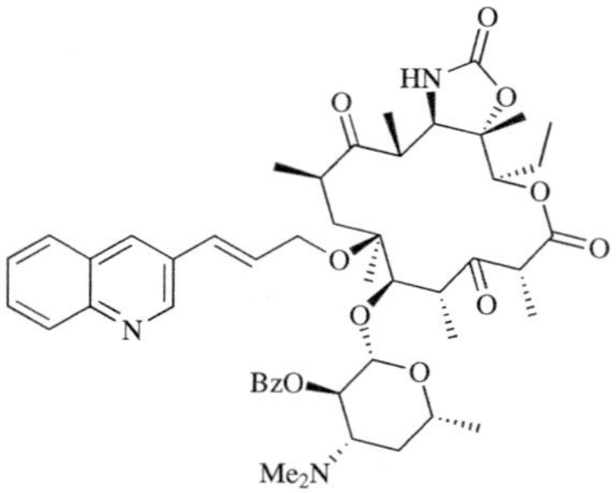 O HN O O O O N O O BzO O Me$_2$N (39) (*E*)/(*Z*) > 95:5	584
x eq	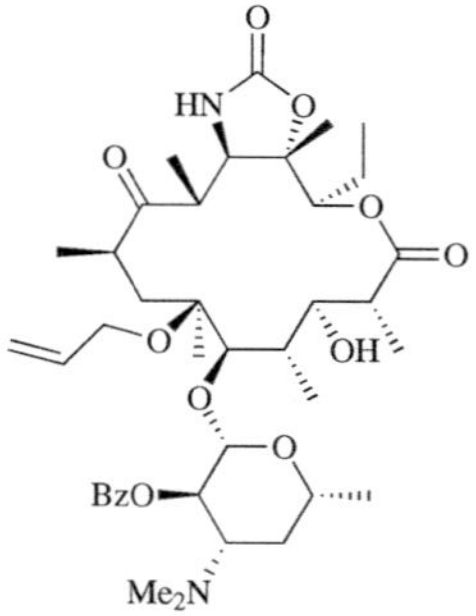 O HN O O O O OH O BzO O Me$_2$N *y* eq	Gr I (*z* eq), DCM	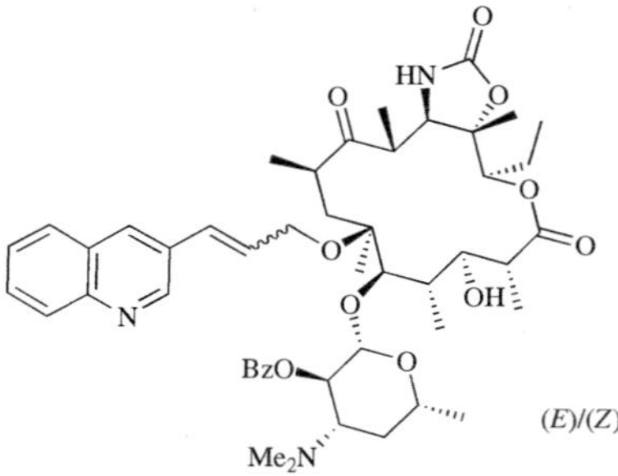 O HN O O O O N OH O BzO O Me$_2$N (*E*)/(*Z*) > 95:5	584

x	y	z	Temp (°)	Time (h)	Conv. (%)[a]
2	1	0.1	rt	20	5
2	1	0.1	40	20	35
2	1	0.1	40	65	59
2	1	0.1	40	91	65
2	1	0.1	40	168	71
2	1	0.05	40	20	27
2	1	0.25	40	20	49
2	1	0.25	40	65	75
1	1	0.25	40	20	45
1	3	0.1	40	20	64
1	3	0.1	40	65	79
5	1	0.1	40	20	23

3 eq

Gr I (0.1 eq), DCM, 40°, 20 h

(86) $(E)/(Z) > 95:5$

584

TABLE 9. CROSS-METATHESIS OF VINYL-SUBSTITUTED HETEROAROMATIC COMPOUNDS (*Continued*)

Heteroaromatic Compound	Cross Partner	Conditions	Product(s) and Yield(s) (%)	Refs.
Please refer to the charts preceding the tables for the catalyst structures.				
C_{11}				
3 eq	BzO, Me_2N, O, HN, OMe, OBz	Gr I (0.1 eq), DCM, 40°, 20 h	(64) (*E*)/(*Z*) > 95:5	584
C_{12}				
TBSO, OTBS, NHPAC	5 eq	Gr II (0.1 eq), DCM, 45°, 2 h	(22) (*E*) only	640

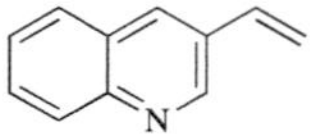

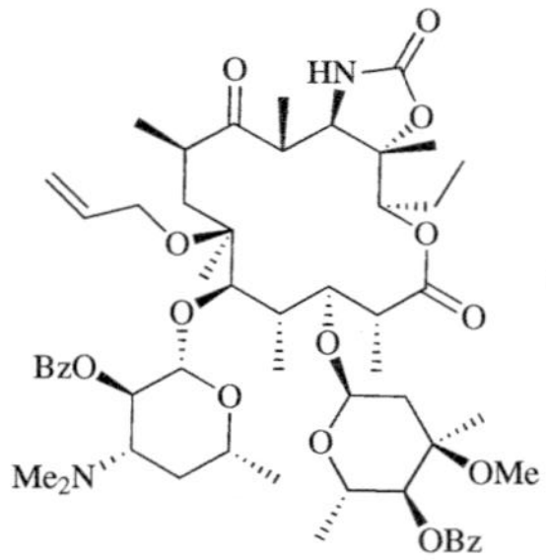

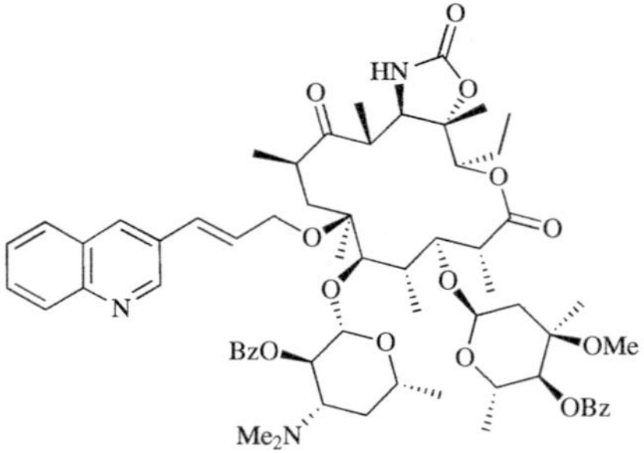

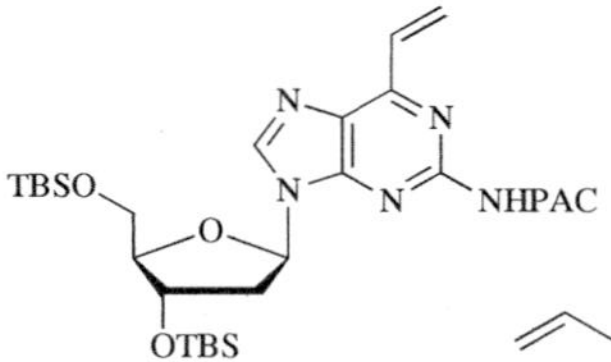

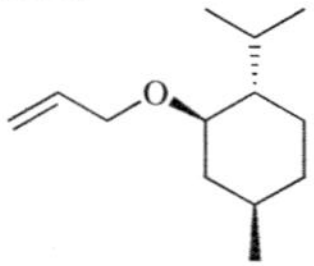

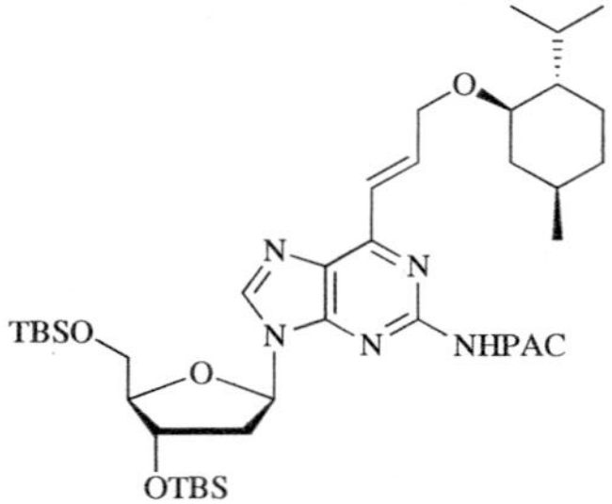

C_{33}		Gr II (0.25 eq), DCM, reflux		(0)	575
C_{34}	4 eq	Gr II (0.25 eq), DCM, reflux, 16 h		(40)	575

TABLE 9. CROSS-METATHESIS OF VINYL-SUBSTITUTED HETEROAROMATIC COMPOUNDS (*Continued*)

Heteroaromatic Compound	Cross Partner	Conditions	Product(s) and Yield(s) (%)	Refs.
Please refer to the charts preceding the tables for the catalyst structures.				
C_{34}				
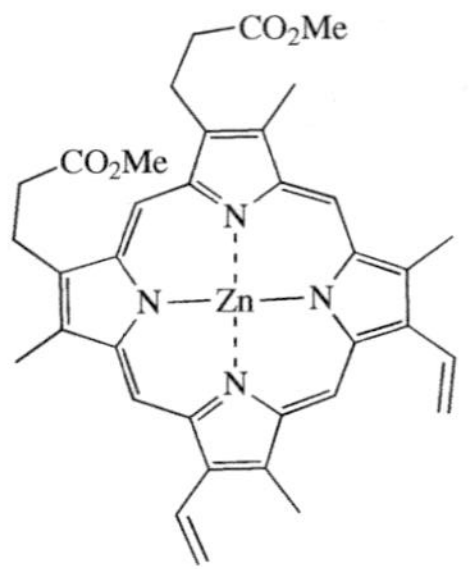 CO_2Me, CO_2Me, N, N, Zn, N, N	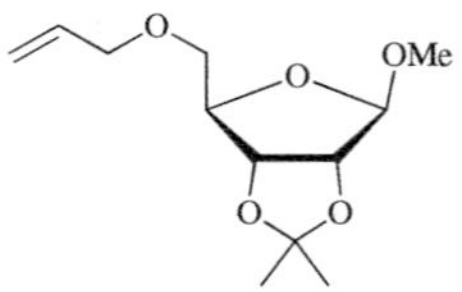 O, O, OMe, O, O 2 eq	Gr II (0.25 eq), DCM, reflux, 8 h	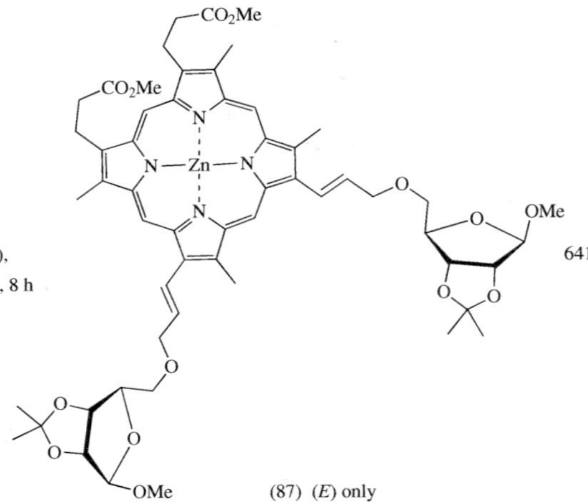 CO_2Me, CO_2Me, N, N, Zn, N, N, O, O, OMe, O, O, O, O, O, O, OMe (87) (*E*) only	641

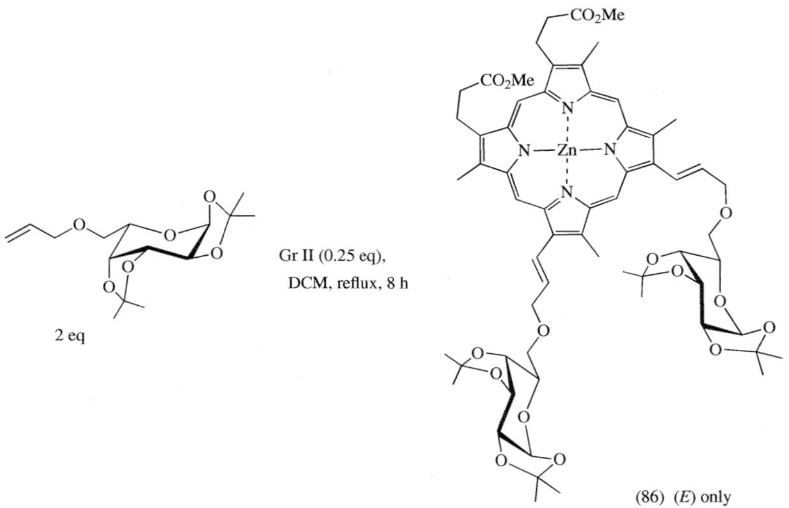

2 eq

Gr II (0.25 eq),
DCM, reflux, 8 h

CO_2Me

CO_2Me

N

N—Zn—N

N

O O O O O

O O O O O O

(86) (*E*) only

641

TABLE 9. CROSS-METATHESIS OF VINYL-SUBSTITUTED HETEROAROMATIC COMPOUNDS (*Continued*)

Heteroaromatic Compound	Cross Partner	Conditions	Product(s) and Yield(s) (%)	Refs.
Please refer to the charts preceding the tables for the catalyst structures.				
C_{34}				
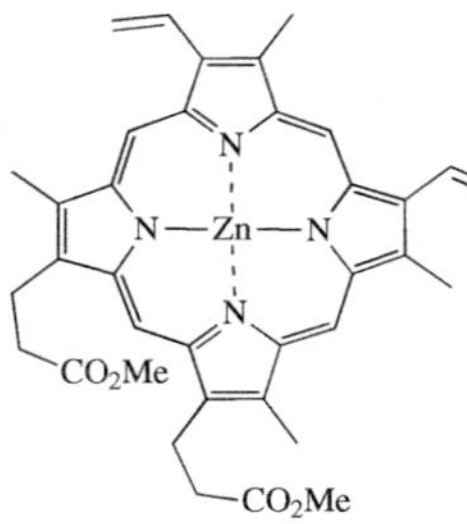 N—Zn—N, CO_2Me, CO_2Me	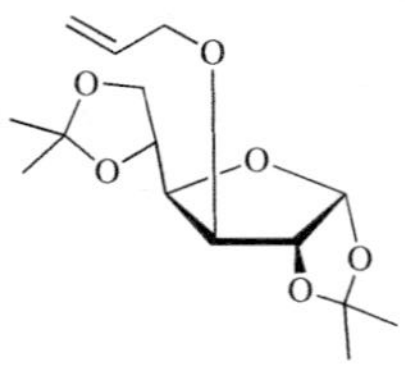 2 eq	Gr II (0.25 eq), DCM, reflux, 8 h	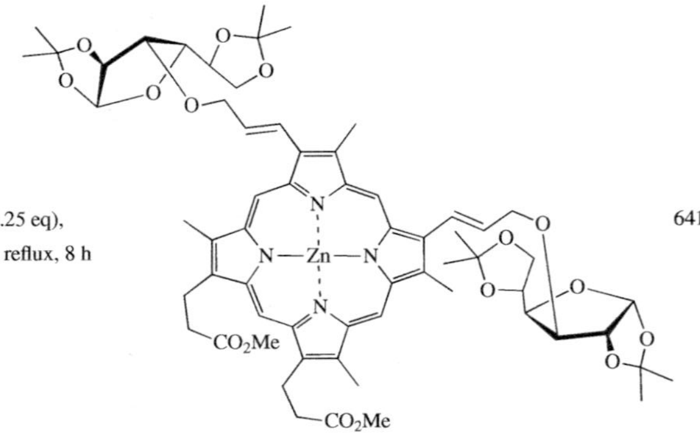 N—Zn—N, CO_2Me, CO_2Me (84) (*E*) only	641

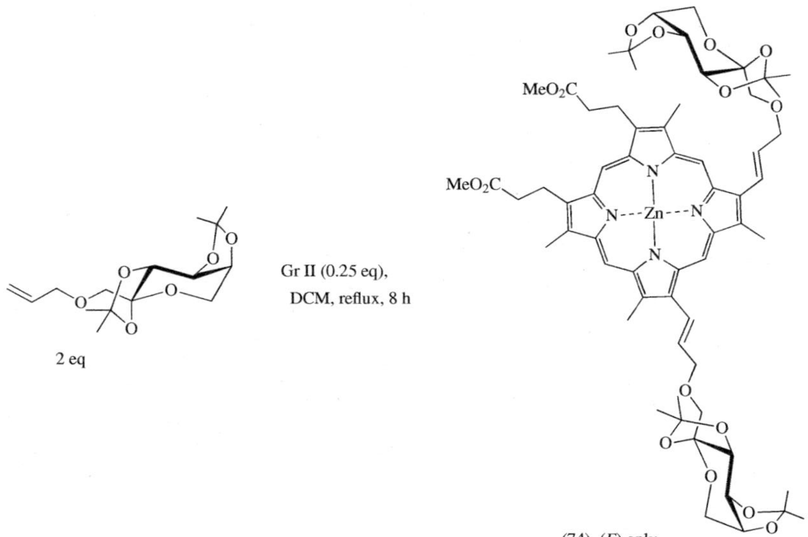

2 eq

Gr II (0.25 eq),
DCM, reflux, 8 h

MeO_2C

MeO_2C

N

N

Zn

N

N

(74) (*E*) only

641

TABLE 9. CROSS-METATHESIS OF VINYL-SUBSTITUTED HETEROAROMATIC COMPOUNDS (*Continued*)

Heteroaromatic Compound	Cross Partner	Conditions	Product(s) and Yield(s) (%)	Refs.
Please refer to the charts preceding the tables for the catalyst structures.				
C_{34}				
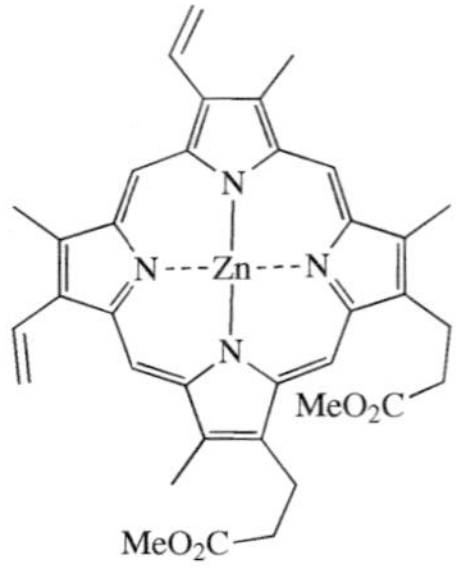 N---Zn---N, MeO_2C, MeO_2C	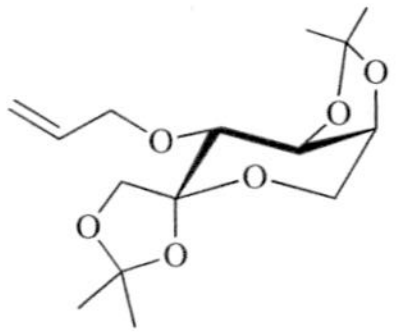 2 eq	Gr II (0.25 eq), DCM, reflux, 8 h	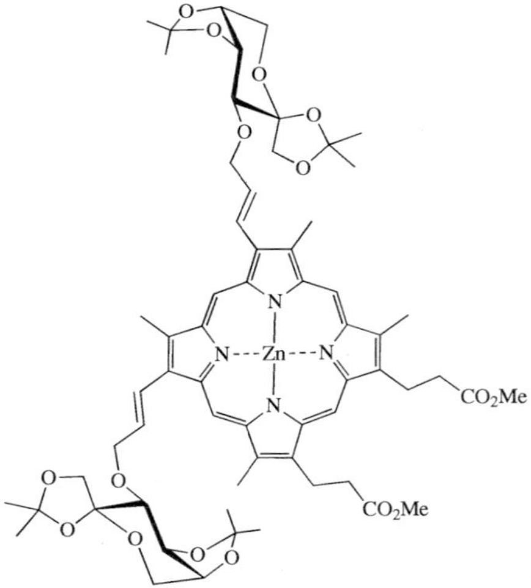 N---Zn---N, CO_2Me, CO_2Me (93) (*E*) only	641
C_{46}				
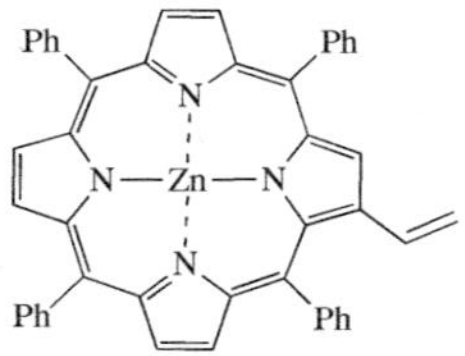 Ph, Ph, N—Zn—N, Ph, Ph	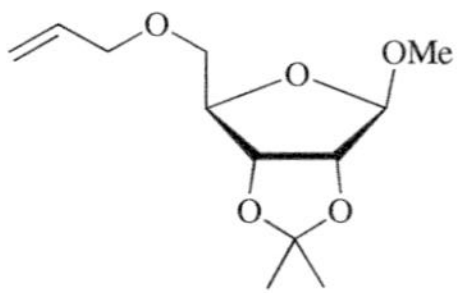 OMe	Gr II (0.25 eq), DCM, reflux, 4 h	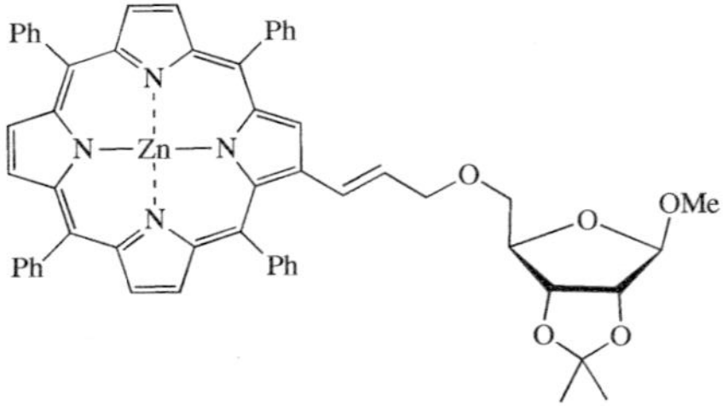 Ph, Ph, N—Zn—N, Ph, Ph, OMe (98) (*E*) only	641

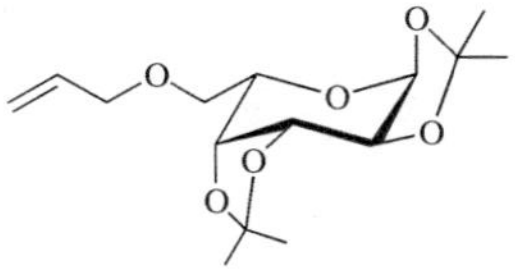

Gr II (0.25 eq),
DCM, reflux, 4 h

Ph Ph N N Zn N N Ph Ph O O O O O O

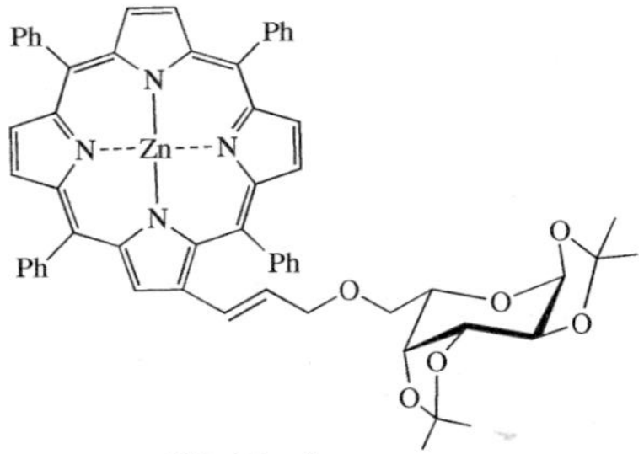

(95) (E) only

641

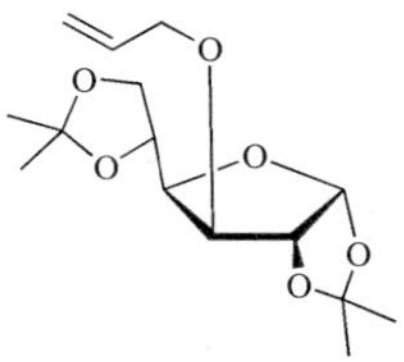

Gr II (0.25 eq),
DCM, reflux, 4 h

Ph Ph N N Zn N N Ph Ph O O O O O O

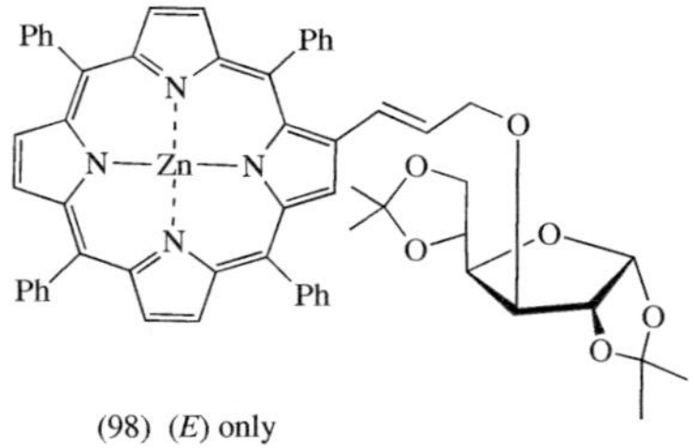

(98) (E) only

641

TABLE 9. CROSS-METATHESIS OF VINYL-SUBSTITUTED HETEROAROMATIC COMPOUNDS (*Continued*)

Heteroaromatic Compound	Cross Partner	Conditions	Product(s) and Yield(s) (%)	Refs.
Please refer to the charts preceding the tables for the catalyst structures.				
C_{46}				
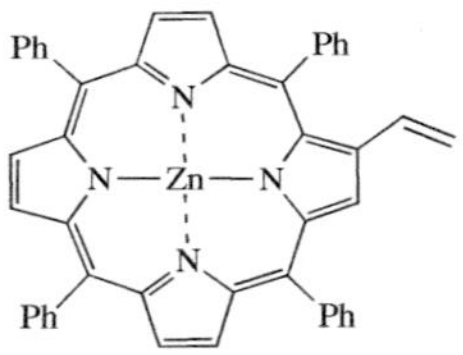 Ph, Ph, N, N—Zn—N, N, Ph, Ph	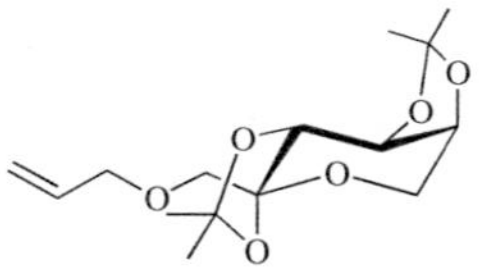 O, O, O, O, O	Gr II (0.25 eq), DCM, reflux, 4 h	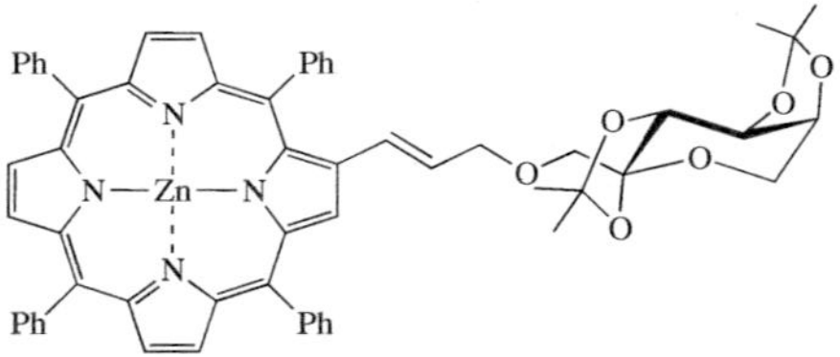 Ph, Ph, N, N—Zn—N, N, Ph, Ph, O, O, O, O, O (97) (*E*) only	641
	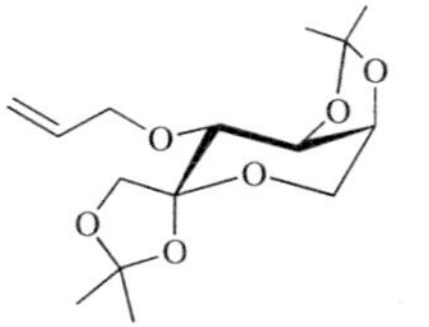 O, O, O, O, O	Gr II (0.25 eq), DCM, reflux, 4 h	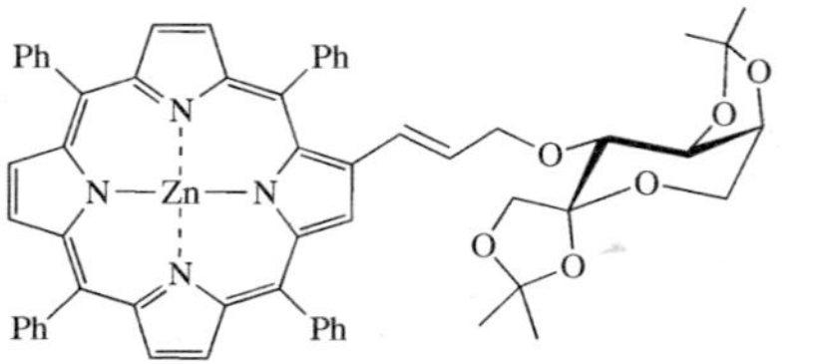 Ph, Ph, N, N—Zn—N, N, Ph, Ph, O, O, O, O, O (95) (*E*) only	641

[a] The conversion was determined by HPLC analysis.

TABLE 10. CROSS-METATHESIS OF VINYL-SUBSTITUTED AROMATIC COMPOUNDS

Aromatic Compound	Cross Partner	Conditions	Product(s) and Yield(s) (%)	Refs.

Please refer to the charts preceding the tables for the catalyst structures.

C_8

Aromatic Compound	Cross Partner	Conditions	Product(s) and Yield(s) (%)	Refs.
Ph-CH=CH$_2$ (2 eq)	CH$_2$=CHCH$_2$OBz	Catalyst (0.05 eq), toluene, 100°, 12 h	PhCH=CHCH$_2$OBz, (*E*)/(*Z*) > 95:5; Catalyst: Ru cat **77** (10), Ru cat **78** (8)	642
	CH$_2$=CHCH$_2$OPh	Catalyst (*x* eq), DCM, 35°, 24 h	PhCH=CHCH$_2$OPh, (*E*)/(*Z*) = 95:5	643
(3 eq)	CH$_2$=CHCH$_2$P(R)(Ph)·BH$_3$	Gr II (*x* eq), DCM, reflux	PhCH=CHCH$_2$P(R)(Ph)·BH$_3$	213

Catalyst	*x*	
Ru cat **118**	0.01	(94)
Ru cat **119**	0.01	(93)
Ru cat **120**	0.01	(94)
Ru cat **121**	0.0005	(94)
Ru cat **132**	0.01	(94)
Ru cat **133**	0.01	(93)
Ru cat **134**	0.01	(93)
Ru cat **135**	0.01	(92)

R	*x*	Time (h)		(*E*)/(*Z*)
MeO	0.02	12	(43)	91:9
Ph	0.06	36	(40)	(*E*) only

TABLE 10. CROSS-METATHESIS OF VINYL-SUBSTITUTED AROMATIC COMPOUNDS (*Continued*)

Aromatic Compound	Cross Partner	Conditions	Product(s) and Yield(s) (%)	Refs.
Please refer to the charts preceding the tables for the catalyst structures.				
C_{8-9}				
R[2], R[1]-substituted styrene	allyl-TMS, 2 eq	Mo cat **1** (0.02 eq), PPh_3 (x eq), solvent, rt, 4 h	R[2], R[1]-substituted aryl–CH=CH–CH$_2$–TMS (*E*) only	208
C_8				
Ph–CH=CH$_2$	allyl-CN, 2 eq	Gr II (0.05 eq), DCM, reflux, 2 h	Ph–CH=CH–CH$_2$–CN (10) (*E*) only	182
Ph–CH=CH$_2$, 2 eq	CH$_2$=CH–CH$_2$–(CH$_2$)$_n$–Br	Mo cat **1** (x eq), DCM, 40°, 12 h	Ph–CH=CH–CH$_2$–(CH$_2$)$_n$–Br **I** (*E*) only + Br–CH$_2$CH$_2$–CH=CH–CH$_2$–Br **II**	

R^1	R^2	x	Solvent	
H	H	—	DME	(85)
H	H	0.04	Et_2O	(92)
H	Me	—	DME	(83)
H	MeO	—	DME	(61)
O_2N	H	—	DME	(40)

n	x	**I**	**II**	Refs.
1	0.01	(50)	(42)	645
2	0.01	(90)	(4)	645
2	0.05	(90)	(0)	36

Gr II (0.1 eq), DCM, reflux, 18 h

R		
H	(51)	644
O_2N	(40)	

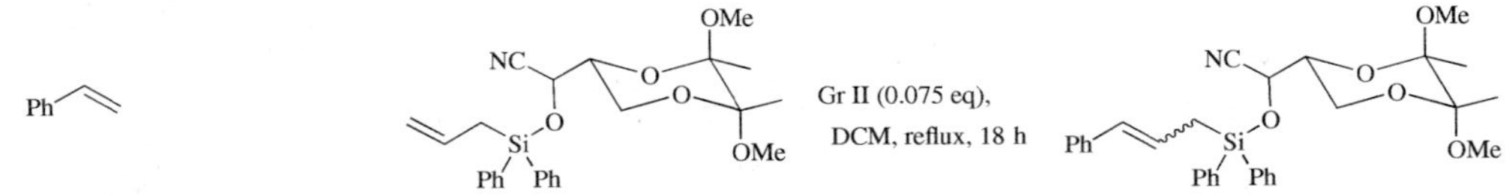

Gr II (0.075 eq), DCM, reflux, 18 h

(51) 644

C_{8-9}

2 eq

Catalyst (x eq), DCM

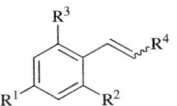

R^1	R^2	R^3	R^4	R^5	Catalyst	x	Temp (°)	Time (h)		(*E*)/(*Z*)	
H	H	H	H	AcO	Ru cat **158**	0.02	25	2	(84)	>99:1	590
H	H	H	H	AcO	Ru cat **157**	0.005	25	2	(92)	>99:1	590
H	H	H	H	AcO	Gr II	0.025	20	3	(81)	98:2	565
H	H	H	H	AcO	Hov II	0.025	20	3	(80)	98:2	565
H	H	H	H	AcO	Ru cat **155**	0.025	20	3	(76)	97:3	565
H	H	H	H	Cl	Gr II	0.05	reflux	12	(93)	>95:5	145
Br	H	H	H	AcO	Gr II	0.05	reflux	12	(93)	>95:5	145
MeO	H	H	H	AcO	Gr II	0.05	reflux	12	(83)	>95:5	145
H	TBSO	H	Me	AcO	Gr II	0.05	reflux	12	(94)	>95:5	145
H	F	H	H	AcO	Gr II	0.05	40	12	(98)	(*E*) only	36
H	F	F	H	AcO	Gr II	0.05	40	12	(50)	(*E*) only	36

TABLE 10. CROSS-METATHESIS OF VINYL-SUBSTITUTED AROMATIC COMPOUNDS (*Continued*)

Aromatic Compound	Cross Partner	Conditions	Product(s) and Yield(s) (%)	Refs.
Please refer to the charts preceding the tables for the catalyst structures.				
C_8				
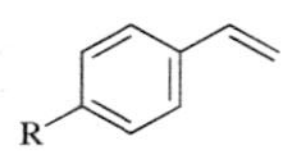 R	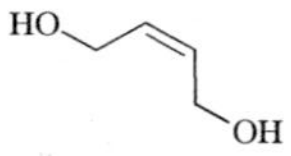 HO, OH; 2 eq	Catalyst (0.05 eq) THF, 22°, 4 h	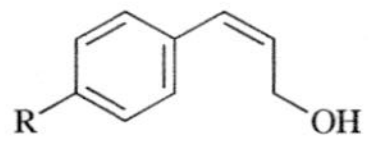 R, OH	183
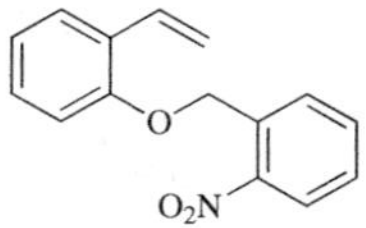 O, O_2N	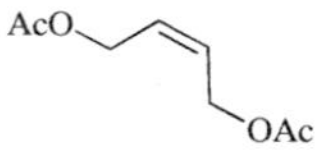 AcO, OAc; 5 eq	Ru cat **353** (0.05 eq), 1-pyrene-carboxaldehyde (7 eq), DCM, 380 nm, 90 h	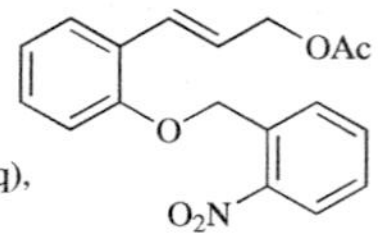 OAc, O, O_2N (78) $(E)/(Z)$ = 95:5	646
C_{8-14}				
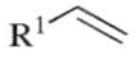 R^1; *x* eq	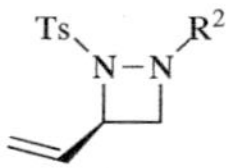 Ts, N–N, R^2	Hov II (*y* eq) DCM, reflux	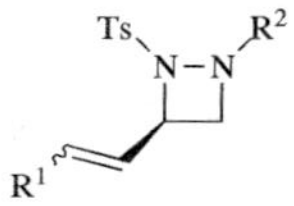 Ts, N–N, R^2, R^1	647

R	Catalyst	Conv. (%)[a]		$(E)/(Z)$
H	Ru cat **153**	57	(53)	6:94
CF_3	Ru cat **152**	60	(55)	7:93
CF_3	Ru cat **153**	66	(60)	7:93
CF_3	Ru cat **215**	19	(—)	—

R^1	x	R^2	y	Time (h)	
Ph	3	Boc	0.05	72	(83)
2-ClC_6H_4	3 + 3	Boc	0.05 + 0.05	48 + 24	(65)
Ph	3	Cbz	0.05	72	(85)
4-FC_6H_4	3 + 3	Cbz	0.05 + 0.05	48 + 24	(82)
2-ClC_6H_4	3 + 3	Cbz	0.05 + 0.05	48 + 24	(65)
3-MeC_6H_4	3	Cbz	0.05	72	(69)

C_8

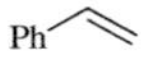

Ph (2 eq) — OH — Gr II (0.05 eq), DCM, reflux, 2 h — Ph, OH (20) (*E*) only — 182

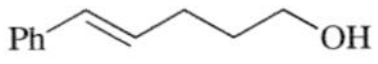

2 eq — OBn — Mo cat **1** (0.01 eq) — Ph, OBn (85) (*E*) only + BnO, OBn (2) — 645

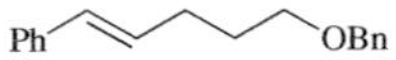

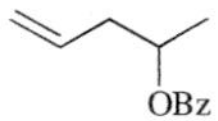

OBz (2 eq) — Gr II (0.05 eq), DCM, reflux, 12 h — Ph, OBz (81) (*E*)/(*Z*) > 95:5 — 145

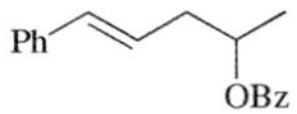

2 eq — CO_2Bn — Mo cat **1** (0.01 eq) — Ph, CO_2Bn (67) (*E*) only + BnO_2C, CO_2Bn (20) — 645

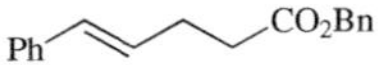

2 eq — Ph, Ph, OMe, O, B, O, OMe, Ph, Ph — Gr I (0.15 eq), DCM, reflux — Ph, Ph, Ph, OMe, O, B, O, OMe, Ph, Ph (89) — 648

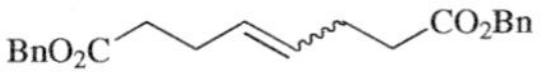

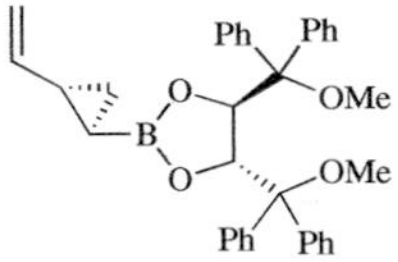

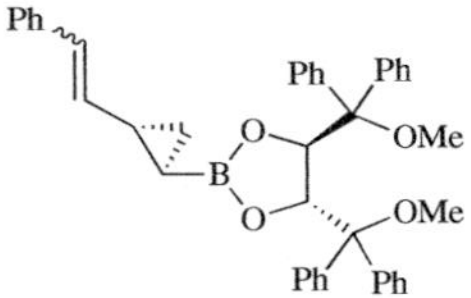

TABLE 10. CROSS-METATHESIS OF VINYL-SUBSTITUTED AROMATIC COMPOUNDS (*Continued*)

Aromatic Compound	Cross Partner	Conditions	Product(s) and Yield(s) (%)	Refs.
Please refer to the charts preceding the tables for the catalyst structures.				
C_8				
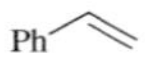 Ph (styrene), *x* eq	 OR	Catalyst (*y* eq), DCM	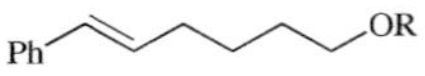 Ph … OR	
			x / R / Catalyst / *y* / Temp (°) / Time (h) / / (*E*)/(*Z*)	
			2 / TBS / Ru cat **158** / 0.02 / 25 / 5 / (83) / >99:1	590
			2 / TBS / Ru cat **157** / 0.005 / 25 / 5 / (85) / >99:1	590
			1 / THP / Gr II / 0.05 / 40 / 12 / (47) / (*E*) only	36
			4 / THP / Gr II / 0.05 / 40 / 12 / (71) / (*E*) only	36
2 eq	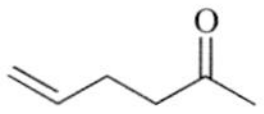 O	Mo cat **1** (*x* eq)	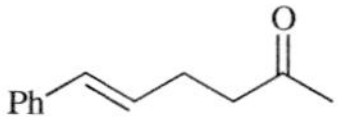 Ph … O *x* / / (*E*)/(*Z*) 0.01 / (0) / — 0.05 / (11) / (*E*) only	645
2 eq	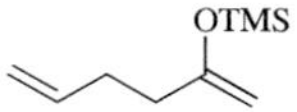 OTMS	Mo cat **1** (*x* eq)	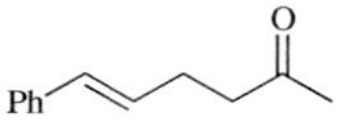 Ph … O *x* 0.01 / (22) 0.05 / (66) (*E*) only	645
2 eq	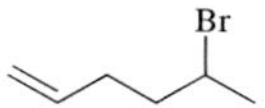 Br	Mo cat **1** (0.01 eq)	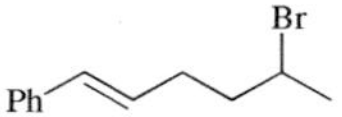 Ph … Br (84) (*E*) only 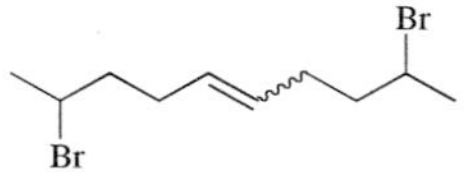 + Br … Br (8)	645
5 eq	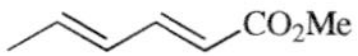 CO_2Me	Hov II (0.05 eq), DCM, rt, 48 h	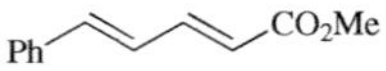 Ph … CO_2Me (35) (*E*) only	549

1.9 eq

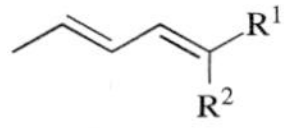

Gr II (2 x 0.05 eq), DCM, 40°, 12 + 12 h

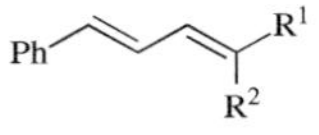

Ph ... R^1, R^2

(*E*)/(*Z*) > 99:1

R^1	R^2	
Br	NC–	(39)
NC–	Br	(66)

649

4 eq

CO_2Me, NHR

Gr II (0.05 eq), toluene, rt, 17 h

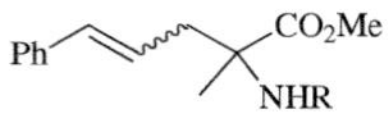

R	
Boc	(64)
Ts	(42)
Ac	(46)
OHC	(49)

457

2 eq

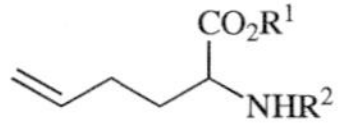

CO_2R^1, NHR^2

Gr I (0.05 eq), DCE, rt, 30 h

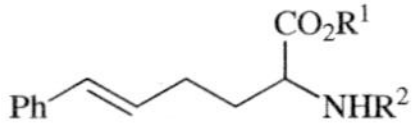

I (*E*)/(*Z*) > 91:9

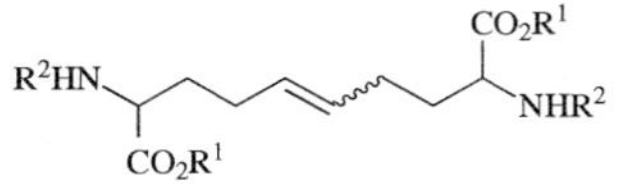

+ **II**

R^1	R^2	**I**	**II**
Me	Boc	(52)	(40)
Me	Phth	(55)	(35)
Me	Ac	(43)	(48)
Bn	Boc	(53)	(44)
t-Bu	Boc	(55)	(45)

434

TABLE 10. CROSS-METATHESIS OF VINYL-SUBSTITUTED AROMATIC COMPOUNDS (*Continued*)

Aromatic Compound	Cross Partner	Conditions	Product(s) and Yield(s) (%)	Refs.
Please refer to the charts preceding the tables for the catalyst structures.				
C_8				
Ph 10 eq	R, O, R, O	Gr I (0.05 eq), DCM, 40°, 24 h	Ph, R, O, R, O; R: Me (15), EtO (20); $(E)/(Z) = 91:9$	492
C_{8-9}				
R 1.5 eq	O, N, CO_2Et	Hov II (0.1 eq), DCM, 40°	R, O, N, CO_2Et	650
C_8				
Ph 2 eq		Mo cat **1** (0.01 eq), DCM, 23°, 1 h	Ph (86) (*E*) only + (2)	645, 36
	CO_2Et 3 eq	Gr II (0.1 eq), toluene, 50°, 16 h	Ph CO_2Et (56) (*E*) only	542

R	Time (h)	Conv. (%)[a]	(*E*)/(*Z*)
H	36	37	>95:5
F	24	80	94:6
MeO	36	26	92:8

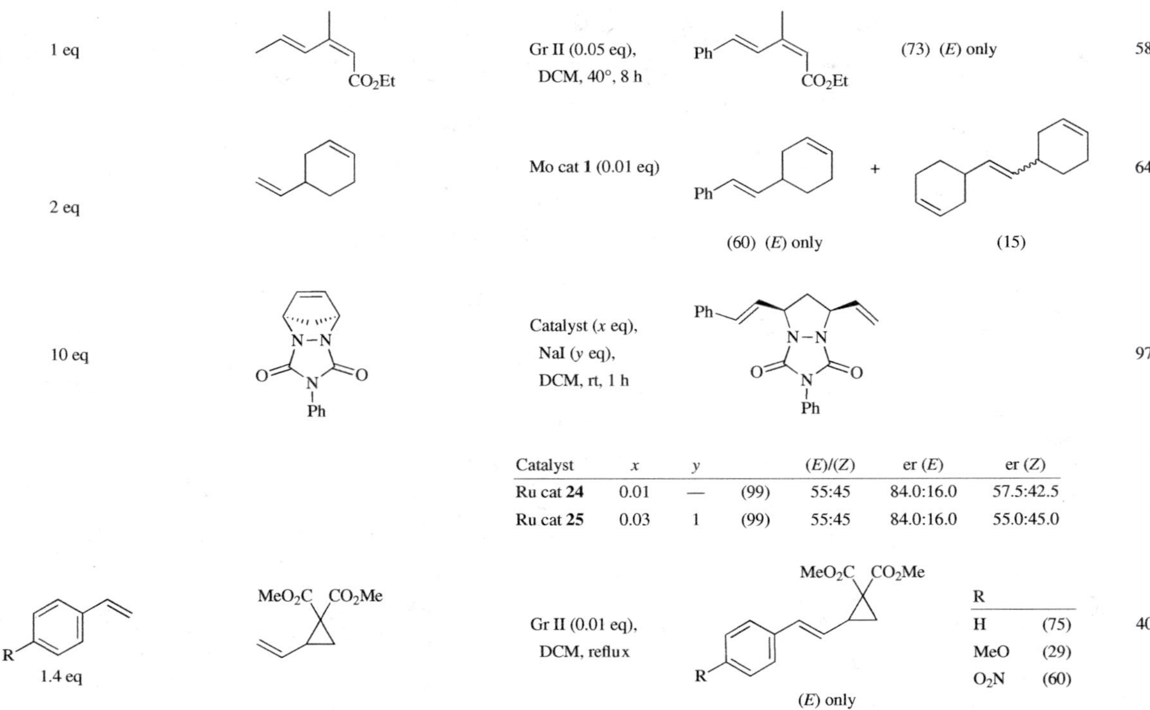

Catalyst	*x*	*y*		(*E*)/(*Z*)	er (*E*)	er (*Z*)
Ru cat **24**	0.01	—	(99)	55:45	84.0:16.0	57.5:42.5
Ru cat **25**	0.03	1	(99)	55:45	84.0:16.0	55.0:45.0

R	
H	(75)
MeO	(29)
O_2N	(60)

TABLE 10. CROSS-METATHESIS OF VINYL-SUBSTITUTED AROMATIC COMPOUNDS (*Continued*)

Aromatic Compound	Cross Partner	Conditions	Product(s) and Yield(s) (%)	Refs.

Please refer to the charts preceding the tables for the catalyst structures.

C_8

Aromatic Compound	Cross Partner	Conditions	Product(s) and Yield(s) (%)	Refs.
3-R^2-4-R^1-styrene (2 eq)	CH_2=CH-n-C_6H_{13}	Mo cat **1** (0.01 eq), DCM, rt, 1 h	**I** (*E*) only + n-C_6H_{13}CH=CH-n-C_6H_{13} **II**	645

R^1	R^2	**I**	**II**
H	H	(89)	(2)
H	O_2N	(48)	(39)
MeO	H	(88)	(2)

Aromatic Compound	Cross Partner	Conditions	Product(s) and Yield(s) (%)	Refs.
PhCH=CH_2	CH_2=CHPh	Catalyst (x eq), solvent	PhCH=CHPh	

Catalyst	x	Solvent	Temp (°)	Time (h)	Gas	Conv. (%)[b]		Refs.
Mo cat **1**	0.005	DME	—	3–24	—	—	(35)	651
Mo cat **1**	0.0002	DME	—	4	—	—	(4)	651
Mo cat **2**	0.00027	DME	—	2.5	—	—	(36)	651
Mo cat **2**	0.001	DME	—	10	—	—	(95)	651
Gr I	0.02	toluene	40	17	O_2	5	(—)	636
Gr I	0.02	toluene	40	17	N_2	19	(—)	636
Mo cat **1**	0.01	toluene	40	17	N_2	48	(—)	636

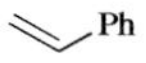

Ph

Catalyst (x eq), milling, neat

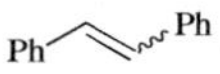

Ph ~ Ph

343

Catalyst	x	Mill	Time (h)	
Gr I	0.05	steel	1	(—)
Gr II	0.01	steel	0.5	(92)
Ru cat **67**	0.02	steel	0.5	(80)
Hov II	0.005	steel	0.5	(90)
Hov II	0.0075	steel	0.5	(95)
Hov II	0.0075	steel	0.5	(96)
Hov II	0.0075	steel	0.5	(95)
Hov II	0.0075	alumina	0.5	(96)

TABLE 10. CROSS-METATHESIS OF VINYL-SUBSTITUTED AROMATIC COMPOUNDS (*Continued*)

Aromatic Compound	Cross Partner	Conditions	Product(s) and Yield(s) (%)	Refs.

Please refer to the charts preceding the tables for the catalyst structures.

C_{8-9}

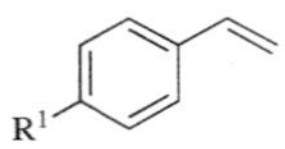

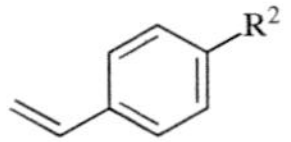

R^1 (4-R¹-styrene) + R^2 (4-R²-styrene); Catalyst (x eq), solvent 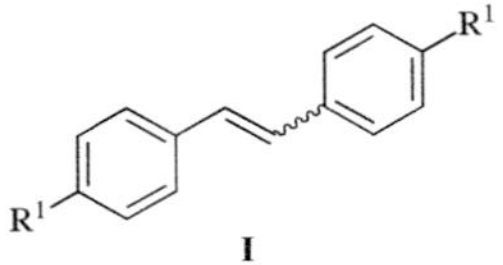 → I 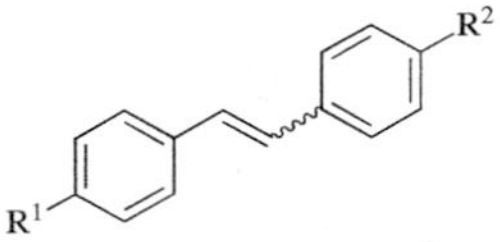 + II 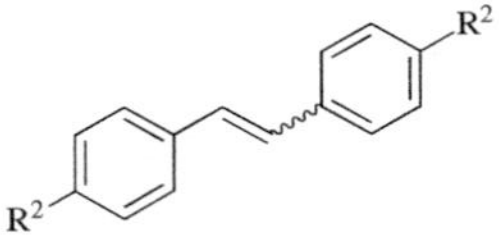 + III

R^1	R^2	Catalyst	x	Solvent	Temp (°)	Time (h)		I/II/III	Refs.
H	H	Ru cat **35**	0.05	DCM	40	2	(98)	—	652
Cl	Cl	Ru cat **35**	0.05	DCM	40	2	(95)	—	652
CF_3	CF_3	Ru cat **35**	0.05	DCM	40	2	(91)	—	652
AcO	AcO	Ru cat **35**	0.05	DCM	40	2	(98)	—	652
MeO	MeO	Ru cat **35**	0.05	DCM	40	2	(94)	—	652
H	Cl	Ru cat **35**	0.05	DCM	40	2	(88)	31:33:36	652
H	MeO	Ru cat **35**	0.05	DCM	40	2	(95)	30:45:25	652
MeO	AcO	Ru cat **35**	0.05	DCM	40	2	(77)	26:36:38	652
Cl	AcO	Ru cat **35**	0.05	DCM	40	2	(91)	31:28:41	652
HO	MeO	Gr II	0.05	toluene	reflux	0.5	(100)	22:65:13	653
MeO	Me	Gr II	0.02	DCM	40	1.5	(75)	7:81:11	143
MeO	AcO	Gr II	0.02	DCM	40	1.5	(60)	20:62:18	143
AcO	Me	Gr II	0.02	DCM	40	1.5	(65)	5:77:17	143

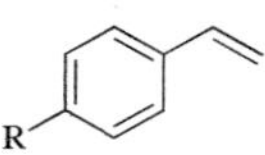

Catalyst (0.05 eq),
DCM, rt, 5 h

654

R	Catalyst	Conv. (%)[a]	R	Cat	Conv. (%)[a]
H	Gr I	12	H	Ru cat **37**	57
Me	Gr I	15	Me	Ru cat **37**	73
Cl	Gr I	5	Cl	Ru cat **37**	51
MeO	Gr I	16	MeO	Ru cat **37**	24
H	Ru cat **4**	4	H	Gr II	63
Me	Ru cat **4**	8	Me	Gr II	88
Cl	Ru cat **4**	6	Cl	Gr II	51
MeO	Ru cat **4**	8	MeO	Gr II	16
H	Ru cat **35**	30	H	Ru cat **12**	73
Me	Ru cat **35**	25	Me	Ru cat **12**	90
Cl	Ru cat **35**	38	Cl	Ru cat **12**	80
MeO	Ru cat **35**	14	MeO	Ru cat **12**	44

TABLE 10. CROSS-METATHESIS OF VINYL-SUBSTITUTED AROMATIC COMPOUNDS (*Continued*)

Aromatic Compound	Cross Partner	Conditions	Product(s) and Yield(s) (%)	Refs.

Please refer to the charts preceding the tables for the catalyst structures.

C_8

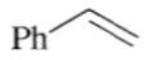

Ph (10 eq); cross partner RO, OR; Catalyst (x eq), NaI (y eq), DCM, rt, 1 h; product Ph, RO, OR; Refs. 97

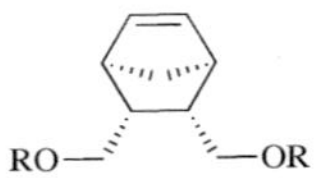

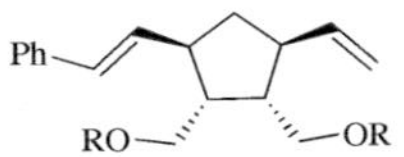

R	Catalyst	x	y		$(E)/(Z)$	er (E)	er (Z)
H	Ru cat **24**	0.01	—	(10)	—	66.5:33.5	64.5:35.5
Ac	Ru cat **24**	0.01	—	(99)	58:42	80.0:20.0	55.0:45.0
Ac	Ru cat **25**	0.03	1	(78)	58:42	86.0:14.0	70.0:30.0

5 eq; cross partner RO, OR; Ru cat **240** (0.01 eq), DCM, –10°, 12 h; product Ph, RO, OR; Refs. 95

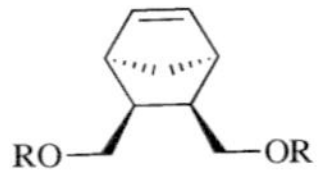

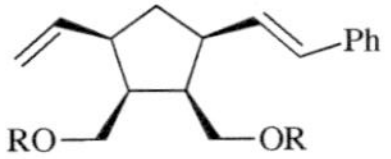

R	Conv. (%)[a]	$(E)/(Z)$	er
Ac	>98	95:5	80.0:20.0
TBS	>98	95:5	85.0:15.0

x eq

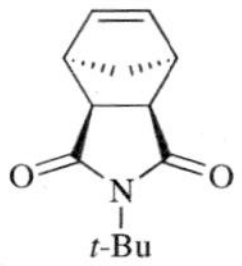

O=C–N(t-Bu)–C=O

Catalyst (y eq),
NaI (z eq),
DCM, rt, 1 h

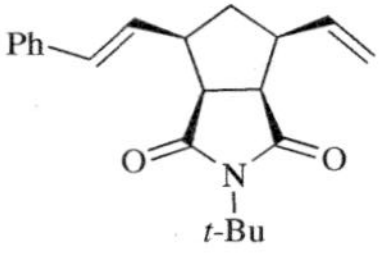

Ph

O N O
t-Bu

x	Catalyst	y	z	Temp (°)	Time (h)	Conv. (%)[a]		(E)/(Z)	er (E)	er (Z)	
10	Ru cat **24**	0.01	—	rt	1	—	(99)	58:42	78.5:21.5	66.5:33.5	97
10	Ru cat **25**	0.03	1	rt	1	—	(99)	55:45	87.5:12.5	75.0:25.0	97
5	Ru cat **240**	0.01	—	25	20	>98	(—)	96:4	91.0:9.0	—	95
5	Ru cat **240**	0.01	—	–10	72	>98	(—)	93:7	96.0:4.0	—	95
5	Ru cat **24**	0.01	—	25	1	>96	(—)	50:50	78.5:21.5	—	96
5	Ru cat **229**	0.025	—	25	72	98	(—)	92:8	95.5:4.5	—	96
5	Ru cat **32**	0.025	—	40	18	98	(—)	90:10	96.0:4.0	—	96

20 eq

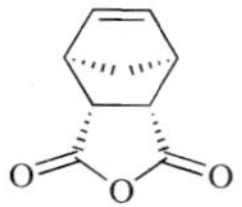

Catalyst (0.025 eq),
THF, 50°, 2 h

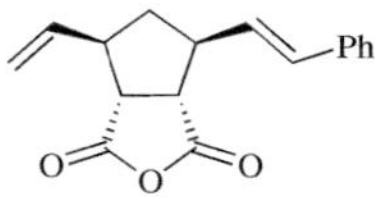

Ph

655

Catalyst		(E)/(Z)	er
Ru cat **58**	(38)	57:43	52.5:47.5
Ru cat **59**	(78)	50:50	51.0:49.0
Ru cat **60**	(9)	77:23	54.5:45.5

TABLE 10. CROSS-METATHESIS OF VINYL-SUBSTITUTED AROMATIC COMPOUNDS (*Continued*)

Aromatic Compound	Cross Partner	Conditions	Product(s) and Yield(s) (%)	Refs.

Please refer to the charts preceding the tables for the catalyst structures.

C_8

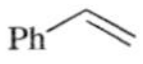

Ph (10 eq)

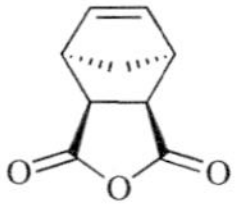

Catalyst (x eq), NaI (y eq), DCM, rt, 1 h

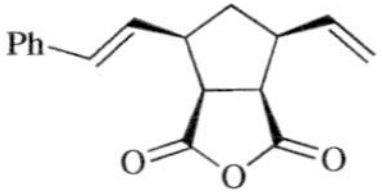

Ph

Refs. 97

Catalyst	x	y	Conv. (%)[a]	$(E)/(Z)$	er (E)	er (Z)
Ru cat **18**	0.01	—	>95	50:50	73.5:26.5	—
Ru cat **16**	0.01	—	>95	50:50	64.5:35.5	—
Ru cat **20**	0.01	—	>95	50:50	81.0:19.0	—
Ru cat **24**	0.01	—	95	50:50	88.0:12.0	52.0:48.0
Ru cat **25**	0.03	1	96	50:50	90.0:10.0	—

C_{8-9}

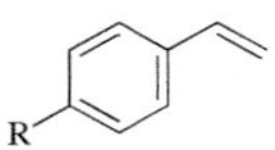

R (5 eq)

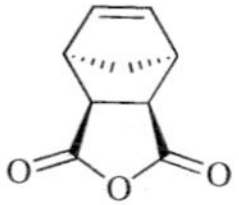

Catalyst (x eq), DCM

R

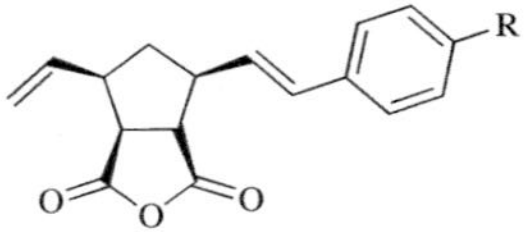

R	Catalyst	x	Temp (°)	Time (h)	Conv. (%)[a]	$(E)/(Z)$	er	Refs.
H	Ru cat **32**	0.07	40	16	78	94:6	93.0:7.0	96
H	Ru cat **229**	0.025	25	16	83	93:7	87.5:12.5	96
H	Ru cat **238**	0.01	25	1	>98	95:5	85.5:14.5	95
H	Ru cat **239**	0.01	25	1	>98	95:5	91.5:8.5	95
H	Ru cat **240**	0.01	25	1	>98	>97:3	94.0:6.0	95
H	Ru cat **240**	0.0005	25	15	>98	>97:3	94.0:6.0	95
H	Ru cat **240**	0.01	–10	12	>98	>97:3	96.5:3.5	95
H	Ru cat **24**	0.01	25	1	>98	50:50	88.0:12.0	95
MeO	Ru cat **240**	0.01	25	1	>98	93:7	90.5:9.5	95
CF_3	Ru cat **240**	0.01	25	1	>98	97:3	84.0:16.0	95
O_2N	Ru cat **240**	0.01	25	1	>98	—	86.0:14.0	95
MeO_2C	Ru cat **240**	0.01	25	1	>98	>97:3	89.5:10.5	95

C_8

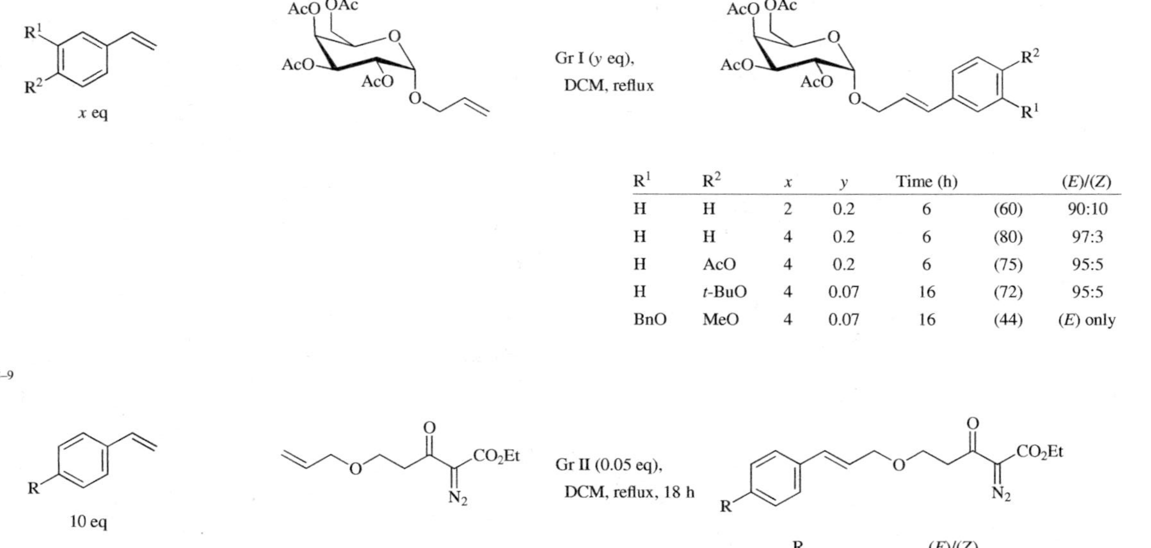

R^1	R^2	x	y	Time (h)		$(E)/(Z)$	
H	H	2	0.2	6	(60)	90:10	571
H	H	4	0.2	6	(80)	97:3	571
H	AcO	4	0.2	6	(75)	95:5	571
H	*t*-BuO	4	0.07	16	(72)	95:5	332
BnO	MeO	4	0.07	16	(44)	(*E*) only	332

$C_{8–9}$

10 eq

Gr II (0.05 eq), DCM, reflux, 18 h

569

R		$(E)/(Z)$
H	(47)	94:6
Me	(65)	98:2
Cl	(54)	97:3

TABLE 10. CROSS-METATHESIS OF VINYL-SUBSTITUTED AROMATIC COMPOUNDS (*Continued*)

Aromatic Compound	Cross Partner	Conditions	Product(s) and Yield(s) (%)	Refs.
Please refer to the charts preceding the tables for the catalyst structures.				
C_{8-9}				
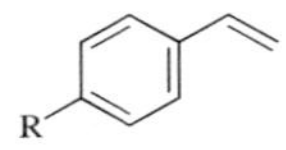 R, 4 eq	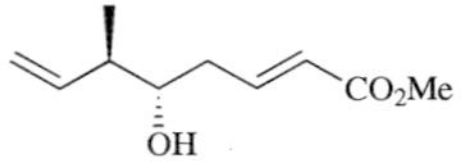 OH, CO_2Me	Hov II (0.05 eq), DCM, reflux, 2 h	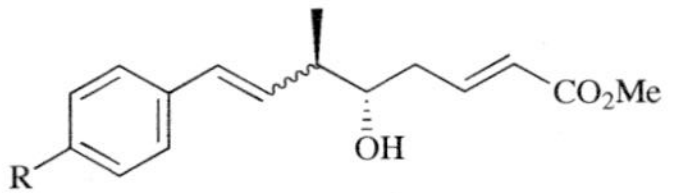 R, OH, CO_2Me R: H (30); CF_3 (23); MeO (26); NC– (8)	656
C_8				
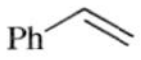 Ph, 3 eq	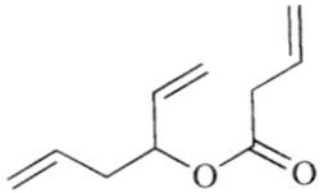 O, O	Gr II (0.025 eq), DCM, reflux, 16 h	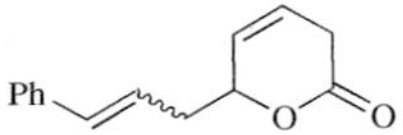 Ph, O, O (62)	394
1.5 eq	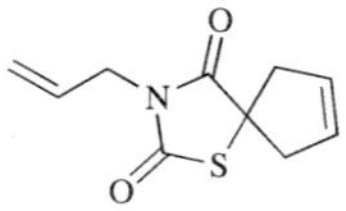 O, N, S, O	Hov II (0.03 eq), DCM, 40°, 12 h	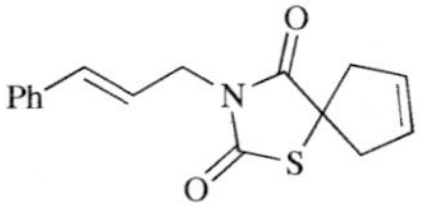 Ph, O, N, S, O (53) (*E*)/(*Z*) > 99:1	657

Olefin	Substrate	Conditions	Product(s) and Yield(s) (%)	Refs.
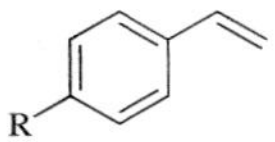 R-C6H4-CH=CH2, 10 eq	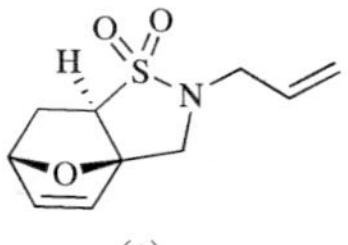 (±)	Gr II (0.1 eq), DCM, 45°, 3 h	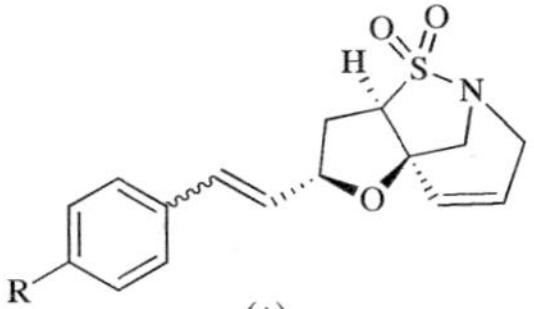 (±); R: H (81), Br (80)	359
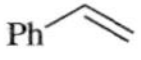 Ph-CH=CH2	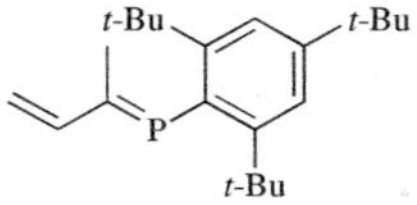 t-Bu, t-Bu, t-Bu, P	Gr II (1 eq), DCM, 35°, 48 h	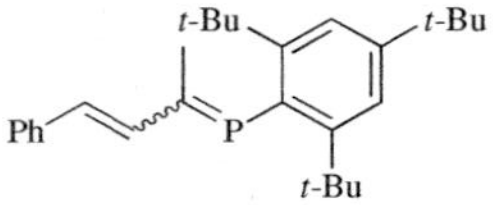 t-Bu, t-Bu, t-Bu, P, Ph (23)	658
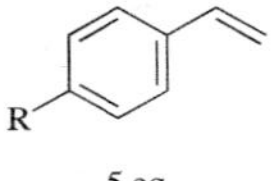 R-C6H4-CH=CH2, 5 eq	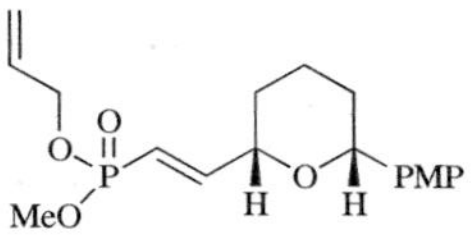 MeO, O, P, H, H, PMP	Gr II (0.1 eq), CuI (0.1 eq), DCM, reflux, 2 h	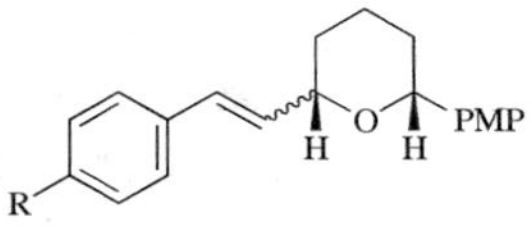 **I** + **II** (~20); R / **I**: H (82), F (53), BnO (50)	220

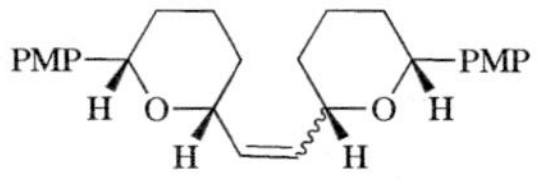

TABLE 10. CROSS-METATHESIS OF VINYL-SUBSTITUTED AROMATIC COMPOUNDS (*Continued*)

Aromatic Compound	Cross Partner	Conditions	Product(s) and Yield(s) (%)	Refs.

Please refer to the charts preceding the tables for the catalyst structures.

C_8

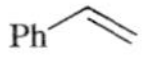

5 eq

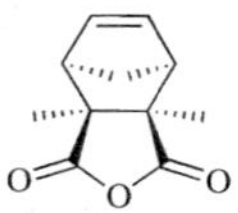

Catalyst (x eq), DCM

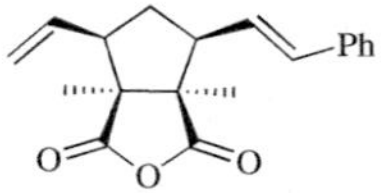

Catalyst	x	Temp (°)	Time (h)	Conv. (%)[a]	(*E*)/(*Z*)	er	Refs.
Ru cat **240**	0.01	25	3.5	>98	95:5	91.0:9.0	95
Ru cat **240**	0.01	–10	48	>98	96:4	95.0:5.0	95
Ru cat **32**	0.05	40	24	97	89:11	90.0:10.0	96
Ru cat **229**	0.05	40	24	96	92:8	85.5:14.5	96

C_{8-9}

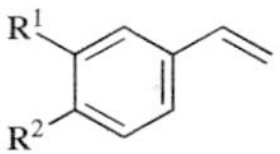

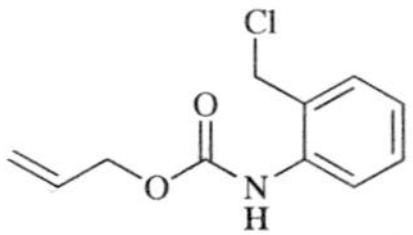

Hov II
(0.05 + 0.02 eq),
DCM, 40°, 16 h

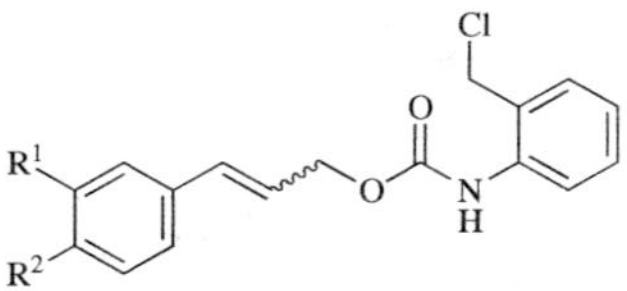

R^1	R^2	
H	H	(37)
H	CF_3	(32)
CF_3	H	(63)

618

C_8

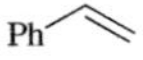

Ph 1.5 eq

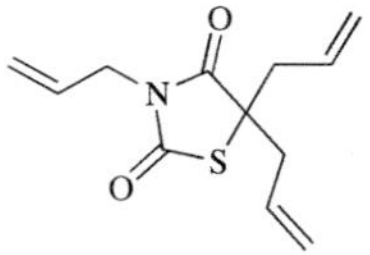

1. Hov II (0.02 eq), DCM, rt, 8 h
2. Hov II (0.03 eq), styrene, 40°, 12 h

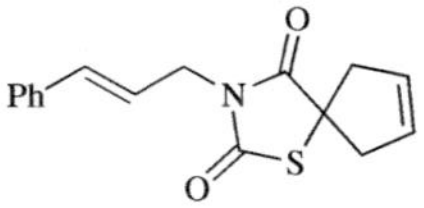

(47) (*E*)/(*Z*) > 99:1 657

C_{8-9}

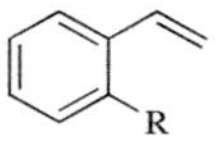

7.5 eq

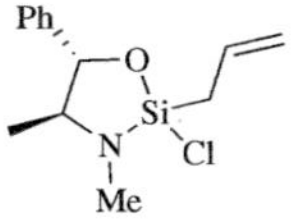

1.5 eq + 1 eq

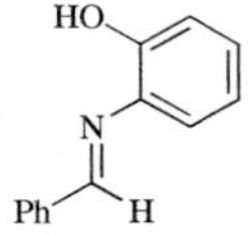

1. Gr II (0.026 eq), DCE, reflux, 3.5 h
2. Imine, reflux, 14 h

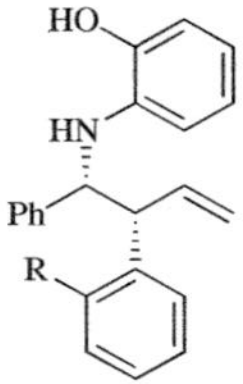

R		dr	er	
H	(68)	87.5:12.5	98.0:2.0	211
Me	(55)	83:17	97.0:3.0	

TABLE 10. CROSS-METATHESIS OF VINYL-SUBSTITUTED AROMATIC COMPOUNDS (*Continued*)

Aromatic Compound	Cross Partner	Conditions	Product(s) and Yield(s) (%)	Refs.

Please refer to the charts preceding the tables for the catalyst structures.

C_8

Aromatic Compound	Cross Partner	Conditions	Product(s) and Yield(s) (%)	Refs.
Ph (7.5 eq)	1.5 eq + 1 eq	1. Gr II (0.026 eq), DCE, reflux, 3.5 h 2. Imine, reflux, 14 h	(54) dr 89:11, er 97.0:3.0	211

C_{8-9}

Aromatic Compound	Cross Partner	Conditions	Product(s) and Yield(s) (%)	Refs.
7.5 eq	1.5 eq + 1 eq	1. Gr II (0.026 eq), solvent, reflux, time 2. Imine, reflux, 14 h		211

R^1	R^2	R^3	Solvent	Time (h)		dr	er
H	H	H	DCE	3.5	(64)	>95:5	98.0:2.0
H	H	Me	$CHCl_3$	5	(65)	>95:5	98.0:2.0
H	O_2N	H	$CHCl_3$	5	(57)	>95:5	99.0:1.0
F	H	H	$CHCl_3$	5	(65)	>95:5	98.0:2.0

C_8

Substrate	Reagents	Conditions	Product(s) and Yield(s) (%)	Refs.
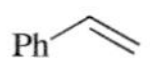 Ph (styrene), 7.5 eq	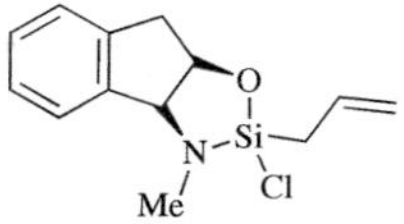 Me, N, Si, O, Cl; 1.5 eq 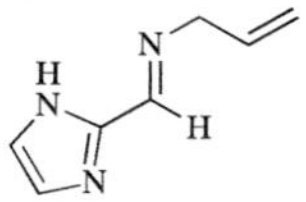 + H, N, N, N, H; 1 eq	1. Hov II (0.036 eq), benzene, reflux, 2 h 2. Imine, rt, 14 h	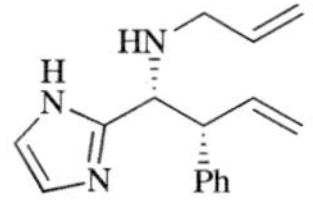 H, N, N, HN, Ph (62) dr > 95:5, er 96.0:4.0	211
7.5 eq	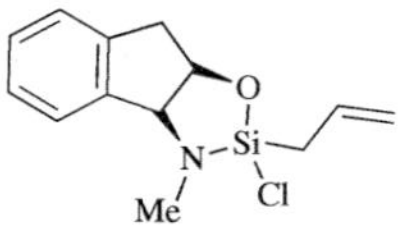 Me, N, Si, O, Cl; 2.5 eq 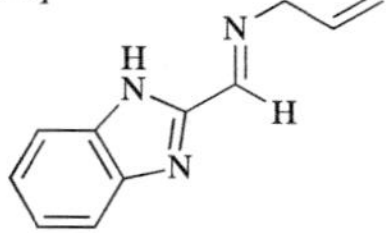 + H, N, N, N, H; 1 eq	1. Hov II (0.06 eq), toluene, reflux, 2 h 2. Imine, 23°, 18 h 3. Hov II (0.1 eq), 45°, 24 h	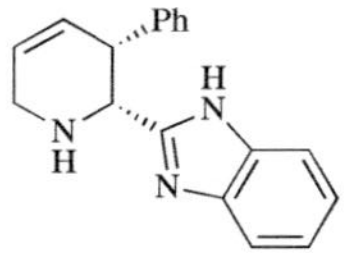 Ph, N, H, H, N, N (48) dr > 95:5, er 94.5:5.5	211

TABLE 10. CROSS-METATHESIS OF VINYL-SUBSTITUTED AROMATIC COMPOUNDS (*Continued*)

Aromatic Compound	Cross Partner	Conditions	Product(s) and Yield(s) (%)	Refs.

Please refer to the charts preceding the tables for the catalyst structures.

C_{8-14}

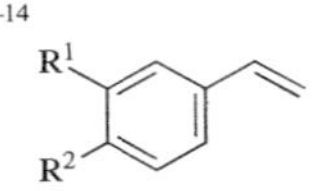

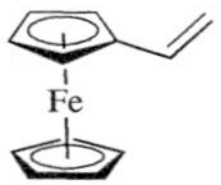

R[1], R[2]; Fe

Mo cat **1** (0.01 eq), solvent

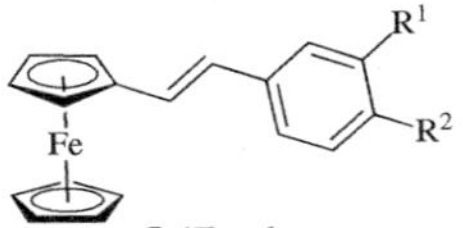

R[1], R[2], Fe

I (*E*) only

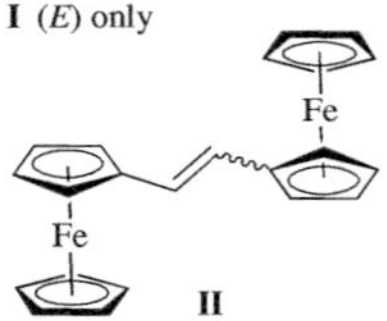

\+ Fe, Fe **II**

\+ R[1], R[2], R[2], R[1] **III**

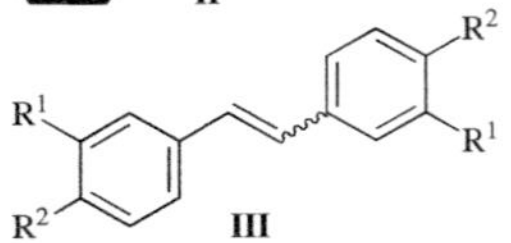

637, 659

R[1]	R[2]	Solvent	Temp (°)	Time (h)	I	I/II/III
H	H	DCM	20	17	(55)	73:8:19
H	H	$CHCl_3$	20	17	(35)	89:3:8
H	H	*n*-hexane	20	17	(41)	64:16:20
H	H	toluene	20	17	(39)	78:3:19
H	H	toluene	–15	17	(42)	92:0:8
H	H	toluene	50	17	(59)	57:11:32
H	H	toluene	rt	15	(62)	87:4:9
H	Ph	toluene	20	17	(42)	88:9:3
H	Ph	toluene	rt	15	(79)	88:9:3
H	Me	toluene	20	17	(59)	81:13:6
H	Me	toluene	rt	15	(55)	81:13:6
H	H	benzene-d_6	20	1.66	(38)	89:2:9
H	Cl	benzene-d_6	20	1.66	(40)	71:0:29
H	*t*-Bu	benzene-d_6	20	1.66	(28)	87:2:11
Me	H	benzene-d_6	20	1.66	(39)	86:3:11

x eq

Catalyst (y eq),
DCM, 24 h

R^1	x	R^2	Catalyst	y	Temp (°)		$(E)/(Z)$	
H	10	Ac	Gr II	0.08	25	(34)	(E) only	660
H	10	Ac	Gr II	0.05	40	(92)	(E) only	660
H	10	Ac	Hov II	0.16	25	(17)	(E) only	660
H	10	Ac	Hov II	0.24	40	(94)	(E) only	660
H	10	Bn	Gr II	0.13	reflux	(78)	(E) only	661
H	10	Ac	Hov II	0.24	reflux	(94)	(E) only	661
O_2N	9	Ac	Hov II	0.12	reflux	(100)	83:17	661
MeO	8.6	Ac	Hov II	0.086	reflux	(47)	(E) only	661

10 eq

Gr II (0.12 eq),
DCM, 40°, 24 h

(100) 660

TABLE 10. CROSS-METATHESIS OF VINYL-SUBSTITUTED AROMATIC COMPOUNDS (*Continued*)

Aromatic Compound	Cross Partner	Conditions	Product(s) and Yield(s) (%)	Refs.
Please refer to the charts preceding the tables for the catalyst structures.				
C_{8-9}				
R[1], R[2]; 2 eq	OH, O, Ph	Gr II (0.05 eq), DCM, 40°, 8 h	R[2], R[1], OH, O, Ph	662
			R[1] / R[2]: H / H (68); H / F (53); H / CF_3 (48); CF_3 / H (58)	
2 eq	OH, O, Ph	Gr II (0.07 eq), DCM, 40°, 8 h	R[2], R[1], OH, O, Ph	662
			R[1] / R[2]: H / H (55); H / F (45); H / CF_3 (59); CF_3 / H (75)	

C_8

Olefin	Substrate	Conditions	Product	Yield (%), (E)/(Z)	Refs.
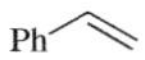 Ph (10 eq)	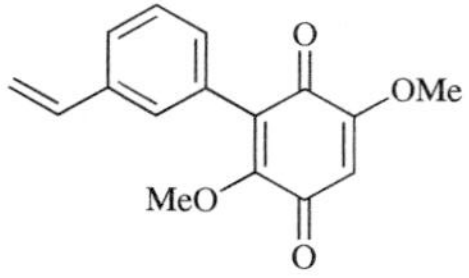 OMe, MeO, O, O	Gr II (0.1 eq), DCM, reflux, 18 h	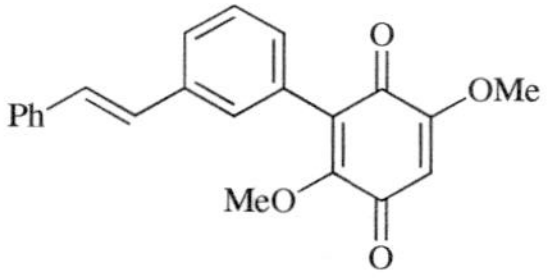 Ph, OMe, MeO, O, O	(60) (E) only	663
2 eq	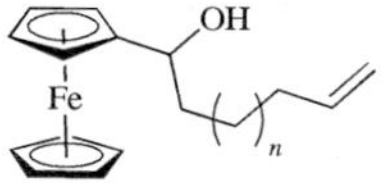 OH, Fe, n	Gr I (0.19 eq), DCM, reflux, 17–24 h	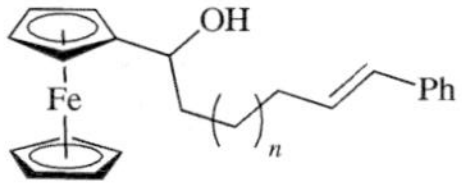 OH, Fe, n, Ph	n = 1 (56) 94:6; n = 2 (54) 94:6	587
2 eq	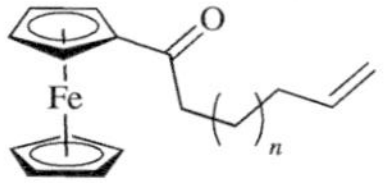 O, Fe, n	Gr I (0.19 eq), DCM, reflux, 17–24 h	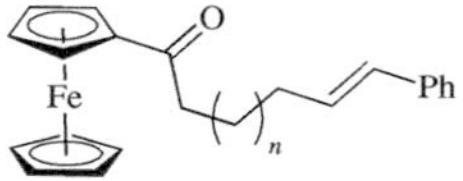 O, Fe, n, Ph	n = 1 (65) 94:6; n = 2 (52) 96:4	587
5 + 5 eq	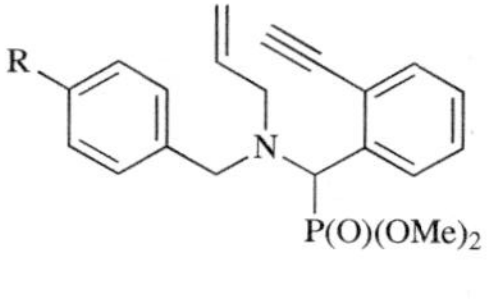 R, N, $P(O)(OMe)_2$	Gr II (0.1 eq), DCM, reflux, 2 + 4 h	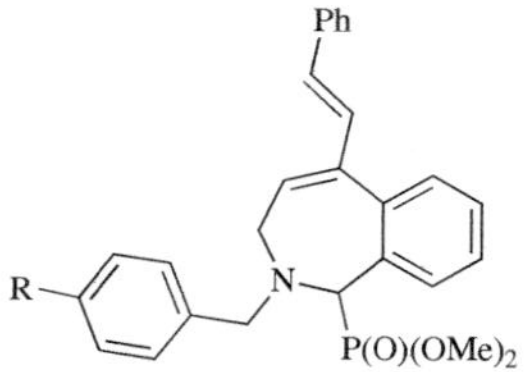 Ph, R, N, $P(O)(OMe)_2$; (E) only	R = H (78); R = Br (69)	95

TABLE 10. CROSS-METATHESIS OF VINYL-SUBSTITUTED AROMATIC COMPOUNDS (*Continued*)

Aromatic Compound	Cross Partner	Conditions	Product(s) and Yield(s) (%)	Refs.
Please refer to the charts preceding the tables for the catalyst structures.				
C_8				
Ph (vinyl) 10 eq	BnHNOC, H, Bn, N, O, O, H, H, N, O, Br	Gr II (0.1 eq), DCM, rt, 12 h	H, Bn, BnHNOC, N, O, Ph, H, H, O, O, H, N, Br (81) $(E)/(Z) > 95{:}5$	664
10 eq	BnHNOC, H, Bn, N, O, O, H, H, N, O, Br	Gr II (0.1 eq), DCM, rt, 12 h	H, Bn, BnHNOC, N, O, Ph, H, H, O, O, H, N, Br (81) $(E)/(Z) > 95{:}5$	557
10 eq	Br, O, N, H, Bn, O, N, O, H, H, O, NHBn	Gr II (0.1 eq), DCM, rt, 3 h	Br, O, H, Bn, N, N, O, O, H, H, Ph, H, O, NHBn (65) $(E)/(Z) > 95{:}5$	557

2 eq

Catalyst (0.03 eq), solvent

(*E*) only 163

Catalyst	Solvent	Temp (°)	Time (h)	
Gr II	DCM	40	2.5	(76)
Ru cat **68**	DCE	80	5.5	(55)
Ru cat **125**	DCM	40	5.5	(74)

C_{8-14}

2 eq

Gr I (0.05 eq), DCM, reflux, 18–40 h

582

R		(*E*)/(*Z*)
H	(74)	>95:5
MeO	(79)	87:13
Ph	(73)	>95:5

TABLE 10. CROSS-METATHESIS OF VINYL-SUBSTITUTED AROMATIC COMPOUNDS (*Continued*)

Aromatic Compound	Cross Partner	Conditions	Product(s) and Yield(s) (%)	Refs.

Please refer to the charts preceding the tables for the catalyst structures.

C_{8-9}

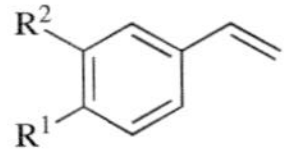

R2, R1

3 eq

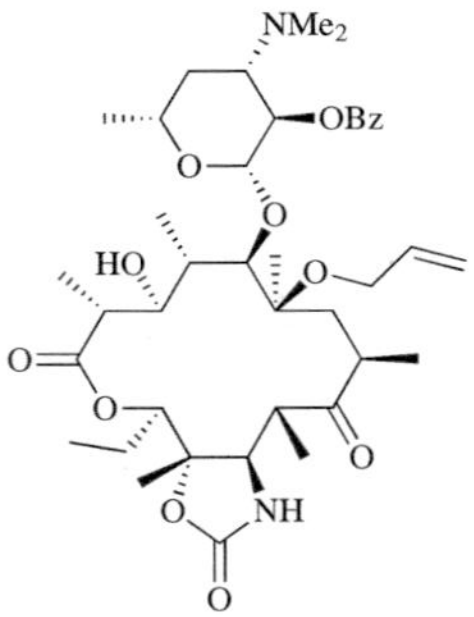

NMe2, OBz, HO, O, NH

Gr I (x eq), DCM, reflux

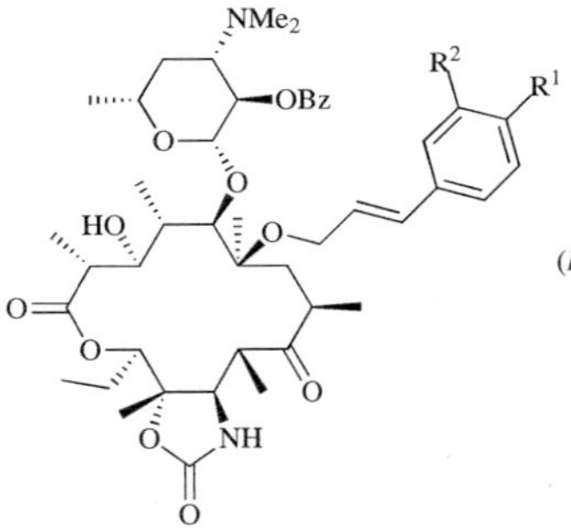

NMe2, OBz, HO, O, NH, R2, R1

$(E)/(Z) > 95:5$ 584

R^1	R^2	x	Time (h)	
H	H	0.1	26	(81)
H	Cl	0.1 + 0.1	65 + 20	(81)
Me	H	0.1 + 0.1	65 + 20	(91)

C_8

Reactant 1	Reactant 2	Conditions	Product	Yield/notes	Ref.
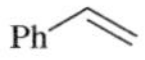 Ph, 2 eq	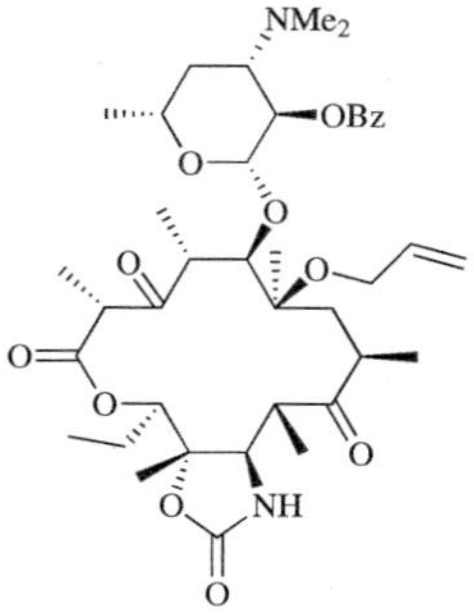 NMe$_2$, OBz, O, NH	Gr I (3 x 0.1 eq), DCM, reflux, 24 + 24 + 6 h	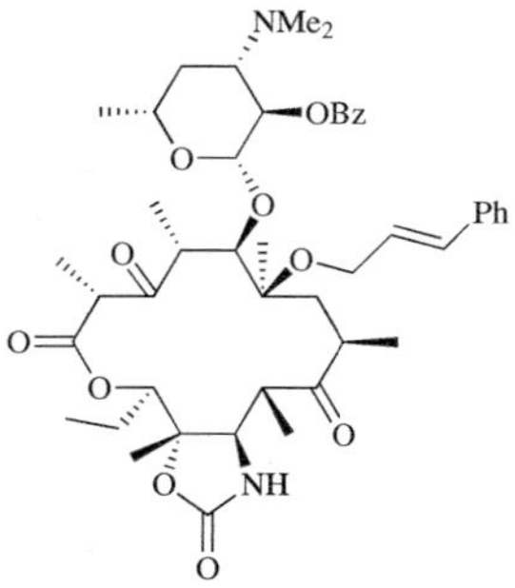 NMe$_2$, OBz, Ph, O, NH	(52) $(E)/(Z)$ > 95:5	584
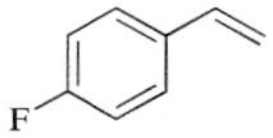 F	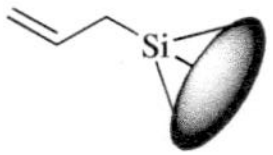 Si	Gr I, THF, 70°, 72 h	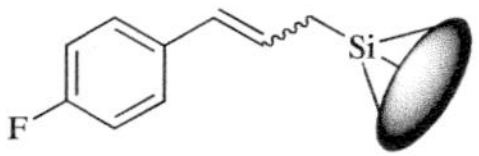 F, Si	(53)	665
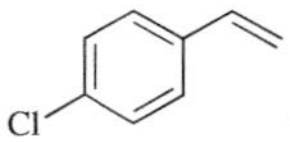 Cl	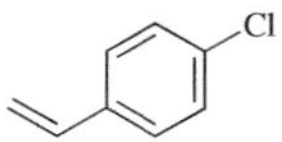 Cl	Hov I (0.05 eq), toluene, 60°, 24 h	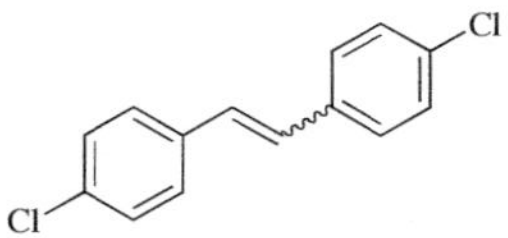 Cl, Cl	26% conv.[a]	654
2 eq	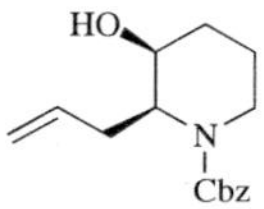 HO, N, Cbz	Gr II (0.1 eq), DCM, reflux, overnight	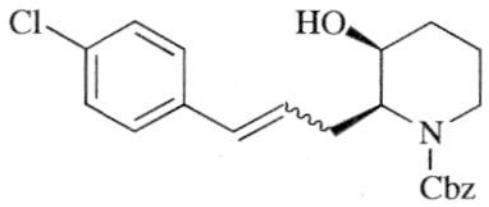 Cl, HO, N, Cbz	(55)	441

TABLE 10. CROSS-METATHESIS OF VINYL-SUBSTITUTED AROMATIC COMPOUNDS (*Continued*)

Aromatic Compound	Cross Partner	Conditions	Product(s) and Yield(s) (%)	Refs.
Please refer to the charts preceding the tables for the catalyst structures.				
C_8				
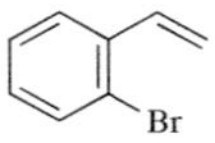 Br; *x* eq	 OAc	Gr II (0.05 eq), DCM, 40°, 12 h	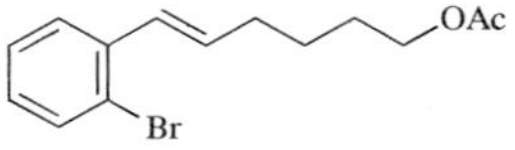 OAc, Br; (*E*) only; *x* 1 (80), 3 (98)	36
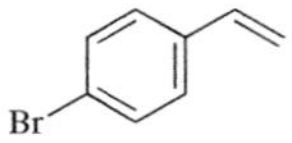 Br; 3 eq	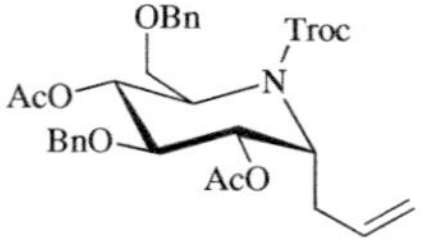 OBn, Troc, N, AcO, BnO, AcO	Gr II (0.1 eq), DCM, reflux, 16–24 h	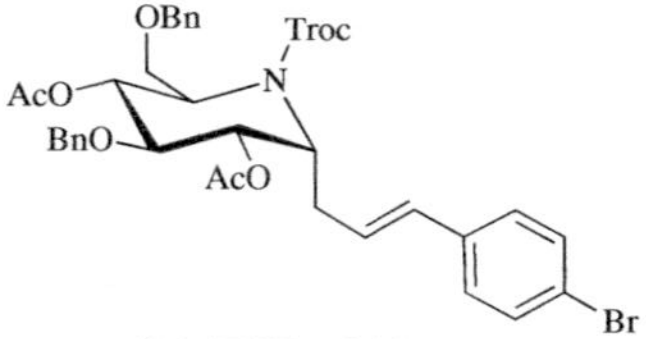 OBn, Troc, N, AcO, BnO, AcO, Br; (45) (*E*)/(*Z*) > 95:5	666
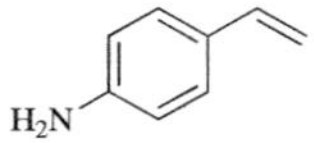 H_2N; 20 eq	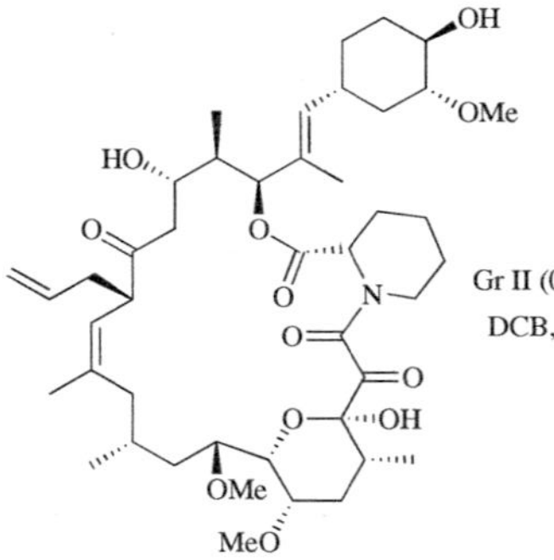 OH, OMe, HO, O, O, O, N, O, O, O, OH, OMe, MeO	Gr II (0.1 eq), DCB, MW	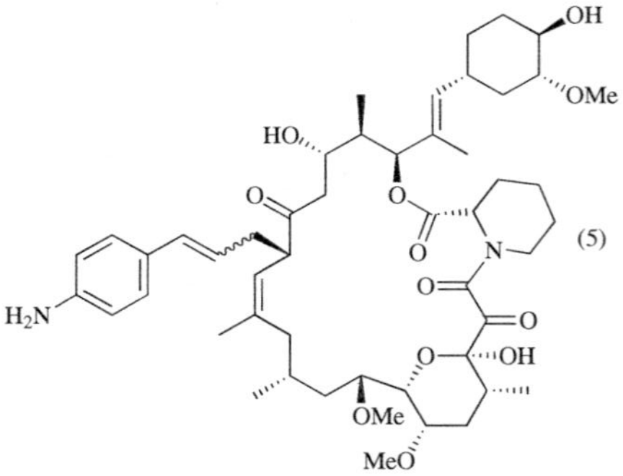 OH, OMe, HO, O, O, O, N, O, O, O, OH, OMe, MeO, H_2N; (5)	432

		Conditions	Product(s) and Yield(s) (%)	Refs.
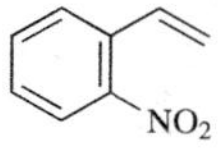 NO_2 2 eq	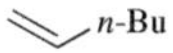 n-Bu	Mo cat **1** (0.05 eq), DCM, 40°, 12 h	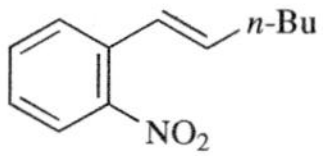 n-Bu, NO_2 (48) (E) only	36
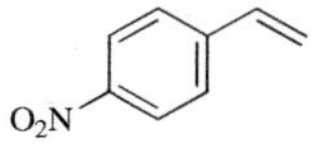 O_2N 3 eq	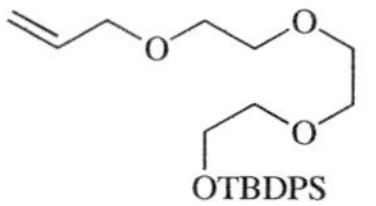 O, O, O, OTBDPS	Gr II (0.05 eq), toluene, 70°, 5 h	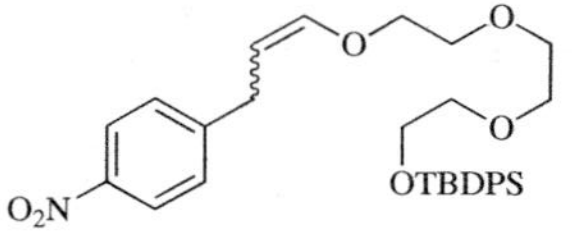 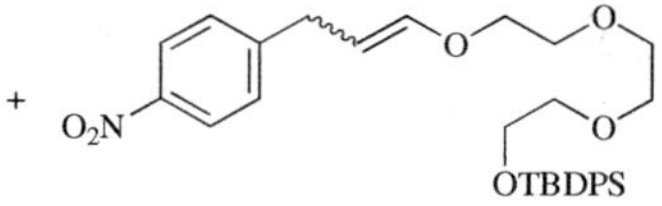 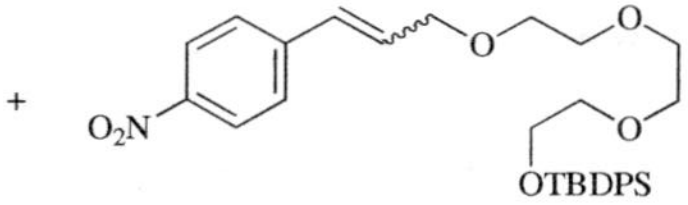 O_2N, O, O, O, OTBDPS **I** + O_2N, O, O, O, OTBDPS **II** + O_2N, O, O, O, OTBDPS **III** **I** + **II** + **III** (65)	667

TABLE 10. CROSS-METATHESIS OF VINYL-SUBSTITUTED AROMATIC COMPOUNDS (*Continued*)

Aromatic Compound	Cross Partner	Conditions	Product(s) and Yield(s) (%)	Refs.
Please refer to the charts preceding the tables for the catalyst structures.				
C_8				
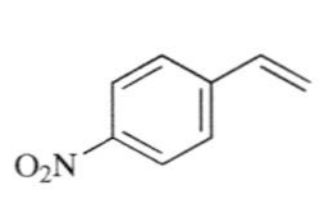 O_2N 3 eq	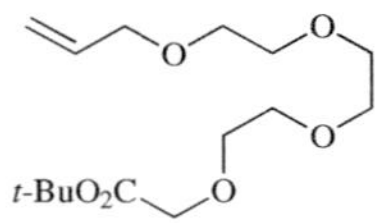 *t*-BuO_2C	Gr II (0.05 eq), DCM, reflux, 6 h	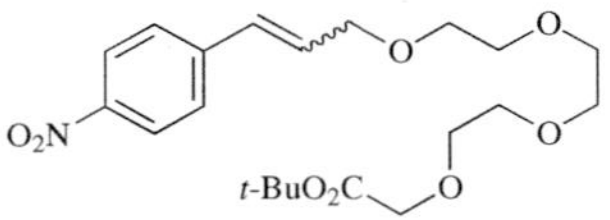 O_2N *t*-BuO_2C (80)	667
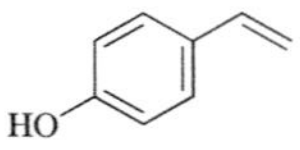 HO 2.5 eq	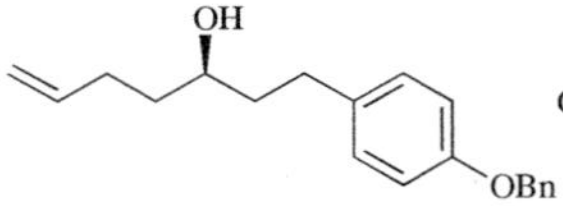 OH OBn	Gr II (0.032 eq), DCM, 50°, 4 h	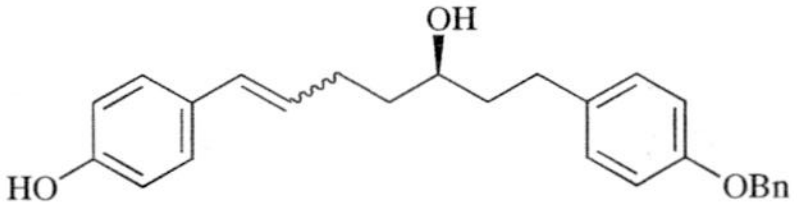 HO OH OBn (68)	668
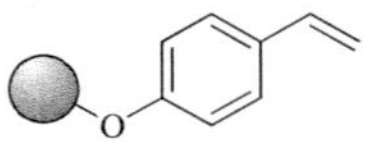 O	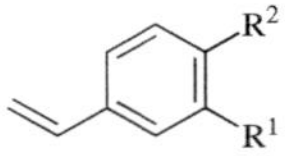 R^2 R^1 *x* eq	1. Ru cat **35** (0.02 eq), benzene, 80°, 12 h 2. 20%TFA, DCM, rt, 0.5 h	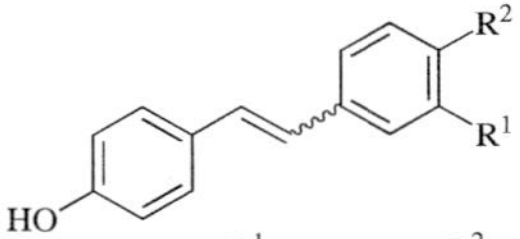 HO R^2 R^1	652

R^1	R^2	*x*	
H	H	10	(71)
H	H	4	(63)
H	Cl	10	(81)
H	CF_3	10	(68)
H	AcO	10	(72)
H	Me	10	(61)
H	MeO	10	(68)
AcO	AcO	10	(61)
$MeOCH_2$	$MeOCH_2$	10	(54)

R, 3 eq + OH, Cl, N

Hov II (0.1 eq), DCM, 50°, 24 h

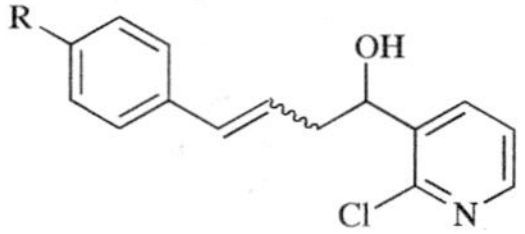

R		
MeO	(37)	603
F	(0)	

R, HO, *x* eq + OH, OH

Gr II (0.03 eq), THF, reflux, 1 h

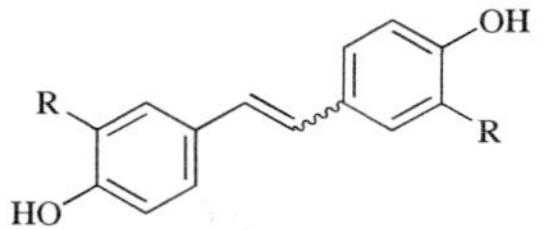

I 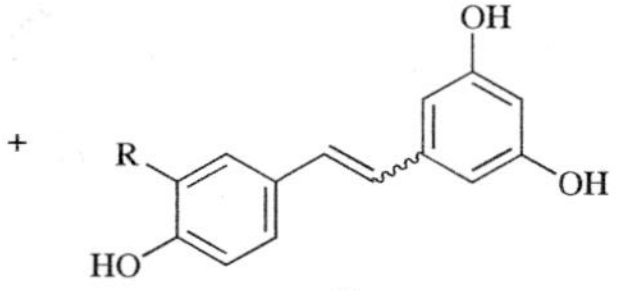 + **II** 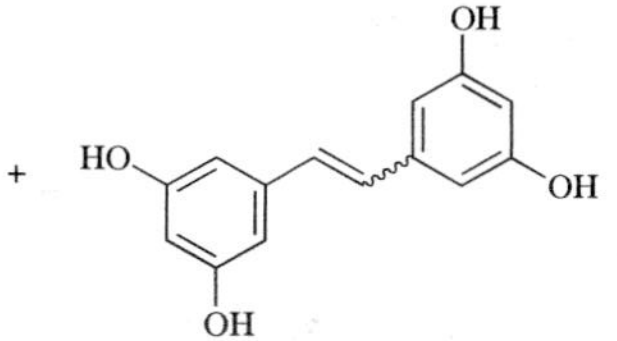 + **III**

653

R	*x*	**I + II + III**	**I/II/III**
H	1	(87)	23:60:17
H	10	(94)	0:95:5
HO	1	(86)	26:52:22

TABLE 10. CROSS-METATHESIS OF VINYL-SUBSTITUTED AROMATIC COMPOUNDS (*Continued*)

Aromatic Compound	Cross Partner	Conditions	Product(s) and Yield(s) (%)	Refs.
Please refer to the charts preceding the tables for the catalyst structures.				
C_8				
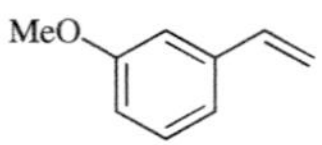 MeO	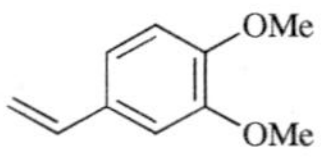 OMe, OMe 1.5 eq	Gr II (0.02 eq), DCM, 40°, 1.5 h	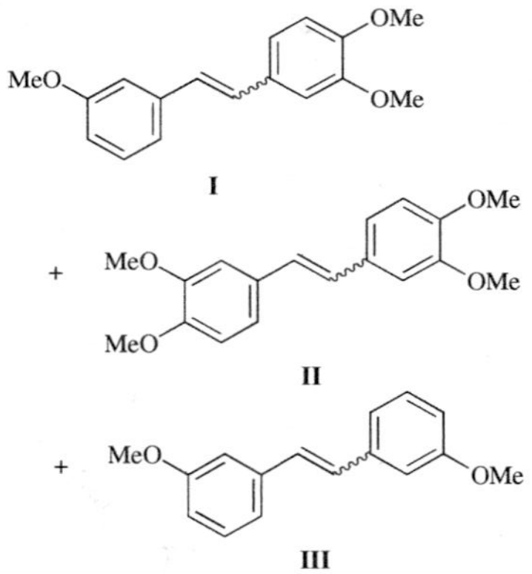 MeO, OMe, OMe **I** + MeO, MeO, OMe, OMe **II** + MeO, OMe **III** **I** + **II** + **III** (67), **I/II/III** = 69:17:14	143

OMe
OMe
1.5 eq

Gr II (0.02 eq),
DCM, 40°, 1.5 h

MeO
OMe
OMe
I

+ MeO
OMe
OMe
OMe
II

+ MeO
OMe
III

I + II + III (77), **I/II/III** = 79:13:7

143

TABLE 10. CROSS-METATHESIS OF VINYL-SUBSTITUTED AROMATIC COMPOUNDS (*Continued*)

Aromatic Compound	Cross Partner	Conditions	Product(s) and Yield(s) (%)	Refs.

Please refer to the charts preceding the tables for the catalyst structures.

C_{8-9}

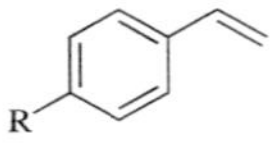

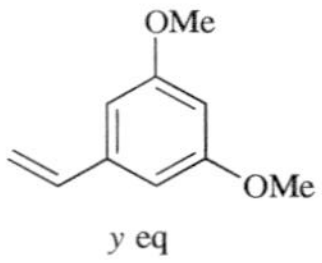

R (x eq) | OMe, OMe (y eq) | Catalyst (z eq), solvent, reflux | R, OMe, OMe I + R, R II + MeO, OMe, OMe, OMe III

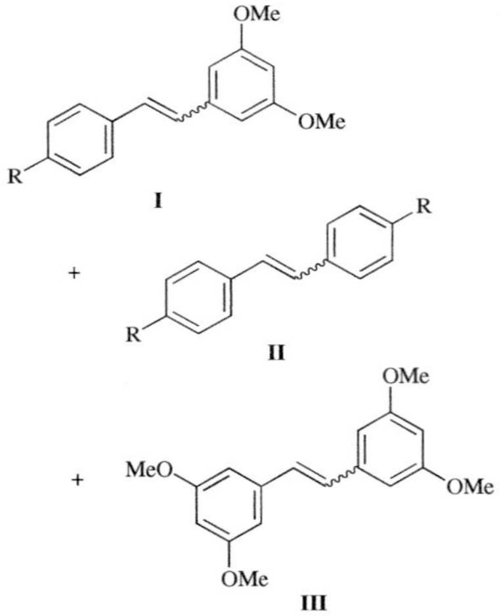

R	x	y	Catalyst	z	Solvent	Time (h)	I + II + III	I/II/III	Refs.
MeO	1	1	Gr II	0.05	toluene	1	(83)	63:19:18	653
MeO	1	1	Gr I	0.05	toluene	1	(78)	62:22:16	653
MeO	1	1	Hov II	0.05	toluene	1	(81)	52:20:28	653
MeO	2	1	Gr II	0.02	DCM	1.5	(86)	54:46:0	143
MeO	1.5	1	Gr II	0.02	DCM	1.5	(86)	47:53:0	143
MeO	1	1	Gr II	0.02	DCM	1.5	(86)	72:28:0	143
MeO	1	1.5	Gr II	0.02	DCM	1.5	(86)	88:5:7	143
MeO	1	2	Gr II	0.02	DCM	1.5	(76)	57:17:21	143
AcO	1	1.5	Gr II	0.02	DCM	1.5	(74)	60:1:39	143
Me	1	1.5	Gr II	0.02	DCM	1.5	(79)	80:9:11	143

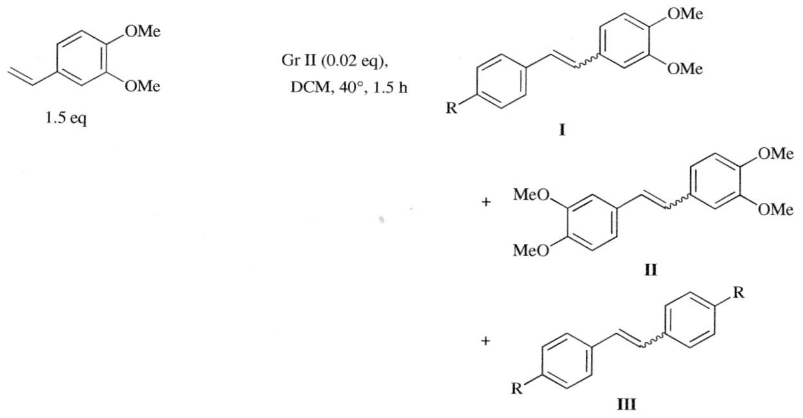

OMe
OMe
1.5 eq

Gr II (0.02 eq),
DCM, 40°, 1.5 h

R, OMe, OMe
I

+ MeO, MeO, OMe, OMe
II

+ R, R
III

143

R	**I + II + III**	**I/II/II**
MeO	(65)	64:24:11
Me	(75)	67:20:13
AcO	(67)	64:31:5

TABLE 10. CROSS-METATHESIS OF VINYL-SUBSTITUTED AROMATIC COMPOUNDS (*Continued*)

Aromatic Compound	Cross Partner	Conditions	Product(s) and Yield(s) (%)	Refs.
Please refer to the charts preceding the tables for the catalyst structures.				
C_8				
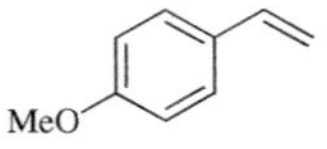 MeO; 5 eq	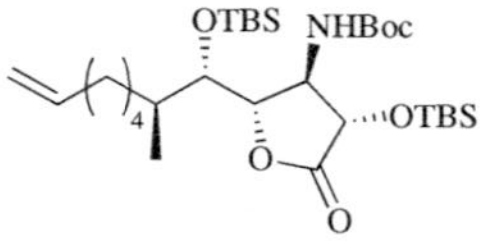 OTBS NHBoc, O, OTBS, 4, O	Gr I (3 x 0.05 eq), DCM, rt, 144 h	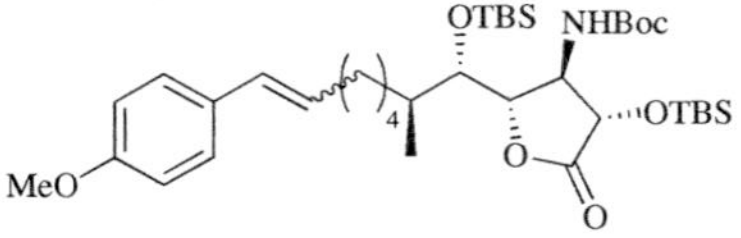 MeO, OTBS NHBoc, 4, O, OTBS, O (88)	669
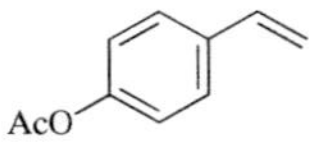 AcO; 2.5 eq	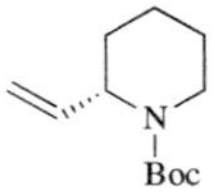 N, Boc	Gr II (0.1 eq), DCM, rt, 1.5 h	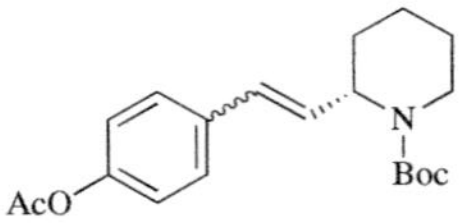 AcO, N, Boc (76)	670
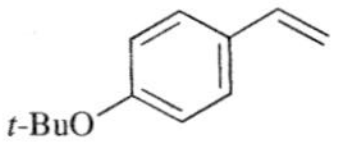 t-BuO; 3.1 eq	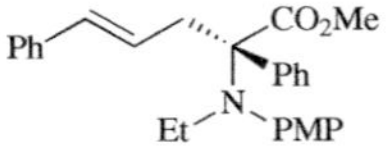 Ph, CO_2Me, Ph, Et, N, PMP; er 90.5:9.5	Hov II (0.2 eq), toluene, 90°, 20 h	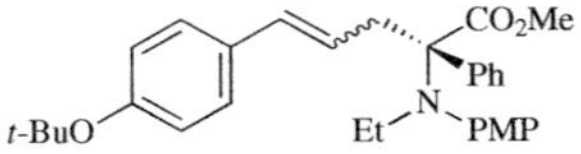 t-BuO, CO_2Me, Ph, Et, N, PMP (74)	368
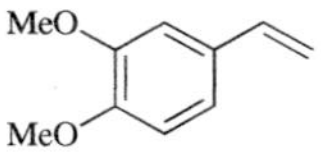 MeO, MeO	 CO_2H; 10 eq	Gr II (0.1 eq), DCM, 40°, 15 h	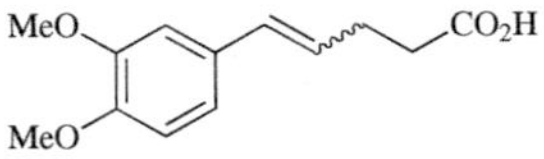 MeO, MeO, CO_2H (41)	671

C_{8-9}

I

3 eq

1. Ru cat **152** (0.01 eq), (*Z*)-butene (*x* eq), THF, 22°, 1 h
2. Ru cat **152** (0.04 eq), **I** (1 eq), THF, 100 torr, 22°, time

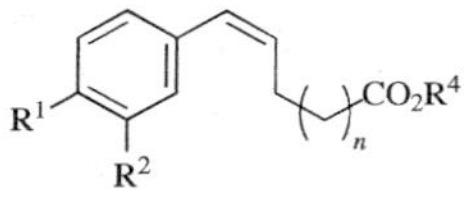

395

R^1	R^2	R^3	R^4	*n*	*x*	Time (h)		(*Z*)/(*E*)
H	H	Me	Bn	1	—	8	(48)	96:4
MeO	MeO	H	Bn	1	20	4	(50)	92:8
H	OHC	H	Me	2	20	4	(64)	96:4

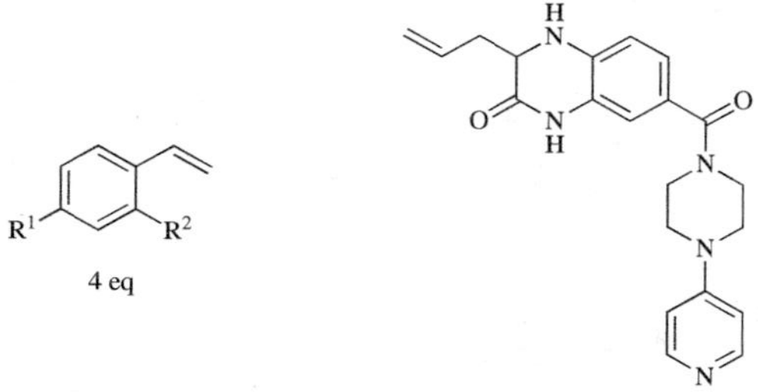

4 eq

1. Gr II (0.1 eq), DCM/MeOH (3:1), 50°
2. H_2, 9 h

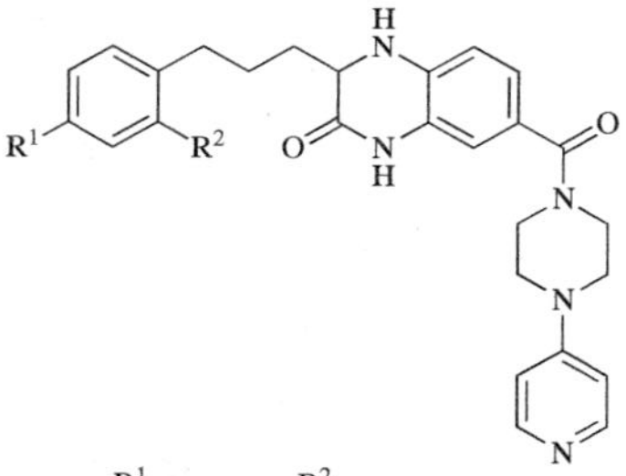

672

R^1	R^2	
H_2NO_2S	H	(49)
H	Me	(69)
H_2NCO	H	(52)

TABLE 10. CROSS-METATHESIS OF VINYL-SUBSTITUTED AROMATIC COMPOUNDS (*Continued*)

Aromatic Compound	Cross Partner	Conditions	Product(s) and Yield(s) (%)	Refs.

Please refer to the charts preceding the tables for the catalyst structures.

C_8

R^1O, R^2O; *x* eq — R^4, R^3 — Gr II (*y* eq), solvent, reflux — R^1O, R^2O, O, R^4, R^3

R^1	R^2	*x*	R^3	R^4	*y*	Solvent	Time (h)		Refs.
H	Me	2	H	Br	0.067	toluene	10	(48)	673
H	Me	—	Br	MeO	0.1	toluene	2	(53)	673
H	Me	—	Br	MOMO	0.1	toluene	2	(30)	674
Me	Bn	—	Br	MeO	0.1	DCM	4	(30)	674
Me	Bn	—	Br	MOMO	0.1	DCM	4	(26)	674

MeO, MeO — OMe, OMe; 1.5 eq — Gr II (0.02 eq), DCM, 40°, 1.5 h — MeO, MeO, OMe, OMe **I** + MeO, MeO, OMe, OMe **II** + MeO, OMe, OMe, OMe **III** — 143

I + **II** + **III** (62), **I**/**II**/**III** = 73:17:10

	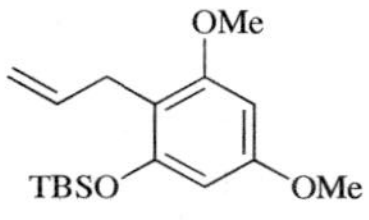 TBSO, OMe, OMe (allyl arene) 2 eq	Gr II (0.05 eq), DCM, reflux, 12 h	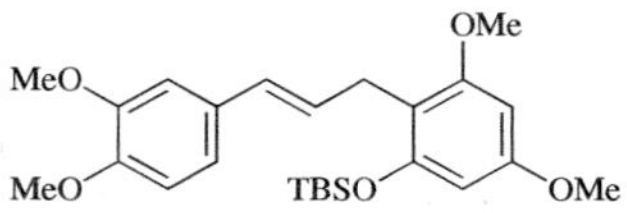 MeO, MeO, TBSO, OMe, OMe (82) $(E)/(Z) > 95:5$	145
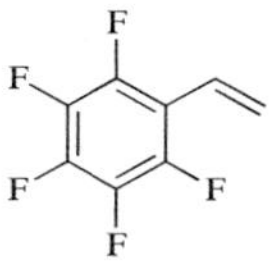 F, F, F, F, F (pentafluorostyrene) 2 eq	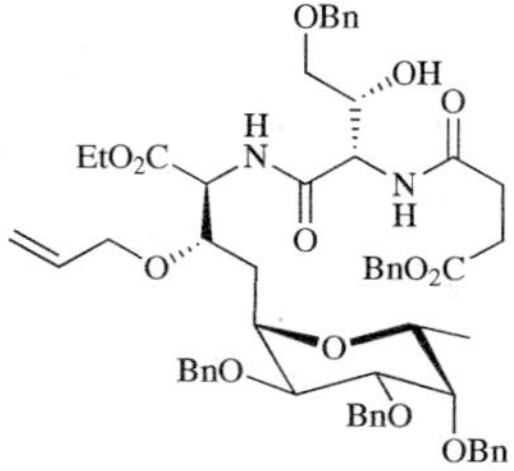 EtO_2C, H, N, OBn, OH, O, N, H, O, BnO_2C, O, BnO, O, BnO, OBn	Gr I (0.05 eq), DCM, reflux, 18–40 h	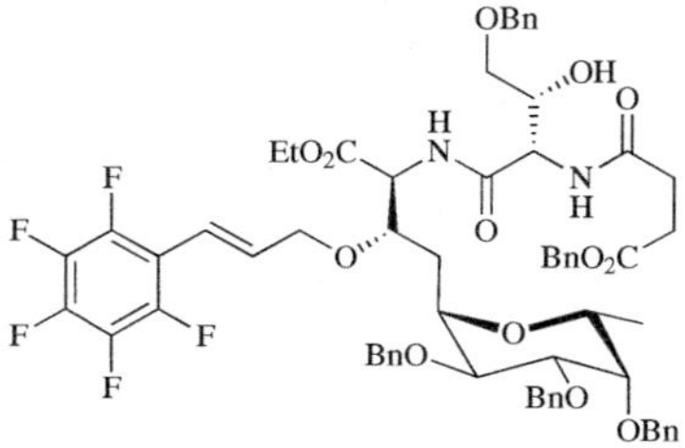 F, F, F, F, F, EtO_2C, H, N, OBn, OH, O, N, H, O, BnO_2C, O, BnO, O, BnO, OBn (17) $(E)/(Z) = 70:30$	582

TABLE 10. CROSS-METATHESIS OF VINYL-SUBSTITUTED AROMATIC COMPOUNDS (*Continued*)

Aromatic Compound	Cross Partner	Conditions	Product(s) and Yield(s) (%)	Refs.
Please refer to the charts preceding the tables for the catalyst structures.				
C_{8-14}				
R-CH=CH$_2$ (8 eq)		Hov II (0.1 + 0.1 eq), DCE, 80°, 5 + 5 h		675
			R: Ph (81); 2-FC_6H_4 (65); 3-FC_6H_4 (73); 4-FC_6H_4 (75); 2-ClC_6H_4 (64); 3-ClC_6H_4 (70); 4-ClC_6H_4 (75); 2-BrC_6H_4 (59); 3-BrC_6H_4 (66); 4-BrC_6H_4 (73); 2-MeC_6H_4 (65); 3-MeC_6H_4 (69); 4-MeC_6H_4 (71); 2-$CF_3C_6H_4$ (57); 3-$CF_3C_6H_4$ (66); 4-$CF_3C_6H_4$ (71); 2,4-$Me_2C_6H_3$ (61); 3,5-$Me_2C_6H_3$ (58); 4-*t*-BuC_6H_4 (60); 4-PhC_6H_4 (32)	
C_9				
4-$CF_3C_6H_4$CH=CH$_2$ (5 eq)	F, OBz	Hov II (0.03 eq), DCM, 100°, 16 h, sealed tube	F, OBz, CF_3 (70) (*E*)/(*Z*) > 95:5	142

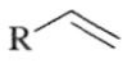

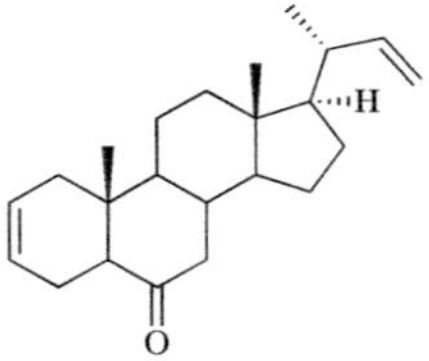

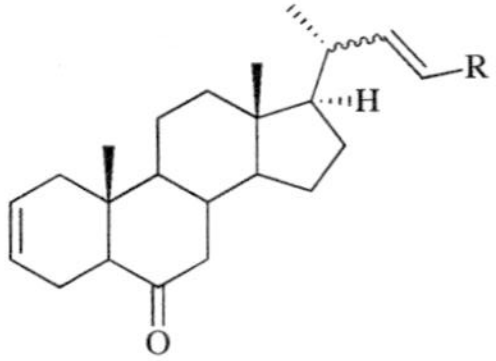

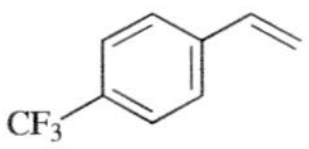

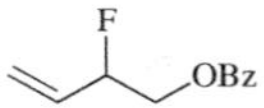

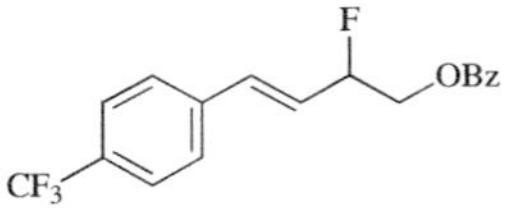

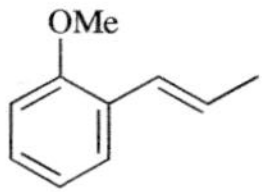 OMe	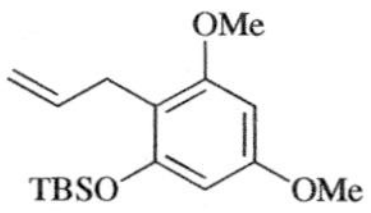 OMe, TBSO, OMe; 2 eq	Gr II (0.05 eq), DCM, reflux, 12 h	OMe, OMe, TBSO, OMe	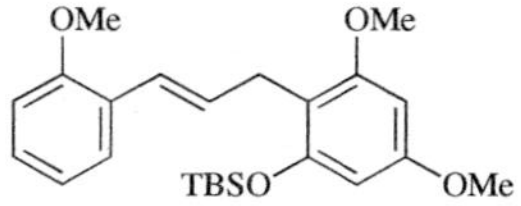 (97) (*E*)/(*Z*) > 95:5	145
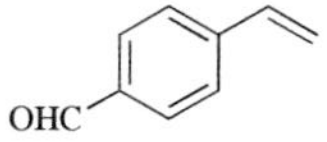 OHC; 2 eq	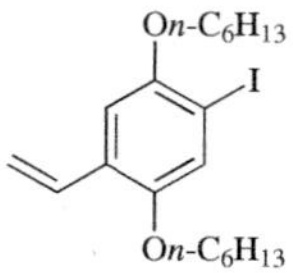 O*n*-C_6H_{13}, I, O*n*-C_6H_{13}	Gr II (0.024 eq), DCM, reflux, 40 h	O*n*-C_6H_{13}, I, OHC, O*n*-C_6H_{13}	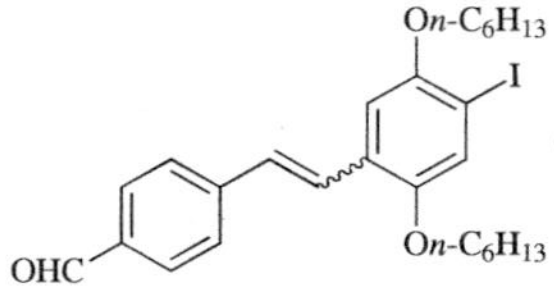 (75)	676
3 eq	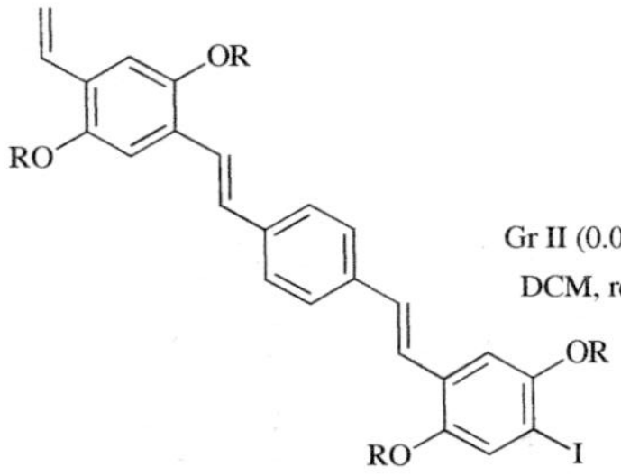 OR, RO, OR, RO, I; R = *n*-C_6H_{13}	Gr II (0.028 eq), DCM, reflux, 40 h	OHC, OR, RO, OR, RO, I	(89)	676

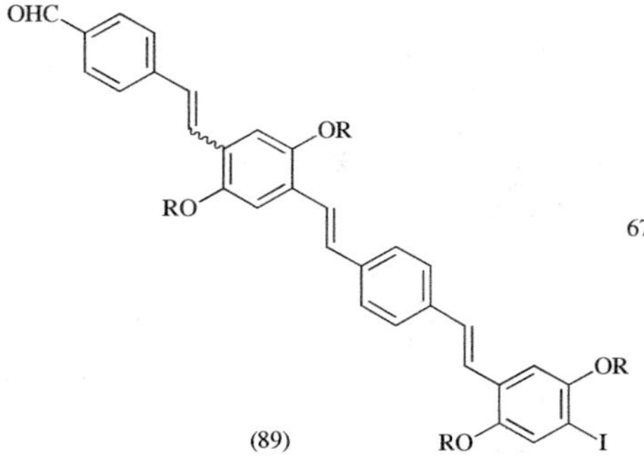

TABLE 10. CROSS-METATHESIS OF VINYL-SUBSTITUTED AROMATIC COMPOUNDS (*Continued*)

Aromatic Compound	Cross Partner	Conditions	Product(s) and Yield(s) (%)	Refs.

Please refer to the charts preceding the tables for the catalyst structures.

C_9

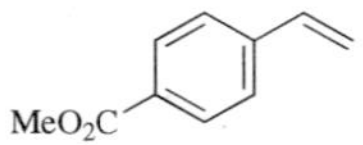

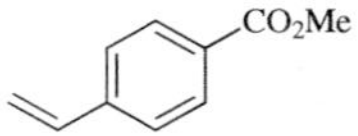

MeO_2C | CO_2Me | Catalyst (*x* eq), milling, additive (3.9 eq), THF (*y* μL) | MeO_2C CO_2Me | 343

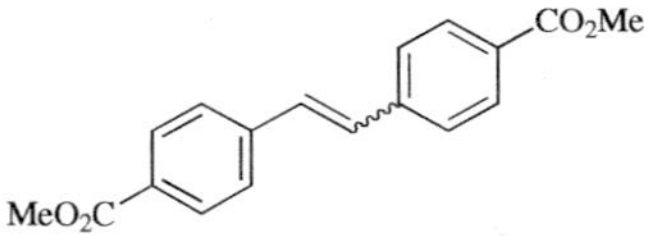

Catalyst	*x*	Mill	Additive	y	Time (h)	
Gr I	0.05	steel	—	—	1	(—)
Gr II	0.05	steel	—	—	1.5	(16)
Gr II	0.05	steel	—	—	1.5	(31)
Ru cat **67**	0.02	steel	—	—	1.5	(0)
Ru cat **67**	0.02	steel	—	—	1.5	(27)
Hov II	0.01	steel	—	—	1.5	(19)
Hov II	0.01	steel	—	—	1.5	(30)
Gr I	0.05	steel	—	50	1	(—)
Gr II	0.05	steel	—	50	1.5	(45)
Ru cat **67**	0.02	steel	—	50	1.5	(35)
Hov II	0.01	steel	—	50	1.5	(40)
Hov II	0.01	steel	—	—	1.5	(30)
Hov II	0.01	steel	NaCl	—	1.5	(93)
Hov II	0.01	steel	NaBr	—	1.5	(92)
Hov II	0.01	steel	NaI	—	1.5	(92)
Hov II	0.01	steel	KCl	—	1.5	(91)
Hov II	0.01	steel	K_2SO_4	—	1.5	(92)
Hov II	0.015	steel	—	—	1	(89)
Hov II	0.015	alumina	—	—	1.5	(91)

HO_2C CO_2H

Catalyst (x eq),
milling,
additive (3.9 eq),
solvent (y µL)

HO_2C CO_2H

343

Catalyst	x	Mill	Additive	Solvent	y	
Gr I	0.05	steel	—	—	—	(—)
Gr II	0.05	steel	—	—	—	(—)
Ru cat **67**	0.05	steel	—	—	—	(—)
Hov II	0.02	steel	—	—	—	(—)
Gr I	0.05	steel	—	THF	50	(—)
Gr II	0.05	steel	—	THF	50	(15)
Gr II	0.05	steel	—	THF	100	(37)
Ru cat **67**	0.05	steel	—	THF	50	(45)
Hov II	0.02	steel	—	THF	50	(49)
Hov II	0.02	steel	NaCl	EtOAc	75	(73)
Hov II	0.02	steel	NaBr	EtOAc	75	(70)
Hov II	0.02	steel	NaI	EtOAc	75	(71)
Hov II	0.02	steel	KCl	EtOAc	75	(74)
Hov II	0.02	steel	K_2SO_4	EtOAc	75	(71)
Hov II	0.03	steel	—	EtOAc	500	(67)
Hov II	0.03	alumina	—	EtOAc	250	(70)

TABLE 10. CROSS-METATHESIS OF VINYL-SUBSTITUTED AROMATIC COMPOUNDS (*Continued*)

Aromatic Compound	Cross Partner	Conditions	Product(s) and Yield(s) (%)	Refs.
Please refer to the charts preceding the tables for the catalyst structures.				
C_9				
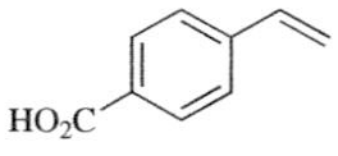 HO_2C 20 eq	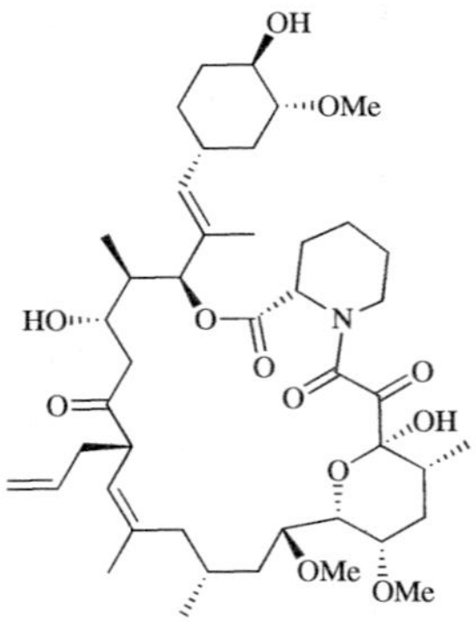 HO, OH, OMe, O, N, O, OH, O, OMe, OMe	Gr II (0.1 eq), solvent, MW	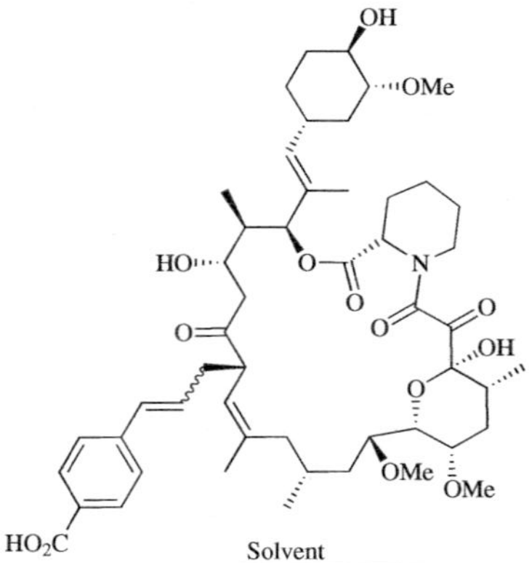 HO_2C, HO, OH, OMe, O, N, O, OH, O, OMe, OMe Solvent: DCB (79); *i*-PrOH (9)	432
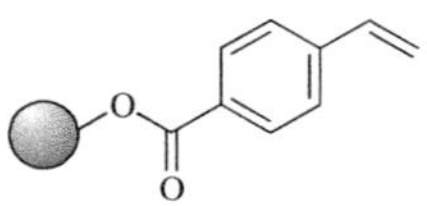 O, O	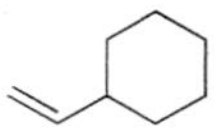 5 eq	1. Gr II (0.05 eq), toluene, 75° (MW), 25 min 2. TFA (0.1 eq), DCM, 1 h 3. CH_2N_2, DCM, 0°, 0.5 h	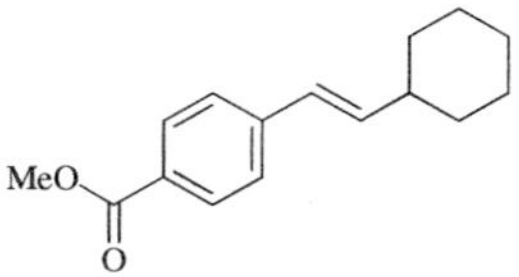 MeO, O (82) (*E*) only	418

R[1] R[2]

5 eq

1. Catalyst (0.05 eq), solvent
2. TFA (0.1 eq), DCM, 1 h
3. CH_2N_2, DCM, 0°, 0.5 h

MeO, O, R[1], R[2]

(*E*) only 418

R^1	R^2	Catalyst	Solvent	Temp (°)	Time (h)	
H	Br	Gr II	DCM	40	20	(43)
H	Br	Hov II	DCM	40	20	(43)
H	Br	Gr II	toluene	75 (MW)	0.42	(80)
AcO	H	Gr II	toluene	75 (MW)	0.42	(85)
Me	H	Gr II	DCM	40	20	(31)
Me	H	Gr II	toluene	75 (MW)	0.42	(30)
Me	H	Hov II	DCM	40	20	(32)
$ClCH_2$	H	Gr II	DCM	40	20	(80)
$ClCH_2$	H	Gr II	toluene	75 (MW)	0.42	(90)
$ClCH_2$	H	Hov II	DCM	40	20	(28)

TABLE 10. CROSS-METATHESIS OF VINYL-SUBSTITUTED AROMATIC COMPOUNDS (*Continued*)

Aromatic Compound	Cross Partner	Conditions	Product(s) and Yield(s) (%)	Refs.

Please refer to the charts preceding the tables for the catalyst structures.

C_9

Aromatic Compound	Cross Partner	Conditions	Product(s) and Yield(s) (%)	Refs.
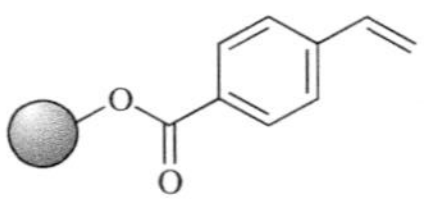	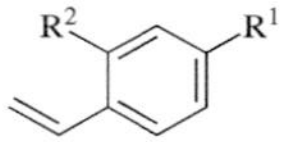 5 eq	1. Cat (0.05 eq), DCM, reflux, 20 h 2. TFA (0.1 eq), DCM, 1 h 3. CH_2N_2, DCM, 0°, 0.5 h	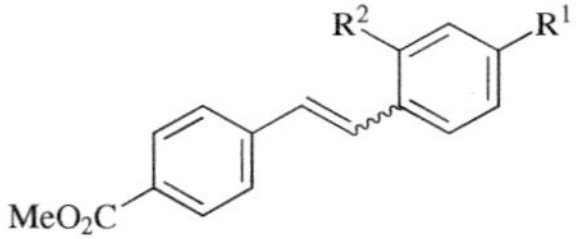 MeO_2C	677

R^1	R^2	Catalyst	
H	Br	Gr II	(10)
H	Br	Hov II	(16)
Me	H	Gr II	(31)
Me	H	Hov II	(32)
$ClCH_2$	H	Gr II	(80)

Aromatic Compound	Cross Partner	Conditions	Product(s) and Yield(s) (%)	Refs.
	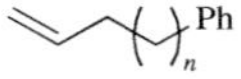 5 eq	1. Catalyst (0.05 eq), solvent 2. TFA (0.1 eq), DCM, 1 h 3. CH_2N_2, DCM, 0°, 0.5 h	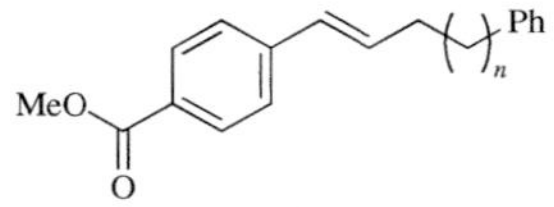 (*E*) only	418

n	Catalyst	Solvent	Temp (°)	Time (h)	
0	Gr I	DCM	40	20	(22)
0	Gr II	DCM	40	20	(86)
0	Gr II	toluene	75 (MW)	0.42	(79)
0	Hov II	DCM	40	20	(84)
1	Gr II	DCM	40	20	(57)
1	Gr II	toluene	75 (MW)	0.42	(86)

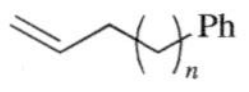

Ph, 5 eq

1. Gr II (0.05 eq), DCM, reflux, 20 h
2. TFA (0.1 eq), DCM, 1 h
3. CH_2N_2, DCM, 0°, 0.5 h

MeO_2C ... Ph

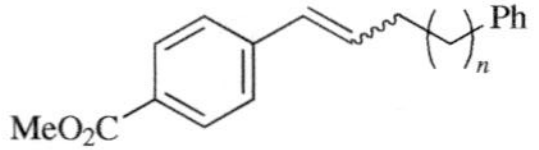

n	
0	(86)
1	(57)

677

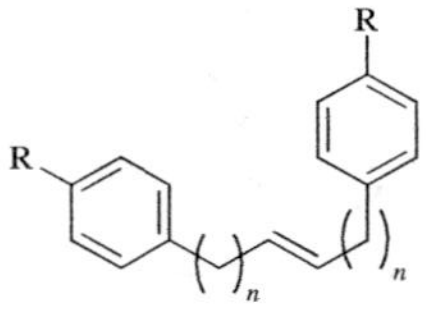

R, R, 5 eq

1. Gr II (0.05 eq), solvent
2. TFA (0.1 eq), DCM, 1 h
3. CH_2N_2, DCM, 0°, 0.5 h

MeO, O, R

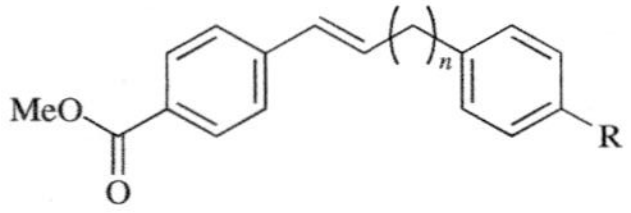

(*E*) only

418, 677

n	R	Solvent	Temp (°)	Time (h)	
0	Me	toluene	75 (MW)	0.5	(15)
0	Me	DCM	reflux	20	(0)
0	$ClCH_2$	DCM	reflux	20	(58)
1	H	DCM	reflux	20	(72)
2	H	DCM	reflux	20	(39)

TABLE 10. CROSS-METATHESIS OF VINYL-SUBSTITUTED AROMATIC COMPOUNDS (*Continued*)

Aromatic Compound	Cross Partner	Conditions	Product(s) and Yield(s) (%)	Refs.
Please refer to the charts preceding the tables for the catalyst structures.				
C_9				
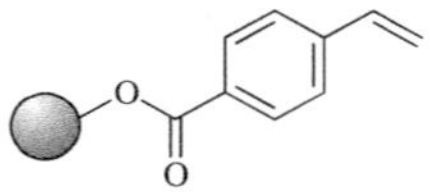	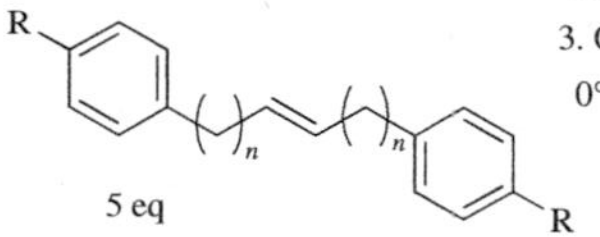 5 eq	1. Gr II (0.05 eq), DCM, reflux, 20 h 2. TFA (0.1 eq), DCM, 1 h 3. CH_2N_2, DCM, 0°, 0.5 h	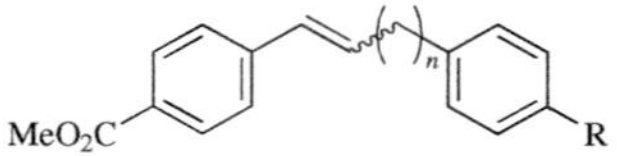 MeO_2C ... R	677
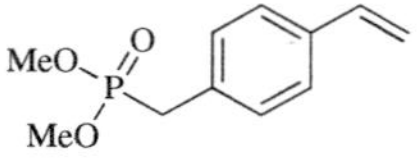 MeO, MeO, P, O *x* eq	 CO_2Me, 7	Catalyst (*y* eq), solvent, 40°	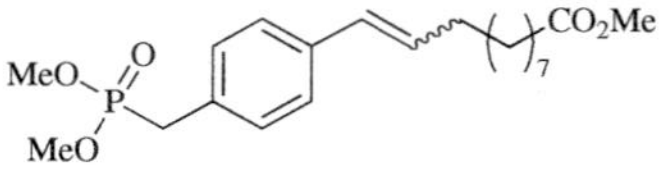 MeO, MeO, P, O, CO_2Me, 7	678

R	*n*	
Me	0	(0)
H	1	(72)
H	2	(39)
$ClCH_2$	0	(58)

x	Catalyst	*y*	Solvent	Time (h)	
2	Gr I	0.02	DCM	12	(29)
2	Gr I	0.02	—	12	(38)
1.4	Gr I	0.02	DCM	24	(44)
1	Gr I	0.02	DCM	12	(38)
1	Gr II	0.01	—	12	(43)
1	Gr II	0.02	—	12	(60)
2	Gr II	0.02	DCM	12	(41)

Substrate 1	Substrate 2	Conditions	Product(s) (yield)	Ref.
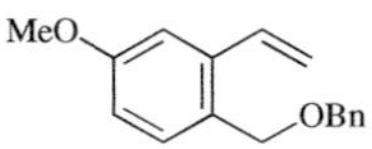 MeO, OBn	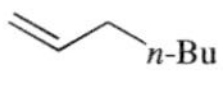 n-Bu; 2 eq	Gr II (0.1 eq), DCM, reflux, 12 h	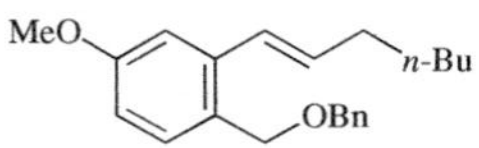 MeO, n-Bu, OBn (80) (*E*) only 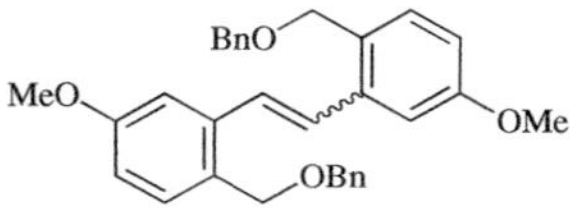 + BnO, MeO, OMe, OBn (7)	678a
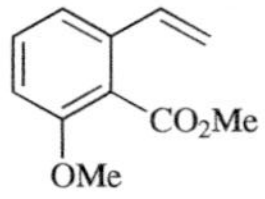 CO_2Me, OMe; 2 eq	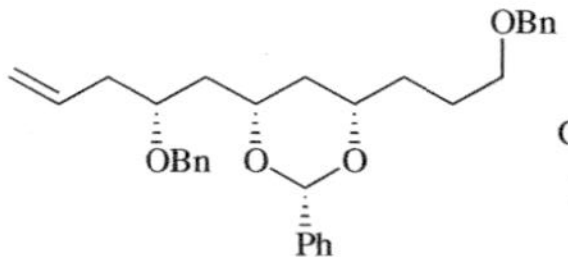 OBn, OBn, O, O, Ph	Gr II (0.05 eq), DCM, reflux, 24 h	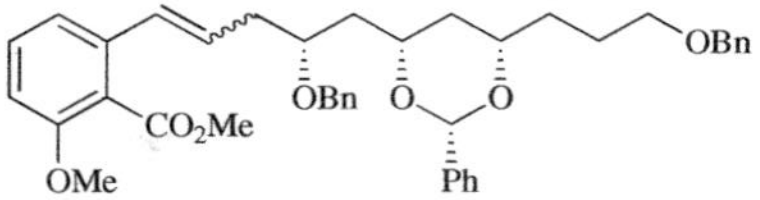 OBn, CO_2Me, OMe, OBn, O, O, Ph (86)	679
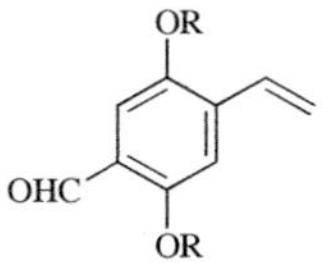 OR, OHC, OR; 4 eq; R = n-C_6H_{13}	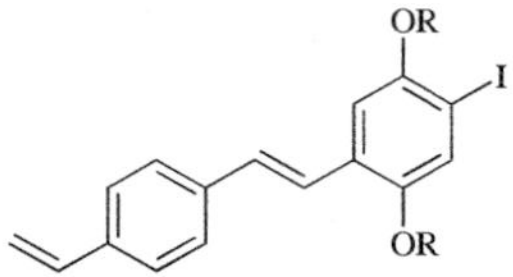 OR, I, OR	Gr II (0.06 eq), DCM, reflux, 40 h	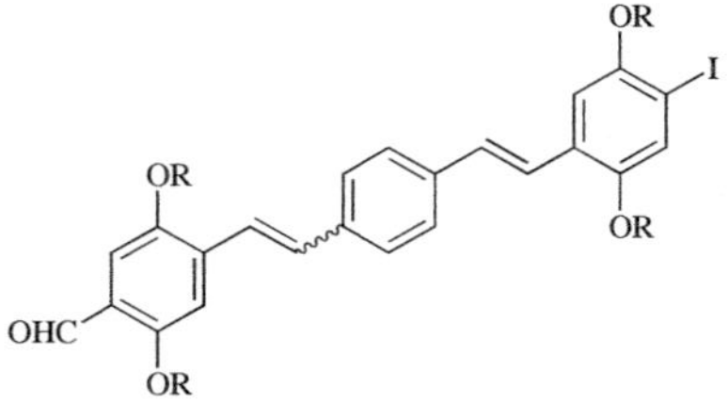 OR, I, OR, OR, OHC, OR (72)	676
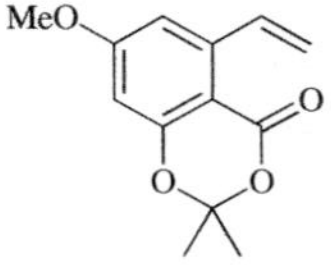 MeO, O, O, O; 1.9 eq	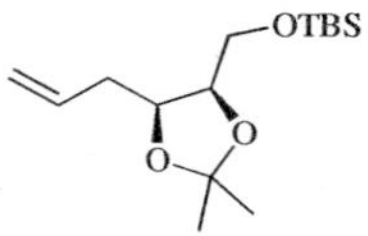 OTBS, O, O	Hov II (0.08 eq), DCM, 40°, 16 h	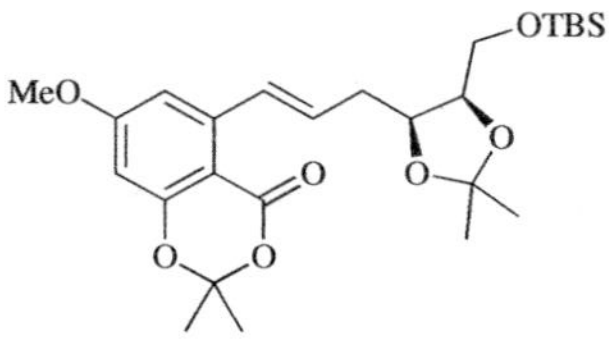 OTBS, MeO, O, O, O, O, O (86) (*E*) only	681

TABLE 10. CROSS-METATHESIS OF VINYL-SUBSTITUTED AROMATIC COMPOUNDS (*Continued*)

Aromatic Compound	Cross Partner	Conditions	Product(s) and Yield(s) (%)	Refs.
Please refer to the charts preceding the tables for the catalyst structures.				
C_9				
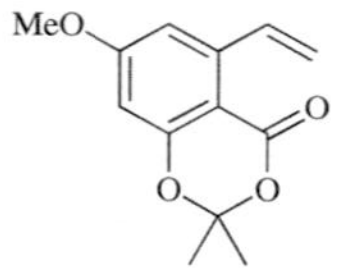 MeO, O, O, O	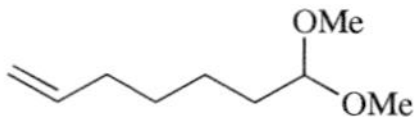 OMe, OMe 3 x 1.25 eq	Gr II (98 x 0.025 eq), CuI (98 x 0.029 eq), Et_2O, reflux, 14 d	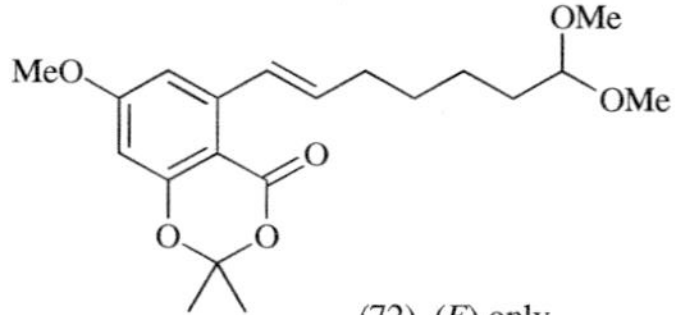 MeO, OMe, OMe, O, O, O (72) (*E*) only	680
2 eq	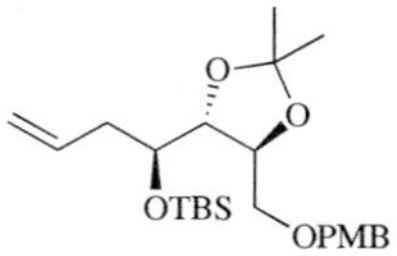 O, O, OTBS, OPMB	Hov II (0.06 eq), DCM, reflux, 2 d	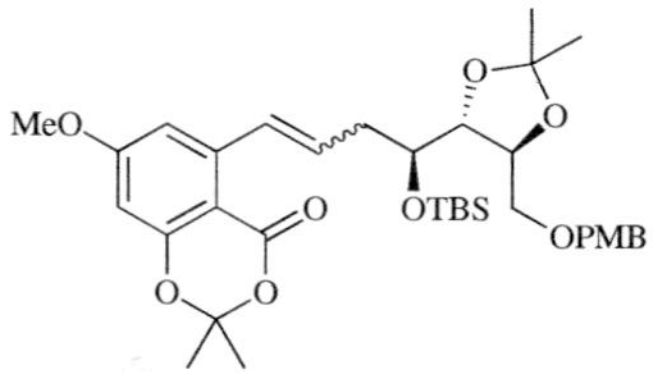 MeO, O, O, O, O, O, OTBS, OPMB (72)	682
C_{10}				
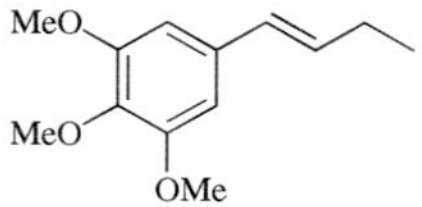 MeO, MeO, OMe	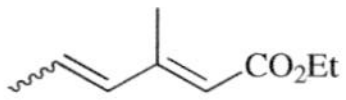 CO_2Et 3 eq	Gr II (0.1 eq), toluene, 50°	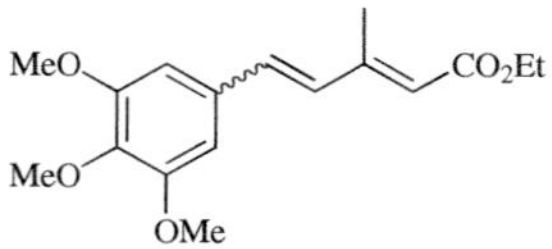 MeO, MeO, OMe, CO_2Et (64)	542
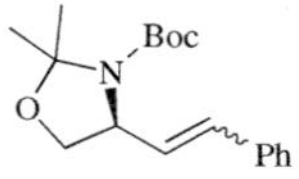 Boc, N, O, Ph 5 eq	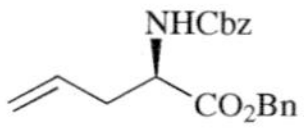 NHCbz, CO_2Bn	Gr II (0.05 eq), toluene, reflux, 3 h	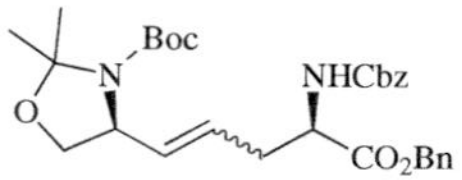 Boc, N, O, NHCbz, CO_2Bn (10)	563

C_{12}

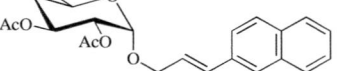

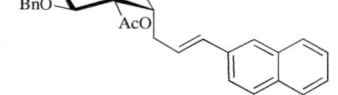

Olefin	Cross partner	Conditions	Product(s) and Yield(s) (%)		Refs.
t-Bu	AcO, OAc	Ru cat **162** (0.02 eq), toluene, 80°, 1 h	t-Bu, OAc	(53)	449
	AcO, OAc 2 eq	Gr II (0.05 eq), DCM, reflux, 12 h	OAc	(81) $(E)/(Z) > 95:5$	145
4 eq	AcO, OAc, O, AcO, AcO, O	Gr I (0.07 eq), DCM, reflux, 16 h	AcO, OAc, O, AcO, AcO, O (53) $(E)/(Z) = 81:19$		332
5 eq	OBn, Troc, N, AcO, BnO, AcO	Gr II (0.2 eq), DCM, reflux, 16–24 h	OBn, Troc, N, AcO, BnO, AcO (90) $(E)/(Z) > 95:5$		666

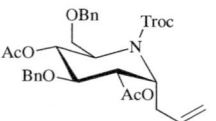

TABLE 10. CROSS-METATHESIS OF VINYL-SUBSTITUTED AROMATIC COMPOUNDS (*Continued*)

Aromatic Compound	Cross Partner	Conditions	Product(s) and Yield(s) (%)	Refs.
Please refer to the charts preceding the tables for the catalyst structures.				
C_{12}				
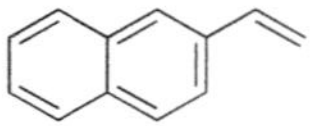	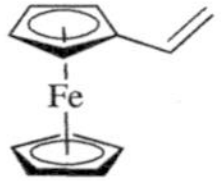 Fe	Mo cat **1** (*x* eq), solvent, rt	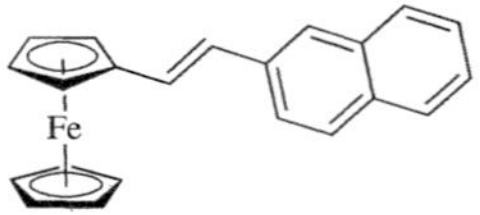 Fe; (*E*) only	
			x / Solvent / Time: 0.01 / benzene d_6 / 100 min (32)	659
			x / Solvent / Time: 0.017–0.01 / toluene / 15 h (45)	637
3.1 eq	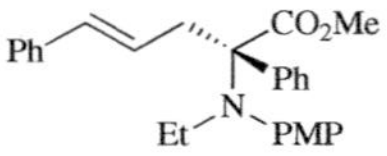 Ph, CO_2Me, Ph, N, Et, PMP; er 90.5:9.5	Hov II (0.2 eq), DCE, 70°, 20 h	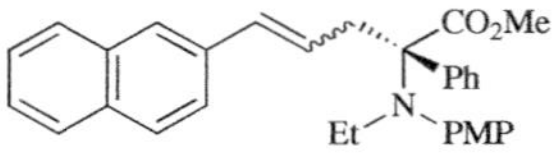 CO_2Me, Ph, N, Et, PMP; (68) er 90.5:9.5	368
2 eq	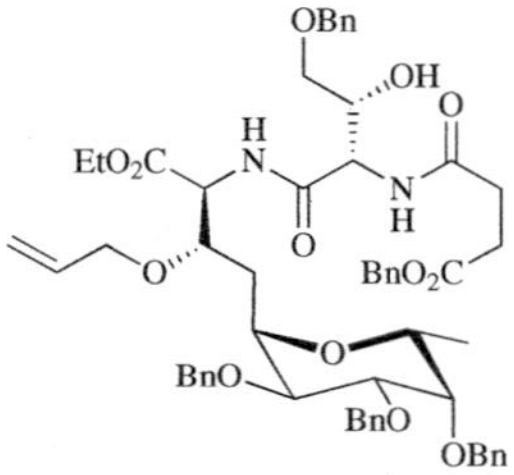 OBn, OH, O, EtO_2C, H, N, N, H, O, BnO_2C, O, O, BnO, BnO, OBn	Gr I (0.05 eq), DCM, reflux, 18–40 h	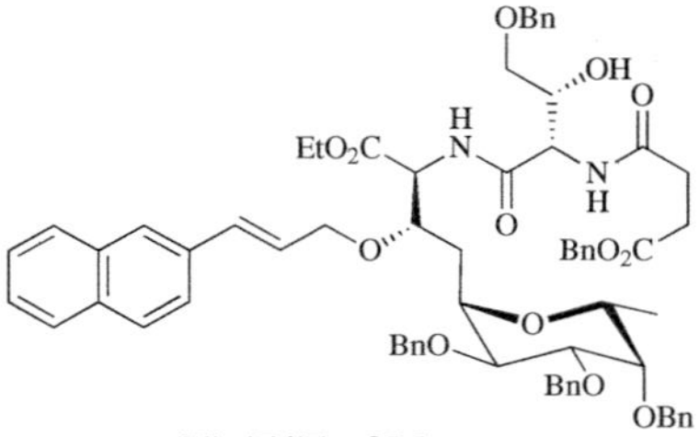 OBn, OH, O, EtO_2C, H, N, N, H, O, BnO_2C, O, O, BnO, BnO, OBn; (71) (*E*)/(*Z*) > 95:5	582

C_{13}

MeO, OMe, O, O, CHO	O, PO(OR)$_2$ 1.5 eq	Hov II (0.1 eq), DCM, 40°, 5 h	OMe, O, CHO, MeO, O, PO(OR)$_2$ R: CF_3CH_2 (65); Ph (69)	683

R	
CF_3CH_2	(65)
Ph	(69)

C_{14}

Ph *x* eq	O, N, R	Gr II (0.05 eq), DCM, reflux, 14 h	Ph, O, N, R	422
n-C_5H_{11}, OH **I**	CO_2Me 3 eq	1. Ru cat **215** (20 eq), (*Z*)-butene (0.01 eq), THF, 22°, 1 h 2. Ru cat **215** (0.05 eq), **I** (1 eq), THF, 100 torr, 35°, 4 h	*n*-C_5H_{11}, OH, CO_2Bn (35) (*Z*)/(*E*) = 71:29	395

x	R	
2.5	H	(38)
2	MeS	(43)

TABLE 10. CROSS-METATHESIS OF VINYL-SUBSTITUTED AROMATIC COMPOUNDS (*Continued*)

Aromatic Compound	Cross Partner	Conditions	Product(s) and Yield(s) (%)	Refs.
Please refer to the charts preceding the tables for the catalyst structures.				
C_{14}				
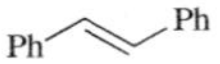 Ph, Ph	 TMS, 2 eq	Mo cat **1** (0.02 eq), DME, 4 h	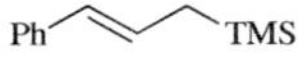 Ph, TMS (80) (*E*)/(*Z*) = 81:19	208
	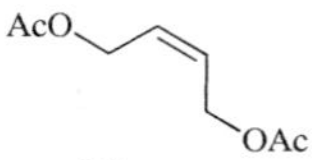 AcO, OAc, 1.2 eq	Gr II (0.05 eq), DCM, 40°, 12 h	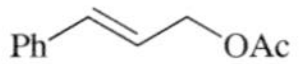 Ph, OAc (51) (*E*) only	36
4 eq	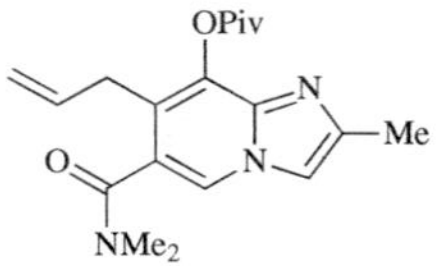 OPiv, N, N, Me, O, NMe_2	Gr II (0.04 eq), DCM, 40°, 18 h	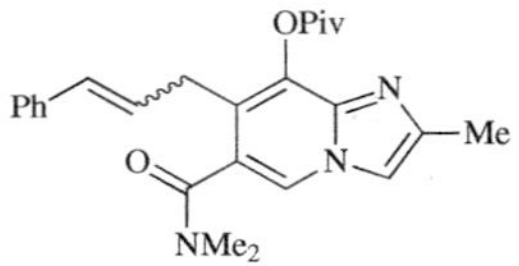 Ph, OPiv, N, N, Me, O, NMe_2 (53)	684
2 eq	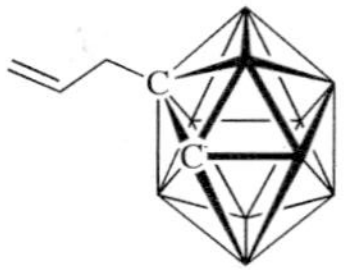 C, C	Hov II (0.007 eq), DCM, reflux, 48 h	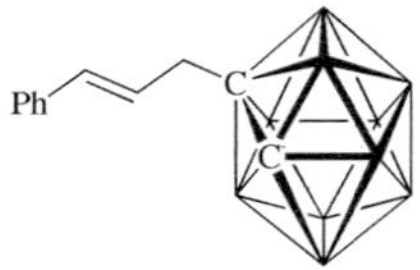 Ph, C, C (81) (*E*)/(*Z*) = 75:25	436
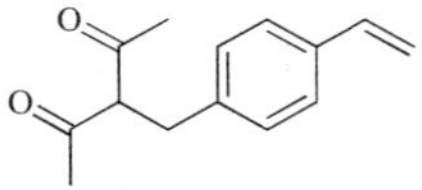 O, O	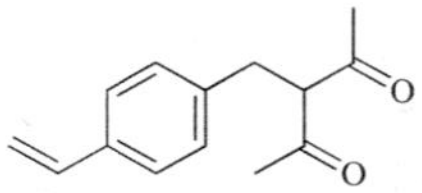 O, O	Gr II (0.05 eq), DCM, 40°, 24 h	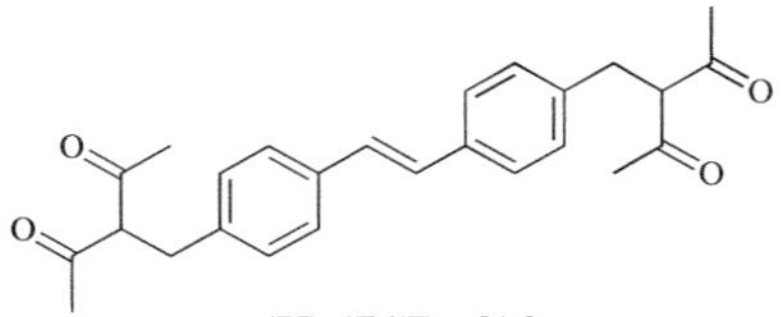 O, O, O, O (75) (*E*)/(*Z*) = 91:9	492

C_{15}	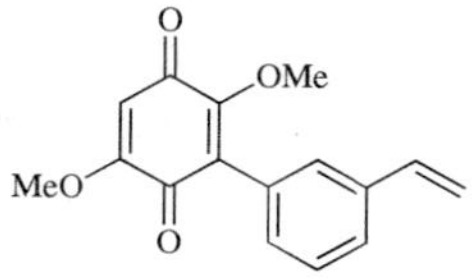	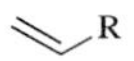 10 eq	Gr II (0.1 eq), DCM, reflux, 18 h	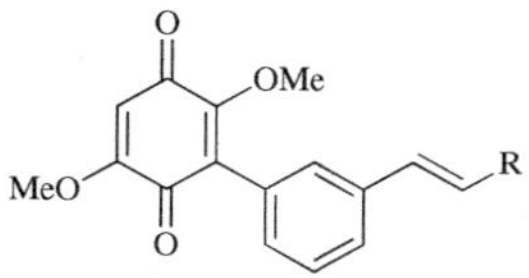 (*E*) only R: *n*-C_5H_{11} (66); *n*-C_6H_{13} (53)	663
	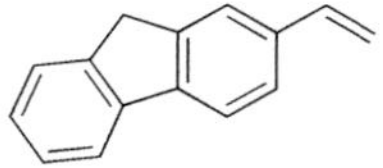 2 eq	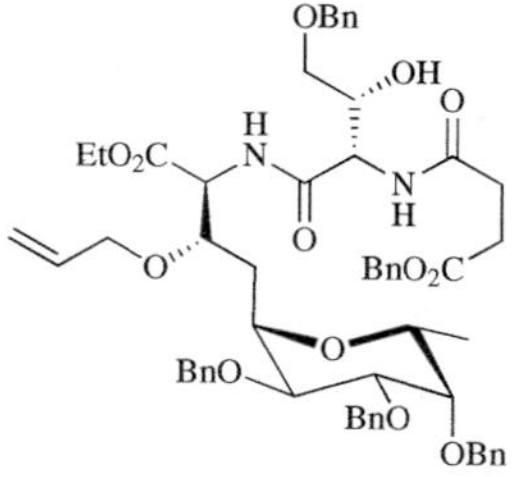	Gr I (0.05 eq), DCM, reflux, 18–40 h	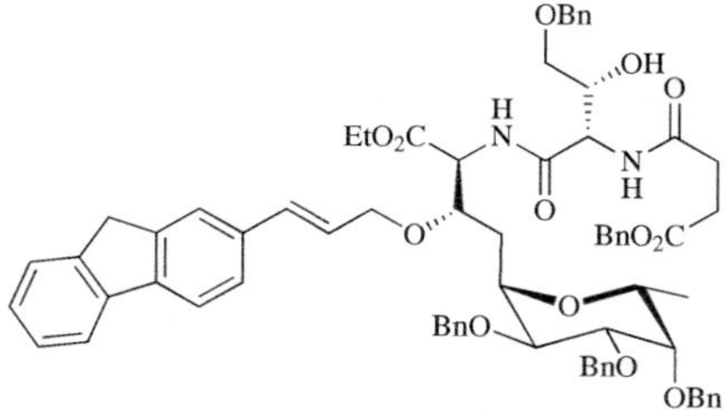 (70) (*E*)/(*Z*) > 95:5	582
C_{18}	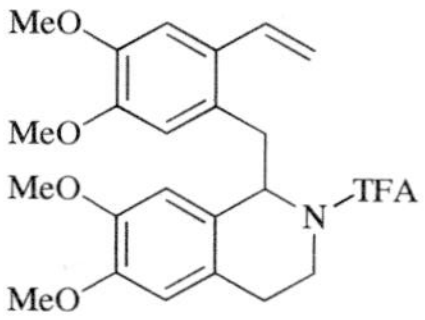	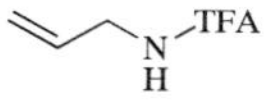 1.4 eq	Gr I (0.1 eq), DCM, 24 h	(30) + 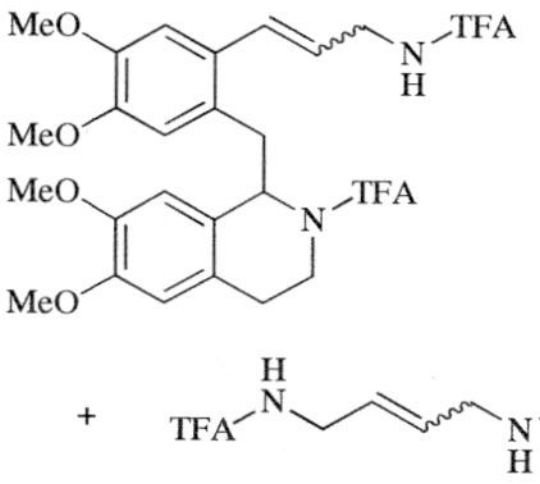 (15)	685

TABLE 10. CROSS-METATHESIS OF VINYL-SUBSTITUTED AROMATIC COMPOUNDS (*Continued*)

Aromatic Compound	Cross Partner	Conditions	Product(s) and Yield(s) (%)	Refs.
Please refer to the charts preceding the tables for the catalyst structures.				
C_{18}				
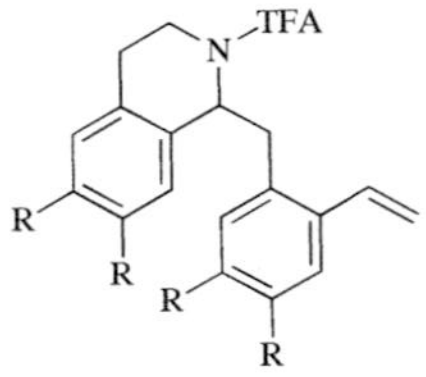 TFA; R; R; R; R; R = OMe	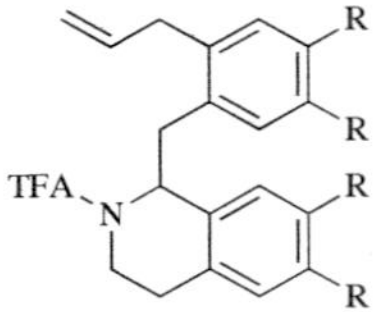 TFA; R; R; R; R; 1.8 eq	Gr I (0.1 eq), DCM, reflux, 48 h	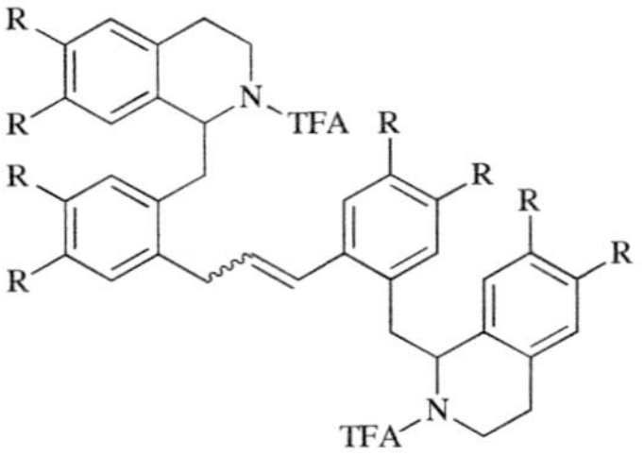 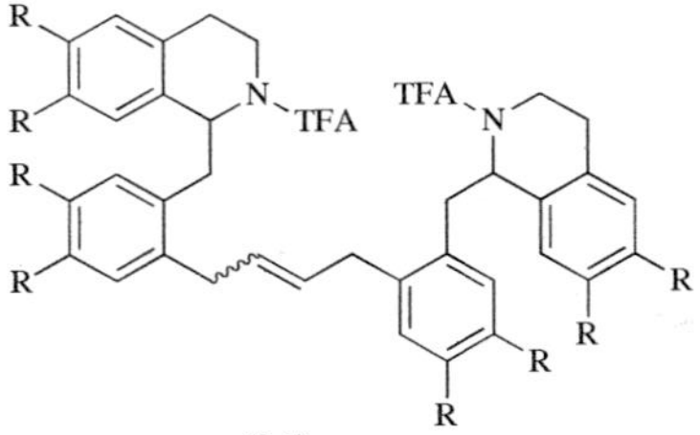 R; R; TFA; R; R; R; R; R; TFA (7.5) + R; R; TFA; R; R; TFA; R; R; R; R (7.5)	685

C_{22}		Hov II (0.13 eq), benzene, 80°, 4.5 h	(83)	686

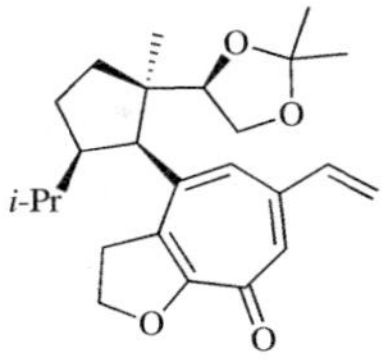

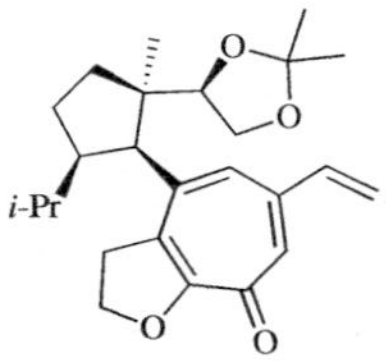

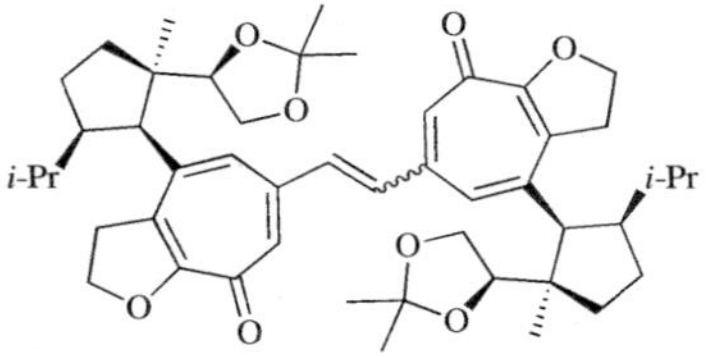

[a] The conversion was determined by ^{1}H NMR analysis.

[b] The conversion was determined by GC analysis.

TABLE 11. CROSS-METATHESIS OF VINYL BORONIC COMPOUNDS

A. VINYL BORONIC ACIDS

Vinyl Boronic Acid	Cross Partner	Conditions	Product(s) and Yield(s) (%)	Refs.
Please refer to the charts preceding the tables for the catalyst structures.				
C_3				
HO–B(OH) (vinyl, wavy bond), 1 eq	TIPS	Gr II (0.05 eq), DCM, reflux	HO–B(OH) TIPS (55) (*E*)/(*Z*) > 95:5	687
1 eq	n-C_6H_{13}	Gr II (0.05 eq), DCM, reflux	HO–B(OH) n-C_6H_{13} (31) (*E*)/(*Z*) > 95:5	687

TABLE 11. CROSS-METATHESIS OF VINYL BORONIC COMPOUNDS (*Continued*)

B. VINYL BORONIC ESTERS

Vinyl Boronic Ester	Cross Partner	Conditions	Product(s) and Yield(s) (%)	Refs.
Please refer to the charts preceding the tables for the catalyst structures.				
C_2				
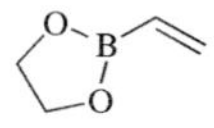	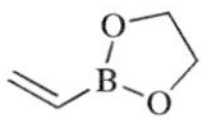	Gr I (0.05 eq), DCM, 40°, 3 h	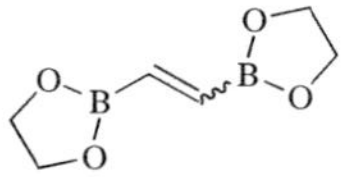 (80)	688
	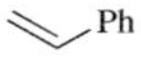 Ph	Gr I (0.05 eq), DCM, 40°, 3 h	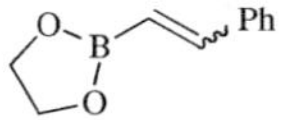 Ph (59)  + Ph Ph (11)	688
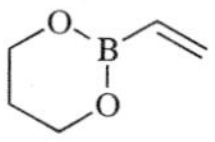	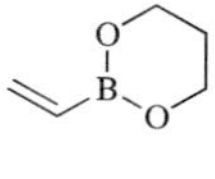	Gr I (0.05 eq), DCM, 40°, 3 h	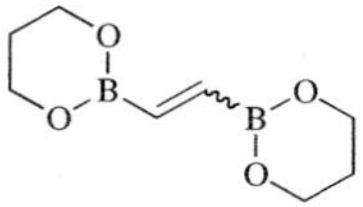 (60)	688
1 eq	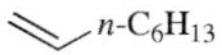 n-C_6H_{13}	Gr I (0.05 eq), DCM, 40°, 3 h	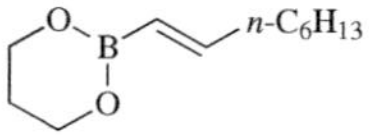 n-C_6H_{13} (75) $(E)/(Z) > 96:4$ 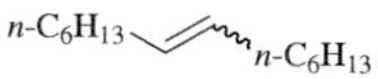 + n-C_6H_{13} n-C_6H_{13} (15) 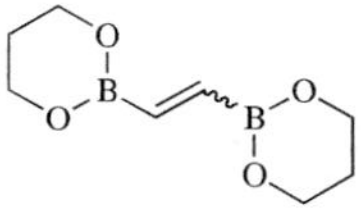 + (8)	688

TABLE 11. CROSS-METATHESIS OF VINYL BORONIC COMPOUNDS (*Continued*)

B. VINYL BORONIC ESTERS (*Continued*)

Vinyl Boronic Ester	Cross Partner	Conditions	Product(s) and Yield(s) (%)	Refs.
Please refer to the charts preceding the tables for the catalyst structures.				
C_2				
1 eq		Gr I (0.05 eq), DCM, 40°, 3 h	**I** (*E*) only + **II** R / **I** / **II**: H (85) (5); Cl (90) (—); Me (85) (—)	688
C_{2-3}				
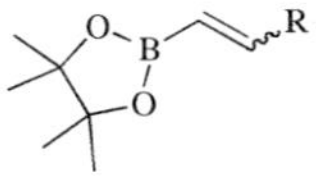 1 eq	 TIPS	Gr II (0.05 eq), DCM, reflux	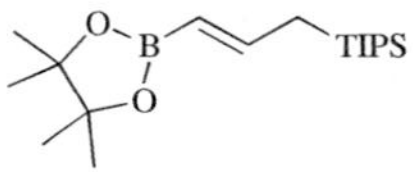 TIPS R / (*E*)/(*Z*): H (86) 88:12; Me (99) 91:9	687
C_2				
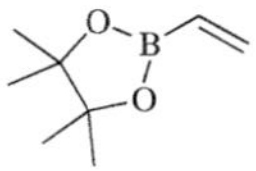 5 eq	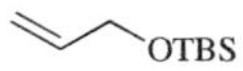 OTBS	Mo cat **5** (0.05 eq), benzene, 22°, 24 h	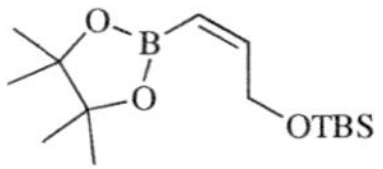 OTBS (70) (*Z*)/(*E*) = 96:4	547

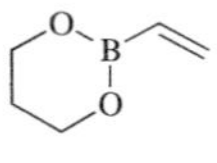

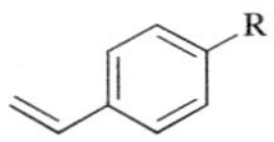

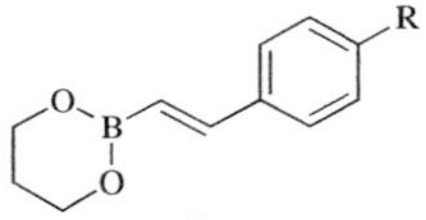

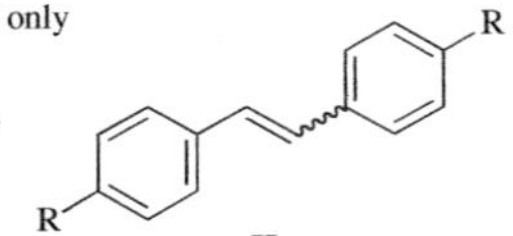

5 eq	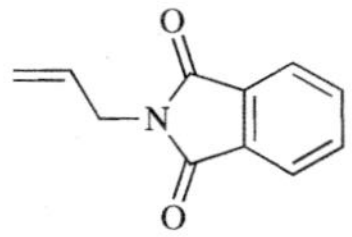 Mo cat **5** (0.05 eq), benzene, 22°, 24 h	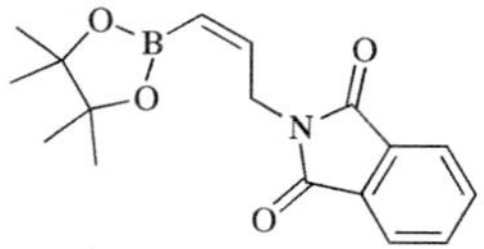	(71) (*Z*)/(*E*) = 93:7	547
5 eq	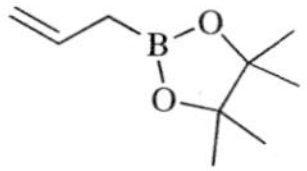 Mo cat **5** (0.05 eq), benzene, 22°, 24 h	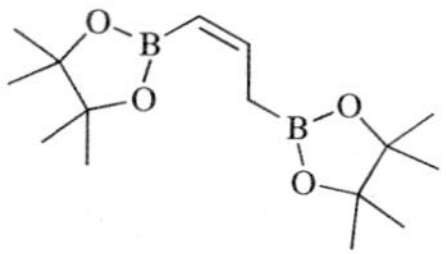	(60) (*Z*)/(*E*) = 97:3	547
1–4 eq	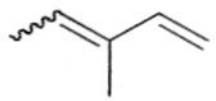 Gr II (0.05 eq), DCM, reflux, 12 h	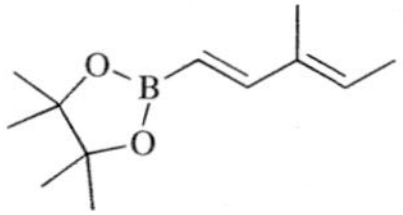	(80) (*E*) only	132

C_{2-3}

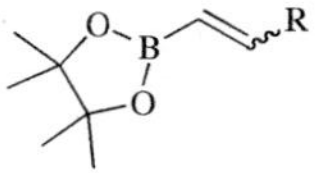 1 eq	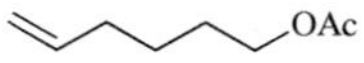 Gr II (0.05 eq), DCM, reflux	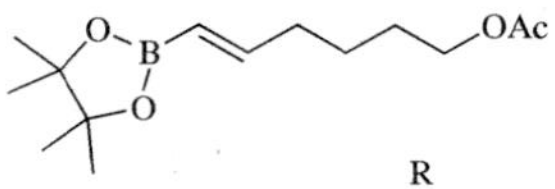 OAc		687

R		(*E*)/(*Z*)
H	(60)	91:9
Me	(65)	93:7

TABLE 11. CROSS-METATHESIS OF VINYL BORONIC COMPOUNDS (*Continued*)

B. VINYL BORONIC ESTERS (*Continued*)

Vinyl Boronic Ester	Cross Partner	Conditions	Product(s) and Yield(s) (%)	Refs.
Please refer to the charts preceding the tables for the catalyst structures.				
C_2				
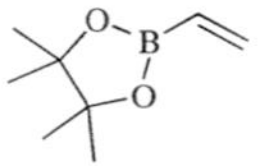 O, B, O 3 eq	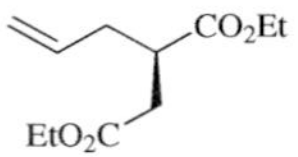 CO_2Et, EtO_2C	Gr I (0.1 eq), DCM, reflux, overnight	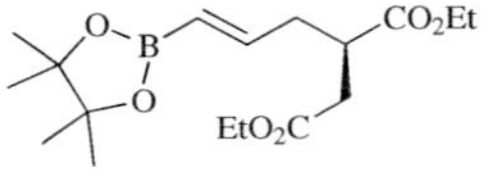 O, B, O, CO_2Et, EtO_2C (92) (*E*)/(*Z*) = 86:14	333
3 eq	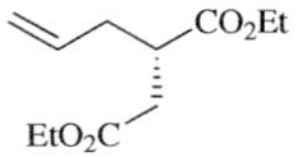 CO_2Et, EtO_2C	Gr I (0.1 eq), DCM, reflux, overnight	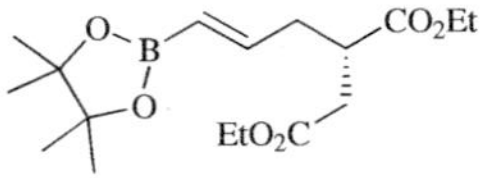 O, B, O, CO_2Et, EtO_2C (90) (*E*)/(*Z*) = 83:17	333
3 eq	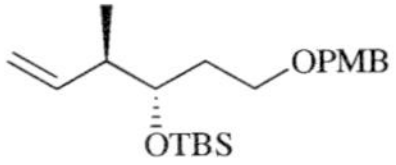 OPMB, OTBS	Hov II (0.05 eq), DCM, reflux, 18 h	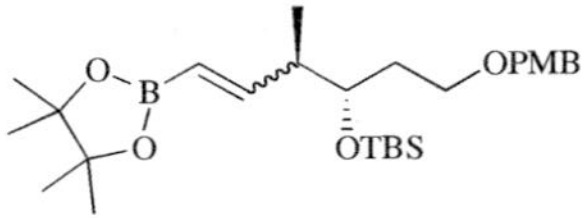 O, B, O, OPMB, OTBS (74)	689
5 eq	 OPMB, 5	Mo cat **5** (0.05 eq), benzene, 22°, 24 h	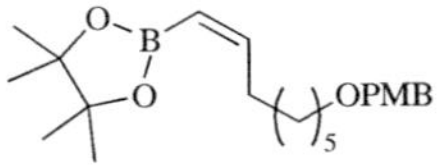 O, B, O, OPMB, 5 (92) (*Z*)/(*E*) = 97:3	547
5 eq	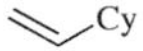 Cy	Mo cat **5** (0.05 eq), benzene, 22°, 24 h	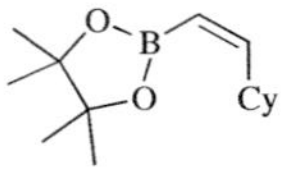 O, B, O, Cy (51) (*Z*)/(*E*) = 93:7	547
5 eq	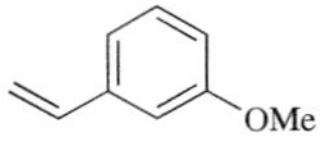 OMe	Mo cat **5** (0.05 eq), benzene, 22°, 24 h	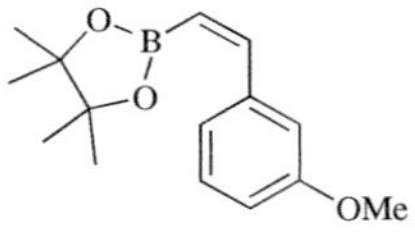 O, B, O, OMe (69) (*Z*)/(*E*) = 95:5	547

	Reactant	Conditions	Product(s) and Yield(s) (%)	Refs.
5 eq	R	Mo cat **5** (0.05 eq), benzene, 22°, 24 h	R	547

R		$(Z)/(E)$
MeO	(69)	93:7
CF_3	(93)	95:5

5 eq	CO_2Me, OMe, N, O	Mo cat **7** (0.1 eq), benzene, 22°, 18 h, 100 torr	CO_2Me, O, B, O, OMe, N, O (76) $(Z)/(E)$ = 92:8	690
2 eq	CO_2Me, OH	Hov II (0.05 eq), DCM, reflux, 4 h	O, B, O, CO_2Me, OH (50)	656
5 eq	Ph	Catalyst (x eq), benzene, 22°	O, B, O, Ph	547

Catalyst	x	Conv. (%)[a]		$(Z)/(E)$
Mo cat **1**	0.05	33	(15)	68:32
Hov II	0.05	98	(78)	9:91
Gr II	0.05	86	(65)	11:89
Mo cat **18**	0.03	56	(—)	96:4
Mo cat **19**	0.03	50	(—)	93:7
Mo cat **5**	0.05	95	(68)	93:7
Mo cat **7**	0.05	98	(69)	93:7
W cat **3**	0.05	66	(—)	86:14

TABLE 11. CROSS-METATHESIS OF VINYL BORONIC COMPOUNDS (*Continued*)

B. VINYL BORONIC ESTERS (*Continued*)

Vinyl Boronic Ester	Cross Partner	Conditions	Product(s) and Yield(s) (%)	Refs.
Please refer to the charts preceding the tables for the catalyst structures.				
C_2				
2 eq	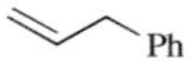	1. Gr I (0.03 eq) DCM, reflux, 18 h 2. RCHO (1 eq), $[Ir(cod)(PPh_3Me_2)]PF_6$ (0.03 eq), THF, rt, 20 h	*anti/syn* > 98:2	691
2 eq		Catalyst (0.03 eq), BQ (*x* eq), solvent, 18 h		692

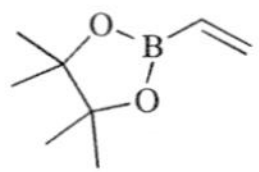

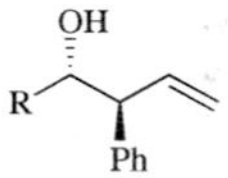

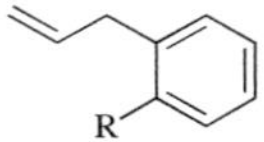

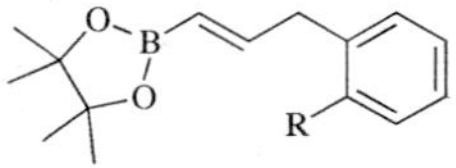

R	
n-C_5H_{11}	(70)
Ph	(60)
4-BrC_6H_4	(66)
4-$O_2NC_6H_4$	(63)
3-F-4-$O_2NC_6H_3$	(65)

R	Catalyst	*x*	Solvent	Temp		(*E*)/(*Z*)
BnO	Gr I	0	DCM	reflux	(73)	90:10
MeO	Gr I	0	DCM	reflux	(89)	87:13
NsHN	Gr I	0	DCM	reflux	(62)	70:30
AcO	Ru cat **170**	0.5	toluene	60°	(55)	85:15

2 eq

Catalyst (0.03 eq),
BQ (x eq),
solvent, 18 h

692

R^1	R^2	Catalyst	x	Solvent	Temp		(*E*)/(*Z*)
MeO	MeO	Ru cat **170**	0.5	toluene	60°	(60)	80:20
HO	MeO	Gr I	0	DCM	reflux	(65)	83:17
BnO	MeO	Gr I	0	DCM	reflux	(56)	75:25
H	BnO	Gr I	0	DCM	reflux	(71)	86:14
MeO	H	Gr I	0	DCM	reflux	(67)	85:15
Me	H	Gr I	0	DCM	reflux	(70)	84:16
MeS	H	Ru cat **170**	0.5	toluene	60°	(42)	70:30
H	H	Gr I	0	DCM	reflux	(70)	85:15
F	H	Gr I	0	DCM	reflux	(76)	93:7
CF_3	H	Gr I	0	DCM	reflux	(70)	92:8

2 eq

Catalyst (0.03 eq),
solvent, reflux

Catalyst	Solvent	Time (h)		(*E*)/(*Z*)	
Gr I	DCM	18	(79)	88:12	692
Ru cat **170**	toluene	1	(65)	(*E*) only	693

TABLE 11. CROSS-METATHESIS OF VINYL BORONIC COMPOUNDS (*Continued*)

B. VINYL BORONIC ESTERS (*Continued*)

Vinyl Boronic Ester	Cross Partner	Conditions	Product(s) and Yield(s) (%)	Refs.

Please refer to the charts preceding the tables for the catalyst structures.

C_2

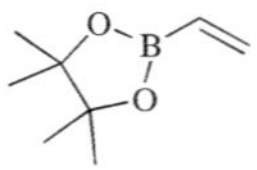

Vinyl Boronic Ester	Cross Partner	Conditions	Product(s) and Yield(s) (%)	Refs.
O-B, O (2 eq)	OMe, OMe	Catalyst (0.03 eq), toluene, 1 h	O-B, O, OMe, OMe	693

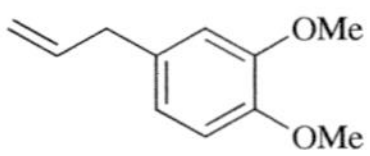

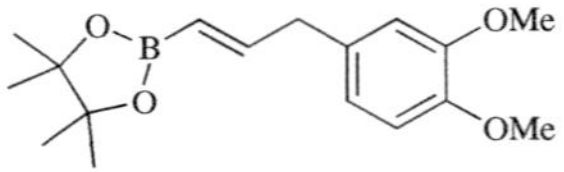

Catalyst	Temp	
Ru cat **170**	50°	(60)
Ru cat **170**	70°	(72)
Ru cat **170**	reflux	(84)
Gr II	reflux	(—)
Hov II	reflux	(—)
Ru cat **187**	reflux	(78)
Ru cat **93**	reflux	(—)

Vinyl Boronic Ester	Cross Partner	Conditions	Product(s) and Yield(s) (%)	Refs.
2 eq	R	Ru cat **170** (0.03 eq), toluene, reflux, 1 h	O-B, O, R (*E*) only	693

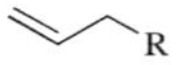

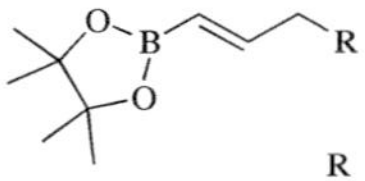

R	
Ph	(65)
4-$MeOC_6H_4$	(65)
3-$BnOC_6H_4$	(58)
4-MeC_6H_4	(65)
4-FC_6H_4	(47)
4-$CF_3C_6H_4$	(66)
3,4-$(MeO)_2C_6H_3$	(62)
3-MeO-4-$BnOC_6H_3$	(80)
3-MeO-4-HOC_6H_3	(49)

2 eq	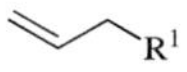 R^1	1. Gr I (0.03 eq) DCM, reflux, 18 h 2. R^{2}CHO (1 eq), [Ir(cod)(PPh$_3$Me$_2$)]PF$_6$ (0.03 eq), THF, rt, 20 h	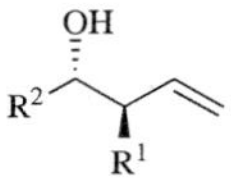 OH, R^2, R^1 *anti/syn* > 98:2	691

R^1	R^2	
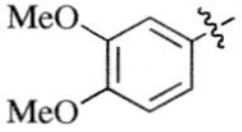 MeO, MeO	4-BrC$_6$H$_4$	(40)
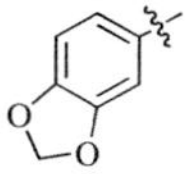 O, O	4-O$_2$NC$_6$H$_4$	(57)
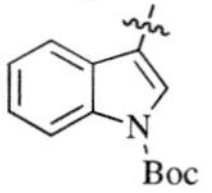 N, Boc	Ph	(31)

TABLE 11. CROSS-METATHESIS OF VINYL BORONIC COMPOUNDS (*Continued*)

B. VINYL BORONIC ESTERS (*Continued*)

Vinyl Boronic Ester	Cross Partner	Conditions	Product(s) and Yield(s) (%)	Refs.
Please refer to the charts preceding the tables for the catalyst structures.				
C_2				
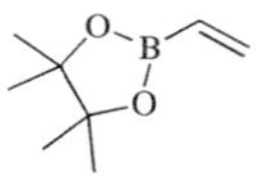 2 eq	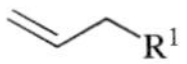	1. Gr I (0.03 eq) DCM, reflux, 18 h 2. R^2CHO (1 eq), $[Ir(cod)(PPh_3Me_2)]PF_6$ (0.03 eq), THF, rt, 20 h	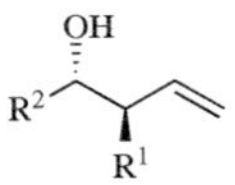 *anti/syn* > 98:2	691

R^1	R^2	
4-MeC_6H_4	4-$O_2NC_6H_4$	(76)
4-MeC_6H_4	Ph	(68)
4-$MeOC_6H_4$	4-$O_2NC_6H_4$	(85)
4-$MeOC_6H_4$	4-BrC_6H_4	(55)
4-$MeOC_6H_4$	3-MeO-4-$PivOC_6H_3$	(42)
4-$MeOC_6H_4$	PhCH=CH	(49)
4-$MeOC_6H_4$	*n*-C_5H_{11}	(49)
2-$MeOC_6H_4$	4-$O_2NC_6H_4$	(33)
2-$BnOC_6H_4$	4-$O_2NC_6H_4$	(30)
2-$AcOC_6H_4$	4-$O_2NC_6H_4$	(57)
2-$AcOC_6H_4$	4-$MeOC_6H_4$	(42)
4-FC_6H_4	4-$O_2NC_6H_4$	(72)
4-$CF_3C_6H_4$	4-ClC_6H_4	(58)
4-$CF_3C_6H_4$	4-$O_2NC_6H_4$	(61)
3-$BnOC_6H_4$	2-Br-4-FC_6H_3	(62)
3-MeO-4-HOC_6H_3	2-Cl-5-$O_2NC_6H_3$	(40)

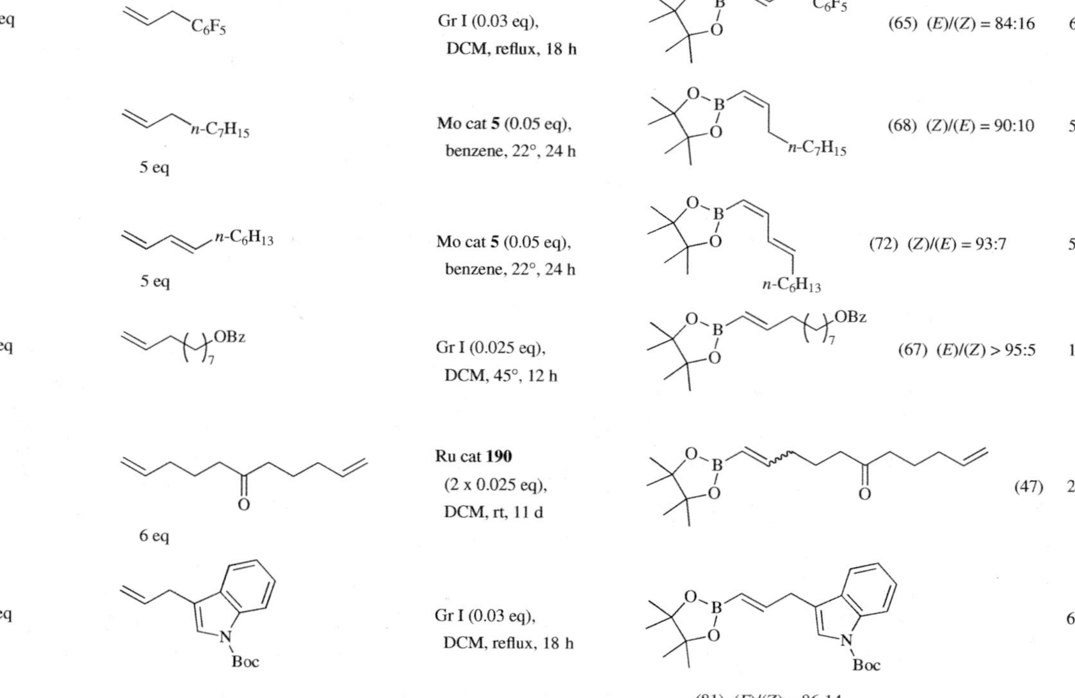

		Conditions	Product(s) and Yield(s) (%)	Refs.
2 eq	C_6F_5	Gr I (0.03 eq), DCM, reflux, 18 h	C_6F_5 (65) (*E*)/(*Z*) = 84:16	692
	n-C_7H_{15} 5 eq	Mo cat **5** (0.05 eq), benzene, 22°, 24 h	*n*-C_7H_{15} (68) (*Z*)/(*E*) = 90:10	547
	n-C_6H_{13} 5 eq	Mo cat **5** (0.05 eq), benzene, 22°, 24 h	*n*-C_6H_{13} (72) (*Z*)/(*E*) = 93:7	547
2 eq	OBz, 7	Gr I (0.025 eq), DCM, 45°, 12 h	OBz, 7 (67) (*E*)/(*Z*) > 95:5	124
	O 6 eq	Ru cat **190** (2 x 0.025 eq), DCM, rt, 11 d	O (47)	231
2 eq	N Boc	Gr I (0.03 eq), DCM, reflux, 18 h	N Boc (81) (*E*)/(*Z*) = 86:14	692

TABLE 11. CROSS-METATHESIS OF VINYL BORONIC COMPOUNDS (*Continued*)

B. VINYL BORONIC ESTERS (*Continued*)

Vinyl Boronic Ester	Cross Partner	Conditions	Product(s) and Yield(s) (%)	Refs.
Please refer to the charts preceding the tables for the catalyst structures.				
C_2				
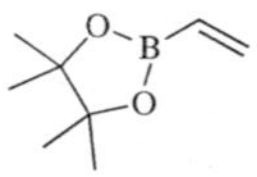 I	 CHO, 7	1. Ru cat **152** (0.01 eq), (*Z*)-butene (20 eq), THF, 22°, 1 h 2. Ru cat **152** (0.04 eq), **I** (3 eq), THF, 22°, 1 h, 100 torr	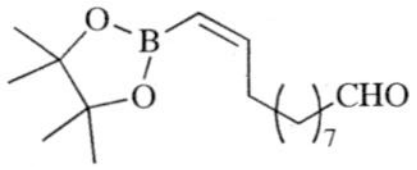 O, B, O, CHO, 7 (54) (*Z*)/(*E*) = 95:5	395
5 eq	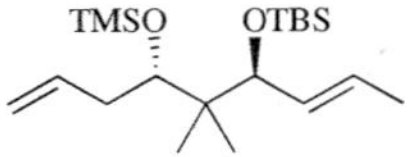 TMSO, OTBS	Mo cat **7** (0.03 eq), benzene, 22°, 20 h, 100 torr	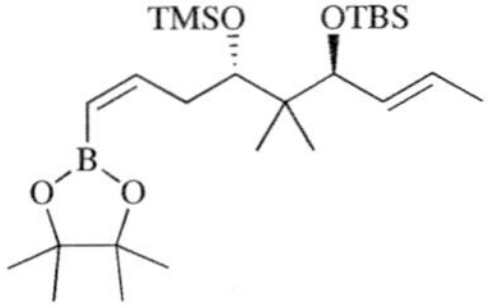 TMSO, OTBS, B, O, O (72) (*Z*)/(*E*) = 98:2	690
7 eq	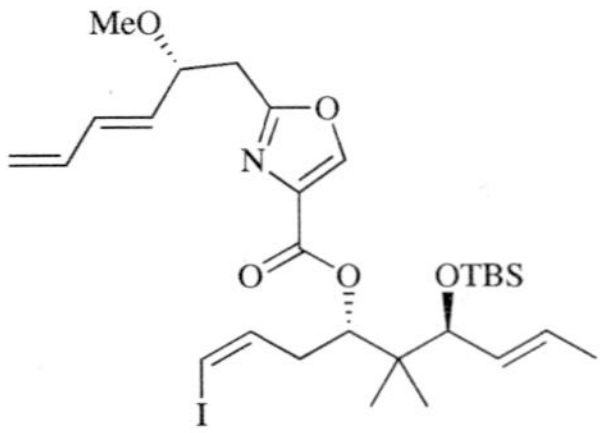 MeO, O, N, O, O, OTBS, I	Mo cat **5** (0.1eq), benzene, 22°, 4 h, 100 torr	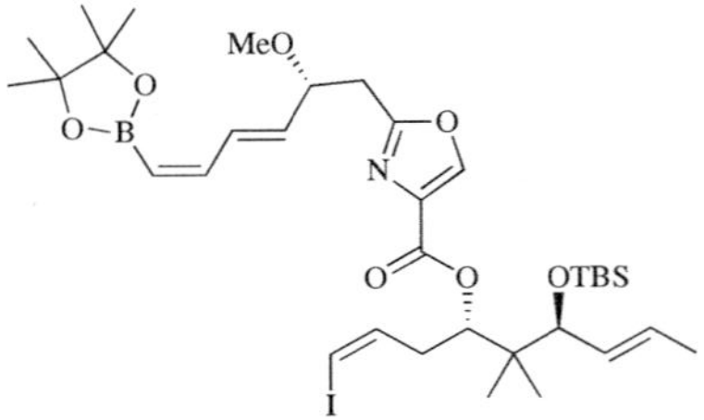 O, B, O, MeO, O, N, O, O, OTBS, I (91) (*Z*)/(*E*) = 98:2	690

	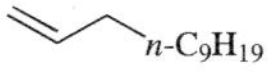 n-C_9H_{19} 4 eq	Ru cat **216** (0.02 eq), THF, 35°, 5 h	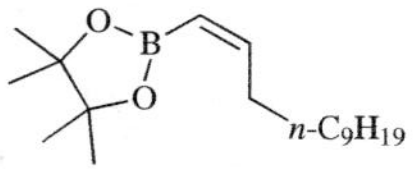 n-C_9H_{19}	(81) (*Z*)/(*E*) = 92:8	461
3 eq	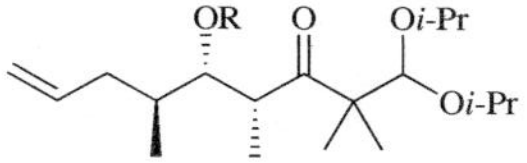 OR O O*i*-Pr O*i*-Pr	Gr I (0.1 eq), DCM, reflux, 8 h	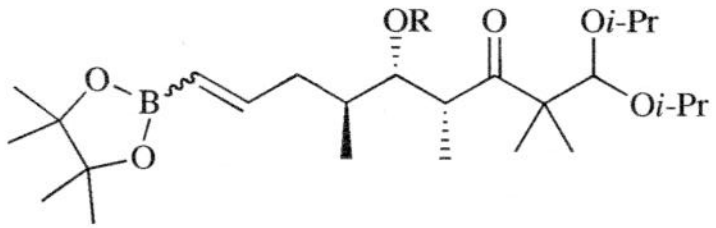 OR O O*i*-Pr O*i*-Pr		694

R	
H	(88)
Troc	(94)

3 eq	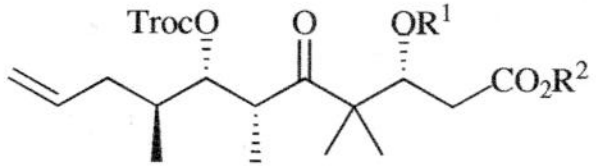 TrocO O OR^1 CO_2R^2	Gr I (0.1 eq), DCM, reflux, 8 h 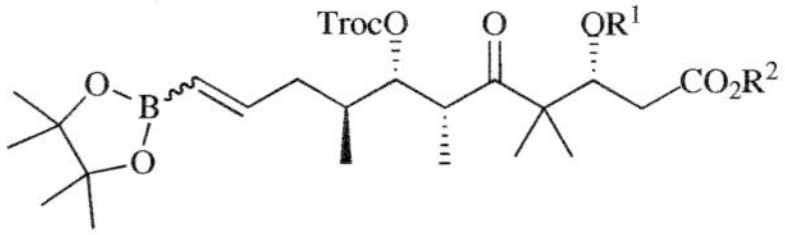	TrocO O OR^1 CO_2R^2	694

R^1	R^2	
H	*t*-Bu	(84)
TES	H	(94)
TES	*t*-Bu	(86)

TABLE 11. CROSS-METATHESIS OF VINYL BORONIC COMPOUNDS (*Continued*)

B. VINYL BORONIC ESTERS (*Continued*)

Vinyl Boronic Ester	Cross Partner	Conditions	Product(s) and Yield(s) (%)	Refs.
Please refer to the charts preceding the tables for the catalyst structures.				
C_2				
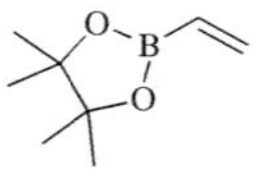 1.5 eq	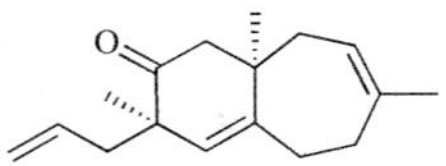	1. Ru cat **271** (0.1 eq) toluene, 60° 2. $NaBO_3$, THF, H_2O	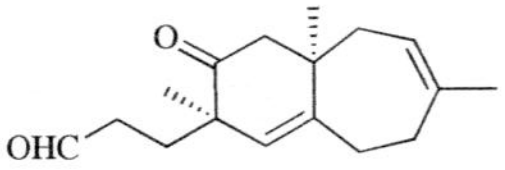 OHC (56)	695
1.5 eq	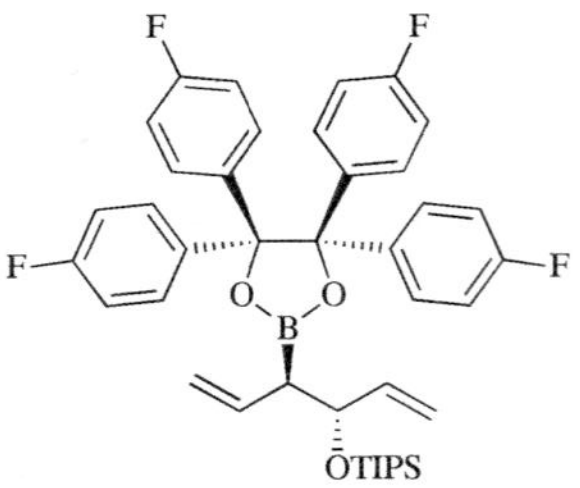 F F F F OTIPS	Hov II (0.1 eq), F_4BQ (0.1 eq), toluene, 80°, 24 h	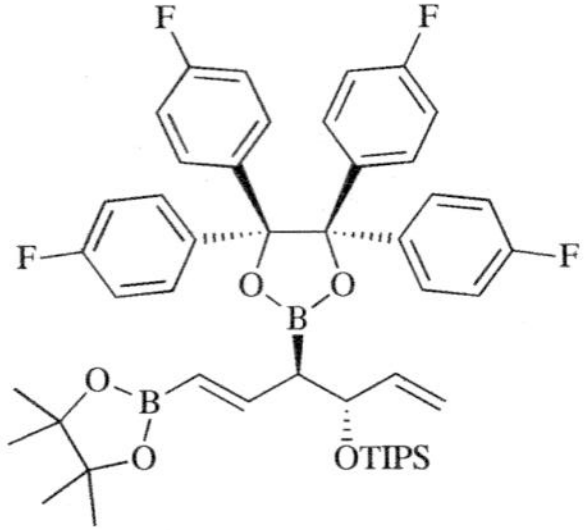 F F F F OTIPS (67) (*E*)/(*Z*) > 95:5	696
1.5 eq	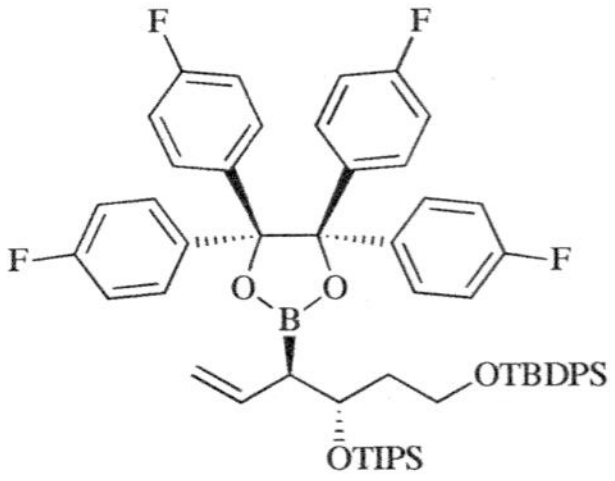 F F F F OTBDPS OTIPS	Hov II (0.1 eq), F_4BQ (0.1 eq), toluene, 80°, 24 h	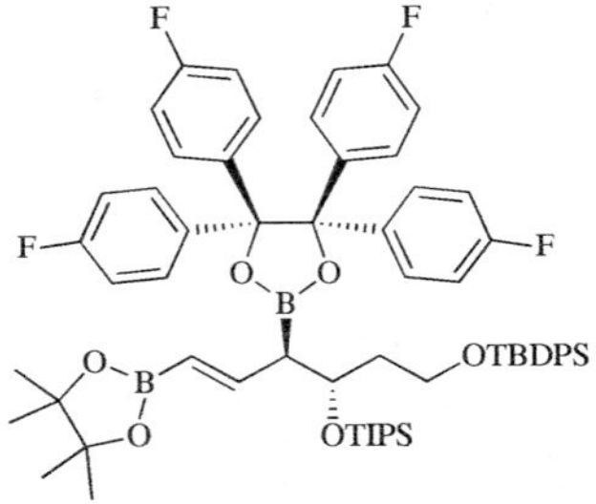 F F F F OTBDPS OTIPS (69) (*E*)/(*Z*) > 95:5	696

1.5 eq

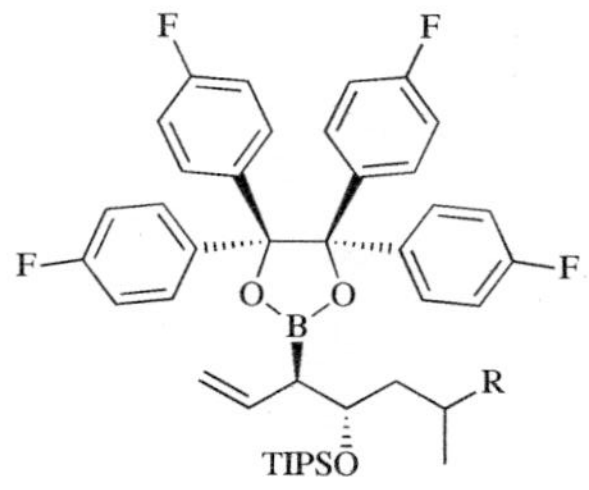

Hov II (0.1 eq),
F_4BQ (x eq),
toluene, 80°, 24 h

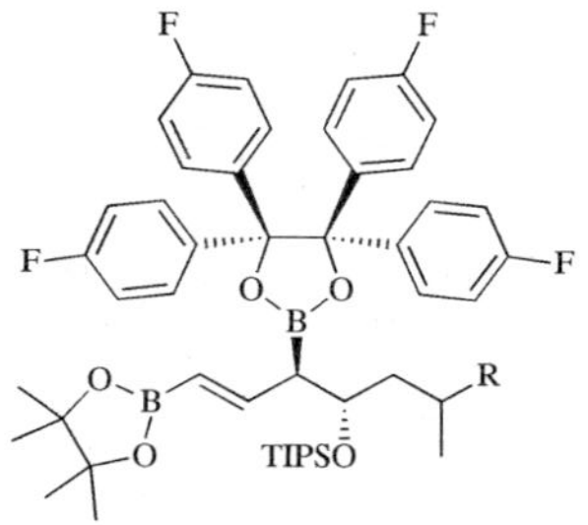

(*E*)/(*Z*) > 95:5 696

R	x	
H	0.08	(55)
H	—	(67)
Me	0.1	(45)
Me	—	(43)

1.5 eq

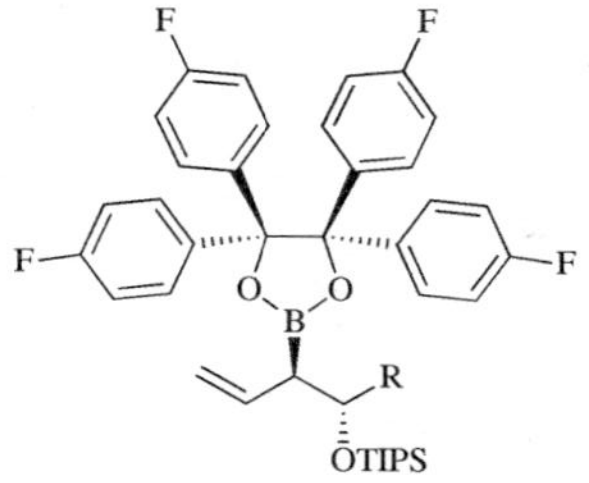

Hov II (0.1 eq),
F_4BQ (x eq),
toluene, 80°, 24 h

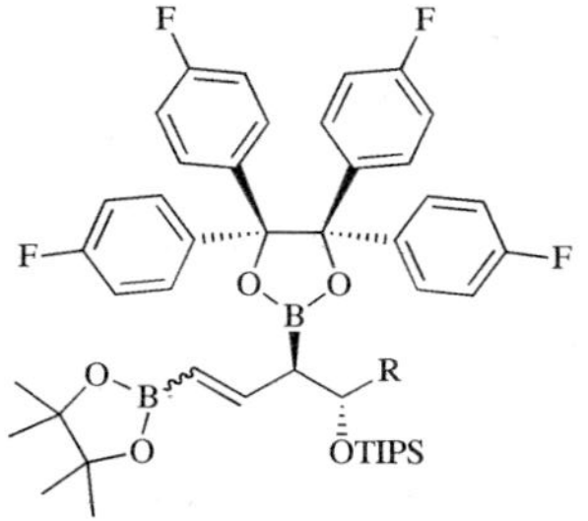

(*E*)/(*Z*) > 95:5 696

R	x	
Cy	0.1	(46)
Cy	—	(88)
Ph	—	(52)

TABLE 11. CROSS-METATHESIS OF VINYL BORONIC COMPOUNDS (*Continued*)

B. VINYL BORONIC ESTERS (*Continued*)

Vinyl Boronic Ester	Cross Partner	Conditions	Product(s) and Yield(s) (%)	Refs.
Please refer to the charts preceding the tables for the catalyst structures.				
C_2				
1.5 eq		Hov II (*x* eq), F_4BQ (*y* eq), toluene, 80°, 24 h	(*E*)/(*Z*) > 95:5	696
1.5 eq		Hov II (0.1 eq), F_4BQ (0.1 eq), toluene, 80°, 24 h	(50) (*E*)/(*Z*) > 95:5	696

R	*x*	*y*	
H	0.1	0.1	(56)
H	0.1	—	(51)
F	0.1	0.1	(61)
F	0.12	—	(74)

1.5 eq

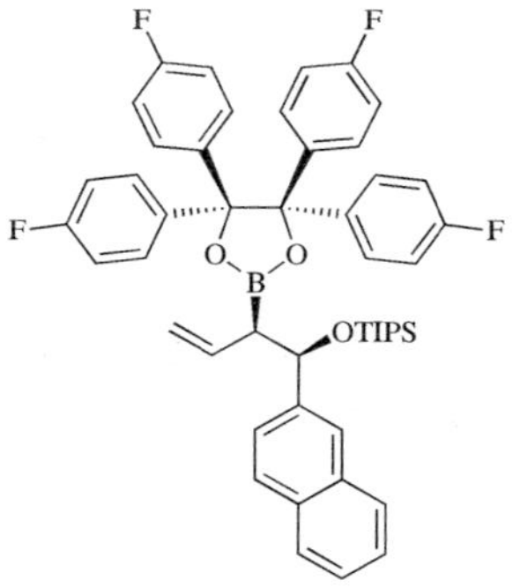

F F F F O B O OTIPS

Hov II (0.1 eq),
toluene, 80°, 24 h

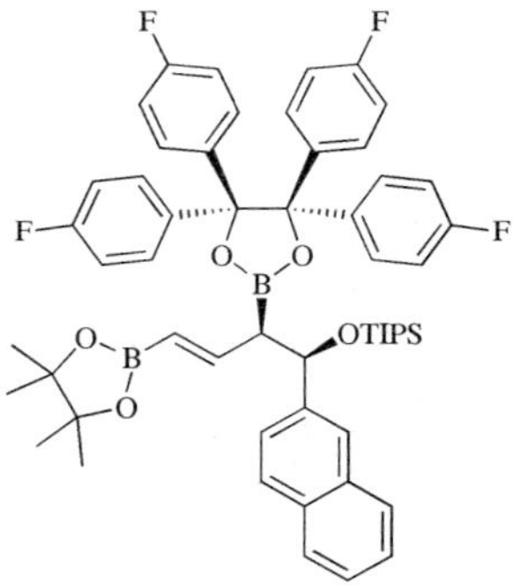

F F F F O B O O B O OTIPS

(67) (*E*)/(*Z*) > 95:5

696

1.5 eq

F F F F O B O TIPSO $(\ )_3$OBn

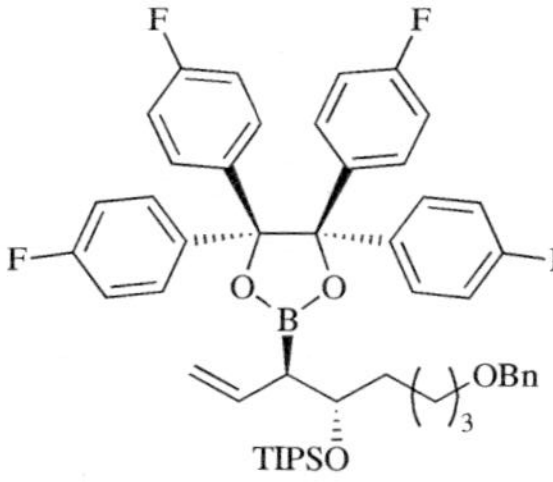

Hov II (*x* eq),
F_4BQ (*y* eq),
toluene, 80°, 24 h

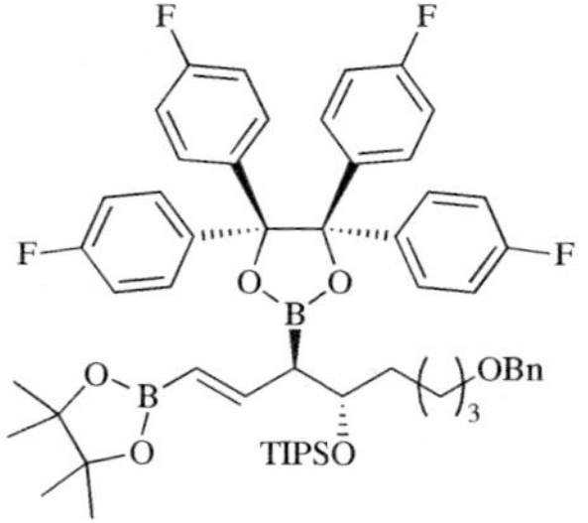

F F F F O B O O B O TIPSO $(\ )_3$OBn

(*E*)/(*Z*) > 95:5

696

x	*y*	
0.1	0.1	(74)
0.15	—	(75)

TABLE 11. CROSS-METATHESIS OF VINYL BORONIC COMPOUNDS (*Continued*)
B. VINYL BORONIC ESTERS (*Continued*)

Vinyl Boronic Ester	Cross Partner	Conditions	Product(s) and Yield(s) (%)	Refs.
Please refer to the charts preceding the tables for the catalyst structures.				
C_2				
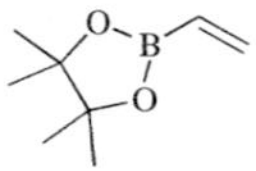 10–20 eq	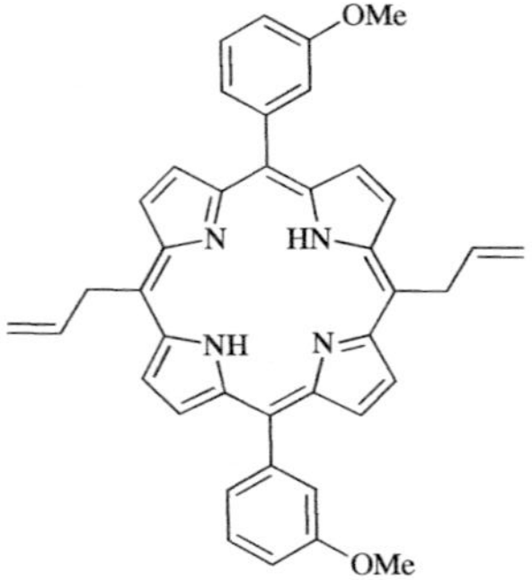	Gr I or Gr II (0.2 eq), DCM, reflux, 12–24 h	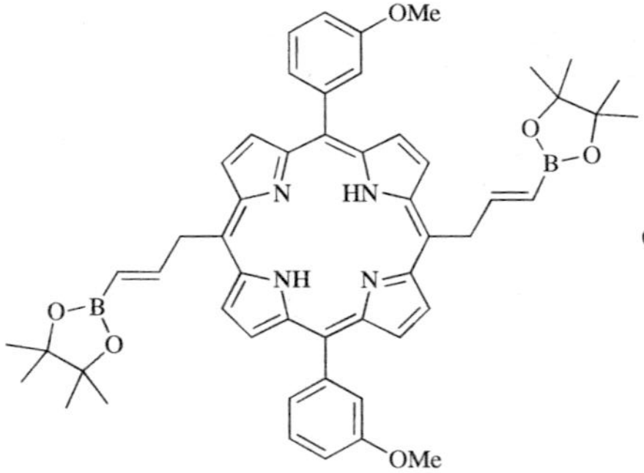 (50) (*E*) only	697
10–20 eq	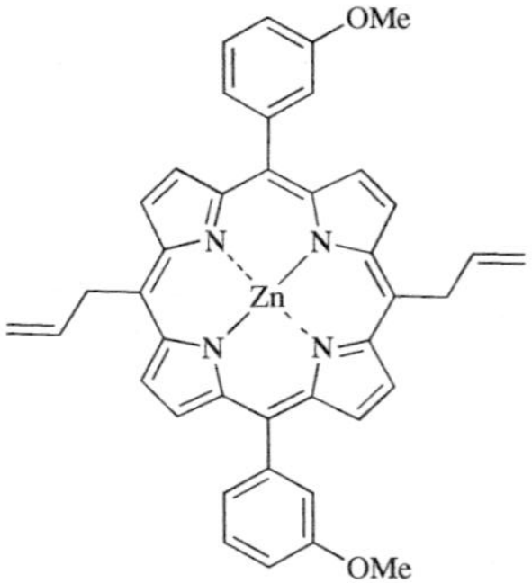	Gr I or Gr II (0.2 eq), DCM, reflux, 12–24 h	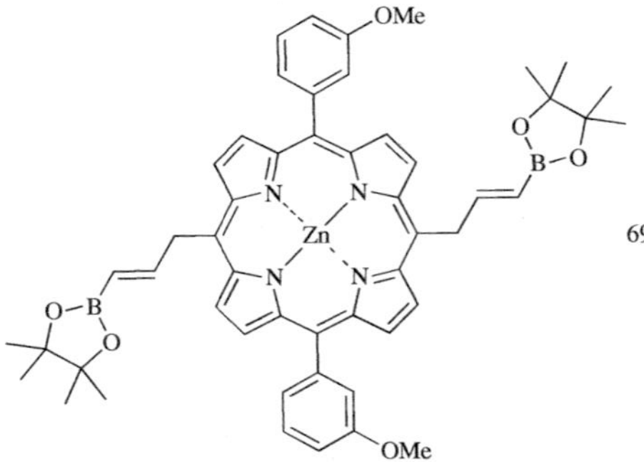 (65) (*E*) only	697

10–20 eq	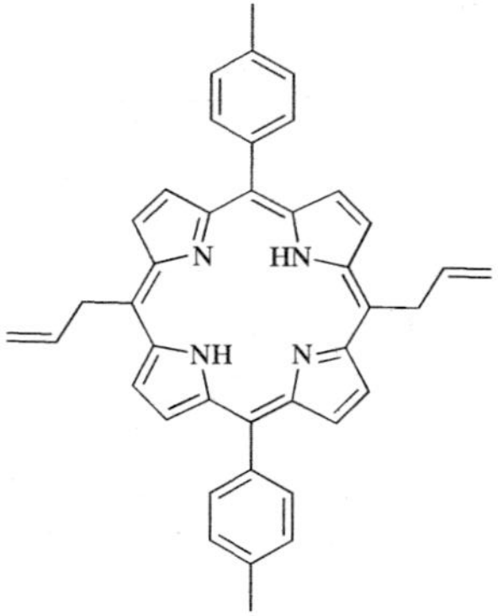	Gr I or Gr II (0.2 eq), DCM, reflux, 12–24 h	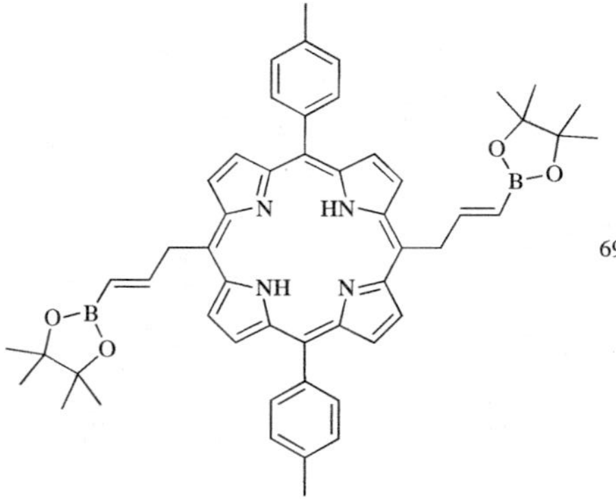 (52) (*E*) only	697
10–20 eq	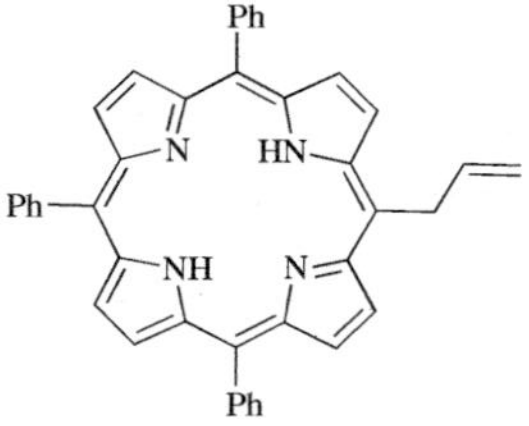	Gr I or Gr II (0.2 eq), DCM, reflux, 12–24 h	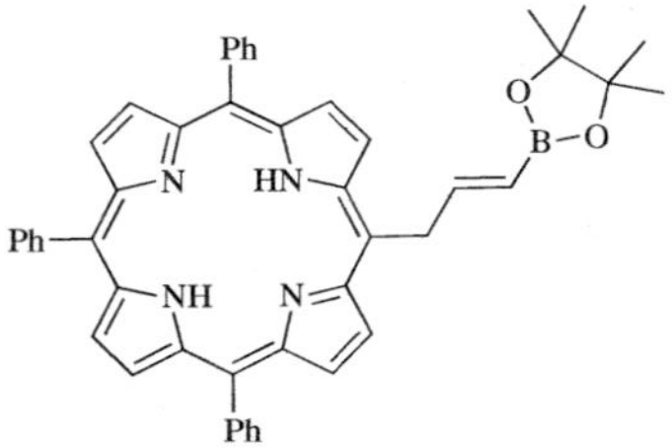 (66) (*E*) only	697
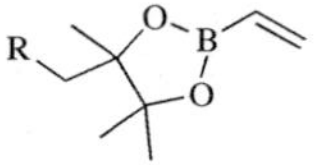 R = Si($CH_2CH_2C_6F_{13}$)$_3$	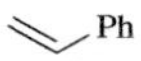 2.13 eq	Ru cat **35** (0.02 eq), $C_6H_5CF_3$, reflux, overnight	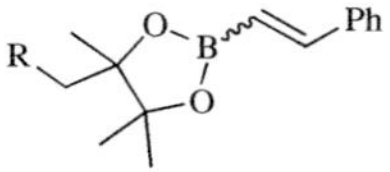 (78)	698

TABLE 11. CROSS-METATHESIS OF VINYL BORONIC COMPOUNDS (*Continued*)
B. VINYL BORONIC ESTERS (*Continued*)

Vinyl Boronic Ester	Cross Partner	Conditions	Product(s) and Yield(s) (%)	Refs.

Please refer to the charts preceding the tables for the catalyst structures.

C_2

Vinyl Boronic Ester	Cross Partner	Conditions	Product(s) and Yield(s) (%)	Refs.
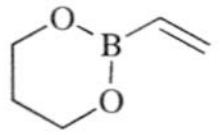	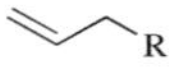 R (5 eq)	Gr I (0.05 eq), DCM, 24 h	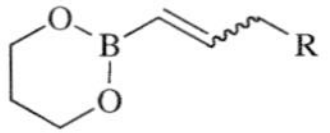	699

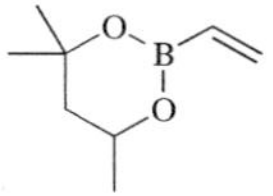

R	Temp (°)		(E)/(Z)
$(EtO)_3Si$	40	(90)	83:17
$(MeO)_3Si$	40	(60)	88:12
$(MeO)_3Si$	rt	(55)	(*E*) only
TMS	40	(88)	(*E*) only

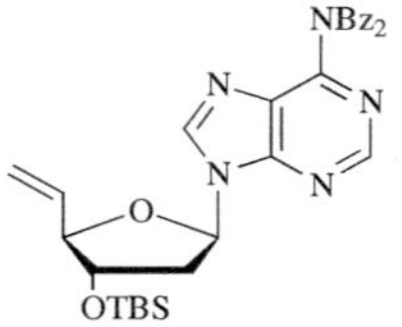

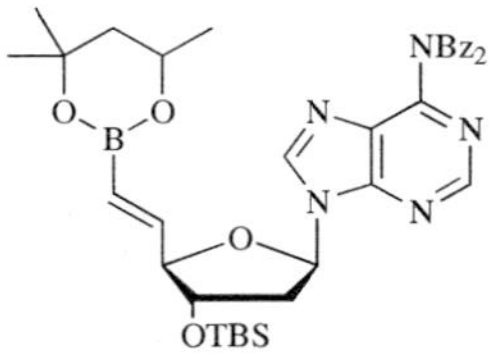

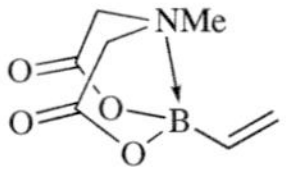

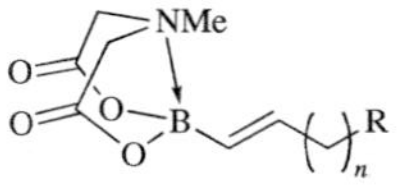

Vinyl Boronic Ester	Cross Partner	Conditions	Product(s) and Yield(s) (%)	Refs.
3 eq	OTBS, NBz_2	Hov II (0.1 eq), DCM, rt, 41 h	OTBS, NBz_2 (44) (*E*) only	700
NMe	R, *n* (*x* eq)	Gr II (0.1 eq), DCM, 40°, 12 h	NMe, R, *n* (*E*)/(*Z*) > 95:5	701

R	*n*	*x*	
Cy	0	1.5	(96)
TIPS	1	2.5	(85)
Me	5	2.5	(80)

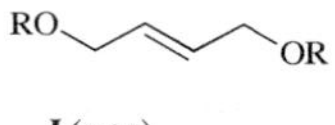

I (x eq)

Gr II (0.1 eq),
DCM, 40°, 12 h

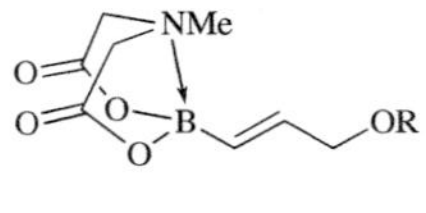

$(E)/(Z) > 95:5$ 701

R	x	$(E)/(Z)$ **I**	
Ac	2.5	(E) only	(84)
Bz	1.5	50:50	(98)

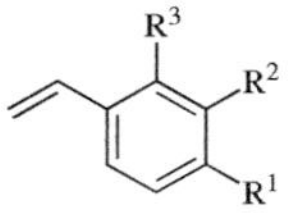

x eq

Gr II (0.1 eq),
DCM, 40°, 12 h

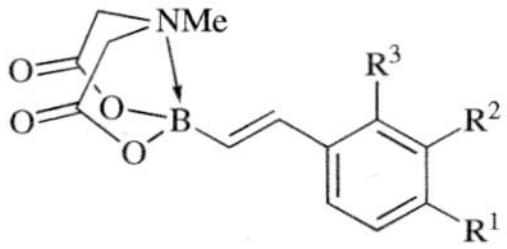

$(E)/(Z) > 95:5$ 701

R^1	R^2	R^3	x	
H	H	H	2.5	(93)
H	H	Br	1.5	(81)
H	Br	H	1.5	(91)
Br	H	H	1.5	(89)

C_3

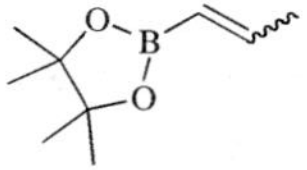

2 eq

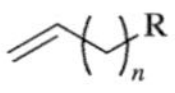

Gr II (0.05 eq),
DCM, reflux

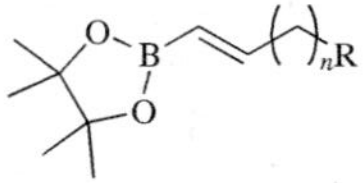

687

R	n		$(E)/(Z)$
BzO	2	(66)	89:11
Me	5	(83)	90:10
c-C_5H_9	0	(80)	>95:5
Cy	0	(80)	>95:5

TABLE 11. CROSS-METATHESIS OF VINYL BORONIC COMPOUNDS (*Continued*)

B. VINYL BORONIC ESTERS (*Continued*)

Vinyl Boronic Ester	Cross Partner	Conditions	Product(s) and Yield(s) (%)	Refs.
Please refer to the charts preceding the tables for the catalyst structures.				
C_3				
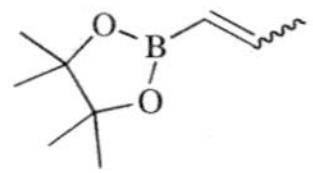 2.5 eq	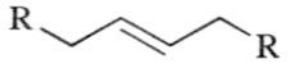 I	Gr II (0.05 eq), DCM, reflux	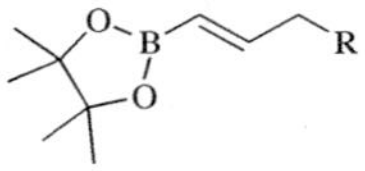	687
1 eq	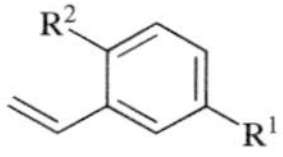	Gr II (0.05 eq), DCM, reflux	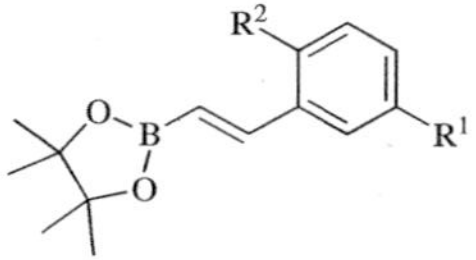 (*E*)/(*Z*) > 95:5	687

R	(*E*)/(*Z*) **I**		(*E*)/(*Z*)
BzO	(*E*) only	(58)	91:9
PhthN	6:94	(65)	94:6

R^1	R^2	
H	H	(92)
H	Br	(80)
O_2N	H	(68)

TABLE 12. CROSS-METATHESIS OF ALLYLIC AMINES (AMINES SUBSTITUTED ON THE α-CARBON)

A. PRIMARY AMINES

Primary Amine	Cross Partner	Conditions	Product(s) and Yield(s) (%)	Refs.
Please refer to the charts preceding the tables for the catalyst structures.				
C_4				
CO_2Me, NHCbz; 2 eq	OTBS	Catalyst (*x* eq), DCM, 25°, 5 h	MeO_2C, NHCbz, OTBS	590
			Catalyst / *x* / / (*E*)/(*Z*)	
			Ru cat **158** / 0.02 / (62) / >99:1	
			Ru cat **157** / 0.005 / (65) / >99:1	
NHCbz, CO_2Me; 2 eq	R^2 R^1	Gr II (0.086 eq), DCM, 35°, 5 h	CO_2Me, CbzHN, R^2 R^1	702
			R^1 / R^2 /	
			H / HO / (83)	
			HO / H / (77)	
CO_2Me, NHBoc; 5 eq	OPMB, R^2 R^1, O, O	Gr II (0.06 eq), DCM, rt, 24 h	NHBoc, MeO_2C, OPMB, R^2 R^1, O, O	703
			R^1 / R^2 /	
			H / HO / (75)	
			HO / H / (78)	

TABLE 12. CROSS-METATHESIS OF ALLYLIC AMINES (AMINES SUBSTITUTED ON THE α-CARBON) (*Continued*)

A. PRIMARY AMINES (*Continued*)

Primary Amine	Cross Partner	Conditions	Product(s) and Yield(s) (%)	Refs.

Please refer to the charts preceding the tables for the catalyst structures.

C_4

Primary Amine	Cross Partner	Conditions	Product(s) and Yield(s) (%)	Refs.
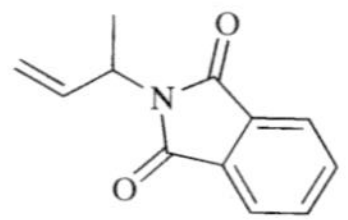	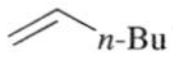 n-Bu, 5 eq	Hov II (0.05 eq), DCM, 40°	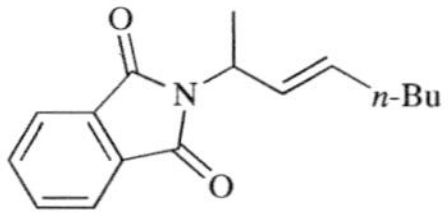 n-Bu (78) (E)/(Z) = 91:9	551
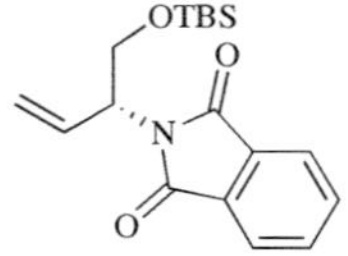 OTBS	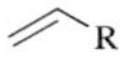 R, 3 eq	Mo cat **18** (0.03 eq), benzene, 22°, 5 h, 7.0 torr	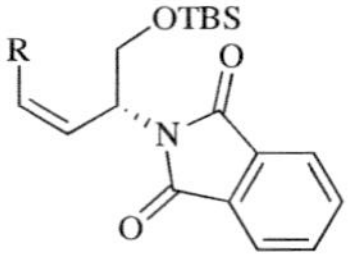 R, OTBS	5

R		(Z)/(E)
$PhO_2C(CH_2)_2$	(63)	96:4
Cy	(65)	97:3
$PMBO(CH_2)_6$	(97)	93:7
$Br(CH_2)_6$	(93)	94:6

Primary Amine	Cross Partner	Conditions	Product(s) and Yield(s) (%)	Refs.
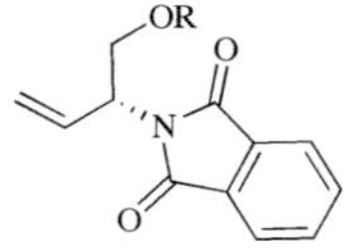 OR	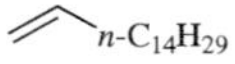 n-$C_{14}H_{29}$, x eq	Catalyst (y eq), solvent	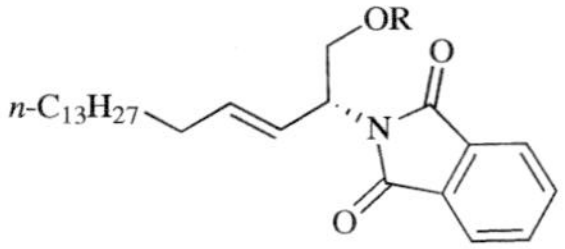 n-$C_{13}H_{27}$, OR	

R	x	Catalyst	y	Solvent	Temp (°)	Time (h)		(E)/(Z)	Refs.
H	4	Gr II	0.05	DCM	40	12	(99)	(E) only	704
TBS	3	Mo cat **5**	0.03	benzene	22	5	(35)	4:96	5
TBS	3	Mo cat **6**	0.03	benzene	22	5	(21)	3:97	5
TBS	3	Mo cat **18**	0.03	benzene	22	5	(88)	3:97	5
TBS	3	Mo cat **8**	0.03	benzene	22	5	(6)	79:21	5
TBS	3	Mo cat **1**	0.03	benzene	22	5	(68)	88:12	5
TBS	3	Hov II	0.03	benzene	22	5	(64)	89:11	5

	Reactant 1	Reactant 2	Conditions	Product(s) and Yield(s) (%)		Refs.
C_5						
	OTBDPS, NHBoc 2 eq	OH, OBn	Gr II (0.1 eq), toluene, 110°, 18 h	TBDPSO, NHBoc, OH, OBn (60)		705
C_6						
	CO_2Me, HN, CCl_3, O	OMe, Ph 1.5 eq	Hov II (0.07 eq), THF, 110°, 24 h	MeO_2C, HN, O, CCl_3, OMe, Ph	(90)	706
C_7						
	NHFmoc, *i*-Pr 1.25 eq	H, N, S, O, O, O, N, H, O*t*-Bu	Hov II (0.5 eq), toluene, 80°, 16 h, Cy_2BCl	FmocHN, *i*-Pr, S, O, O, H, N, O, N, H, O*t*-Bu Loading 20%		154
	NHFmoc, *i*-Pr 1.25 eq	O, H, N, O, N, H, O*t*-Bu	Hov II (0.5 eq), toluene, 80°, 16 h, Cy_2BCl	FmocHN, *i*-Pr, O, H, N, O, N, H, O*t*-Bu Loading 10%		154

TABLE 12. CROSS-METATHESIS OF ALLYLIC AMINES (AMINES SUBSTITUTED ON THE α-CARBON) (*Continued*)

A. PRIMARY AMINES (*Continued*)

Primary Amine	Cross Partner	Conditions	Product(s) and Yield(s) (%)	Refs.
Please refer to the charts preceding the tables for the catalyst structures.				
C_9				
O, N, H, OH, CbzHN	Br 2 x 5 eq	Gr II (0.08 eq), DCM, reflux	HO, H, N, NHCbz, O, Br (69)	707
C_{10}				
NHBoc, Bn 2 eq	O, O, S, NHBn	Gr II, DCM, reflux	BnHN, S, O, O, NHBoc, Bn (39)	154
OBn, OBn, OBn, BnO, O, N(Boc)$_2$, O	n-$C_{14}H_{29}$ 5 eq	Mo cat **18** (0.08 eq), benzene, 22°, 5 h, 1.0 torr	OBn, OBn, OBn, BnO, O, N(Boc)$_2$, O, n-$C_{14}H_{29}$ (85) (*Z*)/(*E*) = 96:4	5

TABLE 12. CROSS-METATHESIS OF ALLYLIC AMINES (AMINES SUBSTITUTED ON THE α-CARBON) (*Continued*)

B. SECONDARY AMINES

Secondary Amine	Cross Partner	Conditions	Product(s) and Yield(s) (%)	Refs.
Please refer to the charts preceding the tables for the catalyst structures.				
C_4				
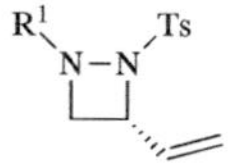 R[1], Ts	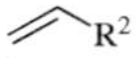 R[2]; *x* eq	Hov II (*y* eq) DCM, reflux	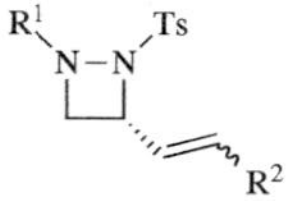 R[1], Ts, R[2]	647
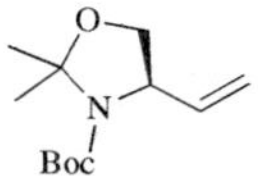 O, N, Boc	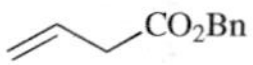 CO_2Bn; 1.2 eq	Gr II (0.004 eq), DCM, reflux, 12 h	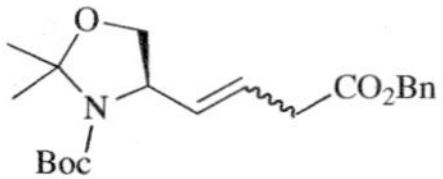 O, N, Boc, CO_2Bn (58)	708
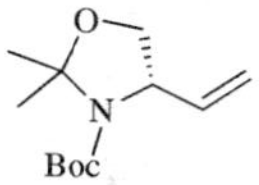 O, N, Boc; 5 eq	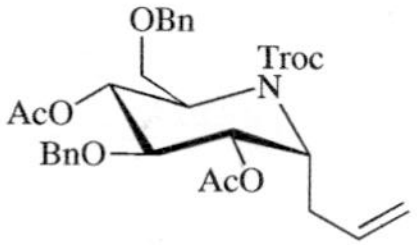 OBn, Troc, N, AcO, BnO, AcO	Gr II (0.2 eq), DCM, reflux, 16–24 h	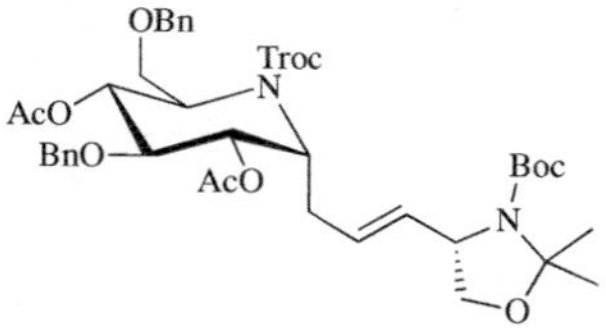 OBn, Troc, N, AcO, BnO, AcO, Boc, N, O; (87) (*E*)/(*Z*) > 95:5	666

R^1	R^2	*x*	*y*	Time (h)	
Cbz	$Br(CH_2)_3$	3	0.05	72	(84)
Cbz	n-C_6H_{13}	3 + 3	0.05 + 0.05	48 + 24	(87)
Boc	n-C_6H_{13}	3	0.05	72	(95)

TABLE 12. CROSS-METATHESIS OF ALLYLIC AMINES (AMINES SUBSTITUTED ON THE α-CARBON) (*Continued*)

B. SECONDARY AMINES (*Continued*)

Secondary Amine	Cross Partner	Conditions	Product(s) and Yield(s) (%)	Refs.

Please refer to the charts preceding the tables for the catalyst structures.

C_{4-10}

Secondary Amine	Cross Partner	Conditions	Product(s) and Yield(s) (%)	Refs.
O, N, Boc, R 5 eq	NHCbz, CO_2Bn	Catalyst (0.05 eq), solvent, reflux	O, N, Boc, NHCbz, CO_2Bn	563

R	Catalyst	Solvent	Time (h)	
H	Gr II	DCM	7	(56)
H	Ru cat **13**	DCM	4.5	(28)
H	Ru cat **271**	DCM	36	(33)
H	Gr II	toluene	3	(64)
Me	Gr II	toluene	4	(76)
Me	Hov II	toluene	3	(56)
Me	Ru cat **112**	toluene	4	(9)
MeO_2C	Gr II	toluene	4	(tr)
$HOCH_2$	Gr II	toluene	3	(9)
Ph	Gr II	toluene	3	(10)

C_4

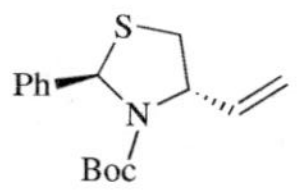

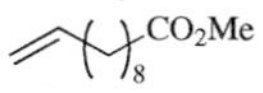

Ph S N Boc

CO_2Me 8

x eq

Catalyst (*y* eq), DCM, reflux

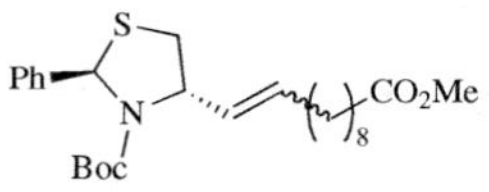

Ph S N Boc CO_2Me 8

x	Catalyst	*y*	Concn (M)	
10	Gr I	0.2	0.04	(0)
10	Gr II	0.2	0.04	(26)
3	Gr II	0.04	0.04	(23)
3	Hov II	0.05	0.03	(81)
1.5	Hov II	0.05	0.03	(53)
3	Hov II	0.025	0.03	(44)
3	Hov II	0.05	0.04	(73)
2.2	Hov II	0.05	0.2	(82)

709

C_5

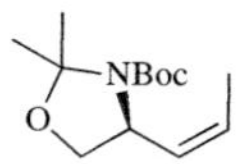

NBoc O

5 eq

(*Z*)/(*E*) = 91:9

NHCbz CO_2Bn

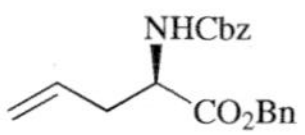

Gr II (0.05 eq), reflux, toluene, 4 h

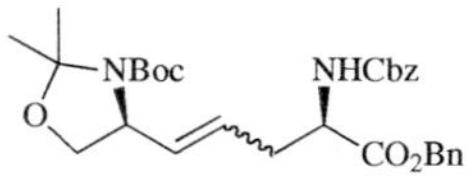

NBoc O NHCbz CO_2Bn

(75)

563

TABLE 12. CROSS-METATHESIS OF ALLYLIC AMINES (AMINES SUBSTITUTED ON THE α-CARBON) (*Continued*)

B. SECONDARY AMINES (*Continued*)

Secondary Amine	Cross Partner	Conditions	Product(s) and Yield(s) (%)	Refs.
Please refer to the charts preceding the tables for the catalyst structures.				
C_6				
Boc, N; 2 eq	O, O, S, NHBn	Gr II, DCM, reflux	Boc, N, O, O, S, NHBn (64)	154
C_7				
Boc, N, MeO_2C	R; 4 eq	Gr II (0.08 eq), $CHCl_3$, 55°	R, N, Boc, CO_2Me **I** + Boc, N, CO_2Me, R **II**	710
R, Cbz, N	Ph; *x* eq	Catalyst (0.1 eq), benzene, 22°, 24 h	R, N, Cbz, Ph; (*E*)/(*Z*) > 98:2	102

R	I + II	I/II
H	(82)	36:64
F	(81)	67:33
MeO	(79)	22:78
MeO_2C	(71)	48:52

R	*x*	Catalyst		er
H	10	Mo cat **12**	(80)	91.0:9.0
H	20	Ru cat **210**	(70)	5.0:95.0
TBSO	10	Mo cat **12**	(80)	92.5:7.5

OTBS, Cbz-N bicyclic substrate + 10 eq styrene (R^1, R^2) → Mo cat **12** (0.1 eq), benzene, 22°, 12 h → OTBS, N-Cbz piperidine product; 102

$(E)/(Z) > 98:2$

R^1	R^2		er
H	H	(93)	95.0:5.0
H	MeO	(65)	90.0:10.0
Me	H	(92)	93.0:7.0

OR^1, Bn-N bicyclic substrate + x eq styrene (R^2, R^3) → Catalyst (0.05 eq), solvent, 22° → OR^1, N-Bn piperidine product; 91

R^1	R^2	R^3	x	Catalyst	Solvent	Time (h)	Conv. (%)[a]		er
H	H	H	20	Ru cat **210**	—	24	>98	(82)	>99.0:1.0
H	H	H	20	Ru cat **212**	—	24	80	(65)	97.0:3.0
H	Br	H	20	Ru cat **210**	—	24	63	(55)	98.5:1.5
H	Br	H	20	Ru cat **212**	—	24	27	(—)	—
H	Me	H	20	Ru cat **210**	—	24	>98	(81)	99.0:1.0
H	Me	H	20	Ru cat **212**	—	24	95	(60)	99.0:1.0
H	F	H	20	Ru cat **210**	—	24	87	(70)	>99.0:1.0
H	H	MeO	20	Ru cat **210**	—	24	87	(67)	>99.0:1.0
H	H	MeO	20	Ru cat **212**	—	24	50	(38)	98.0:2.0
H	H	CF_3	5	Ru cat **210**	—	24	>98	(78)	98.5:1.5
TBS	H	H	20	Ru cat **210**	—	24	>98	(80)	>99.0:1.0
TBS	H	H	20	Ru cat **212**	—	24	70	(—)	98.0:2.0
TBS	H	H	20	Ru cat **213**	—	24	<5	(—)	—
TBS	H	H	10	Mo cat **12**	benzene	1	>98	(85)	60.5:39.5

TABLE 12. CROSS-METATHESIS OF ALLYLIC AMINES (AMINES SUBSTITUTED ON THE α-CARBON) (*Continued*)

B. SECONDARY AMINES (*Continued*)

Secondary Amine	Cross Partner	Conditions	Product(s) and Yield(s) (%)	Refs.
Please refer to the charts preceding the tables for the catalyst structures.				
C_8				
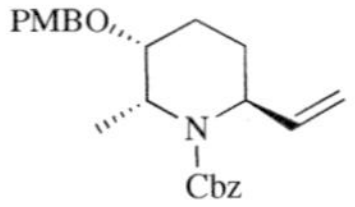 PMBO, N, Cbz	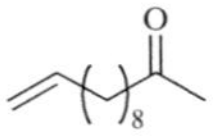 O, 8	Gr II, DCM, reflux	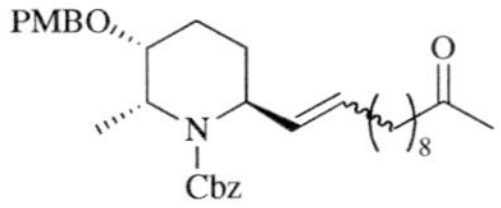 PMBO, N, Cbz, O, 8 (62)	711
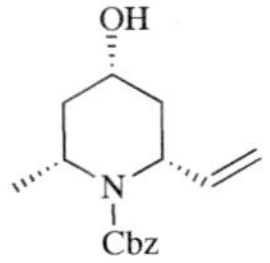 OH, N, Cbz	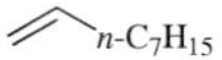 n-C_7H_{15} 10 eq	Gr II (0.1 eq), DCM, reflux, 7 h	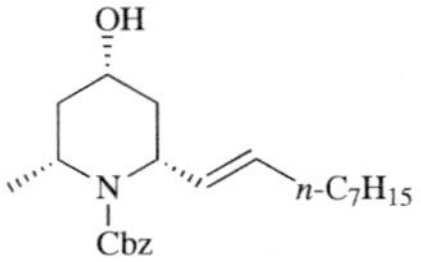 OH, N, Cbz, n-C_7H_{15} (71) (*E*) only	712
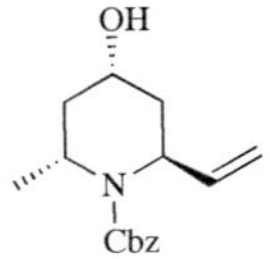 OH, N, Cbz	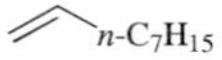 n-C_7H_{15} 10 eq	Gr II (0.1 eq), DCM, reflux, 8 h	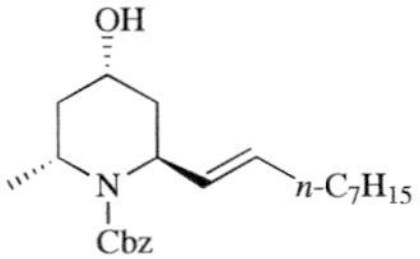 OH, N, Cbz, n-C_7H_{15} (46) (*E*) only	712
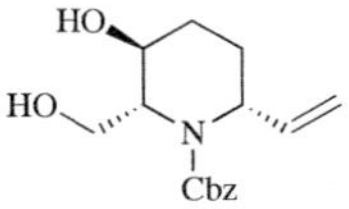 HO, HO, N, Cbz	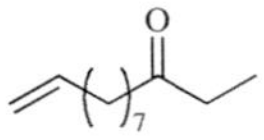 O, 7 5 eq	Hov II (0.12 eq), DCM, 40°, 25 h	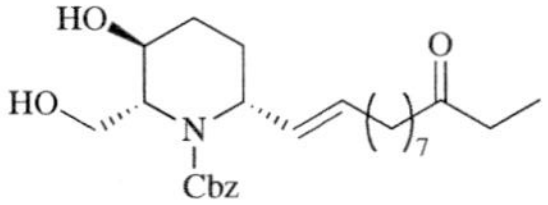 HO, HO, N, Cbz, O, 7 (58) (*E*) only	712

Substrate	Olefin	Conditions	Product(s) and Yield(s) (%)	Refs.
	5 eq	Hov II (0.09 eq), DCM, 40°, 24 h	(62) (*E*) only	712
	3 eq	Hov II (0.05 eq), DCM, 40°, 1 h	(30) (*E*) only	712
	5 eq	Hov II (0.07 eq), DCM, 40°, 1 h	(57) (*E*) only	712
C_9				
	3 eq	Catalyst (0.05 eq), DCM, reflux, 6 h	(*E*)/(*Z*) > 95:5 Catalyst: Hov II (19); Ru cat **271** (30)	177

[a] The conversion was determined by ^{1}H NMR analysis.

TABLE 12. CROSS-METATHESIS OF ALLYLIC AMINES (AMINES SUBSTITUTED ON THE α-CARBON) (*Continued*)

C. TERTIARY AMINES

Tertiary Amine	Cross Partner	Conditions	Product(s) and Yield(s) (%)	Refs.
Please refer to the charts preceding the tables for the catalyst structures.				
C_7				
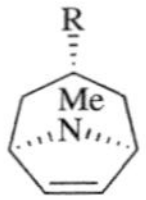 R; Me; N	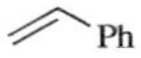 Ph; *x* eq	Mo cat **12** (0.05 eq), benzene, 22°, 1 h	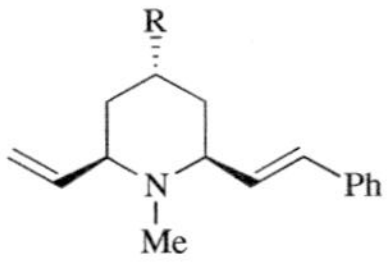 R; N; Me; Ph; (*E*)/(*Z*) > 98:2	102
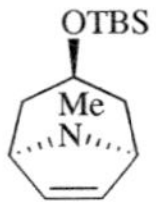 OTBS; Me; N	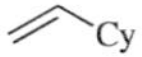 Cy; 10 eq	Mo cat **12** (0.05 eq), benzene, 22°, 12 h	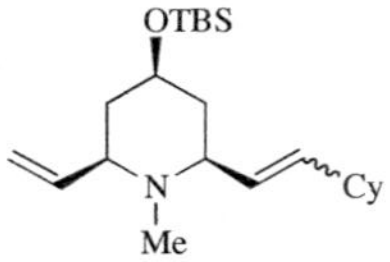 OTBS; N; Me; Cy (43) er 84.0:16.0 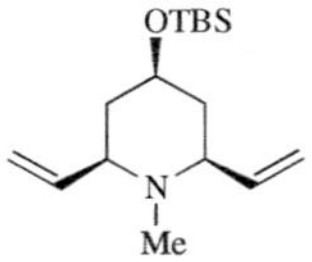 + OTBS; N; Me (25) 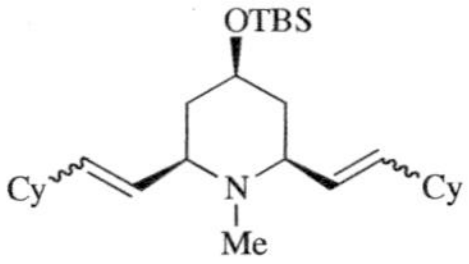 + OTBS; Cy; N; Me; Cy (25)	102

R	*x*		er
H	2	(64)	90.0:10.0
TBSO	10	(90)	92.5:7.5

Ph
x eq

Catalyst (0.05 eq),
solvent, 22°

OTBS
N
Me
Ph

(*E*)/(*Z*) > 98:2

x	Catalyst	Solvent	Time (h)	Conv (%)[a]		er	
20	Ru cat **210**	—	36	30	(—)	66.5:33.5	91
20	Ru cat **212**	—	36	<5	(—)	—	91
20	Ru cat **213**	—	36	<4	(—)	—	91
10	Mo cat **12**	benzene	1	>98	(95)	97.0:3.0	91
2	Ru cat **210**	benzene	12	<2	(—)	—	102
2	Ru cat **212**	benzene	12	<2	(—)	—	102
2	Ru cat **213**	benzene	12	<2	(—)	—	102
2	Mo cat **8**	benzene	12	<2	(—)	—	102
2	Mo cat **9**	benzene	12	>98	(—)	<51.0:49.0	102
2	Mo cat **10**	benzene	12	<2	(—)	—	102
2	Mo cat **12**	benzene	12	>98	(—)	97.0:3.0	102

TABLE 12. CROSS-METATHESIS OF ALLYLIC AMINES (AMINES SUBSTITUTED ON THE α-CARBON) (*Continued*)

C. TERTIARY AMINES (*Continued*)

Tertiary Amine	Cross Partner	Conditions	Product(s) and Yield(s) (%)	Refs.
Please refer to the charts preceding the tables for the catalyst structures.				
C_7				
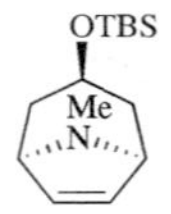 OTBS; Me; N	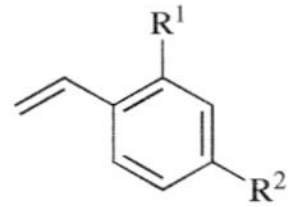 R^1; R^2; 10 eq	Mo cat **12** (0.05 eq), benzene, 22°, 12 h	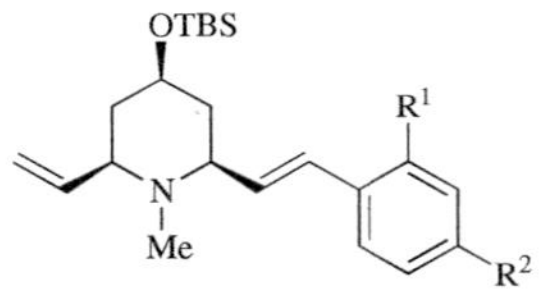 OTBS; N; Me; R^1; R^2; (*E*)/(*Z*) > 98:2	102
C_{11}				
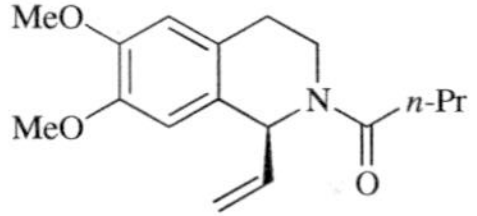 MeO; MeO; N; *n*-Pr; O	 OTBS; 3 eq	Hov II (0.2 eq), BQ, toluene, reflux, 24 h	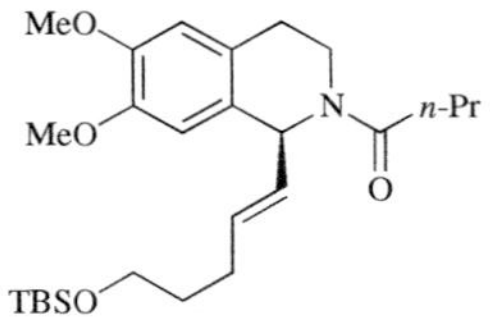 MeO; MeO; N; *n*-Pr; O; TBSO; (52) (*E*)/(*Z*) = 91:9	713

R^1	R^2		er
H	H	(95)	97.0:3.0
H	MeO	(92)	94.5:5.5
H	CF_3	(88)	82.0:18.0
Br	H	(86)	94.0:6.0
Me	H	(91)	99.0:1.0

[a] The conversion was determined by ^1H NMR analysis.

TABLE 13. CROSS-METATHESIS OF SECONDARY ALLYLIC ALCOHOLS AND DERIVATIVES

A. ALLYLIC ALCOHOLS

Allylic Alcohol	Cross Partner	Conditions	Product(s) and Yield(s) (%)	Refs.

Please refer to the charts preceding the tables for the catalyst structures.

C_{3-9}

Allylic Alcohol	Cross Partner	Conditions	Product(s) and Yield(s) (%)	Refs.
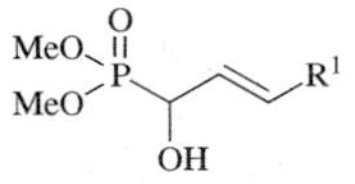 MeO, MeO, P=O, OH, R^1	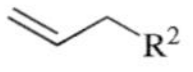 R^2, 5 eq	Gr II (0.05 eq), solvent, 40°	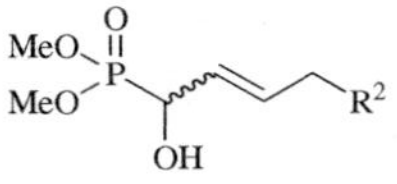 MeO, MeO, P=O, OH, R^2	221

R^1	R^2	Solvent	Time (h)	
H	AcO	$CHCl_3$	—	(<30)
H	TMS	$CHCl_3$	—	(82)
Me	AcO	DCM	12	(46)
Ph	AcO	DCM	96	(0)

C_3

Allylic Alcohol	Cross Partner	Conditions	Product(s) and Yield(s) (%)	Refs.
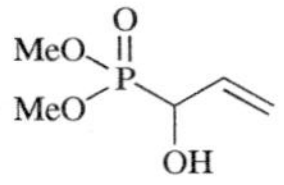 MeO, MeO, P=O, OH	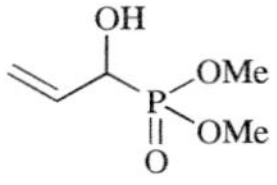 OH, P=O, OMe, OMe	Gr II (0.05 eq), DCM, 40°	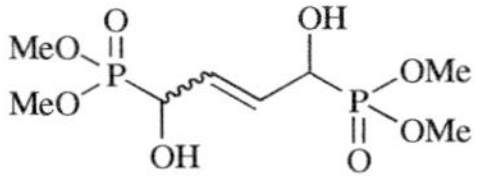 MeO, MeO, P=O, OH, OH, P=O, OMe, OMe (76)	221
	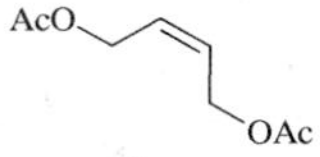 AcO, OAc, 5 eq	Gr II (0.05 eq), $CHCl_3$, 40°	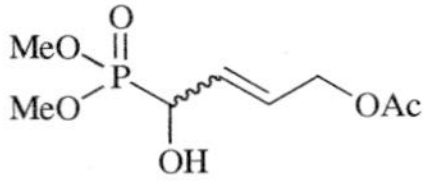 MeO, MeO, P=O, OH, OAc (61)	221

C_{3-4}

Allylic Alcohol	Cross Partner	Conditions	Product(s) and Yield(s) (%)	Refs.
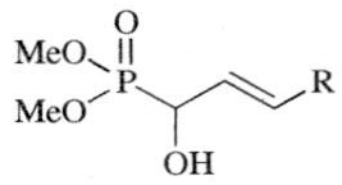 MeO, MeO, P=O, OH, R	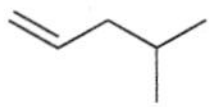 *x* eq	Gr II (*y* eq), solvent, 40°	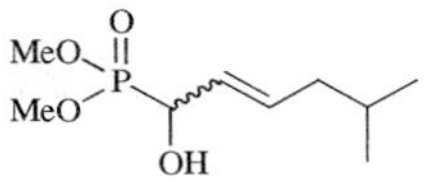 MeO, MeO, P=O, OH	221

R	*x*	*y*	Solvent	Time (h)	
H	5	0.05	$CHCl_3$	—	(57)
Me	3.75	0.0375	DCM	12	(48)

TABLE 13. CROSS-METATHESIS OF SECONDARY ALLYLIC ALCOHOLS AND DERIVATIVES (*Continued*)

A. ALLYLIC ALCOHOLS (*Continued*)

Allylic Alcohol	Cross Partner	Conditions	Product(s) and Yield(s) (%)	Refs.
Please refer to the charts preceding the tables for the catalyst structures.				
C_3				
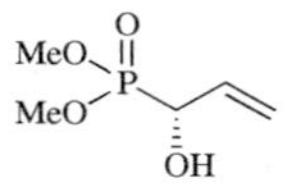 MeO, MeO, P, O, OH; er 85.0:15.0	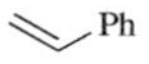 Ph; 5 eq	Gr II (0.05 eq), $CHCl_3$, 40°	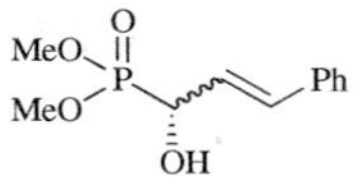 MeO, MeO, P, O, OH, Ph (51) er 85.0:15.0	221
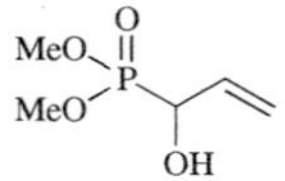 MeO, MeO, P, O, OH	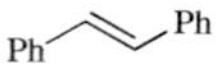 Ph, Ph; 5 eq	Gr II (0.05 eq), $CHCl_3$, 40°	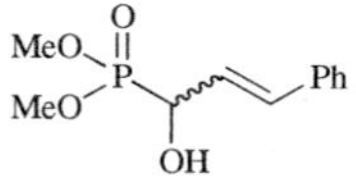 MeO, MeO, P, O, OH, Ph (31)	221
C_4				
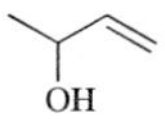 OH	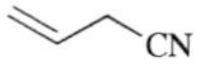 CN; 2 eq	Gr II (0.05 eq), DCM, reflux, 2 h	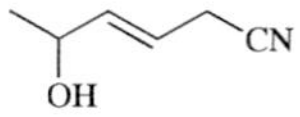 OH, CN (21) (*E*)/(*Z*) = 83:17	182
C_{4-9}				
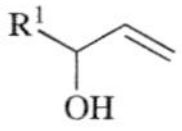 R^1, OH; 2 eq	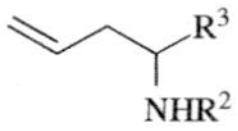 R^3, NHR^2	Gr II (*x* eq), DCM, 45°	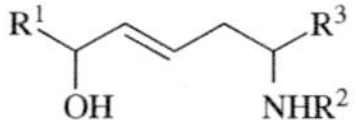 R^1, OH, R^3, NHR^2	714

R^1	R^2	R^3	*x*		(*E*)/(*Z*)
Me	Troc	MeO_2C	0.03	(80)	83:17
Me	Cbz	MeO_2C	0.03	(74)	83:17
Me	Fmoc	MeO_2C	0.03	(72)	83:17
Me	Tos	MeO_2C	0.03	(75)	50:50
Me	*p*-Nos	MeO_2C	0.03	(78)	29:71
Me	Boc	MeO_2C	0.015	(85)	55:45
Me	Troc	$AcOCH_2$	0.04	(90)	(*E*) only
Ph	Troc	MeO_2C	0.04	(88)	(*E*) only

C_4

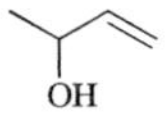

OH

3.2 eq

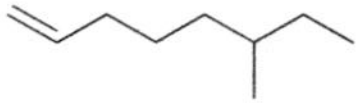

Gr **II** (0.0063 eq), DCM, 0.5 h, 45°

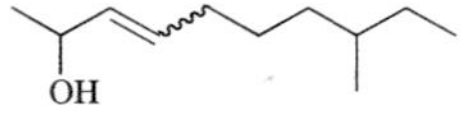

OH

(67) 443

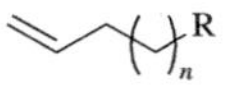

R

n

x eq

Gr **II** (y eq), solvent

R

n

OH

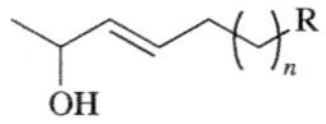

R	n	x	y	Solvent	Temp (°)	Time (h)		$(E)/(Z)$	
AcO	3	2	0.03	DCM	40	12	(92)	93:7	36
AcO	3	1	0.03	DCM	40	12	(50)	93:7	36
AcO	3	1	0.03	DCM	23	12	(62)	93:7	36
MeO_2C	22	3	0.08	DCE	60	5	(65)	>95:5	715

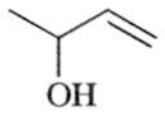

OH

I

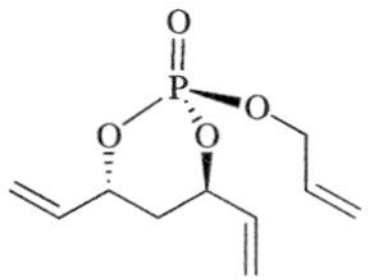

1. Hov II (0.06 eq), BQ, DCM, reflux, 0.75 h
2. Hov II (0.04 eq), **I** (2–4 eq), CuI, DCE, reflux, 4–6 h
3. *o*-NBSH (15 eq), Et_3N, rt, overnight

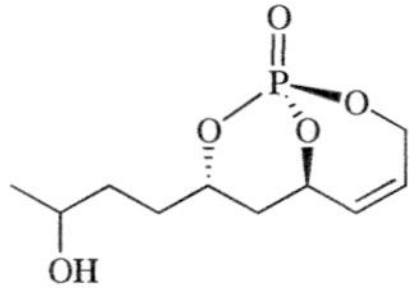

OH

(48) 716

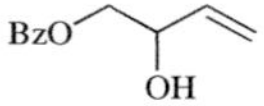

BzO

OH

2 eq

N

Cbz

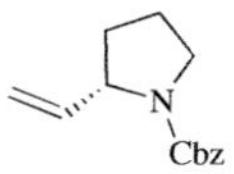

Hov II (0.05 eq), DCM, reflux, 60 h

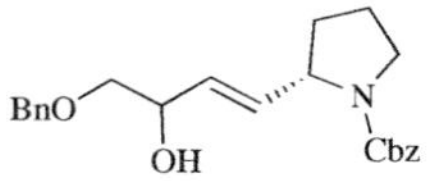

BnO

OH

N

Cbz

(60) $(E)/(Z)$ = 50:50 717

TABLE 13. CROSS-METATHESIS OF SECONDARY ALLYLIC ALCOHOLS AND DERIVATIVES (*Continued*)

A. ALLYLIC ALCOHOLS (*Continued*)

Allylic Alcohol	Cross Partner	Conditions	Product(s) and Yield(s) (%)	Refs.
Please refer to the charts preceding the tables for the catalyst structures.				
C_4				
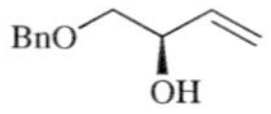 BnO, OH; 4–5 eq	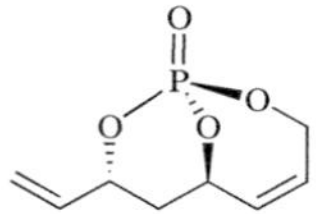	Hov II (0.09 eq), DCM, 40°, 3 h	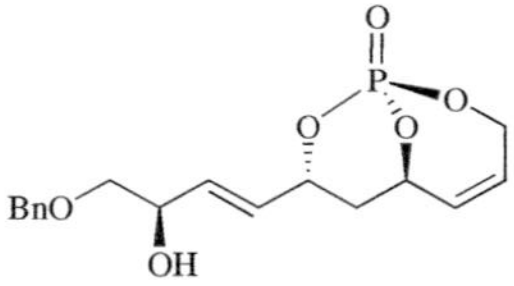 BnO, OH (72) (*E*)/(*Z*) > 99:1	719
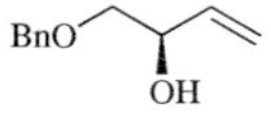 BnO, OH; **I**	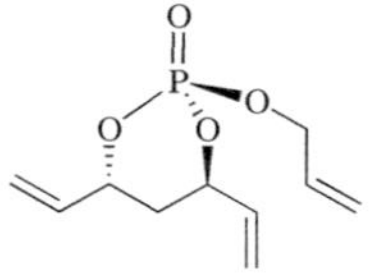	1. Hov II (0.06 eq), BQ, DCM, reflux, 0.75 h, 2. Hov II (0.04 eq), **I** (2–4 eq), CuI, DCE, reflux, 4–6 h 3. *o*-NBSH (15 eq), Et_3N, rt, overnight	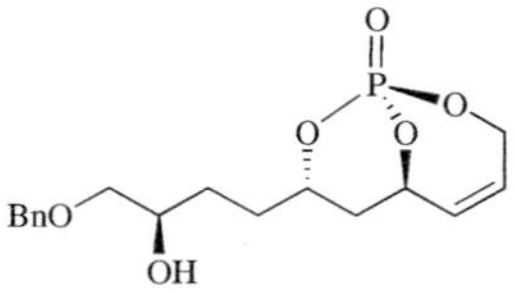 BnO, OH (41)	716
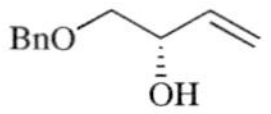 BnO, OH; **I**	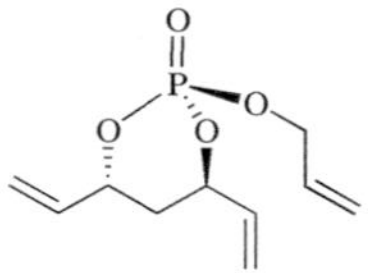	1. Hov II (0.06 eq), BQ, DCM, reflux, 0.75 h, 2. Hov II (0.04 eq), **I** (2–4 eq), CuI, DCE, reflux, 4–6 h 3. *o*-NBSH (15 eq), Et_3N, rt, overnight	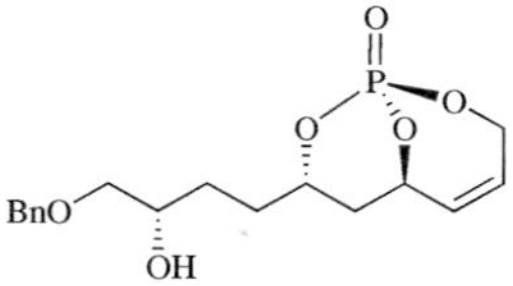 BnO, OH (35)	716
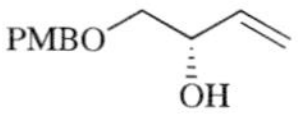 PMBO, OH	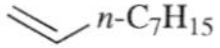 *n*-C_7H_{15}; 5 eq	Gr II (0.1 eq), DCM, reflux, 24 h	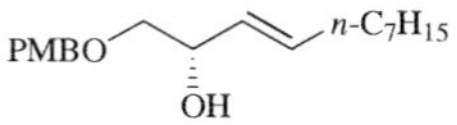 PMBO, OH, *n*-C_7H_{15} (81) (*E*)/(*Z*) = 95:5	718

	Substrate 1	Substrate 2	Conditions	Product(s) and Yield(s) (%)	Refs.
C_5	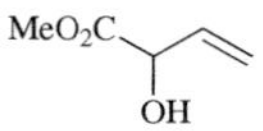 MeO_2C, OH; 3 eq	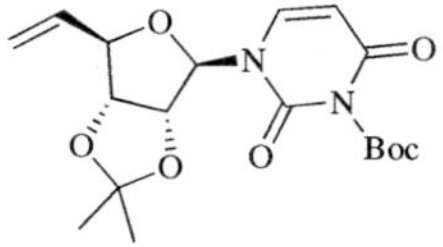 O, N, O, N, O, Boc, O, O	Gr II (0.14 eq), DCM, reflux, 24 h	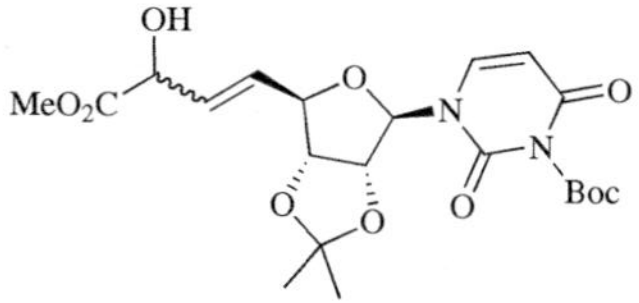 OH, MeO_2C, O, N, O, N, O, Boc, O, O (85)	721
	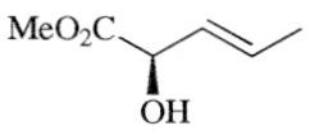 MeO_2C, OH	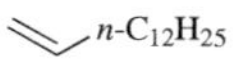 n-$C_{12}H_{25}$; 2 eq	Gr II (0.05 eq), AcOH (0.5 eq), DCM, reflux, 60 h	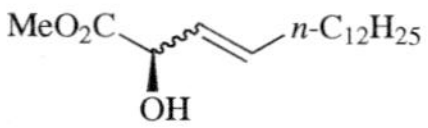 MeO_2C, n-$C_{12}H_{25}$, OH (73)	722
	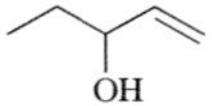 OH; 1 eq	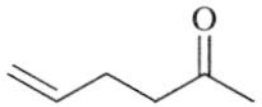 O	Hov II (0.006 eq), DCM, reflux	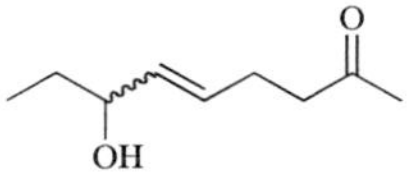 O, OH (48)	723
	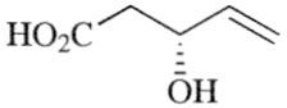 HO_2C, OH	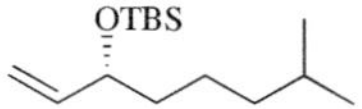 OTBS; 1.2 eq	Gr II (0.05 eq), DCM, reflux, 1 h	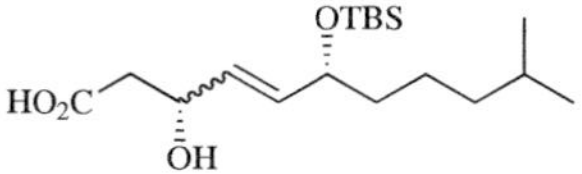 OTBS, HO_2C, OH (72)	724
	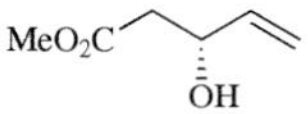 MeO_2C, OH	 OPMB; 2 eq	Gr II (0.05 eq), DCM, 40°, 17.5 h	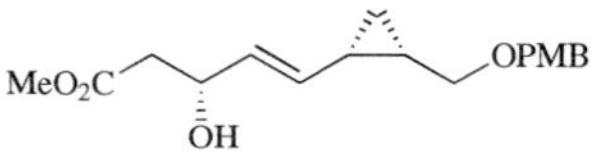 MeO_2C, OPMB, OH (70) $(E)/(Z)$ = 91:9	725

TABLE 13. CROSS-METATHESIS OF SECONDARY ALLYLIC ALCOHOLS AND DERIVATIVES (*Continued*)

A. ALLYLIC ALCOHOLS (*Continued*)

Allylic Alcohol	Cross Partner	Conditions	Product(s) and Yield(s) (%)	Refs.
Please refer to the charts preceding the tables for the catalyst structures.				
C_5				
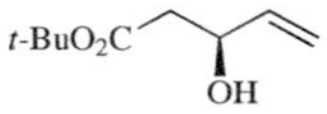 t-BuO_2C … OH	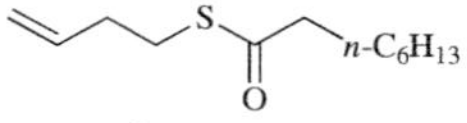 … S … O … n-C_6H_{13} (2 eq)	Gr II (0.03 eq), DCM, 40°, 12 h	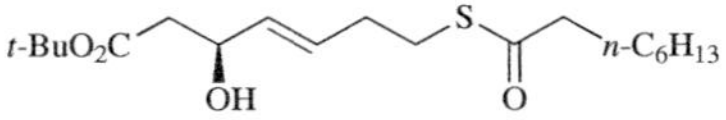 t-BuO_2C … OH … S … O … n-C_6H_{13} (67) (*E*) only	726
$C_{5–6}$				
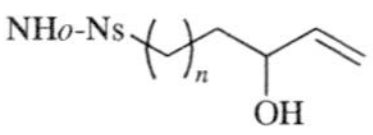 NHo-Ns ( )$_n$ … OH	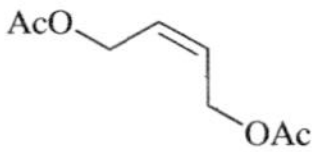 AcO … OAc	Gr II (x eq), DCM, reflux	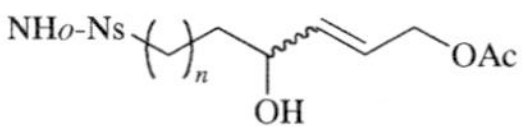 NHo-Ns ( )$_n$ … OH … OAc n / x: 1 / 0.03 (36); 2 / 0.05 (60)	727
C_5				
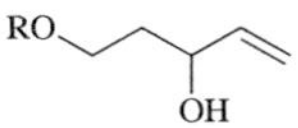 RO … OH (9.8 eq)	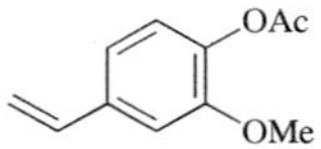 OAc, OMe (styrene)	Hov II (x eq), DCM, rt, 16 h	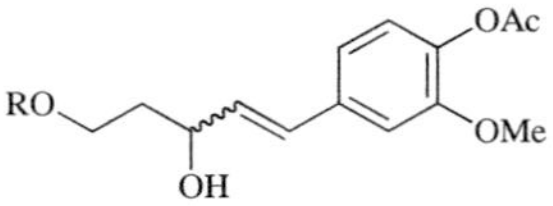 RO … OH … OAc, OMe R / x: H / 0.07 (41); TBS / 0.1 (63)	728
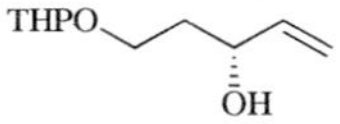 THPO … OH	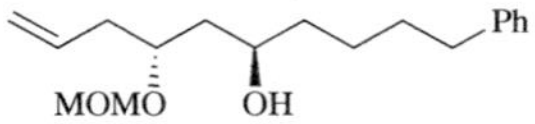 MOMO … OH … Ph (2 eq)	Gr II (0.10 eq), DCM, 0°, 1 h	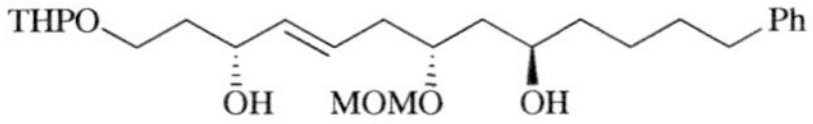 THPO … OH … MOMO … OH … Ph (68) (*E*)/(*Z*) = 95:5	720

PivO OH 2 eq | OMe I | Gr II (0.1 eq), toluene, 80°, 24 h | PivO OH OMe I | (44) | 729

TBSO NHBoc OH | R ( )$_n$ 4 eq | Gr II (0.03 eq), DCM, reflux, 2 h | TBSO NHBoc OH ( )$_n$ R | | 730

R	n	
HO	8	(76)
Me	5	(55)

( )$_n$ I 2 eq | Gr II (0.03 eq), DCM, reflux, 2 h | TBSO NHBoc OH ( )$_n$ I | (*E*) only | 731

n	
5	(51)
7	(47)
8	(54)
9	(52)
10	(51)
12	(57)

TABLE 13. CROSS-METATHESIS OF SECONDARY ALLYLIC ALCOHOLS AND DERIVATIVES (*Continued*)

A. ALLYLIC ALCOHOLS (*Continued*)

Allylic Alcohol	Cross Partner	Conditions	Product(s) and Yield(s) (%)	Refs.

Please refer to the charts preceding the tables for the catalyst structures.

C_5

Allylic Alcohol	Cross Partner	Conditions	Product(s) and Yield(s) (%)	Refs.
RO, NHBoc, OH	n-$C_{13}H_{27}$ (*x* eq)	Hov II (0.1 eq), DCM, 21°, 16 h	RO, NHBoc, OH, n-$C_{13}H_{27}$	732
R^2O, NHR^1, OH	n-$C_{13}H_{27}$ (4 eq)	Gr II (*x* eq), DCM, reflux, 2 h	R^2O, NHR^1, OH, n-$C_{13}H_{27}$	730

R	*x*		(*E*)/(*Z*)
TIPS	3	(80)	(*E*) only
Bn	2	(84)	—

R^1	R^2	*x*		(*E*)/(*Z*)
Boc	TBS	0.03	(72)	—
Boc	TBS	0.01	(75)	94:6
Boc	H	0.03	(58)	—
n-$C_{15}H_{31}CO$	H	0.03	(56)	—
Boc	$(MeO)_2PO$	0.03	(76)	—
Boc	$(Br(CH_2)_2O)(MeO)PO$	0.03	(70)	—
n-$C_{15}H_{31}CO$	$(Br(CH_2)_2O)(MeO)PO$	0.03	(55)	—

Substrate	Partner	Conditions	Product(s) and Yield(s) (%)	Refs.
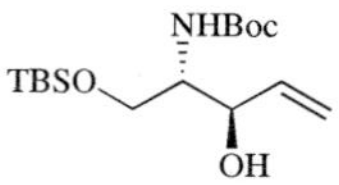 TBSO, NHBoc, OH	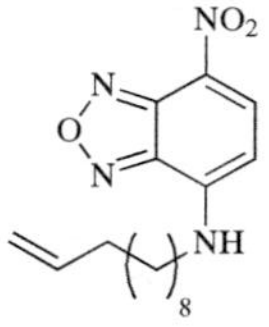 NO_2, N, O, N, NH, 8; 4 eq	Gr II (0.03 eq), DCM, reflux, 2 h	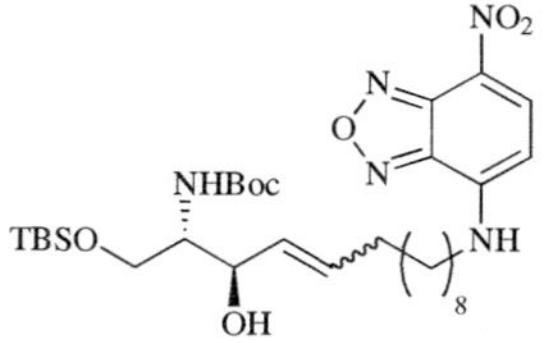 TBSO, NHBoc, OH, NO_2, N, O, N, NH, 8 (57)	730
	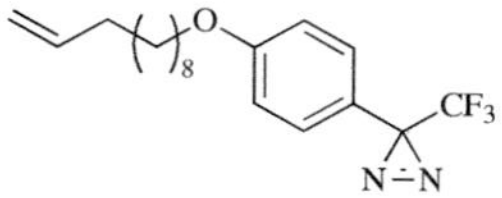 O, 8, CF_3, N=N; 4 eq	Gr II (0.03 eq), DCM, rt, 0.5 h	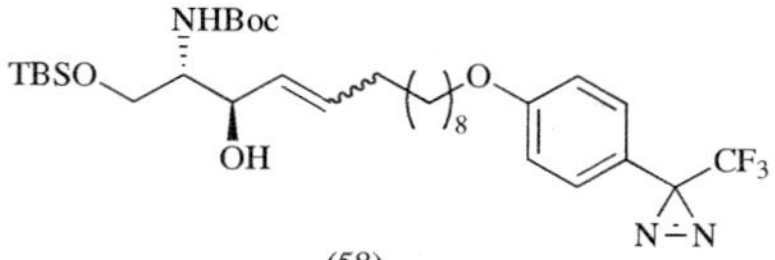 TBSO, NHBoc, OH, O, 8, CF_3, N=N (58)	730
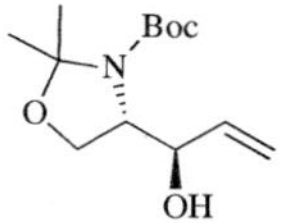 Boc, N, O, OH	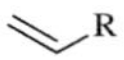 R; 4 eq	Gr II (0.03 eq), DCM, reflux, 5 h	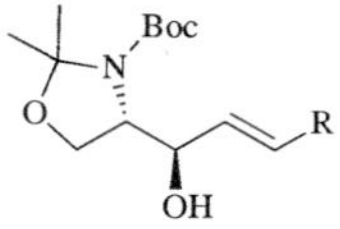 Boc, N, O, OH, R; (*E*) only R: Ph (65); n-C_9H_{19} (69)	733
	 NHBoc, 8; 4 eq	Gr II (0.10 eq), DCM, reflux, 4 h	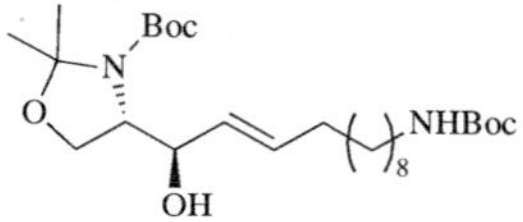 Boc, N, O, OH, NHBoc, 8 (86) (*E*) only	734

TABLE 13. CROSS-METATHESIS OF SECONDARY ALLYLIC ALCOHOLS AND DERIVATIVES (*Continued*)

A. ALLYLIC ALCOHOLS (*Continued*)

Allylic Alcohol	Cross Partner	Conditions	Product(s) and Yield(s) (%)	Refs.
Please refer to the charts preceding the tables for the catalyst structures.				
C_5				
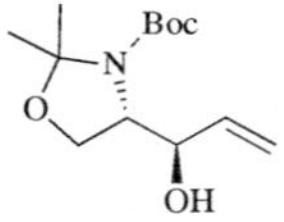 Boc, N, O, OH	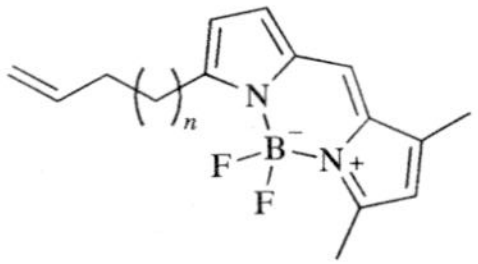 n, N, B, F, N+, F; 2 eq	Gr II (0.1 eq), DCM, reflux, 5 h	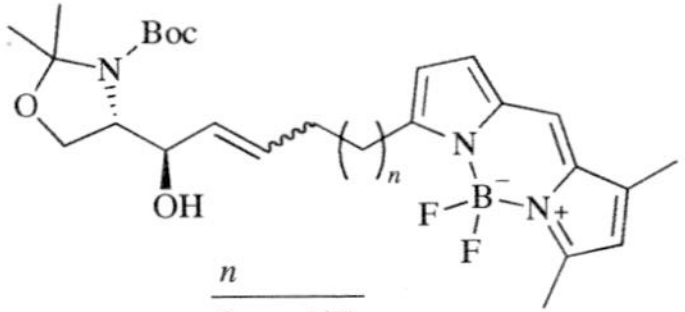 Boc, N, O, OH, n, N, B, F, N+, F; n = 3 (67); n = 8 (57)	735
	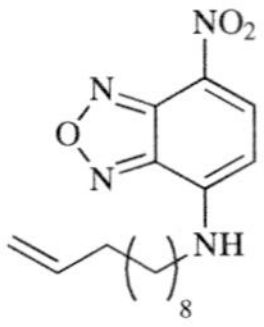 NO_2, N, O, N, NH, 8; 2 eq	Gr II (0.1 eq), DCM, reflux, 5 h	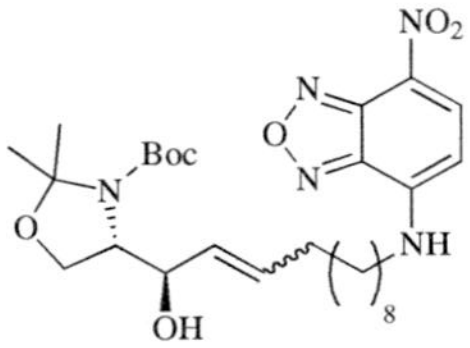 Boc, N, O, OH, NO_2, N, O, N, NH, 8 (66)	735
	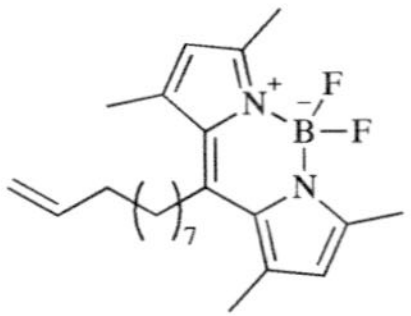 N+, F, B, F, N, 7; 2 eq	Gr II (0.1 eq), DCM, reflux, 5 h	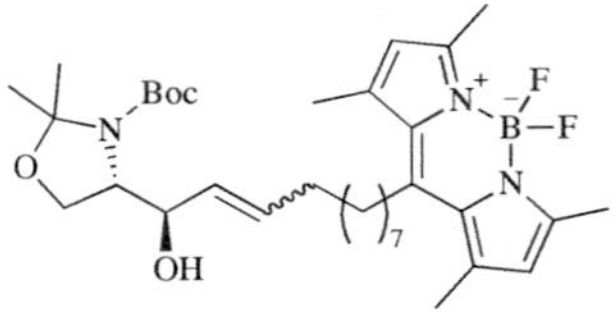 Boc, N, O, OH, N+, F, B, F, N, 7 (79)	735

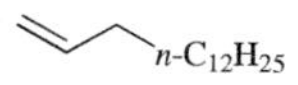

n-$C_{12}H_{25}$

2 eq

Catalyst (x eq), additive, solvent

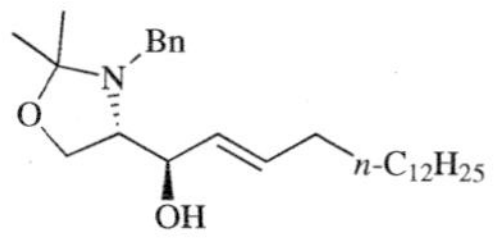

Bn N O OH n-$C_{12}H_{25}$

736

Catalyst	x	Additive	Solvent	Temp (°)	Time (h)		$(E)/(Z)$
Gr I	0.05	—	DCM	40	43	(0)	—
Gr II	0.05	—	DCM	rt	72	(35)	93:7
Gr II	0.05	—	DCM	40	24	(49)	83:17
Gr II	0.1	—	toluene	50	1.5	(7)	92:8
Gr I	0.05	$Ti(Oi\text{-}Pr)_4$	DCM	40	40	(7)	—
Gr II	0.1	$Ti(Oi\text{-}Pr)_4$	DCM	40	22	(30)	—
Gr II	0.1	$Ti(Oi\text{-}Pr)_4$	DCM	40	0.75	(52)	94:6
Ru cat **67**	0.1	—	DCM	rt	48	(53)	93:7
Ru cat **67**	0.1	—	DCM	40	1.5	(59)	94:6

C_6

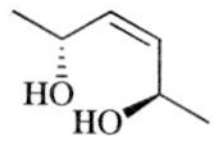

HO HO

3 eq

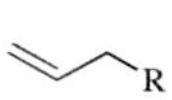

R

Hov II (0.005 eq), DCM, 200°, 7 h

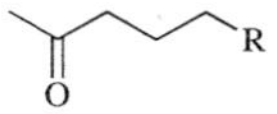

O R

R	
BnO	(52)
Ph	(63)

737

TABLE 13. CROSS-METATHESIS OF SECONDARY ALLYLIC ALCOHOLS AND DERIVATIVES (*Continued*)

A. ALLYLIC ALCOHOLS (*Continued*)

Allylic Alcohol	Cross Partner	Conditions	Product(s) and Yield(s) (%)	Refs.
Please refer to the charts preceding the tables for the catalyst structures.				
C_6				
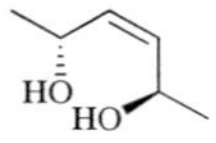 HO, HO; 3 eq	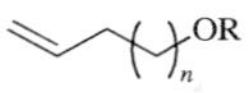 OR, n	Hov II (0.005 eq), DCM	 O, OR, n	737
			R / n / Temp (°) / Time (h): Ac / 3 / 200 / 7 (71); Bn / 4 / rt, 200 / 1 + 7 (57)	
3 eq	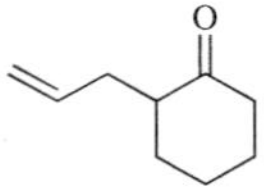 O	1. Hov II (0.005 eq), DCM, rt, 2 h 2. 200°, 7 h	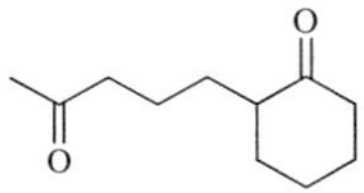 O, O (51)	737
3 eq	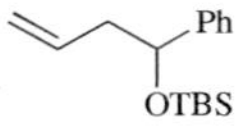 Ph, OTBS	1. Hov II (0.005 eq), DCM, rt, 2 h 2. 200°, 7 h	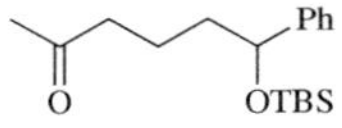 O, Ph, OTBS (69)	737
3 eq	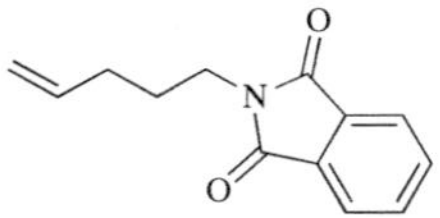 N, O, O	1. Hov II (0.005 eq), DCM, rt, 1 h 2. 200°, 7 h	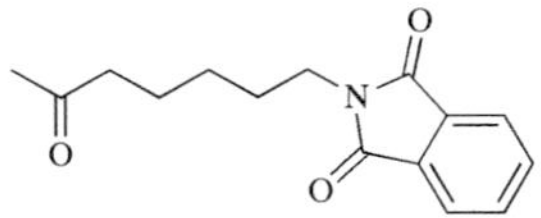 O, N, O, O (69)	737
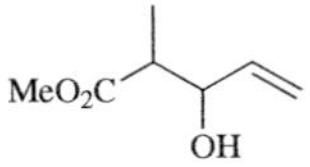 MeO_2C, OH; 2 eq	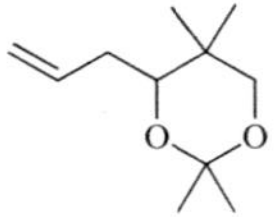 O, O	Gr II (0.025 eq), DCM, reflux, 14 h	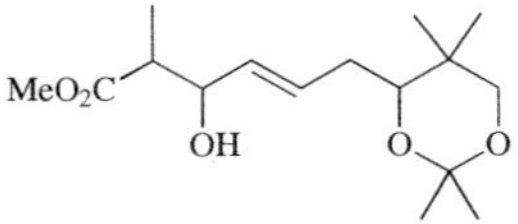 MeO_2C, OH, O, O (80)	738

Alkene 1	Alkene 2	Conditions	Product(s) and Yield(s) (%)	Refs.
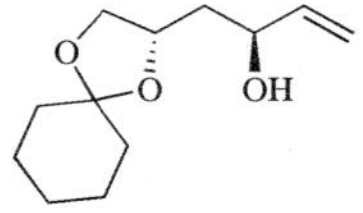 (structure: OH)	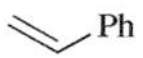 Ph (1.5 eq)	Gr II (0.05 eq), benzene, reflux, 6 h	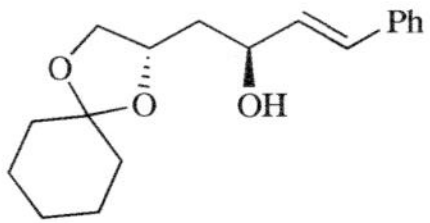 (structure: Ph, OH) (84) $(E)/(Z) = 95{:}5$	739
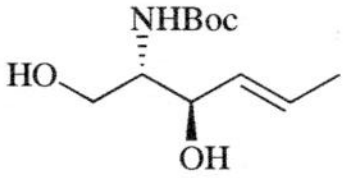 (structure: NHBoc, HO, OH)	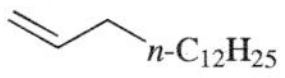 n-$C_{12}H_{25}$ (x eq)	Gr II (0.08 eq), benzene, 55°	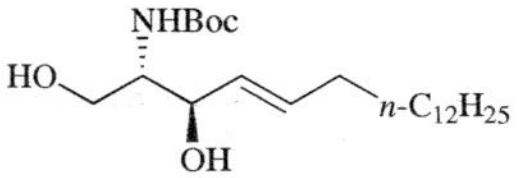 (structure: NHBoc, HO, OH, n-$C_{12}H_{25}$) (E) only; x = 1 (62); x = 2 (30)	740
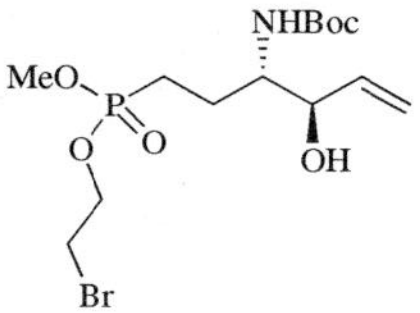 (structure: MeO, P, O, NHBoc, OH, Br)	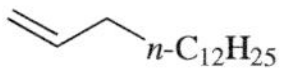 n-$C_{12}H_{25}$ (4 eq)	Gr II (0.03 eq), DCM, reflux, 2 h	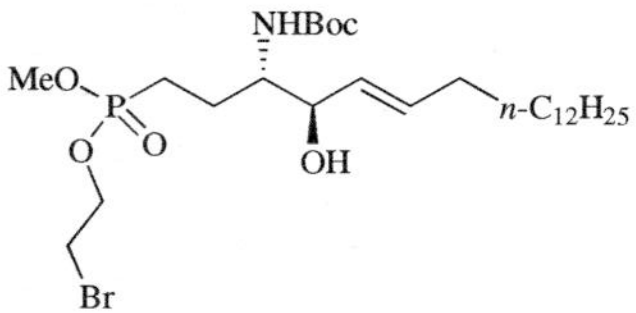 (structure: MeO, P, O, NHBoc, OH, n-$C_{12}H_{25}$, Br) (60) $(E)/(Z) = 95{:}5$	741
	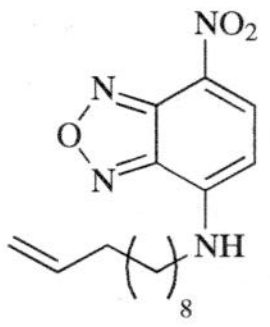 (structure: NO_2, N, O, N, NH, 8) (4 eq)	Gr II (0.03 eq), DCM, reflux, 2 h	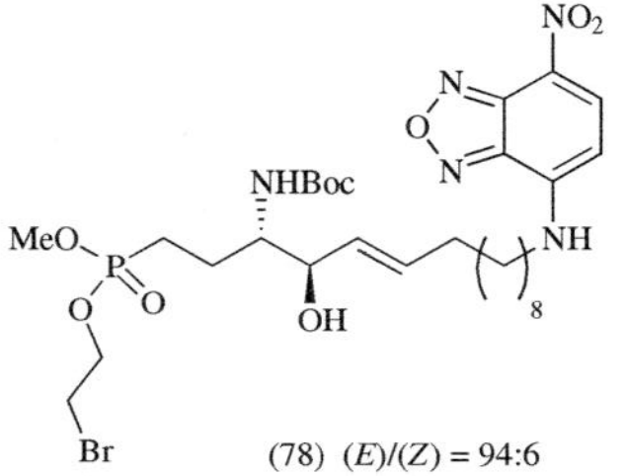 (structure: MeO, P, O, NHBoc, OH, NO_2, NH, 8, Br) (78) $(E)/(Z) = 94{:}6$	741

TABLE 13. CROSS-METATHESIS OF SECONDARY ALLYLIC ALCOHOLS AND DERIVATIVES (*Continued*)

A. ALLYLIC ALCOHOLS (*Continued*)

Allylic Alcohol	Cross Partner	Conditions	Product(s) and Yield(s) (%)	Refs.
Please refer to the charts preceding the tables for the catalyst structures.				
C_6				
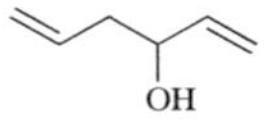 OH	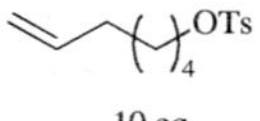 OTs, 4; 10 eq	Gr II (0.075 eq), DCM, 20°, 5 h	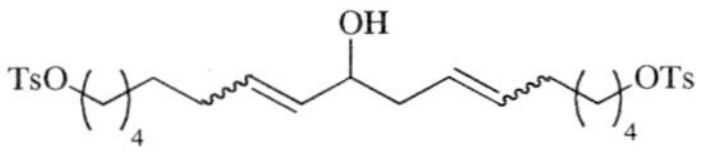 TsO, 4, OH, OTs, 4 (99)	403
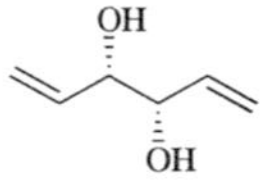 OH, OH	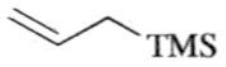 TMS; 3 eq	Gr II (0.05 eq), DCM, 40°, 12 h	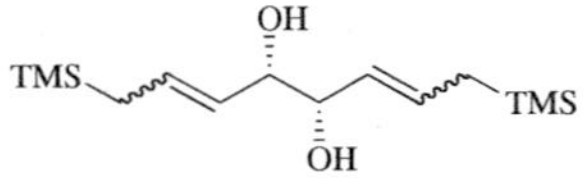 TMS, OH, OH, TMS (82)	209
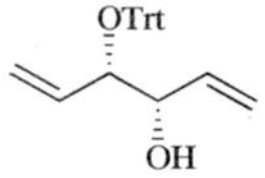 OTrt, OH	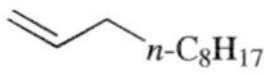 n-C_8H_{17}; 8 eq	Ru cat **139** (0.01 eq), DCM, 20°, 6 h	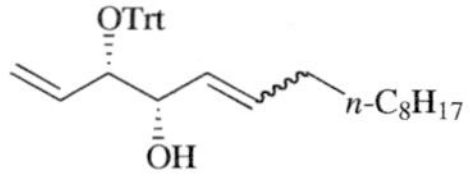 OTrt, OH, n-C_8H_{17} (92)	742
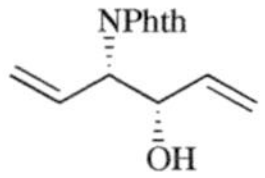 NPhth, OH	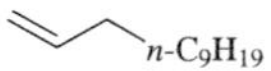 n-C_9H_{19}; 2.5 eq	Gr II (0.05 eq), DCM, reflux, 1 h	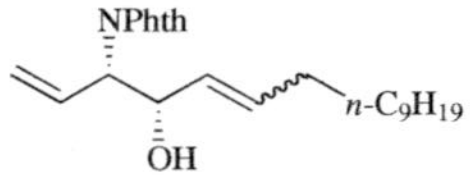 NPhth, OH, n-C_9H_{19} (76)	743

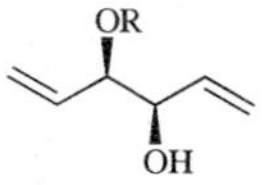

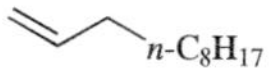

OR OH n-C_8H_{17} x eq Catalyst (y eq), solvent OR OH n-C_8H_{17} 742

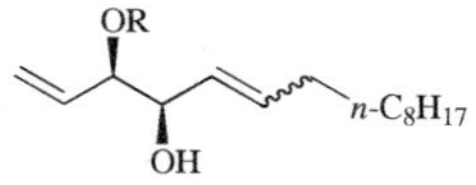

R	x	Catalyst	y	Solvent	Temp (°)	
TBS	5	Gr II	0.02	toluene	80	(51)
TBS	5	Gr II	0.02	DCM	40	(53)
TBS	5	Gr II	0.02	DCM	40	(58)
TBS	5	Gr II	0.02	DCM	20	(66)
TBS	2 + 3	Gr II	0.02	DCM	20	(73)
Tr	5	Gr II	0.02	toluene	80	(53)
Tr	5	Gr II	0.02	DCM	40	(58)
Tr	5 + 5	Ind II	0.02	DCM	40	(83)
Tr	5	Ru cat **139**	0.01	DCM	20	(74)
Tr	8	Ru cat **139**	0.01	DCM	20	(92)

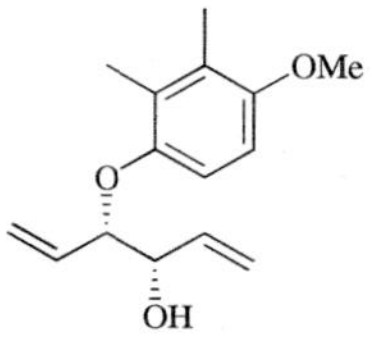

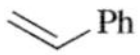

OMe O OH Ph 10 eq Gr II (0.1 eq), DCM, reflux, 2.5 h OMe O OH Ph (72) 743

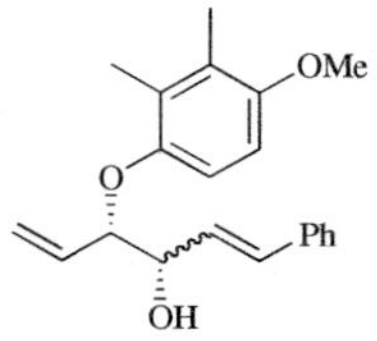

TABLE 13. CROSS-METATHESIS OF SECONDARY ALLYLIC ALCOHOLS AND DERIVATIVES (*Continued*)

A. ALLYLIC ALCOHOLS (*Continued*)

Allylic Alcohol	Cross Partner	Conditions	Product(s) and Yield(s) (%)	Refs.
Please refer to the charts preceding the tables for the catalyst structures.				
C_7				
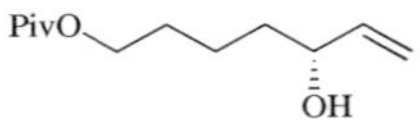 PivO, OH; 3 eq	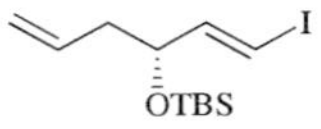 I, OTBS	Gr II (0.03 eq), DCM, reflux, 3 h	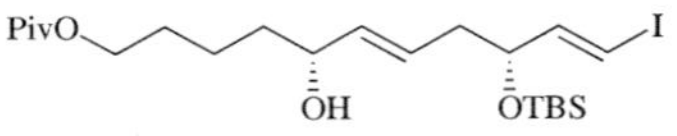 PivO, OH, OTBS, I (70) (*E*)/(*Z*) = 91:9	744
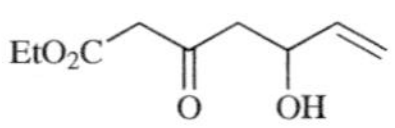 EtO_2C, O, OH	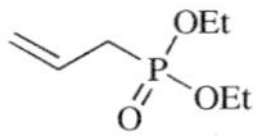 OEt, P, O, OEt; 5 eq	Gr II (0.05 eq), DCM, 40°, 16 h	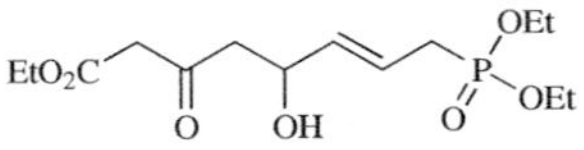 EtO_2C, O, OH, OEt, P, O, OEt (88) (*E*) only	745
	RO, OR; 2 eq	Gr II (0.05 eq), DCM, 40°	EtO_2C, O, OH, OR (*E*) only	745
			R: Ac, Time (h) 20, (70); R: Bn, Time (h) 18, (90)	
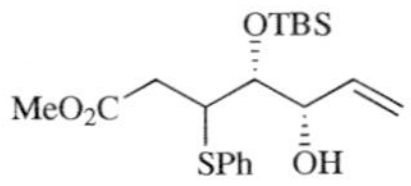 MeO_2C, OTBS, SPh, OH	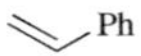 Ph; 5 eq	Gr II (0.05 eq), DCM, reflux, 16 h	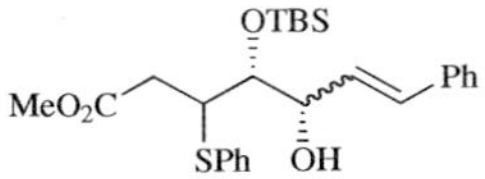 MeO_2C, OTBS, SPh, OH, Ph (90)	743
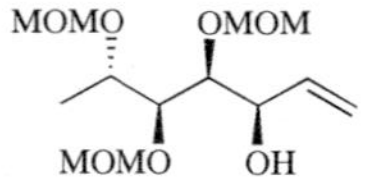 MOMO, OMOM, MOMO, OH	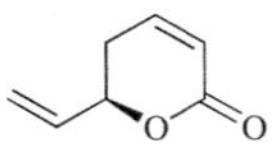 O, O	Gr II, DCM, reflux, 0.5 h	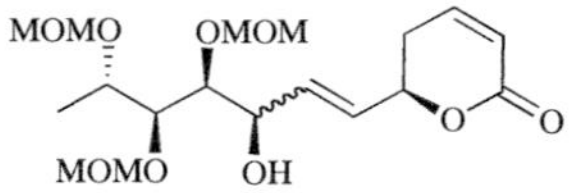 MOMO, OMOM, MOMO, OH, O, O (86)	746

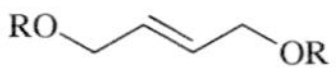

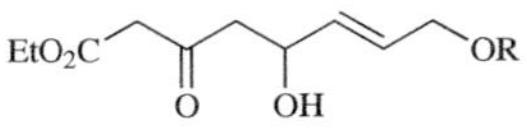

Substrate	Partner	Conditions	Product (Yield, %)	Ref.
	2 eq	Gr II (0.1 eq), DCM, reflux, 5 h	(86)	747
		Gr II (0.03 eq), DCM, reflux, 4 d	(95)	748
	n-C_9H_{19}, 10 eq	Gr II (0.025 eq), DCM, reflux, 12 h	(95)	749
	5 eq	Gr II (0.1 eq), DCM, reflux, 16 h	(59)	743

TABLE 13. CROSS-METATHESIS OF SECONDARY ALLYLIC ALCOHOLS AND DERIVATIVES (*Continued*)
A. ALLYLIC ALCOHOLS (*Continued*)

Allylic Alcohol	Cross Partner	Conditions	Product(s) and Yield(s) (%)	Refs.
Please refer to the charts preceding the tables for the catalyst structures.				
C_7				
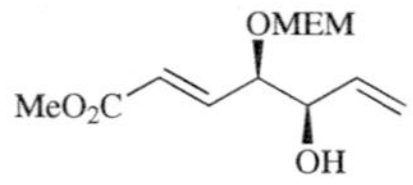 MeO$_2$C, OMEM, OH	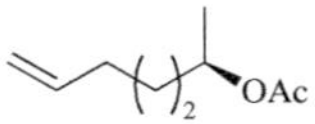 OAc, 1.5 eq	Gr II (0.05 eq), DCM, reflux, 16 h, *p*-cresol (0.5 eq)	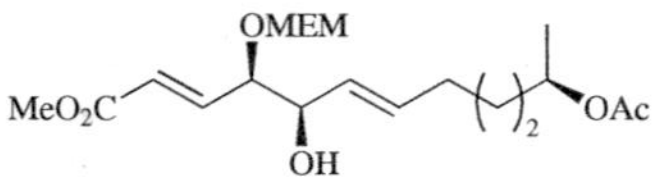 MeO$_2$C, OMEM, OH, OAc (60) (*E*) only	750
	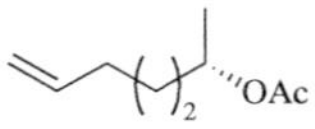 OAc, 1.5 eq	Gr II (0.05 eq), DCM, reflux, 16 h, *p*-cresol (0.5 eq)	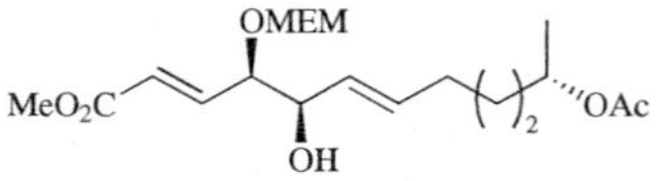 MeO$_2$C, OMEM, OH, OAc (60) (*E*) only	750
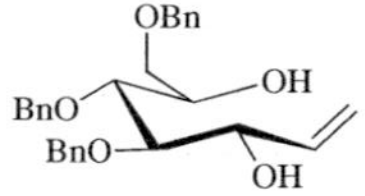 BnO, BnO, OBn, OH, OH; *x* eq	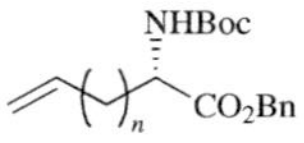 NHBoc, CO$_2$Bn	Gr II (*y* eq), DCM, reflux	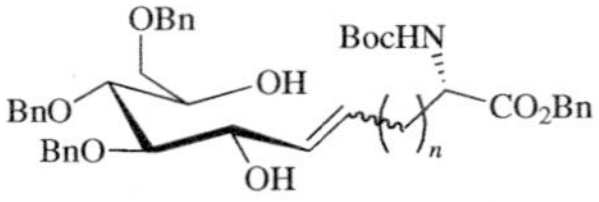 **I** 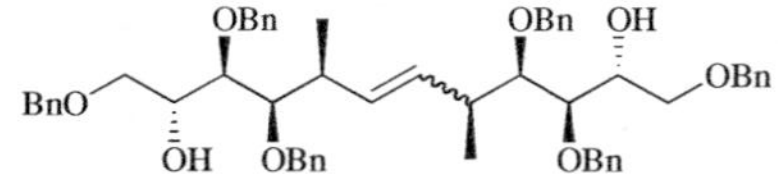 + **II**	426

x	*n*	*y*	Time (h)	I	II
0.5	0	0.075	20	(68)	(—)
0.5	1	0.15	15	(48)	(—)
2	1	0.15	15	(54)	(39)
4	1	0.15	15	(70)	(64)
2	1	0.1	15	(54)	(39)

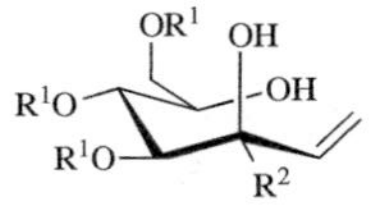

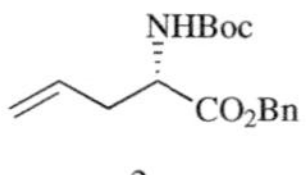

	2 eq	Gr II (x eq), DCM, reflux, 16 h			426

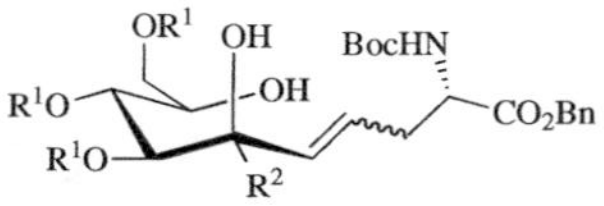

R^1	R^2	x	
TBS	H	0.1	(51)
Bn	AcHN	0.2	(16)

C_8

n-Bu, OH	OH	Ru cat **170** (0.01 eq), DCM, 50°, 24 h	n-Bu, OH, OH	(43)	751
n-C_5H_{11}, OR^1 (2 eq)	R^2O, O, O	Catalyst (0.06 eq), DCM, 40°, 12 h	n-C_5H_{11}, OR^1, R^2O, O, O		752

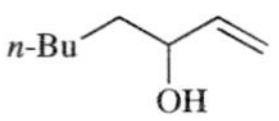

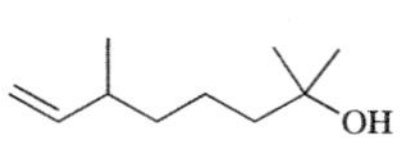

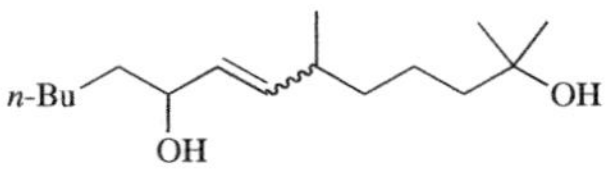

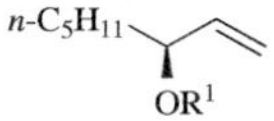

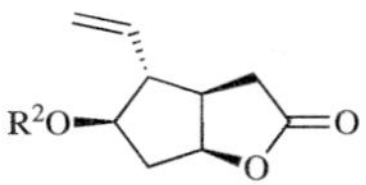

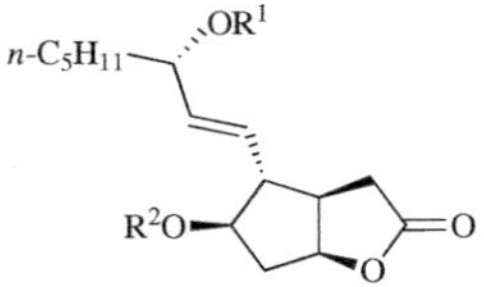

R^1	R^2	Catalyst		$(E)/(Z)$
H	TBS	Gr II	(60)	92:8
TBS	H	Gr II	(70)	93:7
TBS	Bz	Gr II	(40)	88:12
TBS	Bz	Hov II	(25)	88:12
TBS	TBS	Gr II	(76)	86:14
TBS	TBS	Gr II	(84)	94:6

TABLE 13. CROSS-METATHESIS OF SECONDARY ALLYLIC ALCOHOLS AND DERIVATIVES (*Continued*)

A. ALLYLIC ALCOHOLS (*Continued*)

Allylic Alcohol	Cross Partner	Conditions	Product(s) and Yield(s) (%)	Refs.
Please refer to the charts preceding the tables for the catalyst structures.				
C_8				
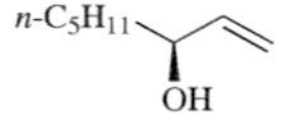 n-C_5H_{11}, OH 5 eq	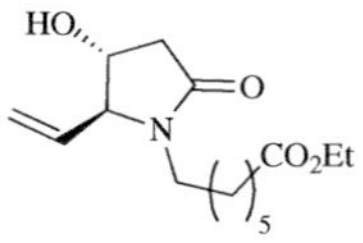 HO, O, N, CO_2Et, 5	Gr II (0.2 eq), DCM, 40°, 24 h	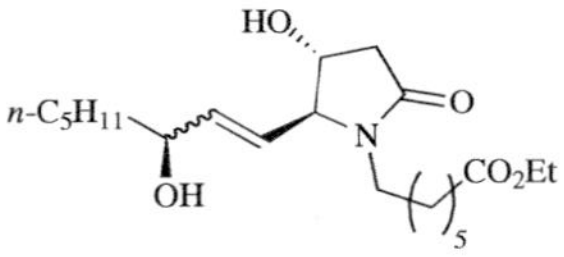 n-C_5H_{11}, OH, HO, O, N, CO_2Et, 5 (53)	255
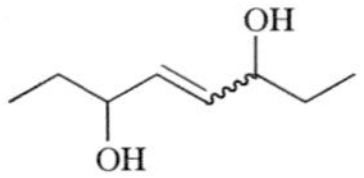 OH, OH	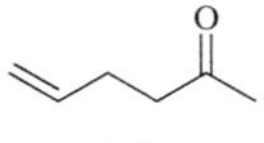 O 0.5 eq	Hov II (0.001 eq), DCM, reflux	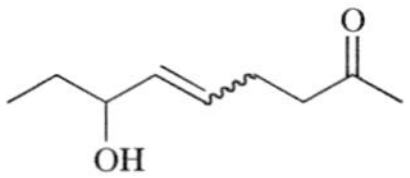 O, OH (72)	723
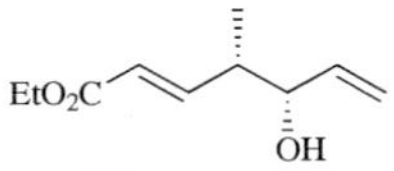 EtO_2C, OH 1.5 eq	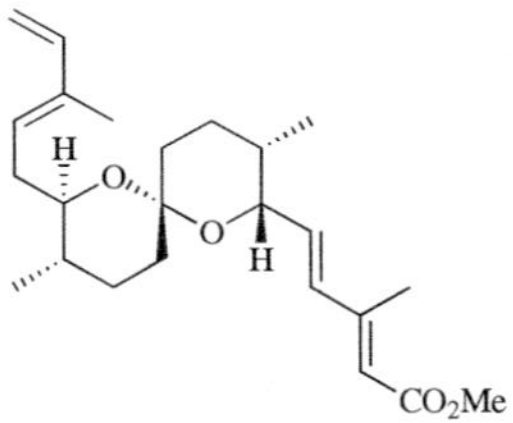 H, O, O, H, CO_2Me	Gr II (2 x 0.2 eq), DCM, 18 h	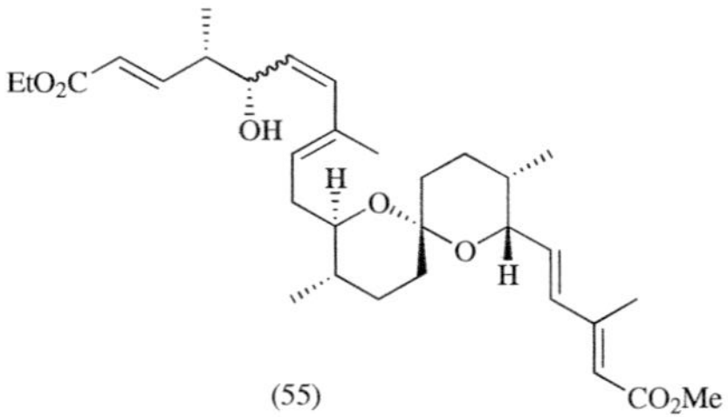 EtO_2C, OH, H, O, O, H, CO_2Me (55)	753

C_9

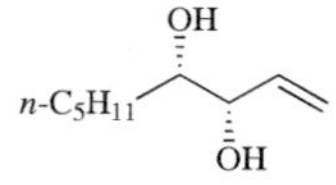
OH
n-C5H11
OH

x eq

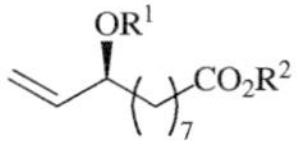
OR1
CO2R2
7

Gr II (y eq), DCM

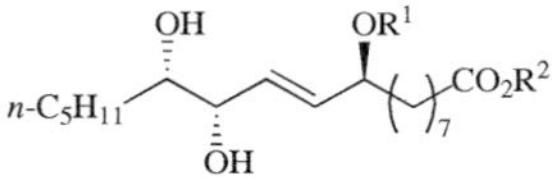
OH
OR1
n-C5H11
CO2R2
7
OH

x	R^1	R^2	y	Temp (°)	Time (h)		(*E*)/(*Z*)	
1.7	H	H	—	25	18	(78)	(*E*) only	754
3	H	Me	0.05	40	5	(35)	—	755
3	TBS	Me	0.05	40	5	(24)	—	755

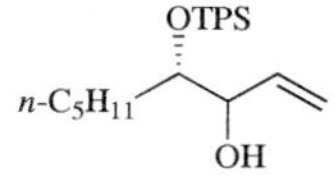
OTPS
n-C5H11
OH

1.7 eq

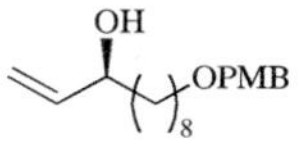
OH
OPMB
8

Gr II, DCM, 25°, 18 h

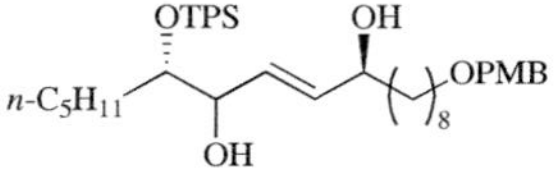
OTPS
OH
n-C5H11
OPMB
8
OH

(78) (*E*) only 754

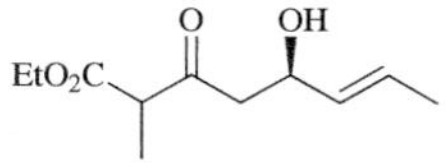
O
OH
EtO2C

2.5 eq

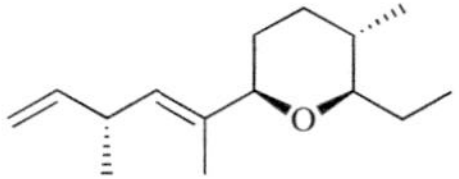
O

Ru cat **88**
(0.08 + 0.05 eq),
$C_6F_5CF_3$, 62°, 4 h

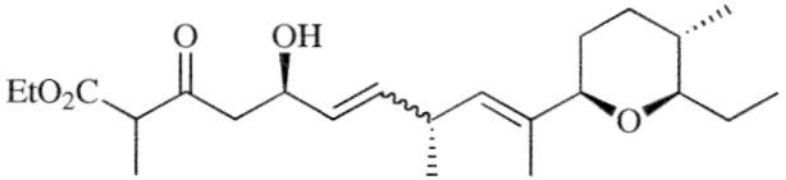
O
OH
EtO2C
O

(86)

529

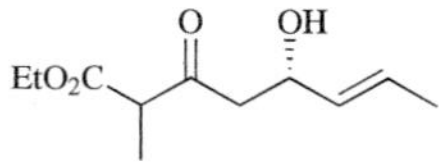
O
OH
EtO2C

4.67 eq

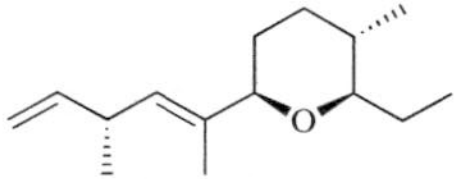
O

Ru cat **88**
(0.06 + 0.09 eq),
$C_6F_5CF_3$, 70°, 4 h

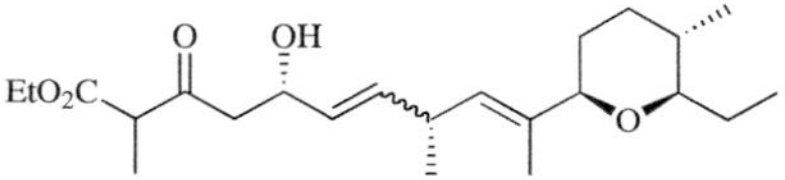
O
OH
EtO2C
O

(75)

529

TABLE 13. CROSS-METATHESIS OF SECONDARY ALLYLIC ALCOHOLS AND DERIVATIVES (*Continued*)

A. ALLYLIC ALCOHOLS (*Continued*)

Allylic Alcohol	Cross Partner	Conditions	Product(s) and Yield(s) (%)	Refs.
Please refer to the charts preceding the tables for the catalyst structures.				
C_9				
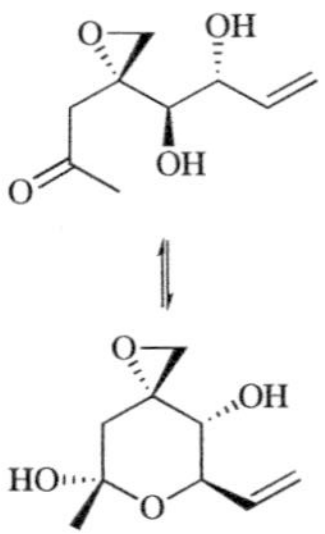 OH, OH ⇌ HO, OH	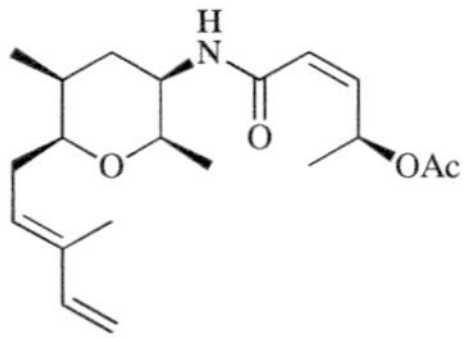 1.8 eq	Ru cat **88** (0.1 eq), DCE, 23°, 10 h	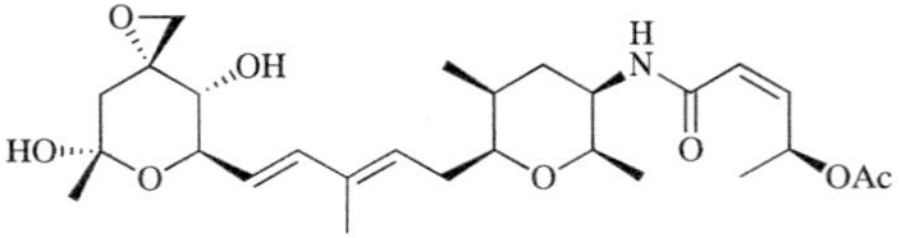 (40) (*E*) only	756
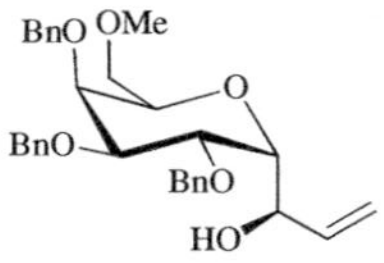 BnO, OMe, BnO, BnO, HO	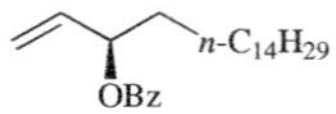 n-$C_{14}H_{29}$, OBz 3.2 eq	Gr II (0.15 eq), DCM, reflux, 2 h	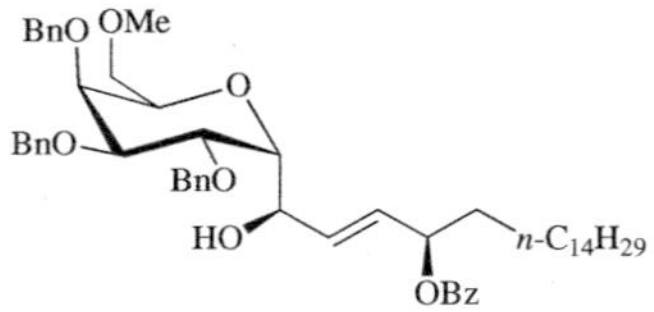 (70) (*E*)/(*Z*) = 95:5	757
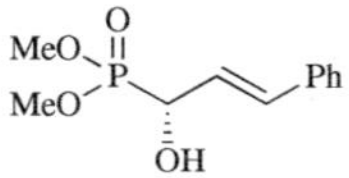 MeO, MeO, P, O, OH, Ph er 97.5:2.5	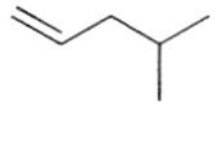 5 eq	Gr II (0.05 eq), DCM, 40°, 36 h	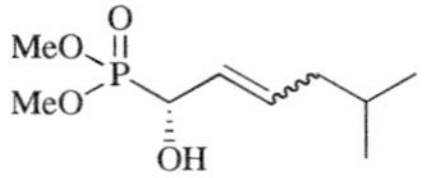 (74) er 97.5:2.5	221

er 97.5:2.5

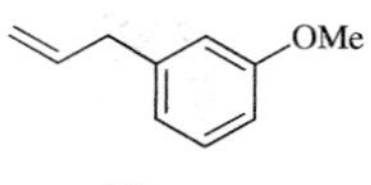

MeO, OMe

1.7 eq

Gr II (0.025 eq), DCM, 40°, 12 h

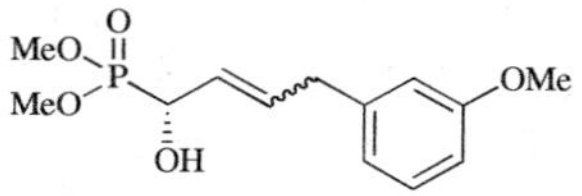

MeO, MeO, P, O, OH, OMe

50% conv., er 96.5:3.5

221

C_{10}

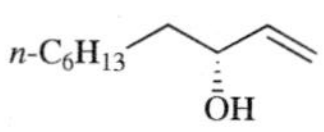

n-C_6H_{13}, OH

x eq

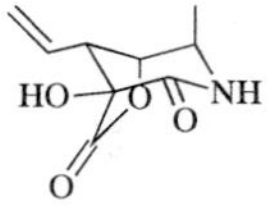

HO, O, O, NH, O

Gr II (*y* eq), DCM, 6 h

n-C_6H_{13}, OH, HO, O, O, NH, O

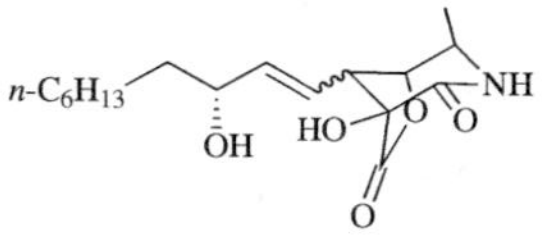

x	*y*	Temp (°)		
2	0.1	rt	(64)	758
5	0.2	35	(58)	759

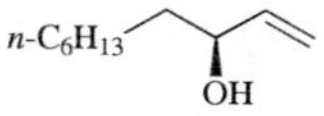

n-C_6H_{13}, OH

5 eq

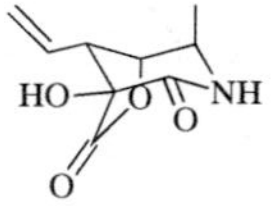

HO, O, O, NH, O

Gr II (0.2 eq), DCM, 35°, 6 h

n-C_6H_{13}, OH, HO, O, O, NH, O

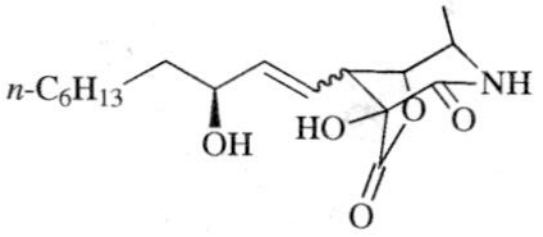

(59) 759

TABLE 13. CROSS-METATHESIS OF SECONDARY ALLYLIC ALCOHOLS AND DERIVATIVES (*Continued*)

A. ALLYLIC ALCOHOLS (*Continued*)

Allylic Alcohol	Cross Partner	Conditions	Product(s) and Yield(s) (%)	Refs.

Please refer to the charts preceding the tables for the catalyst structures.

C_{10}

Allylic Alcohol	Cross Partner	Conditions	Product(s) and Yield(s) (%)	Refs.
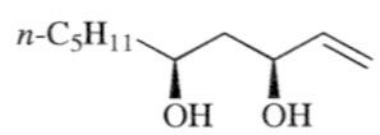 n-C_5H_{11} (diol, OH OH)	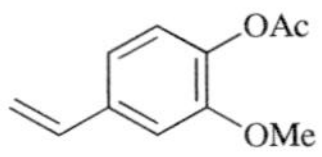 OAc, OMe styrene; x eq	Hov II (y eq), PhOH (z eq), solvent, rt, 16 h	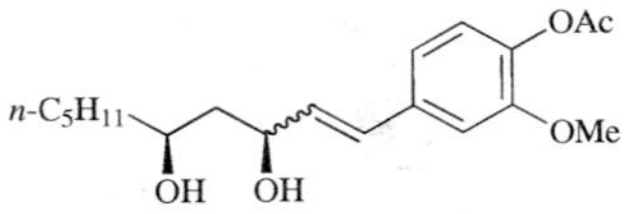 I 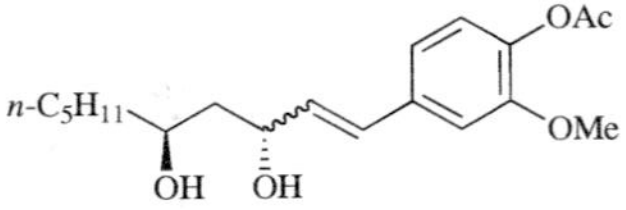 + II 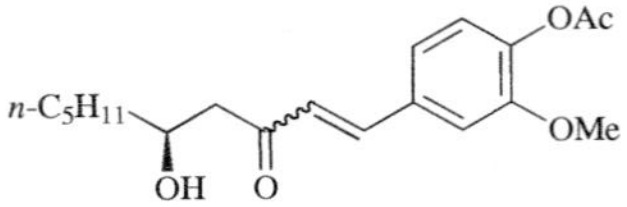 + III	728
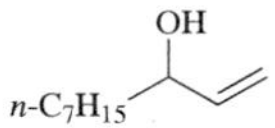 n-C_7H_{15} (OH)	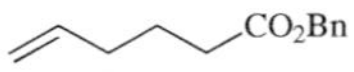 CO_2Bn; 3 eq	1. Ru cat **152** (0.01 eq), (Z)-butene (20 eq), THF, 22°, 1 h 2. Ru cat **152** (0.04 eq), THF, 100 torr, 22°, 8 h	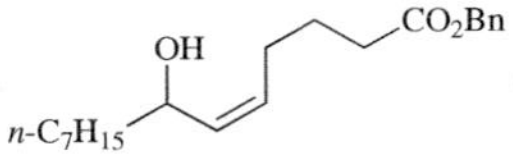 n-C_7H_{15} (OH) CO_2Bn (47) (Z)/(E) = 90:10	395

I

II

III

x	y	z	Solvent	**I + II**	**I/II**	**III**
4.1	0.11	—	toluene	(24)	**I** only	(tr)
5.2	0.11	0.15	DCM	(55)	80:20	(9)
5.6	0.12	0.7	toluene	(63)	**I** only	(9)
9.8	0.1	—	DCM	(21)	80:20	(7)

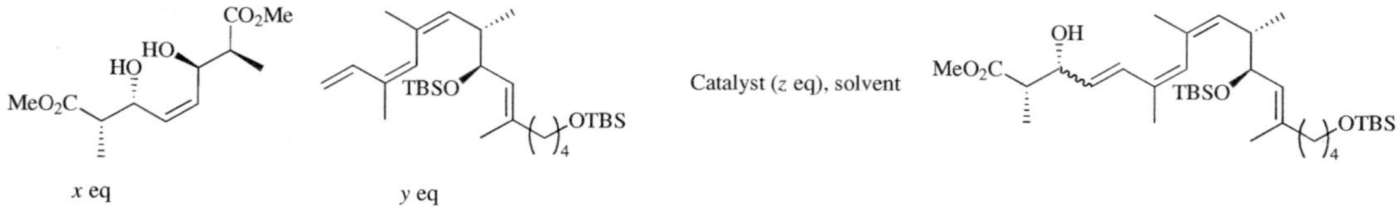

MeO_2C … HO HO … CO_2Me (x eq) + TBSO … OTBS (y eq) → Catalyst (z eq), solvent → MeO_2C … OH … TBSO … OTBS | 760

x	y	Catalyst	z	Solvent	Temp (°)	Time (h)	
1	1	Hov II	0.05	toluene	50	24	(10)
1	3	Hov II	0.05	toluene	50	24	(18)
1	3	Gr II	0.05	toluene	50	24	(tr)
1	3	Hov II	0.1	toluene	50	24	(19)
1	3	Ru cat **67**	0.05	toluene	50	24	(—)
1	3	Ru cat **271**	0.05	toluene	21	24	(—)
1	3	Ru cat **271**	0.05	toluene	50	24	(—)
1	3	Hov II	0.15	toluene	50	72	(26)
3 x 0.2	1	Hov II	3 x 0.025	C_6D_6	50	36	(78)

C_{11}

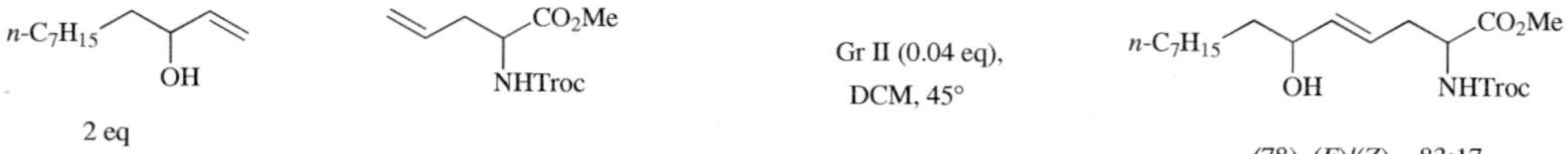

n-C_7H_{15} … OH (2 eq) + CO_2Me NHTroc → Gr II (0.04 eq), DCM, 45° → n-C_7H_{15} … OH … CO_2Me NHTroc | 714

(78) $(E)/(Z) = 83:17$

TABLE 13. CROSS-METATHESIS OF SECONDARY ALLYLIC ALCOHOLS AND DERIVATIVES (*Continued*)

A. ALLYLIC ALCOHOLS (*Continued*)

Allylic Alcohol	Cross Partner	Conditions	Product(s) and Yield(s) (%)	Refs.
Please refer to the charts preceding the tables for the catalyst structures.				
C_{11}				
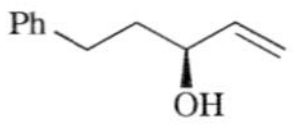 Ph, OH	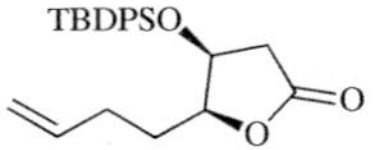 TBDPSO, O	Gr II (0.1 eq), $CHCl_3$, 40°, 2 h	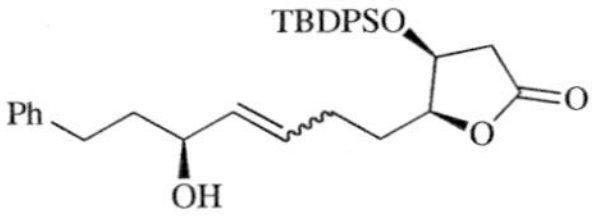 Ph, OH, TBDPSO, O (63)	761
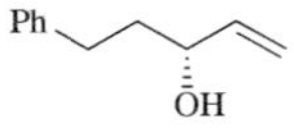 Ph, OH	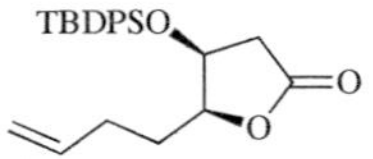 TBDPSO, O	Gr II (0.1 eq), $CHCl_3$, 40°, 2 h	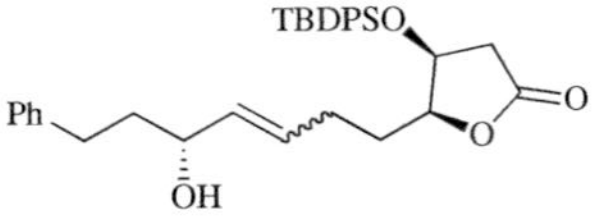 Ph, OH, TBDPSO, O (66)	761
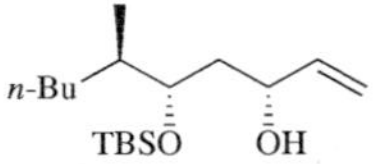 *n*-Bu, TBSO, OH	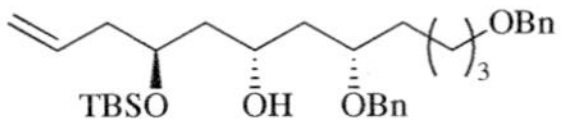 TBSO, OH, OBn, OBn, 3	Gr II, DCM, reflux, 6 h	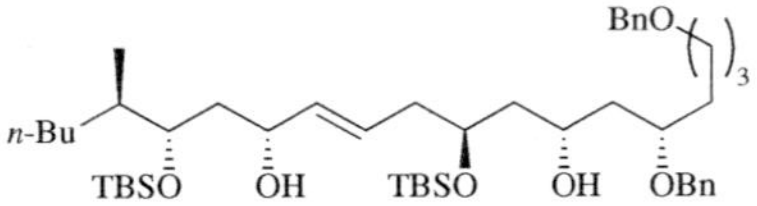 *n*-Bu, TBSO, OH, TBSO, OH, OBn, BnO, 3 (72) (*E*)/(*Z*) = 80:20	762
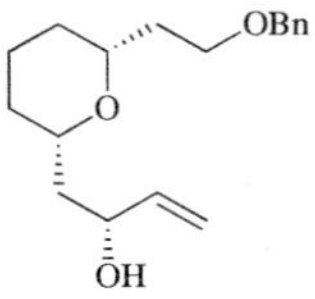 O, OBn, OH	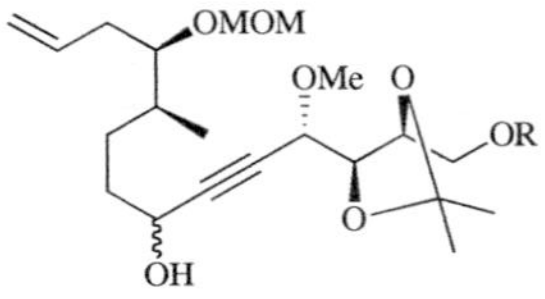 OMOM, OMe, O, OR, OH, R = TBDPS	Gr II (0.1 eq), DCM, rt	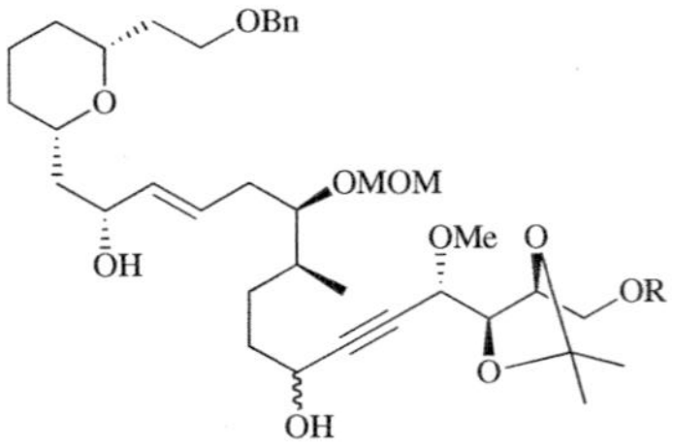 O, OBn, OH, OMOM, OMe, O, OR, OH (65) (*E*) only	763

Reactant		Conditions	Product(s) and Yield(s) (%)	Refs.
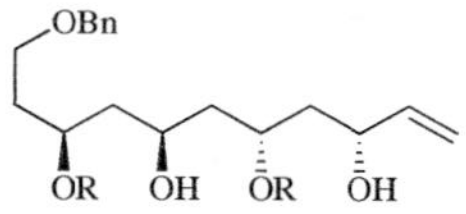 R = TBDPS	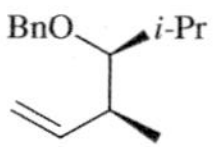 2 eq	Hov II (0.05 eq), DCM, reflux, 6 h	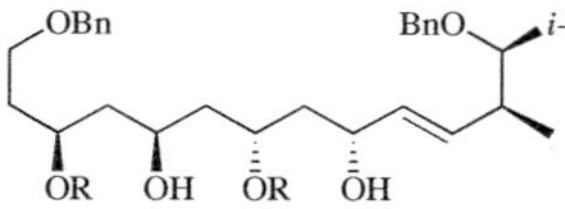 (52) (*E*)/(*Z*) = 90:10	764
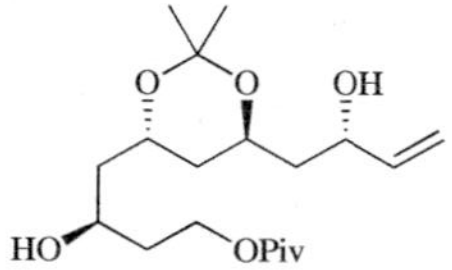	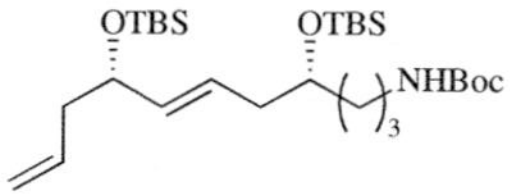 2 eq	Hov II (0.05 eq), DCM, rt	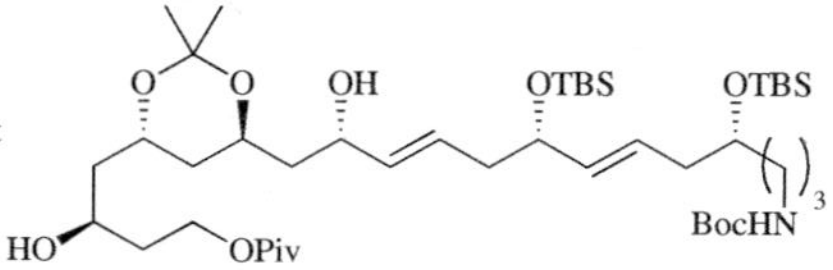 (68) (*E*) only	765
C_{12}				
1.5 eq		Gr II (0.05 eq), DCM, rt, 22 h	(63)	766
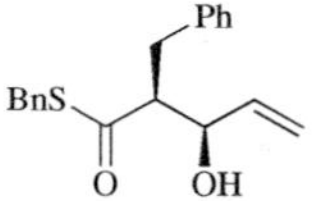 5 eq	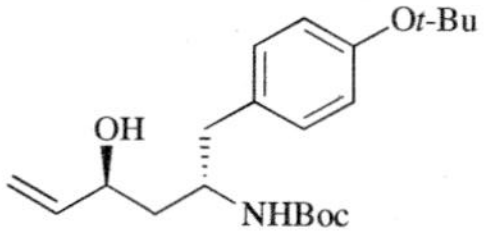	Gr II, DCM, 40°	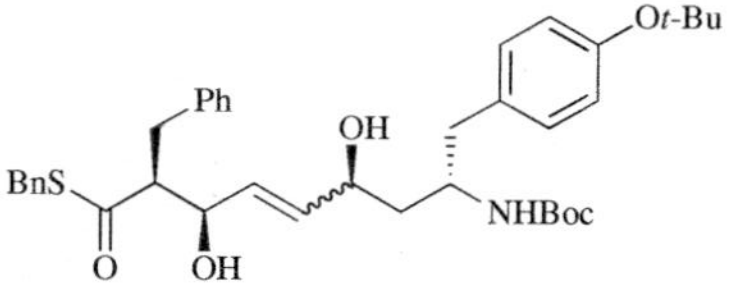 (53)	767

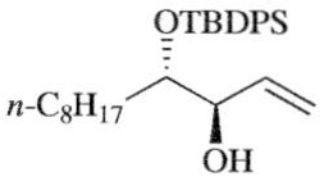

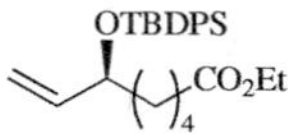

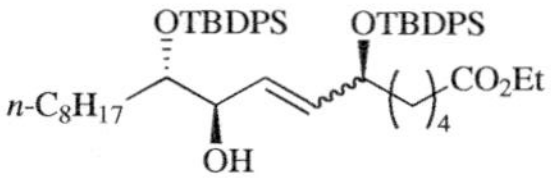

TABLE 13. CROSS-METATHESIS OF SECONDARY ALLYLIC ALCOHOLS AND DERIVATIVES (*Continued*)

A. ALLYLIC ALCOHOLS (*Continued*)

Allylic Alcohol	Cross Partner	Conditions	Product(s) and Yield(s) (%)	Refs.
Please refer to the charts preceding the tables for the catalyst structures.				
C_{12}				
BnS, O, Ph, OH; 5 eq	OH, NH, Boc, Ot-Bu	Gr II, DCM, 40°	BnS, O, Ph, OH, OH, NH, Boc, Ot-Bu (54)	768
OMOM, O, OH, OH, OBn	OTBS; 6 eq	Gr II (0.05 eq), DCM, rt, 12 h	OMOM, O, OH, OH, OBn, OTBS (65)	769
C_{13}				
O, O, N, H, OH, O	( )$_n$; 1.2 eq	Gr II (0.05 eq), DCM, rt, 24 h	O, O, N, H, OH, O, ( )$_n$ *n* = 2 (88) *n* = 4 (86) *n* = 6 (88) *n* = 8 (89)	770
	R; 1.2 eq	Gr II (0.05 eq), DCM, rt, 24 h	O, O, N, H, OH, O, R R = *c*-C_5H_9 (78) R = *c*-C_6H_{11} (75) R = $PhCH_2$ (82)	770

C_{14}

Alkene 1	Alkene 2	Conditions	Product(s) and Yield(s) (%)	Refs.
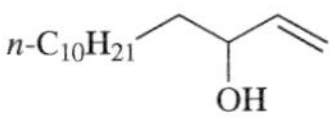 n-$C_{10}H_{21}$, OH; 2.5 eq	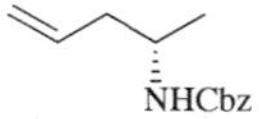 NHCbz	Gr II (0.05 eq), DCM, 45°	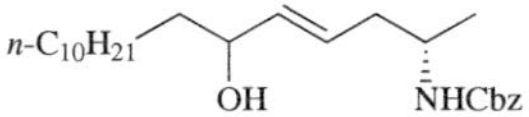 n-$C_{10}H_{21}$, OH, NHCbz (75) (*E*) only	714
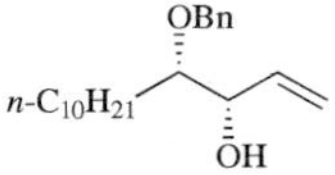 OBn, n-$C_{10}H_{21}$, OH; 1.5 eq	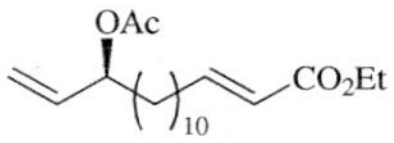 OAc, CO_2Et, 10	Gr II (0.04 eq), DCM, 25°, 22 h	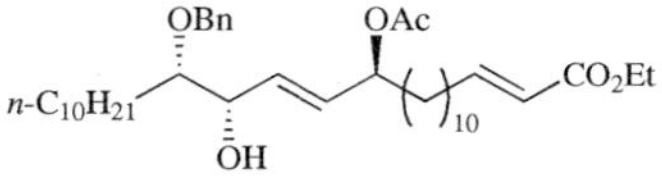 OBn, OAc, n-$C_{10}H_{21}$, OH, CO_2Et, 10 (61) (*E*) only	772
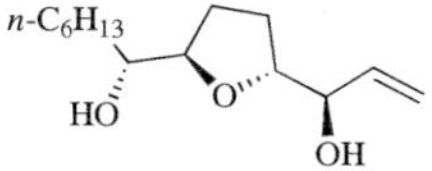 n-C_6H_{13}, HO, O, OH	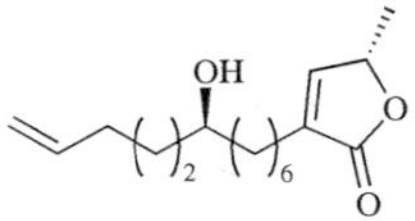 OH, 2, 6, O, O; 2 eq	Gr II (0.1 eq), DCM, rt, 20 h	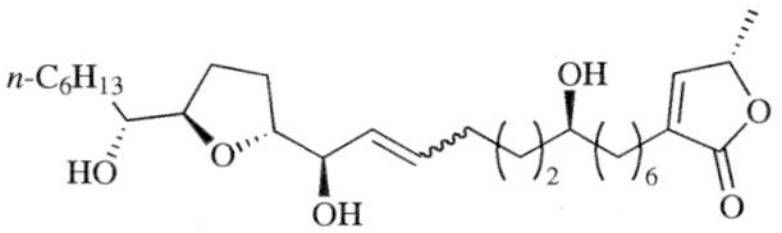 n-C_6H_{13}, HO, O, OH, OH, 2, 6, O, O (73)	771
	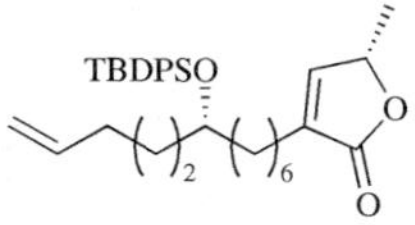 TBDPSO, 2, 6, O, O; 2 eq	Gr II (0.1 eq), DCM, rt, 20 h	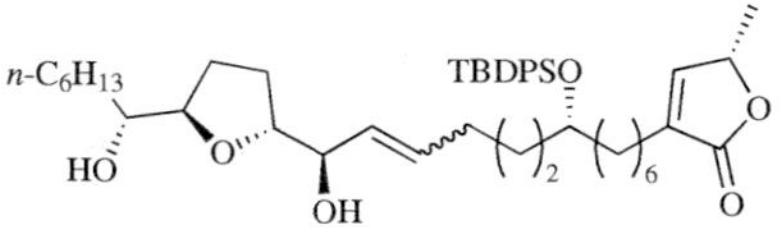 n-C_6H_{13}, HO, O, OH, TBDPSO, 2, 6, O, O (73)	771

TABLE 13. CROSS-METATHESIS OF SECONDARY ALLYLIC ALCOHOLS AND DERIVATIVES (*Continued*)

A. ALLYLIC ALCOHOLS (*Continued*)

Allylic Alcohol	Cross Partner	Conditions	Product(s) and Yield(s) (%)	Refs.
Please refer to the charts preceding the tables for the catalyst structures.				
C_{15}				
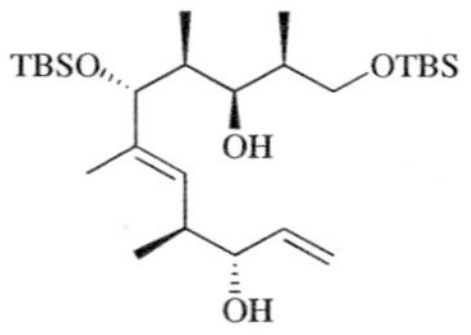 TBSO, OTBS, OH, OH	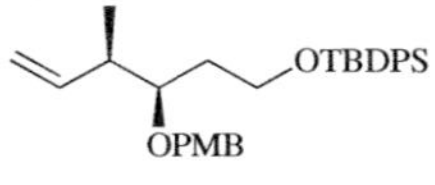 OTBDPS, OPMB 3 eq	Gr II (0.1 eq), DCM, reflux	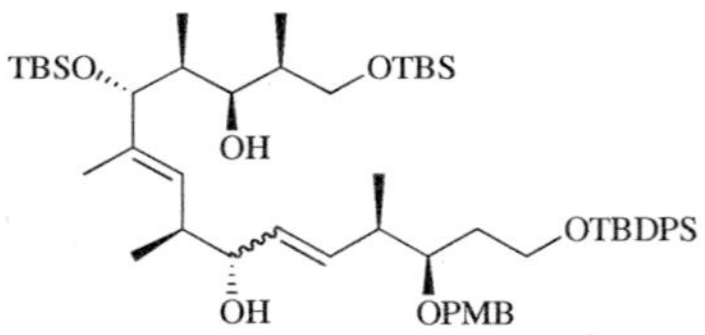 TBSO, OTBS, OH, OH, OTBDPS, OPMB (60)	773
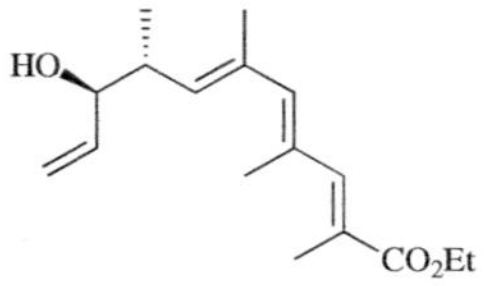 HO, CO_2Et 2.2 eq	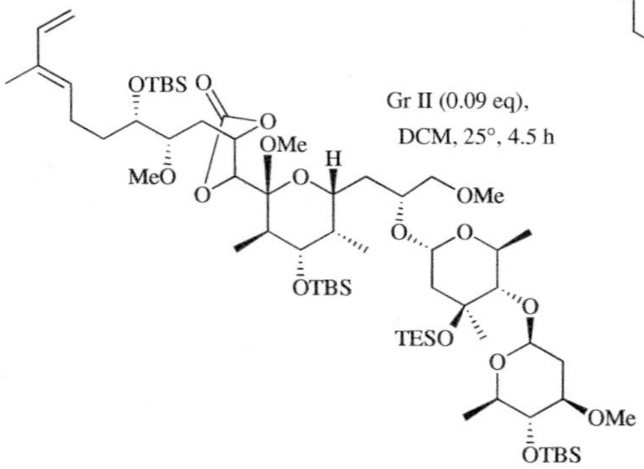 OTBS O, O, OMe, H, MeO, O, OMe, OTBS, O, O, TESO, O, O, OMe, OTBS	Gr II (0.09 eq), DCM, 25°, 4.5 h	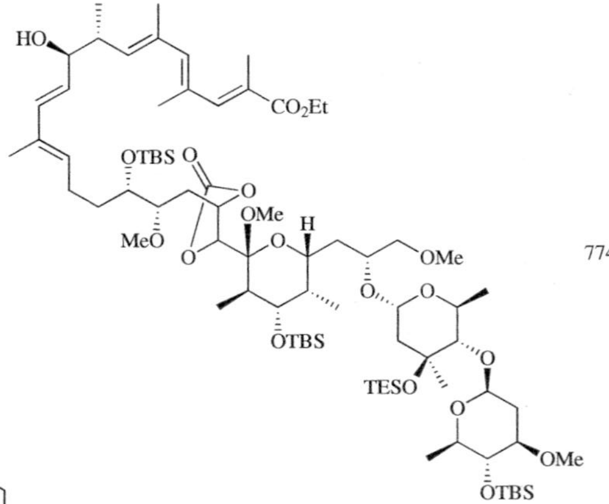 HO, CO_2Et, OTBS O, O, OMe, H, MeO, O, OMe, OTBS, O, O, TESO, O, O, OMe, OTBS (60) $(E)/(Z) = 95:5$	774

C_{17}

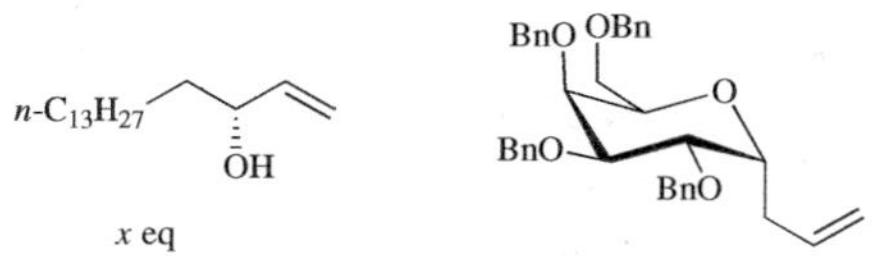

n-$C_{13}H_{27}$ OH

x eq

BnO OBn BnO BnO

Catalyst (y eq), DCM, 45°

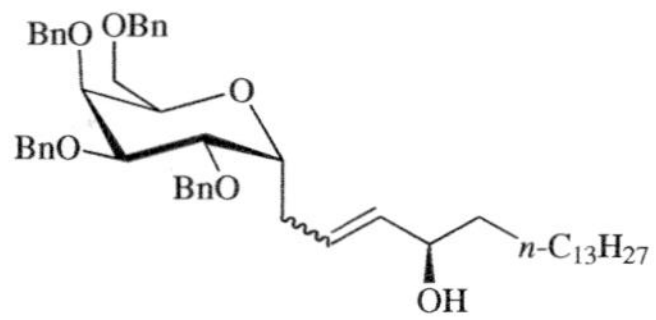

BnO OBn BnO BnO OH n-$C_{13}H_{27}$

x	Catalyst	y	Time (h)	
4	Gr II	0.17	21	(38)
3	Hov II	0.10	3	(52)
0.33	Hov II	0.10	3	(63)

775

C_{18}

HO n-$C_{10}H_{21}$ OH

AcO OPiv

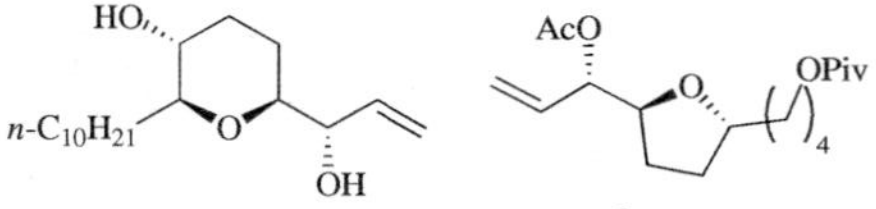

3 eq

Gr II (2 x 0.1 eq), DCM, rt, 2 x 18 h

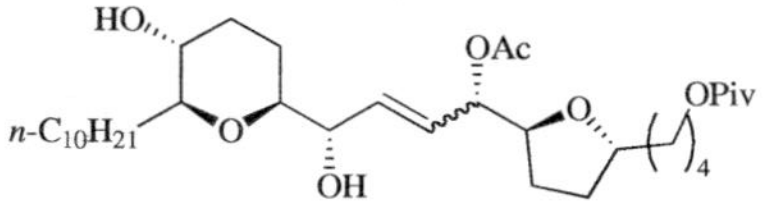

HO n-$C_{10}H_{21}$ OH OAc OPiv

(51)

257

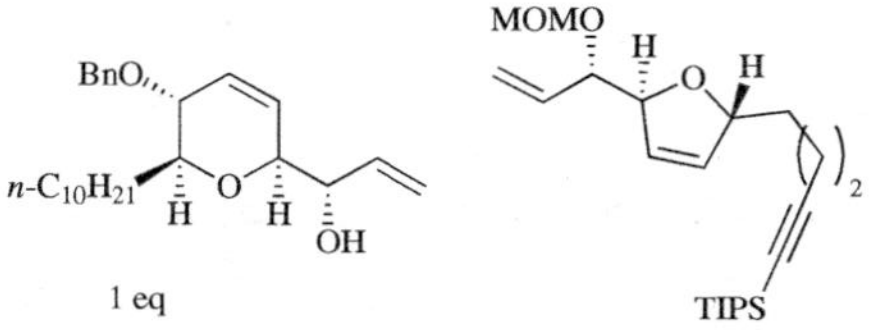

BnO n-$C_{10}H_{21}$ H H OH

1 eq

MOMO H H TIPS

Hov II (2 x 0.01 eq), DCM, reflux

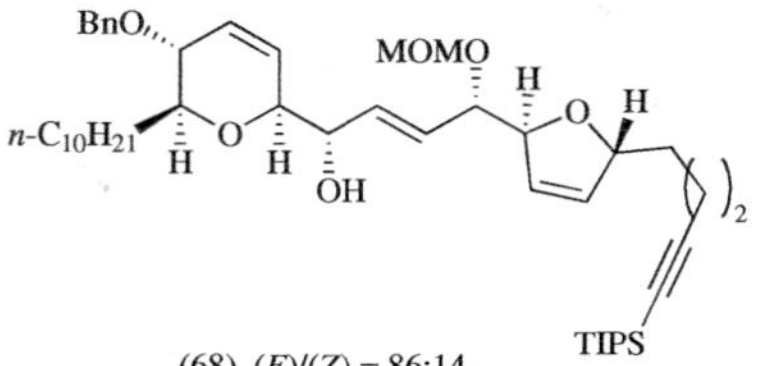

BnO n-$C_{10}H_{21}$ H H OH MOMO H H TIPS

(68) (E)/(Z) = 86:14

776

TABLE 13. CROSS-METATHESIS OF SECONDARY ALLYLIC ALCOHOLS AND DERIVATIVES (*Continued*)

A. ALLYLIC ALCOHOLS (*Continued*)

Allylic Alcohol	Cross Partner	Conditions	Product(s) and Yield(s) (%)	Refs.
Please refer to the charts preceding the tables for the catalyst structures.				
C_{20}				
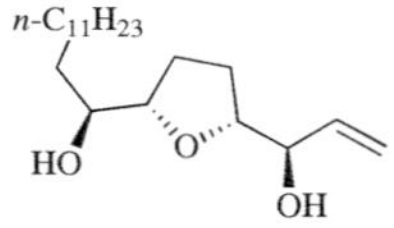 n-$C_{11}H_{23}$, HO, OH	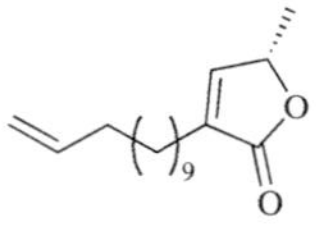 2 eq	Gr II (0.3 eq), DCM, 40°, 12 h	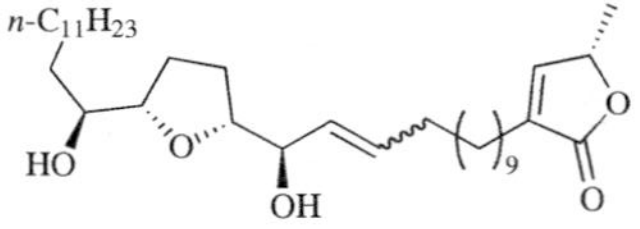 n-$C_{11}H_{23}$, HO, OH (52)	777, 778
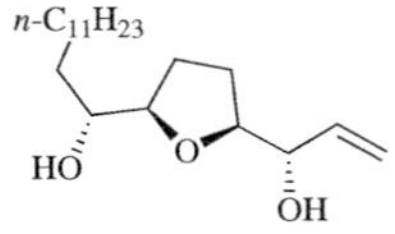 n-$C_{11}H_{23}$, HO, OH	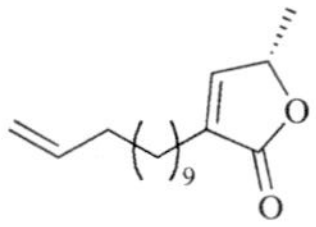 2 eq	Gr II (0.3 eq), DCM, 40°, 12 h	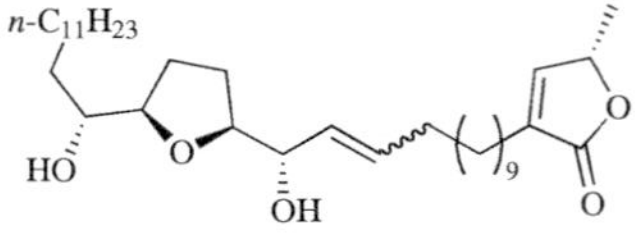 n-$C_{11}H_{23}$, HO, OH (64)	778
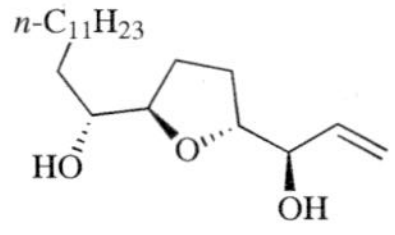 n-$C_{11}H_{23}$, HO, OH	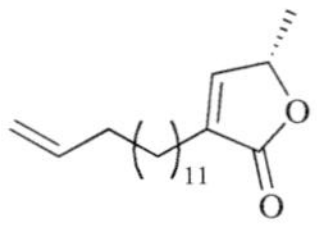 2 eq	Gr II (0.3 eq), DCM, 40°, 12 h	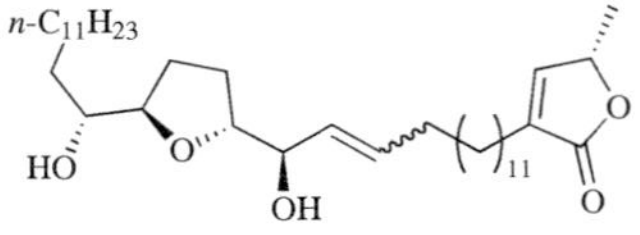 n-$C_{11}H_{23}$, HO, OH (32)	778

C_{24}

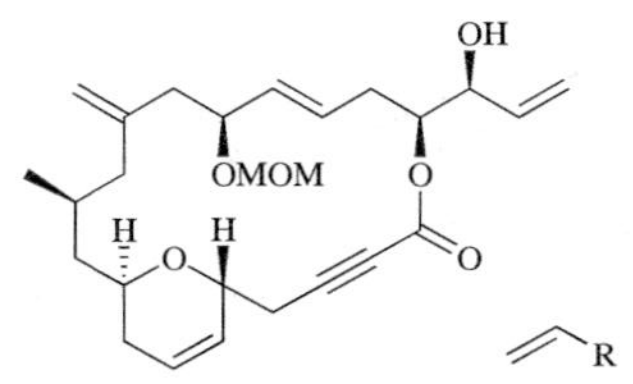

OH OMOM O O

R

x eq

Gr II (0.1 eq), DCM, rt, 24 h

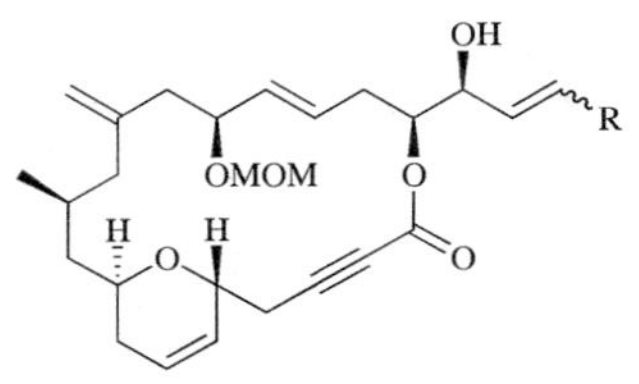

OH OMOM O O R

R	x	
$MeOCH_2$	20	(36)
c-C_6H_{11}	20	(68)
3-MeC_6H_4	50	(40)

779

C_{25}

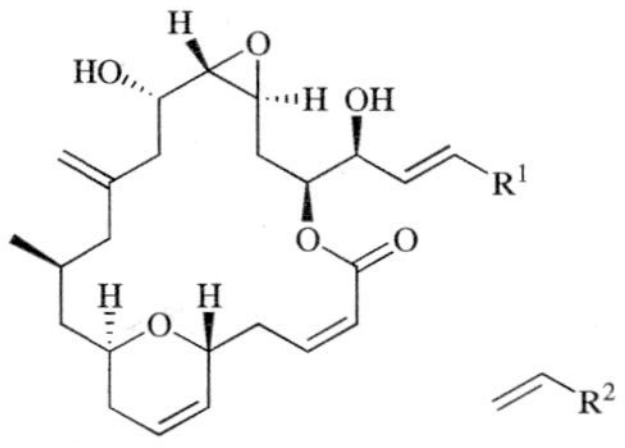

HO H O H OH R^1 O O

R^2

50 eq

Gr II (x eq), DCM, rt, 18 h

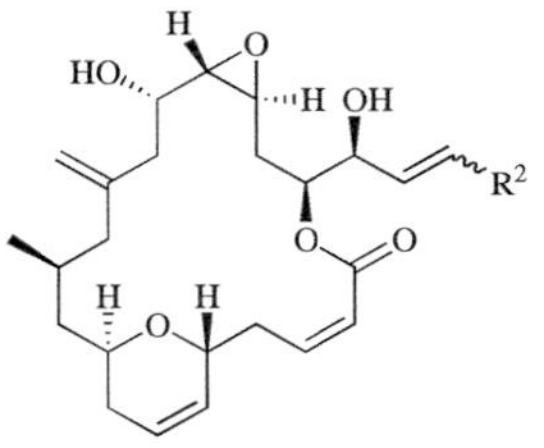

HO H O H OH R^2 O O

R^1	R^2	x	
$MeOCH_2$	Cy	0.1	(50)
Cy	3-MeC_6H_4	0.2	(50)
Cy	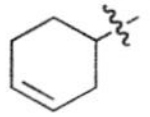 [cyclohex-3-enyl]	0.3	(55)

780

TABLE 13. CROSS-METATHESIS OF SECONDARY ALLYLIC ALCOHOLS AND DERIVATIVES (*Continued*)

B. ALLYLIC ETHERS

Allylic Ester	Cross Partner	Conditions	Product(s) and Yield(s) (%)	Refs.
Please refer to the charts preceding the tables for the catalyst structures.				
C_4				
BnO, OBn	R; 10 eq	Ru cat **151** (0.05 eq), DCM, 40°, 12 h	OBn, OBn, R R / (Z)/(E) MeO (88) 98:2 n-C_6H_{13} (60) 98:2	83
	Ph; 10 eq	Ru cat **151** (0.05 eq), DCM, 40°, 12 h	OBn, OBn, Ph (58) (Z)/(E) = 98:2	83
OBn; 2 eq	OTBDPS, OMOM	Gr II (0.05 eq), DCM, reflux, 6 h	OBn, OMOM, OTBDPS (80)	781
OTBDPS	AcO, OAc; 2 eq	Gr I (0.05 eq), DCM, 45°	OTBDPS, OAc (23) (E)/(Z) = 88:12	124
	OAc, n; x eq	Catalyst (y eq), DCM, 40°	OTBDPS, OAc	

n	x	Catalyst	y	Time (h)		(E)/(Z)	Refs.
2	3	Hov II	0.05	6	(68)	>95:5	177
2	3	Ru cat **271**	0.05	6	(91)	>95:5	177
3	0.5	Gr II	0.03	12	(53)	87:13	36

Reactant 1	Reactant 2	Conditions	Product(s) and Yield(s) (%)	Refs.
OTBS **I**		1. Hov II (0.06 eq), BQ, DCM, reflux, 0.75 h 2. Hov II (0.04 eq), **I** (2–4 eq), CuI, DCE, reflux, 4–6 h 3. *o*-NBSH (15 eq), Et_3N, rt, overnight	OTBS (69)	716
RO, OMe	OMe, OR	Gr I (0.05 eq), DCM, reflux	RO, OMe, OMe, OR R: Me (7); Ac (18)	782
TBSO, PMBO	R, *x* eq	Mo cat **18** (0.03 eq), benzene, 7 torr, 22°, 8 h	TBSO, PMBO, R	184
HO	CO_2Ph, 3 eq	1. Mo cat **18** (0.03 eq), benzene, 7 torr, 22°, 8 h 2. DDQ (2 eq), DCM, H_2O, 0°, 1 h	HO, PMBO, CO_2Ph 89% conv.[a], (87) $(Z)/(E)$ = 90:10	184

R	*x*	Conv. (%)[a]		$(Z)/(E)$
TESO	3	91	(72)	92:8
$Br(CH_2)_5$	2	82	(70)	90:10

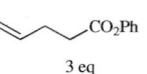

TABLE 13. CROSS-METATHESIS OF SECONDARY ALLYLIC ALCOHOLS AND DERIVATIVES (*Continued*)

B. ALLYLIC ETHERS (*Continued*)

Allylic Ester	Cross Partner	Conditions	Product(s) and Yield(s) (%)	Refs.
Please refer to the charts preceding the tables for the catalyst structures.				
C_4				
TBSO, PMBO	n-C_7H_{15}, 3 eq	1. Mo cat **18** (0.03 eq), benzene, 7 torr, 22°, 8 h 2. $(n\text{-Bu})_4NF$ (2 eq), THF, 22°, 1 h	HO, PMBO, n-C_7H_{15} 93% conv.[a], (87) $(Z)/(E) > 98{:}2$	184
PMBO, OMOM, *x* eq	OR (subscript 8), *y* eq	Gr II (*z* eq), DCM, rt, 10 h	PMBO, OMOM, OR (subscript 8)	783
TBDPSO, OR, 8 eq	HO, H, H, H, O, OBn, O, H, H, H, OBn, OBn	Gr II (0.2 eq), benzene, reflux, 6.5 h	H, HO, H, H, O, OBn, O, H, H, H, OBn, OBn, TBDPSO, OR (*E*) only	784

x	R	*y*	*z*		(*E*)/(*Z*)
1	H	1.5	0.1	(60)	90:10
1	H	1	0.1	(65)	90:10
1.5	H	1	0.1	(76)	90:10
1.5	H	1	0.05 + 0.03	(76)	90:10
1	TBS	1.5	0.1	(0)	—

R	
TBS	(23)
PMB	(34)

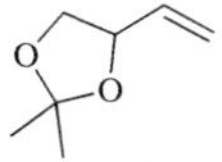

Reactant 1	Reactant 2	Conditions	Product(s) and Yield(s) (%)	Refs.
x eq	nR	Gr I (0.05 eq), DCM, reflux, 1 h	nR	782
	R	Gr I (0.05 eq), DCM, reflux	R	782
	R[1], R[2]	Ru cat **35**, DCM, 40°	R[1], R[2]	785
5 eq	NH, R	Ru cat **35** (0.1 eq), DCM, 40°, 5 h	NH, R; (*E*) only	628

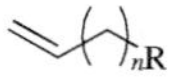

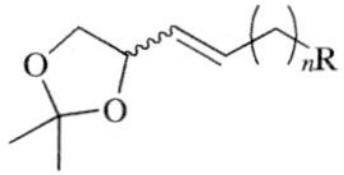

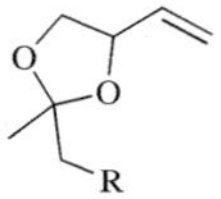

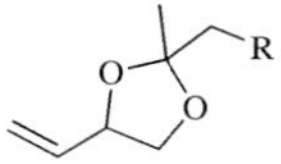

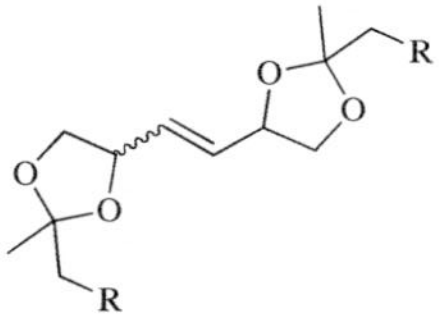

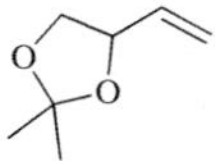

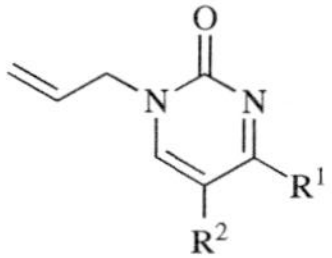

x	R	n	
3	TMS	1	(63)
2	Me	5	(68)
0.5	Me	5	(36)
2	Me	7	(72)
3	Ph	0	(75)

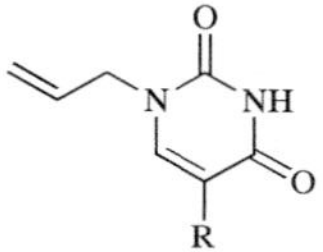

R	
H	(83)
Me	(81)

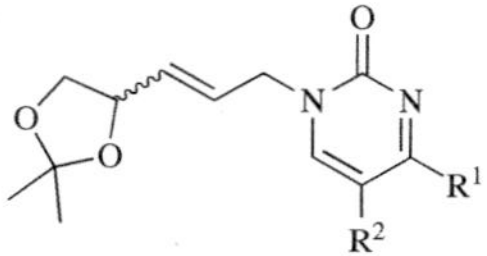

R[1]	R[2]	
HO	H	(78)
HO	Me	(50)
HO	I	(56)
BzHN	H	(57)

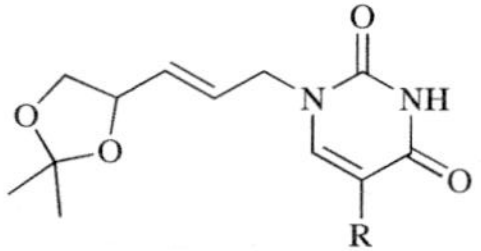

R	
H	(78)
F	(62)
I	(56)
Me	(50)

TABLE 13. CROSS-METATHESIS OF SECONDARY ALLYLIC ALCOHOLS AND DERIVATIVES (*Continued*)

B. ALLYLIC ETHERS (*Continued*)

Allylic Ester	Cross Partner	Conditions	Product(s) and Yield(s) (%)	Refs.
Please refer to the charts preceding the tables for the catalyst structures.				
C_4				
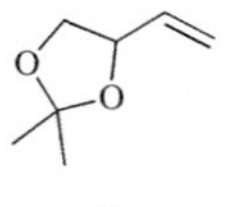 5.12 eq	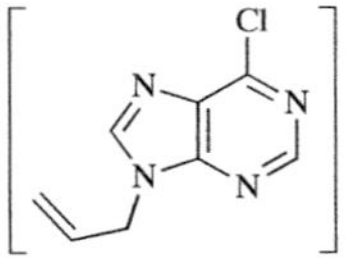 • HCl	Ru cat **35**, DCM, 40°	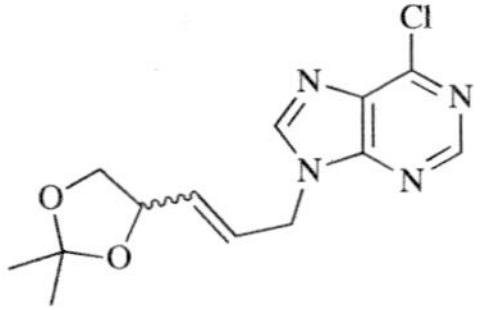 (14)	785
3 eq	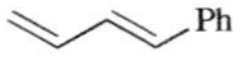 Ph	Gr I (0.05 eq), DCM, reflux, 2 h	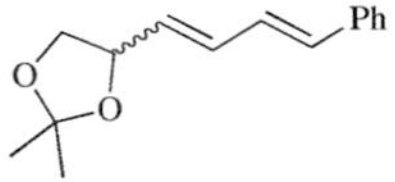 Ph (64)	782
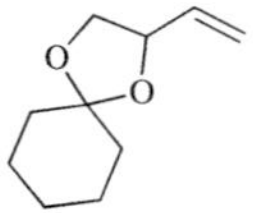	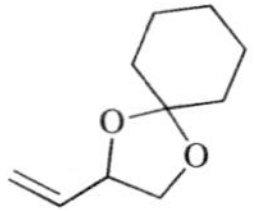	Gr I (0.05 eq), DCM, reflux	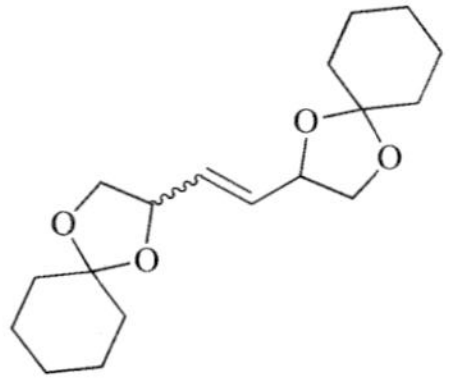 (80)	782

Alkene 1	Alkene 2	Conditions	Product(s) and Yield(s) (%)	Refs.
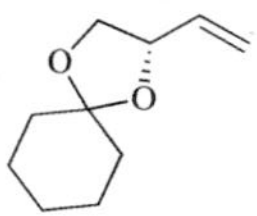 5 eq	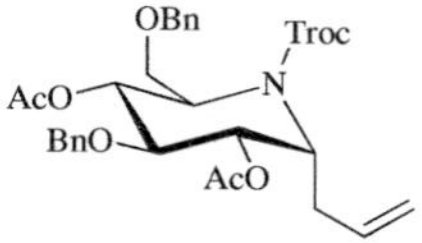 OBn, AcO, BnO, AcO, N, Troc	Gr II (0.2 eq), DCM, reflux, 16–24 h	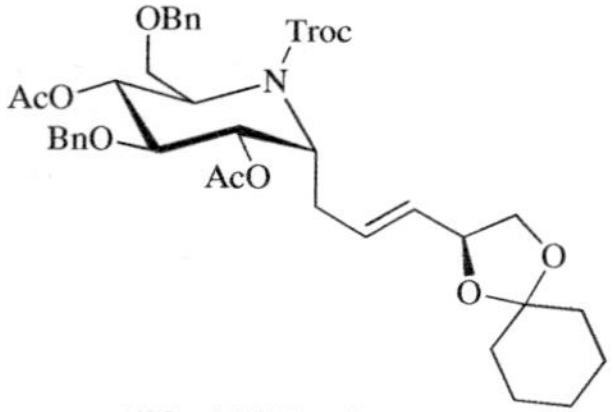 OBn, Troc, N, AcO, BnO, AcO, O, O (88) $(E)/(Z) > 95:5$	666
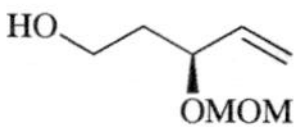 HO, OMOM 4 eq	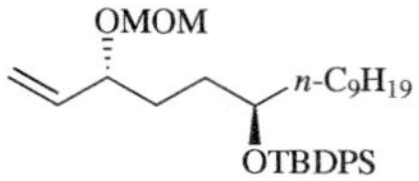 OMOM, n-C_9H_{19}, OTBDPS	Hov II (0.05 eq), DCM, reflux, 24 h	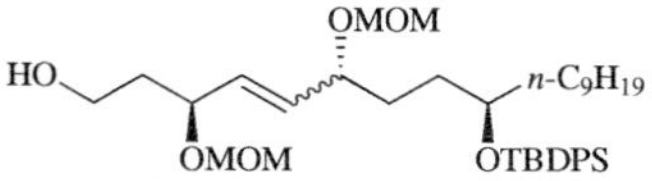 HO, OMOM, OMOM, n-C_9H_{19}, OTBDPS (75)	786
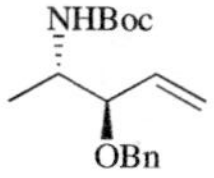 NHBoc, OBn	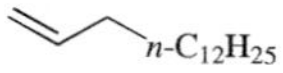 n-$C_{12}H_{25}$ 3 eq	Gr II (0.05 eq), DCM, reflux, 18 h	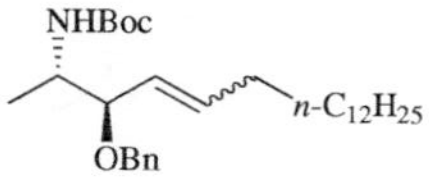 NHBoc, OBn, n-$C_{12}H_{25}$ (87)	787

TABLE 13. CROSS-METATHESIS OF SECONDARY ALLYLIC ALCOHOLS AND DERIVATIVES (*Continued*)
B. ALLYLIC ETHERS (*Continued*)

Allylic Ester	Cross Partner	Conditions	Product(s) and Yield(s) (%)	Refs.
Please refer to the charts preceding the tables for the catalyst structures.				
C_5				
R^2, OR^1; 4 eq	OTBS, n-C_9H_{19}	Catalyst (x eq), DCM, 40°	R^2, OR^1, OTBS, n-C_9H_{19}	786

R^1	R^2	Catalyst	x	Time (h)		$(E)/(Z)$
H	$PMBOCH_2$	Gr II	0.05	24	(30)	50:50
H	$PMBOCH_2$	Hov I	0.05	24	(—)	—
H	$PMBOCH_2$	Hov II	0.05	24	(70)	92:8
H	$PMBOCH_2$	Ru cat **271**	0.05	24	(10)	—
H	$PMBOCH_2$	Ru cat **271**	0.2	80	(40)	89:11
H	$PMBOCH_2$	Ru cat **1**	0.05	24	(25)	50:50
MOM	$PMBOCH_2$	Gr II	0.05	24	(—)	—
MOM	$PMBOCH_2$	Hov I	0.05	24	(—)	—
MOM	$PMBOCH_2$	Hov II	0.05	24	(—)	—
MOM	$PMBOCH_2$	Ru cat **271**	0.05	24	(—)	—
MOM	$PMBOCH_2$	Ru cat **1**	0.05	24	(—)	—
MOM	$HOCH_2$	Hov II	0.05	24	(40)	90:10
MOM	HO_2C	Gr II	0.05	24	(—)	—
MOM	HO_2C	Hov I	0.05	24	(—)	—
MOM	HO_2C	Hov II	0.05	24	(—)	—
MOM	HO_2C	Ru cat **271**	0.05	24	(—)	—
MOM	HO_2C	Ru cat **1**	0.05	24	(—)	—

Substrate	Partner	Conditions	Product(s) and Yield(s) (%)			Refs.
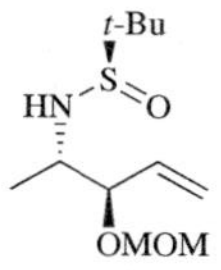 t-Bu, HN, S, O, OMOM	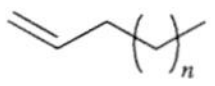 n, 4 eq	Gr II (3 x 0.04 eq), DCM, reflux, 20 h	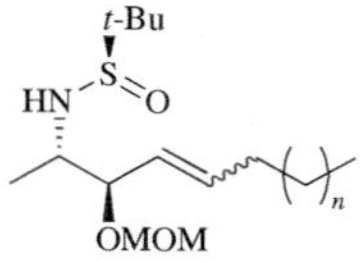 t-Bu, HN, S, O, OMOM, n	*n*: 10 (67); 11 (72)		788
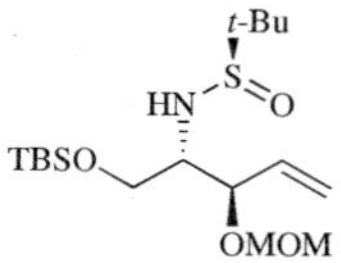 t-Bu, HN, S, O, TBSO, OMOM	 n, 4 eq + 3 x 2 eq	Hov II (4 x 0.04 eq), DCM, reflux, 26 h	t-Bu, HN, S, O, TBSO, OMOM, n	*n*: 5 (78); 11 (78)	(*E*)/(*Z*): 86:14; 90:10	788
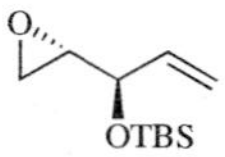 O, OTBS	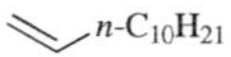 *n*-$C_{10}H_{21}$, 6 eq	1. Gr II (0.05 eq), benzene, 50°, 10 h 2. H_2, Pd/C, EtOAc	O, OTBS, *n*-$C_{10}H_{21}$	(73)		521
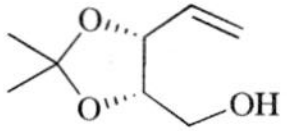 O, O, OH	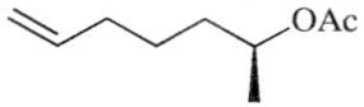 OAc, 2.5 eq	Gr II (0.05 eq), DCM, 40°, 4 h	O, O, OH, OAc	(76)		250
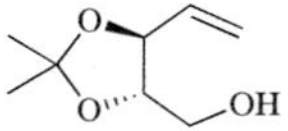 O, O, OH	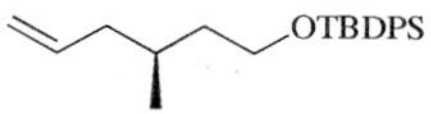 OTBDPS, 1.5 eq	Gr II (0.03 eq), DCM, reflux, 12 h	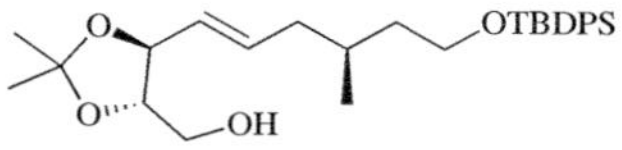 O, O, OH, OTBDPS	(72)	(*E*) only	789

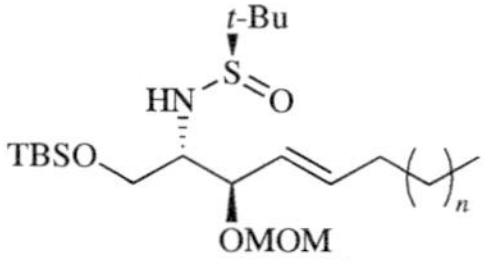

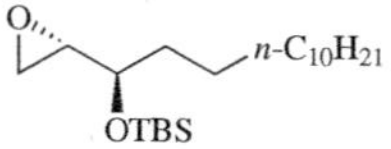

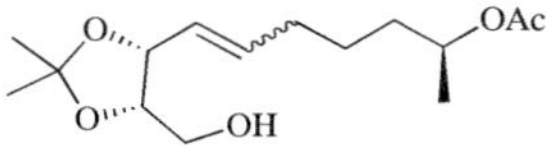

TABLE 13. CROSS-METATHESIS OF SECONDARY ALLYLIC ALCOHOLS AND DERIVATIVES (*Continued*)

B. ALLYLIC ETHERS (*Continued*)

Allylic Ester	Cross Partner	Conditions	Product(s) and Yield(s) (%)	Refs.
Please refer to the charts preceding the tables for the catalyst structures.				
C_5				
MeO_2C	OH 1.2 eq	Gr II (0.1 eq), DCM, rt, 10 h	MeO_2C … OH (75)	790
MeO_2C 1.5 eq	OH	Gr II (0.1 eq), DCM, 40°, 7 h	MeO_2C … OH (73) (*E*) only	791
O *x* eq	R, CO_2t-Bu	Hov II (*y* eq), solvent	O, R, CO_2t-Bu	792

x	R	*y*	Solvent	Temp (°)	Time (h)	
1.5	H	0.05	DCM	40	2.5	(82)
1.5	Me	0.05	DCM	40	2.5	(0)
1.5	Me	0.01	$C_6F_5CF_3$	70	16	(43)
3	Me	0.01	$C_6F_5CF_3$	70	2	(68)
3	Bn	0.01	$C_6F_5CF_3$	70	2	(61)

Reactant 1	Reactant 2	Conditions	Product(s) and Yield(s) (%)	Refs.
1 eq	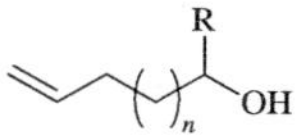 R, OH, n	Hov II (0.05 eq), DCM, 40°, 24 h	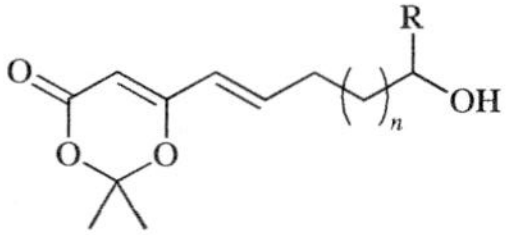 O, O, O, R, OH, n; $(E)/(Z) > 95{:}5$; R / n: H 6 (71); H 8 (75); Me 9 (78); H 10 (77)	793
1.5 eq	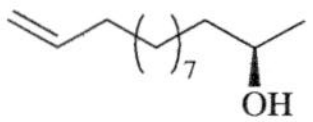 OH, 7	Hov II (0.05 eq), DCM, 40°, 24 h	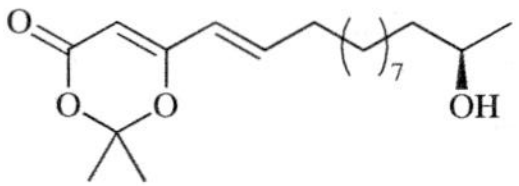 O, O, O, OH, 7; (88) $(E)/(Z) > 95{:}5$	793
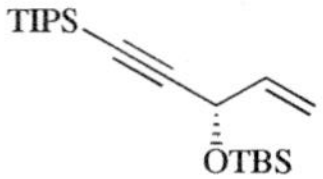 TIPS, OTBS	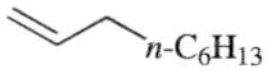 n-C_6H_{13}; 10 eq	1. Mo cat **18** (0.045 eq), benzene, 7 torr, 22°, 4 h; 2. $(n\text{-Bu})_4NF$ (3 eq), THF, 22°, 1 h	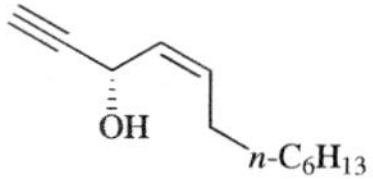 OH, n-C_6H_{13}; (94) $(Z)/(E) = 92{:}8$	184
	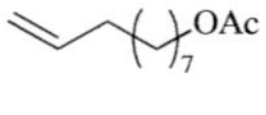 OAc, 7; 3 eq	Mo cat **18** (0.03 eq), benzene, 7 torr, 22°, 8 h	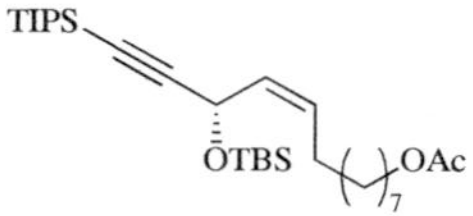 TIPS, OTBS, OAc, 7; (92) $(Z)/(E) = 92{:}8$	184

TABLE 13. CROSS-METATHESIS OF SECONDARY ALLYLIC ALCOHOLS AND DERIVATIVES (*Continued*)

B. ALLYLIC ETHERS (*Continued*)

Allylic Ester	Cross Partner	Conditions	Product(s) and Yield(s) (%)	Refs.
Please refer to the charts preceding the tables for the catalyst structures.				
C_5				
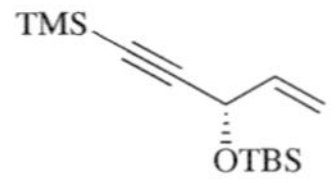 TMS, OTBS; 3 eq	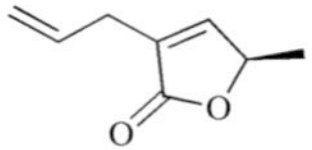	Mo cat **18** (0.06 eq), PhCl, 100 torr, 22°, 5.5 h	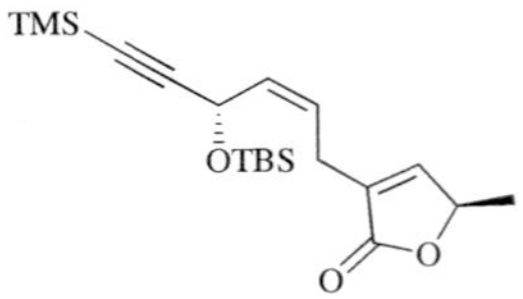 TMS, OTBS; (56) (Z)/(E) = 91:9	184
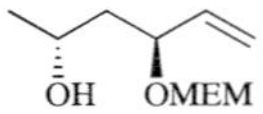 OH, OMEM; 2 eq	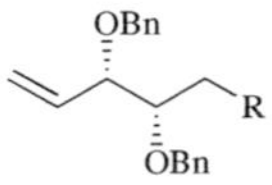 OBn, R, OBn	Gr II (0.2 eq), DCM, rt, 2 d	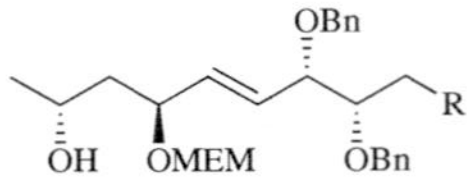 OH, OMEM, OBn, R, OBn; (E)/(Z) = 83:17 R: $HOCH_2$ (38); HO_2C (54)	794
C_6				
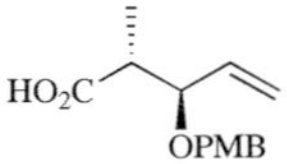 HO_2C, OPMB; 1.25 eq	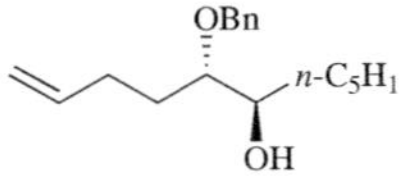 OBn, n-C_5H_1, OH	Gr II (0.1 eq), DCM, reflux, 12 h	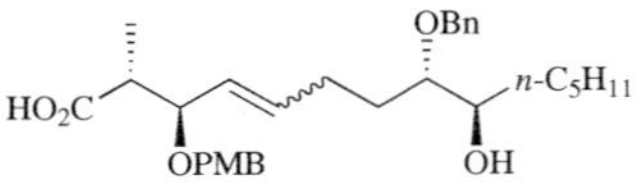 HO_2C, OPMB, OBn, n-C_5H_{11}, OH (70)	795
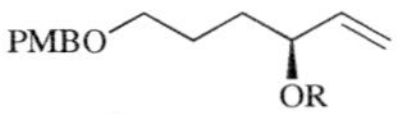 PMBO, OR; 2 eq	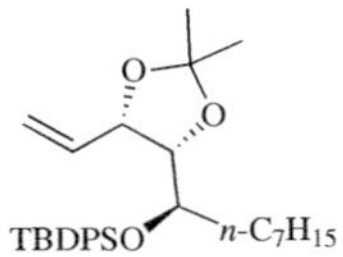 TBDPSO, n-C_7H_{15}	Hov II (0.05 eq), DCM, 40°, 24 h	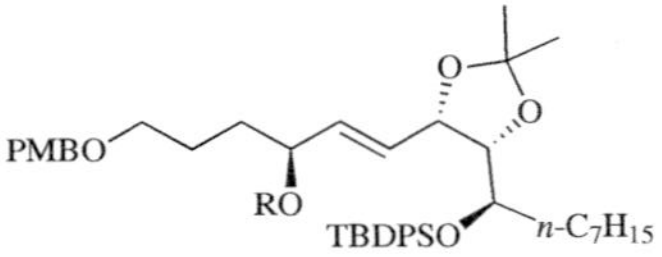 PMBO, RO, TBDPSO, n-C_7H_{15} R: H (65) (E)/(Z) 94:6; MOM (32) (E)/(Z) 94:6	796

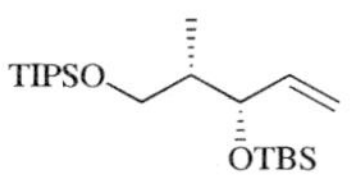

TIPSO OTBS

2 eq

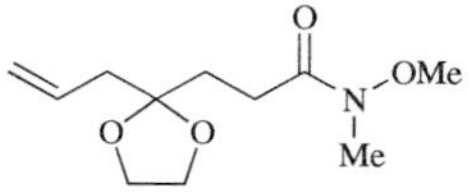

O N OMe Me (ketal)

Hov II (0.09 eq), DCM, reflux, 7 d

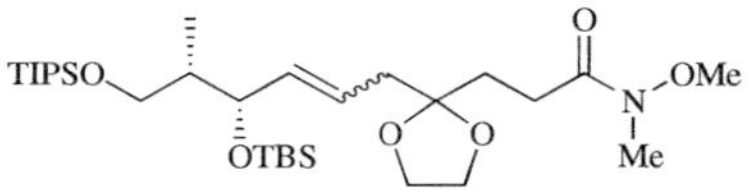

TIPSO OTBS O N OMe Me (85) 797

RO NHBoc OTBS

n-$C_{13}H_{27}$

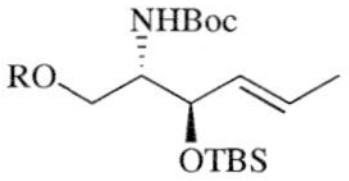

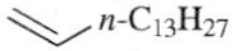

1 eq

Gr II (6 × 0.05 eq), benzene, 55°

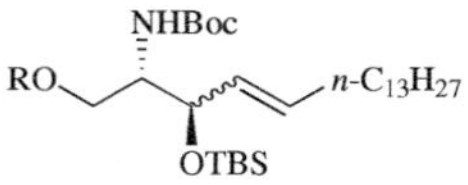

RO NHBoc OTBS n-$C_{13}H_{27}$

R	
H	(18)
TBS	(—)

740

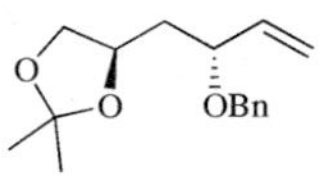

OBn (acetonide)

x eq

OTBS OPMB

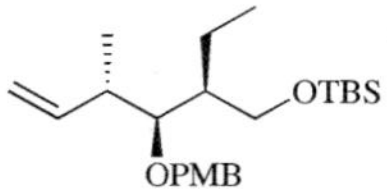

y eq

Catalyst (0.1 eq), solvent, reflux

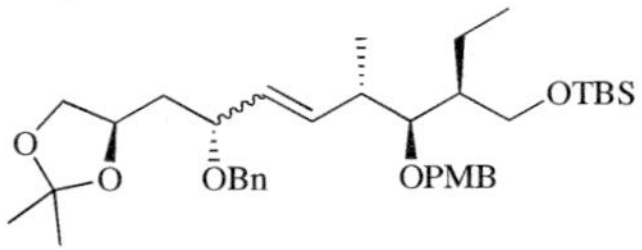

OBn OPMB OTBS 798

x	y	Catalyst	Solvent	Time (h)	
1	1	Gr I	DCM	24	(10)
1	1	Gr II	DCM	24	(29)
1.5	1	Gr II	DCM	24	(35)
1.5	1	Gr II	DCM	48	(47)
1	1.5	Gr II	DCM	24	(40)
1.5	1	Gr II	toluene	24	(32)
1.5	1	Hov I	DCM	24	(0)

TABLE 13. CROSS-METATHESIS OF SECONDARY ALLYLIC ALCOHOLS AND DERIVATIVES (*Continued*)

B. ALLYLIC ETHERS (*Continued*)

Allylic Ester	Cross Partner	Conditions	Product(s) and Yield(s) (%)	Refs.
Please refer to the charts preceding the tables for the catalyst structures.				
C_6				
	OH; 2.5 eq	Gr II (0.05 eq), DCM, reflux	OH; (45) $(E)/(Z) = 90:10$	799
O; O; O	ŌH; 1.5 eq	Gr II (0.02 eq), DCM, reflux, 24 h	O; O; O; ŌH; (85)	334
R^1; R^2; O; O	n-$C_{11}H_{23}$; x eq	Gr II (y eq), DCM, reflux	R^1; R^2; O; O; n-$C_{11}H_{23}$	

R^1	R^2	x	y	Time (h)		$(E)/(Z)$	Refs.
HO	AcHN	5	0.2	18	(82)	94:6	800
HO	BocHN	5	0.2	18	(77)	92:8	800
HO	TFAHN	5	0.2	18	(83)	100:0	800
TBSO	HO	4	0.04	24	(60)	—	801
PhthN	TBSO	4	0.04	24	(88)	—	801

Substrate	Alkene	Conditions	Product(s) and Yield(s) (%)			Refs.
R^1, R^2	n-$C_{11}H_{23}$ 4 eq	Gr II (0.04 eq), DCM, reflux, 24 h	R^1, R^2, n-$C_{11}H_{23}$			
			R^1	R^2		
			TBSO	HO	(57)	801
			PhthN	TBSO	(87)	801
			MOMO	PhthN	(75)	802
TsO, OH	CO_2Bn, 10 1.35 eq	1. Gr II (0.02 eq), DCM, reflux, 14 h 2. TBAF, THF, rt, 4 h	CO_2Bn, 10		(72)	803
MeO, RO, OR 2 eq	N, O, N	Gr I (0.1 eq), Ti(Oi-Pr)$_4$ (0.3 eq), DCM, 40°, 18 h	MeO, RO, OR, N, O, N	R		
				H	(78)	804
				Ac	(73)	
				Me	(81)	
OBn	OH	Gr II (0.074 eq), Triton X 100, H_2O, 25°, 48 h	OH, OBn		(62)	805

TABLE 13. CROSS-METATHESIS OF SECONDARY ALLYLIC ALCOHOLS AND DERIVATIVES (*Continued*)

B. ALLYLIC ETHERS (*Continued*)

Allylic Ester	Cross Partner	Conditions	Product(s) and Yield(s) (%)	Refs.
Please refer to the charts preceding the tables for the catalyst structures.				
C_6				
OBn	Br, 2 eq	Gr II (0.1 eq), DCM, rt, 24 h	Br, OBn (71)	806
OH, 1 eq	OMe, R	Gr II (0.05 eq), DCM, reflux, 18 h	OMe, R, OH (*E*) only; R: H (61), MeO (53)	807
OR, 2 eq	N, O, N	Gr I (0.1 eq), Ti(O*i*-Pr)$_4$ (0.3 eq), DCM, 40°, 18 h	OR, N, O, N; R: H (79), Ac (75), Me (83)	804
MeO, 1 eq	R[1], R[2], R[3]	Gr II (0.05 eq), DCM, 40°, 18 h	MeO, R[1], R[2], R[3] (*E*) only	807

R[1]	R[2]	R[3]	
H	MeO	H	(66)
MeO	H	H	(65)
MeO	HO	H	(58)
MeO	H	MeO	(51)

1 eq	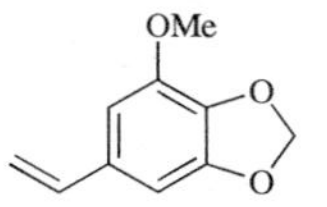	Gr II (0.05 eq), DCM, 40°, 18 h	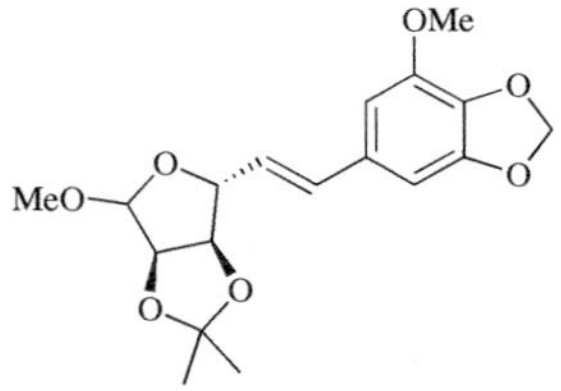	(67) (*E*) only	807
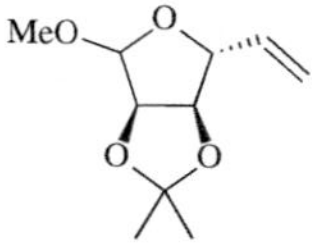 2 eq	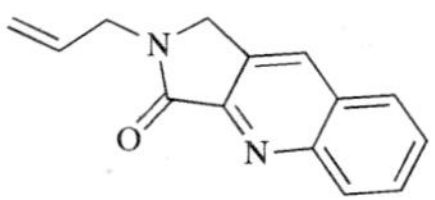	Gr II (0.1 eq), additive (*x* eq), DCM, 40°	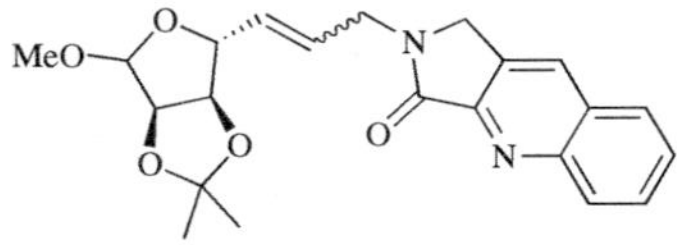 **I** + 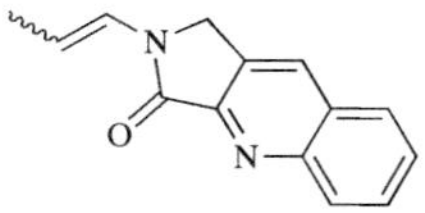 **II**		804

Additive	*x*	Time (h)	**I**	**II**
—	—	18	(31)	(27)
$AlCl_3$	1	2	(0)	(—)
$Sc(OTf)_3$	1	4	(0)	(—)
$TiCl_4$	1	2	(0)	(—)
$Ti(Oi\text{-}Pr)_4$	1	12	(72)	(—)
$Ti(Oi\text{-}Pr)_4$	0.5	16	(78)	(—)
$Ti(Oi\text{-}Pr)_4$	0.3	18	(86)	(—)
$Ti(Oi\text{-}Pr)_4$	0.1	24	(81)	(—)

TABLE 13. CROSS-METATHESIS OF SECONDARY ALLYLIC ALCOHOLS AND DERIVATIVES (*Continued*)

B. ALLYLIC ETHERS (*Continued*)

Allylic Ester	Cross Partner	Conditions	Product(s) and Yield(s) (%)	Refs.
Please refer to the charts preceding the tables for the catalyst structures.				
C_6				
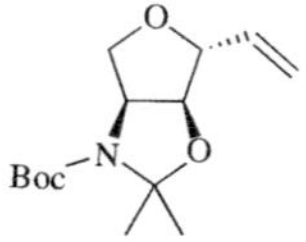 Boc–N, O	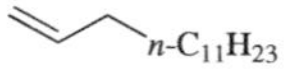 n-$C_{11}H_{23}$ 2.5 eq	Gr II (0.05 + 0.03 eq), DCM, 45°, 8 h	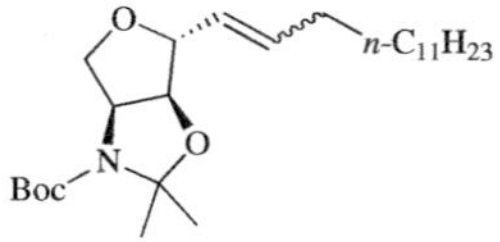 O, Boc–N, O, n-$C_{11}H_{23}$ (80)	470
C_7				
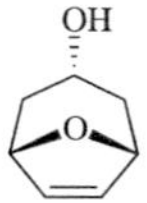 OH, O	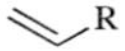 R 20 eq	Ru cat **213** (0.05 eq), THF, 22°, 2.5 h	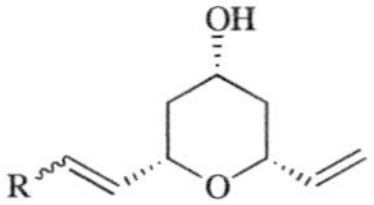 OH, R, O	808
			R: Ph; Conv. (%)[a]: >98; (68); er: 96.0:4.0	
			R: Cy; Conv. (%)[a]: >98; (58); er: 59.0:41.0	
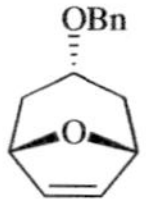 OBn, O	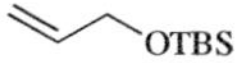 OTBS	**Mo cat 18** (0.05 eq), 22°, 1 h	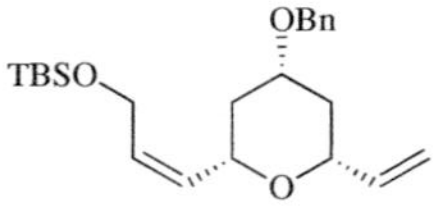 OBn, TBSO, O (45) $(Z)/(E)$ = 95:5, er 86.0:14.0	809

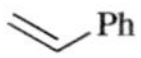

Ph

Catalyst (0.01 eq), 22°, 1 h

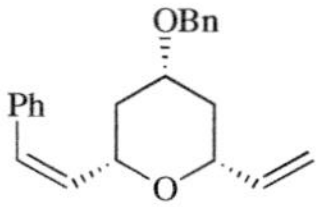

OBn
Ph
O

809

Catalyst		(*Z*)/(*E*)	er
Mo cat **12**	(80)	>98:2	95.0:5.0
Ru cat **212**	(85)	>98:2	98.5:1.5
Ru cat **213**	(87)	95:5	98.0:2.0

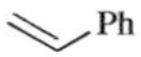

OBn
O

Ph

Mo cat **18** (0.02 eq), 22°

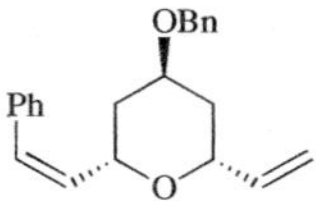

OBn
Ph
O

(—) (*Z*)/(*E*) = 95:5, er 91.0:9.0

809

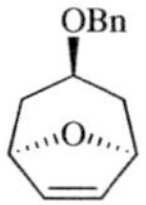

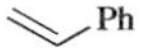

OBn
O

Ph
5 eq

Catalyst (*x* eq), solvent, 1 h

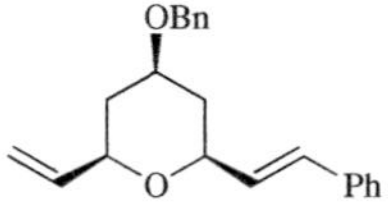

OBn
O
Ph

(*E*)/(*Z*) > 98:2

91

Catalyst	*x*	Solvent	Temp (°)		er
Ru cat **210**	0.02	—	22	(70)	98.0:2.0
Ru cat **213**	0.02	—	22	(90)	94.0:6.0
Ru cat **213**	0.02	—	–15	(87)	95.5:4.5
Mo cat **12**	0.05	benzene	22	(—)	—

TABLE 13. CROSS-METATHESIS OF SECONDARY ALLYLIC ALCOHOLS AND DERIVATIVES (*Continued*)

B. ALLYLIC ETHERS (*Continued*)

Allylic Ester	Cross Partner	Conditions	Product(s) and Yield(s) (%)	Refs.

Please refer to the charts preceding the tables for the catalyst structures.

C_7

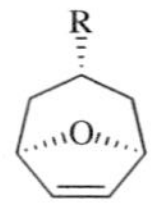

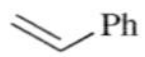

Allylic Ester	Cross Partner	Conditions	Product(s) and Yield(s) (%)	Refs.
R	Ph, *x* eq	Catalyst (*y* eq), solvent	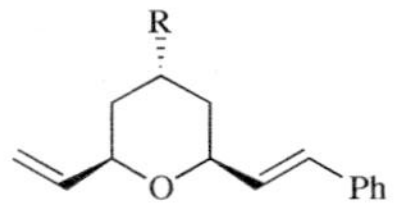 R, O, Ph; (*E*)/(*Z*) > 98:2	91

R	*x*	Catalyst	*y*	Solvent	Temp (°)	Time (h)		er
BnO	2	Ru cat **211**	0.02	—	22	36	(85)	95.0:5.0
BnO	2	Ru cat **213**	0.02	—	22	15	(70)	97.0:3.0
BnO	5	Ru cat **213**	0.02	—	–15	15	(84)	>99.0:1.0
BnO	10	Mo cat **12**	0.02	benzene	22	1	(81)	95.0:5.0
I	2	Ru cat **211**	0.02	—	22	5	(65)	96.5:3.5
I	2	Ru cat **213**	0.02	—	22	5	(70)	91.0:9.0
I	10	Mo cat **12**	0.05	benzene	22	1	(76)	98.5:1.5

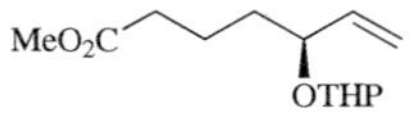

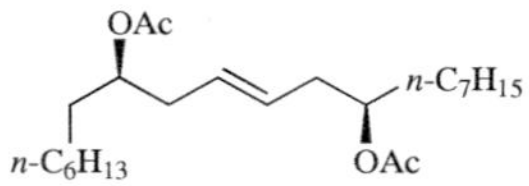

Allylic Ester	Cross Partner	Conditions	Product(s) and Yield(s) (%)	Refs.
MeO_2C, OTHP	OAc, n-C_7H_{15}, n-C_6H_{13}, OAc; 1.6 eq	Ru cat **35** (0.15 eq), DCM, 100° (MW), 40 min	MeO_2C, n-C_7H_{15}, OTHP, OAc (66)	505

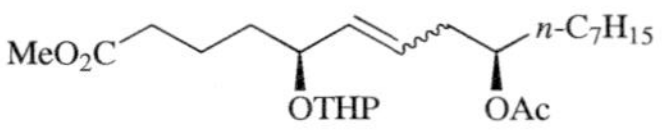

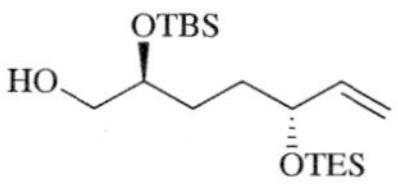 HO, OTBS, OTES	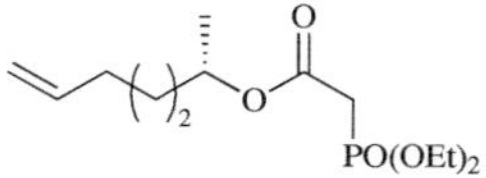 PO(OEt)$_2$, ( )$_2$, O 3 eq	Gr II (0.05 eq), DCM, reflux, 9 h	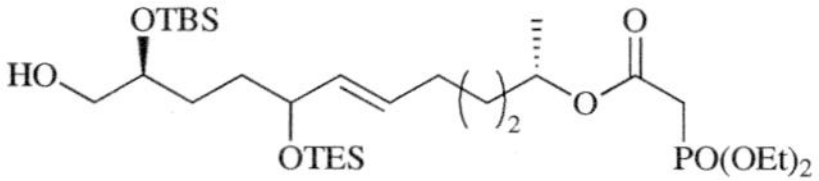 HO, OTBS, OTES, ( )$_2$, O, PO(OEt)$_2$ (98) (*E*)/(*Z*) = 95:5	810
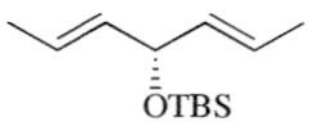 OTBS	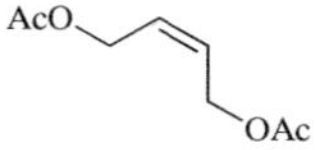 AcO, OAc 5 eq	Ru cat **20** (0.05 eq), DCM, 40°, 6 h	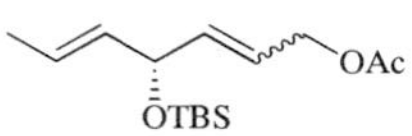 OTBS, OAc (23) er 52.0:48.0	97
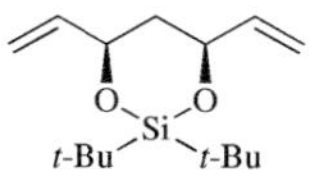 O, Si, O, *t*-Bu, *t*-Bu	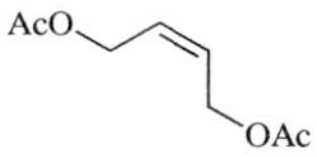 AcO, OAc 5 eq	Ru cat **20** (0.05 eq), DCM, 40°, 6 h	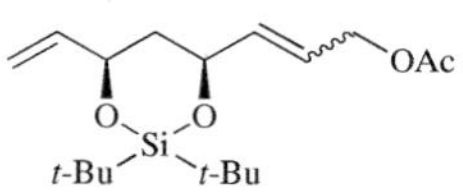 O, Si, O, *t*-Bu, *t*-Bu, OAc (48) er 68.5:31.5	97
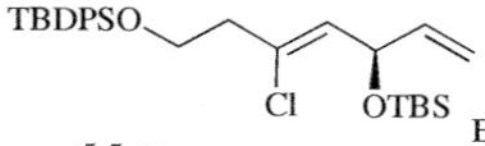 TBDPSO, Cl, OTBS 5.5 eq	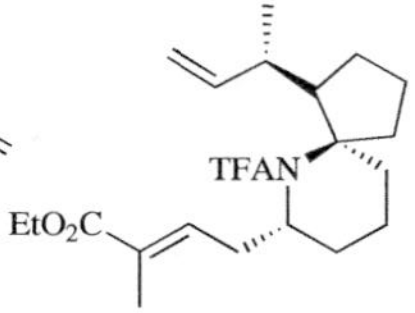 TFAN, EtO_2C	Gr II (0.8 eq), toluene, 90°	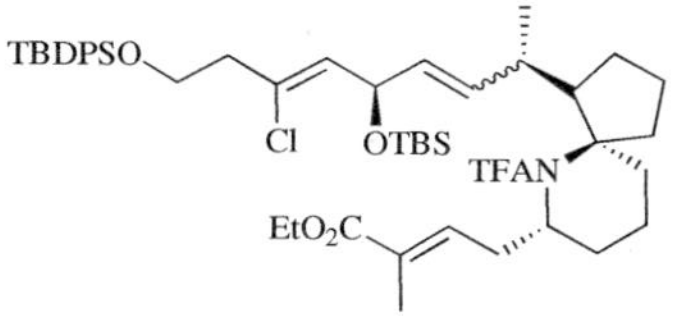 TBDPSO, Cl, OTBS, TFAN, EtO_2C (40)	811

TABLE 13. CROSS-METATHESIS OF SECONDARY ALLYLIC ALCOHOLS AND DERIVATIVES (*Continued*)

B. ALLYLIC ETHERS (*Continued*)

Allylic Ester	Cross Partner	Conditions	Product(s) and Yield(s) (%)	Refs.
Please refer to the charts preceding the tables for the catalyst structures.				
C_7				
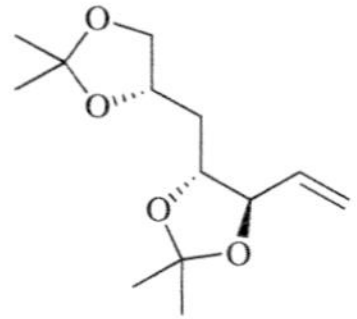	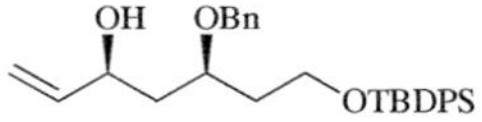 OH OBn OTBDPS	Gr II, DCM, rt, 6 h	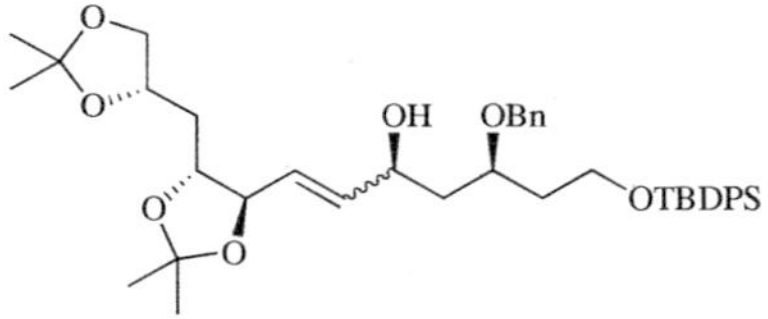 OH OBn OTBDPS (63)	812
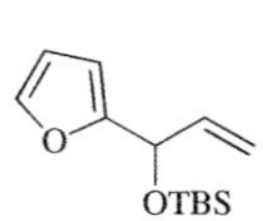 OTBS	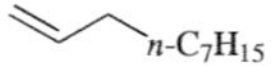 n-C_7H_{15} 3 eq	1. Mo cat **18** (0.03 eq), benzene, 7 torr, 22°, 8 h 2. $(n\text{-}Bu)_4NF$ (2 eq), THF, 22°, 1 h	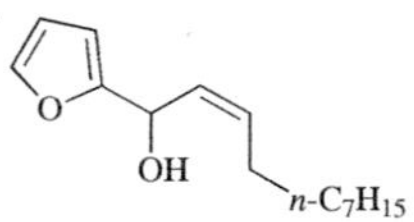 OH n-C_7H_{15} (80) $(Z)/(E)$ = 95:5	184
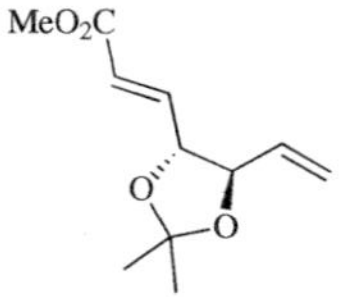 MeO_2C	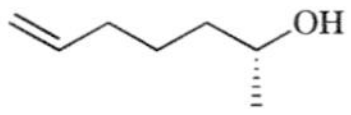 OH 1.5 eq	Gr II (0.1 eq), DCM, reflux	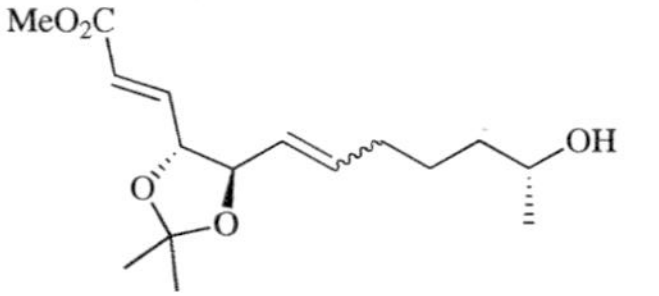 MeO_2C OH (60)	799
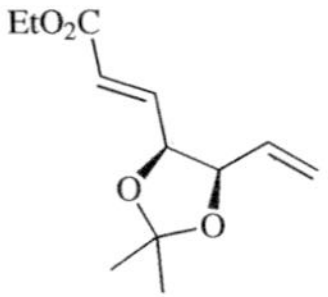 EtO_2C	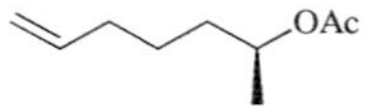 OAc 2.5 eq	Gr II (0.05 eq), DCM, reflux, 4 h	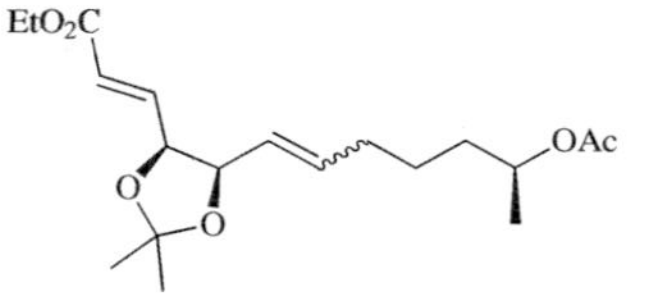 EtO_2C OAc (75)	250

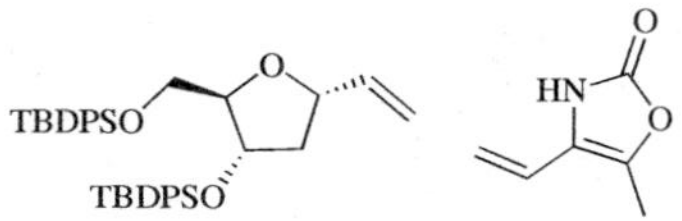

TBDPSO TBDPSO 1 eq HN O O

Catalyst (0.05 eq), DCM, 40°, 8 h

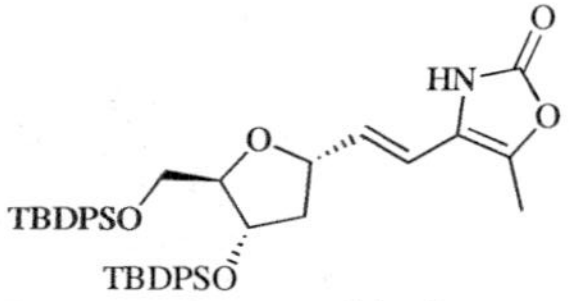

TBDPSO TBDPSO HN O O (*E*) only

Catalyst		
Hov I	(67)	813
Hov II	(15)	
Gr I	(8)	
Gr II	(12)	

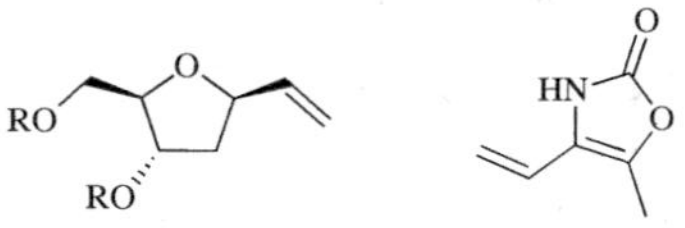

RO RO 1 eq HN O O

Catalyst (0.05 eq), solvent, 8 h

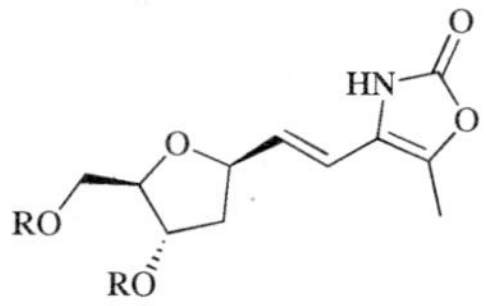

RO RO HN O O (*E*) only 813

R	Catalyst	Solvent	Temp (°)	
H	Hov I	DCM	40	—
H	Hov II	DCM	40	—
H	Gr I	DCM	40	—
H	Gr II	DCM	40	—
TBDPS	Hov I	DCM	40	(63)
TBDPS	Hov II	DCM	40	(18)
TBDPS	Gr I	DCM	40	(10)
TBDPS	Gr I	toluene	110	(5)
TBDPS	Gr II	DCM	40	(14)
TBDPS	Gr II	toluene	110	(5)

TABLE 13. CROSS-METATHESIS OF SECONDARY ALLYLIC ALCOHOLS AND DERIVATIVES (*Continued*)

B. ALLYLIC ETHERS (*Continued*)

Allylic Ester	Cross Partner	Conditions	Product(s) and Yield(s) (%)	Refs.
Please refer to the charts preceding the tables for the catalyst structures.				
C_7				
	R, OMe; 1.5 eq	Gr II (*x* eq), DCM, reflux	R, OMe	
			R / *x* / Time (h) / / (*E*)/(*Z*)	
			$HOCH_2$ / 0.06 / 32 / (62) / (*E*) only	814
			MeO_2C / 0.05 / 12 / (56) / —	815
	CO_2Me, OMe; 1.5 eq	Gr II (0.05 eq), DCM, reflux, 12 h	OMe, CO_2Me (62)	815

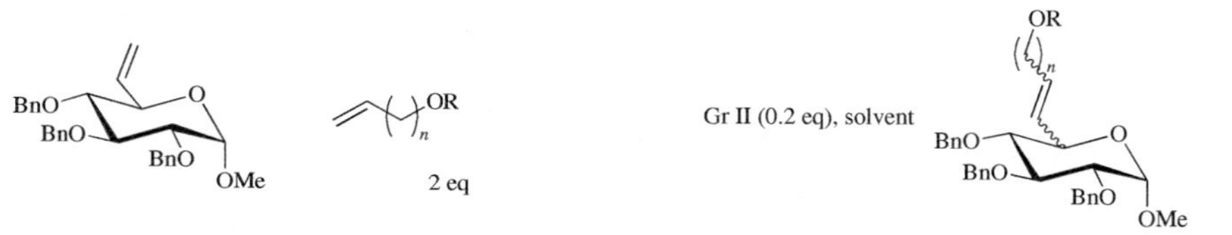

2 eq

Gr II (0.2 eq), solvent

816

R	n	Solvent	Temp (°)	
H	1	DCM	23	(13)
H	1	DCM	40	(15)
H	3	DCE	83	(7)
Ac	3	DCM	23	(17)
Ac	4	DCM	40	(66)
Ac	3	DCM	40	(56)
Ac	2	DCM	40	(49)
Ac	1	DCM	40	(69)

6.25 eq

Gr II (0.23 eq), DCM, reflux

660

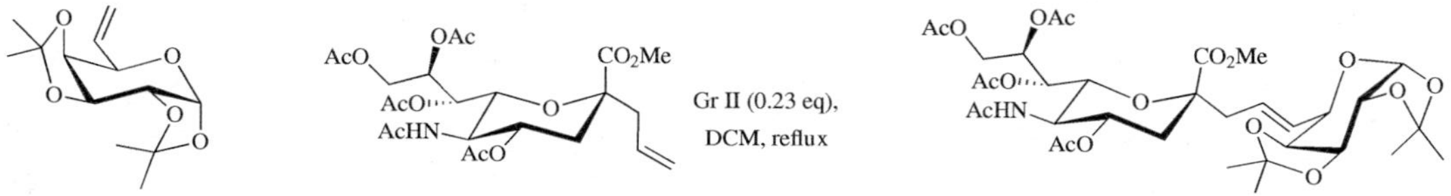

(95) $(E)/(Z)$ = 75:25

TABLE 13. CROSS-METATHESIS OF SECONDARY ALLYLIC ALCOHOLS AND DERIVATIVES (*Continued*)

B. ALLYLIC ETHERS (*Continued*)

Allylic Ester	Cross Partner	Conditions	Product(s) and Yield(s) (%)	Refs.
Please refer to the charts preceding the tables for the catalyst structures.				
C_8				
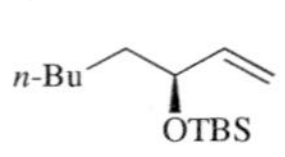 n-Bu, OTBS; 2 eq	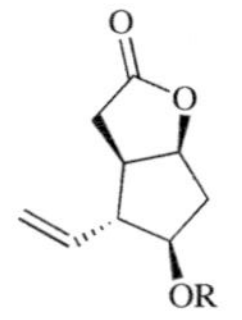 O, O, OR	Catalyst (0.06 eq), DCM, 40°, 12 h	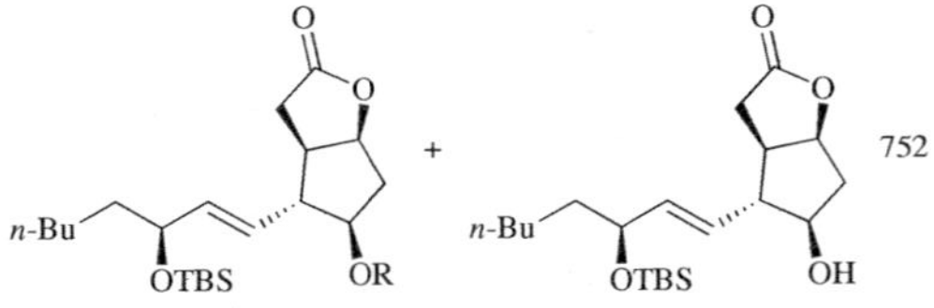 n-Bu, OTBS, O, O, OR **I** + n-Bu, OTBS, O, O, OH **II**	752
5 eq	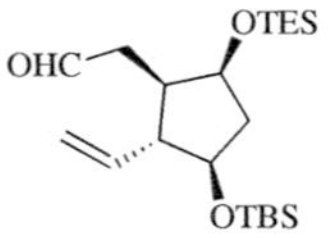 OHC, OTES, OTBS	Hov **II** (0.1 eq), DCE, reflux, 12 h	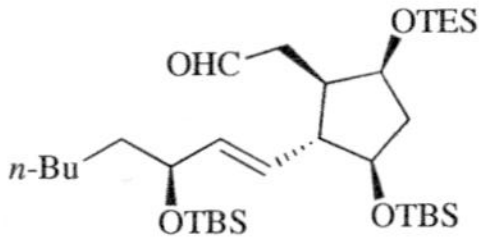 OHC, OTES, n-Bu, OTBS, OTBS (64) (*E*)/(*Z*) > 95:5	817
5 eq	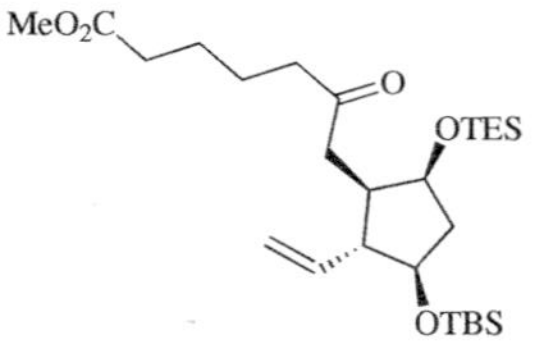 MeO_2C, O, OTES, OTBS	Hov **II** (0.1 eq), DCM, reflux, 14 h	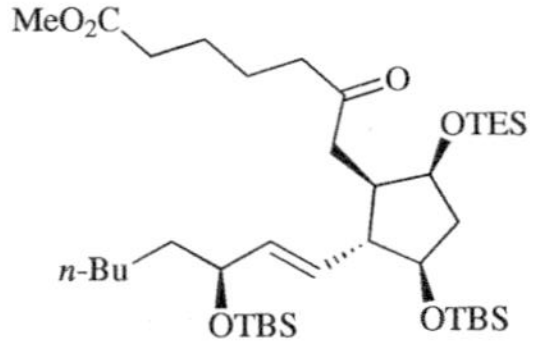 MeO_2C, O, OTES, n-Bu, OTBS, OTBS (64) (*E*)/(*Z*) > 95:5	817

R	Catalyst	**I**	(*E*)/(*Z*) **I**	**II**	(*E*)/(*Z*) **II**
H	Gr II	(70)	93:7	(—)	—
TBS	Gr II	(76)	86:14	(—)	—
TBS	Gr II	(84)	94:6	(—)	—
Bz	Gr II	(40)	88:12	(25)	86:14
Bz	Hov II	(25)	88:12	(15)	86:14

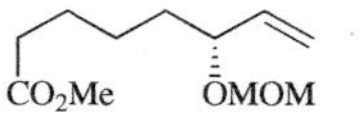

1.25 eq

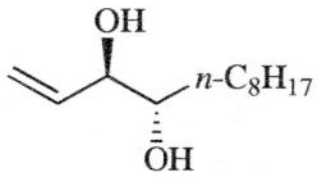

Gr II (0.05 eq), DCM, reflux, 12 h

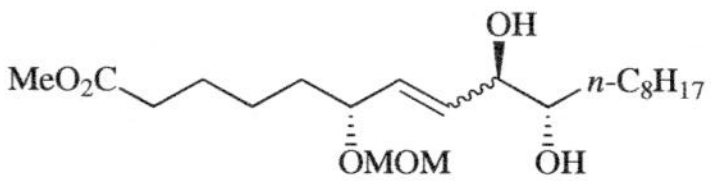

(70) 818

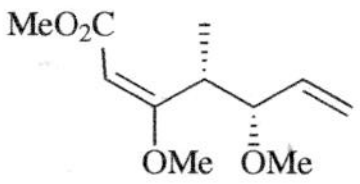

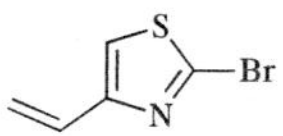

5 x 0.2 eq

Gr II (0.2 + 0.1 eq), DCM, 100° (MW), 5 x 1 h + 0.5 h

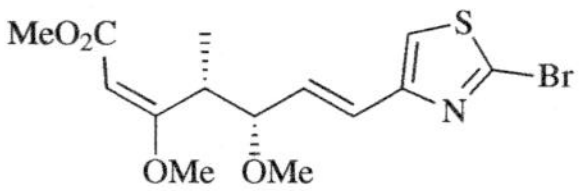

(55) (*E*)/(*Z*) = 95:5 819

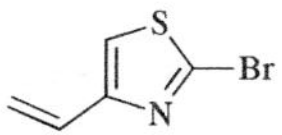

x eq

Catalyst (*y* eq), solvent

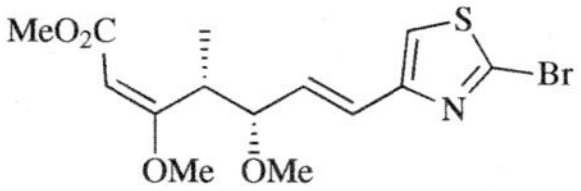

820

x	Catalyst	*y*	Solvent	Temp (°)	Time (h)	Conv. (%)[a]		(*E*)/(*Z*)
1	Gr I	0.1	DCM	40	48	0	(—)	—
1	Gr II	0.1	DCM	40	48	27	(—)	>95:5
1	Hov II	0.1	DCM	40	48	18	(—)	>95:5
2	Gr II	0.2	DCM	40	60	50	(—)	>95:5
2	Gr II	0.2	benzene	60	60	26	(—)	>95:5
2	Gr II	0.2	DCM	20	72	5	(—)	>95:5
2	Gr II	0.3	DCM	40	60	68	(56)	>95:5
2	Gr II	0.4	DCM	40	60	72	(—)	>95:5

TABLE 13. CROSS-METATHESIS OF SECONDARY ALLYLIC ALCOHOLS AND DERIVATIVES (*Continued*)

B. ALLYLIC ETHERS (*Continued*)

Allylic Ester	Cross Partner	Conditions	Product(s) and Yield(s) (%)	Refs.
Please refer to the charts preceding the tables for the catalyst structures.				
C_8				
MeO$_2$C, OMe, OMe	S, N, O; *x* eq	Catalyst (*y* eq), solvent	MeO$_2$C, S, N, O, OMe, OMe	821
	S, N, N, S; 5 x 0.2 eq	Gr II (0.2 +0.1 eq), CD_2Cl_2, MW	MeO$_2$C, S, N, N, S, OMe, OMe (25)	820

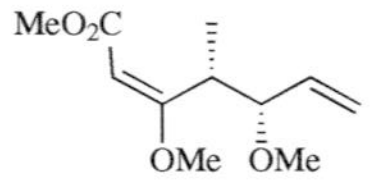

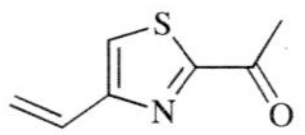

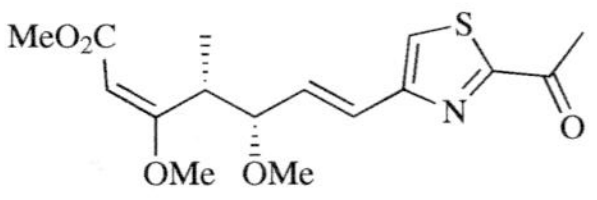

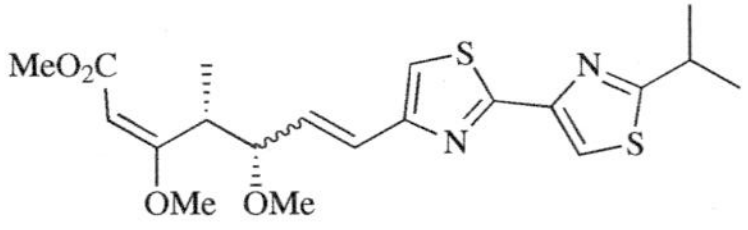

x	Catalyst	*y*	Solvent	Temp (°)	Time (h)	Conv. (%)[a]		(*E*)/(*Z*)
1	Gr I	0.1	DCM	40	48	0	(—)	—
1	Gr II	0.1	DCM	40	48	27	(—)	95:5
1	Hov II	0.1	DCM	40	48	18	(—)	95:5
2	Gr II	0.2	DCM	40	60	50	(—)	95:5
2	Gr II	0.2	benzene	60	60	26	(—)	95:5
2	Gr II	0.2	DCM	20	72	5	(—)	95:5
2	Gr II	0.3	DCM	40	60	68	(56)	95:5
2	Gr II	0.4	DCM	40	60	72	(—)	95:5

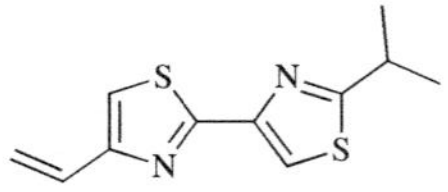

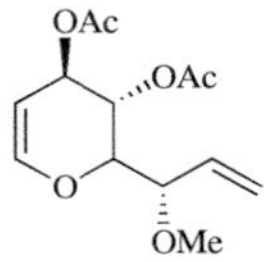

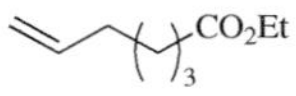

OAc
OAc
OMe

CO_2Et
3.6 eq

Gr II (0.05 eq),
DCM, reflux, 48 h

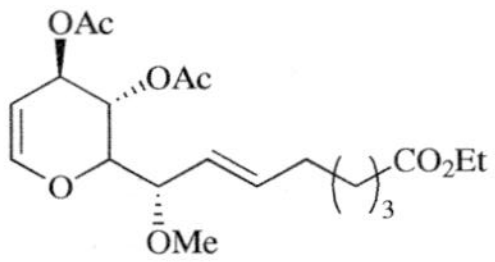

OAc
OAc
CO_2Et
OMe

(55) (*E*) only

822

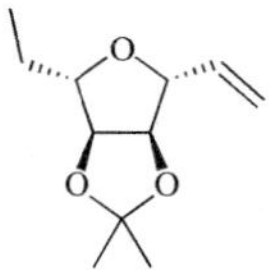

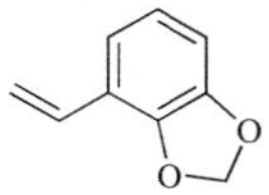

4 eq

1. Gr II (0.05 eq),
DCM, reflux, 24 h
2. HCl (2N), 30°, 12 h

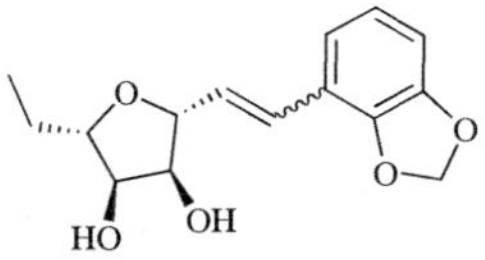

HO OH

(44)

814

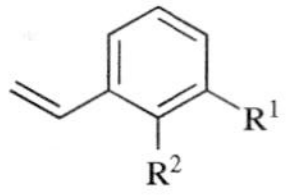

R^1
R^2
x eq

Gr II (0.05 eq),
DCM, reflux

R^1
R^2

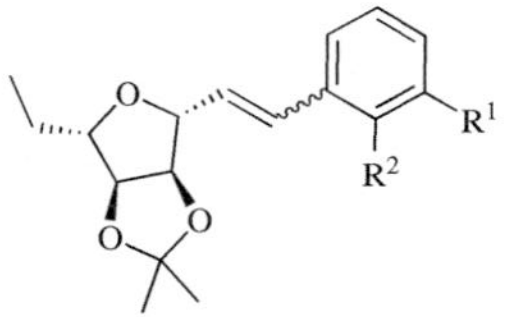

814

R^1	R^2	*x*	Time (h)	
MeO	$HOCH_2$	0.67	12	(70)
EtO	MeO	4	18	(74)
EtO	EtO	4	18	(64)
Cl	Cl	4	24	(90)

TABLE 13. CROSS-METATHESIS OF SECONDARY ALLYLIC ALCOHOLS AND DERIVATIVES (*Continued*)

B. ALLYLIC ETHERS (*Continued*)

Allylic Ester	Cross Partner	Conditions	Product(s) and Yield(s) (%)	Refs.
Please refer to the charts preceding the tables for the catalyst structures.				
C_8				
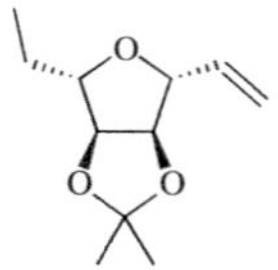	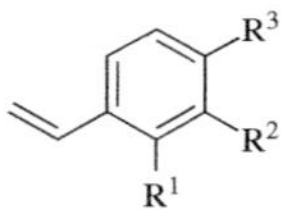 R^1, R^2, R^3; *x* eq	1. Gr II (0.05 eq), DCM, reflux, time 1 2. HCl (2N), 30°, time 2	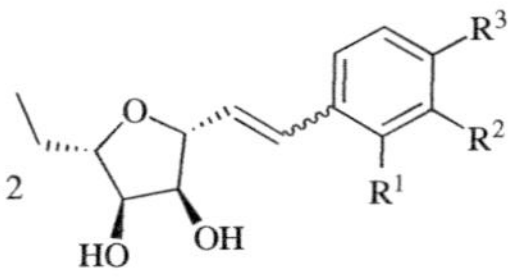 R^1, R^2, R^3, HO, OH	814
	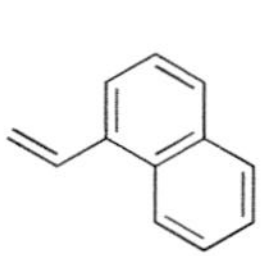 4 eq	Gr II (0.05 eq), DCM, reflux, 24 h	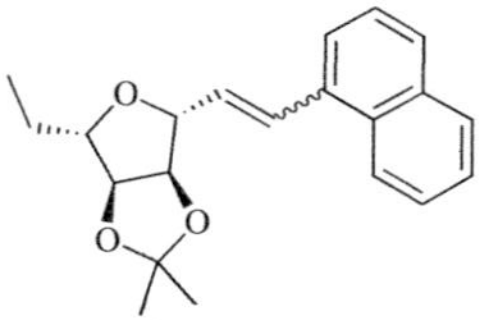 (81)	814
	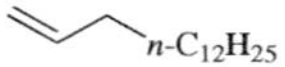 *n*-$C_{12}H_{25}$; 6 eq	Gr II (0.05 eq), DCM, reflux, 6 h	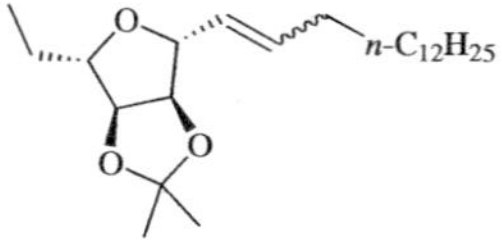 *n*-$C_{12}H_{25}$ (81)	814

R^1	R^2	R^3	*x*	Time 1 (h)	Time 2 (h)	
MeO	MeO	MeO	4	18	12	(40)
MeO	MeO	H	4	24	8	(44)
$MeOCH_2$	MeO	H	0.67	12	8	(43)

Ph

x eq

Catalyst (y eq),
solvent, 22°

Ph

91

x	Catalyst	y	Solvent	Time (h)	Conv. (%)[a]		(E)/(Z)	er
5	Ru cat **211**	0.02	—	36	—	(50–60)	>98:2	97.0:3.0
5	Ru cat **213**	0.02	—	15	—	(50–60)	>98:2	>99.0:1.0
2	Mo cat **8**	0.05	pentane	4	>98	—	—	—
2	Mo cat **9**	0.05	pentane	4	<5	—	—	—
2	Mo cat **10**	0.05	pentane	4	50	—	—	—
2	Mo cat **12**	0.05	pentane	4	>98	—	>98:2	97.0:3.0
5	Mo cat **12**	0.05	benzene	1	—	(50–60)	>98:2	98.5:1.5
2	Mo cat **15**	0.05	pentane	4	<5	—	—	—

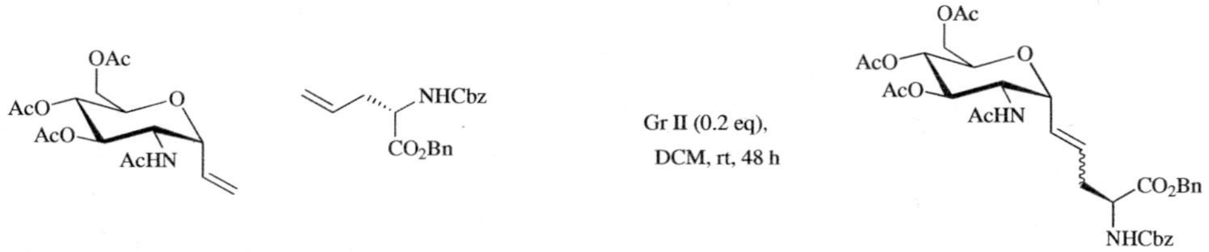

OAc
AcO
AcO
O
AcHN

NHCbz
CO_2Bn

Gr II (0.2 eq),
DCM, rt, 48 h

OAc
AcO
AcO
O
AcHN
CO_2Bn
NHCbz

(77)

423

TABLE 13. CROSS-METATHESIS OF SECONDARY ALLYLIC ALCOHOLS AND DERIVATIVES (*Continued*)
B. ALLYLIC ETHERS (*Continued*)

Allylic Ester	Cross Partner	Conditions	Product(s) and Yield(s) (%)	Refs.
Please refer to the charts preceding the tables for the catalyst structures.				
C_9				
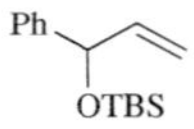 Ph, OTBS	 CO_2Ph, 2 eq	Mo cat **18** (0.03 eq), benzene, 7 torr, 22°, 8 h	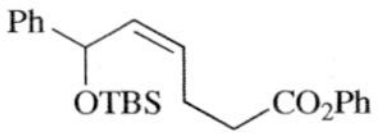 Ph, OTBS, CO_2Ph (72) (*Z*)/(*E*) = 95:5	184
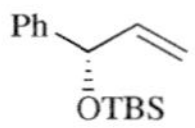 Ph, OTBS, 1.5 eq	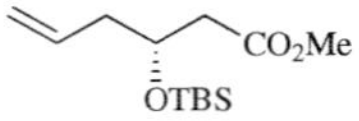 OTBS, CO_2Me	Gr II (0.1 eq), DCM, rt, 24 h	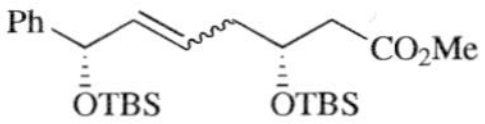 Ph, OTBS, OTBS, CO_2Me (60)	823
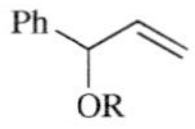 Ph, OR	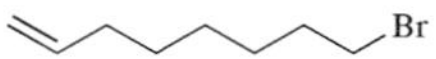 Br, *x* eq	Catalyst (*y* eq), benzene, 7 torr, 22°, 8 h	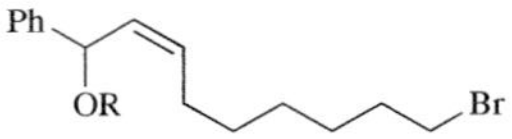 Ph, OR, Br	184

R	*x*	Catalyst	*y*	Conv. (%)[a]		(*Z*)/(*E*)
TBS	1	Mo cat **18**	0.03	56	(45)	91:9
TBS	1.5	Mo cat **18**	0.03	67	(65)	90:10
TBS	2	Mo cat **18**	0.03	79	(69)	95:5
TBS	3	Mo cat **18**	0.03	72	(65)	95:5
TBS	10	Mo cat **18**	0.03	<2	(—)	—
TBS	2	Mo cat **1**	0.05	95	(80)	5:95
TBS	2	Hov II	0.05	98	(76)	5:95
TBS	2	Mo cat **5**	0.05	64	(62)	81:19
TBS	2	W cat **3**	0.05	25	(20)	>95:5
PMB	2	Mo cat **18**	0.03	90	(85)	>98:2

Substrate 1	Substrate 2	Conditions	Product(s) and Yield(s) (%)	Refs.
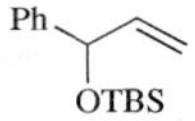 Ph, OTBS	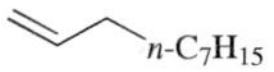 n-C_7H_{15} (3 eq)	Mo cat **18** (0.03 eq), benzene, 7 torr, 22°, 8 h	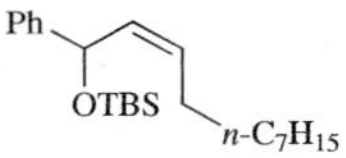 Ph, OTBS, n-C_7H_{15} (86) $(Z)/(E) = 95:5$	184
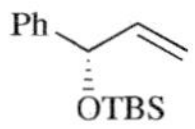 Ph, OTBS (1.5 eq)	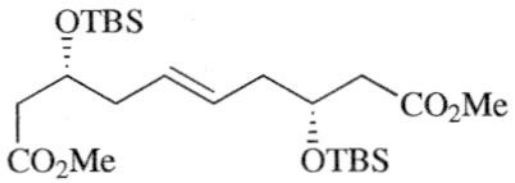 OTBS, CO_2Me, OTBS, CO_2Me	Gr II (0.1 eq), DCM, rt, 24 h	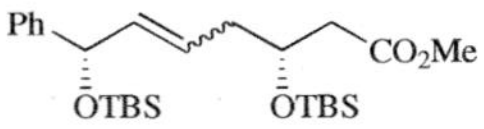 Ph, OTBS, OTBS, CO_2Me (60)	823
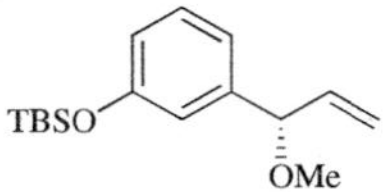 TBSO, OMe (5 eq)	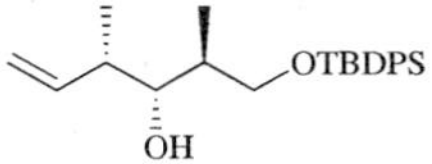 OH, OTBDPS	Gr II (0.1 eq), DCM, reflux, 24 h	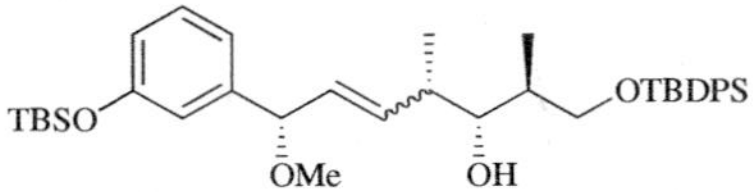 TBSO, OMe, OH, OTBDPS (94)	824
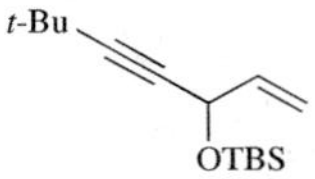 t-Bu, OTBS	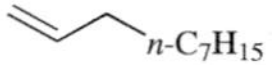 n-C_7H_{15} (3 eq)	1. Mo cat **18** (0.015 eq), benzene, 7 torr, 22°, 8 h 2. $(n\text{-Bu})_4NF$ (2 eq), THF, 22°, 1 h	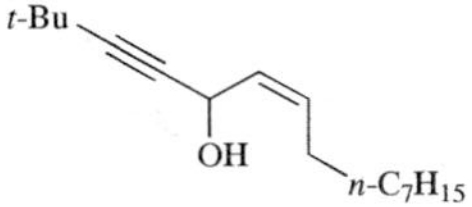 t-Bu, OH, n-C_7H_{15} (76), 84% conv.[a] $(Z)/(E) = 90:10$	184

TABLE 13. CROSS-METATHESIS OF SECONDARY ALLYLIC ALCOHOLS AND DERIVATIVES (*Continued*)

B. ALLYLIC ETHERS (*Continued*)

Allylic Ester	Cross Partner	Conditions	Product(s) and Yield(s) (%)	Refs.
Please refer to the charts preceding the tables for the catalyst structures.				
C_9				
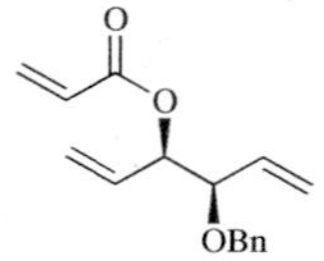 OBn	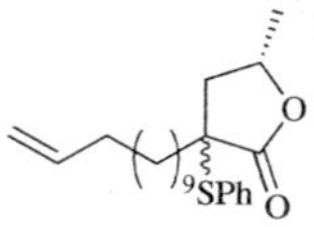 SPh, 3 eq	Catalyst (0.1 eq), solvent, 20 h	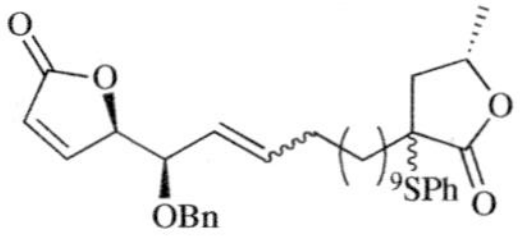 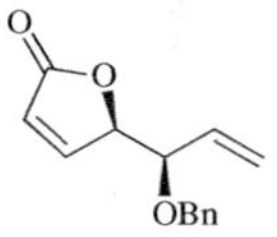 OBn, SPh (I) + OBn (II)	825
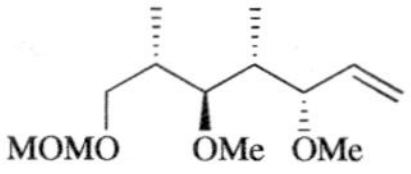 MOMO, OMe, OMe	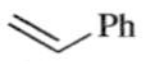 Ph	Gr II, benzene, 50°, 12 h	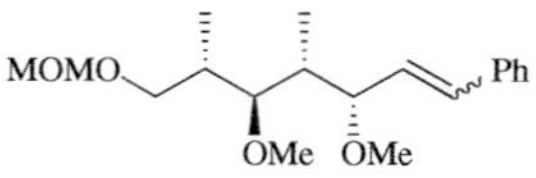 MOMO, Ph, OMe, OMe (92)	826

Catalyst	Solvent	Temp (°)	I	II
Gr I	toluene	80	(0)	(45)
Gr II	toluene	80	(53)	(12)
Gr II	DCM	40	(60)	(19)
Hov I	toluene	80	(66)	(16)
Hov II	toluene	80	(68)	(0)
Hov II	DCM	40	(77)	(0)

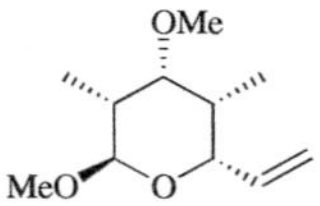	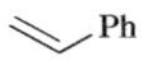	Gr II (0.1 eq), benzene, 55°	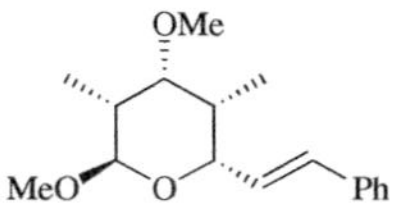 (81) (*E*)/(*Z*) = 98:2	827
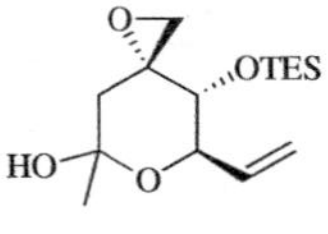 1 eq	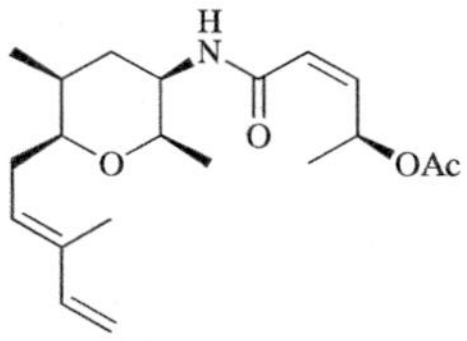	Gr II (0.2 eq), THF, 40°, 5 h	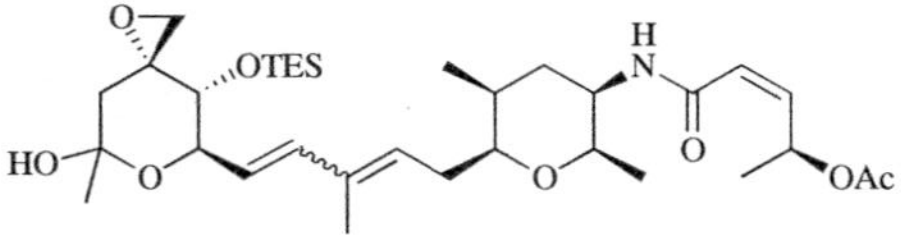 (22)	828
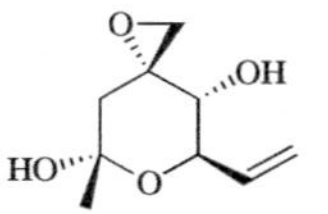 1.8 eq	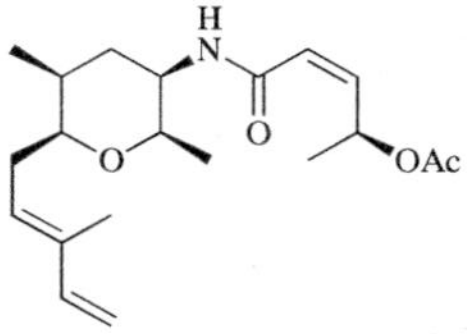	Catalyst (3 x 0.025 eq), DCE, 10 h	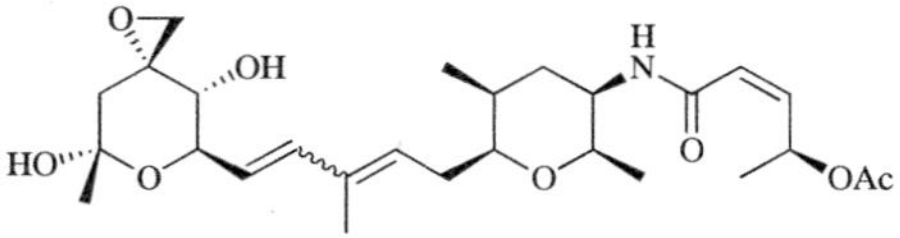	828

Catalyst	Temp (°)	
Gr II	43	(12)
Hov II	42	(13)
Ru cat **88**	23	(28)

TABLE 13. CROSS-METATHESIS OF SECONDARY ALLYLIC ALCOHOLS AND DERIVATIVES (*Continued*)

B. ALLYLIC ETHERS (*Continued*)

Allylic Ester	Cross Partner	Conditions	Product(s) and Yield(s) (%)	Refs.

Please refer to the charts preceding the tables for the catalyst structures.

C_9

Allylic Ester	Cross Partner	Conditions	Product(s) and Yield(s) (%)	Refs.
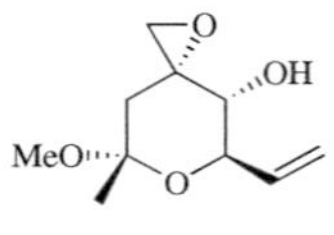 MeO, OH, O; 1.25 eq	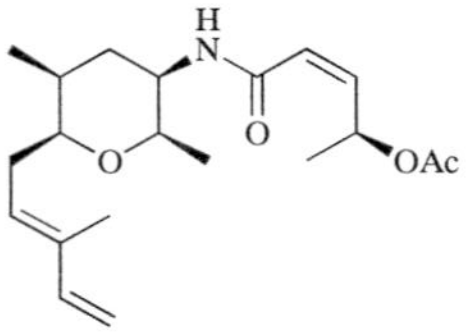 H N, O, OAc	Gr II (0.077 eq), DCM, reflux, 5 h	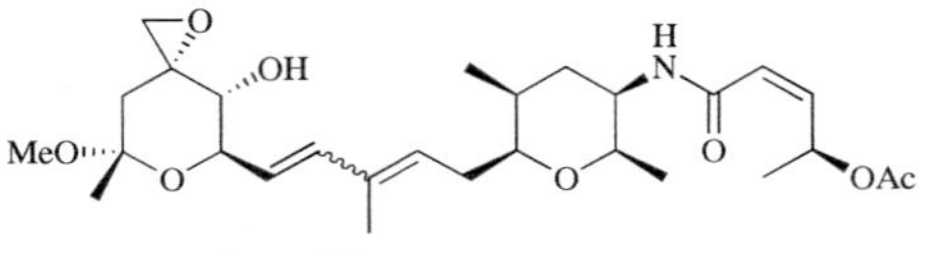 MeO, OH, O, H N, O, OAc (57)	829
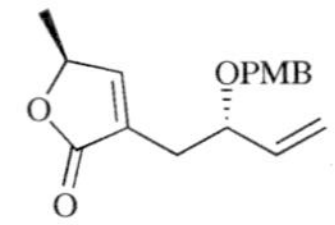 O, O, OPMB; 1 eq	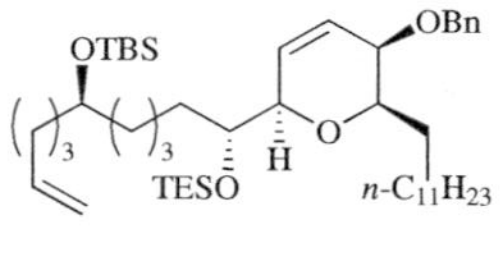 OTBS, ( )$_3$, ( )$_3$, TESO, H, O, OBn, n-$C_{11}H_{23}$	Ru cat **190** (2 x 0.1 eq), DCM, 40°, 24 h	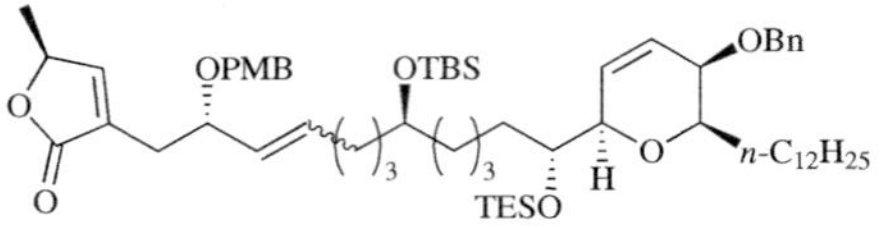 O, O, OPMB, OTBS, ( )$_3$, ( )$_3$, TESO, H, O, OBn, n-$C_{12}H_{25}$ (58)	830
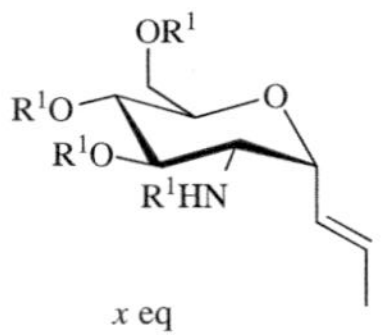 OR^1, R^1O, R^1O, R^1HN, O; x eq	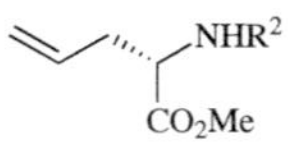 NHR^2, CO_2Me	Catalyst (y eq), DCM, reflux	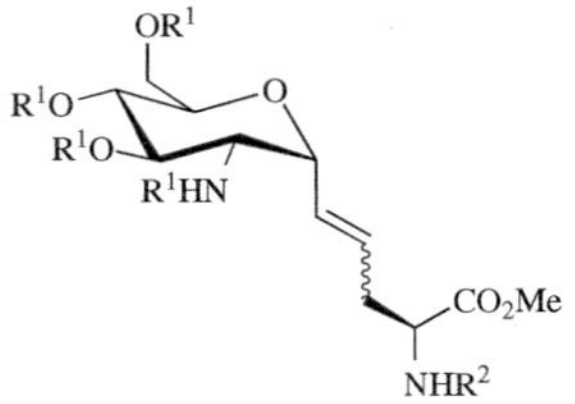 OR^1, R^1O, R^1O, R^1HN, O, CO_2Me, NHR^2	427

R^1	x	R^2	Catalyst	y	Time (h)	
Bn	2	Fmoc	Gr I	0.2	48	(—)
Bn	2	Boc	Gr I	0.2	48	(—)
Bn	2	Fmoc	Gr II	0.1	24	(82)
Bn	2	Boc	Gr II	0.1	48	(77)
Bn	1	Fmoc	Gr II	0.1	24	(68)
Ac	2	Fmoc	Gr II	0.2	48	(69)
Ac	2	Boc	Gr II	0.2	48	(77)

C_{10}

Reactant	Olefin partner	Conditions	Product(s) and Yield(s) (%)	Refs.
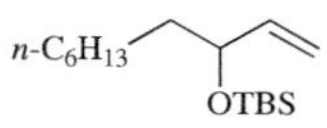 n-C_6H_{13}, OTBS	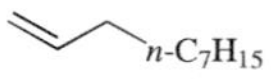 n-C_7H_{15} (3 eq)	1. Mo cat **18** (0.03 eq), benzene, 7 torr, 22°, 8 h 2. (n-Bu)$_4$NF (2 eq), THF, 22°, 1 h	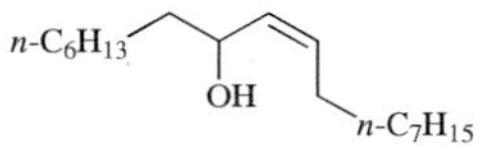 n-C_6H_{13}, OH, n-C_7H_{15} (61) (Z)/(E) = 78:22	184
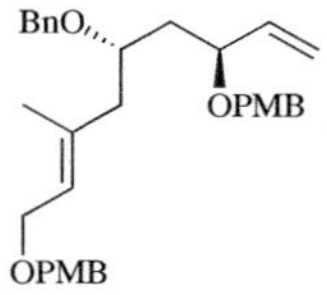 BnO, OPMB, OPMB	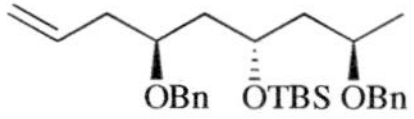 OBn, OTBS, OBn (1.2 eq)	Hov II (0.1 eq), DCM, rt, 6 h	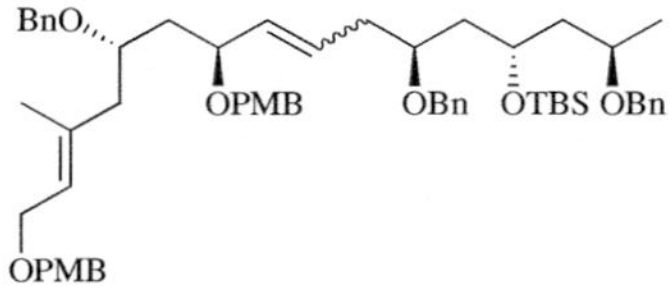 BnO, OPMB, OBn, OTBS, OBn, OPMB (57)	831
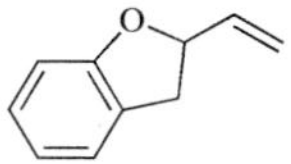 O (2-vinyl-2,3-dihydrobenzofuran)	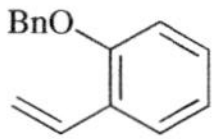 BnO (4.2 eq)	Gr II (0.026 eq), DCM, reflux, 46 h	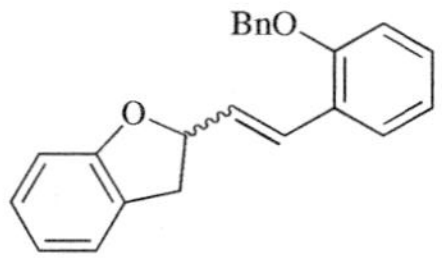 BnO, O (71)	832
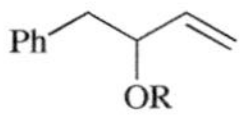 Ph, OR	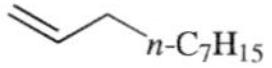 n-C_7H_{15} (3 eq)	1. Mo cat **18** (0.03 eq), benzene, 7 torr, 22°, 8 h 2. (n-Bu)$_4$NF (2 eq), THF, 22°, 1 h	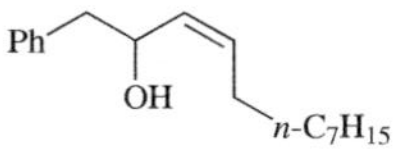 Ph, OH, n-C_7H_{15}	184

R	Conv. (%)[a]		(Z)/(E)
TBS	43	(37)	86:14
PMB	43	(39)	>98:2

TABLE 13. CROSS-METATHESIS OF SECONDARY ALLYLIC ALCOHOLS AND DERIVATIVES (*Continued*)

B. ALLYLIC ETHERS (*Continued*)

Allylic Ester	Cross Partner	Conditions	Product(s) and Yield(s) (%)	Refs.
Please refer to the charts preceding the tables for the catalyst structures.				
C_{10}				
H, H, H, PMBO 1.5 eq	OTBS, H, H, TESO, OTES	Gr II (0.1 eq), toluene, rt, 18 h	H, H, H, PMBO, OTBS, H, H, TESO, OTES (77)	522
TBDPSO 2.5 eq	TBSO, H, H, TBSO, OTBS, PivO	1. Gr II (0.084 eq), toluene, rt, 36 h 2. DIBAL-H, DCM, –78°, 0.5 h	TBSO, H, H, TBSO, OTBS, PivO, TBDPSO (60)	833
OH 2.5 eq	H N, O, OAc	Ru cat **88** (0.1 eq), BQ (1.9 eq), DCE, 40°, 9 h	OH, H N, O, OAc (45)	828

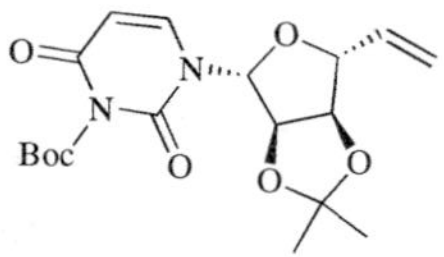

	Reactant	Alkene	Conditions	Product(s) and Yield(s) (%)		Refs.
		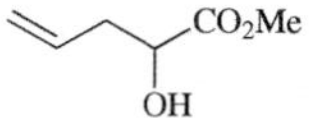 2.5 eq	Gr II (0.1 eq), DCM, reflux, 24.5 h	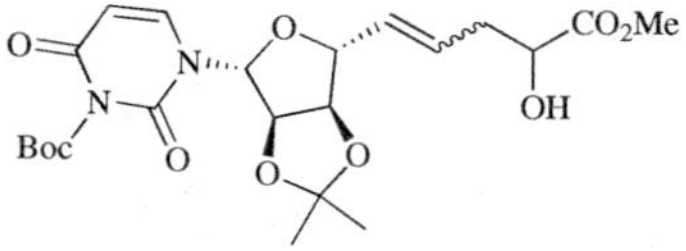	(85)	721
		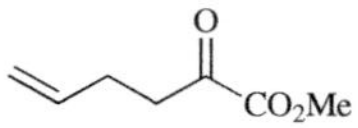 2.5 eq	Gr II (0.15 + 0.05 eq), DCM, reflux, 24 h	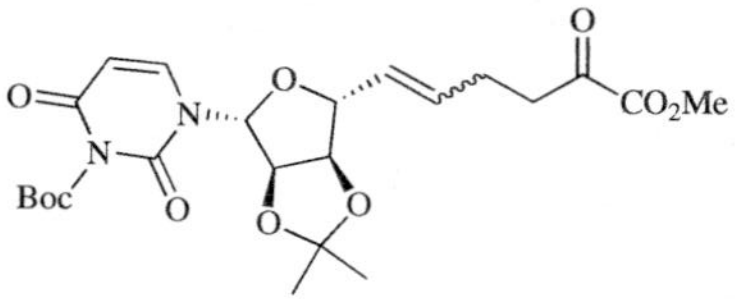	(40)	721
C_{11}	OPMB	3 eq	1. Mo cat **18** (0.03 eq), benzene, 7 torr, 22°, 8 h 2. DDQ (2 eq), DCM, H_2O, 22°, 1 h		(60), 66% conv.[a] $(Z)/(E) > 98:2$	184
C_{11-12}	OTBS	3 eq	1. Mo cat **18** (0.03 eq), benzene, 7 torr, 22°, 8 h 2. $(n\text{-Bu})_4NF$ (2 eq), THF, 22°, 1 h		$(Z)/(E) > 98:2$	184

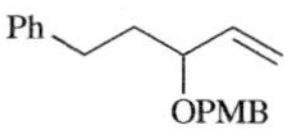

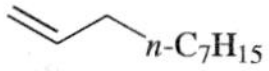

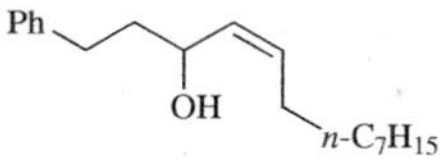

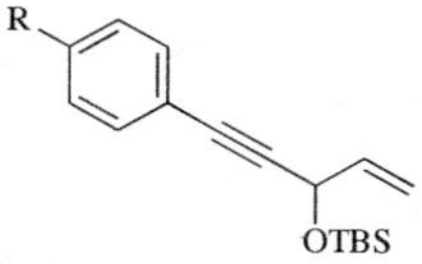

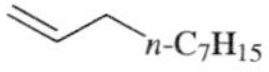

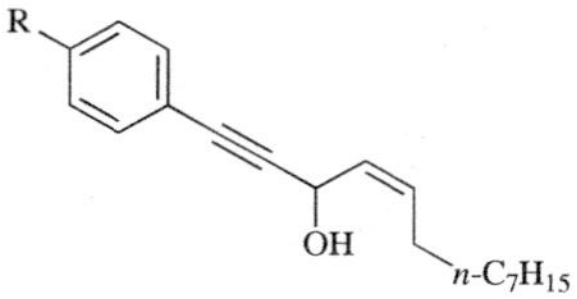

R	Conv. (%)[a]	
H	72	(68)
MeO	66	(60)
CF_3	73	(64)

TABLE 13. CROSS-METATHESIS OF SECONDARY ALLYLIC ALCOHOLS AND DERIVATIVES (*Continued*)
B. ALLYLIC ETHERS (*Continued*)

Allylic Ester	Cross Partner	Conditions	Product(s) and Yield(s) (%)	Refs.
Please refer to the charts preceding the tables for the catalyst structures.				
C_{11}				
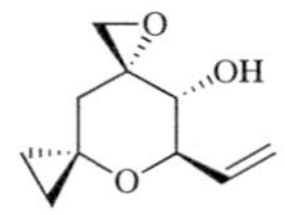 OH	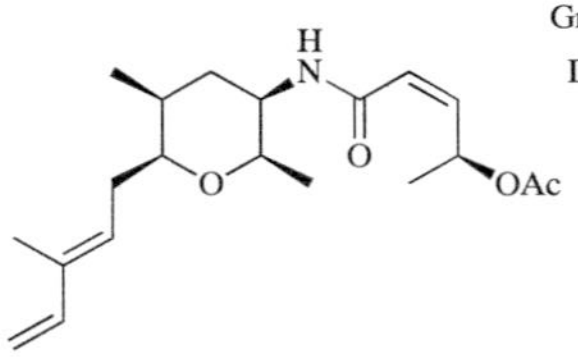 H N O OAc	Gr II (0.15 eq), DCM, 45°, 8 h	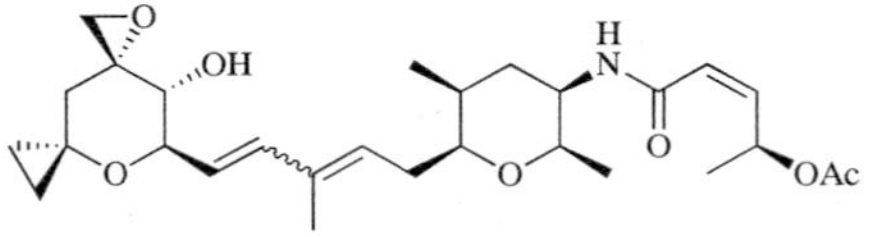 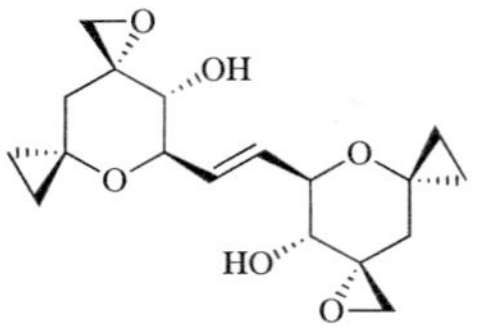 OH H N O OAc (27) 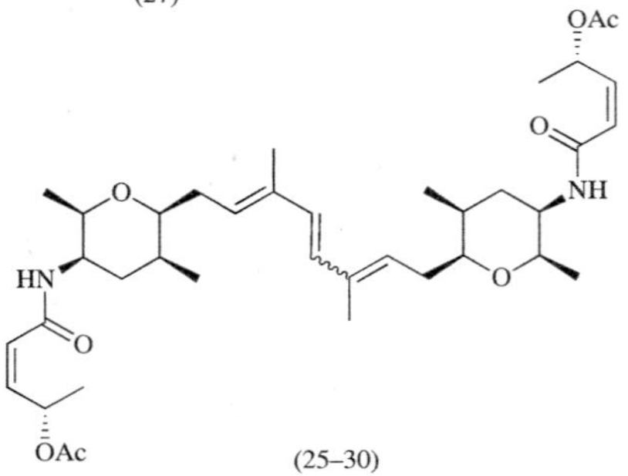 + OAc O NH HN O OAc (25–30) + OH HO (25–30)	834

EtO$_2$C, OR^2, OR^1 + OTIPS (2 eq) — Gr II, DCM, 45° — EtO$_2$C, OR^2, OR^1, OTIPS — 835

R^1	R^2		(*E*)/(*Z*)
TIPS	H	(25)	91:9
i-Pr	*i*-Pr	(73)	67:33

PivO( )$_4$, OTBS, OTBS + OH, OTBDPS (3 eq) — Gr II (0.1 eq), DCM, reflux, 3 h — PivO( )$_4$, OTBS, OTBS, OH, OTBDPS (78) — 744

OH, OTBDPS (3 eq) — Gr II (0.1 eq), DCM, reflux, 3 h — PivO( )$_4$, OTBS, OTBS, OH, OTBDPS (67) — 744

TABLE 13. CROSS-METATHESIS OF SECONDARY ALLYLIC ALCOHOLS AND DERIVATIVES (*Continued*)

B. ALLYLIC ETHERS (*Continued*)

Allylic Ester	Cross Partner	Conditions	Product(s) and Yield(s) (%)	Refs.
Please refer to the charts preceding the tables for the catalyst structures.				
C_{11}				
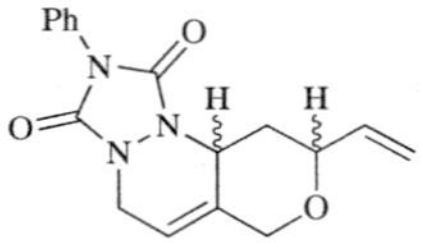	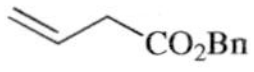 CO_2Bn, 2 eq	Gr II (0.05 eq), DCM, 40°, 6 h	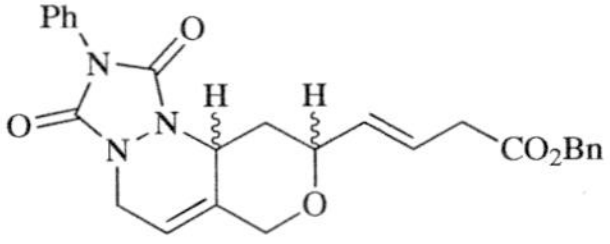 CO_2Bn (66) (*E*) only	836
	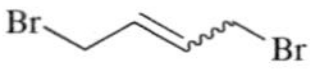 Br, Br, 2 eq	Gr II (0.05 eq), DCM, 40°, 6 h	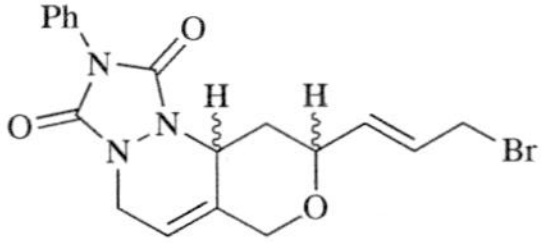 Br (52) (*E*) only	836
	 OTBS, 2 eq	Gr II (0.05 eq), DCM, 40°, 6 h	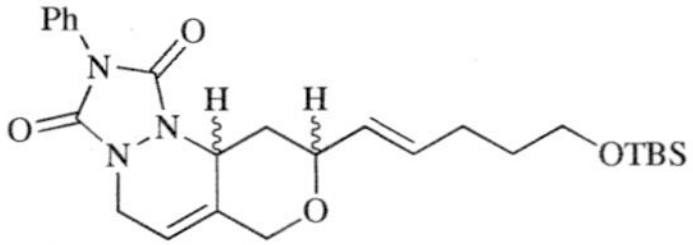 OTBS (74) (*E*) only	836
	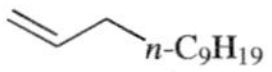 n-C_9H_{19}, 2 eq	Catalyst (0.05 eq), Ti(O*i*-Pr)$_4$ (*x* eq), DCM, 40°, 6 h	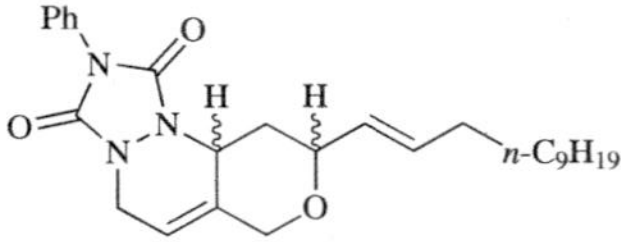 n-C_9H_{19} (*E*) only	836

Catalyst	*x*	
Gr I	—	(56)
Gr I	0.3	(49)
Gr II	—	(81)

1 eq

Hov II (0.05 eq), DCM, 65°, overnight

R^1	R^2	
H	Bz	(51)
H	Boc	(60)
Bz	Bz	(61)
Bz	Boc	(76)

837

1 eq

Hov II (0.05 eq), DCM, 65°, overnight

(77) 837

TABLE 13. CROSS-METATHESIS OF SECONDARY ALLYLIC ALCOHOLS AND DERIVATIVES (*Continued*)

B. ALLYLIC ETHERS (*Continued*)

Allylic Ester	Cross Partner	Conditions	Product(s) and Yield(s) (%)	Refs.

Please refer to the charts preceding the tables for the catalyst structures.

C_{11}

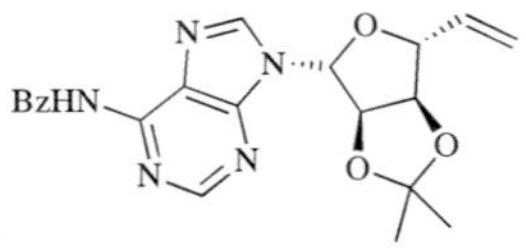

BzHN

1 eq

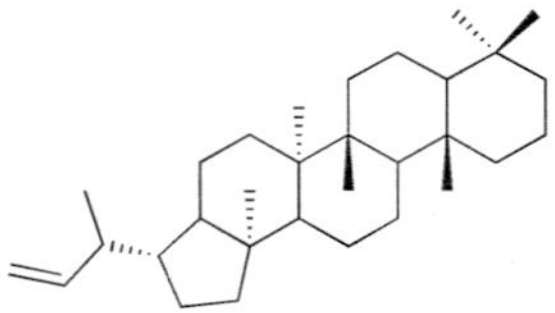

x eq

Catalyst (0.1 eq), solvent

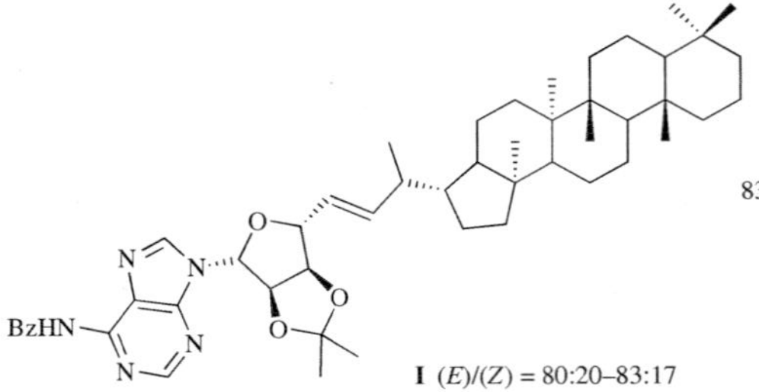

BzHN

I $(E)/(Z)$ = 80:20–83:17

838

\+

BzHN

NHBz

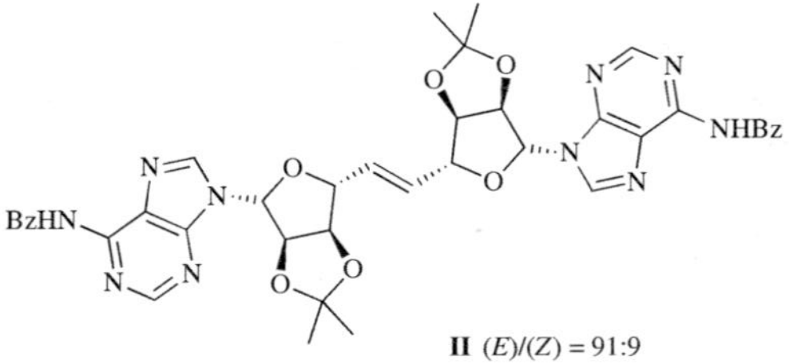

II $(E)/(Z)$ = 91:9

x	Catalyst	Solvent	Temp (°)	Time (h)	I	II
1	Hov II	DCM	40	24	(0)	(0)
1	Hov II	DCM	75 (MW)	3	(13)	(43)
1	Ru cat **89**	DCM	75 (MW)	3	(12)	(45)
1	Gr II	DCM	75 (MW)	3	(6)	(0)
10	Ru cat **89**	DCM	75 (MW)	3	(59)	(40)
10	Hov II	DCM	75 (sealed tube)	overnight	(51)	(46)
10	Ru cat **89**	DCM	75 (sealed tube)	overnight	(52)	(43)
1	Hov II	C_6F_6	75 (MW)	3	(6)	(78)

Substrate	Alkene partner	Conditions	Product(s) and Yield(s) (%)	Refs.
C_{12}				
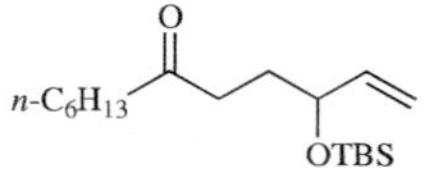 n-C_6H_{13}, O, OTBS 1.63 eq	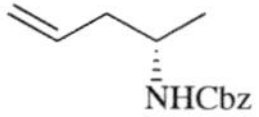 NHCbz	Hov II (0.05 eq), benzene, 75°	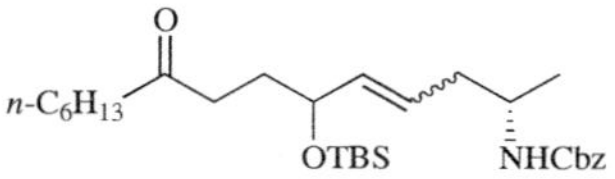 n-C_6H_{13}, O, OTBS, NHCbz (60)	714
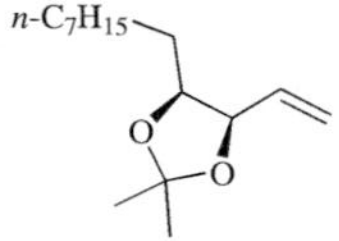 n-C_7H_{15}, O, O	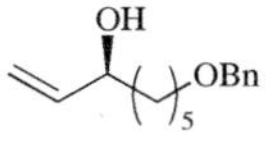 OH, OBn, ( )$_5$ 1.05 eq	Gr II (0.05 eq), DCM, rt, 0.5 h	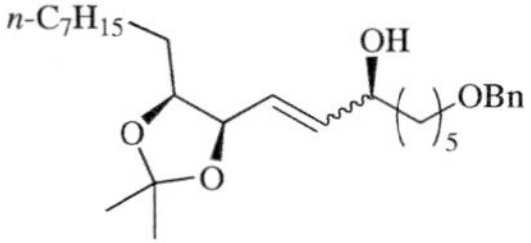 n-C_7H_{15}, O, O, OH, OBn, ( )$_5$ (70)	839
C_{13}				
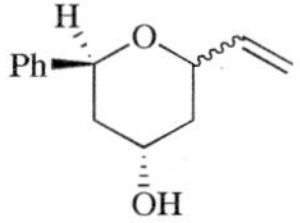 H, Ph, O, OH dr (*cis*/*trans*) 83:17	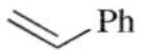 Ph 2 eq	Gr II (0.1 eq), DCM, reflux, 8 h	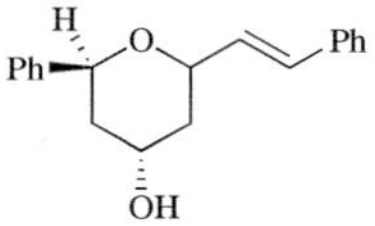 H, Ph, O, OH, Ph (68) (*E*) only	621
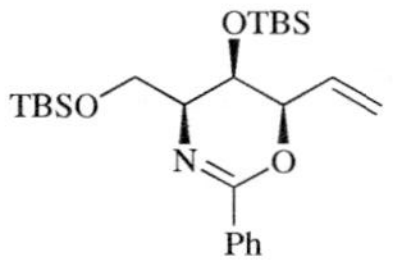 TBSO, OTBS, N, O, Ph	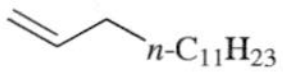 n-$C_{11}H_{23}$ 2 eq	Gr II (0.05 eq), DCM, reflux, 8 h	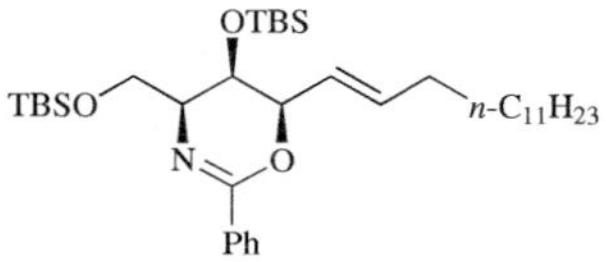 TBSO, OTBS, N, O, Ph, n-$C_{11}H_{23}$ (92) (*E*)/(*Z*) > 95:5	840

TABLE 13. CROSS-METATHESIS OF SECONDARY ALLYLIC ALCOHOLS AND DERIVATIVES (*Continued*)

B. ALLYLIC ETHERS (*Continued*)

Allylic Ester	Cross Partner	Conditions	Product(s) and Yield(s) (%)	Refs.
Please refer to the charts preceding the tables for the catalyst structures.				
C_{13}				
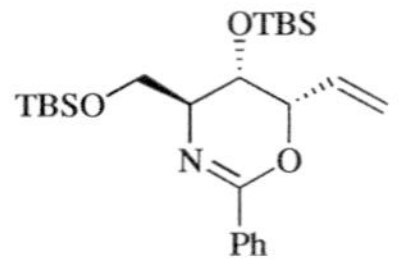 TBSO, OTBS, N, O, Ph	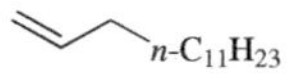 n-$C_{11}H_{23}$ 2 eq	Gr II (0.05 eq), DCM, reflux, 12 h	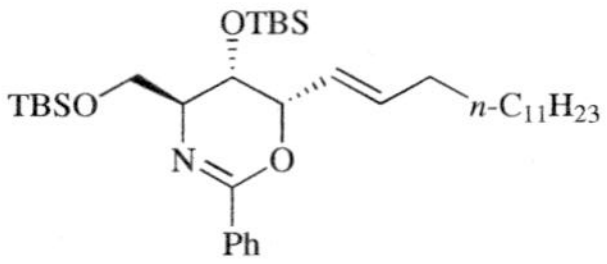 TBSO, OTBS, n-$C_{11}H_{23}$, N, O, Ph (94) $(E)/(Z) > 95:5$	841
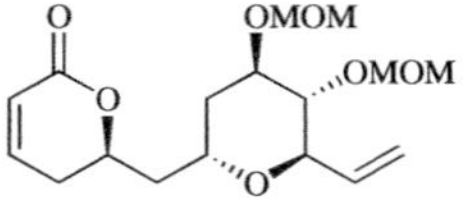 O, O, OMOM, OMOM, O	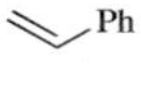 Ph	Gr II (0.1 eq), benzene, 8 h, 55°	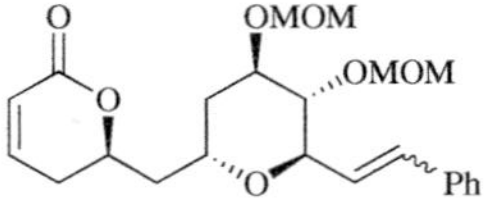 O, O, OMOM, OMOM, O, Ph (87)	842
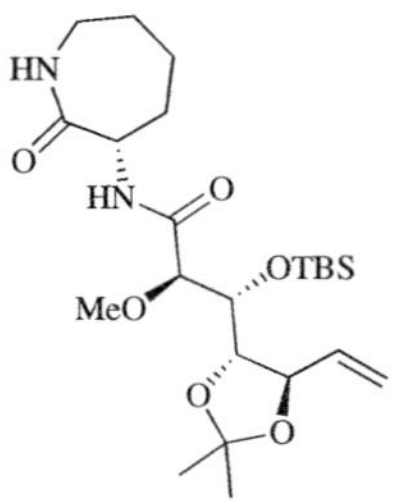 HN, O, HN, O, MeO, OTBS, O, O	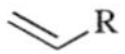 R	Hov II (0.3 eq), DCM, 40°, 4 d	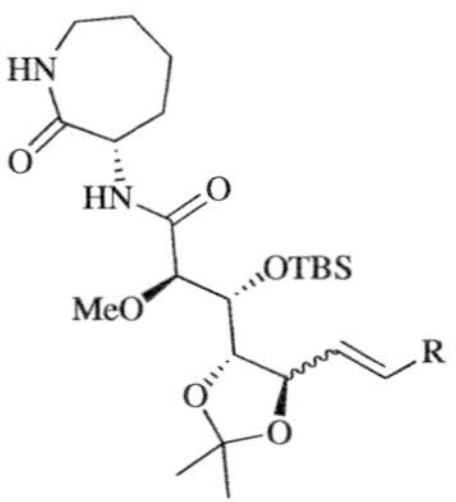 HN, O, HN, O, MeO, OTBS, R, O, O R: *i*-Pr (89); *t*-Bu (90); *s*-Bu (26); *i*-Bu (80); *t*-$BuCH_2$ (74); Ph (94); Cy (77)	843

C_{14}

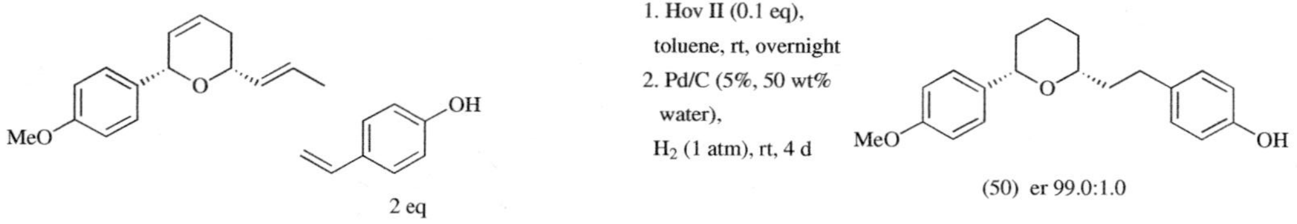

Substrates	Conditions	Product (yield)	Ref.
MeO; OH; 2 eq	1. Hov II (0.1 eq), toluene, rt, overnight 2. Pd/C (5%, 50 wt% water), H_2 (1 atm), rt, 4 d	MeO; O; OH (50) er 99.0:1.0	844
MOMO; H; O; H; H; O; H; OMOM TBSO; O; O; 7 2.05 eq	Gr II (0.05 eq), DCM, reflux, 16 h	TBSO; O; O; 7; TBSO; O; O; 7; MOMO; H; O; H; H; O; H; OMOM (85)	845

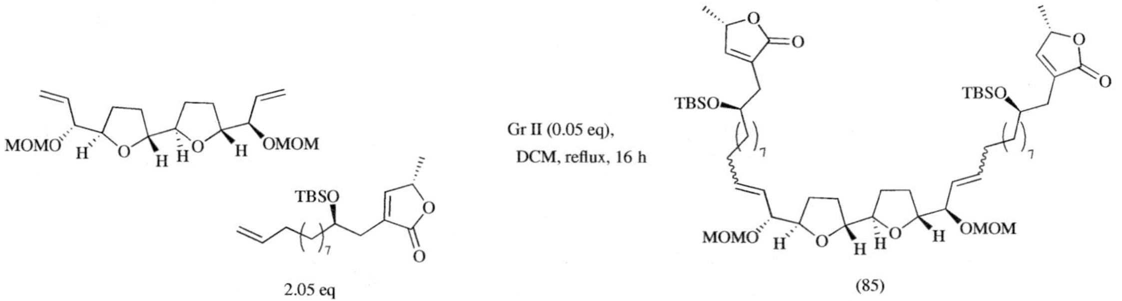

TABLE 13. CROSS-METATHESIS OF SECONDARY ALLYLIC ALCOHOLS AND DERIVATIVES (*Continued*)

B. ALLYLIC ETHERS (*Continued*)

Allylic Ester	Cross Partner	Conditions	Product(s) and Yield(s) (%)	Refs.
Please refer to the charts preceding the tables for the catalyst structures.				
C_{15}				
n-$C_{12}H_{25}$, OR	OH, OH, NHBoc 3 eq	Gr II (0.2 eq), CuI (0.03 eq), solvent, 48 h	n-$C_{12}H_{25}$, OR, OH, OH, NHBoc	722
			R / Solvent / Temp (°) H / DCM / 40 / (23) H / DCE / 80 / (—) Bz / DCM / 40 / (—) TBS / DCM / 40 / (—)	
O, O, O, O, O, O, OH	PMBO, i-Pr 3 eq	Hov II (0.1 eq), DCM, 40°, 24 h	PMBO, i-Pr, O, O, O, O, O, O, OH (53)	846

C_{17}

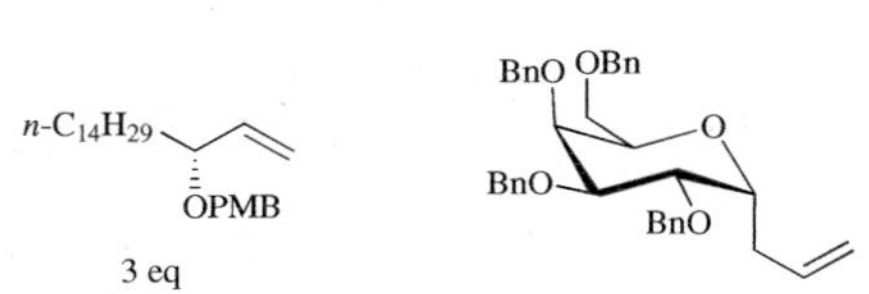

n-$C_{14}H_{29}$ OPMB

3 eq

BnO OBn O BnO BnO

Gr II (0.25 eq), DCM, 45°, 27 h

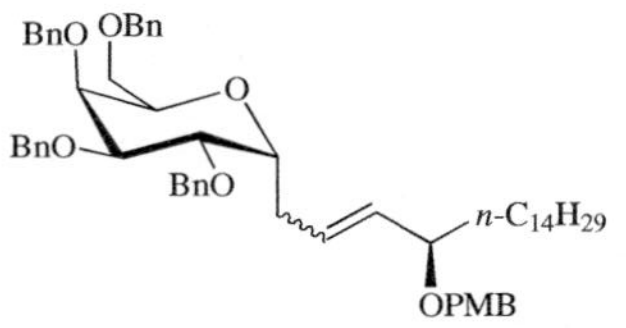

BnO OBn O BnO BnO n-$C_{14}H_{29}$ OPMB

(34) 775

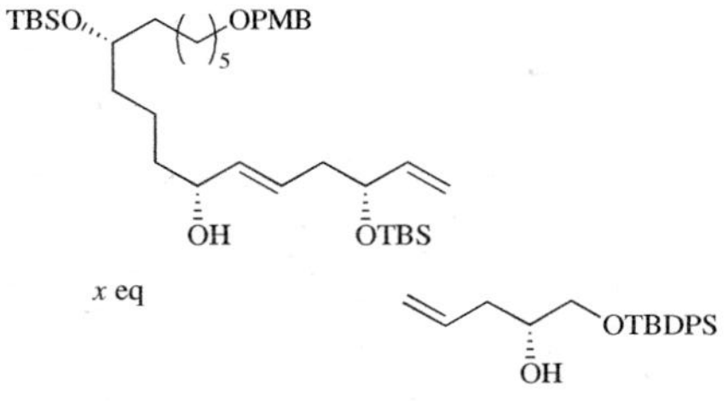

TBSO OPMB 5 OH OTBS

x eq

OTBDPS OH

y eq

Catalyst (0.1 eq), DCM, 45°, 3 h

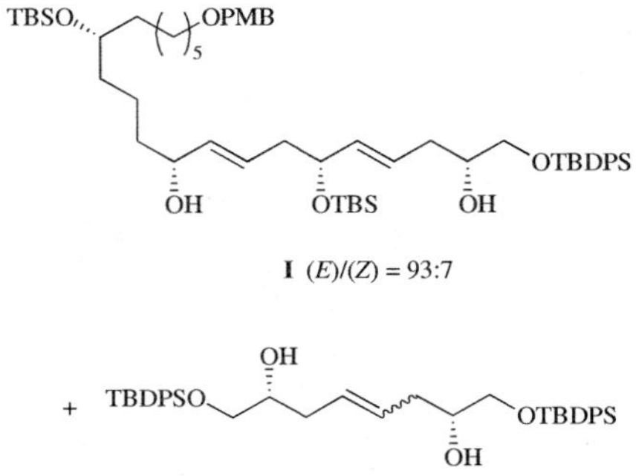

TBSO OPMB 5 OH OTBS OH OTBDPS

I (E)/(Z) = 93:7

+ TBDPSO OH OTBDPS OH

II

847

x	y	Catalyst	**I**	**II**
1	3	Gr II	(38)	(31)
1	3	Ru cat **89**	(26)	(34)
1.3	1	Gr II	(40)	(—)
3	1	Gr II	(70)	(—)

TABLE 13. CROSS-METATHESIS OF SECONDARY ALLYLIC ALCOHOLS AND DERIVATIVES (*Continued*)

B. ALLYLIC ETHERS (*Continued*)

Allylic Ester	Cross Partner	Conditions	Product(s) and Yield(s) (%)	Refs.
Please refer to the charts preceding the tables for the catalyst structures.				
C_{18}				
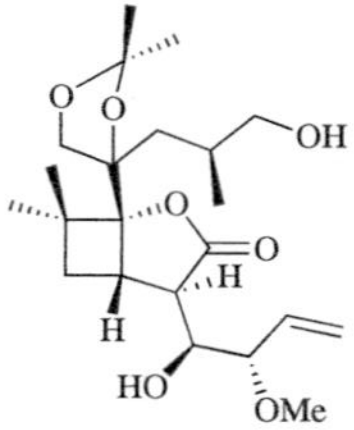 HO, OMe, OH, O, H	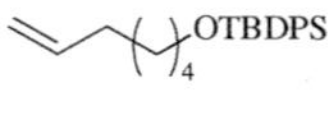 OTBDPS (4) 2.6 eq	Ru cat **89** (0.2 eq), toluene, 45°, 48 h	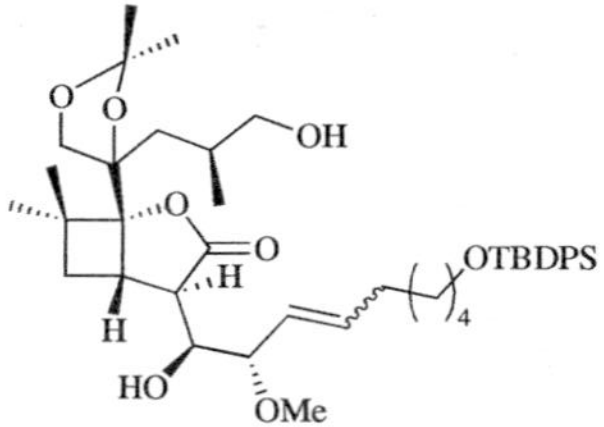 OTBDPS, OH, HO, OMe (67)	848
C_{22}				
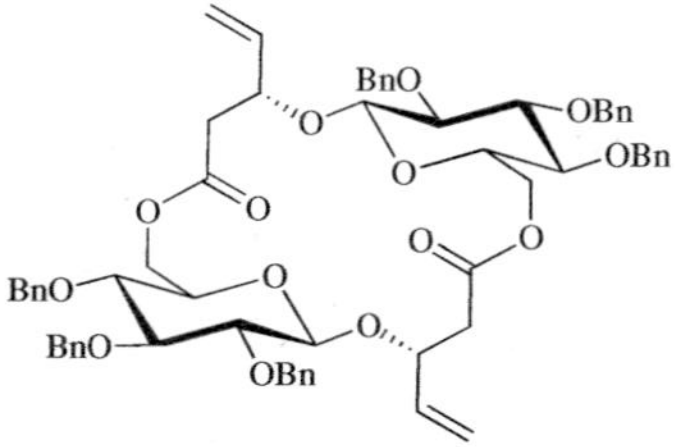 BnO, OBn	 4 x 10 eq	Gr II (4 x 0.02 eq), DCM, 40°, 4 d	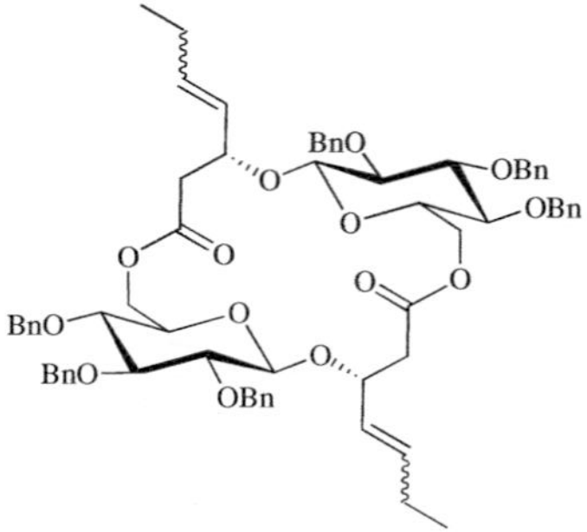 BnO, OBn (89)	849

C_{25-29}

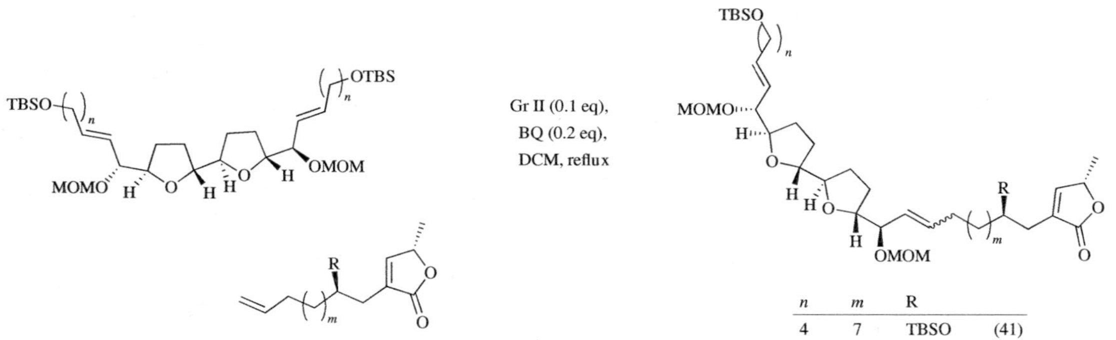

TBSO MOMO H H H H OMOM OTBS

1.1 eq

Gr II (0.1 eq), BQ (0.2 eq), DCM, reflux

537

n	m	R	
4	7	TBSO	(41)
9	7	TBSO	(40)
9	6	H	(41)

[a] The conversion was determined by ^{1}H NMR analysis.

TABLE 13. CROSS-METATHESIS OF SECONDARY ALLYLIC ALCOHOLS AND DERIVATIVES (*Continued*)

C. ALLYLIC ESTERS

Allylic Ester	Cross Partner	Conditions	Product(s) and Yield(s) (%)	Refs.
Please refer to the charts preceding the tables for the catalyst structures.				
C_3				
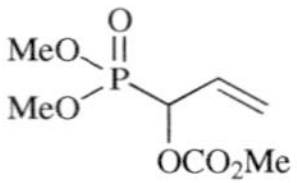 MeO, MeO P(O), OCO_2Me	 R, n; x eq	Gr II (0.05 eq), solvent, 40°	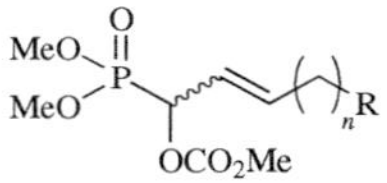 MeO, MeO P(O), OCO_2Me, R, n	221
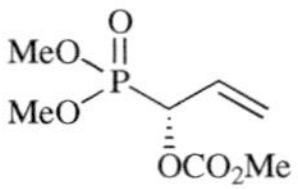 er 84.0:16.0 MeO, MeO P(O), OCO_2Me	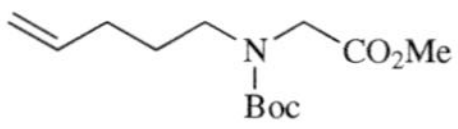 N, Boc, CO_2Me; 2 eq	Gr II (3 x 0.025 eq), DCM, 40°	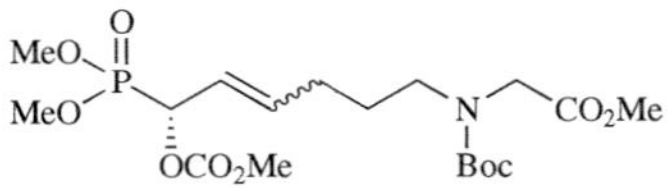 MeO, MeO P(O), OCO_2Me, N, Boc, CO_2Me (70) er 84.0:16.0	221
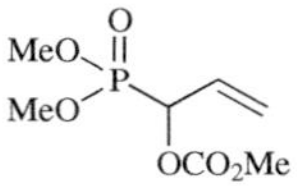 MeO, MeO P(O), OCO_2Me	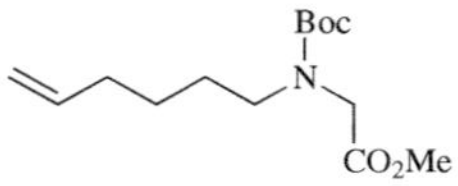 Boc, N, CO_2Me; 2 eq	Gr II (3 x 0.025 eq), DCM, 40°	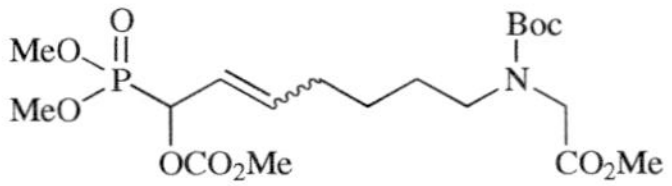 MeO, MeO P(O), OCO_2Me, Boc, N, CO_2Me (94)	221

R	n	x	Solvent	
BocHN	1	—	DCM	(10)
AcO	1	—	DCM	(0)
TMS	1	5	toluene	(60)
Me	3	10	toluene	(97)
Br	3	5	DCM	(62)
HO	4	2	DCM	(62)

C$_3$

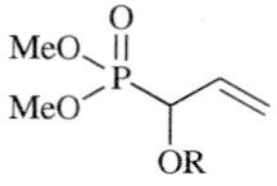 MeO, MeO, O, P, OR	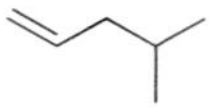 5 eq	Gr II (0.05 eq), DCM, 40° 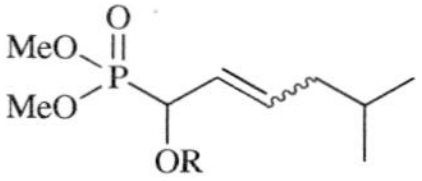	MeO, MeO, O, P, OR	221
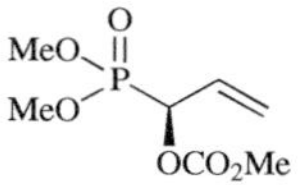 MeO, MeO, O, P, OCO_2Me	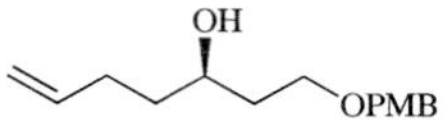 OH, OPMB; 1 eq	Gr II (0.05 eq), CuI (0.1 eq), DCM, 40°, 10 h 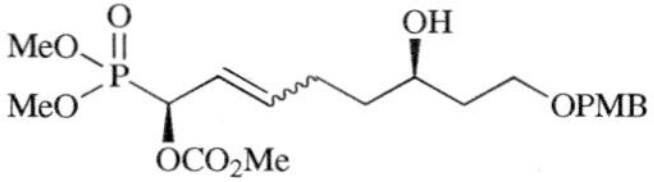	MeO, MeO, O, P, OCO_2Me, OH, OPMB (78)	850
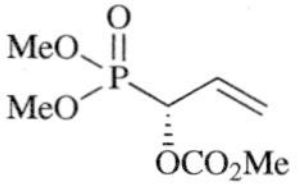 MeO, MeO, O, P, OCO_2Me; er ~ 100:0	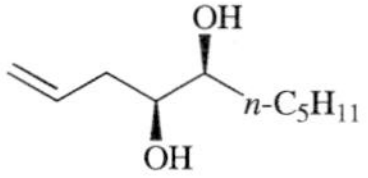 OH, OH, n-C_5H_{11}; 1 eq	Gr II (0.05 eq), DCM, 40°, 3 h 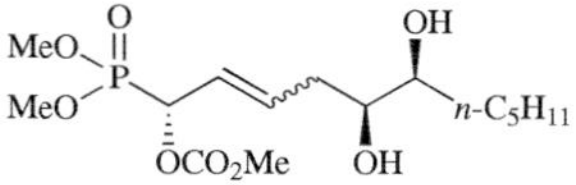	MeO, MeO, O, P, OCO_2Me, OH, OH, n-C_5H_{11} (74)	851
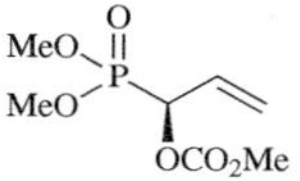 MeO, MeO, O, P, OCO_2Me; er > 97.5:2.5	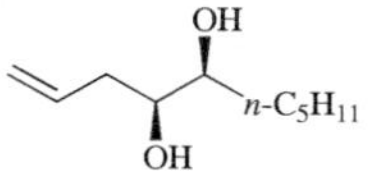 OH, OH, n-C_5H_{11}; 1 eq	Catalyst (0.05 eq), CuI (x eq) DCM, 40°, 5 h 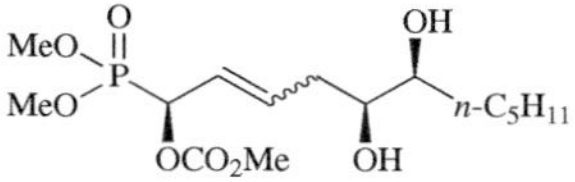	MeO, MeO, O, P, OCO_2Me, OH, OH, n-C_5H_{11}	851

R	
MeO_2C	(92)
Ac	(90)
TsHNCO	(88)
TBS	(89)
$AcCH_2CO$	(82)

Catalyst	x	
Gr II	0.1	(20)
Hov II	—	(68)
Gr II	—	(74)

TABLE 13. CROSS-METATHESIS OF SECONDARY ALLYLIC ALCOHOLS AND DERIVATIVES (*Continued*)

C. ALLYLIC ESTERS (*Continued*)

Allylic Ester	Cross Partner	Conditions	Product(s) and Yield(s) (%)	Refs.
Please refer to the charts preceding the tables for the catalyst structures.				
C_{3-9}				
MeO, MeO, P(=O), R, OCO_2Me; er 84.0:16.0	OMe; 1.5 eq	Gr II (0.05 eq), DCM, 40°	MeO, MeO, P(=O), OMe, OCO_2Me; er 84.0:16.0; R: H (69), Ph (0)	221
C_3				
MeO, MeO, P(=O), OCO_2Me	R; *x* eq	Gr II (0.05 eq), CuI (0.07 eq) DCM, 45°	MeO, MeO, P(=O), R, OCO_2Me; R, *x*: *n*-$C_{10}H_{21}$, 10 (79); *n*-$C_{12}H_{25}$, 10 (76); *n*-$C_{16}H_{33}$, 5 (70); *n*-$C_{18}H_{37}$, 5 (70)	852
3 eq	OAc, AcO, AcO, O, O, NPhth	Gr II (0.05 eq), DCM, 40°	OAc, AcO, AcO, O, O, NPhth, MeO_2CO, P=O, OMe, OMe (87)	221

C_4

OR

x eq

OAc

Catalyst (y eq), DCM, reflux

OR

OAc

R	n	x	Catalyst	y	Time (h)	Conv. (%)[a]		(E)/(Z)	
Ac	3	—	Ru cat **149**	0.03	12	99	(—)	(E) only	853
Ac	4	—	Gr II	0.03	—	>95	(—)	95:5	854
Ac	4	—	Ru cat **33**	0.03	—	>95	(—)	75:25	854
Ac	4	—	Hov II	0.03	—	>95	(—)	95:5	854
Ac	4	—	Ru cat **223**	0.03	—	>95	(—)	95:5	854
Bz	2	3	Hov II	0.05	6	—	(82)	>95:5	177
Bz	2	3	Ru cat **271**	0.05	6	—	(93)	>95:5	177
Bz	3	2	Gr II	0.03	12	—	(82)	91:9	36

OBz

AcO

OAc

x eq

Catalyst (y eq), solvent

OBz

OAc

x	Catalyst	y	Solvent	Temp (°)	Time (h)		(E)/(Z)	
1.8	Gr II	0.03	DCM	40	12	(38)	94:6	36
1.8	Gr II	0.03	benzene	reflux	6	(38)	—	177
1.8	Hov II	0.03	benzene	reflux	6	(59)	—	177
1.8	Ru cat **271**	0.03	benzene	reflux	6	(87)	—	177
2	Gr I	0.05	DCM	45	—	(30)	94:6	124

TABLE 13. CROSS-METATHESIS OF SECONDARY ALLYLIC ALCOHOLS AND DERIVATIVES (*Continued*)

C. ALLYLIC ESTERS (*Continued*)

Allylic Ester	Cross Partner	Conditions	Product(s) and Yield(s) (%)	Refs.
Please refer to the charts preceding the tables for the catalyst structures.				
C_4				
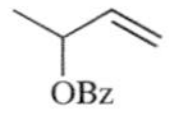 OBz	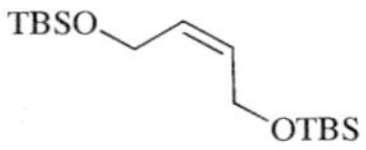 TBSO, OTBS; 2 eq	1. Gr I (0.05 eq), DCM, 45° 2. TBAF, THF	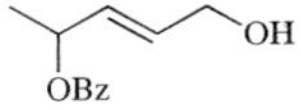 OBz, OH (54) $(E)/(Z) = 98:2$	124
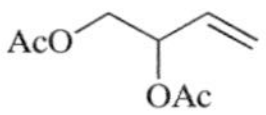 AcO, OAc	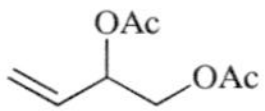 OAc, OAc	Gr I (0.05 eq), DCM, reflux	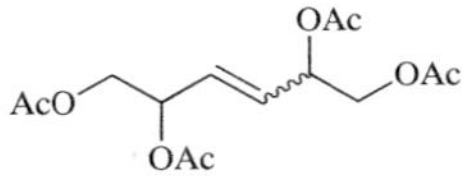 AcO, OAc, OAc, OAc (27)	782
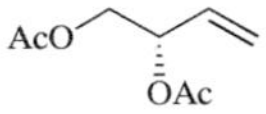 AcO, OAc; 2 eq	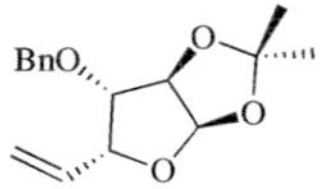 BnO, O, O, O	Gr II (0.036 eq), benzene, 16 h, reflux	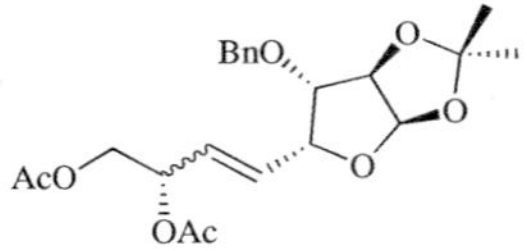 BnO, O, O, O, AcO, OAc (64)	805
3.4 eq	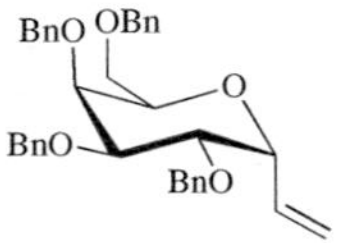 BnO, OBn, BnO, BnO, O	Gr II (0.1 eq), DCM, 45°, 4 d	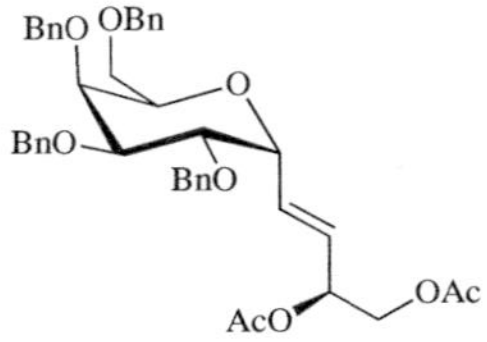 BnO, OBn, BnO, BnO, O, AcO, OAc (82) (E) only	855
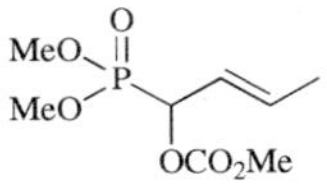 MeO, MeO, P, O, OCO_2Me	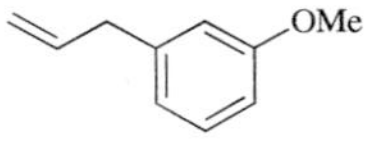 OMe; 2 eq	Gr II (0.05 eq), DCM, 40°, 24 h	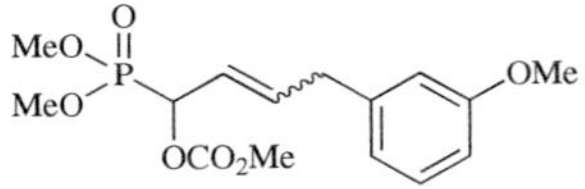 MeO, MeO, P, O, OCO_2Me, OMe (36)	221

C_5

OAc, OAc		Gr II, DCM, reflux, 6 h	OAc, OAc	(67)	856
EtO_2C, OPiv	OR, OH	Gr II (*x* eq), DCM, MW	EtO_2C, OPiv, OH, OR		857

R	*x*	Temp (°)	Time (h)		(*E*)/(*Z*)
Tr	0.15	90	1	(—)	(*E*) only
TBS	0.20	70	2	(60)	—

t-BuO_2C, OAc	4 eq	Gr II (0.05 eq), DCM, reflux, 24 h	t-BuO_2C, OAc	(90)	858
AcO, OBn, OAc	BnO	Gr II (0.035 eq), benzene, reflux, 37 h	AcO, OBn, OAc, BnO	(62) (*E*) only	805

TABLE 13. CROSS-METATHESIS OF SECONDARY ALLYLIC ALCOHOLS AND DERIVATIVES (*Continued*)

C. ALLYLIC ESTERS (*Continued*)

Allylic Ester	Cross Partner	Conditions	Product(s) and Yield(s) (%)	Refs.
Please refer to the charts preceding the tables for the catalyst structures.				
C_5				
NsHN, NHBoc, OBoc	n-$C_{12}H_{25}$ 4 eq	Gr II (0.03 eq), DCM, reflux, 7 h	NsHN, NHBoc, OBoc, n-$C_{12}H_{25}$ (72) (*E*) only	741
NHBoc, O, O, O 3 eq	OR, n-C_6H_{13}	Catalyst (0.2 eq), CuI (x eq), solvent	NHBoc, OR, O, O, O, n-C_6H_{13}	722

R	Catalyst	x	Solvent	Temp (°)	Time (h)	
H	Gr II	—	DCE	80	48	(—)
H	Hov II	—	DCM	40	70	(48)
H	Gr II	0.3	toluene	40	48	(78)
H	Gr II	0.3	Et_2O	rt	48	(83)
H	Gr II	—	DCM	40	48	(48)
Bz	Gr II	—	DCM	40	48	(—)
TBS	Gr II	—	DCM	40	48	(—)
PMP	Gr II	—	DCM	40	48	(37)
MOM	Gr II	—	DCM	40	66	(43)

Allylic Ester	Cross Partner	Conditions	Product(s) and Yield(s) (%)	Refs.
C_6				
MeO, MeO, P, O, NHBoc, OCbz	n-$C_{12}H_{25}$ 4 eq	Gr II (0.03 eq), DCM, reflux, 7 h	MeO, MeO, P, O, NHBoc, OCbz, n-$C_{12}H_{25}$ (68) (*E*) only	741

Substrate	Partner	Conditions	Product(s) and Yield(s) (%)		Refs.
AcO, OBn, OAc, OAc	BnO	Gr II (0.07 eq), benzene, reflux, 48 h	AcO, OBn, OAc, OAc, BnO	(51) (*E*) only	805
AcO, OAc, NHAc, OAc	n-$C_{11}H_{23}$ 5 eq	Gr II (0.2 eq), DCM, reflux, 18 h	AcO, OAc, NHAc, OAc, n-$C_{11}H_{23}$	(73)	800
R^1O, OR^2, AcHN, OR^2	n-$C_{11}H_{23}$ 5 eq	Gr II (0.2 eq), DCM, reflux, 18 h	R^1O, OR^2, AcHN, OR^2, n-$C_{11}H_{23}$		800

R^1	R^2		(*E*)/(*Z*)
H	Bz	(91)	93:7
Ac	Ac	(86)	—

Substrate	Partner	Conditions	Product(s) and Yield(s) (%)		Refs.
HO, O, O	OH 1.5 eq	Gr II (0.02 eq), DCM, reflux, 48 h	HO, OH, O, O	(78)	859

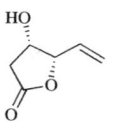

TABLE 13. CROSS-METATHESIS OF SECONDARY ALLYLIC ALCOHOLS AND DERIVATIVES (*Continued*)

C. ALLYLIC ESTERS (*Continued*)

Allylic Ester	Cross Partner	Conditions	Product(s) and Yield(s) (%)	Refs.
Please refer to the charts preceding the tables for the catalyst structures.				
C_6				
RO	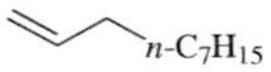 n-C_7H_{15} x eq	Catalyst (0.02 eq), DCM, reflux, 72 h	RO, n-C_7H_{15}	860
			R / x / Catalyst: H / 1.5 / Gr II / (90)	
			R / x / Catalyst: TBS / 2 / Hov II / (45)	
C_7				
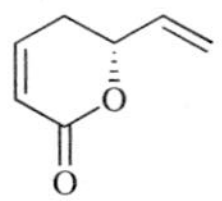 x eq	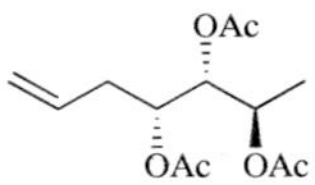 OAc, OAc, OAc y eq	Gr II (z eq), DCM, reflux, 4 h	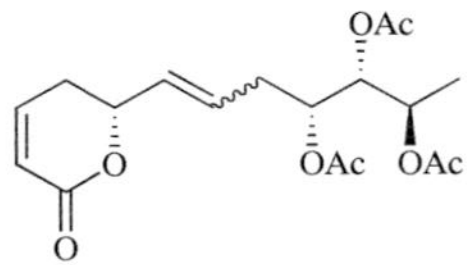 OAc, OAc, OAc	
			x / y / z: 1 / 5 / 0.32 / (66)	1215
			x / y / z: 4 / 1 / 0.57 / (55)	862

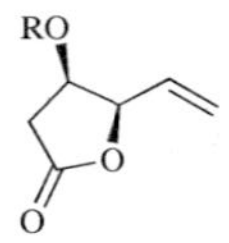

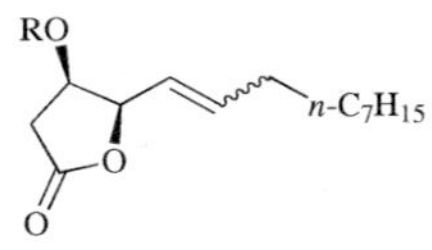

4.86 eq		Gr II (0.46 eq), DCM, reflux, 2 h		(67)	1215
4 eq		Gr II (0.5 eq), DCM, reflux, 4 h		(68)	863
4 eq		Gr II (0.5 eq), DCM, reflux, 4 h		(70)	863
5 eq		Gr II (0.1 eq), DCM, reflux, 8 h		(62)	864

TABLE 13. CROSS-METATHESIS OF SECONDARY ALLYLIC ALCOHOLS AND DERIVATIVES (*Continued*)

C. ALLYLIC ESTERS (*Continued*)

Allylic Ester	Cross Partner	Conditions	Product(s) and Yield(s) (%)	Refs.
Please refer to the charts preceding the tables for the catalyst structures.				
C_7				
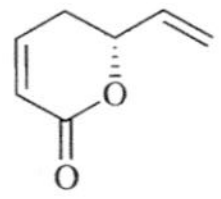 OAc OAc OAc OAc 2 eq	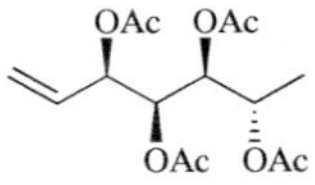	Gr II (0.02 eq), DCM, reflux, 5 h	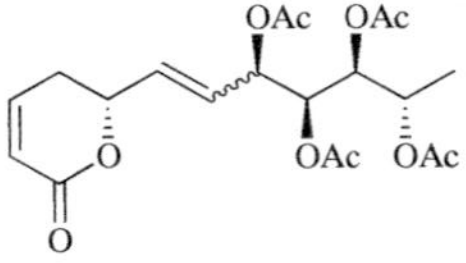 OAc OAc OAc OAc (68)	865
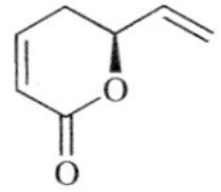 2 eq	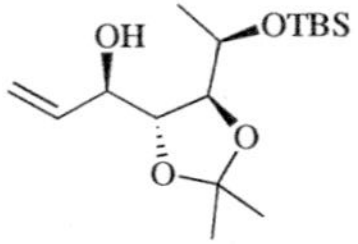 OH OTBS	Gr II (0.02 eq), DCM, reflux, 5 h	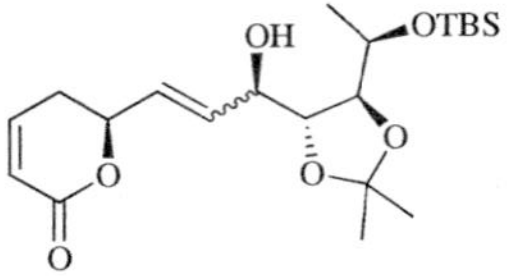 OH OTBS (82)	865
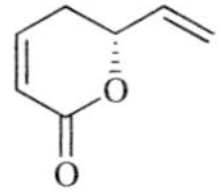	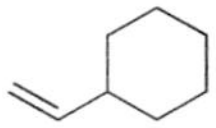 7.5 eq	Hov II (0.03 eq), DCM, 40°, 36 h	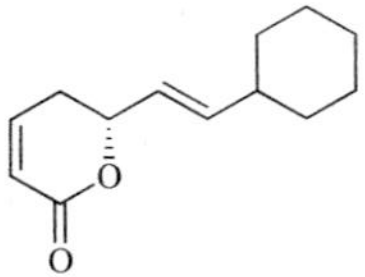 (89) (*E*)/(*Z*) > 95:5	866
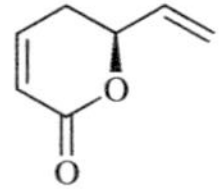	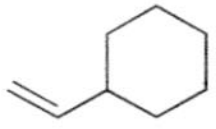 7.5 eq	Catalyst (0.03 eq), DCM, 40°, 36 h	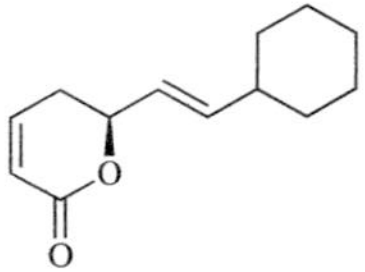 (*E*)/(*Z*) > 95:5 Catalyst: Gr II (88); Hov II (93)	866

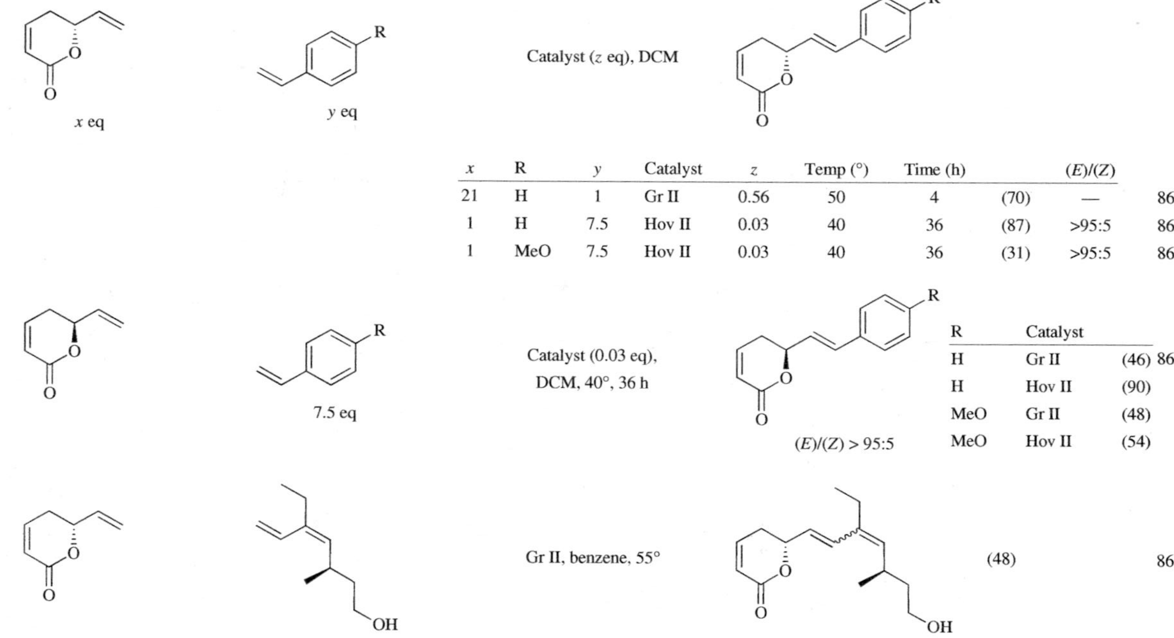

x	R	y	Catalyst	z	Temp (°)	Time (h)		(E)/(Z)	
21	H	1	Gr II	0.56	50	4	(70)	—	867
1	H	7.5	Hov II	0.03	40	36	(87)	>95:5	866
1	MeO	7.5	Hov II	0.03	40	36	(31)	>95:5	866

R	Catalyst		
H	Gr II	(46)	866
H	Hov II	(90)	
MeO	Gr II	(48)	
MeO	Hov II	(54)	

TABLE 13. CROSS-METATHESIS OF SECONDARY ALLYLIC ALCOHOLS AND DERIVATIVES (*Continued*)

C. ALLYLIC ESTERS (*Continued*)

Allylic Ester	Cross Partner	Conditions	Product(s) and Yield(s) (%)	Refs.
Please refer to the charts preceding the tables for the catalyst structures.				
C_7				
	n-C_8H_{17} 7.5 eq	Catalyst (0.03 eq), DCM, 40°, 36 h	n-C_8H_{17} Catalyst / (E)/(Z): Gr II (85) 89:11 Hov II (72) 50:50	866
x eq	Ph, OH, OH	Catalyst (*y* eq), DCM	OH OH Ph	

x	Catalyst	*y*	Temp (°)	Time (h)		(*E*)/(*Z*)	Refs.
5	Gr II	0.15	50	4	(63)	(*E*) only	720
5	Gr II	0.15	50	4	(66)	(*E*) only	869
0.25	Gr II	0.05	50	3	(55)	(*E*) only	870
0.5	Hov II	0.05	40	1.5	(77)	—	871

Allylic Ester	Cross Partner	Conditions	Product(s) and Yield(s) (%)	Refs.
1.5 eq	Ph, OH, OH	Gr II (0.05 eq), DCM, reflux, 10 h	OH OH Ph (65)	739
	OH OH Ph 2 eq	Gr II (0.004 eq), DCM, 50°, 3 h	Ph (82)	872

2 eq | OH OAc Ph | Gr II (0.01 eq), DCM, 50°, 3 h | O O OH OAc Ph (71) | 867

1.5 eq | OR OR Ph | Gr II (0.1 eq), DCM, reflux | O O Ph + O O O O | 873

I II

R	Time (h)	**I**	(*E*)/(*Z*) **I**	**II**
H	0.25	(0)	—	(90)
Ac	4	(80)	(*E*) only	(0)

1.2 eq | OH OBn Ph | Gr II (0.03 eq), DCM, reflux, 8 h | O O OH OBn Ph (74) | 874

TABLE 13. CROSS-METATHESIS OF SECONDARY ALLYLIC ALCOHOLS AND DERIVATIVES (*Continued*)

C. ALLYLIC ESTERS (*Continued*)

Allylic Ester	Cross Partner	Conditions	Product(s) and Yield(s) (%)	Refs.
Please refer to the charts preceding the tables for the catalyst structures.				
C_7				
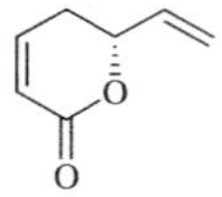	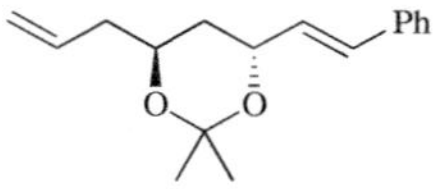 1.4 eq	Gr II (0.1 eq), DCM, reflux, 4 h	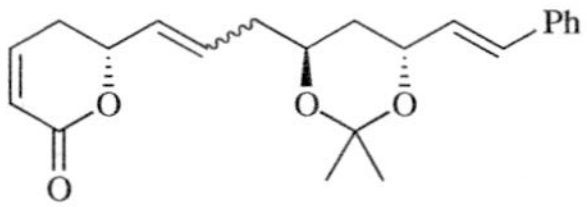 (69)	875
	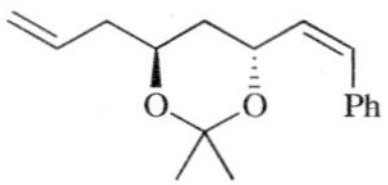 2 eq	Gr II (0.004 eq), DCM, 50°, 4 h	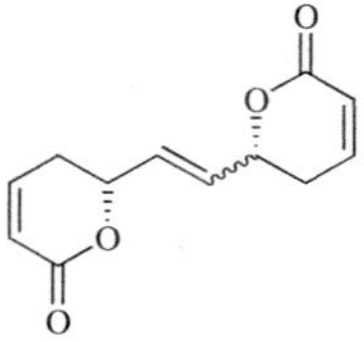 (70)	872
	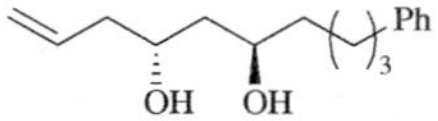 5 eq	Gr II (0.15 eq), DCM, 50°, 4 h	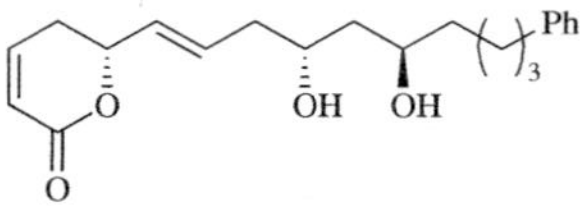 (68) (*E*)/(*Z*) > 95:5	869
	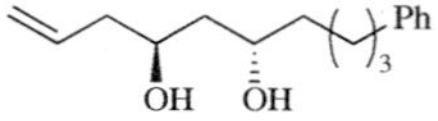	Hov II (0.06 eq), DCM, 40°, 1.5 h	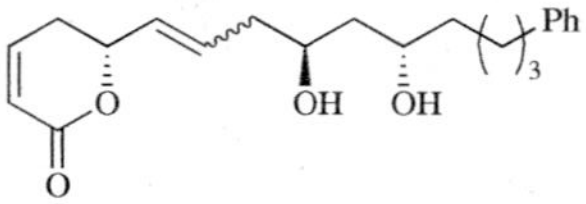 (73)	871

Reactant	Partner	Conditions	Product(s) and Yield(s) (%)	Refs.
3 eq	OH OMOM Ph	Gr II (0.05 eq), DCM, 40°, 12 h	O O OH OMOM Ph (74)	876
1.5 eq	O Ph OMOM OMOM	Gr II (0.07 eq), DCM, reflux, 6 h	O O O Ph OMOM OMOM (87)	877
OH O O 2 eq	OH OH	1. Gr II (0.005 eq), DCM, 40° 2. PPTS, 2,2-DMP (5 eq), 0°, 1 h	OH O O O O (80)	380
MOMO OMOM O O	Ph 2.96 eq	Gr II (0.1 eq), benzene, 55°, 5 h	MOMO OMOM Ph O O (87)	877

TABLE 13. CROSS-METATHESIS OF SECONDARY ALLYLIC ALCOHOLS AND DERIVATIVES (*Continued*)
C. ALLYLIC ESTERS (*Continued*)

Allylic Ester	Cross Partner	Conditions	Product(s) and Yield(s) (%)	Refs.
Please refer to the charts preceding the tables for the catalyst structures.				
C_8				
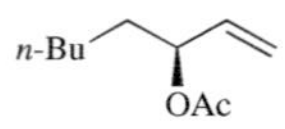 n-Bu, OAc; 2 eq	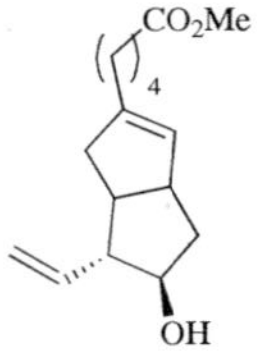 CO_2Me, 4, OH, (±)	Hov II (0.1 eq), DCM, rt, 2 h	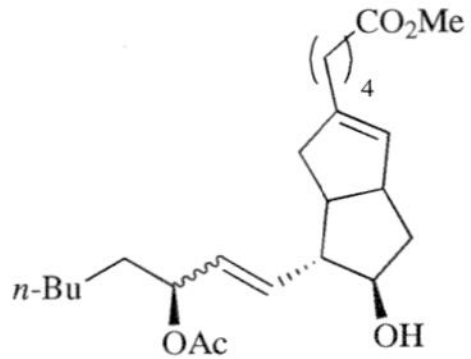 n-Bu, OAc, CO_2Me, 4, OH, (±) (75)	878
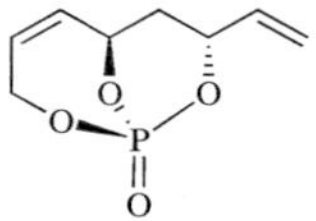 O, O, P, O	 R, n; *x* eq	Hov II (*y* eq), solvent	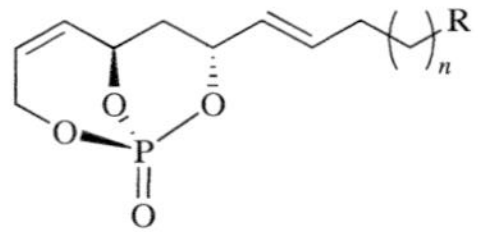 O, O, P, O, R, n	

x	R	*n*	*y*	Solvent	Temp (°)	Time (h)		(*E*)/(*Z*)	
1.1	BocHN	0	0.1	DCM	reflux	3–6	(69)	99:1	719
1.1	HO	0	0.1	DCM	reflux	3–6	(86)	99:1	719
1.1	TBSO	0	0.1	DCM	reflux	3–6	(87)	99:1	719
1.1	$(MeO)_2P(O)$	0	0.1	DCM	reflux	3–6	(80)	67:33	719
1.5	PMBO	2	0.06	DCE	90	2	(82)	>95:5	879, 880
1.1	Me	7	0.1	DCM	reflux	2	(72)	>99:1	881

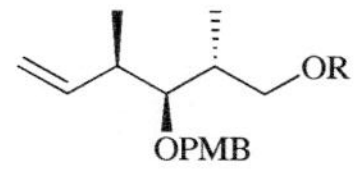

OR
OPMB
2 eq

Hov II (x eq), solvent, 6 h

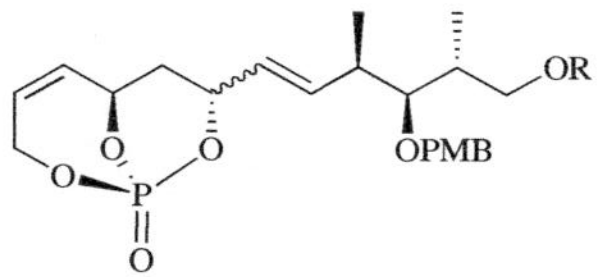

OR
OPMB

R	x	Solvent	Temp (°)		
TBDPS	0.06	toluene	60	(—)	882
TBDPS	0.06	toluene	90	(28)	882
TBDPS	0.12	toluene	90	(31)	882
TBS	0.06	DCM	50	(42)	882
TBS	0.06	toluene	90	(45)	882
TBS	0.12	toluene	90	(60)	882
TBS	0.06	DCE	90	(72)	879

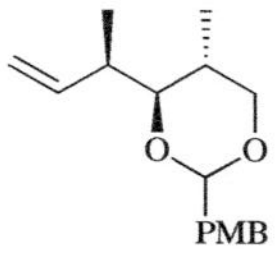

PMB
2 eq

Hov II (0.06 eq), solvent, 6 h

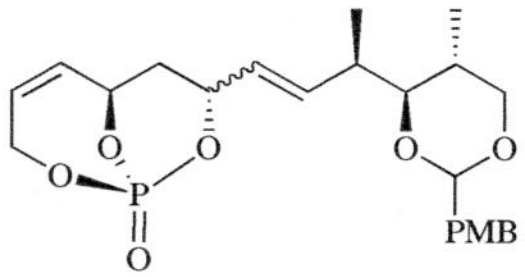

PMB

882

Solvent	Temp (°)	
DCM	50	(32)
toluene	90	(63)
DCE	90	(73)

TABLE 13. CROSS-METATHESIS OF SECONDARY ALLYLIC ALCOHOLS AND DERIVATIVES (*Continued*)

C. ALLYLIC ESTERS (*Continued*)

Allylic Ester	Cross Partner	Conditions	Product(s) and Yield(s) (%)	Refs.
Please refer to the charts preceding the tables for the catalyst structures.				
C_8				
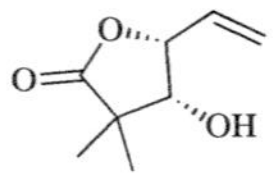 OH	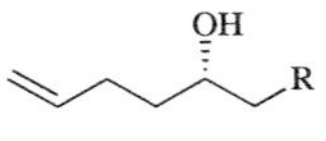 OH, R; 1.3 eq	Gr II (0.02 eq), DCM, reflux, 12 h	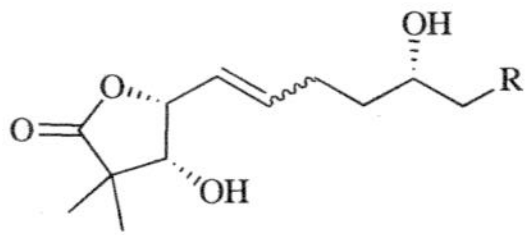 OH, OH, R R: TsO (89); n-C_6H_{13} (77)	883
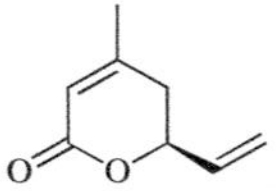 1.5 eq	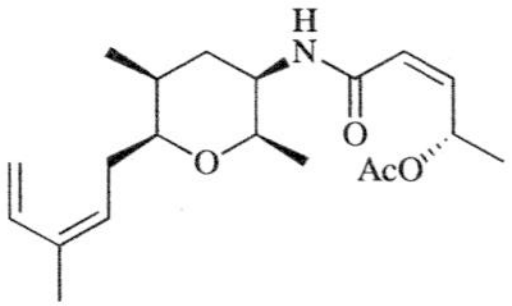 H N, AcO	Gr II (0.1 eq), DCM, 40°, 5 h	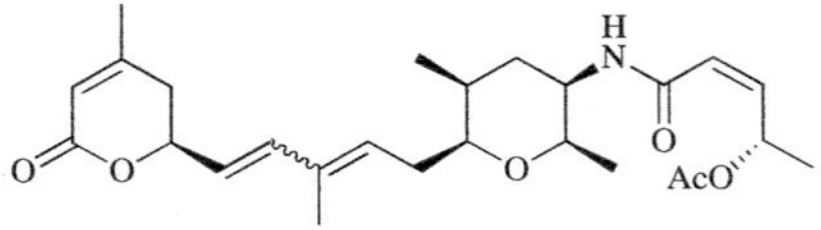 H N, AcO (71)	884
C_9				
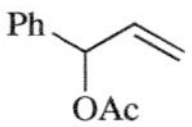 Ph, OAc; 3 + 2 eq	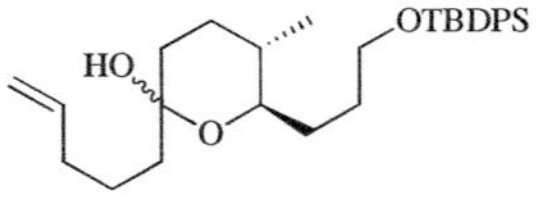 HO, OTBDPS	1. Hov II (0.15 eq), DCM, 80° (MW), 105 min 2. CSA, THF, rt, 2 h	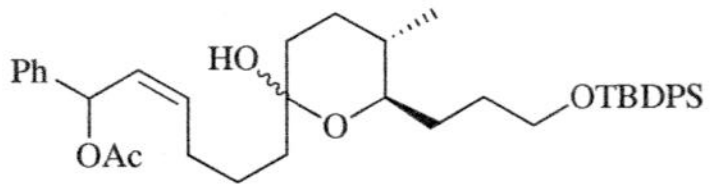 Ph, OAc, HO, OTBDPS (88)	885
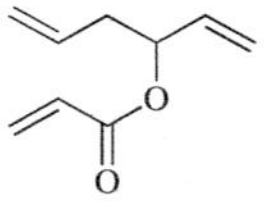	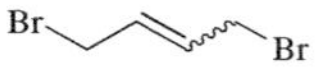 Br, Br; 5 eq	Gr II (0.05 eq), DCM, 50°, 8 h	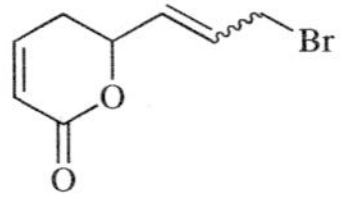 Br (40)	886

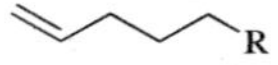

Reactant	Conditions	Product(s) and Yield(s) (%)	Refs.
R (5 eq)	Catalyst (0.05 eq), DCM, 50°, 8 h	R	886

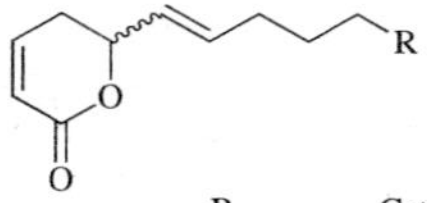

R	Catalyst	
Br	Gr I	(9)
Br	Gr II	(78)
TBSO	Gr II	(69)

Reactant	Conditions	Product(s) and Yield(s) (%)	Refs.
n-Bu (5 eq)	Gr II (0.05 eq), DCM, 50°, 8 h	*n*-Bu (17) + *n*-Bu (52)	886
Ph (5 eq)	Gr II (0.05 eq), DCM, 50°, 8 h	Ph (61)	886
n-$C_{10}H_{21}$ (5 eq)	Gr II (0.05 eq), DCM, 50°, 8 h	*n*-$C_{10}H_{21}$ (73)	886

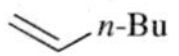

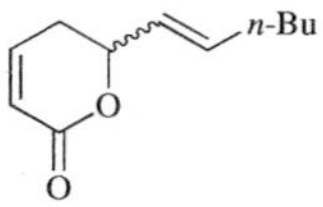

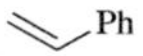

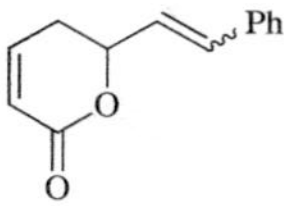

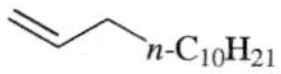

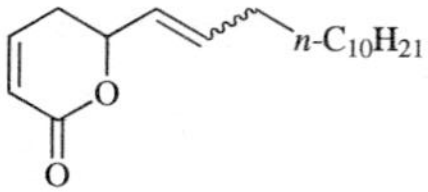

TABLE 13. CROSS-METATHESIS OF SECONDARY ALLYLIC ALCOHOLS AND DERIVATIVES (*Continued*)

C. ALLYLIC ESTERS (*Continued*)

Allylic Ester	Cross Partner	Conditions	Product(s) and Yield(s) (%)	Refs.
Please refer to the charts preceding the tables for the catalyst structures.				
C_9				
OMe; 3.38 eq		Ru cat **88** (2 × 0.06 eq), $C_6F_5CF_3$, 88°, 4 h	OMe (98)	529
3 eq		Ru cat **88** (0.04 + 0.11 eq), $C_6F_5CF_3$, 70°, 4 h	OMe (76)	529
C_{10}				
I	R, *n*; 1–1.5 eq	1. Hov II (0.06 eq), BQ, DCM, reflux, 0.75 h, 2. Hov II (0.04 eq), **I**, CuI, DCM, reflux, 2–3 h 3. *o*-NBSH (15 eq), Et_3N, rt, overnight	R, *n* (see below)	716

R	*n*	
BocHN	0	(59)
TBSO	0	(52)
$(PhO)_2P(O)$	0	(40)
PMBO	3	(43)
Me	6	(65)

Substrate	Olefin	Conditions	Product(s) and Yield(s) (%)	Refs.
I	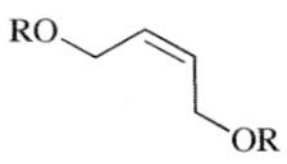 RO–CH$_2$CH=CHCH$_2$–OR 1–1.5 eq	1. Hov II (0.06 eq), BQ, DCM, reflux, 0.75 h, 2. Hov II (0.04 eq), **I**, CuI, DCM, reflux, 2–3 h 3. *o*-NBSH (15 eq), Et_3N, rt, overnight	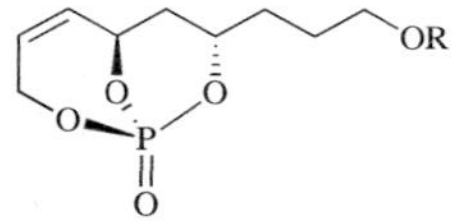 OR R: H (64); Ac (56)	716
I	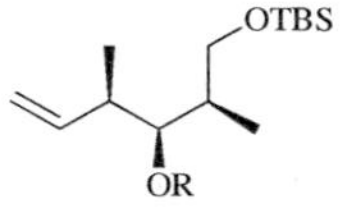 OTBS, OR 2–4 eq	1. Hov II (0.06 eq), BQ, solvent, reflux, 0.75 h, 2. Hov II (0.04 eq), **I**, CuI, solvent, reflux, 4–6 h 3. *o*-NBSH (15 eq), Et_3N, rt, overnight	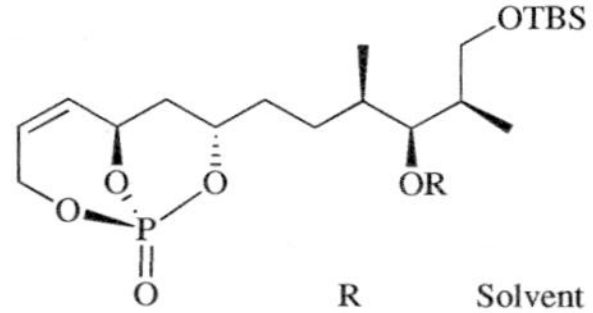 OTBS, OR R / Solvent: H, DCE (54); PMB, DCM (30)	716
I	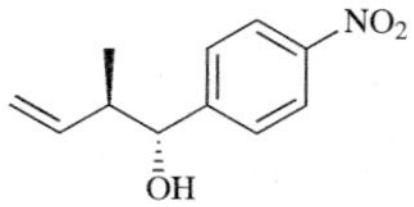 NO_2, OH 2–4 eq	1. Hov II (0.06 eq), BQ, DCM, reflux, 0.75 h, 2. Hov II (0.04 eq), **I**, CuI, DCM, reflux, 4–6 h 3. *o*-NBSH (15 eq), Et_3N, rt, overnight	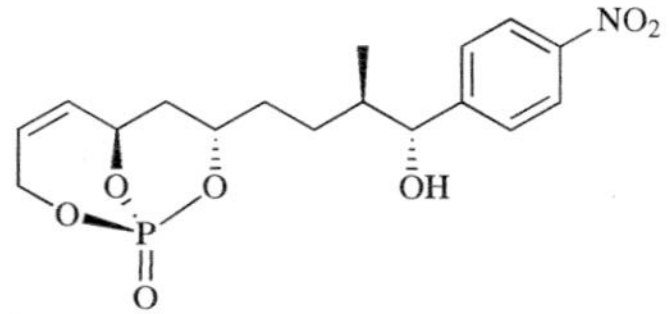 NO_2, OH (85)	716

TABLE 13. CROSS-METATHESIS OF SECONDARY ALLYLIC ALCOHOLS AND DERIVATIVES (*Continued*)

C. ALLYLIC ESTERS (*Continued*)

Allylic Ester	Cross Partner	Conditions	Product(s) and Yield(s) (%)	Refs.
Please refer to the charts preceding the tables for the catalyst structures.				
C_{10}				
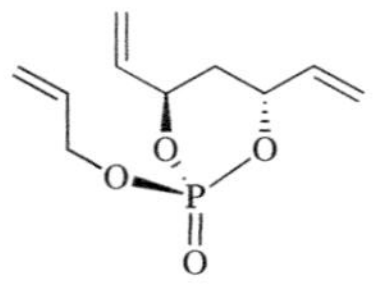 I	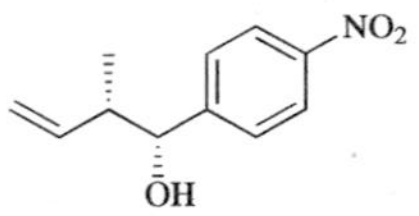 NO_2, OH; 2–4 eq	1. Hov II (0.06 eq), BQ, DCM, reflux, 0.75 h, 2. Hov II (0.04 eq), **I**, CuI, DCM, reflux, 4–6 h 3. *o*-NBSH (15 eq), Et_3N, rt, overnight	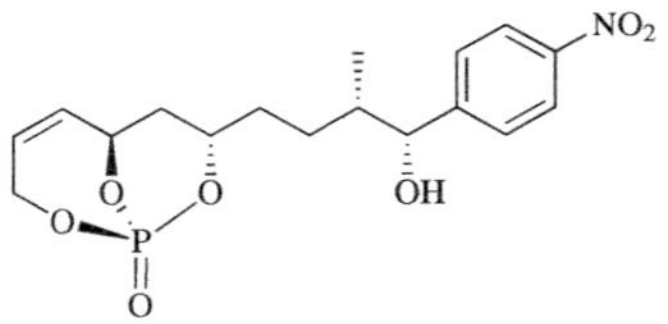 NO_2, OH (79)	716
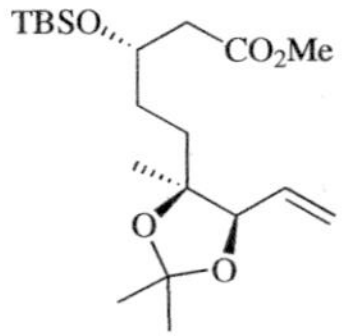 TBSO, CO_2Me	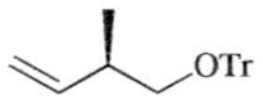 OTr; 4 eq	Hov II (0.2 eq), DCM, 40°	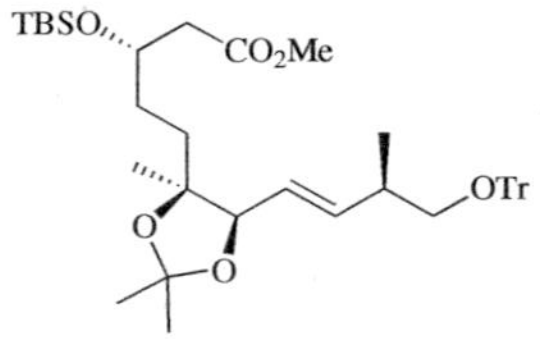 TBSO, CO_2Me, OTr (76) (*E*) only	887
C_{11}				
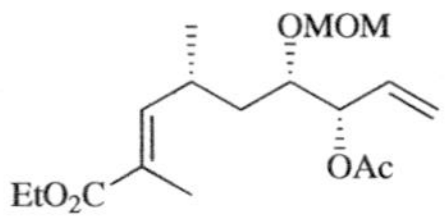 OMOM, OAc, EtO_2C	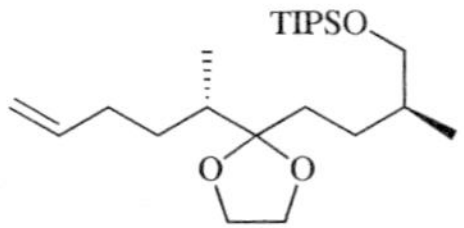 TIPSO; 2.1 eq	Gr II (0.09 eq), DCM, 45°	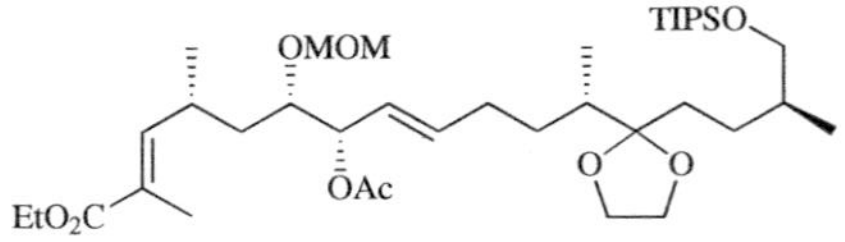 OMOM, TIPSO, OAc, EtO_2C (86) (*E*)/(*Z*) = 83:17	835

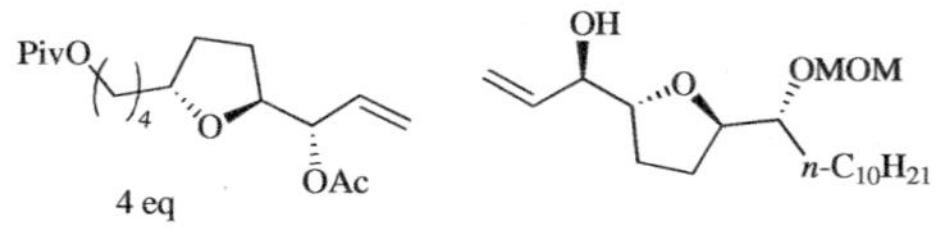

4 eq

1. Gr II (0.1 eq), DCM, rt, 18 h
2. Gr II (0.1 eq), reflux, 4 h

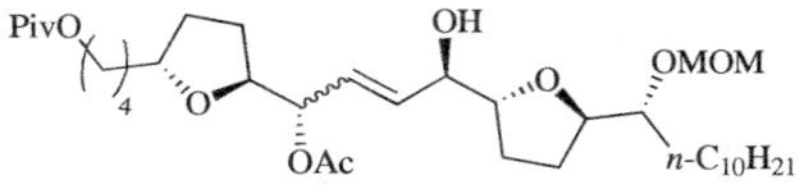

(98) 888

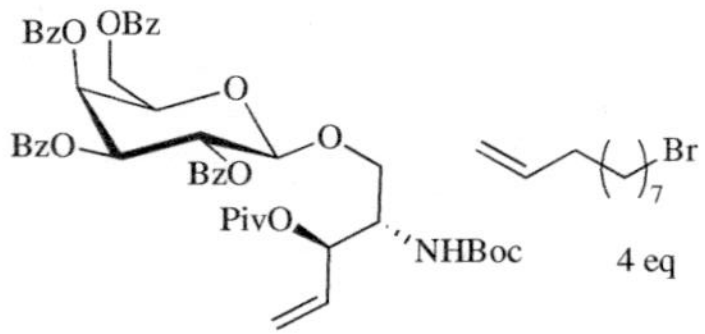

4 eq

Gr II (0.3 eq), DCM, reflux, 16 h

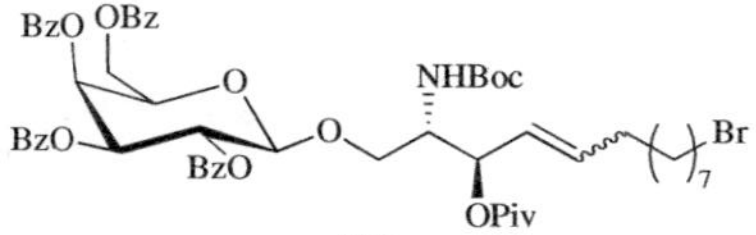

(63) 889

C_{12}

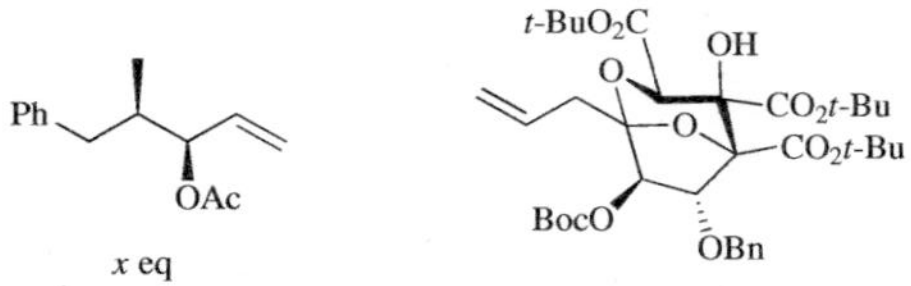

x eq

Catalyst (y eq), benzene

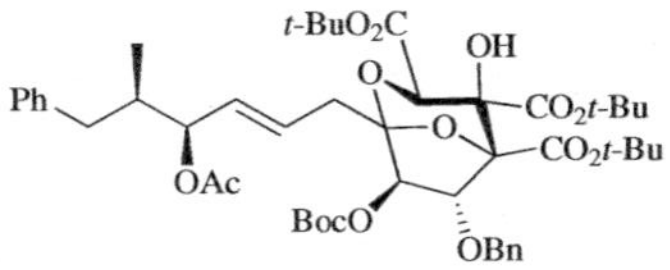

890

x	Catalyst	y	Temp (°)		(E)/(Z)
1.2	Gr II	0.05	70	(67)	(E) only
1.2	Gr II	0.1	70	(60)	(E) only
2	Gr II	0.1	70	(60)	(E) only
1.5	Hov II	0.05	70	(62)	(E) only
1.2	Ru cat **112**	0.05	60	(48)	89:11
1.2	Ru cat **112**	0.2	60	(46)	89:11
2	Ru cat **112**	0.2	60	(90)	89:11

TABLE 13. CROSS-METATHESIS OF SECONDARY ALLYLIC ALCOHOLS AND DERIVATIVES (*Continued*)

C. ALLYLIC ESTERS (*Continued*)

Allylic Ester	Cross Partner	Conditions	Product(s) and Yield(s) (%)	Refs.
Please refer to the charts preceding the tables for the catalyst structures.				
C_{17}				
n-$C_{13}H_{27}$, OAc	OBn 1.2 eq	Gr II (0.04 eq), DCM, 40°, 10 h	n-$C_{13}H_{27}$, OAc, OBn (78) (*E*) only	891
C_{19}				
H N, O, O, O, S, N, N, HN, S, O	O, STr, n *x* eq	Catalyst (*y* eq), solvent	H N, O, O, O, STr, n, S, N, N, HN, S, O	

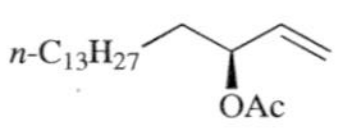

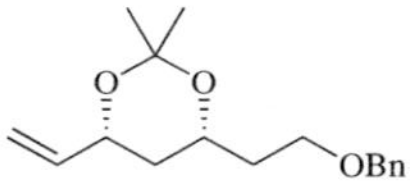

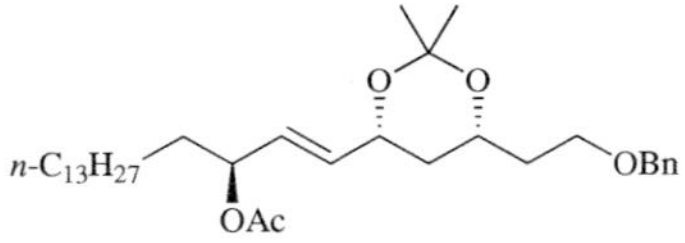

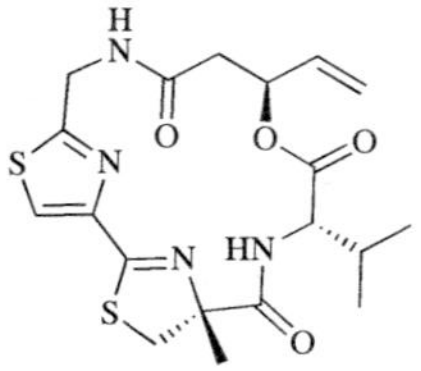

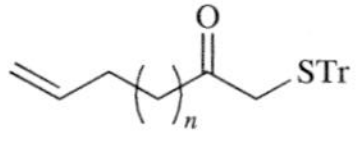

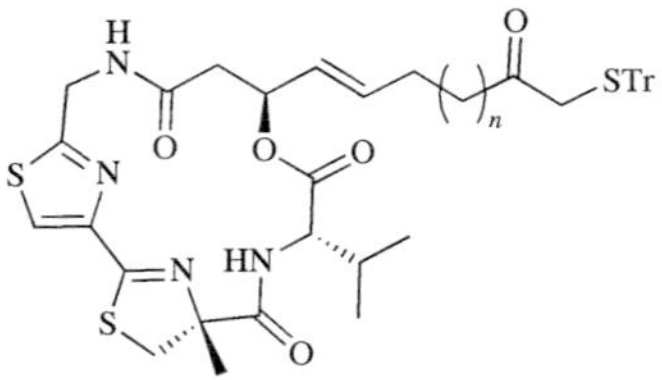

n	*x*	Catalyst	*y*	Solvent	Temp	Time (h)		Refs.
1	4 x 2	Hov II	4 x 0.2	toluene	reflux	4	(51)	893
2	4 x 2	Hov II	4 x 0.2	toluene	reflux	4	(62)	893
2	3.5	Ru cat **88**	0.15	DCE	90°	15	(52)	892

Reactant	Olefin partner	Conditions	Product(s) and Conversion(s) (%)[a]	Refs.
	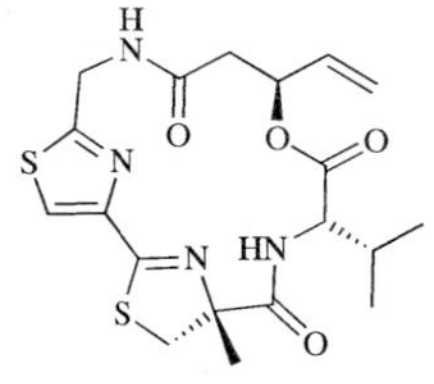 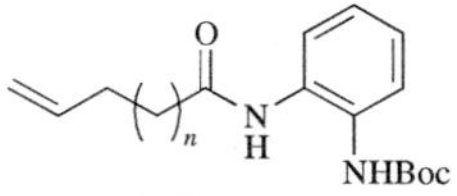 4 x 2 eq	Gr II (4×0.2 eq), toluene, reflux, 4 h	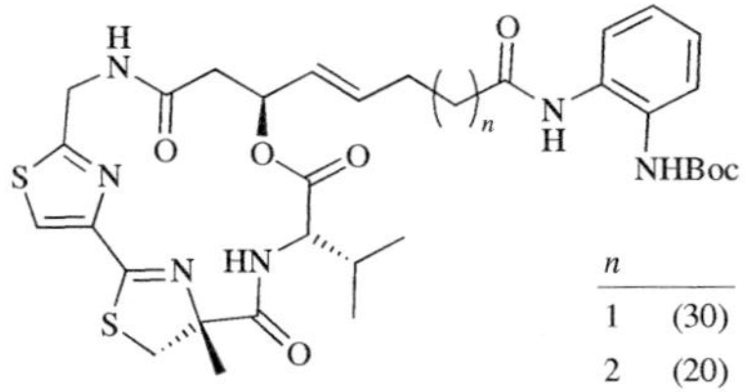 *n* 1 (30); 2 (20)	893
	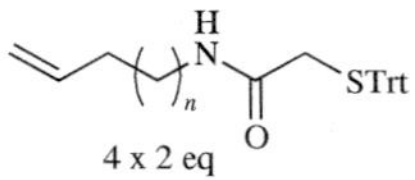 4 x 2 eq	Catalyst (4×0.2 eq), toluene, reflux, 4 h	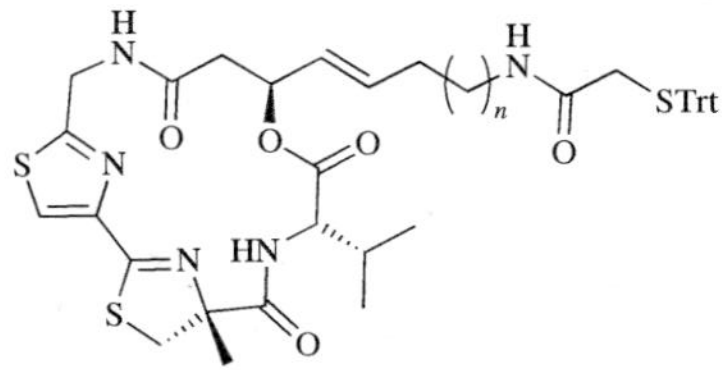 *n* Catalyst: 0 Hov II (42); 1 Gr II (15)	893

[a] The conversion was determined by ^{1}H NMR analysis.

TABLE 13. CROSS-METATHESIS OF SECONDARY ALLYLIC ALCOHOLS AND DERIVATIVES (*Continued*)

D. ACETALS

Acetal	Cross Partner	Conditions	Product(s) and Yield(s) (%)	Refs.
Please refer to the charts preceding the tables for the catalyst structures.				
C_3				
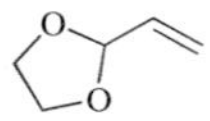	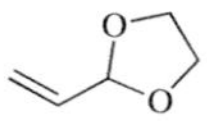	Gr I (0.05 eq), DCM, 45°, 17 h	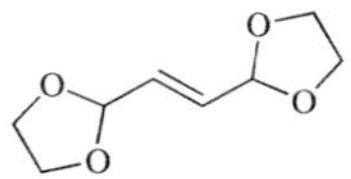 (52) (*E*)/(*Z*) = 90:10	124
	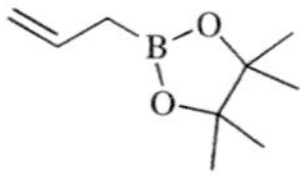 4 eq	1. Ru cat **216** (0.02 eq), THF, 35°, 5 h 2. NaOH, H_2O_2	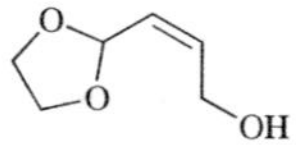 OH (63) (*Z*)/(*E*) = 95:5	461
3 eq		1.Gr II (*x* eq), DCM, reflux, 2–12 h 2. PhCHO (1.5 eq), 23°, 12–48 h	OH, Ph	126
	R 4 eq	Ru cat **216** (0.02 eq), THF (0.5 M), 35°	R	461

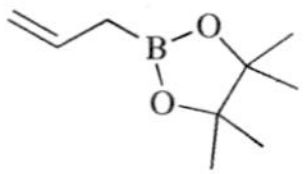

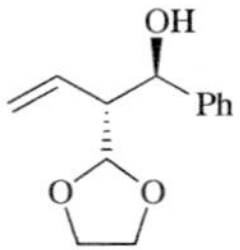

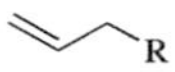

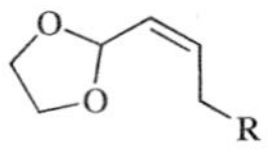

x		*anti/syn*[a]
0.05	(68)	>95:5
0.02	(69)	>95:5

R	Time (h)		(*Z*)/(*E*)
BnPhN	5	(72)	89:11
$HO(CH_2)_2$	5	(74)	95:5
$Br(CH_2)_2$	5	(83)	95:5
Ph	3	(88)	95:5
$MeO_2C(CH_2)_7$	5	(84)	95:5
n-C_9H_{19}	5	(85)	95:5

	Reactant	Conditions	Product(s) and Yield(s) (%)	Refs.
5 eq	OH, OMe, P, OMe, O er 85.0:15.0	Gr II (0.05 eq), $CHCl_3$, 40°	O, O, OH, P, OMe, OMe, O (30) er 85.0:15.0	221
5 eq	OCO_2Me, OMe, P, OMe, O	Gr II (0.05 eq), DCM, 40°	O, O, OCO_2Me, P, OMe, OMe, O (32)	221
	CN 2 eq	Gr II (0.05 eq), DCM, reflux, 2 h	O, O, CN (26) $(E)/(Z) = 71:29$	154
4 eq	R^1HN, COR^2, n	Gr II (0.05 eq), toluene, rt, 17 h	O, O, R^1HN, COR^2, n	457

R^1	R^2	n	
Boc	H_2N	0	(74)
Boc	MeO	0	(45)
OHC	MeO	0	(49)
Boc	H_2N	1	(71)
Boc	MeO	1	(52)
Boc	MeO	2	(50)

TABLE 13. CROSS-METATHESIS OF SECONDARY ALLYLIC ALCOHOLS AND DERIVATIVES (*Continued*)

D. ACETALS (*Continued*)

Acetal	Cross Partner	Conditions	Product(s) and Yield(s) (%)	Refs.
Please refer to the charts preceding the tables for the catalyst structures.				
C_3				
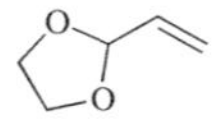 3 eq	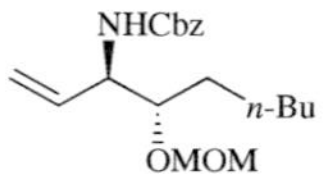 NHCbz, n-Bu, OMOM	Gr II (0.1 eq), DCM, reflux, overnight	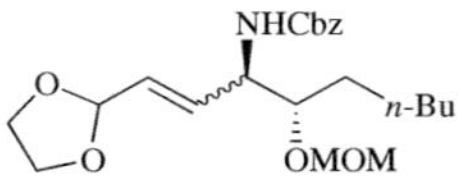 NHCbz, n-Bu, OMOM (87)	894
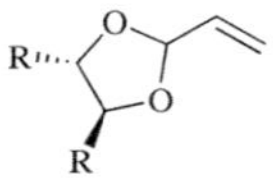 R, 2 eq	 OBz, 7	Gr I (*x* eq), DCM, 45°, 12 h	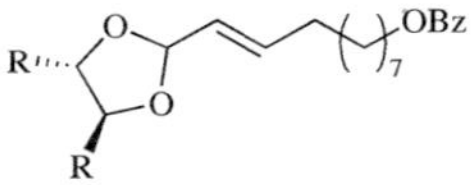 R, OBz, 7; $(E)/(Z) = 88{:}12$	124, 897
			R = H, *x* = 0.01 (74)	
			R = H, *x* = 0.02 (87)	
			R = H, *x* = 0.025 (93)	
			R = H, *x* = 0.05 (91)	
			R = EtO_2C, *x* = 0.025 (86)	
10 eq	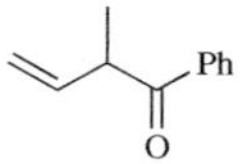 Ph	Ru cat **89** (0.05 eq), DCM, reflux, 48 h	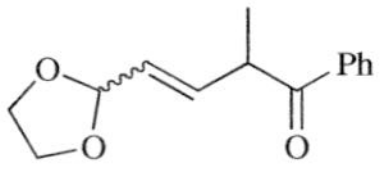 Ph (68)	895

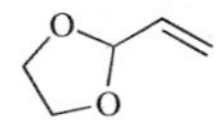

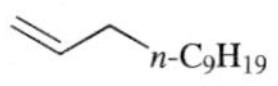

n-C_9H_{19} x eq

Ru cat **216** (y eq), THF, 35° n-C_9H_{19} 461

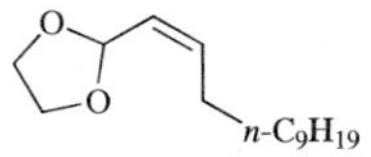

x	y	Conc (M)	Time (h)		$(Z)/(E)$
6	0.05	0.5	3	(84)	93:7
4	0.05	0.5	3	(87)	94:6
2	0.05	0.5	3	(80)	95:5
4	0.02	0.5	3	(83)	95:5
4	0.02	0.3	3	(92)	94:6
2	0.02	0.5	3	(80)	95:5
2	0.02	1.0	3	(66)	95:5
2	0.02	0.3	3	(82)	95:5
2	0.01	0.5	3	(74)	95:5
2	0.02	0.3	7	(87)	94:6

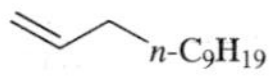

n-C_9H_{19} x eq

Ru cat **216** (0.02 eq), THF (0.5 M), 35° n-C_9H_{19} 461

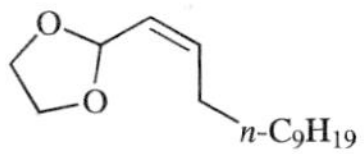

x	Time (h)		$(Z)/(E)$
2	3	(63)	93:7
4	3	(65)	91:9
0.5	3	(82)	95:5
0.25	3	(92)	94:6
2	7	(84)	92:8
4	7	(94)	91:9

TABLE 13. CROSS-METATHESIS OF SECONDARY ALLYLIC ALCOHOLS AND DERIVATIVES (*Continued*)

D. ACETALS (*Continued*)

Acetal	Cross Partner	Conditions	Product(s) and Yield(s) (%)	Refs.
Please refer to the charts preceding the tables for the catalyst structures.				
C_3				
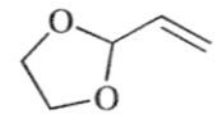	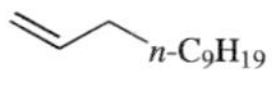 n-C_9H_{19} 4 eq	Catalyst (0.02 eq), THF (0.3 M), 35°, 3 h	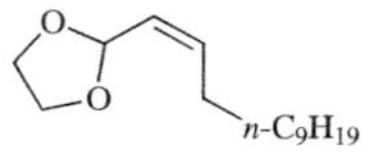 n-C_9H_{19}	461
			Catalyst / (Z)/(E) Ru cat **215** (87) 76:24 Ru cat **216** (92) 94:6 Gr I (96) 10:90 Hov II (92) 5:95	
3 eq	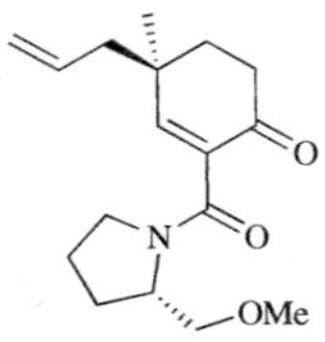 OMe	Hov II (0.025 eq), DCM, reflux, 7 h	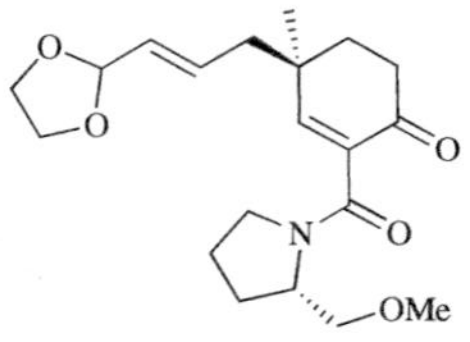 OMe (86) (*E*) only	896
20 eq	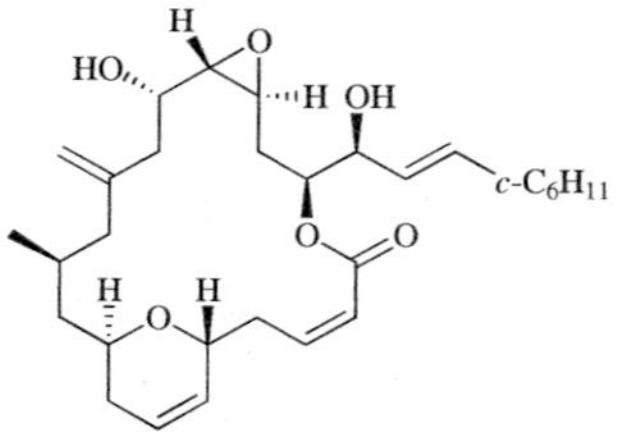 c-C_6H_{11}	Gr II (0.3 eq), DCM, rt, 4 h	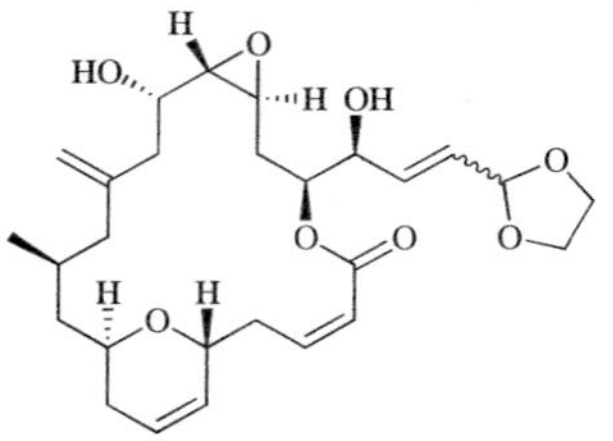 (41)	780

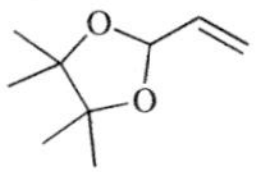	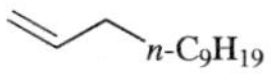 n-C_9H_{19} 4 eq	Ru cat **216** (0.02 eq), THF, 35°, 5 h	n-C_9H_{19}	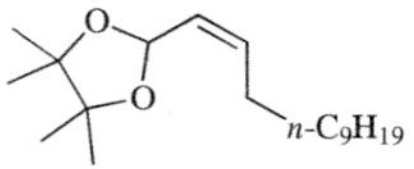 (84) (*Z*)/(*E*) = 95:5	461
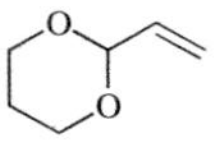	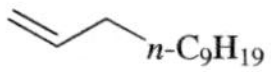 n-C_9H_{19} 4 eq	Ru cat **216** (0.02 eq), THF, 35°, 5 h	n-C_9H_{19}	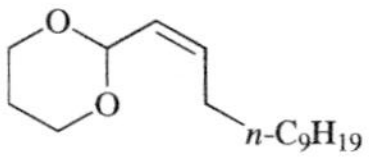 (79) (*Z*)/(*E*) = 95:5	461
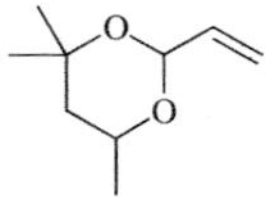	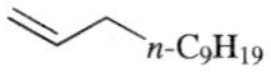 n-C_9H_{19} 4 eq	Ru cat **216** (0.02 eq), THF, 35°, 5 h	n-C_9H_{19}	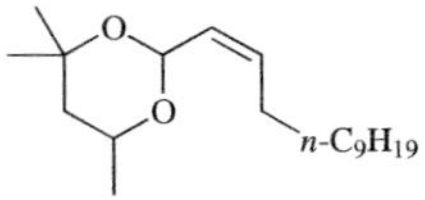 (85) (*Z*)/(*E*) = 95:5	461
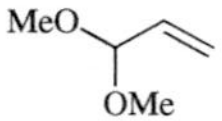 MeO, OMe 3 eq	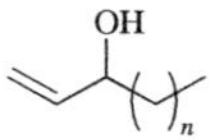 OH, n	Hov II (2 x 0.0125 eq), DCM, rt, 20 h	MeO, OMe, OH, n	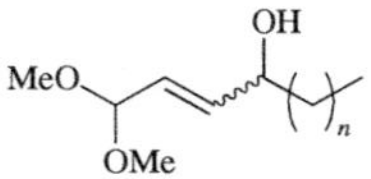 *n*: 1 (55); 4 (65); 7 (50)	898

TABLE 13. CROSS-METATHESIS OF SECONDARY ALLYLIC ALCOHOLS AND DERIVATIVES (*Continued*)

D. ACETALS (*Continued*)

Acetal	Cross Partner	Conditions	Product(s) and Yield(s) (%)	Refs.
Please refer to the charts preceding the tables for the catalyst structures.				
C_3				
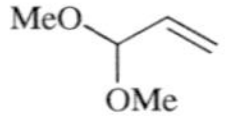 MeO, OMe; 2 eq	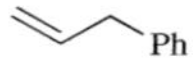 Ph	Gr I (0.05 eq), DCM, reflux, 16 h	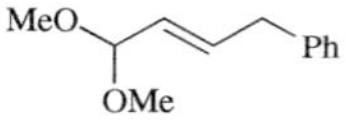 MeO, OMe, Ph (59) $(E)/(Z) > 95:5$	156
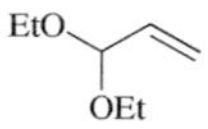 EtO, OEt; 5 eq	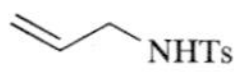 NHTs	Hov II (0.05 eq), $(PhO)_3B$ (0.1 eq), toluene, 80°, 0.5 h	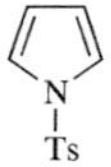 N, Ts (60)	899
4 eq	 BocHN, CO_2Me	Gr II (0.05 eq), toluene, rt, 17 h	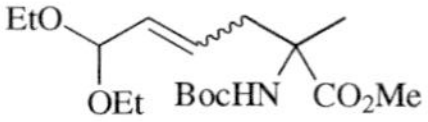 EtO, OEt, BocHN, CO_2Me (66)	457
10 eq	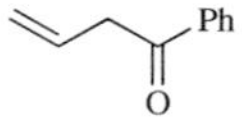 Ph, O	Ru cat **89** (0.05 eq), DCM, 40°, 48 h	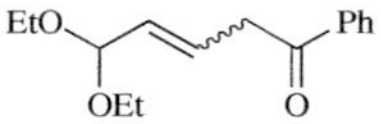 EtO, OEt, Ph, O (50)	895
x eq	 OBz, 7	1. Gr I (0.05 eq), solvent, 45°, 12 h 2. HCO_2H, DCM	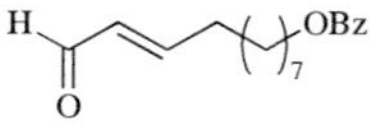 H, O, OBz, 7 $(E)/(Z) = 96:4$	897

x	Solvent	
4	DCM	(72)
2	DCM	(75)
65	—	(<10)

Reactant	Reactant	Conditions	Product(s) and Yield(s) (%)	Refs.
3 eq	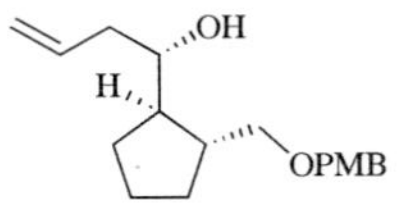	1. Hov II (0.05 eq), DCM, 45°, 18 h 2. Acidic work-up	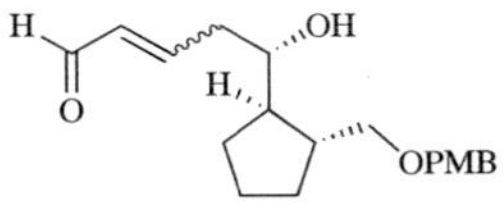 (75)	900
	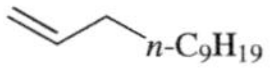 4 eq	Ru cat **216** (0.02 eq), THF, 35°, 5 h	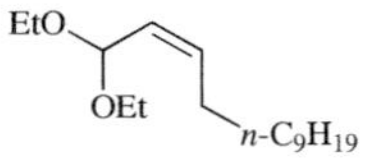 (70) (*Z*)/(*E*) = 95:5	461
21 eq	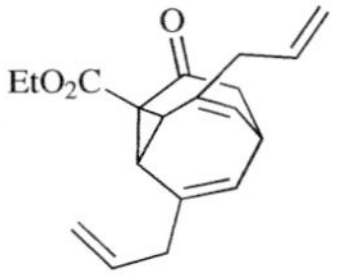	1. Gr II (0.15 eq), DCM, 40°, 12 h 2. HCl, THF	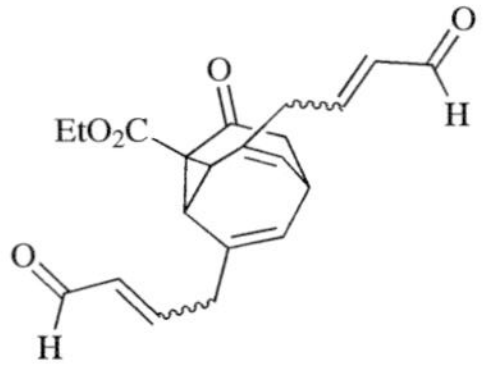 (46)	901
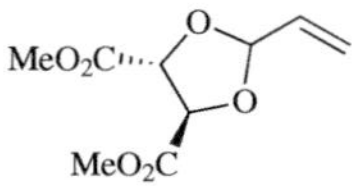 2 eq		Gr I (0.05 eq), DCM, 45°, 12 h	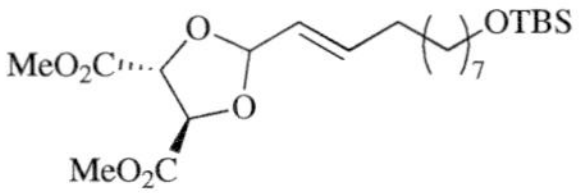 (94) (*E*)/(*Z*) = 86:14	124

TABLE 13. CROSS-METATHESIS OF SECONDARY ALLYLIC ALCOHOLS AND DERIVATIVES (*Continued*)

D. ACETALS (*Continued*)

Acetal	Cross Partner	Conditions	Product(s) and Yield(s) (%)	Refs.
Please refer to the charts preceding the tables for the catalyst structures.				
C_4				
	OAc	Gr II (*x* eq), DCM, 40°, 12 h	OAc	
			x / (*E*)/(*Z*)	
			0.037 (91) (*E*) only	36
			0.02 (95) —	902
2 eq	Ph	Gr II (*x* eq), DCM, 12 h	Ph	
			x / Temp (°) / (*E*)/(*Z*)	
			0.03 40 (70) (*E*) only	36
			0.05 reflux (71) >95:5	145
6 eq	H, H, N, O, O, *n*-Pr	Hov II (2 x 0.5 eq), toluene, 80° (MW), 2 x 0.5 h	H, H, N, O, O, *n*-Pr (80) dr 10:1	525
2 eq	OBz, 7	Gr I (0.05 eq), DCM, 45°, 17 h	OBz, 7 (44) (*E*)/(*Z*) = 91:9	124

[a] The ratio was determined by ^{1}H NMR analysis.

TABLE 14. CROSS-METATHESIS OF TERTIARY ALLYLIC ALCOHOLS AND DERIVATIVES

Tertiary Alcohol	Cross Partner	Conditions	Product(s) and Yield(s) (%)	Refs.
Please refer to the charts preceding the tables for the catalyst structures.				
C_5				
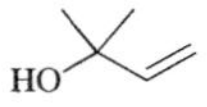 HO (structure) 2.5 eq	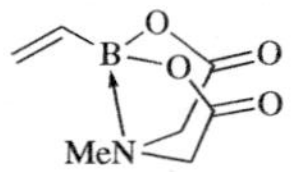 (structure) B, O, O, O, MeN	Gr II (0.1 eq), DCM, 40°, 12 h	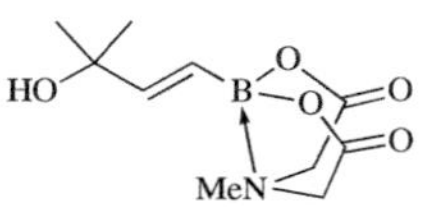 HO (structure) B, O, O, O, MeN (94) (*E*)/(*Z*) > 95:5	701
	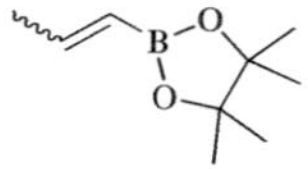 (structure) B, O, O 2 eq	Gr II (0.05 eq), DCM, reflux	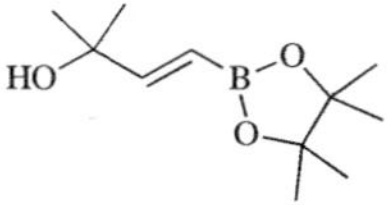 HO (structure) B, O, O (61) (*E*)/(*Z*) > 95:5	687
2.2 eq	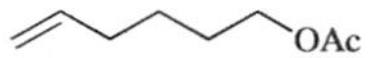 (structure) OAc	Gr II (0.06 eq), DCM, 40°, 12 h	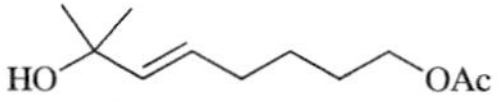 HO (structure) OAc (58) (*E*) only	36
2–4 eq	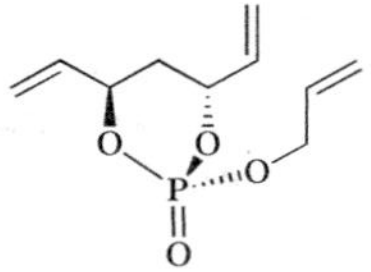 (structure) O, O, P, O, O	1. Hov II (0.06 eq), BQ, DCM, reflux, 0.75 h 2. Hov II (0.04 eq), allylic alcohol, CuI, DCE, reflux, 4–6 h 3. *o*-NBSH (15 eq), Et_3N, rt, overnight	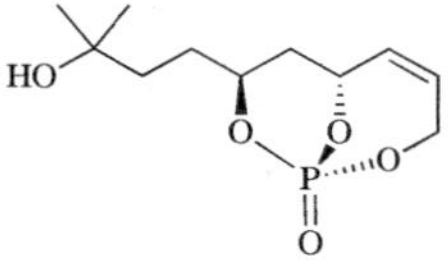 HO (structure) O, O, P, O, O (72)	716

TABLE 14. CROSS-METATHESIS OF TERTIARY ALLYLIC ALCOHOLS AND DERIVATIVES (*Continued*)

Tertiary Alcohol	Cross Partner	Conditions	Product(s) and Yield(s) (%)	Refs.
Please refer to the charts preceding the tables for the catalyst structures.				
C_5				
 HO 90 eq	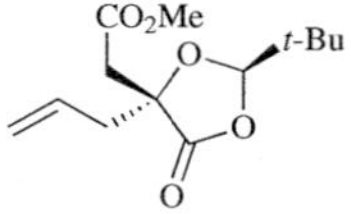 CO_2Me, O, O, O, t-Bu	Gr II (0.05 eq), 45°, 18 h	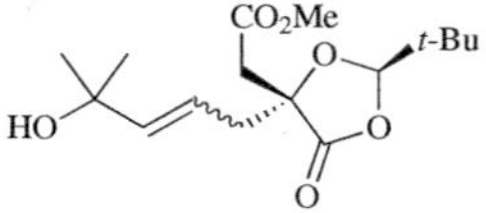 HO, CO_2Me, O, O, O, t-Bu (80)	903
5 eq	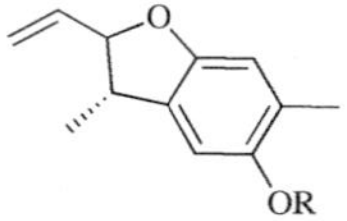 O, OR	Gr II (0.1 eq), DCM, 40°, 20 h	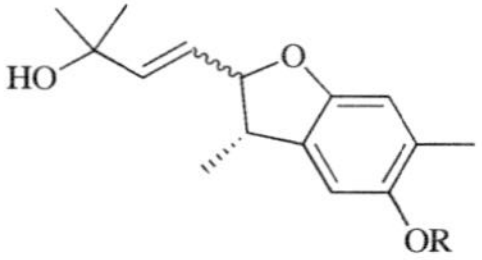 HO, O, OR R: H (32); TBS (98)	904
5 eq	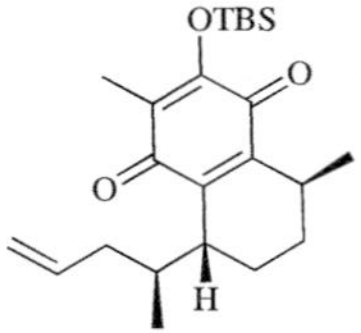 OTBS, O, O, H	1. Gr II (0.1 eq), DCM, reflux 2. TBAF	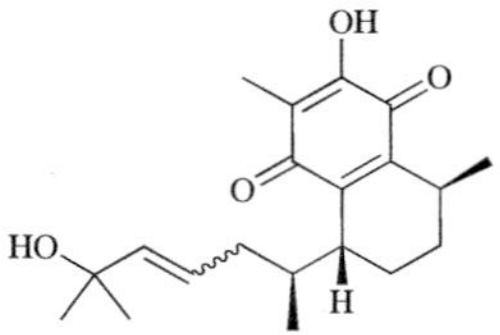 HO, OH, O, O, H (60)	905
20 + 14 eq	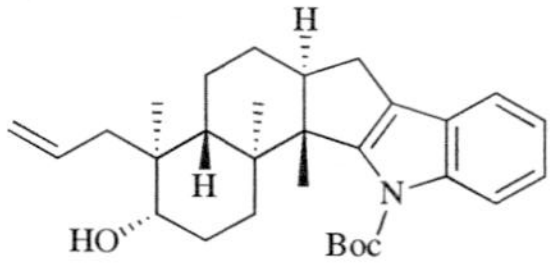 H, H, HO, N, Boc	Gr II (0.27 + 0.14 eq), DCM, reflux, 1.5 h	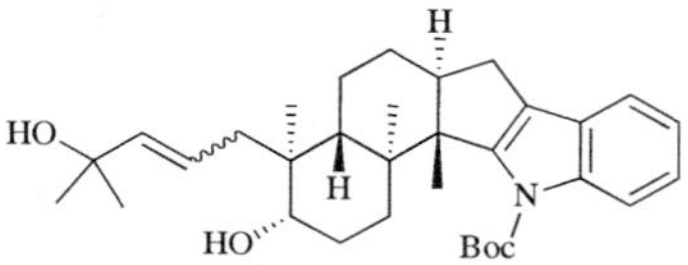 HO, H, H, HO, N, Boc (71)	906

	 TBSO; 2.2 eq	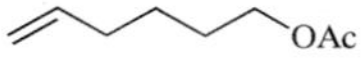 OAc	Gr II (0.06 eq), DCM, 40°, 12 h	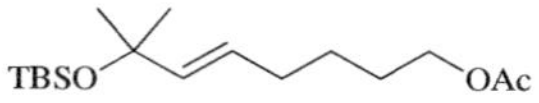 TBSO; OAc	(97) (*E*) only	36
	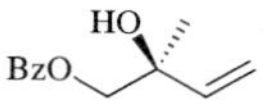 BzO; HO	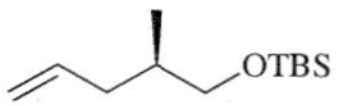 OTBS; 1.2 eq	Hov II (0.06 eq), $PhCF_3$, 85°, 2 h	 BzO; HO; OTBS	(62)	342
C_6						
	 HO; 2 eq	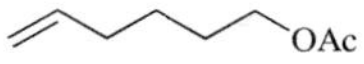 OAc	Gr II (0.048 eq), DCM, 40°, 12 h	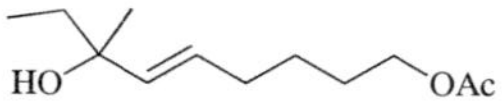 HO; OAc	(93) (*E*) only	36
	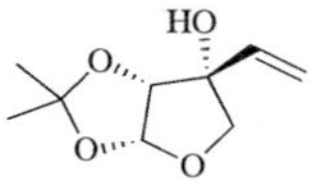 HO; O; O; O	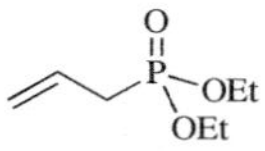 O; P; OEt; OEt; 3 eq	Gr II (0.05 eq), DCM, reflux, overnight	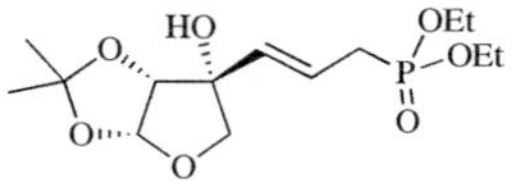 HO; O; O; O; OEt; P; OEt; O	(93) (*E*) only	907

TABLE 14. CROSS-METATHESIS OF TERTIARY ALLYLIC ALCOHOLS AND DERIVATIVES (*Continued*)

Tertiary Alcohol	Cross Partner	Conditions	Product(s) and Yield(s) (%)	Refs.
Please refer to the charts preceding the tables for the catalyst structures.				
C_7				
OH, OAc; *x* eq	OTBDPS	Hov II (0.1 eq), solvent	OH, OAc, OTBDPS + OH, AcO, HO, OAc; **I** (60), (*E*) only; **II**	908
C_{7-8}				
O, R, SO_2, Cl	TMS; *x* eq	Hov II (*y* eq), additive, DCE, 12 h	Cl, O_2, S, O, R, TMS; **I** + Cl, O_2, S, O, R, TMS; **II**	909

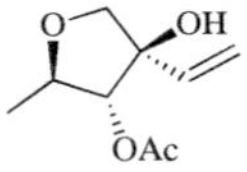

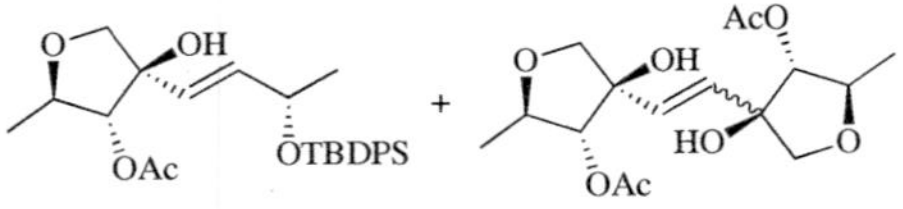

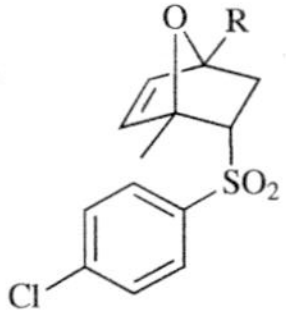

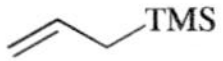

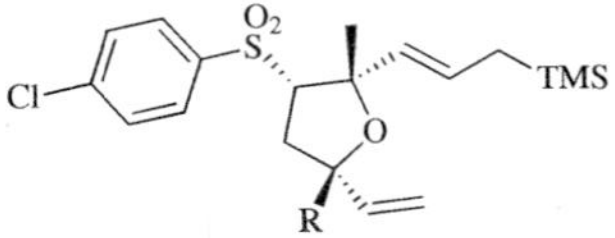

x	Solvent	Temp (°)	Time (h)	II
2	DCM	40	20	(10)
2	toluene	110	1	(10)
0.5	DCM	40	20	(5)
0.5	toluene	110	1	(5)

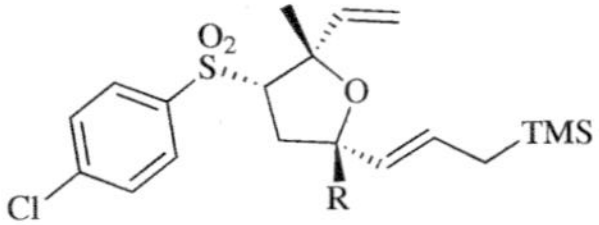

R	*x*	*y*	Additive	Temp (°)	I + II	I/II
H	10	0.01	—	70	(76)	92:8
Me	6	0.05	ethylene	0	(89)	**I** only
Me	10	0.025	—	25	(94)	**I** only

	Substrate	Partner	Conditions	Product(s) and Yield(s) (%)	Refs.
C_7					
	Cl, SO_2, O	O 6 eq	Hov II (0.01 eq), DCE, 84°	Cl, O_2, S, O, H, O (59)	909
C_9					
	OH, n-C_5H_{11} 5 eq	OBz, O, O	Gr II (0.1 eq), DCM, rt, 16 h	OH, OBz, n-C_5H_{11}, O, O (60) (E) only	910
C_{10}					
	OH, Ph	CO_2Me 10 eq	1. Ru cat **163** (0.03 eq), DCM, rt, hν (350 nm) 2. Phenanthrene (2.3 eq), DCM, rt, hν (254 nm)	Ph, O, O (52)	911
	OH, OTBS	TESO, $()_2$, OTBDPS 1.25 eq	Hov II (0.06 eq), $PhCF_3$, 85°, 2 h	OH, $()_3$, OTBS, TESO, $()_2$, OTBDPS (60)	342

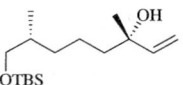

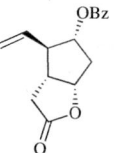

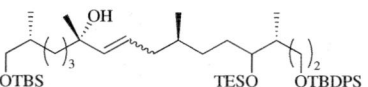

TABLE 14. CROSS-METATHESIS OF TERTIARY ALLYLIC ALCOHOLS AND DERIVATIVES (*Continued*)

Tertiary Alcohol	Cross Partner	Conditions	Product(s) and Yield(s) (%)	Refs.

Please refer to the charts preceding the tables for the catalyst structures.

C_{11}

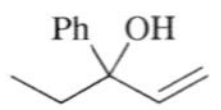

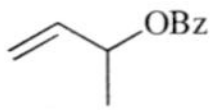

Tertiary Alcohol	Cross Partner	Conditions	Product(s) and Yield(s) (%)	Refs.
Ph OH	OBz 3 eq	Catalyst (0.05 eq), DCM, reflux, 6 h	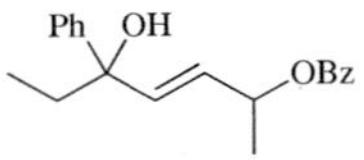 Ph OH, OBz (*E*)/(*Z*) = 60:40 Catalyst: Hov II (50); Ru cat **271** (66)	177
	BzO, OBz 3 eq	Catalyst (0.05 eq), DCM, reflux, 6 h	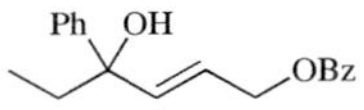 Ph OH, OBz (*E*)/(*Z*) > 95:5 Catalyst: Hov II (81); Ru cat **271** (98)	177
	 OAc 3 eq	Catalyst (x eq), solvent	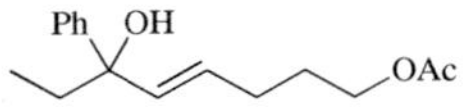 Ph OH, OAc	

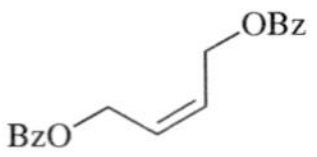

Catalyst	x	Solvent	Temp (°)	Time	Conv. (%)[a]		(*E*)/(*Z*)	Refs.
Hov II	0.05	DCM	40	6 h	—	(70)	60:40	177
Ru cat **271**	0.05	DCM	40	6 h	—	(89)	60:40	177
Ind II	3 x 0.02	$C_6F_5CF_3$	120	3 x 10 min	90	(—)	>95:5	40
Ind II	3 x 0.02	$C_6F_5CF_3$	120	3 x 10 min (MW)	95	(86)	>95:5	40

3 eq

Catalyst (x eq), solvent

Ph OH Ph

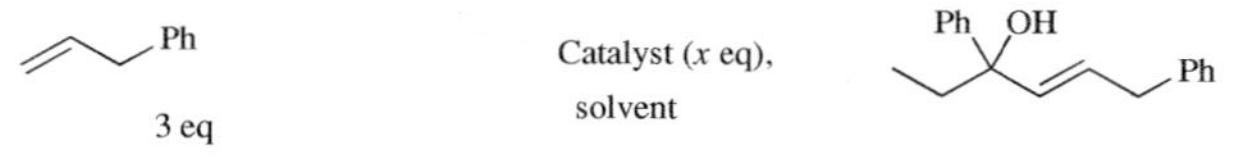

Catalyst	x	Solvent	Temp (°)	Time	Conv. (%)[a]		(E)/(Z)	
Hov II	0.05	DCM	40	6 h	—	(64)	60:40	177
Ru cat **271**	0.05	DCM	40	6 h	—	(91)	60:40	177
Ind II	3 x 0.02	$C_6F_5CF_3$	120	3 x 10 min	69	(—)	>95:5	40
Ind II	3 x 0.02	$C_6F_5CF_3$	120	3 x 10 min (MW)	91	(82)	>95:5	40

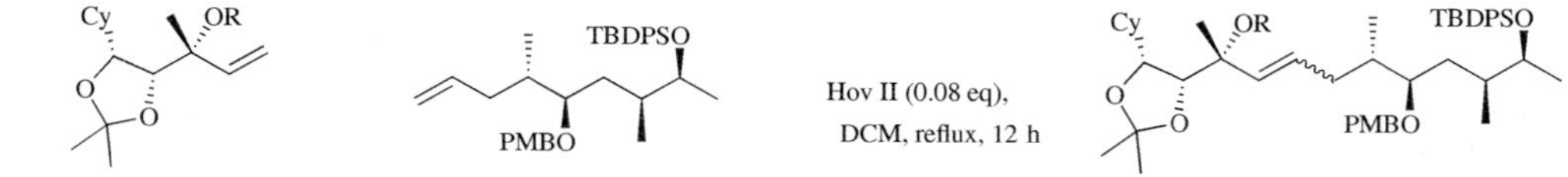

C_{12}

Cy OR O O

2 eq

TBDPSO PMBO

Hov II (0.08 eq), DCM, reflux, 12 h

Cy OR O O TBDPSO PMBO

R		
H	(38)	912
PMB	(0)	

OH O OH

2 eq

O O OH OH OAc

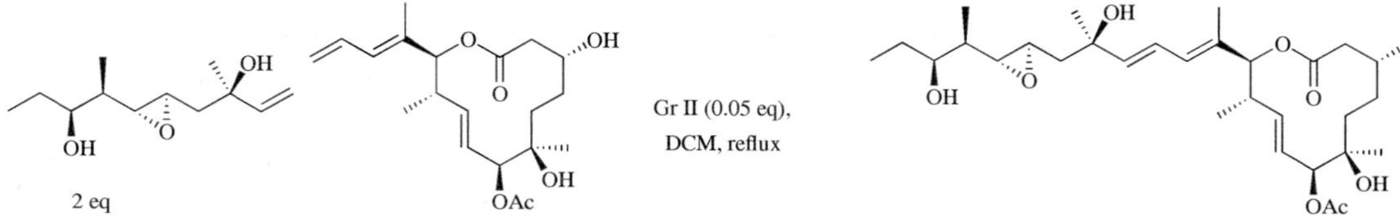

Gr II (0.05 eq), DCM, reflux

OH O OH O O OH OH OAc

(64) (E) only

913

TABLE 14. CROSS-METATHESIS OF TERTIARY ALLYLIC ALCOHOLS AND DERIVATIVES (*Continued*)

Tertiary Alcohol	Cross Partner	Conditions	Product(s) and Yield(s) (%)	Refs.
Please refer to the charts preceding the tables for the catalyst structures.				
C_{20}				
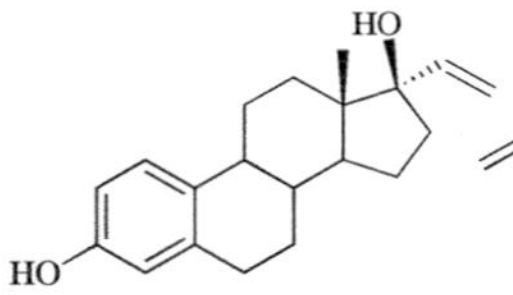 HO, HO; 1 eq	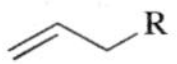 R	Hov II (0.1 eq), DCM, reflux, 12 h	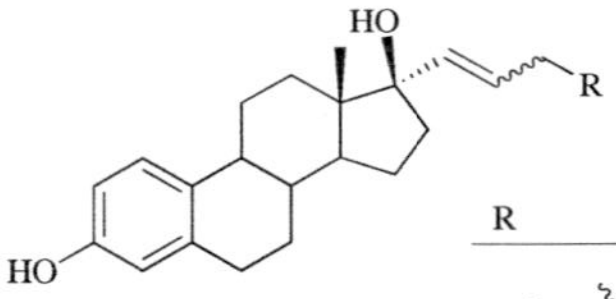 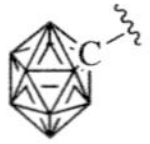  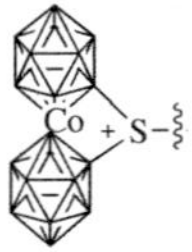 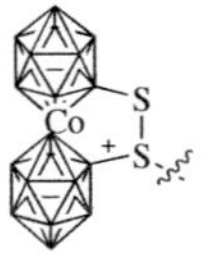 HO, HO, R; R: C (37); C (38); Co, + S (41); Co, S, + S (29)	523

[a] The conversion was determined by ^{1}H NMR analysis.

TABLE 15. CROSS-METATHESIS OF VINYL EPOXIDES

Please refer to the charts preceding the tables for the catalyst structures.

Vinyl Epoxide	Cross Partner	Conditions	Product(s) and Yield(s) (%)	Refs.
C_4				
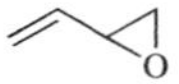	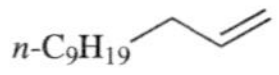 n-C_9H_{19} 4 eq	Ru cat **216** (0.02 eq), THF (0.5 M), 35°, 5 h	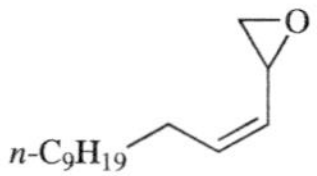 n-C_9H_{19} (40) $(Z)/(E)$ = 95:5	461
20 eq	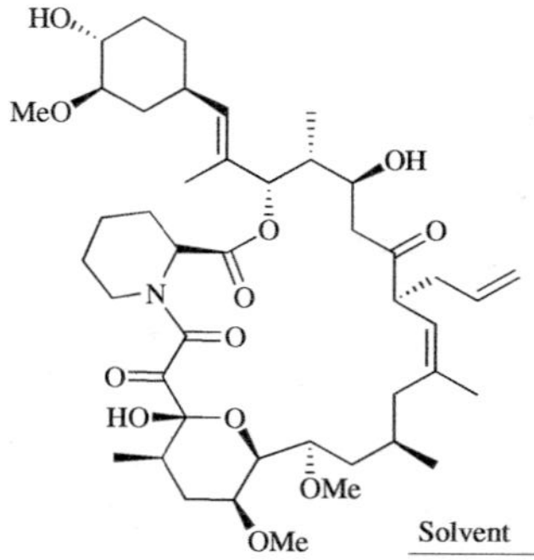 HO, MeO, OH, O, N, O, O, O, HO, O, OMe, OMe	Gr II (0.1 eq), solvent, 150° (MW), 5 min	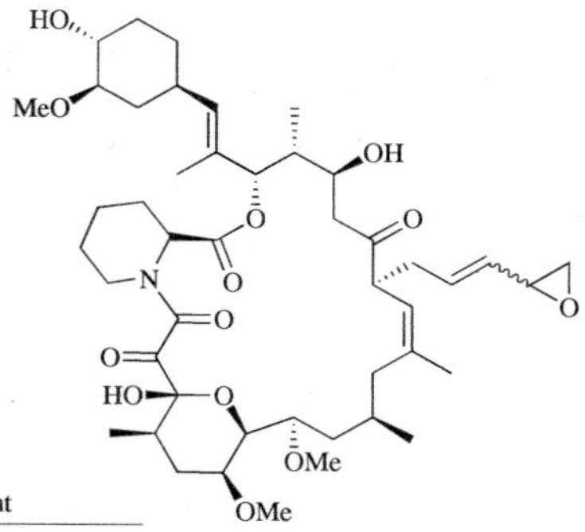 HO, MeO, OH, O, N, O, O, O, HO, O, OMe, OMe	432

Solvent		Solvent	
1,2-DCB	(64)	*i*-PrOH	(42)
DCE	(61)	butan-1-ol	(18)
DCM	(tr)	EtOH	(5)
DMF	(20)	MeOH	(4)
MeCN	(3)	DMSO	(2)

TABLE 15. CROSS-METATHESIS OF VINYL EPOXIDES (*Continued*)

Vinyl Epoxide	Cross Partner	Conditions	Product(s) and Yield(s) (%)	Refs.
Please refer to the charts preceding the tables for the catalyst structures.				
C_6				
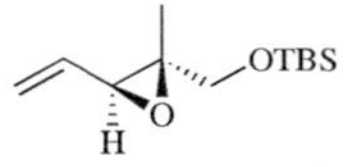 OTBS, H, O; 3 eq	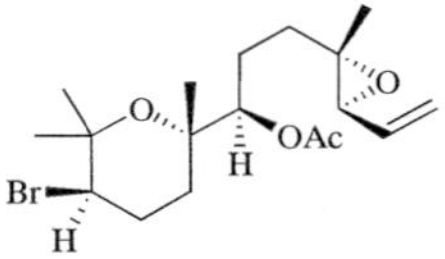 Br, H, O, H, OAc, O	Gr II (0.1 eq), DCM, 30°, 1 d	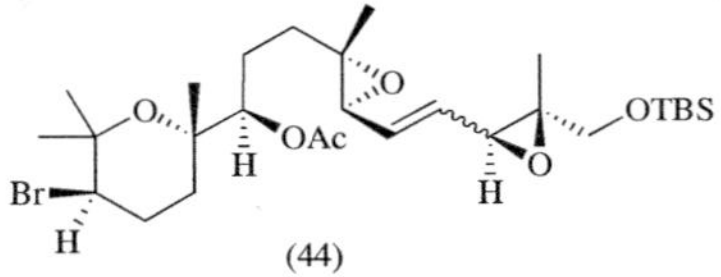 Br, H, O, H, OAc, O, OTBS, H, O (44)	914
C_7				
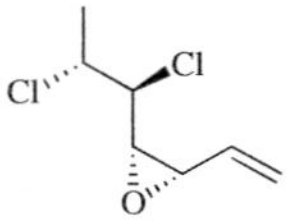 Cl, Cl, O	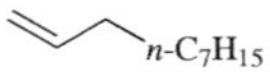 n-C_7H_{15}; 3 eq	Ru cat **215** (0.3 eq), DCE/DCM, 35°	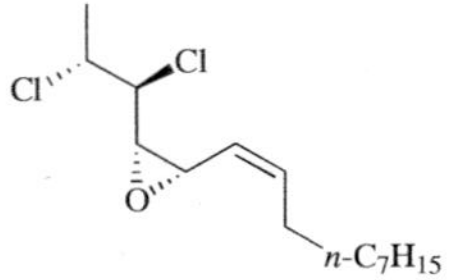 Cl, Cl, O, n-C_7H_{15} (83) $(Z)/(E) > 95:5$	191
	 Cl; x eq	Ru cat **215** (y eq), solvent, 35°	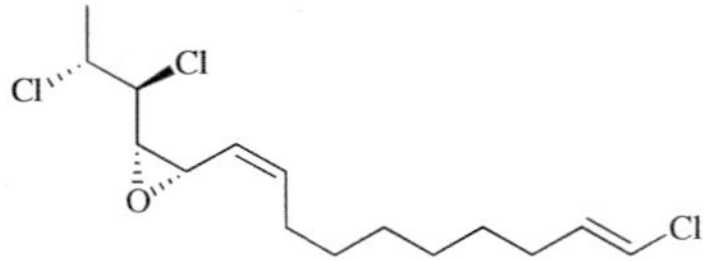 Cl, Cl, O, Cl	915, 191

x	y	Solvent	Time (h)		$(Z)/(E)$
3	0.1	DCE/DCM	2	(10)	95:5
3	0.3	DCE/DCM	3	(32)	95:5
3	0.3	$PhCF_3$/DCM	3	(33)	95:5
3	0.5	$PhCF_3$/DCM	4	(34)	95:5
5	1	DCE	3	(39)	95:5

C_9

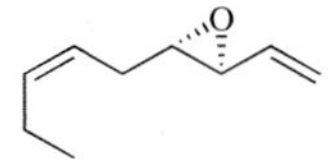

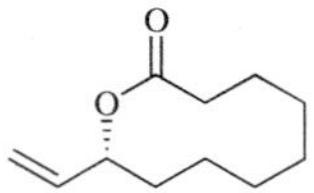

Gr II, DCM, 25°, 2 h

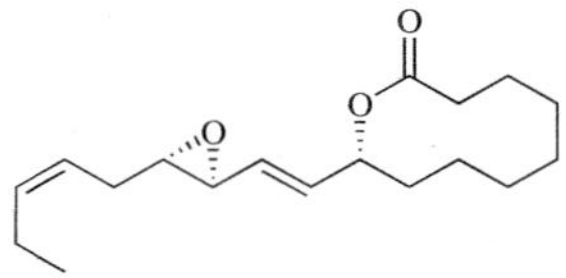

(40) (*E*) only 190

C_{11}

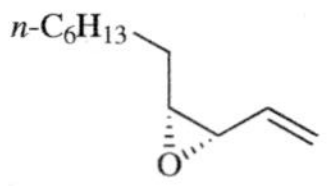

n-C_6H_{13}

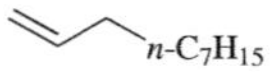

n-C_7H_{15}

5 eq

Ru cat **215** (*x* eq), DCE, 35°, 3 h

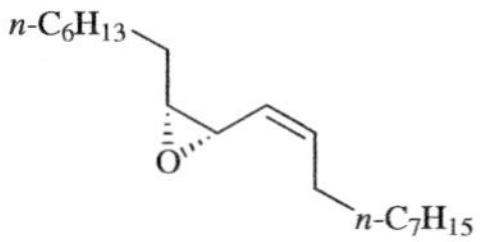

n-C_6H_{13} n-C_7H_{15}

(*Z*) only

x	
0.1	(83)
0.01	(43)

915

C_{12}

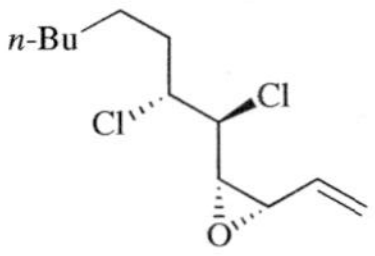

n-Bu Cl Cl

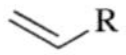

R

10 eq

Catalyst (0.1 eq), solvent, rt

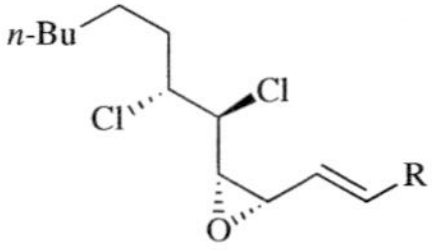

n-Bu Cl Cl R

915

R	Catalyst	Solvent	Time (h)		(*E*)/(*Z*)
n-Bu	Hov II	toluene	19	(—)	—
n-C_8H_{17}	Gr II	DCM	6	(57)	82:18

TABLE 15. CROSS-METATHESIS OF VINYL EPOXIDES (*Continued*)

Vinyl Epoxide	Cross Partner	Conditions	Product(s) and Yield(s) (%)	Refs.

Please refer to the charts preceding the tables for the catalyst structures.

C_{12}

Vinyl Epoxide	Cross Partner	Conditions	Product(s) and Yield(s) (%)	Refs.
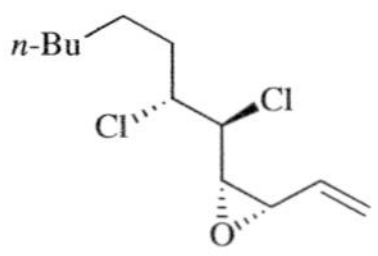 n-Bu, Cl, Cl, O	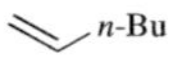 n-Bu 10 eq	Ru cat **215** (*x* eq), solvent, 35°, 1 h	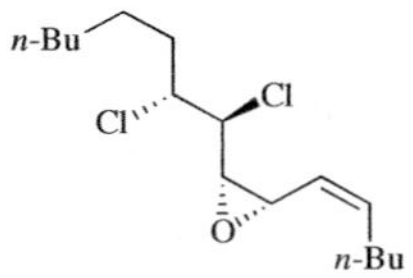 n-Bu, Cl, Cl, O, n-Bu (*Z*) only	915
	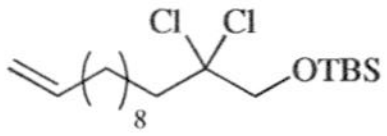 Cl, Cl, OTBS, 8 2 eq	Ru cat **215** (*x* eq), additive, DCE, 35°	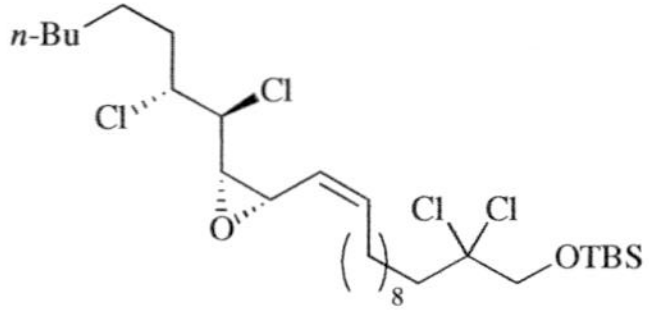 n-Bu, Cl, Cl, O, Cl, Cl, OTBS, 8 (*Z*) only	915
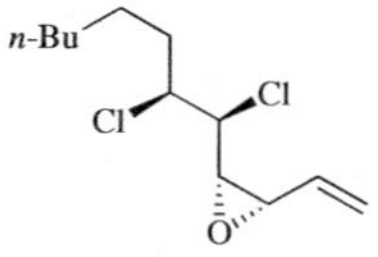 n-Bu, Cl, Cl, O	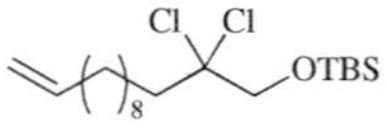 Cl, Cl, OTBS, 8 3 eq	Ru cat **215** (0.3 eq), DCE, DCM, 35°, 3 h	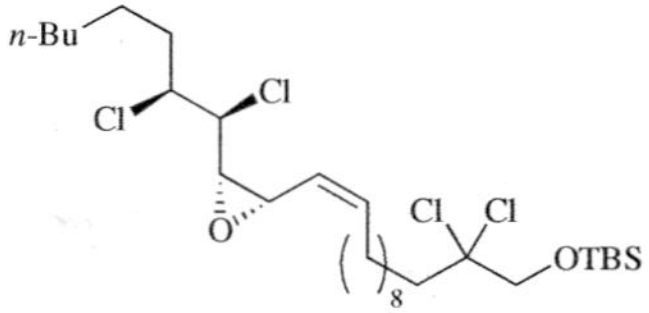 n-Bu, Cl, Cl, O, Cl, Cl, OTBS, 8 (19) (*Z*)/(*E*) > 95:5	915

x	Solvent	Conv. (%)[a]
0.01	THF	12
0.1	DCE	50–60

x	Additive	Time (h)	
0.1	—	1	(19)
0.3	—	2	(29)
0.2	static vacuum, 4 Å MS	0.5	(16)

Carbons	Reactants	Conditions	Product(s) and Yield(s) (%)	Refs.
C_{16}	4.7 eq	Gr II (0.1 eq), DCM, 40°, 3 h	(87) (*E*) only	916
C_{18}	TBDPSO, OTBS; OTBS, OH, 5 eq	Gr II (0.05 eq), toluene, 50°, 3 h	**I** + **II**; **I + II** (49)	917

[a] The conversion was determined by ^{1}H NMR analysis.

TABLE 16. CROSS-METATHESIS OF α,β-UNSATURATED ALDEHYDES

Aldehyde	Cross Partner	Conditions	Product(s) and Yield(s) (%)	Refs.

Please refer to the charts preceding the tables for the catalyst structures.

C_3

Aldehyde	Cross Partner	Conditions	Product(s) and Yield(s) (%)	Refs.
OHC 5 eq	NHTs	Hov II (0.05 eq), $(PhO)_3B$ (0.1 eq), toluene, 80°, 0.5 h	N Ts (90)	899
x eq	OR	Catalyst (*y* eq), DCM, rt	OHC OR	

x	R	Catalyst	*y*	Time (h)		Refs.
4	H	Gr II	0.0078	24	(100)	918
3	TBDPS	Hov II	0.05	12	(85)	919

Aldehyde	Cross Partner	Conditions	Product(s) and Yield(s) (%)	Refs.
3 eq	Y NHBoc *n*	Hov II (0.05 eq), DCM, rt, 15 h	OHC Y NHBoc *n*	920

n	Y	
1	O	(56)
1	CbzN	(43)
1	S	(87)
2	O	(84)
2	CbzN	(44)
2	S	(60)

Aldehyde	Cross Partner	Conditions	Product(s) and Yield(s) (%)	Refs.
3 eq	OH	Hov II (0.05 eq), DCM, rt, 2 h	OHC OH (85)	921
3 eq	NHR	Hov II (0.05 eq), DCM, rt	OHC NHR	920

R	Time (h)	
Boc	7	(81)
Cbz	15	(83)

x eq

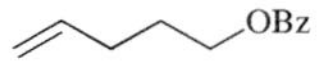

OBz

Catalyst (0.05 eq), solvent

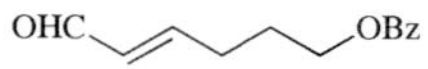

OHC OBz

x	Catalyst	Solvent	Temp (°)	Time (h)		(*E*)/(*Z*)	
2	Hov II	DCM	40	—	(93)	50:50	922
2	Ru cat **191**	DCM	45	12	(86)	50:50	923
2	Ru cat **243**	DCM	45	12	(0)	—	923
—	Ru cat **185**	BMI•PF_6/toluene	rt	6	(92)	>95:5	924

3 eq

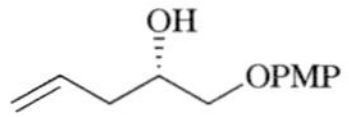

OH OPMP

Hov II (0.025 eq), DCM, 25°, 36 h

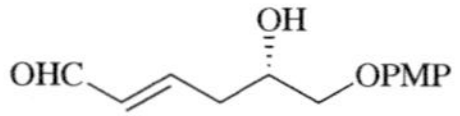

OHC OH OPMP

(41) (*E*)/(*Z*) > 95:5

157

3 eq

OH OPMP

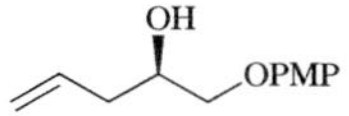

Hov II (0.025 eq), DCM, rt, 36 h

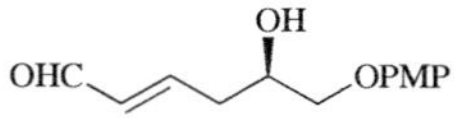

OHC OH OPMP

(75) (*E*)/(*Z*) > 98:2

626

5 eq

OH OBn

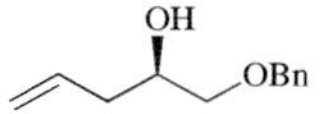

1. Hov II (0.015 eq), DCM, reflux, 4 h
2. 2,6-Lutidine (2.5 eq), TESOTf (1.95 eq), –78°, 0.17 h

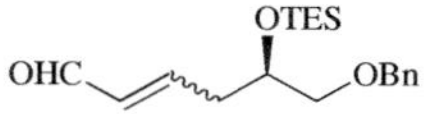

OHC OTES OBn (98)

925

TABLE 16. CROSS-METATHESIS OF α,β-UNSATURATED ALDEHYDES (*Continued*)

Aldehyde	Cross Partner	Conditions	Product(s) and Yield(s) (%)	Refs.

Please refer to the charts preceding the tables for the catalyst structures.

C_3

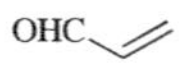

Aldehyde	Cross Partner	Conditions	Product(s) and Yield(s) (%)	Refs.
OHC (3 eq)	OTES, OTBDPS	Hov II (0.05 eq), DCM, 35°, 5 h	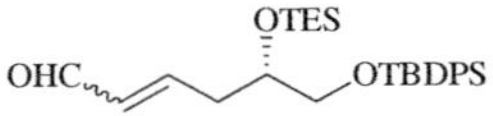 OHC, OTES, OTBDPS (72)	847
3 eq		Ru cat **93** (4 x 0.01 eq), $CDCl_3$, 100° (MW), 2.5 h	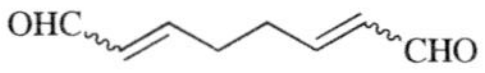 OHC, CHO (50)	926
x eq	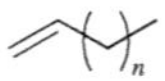 *n*	Catalyst (*y* eq), solvent, rt, 8 h	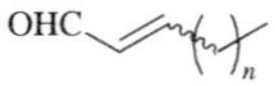 OHC, *n*	927

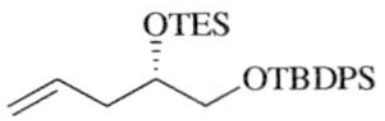

x	*n*	Catalyst	*y*	Solvent	
1	3	HovII	0.2	DCM	(60)
1	5	Hov II	0.2	DCM	(75)
1	7	Gr I	0.1	DCM	(3)
1	7	Gr II	0.1	DCM	(48)
1	7	Hov II	0.1	DCM	(70)
1	7	Ru cat **126**	0.1	DCM	(59)
1	7	HovII	0.14	DCM	(80)
1	7	Hov II	0.07	DCM	(71)
1	7	Hov II	0.035	DCM	(69)
1	7	Hov II	0.014	DCM	(68)
1	7	Hov II	0.01	DCM	(63)
1	7	Hov II	0.0035	DCM	(39)

x	*n*	Catalyst	*y*	Solvent	
1	7	Hov II	0.035	DCE	(67)
1	7	Hov II	0.035	EtOAc	(56)
1	7	Hov II	0.035	pentane	(40)
1	7	Hov II	0.035	THF	(37)
1	7	Hov II	0.035	MeOH	(5)
1	7	Hov II	0.035	*i*-PrOH	(12)
2	7	Hov II	0.035	DCM	(72)
5	7	Hov II	0.035	DCM	(74)
1	7	Hov II	0.035	EtOAc	(60)
1	7	Hov II	0.035	EtOAc	(49)
1	7	Hov II	0.035	EtOAc	(42)

x eq R (alkene) → Catalyst (y eq), HCl/ether (z eq), solvent → OHC R

x	R	Catalyst	y	z	Solvent	Temp (°)	Time (h)		$(E)/(Z)$	
3	BocHN	Hov II	0.05	—	DCM	rt	7	(60)	—	920
3	CbzHN	Hov II	0.05	—	DCM	rt	14	(70)	—	920
3	AcO	Hov II	0.025	—	DCM	rt	24	(80)	>98:2	626
2.6	AcO	Gr II	0.05	0.25	benzene/THF	45	14	(52)	>95:5	928
2	TBSO	Ru cat **286**	0.02	—	DCM	40	8	(48)	>95:5	161
2	TBSO	Ru cat **41**	0.01	—	DCM	rt	5	(12)	>95:5	562
2	TBSO	Ru cat **44**	0.01	—	DCM	rt	5	(31)	>95:5	562
2	TBSO	Ru cat **45**	0.01	—	DCM	rt	5	(25)	>95:5	562
2	TBSO	Ru cat **316**	0.02	—	toluene	80	2	(57)	>95:5	591

3 eq OR (diene) → Hov II (0.05 eq), DCM → OHC OR

R	Temp (°)	Time (h)		$(E)/(Z)$	
Ac	rt	12	(73)	>95:5	157
TBDPS	40	—	(56)	95:5	169
TBS	40	—	(52)	95:5	169

TABLE 16. CROSS-METATHESIS OF α,β-UNSATURATED ALDEHYDES (*Continued*)

Aldehyde	Cross Partner	Conditions	Product(s) and Yield(s) (%)	Refs.
Please refer to the charts preceding the tables for the catalyst structures.				
C_3				
 OHC 3 eq	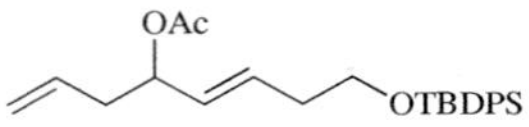 OAc, OTBDPS	Hov II (0.05 eq), DCM, rt, 12 h	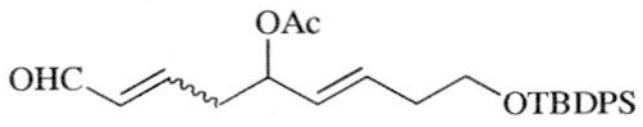 OHC, OAc, OTBDPS (62)	919
5 eq	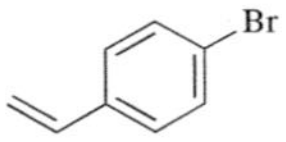 Br	Ru cat **88** (5 x 0.024 eq), $C_6F_5CF_3$, 120° (MW), 50 min	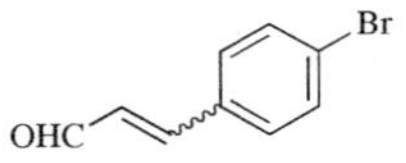 OHC, Br (65)	929
3 eq	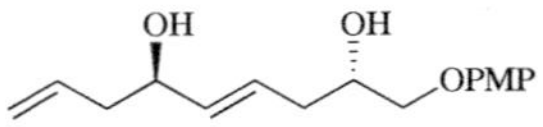 OH, OH, OPMP	Hov II (0.05 eq), DCM, 25°, 12 h	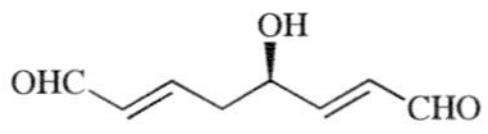 OHC, OH, CHO (43) (*E*)/(*Z*) > 95:5 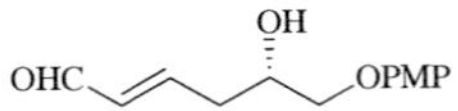 + OHC, OH, OPMP (31) (*E*)/(*Z*) > 95:5	157
3 eq	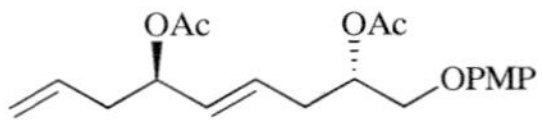 OAc, OAc, OPMP	Hov II (0.05 eq), DCM, 25°, 12 h	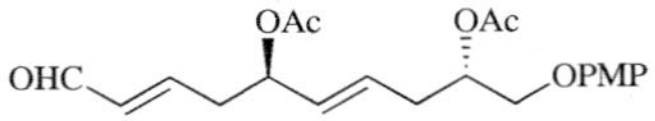 OHC, OAc, OAc, OPMP (63) (*E*)/(*Z*) = >95:5	157

2 eq

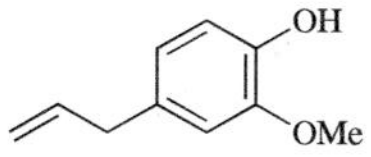

Catalyst (*x* eq), solvent

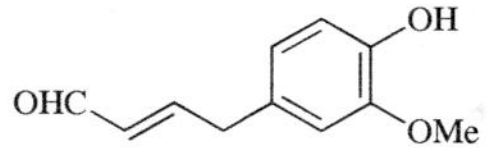

OHC / OH / OMe

Catalyst	*x*	Solvent	Temp (°)	Time (h)		(*E*)/(*Z*)	
Ru cat **286**	0.05	DCM	40	8	(37)	>91:9	161
Ru cat **316**	0.02	toluene	80	5	(62)	86:14	591

2 eq

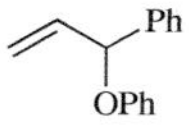

Hov II (0.05 eq), toluene, 80°, 6 h

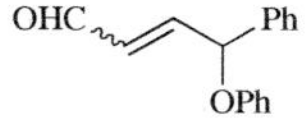

OHC / Ph / OPh (63) 930

3 eq

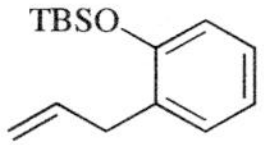

Gr II (0.02 eq), $KHSO_4$ (*x* eq) solvent, 22°, 8 h,

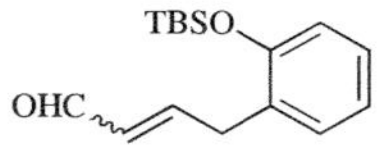

TBSO / OHC 931

Solvent	*x*	
DCM	—	(9)
PTS (2.5% aq.)	—	(43)
PTS (2.5% aq.)	0.03	(70)

3 eq

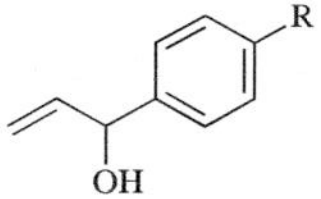

Hov II (0.05 eq), DCM, rt

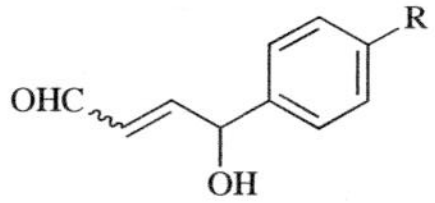

OHC / R / OH 919

R	Time (h)	
H	16	(90)
MeO	12	(90)

TABLE 16. CROSS-METATHESIS OF α,β-UNSATURATED ALDEHYDES (*Continued*)

Aldehyde	Cross Partner	Conditions	Product(s) and Yield(s) (%)	Refs.
Please refer to the charts preceding the tables for the catalyst structures.				
C_3				
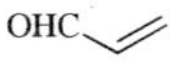 OHC 3 eq	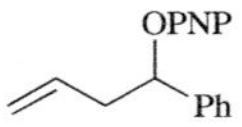 OPNP Ph	Hov II (0.02 eq), DCM, 40°, 15 h	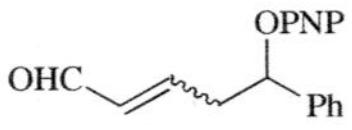 OHC OPNP Ph (55)	932
6 eq	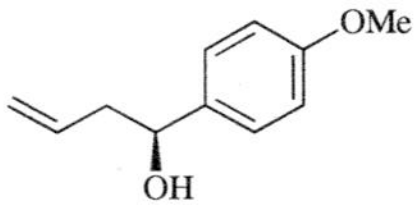 OMe OH	Hov I (0.05 eq), DCM, rt, 2 h	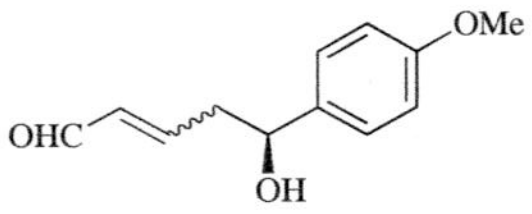 OHC OMe OH (84)	933
6 eq	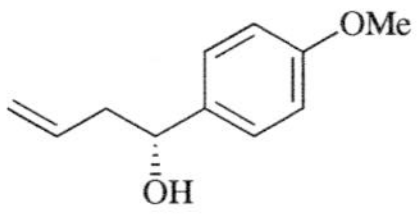 OMe OH	Hov I (0.05 eq), DCM, rt, 2 h	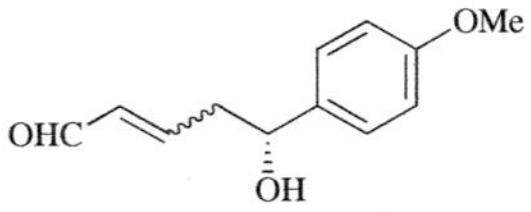 OHC OMe OH (86)	933
$C_{3\text{–}4}$				
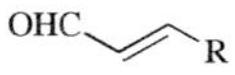 OHC R *x* eq	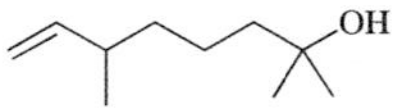 OH	Ru cat **170** (*y* eq), EtOAc, 60°	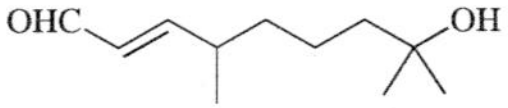 OHC OH	751
C_3				
OHC 5 eq	Ph NHCbz	Hov II (0.1 eq), DCM, 40°, 48 h	OHC Ph NHCbz (56)	934

R	*x*	*y*	Time (h)		(*E*)/(*Z*)
H	1	0.005	16	(80)	94:6
Me	1.2	0.01	23	(53)	95:5

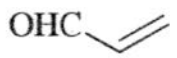

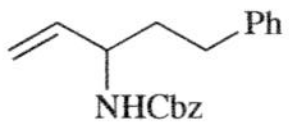

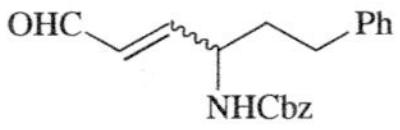

	Reagent	Substrate	Conditions	Product(s) and Yield(s) (%)	Refs.
	5 eq	(O)	Ru cat **190** (2 x 0.025 eq), DCM, rt, 120 h	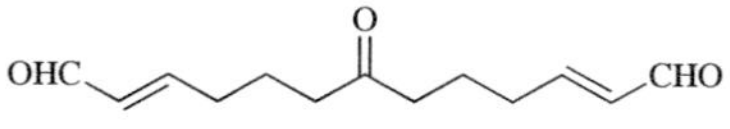 OHC, (O), CHO (63) (*E*) only	231
C_{3-4}	OHC, R 1 eq	(O)	Hov II (0.02 eq), DCM, rt, 8 h	OHC, (O) R: H (33); Me (15)	927
C_3	OHC 3 eq	(OH, O, Cl, Cl, Cl)	Hov II (0.025 eq), DCM, rt, 36 h	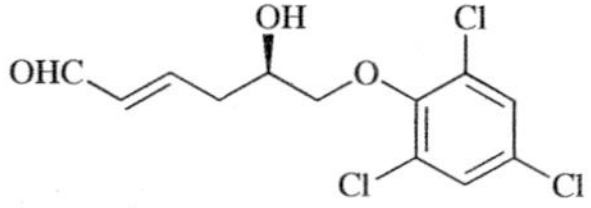 OHC, (OH, O, Cl, Cl, Cl) (90) (*E*)/(*Z*) > 98:2	626
	2 eq	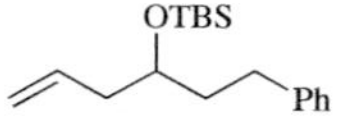 (OTBS, Ph)	Catalyst (0.1 eq), DCM, reflux	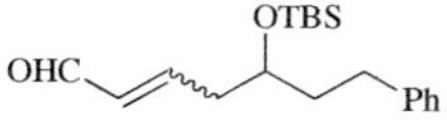 OHC, (OTBS, Ph) Catalyst / Conv. (%): Gr I — (0); Gr II 54 (—); Hov II — (75)	541

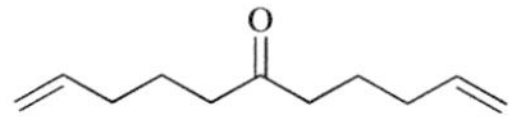

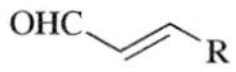

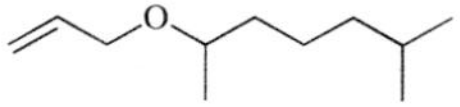

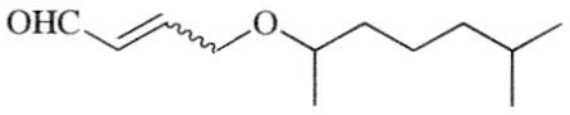

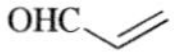

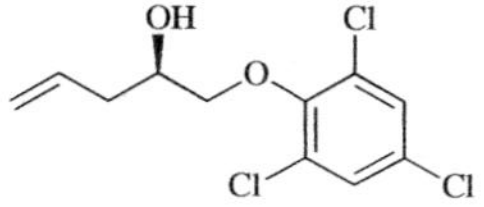

TABLE 16. CROSS-METATHESIS OF α,β-UNSATURATED ALDEHYDES (*Continued*)

Aldehyde	Cross Partner	Conditions	Product(s) and Yield(s) (%)	Refs.
Please refer to the charts preceding the tables for the catalyst structures.				
C_3				
OHC (acrolein), 3 eq	OR, Ph	Hov II (0.05 eq), DCM, 25°, 12 h	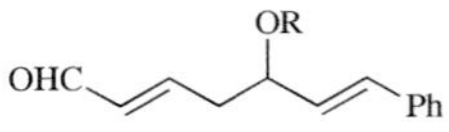 OHC, OR, Ph; **I** (*E*)/(*Z*) > 95:5 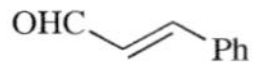 + OHC, Ph **II** R | **I** | **II** | (*E*)/(*Z*) **II** H | (41) | (37) | >95:5 Ac | (71) | (—) | —	157
3 eq	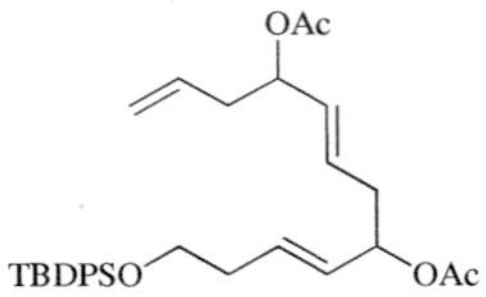 OAc, TBDPSO, OAc	Hov II (0.05 eq), DCM, rt, 12 h	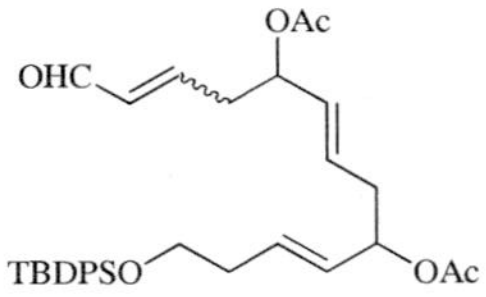 OHC, OAc, TBDPSO, OAc (80)	919
4 eq	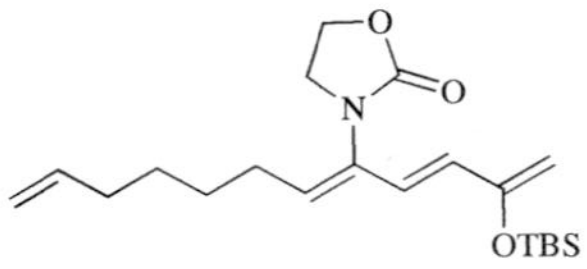 O, O, N, OTBS	Hov II (0.1 eq), DCM, 23°, 10 min	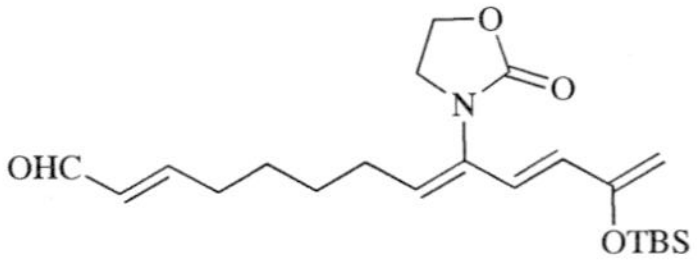 OHC, O, O, N, OTBS (85) (*E*)/(*Z*) > 95:5	938
3 eq	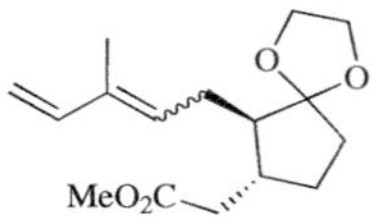 O, O, MeO_2C	Hov II (0.05 eq), DCM, 45°, 3 h	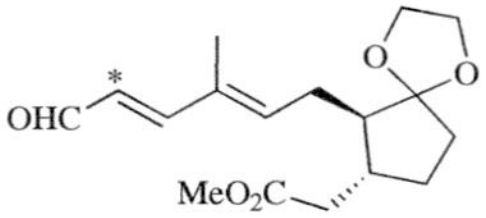 OHC, *, O, O, MeO_2C (23) *(*E*)/(*Z*) = 80:20	939

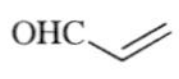

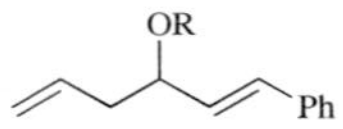

Reactant		Conditions	Product(s) and Yield(s) (%)	Refs.
2 eq	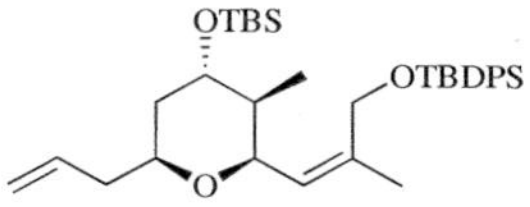 OTBS, OTBDPS	Gr II (0.05 eq), DCM, 50°, 16 h	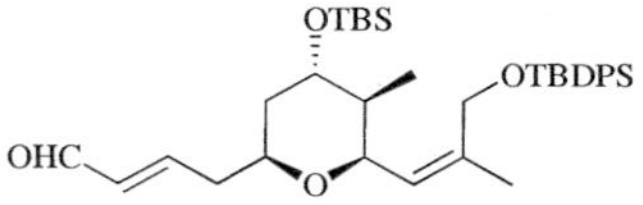 OHC, OTBS, OTBDPS (72) $(E)/(Z) = 94:6$	935
3 eq	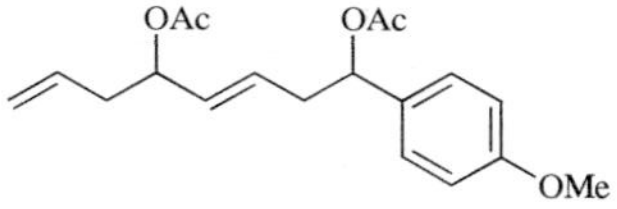 OAc, OAc, OMe	Hov II (0.05 eq), DCM, rt, 16 h	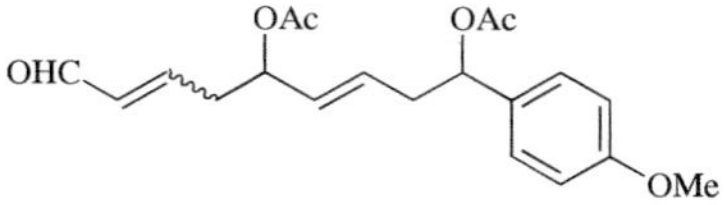 OHC, OAc, OAc, OMe (61)	919
3 eq	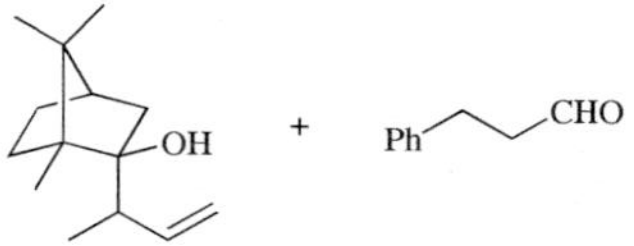 OH (3 eq) + Ph–CHO	1. CSA (0.1 eq), DCM, 25°, 144 h 2. Gr II (0.2 eq), DCM, 4 h	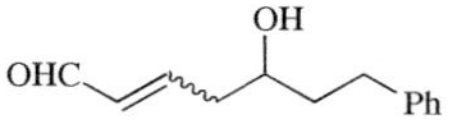 OHC, OH, Ph (46) er 97.0:3.0	936
2 x 3 eq	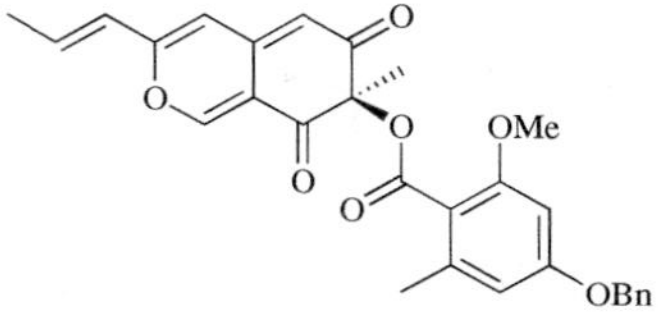 O, O, O, OMe, OBn	Hov II (2 x 0.1 eq), DCM, reflux, 2 x 12 h	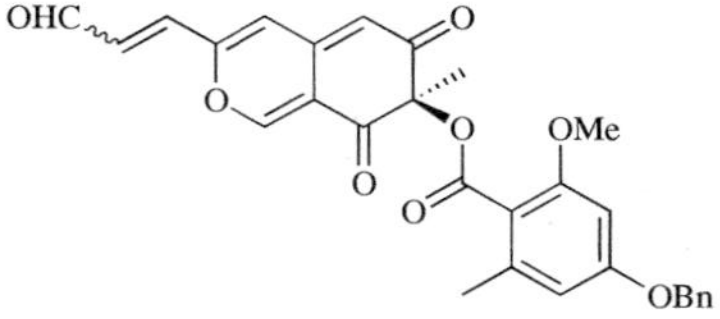 OHC, O, O, O, OMe, OBn (76)	937

TABLE 16. CROSS-METATHESIS OF α,β-UNSATURATED ALDEHYDES (*Continued*)

Aldehyde	Cross Partner	Conditions	Product(s) and Yield(s) (%)	Refs.
Please refer to the charts preceding the tables for the catalyst structures.				
C_3				
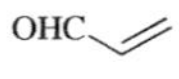 OHC (5 eq)	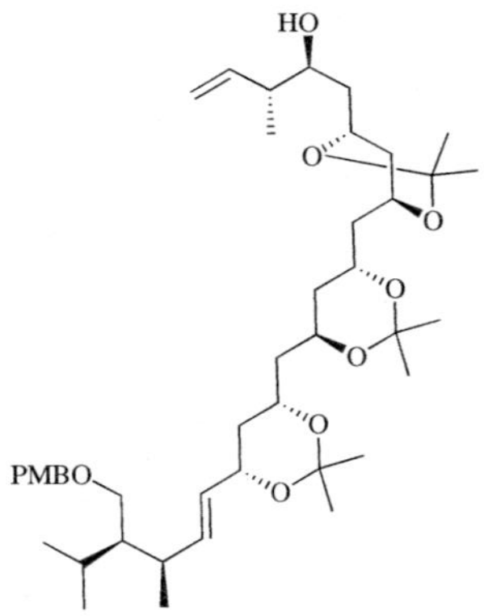 HO, O, O, O, O, O, O, PMBO	Hov II (0.075 eq), DCE, 60°, 24 h	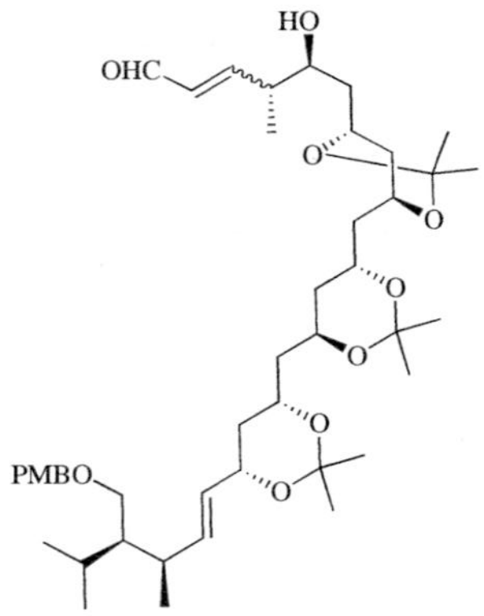 OHC, HO, O, O, O, O, O, O, PMBO (52)	846

C_4

OHC (crotonaldehyde), 5 eq + NHR (allylamine) → Catalyst (x eq), additive (y eq), solvent → N–R pyrrole

R	Catalyst	x	Additive	y	Solvent	Temp (°)	Time (h)	
Ts	Gr II	0.1	—	—	DCM	40	48	(5)
Ts	Ru cat **88**	0.05	—	—	DCM	40	36	(32)
Ts	Hov II	0.05	—	—	DCM	40	36	(60)
Ts	Gr I	0.1	—	—	toluene	80	48	(5)
Ts	Gr II	0.1	—	—	toluene	80	48	(8)
Ts	Ind I	0.1	—	—	toluene	80	48	(6)
Ts	Ind II	0.1	—	—	toluene	80	48	(16)
Ts	Ru cat **88**	0.05	—	—	toluene	80	36	(35)
Ts	Hov II	0.05	—	—	toluene	80	36	(60)
Ts	Hov II	0.05	LiCl	1	toluene	80	2	(5)
Ts	Hov II	0.05	$AlCl_3$	1	toluene	80	2	(5)
Ts	Hov II	0.05	$Ti(O\textit{i}\text{-}Pr)_4$	1	toluene	80	2	(20)
Ts	Hov II	0.05	$Zn(OTf)_2$	1	toluene	80	1	(60)
Ts	Hov II	0.05	$RuCl_3{\bullet}(H_2O)_n$	1	toluene	80	0.75	(60)
Ts	Hov II	0.05	ClBcat	1	toluene	80	1.5	(60)
Ts	Hov II	0.05	$PhBCl_2$	1	toluene	80	1.5	(70)
Ts	Hov II	0.05	$(PhO)_3B$	1	toluene	80	0.5	(93)
Ts	Hov II	0.05	$(PhO)_3B$	0.5	toluene	80	0.5	(93)
Ts	Hov II	0.05	$(PhO)_3B$	0.2	toluene	80	0.5	(92)
Ts	Hov II	0.05	$(PhO)_3B$	0.1	toluene	80	0.5	(92)
Boc	Hov II	0.05	$(PhO)_3B$	0.1	toluene	80	0.5	(60)

899

TABLE 16. CROSS-METATHESIS OF α,β-UNSATURATED ALDEHYDES (*Continued*)

Aldehyde	Cross Partner	Conditions	Product(s) and Yield(s) (%)	Refs.

Please refer to the charts preceding the tables for the catalyst structures.

C_4

Aldehyde	Cross Partner	Conditions	Product(s) and Yield(s) (%)	Refs.
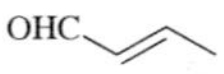 OHC (crotonaldehyde), 5 eq	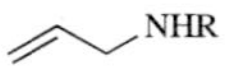 allyl-NHR	Hov II (0.05 eq), $(PhO)_3B$ (0.1 eq), solvent, 80°	 N-R pyrrole	899

R	Solvent	Time (h)	Yield
$MeOC_6H_4O_2S$	toluene	0.5	(91)
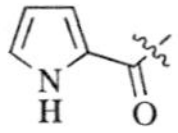 pyrrole-2-carbonyl (N-H)	toluene	1.25	(60)
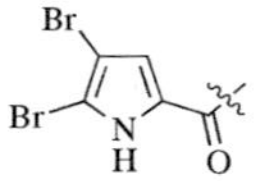 4,5-dibromopyrrole-2-carbonyl (Br, Br, N-H)	toluene	1.25	(60)
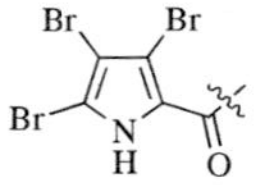 3,4,5-tribromopyrrole-2-carbonyl (Br, Br, Br, N-H)	toluene	1.5	(43)
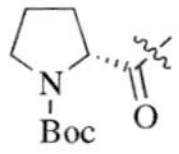 N-Boc-prolyl (Boc)	toluene	0.85	(57)
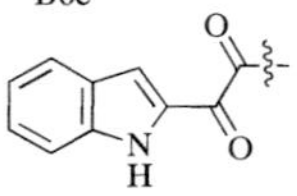 2-(indol-2-yl)-2-oxoacetyl (N-H)	DCE	1.5	(15)

	Reactants	Conditions	Product(s) and Yield(s) (%)	Refs.
3 eq	OBz + Ph_3P=CO$_2$Me **I**	1. Gr II (0.05 eq), DCM, 40°, 3 h 2. **I** (1.2 eq), DCM, rt to 40°, 15 + 1 h	MeO$_2$C ... * ... OBz (60) *(*E*)/(*Z*) = 89:11	940
3 eq	OR + **I** (O, O, N, O, Bn)	1. Hov II (0.05 eq), DCM, 40°, 12 h 2. **I** (1 eq), Bu_2OTf (1.1 eq), Et_3N (1.4 eq), DCM, –78 to 0°, 2 h	O, O, OH, N, O, OR, Bn R / dr: TBS (48) >95:5; PMB (64) >95:5	940
3 eq	OBn + **I** (O, O, N, O, Bn)	1. Hov II (0.05 eq), DCM, 40°, 12 h 2. **I** (1 eq), Bu_2OTf (1.1 eq), Et_3N (1.4 eq), DCM, –78 to 0°, 2 h	O, O, OH, N, O, OBn, Bn (52) dr > 95:5	940

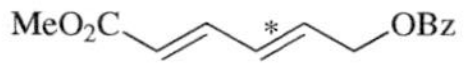

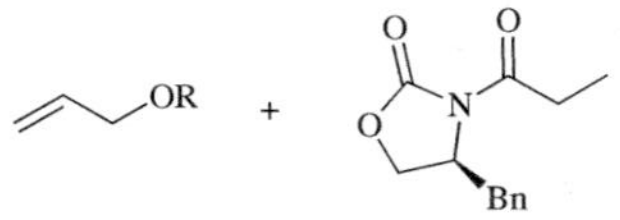

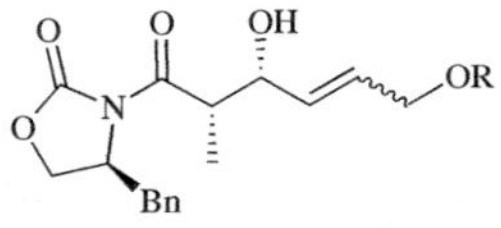

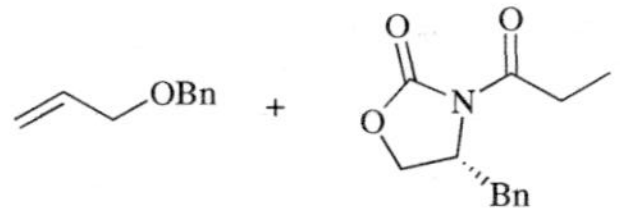

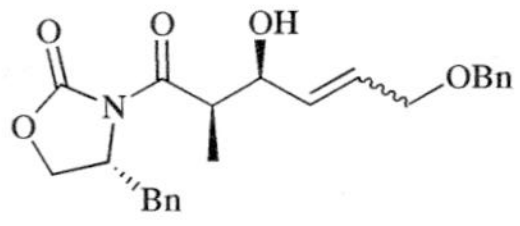

R		dr
TBS	(48)	>95:5
PMB	(64)	>95:5

TABLE 16. CROSS-METATHESIS OF α,β-UNSATURATED ALDEHYDES (*Continued*)

Aldehyde	Cross Partner	Conditions	Product(s) and Yield(s) (%)	Refs.
Please refer to the charts preceding the tables for the catalyst structures.				
C_4				
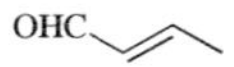 OHC (3 eq)	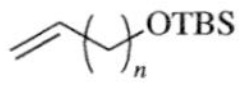 OTBS	1. Gr II (0.05 eq), DCM, reflux, 3 h 2. DIBAL-H, DCM, –78°, 2 h	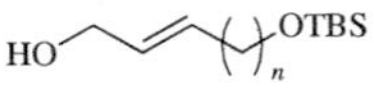 HO, OTBS; (*E*)/(*Z*) > 95:5 *n* = 2 (72); *n* = 3 (74)	941
3 eq	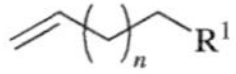 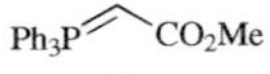 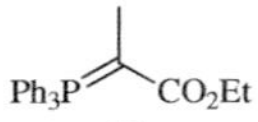 R^1 + Ph_3P=CO$_2$Me (I); Ph_3P=CO$_2$Et (II)	1. Gr II (0.05 eq), DCM, 40°, 3 h 2. Additive (1.2 eq), DCM, rt to 40°, 15 + 1 h	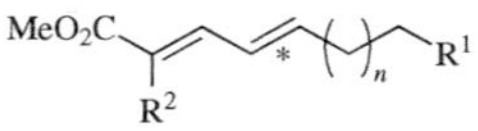 MeO_2C, R^2, *, R^1	940
3 eq	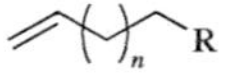 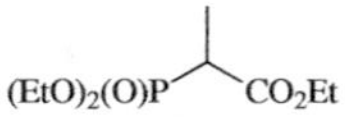 R + $(EtO)_2(O)P$ CO$_2$Et (I)	1. Gr II (0.05 eq), DCM, 40°, 3 h 2. I (1.4 eq), NaH (1.4 eq), THF, 0° to rt, 0.5 h + 15 h	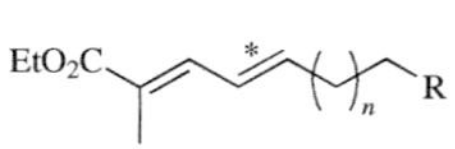 EtO_2C, *, R	940
10 eq	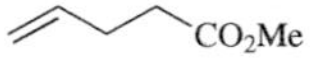 CO$_2$Me	Hov II (0.03 eq), DCM, 40°, 2 h	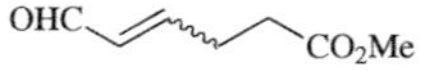 OHC, CO$_2$Me (93)	942

R^1	*n*	Additive	R^2		*(*E*)/(*Z*)
MeO_2C	1	**I**	H	(57)	90:10
MeO_2C	1	**II**	Me	(50)	83:17
TBSO	2	**I**	H	(77)	90:10

n	R		*(*E*)/(*Z*)
1	MeO_2C	(69)	83:17
2	TBSO	(83)	95:5

		Conditions	Product(s) and Yield(s) (%)		Refs.
4 eq	CO_2Me NH O_2S NO_2	Gr II (0.1 eq), DCM, reflux, 0.5 h	OHC CO_2Me NH O_2S NO_2	(—)	943
1 eq	n	Hov II (0.02 eq), DCM, rt, 8 h	OHC n		927

n	
3	(55)
5	(51)
7	(71)

		Conditions	Product(s) and Yield(s) (%)		Refs.
5 eq	NHCbz	Gr II (0.05 eq), DCM, 45°, 48 h	OHC NHCbz	(78)	944
5 eq	m n NHCbz	Gr II (0.05 eq), DCM, 45°, 48 h	OHC m n NHCbz		944

m	n	
1	1	(72)
2	0	(65)
2	1	(81)
3	0	(72)

TABLE 16. CROSS-METATHESIS OF α,β-UNSATURATED ALDEHYDES (*Continued*)

Aldehyde	Cross Partner	Conditions	Product(s) and Yield(s) (%)	Refs.
Please refer to the charts preceding the tables for the catalyst structures.				
C_4				
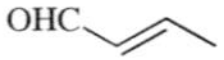 OHC (crotonaldehyde), 1.5 eq	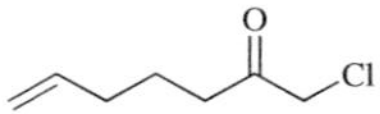 6-chlorohex-... (Cl ketone structure)	Gr II (0.025 eq), DCM, 45°, 9 h	 OHC ... Cl (ketone) (80)	945
1.4 eq	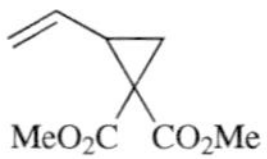 vinylcyclopropane, MeO_2C CO_2Me	Gr II (0.01 eq), DCM, reflux	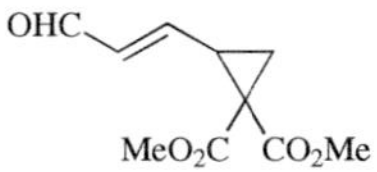 OHC ..., MeO_2C CO_2Me (69) (*E*) only	407
3 eq	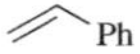 Ph (styrene)	1. Gr II (0.05 eq), DCM, reflux, 3 h 2. DIBAL-H, DCM, –78°, 2 h	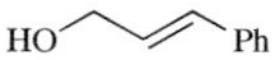 HO ... Ph (70) (*E*)/(*Z*) > 95:5	941
3 eq	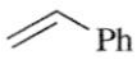 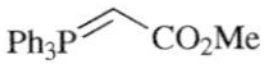 Ph + Ph_3P=CO_2Me **I**	1. Gr II (0.05 eq), DCM, 40°, 3 h 2. **I** (1.2 eq), DCM, rt to 40°, 15 + 1 h	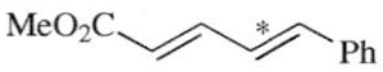 MeO_2C ... * ... Ph (48) *(*E*)/(*Z*) = 92:8	940
3 eq	Ph + $(MeO)_2OP$ CO_2Me **I** $(CF_3CH_2O)_2OP$ CO_2Me **II**	1. Gr II (0.05 eq), DCM, 40°, 3 h 2. **I** or **II** (1.4 eq), NaH (1.4 eq), THF, 0° to rt, 0.5 + 15 h	MeO_2C ... * ... Ph	940

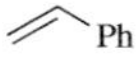

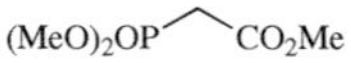

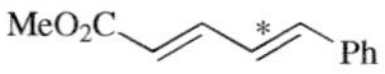

Additive		*(*E*)/(*Z*)
I	(55)	95:5
II	(63)	13:87

Olefin	Partner	Conditions	Product(s) and Yield(s) (%)	Refs.
2 eq, R	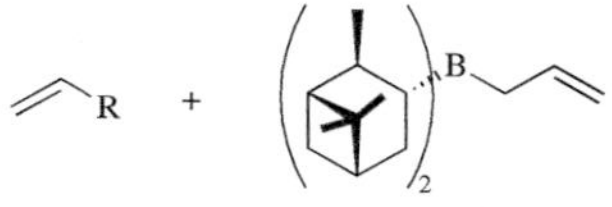 I	1. Hov II (0.05 eq), DCM, 40°, 12 h 2. I (1.9 eq), Et_2O, –78°, 1.17 h	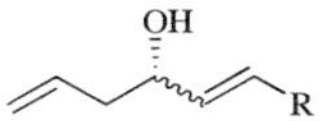 OH, R R (yield) er: Ph (74) 91.5:8.5; 4-FC_6H_4 (71) 91.5:8.5; Cy (72) 91.0:9.0	940
2 eq, Ph	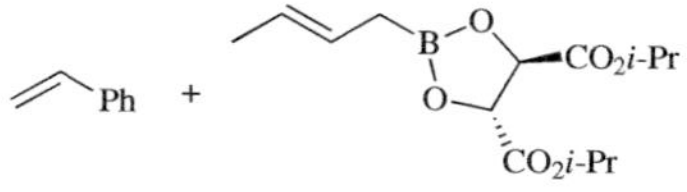 B, O, CO_2i-Pr, CO_2i-Pr, I	1. Hov II (0.05 eq), DCM, 40°, 12 h 2. I (1.5 eq), toluene, –78°, 5 h	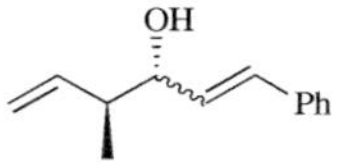 OH, Ph (80) er 83.5:16.5, dr > 95:5	940
3 eq, Ph	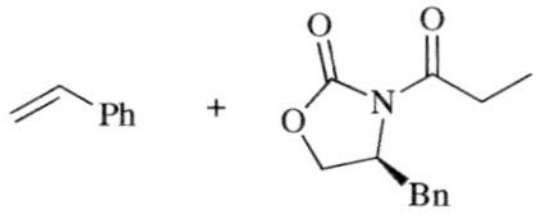 O, O, N, O, Bn, I	1. Hov II (0.05 eq), DCM, 40°, 12 h 2. I (1 eq), Bu_2OTf (1.1 eq), Et_3N (1.4 eq), DCM, –78 to 0°, 2 h	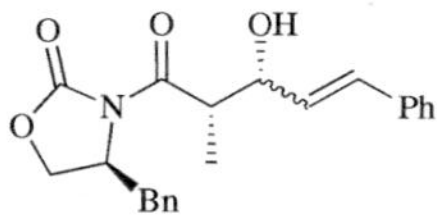 O, O, OH, N, O, Bn, Ph (60) dr > 95:5	940
5 eq, O, O, OTBS	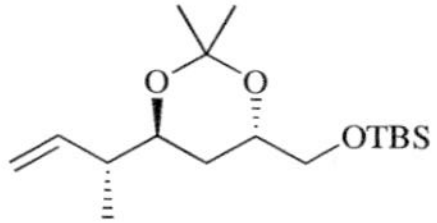	Gr II (0.15 eq) toluene, 60°, 6 h	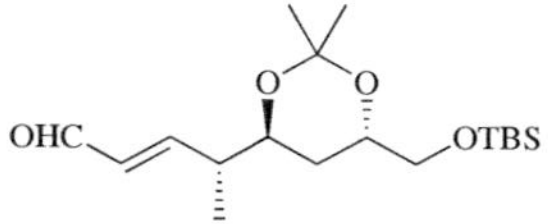 OHC, O, O, OTBS (60) (*E*) only	946

TABLE 16. CROSS-METATHESIS OF α,β-UNSATURATED ALDEHYDES (*Continued*)

Aldehyde	Cross Partner	Conditions	Product(s) and Yield(s) (%)	Refs.
Please refer to the charts preceding the tables for the catalyst structures.				
C_4				
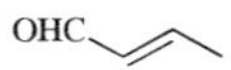 OHC (crotonaldehyde), 3 eq	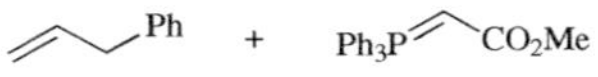 Ph (allylbenzene) + Ph_3P=CHCO$_2$Me **I**	1. Gr II (0.05 eq), DCM, 40°, 3 h 2. **I** (1.2 eq), DCM, rt to 40°, 15 + 1 h	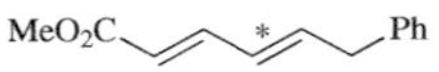 MeO_2C ... Ph (73) *(*E*)/(*Z*) = 90:10	940
2 eq	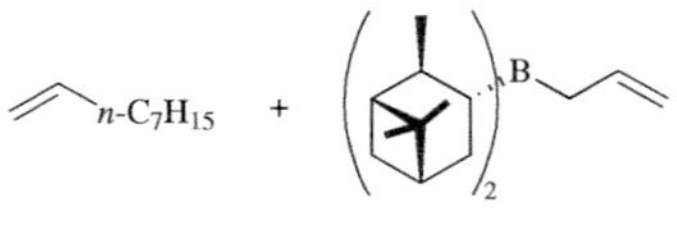 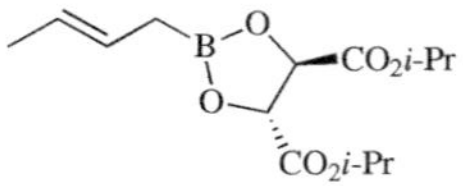 n-C_7H_{15} (alkene) + **I** (diisopinocampheyl allylborane) or **II** (crotylboronate, CO_2i-Pr, CO_2i-Pr)	1. Hov II (0.05 eq), DCM, 40°, 12 h 2. **I** or **II** (x eq), solvent, –78°, time	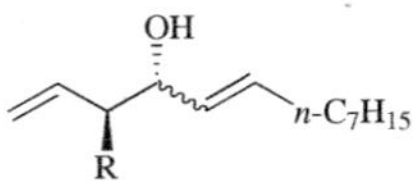 OH, R, n-C_7H_{15} (see below)	940
3 eq	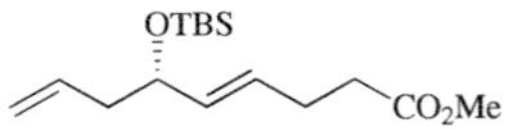 OTBS ... CO_2Me	Hov II (0.05 eq), DCM, 40°, 2 h	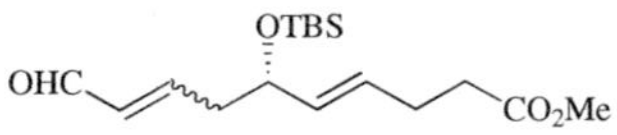 OHC ... OTBS ... CO_2Me (86)	942

Additive	x	Solvent	Time (h)	R		dr	er
I	1.9	Et_2O	1.17	H	(75)	94:6	—
II	1.5	toluene	5	Me	(82)	>95:5	93.0:7.0

OHC (3–10 eq) | Hov II (0.1 eq), additive (0.03–0.1 eq), solvent | 947

R^1	R^2	Additive	Solvent	Temp (°)	Time (h)		(E)/(Z)
H	HO	—	toluene	80–100	11–24	(60)	86:14
H	HO	CSA	DCM	25–35	—	(70)	>95:5
HO	H	—	toluene	80–100	11–24	(54)	94:6
HO	H	CSA	DCM	25–35	—	(48)	80:20

3 eq | CbzHN | Gr II (0.05 eq), DCM, 45°, 48 h | OHC, CbzHN (52) | 944

3 eq | TMSO, CO_2Me | Hov II (0.05 eq), DCM, 40°, 2 h | OHC, TMSO, CO_2Me (86) (*E*) only | 942

5 eq | HO, OBn | Hov II (0.05 eq), toluene, 110°, 18 h | OHC, OBn **I** + OHC, OBn **II** | 948

I + II (60–77), **I/II** = 4–5:1

TABLE 16. CROSS-METATHESIS OF α,β-UNSATURATED ALDEHYDES (*Continued*)

Aldehyde	Cross Partner	Conditions	Product(s) and Yield(s) (%)	Refs.
Please refer to the charts preceding the tables for the catalyst structures.				
C_4				
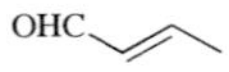 OHC 2 eq	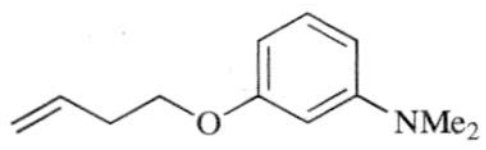 NMe$_2$	Gr II (0.03 eq), DCM, 40°, 82 h	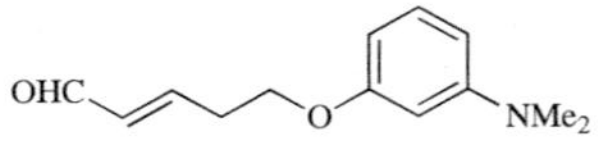 OHC NMe$_2$ (88) (*E*) only	949
10 eq	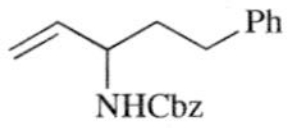 Ph NHCbz	Hov II (0.1 eq), DCM, 55°, 48 h	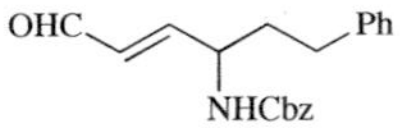 OHC Ph NHCbz (58) (*E*) only	934
2 eq	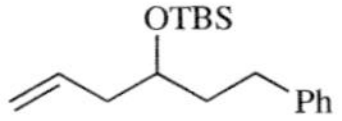 OTBS Ph	Catalyst (0.1 eq), DCM, reflux	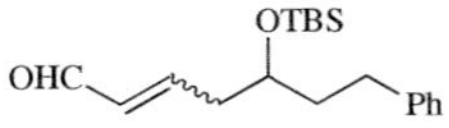 OHC OTBS Ph Catalyst: Gr I (16); Gr II (91); Hov II (94)	541
	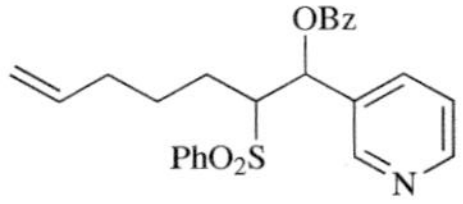 OBz PhO$_2$S N	Gr II, toluene, 60°, 3 h	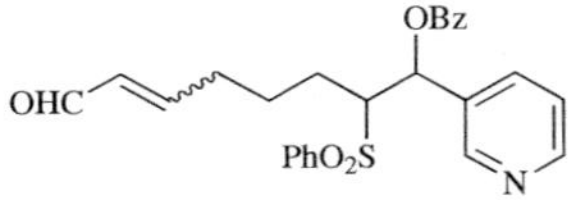 OHC OBz PhO$_2$S N (78)	638

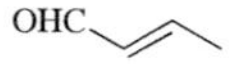

OHC 10 eq

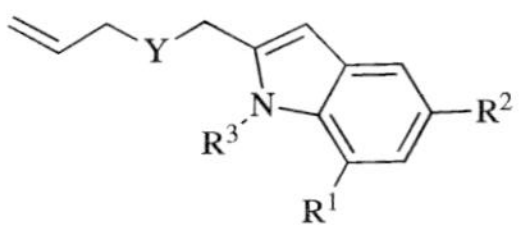

Gr II (0.03 eq), DCM, 40°

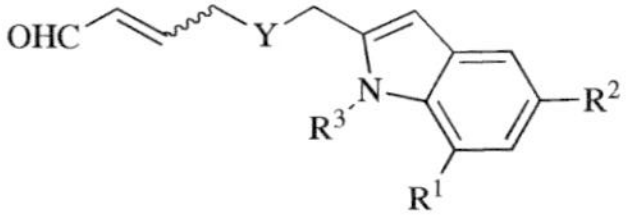

R^2 (*E*) only 949

R^1	R^2	R^3	Y	Time (h)	
H	H	Me	O	18	(55)
H	H	Bn	O	17	(57)
H	F	Me	O	36	(43)
H	F	Bn	O	18	(53)
H	Cl	Me	O	39	(33)
H	Br	Me	O	39	(44)
H	MeO	Bn	O	17	(51)
H	Me	Bn	O	41	(54)
H	H	Bn	TsN	82	(50)
Cl	H	Me	O	39	(56)

5 eq

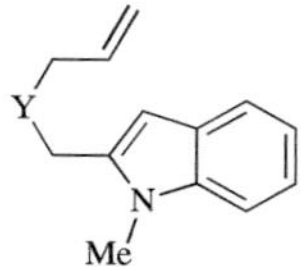

Hov II (0.03 eq), DCE, reflux

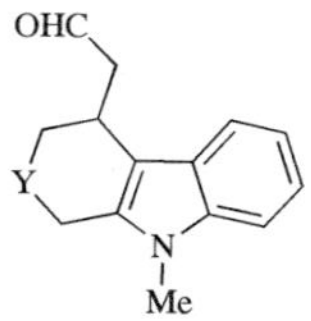

Y	Time (h)		
O	48	(74)	950
TsN	24	(85)	

TABLE 16. CROSS-METATHESIS OF α,β-UNSATURATED ALDEHYDES (*Continued*)

Aldehyde	Cross Partner	Conditions	Product(s) and Yield(s) (%)	Refs.

Please refer to the charts preceding the tables for the catalyst structures.

C_4

OHC–CH=CH–CH3 (x eq) + 2-(pent-4-enyl)indole (N-R[1], R[2], R[3], R[4], R[5]) → Catalyst (y eq), solvent → OHC-CH2-tetrahydrocarbazole (N-R[1], R[2]–R[5]) — Refs. 950, 949

x	R[1]	R[2]	R[3]	R[4]	R[5]	Catalyst	y	Solvent	Temp (°)	Time (min)	Yield (%)
10	Me	H	H	H	H	Gr I	0.03	DCM	40	60	(5)
10	Me	H	H	H	H	Hov I	0.03	DCM	40	60	(15)
10	Me	H	H	H	H	Hov II	0.03	DCM	40	60	(90)
10	Me	H	H	H	H	Gr II	0.03	DCM	40	69	(79)
10	Me	H	H	H	H	Hov II	0.03	toluene	110	40	(90)
10	Me	H	H	H	H	Hov II	0.03	DCE	80	30	(97)
5	Me	H	H	H	H	Hov II	0.03	DCE	80	40	(93)
10	Me	H	H	H	H	Hov II	0.01	DCE	80	720	(63)
5	H	H	H	H	H	Hov II	0.03	DCE	80	70	(82)
5	Me	H	H	H	H	Hov II	0.03	DCE	80	40	(93)
5	Me	Me	H	H	H	Hov II	0.03	DCE	80	40	(95)
5	H	Me	H	H	H	Hov II	0.03	DCE	80	40	(82)
5	Me	Cl	H	H	H	Hov II	0.03	DCE	80	90	(86)
5	Me	F	H	H	H	Hov II	0.03	DCE	80	90	(90)
5	Me	H	Me	H	H	Hov II	0.03	DCE	80	40	(90)
5	Me	H	MeO	H	H	Hov II	0.03	DCE	80	90	(88)
5	Me	H	F	H	H	Hov II	0.03	DCE	80	90	(95)
5	Me	H	H	Me	H	Hov II	0.03	DCE	80	30	(88)
5	Me	H	H	Cl	H	Hov II	0.03	DCE	80	55	(81)
5	Me	H	H	H	Me	Hov II	0.03	DCE	80	30	(96)
5	Me	H	H	H	Et	Hov II	0.03	DCE	80	60	(80)
5	Me	H	Me	H	Me	Hov II	0.03	DCE	80	30	(91)
5	Me	H	Me	H	Cl	Hov II	0.03	DCE	80	40	(99)
10	Bn	H	H	H	H	Gr II	0.03	DCM	40	90	(61)

		Conditions	Product(s) and Yield(s) (%)	Refs.
5 eq	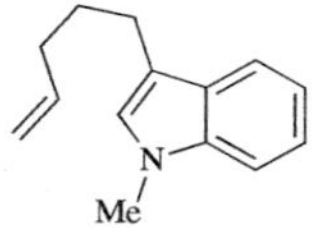	Hov II (0.1 eq), DCE, reflux, 24 h	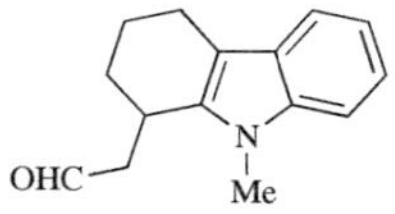 OHC; N; Me (80)	950
5 eq	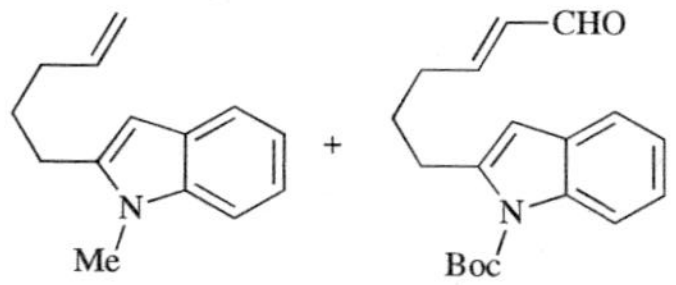 CHO; N; Me; + N; Boc	Hov II (0.03 eq), DCE, reflux, 4 h	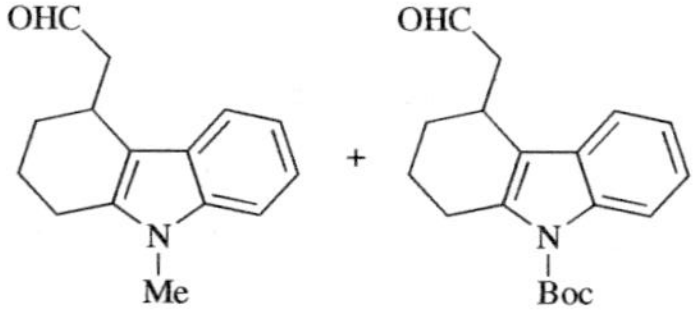 OHC; N; Me (89) + OHC; N; Boc (51)	950
n eq	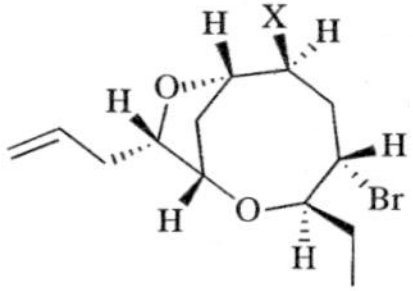 H; X; H; O; H; H; Br; H; O; H	1. Gr II (x eq), DCM, 40°, time 2. DMSO, rt, 12 h	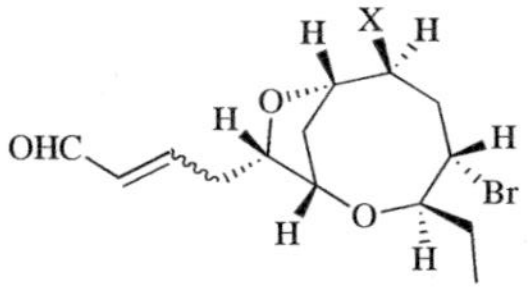 OHC; H; X; H; O; H; H; Br; H; O; H	570

n	X	*x*	Time (h)	
20	Cl	0.1	1	(83)
10	Br	0.29	1.5	(86)

TABLE 16. CROSS-METATHESIS OF α,β-UNSATURATED ALDEHYDES (*Continued*)

Aldehyde	Cross Partner	Conditions	Product(s) and Yield(s) (%)	Refs.
Please refer to the charts preceding the tables for the catalyst structures.				
C_4				
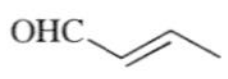 OHC 6 eq	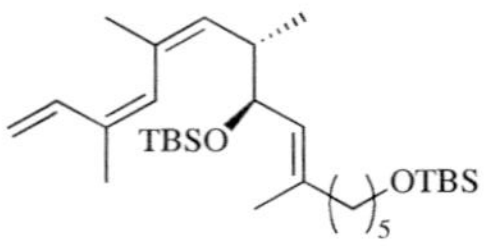 TBSO OTBS ($)_5$	Catalyst, toluene, 60°	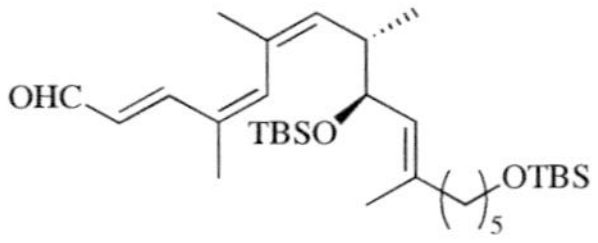 OHC TBSO OTBS ($)_5$ Catalyst / $(E)/(Z)$ Gr II / 50:50 Hov II / 50.50	760
30 eq	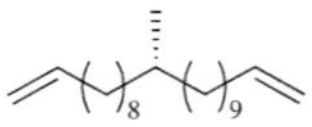 ($)_8$ ($)_9$	Hov II (0.06 eq), DCM, rt, 0.5 h; then 60°, 2.5 h	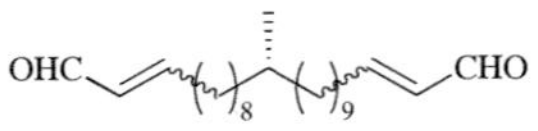 OHC ($)_8$ ($)_9$ CHO (98)	951

TABLE 17. CROSS-METATHESIS OF α,β-UNSATURATED KETONES

Ketone	Cross Partner	Conditions	Product(s) and Yield(s) (%)	Refs.

Please refer to the charts preceding the tables for the catalyst structures.

C_{4-7}

Ketone	Cross Partner	Conditions	Product(s) and Yield(s) (%)	Refs.
R–C(O)–CH=CH₂ (2.65 eq)	CH₂=CHCH₂NHTs	Hov II (0.05 eq), $B(OPh)_3$ (0.1 eq), toluene, 80°	N-Ts-2-R-pyrrole	899

R	Time (min)	Yield (%)
Me	45	(60)
n-Pr	45	(54)
2-furyl	60	(30)

C_4

Ketone	Cross Partner	Conditions	Product(s) and Yield(s) (%)	Refs.
$CH_3C(O)CH=CHR$ (1 eq)	CH₂=CHCH₂NHBoc	Hov II (0.05 eq), additive (x eq), toluene, 80°, 12 h	$CH_3C(O)CH=CHCH_2NHBoc$	952

R	Additive	x	Yield (%)
H	—	—	(27)
H	ClBcat	0.1	(86)
H	ClBcat	0.4	(16)
H	ClBcat	1	(—)
Me	—	—	(55)
Me	$PhBCl_2$	0.2	(64)
Me	Cy_2BCl	0.1	(77)

TABLE 17. CROSS-METATHESIS OF α,β-UNSATURATED KETONES (*Continued*)

Ketone	Cross Partner	Conditions	Product(s) and Yield(s) (%)	Refs.

Please refer to the charts preceding the tables for the catalyst structures.

C_4

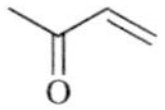

Ketone: x eq; Cross Partner (R); Conditions: Catalyst (y eq), solvent

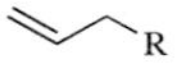

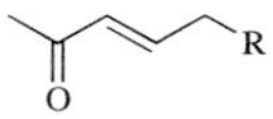

x	R	Catalyst	y	Solvent	Temp (°)	Time (h)		(E)/(Z)	Refs.
8	CbzHN	Hov II	0.1	DCM	40	48	(61)	—	934
2	BzHN	Ru cat **41**	0.01	DCM	rt	6	(8)	>95:5	562
2	BzHN	Ru cat **44**	0.01	DCM	rt	6	(17)	>95:5	562
2	BzHN	Ru cat **45**	0.01	DCM	rt	6	(15)	>95:5	562
3	Ts	Hov II	0.03	DCE	reflux	0.5–1	(67)	—	953
3	Ph_2PO	Gr II	0.02–0.06	DCM	40	20–48	(46)	100:0	214

Ketone: 3 eq; Cross Partner (OR); Conditions: Ru cat **129** (x eq), CuI (0.03 eq), Et_2O, 35°, 16 h; Refs. 954

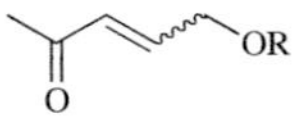

R	x	
H	0.001	(0)
H	0.0025	(5)
Ac	0.001	(95)
Ac	0.0025	(>99)

1 eq	NHBn; 1 eq; + N_2, O, O	1. DCM, 300 W, 0.25 h 2. Hov II (0.04 eq), 100°, 0.5 h 3. IPr (0.2 eq), rt, 20 h	O, O, N, Bn, O	(77) (*E*)/(*Z*) > 95:5	955
3 eq	Ts, N, Bn	1. Hov II (0.02 eq), DCM, 40°, 15 h 2. NaH (2 eq), DMF, 22°, 1 h	H, Ts, N, Bn	(96)	956
3 eq	O, N, O	1. Hov II (0.02 eq), DCM, 40°, 15 h 2. NaH (2 eq), DMF, 22°, 1 h	O, HN, O	(95)	956

TABLE 17. CROSS-METATHESIS OF α,β-UNSATURATED KETONES (*Continued*)

Ketone	Cross Partner	Conditions	Product(s) and Yield(s) (%)	Refs.

Please refer to the charts preceding the tables for the catalyst structures.

C_4

Ketone: *x* eq; Cross Partner: OBz

Conditions: Catalyst (*y* eq), DCM

Products: OBz + BzO OBz

I (*E*)/(*Z*) > 95:5 **II**

x	Catalyst	*y*	Temp (°)	Time (h)	**I**	**II**	(*E*)/(*Z*) **II**	Refs.
2	Ru cat **93**	0.01	25	0.5	(76)	(—)	—	957
2	Ru cat **169**	0.01	25	0.5	(74)	(—)	—	957
2	Ru cat **170**	0.01	25	0.5	(64)	(34)	80:20	957
2	Ru cat **171**	0.01	25	0.5	(62)	(31)	80:20	957
5	Ru cat **170**	0.01	25	1	(92)	(—)	—	957
2	Ru cat **286**	0.02	40	8	(57)	(—)	—	161

C_{4-5}

Ketone: R, 3 eq; Cross Partner: OPNP

Conditions: Catalyst (*x* eq), DCM, 15 h

Product: R OPNP (*E*) only — Refs. 932

R	Catalyst	*x*	Temp (°)	
Me	Hov II	0.02	40	(92)
Et	Hov II	0.05	40	(99)
Et	Hov II	0.02	40	(95)
Et	Hov II	0.02	22	(72)
Et	Ru cat **1**	0.05	40	(88)
Et	Gr II	0.05	40	(38)
Et	Gr II	0.05	20	(29)
Et	Gr II	0.05	40	(68)
Et	Hov II	0.02	40	(95)

C_4

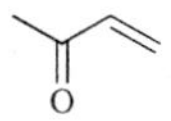

3 eq

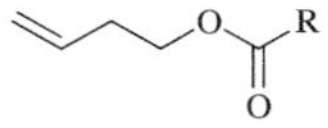

1. Hov II (0.02 eq), DCM, 40°, 15 h
2. DBU (2 eq), 22°, 1 h

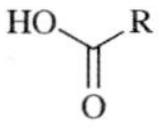

R		
Ph	(95)	956
4-MeC_6H_4	(94)	
Ph$(CH_2)_2$	(91)	
n-$C_{11}H_{23}$	(92)	
1-Naph	(90)	
(Z)-HNCH=CHBn	(90)	

C_{4-8}

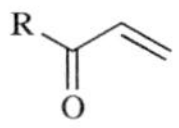

x eq

NHCbz

Catalyst (y eq), additive (0.01 eq), solvent

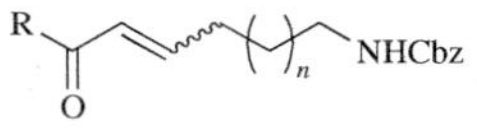

NHCbz + Cbz

I 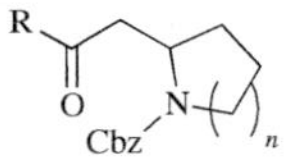 II

958

R	x	n	Catalyst	y	Additive	Solvent	Temp (°)	Time (h)	I	II
Me	1.64	1	Gr II	0.1	—	DCM	reflux	5	(45)	(—)
Me	1.64	1	Hov II	0.05	—	DCM	reflux	5	(71)	(3)
Me	1.64	1	Hov II	0.05	—	toluene	reflux	5	(35)	(21)
Me	1.64	1	Hov II	0.05	BF_3•Et_2O	DCM	45	96	(—)	(99)
Me	1.64	1	Hov II	0.05	BF_3•Et_2O	DCM	100 (MW)	0.33	(—)	(96)
Me	2	1	Hov II	0.05	BF_3•Et_2O	DCM	45	96	(—)	(99)
Me	2	1	Hov II	0.05	BF_3•Et_2O	DCM	100 (MW)	0.33	(—)	(96)
Me	2	2	Hov II	0.05	BF_3•Et_2O	DCM	45	96	(—)	(82)
Me	2	2	Hov II	0.05	BF_3•Et_2O	DCM	100 (MW)	0.33	(—)	(93)
n-Pr	2	1	Hov II	0.05	BF_3•Et_2O	DCM	45	96	(—)	(73)
n-Pr	2	1	Hov II	0.05	BF_3•Et_2O	DCM	100 (MW)	0.33	(—)	(65)
n-Pr	2	2	Hov II	0.05	BF_3•Et_2O	DCM	45	96	(—)	(81)
n-Pr	2	2	Hov II	0.05	BF_3•Et_2O	DCM	100 (MW)	0.33	(—)	(70)
n-Pn	2	1	Hov II	0.05	BF_3•Et_2O	DCM	45	96	(—)	(79)
n-Pn	2	1	Hov II	0.05	BF_3•Et_2O	DCM	100 (MW)	0.33	(—)	(72)
n-Pn	2	2	Hov II	0.05	BF_3•Et_2O	DCM	45	96	(—)	(83)
n-Pn	2	2	Hov II	0.05	BF_3•Et_2O	DCM	100 (MW)	0.33	(—)	(61)

TABLE 17. CROSS-METATHESIS OF α,β-UNSATURATED KETONES (*Continued*)

Ketone	Cross Partner	Conditions	Product(s) and Yield(s) (%)	Refs.

Please refer to the charts preceding the tables for the catalyst structures.

C_4

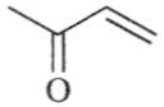

5 eq

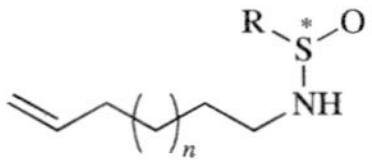

Catalyst (0.1 eq),
Ti(O*i*-Pr)$_4$ (0.1 eq),
solvent

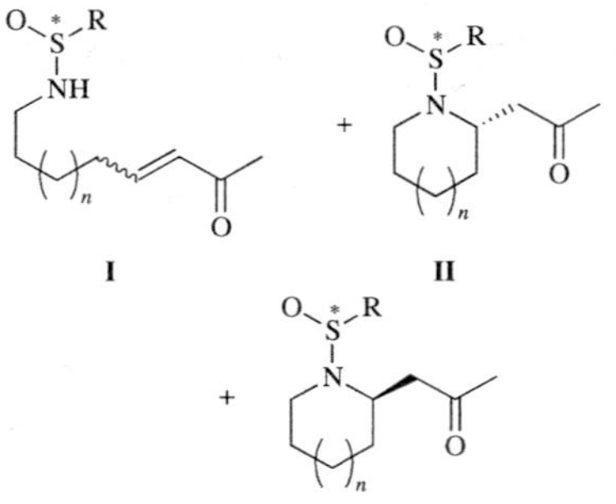

I + II + III

959

R	*n*	Catalyst	Solvent	Temp (°)	Time (h)	I	II + III	dr II/III
(S_s) *p*-Tol	0	Ru cat **35**	DCM	reflux	2	(70)	(12)	—
(R_s) *t*-Bu	0	Ru cat **35**	DCM	reflux	2	(>99)	(—)	—
(S_s) *p*-Tol	0	Ru cat **35**	DCM	reflux	6	(50)	(28)	85:15
(S_s) *p*-Tol	0	Ru cat **35**	toluene	reflux	48	(10)	(70)	88:12
(S_s) *p*-Tol	0	Ru cat **190**	DCM	reflux	48	(—)	(95)	89:11
(S_s) *p*-Tol	0	Ru cat **190**	DCM	reflux	12	(23)	(10)	80:20
(S_s) *p*-Tol	0	Ru cat **190**	DCM	100 (MW)	1	(21)	(20)	81:19
(S_s) *p*-Tol	0	Ru cat **190**	DCM	60 (MW)	7	(14)	(60)	88:12
(R_s) *t*-Bu	0	Ru cat **35**	DCM	reflux	48	(9)	(74)	9:91
(R_s) *t*-Bu	0	Ru cat **190**	DCM	reflux	48	(—)	(92)	8:92
(S_s) *p*-Tol	1	Ru cat **35**	DCM	reflux	2	(72)	(—)	—
(R_s) *t*-Bu	1	Ru cat **35**	DCM	reflux	2	(>99)	(—)	—
(S_s) *p*-Tol	1	Ru cat **190**	DCM	reflux	48	(54)	(30)	70:30
(S_s) *p*-Tol	1	Ru cat **190**	DCE	reflux	48	(41)	(25)	71:29
(R_s) *t*-Bu	1	Ru cat **190**	DCM	reflux	48	(63)	(22)	24:76
(R_s) *t*-Bu	1	Ru cat **190**	DCE	reflux	48	(77)	(15)	24:76

2 eq

NHCbz, F, F

Hov II (0.05 eq), $BF_3{\bullet}Et_2O$ (0.01 eq), DCM

O, N, Cbz, F, F

Temp (°)	Time (h)		
100 (MW)	0.33	(55)	958
45	96	(60)	

3 eq

F, F, NHR, *n*, O

Hov II (0.05 eq), $Ti(O\textit{i}\text{-}Pr)_4$ (0.1 eq), DCM, reflux

O, R, N, *n*, F, F, O

n	R	Time (h)		
0	PMP	8	(61)	960
0	MeO	10	(58)	
0	SAMP	8	(58)	
1	PMP	8	(51)	
1	MeO	10	(49)	
1	SAMP	8	(62)	

x eq

OBz

Catalyst (*y* eq), solvent

O, OBz

x	Catalyst	*y*	Solvent	Temp (°)	Time (h)		(*E*)/(*Z*)	
2	Hov II	0.05	DCM	40	—	(85)	>95:5	922
2	Ru cat **243**	0.05	DCM	45	12	(68)	>95:5	923
2	Ru cat **191**	0.05	DCM	45	12	(97)	>95:5	923
—	Ru cat **185**	0.05	$BMI{\bullet}PF_6$/toluene	rt	6	(88)	>95:5	924
2	Ru cat **170**	0.01	DCM	25	1	(95)	>95:5	957
2	Ru cat **110**	0.01	DCM	rt	0.33	(98)	(*E*) only	961
2	Ru cat **112**	0.01	DCM	rt	0.33	(82)	99:1	962

TABLE 17. CROSS-METATHESIS OF α,β-UNSATURATED KETONES (*Continued*)

Ketone	Cross Partner	Conditions	Product(s) and Yield(s) (%)	Refs.
Please refer to the charts preceding the tables for the catalyst structures.				
C_4				
3 eq	CO_2Me	Hov II (0.03 eq), DCE, reflux, 0.5–1 h	CO_2Me (63)	953
5 eq	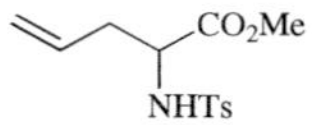 CO_2Me, NHTs	Hov II (0.075 eq) DCM, 55°, 48 h	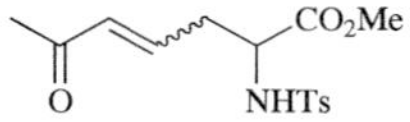 CO_2Me, NHTs (88)	335
	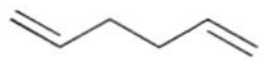 2 eq	1. Gr II (0.05 eq), DCM, 40°, 3 h 2. Gr II (0.05 eq), styrene (3 eq), DCM, 40°, 12 h	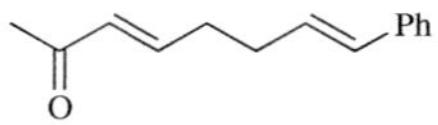 Ph (47) (*E,E*) only	36
C_{4-9}				
R, 3 eq		Ru cat **93** (0.025 eq), $CDCl_3$, 100°, (MW), 1 h	R, R R: Me (85); Ph (70)	926
C_4				
3 eq	*n*-Bu	Catalyst (*x* eq), CuI (0.03 eq), Et_2O, 35°, 16 h	*n*-Bu (*E*) only	963

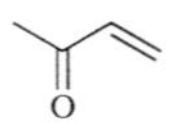

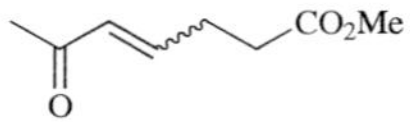

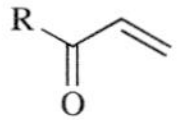

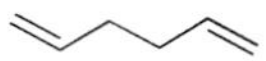

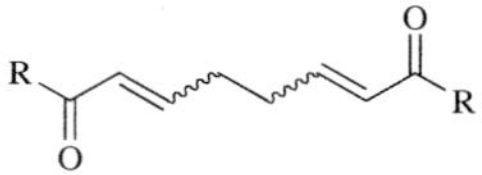

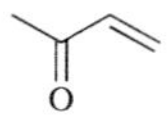

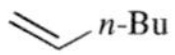

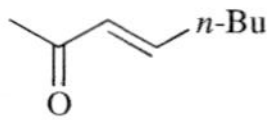

Catalyst	*x*	
Ru cat **129**	0.005	(>99)
Ru cat **131**	0.005	(<2)
Ru cat **139**	0.005	(45)
Hov II	0.005	(48)
Ind II	0.005	(>99)
Ind II	0.0025	(96)
Ru cat **129**	0.0025	(>99)
Ru cat **129**	0.001	(>99)
Ru cat **129**	0.0005	(>98)
Ru cat **129**	0.00025	(73)

3 eq	n-Bu	Catalyst (x eq), additive (0.03 eq), Et_2O, 35°, 16 h	n-Bu (enone product)	954

Catalyst	x	Additive	Conc (M)	Conv (%)[a]
Ru cat **139**	0.01	CuI	0.05	>99
Ru cat **139**	0.01	CuI	0.1	>99
Ru cat **139**	0.01	CuI	0.2	>99
Ru cat **139**	0.01	CuI	0.5	66
Ru cat **139**	0.01	CuCl	0.2	80
Ru cat **139**	0.01	—	0.2	33
Gr II	0.005	CuI	0.2	>99
Ind II	0.005	CuI	0.2	>99
Hov II	0.005	CuI	0.2	48
Ru cat **139**	0.005	CuI	0.2	45
Ru cat **285**	0.005	CuI	0.2	80
Ru cat **173**	0.005	CuI	0.2	>99
Ru cat **129**	0.005	CuI	0.2	99
Gr II	0.0025	CuI	0.2	73
Ind II	0.0025	CuI	0.2	96
Ru cat **173**	0.0025	CuI	0.2	98
Ru cat **129**	0.0025	CuI	0.2	>99
Ind II	0.001	CuI	0.2	88
Ru cat **173**	0.001	CuI	0.2	71
Ru cat **129**	0.001	CuI	0.2	>99
Ru cat **129**	0.0005	CuI	0.2	98
Ru cat **129**	0.0005	—	0.2	48
Ru cat **129**	0.00025	CuI	0.2	76

TABLE 17. CROSS-METATHESIS OF α,β-UNSATURATED KETONES (*Continued*)

Ketone	Cross Partner	Conditions	Product(s) and Yield(s) (%)	Refs.
Please refer to the charts preceding the tables for the catalyst structures.				
C_4				
3 eq	Ts, N	Hov II (0.05 eq), DCM, 45°, 3 h	Ts, N, O (85)	939
C_{4-10}				
R, O; 3 eq	NHCbz + SiPh₃, O, OH, P, O, O, SiPh₃; 0.1 eq	Hov II (0.05 eq), DCM, 0°	R, O, N, Cbz	964
			R / Time (h) / / er: Me 14 (74) 95.0:5.0; Et 18 (76) 97.0:3.0; Bn 20 (61) 96.0:4.0	
R^1, O; 3 eq	NHR^2	Hov II (0.05 eq), DCM, rt, 12 h	R^1, O, NHR^2	964

R	Time (h)		er
Me	14	(74)	95.0:5.0
Et	18	(76)	97.0:3.0
Bn	20	(61)	96.0:4.0

R^1	R^2	
Me	Boc	(69)
Me	Cbz	(89)
Et	Cbz	(85)
Bn	Cbz	(68)

C_4

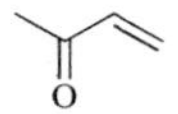

2 eq

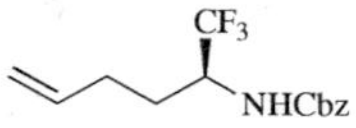

CF_3 NHCbz

Hov II (0.05 eq), $BF_3{\cdot}Et_2O$ (0.01 eq), DCM

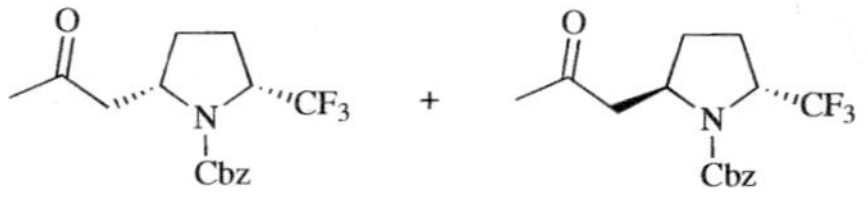

N Cbz CF_3 + N Cbz CF_3 958

I II

Temp (°)	Time (h)		I/II
45	96	(76)	5:1
100 (MW)	0.33	(97)	1:3

2.6 eq

OAc

Gr II (0.05 eq), additive (0.25 eq), benzene, THF, 45°

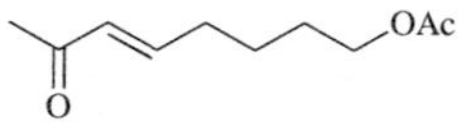

OAc

(*E*)/(*Z*) = 93:7 928

Additive	Time (h)	
—	48	(92)
HCl/Et_2O	14	(90)
$CuCl_2$	14	(12)
CuCl	24	(42)
$B(C_6F_5)_3$	14	(47)
$Ni(COD)_2$	<1	(0)
$AlCl_3$	14	(67)

TABLE 17. CROSS-METATHESIS OF α,β-UNSATURATED KETONES (*Continued*)

Ketone	Cross Partner	Conditions	Product(s) and Yield(s) (%)	Refs.

Please refer to the charts preceding the tables for the catalyst structures.

C_4

Ketone: x eq; Cross Partner: OTBS; Conditions: Catalyst (*y* eq), DCM; Product: OTBS

x	Cat	*y*	Temp (°)	Time (h)	Yield (%)	(*E*)/(*Z*)	Refs.
2	Ru cat **112**	0.025	20	0.66	(91)	99:1	962
2	Ru cat **93**	0.01	25	0.5	(86)	>95:5	957
2	Ru cat **169**	0.01	25	0.5	(48)	>95:5	957
2	Ru cat **170**	0.01	25	0.5	(>98)	>95:5	957
2	Ru cat **171**	0.01	25	0.5	(>98)	>95:5	957
—	Ru cat **93**	0.01	rt	0.5	(86)	>95:5	550
2	Ru cat **286**	0.02	25	5	(59)	>95:5	161
2	Ru cat **286**	0.02	40	8	(77)	>95:5	161
2	Ru cat **41**	0.01	rt	3	(70)	>95:5	562
2	Ru cat **44**	0.01	rt	3	(85)	>95:5	562
2	Ru cat **45**	0.01	rt	3	(80)	>95:5	562
2	Ru cat **112**	0.01	20	0.67	(90)[b]	(*E*) only	609
2	Ru cat **88**	0.025	25	0.5	(82)	(*E*) only	609
2	Ru cat **138**	0.01	25	5	(57)[b]	(*E*) only	609
2	Ru cat **115**	0.01	25	0.5	(50)[b]	(*E*) only	609
2	Ru cat **115**	0.01	25	1	(82)	(*E*) only	609
2	Ru cat **116**	0.01	25	0.67	(65)[b]	(*E*) only	609
—	Ru cat **106**	0.025	20	3	(93)	99:1	216
—	Ru cat **112**	0.01	25	0.67	(90)	99:1	216

—	Ru cat **88**	0.025	25	3	(95)	99:1	216
2	Ru cat **98**	0.01	rt	0.5	(81)	95:5	965
2	Ru cat **99**	0.01	rt	0.5	(88)	95:5	965
2	Ru cat **100**	0.01	rt	0.5	(86)	95:5	965
2	Ru cat **101**	0.01	rt	0.5	(83)	95:5	965
2	Ru cat **102**	0.01	rt	0.5	(80)	95:5	965
2	Ru cat **88**	0.025	25	0.5	(82)	99:1	171
2	Ru cat **88**	0.025	25	3	(95)[b]	99:1	171
2	Ru cat **88**	0.025	rt	0.5	(82)	95:5	160
2	Ru cat **112**	0.01	rt	0.67	(90)	95:5	160

2 eq (alkene bearing R^3, R^2, OR^1); Gr II (0.05 eq), DCM, reflux, 6 h; product: enone bearing R^3, R^2, OR^1

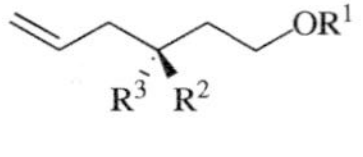

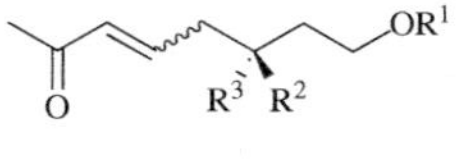

R^1	R^2	R^3		
TBDPS	H	MOMO	(81)	781
Bn	MeO	H	(86)	966

3 eq	(CO_2Et, CO_2Et)	Hov II (0.03 eq) DCE, reflux, 0.5–1 h	(CO_2Et, CO_2Et) (70)	953
3 eq	(dioxolane)	Hov II (0.03 eq) DCE, reflux, 0.5–1 h	(dioxolane) (90)	953

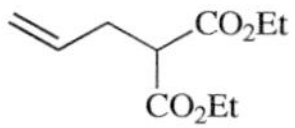

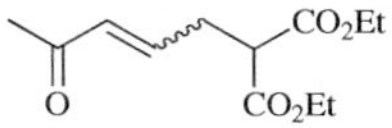

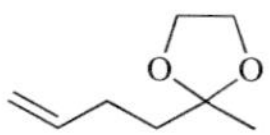

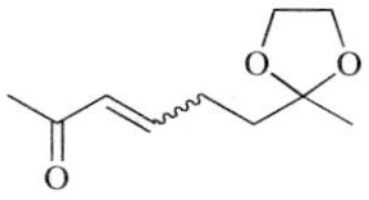

TABLE 17. CROSS-METATHESIS OF α,β-UNSATURATED KETONES (*Continued*)

Ketone	Cross Partner	Conditions	Product(s) and Yield(s) (%)	Refs.

Please refer to the charts preceding the tables for the catalyst structures.

C_4

Ketone	Cross Partner	Conditions	Product(s) and Yield(s) (%)	Refs.
x eq	CO_2R (*)	Hov **II** (0.05 eq), DCM	CO_2R **I** + RO_2C ~ CO_2R **II**	

x	R	*(*E*)/(*Z*)	Temp (°)	Time (h)	**I**	(*E*)/(*Z*) **I**	**II**	Refs.
5	Me	(*E*) only	rt	48	(13)	(*E*) only	(—)	549
3	Et	50:50	45	3	(24)	—	(6)	939

Ketone	Cross Partner	Conditions	Product(s) and Yield(s) (%)	Refs.
6 eq	n	Ru cat **129** (x eq), CuI (0.03 eq), Et_2O, 35°, 16 h	O, n, O	954

n	x	Conv. %
2	0.001	80
2	0.0025	>99
3	0.001	5
3	0.02	>99
5	0.0005	94
5	0.001	>99

C_{4-8}

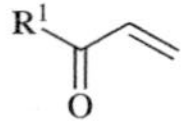

5 eq

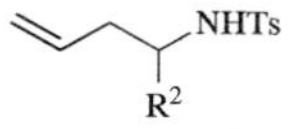

Hov II (0.075 eq), DCM, 55°, 48 h

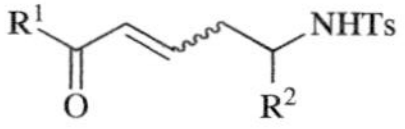

R^1	R^2	
Me	*i*-Pr	(85)
Me	4-$MeOC_6H_4$	(95)
Me	*c*-C_6H_{11}	(76)
Me	4-$CF_3C_6H_4$	(88)
c-C_5H_{11}	Ph	(76)

335

C_{4-10}

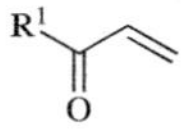

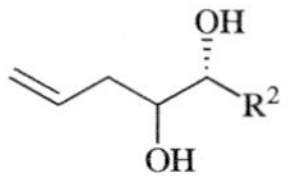

Hov II (0.025 eq), DCM, rt, 6 h

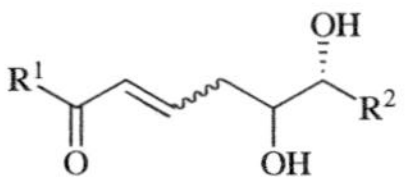

R^1	R^2	
Me	Et	(80)
Me	Ph	(62)
Me	Bn	(75)
Et	Et	(76)
Bn	Et	(95)

967

C_4

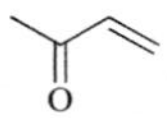

5 eq

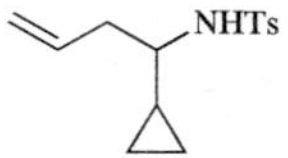

Hov II (0.075 eq) DCM, 55°, 72 h

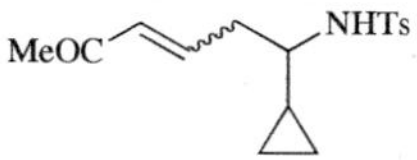

MeOC NHTs (94)

335

TABLE 17. CROSS-METATHESIS OF α,β-UNSATURATED KETONES (*Continued*)

Ketone	Cross Partner	Conditions	Product(s) and Yield(s) (%)	Refs.
Please refer to the charts preceding the tables for the catalyst structures.				
C_4				
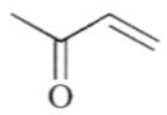 1.4 eq	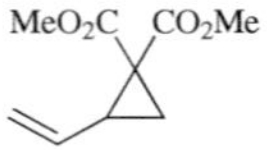 MeO_2C, CO_2Me	Gr II (0.01 eq), DCM, reflux	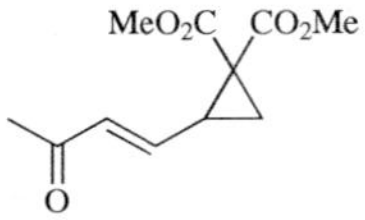 MeO_2C, CO_2Me (82) (*E*) only	407
1 eq	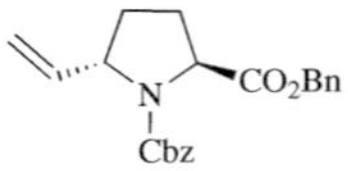 CO_2Bn, N, Cbz	Hov II (0.05 eq), toluene, 80°, 4 h	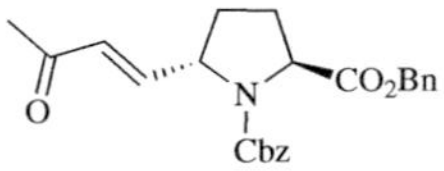 CO_2Bn, N, Cbz (61) (*E*) only	968
3.3 eq	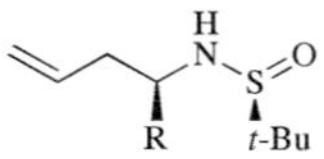 H, N, S=O, R, *t*-Bu	Hov II (0.1 eq), DCM, 45°, 60 h	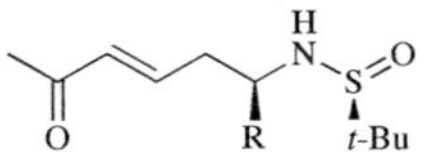 H, N, S=O, R, *t*-Bu (*E*) only R: *n*-Pr (85); *n*-C_9H_{19} (72); *n*-$C_{11}H_{23}$ (75); Ph (85)	969
5 eq	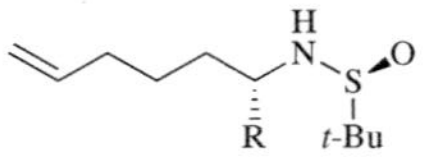 H, N, S, O, R, *t*-Bu	Ru cat **190** (0.1 eq), Ti(O*i*-Pr)$_4$ (0.1 eq), DCM, reflux, 2 h	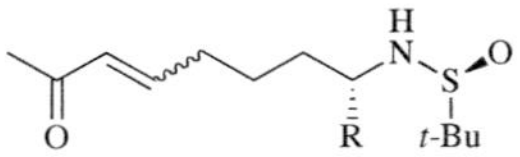 H, N, S, O, R, *t*-Bu R: Me (99); CF_3 (99)	959

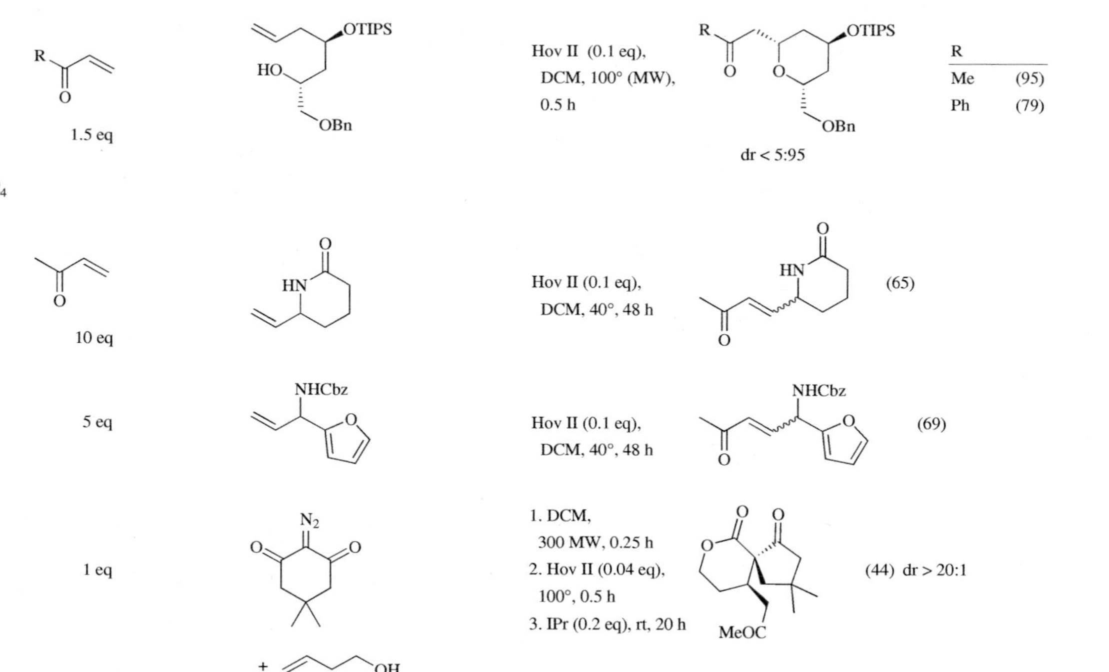

	Reactant	Conditions	Product(s) and Yield(s) (%)	Refs.
C_{4-9}				
1.5 eq		Hov II (0.1 eq), DCM, 100° (MW), 0.5 h	dr < 5:95; R = Me (95), Ph (79)	970
C_4				
10 eq		Hov II (0.1 eq), DCM, 40°, 48 h	(65)	934
5 eq		Hov II (0.1 eq), DCM, 40°, 48 h	(69)	934
1 eq		1. DCM, 300 MW, 0.25 h 2. Hov II (0.04 eq), 100°, 0.5 h 3. IPr (0.2 eq), rt, 20 h	(44) dr > 20:1	955

TABLE 17. CROSS-METATHESIS OF α,β-UNSATURATED KETONES (*Continued*)

Ketone	Cross Partner	Conditions	Product(s) and Yield(s) (%)	Refs.
Please refer to the charts preceding the tables for the catalyst structures.				
C_4				
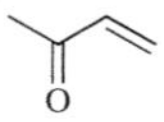 2 eq	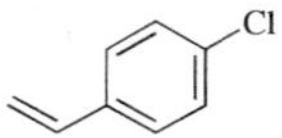	Ru cat **286** (0.02 eq), DCM, 40°, 8 h	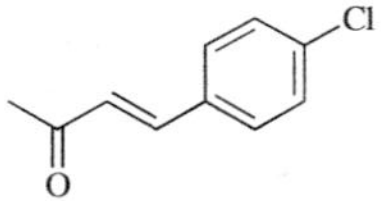 (58) (*E*)/(*Z*) > 95:5	161
6 eq	 CO_2Me	Ru cat **129** (0.005 eq), CuI (0.03 eq), Et_2O, 35°, 16 h	 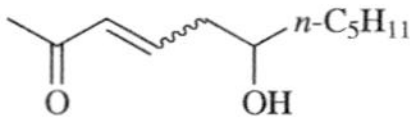 CO_2Me + n-C_5H_{11} OH **I** (>98), **I/II** = 1:1 **II**	954
3 eq	 OH	Hov II (0.03 eq) DCE, reflux, 0.5–1 h	 (60)	953
1 eq	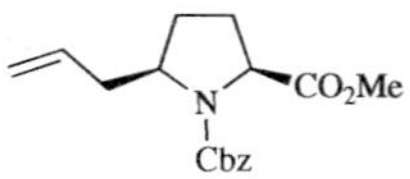 CO_2Me N Cbz	Hov II (0.05 eq), DCM, rt, 12 h	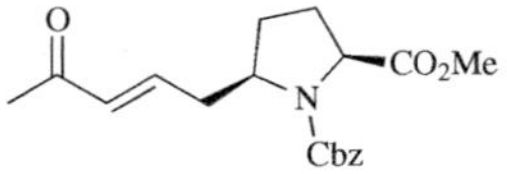 (81) (*E*) only	968

2 eq

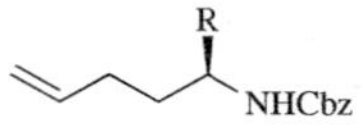

NHCbz

Hov II (0.05 eq), $BF_3 \cdot Et_2O$ (0.01 eq), DCM

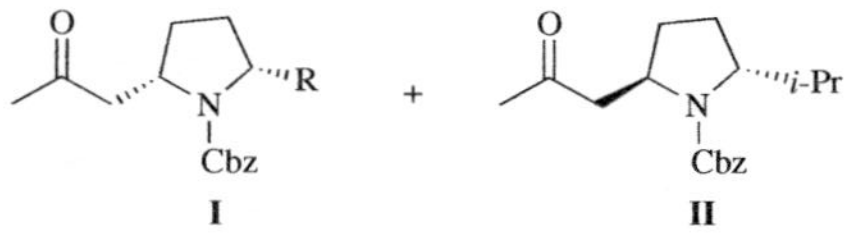

O N Cbz **I** + O N Cbz *i*-Pr **II**

958

R	Temp (°)	Time (h)	**I + II**	**I/II**
i-Pr	45	96	(97)	3:1
i-Pr	100 (MW)	0.33	(81)	1:4
Ph	45	96	(98)	6:1
Ph	100 (MW)	0.33	(86)	1:2
PMP	45	96	(78)	4:1
PMP	100 (MW)	0.33	(97)	1:2

5 eq

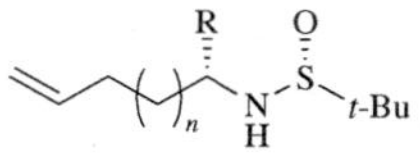

Ru cat **190** (0.1 eq), $Ti(Oi\text{-}Pr)_4$ (0.1 eq), DCM, reflux, 2 h

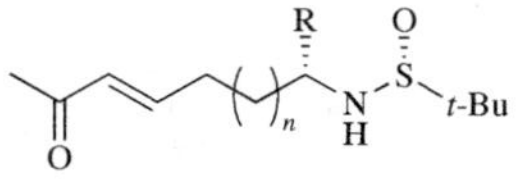

(*E*) only

R	*n*	
i-Pr	1	(>99)
i-Pr	2	(93)
Ph	1	(82)
Ph	2	(84)

959

TABLE 17. CROSS-METATHESIS OF α,β-UNSATURATED KETONES (*Continued*)

Ketone	Cross Partner	Conditions	Product(s) and Yield(s) (%)	Refs.

Please refer to the charts preceding the tables for the catalyst structures.

C_4

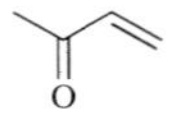

Ketone	Cross Partner	Conditions	Product(s) and Yield(s) (%)	Refs.
1.5 eq	OTIPS, HO, OTBDPS	Catalyst (0.1 eq), 2,6-Cl_2-BQ (x eq), solvent	OTIPS, O, HO, OTBDPS **I** + O, O, OTIPS, OTBDPS **II**	970

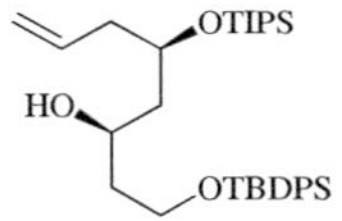

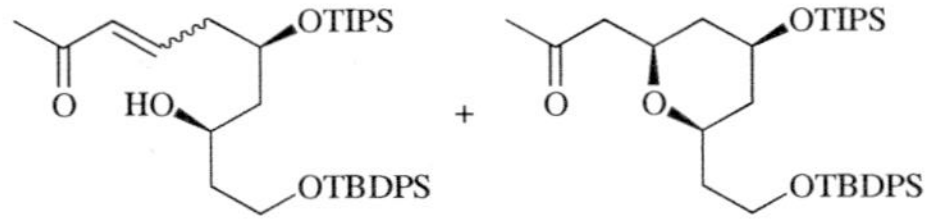

Catalyst	x	Solvent	Temp (°)	Time (h)	I	II	*cis/trans* II
Hov II	—	DCM	35	12	(88)	(0)	—
Hov II	—	DCE	80	15	(0)	(80)	90:10
Gr II	—	DCE	80	15	(19)	(31)	40:60
Hov II	—	DCM	100 (MW)	0.5	(0)	(94)	88:12
Hov II	0.2	DCM	100 (MW)	0.5	(17)	(73)	—

Ketone	Cross Partner	Conditions	Product(s) and Yield(s) (%)	Refs.
x eq	O, P=O, O, O	Catalyst (0.01 eq), DCM, reflux	O, O, P=O, O, O	719

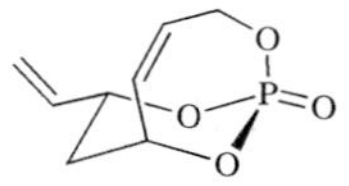

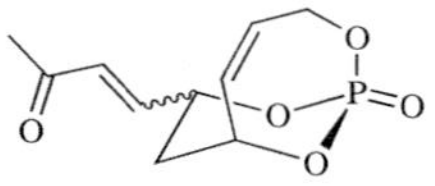

x	Catalyst	Conc (M)	
3	Gr I	0.05	(<5)[c]
3	Gr II	0.05	(25)[c]
3	Gr II	0.1	(26)[c]
3	Hov II	0.05	(75)
4	Hov II	0.1	(71)

5 eq

1. Ru cat **89** (x eq), DCM, 55°, 48 h
2. DMP, 0°

362

R^1	R^2	x	
H	Ph	0.05	(78)
Me	Ph	0.02	(73)
EtO	Ph	0.05	(45)
H	$Ph(CH_2)_2$	0.02	(61)
H	$4\text{-}MeOC_6H_4$	0.05	(62)
Me	2-furyl	0.05	(54)
Me	$3\text{-}BrC_6H_4$	0.02	(60)
H	$3\text{-}CF_3C_6H_4$	0.05	(45)

x eq

Catalyst (y eq), solvent, reflux

(E) only

x	Catalyst	y	Solvent	Time (h)		
3	Hov II	0.03	DCE	0.5–1	(81)	953
0.4	Gr II	0.025	DCM	12	(99)	971
0.4	Hov II	0.025	DCM	12	(99)	971
0.4	Ru cat **86**	0.025	DCM	12	(99)	971

TABLE 17. CROSS-METATHESIS OF α,β-UNSATURATED KETONES (*Continued*)

Ketone	Cross Partner	Conditions	Product(s) and Yield(s) (%)	Refs.

Please refer to the charts preceding the tables for the catalyst structures.

C_4

3 eq

TBSO

Catalyst (*x* eq), additive (y eq), solvent

TBSO

Catalyst	*x*	Additive	*y*	Solvent	Temp (°)	Time (h)	Conv. (%)	Yield (%)	(*E*)/(*Z*)	Refs.
Gr II	0.02	—	—	2.5% aq. PTS	22	4	64	(—)	—	972
Gr II	0.02	—	—	2.5% aq. TPGS-550-M	22	4	50	(—)	—	972
Gr II	0.02	—	—	2.5% aq. TPGS-600	22	4	60	(—)	—	972
Gr II	0.02	—	—	2.5% aq. TPGS-750-W	22	4	74	(—)	—	972
Gr II	0.02	—	—	2.5% aq. TPGS-1000	22	4	35	(—)	—	972
Gr II	0.02	—	—	2.5% aq. PTS	22	12	—	(70)	—	972
Gr II	0.02	—	—	2.5% aq. TPGS-750-W	22	12	—	(74)	—	972
Gr II	0.02	$KHSO_4$	0.04	2.5% aq. TPGS-750-W	22	4	—	(93)	—	972
Gr II	0.02	$KHSO_4$	0.04	2.5% aq. PTS	22	4	—	(91)	—	972
Gr II	0.02	CuCN	0.03	DCM	22	15	24	(—)	—	167
Gr II	0.02	CuOAc	0.03	DCM	22	15	<5	(—)	—	167
Gr II	0.02	$Cu(OTf)_2$	0.03	DCM	22	15	0	(—)	—	167
Gr II	0.02	$Cu(BF_4)_2$	0.03	DCM	22	15	15	(—)	—	167
Gr II	0.02	$Cu(NO_3)_2$	0.03	DCM	22	15	0	(—)	—	167
Gr II	0.02	$Cu(ClO_4)_2$	0.03	DCM	22	15	22	(—)	—	167
Gr II	0.02	$Cu(CF_3CO_2)_2$	0.03	DCM	22	15	0	(—)	—	167
Gr II	0.02	$Cu(CH_3CN)_4PF_6$	0.03	DCM	22	15	27	(—)	—	167
Gr II	0.02	CuCl	0.03	DCM	22	15	35	(—)	—	167
Gr II	0.02	CuBr	0.03	DCM	22	15	43	(—)	—	167
Gr II	0.02	CuI	0.03	DCM	22	15	64	(—)	—	167
Gr II	0.02	—	0.03	DCM	22	15	45	(—)	—	167

Catalyst	x	Additive	y	Solvent	Temp (°)	Time (h)	Conv. (%)		$(E)/(Z)$	
Gr II	0.02	CuI	0.06	DCM	22	15	63	(—)	—	167
Gr II	0.03	CuI	0.06	DCM	22	15	68	(—)	—	167
Gr II	0.02	CuI	0.03	toluene	22	15	54	(—)	—	167
Gr II	0.02	CuI	0.03	THF	22	15	70	(—)	—	167
Gr II	0.02	CuI	0.03	Et_2O	22	15	71	(—)	—	167
Gr II	0.02	—	—	Et_2O	22	15	43	(—)	—	167
Gr II	0.02	—	—	Et_2O	35	3	57	(—)	—	167
Gr II	0.02	CuI	0.03	Et_2O	35	3	99	(98)	>95:5	167
Gr II	0.02	NaI	0.03	Et_2O	35	3	100	(—)	—	167
Ru cat **13**	0.02	CuI	0.03	Et_2O	22	15	59	(—)	—	167
Ru cat **13**	0.02	—	—	Et_2O	22	15	22	(—)	—	167
Ind II	0.02	CuI	0.03	Et_2O	22	15	71	(—)	—	167
Ind II	0.02	—	—	Et_2O	22	15	32	(—)	—	167
Gr II	0.02	CuI	0.03	2.5% aq. TPGS-750-W	22	12	—	(93)	—	167
Gr II	0.02	—	—	2.5% aq. TPGS-750-W	22	12	—	(74)	—	167
Gr II	0.02	—	—	2.5% aq. PTS	22	4	64	(—)	—	931
Gr II	0.02	NaCl	0.03	2.5% aq. PTS	22	4	75	(—)	—	931
Gr II	0.02	—	—	2.5% PTS, H_2O:MeOH 19:1	22	4	59	(—)	—	931
Gr II	0.02	—	—	2.5% PTS, H_2O:EtOH 19:1	22	4	54	(—)	—	931
Gr II	0.02	—	—	2.5% aq. solutol	22	4	33	(—)	—	931
Gr II	0.02	—	—	95% EtOH	22	4	22	(—)	—	931
Gr II	0.02	—	—	AcOH	22	4	20	(—)	—	931
Gr II	0.02	—	—	DCM	22	4	30	(—)	—	931
Gr II	0.02	—	—	10% PTSA in 95% EtOH	22	4	<5	(—)	—	931

TABLE 17. CROSS-METATHESIS OF α,β-UNSATURATED KETONES (*Continued*)

Ketone	Cross Partner	Conditions	Product(s) and Yield(s) (%)	Refs.

Please refer to the charts preceding the tables for the catalyst structures.

C_4

TBSO

Catalyst (*x* eq),
additive (y eq),
solvent

O
TBSO

3 eq

Catalyst	*x*	Additive	*y*	Solvent	Temp (°)	Time (h)	Conv. (%)		(*E*)/(*Z*)	Refs.
Gr II	0.02	—	—	2.5% PTS in 95% EtOH	22	4	28	(—)	—	931
Gr II	0.02	$KHSO_4$	0.04	2.5% aq. PTS	22	4	95	(—)	—	931
Ind II	0.02	$KHSO_4$	0.04	2.5% aq. PTS	22	4	94	(—)	—	931
Gr II	0.02	$KHSO_4$	0.04	2.5% PTS in 95% EtOH	22	4	<5	(—)	—	931
Gr II	0.02	—	—	2.5% aq. PTS	22	12	—	(74)	—	931
Gr II	0.02	CuI	0.03	2.5% aq. PTS	22	12	—	(92)	—	931
Ru cat **80**	0.02	$KHSO_4$	0.04	2.5% aq. PTS	22	4	0	(—)	—	931
Ru cat **80**	0.02	—	—	2.5% aq. PTS	22	4	0	(—)	—	931
Ru cat **89**	0.02	$KHSO_4$	0.04	2.5% aq. PTS	22	4	88	(—)	—	931
Ru cat **89**	0.02	—	—	2.5% aq. PTS	22	4	90	(—)	>95:5	931
Ind II	0.02	—	—	2.5% aq. PTS	22	4	64	(—)	—	931
Hov II	0.02	$KHSO_4$	0.04	2.5% aq. PTS	22	4	61	(—)	—	931
Hov II	0.02	—	—	2.5% aq. PTS	22	4	62	(—)	—	931

C_{4-6}

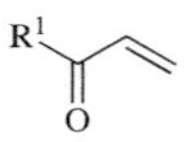

R^1 O 1.5 eq

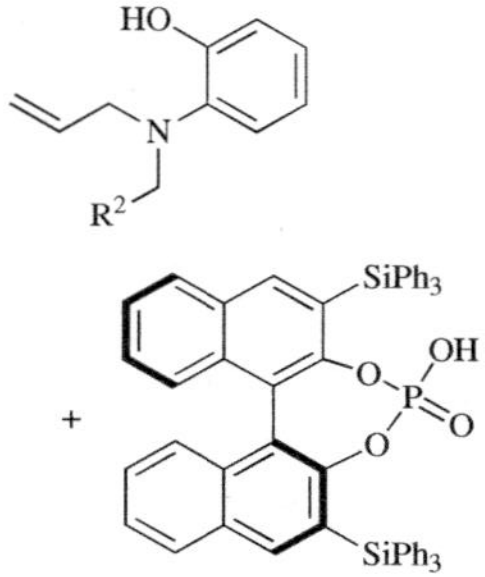

HO N R^2

\+ $SiPh_3$ OH O P O O $SiPh_3$ 0.05 eq

Hov II (0.05 eq), DCM, 40°

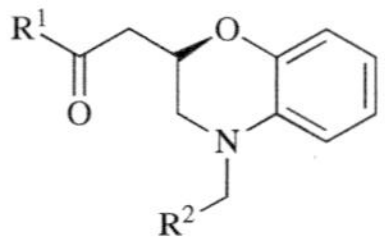

R^1 O O N R^2

R^1	R^2	Time (h)		er
Me	Ph	5	(76)	88.0:12.0
Et	Ph	12	(81)	89.0:11.0
n-Pr	Ph	12	(67)	91.0:9.0
Me	1-Naph	12	(58)	89.5:10.5
Me	2-Naph	14	(63)	82.0:18.0

973

C_4

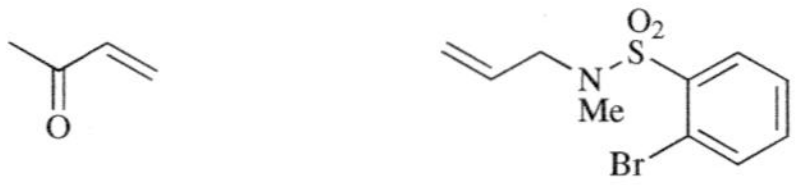

O 3 eq

N Me O_2 S Br

Catalyst (0.05 eq), DCM

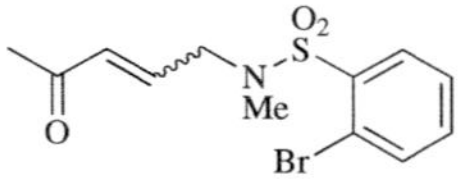

O N Me O_2 S Br

Catalyst	Temp (°)	Time (h)	
Gr II	40	3	(76)
Gr II	110 (MW)	4	(46)
Hov II	110 (MW)	2	(57)

634

TABLE 17. CROSS-METATHESIS OF α,β-UNSATURATED KETONES (*Continued*)

Ketone	Cross Partner	Conditions	Product(s) and Yield(s) (%)	Refs.
Please refer to the charts preceding the tables for the catalyst structures.				
C_{4-9}				
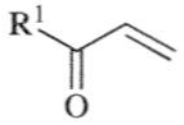 5 eq	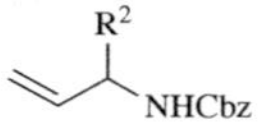 NHCbz, R^2	Hov II (0.1 eq), DCM, 40°, 48 h	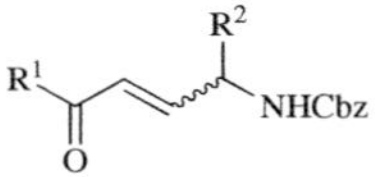 R^1, R^2, NHCbz R^1 / R^2 Me / Cy (56) Me / Ph (69) Ph / Ph (55)	934
C_4				
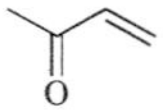 5 eq	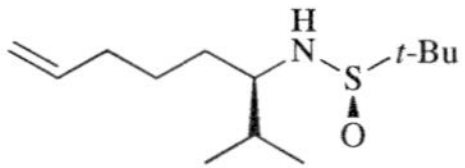 H, N, S, *t*-Bu, O	Ru cat **190** (0.1 eq), Ti(O*i*-Pr)$_4$ (0.1 eq), DCM, reflux, 2 h	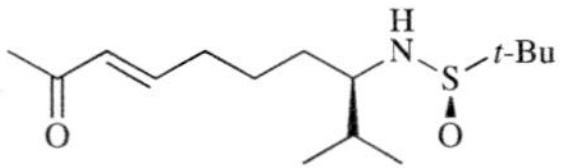 H, N, S, *t*-Bu, O (93) (*E*) only	959
5 eq	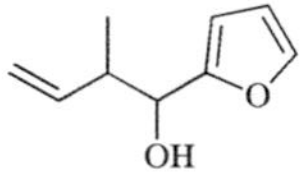 OH, O	1. Ru cat **89** (0.05 eq), DCM, 40°, 48 h 2. DMP (3.0 eq), DCM, H_2O	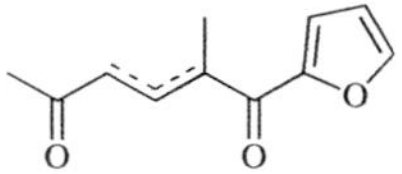 O, O (71) rr = 67:33	895
3 eq	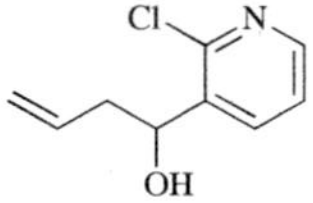 Cl, N, OH	Hov II (0.1 eq), DCM, 50°, 24 h	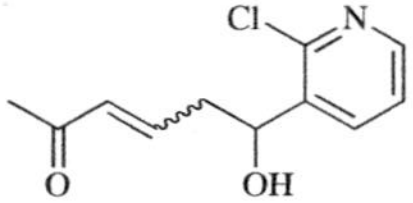 Cl, N, O, OH (78)	603
5 eq	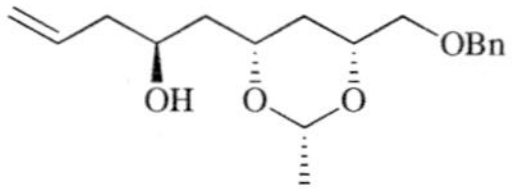 OH, O, O, OBn	Hov II (0.01 eq), DCM, 40°, 2 h	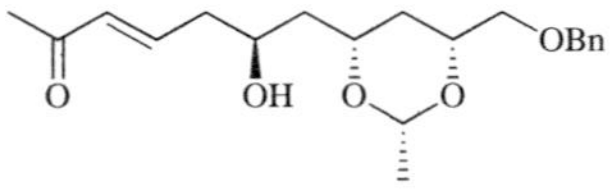 O, OH, O, O, OBn (87) (*E*)/(*Z*) > 95:5	925

Reactant 1	Reactant 2	Conditions	Product(s) and Yield(s) (%)	Refs.
R(C=O)CH=CH₂ 3 eq	OH, OTBS, OBn diene	Hov II (0.27 + 0.27 eq), DCE, reflux, 12 + 16 h	R = Me (94); TBSOCH$_2$ (91)	974
5 eq	Br, Ph	1. MeMgBr (1.2 eq), (*R*,*R*)-Taniaphos (0.012 eq), CuBr•SMe$_2$ (0.01 eq), DCM, –75° 2. Hov II (0.02 eq), DCM, rt	Ph + Ph; **I** (80), er 99.5:0.5 **I/II** > 97:3 **II**	975
x eq	CO_2R, *n*	Ru cat **129** (*y* eq), CuI (0.03 eq), Et_2O, 35°, 16 h	CO_2R, *n*	954

x	R	*n*	*y*	
6	Me	6	0.002	(25)
6	Me	6	0.005	(98)
3	Bu	7	0.001	(99)

TABLE 17. CROSS-METATHESIS OF α,β-UNSATURATED KETONES (*Continued*)

Ketone	Cross Partner	Conditions	Product(s) and Yield(s) (%)	Refs.

Please refer to the charts preceding the tables for the catalyst structures.

C_4

Ketone	Cross Partner	Conditions	Product(s) and Yield(s) (%)	Refs.
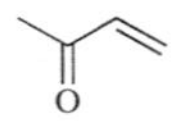 3 eq	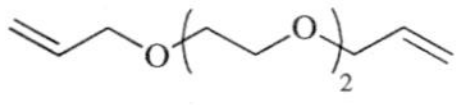	Hov II (0.03 eq) DCE, reflux, 0.5–1 h	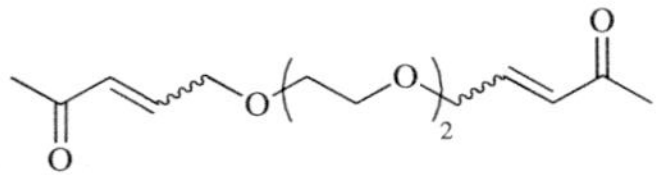 (63)	953
3 eq	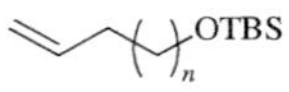 OTBS, n	Catalyst (0.02 eq), CuI (*x* eq), solvent, 22°, 12 h	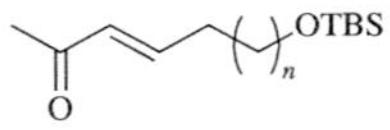 OTBS, n	

n	Catalyst	*x*	Solvent		(*E*)/(*Z*)	Refs.
7	Gr II	—	2.5 % aq. PTS	(89)	(*E*) only	976
8	Ru cat **109**	—	H_2O	(80)	(*E*) only	341
8	Gr II	—	2.5 % aq TPGS-750-M	(76)	—	167
8	Gr II	0.03	2.5 % aq TPGS-750-M	(90)	—	167

Ketone	Cross Partner	Conditions	Product(s) and Yield(s) (%)	Refs.
5 eq	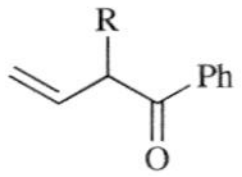 R, Ph	Ru cat **89** (0.05 eq), DCM, 55°	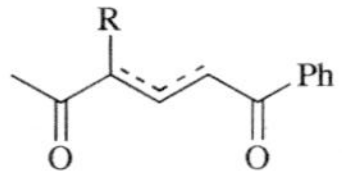 R, Ph	895

R	Time (h)		rr
H	96	(63)	77:23
Me	48	(14)	83:17

C_{4-5}

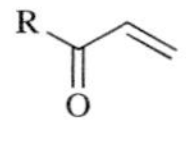

R, O, 3 eq; OPNP, Ph

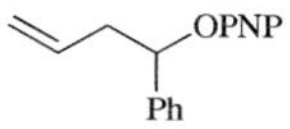

Catalyst (x eq),
KHSO$_4$ (y eq),
solvent

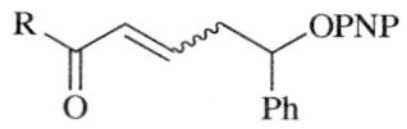

R, O, OPNP, Ph

R	Catalyst	x	y	Solvent	Temp (°)	Time (h)		
Me	Hov II	0.02	—	DCM	40	15	(95)	932
Et	Hov II	0.02	—	DCM	40	15	(97)	932
Et	Gr II	0.04	0.04	2.5 % aq. PTS	22	8	(75)	931

C_{4-9}

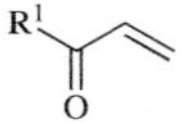

R[1], O, 3 eq; R[2], R[3], OH

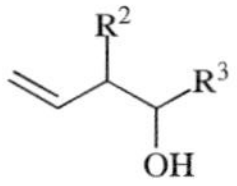

1. Ru cat **89** (0.05 eq),
DCM, 55°, 48 h
2. DMP (3.0 eq),
DCM, H_2O

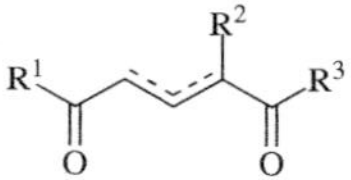

R[1], R[2], R[3], O, O — 895

R[1]	R[2]	R[3]		rr
Me	H	Ph	(75)	77:23
Me	Me	Ph	(83)	83:17
Et	EtO	4-BrC$_6$H$_4$	(44)	91:9
Ph	Me	Ph	(54)	91:9

C_4

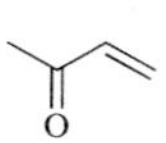

O, 2.5 eq; O

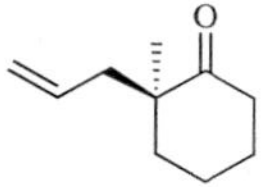

Gr II (0.05 eq),
DCM, 40°, 18 h

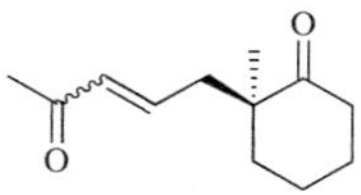

O, O — (78) — 977

TABLE 17. CROSS-METATHESIS OF α,β-UNSATURATED KETONES (*Continued*)

Ketone	Cross Partner	Conditions	Product(s) and Yield(s) (%)		Refs.
Please refer to the charts preceding the tables for the catalyst structures.					
C_4					
3 eq	OAr	1. Hov II (0.02 eq), DCM, 40°, 15 h 2. DBU (2 eq), 22°, 1 h	ArOH Ar		956
			Ph	(98)	
			4-$MeOC_6H_4$	(92)	
			4-$O_2NC_6H_4$	(91)	
			4-IC_6H_4	(92)	
			2-MeO-4-MeC_6H_3	(98)	
			3-$MeOC_6H_4$	(93)	
			4-BzC_6H_4	(90)	
			4-$OHCC_6H_4$	(86)	
			4-PhC_6H_4	(97)	
			4-*t*-BuC_6H_4	(90)	
			4-NCC_6H_4	(40)	
			4-$TBSOC_6H_4$	(92)	
			4-$AcOC_6H_4$	(85)	
			4-$BnOC_6H_4$	(94)	
1 eq	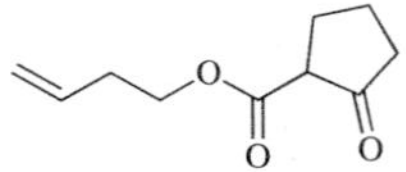	Hov II (0.03 + 0.01 eq), DCM, 100° (MW), 0.33 + 0.17 h		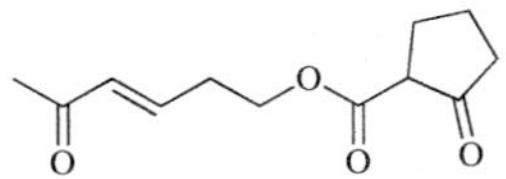 (66) (*E*)/(*Z*) = 91:9	955

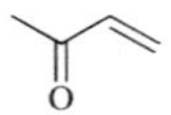

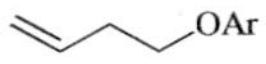

C_{4-9}

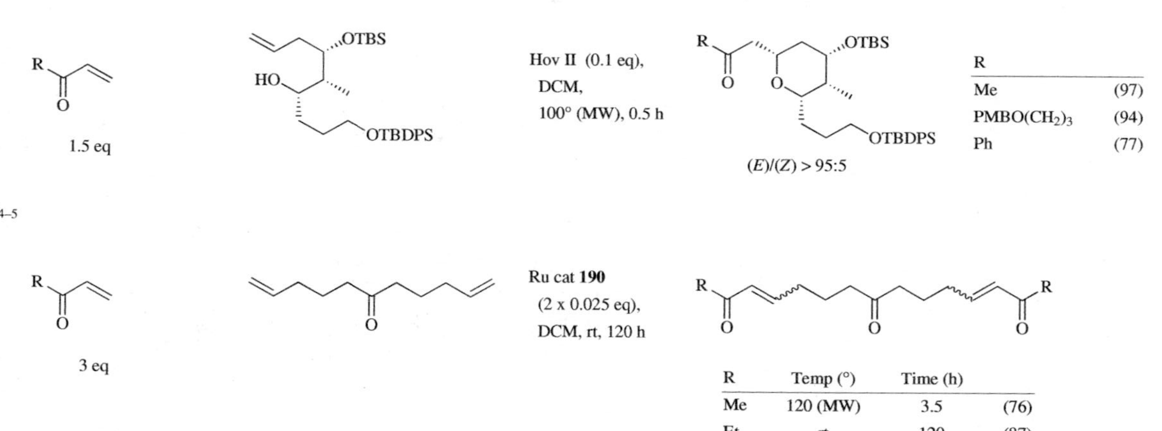

1.5 eq

Hov II (0.1 eq),
DCM,
100° (MW), 0.5 h

$(E)/(Z) > 95{:}5$

R		Ref.
Me	(97)	970
PMBO(CH$_2$)$_3$	(94)	
Ph	(77)	

C_{4-5}

3 eq

Ru cat **190**
(2 x 0.025 eq),
DCM, rt, 120 h

R	Temp (°)	Time (h)		Ref.
Me	120 (MW)	3.5	(76)	231
Et	rt	120	(87)	

TABLE 17. CROSS-METATHESIS OF α,β-UNSATURATED KETONES (*Continued*)

Ketone	Cross Partner	Conditions	Product(s) and Yield(s) (%)	Refs.
Please refer to the charts preceding the tables for the catalyst structures.				
C_4				
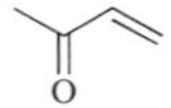 3 eq	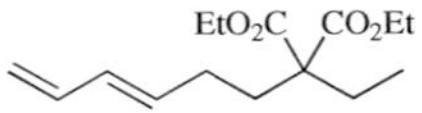 EtO_2C CO_2Et (*E*)/(*Z*) = 50:50	Hov II (0.05 eq), DCM, 45°, 3 h	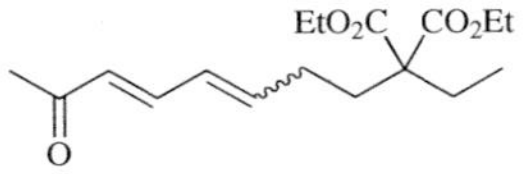 EtO_2C CO_2Et (38) 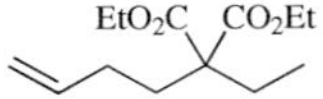 + EtO_2C CO_2Et (11) 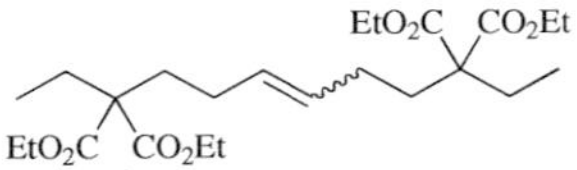 + EtO_2C CO_2Et EtO_2C CO_2Et (13)	939
3 eq	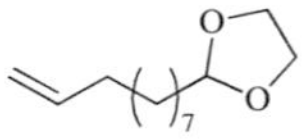 O O ( )$_7$	Gr II (0.02 eq), $KHSO_4$ (0.03 eq), 2.5 % aq PTS, 22°, 4 h	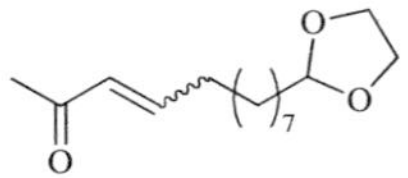 O O ( )$_7$ (90)	931
C_{4-9}				
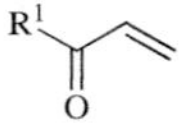 R[1] O 5 eq	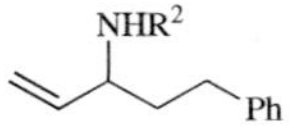 NHR^2 Ph	Hov II (0.1 eq), DCM, 40°, 48 h	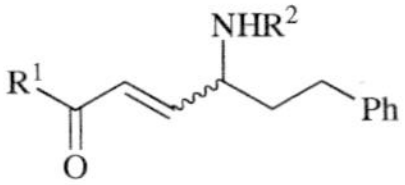 R[1] O NHR^2 Ph	934

R^1	R^2	
Me	Cbz	(72)
Me	Ts	(73)
Me	Boc	(73)
Me	$COCF_3$	(60)
Et	Cbz	(71)
Ph	Cbz	(59)

C_4

1 eq

Hov **II** (x eq),
additive (0.1 eq),
solvent, 12 h

952

x	Additive	Solvent	Temp (°)	
0.05	—	toluene	80	(40)
0.05	ClBcat	toluene	80	(42)
0.05	Cy_2BCl	toluene	80	(99)
0.01	Cy_2BCl	toluene	80	(70)
0.05	Cy_2BCl	DCM	40	(99)

3 eq

Ts
N

Hov **II** (0.05 eq),
DCM, 45°, 3 h

O Ts N O

(84) (E)/(Z) > 95:5 939

TABLE 17. CROSS-METATHESIS OF α,β-UNSATURATED KETONES (*Continued*)

Ketone	Cross Partner	Conditions	Product(s) and Yield(s) (%)	Refs.
Please refer to the charts preceding the tables for the catalyst structures.				
C_4				
1 eq		Hov II (0.05 + 0.02 eq), DCM, 40°, 3 h + overnight	(80) (*E*) only	618
3 eq	R	Hov II (0.03 eq), DCE, reflux, 0.5–1 h	R: H (92); HO (86)	953
C_{4-9}				
R^1, 6 eq	R^2	Gr II, DCM, 40°, 16 h	R^1, R^2	978

R^1	R^2		(*E*)/(*Z*)
Me	Cbz	(34)	>98:2
Me	MeO_2C	(50)	>98:2
Me	CF_3CO	(41)	>98:2
Me	Bn	—	—
Ph	Boc	(38)	>98:2
Me	Boc	(65)	>98:2

Cross partner	Alkene	Conditions	Product(s) and Yield(s) (%)	Refs.
C_4				
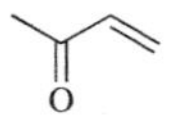 1 eq	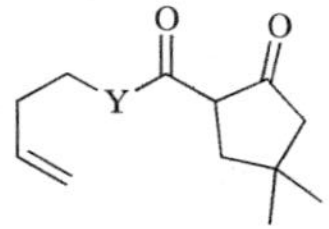	1. Hov II (0.03 + 0.01 eq), DCM, 100° (MW), 20 + 10 min 2. IPr (0.2 eq), 24°, 20 h	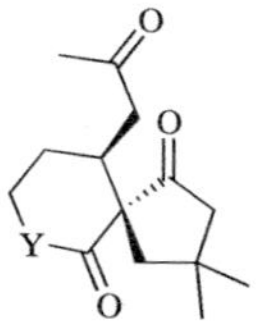 (E)/(Z) > 95:5 Y: BnN (79); O (44)	979
3 eq	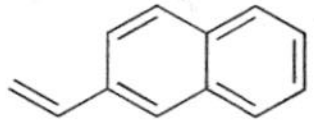	Hov II (0.03 eq), DCE, reflux, 0.5–1 h	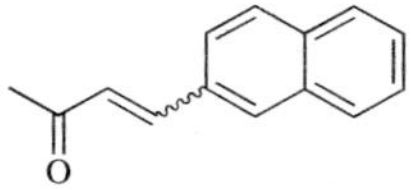 (81)	953
C_{4-5}				
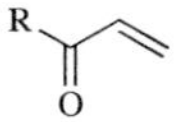 5 eq	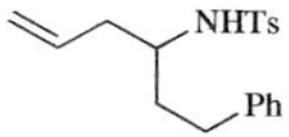 NHTs, Ph	Hov II (0.075 eq), DCM, 55°, 48 h	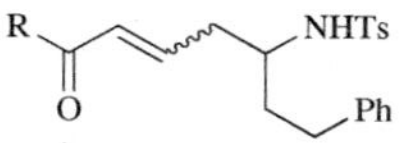 NHTs, Ph R: Me (79); Et (71)	335
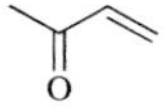 5 eq	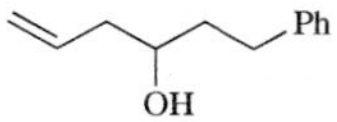 OH, Ph	1. Ru cat **89** (0.05 eq), DCM, 40°, 48 h 2. DMP (3 eq), DCM, H_2O	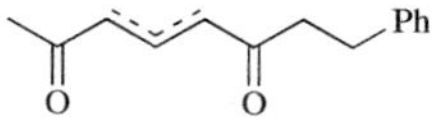 Ph (71) rr = 53:47	895

TABLE 17. CROSS-METATHESIS OF α,β-UNSATURATED KETONES (*Continued*)

Ketone	Cross Partner	Conditions	Product(s) and Yield(s) (%)	Refs.
Please refer to the charts preceding the tables for the catalyst structures.				
C_4				
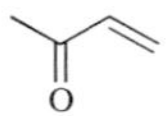 O; 2 eq	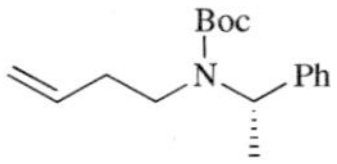 Boc, N, Ph	Ru cat **185** (0.05 eq), solvent	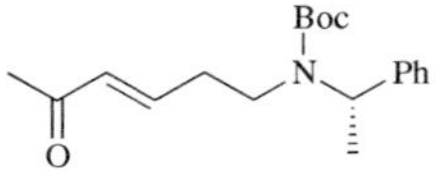 Boc, N, Ph, O; (*E*)/(*Z*) > 95:5	924
3 eq	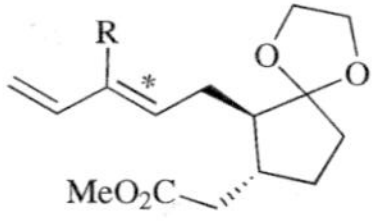 R, *, O, O, MeO_2C	Hov II (0.05 eq), DCM, 45°, 3 h	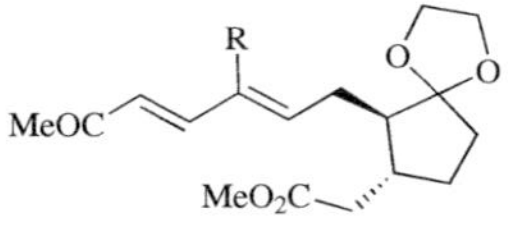 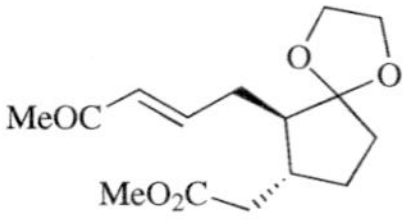 MeOC, R, O, O, MeO_2C **I** + MeOC, O, O, MeO_2C **II**	939
5 eq	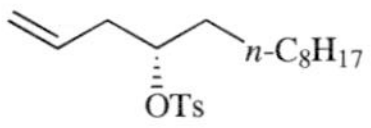 n-C_8H_{17}, OTs	Gr II (0.05 eq), DCM, 40°, 6 h	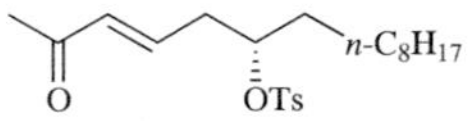 O, OTs, n-C_8H_{17}; (84) (*E*) only	980

Solvent	Temp (°)	Time (h)	
BMI•PF_6/toluene	rt	6	(80)
BMI•PF_6/DCM	40	3	(94)

R	*(*E*)/(*Z*)	**I**	(*E*)/(*Z*) **I**	**II**	(*E*)/(*Z*) **II**
H	50:50	(35)	50:50	(38)	50:50
Me	80:20	(73)	80:20	(0)	—

Eq	Substrate	Conditions	Product(s) and Yield(s) (%)		Refs.
5 eq	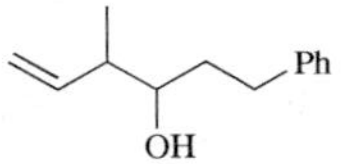 Ph, OH	1. Ru cat **89** (0.05 eq), DCM, 40°, 48 h 2. DMP (3 eq), DCM, H_2O	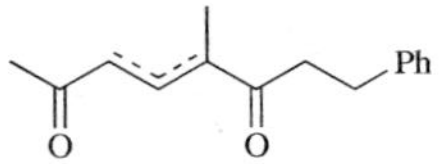 O, O, Ph	(73) dr 50:50	895
3.3 eq	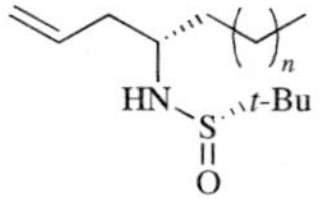 HN, S, O, *t*-Bu, *n*	Hov II (0.1 eq), DCM, 45°, 48 h	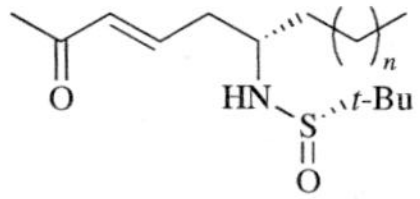 O, HN, S, O, *t*-Bu, *n* (*E*) only	*n* 7 (72); 9 (75)	969
5 eq	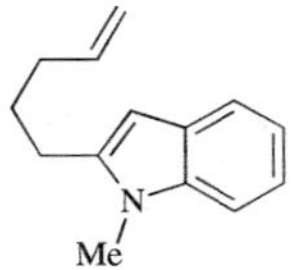 N, Me	Hov II (0.03 eq), DCE, reflux, 0.5 h	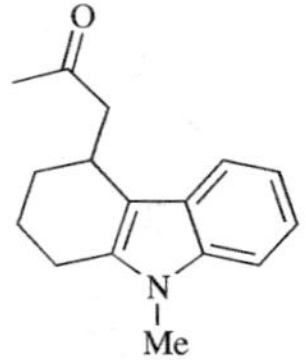 O, N, Me	(98)	950
5 eq	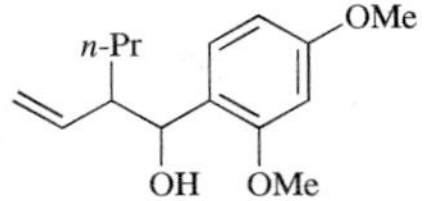 *n*-Pr, OH, OMe, OMe	1. Ru cat **89** (0.05 eq), DCM, 40°, 48 h 2. DMP (3 eq), DCM, H_2O	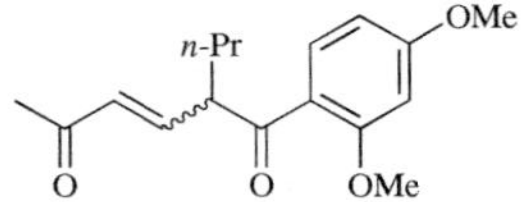 *n*-Pr, O, O, OMe, OMe	(52)	895

TABLE 17. CROSS-METATHESIS OF α,β-UNSATURATED KETONES (*Continued*)

Ketone	Cross Partner	Conditions	Product(s) and Yield(s) (%)	Refs.
Please refer to the charts preceding the tables for the catalyst structures.				
C_4				
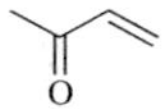 6 eq	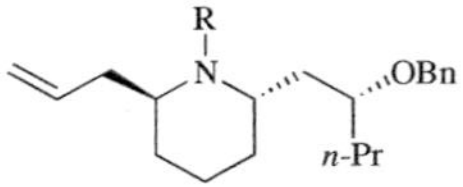 R, N, OBn, *n*-Pr	Hov II (*x* eq), additive, toluene, 80°	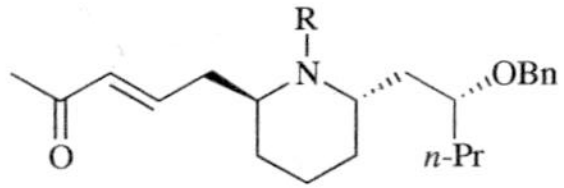 R, N, O, OBn, *n*-Pr	389
			R | *x* | Additive | Time (h) | | (*E*)/(*Z*) Bn | 0.1 | Ti(O*i*-Pr)$_4$ | — | (0) | — MeO$_2$C | 2 x 0.15 | — | 2 x 0.5 | (41) | 100:0	
10 eq	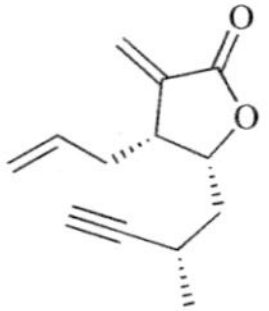 O, O	Hov II (0.2 eq), DCM, 45°, 13 h	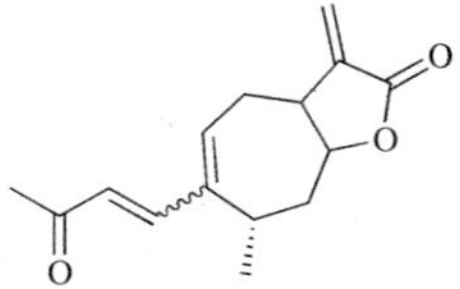 O, O, O (83)	531
	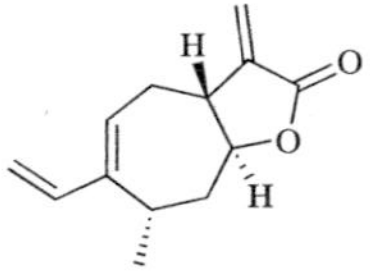 H, O, O, H	Hov II, DCM, reflux	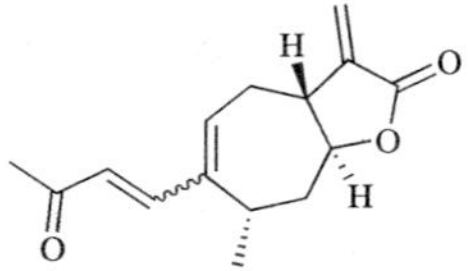 H, O, O, H, O (49)	981
3 eq	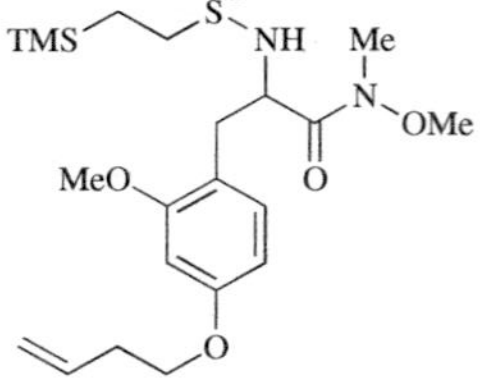 TMS, O$_2$, S, NH, Me, N, OMe, O, MeO, O	1. Hov II (0.02 eq), DCM, 40°, 15 h 2. DBU (2 eq), 22°, 1 h	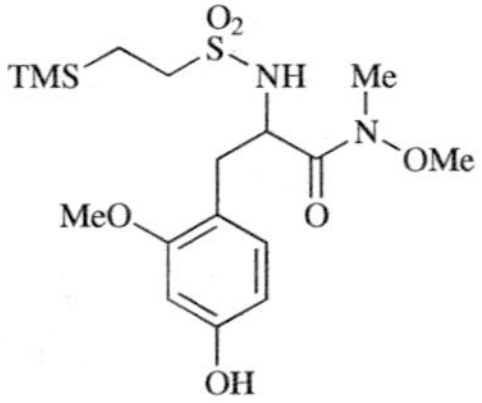 TMS, O$_2$, S, NH, Me, N, OMe, O, MeO, OH (90)	956

C_{4-13}

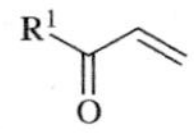

1.5 eq

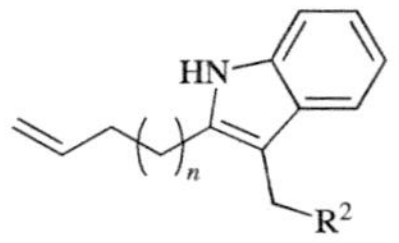

Ru cat **89** (0.03 eq), DCM, 45°

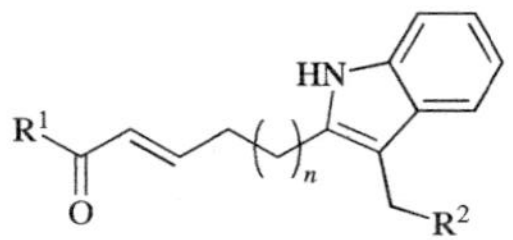

(E)/(Z) > 98:2

982

R^1	R^2	*n*	
Me	H	2	(94)
Ph	H	1	(36)
Ph	H	2	(45)
Ph	Me	2	(68)
Ph	$HOCH_2$	2	(79)
Ph	Et	2	(73)
Ph	$AcCH_2$	2	(67)
Ph	$BzCH_2$	2	(79)
Ph	Ph	2	(81)
4-BrC_6H_4	H	2	(36)
4-ClC_6H_4	H	2	(57)
4-MeC_6H_4	H	2	(55)
4-$MeOC_6H_4$	H	2	(46)
1-Naph	H	2	(53)

TABLE 17. CROSS-METATHESIS OF α,β-UNSATURATED KETONES (*Continued*)

Ketone	Cross Partner	Conditions	Product(s) and Yield(s) (%)	Refs.

Please refer to the charts preceding the tables for the catalyst structures.

C_{4-9}

Ketone	Cross Partner	Conditions	Product(s) and Yield(s) (%)	Refs.
R^1, O; *x* eq	N, Boc, R^2, H, N, R^3	Catalyst (*y* eq), DCM	R^1, O, N, Boc, R^2, H, N, R^3	

R^1	*x*	R^2	R^3	Catalyst	*y*	Temp (°)	Time (h)		(*E*)/(*Z*)	Refs.
Me	—	Ph	H	Gr II	—	40	16	(43)	—	978
Ph	1.5	H	Me	Ru cat **89**	0.03	45	—	(69)	>98:2	982
Ph	1.5	H	Ph	Ru cat **89**	0.03	45	—	(43)	>98:2	982

C_4

Ketone	Cross Partner	Conditions	Product(s) and Yield(s) (%)	Refs.
O; 3 eq	O	1. Hov II (0.02 eq), DCM, 40°, 15 h 2. DBU (2 eq), 22°, 1 h	HO (93)	956
3 eq	Ph, OAc, OAc	1. Hov II (0.05 eq), DCM, rt, 16 h 2. TBAF, THF, rt, 24 h	O, Ph (51)	919

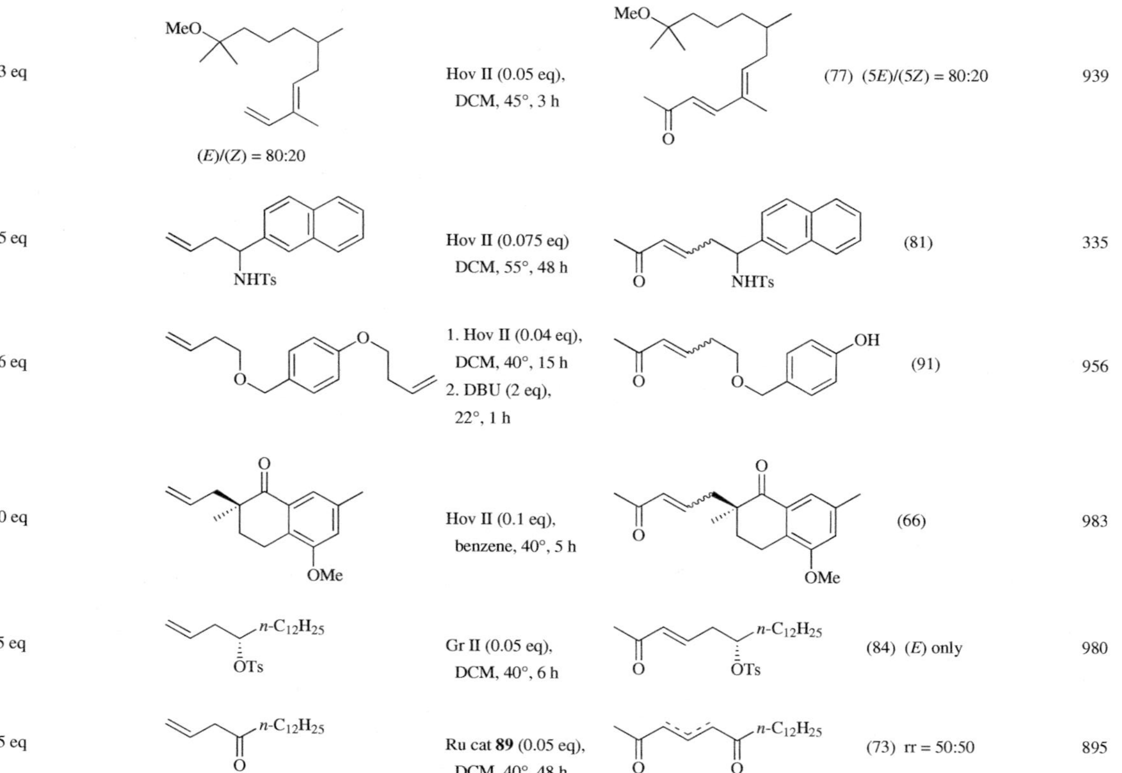

Equiv	Substrate	Conditions	Product	Yield (%), notes	Ref.
3 eq	MeO; $(E)/(Z) = 80{:}20$	Hov II (0.05 eq), DCM, 45°, 3 h	MeO; O	(77) $(5E)/(5Z) = 80{:}20$	939
5 eq	NHTs	Hov II (0.075 eq) DCM, 55°, 48 h	O; NHTs	(81)	335
6 eq	O; O	1. Hov II (0.04 eq), DCM, 40°, 15 h 2. DBU (2 eq), 22°, 1 h	O; O; OH	(91)	956
10 eq	O; OMe	Hov II (0.1 eq), benzene, 40°, 5 h	O; O; OMe	(66)	983
5 eq	n-$C_{12}H_{25}$; OTs	Gr II (0.05 eq), DCM, 40°, 6 h	O; n-$C_{12}H_{25}$; OTs	(84) (*E*) only	980
5 eq	n-$C_{12}H_{25}$; O	Ru cat **89** (0.05 eq), DCM, 40°, 48 h	O; n-$C_{12}H_{25}$; O	(73) rr = 50:50	895

TABLE 17. CROSS-METATHESIS OF α,β-UNSATURATED KETONES (*Continued*)

Ketone	Cross Partner	Conditions	Product(s) and Yield(s) (%)	Refs.
Please refer to the charts preceding the tables for the catalyst structures.				
C_4				
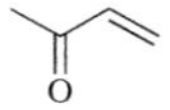 O; 10 eq	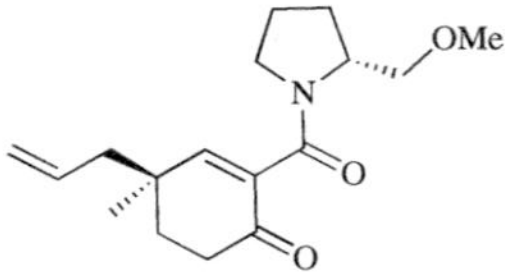 OMe, N, O, O	Hov II (0.025 eq), DCM, reflux, 7 h	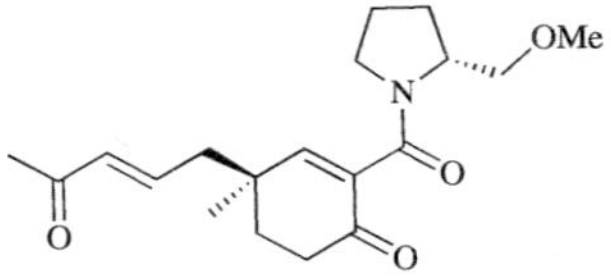 O, OMe, N, O, O (87) (E)/(Z) = 89:11	896
6 eq	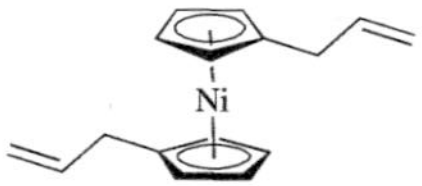 Ni	Gr II (0.06 eq), DCM, reflux, 3 h	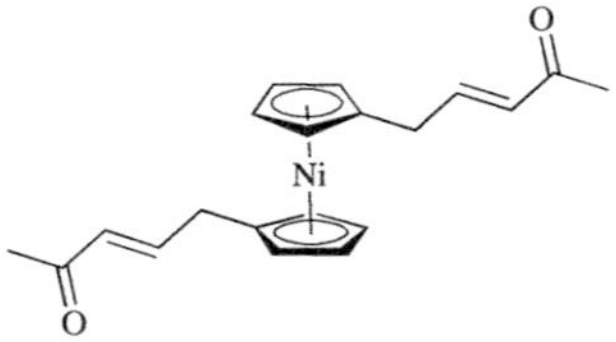 O, Ni, O (82) (E) only	984
10 eq	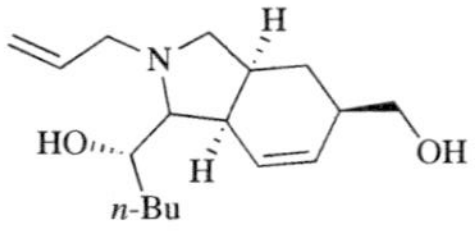 N, H, HO, H, n-Bu, OH	1. Hov II (0.1 eq), TsOH (1 eq), DCM 2. 100° (MW), 4 h	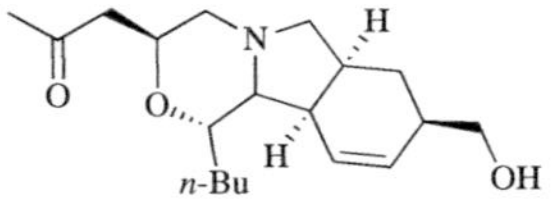 O, N, H, O, H, n-Bu, OH (64)	985

Reactant		Conditions	Product(s) and Yield(s) (%)		Refs.
10 eq	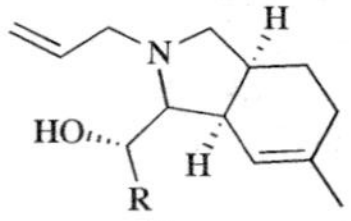 R	1. Hov II (0.1 eq), TsOH (1 eq), DCM 2. 100° (MW), 4 h	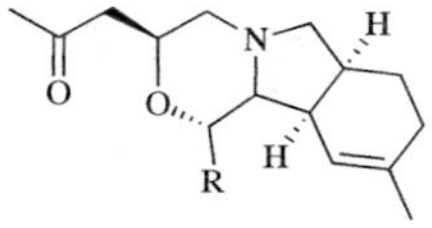 R	**R** *i*-Bu (82) *n*-C_5H_{11} (55) Bn (84)	985
10 eq	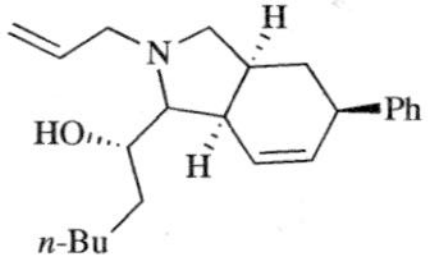 *n*-Bu, Ph	1. Hov II (0.1 eq), TsOH (1 eq), DCM 2. 100° (MW), 4 h	*n*-Bu, Ph	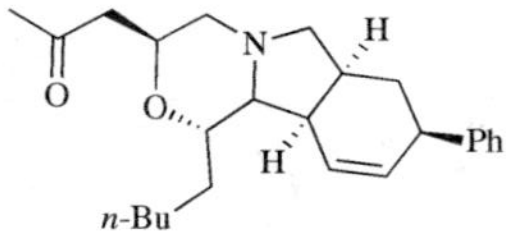 (71)	985
10 eq	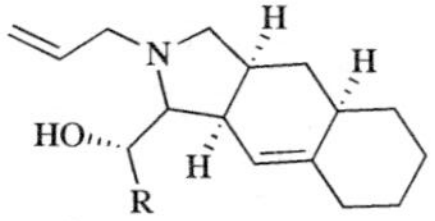 R	1. Hov II (0.1 eq), TsOH (1 eq), DCM 2. 100° (MW), 4 h	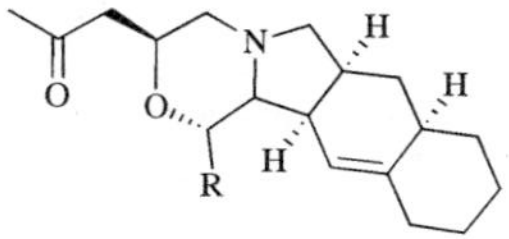 R	**R** *i*-Bu (65) *n*-C_5H_{11} (56)	985

TABLE 17. CROSS-METATHESIS OF α,β-UNSATURATED KETONES (*Continued*)

Ketone	Cross Partner	Conditions	Product(s) and Yield(s) (%)	Refs.

Please refer to the charts preceding the tables for the catalyst structures.

C_4

Ketone	Cross Partner	Conditions	Product(s) and Yield(s) (%)	Refs.
10 eq	CF3, H N, O, Cl, O, N, R2, R1, N H, O	Hov II (*x* eq), DCM, 100° (MW)	CF3, H N, O, Cl, O, N, R2, R1, N H, O, O; (*E*) only	

R^1	R^2	*x*	Time (h)	Yield (%)	Refs.
H	Me	0.1	2	(80)	986
H	Ph	0.2	1	(70)	987
$Ph(CH_2)_2$	Ph	0.2	1	(99)	987

Ketone	Cross Partner	Conditions	Product(s) and Yield(s) (%)	Refs.
5 eq	OH, OMOM, O, O, Cl	Ru cat **363** (0.051 eq), DCM, 60°, 41 h	OH, OMOM, O, O, Cl, O (65)	988

5 eq	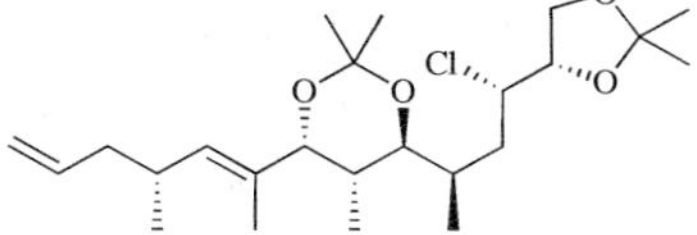	Ru cat **363** (0.05 eq), DCM, 60°, 42.5 h	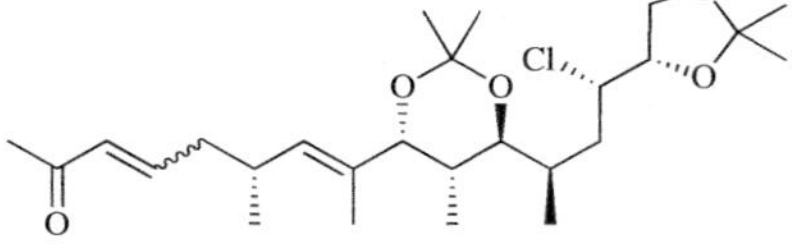	(90)	988
3 eq	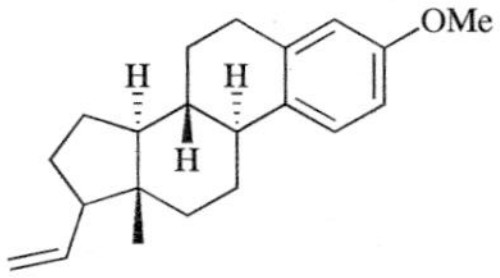	Hov II (0.05 eq), DCM, 45°, 24 h	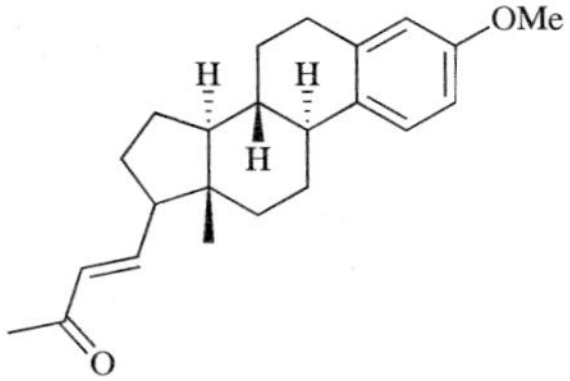	(39) (*E*)/(*Z*) = 95:5	939
5 eq	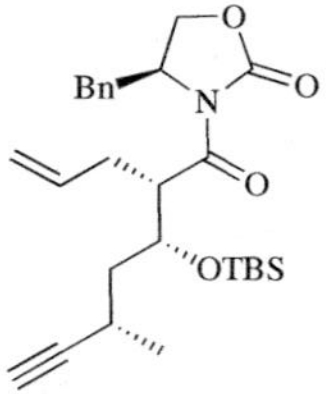	Hov II (0.2 eq), DCM, 45°	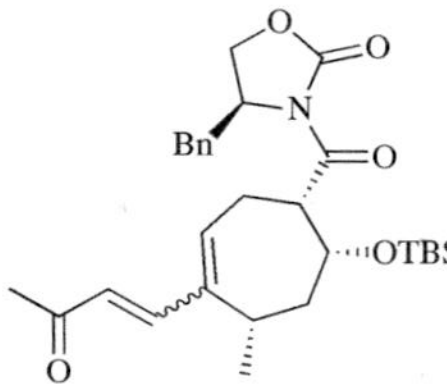	(62)	531

TABLE 17. CROSS-METATHESIS OF α,β-UNSATURATED KETONES (*Continued*)

Ketone	Cross Partner	Conditions	Product(s) and Yield(s) (%)	Refs.
Please refer to the charts preceding the tables for the catalyst structures.				
C_4				
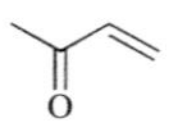 3 eq	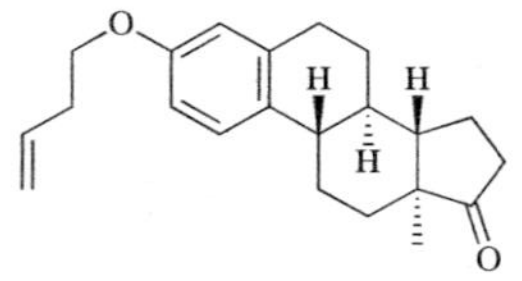	1. Hov II (0.02 eq), DCM, 40°, 15 h 2. DBU (2 eq), 22°, 1 h	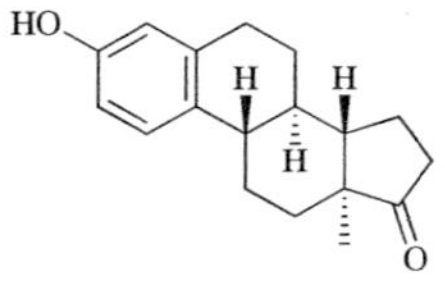 (88)	956
3 eq	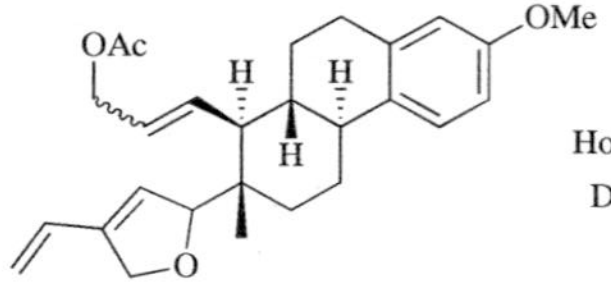	Hov II (0.05 eq), DCM, 45°, 24 h	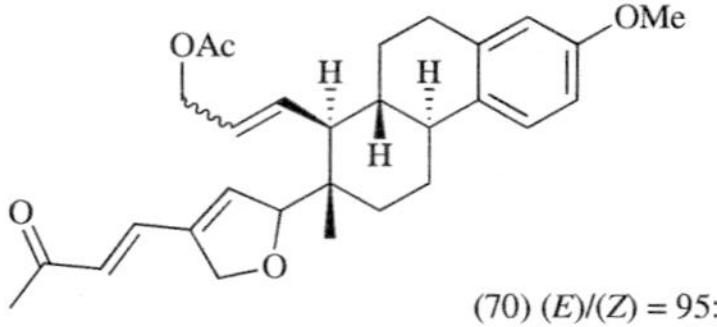 (70) (*E*)/(*Z*) = 95:5	939
10 eq	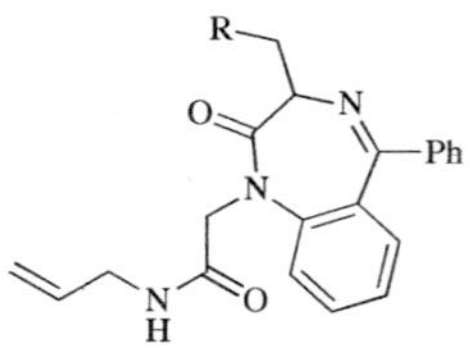	Hov II (0.1 eq), DCM, 100° (MW), 2 h	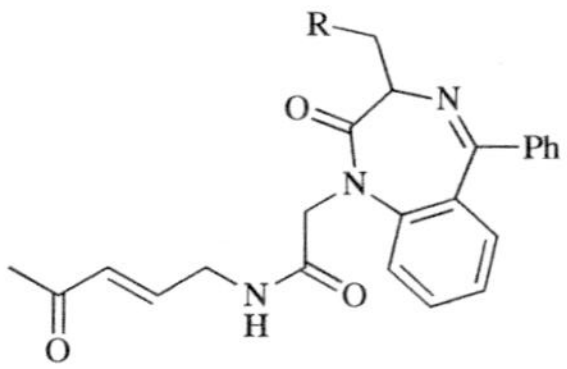 (*E*) only R: 4-morpholinyl (85); 1-piperidinyl (83); *i*-BuHN (87)	986

	Olefin 1	Olefin 2	Conditions	Product(s) and Yield(s) (%)		Refs.
	BnO, O; *x* eq	Bz, N, R¹, R²	Hov II (0.05 eq), DCM, reflux, 60 h	BnO, O, Bz, N, R¹, R²	(*E*) only	717

x	R^1	R^2	
2	H	H	(80)
4	H	BnO	(57)
4	BnO	H	(76)

	Olefin 1	Olefin 2	Conditions	Product(s) and Yield(s) (%)		Refs.
	10 eq	OH, OTBS, OTBS, H, H, O, O, H, CO_2Me	Hov II (0.05 eq), toluene, 80°	BnO, O, OH, OTBS, OTBS, H, H, O, O, H, CO_2Me	(81)	989
C_5	O; 3 eq	OTBS	1. Gr II (0.1 eq), DCM, reflux, 3 h 2. DIBAL-H (3.75 eq) DCM, –78°, 2 h	OH, OTBS	(72) (*E*)/(*Z*) > 95:5	940
	5 eq	NHBoc, CO_2Me	Gr II (0.05 eq), DCM, reflux, 16 h	O, NHBoc, CO_2Me	(—)	943

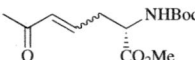

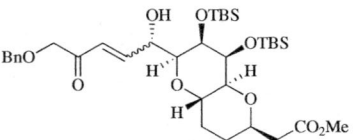

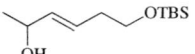

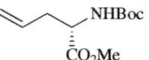

TABLE 17. CROSS-METATHESIS OF α,β-UNSATURATED KETONES (*Continued*)

Ketone	Cross Partner	Conditions	Product(s) and Yield(s) (%)	Refs.
Please refer to the charts preceding the tables for the catalyst structures.				
C_5				
O; 5 eq	NHCbz	Gr II (0.05 eq), DCM, 45°, 48 h	O; NHCbz (67)	944
C_{5-11}				
R[1], R[2], O; 1.5 eq	O; Cl	Gr II (0.025 eq), DCM, 45°, 12 h	R[1], O, O, Cl R[1] / R[2] Me / Me (82) $Ph(CH_2)_2$ / H (83)	945
C_5				
O; 3 eq	Ph	1. Gr II (0.1 eq), DCM, reflux, 3 h 2. DIBAL-H (3.75 eq) DCM, –78°, 2 h	Ph, OH (70) (*E*)/(*Z*) = 95:5	940

3–10 eq

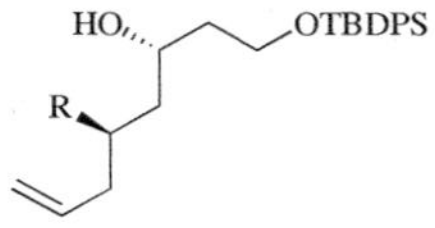

HO, OTBDPS, R

Hov II (0.1 eq), additive (0.03–0.1 eq), solvent

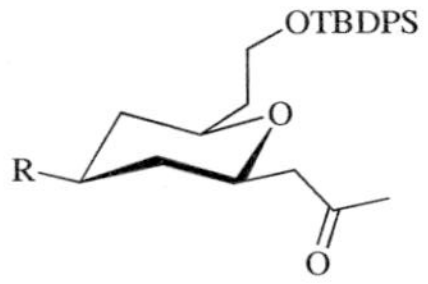

OTBDPS, O, R, O

947

R	Additive	Solvent	Temp (°)	Time (h)		(E)/(Z)
H	—	toluene	80–100	11–24	(73)	75:15
H	CSA	DCM	25–35	—	(19)	>95:5
TIPSO	—	DCM	100 (MW)	0.5	(73)	83:17
TIPSO	—	toluene	80–100	11–24	(80)	71:29
TIPSO	CSA	DCM	25–35	—	(73)	>95:5

x eq

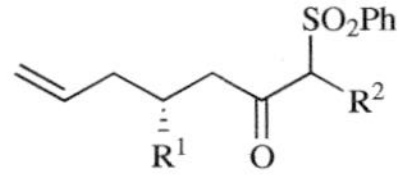

SO_2Ph, R^2, R^1, O

Hov II (y eq), DCM, rt

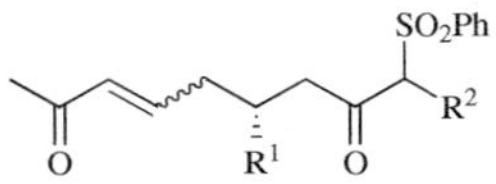

SO_2Ph, R^2, O, R^1, O

990

x	R^1	R^2	y	Time (h)	
3	H	Me	0.1	24	(70)
5	H	$TBSO(CH_2)_2$	0.05	24	(57)
1.5	H	$N_3(CH_2)_3$	0.05	3.5	(63)
3	H	$PhCH_2$	0.05	24	(65)
1.5	Me	$N_3(CH_2)_3$	0.05	2	(63)

TABLE 17. CROSS-METATHESIS OF α,β-UNSATURATED KETONES (*Continued*)

Ketone	Cross Partner	Conditions	Product(s) and Yield(s) (%)	Refs.
Please refer to the charts preceding the tables for the catalyst structures.				
C_5				
1.5 eq	SO₂Ph	Hov II (0.05 eq), DCM, rt, 36 h	SO₂Ph (88)	991
		Catalyst (0.05 eq), DCM		
			Catalyst: Gr II; Temp (°): 40; Time (h): 3; (94); (*E*)/(*Z*): (*E*) only	170
			Catalyst: Hov II; Temp (°): 100 (MW); Time (h): 0.25; (>99); (*E*)/(*Z*): —	165
3 eq	OR	Hov II (0.05 eq), DCM, 40°	OR; (*E*)/(*Z*) = 95:5; R: TBDPS (76), TBS (80)	169
— eq	OH, OH	Hov II (0.025 eq), DCM, rt, 6 h	OH, OH (60)	967

Reactant	Reactant	Conditions	Product(s) and Yield(s) (%)		Refs.
1.5 eq		Catalyst (0.1 eq), DCM, 40°, 24 h	CO_2Et	Catalyst: Gr II (46); Hov II (65)	650
			(E)/(Z) = 91:9		
3 eq	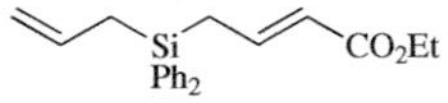 Si Ph$_2$ CO_2Et	Hov II (0.025 eq), DCM, rt, 12 h	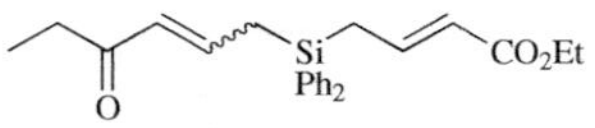 Si Ph$_2$ CO_2Et	(91)	992
	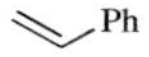 Ph 3 eq 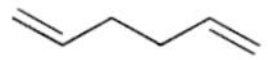 +	Gr II (0.07 eq), DCM, 23°, 12 h	Ph	(47) (E) only	36
3 eq	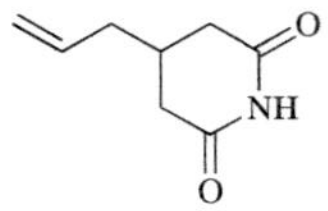 NH	Hov II (0.05 eq), DCM, rt	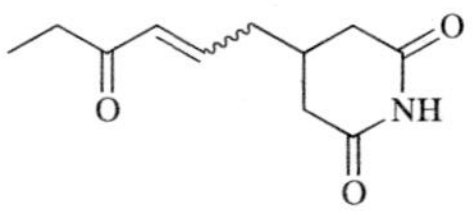 NH	(60)	993
C_{5-10}					
R 3 eq	TBSO	Gr II (0.02 eq), $KHSO_4$ (0.04 eq), 2.5 % aq. PTS, 22°, 4 h	R TBSO	R: Et (88); $Ph(CH_2)_2$ (81)	931

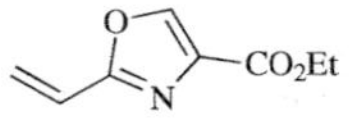

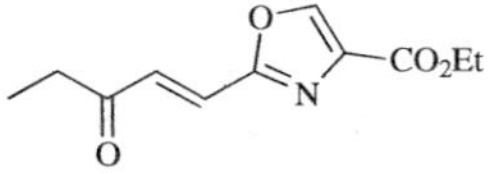

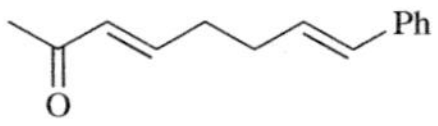

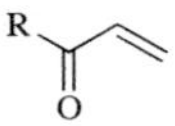

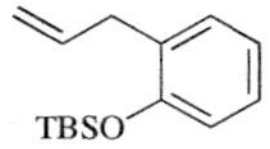

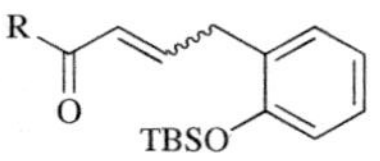

TABLE 17. CROSS-METATHESIS OF α,β-UNSATURATED KETONES (*Continued*)

Ketone	Cross Partner	Conditions	Product(s) and Yield(s) (%)	Refs.
Please refer to the charts preceding the tables for the catalyst structures.				
C_5				
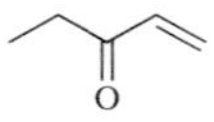 6 eq	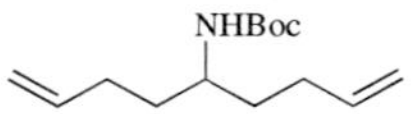 NHBoc	Hov II (0.074 eq), DCM, rt, 144 h	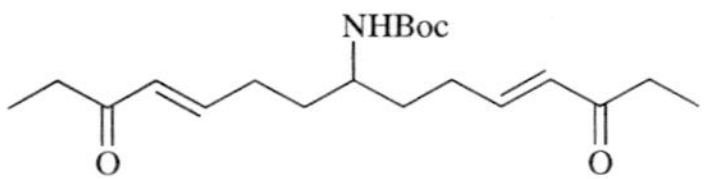 NHBoc (34) (*E*) only 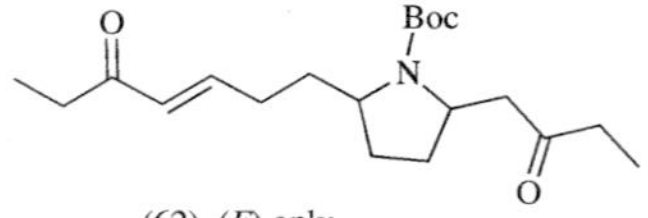 + Boc N (62) (*E*) only	994
1.5 eq	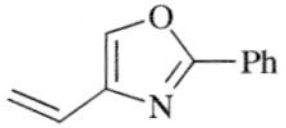 O N Ph	Hov II (0.1 eq), DCM, 40°, 36 h	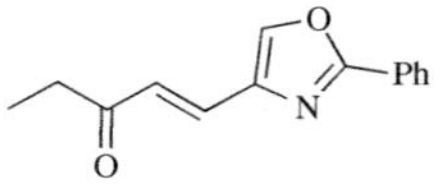 O N Ph (72) (*E*)/(*Z*) > 95:5	650
2000 eq	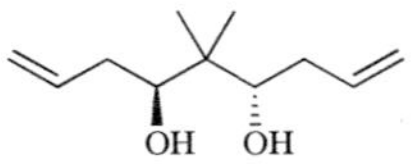 OH OH	Hov II (0.1 eq), DCE, 80°	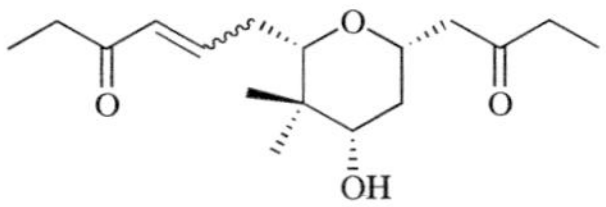 O OH (76) dr 91:9	369

Alkene (eq)	Substrate	Conditions	Product	Yield (%)	Ref.
10 eq	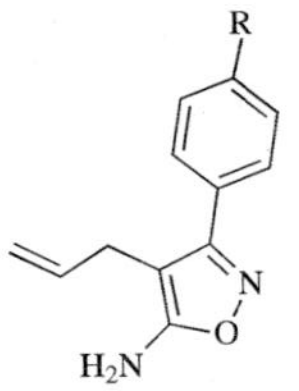 R; H_2N; N; O	Hov II (2 x 0.05 eq), DCM, rt, 48 h, green LED (36 W)	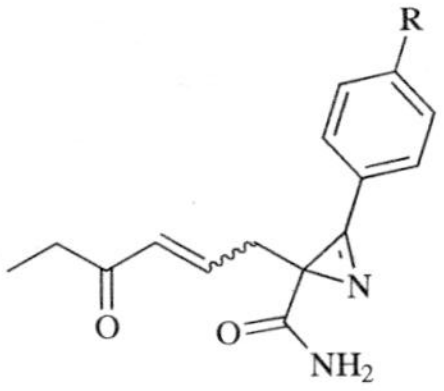 R; N; O; O; NH_2	R: H (72), Cl (65), Br (33), MeO (49), Me (51)	995
10 eq	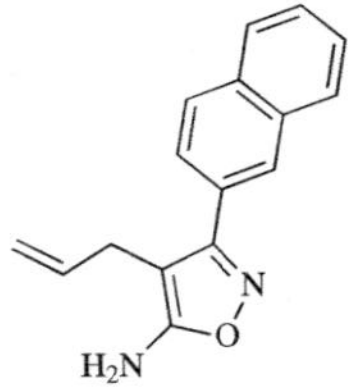 H_2N; N; O	Hov II (2 x 0.05 eq), DCM, rt, 48 h, green LED (36 W)	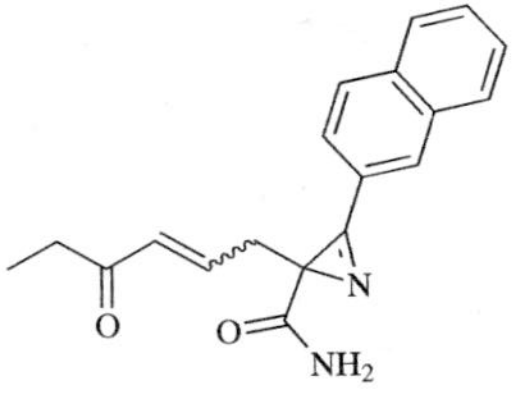 N; O; O; NH_2	(62)	995
4 eq	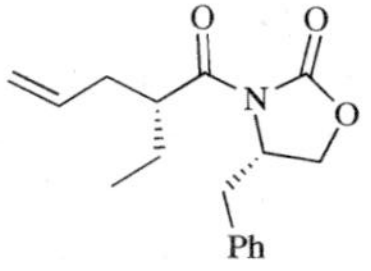 O; O; N; O; Ph	Hov II (0.05 eq), DCM, rt, 3 h	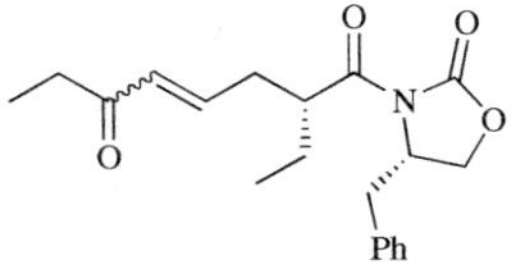 O; O; O; N; O; Ph	(98)	996

TABLE 17. CROSS-METATHESIS OF α,β-UNSATURATED KETONES (*Continued*)

Ketone	Cross Partner	Conditions	Product(s) and Yield(s) (%)	Refs.
Please refer to the charts preceding the tables for the catalyst structures.				
C_5				
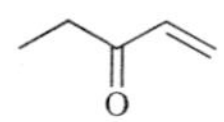 O; 2 eq	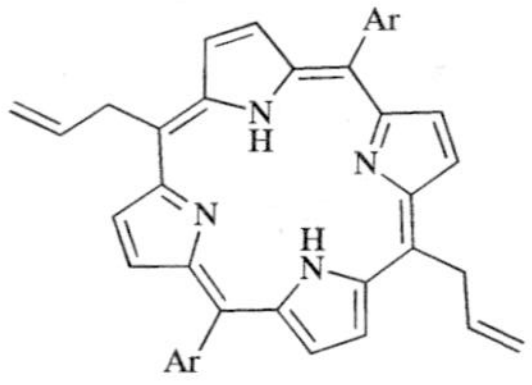 Ar, N, H, N, H, N, N, Ar	Gr II (0.1 eq), DCM, reflux	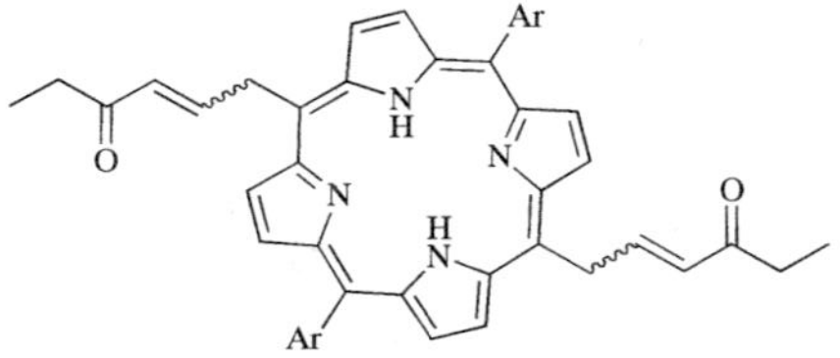 O, Ar, N, H, N, H, N, N, O, Ar Ar: Ph (94); 3-MeOC$_6$H$_4$ (94)	697
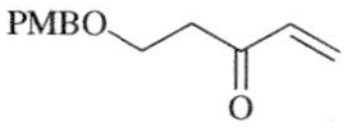 PMBO, O; 1.5 eq	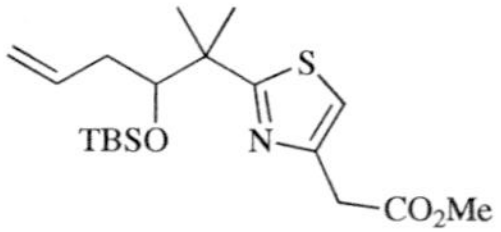 TBSO, S, N, CO$_2$Me	Gr II (5 x 0.005 eq) DCM, 50° (MW), 5 x 0.25 h	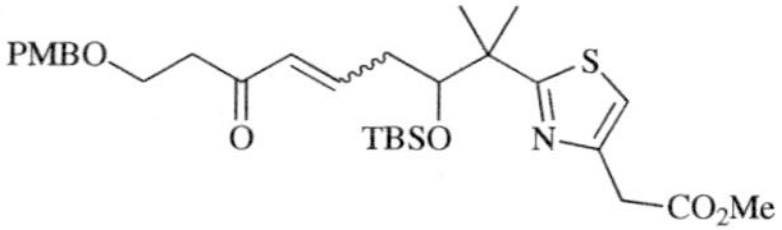 PMBO, O, TBSO, S, N, CO$_2$Me (50)	997
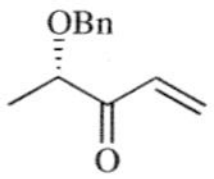 OBn, O; 1.5 eq	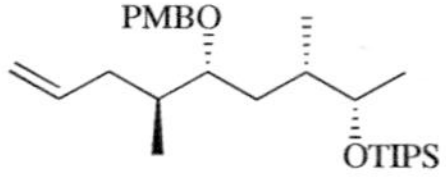 PMBO, OTIPS	Hov II (0.1 eq), DCM, 40°, overnight	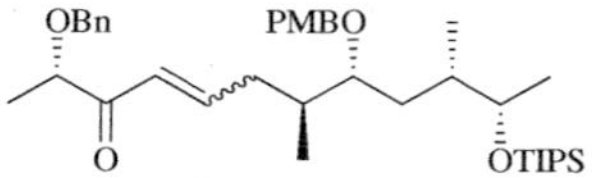 OBn, O, PMBO, OTIPS (70)	998
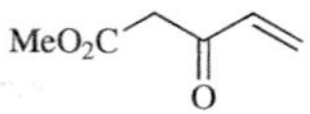 MeO$_2$C, O; 2 eq	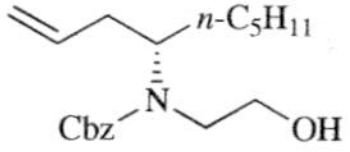 n-C$_5$H$_{11}$, Cbz, N, OH	Hov II (0.075 eq), DCM, reflux, 20 h	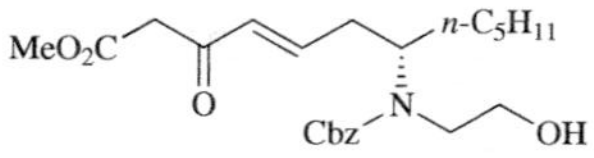 MeO$_2$C, O, n-C$_5$H$_{11}$, Cbz, N, OH (70) (E) only	999

MeO,,P(=O)(MeO) ... O   x eq   R^2 ... R^1 (n)

Catalyst (y eq), DCM

MeO,,P(=O)(MeO) ... O ... (n) R^2 R^1 1000

x	R^1	R^2	n	Catalyst	y	Temp (°)	Time (h)	
3	HO	H	0	Gr II	0.05	rt	16	(10)
3	HO	H	0	Gr II	0.1	40	4	(70)
1	HO	H	0	Hov II	0.05	rt	16	(0)
3	HO	H	0	Hov II	0.05	rt	16	(75)
3	HO	Me	0	Hov II	0.05	rt	16	(14)
3	HO	Ph	0	Hov II	0.05	rt	16	(0)
3	HO	H	1	Hov II	0.05	rt	16	(15)
3	HO	H	2	Hov II	0.05	rt	16	(20)
3	TsHN	H	2	Hov II	2 x 0.05	100 (MW)	2 x 0.25	(72)
3	TsHN	*i*-Pr	2	Hov II	2 x 0.05	100 (MW)	2 x 0.25	(80)
3	TsHN	Ph	2	Hov II	2 x 0.05	100 (MW)	2 x 0.25	(88)

TABLE 17. CROSS-METATHESIS OF α,β-UNSATURATED KETONES (*Continued*)

Ketone	Cross Partner	Conditions	Product(s) and Yield(s) (%)	Refs.

Please refer to the charts preceding the tables for the catalyst structures.

C_5

Ketone	Cross Partner	Conditions	Product(s) and Yield(s) (%)	Refs.
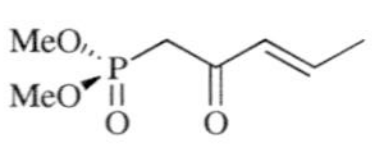 MeO, MeO, P, O, O; 3 eq	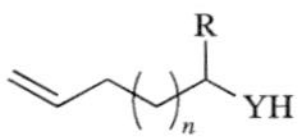 R, YH, n	Hov II (2 x 0.05 eq.), DCM, 100° (MW), 2 x 0.25 h	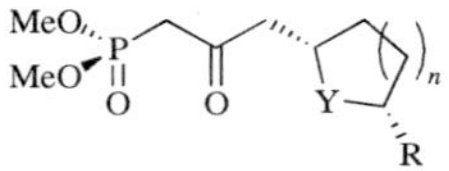 MeO, MeO, P, O, O, Y, R, n	1000

R	Y	*n*		dr
H	TsN	1	(55)	—
H	BocN	1	(80)	—
H	O	1	(54)	—
H	O	2	(46)	—
n-Bu	BocN	1	(43)	70:30
n-Bu	O	1	(77)	50:50
n-Bu	O	2	(80)	>98:2
Ph	BocN	1	(65)	70:30
Ph	TsN	1	(60)	80:20
Ph	O	1	(>99)	60:40
Ph	O	2	(90)	>98:2

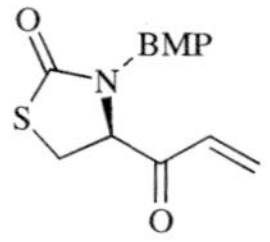

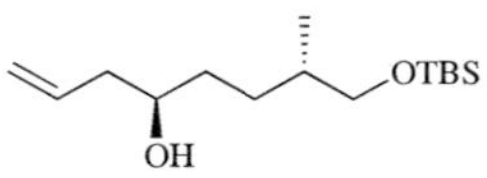

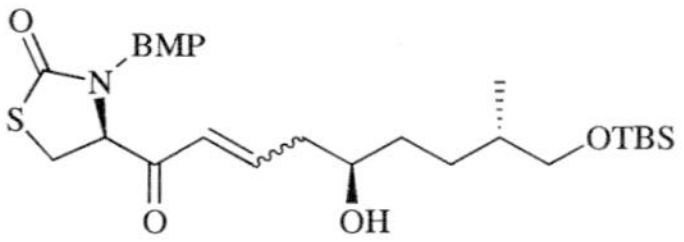

Ketone	Cross Partner	Conditions	Product(s) and Yield(s) (%)	Refs.
O, BMP, N, S, O; 3 eq	OH, OTBS	Hov II (0.1 eq), DCE, 50°, 5 h	O, BMP, N, S, O, OH, OTBS (70)	1001

C_6

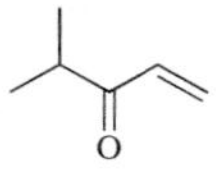

5 eq

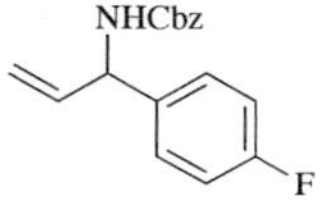

NHCbz

F

Hov II (0.1 eq), DCM, 40°, 96 h

NHCbz

F

(50)

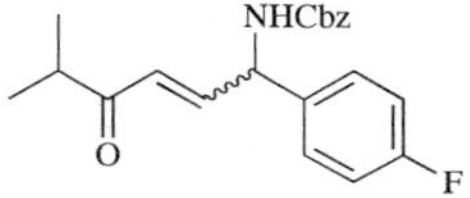

934

C_{6-9}

R

O

1.5 eq

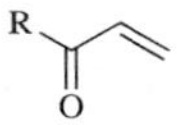

OTIPS

HO

OTBDPS

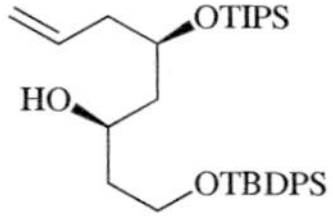

Hov II (0.1 eq), DCM, 100° (MW), 0.5 h

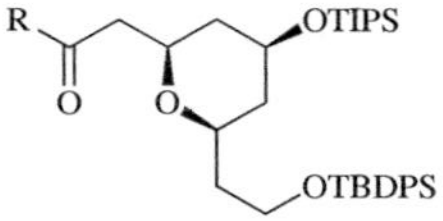

R

O

O

OTIPS

OTBDPS

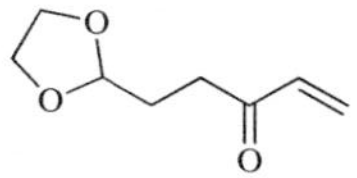

970

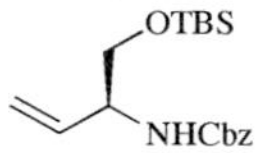

R		(E)/(Z)
PMBO(CH_2)$_3$	(84)	91:9
Ph	(79)	93:7

C_6

O

O

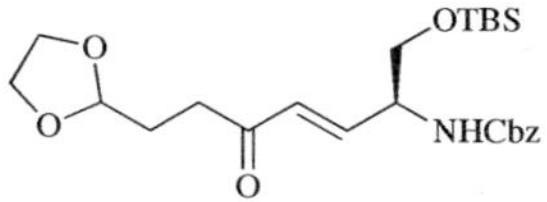

O

2 eq

OTBS

NHCbz

Hov II (0.1 eq), DCM, 40°, 72 h

O

O

O

OTBS

NHCbz

(73) (*E*) only

1002

TABLE 17. CROSS-METATHESIS OF α,β-UNSATURATED KETONES (*Continued*)

Ketone	Cross Partner	Conditions	Product(s) and Yield(s) (%)	Refs.
Please refer to the charts preceding the tables for the catalyst structures.				
C_6				
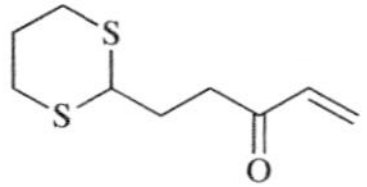 1.5 eq	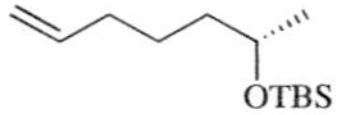 OTBS	Gr II (0.05 eq), DCM, 45°, 2 h	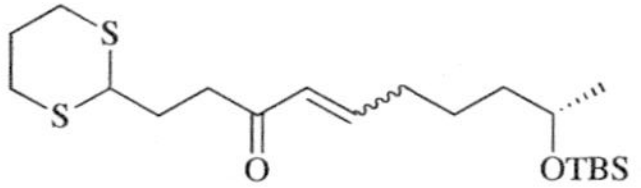 OTBS (76)	861
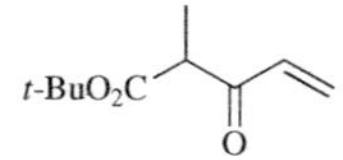 t-BuO_2C — x eq	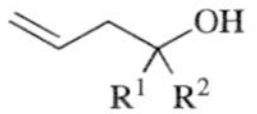 OH, R^1 R^2	Catalyst (y eq), DCM, 40°	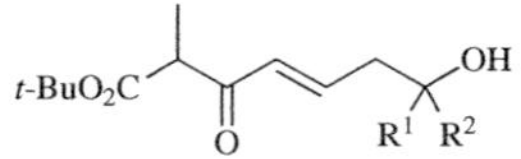 t-BuO_2C, OH, R^1 R^2 (*E*) only	738
C_7				
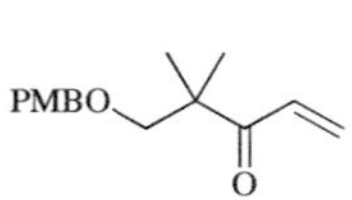 PMBO — 1.5 eq	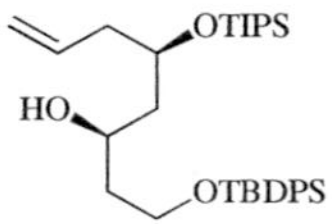 OTIPS, HO, OTBDPS	Hov II (0.1 eq), DCM, 100° (MW), 0.5 h	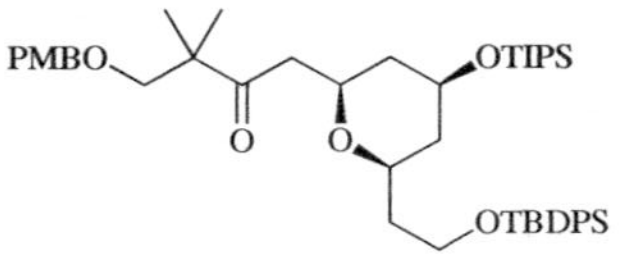 PMBO, OTIPS, OTBDPS (70) *cis*/*trans* = 91:9	970

x	R^1	R^2	Catalyst	y	Time (h)	
1.5	H	H	Hov II	0.1	36	(61)
1.5	*i*-Pr	H	Hov II	0.1	36	(81)
1.5	c-C_6H_{11}	H	Hov II	0.1	36	(78)
1.5	Ph	H	Hov II	0.1	36	(80)
3	n-C_7H_{15}	H	Gr II	0.025	14	(82)
3	Ph	Ph	Gr II	0.025	14	(78)

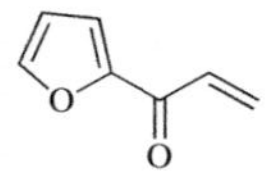

Carbon	Reactant	Reactant	Conditions	Product(s) and Yield(s) (%)	Refs.
	3 eq	NHCbz	Hov II (0.05 eq), DCM, rt, 12 h	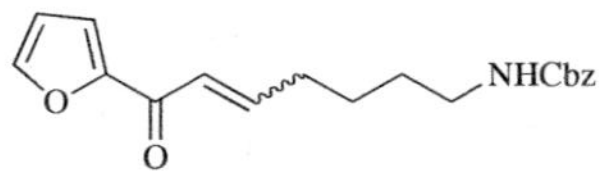 NHCbz (62)	964
	3 eq	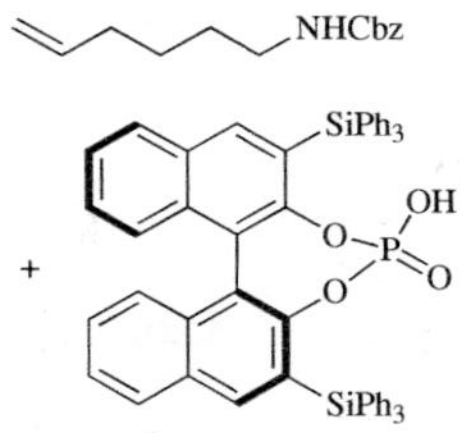 NHCbz + SiPh$_3$, OH, O, P, O, O, SiPh$_3$ 0.1 eq	Hov II (0.05 eq), DCM, 0°, 15 h	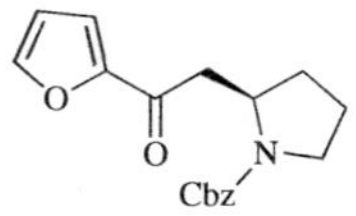 O, N, Cbz (52) er 92.0:8.0	964
C_8	n-C_5H_{11}, O 1.5 eq	OTBS, OH	Hov II (0.05 eq), DCM, 40°, 5 h	n-C_5H_{11}, O, OTBS, OH (74)	1003

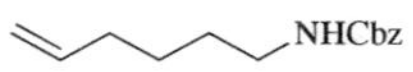

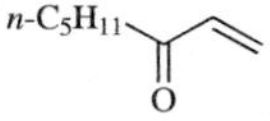

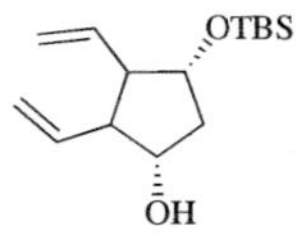

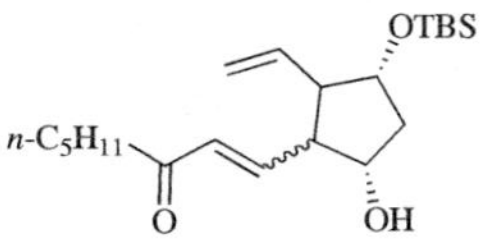

TABLE 17. CROSS-METATHESIS OF α,β-UNSATURATED KETONES (*Continued*)

Ketone	Cross Partner	Conditions	Product(s) and Yield(s) (%)	Refs.
Please refer to the charts preceding the tables for the catalyst structures.				
C_8				
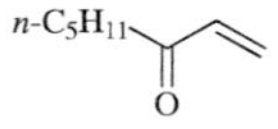 n-C_5H_{11} (vinyl ketone), 3 eq	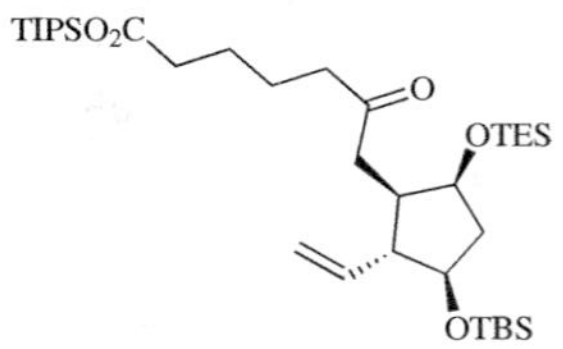 $TIPSO_2C$ … OTES, OTBS	Ru cat **89** (0.05 eq), DCM, reflux, 11 h	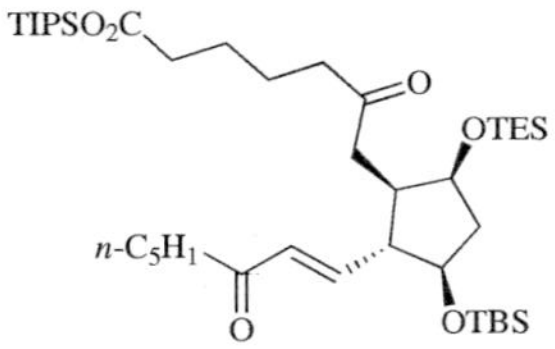 $TIPSO_2C$ … OTES, OTBS, n-C_5H_1 (65) (*E*)/(*Z*) > 95:5	817
C_{8-10}				
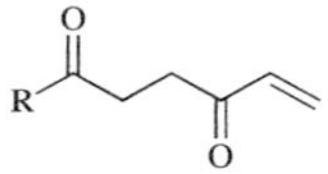 R (dione, vinyl), 1 eq	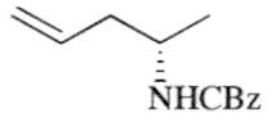 NHCBz	Hov II (0.05 eq) DCM, reflux, 4 h	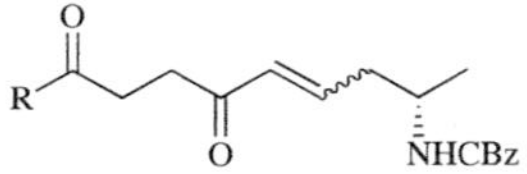 R … NHCBz; R: n-Bu (89), Et (87)	175
C_9				
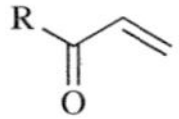 R (vinyl ketone)	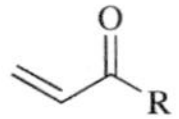 R (vinyl ketone)	Gr II (0.05 eq), DCM, 40°, 3 h	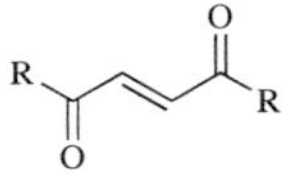 R … R, (*E*) only; R: c-C_6H_{11} (95), n-C_6H_{13} (77)	170

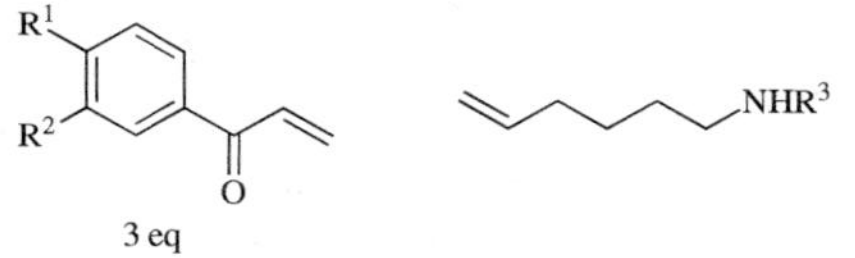

3 eq

NHR^3

Hov II (0.05 eq), DCM, rt, 12 h

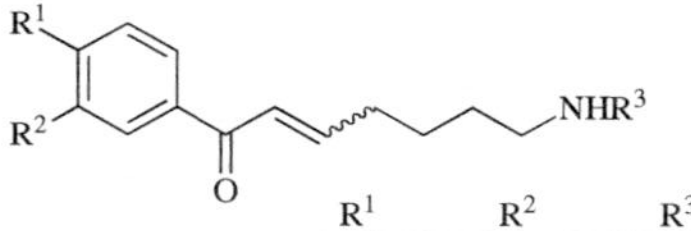

NHR^3

R^1	R^2	R^3	
H	H	Cbz	(75)
MeO	H	Cbz	(68)
O_2N	H	Cbz	(75)
Cl	H	Cbz	(64)
MeO	MeO	Boc	(79)
MeO	MeO	Cbz	(78)

964

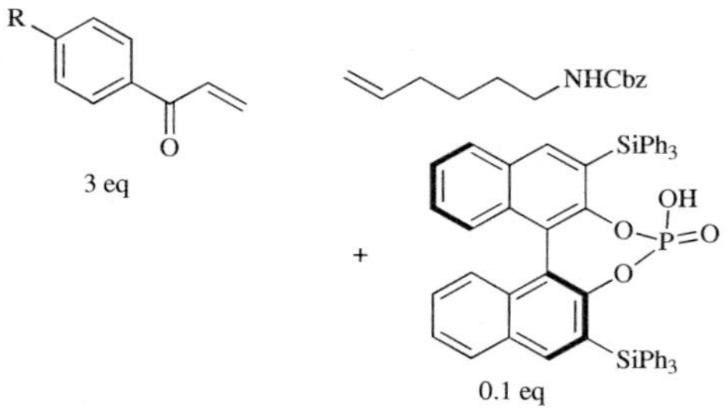

3 eq

NHCbz

+ $SiPh_3$ OH O–P=O O $SiPh_3$

0.1 eq

Hov II (0.05 eq), DCM, 0°

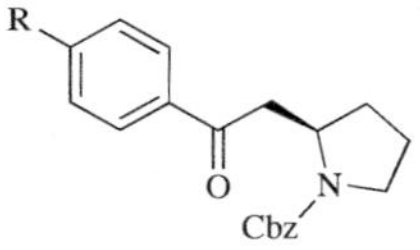

Cbz

R	Time (h)		er
H	11	(76)	97.0:3.0
MeO	18	(59)	98.5:1.5
O_2N	36	(52)	93.5:6.5
Cl	24	(60)	97.0:3.0

964

TABLE 17. CROSS-METATHESIS OF α,β-UNSATURATED KETONES (*Continued*)

Ketone	Cross Partner	Conditions	Product(s) and Yield(s) (%)	Refs.

Please refer to the charts preceding the tables for the catalyst structures.

C_9

Ketone	Cross Partner	Conditions	Product(s) and Yield(s) (%)	Refs.
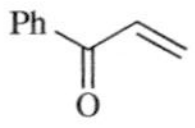 Ph, O; 5 eq	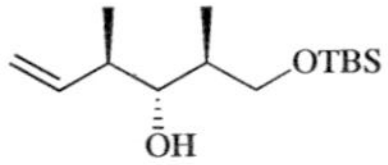 OTBS, OH	Catalyst (0.05–0.1 eq) DCM, 48 h	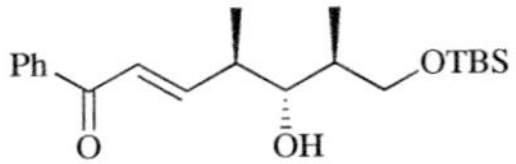 Ph, O, OH, OTBS	1004

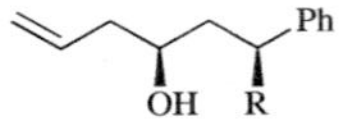

Catalyst	Temp (°)		(E)/(Z)
Hov II	50	(21)	—
Hov II	40	(43)	—
Hov II	40 (MW)	(40)	—
Ru cat **89**	40	(65)	(E) only

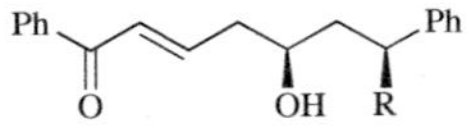

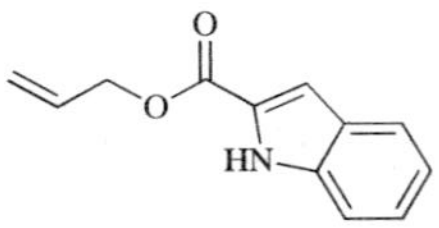

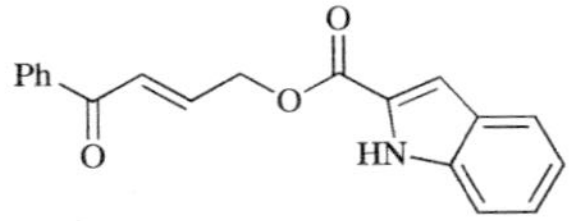

Ketone	Cross Partner	Conditions	Product(s) and Yield(s) (%)	Refs.
x eq	Ph, OH, R	Gr II (y eq), DCM	Ph, O, OH, R, Ph	

x	R	y	Temp (°)	Time (h)		(E)/(Z)	Refs.
3.3	BocHN	0.2	rt	1	(70)	—	1005
5	TBSO	0.05	reflux	—	(75)	95:5	1006
3	TBSO	0.05	reflux	3	(94)	(E) only	1007

Ketone	Cross Partner	Conditions	Product(s) and Yield(s) (%)	Refs.
1.5 eq	O, O, HN	Ru cat **89** (0.05 eq), DCM, 45°	Ph, O, O, O, HN (20) (E)/(Z) > 98:2	982

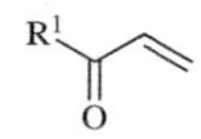

1.5 eq

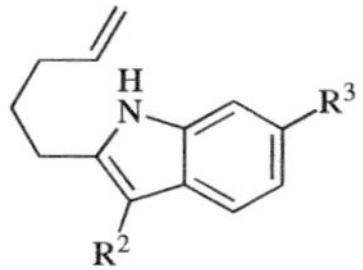

+

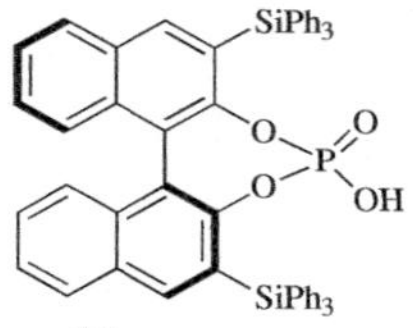

0.1 eq

Ru cat **89** (0.03 eq), 4 Å MS, toluene, 50°

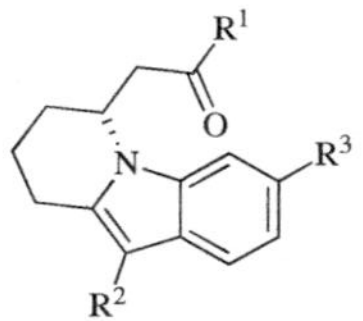

982

R^1	R^2	R^3		er
Ph	Me	H	(89)	95.0:5.0
4-BrC_6H_4	Me	H	(80)	94.0:6.0
Ph	Me	Me	(81)	97.0:3.0
Ph	$HO(CH_2)_2$	H	(45)	94.0:6.0
Ph	$Ac(CH_2)_2$	H	(92)	95.0:5.0
Ph	Bn	H	(88)	95.0:5.0
Ph	$Bz(CH_2)_2$	H	(94)	95.0:5.0
4-BrC_6H_4	$PhthN(CH_2)_2$	H	(96)	95.0:5.0

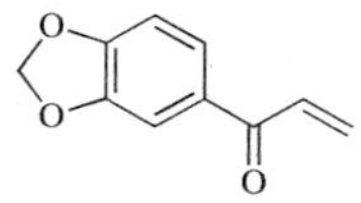

3 eq

Hov II (0.05 eq), DCM, rt, 12 h

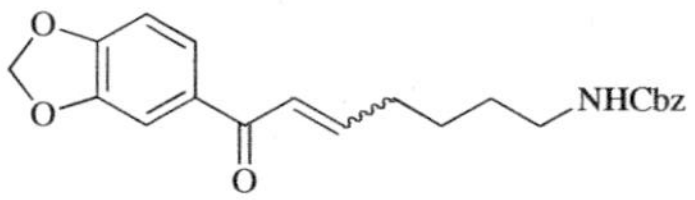

(78) 964

3 eq

+

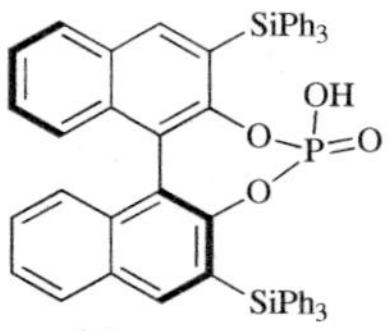

0.1 eq

Hov II (0.05 eq), DCM, 0°, 12 h

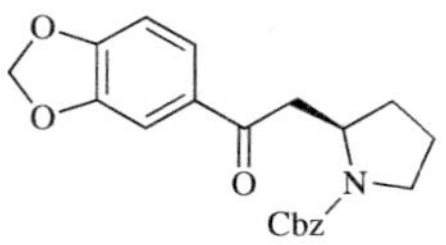

(62) er 97.0:3.0 964

TABLE 17. CROSS-METATHESIS OF α,β-UNSATURATED KETONES (*Continued*)

Ketone	Cross Partner	Conditions	Product(s) and Yield(s) (%)	Refs.

Please refer to the charts preceding the tables for the catalyst structures.

C_9

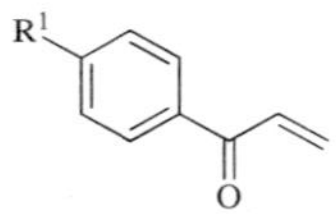

R^1 · 1.5 eq

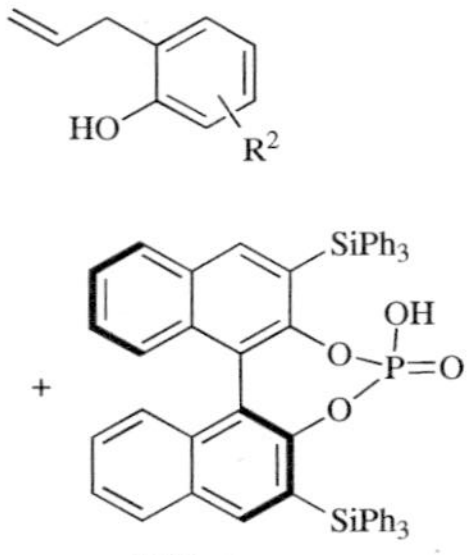

HO · R^2 · + · $SiPh_3$ · OH · O–P=O · O · $SiPh_3$ · 0.05 eq

Hov II (0.05 eq), DCM, 40°

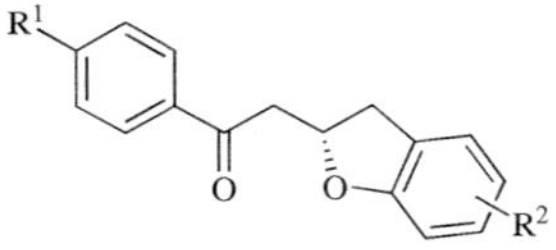

R^1 · O · O · R^2

964

R^1	R^2	Time (h)		er
H	H	48	(56)	89.0:11.0
H	4-MeO	12	(52)	87.0:13.0
H	5-MeO	24	(46)	90.0:10.0
H	6-MeO	12	(59)	83.0:17.0
H	4-Br	24	(40)	89.5:10.5
H	4-Me	12	(49)	89.5:10.5
H	4-*i*-Pr	12	(72)	89.0:11.0
H	4-*t*-Bu	23	(64)	88.0:12.0
H	4-Ph	12	(53)	88.0:12.0
Me	H	12	(49)	90.5:9.5
MeO	H	12	(38)	88.0:12.0
Cl	H	12	(51)	89.5:10.5
Br	H	12	(61)	88.5:11.5
O_2N	H	12	(75)	75.0:25.0

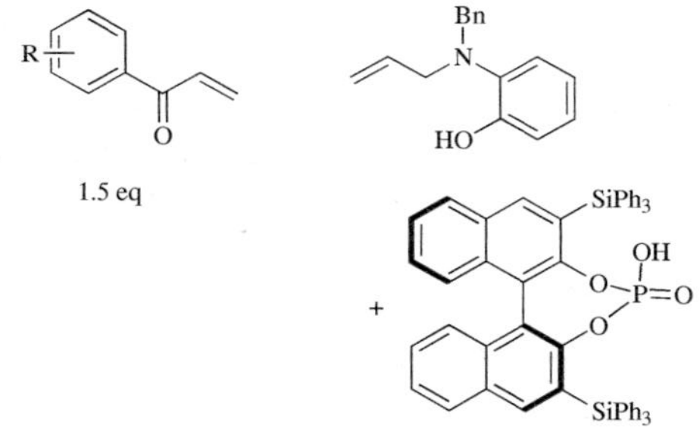

1.5 eq

+

0.05 eq

Hov II (0.05 eq),
DCM, 40°

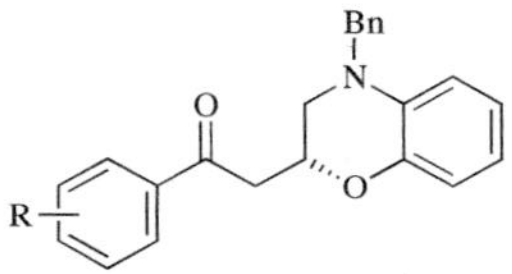

R	Time (h)		er
H	24	(65)	88.5:11.5
2-Br	11	(44)	84.5:15.5
4-Br	11	(52)	88.0:12.0
4-Cl	8	(48)	86.5:13.5
4-O_2N	5	(52)	88.0:12.0
4-MeO	24	(64)	83.5:16.5
4-Me	10	(54)	87.0:13.0

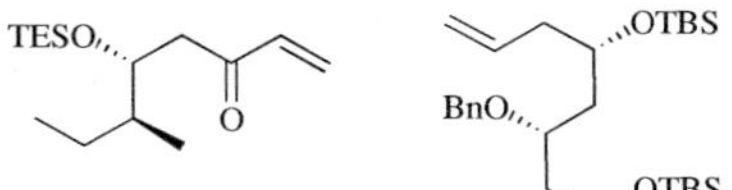

973

2 eq

Hov II (0.1 eq),
DCM, 40°, 24 h

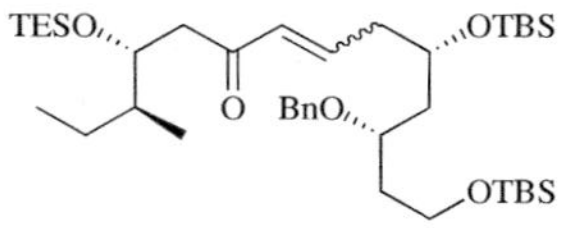

(75)

1008

TABLE 17. CROSS-METATHESIS OF α,β-UNSATURATED KETONES (*Continued*)

Ketone	Cross Partner	Conditions	Product(s) and Yield(s) (%)	Refs.
Please refer to the charts preceding the tables for the catalyst structures.				
C_9				
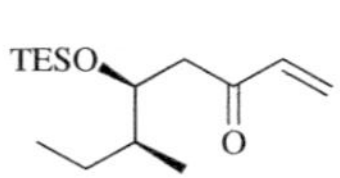 TESO, O; 2 eq	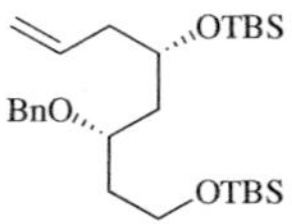 OTBS, BnO, OTBS	Hov II (0.1 eq), DCM, 40°, 24 h	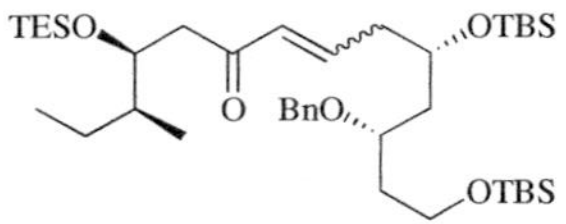 TESO, O, BnO, OTBS, OTBS (77)	1008
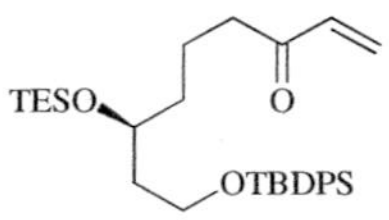 TESO, O, OTBDPS; 1.5 eq	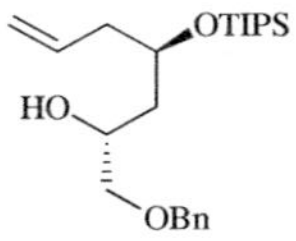 OTIPS, HO, OBn	Hov II (0.1 eq), DCM, 100° (MW), 0.5 h	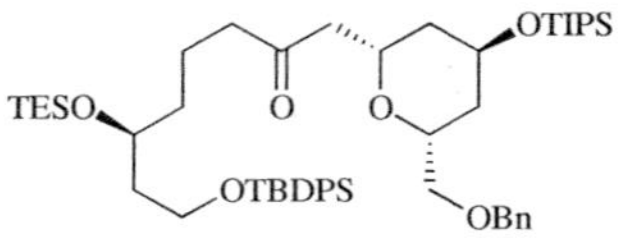 TESO, O, OTBDPS, O, OTIPS, OBn (90)	970
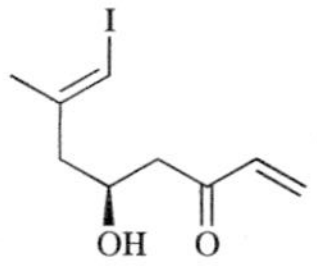 I, OH, O; 1 eq	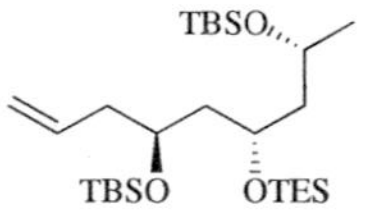 TBSO, TBSO, OTES	Hov II (0.1 eq), DCM, 30°, 15 h	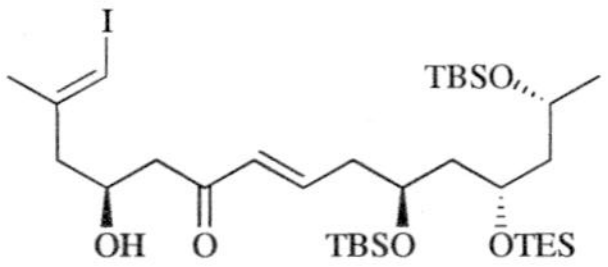 I, OH, O, TBSO, TBSO, OTES (66) (*E*) only	1009, 1010
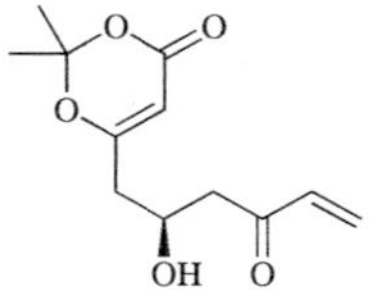 O, O, O, OH, O; 1.2 eq	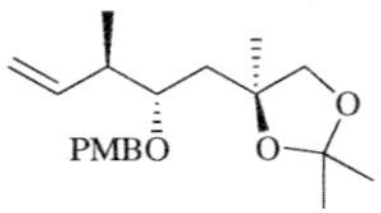 PMBO, O, O	1. Hov II (0.2 eq), DCM, 40°, 15 h 2. H_2, Pd/C, AcOEt, rt	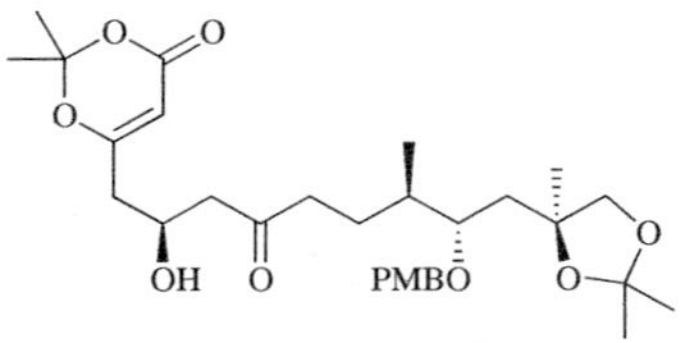 O, O, O, OH, O, PMBO, O, O (79)	1011

C_{10}

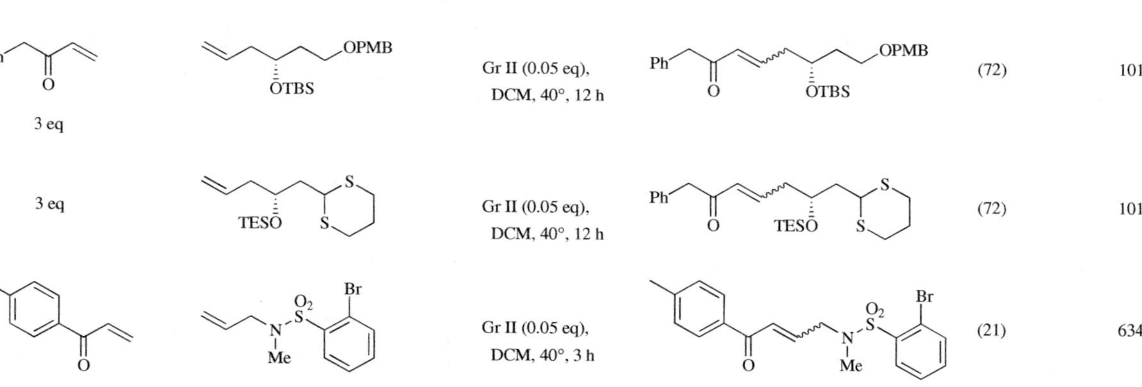

R	x	
TES	0.14	(81)
$PMPCH_2$	0.2	(78)

TABLE 17. CROSS-METATHESIS OF α,β-UNSATURATED KETONES (*Continued*)

Ketone	Cross Partner	Conditions	Product(s) and Yield(s) (%)	Refs.
Please refer to the charts preceding the tables for the catalyst structures.				
C_{10}				
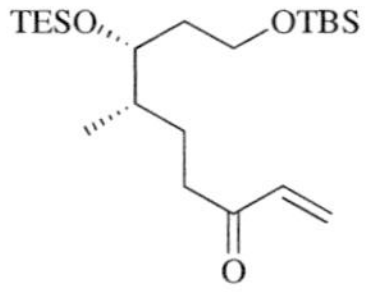 TESO, OTBS, O	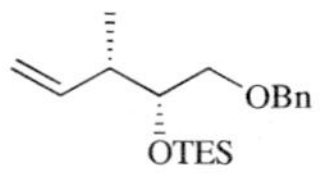 OBn, OTES; 1.2 eq	Gr II (0.025 eq), DCM, reflux, 12 h	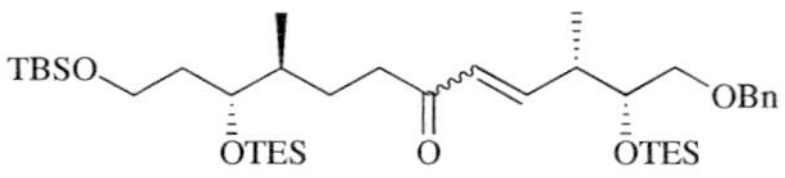 TBSO, OTES, O, OTES, OBn (86)	1014
C_{11}				
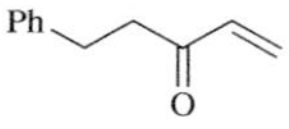 Ph, O; 0.83 eq	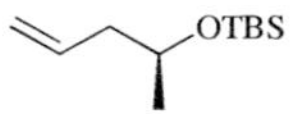 OTBS	Gr II (0.04 eq) DCM, reflux, 22 h	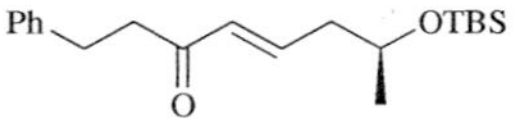 Ph, O, OTBS (89) (*E*) only	1015
x eq	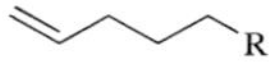 R	Gr II (0.02 eq), DCM, reflux, 22 h	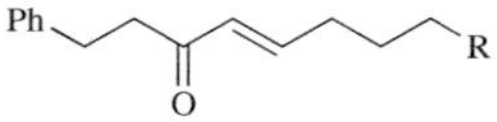 Ph, O, R; (*E*) only; R / *x*: Me 0.5 (75); NC– 0.57 (53)	1015
	 *t*-Bu; 2 eq	Gr II (0.04 eq), DCM, reflux, 22 h	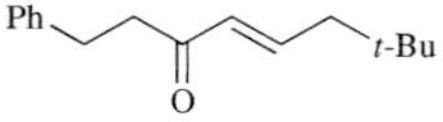 Ph, O, *t*-Bu (58) (*E*) only	1015
	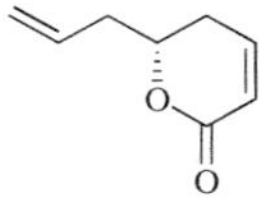 O, O	Gr II (0.05 eq) DCM, 40°, 12 h	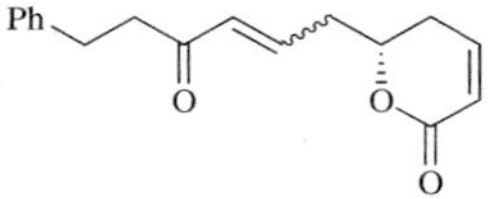 Ph, O, O, O (74)	1016

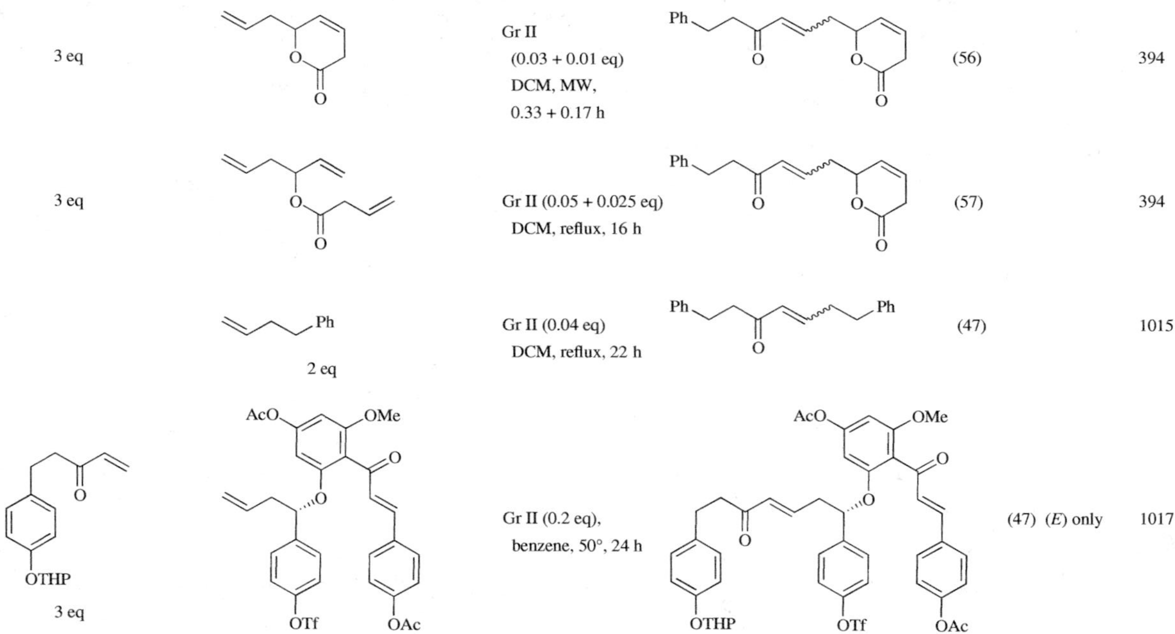

3 eq
Gr II
(0.03 + 0.01 eq)
DCM, MW,
0.33 + 0.17 h
Ph
(56)
394
3 eq
Gr II (0.05 + 0.025 eq)
DCM, reflux, 16 h
Ph
(57)
394
Ph
2 eq
Gr II (0.04 eq)
DCM, reflux, 22 h
Ph
Ph
(47)
1015
OTHP
3 eq
AcO
OMe
OTf
OAc
Gr II (0.2 eq),
benzene, 50°, 24 h
AcO
OMe
OTHP
OTf
OAc
(47) (E) only
1017

TABLE 17. CROSS-METATHESIS OF α,β-UNSATURATED KETONES (*Continued*)

Ketone	Cross Partner	Conditions	Product(s) and Yield(s) (%)	Refs.
Please refer to the charts preceding the tables for the catalyst structures.				
C_{11}				
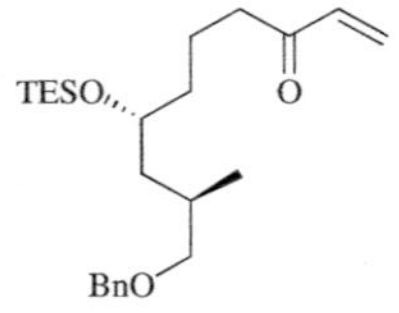 TESO, O, BnO	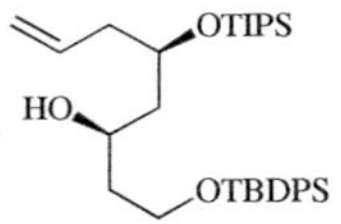 OTIPS, HO, OTBDPS 1.5 eq	Hov II (0.1 eq), DCM, 35°, 21 h	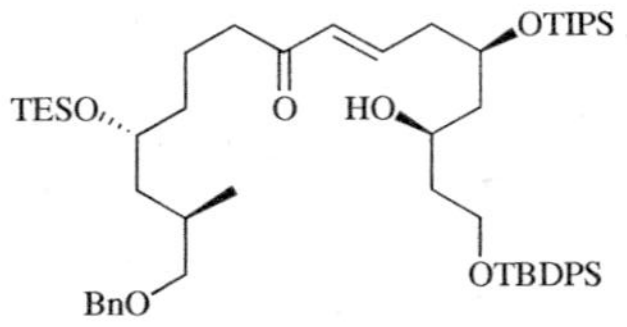 TESO, O, HO, OTIPS, OTBDPS, BnO (93) (*E*)/(*Z*) > 95:5	1018, 1019
	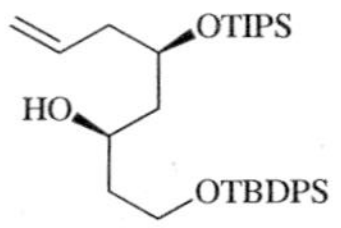 OTIPS, HO, OTBDPS 1.5 eq	Hov II (0.1 eq), DCM, 100° (MW), 0.5 h	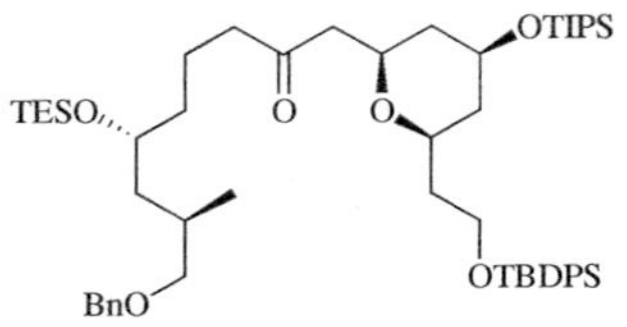 TESO, O, O, OTIPS, OTBDPS, BnO (94) (*E*)/(*Z*) = 89:11	970
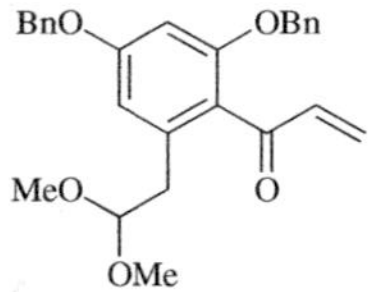 BnO, OBn, MeO, O, OMe	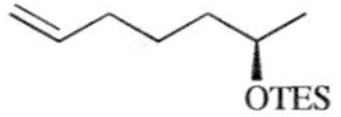 OTES 3 eq	Gr II (0.15 eq) DCM, 45°, 1.5 h	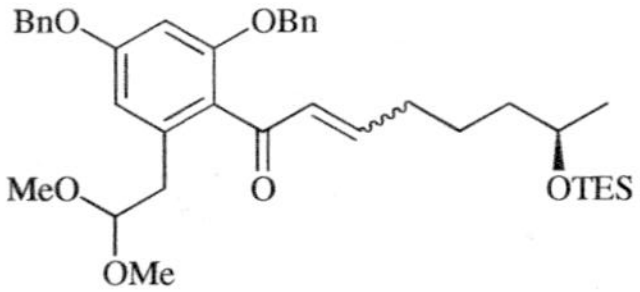 BnO, OBn, MeO, O, OMe, OTES (78)	260

BnO OBn MeO OMe O OBn 3 eq Gr II (0.15 eq) DCM, 45°, 1.5 h BnO OBn MeO OMe O OBn (84) 260

MeO CO$_2$Me OMe O OTBS Gr II (0.2 eq), DCM, reflux, 24 h MeO CO$_2$Me OMe O OTBS (80) 1020

OTBS Gr II (0.2 eq), DCM, reflux, 24 h MeO CO$_2$Me OMe O OTBS (85) 1020

C_{11-12}

OMe R^{1}O R^2 O Ph 2 eq Hov II (0.05 eq), DCM, 18–24 h OMe R^{1}O Ph O (*E*) only 1021

R^1	R^2	Temp	
H	Me	rt	(66)
H	H	rt	(70)
TBS	Me	reflux	(80)

TABLE 17. CROSS-METATHESIS OF α,β-UNSATURATED KETONES (*Continued*)

Ketone	Cross Partner	Conditions	Product(s) and Yield(s) (%)	Refs.
Please refer to the charts preceding the tables for the catalyst structures.				
C_{11-12}				
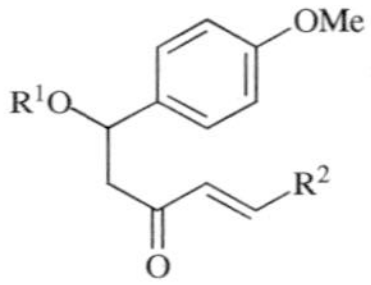 R¹O, OMe, R^2, O	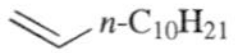 n-$C_{10}H_{21}$ 2 eq	Hov II (0.05 eq), DCM, 18–24 h	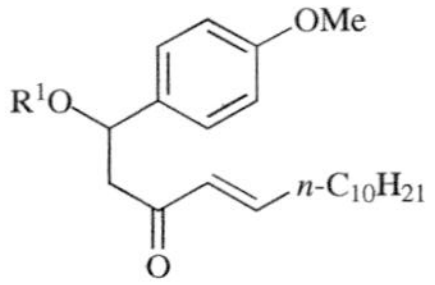 R¹O, OMe, n-$C_{10}H_{21}$, O (*E*) only R¹ / R² / Temp H / Me / rt (85) H / H / rt (76) TBS / Me / reflux (91)	1021
C_{12}				
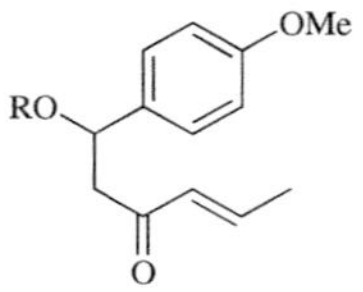 RO, OMe, O	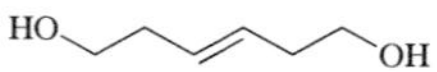 HO, OH 3 eq	Hov II (0.05 eq), DCM, reflux, 18–24 h	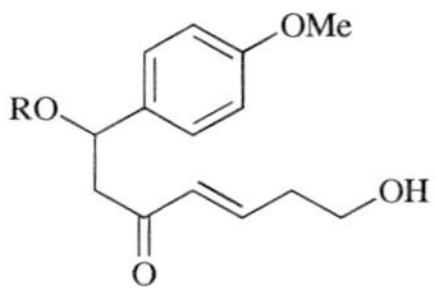 RO, OMe, OH, O (*E*) only R H (24) TBS (50)	1021
C_{13}				
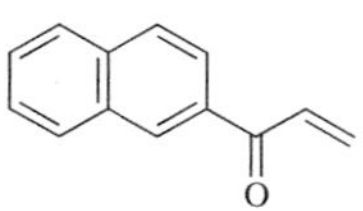 O 1.5 eq	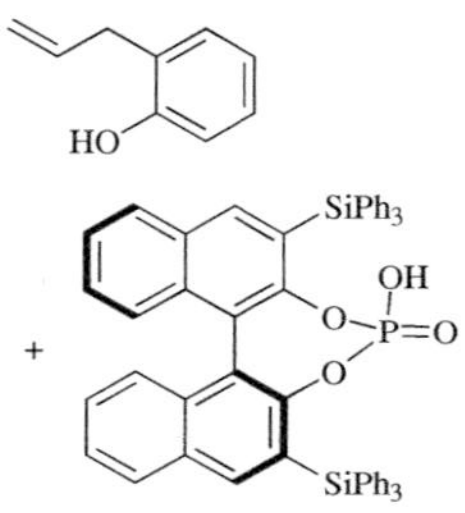 HO + SiPh₃, OH, O, P=O, O, SiPh₃ 0.05 eq	Hov II (0.05 eq), DCM, 40°, 12 h	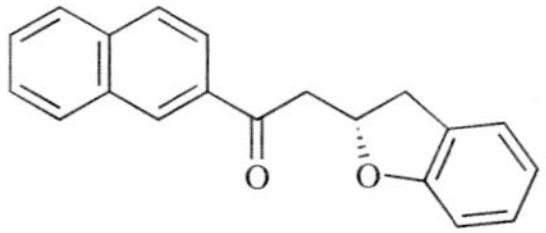 O, O (49) er 89.5:10.5	973

Reactant	Reagent	Conditions	Product(s) and Yield(s) (%)	Refs.
1.5 eq	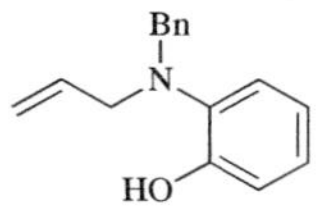 + 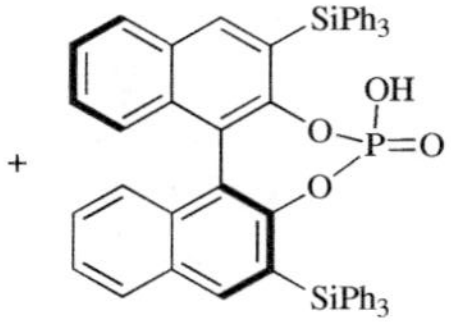 0.05 eq	Hov II (0.05 eq), DCM, 40°, 12 h	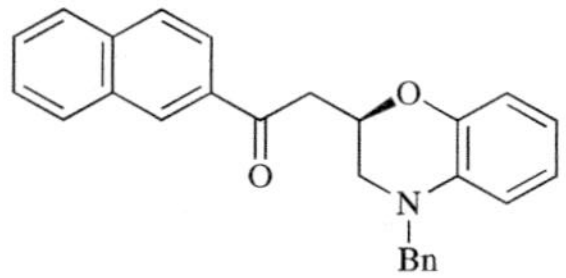 (51) er 88.0:12.0	973
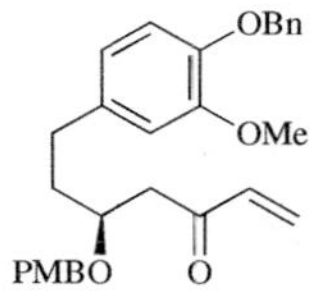	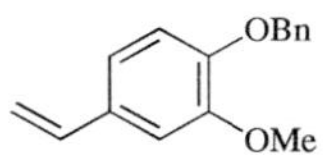 1 eq	Gr II (0.1 eq), DCM, 50°, 48 h	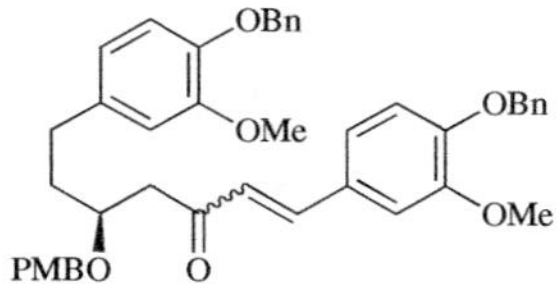 (70)	1022

TABLE 17. CROSS-METATHESIS OF α,β-UNSATURATED KETONES (*Continued*)

Ketone	Cross Partner	Conditions	Product(s) and Yield(s) (%)		Refs.
Please refer to the charts preceding the tables for the catalyst structures.					
C_{14}					
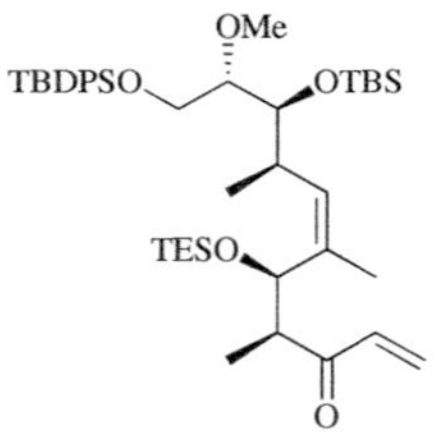 3 eq	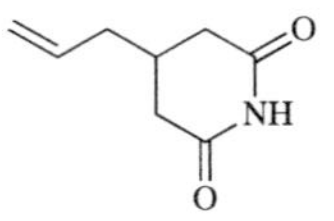	Hov II (0.3 eq), DCM, rt, 72 h		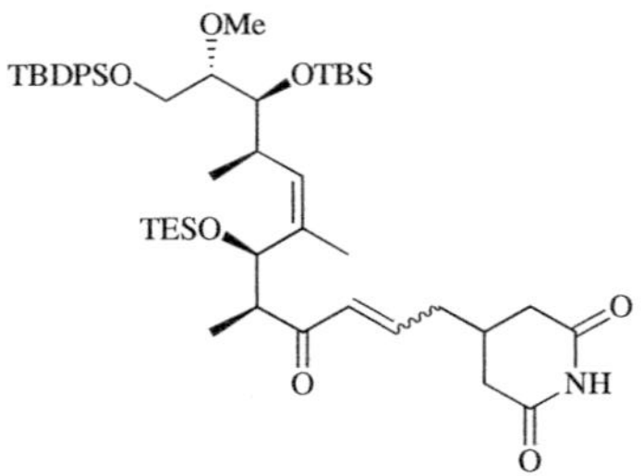 (32)	1023, 993
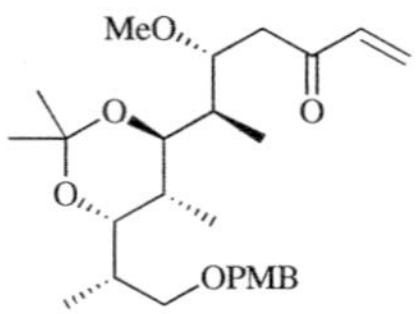	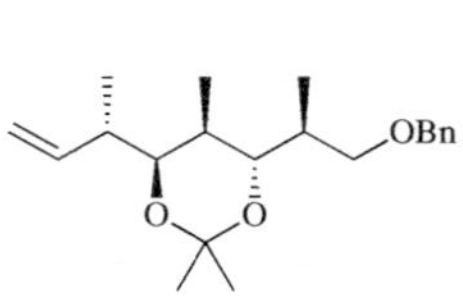 1.5 eq	Ru cat **88** (0.12 eq), DCM, 50°, 24 h	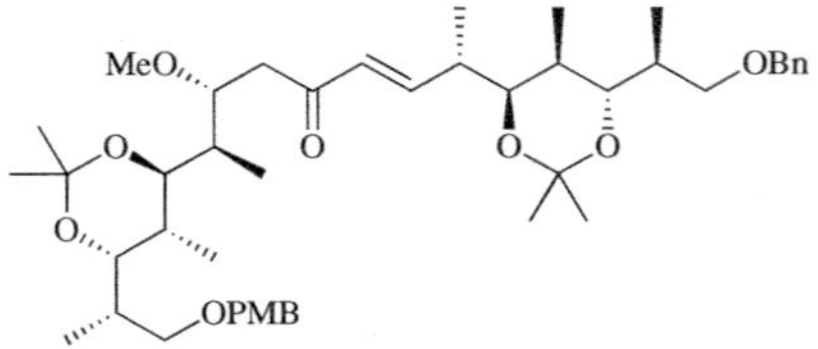 (80) (*E*) only		1024

C_{15}

Reactant 1	Reactant 2	Conditions	Product(s) and Yield(s) (%)	Refs.
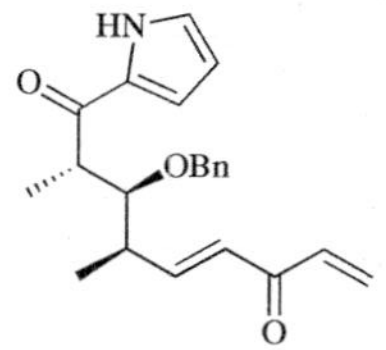 HN, O, OBn, O; 1.8 eq	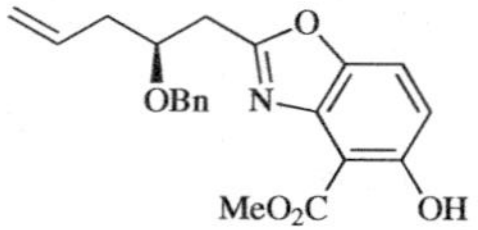 OBn, N, O, MeO_2C, OH	Gr II (0.15 eq), THF, rt, 6 h	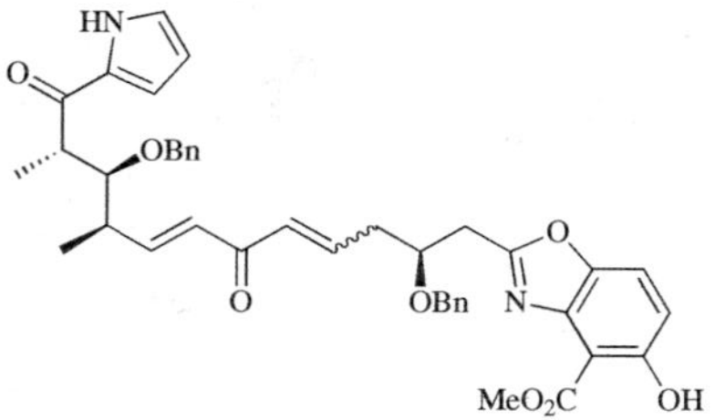 HN, O, OBn, O, OBn, N, O, MeO_2C, OH (59)	340
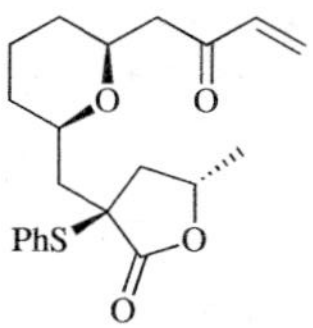 O, O, PhS, O, O	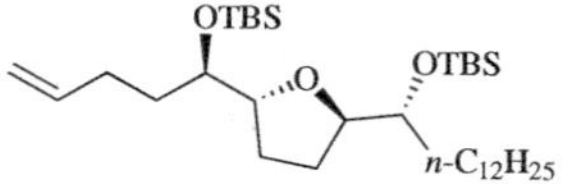 OTBS, O, OTBS, n-$C_{12}H_{25}$; 1.6 eq	Gr II (0.09 eq), DCM, 45°, 2 h	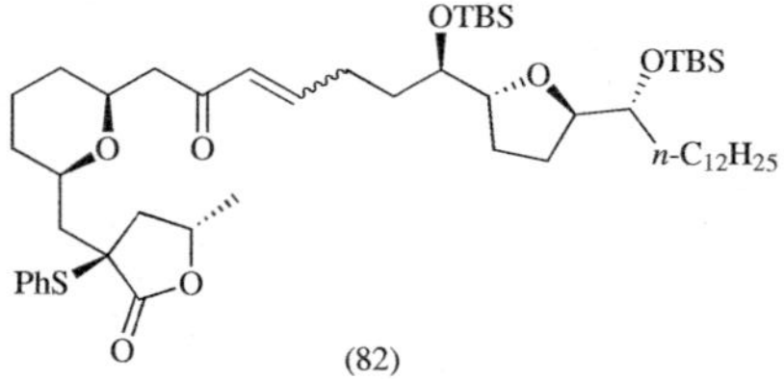 O, O, OTBS, O, OTBS, n-$C_{12}H_{25}$, PhS, O, O (82)	1025
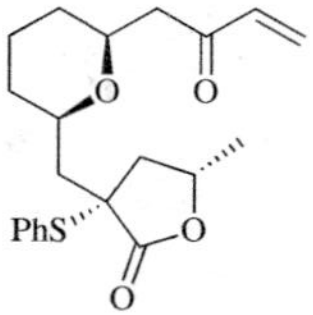 O, O, PhS, O, O	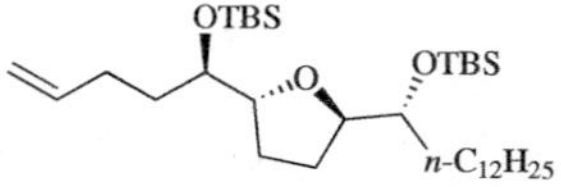 OTBS, O, OTBS, n-$C_{12}H_{25}$; 1.6 eq	Gr II (0.14 eq), DCM, 45°, 2 h	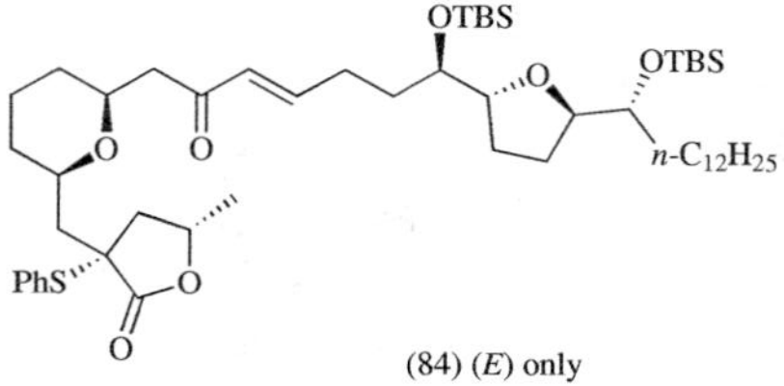 O, O, OTBS, O, OTBS, n-$C_{12}H_{25}$, PhS, O, O (84) (E) only	1025

TABLE 17. CROSS-METATHESIS OF α,β-UNSATURATED KETONES (*Continued*)

Ketone	Cross Partner	Conditions	Product(s) and Yield(s) (%)	Refs.
Please refer to the charts preceding the tables for the catalyst structures.				
C_{15}				
PhS, O, O, O	OTBS, O, OTBS, n-$C_{12}H_{25}$; 1.6 eq	Gr II (0.18 eq), DCM, 45°, 2 h	OTBS, O, OTBS, n-$C_{12}H_{25}$, O, O, PhS, O, O; (83) (*E*) only	1026
PhS, O, O, O	OTBS, O, OTBS, n-$C_{12}H_{25}$; 1.6 eq	Gr II (0.18 eq), DCM, 45°, 2 h	OTBS, O, OTBS, n-$C_{12}H_{25}$, O, O, PhS, O, O; (84) (*E*) only	1026
	OTBS, O, OTBS, n-$C_{12}H_{25}$; 1.6 eq	Gr II (0.13 eq), DCM, 45°, 2 h	OTBS, O, OTBS, n-$C_{12}H_{25}$, O, O, PhS, O, O; (76)	1025

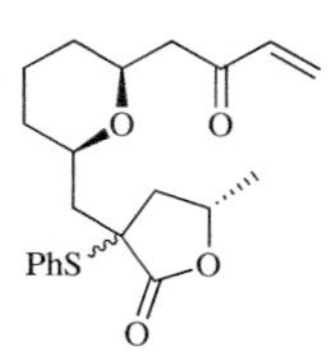

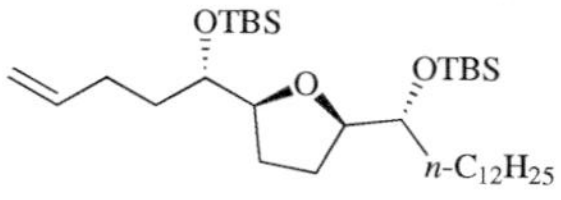

1.6 eq

Gr II (0.18 eq),
DCM, 45°, 2 h

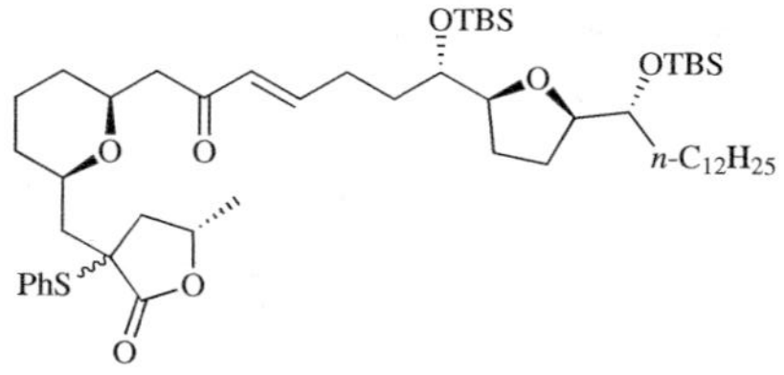

(81) (*E*) only

1026

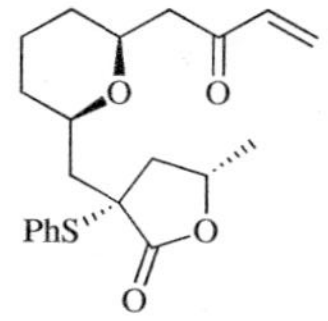

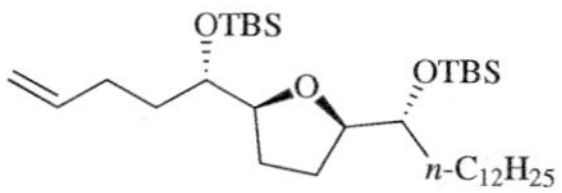

1.6 eq

Gr II (0.18 eq),
DCM, 45°, 2 h

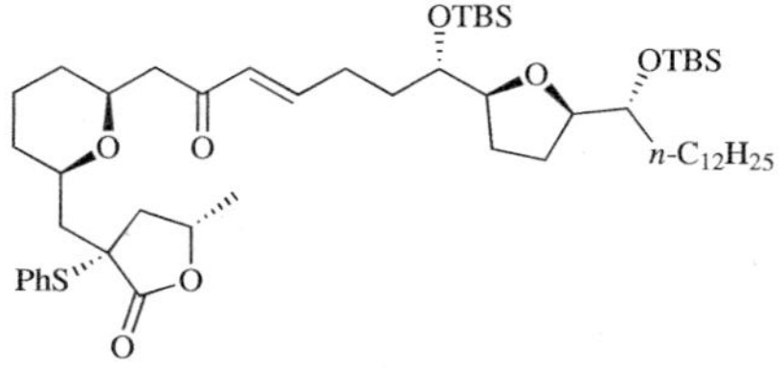

(85) (*E*) only

1026

C_{16}

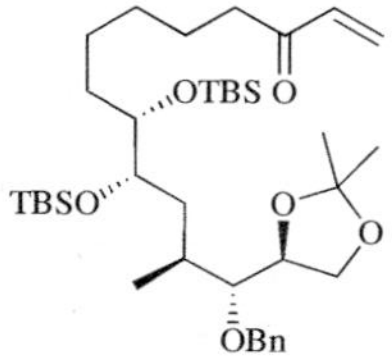

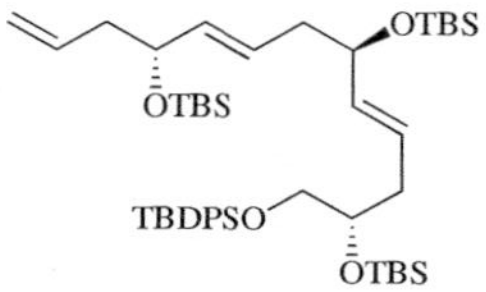

1 eq

Hov II (0.1 eq),
DCM, rt, 47.5 h

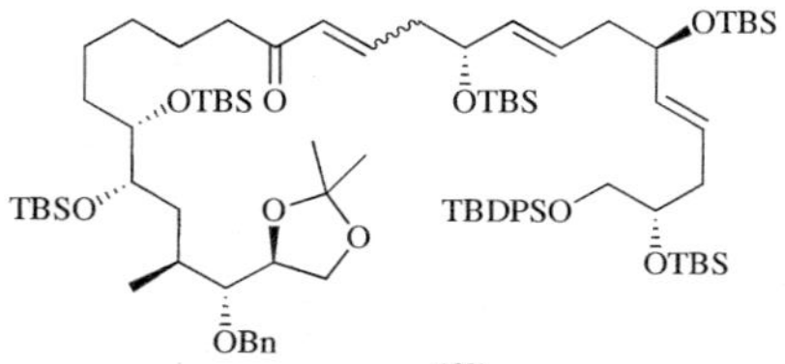

(69)

1027

TABLE 17. CROSS-METATHESIS OF α,β-UNSATURATED KETONES (*Continued*)

Ketone	Cross Partner	Conditions	Product(s) and Yield(s) (%)	Refs.
Please refer to the charts preceding the tables for the catalyst structures.				
C_{31}				
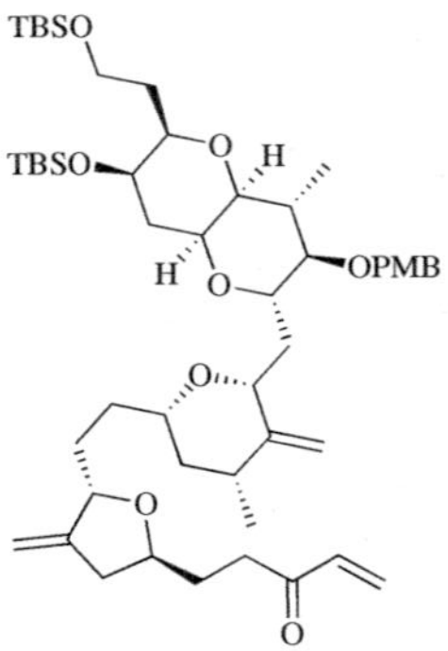 TBSO; TBSO; H; H; OPMB; O; O; O; O	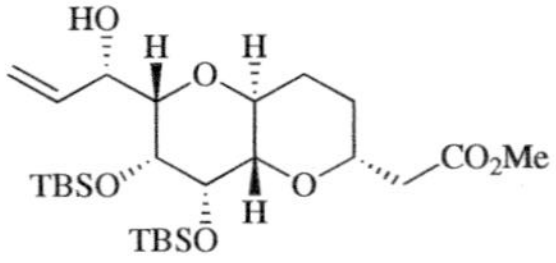 HO; H; H; O; O; TBSO; TBSO; H; CO_2Me; 2 eq	Ru cat **271** (0.2 eq), toluene, 80°, 2 h	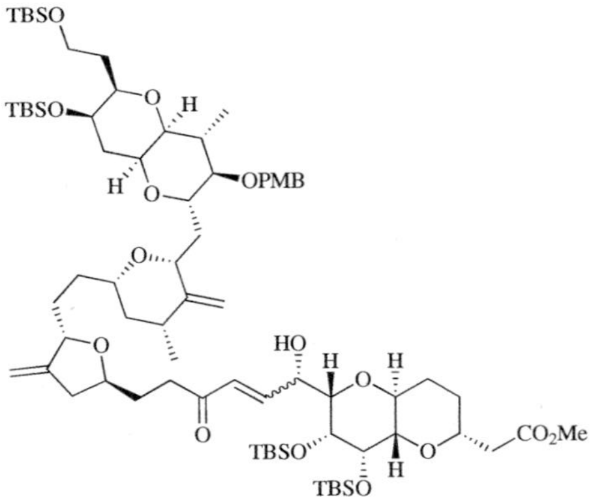 TBSO; TBSO; H; H; OPMB; O; O; O; O; HO; H; H; O; O; TBSO; TBSO; H; CO_2Me; (62)	1028

[a] The conversion was determined by GC analysis.

[b] The yield was determined by GC analysis.

[c] The yield was determined by ^{31}P NMR analysis.

TABLE 18. CROSS-METATHESIS OF α,β-UNSATURATED ACRYLIC ACIDS

Acrylic Acid	Cross Partner	Conditions	Product(s) and Yield(s) (%)	Refs.

Please refer to the charts preceding the tables for the catalyst structures.

C_3

Acrylic Acid	Cross Partner	Conditions	Product(s) and Yield(s) (%)	Refs.
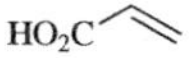 HO_2C (20 eq)	 R, NH_3^+ Y^-, n	Hov II (0.01 eq), EtOAc, reflux, 1 h	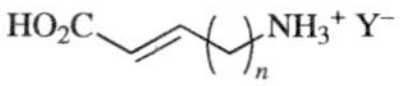 HO_2C, NH_3^+ Y^-, n	548
3 eq	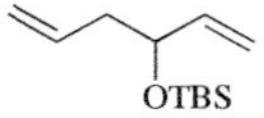 OTBS	Hov II (0.05 eq), DCM, 40°	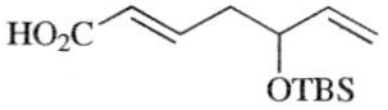 HO_2C, OTBS (81) (E)/(Z) = 95:5	169
3 eq	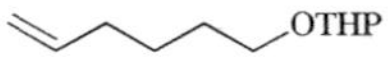 OTHP	Gr II (0.05 eq), DCM, 40°, 15 h	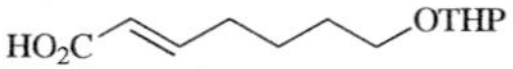 HO_2C, OTHP (100) (E) only	166
3 eq	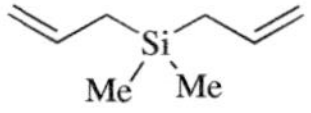 Si, Me, Me	Hov II (0.025 eq), DCM, rt, 2 h	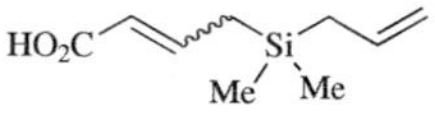 HO_2C, Si, Me, Me (44)	992
3 eq	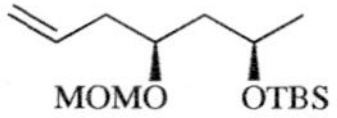 MOMO, OTBS	Hov II (0.05 eq), DCM, 25°, 16 h	 HO_2C, MOMO, OTBS (82) (E)/(Z) = 95:5	1029

n	R	Y	Conv. (%)[a]		(E)/(Z)
1	H	TfO	>95	(—)	>75:25
2	H	BF_4	>99	(—)	87:13
3	H	BF_4	>99	(70)	>95:5
3	H	TfO	>99	(95)	>95:5
3	H	TsO	>99	(—)	>95:5
8	Me	TsO	>99	(—)	>95:5

TABLE 18. CROSS-METATHESIS OF α,β-UNSATURATED ACRYLIC ACIDS (*Continued*)

Acrylic Acid	Cross Partner	Conditions	Product(s) and Yield(s) (%)	Refs.
Please refer to the charts preceding the tables for the catalyst structures.				
C_3				
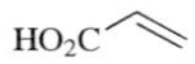 HO_2C (acrylic acid)	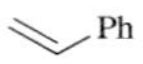 styrene (Ph), *x* eq	Catalyst (0.05 eq), DCM	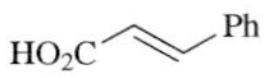 HO_2C–CH=CH–Ph, (*E*) only	
			x 2.5, Gr II, 100 (MW), 0.5 h (99)	971
			x 2.5, Hov II, 100 (MW), 0.5 h (99)	971
			x 2.5, Ru cat **86**, 100 (MW), 0.5 h (99)	971
			x 1.9, Gr II, 40, 15 h (63)	166
3 eq	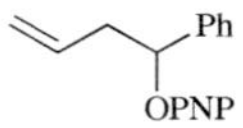 Ph, OPNP	Hov II (0.02 eq), DCM, 40°, 15 h	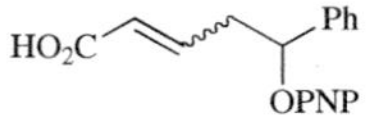 HO_2C, Ph, OPNP (75)	932
3 eq	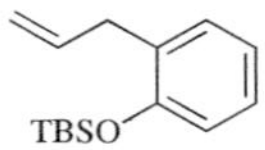 TBSO	Gr II (0.02 eq), additive (0.03 eq), Et_2O, 35°, 3 h	 HO_2C, TBSO	167
			Additive NaI (64) (*E*)/(*Z*) —	
			Additive CuI (82) (*E*)/(*Z*) >95:5	
3 eq	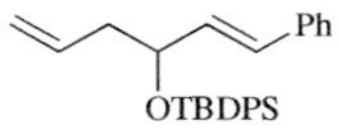 Ph, OTBDPS	Hov II (0.05 eq), DCM, 40°	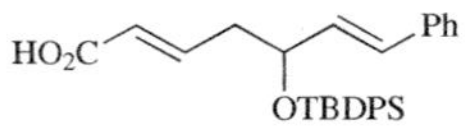 HO_2C, Ph, OTBDPS (83) (*E*)/(*Z*) = 95:5	169

Equiv.	Substrate	Conditions	Product (Yield %)	Ref.
3 eq	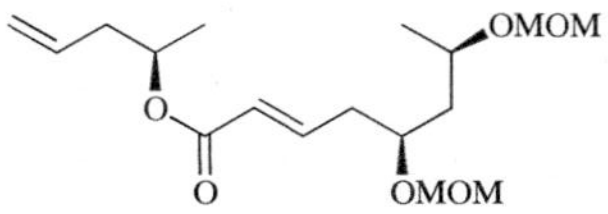 OMOM, OMOM, O	Hov II (0.05 eq), DCM, 25°, 16 h	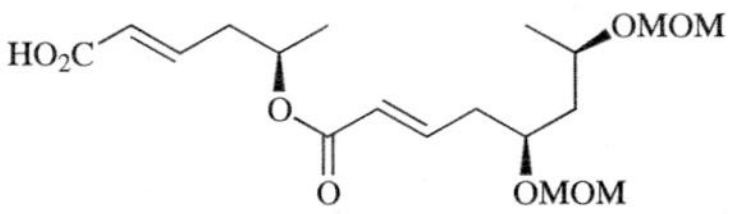 HO_2C, OMOM, OMOM, O (78) (*E*)/(*Z*) = 95:5	1029
3 eq	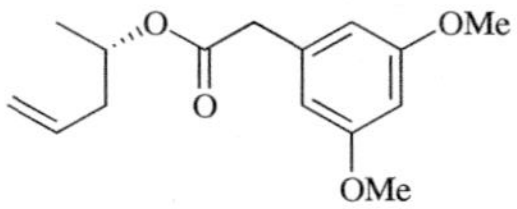 OMe, OMe, O	Hov II (0.05 eq), DCM, 25°, 16 h	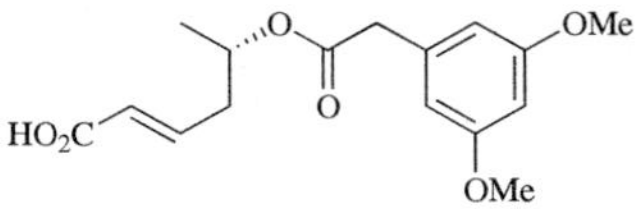 HO_2C, OMe, OMe, O (82) (*E*)/(*Z*) = 95:5	1030
3–20 eq	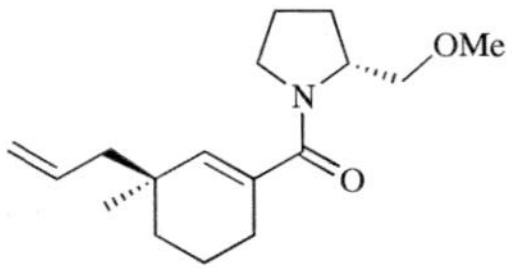 N, OMe, O	Hov II (0.025 eq), DCM, reflux, 7 h	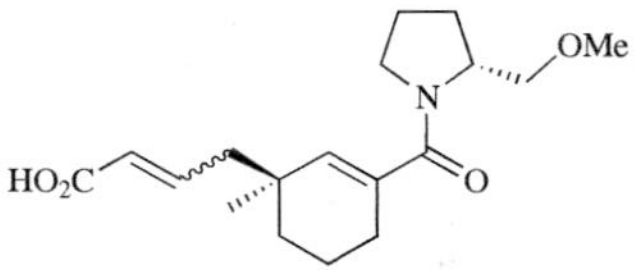 HO_2C, N, OMe, O (86)	896

TABLE 18. CROSS-METATHESIS OF α,β-UNSATURATED ACRYLIC ACIDS (*Continued*)

Acrylic Acid	Cross Partner	Conditions	Product(s) and Yield(s) (%)	Refs.

Please refer to the charts preceding the tables for the catalyst structures.

C_3

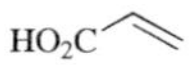

Acrylic Acid	Cross Partner	Conditions	Product(s) and Yield(s) (%)	Refs.
HO_2C (2 eq)	OBn, H, H, H, O	Catalyst (0.03–0.05 eq), solvent	HO_2C, OBn, H, H, H, O	163

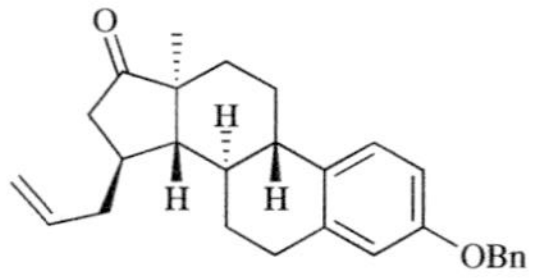

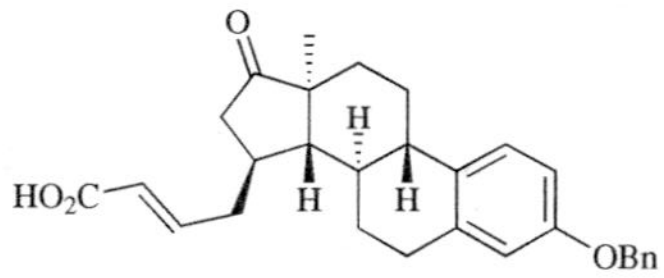

Catalyst	Solvent	Temp (°)	Time (h)		(*E*)/(*Z*)
Gr II	DCM	40	2.5	(88)	(*E*) only
Ru cat **68**	DCE	80	5.5	(0)	—
Ru cat **125**	DCM	40	5.5	(99)	(*E*) only

Acrylic Acid	Cross Partner	Conditions	Product(s) and Yield(s) (%)	Refs.
4 eq	O, O, O	Catalyst (0.05 eq), toluene, 80°, 12 h	HO_2C, O, O, O	1031

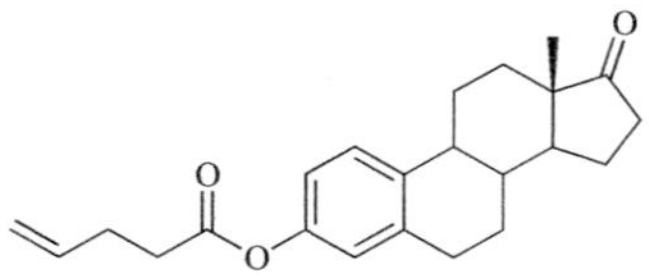

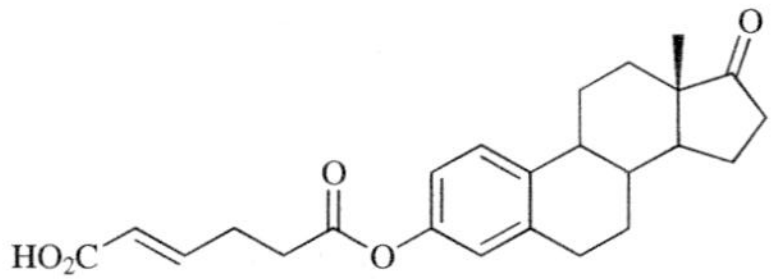

Catalyst		(*E*)/(*Z*)
Ru cat **266**	(90)	78:22
Ru cat **267**	(93)	79:21

18 eq Gr II (0.1 eq), DCM, rt, 48 h (36) 1032

20 eq Gr II (0.1 eq), DCB, MW (20 W) (99) 432

TABLE 18. CROSS-METATHESIS OF α,β-UNSATURATED ACRYLIC ACIDS (*Continued*)

Acrylic Acid	Cross Partner	Conditions	Product(s) and Yield(s) (%)	Refs.

Please refer to the charts preceding the tables for the catalyst structures.

C_3

Acrylic Acid	Cross Partner	Conditions	Product(s) and Yield(s) (%)	Refs.
HO_2C (acrylic acid) 40 eq	H–(O–sugar(HO, OEt, OH)–O–$(CH_2)_m$–CH=CH₂)$_n$	Hov II (x eq), solvent, 37°	H–(O–sugar(HO, OEt, OH)–O–$(CH_2)_m$–CH=CH–CO_2H)$_n$	1033

m	x	Solvent	Time (h)	Conv. (%)[a]
0	0.05	THF	2	31
0	0.05	DCM	2	40
0	0.05	acetic acid	2	23
2	0.05	THF	2	60
2	0.05	DCM	2	45
2	0.05	acetic acid	2	72
2	0.05	THF	2	87
2	0.1	DCM	2	100
2	0.1	acetic acid	2	100
4	0.1	DCM	2	>95
8	0.05	DCM	2	44
8	0.05	acetic acid	2	38
8	0.1	acetic acid	2	59
8	0.05	DCM	12	83
8	0.1	DCM	12	100

20 eq

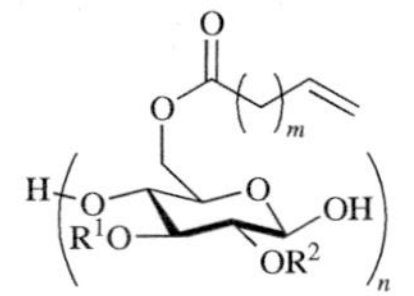

Catalyst (x eq),
BHT (y eq),
THF, 1 h

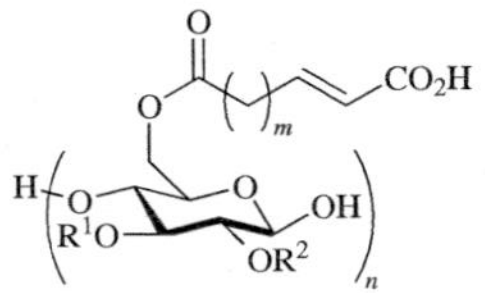

R^1	R^2	m	Catalyst	x	y	Temp (°)	Conv. (%)[a]		$(E)/(Z)$	
H	Ac	—	Hov II	0.03	0.12	30	100	(—)	—	1034
H	H	2	Gr II	0.05	0	rt	—	(94)	94:6	1035
Me	Me	8	Gr II	0.05	0	rt	—	(93)	94:6	1035
H	H	8	Hov II	0.03	0	30	—	(93)	—	1036
Ac	Ac	8	Hov II	0.03	0	30	—	(71)	—	1036
EtCO	EtCO	8	Hov II	0.03	0	30	—	(85)	—	1036
n-PrCO	n-PrCO	8	Hov II	0.03	0	30	—	(95)	—	1036

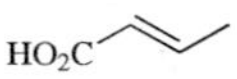

C_4

HO_2C 5 eq

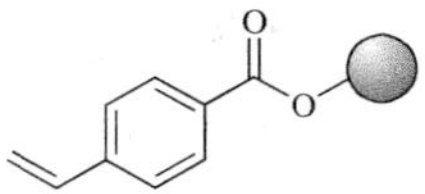

1. Gr II (0.05 eq), DCM, reflux, 20 h
2. TFA (10%), DCM, 1 h
3. CH_2N_2, DCM, 0°, 0.5 h

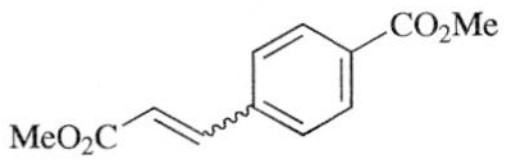

MeO_2C CO_2Me (97) 677

TABLE 18. CROSS-METATHESIS OF α,β-UNSATURATED ACRYLIC ACIDS (*Continued*)

Acrylic Acid	Cross Partner	Conditions	Product(s) and Yield(s) (%)	Refs.
Please refer to the charts preceding the tables for the catalyst structures.				
C_4				
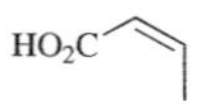 HO_2C **I**	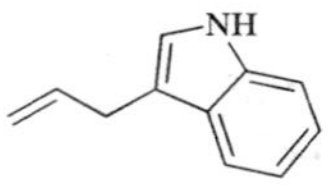 NH	1. Ru cat **364** (0.01 eq), (*Z*)-2-butene (5 eq), THF, 22°, 1 h 2. Ru cat **364** (0.04 eq), **I** (5 eq), THF, 100 torr, 22°, 1 h; then ambient pressure, 7 h 3. Ru cat **364** (0.04 eq), THF, 100 torr, 22°, 1 h; then ambient pressure, 11 h	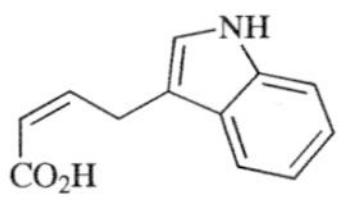 NH CO_2H (52) (*Z*)/(*E*) = >98:2	1037
I	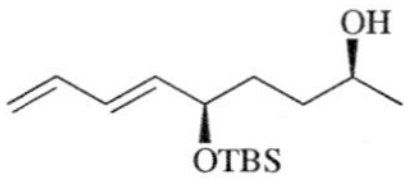 OH OTBS	1. Ru cat **364** (0.02 eq), (*Z*)-2-butene (5 eq), THF, 22°, 1 h 2. Ru cat **364** (0.04 eq), **I** (5 eq), THF, 100 torr, 22°, 1 h; then ambient pressure, 7 h 3. Ru cat **364** (0.04 eq), THF, 100 torr, 22°, 1 h; then ambient pressure, 11 h	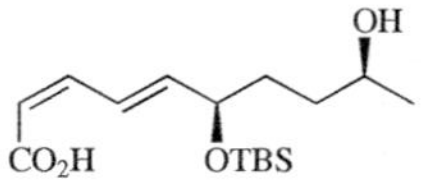 OH CO_2H OTBS (53) (*Z*)/(*E*) > 98:2	1037

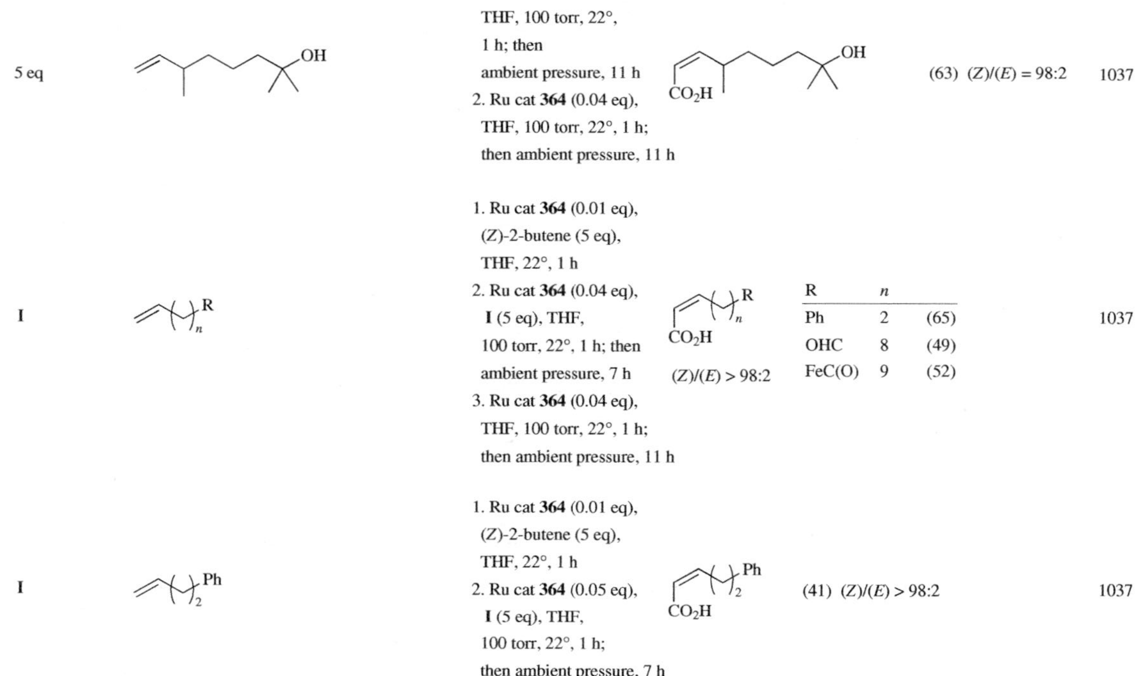

Reactant		Conditions	Product(s) and Yield(s) (%)	Refs.
5 eq		1. Ru cat **364** (0.04 eq), THF, 100 torr, 22°, 1 h; then ambient pressure, 11 h 2. Ru cat **364** (0.04 eq), THF, 100 torr, 22°, 1 h; then ambient pressure, 11 h	(63) (*Z*)/(*E*) = 98:2	1037
I		1. Ru cat **364** (0.01 eq), (*Z*)-2-butene (5 eq), THF, 22°, 1 h 2. Ru cat **364** (0.04 eq), **I** (5 eq), THF, 100 torr, 22°, 1 h; then ambient pressure, 7 h 3. Ru cat **364** (0.04 eq), THF, 100 torr, 22°, 1 h; then ambient pressure, 11 h	(*Z*)/(*E*) > 98:2 R / *n* Ph / 2 / (65) OHC / 8 / (49) FeC(O) / 9 / (52)	1037
I		1. Ru cat **364** (0.01 eq), (*Z*)-2-butene (5 eq), THF, 22°, 1 h 2. Ru cat **364** (0.05 eq), **I** (5 eq), THF, 100 torr, 22°, 1 h; then ambient pressure, 7 h	(41) (*Z*)/(*E*) > 98:2	1037

[a] The conversion was determined by ^{1}H NMR analysis.

TABLE 19. CROSS-METATHESIS OF ACRYLOYL CHLORIDE

Acryloyl Chloride	Cross Partner	Conditions	Product(s) and Yield(s) (%)	Refs.
Please refer to the charts preceding the tables for the catalyst structures.				
C_3				
1.5 eq		1. Hov II (0.05 eq), DCM, 16 h 2. Piperidine (6 eq), 1 h	R: $TMSCH_2$ (87); $PhCH_2$ (86); n-C_6H_{13} (86); Ph (54)	174
1.5 eq	OTBS	1. Hov II (0.05 eq), toluene, 16 h 2. NaN_3 (3 eq), MeCN, 2 h	N_3 OTBS (63)	174
1.5 eq	OTBS	1. Catalyst (0.05 eq), DCM, 16 h 2. RH (6 eq), 1 h	R OTBS	174

Catalyst	Temp	RH	
Gr I	rt	Me_2NH (33% in EtOH)	(tr)
Gr II	rt	Me_2NH (33% in EtOH)	(19)
Gr II	40°	Me_2NH (33% in EtOH)	(74)
Hov II	rt	Me_2NH (33% in EtOH)	(73)
Ru cat **88**	rt	Me_2NH (33% in EtOH)	(54)
Hov II	rt	$MeNH_2$ (33% in EtOH)	(80)
Hov II	rt	NH_3 (28% in H_2O)	(81)

1.5 eq — OTBS (alkene substrate)

1. Hov II (0.05 eq), DCM, 16 h
2. RH (x eq), K_3PO_4 (y eq), 1 h

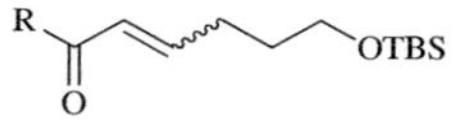

R–C(=O)–CH=CH∼∼–OTBS

174

RH	x	y	
 piperidine (NH)	1.6	3.8	(75)
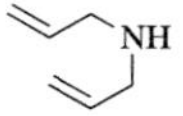 diallylamine (NH)	1.6	3.8	(65)
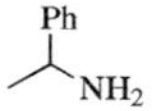 Ph–CH(Me)–NH$_2$	6	—	(73)
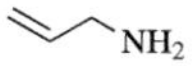 allyl–NH$_2$	6	—	(76)
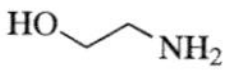 HO–CH$_2$CH$_2$–NH$_2$	1.6	3.8	(68)
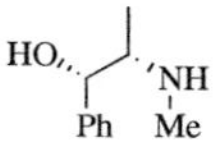 HO, Ph, NH–Me (pseudoephedrine-type)	1.6	3.8	(70)

TABLE 19. CROSS-METATHESIS OF ACRYLOYL CHLORIDE (*Continued*)

Acryloyl Chloride	Cross Partner	Conditions	Product(s) and Yield(s) (%)	Refs.
Please refer to the charts preceding the tables for the catalyst structures.				
C_3				
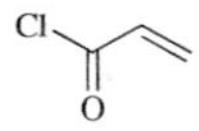 Cl, O; 1.5 eq	 OTBS	1. Hov II (0.05 eq), DCM, 16 h 2. RH•HCl (1.6 eq), NMM (6 eq), 1 h	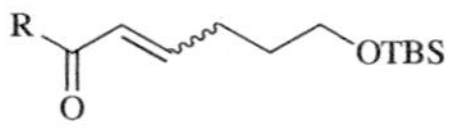 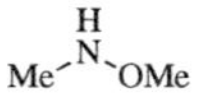 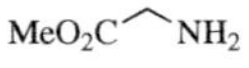 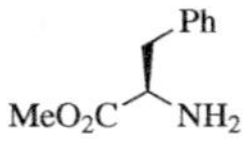 R, O, OTBS RH: Me–NH–OMe (69); MeO_2C ⁀ NH_2 (74); MeO_2C, Ph, NH_2 (78)	174
1.5 eq	 OTBS	1. Hov II (0.05 eq), DCM, 16 h 2. RCH_2OH (3 eq), pyridine (3 eq), 0°, 1 h	 R, O, OTBS R: CH_2=CH (63); HC≡C (65)	174

1.5 eq	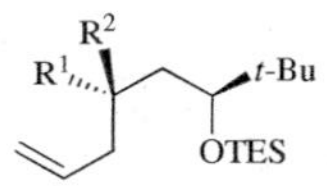	1. Hov II (0.05 eq), DCM, rt, overnight 2. NH_4OH, 1 h		1038

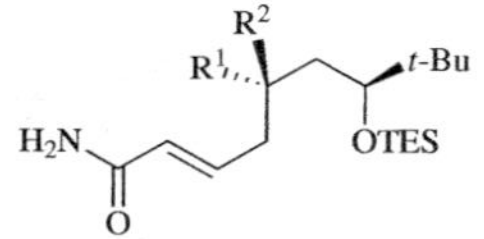

R^1	R^2		(E)/(Z)
H	Me	(82)	96:4
Me	H	(58)	—

20 eq	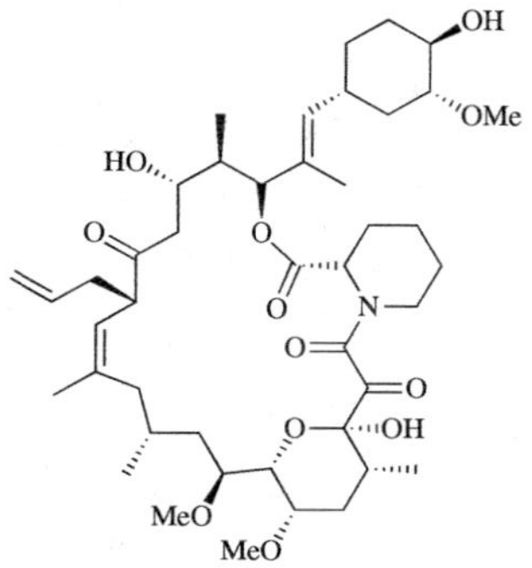	Gr II (0.1 eq), solvent, MW (20 W)		432

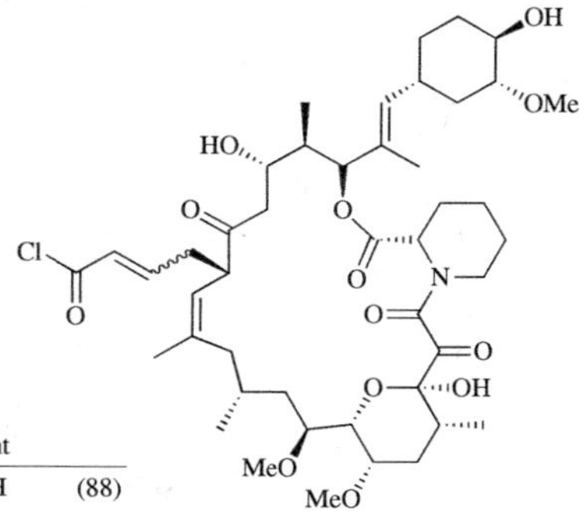

Solvent	
i-PrOH	(88)
DCB	(6)

TABLE 20. CROSS-METATHESIS OF α,β-UNSATURATED ESTERS AND THIOESTERS

A. ESTERS

Ester	Cross Partner	Conditions	Product(s) and Yield(s) (%)	Refs.
Please refer to the charts preceding the tables for the catalyst structures.				
C_3				
MeO_2C–CH=CH$_2$ (2 eq)	CH$_2$=CHCH$_2$OBz	Ru cat **286** (*x* eq), DCM, 8 h	MeO_2C–CH=CH–CH$_2$OBz, (*E*)/(*Z*) > 95:5 *x* / Temp (°): 0.02 / 25 (30) 0.02 / 40 (34) 0.05 / 40 (37)	161
(1 eq)	CH$_2$=CHCH$_2$NHBoc	Hov II (0.05 eq), additive (0.1 eq), toluene, 80°, 12 h	MeO_2C–CH=CH–CH$_2$NHBoc (wavy bond) Additive: — (28) Ti(O*i*-Pr)$_4$ (0) Me_2AlCl (20) ClBcat (50) $PhBCl_2$ (77) Cy_2BCl (84)	952
t-BuO_2C–CH=CH$_2$ (2 eq)	CH$_2$=CHCH$_2$TMS	Gr II (0.02 eq), 2.5% aq. PTS, 22°, 12 h	*t*-BuO_2C–CH=CH–CH$_2$TMS (91) (*E*)/(*Z*) = 95:5	976

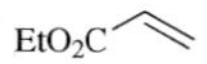

EtO_2C (x eq) | O=C(R) vinyl ketone | Hov II (0.1 eq), DCM, 100° (MW), 0.25 h | EtO_2C–CH=CH–C(O)R **I** + ROC–CH=CH–COR **II** + EtO_2C–CH=CH–CO_2Et **III** | 165

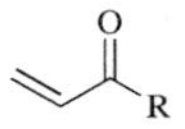

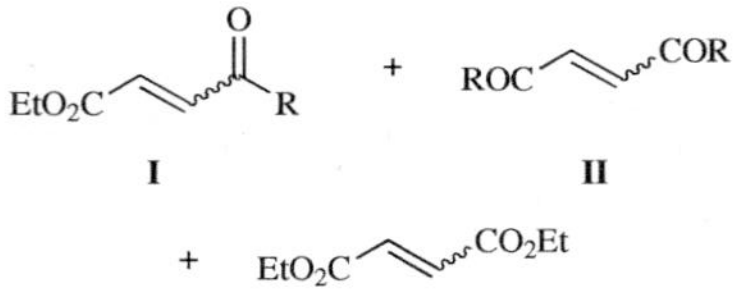

x	R	I	II	III
—	t-BuO	(32)	(9)	(11)
—	H_2N	(0)	(37)	(35)
0.0025	Et	(45)	(8)	(20)
0.01	Et	(36)	(7)	(16)

R^1O_2C (2 eq) | O=C(R^2) vinyl ketone | Gr II (0.05 eq), DCM, 3 h | R^1O_2C–CH=CH–C(O)R^2 (E) only | 170

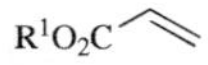

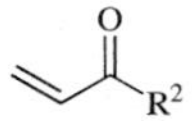

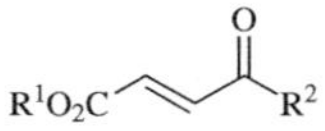

R^1	R^2	
Me	Me	(41)
t-Bu	Et	(41)

TABLE 20. CROSS-METATHESIS OF α,β-UNSATURATED ESTERS AND THIOESTERS (*Continued*)

A. ESTERS (*Continued*)

Ester	Cross Partner	Conditions	Product(s) and Yield(s) (%)	Refs.

Please refer to the charts preceding the tables for the catalyst structures.

C_3

RO_2C–CH=CH₂ (Ester); CH₂=CH–CO_2R (Cross Partner); Catalyst (*x* eq), DCM; RO_2C–CH=CH–CO_2R (Product)

R	Catalyst	*x*	Temp (°)	Time (h)	Yield (%)	(*E*)/(*Z*)	Refs.
n-Bu	Gr II	0.05	40	3	(87)	100:0	170
Cy	Gr II	0.05	40	3	(75)	100:0	170
t-Bu	Gr II	0.05	40	3	(94)	100:0	170
Ad	Gr II	0.05	40	3	(80)	100:0	170
t-Bu	Ru cat **108**	0.025	40	6	(88)	—	1039
n-Bu	Ru cat **108**	0.025	40	6	(91)	—	1039
t-Bu	Hov II	0.05	100 (MW)	0.25	(87)	—	165
Et	Hov II	0.05	100 (MW)	0.25	(66)	—	165
Et	Gr II	0.05	100 (MW)	0.25	(19)	—	165
Et	Gr II	0.05	rt	24	(20)	—	165
Et	Gr II	0.05	100 (MW)	1	(25)	—	165
Et	Hov II	0.05	rt	24	(18)	—	165
Et	Hov II	0.025	100 (MW)	0.25	(63)	—	165
Et	Hov II	0.005	100 (MW)	0.25	(23)	—	165
Et	Hov II	0.05	100 (MW)	0.17	(20)	—	165
2-hydroxyethyl	Ru cat **85**	0.025	40	48	(>99)	—	1040
2-hydroxyethyl	Hov II	0.025	40	2	(>99)	—	1040

Reactant 1	Reactant 2	Conditions	Product(s) and Yield(s) (%)	Refs.
t-BuO_2C (3 eq)	EtO, P(=O), OEt, OEt	Gr II (0.05 eq), DCM, 40°, 22 h	t-BuO_2C, EtO, P(=O), OEt, OEt (73)	1041
MeO_2C (2 eq)	OBz	Ru cat **286** (x eq), DCM, 40°, 8 h	MeO_2C, OBz; $(E)/(Z)$ > 95:5; x = 0.02 (22), 0.05 (28)	161
1 eq	OH (1 eq) + N_2, O, O (1 eq)	1. DCM, 300 W, 0.25 h 2. Hov II (0.04 eq), 100°, 0.5 h 3. IPr (0.2 eq), rt, 20 h	O, O, O, CO_2Me (55) $(E)/(Z)$ > 95:5	955

TABLE 20. CROSS-METATHESIS OF α,β-UNSATURATED ESTERS AND THIOESTERS (*Continued*)

A. ESTERS (*Continued*)

Ester	Cross Partner	Conditions	Product(s) and Yield(s) (%)	Refs.

Please refer to the charts preceding the tables for the catalyst structures.

C_3

MeO_2C (x eq) + OBz → Catalyst (y eq), solvent → MeO_2C ... OBz **I** + BzO ... OBz **II**

x	Catalyst	y	Solvent	Temp (°)	Time (h)	I	(E)/(Z) I	II	Refs.
2	Hov II	0.01	DCM	20	24	(30)	—	(52)	1040
2	Ru cat **85**	0.01	DCM	20	24	(69)	—	(9)	1040
2	Ind II	0.01	DCM	rt	5	(26)	94:6	(3)	1042
2	Ru cat **294**	0.01	DCM	rt	5	(73)	>95:5	(7)	1042
2	Ru cat **320**	0.01	DCM	rt	5	(5)	88:12	(3)	1042
2	Ru cat **285**	0.01	DCM	rt	5	(57)	94:6	(13)	1042
2	Ru cat **297**	0.01	DCM	rt	5	(79)	94:6	(11)	1042
2	Ru cat **322**	0.01	DCM	rt	5	(31)	>95:5	(25)	1042
2	Ru cat **41**	0.01	DCM	rt	5	(79)	>95:5	—	562
2	Ru cat **44**	0.01	DCM	rt	5	(85)	>95:5	—	562
2	Ru cat **45**	0.01	DCM	rt	5	(91)	>95:5	—	562
2	Ru cat **248**	0.01	DCM	rt	5	(69)	>95:5	—	562

5	Ru cat **316**	0.002	MTBE	50	8	(72)	>95:5	—	1043
5	Gr II	0.002	MTBE	50	8	(72)	>95:5	—	1043
2	Ru cat **139**	0.02	DCM	30	18	(75)	>95:5	—	1044
2	Ru cat **324**	0.01	DCM	rt	5	(50)	94:6	(15)	1045
2	Ru cat **325**	0.01	DCM	rt	5	(47)	94:6	(18)	1045
2	Ru cat **298**	0.01	DCM	rt	5	(81)	>95:5	(5)	1045
2	Ru cat **323**	0.01	DCM	rt	5	(23)	93:7	(26)	1045
2	Ru cat **316**	0.02	toluene	80	2.5	(75)	>95:5	—	591
4	Ru cat **209**	0.025	toluene	80	1.5	(72)	90:10	(9)	1046
—	Ru cat **93**	0.01	DCM	rt	0.5	(76)	>95:5	—	550
2	Ru cat **93**	0.01	DCM	25	0.5	(76)	>95:5	—	957
2	Ru cat **169**	0.01	DCM	25	0.5	(74)	>95:5	—	957
2	Ru cat **170**	0.01	DCM	25	0.5	(64)	>95:5	(34)	957
2	Ru cat **171**	0.01	DCM	25	0.5	(62)	>95:5	(31)	957
5	Ru cat **170**	0.01	DCM	25	1	(92)	>95:5	—	957
2	Ru cat **286**	0.02	DCM	25	2	(>99)	>95:5	—	957

n-BuO$_2$C (acrylate), 4 eq + OBz (butenyl benzoate) — Ru cat **208** (0.005 eq), DCM, 20°, 3.5 h → *n*-BuO$_2$C~~OBz (70) (*E*)/(*Z*) > 95:5 + BzO~~OBz (10) — 1046

TABLE 20. CROSS-METATHESIS OF α,β-UNSATURATED ESTERS AND THIOESTERS (*Continued*)

A. ESTERS (*Continued*)

Ester	Cross Partner	Conditions	Product(s) and Yield(s) (%)	Refs.

Please refer to the charts preceding the tables for the catalyst structures.

C_3

Ester	Cross Partner	Conditions	Product(s) and Yield(s) (%)
R^1O_2C–CH=CH$_2$ (*x* eq)	CH$_2$=CH–CH$_2$–CH$_2$–OR^2	Catalyst (*y* eq), solvent	R^1O_2C–CH=CH–CH$_2$–CH$_2$–OR^2

R^1	*x*	R^2	Catalyst	*y*	Solvent	Temp (°)	Time (h)	Yield (%)	(*E*)/(*Z*)	Refs.
Me	4	PMP	Ru cat **209**	0.025	toluene	80	1.5	(72)	90:10	1046
Me	2	PNP	Hov II	0.02	DCM	40	15	(89)	(*E*) only	932
t-Bu	2	PNP	Hov II	0.02	DCM	40	15	(96)	(*E*) only	932
2-ethylhexyl	2	PNP	Hov II	0.02	DCM	40	15	(95)	(*E*) only	932

Ester	Cross Partner	Conditions	Product(s) and Yield(s) (%)	Refs.
MeO_2C–CH=CH$_2$ (10 eq)	CH$_2$=CH–CH(OH)–R	1. Catalyst (0.03 eq), DCM, rt, 350 nm 2. *t*-BuOH/DCM (1:4), rt, h*ν* (254 nm)	MeO_2C–CH$_2$–CH$_2$–C(=O)–R	911

R	Catalyst	Yield (%)
Me	Ru cat **163**	(66)
$MeO_2C(CH_2)_2$	Ru cat **353**	(62)
4-BrC_6H_4	Ru cat **163**	(70)
3-BrC_6H_4	Ru cat **163**	(68)
4-ClC_6H_4	Ru cat **163**	(71)
4-$MeOC_6H_4$	Ru cat **163**	(76)
3-$MeOC_6H_4$	Ru cat **163**	(76)
n-C_9H_{19}	Ru cat **163**	(62)

10 eq

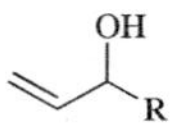

OH, R

1. Catalyst (0.03 eq), DCM, rt, 350 nm
2. Phenanthrene (x eq), DCM, rt, hν (254 nm)

O, O, R

R	Catalyst	x	
Me	Ru cat **163**	0.3	(62)
$MeO_2C(CH_2)_2$	Ru cat **353**	0.3	(53)
4-BrC_6H_4	Ru cat **163**	2.1	(70)
3-BrC_6H_4	Ru cat **163**	2.1	(67)
4-ClC_6H_4	Ru cat **163**	2.1	(72)
4-$MeOC_6H_4$	Ru cat **163**	2.1	(0)
3-$MeOC_6H_4$	Ru cat **163**	2.1	(63)
n-C_9H_{19}	Ru cat **163**	0.3	(70)

911

3 + 3 eq

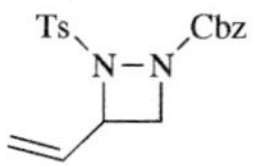

Ts, N–N, Cbz

Gr II (0.05 + 0.05 eq)
DCM, reflux, 6 + 24 h

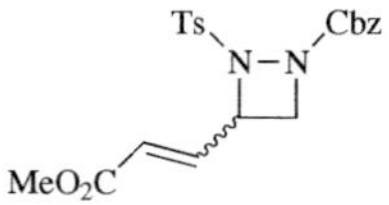

Ts, N–N, Cbz, MeO_2C

(83)

647

TABLE 20. CROSS-METATHESIS OF α,β-UNSATURATED ESTERS AND THIOESTERS (*Continued*)

A. ESTERS (*Continued*)

Ester	Cross Partner	Conditions	Product(s) and Yield(s) (%)	Refs.

Please refer to the charts preceding the tables for the catalyst structures.

C_3

Ester	Cross Partner	Conditions	Product(s) and Yield(s) (%)	Refs.
MeO_2C (20 eq)	R, NH_3^+ Y^- (n)	Hov II (*x* eq), EtOAc, reflux, 1 h	MeO_2C, NH_3^+ Y^- (n)	548

R	Y	*n*	*x*	Conv. (%)[a]	(*E*)/(*Z*)
H	TfO	2	0.05	>99	86:14
H	BF_4	2	0.01	>99	87:13
H	TfO	3	0.01	>95	87:13
H	BF_4	3	0.01	>99	91:9
H	TsO	3	0.01	>99	91:9
Me	BF_4	8	0.01	>99	91:9
Me	TsO	8	0.01	>99	95:5

Ester	Cross Partner	Conditions	Product(s) and Yield(s) (%)	Refs.
t-BuO_2C (10 eq)	OH, OH	Ru cat **353** (0.06 eq), DCM, rt, h*ν* (350 nm)	*t*-BuO_2C, OH, OH (46) (*E*) only	911
R^1O_2C (2 eq)	R^2, OH	Gr II (*x* eq), benzene, rt	R^1O_2C, R^2, OH	336

R^1	R^2	*x*	Time (h)		(*E*)/(*Z*)
Me	Ph	0.02	24	(83)	>91:9
Me	*n*-$C_{15}H_{31}$	0.02	16	(65)	95:5
Et	Me	0.05	24	(75)	>95:5
Et	*n*-$C_{15}H_{31}$	0.05	24	(55)	>95:5

EtO_2C (2.5 eq)	OTBS		Gr II (0.02 eq), benzene, rt, 20 h		EtO_2C … OTBS		(80)			794
R^1O_2C (2 eq)	OR^2		Catalyst (x eq), solvent		R^1O_2C … OR^2					
R^1	R^2	Catalyst	x	Solvent	Temp (°)	Time (h)		(E)/(Z)		
Me	TBS	Ru cat **324**	0.01	DCM	rt	0.5	(77)	>95:5	1045	
Me	TBS	Ru cat **325**	0.01	DCM	rt	0.5	(78)	>95:5	1045	
Me	TBS	Ru cat **286**	0.01	DCM	rt	2.5	(91)	>95:5	1045	
Me	TBS	Ru cat **323**	0.01	DCM	rt	15	(50)	>95:5	1045	
Me	TBS	Ru cat **326**	0.02	toluene	rt	18	(91)	—	1047	
Me	Bz	Ru cat **106**	0.01	DCM	20	0.5	(99)	—	216	
Me	Bz	Ru cat **112**	0.01	DCM	20	0.5	(99)	>97:3	216	
Me	Bz	Ru cat **110**	0.01	DCM	rt	0.33	(95)	98:2	961	
Me	Bz	Gr II	0.01	DCM	20	3	(86)	97:3	962	
Me	Bz	Ru cat **112**	0.01	DCM	20	0.25	(93)	97:3	962	
Me	Ac	Ru cat **109**	0.02	H_2O	22	12	(82)	>95:5	341	
Me	Ac	Hov II	0.02	2.5 % aq. PTS	22	12	(83)	>95:5	976	
t-Bu	Ac	Hov II	0.02	2.5 % aq. PTS	22	12	(87)	95:5	976	

TABLE 20. CROSS-METATHESIS OF α,β-UNSATURATED ESTERS AND THIOESTERS (*Continued*)

A. ESTERS (*Continued*)

Ester	Cross Partner	Conditions	Product(s) and Yield(s) (%)	Refs.

Please refer to the charts preceding the tables for the catalyst structures.

C_3

MeO_2C-CH=CH$_2$ (x eq) + CH$_2$=CH(CH$_2$)$_3$OBz → Catalyst (y eq), solvent → MeO_2C-CH=CH(CH$_2$)$_3$OBz

x	Catalyst	y	Solvent	Temp (°)	Time (h)	Yield (%)	(E)/(Z)	Refs.
2	Ru cat **98**	0.01	DCM	rt	0.5	(69)	95:5	965
2	Ru cat **99**	0.01	DCM	rt	0.5	(50)	95:5	965
2	Ru cat **100**	0.01	DCM	rt	0.5	(71)	95:5	965
2	Ru cat **101**	0.01	DCM	rt	0.5	(73)	95:5	965
2	Ru cat **102**	0.01	DCM	rt	0.5	(62)	95:5	965
2	Ru cat **316**	0.02	toluene	80	2.5	(85)	>95:5	591
2	Ru cat **316**	0.002	toluene	120	15	(81)	>95:5	591
2	Ru cat **191**	0.05	DCM	45	12	(97)	>95:5	923
2	Ru cat **243**	0.05	DCM	45	12	(96)	>95:5	923
2	Hov II	0.05	DCM	40	—	(87)	>95:5	922
—	Ru cat **185**	0.05	BMI•PF_6/toluene	rt	3	(80)	>95:5	924
2	Ru cat **286**	0.02	DCM	25	2	(>99)	>95:5	161
2	Ru cat **185**	0.05	BMI•PF_6/toluene	rt	3	(80)	>99:1	1048
2	Ru cat **186**	0.05	BMI•PF_6/toluene	rt	3	(79)	>99:1	1048
1	Ru cat **186**	0.05	BMI•PF_6/toluene	rt	3	(78)	>99:1	1048
5	Ru cat **186**	0.025	toluene	rt	24	(11)	—	1048
5	Ru cat **186**	0.025	MTBE	rt	24	(<5)	—	1048
5	Ru cat **186**	0.025	DCM	rt	2	(87)	—	1048
5	Ru cat **186**	0.05	BMI•PF_6/toluene	rt	3	(98)	>99:1	1048
5	Ru cat **186**	0.05	BMI•PF_6/benzene	rt	2.5	(93)	>99:1	1048
5	Ru cat **186**	0.05	BMI•PF_6/MTBE	rt	6	(56)	94:6	1048
5	Ru cat **186**	0.05	BMI•PF_6/toluene	rt	3	(98)	>99:1	1048
5	Ru cat **186**	0.05	BMI•PF_6/toluene	50	0.5	(97)	>99:1	1048
2	Ru cat **170**	0.01	DCM	25	1	(91)	>95:5	1048

Olefin 1	Olefin 2	Conditions	Product(s)	Yield (%) / ratio	Ref.
2 eq	OBz, 1 eq + Ph, 1 eq	Catalyst (0.01 eq), [bdmim][PF_6]/toluene, rt, 4 h	MeO_2C … OBz (**I**) + MeO_2C … Ph (**II**)		1050
EtO_2C, 2 eq	OPMP, OH	Hov II (0.025 eq), DCM, rt, 15 h	EtO_2C … OPMP, OH	(70) (*E*)/(*Z*) = 98:2	626
MeO_2C	OPMP, OH	Gr II (0.01 eq), DCM, reflux, 3 h	MeO_2C … OPMP, OH	(87) (*E*)/(*Z*) = 99:1	566
3 eq	OH, OR, *n*	Gr II (0.01 eq), DCM, reflux, 3 h	MeO_2C … OH, OR, *n*		1051

Catalyst	I	(*E*)/(*Z*) **I**	II	(*E*)/(*Z*) **II**
Ru cat **142**	(92)	95:5	(90)	95:5
Ru cat **141**	(90)	94:6	(89)	95:5

R	*n*		(*E*)/(*Z*)
Bn	0	(94)	>99:1
PMB	0	(84)	99:1
Bn	1	(92)	>99:1
Bn	2	(89)	>99:1
Bn	3	(96)	>99:1
TBS	3	(66)	99:1

TABLE 20. CROSS-METATHESIS OF α,β-UNSATURATED ESTERS AND THIOESTERS (*Continued*)

A. ESTERS (*Continued*)

Ester	Cross Partner	Conditions	Product(s) and Yield(s) (%)	Refs.

Please refer to the charts preceding the tables for the catalyst structures.

C_3

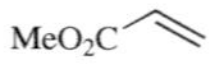

Ester	Cross Partner	Conditions	Product(s) and Yield(s) (%)	Refs.
MeO_2C (10 eq)	OH, OTBS	Gr II (*x* eq), phenol (*y* eq), toluene	MeO_2C … OH, OTBS (I) + MeO_2C … OH, OTBS (II)	1049

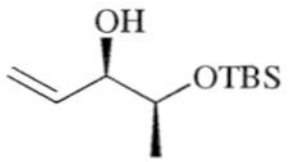

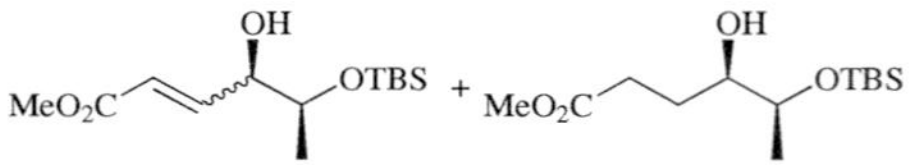

x	C/mol L	*y*	Temp (°)	Time (h)	I	II
0.05	1	—	110	12	(62)	(28)
0.05	0.5	—	110	2.5	(80)	(12)
0.05	0.5	—	110	0.5	(83)	(—)
0.05	0.5	—	80	0.5	(73)	(—)
0.05	1	—	110	0.5	(80)	(—)
0.05	neat	—	80	0.5	(83)	(—)
0.025	0.5	—	110	0.5	(75)	(—)
0.025	0.5	0.5	110	0.5	(86)	(—)
0.025	1	0.5	110	0.5	(98)	(—)
0.025	neat	0.5	80	0.5	(83)	(—)
0.05	0.5	0.5	110	2.5	(77)	(17)
0.025	1	0.5	110	12	(48)	(20)

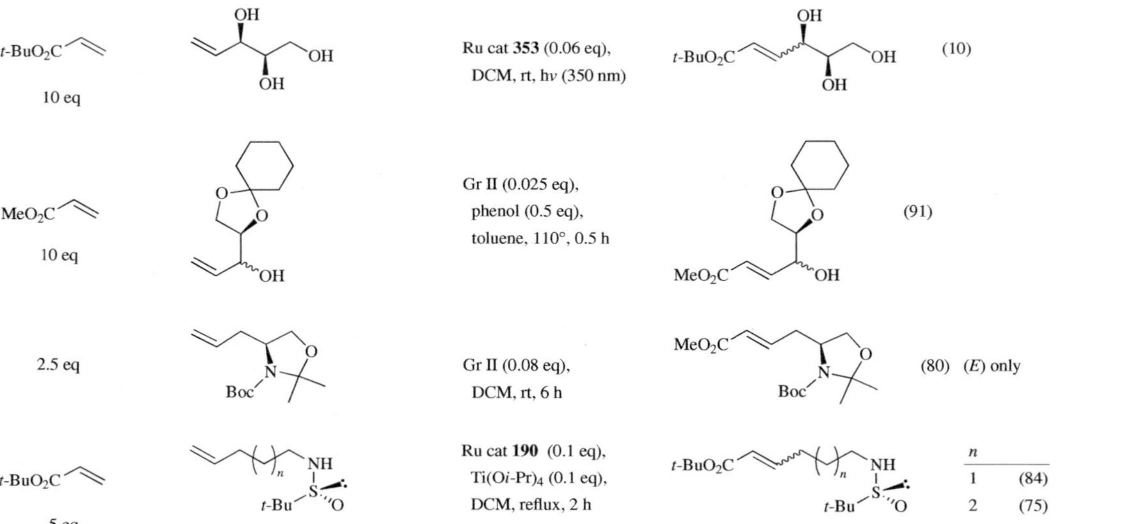

Alkene partner	Substrate	Conditions	Product(s) and Yield(s) (%)		Refs.
t-BuO$_2$C (10 eq)		Ru cat **353** (0.06 eq), DCM, rt, hν (350 nm)	t-BuO$_2$C, OH, OH, OH	(10)	911
MeO$_2$C (10 eq)	O, O, OH	Gr II (0.025 eq), phenol (0.5 eq), toluene, 110°, 0.5 h	MeO$_2$C, O, O, OH	(91)	1049
2.5 eq	Boc, N, O	Gr II (0.08 eq), DCM, rt, 6 h	MeO$_2$C, Boc, N, O	(80) (*E*) only	670
t-BuO$_2$C (5 eq)	NH, n, t-Bu, S, O	Ru cat **190** (0.1 eq), Ti(O*i*-Pr)$_4$ (0.1 eq), DCM, reflux, 2 h	t-BuO$_2$C, NH, n, t-Bu, S, O	*n* 1 (84); 2 (75)	959

TABLE 20. CROSS-METATHESIS OF α,β-UNSATURATED ESTERS AND THIOESTERS (*Continued*)

A. ESTERS (*Continued*)

Ester	Cross Partner	Conditions	Product(s) and Yield(s) (%)	Refs.
Please refer to the charts preceding the tables for the catalyst structures.				
C_3				
R^1O_2C (3 eq)	F, F, NHR^2, O, n	Hov II (0.05 eq), additive (0.1 eq), DCM, reflux	R^1O_2C, F, F, NHR^2, O, n	960

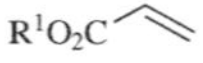

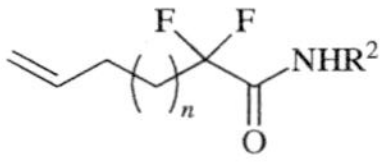

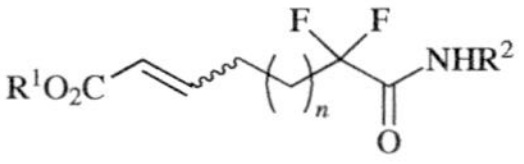

R^1	R^2	n	Additive	Time (h)	
Et	PMP	0	$Ti(Oi\text{-}Pr)_4$	5	(90)
Et	PMP	0	$BF_3{\bullet}OEt_2$	5	(80)
Et	MeO	0	$Ti(Oi\text{-}Pr)_4$	5	(71)
Et	MeO	0	$BF_3{\bullet}OEt_2$	2	(76)
Et	SAMP	0	$Ti(Oi\text{-}Pr)_4$	3	(75)
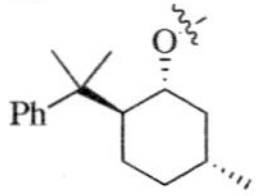 Ph, O	PMP	0	$Ti(Oi\text{-}Pr)_4$	5	(87)
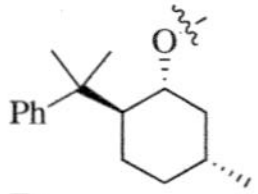 Ph, O	MeO	0	$Ti(Oi\text{-}Pr)_4$	5	(46)
Et	PMP	1	$Ti(Oi\text{-}Pr)_4$	5	(89)
Et	MeO	1	$Ti(Oi\text{-}Pr)_4$	6	(75)
Et	SAMP	1	$Ti(Oi\text{-}Pr)_4$	4	(64)
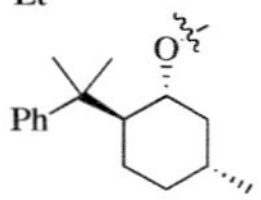 Ph, O	PMP	1	$Ti(Oi\text{-}Pr)_4$	5	(89)
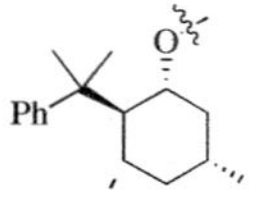 Ph, O	MeO	1	$Ti(Oi\text{-}Pr)_4$	5	(38)

x eq | Gr II (*y* eq), solvent | I + II | 1052, 392

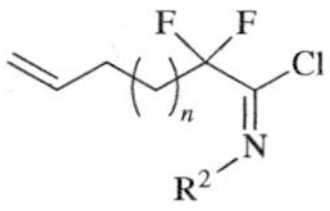

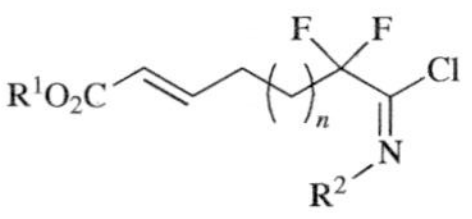

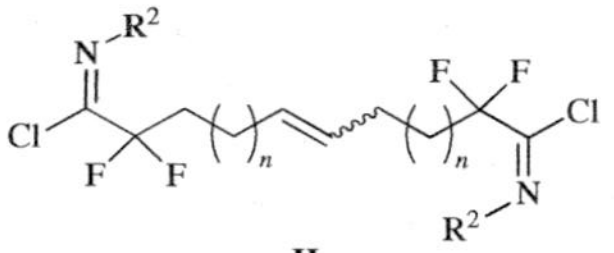

R^1	*x*	R^2	*n*	*y*	Solvent	Temp (°)	Time (h)	I	(*E*)/(*Z*) I	II
Et	5	PMP	0	0.05	toluene	95	15	(95)	96:4	(0)
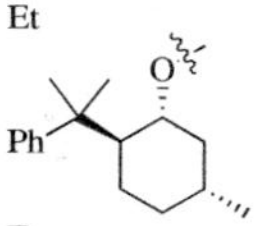 Ph	2	PMP	0	0.05	toluene	95	15	(76)	92:8	(0)
Et	2	PMP	0	0.1	DCM	40	5	(40)	>95:5	(16)
Et	2	PMP	0	0.1	DCM	40	15	(60)	>95:5	(8)
Et	3	PMP	0	0.1	DCM	40	48	(70)	>95:5	(6)
Et	5	PMP	0	0.1	toluene	95	15	(95)	>95:5	(0)
Et	2	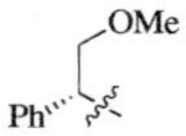 OMe, Ph	0	0.05	toluene	95	15	(71)	90:10	(0)
Et	5	PMP	1	0.05	toluene	95	15	(94)	98:2	(0)
Et	3	PMP	1	0.1	toluene	95	15	(94)	>95:5	(0)
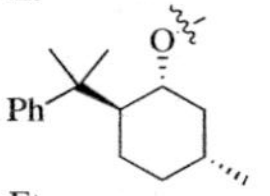 Ph	2	PMP	1	0.1	toluene	95	15	(80)	95:5	(0)
Et	5	PMP	2	0.05	toluene	95	15	(93)	92:8	(0)

TABLE 20. CROSS-METATHESIS OF α,β-UNSATURATED ESTERS AND THIOESTERS (*Continued*)

A. ESTERS (*Continued*)

Ester	Cross Partner	Conditions	Product(s) and Yield(s) (%)	Refs.
Please refer to the charts preceding the tables for the catalyst structures.				
C_3				
MeO_2C (2 eq)	Ph, Ph, OMe, Ph, Ph, OMe, O, B, O	Gr I (0.15 eq), DCM, reflux	MeO_2C … Ph, Ph, OMe, Ph, Ph, OMe, O, B, O (52) + Ph, Ph, MeO, MeO, Ph, Ph, O, B, O … Ph, Ph, OMe, OMe, Ph, Ph, O, B, O (14)	648
BnO_2C (2.5 eq)	*n*-Pr	Gr II (0.025 eq), TPGS-750-M in H_2O (2 wt %) $KHSO_4$ (0.02 M), rt	BnO_2C … *n*-Pr 100[b]	1053
RO_2C (3 eq)		Ru cat **93** (0.025 eq), $CDCl_3$, 100° (MW), 1 h	RO_2C … CO_2R — R: Et (75); *t*-Bu (88)	926

x eq | Hov II (y eq), 4-methoxyphenol (z eq), solvent | RO_2C ... CO_2R | 1054

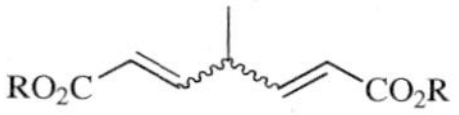

R	x	y	z	Solvent	Temp (°)	Time	
Me, t-Bu	—	0.1	—	DCM	reflux	—	(80–90)
t-Bu	2.5 + 4 x 1.25	0.005 + 7 x 0.0025	—	DCM	reflux	7 d	(80)
t-Bu	20	3 x 0.23	0.0047	—	50	5 h	(58)

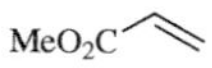

MeO_2C (5 eq) + OBz | Catalyst (0.00025 eq), DCM, 30°, 2 h | MeO_2C ... OBz, $(E)/(Z) > 95:5$

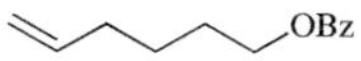

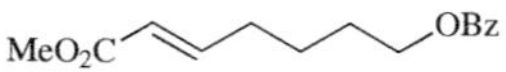

Catalyst		
Ru cat **321**	(92)	1055
Ru cat **322**	(82)	
Ru cat **285**	(100)	

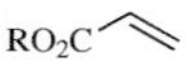

RO_2C (x eq) + OTBS | Catalyst (y eq), solvent | RO_2C ... OTBS

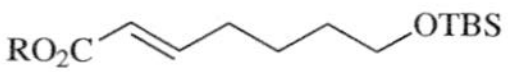

R	x	Catalyst	y	Solvent	Temp (°)	Time (h)		$(E)/(Z)$	
Me	5	Ru cat **316**	0.05	MTBE	50	8	(83)	>95:5	1043
Me	2	Ru cat **112**	0.025	DCM	20	0.33	(91)	97:3	609
Me	2	Ru cat **88**	0.01	DCM	25	1	(95)	97:3	609
Me	2	Ru cat **138**	0.01	DCM	25	2	(68)	97:3	609

TABLE 20. CROSS-METATHESIS OF α,β-UNSATURATED ESTERS AND THIOESTERS (*Continued*)

A. ESTERS (*Continued*)

Ester	Cross Partner	Conditions	Product(s) and Yield(s) (%)	Refs.

Please refer to the charts preceding the tables for the catalyst structures.

C_3

RO_2C–CH=CH$_2$ (x eq) + CH$_2$=CH(CH$_2$)$_4$OTBS → Catalyst (y eq), solvent → RO_2C–CH=CH(CH$_2$)$_4$OTBS

R	x	Catalyst	y	Solvent	Temp (°)	Time (h)	Yield (%)	(E)/(Z)	Refs.
Me	2	Ru cat **115**	0.01	DCM	25	0.33	(59)	97:3	609
Me	2	Ru cat **115**	0.01	DCM	25	5	(85)	97:3	609
Me	2	Ru cat **115**	0.01	DCM	25	11.5	(89)	97:3	609
Me	2	Ru cat **116**	0.01	DCM	25	0.33	(74)	97:3	609
Me	2	Ru cat **116**	0.01	DCM	25	5	(78)	97:3	609
Me	2	Ru cat **139**	0.01	DCM	30	18	(85)	95:5	1044
Me	2	Ru cat **158**	0.01	DCM	25	2	(92)	—	590
Me	2	Ru cat **157**	0.005	DCM	25	2	(90)	—	590
Me	2	Ru cat **316**	0.02	toluene	80	1.75	(97)	—	591
Me	2	Ru cat **316**	0.02	toluene	80	0.5	(90)	—	591
Me	2	Ru cat **314**	0.02	toluene	80	1.75	(38)	—	591
Me	2	Ru cat **314**	0.02	toluene	80	0.5	(6)	—	591
Me	2	Ru cat **315**	0.02	toluene	80	1.75	(67)	—	591
Me	2	Ru cat **315**	0.02	toluene	80	0.5	(13)	—	591
Me	2	Ru cat **317**	0.02	toluene	80	1.75	(94)	—	591
Me	2	Ru cat **317**	0.02	toluene	80	0.5	(60)	—	591
Me	2	Ru cat **316**	0.02	toluene	80	2	(81)	>95:5	591
Me	2	Ind I	0.02	DCM	25	8	(19)	>95:5	161
Me	2	Ind I	0.02	DCM	40	8	(55)	>95:5	161
Me	2	Ru cat 282	0.02	DCM	40	8	(26)	>95:5	161
Me	2	Ru cat **282**	0.02	DCM	25	8	(0)	—	161
Me	2	Ind II	0.02	DCM	25	8	(65)	>95:5	161
Me	2	Ind II	0.02	DCM	40	8	(81)	>95:5	161

	Me	2	Ru cat **286**	0.02	DCM	25	1	(60)	>95:5	161
	Me	2	Ru cat **286**	0.02	DCM	25	3	(88)	>95:5	161
	Me	2	Ru cat **286**	0.02	DCM	25	8	(98)	>95:5	161
	Me	2	Ru cat **286**	0.02	DCM	40	1	(55)	>95:5	161
	Me	2	Ru cat **286**	0.02	DCM	40	8	(98)	>95:5	161
	Me	3	Ru cat **123**	0.05	DCM	rt	1	(97)	—	1207
	Me	—	Ru cat **106**	0.01	DCM	25	0.42	(99)	78:22	216
	Me	—	Ru cat **89**	0.01	DCM	25	0.5	(95)	95:5	216
	Me	2	Ru cat **112**	0.025	DCM	25	0.3	(91)	—	1061
	Me	2	Ru cat **88**	0.01	DCM	25	16	(95)	—	1061
	Me	2	Ru cat **115**	0.01	DCM	25	2	(90)	—	1061
	Me	2	Ru cat **115**	0.01	DCM	25	16	(89)	—	1061
	Me	2	Ru cat **116**	0.01	DCM	25	1	(80)	—	1061
	Me	2	Ru cat **88**	0.01	DCM	25	0.5	(95)	95:5	171
	Me	2	Ru cat **88**	0.01	DCM	rt	0.5	(95)	95:5	160
	Me	2	Ru cat **112**	0.025	DCM	rt	0.33	(91)	95:5	160
	Me	2	Ru cat **112**	0.025	DCM	20	0.33	(91)	99:1	962
	Me	—	Ru cat **268**	0.01	DCM	40	1.3	(97)	—	329
	Me	—	Ru cat **269**	0.01	DCM	40	0.5	(98)	—	329
	Me	—	Ru cat **270**	0.01	DCM	40	0.3	(90)	—	329
	Me	3	Gr II	0.05	H_2O	40	5	(65)	—	1083
	Me	3	Ru cat **282**	0.05	H_2O	40	5	(56)	—	1083
	t-Bu	3	Gr II	0.05	H_2O	40	5	(81)	—	1083
	t-Bu	4	Ru cat **143**	0.05	H_2O	30	12	(78)	100:0	588

MeO_2C (alkene) 4 eq + (alkene)$(CH_2)_3$R —Ru cat **245** (0.25 eq), toluene, 80°, 1 h→ MeO_2C (alkene)$(CH_2)_3$R, (*E*)/(Z) = 95:5

R	Conv. (%)[b]	
TBSO	91	1056
$MeO_2C(CH_2)_4$	98	

TABLE 20. CROSS-METATHESIS OF α,β-UNSATURATED ESTERS AND THIOESTERS (*Continued*)

A. ESTERS (*Continued*)

Ester	Cross Partner	Conditions	Product(s) and Yield(s) (%)	Refs.

Please refer to the charts preceding the tables for the catalyst structures.

C_3

RO_2C–CH=CH₂ (*x* eq) + CH₂=CH(CH₂)₄OAc → Catalyst (*y* eq), solvent → RO_2C–CH=CH(CH₂)₄OAc

R	*x*	Catalyst	*y*	Solvent	Temp (°)	Time (h)	Conv. (%)	(*E*)/(*Z*)	Refs.
Me	2.6	Gr II + HCl	0.05	benzene, THF	45	14	98	>95:5	711
Me	1	Gr I	0.025	CD_2Cl_2	35	3	10	—	1208
Me	1	Hov I	0.025	CD_2Cl_2	35	3	6	—	1208
Me	1	Gr II	0.025	CD_2Cl_2	35	3	100	—	1208
Me	1	Hov II	0.025	CD_2Cl_2	35	3	92	—	1208
Me	1	Ru cat **35**	0.025	CD_2Cl_2	35	3	84	—	1208
Me	1	Ru cat **187**	0.025	CD_2Cl_2	35	3	90	—	1208
Me	1	Ru cat **15**	0.025	CD_2Cl_2	35	3	38	—	1208
Me	1	Gr II	0.025	DCM	35	2	85	—	1209
Me	1	Ru cat **277**	0.025	DCM	35	60	60	—	1209
Me	1	Ru cat **278**	0.025	DCM	35	60	18	—	1209
Me	2	Ru cat **244**	0.01	toluene	50	2	84	—	1210
Me	1	Hov I	0.025	DCE	80	24	62	—	1211
Me	1	Ru cat **299**	0.025	toluene	80	1	24	—	1211
Me	1	Ru cat **300**	0.025	DCE	40	3	32	—	1211
Me	1	Ru cat **301**	0.025	toluene	80	24	53	—	1211
Me	1	Ru cat **302**	0.025	toluene	60	24	71	—	1211
Me	1	Ru cat **303**	0.025	toluene	60	1	42	—	1211
Me	1	Ru cat **304**	0.025	toluene	80	24	26	—	1211
Me	2	Ru cat **31**	0.004	toluene	80	2	95	—	1212
Et	1	Gr II	0.02	DCM	40	12	98	—	902
n-Bu	4	Ru cat **208**	0.005	DCM	20	3.5	82	>95:5	1046

MeO_2C–CH=CH$_2$ (2 eq)	CH$_2$=CH(CH$_2$)$_4$OAc	Catalyst (0.02 eq), solvent	MeO_2CCH=CH(CH$_2$)$_4$OAc	1057

Catalyst	Solvent	Temp (°)	Time (h)	Conv. (%)
Gr I	CD_2Cl_2	45	24	22
Ru cat **46**	CD_2Cl_2	45	24	7
Ru cat **46**	DMSO-d_6	80	24	0
Ru cat **47**	CD_2Cl_2	45	24	0
Ru cat **47**	DMSO-d_6	80	24	0
Ru cat **48**	CD_2Cl_2	45	24	15
Ru cat **48**	CD_2Cl_2	45	72	18
Ru cat **48**	DMSO-d_6	80	24	12
Ru cat **48**	DMSO-d_6	80	72	17

10 eq	CH$_2$=CH(CH$_2$)$_4$Cl	Gr II (0.05 eq), *p*-cresol (0.5 eq), toluene, 120°, 3 h	MeO_2CCH=CH(CH$_2$)$_4$Cl (80) (*E*) only	1058

TABLE 20. CROSS-METATHESIS OF α,β-UNSATURATED ESTERS AND THIOESTERS (*Continued*)

A. ESTERS (*Continued*)

Ester	Cross Partner	Conditions	Product(s) and Yield(s) (%)	Refs.
Please refer to the charts preceding the tables for the catalyst structures.				
C_3				
MeO_2C (vinyl), 10 eq	OH, R	Gr II (0.025 eq), phenol (0.5 eq), toluene, 110°, 0.5 h	MeO_2C ... OH, R R: *n*-Pr (87); *n*-C_5H_{11} (93)	1058
R^1O_2C (vinyl), 5 eq	OH, R^2	Gr II (0.02–0.05 eq), CuI (0.03–0.06 eq), Et_2O, 40°, 12 h	R^1O_2C ... OH, R^2 R^1 / R^2: Me / *i*-Pr (75); Me / Bn (87); Me / 4-BrC_6H_4 (91); *t*-Bu / 4-BrC_6H_4 (85); *t*-Bu / Bn (78); Bn / 4-BrC_6H_4 (86); Bn / Bn (80)	1059
EtO_2C (vinyl), 3 eq	OR	Hov II (0.05 eq), DCM, 40°	EtO_2C ... OR R: TBDPS (71); TBS (77) (*E*)/(*Z*) = 95:5	169

MeO_2C (3 eq)	OBz	1. Hov II (0.1 eq), DCM, reflux, 3 h 2. DIBALH (3.75 eq) DCM, –78 to –45°, 2 h	HO, OBz	(59) $(E)/(Z) > 95:5$	940
3 eq	OBn	1. Hov II (0.1 eq), DCM, reflux, 3 h 2. DIBALH, DCM, –78 to –45°, 2 h	HO, OBn	(59) $(E)/(Z) > 95:5$	941
t-BuO_2C	St-Bu, O (2 eq)	Catalyst (0.05 eq), toluene, 80°	t-BuO_2C, St-Bu, O		1060

Catalyst	Time (h)	
Ru cat **166**	8	(84)
Ru cat **168**	6	(80)

TABLE 20. CROSS-METATHESIS OF α,β-UNSATURATED ESTERS AND THIOESTERS (*Continued*)

A. ESTERS (*Continued*)

Ester	Cross Partner	Conditions	Product(s) and Yield(s) (%)	Refs.
Please refer to the charts preceding the tables for the catalyst structures.				
C_3				
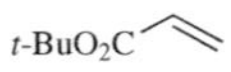 $t\text{-BuO}_2\text{C}$ (5 eq)	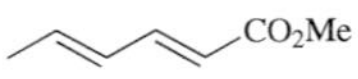 CO_2Me	Hov II (0.05 eq), DCM, rt, 48 h	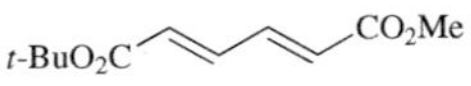 $t\text{-BuO}_2\text{C}$ … CO_2Me (30) (*E,E*)/(*Z,Z*) = 98:2	549
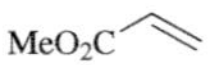 MeO_2C (1.9 eq)	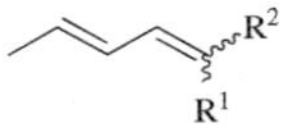 R^2, R^1	Gr II (2 x 0.05 eq), DCM, 40°, 24 h	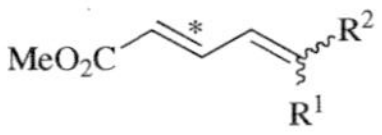 MeO_2C … R^2, R^1; *(*E*)/(*Z*) > 99:1 R^1 / R^2: Br / NC– (4); NC– / Br (18)	649
	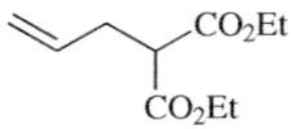 CO_2Et, CO_2Et	Gr II (0.071 eq), benzoic acid (0.075 eq), DCM, paraffin, rt, 22.5 h	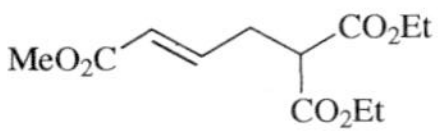 MeO_2C … CO_2Et, CO_2Et (77) (*E*)/(*Z*) = 95:5	511
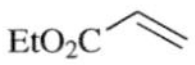 EtO_2C (4 eq)	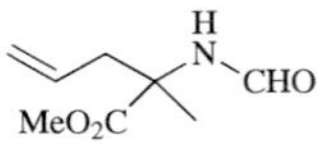 H N CHO, MeO_2C	Gr II (0.05 eq), toluene, rt, 17 h	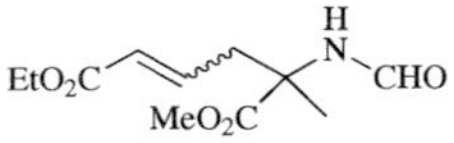 EtO_2C … H N CHO, MeO_2C (70)	457
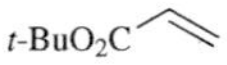 $t\text{-BuO}_2\text{C}$ (10 eq)	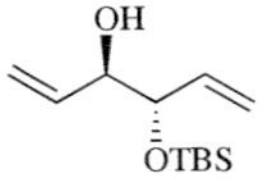 OH, OTBS	Gr II (0.05 eq), BHT (0.5 eq), toluene, reflux, 3 h	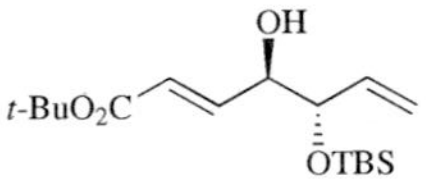 $t\text{-BuO}_2\text{C}$ … OH, OTBS (66) (*E*) only	750

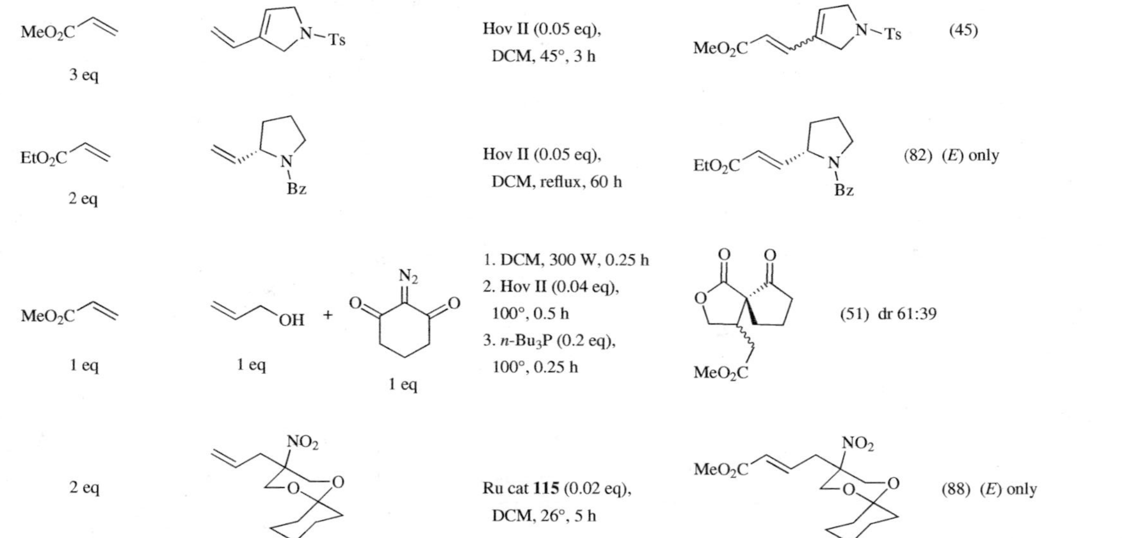

Alkene 1	Alkene 2	Conditions	Product(s) and Yield(s) (%)	Refs.
MeO_2C–CH=CH$_2$ (3 eq)		Hov II (0.05 eq), DCM, 45°, 3 h	(45)	939
EtO_2C–CH=CH$_2$ (2 eq)		Hov II (0.05 eq), DCM, reflux, 60 h	(82) (*E*) only	717
MeO_2C–CH=CH$_2$ (1 eq)	(1 eq) + (1 eq)	1. DCM, 300 W, 0.25 h 2. Hov II (0.04 eq), 100°, 0.5 h 3. *n*-Bu_3P (0.2 eq), 100°, 0.25 h	(51) dr 61:39	955
(2 eq)		Ru cat **115** (0.02 eq), DCM, 26°, 5 h	(88) (*E*) only	609, 1061

TABLE 20. CROSS-METATHESIS OF α,β-UNSATURATED ESTERS AND THIOESTERS (*Continued*)

A. ESTERS (*Continued*)

Ester	Cross Partner	Conditions	Product(s) and Yield(s) (%)	Refs.
Please refer to the charts preceding the tables for the catalyst structures.				
C_3				
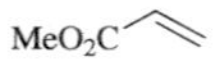 MeO_2C (acrylate), *x* eq	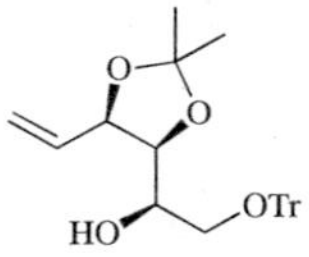 HO, OTr	Gr II (*y* eq), Ti(O*i*-Pr)$_4$ (*z* eq), solvent, 13 h	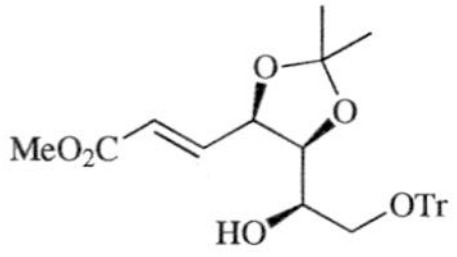 MeO_2C, HO, OTr; (*E*) only	406
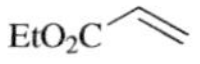 EtO_2C (acrylate), *x* eq	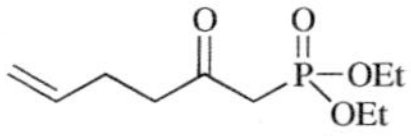 P–OEt, OEt	Catalyst (*y* eq), solvent	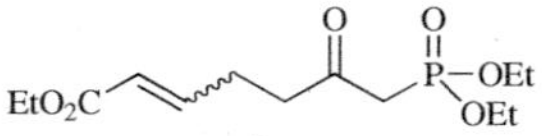 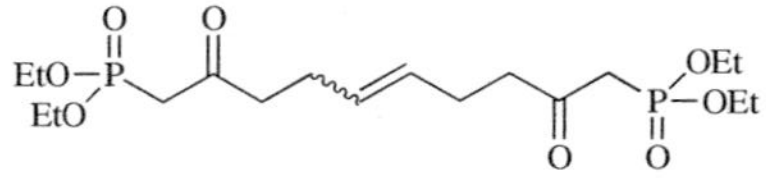 EtO_2C, P–OEt, OEt **I** + EtO–P, EtO, OEt, P–OEt **II**	165

x	*y*	*z*	Solvent	Temp (°)	
2	0.05	—	DCM	rt	(20)
20	0.05	—	DCM	rt	(31)
2	0.1	—	DCM	reflux	(23)
10	0.1	—	DCM	reflux	(37)
2	0.1	0.3	toluene	80	(25)
10	0.1	0.3	toluene	reflux	(15)
10	0.1	0.3	toluene	75	(43)

x eq

Hov II (0.025 eq), DCM, rt, 4 h

EtO_2C — Si(R)(R) —

I (*E*) only

\+ EtO_2C — Si(R)(R) — CO_2Et

II

x	Catalyst	*y*	Solvent	Temp (°)	Time (h)	**I**	**II**
4	Gr II	0.1	DCM	100 (MW)	0.25	(70)	—
4	Gr II	0.1	DCM	rt	24	(46)	—
4	Gr II	0.1	DCM	40	2	(70)	—
4	Hov II	0.1	DCM	100 (MW)	0.25	(79)	—
4	Hov II	0.1	DCM	rt	24	(30)	—
4	Hov II	0.1	DCM	40	6	(80)	—
1.1	Gr II	0.1	DCM	100 (MW)	0.25	(30)	—
1.1	Hov II	0.1	DCM	100 (MW)	0.25	(65)	—
1.1	Gr II	0.1	DCM	100 (MW)	0.02	(30)	—
1.1	Hov II	0.1	DCM	100 (MW)	0.02	(66)	—
1.1	Hov II	0.1	DCM	100 (MW)	0.003	(62)	—
1.1	Hov II	0.05	DCM	100 (MW)	0.25	(81)	—
1.1	Hov II	0.01	DCM	100 (MW)	0.25	(50)	(10)
1.1	Hov II	0.1	DCE	100 (MW)	0.0003	(72)	—
1.1	Hov II	0.1	DCM	60 (MW)	0.25	(51)	(8)

x	R	**I**	**II**	(*E,E*)/(*E,Z*) **II**
1.5	Me	(60)	(12)	83:17
3	Me	(53)	(16)	83:17
6	Me	(40)	(53)	88:12
3	Ph	(5)	(87)	98:2

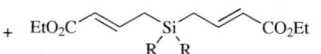

992

TABLE 20. CROSS-METATHESIS OF α,β-UNSATURATED ESTERS AND THIOESTERS (*Continued*)

A. ESTERS (*Continued*)

Ester	Cross Partner	Conditions	Product(s) and Yield(s) (%)	Refs.

Please refer to the charts preceding the tables for the catalyst structures.

C_3

Ester	Cross Partner	Conditions	Product(s) and Yield(s) (%)	Refs.
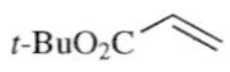 t-BuO_2C (vinyl), x eq	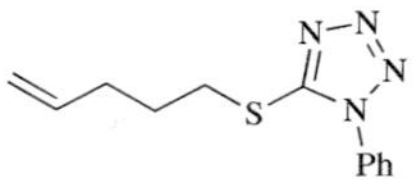 (pent-4-enyl)thio-1-phenyltetrazole (N–N, N, N, S, Ph)	Gr II (0.02 eq), additive (0.03 eq), solvent	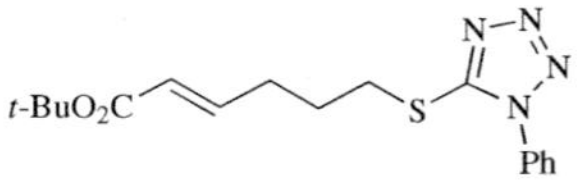 t-BuO_2C–CH=CH–(CH₂)₃–S–(1-phenyltetrazol-5-yl) (N–N, N, N, S, Ph)	

x	Additive	Solvent	Temp (°)	Time (h)		$(E)/(Z)$	Refs.
1	—	2.5 % aq. PTS	22	12	(55)	86:14	976
3	CuI	Et_2O	35	3	(84)	>95:5	167
2	—	2.5 % aq. TPGS-750-M	22	20	(55)	—	167
2	CuI	2.5 % aq. TPGS-750-M	22	20	(80)	—	931
2	—	2.5 % aq. PTS	22	20	(55)	—	931
2	CuI	2.5 % aq. PTS	22	20	(80)	—	931

Ester	Cross Partner	Conditions	Product(s) and Yield(s) (%)	Refs.
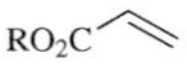 RO_2C (vinyl), 1.5 eq	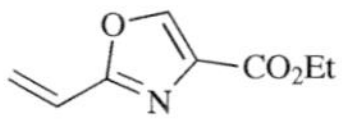 2-vinyloxazole-4-CO_2Et (O, N, CO_2Et)	Hov II (0.1 eq), DCM, 40°, 48 h	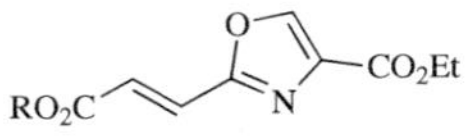 RO_2C–CH=CH–oxazole–CO_2Et (O, N); $(E)/(Z) = 75:25$	650

R	
Me	(52)
t-Bu	(24)

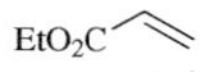

EtO_2C (acrylate), 4 eq	CO_2Et	Hov II (0.1 eq), DCM, rt, 2 h	EtO_2C, CO_2Et (83)	165
x eq	O, CO_2Et	Hov II (*y* eq), DCM	EtO_2C, O, CO_2Et	165

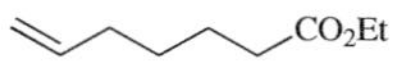

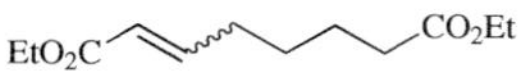

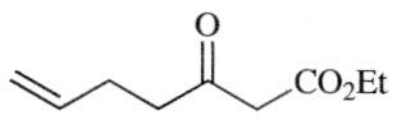

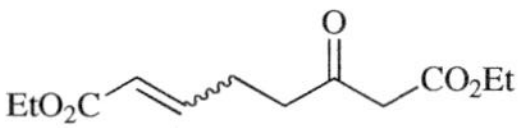

x	*y*	Temp (°)	Time (h)	
4	0.1	rt	24	(33)
4	0.1	40	2	(64)
1.1	0.1	100 (MW)	0.25	(75)
1.1	0.01	100 (MW)	0.25	(59)
1.1	0.1	100 (MW)	0.003	(71)

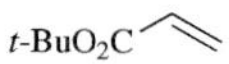

t-BuO_2C (acrylate), 1.5 eq	O, Cl	Gr II (0.025 eq), DCM, 45°, 18 h	*t*-BuO_2C, O, Cl (85)	945

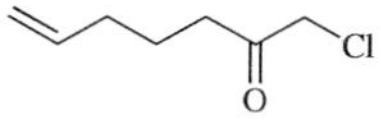

TABLE 20. CROSS-METATHESIS OF α,β-UNSATURATED ESTERS AND THIOESTERS (*Continued*)

A. ESTERS (*Continued*)

Ester	Cross Partner	Conditions	Product(s) and Yield(s) (%)	Refs.

Please refer to the charts preceding the tables for the catalyst structures.

C_3

Ester	Cross Partner	Conditions	Product(s) and Yield(s) (%)	Refs.
EtO_2C (6 eq)		Hov II (*x* eq), DCM, rt	EtO_2C … CO_2Et	

n	*x*	Time (h)		(*E*)/(*Z*)	Refs.
0	2 x 0.025	89	(8)	—	231
1	0.086	190	(66)	—	231
1	0.043	216	(66)	100:0	994

Ester	Cross Partner	Conditions	Product(s) and Yield(s) (%)	Refs.
x eq	R	Hov II (*y* eq), DCM, rt	EtO_2C … R … CO_2Et	231

x	R	*n*	*y*	Time (h)	
12	HO	0	2 x 0.025	89	(44)
6	HO	1	0.05	120	(14)
6	TBSO	1	0.05	120	(57)
6	SESHN	1	0.05	48	(77)

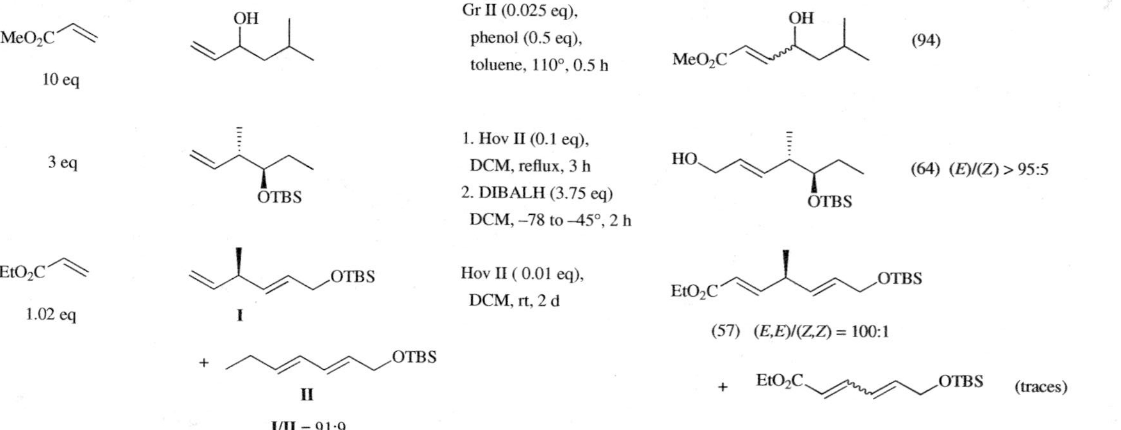

1049

940, 914

1062

TABLE 20. CROSS-METATHESIS OF α,β-UNSATURATED ESTERS AND THIOESTERS (*Continued*)

A. ESTERS (*Continued*)

Ester	Cross Partner	Conditions	Product(s) and Yield(s) (%)	Refs.
Please refer to the charts preceding the tables for the catalyst structures.				
C_3				
MeO_2C (acrylate)	OH, OBn	Gr II, DCM, reflux	MeO_2C … OH, OBn (94) (*E*)/(*Z*) > 99:1	1063
	OR^1, OR^2	Gr II, DCM, 40°	MeO_2C … OR^1, OR^2	
			R^1 / R^2 / Time (h): H / TBDPS / — (95)	1064
			PMB / H / 4 (81)	1065
EtO_2C (acrylate), 2 eq	OH, OH, OTr	Hov II (0.025 eq), DCM, rt, 36 h	EtO_2C … OH, OH, OTr (80) (*E*)/(*Z*) > 98:2	626
t-BuO_2C (acrylate), 3 eq	OPMP, epoxide	Gr II (0.1 eq), DCM, reflux, 8 h	*t*-BuO_2C … OPMP, epoxide (91) (*E*)/(*Z*) = 95:5	1066
RO_2C (acrylate), 1.5 eq	NHBoc, COOMe	Gr II (0.025 eq) DCM, reflux, 2 h	ROOC … NHBoc, COOMe; R: *n*-Bu (90), *t*-Bu (83)	1067

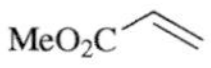 MeO_2C (1.4 eq)	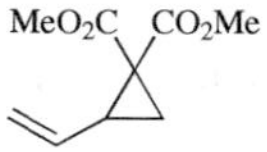 MeO_2C CO_2Me	Gr II (0.01 eq), DCM, reflux	MeO_2C CO_2Me MeO_2C	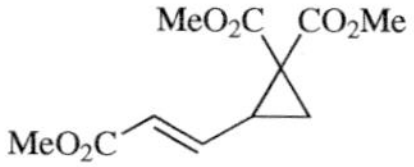 (92) $(E)/(Z)$ = 89:11	407
10 eq	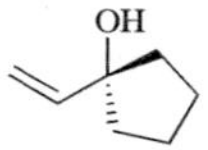 OH	1. Ru cat **163** (0.03 eq), DCM, rt, hν (350 nm) 2. Phenanthrene (0.3 eq), DCM, rt, hν (254 nm)	O O	(52)	911
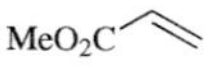 MeO_2C **I** 3 eq	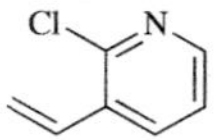 Cl N	Hov II (0.1 eq), DCM, 50°, 24 h	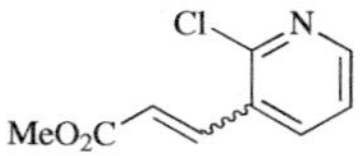 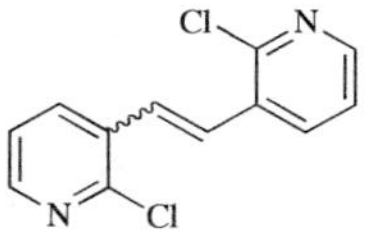 MeO_2C Cl N **II** + Cl N N Cl **III**	(31) **I/II/III** = 40:40:20	603

TABLE 20. CROSS-METATHESIS OF α,β-UNSATURATED ESTERS AND THIOESTERS (*Continued*)

A. ESTERS (*Continued*)

Ester	Cross Partner	Conditions	Product(s) and Yield(s) (%)	Refs.
Please refer to the charts preceding the tables for the catalyst structures.				
C_3				
MeO_2C (3 eq)	R, OH	Hov II (0.1 eq), DCM, 50°, 24 h	MeO_2C, R, OH R: (53); (70); (82); (21); (71); (82)	603
9.5 eq	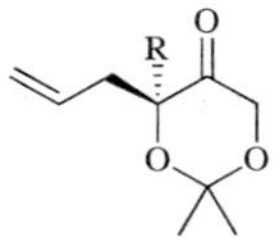 R, O, O, O	Gr II (0.02 eq), DCM, 37°, 40 h	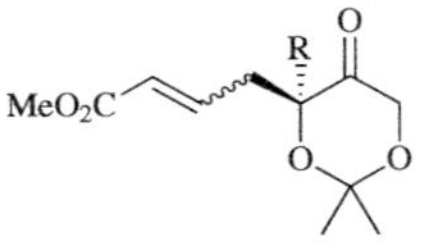 MeO_2C, R, O, O, O R: Me (83); Et (89); Bn (75)	977

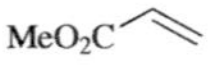

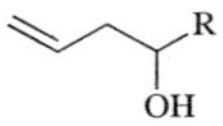

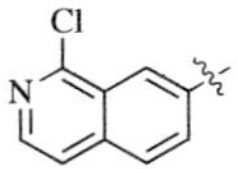

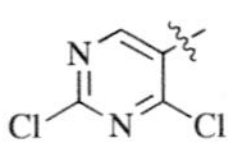

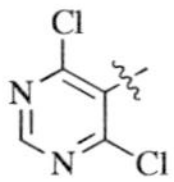

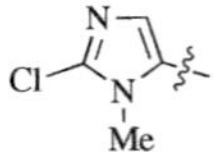

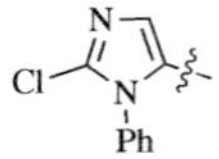

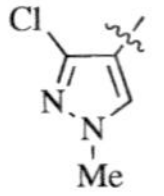

Reactant	Reactant	Conditions	Product(s) and Yield(s) (%)	Refs.
4 eq	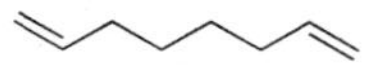	Gr II (0.00025 eq), *p*-cresol (*x* eq), 50°, 2 h	MeO_2C … (**I**) + (**II**)	130
EtO_2C 1 eq	Ph (3 eq) + (1 eq)	Gr II (0.052 eq), DCM, 23°, 12 h	EtO_2C … Ph (34) (*E*) only	36
5 eq	Cl	Ru cat **88** (5 x 0.024 eq), $C_6F_5CF_3$, MW, 120°, 50 min	EtO_2C … Cl (65)	929
R^1O_2C 10 eq	R^2, O, NO_2	Ru cat **353** (0.05 eq), 1-pyrenecarboxaldehyde (7 eq), DCM, h*v* (380 nm), 90 h	R^1O_2C … R^2, O, NO_2; (*E*)/(*Z*) > 99:1	646

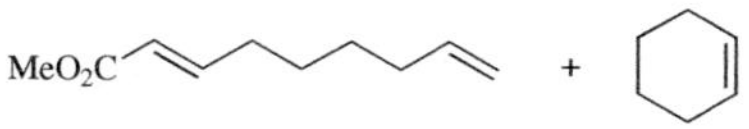

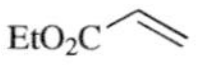

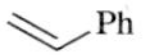

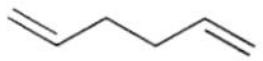

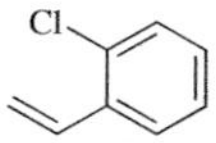

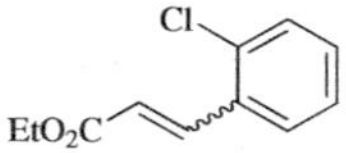

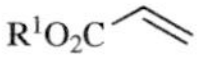

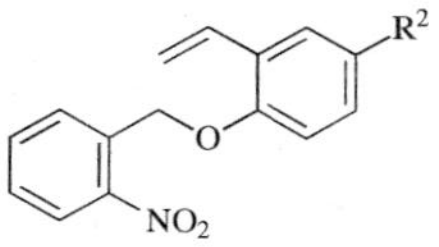

x	**I**	(*E*)/(*Z*) **I**	**II**
—	5	92:8	(29)
500	24	93:7	(66)

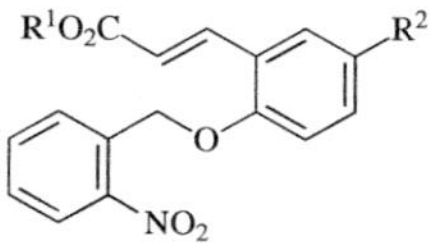

R^1	R^2	
Me	Me	(57)
t-Bu	Cl	(60)
t-Bu	O_2N	(65)

TABLE 20. CROSS-METATHESIS OF α,β-UNSATURATED ESTERS AND THIOESTERS (*Continued*)

A. ESTERS (*Continued*)

Ester	Cross Partner	Conditions	Product(s) and Yield(s) (%)	Refs.

Please refer to the charts preceding the tables for the catalyst structures.

C_3

Ester	Cross Partner	Conditions	Product(s) and Yield(s) (%)	Refs.
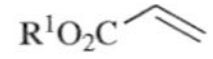 R^1O_2C (vinyl); 10 eq		1. Ru cat **353** (0.05 eq), 1-pyrenecarboxaldehyde (7 eq), DCM, hν (380 nm), 80–100 h 2. hν (254 nm), lamp, 10–30 h		646
10 eq	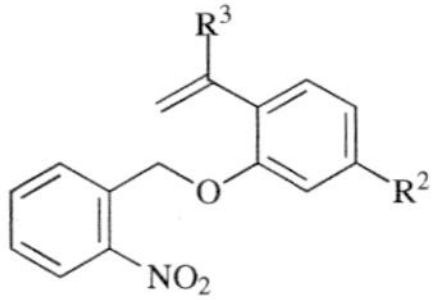	Ru cat **353** (0.05 eq), 1-pyrenecarboxaldehyde (7 eq), DCM, hν (380 nm), 90 h	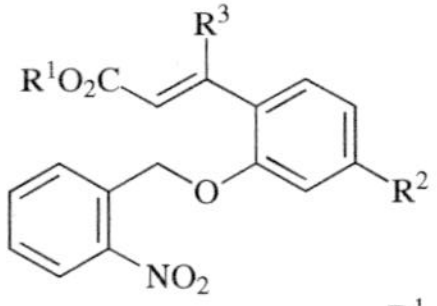 R^1O_2C …; (E)/(Z) > 99:1	646
x eq	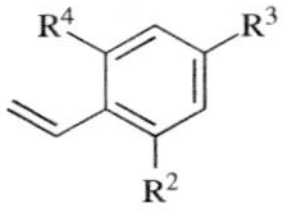	Catalyst (*y* eq), *p*-cresol (*z* eq), solvent	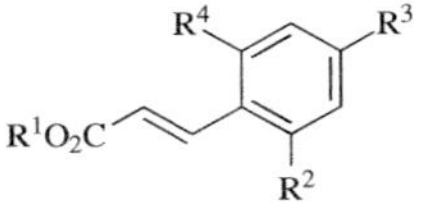 I 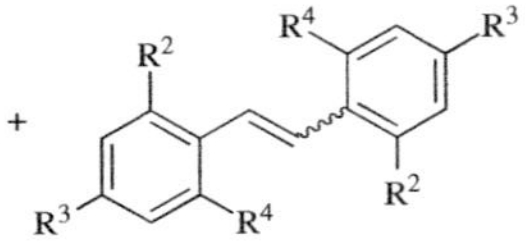 + II	

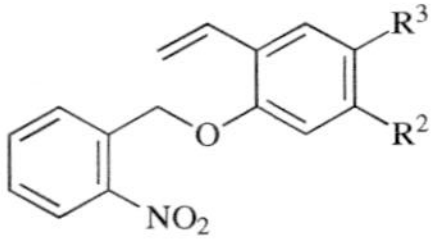

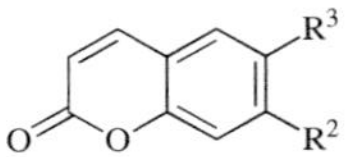

R^1	R^2	R^3	
Me	H	H	(65)
Me	Br	H	(51)
Me	H	Me	(46)
t-Bu	H	Cl	(56)
t-Bu	H	O_2N	(55)
t-Bu	Br	H	(57)
t-Bu	Ph	H	(58)

R^1	R^2	R^3	
Me	H	H	(75)
Me	H	Me	(0)
t-Bu	Br	H	(68)
t-Bu	Ph	H	(63)

R^1	x	R^2	R^3	R^4	Catalyst	y	z	Solvent	Temp (°)	Time (h)	**I**	(*E*)/(*Z*) **I**	**II**	
Me	0.5	H	H	H	Gr II	0.00025	—	neat	50	2	(24)	>98:2	(0)	130
Me	0.5	H	H	H	Gr II	0.00025	500	neat	50	2	(66)	>98:2	(0)	130
Me	0.5	H	H	H	Gr II	0.0000625	—	neat	50	2	(16)	>98:2	(0)	130
Me	0.5	H	H	H	Gr II	0.0000625	500	neat	50	2	(57)	>98:2	(0)	130
Me	2	H	H	H	Gr II	0.00025	—	neat	50	2	(8)	>98:2	(0)	130
Me	2	H	H	H	Gr II	0.00025	500	neat	50	2	(40)	>98:2	(0)	130
Me	2	H	H	H	Gr II	0.05	—	DCM	40	12	(92)	(*E*) only	(0)	36
Me	2	H	O_2N	H	Gr II	0.05	—	DCM	40	12	(89)	(*E*) only	(0)	36
Et	2	F	H	H	Gr II	0.05	—	DCM	40	12	(72)	(*E*) only	(0)	36
Et	2	Cl	H	H	Gr II	0.05	—	DCM	40	12	(62)	(*E*) only	(0)	36
Et	2	Br	H	H	Gr II	0.05	—	DCM	40	12	(49)	(*E*) only	(0)	36
Et	2	F	H	F	Gr II	0.05	—	DCM	40	12	(15)	(*E*) only	(0)	36
Me	2	Cl	H	H	Ru cat **41**	0.01	—	DCM	rt	3	(72)	>95:5	(0)	562
Me	2	Cl	H	H	Ru cat **44**	0.01	—	DCM	rt	3	(83)	>95:5	(0)	562
Me	2	Cl	H	H	Ru cat **45**	0.01	—	DCM	rt	3	(77)	>95:5	(0)	562
Me	5	Cl	H	H	Ru cat **258**	0.00025	—	DCM	30	2	(28)	>95:5	(30)	1055
Me	5	Cl	H	H	Ru cat **321**	0.00025	—	DCM	30	2	(35)	>95:5	(33)	1055
Me	5	Cl	H	H	Ru cat **322**	0.00025	—	DCM	30	2	(38)	>95:5	(33)	1055
Me	3	Cl	H	H	Ru cat **316**	0.02	—	toluene	80	2	(81)	>95:5	(0)	591
Me	5	Cl	H	H	Ru cat **316**	0.001	—	toluene	80	15	(59)	>95:5	(0)	591
Me	2	Cl	H	H	Ru cat **286**	0.02	—	DCM	40	8	(78)	>95:5	(0)	161

TABLE 20. CROSS-METATHESIS OF α,β-UNSATURATED ESTERS AND THIOESTERS (*Continued*)

A. ESTERS (*Continued*)

Ester	Cross Partner	Conditions	Product(s) and Yield(s) (%)	Refs.
Please refer to the charts preceding the tables for the catalyst structures.				
C_3				
RO_2C (2 eq)	OH	Gr II (*x* eq), benzene, rt, 16 h	RO_2C … OH R / *x* / yield / (*E*)/(*Z*) Me / 0.05 / (90) / >96:4 Et / 0.02 / (63) / >95:5	336
t-BuO_2C (3.14 eq)	O	Hov II (0.02 eq), 2.5% aq. PTS, 22°, 12 h	*t*-BuO_2C … O (76) (*E*)/(*Z*) = 95:5	976
MeO_2C (10 eq)	EtO, N, *n*, O	Gr II (0.02 eq), toluene, 70°	MeO_2C … EtO, N, *n*, O *n* / Time (h) / yield 1 / 3 / (83) 2 / 2 / (86) 3 / 2 / (96)	1068
2 eq	Cbz, N, HO	Gr II (0.1 eq), DCM, reflux, overnight	MeO_2C … Cbz, N, HO (62)	441

Olefin	Substrate	Conditions	Product(s) and Yield(s) (%)	Refs.
RO_2C–CH=CH$_2$ (10 eq)	allyl ether diazo keto ester (CO_2Et, N_2)	Gr II (0.05 eq), DCM, 40°, 18 h	RO_2C-substituted diazo keto ester (CO_2Et, N_2); R = Me (86) (E)/(Z) 95:5; R = t-Bu (78) (E)/(Z) >99:1	569
MeO_2C–CH=CH$_2$ (2 eq)	homoallylic alcohol (OH)	Gr II (0.02 eq), benzene, rt, 16 h	MeO_2C product (OH) (70) (E)/(Z) = 95:5	336
EtO_2C–CH=CH$_2$ (3 eq)	OAc, OTBDPS diene	Hov II (0.05 eq) DCM, rt, 12 h	EtO_2C product (OAc, OTBDPS) (69)	919
MeO_2C–CH=CH$_2$ (2 eq)	R, CO_2Me, NHBoc olefin	Gr II (0.1 eq), DCM, reflux, 3 h	MeO_2C product (R, CO_2Me, NHBoc), (E) only; R = Me (93); R = i-Bu (95)	1069
x eq	OH, OTBDPS olefin	Gr II (y eq), DCM	MeO_2C product (OH, OTBDPS)	

x	y	Temp (°)	Time (h)		(E)/(Z)	Refs.
—	0.005	—	—	(91)	—	1070
5	0.025	40	3	(99)	97:3	1071

TABLE 20. CROSS-METATHESIS OF α,β-UNSATURATED ESTERS AND THIOESTERS (*Continued*)

A. ESTERS (*Continued*)

Ester	Cross Partner	Conditions	Product(s) and Yield(s) (%)	Refs.
Please refer to the charts preceding the tables for the catalyst structures.				
C_3				
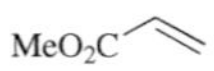 MeO_2C (acrylate), 15 eq	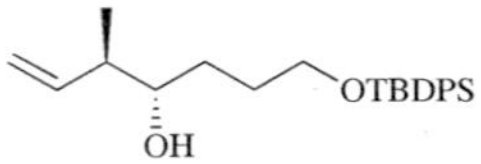 OTBDPS, OH	Gr II (0.075 eq), DCM, 80° (MW), 2 h	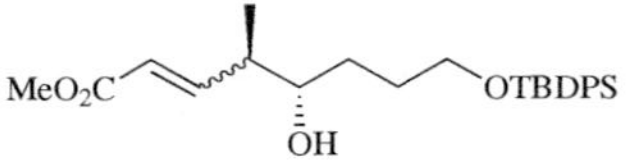 MeO_2C … OTBDPS, OH (83)	885
12 eq	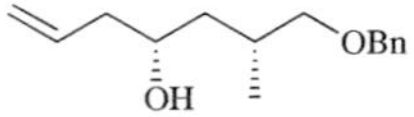 OBn, OH	Gr II (0.02 eq), DCM, 35°, 2 h	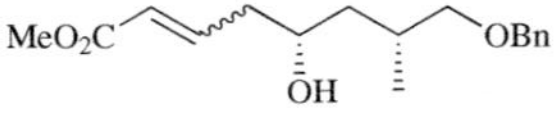 MeO_2C … OBn, OH (72)	1018, 1019
3–10 eq	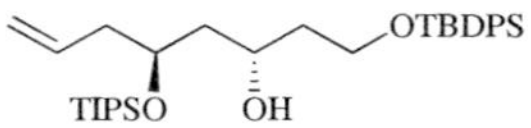 OTBDPS, TIPSO, OH	Hov II (0.1 eq), DCM, 100° (MW), 0.5 h	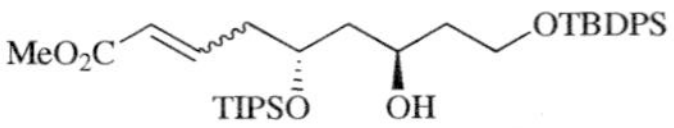 MeO_2C … OTBDPS, TIPSO, OH (98)	947
	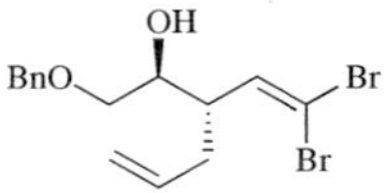 BnO, OH, Br, Br	Gr II (0.1 eq), DCE, 60°	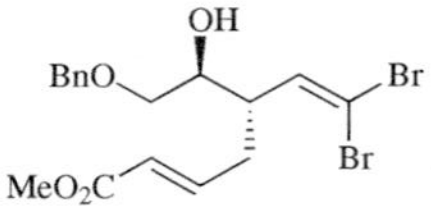 BnO, OH, Br, Br, MeO_2C (68) (*E*) only	1072
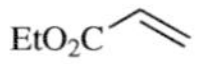 EtO_2C (acrylate), 3 eq	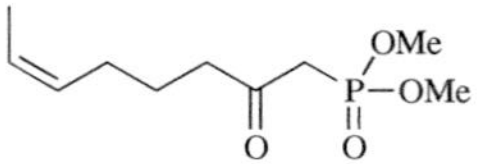 OMe, P–OMe, O, O	Gr II (0.03 + 0.01 eq), Ti(O*i*-Pr)$_4$ (0.15 eq), DCM, 100°, 20 min	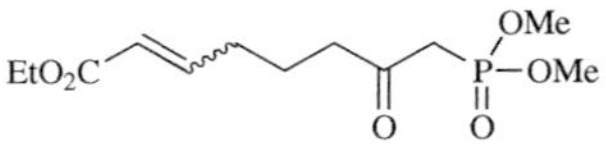 EtO_2C … OMe, P–OMe, O, O (82)	1073
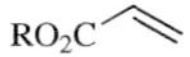 RO_2C (acrylate), 3–5 eq	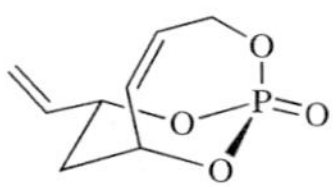 O, P=O, O, O	Gr II (0.01 eq), DCM, reflux	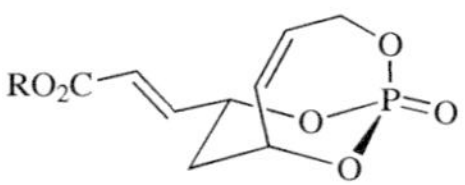 RO_2C … O, P=O, O, O	719

R		(*E*)/(*Z*)
Me	(78)	89:11
t-Bu	(60)	83:17

x eq Ph — Catalyst (*y* eq), solvent, 12 h → RO_2C Ph

R	*x*	Catalyst	*y*	Solvent	Temp (°)		(*E*)/(*Z*)	
Me	0.4	Gr II	0.025	DCM	40	(99)	(*E*) only	971
Me	0.4	Hov II	0.025	DCM	40	(99)	(*E*) only	971
Me	0.4	Ru cat **86**	0.025	DCM	40	(99)	(*E*) only	971
2-ethylhexyl	2	Ru cat **109**	0.02	DCM	22	(94)	90:10	341
t-Bu	2	Gr II	0.02	2.5% aq. PTS	rt	(97)	—	976
t-Bu	1.3	Gr II	0.02	2.5% aq. PTS	rt	(93)	—	976
t-Bu	2	Gr II	0.02	2.5% aq. TPGS	rt	(67)	—	976
t-Bu	1.3	Gr II	0.02	2.3% aq. TPGS	rt	(61)	—	976
t-Bu	2	Gr II	0.02	2.5% aq. PSS	rt	(78)	—	976
t-Bu	1.3	Gr II	0.02	2.5% aq. PSS	rt	(70)	—	976
t-Bu	2	Gr II	0.02	2.5% aq. Triton X-100	rt	(69)	—	976
t-Bu	1.3	Gr II	0.02	2.5% aq. Triton X-100	rt	(63)	—	976
t-Bu	2	Gr II	0.02	2.5% aq. Brij 30	rt	(63)	—	976
t-Bu	1.3	Gr II	0.02	2.5% aq. Brij 30	rt	(55)	—	976
t-Bu	2	Gr II	0.02	2.5% aq. PEG 600	rt	(60)	—	976
t-Bu	1.3	Gr II	0.02	2.3% aq. PEG 600	rt	(47)	—	976
t-Bu	2	Gr II	0.02	2.5% aq. SDS	rt	(68)	>95:5	976
t-Bu	1.3	Gr II	0.02	2.5% aq. SDS	rt	(64)	(*E*) only	976
t-Bu	2	Gr II	0.02	H_2O	rt	(71)	(*E*) only	976
t-Bu	1.3	Gr II	0.02	H_2O	rt	(62)	—	976
t-Bu	2	Gr II	0.02	2.5% aq. PTS	22	(96)	>95:5	976
2-ethylhexyl	2	Gr II	0.02	2.5% aq. PTS	22	(92)	90:10	976

TABLE 20. CROSS-METATHESIS OF α,β-UNSATURATED ESTERS AND THIOESTERS (*Continued*)

A. ESTERS (*Continued*)

Ester	Cross Partner	Conditions	Product(s) and Yield(s) (%)	Refs.

Please refer to the charts preceding the tables for the catalyst structures.

C_3

R^1O_2C–CH=CH$_2$ (2 eq) + allylbenzene bearing R^2 and R^3 → Catalyst (*x* eq), BQ (*y* eq), solvent → R^1O_2C–CH=CH–CH$_2$–Ar(R^2, R^3)

R^1	R^2	R^3	Catalyst	*x*	*y*	Solvent	Temp (°)	Time (h)	Yield (%)	(*E*)/(*Z*)	Refs.
Me	MeO	HO	Gr II	0.02	0	DMC	80	3	(0)	—	1074
Me	MeO	HO	Hov II	0.02	0	DMC	80	3	(65)	—	1074
Me	MeO	HO	Hov II	0.02	0	DMC	80	3	(50)	—	1074
Me	MeO	HO	Hov II	0.02	0	DMC	80	24	(62)	—	1074
Me	MeO	HO	Hov II	0.005	0	DMC	80	3	(0)	—	1074
Me	MeO	HO	Ru cat **89**	0.02	0	DMC	80	3	(0)	—	1074
Me	MeO	HO	Ru cat **139**	0.02	0	DMC	80	3	(80)	—	1074
Me	MeO	HO	Hov II	0.01	0.05	DMC	80	16	(0)	—	1074
Me	MeO	HO	Ru cat **139**	0.01	0.05	DMC	80	8	(73)	—	1074
Me	MeO	HO	Ru cat **139**	0.01	0.05	DMC	80	16	(78)	—	1074
Me	MeO	HO	Ru cat **139**	0.005	0.05	DMC	80	3	(0)	—	1074
Me	MeO	*i*-PrO	Hov II	0.02	0.05	DMC	80	3	(75)	—	1074
Me	MeO	*i*-PrO	Hov II	0.02	0.05	DMC	80	8	(85)	—	1074
Me	MeO	AcO	Hov II	0.02	0.05	DMC	80	8	(81)	—	1074
Et	MeO	HO	Ru cat **41**	0.01	0	DCM	rt	5	(68)	>95:5	562
Et	MeO	HO	Ru cat **44**	0.01	0	DCM	rt	5	(89)	>95:5	562
Et	MeO	HO	Ru cat **45**	0.01	0	DCM	rt	5	(92)	>95:5	562
Ad	H	MeO	Ru cat **109**	0.02	0	DCM	22	12	(73)	(*E*) only	341

x eq (allylarene: R^2, R^3) → Catalyst (0.02 eq), additive (0.03 eq), solvent → R^1O_2C–CH=CH–CH$_2$–Ar (R^2, R^3)

R^1	*x*	R^2	R^3	Catalyst	Additive	Solvent	Temp (°)	Time (h)		(*E*)/(*Z*)	
t-Bu	2	TBSO	H	Gr II	—	2.5% aq. PTS	22	12	(88)	—	972
t-Bu	2	TBSO	H	Gr II	—	2.5% aq. TPGS-750-W	22	12	(91)	—	972
Ad	2	H	MeO	Gr II	—	2.5% aq. PTS	22	12	(78)	—	972
Ad	2	H	MeO	Gr II	—	2.5% aq. TPGS-750-W	22	12	(82)	—	972
Me	3	TBSO	H	Gr II	NaI	Et_2O	35	3	(100)	—	167
Me	3	TBSO	H	Gr II	CuI	Et_2O	35	3	(100)	—	167
Me	3	TBSO	H	Gr II	NaI	DME	35	3	(67)	—	167
Me	3	TBSO	H	Gr II	CuI	DME	35	3	(48)	—	167
t-Bu	3	TBSO	H	Gr II	CuI	Et_2O	35	3	(93)	>95:5	167
Me	3	TBSO	H	Gr II	CuI	Et_2O	35	3	(93)	(*E*) only	167
Me	3	H	MeO	Gr II	CuI	Et_2O	35	3	(94)	(*E*) only	167
Ad	2	H	MeO	Gr II	—	2.5% aq. TPGS-750-W	22	12	(82)	—	167
Ad	2	H	MeO	Gr II	CuI	2.5% aq. TPGS-750-W	22	12	(89)	—	167
t-Bu	2	TBSO	H	Ru cat **290**	—	2.5% aq. PTS	22	12	(27)	—	931
t-Bu	2	TBSO	H	Ru cat **301**	—	2.5% aq. PTS	22	12	(0)	—	931
t-Bu	2	TBSO	H	Ru cat **299**	—	2.5% aq. PTS	22	12	(0)	—	931
t-Bu	2	TBSO	H	Ru cat **139**	—	2.5% aq. PTS	22	12	(80)	—	931
t-Bu	2	TBSO	H	Ind II	—	2.5% aq. PTS	22	12	(99)	—	931
t-Bu	2	TBSO	H	Gr II	NaI	2.5% aq. PTS	22	4	(99)	—	931
Me	2	H	MeO	Gr II	—	2.5% aq. PTS	22	12	(70)	(*E*) only	976
Ad	2	H	MeO	Gr II	—	2.5% aq. PTS	22	12	(78)	(*E*) only	976
t-Bu	2	H	MeO	Gr II	—	2.5% aq. PTS	22	12	(93)	>95:5	976
t-Bu	2	TBSO	H	Gr II	—	2.5% aq. PTS	22	12	(88)	(*E*) only	976

TABLE 20. CROSS-METATHESIS OF α,β-UNSATURATED ESTERS AND THIOESTERS (*Continued*)

A. ESTERS (*Continued*)

Ester	Cross Partner	Conditions	Product(s) and Yield(s) (%)	Refs.
Please refer to the charts preceding the tables for the catalyst structures.				
C_3				
R^1O_2C (6 eq)	OR^2, R^3	1. Ru cat **352** (0.001 eq), toluene, rt, 0.5 h 2. Hov II (0.005 eq), toluene, 70°, 6 h	R^1O_2C OR^2, R^3	1075
RO_2C (6 eq)	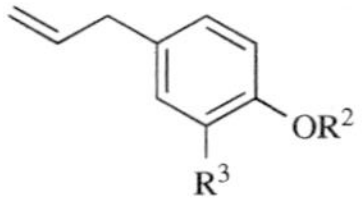	1. Ru cat **352** (0.001 eq), toluene, rt, 0.5 h 2. Hov II (0.005 eq), toluene, 70°, 6 h	RO_2C	1075

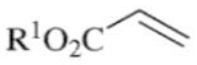

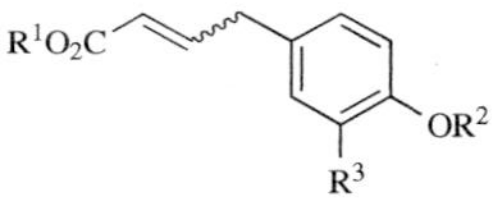

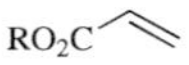

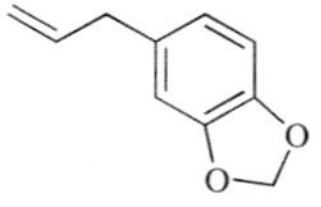

R^1	R^2	R^3	
Me	Me	H	(100)
Me	H	MeO	(>99)
2-ethyhexyl	Me	H	(>99)
2-ethyhexyl	H	MeO	(>99)

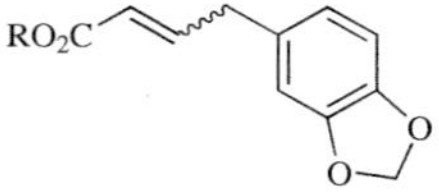

R	
Me	(>99)
2-ethyhexyl	(>99)

R^1O_2C–CH=CH$_2$ (x eq) + substituted (E)-propenylbenzene (R^2, R^3) → Hov II (y eq), DCE, 70°, 6 h → R^1O_2C–CH=CH–Ar (R^2, R^3) 1076

R^1	x	R^2	R^3	y		(E)/(Z)
Me	4	H	MeO	0.02	(99)	97:3
Me	4	H	MeO	0.001	(78)	>99:1
Me	4	H	MeO	0.005	(97)	99:1
Me	2	H	MeO	0.005	(90)	99:1
Me	10	H	MeO	0.005	(93)	98:2
Me	20	H	MeO	0.005	(89)	99:1
Me	6	H	MeO	0.005	(99)	99:1
Me	6	H	Me	0.005	(83)	99:1
Me	6	MeO	H	0.005	(90)	98:2
Et	6	H	Me	0.005	(85)	>99:1
Et	6	MeO	H	0.005	(95)	>99:1
2-ethylhexyl	6	H	Me	0.005	(89)	>99:1
2-ethylhexyl	6	MeO	H	0.005	(98)	>99:1

EtO_2C–CH=CH$_2$ (2 eq) + substituted styrene (R^1, R^2) → Gr II (0.02–0.05 eq), DCM, 40°, 12 h → EtO_2C–CH=CH–Ar (R^1, R^2), (E) only 36

R^1	R^2	
H	CF_3	(44)
OHC	H	(83)

TABLE 20. CROSS-METATHESIS OF α,β-UNSATURATED ESTERS AND THIOESTERS (*Continued*)

A. ESTERS (*Continued*)

Ester	Cross Partner	Conditions	Product(s) and Yield(s) (%)	Refs.
Please refer to the charts preceding the tables for the catalyst structures.				
C_3				
MeO_2C (10 eq)	OH, R^2, R^1	Gr II (0.025 eq), phenol (0.5 eq), toluene, 110°, 0.5 h	MeO_2C, OH, R^2, R^1 R^1 / R^2: H / H (91); Br / H (94); MeO / H (91); TBSO / H (92); BnO / H (88); MOMO / MeO (85)	1049
10 eq	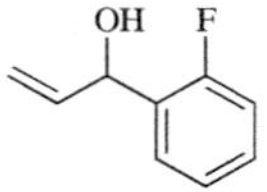 OH, F	Gr II (0.025 eq), phenol (0.5 eq), toluene, 110°, 0.5 h	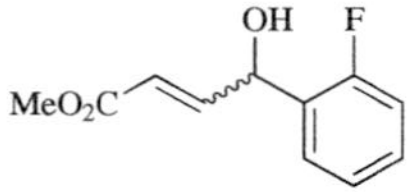 MeO_2C, OH, F (94)	1049
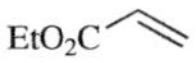 EtO_2C (2 eq)	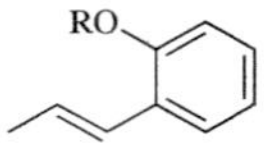 RO	Gr II (0.05 eq), DCM, 40°, 12 h	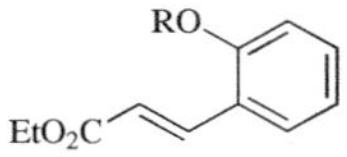 EtO_2C, RO, (*E*) only R: TBS (30); Ac (87)	36 145

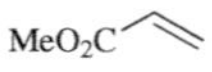

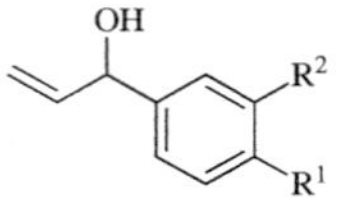

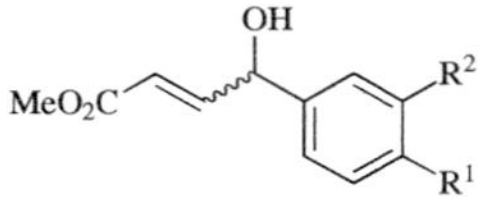

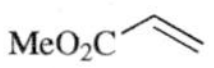

MeO_2C 3 eq 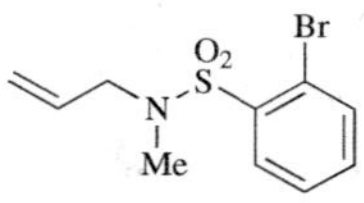 | Gr II (0.05 eq), DCM, 40°, 3 h | (96) | 634

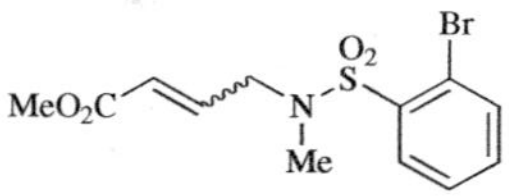

3 eq | Catalyst (0.1 eq), solvent, 24 h | 603

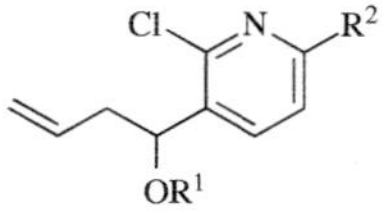

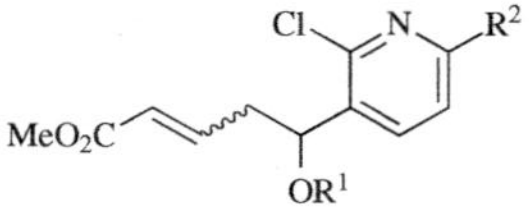

R^1	R^2	Catalyst	Solvent	Temp (°)	Concn (M)	
H	H	Hov II	DCM	50	0.1	(64)
H	H	Hov II	DCM	50	1	(84)
H	H	Hov II	toluene	80	1	(35)
H	H	Gr II	DCM	50	1	(65)
H	Br	Hov II	DCM	50	1	(73)
H	Cl	Hov II	DCM	50	1	(67)
H	F	Hov II	DCM	50	1	(68)
H	MeO	Hov II	DCM	50	1	(51)
H	CF_3	Hov II	DCM	50	1	(73)
H	*t*-Bu	Hov II	DCM	50	1	(52)
Ac	TfO	Hov II	DCM	50	1	(85)

TABLE 20. CROSS-METATHESIS OF α,β-UNSATURATED ESTERS AND THIOESTERS (*Continued*)

A. ESTERS (*Continued*)

Ester	Cross Partner	Conditions	Product(s) and Yield(s) (%)	Refs.
Please refer to the charts preceding the tables for the catalyst structures.				
C_3				
MeO_2C–CH=CH₂ (2 eq)	CH₂=CH–CH(OPh)Ph	Hov II (0.05 eq), toluene, 80°, 6 h	MeO_2C–CH=CH–CH(OPh)Ph (65)	930
R^1O_2C–CH=CH₂ (10 eq)	CH₂=CH–CH(OH)R^2	Catalyst (0.03 eq), DCM, rt, hν (350 nm)	R^1O_2C–CH=CH–CH(OH)R^2	911

R^1	R^2	Catalyst	
Me	Ph	Ru cat **163**	(89)
t-Bu	Ph	Ru cat **163**	(94)
n-Bu	Ph	Ru cat **353**	(77)
Bn	Ph	Ru cat **353**	(82)
Bn	*n*-C_9H_{19}	Ru cat **353**	(78)
4-*t*-BuC_6H_4	Ph	Ru cat **353**	(74)
4-*t*-BuC_6H_4	*n*-C_9H_{19}	Ru cat **353**	(75)

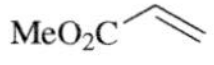

MeO_2C (structure) **I**

Br (structure) Ph

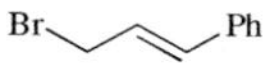

1. RMgBr (1.2 eq), (*R*,*R*)-Taniaphos (0.012 eq), $CuBr•SMe_2$ (0.01 eq), DCM, –75°
2. **I** (5 eq), catalyst (*x* eq), DCM, rt

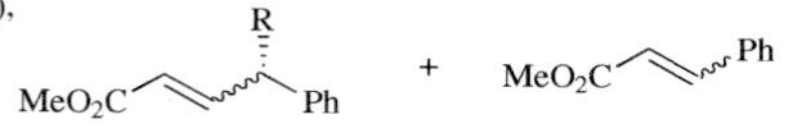

MeO_2C (structure) R, Ph **I** + MeO_2C (structure) Ph **II**

975

R	Catalyst	*x*	**I**	er **I**	**I/II**
Me	Hov II	0.02	(66)	—	>97:3
Et	Hov II	0.02	(67)	99:1	80:20
Et	Gr II	0.015	(49)	99:1	90:10

20 eq

MeO, P, O, O (structure)

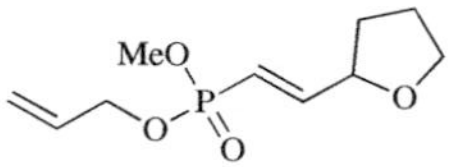

Gr II (0.1 eq), CuI (0.13 eq), DCM, reflux, 2 h

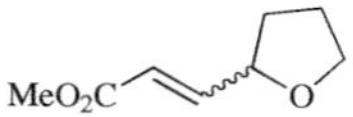

MeO_2C (structure) O (78)

220

1 eq

O, O, O (structure)

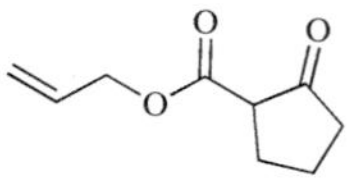

Hov II (0.03 + 0.01 eq), DCM, 100° (MW), 0.33 + 0.17 h

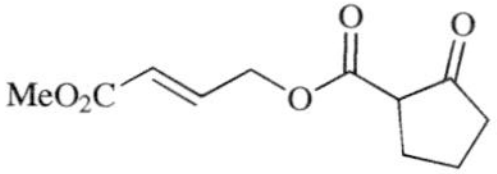

MeO_2C (structure) O, O, O (53) (*E*)/(*Z*) > 91:9

955

TABLE 20. CROSS-METATHESIS OF α,β-UNSATURATED ESTERS AND THIOESTERS (*Continued*)

A. ESTERS (*Continued*)

Ester	Cross Partner	Conditions	Product(s) and Yield(s) (%)	Refs.
Please refer to the charts preceding the tables for the catalyst structures.				
C_3				
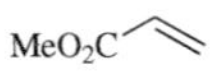 MeO_2C (1 eq)	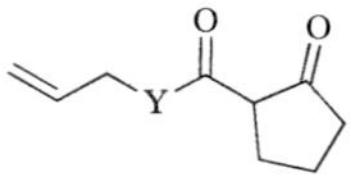	1. Hov II (0.04 eq), DCM, 100° (MW), 0.5 h 2. Alumina basic (0.8 g), MeOH, 50°, 24 h	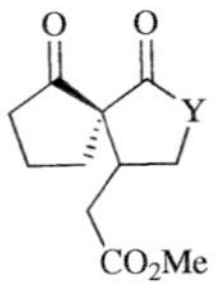 CO_2Me Y: O (65) dr 60:40; MeN (56) dr 57:43	955
1 eq	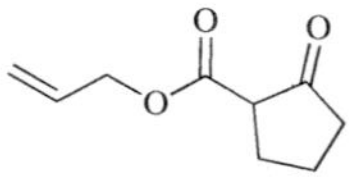	1. Hov II (0.03 + 0.01 eq), DCM, 100° (MW), 20 + 10 min 2. $P(n\text{-}Bu)_3$ (0.2 eq), 100° (MW), 20 min	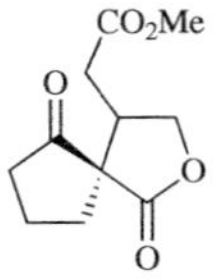 CO_2Me (53) dr = 62:38	979
1 eq	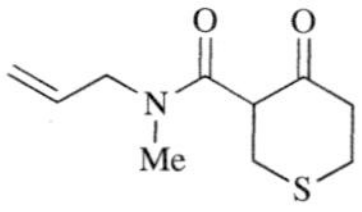 N, Me, S	1. Hov II (0.04 eq), DCM, 100° (MW), 0.5 h 2. Alumina basic (0.8 g), MeOH, 50°, 24 h	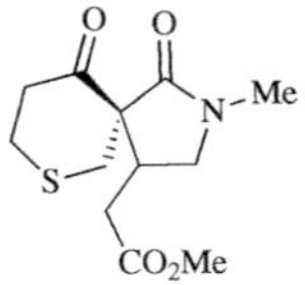 N-Me, S, CO_2Me (43) dr = 56:44	955

		Conditions	Product(s) and Yield(s) (%)		Refs.
1 eq	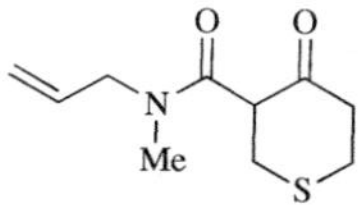	1. Hov II (0.04 eq), DCM, 100° (MW), 0.5 h 2. n-BuP_3 (0.2 eq), 100° (MW), 20 min	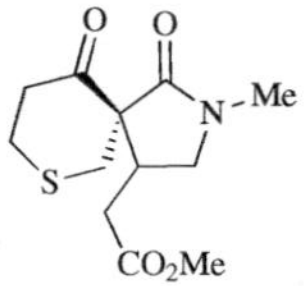	(35) dr = 54:46	955
2.5 eq		Hov II (0.05 eq), DCM, 40°, 14 h		**I** (79) (*E*) only	1077
			+ **II**	**I/II** = 15:1	
	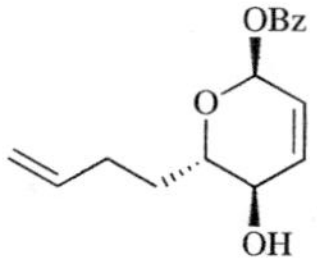	Hov II (0.05 eq), DCM	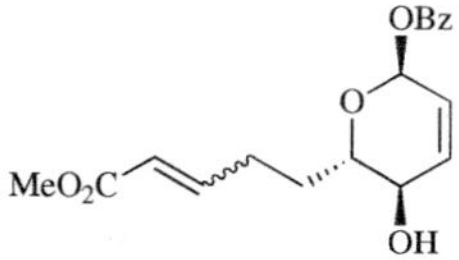	(82)	989

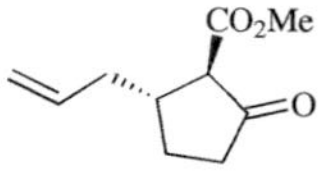

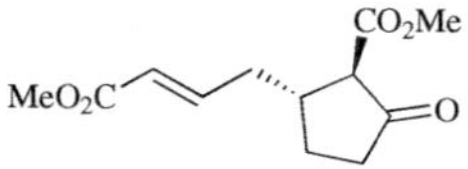

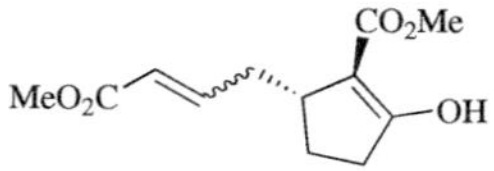

TABLE 20. CROSS-METATHESIS OF α,β-UNSATURATED ESTERS AND THIOESTERS (*Continued*)

A. ESTERS (*Continued*)

Ester	Cross Partner	Conditions	Product(s) and Yield(s) (%)	Refs.

Please refer to the charts preceding the tables for the catalyst structures.

C_3

Ester	Cross Partner	Conditions	Product(s) and Yield(s) (%)	Refs.
MeO_2C (alkene), 10 eq	SPh, N, Ts, O, n	Catalyst (0.05 eq), *p*-cresol (*x* eq), toluene	MeO_2C, SPh, N, Ts, O, n	1078
EtO_2C (alkene), 6 eq	O, N, PMB	Hov II (2 x 0.025 eq), DCM, rt, 144 h	EtO_2C, O, N, PMB, CO_2Et (79)	231

n	Catalyst	*x*	Temp (°)	Time (h)	
1	Gr II	0.5	150 (MW)	0.5	(91)
1	Gr II	0.5	80	3.5	(85)
2	Gr I	0.5	150 (MW)	3.5	(16)
2	Gr II	—	150 (MW)	0.5	(88)
2	Gr II	0.5	150 (MW)	0.5	(96)
2	Gr II	0.5	reflux	3.5	(91)
2	Gr I	—	150 (MW)	0.5	(0)
3	Gr II	0.5	150 (MW)	0.5	(94)
3	Gr II	0.5	80	3.5	(88)

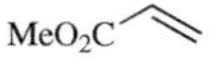 MeO_2C 10 eq	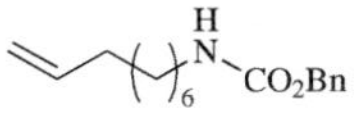 CO_2Bn	Hov II (0.005 eq), rt, 50°, 16.3 h	 MeO_2C CO_2Bn (90)	1079
20 eq	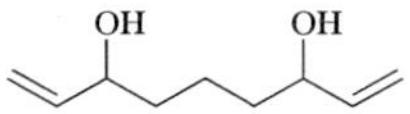 OH OH	Gr II (0.1 eq), phenol (0.5 eq), toluene, 110°, 0.5 h	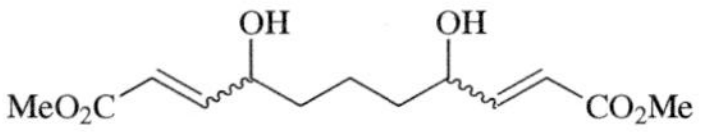 MeO_2C OH OH CO_2Me (82)	1049
3 eq	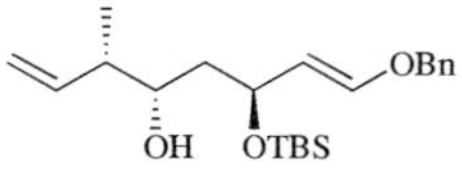 OH OTBS OBn	Hov II (0.27 + 0.27 eq), DCE, reflux, 12 + 16 h	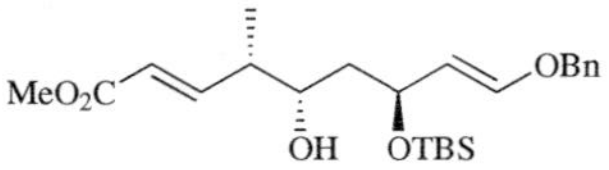 MeO_2C OH OTBS OBn (95) (*E*) only	974
2 eq	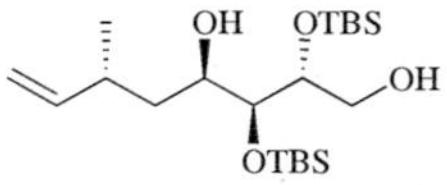 OH OTBS OTBS OH	Gr II (0.07 eq), CuI (0.1 eq), DCM, 40°, 4 h 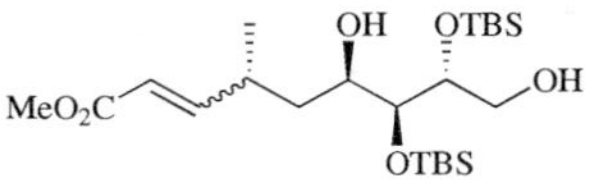	MeO_2C OH OTBS OTBS OH (60)	1080

TABLE 20. CROSS-METATHESIS OF α,β-UNSATURATED ESTERS AND THIOESTERS (*Continued*)

A. ESTERS (*Continued*)

Ester	Cross Partner	Conditions	Product(s) and Yield(s) (%)	Refs.
Please refer to the charts preceding the tables for the catalyst structures.				
C_3				
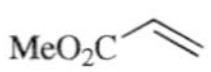 MeO_2C (2 eq)	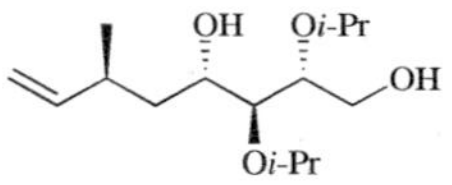 OH, O*i*-Pr, OH, O*i*-Pr	Gr II (0.07 eq), CuI (0.1 eq), DCM, 40°, 4 h	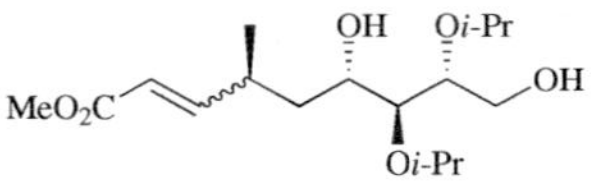 MeO_2C, OH, O*i*-Pr, OH, O*i*-Pr (62)	1080
(3 eq)	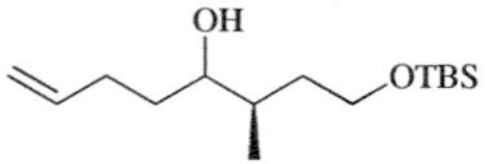 OH, OTBS	Hov II (0.02 eq), DCM, 40°, 2 h	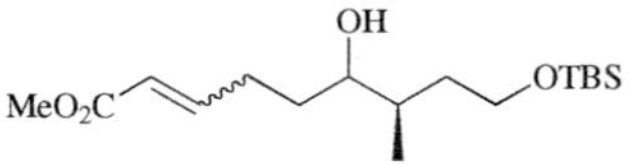 MeO_2C, OH, OTBS (80)	342
EtO_2C (3 eq)	OBn, R, N, AcO, BnO, AcO	Gr II (*x* eq), DCM, 40°, 16–24 h	OBn, R, N, AcO, BnO, AcO, CO_2Et (*E*)/(*Z*) > 95:5	666

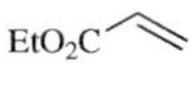

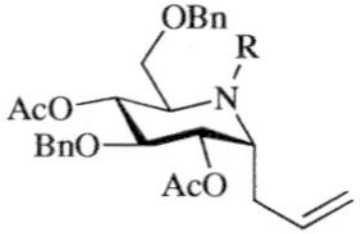

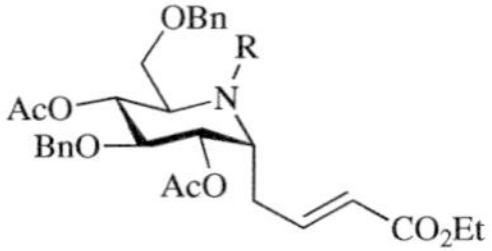

R	*x*	
Tr	0.05	(92)
OHC	0.1	(96)

Reactant 1	Reactant 2	Conditions	Product(s) and Yield(s) (%)	Refs.
RO_2C–CH=CH$_2$ (2 eq)	CH$_2$=CHCH$_2$-n-C_7H_{15}	Gr II (x eq), p-cresol (y eq), 50°, 2 h	RO_2CCH=CHCH$_2$-n-C_7H_{15}	130
MeO_2C–CH=CH$_2$	CH$_2$=CHCH$_2$-n-C_7H_{15}	Catalyst (0.025 eq), solvent, 35°, 3 h	MeO_2CCH=CHCH$_2$-n-C_7H_{15} (E) only	384

R	x	y		(E)/(Z)
Me	0.001	—	(24)	93:7
Me	0.001	500	(100)	97:3
Me	0.0005	—	(56)	94:6
Me	0.0005	500	(97)	96:4
Me	0.00025	—	(58)	93:7
Me	0.00025	500	(96)	96:4
Me	0.0000625	—	(47)	94:6
Me	0.0000625	500	(71)	94:6
Et	0.0005	—	(43)	93:7
Et	0.0005	500	(95)	96:4

Catalyst	Solvent	Conv. (%)[b]	
Gr II	DCM	94	(62)
Gr II	DMC	84	(56)
Hov II	DCM	95	(69)
Hov II	DMC	87	(58)

TABLE 20. CROSS-METATHESIS OF α,β-UNSATURATED ESTERS AND THIOESTERS (*Continued*)

A. ESTERS (*Continued*)

Ester	Cross Partner	Conditions	Product(s) and Yield(s) (%)	Refs.

Please refer to the charts preceding the tables for the catalyst structures.

C_3

Ester	Cross Partner	Conditions	Product(s) and Yield(s) (%)	Refs.
EtO_2C–CH=CH$_2$ (5 eq)	CH$_2$=CH–CH$_2$–(CH$_2$)$_4$–CH$_2$–CH=CH$_2$	Catalyst (*x* eq)	EtO_2C–CH=CH–CH$_2$–(CH$_2$)$_4$–CH$_2$–CH=CH–CO_2Et	1081

Catalyst	*x*	Temp (°)	Time (h)	Yield (%)	(*E*)/(*Z*)
Gr II	0.001	50	0.5	(76)	95:5
Gr II	0.001	50	4	(76)	95:5
Ind II	0.001	50	0.5	(75)	92:8
Ind II	0.001	50	4	(85)	94:6
Ind II	0.001	50	16	(88)	94:6
Hov II	0.001	50	0.5	(100)	93:7
Ru cat **84**	0.001	50	0.5	(84)	93:7
Ru cat **84**	0.001	50	4	(85)	92:8
Ru cat **85**	0.001	50	0.5	(45)	93:7
Ru cat **85**	0.001	50	16	(84)	93:7
Ru cat **139**	0.001	50	0.5	(92)	93:7
Ru cat **139**	0.001	50	4	(93)	92:8
Ru cat **131**	0.001	50	0.5	(55)	91:9
Ru cat **131**	0.001	50	16	(95)	92:8
Hov II	0.0005	50	0.5	(86)	94:6
Hov II	0.0005	50	16	(99)	93:7
Ru cat **131**	0.0005	50	16	(85)	91:9
Hov II	0.0003	50	16	(98)	94:6
Ru cat **131**	0.0003	50	16	(55)	91:9
Hov II	0.001	50	16	(54)	93:7
Hov II	0.0005	80	0.5	(99)	93:7
Hov II	0.0003	80	0.5	(99)	93:7
Hov II	0.0002	80	0.5	(98)	93:7
Gr II	0.0002	80	0.5	(50)	92:8

Ind II	0.0002	80	0.5	(47)	89:11
Ru cat **84**	0.0002	80	0.5	(82)	91:9
Ru cat **139**	0.0002	80	0.5	(82)	93:7
Ru cat **131**	0.0002	80	0.5	(93)	94:6
Ru cat **131**	0.0002	80	16	(100)	93:7
Hov II	0.00016	80	0.5	(92)	92:8
Hov II	0.00016	80	16	(95)	92:8
Ru cat **131**	0.00016	80	32	(99)	92:8
Hov II	0.0001	100	0.5	(96)	92:8
Ru cat **131**	0.0001	100	0.5	(89)	92:8
Ru cat **131**	0.0001	100	4	(98)	92:8

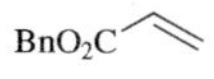

BnO_2C (2 eq) + Ph-alkene; Catalyst (0.02 eq), DCM, 55°, 2 h → BnO_2C … Ph

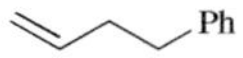

Catalyst	
Ru cat **81**	(82)
Ru cat **91**	(88)

1082

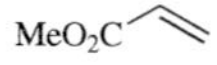

MeO_2C (10 eq) + OH, Ph, n; Gr II (0.025 eq), phenol (0.5 eq), toluene, 110°, 0.5 h → MeO_2C … OH, Ph, n

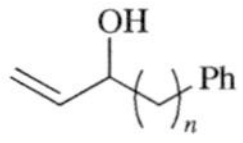

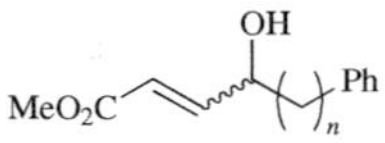

n	
1	(88)
2	(90)

1049

TABLE 20. CROSS-METATHESIS OF α,β-UNSATURATED ESTERS AND THIOESTERS (*Continued*)

A. ESTERS (*Continued*)

Ester	Cross Partner	Conditions	Product(s) and Yield(s) (%)	Refs.

Please refer to the charts preceding the tables for the catalyst structures.

C_3

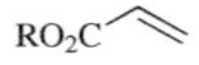

Ester	Cross Partner	Conditions	Product(s) and Yield(s) (%)	Refs.
RO_2C (x eq)	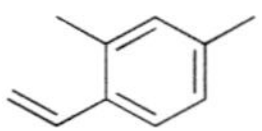	Gr II (0.02 eq), additive (0.03 eq)	RO_2C	

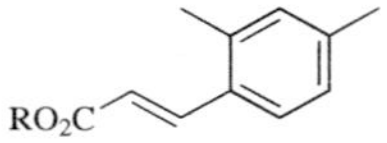

R	x	Additive	Solvent	Temp (°)	Time (h)		(E)/(Z)	Refs.
Et	2	—	DCM	40	12	(87)	(E) only	36
t-Bu	2	—	2.5% aq. PTS	22	12	(72)	—	972
t-Bu	2	—	2.5% aq. TPGS-750-W	22	12	(74)	—	972
t-Bu	2	—	2.5% aq. TPGS-750-W	22	15	(74)	—	167
t-Bu	2	CuI	2.5% aq. TPGS-750-W	22	15	(84)	—	167
t-Bu	2	—	2.5% aq. PTS	22	15	(72)	—	931
t-Bu	2	CuI	2.5% aq. PTS	22	15	(81)	—	931
t-Bu	1	—	2.5% aq. PTS	22	12	(72)	90:10	976

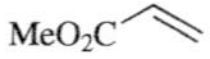

Ester	Cross Partner	Conditions	Product(s) and Yield(s) (%)	Refs.
MeO_2C (1 eq)	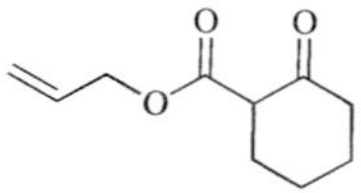	1. Hov II (0.04 eq), DCM, 100° (MW), 0.5 h 2. Alumina basic (0.8 g), MeOH, 50°, 24 h	MeO_2C (52) dr 50:50	955
3 eq	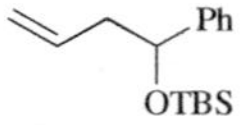 Ph, OTBS	1. Hov II (0.1 eq), DCM, reflux, 3 h 2. DIBALH (3.75 eq) DCM, –78 to –45°, 2 h	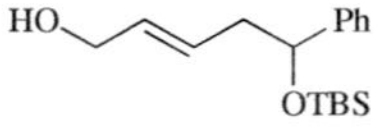 HO, Ph, OTBS (56) (E)/(Z) > 95:5	940

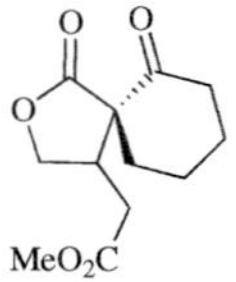

Reactant	Olefin	Conditions	Product(s) and Yield(s) (%)	Refs.
RO_2C–CH=CH$_2$ (x eq)	CH$_2$=CHCH$_2$CH(Ph)OPNP	Hov II (0.02 eq), DCM, 40°, 15 h	RO_2CCH=CHCH$_2$CH(Ph)OPNP	932

R	x	
Me	3	(93)
t-Bu	2	(93)
2-ethylhexyl	2	(96)

Reactant	Olefin	Conditions	Product(s) and Yield(s) (%)	Refs.
MeO_2C–CH=CH$_2$ (3 eq)	CH$_2$=CHCH(R)CH(Ph)OTBS	1. Hov II (0.1 eq), DCM, reflux, 3 h 2. DIBALH (3.75 eq), DCM, –78 to –45°, 2 h	HOCH$_2$CH=CHCH(R)CH(Ph)OTBS (E)/(Z) > 95:5	

R		Refs.
H	(56)	940
Me	(57)	941

Reactant	Olefin	Conditions	Product(s) and Yield(s) (%)
R^1O_2C–CH=CH$_2$ (x eq)	CH$_2$=CHCH$_2$(CH$_2$)$_7$OR^2	Catalyst (y eq), solvent	R^1O_2CCH=CHCH$_2$(CH$_2$)$_7$OR^2

R^1	x	R^2	Catalyst	y	Solvent	Temp (°)	Time (h)		(E)/(Z)	Refs.
Me	3	H	Gr II	0.05	H_2O	40	5	(99)	—	1083
Me	3	H	Ind II	0.05	H_2O	40	5	(90)	—	1083
t-Bu	2	TBS	Gr II	0.02	2.5% aq. PTS	22	12	(95)	—	972
t-Bu	2	TBS	Gr II	0.02	2.5% aq. TPGS-750-W	22	12	(95)	—	972
t-Bu	1	TBS	Gr II	0.02	2.5% aq. PTS	22	12	(95)	(E) only	976

Reactant	Olefin	Conditions	Product(s) and Yield(s) (%)	Refs.
MeO_2C–CH=CH$_2$ (2 eq)	CH$_2$=CHCH$_2$O–P(O)(OMe)–CH=CH–n-C_5H_{11}	Gr II (0.1 eq), CuI (0.1 eq), DCM, reflux, 16 h	MeO_2CCH=CH–n-C_5H_{11} (78) (E) only	220

TABLE 20. CROSS-METATHESIS OF α,β-UNSATURATED ESTERS AND THIOESTERS (*Continued*)

A. ESTERS (*Continued*)

Ester	Cross Partner	Conditions	Product(s) and Yield(s) (%)	Refs.

Please refer to the charts preceding the tables for the catalyst structures.

C_3

Ester	Cross Partner	Conditions	Product(s) and Yield(s) (%)	Refs.
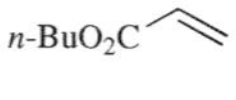 n-BuO_2C (2 eq)	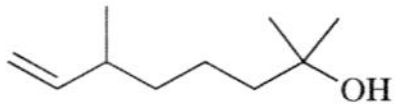 OH	Catalyst (x eq), AcOEt, 60°	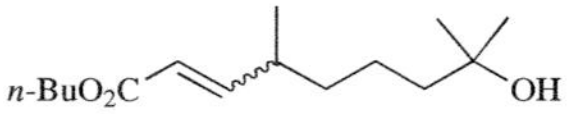 n-BuO_2C … OH	751
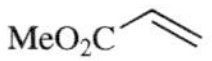 MeO_2C (4 eq)	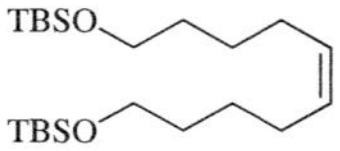 TBSO, TBSO	Catalyst (0.04 eq), DCM, 40°, 8 h	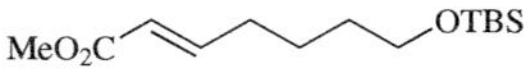 MeO_2C … OTBS	161

Product yields for the n-BuO_2C entry (ref. 751):

Catalyst	x	Time (h)	
Ru cat **170**	0.01	17	(71)
Ru cat **93**	0.01	17	(63)
Ru cat **170**	0.01	17	(73)
Ru cat **172**	0.01	17	(67)
Ru cat **94**	0.01	17	(68)
Ru cat **170**	0.01	18	(75)
Ru cat **170**	0.005	17	(47)
Ru cat **170**	0.002	17	(28)
Gr II	0.01	17	(25)

Product yields for the MeO_2C entry (ref. 161):

Catalyst		(*E*)/(*Z*)
Ind I	(<2)	—
Ru cat **282**	(4)	80:20
Ind II	(35)	75:25
Ru cat **286**	(51)	75:25

Olefin	Substrate	Conditions	Product(s) and Yield(s) (%)	Refs.
EtO_2C (3 eq)	TBDPSO, OAc, OAc	Hov II (0.05 eq), DCM, rt, 12 h	EtO_2C, OAc, OAc, OTBDPS (72)	919
(3 eq)	TBDPSO, MOMO, OTBS, OMe	Hov II (0.025 eq), DCM, rt, 24 h	EtO_2C, MOMO, OTBS, OMe, OTBDPS (85)	1115
MeO_2C (5 eq)	N, O, N, Bn, O	Hov II (0.03 eq), DCM, 45°, 12 h	MeO_2C, N, O, N, Bn, O (62) (*E*) only	1098
(3 eq)	O, N, O, S	Hov II (0.03 eq), DCM, 40°, 12 h	MeO_2C, O, N, O, S (60) (*E*) only	578
RO_2C (10 eq)	O_2, N, S, H, O (±)	Ru cat **35** (0.1 eq), DCM, 45°, 3 h	RO_2C, H, O_2, S, N, O (±); R: Me (56), Et (65), *t*-Bu (78)	359

TABLE 20. CROSS-METATHESIS OF α,β-UNSATURATED ESTERS AND THIOESTERS (*Continued*)

A. ESTERS (*Continued*)

Ester	Cross Partner	Conditions	Product(s) and Yield(s) (%)	Refs.
Please refer to the charts preceding the tables for the catalyst structures.				
C_3				
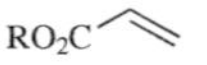 RO_2C 100 eq	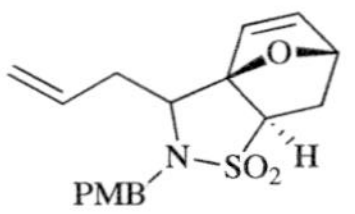 PMB, N–SO_2, H, O (±)	Ru cat **35** (0.1 eq), DCM, 40°, 3 h	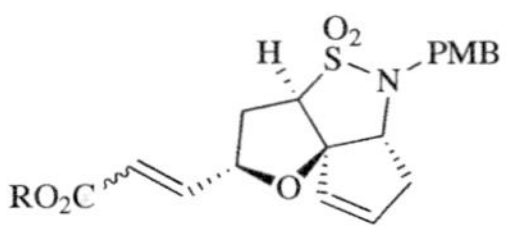 RO_2C, H, O_2 S, N, PMB, O (±); R: Et (52), *t*-Bu (49)	359
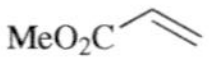 MeO_2C 1.5 eq	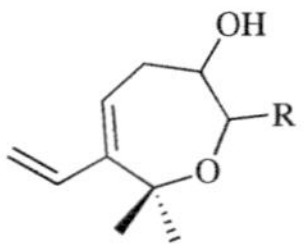 OH, R, O	Gr II (0.15 eq), DCM, reflux, 18 h	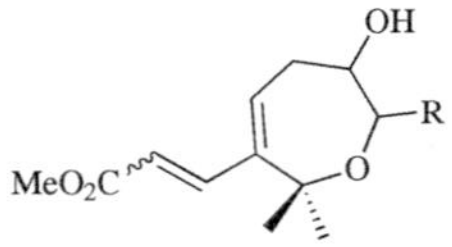 MeO_2C, OH, R, O; R / Conv. (%)[d]: H 100, MeO_2C 100	1084
10 eq	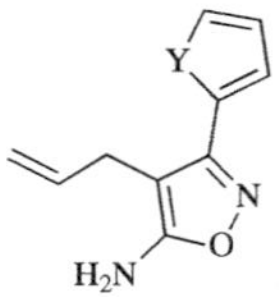 Y, N, O, H_2N	Hov II (2 x 0.05 eq), DCM, rt, 48 h, green LED (36 W)	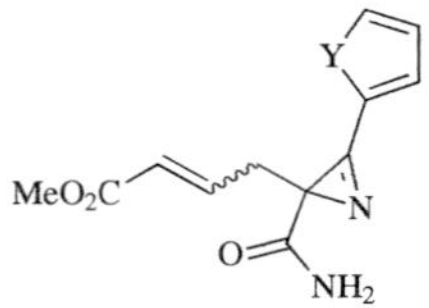 MeO_2C, Y, N, O, NH_2; Y: O (61), S (58)	995
10 eq	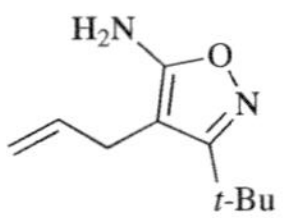 H_2N, O, N, *t*-Bu	Hov II (2 x 0.05 eq), DCM, rt, 48 h, green LED (36 W)	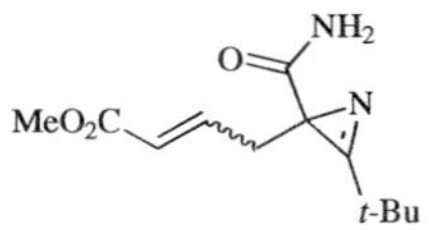 MeO_2C, NH_2, O, N, *t*-Bu (82)	995
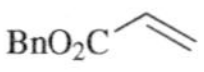 BnO_2C 1.9 eq	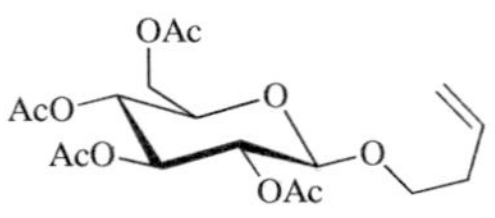 OAc, AcO, AcO, O, OAc, O	Gr II (0.043 eq), DCM, 45°, 8 h	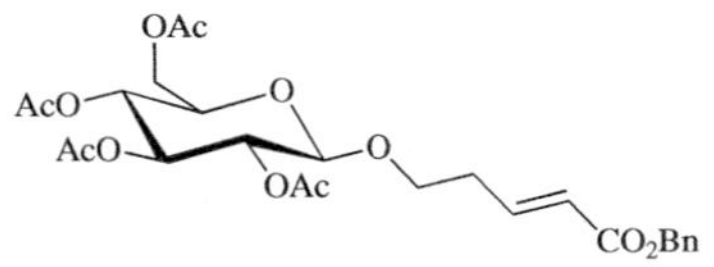 OAc, AcO, AcO, O, OAc, O, CO_2Bn (82) (*E*)/(*Z*) = 93:7	1085

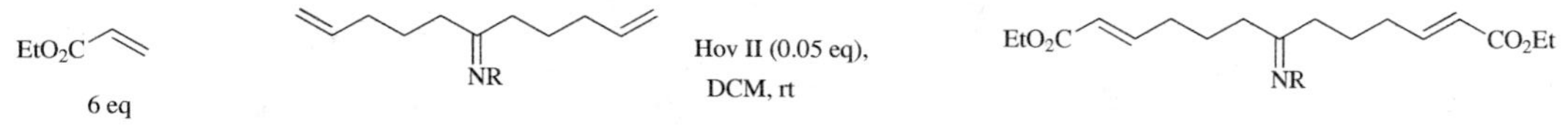

EtO_2C (6 eq) + NR-diene; Hov II (0.05 eq), DCM, rt → EtO_2C…NR…CO_2Et

R	Time (h)		(E,E)/(Z,Z)	
Phth	144	(89)	—	231
HO	144	(0)	—	231
Phth	120	(88)	100:0	1086

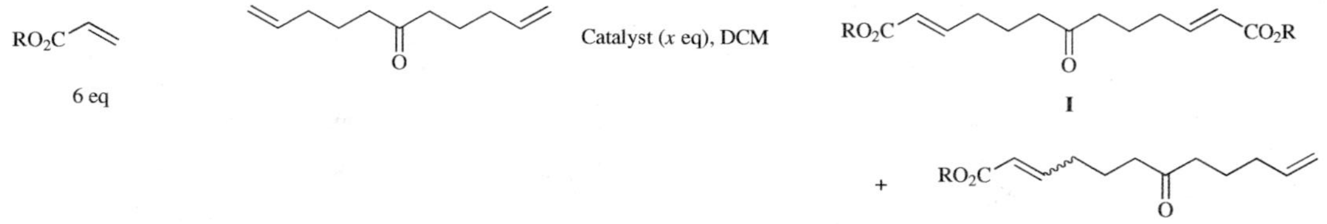

RO_2C (6 eq) + ketone diene; Catalyst (x eq), DCM → RO_2C…O…CO_2R **I** + RO_2C…O… **II**

R	Catalyst	x	Temp (°)	Time (h)	I	(E,E)/(Z,Z) I	II	
Et	Ru cat **35**	0.05	20	120	(31)	—	(44)	231
Et	Ru cat **190**	0.05	20	120	(90)	—	—	231
Et	Ru cat **190**	0.04	20	120	(83)	—	—	231
Et	Ru cat **190**	0.01	20	120	(67)	—	—	231
Et	Ru cat **190**	0.05	40	60	(79)	—	—	231
Me	Ru cat **190**	0.05	rt	120	(78)	—	—	231
Bn	Ru cat **190**	0.05	rt	120	(85)	—	—	231
t-Bu	Ru cat **190**	0.05	rt	120	(78)	—	—	231
Et	Hov II	0.05	rt	120	(90)	—	—	231
Et	Hov II	0.05	rt	72	(89)	100:0	—	1086

TABLE 20. CROSS-METATHESIS OF α,β-UNSATURATED ESTERS AND THIOESTERS (*Continued*)
A. ESTERS (*Continued*)

Ester	Cross Partner	Conditions	Product(s) and Yield(s) (%)	Refs.

Please refer to the charts preceding the tables for the catalyst structures.

C_3

Ester	Cross Partner	Conditions	Product(s) and Yield(s) (%)	Refs.
EtO_2C (6 eq)	R	Hov II (0.05 eq), DCM, rt	EtO_2C … CO_2Et (R) **I** + EtO_2C … (R) **II**	231
t-BuO_2C (x eq)	OR (8)	Catalyst (0.02 eq), solvent, 12 h	t-BuO_2C … OR (8)	

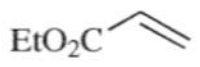

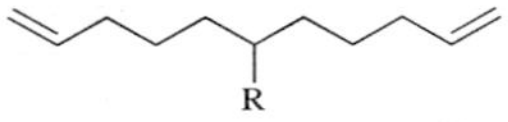

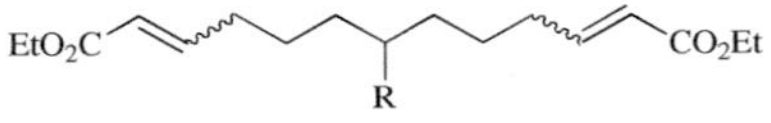

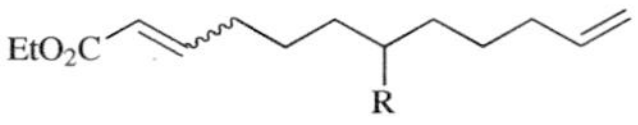

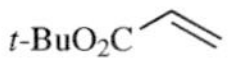

R	Time (h)	I	II
HO	96	(—)	(35)
TBSO	120	(67)	(—)
H_2N	144	(0)	(—)

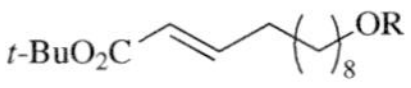

R	x	Catalyst	Solvent	Temp (°)		$(E)/(Z)$	Refs.
H	1	Gr II	2.5% aq. PTS	22	(82)	>95:5	976
H	2	Ru cat **109**	H_2O	22	(85)	(E) only	341
TBS	2	Ru cat **109**	H_2O	22	(94)	(E) only	341
TBS	2	Gr II	H_2O	rt	(70)	—	341
TBS	2	Hov II	H_2O	rt	(56)	—	341

MeO_2C (2 eq) + 8-nonenenitrile analogue ($(CH_2)_7CN$)

1. Catalyst (x eq), toluene, 1 h, temp
2. H_2 (20 bar), t-BuOK (y eq), 80°, 40 h

MeO_2C–…–$(CH_2)_7$–CN **I** + MeO_2C–…–$(CH_2)_8$–CN **II** 1087

Catalyst	x	Temp (°)	y	**I**	**II**
Gr II	0.01	100	0.3	(97)	(—)
Ru cat **139**	0.03	100	0.15	(43)	(50)
Ru cat **139**	0.03	100	0.3	(—)	(96)
Gr II	0.01	rt	0.3	(96)	(—)
Ru cat **139**	0.03	rt	0.2	(—)	(95)
Ru cat **139**	0.03	rt	0.3	(—)	(97)

TABLE 20. CROSS-METATHESIS OF α,β-UNSATURATED ESTERS AND THIOESTERS (*Continued*)

A. ESTERS (*Continued*)

Ester	Cross Partner	Conditions	Product(s) and Yield(s) (%)	Refs.

Please refer to the charts preceding the tables for the catalyst structures.

C_3

MeO_2C (x eq) + alkene-$(CH_2)_7$-R → Catalyst (y eq), solvent → MeO_2C-CH=CH-$(CH_2)_7$-R

R	x	Catalyst	y	Solvent	Temp (°)	Time (h)	Yield (%)	(E)/(Z)	Refs.
OHC	2	Hov II	0.005	toluene	50	0.5	(98)	92:8	1088
OHC	2	Hov II	0.005	toluene	rt	0.5	(98)	92:8	1088
OHC	2	Ru cat **144**	0.005	toluene	50	0.5	(95)	7:93	1088
OHC	2	Ru cat **144**	0.005	toluene	rt	1.5	(85)	6:94	1088
NC–	2	Hov II	0.005	toluene	50	1	(98)	93:7	1088
NC–	2	Hov II	0.005	toluene	25	2	(99)	93:7	1088
NC–	2	Ru cat **79**	0.005	toluene	50	2	(98)	93:7	1088
NC–	2	Ru cat **79**	0.005	toluene	25	2	(95)	93:7	1088
NC–	2	Ru cat **89**	0.005	toluene	50	1	(99)	90:10	1087
NC–	2	Ru cat **89**	0.005	toluene	25	1	(96)	91:9	1087
NC–	2	Hov II	0.01	neat	50	22	(41)	93:7	1087
NC–	5	Hov II	0.01	neat	50	19	(63)	88:12	1087
NC–	10	Hov II	0.01	neat	50	23	(73)	90:10	1087
NC–	10	Hov II	0.01	neat	100	23	(74)	91:9	1087
NC–	20	Hov II	0.01	neat	50	20	(78)	92:8	1087
NC–	4	Hov II	0.0005	toluene	100	7.67	(98)	89:11	1087
NC–	4	Hov II	0.0005	toluene	50	7.67	(96)	94:6	1087
NC–	4	Hov II	0.0005	toluene	25	7.67	(36)	92:8	1087
NC–	4	Hov II	0.00025	toluene	100	7.67	(98)	90:10	1087
NC–	4	Hov II	0.0001	toluene	100	7.67	(92)	89:11	1087
NC–	4	Hov II	0.000075	toluene	100	7.67	(66)	88:12	1087
NC–	4	Hov II	0.000075	toluene	100	7.67	(80)	89:11	1087
NC–	4	Hov II	0.00005	toluene	100	7.67	(61)	87:13	1087
NC–	4	Hov II	0.00005	toluene	100	7.67	(63)	89:11	1087

2 eq $CH_2=CH(CH_2)_7CO_2Me$ Catalyst (x eq), toluene MeO_2C–CH=CH–$(CH_2)_7CO_2Me$

Catalyst	x	Temp (°)	Time (h)		$(E)/(Z)$	
Hov II	0.005	50	0.5	(99)	92:8	1088
Hov II	0.005	rt	0.5	(98)	92:8	1088
Ru cat **144**	0.005	50	0.5	(92)	92:8	1088
Ru cat **144**	0.005	rt	1.5	(89)	92:8	1088
Hov II	0.001	100	3	(82)	92:8	1089
Hov II	0.001	50	3	(82)	93:7	1089
Hov II	0.005	50	3	(99)	92:8	1089
Hov II	0.001	50	3	(92)	93:7	1089
Hov II	0.001	rt	3	(85)	92:8	1089
Hov II	0.0005	100	5	(98)	93:7	1089
Hov II	0.00025	100	5	(95)	90:10	1089
Hov II	0.00025	100	5	(96)	90:10	1089
Hov II	0.000125	100	5	(93)	91:9	1089
Hov II	0.000125	50	5	(84)	91:9	1089
Hov II	0.000123	rt	5	(37)	93:7	1089

TABLE 20. CROSS-METATHESIS OF α,β-UNSATURATED ESTERS AND THIOESTERS (*Continued*)

A. ESTERS (*Continued*)

Ester	Cross Partner	Conditions	Product(s) and Yield(s) (%)	Refs.

Please refer to the charts preceding the tables for the catalyst structures.

C_3

Ester	Cross Partner	Conditions	Product(s) and Yield(s) (%)	Refs.
MeO_2C–CH=CH₂ *x* eq	CH₂=CH–CH₂–$(CH_2)_7$–CO_2Me	Catalyst (*y* eq), solvent	MeO_2C–CH=CH–CH₂–$(CH_2)_7$–CO_2Me	

x	Catalyst	*y*	Solvent	Temp (°)	Time (h)	Conv. (%)[b]	(*E*)/(*Z*)	Refs.
2	Ru cat **88**	0.005	PC	80	3	97	89:11	1090
2	Ru cat **88**	0.0025	PC	80	2+1	93	90:10	1090
3	Ru cat **88**	0.0005	PC	50	3	9	88:12	1090
3	Ru cat **88**	0.0005	PC	80	3	46	86:14	1090
3	Hov II	0.001	PC	100	3	96	88:12	1090
3	Ru cat **88**	0.001	PC	100	3	97	87:13	1090
3	Hov II	0.005	PC	80	15	98	89:11	1090
8	Ru cat **246**	0.0005	toluene	60	0.5	84	93:7	606

Ester	Cross Partner	Conditions	Product(s) and Yield(s) (%)	Refs.
5 eq	CH₂=CH–CH₂–$(CH_2)_7$–CO_2Me **I**	Ind II (0.005 eq), toluene, 80°, 0.5 h	MeO_2C–CH=CH~CH₂–$(CH_2)_7$–CO_2Me **II**	1091

I/toluene (wt %)	Conv. **I** (%)[b]	
1:9	94	(87)
1:4	97	(97)
3:7	>99	(>99)
2:3	>99	(>99)
1:1	>99	(>99)

Olefin 1	Olefin 2	Conditions	Product(s) and Yield(s) (%)	Refs.
1.5 eq	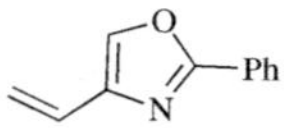	Hov II (0.1 eq), DCM, 40°, 36 h	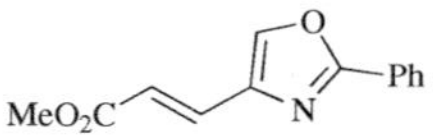 (60) $(E)/(Z) > 95:5$	650
2 eq	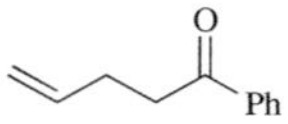	Ru cat **316** (0.004 eq), xylene, 140°, 0.25 h	 (72)	1092
13 eq	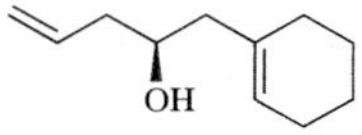	Gr II (0.05 eq), neat, 19°, 50 min	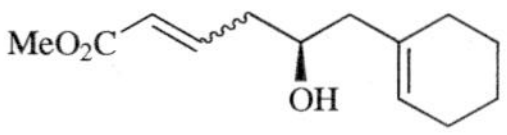 (54) er 98.0:2.0	1093
2 eq	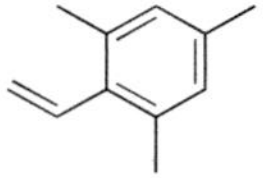	Gr II (0.05 eq), DCM, 40°, 12 h	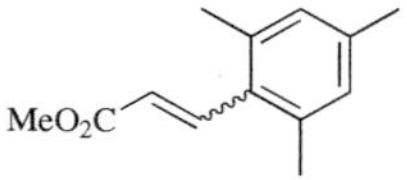 (5)	36
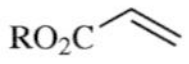 RO_2C 6 eq	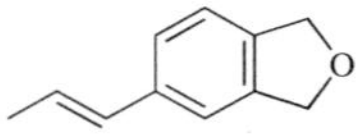	Hov II (0.005 eq), DCE, 70°, 6 h	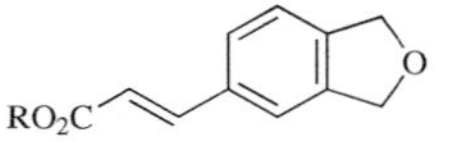	1076
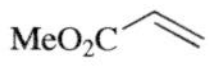 MeO_2C 1 eq	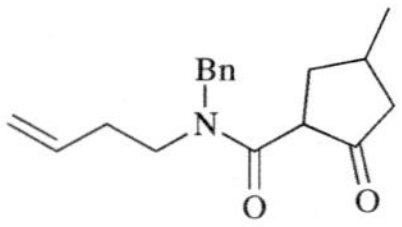	1. Hov II (0.03 + 0.01 eq), DCM, 100° (MW), 20 + 10 min 2. IPr (0.2 eq), 24°, 20 h	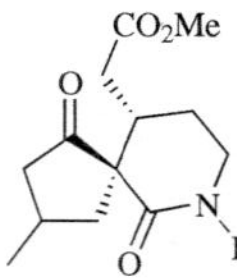 (54) *cis*/*trans* = 74:26	979

R		$(E)/(Z)$
Me	(85)	99:1
Et	(90)	99:1
2-ethylhexyl	(81)	>99:1

TABLE 20. CROSS-METATHESIS OF α,β-UNSATURATED ESTERS AND THIOESTERS (*Continued*)

A. ESTERS (*Continued*)

Ester	Cross Partner	Conditions	Product(s) and Yield(s) (%)	Refs.
Please refer to the charts preceding the tables for the catalyst structures.				
C_3				
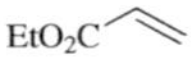 EtO_2C (5 eq)	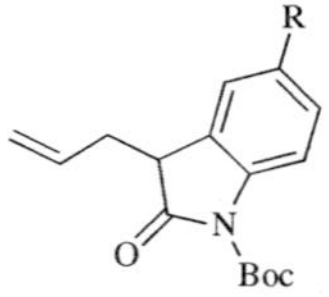 R; O; N; Boc	Ru cat **89** (0.05 eq), DCM, reflux, 5 h	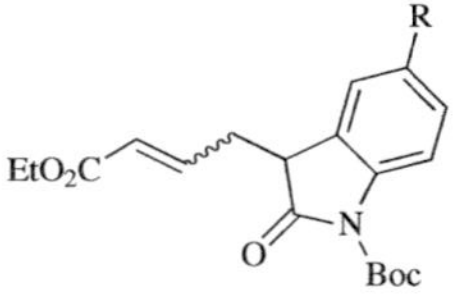 EtO_2C; R; O; N; Boc — R: H (71); MeO (80)	1094
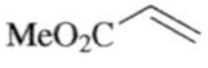 MeO_2C (2.17 eq)	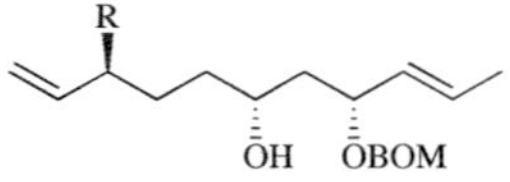 R; OH; OBOM	Catalyst (0.067 eq), additive, solvent, rt	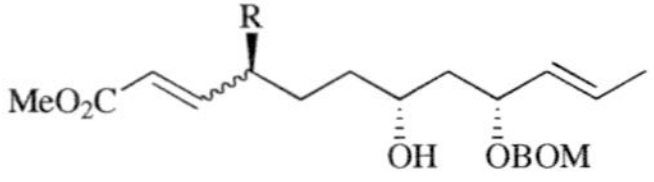 MeO_2C; R; OH; OBOM	
t-BuO_2C (10 eq)	OH; OH; OH	Ru cat **353** (0.05 eq), DCM, hν (350 nm), 30 h	t-BuO_2C; OH; OH; OH (84)	911

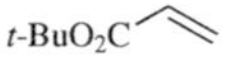

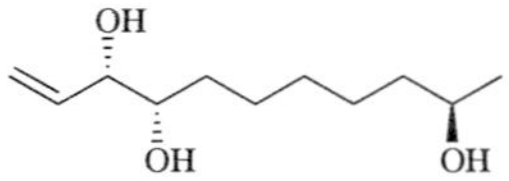

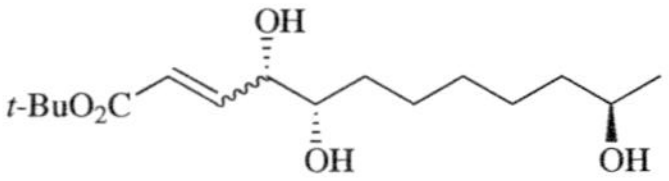

R	Catalyst	Additive	Solvent	Time (h)		Ratio Epimer	Refs.
H	Gr II	none	DCM	7	(76)	—	1095
Me	Gr II	none	DCM	7	(81)	7:1	330
Me	Hov II	none	DCM	7	(76)	4:1	330
Me	Hov II	BQ	DCM	7	(60)	4:1	330
Me	Hov II	none	toluene	3	(71)	—	330
Me	Gr II	phenol (0.5 eq)	DCM	6	(67)	—	330
Me	Hov II	phenol (0.5 eq)	DCM	>6	(75)	—	330

MeO_2C (2 eq) + diene (OH, OTBS, OTBDPS) | Hov II (0.025 eq), DCM, rt, 15 h | MeO_2C product (OH, OTBS, OTBDPS) (75) (*E*)/(*Z*) > 98:2 | 626

RO_2C (*x* eq) + HO-[poly, N]$_{50}$ alkene amide ($()_5$) | Gr II (*y* eq), DCM, 40°, 2 h | RO_2C product | 614

R	*x*	*y*	
Me	7	0.04	(92)
Et	7	0.04	(89)
Bu	7	0.04	(84)
n-C_6H_{13}	7	0.04	(72)
2-hydroxyethyl	12	0.04	(80)
PEG_{480}	7	0.05	(87)
n-$C_8F_{17}(CH_2)_2$	10	0.05	(60)
t-Bu	7	0.04	(87)

EtO_2C (3 eq) + OTBDPS, Ph diene | Hov II (0.05 eq), DCM, 40° | EtO_2C product (OTBDPS, Ph) (78) (*E*)/(*Z*) = 95:5 | 169

TABLE 20. CROSS-METATHESIS OF α,β-UNSATURATED ESTERS AND THIOESTERS (*Continued*)

A. ESTERS (*Continued*)

Ester	Cross Partner	Conditions	Product(s) and Yield(s) (%)	Refs.
Please refer to the charts preceding the tables for the catalyst structures.				
C_3				
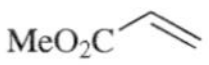 MeO_2C 3 eq	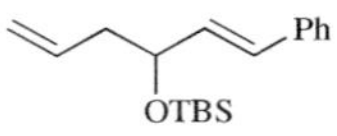 OTBS, Ph	1. Hov II (0.1 eq), DCM, reflux, 3 h 2. DIBALH (3.75 eq), DCM, –78 to –45°, 2 h	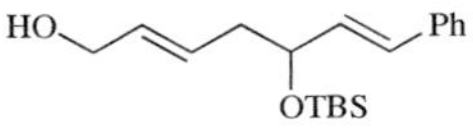 HO, OTBS, Ph (56) (*E*)/(*Z*) > 95:5	940
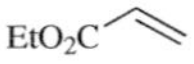 EtO_2C 1.3 eq	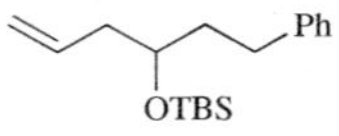 OTBS, Ph	Catalyst (0.1 eq), DCM, reflux	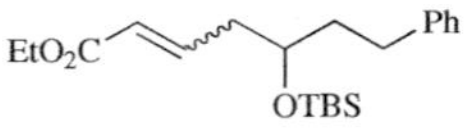 EtO_2C, OTBS, Ph Catalyst: Gr I (26); Gr II (86); Hov II (92)	941
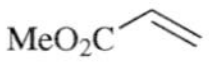 MeO_2C 3 eq	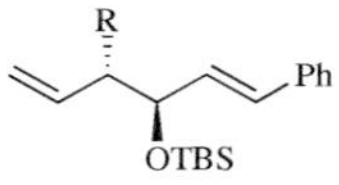 R, OTBS, Ph	1. Hov II (0.1 eq), DCM, reflux, 3 h 2. DIBALH (3.75 eq), DCM, –78 to –45°, 2 h	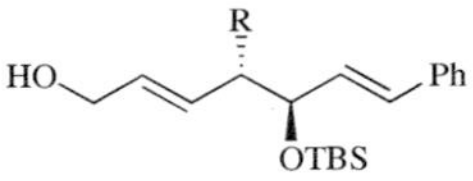 HO, R, OTBS, Ph (*E*)/(*Z*) > 95:5 R: H (54); Me (74)	940 941
2 eq	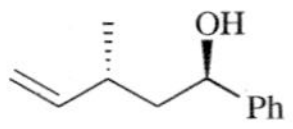 OH, Ph	Gr II (0.07 eq), CuI (0.1 eq), DCM, 40°, 4 h	 MeO_2C, OH, Ph (79)	1080
3 eq	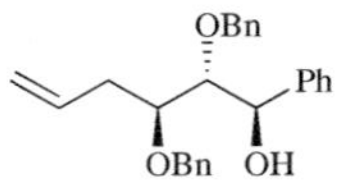 OBn, Ph, OBn, OH	Gr II (0.05 eq), DCM, 40°, 4 h	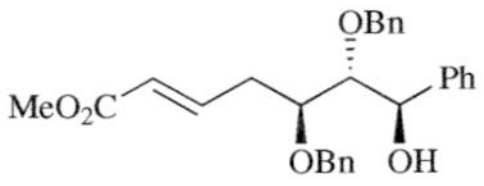 MeO_2C, OBn, Ph, OBn, OH (91) (*E*) only	1096

Reactant	Substrate	Conditions	Product(s) and Yield(s) (%)	Refs.
10 eq	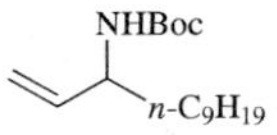 NHBoc, n-C_9H_{19}	1. Hov II (0.03 eq), toluene, reflux, overnight 2. 3N HCl, acetone, overnight	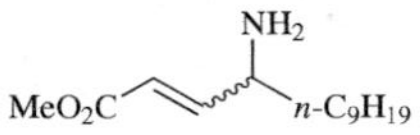 MeO_2C, NH_2, n-C_9H_{19} (67)	911
1 eq	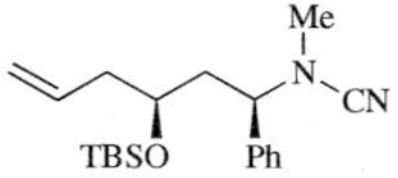 Me, N, CN, TBSO, Ph	Gr II (0.076 eq), DCM, 40°, 2 h	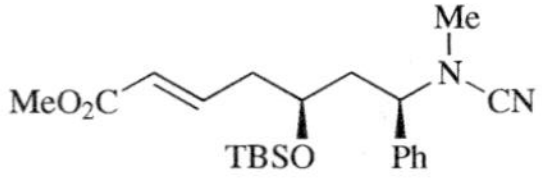 MeO_2C, Me, N, CN, TBSO, Ph (70) (*E*) only	1097
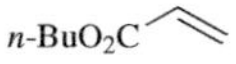 n-BuO_2C (4 eq)	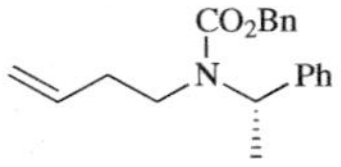 CO_2Bn, N, Ph	Catalyst (0.05 eq), toluene, 80°, 12 h	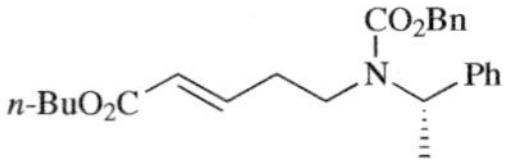 n-BuO_2C, CO_2Bn, N, Ph Catalyst / (*E*)/(*Z*) Ru cat **266** (90) 71:29 Ru cat **267** (93) 73:27	1031
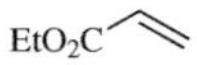 EtO_2C (3 eq)	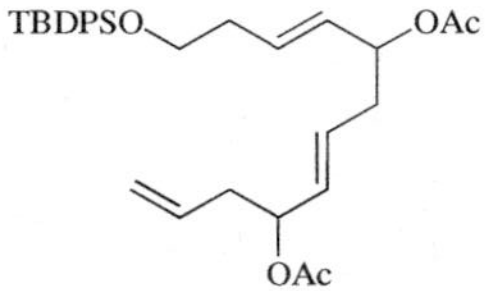 TBDPSO, OAc, OAc	Hov II (0.05 eq), DCM, rt, 12 h	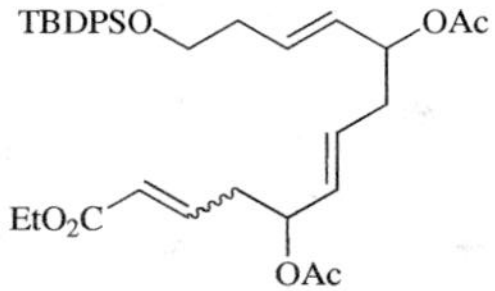 TBDPSO, OAc, EtO_2C, OAc (69)	919

TABLE 20. CROSS-METATHESIS OF α,β-UNSATURATED ESTERS AND THIOESTERS (*Continued*)

A. ESTERS (*Continued*)

Ester	Cross Partner	Conditions	Product(s) and Yield(s) (%)	Refs.
Please refer to the charts preceding the tables for the catalyst structures.				
C_3				
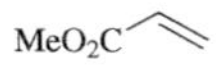 MeO_2C (acrylate), 1 eq	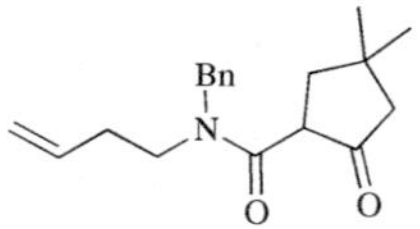 Bn, N, O, O	Hov II (0.03 + 0.01 eq), DCM, 100° (MW), 0.33 + 0.17 h	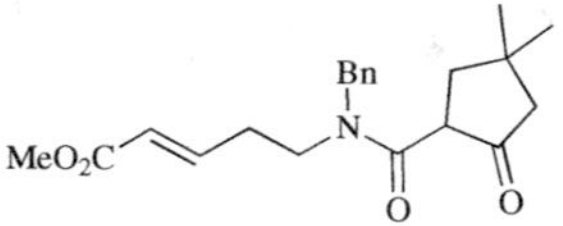 MeO_2C, Bn, N, O, O (73) (*E*)/(*Z*) = 91:9	955
1 eq	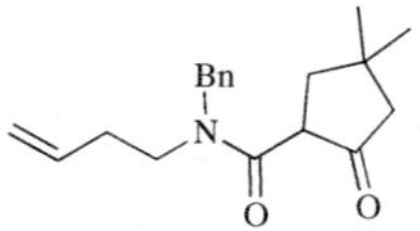 Bn, N, O, O	1. Hov II (0.04 eq), DCM, 100° (MW), 0.5 h 2. Dowex 550A, MeOH, 100° (MW), 0.33 h	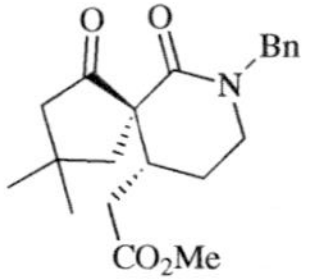 O, O, N, Bn, CO_2Me (68) dr > 95:5	955
1 eq	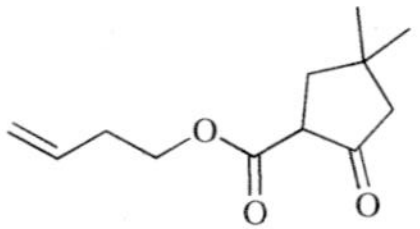 O, O, O	Hov II (0.03 + 0.01 eq), DCM, 100° (MW), 20 + 10 min	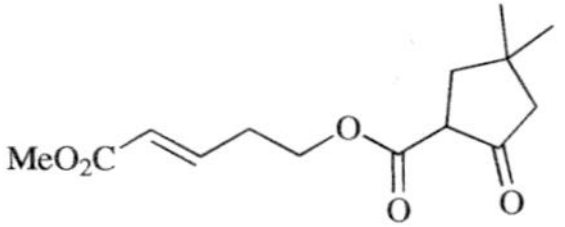 MeO_2C, O, O, O (78) (*E*)/(*Z*) = 91:9	955
10 eq	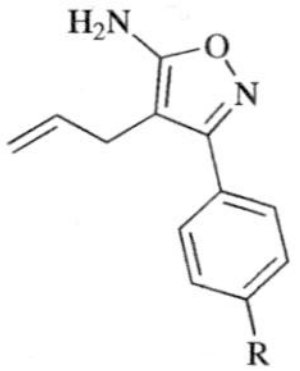 H_2N, O, N, R	Hov II (2 x 0.05 eq), DCM, rt, 48 h, green LED (36 W)	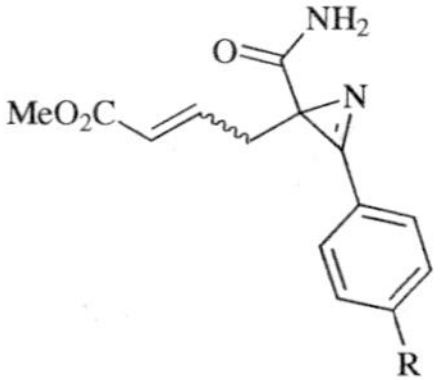 MeO_2C, O, NH_2, N, R R: H (82); F (52); Cl (70)	995

10 eq	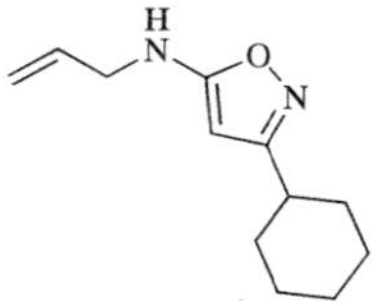	Hov II (2 x 0.05 eq), DCM, rt, 48 h, green LED (36 W)	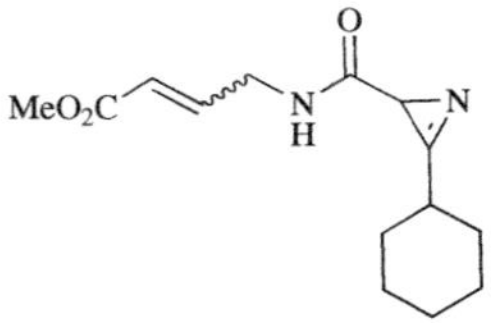 MeO_2C	(59)	995
5 eq	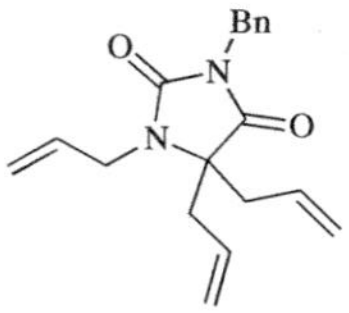	Hov II (0.05), DCM, 45°, 24 h	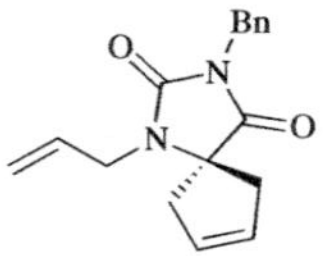	(53)	1098
	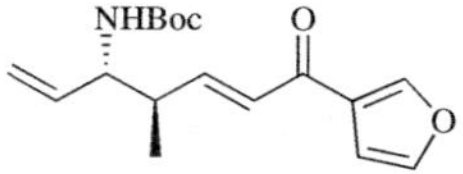	Hov II (0.05 eq), toluene, 70°	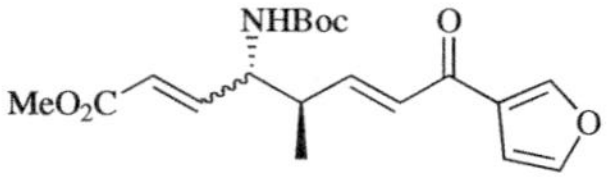 MeO_2C	(75)	1099
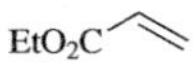 EtO_2C 3 eq	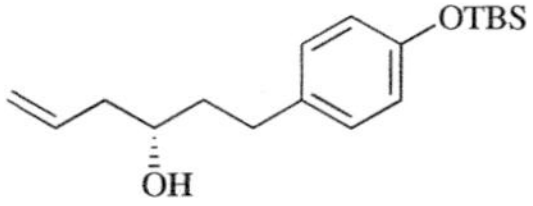	Hov II (0.05 eq) DCM, rt, 48 h	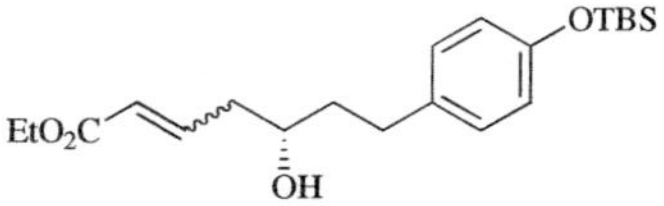 EtO_2C	(80)	1100

TABLE 20. CROSS-METATHESIS OF α,β-UNSATURATED ESTERS AND THIOESTERS (*Continued*)

A. ESTERS (*Continued*)

Ester	Cross Partner	Conditions	Product(s) and Yield(s) (%)	Refs.
Please refer to the charts preceding the tables for the catalyst structures.				
C_3				
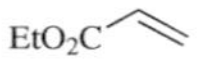 EtO_2C (3 eq)	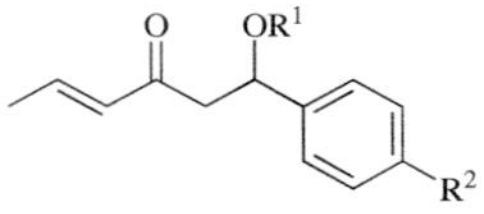 O, OR^1, R^2	Hov II (0.05 eq), DCM, reflux, 18–24 h	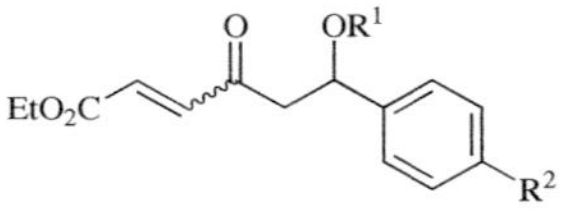 EtO_2C O OR^1 R^2; R^1 / R^2: H / MeO (24); H / O_2N (13); TBS / MeO (62)	1021
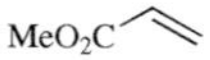 MeO_2C (20 eq)	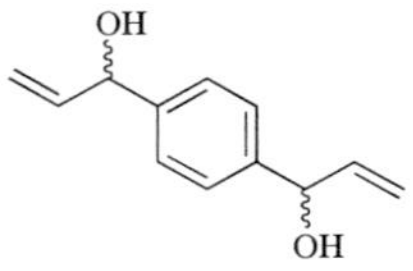 OH, OH	Gr II (0.1 eq), phenol (0.5 eq), toluene, 110°, 0.5 h	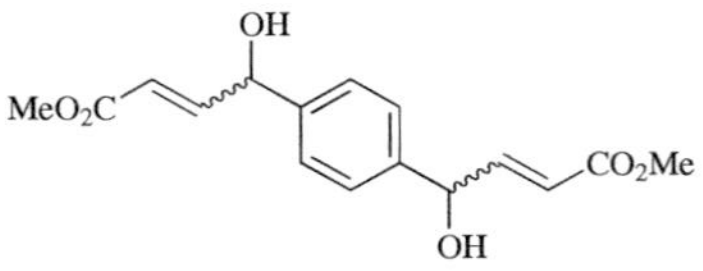 MeO_2C OH, OH, CO_2Me (83)	1049
(2 eq)	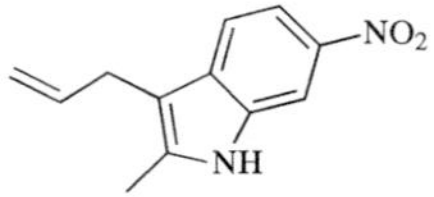 NO_2, NH	Ru cat **88** (0.05 eq), DCM, 25°, 2 h	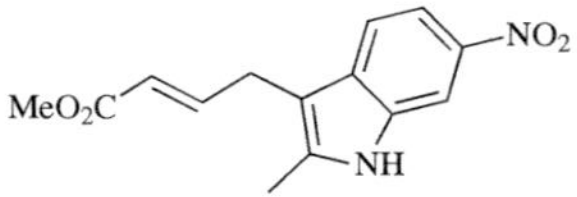 MeO_2C NO_2 NH (91) (*E*) only	171
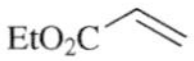 EtO_2C	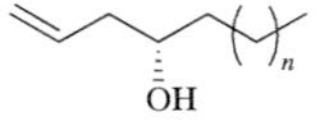 OH, *n*	Gr II (0.05 eq), DCM, rt, 3 h	 EtO_2C OH *n*; *n* = 7 (94); 9 (94)	720
	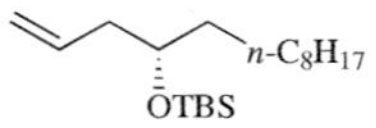 OTBS, *n*-C_8H_{17}	Gr I (0.03 eq), DCM, rt, 1.5 h	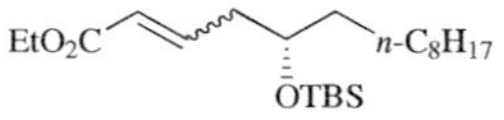 EtO_2C OTBS *n*-C_8H_{17} (92)	1101

Reactant	Substrate	Conditions	Product(s) and Yield(s) (%)		Refs.
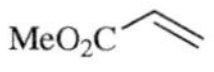 MeO_2C 3 eq	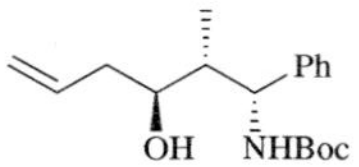 OH NHBoc Ph	1. Gr II (0.05 eq), rt, 5 h 2. HCl, THF, rt, 6 h 3. aq NH_3, MeOH	MeO_2C H N Ph OH	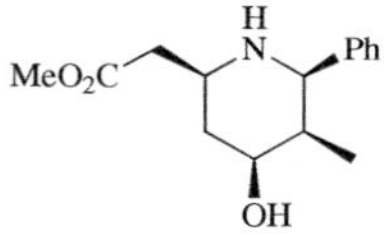 (85) dr > 99:1	1102
3 eq	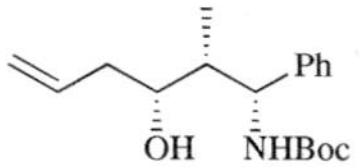 OH NHBoc Ph	1. Gr II (0.05 eq), rt, 5 h 2. HCl, THF, rt, 6 h 3. aq NH_3, MeOH	MeO_2C H N Ph OH	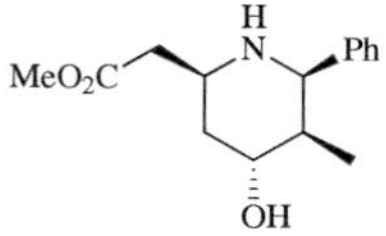 (83) dr > 99:1	1102
3 eq	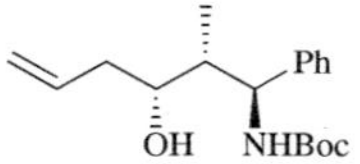 OH NHBoc Ph	1. Gr II (0.05 eq), rt, 5 h 2. HCl, THF, rt, 6 h 3. aq NH_3, MeOH	MeO_2C H N Ph OH	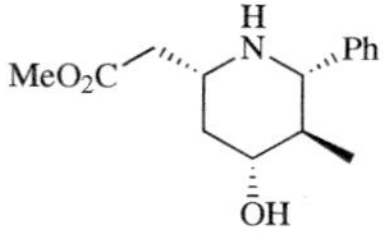 (80) dr > 99:1	1102
3 eq	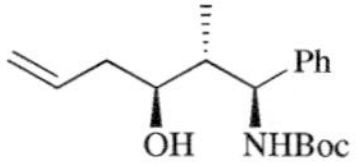 OH NHBoc Ph	1. Gr II (0.05 eq), rt, 5 h 2. HCl, THF, rt, 6 h 3. aq NH_3, MeOH	MeO_2C H N Ph OH	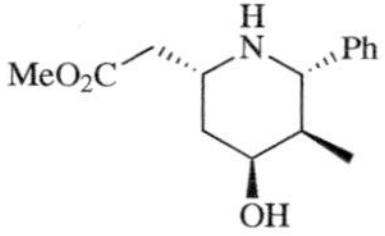 (83) dr > 99:1	1102
3 eq	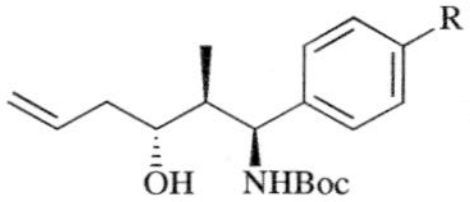 OH NHBoc R	1. Gr II (0.05 eq), rt, 5 h 2. HCl, THF, rt, 6 h 3. aq NH_3, MeOH	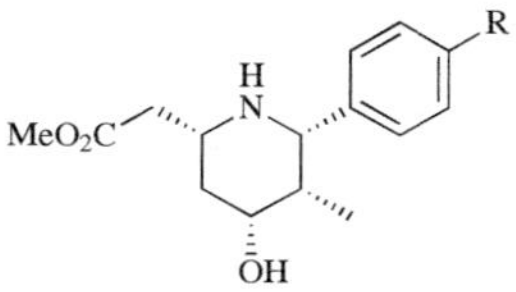 MeO_2C H N R OH dr > 99:1	R: F (75); Me (80)	1102

TABLE 20. CROSS-METATHESIS OF α,β-UNSATURATED ESTERS AND THIOESTERS (*Continued*)

A. ESTERS (*Continued*)

Ester	Cross Partner	Conditions	Product(s) and Yield(s) (%)	Refs.
Please refer to the charts preceding the tables for the catalyst structures.				
C_3				
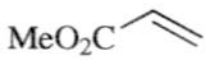 MeO_2C (3 eq)	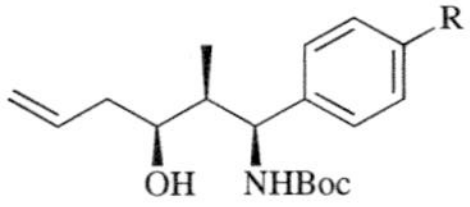 OH, NHBoc, R	1. Gr II (0.05 eq), rt, 5 h 2. HCl, THF, rt, 6 h 3. aq NH_3, MeOH	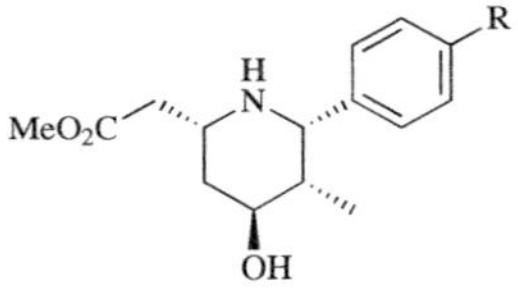 MeO_2C, H N, OH, R; dr > 99:1 R: F (72); Me (81)	1102
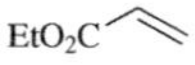 EtO_2C (3 eq)	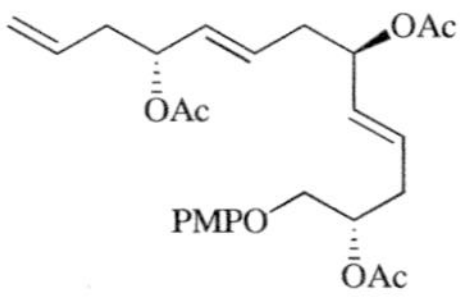 OAc, OAc, PMPO, OAc	Hov II (0.05 eq), DCM, 25°, 24 h	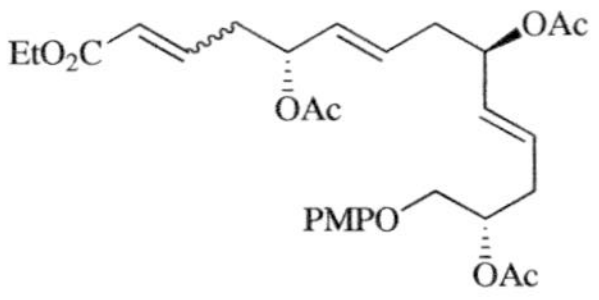 EtO_2C, OAc, OAc, PMPO, OAc (61) (*E*)/(*Z*) > 95:5	157
5 eq	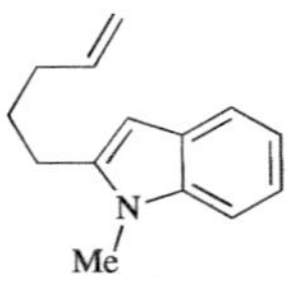 N, Me	Hov II (0.03 eq), BF_3•Et_2O (0.1 eq), DCE, reflux, 1.5 h	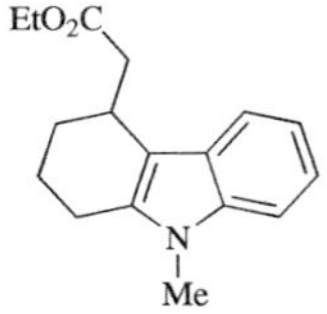 EtO_2C, N, Me (95)	950

Reactant	Substrate	Conditions	Product(s) and Yield(s) (%)	Refs.
R^1O_2CCH=CH$_2$ (13 eq)		Gr II (0.05 eq), neat, 19°, 50 min	R^1O_2C	1093
MeO_2CCH=CH$_2$ (3 eq)		Gr II (0.05 eq), DCM, 40°	MeO_2C, RO	1103
2 eq		Gr II (0.1 eq), CuI (0.1 eq), DCM, reflux, 16 h	MeO_2C *n*-Bu (45)	220

R^1	R^2	R^3		er
Me	H	H	(73)	—
Me	H	MeO	(43)	97.0:3.0
Me	O_2N	H	(73)	98.0:2.0
Me	MeO	H	(35)	—
t-Bu	H	H	(64)	—

R	Time (h)	
H	0.5	(87)
TMS	1	(57)

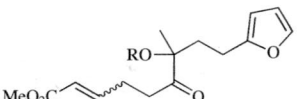

TABLE 20. CROSS-METATHESIS OF α,β-UNSATURATED ESTERS AND THIOESTERS (*Continued*)

A. ESTERS (*Continued*)

Ester	Cross Partner	Conditions	Product(s) and Yield(s) (%)	Refs.
Please refer to the charts preceding the tables for the catalyst structures.				
C_3				
MeO_2C 7 eq	O, O, P, O, *n*-Bu 1 eq + MeO, O, P, O, *n*-Bu 2.5 eq	Gr II (0.35 eq), CuI (0.35 eq), DCM, reflux, 16 h	 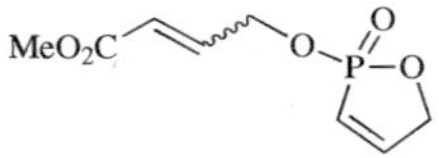 MeO_2C, *n*-Bu **I** + MeO_2C, O, P, O, O **II**	**I** + **II** (86)	220
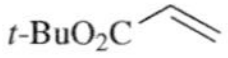 *t*-BuO_2C	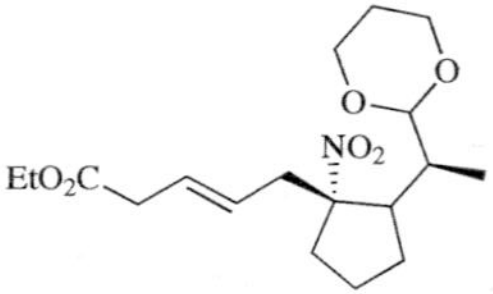 EtO_2C, NO_2, O, O	Hov II (0.05 eq), DCE, reflux, 16 h	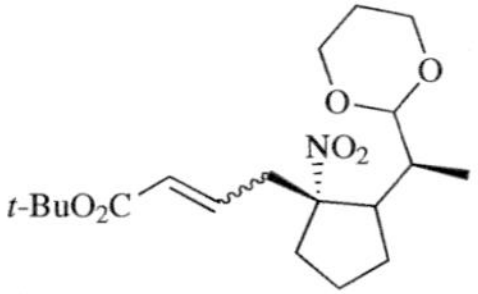 *t*-BuO_2C, NO_2, O, O (68)	1104
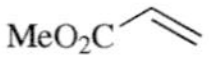 MeO_2C 3 eq	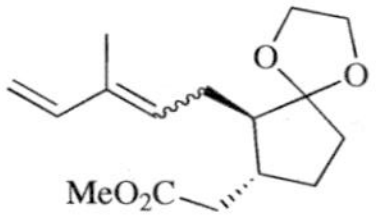 O, O, MeO_2C (*E*)/(*Z*) = 80:20	Hov II (0.05 eq), DCM, 45°, 3 h	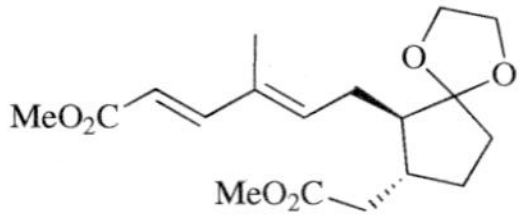 MeO_2C, O, O, MeO_2C (37) (*E*)/(*Z*) = 80:20	939

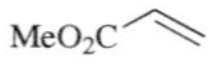

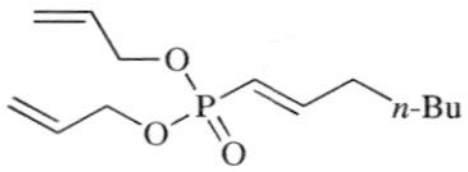

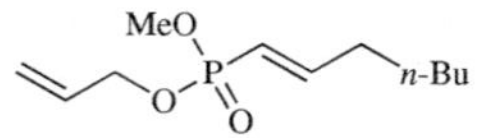

Cross-Partner	Olefin	Conditions	Product(s) and Yield(s) (%)	Refs.
1.5 eq	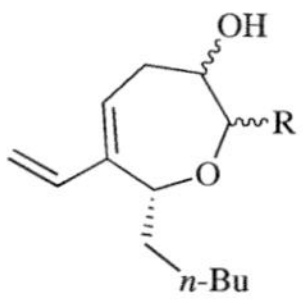 OH, R, O, n-Bu	Gr II (0.15 eq), DCM, reflux, 18 h 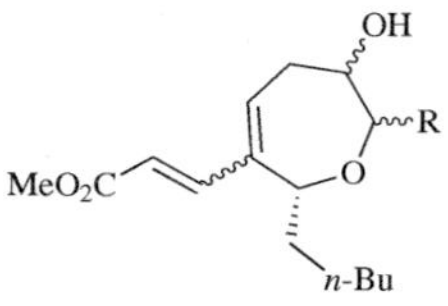	MeO_2C, OH, R, O, n-Bu; R: H, Conv. (%)[d] 100; MeO_2C, 100	1084
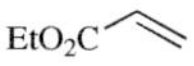 EtO_2C 1 eq	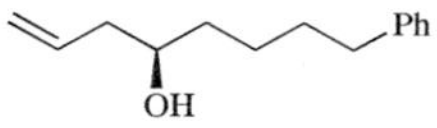 OH, Ph	Gr II (0.05 eq), DCM, reflux, 12 h	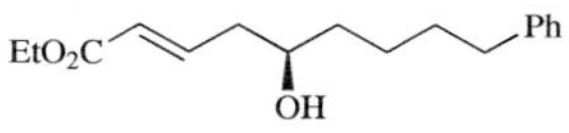 EtO_2C, OH, Ph (85) (*E*) only	1105
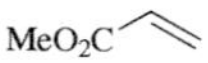 MeO_2C 13 eq	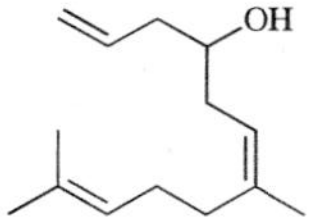 OH	Gr II (0.05 eq) neat, 19°, 50 min 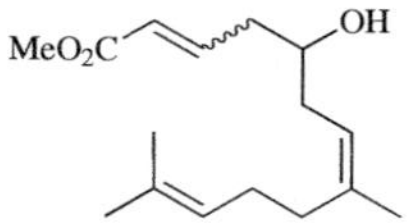	MeO_2C, OH (68)	1093
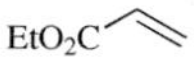 EtO_2C 3 eq	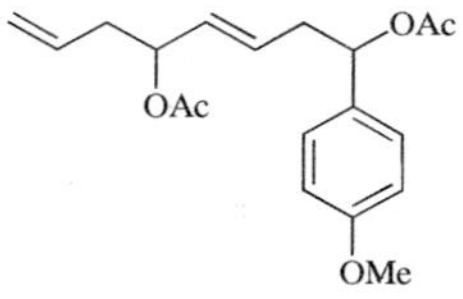 OAc, OAc, OMe	Hov II (0.05 eq), DCM, rt, 16 h 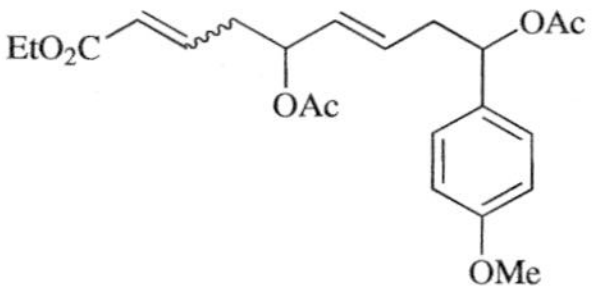	EtO_2C, OAc, OAc, OMe (79)	919

TABLE 20. CROSS-METATHESIS OF α,β-UNSATURATED ESTERS AND THIOESTERS (*Continued*)

A. ESTERS (*Continued*)

Ester	Cross Partner	Conditions	Product(s) and Yield(s) (%)	Refs.
Please refer to the charts preceding the tables for the catalyst structures.				
C_3				
MeO_2C (x eq)	TBSO, Ph, OR	Gr II (0.05 eq), solvent, rt, 4.5 h	MeO_2C, TBSO, Ph, OR (I) + OTBS, OR (II)	1106
71 eq	OH, OTBS, Ph	Gr II (0.5 eq), toluene, rt, 2 h	MeO_2C, OH, OTBS, Ph (90) $(E)/(Z) = 95:5$	1107, 1108
t-BuO_2C (13 eq)	R^2, OH, R^1, Ph	Gr II (0.05 eq), neat, 19°	t-BuO_2C, R^2, OH, R^1, Ph	1093

x	R	Solvent	I	II
excess	H	DCM	(77)	(7)
30	H	DCM	(82)	(7)
20	BOM	DCM	(25)	(71)
excess	BOM	toluene	(51)	(46)

R^1	R^2	Time (min)		dr
H	Me	50	(60)	~55:45
Me	H	60	(80)	96:4

5.6 eq

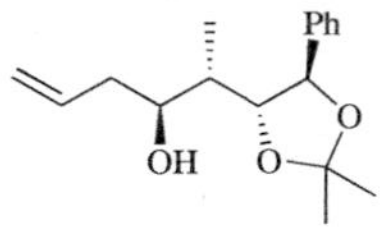

Gr II (0.04 eq), DCM, rt, 19 h

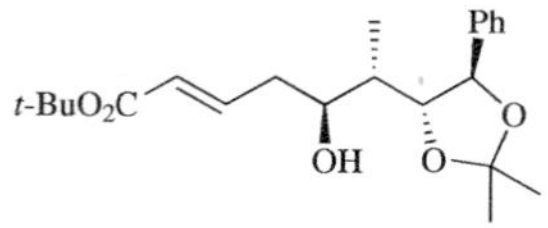

t-BuO_2C

OH Ph

(71) (*E*) only

1109

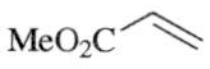

MeO_2C 8 eq

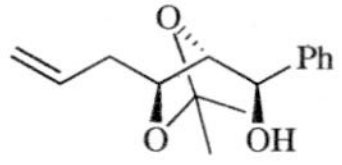

Gr II (0.05 eq), DCM, rt, 3 h

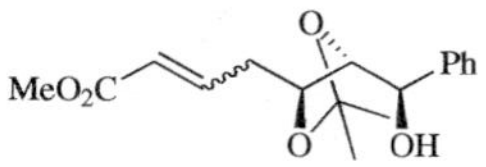

MeO_2C

Ph OH

(97)

1110

3 eq

OH + OHC R *n* 3 eq

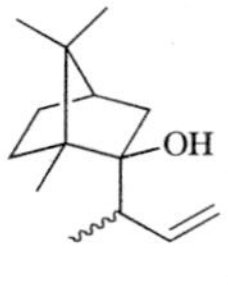

1. CSA (0.1 eq), DCM, 25°
2. Gr II (0.2 eq), DCM, 4 h

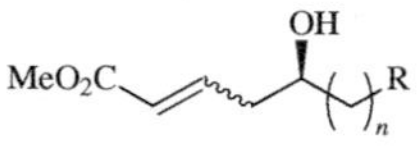

MeO_2C OH R *n*

936

R	*n*	Time (h)		er
H	1	144	(62)	96.0:4.0
BnO	2	144	(46)	96.0:4.0
BnO	3	150	(52)	97.0:3.0
BnO	4	126	(50)	97.0:3.0
Me	7	126	(58)	97.0:3.0
Ph	2	122	(54)	97.0:3.0

TABLE 20. CROSS-METATHESIS OF α,β-UNSATURATED ESTERS AND THIOESTERS (*Continued*)

A. ESTERS (*Continued*)

Ester	Cross Partner	Conditions	Product(s) and Yield(s) (%)	Refs.
Please refer to the charts preceding the tables for the catalyst structures.				
C_3				
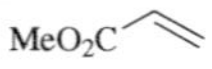 MeO_2C (3 eq)	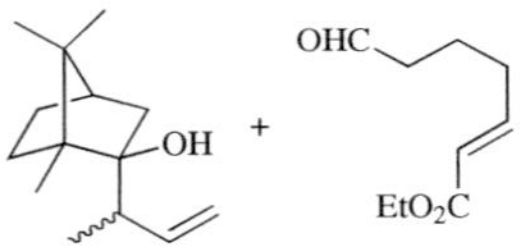 OH + OHC, EtO_2C (3 eq)	1. CSA (0.1 eq), DCM, 25°, 126 h 2. Gr II (0.2 eq), DCM, 4 h	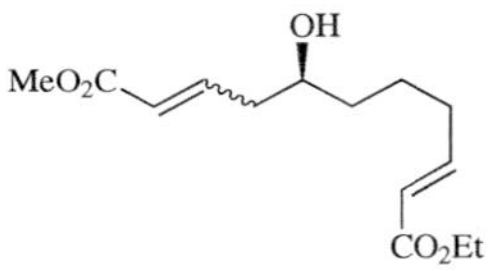 MeO_2C, OH, CO_2Et (46) er = 98:2	936
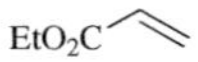 EtO_2C (*x* eq)	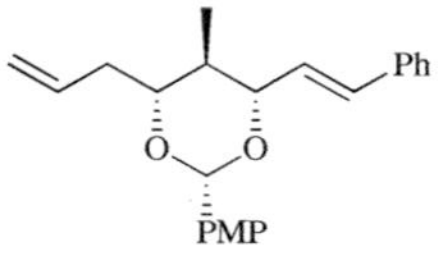 Ph, O, O, PMP	Catalyst (*y* eq), DCM	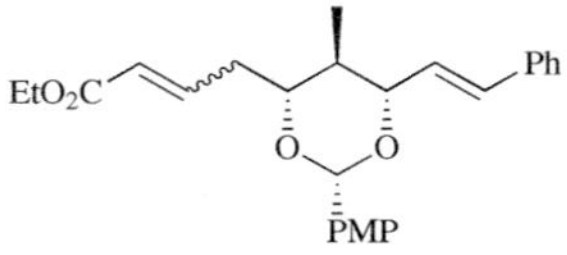 EtO_2C, Ph, O, O, PMP	1111

x	Catalyst	*y*	Temp (°)	Time (h)	
2	Gr II	0.05	rt	14	(61)
2	Gr II	0.05	40	3	(37)
2	Gr II	0.075	rt	14	(53)
5	Gr I	0.1	40	4	(20)
4	Gr I	0.05	40	14	(0)
10	Gr II	0.05	rt	14	(71)
20	Gr II	0.05	rt	14	(95)

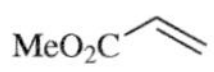

MeO_2C (acrylate)

5 eq

OMe, TBSO, OMe

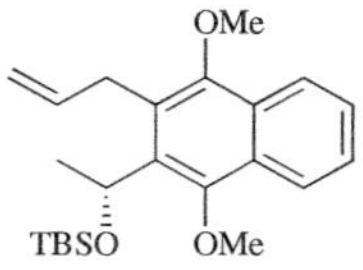

Hov II (2 x 0.05 eq), DCM, rt, 8 h + overnight

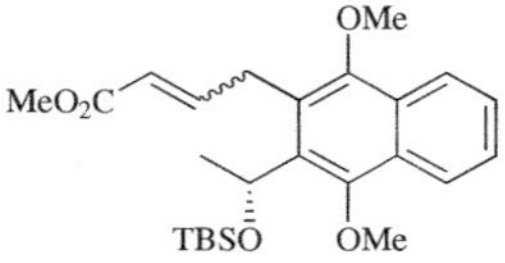

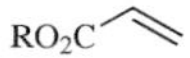

MeO_2C, OMe, TBSO, OMe

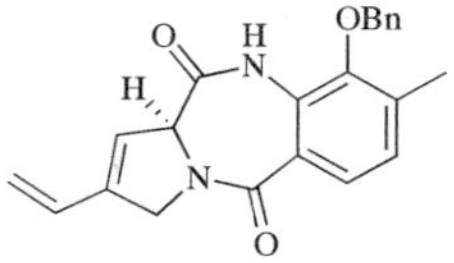

(80)

1112

RO_2C (acrylate)

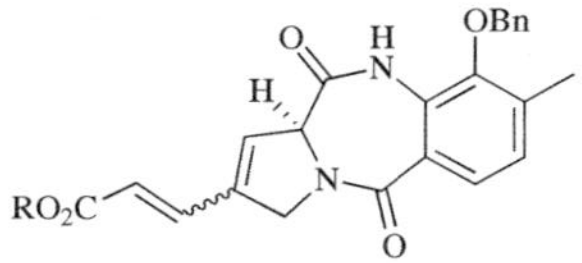

x eq

O, H, N, H, OBn, H, N, O

Catalyst (0.1 eq), DCM

O, H, N, H, OBn, H, RO_2C, N, O

R	*x*	Catalyst	Temp (°)	Time (h)	
Me	10	Gr II	reflux	24	(41)
Me	10	Ru cat **112**	rt	17	(63)
Et	10,000	Ru cat **112**	rt	17	(60)

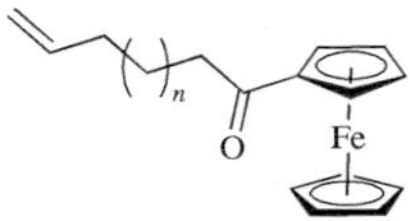

608

10 eq

n, O, Fe

Gr II (0.2 eq), DCM, reflux

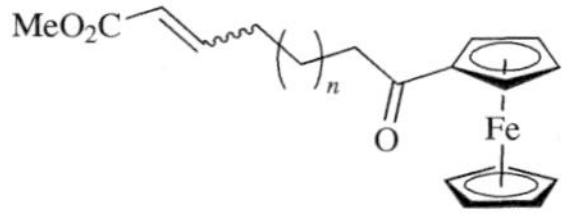

MeO_2C, *n*, O, Fe

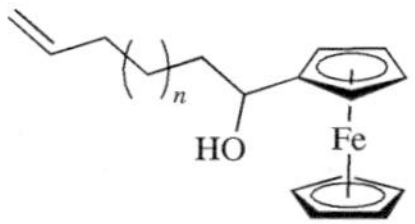

(*E*) only

n	
1	(38)
2	(37)

587

10 eq

n, HO, Fe

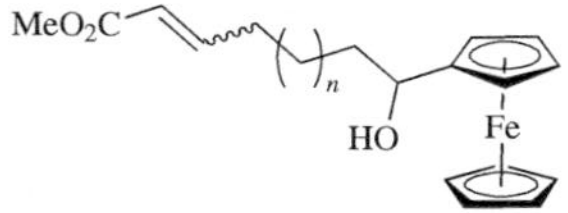

Gr II (0.2 eq), DCM, reflux

MeO_2C, *n*, HO, Fe

(*E*) only

n	
1	(26)
2	(42)

587

6 eq

Ni

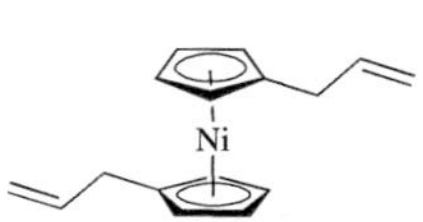

Gr II (0.06 eq), DCM, reflux, 3 h

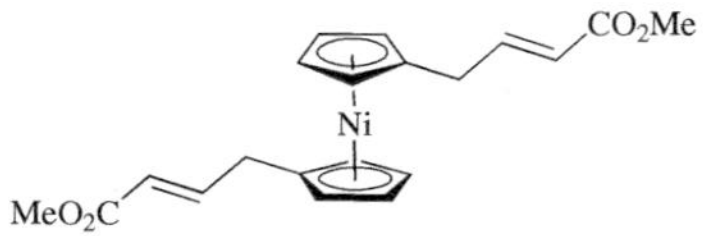

CO_2Me, Ni, MeO_2C

(84) (*E*) only 984

TABLE 20. CROSS-METATHESIS OF α,β-UNSATURATED ESTERS AND THIOESTERS (*Continued*)

A. ESTERS (*Continued*)

Ester	Cross Partner	Conditions	Product(s) and Yield(s) (%)	Refs.
Please refer to the charts preceding the tables for the catalyst structures.				
C_3				
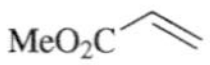 MeO_2C 1.5 eq	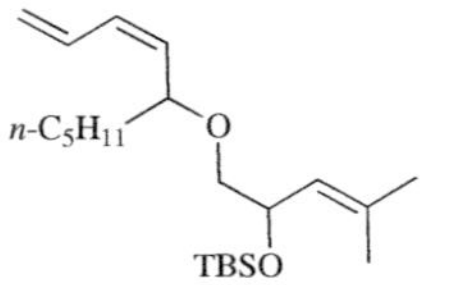 n-C_5H_{11}, O, TBSO	Gr II (0.1 eq), DCM, reflux, 18 h	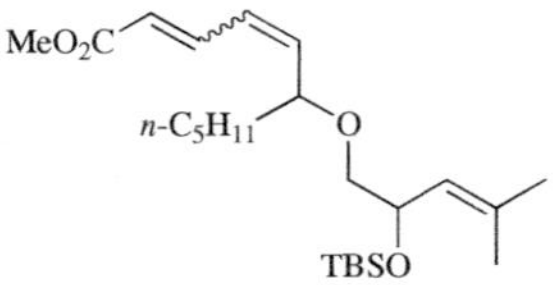 MeO_2C, n-C_5H_{11}, O, TBSO (60)	1113
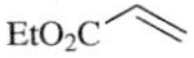 EtO_2C 3 eq	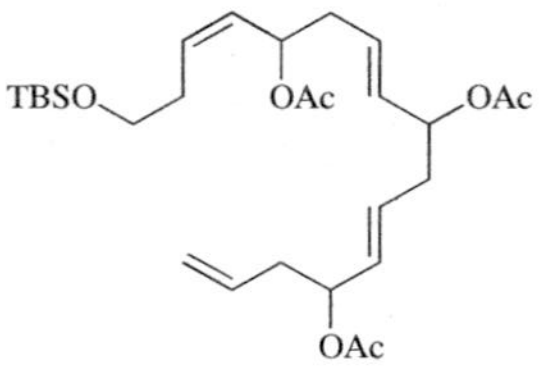 TBSO, OAc, OAc, OAc	Hov II (0.05 eq), DCM, rt, 12 h	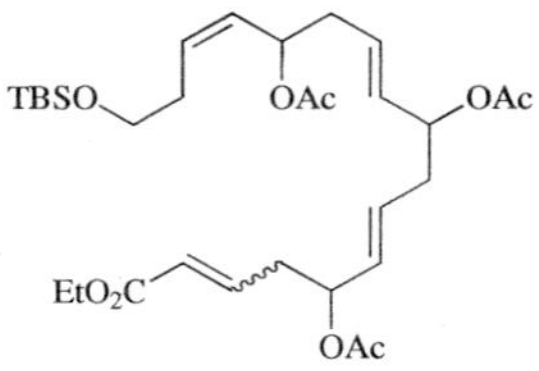 TBSO, OAc, OAc, EtO_2C, OAc (60)	919
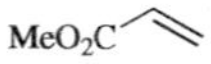 MeO_2C 2 eq	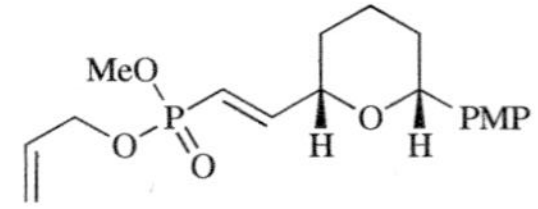 MeO, P, O, O, H, O, H, PMP	Gr II (0.1 eq), CuI (0.1 eq), DCM, reflux, 2 h	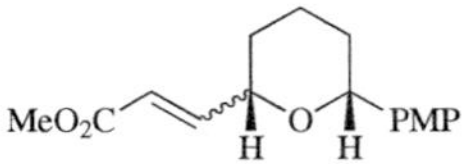 MeO_2C, H, O, H, PMP (73)	220
20 eq	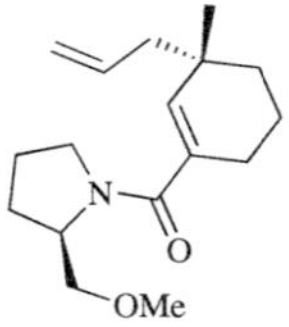 N, O, OMe	Hov II (0.025 eq), DCM, reflux, 7 h	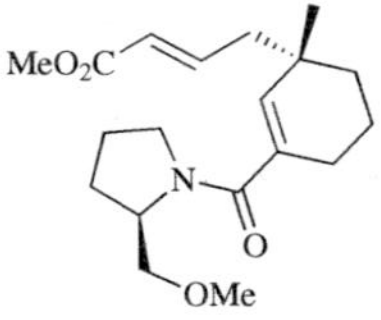 MeO_2C, N, O, OMe (82) $(E)/(Z)$ = 97:3	896

Reagent	Substrate	Conditions	Product	Yield (%)	Ref.
4 eq	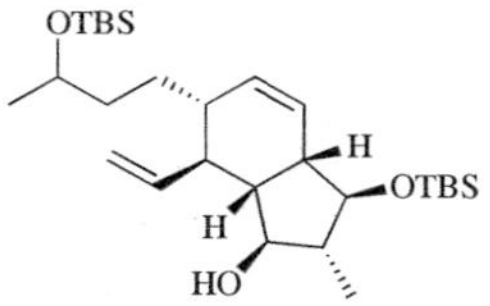 OTBS, H, OTBS, H, HO	Hov II (0.05 eq), DCM, reflux, 18 h	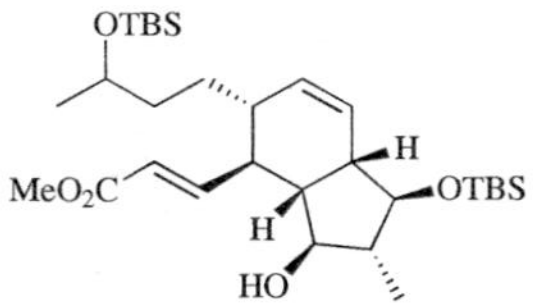 OTBS, MeO_2C, H, OTBS, H, HO	(80) (*E*)/(*Z*) = 84:16	1114
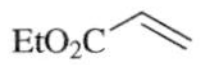 EtO_2C 6 eq	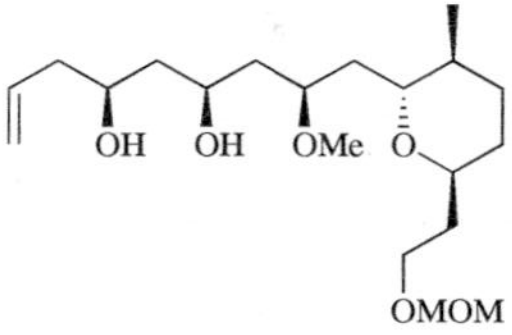 OH OH OMe O, OMOM	Hov II (0.05 eq), DCM, rt, 1.5 h	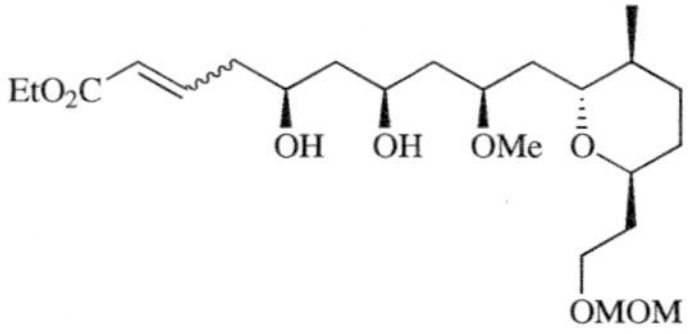 EtO_2C, OH OH OMe O, OMOM	(89)	1115
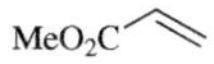 MeO_2C 10 eq	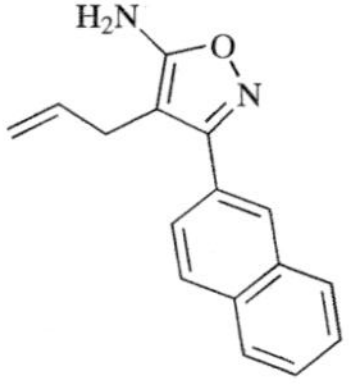 H_2N, O, N	Hov II (2 x 0.05 eq), DCM, rt, 48 h, green LED (36 W)	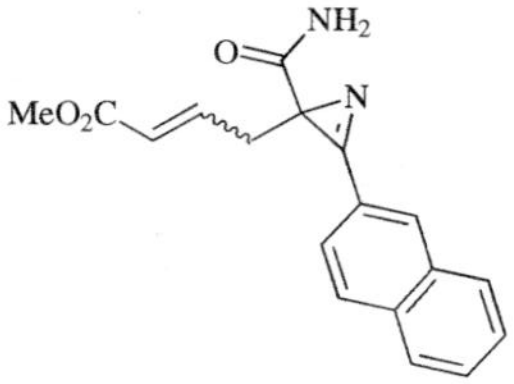 MeO_2C, O, NH_2, N	(70)	995
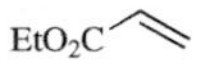 EtO_2C 3 eq	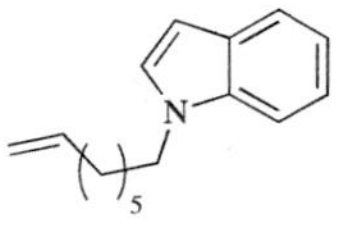 N, 5	Ru cat **360** (0.01 eq), toluene, 80°, 8 h	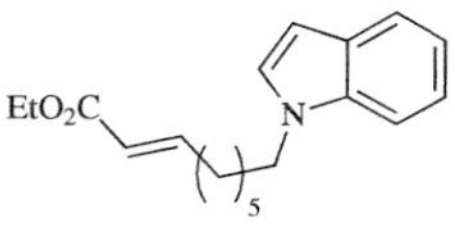 EtO_2C, N, 5	(89) (*E*)/(*Z*) = 90:10	491

TABLE 20. CROSS-METATHESIS OF α,β-UNSATURATED ESTERS AND THIOESTERS (*Continued*)

A. ESTERS (*Continued*)

Ester	Cross Partner	Conditions	Product(s) and Yield(s) (%)	Refs.
Please refer to the charts preceding the tables for the catalyst structures.				
C_3				
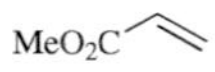 MeO_2C (3 eq)	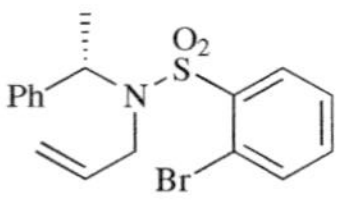 Ph, N, O_2S, Br	Gr II (0.05 eq), DCM, 40°, 3 h	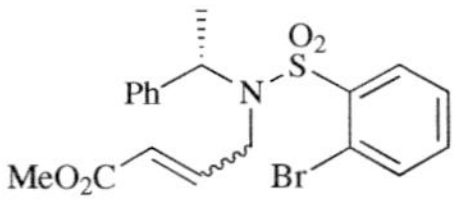 Ph, N, O_2S, MeO_2C, Br (60)	634
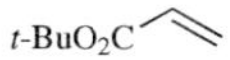 t-BuO_2C (15 eq)	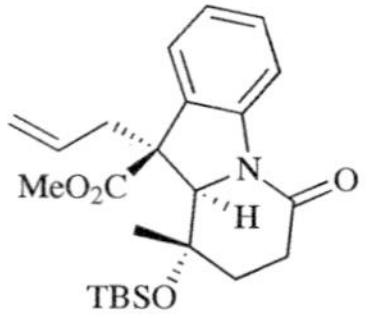 MeO_2C, N, O, H, TBSO	Gr II (0.15 eq), DCM, 40°, overnight	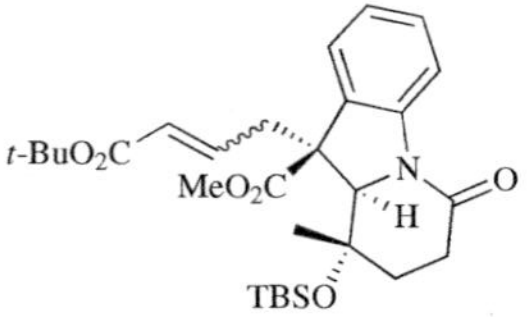 t-BuO_2C, MeO_2C, N, O, H, TBSO (89)	1116
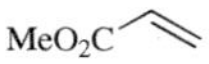 MeO_2C (10 eq)	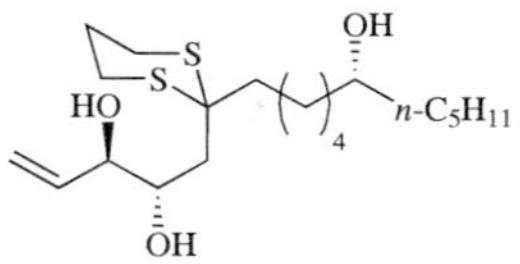 S, S, HO, OH, OH, 4, n-C_5H_{11}	Gr II (0.15 eq), DCM, reflux, 48 h	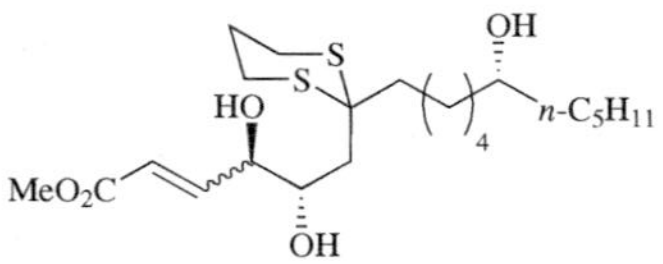 S, S, HO, OH, OH, 4, n-C_5H_{11}, MeO_2C (74)	1117

EtO_2C (10 eq) + n-C_7H_{15} ... NHAc (7) — Catalyst (x eq), solvent → EtO_2C ... NHAc (7) **I** + n-C_7H_{15} ... CO_2Et **II** 1118

Catalyst	x	Solvent	Temp (°)	Time (h)	**I + II**
Hov II	0.001	—	80	16	(12)
Hov II	0.005	—	80	16	(36)
Hov II	0.01	—	80	16	(58)
Ru cat **139**	0.005	—	80	16	(33)
Ru cat **139**	0.01	—	80	16	(48)
Hov II	0.01	DCM	40	16	(33)
Hov II	0.01	toluene	80	4	(88)
Ru cat **131**	0.01	toluene	80	4	(96)

MeO_2C (10 eq) + n-C_7H_{15} ... (n) NH CO_2Bn — Hov II (0.005 eq), 50°, 16 h → EtO_2C ... (n) NH CO_2Bn 1079

n	
7	(91)
13	(90)

TABLE 20. CROSS-METATHESIS OF α,β-UNSATURATED ESTERS AND THIOESTERS (*Continued*)

A. ESTERS (*Continued*)

Ester	Cross Partner	Conditions	Product(s) and Yield(s) (%)	Refs.
Please refer to the charts preceding the tables for the catalyst structures.				
C_3				
EtO_2C–CH=CH$_2$ 10 eq	n-C_7H_{15} (Z)-alkene $(CH_2)_7CO_2Me$	Catalyst (x eq), toluene, 100°	EtO_2C–CH=CH–$(CH_2)_7CO_2Me$ (I) + n-C_7H_{15}–CH=CH–CO_2Et (II)	1118

Catalyst	x	Time (h)	**I + II**
Hov II	0.001	4	(55)
Hov II	0.001	16	(—)
Ru cat **139**	0.001	4	(22)
Ru cat **139**	0.001	16	(—)
Ru cat **131**	0.001	16	(58)
Hov II	0.003	0.5	(90)
Hov II	0.003	4	(92)
Ru cat **139**	0.003	4	(88)
Ru cat **131**	0.003	4	(35)
Ru cat **131**	0.003	16	(>97)

MeO_2C (4 eq) + n-C_7H_{15}–CH=CH–$(CH_2)_7$–CO_2Me → Catalyst (x eq), toluene, 60°, 4 h → MeO_2C–CH=CH–$(CH_2)_7$–CO_2Me (**I**) + n-C_7H_{15}–CH=CH–CO_2Me (**II**) + n-C_7H_{15}–CH=CH–n-C_7H_{15} (**III**) + MeO_2C–$(CH_2)_7$–CH=CH–$(CH_2)_6$–CO_2Me (**IV**) 1119

Catalyst	x	(**I** + **II**)^d	Selectivity (**I** + **II**)^d	(**III** + **IV**)^d	TON	Productive TON
Ru cat **139**	0.26	(99)	96	(4)	380	365
Ru cat **139**	0.026	(96)	92	(8)	3690	3400
Ru cat **139**	0.0026	(69)	52	(48)	26,500	13,800
Ru cat **139**	0.0003	(33)	15	(85)	127,000	19,000
Ru cat **320**	0.0029	(74)	63	(37)	260	160
Ru cat **320**	0.029	(63)	40	(60)	21,170	870
Ru cat **320**	0.0029	(58)	30	(70)	20,000	6000
Ru cat **320**	0.0003	(11)	13	(87)	36,600	4770

TABLE 20. CROSS-METATHESIS OF α,β-UNSATURATED ESTERS AND THIOESTERS (*Continued*)

A. ESTERS (*Continued*)

Ester	Cross Partner	Conditions	Product(s) and Yield(s) (%)	Refs.

Please refer to the charts preceding the tables for the catalyst structures.

C_3

Ester	Cross Partner	Conditions	Product(s) and Yield(s) (%)	Refs.
MeO_2C–CH=CH$_2$ (4 eq)	$MeO_2C(CH_2)_7$CH=CH$(CH_2)_7CO_2Me$	Hov **II** (*x* eq), toluene, 100°	MeO_2CCH=CH$(CH_2)_7CO_2Me$	1089

x	Concn (mol/L)	Time (h)		(*E*)/(*Z*)
0.005	0.05	2	(93)	93:7
0.00025	0.05	6	(87)	90:10
0.00025	0.1	6	(90)	—
0.00025	0.5	6	(30)	90:10

Ester	Cross Partner	Conditions	Product(s) and Yield(s) (%)	Refs.
4 eq	$MeO_2C(CH_2)_7$CH=CHCH$_2$CH(OH)-*n*-C_6H_{13}	Hov **II** (*x* eq), toluene, 100°	MeO_2CCH=CH$(CH_2)_7CO_2Me$ **I** + MeO_2CCH=CHCH$_2$CH(OH)-*n*-C_6H_{13} **II**	1089

x	Time (h)	**I**	(*E*)/(*Z*) **I**	**II**	(*E*)/(*Z*) **II**
0.005	8	(89)	88:12	(85)	96:4
0.001	6	(45)	82:18	(78)	92:8

3 eq

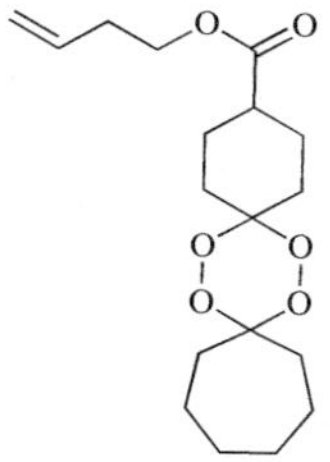

Ru cat **88** (0.03 eq), DCM, 45°, 2 h

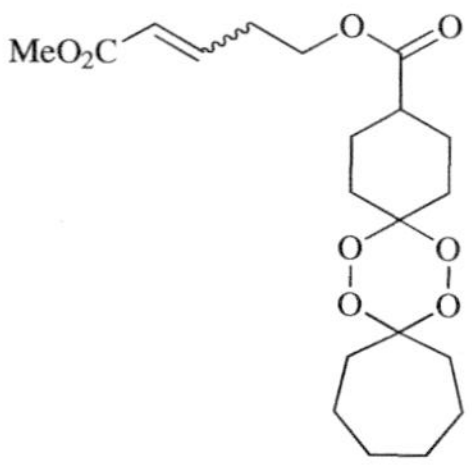

MeO_2C

(89) (*E*) only 193

EtO_2C

4 eq

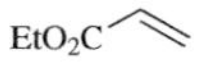

R[2] OMe R[1]

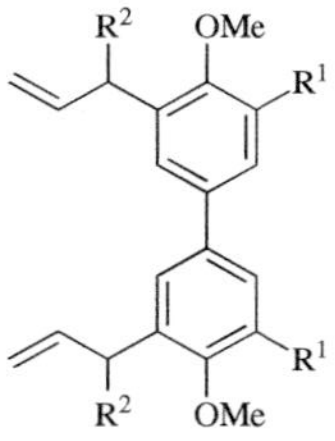

1. Gr II (0.028 eq), DCE, reflux, 10 h
2. 10 % Pd/C, H_2, AcOEt, rt, 10 h

EtO_2C R[2] OMe R[1]

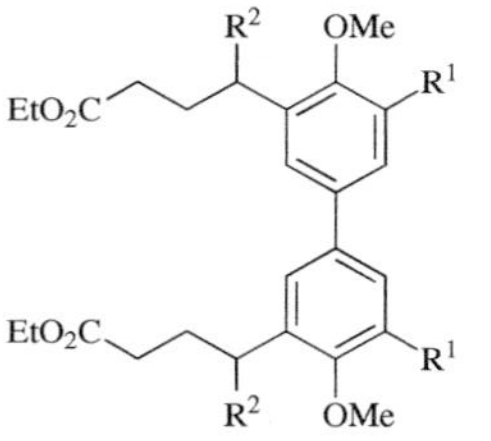

R^1	R^2		
H	H	(70)	1120
H	Me	(72)	
n-Pr	H	(75)	
n-Pr	Me	(66)	

4 eq

OMe *n*-Pr O*n*-Pr

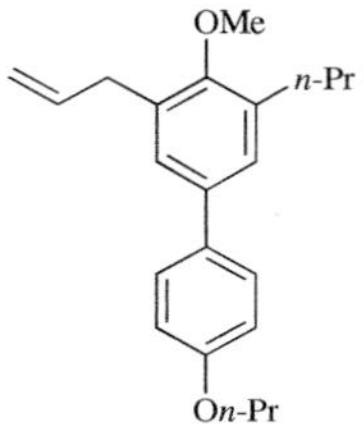

1. Gr II (0.028 eq), DCE, reflux, 10 h
2. 10 % Pd/C, H_2, AcOEt, rt, 10 h

EtO_2C OMe *n*-Pr O*n*-Pr

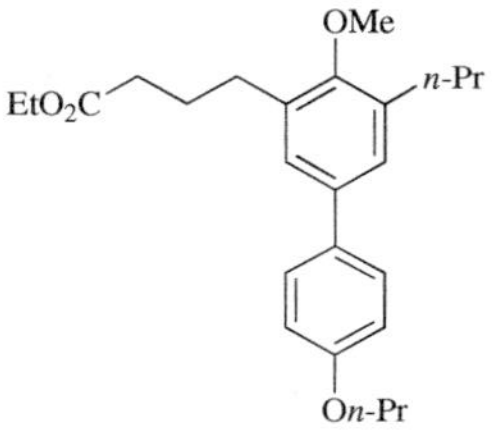

(79) 1120

TABLE 20. CROSS-METATHESIS OF α,β-UNSATURATED ESTERS AND THIOESTERS (*Continued*)

A. ESTERS (*Continued*)

Ester	Cross Partner	Conditions	Product(s) and Yield(s) (%)	Refs.
Please refer to the charts preceding the tables for the catalyst structures.				
C_3				
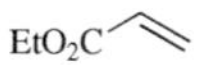 EtO_2C 3 eq	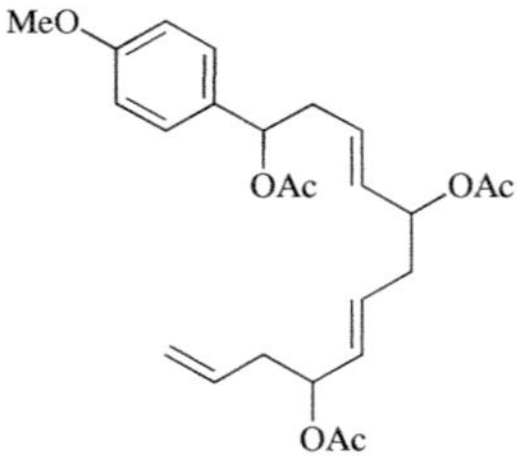 MeO, OAc, OAc, OAc	1. Hov II (0.05 eq), DCM, rt, 16 h 2. TBAF, THF, rt, 24 h	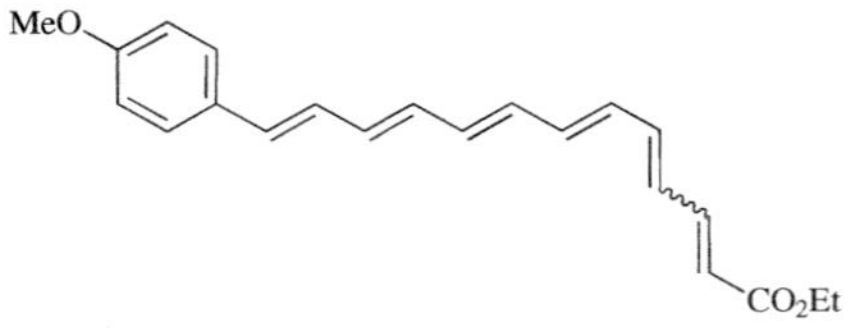 MeO, CO_2Et (55)	919
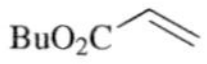 BuO_2C 16 eq	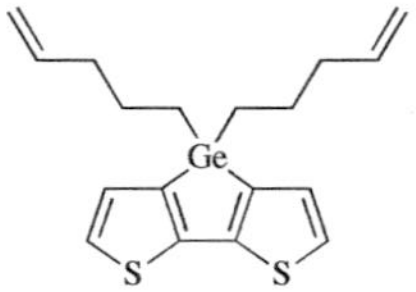 Ge, S, S	1. Gr II (0.06 eq), DCM, reflux, 24 h 2. Wilkinson's catalyst (0.1 eq), H_2 (200 psi), toluene, 50 °, 2 d	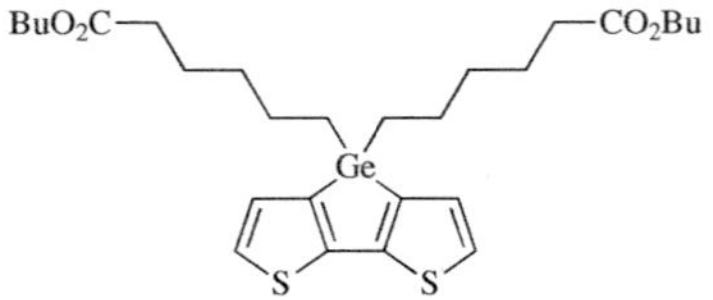 BuO_2C, CO_2Bu, Ge, S, S (81)	472
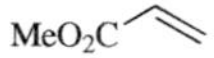 MeO_2C 15 eq	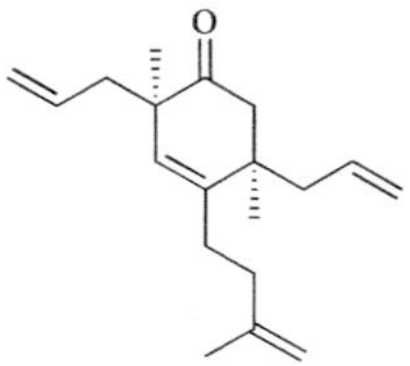 O	Ru cat **271** (0.05 eq), DCM, 30°, 4 h	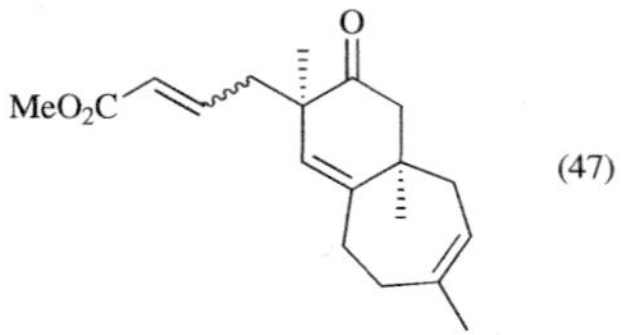 MeO_2C, O (47)	695

2 eq

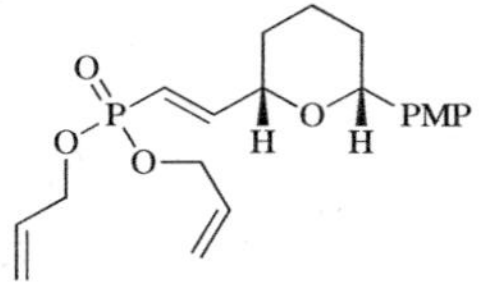

O P O O PMP H H

Gr II (0.1 eq), CuI (0.1 eq), DCM, reflux, 2 h

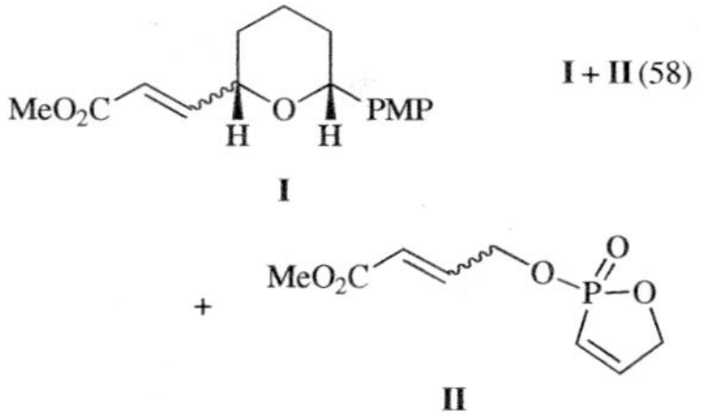

MeO_2C O PMP H H

I

\+ MeO_2C O P O O

II

I + **II** (58)

220

13 eq

Ph Ph OH

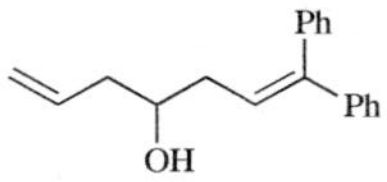

Gr II (0.05 eq), neat, 19°, 50 min

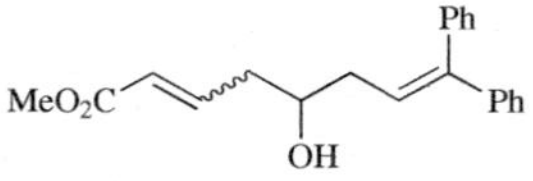

MeO_2C Ph Ph OH

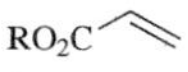

(84)

1093

RO_2C

x eq

Bn O NHCbz O $(\;)_8$

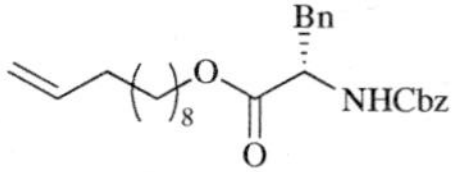

Catalyst (0.02 eq), solvent, 22°, 12 h

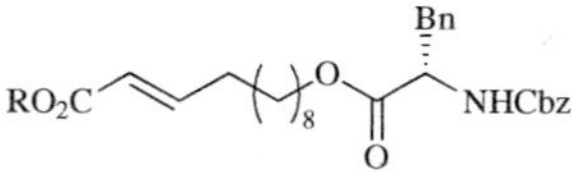

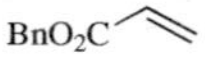

RO_2C Bn O NHCbz O $(\;)_8$

(*E*) only

R	*x*	Catalyst	Solvent		
t-Bu	2	Ru cat **109**	H_2O	(97)	341
t-Bu	1	Gr II	2.5 % aq. PTS	(96)	976
Ad	1	Gr II	2.5 % aq. PTS	(94)	976

BnO_2C

1.3 eq

R O n-C_5H_{11} $(\;)_8$

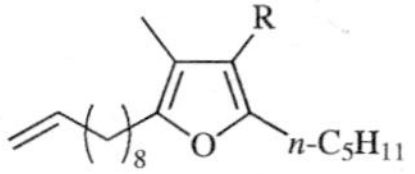

Hov II (0.02 eq), DCM, reflux, 2 h

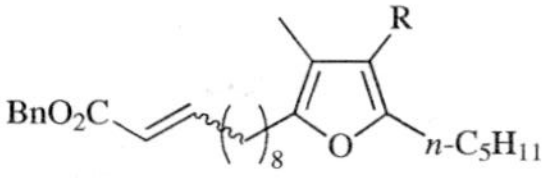

BnO_2C R O n-C_5H_{11} $(\;)_8$

R		
H	(89)	1121
Me	(89)	
CD_3	(87)	

TABLE 20. CROSS-METATHESIS OF α,β-UNSATURATED ESTERS AND THIOESTERS (*Continued*)

A. ESTERS (*Continued*)

Ester	Cross Partner	Conditions	Product(s) and Yield(s) (%)	Refs.
Please refer to the charts preceding the tables for the catalyst structures.				
C_3				
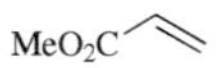 MeO_2C (10 eq)	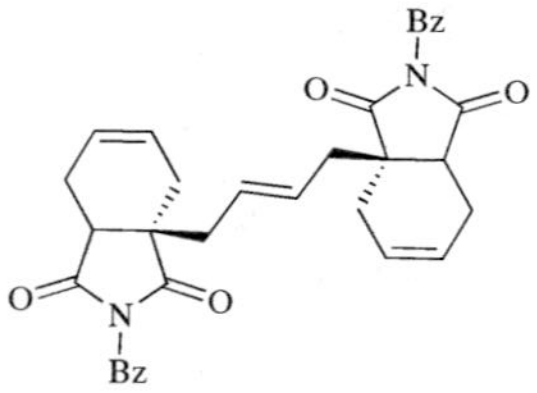 Bz, N, O, O, N, Bz, O, O	Hov II (0.05 eq), DCM, 45°, 72 h	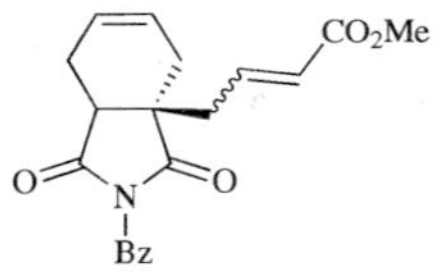 CO_2Me, O, N, Bz, O (79)	1098
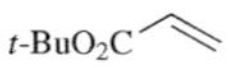 t-BuO_2C (3 eq)	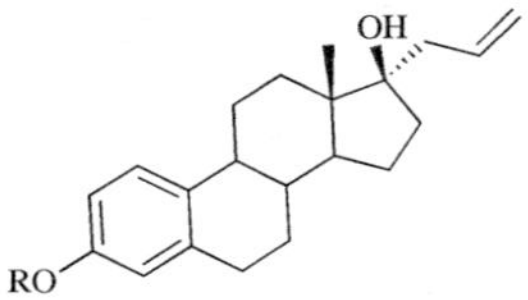 OH, RO	Gr II (0.01 eq), solvent, 6 h	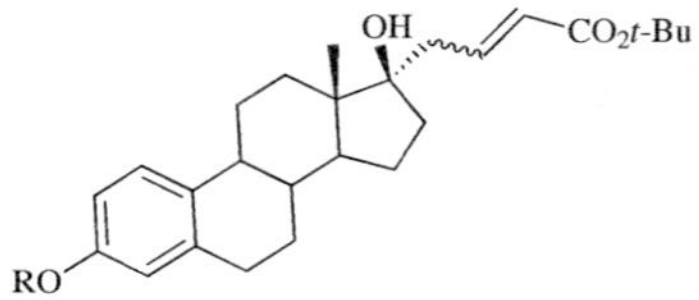 OH, CO_2t-Bu, RO	140

R	Solvent	Temp (°)	
H	DCM	40	(32)
n-$C_8F_{17}(CH_2)_2(i$-$Pr)_2Si$	$C_6F_5CF_3$	70	(93)

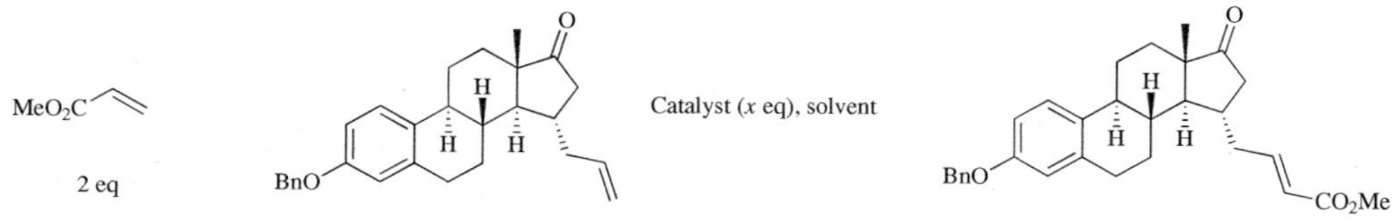

MeO_2C (2 eq) + BnO-steroid (allyl) → Catalyst (x eq), solvent → BnO-steroid (CO_2Me) (*E*) only 163

Catalyst	x	Solvent	Temp (°)	Time (h)	
Gr II	0.05	DCM	40	2.5	(97)
Ru cat **68**	0.05	DCE	80	5.5	(95)
Ru cat **125**	0.05	DCM	40	5.5	(95)
Hov II	0.05	DCM	40	4	(90)
Ru cat **67**	0.05	DCM	40	5.5	(34)
Gr II	0.02	DCM	40	2.5	(97)
Gr II	0.01	DCM	40	2.5	(9)
Gr II	0.005	DCM	40	2.5	(<2)
Gr II	0.0008	DCM	40	2.5	(0)
Ru cat **125**	0.02	DCM	40	8	(>99)
Ru cat **125**	0.01	DCM	40	8	(>99)
Ru cat **125**	0.005	DCM	40	8	(>99)
Ru cat **125**	0.0034	DCM	40	8	(95)
Ru cat **125**	0.0025	DCM	40	8	(72)
Ru cat **125**	0.0017	DCM	40	8	(11)
Ru cat **125**	0.0008	DCM	40	8	(3)

TABLE 20. CROSS-METATHESIS OF α,β-UNSATURATED ESTERS AND THIOESTERS (*Continued*)

A. ESTERS (*Continued*)

Ester	Cross Partner	Conditions	Product(s) and Yield(s) (%)	Refs.
Please refer to the charts preceding the tables for the catalyst structures.				
C_3				
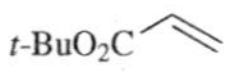 t-BuO_2C (4 eq)	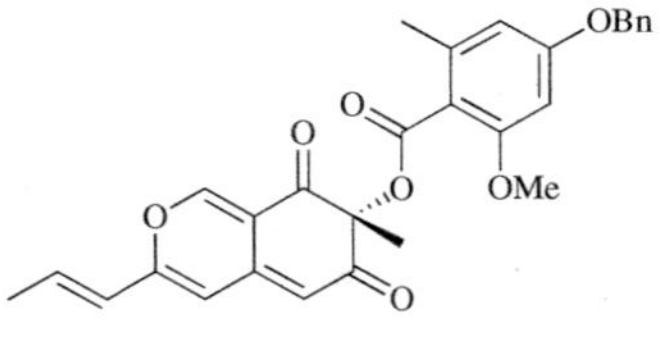 OBn, OMe, O	Hov II (2 x 0.1 eq), DCM, reflux, 2 x 12 h	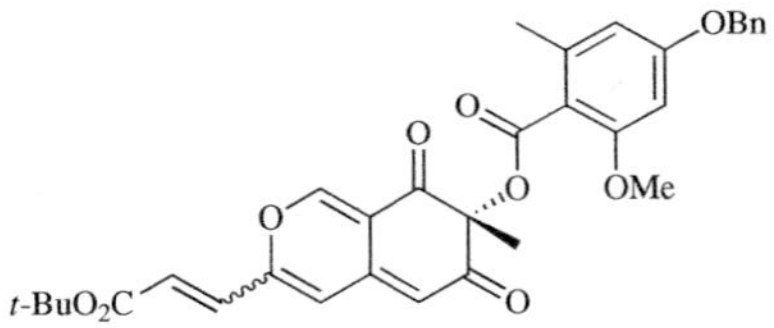 t-BuO_2C, OBn, OMe, O (90)	937
3 eq	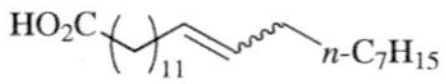 HO_2C ( )$_{11}$ n-C_7H_{15}	Catalyst (x eq), DCM, 40°	 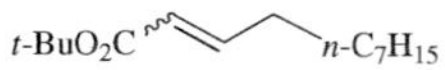 HO_2C ( )$_{11}$ CO_2t-Bu **I** + t-BuO_2C n-C_7H_{15} **II**	1040

Catalyst	x	Time (h)	**I + II**
Ru cat **85**	0.025	22	(>99)
Hov II	0.025	22	(>99)
Ru cat **85**	0.005	17	(75)
Hov II	0.005	17	(81)

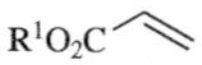

R^1O_2C

10 eq

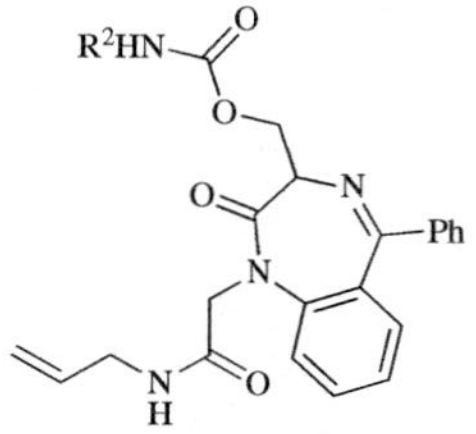

R^2HN O O O N N Ph N H O

Hov II (0.2 eq),
DCM, 100° (MW), 1 h

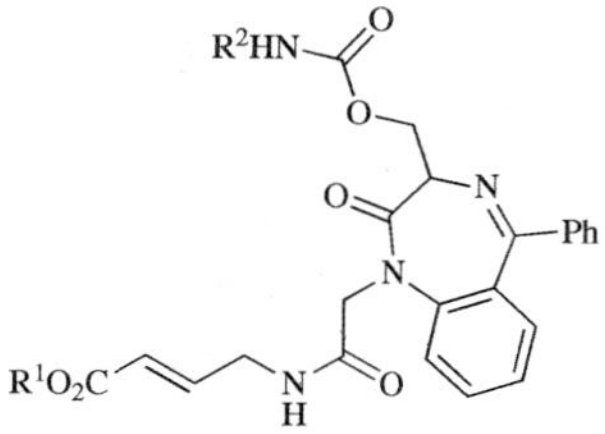

R^2HN O O O N N Ph R^1O_2C N H O

R^1	R^2	
Me	4-Cl-2-$CF_3C_6H_3$	(99)
Et	4-Cl-2-$CF_3C_6H_3$	(99)
i-Bu	4-Cl-2-$CF_3C_6H_3$	(99)
Bn	4-Cl-2-$CF_3C_6H_3$	(99)
Et	4-$MeOC_6H_4$	(77)
i-Bu	4-$MeOC_6H_4$	(67)
Et	1-Naph	(74)
i-Bu	1-Naph	(73)
Et	*n*-Bu	(69)
i-Bu	*n*-Bu	(63)
Et	*n*-C_6H_{13}	(70)
i-Bu	*n*-C_6H_{13}	(62)

(*E*) only

987, 1122

TABLE 20. CROSS-METATHESIS OF α,β-UNSATURATED ESTERS AND THIOESTERS (*Continued*)

A. ESTERS (*Continued*)

Ester	Cross Partner	Conditions	Product(s) and Yield(s) (%)	Refs.

Please refer to the charts preceding the tables for the catalyst structures.

C_3

Ester	Cross Partner	Conditions	Product(s) and Yield(s) (%)	Refs.
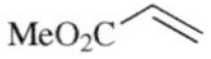 MeO_2C (5 eq)	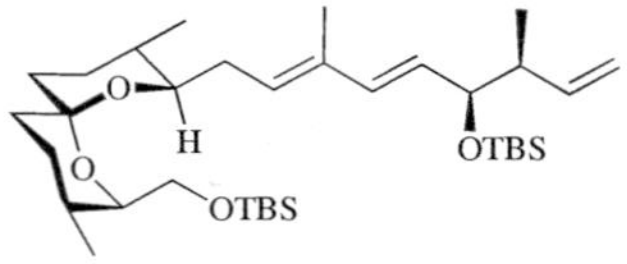 OTBS, H, O, O, OTBS	Gr II (0.1 eq), neat, 90°, 5 h	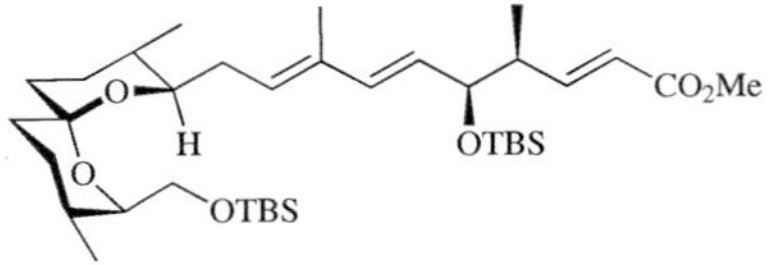 CO_2Me, OTBS, H, O, O, OTBS (65) (*E*) only	1123
(5 eq)	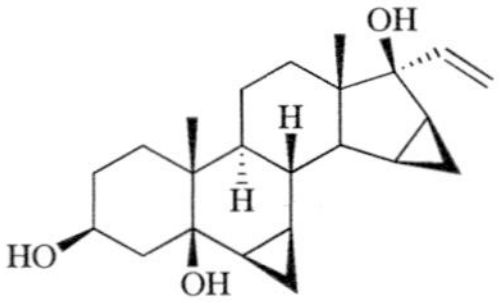 OH, H, H, HO, OH	Catalyst (*x* eq), DCM, reflux	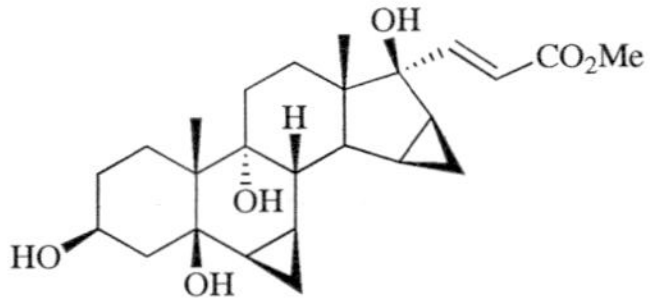 OH, CO_2Me, H, OH, HO, OH	1124

Catalyst	*x*	Time (h)		(*E*)/(*Z*)
Gr I	0.12	24	(0)	—
Gr II	0.12	24	(55)	>95:5
Hov II	0.12	3	(82)	>95:5
Hov II	0.06	3	(80)	>95:5

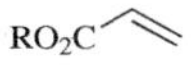

RO_2C (x eq) + steroid ester; Catalyst (y eq), solvent → CO_2R product

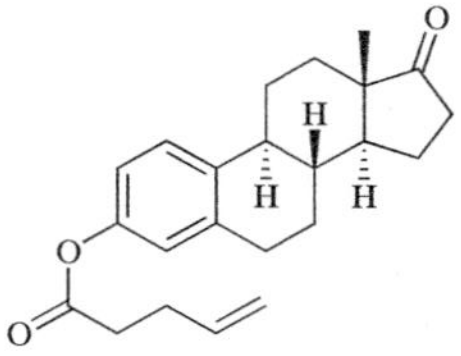

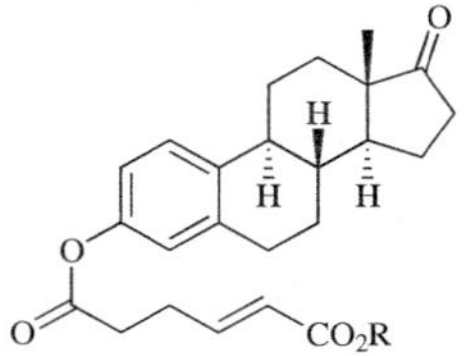

R	x	Catalyst	y	Solvent	Temp (°)	Time (h)		(E)/(Z)	
Me	4	Ru cat **245**	0.25	toluene	80	1	(95)	95:5	1056
Me	—	Ru cat **269**	0.01	DCM	40	1	(88)	—	329
Me	—	Ru cat **270**	0.01	DCM	40	1	(84)	—	329
t-Bu	4	Ru cat **143**	0.05	H_2O	30	12	(74)	(E) only	588
t-Bu	3	Ru cat **139**	0.01	DCM	40	3	(95)	—	1044
t-Bu	3	Ru cat **137**	0.02	DCM	40	3	(52)	—	1044
t-Bu	2	Ru cat **107**	0.01	toluene	80	24	(91)	90:10	596

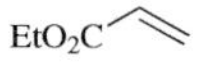

EtO_2C (3 eq) 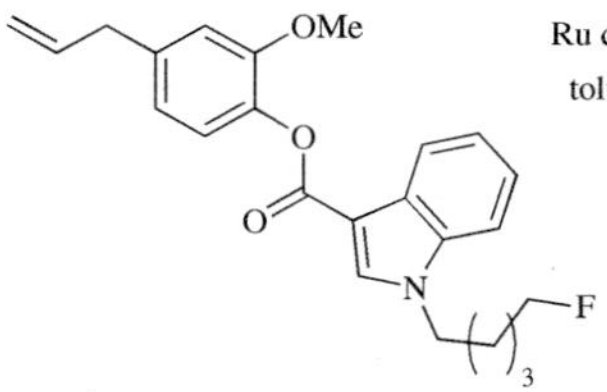 — Ru cat **360** (0.01 eq), toluene, 80°, 8 h 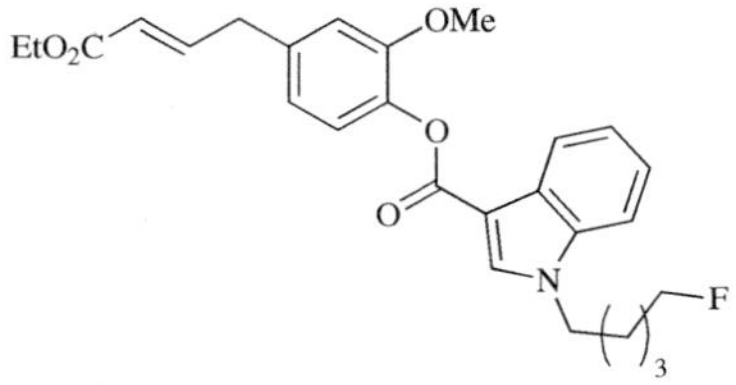 — (91) (E)/(Z) = 67:33 — 491

3 eq — Ru cat **360** (0.01 eq), toluene, 80°, 8 h — EtO_2C — (83) (E)/(Z) = 91:9 — 491

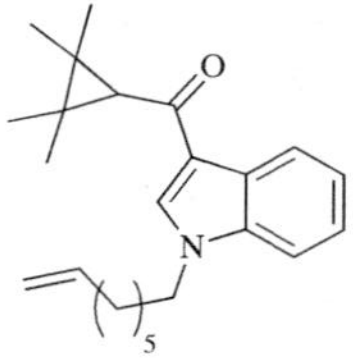

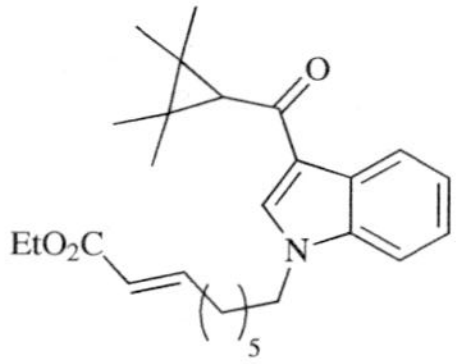

TABLE 20. CROSS-METATHESIS OF α,β-UNSATURATED ESTERS AND THIOESTERS (*Continued*)

A. ESTERS (*Continued*)

Ester	Cross Partner	Conditions	Product(s) and Yield(s) (%)	Refs.
Please refer to the charts preceding the tables for the catalyst structures.				
C_3				
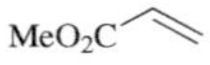 MeO_2C (vinyl) 10 eq	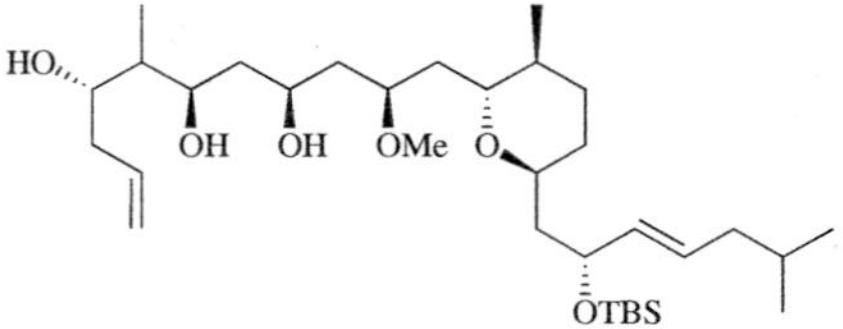 HO, OH, OH, OMe, O, OTBS	Hov II (0.15 eq), DCM, rt, 18 h	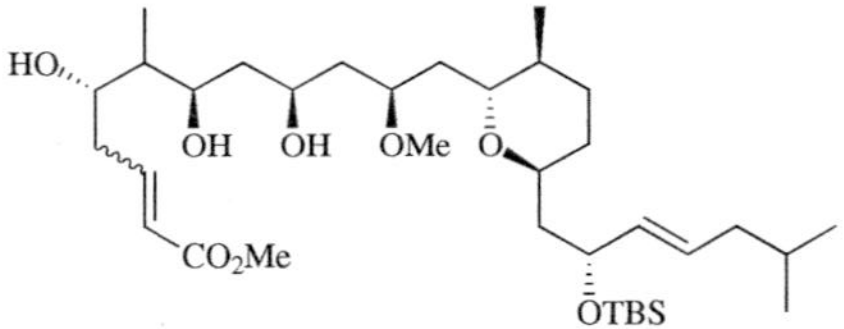 HO, OH, OH, OMe, O, CO_2Me, OTBS (84)	1115
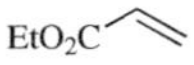 EtO_2C (vinyl) 3 eq	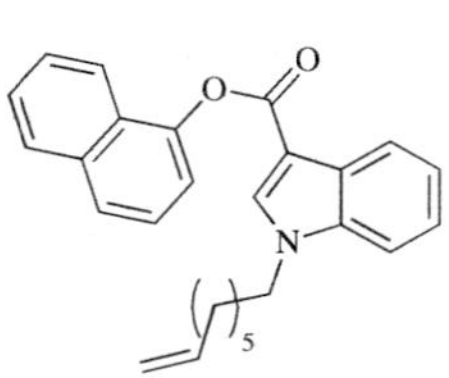 O, O, N, 5	Ru cat **360** (0.01 eq), toluene, 80°, 8 h	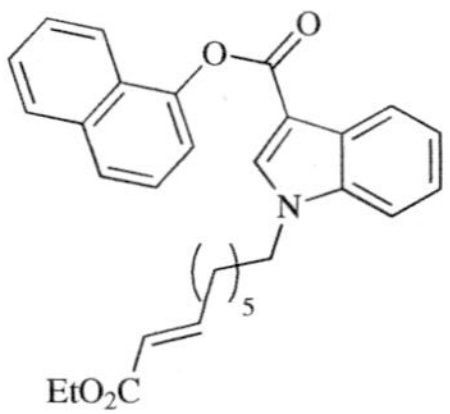 O, O, N, 5, EtO_2C (87) (*E*)/(*Z*) = 91:9	491
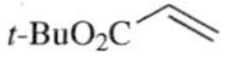 *t*-BuO_2C (vinyl) 40 eq	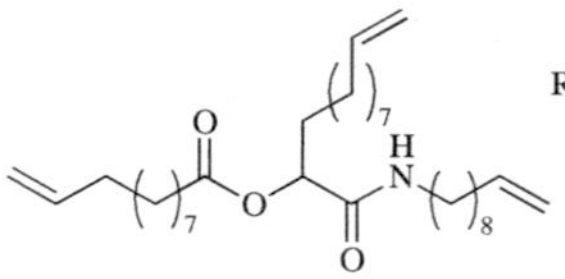 O, O, 7, 7, H, N, O, 8	Ru cat **89** (0.02 eq), BQ (0.06 eq), DCM, 40°, 1 h	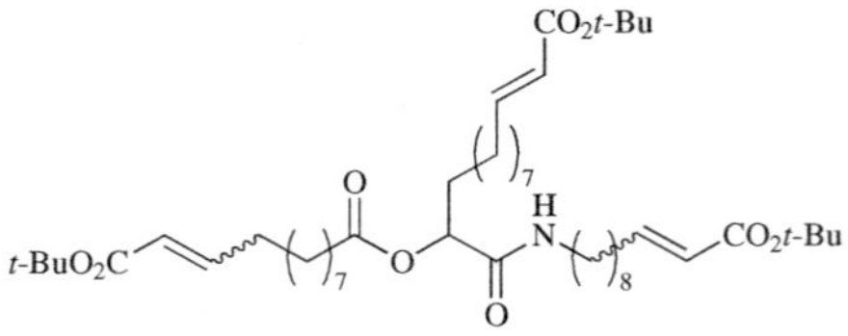 *t*-BuO_2C, O, O, 7, 7, CO_2t-Bu, H, N, O, 8, CO_2t-Bu (88)	1125

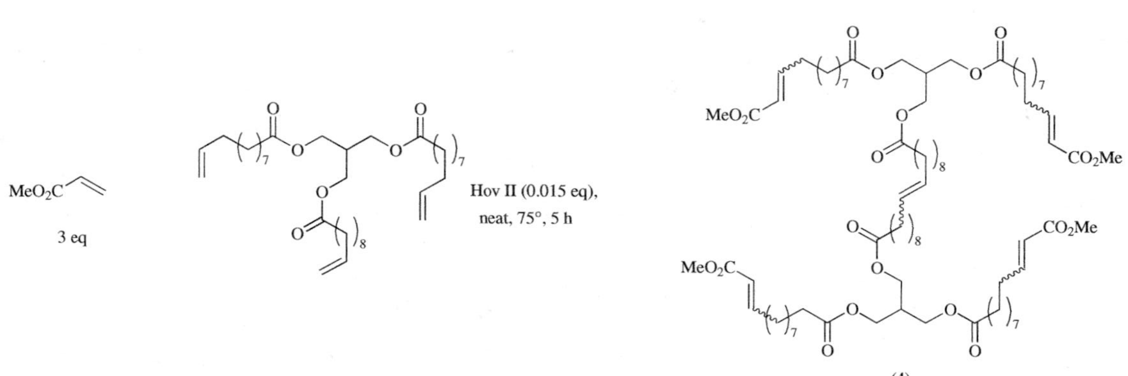

MeO2C
3 eq
Hov II (0.015 eq),
neat, 75°, 5 h
MeO2C
CO2Me
CO2Me
MeO2C

(4) 515

10 eq 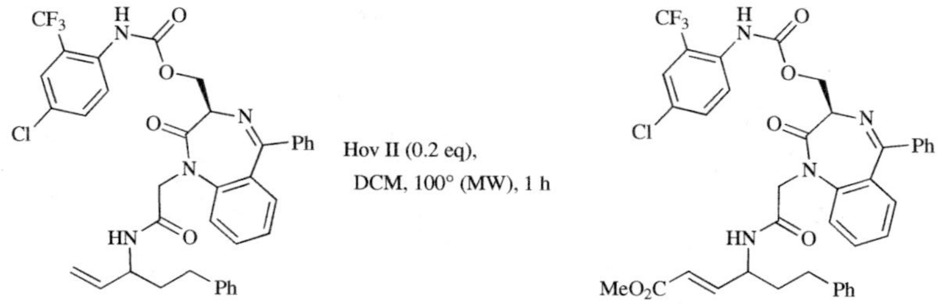

CF3
Cl
Ph
Hov II (0.2 eq),
DCM, 100° (MW), 1 h
CF3
Cl
Ph
MeO2C

(99) (*E*) only 987

TABLE 20. CROSS-METATHESIS OF α,β-UNSATURATED ESTERS AND THIOESTERS (*Continued*)

A. ESTERS (*Continued*)

Ester	Cross Partner	Conditions	Product(s) and Yield(s) (%)	Refs.
Please refer to the charts preceding the tables for the catalyst structures.				
C_3				
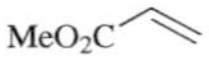 MeO_2C (20 eq)	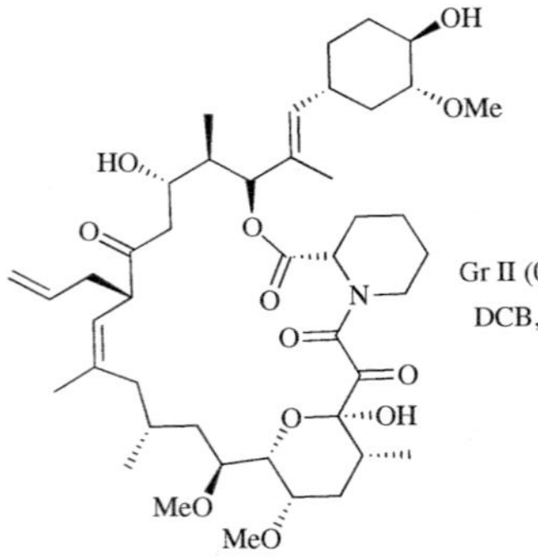 HO, OH, OMe, O, MeO, MeO, N, OH	Gr II (0.1 eq), DCB, MW	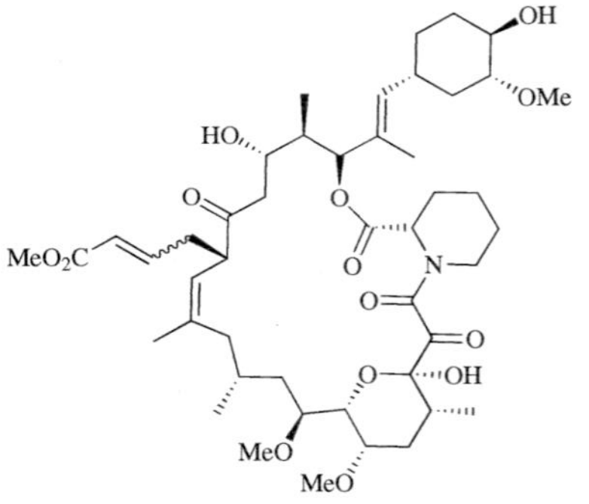 MeO_2C, HO, OH, OMe, O, MeO, MeO, N, OH (97)	432
4.6 eq	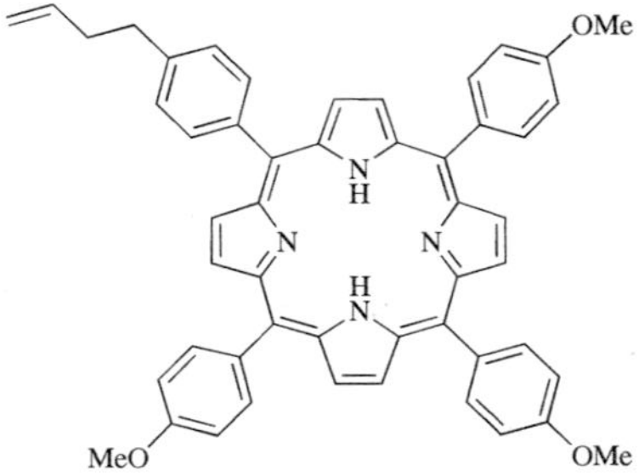 OMe, MeO, OMe, N, NH	Gr II (0.1 eq), DCM, rt, 24 h	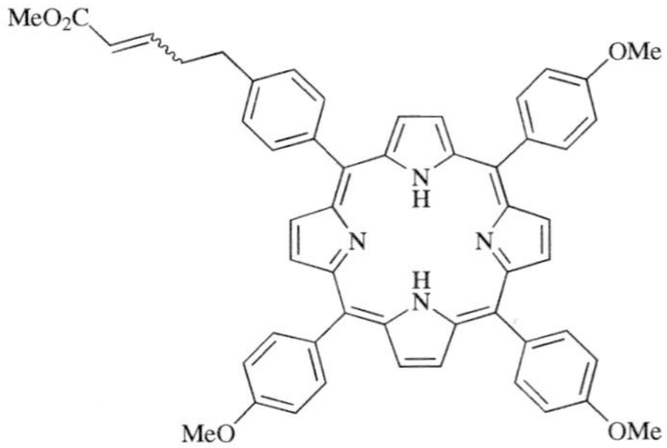 MeO_2C, OMe, MeO, OMe, N, NH (54)	1032

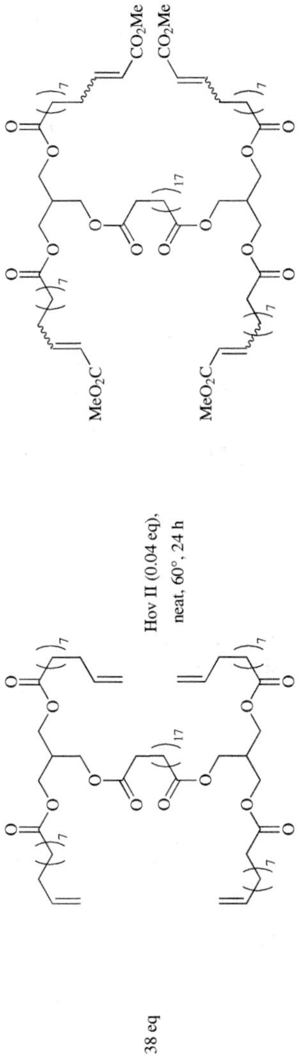

38 eq

Hov II (0.04 eq), neat, 60°, 24 h

MeO_2C CO_2Me CO_2Me MeO_2C

515

(51)

TABLE 20. CROSS-METATHESIS OF α,β-UNSATURATED ESTERS AND THIOESTERS (*Continued*)

A. ESTERS (*Continued*)

Ester	Cross Partner	Conditions	Product(s) and Yield(s) (%)	Refs.

Please refer to the charts preceding the tables for the catalyst structures.

C_3

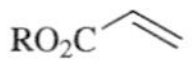

Ester	Cross Partner	Conditions	Product(s) and Yield(s) (%)	Refs.
RO_2C (40 eq)	(polysaccharide structure, m, n)	Hov II (x eq), solvent	(polysaccharide product with CO_2R, m, n)	1033

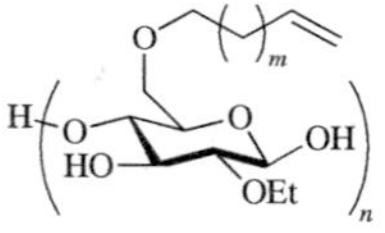

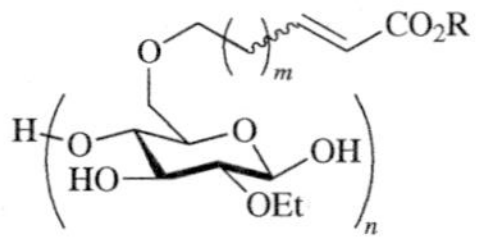

R	m	x	Solvent	Temp (°)	Time (h)	Conv. (%)[a]
Me	0	0.05	THF	37	2	47
Me	0	0.05	DCM	37	2	49
Me	0	0.1	DCM	37	2	70
Me	2	0.05	THF	37	2	35
Me	2	0.05	DMI	70	2	38
Me	2	0.05	DCM	37	2	1
Me	2	0.1	THF	37	2	51
Me	2	0.1	DCM	37	2	100
Me	4	0.1	DCM	37	2	>90
Me	8	0.05	DCM	37	2	33
Me	8	0.05	THF	37	2	29
Me	8	0.05	DMI	70	2	46
Me	8	0.05	DCM	37	12	95
Me	8	0.1	DCM	37	12	100
$HO(CH_2)_2$	2	0.1	DCM	37	2	100

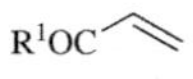

R^1OC (vinyl), 20 eq + sugar polyester (terminal alkene, m, n)

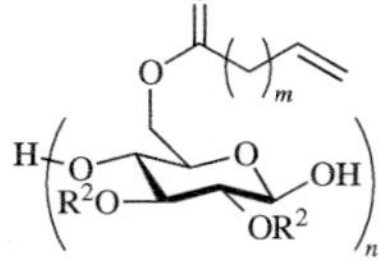

Gr II (0.05 eq), THF, rt, 1 h

Product: $\ldots COR^1$ — 1035

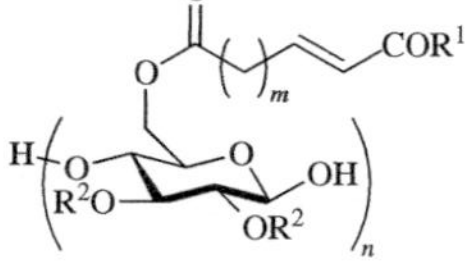

R^1	R^2	m		$(E)/(Z)$
MeO	Me	8	(94)	94:6
$MeO(CH_2CH_2O)_x$	H	2	(88)	89:11
$MeO(CH_2CH_2O)_x$	Me	8	(84)	97:3

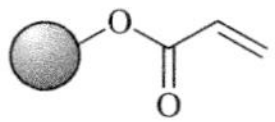

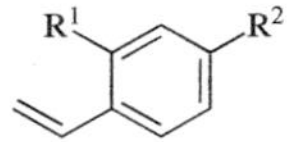

Resin-bound acrylate + styrene (R^1, R^2), 5 eq

1. Catalyst (0.05 eq), DCM, reflux, 20 h
2. TFA (10%), DCM, 1 h
3. CH_2N_2, DCM, 0°, 0.5 h

Product: MeO_2C (cinnamate, R^1, R^2)

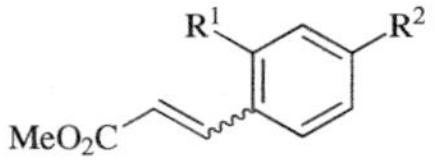

R^1	R^2	Catalyst		
H	$ClCH_2$	Hov II	(23)	418
H	$ClCH_2$	Gr II	(11)	418
Br	H	Hov II	(16)	418, 677
Br	H	Gr II	(10)	418, 677

Alkenyl–Ph (n), 5 eq

1. Catalyst (0.05 eq), DCM, reflux, 20 h
2. TFA (10%), DCM, 1 h
3. CH_2N_2, DCM, 0°, 0.5 h

Product: MeO_2C…Ph (n)

n	Catalyst		
0	Hov II	(59)	418
1	Hov II	(68)	418
0	Gr II	(35)	418, 677
1	Gr II	(26)	418, 677

TABLE 20. CROSS-METATHESIS OF α,β-UNSATURATED ESTERS AND THIOESTERS (*Continued*)

A. ESTERS (*Continued*)

Ester	Cross Partner	Conditions	Product(s) and Yield(s) (%)	Refs.
Please refer to the charts preceding the tables for the catalyst structures.				
C_3				
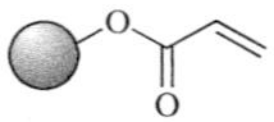	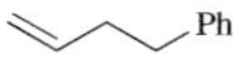 Ph; 5 eq	1. Gr II (0.05 eq), DCM, 50°, 20 h 2. Gr II (0.05 eq), Et_3SiH (5.0 eq), DCM, 150° (MW), 0.5 h 3. TFA (10%), DCM, 1 h 4. CH_2N_2, DCM, 0°, 0.5 h	 MeO_2C Ph (62)	1126
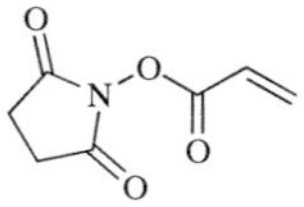 N O O O O; 20 eq	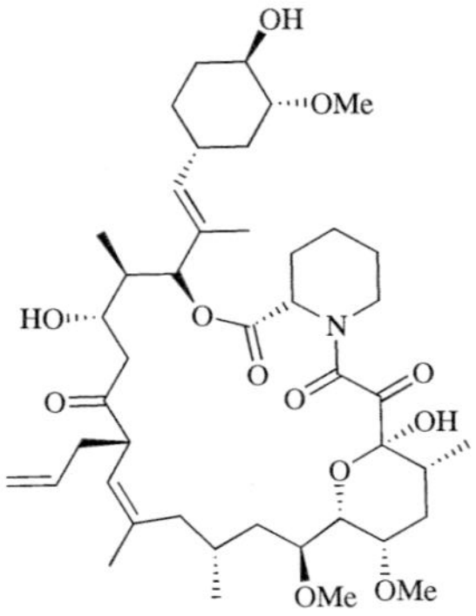 OH OMe HO O O N O O OH O OMe OMe	Gr II (0.1 eq), DCB, MW (20 W)	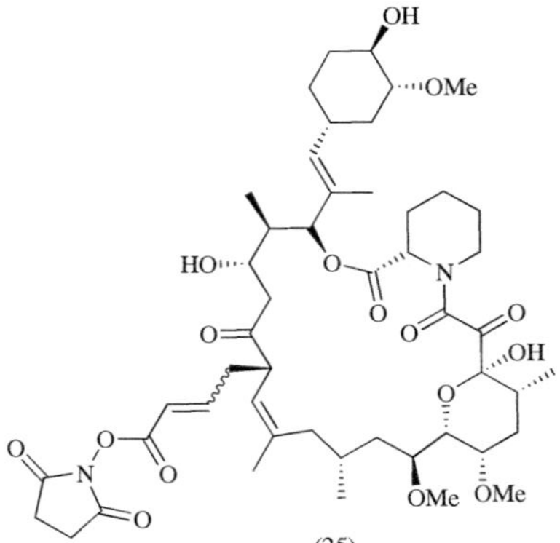 OH OMe HO O O N O O OH O O N O O O OMe OMe (25)	432
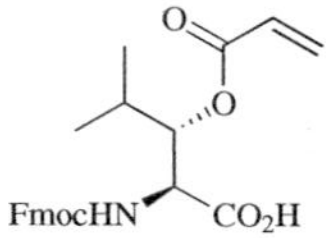 O O FmocHN CO_2H	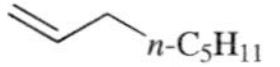 n-C_5H_{11}; 2.3 eq	Gr II (0.1 eq), DCM, reflux, 48 h	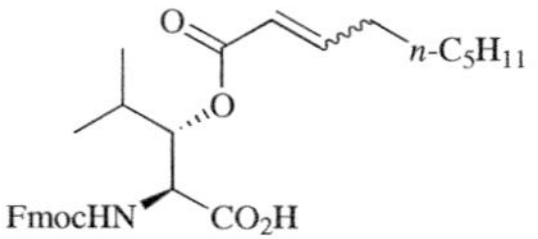 O O n-C_5H_{11} FmocHN CO_2H (63)	372

Reactant(s)	Conditions	Yield (%)	Refs.
2.05 eq	Gr II (0.1 eq), DCE, 85°, 24 h	(80)	1127
RO_2C; PF_6^-, 1.7 eq	Gr II (0.05 eq), DB24C8 (1.7 eq) DCM, reflux, 15 h	PF_6^-, CO_2R	1128, 1129

R	
3,5-dimethylphenyl	(72)
ferrocenyl (Fe)	(35)

TABLE 20. CROSS-METATHESIS OF α,β-UNSATURATED ESTERS AND THIOESTERS (*Continued*)

A. ESTERS (*Continued*)

Ester	Cross Partner	Conditions	Product(s) and Yield(s) (%)	Refs.
Please refer to the charts preceding the tables for the catalyst structures.				
C_3				
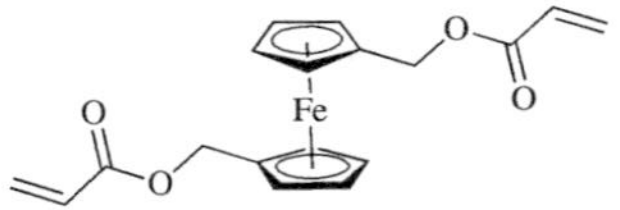 Fe	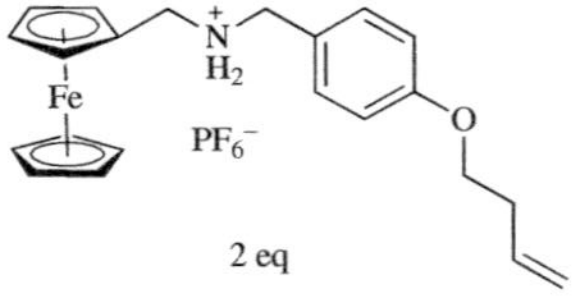 Fe, $\overset{+}{N}H_2$, PF_6^-; 2 eq	Gr II (0.1 eq), DB24C8 (2.7 eq) DCM, reflux, 15 h	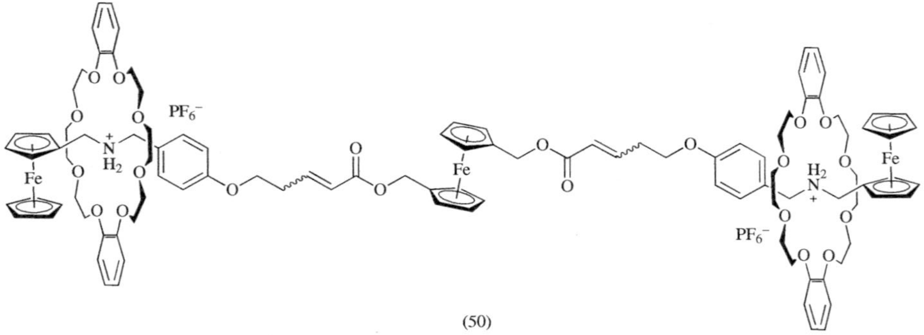 Fe, $\overset{+}{N}H_2$, PF_6^- (50)	1129

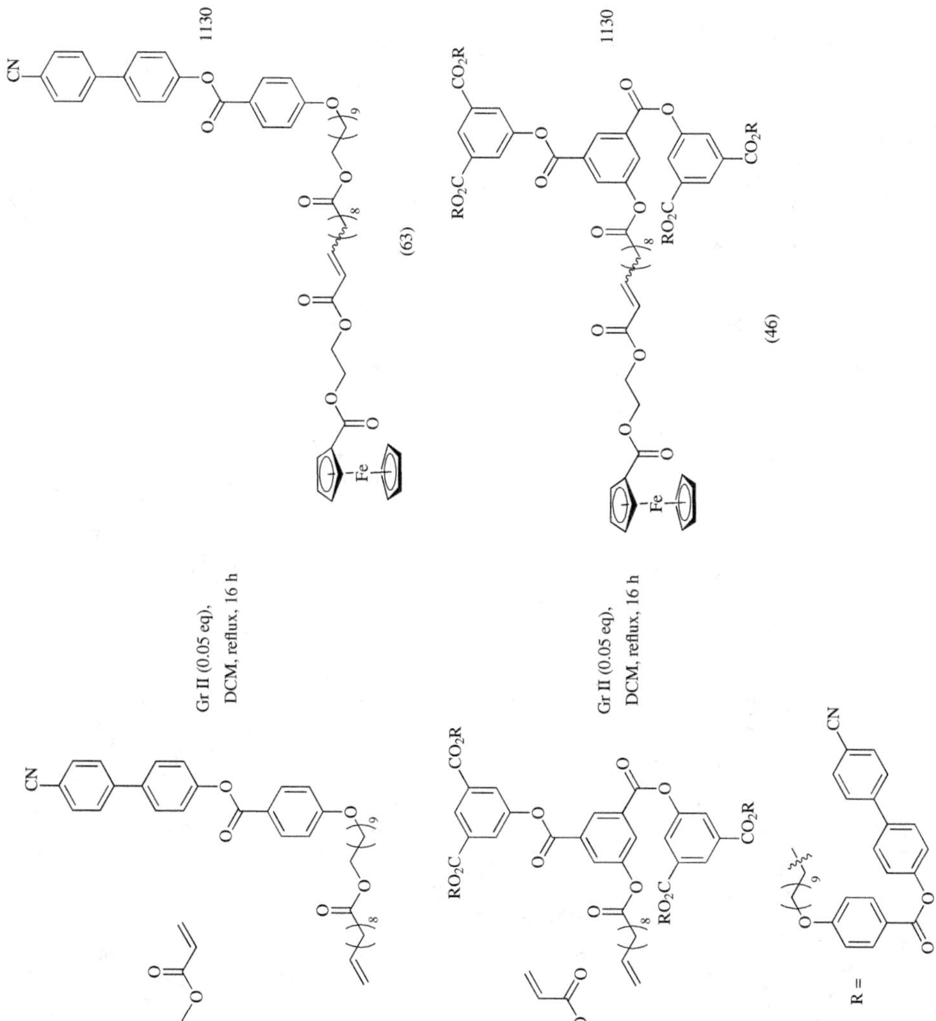

1.2 eq

Gr II (0.05 eq),
DCM, reflux, 16 h

(63) 1130

1.2 eq

Gr II (0.05 eq),
DCM, reflux, 16 h

(46) 1130

R =

TABLE 20. CROSS-METATHESIS OF α,β-UNSATURATED ESTERS AND THIOESTERS (*Continued*)

A. ESTERS (*Continued*)

Ester	Cross Partner	Conditions	Product(s) and Yield(s) (%)	Refs.
Please refer to the charts preceding the tables for the catalyst structures.				
C_3				
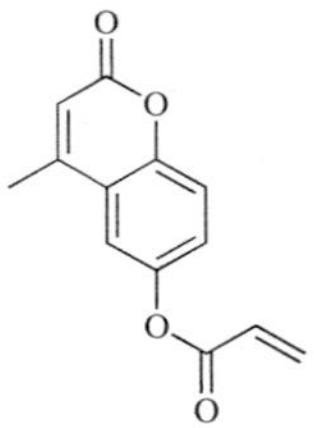	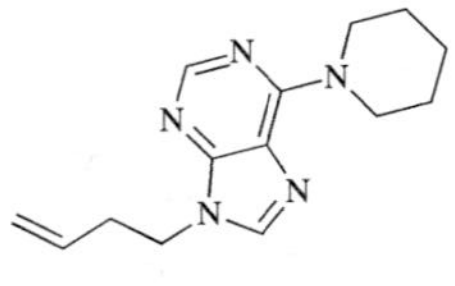 3 eq	Gr II (0.1 + 0.04 eq), DCM, 100° (MW), 2 + 1 h	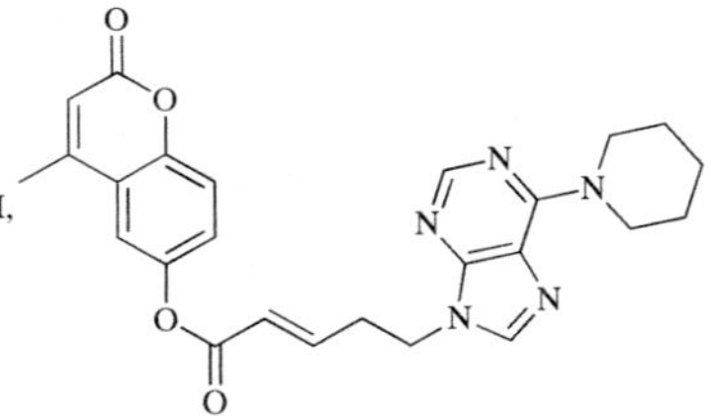 (50) (*E*) only	499
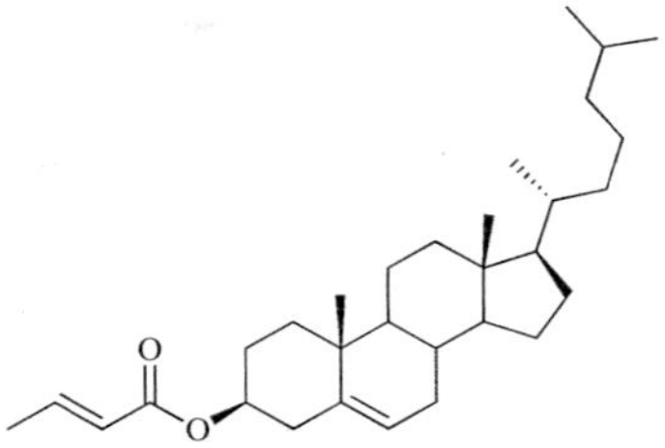 4 eq	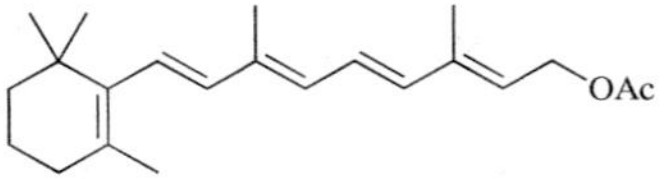 OAc	Hov II (0.15 eq), toluene, rt, 96 h	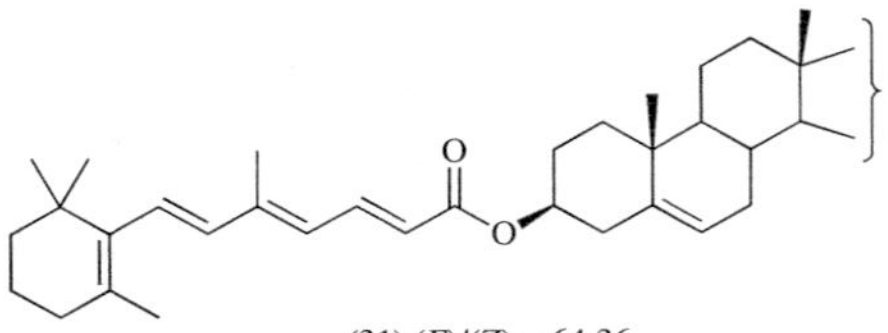 (31) (*E*)/(*Z*) = 64:36	539

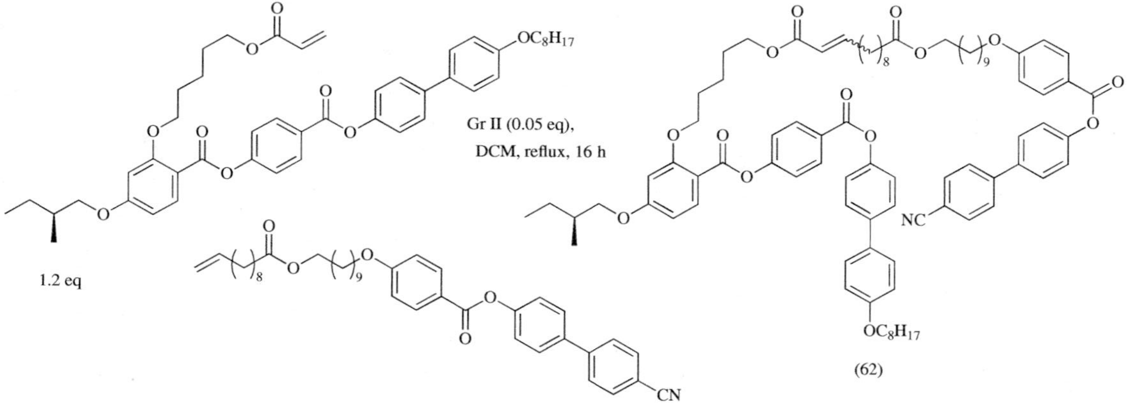

1.2 eq

Gr II (0.05 eq),
DCM, reflux, 16 h

(62) 1130

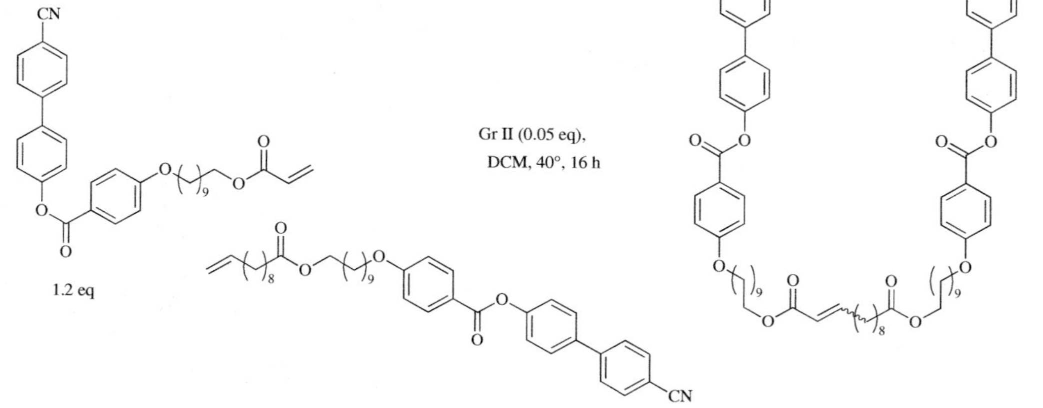

1.2 eq

Gr II (0.05 eq),
DCM, 40°, 16 h

(67) 1130

TABLE 20. CROSS-METATHESIS OF α,β-UNSATURATED ESTERS AND THIOESTERS (*Continued*)

A. ESTERS (*Continued*)

Ester	Cross Partner	Conditions	Product(s) and Yield(s) (%)	Refs.
Please refer to the charts preceding the tables for the catalyst structures.				
C_3				
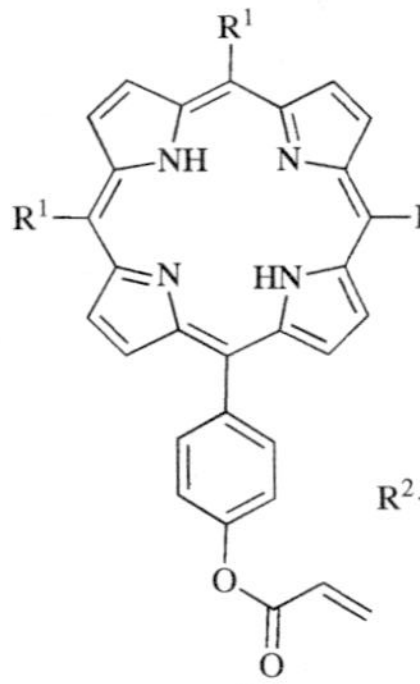 4 eq R^1 = 4-MeC_6H_4	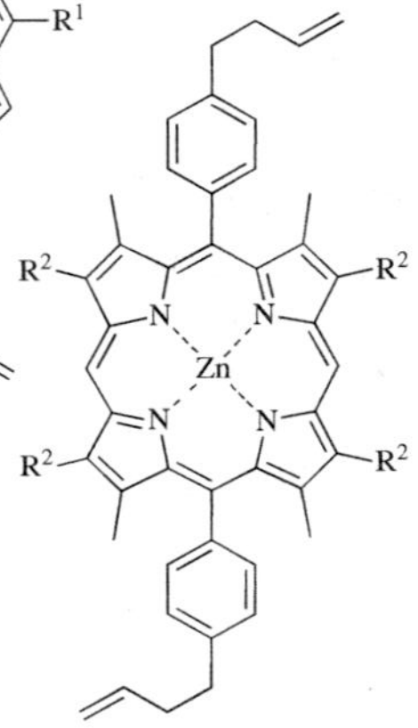 R^2 = *n*-C_7H_{15}	Gr II (1.0 eq), DCM, rt, 72 h	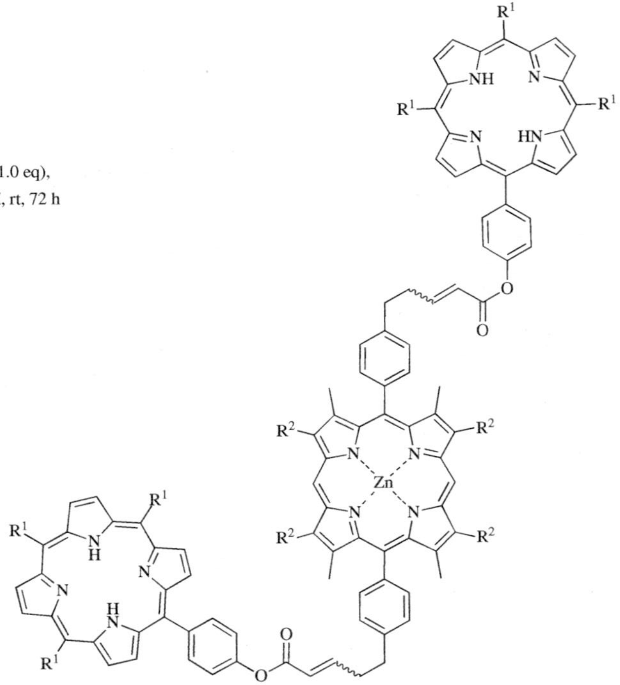 (48)	1032

2 eq

R = 4-MeC_6H_4

Gr II (0.2 eq),
DCM, rt, 24 h

(43) 1032

TABLE 20. CROSS-METATHESIS OF α,β-UNSATURATED ESTERS AND THIOESTERS (*Continued*)

A. ESTERS (*Continued*)

Ester	Cross Partner	Conditions	Product(s) and Yield(s) (%)	Refs.
Please refer to the charts preceding the tables for the catalyst structures.				
C_3				
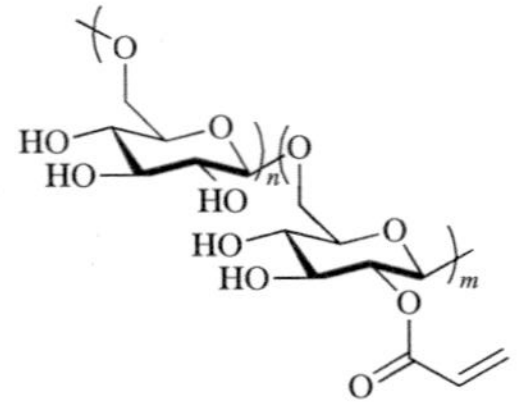 HO, HO, HO, HO, HO, O, *n*, *m*	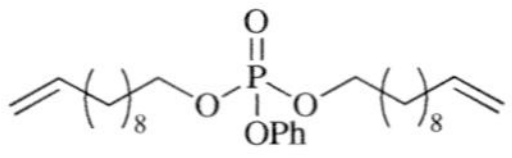 O, P, OPh, 8, 8; 1.05 eq	Hov **II** (0.1 eq), toluene, 40°, 16 h	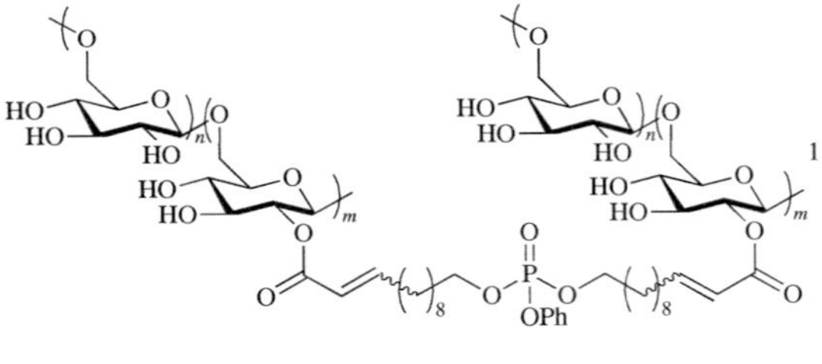 HO, HO, HO, HO, HO, O, *n*, *m*, 8, P, OPh (—)	1131
C_4				
 CO_2t-Bu **I**	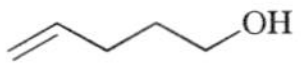 OH	1. Ru cat **364** (0.01 eq), (*Z*)-2-butene (5 eq), THF, 22°, 1 h 2. Ru cat **364** (0.04 eq), **I** (5 eq), THF, 100 torr, 22°, 1 h, ambient pressure, 7 h 3. Ru cat **364** (0.04 eq), THF, 100 torr, 22°, 1 h, ambient pressure, 11 h	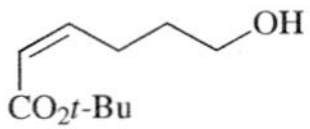 OH, CO_2t-Bu (55) (*Z*)/(*E*) > 98:2	1037

Reactant	Olefin	Conditions	Product(s) and Yield(s) (%)	Refs.
 CO_2Bn **I**	 R-substituted terminal olefin (n)	1. Ru cat **364** (0.01 eq), (*Z*)-2-butene (5 eq), THF, 22°, 1 h 2. Ru cat **364** (0.04 eq), **I** (5 eq), THF, 100 torr, 22°, 1 h, ambient pressure, 7 h 3. Ru cat **364** (0.04 eq), THF, 100 torr, 22°, 1 h, ambient pressure, 11 h	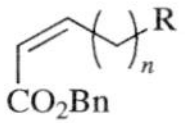 CO_2Bn (R, n) R = HO_2C, n = 2 (57) (*Z*)/(*E*) >98:2 R = Br, n = 6 (63) (*Z*)/(*E*) 98:2 R = HO, n = 7 (64) (*Z*)/(*E*) >98:2 R = Ph, n = 2 (66) (*Z*)/(*E*) 98:2 R = OHC, n = 8 (69) (*Z*)/(*E*) >98:2	1037
I	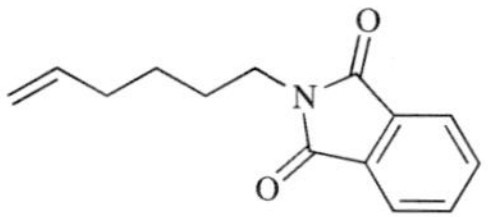 N-(hex-5-enyl)phthalimide	1. Ru cat **364** (0.01 eq), (*Z*)-2-butene (5 eq), THF, 22°, 1 h 2. Ru cat **364** (0.04 eq), **I** (5 eq), THF, 100 torr, 22°, 1 h, ambient pressure, 7 h 3. Ru cat **364** (0.04 eq), THF, 100 torr, 22°, 1 h, ambient pressure, 11 h	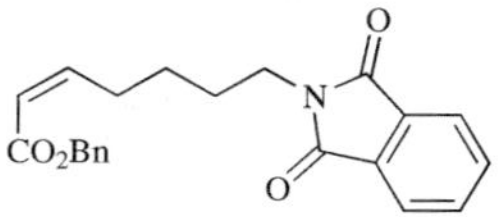 CO_2Bn, N-phthalimide product (52) (*Z*)/(*E*) = 96:4	1037
I	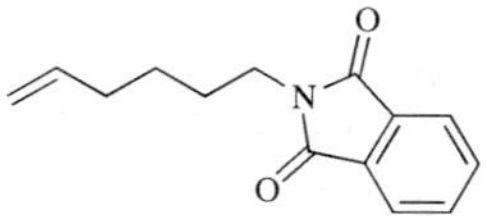 N-(hex-5-enyl)phthalimide	1. Ru cat **364** (0.01 eq), (*Z*)-2-butene (5 eq), THF, 22°, 1 h 2. Ru cat **364** (0.05 eq), **I** (5 eq), THF, 100 torr, 22°, 1 h, ambient pressure, 7 h	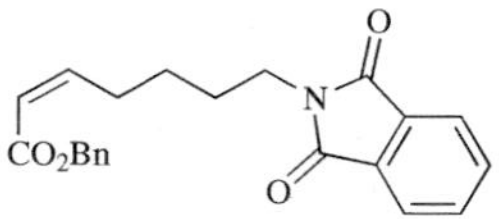 CO_2Bn, N-phthalimide product (45) (*Z*)/(*E*) = 98:2	1037

TABLE 20. CROSS-METATHESIS OF α,β-UNSATURATED ESTERS AND THIOESTERS (*Continued*)

A. ESTERS (*Continued*)

Ester	Cross Partner	Conditions	Product(s) and Yield(s) (%)	Refs.
Please refer to the charts preceding the tables for the catalyst structures.				
C_4				
CO_2Bn **I**	R, *n*	1. Ru cat **364** (0.01 eq), (*Z*)-2-butene (5 eq), THF, 22°, 1 h 2. Ru cat **364** (0.05 eq), **I** (5 eq), THF, 100 torr, 22°, 1 h, ambient pressure, 7 h	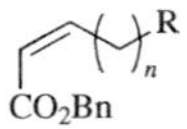 CO_2Bn, R, *n* (*Z*)/(*E*) = 98:2 R / *n* Ph / 2 / (53) Br / 6 / (55)	1037
I	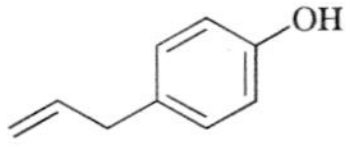 OH	1. Ru cat **364** (0.01 eq), (*Z*)-2-butene (5 eq), THF, 22°, 1 h 2. Ru cat **364** (0.05 eq), **I** (5 eq), THF, 100 torr, 22°, 1 h, ambient pressure, 7 h	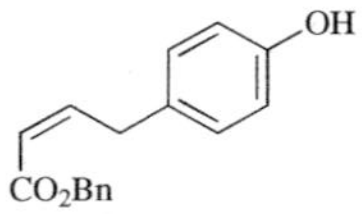 CO_2Bn, OH (46) (*Z*)/(*E*) = 98:2	1037
I	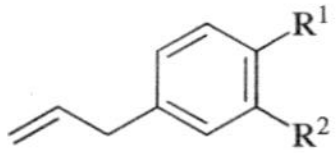 R^1, R^2	1. Ru cat **364** (0.01 eq), (*Z*)-2-butene (5 eq), THF, 22°, 1 h 2. Ru cat **364** (0.04 eq), **I** (5 eq), THF, 100 torr, 22°, 1 h, ambient pressure, 7 h 3. Ru cat **364** (0.04 eq), THF, 100 torr, 22°, 1 h, ambient pressure, 11 h	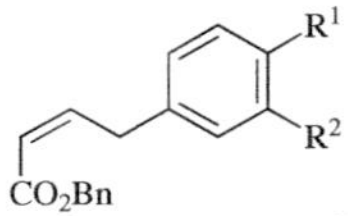 CO_2Bn, R^1, R^2 R^1 / R^2 / / (*Z*)/(*E*) HO / H / (59) / 98:2 MeO / MeO / (67) / >98:2	1037

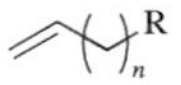

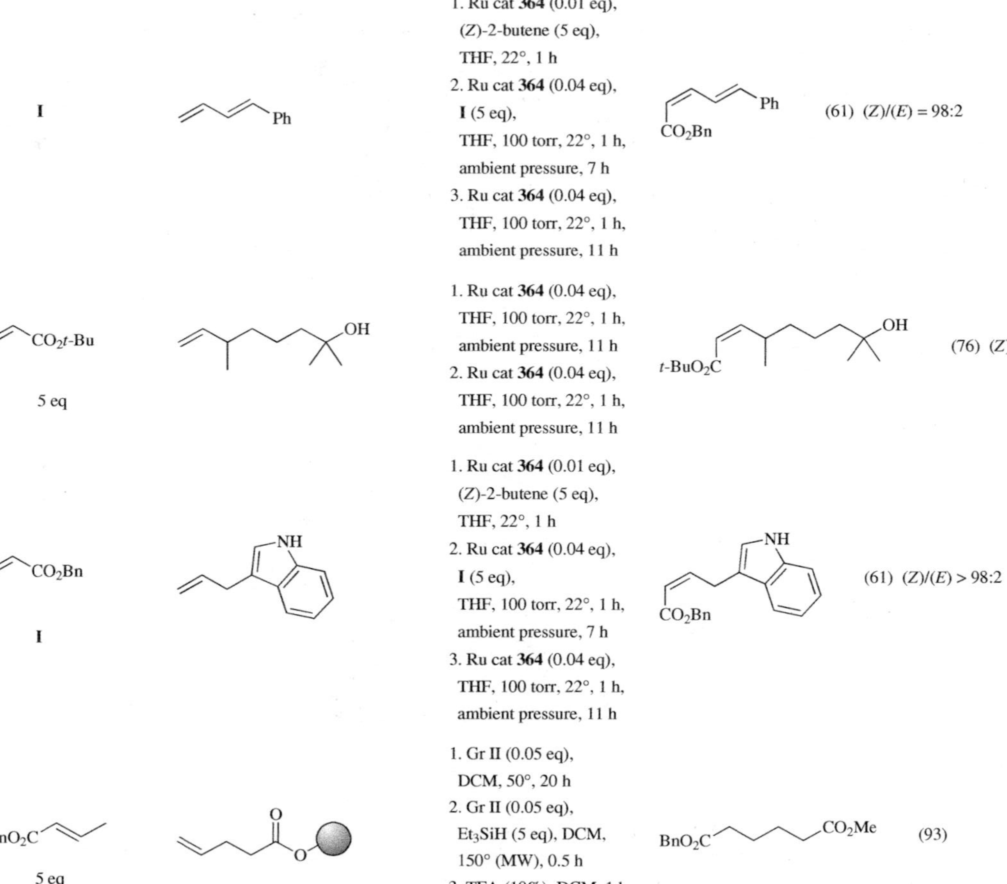

Reactant	Reactant	Conditions	Product	Yield (%)	Refs.
I		1. Ru cat **364** (0.01 eq), (*Z*)-2-butene (5 eq), THF, 22°, 1 h 2. Ru cat **364** (0.04 eq), **I** (5 eq), THF, 100 torr, 22°, 1 h, ambient pressure, 7 h 3. Ru cat **364** (0.04 eq), THF, 100 torr, 22°, 1 h, ambient pressure, 11 h		(61) (*Z*)/(*E*) = 98:2	1037
5 eq		1. Ru cat **364** (0.04 eq), THF, 100 torr, 22°, 1 h, ambient pressure, 11 h 2. Ru cat **364** (0.04 eq), THF, 100 torr, 22°, 1 h, ambient pressure, 11 h		(76) (*Z*)/(*E*) > 98:2	1037
I		1. Ru cat **364** (0.01 eq), (*Z*)-2-butene (5 eq), THF, 22°, 1 h 2. Ru cat **364** (0.04 eq), **I** (5 eq), THF, 100 torr, 22°, 1 h, ambient pressure, 7 h 3. Ru cat **364** (0.04 eq), THF, 100 torr, 22°, 1 h, ambient pressure, 11 h		(61) (*Z*)/(*E*) > 98:2	1037
5 eq		1. Gr II (0.05 eq), DCM, 50°, 20 h 2. Gr II (0.05 eq), Et_3SiH (5 eq), DCM, 150° (MW), 0.5 h 3. TFA (10%), DCM, 1 h 4. CH_2N_2, DCM, 0°, 0.5 h		(93)	636

TABLE 20. CROSS-METATHESIS OF α,β-UNSATURATED ESTERS AND THIOESTERS (*Continued*)

A. ESTERS (*Continued*)

Ester	Cross Partner	Conditions	Product(s) and Yield(s) (%)	Refs.
Please refer to the charts preceding the tables for the catalyst structures.				
C_4				
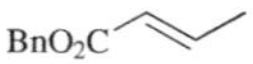 BnO_2C 5 eq	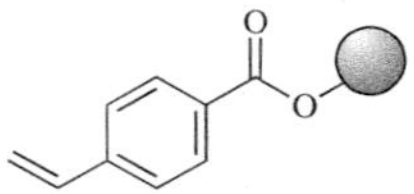	1. Gr II (0.05 eq), DCM, 50°, 20 h 2. Gr II (0.05 eq), Et_3SiH (5 eq), DCM, 150° (MW), 0.5 h 3. TFA (10%), DCM, 1 h 4. CH_2N_2, DCM, 0°, 0.5 h	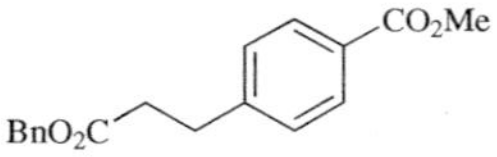 BnO_2C … CO_2Me (51)	636
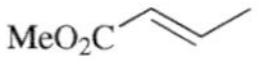 MeO_2C 2.4 eq	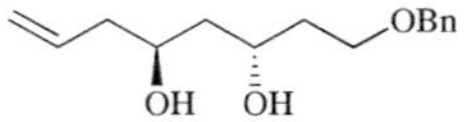 OH OH OBn	Gr II (0.04 eq), DCM, 45°, 3 h	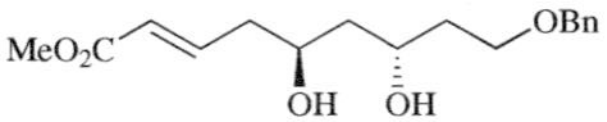 MeO_2C … OH OH OBn (89) (*E*) only	974
5 eq	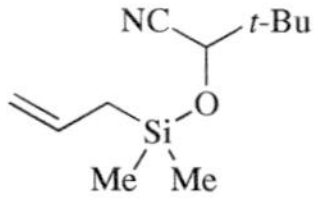 NC *t*-Bu O Si Me Me	Gr II (0.1 eq), DCM, 40°, 18 h	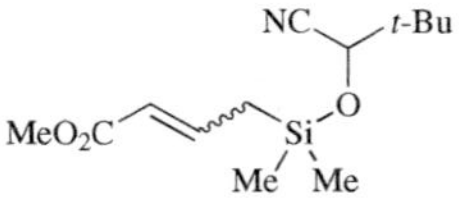 MeO_2C … NC *t*-Bu O Si Me Me (44)	644
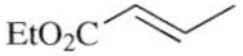 EtO_2C 2 eq	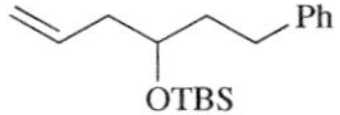 OTBS Ph	Catalyst (0.1 eq), DCM, reflux	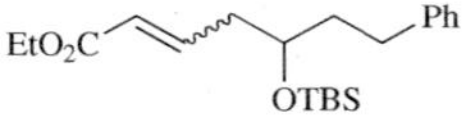 EtO_2C … OTBS Ph Catalyst: Gr I (8); Gr II (92); Hov II (70)	541
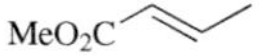 MeO_2C 2 eq	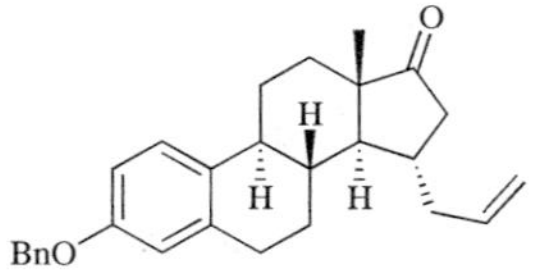 BnO H H H O	Catalyst (0.03–0.05 eq), solvent	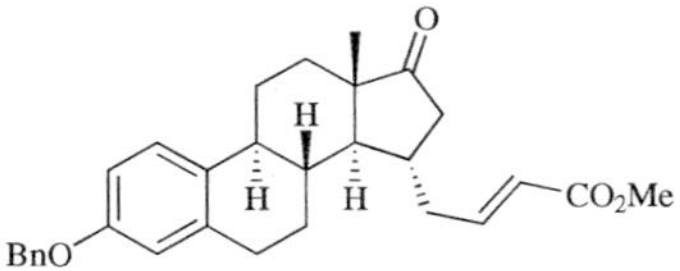 BnO H H H O … CO_2Me (*E*) only	163

Catalyst	Solvent	Temp (°)	Time (h)	
Gr II	DCM	40	2.5	(94)
Ru cat **68**	DCE	80	5.5	(58)
Ru cat **125**	DCM	40	5.5	(73)

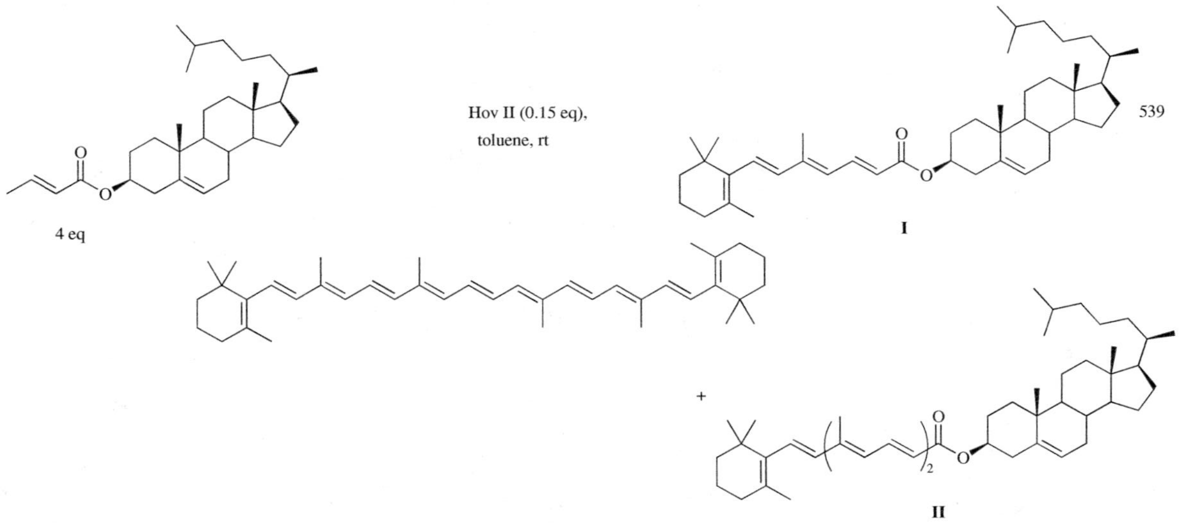

4 eq

Hov II (0.15 eq), toluene, rt

I

539

+

II

Time (h)	**I**	(*E*)/(*Z*) **I**	**II**	(*E*)/(*Z*) **II**
24	(9)	86:14	(30)	91:9
96	(12)	83:17	(8)	89:11

TABLE 20. CROSS-METATHESIS OF α,β-UNSATURATED ESTERS AND THIOESTERS (*Continued*)

A. ESTERS (*Continued*)

Ester	Cross Partner	Conditions	Product(s) and Yield(s) (%)	Refs.

Please refer to the charts preceding the tables for the catalyst structures.

C_4

Ester	Cross Partner	Conditions	Product(s) and Yield(s) (%)	Refs.
MeO_2C–CH=CH–CO_2Me (x eq)	CH_2=CH–CH_2–$(CH_2)_7$–CO_2Me	Catalyst (0.01 eq), $PhSiCl_3$ (y eq), toluene	MeO_2C–CH=CH–$(CH_2)_7$–CO_2Me **I** + MeO_2C–$(CH_2)_8$–CH=CH–$(CH_2)_7$–CO_2Me **II**	1091

x	Concn (M)	Catalyst	y	Temp (°)	Time (h)	Conv. (%)[b]	I	II
5	0.1	Ind II	—	80	1	44	(8)	(18)
5	0.25	Ind II	—	80	1	56	(16)	(19)
5	0.5	Ind II	—	80	1	81	(51)	(14)
5	0.75	Ind II	—	80	1	>99	(99)	(1)
5	1	Ind II	—	80	1	>99	(99)	(1)
10	1	Gr II	—	50	3	71	(32)	(13)
10	1	Ind II	—	50	3	92	(50)	(17)
10	1	Ru cat **313**	100	50	3	49	(26)	(0)
10	1	Ru cat **305**	100	50	3	>99	(99)	(0)
10	1	Ru cat **301**	100	50	3	49	(18)	(1)
10	0.1	Ind II	100	50	3	11	(3)	(7)
10	0.25	Ind II	100	50	3	34	(10)	(14)
10	0.5	Ind II	100	50	3	59	(26)	(19)
10	0.75	Ind II	100	50	3	80	(50)	(19)
10	1.5	Ind II	100	50	3	96	(70)	(14)
10	2	Ind II	100	50	3	>99	(99)	(2)

[a] The conversion was determined by ^1H NMR analysis.

[b] The conversion was determined by GC analysis.

[c] The conversion was determined by TLC analysis.

[d] The yield and selectivity were determined by GC analysis.

TABLE 20. CROSS-METATHESIS OF α,β-UNSATURATED ESTERS AND THIOESTERS (*Continued*)

B. THIOESTERS

Thioester	Cross Partner	Conditions	Product(s) and Yield(s) (%)	Refs.

Please refer to the charts preceding the tables for the catalyst structures.

C_3

Thioester	Cross Partner	Conditions	Product(s) and Yield(s) (%)	Refs.
EtS(C=O)CH=CH$_2$	CH$_2$=CHCH$_2$(CH$_2$)$_n$R (2.5 eq)	Catalyst (x eq), DCM	EtS(C=O)CH=CH~(CH$_2$)$_n$R	1132

R	n	Catalyst	x	Temp (°)	Time (h)	
TMS	0	Hov II	0.02	45	2	(92)
TsHN	0	Hov II	2 x 0.02	45	24	(59)
OHC	1	Hov II	2 x 0.02	45	24	(75)
HO_2C	1	Hov II	2 x 0.02	45	24	(83)
MeO_2C	1	Hov II	0.02	45	4	(91)
AcO	2	Hov II	0.02	45	4	(86)
HO	3	Hov II	2 x 0.02	45	24	(93)
Br	3	Hov II	2 x 0.02	45	24	(75)
Me	4	Gr I	0.02	rt	20	(0)
Me	4	Hov I	0.02	rt	20	(0)
Me	4	Gr II	0.02	rt	20	(76)
Me	4	Hov II	0.02	rt	20	(93)
Me	4	Ru cat **88**	0.02	rt	20	(72)
Me	4	Hov II	0.02	reflux	1	(94)
Me	4	Hov II	0.01	reflux	2	(94)
Me	4	Hov II	0.02	reflux	24	(42)

TABLE 20. CROSS-METATHESIS OF α,β-UNSATURATED ESTERS AND THIOESTERS (*Continued*)
B. THIOESTERS (*Continued*)

Thioester	Cross Partner	Conditions	Product(s) and Yield(s) (%)	Refs.
Please refer to the charts preceding the tables for the catalyst structures.				
C_3				
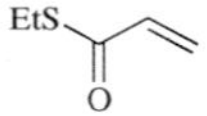 EtS, O	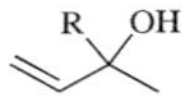 R, OH 2.5 eq	Hov II (2 x 0.02 eq), DCM, 45°, 24 h	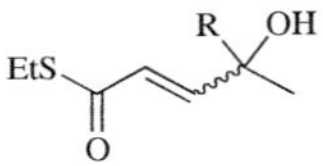 EtS, O, R, OH R: H (66); Me (71)	1132
	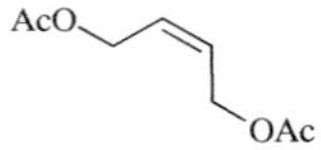 AcO, OAc 2.5 eq	Hov II (2 x 0.02 eq), DCM, 45°, 24 h	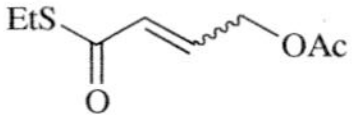 EtS, O, OAc (65)	1132
	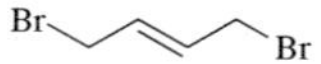 Br, Br 1.5 eq	Hov II (2 x 0.02 eq), DCM, 45°, 24 h	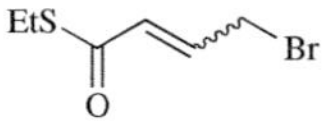 EtS, O, Br (64)	1132

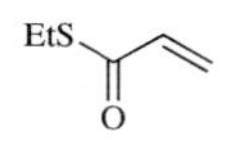

2 eq

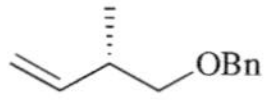

er 97.0:3.0

Hov II (0.1 eq),
DCM, reflux, 6 h

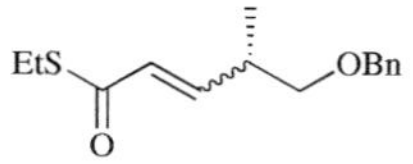

(83)

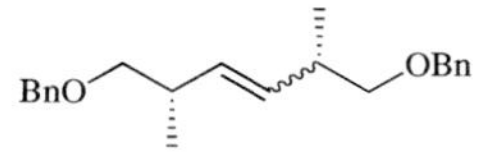

+ BnO ... OBn (5)

975

2.5 eq

Hov II (0.02 eq),
DCM, 45°

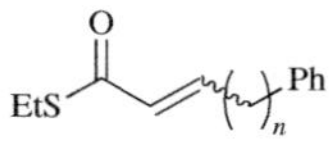

n	Time (h)	
0	18	(72)
1	1	(95)

1132

2 eq

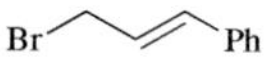

1. RMgBr (1.2 eq),
(*R*,*R*)-Taniaphos (0.012 eq),
$CuBr•SMe_2$ (0.01 eq),
DCM, –75°
2. Hov II (2 x 0.05 eq), DCM, rt

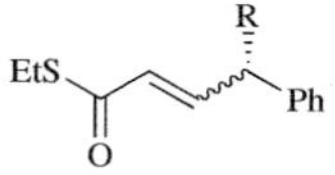

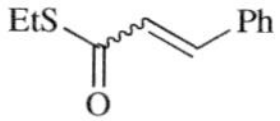

I er = 98.5:1.5 + **II**

R	I	I/II
Me	(74)	>97:3
Et	(67)	80:20

975

TABLE 20. CROSS-METATHESIS OF α,β-UNSATURATED ESTERS AND THIOESTERS (*Continued*)

B. THIOESTERS (*Continued*)

Thioester	Cross Partner	Conditions	Product(s) and Yield(s) (%)	Refs.

Please refer to the charts preceding the tables for the catalyst structures.

C_3

Thioester	Cross Partner	Conditions	Product(s) and Yield(s) (%)	Refs.
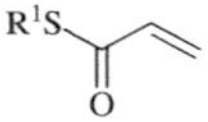 R¹S-acryloyl thioester, 3 eq	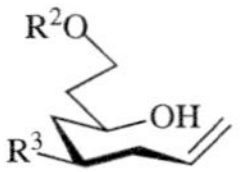 R²O, OH, R³ (alkene)	Hov II (0.1 eq), DCM, 35°, 18 h	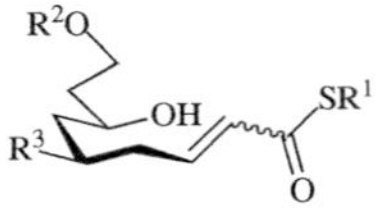 R²O, OH, SR¹, O, R³ (product)	1134
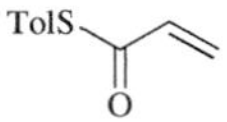 TolS-acryloyl thioester, 3 eq	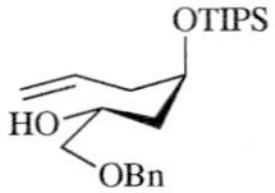 OTIPS, HO, OBn (alkene)	Hov II (0.1 eq), DCM, 35°, 18 h	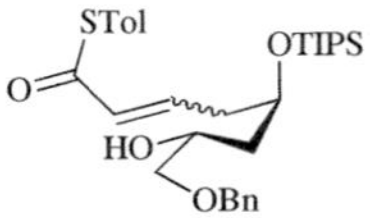 STol, O, OTIPS, HO, OBn (product) (82)	1134

R¹	R²	R³	
Et	MPM	H	(93)
Ph	MPM	H	(94)
4-MeC_6H_4	MPM	H	(92)
4-$MeOC_6H_4$	MPM	H	(82)
4-$O_2NC_6H_4$	MPM	H	(94)
1-Naph	MPM	H	(87)
Tol	TBS	H	(90)
Tol	TBS	TIPSO	(84)

Reactant	Substrate	Conditions	Product(s) and Yield(s) (%)				Refs.
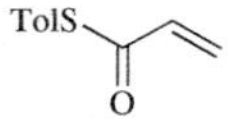 TolS(C=O)CH=CH2, 3 eq	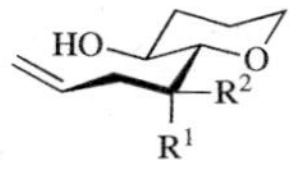 HO, O, R^1, R^2	Hov II (0.1 eq), DCM, 35°, 18 h	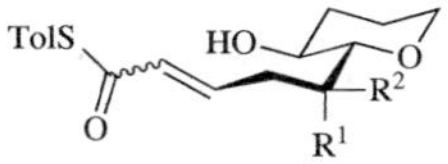 TolS, O, HO, O, R^1, R^2	R^1	R^2		1134
				H	H	(93)	
				H	HO	(94)	
				HO	H	(82)	
3 eq	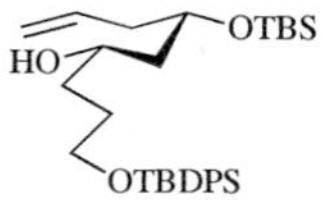 HO, OTBS, OTBDPS	Hov II (0.1 eq), DCM, 35°, 18 h	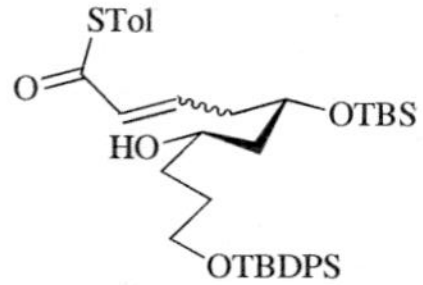 STol, O, HO, OTBS, OTBDPS (79)				1134

TABLE 21. CROSS-METATHESIS OF α,β-UNSATURATED AMIDES

Amide	Cross Partner	Conditions	Product(s) and Yield(s) (%)	Refs.

Please refer to the charts preceding the tables for the catalyst structures.

C_3

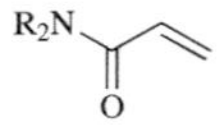

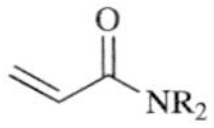

Amide	Cross Partner	Conditions	Product(s) and Yield(s) (%)	Refs.
R_2N	NR_2	Hov II (x eq), DCM, 100° (MW), 15 min	R_2N ... NR_2	165
RHN	NHR	Catalyst (0.025 eq), DCM, 40°, 4 h	RHN ... NHR	1135
Ph ... N H	N H ... Ph	Ru cat **35** (0.025 eq), DCM, 40°, 4 h	Ph ... N H ... N H ... Ph (94) (*E*) only	1135

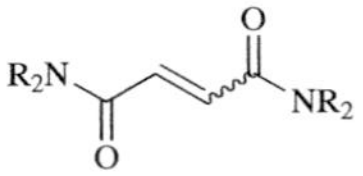

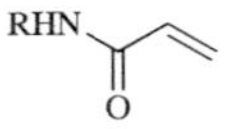

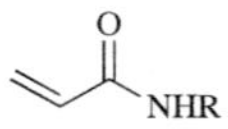

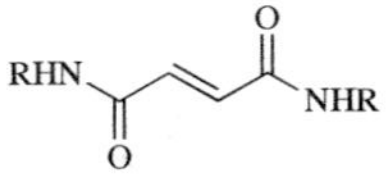

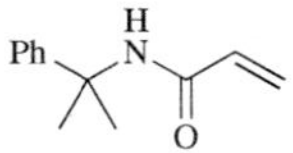

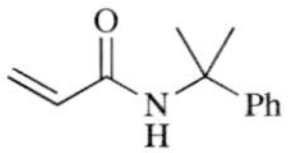

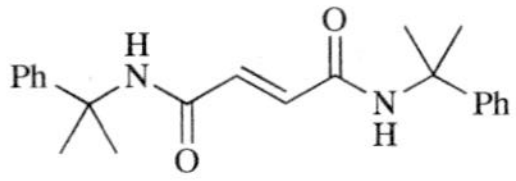

R	x	
H	0.05	(87)
H	0.025	(80)
H	0.005	(63)
Me	0.05	(<5)

R	Catalyst		(*E*)/(*Z*)
t-Bu	Ru cat **35**	(<10)	—
n-C_6H_{13}	Ru cat **35**	(96)	(*E*) only
Bn	Gr I	(15)	(*E*) only
Bn	Ru cat **35**	(98)	(*E*) only
n-C_8H_{17}	Ru cat **35**	(93)	(*E*) only

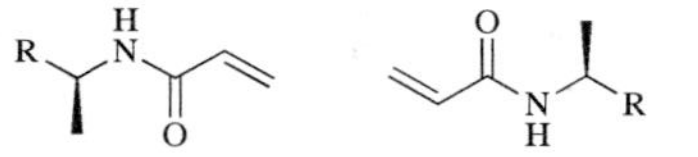

Ru cat **35** (*x* eq), DCM, 40°, 4 h

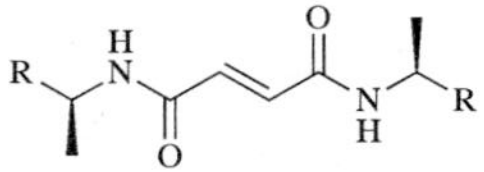

(*E*) only 1135

R	*x*	
Ph	0.025	(95)
Ph	0.000375	(94)
1-Naph	0.025	(95)

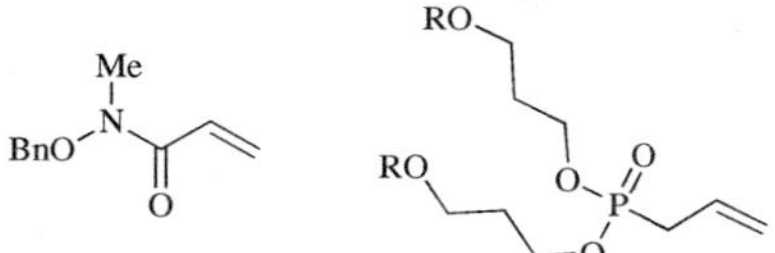

4 eq

Hov II (0.05 eq), toluene, 70°, overnight

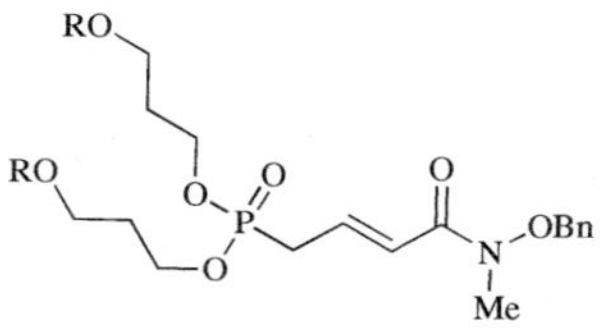

1136

R		(*E*)/(*Z*)
n-C_4H_9	(67)	>99:1
n-C_8H_{17}	(76)	>99:1
n-$C_{12}H_{25}$	(59)	>99:1
n-$C_{16}H_{33}$	(55)	>99:1

TABLE 21. CROSS-METATHESIS OF α,β-UNSATURATED AMIDES (*Continued*)

Amide	Cross Partner	Conditions	Product(s) and Yield(s) (%)	Refs.
Please refer to the charts preceding the tables for the catalyst structures.				
C_3				
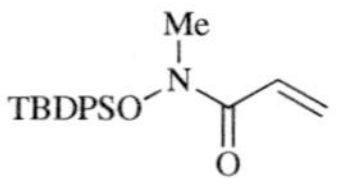 TBDPSO, N, Me, O 4 eq	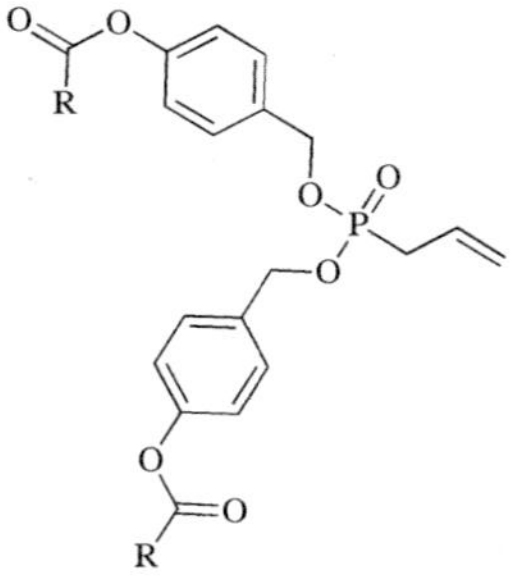 O, O, R, O, P, O, O, O, O, R	Hov II (0.05 eq), toluene, 70°, overnight	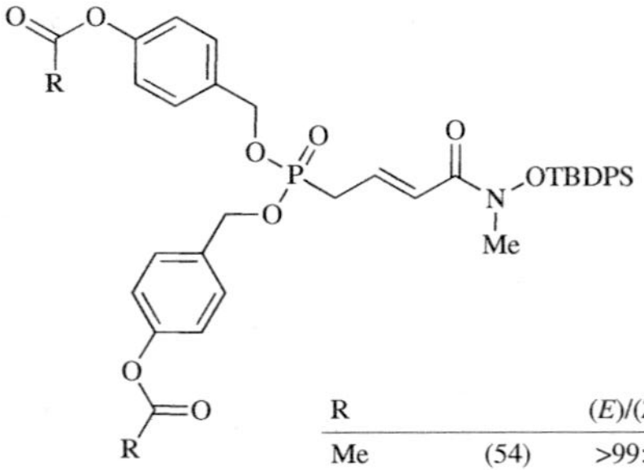 O, O, R, O, P, O, O, N, OTBDPS, Me, O, O, R R / (E)/(Z): Me (54) >99:1 *i*-Pr (54) >99:1 *t*-Bu (60) >99:1 Ph (51) >99:1 *n*-C_8H_{17} (64) >99:1	1136
Bn, O, N, O, O 1.3 eq 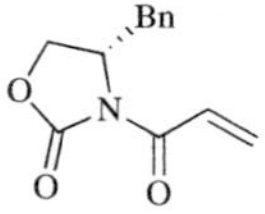	 OTIPS	Gr II (0.1 eq), DCM, 40°, 24 h	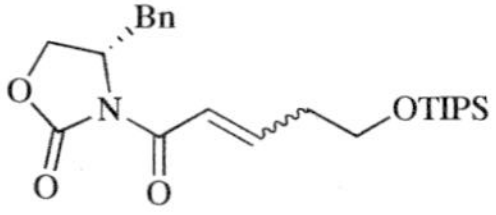 Bn, O, N, OTIPS, O, O (59)	268

2 eq

Gr II (0.02 eq), DCM, reflux, 3 h

(72) 1133

x eq

Catalyst (0.05 eq), solvent

R^1	*x*	R^2	Catalyst	Solvent	Temp (°)	Time (h)		(*E*)/(*Z*)	
Me_2N	0.5	Bz	Hov II	DCM	40	—	(>98)	>95:5	1135
Me_2N	0.5	Bz	Ru cat **243**	DCM	45	12	(40)	95:5	923
Me_2N	0.5	Bz	Ru cat **191**	DCM	45	12	(42)	95:5	923
i-PrHN	0.5	Bz	Ru cat **243**	DCM	45	12	(96)	95:5	923
i-PrHN	0.5	Bz	Ru cat **191**	DCM	45	12	(0)	—	924
Me_2N	—	Bz	Ru cat **185**	BMI•PF_6/toluene	rt	6	(0)	—	924
H_2N	1.5	TBS	Hov II	DCM	rt	16	(71)	—	174
MeHN	1.5	TBS	Hov II	DCM	rt	16	(11)	—	174
Me_2N	1.5	TBS	Hov II	DCM	rt	16	(tr)	—	174

TABLE 21. CROSS-METATHESIS OF α,β-UNSATURATED AMIDES (*Continued*)

Amide	Cross Partner	Conditions	Product(s) and Yield(s) (%)	Refs.
Please refer to the charts preceding the tables for the catalyst structures.				
C_3				
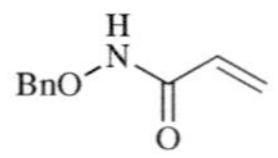 BnO–NH (2 eq)	 CO_2Bn	Gr II (0.02 eq), DCM, reflux, 3 h	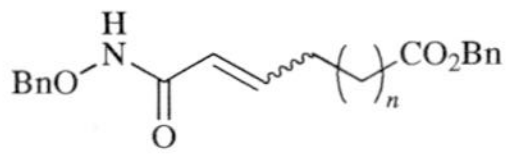 BnO–NH, CO_2Bn; *n* 1 (79), 2 (73), 3 (70)	1133
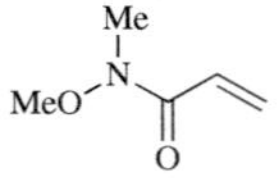 MeO–N(Me) (3 eq)	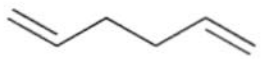	Catalyst (2 x 0.025 eq), $CDCl_3$, 100° (MW), 3 h	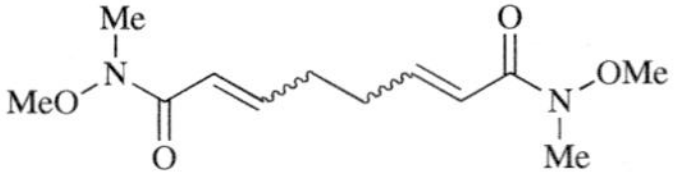 MeO–N(Me), O, N–OMe, Me; Catalyst / Conv. (%): Gr II 72; Hov II 100; Ind II 100; Ru cat **93** 87; Ru cat **170** 94; Ru cat **172** 100	926
3 eq	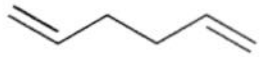	1. Ind II (2 x 0.025 eq), $CDCl_3$, 100° (MW), 2 h 2. CH_3NH_2 (3 eq), THF, 100° (MW), 2 h	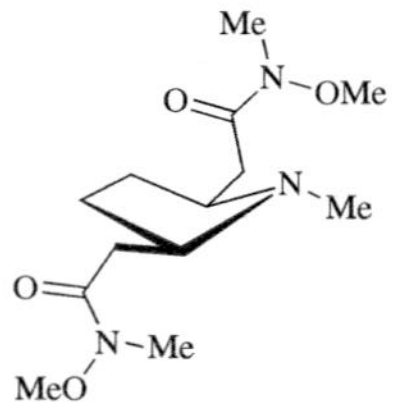 Me, N–OMe, O, N–Me, O, N–Me, MeO (65)	926

3 eq

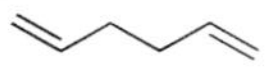

1. Ind II (2 x 0.025 eq), $CDCl_3$, 100° (MW), 2 h
2. (1*R*)-phenylethylamine (3 eq), THF, 100° (MW), 2 h

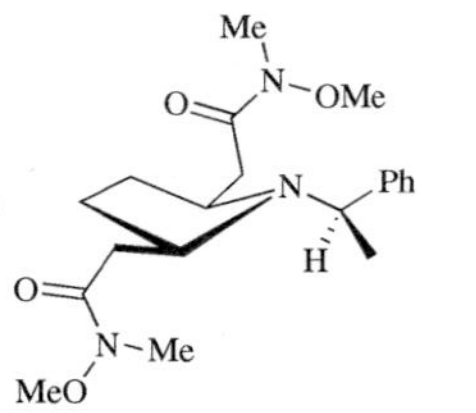

(67) 926

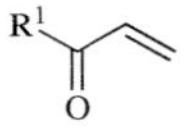

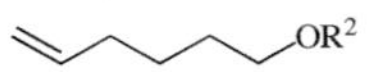

1.25 eq

Gr II (0.05 eq), DCM, 40°, 15 h

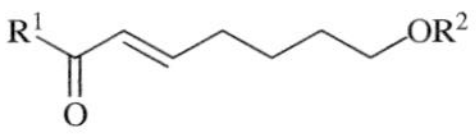

166

R^1	R^2		(*E*)/(*Z*)
H_2N	Ac	(89)	(*E*) only
i-PrHN	THP	(80)	(*E*) only
(MeO)MeN	THP	(89)	98:2
PhHN	THP	(90)	(*E*) only
PhMeN	THP	(97)	97:3
Ph_2N	THP	(87)	98:2

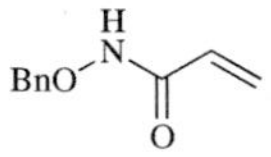

2 eq

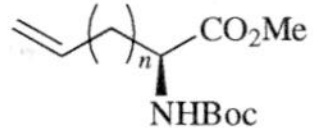

Gr II (0.02 eq), DCM, reflux, 3 h

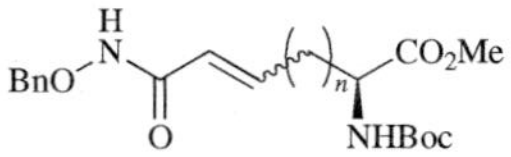

n	
2	(78)
3	(78)

1133

TABLE 21. CROSS-METATHESIS OF α,β-UNSATURATED AMIDES (*Continued*)

Amide	Cross Partner	Conditions	Product(s) and Yield(s) (%)	Refs.
Please refer to the charts preceding the tables for the catalyst structures.				
C_3				
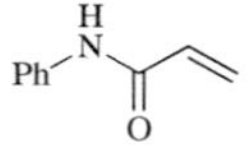 Ph–NH–C(O)CH=CH₂ 2 eq	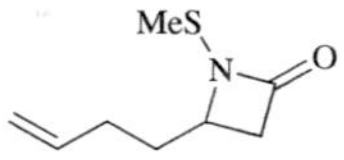 MeS N O	Gr II (0.05 eq), DCM, 40°, 14 h	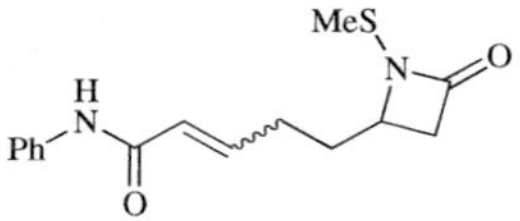 Ph H N O MeS N O (51)	422
1.5 eq	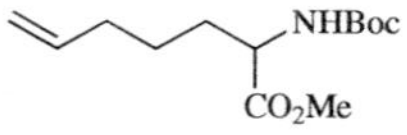 NHBoc CO_2Me	Gr II (0.025 eq) DCM, reflux, 2 h	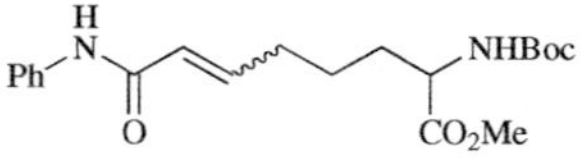 Ph H N O NHBoc CO_2Me (88)	1067
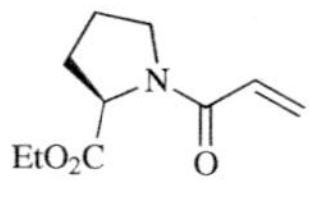 EtO_2C N O 1.2 eq	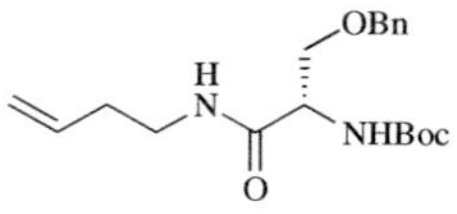 H N O OBn NHBoc	Gr II (0.19 eq), DCM, 80° (MW), 2 h	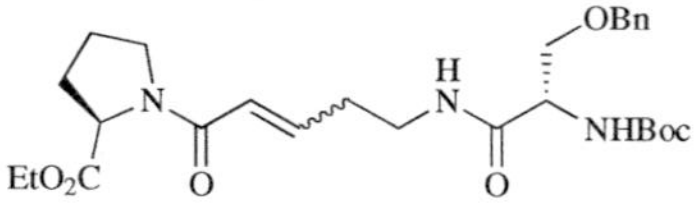 EtO_2C N O H N O OBn NHBoc (50)	1137
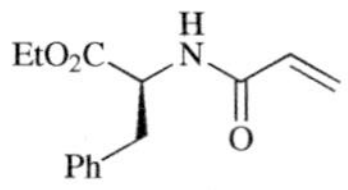 EtO_2C H N O Ph 1.1 eq	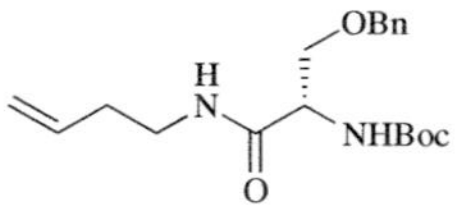 H N O OBn NHBoc	Gr II (0.19 eq), DCM, 90° (MW), 0.5 h	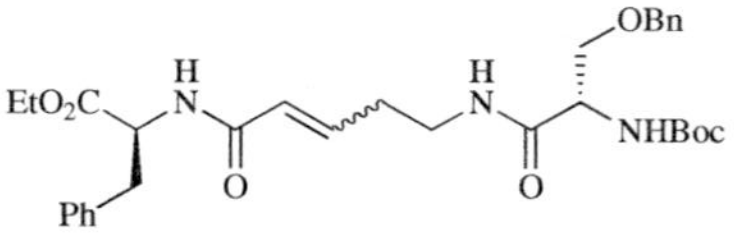 EtO_2C H N O Ph H N O OBn NHBoc (73)	1137

1.3 eq

Gr II (0.19 eq), DMF (3 drops), DCM, 90° (MW), 0.5 h

NHBoc S EtO$_2$C Ph NHBoc (51) 1137

R O Ph x eq

Catalyst (0.05 eq), DCM

R O Ph (*E*) only

R	x	Catalyst	Temp (°)	Time (h)		
Me$_2$N	2.5	Gr II	100 (MW)	0.5	(30)	971
Me$_2$N	2.5	Hov II	100 (MW)	0.5	(65)	971
Me$_2$N	2.5	Ru cat **86**	100 (MW)	0.5	(45)	971
Me$_2$N	1.9	Gr II	40	15	(25)	166
Cy$_2$N	1.9	Gr II	40	15	(57)	166
i-PrHN	1.9	Gr II	40	15	(62)	166
(MeO)MeN	1.9	Gr II	40	15	(66)	166
H$_2$N	1.9	Gr II	40	15	(69)	166
PhHN	1.9	Gr II	40	15	(69)	166
MePhN	1.9	Gr II	40	15	(83)	166
Ph$_2$N	1.9	Gr II	40	15	(87)	166
O N O (oxazolidinone)	1.9	Gr II	40	15	(40)	166

TABLE 21. CROSS-METATHESIS OF α,β-UNSATURATED AMIDES (*Continued*)

Amide	Cross Partner	Conditions	Product(s) and Yield(s) (%)	Refs.
Please refer to the charts preceding the tables for the catalyst structures.				
C_3				
BnO–NH–C(O)–CH=CH₂, 2 eq	4-MeO-styrene	Gr II (0.02 eq), DCM, reflux, 3 h	BnO–NH–C(O)–CH=CH~C₆H₄-4-OMe (57)	1133
camphorsultam N-acryloyl amide	$CH_2=CHCH_2$-n-C_5H_{11}, 1.2 eq	Catalyst (0.025 eq), EtOAc, reflux	I + n-C_6H_{13}CH=CH~n-C_5H_{11} (II)	1138

Catalyst	Time (h)	I	I/II
Gr I	24	(0)	**II** only
Gr II	24	(28)	26:74
Hov I	24	(0)	**II** only
Hov II	12	(89)	91:9
Ru cat **86**	10	(90)	91:9

R^1(C=O)CH=CH$_2$ 3 eq + R^2O…OH

Hov **II** (0.1 eq), DCM, 35°

1134

R^1	R^2	Time (h)	
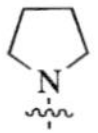 N	TBS	25	(96)
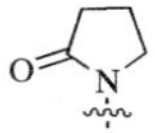 O, N	TBS	6.5	(62)
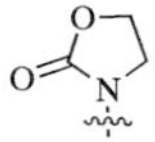 O, O, N	TBS	26	(86)
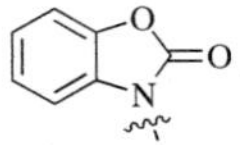 O, O, N	TBS	40	(91)
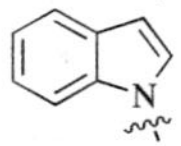 N	TBS	25	(85)

R^1	R^2	Time (h)	
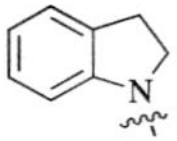 N	TBS	25	(76)
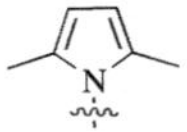 N	TBS	24	(76)
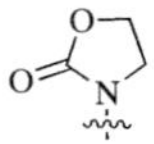 O, O, N	MPM	18	(84)
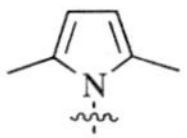 N	MPM	18	(66)

TABLE 21. CROSS-METATHESIS OF α,β-UNSATURATED AMIDES (*Continued*)

Amide	Cross Partner	Conditions	Product(s) and Yield(s) (%)	Refs.

Please refer to the charts preceding the tables for the catalyst structures.

C_3

Amide	Cross Partner	Conditions	Product(s) and Yield(s) (%)	Refs.
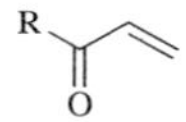 R, O; 3 eq	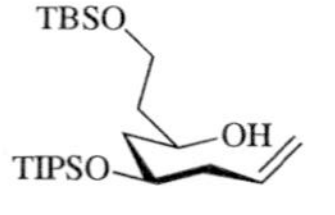 TBSO, OH, TIPSO	Hov II (0.1 eq), DCM, 35°, 18 h	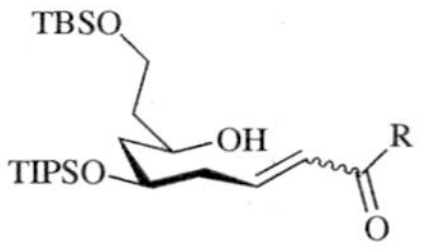 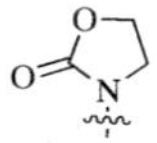 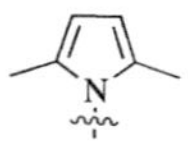 TBSO, OH, TIPSO, R, O; R = oxazolidinone (O, N) (81); R = 2,5-dimethylpyrrolyl (N) (70)	1134
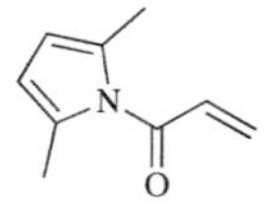 N, O; 3–10 eq	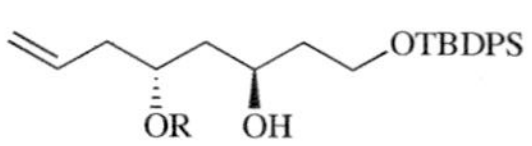 OTBDPS, OR, OH	Hov II (0.1 eq), additive (0.03–0.1 eq), solvent	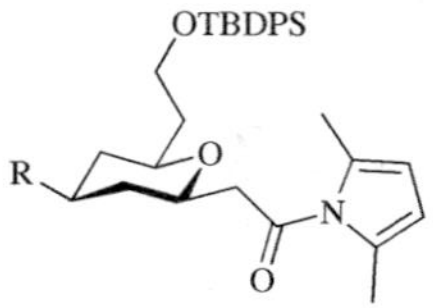 OTBDPS, R, O, N, O	947

R	Additive	Solvent	Temp (°)	Time (h)		(*E*)/(*Z*)
H	—	toluene	80–100	11–24	(36)	88:12
H	CSA	DCM	25–35	—	(72)	>95:5
TIPS	—	DCM	100 (MW)	0.5	(23)	75:25
TIPS	—	toluene	80–100	11–24	(10)	91:9
TIPS	CSA	DCM	25–35	—	(73)	>95:5

3 eq

Gr II (0.2 eq), DCM, 40°, 6 h

R		
$MeO_2C(CH_2)_8$	(69)	429
(glycoside)	(55)	

2.5 eq

Ru cat **353** (0.06 eq), DCM, hν (350 nm), 3.5 h

R		
H	(55)	911
i-Pr	(57)	

x eq

Hov II (0.02 eq), BQ (0.05 eq), DCM, 80°

1074

R^1	R^2	*x*	Time (h)	
H_2N	H	1	4	(40)
H_2N	H	1	4	(60)
H_2N	Ac	1.25	17	(65)
H_2N	*i*-Pr	1.25	17	(57)
i-PrHN	H	1.25	17	(70)
i-PrHN	Ac	1.25	5	(53)
i-PrHN	*i*-Pr	1.25	17	(53)

TABLE 21. CROSS-METATHESIS OF α,β-UNSATURATED AMIDES (*Continued*)

Amide	Cross Partner	Conditions	Product(s) and Yield(s) (%)	Refs.
Please refer to the charts preceding the tables for the catalyst structures.				
C_3				
MeO, Me, N, O; 3 eq	Br, O_2, S, N, Me	Gr II (0.05 eq), DCM, 40°	Me, MeO, N, O, O_2, S, N, Me, Br (43)	634
N, O; 3–10 eq	HO, O, R^2, R^1	Hov II (0.1 eq), additive (0.03–0.1 eq), solvent	N, O, O, O, R^2, R^1	947

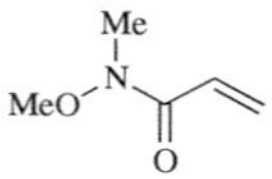

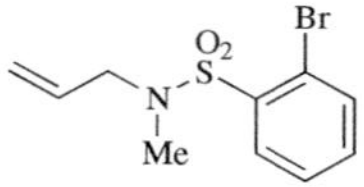

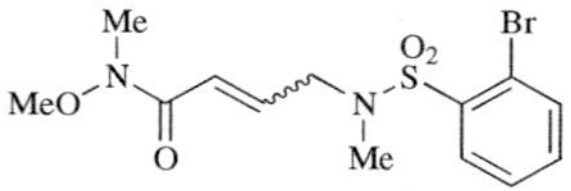

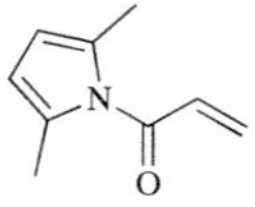

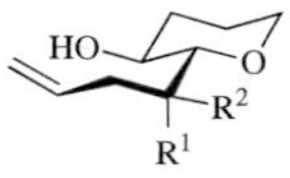

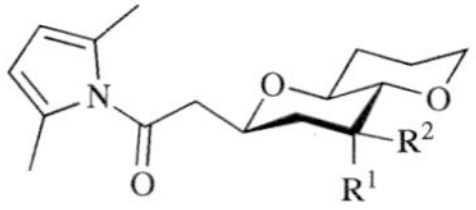

R^1	R^2	Additive	Solvent	Temp (°)	Time (h)		(E)/(Z)
H	HO	—	toluene	80–100	11–24	(71)	88:12
H	HO	CSA	DCM	25–35	—	(63)	>95:5
HO	H	—	toluene	80–100	11–24	(52)	>95:5
HO	H	CSA	DCM	25–35	—	(56)	>95:5

R^1 (3 eq) + substrate; Hov II (0.1 eq), DCM, 35°, 18 h

1134

R^1	R^2	R^3	
O=C(O)N (oxazolidinone)	H	HO	(85)
2,5-dimethylpyrrol-1-yl	H	HO	(71)
O=C(O)N (oxazolidinone)	HO	H	(81)
2,5-dimethylpyrrol-1-yl	HO	H	(64)
2,5-dimethylpyrrol-1-yl	H	H	(68)

TABLE 21. CROSS-METATHESIS OF α,β-UNSATURATED AMIDES (*Continued*)

Amide	Cross Partner	Conditions	Product(s) and Yield(s) (%)	Refs.
Please refer to the charts preceding the tables for the catalyst structures.				
C_3				
R(C=O)CH=CH$_2$ 3 eq	HO, OTBS, OTBDPS	Hov II (0.1 eq), DCM, 35°, 18 h	R(C=O), HO, OTBS, OTBDPS; R = 2-oxo-oxazolidin-3-yl (85); R = 2,5-dimethylpyrrol-1-yl (68)	1134
BnO–NH–C(=O)CH=CH$_2$ 2 eq	CH$_2$=CHCH$_2$Ph	Gr II (0.02 eq), DCM, reflux, 3 h	BnO–NH–C(=O)CH=CHCH$_2$Ph (77)	1133
2 eq	2-MeO-C$_6$H$_4$CH$_2$CH=CH$_2$	Gr II (0.02 eq), DCM, reflux, 3 h	BnO–NH–C(=O)CH=CHCH$_2$(2-MeO-C$_6$H$_4$) (72)	1133

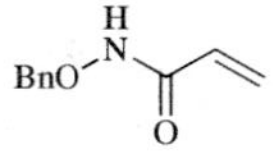

BnO–NH–C(O)–CH=CH₂ (2 eq) + CH₂=CH–CH₂–(CH₂)$_n$–CH₃

Gr II (0.02 eq), DCM, reflux, 3 h

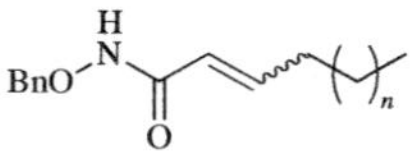

BnO–NH–C(O)–CH=CH–CH₂–(CH₂)$_n$–CH₃

n	
6	(81)
8	(85)

1133

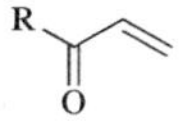

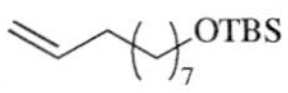

R–C(O)–CH=CH₂ + CH₂=CH–CH₂–(CH₂)$_7$–OTBS (x eq)

Ru cat **35** (y eq), DCM, rt, 12 h

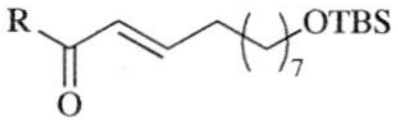

R–C(O)–CH=CH–CH₂–(CH₂)$_7$–OTBS

166

R	x	y		(E)/(Z)
Me_2N	1.25	0.05	(39)	96:4
Me_2N	1.5	0.1	(83)	96:4
$(n\text{-}C_6H_{13})_2N$	1.25	0.05	(77)	—
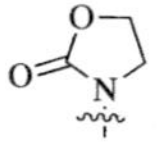 2-oxooxazolidin-3-yl	1.25	0.05	(87)	98:2

EtO₂C, BnO, R, H, N, O (1.1 eq) + H, N, O, NHBoc

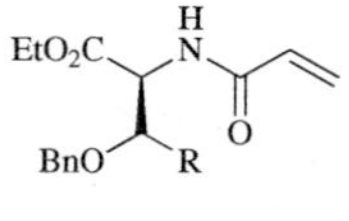

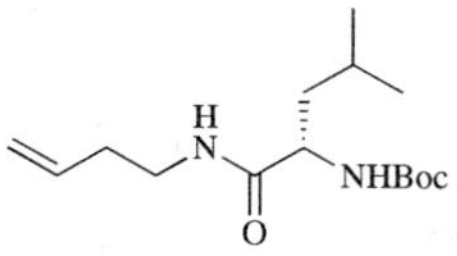

Gr II (0.19 eq), DCM, 90° (MW), 0.5 h

EtO₂C, BnO, R, H, N, O, NHBoc

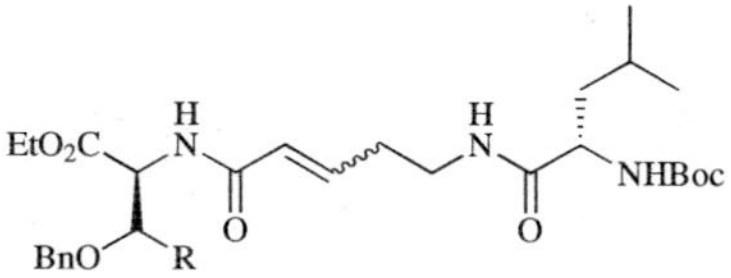

R	
H	(75)
Me	(61)

1137

TABLE 21. CROSS-METATHESIS OF α,β-UNSATURATED AMIDES (*Continued*)

Amide	Cross Partner	Conditions	Product(s) and Yield(s) (%)	Refs.
Please refer to the charts preceding the tables for the catalyst structures.				
C_3				
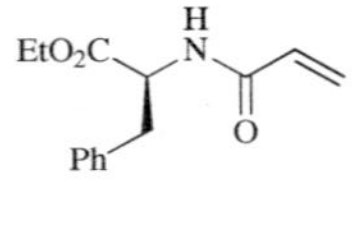 EtO_2C, Ph, NH, O	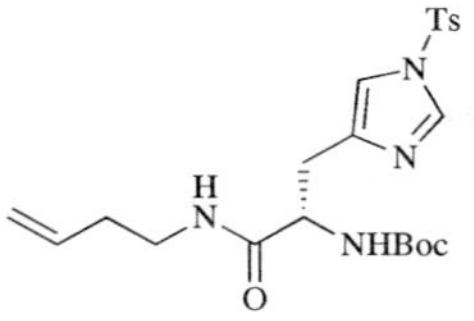 NHBoc, Ts, N, O; 1.3 eq	Gr II (0.19 eq), DMF (3 drops), DCM, 90° (MW), 0.5 h	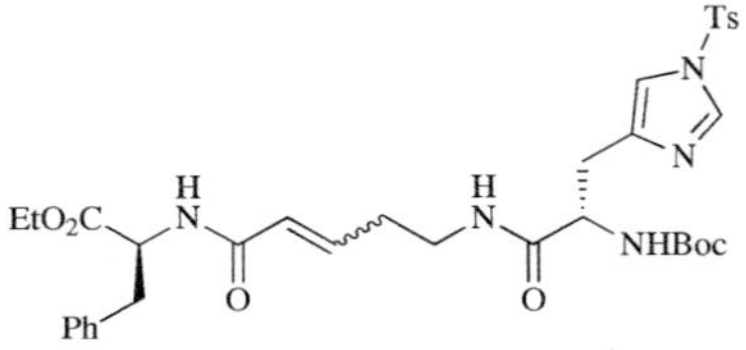 EtO_2C, Ph, NHBoc, Ts (41)	1137
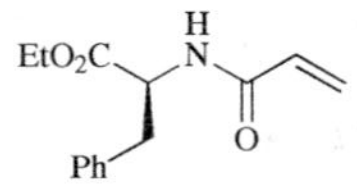 EtO_2C, Ph, NH, O; 1.1 eq	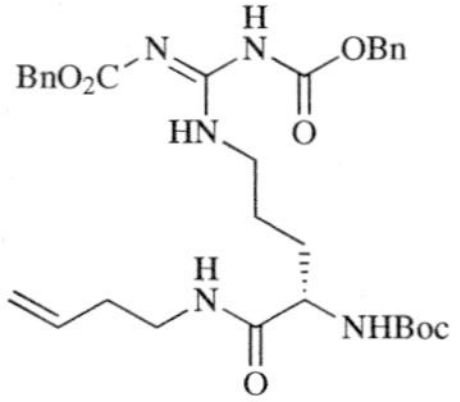 BnO_2C, OBn, HN, NHBoc	Gr II (0.19 eq), DCM	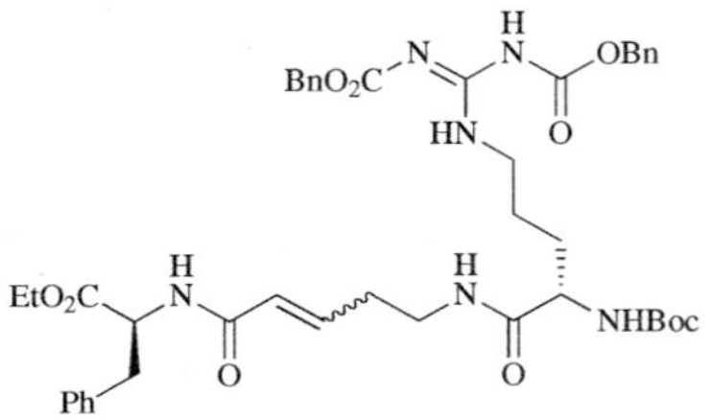 EtO_2C, Ph, BnO_2C, OBn, NHBoc Temp (°) / Time (h): reflux / 96 (70); 90 (MW) / — (43)	1137
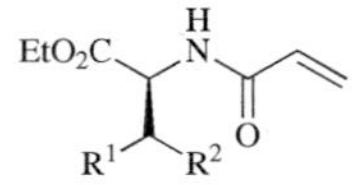 EtO_2C, R^1, R^2, NH, O	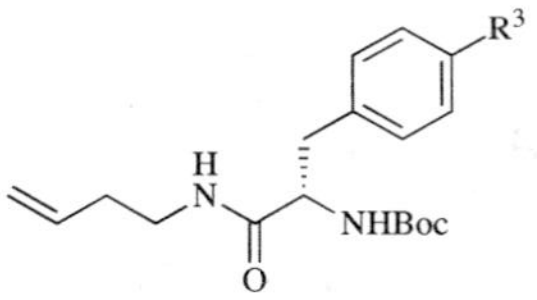 R^3, NHBoc; *x* eq	Gr II (0.19 eq), DCM, 80° (MW), 2 h	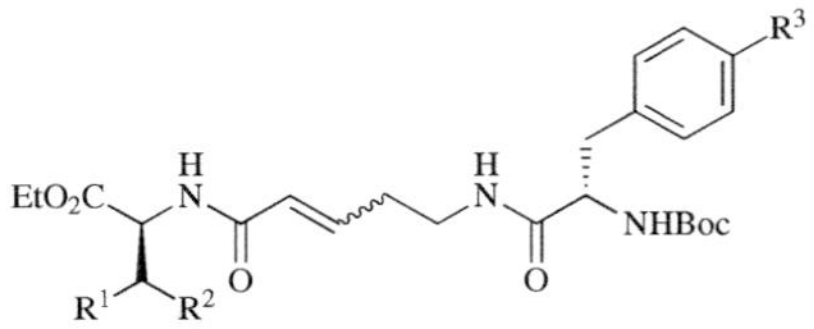 EtO_2C, R^1, R^2, R^3, NHBoc R^1 / R^2 / R^3 / *x*: H / Ph / H / 1.3 (86); H / Ph / BnO / 1.3 (82); Me / Me / H / 1.1 (34)	1137

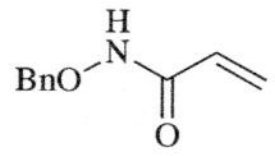

2 eq

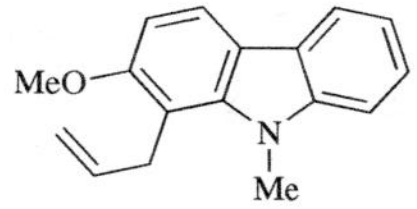

Gr II (0.02 eq), DCM, reflux, 3 h

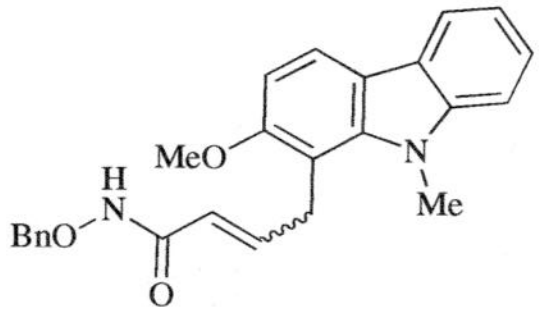

(78) 1133

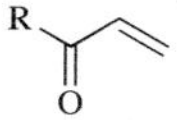

2 eq

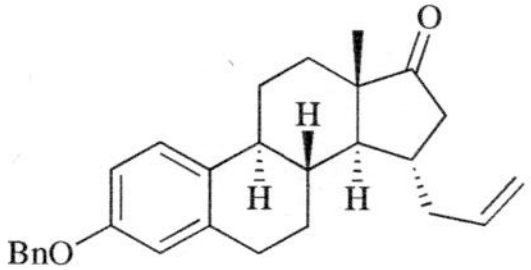

Catalyst (0.03–0.05 eq), solvent

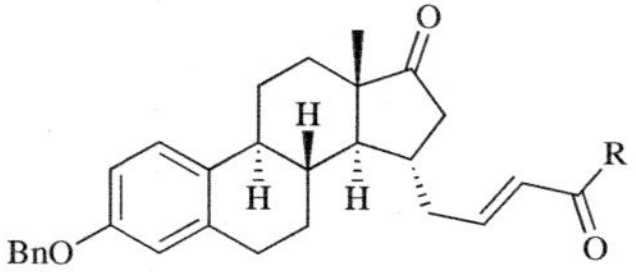

163

R	Catalyst	Solvent	Temp (°)	Time (h)		(*E*)/(*Z*)
H_2N	Gr II	DCM	40	2.5	(61)	100:0
H_2N	Ru cat **68**	DCE	80	5.5	(0)	—
H_2N	Ru cat **125**	DCM	40	5.5	(12)	100:0
PhHN	Gr II	DCM	40	2.5	(62)	100:0
PhHN	Ru cat **68**	DCE	80	5.5	(58)	100:0
PhHN	Ru cat **125**	DCM	40	5.5	(85)	100:0
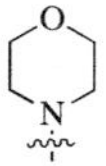 morpholino (O, N)	Gr II	DCM	40	2.5	(51)	100:0
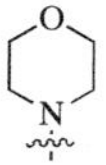 morpholino (O, N)	Ru cat **68**	DCE	80	5.5	(1)	100:0
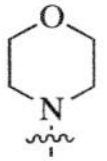 morpholino (O, N)	Ru cat **125**	DCM	40	5.5	(3)	100:0

TABLE 21. CROSS-METATHESIS OF α,β-UNSATURATED AMIDES (*Continued*)

Amide	Cross Partner	Conditions	Product(s) and Yield(s) (%)	Refs.
Please refer to the charts preceding the tables for the catalyst structures.				
C_3				
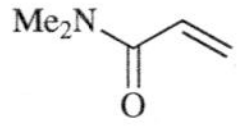 Me_2N (C=O) acrylamide; 10 eq	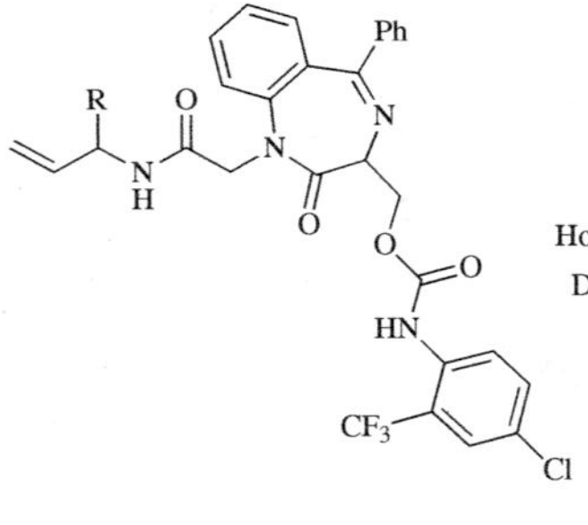 R; O; N H; N; O; Ph; N; O; O; HN; CF_3; Cl	Hov II (0.2 eq), DCM, 100° (MW), 1 h	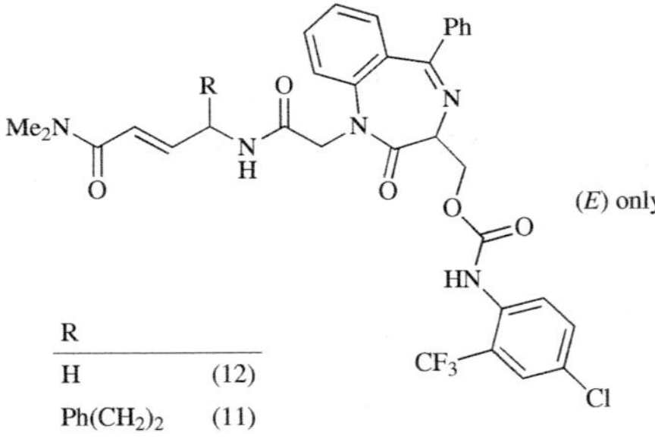 Me_2N; O; R; O; N H; N; O; Ph; N; O; O; HN; CF_3; Cl R: H (12); $Ph(CH_2)_2$ (11) (*E*) only	987
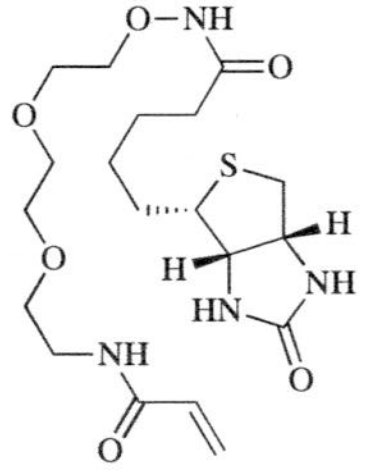 O–NH; O; O; O; NH; O; S; H; H; HN; NH; O; 10 eq	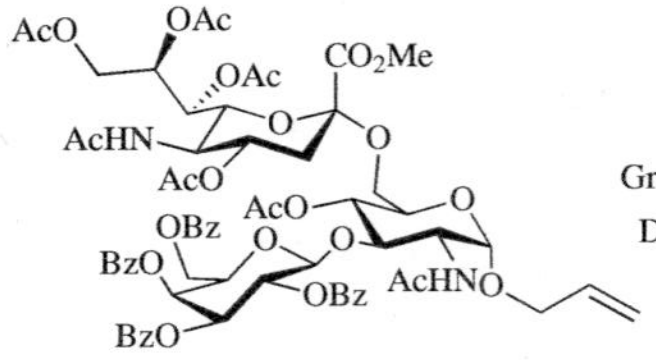 AcO; OAc; OAc; CO_2Me; AcHN; O; O; AcO; AcO; O; OBz; O; O; BzO; OBz; AcHNO; BzO	Gr II (0.1 eq), DCM, 50°, 12 h	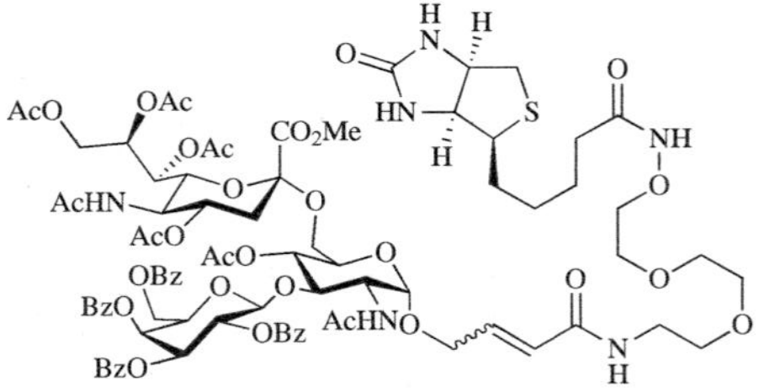 AcO; OAc; OAc; CO_2Me; AcHN; O; O; AcO; AcO; O; OBz; O; O; BzO; OBz; AcHNO; BzO; H; H; O; N; HN; H; S; O; NH; O; O; O; O; N H (>99)	1139

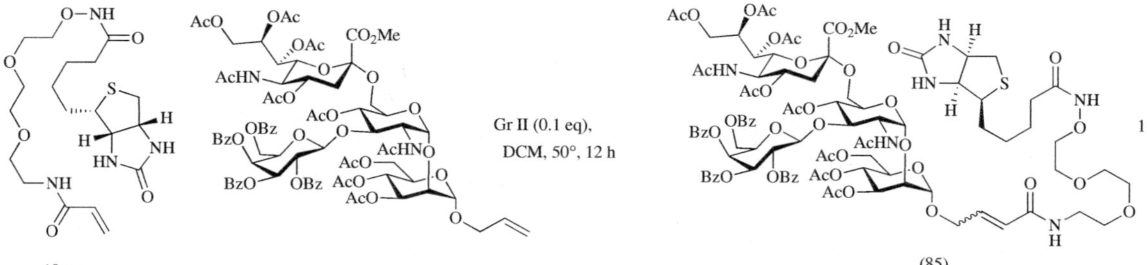

10 eq

Gr II (0.1 eq),
DCM, 50°, 12 h

(85)

1139

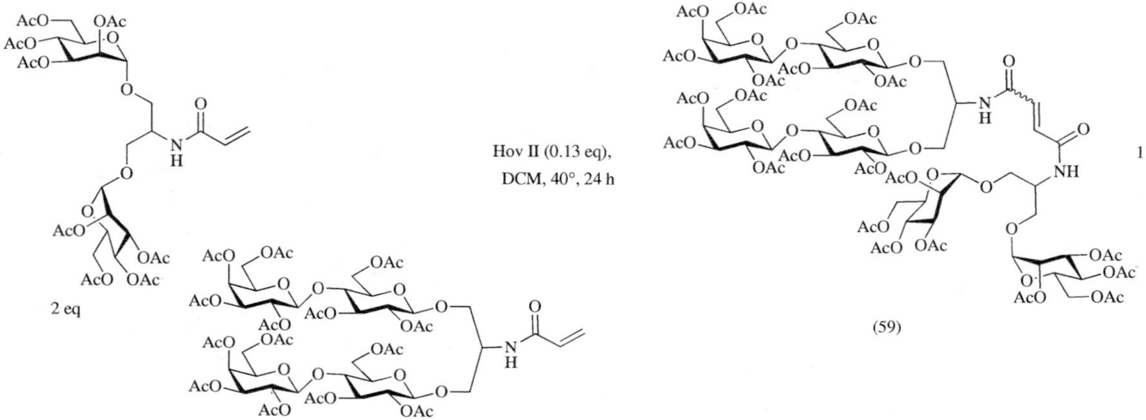

2 eq

Hov II (0.13 eq),
DCM, 40°, 24 h

(59)

1140

TABLE 21. CROSS-METATHESIS OF α,β-UNSATURATED AMIDES (*Continued*)

Amide	Cross Partner	Conditions	Product(s) and Yield(s) (%)	Refs.

Please refer to the charts preceding the tables for the catalyst structures.

C_3

Amide	Cross Partner	Conditions	Product(s) and Yield(s) (%)	Refs.
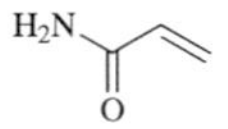 H_2N–C(O)CH=CH₂, 20 eq	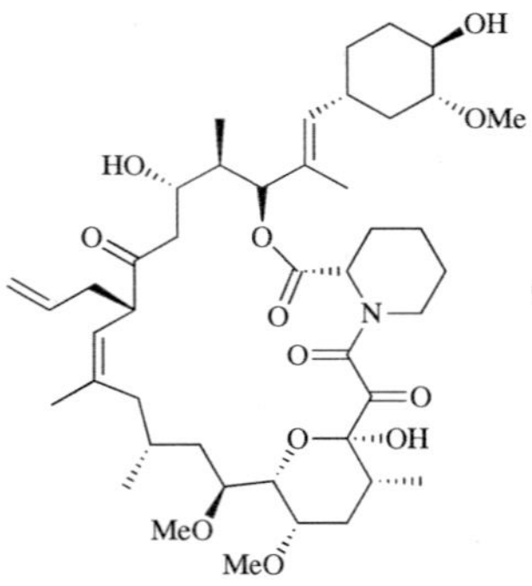 HO, OH, OMe, O, N, OH, MeO, MeO	Gr II (0.1 eq), DCB, MW (20 W)	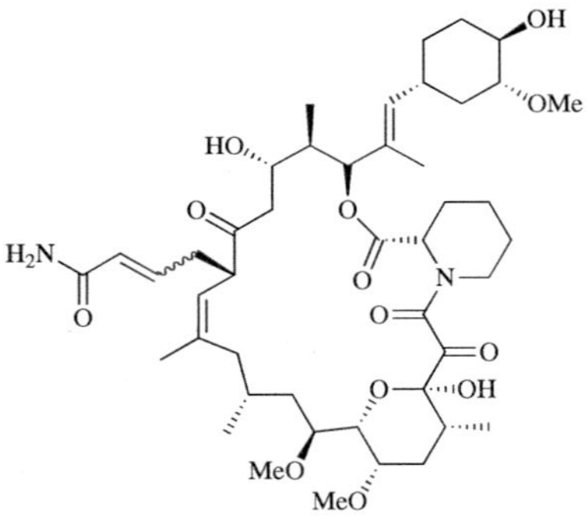 H_2N, O, HO, OH, OMe, N, OH, MeO, MeO (99)	432
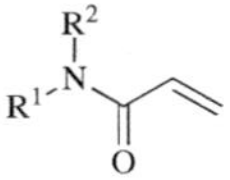 R^1R^2N–C(O)CH=CH₂, 40 eq	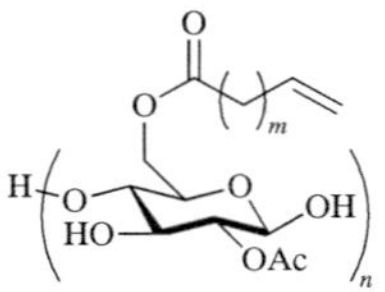 H, O, HO, OH, OAc, O, *m*, *n*	Hov II (*x* eq), BHT (0.12 eq), solvent, 30°	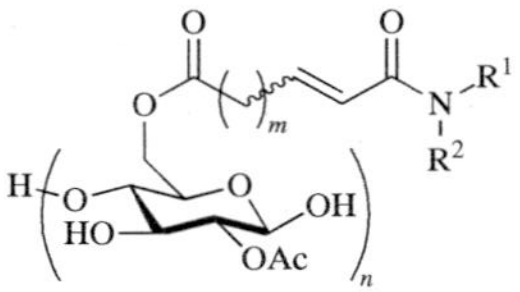 H, O, HO, OH, OAc, *m*, *n*, N, R^1, R^2	1034

R^1	R^2	*x*	Solvent	Time (h)	Conv. (%)[a]
H	H	0.06 + 0.06	THF	1 + 1	50
H	H	0.06 + 0.06	AcOH	1 + 1	95
Me	Me	0.03	THF	1	39
Me	Me	0.03	AcOH	1	68
i-Pr	H	0.06 + 0.06	THF	1 + 1	82
i-Pr	H	0.06 + 0.06	AcOH	1 + 1	73
Ph	H	0.03	THF	1	50
Ph	H	0.03	AcOH	1	37

C_4

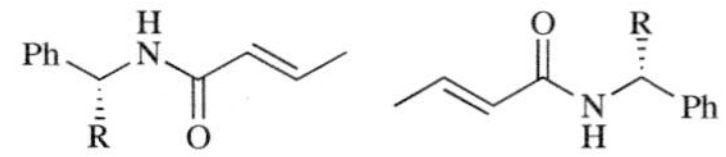

Ru cat **35** (0.025 eq), DCM, 40°, 4 h

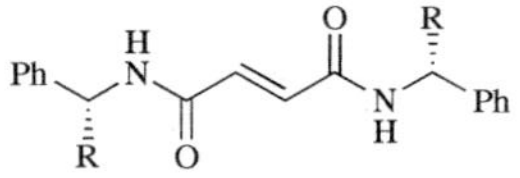

(*E*) only

R		
H	(92)	1135
Me	(88)	

I

1. Ru cat **364** (0.01 eq), (*Z*)-2-butene (5 eq), THF, 22°, 1 h
2. Ru cat **364** (0.05 eq), **I** (5 eq), THF, 100 torr, 22°, 1 h; then ambient pressure, 7 h

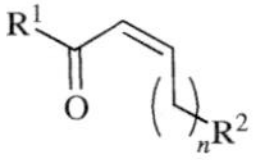

(*Z*)/(*E*) > 98:2

1037

R^1	R^2	*n*	
PMBHN	HO	3	(55)
BnHN	BnO_2C	2	(49)
MeO(Me)N	Ph	2	(49)

TABLE 21. CROSS-METATHESIS OF α,β-UNSATURATED AMIDES (*Continued*)

Amide	Cross Partner	Conditions	Product(s) and Yield(s) (%)	Refs.
Please refer to the charts preceding the tables for the catalyst structures.				
C_4				
R[1] … O **I**	R[2], *n*	1. Ru cat **364** (0.01 eq), (*Z*)-2-butene (5 eq), THF, 22°, 1 h 2. Ru cat **364** (0.04 eq), **I** (5 eq), THF, 100 torr, 22°, 1 h; then ambient pressure, 7 h 3. Ru cat **364** (0.04 eq), THF, 100 torr, 22°, 1 h; then ambient pressure, 11 h	R[1] … O, *n*, R[2] (*Z*)/(*E*) > 98:2	1037

R^1	R^2	*n*	
PMBHN	HO	3	(76)
BnHN	HO	3	(70)
BnHN	BnO_2C	2	(61)
PMBHN	Ph	2	(61)
MeO(Me)N	Ph	2	(82)
MeO(Me)N	Me	8	(80)
MeO(Me)N	OHC	8	(58)
MeO(Me)N	HO_2C	8	(54)

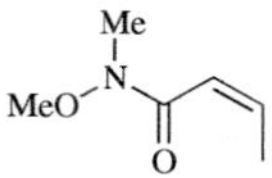

I

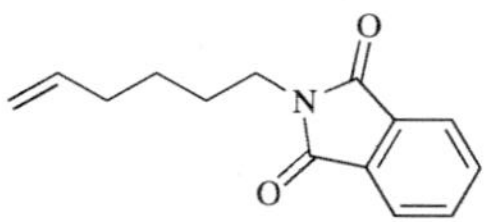

1. Ru cat **364** (0.01 eq),
(*Z*)-2-butene (5 eq),
THF, 22°, 1 h
2. Ru cat **364** (0.04 eq),
I (5 eq), THF,
100 torr, 22°, 1 h; then
ambient pressure, 7 h
3. Ru cat **364** (0.04 eq),
THF, 100 torr, 22°, 1 h;
then ambient pressure, 11 h

R O N O O

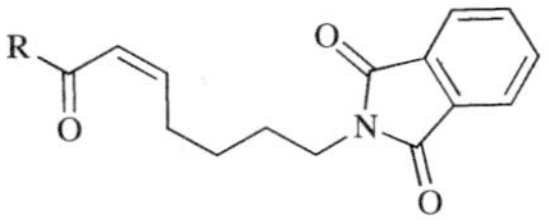

R	
BnHN	(55)
MeO(Me)N	(69)

(*Z*)/(*E*) > 98:2 1037

Me N MeO O

I

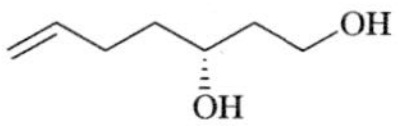

OH OH

1. Ru cat **364** (0.01 eq),
(*Z*)-2-butene (5 eq),
THF, 22°, 1 h
2. Ru cat **364** (0.04 eq),
I (5 eq), THF,
100 torr, 22°, 1 h; then
ambient pressure, 7 h
3. Ru cat **364** (0.04 eq),
THF, 100 torr, 22°, 1 h;
then ambient pressure, 11 h

Me MeO N O OH OH

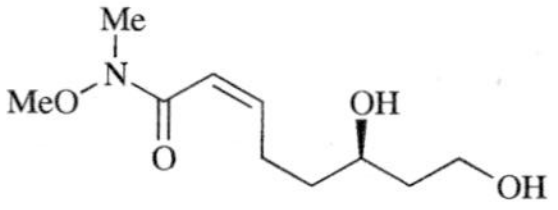

(73) (*Z*)/(*E*) > 98:2 1037

TABLE 21. CROSS-METATHESIS OF α,β-UNSATURATED AMIDES (*Continued*)

Amide	Cross Partner	Conditions	Product(s) and Yield(s) (%)	Refs.

Please refer to the charts preceding the tables for the catalyst structures.

C_4

Amide	Cross Partner	Conditions	Product(s) and Yield(s) (%)	Refs.
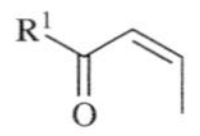 R^1 (Z)-crotonamide **I**	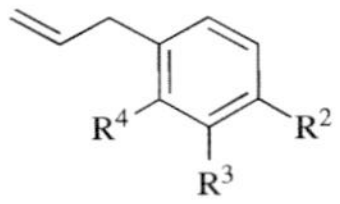 allylarene (R^2, R^3, R^4)	1. Ru cat **364** (0.01 eq), (Z)-2-butene (5 eq), THF, 22°, 1 h 2. Ru cat **364** (0.04 eq), **I** (5 eq), THF, 100 torr, 22°, 1 h; then ambient pressure, 7 h 3. Ru cat **364** (0.04 eq), THF, 100 torr, 22°, 1 h; then ambient pressure, 11 h	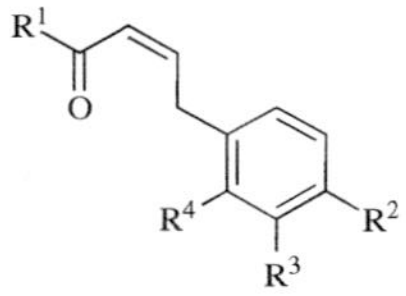 (see below)	1037

R^1	R^2	R^3	R^4		(Z)/(E)
H_2N	H	H	HO	(67)	98:2
H_2N	HO	H	H	(61)	98:2
BnHN	MeO	MeO	H	(59)	98:2
PMBHN	MeO	MeO	H	(63)	>98:2
i-BuHN	MeO	MeO	H	(69)	97:3
MeO(Me)N	MeO	MeO	H	(67)	>98:2

Amide	Cross Partner	Conditions	Product(s) and Yield(s) (%)	Refs.
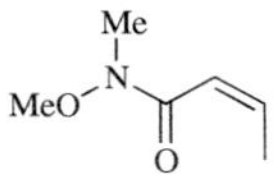 MeO(Me)N (Z)-crotonamide **I**	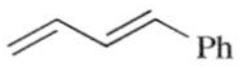 1-phenylbutadiene (Ph)	1. Ru cat **364** (0.01 eq), (Z)-2-butene (5 eq), THF, 22°, 1 h 2. Ru cat **364** (0.04 eq), **I** (5 eq), THF, 100 torr, 22°, 1 h; then ambient pressure, 7 h 3. Ru cat **364** (0.04 eq), THF, 100 torr, 22°, 1 h; then ambient pressure, 11 h	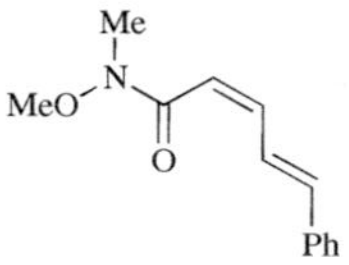 (58) (Z)/(E) > 98:2	1037

5 eq	OH	1. Ru cat **364** (0.04 eq), THF, 100 torr, 22°, 1 h, ambient pressure, 11 h 2. Ru cat **364** (0.04 eq), THF, 100 torr, 22°, 1 h, ambient pressure, 11 h	Me, MeO N, O, OH (80) (*Z*)/(*E*) > 98:2	1037
PhHN, O 1.1 eq	( )$_2$	Hov II (0.15 eq), toluene, DCM, rt, 72 h	O, NHPh (21) (*E*)/(*Z*) = 75:25 + O, NHPh (19) (*E*)/(*Z*) = 75:25	539

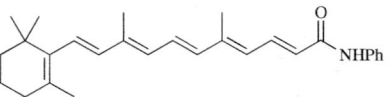

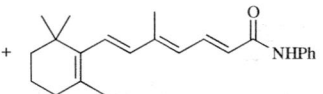

[a] The conversion was determined by ^{1}H NMR analysis.

TABLE 22. CROSS-METATHESIS OF SECONDARY ALLYLIC HALIDES

Allylic Halide	Cross Partner	Conditions	Product(s) and Yield(s) (%)	Refs.
Please refer to the charts preceding the tables for the catalyst structures.				
C_4				
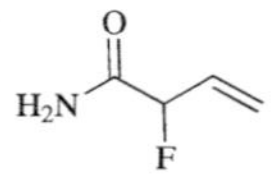 H_2N, O, F	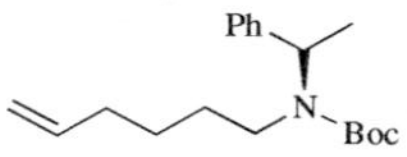 Ph, N, Boc	Hov II (0.05 eq), DCM, 40°	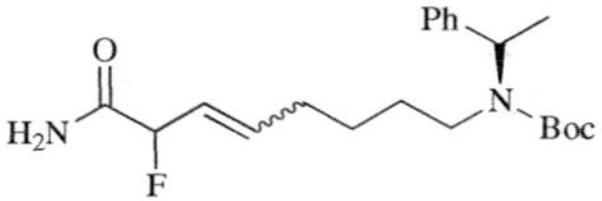 O, H_2N, F, Ph, N, Boc (81)	1141
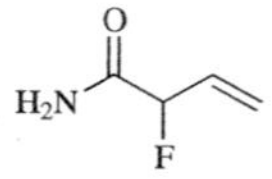 H_2N, O, F	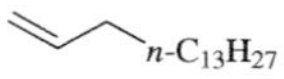 n-$C_{13}H_{27}$ 1.5 eq	Hov II (0.07 eq), DCM, 25°, 24 h	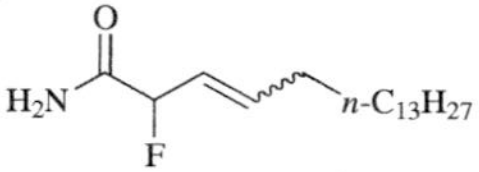 O, H_2N, F, n-$C_{13}H_{27}$ (43)	1142
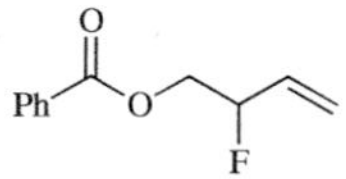 Ph, O, O, F	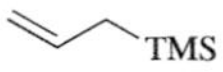 TMS 5 eq	Hov II (0.02 eq), DCM, 100°, 16 h	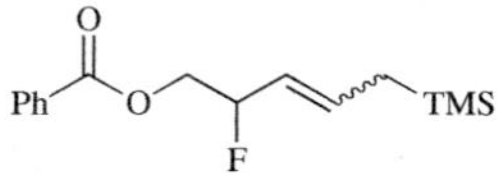 Ph, O, O, F, TMS (0) 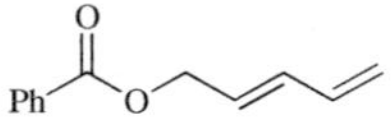 + Ph, O, O (40) (E)/(Z) = 95:5	142

5 eq

Hov II (x eq), DCM, 100°

Ph O F R

142

R	x	Time (h)		(E)/(Z)
$BocMeNCH_2$	0.04	48	(53)	95:5
$Br(CH_2)_2$	0.02	12	(85)	90:10
$BzO(CH_2)_2$	0.04	24	(67)	67:33

x eq

Catalyst (y eq), DCM, 12 h

Ph O F R

142

R	x	Catalyst	y	Temp (°)	Time (h)		(E)/(Z)
H	5	Gr II	0.02	40	12	(30)	(E) only
H	5	Hov II	0.02	25	12	(30)	(E) only
H	5	Gr II	0.02	100	12	(69)	(E) only
H	1	Gr II	0.02	100	12	(43)	(E) only
H	5	Hov II	0.02	100	12	(69)	95:5
CF_3	5	Hov II	0.03	100	16	(70)	95:5
MeO	5	Hov II	0.06	100	48	(0)	—

TABLE 22. CROSS-METATHESIS OF SECONDARY ALLYLIC HALIDES (*Continued*)

Allylic Halide	Cross Partner	Conditions	Product(s) and Yield(s) (%)	Refs.

Please refer to the charts preceding the tables for the catalyst structures.

C_4

Allylic Halide	Cross Partner	Conditions	Product(s) and Yield(s) (%)	Refs.
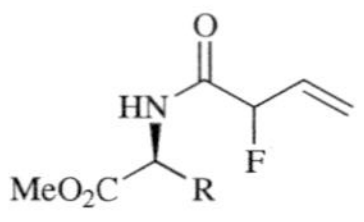 O, HN, F, MeO_2C, R	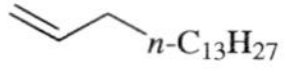 n-$C_{13}H_{27}$, 1.5 eq	Catalyst (x eq), DCM, 25°	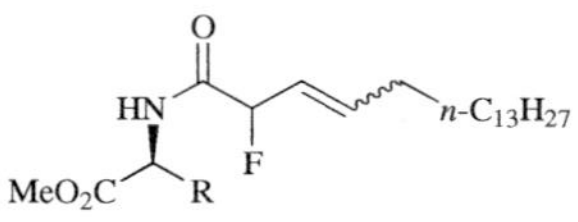 O, HN, F, n-$C_{13}H_{27}$, MeO_2C, R	1142

R	Catalyst	x	Time (h)	
$TBSOCH_2$	Hov II	0.07	24	(53)
MeO_2CCH_2	Hov II	0.07	24	(54)
i-Pr	Hov II	0.07	24	(48)
Ph	Hov II	0.07	24	(55)
$PhCH_2$	Hov II	0.07	24	(4)
$PhCH_2$	Gr II	0.07	24	(15)
$PhCH_2$	Ru cat **199**	0.07	24	(28)
$PhCH_2$	Ru cat **93**	0.07	24	(45)
$PhCH_2$	Hov II	0.017	24	(65)
$PhCH_2$	Hov II	0.025	24	(38)
$PhCH_2$	Hov II	0.15	24	(63)
$PhCH_2$	Hov II	0.07	2	(57)
$PhCH_2$	Hov II	0.07	6	(61)
$PhCH_2$	Hov II	0.07	10	(64)

	Substrate	Olefin	Conditions	Product(s) and Yield(s) (%)	Refs.
	MeO$_2$C, H N, HN, O, F, O, Ph	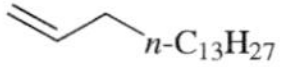 n-$C_{13}H_{27}$ 1.5 eq	Hov II (0.07 eq), DCM, 25°, 24 h	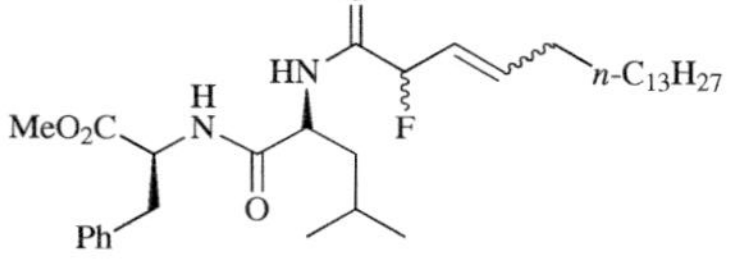 MeO$_2$C, H N, HN, O, F, n-$C_{13}H_{27}$, O, Ph (54)	1142
	O, N, F, O	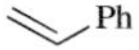 Ph 5 eq	Hov II (0.04 eq), DCM, 100°, 24 h	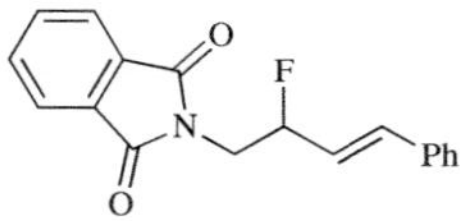 O, F, N, Ph, O (87) $(E)/(Z)$ = 95:5	142
C_{12}					
	Cl, OH, n-C_6H_{13}, Cl, Cl	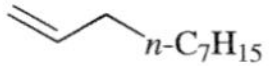 n-C_7H_{15} 10 eq	Catalyst (0.1 eq), toluene, rt, 40 h	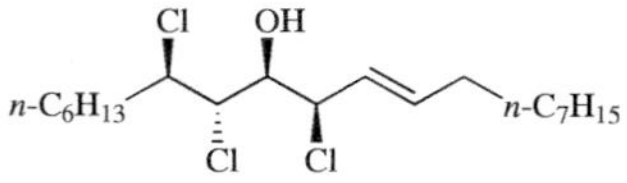 Cl, OH, n-C_6H_{13}, n-C_7H_{15}, Cl, Cl	915

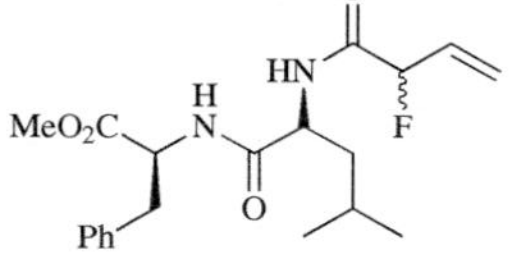

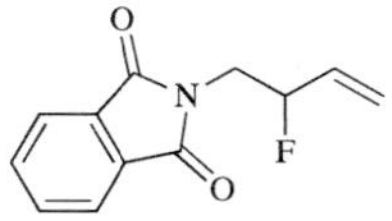

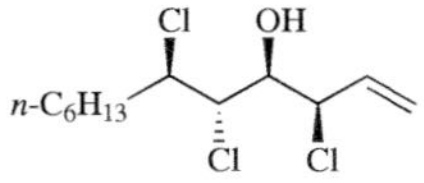

Catalyst		$(E)/(Z)$
Gr II	(43)	(E) only
Hov II	(60)	84:16

TABLE 23. CROSS-METATHESIS OF ALLYLSTANNANES

Allylstannane	Cross Partner	Conditions	Product(s) and Yield(s) (%)	Refs.
Please refer to the charts preceding the tables for the catalyst structures.				
C_3				
R_3Sn (allyl), 2 eq	CN	Mo cat **1** (0.05 eq), DCM, reflux	R_3Sn CN R / (E)/(Z) *n*-Bu (72) 50:50 Ph (22) 64:35	212
2 eq	CO_2Me	Mo cat **1** (0.05 eq), DCM, reflux	R_3Sn CO_2Me R / (E)/(Z) *n*-Bu (74) 69:31 Ph (78) 73:27	212
2 eq	CO_2Et, CO_2Et	Mo cat **1** (0.05 eq), DCM, reflux	R_3Sn CO_2Et, CO_2Et R / (E)/(Z) *n*-Bu (15) 60:40 Ph (31) 63:37	212
2 eq	CO_2Me, CO_2Me	Mo cat **1** (0.05 eq), DCM, reflux	R_3Sn CO_2Me, CO_2Me R / (E)/(Z) *n*-Bu (41) 67:33 Ph (49) (*E*) only	212
2 eq	H N, O, Ar, O, AcO Ar = 9-fluorenyl	Mo cat **1** (0.05 eq), DCM, reflux	R_3Sn H N, O, Ar, O, AcO R / (E)/(Z) *n*-Bu (13) 53:47 Ph (50) 58:42	212

Equiv	Substrate	Conditions	Product(s) and Yield(s) (%)			Refs.
2 eq	Ph	Mo cat **1** (0.05 eq), DCM, reflux	R_3Sn, Ph			212
			R		(*E*)/(*Z*)	
			n-Bu	(55)	58:42	
			Ph	(74)	55:45	
2 eq	OAc, AcO, AcO, O, O, OAc	Mo cat **1** (0.05 eq), DCM, reflux	OAc, AcO, AcO, O, O, OAc, SnR_3			212
			R		(*E*)/(*Z*)	
			n-Bu	(21)	(*Z*) only	
			Ph	(67)	70:30	
2 eq	O, O	Mo cat **1** (0.05 eq), DCM, reflux	R_3Sn, O, O			212
			R		(*E*)/(*Z*)	
			n-Bu	(21)	57:43	
			Ph	(71)	63:37	

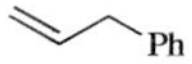

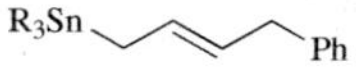

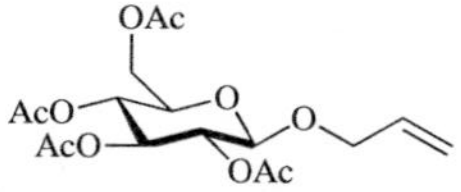

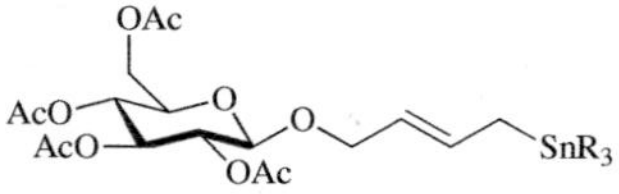

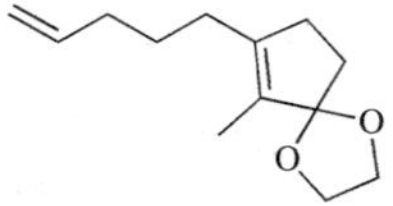

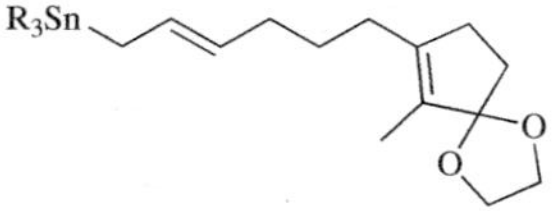

TABLE 24. CROSS-METATHESIS OF ALKENES WITH QUATERNARY ALLYLIC CARBON CENTERS

Alkene	Cross Partner	Conditions	Product(s) and Yield(s) (%)	Refs.
Please refer to the charts preceding the tables for the catalyst structures.				
C_5				
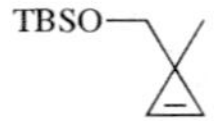 TBSO	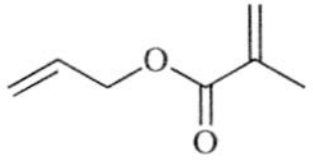 2.05 eq	Ru cat **213** (0.05 eq), THF, 0°, 48 h	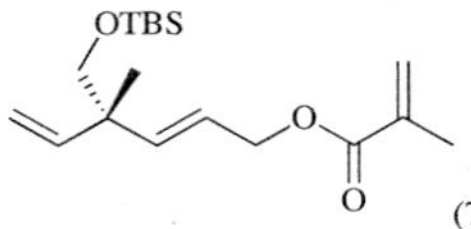 OTBS, O, O (76) (*E*)/(*Z*) = 80:20, er (*E*) 93.0:7.0	1143
C_6				
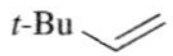 *t*-Bu, excess	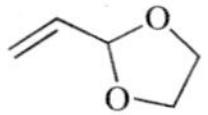 O, O	Gr II (0.03–0.05 eq), 40°, 12 h	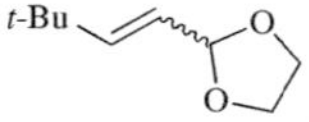 *t*-Bu, O, O (75)	36
excess	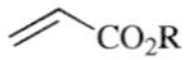 CO_2R	Gr II (0.03–0.05 eq), 40°, 12 h	 *t*-Bu CO_2R; R: H (73); *t*-Bu (73)	36
	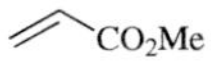 CO_2Me, 10 eq	Gr II (0.001 eq), *p*-cresol (*x* eq), 50°, 2 h	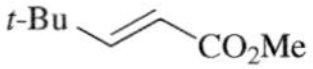 *t*-Bu CO_2Me, (*E*)/(*Z*) = 98:2; *x* / Conv. (%)[a]: — / 11; 500 / 18	130
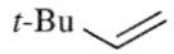 *t*-Bu, excess, + 	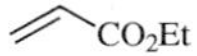 CO_2Et, 1 eq	Gr II (0.075 eq), neat, 23°, 12 h	 *t*-Bu CO_2Et (67) (*E*) only	36

Alkene	Substrate	Conditions	Product(s) and Yield(s) (%)	Refs.
excess		Gr II (0.03–0.05 eq), DCM, 40°, 12 h	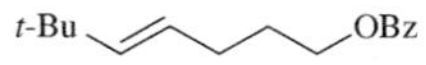 (99) (*E*) only	36
1.9 eq	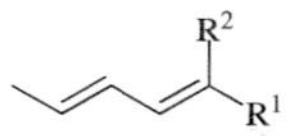	Gr II (2 x 0.05 eq), DCM, 40°, 24 h	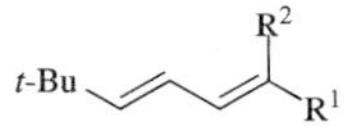	649
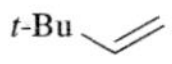 t-Bu (10 eq)	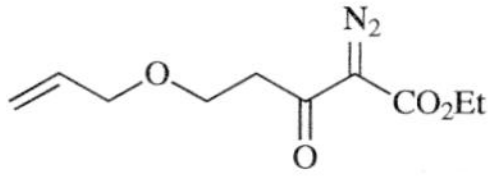	Gr II (0.05 eq), DCM, 18 h, reflux	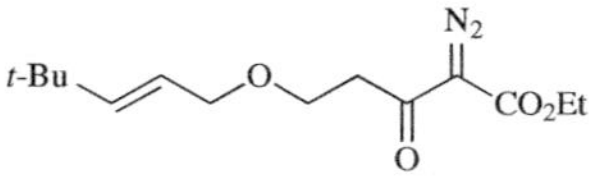 (50) (*E*)/(*Z*) > 99:1	569
	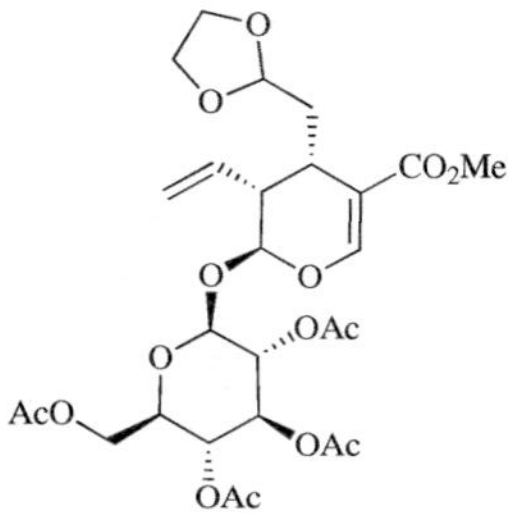	Gr II (0.1 eq), toluene, 100°	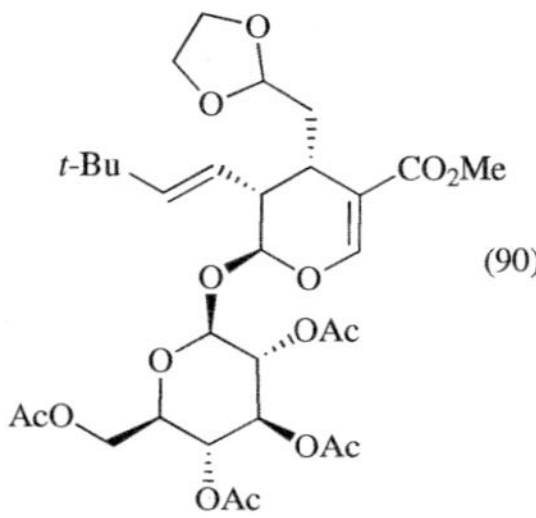 (90) (*E*)/(*Z*) = 95:5	410

R^1	R^2		(*E*)/(*Z*)
Br	NC–	(16)	82:18
NC–	Br	(19)	80:20

TABLE 24. CROSS-METATHESIS OF ALKENES WITH QUATERNARY ALLYLIC CARBON CENTERS (*Continued*)

Alkene	Cross Partner	Conditions	Product(s) and Yield(s) (%)	Refs.
Please refer to the charts preceding the tables for the catalyst structures.				
C_6				
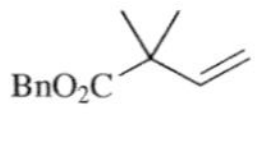 BnO_2C 3.95 eq	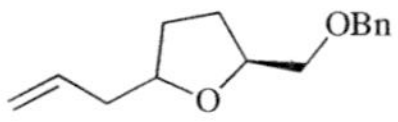 OBn	Gr II (0.1 eq), Cy_2BCl (0.1 eq), toluene, 90° (MW), 9 h	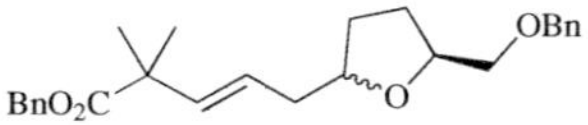 BnO_2C OBn (22) (*E*) only	1144
C_7				
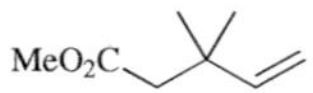 MeO_2C 3.6 eq	 NO_2	Gr II (0.052 eq), DCM, reflux, 18 h	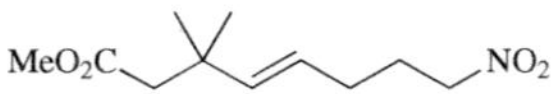 MeO_2C NO_2 (73) (*E*) only	146
C_8				
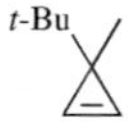 *t*-Bu	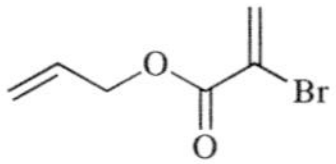 Br 2.05 eq	Ru cat **213** (0.05 eq), THF, 22°, 16 h	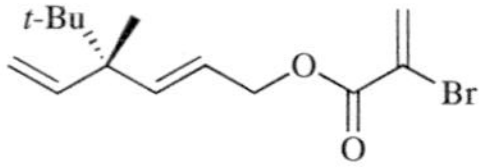 *t*-Bu Br (61) (*E*)/(*Z*) = 67:33, er (*E*) 95.0:5.0	1143
C_9				
 *n*-Bu 2 eq	 OAc	Gr II (0.03–0.05 eq), DCM, 40°, 12 h	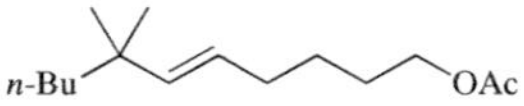 *n*-Bu OAc (90) (*E*) only	36
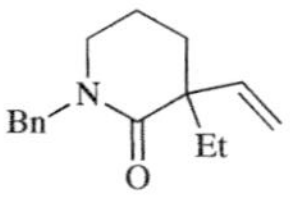 Bn N Et 10 eq	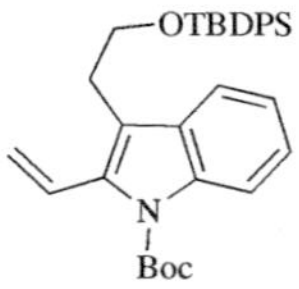 OTBDPS N Boc	Hov II (0.3 eq), 140°, 3 h	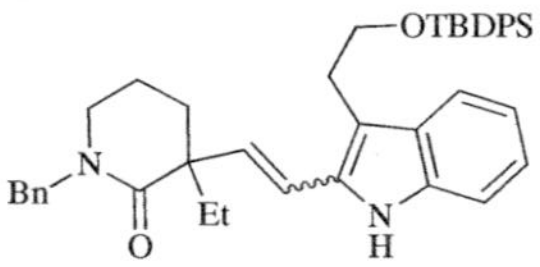 OTBDPS Bn N Et N H (42)	1145

C_{10}

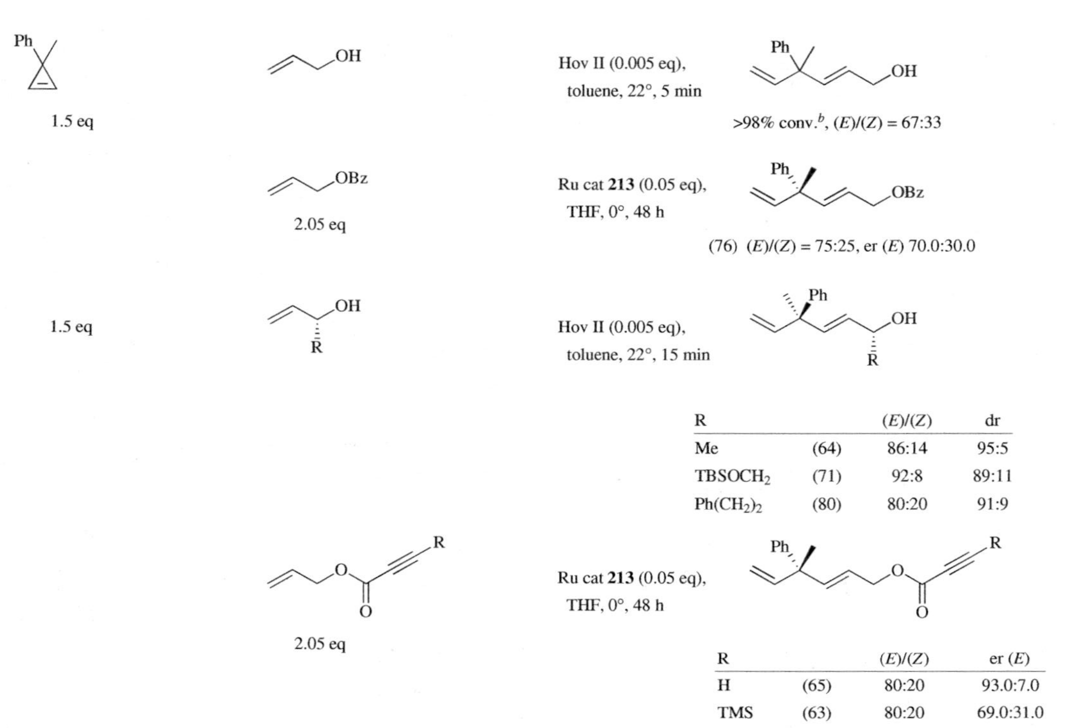

R		(E)/(Z)	dr
Me	(64)	86:14	95:5
$TBSOCH_2$	(71)	92:8	89:11
$Ph(CH_2)_2$	(80)	80:20	91:9

R		(E)/(Z)	er (E)
H	(65)	80:20	93.0:7.0
TMS	(63)	80:20	69.0:31.0
Me	(72)	67:33	86.0:14.0

TABLE 24. CROSS-METATHESIS OF ALKENES WITH QUATERNARY ALLYLIC CARBON CENTERS (*Continued*)

Alkene	Cross Partner	Conditions	Product(s) and Yield(s) (%)	Refs.
Please refer to the charts preceding the tables for the catalyst structures.				
C_{10}				
Ph	R²; R¹; O; O; 2.05 eq	Ru cat **213** (0.05 eq), THF, 0°, 48 h	Ph; R²; R¹; O; O	1143
	O; 2.05 eq	Ru cat **213** (0.05 eq), THF, 0°, 48 h	Ph; O; (76) (*E*)/(*Z*) = 67:33, er (*E*) 64.0:36.0	1143
	O; O; 2.05 eq	Ru cat **213** (0.05 eq), THF, 0°, 48 h	Ph; O; O; (59) (*E*)/(*Z*) = 75:25, er (*E*) 35.0:65.0	1143

R^1	R^2		(*E*)/(*Z*)	er (*E*)	er (*Z*)
H	H	(71)	82:18	8.0:92.0	—
H	Me	(77)	83:17	8.0:92.0	—
H	CF_3	(63)	86:14	9.0:91.0	65.0:34.0
H	Br	(80)	88:12	1.0:99.0	—
Br	H	(69)	86:14	1.0:99.0	75.0:25.0

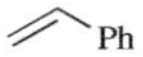

1.5 eq

Ph

2.05 eq

Ru cat **213** (0.05 eq), THF, 0°, 48 h

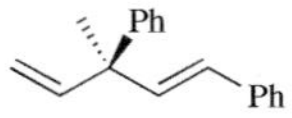

Ph Ph

(90) (*E*)/(*Z*) > 95:5, er (*E*) 97.0:3.0 | 1143

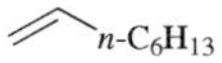

n-C_6H_{13}

Catalyst (x eq), solvent

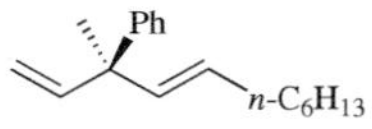

Ph n-C_6H_{13}

Catalyst	x	Solvent	Temp (°)	Time (h)	Conv. (%)[b]		(*E*)/(*Z*)	er (*E*)	
Ru cat **213**	0.05	THF	0	48	—	(59)	60:40	60.0:40.0	1143
Hov II	0.005	toluene	22	4	83	(—)	75:25	—	624

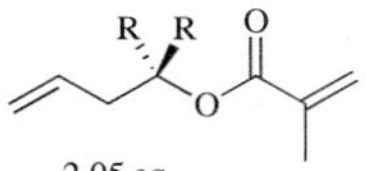

R R O

2.05 eq

Ru cat **213** (0.05 eq), THF, 0°, 48 h

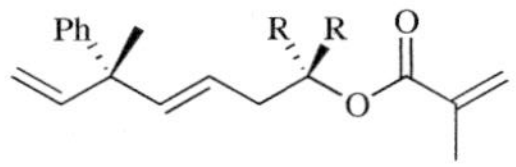

Ph R R O

R		(*E*)/(*Z*)	er (*E*)
H	(71)	80:20	80.0:20.0
Me	(60)	83:17	90.0:10.0

1143

TABLE 24. CROSS-METATHESIS OF ALKENES WITH QUATERNARY ALLYLIC CARBON CENTERS (*Continued*)

Alkene	Cross Partner	Conditions	Product(s) and Yield(s) (%)	Refs.
Please refer to the charts preceding the tables for the catalyst structures.				
C_{10}				
 Ph	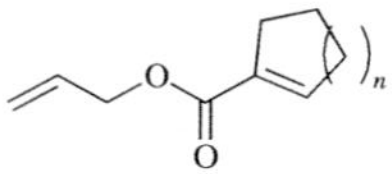 O, O, n 2.05 eq	Ru cat **213** (0.05 eq), THF, 0°, 48 h	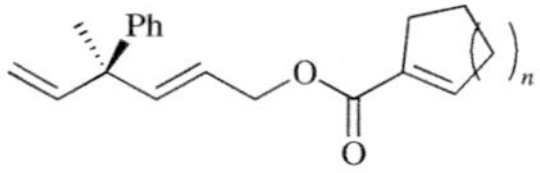 Ph, O, O, n	1143
			n — (*E*)/(*Z*) — er (*E*) — er (*Z*) 1 — (72) — 83:17 — 5.0:95.0 — 74.0:26.0 2 — (83) — 75:25 — 6.0:94.0 — 70.0:30.0	
	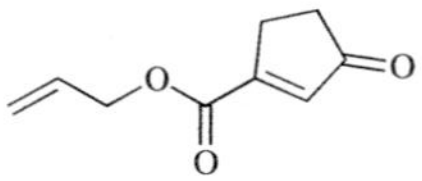 O, O, O 2.05 eq	Ru cat **213** (0.05 eq), THF, 0°, 48 h	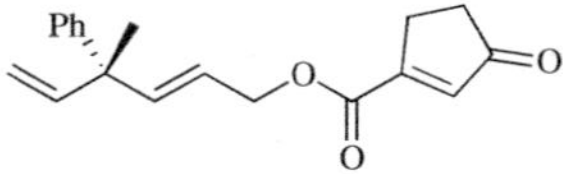 Ph, O, O, O (63) (*E*)/(*Z*) = 67:33, er (*E*) 83.0:17.0	1143
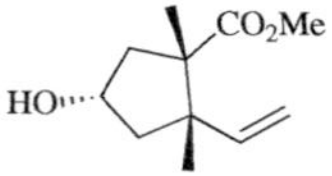 HO, CO_2Me	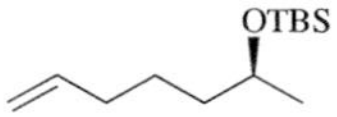 OTBS 3 eq	Gr II (0.05 eq), DCM, reflux	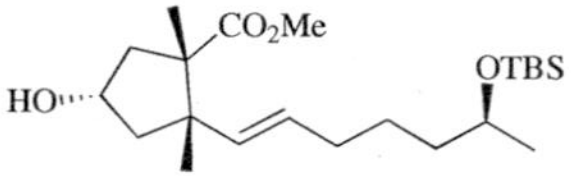 HO, CO_2Me, OTBS (24) (*E*)/(*Z*) = 80:20	462

	Substrate 1	Substrate 2	Conditions	Product (Yield, %)	Ref.
C_{15}	TBSO (2 eq)	F, F–B–N, N	Gr II (0.1 eq), DCM, reflux, 6 h	TBSO; F, F–B–N, N; n = 3 (65); 4 (53); 5 (48)	1146
C_{17}	Ph, O, N, O	OTBDPS, N, Boc (3.5 eq)	Hov II (0.06 eq), xylene, 140°, 3 h	Ph, O, N, O; OTBDPS, N, H (65)	1145

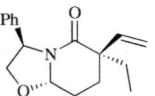

[a] The conversion was determined by GC/GC-MS analysis.

[b] The conversion was determined by ^{1}H NMR analysis.

TABLE 25. CROSS-METATHESIS OF GEMINAL–DISUBSTITUTED COMPOUNDS

A. ALKENES WITH FUNCTIONAL GROUPS IN ALLYLIC AND MORE DISTANT POSITIONS

Alkene	Cross Partner	Conditions	Product(s) and Yield(s) (%)	Refs.
Please refer to the charts preceding the tables for the catalyst structures.				
C_4				
 excess	 CO_2Ad	Gr II (0.01 eq), neat, 40°, 12 h	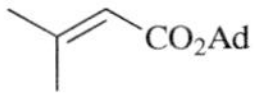 CO_2Ad (83)	129
 excess 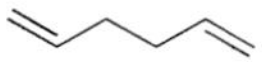 + 3 eq	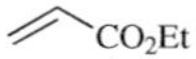 CO_2Et	Gr II (0.05 eq), neat, 40°, 12 h	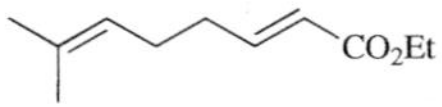 CO_2Et (60) (*E*) only	36
 excess 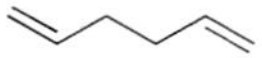 + 3 eq	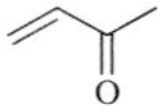 O	Gr II (0.07 eq), neat, 40°, 12 h	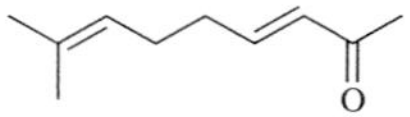 O (89) (*E*) only	36
 excess	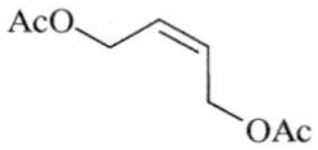 AcO, OAc	Gr II (0.01 eq), neat, 40°, 12 h	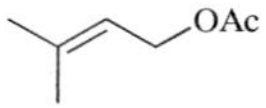 OAc (88)	129
excess	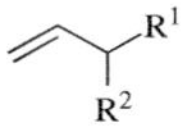 R^1, R^2	Gr II (0.01 eq), neat, 40°, 12 h	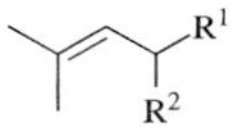 R^1, R^2 R^1 / R^2 BzO / Me (99) $TBSO(CH_2)_2$ / Me (72) $AcO(CH_2)_3$ / H (97)	129

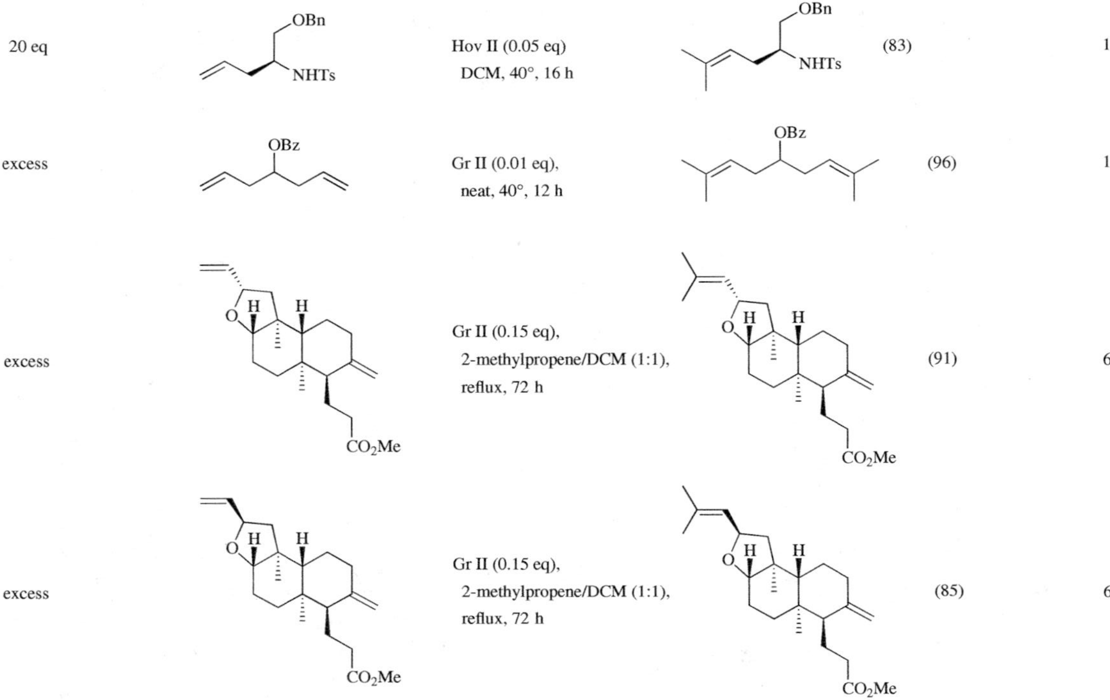

20 eq		Hov II (0.05 eq) DCM, 40°, 16 h		(83)	1147
excess		Gr II (0.01 eq), neat, 40°, 12 h		(96)	1147
excess		Gr II (0.15 eq), 2-methylpropene/DCM (1:1), reflux, 72 h		(91)	622
excess		Gr II (0.15 eq), 2-methylpropene/DCM (1:1), reflux, 72 h		(85)	622

TABLE 25. CROSS-METATHESIS OF GEMINAL–DISUBSTITUTED COMPOUNDS (*Continued*)

A. ALKENES WITH FUNCTIONAL GROUPS IN ALLYLIC AND MORE DISTANT POSITIONS (*Continued*)

Alkene	Cross Partner	Conditions	Product(s) and Yield(s) (%)	Refs.
Please refer to the charts preceding the tables for the catalyst structures.				
C_4				
Cl 6 eq	R	Ru cat **139** (0.02 eq), DCM, reflux, 5 h	R, Cl	620
			R / / (E)/(Z) n-C_7H_{15} (79) 75:25 $NC(CH_2)_7$ (67) 75:25 $MeO_2C(CH_2)_7$ (72) 74:26 3-MeO-4-$AcOC_6H_3$ (70) 75:25	
HO 20 eq	N, R	Gr II (0.1 + 0.1 + 0.05 eq) DCE, 40°, 6 + 6 + 12 h	HO, N, R R / / H (52) MeO (56)	1148
C_{4-5}				
HO, R 5 eq	CO_2Bn	Hov II (0.05 eq) DCM, 40°, 12 h	HO, CO_2Bn	1149
			R / / (E)/(Z) H (61) 87:13 (Z)-Me (64) 88:12 (E)-Me (71) 91:9	

C_4

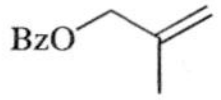

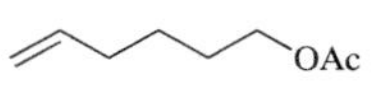

BzO (2-methylallyl benzoate) + hex-5-enyl acetate (2 eq); Gr II (0.03–0.05 eq), DCM, 40°, 12 h

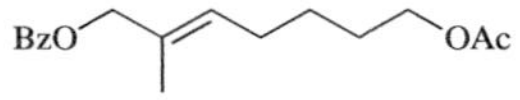

BzO ... OAc

(80) (E)/(Z) = 80:20

36, 241

C_{4-5}

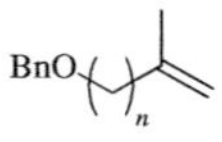

BnO (n), x eq; 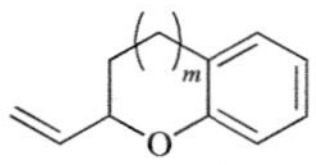 Gr II (0.05 eq), neat, 60°

BnO (n) ... O (m)

832

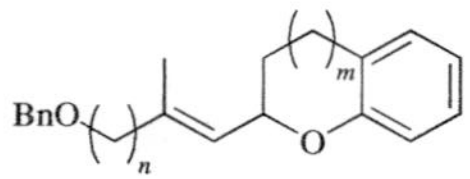

x	n	m	Time (h)		(E)/(Z)
4	1	0	37	(57)	80:20
6	2	0	48	(58)	68:32
4	1	1	70	(51)	78:22
6	2	1	14	(38)	52:48

C_{4-9}

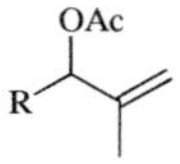

OAc, R, 4 eq; CO_2Me; Gr II (0.04 eq), DCM, reflux, 15 h

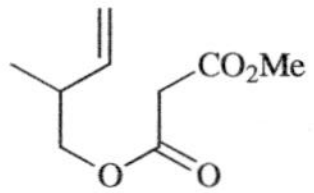

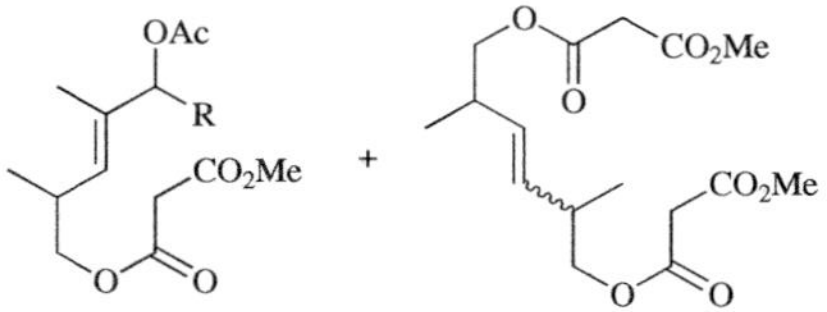

I (OAc, R, CO_2Me) + **II** (CO_2Me, CO_2Me)

586

R	I	II	(E)/(Z) I
H	(43)	(45)	80:20
n-Bu	(28)	(68)	(E) only
c-C_5H_9	(0)	(53)	—

TABLE 25. CROSS-METATHESIS OF GEMINAL–DISUBSTITUTED COMPOUNDS (*Continued*)

A. ALKENES WITH FUNCTIONAL GROUPS IN ALLYLIC AND MORE DISTANT POSITIONS (*Continued*)

Alkene	Cross Partner	Conditions	Product(s) and Yield(s) (%)	Refs.

Please refer to the charts preceding the tables for the catalyst structures.

C_{4-14}

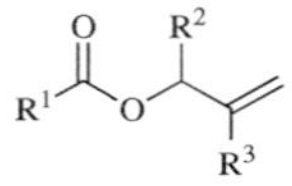

Alkene	Cross Partner	Conditions	Product(s) and Yield(s) (%)	Refs.
$R^1C(O)OCH(R^2)C(R^3)=CH_2$, *x* eq	$CH_2=CHCH_2CH_2OC(O)CH_2CO_2Me$, *y* eq	Gr II (*z* eq), DCM, reflux	I + II	586

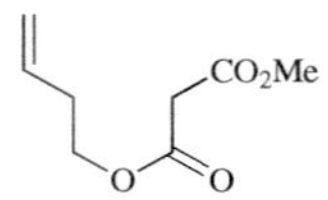

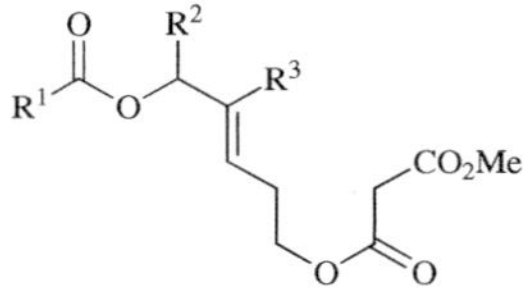

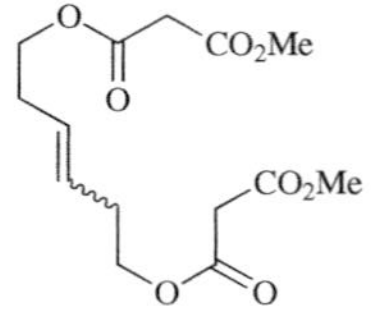

R^1	R^2	R^3	*x*	*y*	*z*	Time (h)	I	(*E*)/(*Z*) I	II
Me	H	Me	1	3.3	0.03	15	(28)	77:23	(0)
Me	H	Me	1	4	0.04	15	(78)	83:17	(tr)
Me	H	Me	2	1	0.04	15	(28)	77:23	(57)
EtO	H	Me	1	3.1	0.03	22	(24)	78:22	(0)
EtO	H	Me	1	4	0.04	15	(58)	78:22	(15)
Ph	H	Me	1	3.4	0.03	15	(20)	77:23	(—)
Ph	H	Me	2.5	1	0.04	15	(43)	79:21	(—)
BnO	H	Me	2	1	0.04	15	(40)	78:22	(42)
Me	H	Et	4	1	0.04	15	(59)	50:50	(0)
Me	H	*n*-Bu	4	1	0.04	15	(36)	50:50	(51)
Me	H	*c*-C_5H_9	4	1	0.04	15	(0)	—	(63)
Me	Et	Me	4	1	0.04	15	(31)	95:5	(39)
Me	*n*-Bu	Me	4	1	0.04	15	(38)	95:5	(42)
Me	*n*-$C_{10}H_{21}$	Me	4	1	0.04	15	(38)	95:5	(30)
BnO	*n*-Bu	Me	4	1	0.04	15	(34)	96:4	(44)
Me	*i*-Pr	Me	4	1	0.04	15	(11)	100:0	(74)
Me	*c*-C_5H_9	Me	4	1	0.04	15	(7)	94:6	(80)

C_4

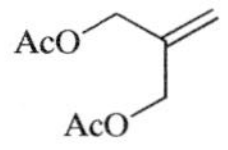

AcO AcO 2 eq

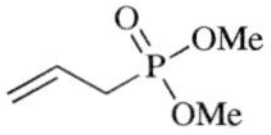

OMe OMe

Catalyst (x eq), solvent

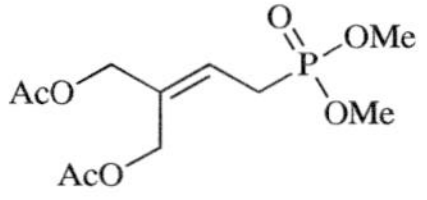

AcO AcO OMe OMe

1150

Catalyst	x	Solvent	Temp (°)	Time (h)	
Ru cat **35**	0.05	DCM	40	24	(9)
Hov II	0.05	DCM	40	24	(16)
Ru cat **35**	3 x 0.05	DCM	40	24	(23)
Ru cat **35**	3 x 0.05	DCE	40	24	(32)
Ru cat **35**	0.1	DCM	110 (MW)	4	(38)
Hov II	0.1	DCM	110 (MW)	4	(70)
Ru cat **35**	3 x 0.05	DCM	55 (U/S)	20	(92)
Ru cat **35**	3 x 0.03	DCM	55 (U/S)	20	(90)
Ru cat **35**	3 x 0.02	DCM	55 (U/S)	20	(86)
Hov II	3 x 0.02	DCM	55 (U/S)	20	(93)
Ru cat **35**	0.05	DCM	55 (U/S)	20	(49)
Ru cat **35**	3 x 0.01	DCM	55 (U/S)	20	(76)

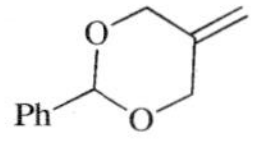

Ph 5 eq

R NH

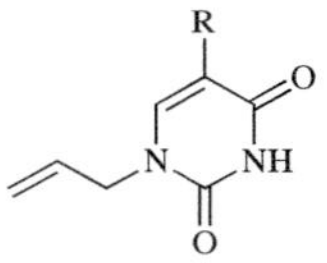

Ru cat **35** (0.1 eq), DCM, 40°, 5 h

Ph R NH

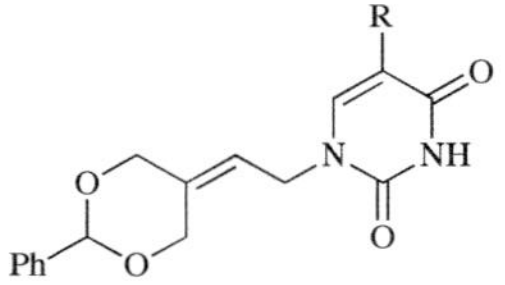

(E) only

R	
H	(53)
F	(70)
I	(57)
Me	(64)

628

TABLE 25. CROSS-METATHESIS OF GEMINAL–DISUBSTITUTED COMPOUNDS (*Continued*)
A. ALKENES WITH FUNCTIONAL GROUPS IN ALLYLIC AND MORE DISTANT POSITIONS (*Continued*)

Alkene	Cross Partner	Conditions	Product(s) and Yield(s) (%)	Refs.

Please refer to the charts preceding the tables for the catalyst structures.

C_4

BzO, BzO — *x* eq | OAc, *n* — *y* eq | Catalyst (*z* eq), solvent | BzO, BzO, OAc, *n*

x	*n*	*y*	Catalyst	*z*	Solvent	Temp (°)	Time	Yield	Refs.
1	3	3	Hov II	0.05	DCM	reflux	24 h	(73)	177
1	3	3	Ru cat **271**	0.05	DCM	reflux	24 h	(54)	177
1	3	3	Ru cat **187**	0.05	DCM	reflux	24 h	(96)	177
3	3	1	Ru cat **98**	0.05	DCM	40	24 h	(32)	957
3	3	1	Ru cat **170**	0.05	DCM	40	24 h	(10)	957
3	3	1	Ru cat **170**	0.05	DCM	40	10 h	(3)	957
1	3	3	Ru cat **170**	0.05	DCM	40	10 h	(43)	957
1	3	3	Hov II	0.05	DCM	40	24 h	(32)	957
1	3	3	Ind II	4 x 0.02	$C_6F_5CF_3$	120	4 x 10 min	(41)	40
1	3	3	Ind II	4 x 0.02	$C_6F_5CF_3$	120 (MW)	4 x 10 min	(62)	40
1	4	2	Gr II	0.03	DCM	40	12 h	(48)	129

AcO, 2, 2, OAc — *x* eq — (*E*)/(*Z*) = 75:25 | Catalyst (0.05 eq), DCM, 40° | BzO, BzO, OAc | 957

x	Catalyst	Time (h)	Yield
3	Ru cat **170**	10	(54)
2	Ru cat **98**	24	(32)
2	Ru cat **170**	24	(33)

C_5

Olefin	Substrate	Conditions	Product(s) and Yield(s) (%)	Refs.
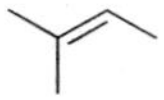 x eq	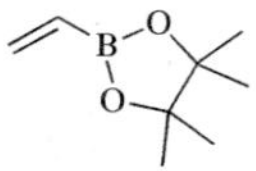	Gr II (0.05 eq), DCM, reflux, 12 h	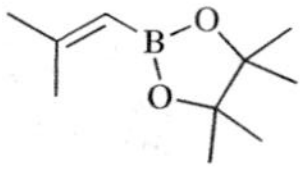 x: 1 (26); 0.5 (22); 2 (48)	132
excess	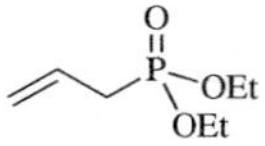 O=P(OEt)(OEt)	Gr II (0.01 eq), neat, 23°, 12 h	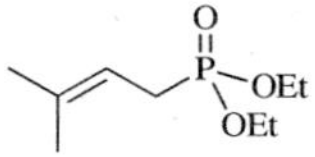 O=P(OEt)(OEt) (97)	129
excess	 OAc	Gr II (0.01 eq), neat, 23°, 12 h	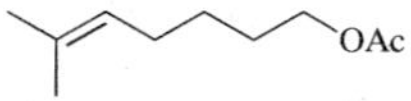 OAc (97)	129
excess	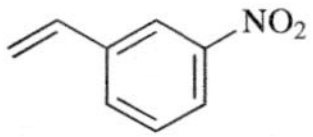 NO_2	Gr II (0.01 eq), neat, 23°, 12 h	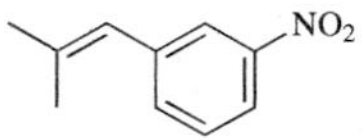 NO_2 (96)	129
excess	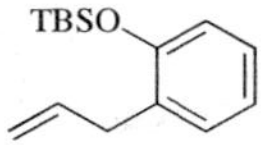 TBSO	Gr II (0.01 eq), neat, 23°, 12 h	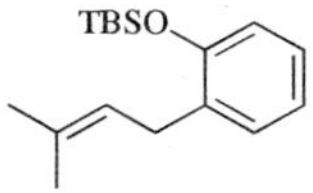 TBSO (99)	129
excess	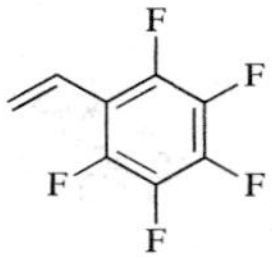 F F F F F	Gr II (0.01 eq), 23°, 12 h 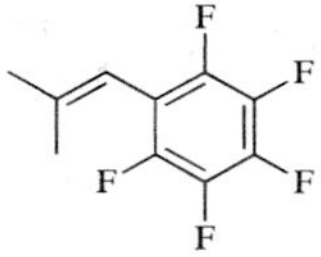	F F F F F (91)	129

TABLE 25. CROSS-METATHESIS OF GEMINAL–DISUBSTITUTED COMPOUNDS (*Continued*)

A. ALKENES WITH FUNCTIONAL GROUPS IN ALLYLIC AND MORE DISTANT POSITIONS (*Continued*)

Alkene	Cross Partner	Conditions	Product(s) and Yield(s) (%)	Refs.
Please refer to the charts preceding the tables for the catalyst structures.				
C_5				
10 eq	N_2 CO_2Et O	Gr II (0.05 eq), DCM, reflux, 18 h	N_2 CO_2Et O (82)	569
excess	OH 9	Ru cat **88** (0.01 eq), neat, 25°, 1 h	OH 9 (99)	171
excess	CHO 8	Gr II (0.01 eq), 23°, 12 h	CHO 8 (91)	129
excess	OH 4	Ru cat **88** (0.01 eq), neat, 25°, 1 h	OH 4 (99)	171
107 eq	BzO OBz OBz OBz OBz O	Gr II (0.03 eq), DCM, rt, 48 h	BzO OBz OBz OBz OBz O (88)	1151

excess	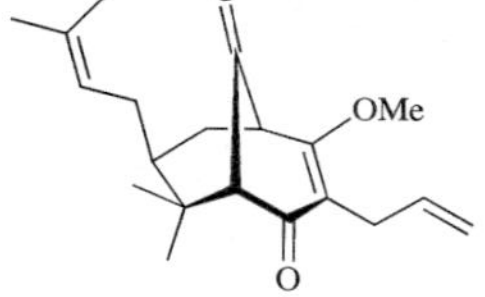 Gr II (0.1 eq), 38°, 2 h	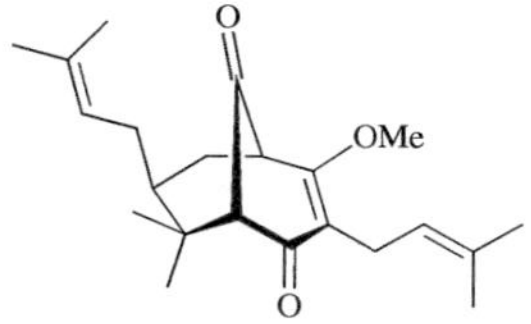	(88)	1152
20 eq	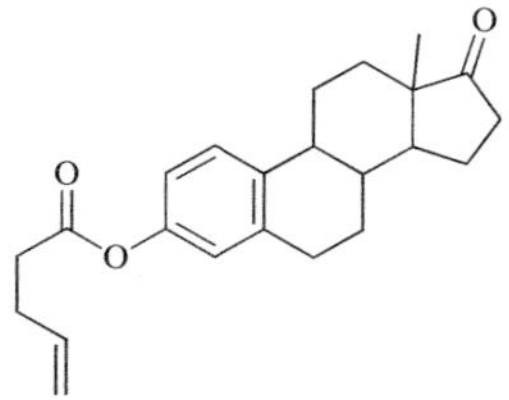 Catalyst (0.01 eq), TMSCl (0.1 eq) DCM, 40°, 2 h	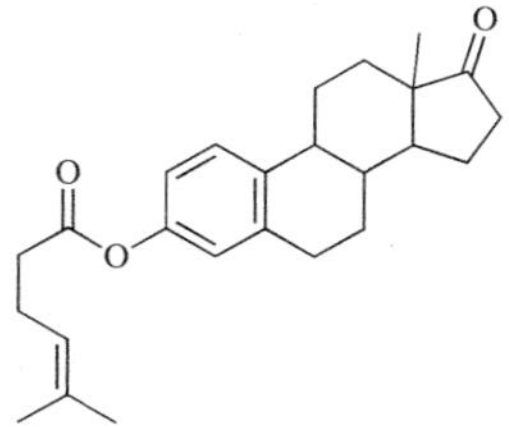 Catalyst Ru cat **194** Ru cat **196**	 (96) (83)	526
60 eq	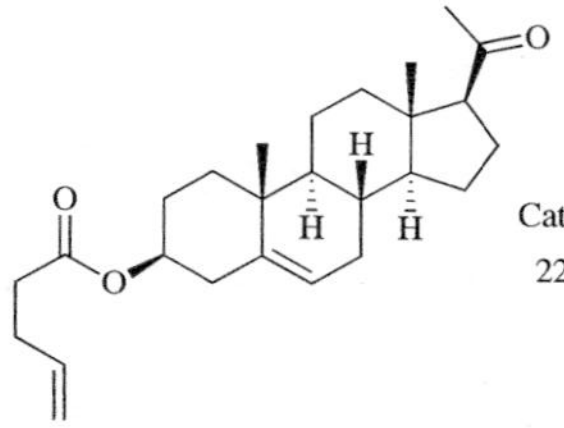 Catalyst (0.025 eq), 22°, 2 h	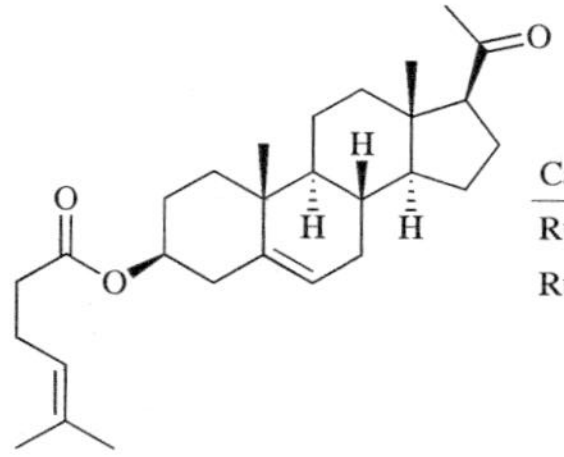 Catalyst Ru cat **327** Ru cat **328**	 (81) (78)	593

TABLE 25. CROSS-METATHESIS OF GEMINAL–DISUBSTITUTED COMPOUNDS (*Continued*)

A. ALKENES WITH FUNCTIONAL GROUPS IN ALLYLIC AND MORE DISTANT POSITIONS (*Continued*)

Alkene	Cross Partner	Conditions	Product(s) and Yield(s) (%)	Refs.
Please refer to the charts preceding the tables for the catalyst structures.				
C_5				
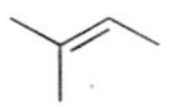	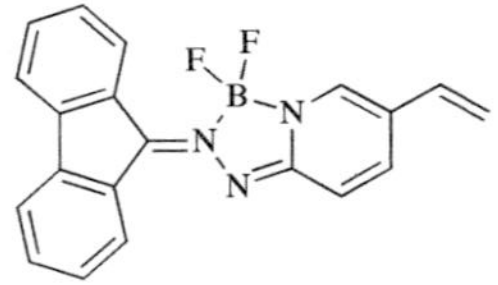	Hov **II** (0.05 eq), DCM, 40°	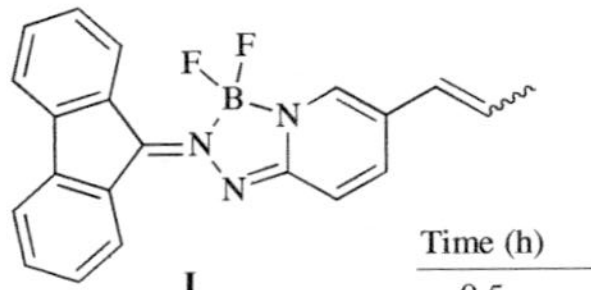 **I** 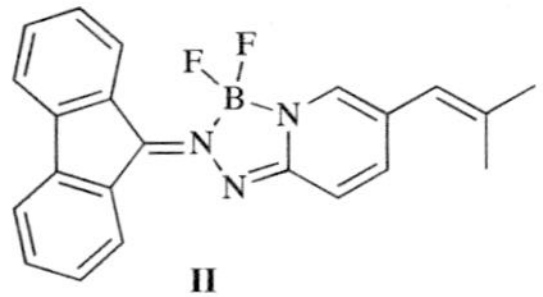 + **II** Time (h) / **I** / **II** 0.5 / (66) / (27) 8 / (52) / (40)	1153
	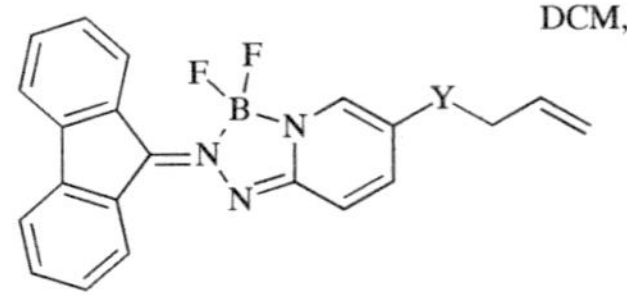	Hov **II** (0.05 eq), DCM, 40°	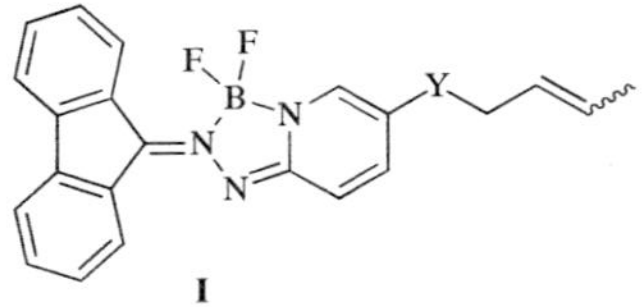 **I** + 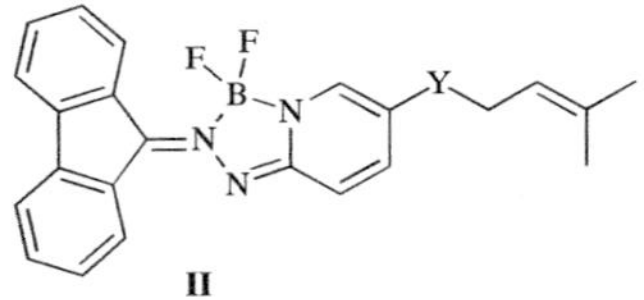 **II**	1153

Y	Time (h)	**I**	**II**
S	0.5	(24)	(57)
S	8	(17	(61)
SO_2	0.5	(25)	(72)
SO_2	8	(1)	(98)

R^1O — *x* eq — R^2; Catalyst (*y* eq), solvent

R^1	R^2	*x*	Catalyst	*y*	Solvent	Temp (°)	Time (h)		(*E*)/(*Z*)	
PNP	Et	0.5	Hov II	0.04	DCM	40	15	(65)	56:44	932
PNP	MeO	1	Hov II	0.04	DCM	40	15	(50)	50:50	932
Bz	MeO	2	Ind II	0.05	DCM	40	8	(0)	—	161
Bz	MeO	2	Ru cat **286**	0.05	DCE	80	8	(0)	—	161
Bz	MeO	0.5	Ind II	0.05	DCM	40	8	(0)	—	161
Bz	MeO	0.5	Ru cat **286**	0.05	DCE	80	8	(0)	—	161
Bz	MeO	2	Hov II	0.01	DCM	20	24	(0)	—	1040
Bz	MeO	2	Ru cat **85**	0.01	DCM	20	24	(0)	—	1040

R^1O **I** — R^2

1. Ru cat **354** (0.01 eq), (*Z*)-butene (5 eq), THF, 22°, 1 h
2. Ru cat **354** (0.05 eq), **I** (5 eq), THF, 100 torr, 22°, 1 h
3. Ambient pressure, 15 h

R^1O — R^2 (*Z*)/(*E*) > 98:2

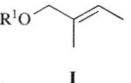

R^1	R^2		
H	$BnO_2C(CH_2)_2$	(75)	1149
H	$Br(CH_2)_6$	(63)	
H	$OHC(CH_2)_8$	(60)	
PMB	$HO(CH_2)_6$	(50)	
PMB	$Br(CH_2)_6$	(54)	

TABLE 25. CROSS-METATHESIS OF GEMINAL–DISUBSTITUTED COMPOUNDS (*Continued*)
A. ALKENES WITH FUNCTIONAL GROUPS IN ALLYLIC AND MORE DISTANT POSITIONS (*Continued*)

Alkene	Cross Partner	Conditions	Product(s) and Yield(s) (%)	Refs.

Please refer to the charts preceding the tables for the catalyst structures.

C_5

Alkene	Cross Partner	Conditions	Product(s) and Yield(s) (%)	Refs.
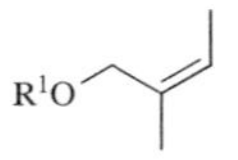 R^1O (I)	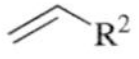 R^2	1. Ru cat **354** (0.01 eq), (*Z*)-butene (*x* eq), THF, 22°, 1 h 2. Ru cat **354** (0.05 eq), **I** (*y* eq), THF, 100 torr, 22°, 1 h 3. Ambient pressure, 15 h	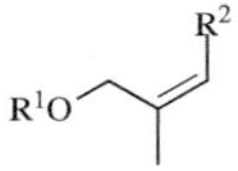 R^1O, R^2 (*Z*)/(*E*) > 98:2	1149

R^1	R^2	*x*	*y*	
H	$BnO_2C(CH_2)_2$	5	5	(74)
H	$HO(CH_2)_6$	5	10	(70)
H	$PMBO(CH_2)_7$	5	5	(78)
H	$Br(CH_2)_6$	10	5	(68)
H	$CF_3O_2C(CH_2)_9$	5	5	(55)
H	$OHC(CH_2)_8$	10	5	(70)
H	$Ph(CH_2)_2$	5	5	(63)
H	$PhCH(OBOM)CH_2$	5	5	(40)
Bn	$BnO_2C(CH_2)_2$	5	5	(72)
Bn	$HO_2C(CH_2)_3$	5	10	(38)
PMB	$HO(CH_2)_6$	5	5	(50)
PMB	$BocHN(CH_2)_2$	5	5	40
AcO	$HO(CH_2)_6$	5	20	(48)

Alkene	Cross Partner	Conditions	Product(s) and Yield(s) (%)	Refs.
BzO	OAc 2 eq	Gr II (0.03–0.05 eq), DCM, 40°, 12 h	BzO, OAc (80) (*E*)/(*Z*) = 74:26	241

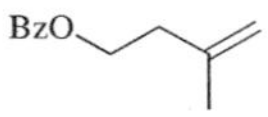

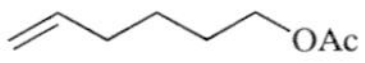

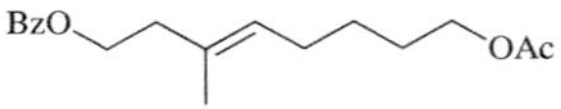

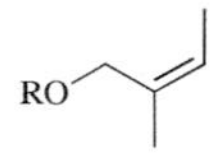

RO I

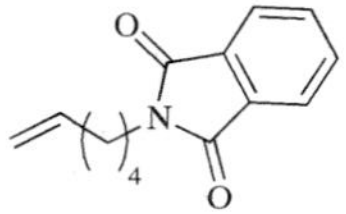

1. Ru cat **354** (0.01 eq), (Z)-butene (5 eq), THF, 22°, 1 h
2. Ru cat **354** (0.05 eq), **I** (5 eq), THF, 100 torr, 22°, 1 h
3. Ambient pressure, 15 h

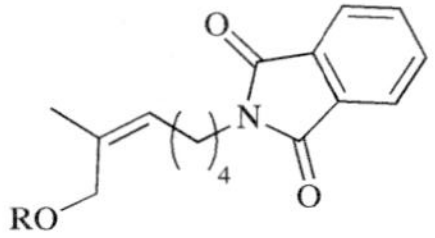

RO O N O 4

(Z)/(E) > 98:2

R	
H	(77)
PMB	(51)

1149

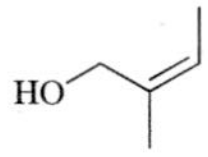

HO I

OBn

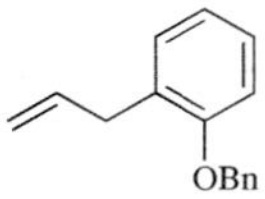

1. Ru cat **354** (0.01 eq), (Z)-butene (5 eq), THF, 22°, 1 h
2. Ru cat **354** (0.03 eq), **I** (10 eq), THF, 100 torr, 22°, 1 h
3. Ambient pressure, 15 h

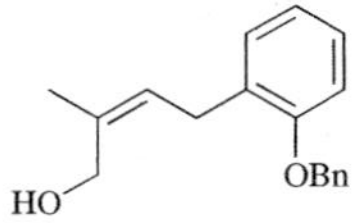

HO OBn

(62) (Z)/(E) > 98:2

1149

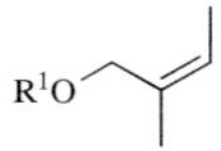

R^1O I

OR^2 R^3

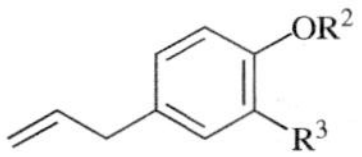

1. Ru cat **354** (0.01 eq), (Z)-butene (5 eq), THF, 22°, 1 h
2. Ru cat **354** (0.05 eq), **I** (x eq), THF, 100 torr, 22°, 1 h
3. Ambient pressure, 15 h

R^1O OR^2 R^3

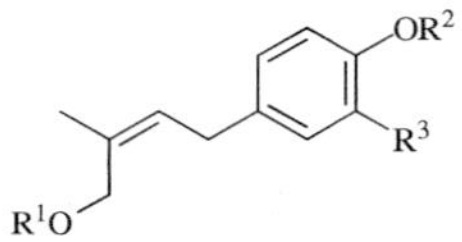

(Z)/(E) > 98:2

R^1	R^2	R^3	x	
H	H	H	10	(55)
H	Me	MeO	5	(81)
Bn	Me	MeO	5	(55)

1149

TABLE 25. CROSS-METATHESIS OF GEMINAL–DISUBSTITUTED COMPOUNDS (*Continued*)

A. ALKENES WITH FUNCTIONAL GROUPS IN ALLYLIC AND MORE DISTANT POSITIONS (*Continued*)

Alkene	Cross Partner	Conditions	Product(s) and Yield(s) (%)	Refs.
Please refer to the charts preceding the tables for the catalyst structures.				
C_5				
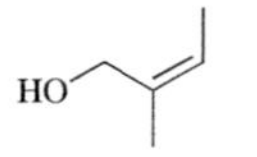 HO **I**	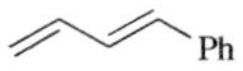 Ph	1. Ru cat **354** (0.01 eq), (Z)-butene (5 eq), THF, 22°, 1 h 2. Ru cat **354** (0.05 eq), **I** (10 eq), THF, 100 torr, 22°, 1 h 3. Ambient pressure, 15 h	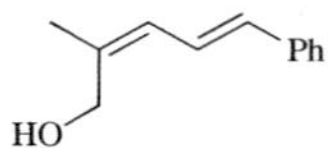 HO Ph (58) (Z)/(E) > 98:2	1149
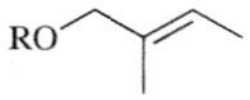 RO **I**	 Ph	1. Ru cat **354** (0.01 eq), (Z)-butene (5 eq), THF, 22°, 1 h 2. Ru cat **354** (0.05 eq), **I** (5 eq), THF, 100 torr, 22°, 1 h 3. Ambient pressure, 15 h	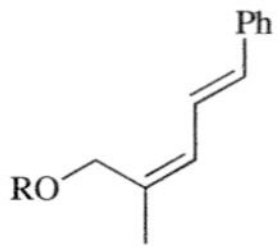 RO Ph (Z)/(E) > 98:2 R — H (52); PMB (41)	1149
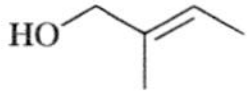 HO **I**	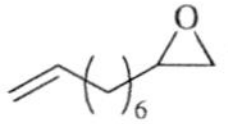 O ( )$_6$	1. Ru cat **354** (0.01 eq), (Z)-butene (5 eq), THF, 22°, 1 h 2. Ru cat **354** (0.05 eq), **I** (5 eq), THF, 100 torr, 22°, 1 h 3. Ambient pressure, 15 h	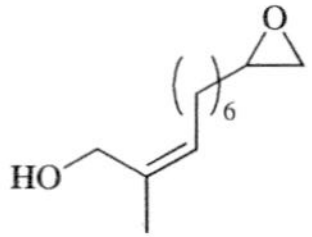 HO O ( )$_6$ (77) (Z)/(E) > 98:2	1149

	Reactant	Partner	Conditions	Product	Yield (%)	Refs.
	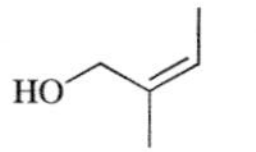 HO (I), alkene structure	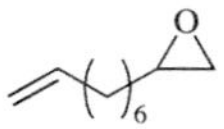 epoxide with $(CH_2)_6$ chain	1. Ru cat **354** (0.01 eq), (*Z*)-butene (5 eq), THF, 22°, 1 h 2. Ru cat **354** (0.05 eq), **I** (5 eq), THF, 100 torr, 22°, 1 h 3. Ambient pressure, 15 h	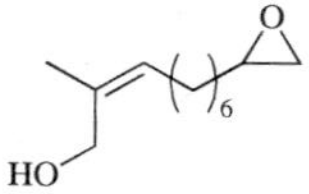 HO, epoxide product with $(CH_2)_6$ chain	(75) (*Z*)/(*E*) > 98:2	1149
	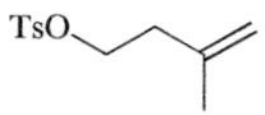 TsO, 10 eq	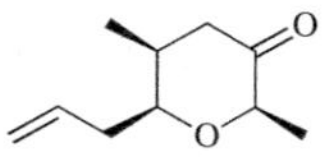	1. Gr II (0.1 eq), DCM, reflux, 12 h 2. *t*-BuOK, DMSO, 75°	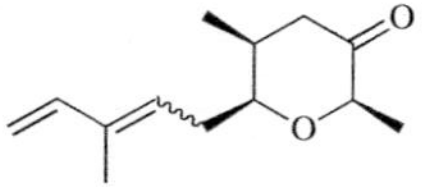	(41)	829, 546
C_6	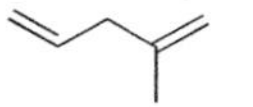 1.4 eq	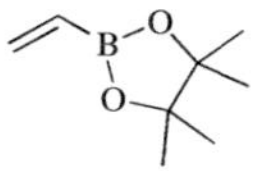	Hov II (0.02 eq), DCM, reflux, 20 h	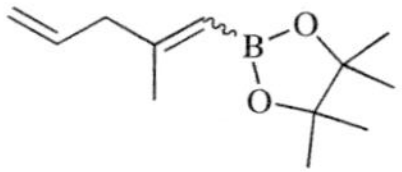	(80)	1154
	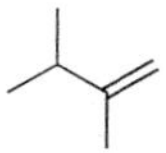 30 eq	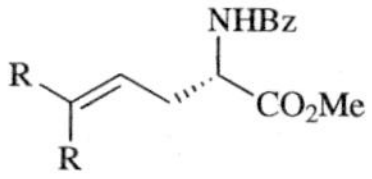 NHBz, CO_2Me, R	Hov II (0.05 eq), benzene, 100°, 14 h	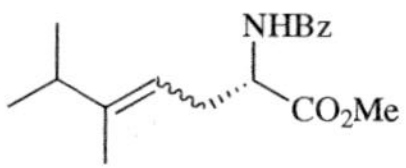 NHBz, CO_2Me	R: H (0); Me (71)	337

TABLE 25. CROSS-METATHESIS OF GEMINAL–DISUBSTITUTED COMPOUNDS (*Continued*)

A. ALKENES WITH FUNCTIONAL GROUPS IN ALLYLIC AND MORE DISTANT POSITIONS (*Continued*)

Alkene	Cross Partner	Conditions	Product(s) and Yield(s) (%)	Refs.
Please refer to the charts preceding the tables for the catalyst structures.				
C_6				
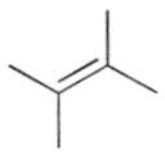	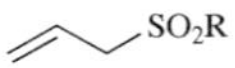 SO_2R	Hov II (0.05 eq), DCE, 60°, 4 h	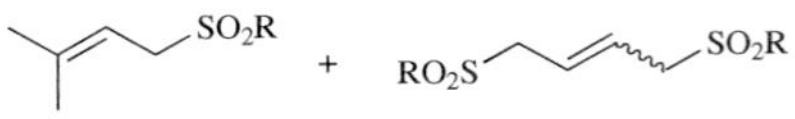 SO_2R (I) + RO_2S … SO_2R (II)	1153
			R: Me, I (45), II (38); Ph, I (51), II (40)	
	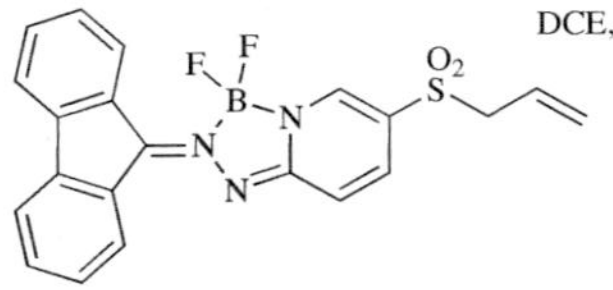 F, F, B, N, N, N, O_2, S	Hov II (0.05 eq), DCE, 60°, 4 h	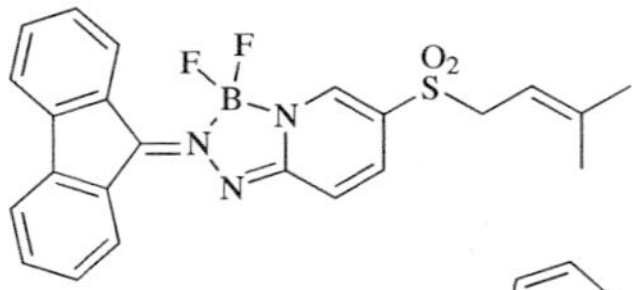 F, F, B, N, N, N, O_2, S (59)	1153
			+ F, F, B, N, N, N, O_2, S, SO_2 (31)	

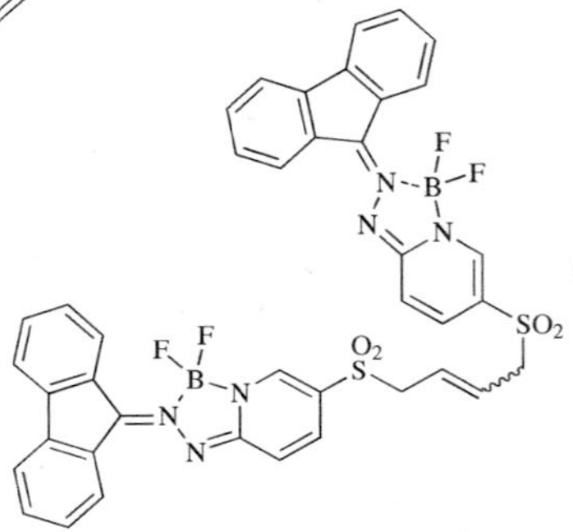

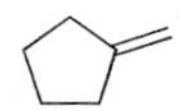

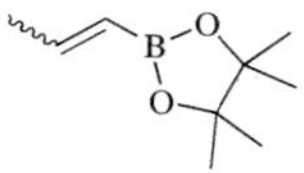

C_{6-8}

Gr II (0.05 eq), DCM, reflux (80) 687

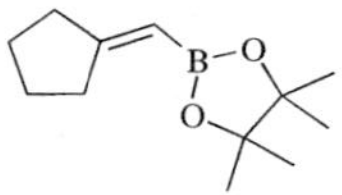

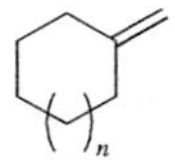

10 eq

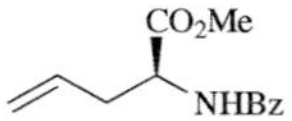

CO_2Me NHBz

Catalyst (x eq), DCM

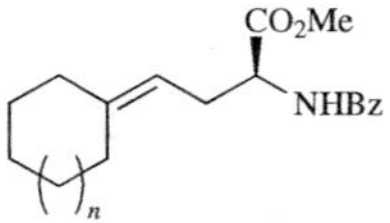

CO_2Me NHBz

1155

n	Catalyst	x	Temp (°)	Time (h)	Conv. (%)[a]	
0	Hov II	0.1	100 (MW)	4	—	(95)
1	Hov II	0.1	100 (MW)	4	—	(96)
2	Hov II	0.1	100 (MW)	4	—	(79)
1	Gr II	0.05	reflux	48	40	(—)
1	Gr II	0.05	100 (MW)	2	60	(—)
1	Hov II	0.05	100 (MW)	2	76	(—)
1	Hov II	0.05	100 (MW)	4	79	(—)
1	Hov II	0.05	100 (MW)	2 + 2	96	(—)

TABLE 25. CROSS-METATHESIS OF GEMINAL–DISUBSTITUTED COMPOUNDS (*Continued*)
A. ALKENES WITH FUNCTIONAL GROUPS IN ALLYLIC AND MORE DISTANT POSITIONS (*Continued*)

Alkene	Cross Partner	Conditions	Product(s) and Yield(s) (%)	Refs.

Please refer to the charts preceding the tables for the catalyst structures.

C_6

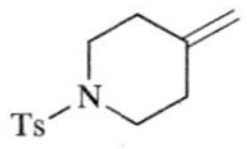

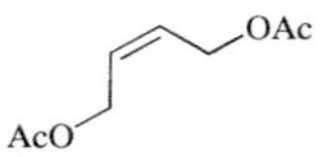

Alkene	Cross Partner	Conditions	Product(s) and Yield(s) (%)	Refs.
Ts–N (piperidine)	AcO, OAc (2 eq)	Catalyst (x eq), DCM, 40°	Ts–N, OAc	

Catalyst	x	Time (h)	Yield (%)	Refs.
Ru cat **88**	0.05	16	(47)	171
Hov II	0.05	16	(54)	171
Ru cat **166**	0.1	7	(63)	1060
Ru cat **167**	0.1	7	(62)	1060

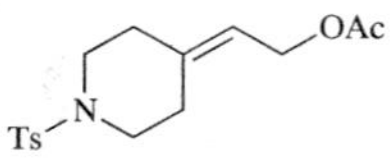

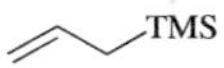

Alkene	Cross Partner	Conditions	Product(s) and Yield(s) (%)	Refs.
	TMS (2 eq)	Catalyst (0.05 eq), DCM, 40°, 16 h	Ts–N, TMS	

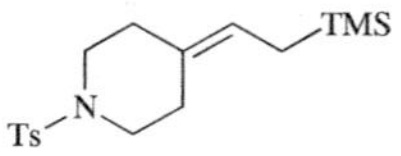

Catalyst	Yield (%)	Refs.
Ru cat **88**	(5)	171
Gr II	(0)	

C_{6-12}

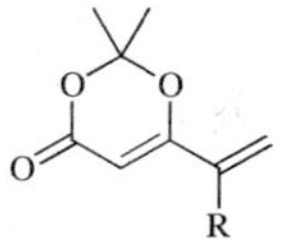

Alkene	Cross Partner	Conditions	Product(s) and Yield(s) (%)	Refs.
O, O, O, R (x eq)	CO_2t-Bu	Hov II (y eq), solvent	O, O, O, R, CO_2t-Bu	792

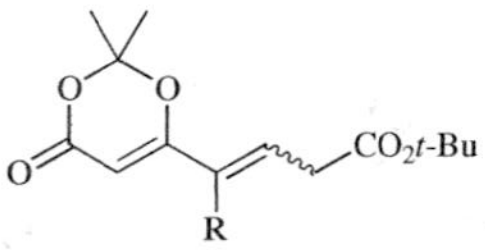

R	x	y	Solvent	Time (h)	Temp (°)	Yield (%)
Me	1.5	0.05	DCM	2.5	40	(0)
Me	1.5	0.05	$C_6F_5CF_3$	16	70	(43)
Me	3	0.1	$C_6F_5CF_3$	2	70	(68)
Bn	3	0.1	$C_6F_5CF_3$	2	70	(61)

C_6

Olefin 1	Olefin 2	Conditions	Product(s) and Yield(s) (%)	Refs.
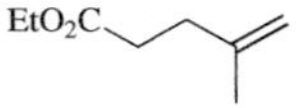 EtO_2C	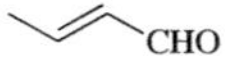 CHO 10 eq	Hov II (0.05 eq), DCM, 40°, 4 h	 EtO_2C … CHO (32)	942
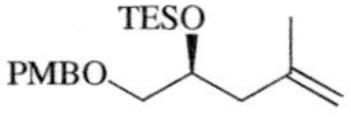 PMBO, TESO 10 eq	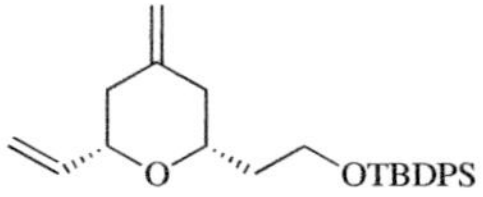 OTBDPS	Gr II (0.01 eq), DCM, reflux, 9 h	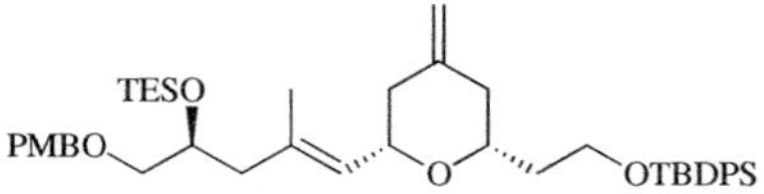 PMBO, TESO, OTBDPS (57) $(E)/(Z)$ = 63:37	1156

C_7

Olefin 1	Olefin 2	Conditions	Product(s) and Yield(s) (%)	Refs.
	R^1, R^2, O 2 eq	Gr II (0.05 eq), DCM, reflux, 3 h	R^2, O	170
	B, O, O	Gr II (0.05 eq), DCM, reflux	B, O, O (91)	687

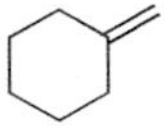

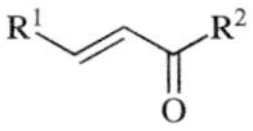

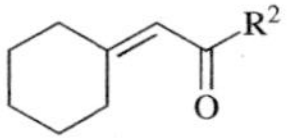

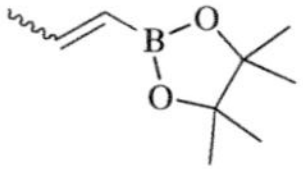

R^1	R^2	
H	*t*-BuO	(75)
H	HO	(83)
Me	Et	(99)

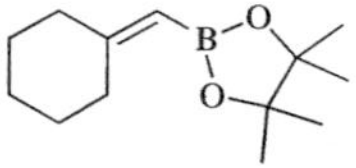

TABLE 25. CROSS-METATHESIS OF GEMINAL–DISUBSTITUTED COMPOUNDS (*Continued*)
A. ALKENES WITH FUNCTIONAL GROUPS IN ALLYLIC AND MORE DISTANT POSITIONS (*Continued*)

Alkene	Cross Partner	Conditions	Product(s) and Yield(s) (%)	Refs.

Please refer to the charts preceding the tables for the catalyst structures.

C_7

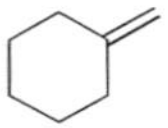

Alkene: methylenecyclohexane. Cross partner: OAc-terminated alkene (n), x eq. Conditions: Catalyst (y eq), solvent. Product: cyclohexylidene–(n)–OAc.

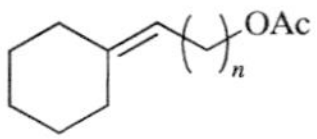

n	x	Catalyst	y	Solvent	Temp (°)	Time	Yield (%)	Refs.
3	3	Hov II	0.05	DCM	reflux	24 h	(78)	177
3	3	Ru cat **271**	0.05	DCM	reflux	24 h	(60)	177
3	3	Ru cat **187**	0.05	DCM	reflux	24 h	(98)	177
3	3	Hov II	0.05	DCM	40	24 h	(78)	845
3	3	Ind II	4 x 0.02	$C_6F_5CF_3$	120	4 x 10 min	(50)	845
3	3	Ind II	4 x 0.02	$C_6F_5CF_3$	120 (MW)	4 x 10 min	(56)	845
4	2	Gr II	0.03	DCM	40	12	(65)	129

Alkene: 2 eq. Cross partner: R^1–CH=CH–C(=O)–CH$_2$–CH(OR^2)–C$_6$H$_4$–R^3. Conditions: Hov II (0.05 eq), DCM, reflux, 18–24 h. Product: cyclohexylidene–CH–C(=O)–CH$_2$–CH(OR^2)–C$_6$H$_4$–R^3. Refs. 1021

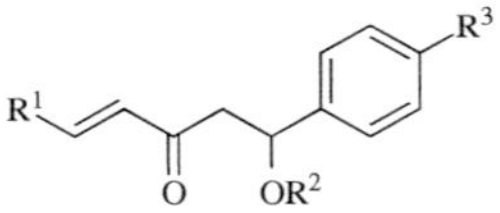

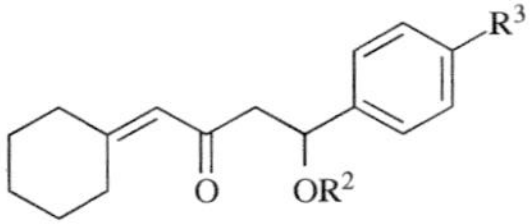

R^1	R^2	R^3	Yield (%)
H	H	MeO	(50)
Me	H	MeO	(42)
Me	H	O_2N	(55)
Me	TBS	MeO	(90)

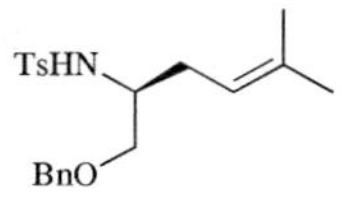

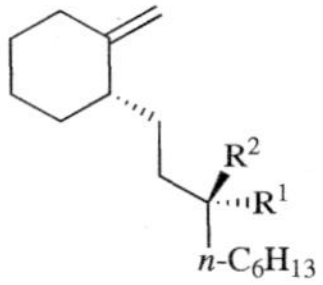

TsHN, BnO

2 eq

Hov II (0.25 eq)
toluene, 120°, 48 h

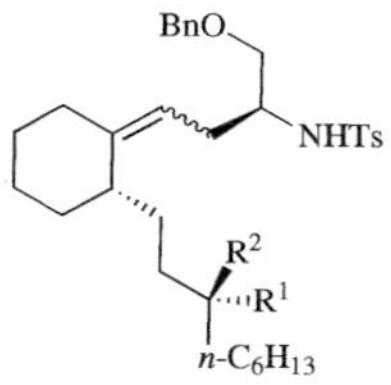

BnO, NHTs

n-C_6H_{13}

R^1	R^2	
H	BzO	(74)
BzO	H	(70)

1147

C_8

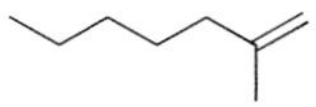

4 eq

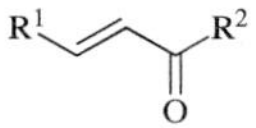

Gr II (0.05 eq),
DCM, 40°

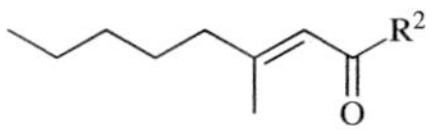

(*E*)/(*Z*) = 67:33

36, 170

R^1	R^2	Time (h)	
H	HO	3	(83)
H	Et	12	(26)
H	EtO	12	(55)
Me	Et	3	(68)
Me	EtO	3	(83)

TABLE 25. CROSS-METATHESIS OF GEMINAL–DISUBSTITUTED COMPOUNDS (*Continued*)
A. ALKENES WITH FUNCTIONAL GROUPS IN ALLYLIC AND MORE DISTANT POSITIONS (*Continued*)

Alkene	Cross Partner	Conditions	Product(s) and Yield(s) (%)	Refs.

Please refer to the charts preceding the tables for the catalyst structures.

C_8

Alkene	Cross Partner	Conditions	Product(s) and Yield(s) (%)	Refs.
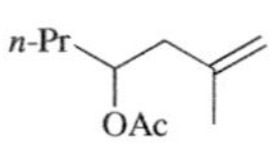 n-Pr, OAc; x eq	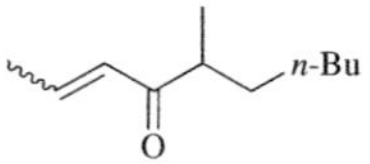 n-Bu, O	Gr II (0.05 eq), DCM, 40°, overnight	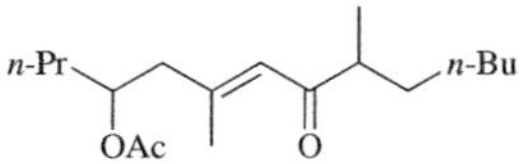 n-Pr, OAc, O, n-Bu	1157

x		(E)/(Z)
2	(15)	50:50
4	(28)	75:25

Alkene	Cross Partner	Conditions	Product(s) and Yield(s) (%)	Refs.
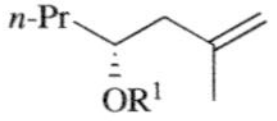 n-Pr, OR¹	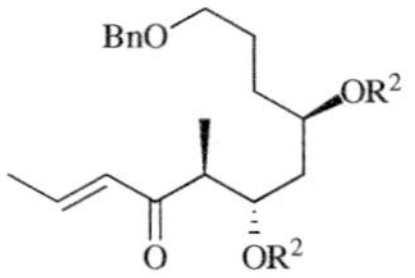 BnO, OR², O, OR²; 6 eq	Catalyst (x eq), DCM, 40°, 72 h	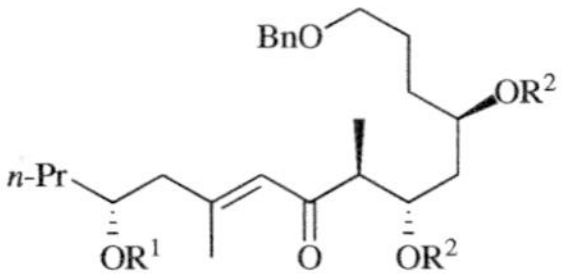 BnO, OR², n-Pr, OR¹, O, OR²	1157

R^1	R^2	Catalyst	x		(E)/(Z)
TBS	Ac	Hov II	0.05	(—)	—
TBS	Ac	Hov II	0.15	(26)	79:21
TBS	Ac	Gr II	0.3	(39)	80:20
TBS	TES	Hov II	0.05	(—)	—
TBS	TES	Hov II	0.3	(46)	(E) only
TBS	TES	Hov II	0.45	(47)	(E) only
PMB	TBS	Hov II	0.3	(45)	(E) only

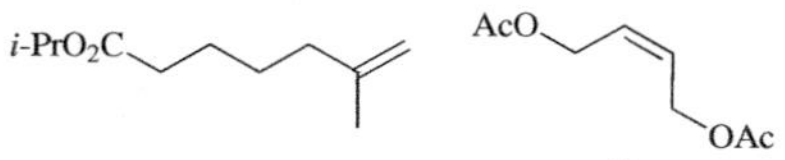

$i\text{-PrO}_2C$ (chain with terminal 1,1-disubstituted alkene) + AcO–CH₂CH=CHCH₂–OAc (2 eq) → Catalyst (0.05 eq), solvent → $i\text{-PrO}_2C$ (chain with trisubstituted alkene)–OAc   171, 140

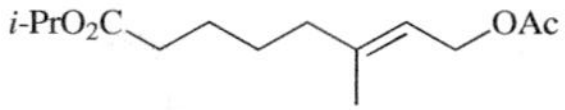

Catalyst	Solvent	Temp (°)	Time (h)		(E)/(Z)
Ru cat **115**	DCE	70	3	(24)	75:25
Ru cat **115**	toluene	70	3	(46)	75:25
Ru cat **115**	perfluorobenzene	70	3	(52)	75:25
Hov II	DCE	70	3	(12)	75:25
Hov II	toluene	70	3	(29)	75:25
Hov II	perfluorobenzene	70	3	(42)	75:25
Gr II	DCE	70	3	(31)	75:25
Gr II	toluene	70	3	(40)	75:25
Gr II	perfluorobenzene	70	3	(50)	75:25
Ind II	DCE	70	3	(27)	75:25
Ind II	toluene	70	3	(40)	75:25
Ind II	perfluorobenzene	70	3	(0)	—
Ru cat **282**	DCE	70	3	(29)	75:25
Ru cat **282**	toluene	70	3	(40)	75:25
Ru cat **282**	perfluorobenzene	70	3	(40)	75:25
Ru cat **88**	DCM	40	16	(37)	70:30
Hov II	DCM	40	16	(17)	70:30

TABLE 25. CROSS-METATHESIS OF GEMINAL–DISUBSTITUTED COMPOUNDS (*Continued*)

A. ALKENES WITH FUNCTIONAL GROUPS IN ALLYLIC AND MORE DISTANT POSITIONS (*Continued*)

Alkene	Cross Partner	Conditions	Product(s) and Yield(s) (%)	Refs.
Please refer to the charts preceding the tables for the catalyst structures.				
C_8				
30 eq	R, R; NHBz; CO_2Me	Hov II (0.05 eq), benzene, 100°, 14 h	I: NHBz, CO_2Me + II: MeO_2C, BzHN, NHBz, CO_2Me R / I / II: H (7) (55); Me (55) (0)	337
C_9				
Ph; 30 eq	R, R; NHBz; CO_2Me	Hov II (0.05 eq), benzene, 100°, 14 h	Ph; NHBz; CO_2Me R: H (0); Me (97)	337
TIPSO; H; O	OAc; 1.5 eq	Catalyst (*x* eq), solvent, reflux, 4 h	I: TIPSO; H; O; OAc + II: TIPSO; O; OAc **II** (*E*)/(*Z*) = 83:17	1158

Catalyst	x	Solvent	**I**	(*E*)/(*Z*) **I**	**II**
Gr II	0.1	DCM	(44)	83:17	(27)
Hov II	0.05	$CHCl_3$	(0)	—	(75)

t-BuO_2C ... CO_2t-Bu + AcO ... OAc (2 eq)

Catalyst (0.05 eq), solvent, 70°, 3 h

t-BuO_2C ... OAc, CO_2t-Bu (*E*)/(*Z*) = 75:25 140, 141

Catalyst	Solvent	
Ru cat **115**	DCE	(13)
Ru cat **115**	toluene	(42)
Ru cat **115**	perfluorobenzene	(47)
Ru cat **115**	perfluorotoluene	(50)
Hov II	DCE	(7)
Hov II	toluene	(20)
Hov II	perfluorobenzene	(39)
Hov II	perfluorotoluene	(44)
Gr II	DCE	(27)
Gr II	toluene	(37)

Catalyst	Solvent	
Gr II	perfluorobenzene	(45)
Gr II	perfluorotoluene	(50)
Ind II	DCE	(25)
Ind II	toluene	(38)
Ind II	perfluorobenzene	(47)
Ind II	perfluorotoluene	(51)
Ru cat **282**	DCE	(24)
Ru cat **282**	toluene	(39)
Ru cat **282**	perfluorobenzene	(32)
Ru cat **282**	perfluorotoluene	(13)

TABLE 25. CROSS-METATHESIS OF GEMINAL–DISUBSTITUTED COMPOUNDS (*Continued*)

A. ALKENES WITH FUNCTIONAL GROUPS IN ALLYLIC AND MORE DISTANT POSITIONS (*Continued*)

Alkene	Cross Partner	Conditions	Product(s) and Yield(s) (%)	Refs.

Please refer to the charts preceding the tables for the catalyst structures.

C_9

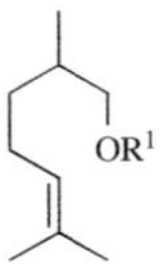

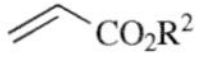

OR[1] | CO_2R^2 (2 eq) | Catalyst (x eq), additive (0.03 eq), solvent | OR[1], CO_2R^2

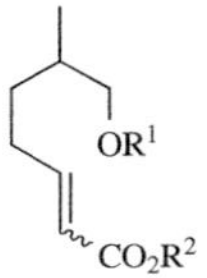

R^1	R^2	Catalyst	x	Additive	Solvent	Temp (°)	Time (h)		Refs.
H	Me	Hov II	0.02	—	DCM	35	15	(0)	1159
H	Me	Hov II	0.02	—	toluene	40	15	(0)	1159
H	Me	Hov II	0.02	—	DMC	40	15	(0)	1159
H	Me	Hov II	0.02	—	DMC	60	15	(68)	1159
H	Me	Hov II	0.02	—	toluene	60	15	(50)	1159
H	Me	Hov II	0.02	—	DMC	80	15	(60)	1159
H	Me	Hov II	0.005	—	DMC	80	24	(53)	1159
TBS	Et	Gr II	0.03	—	2.5 % aq. PTS	22	12	(82)	931
TBS	*t*-Bu	Gr II	0.03	—	2.5 % aq. PTS	22	12	(84)	931
TBS	*t*-Bu	Gr II	0.03	NaCl	2.5 % aq. PTS	22	12	(86)	931
Ac	*t*-Bu	Gr II	0.03	—	2.5 % aq. PTS	22	12	(80)	931
Ac	2-ethylhexyl	Gr II	0.03	—	2.5 % aq. PTS	22	12	(78)	931
Ac	*n*-Bu	Ru cat **170**	0.01	—	AcOEt	80	1	(43)	751

C_{10}

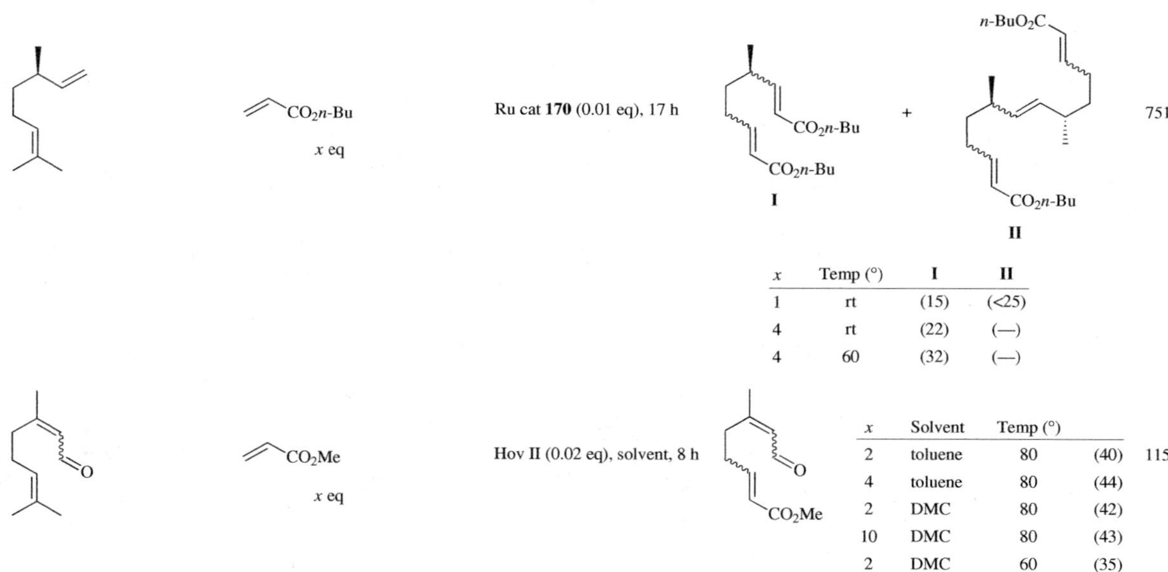

x	Temp (°)	I	II
1	rt	(15)	(<25)
4	rt	(22)	(—)
4	60	(32)	(—)

x	Solvent	Temp (°)	
2	toluene	80	(40)
4	toluene	80	(44)
2	DMC	80	(42)
10	DMC	80	(43)
2	DMC	60	(35)

TABLE 25. CROSS-METATHESIS OF GEMINAL–DISUBSTITUTED COMPOUNDS (*Continued*)

A. ALKENES WITH FUNCTIONAL GROUPS IN ALLYLIC AND MORE DISTANT POSITIONS (*Continued*)

Alkene	Cross Partner	Conditions	Product(s) and Yield(s) (%)	Refs.

Please refer to the charts preceding the tables for the catalyst structures.

C_{10}

Alkene	Cross Partner	Conditions	Product(s) and Yield(s) (%)	Refs.
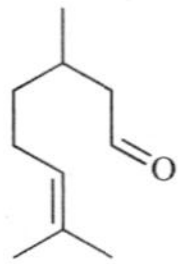	 CO_2Me, *x* eq	Catalyst (*y* eq), solvent	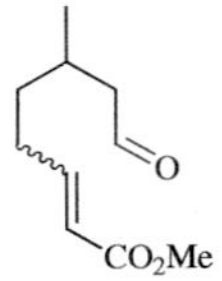 CO_2Me, O	1159

x	Catalyst	*y*	Solvent	Temp (°)	Time (h)	
2	Hov II	0.02	DCM	35	15	(54)
2	Hov II	0.02	toluene	40	15	(70)
2	Hov II	0.02	DMC	40	15	(74)
2	Hov II	0.02	DMC	60	15	(68)
2	Hov II	0.005	DMC	80	3	(70)
4	Hov II	0.005	DMC	80	3	(70)
10	Hov I	0.005	DMC	80	3	(45)
2	Gr II	0.005	DMC	80	3	(20)
2	Ind II	0.005	DMC	80	3	(10)
2	Ru cat **89**	0.005	DMC	80	3	(80)

Alkene	Cross Partner	Conditions	Product(s) and Yield(s) (%)	Refs.
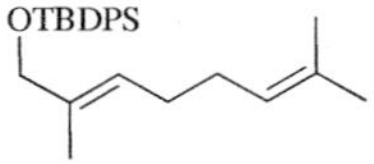 OTBDPS	 CO_2Et, 1 eq	Gr II (0.05 eq), DCM, 23°, 12 h	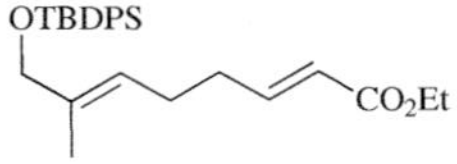 OTBDPS, CO_2Et (61) (*E*) only	902

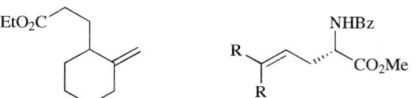

EtO_2C 2 eq | CO_2t-Bu 2 eq | Gr II (0.03 eq), 2.5 % aq. PTS, 22°, 12 h | EtO_2C CO_2t-Bu (50) | 931

AcO OAc | Ind II (4 x 0.05 eq), solvent, 120° | EtO_2C OAc | (*E*)/(*Z*) = 82:18 | 40

Solvent	Time	MW	
$C_6D_5CD_3$	24 h	—	(10)
$C_6F_5CF_3$	4 x 10 min	200 W	(51)

EtO_2C 30 eq | R R NHBz CO_2Me | Hov II (0.05 eq), benzene, 100°, 14 h | EtO_2C NHBz CO_2Me | 337

R	
H	(0)
Me	(57)

O 2 eq | AcO OAc | Ind II (*x* eq), solvent | O OAc | (*E*) only | 40

x	Solvent	Temp (°)	Time	
0.2	CD_2Cl_2	40	6 h	(10)
4 x 0.05	$C_6F_5CF_3$	120	4 x 10 min	(35)
4 x 0.05	$C_6F_5CF_3$	120 (MW)	4 x 10 min	(46)

TABLE 25. CROSS-METATHESIS OF GEMINAL–DISUBSTITUTED COMPOUNDS (*Continued*)

A. ALKENES WITH FUNCTIONAL GROUPS IN ALLYLIC AND MORE DISTANT POSITIONS (*Continued*)

Alkene	Cross Partner	Conditions	Product(s) and Yield(s) (%)	Refs.
Please refer to the charts preceding the tables for the catalyst structures.				
C_{10}				
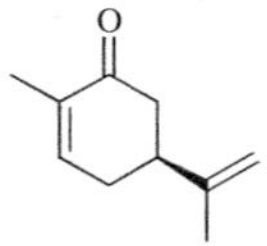 O; 30 eq	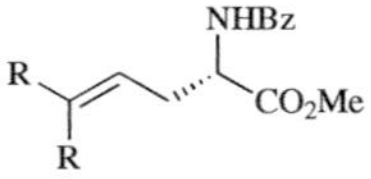 R; R; NHBz; CO_2Me	Hov II (0.05 eq), benzene, 100°, 14 h	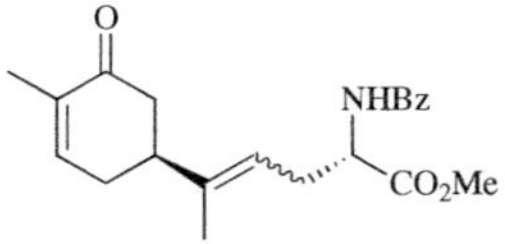 O; NHBz; CO_2Me; R: H (0), Me (43)	337
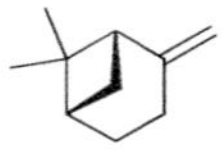 30 eq	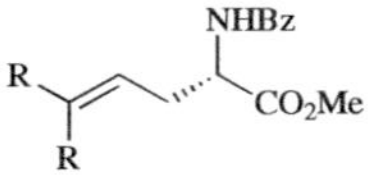 R; R; NHBz; CO_2Me	Hov II (0.05 eq), benzene, 100°, 14 h	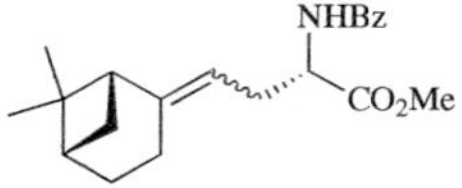 NHBz; CO_2Me; R: H (0), Me (36)	337
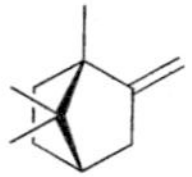 30 eq	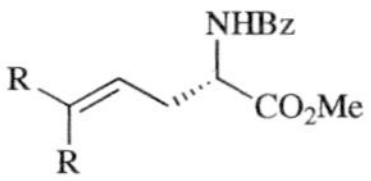 R; R; NHBz; CO_2Me	Hov II (0.05 eq), benzene, 100°, 14 h	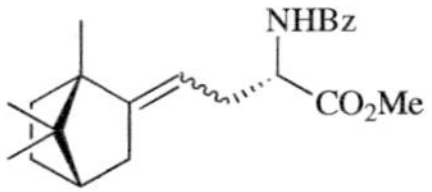 NHBz; CO_2Me; R: H (0), Me (24)	337
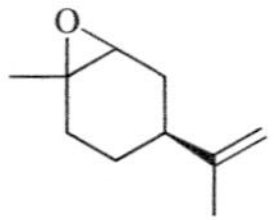 O; 30 eq	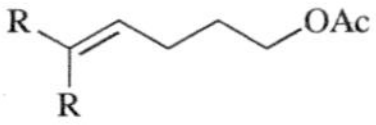 R; R; OAc	Hov II (0.05 eq), benzene, 100°, 14 h	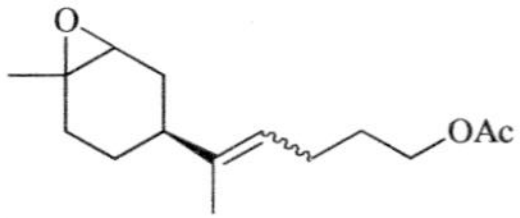 O; OAc; R: H (0), Me (87)	337

C_{10-11}

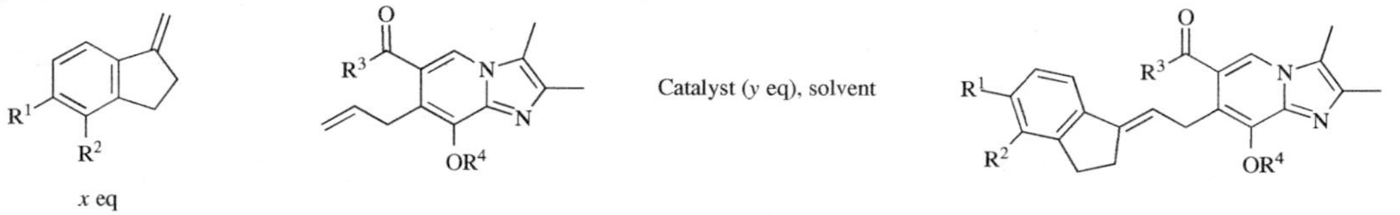

R^1 R^2 x eq

R^3 O OR^4

Catalyst (y eq), solvent

R^1 R^2 R^3 O OR^4

502

R^1	R^2	x	R^3	R^4	Catalyst	y	Solvent	Temp (°)	Time (h)		$(E)/(Z)$
H	H	3.7	Me_2N	Piv	Gr II	0.08	DCE	80	2	(25)	(E) only
H	H	3	Me_2N	TDS	Gr II	0.09	DCM	40	18	(72)	(E) only
H	H	3.7	EtO	TDS	Gr II	0.08	DCM	40	18	(68)	—
H	H	2.2	EtO	TDS	Hov II	0.056	DCM	40	24	(40)	(E) only
F	H	2.8	EtO	TDS	Hov II	0.07	DCM	40	24	(34)	—
H	Me	2.2	EtO	TDS	Hov II	0.056	DCM	40	17	(43)	—

C_{10}

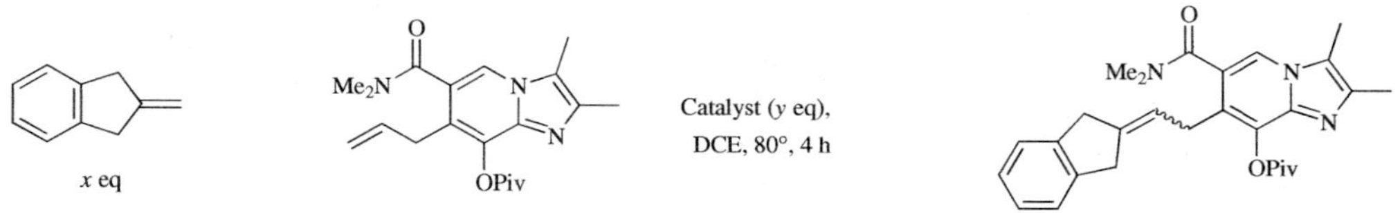

x eq

Me_2N O OPiv

Catalyst (y eq),
DCE, 80°, 4 h

Me_2N O OPiv

502

x	Catalyst	y	
4	Gr II	0.08	(44)
5	Hov II	0.082	(27)

TABLE 25. CROSS-METATHESIS OF GEMINAL–DISUBSTITUTED COMPOUNDS (*Continued*)
A. ALKENES WITH FUNCTIONAL GROUPS IN ALLYLIC AND MORE DISTANT POSITIONS (*Continued*)

Alkene	Cross Partner	Conditions	Product(s) and Yield(s) (%)	Refs.
Please refer to the charts preceding the tables for the catalyst structures.				
C_{10}				
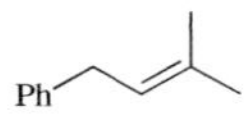 Ph	 CO_2R (2 eq)	Gr II (0.03 eq), 2.5 % aq. PTS, 22°, 12 h	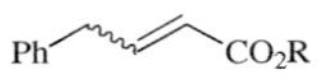 Ph CO_2R R: *t*-Bu (88); 2-ethylhexyl (88)	931
2 eq	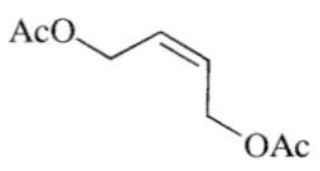 AcO OAc	Gr II (0.03 eq), 2.5 % aq. PTS, 22°, 12 h	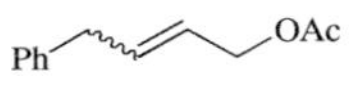 Ph OAc (84)	931
C_{11-12}				
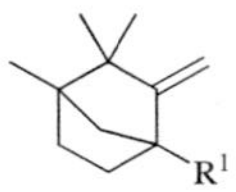 R^1 (30 eq)	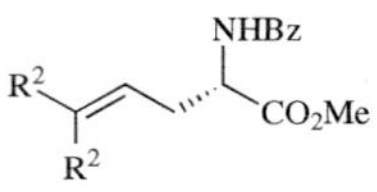 R^2 R^2 NHBz CO_2Me	Hov II (0.05 eq), benzene, 100°, 14 h	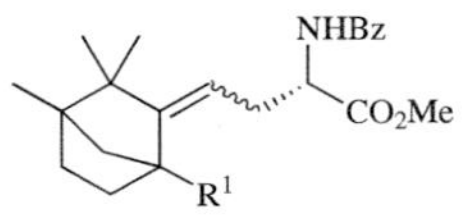 NHBz CO_2Me R^1 R^1 = H, R^2 = H (0) R^1 = H, R^2 = Me (34) R^1 = Me, R^2 = Me (0)	337

C_{11}

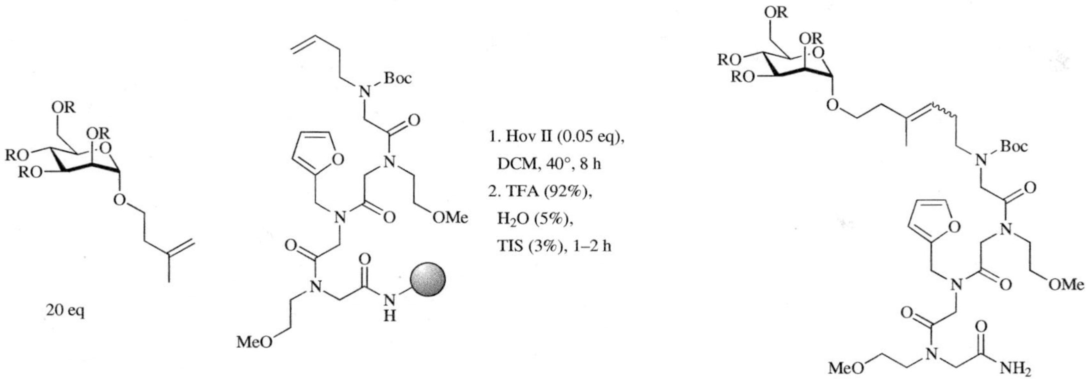

OR
OR
RO
RO
O

20 eq

N
Boc
O
O
N
O
N
O
OMe
O
O
N
N
H
MeO

1. Hov II (0.05 eq),
DCM, 40°, 8 h
2. TFA (92%),
H_2O (5%),
TIS (3%), 1–2 h

OR
OR
RO
RO
O
O
N
Boc
O
O
N
O
N
O
OMe
O
O
N
NH_2
MeO

R	
Bz	(<5)
Ac	(<5)

12

TABLE 25. CROSS-METATHESIS OF GEMINAL–DISUBSTITUTED COMPOUNDS (*Continued*)

A. ALKENES WITH FUNCTIONAL GROUPS IN ALLYLIC AND MORE DISTANT POSITIONS (*Continued*)

Alkene	Cross Partner	Conditions	Product(s) and Yield(s) (%)	Refs.
Please refer to the charts preceding the tables for the catalyst structures.				
C_{11}				
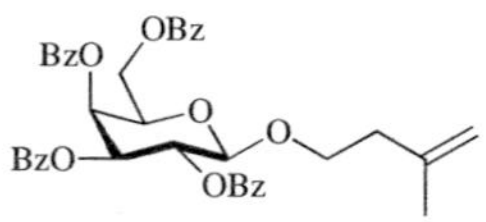 BzO, OBz, BzO, OBz, O, O 20 eq	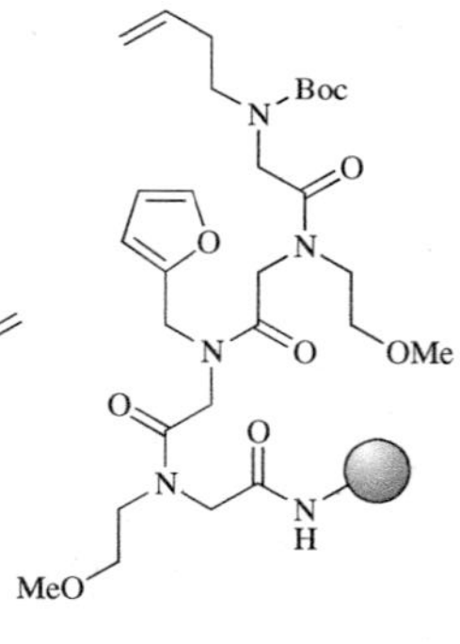 Boc, N, O, O, N, OMe, N, O, O, N, O, N, H, MeO	1. Hov II (0.05 eq), DCM, 40°, 8 h 2. TFA (92%), H_2O (5%), TIS (3%), 1–2 h	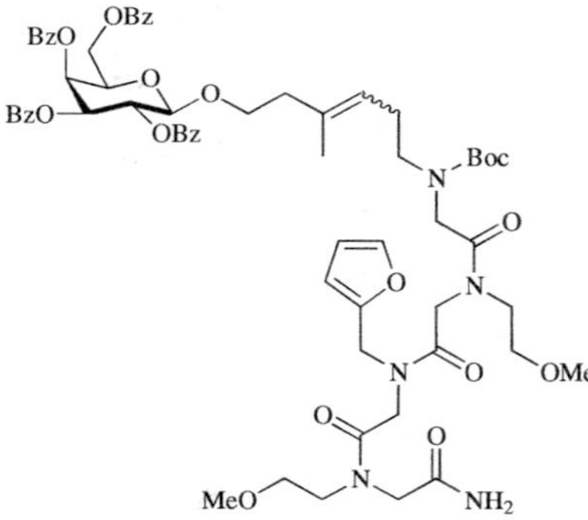 BzO, OBz, BzO, O, O, OBz, Boc, N, O, O, N, O, N, O, OMe, O, O, N, MeO, NH_2 (<5)	12
C_{12}				
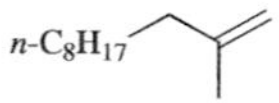 n-C_8H_{17} 3 eq	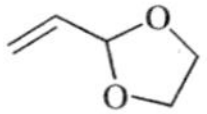 O, O	Gr II (0.03–0.05 eq), DCM, 40°, 12 h	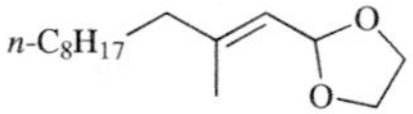 n-C_8H_{17}, O, O (67) $(E)/(Z) = 75:25$	36, 241

AcO–CH₂CH=CHCH₂–OAc

2 eq

Gr II (0.05 eq),
DCM, 40°, 12 h

n-C_8H_{17} … OAc

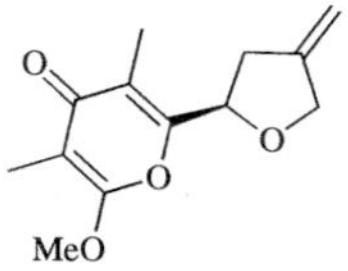

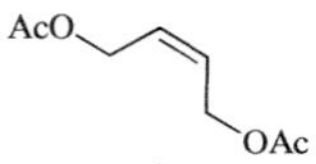

(53) (*E*)/(*Z*) = 71:29 241

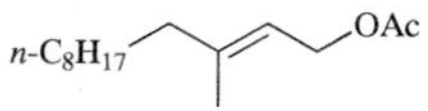

R

2 eq

Gr II (0.05 eq),
DCM, 40°, 12 h

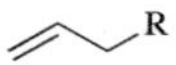

n-C_8H_{17} … R

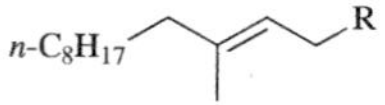

241

R		(*E*)/(*Z*)
PhO_2S	(87)	77:23
$AcO(CH_2)_3$	(60)	70:30

O O

MeO O O

2 eq

Gr II (0.05 eq),
DCM, reflux

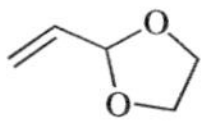

(65) (*Z*)/(*E*) = 55:45 1160

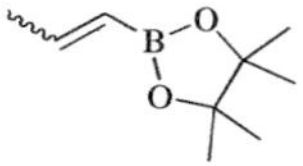

B O O

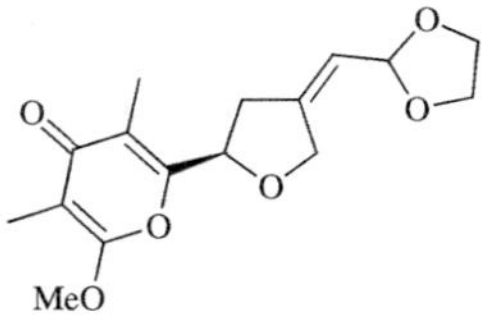

2 eq

Gr II (0.05 eq),
DCM, reflux

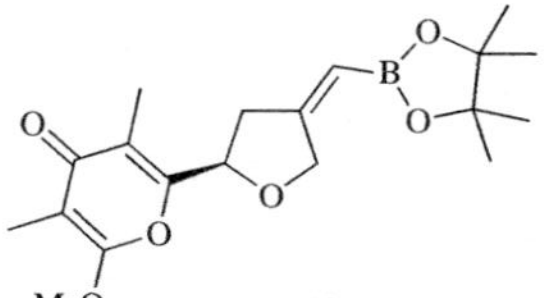

(98) (*Z*)/(*E*) = 55:45

1160

TABLE 25. CROSS-METATHESIS OF GEMINAL–DISUBSTITUTED COMPOUNDS (*Continued*)

A. ALKENES WITH FUNCTIONAL GROUPS IN ALLYLIC AND MORE DISTANT POSITIONS (*Continued*)

Alkene	Cross Partner	Conditions	Product(s) and Yield(s) (%)	Refs.
Please refer to the charts preceding the tables for the catalyst structures.				
C_{13}				
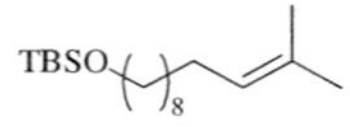 TBSO	 CO_2t-Bu 2 eq	Gr II (0.03 eq), 2.5 % aq. PTS, 22°, 12 h	 TBSO CO_2t-Bu (92)	931
C_{14}				
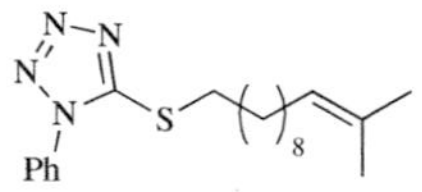 N–N N N Ph S	 CO_2t-Bu 2 eq	Gr II (0.03 eq) 2.5% aq PTS, 22°, 12 h	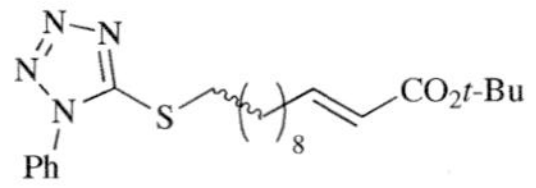 N–N N N Ph S CO_2t-Bu (55)	931
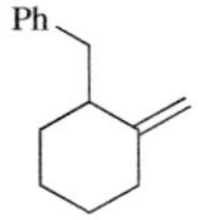 Ph	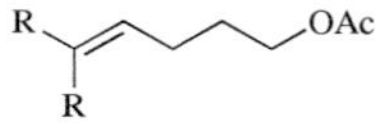 R R OAc *x* eq	Catalyst (0.05 eq), solvent	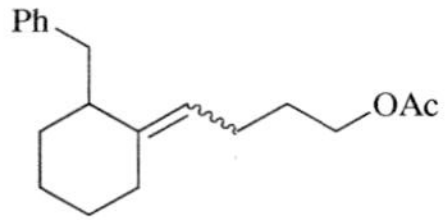 Ph OAc	

R	*x*	Catalyst	Solvent	Temp (°)	Time (h)		Refs.
H	3	Hov II	DCM	reflux	24	(17)	177
H	3	Ru cat **271**	DCM	reflux	24	(0)	177
H	3	Ru cat **187**	DCM	reflux	24	(7)	177
H	0.03	Hov II	benzene	100	14	(17)	337
H	0.03	Ru cat **271**	benzene	100	14	(0)	337
Me	0.03	Hov II	benzene	100	14	(90)	337

Reactant	Alkene	Conditions	Product(s) and Yield(s) (%)		Refs.
30 eq	OAc	Hov II (0.05 eq), benzene, 100°, 14 h	Ph, OAc	(53)	337
30 eq	R, R, NHBz, CO_2Me	Hov II (0.05 eq), benzene, 100°, 14 h	Ph, NHBz, CO_2Me	R: H (0); Me (50)	337
	CO_2Et (3 eq)	Hov II (0.1 eq), toluene, 50°, 16 h	CO_2Et (E)/(Z) = 52:48 + CO_2Et (E)/(Z) = 81:19	~5% conv.[a]	542

TABLE 25. CROSS-METATHESIS OF GEMINAL–DISUBSTITUTED COMPOUNDS (*Continued*)
A. ALKENES WITH FUNCTIONAL GROUPS IN ALLYLIC AND MORE DISTANT POSITIONS (*Continued*)

Alkene	Cross Partner	Conditions	Product(s) and Yield(s) (%)	Refs.

Please refer to the charts preceding the tables for the catalyst structures.

C_{14}

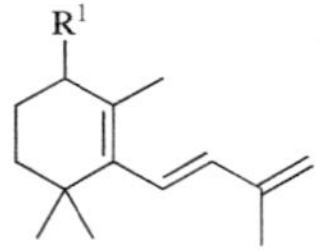

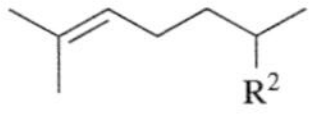

Alkene	Cross Partner	Conditions	Product(s) and Yield(s) (%)	Refs.
R^1	R^2, x eq	Gr II (0.05 eq), DCM, 35°, 18 h	R^1, R^2	337

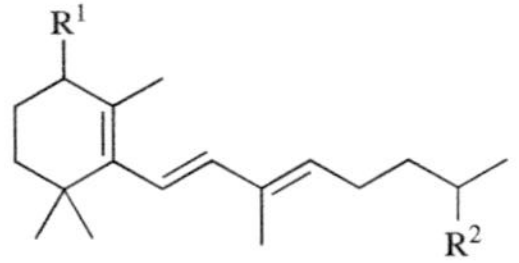

R^1	R^2	x		$(E)/(Z)$
H	H	2.1	(95)	64:36
H	TBSO	2.1	(75)	64:36
TBSO	H	2.1	(95)	73:27
TBSO	TBSO	1.5	(75)	76:23

Alkene	Cross Partner	Conditions	Product(s) and Yield(s) (%)	Refs.
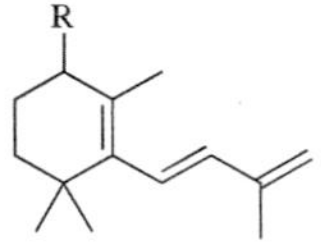 R	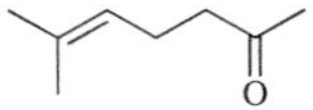 O, x eq	Gr II (0.05 eq), DCM, 35°, 18 h	R, O	337

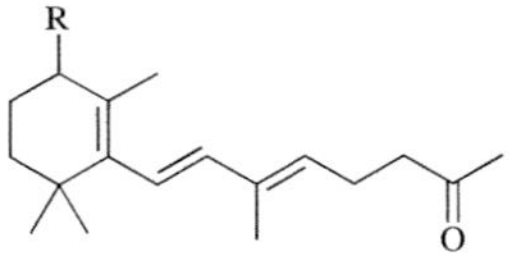

R	x		$(E)/(Z)$
H	2.1	(75)	55:45
TBSO	1.5	(70)	71:29

Reactant	Olefin	Conditions	Product(s) and Yield(s) (%)	Refs.
	2.1 eq	Gr II (0.05 eq), DCM, 35°, 18 h		337

R		$(E)/(Z)$
H	(69)	77:23
TBSO	(89)	76:24

Reactant	Olefin	Conditions	Product(s) and Yield(s) (%)	Refs.
	2.1 eq	Gr II (0.05 eq), DCM, 35°, 18 h	(57) $(E)/(Z)$ = 64:36	337
C_{15} OMe	CO_2t-Bu, 3 eq	Gr II (0.03 eq), CuI (0.04 eq), Et_2O, 35°, 24 h	OMe, CO_2t-Bu (81)	167
OR, OH	3 eq	Catalyst (0.1 eq), DCM, reflux, 48 h	OH	578

R	Catalyst		$(E)/(Z)$
H	Gr I	(0)	—
H	Gr II	(0)	—
Ac	Gr II	(19)	60:40

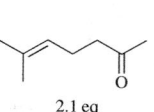

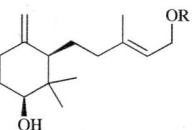

TABLE 25. CROSS-METATHESIS OF GEMINAL–DISUBSTITUTED COMPOUNDS (*Continued*)
A. ALKENES WITH FUNCTIONAL GROUPS IN ALLYLIC AND MORE DISTANT POSITIONS (*Continued*)

Alkene	Cross Partner	Conditions	Product(s) and Yield(s) (%)	Refs.
Please refer to the charts preceding the tables for the catalyst structures.				
C_{21}				
CbzHN, Bn, O, O, 7	CO_2t-Bu 2 eq	Gr II (0.03 eq) 2.5% aq PTS, 22°, 12 h	CbzHN, Bn, O, O, 7, CO_2t-Bu (97)	931
C_{28}				
C_4F_8H, O, O, HO, OAc, HO	AcO, OAc 3 eq	Gr II (0.05 eq), solvent, 6 h	AcO, C_4F_8H, O, O, HO, OAc, HO (*E*) only	140

Solvent	Temp (°)	Conv. (%)[a]	
DCM	40	28	(—)
$C_6F_5CF_3$	70	92	(71)

[a] The conversion was determined by ^1H NMR analysis

TABLE 25. CROSS-METATHESIS OF GEMINAL–DISUBSTITUTED COMPOUNDS (*Continued*)

B. BORONIC ESTERS

Boronic Ester	Cross Partner	Conditions	Product(s) and Yield(s) (%)	Refs.

Please refer to the charts preceding the tables for the catalyst structures.

C_3

Boronic Ester	Cross Partner	Conditions	Product(s) and Yield(s) (%)	Refs.
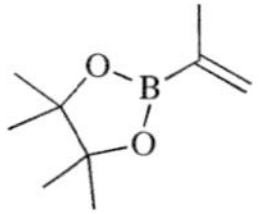	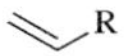 R *x* eq	Gr II (0.05 eq), DCM, reflux, 12 h	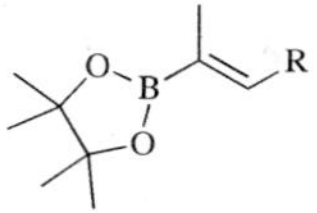 R	

R	*x*		(*E*)/(*Z*)	Refs.
$TIPSCH_2$	1	(59)	>95:5	125
$TMSCH_2$	2	(59)	>95:5	125
Ph	2	(30)	>95:5	125
$BzO(CH_2)_2$	1	(46)	>95:5	125
$AcO(CH_2)_4$	0.5	(58)	>95:5	125
$AcO(CH_2)_4$	—	(64)	—	902

Boronic Ester	Cross Partner	Conditions	Product(s) and Yield(s) (%)	Refs.
2 eq	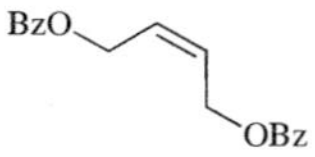 BzO, OBz (*Z*)/(*E*) = 94:6	Gr II (0.05 eq), DCM, reflux, 12 h	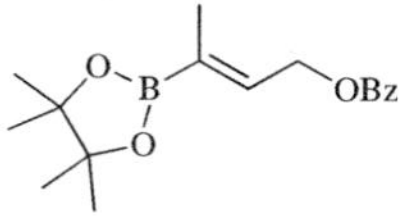 OBz (30) (*E*)/(*Z*) = 67:33	902

TABLE 25. CROSS-METATHESIS OF GEMINAL–DISUBSTITUTED COMPOUNDS (*Continued*)

B. BORONIC ESTERS (*Continued*)

Boronic Ester	Cross Partner	Conditions	Product(s) and Yield(s) (%)	Refs.
Please refer to the charts preceding the tables for the catalyst structures.				
C_{3-8}				
R, O, B, O; *x* eq	OAc, 3	Gr II (*y* eq), DCM, reflux, 12 h	R, O, B, O, OAc, 3	125

R	*x*	*y*		(*E*)/(*Z*)
Me	2	0.05	(58)	>95:5
$HOCH_2$	2	0.05	(0)	—
$AcOCH_2$	2	0.05	(40)	50:50
$TBSOCH_2$	2	0.05	(0)	—
$HO(CH_2)_2$	2	0.05	(0)	—
$AcO(CH_2)_2$	2	0.05	(0)	—
$TBSO(CH_2)_2$	1.3	0.05	(17)	83:17
$TBSO(CH_2)_3$	1.2	0.05	(35)	80:20
n-C_6H_{13}	2	0.05	(40)	75:25
n-C_6H_{13}	2	0.2	(54)	80:20

TABLE 25. CROSS-METATHESIS OF GEMINAL–DISUBSTITUTED COMPOUNDS (*Continued*)
C. LACTONES WITH AN EXOCYCLIC C–C DOUBLE BOND

Lactone	Cross Partner	Conditions	Product(s) and Yield(s) (%)	Refs.
Please refer to the charts preceding the tables for the catalyst structures.				
C_5				
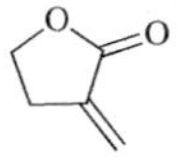	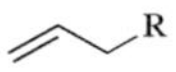 R 1.5 eq	Gr II (*x* eq), ClBcat (0.05 eq), DCM, 40°, 14 h	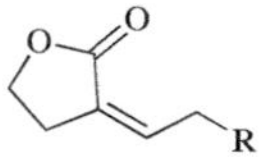 R (*Z*)/(*E*) = 83:17 R / *x* TMS 0.025 (64) TMS 2 x 0.025 (81) $(EtO)_2OP$ 0.025 (45) $(EtO)_2OP$ 2 x 0.025 (59)	173
	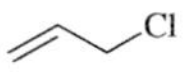 Cl 1.5 eq	Hov II (2 x 0.05 eq), 2,6-Cl_2-BQ (0.1 eq), DCM, reflux, 12 h	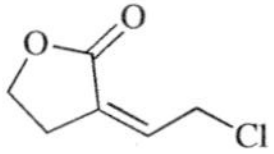 Cl (44) (*Z*)/(*E*) > 95:5	1162
	 OAc 1.5 eq	Gr II (*x* eq), ClBcat (0.05 eq), DCM, 40°, 14 h	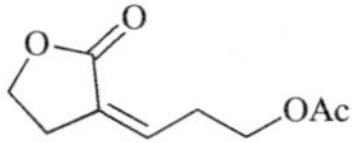 OAc (*Z*)/(*E*) = 90:10 *x* 0.025 (28) 2 x 0.025 (68)	173
	 OAc, *n* 1.5 eq	Hov II (2 x 0.05 eq), 2,6-Cl_2-BQ (0.1 eq), DCM, reflux, 12 h	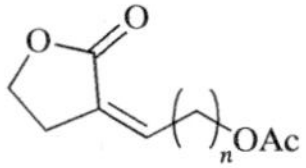 OAc, *n* *n* / (*Z*)/(*E*) 2 (75) 91:9 8 (80) 94:6	1162
	 Br 1.5 eq	Hov II (2 x 0.05 eq), 2,6-Cl_2-BQ (0.1 eq), DCM, reflux, 12 h	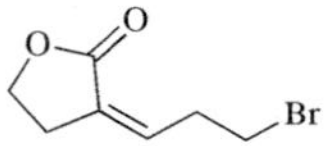 Br (43) (*Z*)/(*E*) > 95:5	1162

TABLE 25. CROSS-METATHESIS OF GEMINAL–DISUBSTITUTED COMPOUNDS (*Continued*)
C. LACTONES WITH AN EXOCYCLIC C–C DOUBLE BOND (*Continued*)

Lactone	Cross Partner	Conditions	Product(s) and Yield(s) (%)	Refs.
Please refer to the charts preceding the tables for the catalyst structures.				
C_5				
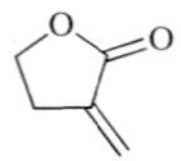	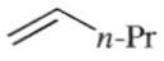 *n*-Pr 1.5 eq	Hov II (2 x 0.05 eq), 2,6-Cl_2-BQ (0.1 eq), DCM, reflux, 12 h	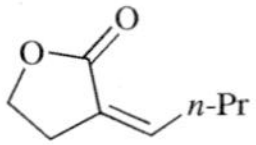 *n*-Pr (83) (*Z*)/(*E*) > 95:5	1162
	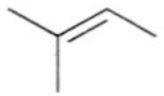 10 eq	Hov II (2 x 0.05 eq), 2,6-Cl_2-BQ (0.1 eq), DCM, reflux, 12 h	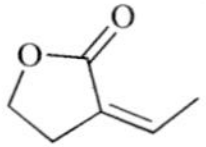 (93) (*Z*)/(*E*) = 86:14	1162
	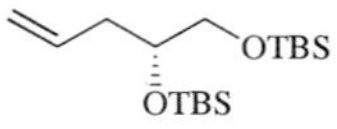 OTBS OTBS 1.5 eq	Gr II (0.025 eq), ClBcat (0.05 eq), DCM, 40°, 14 h	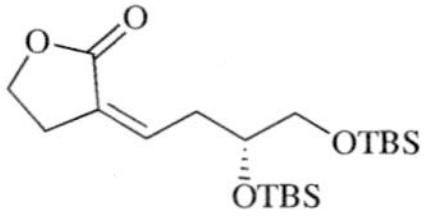 OTBS OTBS (72) (*Z*)/(*E*) = 89:11	173, 1163
I	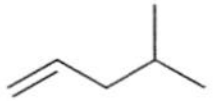 1.5 eq	Catalyst (*x* eq), CD_2Cl_2, 40°, 20 h	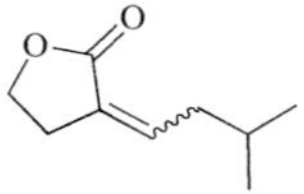 **II** + 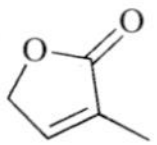 **III**	173

Catalyst	*x*	**I/II/III**[a]
Gr I	0.1	**I** only
Gr II	0.1	10:42:48
Gr II	0.05	9:55:36
Gr II	0.025	31:50:19
Gr II	0.01	46:44:10
Hov II	0.1	**I** only

I

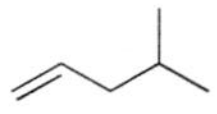

1.5 eq

Gr II (0.025 eq),
additive (0.05 eq),
CD_2Cl_2, 40°, 14 h

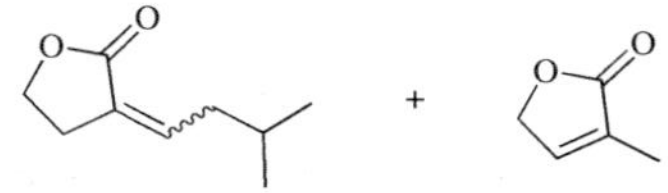

II + **III**

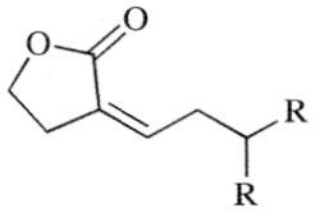

173

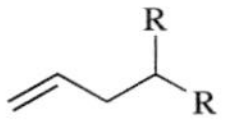

Additive	**II**	**I/II/III**[a]
—	(48)	31:50:19
Cy_2PCl	(14)	0:16:84
Ph_3PO	(56)	0:60:40
Cy_3PO	(64)	0:67:33
Ph_3As	(63)	0:69:31
Ph_2PCl	(71)	0:88:12
2,6-Cl_2-BQ	(77)	14:86:0
ClBcat	(87)	0:91:9

1.5 eq

Gr II (*x* eq),
ClBcat (0.05 eq),
DCM, 40°, 14 h

173

R	*x*		(*Z*)/(*E*)
Me	0.025	(87)	>95:5
MeO_2C	0.025	(24)	86:14
MeO_2C	2 x 0.025	(41)	86:14

TABLE 25. CROSS-METATHESIS OF GEMINAL–DISUBSTITUTED COMPOUNDS (*Continued*)
C. LACTONES WITH AN EXOCYCLIC C–C DOUBLE BOND (*Continued*)

Lactone	Cross Partner	Conditions	Product(s) and Yield(s) (%)	Refs.
Please refer to the charts preceding the tables for the catalyst structures.				
C_5				
	R; 1.5 eq	Gr II (0.025 eq), ClBcat (0.05 eq), DCM, 40°, 14 h	R; (Z)/(E) > 95:5	173
	R^1, R^2; 1.5 eq	Gr II (*x* eq), ClBcat (0.05 eq), DCM, 40°, 14 h	R^1, R^2	173
	OR, Ph; 1.5 eq	Gr II (*x* eq), ClBcat (0.05 eq), DCM, 40°, 14 h	OR, Ph	173

R	
H	(67)
F	(65)
MeO	(85)

R^1	R^2	*x*		(Z)/(E)
HO	MeO	0.025	(86)	>95:5
MeO	H	0.025	(54)	91:9
MeO	H	2 x 0.025	(70)	91:9

R	*x*		(Z)/(E)
MOM	0.025	(51)	>95:5
MOM	2 x 0.025	(88)	>95:5
TBS	0.025	(94)	91:9

C_7

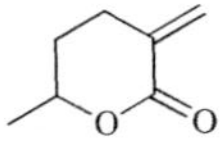

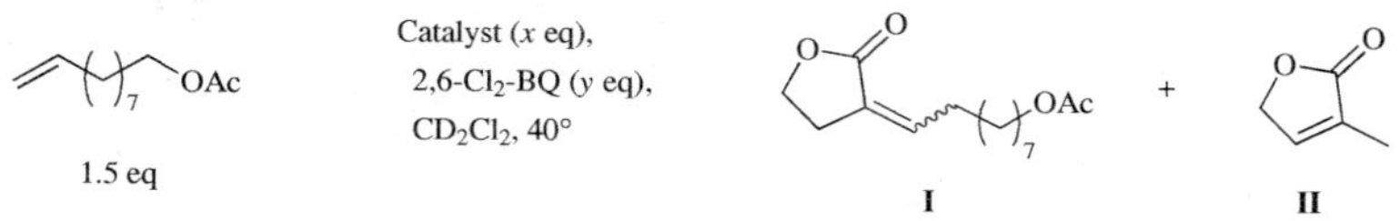

OAc (7) 1.5 eq

Catalyst (x eq), 2,6-Cl_2-BQ (y eq), CD_2Cl_2, 40°

I + II

1162

Catalyst	x	y	I/II[a]
Gr II	0.05	—	0:92
Hov II	0.05	—	0:87
Gr II	0.05	10	23:25
Hov II	0.05	10	67:0
Hov II	2 x 0.05	10	98:0

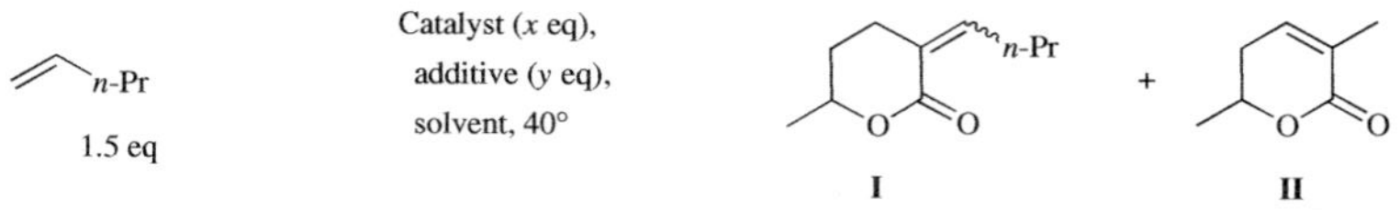

n-Pr 1.5 eq

Catalyst (x eq), additive (y eq), solvent, 40°

I + II

1162

Catalyst	x	Additive	y	Solvent	Time (h)	I/II[a]
Gr II	0.05	—	—	CD_2Cl_2	48	19:0
Hov II	0.05	—	—	CD_2Cl_2	48	11:0
Gr II	0.05	—	—	tol-d_8	48	0:66
Gr II	0.05	2,6-Cl_2-BQ	0.1	tol-d_8	48	0:26
Gr II	0.05	Ti(Oi-Pr)$_4$	0.2	CD_2Cl_2	48	0:0
Gr II	0.05	Ti(Oi-Pr)$_4$	0.2	tol-d_8	8	**II** only
Gr II	0.05	2,6-Cl_2-BQ, Ti(Oi-Pr)$_4$	0.1, 0.2	tol-d_8	8	**II** only
Gr II	2 x 0.05	HOAc	1.5	CD_2Cl_2	36	54:tr

TABLE 25. CROSS-METATHESIS OF GEMINAL–DISUBSTITUTED COMPOUNDS (*Continued*)
C. LACTONES WITH AN EXOCYCLIC C–C DOUBLE BOND (*Continued*)

Please refer to the charts preceding the tables for the catalyst structures.

Lactone	Cross Partner	Conditions	Product(s) and Yield(s) (%)	Refs.
C_{12}				
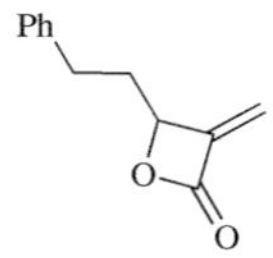 Ph	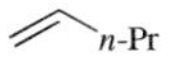 n-Pr, 1.5 eq	Gr II (0.05 eq), DCM, 40–55°, 12–24 h	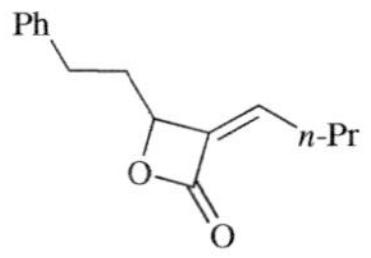 Ph, n-Pr (76) (Z)/(E) = 82:18	172
C_{17}				
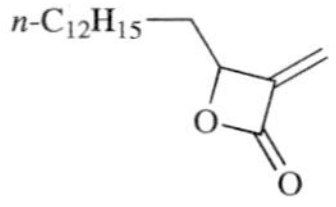 n-$C_{12}H_{15}$	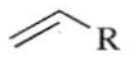 R, 2 eq	Gr II (0.05 eq), DCM, reflux, overnight	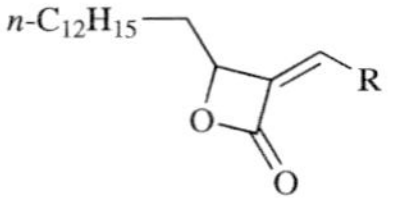 n-$C_{12}H_{15}$, R	1164
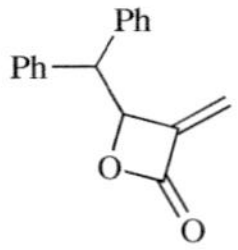 Ph, Ph	 OAc, *n*, 1.5 eq	Gr II (0.05 eq), DCM, 40–55°, 12–24 h	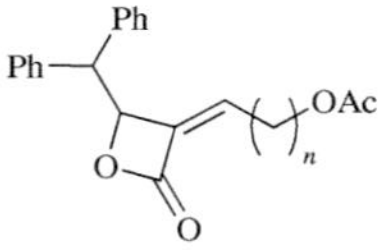 Ph, Ph, OAc, *n*	172

R		(Z)/(E)
n-C_6H_{13}	(97)	76:24
Ph	(40)	(Z) only
$Ph(CH_2)_3$	(51)	(Z) only
$Ph(CH_2)_4$	(68)	(Z) only
n-$C_{18}H_{37}$	(74)	69:31

n		(Z)/(E)
1	(84)	>95:5
2	(90)	>95:5
3	(85)	93:7
7	(84)	>95:5

Reactant	Conditions	Product(s) and Yield(s) (%)			Refs.
 Cl (1.5 eq)	Hov II (0.05 eq), DCM, 40–55°, 12–24 h	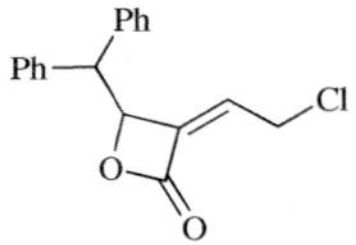 (80) (*Z*)/(*E*) = 92:8			172
 R (1.5 eq)	Gr II (0.05 eq), DCM, 40–55°, 12–24 h	R		(*Z*)/(*E*)	172
		TBSO	(88)	90:10	
		Br	(93)	>95:5	
		Me	(94)	92:8	
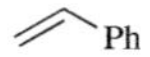 Ph (1.5 eq)	Gr II (0.05 eq), DCM, 40–55°, 12–24 h	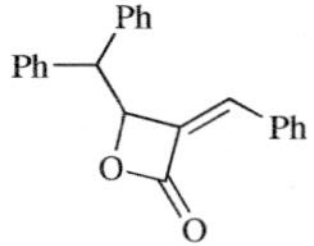 (55) (*Z*)/(*E*) > 95:5			172

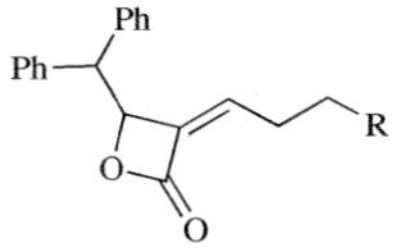

[a] The ratio was determined by ^{1}H NMR analysis.

TABLE 25. CROSS-METATHESIS OF GEMINAL–DISUBSTITUTED COMPOUNDS (*Continued*)
D. LACTAMS WITH AN EXOCYCLIC C–C DOUBLE BOND

Lactam	Cross Partner	Conditions	Product(s) and Yield(s) (%)	Refs.

Please refer to the charts preceding the tables for the catalyst structures.

C_4

Lactam	Cross Partner	Conditions	Product(s) and Yield(s) (%)	Refs.
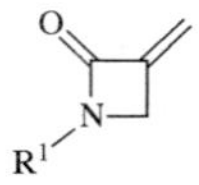 R^1–N (3-methylene-2-azetidinone)	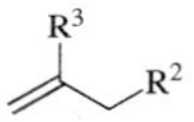 R^3C(=CH$_2$)CH$_2R^2$, 2 eq	Catalyst (x eq), DCM, reflux	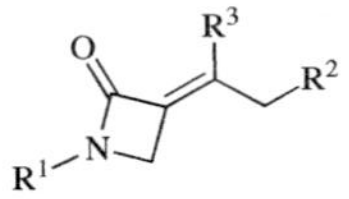 R^1–N 2-azetidinone with exocyclic =C(R^3)CH$_2R^2$	1165

R^1	R^2	R^3	Catalyst	x		$(E)/(Z)$
H	Cl	H	Hov II	0.06	(25)	52:48
H	Cl	H	Hov II	0.09	(60)	52:48
Boc	Cl	CH_3	Hov II	0.1	(58)	39:61
Boc	TMS	CH_3	Hov II	0.1	(73)	39:61
Boc	HO	CH_3	Hov II	0.1	(86)	39:61
Boc	Cl	H	Gr II	0.15	(0)	—
Boc	Cl	H	Hov II	0.06	(78)	55:45
Ph	Cl	H	Hov II	0.06	(54)	52:48
4-MeOC_6H_4	Cl	H	Hov II	0.06	(48)	52:48
Bn	Cl	H	Hov II	0.06	(58)	52:48

Catalyst (x eq),
DCM, reflux

2 eq

1165

R^1	R^2	R^3	Catalyst	x		$(E)/(Z)$
Boc	$AcO(CH_2)_2$	H	Gr II	0.1	(57)	67:33
Boc	$AcO(CH_2)_2$	H	Hov II	0.03	(90)	67:33
H	$AcO(CH_2)_2$	H	Hov II	0.03	(90)	67:33
Bn	$AcO(CH_2)_2$	H	Hov II	0.03	(77)	60:40
Ph	$AcO(CH_2)_2$	H	Hov II	0.03	(76)	67:33
4-$MeOC_6H_4$	$AcO(CH_2)_2$	H	Hov II	0.03	(57)	60:40
4-$MeOC_6H_4$	$AcO(CH_2)_2$	H	Hov II	0.1	(72)	60:40
Boc	*n*-Pr	H	Gr II	0.02	(84)	71:29
H	*n*-Pr	H	Gr II	0.06	(61)	67:33
H	*n*-Pr	H	Hov II	0.03	(89)	67:33
Bn	*n*-Pr	H	Gr II	0.04	(89)	67:33
Ph	*n*-Pr	H	Gr II	0.04	(82)	75:25
4-$MeOC_6H_4$	*n*-Pr	H	Gr II	0.04	(80)	71:29
Boc	*i*-Pr	Me	Hov II	0.1	(0)	—
Boc	*n*-Pr	Me	Hov II	0.1	(85)	50:50
Bn	*n*-Pr	Me	Hov II	0.1	(74)	55:45
Boc	Et	Et	Hov II	0.1	(65)	—

TABLE 25. CROSS-METATHESIS OF GEMINAL–DISUBSTITUTED COMPOUNDS (*Continued*)
D. LACTAMS WITH AN EXOCYCLIC C–C DOUBLE BOND (*Continued*)

Lactam	Cross Partner	Conditions	Product(s) and Yield(s) (%)	Refs.
Please refer to the charts preceding the tables for the catalyst structures.				
C_4				
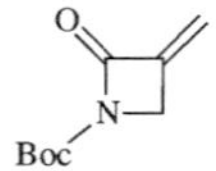 O, N, Boc	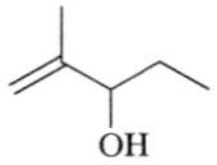 OH 2 eq	Hov II (x eq), DCM, reflux	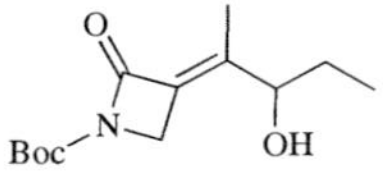 O, N, Boc, OH $(E)/(Z) = 71{:}29$ x — 0.1 (40); 0.15 (69)	1165
	R^1, R^2 2 eq	Hov II (x eq), DCM, reflux	O, N, Boc, R^1, R^2	1165
	Ph 2 eq	Hov II (0.1 eq), DCM, reflux, 12 h	O, N, Boc, Ph (81) $(E)/(Z) = 60{:}40$	1165

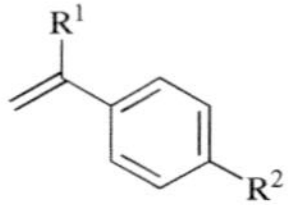

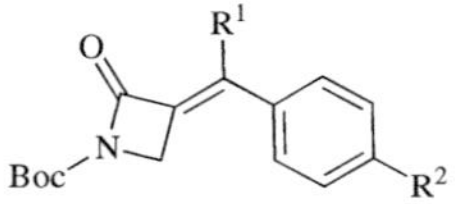

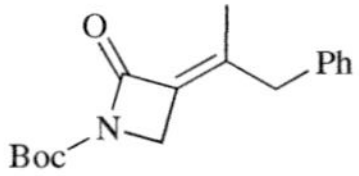

R^1	R^2	x		$(E)/(Z)$
H	MeO	0.03	(87)	83:17
H	F	0.03	(90)	75:25
Me	H	0.1	(0)	—

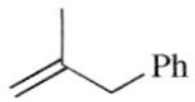

C_{16}

R[1], R[2], O, N; R[3]

Hov II (3 x 0.03 eq),
solvent, 100°, 40 h

R[1], R[2], R[3], O, N

1166

R^1	R^2	R^3	Solvent	
H	H	Ph	$C_6F_5CF_3$	(88)
H	H	4-$O_2NC_6H_4$	toluene	(66)
H	H	4-$O_2NC_6H_4$	$C_6F_5CF_3$	(51)
H	H	4-NCC_6H_4	$C_6F_5CF_3$	(83)
H	H	4-$CF_3C_6H_4$	$C_6F_5CF_3$	(62)
H	H	4-ClC_6H_4	$C_6F_5CF_3$	(41)
H	H	4-$MeO_2CC_6H_4$	$C_6F_5CF_3$	(51)
H	H	4-$AcOC_6H_4$	$C_6F_5CF_3$	(52)
H	H	4-$MeOC_6H_4$	$C_6F_5CF_3$	(0)
H	H	2-naphthyl	$C_6F_5CF_3$	(0)
H	H	Bn	toluene	(43)
H	H	Bn	$C_6F_5CF_3$	(43)
H	H	*rac*-PhCH(HO)CH_2	toluene or $C_6F_5CF_3$	(0)
H	H	*rac*-PhCH(AcO)CH_2	toluene or $C_6F_5CF_3$	(0)
H	H	*rac*-PhCH(BnO)CH_2	toluene or $C_6F_5CF_3$	(0)
H	H	*rac*-PhCH(TBSO)CH_2	toluene	(72)
H	H	*rac*-PhCH(TBSO)CH_2	$C_6F_5CF_3$	(57)
F	TBSO (*rac*)	(*S*)-4-FC_6H_4CH(TBSO)CH_2	toluene	(67)
F	TBSO (*R*)	(*S*)-4-FC_6H_4CH(TBSO)CH_2	toluene	(65)

TABLE 25. CROSS-METATHESIS OF GEMINAL–DISUBSTITUTED COMPOUNDS (*Continued*)

E. α-METHACRYLATES

Acrylate	Cross Partner	Conditions	Product(s) and Yield(s) (%)	Refs.
Please refer to the charts preceding the tables for the catalyst structures.				
C_4				
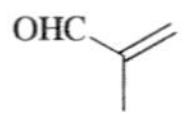 OHC 5 eq	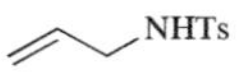 NHTs	Hov II (0.05 eq), $(PhO)_3B$ (0.1 eq), toluene, 80°, 0.75 h	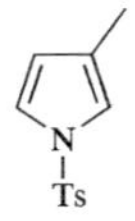 N Ts (70)	899
3 eq	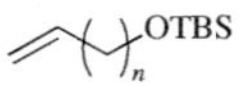 OTBS *n*	1. Gr II (0.05 eq), DCM, reflux, 3 h 2. DIBAL-H (3.75 eq), DCM, –78°, 2 h	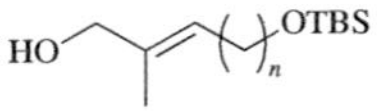 HO OTBS *n* (*E*)/(*Z*) > 95:5; *n* = 2 (56), 3 (69), 4 (59)	940, 941
3 eq	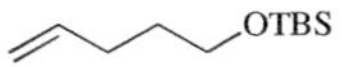 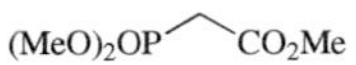 OTBS + $(MeO)_2OP$ CO_2Me **I**	1. Gr II (0.05 eq), DCM, 40°, 3 h 2. **I** (1.4 eq), NaH (1.4 eq), THF, 0° to rt, 0.5 + 15 h	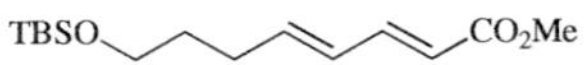 TBSO CO_2Me (81) (*E*)/(*Z*) = 95:5	940
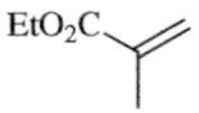 EtO_2C excess	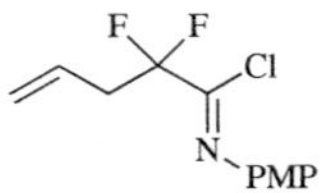 F F Cl N PMP	Gr II (0.1 eq), neat, reflux, 96 h	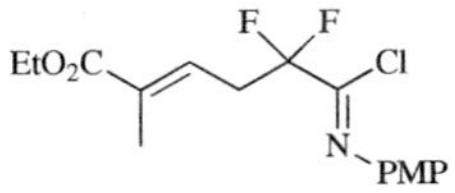 EtO_2C F F Cl N PMP (85) (*E*)/(*Z*) > 95:5	392

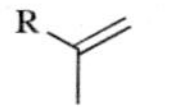 R (alkene), 3 eq	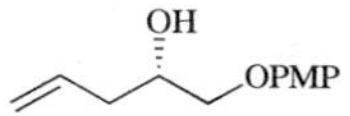 OH, OPMP	Hov II (0.025 eq), DCM, rt, 36 h	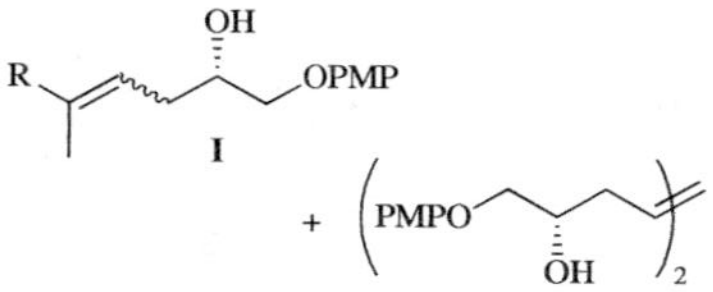 R, OH, OPMP **I** + (PMPO, OH)$_2$ **II** (*E*)/(*Z*) = 80:20	626
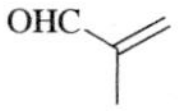 OHC, 2 eq	 R	Hov II (0.02 eq), rt, 8 h	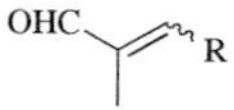 OHC, R	927
2.6 eq	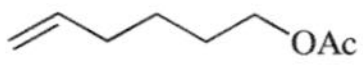 OAc	Gr II (0.05 eq), HCl/Et_2O (0.25 eq), benzene/THF, 45°, 14 h	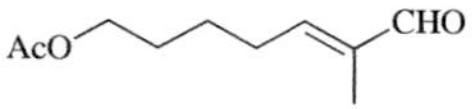 AcO, CHO (82) (*E*)/(*Z*) > 95:5	928
3 eq	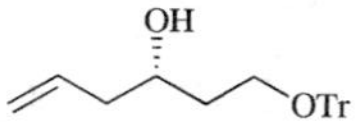 OH, OTr	Hov II (0.025 eq), DCM, rt, 36 h	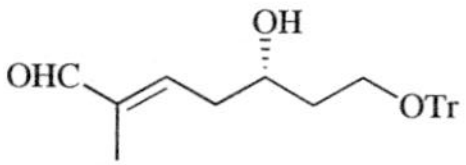 OHC, OH, OTr (2) (*E*)/(*Z*) >97:3 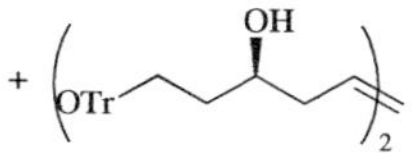 + (OTr, OH)$_2$ (25) (*E*)/(*Z*) = 80:20	626

R	I	II
OHC	(2)	(40)
MeO_2C	(0)	(51)

R	
n-Bu	(25)
n-C_6H_{13}	(35)
n-C_8H_{17}	(30)

TABLE 25. CROSS-METATHESIS OF GEMINAL–DISUBSTITUTED COMPOUNDS (*Continued*)

E. α-METHACRYLATES (*Continued*)

Acrylate	Cross Partner	Conditions	Product(s) and Yield(s) (%)	Refs.
Please refer to the charts preceding the tables for the catalyst structures.				
C_4				
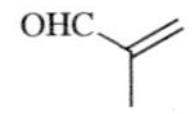 OHC; 36 eq	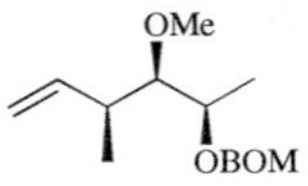 OMe; OBOM	Hov II (0.02 eq), DCM, 45°, 16 h	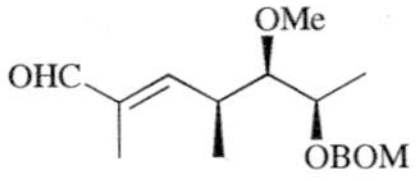 OHC; OMe; OBOM (58) (*E*)/(*Z*) = 85:15	1167
— eq	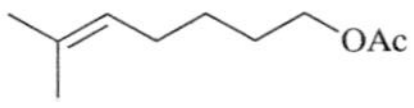 OAc	Gr II (0.05 eq), DCM, 40°, 12 h	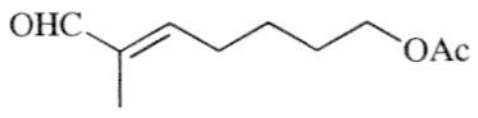 OHC; OAc (97) (*E*) only	36, 902
3 eq	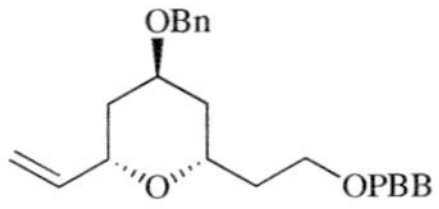  OBn; O; OPBB + $(EtO)_2OP$ CO_2Et **I**	1. Hov II (0.05 eq), DCM, reflux, 12 h 2. **I** (1.43 eq), NaH (1.43 eq), THF, rt, 1 h	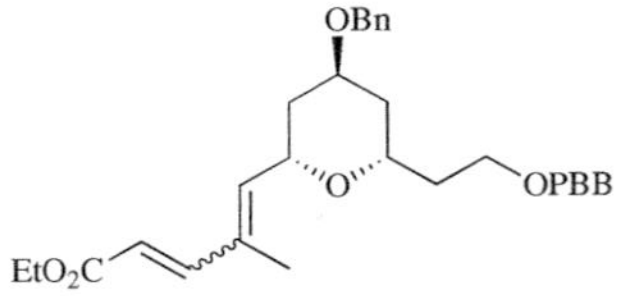 EtO_2C; OBn; O; OPBB (80)	1168
MeO_2C; *x* eq	OR; OMe	Gr II (0.05 eq), neat, 90°, 16 h	MeO_2C; OR; OMe	1074

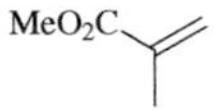

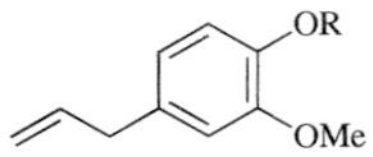

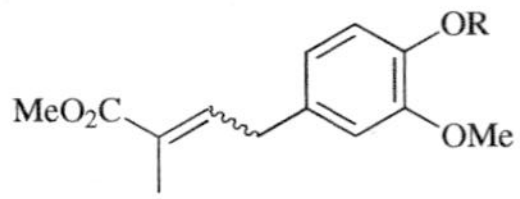

R	*x*	
H	31	(60)
Ac	39	(72)
i-Pr	39	(78)

Reactant 1	Reactant 2	Conditions	Product	Yield (%)	Refs.
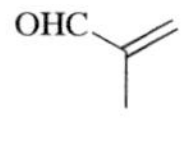 OHC (3 eq)	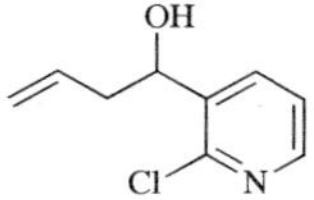 OH, Cl, N	Hov II (0.1 eq), DCM, 50°, 24 h	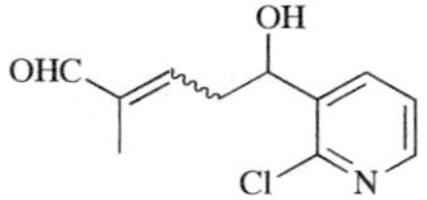 OHC, OH, Cl, N	(52)	603
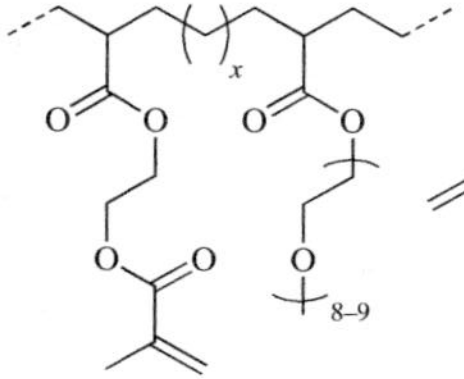 O, O, O, O, O, x, 8–9	 Ph (10 eq)	Gr II (0.1 eq), DCE, 70°, 12 h	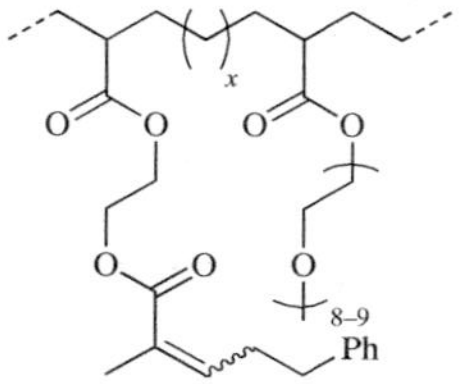 O, O, O, O, O, x, 8–9, Ph	(15)	1169
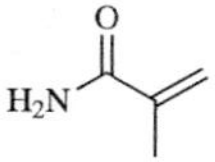 H_2N, O	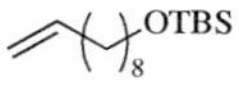 OTBS, 8 (1.2 eq)	Gr II (0.05 eq), DCM, reflux, 12 h	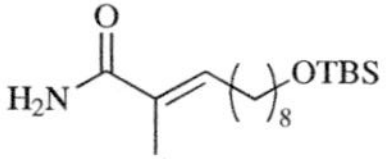 H_2N, O, OTBS, 8	(71) $(E)/(Z) > 95:5$	36
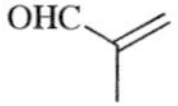 OHC (10 eq)	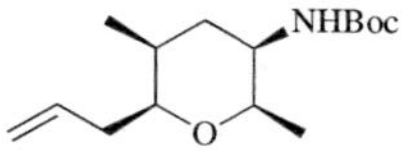 NHBoc, O	Gr II (0.05 eq), DCE, 40°	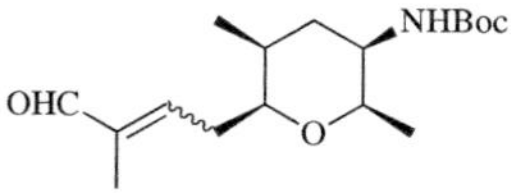 OHC, NHBoc, O	(67)	828

TABLE 25. CROSS-METATHESIS OF GEMINAL–DISUBSTITUTED COMPOUNDS (*Continued*)

E. α-METHACRYLATES (*Continued*)

Acrylate	Cross Partner	Conditions	Product(s) and Yield(s) (%)	Refs.
Please refer to the charts preceding the tables for the catalyst structures.				
C_4				
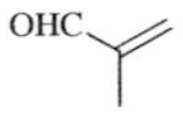 OHC; — eq	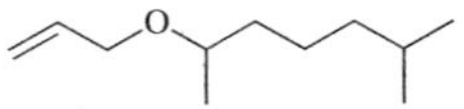	Hov II (0.02 eq), rt, 8 h	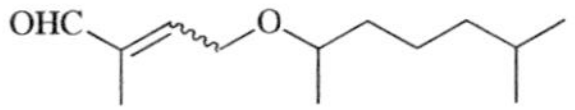 OHC; (15)	927
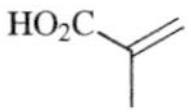 HO_2C	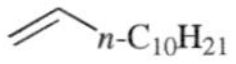 n-$C_{10}H_{21}$; 1.1 eq	Gr II (0.05 eq), DCM, 40°, 12 h	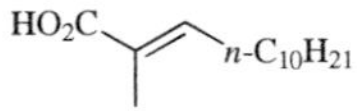 HO_2C, n-$C_{10}H_{21}$ (23) $(E)/(Z)$ = 80:20	36
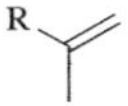 R; 3 eq	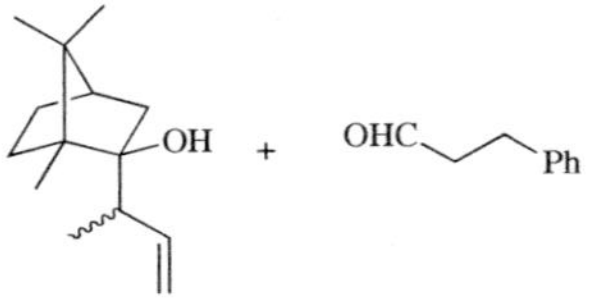 OH + OHC, Ph; 3 eq	1. CSA (0.1 eq), DCM, 25°, 122 h 2. Gr II (0.2 eq), DCM, 4 h	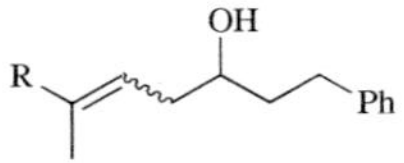 R, OH, Ph R: MeO_2C (0), er — R: OHC (30), er 97.0:3.0	936
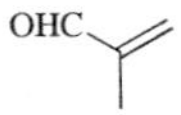 OHC; 4 eq	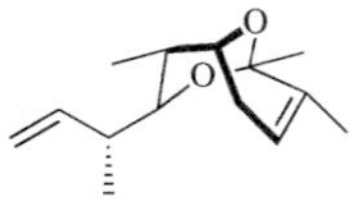 O, O	Ru cat **271** (0.1 eq), DCM, 40°	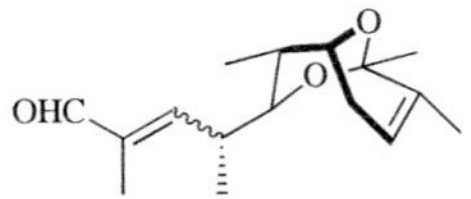 OHC, O, O (61)	1170
20 eq	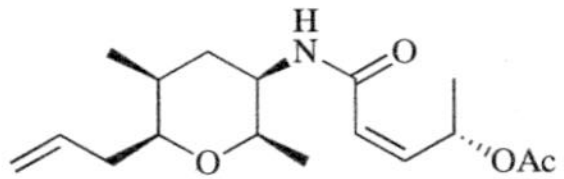 H N O, O, OAc	Ru cat **88** (0.05 eq), DCM, 23°	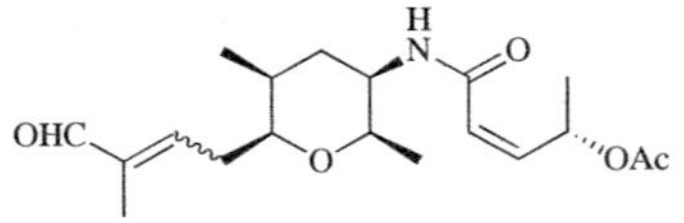 OHC, H N O, O, OAc (57)	756, 828

EtO_2C 20 eq	TBDPSO, N, TFA	Hov II (0.1 eq), neat, reflux, 13 h	EtO_2C, TBDPSO, N, TFA (74) (*E*) only + TBDPSO, N, TFA (dimer,)$_2$ (14) (*E*) only	811
4 eq	OMe, OMe	1. Gr II (0.028 eq), DCE, reflux, 10 h 2. 10% Pd/C, H_2, AcOEt, rt, 10 h	EtO_2C, EtO_2C, OMe, OMe (60)	1120

TABLE 25. CROSS-METATHESIS OF GEMINAL–DISUBSTITUTED COMPOUNDS (*Continued*)

E. α-METHACRYLATES (*Continued*)

Acrylate	Cross Partner	Conditions	Product(s) and Yield(s) (%)	Refs.
Please refer to the charts preceding the tables for the catalyst structures.				
C_4				
MeO_2C 4 eq	n-C_7H_{15} CO_2Me 7	Catalyst (x eq), toluene, 60°, 4 h	MeO_2C CO_2Me 7 **I** + n-C_7H_{15} CO_2Me **II** + n-C_7H_{15} n-C_7H_{15} **III** + MeO_2C 7 6 CO_2Me **IV**	1119

Catalyst	x	**I** + **II**[a]	Selectivity **I** + **II**[a]	**III** + **IV**[a]	TON	Productive TON
Ru cat **139**	0.26	(96)	97	(3)	420	360
Ru cat **139**	0.026	(95)	99	(1)	3640	3600
Ru cat **139**	0.0026	(96)	96	(4)	37000	35450
Ru cat **139**	0.0003	(53)	16	(84)	177700	28300
Ru cat **320**	0.0029	(96)	97	(3)	330	320
Ru cat **320**	0.029	(97)	97	(3)	3300	3200
Ru cat **320**	0.0029	(96)	97	(3)	33000	31100
Ru cat **320**	0.0003	(26)	4	(96)	88500	3470

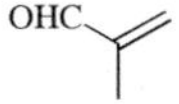 OHC (methacrolein), 5 eq	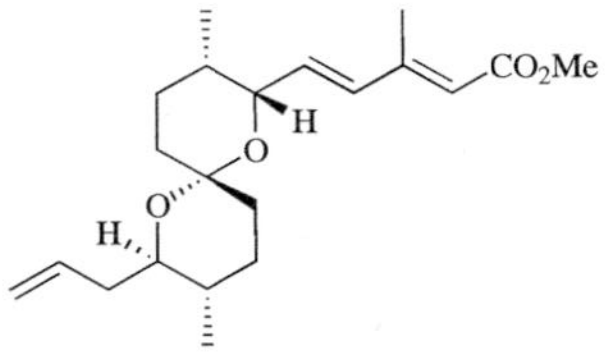 CO_2Me, H, O, O, H	Ru cat **35** (0.05 eq), DCM, reflux	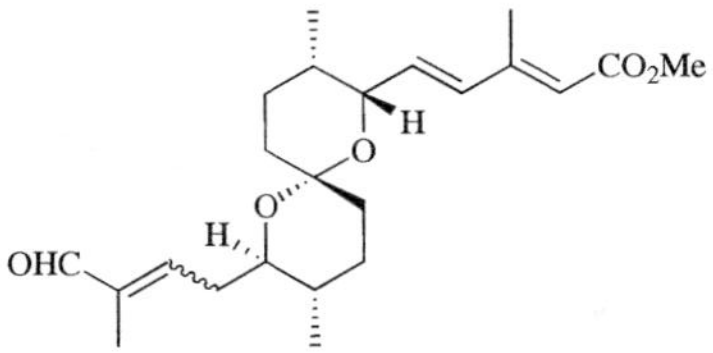 OHC, CO_2Me, H, O, O, H (74)	753
5 eq	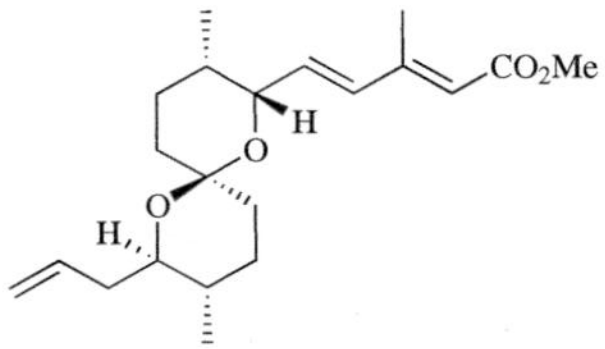 CO_2Me, H, O, O, H	Ru cat **35** (0.05 eq), DCM, reflux	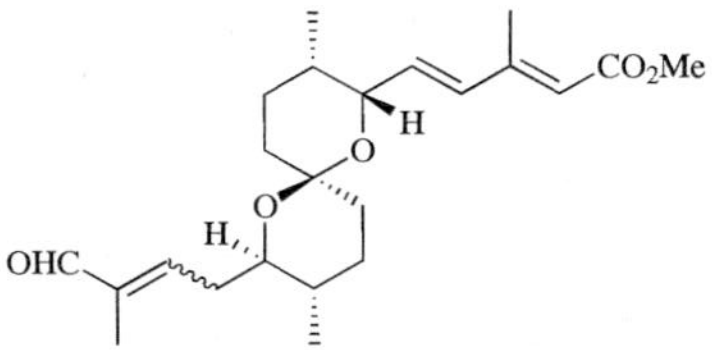 OHC, CO_2Me, H, O, O, H (74)	753

TABLE 25. CROSS-METATHESIS OF GEMINAL–DISUBSTITUTED COMPOUNDS (*Continued*)

E. α-METHACRYLATES (*Continued*)

Acrylate	Cross Partner	Conditions	Product(s) and Yield(s) (%)	Refs.
Please refer to the charts preceding the tables for the catalyst structures.				
C_4				
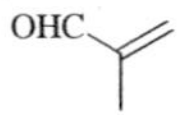 OHC; 12.5 eq	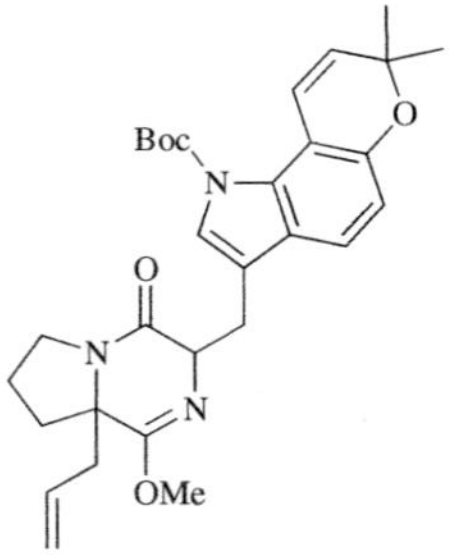 Boc, N, O, N, OMe	Catalyst (*x* eq), DCM	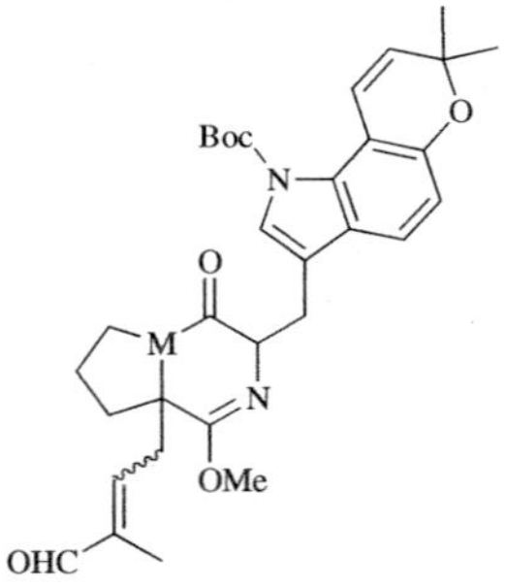 Boc, N, O, M, N, OMe, OHC	1171
$CO(OCH_2)_nOMe$; 40 eq	O, H, O, HO, O, OH, OEt, 2, *m*	Hov II (0.1 eq), DCM, 37°	$CO(OCH_2)_nOMe$, O, H, O, HO, O, OH, OEt, 2, *m*	1033

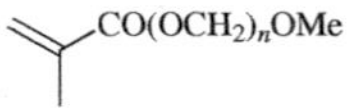

Catalyst	*x*	Temp (°)	Time (h)	
Gr II	0.2	reflux	24	(65)
Hov II	0.05	reflux	48	(71)
Hov II	0.05	100 (MW)	0.33	(73)

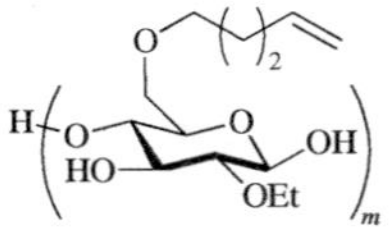

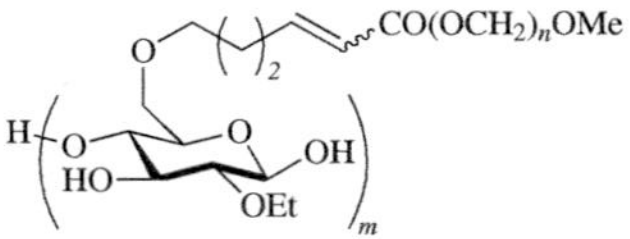

Time (h)	Conv. (%)[b]
2	85
12	100

[a] The yield and selectivity were determined by GC analysis.

[b] The conversion was determined by ^1H NMR analysis.

TABLE 26. CROSS-METATHESIS OF ALLYLIC NITRO COMPOUNDS

Nitro Compound	Cross Partner	Conditions	Product(s) and Yield(s) (%)	Refs.
Please refer to the charts preceding the tables for the catalyst structures.				
C_3				
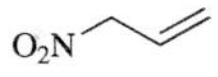 O_2N (allyl) 2 eq	 R (n)	Ru cat **88** (0.05 eq), $B(OPh)_3$ (0.25 eq), DCM, 40°, 5 h	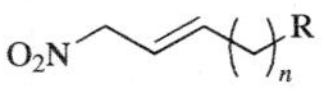 O_2N … R (n)	179

n	R		(*E*)/(*Z*)
1	Ph	(66)	93:7
1	$(EtCO_2)_2CH$	(52)	84:16
2	Ac	(71)	94:6
2	 O (oxiranyl)	(43)	89:11
3	MeO_2C	(62)	93:7
8	Br	(70)	88:12
10	Me	(73)	84:16

TABLE 26. CROSS-METATHESIS OF ALLYLIC NITRO COMPOUNDS (*Continued*)

Nitro Compound	Cross Partner	Conditions	Product(s) and Yield(s) (%)	Refs.

Please refer to the charts preceding the tables for the catalyst structures.

C_3

Nitro Compound	Cross Partner	Conditions	Product(s) and Yield(s) (%)	Refs.
O_2N (allyl), 2 eq	OTBS (alkene)	Catalyst (*x* eq), additive (*y* eq), solvent, 5 h	O_2N ~ OTBS	179

Catalyst	*x*	Additive	*y*	Solvent	Temp (°)	Conv. (%)[a]	
Gr II	0.05	—	—	DCM	40	1	(—)
Hov II	0.05	—	—	DCM	40	37	(—)
Ru cat **88**	0.05	—	—	DCM	40	44	(—)
Ru cat **88**	0.1	—	—	DCM	40	51	(—)
Ru cat **88**	0.05	—	—	toluene	70	36	(—)
Ru cat **88**	0.05	—	—	$C_6F_5CF_3$	70	60	(—)
Ru cat **88**	0.05	Ti(O*i*-Pr)$_4$	0.1	DCM	40	19	(16)
Ru cat **88**	0.05	Ph_2SnCl_2	0.1	DCM	40	62	(58)
Ru cat **88**	0.05	BF_3•Et_2O	0.1	DCM	40	86	(58)
Ru cat **88**	0.05	B(OMe)$_3$	0.1	DCM	40	26	(26)
Ru cat **88**	0.05	B(OPh)$_3$	0.1	DCM	40	68	(68)
Ru cat **88**	0.05	B(OPh)$_3$	0.05	DCM	40	70	(65)
Ru cat **88**	0.05	B(OPh)$_3$	0.25	DCM	40	80	(74)
Ru cat **88**	0.05	B(OPh)$_3$	0.5	DCM	40	83	(74)
Ru cat **88**	0.05	B(OPh)$_3$	1.0	DCM	40	81	(78)
Ru cat **88**	0.05	PhO_2BCl	0.01	DCM	40	54	(54)

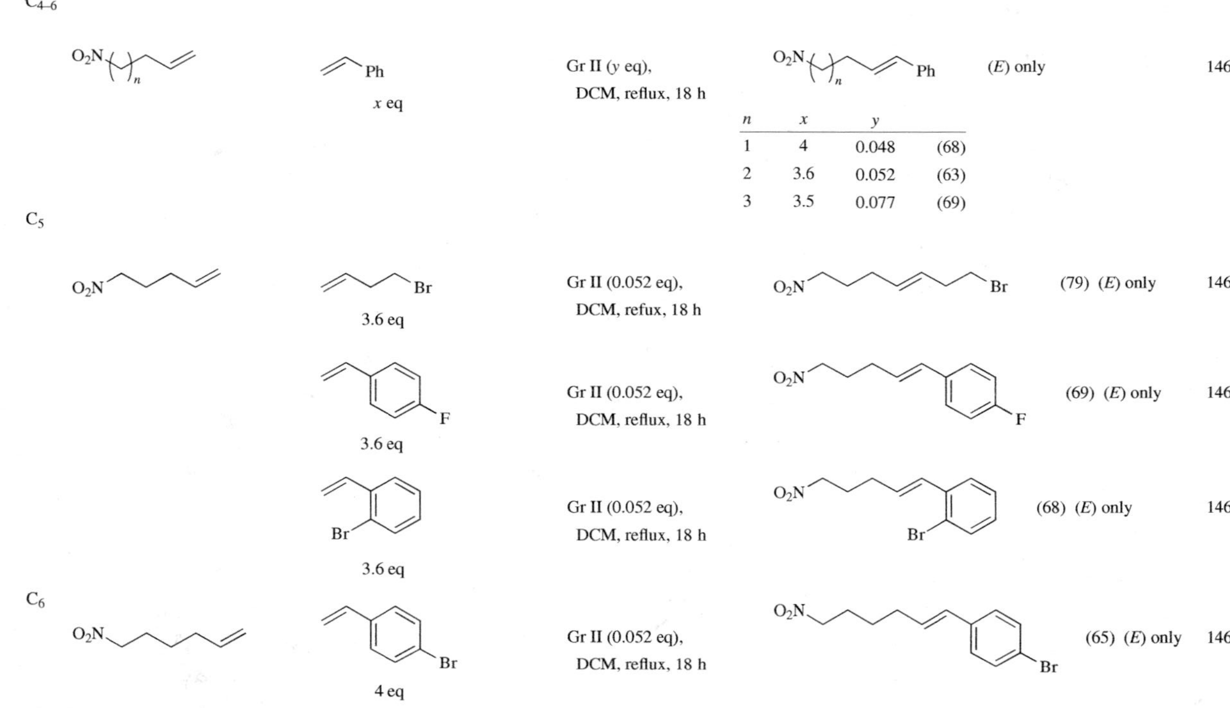

	Reagent	Conditions	Product		Refs.
C_{4-6}	Ph, x eq	Gr II (y eq), DCM, reflux, 18 h	O_2N, Ph	(E) only	146
C_5	Br, 3.6 eq	Gr II (0.052 eq), DCM, refux, 18 h	O_2N, Br	(79) (E) only	146
	F, 3.6 eq	Gr II (0.052 eq), DCM, reflux, 18 h	O_2N, F	(69) (E) only	146
	Br, 3.6 eq	Gr II (0.052 eq), DCM, reflux, 18 h	O_2N, Br	(68) (E) only	146
C_6	Br, 4 eq	Gr II (0.052 eq), DCM, reflux, 18 h	O_2N, Br	(65) (E) only	146

n	x	y	
1	4	0.048	(68)
2	3.6	0.052	(63)
3	3.5	0.077	(69)

[a] The conversion was determined by [1]NMR analysis.

TABLE 27. CROSS-METATHESIS OF ACRYLONITRILE AND METHACRYLONITRILE

Nitrile	Cross Partner	Conditions	Product(s) and Yield(s) (%)	Refs.

Please refer to the charts preceding the tables for the catalyst structures.

C_3

Nitrile: NC–CH=CH₂ (acrylonitrile); Cross Partner: CH₂=CH–CN; Conditions: Catalyst (x eq), solvent; Product: NC–CH=CH–CN

Catalyst	x	Solvent	Temp (°)	Time (h)	Yield (%)	(Z)/(E)	Refs.
Gr I	0.02	DCM	45	12	(0)	—	151
Gr I	0.1	DCM	45	24	(0)	—	151
Gr II	0.02	DCM	45	12	(2)	—	151
Gr II	0.1	DCM	45	24	(12)	—	151
Hov II	0.02	DCM	45	24	(21)	—	151
Ru cat **30**	0.1	DCM	45	24	(15)	—	151
Ru cat **65**	0.02	DCM	45	24	(39)	—	151
Ru cat **65**	0.1	toluene	110	12	(38)	—	151
Ru cat **69**	0.02	DCM	45	12	(12)	75:25	1172
Ru cat **69**	0.1	DCM	45	12	(36)	75:25	1172

Cross Partner: 2 eq CH₂=CH–R; Conditions: Ru cat **69** (x eq), DCM, 45°, 12 h; Products: NC–CH=CH–R (**I**) + R–CH=CH–R (**II**); Ref. 1172

R	x	**I**	(Z)/(E) **I**	**II**
$HOCH_2$	0.02	(56)	75:25	(5)
$HOCH_2$	0.1	(68)	75:25	(3)
MeO_2C	0.02	(60)	80:20	(—)
MeO_2C	0.1	(71)	80:20	(—)
OHC	0.02	(61)	80:20	(—)
OHC	0.1	(77)	80:20	(—)
HO_2C	0.02	(35)	75:25	(11)
HO_2C	0.1	(55)	75:25	(7)

x eq

Catalyst (y eq), solvent

I + II

R	x	Catalyst	y	Solvent	Temp (°)	Time (h)	I	(Z)/(E) I	II	Refs.
Br	2	Mo cat **1**	0.05	DCM	rt	3	(18)	90:10	(20)	149
BnO	2	Mo cat **1**	0.05	DCM	rt	3	(40)	80:20	(—)	149
TBSO	2	Mo cat **1**	0.05	DCM	rt	3	(73)	84:16	(10)	149
TMS	2	Mo cat **1**	0.05	DCM	rt	3	(76)	75:25	(—)	149
BnO_2C	2	Mo cat **1**	0.05	DCM	rt	3	(44)	85:15	(6)	149
TMS	2	Mo cat **1**	0.05	DME	rt	4	(76)	82:18	(—)	208
BzO	2	Ru cat **286**	0.02	DCM	40	24	(0)	—	(—)	161
BzO	—	Ru cat **98**	0.02	DCM	40	24	(40)	80:20	(—)	965
BzO	2	Ru cat **99**	0.02	DCM	40	24	(48)	80:20	(—)	965
BzO	2	Ru cat **189**	0.02	DCM	40	24	(84)	80:20	(—)	965
OHC	2	Hov II	0.05	DCM	45	2	(91)	80:20	(—)	153

2 eq

Hov II (0.05 eq), DCM, reflux, 3 h

182

Concn (M)		(Z)/(E)
0.5	(35)	67:33
0.07	(74)	71:29
0.24	(60)	67:33

TABLE 27. CROSS-METATHESIS OF ACRYLONITRILE AND METHACRYLONITRILE (*Continued*)

Nitrile	Cross Partner	Conditions	Product(s) and Yield(s) (%)	Refs.

Please refer to the charts preceding the tables for the catalyst structures.

C_3

NC (2 eq) + cross partner (R) — Catalyst (x eq), DCM → I + II

R	Catalyst	x	Temp (°)	Time (h)	I	(Z)/(E) I	II	Refs.
Br	Mo cat **1**	0.05	rt	3	(45)	88:12	(11)	149
BnO	Mo cat **1**	0.05	rt	3	(60)	88:12	(<1)	149
TBSO	Mo cat **1**	0.05	rt	3	(68)	85:15	(—)	149
HO	Hov II	0.05	45	2	(81)	75:25	(—)	153
BzO	Hov II	0.05	45	2	(88)	80:20	(—)	153
BzO	Ru cat **243**	0.05	45	12	(15)	—	(—)	923
BzO	Ru cat **191**	0.05	45	12	(98)	—	(—)	923
BzO	Ru cat **98**	0.02	40	24	(39)	80:20	(—)	965
BzO	Ru cat **99**	0.02	40	24	(42)	80:20	(—)	965
BzO	Ru cat **100**	0.02	40	24	(24)	80:20	(—)	965
BzO	Ru cat **101**	0.02	40	24	(47)	80:20	(—)	965
BzO	Ru cat **102**	0.02	40	24	(49)	80:20	(—)	965

x eq	OR	Gr II (0.01 eq), CuCl (y eq), DCM, 40°	NC OR	150

x	R	y	Time (h)	Conv. (%)[a]
1	H	0.04	16	53
1	H	—	16	29
1	Bz	—	3	40
1	Bz	0.04	3	65
1	Bz	0.08	3	63
1	Bz	0.16	3	46
4	Bz	0.04	3	30
1	Bz	0.04	16	65
1	Bz	—	16	45

2 eq	CO_2Me, NHBoc	Hov II (0.08 eq), DCM, 45°, 6 h	CN, CO_2Me, NHBoc (83) (Z)/(E) = 75:25	153
3 eq	O, O	Hov II (0.05 eq), DCM, 45°, 2 h	CN, O, O (60) (Z)/(E) = 80:20	153
2 eq	OPMP, OH	Hov II (0.025 eq), DCM, rt, 15 h	(OPMP, OH)$_2$ (49) (E)/(Z) = 80:20 + CN, OPMP, OH (20) (Z) only	626
3 eq	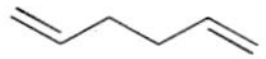	Ru cat **93** (4 x 0.01 eq), $CDCl_3$, 100° (MW), 4 h	CN, CN (55)	926

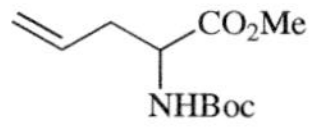

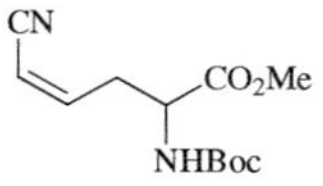

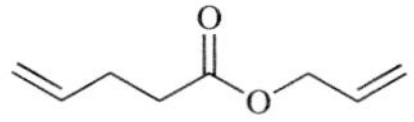

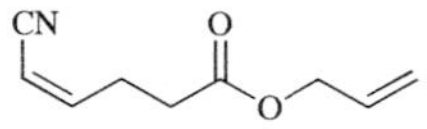

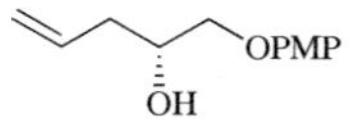

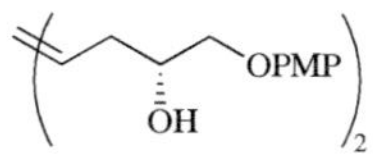

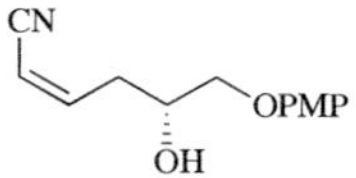

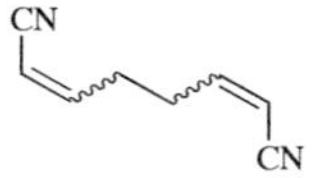

TABLE 27. CROSS-METATHESIS OF ACRYLONITRILE AND METHACRYLONITRILE (*Continued*)

Nitrile	Cross Partner	Conditions	Product(s) and Yield(s) (%)	Refs.

Please refer to the charts preceding the tables for the catalyst structures.

C_3

NC⁄⁄ + ⁄⁄⁄R — Hov **II** (0.02 eq), rt, 8 h — NC⁄⁄∿R

R		Refs.
n-Pr	(48)	927
n-C_5H_{11}	(65)	
n-C_7H_{15}	(80)	

2 eq + ⁄⁄⁄⁄⁄R — Catalyst (*x* eq), DCM — **I** (CN) + **II** (dimer)$_2$

R	Catalyst	*x*	Temp (°)	Time (h)	I	(*Z*)/(*E*) I	II	Refs.
BnO	Mo cat **1**	0.05	rt	3	(77)	90:10	—	149
TBSO	Mo cat **1**	0.05	rt	3	(90)	87:13	(7)	149
TBSO	Ru cat **88**	0.05	25	2	(83)	73:27	—	171
TBSO	Ru cat **122**	0.05	40	8	(91)	73:27	—	171
H	Ru cat **65**	0.02	45	12	(37)	25:75	(7)	1172
H	Ru cat **65**	0.1	45	12	(51)	25:75	(5)	1172
TBSO	Ru cat **158**	0.01	25	2	(65)	7:93	—	590
TBSO	Ru cat **157**	0.005	25	2	(57)	7:93	—	590
TBSO	Ru cat **138**	0.03	25	5	(95)	75:25	—	590
TBSO	Ru cat **115**	0.01	25	24	(59)	75:25	—	609
TBSO	Ru cat **130**	0.03	25	24	(48)	75:25	—	609
TBSO	Ru cat **116**	0.01	25	5	(45)	75:25	—	609
TBSO	Ru cat **116**	0.01	25	24	(51)	75:25	—	609
TBSO	Ru cat **115**	0.03	25	5	(94)	75:25	—	1061
TBSO	Ru cat **286**	0.02	25	5	(13)	—	—	161
TBSO	Ru cat **286**	0.02	40	8	(23)	74:26	—	161
TBSO	Ru cat **286**	0.05	40	8	(42)	73:27	—	161
TBSO	Ru cat **106**	0.05	20	3	(77)	72:28	—	216

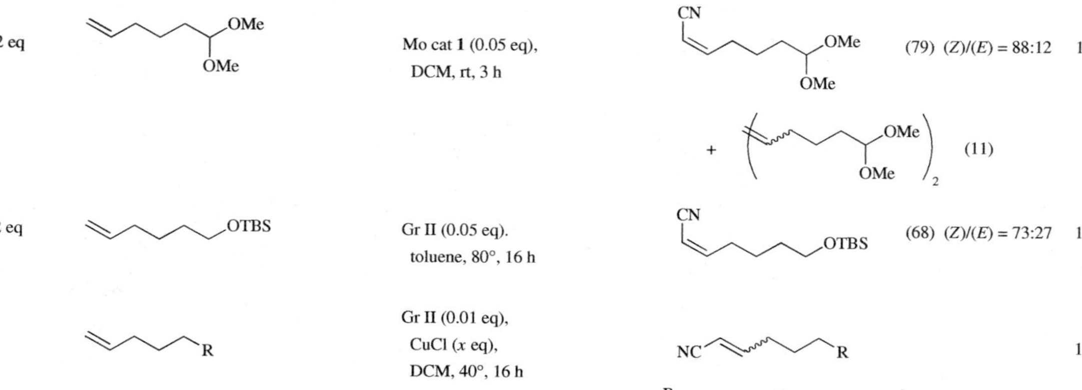

R	*x*	
HO_2C	4	(34)
HO_2C	—	(10)
$AcOCH_2$	4	(64)
$AcOCH_2$	—	(52)

TABLE 27. CROSS-METATHESIS OF ACRYLONITRILE AND METHACRYLONITRILE (*Continued*)

Nitrile	Cross Partner	Conditions	Product(s) and Yield(s) (%)	Refs.
Please refer to the charts preceding the tables for the catalyst structures.				
C_3				
NC (2 eq)	St-Bu, O	Catalyst (0.05 eq), toluene, 80°, 6 h	NC, St-Bu, O	1060
2 eq	CO_2Et, CO_2Et	Catalyst (*x* eq), DCM	CN, CO_2Et, CO_2Et	

Catalyst	
Ru cat **166**	(70)
Ru cat **167**	(72)
Ru cat **168**	(69)

Catalyst	*x*	Temp (°)	Time (h)		(*Z*)/(*E*)	Refs.
Hov II	0.08	45	6	(79)	75:25	153
Ru cat **88**	0.05	rt	0.5	(87)	67:33	160
Ru cat **114**	0.05	40	6	(79)	75:25	1173
Ru cat **88**	0.05	25	3	(85)	67:33	171

2 eq

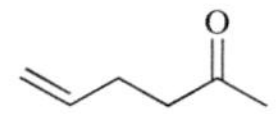

O

Catalyst (x eq),
CuCl (y eq), DCM

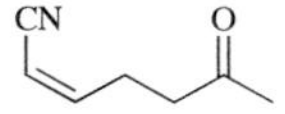

CN O

Catalyst	x	y	Temp (°)	Time (h)		$(Z)/(E)$	
Ru cat **149**	0.05	—	45	3	(73)	80:20	853
Hov II	0.01	4	40	16	(27)	—	150
Hov II	0.01	—	40	16	(16)	—	150

3 eq

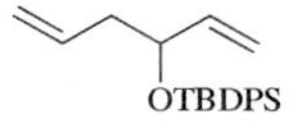

OTBDPS

Hov II (0.05 eq),
DCM, 40°

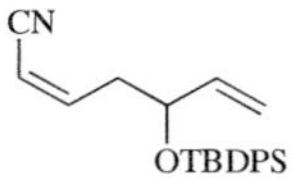

CN

OTBDPS

(23) $(Z)/(E)$ = 75:25 169

3 eq

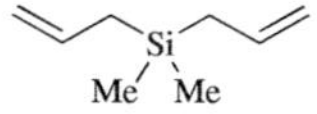

Si
Me Me

Hov II (0.025 eq),
DCM, rt, 4 h

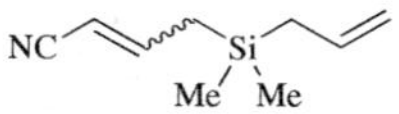

NC Si
Me Me

(<1) 992

1 eq

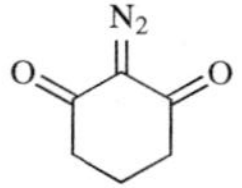

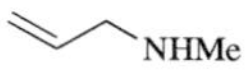

N_2
O O

+ NHMe
1 eq

1. DCM, MW (300 W), 0.25 h
2. Hov II (0.04 eq), 100°, 0.5 h
3. n-Bu_3P (0.2 eq), 100°, 0.25 h

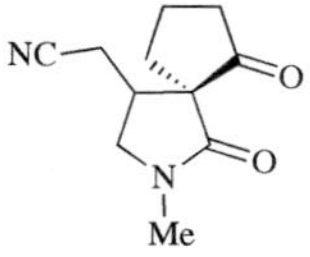

NC
O
O
N
Me

(34) dr 55:45 955

Nitrile	Cross Partner	Conditions	Product(s) and Yield(s) (%)	Refs.

Please refer to the charts preceding the tables for the catalyst structures.

C_3

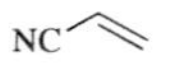

NC (1 eq) + N_2 (R[1], R[2]) + OH (1 eq)

1. DCM, MW (300 W), 0.25 h
2. Hov II (0.04 eq), 100°, 0.5 h
3. Additive (0.2 eq), temp, time

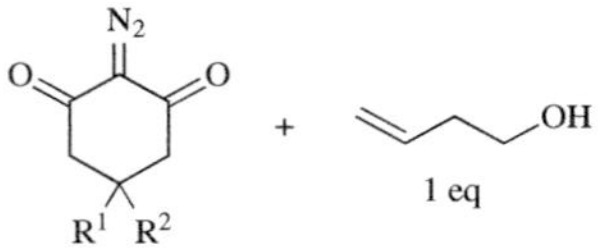

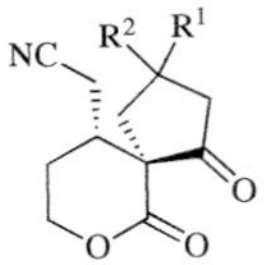

NC, R[2], R[1], O, O, O — 955

R^1	R^2	Additive	Temp (°)	Time (h)		dr
H	H	n-Bu_3P	100	0.25	(60)	>95:5
H	H	IPr	rt	20	(54)	>95:5
Me	H	n-Bu_3P	100	0.25	(41)	>74:21:4:1
Me	H	IPr	rt	20	(41)	>76:19:4:1
Me	Me	n-Bu_3P	100	0.25	(40)	>95:5
Me	Me	IPr	rt	20	(60)	>95:5

(1 eq) + N_2 + OH (1 eq)

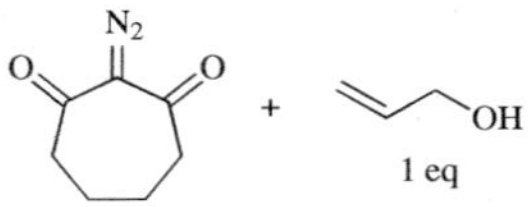

1. DCM, MW (300 W), 0.25 h
2. Hov II (0.04 eq), 100°, 0.5 h
3. Additive (0.2 eq), temp, time

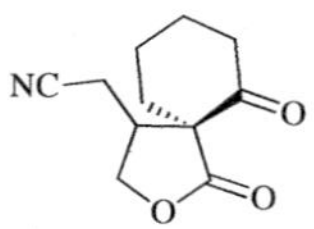

NC, O, O, O — 955

Additive	Temp (°)	Time (h)		dr
n-Bu_3P	100	0.25	(49)	64:36
IPr	rt	20	(50)	73:27

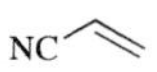

NC 2 eq | CO_2H | Hov II (0.05 eq), DCM, 45°, 2 h | CN, CO_2H | (76) (Z)/(E) = 75:25 | 153

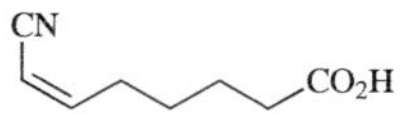

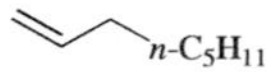

x eq | n-C_5H_{11} | Catalyst (y eq), solvent | CN, n-C_5H_{11} **I** + n-C_6H_{13}, n-C_5H_{11} **II**

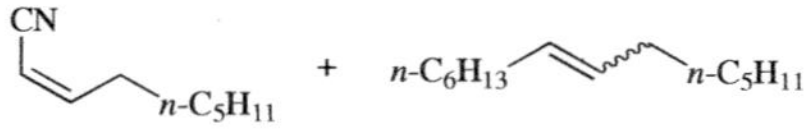

x	Cat	y	Solvent	Temp (°)	Time (h)	**I**	(Z)/(E) **I**	**II**	
2	Mo cat **1**	0.02	DCM	rt	3	(56)	90:10	(4)	149
2	Mo cat **1**	0.02	Et_2O	rt	3	(75)	90:10	(4)	149
0.5	Ru cat **54**	0.01	DCM	40	12	(56)	76:24	(—)	1174
0.5	Ru cat **55**	0.01	DCM	40	12	(75)	67:33	(—)	1174
0.5	Ru cat **56**	0.01	DCM	40	12	(81)	74:26	(—)	1174
0.5	Ru cat **57**	0.01	DCM	40	12	(70)	75:25	(—)	1174
0.5	Ru cat **70**	0.01	DCM	40	12	(57)	73:27	(—)	1174

2 eq | OMe, OMe | Gr II (0.02 eq), DCM, rt, 12 h | CN, OMe, OMe | (64) (Z)/(E) = 75:25 | 149

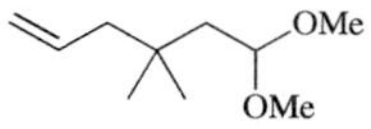

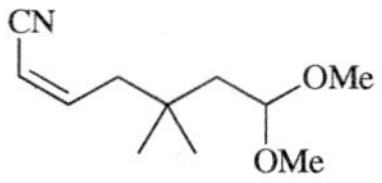

TABLE 27. CROSS-METATHESIS OF ACRYLONITRILE AND METHACRYLONITRILE (*Continued*)

Nitrile	Cross Partner	Conditions	Product(s) and Yield(s) (%)	Refs.

Please refer to the charts preceding the tables for the catalyst structures.

C_3

Nitrile	Cross Partner	Conditions	Product(s) and Yield(s) (%)	Refs.
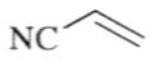 NC (1 eq)	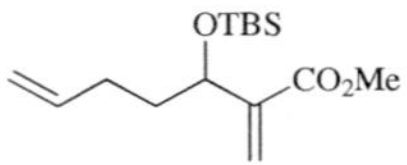 OTBS, CO_2Me	Gr II (0.01 eq), CuCl (*x* eq), DCM, 40°, overnight	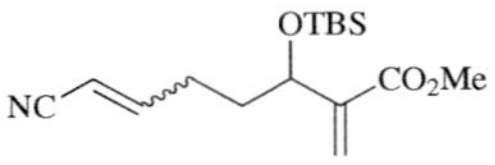 NC, OTBS, CO_2Me; *x* = 4 (52); *x* = — (42)	150
x eq	 Ph	Catalyst (*y* eq), solvent, reflux	NC, Ph	

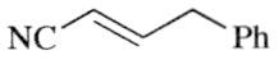

x	Catalyst	*y*	Solvent	Time (h)		(*E*)/(*Z*)	Refs.
2	Gr I	0.05	DCM	16	(60)	40:60	156
—	Gr II	0.03	DCM	—	(67)	37:63	853
—	Ru cat **33**	0.03	DCM	—	(7)	71:29	853
—	Hov II	0.03	DCM	—	(95)	44:56	853
—	Ru cat **223**	0.03	DCM	—	(20)	64:36	853
0.4	Gr II	0.01	DCM	5	(51)	75:25	971
0.4	Hov II	0.01	DCM	5	(67)	75:25	971
0.4	Ru cat **86**	0.01	DCM	5	(59)	75:25	971
1.1	Ru cat **227**	0.05	$CDCl_3$	—	(30)	55:45	600
1.1	Ru cat **228**	0.05	$CDCl_3$	—	(32)	38:62	600
1.1	Ru cat **223**	0.05	$CDCl_3$	—	(<33)	74:26	600
1.1	Hov II	0.05	$CDCl_3$	—	(<33)	64:36	600

2 eq

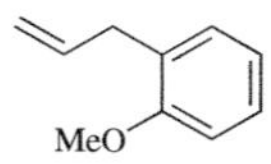

Mo cat **1** (0.05 eq), DCM, rt 3 h

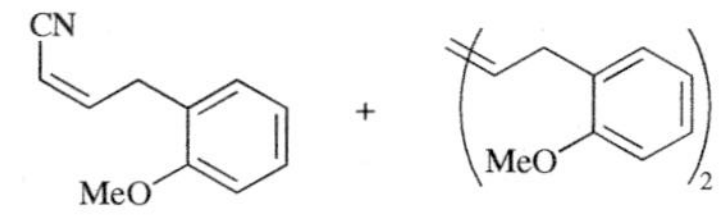

MeO CN + MeO ($_2$)

(72) (*Z*)/(*E*) = 90:10 (9)

149

1 eq

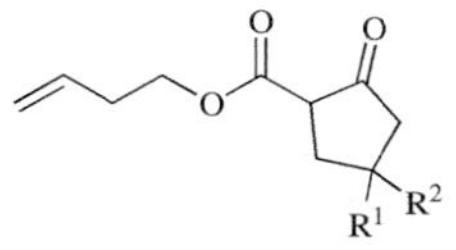

1. Hov II (0.03 + 0.01 eq), DCM, 100° (MW), 20 + 10 min
2. Additive (*x* eq), temp, time

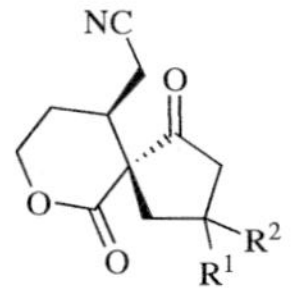

979

R^1	R^2	Additive	*x*	Temp (°)	Time		dr
H	H	*n*-Bu_3P	0.1	100 (MW)	10 min	(60)	>95:5
H	H	IPr	0.2	24	20 h	(40)	>95:5
Me	H	IPr	0.2	24	20 h	(44)	80:20
Me	Me	*n*-Bu_3P	0.2	100 (MW)	10 min	(41)	>95:5
Me	Me	IPr	0.2	24	20 h	(40)	>95:5

1 eq

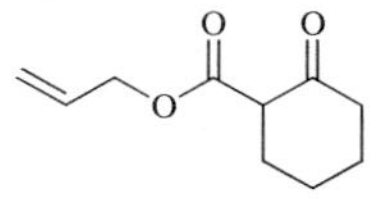

1. Hov II (0.03 + 0.01 eq), DCM, 100° (MW), 20 + 10 min
2. Additive (0.2 eq), temp, time

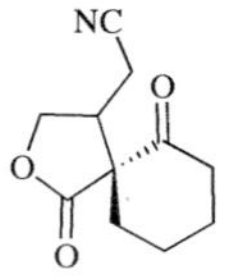

979

Additive	Temp (°)	Time		*cis*/*trans*
n-Bu_3P	100 (MW)	10 min	(52)	64:36
IPr	24	20 h	(50)	73:27

TABLE 27. CROSS-METATHESIS OF ACRYLONITRILE AND METHACRYLONITRILE (*Continued*)

Nitrile	Cross Partner	Conditions	Product(s) and Yield(s) (%)	Refs.

Please refer to the charts preceding the tables for the catalyst structures.

C_3

Nitrile	Cross Partner	Conditions	Product(s) and Yield(s) (%)	Refs.
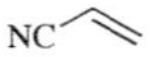 NC, 10 eq	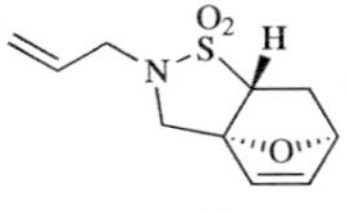 O_2 S N H O (±)	Ru cat **35** (0.1 eq), DCM, 45°, 3h	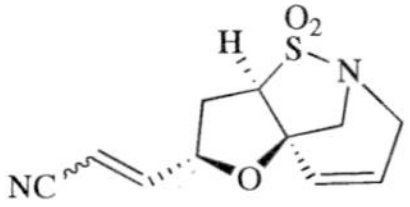 NC H O_2 S N O (±) (67)	359
2 eq	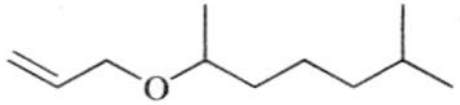 O	Hov II (0.02 eq), rt, 8 h	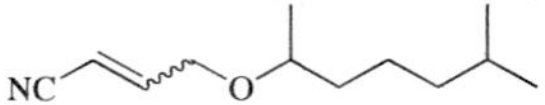 NC O (5)	927
3 eq	 CO_2Me 7	Catalyst (*x* eq), PC/EA/cyclohexane (1:1:1), 3 h	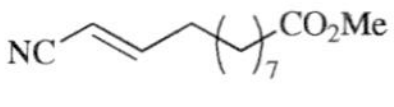 NC CO_2Me 7	1090

Catalyst	Temp (°)	Concn (mol/L)	Conv. (%)[b]	(*E*)/(*Z*)
Hov II	80	0.05	20	72:28
Ru cat **89**	80	0.05	27	72:28
Ru cat **89**	80	0.025	57	81:19
Ru cat **89**	80	0.01	79	90:10
Ru cat **89**	100	0.01	87	73:27
Ru cat **89**	100	0.01	34	86:14

Nitrile	Cross Partner	Conditions	Product(s) and Yield(s) (%)	Refs.
2 eq	 CO_2Me 7	Catalyst (*x* eq), toluene, 70°, 2 h	 NC CO_2Me 7	152

Catalyst	*x*	Concn (mol/L)	Conv. (%)[b]
Ru cat **356**	0.0003	0.1	86
Ru cat **357**	0.0003	0.1	87
Ru cat **356**	0.000025	0.25	63
Ru cat **357**	0.000025	0.25	73

	1 eq	$CH_2=CHCH_2O_2C$-(benzosuberone)	1. Hov II (0.04 eq), DCM, 100° (MW), 0.5 h 2. n-Bu_3P (0.2 eq), 100° (MW), 0.33 h	NC–spirolactone	(30) dr 54:46	955
	5 eq	n-$C_8H_{17}CH=CH(CH_2)_7CO_2Me$	Hov II (0.01 eq), toluene, 95°, 2 h	$NCCH=CH(CH_2)_7CO_2Me$	>99% conv.[b]	536
C_4						
NC(Me)C=CH2		$CH_2=CH(CH_2)_3OBz$	Hov II (0.05 eq), DCE, 70°, 4 h	$MeC(CN)=CH(CH_2)_3OBz$	(68) $(Z)/(E)$ = 67:33	153
	4 eq	$CH_2=CH(CH_2)_4OTBS$	Catalyst (0.05 eq), solvent	$MeC(CN)=CH(CH_2)_4OTBS$		

Catalyst	Solvent	Temp (°)	Time (h)		$(Z)/(E)$	
Ru cat **88**	DCM	40	16	(58)	67:33	160, 609
Ru cat **122**	DCM	40	16	(40)	67:33	171
Gr II	toluene	80	24	(0)	—	171
Ru cat **115**	DCM	40	16	(56)	67:33	1061
Gr II	DCM	40	24	(30)	67:33	609
Hov II	DCM	40	16	(40)	67:33	609

[a] The conversion was determined by ^1H NMR analysis.

[b] The conversion was determined by GC analysis.

TABLE 28. CROSS-METATHESIS OF VINYL ETHERS AND ESTERS

A. VINYL ETHERS

Vinyl Ether	Cross Partner	Conditions	Product(s) and Yield(s) (%)	Refs.
Please refer to the charts preceding the tables for the catalyst structures.				
C_2				
BnO (5 eq)		Mo cat **5** (0.05 eq), benzene, 22°, 24 h	BnO O B O (80) $(Z)/(E)$ = 93:7	547
n-$C_{11}H_{23}$ O	F F F F (2 eq)	Catalyst (0.02 eq), C_6D_6, 60°, 1 h	n-$C_{11}H_{23}$ O F F	1177
	F R F F (10 eq)	Ru cat **172** (0.1 eq), C_6D_6, 60°, 1 h	n-$C_{11}H_{23}$ O F R **I** + n-$C_{11}H_{23}$ O F F **II**	1177

Catalyst		TON
Hov II	(25)	12.5
Ru cat **139**	(27)	13.4
Ru cat **172**	(23)	11.7
Gr II	(6)	3.2
Ru cat **67**	(2)	1.1
Ru cat **271**	(2)	0.8

R	**I**	$(Z)/(E)$ **I**	**II**	TON
F	(64)	—	(64)	6.9
Cl	(51)	61:39	(11)	6.2
CF_3	(22)	75:25	(—)	2.2
H	(72)	80:20	(—)	7.2

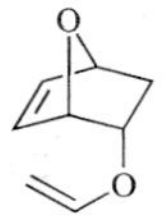

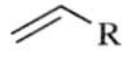

	5 eq	Hov II (0.1 eq), DCM, 40°, 24 h		1176
n-BuO, 10 eq		Mo cat **5** (x eq), benzene, 22°, 2 h	n-BuO ... R	5
10 eq		Mo cat **5** (x eq), benzene, 22°, 2 h	n-BuO ... R	5
20 eq	R^1O, R^2	Ru cat **210** (0.05 eq), benzene, 22°, 24 h	R^1O, R^2, On-Bu	808, 1175

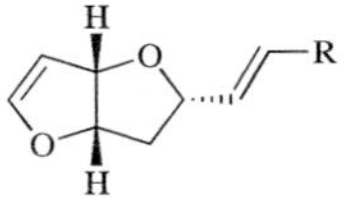

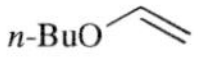

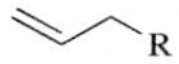

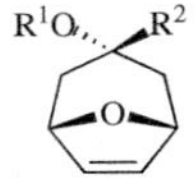

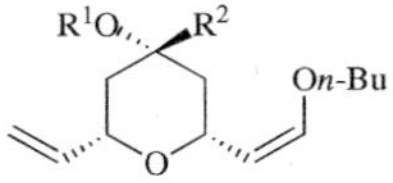

R		(E)/(Z)
NC–	(72)	41:59
MeO_2C	(72)	(E) only
$AcOCH_2$	(85)	—

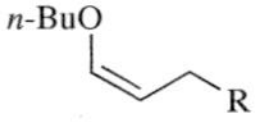

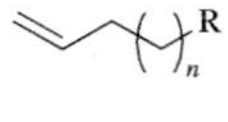

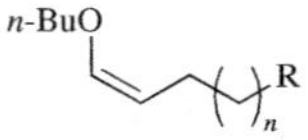

R	x		(Z)/(E)
TIPSO	0.025	(77)	94:6
PhHN	0.025	(51)	98:2
Ph	0.012	(57)	>98:2

n	R	x		(Z)/(E)
2	PhO_2C	0.025	(73)	98:2
5	Me	0.025	(68)	98:2
5	Br	0.012	(70)	>98:2
5	TMS—≡	0.05	(59)	>98:2

R^1	R^2		(Z)/(E)	er
TBS	H	(71)	87:13	93:7
TBS	Me	(72)	92:8	95:5
Bn	H	(62)	94:6	91:9

TABLE 28. CROSS-METATHESIS OF VINYL ETHERS AND ESTERS (*Continued*)

A. VINYL ETHERS (*Continued*)

Vinyl Ether	Cross Partner	Conditions	Product(s) and Yield(s) (%)	Refs.
Please refer to the charts preceding the tables for the catalyst structures.				
C_2				
x eq	OR	Mo cat **18** (0.006 eq), benzene, 22°	OR, O*n*-Bu (*Z*)/(*E*) > 98:2	339

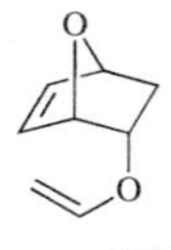

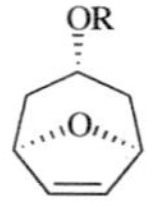

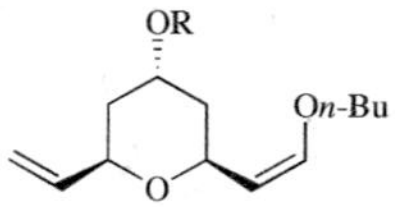

x	R	Time (min)	Conv. (%)[a]		er
1.1	TBS	30	>98	(53)	52.0:48.0
5	TBS	0.5	97	(83)	95.0:5.0
5	TBS	1	>98	(89)	90.0:10.0
5	TBS	3	>98	(96)	85.5:14.5
5	TBS	30	>98	(80)	62.5:37.5
10	TBS	30	>98	(80)	73.0:27.0
20	TBS	30	>98	(85)	84.5:15.5
30	TBS	30	>98	(88)	95.0:5.0
1.1	Bn	30	96	(54)	57.5:42.5
30	Bn	30	>98	(88)	95.0:5.0

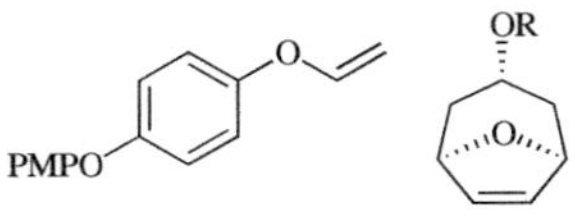

PMPO OR
x eq

Mo cat **18** (0.006 eq), benzene, 22°

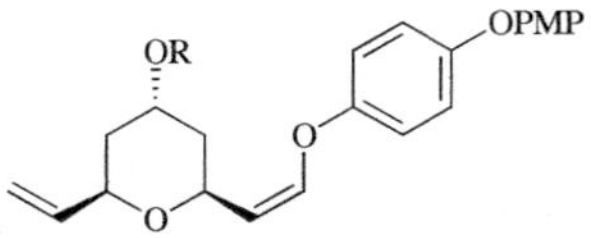

OR OPMP

$(Z)/(E) > 98:2$

339

x	R	Time (min)	Conv. (%)[a]		er
1.1	TBS	30	97	(86)	56.5:43.5
5	TBS	0.5	97	(86)	90.0:10.0
5	TBS	1	97	(82)	87.0:13.0
5	TBS	3	>98	(83)	86.0:14.0
5	TBS	30	>98	(87)	84.5:15.5
10	TBS	30	>98	(90)	91.5:8.5
20	TBS	30	>98	(87)	92.0:8.0
30	TBS	30	>98	(82)	91.5:8.5
1.1	Bn	30	94	(82)	82.0:18.0
20	Bn	30	>98	(87)	96.5:3.5

TABLE 28. CROSS-METATHESIS OF VINYL ETHERS AND ESTERS (*Continued*)

A. VINYL ETHERS (*Continued*)

Vinyl Ether	Cross Partner	Conditions	Product(s) and Yield(s) (%)	Refs.

Please refer to the charts preceding the tables for the catalyst structures.

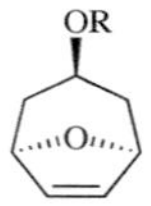

C_2

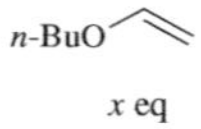

n-BuO (x eq) | OR | Catalyst (y eq), solvent, 22° | OR, On-Bu | 339

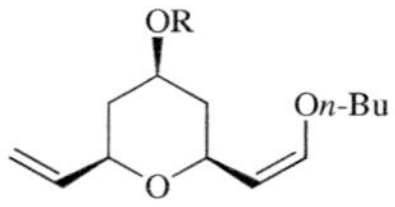

x	R	Catalyst	y	Solvent	Time (h)	Conv (%)[a]		(Z)/(E)	er
1.1	TBS	Hov II	0.01	benzene	0.5	69	(32)	25:75	—
1.1	TBS	Mo cat **1**	0.01	benzene	0.5	98	(10)	50:50	—
1.1	TBS	Mo cat **8**	0.01	benzene	12	<2	(0)	—	—
1.1	TBS	Mo cat **3**	0.01	benzene	0.5	>98	(48)	80:20	—
1.1	TBS	Mo cat **5**	0.01	benzene	0.5	85	(68)	>98:2	99.0:1.0
1.1	TBS	Mo cat **6**	0.01	benzene	0.5	<5	(0)	—	—
1.1	TBS	Mo cat **16**	0.0007	benzene	0.5	>98	(87)	>98:2	97.0:3.0
1.1	TBS	Mo cat **17**	0.005	benzene	0.5	>98	(87)	>98:2	98.0:2.0
1.1	TBS	Mo cat **18**	0.006	benzene	0.5	>98	(89)	>98:2	99.0:1.0
5	TBS	Mo cat **18**	0.0015	—	0.17	>98	(76)	>98:2	99.0:1.0
1.1	TBS	Mo cat **19**	0.007	benzene	0.5	>98	(87)	>98:2	98.0:2.0
1.1	Bn	Mo cat **18**	0.006	benzene	0.5	>98	(75)	>98:2	99.0:1.0
5	Bn	Mo cat **18**	0.0015	—	0.17	>98	(74)	>98:2	99.0:1.0

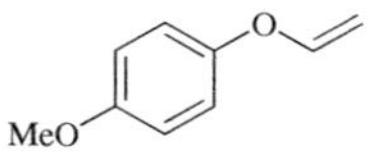

MeO, O (x eq) | OR | Catalyst (y eq), solvent, 22° | OR, O, OMe | 339

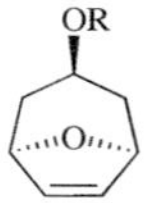

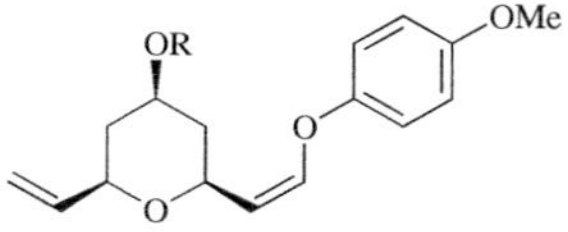

x	R	Catalyst	y	Solvent	Time (h)	Conv. (%)[a]		$(Z)/(E)$	er
1.1	TBS	Hov II	0.01	benzene	0.5	95	(<5)	—	—
1.1	TBS	Mo cat **1**	0.01	benzene	0.5	98	(14)	80:20	—
1.1	TBS	Mo cat **8**	0.01	benzene	12	<2	(0)	—	—
1.1	TBS	Mo cat **3**	0.01	benzene	0.5	>98	(75)	82:18	—
1.1	TBS	Mo cat **5**	0.01	benzene	0.5	86	(47)	>98:2	99.0:1.0
1.1	TBS	Mo cat **6**	0.01	benzene	0.5	<5	(0)	—	—
1.1	TBS	Mo cat **16**	0.0007	benzene	0.5	60	(41)	>98:2	88.5:11.5
1.1	TBS	Mo cat **17**	0.005	benzene	0.5	>98	(80)	>98:2	92.0:8.0
1.1	TBS	Mo cat **18**	0.006	benzene	0.5	>98	(90)	>98:2	98.5:1.5
5	TBS	Mo cat **18**	0.0015	—	0.17	>98	(80)	>98:2	99.0:1.0
1.1	TBS	Mo cat **19**	0.007	benzene	0.5	>98	(85)	>98:2	99.0:1.0
1.1	Bn	Mo cat **18**	0.006	benzene	0.5	>98	(80)	>98:2	97.0:3.0
5	Bn	Mo cat **18**	0.0015	—	0.17	>98	(77)	>98:2	99.0:1.0

10 eq Cy Mo cat **5** (0.05 eq), benzene, 22°, 2 h MeO O Cy (75) $(Z)/(E) > 98{:}2$ 5

TABLE 28. CROSS-METATHESIS OF VINYL ETHERS AND ESTERS (*Continued*)

A. VINYL ETHERS (*Continued*)

Vinyl Ether	Cross Partner	Conditions	Product(s) and Yield(s) (%)	Refs.
Please refer to the charts preceding the tables for the catalyst structures.				
C_2				
HO O (20 eq)		Gr II (0.05 eq), THF, rt, 1 h	*m* R (*E*)/(*Z*): 2 H (96) 91:9; 8 Me (97) 96:4	1035
C_3				
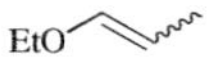 EtO	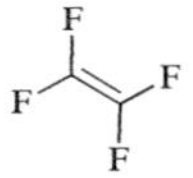 F F F F (2 eq)	Ru cat **172** (0.1 eq), C_6D_6, 60°, 1 h	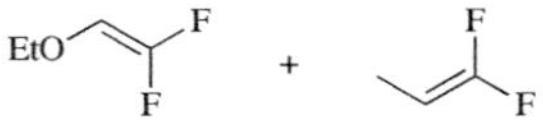 EtO F F (27) TON = 2.7 + F F (17)	1177

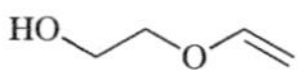

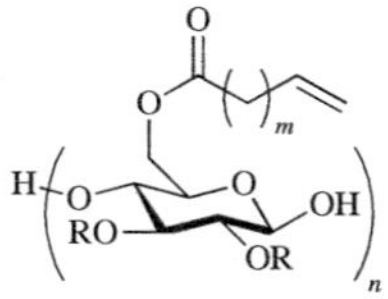

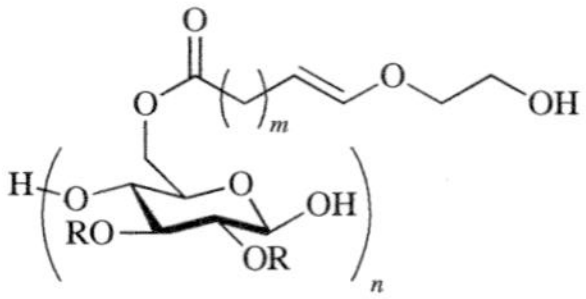

m	R		(*E*)/(*Z*)
2	H	(96)	91:9
8	Me	(97)	96:4

[a] The conversion was determined by ^{1}H NMR analysis.

TABLE 28. CROSS-METATHESIS OF VINYL ETHERS AND ESTERS (*Continued*)

B. VINYL ESTERS

Vinyl Ester	Cross Partner	Conditions	Product(s) and Yield(s) (%)	Refs.
Please refer to the charts preceding the tables for the catalyst structures.				
C_2				
AcO (5 eq)	PMB, NHBn, O, H, I, H, H	Hov II (0.02 eq), benzene, 60°, 4 h	PMB, NHBn, OAc, O, H, I; (99) (*E*)/(*Z*) = 80:20	1178
5 eq	R, NHBn, O, H, H, H, I	Hov II (*x* eq), solvent	R, NHBn, OAc, O, H, I	1178

R	*x*	Solvent	Temp (°)	Time (h)		(*E*)/(*Z*)
PMB	0.02	benzene	60	4	(87)	93:7
$TBSO(CH_2)_2$	0.15	benzene, DCE	45	9	(74)	84:16

Vinyl Ester	Cross Partner	Conditions	Product(s) and Yield(s) (%)	Refs.
5 eq	PMB, H, NHBn, O, I, H, H	Hov II (0.01 eq), benzene, rt, 14 h	PMB, H, NHBn, OAc, O, H, I, H; (87) (*E*)/(*Z*) = 93:7	664, 1179

TABLE 28. CROSS-METATHESIS OF VINYL ETHERS AND ESTERS (*Continued*)

B. VINYL ESTERS (*Continued*)

Vinyl Ester	Cross Partner	Conditions	Product(s) and Yield(s) (%)	Refs.
Please refer to the charts preceding the tables for the catalyst structures.				
C_2				
AcO 5 eq	PMB N O O H H O	Hov II (0.005 eq), benzene, rt, 8.5 h	PMB N OAc O O H H H O (88) $(E)/(Z)$ = 55:45	664, 1179
5 eq	PMB H O NHBn N O O H H Y	1. Hov II (x eq), vinyl acetate (y eq), benzene, temp 1, time 1 2. Hov II (z eq), solvent, temp 2, time 2	PMB H O NHBn N OAc O O H H H Y	664, 1179

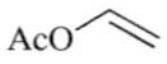

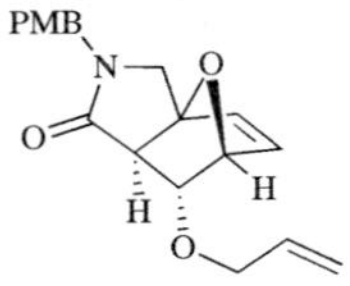

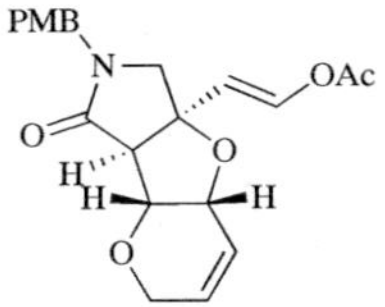

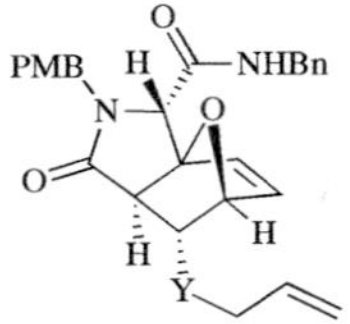

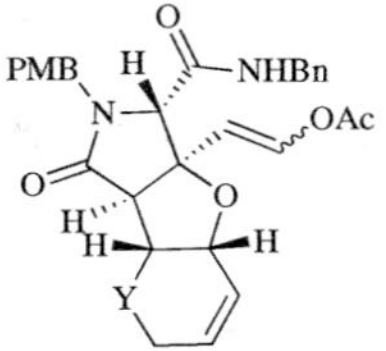

Y	x	y	Temp 1 (°)	Time 1 (h)	z	Solvent	Temp 2 (°)	Time 2 (h)	
NsN	0.02	—	rt	11	—	—	—	—	(97)
NsN	0.01	5	69	25	0.01	benzene	80	10	(90)
TFAN	0.01	5	69	19	0.01	benzene	69	11	(85)
O	0.005	—	rt	4	—	benzene	—	—	(100)
O	0.01	5	69	46	0.01	benzene	69	21	(84)

2 eq

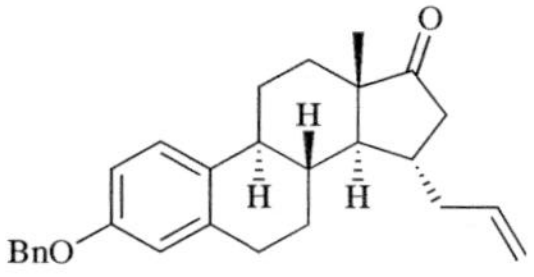

BnO

Catalyst (0.03 eq),
solvent

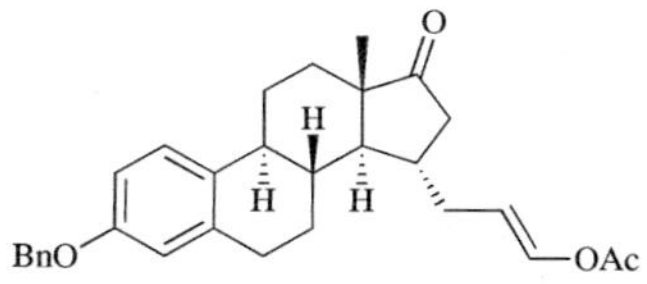

BnO OAc

163

Catalyst	Solvent	Temp (°)	Time (h)		(*E*)/(*Z*)
Gr II	DCM	40	2.5	(86)	(*E*) only
Ru cat **68**	DCE	50	5.5	(0)	—
Ru cat **125**	DCM	40	5.5	(32)	(*E*) only

C_8

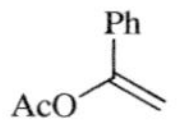

Ph AcO

3 eq

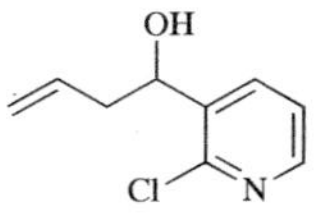

OH Cl N

Hov II (0.1 eq),
DCM, 50°, 24 h

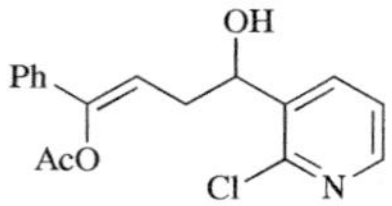

Ph AcO OH Cl N

(68) (*E*)/(*Z*) = 50:50

603

TABLE 29. CROSS-METATHESIS OF VINYLIC COMPOUNDS WITH A PERFLUORINATED CARBON CHAIN

Perfluorinated Alkene	Cross Partner	Conditions	Product(s) and Yield(s) (%)	Refs.

Please refer to the charts preceding the tables for the catalyst structures.

C_{3-6}

R^1-CH=CH$_2$ (10 eq) + CH$_2$=CH-CH$_2$-(CH$_2$)$_n$-R^2; Catalyst (x eq), $C_6F_5CF_3$ → R^1-CH=CH-CH$_2$-(CH$_2$)$_n$-R^2 **I** $(E)/(Z) > 95:5$ + R^1-CH=CH-R^1 **II** — Ref. 137

R^1	R^2	n	Catalyst	x	Temp (°)	Time (h)	I	II
CF_3	BzO	2	Hov II	0.05	45	3	(>95)	(0)
CF_3	BzO	2	Gr II	0.05	60	4	(85)	(5)
CF_3	HO_2C	2	Hov II	0.05	45	3	(70)	(19)
CF_3	HO_2C	2	Gr II	0.05	60	4	(58)	(23)
CF_3	HO	3	Hov II	0.05	45	3	(75)	(18)
CF_3	HO	3	Gr II	0.05	60	4	(50)	(50)
n-C_4F_9	BzO	2	Hov II	0.1	45	3	(>95)	(0)
n-C_4F_9	BzO	2	Gr II	0.1	60	4	(>95)	(0)
n-C_4F_9	HO_2C	2	Hov II	0.1	45	3	(59)	(41)
n-C_4F_9	HO_2C	2	Gr II	0.1	60	4	(42)	(56)
n-C_4F_9	HO	3	Hov II	0.1	45	3	(7)	(79)
n-C_4F_9	HO	3	Gr II	0.1	60	4	(21)	(73)

R 10 eq

Catalyst (x eq), $C_6F_5CF_3$

I $(E)/(Z) > 95:5$ + **II** 137

R	n	Catalyst	x	Temp (°)	Time (h)	**I**	**II**
CF_3	1	Hov II	0.05	45	3	(50)	(0)
CF_3	1	Gr II	0.05	60	4	(16)	(19)
CF_3	7	Hov II	0.05	45	3	(>95)	(0)
CF_3	7	Gr II	0.05	60	4	(>95)	(0)
n-C_4F_9	1	Hov II	0.1	45	3	(57)	(0)
n-C_4F_9	1	Gr II	0.1	60	4	(13)	(47)
n-C_4F_9	7	Hov II	0.1	45	3	(>95)	(0)
n-C_4F_9	7	Gr II	0.1	60	4	(24)	(35)

TABLE 29. CROSS-METATHESIS OF VINYLIC COMPOUNDS WITH A PERFLUORINATED CARBON CHAIN (*Continued*)

Perfluorinated Alkene	Cross Partner	Conditions	Product(s) and Yield(s) (%)	Refs.

Please refer to the charts preceding the tables for the catalyst structures.

C_{3-6}

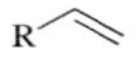

R–CH=CH$_2$, 10 eq | Cross partner: CO_2Me | Catalyst (x eq), $C_6F_5CF_3$ | R + R (II) | 137

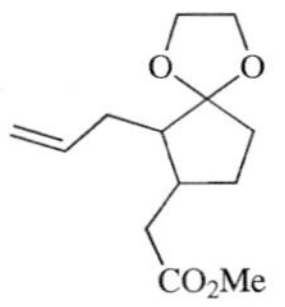

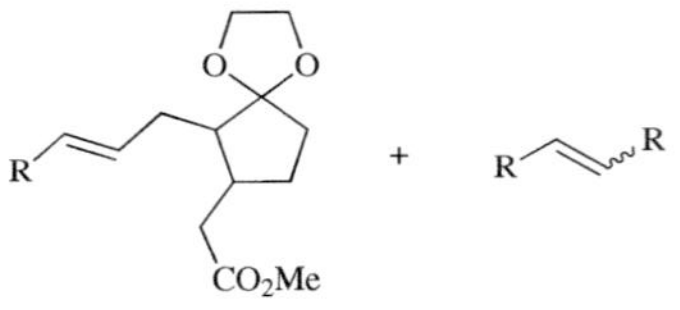

I $(E)/(Z) > 95{:}5$ II

R	Catalyst	x	Temp (°)	Time (h)	I	II
CF_3	Hov II	0.05	45	3	(>95)	(0)
CF_3	Gr II	0.05	60	4	(77)	(10)
n-C_4F_9	Hov II	0.1	45	3	(79)	(21)
n-C_4F_9	Gr II	0.1	60	4	(38)	(42)

10 eq | Cross partner: CO_2Et | Catalyst (x eq), $C_6F_5CF_3$ | R + R (II) | 137

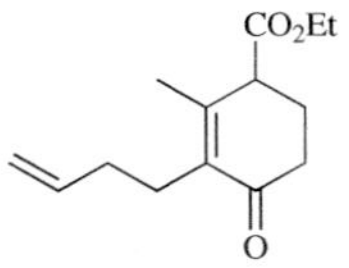

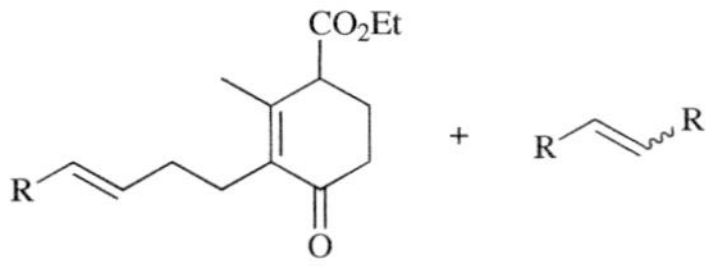

I $(E)/(Z) > 95{:}5$ II

R	Catalyst	x	Temp (°)	Time (h)	I	II
CF_3	Hov II	0.05	45	3	(>95)	(0)
CF_3	Gr II	0.05	60	4	(41)	(33)
n-C_4F_9	Hov II	0.1	45	3	(90)	(10)
n-C_4F_9	Gr II	0.1	60	4	(40)	(45)

C_4

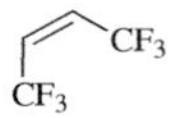

$CF_3CH=CHCF_3$

5 eq

(Z)/(E) > 98:2

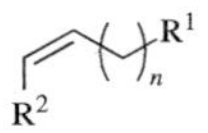

Mo cat **31** (x eq), benzene, 22°

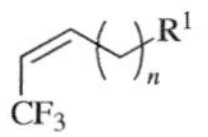

1180

R^1	R^2	n	x	Time (h)		(Z)/(E)
PMBO	Et	1	0.05	4	(77)	>98:2
TsO	Et	2	0.02	2	(77)	>98:2
Bn_2N	Et	2	0.03	2	(92)	>98:2
PhS	Et	2	0.03	2	(69)	95:5
B(pin)	Me	1	0.05	2	(70)	97:3
TES	Me	1	0.03	2	(62)	94:6

5 eq

(Z)/(E) > 98:2

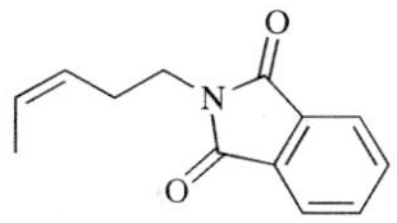

Mo cat **31** (0.03 eq), benzene, 22°, 2 h

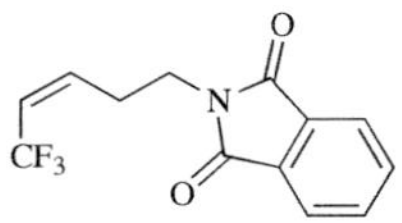

(80) (Z)/(E) > 98:2 1180

OTBDPS

10 eq

(Z)/(E) > 98:2

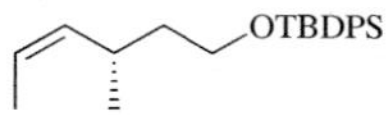

Mo cat **31** (0.05 eq), benzene, 22°, 4 h

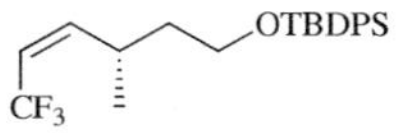

OTBDPS

(85) (Z)/(E) > 98:2 1180

TABLE 29. CROSS-METATHESIS OF VINYLIC COMPOUNDS WITH A PERFLUORINATED CARBON CHAIN (*Continued*)

Perfluorinated Alkene	Cross Partner	Conditions	Product(s) and Yield(s) (%)	Refs.
Please refer to the charts preceding the tables for the catalyst structures.				
C_4				
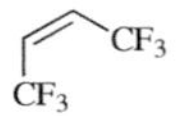 CF_3, CF_3 10 eq (*Z*)/(*E*) > 98:2	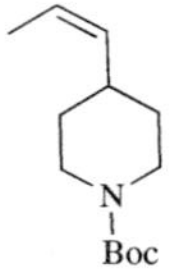 Boc, N	Mo cat **31** (0.05 eq), benzene, 22°, 4 h	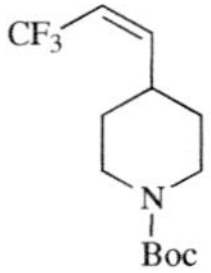 CF_3, N, Boc (88) (*Z*)/(*E*) > 98:2	1180
10 eq (*Z*)/(*E*) > 98:2	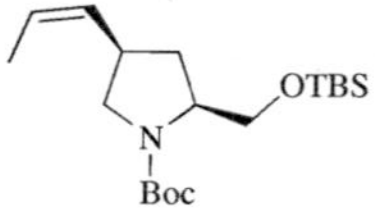 OTBS, N, Boc	Mo cat **31** (0.05 eq), benzene, 22°, 4 h	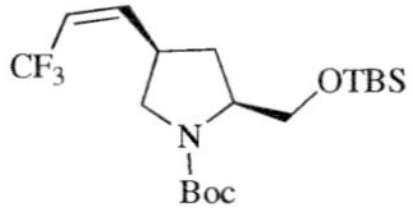 CF_3, OTBS, N, Boc (84) (*Z*)/(*E*) > 98:2	1180
5 eq (*Z*)/(*E*) > 98:2	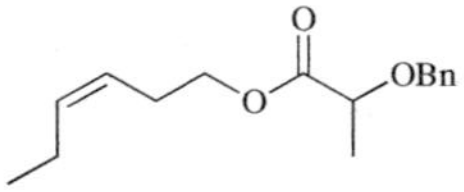 O, O, OBn	Mo cat **31** (0.02 eq), benzene, 22°, 2 h	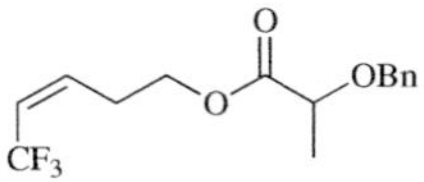 O, O, OBn, CF_3 (70) (*Z*)/(*E*) = 97:3	1180
5 eq (*Z*)/(*E*) > 98:2	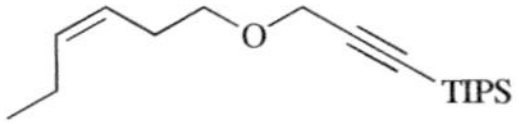 O, TIPS	Mo cat **31** (0.05 eq), benzene, 22°, 2 h	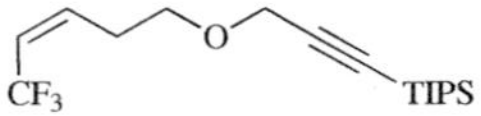 O, CF_3, TIPS (82) (*Z*)/(*E*) > 98:2	1180

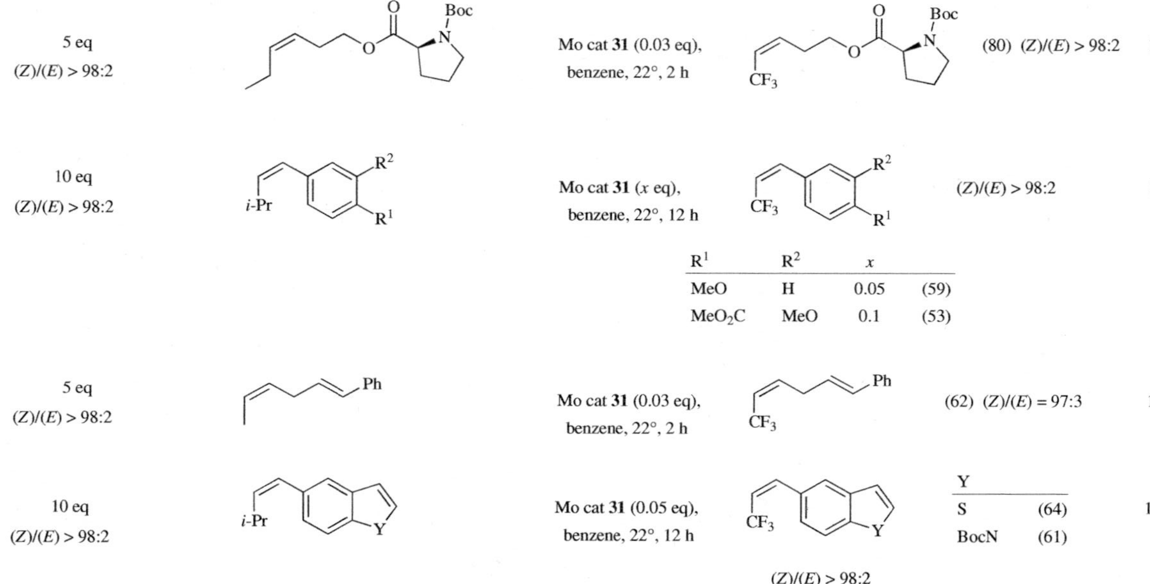
5 eq
(Z)/(E) > 98:2
O
Boc
N
O
Mo cat **31** (0.03 eq),
benzene, 22°, 2 h
CF3
(80) (Z)/(E) > 98:2
1180
10 eq
(Z)/(E) > 98:2
i-Pr
R2
R1
Mo cat **31** (x eq),
benzene, 22°, 12 h
CF3
(Z)/(E) > 98:2
1180
R1 R2 x
MeO H 0.05 (59)
MeO2C MeO 0.1 (53)
5 eq
(Z)/(E) > 98:2
Ph
Mo cat **31** (0.03 eq),
benzene, 22°, 2 h
CF3
(62) (Z)/(E) = 97:3
1180
10 eq
(Z)/(E) > 98:2
i-Pr
Y
Mo cat **31** (0.05 eq),
benzene, 22°, 12 h
CF3
Y
S (64)
BocN (61)
1180
(Z)/(E) > 98:2

TABLE 29. CROSS-METATHESIS OF VINYLIC COMPOUNDS WITH A PERFLUORINATED CARBON CHAIN (*Continued*)

Perfluorinated Alkene	Cross Partner	Conditions	Product(s) and Yield(s) (%)	Refs.
Please refer to the charts preceding the tables for the catalyst structures.				
C_4				
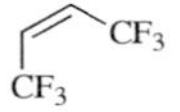 CF_3, CF_3; 5 eq; (*Z*)/(*E*) > 98:2	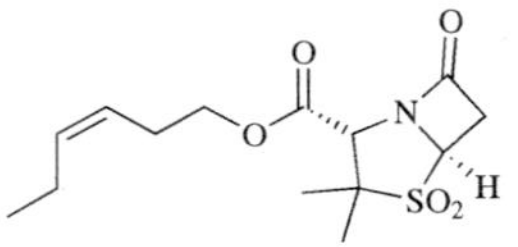 O, O, N, SO_2, H	Mo cat **31** (0.05 eq), benzene, 22°, 2 h	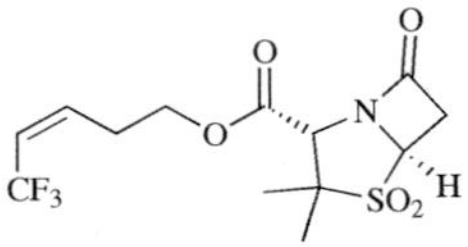 CF_3, O, O, N, SO_2, H; (83) (*Z*)/(*E*) > 98:2	1180
10 eq; (*Z*)/(*E*) > 98:2	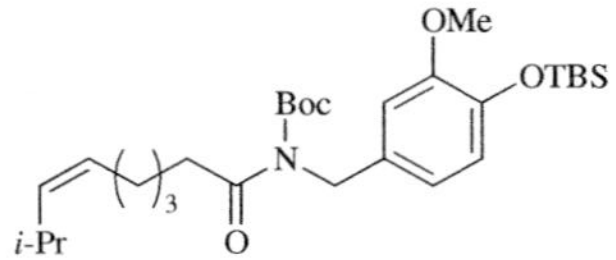 *i*-Pr, 3, O, Boc, N, OMe, OTBS	Mo cat **31** (0.05 eq), benzene, 22°, 4 h	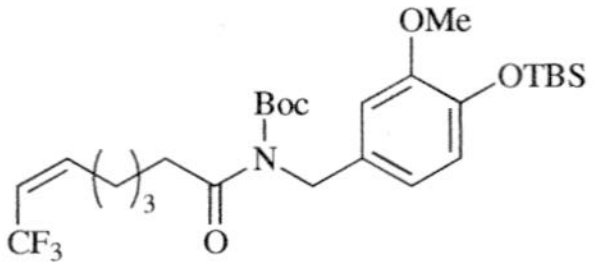 CF_3, 3, O, Boc, N, OMe, OTBS; (86) (*Z*)/(*E*) > 98:2	1180
x eq; (*Z*)/(*E*) > 98:2	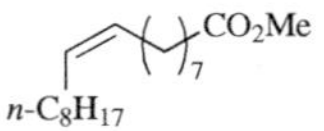 *n*-C_8H_{17}, 7, CO_2Me	Catalyst (*y* eq), benzene, 22°	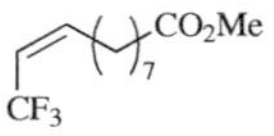 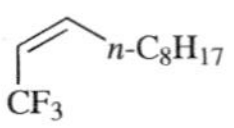 CF_3, 7, CO_2Me **I** + CF_3, *n*-C_8H_{17} **II**	1180

x	Catalyst	*y*	Time (h)	**I**	(*Z*)/(*E*) **I**	**II**	(*Z*)/(*E*) **II**
20	Mo cat **28**	0.05	4	(66)	91:9	(36)	91:9
20	Mo cat **30**	0.05	4	(64)	81:19	(36)	81:19
20	Mo cat **31**	0.05	4	(90)	98:2	(65)	98:2
5	Mo cat **31**	0.02	0.25	(95)	>98:2	(67)	>98:2

Reactant	Substrate	Conditions	Product(s) and Yield(s) (%)	Refs.
5 eq (*Z*)/(*E*) > 98:2	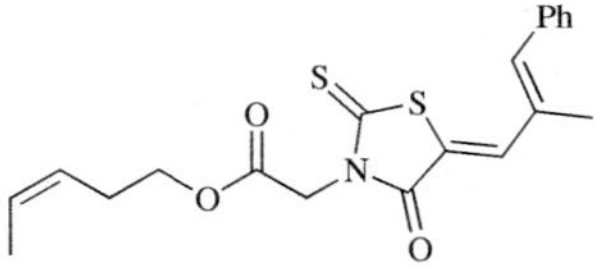	Mo cat **31** (0.05 eq), benzene, 22°, 4 h	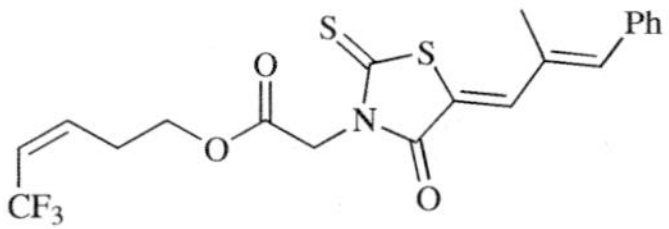 CF_3 (92) (*Z*)/(*E*) > 98:2	1180
5 eq (*Z*)/(*E*) > 98:2	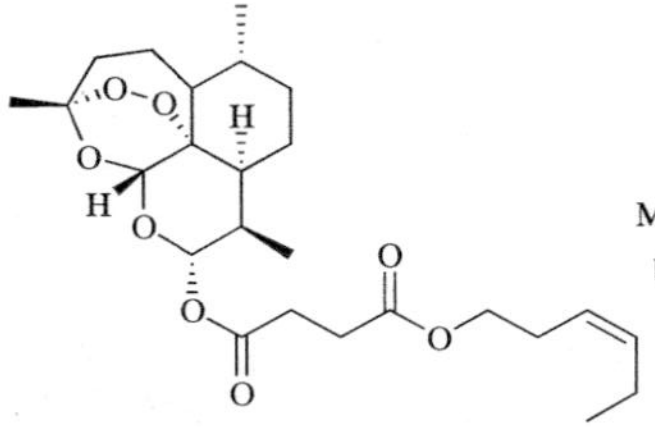	Mo cat **31** (0.05 eq), benzene, 22°, 4 h	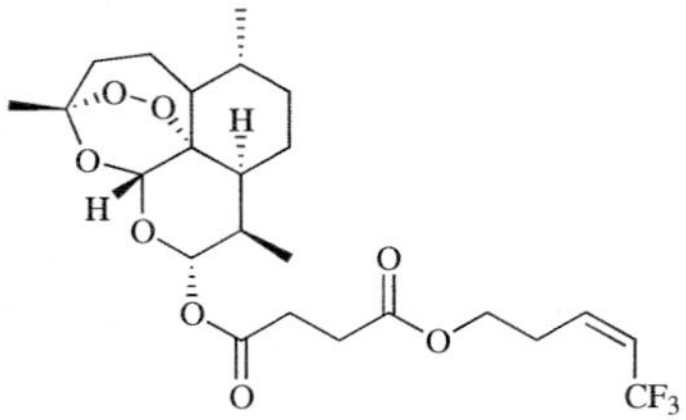 CF_3 (81) (*Z*)/(*E*) > 98:2	1180
C_6				
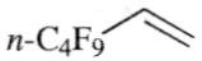 *n*-C_4F_9 2 eq	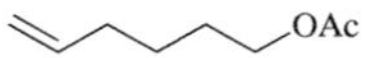 OAc	Ru cat **11** (0.042 eq), DCM, reflux, 12 h	 *n*-C_4F_9 … OAc (34) (*E*)/(*Z*) = 70:30	139

TABLE 29. CROSS-METATHESIS OF VINYLIC COMPOUNDS WITH A PERFLUORINATED CARBON CHAIN (*Continued*)

Perfluorinated Alkene	Cross Partner	Conditions	Product(s) and Yield(s) (%)	Refs.
Please refer to the charts preceding the tables for the catalyst structures.				
C_{6-9}				
R[1] (allyl); 2 eq	R[2]O, OH, vinyl steroid	Hov II (0.1 eq), DCM, 40°, 4 h	R[2]O, OH, R[1] steroid product	134
2 eq	R[2]O, HO, allyl steroid	Hov II (0.1 eq), DCM, 40°, 4 h	R[2]O, HO, R[1] steroid product	134

First entry:

R^1	R^2	
n-C_3F_7	H	(25)
n-C_3F_7	Me	(29)
i-C_3F_7	H	(12)
n-C_6F_{13}	H	(39)
n-C_6F_{13}	Me	(36)

Second entry:

R^1	R^2	
n-C_3F_7	H	(58)
n-C_3F_7	Me	(68)
i-C_3F_7	H	(67)
n-C_6F_{13}	H	(53)
n-C_6F_{13}	Me	(62)

2 eq

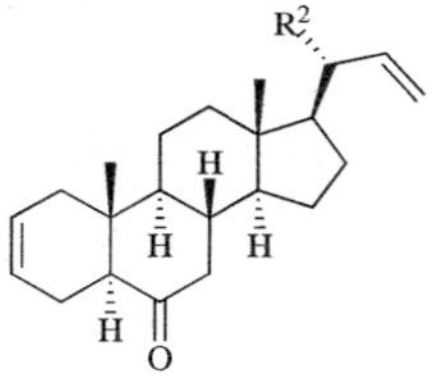

Hov II (0.1 eq), DCM, reflux, 4 h

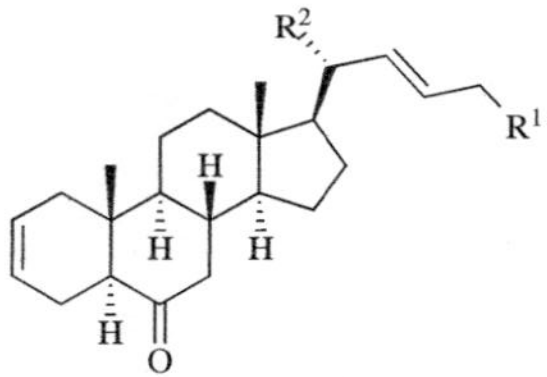

R^1	R^2		(*E*)/(*Z*)
i-C_3F_7	H	(59)	95:5
n-C_3F_7	Me	(71)	(*E*) only
n-C_6F_{13}	Me	(67)	(*E*) only

136

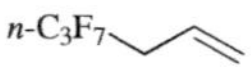

C_6

n-C_3F_7

2 eq

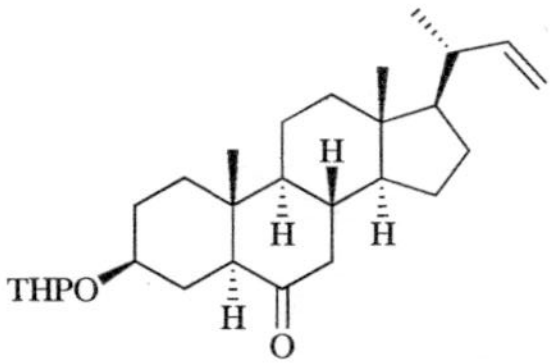

Hov II (0.1 eq), DCM, reflux, 4 h

THPO

THPO

n-C_3F_7

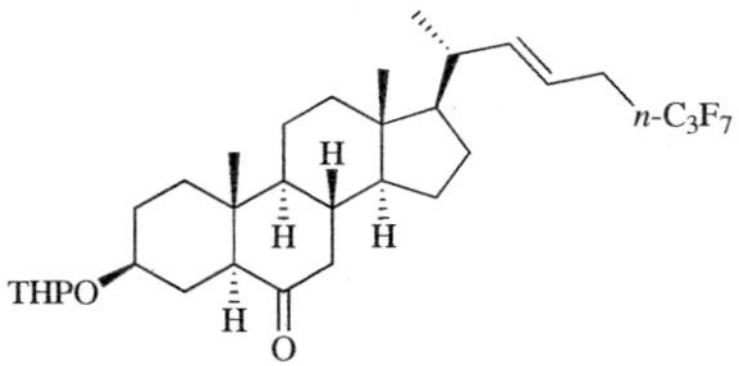

(81) (*E*) only

136

TABLE 29. CROSS-METATHESIS OF VINYLIC COMPOUNDS WITH A PERFLUORINATED CARBON CHAIN (*Continued*)

Perfluorinated Alkene	Cross Partner	Conditions	Product(s) and Yield(s) (%)	Refs.
Please refer to the charts preceding the tables for the catalyst structures.				
C_{6-9}				
R^1 (allyl), 2 eq	R^2	Catalyst (0.1 eq), DCM, reflux, 3 h	R^2, R^1	135
R (allyl), 2 eq	H C, C	Hov II (0.1 eq), DCM, reflux, 6 h	H C, C, R	612

R^1	R^2	Catalyst		$(E)/(Z)$
n-C_3F_7	H	Hov II	(75)	60:40
n-C_3F_7	THPO	Hov II	(81)	88:12
i-C_3F_7	HO	Hov II	(75)	60:40
i-C_3F_7	TBSO	Hov II	(63)	88:12
n-C_6F_{13}	H	Hov II	(64)	80:20
n-C_6F_{13}	HO	Hov II	(70)	67:33
n-C_6F_{13}	TBSO	Hov II	(75)	(*E*) only
n-C_6F_{13}	TBSO	Gr II	(59)	(*E*) only

R	
n-C_3F_7	(34)
i-C_3F_7	(32)
n-C_6F_{13}	(34)

2 eq	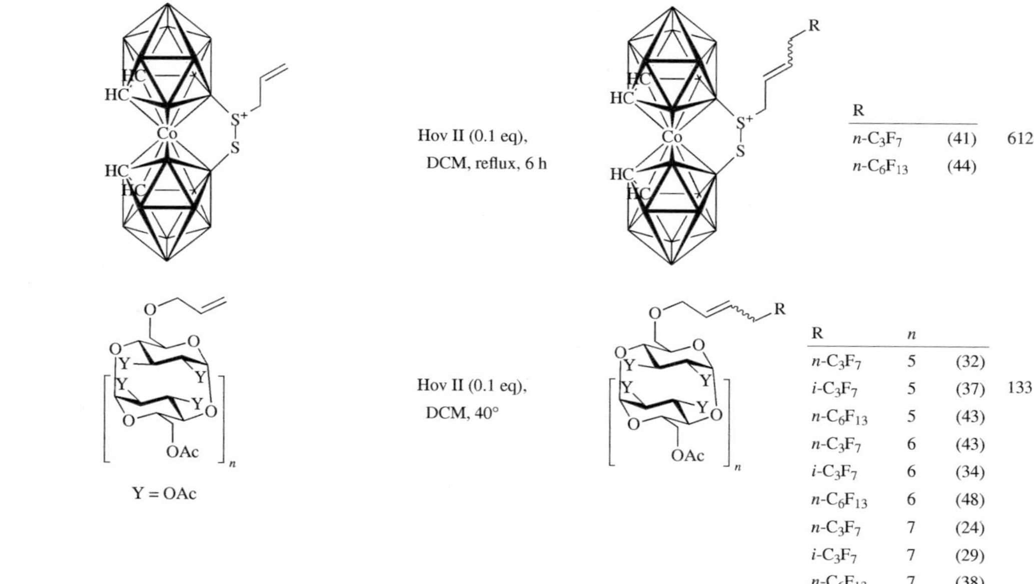	Hov II (0.1 eq), DCM, reflux, 6 h	R: n-C_3F_7 (41); n-C_6F_{13} (44)		612
5 eq	Y = OAc	Hov II (0.1 eq), DCM, 40°			133

R	n	
n-C_3F_7	5	(32)
i-C_3F_7	5	(37)
n-C_6F_{13}	5	(43)
n-C_3F_7	6	(43)
i-C_3F_7	6	(34)
n-C_6F_{13}	6	(48)
n-C_3F_7	7	(24)
i-C_3F_7	7	(29)
n-C_6F_{13}	7	(38)

TABLE 29. CROSS-METATHESIS OF VINYLIC COMPOUNDS WITH A PERFLUORINATED CARBON CHAIN (*Continued*)

Perfluorinated Alkene	Cross Partner	Conditions	Product(s) and Yield(s) (%)	Refs.

Please refer to the charts preceding the tables for the catalyst structures.

C_{8-10}

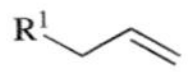

Perfluorinated Alkene	Cross Partner	Conditions	Product(s) and Yield(s) (%)	Refs.
R^1CH$_2$CH=CH$_2$ (x eq)	OH, OR^2, NHBoc	Hov II (2 x 0.05 eq), DCM, 42° (MW), 25 min	R^1, OH, OR^2, NHBoc	138

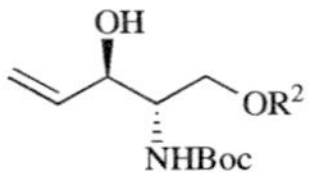

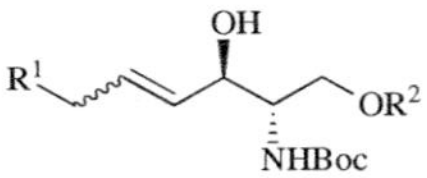

R^1	x	R^2	
n-C_5F_{11}	2 + 1	H	(34)
n-C_6F_{13}	2 + 1	H	(36)
n-C_7F_{15}	2 + 1	H	(32)
n-C_5F_{11}	2	H	(50)
n-C_6F_{13}	2	H	(60)
n-C_7F_{15}	2	H	(50)
n-C_5F_{11}	2 + 1	TBS	(38)
n-C_6F_{13}	2 + 1	TBS	(46)
n-C_7F_{15}	2 + 1	TBS	(47)
n-C_5F_{11}	2	TBS	(58)
n-C_6F_{13}	2	TBS	(65)
n-C_7F_{15}	2	TBS	(50)

C_8

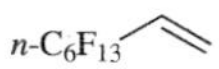

n-C_6F_{13} (5 eq)

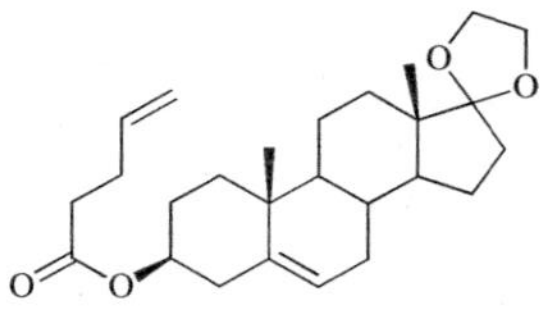

Ind II (0.05 eq), solvent, 60°, 20 h

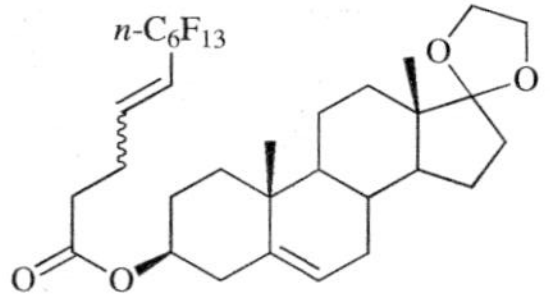

I

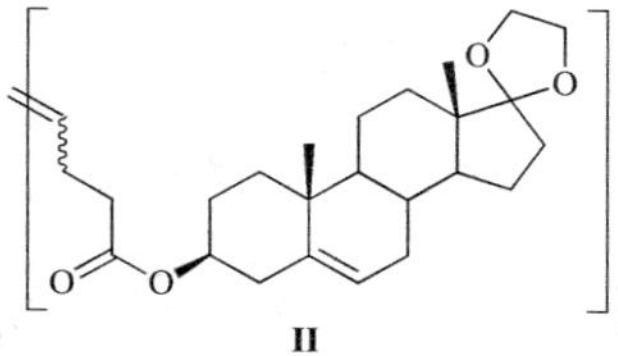

+ [ ]$_2$

II

Solvent	I	II
DCE	(12)	(23)
C_6F_6	(90)	(6)

141

C_9

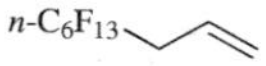

n-C_6F_{13} (2 eq)

OAc AcO AcO AcO

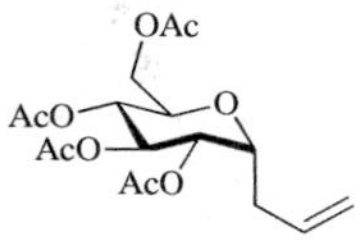

Hov II (0.1 eq), DCM, reflux, 3 h

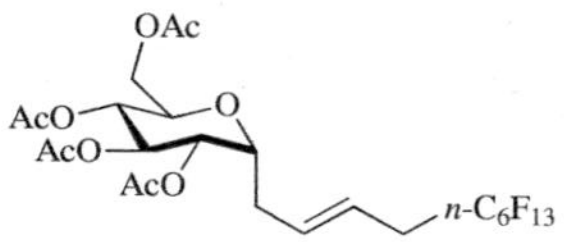

OAc AcO AcO AcO n-C_6F_{13}

(64) (*E*) only 135

TABLE 29. CROSS-METATHESIS OF VINYLIC COMPOUNDS WITH A PERFLUORINATED CARBON CHAIN (*Continued*)

Perfluorinated Alkene	Cross Partner	Conditions	Product(s) and Yield(s) (%)	Refs.
Please refer to the charts preceding the tables for the catalyst structures.				
C_9				
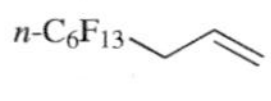 n-C_6F_{13} 2 eq	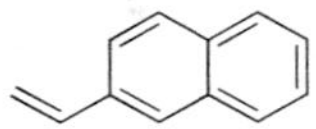	Hov II (0.1 eq), DCM, reflux, 3 h	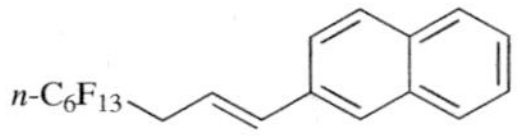 n-C_6F_{13} (17) (*E*) only	135
2 eq	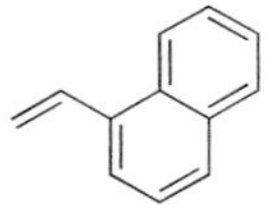	Hov II (0.1 eq), DCM, reflux, 3 h	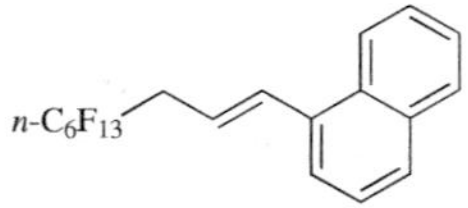 n-C_6F_{13} (15) (*E*) only	135
2 eq	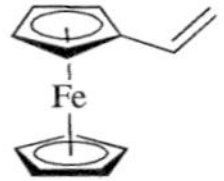 Fe	Hov II (0.1 eq), DCM, reflux, 3 h	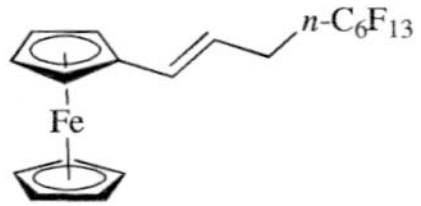 n-C_6F_{13} Fe (48) (*E*) only	135
	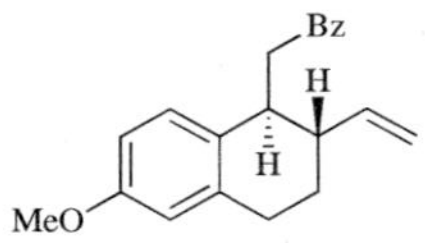 Bz H H MeO 2 eq	Hov II (0.1 eq), DCM, reflux, 3 h	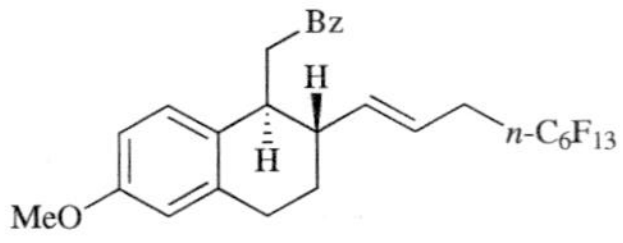 Bz H H n-C_6F_{13} MeO (66) (*E*) only	135

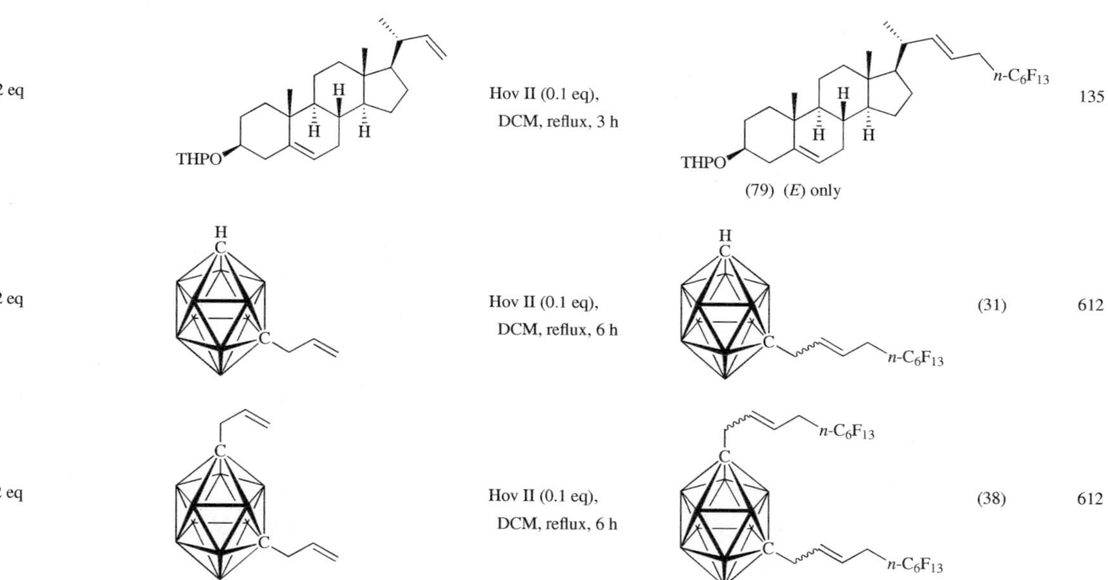

	Conditions	Product	Yield (%)	Ref.
2 eq	Hov II (0.1 eq), DCM, reflux, 3 h	(79) (*E*) only		135
2 eq	Hov II (0.1 eq), DCM, reflux, 6 h		(31)	612
2 eq	Hov II (0.1 eq), DCM, reflux, 6 h		(38)	612

TABLE 29. CROSS-METATHESIS OF VINYLIC COMPOUNDS WITH A PERFLUORINATED CARBON CHAIN (*Continued*)

Perfluorinated Alkene	Cross Partner	Conditions	Product(s) and Yield(s) (%)	Refs.
Please refer to the charts preceding the tables for the catalyst structures.				
C_9				
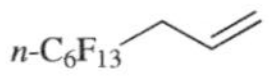 n-C_6F_{13} (allyl) 2 eq	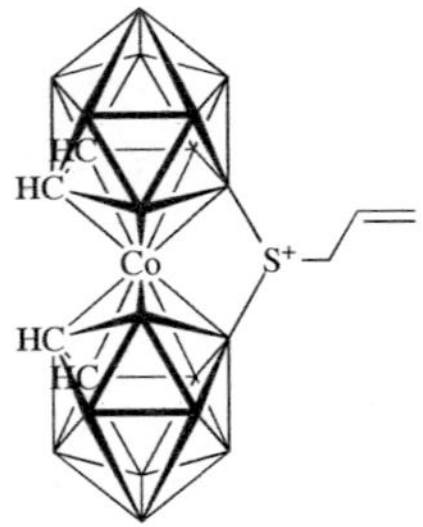 HC, HC, Co, HC, HC, S^+	Hov II (0.1 eq), DCM, reflux, 6 h	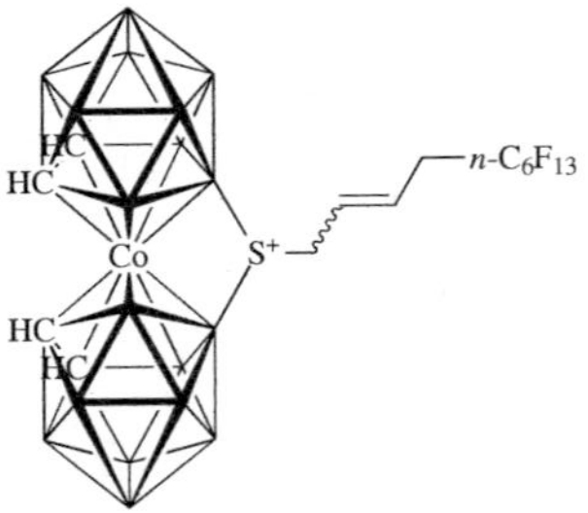 HC, HC, Co, HC, HC, S^+, n-C_6F_{13} (53)	612
5 eq	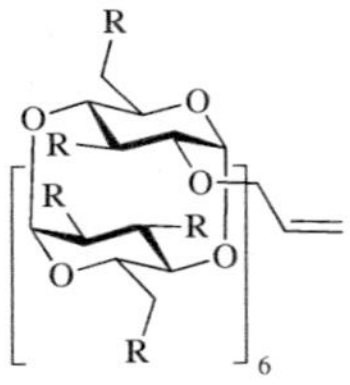 R, O, R, O, R, R, O, O, R, 6; R = OAc	Hov II (0.1 eq), DCM, 40°	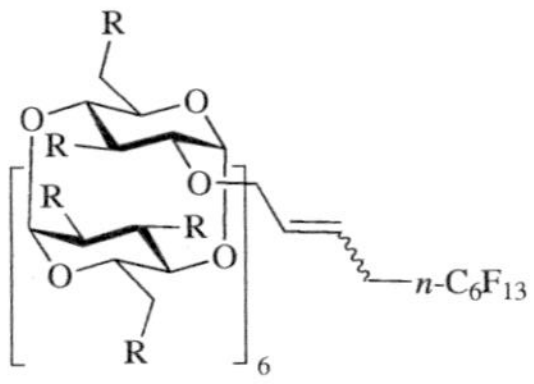 R, O, R, O, R, R, O, O, R, 6, n-C_6F_{13} (50)	133

C_{13}

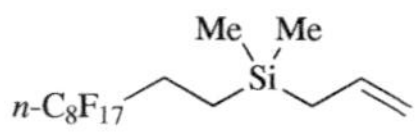

n-C_8F_{17} Me Me Si

3 eq

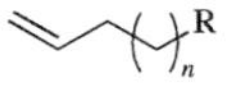

Catalyst (0.05 eq),
BQ (0.5 eq),
DCM, reflux, 12 h

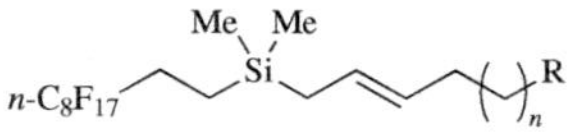

n-C_8F_{17} Me Me Si R

1181

R	n	Catalyst		$(E)/(Z)$
BnO	0	Gr II	(42)	83:17
BzO	0	Hov II	(36)	83:17
BzO	3	Gr II	(40)	—
BnO_2C	0	Gr II	(39)	75:25

3 eq

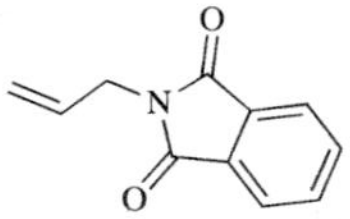

Gr II (0.05 eq),
BQ (x eq),
DCM, reflux

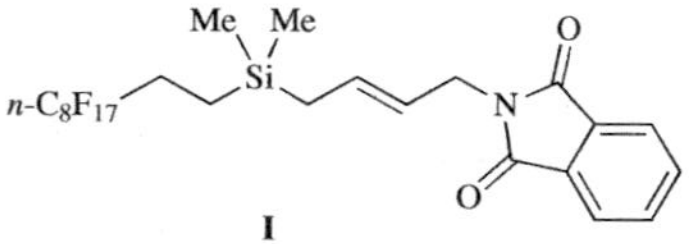

n-C_8F_{17} Me Me Si N O O

I

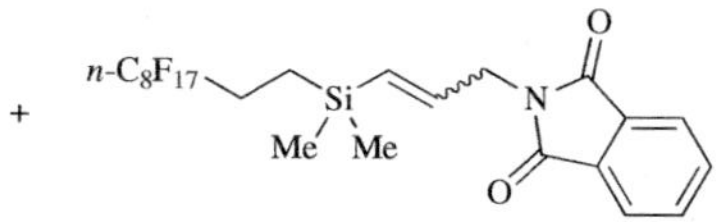

1181

\+ n-C_8F_{17} Si Me Me N O O

II

x	Time (h)	**I**	**I/II**	$(E)/(Z)$ **I**
—	48	(60)	80:20	75:25
0.5	12	(45)	**I** only	80:20
0.5	48	(43)	89:11	80:20

TABLE 30. CROSS-METATHESIS OF VINYL HALIDES

Vinyl Halide	Cross Partner	Conditions	Product(s) and Yield(s) (%)	Refs.

Please refer to the charts preceding the tables for the catalyst structures.

C_2

Vinyl Halide	Cross Partner	Conditions	Product(s) and Yield(s) (%)	Refs.
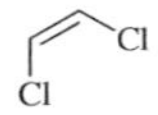 Cl, Cl (1.06 eq)	 OPh	Hov II (0.05 eq), C_6D_6/CD_2Cl_2 (10:1), 50°, 16 h	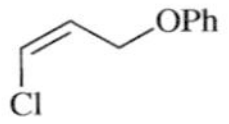 Cl, OPh (20) (Z)/(E) = 56:44	104
5 eq	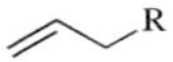 R	Mo cat **26** (*x* eq), benzene, 22°, 4 h	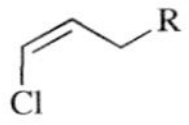 Cl, R	106

R	*x*	Conv. (%)[a]		(Z)/(E)
B(pin)	0.05	90	(66)	>98:2
TIPS	0.05	90	(84)	96:4
Ph_3Sn	0.03	97	(89)	>98:2
$MeO_2C(CH_2)_2$	0.03	90	(65)	>98:2

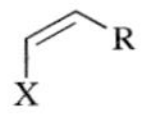

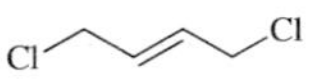

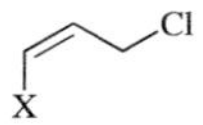

Vinyl Halide	Cross Partner	Conditions	Product(s) and Yield(s) (%)	Refs.
X, R (*x* eq)	Cl, Cl	Hov II (*y* eq), C_6D_6	X, Cl	104

X	R	*x*	*y*	Temp (°)	Time (h)		(Z)/(E)
F	H	2.5	0.05	50	24	(11)	56:44
Cl	Cl	5.6	0.1	50	3	(78)	50:50
Cl	Cl	6.4	0.05	45	3	(53)	56:44

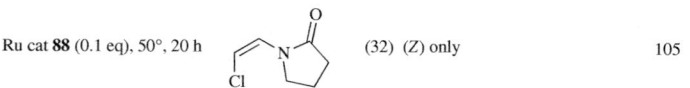

Cl, Cl (100 eq) | N-vinylpyrrolidinone | Ru cat **88** (0.1 eq), 50°, 20 h | (32) (Z) only | 105

X, X (x eq) | n-Bu alkene | Catalyst (0.05 eq), solvent, 3 h | 104

X	x	Catalyst	Solvent	Temp (°)		(Z)/(E)
Cl	5	Hov II	C_6D_6	50	(>90)	60:40
Cl	2.7	Hov II	C_6D_6	45	(60)	56:44
Cl	5	Ru cat **71**	C_6D_6/CD_2Cl_2 (10:1)	50	(43)	62:38
Br	1	Ru cat **71**	C_6D_6/CD_2Cl_2 (10:1)	50	(16)	33:67

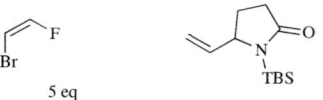

Br, F (5 eq) | N-TBS lactam | 1. Mo cat **26** (0.05 eq), benzene, 22°, 4 h 2. HCl, MeOH, 22°, 2 h | 60% conv.[a], (55) (Z)/(E) > 98:2 | 106

TABLE 30. CROSS-METATHESIS OF VINYL HALIDES (*Continued*)

Vinyl Halide	Cross Partner	Conditions	Product(s) and Yield(s) (%)	Refs.
Please refer to the charts preceding the tables for the catalyst structures.				
C_2				
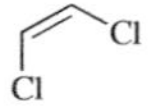 Cl, Cl; 5 eq	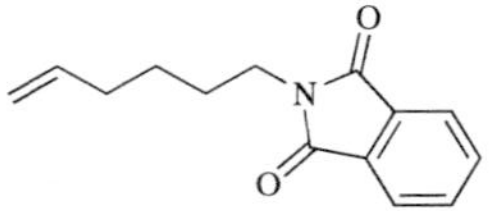	Mo cat **26** (0.03 eq), benzene, 22°, 4 h	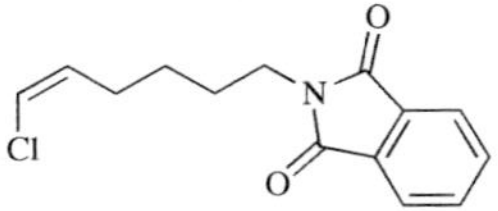 Cl; 81% conv.[a] (65) (Z)/(E) > 98:2	106
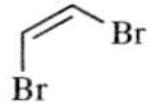 Br, Br; 8 eq; (*Z*)/(*E*) = 64:36	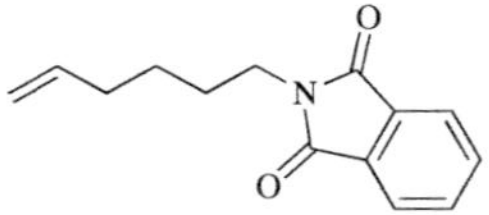	Mo cat **26** (0.05 eq), benzene, 22°, 4 h	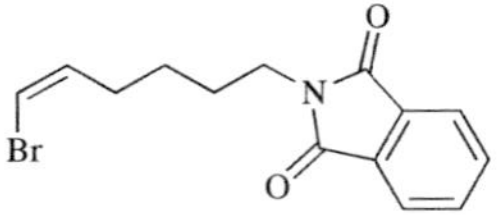 Br; 92% conv.[a] (67) (*Z*)/(*E*) = 87:13	106
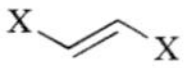 X, X; 100 eq	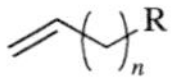 R, *n*	Ru cat **88** (*x* eq), 20 h	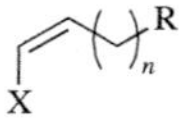 X, R, *n*	105

X	R	*n*	*x*	Temp (°)		(*Z*)/(*E*)
Cl	TBSO	4	0.15	50	(76)	60:40
Cl	Br	8	0.05	50	(41)	58:42
Cl	Br	8	0.1	50	(87)	58:42
Br	Br	8	0.15	80	(34)	36:64

Cl Cl (5 eq) + Br — Catalyst (x eq) → Cl Br 106

Catalyst	x	Temp (°)	Time (h)	Conv. (%)[a]		(Z)/(E)
Hov II	0.05	50	4	82	(59)	58:42
Ru cat **215**	0.05	50	4	10	(<5)	—
Ru cat **152**	0.05	50	4	<10	(<5)	—
Mo cat **1**	0.05	22	4	67	(<5)	—
W cat **4**	0.05	22	4	45	(<10)	—
Mo cat **5**	0.05	22	4	43	(<5)	—
Mo cat **18**	0.05	22	4	60	(27)	>98:2
Mo cat **24**	0.05	22	4	87	(60)	>98:2
Mo cat **25**	0.05	22	4	62	(40)	98:2
Mo cat **25**	0.05	22	12	95	(84)	93:7
Mo cat **26**	0.03	22	4	90	(75)	>98:2

TABLE 30. CROSS-METATHESIS OF VINYL HALIDES (*Continued*)

Vinyl Halide	Cross Partner	Conditions	Product(s) and Yield(s) (%)	Refs.

Please refer to the charts preceding the tables for the catalyst structures.

C_2

Vinyl Halide	Cross Partner	Conditions	Product(s) and Yield(s) (%)	Refs.
X^1, X^2 (5 eq)	R	Mo cat **26** (*x* eq), benzene, 22°, 4 h	X^2, R; (*Z*)/(*E*) > 98:2	106

X^1	X^2	R	*x*	Conv. (%)[a]	
Cl	Cl	TBSO	0.03	88	(80)
Cl	Cl	BnS	0.05	94	(71)
Cl	Cl	TIPS—≡—ξ—	0.03	97	(89)
Br	F	TIPS—≡—ξ—	0.05	>98	(64)
Cl	Cl	oxiranyl (O)	0.05	91	(68)

Vinyl Halide	Cross Partner	Conditions	Product(s) and Yield(s) (%)	Refs.
Br, Br (8 eq); (*Z*)/(*E*) = 64:36	R	Mo cat **26** (0.05 eq), benzene, 22°, 4 h	Br, R	106

R	Conv. (%)[a]		(*Z*)/(*E*)
TBSO	95	(63)	88:12
$Bn_2N(CH_2)_3$	90	(66)	88:12
TIPS—≡—ξ—	98	(83)	87:13
N-Boc prolinoyl-O (N, Boc, O, O)	84	(66)	87:13

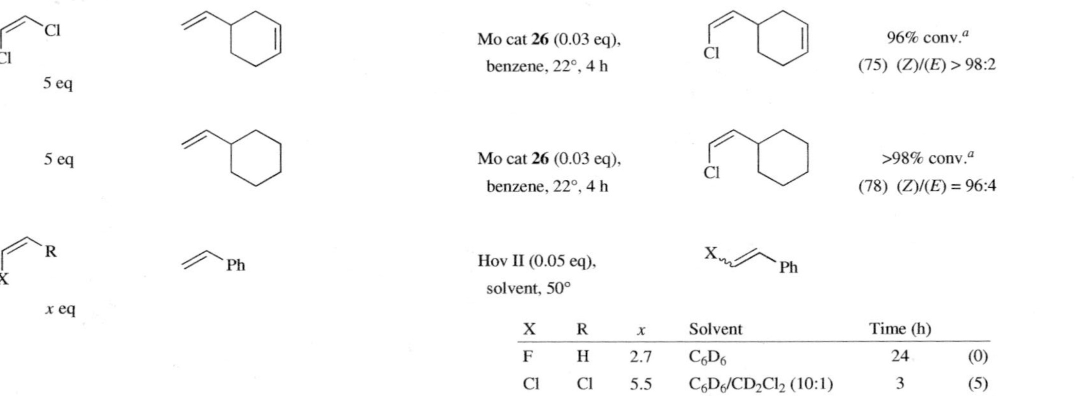

Mo cat **26** (0.03 eq), benzene, 22°, 4 h — 96% conv.[a] (75) (Z)/(E) > 98:2 — 106

Mo cat **26** (0.03 eq), benzene, 22°, 4 h — >98% conv.[a] (78) (Z)/(E) = 96:4 — 106

Hov II (0.05 eq), solvent, 50° — 104

X	R	x	Solvent	Time (h)	
F	H	2.7	C_6D_6	24	(0)
Cl	Cl	5.5	C_6D_6/CD_2Cl_2 (10:1)	3	(5)
Cl	H	12	C_6D_6/CD_2Cl_2 (10:1)	3	(5)

TABLE 30. CROSS-METATHESIS OF VINYL HALIDES (*Continued*)

Vinyl Halide	Cross Partner	Conditions	Product(s) and Yield(s) (%)	Refs.

Please refer to the charts preceding the tables for the catalyst structures.

C_2

Vinyl Halide	Cross Partner	Conditions	Product(s) and Yield(s) (%)	Refs.
Cl, Cl (E); 100 eq	OMe	Catalyst (*x* eq), CuCl (*y* eq), reflux	Cl, OMe; (*Z*)/(*E*) = 83:17	105
Cl, Cl (Z); 10 eq; (*Z*)/(*E*) > 98:2	OMe	Catalyst (0.05 eq), benzene, 22°, 4 h	Cl, OMe	1180

First entry (ref. 105):

Catalyst	*x*	*y*	Time (h)	
Gr II	0.05	—	20	(15)
Gr II	0.05	0.05	20	(32)
Hov II	0.05	—	20	(24)
Ru cat **88**	0.05	—	20	(54)
Ru cat **88**	0.01	—	20	(21)
Ru cat **115**	0.05	—	20	(24)
Ru cat **115**	0.15	—	6	(91)

Second entry (ref. 1180):

Catalyst		(*Z*)/(*E*)
Mo cat **27**	(34)	52:48
Mo cat **29**	(80)	>98:2

5 eq

Mo cat **26** (0.05 eq), benzene, 22°, 4 h

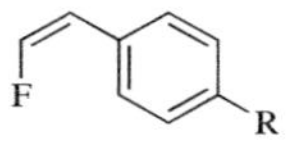

106

R	Conv. (%)[a]		$(Z)/(E)$
MeO	>98	(71)	95:5
AcO	84	(72)	94:6
Cl	78	(64)	97:3
Br	88	(66)	93:7

100 eq

Ru cat **88** (0.1 eq), 50°, 20 h

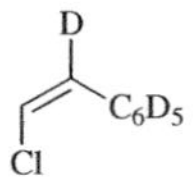

(52) $(Z)/(E)$ = 72:28

105

x eq

Catalyst (*y* eq), solvent

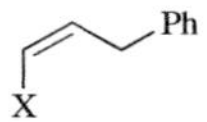

104

X	R	*x*	Catalyst	*y*	Solvent	Temp (°)	Time (h)		$(Z)/(E)$
F	H	0.31	Hov II	0.1	C_6D_6/CD_2Cl_2 (10:1)	50	24	(<10)	—
Cl	Cl	2.5	Ru cat **71**	0.1	C_6D_6/CD_2Cl_2 (10:1)	50	3	(58)	50:50
Cl	Cl	3.4	Hov II	0.1	C_6D_6	50	3	(50)	50:50
Cl	Cl	3.2	Hov II	0.05	C_6D_6	43	3	(50)	50:50
Cl	H	7.5	Hov II	0.1	C_6D_6/CD_2Cl_2 (10:1)	50	24	(20)	75:25

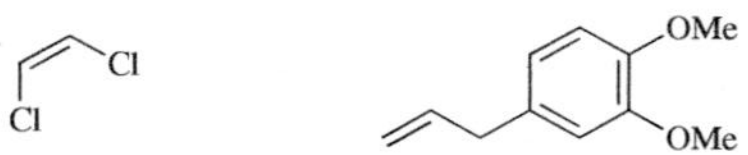

5 eq

Mo cat **26** (0.03 eq), benzene, 22°, 4 h

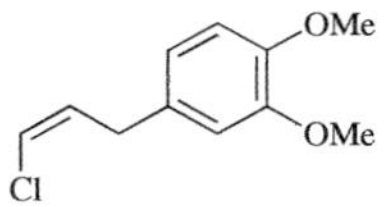

86% conv.[a]
(50) $(Z)/(E)$ = 95:5

106

TABLE 30. CROSS-METATHESIS OF VINYL HALIDES (*Continued*)

Vinyl Halide	Cross Partner	Conditions	Product(s) and Yield(s) (%)	Refs.
Please refer to the charts preceding the tables for the catalyst structures.				
C_2				
X, R (x eq)	n-Bu, n-Bu	Catalyst (y eq), solvent	X, n-Bu	104
Br, Br; 8 eq; (Z)/(E) = 64:36	OTBS	Mo cat **26** (0.05 eq), benzene, 22°, 4 h	Br, OTBS; 74% conv.[a]; (57) (Z)/(E) = 91:9	106

X	R	x	Catalyst	y	Solvent	Temp (°)	Time (h)		(Z)/(E)
F	H	8.5	Hov II	0.05	CD_2Cl_2	23/50	0.5/3	(9)	79:21
Cl	Cl	5	Hov II	0.05	C_6D_6	23	16	(100)	71:29
Cl	Cl	3	Hov II	0.04	C_6D_6	23	40	(>98)	69:31
Cl	Cl	4.1	Hov II	0.05	C_6D_6	45	3	(85)	67:33
Cl	Cl	1.6	Ru cat **67**	0.05	C_6D_6/CD_2Cl_2 (10:1)	50	3	(<5)	—
Cl	Cl	0.93	Ru cat **71**	0.05	C_6D_6/CD_2Cl_2 (10:1)	50	24	(0)	—
Cl	H	8	Hov II	0.05	C_6D_6/CD_2Cl_2 (10:1)	23	24	(29)	(Z) only
Cl	H	8	Ru cat **71**	0.05	C_6D_6/CD_2Cl_2 (10:1)	50	3	(25)	88:12
Cl	H	5.8	Ru cat **67**	0.05	C_6D_6/CD_2Cl_2 (10:1)	50	3	(0)	—
Br	Br	2	Hov II	0.05	C_6D_6/CD_2Cl_2 (10:1)	50	24	(22)	—
Br	H	8	Hov II	0.05	C_6D_6/CD_2Cl_2 (10:1)	50	3	(5)	—

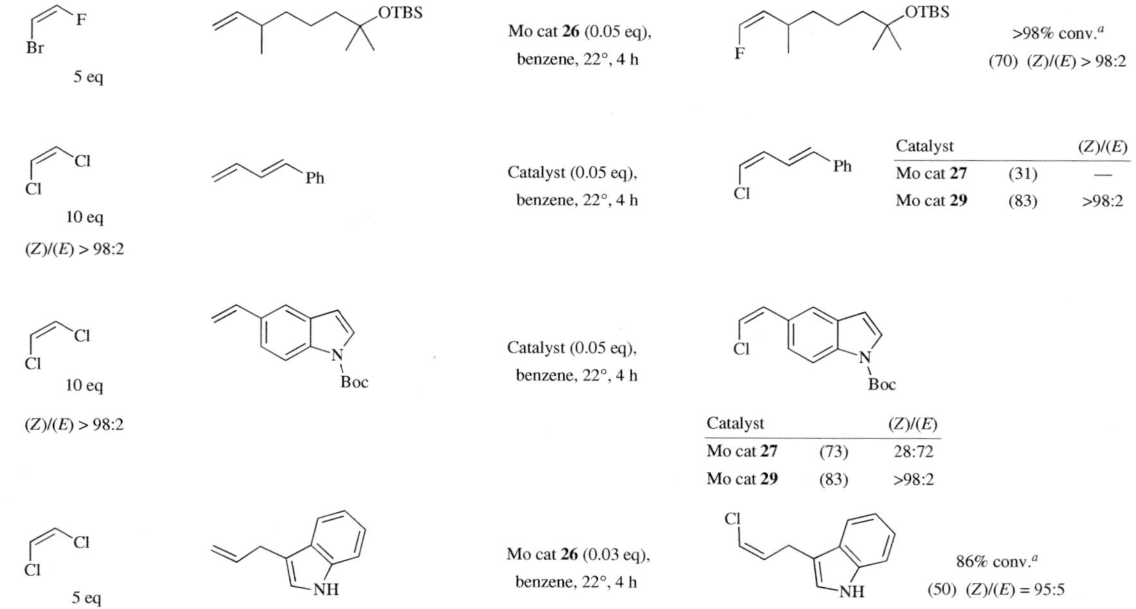

F
Br
5 eq
OTBS
Mo cat **26** (0.05 eq),
benzene, 22°, 4 h
OTBS
F
>98% conv.[a]
(70) (Z)/(E) > 98:2
106
Cl
Cl
10 eq
(Z)/(E) > 98:2
Ph
Catalyst (0.05 eq),
benzene, 22°, 4 h
Ph
Cl
Catalyst | | (Z)/(E)
Mo cat **27** | (31) | —
Mo cat **29** | (83) | >98:2
1180
Cl
Cl
10 eq
(Z)/(E) > 98:2
N
Boc
Catalyst (0.05 eq),
benzene, 22°, 4 h
Cl
N
Boc
1180
Catalyst | | (Z)/(E)
Mo cat **27** | (73) | 28:72
Mo cat **29** | (83) | >98:2
Cl
Cl
5 eq
NH
Mo cat **26** (0.03 eq),
benzene, 22°, 4 h
Cl
NH
86% conv.[a]
(50) (Z)/(E) = 95:5
106

TABLE 30. CROSS-METATHESIS OF VINYL HALIDES (*Continued*)

Vinyl Halide	Cross Partner	Conditions	Product(s) and Yield(s) (%)	Refs.
Please refer to the charts preceding the tables for the catalyst structures.				
C_2				
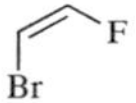 Br, F; 5 eq	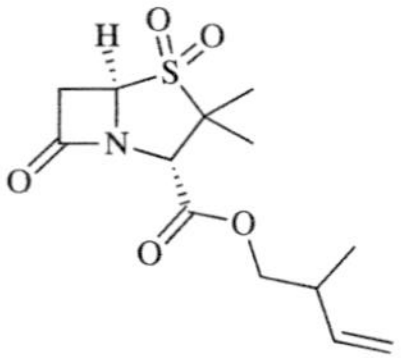 H, O, O, S, N, O, O, O	Mo cat **26** (0.05 eq), benzene, 22°, 4 h	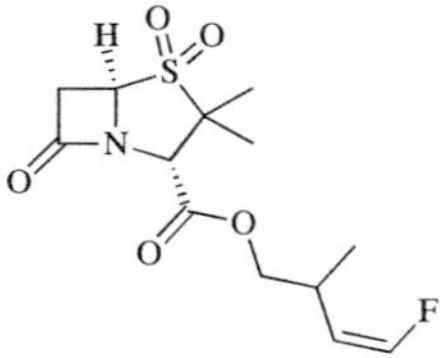 H, O, O, S, N, O, O, O, F; 86% conv.[a] (80) (Z)/(E) > 98:2	106
5 eq	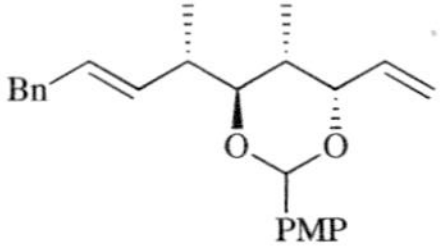 Bn, O, O, PMP	Mo cat **26** (0.05 eq), benzene, 22°, 4 h	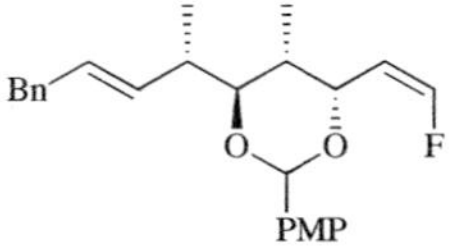 Bn, O, O, PMP, F; 80% conv.[a] (63) (Z)/(E) > 98:2	106
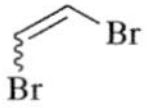 Br, Br; 8 eq; (Z)/(E) = 64:36	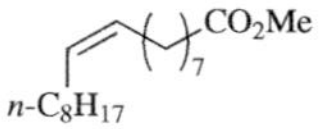 n-C_8H_{17}, CO_2Me, 7	Catalyst (x eq), benzene, 22°, 4 h	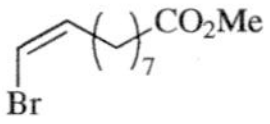 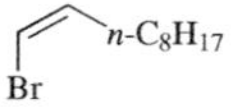 Br, CO_2Me, 7 **I** + Br, n-C_8H_{17} **II**	1180

Catalyst	x	**I**	(Z)/(E) **I**	**II**	(Z)/(E) **II**
Mo cat **27**	0.05	(45)	89:11	(45)	89:11
Mo cat **29**	0.03	(90)	>98:2	(85)	>98:2

Reagent	Substrate	Conditions	Product(s) and Yield(s) (%)	Refs.
Br, F (5 eq)	MeO$_2$C, H, H	Mo cat **26** (0.05 eq), benzene, 22°, 4 h	MeO$_2$C, H, H, F 83% conv.[a] (70) (Z)/(E) = 96:4	106
Cl, Cl (5 eq)	OTBS, H, H, H, H	Mo cat **26** (0.05 eq), benzene, 22°, 4 h	OTBS, Cl, H, H, H, H 75% conv.[a] (70) (Z)/(E) = 95:5	106
Br, F (5 eq)	N, N, O, O, N, S, Cl	Mo cat **26** (0.05 eq), benzene, 22°, 4 h	N, N, O, O, F, N, S, Cl 88% conv.[a] (78) (Z)/(E) > 98:2	106

[a] The conversion was determined by ^{1}H NMR analysis.

TABLE 31. CROSS-METATHESIS OF VINYL SILANES AND SILOXANES
A. VINYL SILANES

Vinyl Silane	Cross Partner	Conditions	Product(s) and Yield(s) (%)	Refs.

Please refer to the charts preceding the tables for the catalyst structures.

C_2

Vinyl Silane	Cross Partner	Conditions	Product(s) and Yield(s) (%)	Refs.
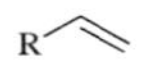 R (vinyl), x eq	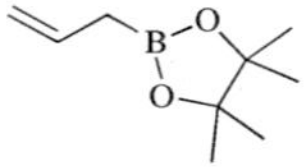	Catalyst (0.05 eq), DCM, 40°, 3 h	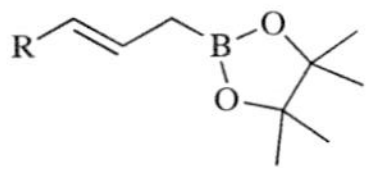 R	699

R	x	Catalyst		(E)/(Z)
TMS	5	Hov II	(0)	—
$PhMe_2Si$	5	Hov II	(0)	—
Cl_3Si	3	Gr II	(92)	(E) only
Cl_2PhSi	3	Hov II	(83)	(E) only

Vinyl Silane	Cross Partner	Conditions	Product(s) and Yield(s) (%)	Refs.
x eq	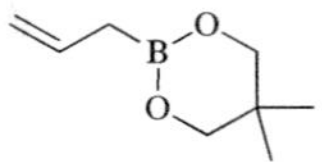	Catalyst (0.05 eq), DCM, 40°, 3 h	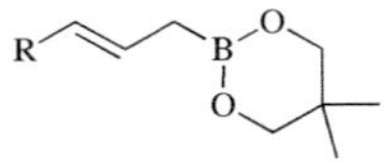 R	699

R	x	Catalyst		(E)/(Z)
TMS	5	Hov II	(0)	—
$PhMe_2Si$	5	Hov II	(0)	—
Cl_3Si	3	Gr II	(95)	(E) only
Cl_2PhSi	3	Hov II	(93)	(E) only

Cl_3Si–CH=CH$_2$ + CH$_2$=CH–R (x eq) → Catalyst (y eq), DCM, reflux → Cl_3Si–CH=CH–R 200

R	x	Catalyst	y	Time (h)	Conv. (%)[a]	[b]	(*E*)/(*Z*)
n-BuOCH$_2$	1	Gr II	0.05	1	100	(92)	89:11
TMSCH$_2$	0.2	Gr II	0.005	1	100	(100)	96:4
n-Bu	0.2	Gr I	0.05	3	0	(—)	—
n-Bu	0.2	Gr II	0.05	3	100	(100)	95:5
n-Bu	1	Hov II	0.05	1	99	(97)	95:5
Ph	0.07	Gr I	0.005	1	0	(—)	—
Ph	0.07	Gr II	0.005	1	85	(83)	(*E*) only
Ph	1	Hov II	0.05	1	100	(85)	(*E*) only

$MeCl_2Si$–CH=CH$_2$ + CH$_2$=CH–R (x eq) → Catalyst (y eq), DCM, reflux → $MeCl_2Si$–CH=CH–R 200

R	x	Catalyst	y	Time (h)	Conv. (%)[a]	[b]	(*E*)/(*Z*)
n-BuOCH$_2$	1	Gr II	0.05	1	100	(98)	96:4
TMSCH$_2$	0.2	Gr II	0.005	1	100	(100)	95:5
n-Bu	0.2	Gr I	0.005	1	0	(—)	—
n-Bu	0.2	Gr II	0.005	1	100	(100)	95:5
n-Bu	1	Gr II	0.05	20	traces	(—)	—
n-Bu	1	Hov II	0.05	6	90	(88)	(*E*) only
Ph	0.07	Gr II	0.005	1	80	(78)	(*E*) only
Ph	1	Hov II	0.02	1	100	(70)	(*E*) only
4-MeC$_6$H$_4$	1	Gr II	0.05	1	100	(95)	(*E*) only
4-ClC$_6$H$_4$	1	Gr II	0.05	1	100	(90)	(*E*) only
4-ClCH$_2$C$_6$H$_4$	1	Gr II	0.05	1	100	(98)	(*E*) only

TABLE 31. CROSS-METATHESIS OF VINYL SILANES AND SILOXANES (*Continued*)

A. VINYL SILANES (*Continued*)

Vinyl Silane	Cross Partner	Conditions	Product(s) and Yield(s) (%)	Refs.

Please refer to the charts preceding the tables for the catalyst structures.

C_2

Vinyl Silane	Cross Partner	Conditions	Product(s) and Yield(s) (%)	Refs.
SiX_2R^1-CH=CH$_2$	CH$_2$=CH-R^2 (*x* eq)	Catalyst (0.05 eq), DCM, reflux	SiX_2R^1-CH=CH-R^2	200

R^1	X	R^2	*x*	Catalyst	Time (h)	Conv. (%)[a]	[b]	(*E*)/(*Z*)
Ph	F	Ph	1	Gr I	20	100	(99)	(*E*) only
Ph	F	Ph	1	Gr II	20	80	(50)	(*E*) only
Ph	Cl	*n*-BuOCH$_2$	1	Gr II	1	100	(90)	(*E*) only
Ph	Cl	*n*-Bu	1	Gr II	1	100	(95)	96:4
Ph	Cl	Ph	1	Gr II	1	100	(98)	(*E*) only
Ph	Cl	4-ClC$_6$H$_4$	1	Gr II	1	100	(95)	(*E*) only
4-CF$_3$C$_6$H$_4$	Cl	*n*-Bu	0.25	Gr II	3	100	(95)	96:4
4-MeC$_6$H$_4$	Cl	*n*-Bu	1	Gr II	1	100	(99)	96:4
4-CF$_3$C$_6$H$_4$	Cl	Ph	0.25	Gr II	1	100	(95)	(*E*) only
4-MeC$_6$H$_4$	Cl	Ph	1	Gr II	2	100	(95)	(*E*) only

Vinyl Silane	Cross Partner	Conditions	Product(s) and Yield(s) (%)	Refs.
Ph_2XSi-CH=CH$_2$ (1 eq)	CH$_2$=CH-R	Gr II (0.05 eq), DCM, reflux	Ph_2XSi-CH=CH-R	200

X	R	Time (h)	Conv. (%)[a]	[b]	(*E*)/(*Z*)
Cl	*n*-Bu	2	100	(90)	96:4
F	Ph	20	100	(20)	(*E*) only
Cl	Ph	1	100	(99)	(*E*) only
Cl	4-ClC$_6$H$_4$	1	100	(97)	(*E*) only

TMS⁄⁄ + ⁄⁄R (x eq) → Catalyst (y eq), solvent → TMS⁄⁄R

R	x	Catalyst	y	Solvent	Time (h)	Temp (°)	Conv. (%)[a]	[b]	(E)/(Z)	Refs.
Ph	3	Gr I	0.02	DCM	6	rt	0	(—)	—	200
Ph	0.2	Gr II	0.02	DCM	6	reflux	0	(—)	—	200
Ph	0.2	Hov II	0.02	DCM	6	reflux	90	(85)	(E) only	200
n-Bu	0.2	$RuCl_2(PPh_3)_3$	0.1	—	10	105	55	(45)	—	1182
n-C_5H_{11}	0.2	$RuCl_2(PPh_3)_3$	0.1	—	10	105	60	(55)	—	1182
n-C_6H_{13}	0.2	$RuCl_2(PPh_3)_3$	0.1	—	10	105	55	(50)	—	1182
n-C_7H_{15}	0.2	$RuCl_2(PPh_3)_3$	0.1	—	10	105	45	(45)	—	1182
n-C_8H_{17}	0.2	$RuCl_2(PPh_3)_3$	0.1	—	10	105	50	(45)	—	1182
n-$C_{16}H_{33}$	0.5	$RuCl_2(PPh_3)_3$	0.1	—	10	105	40	(40)	—	1182

R_3Si⁄⁄ (5 eq) + ⁄⁄n-Bu → Gr II (0.02 eq), DCM, reflux, 3 h → R_3Si⁄⁄n-Bu 200

R	Conv. (%)[a]	[b]	(E)/(Z)
Ph	0	(—)	—
4-MeC_6H_4	0	(—)	—
4-$CF_3C_6H_4$	100	(97)	96:4

TABLE 31. CROSS-METATHESIS OF VINYL SILANES AND SILOXANES (*Continued*)

A. VINYL SILANES (*Continued*)

Vinyl Silane	Cross Partner	Conditions	Product(s) and Yield(s) (%)	Refs.

Please refer to the charts preceding the tables for the catalyst structures.

C_2

Vinyl Silane	Cross Partner	Conditions	Product(s) and Yield(s) (%)	Refs.
R^1-CH=CH$_2$	CH$_2$=CH-R^2 (10 eq)	Catalyst (0.01 eq), solvent	R^1-CH=CH-R^2	1183

R^1	R^2	Catalyst	Solvent	Temp (°)	Time (h)	*b*	(*E*)/(*Z*)
$PhMe_2Si$	n-C_5H_{11}	$RuCl_2(PPh_3)_3$	benzene	reflux	48	(60)	98:2
$PhMe_2Si$	n-C_5H_{11}	$RuCl_2(PPh_3)_3$	benzene	130	48	(70)	—
$PhMe_2Si$	n-C_5H_{11}	$RuCl_2(PPh_3)_3$	—	130	48	(60)	—
$PhMe_2Si$	n-C_6H_{13}	$RuCl_2(PPh_3)_3$	benzene	reflux	4	(70)	95:5
$PhMe_2Si$	n-C_6H_{13}	$RuCl_2(PPh_3)_3$	—	130	48	(80)	—
$PhMe_2Si$	n-C_8H_{17}	$RuCl_2(PPh_3)_3$	benzene	130	2	(70)	98:2
$PhMe_2Si$	n-C_8H_{17}	$RuCl_2(PPh_3)_3$	—	130	48	(85)	—
$PhMe_2Si$	n-C_8H_{17}	$[RuCl_2(CO)_3]_2$	—	130	48	(85)	—
$PhMe_2Si$	n-C_8H_{17}	$Ru_2(acac)_3$	—	130	48	(50)	—
$PhMe_2Si$	n-C_8H_{17}	$RhCl(PPh_3)_3$	—	130	48	(40)	—
$PhMe_2Si$	n-$C_{10}H_{21}$	$RuCl_2(PPh_3)_3$	benzene	130	4	(65)	94:6
$PhMe_2Si$	n-$C_{10}H_{21}$	$RuCl_2(PPh_3)_3$	—	130	48	(80)	—
$PhMe_2Si$	n-$C_{12}H_{25}$	$RuCl_2(PPh_3)_3$	benzene	130	6	(65)	95:5
$PhMe_2Si$	n-$C_{12}H_{25}$	$RuCl_2(PPh_3)_3$	—	130	48	(70)	—
$PhMe_2Si$	n-$C_{16}H_{33}$	$RuCl_2(PPh_3)_3$	benzene	130	6	(70)	—
$PhMe_2Si$	n-$C_{16}H_{33}$	$RuCl_2(PPh_3)_3$	—	130	48	(75)	—
$PhMe_2Si$	Ph	$[RuCl_2(CO)_3]_2$	—	120	0.3	(85)	97:3
$PhMe_2Si$	Ph	$RuCl_2(PPh_3)_3$	benzene	120	0.3	(85)	—
Ph_2MeSi	n-C_8H_{17}	$RuCl_2(PPh_3)_3$	benzene	130	6	(70)	98:2
Ph_3Si	n-C_8H_{17}	$RuCl_2(PPh_3)_3$	benzene	150	24	(30)	—

C_4

TMS (1.1 eq)

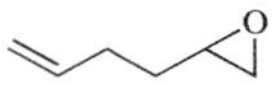

4 eq

Gr I (0.05 eq), DCM, 40°, 8 h  (17) 1184

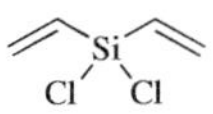

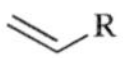

Cl Si Cl

Gr II (0.05 eq), DCM, reflux 203

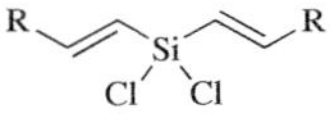

R	Time (h)		$(E)/(Z)$
4-ClC_6H_4	3	(98)	(E) only
n-C_8H_{17}	1	(99)	94:6

CF_3 Si CF_3

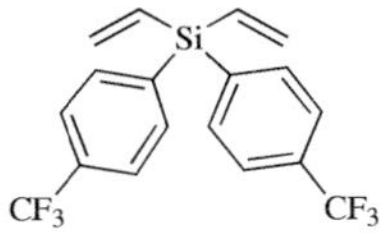

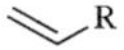

x eq

Catalyst (y eq), DCM, reflux 203

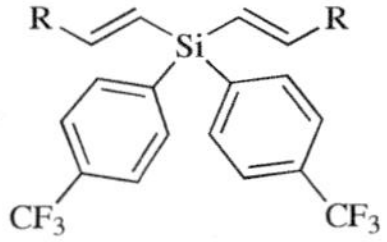

R	x	Catalyst	y	Time (h)		$(E)/(Z)$
4-ClC_6H_4	4	Gr I	0.05	5	(0)	—
4-ClC_6H_4	4	Gr II	0.05	5	(14)	(E) only
4-ClC_6H_4	10	Gr II	0.1	5	(64)	(E) only
4-ClC_6H_4	10	Hov II	0.1	3	(96)	(E) only
4-$MeOC_6H_4$	10	Gr II	0.1	5	(70)	(E) only
n-C_8H_{17}	4	Gr I	0.05	5	(0)	—
n-C_8H_{17}	4	Gr II	0.05	3	(57)	94:6
n-C_8H_{17}	10	Gr II	0.05	3	(96)	94:6
n-C_8H_{17}	10	Hov II	0.05	3	(56)	94:6

TABLE 31. CROSS-METATHESIS OF VINYL SILANES AND SILOXANES (*Continued*)

A. VINYL SILANES (*Continued*)

Vinyl Silane	Cross Partner	Conditions	Product(s) and Yield(s) (%)	Refs.
Please refer to the charts preceding the tables for the catalyst structures.				
C_4				
$(CH_2{=}CH)_2Si(C_6F_5)_2$	$CH_2{=}CHR$ (*x* eq)	Catalyst (*y* eq), DCM, reflux	$(RCH{=}CH)_2Si(C_6F_5)_2$	203

R	*x*	Catalyst	*y*	Time (h)		(*E*)/(*Z*)
4-ClC_6H_4	10	Gr II	0.1	5	(72)	(*E*) only
4-ClC_6H_4	10	Hov II	0.1	3	(98)	(*E*) only
4-$MeOC_6H_4$	10	Gr II	0.1	5	(74)	(*E*) only
n-C_8H_{17}	4	Gr I	0.05	5	(0)	—
n-C_8H_{17}	10	Gr II	0.05	3	(98)	94:6

[a] The conversion was determined by GC analysis.

[b] The yield was determined by GC analysis.

TABLE 31. CROSS-METATHESIS OF VINYL SILANES AND SILOXANES (*Continued*)

B. VINYL SILOXANES

Vinyl Siloxane	Cross Partner	Conditions	Product(s) and Yield(s) (%)	Refs.

Please refer to the charts preceding the tables for the catalyst structures.

C_2

R (5 eq) 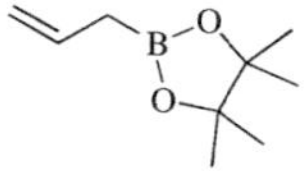 — Catalyst (0.05 eq), DCM, 40° 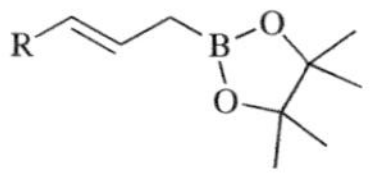 — 699

R	Catalyst	Time (h)		(E)/(Z)
$(MeO)_3Si$	Gr I	3	(99)	91:9
$(MeO)_3Si$	Gr II	3	(94)	(*E*) only
$(MeO)_3Si$	Hov II	1	(99)	(*E*) only
$(EtO)_3Si$	Gr I	3	(98)	(*E*) only
$(EtO)_3Si$	Gr II	3	(87)	(*E*) only
$(EtO)_3Si$	Hov II	1	(99)	(*E*) only

5 eq — Catalyst (0.05 eq), DCM, 40° 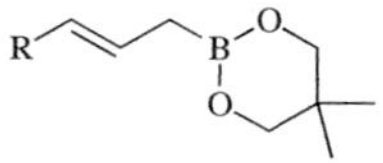 — 699

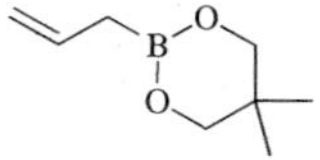

R	Catalyst	Time (h)		(E)/(Z)
$(MeO)_3Si$	Gr I	3	(78)	75:25
$(MeO)_3Si$	Gr II	3	(98)	(*E*) only
$(MeO)_3Si$	Hov II	1	(99)	(*E*) only
$(EtO)_3Si$	Gr I	3	(92)	(*E*) only
$(EtO)_3Si$	Gr II	3	(82)	(*E*) only
$(EtO)_3Si$	Hov II	1	(99)	(*E*) only

TABLE 31. CROSS-METATHESIS OF VINYL SILANES AND SILOXANES (*Continued*)

B. VINYL SILOXANES (*Continued*)

Vinyl Siloxane	Cross Partner	Conditions	Product(s) and Yield(s) (%)	Refs.

Please refer to the charts preceding the tables for the catalyst structures.

C_2

R^1-CH=CH$_2$ (x eq) + CH$_2$=CH-R^2 → Catalyst (y eq), DCM → R^1-CH=CH-R^2 — Refs. 200

R^1	x	R^2	Catalyst	y	Temp	Time (h)	Conv. (%)[a]	[b]	$(E)/(Z)$
$(AcO)_3Si$	5	$TMSCH_2$	Gr II	0.005	reflux	1	100	(100)	96:4
$(AcO)_3Si$	5	*n*-Bu	Gr II	0.005	reflux	1	100	(100)	95:5
$(AcO)_3Si$	15	Ph	Gr II	0.005	reflux	1	80	(75)	(*E*) only
$(Me_3SiO)Cl_2Si$	1	*n*-$BuOCH_2$	Gr II	0.05	reflux	1	100	(95)	96:4
$(Me_3SiO)Cl_2Si$	1	*n*-Bu	Gr II	0.05	reflux	1	100	(97)	95:5
$(Me_3SiO)Cl_2Si$	1	Ph	Gr II	0.05	reflux	1	100	(95)	(*E*) only
$Ph(EtO)_2Si$	1	*n*-Bu	Gr II	0.05	reflux	1	90	(75)	96:4
$Ph(EtO)_2Si$	1	*n*-Bu	Hov II	0.05	reflux	20	100	(20)	(*E*) only
$Ph(EtO)_2Si$	1	Ph	Gr I	0.05	reflux	20	40	(40)	(*E*) only
$Ph(EtO)_2Si$	1	Ph	Gr II	0.05	reflux	20	90	(88)	(*E*) only
$Ph(EtO)_2Si$	3	4-ClC_6H_4	Gr II	0.05	reflux	20	95	(90)	(*E*) only
$Me(EtO)_2Si$	1	Ph	Gr I	0.02	rt	6	10	(7)	(*E*) only
$Me(EtO)_2Si$	1	Ph	Gr II	0.02	reflux	3	10	(6)	(*E*) only
$Ph_2(EtO)Si$	1	Ph	Gr I	0.05	reflux	3	0	(0)	—
$Ph_2(EtO)Si$	1	Ph	Gr II	0.05	reflux	3	100	(50)	(*E*) only
$Me_2(EtO)Si$	1	Ph	Gr I	0.02	rt	6	3	(0)	—
$Me_2(EtO)Si$	1	Ph	Gr II	0.05	reflux	3	7	(5)	—

Me_3SiO
$Me_3SiO-Si$
R^1
x eq

R^2

Catalyst (0.05 eq)
DCM, reflux, 5 h

Me_3SiO
$Me_3SiO-Si$ R^2
R^1

202

R^1	x	R^2	Catalyst	Conv. (%)[a]	[b]	(*E*)/(*Z*)
MeO	5	Ph	Gr II	94	(93)	(*E*) only
MeO	5	4-$MeOC_6H_4$	Gr II	95	(94)	91:9
MeO	5	*n*-C_8H_{17}	Gr II	93	(93)	96:4
MeO	5	$TMSCH_2$	Gr II	80	(80)	(*E*) only
CF_3	1	Ph	Gr I	86	(86)	(*E*) only
CF_3	5	Ph	Gr I	97	(97)	(*E*) only
CF_3	1	4-$MeOC_6H_4$	Gr I	94	(94)	(*E*) only
CF_3	1	4-ClC_6H_4	Gr I	80	(80)	(*E*) only
CF_3	5	*n*-C_8H_{17}	Gr I	100	(95)	91:9
CF_3	2	*n*-$BuOCH_2$	Gr I	100	(100)	89:11
CF_3	5	4-$MeOC_6H_4$	Gr II	99	(95)	(*E*) only
CF_3	5	*n*-C_8H_{17}	Gr II	100	(95)	91:9
CF_3	5	$TMSCH_2$	Gr II	100	(100)	96:4
CF_3	5	*n*-$BuOCH_2$	Gr II	100	(95)	87:13

TABLE 31. CROSS-METATHESIS OF VINYL SILANES AND SILOXANES (*Continued*)

B. VINYL SILOXANES (*Continued*)

Vinyl Siloxane	Cross Partner	Conditions	Product(s) and Yield(s) (%)	Refs.

Please refer to the charts preceding the tables for the catalyst structures.

C_2

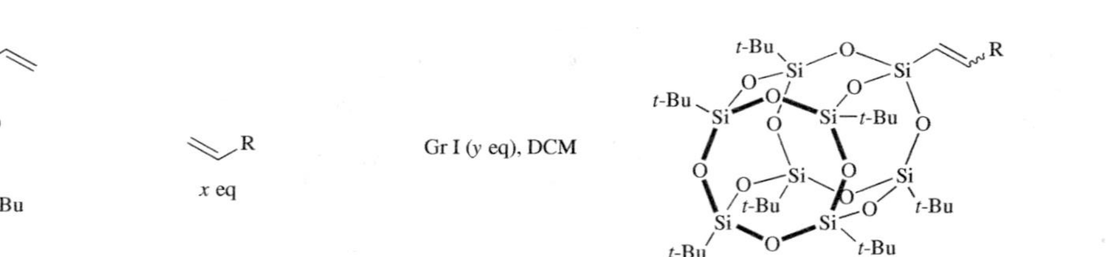

R	x	y	Temp	Time (h)	Yield (%)	Refs.
$TMSCH_2$	3	0.01	reflux	18	(85)	1188
$PhCH_2$	2	0.02	reflux	18	(87)	1188
Ph	2	0.01	reflux	6	(97)	1188
4-MeC_6H_4	3	0.01	reflux	6	(94)	1188
4-$MeOC_6H_4$	1.5	0.01	reflux	6	(97)	1188
4-ClC_6H_4	1.5	0.01	reflux	6	(96)	1188
4-BrC_6H_4	1.5	0.01	reflux	6	(95)	1188
4-BrC_6H_4	3	0.03	30°	72	(68)	338

$(EtO)_3Si$–CH=CH$_2$ (x eq) + CH$_2$=CH–R → Catalyst (y eq), DCM → $(EtO)_3Si$–CH=CH–R

x	R	Catalyst	y	Temp	Time (h)	Conv. (%)[a]	[b]	(*E*)/(*Z*)	
1	Ph	Gr I	0.02	rt	6	95	(95)	(*E*) only	200
1	Ph	Gr II	0.02	reflux	1	100	(80)	(*E*) only	200
3	Ph	Gr II	0.02	reflux	3	100	(95)	(*E*) only	200
1	Ph	Hov II	0.05	reflux	2	100	(85)	(*E*) only	200
5	*n*-Bu	Gr I	0.05	reflux	3	100	(75)	90:10	200
5	*n*-Bu	Gr II	0.05	reflux	3	100	(99)	95:5	200
5	$PhOCH_2$	Gr I	0.05	reflux	3	80	(72)	83:17	1185
5	$PhCH_2$	Gr I	0.05	reflux	3	85	(68)	91:9	1185
5	$(EtO)_3SiCH_2$	Gr I	0.05	reflux	3	75	(71)	94:6	1185
5	$TMSCH_2$	Gr I	0.05	reflux	3	100	(95)	(*E*) only	1185
5	*n*-Bu	Gr I	0.05	reflux	3	100	(75)	90:10	1185
5	*n*-C_8H_{17}	Gr I	0.05	reflux	3	75	(60)	90:10	1185
5	Ph	Gr I	0.05	reflux	3	70	(67)	(*E*) only	1185
0.3	Ph	Gr I	0.05	reflux	3	100	(95)	(*E*) only	1185
5	4-ClC_6H_4	Gr I	0.05	reflux	3	72	(67)	(*E*) only	1185
0.3	4-ClC_6H_4	Gr I	0.05	reflux	3	100	(95)	(*E*) only	1185
5	4-MeC_6H_4	Gr I	0.05	reflux	1	100	(95)	(*E*) only	1185
5	4-$MeOC_6H_4$	Gr I	0.05	reflux	1	100	(95)	(*E*) only	1185
5	$MeCO_2CH_2$	Gr I	0.05	reflux	3	100	(90)	92:8	1186
5	$EtCO_2CH_2$	Gr I	0.05	reflux	3	99	(94)	93:7	1186
5	*n*-$PrCO_2CH_2$	Gr I	0.05	reflux	3	95	(93)	93:7	1186
10	*n*-$PrCO_2CH_2$	Gr I	0.05	reflux	3	75	(75)	89:11	1186
10	*n*-$PrCO_2CH_2$	Gr I	0.05	reflux	4	90	(90)	92:8	1186
5	*n*-$PrCO_2CH_2$	Gr II	0.05	reflux	1	91	(90)	80:20	1186
5	$PhCO_2CH_2$	Gr I	0.05	reflux	3	85	(80)	67:33	1186
5	*c*-$C_6H_{11}(CH_2)_2CO_2CH_2$	Gr I	0.05	reflux	3	95	(90)	(*E*) only	1186
5	—≡—CO_2CH_2	Gr I	0.05	reflux	3	92	(90)	75:25	1186

TABLE 31. CROSS-METATHESIS OF VINYL SILANES AND SILOXANES (*Continued*)
B. VINYL SILOXANES (*Continued*)

Vinyl Siloxane	Cross Partner	Conditions	Product(s) and Yield(s) (%)	Refs.

Please refer to the charts preceding the tables for the catalyst structures.

C_2

(MeO)$_3$Si-CH=CH$_2$ (*x* eq) + CH$_2$=CH-R; Gr I (*y* eq), DCM, reflux → (MeO)$_3$Si-CH=CH-R

x	R	*y*	Time (h)	Conv. (%)[a]	[b]	(*E*)/(*Z*)	Refs.
5	*n*-Bu	0.05	3	80	(60)	86:14	1185
1	Ph	0.02	6	95	(94)	(*E*) only	200
5	Cy(CH$_2$)$_2$CO$_2$CH$_2$	0.05	3	92	(88)	(*E*) only	1186

(Me$_3$SiO)$_3$Si-CH=CH$_2$ (*x* eq) + CH$_2$=CH-R; Catalyst (*y* eq), DCM, reflux → (Me$_3$SiO)$_3$Si-CH=CH-R

x	R	Catalyst	*y*	Time (h)	Conv. (%)[a]	[b]	(*E*)/(*Z*)	Refs.
5	*n*-Bu	Gr I	0.05	3	90	(72)	91:9	200
5	*n*-Bu	Gr II	0.05	3	100	(95)	95:5	1185
1	Ph	Gr I	0.02	6	97	(95)	(*E*) only	1185
5	Cy(CH$_2$)$_2$CO$_2$CH$_2$	Gr I	0.05	3	96	(90)	89:11	1186

R_3Si (vinyl)

6 eq

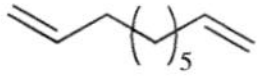

Gr I (0.05 eq),
DCM, reflux

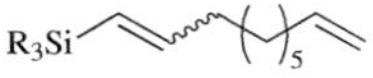

I

+ II

R	Time (h)	Conv. (%)[a]	**I**[b]	**II**[b]
MeO	0.17	84	(60)	(11)
MeO	2	100	(7)	(73)
EtO	0.17	88	(50)	(18)
EtO	2	100	(11)	(58)
Me_3SiO	0.17	83	(40)	(13)
Me_3SiO	2	100	(15)	(45)

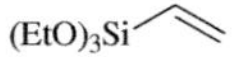

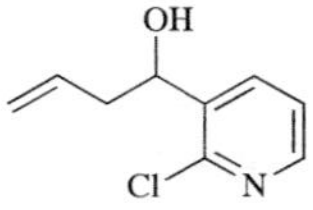

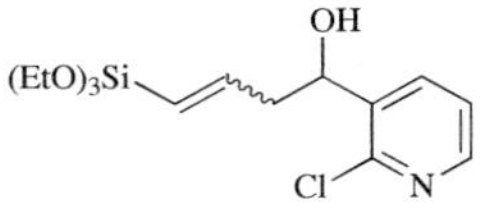

1187

$(EtO)_3Si$ (vinyl)

5 eq

NC, t-Bu, O, Si, Me, Me

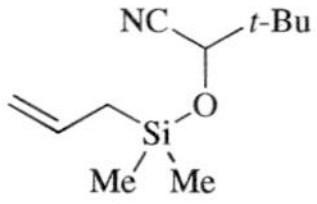

Gr II (0.1 eq),
DCM, reflux, 21 h

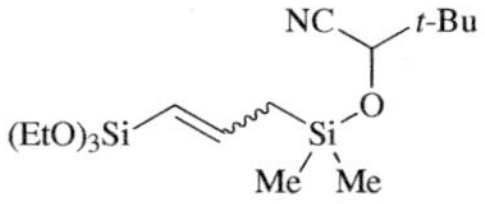

$(EtO)_3Si$, NC, t-Bu, O, Si, Me, Me

(34) 644

3 eq

OH, Cl, N

Hov II (0.1 eq),
DCM, 50°, 24 h

$(EtO)_3Si$, OH, Cl, N

(62) 603

TABLE 31. CROSS-METATHESIS OF VINYL SILANES AND SILOXANES (*Continued*)
B. VINYL SILOXANES (*Continued*)

Vinyl Siloxane	Cross Partner	Conditions	Product(s) and Yield(s) (%)	Refs.
Please refer to the charts preceding the tables for the catalyst structures.				
C_2				
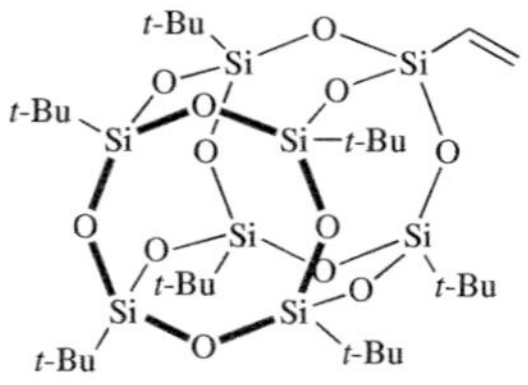	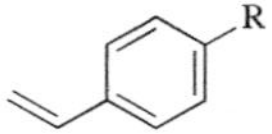	Gr I (0.01 eq), DCM, reflux, 24 h	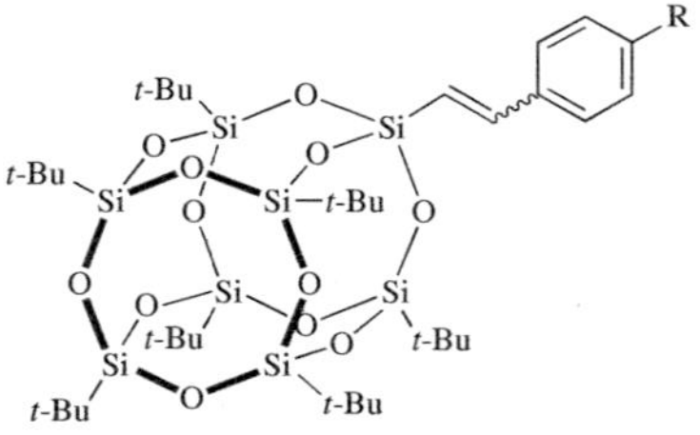	526

R	Conv. (%)[a]	[b]
H	>99	(90)
1-naphthyl	>99	(87)
9-anthracenyl	>99	(88)
2-thienyl	>99	(82)

C_4

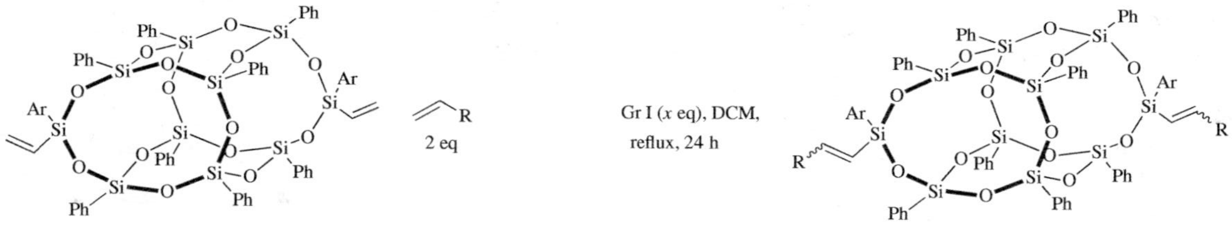

2 eq

Gr I (x eq), DCM, reflux, 24 h

1189

Ar	R	x	Conv. (%)[c]	
Ph	Ph	0.02	—	(94)
Ph	4-BrC_6H_4	0.02	—	(90)
Ph	4-ClC_6H_4	0.02	—	(89)
Ph	4-$MeOC_6H_4$	0.02	99	(—)
Ph	4-$CF_3C_6H_4$	0.02	—	(95)
Ph	4-MeC_6H_4	0.02	—	(92)
4-$MeOC_6H_4$	Ph	0.02	—	(88)
4-$MeOC_6H_4$	4-BrC_6H_4	0.02	—	(89)
4-$MeOC_6H_4$	4-ClC_6H_4	0.02	—	(93)
4-$MeOC_6H_4$	4-$MeOC_6H_4$	0.02	99	(—)
4-$MeOC_6H_4$	4-$CF_3C_6H_4$	0.02	99	(—)
4-$MeOC_6H_4$	4-MeC_6H_4	0.04	58	(—)
4-$MeOC_6H_4$	$TMSCH_2$	0.04	60	(—)
4-$MeOC_6H_4$	$HOCH_2$	0.04	55	(—)

TABLE 31. CROSS-METATHESIS OF VINYL SILANES AND SILOXANES (*Continued*)

B. VINYL SILOXANES (*Continued*)

Vinyl Siloxane	Cross Partner	Conditions	Product(s) and Yield(s) (%)	Refs.

Please refer to the charts preceding the tables for the catalyst structures.

C_4

Vinyl Siloxane	Cross Partner	Conditions	Product(s) and Yield(s) (%)	Refs.
Ph, Si–Ph, O, Si, Ph, Ph	R, 4 eq	Gr II (0.05 eq), DCM, reflux	R, Ph, Si–Ph, R, O, Si, Ph, Ph	203

R	Time (h)	^{b}	(*E*)/(*Z*)
4-ClC_6H_4	5	(4)	(*E*) only
n-C_8H_{17}	3	(12)	(*E*) only

Vinyl Siloxane	Cross Partner	Conditions	Product(s) and Yield(s) (%)	Refs.
CF_3, Si, CF_3, O, Si, CF_3, CF_3	R, *x* eq	Catalyst (*y* eq), DCM, reflux	**I**: CF_3, R, Si, R, O, Si, CF_3, CF_3, CF_3 + **II**: CF_3, R, Si, O, Si, CF_3, CF_3, CF_3	203

R	*x*	Catalyst	*y*	Time (h)	**I**^{b}	(*E*)/(*Z*) **I**	**II**^{b}	(*E*)/(*Z*) **II**
4-ClC_6H_4	10	Gr II	0.1	5	(73)	(*E*) only	(25)	(*E*) only
4-ClC_6H_4	10	Hov II	0.1	5	(83)	(*E*) only	(12)	(*E*) only
4-$MeOC_6H_4$	10	Gr II	0.1	5	(78)	(*E*) only	(20)	(*E*) only
n-C_8H_{17}	4	Gr I	0.05	3	(44)	96:4	(51)	96:4
n-C_8H_{17}	10	Gr I	0.05	3	(66)	96:4	(32)	96:4
n-C_8H_{17}	10	Gr II	0.05	3	(92)	96:4	(8)	96:4
n-C_8H_{17}	10	Hov II	0.05	3	(88)	96:4	(10)	96:4

C_6

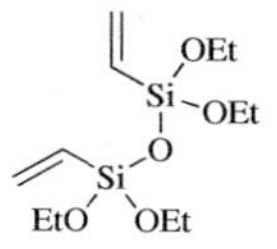

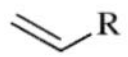

OEt, Si, OEt, O, Si, EtO, OEt

R, x eq

Catalyst (0.05 eq), DCM, reflux

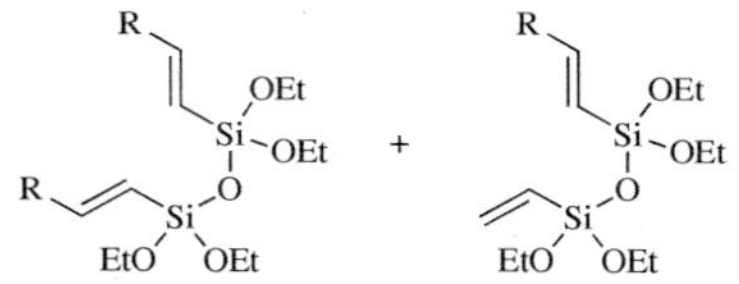

R, OEt, Si, OEt, R, Si, O, EtO, OEt + R, OEt, Si, OEt, Si, O, EtO, OEt

I (*E*) only **II** (*E*) only

203

R	x	Catalyst	Time (h)	**I**[b]	**II**[b]
4-ClC_6H_4	4	Gr I	24	(86)	(10)
4-ClC_6H_4	10	Gr II	3	(95)	(2)
4-ClC_6H_4	10	Hov II	5	(66)	(20)
n-C_8H_{17}	4	Gr I	5	(5)	(54)

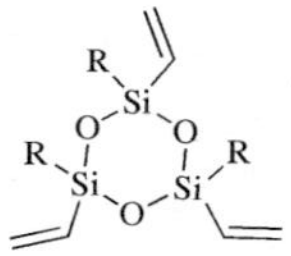

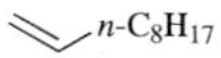

R, Si, O, O, R, R, Si, O, Si

n-C_8H_{17}, x eq

Catalyst (0.05 eq), DCM, 40°

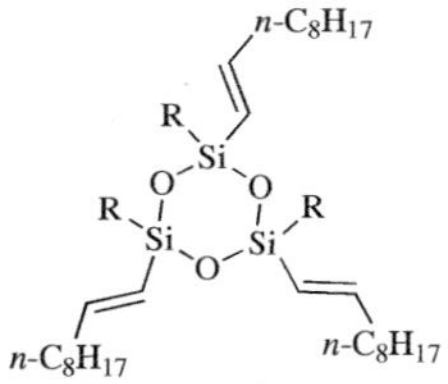

n-C_8H_{17}, R, Si, O, O, R, R, Si, O, Si, n-C_8H_{17}, n-C_8H_{17}

445

R	x	Catalyst	Time (h)		(*E*)/(*Z*)
TMSO	3	Gr I	3	(78)	89:11
TMSO	15	Gr II	2	(82)	89:11
Ph	3	Gr II	5	(90)	89:11
4-$MeOC_6H_4$	3	Gr II	5	(91)	91:9

TABLE 31. CROSS-METATHESIS OF VINYL SILANES AND SILOXANES (*Continued*)

B. VINYL SILOXANES (*Continued*)

Vinyl Siloxane	Cross Partner	Conditions	Product(s) and Yield(s) (%)	Refs.

Please refer to the charts preceding the tables for the catalyst structures.

C_{16}

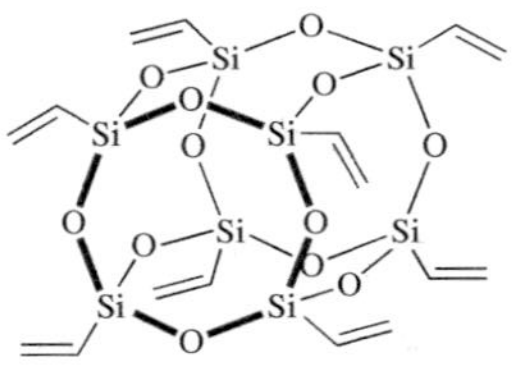

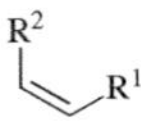

R^2 / R^1 (x eq); Catalyst (y eq), solvent

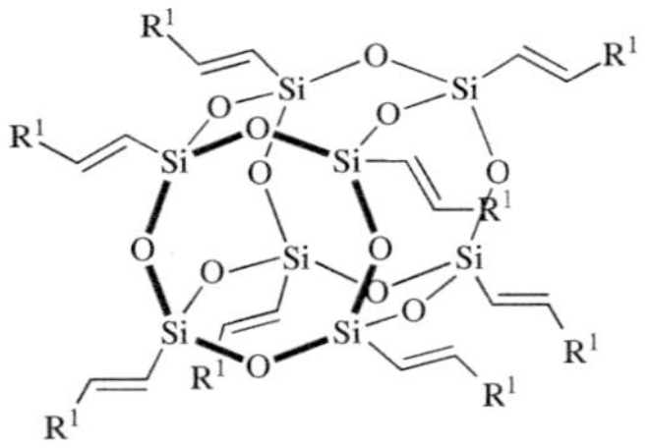

R^1	R^2	*x*	Catalyst	*y*	Solvent	Temp (°)	Time (h)	Conv. (%)[a]		(*E*)/(*Z*)	Refs.
Ph	H	1.5	Gr I	0.005	DCM	rt	72	—	(30-40)	—	1213
4-MeC_6H_4	H	1.5	Gr I	0.005	DCM	rt	72	—	(30-40)	—	1213
4-$MeOC_6H_4$	H	1.5	Gr I	0.005	DCM	rt	72	—	(30-40)	—	1213
4-ClC_6H_4	H	1.5	Gr I	0.005	DCM	rt	72	—	(30-40)	—	1213
4-BrC_6H_4	H	1.5	Gr I	0.005	DCM	rt	72	—	(30-40)	—	1213
3-$O_2NC_6H_4$	H	1.5	Gr I	0.005	DCM	rt	72	—	(30-40)	—	1213
n-Pr	*n*-Pr	3	Mo cat **1**	0.016	benzene	25	100	100	(75)	85:15	204
n-Pr	*n*-Pr	3	Mo cat **1**	0.05	C_6D_6	25	4.5	100	(100)	85:15	204
n-Pr	H	2	Mo cat **1**	0.025	—	25	12	100	(100)	85:15	204
Ph	H	5	Mo cat **1**	0.03	C_6D_6	25	12	100	(81)	(*E*) only	204
$TMSCH_2$	H	1.5	Mo cat **1**	0.05	C_6D_6	25	12	100	(100)	88:12	204
$EtO_2C(CH_2)_8$	H	10	Mo cat **1**	0.06	C_6D_6	25	22	100	(100)	80:20	204
$Me_3Si(CH_2)_6$	H	2	Mo cat **1**	0.06	C_6D_6	25	6	100	(100)	84:16	204
$Br(CH_2)_3$	H	12	Mo cat **1**	0.06	C_6D_6	25	6	100	(76)	94:6	204
$Br(CH_2)_3$	H	3	Gr I	0.06	CD_2Cl_2	38	24	25	—	80:20	204
$HO(CH_2)_3$	H	3	Gr I	0.06	CD_2Cl_2	38	48	10	—	80:20	204

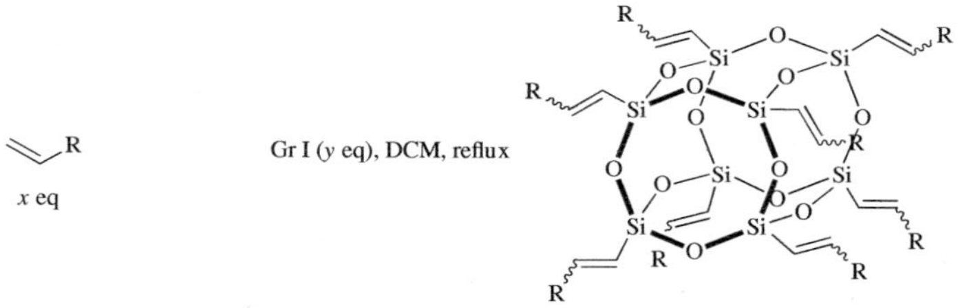

$\diagup\!\!\diagdown R$ (x eq) — Gr I (y eq), DCM, reflux

R	x	y	Time (h)	Conv. (%)[a]	[b]	
Ph	8	0.01	24	>99	(86)	205
4-ClC_6H_4	8	0.01	24	>99	(92)	205
4-BrC_6H_4	8	0.01	24	>99	(93)	205
4-$CF_3C_6H_4$	8	0.01	24	>99	(92)	205
4-(1-naphthyl)C_6H_4	8	0.01	24	>99	(94)	205
4-(9-anthracenyl)C_6H_4	8	0.01	24	>99	(89)	205
4-(2-thienyl)C_6H_4	8	0.01	24	>99	(92)	205
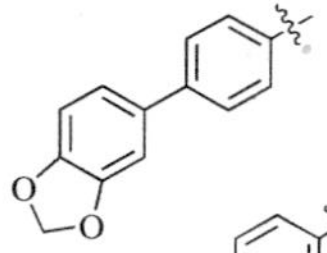 O O	17	0.035	96	—	(67)	206
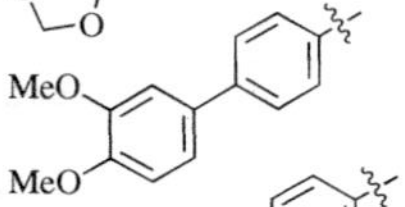 MeO MeO	17	0.035	96	—	(53)	206
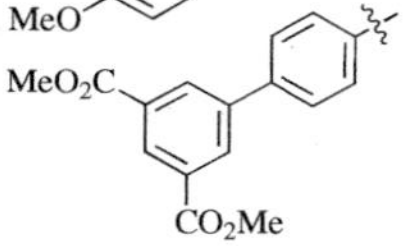 MeO_2C CO_2Me	17	0.035	96	—	(62)	206

TABLE 31. CROSS-METATHESIS OF VINYL SILANES AND SILOXANES (*Continued*)

B. VINYL SILOXANES (*Continued*)

Vinyl Siloxane	Cross Partner	Conditions	Product(s) and Yield(s) (%)	Refs.
Please refer to the charts preceding the tables for the catalyst structures.				
C_{16}				
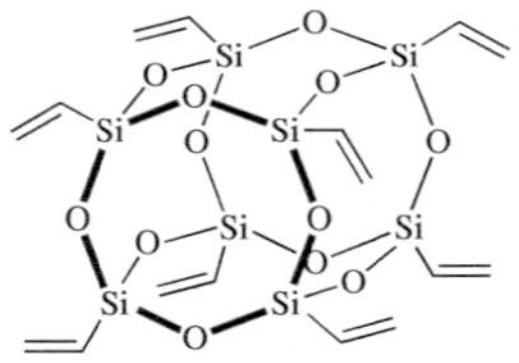	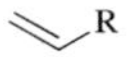 R; 12.5 eq	Gr I (0.45 eq), DCM, reflux, 90 h	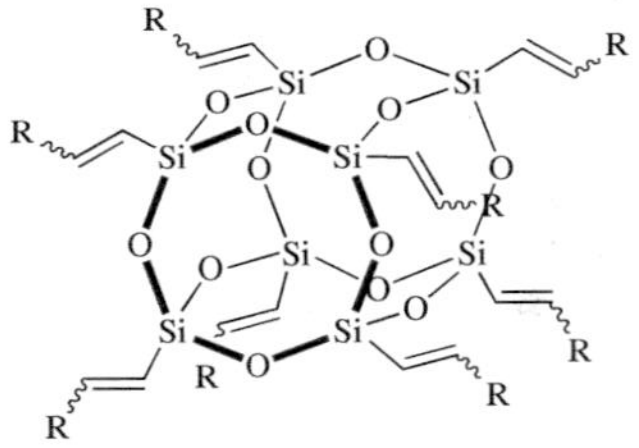 R	207
			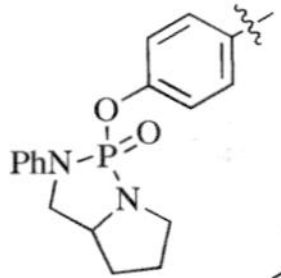 (72)	
			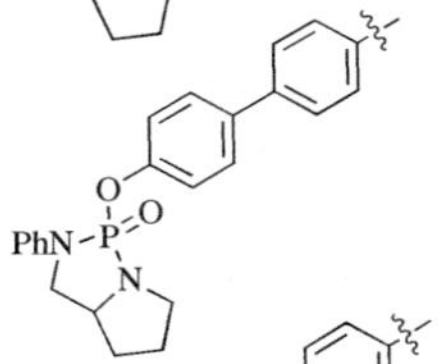 (—)	
			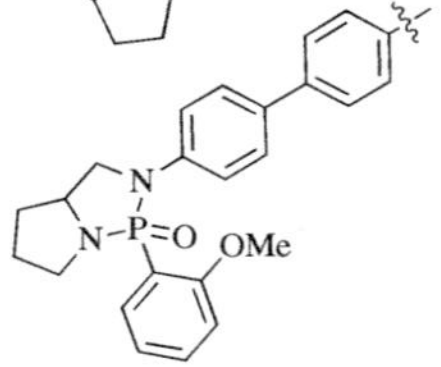 (—)	

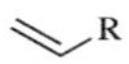 12.5 eq	Gr I (0.45 eq), DCM, reflux, 90 h	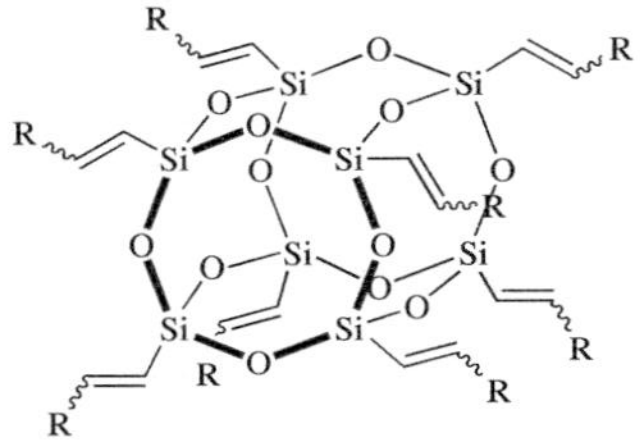		207
		R		
		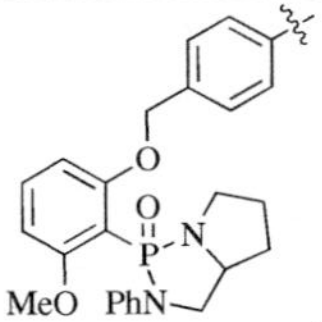	(—)	
		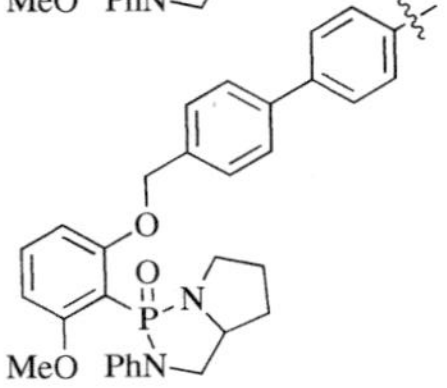	(—)	
		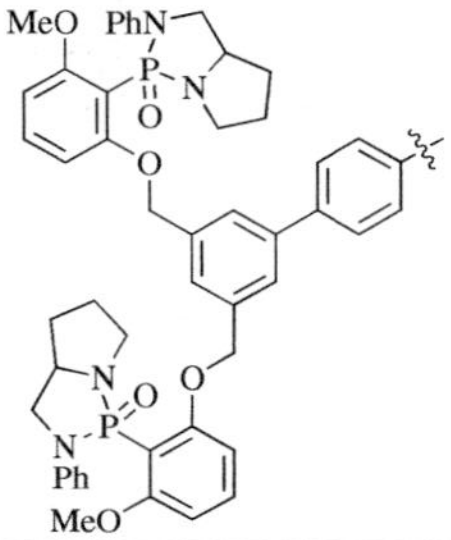	(—)	

[a] The conversion was determined by GC analysis.

[b] The yield was determined by GC analysis.

[c] The conversion was determined by ^{1}H NMR analysis.

TABLE 32. CROSS-METATHESIS OF VINYLIC ORGANOPHOSPHORUS COMPOUNDS

A. VINYL PHOSPHONATES

Vinylphosphonate	Cross Partner	Conditions	Product(s) and Yield(s) (%)	Refs.

Please refer to the charts preceding the tables for the catalyst structures.

C_2

Vinylphosphonate	Cross Partner	Conditions	Product(s) and Yield(s) (%)	Refs.
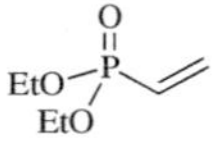 EtO–P(O)(OEt)– vinyl	 x eq	Gr II (0.05 eq), DCM, reflux	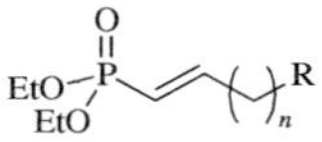	

R	n	x	Time (h)		$(E)/(Z)$	Refs.
BnTsN	1	—	20	(61)	—	1190
Br	2	1.5	12	(82)	>95:5	144
AcO	4	1.5	12	(95)	>95:5	144
OHC	8	1.5	12	(77)	>95:5	144

Vinylphosphonate	Cross Partner	Conditions	Product(s) and Yield(s) (%)	Refs.
6 eq	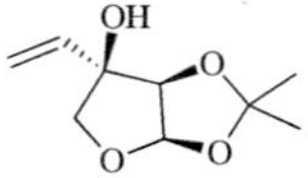	Gr II (0.05 eq), DCM, reflux, overnight	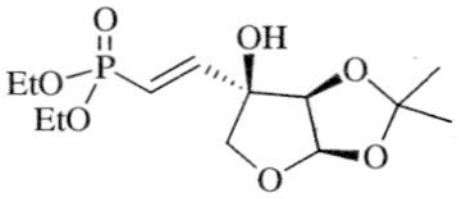 (100) (E) only	907
5 eq	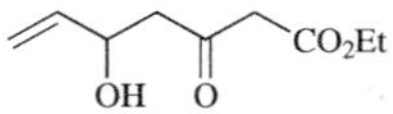 CO_2Et	Gr II (0.05 eq), DCM, 40°, 48 h	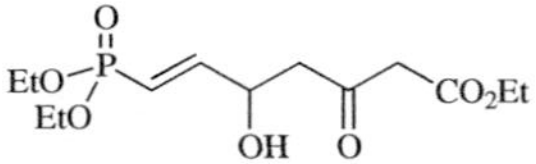 CO_2Et (47) (E) only	745
	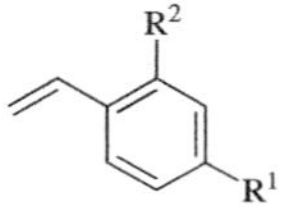 1.5 eq	Gr II (0.05 eq), DCM, 40°, 12 h	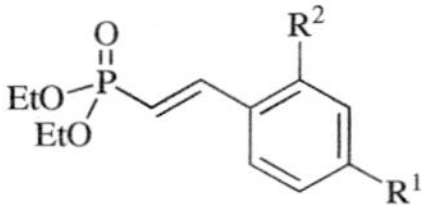 $(E)/(Z) > 95:5$	144

R^1	R^2	
H	H	(97)
MeO	H	(97)
Br	H	(93)
Me	Me	(77)

4 eq	H N O O N R	Gr II (0.05 eq), DCM, reflux, 22 h	O EtO P EtO H N O O N R (*E*) only	R: H (67); PhS (42); Cl (63); Br (37)	669
	Ph 1.5 eq	Gr II (0.05 eq), DCM, 40°, 12 h	O EtO P EtO Ph	(90) (*E*)/(*Z*) > 95:5	144
3 eq	TBSO	Gr II (0.02 eq), solvent, 22° (MW), 4 h	TBSO O EtO P EtO	Solvent: PTS (51); PTS/$KHSO_4$ (65); DCM (39)	931
2 eq	OBn Troc AcO N BnO AcO	Gr II (0.1 eq), DCM, reflux, 16–24 h	OBn Troc AcO N BnO AcO OEt P OEt O	(71) (*E*)/(*Z*) > 95:5	666

TABLE 32. CROSS-METATHESIS OF VINYLIC ORGANOPHOSPHORUS COMPOUNDS (*Continued*)

A. VINYL PHOSPHONATES (*Continued*)

Vinylphosphonate	Cross Partner	Conditions	Product(s) and Yield(s) (%)	Refs.
Please refer to the charts preceding the tables for the catalyst structures.				
C_2				
EtO, EtO, O, P; 2 eq	BnO_2C, N, OBn, OBn, BnO, OBn	Catalyst (0.1 eq), DCM, 40°	EtO, EtO, P, O, BnO_2C, N, OBn, OBn, BnO, OBn; **I** (*E*) only + BnO_2C, BnO, N, BnO, BnO, OBn, BnO, OBn, OBn, OBn, N, BnO_2C; **II**	1191, 1192

Catalyst	Time (h)	**I**	**II**
Ru cat **35**	24	(68)	(8)
Gr II	12	(—)	(—)
Ru cat **35**	12	(68)	(8)
Hov II	12	(75)	(10)

Reactant	Conditions	Product(s) and Yield(s) (%)	Refs.
4 eq	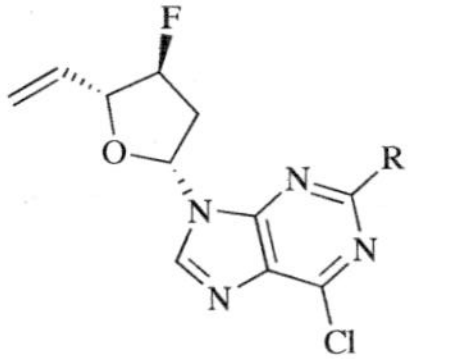 Gr II (0.04 eq), DCM, reflux, 26 h	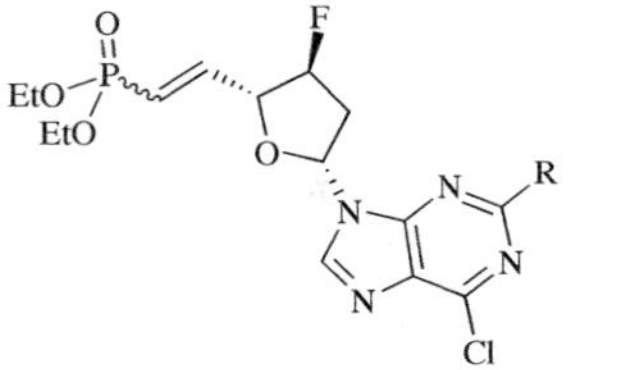 R: H (57); F (60)	1193
5 eq	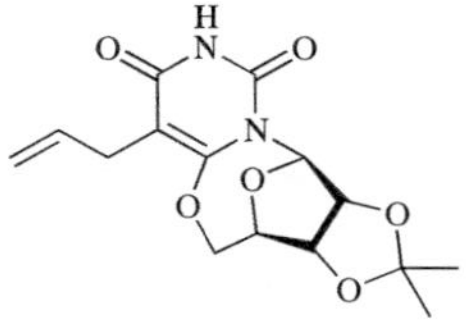 Ru cat **35** (0.06 eq), DCM, reflux, 7 h	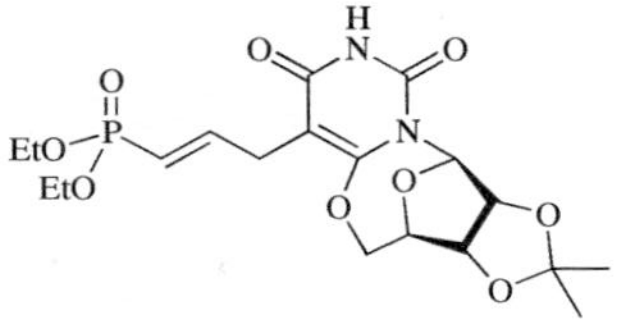 (60) (*E*) only	631
4 eq	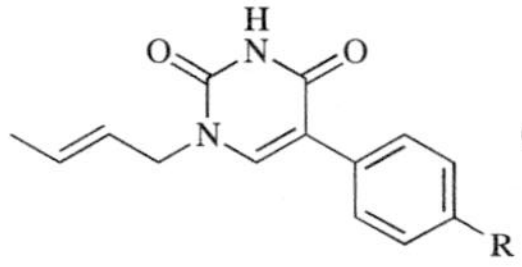 Gr II (0.05 eq), DCM, reflux, 22 h	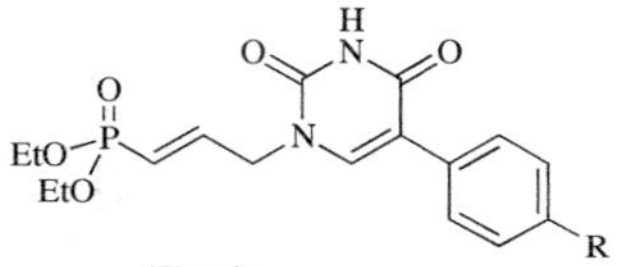 (*E*) only; R: H (67); F (71); MeO (52)	629

TABLE 32. CROSS-METATHESIS OF VINYLIC ORGANOPHOSPHORUS COMPOUNDS (*Continued*)

A. VINYL PHOSPHONATES (*Continued*)

Vinylphosphonate	Cross Partner	Conditions	Product(s) and Yield(s) (%)	Refs.
Please refer to the charts preceding the tables for the catalyst structures.				
C_2				
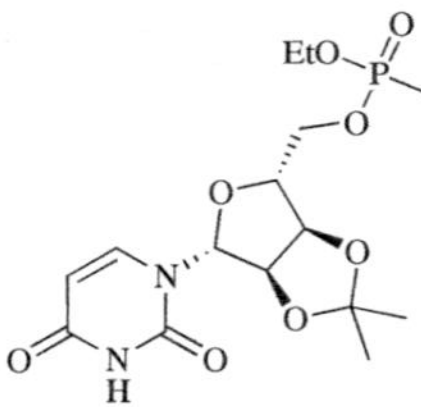 2 eq	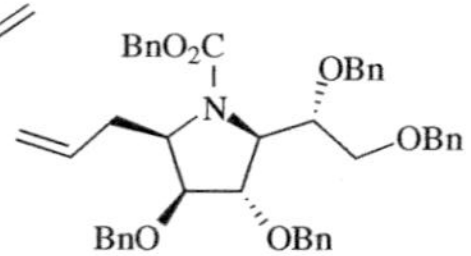	Catalyst (0.1 eq), DCM, 40°	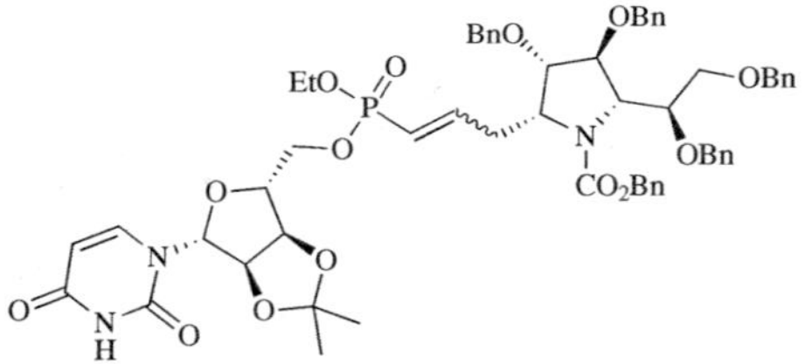 I 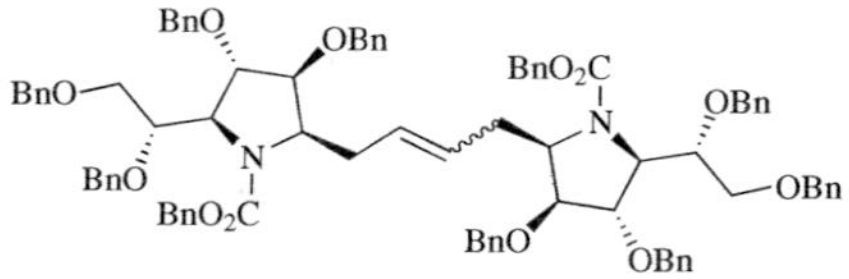 + II	1191, 1192

Catalyst	Time (h)	I	II
Ru cat **35**	44	(51)	(8)
Hov II	24	(61)	(—)

TABLE 32. CROSS-METATHESIS OF VINYLIC ORGANOPHOSPHORUS COMPOUNDS (*Continued*)

B. VINYL PHOSPHINE OXIDES

Vinylphosphine Oxide	Cross Partner	Conditions	Product(s) and Yield(s) (%)	Refs.
Please refer to the charts preceding the tables for the catalyst structures.				
C_6				
	CO_2Me 30 eq	Ru cat **88** (0.05 eq), DCM, reflux, 24 h	CO_2Me (54) (*E*) only	215
		Ru cat **88** (0.05 eq), DCM, reflux, 24 h	(62) (*E*) only	215
3 eq	R	Gr II (0.04 eq), DCM, reflux, 24 h	R: Ph (68); n-$C_{10}H_{21}$ (75)	217
C_9				
Me, Ph	CO_2Me 30 eq	Ru cat **88** (0.05 eq), DCM, reflux, 24 h	Me, Ph, CO_2Me (62) (*E*) only	215
	R, *n*, 2.5 eq	Catalyst (0.05 eq), DCM, reflux, 16 h	Ph, Me, R, *n* (*E*) only	

n	R	Catalyst		Refs.
2	Ac	Ru cat **114**	(85)	164
8	MeO_2C	Gr II	(74)	164
4	Br	Ru cat **88**	(86)	171

TABLE 32. CROSS-METATHESIS OF VINYLIC ORGANOPHOSPHORUS COMPOUNDS (*Continued*)
B. VINYL PHOSPHINE OXIDES (*Continued*)

Vinylphosphine Oxide	Cross Partner	Conditions	Product(s) and Yield(s) (%)	Refs.
Please refer to the charts preceding the tables for the catalyst structures.				
C_9				
Me'''P(=O)(Ph) vinyl	n-C_6F_{13} alkene, 2.5 eq	Ru cat **88** (0.05 eq), DCM, reflux, 16 h	(85) (*E*) only + n-C_6F_{13} product (0)	164
	Ph, Me, P=O vinyl	Catalyst (0.05 eq), DCM, reflux	(*E*) only; Catalyst / Time (h): Gr II 16 (18); Hov II 16 (44); Ru cat **88** 16 (95); Ru cat **88** 24 (80)	171, 215, 164
C_{10}				
Ph, O, P divinyl, 3 eq	alkene with R, n	Gr II (0.02 eq), DCM, reflux, 24 h	n / R: 1 TMS (72); 4 Br (86)	217
3 eq	4-vinylanisole (OMe)	Gr II (0.02 eq), DCM, reflux, 24 h	OMe product (56)	217

C_{12}

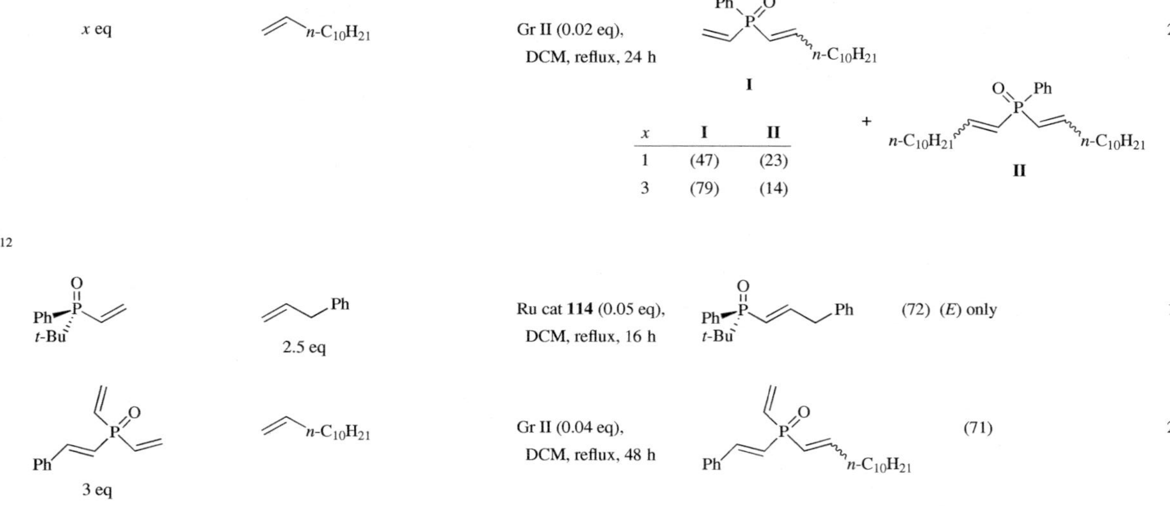

Reactants	Conditions	Product(s) and Yield(s) (%)	Refs.
x eq; n-$C_{10}H_{21}$ alkene	Gr II (0.02 eq), DCM, reflux, 24 h	I + II	217
2.5 eq	Ru cat **114** (0.05 eq), DCM, reflux, 16 h	(72) (*E*) only	164
3 eq; n-$C_{10}H_{21}$ alkene	Gr II (0.04 eq), DCM, reflux, 48 h	(71)	217

x	**I**	**II**
1	(47)	(23)
3	(79)	(14)

Vinylphosphine Oxide	Cross Partner	Conditions	Product(s) and Yield(s) (%)	Refs.

Please refer to the charts preceding the tables for the catalyst structures.

C_{14}

Ph–P(=O)(Ph)–CH=CH₂ | Cross partner: CH₂=CH–(CH₂)$_n$–R, *x* eq | Catalyst (*y* eq), DCM, 40° | Ph–P(=O)(Ph)–CH=CH–(CH₂)$_n$–R (*E*) only

n	R	*x*	Catalyst	*y*	Time (h)	Yield (%)	Refs.
0	Ac	3	Gr II	0.02–0.06	20–48	(0)	214
1	TMS	3	Gr II	0.02–0.06	20–48	(97)	214
3	Ph_2OP	—	Gr II	0.02–0.06	20–48	(93)	214
4	AcO	3	Gr II	0.02–0.06	20–48	(94)	214
4	TBSO	0.5	Gr II	0.05	16	(82)	216
4	TBSO	0.5	Ru cat **88**	0.05	16	(82)	171
4	TBSO	0.5	Ru cat **114**	0.05	16	(78)	164
4	TBSO	0.5	Ru cat **106**	0.05	16	(87)	164
4	Br	2.5	Ru cat **114**	0.05	16	(99)	164
4	Br	3	Gr II	0.02–0.06	20–48	(92)	164
9	Me	3	Gr II	0.02–0.06	20–48	(100)	164

Cross partner: R–CH₂–CH=CH–CH₂–R (*Z*), 2.5 eq | Catalyst (0.05 eq), DCM, reflux, 16 h | Ph–P(=O)(Ph)–CH=CH–CH₂–R (*E*) only

R	Catalyst	Yield (%)	Refs.
AcO	Ru cat **88**	(52)	164
Cl	Ru cat **114**	(65)	

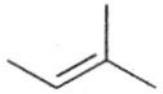

	excess	Ru cat **88** (0.05 eq), solvent, reflux, 16 h	Ph–P(O)(Ph)–CH=CH~ **I** + Ph–P(O)(Ph)–CH=C(CH$_3$)$_2$ **II**	164
			Solvent / I / II: neat (98) (2); DCM (70) (30)	
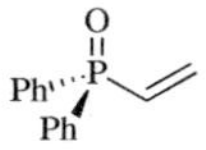 Ph, Ph, O, P	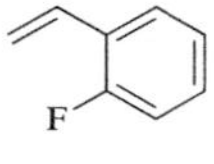 F; 3 eq	Ru cat **88** (0.05 eq), DCM, reflux, 24 h	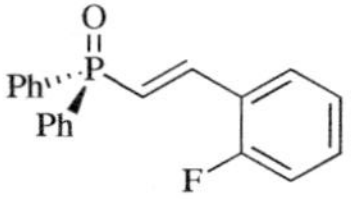 Ph, Ph, O, P, F; (32) (*E*) only	215
	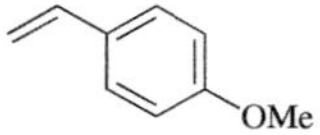 OMe; 3 eq	Catalyst (0.05 eq), DCM, reflux, 24 h	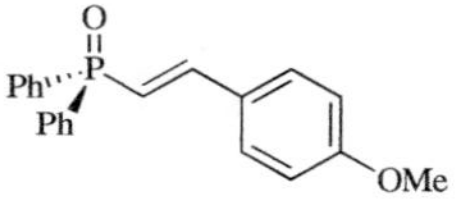 Ph, Ph, O, P, OMe; (*E*) only; Catalyst: Gr II (78), Ru cat **88** (81), Ru cat **106** (83)	214, 216, 215
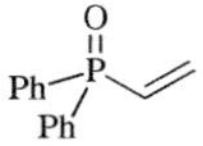 Ph, Ph, O, P; 2 eq	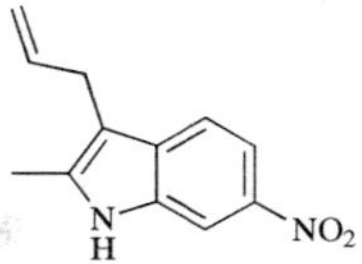 N, H, NO_2	Ru cat **88** (0.05 eq), DCM, reflux, 16 h	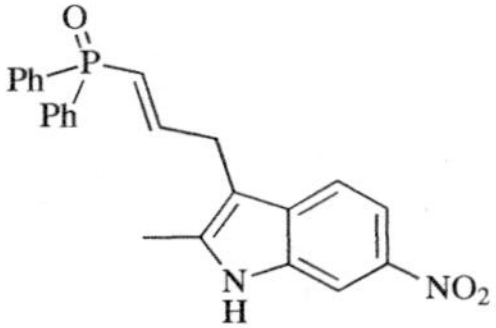 Ph, Ph, O, P, N, H, NO_2; (76) (*E*) only	171, 164

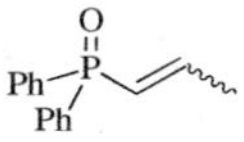

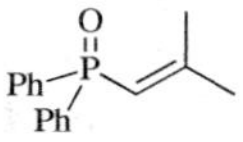

Solvent	I	II
neat	(98)	(2)
DCM	(70)	(30)

Catalyst	
Gr II	(78)
Ru cat **88**	(81)
Ru cat **106**	(83)

TABLE 32. CROSS-METATHESIS OF VINYLIC ORGANOPHOSPHORUS COMPOUNDS (*Continued*)

B. VINYL PHOSPHINE OXIDES (*Continued*)

Vinylphosphine Oxide	Cross Partner	Conditions	Product(s) and Yield(s) (%)	Refs.
Please refer to the charts preceding the tables for the catalyst structures.				
C_{14}				
OMe, O, P, Ph	F; 3 eq	Ru cat **88** (0.05 eq), DCM, reflux, 24 h	OMe, O, P, Ph, F (29) (*E*) only	215
O, R^1–P, R^2	R^2, P–R^1, O	Ru cat **88** (0.05 eq), DCM, reflux, 24 h	O, R^1–P, R^2, R^2, P–R^1, O; (*E*) only R^1 / R^2 n-C_6H_{13} / n-C_6H_{13} (65) Bn / Ph (40) Bn / Bn (46)	215
C_{15}				
O, Bn–P, Ph; 30 eq	CO_2Me	Ru cat **88** (0.05 eq), DCM, reflux, 24 h	O, Bn–P, Ph, CO_2Me (34) (*E*) only	215
C_{16}				
O, P, n-$C_{10}H_{21}$; 3 eq	Ph	Gr II (0.04 eq), DCM, reflux, 48 h	O, P, n-$C_{10}H_{21}$, Ph (87)	217

C_{18}				
R, R, O, P (aryl vinyl phosphine oxide); NO_2, *i*-PrO (styrene)	Ru cat **88** (0.05 eq), DCM, reflux, 24 h	R, R, O, P, NO_2, *i*-PrO (*E*) only	R: H (6); MeO (6)	215
C_{20}				
O, Ph, P, n-$C_{10}H_{21}$ 3 eq; Ph	Gr II (0.02 eq), DCM, reflux, 48 h	**I** (69) + **II** (—) + **III** (—) + **IV** (—) **I/II/III/IV** = 8:1:5:4		217
C_{22}				
Ph, O, P, n-$C_{10}H_{21}$ 3 eq; Ph	Gr II (0.04 eq), DCM, reflux, 24 h	Ph, O, P, n-$C_{10}H_{21}$, Ph	(65)	217

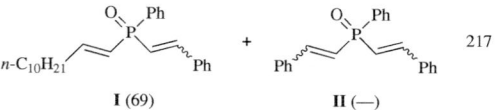

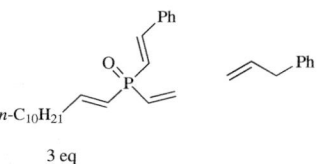

TABLE 32. CROSS-METATHESIS OF VINYLIC ORGANOPHOSPHORUS COMPOUNDS (*Continued*)
C. VINYL PHOSPHINE BORANES

Vinyl Phosphine Borane	Cross Partner	Conditions	Product(s) and Yield(s) (%)	Refs.
Please refer to the charts preceding the tables for the catalyst structures.				
C_8				
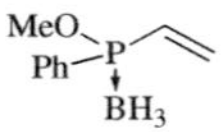 MeO, Ph, P, BH_3	 Br, 3; 3 eq	Gr II (0.06 eq), DCM, reflux, 36 h	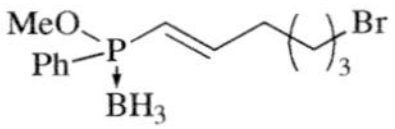 MeO, Ph, P, BH_3, Br, 3 (55) (*E*) only	213
	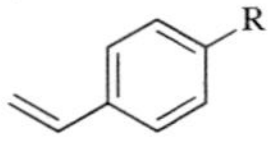 R; 3 eq	Gr II (*x* eq), DCM, reflux	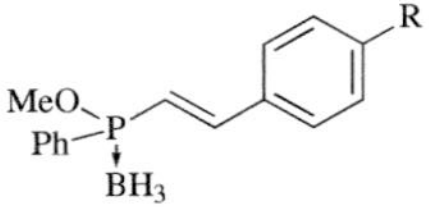 MeO, Ph, P, BH_3, R (*E*) only	213
			R: H, *x*: 0.08, Time (h): 48 (66)	
			R: MeO, *x*: 0.06, Time (h): 36 (46)	
C_{10}				
Ph, P, BH_3	n-C_9H_{19}; 3 eq	Gr II (0.08 eq), DCM, reflux, 48 h	n-C_9H_{19}, Ph, P, BH_3, n-C_9H_{19} (22) (*E*) only	213
			+ Ph, P, BH_3, n-C_9H_{19} (17) (*E*) only	
C_{14}				
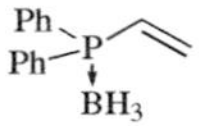 Ph, Ph, P, BH_3	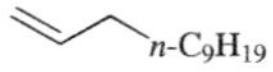 n-C_9H_{19}; 3 eq	Gr II (0.08 eq), DCM, reflux, 48 h	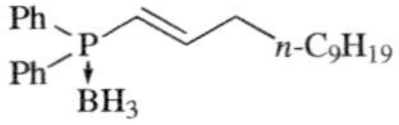 Ph, Ph, P, BH_3, n-C_9H_{19} (35) (*E*) only	213

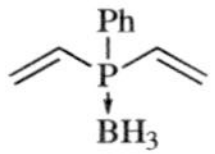

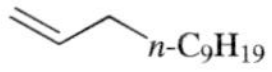

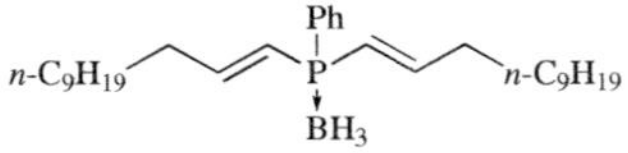

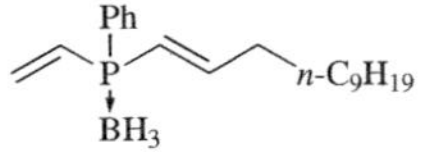

TABLE 33. CROSS-METATHESIS OF VINYLIC ORGANOSULFUR COMPOUNDS

A. VINYL THIOETHERS

Vinyl Thioether	Cross Partner	Conditions	Product(s) and Yield(s) (%)	Refs.
Please refer to the charts preceding the tables for the catalyst structures.				
C_{4-8}				
R^1S-CH=CH$_2$	R^3CH=CHR^2 (x eq)	Hov II (0.05 eq), solvent, 23°	R^1SCH=CHR^2	104
C_{8-9}				
RS-CH=CH$_2$	ClCH=CHCl (100 eq)	Catalyst (x eq), 50°	RSCH=CHCl (Z) only	105

R^1	R^2	R^3	x	Solvent	Time (h)		(Z)/(E)
Et	F	H	—	C_6D_6/CD_2Cl_2 10:1	24	(0)	—
Et	Cl	Cl	5.6	C_6D_6	24	(5)	(Z) only
Ph	F	H	—	C_6D_6/CD_2Cl_2 (10:1)	24	(0)	—
Ph	Cl	Cl	6	C_6D_6	24	(50)	88:12
Ph	Cl	Cl	3.8	C_6D_6	40	(70)	81:19

R	Catalyst	x	Time (h)	
Ph	Ru cat **115**	0.1	20	(85)
Bn	Ru cat **88**	0.05	24	(35)

TABLE 33. CROSS-METATHESIS OF VINYLIC ORGANOSULFUR COMPOUNDS (*Continued*)

B. VINYL SULFONES

Vinyl Sulphone	Cross Partner	Conditions	Product(s) and Yield(s) (%)	Refs.
Please refer to the charts preceding the tables for the catalyst structures.				
C_{3-8}				
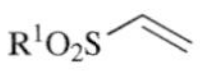 R^1O_2S 5 eq	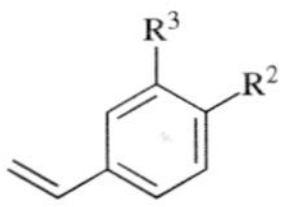	Ru cat **88** (5 x 0.024 eq), $C_6F_5CF_3$, 120° (MW), 50 min	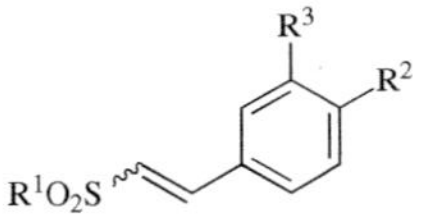 R^1O_2S	929
			R^1 / R^2 / R^3	
			Me / H / H (50)	
			Me / MeO / EtO (57)	
			CH_2=CH / MeO / EtO (45)	
			Et / $CF_3CH_2OCH_2$ / H (43)	
			Ph / MeO / EtO (80)	
C_3				
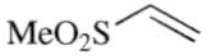 MeO_2S **I**	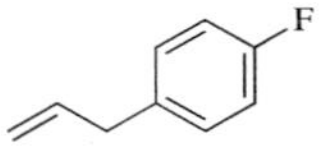	1. Ru cat **352** (0.02 eq), acetone-d_6, rt, 10 min 2. Ru cat **88** (5 x 0.024 eq), **I** (5 eq), $C_6F_5CF_3$, 120° (MW), 50 min	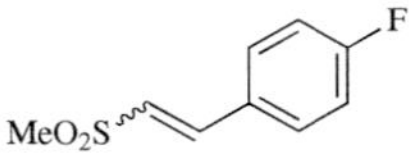 MeO_2S (35)	929

C_{3-8}

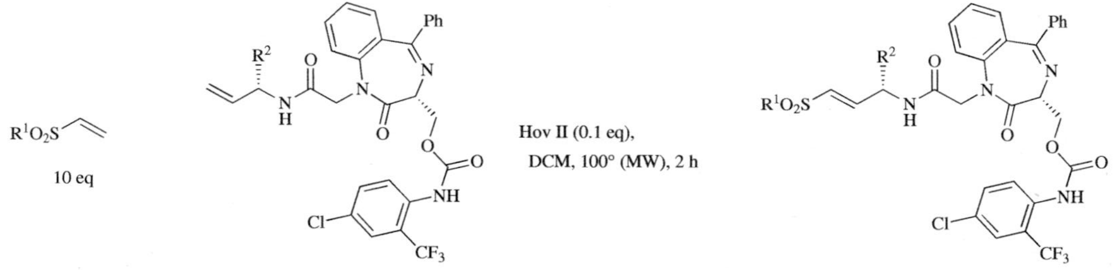

R^1O_2S (vinyl sulfone), 10 eq

Hov II (0.1 eq),
DCM, 100° (MW), 2 h

R^1O_2S … NH … Cl … CF_3

(*E*) only 238

R^1	R^2	
Me	H	(47)
Et	H	(53)
Ph	H	(49)
4-$MeOC_6H_4$	H	(51)
Me	$Ph(CH_2)_2$	(47)
Et	$Ph(CH_2)_2$	(49)
Ph	$Ph(CH_2)_2$	(48)
4-$MeOC_6H_4$	$Ph(CH_2)_2$	(46)

TABLE 33. CROSS-METATHESIS OF VINYLIC ORGANOSULFUR COMPOUNDS (*Continued*)

B. VINYL SULFONES (*Continued*)

Vinyl Sulphone	Cross Partner	Conditions	Product(s) and Yield(s) (%)	Refs.

Please refer to the charts preceding the tables for the catalyst structures.

C_4

3 eq

Catalyst (0.05 eq), DCM, reflux

I

+

II

R	*n*	Catalyst	Time (h)	I	II	Refs.
Ph	0	Gr II	24	(84)	(9)	228
TMS	0	Gr II	24	(66)	(31)	233
TBSO	3	Gr II	16	(69)	(13)	233
Br	7	Ru cat **88**	22	(72)	(19)	233
HO	8	Gr II	24	(71)	(12)	233

3 eq

Catalyst (0.05 eq), DCM, reflux

228, 233

R	*n*	Catalyst	Time (h)	
Br	3	Ru cat **88**	20	(36)
Me	3	Gr II	20	(42)
Me	3	Ru cat **88**	20	(51)
Me	15	Gr II	21	(59)
Me	15	Ru cat **88**	18	(70)

x eq n-C_9H_{19} y eq

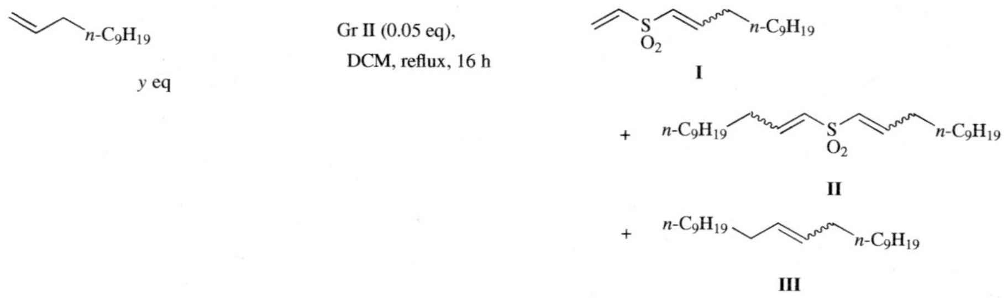

Gr **II** (0.05 eq), DCM, reflux, 16 h

n-C_9H_{19} SO_2 **I**

+ n-C_9H_{19} SO_2 n-C_9H_{19} **II**

+ n-C_9H_{19} n-C_9H_{19} **III**

233

x	y	**I/II/III**[a]
1	1	74:20:6
1	3	41:22:37
3	1	82:7:11

TABLE 33. CROSS-METATHESIS OF VINYLIC ORGANOSULFUR COMPOUNDS *(Continued)*

B. VINYL SULFONES *(Continued)*

Vinyl Sulphone	Cross Partner	Conditions	Product(s) and Yield(s) (%)	Refs.

Please refer to the charts preceding the tables for the catalyst structures.

C_8

Vinyl Sulphone	Cross Partner	Conditions	Product(s) and Yield(s) (%)
PhO_2S (vinyl)	$CH_2{=}CH(CH_2)_nR$ (drawn), x eq	Catalyst (y eq), DCM	PhO_2S–CH=CH–$(CH_2)_n$R (drawn)

R	n	x	Catalyst	y	Temp (°)	Time (h)		(E)/(Z)	Refs.
TMS	0	3	Gr II	0.05	reflux	3–24	(26)	—	171
TMS	0	3	Ru cat **88**	0.05	reflux	3–24	(33)	—	146
Ph	0	0.5	Gr II	0.05	25	16	(68)	—	216
Ph	0	0.5	Ru cat **88**	0.05	25	16	(86)	—	160
Ph	0	2.5	Gr II	0.025	reflux	12	(43)	(*E*) only	971
Ph	0	2.5	Hov II	0.025	reflux	12	(71)	(*E*) only	1173
Ph	0	2.5	Ru cat **86**	0.025	reflux	12	(62)	(*E*) only	232
HO	3	0.5	Gr II	0.1	reflux	3–24	(81)	—	228
TBSO	3	0.5	Gr II	0.05	45	3–24	(85)	—	228
TBSO	3	0.5	Ru cat **88**	0.025	25	16	(90)	—	228
TBSO	3	0.5	Ru cat **106**	0.025	20	16	(89)	—	228
TBSO	3	0.5	Ru cat **114**	0.05	40	16	(80)	—	228
O_2N	3	0.25	Gr II	0.045	reflux	18	(50)	(*E*) only	228
MeO_2C	7	—	Gr II	0.05	reflux	3–24	(76)	—	228
HO	8	0.5	Ru cat **88**	0.05	reflux	16	(96)	—	228

Vinyl Sulphone	Cross Partner	Conditions	Product(s) and Yield(s) (%)	Refs.
3 eq	1,5-hexadiene (drawn)	Ru cat **93** (3 x 0.025 eq), $CDCl_3$, 100° (MW), 2.5 h	PhO_2S–CH=CH–$(CH_2)_2$–CH=CH–SO_2Ph (drawn) (55)	926

x eq OTBS — Catalyst (y eq), solvent → PhO_2S OTBS (I) + TBSO OTBS (II)

x	Catalyst	y	Solvent	Temp (°)	Time (h)	**I**	**II**	
2	Gr II	0.1	toluene	50	3	(86)	(5)	232
2	Gr II	0.1	Et_2O	reflux	3	(80)	(20)	232
2	Gr II	0.1	DCM	reflux	3	(92)	(8)	232
3	Gr II	0.05	DCM	reflux	3	(82)	(0)	232
1	Gr II	0.05	DCM	reflux	3	(57)	(17)	232
0	Gr II	0.05	DCM	reflux	3	(0)	(>95)	232
2	Gr I	0.1	DCM	45	3	(0)	(82)	228
2	Mo cat **1**	0.1	DCM	45	3	(0)	(0)	228
2	Ru cat **35**	0.1	DCM	45	3	(98)	(2)	228
2	Hov II	0.1	DCM	45	3	(87)	(13)	228
2	Ru cat **87**	0.1	DCM	45	3	(80)	(16)	228
2	Ru cat **88**	0.1	DCM	45	3	(95)	(5)	228
2	Ru cat **167**	0.05	toluene	80	0.5	(78)	(—)	1060

TABLE 33. CROSS-METATHESIS OF VINYLIC ORGANOSULFUR COMPOUNDS (*Continued*)

B. VINYL SULFONES (*Continued*)

Vinyl Sulphone	Cross Partner	Conditions	Product(s) and Yield(s) (%)	Refs.
Please refer to the charts preceding the tables for the catalyst structures.				
C_8				
PhO_2S 2 eq	CO_2Et, CO_2Et	Gr II (0.05 eq), DCM, reflux, 3–24 h	PhO_2S, CO_2Et, CO_2Et (74)	232, 228
3 eq	OBn, Troc, N, AcO, BnO, AcO	Gr II (0.1 eq), DCM, reflux, 16–24 h	OBn, Troc, N, AcO, BnO, AcO, SO_2Ph (50) $(E)/(Z) > 95{:}5$	666
2 eq	Ph, OR	Catalyst (x eq), DCM, reflux, 3–24 h	PhO_2S, Ph, OR	232, 228
6 eq	O	Ru cat **190** (0.025 eq), DCM, 50°, 96 h	PhO_2S, O, SO_2Ph (73)	231

R	Catalyst	x	
H	Ru cat **87**	0.05	(71)
TBS	Gr II	0.1	(55)
TBS	Ru cat **35**	0.05	(53)
TBS	Ru cat **87**	0.05	(30)
TBS	Ru cat **88**	0.05	(48)

2 eq

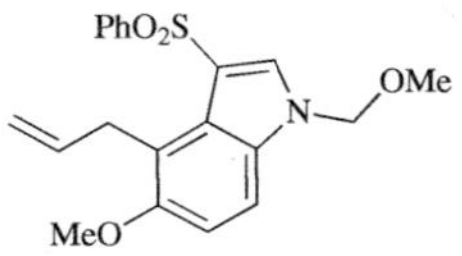

PhO$_2$S, OMe, N, MeO

Catalyst (x eq), DCM, reflux, 3–24 h

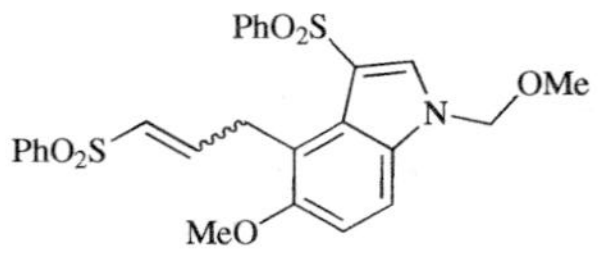

PhO$_2$S, PhO$_2$S, OMe, N, MeO

232, 228, 171

Catalyst	x	
Gr II	0.1	(33)
Ru cat **35**	0.05	(44)
Ru cat **35**	0.1	(54)
Ru cat **87**	0.05	(41)
Ru cat **88**	0.05	(53)
Ru cat **122**	0.05	(54)

x eq

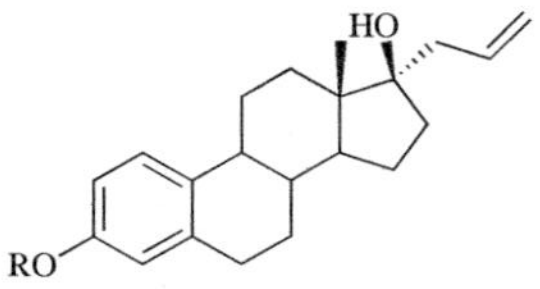

HO, RO

Gr II (0.05 eq), solvent, 6 h

HO, SO$_2$Ph, RO

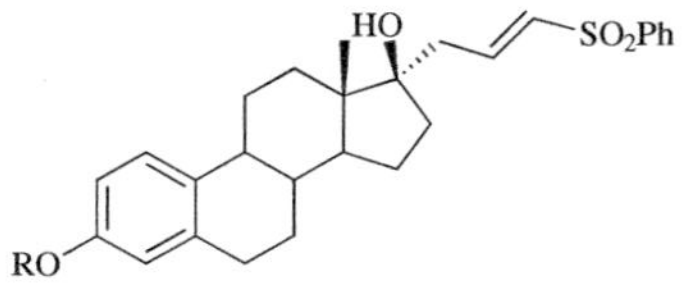

x	R	Solvent	Temp (°)		$(E)/(Z)$	
—	H	DCE/$C_6F_5CF_3$ (1:4)	70	(55)	(E) only	229
3	H	DCM	40	(29)	—	140
3	n-$C_8F_{17}(CH_2)_2(i\text{-}Pr)_2Si$	$C_6F_5CF_3$	70	(69)	—	140

TABLE 33. CROSS-METATHESIS OF VINYLIC ORGANOSULFUR COMPOUNDS (*Continued*)

B. VINYL SULFONES (*Continued*)

Vinyl Sulphone	Cross Partner	Conditions	Product(s) and Yield(s) (%)	Refs.

Please refer to the charts preceding the tables for the catalyst structures.

C_{8-13}

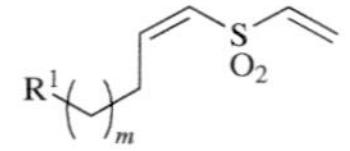

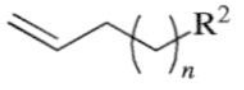

Cross Partner: 4 eq

Conditions: Gr II (x eq), DCM, reflux

Refs.: 233

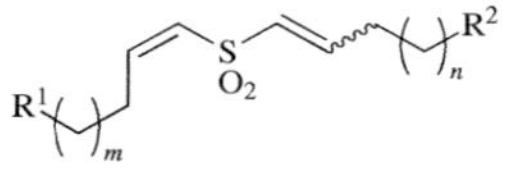

R^1	m	R^2	n	x	Time (h)	
TBSO	3	Me	3	0.1	23	(85)
HO	8	Br	7	0.08	16	(74)

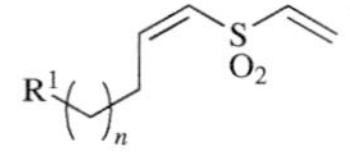

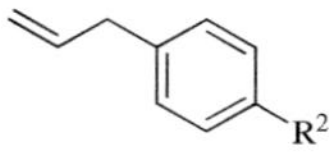

C_{11-21}

Cross Partner: 4 eq

Conditions: Catalyst (x eq), DCM, reflux

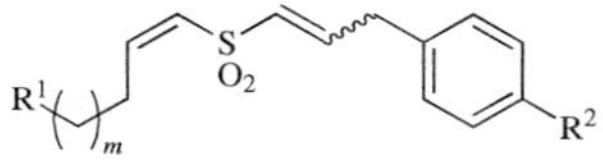

Refs.: 233

R^1	n	R^2	Catalyst	x	Time (h)	
Ph	0	H	Ru cat **88**	0.1	16	(68)
Ph	0	Br	Ru cat **88**	0.08	44	(74)
Me	15	H	Gr II	0.2	24	(54)

[a] The ratio of products was determined by GC analysis.

TABLE 33. CROSS-METATHESIS OF VINYLIC ORGANOSULFUR COMPOUNDS (*Continued*)

C. VINYL SULFONAMIDES

Vinyl Sulfonamide	Cross Partner	Conditions	Product(s) and Yield(s) (%)		Refs.
Please refer to the charts preceding the tables for the catalyst structures.					
C_2					
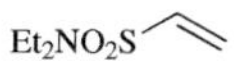 Et_2NO_2S 5 eq	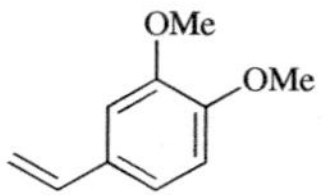 OMe OMe	Ru cat **88** (5 x 0.024 eq), $C_6F_5CF_3$, 120° (MW), 50 min	Et_2NO_2S OMe OMe	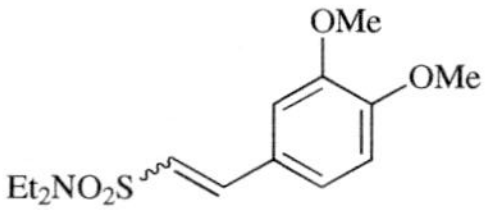 (89)	929
5 eq	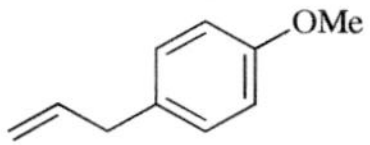 OMe	1. Ru cat **352** (0.02 eq), acetone-d_6, D_2O, rt, 10 min 2. Ru cat **88** (5 × 0.02 eq), **I** (5 eq), $C_6F_5CF_3$, 120° (MW), 50 min	Et_2NO_2S D OMe	(41)	929

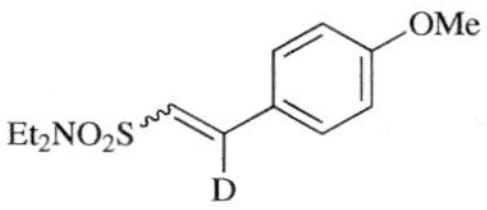

TABLE 34. ASYMMETRIC CROSS-METATHESIS

Prochiral Diene	Cross Partner	Conditions	Product(s) and Yield(s) (%)	Refs.
Please refer to the charts preceding the tables for the catalyst structures.				
C_5				
OTBS	AcO, OAc 5 eq	Catalyst (0.05 eq), DCM, 40°	OTBS, OAc	97
			Catalyst / er Ru cat **16** (15) 61.0:39.0 Ru cat **18** (20) 72.0:28.0 Ru cat **20** (28) 72.0:28.0 Ru cat **22** (20) 67.0:33.0 Ru cat **24** (12) 68.5:31.5	
	AcO, OAc 5 eq	Ru cat **215** (0.05 eq), THF, 35°, 19 h	OTBS, OAc (35) er 75.0:25.0	98
OTIPS 5 eq	AcO, OAc	Ru cat **20** (0.05 eq), 40°, 6 h	OTIPS, OAc (52) er 76.0:24.0	97

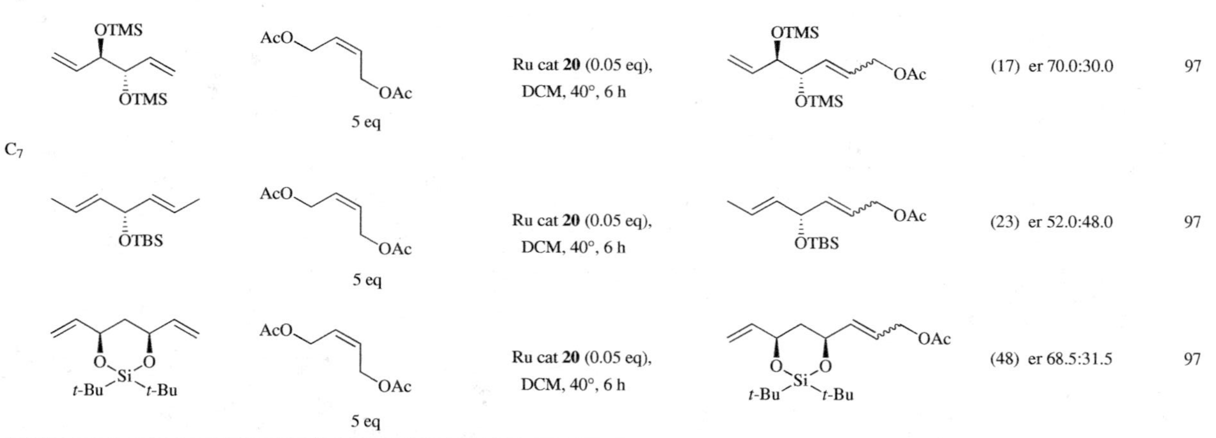

	Reactant	Partner	Conditions	Product	Yield (%) / er	Ref.
C_6	OTMS, OTMS	AcO, OAc, 5 eq	Ru cat **20** (0.05 eq), DCM, 40°, 6 h	OTMS, OTMS, OAc	(17) er 70.0:30.0	97
C_7	OTBS	AcO, OAc, 5 eq	Ru cat **20** (0.05 eq), DCM, 40°, 6 h	OTBS, OAc	(23) er 52.0:48.0	97
	O, Si, O, *t*-Bu, *t*-Bu	AcO, OAc, 5 eq	Ru cat **20** (0.05 eq), DCM, 40°, 6 h	O, Si, O, *t*-Bu, *t*-Bu, OAc	(48) er 68.5:31.5	97

TABLE 35. RING-OPENING CROSS-METATHESIS

Cyclic Alkene	Cross Partner	Conditions	Product(s) and Yield(s) (%)	Refs.
Please refer to the charts preceding the tables for the catalyst structures.				
C_5				
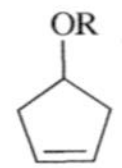 OR	 CO_2Et 6 eq	Hov II (2 x 0.025 eq), DCM, rt	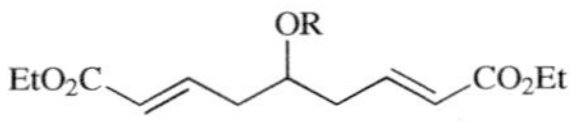 EtO_2C … OR … CO_2Et (*E*) only R / Time (h) H / 72 / (20) TBS / 48 / (66)	1194
 HO…OH	 CO_2Et 6 eq	Hov II (2 x 0.025 eq), DCM, rt, 89 h	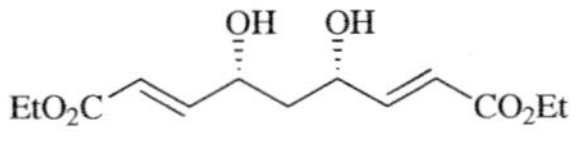 EtO_2C … OH OH … CO_2Et (52) (*E*) only	1194
C_{5-6}				
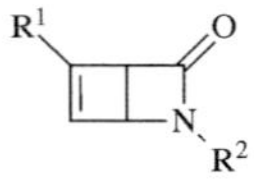 R^1, O, N, R^2	excess	Gr I (0.1 eq), DCM, rt	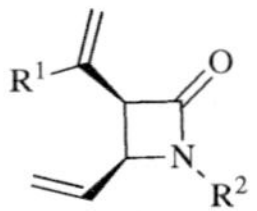 R^1, O, N, R^2 R^1 / R^2 H / H / (50) H / TMS / (0) H / TBDPS / (60) H / TBS / (89) Me / H / (28) Me / TMS / (0) Me / TBDPS / (53) Me / TBS / (90)	119

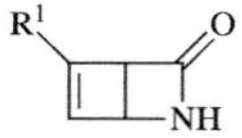

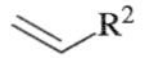

R^1 NH O | R^2 5 eq | Gr I (0.1 eq), DCM, rt | I + II | 119

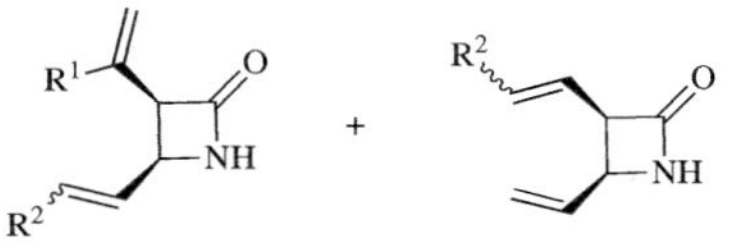

R^1	R^2	I	II
H	*n*-Bu	(9)	(3)
H	Ph	(7)	(5)
Me	*n*-Bu	(10)	(—)
Me	Ph	(15)	(—)

C_6

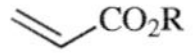

CO_2R *x* eq | Gr II (*y* eq), solvent | RO_2C ... CO_2R (*E*) only

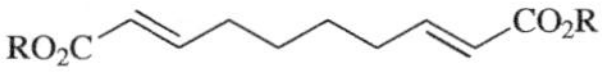

R	*x*	*y*	Solvent	Temp	Time (h)		
H	0.33	0.05	DCM	reflux	3	(94)	170
t-Bu	2	0.05	DCM	rt	3	(88)	170
t-Bu	0.25	0.005	2.5 % aq. PTS	rt	15	(80)	976

TABLE 35. RING-OPENING CROSS-METATHESIS (*Continued*)

Cyclic Alkene	Cross Partner	Conditions	Product(s) and Yield(s) (%)	Refs.

Please refer to the charts preceding the tables for the catalyst structures.

C_6

Cyclic Alkene	Cross Partner	Conditions	Product(s) and Yield(s) (%)	Refs.
	R^1, R^2; *x* eq	**Gr II** (*y* eq), solvent	R^2, O, O, R^2 (*E*) only	

R^1	R^2	*x*	*y*	Solvent	Temp	Time (h)		Refs.
H	Me	2.5	0.02	2 % aq. PTS	rt	15	(68)	976
Me	Et	0.33	0.05	DCM	reflux	3	(72)	170

Cyclic Alkene	Cross Partner	Conditions	Product(s) and Yield(s) (%)	Refs.
O, O	CO_2Et; 6 eq	**Hov II** (2 x 0.025 eq), DCM, rt, 72 h	EtO_2C, O, O, CO_2Et (50) (*E*) only	1194

C_7

Cyclic Alkene	Cross Partner	Conditions	Product(s) and Yield(s) (%)	Refs.
	B, O, O; 10 eq	Ru cat **151** (0.05 eq), DCM, 22°, 2 h	B, O, O (64) (*Z*)/(*E*) = 90:10	83
	R; 10 eq	Ru cat **151** (0.05 eq), DCM, 22°, 2 h	R	83

R		(*Z*)/(*E*)
HO	(68)	88:12
$HOCH_2$	(84)	87:13
n-$BuOCH_2$	(66)	87:13

	Substrate	Olefin	Conditions	Product(s) and Yield(s) (%)		Refs.
C_{7-8}	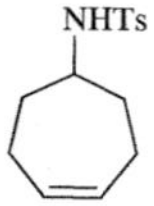 NHTs	CO_2Et 6 eq	Hov II (2 x 0.025 eq), DCM, rt, 48 h	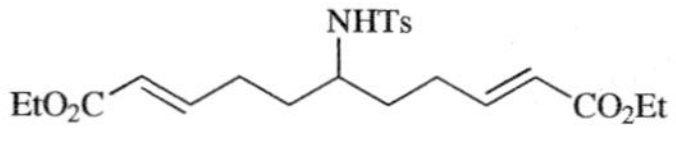 NHTs EtO_2C CO_2Et	(79) (*E*) only	1194
	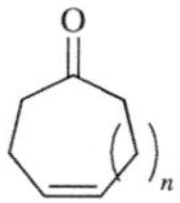 O ( )$_n$	CO_2R 6 eq	Hov II (2 x 0.025 eq), DCM, rt, 48 h	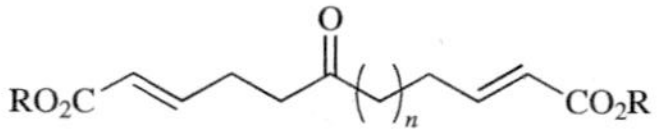 O RO_2C ( )$_n$ CO_2R	(*E*) only	1194

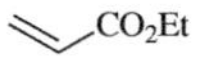

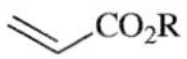

R	*n*	
Et	1	(63)
Et	2	(77)
Bn	2	(72)
t-Bu	2	(60)

	Substrate	Olefin	Conditions	Product(s)		Refs.
C_7	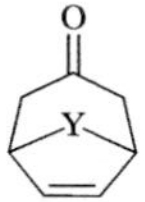 O Y		Gr I (0.01 eq), DCM, rt, 26 h	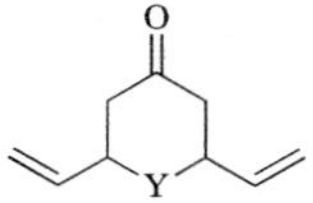 O Y		345

Y	
O	(43)
CH_2	(50)

TABLE 35. RING-OPENING CROSS-METATHESIS (*Continued*)

Cyclic Alkene	Cross Partner	Conditions	Product(s) and Yield(s) (%)	Refs.

Please refer to the charts preceding the tables for the catalyst structures.

C_7

Cyclic Alkene	Cross Partner	Conditions	Product(s) and Yield(s) (%)	Refs.
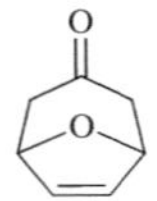	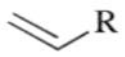 R, bubbled	Gr I (0.01 eq), DCM, rt	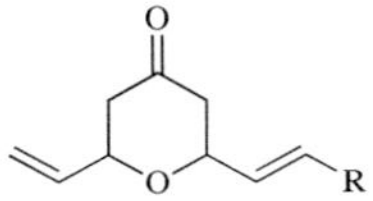	345

R	Time (h)		(*E*)/(*Z*)
Me	4	(40)	78:22
Et	1	(42)	89:11

$C_{7\text{-}10}$

Cyclic Alkene	Cross Partner	Conditions	Product(s) and Yield(s) (%)	Refs.
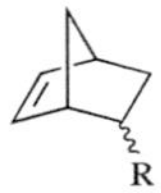	*x* bar	Catalyst (*y* eq), toluene, rt	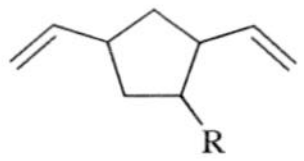	

R	*x*	Catalyst	*y*	Time (h)		Refs.
$(EtO)_3Si$	4	Ru cat **73**	0.00025	24	(99)	346
Me_3SiOMe_2Si	4	Ru cat **73**	0.001	24	(99)	346
$(EtO)_3Si$	3	Gr I	0.00025	24	(97)	347
$CH_2{=}CHCH_2$	2	Gr I	0.0014	24	(88)	347
Ac	2	Gr I	0.001	48	(92)	347

C_7

Cyclic Alkene	Cross Partner	Conditions	Product(s) and Yield(s) (%)	Refs.
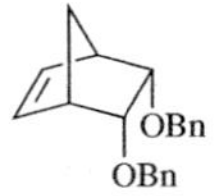 OBn, OBn	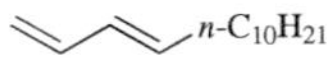 n-$C_{10}H_{21}$, 1.5 eq	Ru cat **216** (0.02 eq), 50°, 24 h	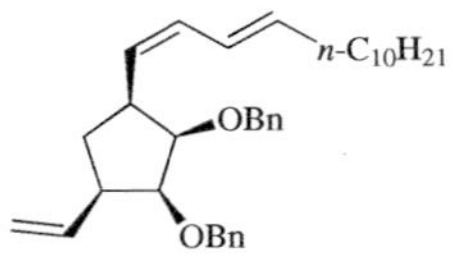 n-$C_{10}H_{21}$, OBn, OBn (>95) (*Z*)/(*E*) > 95:5	538

Boc
N
MeO_2C

SPh
4 eq

Gr II (0.08 eq),
$CHCl_3$, 55°, 16 h

PhS
MeO_2C N Boc

(89) 710

O
R
2 eq

Hov II (0.04 + 0.04 eq),
toluene, 80°, 24 h

MeO_2C N Boc R O
(*E*) only

R		
H	(76)	1195
Me	(87)	
Et	(89)	

CO_2R
x eq

Catalyst (*y* eq), solvent

MeO_2C N Boc CO_2R

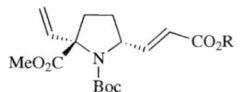

R	*x*	Catalyst	*y*	Solvent	Temp (°)	Time (h)		(*E*)/(*Z*)	
H	10	Hov II	0.04 + 0.04	toluene	80	48	(23)	100:0	710
Me	4	Hov II	0.08	toluene	80	—	(98)	—	1195
Me	10	Gr II	0.08	DCM	rt	144	(15)	100:0	1195
Me	10	Gr II	0.08	$CHCl_3$	55	48	(50)	100:0	1195
Me	10	Gr II	0.04	toluene	80	24	(76)	100:0	1195
Me	10	Hov II	0.04	toluene	80	4	(92)	100:0	1195
Et	10	Hov II	0.04	toluene	80	4	(93)	100:0	1195
i-Bu	10	Hov II	0.04	toluene	80	4	(89)	100:0	1195
t-Bu	10	Hov II	0.04	toluene	80	4	(90)	100:0	1195

TABLE 35. RING-OPENING CROSS-METATHESIS (*Continued*)

Cyclic Alkene	Cross Partner	Conditions	Product(s) and Yield(s) (%)	Refs.

Please refer to the charts preceding the tables for the catalyst structures.

C_7

Cyclic Alkene	Cross Partner	Conditions	Product(s) and Yield(s) (%)	Refs.
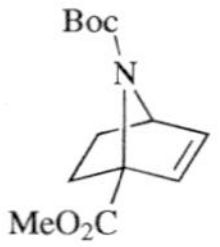 Boc, N, MeO_2C	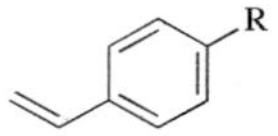 R; 4 eq	Gr II (0.08 eq), $CHCl_3$, 55°	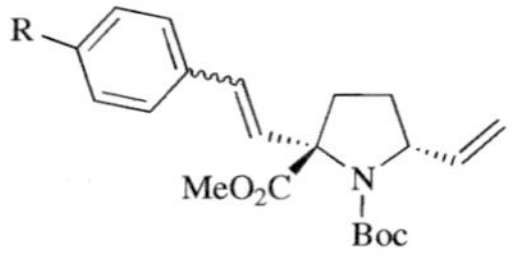 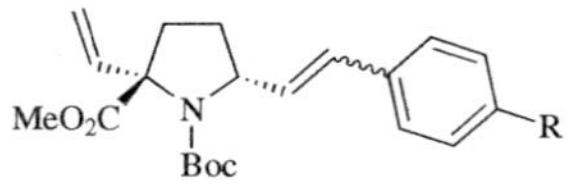 R, MeO_2C, N, Boc (I) + MeO_2C, N, Boc, R (II)	710

R	I + II	I/II
H	(82)	64:36
F	(81)	33:67
MeO	(79)	78:22
MeO_2C	(71)	52:48

C_{7-8}

Cyclic Alkene	Cross Partner	Conditions	Product(s) and Yield(s) (%)	Refs.
Y, CO_2Et, NHBz		Catalyst (0.05 eq), DCM, 20°, 2 h	EtO_2C, NHBz, Y	348

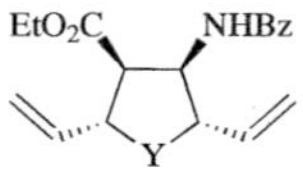

Y	Catalyst	
O	Hov I	(79)
CH_2	Gr I	(37)
CH_2	Gr II	(33)
CH_2	Hov I	(38)
CH_2	Hov II	(41)

Y	Catalyst		Refs.
O	Hov I	(68)	348
CH_2	Gr I	(46)	
CH_2	Gr II	(20)	
CH_2	Hov I	(80)	
CH_2	Hov II	(28)	

Catalyst (0.05 eq), DCM, 20°, 2 h

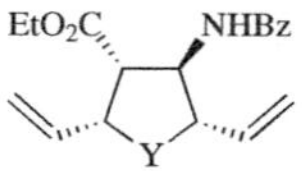

C_7

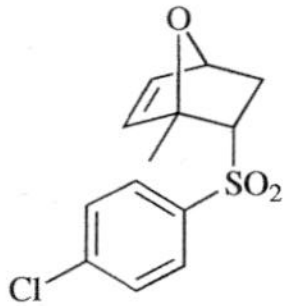

50 eq

Hov II (0.05 eq), DCE, 70°, 12 h

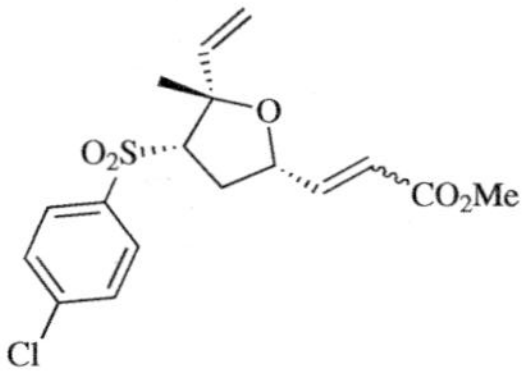

(67) 909

C_{7-8}

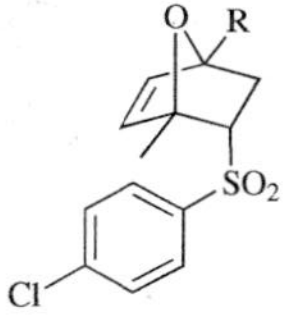

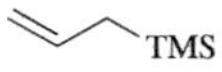

x eq

Hov II (*y* eq), DCE, 12 h

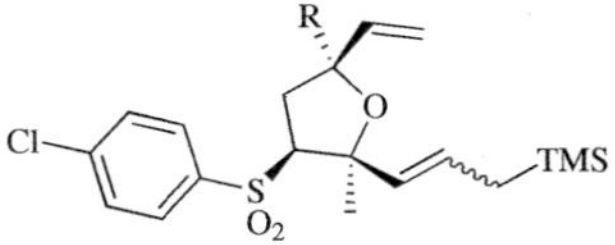

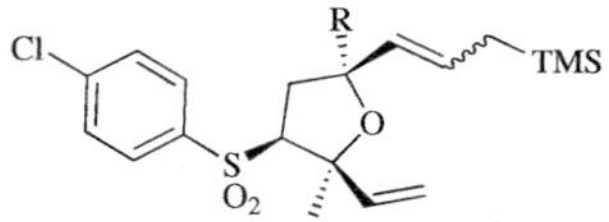

909

R	*x*	*y*	Temp (°)	**I + II**	**I/II**
H	10	0.01	70	(76)	92:8
Me	15	0.025	25	(94)	**I** only

TABLE 35. RING-OPENING CROSS-METATHESIS (*Continued*)

Cyclic Alkene	Cross Partner	Conditions	Product(s) and Yield(s) (%)	Refs.
Please refer to the charts preceding the tables for the catalyst structures.				
C_7				
Cl, SO_2, O	O, 6 eq	Hov II (0.01 eq), DCE, 84°	Cl, O_2, S, O, O (59)	909
H, OH, H, OTBS	O, CO_2Me, 2 eq	1. Gr I (0.05 eq), ethylene, benzene, rt, 12 h 2. Hov II (0.1 eq), reflux, 18 h	MeO_2C, O, OTBS, OH (77)	1196
H, OTBS, H, OTBS	O, CO_2Me, 2 eq	1. Gr I (0.05 eq), ethylene, benzene, rt, 12 h 2. Hov II (0.1 eq), reflux, 18 h	MeO_2C, O, OTBS, OTBS (68)	1196

C_8

x bar

Mo cat **6** (y eq), neat, rt

350

x	y	Time (h)	Conv. (%)[a]	
10	0.0002	16	98	(90)
10	0.0001	20	98	(80)
20	0.0002	20	93	(93)
20	0.0001	16	88	(88)
20	0.0006	20	75	(75)

R^1 R^2

x eq

Hov II (y eq), C_6D_6

R^1 6

104

R^1	R^2	x	y	Temp (°)	Time (h)		$(Z)/(E)$
F	H	3	0.05	23	48	(55)[b]	72:28
F	H	18	0.05	50	24	(0)	—
Cl	H	10.6	0.05	23	24	(93)	—
Cl	Cl	5.5	0.05	23	24	(>95)	68:32
Cl	Cl	5.5	0.05	45	1	(>95)	68:32
Cl	Cl	4	0.03	23	44	(97)	—

TABLE 35. RING-OPENING CROSS-METATHESIS (*Continued*)

Cyclic Alkene	Cross Partner	Conditions	Product(s) and Yield(s) (%)	Refs.
Please refer to the charts preceding the tables for the catalyst structures.				
C_8				
	Cl, Cl; 5 eq	Mo cat **26** (0.05 eq), benzene, 22°, 2 h	Cl, Cl; >98% conv.[a], (75) (*Z*,*Z*)/(*E*,*E*) > 98:2	106
	Br, Br; 8 eq; (*Z*)/(*E*) = 64:36	Mo cat **26** (0.05 eq), benzene, 22°, 4 h	Br, Br; >98% conv.[a], (88) (*Z*,*Z*)/(*E*,*E*) > 98:2	106
	SiR_3; 3 eq	Gr I (0.05 eq), DCM, reflux	R_3Si ... 5 **I** + R_3Si ... 5 ... SiR_3 **II**	1187

R	Time (h)	Conv. (%)[c]	**I**[d]	**II**[d]
MeO	0.17	96	(43)	(35)
MeO	2	100	(8)	(72)
EtO	0.17	81	(47)	(8)
EtO	2	100	(10)	(60)
Me_3SiO	0.17	94	(29)	(35)
Me_3SiO	2	100	(2)	(60)

$C_{8–10}$

C_8

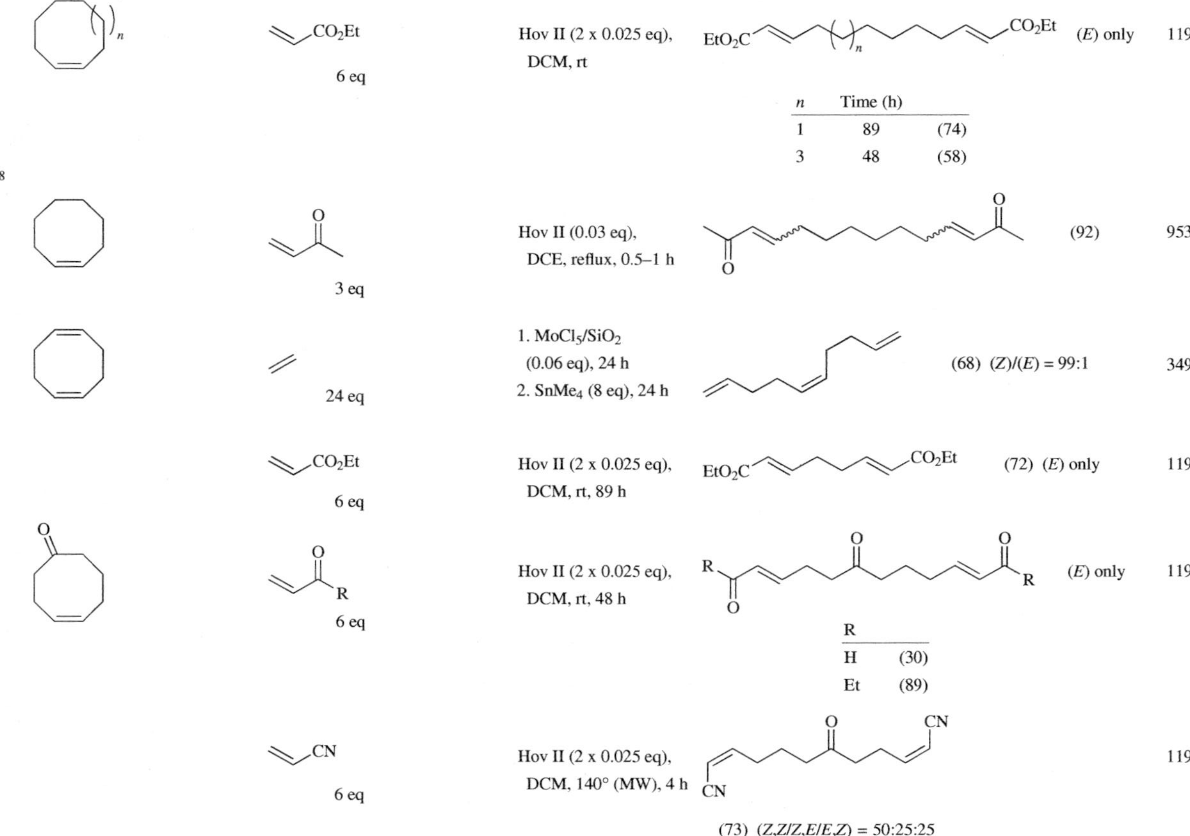

CO2Et
6 eq
Hov II (2 x 0.025 eq),
DCM, rt
EtO2C
CO2Et
(E) only
1194
n Time (h)
1 89 (74)
3 48 (58)
O
3 eq
Hov II (0.03 eq),
DCE, reflux, 0.5–1 h
(92)
953
24 eq
1. MoCl5/SiO2
(0.06 eq), 24 h
2. SnMe4 (8 eq), 24 h
(68) (Z)/(E) = 99:1
349
CO2Et
6 eq
Hov II (2 x 0.025 eq),
DCM, rt, 89 h
EtO2C
CO2Et
(72) (E) only
1194
R
6 eq
Hov II (2 x 0.025 eq),
DCM, rt, 48 h
(E) only
1194
R
H (30)
Et (89)
CN
6 eq
Hov II (2 x 0.025 eq),
DCM, 140° (MW), 4 h
1194
(73) (Z,Z/Z,E/E,Z) = 50:25:25

TABLE 35. RING-OPENING CROSS-METATHESIS (*Continued*)

Cyclic Alkene	Cross Partner	Conditions	Product(s) and Yield(s) (%)	Refs.

Please refer to the charts preceding the tables for the catalyst structures.

C_8

Cyclic Alkene	Cross Partner	Conditions	Product(s) and Yield(s) (%)	Refs.
	TMS 6 eq	Hov II (2 x 0.025 eq), DCM, rt, 48 h	TMS ... TMS (14)	1194
	R [e] n	Catalyst (0.01 eq), DCM, rt	R, n	345
	Ph 5 eq	Gr I (0.01 eq), DCM, rt, 6 h	Ph (61) (*E*) only	345
AcO, OAc, O	HO, HO 6 eq	1. Catalyst (*x* eq), DCM, rt, 4 h 2. 200°, 8.5 h	AcO, OAc, O, O, O	737

R	*n*	Catalyst	Time		(*E*)/(*Z*)
H	1	Gr I	75 min	(40)	78:22
H	2	Gr I	1 h	(65)	86:14
H	4	Gr I	20 min	(50)	88:12
Ph	1	Gr II	19 h	(50)	34:66

Catalyst	*x*	
Gr II	0.1	(72)
Hov II	0.01	(58)

Reactant 1	Reactant 2	Conditions	Product(s) and Yield(s) (%)		Refs.
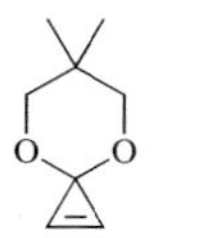	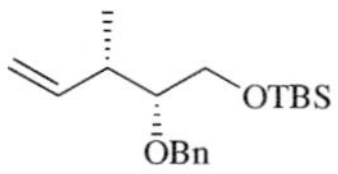 OTBS, OBn	Gr II (0.1 eq), benzene, 90°, 3.5 h	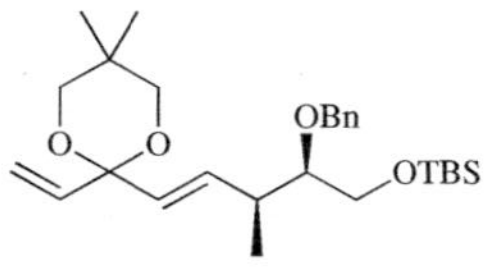 OBn, OTBS	(72) $(E)/(Z) = 60:40$	1123
2 eq	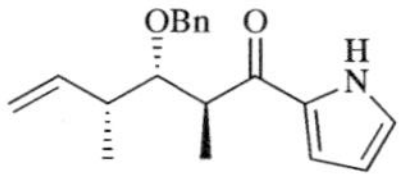 OBn, O, H N	Gr II (0.1 eq), aq $HClO_4$, THF, rt	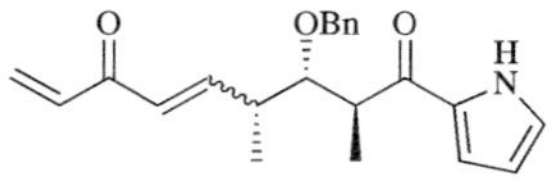 O, OBn, O, H N	(50)	340
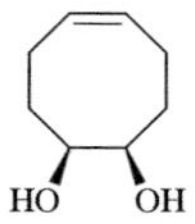 HO, OH	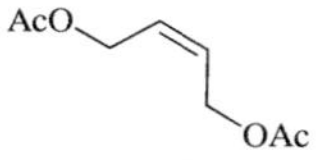 AcO, OAc, 15 eq	Gr II (0.1 eq), DCM, reflux	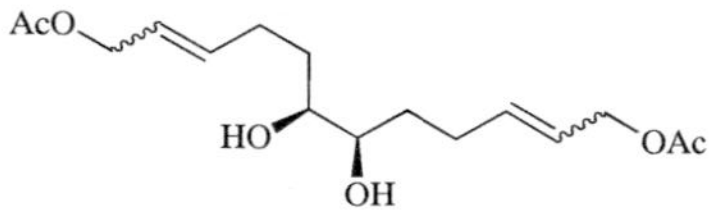 AcO, HO, OH, OAc	(99)	1197
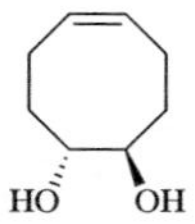 HO, OH	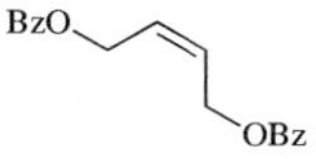 BzO, OBz, 7.7 eq	Gr II (0.05 eq), DCM, reflux	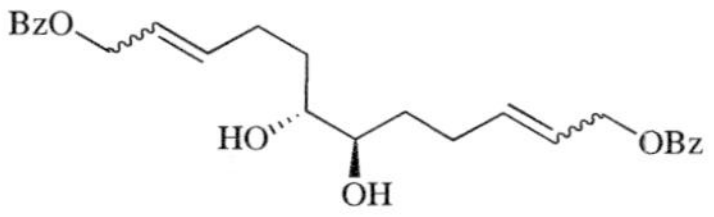 BzO, HO, OH, OBz	(74)	1197

TABLE 35. RING-OPENING CROSS-METATHESIS (*Continued*)

Cyclic Alkene	Cross Partner	Conditions	Product(s) and Yield(s) (%)	Refs.

Please refer to the charts preceding the tables for the catalyst structures.

C_8

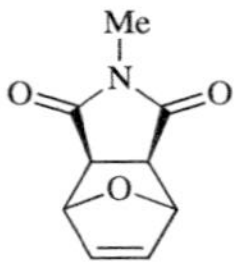

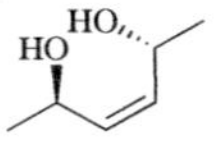

Cyclic Alkene	Cross Partner	Conditions	Product(s) and Yield(s) (%)	Refs.
Me, N, O, O, O	HO, HO; 6 eq	1. Catalyst (x eq), DCM, rt, 4 h 2. 200°, 8.5 h	Me, N, O, O, O, O, O	737

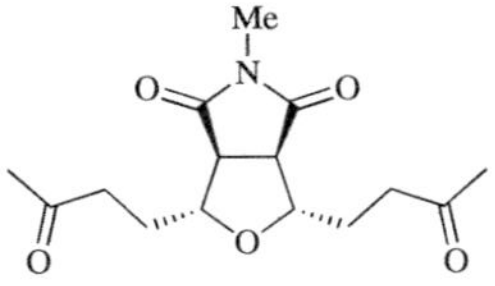

Catalyst	x	
Gr II	0.1	(72)
Hov II	0.01	(63)

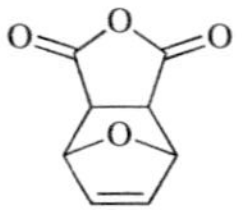

Cyclic Alkene	Cross Partner	Conditions	Product(s) and Yield(s) (%)	Refs.
O, O, O, O	SiR_3	Catalyst (x eq), solvent	O, O, O, O, SiR_3	

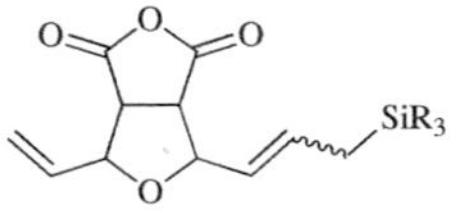

R	Catalyst	x	Solvent	Temp	Time (h)		(E)/(Z)	Refs.
Me	Ru cat **351**	0.02	$CDCl_3$	rt	2	(95)	50:50	1198
Me	Ru cat **347**	0.01	$CHCl_3$	45°	5.5	(>95)	—	1199
Me	Ru cat **348**	0.14	$CHCl_3$	45°	4	(50)	—	1199
Me	Ru cat **349**	0.02	$CDCl_3$	rt	12	(100)	—	1200
Me	Ru cat **350**	0.02	$CDCl_3$	rt	12	(100)	—	1200
Ph	Ru cat **349**	0.02	$CDCl_3$	rt	12	(100)	—	1200
Ph	Ru cat **350**	0.02	$CDCl_3$	rt	12	(100)	—	1200

Reactant		Conditions	Product(s) and Yield(s) (%)		Refs.
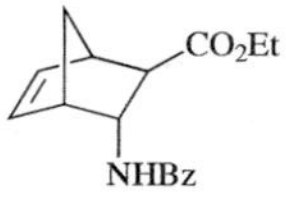 CO_2Et, NHBz	$=$	Catalyst (0.05 eq), DCM, 20°, 2 h	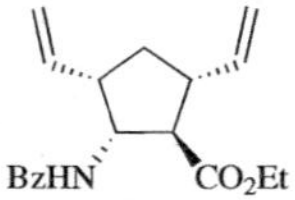 BzHN, CO_2Et	Catalyst Gr I (68) Gr II (29) Hov I (45) Hov II (16)	348
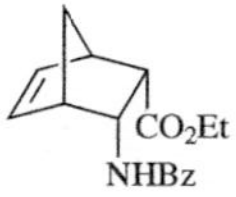 CO_2Et, NHBz	$=$	Catalyst (0.05 eq), DCM, 20°, 2 h	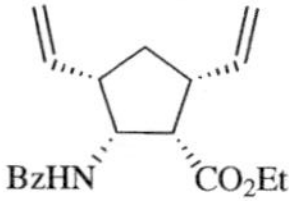 BzHN, CO_2Et	Catalyst Gr I (6) Gr II (26) Hov I (29) Hov II (31)	348
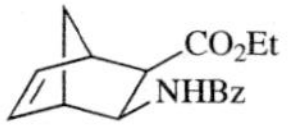 CO_2Et, NHBz	$=$	Hov I (0.05 eq), DCM, 20°, 2 h	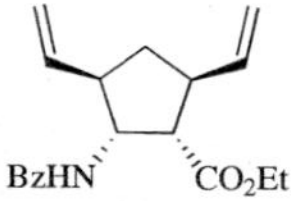 BzHN, CO_2Et	(39)	348
 NHBz, CO_2Et	$=$	Hov I (0.05 eq), DCM, 20°, 2 h	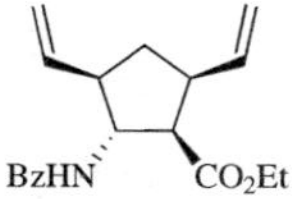 BzHN, CO_2Et	(64)	348

TABLE 35. RING-OPENING CROSS-METATHESIS (*Continued*)

Cyclic Alkene	Cross Partner	Conditions	Product(s) and Yield(s) (%)	Refs.
Please refer to the charts preceding the tables for the catalyst structures.				
C_8				
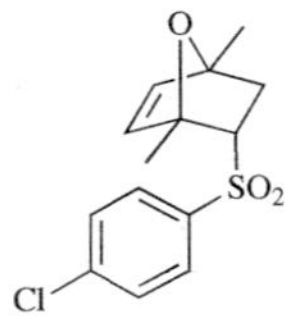 Cl, SO_2, O	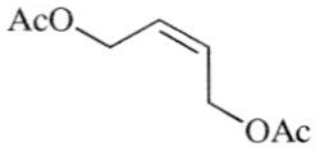 AcO, OAc 6 eq	Hov II (0.1 eq), ethylene, DCE, 25°, 12 h	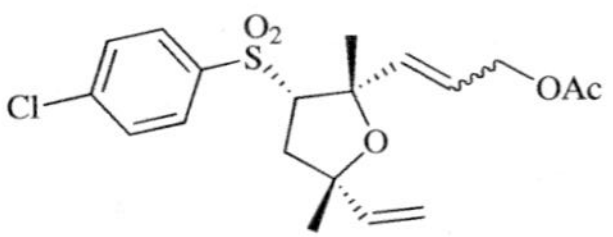 Cl, O_2 S, OAc, O (55)	909
	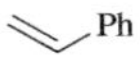 Ph 10 eq	Hov II (0.05 eq), DCE, 25°, 12 h	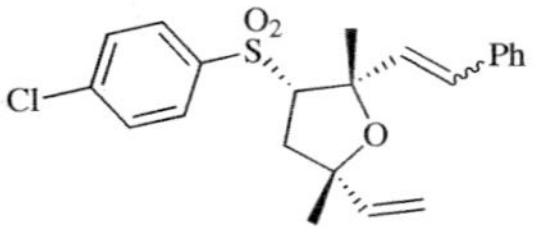 Cl, O_2 S, Ph, O (93)	909
	 Ph 6 eq	Hov II (0.05 eq), ethylene, DCE, 0°, 12 h,	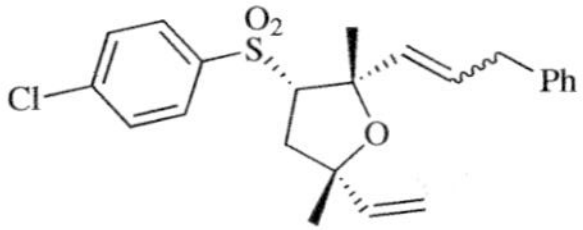 Cl, O_2 S, Ph, O (89)	909

C_9

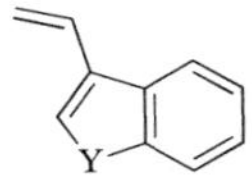

HO OH

10 eq

Ru cat **151** (0.05 eq), DCM, 22°, 2 h

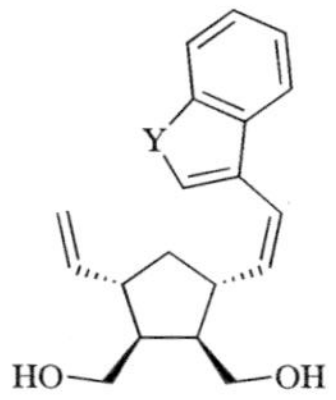

HO OH

Y		(*Z*)/(*E*)
TsN	(93)	93:7
S	(97)	98:2

83

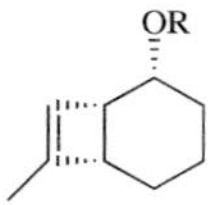

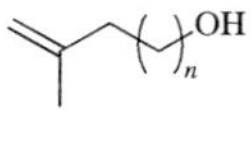

OR

5 eq

Hov II (*x* eq), solvent, 22 h

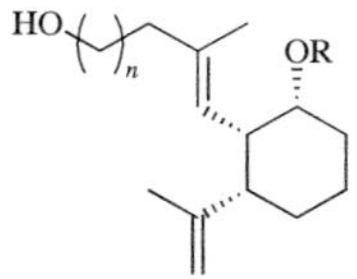

OR

1201

R	*n*	*x*	Solvent	Temp (°)		(E)/(Z)
H	0	0.1	DCM	60	(47)	50:50
H	0	0.1	DCE	90	(40)	67:33
H	0	0.1	DCM	rt	(85)	50:50
TES	0	0.1	DCM	60	(60)	50:50
H	1	0.1	DCM	60	(31)	75:25
H	1	0.1	DCM	rt	(0)	—
TES	1	0.15	DCM	60	(15)	>95:5

TABLE 35. RING-OPENING CROSS-METATHESIS (*Continued*)

Cyclic Alkene	Cross Partner	Conditions	Product(s) and Yield(s) (%)	Refs.

Please refer to the charts preceding the tables for the catalyst structures.

C_9

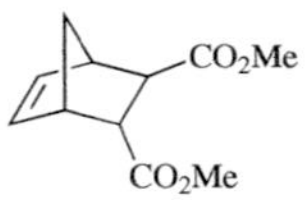

Cyclic Alkene	Cross Partner	Conditions	Product(s) and Yield(s) (%)	Refs.
CO_2Me, CO_2Me	x bar	Catalyst (0.01 eq), benzene	MeO_2C, CO_2Me	353

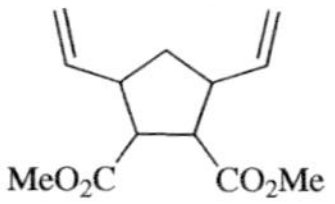

x	Catalyst	Time (h)	Temp (°)	
1	W cat **1**	16	20	(22)
3.7	W cat **1**	16	20	(24)
20	W cat **1**	16	20	(3)
1	W cat **2**	16	20	(25)
3.7	W cat **2**	16	20	(41)
3.7	W cat **2**	22	20	(47)
3.7	W cat **2**	48	20	(53)
20	W cat **2**	16	20	(24)
1	W cat **2**	22	20	(45)
1	W cat **2**	24	20	(63)
1	W cat **2**	21	60	(85)
1	W cat **2**	48	60	(70)
1	W cat **2**	93	60	(>98)
1	W cat **2**	21	80	(52)

Substrate	Alkene	Conditions	Product(s) and Yield(s) (%)	Refs.
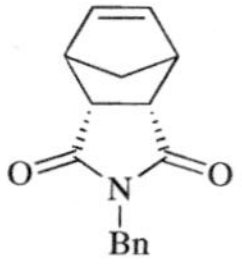 O, N, O, Bn	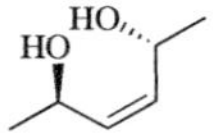 HO, HO; 6 eq	1. Gr II (0.1 eq), DCM, rt, 4 h 2. 200°, 8.5 h	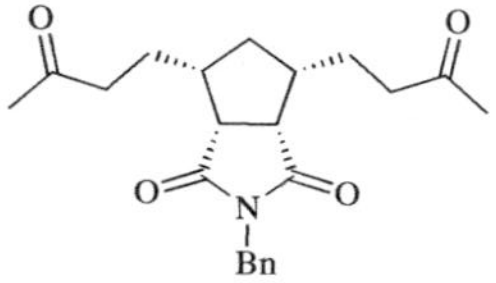 O, O, O, N, O, Bn (41)	737
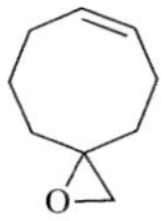 O	 CO_2Et; 6 eq	Hov II (2 x 0.025 eq), DCM, rt, 48 h	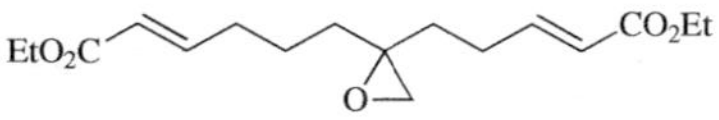 EtO_2C, CO_2Et, O (67) (*E*) only	1194
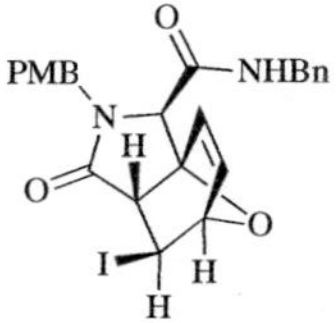 PMB, O, NHBn, N, H, O, O, I, H, H	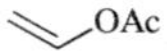 OAc; 5 eq	Hov II (0.02 eq), benzene, 60°, 4 h	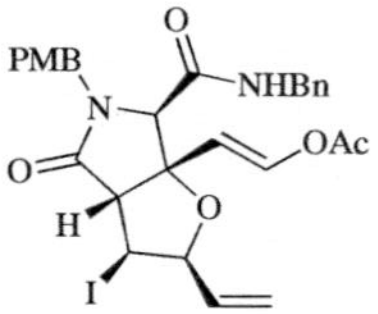 PMB, O, NHBn, N, OAc, O, H, O, I (99) (*E*)/(*Z*) = 80:20	1178

TABLE 35. RING-OPENING CROSS-METATHESIS (*Continued*)

Cyclic Alkene	Cross Partner	Conditions	Product(s) and Yield(s) (%)	Refs.
Please refer to the charts preceding the tables for the catalyst structures.				
C_9				
R, NHBn, H, O, H, H, I	OAc 5 eq	Hov II (x eq), solvent	R, NHBn, OAc, O, H, I	1178
			R / x / Solvent / Temp (°) / Time (h) / (*E*)/(*Z*) PMB / 0.02 / benzene / 60 / 4 / (87) / 93:7 TBSO$(CH_2)_2$ / 0.15 / benzene, DCE / 45 / 9 / (74) / 84:16	
PMB, H, NHBn, O, I, H, H	OAc 5 eq	Hov II (0.01 eq), benzene, rt, 14 h	PMB, H, NHBn, OAc, O, H, I, H (87) (*E*)/(*Z*) = 93:7	664, 1179
C_{10}				
	CO_2i-Pr 2 eq	Gr II (0.0013 eq), DCM, 40°, 16 h	*i*-PrO_2C ( )$_6$ CO_2i-Pr (80–85) (*E*) only	1202
Ph 1.5 eq	OH	Hov II (0.005 eq), toluene, 22°, 5 min	Ph, OH >98% conv.[a], (*E*)/(*Z*) = 67:33	624

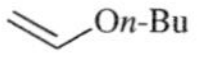

On-Bu

10 eq

Ru cat **151** (0.02 eq), DCM, 22°, 2 h

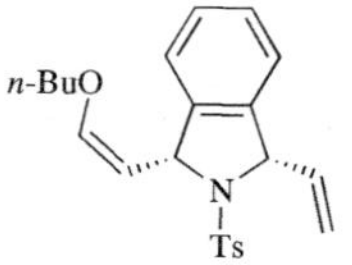

n-BuO

N–Ts

(83) (Z)/(E) = 98:2

83

Boc–N

R

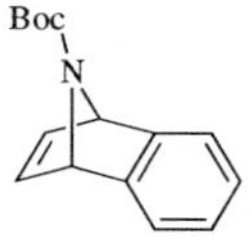

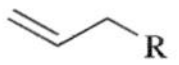

x eq

Catalyst (y eq), DCM, rt, 21 h

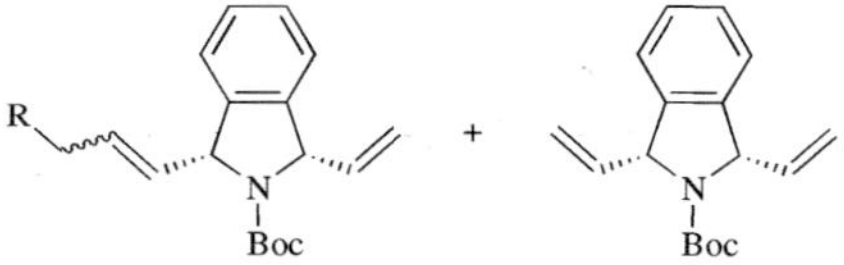

R

N–Boc

I

+

N–Boc

II

1203

R	x	Catalyst	y	**I**	**II**
TMS	1.2	Hov II	0.07	(98)	(—)
TMS	1.2	Gr II	0.03	(100)	(—)
TMS	1.2	Hov II	0.03	(97)	(—)
TMS	1.2	Hov II	0.01	(73)	(—)
MeO_2C	2	Gr II	0.05	(61)	(12)
MeO_2C	2	Hov II	0.03	(62)	(9)

TABLE 35. RING-OPENING CROSS-METATHESIS (*Continued*)

Cyclic Alkene	Cross Partner	Conditions	Product(s) and Yield(s) (%)	Refs.
Please refer to the charts preceding the tables for the catalyst structures.				
C_{14}				
	3 bar	Gr I (0.0003 eq), toluene, rt, 24 h	(99)	347
Me Si O Si Me Me Me (n)	4 bar	Gr I (x eq), toluene, rt, 24 h	Me Si O Si Me Me Me (n)	347
			n / x: 1 / 0.0005 (99); 12 / 0.0007 (99); 13 / 0.0007 (99)	
C_{15}				
		Gr I (0.05 eq), rt, 20 h	(76)	385

	5 eq OAc	Gr II (0.05 eq), 60°, 20 h	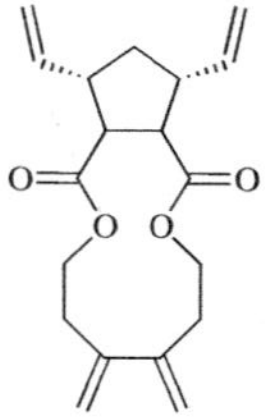	(100)	385
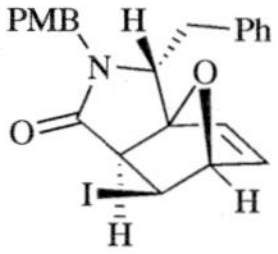 PMB, H, Ph, N, O, O, I, H, H	 OAc 5 eq	Hov II (x eq), benzene, rt	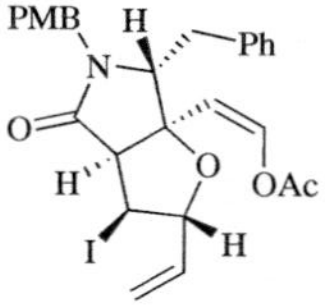 PMB, H, Ph, N, O, H, O, OAc, I, H		664, 1179

x	Time (h)	Concn (M)		$(Z)/(E)$
0.01	3	0.041	(59)	56:44
0.03	5	0.0038	(60)	50:50

[a] The conversion was determined by ^{1}H NMR analysis.

[b] This reaction included the use of ultrasound.

[c] The conversion was determined by GC analysis.

[d] The yield was determined by GC analysis.

[e] The alkene was bubbled through the reaction mixture every 3 h.

TABLE 36. ASYMMETRIC RING-OPENING CROSS-METATHESIS

Cyclic Alkene	Cross Partner	Conditions	Product(s) and Yield(s) (%)	Refs.

Please refer to the charts preceding the tables for the catalyst structures.

C_4

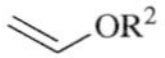

R¹O, OR¹ (cyclobutene) | OR² (x eq) | Mo cat **18** (0.006 eq), benzene, 0.5 h | OR¹ OR² OR¹ | 339

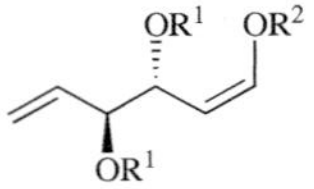

R^1	R^2	x	Temp (°)	Conv. (%)[a]		(Z)/(E)	er
Bn	*n*-Bu	1.1	22	60	(52)	97:3	79.5:20.5
Bn	*n*-Bu	2	22	65	(53)	97:3	80.0:20.0
Bn	*n*-Bu	5	22	97	(89)	97:3	82.5:17.5
Bn	*n*-Bu	10	22	>98	(90)	96:4	86.0:14.0
Bn	*n*-Bu	20	60	>98	(61)	84:16	85.0:15.0
TBS	*n*-Bu	1.1	22	60	(52)	97:3	79.5:20.5
TBS	*n*-Bu	10	22	>98	(90)	96:4	86.0:14.0
Bn	PMP	1.1	22	57	(43)	95:5	68.0:32.0
Bn	PMP	2	22	68	(58)	94:6	75.0:25.0
Bn	PMP	5	22	98	(90)	94:6	81.0:19.0
Bn	PMP	10	22	>98	(90)	94:6	85.0:15.0
Bn	PMP	20	60	>98	(73)	98:2	85.0:15.0
TBS	PMP	1.1	22	57	(34)	95:5	68.0:32.0
TBS	PMP	10	22	>98	(90)	94:6	85.0:15.0

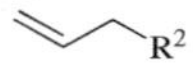

7 eq

Ru cat **215** (0.01 eq), THF, 23°

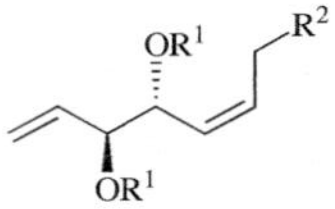

98, 1204

R^1	R^2	Time (h)		(*Z*)/(*E*)	er (*Z*)	er (*E*)
H	BzO	1–1.5	(67)	75:25	95.5:4.5	83.5:16.5
Bz	HO	1–1.5	(69)	75:25	98.0:2.0	91.0:9.0
Bn	HO	1.5	(62)	89:11	96.5:3.5	93.0:7.0
Bn	AcO	1–1.5	(79)	85:15	97.5:2.5	—
Bn	BzO	1.5	(61)	88:12	98.5:1.5	94.0:6.0
Bn	TBSO	1.5	(68)	87:13	94.5:5.5	88.5:11.5
Bn	BnO	1.5	(64)	86:14	95.5:4.5	—
Bn	4-$MeOC_6H_4$	1.5	(76)	90:10	96.5:3.5	89.5:10.5
Bn	$AcCH_2$	1.5	(65)	90:10	96.0:4.0	92.0:8.0
Bn	Bpin	1.5	(50)	—	95.5:4.5	—
TBS	HO	1.5	(66)	88:12	99.5:0.5	—
i-Pr	BzO	1.5	(<5)	—	—	—

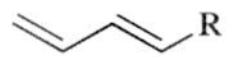

10 eq

Ru cat **151** (0.05 eq), DCM, 40°, 12 h

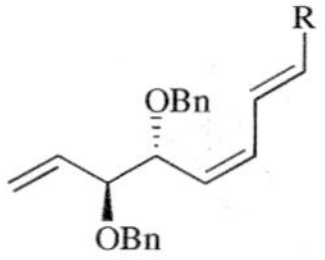

(*Z*)/(*E*) = 98:2

R	
MeO	(88)
n-C_6H_{13}	(60)

83

TABLE 36. ASYMMETRIC RING-OPENING CROSS-METATHESIS (*Continued*)

Cyclic Alkene	Cross Partner	Conditions	Product(s) and Yield(s) (%)	Refs.
Please refer to the charts preceding the tables for the catalyst structures.				
C_4				
BnO, OBn	Ph; 10 eq	Ru cat **151** (0.05 eq), DCM, 40°, 12 h	OBn, Ph, OBn; (58) (Z)/(E) = 98:2	83
TBSO, OTBS	Ph, OH; 1.03 eq	Ru cat **187** (0.02 eq), toluene, 22°, 2 h	OTBS, OH, OTBS, Ph; (51) dr 98:2 + TBSO, Ph, OH, OTBS; (36) dr 98:2	624
BnO, OBn	n-$C_{10}H_{21}$; 1.5 eq	Ru cat **216** (0.02 eq) 50°, 24 h	n-$C_{10}H_{21}$, OBn, OBn; (89) (Z)/(E) > 95:5	538
C_7				
OBn, Me, N	O, OMe; x eq	Mo cat **18** (0.03 eq), benzene, 60°, 3 h	OBn, OMe, O, N, Me	624

x	Conv. (%)[a]		(Z)/(E)	er
2	28	(25)	75:25	72.0:28.0
10	69	(50)	80:20	75.0:25.0
20	98	(90)	92:8	91.0:9.0

Ph
x eq

Catalyst (*y* eq),
benzene, 22°, 24 h

Ph

$(E)/(Z) > 98:2$

102

R^1	R^2	*x*	Catalyst	*y*		er
H	EtO_2C	10	Mo cat **12**	0.05	(72)	95.0:5.0
H	EtO_2C	20	Ru cat **210**	0.1	(66)	3.0:97.0
TBSO	EtO_2C	10	Mo cat **12**	0.1	(60)	90.0:10.0
BnO	Me	10	Mo cat **12**	0.05	(86)	98.0:2.0

TBSO Me

Ph
x eq

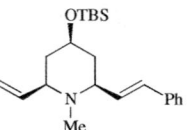

Catalyst (0.05 eq),
solvent, 22°

OTBS

Ph

Me

$(E)/(Z) > 98:2$

x	Catalyst	Solvent	Time (h)	Conv. (%)[a]		er	
20	Ru cat **210**	—	36	30	(—)	66.5:33.5	91
20	Ru cat **212**	—	36	<5	(—)	—	91
20	Ru cat **213**	—	36	<4	(—)	—	91
10	Mo cat **12**	benzene	1	>98	(95)	97.0:3.0	91
2	Ru cat **210**	benzene	12	<2	(—)	—	102
2	Ru cat **212**	benzene	12	<2	(—)	—	102
2	Ru cat **213**	benzene	12	<2	(—)	—	102
2	Mo cat **8**	benzene	12	<2	(—)	—	102
2	Mo cat **9**	benzene	12	>98	(—)	<51.0:49.0	102
2	Mo cat **10**	benzene	12	<2	(—)	—	102
2	Mo cat **12**	benzene	12	>98	(—)	97.0:3.0	102

TABLE 36. ASYMMETRIC RING-OPENING CROSS-METATHESIS (*Continued*)

Cyclic Alkene	Cross Partner	Conditions	Product(s) and Yield(s) (%)	Refs.

Please refer to the charts preceding the tables for the catalyst structures.

C_7

Cyclic Alkene	Cross Partner	Conditions	Product(s) and Yield(s) (%)	Refs.
TBSO, Me (N)	R^1, R^2; 10 eq	Mo cat **12** (0.05 eq), benzene, 22°, 12 h	OTBS, N–Me, R^1, R^2; (*E*)/(*Z*) > 98:2	102

R^1	R^2		er
H	H	(95)	97.0:3.0
H	MeO	(92)	94.5:5.5
H	CF_3	(88)	82.0:18.0
Br	H	(86)	94.0:6.0
Me	H	(91)	99.0:1.0

Cyclic Alkene	Cross Partner	Conditions	Product(s) and Yield(s) (%)	Refs.
TBSO, Cbz (N)	R^1, R^2; 10 eq	Mo cat **12** (0.1 eq), benzene, 22°, 12 h	OTBS, N–Cbz, R^1, R^2; (*E*)/(*Z*) > 98:2	102

R^1	R^2		er
H	H	(93)	95.0:5.0
H	MeO	(65)	90.0:10.0
Me	H	(92)	93.0:7.0

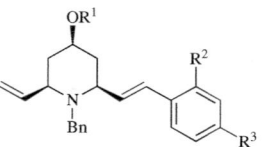

Catalyst (0.05 eq), solvent, 22°

x eq

$(E)/(Z) > 98:2$

91

R^1	R^2	R^3	x	Catalyst	Solvent	Time (h)	Conv. (%)[a]		er
H	H	H	20	Ru cat **210**	—	24	>98	(82)	>99.0:1.0
H	H	H	20	Ru cat **212**	—	24	80	(65)	97.0:3.0
H	Br	H	20	Ru cat **210**	—	24	63	(55)	98.5:1.5
H	Br	H	20	Ru cat **212**	—	24	27	(—)	—
H	Me	H	20	Ru cat **210**	—	24	>98	(81)	99.0:1.0
H	Me	H	20	Ru cat **212**	—	24	95	(60)	99.0:1.0
H	F	H	20	Ru cat **210**	—	24	87	(70)	>99.0:1.0
H	H	MeO	20	Ru cat **210**	—	24	87	(67)	>99.0:1.0
H	H	MeO	20	Ru cat **212**	—	24	50	(38)	98.0:2.0
H	H	CF_3	5	Ru cat **210**	—	24	>98	(78)	98.5:1.5
TBS	H	H	20	Ru cat **210**	—	24	>98	(80)	>99.0:1.0
TBS	H	H	20	Ru cat **212**	—	24	70	(—)	98.0:2.0
TBS	H	H	20	Ru cat **213**	—	24	<5	(—)	—
TBS	H	H	10	Mo cat **12**	benzene	1	>98	(85)	60.5:39.5

TABLE 36. ASYMMETRIC RING-OPENING CROSS-METATHESIS (*Continued*)

Cyclic Alkene	Cross Partner	Conditions	Product(s) and Yield(s) (%)	Refs.

Please refer to the charts preceding the tables for the catalyst structures.

C_7

Cyclic Alkene	Cross Partner	Conditions	Product(s) and Yield(s) (%)	Refs.
TBSO, Me, N	Cy, 10 eq	Mo cat **12** (0.05 eq), benzene, 22°, 12 h	OTBS, N, Me, Cy (43) er 84.0:16.0 + OTBS, N, Me (25) + OTBS, Cy, N, Me, Cy (25)	102
R^2, R^1, N	Ph, *x* eq	Catalyst (*y* eq), benzene, 22°	R^2, N, R^1, Ph; (*E*)/(*Z*) > 98:2	102

R^1	R^2	*x*	Catalyst	*y*	Time (h)		er
Cbz	H	10	Mo cat **12**	0.1	24	(80)	91.0:9.0
Cbz	H	20	Ru cat **210**	0.1	24	(70)	5.0:95.0
Cbz	TBSO	10	Mo cat **12**	0.1	24	(80)	92.5:7.5
Me	H	2	Mo cat **12**	0.05	1	(64)	90.0:10.0
Me	TBSO	10	Mo cat **12**	0.05	1	(90)	92.5:7.5

OR

x eq

Catalyst (y eq),
solvent, 22°

OH

OR

808, 1175

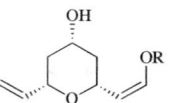

R	x	Catalyst	y	Solvent	Time (h)	Conv. (%)[a]		$(Z)/(E)$	er
n-Bu	20	Ru cat **210**	0.05	benzene	8	>98	(80)	95:5	>98.0:2.0
n-Bu	10	Ru cat **210**	0.005	benzene	8	77	(60)	95:5	>98.0:2.0
n-Bu	20	Ru cat **212**	0.05	benzene	24	85	(58)	92:8	97.0:3.0
n-Bu	20	Hov I	0.05	benzene	24	7	(0)	82:18	—
n-Bu	20	Hov II	0.05	benzene	24	79	(41)	23:77	—
n-C_6H_{13}	20	Ru cat **210**	0.05	benzene	24	>98	(64)	98:2	>98.0:2.0
n-C_6H_{13}	20	Hov II	0.05	benzene	24	>98	(58)	37:63	—
n-C_6H_{13}	20	Ru cat **213**	0.05	THF	2.5	>98	(58)	—	59.0:41.0
Ph	20	Ru cat **213**	0.05	THF	2.5	>98	(68)	—	96.0:4.0
PMP	20	Ru cat **210**	0.05	benzene	24	>98	(67)	95:5	97.0:3.0
PMP	20	Hov II	0.05	benzene	24	>98	(41)	43:57	—
CF_3CH_2	20	Ru cat **210**	0.05	benzene	24	>98	(63)	94:6	96.0:4.0
CF_3CH_2	20	Hov II	0.05	benzene	24	>98	(59)	27:73	—
$Cl(CH_2)_2$	20	Ru cat **210**	0.05	benzene	24	>98	(63)	95:5	>98.0:2.0
$Cl(CH_2)_2$	20	Hov II	0.05	benzene	24	>98	(31)	24:76	—

TABLE 36. ASYMMETRIC RING-OPENING CROSS-METATHESIS (*Continued*)

Cyclic Alkene	Cross Partner	Conditions	Product(s) and Yield(s) (%)	Refs.

Please refer to the charts preceding the tables for the catalyst structures.

C_7

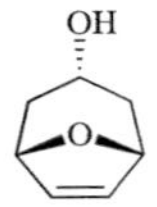

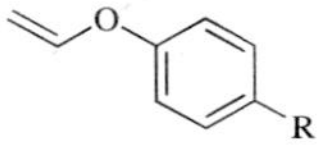

OH · O · 20 eq · R

Ru cat **213** (0.05 eq)
benzene, 22°, 24 h

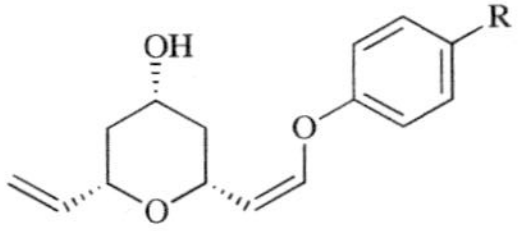

OH · O · R · 1205

R	Conv. (%)[a]		Efficiency[b]	(Z)/(E)	er[c]
n-BuO	66	(64)	0.97	92:8	97.0:3.0
Et	74	(69)	0.93	91:9	97.0:3.0
H	82	(74)	0.90	93:7	96.0:4.0
Br	90	(79)	0.88	93:7	96.0:4.0
Cl	93	(75)	0.81	93:7	95.0:5.0
CF_3	95	(73)	0.77	93:7	96.0:4.0

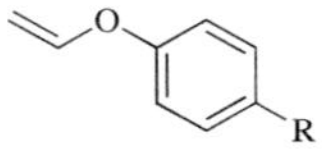

O · 20 eq · R

Catalyst (0.05 eq)
benzene, 22°, 24 h

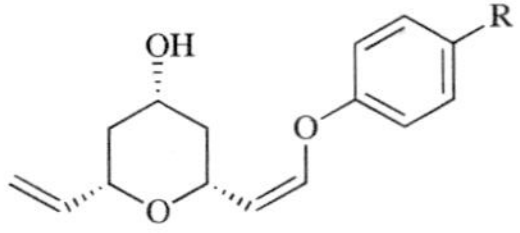

OH · O · R · 1205

R	Catalyst	Conv. (%)[a]		Efficiency[b]	(Z)/(E)
n-BuO	Hov I	62	(59)	0.95	83:17
Et	Hov I	72	(66)	0.92	77:23
H	Hov I	81	(70)	0.86	76:24
Br	Hov I	83	(65)	0.78	79:21

Cl	Hov I	86	(66)	0.77	81:19
CF_3	Hov I	94	(70)	0.74	82:18
n-BuO	Hov II	>98	(60)	0.60	40:60
Et	Hov II	>98	(59)	0.59	45:55
H	Hov II	>98	(63)	0.63	45:55
Br	Hov II	>98	(59)	0.59	42:58
Cl	Hov II	>98	(61)	0.61	43:57
CF_3	Hov II	>98	(58)	0.58	43:57

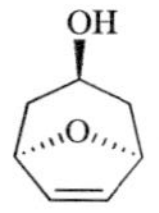

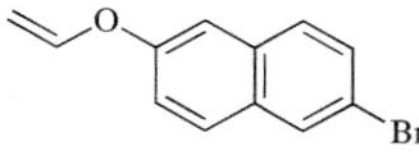

O, Br

20 eq

Ru cat **213** (0.05 eq),
benzene, 22°, 24 h

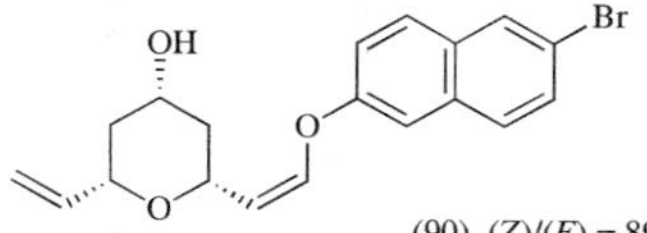

OH, O, O, Br

(90) (*Z*)/(*E*) = 89:11, er 96.0:4.0 1175

SPh

20 eq

Ru cat **213** (0.05 eq),
benzene, 22°, 24 h

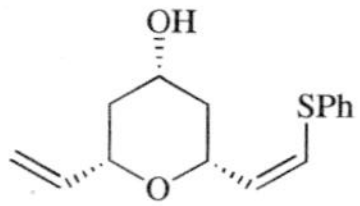

OH, O, SPh

(67) (*Z*)/(*E*) = 91:9, er 96.0:4.0 1175

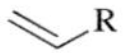

OH, O

R

20 eq

Ru cat **213** (0.05 eq),
THF, 22°, 2.5 h

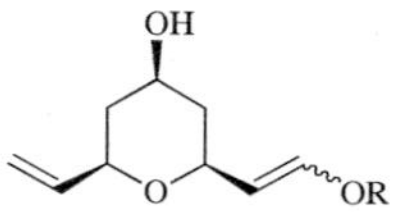

OH, O, OR

808

R	Conv (%)[a]		er
Ph	>98	(68)	96:4
Cy	>98	(58)	59:41

TABLE 36. ASYMMETRIC RING-OPENING CROSS-METATHESIS (*Continued*)

Cyclic Alkene	Cross Partner	Conditions	Product(s) and Yield(s) (%)	Refs.

Please refer to the charts preceding the tables for the catalyst structures.

C_7

Cyclic Alkene	Cross Partner	Conditions	Product(s) and Yield(s) (%)	Refs.
OR^1	OR^2, *x* eq	Mo cat **18** (0.006 eq), benzene, 22°	OR^1, OR^2; (*Z*)/(*E*) > 98:2	339

R^1	R^2	*x*	Time (min)	Conv. (%)[a]		er
TBS	*n*-Bu	1.1	30	>98	(53)	52.0:48.0
TBS	*n*-Bu	5	0.5	97	(83)	95.0:5.0
TBS	*n*-Bu	5	1	>98	(89)	90.0:10.0
TBS	*n*-Bu	5	3	>98	(96)	85.5:14.5
TBS	*n*-Bu	5	30	>98	(80)	62.5:37.5
TBS	*n*-Bu	10	30	>98	(80)	73.0:27.0
TBS	*n*-Bu	20	30	>98	(85)	84.5:15.5
TBS	*n*-Bu	30	30	>98	(88)	95.0:5.0
TBS	*n*-Bu	1.1	30	97	(86)	56.5:43.5
TBS	*n*-Bu	5	30	97	(86)	90.0:10.0
TBS	*n*-Bu	5	30	97	(82)	87.0:13.0
Bn	*n*-Bu	1.1	30	96	(54)	57.5:42.5
Bn	*n*-Bu	30	30	>98	(88)	95.0:5.0
TBS	PMP	5	0.5	>98	(83)	86.0:14.0
TBS	PMP	5	1	>98	(87)	84.5:15.5
TBS	PMP	10	3	>98	(90)	91.5:8.5
TBS	PMP	20	30	>98	(87)	92.0:8.0
TBS	PMP	30	30	>98	(82)	91.5:8.5
Bn	PMP	1.1	30	94	(82)	82.0:18.0
Bn	PMP	20	30	>98	(87)	96.5:3.5

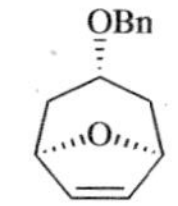

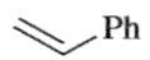

Reactant	Reagent	Conditions	Product(s) and Yield(s) (%)	Refs.
OBn	Ph	Mo cat **18** (0.02 eq), 22°	OBn, O, Ph (—) $(E)/(Z) = 5:95$ er 91.0:9.0	809
R	Ph, x eq	Catalyst (y eq), solvent	R, O, Ph $(E)/(Z) > 98:2$	91

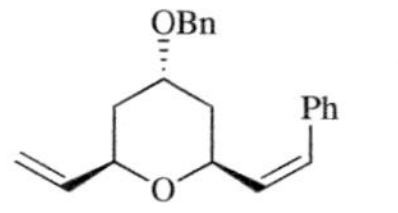

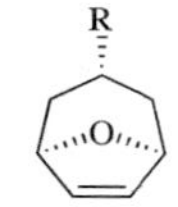

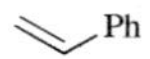

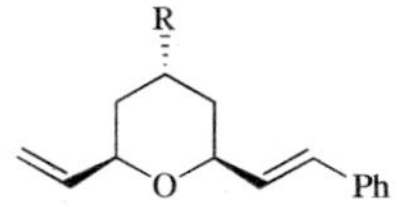

R	x	Catalyst	y	Solvent	Temp (°)	Time (h)		er
BnO	2	Ru cat **211**	0.02	—	22	36	(85)	95.0:5.0
BnO	2	Ru cat **213**	0.02	—	22	15	(70)	97.0:3.0
BnO	5	Ru cat **213**	0.02	—	–15	15	(84)	>99.0:1.0
BnO	10	Mo cat **12**	0.02	benzene	22	1	(81)	95.0:5.0
I	2	Ru cat **211**	0.02	—	22	5	(65)	96.5:3.5
I	2	Ru cat **213**	0.02	—	22	5	(70)	91.0:9.0
I	10	Mo cat **12**	0.05	benzene	22	1	(76)	98.5:1.5

TABLE 36. ASYMMETRIC RING-OPENING CROSS-METATHESIS (*Continued*)

Cyclic Alkene	Cross Partner	Conditions	Product(s) and Yield(s) (%)	Refs.

Please refer to the charts preceding the tables for the catalyst structures.

C_7

OR

On-Bu, *x* eq

Catalyst (*y* eq), benzene, 22°

OR, On-Bu

339

R	*x*	Catalyst	*y*	Time (h)	Conv. (%)[a]		(*Z*)/(*E*)	er
TBS	1.1	Hov II	0.01	0.5	69	(32)	25:75	—
TBS	1.1	Mo cat **1**	0.01	0.5	98	(10)	50:50	—
TBS	1.1	Mo cat **8**	0.01	12	<2	(0)	—	—
TBS	1.1	Mo cat **3**	0.01	0.5	>98	(48)	80:20	—
TBS	1.1	Mo cat **5**	0.01	0.5	85	(68)	>98:2	99.0:1.0
TBS	1.1	Mo cat **6**	0.01	0.5	<5	(0)	—	—
TBS	1.1	Mo cat **16**	0.0007	0.5	>98	(87)	>98:2	97.0:3.0
TBS	1.1	Mo cat **17**	0.005	0.5	>98	(87)	>98:2	98.0:2.0
TBS	1.1	Mo cat **18**	0.006	0.5	>98	(89)	>98:2	99.0:1.0
TBS	5	Mo cat **18**	0.05	0.17	>98	(76)	>98:2	99.0:1.0
TBS	1.1	Mo cat **19**	0.007	0.5	>98	(87)	>98:2	98.0:2.0
Bn	1.1	Mo cat **18**	0.006	0.5	>98	(75)	>98:2	99.0:1.0
Bn	5	Mo cat **18**	0.006	0.17	>98	(74)	>98:2	99.0:1.0

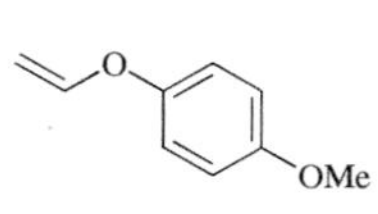

OMe

x eq

Catalyst (y eq), benzene, 22°

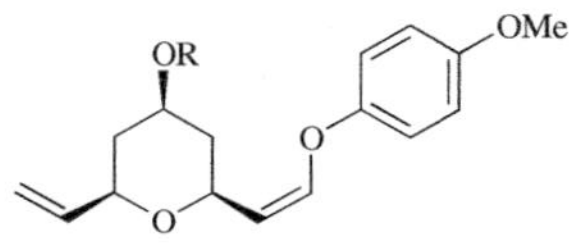

OR OMe O O

339

R	x	Catalyst	y	Time (h)	Conv. (%)[a]		(Z)/(E)	er
TBS	1.1	Hov II	0.01	0.5	95	(<5)	—	—
TBS	1.1	Mo cat **1**	0.01	0.5	98	(14)	80:20	—
TBS	1.1	Mo cat **8**	0.01	12	<2	(0)	—	—
TBS	1.1	Mo cat **3**	0.01	0.5	>98	(75)	82:18	—
TBS	1.1	Mo cat **5**	0.01	0.5	86	(47)	>98:2	99.0:1.0
TBS	1.1	Mo cat **6**	0.01	0.5	<5	(0)	—	—
TBS	1.1	Mo cat **16**	0.0007	0.5	60	(41)	>98:2	86.0:14.0
TBS	1.1	Mo cat **17**	0.005	0.5	>98	(80)	>98:2	92.0:8.0
TBS	1.1	Mo cat **18**	0.006	0.5	>98	(90)	>98:2	99.0:1.0
TBS	5	Mo cat **18**	0.05	0.17	>98	(80)	>98:2	99.0:1.0
TBS	1.1	Mo cat **19**	0.007	0.5	>98	(85)	>98:2	99.0:1.0
Bn	1.1	Mo cat **18**	0.006	0.5	>98	(80)	>98:2	97.0:3.0
Bn	5	Mo cat **18**	0.006	0.17	>98	(77)	>98:2	99.0:1.0

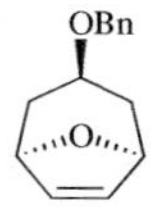

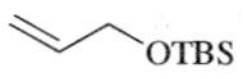

OBn O OTBS

Mo cat **18** (0.05 eq), 22°, 1 h

OBn OTBS O

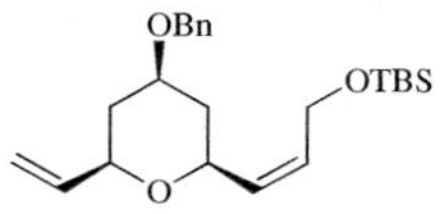

(45) (E)/(Z) = 5:95
er 86.0:14.0

809

TABLE 36. ASYMMETRIC RING-OPENING CROSS-METATHESIS (*Continued*)

Cyclic Alkene	Cross Partner	Conditions	Product(s) and Yield(s) (%)	Refs.

Please refer to the charts preceding the tables for the catalyst structures.

C_7

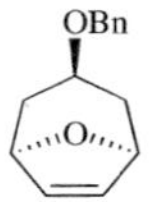

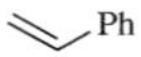

Cyclic Alkene	Cross Partner	Conditions	Product(s) and Yield(s) (%)	Refs.
OBn	Ph	Catalyst (0.01 eq), 22°, 1 h	OBn, O, Ph	809
	Ph, 5 eq	Catalyst (x eq), solvent, 1 h	OBn, O, Ph; $(E)/(Z) > 98:2$	91

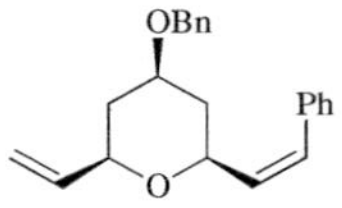

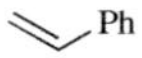

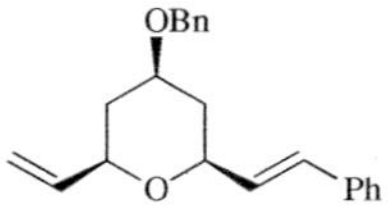

Catalyst		$(E)/(Z)$	er
Mo cat **12**	(80)	>2:98	95.0:5.0
Ru cat **212**	(85)	>2:98	98.5:1.5
Ru cat **213**	(87)	5:95	98.0:2.0

Catalyst	x	Solvent	Temp (°)		er
Ru cat **210**	0.02	—	22	(70)	98.0:2.0
Ru cat **213**	0.02	—	22	(90)	94.0:6.0
Ru cat **213**	0.02	—	–15	(87)	95.5:4.5
Mo cat **12**	0.05	benzene	22	(—)	—

C_{7-8}

R¹O R²

On-Bu

20 eq

Ru cat **210** (0.05 eq),
benzene, 22°, 24 h

R¹O R² On-Bu

808, 1175

R^1	R^2		(Z)/(E)	er
TBS	H	(71)	87:13	93.0:7.0
TBS	Me	(72)	92:8	95.0:5.0
Bn	H	(62)	94:6	91.0:9.0

OBn

OAc

7 eq

Ru cat **215** (0.01 eq),
THF, 23°, 1–1.5 h

OBn OAc

98

(56) (*E*)/(*Z*) = 85:15, er (*E*) 96.5:3.5, er (*Z*) 97.0:3.0

Ph

1 eq

Catalyst (0.05 eq),
C_6D_6, 40 h

OBn Ph + Ph OBn Ph

I **II**

1206

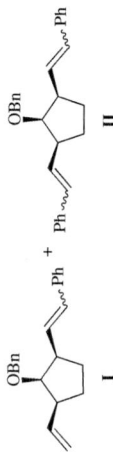

Catalyst	Temp (°)	I	II	er I
Mo cat **12**	25	(56)	(36)	95.5:4.5
Mo cat **20**	25	(36)	(42)	90.0:10.0
Mo cat **22**	25	(<2)	(<2)	—
Mo cat **15**	25	(64)	(24)	2.0:98.0
Mo cat **21**	75	(61)	(14)	1.0:99.0
Mo cat **23**	75	(57)	(14)	1.0:99.0

TABLE 36. ASYMMETRIC RING-OPENING CROSS-METATHESIS (*Continued*)

Cyclic Alkene	Cross Partner	Conditions	Product(s) and Yield(s) (%)	Refs.
Please refer to the charts preceding the tables for the catalyst structures.				
C_7				
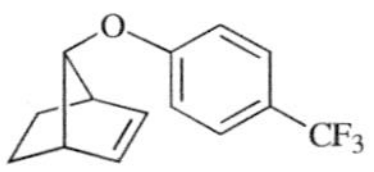 CF_3	 OAc 7 eq	Ru cat **215** (0.01 eq), THF, 23°, 1 h	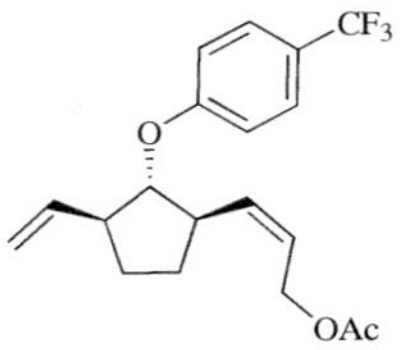 CF_3, O, OAc (40) (*Z*)/(*E*) = 70:30, er (*Z*) 97.5:2.5, er (*E*) 97.5:2.5	98, 417
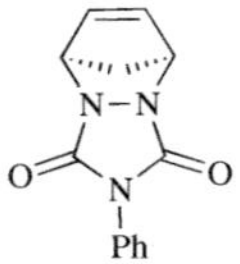 N–N, O, N, Ph	 Ph 10 eq	Catalyst (*x* eq), NaI (*y* eq), DCM, rt, 1 h	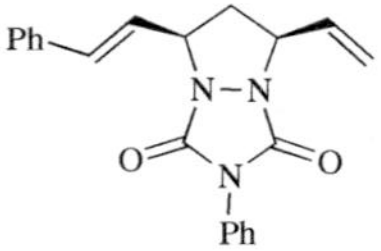 Ph, N–N, O, N, Ph	97

Catalyst	*x*	*y*		(*E*)/(*Z*)	er (*E*)	er (*Z*)
Ru cat **24**	0.01	—	(99)	55:45	84.0:16.0	57.5:42.5
Ru cat **25**	0.03	1	(99)	55:45	84.0:16.0	55.0:45.0

C_8

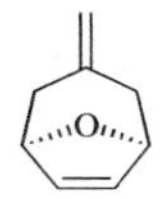

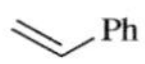

Ph

x eq

Catalyst (y eq), solvent, 22°

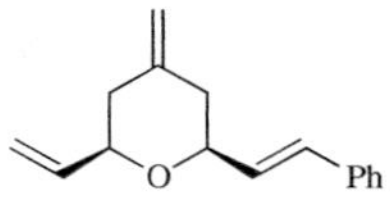

O Ph

91

x	Catalyst	y	Solvent	Time (h)	Conv. (%)[a]		$(E)/(Z)$	er
5	Ru cat **211**	0.02	—	36	—	(50-60)	>98:2	97.0:3.0
5	Ru cat **213**	0.02	—	15	—	(50-60)	>98:2	>99.0:1.0
2	Mo cat **8**	0.05	pentane	4	>98	—	—	—
2	Mo cat **9**	0.05	pentane	4	<5	—	—	—
2	Mo cat **10**	0.05	pentane	4	50	—	—	—
2	Mo cat **12**	0.05	pentane	4	>98	—	>98:2	97.0:3.0
5	Mo cat **12**	0.05	benzene	1	—	(50-60)	>98:2	98.5:1.5
2	Mo cat **15**	0.05	pentane	4	<5	—	—	—

HO OH

0.2 eq

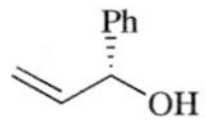

Ph OH

er 96.0:4.0

Ru cat **151** (0.05 eq), DCM, 22°, 8 h

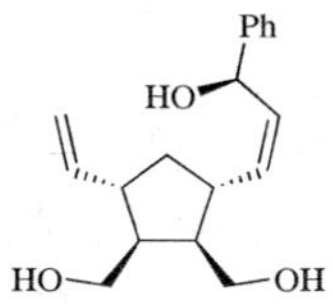

Ph HO HO OH

(67) $(Z)/(E)$ = 98:2, dr 98:2

83

TABLE 36. ASYMMETRIC RING-OPENING CROSS-METATHESIS (*Continued*)

Cyclic Alkene	Cross Partner	Conditions	Product(s) and Yield(s) (%)	Refs.
Please refer to the charts preceding the tables for the catalyst structures.				
C_8				
Me, N, O, O, O	TMS	Ru cat **149** (0.005 eq), DCM, reflux, 2 h	Me, N, O, O, TMS, TMS, O >99% conv.[a]	853
C_9				
HO, OH	On-Bu 20 eq	Catalyst (0.05 eq) 22°, 24 h	n-BuO, HO, OH	1205

Catalyst	Conv. (%)[a]		(Z)/(E)	er[c]
Ru cat **212**	>98	(86)	94:6	81.0:19.0
Ru cat **213**	>98	(81)	85:15	67.0:33.0

Cyclic Alkene	Cross Partner	Conditions	Product(s) and Yield(s) (%)	Refs.
	R 10 eq	Ru cat **151** (0.05 eq), DCM, 22°, 8 h	R, HO, OH (Z)/(E) = 98:2	83

R	
TMS	(89)
4-MeOC$_6$H$_4$	(83)
PhHNOCCH$_2$	(65)
TBSOCH$_2$	(68)
n-C$_7$H$_{15}$	(58)

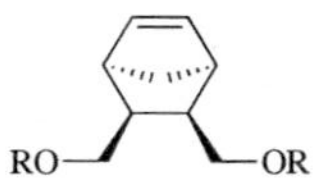

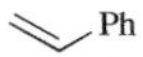

RO— —OR

Ph

5 eq

Ru cat **240** (0.01 eq),
DCM, –10°, 12 h

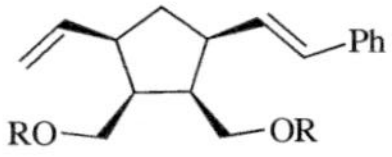

RO— —OR Ph

(*E*)/(*Z*) = 95:5

95

R	Conv. (%)[a]	er
Ac	>98	80.0:20.0
TBS	>98	85.0:15.0

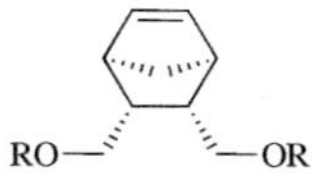

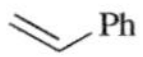

RO— —OR

Ph

10 eq

Catalyst (x eq),
NaI (y eq), DCM, rt, 1 h

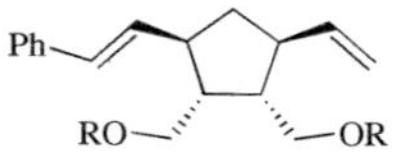

Ph RO— —OR

97

R	Catalyst	x	y		(*E*)/(*Z*)	er (*E*)	er (*Z*)
H	Ru cat **24**	0.01	—	(10)	—	66.5:33.5	64.5:35.5
Ac	Ru cat **24**	0.01	—	(99)	58:42	80.0:20.0	55.0:45.0
Ac	Ru cat **25**	0.03	1	(78)	58:42	86.0:14.0	70.0:30.0

TABLE 36. ASYMMETRIC RING-OPENING CROSS-METATHESIS (*Continued*)

Cyclic Alkene	Cross Partner	Conditions	Product(s) and Yield(s) (%)	Refs.

Please refer to the charts preceding the tables for the catalyst structures.

C_9

Cyclic Alkene	Cross Partner	Conditions	Product(s) and Yield(s) (%)	Refs.
OR[1], OR[1]	R[2], 7 eq	Ru cat **215** (0.01 eq), THF, 23°, 1 h	R_2, R[1]O, OR[1]	417

R^1	R^2		(*Z*)/(*E*)	er
Ac	AcO	(45)	97:3	91.0:9.0
Bn	AcO	(64)	95:5	96.5:3.5
Bn	BocHN	(41)	95:5	97.0:3.0
Bn	EtO_2CCH_2	(65)	95:5	95.6:4.5
Bn	*n*-Pr	(62)	95:5	94.5:5.5

Cyclic Alkene	Cross Partner	Conditions	Product(s) and Yield(s) (%)	Refs.
OBn, OBn	B, O, O; 7 eq	Ru cat **215** (0.01 eq), THF, 23°, 1 h	BnO, BnO, B–O, O; (48) (*Z*)/(*E*) = 95:5, er 87.5:12.5	417

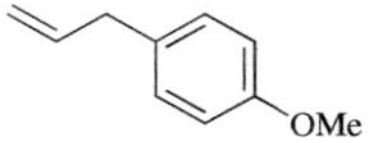

7 eq

Ru cat **215** (0.01 eq), THF, 23°, 1 h

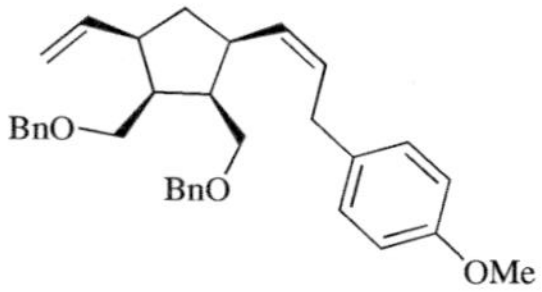

BnO BnO OMe

(51) $(Z)/(E) = 95:5$, er 90.5:9.5

417

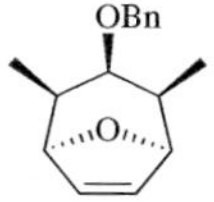

OBn

Ph

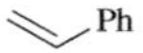

x eq

Catalyst (y eq), solvent

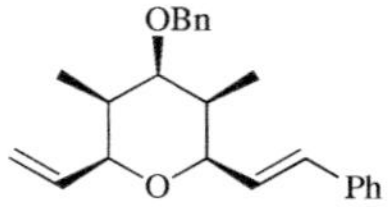

OBn O Ph

$(E)/(Z) > 98:2$

91

x	Catalyst	y	Solvent	Temp (°)	Time (h)		er
4	Ru cat **211**	0.05	—	22	44	(44)	90.0:10.0
4	Ru cat **213**	0.05	—	22	15	(56)	90.5:9.5
8	Ru cat **213**	0.02	—	–15	20	(64)	94.5:5.5
10	Mo cat **12**	0.05	benzene	22	5	(64)	83.5:16.5

TABLE 36. ASYMMETRIC RING-OPENING CROSS-METATHESIS (*Continued*)

Cyclic Alkene	Cross Partner	Conditions	Product(s) and Yield(s) (%)	Refs.

Please refer to the charts preceding the tables for the catalyst structures.

C_9

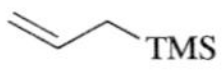

Cyclic Alkene	Cross Partner	Conditions	Product(s) and Yield(s) (%)	Refs.
N–R	TMS, 2 eq	Catalyst (x eq), solvent	TMS, N–R	96

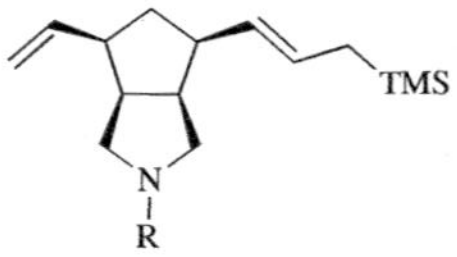

R	Catalyst	x	Solvent	Temp (°)	Time (h)	Conv. (%)[a]	$(E)/(Z)$	er
CF_3CO	Ru cat **32**	0.025	DCM	rt	2	>99	67:33	90.0:10.0
CF_3CO	Ru cat **32**	0.025	DCM	0	6	>99	67:33	91.0:9.0
Tos	Ru cat **32**	0.03	DCM	rt	12	>99	40:60	87.5:12.5
Tos	Ru cat **32**	0.03	MTBE	rt	12	>99	67:33	79.0:21.0
Tos	Ru cat **32**	0.03	THF	rt	12	>99	50:50	81.5:18.5
Tos	Ru cat **32**	0.03	benzene	rt	12	>99	67:33	76.5:23.5
Tos	Ru cat **32**	0.03	toluene	rt	12	>99	67:33	78.5:21.5
Tos	Ru cat **32**	0.03	trifluorotoluene	rt	12	92	60:40	84.5:15.5
Tos	Ru cat **32**	0.025	DCM	40	1	>99	40:60	87.5:12.5
Tos	Ru cat **32**	0.025	DCM	rt	2	>99	40:60	87.5:12.5
Tos	Ru cat **32**	0.025	DCM	0	2	>99	33:67	89.5:10.5
Tos	Ru cat **32**	0.025	DCM	–10	48	>99	33:67	89.5:10.5
Tos	Ru cat **229**	0.025	DCM	rt	18	>99	50:50	86.5:13.5
Tos	Ru cat **24**	0.025	DCM	rt	2	>99	75:24	67.0:33.0

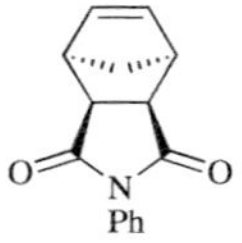

Cyclic Alkene	Cross Partner	Conditions	Product(s) and Yield(s) (%)	Refs.
O, N, Ph, O	OAc, 7 eq	Ru cat **215** (0.01 eq), THF, 23°, 1 h	O, N, Ph, O, OAc (63) $(E)/(Z) = 6:94$ er 80:20	417

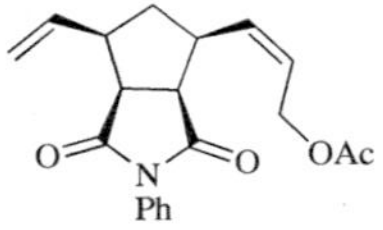

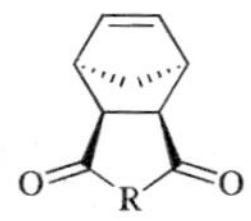

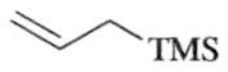

TMS

2 eq

Catalyst (x eq), DCM, rt

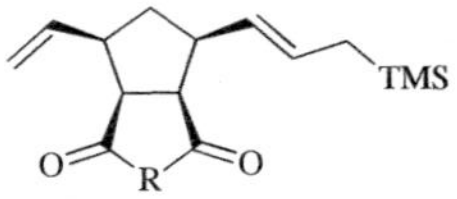

TMS

$(E)/(Z) = 67:33$ 96

R	Catalyst	x	Time (h)	Conv. (%)[a]	er
t-BuN	Ru cat **32**	0.03	96	96	99.0:1.0
t-BuN	Ru cat **229**	0.03	96	>99	97.5:2.5
t-BuN	Ru cat **24**	0.02	18	>99	67.5:32.5
O	Ru cat **32**	0.026	72	87	86.0:14.0

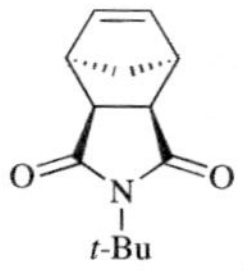

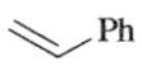

t-Bu

Ph

x eq

Catalyst (y eq),
NaI (z eq), DCM, rt, 1 h

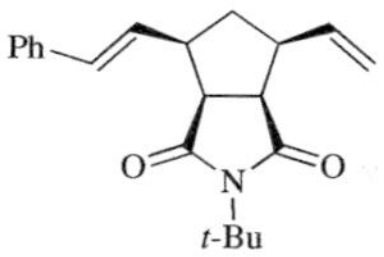

Ph

t-Bu

x	Catalyst	y	z	Temp (°)	Time (h)	Conv. (%)[a]		$(E)/(Z)$	er (E)	er (Z)	
10	Ru cat **24**	0.01	—	rt	1	—	(99)	58:42	78.5:21.5	66.5:33.5	97
10	Ru cat **25**	0.03	1	rt	1	—	(99)	55:45	87.5:12.5	75.0:25.0	97
5	Ru cat **240**	0.01	—	25	20	>98	(—)	96:4	91.0:9.0	—	95
5	Ru cat **240**	0.01	—	–10	72	>98	(—)	93:7	96.0:4.0	—	95
5	Ru cat **24**	0.01	—	25	1	>96	(—)	50:50	78.5:21.5	—	96
5	Ru cat **229**	0.025	—	25	72	98	(—)	92:8	95.5:4.5	—	96
5	Ru cat **32**	0.025	—	40	18	98	(—)	90:10	96.0:4.0	—	96

TABLE 36. ASYMMETRIC RING-OPENING CROSS-METATHESIS (*Continued*)

Cyclic Alkene	Cross Partner	Conditions	Product(s) and Yield(s) (%)	Refs.

Please refer to the charts preceding the tables for the catalyst structures.

C_9

Cyclic Alkene	Cross Partner	Conditions	Product(s) and Yield(s) (%)	Refs.
O, O, O	On-Bu, 20 eq	Ru cat **213** (0.05 eq) solvent, 22°, 24 h	On-Bu, O, O, O	1205

Solvent	Conv. (%)[a]		(Z)/(E)	er[c]
benzene	>98	(80)	81:19	80:20
THF	>98	(44)	85:15	77:23

Cyclic Alkene	Cross Partner	Conditions	Product(s) and Yield(s) (%)	Refs.
R, R, O, O, O	OAc, 7 eq	Ru cat **215** (0.01 eq), THF, 23°, 1 h	R, R, OAc, O, O, O	417

R		(E)/(Z)	er
H	(58)	2:98	87.5:12.5
Me	(65)	4:96	97.5:2.5

Cyclic Alkene	Cross Partner	Conditions	Product(s) and Yield(s) (%)	Refs.
O, O, O	Ph, 20 eq	Catalyst (0.025 eq), THF, 50°, 2 h	Ph, O, O, O	655

Catalyst		(E)/(Z)	er
Ru cat **58**	(38)	57:43	52.5:47.5
Ru cat **59**	(78)	50:50	51.0:49.0
Ru cat **60**	(9)	77:23	54.5:45.5

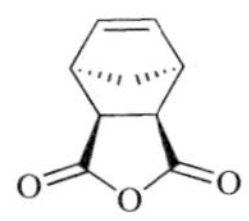

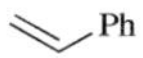

10 eq

Catalyst (x eq),
NaI (y eq), DCM, rt, 1 h

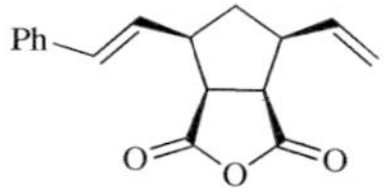

Ph

$(E)/(Z) = 50:50$ 97

Catalyst	x	y	Conv. (%)[a]	er (E)	er (Z)
Ru cat **18**	0.01	—	>95	73.5:26.5	—
Ru cat **16**	0.01	—	>95	64.5:35.5	—
Ru cat **20**	0.01	—	>95	81.0:19.0	—
Ru cat **24**	0.01	—	95	88.0:12.0	52.0:48.0
Ru cat **25**	0.03	1	96	90.0:10.0	—

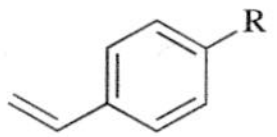

R

5 eq

Catalyst (x eq), DCM

R

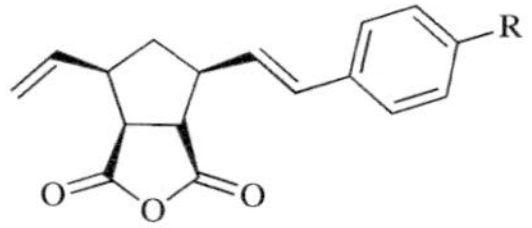

R	Catalyst	x	Temp (°)	Time (h)	Conv. (%)[a]	$(E)/(Z)$	er	
H	Ru cat **238**	0.01	25	1	>98	95:5	85.5:14.5	95
H	Ru cat **239**	0.01	25	1	>98	95:5	91.5:8.5	95
H	Ru cat **240**	0.01	25	1	>98	>97:3	94.0:6.0	95
H	Ru cat **240**	0.0005	25	15	>98	>97:3	94.0:6.0	95
H	Ru cat **240**	0.01	–10	12	>98	>97:3	96.5:3.5	95
H	Ru cat **24**	0.01	25	1	>98	50:50	88.0:12.0	95
MeO	Ru cat **240**	0.01	25	1	>98	93:7	90.5:9.5	95
CF_3	Ru cat **240**	0.01	25	1	>98	97:3	84.0:16.0	95
O_2N	Ru cat **240**	0.01	25	1	>98	—	86.0:14.0	95
MeO_2C	Ru cat **240**	0.01	25	1	>98	>97:3	89.5:10.5	95
H	Ru cat **32**	0.07	40	16	78	94:6	93.0:7.0	96
H	Ru cat **229**	0.025	25	16	83	93:7	87.5:12.5	96

TABLE 36. ASYMMETRIC RING-OPENING CROSS-METATHESIS (*Continued*)

Cyclic Alkene	Cross Partner	Conditions	Product(s) and Yield(s) (%)	Refs.

Please refer to the charts preceding the tables for the catalyst structures.

C_{10}

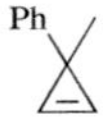

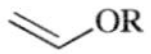

Ph

OR, *x* eq

Catalyst (*y* eq), solvent, 22°

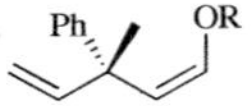

Ph, OR

R	*x*	Catalyst	*y*	Solvent	Time (h)	Conv. (%)[a]		(*Z*)/(*E*)	er	Refs.
n-Bu	10	Ru cat **151**	0.02	DCM	2	—	(79)	88:12	—	83
n-Bu	10	Mo cat **18**	0.03	benzene	0.5	93	(79)	>98:2	95.0:5.0	339
PMP	2	Mo cat **18**	0.03	benzene	0.5	95	(71)	>98:2	97.0:3.0	339

OBz, 2.05 eq

Ru cat **213** (0.05 eq), THF, 0°, 48 h

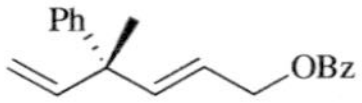

Ph, OBz

(76) (*E*)/(*Z*) = 75:25, er (*E*) 70.0:30.0 — 1143

1.5 eq

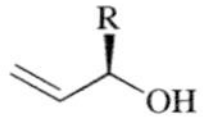

R, OH

Hov II (0.005 eq), toluene, 22°, 0.25 h

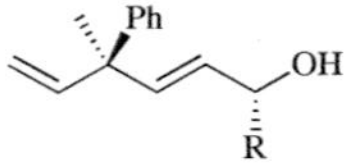

Ph, OH, R — 624

R		(*E*)/(*Z*)	dr
Me	(64)	86:14	95:5
$TBSOCH_2$	(71)	92:8	89:11
$Ph(CH_2)_2$	(80)	80:20	91:9

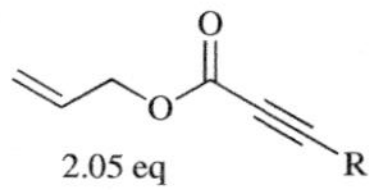

2.05 eq

R

Ru cat **213** (0.05 eq), THF, 0°, 48 h

Ph

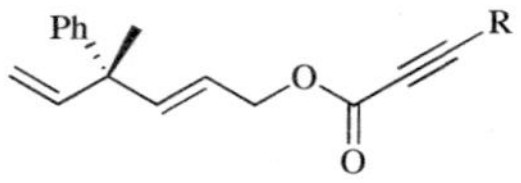

1143

R		$(E)/(Z)$	er (E)
H	(65)	80:20	93.0:7.0
Me	(72)	67:33	86.0:14.0
TMS	(63)	80:20	69.0:31.0

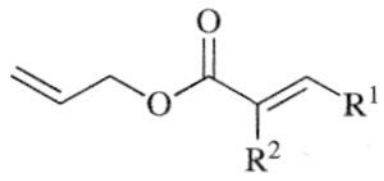

2.05 eq

R^1 R^2

Ru cat **213** (0.05 eq), THF, 0°, 48 h

Ph

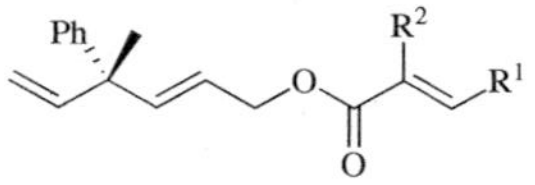

1143

R^1	R^2		$(E)/(Z)$	er (E)	er (Z)
H	H	(71)	82:18	8.0:92.0	—
H	Me	(77)	83:17	8.0:92.0	—
H	CF_3	(63)	86:14	9.0:91.0	65.0:34.0
H	Br	(80)	88:12	1.0:99.0	—
Br	H	(69)	86:14	1.0:99.0	75.0:25.0

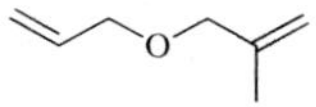

2.05 eq

Ru cat **213** (0.05 eq), THF, 0°, 48 h

Ph

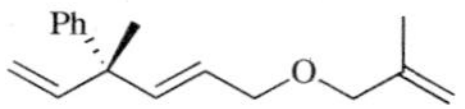

(76) $(E)/(Z)$ = 67:33, er (E) 64.0:36.0

1143

TABLE 36. ASYMMETRIC RING-OPENING CROSS-METATHESIS (*Continued*)

Cyclic Alkene	Cross Partner	Conditions	Product(s) and Yield(s) (%)	Refs.

Please refer to the charts preceding the tables for the catalyst structures.

C_{10}

Cyclic Alkene	Cross Partner	Conditions	Product(s) and Yield(s) (%)	Refs.
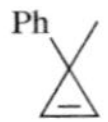 Ph	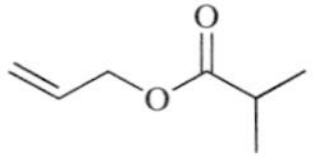 2.05 eq	Ru cat **213** (0.05 eq), THF, 0°, 48 h	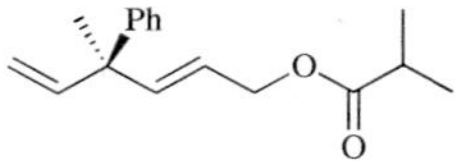 Ph; (59) (*E*)/(*Z*) = 75:25, er (*E*) 35.0:65.0	1143
	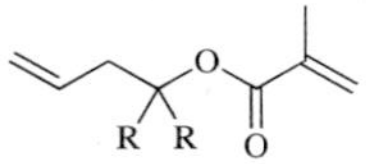 R R O; 2.05 eq	Ru cat **213** (0.05 eq), THF, 0°, 48 h	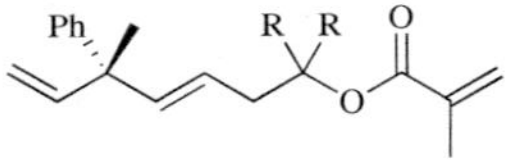 Ph, R R O	1143

R		(*E*)/(*Z*)	er (*E*)
H	(71)	80:20	80.0:20.0
Me	(60)	83:17	90.0:10.0

Cyclic Alkene	Cross Partner	Conditions	Product(s) and Yield(s) (%)	Refs.
1.5 eq	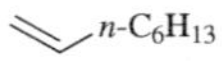 n-C_6H_{13}	Catalyst (*x* eq), solvent	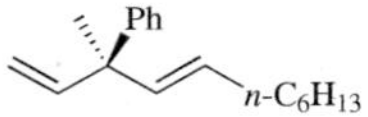 Ph, n-C_6H_{13}	

Catalyst	*x*	Solvent	Temp (°)	Time (h)	Conv. (%)[a]		(*E*)/(*Z*)	er (*E*)	Refs.
Ru cat **213**	0.05	THF	0	48	—	(59)	60:40	60.0:40.0	1143
Hov II	0.005	toluene	22	4	83	(—)	75:25	—	624

Cyclic Alkene	Cross Partner	Conditions	Product(s) and Yield(s) (%)	Refs.
	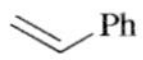 Ph; 2.05 eq	Ru cat **213** (0.05 eq), THF, 0°, 48 h	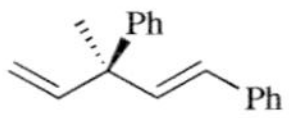 Ph, Ph; (90) (*E*)/(*Z*) > 95:5, er (*E*) 97.0:3.0	1143

2.05 eq

Ru cat **213** (0.05 eq), THF, 0°, 48 h

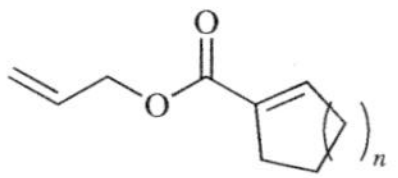

1143

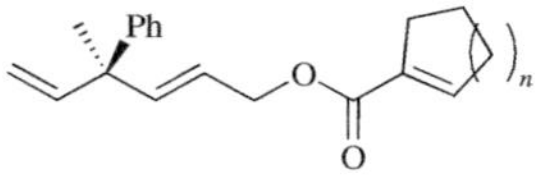

n		$(E)/(Z)$	er (E)	er (Z)
1	(72)	83:17	5.0:95.0	74.0:26.0
2	(83)	75:25	6.0:94.0	70.0:30.0

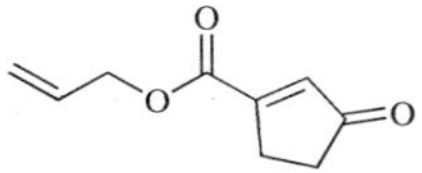

2.05 eq

Ru cat **213** (0.05 eq), THF, 0°, 48 h

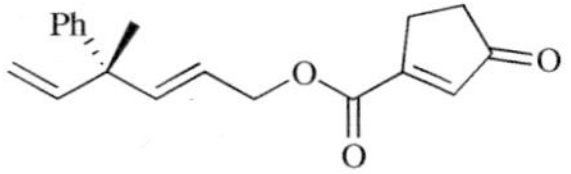

(63) $(E)/(Z)$ = 67:33, er (E) 83.0:17.0

1143

2 eq

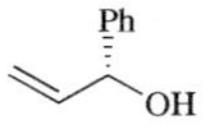

er 96.0:4.0

Ru cat **151** (0.05 eq), DCM, 22°, 8 h

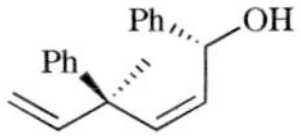

(78) $(Z)/(E)$ = 91:9, dr 93:7

83

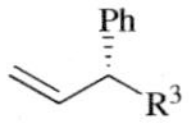

C_{10-14}

1 eq

Hov II (0.005 eq), toluene, 22°

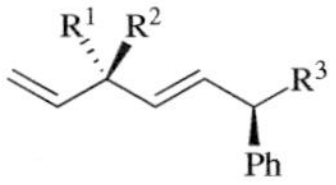

624

R^1	R^2	R^3	Time	Conv. (%)[a]		$(E)/(Z)$	dr
Ph	Me	HO	5 min	—	(87)	91:9	96:4
Ph	Me	MeO	18 h	51	(33)	86:14	79:21
Ph	Me	Me	18 h	56	(55)	90:10	91:9
Ph	$AcOCH_2$	HO	4h	—	(80)	89:11	95:5
2-Naph	Me	HO	5 min	—	(76)	91:9	96:4

TABLE 36. ASYMMETRIC RING-OPENING CROSS-METATHESIS (*Continued*)

Cyclic Alkene	Cross Partner	Conditions	Product(s) and Yield(s) (%)	Refs.

Please refer to the charts preceding the tables for the catalyst structures.

C_{10}

Cyclic Alkene	Cross Partner	Conditions	Product(s) and Yield(s) (%)	Refs.
Ts	OR, 20 eq	Ru cat **213** (0.05 eq) solvent, 22°, 24 h	OR, N, Ts	1205

R	Solvent	Conv. (%)[a]		(*Z*)/(*E*)	er[c]
n-Bu	—	>98	(90)	95:5	93.0:7.0
Cl(CH$_2$)$_2$	—	>98	(89)	>98:2	97.0:3.0
Cy	THF	>98	(93)	95:5	94.0:6.0

C_{11}

Cyclic Alkene	Cross Partner	Conditions	Product(s) and Yield(s) (%)
	Ph, 5 eq	Catalyst (*x* eq), DCM	Ph

Catalyst	*x*	Temp (°)	Time (h)	Conv (%)[a]	(*E*)/(*Z*)	er	Refs.
Ru cat **240**	0.01	25	3.5	>98	95:5	91.0:9.0	95
Ru cat **240**	0.01	–10	48	>98	96:4	95.0:5.0	95
Ru cat **32**	0.05	40	24	97	89:11	90.0:10.0	96
Ru cat **229**	0.05	40	24	96	92:8	85.5:14.5	96

Cyclic Alkene	Cross Partner	Conditions	Product(s) and Yield(s) (%)	Refs.
	OAc, 7 eq	Ru cat **215** (0.01 eq), THF, 23°, 1 h	AcO; (55) (*Z*)/(*E*) = 76:24, er (*E*) > 99.0:1.0, er (*Z*) > 99.0:1.0	98

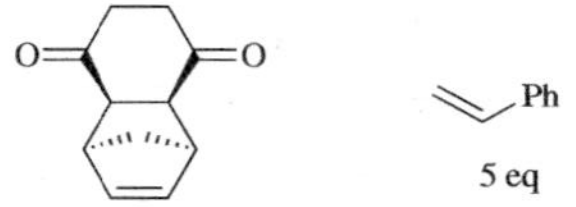

5 eq

Catalyst (x eq), DCM

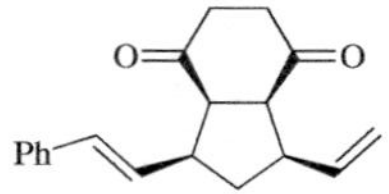

Catalyst	x	Temp (°)	Time (h)	Conv. (%)[a]	(*E*)/(*Z*)	er	
Ru cat **240**	0.01	25	120	61	95:5	93.0:7.0	95
Ru cat **32**	0.06	40	16	36	75:25	92.5:7.5	96

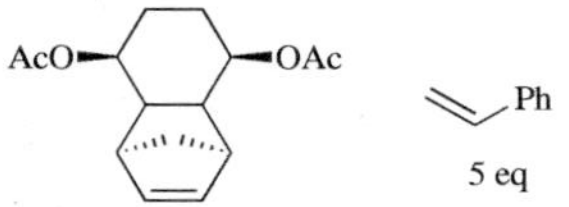

5 eq

Ru cat **240** (0.01 eq), DCM, 25°, 2 h

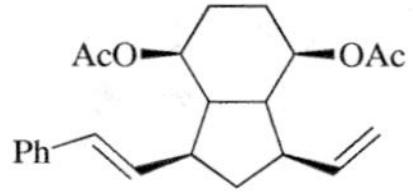

95

Conv. (%)[a] > 98, (*E*)/(*Z*) > 97:3, er 88.0:12.0

C_{14}

1 eq

Hov II (0.005 eq), toluene, 22°, 0.25 h

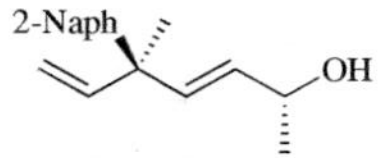

(84) (*E*)/(*Z*) = 86:14, dr 97:3

624

[a] The conversion was determined by ^{1}H NMR analysis.

[b] The efficiency was calculated as yield/conversion.

[c] The enantioselectivity was determined by HPLC analysis.

REFERENCES

[1] Calderon, N.; Chen, H. Y.; Scott, K. W. *Tetrahedron Lett.* **1967**, *8*, 3327.
[2] Astruc, D. *New J. Chem.* **2005**, *29*, 42.
[3] *Handbook of Metathesis,* 2nd ed.; Grubbs, R. H., Wenzel, A. G., O'Leary, D. J., Khosravi, E., Eds.; Wiley: Weinheim, Germany, 2015; Vols. 1–3.
[4] *Olefin Metathesis: Theory and Practice*; Grela, K., Ed.; Wiley: Hoboken, NJ, 2014.
[5] Meek, S. J.; O'Brien, R. V.; Llaveria, J.; Schrock, R. R.; Hoveyda, A. H. *Nature* **2011**, *471*, 461.
[6] Hoveyda, A. H.; Malcolmson, S. J.; Meek, S. J.; Zhugralin, A. R. *Angew. Chem., Int. Ed.* **2010**, *49*, 34.
[7] Nicolaou, K. C.; Bulger, P. G.; Sarlah, D. *Angew. Chem., Int. Ed.* **2005**, *44*, 4490.
[8] Morzycki, J. W. *Steroids* **2011**, *76*, 949.
[9] Lin, Y. A.; Chalker, J. M.; Davis, B. G. *ChemBioChem* **2009**, *10*, 959.
[10] Yet, L. *Chem. Rev.* **2000**, *100*, 2963.
[11] Binder, J. B.; Raines, R. T. *Curr. Opin. Chem. Biol.* **2008**, *12*, 767.
[12] Khan, S. N.; Kim, A.; Grubbs, R. H.; Kwon, Y.-U. *Org. Lett.* **2012**, *14*, 2952.
[13] Franzen, R. G. *Top. Catal.* **2016**, *59*, 1143.
[14] Buchmeiser, M. R. *Chem. Rev.* **2000**, *100*, 1565.
[15] Leitgeb, A.; Wappel, J.; Slugovc, C. *Polymer* **2010**, *51*, 2927.
[16] Bielawski, C. W.; Grubbs, R. H. *Prog. Polym. Sci.* **2007**, *32*, 1.
[17] Mol, J. C. *J. Mol. Catal. A: Chem.* **2004**, *213*, 39.
[18] Požgan, F.; Dixneuf, P. H. In *Metathesis Chemistry: From Nanostructure Design to Synthesis of Advanced Materials*; Imamoglu, Y., Dragutan, V., Karabulut, S., Eds.; Springer: Dorgrecht, Netherlands, 2007; Vol. 243, pp 195–222.
[19] Fürstner, A. *Angew. Chem., Int. Ed.* **2000**, *39*, 3012.
[20] The Nobel Prize in Chemistry 2005. https://www.nobelprize.org/prizes/chemistry/2005/summary/ (accessed Jan 11, 2021).
[21] *Handbook of Metathesis*, 1st ed.; Grubbs, R. H., Ed.; Wiley: Weinheim, Germany, 2003; Vols. 1–3.
[22] Lozano-Vila, A. M.; Monsaert, S.; Bajek, A.; Verpoort, F. *Chem. Rev.* **2010**, *110*, 4865.
[23] Armstrong, S. K. *J. Chem. Soc., Perkin Trans. 1* **1998**, 371.
[24] Felpin, F. X.; Lebreton, J. *Eur. J. Org. Chem.* **2003**, 3693.
[25] Monfette, S.; Fogg, D. E. *Chem. Rev.* **2009**, *109*, 3783.
[26] Yet, L. *Org. React.* **2016**, *89*, 1.
[27] Frenzel, U.; Nuyken, O. *J. Polym. Sci., Part A: Polym. Chem.* **2002**, *40*, 2895.
[28] Brzezinska, K. R.; Wagener, K. B.; Burns, G. T.. *J. Polym Sci., Part A: Polym. Chem.* **1999**, *37*, 849.
[29] Schwendeman, J. E.; Church, A. C.; Wagener, K. B. *Adv. Synth. Catal.* **2002**, *344*, 597.
[30] Hopkins, T. E.; Wagener, K. B. *J. Mol. Catal. A: Chem.* **2004**, *213*, 93.
[31] Mutlu, H.; de Espinosa, L. M.; Meier, M. A. R. *Chem. Soc. Rev.* **2011**, *40*, 1404.
[32] Blackwell, H. E.; O'Leary, D. J.; Chatterjee, A. K.; Washenfelder, R. A.; Bussmann, D. A.; Grubbs, R. H. *J. Am. Chem. Soc.* **2000**, *122*, 58.
[33] Vernall, A. J.; Abell, A. D. *Aldrichimica Acta* **2003**, *36*, 93.
[34] Connon, S. J.; Blechert, S. *Angew. Chem., Int. Ed.* **2003**, *42*, 1900.
[35] Pederson, R. L.; Fellows, I. M.; Ung, T. A.; Ishihara, H.; Hajela, S. P. *Adv. Synth. Catal.* **2002**, *344*, 728.
[36] Chatterjee, A. K.; Choi, T.-L.; Sanders, D. P.; Grubbs, R. H. *J. Am. Chem. Soc.* **2003**, *125*, 11360.
[37] Tallarico, J. A.; Randall, M. L.; Snapper, M. L. *Tetrahedron* **1997**, *53*, 16511.
[38] Wright, D. L.; Usher, L. C.; Estrella-Jimenez, M. *Org. Lett.* **2001**, *3*, 4275.
[39] Debleds, O.; Campagne, J.-M. *J. Am. Chem. Soc.* **2008**, *130*, 1562.
[40] Samojłowicz, C.; Borré, E.; Mauduit, M.; Grela, K. *Adv. Synth. Catal.* **2011**, 1993.
[41] Blechert, S.; Stapper, C. *Eur. J. Org. Chem.* **2002**, 2855.
[42] Choi, T. L.; Grubbs, R. H. *Chem. Commun.* **2001**, 2648.
[43] Wallace, D. J. *Tetrahedron Lett.* **2003**, *44*, 2145.
[44] Li, J.; Lee, D. *Eur. J. Org. Chem.* **2011**, 4269.
[45] Schuster, M.; Blechert, S. *Angew. Chem., Int. Ed. Engl.* **1997**, *36*, 2036.
[46] Grubbs, R. H.; Wenzel, A. G.; Chatterjee, A. K. In *Comprehensive Organometallic Chemistry III*; Crabtree, R. H., Ed.; Elsevier: Oxford, 2007; pp 179–205.
[47] Michalak, M.; Gułajski, Ł.; Grela, K. *Sci. Synth.* **2010**, 327.
[48] Żukowska, K.; Grela, K. In *Comprehensive Organic Synthesis II,* 2nd ed.; Elsevier: Amsterdam, 2014; pp 1257–1301.

[49] Żukowska, K.; Grela, K. In *Olefin Metathesis: Theory and Practice* (ed. Grela, K.), Wiley: Hoboken, NJ, 2014; pp 37–83.
[50] O'Leary, D. J.; O'Neil, G. W. In *Handbook of Metathesis*; Grubbs, R. H., O'Leary, D. J., Eds.; Wiley: Weinheim, Germany, 2015; Vol. 2, pp 171–294.
[51] Deraedt, C.; d'Halluin, M.; Astruc, D. *Eur. J. Inorg. Chem.* **2013**, *4881.*
[52] Fürstner, A. *Chem. Commun.* **2011**, *47*, 6505.
[53] Hong, S. H.; Wenzel, A. G.; Salguero, T. T.; Day, M. W.; Grubbs, R. H. *J. Am. Chem. Soc.* **2007**, *129*, 7961.
[54] Nascimento, D. L.; Fogg, D. E. *J. Am. Chem. Soc.* **2019**, *141*, 19236.
[55] McClennan, W. L.; Rufh, S. A.; Lummiss, J. A. M.; Fogg, D. E. *J. Am. Chem. Soc.* **2016**, *138*, 14668.
[56] Jawiczuk, M.; Marczyk, A.; Trzaskowski, B. *Catalysts* **2020**, *10*, 887.
[57] Mol, J. C.; Moulijn, J. A.; Boelhouw, C. *J. Chem. Soc., Chem. Commun.* **1968**, 633.
[58] Calderon, N.; Ofstead, E. A.; Ward, J. P.; Judy, W. A.; Scott, K. W. *J. Am. Chem. Soc.* **1968**, *90*, 4133.
[59] Schrock, R. R. *Chem. Rev.* **2009**, *109*, 3211.
[60] Nelson, D. J.; Manzini, S.; Urbina-Blanco, C. A.; Nolan, S. P. *Chem. Commun.* **2014**, *50*, 10355.
[61] Dias, E. L.; Nguyen, S. T.; Grubbs, R. H. *J. Am. Chem. Soc.* **1997**, *119*, 3887.
[62] Sanford, M. S.; Love, J. A.; Grubbs, R. H. *J. Am. Chem. Soc.* **2001**, *123*, 6543.
[63] Vyboishchikov, S. F.; Bühl, M.; Thiel, W. *Chem.—Eur. J.* **2002**, *8*, 3962.
[64] Adlhart, C.; Chen, P. *J. Am. Chem. Soc.* **2004**, *126*, 3496.
[65] Kingsbury, J. S.; Hoveyda, A. H. *J. Am. Chem. Soc.* **2005**, *127*, 4510.
[66] Love, J. A.; Morgan, J. P.; Trnka, T. M.; Grubbs, R. H. *Angew. Chem., Int. Ed.* **2002**, *41*, 4035.
[67] Trzaskowski, B.; Grela, K. *Organometallics* **2013**, *32*, 3625.
[68] Thiel, V.; Hendann, M.; Wannowius, K.-J.; Plenio, H. *J. Am. Chem. Soc.* **2012**, *134*, 1104.
[69] Urbina-Blanco, C. A.; Poater, A.; Lebl, T.; Manzini, S.; Slawin, A. M. Z.; Cavallo, L.; Nolan, S. P. *J. Am. Chem. Soc.* **2013**, *135*, 7073.
[70] Ashworth, I. W.; Hillier, I. H.; Nelson, D. J.; Percy, J. M.; Vincent, M. A. *Chem. Commun.* **2011**, *47*, 5428.
[71] Nuñez-Zarur, F.; Solans-Monfort, X.; Rodríguez-Santiago, L.; Sodupe, M. *Organometallics* **2012**, *31*, 4203.
[72] Pazio, A.; Woźniak, K.; Grela, K.; Trzaskowski, B. *Organometallics* **2015**, *34*, 563.
[73] Kingsbury, J. S.; Harrity, J. P. A.; Bonitatebus, P. J.; Hoveyda, A. H. *J. Am. Chem. Soc.* **1999**, *121*, 791.
[74] Bieniek, M.; Michrowska, A.; Usanov, D. L.; Grela, K. *Chem.—Eur. J.* **2008**, *14*, 806.
[75] Vorfalt, T.; Wannowius, K.-J.; Plenio, H. *Angew. Chem., Int. Ed.* **2010**, *49*, 5533.
[76] Bates, J. M.; Lummiss, J. A. M.; Bailey, G. A.; Fogg, D. E. *ACS Catal.* **2014**, *4*, 2387.
[77] Lujan, C.; Nolan, S. P. *Catal. Sci. Technol.* **2012**, *2*, 1027.
[78] Hoveyda, A. H. *J. Org. Chem.* **2014**, *79*, 4763.
[79] Malcolmson, S. J.; Meek, S. J.; Sattely, E. S.; Schrock, R. R.; Hoveyda, A. H. *Nature* **2008**, *456*, 933.
[80] Flook, M. M.; Jiang, A. J.; Schrock, R. R.; Müller, P.; Hoveyda, A. H. *J. Am. Chem. Soc.* **2009**, *131*, 7962.
[81] Jiang, A. J.; Zhao, Y.; Schrock, R. R.; Hoveyda, A. H. *J. Am. Chem. Soc.* **2009**, *131*, 16630.
[82] Keitz, B. K.; Endo, K.; Patel, P. R.; Herbert, M. B.; Grubbs, R. H. *J. Am. Chem. Soc.* **2012**, *134*, 693.
[83] Koh, M. J.; Khan, R. K. M.; Torker, S.; Hoveyda, A. H. *Angew. Chem., Int. Ed.* **2014**, *53*, 1968.
[84] Keitz, B. K.; Endo, K.; Herbert, M. B.; Grubbs, R. H. *J. Am. Chem. Soc.* **2011**, *133*, 9686.
[85] Occhipinti, G.; Hansen, F. R.; Törnroos, K. W.; Jensen, V. R. *J. Am. Chem. Soc.* **2013**, *135*, 3331.
[86] Liu, P.; Xu, X.; Dong, X.; Keitz, B. K.; Herbert, M. B.; Grubbs, R. H.; Houk, K. N. *J. Am. Chem. Soc.* **2012**, *134*, 1464.
[87] Dang, Y.; Wang, Z.-X.; Wang, X. *Organometallics* **2012**, *31*, 7222.
[88] Johns, A. M.; Ahmed, T. S.; Jackson, B. W.; Grubbs, R. H.; Pederson, R. L. *Org. Lett.* **2016**, *18*, 772.
[89] Montgomery, T. P; S., Ahmed, T. S.; Grubbs, R. H. *Angew. Chem., Int. Ed.* **2017**, *56*, 11024.
[90] Stenne, B.; Collins, S. K. In *Olefin Metathesis: Theory and Practice*; Grela, K., Ed.; Wiley: Hoboken, NJ, 2014; pp 233–267.
[91] Cortez, G. A.; Baxter, C. A.; Schrock, R. R.; Hoveyda, A. H. *Org. Lett.* **2007**, *9*, 2871.
[92] Seiders, T. J.; Ward, D. W.; Grubbs, R. H. *Org. Lett.* **2001**, *3*, 3225.
[93] Van Veldhuizen, J. J.; Garber, S. B.; Kingsbury, J. S.; Hoveyda, A. H. J. *Am. Chem. Soc.* **2002**, *124*, 4954.
[94] Fournier, P. A.; Collins, S. K. *Organometallics* **2007**, *26*, 2945.
[95] Tiede, S.; Berger, A.; Schlesiger, D.; Rost, D.; Lühl, A.; Blechert, S. *Angew. Chem., Int. Ed.* **2010**, *49*, 3972.
[96] Kannenberg, A.; Rost, D.; Eibauer, S.; Tiede, S.; Blechert, S. *Angew. Chem., Int. Ed.* **2011**, *50*, 3299.

[97] Berlin, J. M.; Goldberg, S. D.; Grubbs, R. H. *Angew. Chem., Int. Ed.* **2006**, *45*, 7591.
[98] Hartung, J.; Dornan, P. K.; Grubbs, R. H. *J. Am. Chem. Soc.* **2014**, *136*, 13029.
[99] La, D. S.; Sattely, E. S.; Ford, J. G.; Schrock, R. R.; Hoveyda, A. H. *J. Am. Chem. Soc.* **2001**, *123*, 7767.
[100] Gillingham, D. G.; Kataoka, O.; Garber, S. B.; Hoveyda, A. H. *J. Am. Chem. Soc.* **2004**, *126*, 12288.
[101] Gillingham, D. G.; Hoveyda, A. H. *Angew. Chem., Int. Ed.* **2007**, *46*, 3860.
[102] Cortez, G. A.; Schrock, R. R.; Hoveyda, A. H. *Angew. Chem., Int. Ed.* **2007**, *46*, 4534.
[103] Joelle, P. *Curr. Top. Med. Chem.* **2005**, *5*, 1559.
[104] Macnaughtan, M. L.; Gary, J. B.; Gerlach, D. L.; Johnson, M. J. A.; Kampf, J. W. *Organometallics* **2009**, *28*, 2880.
[105] Sashuk, V.; Samojłowicz, C.; Szadkowska, A.; Grela, K. *Chem. Commun.* **2008**, 2468.
[106] Koh, M. J.; Nguyen, T. T.; Zhang, H.; Schrock, R. R.; Hoveyda, A. H. *Nature* **2016**, *531*, 459.
[107] Schrock, R. R.; Murdzek, J. S.; Bazan, G. C.; Robbins, J.; DiMare, M.; O'Regan, M. *J. Am. Chem. Soc.* **1990**, *112*, 3875.
[108] Ondi, L.; Nagy, G. M.; Czirok, J. B.; Bucsai, A.; Frater, G. E. *Org. Process Res. Dev.* **2016**, *20*, 1709.
[109] Heppekausen, J.; Fürstner, A. *Angew. Chem., Int. Ed.* **2011**, *50*, 7829.
[110] Bidange, J.; Fischmeister, C.; Bruneau, C. *Chem.—Eur. J.* **2016**, *22*, 12226.
[111] Spekreijse, J.; Sanders, J. P. M.; Bitter, J. H.; Scott, E. L. *ChemSusChem* **2017**, *10*, 470.
[112] Behr, A.; Krema, S.; Kamper, A. *RSC Adv.* **2012**, *2*, 12775.
[113] Park, C. P.; Van Wingerden, M. M.; Han, S.-Y.; Kim, D.-P.; Grubbs, R. H. *Org. Lett.* **2011**, *13*, 2398.
[114] Biermann, U.; Bornscheuer, U.; Meier, M. A. R.; Metzger, J. O.; Schäfer, H. J. *Angew. Chem., Int. Ed.* **2011**, *50*, 3854.
[115] Gawin, R.; Kozakiewicz, A.; Guńka, P. A.; Dąbrowski, P.; Skowerski, K. *Angew. Chem., Int. Ed.* **2017**, *56*, 910.
[116] Wipf, P.; Rector, S. R.; Takahashi, H. *J. Am. Chem. Soc.* **2002**, *124*, 14848.
[117] Ratnayake, A. S.; Hemscheidt, T. *Org. Lett.* **2002**, *4*, 4667.
[118] Jermacz, I.; Maj, J.; Morzycki, J. W.; Wojtkielewicz, A. *Toxicol. Mech. Methods* **2008**, *18*, 469.
[119] Adamo, M. F. A.; Disetti, P.; Piras, L. *Tetrahedron Lett.* **2009**, *50*, 3580.
[120] Matteson, D. S. *Tetrahedron* **1989**, *45*, 1859.
[121] Brown, H. C.; Hamaoka, T.; Ravindran, N. *J. Am. Chem. Soc.* **1973**, *95*, 5786.
[122] Brown, H. C.; Hamaoka, T.; Ravindran, N. *J. Am. Chem. Soc.* **1973**, *95*, 6456.
[123] Miyaura, N.; Suzuki, A. *Chem. Rev.* **1995**, *95*, 2457.
[124] Blackwell, H. E.; O'Leary, D. J.; Chatterjee, A. K.; Washenfelder, R. A.; Bussmann, D. A.; Grubbs, R. H. J. *Am. Chem. Soc.* **2000**, *122*, 58.
[125] Morrill, C.; Funk, T. W.; Grubbs, R. H. *Tetrahedron Lett.* **2004**, *45*, 7733.
[126] Goldberg, S. D.; Grubbs, R. H. *Angew. Chem., Int. Ed.* **2002**, *41*, 807.
[127] Wang, Z. J.; Spiccia, N. D.; Jackson, W. R.; Robinson, A. J. *J. Pept. Sci.* **2013**, *19*, 470.
[128] Sheddan, N. A.; Mulzer, J. *Org. Lett.* **2006**, *8*, 3101.
[129] Chatterjee, A. K.; Sanders, D. P.; Grubbs, R. H. *Org. Lett.* **2002**, *4*, 1939.
[130] Forman, G. S.; Tooze, R. P. *J. Organomet. Chem.* **2005**, *690*, 5863.
[131] Amans, D.; Bellosta, V.; Cossy, J. *Angew. Chem., Int. Ed.* **2006**, *45*, 5870.
[132] Funk, T. W.; Efskind, J.; Grubbs, R. H. *Org. Lett.* **2004**, *7*, 187.
[133] Řezanka, M.; Eignerová, B.; Jindřich, J.; Kotora, M. *Eur. J. Org. Chem.* **2010**, 6256.
[134] Eignerová, B.; Sedlák, D.; Dračínský, M.; Bartůněk, P.; Kotora, M. *J. Med. Chem.* **2010**, *53*, 6947.
[135] Eignerová, B.; Dračínský, M.; Kotora, M. *Eur. J. Org. Chem.* **2008**, 4493.
[136] Eignerová, B.; Slavíková, B.; Buděšínský, M.; Dračínský, M.; Klepetářová, B.; Šťastná, E.; Kotora, M. *J. Med. Chem.* **2009**, *52*, 5753.
[137] Imhof, S.; Randl, S.; Blechert, S. *Chem. Commun.* **2001**, 1692.
[138] Prchalová, E.; Votruba, I.; Kotora, M. *J. Fluorine Chem.* **2012**, *141*, 49.
[139] Chatterjee, A. K.; Morgan, J. P.; Scholl, M.; Grubbs, R. H. *J. Am. Chem. Soc.* **2000**, *122*, 3783.
[140] Samojłowicz, C.; Bieniek, M.; Pazio, A.; Makal, A.; Woźniak, K.; Poater, A.; Cavallo, L.; Wójcik, J.; Zdanowski, K.; Grela, K. *Chem.—Eur. J.* **2011**, *17*, 12981.
[141] Samojłowicz, C.; Bieniek, M.; Zarecki, A.; Kadyrov, R.; Grela, K. *Chem. Commun.* **2008**, 6282.
[142] Thibaudeau, S.; Fuller, R.; Gouverneur, V. *Org. Biomol. Chem.* **2004**, *2*, 1110.
[143] Velder, J.; Ritter, S.; Lex, J.; Schmalz, H.-G. *Synthesis* **2006**, 273.
[144] Chatterjee, A. K.; Choi, T.-L.; Grubbs, R. H. *Synlett* **2001**, 1034.
[145] Chatterjee, A. K.; Toste, F. D.; Choi, T.-L.; Grubbs, R. H. *Adv. Synth. Catal.* **2002**, *344*, 634.
[146] Marsh, G. P.; Parsons, P. J.; McCarthy, C.; Corniquet, X. G. *Org. Lett.* **2007**, *9*, 2613.
[147] Dash, J.; Arseniyadis, S.; Cossy, J. *Adv. Synth. Catal.* **2007**, *349*, 152.
[148] Lafaye, K.; Bosset, C.; Nicolas, L.; Guérinot, A.; Cossy J. *Beilstein J. Org. Chem.* **2015**, *11*, 2223.

149 Crowe, W. E.; Goldberg, D. R. *J. Am. Chem. Soc.* **1995**, *117*, 5162.
150 Rivard, M.; Blechert, S. *Eur. J. Org. Chem.* **2003**, 2225.
151 Bai, C.-X.; Lu, X.-B.; He, R.; Zhang, W.-Z.; Feng, X.-J. *Org. Biomol. Chem.* **2005**, *3*, 4139.
152 Gawin, R.; Tracz, A.; Chwalba, M.; Kozakiewicz, A.; Trzaskowski, B.; Skowerski, K. *ACS Catal.* **2017**, *7*, 5443.
153 Randl, S.; Gessler, S.; Wakamatsu, H.; Blechert, S. *Synlett* **2001**, 0430.
154 Brouwer, A. J.; Elgersma, R. C.; Jagodzinska, M.; Rijkers, D. T. S.; Liskamp, R. M. J. *Bioorg. Med. Chem. Lett.* **2008**, *18*, 78.
155 Michrowska, A.; List, B. *Nat. Chem.* **2009**, *1*, 225.
156 Blanco, O. M.; Castedo, L. *Synlett* **1999**, 557.
157 BouzBouz, S.; Cossy, J. *Org. Lett.* **2001**, *3*, 1451.
158 Cran, J. W.; Krafft, M. E.; Seibert, K. A.; Haxell, T. F. N.; Wright, J. A.; Hirosawa, C.; Abboud, K. A. *Tetrahedron* **2011**, *67*, 9922.
159 Krafft, M. E.; Haxell, T. F. N. *J. Am. Chem. Soc.* **2005**, *127*, 10168.
160 Grela, K.; Harutyunyan, S.; Michrowska, A. *Angew. Chem., Int. Ed.* **2002**, *41*, 4038.
161 Boeda, F.; Bantreil, X.; Clavier, H.; Nolan, S. P. *Adv. Synth. Catal.* **2008**, *350*, 2959.
162 ElMarrouni, A.; Fukuda, A.; Heras, M.; Arseniyadis, S.; Cossy, J. *J. Org. Chem.* **2010**, *75*, 8478.
163 Kirschning, A.; Harmrolfs, K.; Mennecke, K.; Messinger, J.; Schön, U.; Grela, K. *Tetrahedron Lett.* **2008**, *49*, 3019.
164 Demchuk, O. M.; Pietrusiewicz, K. M.; Michrowska, A.; Grela, K. *Org. Lett.* **2003**, *5*, 3217.
165 Bargiggia, F. C.; Murray, W. V. *J. Org. Chem.* **2005**, *70*, 9636.
166 Choi, T.-L.; Chatterjee, A. K.; Grubbs, R. H. *Angew. Chem., Int. Ed.* **2001**, *40*, 1277.
167 Voigtritter, K.; Ghorai, S.; Lipshutz, B. H. *J. Org. Chem.* **2011**, *76*, 4697.
168 BouzBouz, S.; De Lemos, E.; Cossy, J. *Adv. Synth. Catal.* **2002**, *344*, 627.
169 BouzBouz, S.; Simmons, R.; Cossy, J. *Org. Lett.* **2004**, *6*, 3465.
170 Choi, T.-L.; Lee, C. W.; Chatterjee, A. K.; Grubbs, R. H. *J. Am. Chem. Soc.* **2001**, *123*, 10417.
171 Michrowska, A.; Bujok, R.; Harutyunyan, S.; Sashuk, V.; Dolgonos, G.; Grela, K. *J. Am. Chem. Soc.* **2004**, *126*, 9318.
172 Raju, R.; Howell, A. R. *Org. Lett.* **2006**, *8*, 2139.
173 Moïse, J.; Arseniyadis, S.; Cossy, J. *Org. Lett.* **2007**, *9*, 1695.
174 Ferrié, L.; BouzBouz, S.; Cossy, J. *Org. Lett.* **2009**, *11*, 5446.
175 Randl, S.; Blechert, S. *J. Org. Chem.* **2003**, *68*, 8879.
176 O'Leary, D. J.; Blackwell, H. E.; Washenfelder, R. A.; Grubbs, R. H. *Tetrahedron Lett.* **1998**, *39*, 7427.
177 Stewart, I. C.; Douglas, C. J.; Grubbs, R. H. *Org. Lett.* **2008**, *10*, 441.
178 Kanemitsu, T.; Seeberger, P. H. *Org. Lett.* **2003**, *5*, 4541.
179 Wdowik, T.; Samojłowicz, C.; Jawiczuk, M.; Zarecki, A.; Grela, K. *Synlett* **2010**, *2931*.
180 Barluenga, S.; Dakas, P.-Y.; Ferandin, Y.; Meijer, L.; Winssinger, N. *Angew. Chem., Int. Ed.* **2006**, *45*, 3951.
181 Skowerski, K.; Szczepaniak, G.; Wierzbicka, C.; Gulajski, L.; Bieniek, M.; Grela, K. *Catal. Sci. Technol.* **2012**, *2*, 2424.
182 Hoveyda, H. R.; Vézina, M. *Org. Lett.* **2005**, *7*, 2113.
183 Koh, M. J.; Khan, R. K. M.; Torker, S.; Yu, M.; Mikus, M. S.; Hoveyda, A. H. *Nature* **2015**, *517*, 181.
184 Mann, T. J.; Speed, A. W. H.; Schrock, R. R.; Hoveyda, A. H. *Angew. Chem., Int. Ed.* **2013**, *52*, 8395.
185 van Lierop, B. J.; Lummiss, J. A. M.; Fogg, D. E. In *Olefin Metathesis: Theory and Practice*; Grela, K., Ed.; Wiley: Weinheim, Germany, 2014; pp 85–152.
186 Nicolaou, K. C.; Hughes, R.; Cho, S. Y.; Winssinger, N.; Labischinski, H.; Endermann, R. *Chem.—Eur. J.* **2001**, *7*, 3824.
187 Blackwell, H. E.; O'Leary, D. J.; Chatterjee, A. K.; Washenfelder, R. A.; Bussmann, D. A.; Grubbs, R. H. J. *Am. Chem. Soc.* **2000**, *122*, 58.
188 Rega, M.; Jiménez, C.; Rodríguez, J. *Steroids* **2007**, *72*, 729.
189 Chen, G.-W.; Kirschning, A. *Chem.—Eur. J.* **2002**, *8*, 2717.
190 Yadav, J. S.; Somaiah, R.; Ravindar, K.; Chandraiah, L. *Tetrahedron Lett.* **2008**, *49*, 2848.
191 Chung, W.-J.; Carlson, J. S.; Bedke, D. K.; Vanderwal, C. D. *Angew. Chem., Int. Ed.* **2013**, *52*, 10052.
192 Knight, D. W.; Morgan, I. R.; Proctor, A. J. *Tetrahedron Lett.* **2010**, *51*, 638.
193 Jana, A.; Grela, K. *Org. Lett.* **2017**, *19*, 520.
194 Mauduit, M.; Caijo, F.; Crévisy, C. (French National Centre for Scientific Research; Rennes School of Chemistry) Intl. Patent WO 2010/037786, April 8, 2010.
195 Grellepois, F.; Crousse, B.; Bonnet-Delpon, D.; Bégué, J.-P. *Org. Lett.* **2005**, *7*, 5219.

196 Macnaughtan, M. L.; Johnson, M. J. A.; Kampf, J. W. *J. Am. Chem. Soc.* **2007**, *129*, 7708.
197 Bandini, M.; Cozzi, P. G.; Licciulli, S.; Umani-Ronchi, A. *Synthesis* **2004**, 409.
198 Jacobs, T.; Rybak, A.; Meier, M. A. R. *Appl. Catal., A* **2009**, *353*, 32.
199 Bieniek, M.; Samojłowicz, C.; Sashuk, V.; Bujok, R.; Śledź, P.; Lugan, N.; Lavigne, G.; Arlt, D.; Grela, K. *Organometallics* **2011**, *30*, 4144.
200 Pietraszuk, C.; Fischer, H.; Rogalski, S.; Marciniec, B. *J. Organomet. Chem.* **2005**, *690*, 5912.
201 Marciniec, B. *Coord. Chem. Rev.* **2005**, *249*, 2374.
202 Żak, P.; Pietraszuk, C.; Marciniec, B. *J. Mol. Catal. A: Chem.* **2008**, *289*, 1.
203 Pietraszuk, C.; Rogalski, S.; Majchrzak, M.; Marciniec, B. *J. Organomet. Chem.* **2006**, *691*, 5476.
204 Feher, F. J.; Soulivong, D.; Eklund, A. G.; Wyndham, K. D. *Chem. Commun.* **1997**, 1185.
205 Żak, P.; Marciniec, B.; Majchrzak, M.; Pietraszuk, C. *J. Organomet. Chem.* **2011**, *696*, 887.
206 Vautravers, N. R.; Andre, P.; Slawin, A. M. Z.; Cole-Hamilton, D. J. *Org. Biomol. Chem.* **2009**, *7*, 717.
207 Cheng, G.; Slawin, A. M. Z.; Vautravers, N. R.; Andre, P.; Morris, R. E.; Samuel, I. D. W.; Cole-Hamilton, D. *Org. Biomol. Chem.* **2011**, *9*, 1189.
208 Crowe, W. E.; Goldberg, D. R.; Zhang, Z. J. *Tetrahedron Lett.* **1996**, *37*, 2117.
209 Purser, S.; Claridge, T. D. W.; Odell, B.; Moore, P. R.; Gouverneur, V. *Org. Lett.* **2008**, *10*, 4263.
210 Thibaudeau, S.; Gouverneur, V. *Org. Lett.* **2003**, *5*, 4891.
211 Huber, J. D.; Perl, N. R.; Leighton, J. L. *Angew. Chem., Int. Ed.* **2008**, *47*, 3037.
212 Feng, J.; Schuster, M.; Blechert, S. *Synlett* **1997**, *1*, 129.
213 Dunne, K. S.; Lee, S. E.; Gouverneur, V. *J. Organomet. Chem.* **2006**, *691*, 5246.
214 Bisaro, F.; Gouverneur, V. *Tetrahedron Lett.* **2003**, *44*, 7133.
215 Vinokurov, N.; Michrowska, A.; Szmigielska, A.; Drzazga, Z.; Wójciuk, G.; Demchuk, O. M.; Grela, K.; Pietrusiewicz, K. M.; Butenschön, H. *Adv. Synth. Catal.* **2006**, *348*, 931.
216 Vinokurov, N.; Garabatos-Perera, J. R.; Zhao-Karger, Z.; Wiebcke, M.; Butenschön, H. *Organometallics* **2008**, *27*, 1878.
217 Bisaro, F.; Gouverneur, V. *Tetrahedron* **2005**, *61*, 2395.
218 Lera, M.; Hayes, C. J. *Org. Lett.* **2001**, *3*, 2765.
219 Pradère, U.; Clavier, H.; Roy, V.; Nolan, S. P.; Agrofoglio, L. A. *Eur. J. Org. Chem.* **2011**, 7324.
220 Malla, R. K.; Ridenour, J. N.; Spilling, C. D. *Beilstein J. Org. Chem.* **2014**, *10*, 1933.
221 He, A.; Yan, B.; Thanavaro, A.; Spilling, C. D.; Rath, N. P. *J. Org. Chem.* **2004**, *69*, 8643.
222 Liu, Z.; Rainier, J. D. *Org. Lett.* **2005**, *7*, 131.
223 Katayama, H.; Urushima, H.; Nishioka, T.; Wada, C.; Nagao, M.; Ozawa, F. *Angew. Chem., Int. Ed.* **2000**, *39*, 4513.
224 Spagnol, G.; Heck, M.-P.; Nolan, S. P.; Mioskowski, C. *Org. Lett.* **2002**, *4*, 1767.
225 Lin, Y. A.; Davis B. G. *Beilstein J. Org. Chem.* **2010**, *6*, 1219.
226 Lin, Y. A.; Chalker, J. M.; Davis, B. G. *J. Am. Chem. Soc.* **2010**, *132*, 16805.
227 Hunter, L.; Condie, G. C.; Harding, M. M. *Tetrahedron Lett.* **2010**, *51*, 5064.
228 Michrowska, A.; Bieniek, M.; Kim, M.; Klajn, R.; Grela, K. *Tetrahedron* **2003**, *59*, 4525.
229 Samojłowicz, C.; Grela, K. *ARKIVOC* **2011**, 71.
230 Woźniak, Ł.; Rajkiewicz, A. A.; Monsigny, L.; Kajetanowicz, A.; Grela, K. *Org. Lett.* **2020**, *22*, 4970.
231 Newton, A. F.; Roe, S. J.; Legeay, J.-C.; Aggarwal, P.; Gignoux, C.; Birch, N. J.; Nixon, R.; Alcaraz, M.-L.; Stockman, R. A. *Org. Biomol. Chem.* **2009**, *7*, 2274.
232 Grela, K.; Bieniek, M. *Tetrahedron Lett.* **2001**, *42*, 6425.
233 Bieniek, M.; Kołoda, D.; Grela, K. *Org. Lett.* **2006**, *8*, 5689.
234 Evans, P.; Leffray, M. *Tetrahedron* **2003**, *59*, 7973.
235 Au, C. W. G.; Pyne, S. G. *J. Org. Chem.* **2006**, *71*, 7097.
236 Burrell, A. J. M.; Coldham, I.; Oram, N. *Org. Lett.* **2009**, *11*, 1515.
237 Fujitani, M.; Tsuchiya, M.; Okano, K.; Takasu, K.; Ihara, M.; Tokuyama, H. *Synlett* **2010**, 822.
238 Ettari, R.; Nizi, E.; Di Francesco, M. E.; Dude, M.-A.; Pradel, G.; Vičík, R.; Schirmeister, T.; Micale, N.; Grasso, S.; Zappalà, M. *J. Med. Chem.* **2008**, *51*, 988.
239 Meadows, D. C.; Gervay-Hague, J. *Med. Res. Rev.* **2006**, *26*, 793.
240 Dunny, E.; Doherty, W.; Evans, P.; Malthouse, J. P. G.; Nolan, D.; Knox, A. J. S. *J. Med. Chem.* **2013**, *56*, 6638.
241 Chatterjee, A. K.; Grubbs, R. H. *Org. Lett.* **1999**, *1*, 1751.
242 Ibrahem, I.; Yu, M.; Schrock, R. R.; Hoveyda, A. H. *J. Am. Chem. Soc.* **2009**, *131*, 3844.
243 Hoveyda, A. H.; Liu, Z.; Qin, C.; Koengeter, T.; Mu, Y. *Angew. Chem., Int. Ed.* **2020**, *59*, 22324.
244 Marinescu, S. C.; Schrock, R. R.; Muller, P.; Takase, M. K.; Hoveyda, A. H. *Organometallics* **2011**, *30*, 1780.

245 Rosebrugh, L. E.; Herbert, M. B.; Marx, V. M.; Keitz, B. K.; Grubbs, R. H. *J. Am. Chem. Soc.* **2013**, *135*, 1276.
246 Yu, M.; Schrock, R. R.; Hoveyda, A. H. *Angew. Chem., Int. Ed.* **2015**, *54*, 215.
247 Prunet, J.; Grimaud, L. In *Metathesis in Natural Product Synthesis*; Cossy, J., Arseniyadis, S., Meyer, C., Eds.; Wiley: Weinheim, Germany, 2010; pp 287–312.
248 Yajima, A.; Kawajiri, A.; Mori, A.; Katsuta, R.; Nukada, T. *Tetrahedron Lett.* **2014**, *55*, 4350.
249 Chhor, R. B.; Nosse, B.; Sörgel, S.; Böhm, C.; Seitz, M.; Reiser, O. *Chem.—Eur. J.* **2003**, *9*, 260.
250 Prasad, K. R.; Revu, O. *Synthesis* **2012**, *44*, 2243.
251 Paquette, L. A.; Bailey, S. *J. Org. Chem.* **1995**, *60*, 7849.
252 In *Prostaglandins, Leukotrienes, and Other Eicosanoids: From Biogenesis to Clinical Application*; Marks, F., Fürstenberger, G., Eds.; Wiley: Weinheim, Germany, 1999.
253 Schrader, T. O.; Snapper, M. L. *J. Am. Chem. Soc.* **2002**, *124*, 10998.
254 Jacobo, S. H.; Chang, C.-T.; Lee, G.-J.; Lawson, J. A.; Powell, W. S.; Pratico, D.; FitzGerald, G. A.; Rokach, J. *J. Org. Chem.* **2006**, *71*, 1370.
255 Wang, X.-G.; Wang, A.-E.; Hao, Y.; Ruan, Y.-P.; Huang, P.-Q. *J. Org. Chem.* **2013**, *78*, 9488.
256 Rupprecht, J. K.; Hui, Y.-H.; McLaughlin, J. L. *J. Nat. Prod.* **1990**, *53*, 237.
257 Zhu, L.; Mootoo, D. R. *Org. Biomol. Chem.* **2005**, *3*, 2750.
258 Wender, P. A.; Longcore, K. E. *Org. Lett.* **2007**, *9*, 691.
259 Crimmins, M. T.; Christie, H. S.; Chaudhary, K.; Long, A. *J. Am. Chem. Soc.* **2005**, *127*, 13810.
260 Takahashi, S.; Akita, Y.; Nakamura, T.; Koshino, H. *Tetrahedron: Asymmetry* **2012**, *23*, 952.
261 Hafner, A.; Duthaler, R. O.; Marti, R.; Rihs, G.; Rothe-Streit, P.; Schwarzenbach, F. *J. Am. Chem. Soc.* **1992**, *114*, 2321.
262 Zieliński, G. K.; Grela, K. *Chem.—Eur. J.* **2016**, *22*, 9440.
263 Smith, A. B.; Adams, C. M.; Kozmin, S. A.; Paone, D. V. *J. Am. Chem. Soc.* **2001**, *123*, 5925.
264 Biard, J.-F.; Roussakis, C.; Kornprobst, J.-M.; Gouiffes-Barbin, D.; Verbist, J.-F.; Cotelle, P.; Foster, M. P.; Ireland, C. M.; Debitus, C. *J. Nat. Prod.* **1994**, *57*, 1336.
265 Statsuk, A. V.; Liu, D.; Kozmin, S. A. *J. Am. Chem. Soc.* **2004**, *126*, 9546.
266 Tiong, E. A.; Rivalti, D.; Williams, B. M.; Gleason, J. L. *Angew. Chem., Int. Ed.* **2013**, *52*, 3442.
267 Takao, K.-I.; Yasui, H.; Yamamoto, S.; Sasaki, D.; Kawasaki, S.; Watanabe, G.; Tadano, K.-I. *J. Org. Chem.* **2004**, *69*, 8789.
268 Hart, A. C.; Phillips, A. J. *J. Am. Chem. Soc.* **2006**, *128*, 1094.
269 Limanto, J.; Snapper, M. L. *J. Am. Chem. Soc.* **2000**, *122*, 8071.
270 Bates, R. In *Organic Synthesis Using Transition Metals*; Wiley: Chichester, U.K., 2012; pp 21–88.
271 Bates, R. In *Organic Synthesis Using Transition Metals*; Wiley: Chichester, U.K., 2012; pp 153–190.
272 Bates, R. In *Organic Synthesis Using Transition Metals*; Wiley: Chichester, U.K., 2012; pp 253–323.
273 Parashar, R. K. In *Reaction Mechanisms in Organic Synthesis*; Wiley: Chichester, U.K., 2008; pp 148–190.
274 Parashar, R. K. In *Reaction Mechanisms in Organic Synthesis*; Wiley: Chichester, U.K., 2008; pp 191–223.
275 Baughman, T. W.; Sworen, J. C.; Wagener, K. B. *Tetrahedron* **2004**, *60*, 10943.
276 DePuy, C. H.; King, R. W. *Chem. Rev.* **1960**, *60*, 431.
277 Concellón, J. M.; Rodríguez-Solla, H.; Simal, C.; Huerta, M. *Org. Lett.* **2005**, *7*, 5833.
278 Saytzeff, A. *Liebigs Ann. Chem.* **1875**, *179*, 296.
279 Block, E. *Org. React.* **1984**, *30*, 457.
280 Corey, E. J.; Winter, R. A. E. *J. Am. Chem. Soc.* **1963**, *85*, 2677.
281 Corey, E. J.; Carey, F. A.; Winter, R. A. E. *J. Am. Chem. Soc.* **1965**, *87*, 934.
282 Semmelhack, M. F.; Stauffer, R. D. *Tetrahedron Lett.* **1973**, *14*, 2667.
283 Crank, G.; Eastwood, F. *Aust. J. Chem.* **1964**, *17*, 1392.
284 Paquette, L. A. *Org. React.* **1977**, *25*, 1.
285 Maryanoff, B. E.; Reitz, A. B. *Chem. Rev.* **1989**, *89*, 863.
286 McMurry, J. E. *Chem. Rev.* **1989**, *89*, 1513.
287 McMurry, J. E. *Acc. Chem. Res.* **1983**, *16*, 405.
288 Nicolaou, K. C.; Theodorakis, E. A.; Rutjes, F. P. J. T.; Sato, M.; Tiebes, J.; Xiao, X. Y.; Hwang, C. K.; Duggan, M. E.; Yang, Z. *J. Am. Chem. Soc.* **1995**, *117*, 10239.
289 Barrett, A. G. M.; Peña, M.; Willardsen, J. A. *J. Org. Chem.* **1996**, *61*, 1082.
290 Schmid, R.; Huesmann, P. L.; Johnson, W. S. *J. Am. Chem. Soc.* **1980**, *102*, 5122.
291 Wadsworth, W. S., Jr. *Org. React.* **1977**, *25*, 73.
292 Denmark, S. E.; Middleton, D. S. *J. Org. Chem.* **1998**, *63*, 1604.
293 Still, W. C.; Gennari, C. *Tetrahedron Lett.* **1983**, *24*, 4405.
294 Trost, B. M.; Metz, P.; Hane, J. T. *Tetrahedron Lett.* **1986**, *27*, 5691.

[295] Ager, D. J. *Synthesis* **1984**, 384.
[296] Ager, D. J. *Org. React.* **1990**, *38*, 1.
[297] van Staden, L. F.; Gravestock, D.; Ager, D. J. *Chem. Soc. Rev.* **2002**, *31*, 195.
[298] Julia, M. *Pure Appl. Chem.* **1985**, *57*, 763.
[299] Keck, G. E.; Savin, K. A.; Weglarz, M. A. *J. Org. Chem.* **1995**, *60*, 3194.
[300] Baudin, J. B.; Hareau, G.; Julia, S. A.; Ruel, O. *Tetrahedron Lett.* **1991**, *32*, 1175.
[301] Blakemore, P. R.; Cole, W. J.; Kocieński, P. J.; Morley, A. *Synlett* **1998**, 26.
[302] Tebbe, F. N.; Parshall, G. W.; Reddy, G. S. *J. Am. Chem. Soc.* **1978**, *100*, 3611.
[303] Petasis, N. A.; Bzowej, E. I. *J. Am. Chem. Soc.* **1990**, *112*, 6392.
[304] Petasis, N. A.; Lu, S. P.; Bzowej, E. I.; Fu, D. K.; Staszewski, J. P.; Akritopoulou-Zanze, I.; Patane, M. A.; Hu, Y. H. *Pure Appl. Chem.* **1996**, *68*, 667.
[305] Nicolaou, K. C.; Postema, M. H. D.; Claiborne, C. F. *J. Am. Chem. Soc.* **1996**, *118*, 1565.
[306] Pine, S. H.; Pettit, R. J.; Geib, G. D.; Cruz, S. G.; Gallego, C. H.; Tijerina, T.; Pine, R. D. *J. Org. Chem.* **1985**, *50*, 1212.
[307] Petasis, N. A.; Bzowej, E. I. *J. Org. Chem.* **1992**, *57*, 1327.
[308] Yan, T.-H.; Tsai, C.-C.; Chien, C.-T.; Cho, C.-C.; Huang, P.-C. *Org. Lett.* **2004**, *6*, 4961.
[309] Heck, R. F.; Nolley, J. P. *J. Org. Chem.* **1972**, *37*, 2320.
[310] Mizoroki, T.; Mori, K.; Ozaki, A. *Bull. Chem. Soc. Jpn.* **1971**, *44*, 581.
[311] Cabri, W.; Candiani, I.; Bedeschi, A.; Penco, S.; Santi, R. *J. Org. Chem.* **1992**, *57*, 1481.
[312] Cabri, W.; Candiani, I.; Bedeschi, A.; Santi, R. *J. Org. Chem.* **1992**, *57*, 3558.
[313] Firmansjah, L.; Fu, G. C. *J. Am. Chem. Soc.* **2007**, *129*, 11340.
[314] Bloome, K. S.; McMahen, R. L.; Alexanian, E. J. *J. Am. Chem. Soc.* **2011**, *133*, 20146.
[315] Tamao, K.; Sumitani, K.; Kumada, M. *J. Am. Chem. Soc.* **1972**, *94*, 4374.
[316] Corriu, R. J. P.; Masse, J. P. *J. Chem. Soc., Chem. Commun.* **1972**, 144.
[317] Satyanarayana Reddy, M.; Thirumalai Rajan, S.; Eswaraiah, S.; Venkat Reddy, G.; Rama Subba Reddy, K.; Sahadeva Reddy, M. (MSN Laboratories Limited) Intl. Patent WO 2011/148392 A1, December 1, 2011.
[318] Stille, J. K. *Angew. Chem., Int. Ed. Engl.* **1986**, *25*, 508.
[319] Tang, W.; Prusov, E. V. *Org. Lett.* **2012**, *14*, 4690.
[320] Miyaura, N.; Yamada, K.; Suzuki, A. *Tetrahedron Lett.* **1979**, *20*, 3437.
[321] Baxter, J. M.; Steinhuebel, D.; Palucki, M.; Davies, I. W. *Org. Lett.* **2004**, *7*, 215.
[322] Fujii, S.; Chang, S. Y.; Burke, M. D. *Angew. Chem., Int. Ed.* **2011**, *50*, 7862.
[323] Hatanaka, Y.; Hiyama, T. *J. Org. Chem.* **1988**, *53*, 918.
[324] Denmark, S. E.; Kallemeyn, J. M. *J. Am. Chem. Soc.* **2006**, *128*, 15958.
[325] Denmark, S. E.; Ambrosi, A. *Org. Process Res. Dev.* **2015**, *19*, 982.
[326] King, A. O.; Okukado, N.; Negishi, E.-i. *J. Chem. Soc., Chem. Commun.* **1977**, 683.
[327] Smith, A. B.; Beauchamp, T. J.; LaMarche, M. J.; Kaufman, M. D.; Qiu, Y.; Arimoto, H.; Jones, D. R.; Kobayashi, K. *J. Am. Chem. Soc.* **2000**, *122*, 8654.
[328] Smith, A. B.; Kaufman, M. D.; Beauchamp, T. J.; LaMarche, M. J.; Arimoto, H. *Org. Lett.* **1999**, *1*, 1823.
[329] Skowerski, K.; Wierzbicka, C.; Szczepaniak, G.; Gulajski, L.; Bieniek, M.; Grela, K. *Green Chem.* **2012**, *14*, 3264.
[330] Bates, R. W.; Li, L.; Palani, K.; Phetsang, W.; Loh, J. K. *Asian J. Org. Chem.* **2014**, *3*, 792.
[331] Tadd, A. C.; Meinander, K.; Luthman, K.; Wallén, E. A. A. *J. Org. Chem.* **2010**, *76*, 673.
[332] Plettenburg, O.; Mui, C.; Bodmer-Narkevitch, V.; Wong, C.-H. *Adv. Synth. Catal.* **2002**, *344*, 622.
[333] Kerherve, J.; Botuha, C.; Dubois, J. *Org. Biomol. Chem.* **2009**, *7*, 2214.
[334] Wang, L.; Gao, Y.; Liu, J.; Cai, C.; Du, Y. *Tetrahedron* **2014**, *70*, 2616.
[335] Donohoe, T. J.; Bower, J. F.; Baker, D. B.; Basutto, J. A.; Chan, L. K. M.; Gallagher, P. *Chem. Commun.* **2011**, *47*, 10611.
[336] Smith, C. M.; O'Doherty, G. A. *Org. Lett.* **2003**, *5*, 1959.
[337] Wang, Z. J.; Jackson, W. R.; Robinson, A. J. *Org. Lett.* **2013**, *15*, 3006.
[338] Araki, H.; Naka, K. *J. Polym. Sci., Part A: Polym. Chem.* **2012**, *50*, 4170.
[339] Yu, M.; Ibrahem, I.; Hasegawa, M.; Schrock, R. R.; Hoveyda, A. H. *J. Am. Chem. Soc.* **2012**, *134*, 2788.
[340] Matsumoto, K.; Kozmin, S. A. *Adv. Synth. Catal.* **2008**, *350*, 557.
[341] Lipshutz, B. H.; Ghorai, S. *Tetrahedron* **2010**, *66*, 1057.
[342] Wang, S.-Y.; Song, P.; Chan, L.-Y.; Loh, T.-P. *Org. Lett.* **2010**, *12*, 5166.
[343] Do, J.-L.; Mottillo, C.; Tan, D.; Štrukil, V.; Friščić, T. *J. Am. Chem. Soc.* **2015**, *137*, 2476.
[344] Baader, S.; Ohlmann, D. M.; Gooßen, L. J. *Chem.—Eur. J.* **2013**, *19*, 9807.

345 Mihovilovic, M. D.; Grötzl, B.; Kandioller, W.; Snajdrova, R.; Muskotál, A.; Bianchi, D. A.; Stanetty, P. *Adv. Synth. Catal.* **2006**, *348*, 463.
346 Karlou-Eyrisch, K.; Müller, B. K. M.; Herzig, C.; Nuyken, O. *J. Organomet. Chem.* **2000**, *606*, 3.
347 Eyrisch, K. K.; Müller, B. K. M.; Herzig, C.; Nuyken, O. *Des. Monomers Polym.* **2004**, *7*, 661.
348 Kiss, L.; Kardos, M.; Forró, E.; Fülöp, F. *Eur. J. Org. Chem.* **2015**, 1283.
349 Bykov, V. I.; Butenko, T. A.; Petrova, E. B.; Finkelshtein, E. S. *Tetrahedron* **1999**, *55*, 8249.
350 Marinescu, S. C.; Schrock, R. R.; Müller, P.; Hoveyda, A. H. *J. Am. Chem. Soc.* **2009**, *131*, 10840.
351 Marinescu, S. C.; Levine, D. S.; Zhao, Y.; Schrock, R. R.; Hoveyda, A. H. *J. Am. Chem. Soc.* **2011**, *133*, 11512.
352 Miyazaki, H.; Herbert, M. B.; Liu, P.; Dong, X.; Xu, X.; Keitz, B. K.; Ung, T.; Mkrtumyan, G.; Houk, K. N.; Grubbs, R. H. *J. Am. Chem. Soc.* **2013**, *135*, 5848.
353 Gerber, L. C. H.; Schrock, R. R. *Organometallics* **2013**, *32*, 5573.
354 Arjona, O.; Csákÿ, A. G.; Medel, R.; Plumet, J. *J. Org. Chem.* **2002**, *67*, 1380.
355 Phillips, G. A.; Palmer, C.; Stevens, A. C.; Piotrowski, M. L.; Dekruyf, D. S. R.; Pagenkopf, B. L. *Tetrahedron Lett.* **2015**, *56*, 6052.
356 Spekreijse, J.; Le Notre, J.; van Haveren, J.; Scott, E. L.; Sanders, J. P. M. *Green Chem.* **2012**, *14*, 2747.
357 Miege, F.; Meyer, C.; Cossy, J. *Org. Lett.* **2010**, *12*, 248.
358 Carreras, J.; Avenoza, A.; Busto, J. H.; Peregrina, J. M. *J. Org. Chem.* **2011**, *76*, 3381.
359 Jeon, K. O.; Rayabarapu, D.; Rolfe, A.; Volp, K.; Omar, I.; Hanson, P. R. *Tetrahedron* **2009**, *65*, 4992.
360 Csákÿ, A. G.; Medel, R.; Murcia, M. C.; Plumet, J. *Helv. Chim. Acta* **2005**, *88*, 1387.
361 Yadav, R. N.; Mondal, S.; Ghosh, S. *Tetrahedron Lett.* **2011**, *52*, 1942.
362 Tatton, M. R.; Simpson, I.; Donohoe, T. J. *Org. Lett.* **2014**, *16*, 1920.
363 Harrity, J. P. A.; Visser, M. S.; Gleason, J. D. *J. Am. Chem. Soc.* **1997**, *119*, 1488.
364 Harrity, J. P. A.; La, D. S.; Cefalo, D. R.; Visser, M. S.; Hoveyda, A. H. *J. Am. Chem. Soc.* **1998**, *120*, 2343.
365 Julis, J.; Bartlett, S. A.; Baader, S.; Beresford, N.; Routledge, E. J.; Cazin, C. S. J.; Cole-Hamilton, D. J. *Green Chem.* **2014**, *16*, 2846.
366 Clark, J. R.; French, J. M.; Diver, S. T. *J. Org. Chem.* **2012**, *77*, 1599.
367 van der Klis, F.; Le Nôtre, J.; Blaauw, R.; van Haveren, J.; van Es, D. S. *Eur. J. Lipid Sci. Technol.* **2012**, *114*, 911.
368 Curto, J. M.; Kozlowski, M. C. *J. Org. Chem.* **2014**, *79*, 5359.
369 Waldeck, A. R.; Krische, M. J. *Angew. Chem., Int. Ed.* **2013**, *52*, 4470.
370 Bidange, J.; Dubois, J.-L.; Couturier, J.-L.; Fischmeister, C.; Bruneau, C. *Eur. J. Lipid Sci. Technol.* **2014**, *116*, 1583.
371 Thurier, C.; Fischmeister, C.; Bruneau, C.; Olivier-Bourbigou, H.; Dixneuf, P. H. *ChemSusChem* **2008**, *1*, 118.
372 Ries, O.; Büschleb, M.; Granitzka, M.; Stalke, D.; Ducho, C. *Beilstein J. Org. Chem.* **2014**, *10*, 1135.
373 Schrodi, Y.; Ung, T.; Vargas, A.; Mkrtumyan, G.; Lee, C. W.; Champagne, T. M.; Pederson, R. L.; Hong, S. H. *Clean: Soil, Air, Water* **2008**, *36*, 669.
374 Marx, V. M.; Sullivan, A. H.; Melaimi, M.; Virgil, S. C.; Keitz, B. K.; Weinberger, D. S.; Bertrand, G.; Grubbs, R. H. *Angew. Chem., Int. Ed.* **2015**, *54*, 1919.
375 Anderson, D. R.; Ung, T.; Mkrtumyan, G.; Bertrand, G.; Grubbs, R. H.; Schrodi, Y. *Organometallics* **2008**, *27*, 563.
376 Forman, G. S.; Bellabarba, R. M.; Tooze, R. P.; Slawin, A. M. Z.; Karch, R.; Winde, R. *J. Organomet. Chem.* **2006**, *691*, 5513.
377 Forman, G. S.; McConnell, A. E.; Hanton, M. J.; Slawin, A. M. Z.; Tooze, R. P.; van Rensburg, W. J.; Meyer, W. H.; Dwyer, C.; Kirk, M. M.; Serfontein, D. W. *Organometallics* **2004**, *23*, 4824.
378 Burdett, K. A.; Harris, L. D.; Margl, P.; Maughon, B. R.; Mokhtar-Zadeh, T.; Saucier, P. C.; Wasserman, E. P. *Organometallics* **2004**, *23*, 2027.
379 Öztürk, B. Ö.; Topoğlu, B.; Karabulut Şehitoğlu, S. *Eur. J. Lipid Sci. Technol.* **2015**, *117*, 200.
380 Reddy, D. V.; Sabitha, G.; Yadav, J. S. *Tetrahedron Lett.* **2015**, *56*, 4112.
381 Kadyrov, R.; Azap, C.; Weidlich, S.; Wolf, D. *Top. Catal.* **2012**, *55*, 538.
382 Alexander, K. A.; Paulhus, E. A.; Lazarus, G. M. L.; Leadbeater, N. E. *J. Organomet. Chem.* **2016**, *812*, 74.
383 Thomas, R. M.; Keitz, B. K.; Champagne, T. M.; Grubbs, R. H. *J. Am. Chem. Soc.* **2011**, *133*, 7490.
384 Miao, X.; Fischmeister, C.; Bruneau, C.; Dixneuf, P. H. *ChemSusChem* **2008**, *1*, 813.
385 Groaz, E.; Banti, D.; North, M. *Adv. Synth. Catal.* **2007**, *349*, 142.
386 Enholm, E.; Low, T. *J. Org. Chem.* **2006**, *71*, 2272.

[387] Shinde, T.; Varga, V.; Polášek, M.; Horáček, M.; Žilková, N.; Balcar, H. *Appl. Catal., A* **2014**, *478*, 138.

[388] Wolf, S.; Plenio, H. *Green Chem.* **2011**, *13*, 2008.

[389] Sancibrao, P.; Karila, D.; Kouklovsky, C.; Vincent, G. *J. Org. Chem.* **2010**, *75*, 4333.

[390] Boyle, T. P.; Bremner, J. B.; Coates, J. A.; Keller, P. A.; Pyne, S. G. *Tetrahedron* **2005**, *61*, 7271.

[391] Salim, H.; Piva, O. *Tetrahedron Lett.* **2007**, *48*, 2059.

[392] Fustero, S.; Sánchez-Roselló, M.; Aceña, J. L.; Fernández, B.; Asensio, A.; Sanz-Cervera, J. F.; del Pozo, C. *J. Org. Chem.* **2009**, *74*, 3414.

[393] Hiebel, M.-A.; Pelotier, B.; Goekjian, P.; Piva, O. *Eur. J. Org. Chem.* **2008**, 713.

[394] Cros, F.; Pelotier, B.; Piva, O. *Eur. J. Org. Chem.* **2010**, 5063.

[395] Xu, C.; Shen, X.; Hoveyda, A. H. *J. Am. Chem. Soc.* **2017**, *139*, 10919.

[396] Bourcet, E.; Virolleaud, M.-A.; Fache, F.; Piva, O. *Tetrahedron Lett.* **2008**, *49*, 6816.

[397] Ying, Y.; Taori, K.; Kim, H.; Hong, J.; Luesch, H. *J. Am. Chem. Soc.* **2008**, *130*, 8455.

[398] Czajkowska, D.; Morzycki, J. W. *Tetrahedron Lett.* **2009**, *50*, 2904.

[399] Madelaine, C.; Ouhamou, N.; Chiaroni, A.; Vedrenne, E.; Grimaud, L.; Six, Y. *Tetrahedron* **2008**, *64*, 8878.

[400] McNaughton, B. R.; Bucholtz, K. M.; Camaaño-Moure, A.; Miller, B. L. *Org. Lett.* **2005**, *7*, 733.

[401] Mangold, S. L.; O'Leary, D. J.; Grubbs, R. H. *J. Am. Chem. Soc.* **2014**, *136*, 12469.

[402] Woodward, C. P.; Spiccia, N. D.; Jackson, W. R.; Robinson, A. J. *Chem. Commun.* **2011**, *47*, 779.

[403] Cros, F.; Pelotier, B.; Piva, O. *Synthesis* **2010**, 233.

[404] Nasveschuk, C. G.; Ungermannova, D.; Liu, X.; Phillips, A. *J. Org. Lett.* **2008**, *10*, 3595.

[405] Souto, J. A.; Vaz, E.; Lepore, I.; Pöppler, A.-C.; Franci, G.; Álvarez, R.; Altucci, L.; de Lera, A. R. *J. Med. Chem.* **2010**, *53*, 4654.

[406] Chang, C.-W.; Chen, Y.-N.; Adak, A. K.; Lin, K.-H.; Tzou, D.-L. M.; Lin, C.-C. *Tetrahedron* **2007**, *63*, 4310.

[407] Vriesen, M. R.; Grover, H. K.; Kerr, M. A. *Synlett* **2014**, *25*, 428.

[408] Virolleaud, M.-A.; Piva, O. *Synlett* **2004**, *2087*.

[409] Desire, J.; Mondon, M.; Fontelle, N.; Nakagawa, S.; Hirokami, Y.; Adachi, I.; Iwaki, R.; Fleet, G. W. J.; Alonzi, D. S.; Twigg, G.; Butters, T. D.; Bertrand, J.; Cendret, V.; Becq, F.; Norez, C.; Marrot, J.; Kato, A.; Bleriot, Y. *Org. Biomol. Chem.* **2014**, *12*, 8977.

[410] Galan, M. C.; O'Connor, S. E. *Tetrahedron Lett.* **2006**, *47*, 1563.

[411] Park, S.; Lee, D. *J. Am. Chem. Soc.* **2006**, *128*, 10664.

[412] Kuntiyong, P.; Akkarasamiyo, S.; Piboonsrinakara, N.; Hemmara, C.; Songthammawat, P. *Tetrahedron* **2011**, *67*, 8034.

[413] Ko, H. M.; Lee, D. G.; Kim, M. A.; Kim, H. J.; Park, J.; Lah, M. S.; Lee, E. *Tetrahedron* **2007**, *63*, 5797.

[414] Rival, N.; Hanquet, G.; Bensoussan, C.; Reymond, S.; Cossy, J.; Colobert, F. *Org. Biomol. Chem.* **2013**, *11*, 6829.

[415] Gahalawat, S.; Pandey, S. K. *RSC Adv.* **2015**, *5*, 41013.

[416] Virolleaud, M.-A.; Menant, C.; Fenet, B.; Piva, O. *Tetrahedron Lett.* **2006**, *47*, 5127.

[417] Hartung, J.; Grubbs, R. H. *J. Am. Chem. Soc.* **2013**, *135*, 10183.

[418] Poeylaut-Palena, A. A.; Mata, E. G. *Org. Biomol. Chem.* **2010**, *8*, 3947.

[419] Prasad, K. R.; Gandi, V. R. *Tetrahedron: Asymmetry* **2010**, *21*, 275.

[420] Prasad, K. R.; Gandi, V. R. *Tetrahedron: Asymmetry* **2011**, *22*, 499.

[421] Kaliappan, K. P.; Si, D. *Synlett* **2009**, 2441.

[422] Galletti, P.; Quintavalla, A.; Ventrici, C.; Giannini, G.; Cabri, W.; Penco, S.; Gallo, G.; Vincenti, S.; Giacomini, D. *ChemMedChem* **2009**, *4*, 1991.

[423] Schmidtmann, F. W.; Benedum, T. E.; McGarvey, G. J. *Tetrahedron Lett.* **2005**, *46*, 4677.

[424] McCune, C. D.; Chan, S. J.; Beio, M. L.; Shen, W.; Chung, W. J.; Szczesniak, L. M.; Chai, C.; Koh, S. Q.; Wong, P. T. H.; Berkowitz, D. B. *ACS Cent. Sci.* **2016**, *2*, 242.

[425] Xie, W.; Zou, B.; Pei, D.; Ma, D. *Org. Lett.* **2005**, *7*, 2775.

[426] Nolen, E. G.; Kurish, A. J.; Potter, J. M.; Donahue, L. A.; Orlando, M. D. *Org. Lett.* **2005**, *7*, 3383.

[427] McGarvey, G. J.; Benedum, T. E.; Schmidtmann, F. W. *Org. Lett.* **2002**, *4*, 3591.

[428] Bouchet, S.; Linot, C.; Ruzic, D.; Agbaba, D.; Fouchaq, B.; Roche, J.; Nikolic, K.; Blanquart, C.; Bertrand, P. *ACS Med. Chem. Lett.* **2019**, *10*, 863.

[429] Vernall, A. J.; Abell, A. D. *Org. Biomol. Chem.* **2004**, *2*, 2555.

[430] Maiorana, S.; Licandro, E.; Perdicchia, D.; Baldoli, C.; Vandoni, B.; Giannini, C.; Salmain, M. *J. Mol. Catal. A: Chem.* **2003**, *204–205*, 165.

[431] Enholm, E. J.; Hastings, J. M.; Edwards, C. *Synlett* **2008**, 203.

432 Marinec, P. S.; Evans, C. G.; Gibbons, G. S.; Tarnowski, M. A.; Overbeek, D. L.; Gestwicki, J. E. *Bioorg. Med. Chem.* **2009**, *17*, 5763.
433 Thomas, R. M.; Fedorov, A.; Keitz, B. K.; Grubbs, R. H. *Organometallics* **2011**, *30*, 6713.
434 Gibson, S. E.; Gibson, V. C.; Keen, S. P. *Chem. Commun.* **1997**, 1107.
435 Michaut, A.; Boddaert, T.; Coquerel, Y.; Rodriguez, J. *Synthesis* **2007**, 2867.
436 Wei, X.; Carroll, P. J.; Sneddon, L. G. *Organometallics* **2005**, *25*, 609.
437 Cannon, J. S.; Grubbs, R. H. *Angew. Chem., Int. Ed.* **2013**, *52*, 9001.
438 Feuillastre, S.; Piva, O. *Synlett* **2014**, *25*, 2883.
439 Kim, S.-G. *Synthesis* **2009**, 2418.
440 Prasad, K. R.; Anbarasan, P. *J. Org. Chem.* **2007**, *72*, 3155.
441 Wijdeven, M. A.; van Delft, F. L.; Rutjes, F. P. J. T. *Tetrahedron* **2010**, *66*, 5623.
442 Nookaraju, U.; Kumar, P. *RSC Adv.* **2015**, *5*, 63311.
443 Mori, K. *Biosci., Biotechnol., Biochem.* **2010**, *74*, 595.
444 Shikichi, Y.; Mori, K. *Biosci., Biotechnol., Biochem.* **2012**, *76*, 407.
445 O'Neil, G. W.; Moser, D. J.; Volz, E. O. *Tetrahedron Lett.* **2009**, *50*, 7355.
446 Li, J.; Ahmed, T. S.; Xu, C.; Stoltz, B. M.; Grubbs, R. H. *J. Am. Chem. Soc.* **2019**, *141*, 154.
447 Verbicky, C. A.; Zercher, C. K. *Tetrahedron Lett.* **2000**, *41*, 8723.
448 Bek, D.; Gawin, R.; Grela, K.; Balcar, H. *Catal. Commun.* **2012**, *21*, 42.
449 Żukowska, K.; Szadkowska, A.; Pazio, A. E.; Woźniak, K.; Grela, K. *Organometallics* **2012**, *31*, 462.
450 Loiseau, F.; Kholod, I.; Neier, R. *Eur. J. Org. Chem.* **2010**, 4642.
451 Yadav, J. S.; Pandurangam, T.; Reddy, V. V. B.; Reddy, B. V. S. *Synthesis* **2010**, 4300.
452 Sengupta, S.; Sim, T. *Eur. J. Org. Chem.* **2014**, 5063.
453 van Innis, L.; Plancher, J. M.; Markó, I. E. *Org. Lett.* **2006**, *8*, 6111.
454 Fernández de la Pradilla, R.; Lwoff, N. *Tetrahedron Lett.* **2008**, *49*, 4167.
455 Abbineni, C.; Sasmal, P. K.; Mukkanti, K.; Iqbal, J. *Tetrahedron Lett.* **2007**, *48*, 4259.
456 Sasmal, P. K.; Abbineni, C.; Iqbal, J.; Mukkanti, K. *Tetrahedron* **2010**, *66*, 5000.
457 Storcken, R. P. M.; Panella, L.; van Delft, F. L.; Kaptein, B.; Broxterman, Q. B.; Schoemaker, H. E.; Rutjes, F. P. J. T. *Adv. Synth. Catal.* **2007**, *349*, 161.
458 Garner, A. L.; Koide, K. *Org. Lett.* **2007**, *9*, 5235.
459 Liao, Y.; Fathi, R.; Yang, Z. *J. Comb. Chem.* **2002**, *5*, 79.
460 Lumini, M.; Cordero, F. M.; Pisaneschi, F.; Brandi, A. *Eur. J. Org. Chem.* **2008**, 2817.
461 Quigley, B. L.; Grubbs, R. H. *Chem. Sci.* **2014**, *5*, 501.
462 Seo, S.-Y.; Jung, J.-K.; Paek, S.-M.; Lee, Y.-S.; Kim, S.-H.; Suh, Y.-G. *Tetrahedron Lett.* **2006**, *47*, 6527.
463 Wu, Q.-K.; Kinami, K.; Kato, A.; Li, Y.-X.; Jia, Y.-M.; Fleet, G. W. J.; Yu, C.-Y. *Molecules* **2019**, *24*, 3712.
464 Fujiwara, T.; Hashimoto, K.; Umeda, M.; Murayama, S.; Ohno, Y.; Liu, B.; Nambu, H.; Yakura, T. *Tetrahedron* **2018**, *74*, 4578.
465 O'Doherty, I.; Yim, J. J.; Schmelz, E. A.; Schroeder, F. C. *Org. Lett.* **2011**, *13*, 5900.
466 Ghosal, P.; Kumar, V.; Shaw, A. K. *Tetrahedron* **2010**, *66*, 7504.
467 Wang, Y.; Romo, D. *Org. Lett.* **2002**, *4*, 3231.
468 Pandey, R. K.; Wang, L.; Wallock, N. J.; Lindeman, S.; Donaldson, W. A. *J. Org. Chem.* **2008**, *73*, 7236.
469 Prasad, K. R.; Penchalaiah, K. *Tetrahedron: Asymmetry* **2011**, *22*, 1400.
470 Jana, A. K.; Panda, G. *RSC Adv.* **2013**, *3*, 16795.
471 Duque, R.; Ochsner, E.; Clavier, H.; Caijo, F.; Nolan, S. P.; Mauduit, M.; Cole-Hamilton, D. J. *Green Chem.* **2011**, *13*, 1187.
472 Zong, K.; Deininger, J. J.; Reynolds, J. R. *Org. Lett.* **2013**, *15*, 1032.
473 Wu, W.-J.; Chen, H.-J.; Wu, Y.; Liu, B. *Tetrahedron* **2014**, *70*, 92.
474 Mori, K.; Tashiro, T. *Tetrahedron Lett.* **2009**, *50*, 3266.
475 Saidhareddy, P.; Shaw, A. K. *RSC Adv.* **2015**, *5*, 29114.
476 Gandi, V. R. *Tetrahedron* **2013**, *69*, 6507.
477 Li, L.-L.; Ding, J.-Y.; Gao, L.-X.; Han, F.-S. *Org. Biomol. Chem.* **2015**, *13*, 1133.
478 Tae, H. S.; Hines, J.; Schneekloth, A. R.; Crews, C. M. *Org. Lett.* **2010**, *12*, 4308.
479 Yadav, J. S.; Das, S. K.; Sabitha, G. *J. Org. Chem.* **2012**, *77*, 11109.
480 Crimmins, M. T.; Caussanel, F. *J. Am. Chem. Soc.* **2006**, *128*, 3128.
481 Reddy, C. R.; Jithender, E.; Prasad, K. R. *J. Org. Chem.* **2013**, *78*, 4251.
482 Reddy, C. R.; Jithender, E.; Singh, A.; Ummanni, R. *Synthesis* **2014**, *46*, 822.
483 Goldup, S. M.; Pilkington, C. J.; White, A. J. P.; Burton, A.; Barrett, A. G. M. *J. Org. Chem.* **2006**, *71*, 6185.

[484] Bini, D.; Forcella, M.; Cipolla, L.; Fusi, P.; Matassini, C.; Cardona, F. *Eur. J. Org. Chem.* **2011**, 3995.
[485] Mori, K. *Tetrahedron* **2009**, *65*, 3900.
[486] Fernández, A.; Levine, Z. G.; Baumann, M.; Sulzer-Mossé, S.; Sparr, C.; Schläger, S.; Metzger, A.; Baxendale, I. R.; Ley, S. V. *Synlett* **2013**, *24*, 514.
[487] Sai Baba, V.; Das, P.; Mukkanti, K.; Iqbal, J. *Tetrahedron Lett.* **2006**, *47*, 7927.
[488] Keitz, B. K.; Endo, K.; Patel, P. R.; Herbert, M. B.; Grubbs, R. H. *J. Am. Chem. Soc.* **2011**, *134*, 693.
[489] Luján, C.; Nolan, S. P. *J. Organomet. Chem.* **2011**, *696*, 3935.
[490] Purushotham Reddy, S.; Chinnababu, B.; Venkateswarlu, Y. *Helv. Chim. Acta* **2014**, *97*, 999.
[491] Małecki, P.; Gajda, K.; Gajda, R.; Woźniak, K.; Trzaskowski, B.; Kajetanowicz, A.; Grela, K. *ACS Catal.* **2019**, *9*, 587.
[492] Miętkiewski, M.; Powała, B.; Staniszewski, B.; Kubicki, M.; Urbaniak, W.; Pietraszuk, C. *Cent. Eur. J. Chem.* **2011**, *9*, 728.
[493] Saha, D.; Guchhait, S.; Goswami, R. K. *Org. Lett.* **2020**, *22*, 745.
[494] Ribes, C.; Falomir, E.; Murga, J.; Carda, M.; Alberto Marco, J. *Org. Biomol. Chem.* **2009**, *7*, 1355.
[495] Cossy, J.; Willis, C.; Bellosta, V. *Synlett* **2001**, 1578.
[496] Cossy, J.; Willis, C.; Bellosta, V.; BouzBouz, S. *J. Org. Chem.* **2002**, *67*, 1982.
[497] Dominique, R.; Das, S. K.; Liu, B.; Nahra, J.; Schmor, B.; Gan, Z.; Roy, R. In *Recognition of Carbohydrates in Biological Systems, Part A: General Procedures*; Yuan, C. L., Reiko, T. L., Eds.; Methods in Enzymology; Academic Press: Cambridge, MA, 2003; Vol. 362, pp 17–28.
[498] Ronchi, P.; Vignando, S.; Guglieri, S.; Polito, L.; Lay, L. *Org. Biomol. Chem.* **2009**, *7*, 2635.
[499] Kallitsakis, M. G.; Hadjipavlou-Litina, D. J.; Litinas, K. E. J. *Enzyme Inhib. Med. Chem.* **2013**, *28*, 765.
[500] Mori, K. *Tetrahedron* **2009**, *65*, 2798.
[501] Yadav, J. S.; Ramesh, K.; Subba Reddy, U. V.; Subba Reddy, B. V.; Ghamdi, A. A. K. A. *Tetrahedron Lett.* **2011**, *52*, 2943.
[502] Palmer, A. M.; Chrismann, S.; Münch, G.; Brehm, C.; Zimmermann, P. J.; Buhr, W.; Senn-Bilfinger, J.; Feth, M. P.; Simon, W. A. *Bioorg. Med. Chem.* **2009**, *17*, 368.
[503] Ribes, C.; Falomir, E.; Murga, J.; Carda, M.; Marco, J. A. *Tetrahedron* **2009**, *65*, 10612.
[504] Štefane, B.; Požgan, F. *Monatsh. Chem.* **2013**, *144*, 633.
[505] Gotz, K.; Liermann, J. C.; Thines, E.; Anke, H.; Opatz, T. *Org. Biomol. Chem.* **2010**, *8*, 2123.
[506] Prasad, K. R.; Gandi, V. R. *Synlett* **2009**, 2593.
[507] Hoye, T. R.; Eklov, B. M.; Jeon, J.; Khoroosi, M. *Org. Lett.* **2006**, *8*, 3383.
[508] Berhal, F.; Takechi, S.; Kumagai, N.; Shibasaki, M. *Chem.—Eur. J.* **2011**, *17*, 1915.
[509] Ronchi, P.; Scarponi, C.; Salvi, M.; Fallarini, S.; Polito, L.; Caneva, E.; Bagnoli, L.; Lay, L. *J. Org. Chem.* **2013**, *78*, 5172.
[510] Czaban, J.; Schertzer, B. M.; Grela, K. *Adv. Synth. Catal.* **2013**, *355*, 1997.
[511] Taber, D. F.; Frankowski, K. J. *J. Org. Chem.* **2003**, *68*, 6047.
[512] Yadav, J. S.; Reddy, P. A. N.; Kumar, A. S.; Prasad, A. R.; Reddy, B. V. S.; Al Ghamdi, A. A. *Tetrahedron Lett.* **2014**, *55*, 1395.
[513] Dommisse, A.; Wirtz, J.; Koch, K.; Barthlott, W.; Kolter, T. *Eur. J. Org. Chem.* **2007**, 3508.
[514] Sohn, T.-i.; Kim, M. J.; Kim, D. *J. Am. Chem. Soc.* **2010**, *132*, 12226.
[515] Biermann, U.; Meier, M. A. R.; Butte, W.; Metzger, J. O. *Eur. J. Lipid Sci. Technol.* **2011**, *113*, 39.
[516] Hernández-Torres, G.; Urbano, A.; Carreño, M. C.; Colobert, F. *Org. Lett.* **2009**, *11*, 4930.
[517] Cabrera, J.; Padilla, R.; Dehn, R.; Deuerlein, S.; Gułajski, Ł.; Chomiszczak, E.; Teles, J. H.; Limbach, M.; Grela, K. *Adv. Synth. Catal.* **2012**, *354*, 1043.
[518] Chen, Z.; Sinha, S. C. *Tetrahedron* **2008**, *64*, 1603.
[519] Markowski, T.; Drescher, S.; Meister, A.; Hause, G.; Blume, A.; Dobner, B. *Eur. J. Org. Chem.* **2011**, 5894.
[520] Neto, V.; Granet, R.; Krausz, P. *Tetrahedron* **2010**, *66*, 4633.
[521] Quinn, K. J.; Islamaj, L.; Couvertier, S. M.; Shanley, K. E.; Mackinson, B. L. *Eur. J. Org. Chem.* **2010**, 5943.
[522] Crimmins, M. T.; Haley, M. W.; O'Bryan, E. A. *Org. Lett.* **2011**, *13*, 4712.
[523] Sedlák, D.; Eignerová, B.; Dračínský, M.; Janoušek, Z.; Bartůněk, P.; Kotora, M. *J. Organomet. Chem.* **2013**, *747*, 178.
[524] Bastien, D.; Leblanc, V.; Asselin, É.; Bérubé, G. *Bioorg. Med. Chem. Lett.* **2010**, *20*, 2078.
[525] Vincent, G.; Karila, D.; Khalil, G.; Sancibrao, P.; Gori, D.; Kouklovsky, C. *Chem.—Eur. J.* **2013**, *19*, 9358.
[526] Kozłowska, A.; Dranka, M.; Zachara, J.; Pump, E.; Slugovc, C.; Skowerski, K.; Grela, K. *Chem.—Eur. J.* **2014**, *20*, 14120.
[527] Bosque, I.; Gonzalez-Gomez, J. C.; Loza, M. I.; Brea, J. *J. Org. Chem.* **2014**, *79*, 3982.

528 Li, J.; Stoltz, B. M.; Grubbs, R. H. *Org. Lett.* **2019**, *21*, 10139.
529 Lindner, F.; Friedrich, S.; Hahn, F. *J. Org. Chem.* **2018**, *83*, 14091.
530 Schachner, J. A.; Cabrera, J.; Padilla, R.; Fischer, C.; van der Schaaf, P. A.; Pretot, R.; Rominger, F.; Limbach, M. *ACS Catal.* **2011**, *1*, 872.
531 Kummer, D. A.; Brenneman, J. B.; Martin, S. F. *Tetrahedron* **2006**, *62*, 11437.
532 Zelin, J.; Nieres, P. D.; Trasarti, A. F.; Apesteguía, C. R. *Appl. Catal., A* **2015**, *502*, 410.
533 Jose, J.; Pourfallah, G.; Merkley, D.; Li, S.; Bouzidi, L.; Leao, A. L.; Narine, S. S. *Polym. Chem.* **2014**, *5*, 3203.
534 Kajetanowicz, A.; Sytniczuk, A.; Grela, K. *Green Chem.* **2014**, *16*, 1579.
535 Thomas, P. A.; Marvey, B. B.; Ebenso, E. E. *Int. J. Mol. Sci.* **2011**, *12*, 3989.
536 Abel, G. A.; Nguyen, K. O.; Viamajala, S.; Varanasi, S.; Yamamoto, K. *RSC Adv.* **2014**, *4*, 55622.
537 Marshall, J. A.; Sabatini, J. J.; Valeriote, F. *Bioorg. Med. Chem. Lett.* **2007**, *17*, 2434.
538 Luo, S.-X.; Cannon, J. S.; Taylor, B. L. H.; Engle, K. M.; Houk, K. N.; Grubbs, R. H. *J. Am. Chem. Soc.* **2016**, *138*, 14039.
539 Wojtkielewicz, A.; Maj, J.; Dzieszkowska, A.; Morzycki, J. W. *Tetrahedron* **2011**, *67*, 6868.
540 Glaus, F.; Altmann, K.-H. *Angew. Chem., Int. Ed.* **2015**, *54*, 1937.
541 Zhang, Y.; Arpin, C. C.; Cullen, A. J.; Mitton-Fry, M. J.; Sammakia, T. *J. Org. Chem.* **2011**, *76*, 7641.
542 Maj, J.; Morzycki, J. W.; Rárová, L.; Wasilewski, G.; Wojtkielewicz, A. *Tetrahedron Lett.* **2012**, *53*, 5430.
543 Sabitha, G.; Gurumurthy, C.; Yadav, J. S. *Synthesis* **2014**, *46*, 110.
544 Gui, H.; Liu, J.; Song, L.; Hui, C.; Feng, J.; Xu, Z.; Ye, T. *Synlett* **2014**, *25*, 138.
545 Yadav, J. S.; Reddy, G. M.; Anjum, S. R.; Reddy, B. V. S. *Eur. J. Org. Chem.* **2014**, 4389.
546 Ghosh, A. K.; Chen, Z.-H.; Effenberger, K. A.; Jurica, M. S. *J. Org. Chem.* **2014**, *79*, 5697.
547 Kiesewetter, E. T.; O'Brien, R. V.; Yu, E. C.; Meek, S. J.; Schrock, R. R.; Hoveyda, A. H. *J. Am. Chem. Soc.* **2013**, *135*, 6026.
548 Spiccia, N. D.; Solyom, S.; Woodward, C. P.; Jackson, W. R.; Robinson, A. J. *J. Org. Chem.* **2016**, *81*, 1798.
549 Ferrié, L.; Amans, D.; Reymond, S.; Bellosta, V.; Capdevielle, P.; Cossy, J. *J. Organomet. Chem.* **2006**, *691*, 5456.
550 Rix, D.; Caijo, F.; Laurent, I.; Boeda, F.; Clavier, H.; Nolan, S. P.; Mauduit, M. *J. Org. Chem.* **2008**, *73*, 4225.
551 Morandi, B.; Wickens, Z. K.; Grubbs, R. H. *Angew. Chem., Int. Ed.* **2013**, *52*, 9751.
552 La Ferla, B.; Spinosa, V.; D'Orazio, G.; Palazzo, M.; Balsari, A.; Foppoli, A. A.; Rumio, C.; Nicotra, F. *ChemMedChem* **2010**, *5*, 1677.
553 Poulsen, S.-A.; Bornaghi, L. F. *Tetrahedron Lett.* **2005**, *46*, 7389.
554 Enholm, J. E.; Low, T.; Cooper, D.; Ghivirija, I. *Tetrahedron* **2012**, *68*, 6920.
555 den Hartog, T.; Maciá, B.; Minnaard, A. J.; Feringa, B. L. *Adv. Synth. Catal.* **2010**, *352*, 999.
556 Dieltiens, N.; Stevens, C. V. *Synlett* **2006**, 2771.
557 Ikoma, M.; Oikawa, M.; Sasaki, M. *Eur. J. Org. Chem.* **2009**, *72*.
558 Rivollier, J.; Thuéry, P.; Heck, M.-P. *Org. Lett.* **2013**, *15*, 480.
559 Lin, Y. A.; Chalker, J. M.; Floyd, N.; Bernardes, G. J. L.; Davis, B. G. *J. Am. Chem. Soc.* **2008**, *130*, 9642.
560 Wang, T.; Yu, X.; Zhang, H.; Wu, S.; Guo, W.; Wang, J. *Appl. Organomet. Chem.* **2019**, *33*, e4939.
561 Raghavan, S.; Krishnaiah, V. *J. Org. Chem.* **2009**, *75*, 748.
562 Vieille-Petit, L.; Clavier, H.; Linden, A.; Blumentritt, S.; Nolan, S. P.; Dorta, R. *Organometallics* **2010**, *29*, 775.
563 Saito, Y.; Yoshimura, Y.; Wakamatsu, H.; Takahata, H. *Molecules* **2013**, *18*, 1162.
564 Moulins, J. R.; Burnell, D. *J. Tetrahedron Lett.* **2011**, *52*, 3992.
565 Hryniewicka, A.; Kozłowska, A.; Witkowski, S. *J. Organomet. Chem.* **2012**, *701*, 87.
566 Bonini, C.; Chiummiento, L.; Videtta, V.; Colobert, F.; Solladié, G. *Synlett* **2006**, 2427.
567 Lindhorst, T. K.; Elsner K. *Beilstein J. Org. Chem.* **2014**, *10*, 1482.
568 Park, J. K.; Lackey, H. H.; Ondrusek, B. A.; McQuade, D. T. *J. Am. Chem. Soc.* **2011**, *133*, 2410.
569 Hodgson, D. M.; Angrish, D.; Erickson, S. P.; Kloesges, J.; Lee, C. H. *Org. Lett.* **2008**, *10*, 5553.
570 Kim, B.; Sohn, T.-i.; Kim, D.; Paton, R. S. *Chem.—Eur. J.* **2018**, *24*, 2634.
571 Roy, R.; Dominique, R.; Das, S. K. *J. Org. Chem.* **1999**, *64*, 5408.
572 Biswas, K.; Coltart, D. M.; Danishefsky, S. J. *Tetrahedron Lett.* **2002**, *43*, 6107.
573 Hu, Y.-J.; Roy, R. *Tetrahedron Lett.* **1999**, *40*, 3305.
574 Kopitzki, S.; Jensen, K. J.; Thiem, J. *Chem.—Eur. J.* **2010**, *16*, 7017.
575 Lonin, I. S.; Grin, M. A.; Lakhina, A. A.; Mironov, A. F. *Mendeleev Commun.* **2012**, *22*, 157.
576 Splain, R. A.; Kiessling, L. L. *Bioorg. Med. Chem.* **2010**, *18*, 3753.

577 Berkowitz, D. B.; Maiti, G.; Charette, B. D.; Dreis, C. D.; MacDonald, R. G. *Org. Lett.* **2004**, *6*, 4921.
578 Yadav, J. S.; Satyanarayana, K.; Sreedhar, P.; Srihari, P.; Shaik, T. B.; Kalivendi, S. V. *Bioorg. Med. Chem. Lett.* **2010**, *20*, 3814.
579 Jones, S. A.; Duncan, J.; Aitken, S. G.; Coxon, J. M.; Abell, A. D. *Aust. J. Chem.* **2014**, *67*, 1257.
580 Yun, H.; Sim, J.; An, H.; Lee, J.; Lee, H. S.; Shin, Y. K.; Paek, S.-M.; Suh, Y.-G. *Org. Biomol. Chem.* **2014**, *12*, 7127.
581 Sukkari, H. E.; Gesson, J.-P.; Renoux, B. *Tetrahedron Lett.* **1998**, *39*, 4043.
582 Huwe, C. M.; Woltering, T. J.; Jiricek, J.; Weitz-Schmidt, G.; Wong, C.-H. *Bioorg. Med. Chem.* **1999**, *7*, 773.
583 Faure, R.; Shiao, T. C.; Lagnoux, D.; Giguere, D.; Roy, R. *Org. Biomol. Chem.* **2007**, *5*, 2704.
584 Hsu, M. C.; Junia, A. J.; Haight, A. R.; Zhang, W. *J. Org. Chem.* **2004**, *69*, 3907.
585 Šnajdr, I.; Janoušek, Z.; Jindřich, J.; Kotora M. *Beilstein J. Org. Chem.* **2010**, *6*, 1099.
586 Barile, F.; Bassetti, M.; D'Annibale, A.; Gerometta, R.; Palazzi, M. *Eur. J. Org. Chem.* **2011**, 6519.
587 Seshadri, H.; Lovely, C. J. *Org. Lett.* **2000**, *2*, 327.
588 Gawin, R.; Czarnecka, P.; Grela, K. *Tetrahedron* **2010**, *66*, 1051.
589 Moura-Letts, G.; Curran, D. P. *Org. Lett.* **2006**, *9*, 5.
590 Wu, G.-L.; Cao, S.-L.; Chen, J.; Chen, Z. *Eur. J. Org. Chem.* **2012**, 6777.
591 Bantreil, X.; Poater, A.; Urbina-Blanco, C. A.; Bidal, Y. D.; Falivene, L.; Randall, R. A. M.; Cavallo, L.; Slawin, A. M. Z.; Cazin, C. S. J. *Organometallics* **2012**, *31*, 7415.
592 Ablialimov, O.; Kędziorek, M.; Torborg, C.; Malińska, M.; Woźniak, K.; Grela, K. *Organometallics* **2012**, *31*, 7316.
593 Ablialimov, O.; Kędziorek, M.; Malińska, M.; Woźniak, K.; Grela, K. *Organometallics* **2014**, *33*, 2160.
594 Siano, V.; D'Auria, I.; Grisi, F.; Costabile, C.; Longo, P. *Cent. Eur. J. Chem.* **2011**, *9*, 605.
595 Sauvage, X.; Zaragoza, G.; Demonceau, A.; Delaude, L. *Adv. Synth. Catal.* **2010**, *352*, 1934.
596 Kajetanowicz, A.; Czaban, J.; Krishnan, G. R.; Malińska, M.; Woźniak, K.; Siddique, H.; Peeva, L. G.; Livingston, A. G.; Grela, K. *ChemSusChem* **2013**, *6*, 182.
597 Kosnik, W.; Grela, K. *Dalton Trans.* **2013**, *42*, 7463.
598 Smoleń, M.; Kędziorek, M.; Grela, K. *Catal. Commun.* **2014**, *44*, 80.
599 Teo, P.; Grubbs, R. H. *Organometallics* **2010**, *29*, 6045.
600 Rosen, E. L.; Sung, D. H.; Chen, Z.; Lynch, V. M.; Bielawski, C. W. *Organometallics* **2009**, *29*, 250.
601 Kniese, M.; Meier, M. A. R. *Green Chem.* **2010**, *12*, 169.
602 Wdowik, T.; Samojlowicz, C.; Jawiczuk, M.; Malinska, M.; Wozniak, K.; Grela, K. *Chem. Commun.* **2013**, *49*, 674.
603 Lafaye, K.; Nicolas, L.; Guérinot, A.; Reymond, S.; Cossy, J. *Org. Lett.* **2014**, *16*, 4972.
604 Carter, K. P.; Moser, D. J.; Storvick, J. M.; O'Neil, G. W. *Tetrahedron Lett.* **2011**, *52*, 4494.
605 Spiccia, N. D.; Border, E.; Illesinghe, J.; Jackson, W. R.; Robinson, A. J. Synthesis 2013, *45*, 1683.
606 Skowerski, K.; Pastva, J.; Czarnocki, S. J.; Janoscova, J. *Org. Process Res. Dev.* **2015**, *19*, 872.
607 Kotha, S.; Seema, V.; Singh, K.; Deodhar, K. D. *Tetrahedron Lett.* **2010**, *51*, 2301.
608 Kitamura, T.; Sato, Y.; Mori, M. *Tetrahedron* **2004**, *60*, 9649.
609 Bieniek, M.; Samojłowicz, C.; Sashuk, V.; Bujok, R.; Śledź, P.; Lugan, N.; Lavigne, G.; Arlt, D.; Grela, K. *Organometallics* **2011**, *30*, 4144.
610 Tzur, E.; Szadkowska, A.; Ben-Asuly, A.; Makal, A.; Goldberg, I.; Woźniak, K.; Grela, K.; Lemcoff, N. G. *Chem.—Eur. J.* **2010**, *16*, 8726.
611 Szadkowska, A.; Żukowska, K.; Pazio, A. E.; Woźniak, K.; Kadyrov, R.; Grela, K. *Organometallics* **2011**, *30*, 1130.
612 Eignerová, B.; Janoušek, Z.; Dračínský, M.; Kotora, M. *Synlett* **2010**, *885*.
613 Fukuda, K.; Tojino, M.; Goto, K.; Dohi, H.; Nishida, Y.; Mizuno, M. *Carbohydr. Res.* **2015**, *407*, 122.
614 de Espinosa, L. M.; Kempe, K.; Schubert, U. S.; Hoogenboom, R.; Meier, M. A. R. *Macromol. Rapid Commun.* **2012**, *33*, 2023.
615 Tosatti, P.; Campbell, A. J.; House, D.; Nelson, A.; Marsden, S. P. *J. Org. Chem.* **2011**, *76*, 5495.
616 Paterson, I.; Anderson, E. A.; Dalby, S. M.; Lim, J. H.; Maltas, P. *Org. Biomol. Chem.* **2012**, *10*, 5873.
617 Centrone, C. A.; Lowary, T. L. *J. Org. Chem.* **2002**, *67*, 8862.
618 Feuillastre, S.; Pellet, V.; Piva, O. *Synthesis* **2012**, *44*, 2431.
619 Eustache, J.; Van de Weghe, P.; Nouen, D. L.; Uyehara, H.; Kabuto, C.; Yamamoto, Y. *J. Org. Chem.* **2005**, *70*, 4043.
620 Bilel, H.; Hamdi, N.; Zagrouba, F.; Fischmeister, C.; Bruneau, C. *Catal. Sci. Technol.* **2014**, *4*, 2064.
621 Lee, K.; Kim, H.; Hong, J. *Org. Lett.* **2009**, *11*, 5202.

[622] Kim, H.; Baker, J. B.; Lee, S.-U.; Park, Y.; Bolduc, K. L.; Park, H.-B.; Dickens, M. G.; Lee, D.-S.; Kim, Y.; Kim, S. H.; Hong, J. *J. Am. Chem. Soc.* **2009**, *131*, 3192.
[623] Amans, D.; Bellosta, V.; Cossy, J. *Chem.—Eur. J.* **2009**, *15*, 3457.
[624] Hoveyda, A. H.; Lombardi, P. J.; O'Brien, R. V.; Zhugralin, A. R. *J. Am. Chem. Soc.* **2009**, *131*, 8378.
[625] Kammari, B. R.; Bejjanki, N. K.; Kommu, N. *Tetrahedron: Asymmetry* **2015**, *26*, 296.
[626] Cossy, J.; BouzBouz, S.; Hoveyda, A. H. *J. Organomet. Chem.* **2001**, *624*, 327.
[627] Engelhardt, F. C.; Schmitt, M. J.; Taylor, R. E. *Org. Lett.* **2001**, *3*, 2209.
[628] Amblard, F.; Nolan, S. P.; Schinazi, R. F.; Agrofoglio, L. A. *Tetrahedron* **2005**, *61*, 537.
[629] Kumamoto, H.; Topalis, D.; Broggi, J.; Pradère, U.; Roy, V.; Berteina-Raboin, S.; Nolan, S. P.; Deville-Bonne, D.; Andrei, G.; Snoeck, R.; Garin, D.; Crance, J.-M.; Agrofoglio, L. A. *Tetrahedron* **2008**, *64*, 3517.
[630] Roy, V.; Mieczkowski, A.; Topalis, D.; Berteina-Raboin, S.; Deville-Bonne, D.; Nolan, S. P.; Agrofoglio, L. A. *Synthesis* **2008**, 2127.
[631] Mieczkowski, A.; Blu, J.; Roy, V.; Agrofoglio, L. A. *Tetrahedron* **2009**, *65*, 9791.
[632] Bessières, M.; Hervin, V.; Roy, V.; Chartier, A.; Snoeck, R.; Andrei, G.; Lohier, J.-F.; Agrofoglio, L. A. *Eur. J. Med. Chem.* **2018**, *146*, 678.
[633] Montagu, A.; Pradére, U.; Roy, V.; Nolan, S. P.; Agrofoglio, L. A. *Tetrahedron* **2011**, *67*, 5319.
[634] Feuillastre, S.; Pelotier, B.; Piva, O. *Synthesis* **2013**, *45*, 810.
[635] Samojlowicz, C.; Grela, K. *ARKIVOC* **2011**, 71.
[636] Kawai, T.; Shida, Y.; Yoshida, H.; Abe, J.; Iyoda, T. *J. Mol. Catal. A: Chem.* **2002**, *190*, 33.
[637] Yasuda, T.; Abe, J.; Iyoda, T.; Kawai, T. *Chem. Lett.* **2001**, *30*, 812.
[638] Storvick, J. M.; Ankoudinova, E.; King, B. R.; Van Epps, H.; O'Neil, G. W. *Tetrahedron Lett.* **2011**, *52*, 5858.
[639] Hoffman, T. J.; Kolleth, A.; Rigby, J. H.; Arseniyadis, S.; Cossy, J. *Org. Lett.* **2010**, *12*, 3348.
[640] Hogendorf, W. F. J.; Verhagen, C. P.; Malta, E.; Goosen, N.; Overkleeft, H. S.; Filippov, D. V.; Van der Marel, G. A. *Tetrahedron* **2009**, *65*, 10430.
[641] da Silva, F. D. C.; Ferreira, V. F.; de Souza, M. C. B. V.; Tomé, A. C.; Neves, M. G. P. M. S.; Silva, A. M. S.; Cavaleiro, J. A. S. *Synlett* **2008**, 1205.
[642] Qiao, W.; Shao, M.; Wang, J. *J. Organomet. Chem.* **2012**, *713*, 197.
[643] Zhang, Y.; Shao, M.; Zhang, H.; Li, Y.; Liu, D.; Cheng, Y.; Liu, G.; Wang, J. *J. Organomet. Chem.* **2014**, *756*, 1.
[644] Robertson, J.; Green, S. P.; Hall, M. J.; Tyrrell, A. J.; Unsworth, W. P. *Org. Biomol. Chem.* **2008**, *6*, 2628.
[645] Crowe, W. E.; Zhang, Z. J. *J. Am. Chem. Soc.* **1993**, *115*, 10998.
[646] Eivgi, O.; Sutar, R. L.; Reany, O.; Lemcoff, N. G. *Adv. Synth. Catal.* **2017**, *359*, 2352.
[647] Rajkumar, S.; Clarkson, G. J.; Shipman, M. *Org. Lett.* **2017**, *19*, 2058.
[648] Garcia, P. G.; Hohn, E.; Pietruszka, J. *J. Organomet. Chem.* **2003**, *680*, 281.
[649] Krishna, C. V.; Bhonde, V. R.; Devendar, A.; Maitra, S.; Mukkanti, K.; Iqbal, J. *Tetrahedron Lett.* **2008**, *49*, 2013.
[650] Hoffman, T. J.; Rigby, J. H.; Arseniyadis, S.; Cossy, J. *J. Org. Chem.* **2008**, *73*, 2400.
[651] Ferré-Filmon, K.; Delaude, L.; Demonceau, A.; Noels, A. F. *Coord. Chem. Rev.* **2004**, *248*, 2323.
[652] Chang, S.; Na, Y.; Shin, H. J.; Choi, E.; Jeong, L. S. *Tetrahedron Lett.* **2002**, *43*, 7445.
[653] Ferré-Filmon, K.; Delaude, L.; Demonceau, A.; Noels, A. F. *Eur. J. Org. Chem.* **2005**, 3319.
[654] Jafarpour, L.; Heck, M.-P.; Baylon, C.; Lee, H. M.; Mioskowski, C.; Nolan, S. P. *Organometallics* **2002**, *21*, 671.
[655] Blacquiere, J. M.; McDonald, R.; Fogg, D. E. *Angew. Chem., Int. Ed.* **2010**, *49*, 3807.
[656] Bolduc, K. L.; Larsen, S. D.; Sherman, D. H. *Chem. Commun.* **2012**, *48*, 6414.
[657] Dhara, K.; Paladhi, S.; Midya, G. C.; Dash, J. *Org. Biomol. Chem.* **2011**, *9*, 3801.
[658] Öberg, E.; Orthaber, A.; Santoni, M.-P.; Howard, F.; Ott, S. *Phosphorus, Sulfur Silicon Relat. Elem.* **2012**, *188*, 152.
[659] Yasuda, T.; Abe, J.; Yoshida, H.; Iyoda, T.; Kawai, T. *Adv. Synth. Catal.* **2002**, *344*, 705.
[660] Meinke, S.; Thiem, J. *Carbohydr. Res.* **2008**, *343*, 1824.
[661] Meinke, S.; Schroven, A.; Thiem, J. *Org. Biomol. Chem.* **2011**, *9*, 4487.
[662] Raffier, L.; Izquierdo, F.; Piva, O. *Synthesis* **2011**, 4037.
[663] Viault, G.; Grée, D.; Das, S.; Yadav, J. S.; Grée, R. *Eur. J. Org. Chem.* **2011**, 1233.
[664] Oikawa, M.; Ikoma, M.; Sasaki, M.; Gill, M. B.; Swanson, G. T.; Shimamoto, K.; Sakai, R. *Eur. J. Org. Chem.* **2009**, 5531.
[665] Plass, K. E.; Liu, X.; Brunschwig, B. S.; Lewis, N. S. *Chem. Mater.* **2008**, *20*, 2228.
[666] Godin, G.; Compain, P.; Martin, O. R. *Org. Lett.* **2003**, *5*, 3269.
[667] Sawant, P.; Maier, M. E. *Eur. J. Org. Chem.* **2012**, *6576*.

[668] Kashanna, J.; Jangili, P.; Kumar, R. A.; Das, B. *Helv. Chim. Acta* **2012**, *95*, 1666.
[669] Pullin, R. D. C.; Rathi, A. H.; Melikhova, E. Y.; Winter, C.; Thompson, A. L.; Donohoe, T. J. *Org. Lett.* **2013**, *15*, 5492.
[670] Radha Krishna, P.; Reddy, B. K. *Tetrahedron Lett.* **2010**, *51*, 6262.
[671] Albury, A. M. M.; De Joarder, D.; Jennings, M. P. *Tetrahedron Lett.* **2015**, *56*, 3057.
[672] Mangold, S. L.; Prost, L. R.; Kiessling, L. L. *Chem. Sci.* **2012**, *3*, 772.
[673] Kumar, G. D. K.; Natarajan, A. *Tetrahedron Lett.* **2008**, *49*, 2103.
[674] Bryant, V. C.; Kumar, G. D. K.; Nyong, A. M.; Natarajan, A. *Bioorg. Med. Chem. Lett.* **2012**, *22*, 245.
[675] Korinkova, P.; Bazgier, V.; Oklestkova, J.; Rarova, L.; Strnad, M.; Kvasnica, M. *Steroids* **2017**, *127*, 46.
[676] Norris, B. N.; Pan, T.; Meyer, T. Y. *Org. Lett.* **2010**, *12*, 5514.
[677] Poeylaut-Palena, A. A.; Testero, S. A.; Mata, E. G. *J. Org. Chem.* **2008**, *73*, 2024.
[678] Sacristán, M.; Ronda, J. C.; Galià, M.; Cádiz, V. *J. Appl. Polym. Sci.* **2011**, *122*, 1649.
[678a] Chanti Babu, D.; Bhujanga Rao, C.; Ramesh, D.; Raghavendra Swamy, S.; Venkateswarlu, Y. *Tetrahedron Lett.* **2012**, *53*, 3633.
[679] Li, M.; O'Doherty, G. A. *Org. Lett.* **2006**, *8*, 6087.
[680] Zhang, Y.; Dlugosch, M.; Jübermann, M.; Banwell, M. G.; Ward, J. S. *J. Org. Chem.* **2015**, *80*, 4828.
[681] Bolte, B.; Basutto, J. A.; Bryan, C. S.; Garson, M. J.; Banwell, M. G.; Ward, J. S. *J. Org. Chem.* **2015**, *80*, 460.
[682] Chakraborty, J.; Nanda, S. *Org. Biomol. Chem.* **2019**, *17*, 7369.
[683] Napolitano, C.; McArdle, P.; Murphy, P. V. *J. Org. Chem.* **2010**, *75*, 7404.
[684] Palmer, A. M.; Grobbel, B.; Jecke, C.; Brehm, C.; Zimmermann, P. J.; Buhr, W.; Feth, M. P.; Simon, W.-A.; Kromer, W. *J. Med. Chem.* **2007**, *50*, 6240.
[685] Mbere-Nguyen, U.; Ung, A. T.; Pyne, S. G. *Tetrahedron* **2009**, *65*, 5990.
[686] Tymann, D.; Bednarzick, U.; Iovkova-Berends, L.; Hiersemann, M. *Org. Lett.* **2018**, *20*, 4072.
[687] Morrill, C.; Grubbs, R. H. *J. Org. Chem.* **2003**, *68*, 6031.
[688] Marciniec, B.; Jankowska, M.; Pietraszuk, C. *Chem. Commun.* **2005**, 663.
[689] Schmidt, J. J.; Khatri, Y.; Brody, S. I.; Zhu, C.; Pietraszkiewicz, H.; Valeriote, F. A.; Sherman, D. H. *ACS Chem. Biol.* **2020**, *15*, 524.
[690] Speed, A. W. H.; Mann, T. J.; O'Brien, R. V.; Schrock, R. R.; Hoveyda, A. H. *J. Am. Chem. Soc.* **2014**, *136*, 16136.
[691] Hemelaere, R.; Carreaux, F.; Carboni, B. *Chem.—Eur. J.* **2014**, *20*, 14518.
[692] Hemelaere, R.; Carreaux, F.; Carboni, B. *J. Org. Chem.* **2013**, *78*, 6786.
[693] Hemelaere, R.; Caijo, F.; Mauduit, M.; Carreaux, F.; Carboni, B. *Eur. J. Org. Chem.* **2014**, 3328.
[694] Njardarson, J. T.; Biswas, K.; Danishefsky, S. J. *Chem. Commun.* **2002**, 2759.
[695] Enquist, J. A.; Virgil, S. C.; Stoltz, B. M. *Chem.—Eur. J.* **2011**, *17*, 9957.
[696] Winbush, S. M.; Roush, W. R. *Org. Lett.* **2010**, *12*, 4344.
[697] Sergeeva, N. N.; Pablo, V. L.; Senge, M. O. *J. Organomet. Chem.* **2008**, *693*, 2637.
[698] Huang, Y.; Chen, D.; Qing, F.-L. *Tetrahedron* **2003**, *59*, 7879.
[699] Jankowska, M.; Pietraszuk, C.; Marciniec, B.; Zaidlewicz, M. *Synlett* **2006**, *1695*.
[700] Martin, A. R.; Mohanan, K.; Luvino, D.; Floquet, N.; Baraguey, C.; Smietana, M.; Vasseur, J.-J. *Org. Biomol. Chem.* **2009**, *7*, 4369.
[701] Uno, B. E.; Gillis, E. P.; Burke, M. D. *Tetrahedron* **2009**, *65*, 3130.
[702] Swaroop, P. S.; Tripathy, S.; Jachak, G.; Reddy, D. S. *Tetrahedron Lett.* **2014**, *55*, 4777.
[703] Ghosh, A. K.; Lv, K. *Eur. J. Org. Chem.* **2014**, 6761.
[704] Llaveria, J.; Díaz, Y.; Matheu, M. I.; Castillón, S. *Org. Lett.* **2009**, *11*, 205.
[705] Radha Krishna, P.; Dayaker, G. *Tetrahedron Lett.* **2007**, *48*, 7279.
[706] Meiries, S.; Parkin, A.; Marquez, R. *Tetrahedron* **2009**, *65*, 2951.
[707] Pirrung, M. C.; Biswas, G.; Ibarra-Rivera, T. R. *Org. Lett.* **2010**, *12*, 2402.
[708] Bélanger, D.; Tong, X.; Soumaré, S.; Dory, Y. L.; Zhao, Y. *Chem.—Eur. J.* **2009**, *15*, 4428.
[709] Suyama, T. L.; Gerwick, W. H. *Org. Lett.* **2008**, *10*, 4449.
[710] Carreras, J.; Avenoza, A.; Busto, J. H.; Peregrina, J. M. *J. Org. Chem.* **2009**, *74*, 1736.
[711] Singh, S.; Singh, O. V.; Han, H. *Tetrahedron Lett.* **2007**, *48*, 8270.
[712] Gnamm, C.; Brödner, K.; Krauter, C. M.; Helmchen, G. *Chem.—Eur. J.* **2009**, *15*, 10514.
[713] Moon, H.; An, H.; Sim, J.; Kim, K.; Paek, S.-M.; Suh, Y.-G. *Tetrahedron Lett.* **2015**, *56*, 608.
[714] Mix, S.; Blechert, S. *Adv. Synth. Catal.* **2007**, *349*, 157.
[715] Santhanam, B.; Boons, G.-J. *Org. Lett.* **2004**, *6*, 3333.
[716] Venukadasula, P. K. M.; Chegondi, R.; Suryn, G. M.; Hanson, P. R. *Org. Lett.* **2012**, *14*, 2634.
[717] Koch, D.; Maechling, S.; Blechert, S. *Tetrahedron* **2007**, *63*, 7112.

718 Kumaraswamy, G.; Sadaiah, K. *Tetrahedron* **2012**, *68*, 262.
719 Waetzig, J. D.; Hanson, P. R. *Org. Lett.* **2006**, *8*, 1673.
720 Sabitha, G.; Nayak, S.; Bhikshapathi, M.; Yadav, J. S. *Tetrahedron Lett.* **2009**, *50*, 5428.
721 Boudreau, M. A.; Vederas, J. C. *Org. Biomol. Chem.* **2007**, *5*, 627.
722 Wisse, P.; de Geus, M. A. R.; Cross, G.; van den Nieuwendijk, A. M. C. H.; van Rooden, E. J.; van den Berg, R. J. B. H. N.; Aerts, J. M. F. G.; van der Marel, G. A.; Codée, J. D. C.; Overkleeft, H. S. *J. Org. Chem.* **2015**, *80*, 7258.
723 Emura, C.; Higuchi, R.; Miyamoto, T. *Chem. Lett.* **2010**, *39*, 1002.
724 Sabitha, G.; Reddy, S. S. S.; Bhaskar, V.; Yadav, J. S. *Synthesis* **2010**, 1217.
725 Hopkins, C. D.; Schmitz, J. C.; Chu, E.; Wipf, P. *Org. Lett.* **2011**, *13*, 4088.
726 Ghosh, A. K.; Kulkarni, S. *Org. Lett.* **2008**, *10*, 3907.
727 Bates, R. W.; Dewey, M. R.; Tang, C. H.; Safii, S. b.; Hong, Y.; Hsieh, J. K. H.; Siah, P. S. *Synlett* **2011**, 2053.
728 Wan, Z.-L.; Zhang, G.-L.; Chen, H.-J.; Wu, Y.; Li, Y. *Eur. J. Org. Chem.* **2014**, 2128.
729 Sawant, P.; Maier, M. E. *Synlett* **2011**, 3002.
730 Yamamoto, T.; Hasegawa, H.; Hakogi, T.; Katsumura, S. *Org. Lett.* **2006**, *8*, 5569.
731 Qu, W.; Ploessl, K.; Truong, H.; Kung, M.-P.; Kung, H. F. *Bioorg. Med. Chem. Lett.* **2009**, *19*, 3382.
732 Jung, M. E.; Yi, S. W. *Tetrahedron Lett.* **2012**, *53*, 4216.
733 Kumar, G.; Kaur, S.; Singh, V. *Helv. Chim. Acta* **2011**, *94*, 650.
734 Ullrich, T.; Ghobrial, M.; Peters, C.; Billich, A.; Guerini, D.; Nussbaumer, P. *ChemMedChem* **2008**, *3*, 356.
735 Peters, C.; Billich, A.; Ghobrial, M.; Högenauer, K.; Ullrich, T.; Nussbaumer, P. *J. Org. Chem.* **2007**, *72*, 1842.
736 Torssell, S.; Somfai, P. *Org. Biomol. Chem.* **2004**, *2*, 1643.
737 Finnegan, D.; Seigal, B. A.; Snapper, M. L. *Org. Lett.* **2006**, *8*, 2603.
738 Bourcet, E.; Fache, F.; Piva, O. *Tetrahedron* **2010**, *66*, 1319.
739 Sabitha, G.; Vangala, B.; Reddy, S. S. S.; Yadav, J. S. *Helv. Chim. Acta* **2010**, *93*, 329.
740 Hasegawa, H.; Yamamoto, T.; Hatano, S.; Hakogi, T.; Katsumura, S. *Chem. Lett.* **2004**, *33*, 1592.
741 Yamamoto, T.; Hasegawa, H.; Ishii, S.; Kaji, S.; Masuyama, T.; Harada, S.; Katsumura, S. *Tetrahedron* **2008**, *64*, 11647.
742 Schmidt, B.; Kunz, O.; Petersen, M. H. *J. Org. Chem.* **2012**, *77*, 10897.
743 Trost, B. M.; Aponick, A.; Stanzl, B. N. *Chem.—Eur. J.* **2007**, *13*, 9547.
744 Oishi, T.; Kanemoto, M.; Swasono, R.; Matsumori, N.; Murata, M. *Org. Lett.* **2008**, *10*, 5203.
745 Comin, M. J.; Parrish, D. A.; Deschamps, J. R.; Marquez, V. E. *Org. Lett.* **2006**, *8*, 705.
746 Sabitha, G.; Reddy, C. N.; Gopal, P.; Yadav, J. S. *Tetrahedron Lett.* **2010**, *51*, 5736.
747 Kumar, K. S.; Reddy, C. S. *Org. Biomol. Chem.* **2012**, *10*, 2647.
748 Raghavan, S.; Subramanian, S. G. *Tetrahedron* **2011**, *67*, 7529.
749 Ghosal, P.; Kumar, V.; Shaw, A. K. *Carbohydr. Res.* **2010**, *345*, 41.
750 Lu, K.-J.; Chen, C.-H.; Hou, D.-R. *Tetrahedron* **2009**, *65*, 225.
751 Borré, E.; Dinh, T. H.; Caijo, F.; Crévisy, C.; Mauduit, M. *Synthesis* **2011**, 2125.
752 Sheddan, N. A.; Arion, V. B.; Mulzer, J. *Tetrahedron Lett.* **2006**, *47*, 6689.
753 Crimmins, M. T.; O'Bryan, E. A. *Org. Lett.* **2010**, *12*, 4416.
754 Sharma, A.; Mahato, S.; Chattopadhyay, S. *Tetrahedron Lett.* **2009**, *50*, 4986.
755 Miura, A.; Kuwahara, S. *Tetrahedron* **2009**, *65*, 3364.
756 Albert, B. J.; Sivaramakrishnan, A.; Naka, T.; Koide, K. *J. Am. Chem. Soc.* **2006**, *128*, 2792.
757 Liu, Z.; Bittman, R. *Org. Lett.* **2012**, *14*, 620.
758 Fu, R.; Chen, J.; Guo, L.-C.; Ye, J.-L.; Ruan, Y.-P.; Huang, P.-Q. *Org. Lett.* **2009**, *11*, 5242.
759 Fu, R.; Ye, J.-L.; Dai, X.-J.; Ruan, Y.-P.; Huang, P.-Q. *J. Org. Chem.* **2010**, *75*, 4230.
760 Swick, S. M.; Schaefer, S. L.; O'Neil, G. W. *Tetrahedron Lett.* **2015**, *56*, 4039.
761 Hande, S. M.; Uenishi, J. I. *Tetrahedron Lett.* **2009**, *50*, 189.
762 Yadav, J. S.; Swapnil, N.; Venkatesh, M.; Prasad, A. R. *Tetrahedron Lett.* **2014**, *55*, 1164.
763 Rajesh, A.; Sharma, G. V. M.; Damera, K. *Tetrahedron Lett.* **2014**, *55*, 4067.
764 Fu, F.; Loh, T.-P. *Tetrahedron Lett.* **2009**, *50*, 3530.
765 Kumpulainen, E. T. T.; Kang, B.; Krische, M. J. *Org. Lett.* **2011**, *13*, 2484.
766 Chatterjee, S.; Kanojia, S. V.; Chattopadhyay, S.; Sharma, A. *Tetrahedron: Asymmetry* **2011**, *22*, 367.
767 Shi, Z.; Harrison, B. A.; Verdine, G. L. *Org. Lett.* **2003**, *5*, 633.
768 Harrison, B. A.; Pasternak, G. W.; Verdine, G. L. *J. Med. Chem.* **2003**, *46*, 677.
769 Srihari, P.; Kumaraswamy, B.; Yadav, J. S. *Tetrahedron* **2009**, *65*, 6304.
770 Singh, A.; Ha, H.-J.; Park, J.; Kim, J. H.; Lee, W. K. *Bioorg. Med. Chem.* **2011**, *19*, 6174.
771 Wang, F.; Kawamura, A.; Mootoo, D. R. *Bioorg. Med. Chem.* **2008**, *16*, 8413.

772 Biswas, S.; Chattopadhyay, S.; Sharma, A. *Tetrahedron: Asymmetry* **2010**, *21*, 27.
773 Nyavanandi, V. K.; Nadipalli, P.; Nanduri, S.; Naidu, A.; Iqbal, J. *Tetrahedron Lett.* **2007**, *48*, 6905.
774 Crimmins, M. T.; Christie, H. S.; Long, A.; Chaudhary, K. *Org. Lett.* **2009**, *11*, 831.
775 Liu, Z.; Byun, H.-S.; Bittman, R. *J. Org. Chem.* **2011**, *76*, 8588.
776 Crimmins, M. T.; Zhang, Y.; Diaz, F. A. *Org. Lett.* **2006**, *8*, 2369.
777 Konno, H.; Okuno, Y.; Makabe, H.; Nosaka, K.; Onishi, A.; Abe, Y.; Sugimoto, A.; Akaji, K. *Tetrahedron Lett.* **2008**, *49*, 782.
778 Konno, H.; Makabe, H.; Hattori, Y.; Nosaka, K.; Akaji, K. *Tetrahedron* **2010**, *66*, 7946.
779 Wender, P. A.; Hilinski, M. K.; Skaanderup, P. R.; Soldermann, N. G.; Mooberry, S. L. *Org. Lett.* **2006**, *8*, 4105.
780 Mooberry, S. L.; Hilinski, M. K.; Clark, E. A.; Wender, P. A. *Mol. Pharmaceutics* **2008**, *5*, 829.
781 Reddy, G. V.; Kumar, R. S. C.; Sreedhar, E.; Babu, K. S.; Rao, J. M. *Tetrahedron Lett.* **2010**, *51*, 1723.
782 Maishal, T. K.; Sinha-Mahapatra, D. K.; Paranjape, K.; Sarkar, A. *Tetrahedron Lett.* **2002**, *43*, 2263.
783 Kumaraswamy, G.; Sadaiah, K.; Raghu, N. *Tetrahedron: Asymmetry* **2012**, *23*, 587.
784 Kadota, I.; Abe, T.; Uni, M.; Takamura, H.; Yamamoto, Y. *Tetrahedron Lett.* **2008**, *49*, 3643.
785 Amblard, F.; Nolan, S. P.; Gillaizeau, I.; Agrofoglio, L. A. *Tetrahedron Lett.* **2003**, *44*, 9177.
786 Rej, R. K.; Pal, P.; Nanda, S. *Tetrahedron* **2014**, *70*, 4457.
787 Ghosal, P.; Shaw, A. K. *Tetrahedron Lett.* **2010**, *51*, 4140.
788 Séguin, C.; Ferreira, F.; Botuha, C.; Chemla, F.; Pérez-Luna, A. *J. Org. Chem.* **2009**, *74*, 6986.
789 Reddy, C. R.; Dharmapuri, G.; Rao, N. N. *Org. Lett.* **2009**, *11*, 5730.
790 Sabitha, G.; Shankaraiah, K.; Prasad, M. N.; S. Yadav, *J. Synthesis* **2013**, *45*, 251.
791 Rao, K. S.; Ghosh, S. *Synthesis* **2013**, *45*, 2745.
792 Blencowe, P. S.; Barrett, A. G. M. *Eur. J. Org. Chem.* **2014**, 4844.
793 Gebauer, J.; Blechert, S. *J. Org. Chem.* **2006**, *71*, 2021.
794 Gupta, P.; Kumar, P. *Eur. J. Org. Chem.* **2008**, 1195.
795 Kamal, A.; Balakrishna, M.; Reddy, P. V.; Rahim, A. *Tetrahedron: Asymmetry* **2014**, *25*, 148.
796 Rej, R. K.; Jana, A.; Nanda, S. *Tetrahedron* **2014**, *70*, 2634.
797 Joyasawal, S.; Lotesta, S. D.; Akhmedov, N. G.; Williams, L. *J. Org. Lett.* **2010**, *12*, 988.
798 Dayaker, G.; Krishna, P. R. *Helv. Chim. Acta* **2014**, *97*, 868.
799 Chou, C.-Y.; Hou, D.-R. *J. Org. Chem.* **2006**, *71*, 9887.
800 Cai, Y.; Ling, C.-C.; Bundle, D. R. *Carbohydr. Res.* **2009**, *344*, 2120.
801 Sridhar, P. R.; Suresh, M.; Kumar, P. V.; Seshadri, K.; Rao, C. V. *Carbohydr. Res.* **2012**, *360*, 40.
802 Perali, R. S.; Mandava, S.; Chalapala, S. *Tetrahedron* **2011**, *67*, 9283.
803 Gao, Y.; Shan, Q.; Liu, J.; Wang, L.; Du, Y. *Org. Biomol. Chem.* **2014**, *12*, 2071.
804 Nagarapu, L.; Gaikwad, H. K.; Bantu, R.; Manikonda, S. R.; Kumar, C. G.; Pombala, S. *Tetrahedron Lett.* **2012**, *53*, 1287.
805 Jana, P. K.; Mandal, S. B.; Bhattacharjya, A. *Tetrahedron Lett.* **2011**, *52*, 6767.
806 Kalamkar, N. B.; Puranik, V. G.; Dhavale, D. D. *Tetrahedron* **2011**, *67*, 2773.
807 Nagarapu, L.; Paparaju, V.; Satyender, A. *Bioorg. Med. Chem. Lett.* **2008**, *18*, 2351.
808 Khan, R. K. M.; Zhugralin, A. R.; Torker, S.; O'Brien, R. V.; Lombardi, P. J.; Hoveyda, A. H. *J. Am. Chem. Soc.* **2012**, *134*, 12438.
809 Córdova, A.; Rios, R. *Angew. Chem., Int. Ed.* **2009**, *48*, 8827.
810 Kanematsu, M.; Yoshida, M.; Shishido, K. *Angew. Chem., Int. Ed.* **2011**, 50, 2618.
811 Xu, S.; Arimoto, H.; Uemura, D. *Angew. Chem., Int. Ed.* **2007**, *46*, 5746.
812 Rajesh, A.; Sharma, G. V. M.; Damera, K. *Tetrahedron Lett.* **2014**, *55*, 6474.
813 Rothman, J. H. *J. Org. Chem.* **2008**, *74*, 925.
814 Ghosal, P.; Sharma, D.; Kumar, B.; Meena, S.; Sinha, S.; Shaw, A. K. *Org. Biomol. Chem.* **2011**, *9*, 7372.
815 Vamshikrishna, K.; Srihari, P. *Tetrahedron* **2012**, *68*, 1540.
816 Postema, M. H. D.; Piper, J. L. *Tetrahedron Lett.* **2002**, *43*, 7095.
817 Kim, T.; Lee, S. I.; Kim, S.; Shim, S. Y.; Ryu, D. H. *Tetrahedron* **2019**, *75*, 130593.
818 Salva Reddy, N.; Das, B. *Helv. Chim. Acta* **2015**, *98*, 78.
819 Colon, A.; Hoffman, T. J.; Gebauer, J.; Dash, J.; Rigby, J. H.; Arseniyadis, S.; Cossy, J. *Chem. Commun.* **2012**, *48*, 10508.
820 Gebauer, J.; Arseniyadis, S.; Cossy, J. *Eur. J. Org. Chem.* **2008**, 2701.
821 Gebauer, J.; Arseniyadis, S.; Cossy, J. *Org. Lett.* **2007**, *9*, 3425.
822 Anquetin, G.; Rawe, S. L.; McMahon, K.; Murphy, E. P.; Murphy, P. V. *Chem.—Eur. J.* **2008**, *14*, 1592.
823 Krishna, P. R.; Alivelu, M. *Helv. Chim. Acta* **2011**, *94*, 1102.

824 Kamal, A.; Vangala, S. R. *Org. Biomol. Chem.* **2013**, *11*, 4442.
825 Quinn, K. J.; Smith, A. G.; Cammarano, C. M. *Tetrahedron* **2007**, *63*, 4881.
826 Yadav, J. S.; Reddy, M. S.; Rao, P. P.; Prasad, A. R. *Synlett* **2007**, 2049.
827 Yadav, J. S.; Reddy, P. V.; Chandraiah, L. *Tetrahedron Lett.* **2007**, *48*, 145.
828 Albert, B. J.; Sivaramakrishnan, A.; Naka, T.; Czaicki, N. L.; Koide, K. *J. Am. Chem. Soc.* **2007**, *129*, 2648.
829 Ghosh, A. K.; Chen, Z.-H. *Org. Lett.* **2013**, *15*, 5088.
830 Crimmins, M. T.; Jacobs, D. L. *Org. Lett.* **2009**, *11*, 2695.
831 Rajesh, A.; Sharma, G. V. M.; Damera, K. *Synthesis* **2015**, *47*, 845.
832 Sperry, J.; Liu, Y.-C.; Wilson, Z. E.; Hubert, J. G.; Brimble, M. A. *Synthesis* **2011**, 1383.
833 Nyalata, S.; Raghavan, S. *Org. Lett.* **2019**, *21*, 7778.
834 Ghosh, A. K.; Reddy, G. C.; Kovela, S.; Relitti, N.; Urabe, V. K.; Prichard, B. E.; Jurica, M. S. *Org. Lett.* **2018**, *20*, 7293.
835 Ghosh, A. K.; Gong, G. *J. Org. Chem.* **2006**, *71*, 1085.
836 Virolleaud, M.-A.; Piva, O. *Eur. J. Org. Chem.* **2007**, 1606.
837 Andrei, D.; Wnuk, S. F. *Org. Lett.* **2006**, *8*, 5093.
838 Liu, W.; Bodlenner, A.; Rohmer, M. *Org. Biomol. Chem.* **2015**, *13*, 3393.
839 Yadav, J. S.; Shankar, K. S.; Reddy, A. S.; Reddy, B. V. S. *Helv. Chim. Acta* **2014**, *97*, 546.
840 Mu, Y.; Jin, T.; Kim, G.-W.; Kim, J.-S.; Kim, S.-S.; Tian, Y.-S.; Oh, C.-Y.; Ham, W.-H. *Eur. J. Org. Chem.* **2012**, 2614.
841 Mu, Y.; Kim, J.-Y.; Jin, X.; Park, S.-H.; Joo, J.-E.; Ham, W.-H. *Synthesis* **2012**, *44*, 2340.
842 Sabitha, G.; Reddy, S. S. S.; Yadav, J. S. *Tetrahedron Lett.* **2011**, *52*, 2407.
843 Sarabia, F.; Martín-Gálvez, F.; Chammaa, S.; Martín-Ortiz, L.; Sánchez-Ruiz, A. *J. Org. Chem.* **2010**, *75*, 5526.
844 Bohrsch, V.; Blechert, S. *Chem. Commun.* **2006**, 1968.
845 Marshall, J. A.; Sabatini, J. J. *Org. Lett.* **2006**, *8*, 3557.
846 Han, S. B.; Hassan, A.; Kim, I. S.; Krische, M. J. *J. Am. Chem. Soc.* **2010**, *132*, 15559.
847 Tsuruda, T.; Ebine, M.; Umeda, A.; Oishi, T. *J. Org. Chem.* **2015**, *80*, 859.
848 Paquette, L. A.; Dong, S.; Parker, G. D. *J. Org. Chem.* **2007**, *72*, 7135.
849 Kaji, E.; Komori, T.; Yokoyama, M.; Kato, T.; Nishino, T.; Shirahata, T. *Tetrahedron* **2010**, *66*, 4089.
850 Roy, S.; Spilling, C. D. *Org. Lett.* **2010**, *12*, 5326.
851 Roy, S.; Spilling, C. D. *Org. Lett.* **2012**, *14*, 2230.
852 Point, V.; Malla, R. K.; Diomande, S.; Martin, B. P.; Delorme, V.; Carriere, F.; Canaan, S.; Rath, N. P.; Spilling, C. D.; Cavalier, J. F. *J. Med. Chem.* **2012**, *55*, 10204.
853 Vehlow, K.; Maechling, S.; Köhler, K.; Blechert, S. *Tetrahedron Lett.* **2006**, *47*, 8617.
854 Vehlow, K.; Maechling, S.; Blechert, S. *Organometallics* **2006**, *25*, 25.
855 Kulkarni, S. S.; Gervay-Hague, J. *Org. Lett.* **2006**, *8*, 5765.
856 Konda, S.; Bhaskar, K.; Nagarapu, L.; Akkewar, D. M. *Tetrahedron Lett.* **2014**, *55*, 3087.
857 Morita, A.; Kuwahara, S. *Tetrahedron Lett.* **2007**, *48*, 3163.
858 Wohlrab, A.; Lamer, R.; VanNieuwenhze, M. S. *J. Am. Chem. Soc.* **2007**, *129*, 4175.
859 Chaudhari, D. A.; Kattanguru, P.; Fernandes, R. A. *RSC Adv.* **2015**, *5*, 42131.
860 Bethi, V.; Kattanguru, P.; Fernandes, R. A. *Eur. J. Org. Chem.* **2014**, 3249.
861 Izuchi, Y.; Kanomata, N.; Koshino, H.; Hongo, Y.; Nakata, T.; Takahashi, S. *Tetrahedron: Asymmetry* **2011**, *22*, 246.
862 Kamal, A.; Balakrishna, M.; Reddy, P. V.; Faazil, S. *Tetrahedron: Asymmetry* **2010**, *21*, 2517.
863 Radha Krishna, P.; Nomula, R.; Kunde, R. *Synthesis* **2014**, *46*, 307.
864 Ramesh, P.; Meshram, H. M. *Tetrahedron Lett.* **2012**, *53*, 4008.
865 Karnekanti, R.; Hanumaiah, M.; Sharma, G. V. M. *Synthesis* **2015**, *47*, 2997.
866 Fischer, T.; Pietruszka, J. *Adv. Synth. Catal.* **2012**, *354*, 2521.
867 Das, B.; Nagendra, S.; Reddy, C. R. *Tetrahedron: Asymmetry* **2011**, *22*, 1249.
868 Yadav, J. S.; Yadagiri, K.; Madhuri, C.; Sabitha, G. *Tetrahedron Lett.* **2011**, *52*, 4269.
869 Das, B.; Veeranjaneyulu, B.; Balasubramanyam, P.; Srilatha, M. *Tetrahedron: Asymmetry* **2010**, *21*, 2762.
870 Ghadigaonkar, S.; Koli, M. R.; Gamre, S. S.; Choudhary, M. K.; Chattopadhyay, S.; Sharma, A. *Tetrahedron: Asymmetry* **2012**, *23*, 1093.
871 Jayasinghe, S.; Venukadasula, P. K. M.; Hanson, P. R. *Org. Lett.* **2014**, *16*, 122.
872 Nagendra, S.; Reddy, V. K.; Das, B. *Helv. Chim. Acta* **2015**, *98*, 520.
873 Sabitha, G.; Reddy, S. S. S.; Reddy, D. V.; Bhaskar, V.; Yadav, J. S. *Synthesis* **2010**, 3453.
874 Sabitha, G.; Bhaskar, V.; Reddy, S. S. S.; Yadav, J. S. *Chin. J. Chem.* **2010**, *28*, 2421.
875 Balasubramanyam, P.; Reddy, G. C.; Salvanna, N.; Das, B. *Synthesis* **2011**, 3706.

[876] Rajaram, S.; Ramesh, D.; Ramulu, U.; Prabhakar, P.; Venkateswarlu, Y. *Helv. Chim. Acta* **2014**, *97*, 112.
[877] Sabitha, G.; Reddy, S. S. S.; Yadav, J. S. *Tetrahedron Lett.* **2010**, *51*, 6259.
[878] Nagy, E. E.; Hyatt, I. F. D.; Gettys, K. E.; Yeazell, S. T.; Frempong, S. K.; Croatt, M. P. *Org. Lett.* **2013**, *15*, 586.
[879] Hanson, P. R.; Chegondi, R.; Nguyen, J.; Thomas, C. D.; Waetzig, J. D.; Whitehead, A. *J. Org. Chem.* **2011**, *76*, 4358.
[880] Whitehead, A.; Waetzig, J. D.; Thomas, C. D.; Hanson, P. R. *Org. Lett.* **2008**, *10*, 1421.
[881] Venukadasula, P. K. M.; Chegondi, R.; Maitra, S.; Hanson, P. R. *Org. Lett.* **2010**, *12*, 1556.
[882] Waetzig, J. D.; Hanson, P. R. *Org. Lett.* **2007**, *10*, 109.
[883] Li, J.; Zhao, C.; Liu, J.; Du, Y. *Tetrahedron* **2015**, *71*, 3885.
[884] Ghosh, A. K.; Veitschegger, A. M.; Sheri, V. R.; Effenberger, K. A.; Prichard, B. E.; Jurica, M. S. *Org. Lett.* **2014**, *16*, 6200.
[885] Commandeur, M.; Commandeur, C.; Cossy, J. *Org. Lett.* **2011**, *13*, 6018.
[886] Virolleaud, M.-A.; Piva, O. *Tetrahedron Lett.* **2007**, *48*, 1417.
[887] Skaanderup, P. R.; Jensen, T. *Org. Lett.* **2008**, *10*, 2821.
[888] Zhu, L.; Mootoo, D. R. *Org. Lett.* **2003**, *5*, 3475.
[889] Lankalapalli, R. S.; Baksa, A.; Liliom, K.; Bittman, R. *ChemMedChem* **2010**, *5*, 682.
[890] Hirata, Y.; Nakamura, S.; Watanabe, N.; Kataoka, O.; Kurosaki, T.; Anada, M.; Kitagaki, S.; Shiro, M.; Hashimoto, S. *Chem.—Eur. J.* **2006**, *12*, 8898.
[891] Sabitha, G.; Prasad, M. N.; Shankaraiah, K.; Reddy, N. M.; Yadav, J. S. *Synthesis* **2010**, 3891.
[892] Chen, Q.-Y.; Chaturvedi, P. R.; Luesch, H. *Org. Process Res. Dev.* **2018**, *22*, 190.
[893] Bowers, A. A.; West, N.; Newkirk, T. L.; Troutman-Youngman, A. E.; Schreiber, S. L.; Wiest, O.; Bradner, J. E.; Williams, R. M. *Org. Lett.* **2009**, *11*, 1301.
[894] Zaed, A. M.; Sutherland, A. *Org. Biomol. Chem.* **2010**, *8*, 4394.
[895] Donohoe, T. J.; Basutto, J. A.; Bower, J. F.; Rathi, A. *Org. Lett.* **2011**, *13*, 1036.
[896] Qiao, Y.; Kumar, S.; Malachowski, W. P. *Tetrahedron Lett.* **2010**, *51*, 2636.
[897] O'Leary, D. J.; Blackwell, H. E.; Washenfelder, R. A.; Miura, K.; Grubbs, R. H. *Tetrahedron Lett.* **1999**, *40*, 1091.
[898] Soulère, L.; Queneau, Y.; Doutheau, A. *Chem. Phys. Lipids* **2007**, *150*, 239.
[899] Shafi, S.; Kędziorek, M.; Grela, K. *Synlett* **2011**, *124*.
[900] Madduri, A. V. R.; Minnaard, A. J. *Chem.—Eur. J.* **2010**, *16*, 11726.
[901] Lippert, A. R.; Kaeobamrung, J.; Bode, J. W. *J. Am. Chem. Soc.* **2006**, *128*, 14738.
[902] Chatterjee, A. K.; Grubbs, R. H. *Angew. Chem., Int. Ed.* **2002**, *41*, 3171.
[903] Hue, V. T.; Nhung, N. T. H.; Hung, M. D. *ARKIVOC* **2014**, 206.
[904] Morimoto, S.; Shindo, M.; Yoshida, M.; Shishido, K. *Tetrahedron Lett.* **2006**, *47*, 7353.
[905] Davies, H. M. L.; Dai, X. *Tetrahedron* **2006**, *62*, 10477.
[906] Murokawa, T.; Enomoto, M.; Teranishi, T.; Ogura, Y.; Kuwahara, S. *Tetrahedron Lett.* **2018**, *59*, 4107.
[907] Huang, Q.; Herdewijn, P. *J. Org. Chem.* **2011**, *76*, 3742.
[908] Radha Krishna, P.; Alivelu, M.; Prabhakar Rao, T. *Eur. J. Org. Chem.* **2012**, 616.
[909] Benjamin, N. M.; Martin, S. F. *Org. Lett.* **2011**, *13*, 450.
[910] Patel, P.; Lee, G.-J.; Kim, S.; Grant, G. E.; Powell, W. S.; Rokach, J. *J. Org. Chem.* **2008**, *73*, 7213.
[911] Sutar, R.; Sen, S.; Eivgi, O.; Segalovich, G.; Schapiro, I.; Reany, O.; Lemcoff, N. G. *Chem. Sci.* **2018**, *9*, 1368.
[912] Wang, S.-Y.; Chin, Y.-J.; Loh, T.-P. *Synthesis* **2009**, 3557.
[913] Kanada, R. M.; Itoh, D.; Nagai, M.; Niijima, J.; Asai, N.; Mizui, Y.; Abe, S.; Kotake, Y. *Angew. Chem., Int. Ed.* **2007**, *46*, 4350.
[914] McDonald, F. E.; Wei, X. *Org. Lett.* **2002**, *4*, 593.
[915] Chung, W.-j.; Carlson, J. S.; Vanderwal, C. D. *J. Org. Chem.* **2014**, *79*, 2226.
[916] Morimoto, Y.; Okita, T.; Kambara, H. *Angew. Chem., Int. Ed.* **2009**, *48*, 2538.
[917] Cribiú, R.; Jäger, C.; Nevado, C. *Angew. Chem., Int. Ed.* **2009**, *48*, 8780.
[918] Zhang, W.; Bah, J.; Wohlfarth, A.; Franzén, J. *Chem.—Eur. J.* **2011**, *17*, 13814.
[919] BouzBouz, S.; Roche, C.; Cossy, J. *Synlett* **2009**, 803.
[920] Fustero, S.; Jiménez, D.; Moscardó, J.; Catalán, S.; del Pozo, C. *Org. Lett.* **2007**, *9*, 5283.
[921] Mohapatra, D. K.; Maity, S.; Rao, T. S.; Yadav, J. S.; Sridhar, B. *Eur. J. Org. Chem.* **2013**, 2859.
[922] Gessler, S.; Randl, S.; Blechert, S. *Tetrahedron Lett.* **2000**, *41*, 9973.
[923] Randl, S.; Buschmann, N.; Connon, S. J.; Blechert, S. *Synlett* **2001**, 1547.
[924] Clavier, H.; Audic, N.; Guillemin, J.-C.; Mauduit, M. *J. Organomet. Chem.* **2005**, *690*, 3585.
[925] Evans, P. A.; Grisin, A.; Lawler, M. J. *J. Am. Chem. Soc.* **2012**, *134*, 2856.
[926] Boufroura, H.; Mauduit, M.; Drège, E.; Joseph, D. *J. Org. Chem.* **2013**, *78*, 2346.

927 Rountree, S. M.; Taylor, S. F. R.; Hardacre, C.; Lagunas, M. C.; Davey, P. N. *Appl.Catal. A* **2014**, *486*, 94.
928 Morgan, J. P.; Grubbs, R. H. *Org. Lett.* **2000**, *2*, 3153.
929 Jana, A.; Zieliński, G. K.; Czarnocka-Śniadała, S.; Grudzień, K.; Podwysocka, D.; Szulc, M.; Kajetanowicz, A.; Grela, K. *ChemCatChem* **2019**, *11*, 5808.
930 Ammar, H. B.; Hassine, B. B.; Fischmeister, C.; Dixneuf, P. H.; Bruneau, C. *Eur. J. Inorg. Chem.* **2010**, 4752.
931 Lipshutz, B. H.; Ghorai, S.; Leong, W. W. Y.; Taft, B. R.; Krogstad, D. V. *J. Org. Chem.* **2011**, *76*, 5061.
932 Lipshutz, B. H.; Ghorai, S.; Bošković, Ž. V. *Tetrahedron* **2008**, *64*, 6949.
933 Nagarjuna, B.; Thirupathi, B.; Venkata Rao, C.; Mohapatra, D. K. *Tetrahedron Lett.* **2015**, *56*, 4916.
934 Donohoe, T. J.; Race, N. J.; Bower, J. F.; Callens, C. K. A. *Org. Lett.* **2010**, *12*, 4094.
935 Sawant, K. B.; Ding, F.; Jennings, M. P. *Tetrahedron Lett.* **2006**, *47*, 939.
936 Lee, C.-H. A.; Loh, T.-P. *Tetrahedron Lett.* **2006**, *47*, 809.
937 Zhu, J.; Porco, J. A. *Org. Lett.* **2006**, *8*, 5169.
938 Movassaghi, M.; Hunt, D. K.; Tjandra, M. *J. Am. Chem. Soc.* **2006**, *128*, 8126.
939 Dewi, P.; Randl, S.; Blechert, S. *Tetrahedron Lett.* **2005**, *46*, 577.
940 Sirasani, G.; Paul, T.; Andrade, R. B. *Tetrahedron* **2011**, *67*, 2197.
941 Paul, T.; Sirasani, G.; Andrade, R. B. *Tetrahedron Lett.* **2008**, *49*, 3363.
942 Altendorfer, M.; Raja, A.; Sasse, F.; Irschik, H.; Menche, D. *Org. Biomol. Chem.* **2013**, *11*, 2116.
943 Rajapakse, H. A.; Walji, A. M.; Moore, K. P.; Zhu, H.; Mitra, A. W.; Gregro, A. R.; Tinney, E.; Burlein, C.; Touch, S.; Paton, B. L.; Carroll, S. S.; DiStefano, D. J.; Lai, M.-T.; Grobler, J. A.; Sanchez, R. I.; Williams, T. M.; Vacca, J. P.; Nantermet, P. G. *ChemMedChem* **2011**, *6*, 253.
944 Carlson, E. C.; Rathbone, L. K.; Yang, H.; Collett, N. D.; Carter, R. G. *J. Org. Chem.* **2008**, *73*, 5155.
945 Bremeyer, N.; Smith, S. C.; Ley, S. V.; Gaunt, M. J. *Angew. Chem., Int. Ed.* **2004**, *43*, 2681.
946 Saibaba, V.; Sampath, A.; Mukkanti, K.; Iqbal, J.; Das, P. *Synthesis* **2007**, 2797.
947 Fuwa, H.; Noguchi, T.; Noto, K.; Sasaki, M. *Org. Biomol. Chem.* **2012**, *10*, 8108.
948 Park, H.; Kim, H.; Hong, J. *Org. Lett.* **2011**, *13*, 3742.
949 Li, C.-F.; Liu, H.; Liao, J.; Cao, Y.-J.; Liu, X.-P.; Xiao, W.-J. *Org. Lett.* **2007**, *9*, 1847.
950 Chen, J.-R.; Li, C.-F.; An, X.-L.; Zhang, J.-J.; Zhu, X.-Y.; Xiao, W.-J. *Angew. Chem., Int. Ed.* **2008**, *47*, 2489.
951 Mori, K.; Akasaka, K.; Matsunaga, S. *Tetrahedron* **2014**, *70*, 392.
952 Vedrenne, E.; Dupont, H.; Oualef, S.; Elkaïm, L.; Grimaud, L. *Synlett* **2005**, *670*.
953 Poulhès, F.; Sylvain, R.; Perfetti, P.; Bertrand, M. P.; Gil, G.; Gastaldi, S. *Synthesis* **2010**, 1334.
954 Abbas, M.; Leitgeb, A.; Slugovc, C. *Synlett* **2013**, *24*, 1193.
955 Boddaert, T.; Coquerel, Y.; Rodriguez, J. *Eur. J. Org. Chem.* **2011**, 5061.
956 Lipshutz, B. H.; Ghorai, S.; Leong, W. W. Y. *J. Org. Chem.* **2009**, *74*, 2854.
957 Clavier, H.; Caijo, F.; Borré, E.; Rix, D.; Boeda, F.; Nolan, S. P.; Mauduit, M. *Eur. J. Org. Chem.* **2009**, 4254.
958 Fustero, S.; Jiménez, D.; Sánchez-Roselló, M.; del Pozo, C. *J. Am. Chem. Soc.* **2007**, *129*, 6700.
959 Fustero, S.; Monteagudo, S.; Sánchez-Roselló, M.; Flores, S.; Barrio, P.; del Pozo, C. *Chem.—Eur. J.* **2010**, *16*, 9835.
960 Fustero, S.; Báez, C.; Sánchez-Roselló, M.; Asensio, A.; Miro, J.; del Pozo, C. *Synthesis* **2012**, *44*, 1863.
961 Wakamatsu, H.; Blechert, S. *Angew. Chem., Int. Ed.* **2002**, *41*, 794.
962 Wakamatsu, H.; Blechert, S. *Angew. Chem., Int. Ed.* **2002**, *41*, 2403.
963 Leitgeb, A.; Abbas, M.; Fischer, R. C.; Poater, A.; Cavallo, L.; Slugovc, C. *Catal. Sci. Technol.* **2012**, *2*, 1640.
964 Liu, H.; Zeng, C.; Guo, J.; Zhang, M.; Yu, S. *RSC Adv.* **2013**, *3*, 1666.
965 Borré, E.; Caijo, F.; Crévisy, C.; Mauduit, M. *Beilstein J. Org. Chem.* **2010**, *6*, 1159.
966 Reddy, G. V.; Kumar, R. S. C.; Babu, K. S.; Rao, J. M. *Tetrahedron Lett.* **2009**, *50*, 4117.
967 Kim, S.-G.; Park, T.-H.; Kim, B. J. *Tetrahedron Lett.* **2006**, *47*, 6369.
968 Lesma, G.; Colombo, A.; Sacchetti, A.; Silvani, A. *J. Org. Chem.* **2008**, *74*, 590.
969 González-Gómez, J. C.; Foubelo, F.; Yus, M. *Synlett* **2008**, *2777*.
970 Fuwa, H.; Noto, K.; Sasaki, M. *Org. Lett.* **2010**, *12*, 1636.
971 Ettari, R.; Micale, N. *J. Organomet. Chem.* **2007**, *692*, 3574.
972 Lipshutz, B. H.; Ghorai, S.; Abela, A. R.; Moser, R.; Nishikata, T.; Duplais, C.; Krasovskiy, A.; Gaston, R. D.; Gadwood, R. C. *J. Org. Chem.* **2011**, *76*, 4379.
973 Zhang, J.-W.; Cai, Q.; Gu, Q.; Shi, X.-X.; You, S.-L. *Chem. Commun.* **2013**, *49*, 7750.
974 Bates, R. W.; Song, P. *Synthesis* **2010**, 2935.

975 van Zijl, A. W.; Szymanski, W.; López, F.; Minnaard, A. J.; Feringa, B. L. *J. Org. Chem.* **2008**, *73*, 6994.
976 Lipshutz, B. H.; Aguinaldo, G. T.; Ghorai, S.; Voigtritter, K. *Org. Lett.* **2008**, *10*, 1325.
977 Behenna, D. C.; Mohr, J. T.; Sherden, N. H.; Marinescu, S. C.; Harned, A. M.; Tani, K.; Seto, M.; Ma, S.; Novák, Z.; Krout, M. R.; McFadden, R. M.; Roizen, J. L.; Enquist, J. A.; White, D. E.; Levine, S. R.; Petrova, K. V.; Iwashita, A.; Virgil, S. C.; Stoltz, B. M. *Chem.—Eur. J.* **2011**, *17*, 14199.
978 Angeli, M.; Bandini, M.; Garelli, A.; Piccinelli, F.; Tommasi, S.; Umani-Ronchi, A. *Org. Biomol. Chem.* **2006**, *4*, 3291.
979 Boddaert, T.; Coquerel, Y.; Rodriguez, J. *Adv. Synth. Catal.* **2009**, *351*, 1744.
980 Kumar, R. S. C.; Sreedhar, E.; Reddy, G. V.; Babu, K. S.; Rao, J. M. *Tetrahedron: Asymmetry* **2009**, *20*, 1160.
981 Yokoe, H.; Yoshida, M.; Shishido, K. *Tetrahedron Lett.* **2008**, *49*, 3504.
982 Cai, Q.; Zheng, C.; You, S.-L. *Angew. Chem., Int. Ed.* **2010**, *49*, 8666.
983 Mukherjee, H.; McDougal, N. T.; Virgil, S. C.; Stoltz, B. M. *Org. Lett.* **2011**, *13*, 825.
984 Buchowicz, W.; Szmajda, M. *Organometallics* **2009**, *28*, 6838.
985 Cannillo, A.; Norsikian, S.; Tran Huu Dau, M.-E.; Retailleau, P.; Iorga, B. I.; Beau, J.-M. *Chem.—Eur. J.* **2014**, *20*, 12133.
986 Ettari, R.; Zappalà, M.; Micale, N.; Grazioso, G.; Giofrè, S.; Schirmeister, T.; Grasso, S. *Eur. J. Med. Chem.* **2011**, *46*, 2058.
987 Ettari, R.; Micale, N.; Schirmeister, T.; Gelhaus, C.; Leippe, M.; Nizi, E.; Di Francesco, M. E.; Grasso, S.; Zappalà, M. *J. Med. Chem.* **2009**, *52*, 2157.
988 Klüppel, A.; Gille, A.; Karayel, C. E.; Hiersemann, M. *Org. Lett.* **2019**, *21*, 2421.
989 Jackson, K. L.; Henderson, J. A.; Morris, J. C.; Motoyoshi, H.; Phillips, A. J. *Tetrahedron Lett.* **2008**, *49*, 2939.
990 Yang, H.; Carter, R. G. *J. Org. Chem.* **2010**, *75*, 4929.
991 Saha, M.; Carter, R. G. *Org. Lett.* **2013**, *15*, 736.
992 BouzBouz, S.; Boulard, L.; Cossy, J. *Org. Lett.* **2007**, *9*, 3765.
993 Reymond, S.; Cossy, J. *Tetrahedron* **2007**, *63*, 5918.
994 Legeay, J.-C.; Lewis, W.; Stockman, R. A. *Chem. Commun.* **2009**, 2207.
995 Ge, Y.; Sun, W.; Pei, B.; Ding, J.; Jiang, Y.; Loh, T.-P. *Org. Lett.* **2018**, *20*, 2774.
996 Llàcer, E.; Urpí, F.; Vilarrasa, J. *Org. Lett.* **2009**, *11*, 3198.
997 Batt, F.; Fache, F. *Eur. J. Org. Chem.* **2011**, 6039.
998 Crimmins, M. T.; Dechert, A.-M. R. *Org. Lett.* **2012**, *14*, 2366.
999 Dewi-Wülfing, P.; Gebauer, J.; Blechert, S. *Synlett* **2006**, *487*.
1000 Cochet, T.; Roche, D.; Bellosta, V.; Cossy, J. *Eur. J. Org. Chem.* **2012**, 801.
1001 Williams, B. D.; Smith, A. B., III. *J. Org. Chem.* **2014**, *79*, 9284.
1002 Dewi-Wülfing, P.; Blechert, S. *Eur. J. Org. Chem.* **2006**, 1852.
1003 Shizuka, M.; Snapper, M. L. *Synthesis* **2007**, 2397.
1004 Pasqua, A. E.; Ferrari, F. D.; Hamman, C.; Liu, Y.; Crawford, J. J.; Marquez, R. *J. Org. Chem.* **2012**, *77*, 6989.
1005 Chandrasekhar, S.; Kiran Babu, G. S.; Raji Reddy, C. *Tetrahedron: Asymmetry* **2009**, *20*, 2216.
1006 Bressy, C.; Allais, F.; Cossy, J. *Synlett* **2006**, *3455*.
1007 Bates, R. W.; Song, P. *Tetrahedron* **2007**, *63*, 4497.
1008 De Joarder, D.; Jennings, M. P. *Tetrahedron Lett.* **2011**, *52*, 5124.
1009 Amans, D.; Bellosta, V.; Cossy, J. *Org. Lett.* **2007**, *9*, 1453.
1010 Amans, D.; Bareille, L.; Bellosta, V.; Cossy, J. *J. Org. Chem.* **2009**, *74*, 7665.
1011 Gebauer, J.; Arseniyadis, S.; Cossy, J. *Synlett* **2008**, *712*.
1012 ElMarrouni, A.; Lebeuf, R.; Gebauer, J.; Heras, M.; Arseniyadis, S.; Cossy, J. *Org. Lett.* **2011**, *14*, 314.
1013 Allais, F.; Aouhansou, M.; Majira, A.; Ducrot, P.-H. *Synthesis* **2010**, 2787.
1014 Lynch, J. E.; Zanatta, S. D.; White, J. M.; Rizzacasa, M. A. *Chem.—Eur. J.* **2011**, *17*, 297.
1015 Drummond, L. J.; Sutherland, A. *Tetrahedron* **2010**, *66*, 5349.
1016 Reddy, D. K.; Shekhar, V.; Reddy, T. S.; Reddy, S. P.; Venkateswarlu, Y. *Tetrahedron: Asymmetry* **2009**, *20*, 2315.
1017 Tian, X.; Rychnovsky, S. D. *Org. Lett.* **2007**, *9*, 4955.
1018 Fuwa, H.; Sasaki, M. *Org. Lett.* **2010**, *12*, 584.
1019 Fuwa, H.; Suzuki, T.; Kubo, H.; Yamori, T.; Sasaki, M. *Chem.—Eur. J.* **2011**, *17*, 2678.
1020 Allu, S. R.; Banne, S.; Jiang, J.; Qi, N.; Guo, J.; He, Y. *J. Org. Chem.* **2019**, *84*, 7227.
1021 Silva, F. A.; Gouverneur, V. *Tetrahedron Lett.* **2005**, *46*, 8705.
1022 Sabitha, G.; Chandrashekhar, G.; Yadagiri, K.; Yadav, J. S. *Tetrahedron: Asymmetry* **2011**, *22*, 1729.

1023 Reymond, S.; Cossy, J. *Eur. J. Org. Chem.* **2006**, 4800.
1024 Gregg, C.; Gunawan, C.; Ng, A. W. Y.; Wimala, S.; Wickremasinghe, S.; Rizzacasa, M. A. *Org. Lett.* **2013**, *15*, 516.
1025 Takahashi, S.; Hongo, Y.; Tsukagoshi, Y.; Koshino, H. *Org. Lett.* **2008**, *10*, 4223.
1026 Takahashi, S.; Takahashi, R.; Hongo, Y.; Koshino, H.; Yamaguchi, K.; Miyagi, T. *J. Org. Chem.* **2009**, *74*, 6382.
1027 Huckins, J. R.; de Vicente, J.; Rychnovsky, S. D. *Org. Lett.* **2007**, *9*, 4757.
1028 Jackson, K. L.; Henderson, J. A.; Motoyoshi, H.; Phillips, A. J. *Angew. Chem., Int. Ed.* **2009**, *48*, 2346.
1029 BouzBouz, S.; Cossy, J. *Tetrahedron Lett.* **2006**, *47*, 901.
1030 Yadav, J. S.; Thrimurtulu, N.; Gayathri, K. U.; Reddy, B. V. S.; Prasad, A. R. *Synlett* **2009**, 790.
1031 Kluciar, M.; Grela, K.; Mauduit, M. *Dalton Trans.* **2013**, *42*, 7354.
1032 Langford, S. J.; Latter, M. J.; Woodward, C. P. *Org. Lett.* **2006**, *8*, 2595.
1033 Dong, Y.; Edgar, K. J. *Polym. Chem.* **2015**, *6*, 3816.
1034 Meng, X.; Edgar, K. *J. Carbohydr. Polym.* **2015**, *132*, 565.
1035 Meng, X.; Matson, J. B.; Edgar, K. J. *Polym. Chem.* **2014**, *5*, 7021.
1036 Meng, X.; Matson, J. B.; Edgar, K. J. *Biomacromolecules* **2014**, *15*, 177.
1037 Liu, Z.; Xu, C.; del Pozo, J.; Torker, S.; Hoveyda, A. H. *J. Am. Chem. Soc.* **2019**, *141*, 7137.
1038 Tello-Aburto, R.; Newar, T. D.; Maio, W. A. *J. Org. Chem.* **2012**, *77*, 6271.
1039 Yao, Q.; Rodriguez Motta, A. *Tetrahedron Lett.* **2004**, *45*, 2447.
1040 Wappel, J.; Urbina-Blanco, C. A.; Abbas, M.; Albering, J. H.; Saf, R.; Nolan, S. P.; Slugovc C. *Beilstein J. Org. Chem.* **2010**, *6*, 1091.
1041 Coudray, L.; Montchamp, J.-L. *Eur. J. Org. Chem.* **2009**, 4646.
1042 Urbina-Blanco, C. A.; Manzini, S.; Gomes, J. P.; Doppiu, A.; Nolan, S. P. *Chem. Commun.* **2011**, *47*, 5022.
1043 Schmid, T. E.; Bantreil, X.; Citadelle, C. A.; Slawin, A. M. Z.; Cazin, C. S. J. *Chem. Commun.* **2011**, *47*, 7060.
1044 Guidone, S.; Blondiaux, E.; Samojłowicz, C.; Gułajski, Ł.; Kędziorek, M.; Malińska, M.; Pazio, A.; Woźniak, K.; Grela, K.; Doppiu, A.; Cazin, C. S. J. *Adv. Synth. Catal.* **2012**, *354*, 2734.
1045 Manzini, S.; Urbina Blanco, C. A.; Slawin, A. M. Z.; Nolan, S. P. *Organometallics* **2012**, *31*, 6514.
1046 Borré, E.; Rouen, M.; Laurent, I.; Magrez, M.; Caijo, F.; Crévisy, C.; Solodenko, W.; Toupet, L.; Frankfurter, R.; Vogt, C.; Kirschning, A.; Mauduit, M. *Chem.—Eur. J.* **2012**, *18*, 16369.
1047 Torborg, C.; Szczepaniak, G.; Zielinski, A.; Malinska, M.; Wozniak, K.; Grela, K. *Chem. Commun.* **2013**, *49*, 3188.
1048 Clavier, H.; Nolan, S. P.; Mauduit, M. *Organometallics* **2008**, *27*, 2287.
1049 Schmidt, B.; Hauke, S. *Org. Biomol. Chem.* **2013**, *11*, 4194.
1050 Chen, S.-W.; Kim, J. H.; Ryu, K. Y.; Lee, W.-W.; Hong, J.; Lee, S.-G. *Tetrahedron* **2009**, *65*, 3397.
1051 Bonini, C.; Campaniello, M.; Chiummiento, L.; Videtta, V. *Tetrahedron* **2008**, *64*, 8766.
1052 Fustero, S.; Sánchez-Roselló, M.; Sanz-Cervera, J. F.; Aceña, J. L.; del Pozo, C.; Fernández, B.; Bartolomé, A.; Asensio, A. *Org. Lett.* **2006**, *8*, 4633.
1053 Lipshutz, B. H.; Bošković, Z.; Crowe, C. S.; Davis, V. K.; Whittemore, H. C.; Vosburg, D. A.; Wenzel, A. G. *J. Chem. Educ.* **2013**, *90*, 1514.
1054 Tanis, P. S.; Infantine, J. R.; Leighton, J. L. *Org. Lett.* **2013**, *15*, 5464.
1055 Urbina-Blanco, C. A.; Leitgeb, A.; Slugovc, C.; Bantreil, X.; Clavier, H.; Slawin, A. M. Z.; Nolan, S. P. *Chem.—Eur. J.* **2011**, *17*, 5045.
1056 Pastva, J.; Skowerski, K.; Czarnocki, S. J.; Žilková, N.; Čejka, J.; Bastl, Z.; Balcar, H. *ACS Catal.* **2014**, *4*, 3227.
1057 Lund, C. L.; Sgro, M. J.; Stephan, D. W. *Organometallics* **2012**, *31*, 580.
1058 Chen, L.-J.; Hou, D.-R. *Tetrahedron: Asymmetry* **2008**, *19*, 715.
1059 Nair, R. N.; Bannister, T. D. *J. Org. Chem.* **2014**, *79*, 1467.
1060 Barbasiewicz, M.; Michalak, M.; Grela, K. *Chem.—Eur. J.* **2012**, *18*, 14237.
1061 Bieniek, M.; Bujok, R.; Cabaj, M.; Lugan, N.; Lavigne, G.; Arlt, D.; Grela, K. *J. Am. Chem. Soc.* **2006**, *128*, 13652.
1062 Huang, Y.; Fananas-Mastral, M.; Minnaard, A. J.; Feringa, B. L. *Chem. Commun.* **2013**, *49*, 3309.
1063 Vincent, A.; Prunet, J. *Synlett* **2006**, *2269.*
1064 Hanessian, S.; Auzzas, L.; Giannini, G.; Marzi, M.; Cabri, W.; Barbarino, M.; Vesci, L.; Pisano, C. *Bioorg. Med. Chem. Lett.* **2007**, *17*, 6261.
1065 Kaliappan, K. P.; Gowrisankar, P. *Synlett* **2007**, 1537.
1066 Kim, H.; Kim, M.-Y.; Tae, J. *Synlett* **2009**, 2949.
1067 Chattopadhyay, S. K.; Sil, S.; Mukherjee, J. P. *Beilstein J. Org. Chem.* **2017**, *13*, 2153.

1068 Mentink, G.; van Maarseveen, J. H.; Hiemstra, H. *Org. Lett.* **2002**, *4*, 3497.
1069 Fustero, S.; Albert, L.; Mateu, N.; Chiva, G.; Miró, J.; González, J.; Aceña, J. L. *Chem.—Eur. J.* **2012**, *18*, 3753.
1070 Paterson, I.; Anderson, E. A.; Dalby, S. M.; Genovino, J.; Lim, J. H.; Moessner, C. *Chem. Commun.* **2007**, 1852.
1071 Hiebel, M.-A.; Pelotier, B.; Lhoste, P.; Piva, O. *Synlett* **2008**, 1202.
1072 Yoshimura, F.; Takahashi, M.; Tanino, K.; Miyashita, M. *Tetrahedron Lett.* **2008**, *49*, 6991.
1073 Lisboa, M. P.; Jones, D. M.; Dudley, G. B. *Org. Lett.* **2013**, *15*, 886.
1074 Bilel, H.; Hamdi, N.; Zagrouba, F.; Fischmeister, C.; Bruneau, C. *RSC Adv.* **2012**, *2*, 9584.
1075 Higman, C. S.; de Araujo, M. P.; Fogg, D. E. *Catal. Sci. Technol.* **2016**, *6*, 2077.
1076 Lummiss, J. A. M.; Oliveira, K. C.; Pranckevicius, A. M. T.; Santos, A. G.; dos Santos, E. N.; Fogg, D. E. *J. Am. Chem. Soc.* **2012**, *134*, 18889.
1077 Shizuka, M.; Snapper, M. L. *Angew. Chem., Int. Ed.* **2008**, *47*, 5049.
1078 Chou, S.-S. P.; Huang, J.-L. *Tetrahedron Lett.* **2012**, *53*, 5552.
1079 Winkler, M.; Meier, M. A. R. *Green Chem.* **2014**, *16*, 3335.
1080 Paudyal, M. P.; Rath, N. P.; Spilling, C. D. *Org. Lett.* **2010**, *12*, 2954.
1081 Abbas, M.; Slugovc, C. *Tetrahedron Lett.* **2011**, *52*, 2560.
1082 Matsugi, M.; Curran, D. P. *J. Org. Chem.* **2005**, *70*, 1636.
1083 Gułajski, Ł.; Sledz, P.; Lupa, A.; Grela, K. *Green Chem.* **2008**, *10*, 271.
1084 Basu, S.; Waldmann, H. *Bioorg. Med. Chem.* **2014**, *22*, 4430.
1085 Munasinghe, V. R. N.; Corrie, J. E. T.; Kelly, G.; Martin, S. R. *Bioconjugate Chem.* **2006**, *18*, 231.
1086 Newton, A. F.; Rejzek, M.; Alcaraz, M.-L.; Stockman, R. A. *Beilstein J. Org. Chem.* **2008**, *4*, 4.
1087 Miao, X.; Fischmeister, C.; Dixneuf, P. H.; Bruneau, C.; Dubois, J. L.; Couturier, J. L. *Green Chem.* **2012**, *14*, 2179.
1088 Miao, X.; Blokhin, A.; Pasynskii, A.; Nefedov, S.; Osipov, S. N.; Roisnel, T.; Bruneau, C.; Dixneuf, P. H. *Organometallics* **2010**, *29*, 5257.
1089 Miao, X.; Malacea, R.; Fischmeister, C.; Bruneau, C.; Dixneuf, P. H. *Green Chem.* **2011**, *13*, 2911.
1090 Huang, S.; Bilel, H.; Zagrouba, F.; Hamdi, N.; Bruneau, C.; Fischmeister, C. *Catal. Commun.* **2015**, *63*, 31.
1091 Behr, A.; Toepell, S.; Harmuth, S. *RSC Adv.* **2014**, *4*, 16320.
1092 Songis, O.; Slawin, A. M. Z.; Cazin, C. S. *J. Chem. Commun.* **2012**, *48*, 1266.
1093 Lautens, M.; Maddess, M. L. *Org. Lett.* **2004**, *6*, 1883.
1094 Li, Y.-M.; Li, X.; Peng, F.-Z.; Li, Z.-Q.; Wu, S.-T.; Sun, Z.-W.; Zhang, H.-B.; Shao, Z.-H. *Org. Lett.* **2011**, *13*, 6200.
1095 Bates, R. W.; Palani, K. *Tetrahedron Lett.* **2008**, *49*, 2832.
1096 Venkataiah, M.; Somaiah, P.; Reddipalli, G.; Fadnavis, N. W. *Tetrahedron: Asymmetry* **2009**, *20*, 2230.
1097 McCall, W. S.; Comins, D. L. *Org. Lett.* **2009**, *11*, 2940.
1098 Dhara, K.; Midya, G. C.; Dash, J. *J. Org. Chem.* **2012**, *77*, 8071.
1099 Bates, R. W.; Lim, C. J. *Synlett* **2010**, 866.
1100 Dittoo, A.; Bellosta, V.; Cossy, J. *Synlett* **2008**, *2459.*
1101 Sabitha, G.; Yagundar Reddy, A.; Yadav, J. S. *Tetrahedron Lett.* **2012**, *53*, 5624.
1102 Kumaraswamy, G.; Kumar, R. S.; Sampath, B.; Poornachandra, Y.; Kumar, C. G.; Vemulapalli, S. P. B.; Bharatam, J. *Bioorg. Med. Chem. Lett.* **2014**, *24*, 4439.
1103 Robertson, J.; North, C.; Sadig, J. E. R. *Tetrahedron* **2011**, *67*, 5011.
1104 Stevenson, B.; Lewis, W.; Dowden, J. *Synlett* **2010**, *672.*
1105 Radha Krishna, P.; Srinivas, R. *Tetrahedron: Asymmetry* **2007**, *18*, 2197.
1106 Fuwa, H.; Saito, A.; Sasaki, M. *Angew. Chem., Int. Ed.* **2010**, *49*, 3041.
1107 Fuwa, H.; Yamaguchi, H.; Sasaki, M. *Org. Lett.* **2010**, *12*, 1848.
1108 Fuwa, H.; Yamaguchi, H.; Sasaki, M. *Tetrahedron* **2010**, *66*, 7492.
1109 Eissler, S.; Nahrwold, M.; Neumann, B.; Stammler, H.-G.; Sewald, N. *Org. Lett.* **2007**, *9*, 817.
1110 Thakur, P.; Boyapelly, K.; Gurram, R. R.; Bandichhor, R.; Mukkanti, K. *Tetrahedron: Asymmetry* **2012**, *23*, 659.
1111 Crimmins, M. T.; Haley, M. W. *Org. Lett.* **2006**, *8*, 4223.
1112 Hume, P. A.; Sperry, J.; Brimble, M. A. *Org. Biomol. Chem.* **2011**, *9*, 5423.
1113 Basu, S.; Waldmann, H. *J. Org. Chem.* **2006**, *71*, 3977.
1114 Stellfeld, T.; Bhatt, U.; Kalesse, M. *Org. Lett.* **2004**, *6*, 3889.
1115 Ferrie, L.; Boulard, L.; Pradaux, F.; Bouzbouz, S.; Reymond, S.; Capdevielle, P.; Cossy, J. *J. Org. Chem.* **2008**, *73*, 1864.
1116 Beemelmanns, C.; Lentz, D.; Reissig, H.-U. *Chem.—Eur. J.* **2011**, *17*, 9720.

1117 Ramakrishna, K.; Kaliappan, K. P. *Org. Biomol. Chem.* **2015**, *13*, 234.
1118 Abbas, M.; Slugovc, C. *Monatsh. Chem.* **2012**, *143*, 669.
1119 Vignon, P.; Vancompernolle, T.; Couturier, J.-L.; Dubois, J.-L.; Mortreux, A.; Gauvin, R. M. *ChemSusChem* **2015**, *8*, 1143.
1120 Chang, M.-Y.; Lee, T.-W.; Wu, M.-H. *Org. Lett.* **2012**, *14*, 2198.
1121 Knight, D. W.; Smith, A. W. T. *Tetrahedron* **2015**, *71*, 7436.
1122 Ettari, R.; Micale, N.; Grazioso, G.; Bova, F.; Schirmeister, T.; Grasso, S.; Zappalà, M. *ChemMedChem* **2012**, *7*, 1594.
1123 Marjanovic, J.; Kozmin, S. A. *Angew. Chem., Int. Ed.* **2007**, *46*, 8854.
1124 Bandini, M.; Contento, M.; Garelli, A.; Monari, M.; Tolomelli, A.; Umani-Ronchi, A.; Andriolo, E.; Montorsi, M. *Synthesis* **2008**, 3801.
1125 Kreye, O.; Kugele, D.; Faust, L.; Meier, M. A. R. *Macromol. Rapid Commun.* **2014**, *35*, 317.
1126 Poeylaut-Palena, A. A.; Testero, S. A.; Mata, E. G. *Chem. Commun.* **2011**, *47*, 1565.
1127 Padakanti, S.; Sreekanth, B. R.; Mahendar, V.; Pal, M.; Mukkanti, K.; Iqbal, J.; Das, P. *Synlett* **2008**, *2417*.
1128 Suzaki, Y.; Osakada, K. *Chem. Lett.* **2006**, *35*, 374.
1129 Suzaki, Y.; Osakada, K. *Dalton Trans.* **2007**, 2376.
1130 Trinh, T. M. N.; Nguyen, T. T.; Kopp, C.; Pieper, P.; Russo, V.; Heinrich, B.; Donnio, B.; Nguyen, T. L. A.; Deschenaux, R. *Eur. J. Org. Chem.* **2015**, 6005.
1131 Malzahn, K.; Marsico, F.; Koynov, K.; Landfester, K.; Weiss, C. K.; Wurm, F. R. *ACS Macro Lett.* **2014**, *3*, 40.
1132 van Zijl, A. W.; Minnaard, A. J.; Feringa, B. L. *J. Org. Chem.* **2008**, *73*, 5651.
1133 Chattopadhyay, S. K.; Ghosh, S.; Sil, S. *Beilstein J. Org. Chem.* **2018**, *14*, 3070.
1134 Fuwa, H.; Ichinokawa, N.; Noto, K.; Sasaki, M. *J. Org. Chem.* **2011**, *77*, 2588.
1135 Streuff, J.; Muñiz, K. *J. Organomet. Chem.* **2005**, *690*, 5973.
1136 Courtens, C.; Risseeuw, M.; Caljon, G.; Cos, P.; Van Calenbergh, S. *ACS Med. Chem. Lett.* **2018**, *9*, 986.
1137 Morris, T.; Sandham, D.; Caddick, S. *Org. Biomol. Chem.* **2007**, *5*, 1025.
1138 Neisius, N. M.; Plietker, B. *J. Org. Chem.* **2008**, *73*, 3218.
1139 Bao, G.-m.; Tanaka, K.; Ikenaka, K.; Fukase, K. *Bioorg. Med. Chem.* **2010**, *18*, 3760.
1140 Jiang, S.; Niu, S.; Zhao, Z.-H.; Li, Z.-J.; Li, Q. *Carbohydr. Res.* **2015**, *414*, 39.
1141 Reddy, K. H. v.; Bédier, M.; Bouzbouz, S. *Eur. J. Org. Chem.* **2018**, 1455.
1142 Bédier, M.; BouzBouz, S. *Synlett* **2015**, *26*, 2531.
1143 Giudici, R. E.; Hoveyda, A. H. *J. Am. Chem. Soc.* **2007**, *129*, 3824.
1144 Finch, O. C.; Furkert, D. P.; Brimble, M. A. *Tetrahedron* **2014**, *70*, 590.
1145 Mizutani, M.; Inagaki, F.; Nakanishi, T.; Yanagihara, C.; Tamai, I.; Mukai, C. *Org. Lett.* **2011**, *13*, 1796.
1146 West, R.; Panagabko, C.; Atkinson, J. *J. Org. Chem.* **2010**, *75*, 2883.
1147 Burnley, J.; Wang, Z. J.; Jackson, W. R.; Robinson, A. J. *J. Org. Chem.* **2017**, *82*, 8497.
1148 Tseng, Y.-L.; Liang, M.-C.; Chen, I. C.; Wu, Y.-K. *Synlett* **2018**, *29*, 609.
1149 Xu, C.; Liu, Z.; Torker, S.; Shen, X.; Xu, D.; Hoveyda, A. H. *J. Am. Chem. Soc.* **2017**, *139*, 15640.
1150 Sari, O.; Hamada, M.; Roy, V.; Nolan, S. P.; Agrofoglio, L. A. *Org. Lett.* **2013**, *15*, 4390.
1151 Escobar, Z.; Solano, C.; Larsson, R.; Johansson, M.; Salamanca, E.; Gimenez, A.; Muñoz, E.; Sterner, O. *Tetrahedron* **2014**, *70*, 9052.
1152 Spessard, S. J.; Stoltz, B. M. *Org. Lett.* **2002**, *4*, 1943.
1153 Sapkota, R. R.; Jarvis, J. M.; Schaub, T. M.; Talipov, M. R.; Arterburn, J. B. *ChemistryOpen* **2019**, *8*, 201.
1154 Esteban, J.; Costa, A. M.; Vilarrasa, J. *Org. Lett.* **2008**, *10*, 4843.
1155 Wang, Z. J.; Spiccia, N. D.; Gartshore, C. J.; Illesinghe, J.; Jackson, W. R.; Robinson, A. J. *Synthesis* **2013**, *45*, 3118.
1156 Ghosh, A. K.; Cheng, X. *Org. Lett.* **2011**, *13*, 4108.
1157 Braun, M.-G.; Vincent, A.; Boumediene, M.; Prunet, J. *J. Org. Chem.* **2011**, *76*, 4921.
1158 Plummer, C. W.; Soheili, A.; Leighton, J. L. *Org. Lett.* **2012**, *14*, 2462.
1159 Bilel, H.; Hamdi, N.; Zagrouba, F.; Fischmeister, C.; Bruneau, C. *Green Chem.* **2011**, *13*, 1448.
1160 Jacobsen, M. F.; Moses, J. E.; Adlington, R. M.; Baldwin, J. E. *Tetrahedron* **2006**, *62*, 1675.
1161 López-Sánchez, C.; Hernández-Cervantes, C.; Rosales, A.; Álvarez-Corral, M.; Muñoz-Dorado, M.; Rodríguez-García, I. *Tetrahedron* **2009**, *65*, 9542.
1162 Raju, R.; Allen, L. J.; Le, T.; Taylor, C. D.; Howell, A. R. *Org. Lett.* **2007**, *9*, 1699.
1163 Moïse, J.; Sonawane, R. P.; Corsi, C.; Wendeborn, S. V.; Arseniyadis, S.; Cossy, J. *Synlett* **2008**, 2617.

1164 Camara, K.; Kamat, S. S.; Lasota, C. C.; Cravatt, B. F.; Howell, A. R. *Bioorg. Med. Chem. Lett.* **2015**, *25*, 317.
1165 Liang, Y.; Raju, R.; Le, T.; Taylor, C. D.; Howell, A. R. *Tetrahedron Lett.* **2009**, *50*, 1020.
1166 Humpl, M.; Tauchman, J.; Topolovčan, N.; Kretschmer, J.; Hessler, F.; Císařová, I.; Kotora, M.; Veselý, J. *J. Org. Chem.* **2016**, *81*, 7692.
1167 Murray, T. J.; Forsyth, C. J. *Org. Lett.* **2008**, *10*, 3429.
1168 Wilson, M. R.; Taylor, R. E. *Org. Lett.* **2012**, *14*, 3408.
1169 Rünzi, T.; Guironnet, D.; Göttker-Schnetmann, I.; Mecking, S. *J. Am. Chem. Soc.* **2010**, *132*, 16623.
1170 Chen, M.; Roush, W. R. *Org. Lett.* **2011**, *14*, 426.
1171 Artman, G. D.; Grubbs, A. W.; Williams, R. M. *J. Am. Chem. Soc.* **2007**, *129*, 6336.
1172 Bai, C.-X.; Zhang, W.-Z.; He, R.; Lu, X.-B.; Zhang, Z.-Q. *Tetrahedron Lett.* **2005**, *46*, 7225.
1173 Grela, K.; Kim, M. *Eur. J. Org. Chem.* **2003**, 963.
1174 Zhang, W.-Z.; He, R.; Zhang, R. *Eur. J. Inorg. Chem.* **2007**, 5345.
1175 Khan, R. K. M.; O'Brien, R. V.; Torker, S.; Li, B.; Hoveyda, A. H. *J. Am. Chem. Soc.* **2012**, *134*, 12774.
1176 Quinn, K. J.; Curto, J. M.; Faherty, E. E.; Cammarano, C. M. *Tetrahedron Lett.* **2008**, *49*, 5238.
1177 Takahira, Y.; Morizawa, Y. *J. Am. Chem. Soc.* **2015**, *137*, 7031.
1178 Oikawa, M.; Kasori, Y.; Katayama, L.; Murakami, E.; Oikawa, Y.; Ishikawa, Y. *Synthesis* **2013**, *45*, 3106.
1179 Ikoma, M.; Oikawa, M.; Gill, M. B.; Swanson, G. T.; Sakai, R.; Shimamoto, K.; Sasaki, M. *Eur. J. Org. Chem.* **2008**, 5215.
1180 Koh, M. J.; Nguyen, T. T.; Lam, J. K.; Torker, S.; Hyvl, J.; Schrock, R. R.; Hoveyda, A. H. *Nature* **2017**, *542*, 80.
1181 Boldon, S.; Moore, J. E.; Gouverneur, V. *Chem. Commun.* **2008**, 3622.
1182 Foltynowicz, Z.; Pietraszuk, C.; Marciniec, B. *Appl. Organomet. Chem.* **1993**, *7*, 539.
1183 Marciniec, B.; Pietraszuk, C. *J. Organomet. Chem.* **1993**, *447*, 163.
1184 Langer, P.; Holtz, E. *Synlett* **2002**, 110.
1185 Pietraszuk, C.; Fischer, H.; Kujawa, M.; Marciniec, B. *Tetrahedron Lett.* **2001**, *42*, 1175.
1186 Kujawa-Welten, M.; Marciniec, B. *J. Mol. Catal. A: Chem.* **2002**, *190*, 79.
1187 Pietraszuk, C.; Marciniec, B.; Jankowska, M. *Adv. Synth. Catal.* **2002**, *344*, 789.
1188 Żak, P.; Pietraszuk, C.; Marciniec, B.; Spólnik, G.; Danikiewicz, W. *Adv. Synth. Catal.* **2009**, *351*, 2675.
1189 Żak, P.; Dudziec, B.; Kubicki, M.; Marciniec, B. *Chem.—Eur. J.* **2014**, *20*, 9387.
1190 Rabasso, N.; Fadel, A. *Tetrahedron Lett.* **2010**, *51*, 60.
1191 Liautard, V.; Desvergnes, V.; Martin, O. R. *Org. Lett.* **2006**, *8*, 1299.
1192 Liautard, V.; Desvergnes, V.; Itoh, K.; Liu, H.-W.; Martin, O. R. *J. Org. Chem.* **2008**, *73*, 3103.
1193 Shen, G. H.; Hong, J. H. *Bull. Korean Chem. Soc.* **2013**, *34*, 3621.
1194 Roe, S. J.; Legeay, J.-C.; Robbins, D.; Aggarwal, P.; Stockman, R. A. *Chem. Commun.* **2009**, 4399.
1195 Carreras, J.; Avenoza, A.; Busto, J. H.; Peregrina, J. M. *Org. Lett.* **2007**, *9*, 1235.
1196 Pandya, B. A.; Snapper, M. L. *J. Org. Chem.* **2008**, *73*, 3754.
1197 Wysocki, L. M.; Dodge, M. W.; Voight, E. A.; Burke, S. D. *Org. Lett.* **2006**, *8*, 5637.
1198 Krause, J. O.; Nuyken, O.; Wurst, K.; Buchmeiser, M. R. *Chem.—Eur. J.* **2004**, *10*, 777.
1199 Mayr, M.; Wang, D.; Kröll, R.; Schuler, N.; Prühs, S.; Fürstner, A.; Buchmeiser, M. R. *Adv. Synth. Catal.* **2005**, *347*, 484.
1200 Yang, L.; Mayr, M.; Wurst, K.; Buchmeiser, M. R. *Chem.—Eur. J.* **2004**, *10*, 5761.
1201 Commandeur, M.; Commandeur, C.; Paolis, M. D.; Edmunds, A. J. F.; Maienfisch, P.; Ghosez, L. *Tetrahedron Lett.* **2009**, *50*, 3359.
1202 Griffiths, P. C.; Knight, D. W.; Morgan, I. R.; Ford, A.; Brown, J.; Davies, B.; Heenan, R. K.; King, S. M.; Dalgliesh, R. M.; Tomkinson, J.; Prescott, S.; Schweins, R.; Paul A. *Beilstein J. Org. Chem.* **2010**, *6*, 1079.
1203 Büchert, M.; Meinke, S.; Prenzel, A. H. G. P.; Deppermann, N.; Maison, W. *Org. Lett.* **2006**, *8*, 5553.
1204 Hartung, J.; Grubbs, R. H. *Angew. Chem., Int. Ed.* **2014**, *53*, 3885.
1205 Torker, S.; Koh, M. J.; Khan, R. K. M.; Hoveyda, A. H. *Organometallics* **2016**, *35*, 543.
1206 Pilyugina, T. S.; Schrock, R. R.; Müller, P.; Hoveyda, A. H. *Organometallics* **2007**, *26*, 831.
1207 Kirschning, A.; Gułajski, L.; Mennecke, K.; Meyer, A.; Busch, T.; Grela, K. *Synlett* **2008**, 2692.
1208 Ritter, T.; Hejl, A.; Wenzel, A. G.; Funk, T. W.; Grubbs, R. H. *Organometallics* **2006**, *25*, 5740.
1209 Kumar, P. S.; Wurst, K.; Buchmeiser, M. R. *Organometallics* **2009**, *28*, 1785.
1210 Schoeps, D.; Buhr, K.; Dijkstra, M.; Ebert, K.; Plenio, H. *Chem.–Eur. J.* **2009**, *15*, 2960.
1211 Vila, A. M. L.; Monsaert, S.; Drozdzak, R.; Wolowiec, S.; Verpoort, F. *Adv. Synth. Catal.* **2009**, *351*, 2689.

[1212] Bek, D.; Balcar, H.; Žilková, N.; Zukal, A.; Horáček, M.; Cejka, J. *ACS Catal.* **2011**, *1*, 709.
[1213] Sulaiman, S.; Bhaskar, A.; Zhang, J.; Guda, R.; Goodson, T.; Laine, R. M. *Chem. Mater.* **2008**, *20*, 5563.
[1214] Behr, A.; Pérez Gomes, J. *Beilstein J. Org. Chem.* **2011**, *7*, 1.
[1215] Das, B.; Balasubramanyam, P.; Veeranjaneyulu, B.; Chinna Reddy, G. *Helv. Chim. Acta* **2011**, *94*, 881.

CHAPTER 2

THE CATALYTIC ENANTIOSELECTIVE STETTER REACTION

Darrin M. Flanigan, Kerem E. Ozboya, Alberto Muñoz, and Tomislav Rovis

Department of Chemistry, Columbia University, New York, NY 10027

Subhash D. Tanpure and Paul R. Blakemore

Department of Chemistry, Oregon State University, Corvallis, OR 97331

Edited by Steven M. Weinreb

Dedicated to the memory of Professor Dieter Enders

CONTENTS

tr2504@columbia.edu
How to cite: Flanigan, D.M.; Ozboya, K.E.; Muñoz, A.; Rovis, T.; Tanpure, S.D.; Blakemore, P.R. The Catalytic Enantioselective Stetter Reaction. *Org. React.* **2021**, *106*, 1191–1330.

Doi:10.1002/0471264180.or106.02

ACKNOWLEDGMENTS

The authors gratefully acknowledge the guidance and assistance of the editorial staff of *Organic Reactions* that was provided during the preparation of this chapter, particularly the editorial assistance of Professors P. Andrew Evans and Steven M. Weinreb. D.M.F., K.E.O., A.M., and T.R. wrote the main chapter and compiled tabular survey entries for literature through 2016. P.R.B. edited and updated the chapter for extended coverage through the end of 2019, and S.D.T. compiled tabular survey entries for literature from 2017–2019.

INTRODUCTION

Transformations that proceed by a reversal in the usual electrophilic and nucleophilic properties associated with a particular functional group, a concept referred to by the German noun 'Umpolung' (literally, 'polarity reversal'),[1] allow for the implementation of unconventional disconnections in retrosynthesis and lead directly to motifs that are difficult to access with traditional chemistries (e.g., 1,4- and 1,6-relationships between oxygenated carbon atoms along an alkyl chain). A commonly explored type of Umpolung reactivity involves the generation of acyl anion equivalents wherein the characteristic electrophilicity of the carbonyl group carbon atom is exchanged for nucleophilicity. This scenario can be effected by multistep tactics involving stoichiometric reagents, for example, via the generation and subsequent metalation of dithianes.[2,3] However, a more attractive approach is to generate an acyl anion equivalent in situ by the interaction of a catalyst with the requisite aldehyde. Adding to the allure of this approach, if the catalyst employed is a chiral molecule, then catalytic enantioselective synthesis may be possible. The catalytic generation of acyl anion equivalents is largely limited to transformations of aldehydes mediated by cyanide anion or *N*-heterocyclic carbenes (NHCs).[4,5] Examples of such processes of both artificial and natural origin exist and include a slew of biosynthetic pathways mediated by the enzyme cofactor thiamine, the deprotonated form of which is a thiazolylidene carbene.[6,7]

The coupling of an aldehyde with an electron-deficient alkene via the conjugate addition of a catalytically generated acyl anion equivalent, commonly known as the Stetter reaction, is a singularly powerful method for generating 1,4-dicarbonyl compounds and related systems. Scheme 1 illustrates this process with an NHC as the catalyst. The Stetter reaction exploits the Umpolung reactivity of an aldehyde

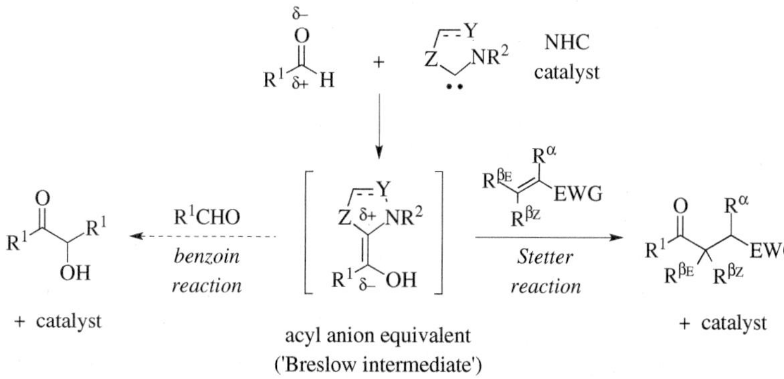

EWG = electron-withdrawing group (e.g., COR^1, CO_2R^1, NO_2 etc.)

Scheme 1

to couple two electrophiles, and it is thus resistant to any uncatalyzed background reaction. The closely related benzoin reaction, the corresponding aldehyde–aldehyde coupling process,[5] proceeds by a similar mechanism and can compete with the Stetter reaction under the same operational conditions. The cyanide anion-[8] and thiazolylidene-catalyzed[9] reactions that now bear his name were first described by Stetter in the early 1970s.[10] The enantioselective variant remained elusive for some time until the first success was charted by Enders in 1996 using a chiral triazolium precatalyst.[11] Building on parallel advances in the design of effective NHC catalysts for the enantioselective benzoin reaction,[12–14] the subsequent refinement of a broadly useful family of fused polycyclic chiral triazolylidene carbenes[15,16] led to further developments that now enable the reliable delivery of Stetter products with generally good-to-excellent levels of enantioselectivity.

The catalytic enantioselective Stetter reaction was initially investigated primarily in the intramolecular manifold. These studies illuminated several factors that impact selectivity, including pre-catalyst structure and acidity, the nature and concentration of the base used to generate the active carbene catalyst, and the propensity of various bases to facilitate product epimerization, which erodes enantioselectivity. Solvent effects that influence reactivity and some fundamental studies on the mechanism were also explored within the construct of the intramolecular reaction. The insights garnered from these investigations enabled the subsequent development of the catalytic enantioselective intermolecular Stetter reaction, which is, in general, more challenging to effect.

This review covers the enantioselective Stetter reaction with an emphasis on small chiral NHCs as the catalysts. Mechanistically distinct processes that achieve the same or closely related overall transformations, such as enzymatic decarboxylative conjugate additions of pyruvate, sila-Stetter reactions, and metal-catalyzed alkene hydroacylation processes, are featured in "Comparison with Other Methods" to provide the reader with alternative strategies. Several reviews focused on the catalytic enantioselective Stetter reaction have been reported,[16–19] and this topic has also been covered in the context of more general accounts on organocatalysis with *N*-heterocyclic carbenes.[4,5,20–22] Non-enantioselective Stetter reactions were previously reviewed in this series.[23]

MECHANISM AND STEREOCHEMISTRY

Mechanism and Operational Considerations

The mechanism of the Stetter reaction has been investigated in both the intra- and inter-molecular modes using experimental and computational methods. The general mechanism of the Stetter reaction is illustrated below for a β-substituted acceptor molecule using a generically-depicted fused polycyclic triazolylidene catalyst (Scheme 2). After generation of the free carbene catalyst **2** by deprotonation of its triazolium salt precursor **1**, the reaction commences, by analogy to the benzoin reaction, with initial addition of the free carbene to an aldehyde to generate tetrahedral intermediate **3**. This species then undergoes proton transfer to afford the neutral enaminol **4**, which is commonly referred to as the 'Breslow intermediate.' Subsequent addition of **4** to an acceptor alkene to form intermediate **5** is the enantio-determining step. Proton transfer then leads to the tetrahedral intermediate **6**, which collapses to release the product with concomitant regeneration of the active catalyst **2** to close the catalytic cycle. Different aspects of this multistep mechanism and operational considerations important for the realization of successful enantioselective Stetter reactions are now elaborated.

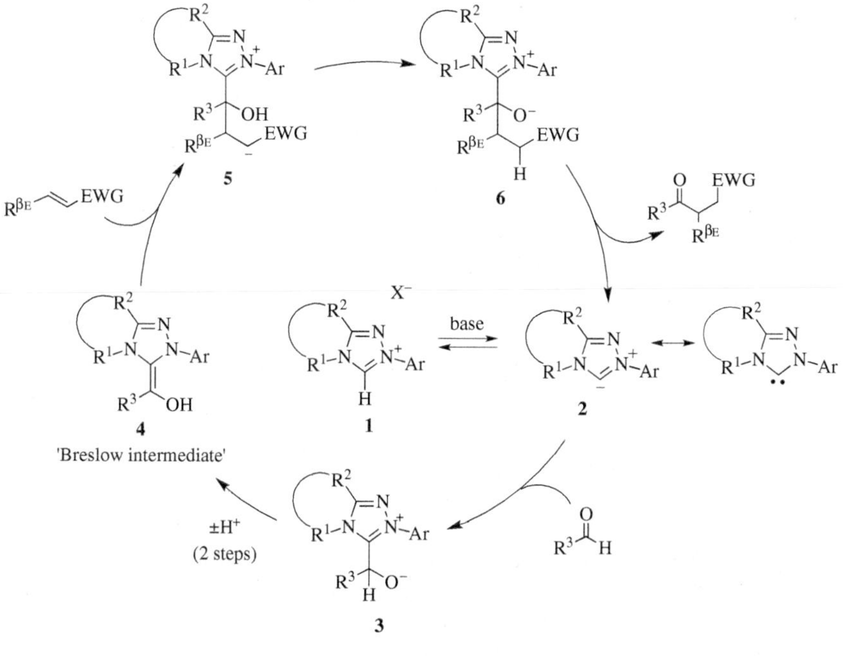

Scheme 2

Generation of *N*-Heterocyclic Carbene Catalysts. As most commonly practiced, the first step in the catalytic enantioselective Stetter reaction is the in situ formation of the free carbene catalyst **2** by deprotonation of its azolium salt precursor **1** with a suitable base. Chiral carbene catalysts that are successful in the enantioselective Stetter reaction have an experimentally determined pK_{aH} in a narrow range of 17.0 to 17.5 in water (Figure 1).[24] As shown, variations to the structure of the commonly used *N*-aryl substituents have the expected effect on acidity. In contrast, such fine control of catalyst basicity is not necessary for non-enantioselective Stetter reactions, which can be satisfactorily effected using carbenes manifesting a significantly greater pK_{aH} range (16.5–18).[25,26] In general, both kinetic and thermodynamic bases are used to generate the active catalyst; some examples include alkyllithiums, metal amides, alkoxides, carbonates, acetates, tertiary amines, and pyridines. Whereas strong bases (pK_{aH} $\geq$30) such as KHMDS and *n*-butyllithium lead to stoichiometric deprotonation of the azolium salt to form the carbene irreversibly, it is noteworthy that weaker bases with conjugate acids that are significantly more acidic than the azolium salt are also effective in promoting the desired net reactivity. The free carbene can be used directly in some cases by simply treating the azolium salt with KHMDS and removing the HMDS and KX (X = counterion of the azolium cation) byproducts via evaporation and filtration, respectively. These conditions can lead to the retention of stereoselectivity in cases where the product ketone is readily enolized in the presence of any exogenous base.[27] The effect of cation structure on the acidity of a wide range of triazolium salts in DMSO solvent has been recently studied,[28] and gas-phase acidities have also been calculated and measured experimentally.[29]

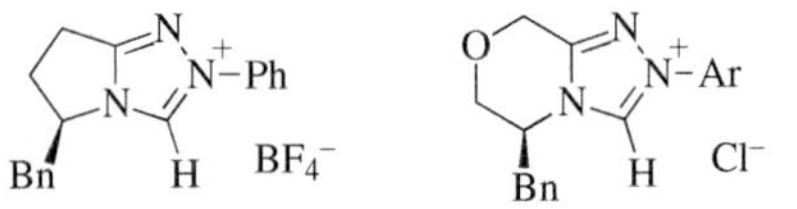

TrP2a•HBF$_4$ 17.5

Precatalyst	Ar	pK_a
TrM2f•HCl	4-FC$_6$H$_4$	17.0
TrM2g•HCl	4-(NC)C$_6$H$_4$	17.1
TrM2a•HCl	Ph	17.2
TrM2c•HCl	2,4,6-Me$_3$C$_6$H$_2$	17.2
TrM2d•HCl	4-MeOC$_6$H$_4$	17.3

Figure 1. Experimentally determined pK$_a$ data (H$_2$O) for a selection of triazolium salts commonly used as precursors to NHC catalysts for the enantioselective Stetter reaction.

Formation and Reactivity of Breslow Intermediates. The aforementioned 'Breslow intermediate' (**4**, Scheme 2) is the putative acyl anion equivalent responsible for the Umpolung reactivity of otherwise electrophilic aldehydes in the Stetter reaction. The intermediacy of such enaminols was originally proposed by Breslow to account for the mechanism of action of the coenzyme thiamine pyrophosphate (a thiazolium salt) in the various processes that it mediates.[6] Many experiments designed either to directly detect or to infer the existence of the Breslow intermediates relevant to Stetter reactions have been conducted. For example, the keto form of a Breslow intermediate was isolated from an equimolar mixture of propionaldehyde and a triazolylidene NHC (Scheme 3).[30] This finding implicates enol-to-keto tautomerization of Breslow intermediates as a potential catalyst deactivation pathway in NHC-mediated Stetter reactions. However, it should be noted that such keto forms have not been detected under any catalytically relevant conditions (i.e., aldehyde/carbene ≥ 10:1).[30] Interestingly, *C*-protonated forms of triazolylidene-derived Breslow intermediates are readily isolated by the addition of triazolium tetrafluoroborate salts to aldehydes in the presence of triethylamine.[31]

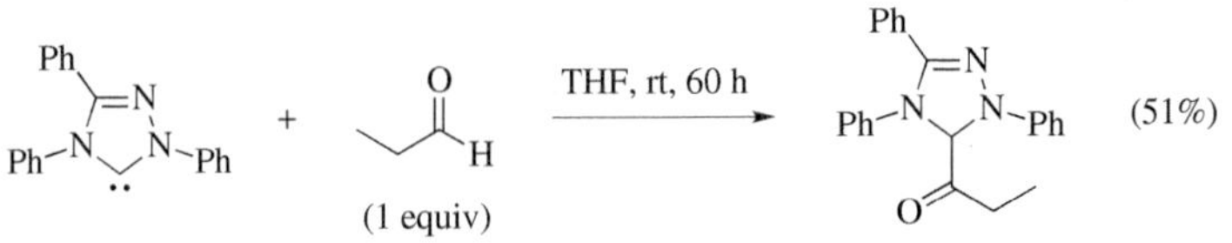

(1 equiv) — THF, rt, 60 h — (51%)

Scheme 3

Much of the work focused on the isolation of Breslow intermediates has been carried out with imidazolidinylidenes and related NHCs. Several different forms of the derived enaminols have been isolated and characterized, and their reactivity has been studied. For example, *O*-methylated Breslow intermediates are available by deprotonation of the corresponding azolium salts (e.g., **7**) in good yields (Scheme 4).[32] Thiazole-derived examples (e.g., **8**) slightly favor the (*Z*)-enol ether geometry, whereas triazole-derived *O*-methylated Breslow intermediates (e.g., **9**) favor the (*E*)-isomer. The relative nucleophilicity of different *O*-methylated Breslow intermediates was compared by measuring their rates of reaction with benzhydrylium cations (Ar_2CH^+).[32] As illustrated in Scheme 5, alkylation occurs at the anticipated site, and triazolylidene **9** reacts significantly more slowly than less sterically encumbered (and more electron-rich) thiazole- and benzimidazole-derived *O*-methylated Breslow intermediates **8** and **12**. Benzofusion results in a dramatic attenuation of nucleophilicity, for thiazolium systems at least, as seen by a comparison of the reactions involving **8** and **10**.

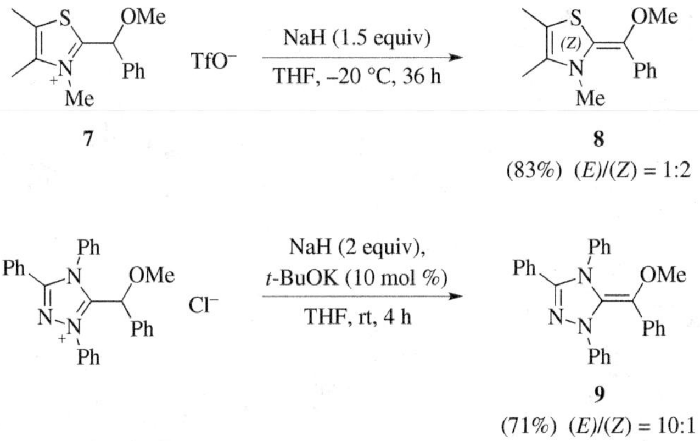

7 (TfO^-) → NaH (1.5 equiv), THF, –20 °C, 36 h → **8** (83%) (*E*)/(*Z*) = 1:2

(Cl^-) → NaH (2 equiv), *t*-BuOK (10 mol %), THF, rt, 4 h → **9** (71%) (*E*)/(*Z*) = 10:1

Scheme 4

10 → $Ar_2CH^+ BF_4^-$ (0.06 equiv), THF, 20 °C, k_{rel} = 1.00 → **11** (BF_4^-)

Ar = 4-[$(CH_2)_4N$]C_6H_4

relative rate of reaction with Ar_2CH^+ compared to above transformation

	9	**8**	**12**
k_{rel} =	68.0	2700	6800

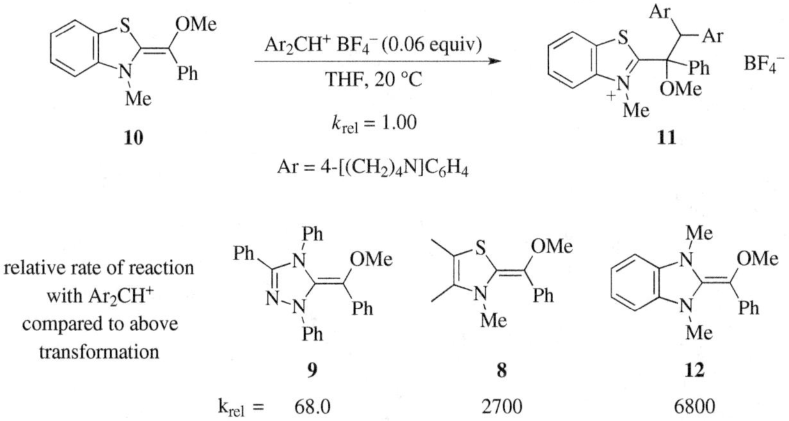

Scheme 5

Several aryl aldehydes, as well as cinnamaldehyde, smoothly generate the expected Breslow intermediates upon mixing with imidazolidinylidene carbenes.[33] ^{1}H NMR spectral analysis shows that addition of SIPr to deuterated benzaldehyde in d_8-THF results in the transfer of the formyl deuterium atom to the enol oxygen of the Breslow intermediate (Scheme 6). This observation aligns with an earlier report in which deuterated benzaldehyde is coupled with *N*,*N*-dimethylacrylamide in a phosphine-catalyzed Stetter-type reaction to afford the γ-keto amide with high levels of deuteration at the α-position of the amide.[34]

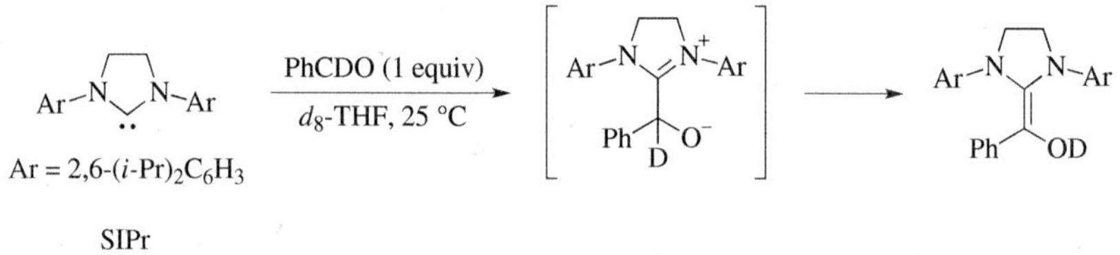

Ar = 2,6-(i-Pr)$_2$C$_6$H$_3$

SIPr

PhCDO (1 equiv)

d_8-THF, 25 °C

Scheme 6

The reversibility of Breslow intermediate formation in the case of imidazolidinylidenes was established by ^{1}H NMR spectral analysis of the benzoin reactions of **13** and **15** (Scheme 7).[33] Addition of excess benzaldehyde to SIMes results in the rapid formation of Breslow intermediate **13** and its slow conversion to traces of the homo-benzoin adduct **17**. When a more electrophilic aldehyde **14** is added following formation of **13** as before, immediate equilibration between Breslow intermediates **13** and **15** occurs, with the latter being overwhelmingly favored. Upon extended incubation of **15** in the presence of benzaldehyde and **14**, no benzoin products derived

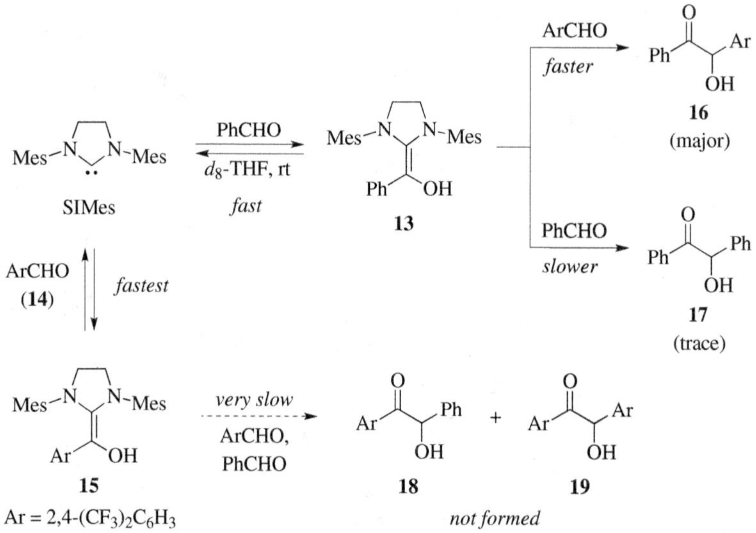

SIMes

PhCHO

d_8-THF, rt

fast

13

ArCHO *faster*

16 (major)

PhCHO *slower*

17 (trace)

ArCHO (**14**) *fastest*

15

Ar = 2,4-(CF$_3$)$_2$C$_6$H$_3$

very slow

ArCHO, PhCHO

18 **19**

not formed

Scheme 7

from **15** (compounds **18** and **19**) are formed but rather, cross-benzoin product **16** (derived from **13**) emerges as the major product. An essentially identical result is realized beginning with an equimolar combination of SIMes and **14** (to give **15**) followed by the subsequent addition of a 1:1 mixture of benzaldehyde and **14**. The behavior described is consistent with a Curtin–Hammett-type situation in which equilibration between intermediates **13** and **15** is rapid, and the thermodynamically favored compound (**15**) is less kinetically competent than its equilibrium partner (**13**) in the ensuing nucleophilic addition reaction to an aldehyde.

The above studies answer several questions about the structure of Breslow intermediates and some of the fundamental steps involved in their formation. However, extrapolating the conclusions to the enantioselective Stetter reaction is tenuous because much of the mechanistic work conducted involves symmetrical imidazolium scaffolds, rather than the more relevant triazolium heterocycles. For the former, enol geometry of the Breslow intermediate is not a consideration, but this attribute for non-symmetrical systems is a critical determinant of the stereochemical outcome in enantioselective Stetter reactions. Unfortunately, and as described above (Scheme 3), to date, only the keto tautomers of Breslow intermediates derived from triazolylidenes and aldehydes have been observable (and isolable).[30]

To gain potentially pertinent mechanistic insights, investigations on the easily isolated aza-analogues of Breslow intermediates derived from iminium ions and triazolium NHC precursors used in enantioselective Stetter reactions were undertaken (Scheme 8).[35] For example, aza-Breslow intermediate **20** is isolated in good yield by the illustrated reaction of a triazolium salt and the iminium ion derived from *N*-methylation of benzylideneaniline. Species such as **20** are surprisingly stable, which permits their characterization in solution (NMR spectroscopy) and in the solid state (X-ray diffraction). Single-crystal X-ray diffraction analyses of **20** and a closely related compound reveal that the (*Z*)-enamine is favored in the solid state, suggesting that these adducts may behave differently relative to the corresponding aldehyde-derived classical Breslow intermediates, which are calculated to favor the (*E*)-isomer.[36] The investigation of the solution-phase structure of **20** by ^{1}H NMR

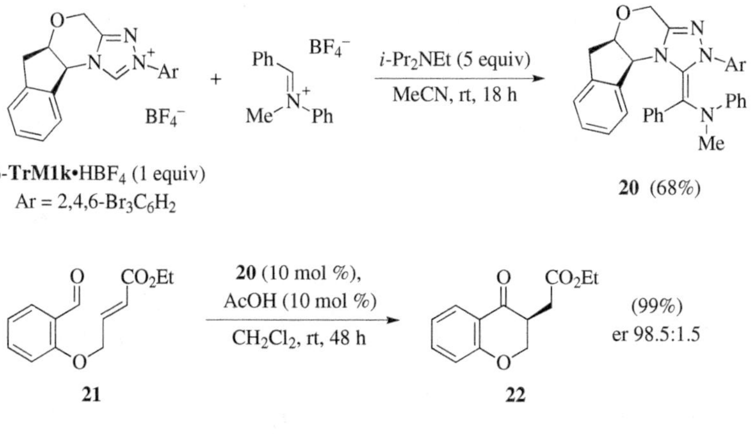

β-**TrM1k**•HBF_4 (1 equiv), Ar = 2,4,6-$Br_3C_6H_2$ + iminium BF_4^- → *i*-Pr_2NEt (5 equiv), MeCN, rt, 18 h → **20** (68%)

21 → **20** (10 mol %), AcOH (10 mol %), CH_2Cl_2, rt, 48 h → **22** (99%), er 98.5:1.5

Scheme 8

spectroscopy indicates the presence of four distinct isomers, which are likely the result of restricted rotation about the carbon–nitrogen and carbon–carbon bonds of the enamine moiety. Cyclic voltammetry studies on **20** reveal that the enamine undergoes a quasi-reversible one-electron oxidation at +0.17 V vs SCE.[35] The corresponding aza-Breslow intermediate derived from cinnamaldehyde shows an irreversible oxidation at +0.49 V vs SCE. In addition to their remarkable stability, these intermediates deliver active catalysts for the intramolecular Stetter reaction, presumably via exchange of the dienamine moiety for the canonical enaminol, as illustrated by the cyclization of **21** to **22** catalyzed by **20**.

Experimental Mechanistic Studies. Limited experimental data are available at this time to support mechanistic details of the Stetter reaction as it is typically practiced in the enantioselective mode. However, in this regard, the intramolecular variant has received more attention than its intermolecular counterpart. In one of the more detailed studies available, a series of kinetic and competition experiments involving the intramolecular Stetter reactions of substrate **21** and related compounds, mediated by the NHC catalyst derived from triazolium salt **TrP2a**•HBF_4, have supported some fundamental aspects of the reaction pathway.[37] A selection of the experiments performed in this study, which also determined that the reaction is first order in both aldehyde and catalyst (or its salt precursor), is illustrated in Scheme 9. The deuterium labeling experiment shown involves isotopic replacement of the formyl hydrogen atom in **21** with a deuterium atom and reveals a significant primary kinetic isotope effect (k_H/k_D = 2.62) for the intramolecular Stetter reaction. This finding suggests that the initial proton transfer event (i.e., **3** to **4**, Scheme 2) is turnover-limiting. In support of this hypothesis, the nature of the acceptor moiety has only a small effect on the rates of conversion of an equimolar mixture of enoate **21** and a closely related enamide analogue (competition experiment 1, Scheme 9). On this basis, and as expected, an analogue of **21** with a more electron-rich benzaldehyde moiety is slower to undergo Stetter cyclization, while a more electron-deficient analogue reacts more quickly (competition experiment 2). In a final experiment, ostensibly at odds with the last pair, an analogue of **21** lacking the arylether linker (presumably making the benzaldehyde moiety less electron-rich than that in **21**) actually reacts more slowly than **21**. One plausible explanation for this curious finding is that the ether oxygen atom in the tether of **21** plays a role in mediating the necessary proton-transfer event from the initial tetrahedral intermediate **3** to the Breslow intermediate **4**. Certainly, some agent is needed to facilitate the conversion of **3** to **4** since the direct concerted intramolecular proton transfer pathway, as an electrophilic rearrangement and formally a four-electron [1,2]-sigmatropy, contravenes the principle of conservation of orbital symmetry.[38,39] The conversion of the *C*-protonated forms of the Breslow intermediate to the neutral enaminol (i.e., the likely second phase of the progression of **3** to **4**) has been studied experimentally by measuring H/D exchange rates in 3-(hydroxybenzyl)azolium salts.[31]

Mechanistic insights of potential generality were garnered from an investigation of the enantioselective intermolecular Stetter reaction between α,β-unsaturated

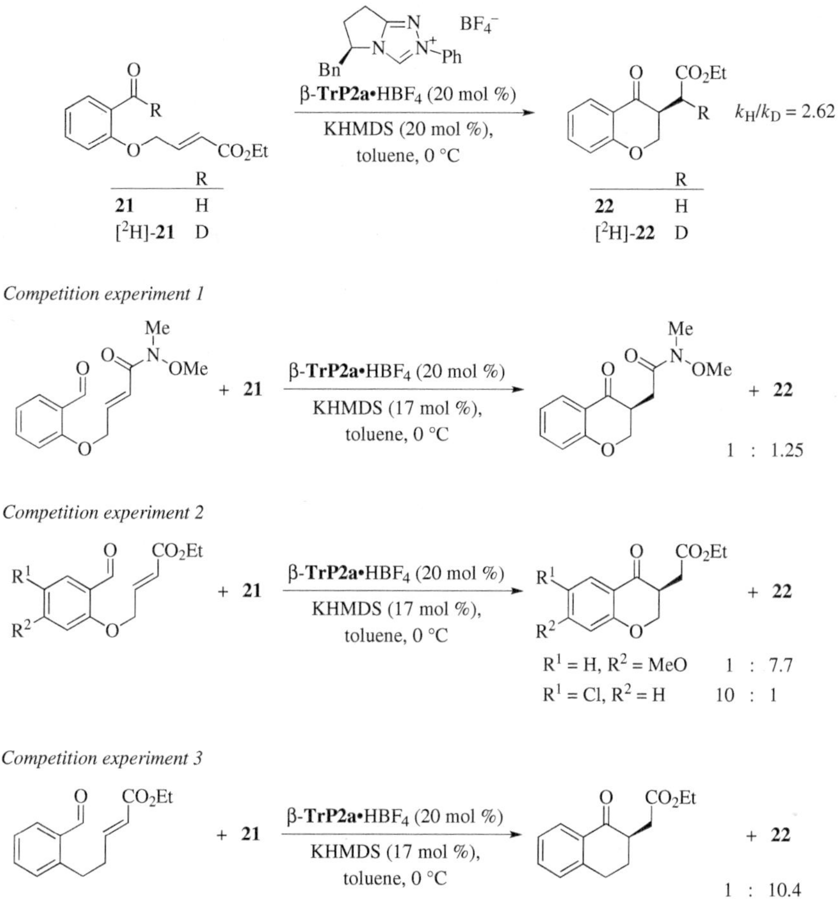

BF_4^- Bn N N N Ph

R	
21	H
[^{2}H]-**21**	D

β-**TrP2a**•HBF_4 (20 mol %)
KHMDS (20 mol %), toluene, 0 °C

CO_2Et R $k_H/k_D = 2.62$

R	
22	H
[^{2}H]-**22**	D

Competition experiment 1

Me N OMe + **21** → β-**TrP2a**•HBF_4 (20 mol %), KHMDS (17 mol %), toluene, 0 °C → Me N OMe + **22**

1 : 1.25

Competition experiment 2

R^1 R^2 CO_2Et + **21** → β-**TrP2a**•HBF_4 (20 mol %), KHMDS (17 mol %), toluene, 0 °C → R^1 R^2 CO_2Et + **22**

R^1 = H, R^2 = MeO 1 : 7.7
R^1 = Cl, R^2 = H 10 : 1

Competition experiment 3

CO_2Et + **21** → β-**TrP2a**•HBF_4 (20 mol %), KHMDS (17 mol %), toluene, 0 °C → CO_2Et + **22**

1 : 10.4

Scheme 9

aldehydes and nitroalkene acceptors (Scheme 10).[40] These experiments were primarily aimed at elucidating the origin of the accelerating effect of catechol on the Stetter process. As seen in the illustrated example involving cinnamaldehyde, the addition of catechol results in a significant yield enhancement without a change in the observed level of enantioselectivity. Related phenolic additives lacking a 1,2-diol moiety do not give the same boost to yield, indicating that the proximity of the hydroxyl groups in catechol is critical. It is speculated, for example as in model **23** (Figure 2), that catechol (pK_a 9.45, and thus present as its catecholate monoanion in the presence of Hünig's base) mediates the crucial aforementioned 1,2-proton transfer necessary for the maturation of tetrahedral intermediate **3** to Breslow intermediate **4**. A series of deuterium-labeling experiments lend credence to this hypothesis: in these experiments, the reaction was conducted as shown with either cinnamaldehyde or its formyl-deuterated isotopologue (PhCH=CHCDO) and with either methanol or d_4-methanol (CD_3OD) as solvent (which effects H/D exchange in catechol to give 1,2-(DO)C_6H_4). Thus, given the turnover-limiting nature of the initial proton transfer event on the entire Stetter process, the reaction involving RCHO/CH_3OH is the fastest, as expected, while that involving RCDO/CD_3OD is the slowest. Other labeling combinations are intermediate in rate (relative rates of reaction: RCHO + CH_3OH = 4.2; RCHO + CD_3OD = 2.3; RCDO + CH_3OH = 1.6; RCDO + CD_3OD = 1.0). Other observations are also in accord with the speculated role for catechol in the proton-transfer step; for example, stereoselectivity is not influenced if catechol is replaced by other phenolic compounds (including related enantioenriched chiral diols), and intramolecular Stetter reactions also go to completion more rapidly and with lower catalyst loadings in the presence of catechol.[40]

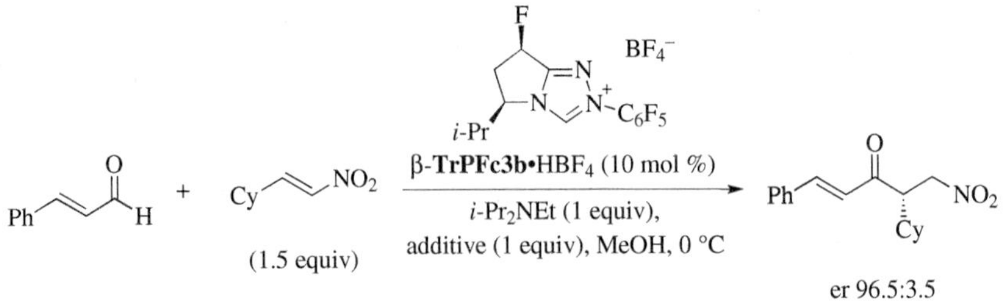

F; BF_4^-; N; N; N; C_6F_5; *i*-Pr

Ph–CH=CH–CHO + Cy–CH=CH–NO_2 (1.5 equiv) → **β-TrPFc3b•HBF₄** (10 mol %), *i*-Pr_2NEt (1 equiv), additive (1 equiv), MeOH, 0 °C → Ph–CH=CH–C(O)–CH(Cy)–CH₂–NO_2 er 96.5:3.5

Additive	Time (h)	Yield (%)
none	8	5
catechol	2	80
phenol	8	8
hydroquinone	8	15
2-methoxyphenol	8	9

Scheme 10

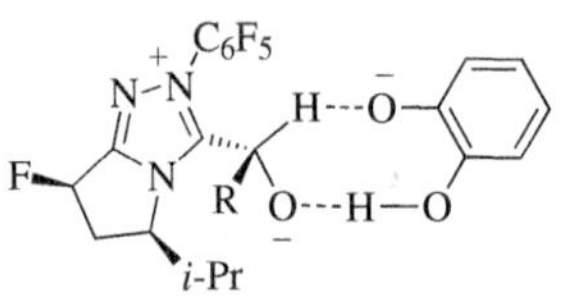

C_6F_5; N–N; H---O; F; N; R; O---H—O; *i*-Pr

23

Figure 2. Putative mediation of 1,2-proton transfer by catechol.

Stereoselectivity

A variety of models have been advanced to explain the origin of stereoselectivity in enantioselective Stetter reactions employing chiral NHC catalysts. As is common in arguments presented to explain the sense of stereoinduction in catalytic enantioselective transformations, a majority of these models should be regarded as highly speculative. Nonetheless, in many cases, including the examples chosen for inclusion here, they are based on sound principles and have predictive and heuristic value.

High-level calculations have been conducted to provide more refined stereocontrol models in some cases (see "Computational Studies").

Salicylaldehyde-derived enoate **24** (and the analogous ethyl ester **21**) has been used as a model substrate in a large number of investigations of the intramolecular Stetter reaction.[41] In the inaugural report of an effective catalytic enantioselective Stetter process, conversion of **24** to chromanone **25** using the chiral triazolylidene catalyst (derived by in situ deprotonation of triazolium salt α-**TrD1a**•$HClO_4$) is proposed to occur via the illustrated transition state (Scheme 11).[11] Here, the dioxanyl-anchored phenyl substituent of the (*S,S*)-configured carbene blocks the *Re*-face of the enaminol, conceivably as depicted due to a π-stacking effect,[42] leaving the *Si*-face open to attack the pendant enoate. The 'strongly simplifying character' of this stereocontrol model was noted by its originators.[11]

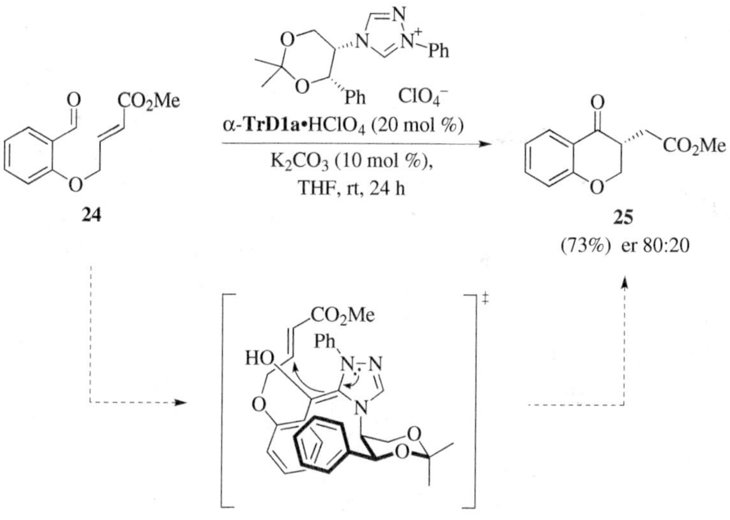

α-**TrD1a**•$HClO_4$ (20 mol %)

K_2CO_3 (10 mol %), THF, rt, 24 h

24 → **25** (73%) er 80:20

Scheme 11

An obstacle to the design of effective NHCs for applications in enantioselective catalysis is the creation of a pervasive and non-promiscuous chiral environment about the reactive carbene center such that high levels of stereocontrol can be realized. In a cleverly designed C_2-symmetric imidazolinium salt NHC precursor (**Im1c**•BF_4), phenyl substituents on stereogenic centers in the carbon backbone induce specific orientations of bulky mesitylmethyl groups residing on the nitrogen atoms that directly flank the C2-position. Remarkably, the precatalyst used in this case (**Im1c**•HBF_4) is capable of inducing good enantioselectivity in the cyclization of a very simple intramolecular Stetter reaction substrate at elevated temperature (Scheme 12).[43] The illustrated model for stereocontrol should be regarded as notional only.

Two stereogenic centers are created from an intramolecular Stetter reaction if the tethered acceptor moiety bears an additional α-substituent. For this to be an effective stratagem, a highly diastereoselective protonation of the incipient enolate

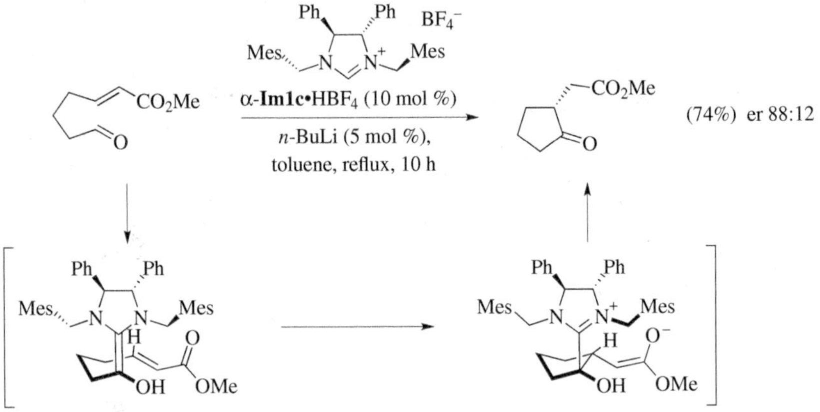

α-**Im1c**•HBF_4 (10 mol %)
n-BuLi (5 mol %), toluene, reflux, 10 h
(74%) er 88:12

Scheme 12

(or the equivalent relevant carbanion) must occur following addition of the Breslow intermediate to the acceptor (cf., **5** to **6**, Scheme 2). Pertinent examples involving cyclizations of (*E*)- and (*Z*)-enoates **26** demonstrate the power of the approach and also illustrate the efficacy of a member of the fused polycyclic triazolylidene catalyst family (Scheme 13).[27] Here, an organized cyclization transition state provides the initially emerging stereogenic center at the β-position via a preferred Breslow intermediate geometry (the (*E*)-enaminol is calculated to be favored from fused triazolylidenes),[36,44] which is followed by an ensuing (and potentially intramolecular) proton-transfer event that leads to the second stereogenic center at the α-position. In the kinetically controlled scenario depicted, the stereochemical integrity of the process is dependent on the relevant proton-transfer event occurring faster than rotation about the indicated bond (in bold) with no epimerization occurring after product formation. The plausibility of this scenario is supported by the observation that (*E*)-**26** leads preferentially to *ul*-**27**, while its geometrical isomer (*Z*)-**26** gives instead the diastereomeric product *lk*-**27** with the same absolute configuration at the α-stereocenter. However, the reduced stereospecificity and lower enantioselectivity obtained in the second reaction reveal that the two starting materials are not equally processed.

High levels of stereoselectivity have likewise been realized from intermolecular Stetter reactions, and appropriate stereocontrol models based on both open and closed transition states have been proposed to account for the observed absolute stereochemical outcomes.[45,46] Notably, in the addition of benzaldehyde to chalcones, the typical triazolylidene NHC catalysts bearing *N*-aryl substituents are ineffective at promoting the Stetter process, but replacement of this group with an *N*-benzyl moiety (e.g., as in **TrP11x**) provides good reactivity (Scheme 14).[45] As above, the putative stereocontrol models for this process (as illustrated) assume that the enaminol intermediate derived from an aldehyde and a triazolylidene-type NHC has an (*E*)-configuration, which is consistent with calculations.[36,44] The reaction protocol shown is also successful for the addition of heteroaryl aldehydes to arylidene malonate acceptors.[46] In these intermolecular Stetter reactions, and others, much of the

(*E*)-**26**

β-**TrP2e** (20 mol %)
toluene, rt, 24 h

ul-**27**
(80%) dr 98:2
er 96:4

intramolecular proton transfer

R = MeO_2CCH_2
(phenyl ring omitted for clarity)

— = slow rotation vs proton transfer

(*Z*)-**26**

β-**TrP2e** (20 mol %)
toluene, rt, 24 h

lk-**27** (70%) dr 86:14 er 69:31

Scheme 13

(1.2 equiv)

α-**TrP11x**•HBF_4 (10 mol %)
Cs_2CO_3 (10 mol %),
THF, rt, 6 h

28
(65%) er 83:17

closed TS

open TS

Scheme 14

aldehyde is initially consumed in a benzoin dimerization at an early stage. However, the benzoin reaction is reversible and its product acts as an effective reservoir of the aldehyde, allowing the subsequent (irreversible) Stetter reaction to proceed largely unimpeded.

The introduction of fluorine atoms to the canonical fused polycyclic triazolylidine catalysts has a significant effect on the level of enantioselectivity that can be attained. As illustrated in the example shown below of an intermolecular Stetter reaction between cinnamaldehyde and a nitroalkene, the addition of a fluorine atom to the pyrrolidine ring of NHC **TrP3b** results in the more effective catalyst **TrPFc3b** (Scheme 15).[40] The effect of catechol on this reaction was previously described above. Remarkably, the NHC catalyst that contains only a fluorine-atom-bearing stereogenic center (**TrPF1b**) is also a powerful stereocontroller, indicating that a strong stereoelectronic effect is operating. The improved selectivity derived from catalyst fluorination was initially ascribed to a conformational effect imparted by the fluorine atom substituent; however, calculations implicate the involvement of an attractive interaction between the developing negative charge on a nitroalkene oxygen atom and the positive charge on the azolium salt.[47] These arguments are described in detail below for a related process (see "Computational Studies").

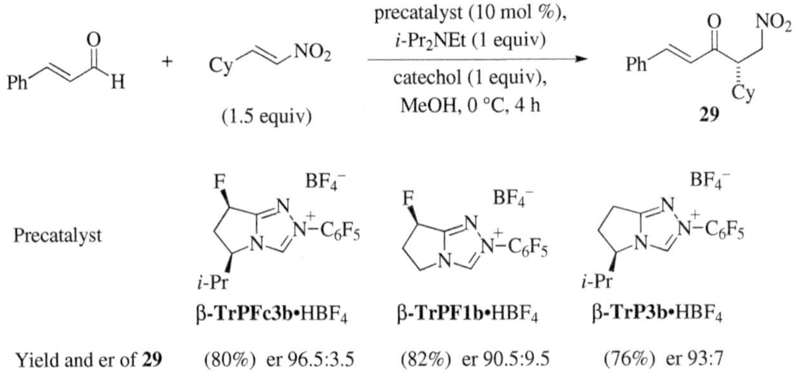

Ph–CH=CH–CHO + Cy–CH=CH–NO_2 (1.5 equiv) → **29**

precatalyst (10 mol %), *i*-Pr_2NEt (1 equiv), catechol (1 equiv), MeOH, 0 °C, 4 h

Precatalyst	β-**TrPFc3b**•HBF_4	β-**TrPF1b**•HBF_4	β-**TrP3b**•HBF_4
Yield and er of **29**	(80%) er 96.5:3.5	(82%) er 90.5:9.5	(76%) er 93:7

Scheme 15

Finally, enantioselectivity has been realized in intermolecular Stetter reactions involving α-substituted acceptors.[48] The situation here is an interesting one because the sole stereogenic center initially created is at the 'acyl anionic' carbon atom. However, this stereocenter is transient because the position later reverts to an sp^2-hybridization state upon catalyst turnover. Thus, following addition of the Breslow intermediate to the acceptor, a second stereoinductive event is required in which stereochemical information at the new γ-position is relayed to the α-position during a diastereocontrolled proton transfer (e.g., **30** to **31**, Scheme 16). The concept has been successfully applied in the enantioselective synthesis of α-amino acid derivatives from α-aminoacrylates.[48,49]

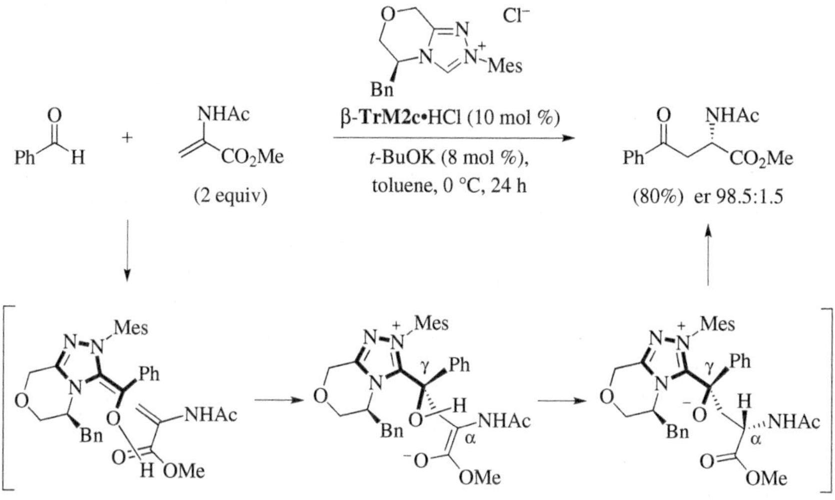

Scheme 16

Computational Studies

A limited number of detailed computational studies on the mechanism of the Stetter reaction exist.[36,47,50] Nevertheless, the insights garnered from these efforts are augmented by theoretical investigations on the related benzoin reaction, since both processes involve the formation of a Breslow intermediate.[44,51,52] Computational studies of complete Stetter reaction cycles using 1,3-dimethyltriazolylidene as a simple model NHC catalyst have been conducted for the addition of acetaldehyde to (*E*)-pent-3-en-2-one [B3LYP/6-31G(d)][36] and for the cyclization of **21** to **22** [B3LYP/6-31G**].[50] These findings support the depiction of the Stetter reaction as shown in Scheme 2; thus, for both intermolecular and intramolecular variants, intermediates and appropriate transition states have been located for a total of six discrete steps. Of note, and as discussed above, a direct 1,2-hydrogen atom transfer to advance from intermediate **3** to **4** contravenes the Woodward–Hoffmann rules, and computed estimates for this concerted process are accordingly high, for example, 42.6 kcal/mol en route from **21** to **22** [B3LYP/6-31G**].[50] The calculated barrier drops considerably for a two-step conversion via initial protonation on the alkoxide oxygen atom of zwitterion **3** followed by loss of a protic hydrogen atom from the adjacent carbon–hydrogen bond to furnish the Breslow intermediate **4**. The computed barrier for the deprotonation step remains quite high (for example, the barrier is 21.4 kcal/mol using triethylamine as base),[50] which is consistent with the conclusions of the experimental work described above (cf., Scheme 9).[37] Theoretical work most relevant to the practice of the enantioselective Stetter reaction, namely determination of the reactive geometry of the key enaminol intermediate and modeling of pertinent transition states for carbon–carbon bond formation, is now considered.

Enaminol Geometry Calculations. A critical determinant of the stereochemical outcome of the enantioselective Stetter reaction is the geometry of the enaminol during the addition step, since (*E*)- and (*Z*)-isomers of the Breslow intermediate are likely to favor opposite enantiomers of the final product. It has been calculated [B3LYP/6-311+G(2d,p)] that Stetter reaction pathways involving (*Z*)-configured triazolylidine-derived Breslow intermediates generally have higher barriers for all the significant steps.[36] Hence, to ensure useful enantioselectivities, the NHC catalyst must be designed to favor the (*E*)-isomer of the enaminol over its (*Z*)-configured counterpart. Fortunately, for a majority of substitution patterns that are likely to be considered in any reasonable catalyst design, (*E*)-isomers are computed to be preferred (Figure 3).[36] Here, the results of calculations broadly agree with expectations based on chemical intuition. Thus, electronic factors that enhance the intrinsic preference for the (*E*)-isomer include electron-withdrawing groups (EWG) on the nitrogen atom (N1) proximal to the hydroxyl group (which alleviates lone-pair/lone-pair destabilization between OH and N) and electron-donating groups (EDG) on the distal nitrogen atom (N3) (which destabilizes the (*Z*)-isomer by exacerbating the same interactions to the opposite side). Electronic factors that reduce N→O lone-pair/lone-pair interactions in the (*Z*)-isomer are to be avoided. For example, if both N3 and C5 are equipped with EWGs, then the (*Z*)-isomer is computed to be favored over the (*E*)-isomer. Steric factors are less influential, albeit a combination of a large N1 substituent with a bulky aldehyde (large R) will enhance the preference for the (*E*)-isomer.[36] The same kind of trends in enaminol geometrical preferences have been calculated for the fluorinated NHC catalysts used in the enantioselective intermolecular Stetter reaction considered in the next section; in these reactions, the (*E*)-isomer of the Breslow intermediate is computed to be 2–3 kcal/mol lower in energy than the corresponding (*Z*)-enaminol [M06-2X/6-31+G(d)].[47]

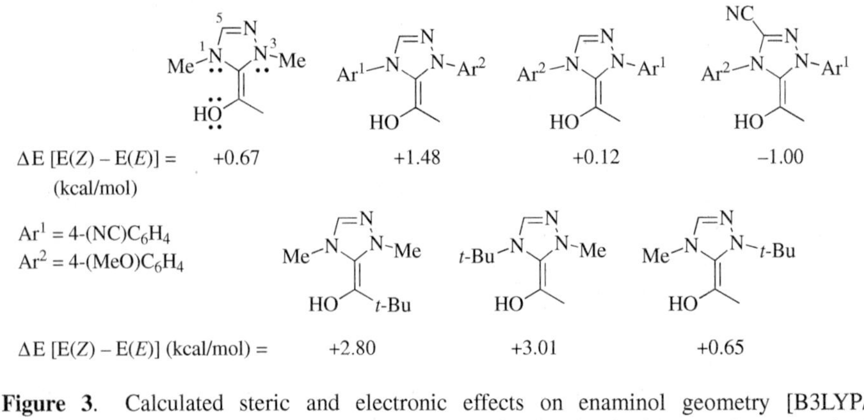

ΔE [E(Z) – E(*E*)] = (kcal/mol) +0.67 +1.48 +0.12 −1.00

Ar^1 = 4-(NC)C_6H_4
Ar^2 = 4-(MeO)C_6H_4

ΔE [E(Z) – E(*E*)] (kcal/mol) = +2.80 +3.01 +0.65

Figure 3. Calculated steric and electronic effects on enaminol geometry [B3LYP/6-311+G(2d,p)]. Energy shown is the computed preference for the illustrated (*E*)-enaminol over the (*Z*)-enaminol; $\Delta E > 0$ indicates (*E*)-isomer favored, $\Delta E < 0$ indicates (*Z*)-isomer favored.

Computed Transition States for Carbon–Carbon Bond Formation. The highest level calculations performed to date for the enantioselective Stetter reaction focus on explaining the kind of fluorine-atom effect highlighted above (see Scheme 15).[40] Nonetheless, these studies also inform more generally on the origin of stereoselectivity for transformations involving fused polycyclic triazolylidene catalysts.[47,53] The calculated transition state structure (gas phase geometries optimized using B3LYP/6-31G(d) and single-point calculations performed using M06-2X/6-31+G(d) with the conductor-like polarizable continuum model (CPCM) of solvation) for the illustrated intermolecular Stetter reaction conveys the main features (Scheme 17).[47] Thus, the acceptor molecule approaches the (*E*)-isomer of the enaminol from its more exposed face with a preferred gauche arrangement of π-bonds in the donor and the acceptor (in agreement with the Seebach topological rule)[54] that minimizes steric interactions. The sp^3-hybridization state of the N3 nitrogen atom (a recurring attribute calculated for Breslow intermediates derived from triazolylidenes) is also noteworthy. The fluorine-atom substituent within the catalyst accentuates the observed selectivity (a yield of 90% and an er of 94:6 are obtained in its absence) for two reasons: (1) the effect of the isopropyl group is maximized because it resides in a pseudoaxial orientation, thereby allowing a stabilizing hyperconjugative interaction that involves the likewise pseudoaxial *cis* carbon–fluorine bond and an adjacent antiperiplanar carbon–hydrogen bond, and (2) the polarization of the carbon–fluorine bond helps to further electrostatically stabilize developing negative charge density on the nitro group. On this basis, it is understandable that a poorer outcome is observed for an otherwise identical reaction using an isomer of **TrPFc3b** that exhibits a *trans* relationship between fluorine and isopropyl substituents (**TrPFt3b**).[47] However, the fact that the *trans* catalyst actually

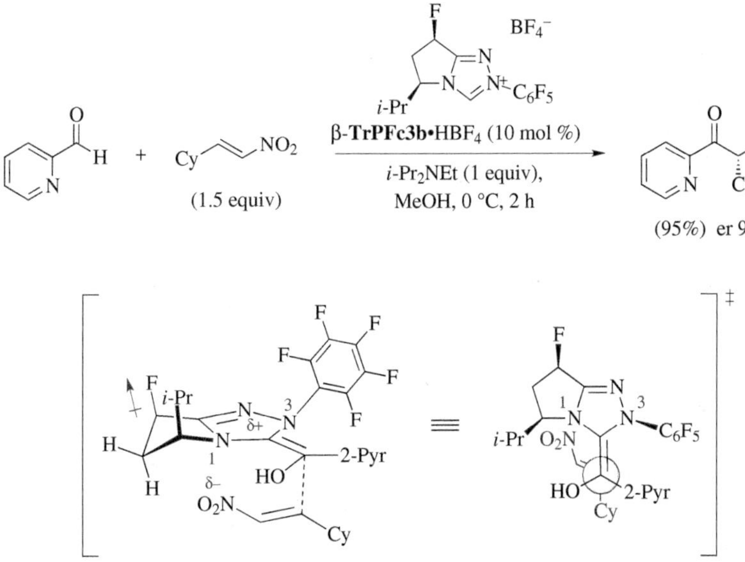

β-**TrPFc3b**•HBF_4 (10 mol %)

i-Pr_2NEt (1 equiv), MeOH, 0 °C, 2 h

(1.5 equiv)

(95%) er 97.5:2.5

Scheme 17

performs better for the addition of aliphatic aldehydes to nitroalkenes indicates that the situation is more complex than suggested by this model.[53]

SCOPE AND LIMITATIONS

The catalytic enantioselective Stetter reaction is an atom-economical process of general utility that can be employed in both intramolecular and intermolecular modes. Although the cyclization and cross-coupling variants proceed through essentially the same fundamental mechanistic steps, there are significant practical differences between the two reactivity modes, and intramolecular Stetter reactions are more flexible with regard to the range of donor aldehydes and acceptor alkenes that are tolerated. Accordingly, in this section, the scope and limitations of intra- and intermolecular Stetter reactions are considered separately with representative and instructive examples for each (a newly introduced aza-Stetter variant is also briefly delineated). First, however, a brief digression on the topic of triazolium salt preparation is in order because some of the most effective Stetter reaction precatalysts are difficult or expensive to procure commercially.

Preparation of Triazolium Salt Precursors to NHC Catalysts

The broader aspects of the typical synthetic route used to prepare the fused polycyclic triazolium salt precursors that provide the most effective NHC catalysts for enantioselective Stetter reactions are encapsulated in the illustrated preparation of precatalyst **TrM1b**•HBF_4 (Scheme 18; see also "Experimental Procedures").[55,56] The conserved part of the approach involves annulation of the triazolium ring onto

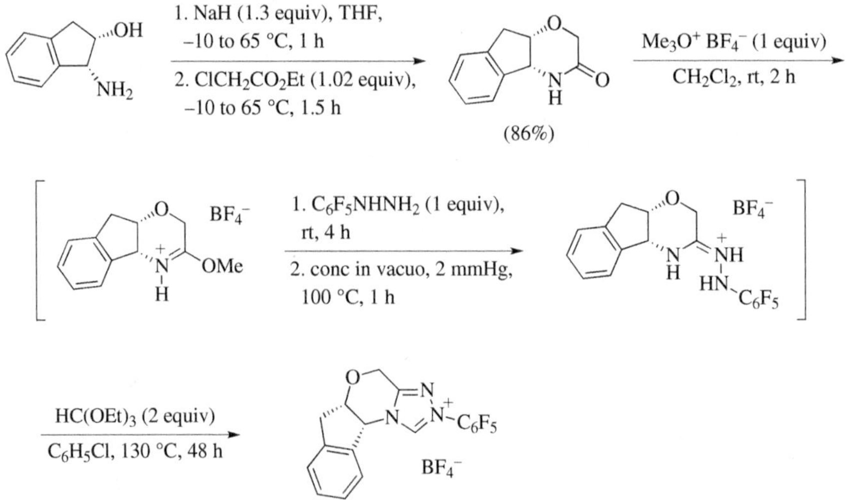

1. NaH (1.3 equiv), THF, −10 to 65 °C, 1 h
2. $ClCH_2CO_2Et$ (1.02 equiv), −10 to 65 °C, 1.5 h

(86%)

$Me_3O^+ BF_4^-$ (1 equiv), CH_2Cl_2, rt, 2 h

1. $C_6F_5NHNH_2$ (1 equiv), rt, 4 h
2. conc in vacuo, 2 mmHg, 100 °C, 1 h

$HC(OEt)_3$ (2 equiv), C_6H_5Cl, 130 °C, 48 h

α-**TrM1b**•HBF_4 (61%)

Scheme 18

an appropriate chiral lactam via imidate formation, nucleophilic substitution with a hydrazine equipped with the desired final *N*-substituent, and then condensation of the resulting imidolyl hydrazone intermediate with ethyl orthoformate.[12] This three-step sequence can be conducted in a one-pot fashion, and it is effective with a wide variety of lactams appropriate for the synthesis of several triazolium salts. Given the simplicity of the triazolium annulation, the overall accessibility of a given NHC catalyst precursor is largely determined by the availability of the requisite chiral lactam. In the case of the popular fused pyrrolidine catalysts **TrP2**, the requisite lactam is obtained in a straightforward five-step sequence (>70% overall yield) from (*S*)-phenylalanine.[57] Other systems are similarly easy to access. For example, the lactam **32** needed to prepare one of the most active precatalysts yet identified, *N*-2,6-dimethoxyphenyl-substituted triazolium **TrP5h**•HBF_4,[58] is obtained as indicated in just three steps from inexpensive (*S*)-pyroglutamic acid (Scheme 19).[58]

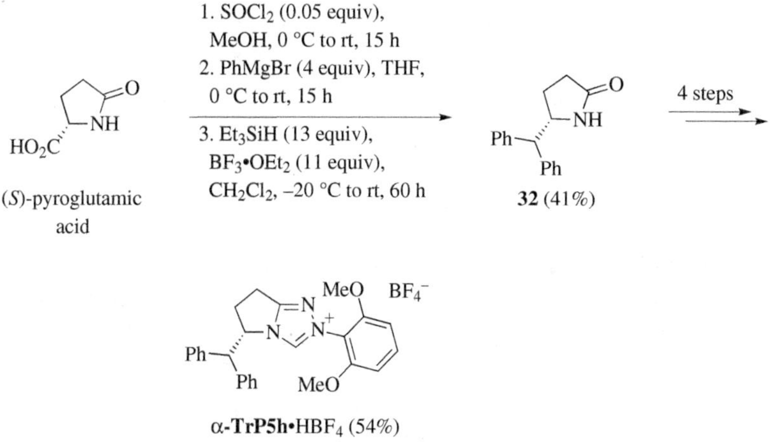

1. $SOCl_2$ (0.05 equiv), MeOH, 0 °C to rt, 15 h
2. PhMgBr (4 equiv), THF, 0 °C to rt, 15 h
3. Et_3SiH (13 equiv), BF_3•OEt_2 (11 equiv), CH_2Cl_2, –20 °C to rt, 60 h

(*S*)-pyroglutamic acid → **32** (41%) → 4 steps → α-**TrP5h**•HBF_4 (54%)

O NH HO_2C Ph Ph MeO BF_4^- N N N + MeO

Scheme 19

Further derivatization of preformed triazolium salts bearing aryl bromides is possible via reduction of the heterocycle to a triazolidine, cross-coupling at the halogenated site, and then oxidation to restore the triazolium salt.[59] Triazolium salts with peripheral azide functionality can be directly functionalized, providing yet another strategy for generating diverse collections of NHC precatalysts for evaluation in Stetter reactions and other reactions of interest.[60,61]

Intramolecular Stetter Reactions

The catalytic enantioselective intramolecular Stetter reaction is broad in scope, and the cyclic ketone products are usually obtained in high yields and with excellent levels of stereoselectivity.[16] A wide variety of aryl aldehydes are competent as donors in this process and, although used less commonly, aliphatic aldehydes can also be successfully employed in many cases. The intramolecular Stetter reaction accommodates a wide range of acceptor moieties, such as enals, enoates, enamides, nitroalkenes, and related unsaturated phosphorus- and sulfur-based functionalities. Furthermore, different substitution patterns within the acceptor are tolerated, enabling access to various combinations of ternary and quaternary stereogenic centers.

Aromatic Aldehyde Donors. Intramolecular reactions between aryl aldehydes and tethered conjugate-addition acceptors are the most commonly encountered examples of enantioselective Stetter chemistry, and various substitution patterns are tolerated. Many types of aryl aldehydes are competent substrates in this reaction, which proceeds with high enantioselectivity regardless of whether electron-withdrawing or electron-donating groups are present on the aromatic nucleus.[62] Electron-withdrawing groups on the aryl ring bearing the aldehyde moiety tend to lead to the products in higher yields. The cyclization of benzaldehyde enoate derivative **21** to chromanone **22** (a reaction prominent in several of the mechanistic studies above) is a benchmark for new NHC designs, and the efficacy of many catalysts has been evaluated against this prototypical intramolecular Stetter reaction.[41] Some representative examples of chiral triazolium and related NHC catalyst precursors that are successful in this transformation are illustrated in Scheme 20.[11,15,62–66] As seen, fused polycyclic triazolylidenes (**TrM** = morpholine fused; **TrP** = pyrrolidine fused) typically afford superior results to other types of NHCs in enantioselective Stetter reactions.[15,62,63] For example, triazolium salt **TrM1c**•$HClO_4$ gives chromanone **22** with excellent yield and in almost perfect enantioselectivity, but an otherwise identical imidazolium precatalyst (**ImM1c**•$HClO_4$) fails to promote the reaction at all under the same reaction conditions (KHMDS, xylenes, room temperature).[63] Slightly improved results, albeit modest, are charted for the non-fused triazolium salt **TrD1a**•$HClO_4$[11] and the macrocyclic imidazolidium salt precatalyst **Im2**•HCl.[66]

Salicylaldehyde-derived substrates, analogous to the aforementioned model **21**, with different types of tethered acceptor alkenes are also successfully converted to chromanones (Scheme 21).[15,62,67,68] In these systems, a wide range of activating EWG groups are tolerated and give satisfactory results from (*E*)-configured acceptors. For example, unsaturated esters, ketones, amides, nitriles, and phosphonates all afford the desired products. In the case of (*E*)-enoates, larger alkoxy groups provide marginally improved stereoselectivity,[43,62] while (*Z*)-enoates lead to products with the same absolute configuration (for a specific catalyst) but with lower enantioselectivity. Poorer outcomes are also noted for acceptors with aryl ketone-, cyano-, and aldehyde-based activating moieties. In the examples given here, using the NHC catalyst with an *N*-pentafluorophenyl substituent (**TrM1b**) is important to obtain optimal results, since **TrM1a**, which bears an *N*-phenyl substituent, is more limited in scope (e.g., the enal substrate does not react in the desired manner). Stetter reactions with nitroalkene- or primary-amide-based acceptors are also unsuccessful with catalyst **TrM1a**; in the latter case, it is postulated that the acidic N–H bonds of the amide deactivate the NHC by protonation.[68] Further variations on the theme of classic substrates like **21** are now considered, including examples with different tethering

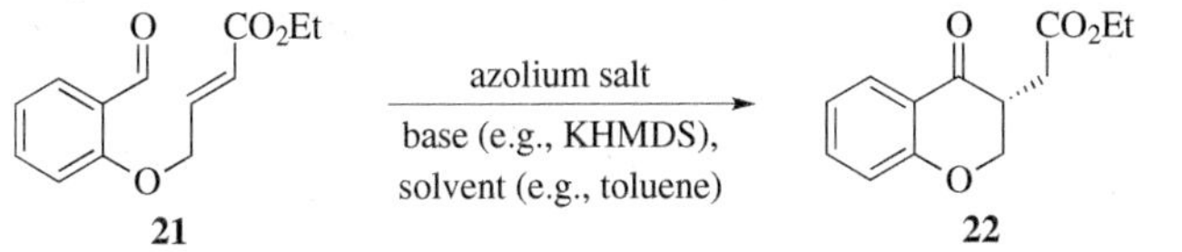

21 → 22: azolium salt; base (e.g., KHMDS), solvent (e.g., toluene)

Yield and er of **22** obtained from **21** using given azolium salt:

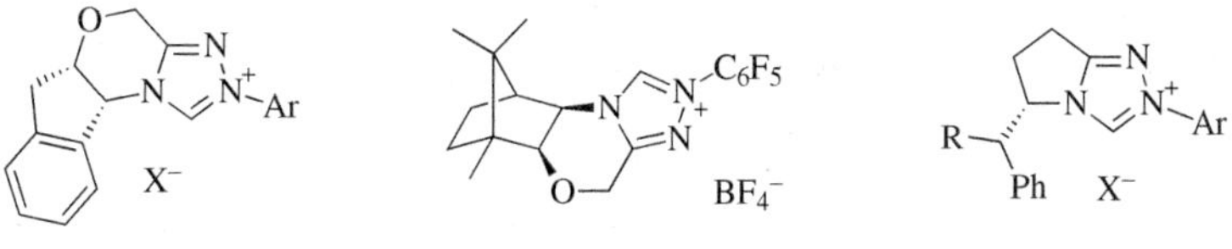

α-**TrM1a**•HBF_4 (Ar = Ph)	(58%) er 97.5:2.5 Ref 15	α-**TrM4b**•HBF_4	(95%) er 96.5:3.5 Ref 64	α-**TrP2a**•HBF_4 (Ar = Ph, R = H)	(94%) er 95:5 Ref 62
α-**TrM1d**•HBF_4 (Ar = PMP)	(94%) er 97:3 Ref 15			α-**TrP2e**•HBF_4 (Ar = 4-$CF_3C_6H_4$, R = H)	(95%) er 96:4 Ref 62
α-**TrM1c**•$HClO_4$ (Ar = Mes)	(94%) er 99:1 Ref 63			α-**TrP5p**•$HClO_4$ (Ar = 2-Pyr, R = Ph)	(85%) er 93:7 Ref 65

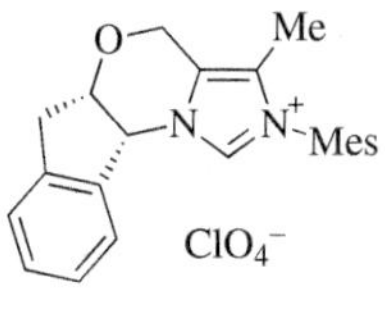

α-**ImM1c**•$HClO_4$
(0%) er —
Ref 63

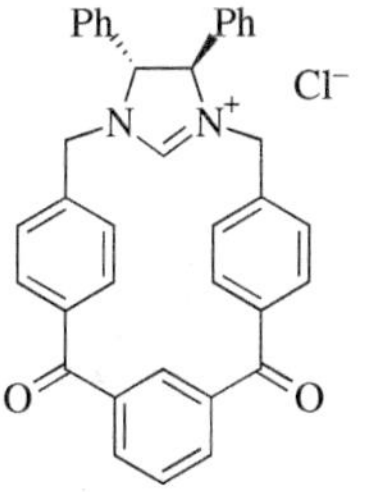

α-**Im2**•HCl
(58%) er 84:16
Ref 66

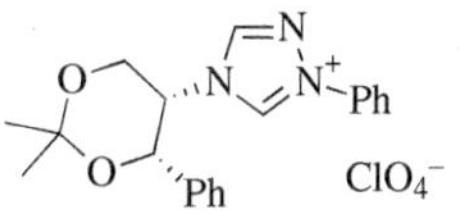

α-**TrD1a**•$HClO_4$
(73%) er 80:20
Ref 11

Scheme 20

segments between donor and acceptor moieties, and with higher-substituted acceptor alkenes, which afford opportunities for the generation of multiple stereogenic centers and/or quaternary carbon atoms in a stereoselective manner.

The linker between the two coupling partners involved in the intramolecular Stetter reactions of aryl aldehydes can be varied. For example, replacement of the ether linker found in the canonical substrate **21** with sulfur-atom, nitrogen-atom, or methylene-group-based linkages gives the expected products with comparable enantioselectivities. However, reaction yields in these cases are typically lower as compared to the oxygenated congener, particularly for the methylene derivative (Scheme 22).[15] As noted above (see "Experimental Mechanistic Studies", Scheme 9),[37] an oxygen atom in the tether may assist in the formation of the Breslow intermediate by mediating the necessary proton transfers en route from **3** to **4** (cf., Scheme 2).

BF_4^-

(20 mol %)

β-TrM1a•HBF$_4$ (Ar = Ph)

or

β-TrM1b•HBF$_4$ (Ar = C_6F_5)

KHMDS (20 mol %), toluene, rt, 1–24 h

EWG	Ar	Yield (%)	er	Ref
MeO_2C	C_6F_5	94	97.5:2.5	62
EtO_2C	Ph	58	97.5:2.5	15
t-BuO_2C	C_6F_5	94	98.5:1.5	62
(Z)-MeO_2C	C_6F_5	80	61:39	62
PhC(O)	C_6F_5	94	89:11	62
EtC(O)	C_6F_5	94	96:4	62
(MeO)MeNC(O)	C_6F_5	94	96:4	62
$H_2NC(O)$	Ph	0	—	68
EtSC(O)	C_6F_5	85	85:15	62
NC–	C_6F_5	80	89:11	62
NC–	Ph	78	87.5:12.5	68
OHC	C_6F_5	50	65:35	62
OHC	Ph	0	—	68
O_2N	Ph	0	—	68
$Ph_2P(O)$	C_6F_5	90	93:7	67
$(EtO)_2P(O)$	C_6F_5	65	90:10	67

Scheme 21

BF_4^-

β-TrM1d•HBF$_4$

(20 mol %)

KHMDS (20 mol %), toluene, rt, 24 h

Y	R	Yield (%)	er
O	Et	94	97:3
MeN	Me	64	91:9
S	Me	63	98:2
CH_2	Me	35	97:3

Scheme 22

A shorter tether permits the synthesis of a benzofuranone in excellent yield. However, in this case (n = 0), the product is obtained essentially as a racemate (Scheme 23).[62] The benzofuranone product has an enantiomeric ratio of 90:10 er at 10% conversion of the substrate, but the level of enantioenrichment is eroded as the transformation progresses due to background racemization likely caused by enolization of the product ketone, which is anticipated to have an α-C–H bond with a relatively low pK_a. The homologue of substrate **21** ($n = 2$) fails to react under the same conditions, and seven-membered rings have yet to be successfully prepared using the intramolecular Stetter reaction of aryl aldehydes.

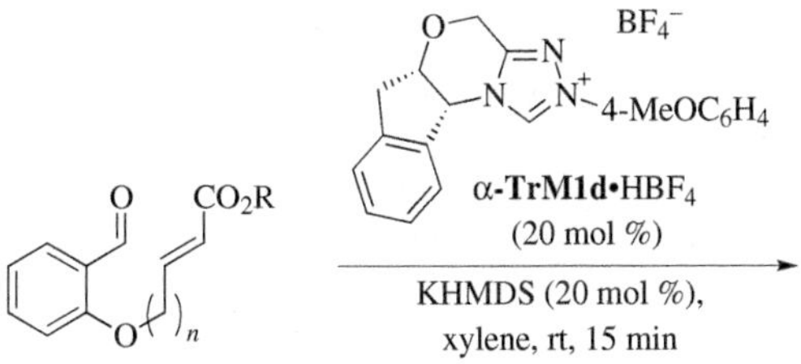

BF_4^- 4-$MeOC_6H_4$

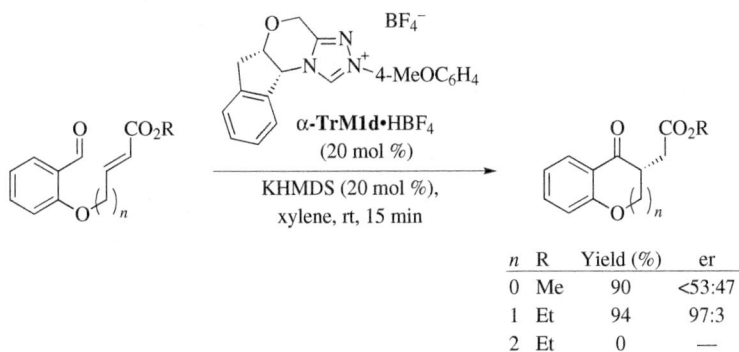

α-**TrM1d**•HBF_4 (20 mol %)

KHMDS (20 mol %), xylene, rt, 15 min

n	R	Yield (%)	er
0	Me	90	<53:47
1	Et	94	97:3
2	Et	0	—

Scheme 23

In an interesting example of five-membered-ring synthesis from a benzaldehyde derivative, a tethered cyclohexadienone acts as the acceptor, and the substrates are enantioselectively desymmetrized during their Stetter cyclization to give indanone products (Scheme 24).[69] The level of enantioselectivity obtained is dependent on the base used to generate the active NHC catalyst (the use of stronger bases such as KHMDS and DBU gives products with <70:30 er) and also on the electronic character of the benzaldehyde moiety, with more electron-rich systems generally giving inferior results. The stereochemical configurations of the products obtained in this study were not determined. The related enantioselective desymmetrizing cyclizations of

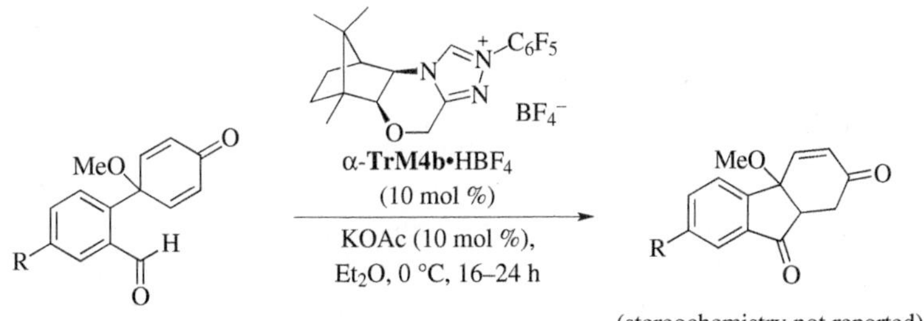

C_6F_5 BF_4^-

α-**TrM4b**•HBF_4 (10 mol %)

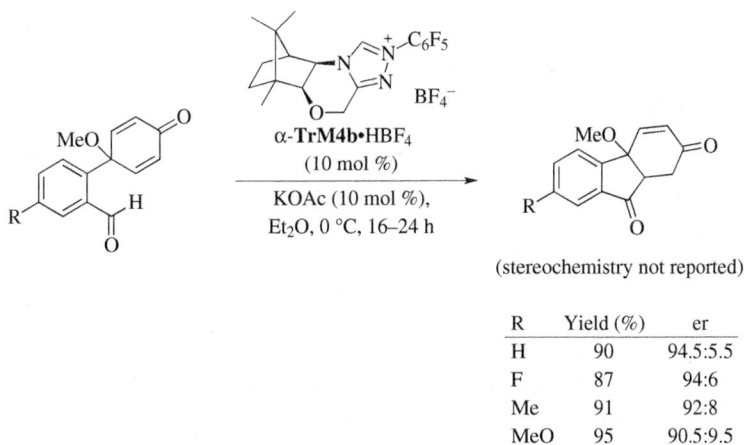

KOAc (10 mol %), Et_2O, 0 °C, 16–24 h

(stereochemistry not reported)

R	Yield (%)	er
H	90	94.5:5.5
F	87	94:6
Me	91	92:8
MeO	95	90.5:9.5

Scheme 24

tethered aryl aldehydes into acyclic geminal bis(enone) moieties have been recently described with comparable results.[70]

Intramolecular Stetter reaction substrates that bear an additional β-substituent in the acceptor moiety (β,β-disubstituted acceptors) are an important class of starting materials because their successful enantioselective cyclization provides quaternary carbon atoms, and stereodefined all-carbon-substituted stereogenic centers are particularly challenging.[71] Conjugated esters and ketones can be used as acceptors in this reaction variant, and substrates decorated similarly to the aforementioned lower-substituted examples are tolerated, as are different types of tethers (Scheme 25).[72] Notably, the five-membered-ring products (e.g., the benzofuranone) are now obtained with excellent enantioselectivities because the lack of an acidic α-C–H bond precludes epimerization via enolization (cf., Scheme 23, $n = 0$). The preparation of six-membered derivatives is more challenging, and this requires the more electrophilic α,β-unsaturated ketones.[73]

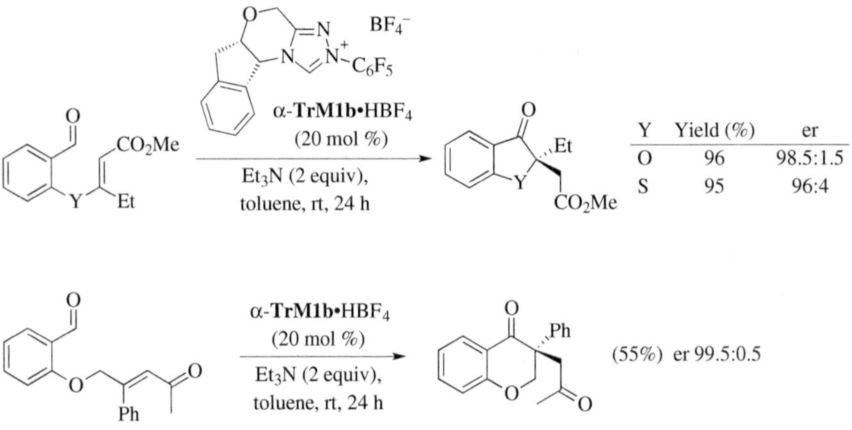

α-**TrM1b**•HBF_4 (20 mol %), Et_3N (2 equiv), toluene, rt, 24 h

Y	Yield (%)	er
O	96	98.5:1.5
S	95	96:4

α-**TrM1b**•HBF_4 (20 mol %), Et_3N (2 equiv), toluene, rt, 24 h — (55%) er 99.5:0.5

Scheme 25

Trisubstituted acceptors bearing substitution at the α- and β-positions lead to Stetter products that contain two contiguous stereogenic centers (Scheme 26).[27] α-Alkyl enoates are competent substrates in this reaction, and they deliver the keto ester products in good yields and with excellent diastereo- and enantioselectivities (see also Scheme 13 in "Stereoselectivity"). In some cases, the base-mediated epimerization of the ketone α-stereocenter leads to compromised diastereoselectivity, but this undesired process can be avoided by using the free carbene catalyst directly in place of the azolium salt precatalyst and an exogenous base. The difference in reactivity implicates the exogenous base rather than the carbene itself as the source of epimerization. In addition to α,β-unsaturated ketones, lactones and cyclopentanones bearing an *exo*-alkylidene moiety are also competent acceptors in this type of reaction (Scheme 27).[27]

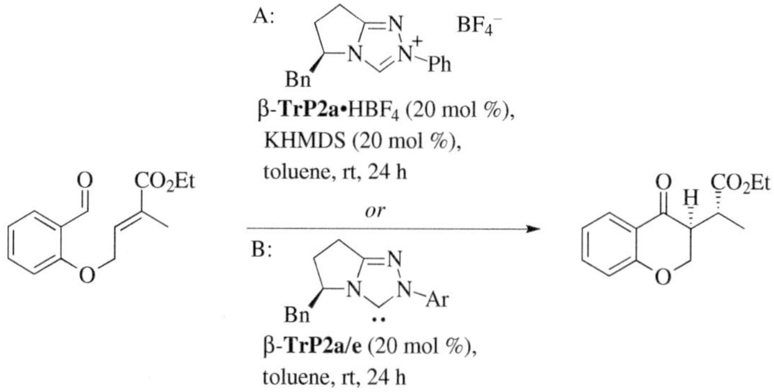

A: β-**TrP2a**•HBF_4 (20 mol %), KHMDS (20 mol %), toluene, rt, 24 h

or

B: β-**TrP2a/e** (20 mol %), toluene, rt, 24 h

Conditions	Catalyst/Precatalyst	Ar	Yield (%)	dr	er
A	β-**TrP2a**•HBF_4	Ph	85	3:1	95:5
B	β-**TrP2a**	Ph	60	12:1	94:6
B	β-**TrP2e**	4-$CF_3C_6H_4$	94	30:1	97.5:2.5

Scheme 26

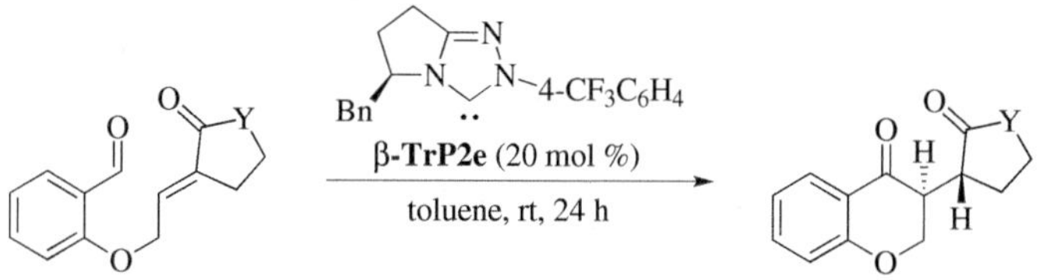

β-**TrP2e** (20 mol %)

toluene, rt, 24 h

Y	Yield (%)	dr	er
O	95	10:1	97:3
CH_2	80	18:1	97.5:2.5

Scheme 27

Aliphatic Aldehyde Donors. Although enolizable aliphatic aldehydes are subject to aldol and other potential side reactions, a majority of the types of enantioselective intramolecular Stetter reactions surveyed above for aryl aldehyde donors are also successful with the analogous aliphatic aldehyde substrates (Scheme 28).[16,62,67] However, there are some limitations: for example, six-membered rings cannot be formed with enoate moieties, albeit cyclohexanones are successfully formed with more electrophilic acceptors such as α,β-unsaturated ketones. Otherwise, the scope of acceptor moieties that can be used in these reactions is broad and also includes unsaturated lactones, amides, and phosphonates, as well as alkylidene malonates.[16,73] Furthermore, heteroatom linkers in the tethering moiety are well tolerated. Interestingly, for a given NHC catalyst/precatalyst, intramolecular Stetter reactions with aliphatic aldehyde donors proceed with the opposite sense of absolute stereoinduction relative to those with aryl aldehyde donors.[62]

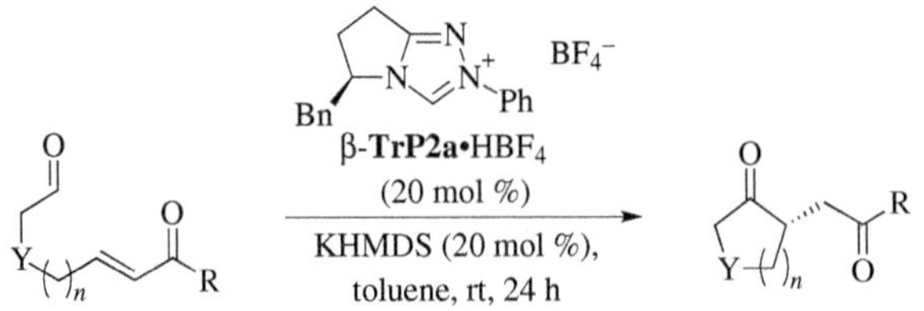

β-TrP2a•HBF$_4$ (20 mol %)
KHMDS (20 mol %), toluene, rt, 24 h

Y	n	R	Yield (%)	er
CH_2	1	EtO	81	97.5:2.5
CbzN	1	EtO	80	99.5:0.5
CH_2	2	EtO	0	—
CH_2	2	Ph	60	71:29

Scheme 28

The influence of preexisting stereocenters within the tether on stereochemical outcome has also been investigated as a means to generate enantioenriched polysubstituted cyclopentanones. In the case of substrates with an alkyl substituent in the β- (not illustrated) or γ-position relative to the aldehyde, the catalyst overrides any intrinsic bias of the substrate, and each enantiomer of a racemic starting material is stereodivergently converted to a highly enantioenriched product within a different diastereomeric series (first example, Scheme 29).[74] However, in the case of an aldehyde with an α-substituent, the catalyst exerts little influence, and essentially racemic products are generated with a diastereomeric bias that largely mimics that obtained with an achiral NHC catalyst (second example, Scheme 29).

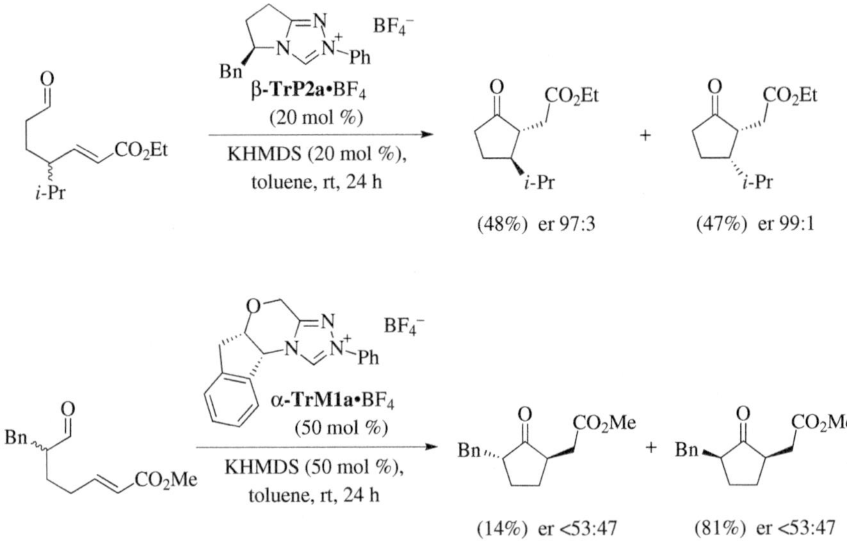

β-TrP2a•BF$_4$ (20 mol %)
KHMDS (20 mol %), toluene, rt, 24 h

(48%) er 97:3 + (47%) er 99:1

α-TrM1a•BF$_4$ (50 mol %)
KHMDS (50 mol %), toluene, rt, 24 h

(14%) er <53:47 + (81%) er <53:47

Scheme 29

More highly substituted acceptors have also been studied in the context of aliphatic aldehyde donors. Here again, opportunities for forming stereodefined quaternary carbon atoms as well as contiguous stereocenters are available. The coupling partners can be tethered with aliphatic chains, and tethers containing nitrogen atoms are also successful substrates (Scheme 30).[72,73] As is generally the case, the (*Z*)-configured acceptor moieties are converted to products with lower efficiency and enantioselectivity compared with the (*E*)-configured examples. In the case of sulfur-linked substrates, the cyclization fails with a thioether, but the transformation is successful with the corresponding sulfone. The difference in reactivity between these last two substrates is likely to be electronic in nature rather than of steric origin. In contrast, the thioether-linked substrates with an aryl aldehyde behave normally in the enantioselective intramolecular Stetter reaction (see Schemes 22 and 25).

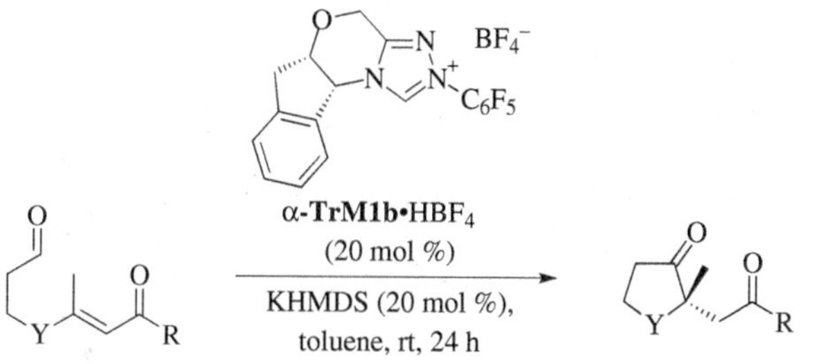

BF_4^- C_6F_5

α-**TrM1b**•HBF_4 (20 mol %)

KHMDS (20 mol %), toluene, rt, 24 h

(*E*) or (*Z*)	Y	R	Yield (%)	er
(*E*)	CH_2	Ph	85	98:2
(*Z*)	CH_2	Ph	50	78:22
(*E*)	AcN	Me	65	98:2

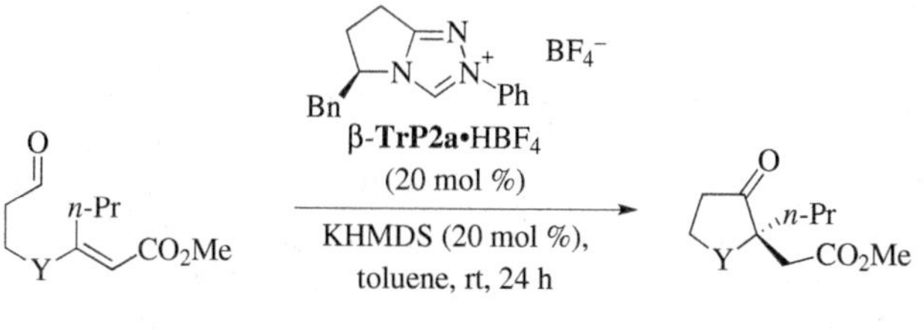

BF_4^- Ph Bn

β-**TrP2a**•HBF_4 (20 mol %)

KHMDS (20 mol %), toluene, rt, 24 h

n-Pr CO_2Me

Y	Yield (%)	er
S	0	—
O_2S	98	90:10

Scheme 30

Finally, the enantioselective desymmetrization of cyclohexadienones by an intramolecular Stetter reaction of aliphatic aldehydes has also been accomplished (Scheme 31).[75] The substrates are easily produced from phenol derivatives by oxidation with (diacetoxyiodo)benzene in the presence of ethylene glycol, followed by conversion of the residual primary alcohol to an aldehyde with Dess–Martin periodinane. Depending on the level of substitution on the dienone, this method reliably generates two or three contiguous stereogenic centers. Analogous substrates incorporating aldehydes linked to the dienone via an alkyl chain tether are likewise successfully converted to highly enantioenriched *cis*-fused bicyclo[4.3.0]nona-3-en-2,7-diones.[75]

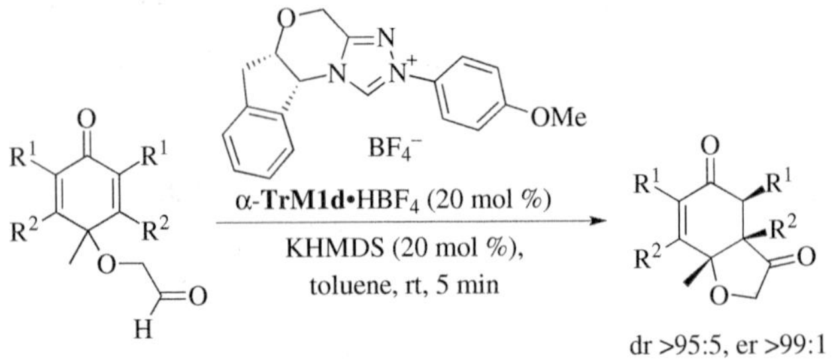

α-**TrM1d**•HBF_4 (20 mol %)

KHMDS (20 mol %), toluene, rt, 5 min

R^1	R^2	Yield (%)
Me	H	86
H	Me	64

dr >95:5, er >99:1

Scheme 31

Intermolecular Stetter Reactions

The enantioselective intermolecular Stetter reaction presents a number of challenges and is more limited in its range of applications in comparison to the aforementioned Stetter cyclization. Typically, a specific type of donor aldehyde must be paired with a certain class of acceptor alkene for a productive outcome, and all intermolecular Stetter reactions are plagued to some extent by benzoin side reactions. Nevertheless, whether or not this competing reaction is detrimental to the desired transformation depends on the degree to which the benzoin adduct can act as an effective reservoir of the free aldehyde via a retro-benzoin reaction.[45,76] Despite the various factors that complicate its execution, several successful and synthetically useful variants of the enantioselective intermolecular Stetter reaction are now available. The usual fused polycyclic triazolylidenes are often the optimal NHC catalysts for these processes, and a variety of aldehyde types are tolerated (special systems such as heterocyclic aldehydes and glyoxylic esters are ideal; enolizable aliphatic aldehydes are more problematic), although enantioselectivities are not uniformly high. The scope of acceptors for the intermolecular Stetter reaction is more limited compared to the intramolecular mode and is generally restricted to highly activated systems (e.g., nitroalkenes, chalcones, and alkylidene malonates). However, there are examples in which less activated systems are used (e.g., simple enoates). Key facets of the enantioselective intermolecular Stetter reaction are now elaborated for different kinds of aldehyde donors and, in each case, the types of alkene acceptors that can be successfully engaged are identified.

Aryl and Heteroaryl Aldehyde Donors. Aryl aldehydes are commonly used in the enantioselective intermolecular Stetter reaction, and these well-behaved donors are successfully coupled with a comparatively broad range of acceptors, including chalcones, arylidene malonates, nitroalkenes, and, to a more limited extent, enoates. Heteroaryl aldehydes that contain a heteroatom *ortho* to the carbonyl group are particularly good substrates, which may be ascribed to a less crowded environment around the formyl functionality, and generally perform better than the corresponding carbocyclic aryl aldehydes.

Simple enoates are generally ineffective as acceptors for the intermolecular addition process due to their comparatively low electrophilicity. However, acrylate derivatives that lack β-substituents, and which result in a less sterically impeded conjugate addition, are viable coupling partners for aryl aldehydes under certain reaction conditions. In such cases, for the Stetter reaction to lead to a chiral nonracemic product, an additional α-substituent is required in the acceptor, and enantioselectivity must be realized by a diastereocontrolled-protonation event that occurs during the penultimate stage of the coupling process (cf. **5** to **6**, Scheme 2). One such scenario has already been described in detail for the addition of aryl aldehydes to α-amidoacrylates to generate α-amino acid derivatives (Scheme 16 in "Stereoselectivity"). The process can be extended to other types of α-substituted acrylates by the use of triazolylidene catalyst **TrP5h**, which bears an *N*-(2,6-dimethoxy)phenyl group and is one of the most effective NHCs for Stetter chemistry yet identified (Scheme 32).[58] The analogous *N*-mesityl catalyst (**TrP5c**) gives low yields of the Stetter products because, in this case, the benzoin dimerization that initially sequesters the aldehyde is ostensibly irreversible. As illustrated, less electron-rich benzaldehyde derivatives (and 2-furaldehyde) give superior results in the context of efficiency, whereas a bulky α-alkyl group on the acrylate improves enantioselectivity. Significantly, the same catalyst mediates an intermolecular Stetter addition to a simple (*Z*)-configured, β-substituted enoate (albeit with only modest enantioselectivity) (Scheme 33).[58] However, the corresponding (*E*)-enoate fails to react under the same conditions. This noteworthy result (the absolute stereochemical outcome of which is not known) remains the current state-of-the-art for an enantioselective, intermolecular Stetter reaction involving a simple enoate acceptor and provides complementary reactivity to the intramolecular variant.

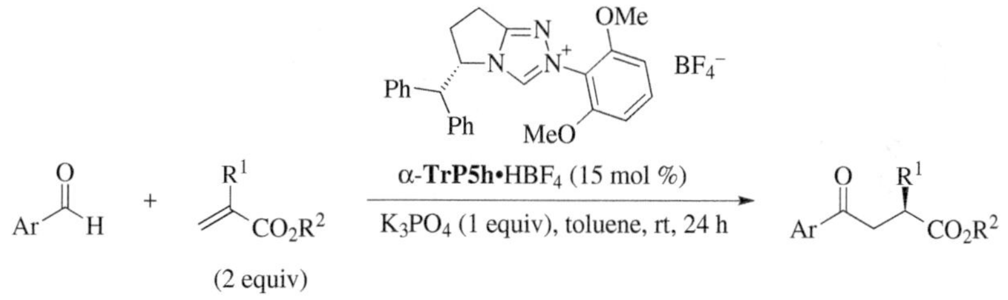

OMe
BF_4^-
Ph
Ph MeO

ArCHO + $CH_2=C(R^1)CO_2R^2$ (2 equiv) → α-**TrP5h**•HBF_4 (15 mol %), K_3PO_4 (1 equiv), toluene, rt, 24 h → $ArCOCH_2CH(R^1)CO_2R^2$

Ar	R^1	R^2	Yield (%)	er
Ph	Me	*n*-Bu	53	96:4
3-MeC_6H_4	Me	*n*-Bu	43	96:4
4-ClC_6H_4	Me	*n*-Bu	90	95.5:4.5
4-$CF_3C_6H_4$	Me	*n*-Bu	86	95.5:4.5
2-furyl	Me	*n*-Bu	97	96.5:3.5
4-ClC_6H_4	Me	Me	82	96.5:3.5
4-ClC_6H_4	Me	*i*-Pr	88	94.5:5.5
4-ClC_6H_4	*i*-Pr	Et	76	98.5:1.5
4-ClC_6H_4	Bn	Me	34	90:10
4-ClC_6H_4	Ph	Me	28	80:20

Scheme 32

The polyactivated nature of arylidenemalonate derivatives makes them suitable substrates for the intermolecular Stetter addition of heteroaryl aldehydes

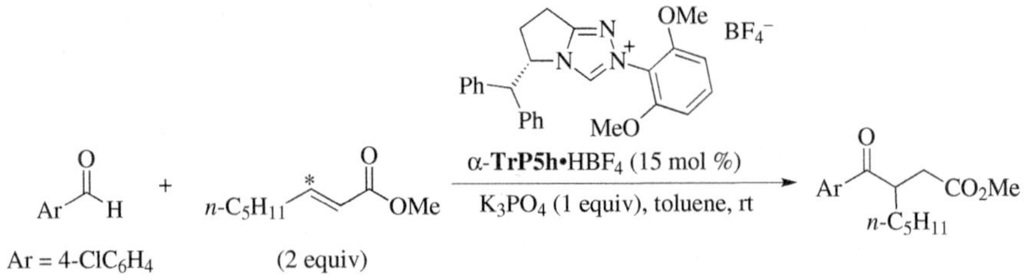

*(*E*) or (*Z*)	Yield (%)	er
(*E*)	0	—
(*Z*)	59	90:10

Scheme 33

(Scheme 34),[46] and as outlined above, the same precatalyst and reaction conditions also effect the addition of aryl aldehydes to chalcones (Scheme 14).[45] Although these processes are reasonably effective, the products obtained are not particularly useful. By contrast, the Stetter reaction between heteroaryl aldehydes and α,β-unsaturated keto esters is high yielding and highly enantioselective (Scheme 35),[77] and affords

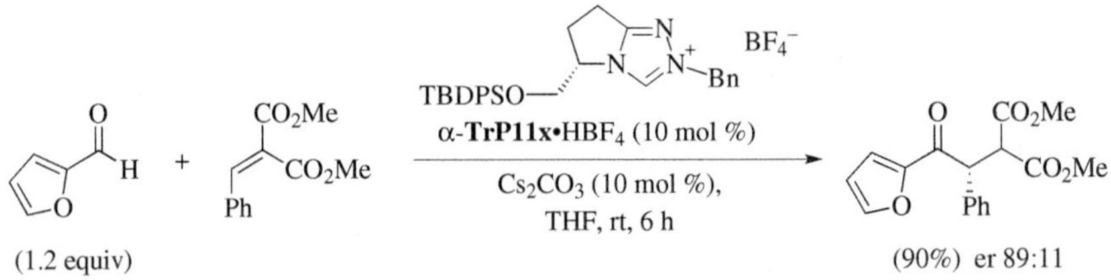

Scheme 34

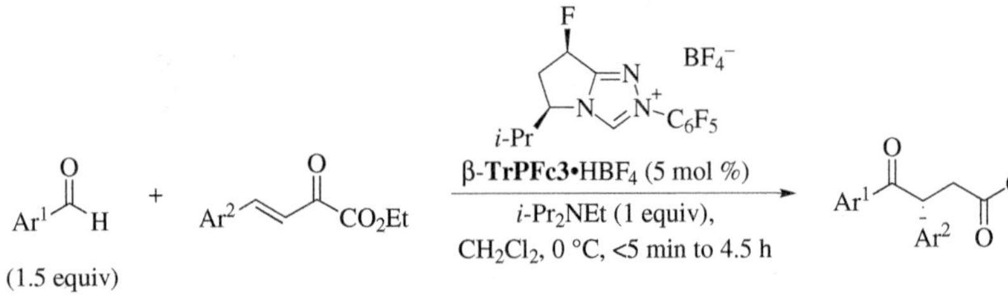

Ar^1	Ar^2	Yield (%)	er
2-furyl	Ph	92	95:5
2-furyl	4-FC_6H_4	80	95:5
2-furyl	3,4-$(MeO)_2C_6H_3$	86	95:5
2-pyridyl	Ph	88	95.5:4:5
2-pyrazinyl	Ph	94	93.5:6.5
4-$(MeO_2C)C_6H_4$	Ph	30	86:14

Scheme 35

products that can be processed into a variety of useful materials. Numerous heteroaryl aldehydes are competent in this reaction, and electron-deficient aryl aldehydes can also be employed, but the reactions are less efficient. Many β-substituents are tolerated in the α,β-unsaturated keto esters, including electron-rich or electron-deficient aryl groups. Nevertheless, aliphatic substituents on the acceptor and β,β-disubstituted keto esters are notably absent from the substrate scope. Moreover, less activated enones bearing aliphatic substituents have not been reported.

A variety of heteroaromatic aldehydes with at least one heteroatom adjacent to the formyl group readily add to aliphatic nitroalkenes with high enantioselectivity using the fluoropyrrolidine-containing catalyst **TrPFc3b** (Scheme 36).[78] The origin of the enhancement effect that the fluorine-atom-bearing stereogenic center has on catalyst activity in this intermolecular Stetter reaction was described previously (see Scheme 17 and "Computational Studies").[47] Enantioselectivity diminishes as the size of the nitroalkene substituent R is reduced, and benzaldehyde does not react under the reaction conditions illustrated.[78]

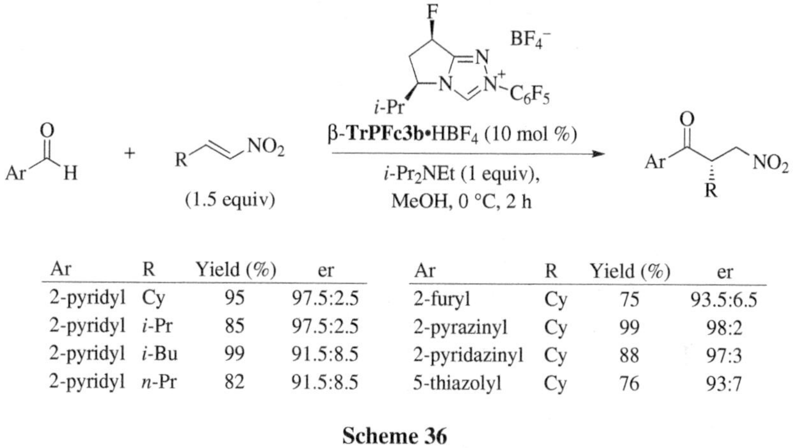

F
BF_4^-
N
N^+
C_6F_5
i-Pr
Ar H (O) + R ⁀ NO_2 (1.5 equiv)
β-**TrPFc3b**•HBF_4 (10 mol %)
i-Pr_2NEt (1 equiv), MeOH, 0 °C, 2 h
Ar (O) R NO_2

Ar	R	Yield (%)	er
2-pyridyl	Cy	95	97.5:2.5
2-pyridyl	*i*-Pr	85	97.5:2.5
2-pyridyl	*i*-Bu	99	91.5:8.5
2-pyridyl	*n*-Pr	82	91.5:8.5
2-furyl	Cy	75	93.5:6.5
2-pyrazinyl	Cy	99	98:2
2-pyridazinyl	Cy	88	97:3
5-thiazolyl	Cy	76	93:7

Scheme 36

Glyoxamide Donors. Glyoxamide donors were the first substrates identified to give intermolecular Stetter reaction products efficiently and with high enantioselectivity.[76,79] The excellent reactivity of this class of substrates was recognized in the original work of Stetter,[80] which is attributed to the facile formation of the Breslow intermediate as a result of the high acidity of the doubly activated α-C–H bond in the preceding tetrahedral intermediate (**3**, R^3 = C(=O)NR_2, Scheme 2). Stetter reactions involving glyoxamides tend to generate the benzoin adduct early in the transformation, but over time, and as is usually the case (vide supra), this compound acts as a reservoir for the aldehyde via a retro-benzoin reaction, and therefore its formation is largely inconsequential. Glyoxamide donors have been paired successfully with acceptor alkylidenemalonate and related systems. In the case of malonate diester acceptors, glyoxylate ester donors give poor enantioselectivities (er <65:35), but glyoxamide donors, and in particular morpholinyl glyoxamides, afford Stetter products with the usual high levels of enantioselectivity expected for polycyclic fused triazolylidine catalysts (Scheme 37).[76] In these cases, stereoselectivity is improved by using bulkier ester alkoxy groups in the acceptor and replacing triethylamine with Hünig's base. Under fully optimized reaction conditions, which include a magnesium sulfate desiccant to scavenge residual water and prolong catalyst activity, a variety of β-substituents are accommodated in the alkylidenemalonate acceptor. When R is a methyl group, partial racemization of the product occurs upon protracted reaction time, and higher enantioselectivity is achieved at the expense of yield by terminating the reaction early.

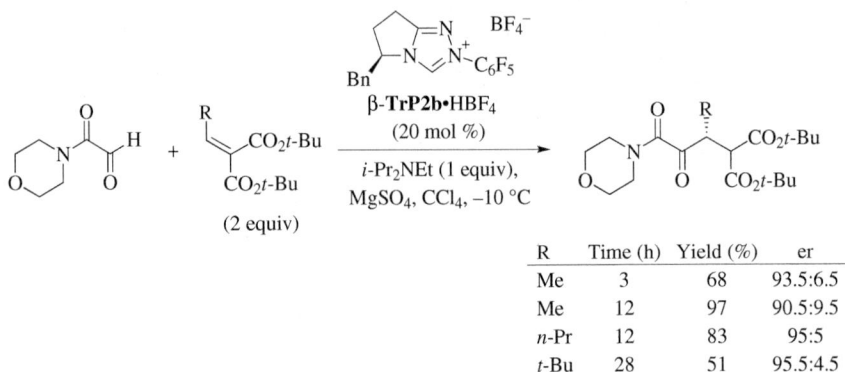

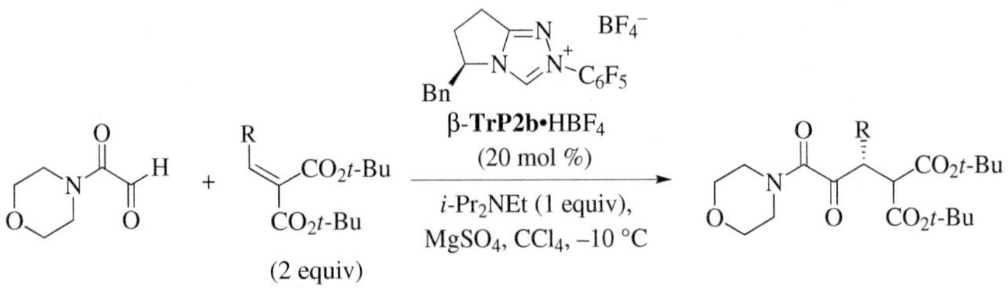

BF_4^- Bn C_6F_5
β-TrP2b•HBF₄ (20 mol %)
CO_2t-Bu, CO_2t-Bu (2 equiv)
i-Pr_2NEt (1 equiv), $MgSO_4$, CCl_4, –10 °C

R	Time (h)	Yield (%)	er
Me	3	68	93.5:6.5
Me	12	97	90.5:9.5
n-Pr	12	83	95:5
t-Bu	28	51	95.5:4.5

Scheme 37

Related, doubly activated α-alkylidene β-ketoamide acceptors, readily obtained as (*Z*)-isomers by piperidine-mediated Knoevenagel condensation,[81] also provide good substrates for the Stetter process (Scheme 38).[79] The products are generated not only with high levels of enantioselectivity, as would be anticipated, but also

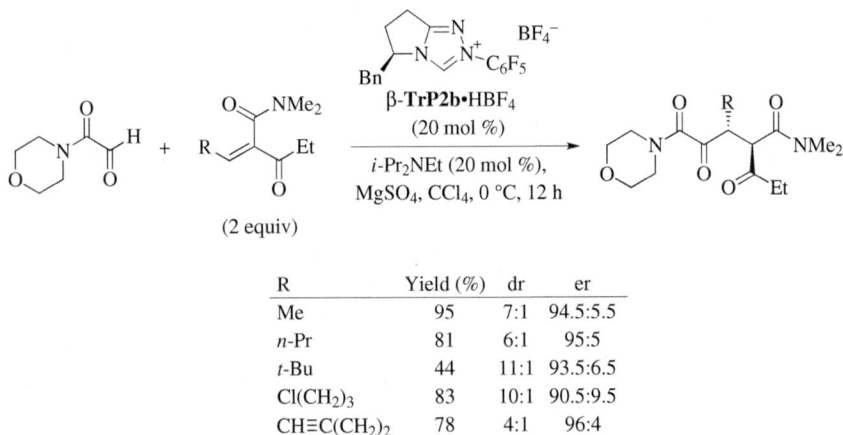

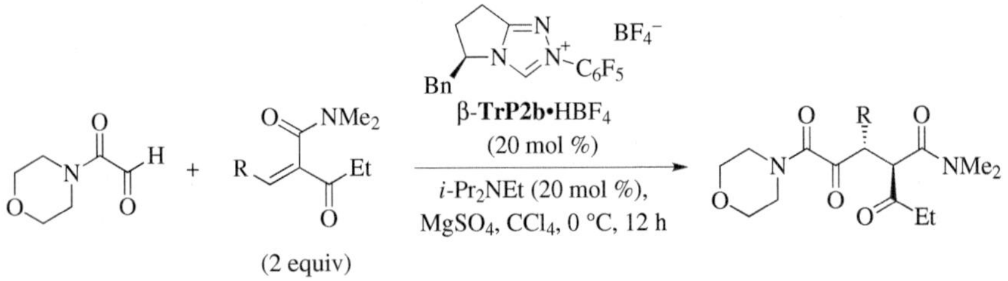

BF_4^- Bn C_6F_5
β-TrP2b•HBF₄ (20 mol %)
NMe_2, Et (2 equiv)
i-Pr_2NEt (20 mol %), $MgSO_4$, CCl_4, 0 °C, 12 h

R	Yield (%)	dr	er
Me	95	7:1	94.5:5.5
n-Pr	81	6:1	95:5
t-Bu	44	11:1	93.5:6.5
$Cl(CH_2)_3$	83	10:1	90.5:9.5
$CH{\equiv}C(CH_2)_2$	78	4:1	96:4

Scheme 38

with modest-to-good diastereoselectivity. The origin of kinetic control in the diastereoselective intermolecular Stetter reactions of acceptors with additional α-substituents was described previously (see Scheme 16, "Stereoselectivity"), but it is surprising that the stereochemical integrity of the new α-stereogenic center is maintained. The ostensible resistance of this position to facile epimerization in the presence of Hünig's base was attributed to 1,3-allylic strain in the enol/enolate, thereby disfavoring enolization.[79]

α,β-Unsaturated Aldehyde Donors. α,β-Unsaturated aldehydes are a challenging class of substrates to employ in Stetter reactions because they can act as either donors or acceptors, and the corresponding Breslow intermediates are ambident nucleophiles that may react with electrophiles at either the d1 or the d3 site. The latter so-called homoenolate behavior has many applications and can be deliberately invoked by the appropriate combination of donor and acceptor, as well as by the choice of NHC catalyst.[5,82] For example, in the illustrated reaction of a β-aryl enal with an aryl aldehyde using an *N,N′*-dimesityl imidazolium salt precatalyst, the Stetter process is not observed, but rather the Breslow intermediate derived from the enal engages with the aryl aldehyde at the distal site (Scheme 39).[83,84]

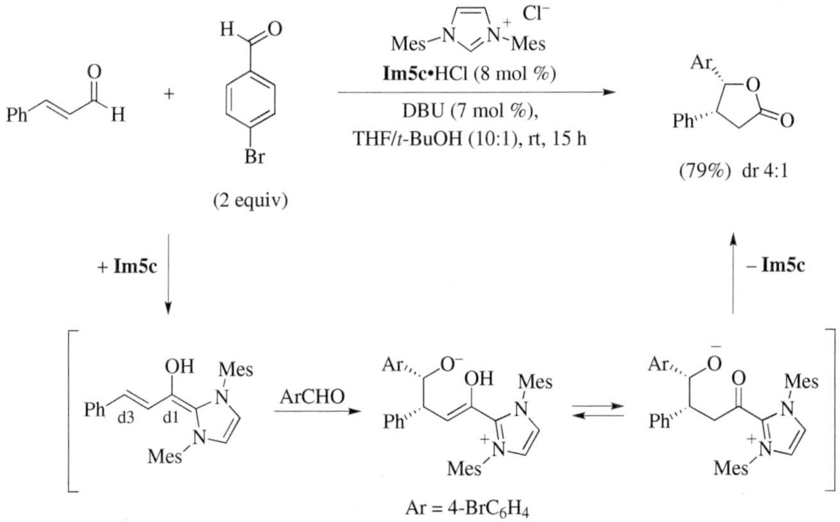

Scheme 39

Notwithstanding the complex reactivity of α,β-unsaturated aldehydes, such enals have been successfully coupled in enantioselective intermolecular Stetter reactions with nitroalkenes (Scheme 40).[40] As illustrated by the selected substrates, the scope of this reaction is broad: a wide range of enal types is accommodated, including cinnamaldehyde (and related other aryl and heteroaryl examples) and extending to simple β-alkyl substituted enals and even conjugated dienals. Various (*E*)-configured β-substituted derivatives of nitroethene are also tolerated, but the reaction fails with 1-nitrocyclohexene.[40] The effect of the fluorine-atom substituent in the pyrrolidine ring in the triazolium salt precatalyst used in this example (see Scheme 15) was discussed previously in "Mechanism and Stereoselectivity", as was a hypothesis relating to the accelerating role of catechol, which is an essential additive for success in this reaction (see Scheme 10). Interestingly, the same kinds of substrates can be coupled via the homoenolate manifold by use of a bulkier triazolium precatalyst and slightly different reaction conditions (Scheme 41).[85] Yet another variation in the catalyst structure leads to the corresponding *anti* products instead of *syn*.[86]

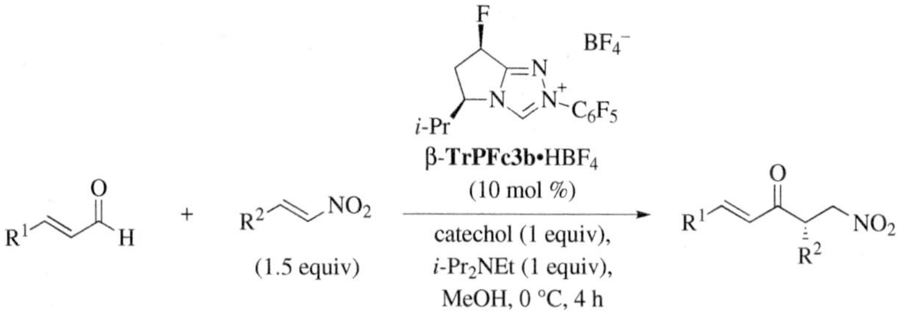

F
BF_4^-
N
N
N^+
C_6F_5
i-Pr
β-TrPFc3b•HBF_4 (10 mol %)
R^1 CH=CH CHO + R^2CH=CHNO_2 (1.5 equiv)
catechol (1 equiv), *i*-Pr_2NEt (1 equiv), MeOH, 0 °C, 4 h
R^1 O R^2 NO_2

R^1	R^2	Yield (%)	er
Ph	Et	66	76:24
Ph	*i*-Bu	84	71.5:28.5
Ph	*i*-Pr	75	95.5:4.5
Ph	Cy	80	96.5:3.5

R^1	R^2	Yield (%)	er
Me	Cy	67	93:7
i-Pr	Cy	70	91.5:8.5
$TIPSOCH_2$	Cy	60	91:9
(*E*)-MeC=CH	Cy	90	97:3

Scheme 40

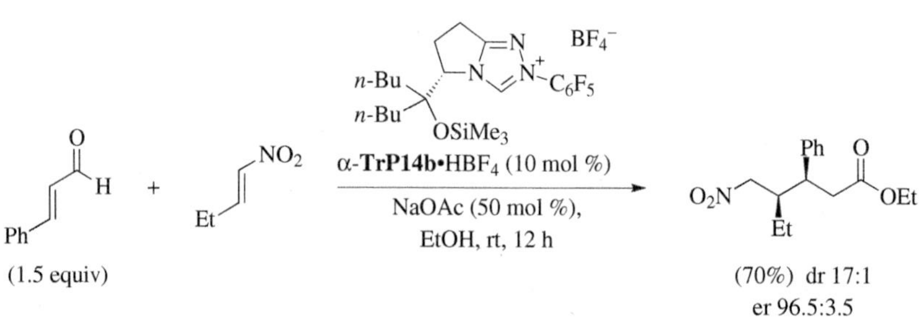

BF_4^-
N
N
N^+
C_6F_5
n-Bu
n-Bu
$OSiMe_3$
α-TrP14b•HBF_4 (10 mol %)
Ph CH=CH CHO (1.5 equiv) + Et CH=CH NO_2
NaOAc (50 mol %), EtOH, rt, 12 h
O_2N Et Ph O OEt
(70%) dr 17:1
er 96.5:3.5

Scheme 41

Finally, enals have also been combined with alkylidene diketones in highly enantioselective intermolecular Stetter reactions. This reaction system successfully employs aryl, alkyl, and even disubstituted enals (Scheme 42).[87] A variety of alkylidene diketones are competent in this reaction, but it should be noted that for electron-rich electrophiles, the use of catechol as an additive is essential for good reactivity.

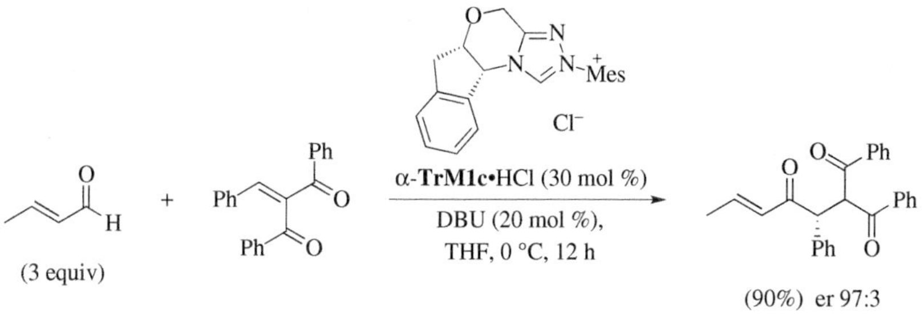

O N N Mes Cl$^-$

Ph O H + Ph O Ph O

α-**TrM1c**•HCl (30 mol %)

DBU (20 mol %), THF, 0 °C, 12 h

(3 equiv)

O O Ph Ph Ph O

(90%) er 97:3

Scheme 42

Aliphatic Aldehyde Donors. Owing to the heightened possibilities for side reactions, aliphatic aldehydes have rarely been successfully employed in the enantioselective intermolecular Stetter reaction. Sterically encumbered aldehydes are generally poor substrates, while linear aldehydes give the best results. Nevertheless, the types of acceptors that can be combined with aliphatic aldehyde donors is limited to chalcones, β-unsubstituted acrylate derivatives, and nitroalkenes. For example, in an early report of such a reaction, the addition of butanal to chalcone proceeds in low yield (ca. 30%) and with modest enantioselectivity (er <70:30) using a chiral thiazolium precatalyst.[88] Acetaldehyde is added to various chalcones with improved results using triazolylidene catalyst **TrM1b** (Scheme 43),[89] although higher enantioselectivity for the same kinds of targets can be realized and a broader range of acceptors tolerated via the enzymatic decarboxylative conjugate addition of pyruvate (see "Comparison with Other Methods").[90] Surprisingly, the highest enantioselectivity yet realized for an intermolecular Stetter reaction with an aliphatic aldehyde donor utilizes methyl α-acetamidoacrylate (Scheme 44).[48]

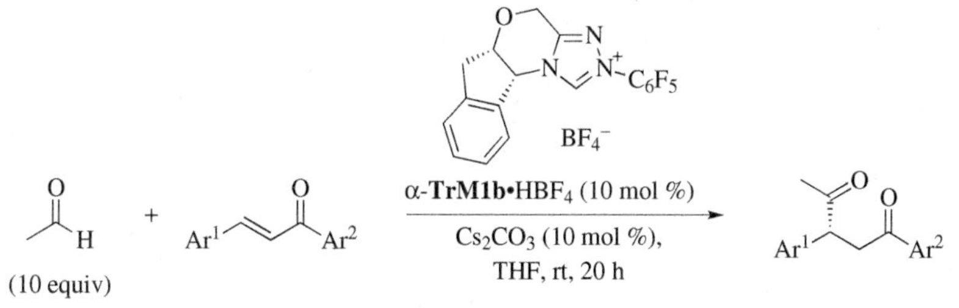

O N N C_6F_5 BF_4^-

O H + Ar^1 O Ar^2

α-**TrM1b**•HBF_4 (10 mol %)

Cs_2CO_3 (10 mol %), THF, rt, 20 h

(10 equiv)

O O Ar^1 Ar^2

Ar^1	Ar^2	Yield (%)	er
Ph	Ph	42	78.5:21.5
Ph	4-ClC_6H_4	78	81:19
Ph	4-$MeOC_6H_4$	43	79:21
2-naphthyl	Ph	62	88:12

Scheme 43

Finally, the enantioselective intermolecular Stetter couplings to β-aryl nitroethene derivatives exhibit arguably the best scope with aliphatic aldehyde donors

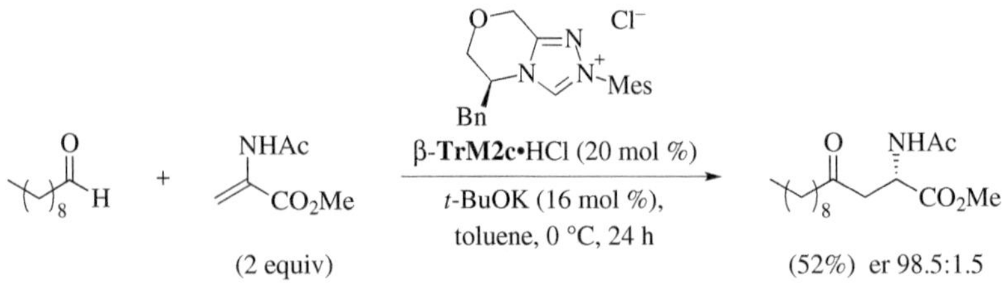

β-**TrM2c**•HCl (20 mol %)
t-BuOK (16 mol %), toluene, 0 °C, 24 h

(2 equiv)

(52%) er 98.5:1.5

Scheme 44

(Scheme 45).[53] Interestingly, in these cases, *trans*-fluoropyrrolidinyl catalyst **TrPFt4b** affords the highest enantioselectivity, whereas an analogous *cis*-fluoropyrrolidinyl catalyst (**TrPFc3b**), which is optimal for comparable reactions involving aryl aldehydes (Scheme 36), delivers the products with significantly lower yields and enantioselectivities. In general, unbranched aldehydes are optimal, and simple functional groups are tolerated in the side chain (e.g., terminal alkene, thioether, TBS ether, alkyl chloride). Aryl-substituted nitroalkenes are competent coupling partners, but alkyl-substituted nitroalkenes lead to diminished yields.

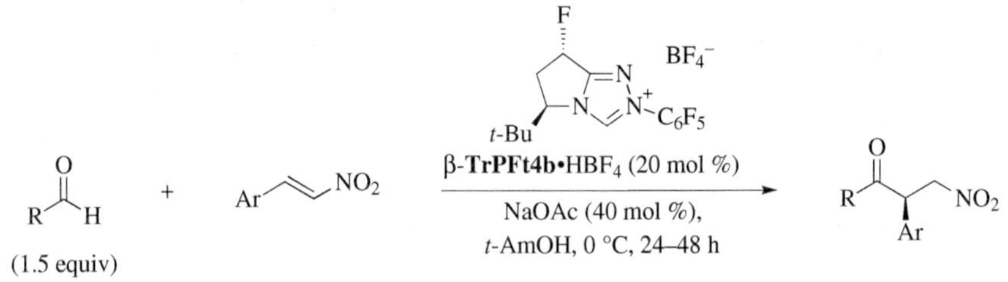

β-**TrPFt4b**•HBF_4 (20 mol %)
NaOAc (40 mol %), *t*-AmOH, 0 °C, 24–48 h

(1.5 equiv)

R	Ar	Yield (%)	er
Me	Ph	71	81:19
n-Pr	Ph	80	96.5:3.5
i-Bu	Ph	32	97.5:2.5
Cy	Ph	<5	—
$Cl(CH_2)_3$	Ph	83	96.5:3.5

R	Ar	Yield (%)	er
n-Pr	2-FC_6H_4	75	96.5:3.5
n-Pr	2-ClC_6H_4	70	95.5:4.5
n-Pr	2-$MeOC_6H_4$	83	97:3
n-Pr	3-$MeOC_6H_4$	63	95.5:4.5

Scheme 45

Aza-Stetter Reactions with Imine Donors

An analogous intermolecular enantioselective aza-Stetter reaction using imine donors in place of the usual aldehydes was recently reported (Scheme 46).[91] The broader aspects of this process are similar to those of the usual transformation involving aldehydes, and the familiar fused polycyclic triazolylidene NHC catalysts again afford the products with remarkably high enantioselectivities. At this early stage of development, the enantioselective aza-Stetter reaction is severely limited in the scope of acceptor alkenes that are tolerated as substrates, in which only the α-methylenated 2-chromanones provide satisfactory results. Acceptors such as chalcone and cyclohex-2-enone fail to react with the aza-Breslow intermediate, and instead, products derived from an aza-benzoin reaction are obtained exclusively. Also, the use of enolizable aliphatic imines is precluded because of their ability to isomerize to the corresponding enamines.

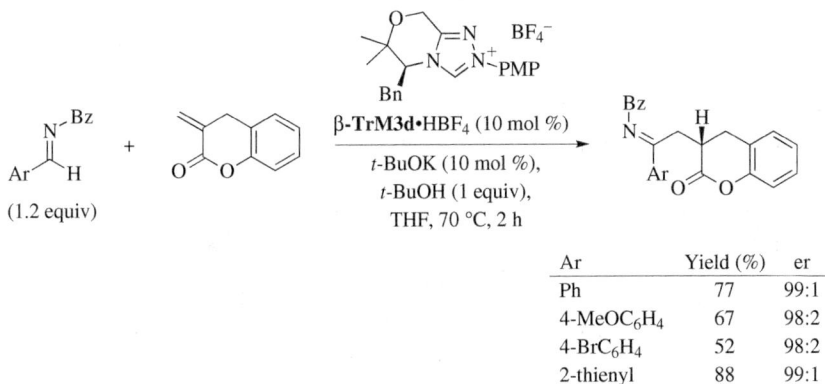

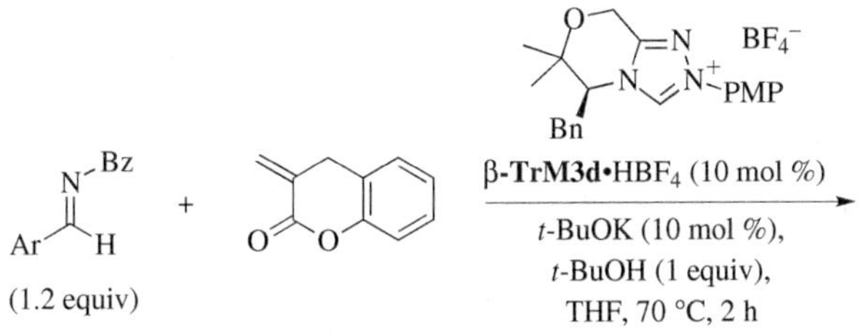

(1.2 equiv)

β-TrM3d•HBF₄ (10 mol %)

t-BuOK (10 mol %), *t*-BuOH (1 equiv), THF, 70 °C, 2 h

Ar	Yield (%)	er
Ph	77	99:1
4-$MeOC_6H_4$	67	98:2
4-BrC_6H_4	52	98:2
2-thienyl	88	99:1

Scheme 46

APPLICATIONS TO SYNTHESIS

The enantioselective Stetter reaction has only rarely been used in target-directed synthesis efforts. Nevertheless, non-enantioselective variants of the process have been instrumental in the efficient syntheses of various biologically active natural product molecules and active pharmaceutical ingredients (API).[92] For example, the intermolecular Stetter reaction has been exploited for syntheses of *cis*-jasmone and dihydrojasmone,[93] haloperidol,[94] and *trans*-sabinene hydrate.[95] In these cases, an unsubstituted acceptor is used, and no new stereocenter is created in the reaction. Diastereoselective intermolecular Stetter reactions have been used for efficient syntheses of roseophilin,[96,97] monomorine,[98] and englerin A.[99] The intramolecular process has proven to be similarly effective since the earliest days of its discovery, as exemplified by syntheses of hirsutic acid[100] and mitchellene B.[101] Surveyed below are selected examples of total syntheses where the Stetter reaction plays an important role in establishing some aspect of the stereochemistry of the target molecule and in which the use of a chiral NHC catalyst may be potentially advantageous. Related studies that demonstrate relevant aspects of Stetter chemistry in target-directed synthesis, such as solving chemoselectivity issues, are also highlighted.

A diastereoselective intramolecular Stetter reaction is a critical component of a synthesis of the racemate of the antibiotic platensimycin (Scheme 47).[102] In this case, treatment of a prochiral, aldehyde-containing cyclohexadienone with a simple, achiral, triazolium salt NHC precursor results in a highly diastereoselective cyclization that correctly installs the desired relative stereochemical relationship between vicinal tertiary and quaternary carbon atoms. Based upon the successful enantioselective desymmetrization of related model dienones (see Scheme 31),[75] the prospects are excellent for rendering this particular intramolecular Stetter reaction enantioselective by using a typical scalemic fused polycyclic triazolylidene catalyst (e.g., **TrM1d**).

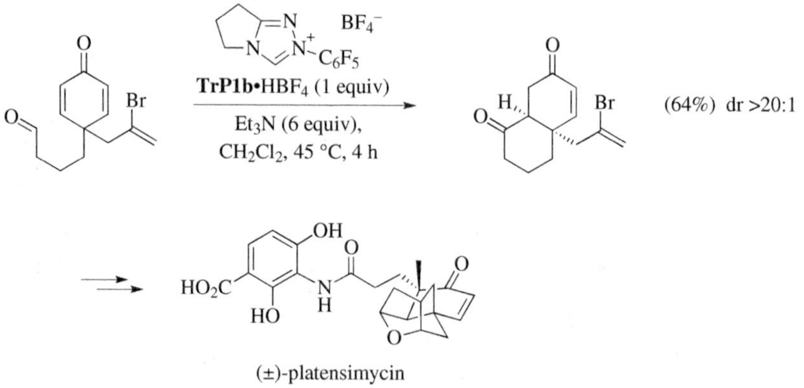

BF_4^- C_6F_5
TrP1b•HBF_4 (1 equiv)
Et_3N (6 equiv), CH_2Cl_2, 45 °C, 4 h
(64%) dr >20:1

(±)-platensimycin

Scheme 47

In a more recent synthesis of the closely related natural-product isolate platencin, an attempt to effect an intramolecular Stetter reaction from a scalemic cyclohexenone containing pendant aldehyde and ketone functionalities resulted in the predominant formation of the unwanted benzoin reaction product (Scheme 48).[103] Hence, the desired 1,4-diketone-bridged product is generated instead from the same substrate using a mercaptan-mediated acyl-radical cyclization.

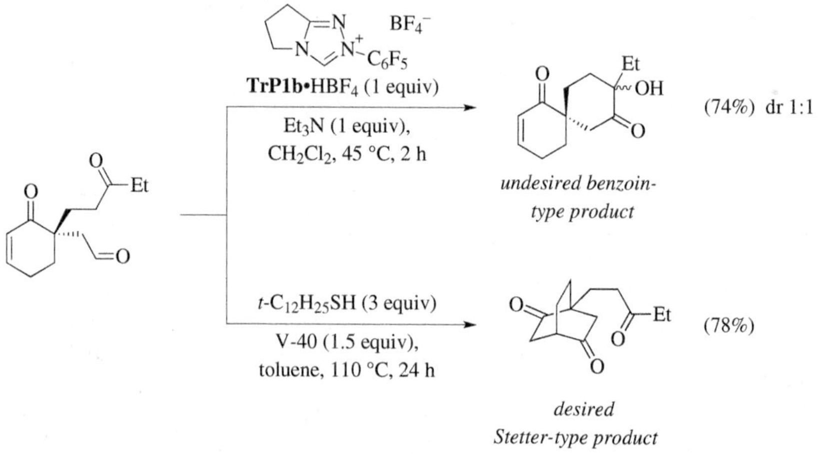

BF_4^- C_6F_5
TrP1b•HBF_4 (1 equiv)
Et_3N (1 equiv), CH_2Cl_2, 45 °C, 2 h
(74%) dr 1:1
undesired benzoin-type product

t-$C_{12}H_{25}SH$ (3 equiv)
V-40 (1.5 equiv), toluene, 110 °C, 24 h
(78%)
desired Stetter-type product

Scheme 48

Two of the stereogenic centers present in the antimalarial agent (–)-dihydroartemisinic acid are generated from the enoate, which is obtained in one step from

(*R*)-citronellal by an intramolecular Stetter reaction mediated by a triazolylidene catalyst (Scheme 49).[104] In this example, the use of an enantiopure chiral catalyst ensures high diastereoselectivity. In related model systems leading to five-membered rings, catalyst stereocontrol dominates in the cyclization of enoate-based substrates that have substituents in the β- or γ-position of the aldehyde (see Scheme 29).[74]

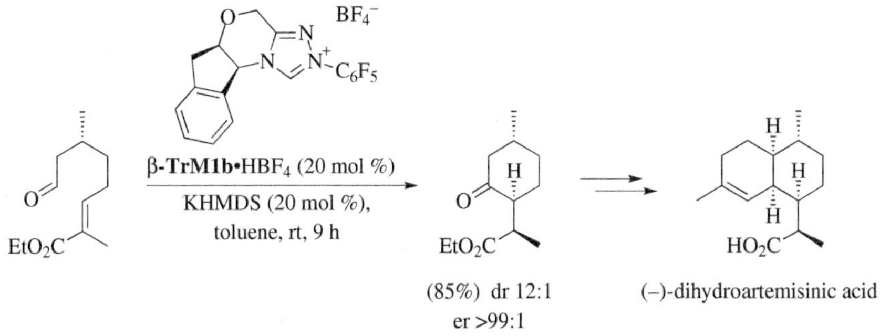

BF_4^- C_6F_5

β-TrM1b•HBF$_4$ (20 mol %)

KHMDS (20 mol %), toluene, rt, 9 h

EtO_2C → EtO_2C → HO_2C

(85%) dr 12:1 er >99:1

(–)-dihydroartemisinic acid

Scheme 49

The enantioselective intramolecular Stetter reaction of suitably linked substrates can provide an efficient entry to otherwise difficult-to-obtain spirocyclic quaternary stereocenters. For example, in the course of a model study directed at the synthesis of spirofuranone-lactam-containing natural products, such as FD-838 and cephalimysin A,[105] the spirocyclization of a maleimide with a tethered aldehyde proceeds under what are now quite standard conditions, using catalyst **TrM1b**, to furnish the expected product in high yield and with excellent enantioselectivity (Scheme 50).[106] Although such reactions are usually successful, occasional failures have been noted. For example, the same catalyst failed to provide a spirocyclic diketone en route to the biologically active tetralone (–)-nidemone. However, the requisite intermediate was successfully generated in racemic form using an achiral thiazolium salt precatalyst and was subsequently advanced to (±)-nidemone (Scheme 51).[107]

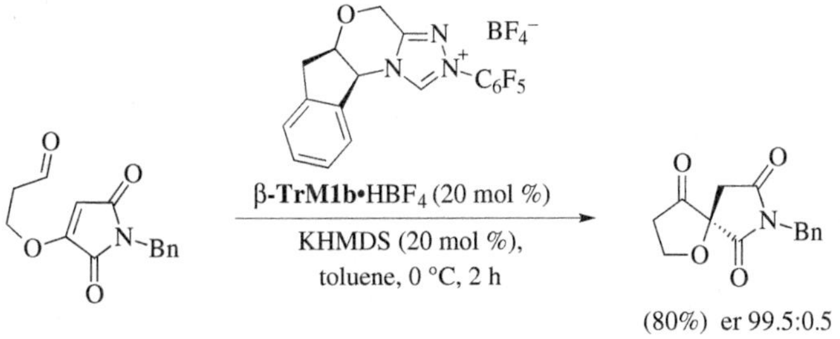

BF_4^- C_6F_5

β-TrM1b•HBF$_4$ (20 mol %)

KHMDS (20 mol %), toluene, 0 °C, 2 h

N–Bn → N–Bn

(80%) er 99.5:0.5

Scheme 50

Finally, building on the promise of the model study in Scheme 50,[106] elaborations of three diastereomers of the potent cytotoxic agent (–)-cephalimysin A

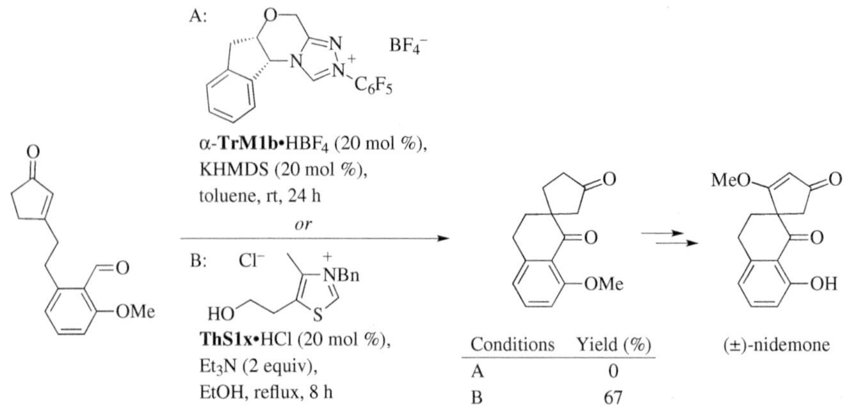

A: α-**TrM1b**•HBF_4 (20 mol %), KHMDS (20 mol %), toluene, rt, 24 h

or

B: **ThS1x**•HCl (20 mol %), Et_3N (2 equiv), EtOH, reflux, 8 h

Conditions	Yield (%)
A	0
B	67

(±)-nidemone

Scheme 51

(including 9-epi-cephalimysin A) were successfully realized using an enantioselective intramolecular Stetter reaction to set the absolute configuration of the key spirocenter (Scheme 52).[108] α-Substituted aldehydes are challenging substrates, and this effort was further complicated by the fact that the requisite donor aldehyde **33**, itself a vinylogous ester, could not be directly accessed in the (*Z*)-configuration required for the Stetter cyclization. To solve these problems, a tandem alkene-photoisomerization–Stetter sequence was developed using an NHC catalyst (**TrM1i**) specifically tailored to more effectively process the transient photoisomerization-generated substrate molecule, (*Z*)-**33**. Interestingly, the presence of an *N*-3,5-bis(trifluoromethyl)phenyl substituent in catalyst β-**TrM1i** leads to a spirocyclic product of opposite absolute configuration to that obtained with an otherwise identical NHC catalyst bearing an *N*-pentafluorophenyl substituent (β-**TrM1b**).

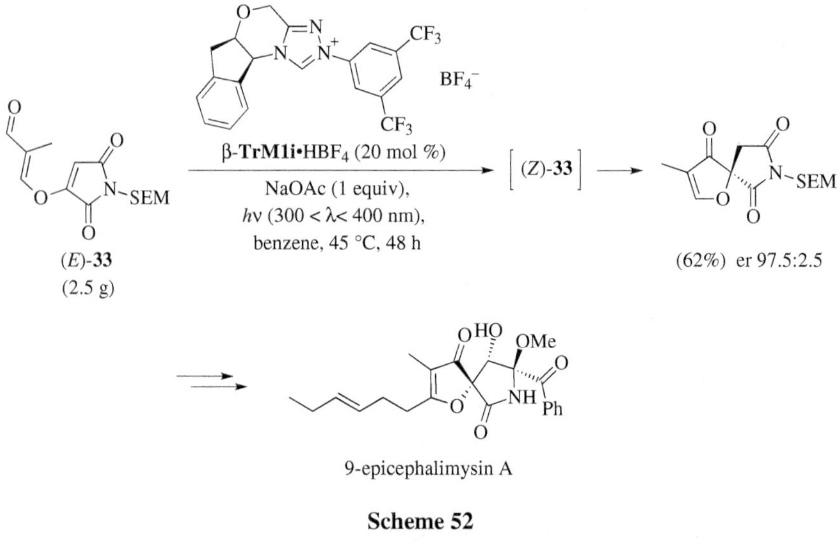

(*E*)-**33** (2.5 g)

β-**TrM1i**•HBF_4 (20 mol %), NaOAc (1 equiv), $h\nu$ ($300 < \lambda < 400$ nm), benzene, 45 °C, 48 h

[(*Z*)-**33**]

(62%) er 97.5:2.5

9-epicephalimysin A

Scheme 52

COMPARISON WITH OTHER METHODS

The enantioselective Stetter reaction is a straightforward and effective way to access scalemic 1,4-dicarbonyl compounds and related systems (such as β-nitroketones, etc.) from simple and readily available starting materials. Nonetheless, the Stetter reaction does not always furnish the desired products due to substrate-specific reactivity issues and, as with any given method, an acceptable level of stereocontrol may not be achieved with a particular transformation. As far as replicating the net outcome of a generic (non-enantioselective) Stetter process is concerned, one can conceive of numerous alternative reagents and reaction sequences that could achieve the same overall transformation. For example, as illustrated above (Scheme 48),[103] acyl radical additions to enones can be applied to some Stetter reaction substrates to furnish the originally intended product. Alternatively, samarium(II) iodide mediated reductive coupling of an aldehyde and an enone, followed by oxidation of the resulting γ-hydroxyketone, will also provide a 1,4-diketone from the standard Stetter starting materials.[109] Absent from many of the perhaps more obvious alternate strategies that can counterfeit the Stetter reaction is a capacity to achieve reagent-controlled enantio- and diastereoselectivity. This section surveys viable alternatives to the Stetter reaction that can likewise be used to access non-racemic 1,4-dicarbonyl compounds from comparable starting materials. Pertinent examples of methods that achieve the hydroacylation of alkenes in a more general sense are also included for comparison.

Enzymatic Stetter-Like Reactions with α-Keto Acid Donors

Although enzyme-catalyzed enantioselective benzoin reactions are well known,[7] true Stetter processes (i.e., with aldehydes as donors) are yet to be documented in a biological pathway, and a biocatalysis variant of the actual Stetter reaction has not been reported to date. Nevertheless, there are close enzymatic analogues of the Stetter reaction that involve α-keto acids and these can be exploited for synthetic purposes. For example, the thiamine diphosphate (ThDP)-dependent enzyme PigD is speculated to catalyze decarboxylation of pyruvate to afford a thiazolylidene-based Breslow intermediate which adds in a 1,4-fashion to a 2-octenoyl-ACP (or CoA) thioester, leading to a 3-acetyloctanoyl system en route to the red pigment prodigiosin.[110] Biocatalysis experiments using the isolated PigD enzyme in the presence of ThDP fail to provide Stetter-type products from pyruvate and conjugated enals; cross-benzoin adducts are obtained instead. However, the desired 1,4-addition process occurs with α,β-unsaturated ketones as acceptors (Scheme 53).[90,111] Enones flanked by either alkyl or aryl substituents are successfully transformed by this biocatalysis platform to give the same 1,4-diketone products that would be derived from a Stetter reaction using acetaldehyde. Although the efficiency of the reaction is low, the high level of enantioselectivity that it offers easily surpasses that delivered by comparable experiments involving NHC-catalyzed additions of acetaldehyde (see Scheme 43).[89] More recently, new 'Stetterase' enzymes have been identified that transfer an acyl moiety from α-ketoglutarate (analogous to the pyruvate reaction with PigD), but which offer increased acceptor-alkene substrate scope.[112] With regard to the range of acyl groups that can be transferred, these Stetter-like biocatalytic reactions remain less versatile than their chiral NHC-catalyzed counterparts.

CH_3COCO_2Na (1.25 equiv) + $R^1CH{=}CHCOR^2$ → PigD (0.066 mol %), ThDP (4 mol %), FAD (0.2 mol %), DMSO–H_2O, 30 °C, 24–40 h → $R^1CH(COCH_3)CH_2COR^2$

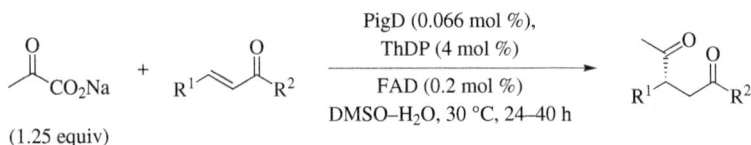

R^1	R^2	Yield (%)	er
Ph	Ph	3	>99:1
2-(HO)C_6H_4	Ph	39	97:3
Ph	Me	7	97.5:2.5
n-C_5H_{11}	Me	38	>99:1

Scheme 53

Stetter-Like Reactions of Acylsilanes

With their facility to undergo Brook rearrangements following the addition of nucleophiles, acylsilanes are natural substrates to explore Umpolung-type reactivity since a carbanionic nucleophile emerges at the site of the initially electrophilic carbonyl group.[113] Accordingly, acylsilanes have found some applications in Stetter-like chemistry, although any such transformations are necessarily more circuitous than actual Stetter reactions because few acylsilanes are commercially available, and some effort is typically required for their preparation. A non-enantioselective intermolecular 'sila-Stetter' process catalyzed by a classic thiazolium salt precatalyst gives good yields of 1,4-diketones from chalcones and aryl acylsilanes in the presence of isopropyl alcohol.[114] While the products emerging from this process are unremarkable, and the use of chiral catalysts has not been reported, it is notable that this transformation is not plagued by competing benzoin-like reactions. An enantioselective 'sila-Stetter' reaction variant catalyzed by lithiated chiral phosphites offers an opportunity to engage acceptor alkenes that are not electrophilic enough to participate in normal intermolecular Stetter reactions (Scheme 54).[115] Thus, β-substituted acrylamides are viable acceptor alkenes, and examples of both aryl- and alkyl-substituted substrates give products in reasonable yields and with high enantioselectivities. The nature of the aryl group in the acylsilane is important, and the electron-rich 4-anisyl group present in the illustrated substrate is crucial for good performance. Consequently, the replacement of the 4-anisyl group with a simple phenyl group results in significantly lower yields (typically by 30–40%).

This interesting reaction presumably proceeds via an α-silyloxy α-lithiated phosphonate that is formed by a [1,2]-Brook rearrangement of the tetrahedral intermediate generated by attack of the metallophosphite on the acylsilane (Scheme 55).[115] Because the lithiated phosphonate is significantly more nucleophilic than the neutral Breslow intermediates of the Stetter reaction, it is capable of adding to the enamide electrophile. Following conjugate addition, a retro-[1,4]-Brook rearrangement furnishes a lithiated *P*,*O*-acetal, which collapses to close the catalytic

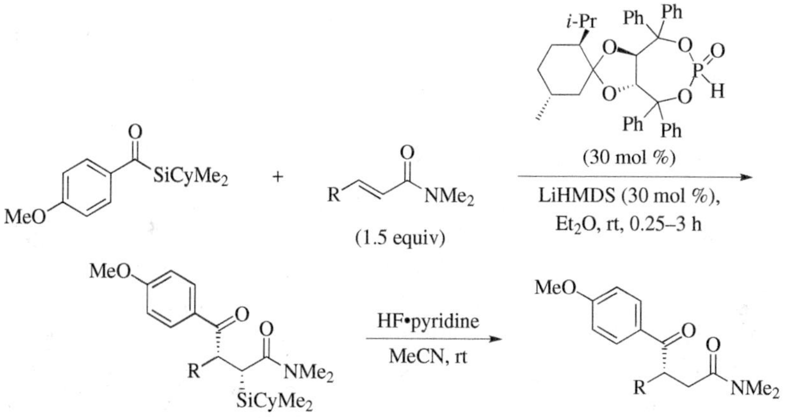

R	Yield (%)	er	R	Yield (%)	er
Ph	68	95:5	2-naphthyl	66	94.5:5.5
4-$MeOC_6H_4$	63	96:4	2-furyl	15	62:38
4-ClC_6H_4	66	97.5:2.5	Me	56	93:7
4-$CF_3C_6H_4$	80	95:5	Et	82	85.5:14.5

Scheme 54

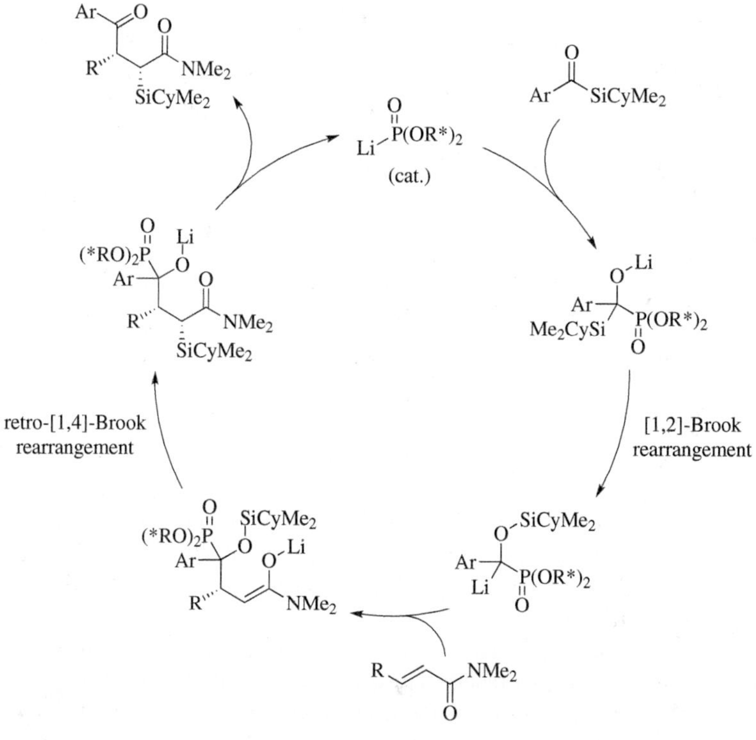

Scheme 55

cycle. As shown, the initial product of this scheme is actually a stereodefined α-silyl amide. However, the silyl group is typically removed by the subsequent treatment with fluoride.

NHC-Catalyzed Hydroacylation of Unactivated Alkenes

The Stetter reaction relies on the conjugate addition of an acyl anion equivalent to an acceptor alkene (Scheme 1). This transformation belongs to a wider collection of alkene hydroacylation processes that involve the apparent or actual insertion of a C=C bond into the formyl C–H bond of an aldehyde (i.e., the formal addition of C and H across C=C). In the case of the Stetter reaction, the alkene must be a conjugate-addition acceptor. However, alternative mechanistic pathways are available in certain scenarios to achieve the same overall result, but without the intervention of a conjugate addition step. Transition-metal catalysis affords one such possibility (see the next section) and, remarkably, NHC catalysis can provide another, as long as the familiar Breslow intermediate engages with the alkene in an unorthodox manner. The reactivity invoked here, which advances the Breslow intermediate directly to the penultimate Stetter intermediate in a single step (cf., **4** to **6** in Scheme 2), is likely a concerted, asynchronous, pericyclic group-transfer reaction (thermally allowed as $\omega 2s + \sigma 2s + \pi 2s$) that has as its closest parallel a retro-Cope elimination (Scheme 56).[116] This interesting behavior was first observed for the intramolecular hydroacylation of enol ethers,[117,118] and it has since been rendered enantioselective[116] and extended to encompass electron-neutral alkenes[119] as well as other electron-rich alkenes.[120] Some intermolecular variants are also known.[120,121] The canonical family of fused polycyclic triazolylidene NHC catalysts again provide for optimal reactivity and enantioselectivity.

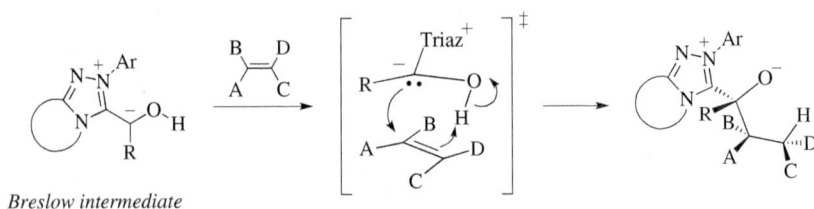

Breslow intermediate

Scheme 56

Accessing the concerted NHC-catalyzed addition pathway generally requires higher reaction temperatures (80–140 °C) than those typically needed for the classical Stetter reactions involving acceptor alkenes (0–40 °C). Nonetheless, the intramolecular variant is a versatile process that can provide excellent enantioselectivities (Scheme 57).[119] The cyclization is successful even with only simple alkyl tethers linking the two reacting unsaturated moieties. Furthermore, five- and six-membered rings are both accessible from either aliphatic or aromatic aldehydes. The only significant limitation is the fact that product ketones with enolizable α-stereogenic centers are racemized by the high temperature and basic reaction conditions (e.g., R^1 = H gives 50:50 er). Thus, quaternary, but not ternary, stereocenters can be established using this chemistry. To date, intermolecular NHC-catalyzed hydroacylation has been achieved with only a small selection of rather unique alkene classes, specifically, cyclopropenes[121] and, as illustrated, α-amidostyrenes (Scheme 58).[120] Surprisingly, the latter process occurs at ambient temperature.

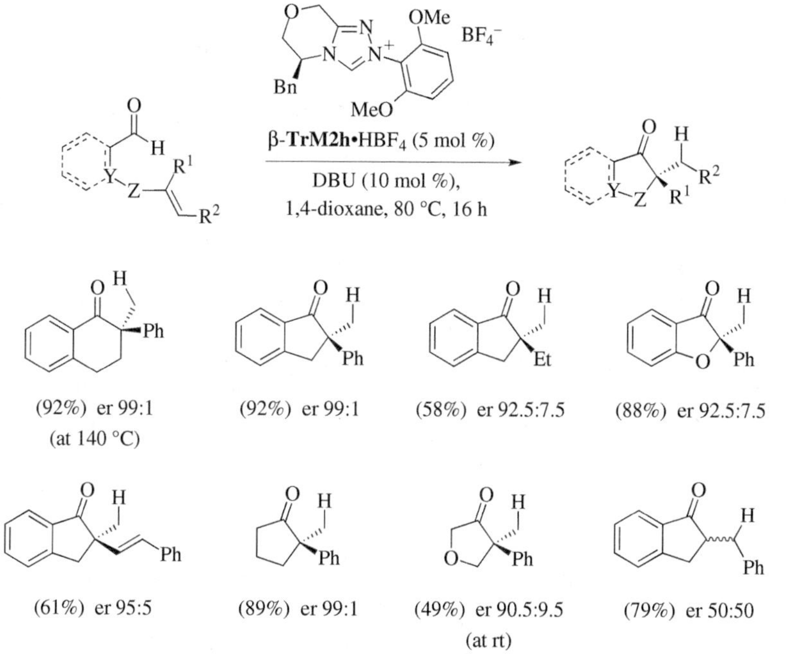

β-**TrM2h**•HBF_4 (5 mol %)

DBU (10 mol %), 1,4-dioxane, 80 °C, 16 h

(92%) er 99:1 (at 140 °C)

(92%) er 99:1

(58%) er 92.5:7.5

(88%) er 92.5:7.5

(61%) er 95:5

(89%) er 99:1

(49%) er 90.5:9.5 (at rt)

(79%) er 50:50

Scheme 57

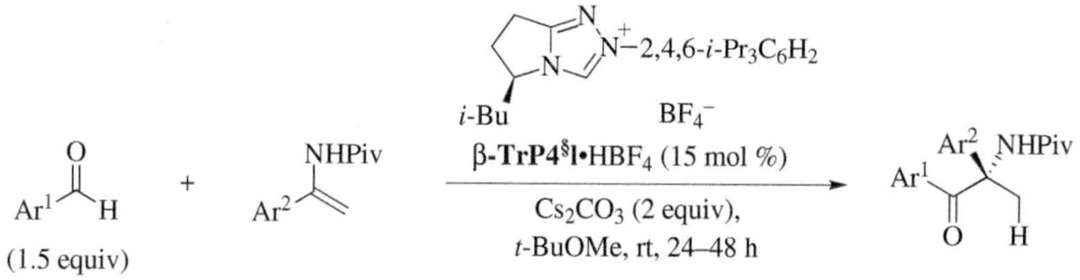

β-**TrP4**[§]**1**•HBF_4 (15 mol %)

Cs_2CO_3 (2 equiv), *t*-BuOMe, rt, 24–48 h

(1.5 equiv)

Ar^1	Ar^2	Yield (%)	er
4-FC_6H_4	Ph	98	95:5
4-$MeOC_6H_4$	Ph	73	95:5
2-furyl	Ph	95	98:2
3-pyridyl	Ph	72	96:4

Ar^1	Ar^2	Yield (%)	er
4-ClC_6H_4	4-FC_6H_4	97	94:6
4-ClC_6H_4	4-$MeOC_6H_4$	98	94.5:5.5
4-ClC_6H_4	2-furyl	98	97:3
4-ClC_6H_4	3-pyridyl	93[a]	97.5:2.5

[a] 2 equivalents of the aldehyde were used.

Scheme 58

Metal-Catalyzed Hydroacylation of Alkenes

A variety of methods based on metal-catalyzed (usually rhodium-catalyzed) hydroacylation of alkenes are available for the enantioselective synthesis of 1,4-dicarbonyl compounds, and these transformations largely complement the Stetter reaction in scope. Because many kinds of coordinated alkenes readily undergo migratory insertion into metal–carbon σ-bonds, the metal-catalyzed process is not restricted to acceptor alkenes. Nonetheless, problems associated with competing decarbonylation can limit the hydroacylation process, particularly in the arena of intermolecular reactions. Herein, the enantioselective metal-catalyzed alkene hydroacylation is briefly compared with similar Stetter reactions; readers requiring a more comprehensive treatment of the topic of hydroacylation are referred to relevant reviews.[122–124]

To date, intramolecular Stetter reactions have only been successfully realized for *exo*-cyclizations that afford five- and six-membered ring systems. These limitations are absent for rhodium-catalyzed hydroacylation, in which both *exo*- and *endo*-type cyclizations have been reported for a wide range of ring sizes (including five-, six-, seven-, and eight-membered rings).[125,126] For example, scalemic indanones can be prepared through the treatment of 2-vinyl benzaldehydes with a rhodium(I)–BINAP complex (Scheme 59).[127] The mechanism for this transformation consists of oxidative addition of the chiral rhodium(I)–BINAP complex into the formyl carbon–hydrogen bond and subsequent hydrometallation of the pendent alkene (i.e., migratory insertion), followed by reductive elimination of the resulting metallocycle to release the product and regenerate the active catalyst. A variety of substituents are tolerated on the alkene, and there is no need for an electron-withdrawing group, although 1,4-dicarbonyl compounds are also accessible (e.g., R = CO_2Et). More recently, enantioselectivity has been achieved in the same process by using a chiral secondary amine organocatalyst as the source of stereoinduction in cooperation with a simple achiral rhodium complex.[128]

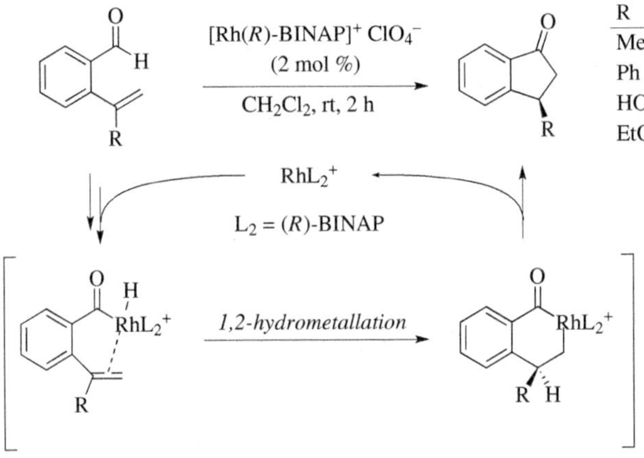

[Rh(*R*)-BINAP]$^+$ ClO_4^- (2 mol %), CH_2Cl_2, rt, 2 h

R	Yield (%)	er
Me	97	>99:1
Ph	98	99:1
$HO(CH_2)_2$	97	98:2
EtO_2C	89	98:2

RhL_2^+

L_2 = (*R*)-BINAP

1,2-hydrometallation

Scheme 59

In the case of the intermolecular rhodium-catalyzed alkene hydroacylation, decarbonylation of the acylrhodium intermediate following the initial oxidative addition event is a major challenge to the wider implementation of this otherwise highly versatile chemistry.[129] The unwanted retro-α-migratory insertion of a CO group, which leads to catalytically inert rhodium carbonyl complexes, can in some cases be circumvented by the incorporation of an additional ligating substituent into one of the substrates. This so called 'chelation assistance' serves to maintain a high level of coordinative saturation about the metal center and thereby disfavors decarbonylation (which causes both the coordination number and the formal electron count of the metal center to increase).

In a pertinent example, α-substituted enamides are hydroacylated with high enantioselectivity by aliphatic aldehydes in the presence of a rhodium–QuinoxP* complex (Scheme 60).[130] The corresponding reaction with aryl aldehydes performs poorly (yields <40%, er <85:15), and this process is also unsatisfactory with differently substituted enamides (e.g., *N,N*-diphenylcrotonamide gives a racemic product). The intermolecular Stetter reaction is broadly complementary in such cases since it gives good results with aryl aldehydes, albeit simple enamide substrates lack the necessary electrophilicity for reactivity and need to be replaced by, for example, unsaturated α-keto esters (see Scheme 35). Notwithstanding the occasionally more limited aldehyde scope, the tolerance of metal-catalyzed hydroacylation for such a wide variety of different alkene types is arguably the main advantage that it offers over comparable Stetter reactions. Furthermore, intermolecular hydroacylations of allenes and alkynes are also successful, providing that chelation assistance is available.[122,131,132]

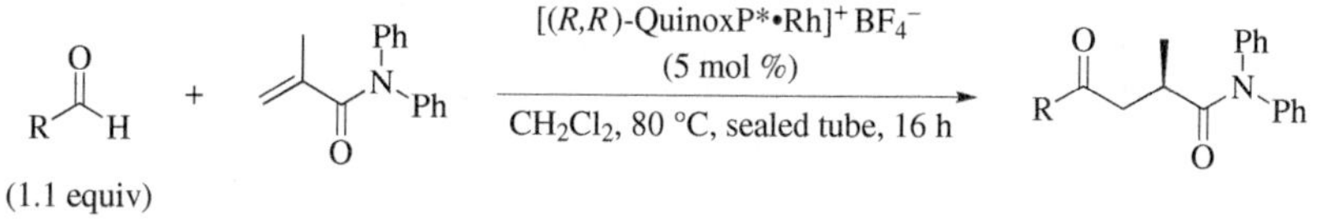

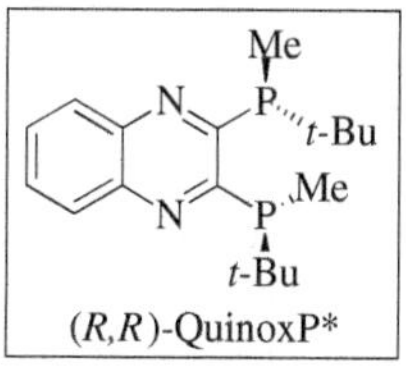

RCHO (1.1 equiv) + $CH_2=C(Me)C(O)NPh_2$ → $[(R,R)\text{-QuinoxP*}\bullet\text{Rh}]^+ BF_4^-$ (5 mol %), CH_2Cl_2, 80 °C, sealed tube, 16 h → $RC(O)CH_2CH(Me)C(O)NPh_2$

(*R,R*)-QuinoxP*

R	Yield (%)	er
n-Pr	79	99:1
$Ph(CH_2)_2$	87	99:1
i-Bu	74	99:1
c-C_6H_{11}	76	99:1
t-Bu	0	—
Ph	12	79:21

Scheme 60

Photo-Mediated Acyl Radical Additions

Advances in methods to generate acyl radicals via visible-light-mediated photochemistry have recently been combined with stereocontrol paradigms from the realms of organocatalysis and metal catalysis to provide new enantioselective approaches to 1,4-dicarbonyl compounds from acceptor alkenes. To date, methods of this kind require at least one specialized component to enable a stereoselective intermolecular Giese-type radical addition reaction. Nonetheless, these methods are potentially more versatile than current state-of-the-art intermolecular enantioselective Stetter reactions. For example, acyl radicals produced via direct visible light irradiation of 4-acyl-1,4-dihydropyridines add to enal acceptors that have been activated in situ as iminium ions by condensation with chiral secondary amine organocatalysts (Scheme 61).[133] Although this process offers somewhat modest levels of enantio-selectivity, it is notable that both alkyl and aryl substituents are tolerated at variable sites on the two substrates. A closely related and similarly efficient process based on cooperative catalysis employs enals and the same organocatalyst, but instead deploys acyl radicals generated by photoredox-catalyst-mediated decarboxylation of α-keto acids.[134]

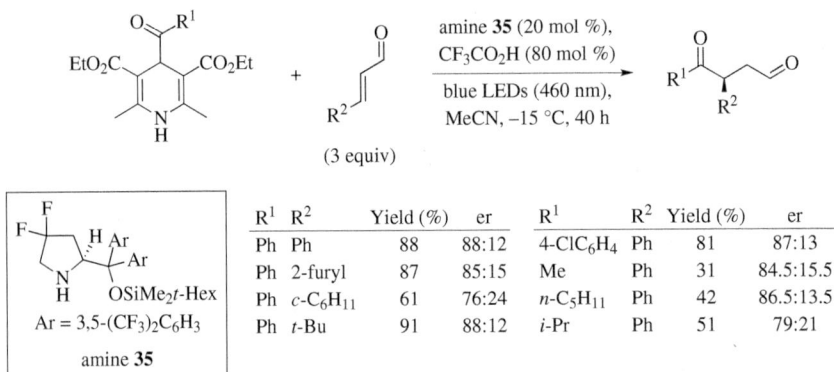

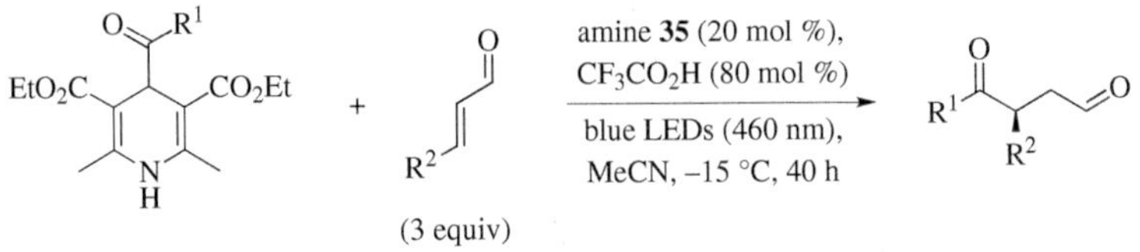

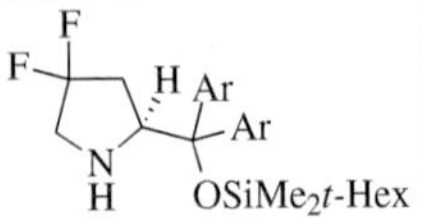

amine **35** (20 mol %), CF_3CO_2H (80 mol %), blue LEDs (460 nm), MeCN, –15 °C, 40 h

(3 equiv)

$Ar = 3,5\text{-}(CF_3)_2C_6H_3$

amine **35**

R^1	R^2	Yield (%)	er	R^1	R^2	Yield (%)	er
Ph	Ph	88	88:12	4-ClC_6H_4	Ph	81	87:13
Ph	2-furyl	87	85:15	Me	Ph	31	84.5:15.5
Ph	*c*-C_6H_{11}	61	76:24	*n*-C_5H_{11}	Ph	42	86.5:13.5
Ph	*t*-Bu	91	88:12	*i*-Pr	Ph	51	79:21

Scheme 61

In a second example of cooperative catalysis, aldehydes are coupled with enam-ides to furnish 1,4-dicarbonyl compounds in modest yields but with excellent levels of enantioselectivity (Scheme 62).[135] This process is particularly notable because the acyl radicals are generated directly from aldehydes by hydrogen-atom transfer using the photoexcited state of the neutral form of the pigment eosin Y. Furthermore, both alkyl and aryl aldehydes are assimilated with comparable facility. A two-point bind-ing interaction is critical between the chiral-at-metal rhodium(III)-catalyst and the acceptor alkene, and hence, it is essential that the illustrated type of pyrazolyl amide is employed. It is doubtful that any of the triazolylidene catalysts currently avail-able would be capable of coupling aldehydes with these types of enamides via an intermolecular Stetter reaction.

Cα-Cβ Disconnection-Based Approaches to 1,4-Dicarbonyl Compounds

The above methods for 1,4-dicarbonyl compound synthesis all rely on the addition of a C1 (aldehyde or aldehyde-type) synthon to a C3 partner. A more convergent entry to 1,4-dicarbonyl compounds involves the combination of two C2 synthons such that the target system is retrosynthetically disconnected at the Cα–Cβ bond.

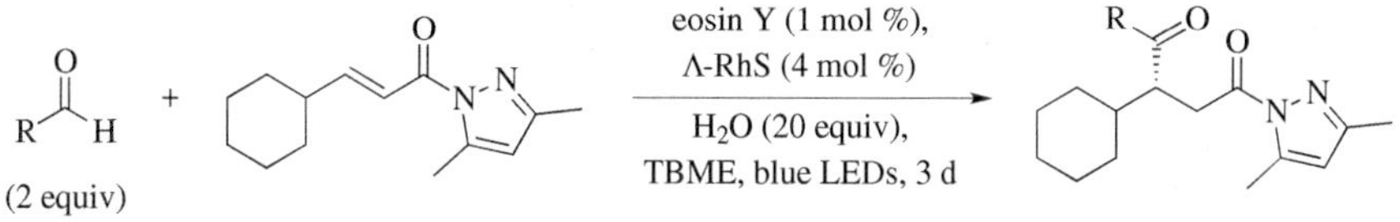

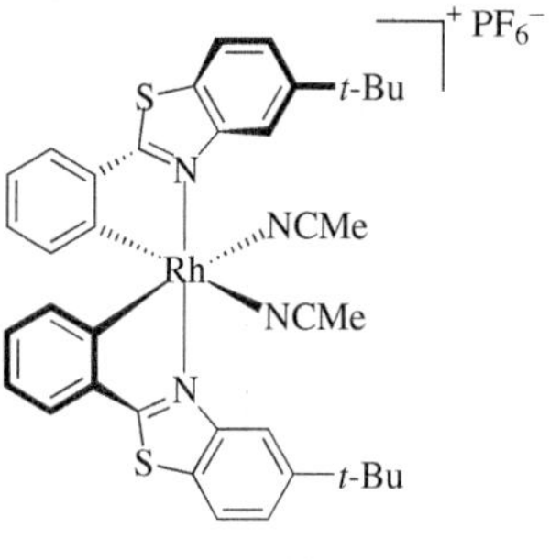

R	Temp (°C)	Yield (%)	er
Me	10	45	91.5:8.5
n-Bu	10	57	95:5
i-Bu	10	53	95:5
i-Pr	10	63	95:5
Ph	30	44	99.5:0.5
4-$MeOC_6H_4$	30	46	95.5:4.5
4-FC_6H_4	30	32	93:7
4-ClC_6H_4	30	32	88:12

Scheme 62

The most commonly explored tactic to implement such a 'C2 + C2' strategy is the oxidative coupling of two enolates, and recent advances have led to such processes being viable alternatives to traditional methods for the construction of 1,4-dicarbonyl compounds.[136] Oxidative enolate coupling was initially limited to homocouplings that form symmetrical diketones and diesters, but the intermolecular cross-coupling of two electronically dissimilar enolates is also possible, and non-symmetrical 1,4-dicarbonyl products are prepared with high efficiency. Most asymmetric methods of this type rely on a chiral auxiliary to control stereoinduction, and not all newly generated stereogenic centers are installed with good selectivity. For example, lithium enolates derived from acylated Evans-type oxazolidinones are efficiently oxidatively cross-coupled with the lithium enolates of esters to afford 1,4-dicarbonyl products with essentially complete control of the α-stereocenter but only modest stereoselectivity at the β-center (Scheme 63).[137–139] Interestingly, the use of an iron(III) oxidant, as depicted, leads to the opposite sense of α-diastereoselectivity as compared to results obtained with a copper(II)-based oxidant. A related chiral-auxiliary-based approach has been recently reported to control quaternary α-stereocenters via oxidative enolate coupling of a cyclic alanine derivative.[140]

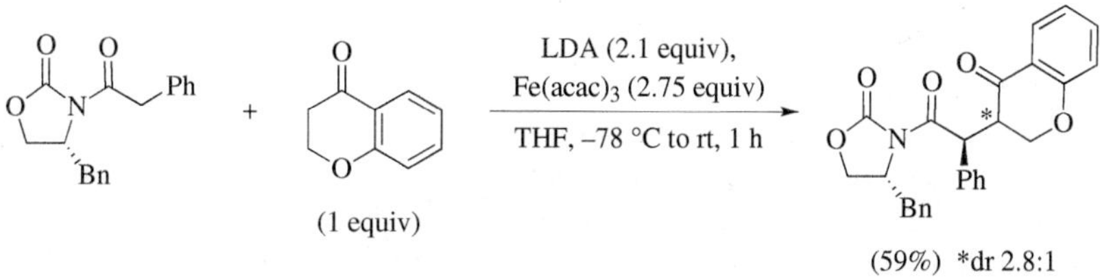

Scheme 63

Despite the progress on diastereoselective, intermolecular, oxidative enolate coupling reactions, the catalytic enantioselective variant has been less studied. Silyl enol ethers can be coupled with enamines generated in situ from aldehyde substrates and a chiral secondary amine organocatalyst to generate substituted γ-keto aldehydes with high enantioselectivity (Scheme 64).[141] This effective organocatalytic enantioselective platform relies on singly occupied molecular orbital (SOMO) activation of the enamine intermediate (i.e., initial single-electron oxidation) such that the resulting radical cation can add to the silyl enol ether in a bond-forming event. More recently, a comparable method for the enantioselective homocoupling of α-aryl propionaldehydes has been reported that provides a synthesis of C_2-symmetric 1,4-dialdehydes bearing contiguous stereodefined quaternary α- and β-position carbon atoms.[142]

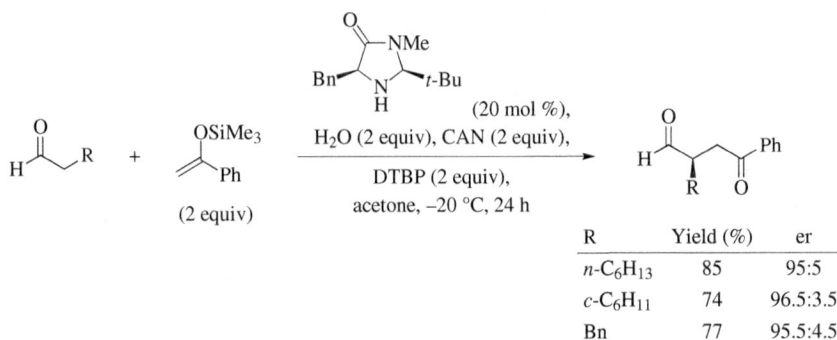

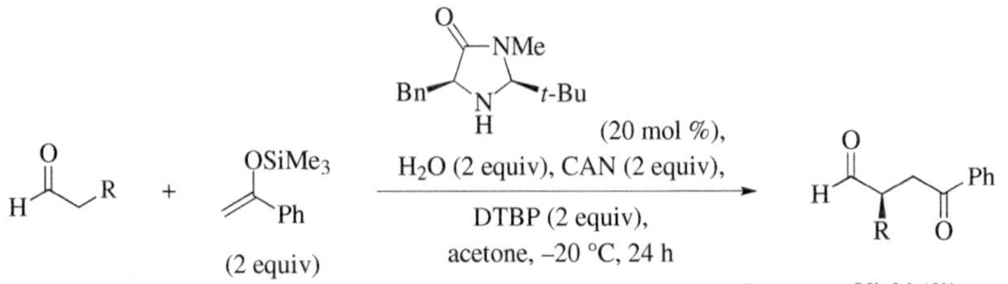

O NMe Bn N t-Bu H (20 mol %), H_2O (2 equiv), CAN (2 equiv), DTBP (2 equiv), acetone, –20 °C, 24 h

$OSiMe_3$ Ph (2 equiv)

R	Yield (%)	er
n-C_6H_{13}	85	95:5
c-C_6H_{11}	74	96.5:3.5
Bn	77	95.5:4.5

Scheme 64

Finally, an unorthodox approach to 1,4-dicarbonyl compounds based on the [3,3]-sigmatropic rearrangement of cationic *S,O*-divinyl sulfenate species has recently emerged that provides for the stereospecific installation of α- and β-position stereogenic centers, one of which can be quaternary (Scheme 65).[143] The pivotal oxasulfonium reactive intermediate **36** is accessed by protic activation of an *N*-acyl ynamide in the presence of a scalemic vinyl sulfoxide. As illustrated, by virtue of the chair-like transition state typical for [3,3]-sigmatropic rearrangements, the absolute and relative configuration of the 1,4-dicarbonyl product is determined by the stereochemical attributes of the sulfoxide (i.e., (*R*)- or (*S*)-configuration at the sulfur atom and (*E*)- or (*Z*)-configuration of the alkene).[143] Good-to-high chemical yields are realized with excellent stereochemical fidelity and high diastereoselectivity toward *syn*- or *anti*-type contiguous stereodiads.

Overall, the methods highlighted in this section are attractive alternatives to the intermolecular Stetter reaction and are particularly useful in scenarios where the latter process would fail because of the necessity to deploy a less activated and/or a bulkier acceptor alkene. Nonetheless, both the enolate oxidative coupling and the sulfoxide-based chemistry described are intrinsically less atom-economical compared to the Stetter processes, and they are operationally more complicated to implement. In addition, precursor synthesis is non-trivial in the case of the latter

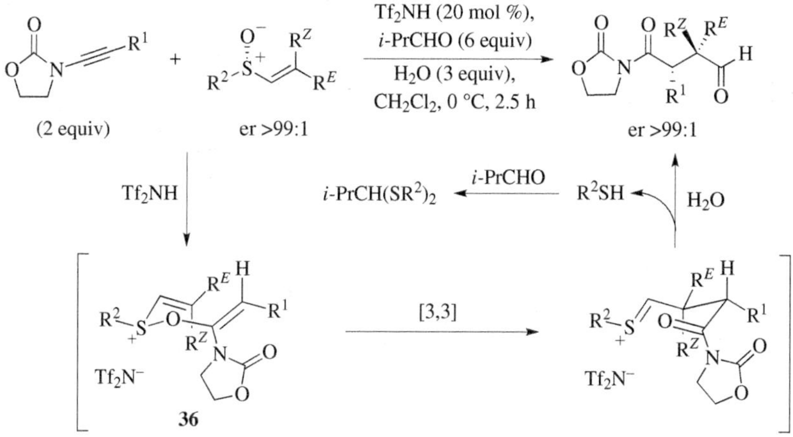

R^1	R^2	R^Z	R^E	Yield (%)	dr
$Ph(CH_2)_3$	4-Tol	H	*n*-Bu	85	8:1
c-C_6H_{11}	4-Tol	H	*n*-Bu	78	8:1
$Ph(CH_2)_3$	*n*-C_8H_{17}	*n*-Bu	H	75	8:1
c-C_6H_{11}	*n*-C_8H_{17}	*n*-Bu	H	80	8:1
n-Bu	*n*-C_8H_{17}	Me	*n*-Bu	60	6:1
n-Bu	*n*-C_8H_{17}	*n*-Bu	Me	79	6:1

Scheme 65

method, and this process relies on stoichiometric enantioenriched starting materials rather than the use of enantioselective catalysis.

EXPERIMENTAL CONDITIONS

The enantioselective Stetter reaction is operationally straightforward to perform: the reactions are usually conducted with standard laboratory equipment and glassware, and no special techniques are required to obtain satisfactory results. Likewise, procedures to prepare the typical azolium salt precatalysts are robust and are not problematic for an experienced synthetic chemist to execute. A variety of base and solvent combinations have been utilized in reported Stetter reaction experimental procedures. However, regardless of the reagents selected, the presence of air is detrimental to chemical yield, and reactions are best conducted under an inert atmosphere of argon gas.[62] A reagent combination of KHMDS as base, which generates the active NHC catalyst irreversibly from its azolium salt precursor, in toluene is the most commonly used process for intramolecular reactions. Low-polarity solvents generally afford the highest enantioselectivities for intramolecular Stetter reactions, albeit protic solvents are sometimes effective (and lead to inversion of selectivities in some cases),[75,144] and solvent-free conditions using Cs_2CO_3 as the base are also viable for some cyclizations.[145] The intermolecular enantioselective Stetter reaction is carried out in a wide variety of different solvents, and special attention should be given to the correct solvent selection. For example, Stetter reactions with glyoxamide donors give optimal stereoselectivity in low-polarity solvents such as CCl_4, whereas MeOH is the solvent of choice for reactions involving nitroalkene acceptors. The only solvent that has been found to have a consistent deleterious effect on the Stetter reaction is water. Thus, for example, toluene saturated with water leads to no reaction in the prototypical cyclization of **21** to **22** using precatalyst **TrP2a**•HBF_4, which is possibly due to the hydration of the NHC.[62]

EXPERIMENTAL PROCEDURES

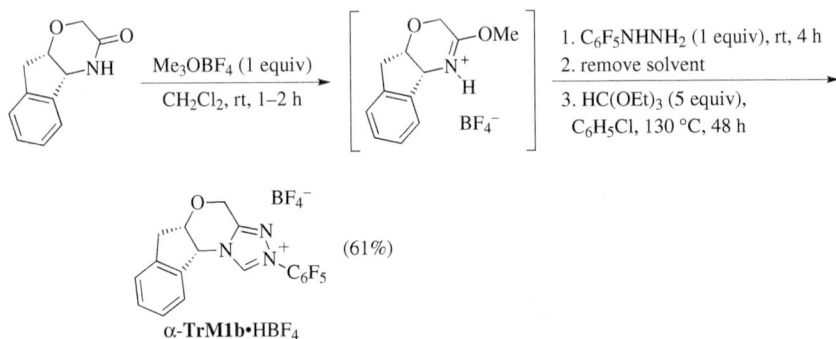

Me_3OBF_4 (1 equiv)
CH_2Cl_2, rt, 1–2 h

1. $C_6F_5NHNH_2$ (1 equiv), rt, 4 h
2. remove solvent
3. $HC(OEt)_3$ (5 equiv), C_6H_5Cl, 130 °C, 48 h

(61%)

α-**TrM1b**•HBF_4

2-Pentafluorophenyl-6,10b-dihydro-4*H*,5a*H*-5-oxa-3,10c-diaza-2-azonia-cyclopenta[*c*]fluorene Tetrafluoroborate (TrM1b•BF_4) [Synthesis of a Precatalyst].[55,56] A flame-dried, 1-L, round-bottomed flask was charged with 4,4a,9,9a-tetrahydro-1-oxa-4-azafluoren-3-one (10.0 g, 52.9 mmol) and a magnetic stirring bar under Ar. Dichloromethane (300 mL) was added, followed by trimethyloxonium tetrafluoroborate (7.82 g, 52.9 mmol). The reaction mixture was stirred at rt until homogeneous (ca. 1–2 h). Pentafluorophenylhydrazine (10.5 g, 52.9 mmol) was added in a single portion and the mixture was stirred for an additional 4 h. The stirring bar was removed, and the solvent was removed in vacuo. The flask was placed under vacuum (2 mm Hg) for 1 h at 100 °C. A magnetic stirring bar was added, followed by chlorobenzene (300 mL) and triethyl orthoformate (19.6 g, 22.0 mL, 132.1 mmol). A reflux condenser was fitted to the flask, and the mixture was heated at 130 °C in an oil bath for 24 h open to the atmosphere. More triethyl orthoformate (19.58 g, 22.0 mL, 132.1 mmol) was added and stirring was continued for a further 24 h at 130 °C. Another portion of triethyl orthoformate (19.6 g, 22.0 mL, 132.1 mmol) was added and stirring was continued for 24 h at the same temperature. The reaction mixture was then cooled to rt, and the solution was poured into a 1-L round-bottomed flask containing toluene (300 mL) and a magnetic stirring bar, resulting in precipitation of the product salt. The reaction flask was rinsed with toluene (50 mL), and the rinse was added to the suspended precipitate. The precipitate slurry was stirred for 10 min, and the product salt was collected by vacuum filtration. The filter cake was washed with toluene (200 mL) and hexane (200 mL). The washed solid was then transferred to a 125-mL Erlenmeyer flask and carefully triturated with a mixture of ethyl acetate (20 mL) and methanol (5 mL) and then stirred for 30 min. The triturated solid was collected by vacuum filtration through a medium fritted funnel, washed with cold ethyl acetate (15 mL), and dried in vacuo (1 h, 2 mm Hg/100 °C) to give the title compound (15.1 g, 32.3 mmol, 61%) as an off-white solid: mp 223–226 °C; TLC R_f 0.22 (CH_2Cl_2/acetone, 3:1); $[\alpha]_D{}^{23}$ + 130.8 (*c* 1.28, MeCN); IR (neat) 3147, 3106, 3028, 2967, 1595, 1530, 1517, 1487, 1461, 1056, 1046, 998 cm^{-1}; ^1H NMR (400 MHz, acetone-d_6) δ 11.09 (br s, 1H), 7.63 (d, J = 7.6 Hz, 1H), 7.43 (q, J = 7.4 Hz, 2H), 7.34 (t, J = 7.3 Hz, 1H), 6.33 (d, J = 4.0 Hz, 1H), 5.39 (d, J = 16.4 Hz, 1H), 5.25 (d, J = 16.4 Hz, 1H), 5.19 (t, J = 4.5 Hz, 1H), 3.55 (dd, J = 17.1, 4.9 Hz, 1H), 3.28 (d, J = 17.2 Hz, 1H); ^{13}C NMR (100 MHz, acetone-d_6) δ 152.5, 147.1, 141.7, 136.2, 130.4, 128.1, 126.4, 125.2, 78.2, 63.5, 60.8, 37.9; HRMS–APCI (*m/z*): M^+ calcd for $C_{18}H_{11}N_3OF_5$, 380.0817; found, 380.0816.

α-**TrM1c**•$HClO_4$ (20 mol %)
KHMDS (17 mol %), xylenes, rt, 24 h

21 → **22** (94%) er 99:1

Ethyl (*R*)-(4-Oxochroman-3-yl)acetate [Enantioselective Intramolecular Stetter Reaction of a Salicylaldehyde-Derived Substrate].[63] An oven-dried, 10-mL, round-bottomed flask equipped with a magnetic stirring bar was charged with triazolium salt precatalyst α-**TrM1c**•$HClO_4$ (8 mg, 0.024 mmol) and xylenes (4.0 mL) and then was purged with N_2 gas. KHMDS (0.040 mL, 0.50 M in toluene, 0.020 mmol) was added, and the mixture was stirred at rt for 15 min. A solution of ethyl (*E*)-4-(2-formylphenoxy)-2-butenoate (28 mg, 0.120 mmol) in xylenes (2.0 mL) was then added, and the resulting solution was stirred at rt for 24 h under N_2. The reaction mixture was directly loaded onto a silica gel chromatography column and eluted (hexanes/*i*-PrOH, 86:14) to afford the chromanone product (26.5 mg, 0.113 mmol, 94%, er 99:1) as a colorless oil: HPLC t_R (major enantiomer) 27.7 min, t_R (minor enantiomer) 39.9 min (Chiracel AD-H, hexanes/*i*-PrOH, 97:3, 0.5 mL/min); $[\alpha]_D{}^{23}$ – 17.0 (*c* 0.12, CH_2Cl_2); ^1H NMR (500 MHz, $CDCl_3$) δ 7.89 (dd, J = 7.9, 1.7 Hz, 1H), 7.49 (m, 1H), 7.04 (m, 1H), 6.98 (dd, J = 8.4, 0.6 Hz, 1H), 4.61 (dd, J = 11.2, 5.2 Hz, 1H), 4.30 (t, J = 11.2 Hz, 1H), 4.20 (m, 2H), 3.35 (m, 1H), 2.95 (dd, J = 17.0, 4.8 Hz, 1H), 2.42 (dd, J = 17.0, 8.2 Hz, 1H), 1.29 (t, J = 7.1 Hz, 3H); ^{13}C NMR (125 MHz, $CDCl_3$) δ 192.8, 171.6, 162.0, 136.2, 127.6, 121.7, 120.7, 118.0, 70.5, 61.2, 42.8, 30.6, 14.4.

KHMDS (20 mol %), toluene, rt, 5 min

– $(Me_3Si)_2NH$
– KBF_4

β-TrP2e•HBF_4 (20 mol %)

β-TrP2e
not isolated

β-TrP2e
toluene, rt, 24 h

(94%) dr 30:1
er 97.5:2.5

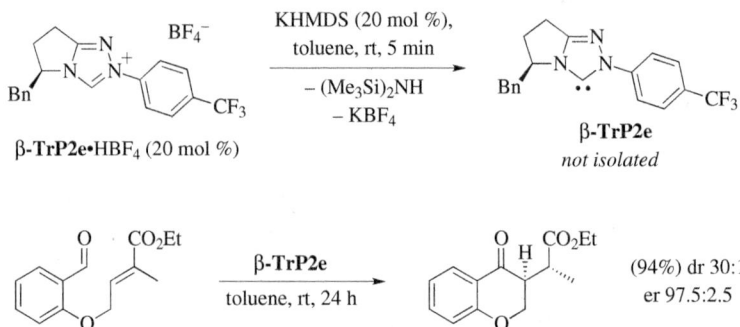

Ethyl (2*R*,3′*S*)-2-(4-Oxochroman-3-yl)propionate [Enantio- and Diastereoselective Intramolecular Stetter Reaction of a Salicylaldehyde-Derived Substrate].[27] Triazolium salt β-**TrP2e**•HBF_4 (13 mg, 0.030 mmol) and KHMDS (0.061 mL, 0.50 M in toluene, 0.031 mmol) were combined in anhydrous toluene (2 mL) at rt under Ar. After stirring the mixture at rt for 5 min, the volatile components were removed in vacuo. (Note: if desired, the KBF_4 byproduct can be removed by microfiltration of the toluene solution through a Gelman 0.45 µm filter prior to removal of the volatile components. However, in this experiment, the presence of KBF_4 did not compromise diastereoselectivity and was not removed.) The residual free carbene β-**TrP2e** (and KBF_4) was then dissolved in anhydrous toluene (3 mL) and added to a solution of ethyl (*E*)-4-(2-formylphenoxy)-2-methyl-2-butenoate (38 mg, 0.148 mmol) in anhydrous toluene (2 mL), and the mixture was stirred at rt for 24 h under Ar. After this time, 15% AcOH in toluene (2 mL) was added, and the reaction mixture was subjected to silica gel column chromatography (EtOAc/hexanes, 50:50) to give the title chromanone (35 mg, 0.140 mmol, 94%, dr 30:1, er 97.5:2.5) as a colorless oil: HPLC t_R (major enantiomer) 54.9 min, t_R (minor enantiomer) 98.1 min (Chiracel OB-H, hexanes/*i*-PrOH, 97:3, 0.3 mL/min); TLC R_f 0.7 (EtOAc/hexanes, 50:50); GC t_R (minor diastereomer) 16.7 min, t_R (major diastereomer) 19.0 min (CP Wax 52CB, 130 °C, 3 mL/min); $[\alpha]_D^{23}$ + 7.9 ($CHCl_3$); IR (neat) 1723, 1701, 1600 cm^{-1}; ^1H NMR (400 MHz, $CDCl_3$) δ 7.86 (d, J = 7.9 Hz, 1H), 7.45 (m, 1H), 6.99 (m, 1H), 6.94 (d, J = 8.3 Hz, 1H), 4.59 (dd, J = 11.3, 5.3 Hz, 1H), 4.34 (t, J = 11.7 Hz, 1H), 4.16 (q, J = 7.0 Hz, 2H), 3.26 (dt, J = 12.2, 5.3 Hz, 1H), 3.10 (qd, J = 7.1, 6.0 Hz, 1H), 1.25 (t, J = 7.1 Hz, 3H), 1.20 (d, J = 7.2 Hz, 3H); ^{13}C NMR (100 MHz, $CDCl_3$) δ 192.6, 174.9, 161.8, 136.1, 127.6, 122.2, 121.7, 117.9, 68.7, 61.1, 47.8, 36.6, 14.4, 13.7; HRMS–FAB (*m*/*z*): M^+ calcd for $C_{14}H_{16}O_4$, 248.1049; found, 248.1119.

β-**TrPFc3b**•HBF_4 (10 mol %)

Cy–CH=CH–NO_2 (1.5 equiv)

i-Pr_2NEt (1 equiv), MeOH, 0 °C, 2 h

(95%) er 97.5:2.5

(S)-2-Cyclohexyl-3-nitro-1-(pyridin-2-yl)propan-1-one [Enantioselective Intermolecular Stetter Reaction of a Heteroaryl Aldehyde with a Nitroalkene Acceptor].[78] A dry, 4-mL glass vial equipped with a magnetic stirring bar was charged with triazolium salt β-**TrPFc3b**•HBF_4 (16 mg, 0.037 mmol), pyridine-2-carboxaldehyde (40 mg, 0.37 mmol), (E)-(2-nitrovinyl)cyclohexane (86 mg, 0.556 mmol) and methanol (1.0 mL). The vial was cooled to 0 °C (ice/water bath), and diisopropylethylamine (0.064 mL, 0.371 mmol) was added dropwise. The reaction mixture was stirred at 0 °C for 2 h, and then glacial AcOH (0.10 mL) was added. The reaction mixture was then concentrated in vacuo, and the residue was subjected to silica gel column chromatography (Et_2O/hexanes, 50:50) to give the title compound (92 mg, 0.351 mmol, 95%, er 97.5:2.5) as a colorless solid: mp 128–130 °C;); HPLC t_R (minor) 6.90 min, t_R (major) 7.74 min (Chiracel OD-H, hexane/i-PrOH, 90:10, 1.0 mL/min); TLC R_f 0.30 (Et_2O/hexanes, 1:1); $[\alpha]_D^{21}$ – 68.0 (c 1.00, CH_2Cl_2); IR (neat) 3070, 3003, 2924, 2856, 1696, 1544, 1448, 1392 cm^{-1}; ^1H NMR (300 MHz, $CDCl_3$) δ 8.72 (dm, J = 4.8 Hz, 1H), 8.08 (d, J = 7.9 Hz, 1H), 7.86 (td, J = 7.8, 1.8 Hz, 1H), 7.50 (ddd, J = 7.8, 4.8, 1.1 Hz, 1H), 5.06 (dd, J = 14.3, 10.9 Hz, 1H), 4.80 (m, 1H), 4.59 (dd, J = 14.3, 3.2 Hz, 1H), 1.85 (m, 1H), 1.65 (m, 5H), 1.15 (m, 4H), 0.93 (m, 1H); ^{13}C NMR (75 MHz, $CDCl_3$) δ 200.9, 152.7, 149.3, 137.3, 127.7, 122.7, 73.9, 47.3, 39.0, 31.5, 29.6, 26.6, 26.5, 26.2; HRMS–ESI (m/z): $[M + H]^+$ calcd for $C_{14}H_{19}N_2O_3$, 263.1390; found, 263.1393.

β-**TrM2c**•HBF_4 (10 mol %)

t-BuOK (8 mol %), toluene, 0 °C, 3 h

(2 equiv)

(93%) er 97.5:2.5

Methyl (*S*)-2-Acetamido-4-(4-chlorophenyl)-4-oxobutanoate [Enantioselective Intermolecular Stetter Reaction of an Aryl Aldehyde with an α-Acetamidoacrylate].[48] A flame-dried, screw-capped test tube equipped with a magnetic stirring bar was charged with *t*-BuOK (4.5 mg, 0.040 mmol) and triazolium salt β-**TrM2c**•HBF_4 (18.5 mg, 0.050 mmol) in a glove box. The vessel was fitted with a septum and removed from the glove box. Toluene (1.8 mL) was added, and the contents were stirred at rt for 30 min under Ar. The reaction mixture was cooled to 0 °C, and 4-chlorobenzaldehyde (70 mg, 0.50 mmol) and methyl 2-acetamidoacrylate (143 mg, 1.00 mmol) were added. After stirring at 0 °C for 3 h, the reaction mixture was diluted with EtOAc (2 mL) and filtered through a thin pad of SiO_2, eluting with EtOAc (10 mL). The filtrate was concentrated in vacuo, and the residue was purified by silica gel column chromatography (EtOAc/pentane, 50:50 to 60:40) to give the title compound (132 mg, 0.465 mmol, 93%, er 97.5:2.5) as a colorless solid: HPLC t_R (major) 8.07 min; t_R (minor) 12.02 min (Chiracel AD-H, hexanes/*i*-PrOH, 70:30, 1.0 mL/min; TLC R_f 0.24 (EtOAc/pentane, 6:4); $[\alpha]_D^{20}$ + 117.0 (*c* 0.93, $CHCl_3$); IR (ATR) 3354, 1733, 1686, 1630, 1587, 1524, 1437, 1405, 1342, 1291, 1239, 1213, 1184, 1088, 1020, 969, 822 cm^{-1}; ^1H NMR (300 MHz, $CDCl_3$) δ 7.82 (d, J = 8.3 Hz, 2H), 7.38 (d, J = 8.3 Hz, 2H), 6.78 (br d, J = 6.9 Hz, NH), 4.93 (m, 1H), 3.70–3.62 (m, 1H), 3.69 (s, 3H), 3.53 (dd, J = 18.4, 4.2 Hz, 1H), 1.97 (s, 3H); ^{13}C NMR (75 MHz, $CDCl_3$) δ 197.6, 171.6, 170.0, 140.3, 134.2, 129.6 (2C), 129.1 (2C), 52.7, 48.2, 40.5, 23.1; HRMS–ESI (*m/z*): $[M + Na]^+$ calcd for $C_{13}H_{14}ClNO_4Na$, 306.0509; found, 306.0504.

β-**TrP2b**•HBF_4 (20 mol %)

i-Pr_2NEt (1 equiv), $MgSO_4$ (1 equiv), CCl_4, –10 °C, 12 h

(2 equiv) (84%) er 95:5

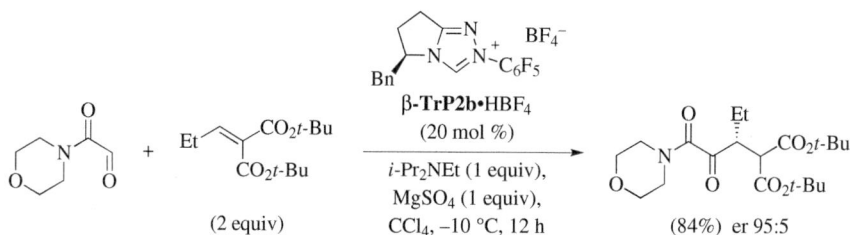

Di-*tert*-butyl (*R*)-2-(1-Morpholino-1,2-dioxopentan-3-yl)malonate [Enantioselective Intermolecular Stetter Reaction of a Glyoxamide with an Alkylidene Malonate Acceptor].[76] A 5-mL, flame-dried test tube equipped with a magnetic stirring bar was charged with triazolium salt β-**TrP2b**•HBF_4 (14.5 mg, 0.032 mmol, 20 mol %), di-*tert*-butyl propylidenemalonate (82 mg, 0.32 mmol), and $MgSO_4$ (20 mg, 0.16 mmol). The vessel was fitted with a septum and flushed with Ar, and then morpholine glyoxamide (23 mg, 0.161 mmol) was added, followed by CCl_4 (0.5 mL). The mixture was cooled to –10 °C (internal temperature), diisopropylethylamine (0.028 mL, 0.16 mmol) was added, and stirring was continued for 12 h at the same temperature. After this time, AcOH (0.1 mL) was added, and the reaction mixture was loaded directly onto a silica gel chromatography column and eluted (hexanes/EtOAc , 67:33) to afford the title compound (53.7 mg, 0.134 mmol, 84%, er 95:5) as a colorless solid: mp 87–88 °C; HPLC t_R (minor) 4.9 min; t_R (major) 5.7 min (Chiracel ADH, hexane/*i*-PrOH, 85:15, 1.0 mL/min); TLC R_f 0.4 (hexanes/EtOAc, 2:1); $[\alpha]_D^{23}$ + 28.6 (*c* 4.90, $CHCl_3$); IR (neat) 2975, 2924, 1720, 1643, 1369, 1116 cm^{-1}; ^1H NMR (300 MHz, $CDCl_3$) δ 3.80–3.45 (m, 10H), 1.85–1.65 (m, 2H), 1.44 (s, 9H), 1.39 (s, 9H), 0.91 (t, J = 7.5 Hz, 3H); ^{13}C NMR (75 MHz, $CDCl_3$) δ 199.7, 168.6, 167.4, 164.7, 82.6, 82.2, 67.1, 66.8, 56.4, 48.3, 46.6, 42.3, 28.0, 23.6, 10.7; HRMS–FAB (m/z): M^+ calcd for $C_{20}H_{33}NO_7$, 399.2257; found, 399.2251.

β-**TrPFc3b**•HBF_4 (10 mol %)

(1.5 equiv)

i-Pr_2NEt (1 equiv), catechol (1 equiv), MeOH, 0 °C, 4 h

(80%) er 96.5:3.5

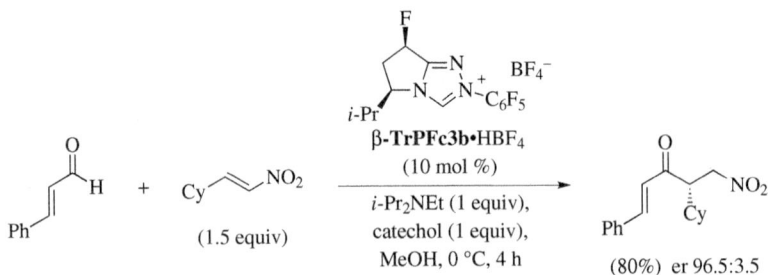

(*S*,*E*)-4-Cyclohexyl-5-nitro-1-phenylpent-1-en-3-one [Enantioselective Intermolecular Stetter Reaction of an α,β-Unsaturated Aldehyde with a Nitroalkene Acceptor].[40] A dry, 4-mL glass vial equipped with a magnetic stirring bar was charged with triazolium salt β-**TrPFc3b**•HBF_4 (16 mg, 0.038 mmol), catechol (42 mg, 0.38 mmol), and methanol (0.8 mL). The vessel was fitted with a septum, flushed with Ar, and then cooled to 0 °C in an ice-water bath. Diisopropylethylamine (0.066 mL, 0.38 mmol) was added, and the mixture was stirred for 5 min at 0 °C before a solution of cinnamaldehyde (50 mg, 0.38 mmol) and (*E*)-(2-nitrovinyl)cyclohexane (88 mg, 0.57 mmol) in methanol (0.2 mL) was added. The reaction mixture was stirred at 0 °C for 4 h, and then glacial AcOH (0.1 mL) was added. The reaction mixture was concentrated in vacuo, and the residue was purified by silica gel column chromatography (hexanes/Et_2O, 80:20) to afford the title compound (87 mg, 0.302 mmol, 80%, er 96.5:3.5) as a colorless solid: mp 94–97 °C; HPLC t_R (minor) 10.66 min; t_R (major) 15.17 min (Chiracel IC, hexanes/*i*-PrOH, 70:30, 1.0 mL/min); TLC R_f 0.29 (hexanes/Et_2O, 4:1); $[\alpha]_D^{21}$ − 106.7 (*c* 0.6, CH_2Cl_2); IR (neat) 2929, 2854, 1685, 1657, 1608, 1576, 1450, 1374 cm^{-1}; ^1H NMR (300 MHz, $CDCl_3$) δ 7.66 (d, J = 16.0 Hz, 1H), 7.61–7.58 (m, 2H), 7.44–7.41 (m, 3H), 6.86 (d, J = 16.0 Hz, 1H), 4.99 (dd, J = 14.6, 10.2 Hz, 1H), 4.45 (dd, J = 14.6, 3.6 Hz, 1H), 3.58 (ddd, J = 10.2, 5.1, 3.7 Hz, 1H), 1.79–1.59 (m, 6H), 1.32–0.93 (m, 5H); ^{13}C NMR (75 MHz, $CDCl_3$) δ 198.9, 144.3, 134.7, 131.3, 129.4, 129.0, 125.7, 73.9, 53.2, 31.7, 30.3, 29.3, 26.8, 26.7, 26.3; HRMS–ESI (*m*/*z*): M^+ calcd for $C_{17}H_{21}NO_3$, 287.1521; found, 287.1515.

TABULAR SURVEY

The Tables are divided into intramolecular (Tables 1–3) and intermolecular examples (Tables 4–7) of the catalytic enantioselective Stetter reaction, followed by examples of the recently introduced imine variant (Table 8). Within each Table, the aldehyde substrate is the primary sorting rubric, and examples appear in order of increasing carbon-atom count. Simple groups on heteroatoms and protecting groups are ignored in the carbon-atom count; however, an exception to this rule is made for intramolecular reactions involving substrates with reactive moieties linked together by N, O, or S atoms. A system of inverse Cahn–Ingold–Prelog priority is used to organize aldehyde substrates containing the same number of carbon atoms and, for intramolecular reactions, the formation of five-membered rings is considered before six-membered rings. Carbon-atom count of the acceptor is used as the secondary organizational rubric for intermolecular examples. Entries within subtables are organized to best emphasize the dependence of reaction outcome upon key substrate characteristics, e.g., linear alkyl substituents are generally listed before branched examples, and aromatic substituents are grouped according to their electronic and steric characteristics. Charts 1 and 2 contain the structures and codes of the *N*-heterocyclic carbene catalysts (or precatalysts) that appear throughout this chapter and in the tabular surveys. The Tables contain all examples of the catalytic enantioselective Stetter reaction published in the primary peer-reviewed literature through to the end of 2019 as retrieved by SciFinder Scholar and Reaxysis substructure, keyword, and citation searches.

Aside from the following abbreviations, all others used in the text and the Tables can be found in *The Journal of Organic Chemistry's* list of "Standard Abbreviations and Acronyms."

API	active pharmaceutical ingredient
Ad	adamantyl
BEMP	2-*tert*-butylimino-2-diethylamino-1,3-dimethylperhydro-1,3,2-diazaphosphorine
Bpin	4,4,5,5-tetramethyl-1,3,2-dioxaborolanyl
CPCM	conductor-like polarizable continuum model
CPME	cyclopentyl methyl ether
DVB	divinyl benzene
EDG	electron-donating groups
EWG	electron-withdrawing group
MMF	monolithic-microreactor flow
P2-Et	1-ethyl-2,2,4,4,4-pentakis(dimethylamino)-$2\lambda^5,4\lambda^5$-catenadi(phosphazene)
PEMP	BEMP PS; polymer-supported 2-*tert*-butylimino-2-diethylamino-1,3-dimethylperhydro-1,3,2-diazaphosphorine
Phth	phthaloyl
TBD	1,5,7-triazabicyclo[4.4.0]dec-5-ene
TBDPS	*tert*-butyldiphenylsilyl

Chart 1. Triazolylidene Catalysts

Key for triazolylidene NHC catalyst codes

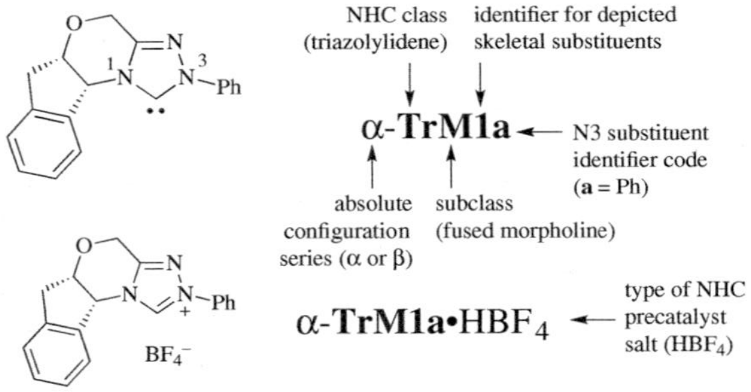

NHC class (triazolylidene) — identifier for depicted skeletal substituents

α-TrM1a ← N3 substituent identifier code (**a** = Ph)

absolute configuration series (α or β) — subclass (fused morpholine)

α-TrM1a•HBF_4 ← type of NHC precatalyst salt (HBF_4)

BF_4^-

N3 substituent codes

	R^1
a	Ph
b	C_6F_5
c	2,4,6-$Me_3C_6H_2$
d	4-$MeOC_6H_4$
e	4-$(CF_3)C_6H_4$
f	4-FC_6H_4
g	4-$(NC)C_6H_4$
h	2,6-$(MeO)_2C_6H_3$
i	3,5-$(CF_3)_2C_6H_3$
i§	3,5-$Me_2C_6H_3$
j	2,4,6-$Cl_3C_6H_2$
k	2,4,6-$Br_3C_6H_2$
l	2,4,6-*i*-$Pr_3C_6H_2$
p	2-pyr
q	6-Me-2-pyr
x	Bn

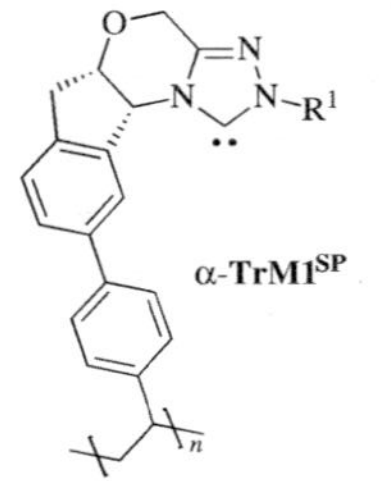

α-TrM1^{SP} as copolymer with styrene and 1,4-divinylbenzene

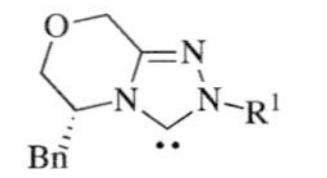

α-TrM2 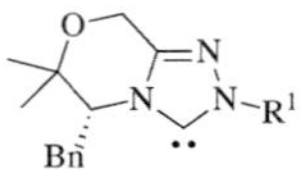 **α-TrM3** 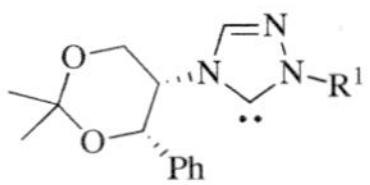 **α-TrD1** 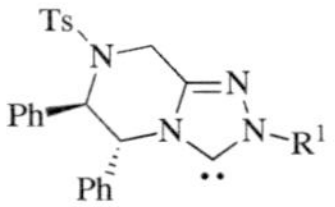 **α-TrY1**

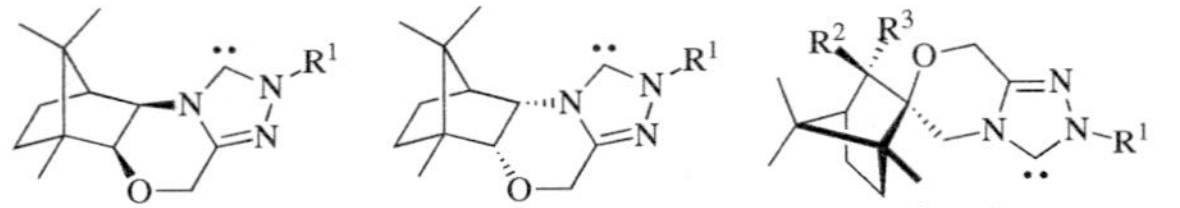

α-TrM4 **β-TrM4II**

α-TrM4III, $R^2 = R^3 = H$

α-TrM4IV, $R^2/R^3 = O(CH_2)_2O$

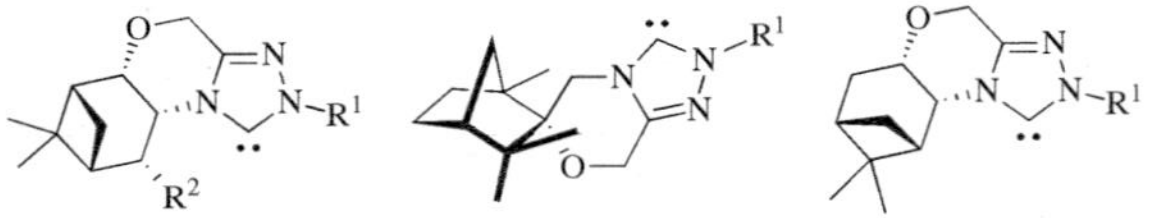

α-TrM4V, $R^2 = H$

α-TrM4VI, $R^2 = Me$

α-TrM4VII **α-TrM4VIII**

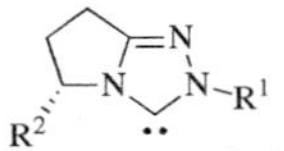

	R^2
TrP1	H
α-TrP2	Bn
α-TrP3	*i*-Pr
α-TrP4	*t*-Bu
α-TrP4§	*i*-Bu

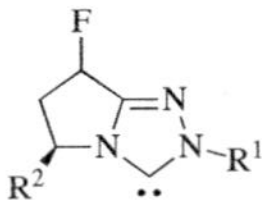

	R^2	C(F) config
β-TrPF1	H	(*R*)
β-TrPFc3	*i*-Pr	(*R*) (*cis*)
β-TrPFc4	*t*-Bu	(*R*) (*cis*)
β-TrPFt3	*i*-Pr	(*S*) (*trans*)
β-TrPFt4	*t*-Bu	(*S*) (*trans*)

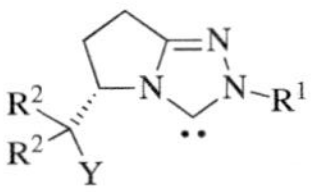

	R^2	Y
α-TrP5	Ph	H
α-TrP6	2-MeC_6H_4	H
α-TrP7	2,4,6-$Me_3C_6H_2$	H
α-TrP8	Ph	HO
α-TrP9	Ph	Me_3SiO
α-TrP10	Ph	Me_2*t*-BuSiO
α-TrP11	H	Ph_2*t*-BuSiO
α-TrP12	H	*i*-Pr_3SiO
α-TrP13	Me	Me_2*t*-BuSiO
α-TrP14	*n*-Bu	Me_3SiO

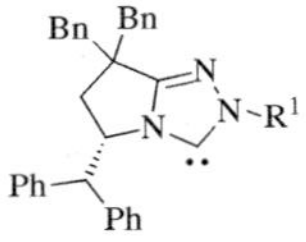

α-TrP5§

Chart 2. Imidazolylidene and Thiazolylidene Catalysts

Imidazolylidene NHC catalyst codes

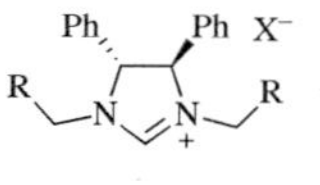

α-**Im1**•HX

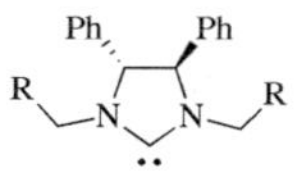

α-**Im1**

	R
a	Ph
c	2,4,6-$Me_3C_6H_2$
k	*t*-$BuCH_2$
l	2,4,6-$Me_3C_6H_2(CH_2)_2$

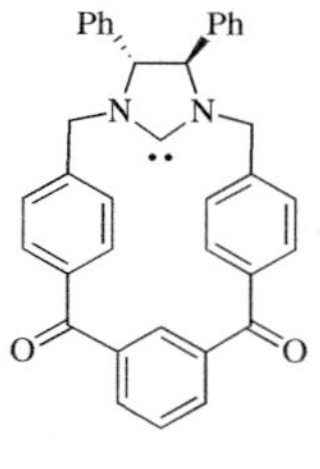

α-**Im2**

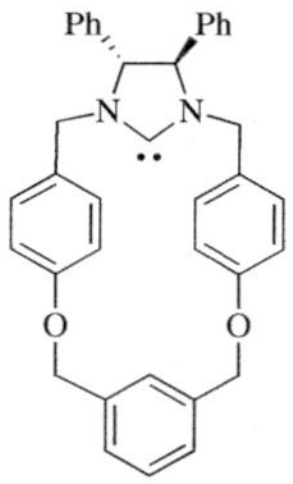

α-**Im3**

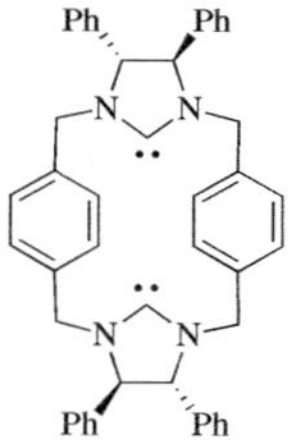

α-**Im4**

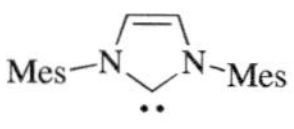

Im5c

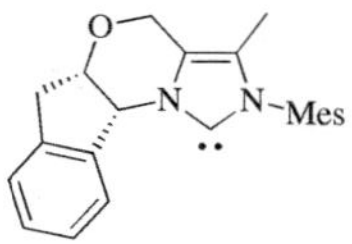

α-**ImM1c**

Thiazolylidene NHC catalyst codes

β-**ThB1**

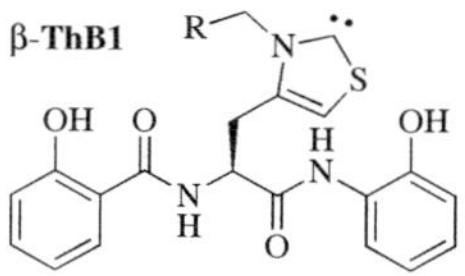

	R
z	H
a	Ph
d	4-(*n*-$C_{10}H_{21}O$)C_6H_4

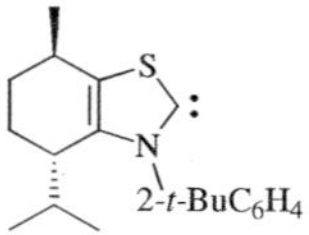

α-**ThMt1**

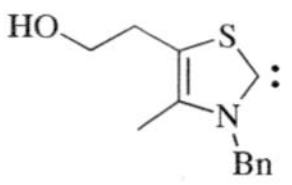

ThS1x

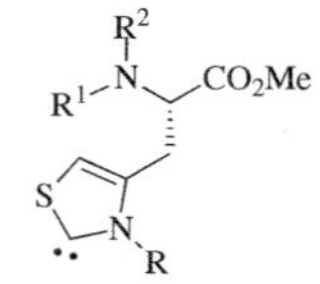

	R	R^1	R^2
α-**ThA1**	Et	Ac	H
α-**ThA2**	Et	Boc	H
α-**ThA3**	Et	Boc	Me
α-**ThA4**	Me	Ts	H
α-**ThA5**	Et	Ts	H
α-**ThA6**	Bn	Ts	H

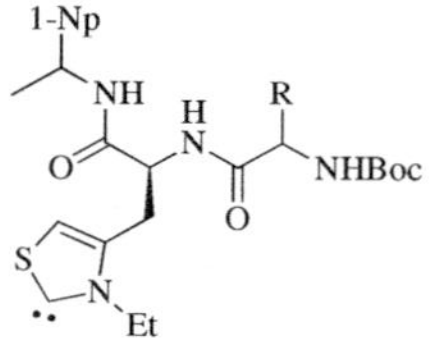

	1-Np	R
α-**ThAF1**	(*R*)	(*S*)-Bn
α-**ThAF2**	(*S*)	(*R*)-Bn
α-**ThAF3**	(*S*)	(*S*)-Bn
α-**ThAV1**	(*S*)	(*S*)-*i*-Pr
α-**ThAT1**	(*S*)	(*S*)-[(*R*)-(BnO)MeHC

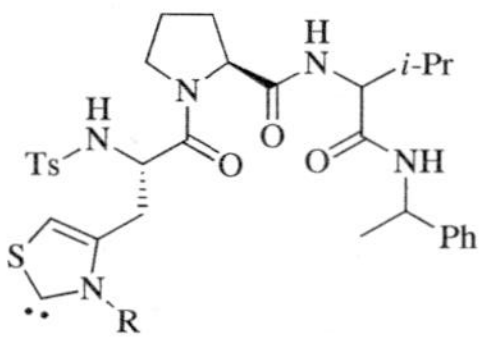

	i-Pr	Ph
α-**ThAPV1**	(*S*)	(*S*)
α-**ThAPV2**	(*R*)	(*S*)
α-**ThAPV3**	(*S*)	(*R*)
α-**ThAPV4**	(*R*)	(*R*)

TABLE 1. INTRAMOLECULAR STETTER REACTIONS OF ARYL ALDEHYDES

Aryl Aldehyde Substrate	Conditions	Product(s), Yield(s) (%), and Stereoselectivity	Refs.

Please refer to the charts preceding the tables for catalyst and precatalyst structures used herein.

C_{10}

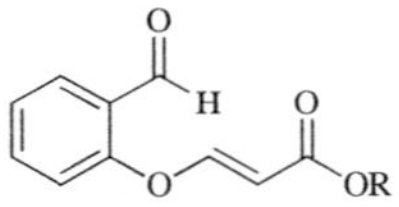

Conditions: Precatalyst (x mol %), base (y mol %), xylene, rt, 24 h

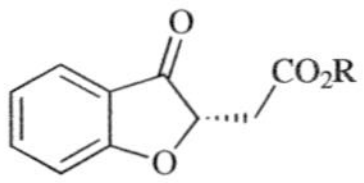

R	Precatalyst	x	Base	y	Time (h)		er	Refs.
Me	β-**TrP2a**•HBF_4	20	KHMDS	20	24	(90)	47:53	15
Me	α-**TrM1d**•HBF_4	20	KHMDS	20	24	(90)	53:47	62
Et	α-**TrY1c**•HBF_4	10	Et_3N	10	66	(59)	50:50	146
Et	α-**TrM4b**•HBF_4	10	*i*-Pr_2NEt	10	(—)	(82)	65:35	64

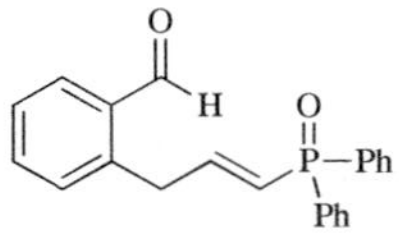

Conditions: α-**TrM1b**•HBF_4 (20 mol %), KHMDS (20 mol %), toluene, rt, 12 h

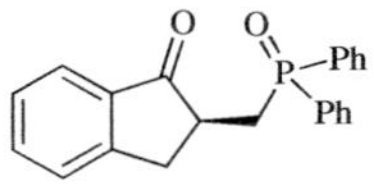

Product: (99) er 95:5 — Refs. 67

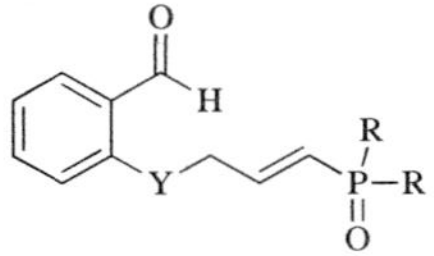

Conditions: Precatalyst (20 mol %), KHMDS (20 mol %), toluene, rt, 12 h 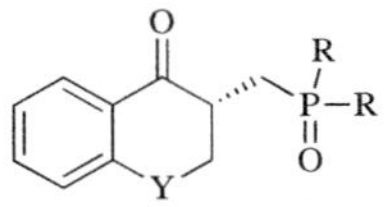 — Refs. 67

R	Y	Precatalyst		er
EtO	O	α-**TrM1b**•HBF_4	(65)	90:10
Ph	O	α-**TrM1b**•HBF_4	(90)	93:7
Ph	O	β-**TrP2b**•HBF_4	(—)	13:87
Ph	S	α-**TrM1b**•HBF_4	(70)	96:4

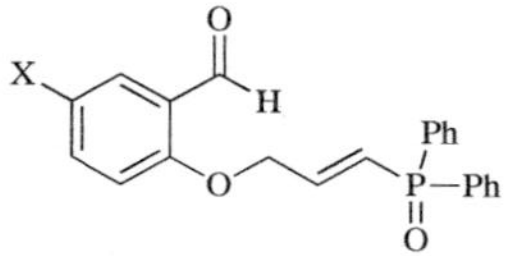

Precatalyst (x mol %),
KHMDS (y mol %),
toluene, rt, 12 h

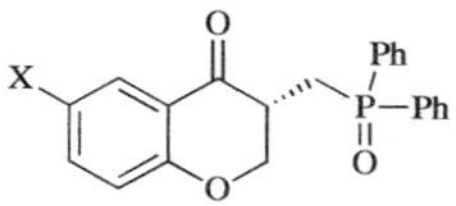

X	Precatalyst	x	y		er
Cl	α-**TrM1b**•HBF_4	20	20	(90)	97:3
Cl	β-**TrP2b**•HBF_4	20	20	(—)	87.5:12.5
Br	α-**TrM1b**•HBF_4	20	20	(88)	98:2
Br	α-**TrM1b**•HBF_4	10	10	(80)	97:3

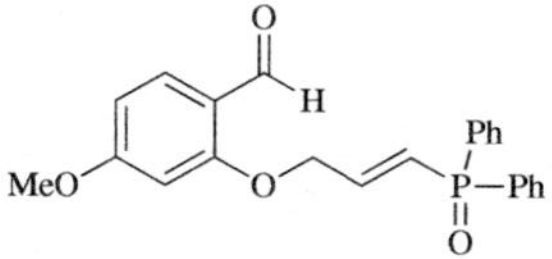

67

α-**TrM1b**•HBF_4 (20 mol %),
KHMDS (20 mol %),
toluene, rt, 12 h

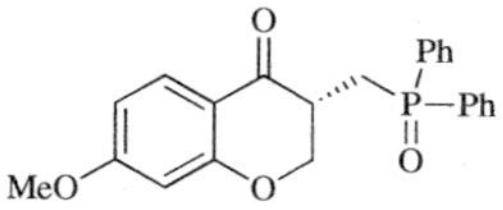

(86) er 96.5:3.5

67

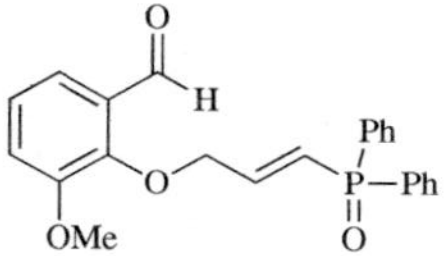

Precatalyst (20 mol %),
KHMDS (20 mol %),
toluene, rt, 12 h

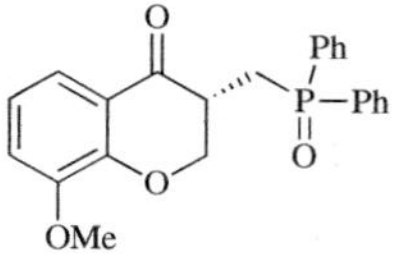

Precatalyst		er
α-**TrM1b**•HBF_4	(75)	93.5:6.5
β-**TrP2b**•HBF_4	(—)	95.5:4.5

67

TABLE 1. INTRAMOLECULAR STETTER REACTIONS OF ARYL ALDEHYDES (*Continued*)

Aryl Aldehyde Substrate	Conditions	Product(s), Yield(s) (%), and Stereoselectivity	Refs.

Please refer to the charts preceding the tables for catalyst and precatalyst structures used herein.

C_{11-18}

R[1] R[2] CO_2R^3 O H O

α-TrM4Vb•HBF_4 (20 mol %), KHMDS (20 mol %), cyclohexane, rt, 3 h

O R[2] CO_2R^3 O R[1]

147

R^1	R^2	R^3		er
H	Me	Me	(96)	98:2
H	Et	Et	(96)	98:2
H	n-Pr	Me	(98)	98.5:1.5
H	n-Pr	Bn	(95)	98:2
H	n-C_5H_{11}	Me	(95)	99:1
H	n-C_6H_{13}	Me	(93)	99:1
H	(c-C_6H_{11})CH_2	Me	(91)	99:1

R^1	R^2	R^3		er
H	Ph$(CH_2)_2$	Me	(92)	99:1
4-Me	Me	Me	(94)	97:3
3-Cl	n-Pr	Me	(95)	98:2
3-Br	n-Pr	Me	(92)	96.5:3.5
3-CF_3O	n-Pr	Me	(94)	98.5:1.5
3-F	n-Pr	Me	(97)	92:8

β-TrM4IIj•$HBPh_4$ (20 mol %), t-BuOK (20 mol %), cyclohexane, rt, 3 h

O CO_2R^3 R[2] O R[1]

147

R^1	R^2	R^3		er
H	Me	Me	(98)	99.5:0.5
H	Et	Et	(97)	95.5:4.5
H	n-Pr	Me	(98)	98.5:1.5
H	n-Pr	Bn	(98)	98.5:1.5
H	n-C_5H_{11}	Me	(95)	99:1
H	n-C_6H_{13}	Me	(93)	99:1
H	(c-C_6H_{11})CH_2	Me	(95)	99.5:0.5

R^1	R^2	R^3		er
H	Ph$(CH_2)_2$	Me	(99)	99.5:0.5
4-Me	Me	Me	(96)	96.5:3.5
3-Cl	n-Pr	Me	(94)	99:1
3-Br	n-Pr	Me	(90)	97:3
3-CF_3O	n-Pr	Me	(92)	92.5:7.5
3-F	n-Pr	Me	(95)	98:2

C_{11-12}

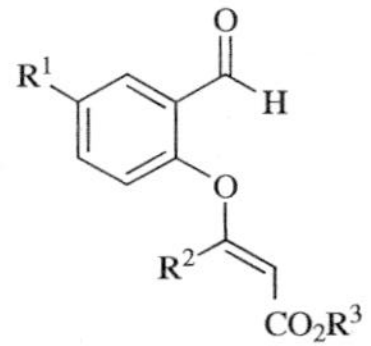

Precatalyst (20 mol %), base (x mol %), toluene, rt, 24 h

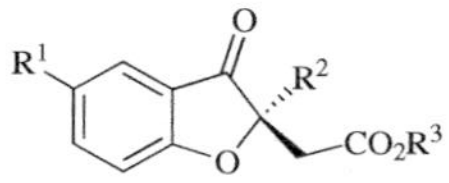

R^1	R^2	R^3	Precatalyst	Base	x	
H	Me	Et	α-**TrM1b**•HBF_4	Et_3N	200	(95)
H	Et	Me	α-**TrM1d**•HBF_4	KHMDS	20	(45)
H	Et	Me	α-**TrM1a**•HBF_4	KHMDS	20	(80)
H	Et	Me	α-**TrM1b**•HBF_4	KHMDS	20	(85)
H	Et	Me	α-**TrM1b**•HBF_4	Et_3N	200	(96)
Br	Et	Me	α-**TrM1b**•HBF_4	Et_3N	200	(92)

C_{11}

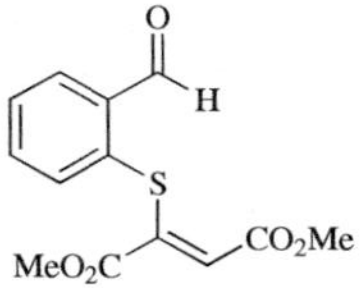

β-**TrM1d**•HBF_4 (20 mol %), KHMDS (20 mol %), toluene, rt, 24 h

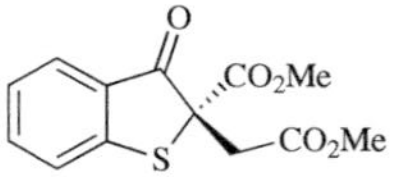

(85) er 95:5 73

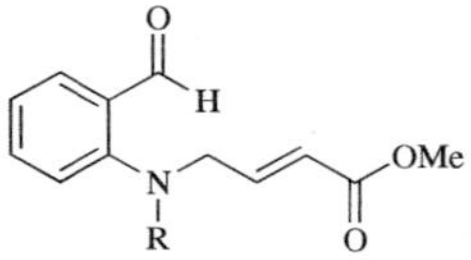

α-**TrM1d**•HBF_4 (20 mol %), KHMDS (20 mol %), xylene, rt, 24 h

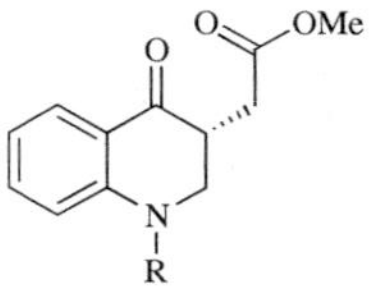

R		er
Me	(64)	91:9
MeO_2C–CH=CH–CH_2–	(72)	92:8

15

TABLE 1. INTRAMOLECULAR STETTER REACTIONS OF ARYL ALDEHYDES (*Continued*)

Aryl Aldehyde Substrate	Conditions	Product(s), Yield(s) (%), and Stereoselectivity	Refs.

Please refer to the charts preceding the tables for catalyst and precatalyst structures used herein.

C_{11}

Aryl Aldehyde Substrate	Conditions	Product(s), Yield(s) (%), and Stereoselectivity	Refs.
O, H, O, H, O	**α-TrM1b**•HBF_4 (20 mol %), KHMDS (20 mol %), toluene, rt, 12 h	O, H, O, O (50) er 65:35	62
O, H, O, *, CN	Precatalyst (20 mol %), KHMDS (20 mol %), toluene, rt, 12 h	O, CN, O	62

*(E)/(Z)	Precatalyst		er
1:1	**β-TrM1b**•HBF_4	(88)	90:10
(*E*) only	**α-TrM1a**•HBF_4	(80)	11:89

Aryl Aldehyde Substrate	Conditions	Product(s), Yield(s) (%), and Stereoselectivity	Refs.
O, H, O, NMe(OMe), O	**α-TrM1b**•HBF_4 (20 mol %), KHMDS (20 mol %), toluene, rt, 12 h	O, NMe(OMe), O, O (94) er 96:4	62
O, H, O, *, CO_2Me	Precatalyst (*x* mol %), base (*y* mol %), solvent	O, CO_2Me, O	

*(*E*) or (*Z*)	Precatalyst	*x*	Base	*y*	Solvent	Temp	Time (h)		er	Refs.
(*E*)	**α-TrD1a**•$HClO_4$	20	K_2CO_3	10	THF	rt	24	(73)	80:20	11
(*E*)	**α-TrP5p**•HBF_4	20	KO*t*-Bu	19	CH_2Cl_2	reflux	12	(93)	90:10	65
(*E*)	**α-TrM1b**•HBF_4	20	KHMDS	20	toluene	rt	24	(95)	97:3	62
(*Z*)	**α-ThMt1**•$HClO_4$	20	Et_3N	20	THF	rt	21	(75)	75:25	148
(*Z*)	**β-TrM1b**•HBF_4	20	KHMDS	20	toluene	rt	24	(80)	39:61	62

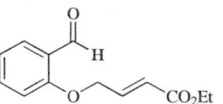

Precatalyst (x mol %), base (y mol %), solvent (0.2 M)

Precatalyst	x	Base	y	Solvent	Temp	Time (h)	Notes		er	
β-**TrM1d**•HBF_4	20	KHMDS	20	xylene	rt	24	—	(94)	97:3	15
α-**TrM1c**•$HClO_4$	20	KHMDS	20	xylene	rt	24	at 0.02 M	(94)	99:1	63
α-**ImM1c**•$HClO_4$	20	KHMDS	20	xylene	rt	24	at 0.02 M	(0)	—	63
α-**TrY1a**•HBF_4	20	KHMDS	20	xylene	rt	36	—	(60)	75.5:24.5	146
α-**TrY1d**•HBF_4	20	KHMDS	20	xylene	rt	36	—	(43)	84:16	146
α-**TrY1c**•HBF_4	20	KHMDS	20	xylene	rt	0.6	—	(95)	92.5:7.5	146
α-**TrY1c**•HBF_4	10	NaHMDS	10	xylene	rt	1.3	—	(97)	92.5:7.5	146
α-**TrY1c**•HBF_4	10	LiHMDS	10	xylene	rt	48	—	(56)	95:5	146
α-**TrY1c**•HBF_4	10	DBU	10	xylene	rt	4	—	(90)	93.5:6.5	146
α-**TrY1c**•HBF_4	10	i-Pr_2NEt	10	xylene	rt	85	—	(90)	97.5:2.5	146
α-**TrY1c**•HBF_4	10	KHMDS	10	xylene	rt	0.6	—	(95)	91:9	146
α-**TrY1c**•HBF_4	10	Et_3N	10	xylene	rt	16	—	(95)	97:3	146
α-**TrY1i**•HBF_4	20	KHMDS	20	xylene	rt	36	—	(17)	55:45	146
α-**TrM4a**•HBF_4	10	KHMDS	10	xylene	rt	24	at 0.05 M	(10)	71.5:28.5	64
α-**TrM4d**•HBF_4	10	KHMDS	10	xylene	rt	24	at 0.05 M	(7)	91:9	64
α-**TrM4c**•HBF_4	10	KHMDS	10	xylene	rt	24	at 0.05 M	(0)	—	64
α-**TrM4b**•HBF_4	10	KHMDS	10	xylene	rt	24	at 0.05 M	(45)	98.5:1.5	64
α-**TrM4b**•HBF_4	10	KHMDS	10	xylene	rt	1	at 0.1 M	(95)	96.5:3.5	64
α-**TrM4b**•HBF_4	10	Et_3N	10	xylene	rt	21	at 0.1 M	(80)	98:2	64
α-**TrM4b**•HBF_4	10	DBU	10	xylene	rt	24	at 0.1 M	(69)	98:2	64
α-**TrM4b**•HBF_4	10	i-Pr_2NEt	10	xylene	rt	4	at 0.1 M	(94)	98:2	64
α-**TrM4b**•HBF_4	10	Cs_2CO_3	10	xylene	rt	24	at 0.1 M	(67)	96:4	64
α-**TrM4b**•HBF_4	10	DBN	10	xylene	rt	20	at 0.1 M	(86)	97.5:2.5	64
α-**TrM4b**•HBF_4	10	DABCO	10	xylene	rt	2.5	at 0.1 M	(88)	96.5:3.5	64
α-**TrM4i**•HBF_4	10	KHMDS	10	xylene	rt	24	at 0.05 M	(5)	93:7	64

TABLE 1. INTRAMOLECULAR STETTER REACTIONS OF ARYL ALDEHYDES (*Continued*)

Aryl Aldehyde Substrate	Conditions	Product(s), Yield(s) (%), and Stereoselectivity	Refs.

Please refer to the charts preceding the tables for catalyst and precatalyst structures used herein.

Continued from previous page

C_{11}

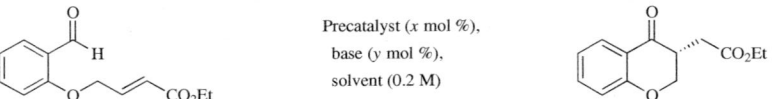

Precatalyst (x mol %), base (y mol %), solvent (0.2 M)

Precatalyst	x	Base	y	Solvent	Temp	Time (h)	Notes	Yield (%)	er	Refs.
α-**Im1a**•HCl	20	NaHMDS	20	xylene	60 °C	48	—	(43)	50:50	66
α-**Im1c**•HBF_4	20	NaHMDS	20	xylene	60 °C	48	—	(38)	54.5:45.5	66
α-**Im4**•2HBr	20	NaHMDS	20	xylene	60 °C	48	—	(33)	58:42	66
α-**Im4**•2HBr	40	NaHMDS	20	xylene	60 °C	48	—	(22)	60:40	66
α-**Im2**•HCl	20	NaHMDS	20	xylene	60 °C	48	—	(58)	84:16	66
α-**Im3**•HCl	20	NaHMDS	20	xylene	60 °C	48	—	(47)	69:31	66
α-**TrY1c**•HBF_4	10	Et_3N	10	toluene	rt	22	—	(95)	94:6	146
α-**TrP5p**•HBF_4	20	KO*t*-Bu	19	toluene	reflux	38	—	(57)	50:50	65
β-**TrP2e**•HBF_4	20	KHMDS	20	toluene	rt	24	—	(95)	4:96	62
α-**TrM1a**•HBF_4	20	KHMDS	20	xylene	rt	24	—	(58)	97.5:2.5	15
β-**TrP2a**•HBF_4	20	KHMDS	20	toluene	rt	24	—	(94)	5:95	62
β-**TrP2a**•HBF_4	20	KHMDS	20	toluene	rt	24	open to air	(42)	5:95	62
β-**TrP2a**•HBF_4	20	KHMDS	20	toluene	rt	24	ACS-grade toluene	(70)	9:91	62
β-**TrP2a**•HBF_4	20	KHMDS	20	toluene	rt	24	wet toluene	(0)	—	62
β-**TrP2a**•HBF_4	20	KHMDS	20	toluene	rt	24	ACS-grade toluene, 4 Å MS	(90)	7:93	62
β-**TrP2a**•HBF_4	20	KHMDS	20	toluene	rt	24	ACS-grade toluene, Na_2SO_4	(80)	9:91	62
β-**TrP2a**•HBF_4	20	KHMDS	20	toluene	rt	24	ACS-grade toluene, $MgSO_4$	(90)	6:94	62
α-**TrD1a**•$HClO_4$	20	K_2CO_3	10	THF	rt	12	—	(69)	78:22	11
α-**TrY1c**•HBF_4	10	Et_3N	10	THF	rt	42	—	(93)	94:6	146

α-**TrM4b**•HBF_4	10	*i*-Pr_2NEt	10	THF	rt	6.5	at 0.1 M	(94)	96.5:3.5	64
α-**TrP5p**•HBF_4	20	KHMDS	19	THF	reflux	16	—	(46)	70.5:29.5	65
α-**TrM4b**•HBF_4	10	*i*-Pr_2NEt	10	EtOH	rt	34	at 0.1 M	(68)	92:8	64
α-**TrY1c**•HBF_4	10	Et_3N	10	Et_2O	rt	52	—	(79)	90:10	146
α-**TrM4b**•HBF_4	10	*i*-Pr_2NEt	10	Et_2O	rt	5	at 0.1 M	(95)	95:5	64
α-**TrP5p**•HBF_4	20	KHMDS	19	Et_2O	reflux	16	—	(86)	59:41	65
α-**TrP5p**•HBF_4	20	KHMDS	19	dioxane	reflux	16	—	(44)	51:49	65
α-**TrP5p**•HBF_4	20	KHMDS	19	DCE	reflux	14	—	(85)	87:13	65
α-**TrY1c**•HBF_4	10	Et_3N	10	CH_2Cl_2	rt	65	—	(82)	87:13	146
α-**TrM4b**•HBF_4	10	*i*-Pr_2NEt	10	CH_2Cl_2	rt	6.5	at 0.1 M	(90)	95:5	64
α-**TrP5p**•HBF4	20	KO*t*-Bu	19	CH_2Cl_2	reflux	25	—	(85)	93:7	65
α-**TrP5p**•HBF4	20	KHMDS	19	CH_2Cl_2	reflux	17	—	(75)	89:11	65
α-**TrP5p**•HBF4	20	*n*-BuLi	19	CH_2Cl_2	reflux	17	—	(0)	—	65
α-**TrP5p**•HBF4	20	LiO*t*-Bu	19	CH_2Cl_2	reflux	13	—	(0)	—	65
α-**TrP5p**•HBF4	20	NaO*t*-Bu	19	CH_2Cl_2	reflux	13	—	(80)	92:8	65
α-**TrP5p**•HBF4	20	K_2CO_3	19	CH_2Cl_2	reflux	20	—	(99)	90:10	65
α-**TrP5p**•HBF4	20	Cs_2CO_3	19	CH_2Cl_2	reflux	12	—	(92)	89.5:10.5	65
α-**TrP5p**•HBF4	20	Et_3N	19	CH_2Cl_2	reflux	12	—	(25)	90:10	65
α-**TrP5p**•HBF4	20	KO*t*-Bu	19	CH_2Cl_2	reflux	16	with HMPA (50 mol %)	(29)	93:7	65
α-**TrP5p**•HBF_4	20	KO*t*-Bu	19	CH_2Cl_2	reflux	16	with KBr (100 mol %)	(75)	93:7	65
α-**TrP5p**•HBF_4	20	KO*t*-Bu	19	CH_2Cl_2	reflux	20	with Ti(O*i*-Pr)$_4$ (100 mol %)	(14)	55:45	65
α-**TrP5p**•HBF_4	20	KO*t*-Bu	19	CH_2Cl_2	reflux	21	with Mg(O*t*-Bu)$_2$ (100 mol %)	(32)	70:30	65
α-**TrP5q**•HBF_4	20	KO*t*-Bu	19	CH_2Cl_2	reflux	15	—	(95)	92:8	65
α-**TrP5§p**•HBF_4	20	KO*t*-Bu	19	CH_2Cl_2	reflux	17	—	(20)	88.5:11.5	65
α-Tr**P6p**•HBF_4	20	KO*t*-Bu	19	CH_2Cl_2	reflux	15	—	(51)	65.5:34.5	65
α-Tr**P8p**•HBF_4	20	KO*t*-Bu	19	CH_2Cl_2	reflux	17	—	(15)	91.5:8.5	65
α-Tr**P9q**•HBF_4	20	KO*t*-Bu	19	CH_2Cl_2	reflux	17	—	(0)	—	65
α-Tr**P5a**•HBF_4	20	KO*t*-Bu	19	CH_2Cl_2	reflux	21	—	(81)	83:17	65

TABLE 1. INTRAMOLECULAR STETTER REACTIONS OF ARYL ALDEHYDES (*Continued*)

Aryl Aldehyde Substrate	Conditions	Product(s), Yield(s) (%), and Stereoselectivity	Refs.

Please refer to the charts preceding the tables for catalyst and precatalyst structures used herein.

Continued from previous page

C_{11}

Precatalyst (x mol %),
base (y mol %),
solvent (0.2 M)

Precatalyst	x	Base	y	Solvent	Temp	Time (h)	Notes		er	Refs.
α-**TrM1a**^{SP}•HBF_4	20	KHMDS	20	toluene	rt	16	—	(95)	96:4	149
α-**TrM1a**^{SP}•HBF_4	10	KHMDS	10	toluene	rt	16	—	(95)	96.5:3.5	149
α-**TrM1a**^{SP}•HBF_4	5	KHMDS	5	toluene	rt	16	—	(63)	96.5:3.5	149
α-**TrM1a**^{SP}•HBF_4	10	Et_3N	10	toluene	rt	16	—	(85)	97:3	149
α-**TrM1a**^{SP}•HBF_4	10	DBU	10	toluene	rt	16	—	(86)	95:5	149
α-**TrM1a**^{SP}•HBF_4	10	KHMDS	10	pentane	rt	16	—	(75)	96:4	149
α-**TrM1a**^{SP}•HBF_4	10	Et_3N	10	EtOH	rt	16	—	(62)	95:5	149
α-**TrM1a**^{SP}•HBF_4	10	Et_3N	10	DMF	rt	16	—	(0)	—	149
α-**TrM1a**^{SP}•HBF_4	10	pH 8.5	10	EtOH	rt	16	aqueous phosphate buffer used	(0)	—	149
α-**TrM1a**^{SP}•HBF_4	10	KHMDS	10	toluene	0 °C	16	—	(35)	97:3	149
α-**TrM1a**^{SP}•HBF_4	5	KHMDS	5	toluene	50 °C	16	—	(95)	95:5	149
α-**TrM1a**^{SP}•HBF_4	10	KHMDS	10	toluene	rt	16	SiO_2-supported catalyst	(45)	96.5:3.5	149
α-**TrM1a**^{SP}•HBF_4	10	KHMDS	10	toluene	rt	16	5th recycle of catalyst	(64)	96:4	149
α-**TrM1a**^{SP}•HBF_4	10	KHMDS	10	toluene	rt	16	10th recycle of catalyst	(38)	95.5:4.5	149
α-**TrM1a**^{SP}•HBF_4	10	KHMDS	10	toluene	rt	16	no styrene/DVB used in polym.	(70)	96:4	149

β-TrM4VIIIa•$HBPh_4$	8	*i*-Pr_2NEt	8	TBME	rt	20	—	(<5)	—	150
β-TrM4VIIIb•$HBPh_4$	8	*i*-Pr_2NEt	8	TBME	rt	20	—	(99)	2:98	150
β-TrM4VIIIb•$HBPh_4$	5	*i*-Pr_2NEt	8	TBME	rt	20	—	(83)	2:98	150
β-TrM4VIIIb•$HBPh_4$	3	*i*-Pr_2NEt	8	TBME	rt	20	—	(71)	2.5:97.5	150
β-TrM4VIIIc•HCl	8	*i*-Pr_2NEt	8	TBME	rt	20	—	(<2)	—	150
β-TrM4VIIIb•$HBPh_4$	8	*c*-Cy_2NEt	8	TBME	rt	20	—	(99)	3:97	150
β-TrM4VIIIb•$HBPh_4$	8	DBU	8	TBME	rt	20	—	(91)	4:96	150
β-TrM4VIIIb•$HBPh_4$	8	Et_3N	8	TBME	rt	20	—	(90)	6:94	150
β-TrM4VIIIb•$HBPh_4$	8	DABCO	8	TBME	rt	20	—	(87)	4:96	150
β-TrM4VIIIb•$HBPh_4$	8	KHMDS	8	TBME	rt	20	—	(93)	8:92	150
β-TrM4VIIIb•$HBPh_4$	8	P2-Et	8	TBME	rt	20	—	(94)	5:95	150
β-TrM4VIIIb•$HBPh_4$	8	*i*-Pr_2NEt	8	*o*-xylene	rt	20	—	(98)	4:96	150
β-TrM4VIIIb•$HBPh_4$	8	*i*-Pr_2NEt	8	THF	rt	20	—	(97)	3:97	150
β-TrM4VIIIb•$HBPh_4$	8	*i*-Pr_2NEt	8	CH_2Cl_2	rt	20	—	(97)	9:91	150
β-TrM4VIIIb•$HBPh_4$	8	*i*-Pr_2NEt	8	*t*-AmOMe	rt	20	—	(97)	3:97	150
β-TrM4VIIIb•$HBPh_4$	8	*i*-Pr_2NEt	8	cyclohexane	rt	20	—	(99)	1:99	150
β-TrM4VIIIb•$HBPh_4$	8	*i*-Pr_2NEt	8	CPME	rt	20	—	(96)	2:98	150

TABLE 1. INTRAMOLECULAR STETTER REACTIONS OF ARYL ALDEHYDES (*Continued*)

Aryl Aldehyde Substrate	Conditions	Product(s), Yield(s) (%), and Stereoselectivity	Refs.

Please refer to the charts preceding the tables for catalyst and precatalyst structures used herein.

Continued from previous page

C_{11}

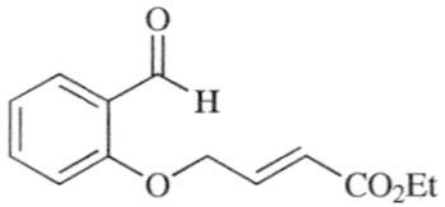

Precatalyst (x mol %), base (y mol %), solvent (0.2 M)

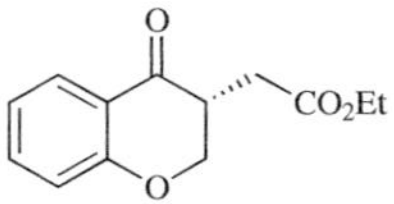

CO_2Et (substrate) → CO_2Et (product)

Precatalyst	x	Base	y	Solvent	Temp	Time (h)	Notes		er	Refs.
β-**TrM4VIa**•HBF_4	10	Et_3N	10	toluene	rt	20	—	(<5)	—	151
β-**TrM4VIb**•$HBPh_4$	10	Et_3N	10	toluene	rt	20	—	(92)	5.5:94.5	151
β-**TrM4VIb**•$HBPh_4$	10	Et_3N	10	toluene	0 °C	20	—	(85)	5.5:94.5	151
β-**TrM4VIb**•$HBPh_4$	3	Et_3N	3	toluene	rt	20	—	(72)	5:95	151
β-**TrM4VIb**•$HBPh_4$	10	i-Pr_2NEt	10	toluene	rt	20	—	(96)	5:95	151
β-**TrM4VIb**•$HBPh_4$	10	c-Cy_2NEt	10	toluene	rt	20	—	(95)	5.5:94.5	151
β-**TrM4VIb**•$HBPh_4$	10	DBU	10	toluene	rt	20	—	(93)	7:93	151
β-**TrM4VIb**•$HBPh_4$	10	DMAP	10	toluene	rt	20	—	(81)	6:94	151
β-**TrM4VIb**•$HBPh_4$	10	DABCO	10	toluene	rt	20	—	(82)	7:93	151
β-**TrM4VIb**•$HBPh_4$	10	TBD	10	toluene	rt	20	—	(85)	6.5:93.5	151
β-**TrM4VIb**•$HBPh_4$	10	P2-Et	10	toluene	rt	20	—	(92)	6:94	151
β-**TrM4VIb**•$HBPh_4$	10	KHMDS	10	toluene	rt	20	—	(94)	5.5:94.5	151
β-**TrM4VIb**•$HBPh_4$	10	t-BuOK	10	toluene	rt	20	—	(92)	5:95	151
β-**TrM4VIb**•$HBPh_4$	10	i-Pr_2NEt	10	o-xylene	rt	20	—	(95)	5.5:94.5	151
β-**TrM4VIb**•$HBPh_4$	10	i-Pr_2NEt	10	THF	rt	20	—	(93)	5.5:94.5	151
β-**TrM4VIb**•$HBPh_4$	10	i-Pr_2NEt	10	CPME	rt	20	—	(92)	5:95	151
β-**TrM4VIb**•$HBPh_4$	10	i-Pr_2NEt	10	TBME	rt	20	—	(81)	5:95	151
β-**TrM4VIb**•$HBPh_4$	10	i-Pr_2NEt	10	t-AmOMe	rt	20	—	(82)	5.5:94.5	151
β-**TrM4VIb**•$HBPh_4$	10	i-Pr_2NEt	10	cyclohexane	rt	20	—	(96)	4.5:95.5	151

Precatalyst (20 mol %),
base (x mol %), solvent

CO_2R CO_2R

R	Precatalyst	Base	x	Solvent	Temp	Time (h)		er	
i-Pr	α-**TrP5p**•HBF_4	KO*t*-Bu	19	CH_2Cl_2	reflux	14	(73)	92:8	65
allyl	α-**TrM1b**•HBF_4	KHMDS	20	toluene	rt	24	(94)	97.5:2.5	62
t-Bu	α-**TrM1b**•HBF_4	KHMDS	20	toluene	rt	24	(94)	98.5:1.5	62
t-Bu	α-**TrP5p**•HBF_4	KO*t*-Bu	19	CH_2Cl_2	reflux	12	(67)	89:11	65
t-Bu	α-**ThA2**•HI	*i*-Pr_2NEt	100	CH_2Cl_2	4 °C	48	(<10)	71:29	152
t-Bu	α-**ThA3**•HI	*i*-Pr_2NEt	100	CH_2Cl_2	4 °C	48	(11)	50:50	152
t-Bu	α-**ThA1**•HI	*i*-Pr_2NEt	100	CH_2Cl_2	4 °C	48	(38)	81.5:18.5	152
t-Bu	α-**ThA5**•HI	*i*-Pr_2NEt	100	CH_2Cl_2	4 °C	48	(10)	85:15	152
t-Bu	α-**ThA4**•HI	*i*-Pr_2NEt	100	CH_2Cl_2	4 °C	48	(<10)	86.5:13.5	152
t-Bu	α-**ThA6**•HBr	*i*-Pr_2NEt	100	CH_2Cl_2	4 °C	48	(40)	90:10	152
t-Bu	α-**ThAPV2**•HI	*i*-Pr_2NEt	100	CH_2Cl_2	4 °C	48	(17)	57.5:42.5	152
t-Bu	α-**ThAPV3**•HI	*i*-Pr_2NEt	100	CH_2Cl_2	4 °C	48	(14)	60.5:39.5	152
t-Bu	α-**ThAPV1**•HI	*i*-Pr_2NEt	100	CH_2Cl_2	4 °C	48	(11)	59:41	152
t-Bu	α-**ThAF1**•HI	*i*-Pr_2NEt	100	CH_2Cl_2	4 °C	48	(20)	77.5:22.5	152
t-Bu	α-**ThAF2**•HI	*i*-Pr_2NEt	100	CH_2Cl_2	4 °C	48	(20)	91:9	152
t-Bu	α-**ThAPV4**•HI	*i*-Pr_2NEt	100	CH_2Cl_2	4 °C	48	(15)	57:43	152
t-Bu	α-**ThAF3**•HI	*i*-Pr_2NEt	100	CH_2Cl_2	4 °C	48	(28)	90:10	152
t-Bu	α-**ThAV1**•HI	*i*-Pr_2NEt	100	CH_2Cl_2	4 °C	48	(22)	82.5:17.5	152
t-Bu	α-**ThAT1**•HI	*i*-Pr_2NEt	100	CH_2Cl_2	4 °C	48	(67)	86.5:13.5	152
1-Ad	α-**TrP5p**•HBF_4	KO*t*-Bu	19	CH_2Cl_2	reflux	12	(94)	90:10	65

TABLE 1. INTRAMOLECULAR STETTER REACTIONS OF ARYL ALDEHYDES (*Continued*)

Aryl Aldehyde Substrate	Conditions	Product(s), Yield(s) (%), and Stereoselectivity	Refs.

Please refer to the charts preceding the tables for catalyst and precatalyst structures used herein.

C_{11}

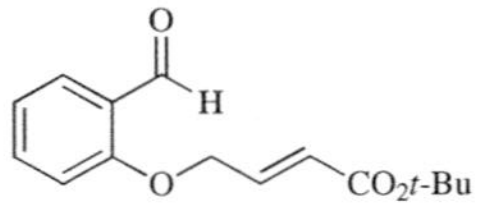

Aryl Aldehyde Substrate	Conditions	Product	Refs.
CO_2t-Bu	Precatalyst (20 mol %), base (*x* mol %), additive (*y* mol %), solvent, 48 h	CO_2t-Bu	153

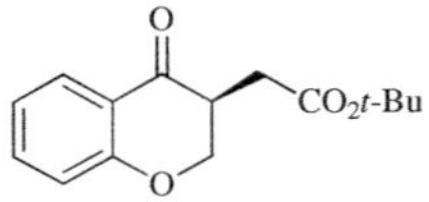

Precatalyst	Base	*x*	Additive	*y*	Solvent	Temp		er
β-**ThB1z**•HI	KHMDS	20	none	—	dioxane	rt	(32)	82:18
β-**ThB1a**•HBr	KHMDS	20	none	—	dioxane	rt	(62)	76.5:23.5
β-**ThB1d**•HBr	KHMDS	20	none	—	dioxane	rt	(21)	83.5:16.5
β-**ThB1d**•HBr	KHMDS	20	none	—	CH_2Cl_2	rt	(<2)	67:33
β-**ThB1d**•HBr	KHMDS	20	none	—	THF	rt	(<2)	76.5:23.5
β-**ThB1d**•HBr	KHMDS	20	none	—	EtOAc	rt	(<2)	74.5:25.5
β-**ThB1d**•HBr	KHMDS	20	none	—	MeCN	rt	(<10)	68:32
β-**ThB1d**•HBr	KHMDS	20	none	—	MeOH	rt	(90)	50:50
β-**ThB1d**•HBr	*t*-BuOK	20	none	—	dioxane	rt	(22)	84:16
β-**ThB1d**•HBr	DBU	20	none	—	dioxane	rt	(<2)	85:15
β-**ThB1d**•HBr	NaOAc	40	none	—	dioxane	rt	(20)	85.5:14.5
β-**ThB1d**•HBr	NaOAc	100	none	—	dioxane	rt	(28)	82:18
β-**ThB1d**•HBr	NaOAc	100	none	—	dioxane/MeOH (1:1)	rt	(92)	52:48
β-**ThB1d**•HBr	NaOAc	100	1,2-$(HO)_2C_6H_4$	100	dioxane	rt	(41)	78:22
β-**ThB1d**•HBr	NaOAc	100	2,6-*t*-$Bu_2C_6H_3OH$	100	dioxane	rt	(36)	83.5:16.5
β-**ThB1d**•HBr	NaOAc	100	4-$(MeO)C_6H_4OH$	100	dioxane	rt	(52)	81.5:18.5
β-**ThB1d**•HBr	NaOAc	100	4-$(MeO)C_6H_4OH$	100	dioxane	60 °C	(93)	66.5:33.5
β-**ThB1d**•HBr	NaOAc	100	$La(NO_3)_3$•$6H_2O$	10	dioxane	rt	(44)	77:23
β-**ThB1d**•HBr	NaOAc	100	$Nd(NO_3)_3$•$6H_2O$	10	dioxane	rt	(52)	79:21

C_{11-12}

β-**ThB1d**•HBr (20 mol %),
NaOAc (1 eq),
4-(MeO)C_6H_4OH (1 eq),
dioxane, rt

153

R^1	R^2	Time (h)		er
H	EtO_2C	48	(50)	75:25
H	$t\text{-}BuO_2C$	48	(53)	81:19
Me	$t\text{-}BuO_2C$	48	(45)	81.5:18.5
MeO	$t\text{-}BuO_2C$	48	(60)	79.5:20.5
Cl	$t\text{-}BuO_2C$	48	(88)	69:31
Cl	$t\text{-}BuO_2C$	21	(50)	79.5:20.5
EtO_2C	$t\text{-}BuO_2C$	24	(54)	77.5:22.5
H	NC–	48	(66)	50:50

A: α-**TrM1a**^{SP}•HBF_4
(10 mol %),
KHMDS (10 mol %),
toluene, rt, time
or
B: α-**TrM1a**^{SP}•HBF_4 in
MMF system (0.63 mmol•g^{-1}),
substrate (0.2 M in toluene),
Et_3N (0.4 M in toluene), rate, rt

149

Conditions	R	Time (h)		er
A	H	16	(95)	96.5:3.5
A	3-Me	16	(95)	97.5:2.5
A	3-MeO	16	(95)	95:5
A	3-Cl	4.5	(95)	96:4
A	3-Br	4.5	(95)	96:4
A	5-Me	16	(95)	90.5:9.5
A	5-MeO	16	(95)	95:5

Conditions	R	Rate (μL•min^{-1})	Productivity[a]	Conv (%)	er
B	H	10	138	(90)	97:3
B	3-Me	10	141	(92)	97.5:2.5
B	3-MeO	10	140	(91)	95:5
B	3-Cl	20	292	(95)	96:4
B	3-Br	20	292	(95)	96:4
B	5-Me	10	138	(90)	90.5:9.5
B	5-MeO	10	142	(92)	95:5

TABLE 1. INTRAMOLECULAR STETTER REACTIONS OF ARYL ALDEHYDES (*Continued*)

Aryl Aldehyde Substrate	Conditions	Product(s), Yield(s) (%), and Stereoselectivity	Refs.

Please refer to the charts preceding the tables for catalyst and precatalyst structures used herein.

C_{11-12}

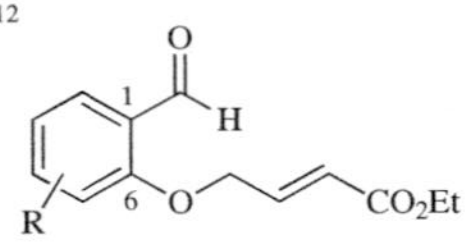

Aryl Aldehyde Substrate	Conditions	Product(s), Yield(s) (%), and Stereoselectivity	Refs.
1 O H 6 O R CO_2Et	β-**TrM4VIb**•$HBPh_4$ (10 mol %), i-Pr_2NEt (10 mol %), cyclohexane, rt, 20 h	O CO_2Et O R	151

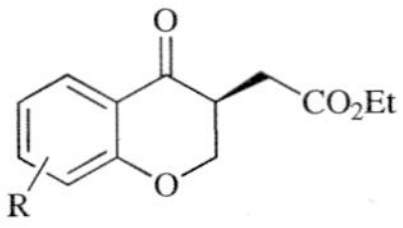

R		er	R		er
5-I	(85)	97:3	3-Me	(93)	95:5
5-F	(97)	94.5:5.5	3-Cl	(99)	97:3
4-MeO	(95)	97:3	3-Br	(95)	97.5:2.5
4-Me	(97)	95.5:4.5	3-F	(96)	96.5:3.5
$3\text{-}CF_3O$	(94)	98:2	$2,4\text{-}Cl_2$	(94)	97:3
3-MeO	(95)	96.5:3.5			

C_{11-14}

Aryl Aldehyde Substrate	Conditions	Product(s), Yield(s) (%), and Stereoselectivity	Refs.
1 O H 6 O R CO_2Et	β-**TrM4VIIIb**•$HBPh_4$ (10 mol %), i-Pr_2NEt (10 mol %), cyclohexane, rt, 20 h	O CO_2Et O R	150

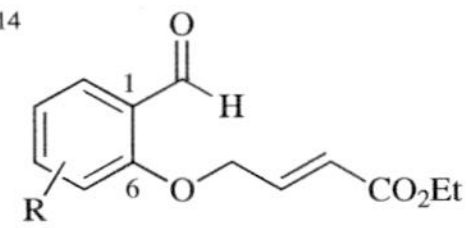

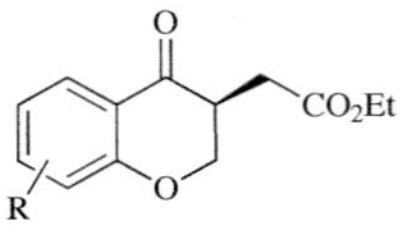

R		er	R		er
5-I	(96)	97:3	3-Cl	(97)	98:2
5-F	(98)	98:2	3-Br	(97)	97:3
4-MeO	(95)	98:2	3-F	(99)	97:3
4-Me	(98)	99:1	$3,5\text{-}Me_2$	(97)	97:3
$3\text{-}CF_3$	(98)	98:2	$2,4,5\text{-}Me_3$	(98)	96.5:3.5
3-MeO	(99)	99:1	$2,4\text{-}Cl_2$	(98)	99.5:0.5
3-Me	(98)	97:3	$3,5\text{-}Cl_2$	(96)	93.5:6.5

C_{11}

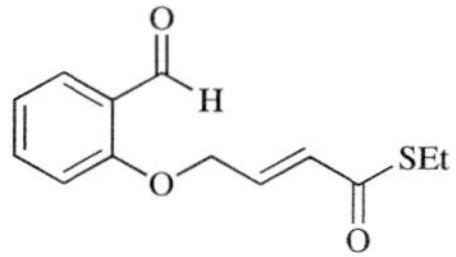

β-TrM1b•HBF_4 (20 mol %), KHMDS (20 mol %), toluene, rt, 24 h — (85) er 85:15 — 62

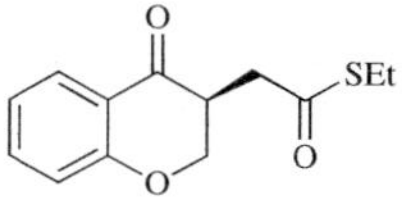

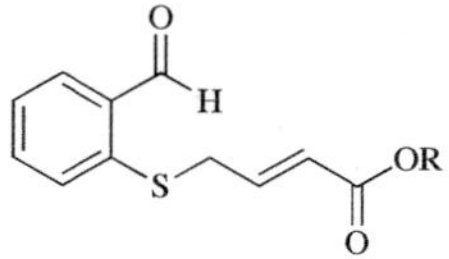

Precatalyst (x mol %), base (y mol %), solvent, rt

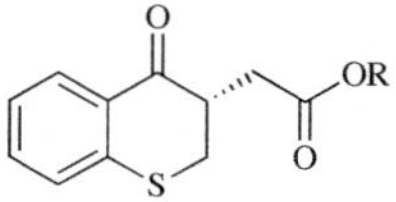

R	Precatalyst	x	Base	y	Solvent	Time (h)		er	
Me	α-**TrM1d**•HBF_4	20	KHMDS	20	xylene	24	(63)	98:2	15
Me	β-**TrP2a**•HBF_4	20	KHMDS	20	toluene	24	(84)	95:5	62
Et	α-**TrY1c**•HBF_4	10	Et_3N	10	xylene	48	(70)	94.5:5.5	146
Et	α-**TrM4b**•HBF_4	20	i-Pr_2NEt	10	xylene	27	(26)	94:6	64

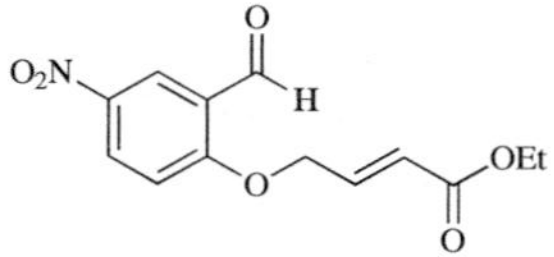

Precatalyst (10 mol %), base (10 mol %), xylene, rt

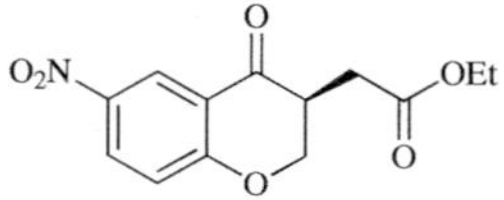

Precatalyst	Base	Time (h)		er	
α-**TrY1c**•HBF_4	Et_3N	15	(90)	50:50	146
α-**TrM4b**•HBF_4	i-Pr_2NEt	10	(86)	16:84	64

TABLE 1. INTRAMOLECULAR STETTER REACTIONS OF ARYL ALDEHYDES (*Continued*)

Aryl Aldehyde Substrate	Conditions	Product(s), Yield(s) (%), and Stereoselectivity	Refs.

Please refer to the charts preceding the tables for catalyst and precatalyst structures used herein.

C_{11}

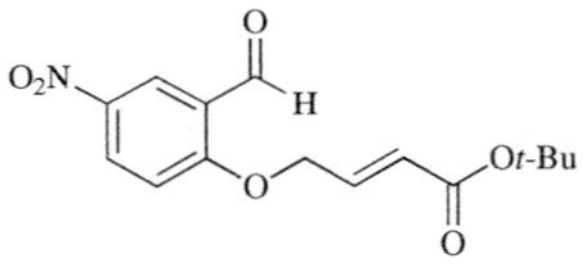

Aryl Aldehyde Substrate	Conditions	Product(s), Yield(s) (%), and Stereoselectivity	Refs.
O_2N, H, O, O*t*-Bu	α-**ThAT1**•HI (20 mol %), *i*-Pr_2NEt (100 mol %), CH_2Cl_2, rt, 48 h	O_2N, O, O*t*-Bu (88) er 50:50	152
MeO, H, O, OR	Precatalyst (*x* mol %), base (*y* mol %), solvent	MeO, O, OR	

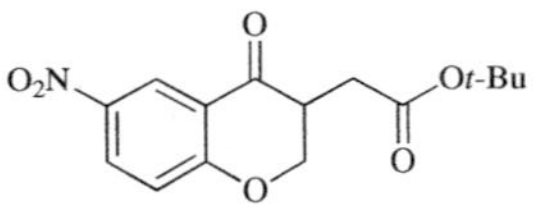

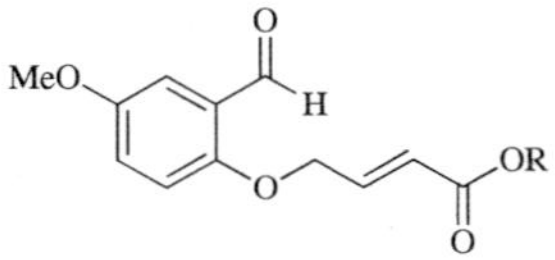

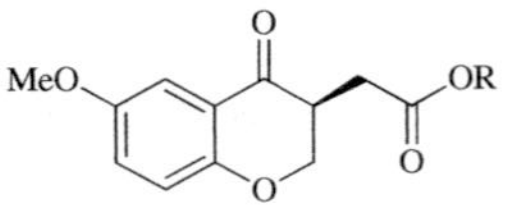

R	Precatalyst	*x*	Base	*y*	Solvent	Temp	Time (h)		er	Refs.
Me	α-**TrD1a**•$HClO_4$	20	K_2CO_3	10	THF	rt	24	(56)	19:81	11
Et	α-**TrP5p**•HBF_4	20	KO*t*-Bu	19	CH_2Cl_2	reflux	15	(71)	8.5:91.5	65
Et	α-**TrY1c**•HBF_4	10	Et_3N	10	xylene	rt	31	(95)	6:94	146
Et	α-**TrM4b**•HBF_4	10	*i*-Pr_2NEt	10	xylene	rt	4	(97)	2:98	64
Et	β-**TrP2a**•HBF_4	20	KHMDS	20	toluene	rt	24	(94)	95:5	62
Et	β-**TrP2e**•HBF_4	20	KHMDS	20	toluene	rt	24	(84)	97:3	62
t-Bu	α-**ThAT1**•HI	20	*i*-Pr_2NEt	100	CH_2Cl_2	rt	48	(47)	18:82	152

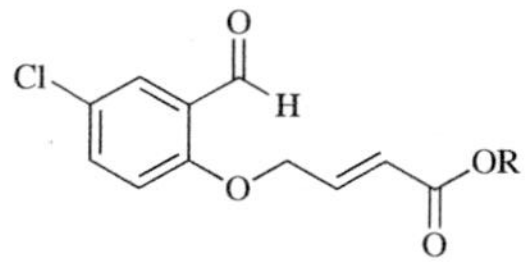

Precatalyst (x mol %),
base (y mol %), solvent

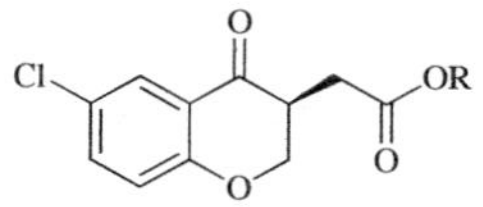

R	Precatalyst	x	Base	y	Solvent	Temp	Time (h)		er	
Me	α-**TrD1a**•$HClO_4$	20	K_2CO_3	10	THF	rt	14	(92)	43:57	11
Et	α-**TrP5p**•HBF_4	20	KO*t*-Bu	19	CH_2Cl_2	reflux	14	(96)	11.5:88.5	65
Et	α-**TrY1c**•HBF_4	10	Et_3N	10	xylene	rt	9	(98)	11:89	146
Et	α-**TrM4b**•HBF_4	10	*i*-Pr_2NEt	10	xylene	rt	4	(94)	3.5:96.5	64
Et	β-**TrP2a**•HBF_4	20	KHMDS	20	toluene	rt	24	(94)	94:6	62
Et	β-**TrP2e**•HBF_4	20	KHMDS	20	toluene	rt	24	(94)	96:4	62

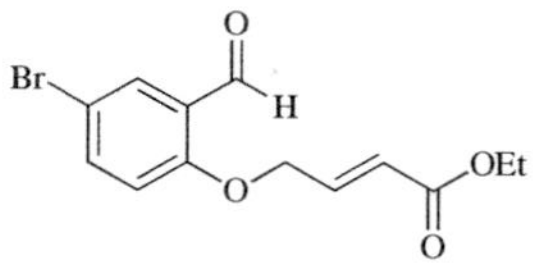

Precatalyst (x mol %),
base (y mol %), solvent

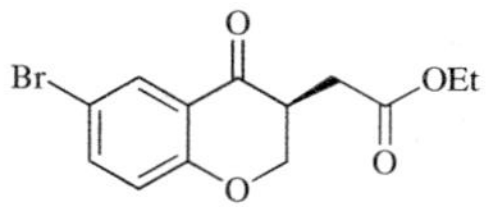

Precatalyst	x	Base	y	Solvent	Temp	Time (h)		er	
α-**TrM1d**•HBF_4	20	KHMDS	20	toluene	rt	24	(68)	21:79	62
α-**TrP5p**•HBF_4	20	KO*t*-Bu	19	CH_2Cl_2	reflux	14	(99)	18:82	65
α-**TrY1c**•HBF_4	10	Et_3N	10	xylene	rt	12	(93)	11:89	146
α-**TrM4b**•HBF_4	10	*i*-Pr_2NEt	10	xylene	rt	4	(94)	3.5:96.5	64
β-**TrP2a**•HBF_4	20	KHMDS	20	toluene	rt	24	(95)	95:5	62
β-**TrP2e**•HBF_4	20	KHMDS	20	toluene	rt	24	(94)	96:4	62

TABLE 1. INTRAMOLECULAR STETTER REACTIONS OF ARYL ALDEHYDES (*Continued*)

Aryl Aldehyde Substrate	Conditions	Product(s), Yield(s) (%), and Stereoselectivity	Refs.

Please refer to the charts preceding the tables for catalyst and precatalyst structures used herein.

C_{11}

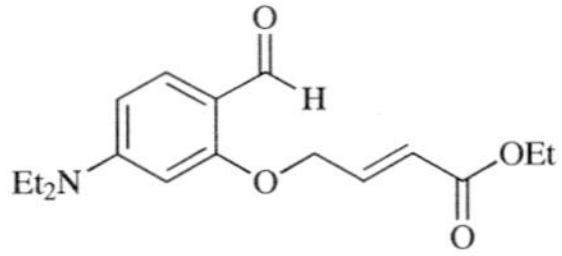

Precatalyst (*x* mol %),
base (*y* mol %), solvent, rt

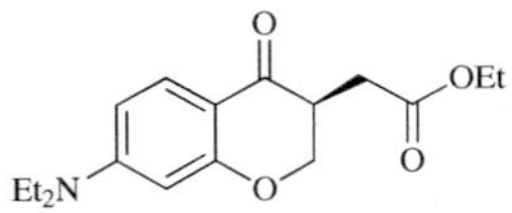

Precatalyst	*x*	Base	*y*	Solvent	Time (h)		er	Refs.
α-**TrY1c**•HBF_4	10	KHMDS	10	xylene	67	(56)	2.5:97.5	146
α-**TrM4b**•HBF_4	20	*i*-Pr_2NEt	10	xylene	24	(90)	1.5:98.5	64
β-**TrP2e**•HBF_4	20	KHMDS	20	toluene	24	(55)	97.5:2.5	62

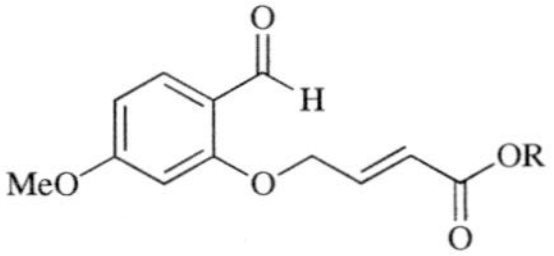

Precatalyst (*x* mol %),
base (*y* mol %), solvent

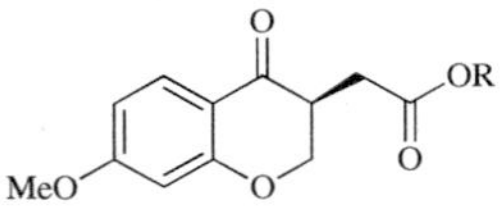

R	Precatalyst	*x*	Base	*y*	Solvent	Temp	Time (h)		er	Refs.
Me	α-**TrD1a**•$HClO_4$	50	K_2CO_3	20	THF	rt	48	(22)	14:86	11
Et	α-**TrP5p**•HBF_4	20	KO*t*-Bu	19	CH_2Cl_2	reflux	15	(15)	11:89	65
Et	α-**TrY1c**•HBF_4	10	Et_3N	10	xylene	rt	26	(87)	1.5:98.5	146
Et	α-**TrM4b**•HBF_4	10	*i*-Pr_2NEt	10	xylene	rt	24	(89)	1.5:98.5	64
Et	β-**TrP2a**•HBF_4	20	KHMDS	20	toluene	rt	24	(45)	95:5	62
Et	β-**TrP2e**•HBF_4	20	KHMDS	20	toluene	rt	24	(94)	95.5:4.5	62
t-Bu	α-**ThAT1**•HI	20	*i*-Pr_2NEt	100	CH_2Cl_2	rt	48	(39)	12:88	152

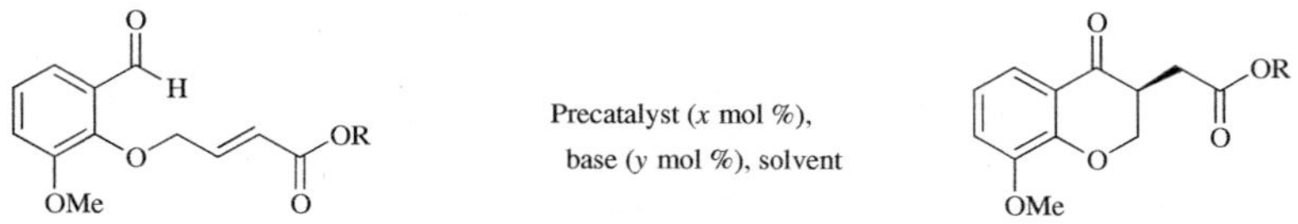

Precatalyst (x mol %), base (y mol %), solvent

R	Precatalyst	x	Base	y	Solvent	Temp	Time (h)		er	
Me	α-**TrD1a**•$HClO_4$	20	K_2CO_3	10	THF	rt	24	(44)	16:84	11
Et	α-**TrD1a**•$HClO_4$	20	K_2CO_3	10	THF	rt	24	(69)	19:81	11
Et	α-**TrP5p**•HBF_4	20	KO*t*-Bu	19	CH_2Cl_2	40 °C	13	(89)	14:86	65
Et	α-**TrY1c**•HBF_4	10	Et_3N	10	xylene	rt	34	(98)	9:91	146
Et	α-**TrM4b**•HBF_4	10	*i*-Pr_2NEt	10	xylene	rt	5	(95)	2:98	64
Et	α-**TrM1d**•HBF_4	10	KHMDS	10	xylene	rt	24	(95)	6:94	15
Et	β-**TrP2a**•HBF_4	20	KHMDS	20	toluene	rt	24	(95)	90:10	62
Et	β-**TrP2e**•HBF_4	20	KHMDS	20	toluene	rt	24	(86)	97.5:2.5	62

TABLE 1. INTRAMOLECULAR STETTER REACTIONS OF ARYL ALDEHYDES (*Continued*)

Aryl Aldehyde Substrate	Conditions	Product(s), Yield(s) (%), and Stereoselectivity	Refs.

Please refer to the charts preceding the tables for catalyst and precatalyst structures used herein.

C_{12-18}

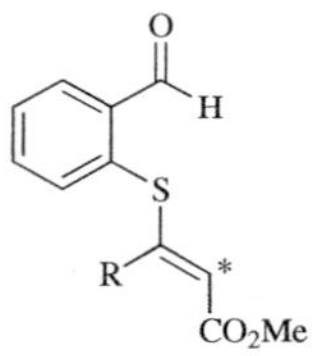

Precatalyst (20 mol %), base (*x* mol %), toluene, rt, 24 h

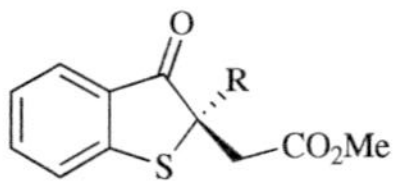

R	*(*E*) or (*Z*)	Precatalyst	Base	*x*		er	Refs.
Et	(*E*)	α-**TrM1b**•HBF_4	Et_3N	200	(95)	96:4	72
Et	(*E*)	α-**TrM1b**•HBF_4	KO*t*-Bu	200	(90)	99:1	73
Et	(*Z*)	α-**TrM1b**•HBF_4	KO*t*-Bu	200	(89)	93:7	73
n-Pr	(*E*)	α-**TrM1b**•HBF_4	Et_3N	200	(54)	94:6	73
n-Pr	(*E*)	α-**TrM1b**•HBF_4	KO*t*-Bu	200	(83)	99:1	73
n-Pr	(*Z*)	α-**TrM1b**•HBF_4	KO*t*-Bu	200	(85)	94.5:5.5	73
n-Pr	(*E*)	β-**TrP2a**•HBF_4	KHMDS	20	(80)	5:95	73
$PhCH_2CH_2$	(*E*)	α-**TrM1b**•HBF_4	Et_3N	200	(33)	94:6	73
$PhCH_2CH_2$	(*E*)	α-**TrM1b**•HBF_4	KO*t*-Bu	200	(91)	99:1	73
$PhCH_2CH_2$	(*Z*)	α-**TrM1b**•HBF_4	KO*t*-Bu	200	(92)	92:8	73
Ph	(*E*)	α-**TrM1b**•HBF_4	Et_3N	200	(11)	91:9	73
Ph	(*E*)	α-**TrM1b**•HBF_4	KO*t*-Bu	200	(15)	91:9	73

C_{12}

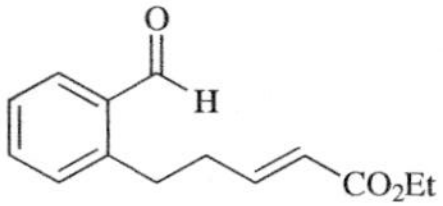

Precatalyst (20 mol %),
KHMDS (20 mol %),
xylenes, rt, 24 h

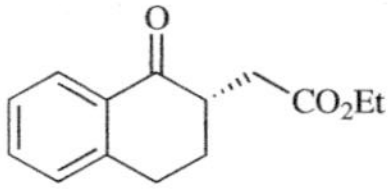

Precatalyst		er	Refs.
α-**TrM1d**•HBF_4	(35)	97:3	15
α-**TrM1a**•HBF_4	(50)	97:3	62
β-**TrP2a**•HBF_4	(90)	4:96	62

C_{12-18}

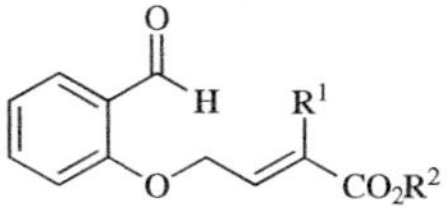

β-**TrP2e**•HBF_4 (20 mol %),
KHMDS (20 mol %),
toluene, rt, 24 h

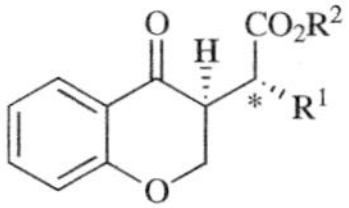

R^1	R^2		*dr	er	Refs.
Me	Et	(94)	30:1	97.5:2.5	
Et	Et	(95)	35:1	96:4	27
allyl	Me	(95)	13:1	92:8	
n-Bu	Et	(53)	12:1	97:3	
Bn	Et	(80)	20:1	92:8	

C_{12}

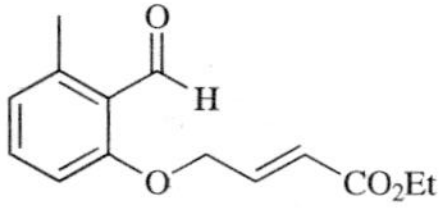

α-**TrP5p**•HBF_4 (20 mol %),
t-BuOK (19 mol %),
CH_2Cl_2, reflux, 14 h

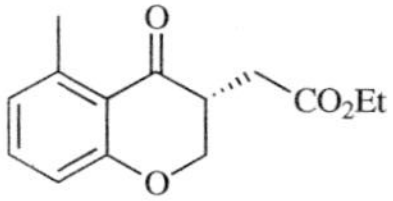

(55) er 87.5:12.5 65

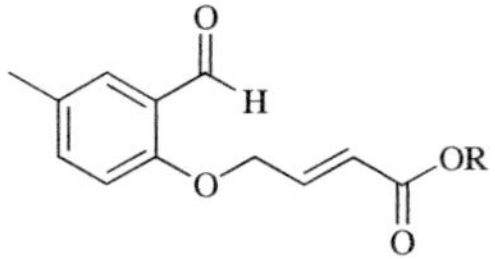

Precatalyst (*x* mol %),
base (*y* mol %), solvent

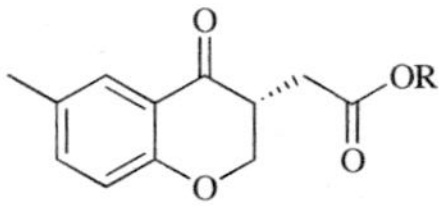

R	Precatalyst	*x*	Base	*y*	Solvent	Temp	Time (h)		er	Refs.
Et	α-**TrP5p**•HBF_4	20	KO*t*-Bu	19	CH_2Cl_2	reflux	14	(65)	89.5:10.5	65
Et	α-**TrY1c**•HBF_4	10	Et_3N	10	xylene	rt	21	(98)	98:2	146
Et	α-**TrM4b**•HBF_4	10	*i*-Pr_2NEt	10	xylene	rt	7	(95)	98.5:1.5	64
Et	α-**TrM1d**•HBF_4	10	KHMDS	10	xylene	rt	24	(80)	98.5:1.5	15
Et	β-**TrP2a**•HBF_4	20	KHMDS	20	toluene	rt	24	(95)	2:98	62
t-Bu	α-**ThAT1**•HI	20	*i*-Pr_2NEt	100	CH_2Cl_2	rt	48	(45)	87.5:12.5	152

TABLE 1. INTRAMOLECULAR STETTER REACTIONS OF ARYL ALDEHYDES (*Continued*)

Aryl Aldehyde Substrate	Conditions	Product(s), Yield(s) (%), and Stereoselectivity	Refs.

Please refer to the charts preceding the tables for catalyst and precatalyst structures used herein.

C_{12}

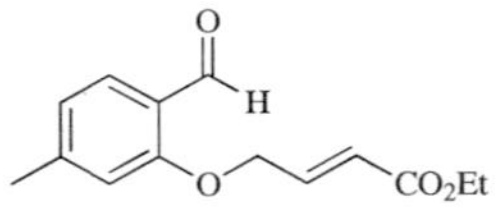

Aryl Aldehyde Substrate	Conditions	Product(s), Yield(s) (%), and Stereoselectivity	Refs.
CO_2Et	α-**TrP5p**•HBF_4 (20 mol %), *t*-BuOK (19 mol %), CH_2Cl_2, reflux, 14 h	CO_2Et (21) er 91:9	65
OR	Precatalyst (*x* mol %), base (*y* mol %), solvent	OR	

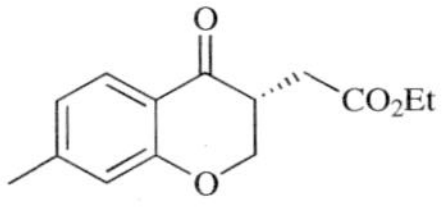

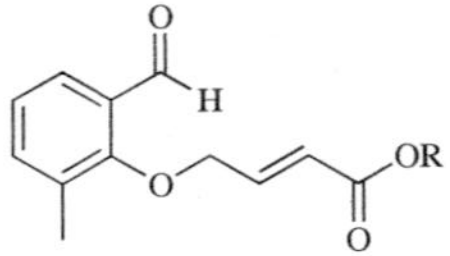

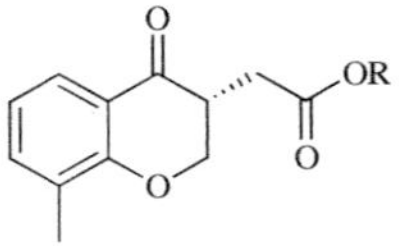

R	Precatalyst	*x*	Base	*y*	Solvent	Temp	Time (h)		er	Refs.
Et	α-**TrP5p**•HBF_4	20	KO*t*-Bu	19	CH_2Cl_2	reflux	15	(76)	88:12	65
Et	α-**TrY1c**•HBF_4	10	Et_3N	10	xylene	rt	8	(98)	90:10	146
Et	α-**TrM4b**•HBF_4	10	*i*-Pr_2NEt	10	xylene	rt	6	(95)	97.5:2.5	64
Et	α-**TrM1d**•HBF_4	20	KHMDS	20	xylene	rt	24	(90)	92:8	15
Et	β-**TrP2a**•HBF_4	20	KHMDS	20	toluene	rt	24	(94)	10:90	62
t-Bu	α-**ThAT1**•HI	20	*i*-Pr_2NEt	100	CH_2Cl_2	rt	48	(45)	87:13	152

C_{13}

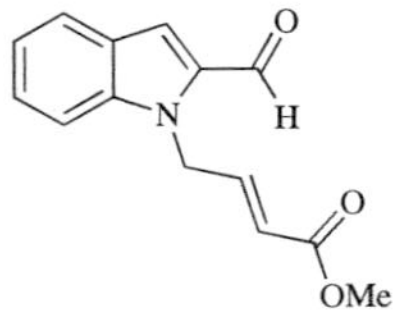

Aryl Aldehyde Substrate	Conditions	Product(s), Yield(s) (%), and Stereoselectivity	Refs.
OMe	β-**TrP2a**•HBF_4 (20 mol %), KHMDS (20 mol %), toluene, rt, 24 h	OMe (80) er 82:18	62

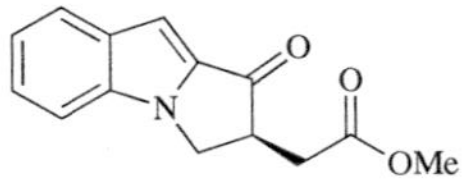

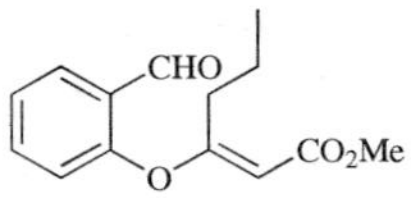

CHO

CO_2Me

Precatalyst (20 mol %),
Et_3N (200 mol %),
toluene, rt, 20 h

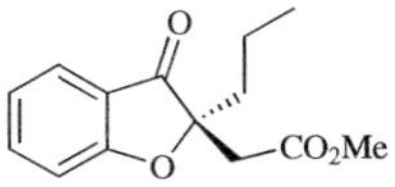

CO_2Me

147

Precatalyst		er
α-**TrM4IIIa**•HBF_4	(<5)	—
α-**TrM4IIIb**•HBF_4	(97)	69:31
α-**TrM4IIIc**•HCl	(37)	74:26
α-**TrM4IVa**•HBF_4	(<5)	—
α-**TrM4IVc**•HCl	(31)	71:29
α-**TrM4VIIa**•HBF_4	(<5)	—
α-**TrM4VIIb**•HBF_4	(98)	60:40
α-**TrM4VIIc**•HCl	(24)	64:36

Precatalyst		er
β-**TrM4IIa**•HBF_4	(<5)	—
β-**TrM4IIb**•HBF_4	(98)	13:87
β-**TrM4IIc**•HCl	(27)	10:90
β-**TrM4IIj**•$HBPh_4$	(98)	8:92
α-**TrM4Va**•HBF_4	(<5)	—
α-**TrM4Vb**•HBF_4	(97)	97:3
α-**TrM4IIc**•HCl	(<5)	—

TABLE 1. INTRAMOLECULAR STETTER REACTIONS OF ARYL ALDEHYDES (*Continued*)

Aryl Aldehyde Substrate	Conditions	Product(s), Yield(s) (%), and Stereoselectivity	Refs.

Please refer to the charts preceding the tables for catalyst and precatalyst structures used herein.

C_{13}

Aryl Aldehyde Substrate	Conditions	Product(s), Yield(s) (%), and Stereoselectivity	Refs.
CHO, CO_2Me	α-**TrM4Vb**•HBF_4 (20 mol %), base (20 mol %), solvent, rt	O, CO_2Me	147

Base	Solvent	Time (h)		er
Et_3N	toluene	20	(96)	97:3
i-Pr_2NEt	toluene	20	(98)	94.5:5.5
c-Cy_2NEt	toluene	20	(97)	96.5:3.5
DMAP	toluene	20	(89)	96:4
TBD	toluene	20	(51)	96:4
P2-Et	toluene	0.5	(99)	96:4
KHMDS	toluene	0.5	(99)	97:3
t-BuOK	toluene	20	(98)	97:3

Base	Solvent	Time (h)		er
BEMP	toluene	0.3	(98)	96:4
PEMP	toluene	20	(98)	96.5:3.5
KHMDS	*o*-xylene	20	(90)	92:8
KHMDS	THF	24	(82)	96:4
KHMDS	CPME	0.5	(99)	97:3
KHMDS	*t*-AmOMe	0.5	(98)	98:2
KHMDS	cyclohexane	0.5	(98)	98.5:1.5
KHMDS	TBME	0.5	(97)	98:2
KHMDS	EtOH	48	(0)	—

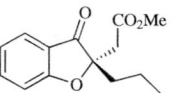

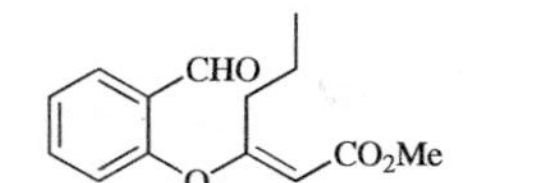

β-**TrM4IIj**•$HBPh_4$ (20 mol %), base (20 mol %), solvent, rt,

147

Base	Solvent	Time (h)		er
Et_3N	toluene	20	(98)	92:8
i-Pr_2NEt	toluene	20	(97)	94.5:5.5
c-Cy_2NEt	toluene	20	(97)	95:5
DMAP	toluene	20	(91)	94:6
TBD	toluene	20	(98)	96:4
P2-Et	toluene	0.5	(99)	96:4
KHMDS	toluene	0.5	(99)	94.5:5.5
t-BuOK	toluene	20	(98)	97:3
BEMP	toluene	0.3	(98)	93:7

Base	Solvent	Time (h)		er
PEMP	toluene	3	(99)	96:4
t-BuOK	*o*-xylene	20	(99)	96:4
t-BuOK	THF	24	(84)	96.5:3.5
t-BuOK	CPME	0.5	(99)	97:3
t-BuOK	*t*-AmOMe	0.5	(98)	96.5:3.5
t-BuOK	cyclohexane	0.5	(98)	98.5:1.5
t-BuOK	TBME	0.5	(96)	97.5:2.5
t-BuOK	EtOH	48	(0)	—

TABLE 1. INTRAMOLECULAR STETTER REACTIONS OF ARYL ALDEHYDES (*Continued*)

Aryl Aldehyde Substrate	Conditions	Product(s), Yield(s) (%), and Stereoselectivity	Refs.

Please refer to the charts preceding the tables for catalyst and precatalyst structures used herein.

C_{13}

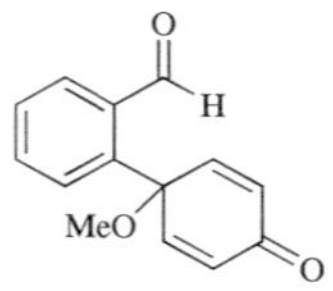

Aryl Aldehyde Substrate	Conditions	Product(s), Yield(s) (%), and Stereoselectivity	Refs.
MeO, H, O, O	Precatalyst (10 mol %), base (10 mol %), solvent (0.2 M), rt	O, MeO, O — *stereochemical configuration not determined*	69

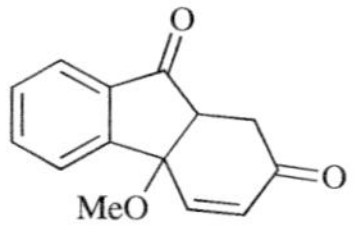

Precatalyst	Base	Solvent	Time (h)		er
α-**TrM4a**•HBF_4	KOAc	Et_2O	30	(15)	66:34
α-**TrM4d**•HBF_4	KOAc	Et_2O	30	(11)	66:34
α-**TrM4c**•HBF_4	KOAc	Et_2O	30	(73)	54:46
α-**TrM4i**•HBF_4	KOAc	Et_2O	30	(<2)	(—)
α-**TrM4b**•HBF_4	KOAc	Et_2O	6	(96)	93:7
α-**TrM4b**•HBF_4	KOAc	Et_2O	24	(90)	94.5:5.5
α-**TrM4b**•HBF_4	KOAc	Et_2O	16	(96)[b]	92:8[b]
α-**TrM4b**•HBF_4	KOAc	Et_2O	20	(96)[c]	92:8[c]
β-**TrM1b**•HBF_4	KOAc	Et_2O	30	(96)	22.5:77.5
β-**TrM1c**•HBF_4	KOAc	Et_2O	30	(74)	71:29
β-**TrM1i**•HBF_4	KOAc	Et_2O	30	(51)	44:56
α-**TrM4b**•HBF_4	KOAc	toluene	6	(96)	89.5:10.5
α-**TrM4b**•HBF_4	KOAc	xylene	1	(98)	91:9
α-**TrM4b**•HBF_4	KOAc	CH_2Cl_2	16	(96)	86.5:13.5
α-**TrM4b**•HBF_4	KOAc	$CHCl_3$	6	(92)	85.5:14.5
α-**TrM4b**•HBF_4	KOAc	CCl_4	3	(94)	87:13

Precatalyst	Base	Solvent	Time (h)		er
α-**TrM4b**•HBF_4	KOAc	cyclohexane	1	(90)	87:13
α-**TrM4b**•HBF_4	KOAc	MeOH	16	(<2)	(—)
α-**TrM4b**•HBF_4	KOAc	THF	16	(76)	91.5:8.5
α-**TrM4b**•HBF_4	KOAc	1,4-dioxane	16	(60)	88.5:11.5
α-**TrM4b**•HBF_4	KOAc	DME	24	(95)	91.5:8.5
α-**TrM4b**•HBF_4	KHMDS	toluene	12	(96)	53:47
α-**TrM4b**•HBF_4	DBU	toluene	48	(45)	68:32
α-**TrM4b**•HBF_4	Et_3N	toluene	48	(45)	75:25
α-**TrM4b**•HBF_4	*i*-Pr_2NEt	toluene	48	(72)	80:20
α-**TrM4b**•HBF_4	NaOAc	toluene	24	(96)	85:15
α-**TrM4b**•HBF_4	LiOAc	toluene	48	(77)	83:17
α-**TrM4b**•HBF_4	CsOAc	toluene	2	(92)	89.5:11.5
α-**TrM4b**•HBF_4	$NaHCO_3$	toluene	48	(79)	75:25
α-**TrM4b**•HBF_4	K_3PO_4	toluene	12	(95)	65.5:34.5
α-**TrM4b**•HBF_4	KO*t*-Bu	toluene	12	(93)	75.5:24.5
α-**TrM4b**•HBF_4	Cs_2CO_3	toluene	48	(64)	67:33

C_{13-14}

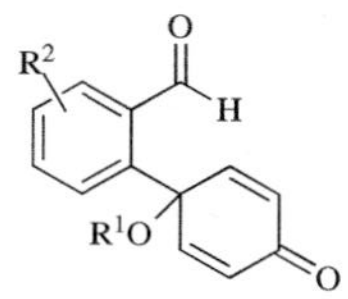

α-**TrM4b**•HBF_4 (10 mol %), KOAc (10 mol %), Et_2O, 0 °C

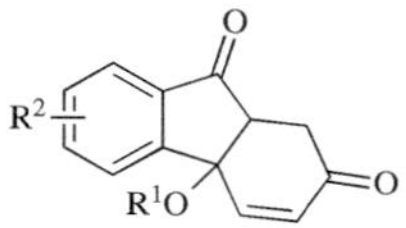

stereochemical configuration not determined

69

R^1	R^2	Time (h)		er
Me	H	24	(90)	94.5:5.5
Me	4-F	21	(91)	93:7
Me	5-F	16	(87)	94:6
Me	6-F	12	(92)	94.5:5.5
Me	5-Cl	3	(96)	93:7

R^1	R^2	Time (h)		er
Me	5-MeO	21	(95)	90.5:9.5
Me	4-Me	16	(95)	91.5:8.5
Me	5-Me	12	(91)	92:8
Et	H	16	(87)	94:6
allyl	H	40	(91)	92.5:7.5

C_{13}

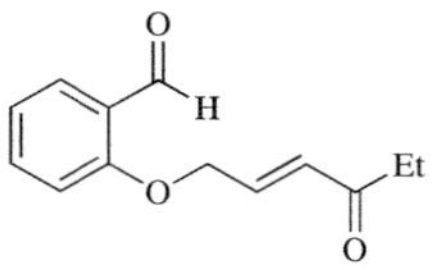

α-**TrM1b**•HBF_4 (20 mol %), KHMDS (20 mol %), toluene, rt, 24 h

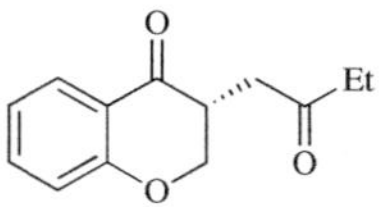

(94) er 96:4

62

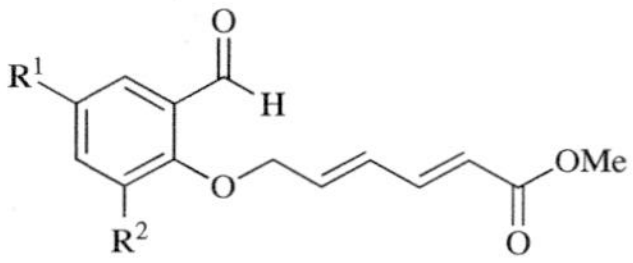

α-**TrM1b**•HBF_4 (10 mol %), DBU (10 mol %), toluene, rt, 18–20 h

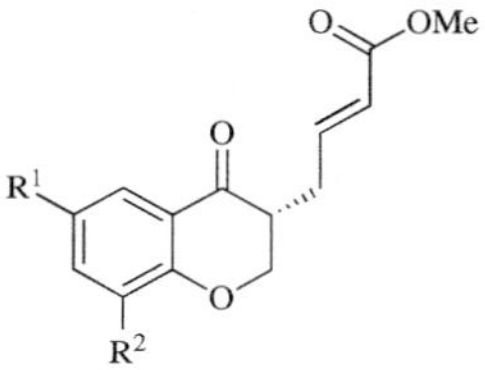

R^1	R^2		er
H	H	(88)	92:8
Br	H	(70)	97:3
H	MeO	(83)	98:2

154

TABLE 1. INTRAMOLECULAR STETTER REACTIONS OF ARYL ALDEHYDES (*Continued*)

Aryl Aldehyde Substrate	Conditions	Product(s), Yield(s) (%), and Stereoselectivity	Refs.

Please refer to the charts preceding the tables for catalyst and precatalyst structures used herein.

C_{13}

Aryl Aldehyde Substrate	Conditions	Product(s), Yield(s) (%), and Stereoselectivity	Refs.
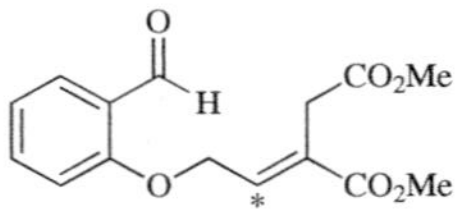 *	β-**TrP2e**•HBF_4 (20 mol %), KHMDS (20 mol %), toluene, rt, 24 h	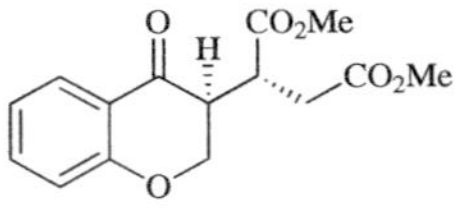 I 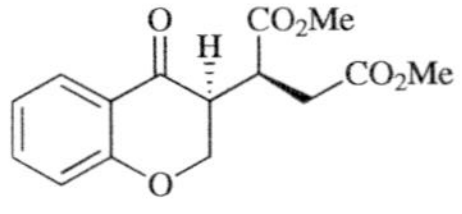 + II	27

*(E) or (Z)		I/II	er
(*E*)	(80)	42:1	96:4
(*Z*)	(70)	1:6	69:31

C_{13-14}

Aryl Aldehyde Substrate	Conditions	Product(s), Yield(s) (%), and Stereoselectivity	Refs.
Y	β-**TrP2e**•HBF_4 (20 mol %), KHMDS (20 mol %), toluene, rt, 24 h	H, O, Y, *	27

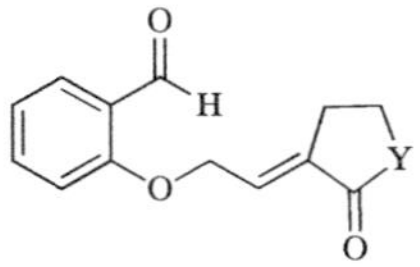

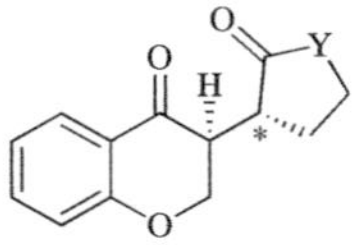

Y		*dr	er
O	(95)	10:1	97:3
CH_2	(80)	18:1	98:2

C_{15-22}

Aryl Aldehyde Substrate	Conditions	Product(s), Yield(s) (%), and Stereoselectivity	Refs.
CHO, R, CO_2Me	Precatalyst (20 mol %), base (20 mol %), cyclohexane, rt, 3 h	R, CO_2Me	147

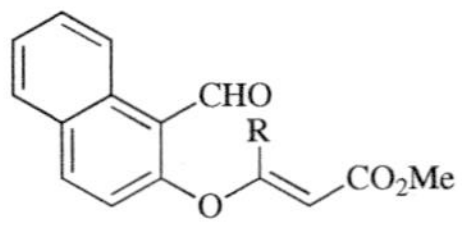

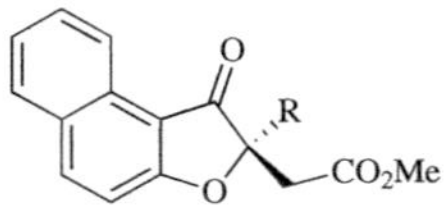

R	Precatalyst	Base		er
Me	α-**TrM4Vb**•HBF_4	KHMDS	(98)	66:34
Me	β-**TrM4IIj**•$HBPh_4$	*t*-BuOK	(96)	48.5:51.5
n-Pr	α-**TrM4Vb**•HBF_4	KHMDS	(95)	75:25
n-Pr	β-**TrM4IIj**•$HBPh_4$	*t*-BuOK	(96)	34:66
$Ph(CH_2)_2$	α-**TrM4Vb**•HBF_4	KHMDS	(94)	64:36
$Ph(CH_2)_2$	β-**TrM4IIj**•$HBPh_4$	*t*-BuOK	(97)	25:75

C_{15}

Reactant	Conditions	Product(s) and Yield(s) (%)	Refs.
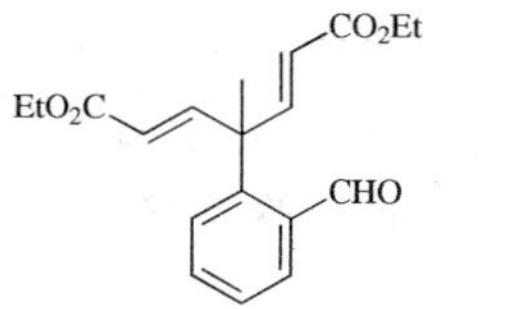 EtO_2C, CO_2Et, CHO	α-**TrM1j**•HBF_4 (20 mol %), *t*-BuOK (20 mol %), toluene, 100 °C, 24 h	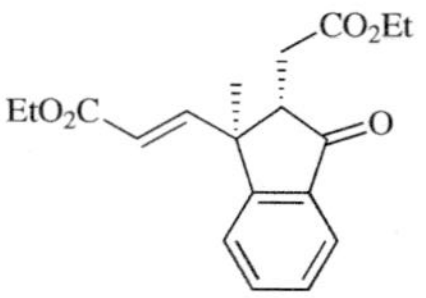 EtO_2C, CO_2Et, O (99) dr 1.3:1, er (major) 62:38, er (minor) 57:43	70
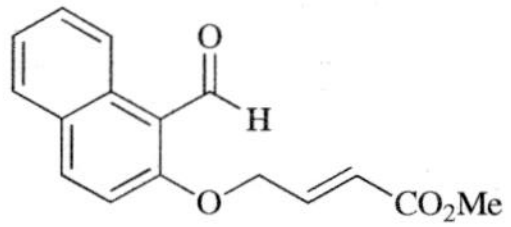 O, H, O, CO_2Me	α-**TrD1a**•$HClO_4$ (20 mol %), K_2CO_3 (10 mol %), THF, rt, 10 h	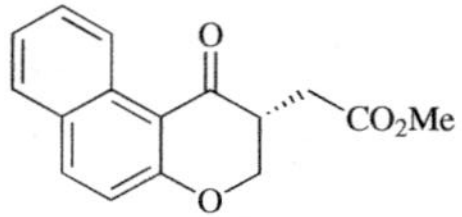 O, CO_2Me, O (51) er 82.5:17.5	11
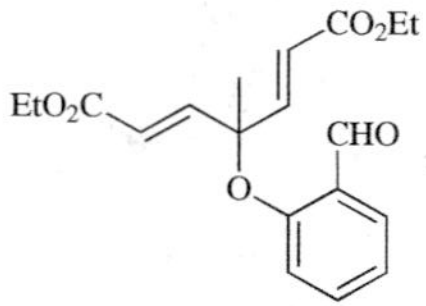 EtO_2C, CO_2Et, CHO, O	α-**TrM1j**•HBF_4 (20 mol %), *t*-BuOK (20 mol %), toluene, –25 °C, 24 h	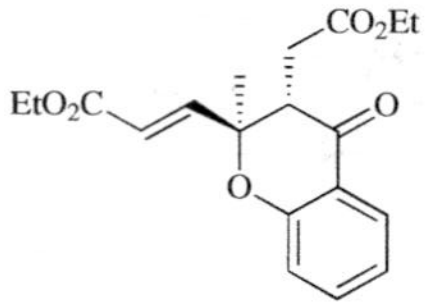 EtO_2C, CO_2Et, O, O (98) dr 2.9:1, er 94:6	70

C_{16}

Reactant	Conditions	Product(s) and Yield(s) (%)	Refs.
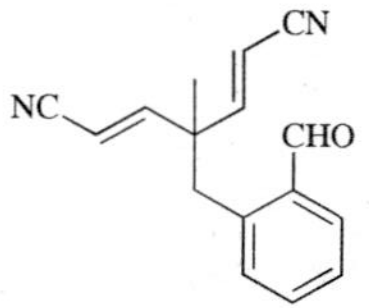 NC, CN, CHO	Precatalyst (20 mol %), *t*-BuOK (20 mol %), solvent, –25 °C, 24 h	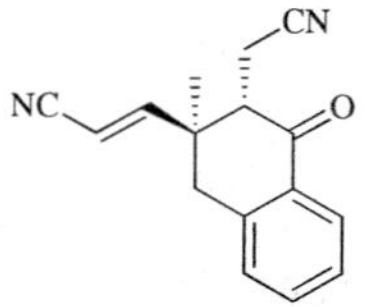 NC, CN, O	70

Precatalyst	Solvent		dr	er
α-**TrM1j**•HBF_4	toluene	(92)	2:1	83:17
α-**TrM1b**•HBF_4	THF	(48)	3.2:1	88:12

TABLE 1. INTRAMOLECULAR STETTER REACTIONS OF ARYL ALDEHYDES (*Continued*)

Aryl Aldehyde Substrate	Conditions	Product(s), Yield(s) (%), and Stereoselectivity	Refs.

Please refer to the charts preceding the tables for catalyst and precatalyst structures used herein.

C_{16}

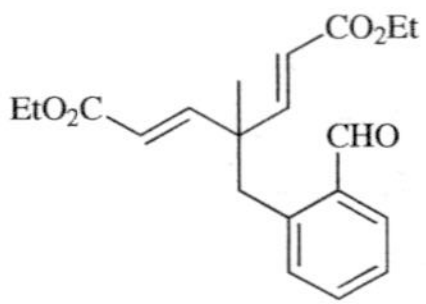

Substrate: EtO_2C, CO_2Et, CHO

Conditions: Precatalyst (20 mol %), base (20 mol %), solvent, 24 h

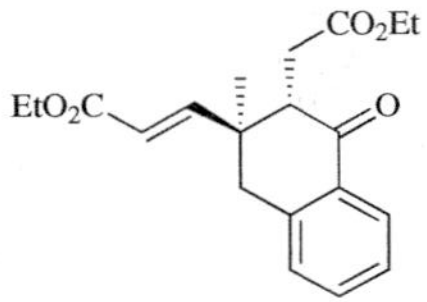

Product: EtO_2C, CO_2Et, O

Refs.: 70

Precatalyst	Base	Solvent	Temp (°C)		dr	er
α-**TrM1a**•HBF_4	KHMDS	toluene	23	(<2)	—	—
α-**TrM1b**•HBF_4	KHMDS	toluene	23	(65)	4.2:1	93.5:6.5
α-**TrM1c**•HBF_4	KHMDS	toluene	23	(49)	5.2:1	99.5:0.5
α-**TrM1j**•HBF_4	KHMDS	toluene	23	(93)	4.3:1	99.5:0.5
α-**TrM1j**•HBF_4	K_2CO_3	toluene	23	(87)	4.3:1	98.5:1.5
α-**TrM1j**•HBF_4	DBU	toluene	23	(75)	4.3:1	99.5:0.5
α-**TrM1j**•HBF_4	*t*-BuOK	toluene	23	(99)	4.3:1	99.5:0.5
α-**TrM1j**•HBF_4	*t*-BuOK	CH_2Cl_2	23	(99)	3.6:1	99:1
α-**TrM1j**•HBF_4	*t*-BuOK	THF	23	(99)	4.4:1	99.5:0.5
α-**TrM1j**•HBF_4	*t*-BuOK	toluene	–25	(99)	7.2:1	99.5:0.5

$C_{16–17}$

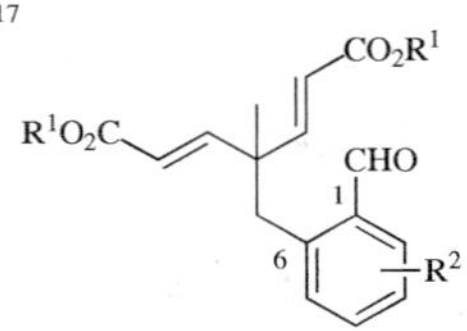

Substrate: R^1O_2C, CO_2R^1, CHO, 1, 6, R^2

Conditions: α-**TrM1j**•HBF_4 (20 mol %), *t*-BuOK (20 mol %), toluene, –25 to 0 °C, 24 h

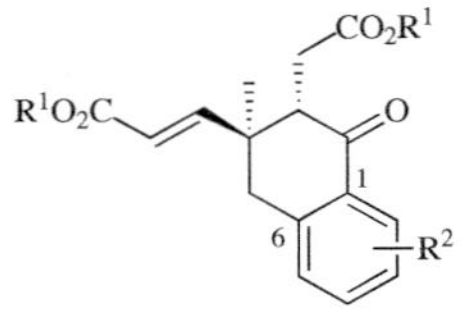

Product: R^1O_2C, CO_2R^1, O, 1, 6, R^2

Refs.: 70

R^1	R^2		dr	er
Me	H	(63)	6:1	99:1
Me	3-Me	(91)	6.6:1	96:4
Me	3-F	(60)	4.8:1	91.5:8.5
n-Bu	3-Me	(81)	4.2:1	98:2
i-$Pr(CH_2)_2$	3-Me	(99)	3.2:1	95:5
Ph	H	(49)	4.5:1	98.5:1.5
4-MeC_6H_4	H	(82)	3.5:1	98.5:1.5
4-FC_6H_4	3-Me	(99)	5.7:1	98:2
4-ClC_6H_4	3-Me	(57)	3.7:1	98.5:1.5
2-furyl	3-Me	(82)	9:1	92.5:7.5
2-thienyl	3-Me	(81)	4.9:1	96:4

Substrate	Conditions	Product(s) and Yield(s) (%)	Refs.
C_{16-22} R^1O_2C … CO_2R^1, R^2, CHO, R^3	α-**TrM1j**•HBF_4 (20 mol %), *t*-BuOK (20 mol %), toluene, –25 °C, 24 h	R^1O_2C … CO_2R^1, R^2, O, R^3	70
C_{17} … COPh	Precatalyst (20 mol %), KHMDS (20 mol %), toluene, rt, 24 h	… COPh	62
C_{18} … Ph, O	α-**TrM1b**•HBF_4 (20 mol %), Et_3N (10 equiv), toluene, rt, 24 h	… Ph, O (55) er 99:1	73

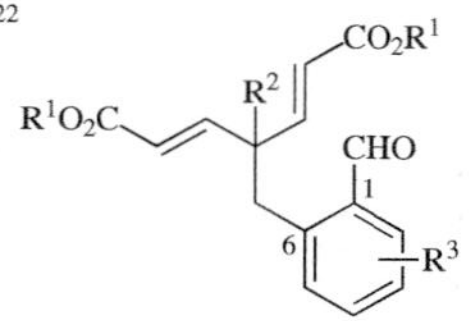

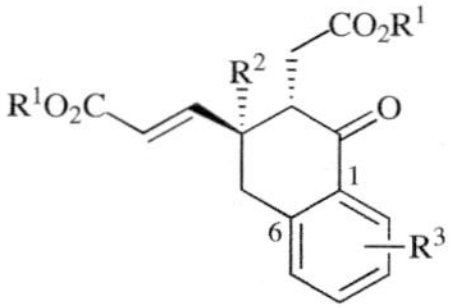

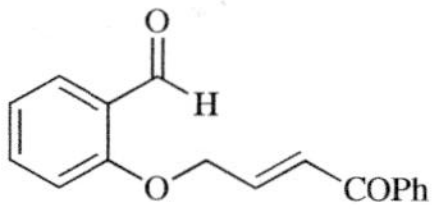

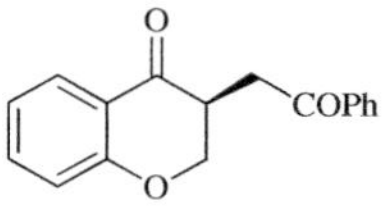

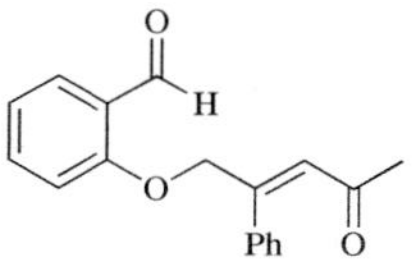

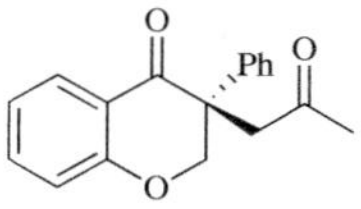

R^1	R^2	R^3		dr	er
Me	Me	H	(88)	3.8:1	99.5:0.5
Et	Et	H	(99)	6:1	99:1
Et	allyl	H	(99)	4.8:1	97:3
Et	Bn	H	(67)	3:1	99:1
Ph	Me	H	(64)	3:1	97.5:2.5
t-Bu	Me	H	(71)	4.4:1	98:2

R^1	R^2	R^3		dr	er
Et	Me	4-F	(75)	3.8:1	95:5
Et	Me	3-Cl	(48)	4.5:1	98.5:1.5
Et	Me	3-F	(86)	5.3:1	96:4
Et	allyl	3-F	(98)	3.7:1	96.5:3.5
Et	Me	3-Me	(70)	6.1:1	99:1

Precatalyst		er
β-**TrM1b**•HBF_4	(94)	89:11
β-**TrP2a**•HBF_4	(85)	84:16

[a] The units are [mmol(product)]•h^{-1}•$[mmol(cat.)]^{-1}$ $\times 10^3$.

[b] The reaction was conducted at 0.05 M.

[c] The reaction was conducted at 0 °C with KOAc (50 mol %).

TABLE 2. INTRAMOLECULAR STETTER REACTIONS OF ALIPHATIC ALDEHYDES

Aliphatic Aldehyde Substrate	Conditions	Product(s), Yield(s) (%), and Stereoselectivity	Refs.

Please refer to the charts preceding the tables for catalyst and precatalyst structures used herein.

C_{5-6}

Substrate	Conditions	Product	Refs.
Y, H, O, P—R, R	α-**TrM1b**•HBF_4 (20 mol %), KHMDS (20 mol %), toluene, rt, 12 h	O, R, P—R, O, Y	67

Y	R		er
O	EtO	(94)	94:6
CH_2	EtO	(66)	87:13
CH_2	Ph	(96)	95:5
CH_2	PhO	(80)	95:5

C_6

Substrate	Conditions	Product	Refs.
O, H, Cbz, N, CO_2Et	β-**TrP2a**•HBF_4 (20 mol %), KHMDS (20 mol %), toluene, rt, 24 h	O, CO_2Et, N, Cbz (80) er 99:1	62

C_7

Substrate	Conditions	Product	Refs.
O, H, CO_2Me	Precatalyst (10 mol %), base (5 mol %), toluene, reflux	O, CO_2Me	43

Precatalyst	Base	Time (h)		er
α-**Im1c**•HBF_4	*n*-BuLi	10	(74)	88:12
α-**Im1c**•HBF_4	KH	16	(85)	60:40
α-**Im1c**•HPF_6	*n*-BuLi	10	(6)	86:14
α-**Im1c**•HPF_6	KH	3	(83)	73:27
α-**Im1c**•HCl	*n*-BuLi	10	(85)	54:46
α-**Im1c**•HBr	*n*-BuLi	10	(0)	—

Catalyst	Base	Time (h)		er
α-**Im1c**•HBr	KH	5	(71)	86:14
α-**Im1k**•HBF_4	*n*-BuLi	10	(5)	84:16
α-**Im1k**•HBr	KH	5	(67)	78:22
α-**Im1l**•HBF_4	*n*-BuLi	10	(91)	60:40
α-**Im1a**•HBF_4	*n*-BuLi	10	(20)	53:47
α-**Im1a**•HBr	KH	5	(58)	54:46

C_8

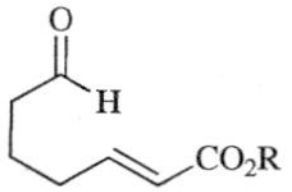

Precatalyst (20 mol %), KHMDS (20 mol %), solvent, rt, 24 h

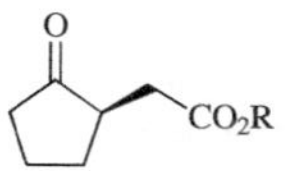

CO_2R

R	Precatalyst	Solvent		er	
Et	α-**TrM1d**•HBF_4	xylene	(80)	91:9	62
Et	β-**TrP2a**•HBF_4	xylene	(81)	97.5:2.5	15
Bn	β-**TrP2a**•HBF_4	toluene	(85)	5:95	62

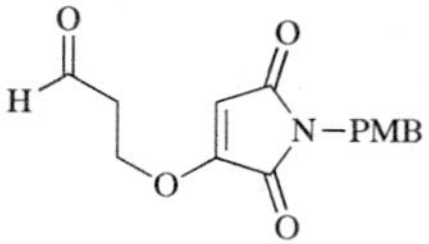

β-**TrM1b**•HBF_4 (20 mol %), KHMDS (20 mol %), toluene, 0 °C

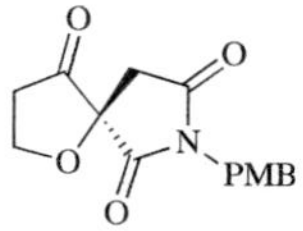

N–PMB

(88) er 99:1 — 106

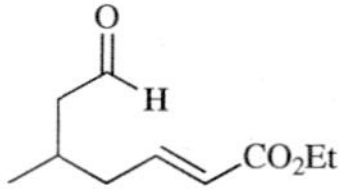

α-**TrM1a**•HBF_4 (20 mol %), KHMDS (20 mol %), toluene, rt, 24 h

CO_2Et + CO_2Et

I er 99:1 — **II** er 97:3

I + II (97)

I/II = 1:1 — 74

α-**TrM1a**•HBF_4 (20 mol %), KHMDS (20 mol %), toluene, rt, 24 h

CO_2Et + CO_2Et

I er 95:5 — **II** er 98:2

I + II (90) **I/II** = 1:1 — 74

TABLE 2. INTRAMOLECULAR STETTER REACTIONS OF ALIPHATIC ALDEHYDES (*Continued*)

Aliphatic Aldehyde Substrate	Conditions	Product(s), Yield(s) (%), and Stereoselectivity	Refs.

Please refer to the charts preceding the tables for catalyst and precatalyst structures used herein.

C_8

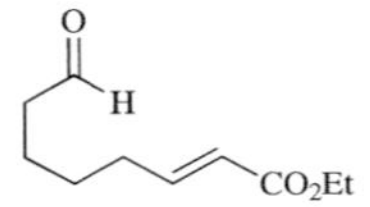

Aliphatic Aldehyde Substrate	Conditions	Product(s), Yield(s) (%), and Stereoselectivity	Refs.
O, H, CO_2Et	**β-TrP2a**•HBF_4 (20 mol %), KHMDS (20 mol %), toluene, rt, 24 h	O, CO_2Et (0)	62
O, H, Ts, N, CO_2Me	Precatalyst (10 mol %), DBU (10 mol %), toluene, rt, 18 h	O, CO_2Me, N, Ts	154

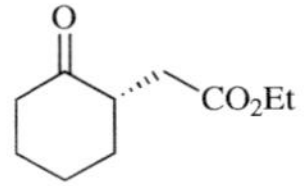

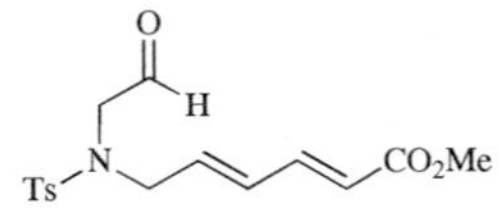

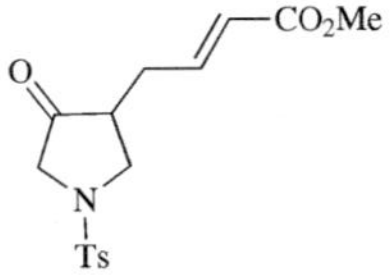

Precatalyst		$[\alpha]_D$	er
β-TrM1b•HBF_4	(72)	– 14.5	—
α-TrM1b•HBF_4	(66)	+ 13.4	—

$C_{8–9}$

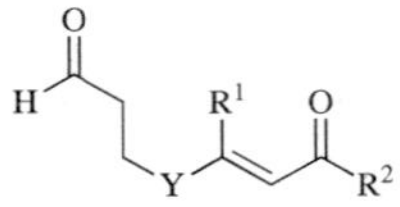

Aliphatic Aldehyde Substrate	Conditions	Product(s), Yield(s) (%), and Stereoselectivity	Refs.
O, H, R^1, O, Y, R^2	Precatalyst (20 mol %), KHMDS (20 mol %), toluene, rt, 24 h	O, R^1, O, Y, R^2	73

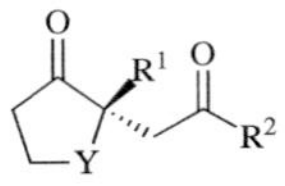

Y	R^1	R^2	Precatalyst		er
AcN	Me	Me	**α-TrM1b**•HBF_4	(65)	97.5:2.5
S	*n*-Pr	MeO	**α-TrM1b**•HBF_4	(—)	(—)
S	*n*-Pr	MeO	**β-TrP2a**•HBF_4	(—)	(—)
O_2S	*n*-Pr	MeO	**β-TrP2a**•HCl	(98)	90:10

Substrate	Conditions	Product(s) and Yield(s) (%)	Refs.
C_8			
(aldehyde, CO_2Me, CO_2Me)	Precatalyst (20 mol %), KHMDS (20 mol %), toluene, rt, 12 h	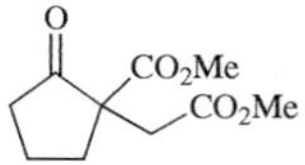 (cyclopentanone, CO_2Me, CO_2Me) Precatalyst / er: α-**TrM1b**•HBF_4 (78) 83:17 β-**TrP2a**•HBF_4 (86) 95:5	72
C_9			
(aldehyde, enone)	α-**TrM1b**•HBF_4 (20 mol %), KHMDS (20 mol %), toluene, rt, 24 h	(81) er 97.5:2.5	72
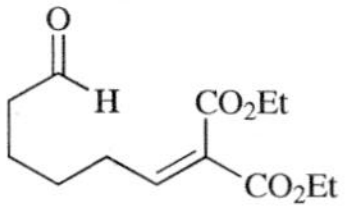 (aldehyde, CO_2Et, CO_2Et)	β-**TrP2a**•HCl (20 mol %), KHMDS (20 mol %), toluene, rt, 36 h	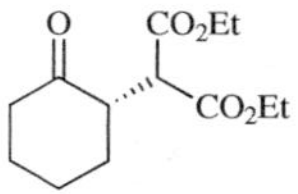 (CO_2Et, CO_2Et) (97) er 91:9	68
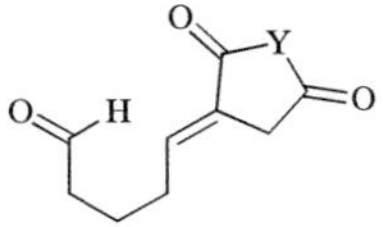 (aldehyde, Y-ring)	β-**TrP2e**•HBF_4 (20 mol %), KHMDS (20 mol %), toluene, rt, 24 h	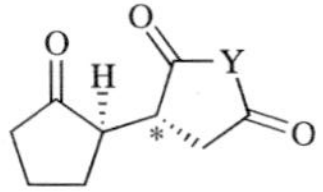 Y / *dr / er: PhN (80) 15:1 94:6 O (94) 50:1 99:1	27

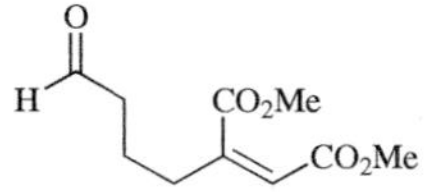

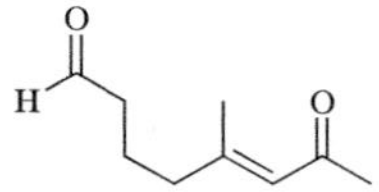

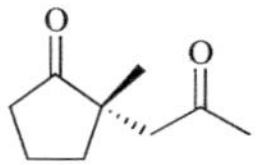

TABLE 2. INTRAMOLECULAR STETTER REACTIONS OF ALIPHATIC ALDEHYDES (*Continued*)

Aliphatic Aldehyde Substrate	Conditions	Product(s), Yield(s) (%), and Stereoselectivity	Refs.

Please refer to the charts preceding the tables for catalyst and precatalyst structures used herein.

C_9

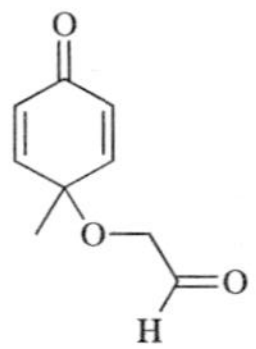

Precatalyst (20 mol %),
base (20 mol %),
solvent (*x* M), rt

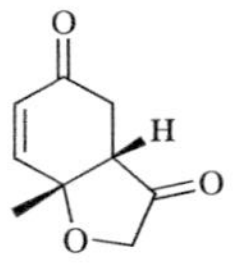

Precatalyst	Base	Solvent	*x*	Time	Yield (%)	er	Refs.
β-**TrP2e**•HBF_4	KHMDS	toluene	0.04	5 min	(83)	78.5:21.5	144
β-**TrP2a**•HBF_4	KHMDS	toluene	0.04	5 min	(84)	74.5:25.5	144
β-**TrP2d**•HBF_4	KHMDS	toluene	0.04	5 min	(79)	67:33	144
α-**TrM1a**•HBF_4	KHMDS	toluene	0.04	5 min	(75)	90:10	75
α-**TrM1b**•HBF_4	KHMDS	toluene	0.04	5 min	(92)	66:34	75
α-**TrM1d**•HBF_4	KHMDS	toluene	0.04	5 min	(90)	94:6	75
α-**TrM1d**•HBF_4	KHMDS	toluene	0.12	5 min	(85)	90:10	144
α-**TrM1d**•HBF_4	KHMDS	toluene	0.013	5 min	(95)	95:5	144
α-**TrM1d**•HBF_4	KHMDS	toluene	0.008	5 min	(90)	95:5	75
α-**TrM1d**•HBF_4	KHMDS	toluene	0.005	5 min	(86)	95:5	75
α-**TrM1d**•HBF_4	*t*-BuOK	toluene	0.04	30 min	(83)	94:6	144
α-**TrM1d**•HBF_4	KH	toluene	0.04	16 h	(76)	88:12	144
α-**TrM1d**•HBF_4	*i*-Pr_2NEt	toluene	0.04	21 h	(62)	85:15	144
α-**TrM1d**•HBF_4	Et_3N	toluene	0.04	30 h	(68)	83:17	144
α-**TrM1d**•HBF_4	KHMDS	xylenes	0.04	5 min	(81)	87:13	144
α-**TrM1d**•HBF_4	KHMDS	benzene	0.04	5 min	(88)	94:6	144
α-**TrM1d**•HBF_4	KHMDS	Et_2O	0.04	5 min	(67)	93:7	144

Precatalyst	Base	Solvent	x	Time		er	
α-**TrM1d**•HBF_4	KHMDS	DMF	0.04	5 min	(71)	87:13	144
α-**TrM1d**•HBF_4	KHMDS	THF	0.04	5 min	(77)	94:6	144
α-**TrM1d**•HBF_4	KHMDS	CH_2Cl_2	0.04	5 min	(16)	84:16	144
α-**TrM1d**•HBF_4	KHMDS	MeOH	0.04	5 min	(29)	44:56	144
α-**TrM1d**•HBF_4	KHMDS	EtOH	0.04	5 min	(63)	29:71	144
α-**TrM1d**•HBF_4	KHMDS	CF_3CH_2OH	0.04	5 min	(0)	—	144
α-**TrM1d**•HBF_4	KHMDS	*i*-PrOH	0.04	5 min	(64)	18:82	144
α-**TrM1d**•HBF_4	KHMDS	*t*-BuOH	0.04	5 min	(30)	21:79	144
α-**TrM1d**•HBF_4	KHMDS	toluene/*i*-PrOH (3:1)	0.04	5 min	(—)	52:48	144
α-**TrM1d**•HBF_4	KHMDS	toluene/*i*-PrOH (1:1)	0.04	5 min	(—)	30:70	144
α-**TrM1d**•HBF_4	KHMDS	toluene/*i*-PrOH (1:3)	0.04	5 min	(—)	19:81	144

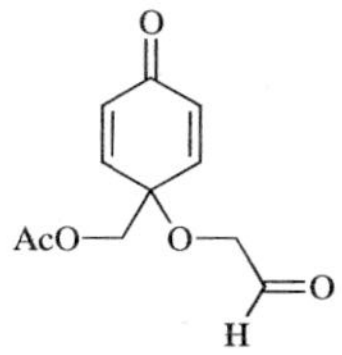

AcO, O, O, H

α-**TrM1d**•HBF_4 (20 mol %),
KHMDS (20 mol %),
toluene (0.008 M), rt, 5 min

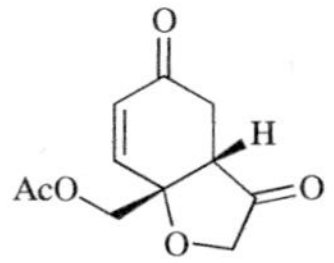

O, H, AcO, O, O

(86) dr >95:5, er 91.5:8.5 — 75

O, MeO, O, H

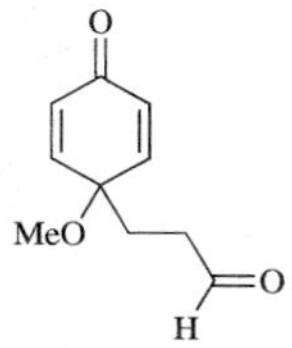

α-**TrM1d**•HBF_4 (10 mol %),
KHMDS (10 mol %),
toluene (0.008 M), rt, 5 min

O, H, MeO, O

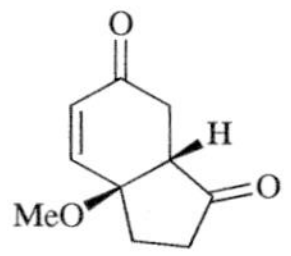

(60) dr >95:5, er 95:5 — 75

TABLE 2. INTRAMOLECULAR STETTER REACTIONS OF ALIPHATIC ALDEHYDES (*Continued*)

Aliphatic Aldehyde Substrate	Conditions	Product(s), Yield(s) (%), and Stereoselectivity	Refs.
Please refer to the charts preceding the tables for catalyst and precatalyst structures used herein.			
C_{10}			
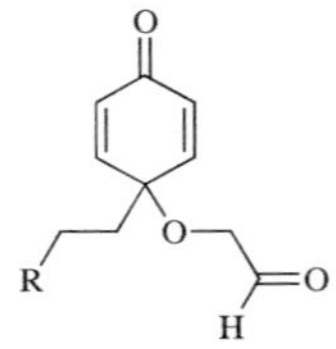 R, O, H, O	α-**TrM1d**•HBF_4 (20 mol %), KHMDS (10 mol %), toluene (0.008 M), rt, 5 min	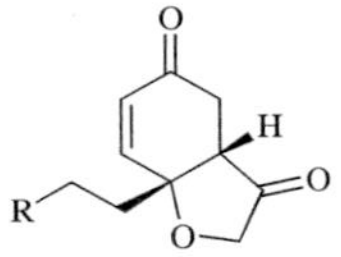 O, H, O, R, O R / er H (86) 97:3 MeO (86) 91:9 BocHN (28) 82:18	75 75 144
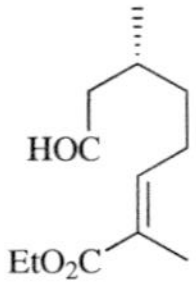 HOC, EtO_2C er >99:1	β-**TrM1b**•HBF_4 (20 mol %), KHMDS (20 mol %), toluene, rt, 24 h	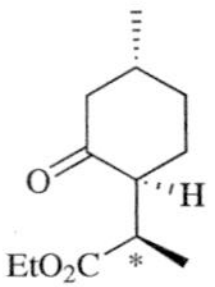 O, H, EtO_2C * (85) *dr 12:1, er >99:1	104
C_{11}			
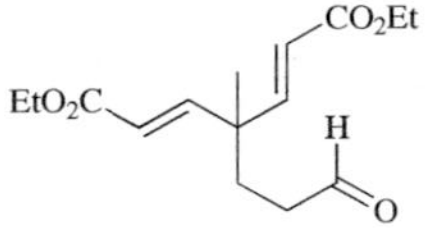 CO_2Et, EtO_2C, H, O	α-**TrM1j**•HBF_4 (20 mol %), *t*-BuOK (20 mol %), toluene, –25 °C, 24 h	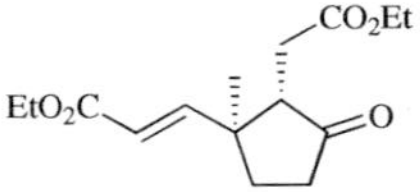 CO_2Et, EtO_2C, O (99) dr 2.3:1, er 61.5:38.5	70

Substrate	Conditions	Product(s) and Yield(s) (%)	Refs.
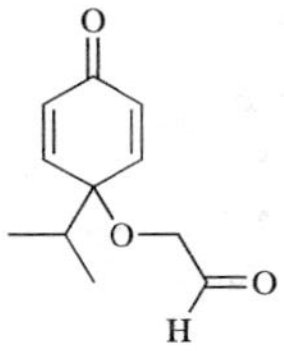	α-**TrM1d**•HBF_4 (10 mol %), KHMDS (10 mol %), toluene (0.008 M), rt, 5 min	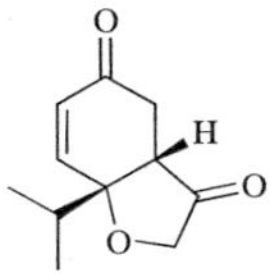 (87) er 97:3	75
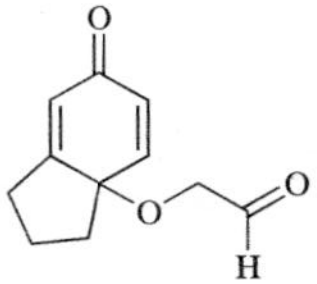	α-**TrM1d**•HBF_4 (20 mol %), KHMDS (20 mol %), toluene (0.008 M), rt, 5 min	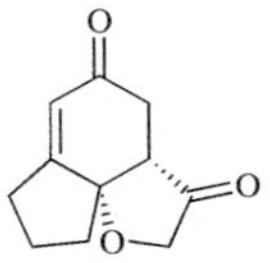 **I** er 94:6 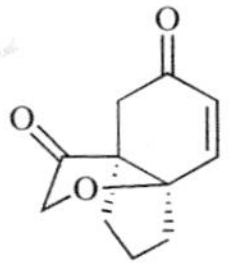 + **II** er 97:3 **I + II** (75) **I/II** 1:1	144
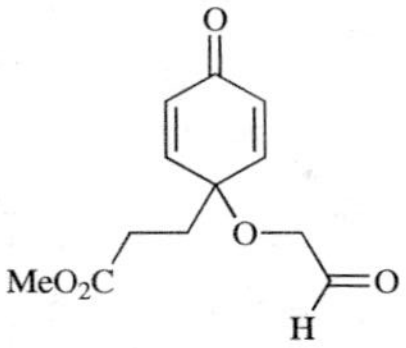 MeO_2C	α-**TrM1d**•HBF_4 (10 mol %), KHMDS (10 mol %), toluene (0.008 M), rt, 5 min	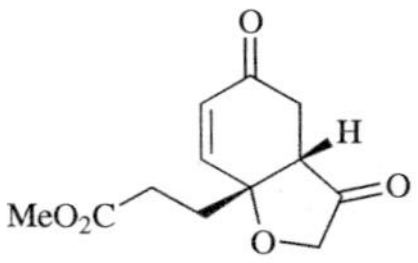 MeO_2C (94) er 94:6	75

TABLE 2. INTRAMOLECULAR STETTER REACTIONS OF ALIPHATIC ALDEHYDES (*Continued*)

Aliphatic Aldehyde Substrate	Conditions	Product(s), Yield(s) (%), and Stereoselectivity	Refs.
Please refer to the charts preceding the tables for catalyst and precatalyst structures used herein.			
C_{11}			
	α-**TrM1d**•HBF_4 (10 mol %), KHMDS (10 mol %), toluene (0.008 M), rt, 5 min	(64) dr >95:5, er 99:1	75
	α-**TrM1d**•HBF_4 (10 mol %), KHMDS (10 mol %), toluene (0.008 M), rt, 5 min	R / dr / er Me (86) >95:5 >99:1 $MeOCH_2$ (71) >95:5 99:1	75
C_{12}			
	α-**TrM1d**•HBF_4 (20 mol %), KHMDS (20 mol %), toluene (0.008 M), rt, 5 min	(86) er 97:3	75
	α-**TrM1d**•HBF_4 (20 mol %), KHMDS (20 mol %), toluene (0.008 M), rt, 5 min	**I** + **II** **I** er 53:47; **II** er >98:2 **I** + **II** (74) **I/II** 34:1	144

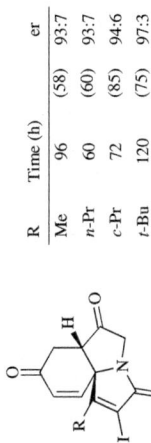

Substrate	Conditions	Product(s) and Yield(s) (%), er	Refs.
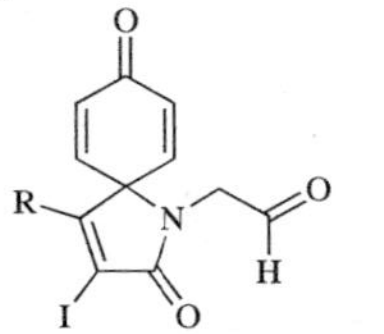 C_{12-15} R = Me, n-Pr, c-Pr, t-Bu	α-**TrM4b**•HBF_4 (10 mol %), i-Pr_2NEt (10 mol %), xylene, rt	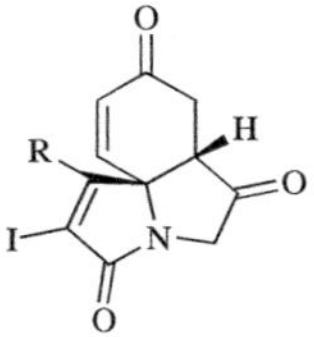 see table below	155
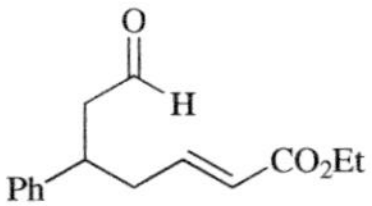 C_{13} Ph, CO_2Et	α-**TrM1a**•HBF_4 (20 mol %), KHMDS (20 mol %), toluene, rt, 24 h	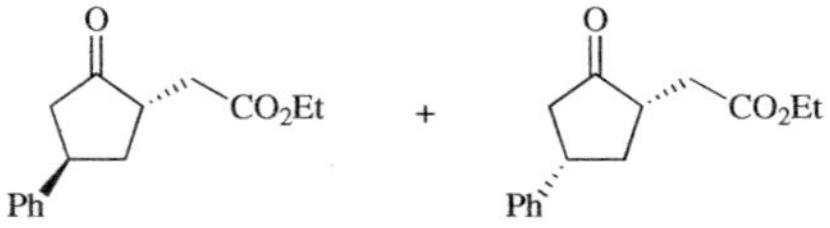 **I** er 99:1 + **II** er 98:2; **I** + **II** (97) **I/II** 1:1	74
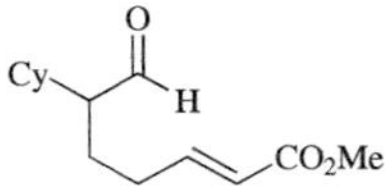 Cy, CO_2Me	β-**TrP2a**•HCl (20 mol %), KHMDS (20 mol %), toluene, rt, 24 h	 **I** er 60:40 + **II** er 53:47; **I** + **II** (74) **I/II** = 4:96	74

R	Time (h)		er
Me	96	(58)	93:7
n-Pr	60	(60)	93:7
c-Pr	72	(85)	94:6
t-Bu	120	(75)	97:3

TABLE 2. INTRAMOLECULAR STETTER REACTIONS OF ALIPHATIC ALDEHYDES (*Continued*)

Aliphatic Aldehyde Substrate	Conditions	Product(s), Yield(s) (%), and Stereoselectivity	Refs.

Please refer to the charts preceding the tables for catalyst and precatalyst structures used herein.

C_{13-17}

Aliphatic Aldehyde Substrate	Conditions	Product(s), Yield(s) (%), and Stereoselectivity	Refs.
	α-**TrM1b**•HBF_4 (20 mol %), KHMDS (20 mol %), toluene, rt, 24 h		

R^1	R^2		er	Refs.
Me	4-pyr	(85)	98:2	72
Me	Ph	(85)	98:2	73
Me	Ph	(50)[a]	78:22[a]	73
Me	4-$(O_2N)C_6H_4$	(90)	92:8	72
Me	$Ph(CH_2)_2$	(63)	99:1	72
n-Bu	Ph	(71)	99:1	72

C_{14}

Aliphatic Aldehyde Substrate	Conditions	Product(s), Yield(s) (%), and Stereoselectivity	Refs.
	β-**TrP2a**•HBF_4 (20 mol %), KHMDS (20 mol %), toluene, rt, 24 h	(60) er 71:29	62
	α-**TrM1a**•HBF_4 (50 mol %), KHMDS (50 mol %), toluene, rt, 24 h	**I** er <52:48 + **II** er <53:47; **I** + **II** (95) **I/II** = 15:85	74
	α-**TrM1d**•HBF_4 (20 mol %), KHMDS (20 mol %), toluene (0.008 M), rt, 5 min		75

Ar		er
Ph	(87)	94:6
4-BrC_6H_4	(78)	93:7

C_{14-23}

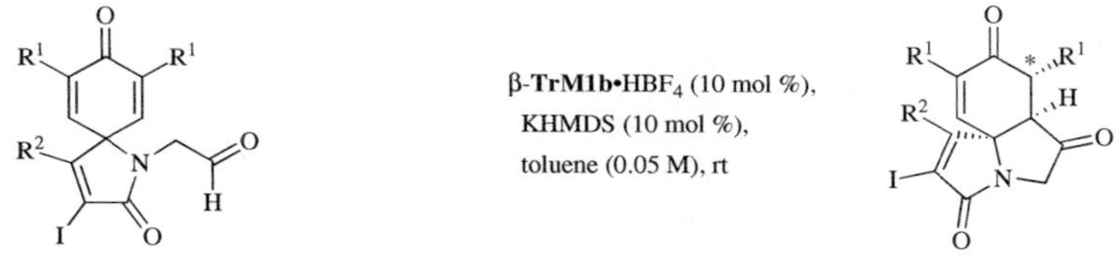

β-**TrM1b**•HBF_4 (10 mol %),
KHMDS (10 mol %),
toluene (0.05 M), rt

156

R^1	R^2	Time (h)		*dr	er
Me	Me	15	(90)	10:1	94:6
Me	*n*-Pr	15	(96)	>19:1	>99:1
Me	*c*-Pr	15	(92)	>19:1	99:1
Me	2-thienyl	15	(91)	>19:1	99:1
Me	4-F-Ph	15	(86)	>19:1	99:1
Me	4-Cl-Ph	15	(86)	>19:1	99:1

R^1	R^2	Time (h)		*dr	er
Me	2-Tol	40	(80)	5:1	99:1
Me	3-Tol	40	(85)	>19:1	99:1
Me	4-Tol	15	(90)	>19:1	99:1
Et	Ph	15	(96)	>19:1	99:1
i-Pr	Ph	60	(70)	>19:1	>99:1

TABLE 2. INTRAMOLECULAR STETTER REACTIONS OF ALIPHATIC ALDEHYDES (*Continued*)

Aliphatic Aldehyde Substrate	Conditions	Product(s), Yield(s) (%), and Stereoselectivity	Refs.

Please refer to the charts preceding the tables for catalyst and precatalyst structures used herein.

C_{17}

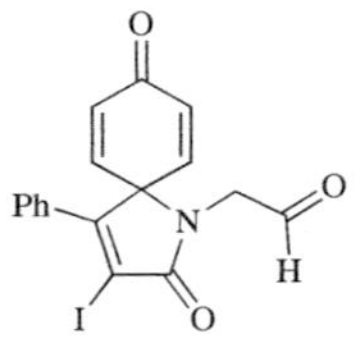

Precatalyst (10 mol %),
base (10 mol %),
solvent, rt, 40 h

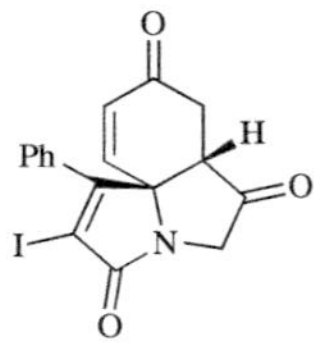

155

Precatalyst	Base	Solvent	Time (h)		er
α-**TrM4a**•HBF_4	i-Pr_2NEt	xylene	40	(0)	—
α-**TrM4d**•HBF_4	i-Pr_2NEt	xylene	40	(0)	—
α-**TrM4c**•HBF_4	i-Pr_2NEt	xylene	40	(0)	—
α-**TrM4i**•HBF_4	i-Pr_2NEt	xylene	40	(<5)	62:38
α-**TrM4b**•HBF_4	i-Pr_2NEt	xylene	40	(81)	96:4
β-**TrM1b**•HBF_4	i-Pr_2NEt	xylene	40	(84)	50:50
α-**TrM4b**•HBF_4	KHMDS	xylene	40	(84)	91:9
α-**TrM4b**•HBF_4	DBU	xylene	40	(74)	95:5
α-**TrM4b**•HBF_4	Et_3N	xylene	40	(67)	90:10
α-**TrM4b**•HBF_4	Cs_2CO_3	xylene	10	(84)	93:7
α-**TrM4b**•HBF_4	i-Pr_2NEt	toluene	40	(50)	94:6
α-**TrM4b**•HBF_4	i-Pr_2NEt	THF	40	(30)	86:14
α-**TrM4b**•HBF_4	i-Pr_2NEt	CH_2Cl_2	40	(64)	75:25
α-**TrM4b**•HBF_4	i-Pr_2NEt	$CHCl_3$	40	(60)	64:36
α-**TrM4b**•HBF_4	i-Pr_2NEt	Et_2O	40	(30)	90:10

C_{17-18}

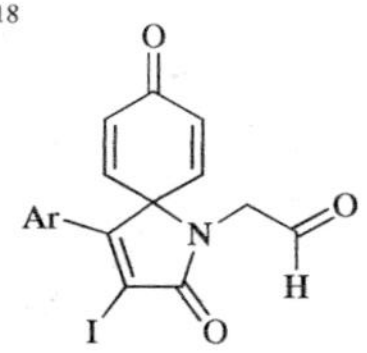

α-**TrM4b**•HBF_4 (10 mol %), base (10 mol %), xylene, rt

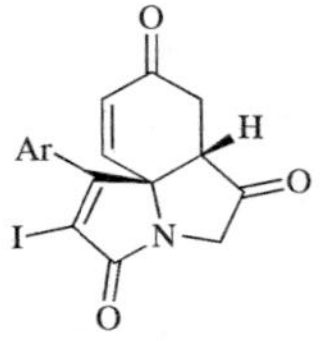

Ar	Base	Time (h)		er	Ref.
4-FC_6H_4	*i*-Pr_2NEt	90	(55)	92:8	
4-ClC_6H_4	Cs_2CO_3	48	(54)	93:7	
4-MeC_6H_4	*i*-Pr_2NEt	50	(65)	79:21	155
4-MeC_6H_4	Cs_2CO_3	24	(55)	90:10	
3-MeC_6H_4	*i*-Pr_2NEt	90	(60)	85:15	
3-MeC_6H_4	Cs_2CO_3	72	(60)	85:15	

C_{17-20}

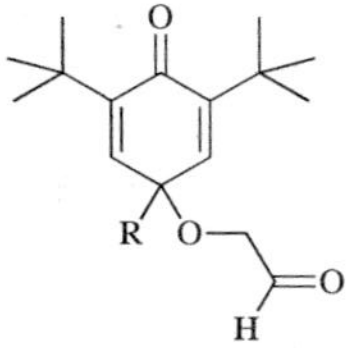

α-**TrM1d**•HBF_4 (20 mol %), KHMDS (20 mol %), toluene (0.008 M), rt, 2 h

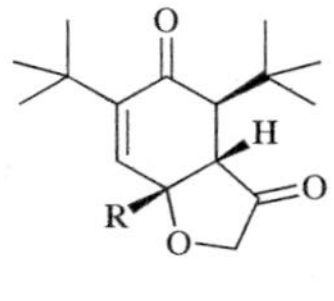

R		dr	er	Ref.
Me	(80)	>19:1	>99:1	75
t-Bu	(62)	>19:1	>99:1	

TABLE 2. INTRAMOLECULAR STETTER REACTIONS OF ALIPHATIC ALDEHYDES (*Continued*)

Aliphatic Aldehyde Substrate	Conditions	Product(s), Yield(s) (%), and Stereoselectivity	Refs.

Please refer to the charts preceding the tables for catalyst and precatalyst structures used herein.

C_{19}

Precatalyst (10 mol %), base (10 mol %), solvent, rt

Refs. 156

Precatalyst	Base	Solvent	Conc (M)	Time (h)		er
α-**TrM4a**•HBF_4	i-Pr_2NEt	o-xylene	0.10	—	(0)	—
α-**TrM4d**•HBF_4	i-Pr_2NEt	o-xylene	0.10	—	(0)	—
α-**TrM4c**•HBF_4	i-Pr_2NEt	o-xylene	0.10	—	(0)	—
α-**TrM4b**•HBF_4	i-Pr_2NEt	o-xylene	0.10	140	(9)	1:99
α-**TrM4i**•HBF_4	i-Pr_2NEt	o-xylene	0.10	—	(<2)	—
β-**TrM1a**•HBF_4	i-Pr_2NEt	o-xylene	0.10	—	(0)	—
β-**TrM1d**•HBF_4	i-Pr_2NEt	o-xylene	0.10	—	(0)	—
β-**TrM1c**•HBF_4	i-Pr_2NEt	o-xylene	0.10	—	(0)	—
β-**TrM1b**•HBF_4	i-Pr_2NEt	o-xylene	0.10	72	(72)	>99:1
β-**TrM1b**•HBF_4	i-Pr_2NEt	toluene	0.10	15	(82)	>99:1
β-**TrM1b**•HBF_4	i-Pr_2NEt	CH_2Cl_2	0.10	15	(71)	98:2
β-**TrM1b**•HBF_4	i-Pr_2NEt	$CHCl_3$	0.10	15	(86)	96:4
β-**TrM1b**•HBF_4	i-Pr_2NEt	THF	0.10	15	(55)	99:1
β-**TrM1b**•HBF_4	i-Pr_2NEt	dioxane	0.10	15	(59)	99:1
β-**TrM1b**•HBF_4	DBU	toluene	0.10	15	(54)	>99:1
β-**TrM1b**•HBF_4	Et_3N	toluene	0.10	15	(55)	>99:1
β-**TrM1b**•HBF_4	Cs_2CO_3	toluene	0.10	15	(70)	>99:1
β-**TrM1b**•HBF_4	KHMDS	toluene	0.10	15	(92)	>99:1
β-**TrM1b**•HBF_4	KHMDS	toluene	0.20	15	(88)	>99:1
β-**TrM1b**•HBF_4	KHMDS	toluene	0.05	15	(95)	>99:1
β-**TrM1b**•HBF_4	KHMDS	toluene	0.025	15	(80)	>99:1
β-**TrM1b**•HBF_4	KHMDS	toluene	0.10	15	(85)[b]	>99:1[b]

[a] The reaction was conducted with the (Z)-alkene.

[b] The reaction was conducted with 5 mol % of the precatalyst and 5 mol % of the base.

TABLE 3. INTRAMOLECULAR STETTER REACTIONS OF MISCELLANEOUS ALDEHYDES

Aldehyde Substrate	Conditions	Product(s), Yield(s) (%), and Stereoselectivity	Refs.

Please refer to the charts preceding the tables for catalyst and precatalyst structures used herein.

C_8

H, O, N—PMB, O, O, O

Precatalyst (30 mol %), NaOAc (1 equiv), *h*ν (300–400 nm), solvent, 45 °C, 24 h

O, O, O, N, PMB, O

Ref. 108

Precatalyst	Solvent		er
α-TrM1b•HBF_4	$CHCl_3$	(35)	64:36
α-TrM1j•HBF_4	$CHCl_3$	(40)	80:20
α-TrM1i•HBF_4	$CHCl_3$	(47)	6:94
β-TrP2b•HBF_4	$CHCl_3$	(37)	75:25
β-TrP2j•HBF_4	$CHCl_3$	(20)	86:14
β-TrP2i•HBF_4	$CHCl_3$	(32)	72:28
α-TrM1i•HBF_4	$Cl(CH_2)_2Cl$	(40)	4:96
α-TrM1i•HBF_4	toluene	(43)	3:97
α-TrM1i•HBF_4	benzene	(58)	3:97

C_{15}

CO_2Me, CHO, Ph

α-TrM1c•HCl (*x* mol %), base (*y* mol %), AcOH (*z* mol %), solvent, rt

CO_2Me, O, Ph

Ref. 157

x	Base	*y*	*z*	Solvent	Time (h)		er
20	KHMDS	20	0	THF	7	(16)	97.5:2.5
20	KO*t*-Bu	20	0	THF	7	(17)	99:1
20	DBU	20	0	THF	7	(0)	(—)
20	*i*-Pr_2NEt	20	0	THF	7	(0)	(—)
20	Et_3N	20	0	THF	7	(0)	(—)
20	K_2CO_3	20	0	THF	7	(52)	98.5:1.5
20	NaOAc	20	0	THF	5	(85)	98.5:1.5
20	NaOBz	20	0	THF	7	(83)	98.5:1.5
10	NaOAc	10	0	THF	7	(79)	98.5:1.5
10	NaOAc	50	0	THF	7	(84)	98.5:1.5
10	NaOAc	100	0	THF	7	(88)	98.5:1.5
10	NaOAc	100	0	Et_2O	2.5	(96)	99:1
5	NaOAc	50	10	Et_2O	4	(96)	99.5:0.5
2.5	NaOAc	25	10	Et_2O	7.5	(96)	99.5:0.5

TABLE 3. INTRAMOLECULAR STETTER REACTIONS OF MISCELLANEOUS ALDEHYDES (*Continued*)

Aldehyde Substrate	Conditions	Product(s), Yield(s) (%), and Stereoselectivity	Refs.

Please refer to the charts preceding the tables for catalyst and precatalyst structures used herein.

C_{15-20}

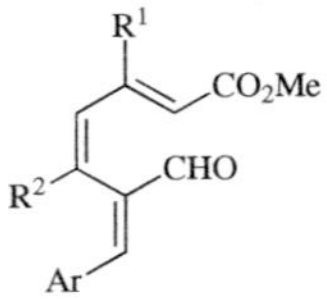

Aldehyde Substrate	Conditions	Product	Refs.
R¹, R², Ar, CO_2Me, CHO	α-**TrM1c**•HCl (x mol %), NaOAc (50 mol %), AcOH (25 mol %), Et_2O, rt, 4–40 h	R¹, R², Ar, CO_2Me, O	157

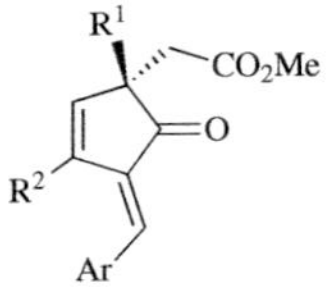

R^1	R^2	Ar	x		er
Me	H	4-BrC_6H_4	5	(92)	99.5:0.5
Me	H	4-ClC_6H_4	5	(93)	99.5:0.5
Me	H	4-MeC_6H_4	5	(95)	99.5:0.5
Me	H	4-(MeO)C_6H_4	5	(75)	99.5:0.5
Me	H	3-ClC_6H_4	5	(94)	99.5:0.5
Me	H	3,4-(OCH_2O)C_6H_3	5	(87)	>99.5:0.5
Me	Me	Ph	5	(80)	>99.5:0.5
Me	Me	4-BrC_6H_4	5	(94)	>99.5:0.5
Me	Me	3-ClC_6H_4	5	(94)	>99.5:0.5
Ph	H	Ph	10	(80)	99.5:0.5
Ph	H	4-BrC_6H_4	10	(71)	>99.5:0.5
Ph	H	4-ClC_6H_4	10	(82)	99.5:0.5
Ph	H	3-ClC_6H_4	10	(74)	>99.5:0.5

TABLE 4. INTERMOLECULAR STETTER REACTIONS OF ARYL AND HETEROARYL ALDEHYDES

Aldehyde Donor	Acceptor	Conditions	Product(s), Yield(s) (%), and Stereoselectivity	Refs.
Please refer to the charts preceding the tables for catalyst and precatalyst structures used herein.				
C_4				
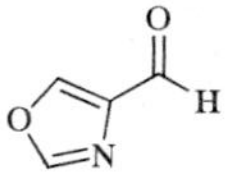 O, H, O, N	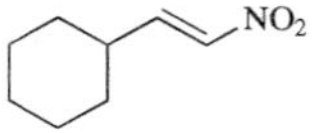 NO_2 1.5 eq	Precatalyst (10 mol %), i-Pr_2NEt (100 mol %), MeOH, 0 °C, 2 h	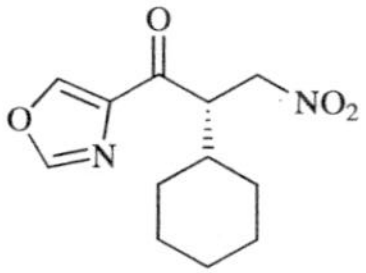 O, O, N, NO_2 Precatalyst / er **β-TrPFc3b**•HBF_4 (76) 93:7 **β-TrP3b**•HBF_4 (—) 89.5:10.5	78
C_5				
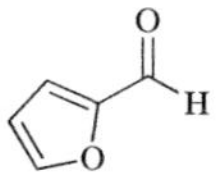 O, H, O	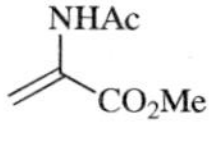 NHAc, CO_2Me 2 eq	**β-TrM2c**•HCl (10 mol %), t-BuOK (8 mol %), toluene, 0 °C, 3 h	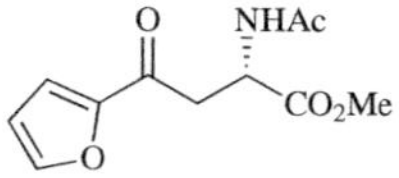 O, NHAc, CO_2Me, O (86) er 99:1	48
	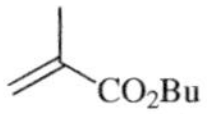 CO_2Bu 2 eq	**α-TrP5h**•HBF_4 (15 mol %), K_3PO_4 (100 mol %), toluene, rt, 24 h	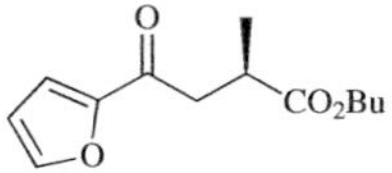 O, CO_2Bu, O (97) er 97:3	58
	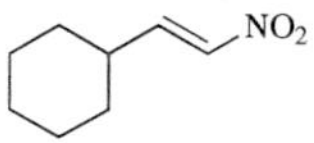 NO_2 1.5 eq	**β-TrPFc3b**•HBF_4 (10 mol %), i-Pr_2NEt (100 mol %), MeOH, 0 °C, 2 h	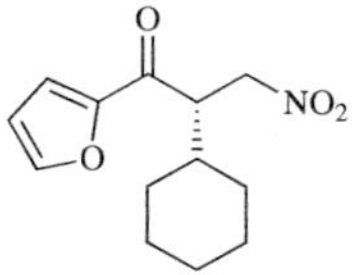 O, NO_2, O (75) er 94:6	78

TABLE 4. INTERMOLECULAR STETTER REACTIONS OF ARYL AND HETEROARYL ALDEHYDES (*Continued*)

Aldehyde Donor	Acceptor	Conditions	Product(s), Yield(s) (%), and Stereoselectivity	Refs.

Please refer to the charts preceding the tables for catalyst and precatalyst structures used herein.

C_5

Aldehyde Donor	Acceptor	Conditions	Product(s), Yield(s) (%), and Stereoselectivity	Refs.
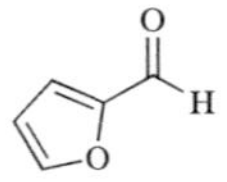 1.2 eq	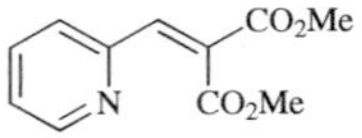 CO_2Me, CO_2Me, N	**α-TrP11x**•HBF_4 (10 mol %), Cs_2CO_3 (10 mol %), THF, rt, 6 h	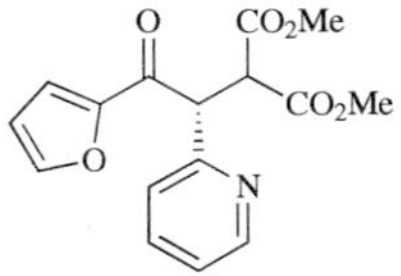 O, CO_2Me, CO_2Me, O, N (98) er 70:30	46
1.2 eq	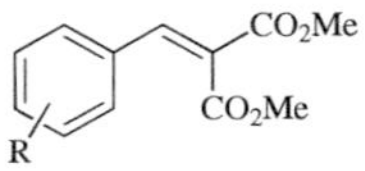 CO_2Me, CO_2Me, R	Precatalyst (10 mol %), base (10 mol %), solvent, rt, 6 h	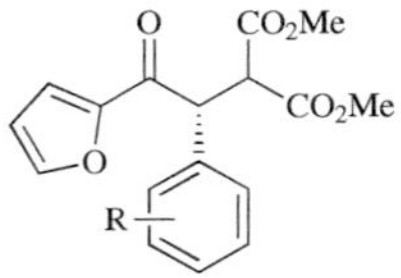 O, CO_2Me, CO_2Me, O, R	46

R	Precatalyst	Base	Solvent		er
H	**α-TrP11x**•HBF_4	DBU	THF	(86)	80:20
H	**α-TrP12a**•HBF_4	DBU	THF	(0)	(—)
H	**α-TrP13x**•HBF_4	DBU	THF	(0)	(—)
H	**β-TrM2x**•HBF_4	DBU	THF	(0)	(—)
H	**α-TrP11x**•HBF_4	Et_3N	THF	(8)	(—)
H	**α-TrP11x**•HBF_4	K_2CO_3	THF	(66)	82.5:17.5
H	α-TrP11x•HBF_4	KO*t*-Bu	THF	(73)	82.5:17.5
H	**α-TrP11x**•HBF_4	Cs_2CO_3	THF	(90)	89:11
H	**α-TrP11x**•HBF_4	Cs_2CO_3	THF	(94)[a]	80:20[a]

R	Precatalyst	Base	Solvent		er
H	**α-TrP11x**•HBF_4	Cs_2CO_3	CH_2Cl_2	(5)	(—)
H	**α-TrP11x**•HBF_4	Cs_2CO_3	toluene	(3)	(—)
H	**α-TrP11x**•HBF_4	Cs_2CO_3	EtOH	(42)	74:26
H	**α-TrP11x**•HBF_4	KHMDS	toluene	(77)	85:15
3-Cl	**α-TrP11x**•HBF_4	Cs_2CO_3	THF	(85)	84:16
4-Cl	**α-TrP11x**•HBF_4	Cs_2CO_3	THF	(92)	81:19
4-Br	**α-TrP11x**•HBF_4	Cs_2CO_3	THF	(88)	85:15
4-Me	**α-TrP11x**•HBF_4	Cs_2CO_3	THF	(84)	86:14

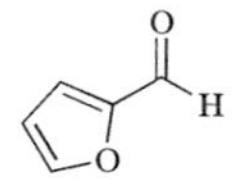

1.5 eq	Ar–CH=CH–CO–CO_2Et	β-**TrPFc3b**•HBF_4 (5 mol %), *i*-Pr_2NEt (100 mol %), CH_2Cl_2, 0 °C, 5 min to 4.5 h		77
1.2 eq	Ph–CH=CH–CO–Ph	α-**TrP11x**•HBF_4 (10 mol %), Cs_2CO_3 (10 mol %), THF, 0 °C, 6 h	(98) er 78:22	45
	1.5 eq	β-**TrPFc3b**•HBF_4 (10 mol %), *i*-Pr_2NEt (100 mol %), MeOH, 0 °C, 2 h	(70) er 98:2	78

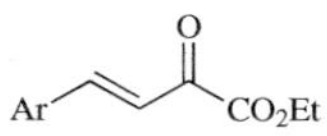

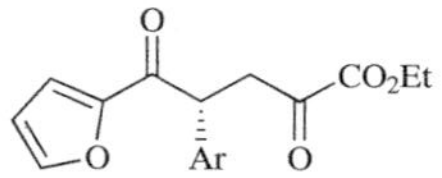

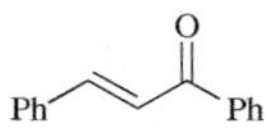

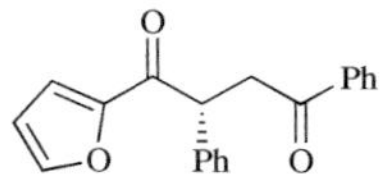

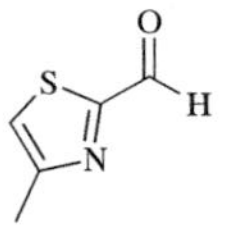

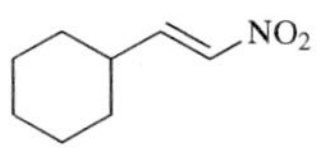

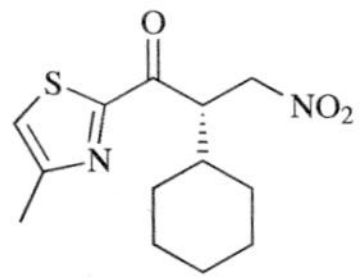

Ar		er	Ar		er
3-furyl	(63)	94:6	4-(MeO)C_6H_4	(96)	95:5
Ph	(92)	95:5	3,4-$(MeO)_2C_6H_3$	(86)	95:5
4-FC_6H_4	(80)	95:5	2-naphthyl	(97)	95:5
4-BrC_6H_4	(90)	95:5			

TABLE 4. INTERMOLECULAR STETTER REACTIONS OF ARYL AND HETEROARYL ALDEHYDES (*Continued*)

Aldehyde Donor	Acceptor	Conditions	Product(s), Yield(s) (%), and Stereoselectivity	Refs.

Please refer to the charts preceding the tables for catalyst and precatalyst structures used herein.

C_5

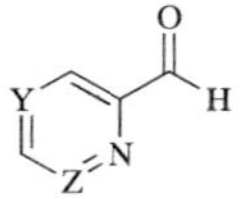

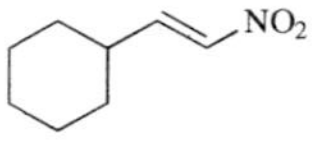

Aldehyde Donor	Acceptor	Conditions	Product(s), Yield(s) (%), and Stereoselectivity	Refs.
Y, Z-heteroaryl aldehyde	1.5 eq	Precatalyst (10 mol %), i-Pr_2NEt (100 mol %), MeOH, 0 °C, 2 h		78

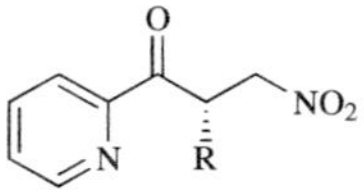

Y	Z	Precatalyst		er
N	CH	**β-TrPFc3b**•HBF_4	(99)	97.5:2.5
N	CH	**β-TrP3b**•HBF_4	(—)	95:5
CH	N	**β-TrPFc3b**•HBF_4	(88)	97:3

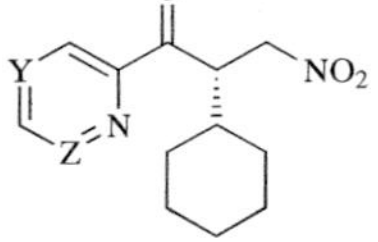

Aldehyde Donor	Acceptor	Conditions	Product(s), Yield(s) (%), and Stereoselectivity	Refs.
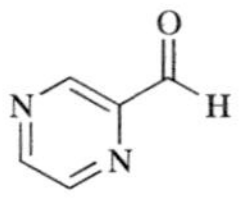 1.5 eq	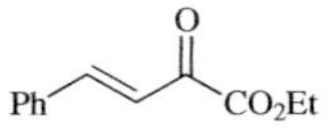	**β-TrPFc3b**•HBF_4 (5 mol %), i-Pr_2NEt (100 mol %), CH_2Cl_2, 0 °C, 4.5 h	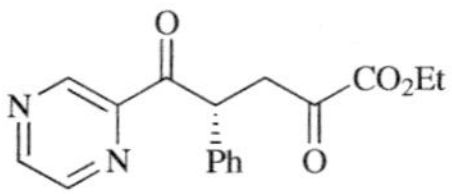 (94) er 93.5:6.5	77

C_6

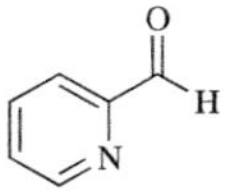

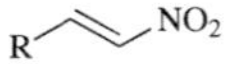

Aldehyde Donor	Acceptor	Conditions	Product(s), Yield(s) (%), and Stereoselectivity	Refs.
	1.5 eq	Precatalyst (10 mol %), i-Pr_2NEt (100 mol %), MeOH, 0 °C, 2 h		78

R	Precatalyst		er
n-Pr	**β-TrPFc3b**•HBF_4	(82)	91.5:8.5
i-Pr	**β-TrPFc3b**•HBF_4	(85)	98.5:1.5
c-Pr	**β-TrPFc3b**•HBF_4	(72)	94:6
i-Bu	**β-TrPFc3b**•HBF_4	(99)	91.5:8.5
c-C_5H_9	**β-TrPFc3b**•HBF_4	(98)	95:5
c-C_6H_{11}	**β-TrPFc3b**•HBF_4	(95)	97.5:2.5
c-C_6H_{11}	**β-TrPF1b**•HBF_4	(93)	83:17

C_{6-10}

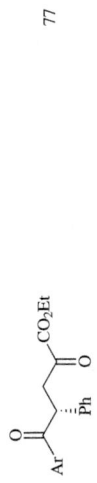

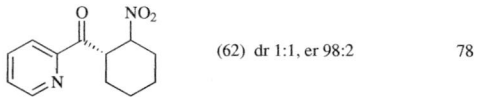

Reactant	Reactant	Conditions	Product(s) and Yield(s) (%)	Refs.
	1-nitrocyclohexene (1.5 eq)	β-**TrPFc3b**•HBF_4 (10 mol %), i-Pr_2NEt (100 mol %), MeOH, 0 °C, 2 h	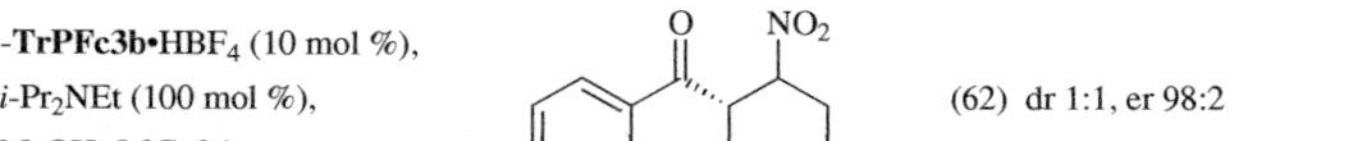 (62) dr 1:1, er 98:2	78
1.2 eq	Ph–CH=C(CO_2Me)$_2$	α-**TrP11x**•HBF_4 (10 mol %), Cs_2CO_3 (10 mol %), THF, 0 °C, 6 h	(94) er 65:35	46
ArCHO (1.5 eq)	Ph–CH=CH–CO–CO_2Et	β-**TrPFc3b**•HBF_4 (5 mol %), i-Pr_2NEt (100 mol %), CH_2Cl_2, 0 °C, 5 min to 4.5 h	Ar–CO–CH(Ph)–CH_2–CO–CO_2Et	77

Ar		er
2-pyridyl	(88)	95.5:4.5
5-Me-2-furyl	(89)	92:8
2-quinolyl	(95)	>99:1

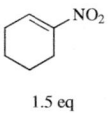

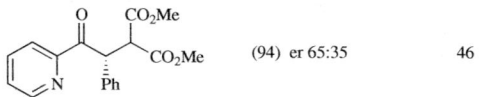

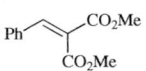

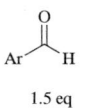

TABLE 4. INTERMOLECULAR STETTER REACTIONS OF ARYL AND HETEROARYL ALDEHYDES (*Continued*)

Aldehyde Donor	Acceptor	Conditions	Product(s), Yield(s) (%), and Stereoselectivity	Refs.

Please refer to the charts preceding the tables for catalyst and precatalyst structures used herein.

C_7

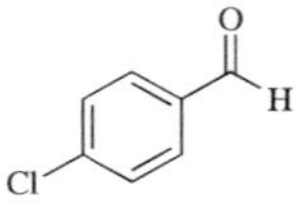

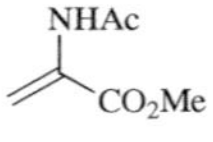

4-ClC₆H₄CHO; NHAc / CO_2Me (2 eq); **β-TrM2c•HCl** (10 mol %), base (10 mol %), solvent; product: O, NHAc, CO_2Me, Cl — 48

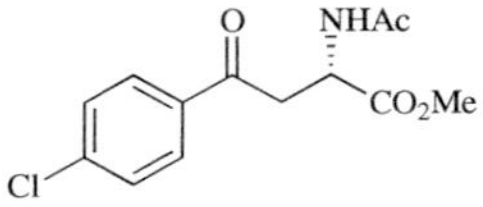

Base	Solvent	Temp (°C)	Time (h)		er
KHMDS	dioxane	25	24	(10)	98.5:1.5
TBD	dioxane	25	24	(22)[b]	—
TBD	THF	25	24	(17)[b]	—
TBD	toluene	25	24	(41)[b]	92:8
DBU	toluene	25	24	(49)[b]	93:7
K_2CO_3	toluene	25	24	(23)[b]	—
K_2CO_3	toluene	25	24	(31)[b]	95:5
KHMDS	toluene	25	24	(67)[b,c]	97.5:2.5
t-BuOK	toluene	25	24	(65)[b,c]	97.5:2.5
t-BuOK	toluene	0	4	(98)[c]	97.5:2.5
t-BuOK	toluene	0	65	(99)[b,c]	97.5:2.5
t-BuOK	toluene	0	3	(93)	97.5:2.5

C_{7-13}

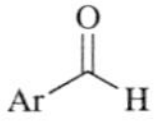

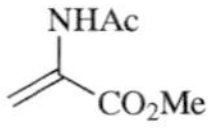

ArCHO; NHAc / CO_2Me (2 eq); **β-TrM2c•HCl** (10 mol %), *t*-BuOK (8 mol %), toluene, 0 °C, 3–24 h; product: Ar, O, NHAc, CO_2Me — 48

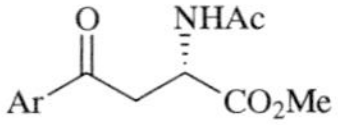

Ar		er
Ph	(80)	98.5:1.5
2-FC_6H_4	(38)	99.5:0.5
3,4-$Cl_2C_6H_3$	(71)	97:3
3-BrC_6H_4	(81)	98.5:1.5
4-BrC_6H_4	(98)	98:2
4-MeC_6H_4	(48)[d]	98.5:1.5[d]
4-PhC_6H_4	(42)[d]	99:1[d]
4-$CF_3C_6H_4$	(73)	96.5:3.5
4-NCC_6H_4	(87)	97.5:2.5
4-(MeO_2C)C_6H_4	(83)	98:2
2-naphthyl	(89)	98.5:1.5
ferrocenyl	(56)[d]	99.5:0.5[d]

C_7

2 eq

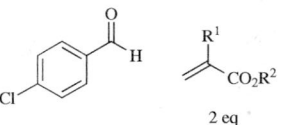

Precatalyst (15 mol %), base (100 mol %), toluene, rt, 24 h

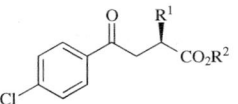

58

Catalyst	Base	R[1]	R[2]		er
β-**TrM2c**•HCl	*t*-BuOK	Me	*n*-Bu	(<5)[e]	—
β-**TrM2h**•HBF_4	*t*-BuOK	Me	*n*-Bu	(15)[e]	8:92[e]
β-**TrM2c**•HBF_4	*t*-BuOK	Me	*n*-Bu	(15)[e]	99:1[e]
α-**TrP5h**•HBF_4	*t*-BuOK	Me	*n*-Bu	(50)[e]	96:4[e]
β-**TrP2h**•HCl	*t*-BuOK	Me	*n*-Bu	(70)[e]	9:91[e]
α-**TrP10h**•HCl	*t*-BuOK	Me	*n*-Bu	(<5)[e]	—
α-**TrP7i**[§]•HCl	*t*-BuOK	Me	*n*-Bu	(59)[e]	93.5:6.5[e]

Catalyst	Base	R[1]	R[2]		er
α-**TrP5h**•HBF_4	K_3PO_4	Me	*n*-Bu	(90)	96:4
α-**TrP5h**•HBF_4	K_3PO_4	Me	Me	(82)	96.5:3.5
α-**TrP5h**•HBF_4	K_3PO_4	Me	*i*-Pr	(88)	95.5:4.5
α-**TrP5h**•HBF_4	K_3PO_4	Et	Et	(83)	96:4
α-**TrP5h**•HBF_4	K_3PO_4	Bu	Et	(79)	96:4
α-**TrP5h**•HBF_4	K_3PO_4	*i*-Pr	Et	(76)	98.5:1.5
α-**TrP5h**•HBF_4	K_3PO_4	Ph	Me	(28)	80:20
α-**TrP5h**•HBF_4	K_3PO_4	Bn	Me	(34)	90:10
α-**TrP5h**•HBF_4	K_3PO_4	$PhthNCH_2$	Me	(70)	93.5:6.5

C_{7-13}

2 eq

α-**TrP5h**•HBF_4 (15 mol %), K_3PO_4 (1 equiv), toluene, rt, 24 h

58

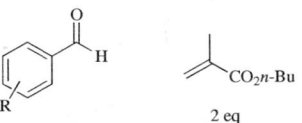

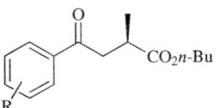

R		er
H	(53)	96:4
2-F	(77)	95.5:4.5
3-Cl	(91)	96:4
3,4-Cl_2	(86)	95:5
4-Br	(89)	95.5:4.5

R		er
3-Me	(43)	96:4
4-*t*-Bu	(31)	96:4
4-Ph	(30)	96.5:3.5
4-CF_3	(86)	95.5:4.5
4-MeO_2C	(86)	96.5:3.5

TABLE 4. INTERMOLECULAR STETTER REACTIONS OF ARYL AND HETEROARYL ALDEHYDES (*Continued*)

Aldehyde Donor	Acceptor	Conditions	Product(s), Yield(s) (%), and Stereoselectivity	Refs.

Please refer to the charts preceding the tables for catalyst and precatalyst structures used herein.

C_7

Aldehyde Donor	Acceptor	Conditions	Product(s), Yield(s) (%), and Stereoselectivity	Refs.
Cl, H, O	CO_2Me, n-C_5H_{11}; 2 eq	α-**TrP5h**•HBF_4 (20 mol %), t-BuOK (16 mol %), toluene, rt, 24 h	O, CO_2Me, Cl, n-C_5H_{11} (59) er 90:10	58

C_{7-11}

Aldehyde Donor	Acceptor	Conditions	Product(s), Yield(s) (%), and Stereoselectivity	Refs.
O, Ar^1, H; 1.2 eq	O, Ar^2, Ph	α-**TrP11x**•HBF_4 (10 mol %), Cs_2CO_3 (10 mol %), THF, 0 °C, 6 h	O, Ar^1, Ph, Ar^2, O	45

Ar^1	Ar^2		er
Ph	Ph	(65)	83:17
Ph	4-MeC_6H_4	(55)	82:18
Ph	4-ClC_6H_4	(57)	77.5:22.5
4-ClC_6H_4	Ph	(55)	83.5:16.5

Ar^1	Ar^2		er
4-BrC_6H_4	Ph	(68)	77.5:22.5
3-MeC_6H_4	Ph	(50)	85:15
4-MeC_6H_4	Ph	(43)	89:11
2-naphthyl	Ph	(65)	85:15

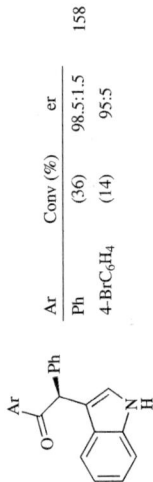

Carbon No.	Reactants		Conditions	Product(s) and Yield(s) (%)	Refs.
C_7	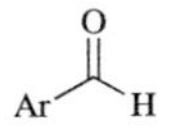 Ar(C=O)H, 1.2 eq	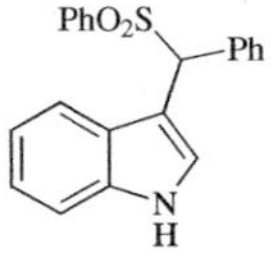 3-[PhO$_2$S(Ph)CH]-1H-indole	β-**TrM1c**•HBF_4 (10 mol %), K_2CO_3 (120 mol %), THF, rt	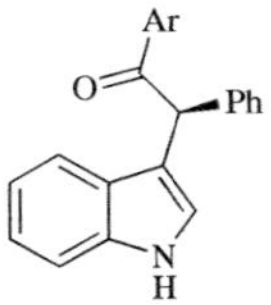 ArC(O)CH(Ph)-(1H-indol-3-yl) Ar — Conv (%) — er Ph — (36) — 98.5:1.5 4-BrC_6H_4 — (14) — 95:5	158
C_8	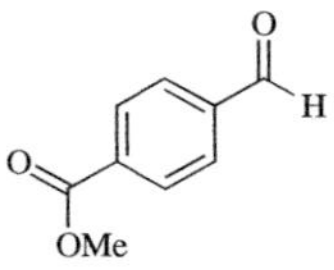 4-$MeO_2CC_6H_4CHO$, 1.5 eq	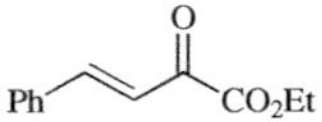 PhCH=CHC(O)CO_2Et	β-**TrPFc3b**•HBF_4 (10 mol %), *i*-Pr_2NEt (100 mol %), CH_2Cl_2, 0 °C, 4.5 h	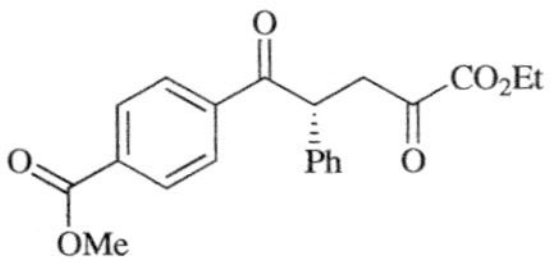 4-$MeO_2CC_6H_4$C(O)CH(Ph)CH_2C(O)CO_2Et (30) er 86:14	77

[a] The reaction was conducted at 40 °C.

[b] The yield was determined against an internal standard by ^{1}H NMR spectral analysis.

[c] The reaction was conducted using 15 mol % precatalyst.

[d] The reaction was conducted with 20 mol % precatalyst and 16 mol % *t*-BuOK.

[e] The reaction was conducted using 10 mol % catalyst and 8 mol % *t*-BuOK, and the yield was determined against an internal standard by ^{1}H NMR spectral analysis.

TABLE 5. INTERMOLECULAR STETTER REACTIONS OF GLYOXLATES AND GLYOXAMIDES

Glyoxyl Donor	Acceptor	Conditions	Product(s), Yield(s) (%), and Stereoselectivity	Refs.

Please refer to the charts preceding the tables for catalyst and precatalyst structures used herein.

C_2

Glyoxyl Donor	Acceptor	Conditions	Product(s), Yield(s) (%), and Stereoselectivity	Refs.
Y–CO–CHO	CO_2Me, CO_2Me 2 eq	Precatalyst (20 mol %), Et_3N (20 mol %), toluene, rt, 8 h	CO_2Me, CO_2Me	76
Morpholine glyoxamide (H)	R, CO_2t-Bu, CO_2t-Bu 2 eq	**β-TrP2b**•HBF_4 (20 mol %), base (*x* mol %), $MgSO_4$, CCl_4	R, CO_2t-Bu, CO_2t-Bu	76
Pyrrolidine glyoxamide (H)	Et, CO_2R, CO_2R 2 eq	**β-TrP2b**•HBF_4 (20 mol %), Et_3N (20 mol %), CCl_4, rt, 8 h	Et, CO_2R, CO_2R	76

Y	Precatalyst		er
$(CH_2)_4N$	**β-TrP2b**•HBF_4	(50)	75.5:24.5
$(CH_2)_4N$	**β-TrP2a**•HBF_4	(<2)	(—)
$(CH_2)_4N$	**β-TrM1b**•HBF_4	(12)	82.5:17.5
EtO	**β-TrP2b**•HBF_4	(100)	61.5:38.5
BnHN	**β-TrP2b**•HBF_4	(51)	<53:47
Me_2N	**β-TrP2b**•HBF_4	(24)	71:29

R	Base	x	Temp (°C)	Time (h)		er
Me	*i*-Pr_2NEt	100	−10	3	(68)	93.5:6.5
Et	Et_3N	20	23	8	(62)	92.5:7.5

R		er
Me	(52)	75.5:24.5
i-Pr	(54)	87.5:12.5
t-Bu	(51)	90.5:9.5

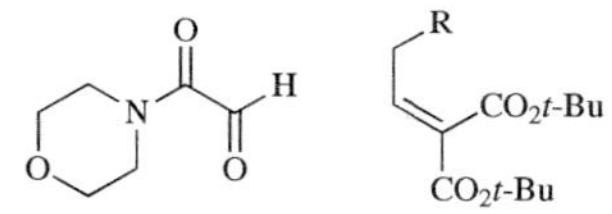

2 eq

β-**TrP2b**•HBF_4 (20 mol %),
i-Pr_2NEt (1 equiv), $MgSO_4$,
CCl_4, −10 °C

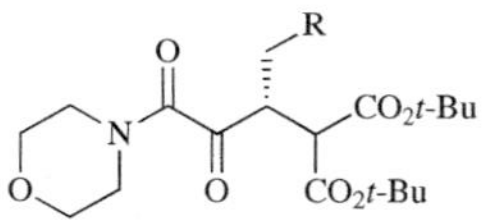

CO_2t-Bu
CO_2t-Bu

76

R	Time (h)		er	R	Time (h)		er
$BnOCH_2$	12	(91)	90:10	i-Pr	28	(51)	95.5:4.5
Et	12	(83)	95:5	allyl	12	(97)	94.5:5.5
$ClCH_2CH_2$	12	(84)	90.5:9.5	Bn	12	(81)	94:6
n-Pr	12	(70)	95:5		12	(88)	92:8

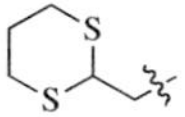

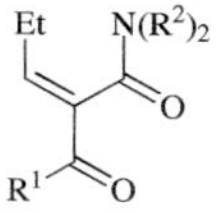

2 eq

β-**TrP2b**•HBF_4 (20 mol %),
i-Pr_2NEt (1 equiv), $MgSO_4$,
CCl_4, 12 h

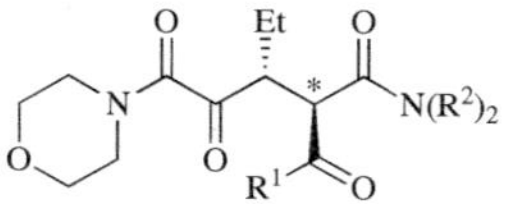

$N(R^2)_2$

79

R^1	R^2	Temp		*dr	er
Me	Me	rt	(68)	6:1	91:9
Me	Et	rt	(66)	14:1	88.5:11.5
Et	Me	rt	(92)	5:1	94.5:5.5
Et	Me	0 °C	(90)	12:1	96:4
Ph	Me	rt	(60)	14:1	53.5:46.5

TABLE 5. INTERMOLECULAR STETTER REACTIONS OF GLYOXLATES AND GLYOXAMIDES (*Continued*)

Glyoxyl Donor	Acceptor	Conditions	Product(s), Yield(s) (%), and Stereoselectivity	Refs.

Please refer to the charts preceding the tables for catalyst and precatalyst structures used herein.

C_2

Acceptor: 2 eq

Conditions: β-**TrP2b**•HBF_4 (20 mol %), *i*-Pr_2NEt (20 mol %), $MgSO_4$, CCl_4, 0 °C

Refs.: 79

R	Time (h)		*dr	er
Me	12	(95)	7:1	94.5:5.5
Et	12	(90)	12:1	96:4
n-Pr	12	(81)	6:1	95:5
n-Bu	12	(71)	12:1	96:4
i-Bu	20	(44)	11:1	93.5:6.5

R	Time (h)		*dr	er
$BnO(CH_2)_2$	12	(87)	11:1	99:1
$Cl(CH_2)_3$	12	(83)	10:1	91:9
$[S(CH_2)_3S]CH(CH_2)_2$	12	(77)	9:1	93:7
$CH_2{=}CH(CH_2)_2$	12	(83)	14:1	95:5
$HC{\equiv}C(CH_2)_2$	12	(78)	4:1	96:4
$Ph(CH_2)_2$	12	(65)	19:1	91.5:8.5

Acceptor: 2 eq

Conditions: β-**TrP2b**•HBF_4 (20 mol %), *i*-Pr_2NEt (20 mol %), $MgSO_4$, CCl_4, 0 °C, 12 h

Refs.: 79

R^1	R^2		*dr	er
n-Pr	H	(92)	11:1	96:4
n-Pr	4-ClC_6H_4	(64)	5:1	97:3
$CH_2{=}CH(CH_2)_2$	H	(94)	9:1	95:5

TABLE 6. INTERMOLECULAR STETTER REACTIONS OF α,β-UNSATURATED ALDEHYDES

Enal Donor	Acceptor	Conditions	Product(s), Yield(s) (%), and Stereoselectivity	Refs.

Please refer to the charts preceding the tables for catalyst and precatalyst structures used herein.

C_4

Enal Donor	Acceptor	Conditions	Product(s), Yield(s) (%), and Stereoselectivity	Refs.
	1.5 eq	**β-TrPFc3b**•HBF_4 (10 mol %), *i*-Pr_2NEt (100 mol %), catechol (100 mol %), MeOH, 0 °C, 4 h		40
3 eq		**α-TrM1c**•HCl (30 mol %), DBU (20 mol %), THF, 0 °C		87

Y		er
H	(67)	93:7
TIPSO	(60)	91:9

Ar^1	Ar^2		er
Ph	Ph	(90)	97:3
Ph	4-FC_6H_4	(74)	96:4
Ph	4-ClC_6H_4	(64)	97:3
Ph	4-BrC_6H_4	(89)	92.5:7.5

Ar^1	Ar^2		er
4-*i*-PrC_6H_4	Ph	(51)[a]	95:5[a]
4-FC_6H_4	Ph	(71)	96:4
4-BrC_6H_4	Ph	(80)	96:4

TABLE 6. INTERMOLECULAR STETTER REACTIONS OF α,β-UNSATURATED ALDEHYDES (*Continued*)

Enal Donor	Acceptor	Conditions	Product(s), Yield(s) (%), and Stereoselectivity	Refs.

Please refer to the charts preceding the tables for catalyst and precatalyst structures used herein.

C_5

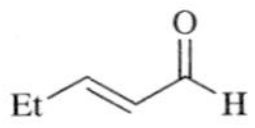

Enal donor: Et–CH=CH–CHO, 3 eq

Acceptor: Ar^1CH=C(COAr²)₂

Conditions: Precatalyst (*x* mol %), base (20 mol %), THF, 12 h

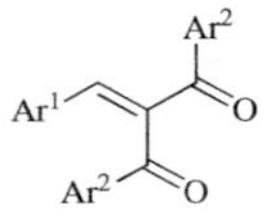

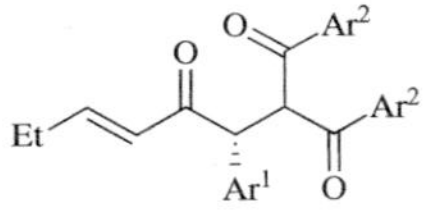

Refs.: 87

Ar^1	Ar^2	Precatalyst	*x*	Base	Temp		er
Ph	Ph	α-**TrM1c**•HCl	20	DBU	rt	(80)	83.5:16.5
Ph	Ph	α-**TrM1c**•HCl	20	Et_3N	rt	(33)	95.5:4.5
Ph	Ph	α-**TrM1c**•HCl	20	*i*-Pr_2NEt	rt	(30)	95:5
Ph	Ph	α-**TrM1c**•HCl	20	DMAP	rt	(40)	96:4
Ph	Ph	α-**TrM1c**•HCl	20	LiHMDS	rt	(22)	92:8
Ph	Ph	β-**TrP2a**•HBF_4	30	DBU	0 °C	(16)	—
Ph	Ph	α-**TrM1a**•HBF_4	30	DBU	0 °C	(22)	—
Ph	Ph	α-**TrM1a**•HBF_4	30	DBU	0 °C	(54)[b]	64.5:35.5[b]
Ph	Ph	β-**TrM2c**•HBF_4	30	DBU	0 °C	(43)	80:20
Ph	Ph	α-**TrM1c**•HCl	30	DBU	0 °C	(86)	95:5
3-$MeOC_6H_4$	Ph	α-**TrM1c**•HCl	30	DBU	0 °C	(68)[a]	95.5:4.5[a]
3-$MeOC_6H_4$	4-BrC_6H_4	α-**TrM1c**•HCl	30	DBU	0 °C	(86)	97:3

C_6

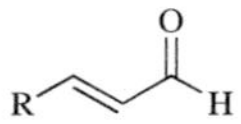

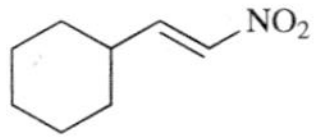

Enal donor: R–CH=CH–CHO

Acceptor: cyclohexyl–CH=CH–NO_2, 1.5 eq

Conditions: β-**TrPFc3b**•HBF_4 (10 mol %), *i*-Pr_2NEt (100 mol %), catechol (100 mol %), MeOH, 0 °C, 4 h

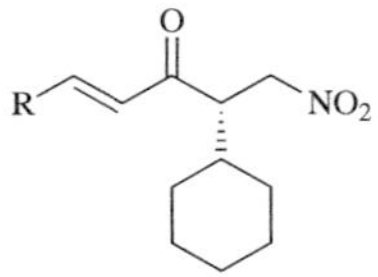

Refs.: 40

R		er
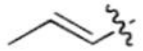 CH₃CH=CH–	(90)	97:3
i-Pr	(70)	91.5:8.5

C_{6-8}

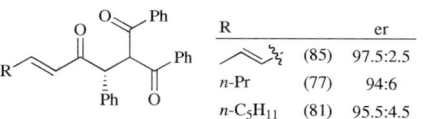

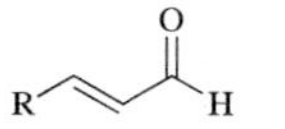

3 eq

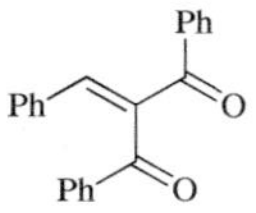

α-**TrM1c**•HCl (30 mol %),
DBU (20 mol %),
THF, 0 °C

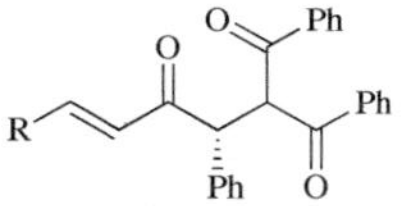

R		er
(CH=CHCH3 wavy)	(85)	97.5:2.5
n-Pr	(77)	94:6
n-C_5H_{11}	(81)	95.5:4.5

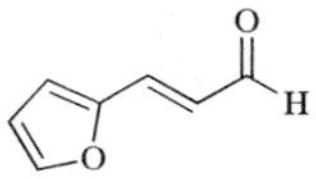

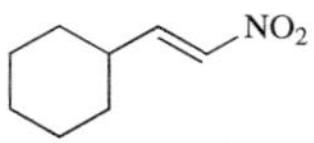

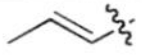

87

C_7

1.5 eq

β-**TrPFc3b**•HBF_4 (10 mol %),
i-Pr_2NEt (1 equiv),
catechol (1 equiv),
MeOH, 0 °C, 4 h

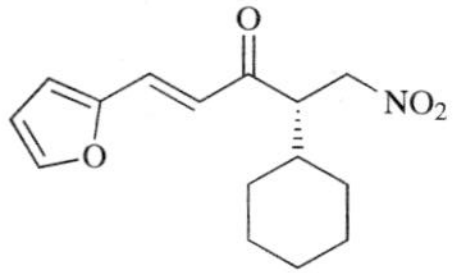

(98) er 96.5:3.5

40

C_9

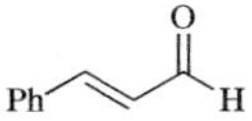

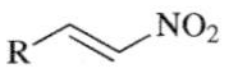

1.5 eq

β-**TrPFc3b**•HBF_4 (10 mol %),
i-Pr_2NEt (1 equiv),
catechol (1 equiv),
MeOH, 0 °C, 4 h

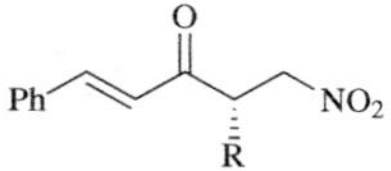

R		er
Et	(66)[c]	76:24[c]
i-Pr	(75)	95.5:4.5
i-Bu	(84)	71.5:28.5
c-C_5H_9	(84)	94:6

R		er
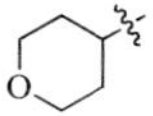 tetrahydropyran-4-yl	(82)	95:5
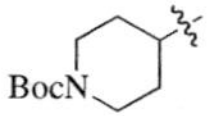 BocN piperidin-4-yl	(78)	95:5

40

TABLE 6. INTERMOLECULAR STETTER REACTIONS OF α,β-UNSATURATED ALDEHYDES (*Continued*)

Enal Donor	Acceptor	Conditions	Product(s), Yield(s) (%), and Stereoselectivity	Refs.

Please refer to the charts preceding the tables for catalyst and precatalyst structures used herein.

C_9

Enal Donor	Acceptor	Conditions	Product(s), Yield(s) (%), and Stereoselectivity	Refs.
Ph, O, H	NO_2; 1.5 eq	**β-TrPFc3b•**HBF_4 (10 mol %), *i*-Pr_2NEt (1 equiv), catechol (1 equiv), MeOH, 0 °C, 4 h	O, NO_2, Ph; (<5) dr —, er —	40
	NO_2; 1.5 eq	Precatalyst (10 mol %), *i*-Pr_2NEt (1 equiv), additive (1 equiv), MeOH, 0 °C	O, Ph, NO_2	40

Precatalyst	Additive	Time (h)		er
β-TrPFc3b•HBF_4	—	8	(5)	95.5:4.5
β-TrPFc3b•HBF_4	C_6H_5OH	8	(8)	95.5:4.5
β-TrPFc3b•HBF_4	1,4-$(HO)_2C_6H_4$	8	(15)	96.5:3.5
β-TrPFc3b•HBF_4	1,2-$(HO)_2C_6H_4$	2	(80)	95.5:4.5
β-TrPFc3b•HBF_4	1,2-$(HO)_2C_6H_4$	0.5	(48)	96.5:3.5
β-TrPF1b•HBF_4	1,2-$(HO)_2C_6H_4$	2	(82)	90.5:9.5
β-TrP3b•HBF_4	1,2-$(HO)_2C_6H_4$	2	(76)	93:7
β-TrPFc3b•HBF_4	1-(HO)-2-(MeO)C_6H_4	8	(9)	96.5:3.5
β-TrPFc3b•HBF_4	1,2-$(HO)_2$-4-*t*-BuC_6H_3	0.5	(55)	96.5:3.5
β-TrPFc3b•HBF_4	1,2-$(HO)_2$-4-(MeO)C_6H_3	0.5	(39)	96.5:3.5
β-TrPFc3b•HBF_4	1,2-$(HO)_2$-4-$(EtO_2C)C_6H_3$	0.5	(40)	96.5:3.5
β-TrPFc3b•HBF_4	1,2-$(HO)_2$-4-(NC)C_6H_3	0.5	(36)	96.5:3.5

Enal Donor	Acceptor	Conditions	Product(s), Yield(s) (%), and Stereoselectivity	Refs.
R, O, H	NO_2; 1.5 eq	**β-TrPFc3b•**HBF_4 (10 mol %), *i*-Pr_2NEt (1 equiv), catechol (1 equiv), MeOH, 0 °C, 2 h	O, NO_2, R	40

R		er
4-Br	(57)	99:1
2-MeO	(97)	96.5:3.5

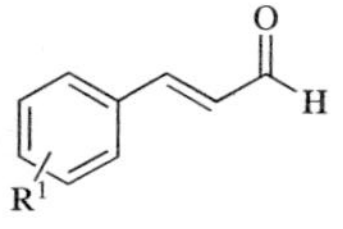

3 eq

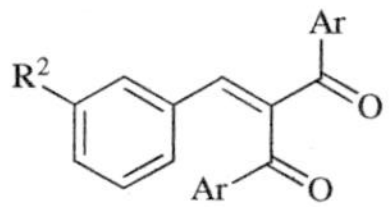

α-**TrM1c**•HCl (30 mol %),
DBU (20 mol %),
toluene, 0 °C

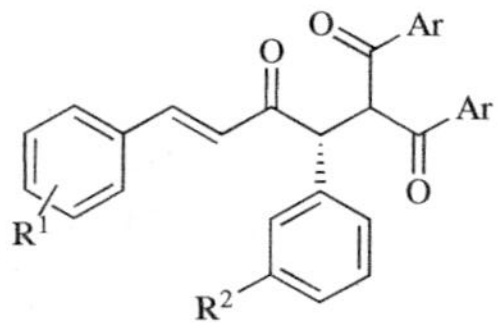

87

R^1	R^2	Ar		er
H	3-MeO	Ph	(40)	92.5:7.5
4-MeO	H	Ph	(47)	94:6
2-MeO	H	4-BrC_6H_4	(49)	96:4
4-Br	H	Ph	(33)	91:9

C_{10}

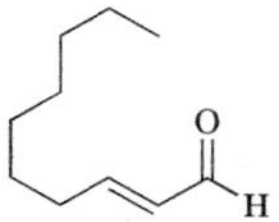

3 eq

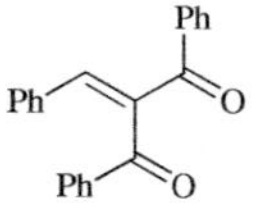

α-**TrM1c**•HCl (30 mol %),
DBU (20 mol %),
THF, 0 °C

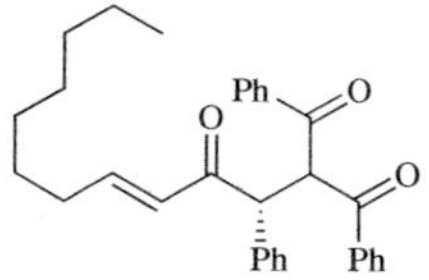

(84) er 95.5:4.5

87

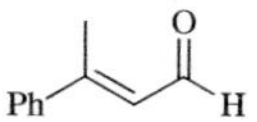

3 eq

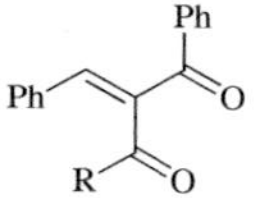

α-**TrM1c**•HCl (30 mol %),
DBU (20 mol %),
toluene, 0 °C

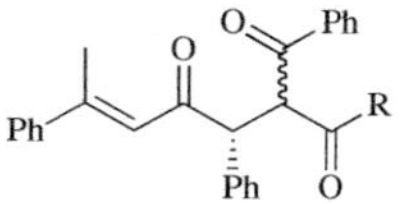

87

R		dr	er (major)	er (minor)
EtO	(65)	2.8:1	98.5:1.5	(—)
Me	(70)	1.3:1	94:6	97:3

TABLE 6. INTERMOLECULAR STETTER REACTIONS OF α,β-UNSATURATED ALDEHYDES (*Continued*)

Enal Donor	Acceptor	Conditions	Product(s), Yield(s) (%), and Stereoselectivity	Refs.
Please refer to the charts preceding the tables for catalyst and precatalyst structures used herein.				
C_{10}				
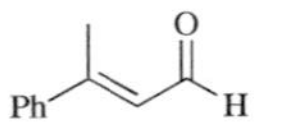 3 eq	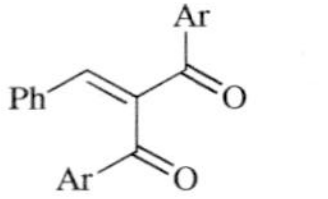	α-**TrM1c**•HCl (30 mol %), DBU (20 mol %), toluene, 0 °C	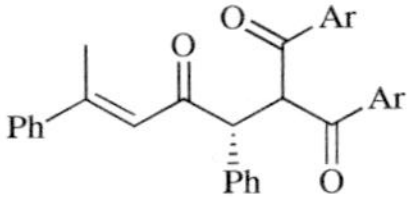 Ar / er: Ph (89) 91.5:8.5; 4-BrC_6H_4 (90) 94:6	87
3 eq	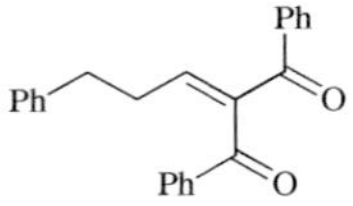	α-**TrM1c**•HCl (30 mol %), DBU (20 mol %), toluene, 0 °C	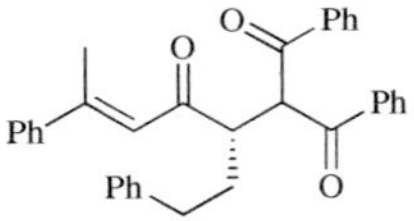 (35) er 78:22	87
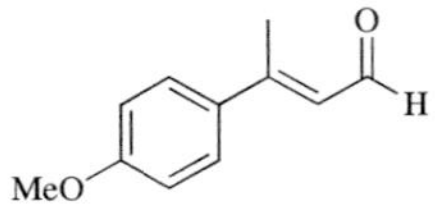 MeO 3 eq	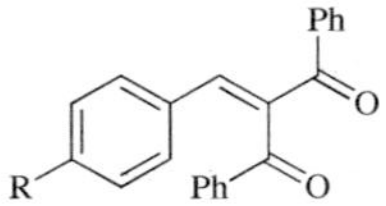 R	α-**TrM1c**•HCl (30 mol %), DBU (20 mol %), toluene, 0 °C	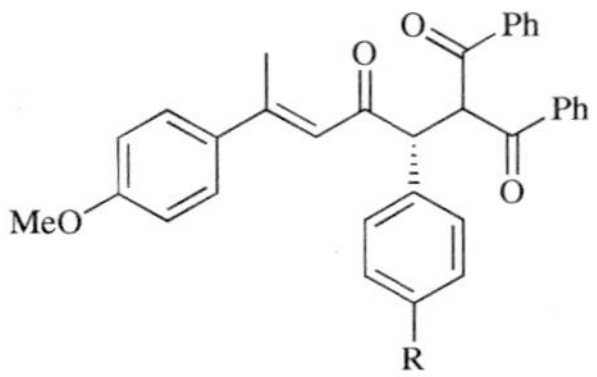 MeO R; R / er: Br (93) 85:15; *i*-Pr (91) 87:13	87
C_{12}				
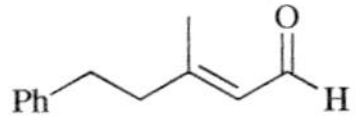 3 eq	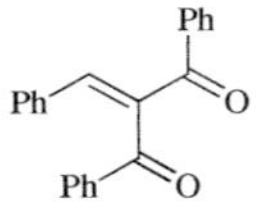	α-**TrM1c**•HCl (30 mol %), DBU (20 mol %), toluene, 0 °C	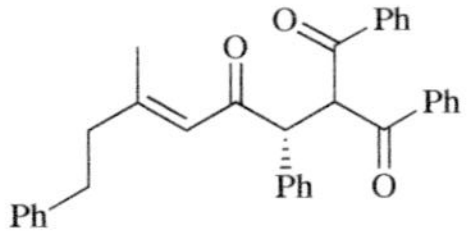 (81) er 71:29	87

C_{13}	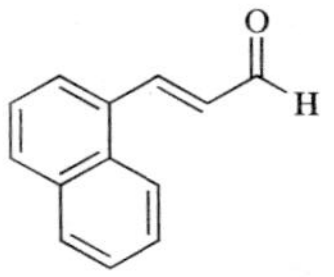 3 eq; Ph, 4-FC_6H_4, O, 4-FC_6H_4, O	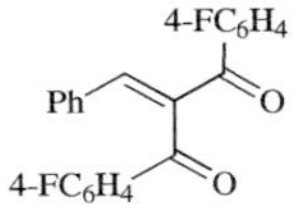 α-**TrM1c**•HCl (30 mol %), DBU (20 mol %), toluene, 0 °C	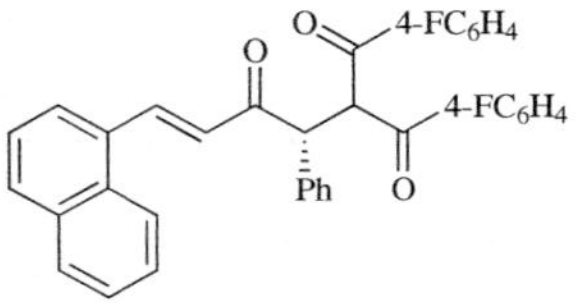 O, O, 4-FC_6H_4, 4-FC_6H_4, Ph, O	(39) er 94:6	87
C_{17}	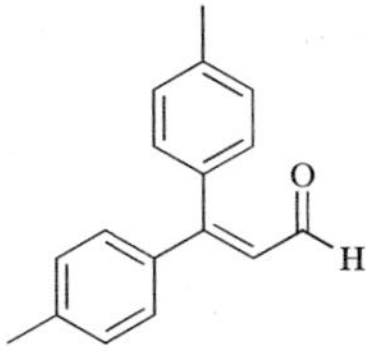 3 eq; Ph, O, Ph, O, OMe	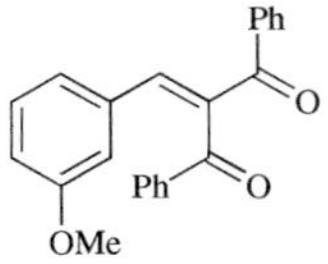 α-**TrM1c**•HCl (30 mol %), DBU (20 mol %), toluene, 0 °C	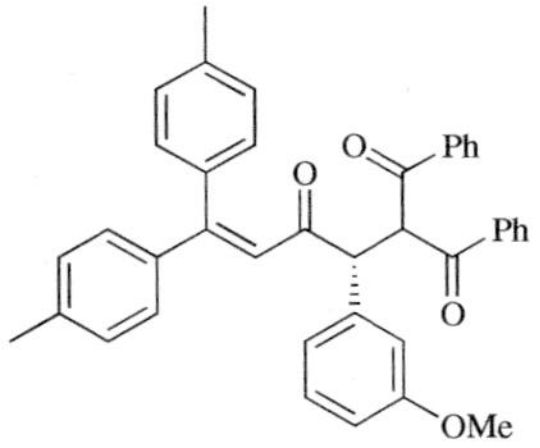 O, O, Ph, Ph, O, OMe	(33) er 82.5:17.5	87

[a] Catechol (1 equiv) was added to the reaction.

[b] The reaction was conducted in toluene.

[c] 4-Ethoxycarbonylcatechol (1 equiv) was used in place of catechol.

TABLE 7. INTERMOLECULAR STETTER REACTIONS OF ALIPHATIC ALDEHYDES

Aldehyde Donor	Acceptor	Conditions	Product(s), Yield(s) (%), and Stereoselectivity	Refs.

Please refer to the charts preceding the tables for catalyst and precatalyst structures used herein.

C_2

Aldehyde Donor	Acceptor	Conditions	Product(s), Yield(s) (%), and Stereoselectivity	Refs.
O / H (1.5 eq)	Ph / NO_2	β-**TrPFt4b**•HBF_4 (20 mol %), NaOAc (40 mol %), *t*-AmOH, 0 °C, 24–48 h	O / NO_2 / Ph (71) er 81:19	53
10 eq	O / Ar^1 / Ar^2	Precatalyst (10 mol %), Cs_2CO_3 (10 mol %), solvent, rt, 24 h	O / Ar^2 / Ar^1 / O	89

Ar^1	Ar^2	Precatalyst	Solvent		er
Ph	Ph	α-**TrM1b**•HBF_4	THF	(42)	78.5:21.5
Ph	4-ClC_6H_4	α-**TrP8a**•HBF_4	THF	(0)	N/A
Ph	4-ClC_6H_4	α-**TrP8b**•HBF_4	THF	(0)	N/A
Ph	4-ClC_6H_4	α-**TrP9a**•HBF_4	THF	(0)	N/A
Ph	4-ClC_6H_4	β-**TrM3c**•HBF_4	THF	(0)	N/A
Ph	4-ClC_6H_4	α-**TrM1c**•HCl	THF	(7)	—
Ph	4-ClC_6H_4	α-**TrM1b**•HBF_4	THF	(80)	81:19
Ph	4-ClC_6H_4	α-**TrM1b**•HBF_4	THF	(76)[a]	78:22[a]
Ph	4-ClC_6H_4	α-**TrM1b**•HBF_4	toluene	(45)	77:23
Ph	4-ClC_6H_4	α-**TrM1b**•HBF_4	$CHCl_3$	(44)	80:20
Ph	3-(MeO)C_6H_4	α-**TrM1b**•HBF_4	THF	(43)	79:21
Ph	4-(NC)C_6H_4	α-**TrM1b**•HBF_4	$CHCl_3$	(85)	80:20
2-naphthyl	Ph	α-**TrM1b**•HBF_4	THF	(62)	88:12

C_3

R, H, O; 1.5 eq

Ph, NO_2

β-**TrPFt4b**•HBF_4 (20 mol %), NaOAc (40 mol %), *t*-AmOH, 0 °C, 24–48 h

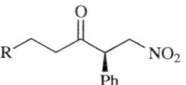

R, O, NO_2, Ph

R		er
H	(87)	96:4
MeS	(67)	96:4

53

C_4

O, H; 1.5 eq

R, NO_2

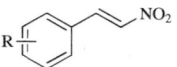

Precatalyst (20 mol %), NaOAc (40 mol %), *t*-AmOH, 0 °C, 24–48 h

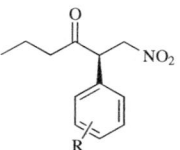

O, NO_2, R

53

R	Precatalyst		er
H	β-**TrPFc4b**•HBF_4	(15)	87:13
H	β-**TrP4b**•HBF_4	(18)	90:10
H	β-**TrP3b**•HBF_4	(49)	75:25
H	β-**TrPFc3b**•HBF_4	(53)	74:26
H	β-**TrPFt3b**•HBF_4	(78)	87:13
H	β-**TrPFt4b**•HBF_4	(80)	97:3
2-Cl	β-**TrPFt4b**•HBF_4	(70)	95.5:4.5

R	Precatalyst		er
2-F	β-**TrPFt4b**•HBF_4	(75)	97:3
2-MeO	β-**TrPFt4b**•HBF_4	(83)	97:3
3-MeO	β-**TrPFt4b**•HBF_4	(63)	95.5:4.5
3-Br	β-**TrPFt4b**•HBF_4	(50)	95.5:4.5
4-Cl	β-**TrPFt4b**•HBF_4	(70)	96:4
4-Me	β-**TrPFt4b**•HBF_4	(81)	96:4
4-Bpin	β-**TrPFt4b**•HBF_4	(62)	95.5:4.5

TABLE 7. INTERMOLECULAR STETTER REACTIONS OF ALIPHATIC ALDEHYDES (*Continued*)

Aldehyde Donor	Acceptor	Conditions	Product(s), Yield(s) (%), and Stereoselectivity	Refs.
Please refer to the charts preceding the tables for catalyst and precatalyst structures used herein.				
C_{4-9}				
R, 1.5 eq	Ph, NO_2	β-**TrPFt4b**•HBF_4 (20 mol %), NaOAc (40 mol %), *t*-AmOH, 0 °C, 24–48 h	R, NO_2, Ph R / er $TBSO(CH_2)_2$ (68) 93.5:6.5 $Cl(CH_2)_2$ (83) 96.5:3.5 *i*-Pr (32) 97.5:2.5 allyl (83) 96.5:3.5 benzyl (76) 96.5:3.5	53
C_7				
H, 1.5 eq	Ph, NO_2	β-**TrPFt4b**•HBF_4 (20 mol %), NaOAc (40 mol %), *t*-AmOH, 0 °C, 24–48 h	NO_2, Ph (<5) er —	53
C_8				
Ph, H	CO_2Bu, 2 eq	α-**TrP5h**•HBF_4 (15 mol %), K_3PO_4 (100 mol %), toluene, rt, 24 h	Ph, CO_2Bu (26) er 93.5:6.5	58
C_{10}				
H	NHAc, CO_2Me, 2 eq	β-**TrM2c**•HCl (10 mol %), *t*-BuOK (8 mol %), toluene, 0 °C, 3 h	NHAc, CO_2Me (52) er 98.5:1.5	48

[a] DBU (10 mol %) was used in place of Cs_2CO_3.

TABLE 8. AZA-STETTER REACTIONS OF IMINES

Imine Donor	Acceptor	Conditions	Product(s), Yield(s) (%), and Stereoselectivity	Refs.

Please refer to the charts preceding the tables for catalyst and precatalyst structures used herein.

C_{5-7}

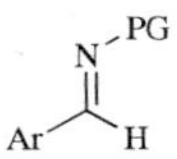

Imine donor: Ar–CH=N–PG, 1.2 eq

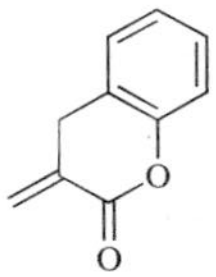

Conditions: β-**TrM3d**•HBF_4 (10 mol %), *t*-BuOK (10 mol %), *t*-BuOH (1 equiv), THF, reflux, 2 h

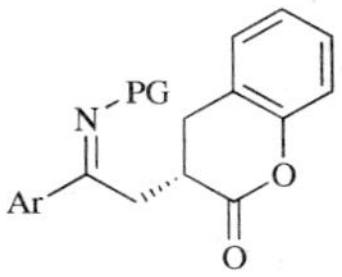

Ref. 91

Ar	PG		er
Ph	Bz	(77)	99:1
4-(MeO)C_6H_4	Bz	(67)	98:2
3-(MeO)C_6H_4	Bz	(80)	99:1
4-BrC_6H_4	Bz	(52)	98:2
2-thienyl	Bz	(88)	99:1

Ar	PG		er
Ph	4-(MeO)C_6H_4C(=O)	(73)	98:2
4-(MeO)C_6H_4	4-(MeO)C_6H_4C(=O)	(59)	99:1
4-BrC_6H_4	4-(MeO)C_6H_4C(=O)	(61)	98:2
Ph	4-ClC_6H_4C(=O)	(60)	98:2

Imine donor: 1.2 eq

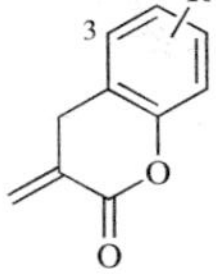

Conditions: β-**TrM3d**•HBF_4 (10 mol %), *t*-BuOK (10 mol %), *t*-BuOH-THF, reflux, 2 h

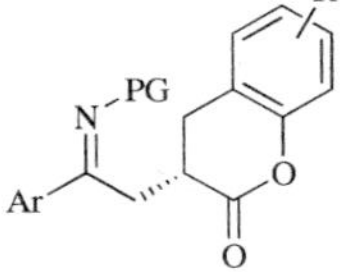

Ref. 91

Ar	PG	R		er
Ph	4-(MeO)C_6H_4C(=O)	4-MeO	(45)	98:2
Ph	4-(MeO)C_6H_4C(=O)	4-Br	(49)	96:4
Ph	4-(MeO)C_6H_4C(=O)	3,5-Me_2	(75)	98:2
4-MeC_6H_4	Bz	4-Me	(61)	99:1
4-MeC_6H_4	4-(MeO)C_6H_4C(=O)	4-Me	(78)	99:1
4-(MeO)C_6H_4	Bz	4-Et	(64)	97:3
2-thienyl	Bz	4-Me	(79)	99:1

TABLE 8. AZA-STETTER REACTIONS OF IMINES (*Continued*)

Imine Donor	Acceptor	Conditions	Product(s), Yield(s) (%), and Stereoselectivity	Refs.

Please refer to the charts preceding the tables for catalyst and precatalyst structures used herein.

C_{5-7}

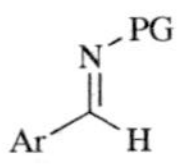

1.2 eq

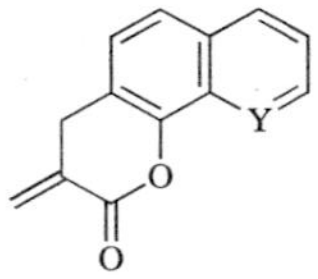

β-**TrM3d**•HBF_4 (10 mol %), *t*-BuOK (10 mol %), *t*-BuOH/THF, reflux, 2 h

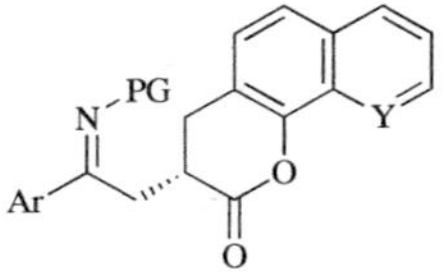

91

Ar	PG	Y		er
Ph	4-(MeO)C_6H_4C(=O)	CH	(63)	96:4
3-(MeO)C_6H_4	Bz	CH	(68)	>99:1
2-thienyl	Bz	CH	(81)	98:2
3-(MeO)C_6H_4	Bz	N	(44)	96:4
2-thienyl	Bz	N	(47)	98:2

REFERENCES

[1] Seebach, D. *Angew. Chem.* **1979**, *91*, 259.
[2] Groebel, B. T.; Seebach, D. *Synthesis* **1977**, 357.
[3] Smith, A. B., III; Adams, C. M. *Acc. Chem. Res.* **2004**, *37*, 365.
[4] Bugaut, X.; Glorius, F. *Chem. Soc. Rev.* **2012**, *41*, 3511.
[5] Flanigan, D. M.; Romanov-Michailidis, F.; White, N. A.; Rovis, T. *Chem. Rev.* **2015**, *115*, 9307.
[6] Breslow, R. *J. Am. Chem. Soc.* **1958**, *80*, 3719.
[7] Frank, R. A. W.; Leeper, F. J.; Luisi, B. F. *Cell. Mol. Life Sci.* **2007**, *64*, 892.
[8] Stetter, H.; Schreckenberg, M. *Angew. Chem., Int. Ed. Engl.* **1973**, *12*, 81.
[9] Stetter, H.; Kuhlmann, H. *Angew. Chem., Int. Ed. Engl.* **1974**, *13*, 539.
[10] Stetter, H. *Angew. Chem., Int. Ed. Engl.* **1976**, *15*, 639.
[11] Enders, D.; Breuer, K.; Runsink, J.; Teles, J. H. *Helv. Chim. Acta* **1996**, *79*, 1899.
[12] Knight, R. L.; Leeper, F. J. *J. Chem. Soc., Perkin Trans. 1* **1998**, 1891.
[13] Dvorak, C. A.; Rawal, V. H. *Tetrahedron Lett.* **1998**, *39*, 2925.
[14] Enders, D.; Kallfass, U. *Angew. Chem., Int. Ed.* **2002**, *41*, 1743.
[15] Kerr, M. S.; Read de Alaniz, J.; Rovis, T. *J. Am. Chem. Soc.* **2002**, *124*, 10298.
[16] Read de Alaniz, J.; Rovis, T. *Synlett* **2009**, *1189*.
[17] Núñez, M. G.; García, P.; Moro, R. F.; Díez, D. *Tetrahedron* **2010**, *66*, 2089.
[18] Dominguez de Maria, P.; Shanmuganathan, S. *Curr. Org. Chem.* **2011**, *15*, 2083.
[19] Gravel, M.; Holmes, J. M. Stetter Reaction. In *Comprehensive Organic Synthesis, 2nd ed.*; Knochel, P., Molander, G. A., Eds.; Elsevier: Amsterdam, 2014; Vol. 4, pp 1384–1406.
[20] Enders, D.; Niemeier, O.; Henseler, A. *Chem. Rev.* **2007**, *107*, 5606.
[21] Haghshenas, P.; Langdon, S. M.; Gravel, M. *Synlett* **2017**, *28*, 542.
[22] Zhao, M.; Zhang, Y.-T.; Chen, J.; Zhou, L. *Asian J. Org. Chem.* **2018**, *7*, 54.
[23] Stetter, H.; Kuhlmann, H. *Org. React.* **1991**, *40*, 407.
[24] Massey, R. S.; Collett, C. J.; Lindsay, A. G.; Smith, A. D.; O'Donoghue, A. C. *J. Am. Chem. Soc.* **2012**, *134*, 20421.
[25] Washabaugh, M. W.; Jencks, W. P. *J. Am. Chem. Soc.* **1989**, *111*, 674.
[26] Wang, N.; Xu, J.; Lee, J. K. *Org. Biomol. Chem.* **2018**, *16*, 6852.
[27] Read de Alaniz, J.; Rovis, T. *J. Am. Chem. Soc.* **2005**, *127*, 6284.
[28] Konstandaras, N.; Dunn, M. H.; Guerry, M. S.; Barnett, C. D.; Cole, M. L.; Harper, J. B. *Org. Biomol. Chem.* **2020**, *18*, 66.
[29] Niu, Y.; Wang, N.; Munoz, A.; Xu, J.; Zeng, H.; Rovis, T.; Lee, J. K. *J. Am. Chem. Soc.* **2017**, *139*, 14917.
[30] Berkessel, A.; Elfert, S.; Etzenbach-Effers, K.; Teles, J. H. *Angew. Chem., Int. Ed.* **2010**, *49*, 7120.
[31] Collett, C. J.; Massey, R. S.; Maguire, O. R.; Batsanov, A. S.; O'Donoghue, A. C.; Smith, A. D. *Chem. Sci.* **2013**, *4*, 1514.
[32] Maji, B.; Mayr, H. *Angew. Chem., Int. Ed.* **2012**, *51*, 10408.
[33] Berkessel, A.; Elfert, S.; Yatham, V. R.; Neudörfl, J.-M.; Schlörer, N. E.; Teles, J. H. *Angew. Chem., Int. Ed.* **2012**, *51*, 12370.
[34] Gong, J. H.; Im, Y. J.; Lee, K. Y.; Kim, J. N. *Tetrahedron Lett.* **2002**, *43*, 1247.
[35] DiRocco, D. A.; Oberg, K. M.; Rovis, T. *J. Am. Chem. Soc.* **2012**, *134*, 6143.
[36] Hawkes, K. J.; Yates, B. F. *Eur. J. Org. Chem.* **2008**, 5563.
[37] Moore, J. L.; Silvestri, A. P.; de Alaniz, J. R.; DiRocco, D. A.; Rovis, T. *Org. Lett.* **2011**, *13*, 1742.
[38] Kemp, D. S. *J. Org. Chem.* **1971**, *36*, 202.
[39] Goldfuss, B.; Schumacher, M. *J. Mol. Model.* **2006**, *12*, 591.
[40] DiRocco, D. A.; Rovis, T. *J. Am. Chem. Soc.* **2011**, *133*, 10402.
[41] Ciganek, E. *Synthesis* **1995**, 1311.
[42] Jones, G. B.; Chapman, B. J. *Synthesis* **1995**, 475.
[43] Matsumoto, Y.; Tomioka, K. *Tetrahedron Lett.* **2006**, *47*, 5843.
[44] Dudding, T.; Houk, K. N. *Proc. Natl. Acad. Sci. U. S. A.* **2004**, *101*, 5770.
[45] Enders, D.; Han, J.; Henseler, A. *Chem. Commun.* **2008**, 3989.
[46] Enders, D.; Han, J. *Synthesis* **2008**, 3864.
[47] Um, J. M.; DiRocco, D. A.; Noey, E. L.; Rovis, T.; Houk, K. N. *J. Am. Chem. Soc.* **2011**, *133*, 11249.
[48] Jousseaume, T.; Wurz, N. E.; Glorius, F. *Angew. Chem., Int. Ed.* **2011**, *50*, **1410**.
[49] Kuniyil, R.; Sunoj, R. B. *Org. Lett.* **2013**, *15*, 5040.
[50] Domingo, L. R.; Zaragozá, R. J.; Saéz, J. A.; Arnó, M. *Molecules* **2012**, *17*, 1335.

[51] Schumacher, M.; Goldfuss, B. *New J. Chem.* **2015**, *39*, 4508.
[52] Langdon, S. M.; Legault, C. Y.; Gravel, M. *J. Org. Chem.* **2015**, *80*, 3597.
[53] DiRocco, D. A.; Noey, E. L.; Houk, K. N.; Rovis, T. *Angew. Chem., Int. Ed.* **2012**, *51*, 2391.
[54] Seebach, D.; Golinski, J. *Helv. Chim. Acta* **1981**, *64*, 1413.
[55] Kerr, M. S.; Read de Alaniz, J.; Rovis, T. *J. Org. Chem.* **2005**, *70*, 5725.
[56] Vora, H. U.; Lathrop, S. P.; Reynolds, N. T.; Kerr, M. S.; Read de Alaniz, J.; Rovis, T. *Org. Synth.* **2010**, *87*, 350.
[57] Smrcina, M.; Majer, P.; Majerova, E.; Guerassina, T. A.; Eissenstat, M. A. *Tetrahedron* **1997**, *53*, 12867.
[58] Wurz, N. E.; Daniliuc, C. G.; Glorius, F. *Chem.—Eur. J.* **2012**, *18*, 16297.
[59] Ozboya, K. E.; Rovis, T. *Synlett* **2014**, *25*, 2665.
[60] Zheng, P.; Gondo, C. A.; Bode, J. W. *Chem.—Asian J.* **2011**, *6*, 614.
[61] Brand, J. P.; Siles, J. I. O.; Waser, J. *Synlett* **2010**, *881*.
[62] Read de Alaniz, J.; Kerr, M. S.; Moore, J. L.; Rovis, T. *J. Org. Chem.* **2008**, *73*, 2033.
[63] Struble, J. R.; Kaeobamrung, J.; Bode, J. W. *Org. Lett.* **2008**, *10*, 957.
[64] Rong, Z.-Q.; Li, Y.; Yang, G.-Q.; You, S.-L. *Synlett* **2011**, 1033.
[65] Soeta, T.; Tabatake, Y.; Ukaji, Y. *Tetrahedron* **2012**, *68*, 10188.
[66] Lu, T.; Gu, L.; Kang, Q.; Zhang, Y. *Tetrahedron Lett.* **2012**, *53*, 6602.
[67] Cullen, S. C.; Rovis, T. *Org. Lett.* **2008**, *10*, 3141.
[68] Kerr, M. S.; Rovis, T. *Synlett* **2003**, 1934.
[69] Jia, M.-Q.; You, S.-L. *Synlett* **2013**, *24*, 1201.
[70] Lin, Q.; Li, Y.; Das, D. K.; Zhang, G.; Zhao, Z.; Yang, S.; Fang, X. *Chem. Commun.* **2016**, *52*, 6459.
[71] Douglas, C. J.; Overman, L. E. *Proc. Natl. Acad. Sci. U. S. A.* **2004**, *101*, 5363.
[72] Kerr, M. S.; Rovis, T. *J. Am. Chem. Soc.* **2004**, *126*, 8876.
[73] Moore, J. L.; Kerr, M. S.; Rovis, T. *Tetrahedron* **2006**, *62*, 11477.
[74] Reynolds, N. T.; Rovis, T. *Tetrahedron* **2005**, *61*, 6368.
[75] Liu, Q.; Rovis, T. *J. Am. Chem. Soc.* **2006**, *128*, 2552.
[76] Liu, Q.; Perreault, S.; Rovis, T. *J. Am. Chem. Soc.* **2008**, *130*, 14066.
[77] Sánchez-Larios, E.; Thai, K.; Bilodeau, F.; Gravel, M. *Org. Lett.* **2011**, *13*, 4942.
[78] DiRocco, D. A.; Oberg, K. M.; Dalton, D. M.; Rovis, T. *J. Am. Chem. Soc.* **2009**, *131*, 10872.
[79] Liu, Q.; Rovis, T. *Org. Lett.* **2009**, *11*, 2856.
[80] Stetter, H.; Skobel, H. *Chem. Ber.* **1987**, *120*, 643.
[81] Inokuchi, T.; Kawafuchi, H. *J. Org. Chem.* **2006**, *71*, 947.
[82] Mahatthananchai, J.; Bode, J. W. *Acc. Chem. Res.* **2014**, *47*, 696.
[83] Sohn, S. S.; Rosen, E. L.; Bode, J. W. *J. Am. Chem. Soc.* **2004**, *126*, 14370.
[84] Burstein, C.; Glorius, F. *Angew. Chem., Int. Ed.* **2004**, *43*, 6205.
[85] White, N. A.; DiRocco, D. A.; Rovis, T. *J. Am. Chem. Soc.* **2013**, *135*, 8504.
[86] Maji, B.; Ji, L.; Wang, S.; Vedachalam, S.; Ganguly, R.; Liu, X.-W. *Angew. Chem., Int. Ed.* **2012**, *51*, 8276.
[87] Fang, X.; Chen, X.; Lv, H.; Chi, Y. R. *Angew. Chem., Int. Ed.* **2011**, *50*, 11782.
[88] Enders, D.; Balensiefer, T. *Acc. Chem. Res.* **2004**, *37*, 534.
[89] Kim, S. M.; Jin, M. Y.; Kim, M. J.; Cui, Y.; Kim, Y. S.; Zhang, L.; Song, C. E.; Ryu, D. H.; Yang, J. W. *Org. Biomol. Chem.* **2011**, *9*, 2069.
[90] Dresen, C.; Richter, M.; Pohl, M.; Ludeke, S.; Muller, M. *Angew. Chem., Int. Ed.* **2010**, *49*, 6600.
[91] Fernando, J. E. M.; Nakano, Y.; Zhang, C.; Lupton, D. W. *Angew. Chem., Int. Ed.* **2019**, *58*, 4007.
[92] Izquierdo, J.; Hutson, G. E.; Cohen, D. T.; Scheidt, K. A. *Angew. Chem., Int. Ed.* **2012**, *51*, 11686.
[93] Stetter, H.; Kuhlmann, H. *Synthesis* **1975**, 379.
[94] Anjaiah, S.; Chandrasekhar, S.; Grée, R. *Adv. Synth. Catal.* **2004**, *346*, 1329.
[95] Galopin, C. C. *Tetrahedron Lett.* **2001**, *42*, 5589.
[96] Harrington, P. E.; Tius, M. A. *Org. Lett.* **1999**, *1*, 649.
[97] Harrington, P. E.; Tius, M. A. *J. Am. Chem. Soc.* **2001**, *123*, 8509.
[98] Randl, S.; Blechert, S. *J. Org. Chem.* **2003**, *68*, 8879.
[99] Ratnayake, R.; Covell, D.; Ransom, T. T.; Gustafson, K. R.; Beutler, J. A. *Org. Lett.* **2009**, *11*, 57.
[100] Trost, B. M.; Shuey, C. D.; DiNinno, F. *J. Am. Chem. Soc.* **1979**, *101*, 1284.
[101] Klepp, J.; Sadgrove, N. J.; Legendre, S. V. A.-M.; Sumby, C. J.; Greatrex, B. W. *J. Org. Chem.* **2019**, *84*, 9637.
[102] Nicolaou, K. C.; Tang, Y.; Wang, J. *Chem. Commun.* **2007**, 1922.
[103] Defieber, C.; Mohr, J. T.; Grabovyi, G. A.; Stoltz, B. M. *Synthesis* **2018**, *50*, 4359.
[104] Rej, R. K.; Acharyya, R. K.; Nanda, S. *Tetrahedron* **2016**, *72*, 4931.

105 Yamada, T.; Kitada, H.; Kajimoto, T.; Numata, A.; Tanaka, R. *J. Org. Chem.* **2010**, *75*, 4146.
106 Orellana, A.; Rovis, T. *Chem. Commun.* **2008**, 730.
107 Hsu, D.-S.; Yeh, J.-Y.; Cheng, C.-Y. *Org. Lett.* **2017**, *19*, 5549.
108 Lathrop, S. P.; Rovis, T. *Chem. Sci.* **2013**, *4*, 1668.
109 Williams, D. R.; Mondal, P. K.; Bawel, S. A.; Nag, P. P. *Org. Lett.* **2014**, *16*, 1956.
110 Hu, D. X.; Withall, D. M.; Challis, G. L.; Thomson, R. J. *Chem. Rev.* **2016**, *116*, 7818.
111 Kasparyan, E.; Richter, M.; Dresen, C.; Walter, L. S.; Fuchs, G.; Leeper, F. J.; Wacker, T.; Andrade, S. L. A.; Kolter, G.; Pohl, M.; Mueller, M. *Appl. Microbiol. Biotechnol.* **2014**, *98*, 9681.
112 Beigi, M.; Waltzer, S.; Zarei, M.; Mueller, M. *J. Biotechnol.* **2014**, *191*, 64.
113 Zhang, H.-J.; Priebbenow, D. L.; Bolm, C. *Chem. Soc. Rev.* **2013**, *42*, 8540.
114 Mattson, A. E.; Bharadwaj, A. R.; Scheidt, K. A. *J. Am. Chem. Soc.* **2004**, *126*, 2314.
115 Nahm, M. R.; Potnick, J. R.; White, P. S.; Johnson, J. S. *J. Am. Chem. Soc.* **2006**, *128*, 2751.
116 Piel, I.; Steinmetz, M.; Hirano, K.; Froehlich, R.; Grimme, S.; Glorius, F. *Angew. Chem., Int. Ed.* **2011**, *50*, 4983.
117 He, J.; Zheng, J.; Liu, J.; She, X.; Pan, X. *Org. Lett.* **2006**, *8*, 4637.
118 He, J.; Tang, S.; Liu, J.; Su, Y.; Pan, X.; She, X. *Tetrahedron* **2008**, *64*, 8797.
119 Janssen-Müller, D.; Schedler, M.; Fleige, M.; Daniliuc, C. G.; Glorius, F. *Angew. Chem., Int. Ed.* **2015**, *54*, 12492.
120 Wu, J.; Zhao, C.; Wang, J. *J. Am. Chem. Soc.* **2016**, *138*, 4706.
121 Liu, F.; Bugaut, X.; Schedler, M.; Fröhlich, R.; Glorius, F. *Angew. Chem., Int. Ed.* **2011**, *50*, 12626.
122 Willis, M. C. Hydroacylation of Alkenes, Alkynes, and Allenes. In *Comprehensive Organic Synthesis, 2nd ed.*; Knochel, P., Molander, G. A., Eds.; Elsevier: Amsterdam, 2014; Vol. 4, pp 961–994.
123 Newton, C. G.; Wang, S.-G.; Oliveira, C. C.; Cramer, N. *Chem. Rev.* **2017**, *117*, 8908.
124 Guo, R.; Zhang, G. *Synlett* **2018**, *29*, 1801.
125 Aloise, A. D.; Layton, M. E.; Shair, M. D. *J. Am. Chem. Soc.* **2000**, *122*, 12610.
126 Beletskiy, E. V.; Sudheer, C.; Douglas, C. J. *J. Org. Chem.* **2012**, *77*, 5884.
127 Kundu, K.; McCullagh, J. V.; Morehead, A. T., Jr. *J. Am. Chem. Soc.* **2005**, *127*, 16042.
128 Rastelli, E. J.; Truong, N. T.; Coltart, D. M. *Org. Lett.* **2016**, *18*, 5588.
129 Leung, J. C.; Krische, M. J. *Chem. Sci.* **2012**, *3*, 2202.
130 Shibata, Y.; Tanaka, K. *J. Am. Chem. Soc.* **2009**, *131*, 12552.
131 Osborne, J. D.; Randell-Sly, H. E.; Currie, G. S.; Cowley, A. R.; Willis, M. C. *J. Am. Chem. Soc.* **2008**, *130*, 17232.
132 Hooper, J. F.; Seo, S.; Truscott, F. R.; Neuhaus, J. D.; Willis, M. C. *J. Am. Chem. Soc.* **2016**, *138*, 1630.
133 Goti, G.; Bieszczad, B.; Vega-Penaloza, A.; Melchiorre, P. *Angew. Chem., Int. Ed.* **2019**, *58*, 1213.
134 Zhao, J.-J.; Zhang, H.-H.; Shen, X.; Yu, S. *Org. Lett.* **2019**, *21*, 913.
135 Kuang, Y.; Wang, K.; Shi, X.; Huang, X.; Meggers, E.; Wu, J. *Angew. Chem., Int. Ed.* **2019**, *58*, 16859.
136 Guo, F.; Clift, M. D.; Thomson, R. J. *Eur. J. Org. Chem.* **2012**, 4881.
137 Baran, P. S.; DeMartino, M. P. *Angew. Chem., Int. Ed.* **2006**, *45*, 7083.
138 DeMartino, M. P.; Chen, K.; Baran, P. S. *J. Am. Chem. Soc.* **2008**, *130*, 11546.
139 Quesnelle, C. A.; Gill, P.; Kim, S.-H.; Chen, L.; Zhao, Y.; Fink, B. E.; Saulnier, M.; Frennesson, D.; DeMartino, M. P.; Baran, P. S.; Gavai, A. V. *Synlett* **2016**, *27*, 2254.
140 Mambrini, A.; Gori, D.; Guillot, R.; Kouklovsky, C.; Alezra, V. *Chem. Commun.* **2018**, *54*, 12742.
141 Jang, H.-Y.; Hong, J.-B.; MacMillan, D. W. C. *J. Am. Chem. Soc.* **2007**, *129*, 7004.
142 Naesborg, L.; Leth, L. A.; Reyes-Rodriguez, G. J.; Palazzo, T. A.; Corti, V.; Jorgensen, K. A. *Chem.—Eur. J.* **2018**, *24*, 14844.
143 Kaldre, D.; Klose, I.; Maulide, N. *Science* **2018**, *361*, 664.
144 Liu, Q.; Rovis, T. *Org. Process Res. Dev.* **2007**, *11*, 598.
145 Ema, T.; Nanjo, Y.; Shiratori, S.; Terao, Y.; Kimura, R. *Org. Lett.* **2016**, *18*, 5764.
146 Jia, M.-Q.; Li, Y.; Rong, Z.-Q.; You, S.-L. *Org. Biomol. Chem.* **2011**, *9*, 2072.
147 Rafinski, Z. *Catalysts* **2019**, *9*, 192.
148 Pesch, J.; Harms, K.; Bach, T. *Eur. J. Org. Chem.* **2004**, 2025.
149 Ragno, D.; Di Carmine, G.; Brandolese, A.; Bortolini, O.; Giovannini, P. P.; Massi, A. *ACS Catal.* **2017**, *7*, 6365.
150 Rafinski, Z. *ChemCatChem* **2016**, *8*, 2599.
151 Rafinski, Z.; Krzeminski, M. P. *Catalysts* **2019**, *9*, 117.
152 Mennen, S. M.; Blank, J. T.; Tran-Dube, M. B.; Imbriglio, J. E.; Miller, S. J. *Chem. Commun.* **2005**, 195.
153 Bao, Y.; Kumagai, N.; Shibasaki, M. *Tetrahedron: Asymmetry* **2014**, *25*, 1401.

[154] Law, K. R.; McErlean, C. S. P. *Chem.—Eur. J.* **2013**, *19*, 15852.
[155] Jia, M.-Q.; You, S.-L. *Chem. Commun.* **2012**, *48*, 6363.
[156] Jia, M.-Q.; Liu, C.; You, S.-L. *J. Org. Chem.* **2012**, *77*, 10996.
[157] Liu, G.; Wilkerson, P. D.; Toth, C. A.; Xu, H. *Org. Lett.* **2012**, *14*, 858.
[158] Li, Y.; Shi, F.-Q.; He, Q.-L.; You, S.-L. *Org. Lett.* **2009**, *11*, 3182.

CUMULATIVE CHAPTER TITLES BY VOLUME

Volume 1 (1942)

1. **The Reformatsky Reaction:** Ralph L. Shriner
2. **The Arndt-Eistert Reaction:** W. E. Bachmann and W. S. Struve
3. **Chloromethylation of Aromatic Compounds:** Reynold C. Fuson and C. H. McKeever
4. **The Amination of Heterocyclic Bases by Alkali Amides:** Marlin T. Leffler
5. **The Bucherer Reaction:** Nathan L. Drake
6. **The Elbs Reaction:** Louis F. Fieser
7. **The Clemmensen Reduction:** Elmore L. Martin
8. **The Perkin Reaction and Related Reactions:** John R. Johnson
9. **The Acetoacetic Ester Condensation and Certain Related Reactions:** Charles R. Hauser and Boyd E. Hudson, Jr.
10. **The Mannich Reaction:** F. F. Blicke
11. **The Fries Reaction:** A. H. Blatt
12. **The Jacobson Reaction:** Lee Irvin Smith

Volume 2 (1944)

1. **The Claisen Rearrangement:** D. Stanley Tarbell
2. **The Preparation of Aliphatic Fluorine Compounds:** Albert L. Henne
3. **The Cannizzaro Reaction:** T. A. Geissman
4. **The Formation of Cyclic Ketones by Intramolecular Acylation:** William S. Johnson
5. **Reduction with Aluminum Alkoxides (The Meerwein-Ponndorf-Verley Reduction):** A. L. Wilds
6. **The Preparation of Unsymmetrical Biaryls by the Diazo Reaction and the Nitrosoacetylamine Reaction:** Werner E. Bachmann and Roger A. Hoffman

7. **Replacement of the Aromatic Primary Amino Group by Hydrogen:** Nathan Kornblum
8. **Periodic Acid Oxidation:** Ernest L. Jackson
9. **The Resolution of Alcohols:** A. W. Ingersoll
10. **The Preparation of Aromatic Arsonic and Arsinic Acids by the Bart, Béchamp, and Rosenmund Reactions:** Cliff S. Hamilton and Jack F. Morgan

Volume 3 (1946)

1. **The Alkylation of Aromatic Compounds by the Friedel-Crafts Method:** Charles C. Price
2. **The Willgerodt Reaction:** Marvin Carmack and M. A. Spielman
3. **Preparation of Ketenes and Ketene Dimers:** W. E. Hanford and John C. Sauer
4. **Direct Sulfonation of Aromatic Hydrocarbons and Their Halogen Derivatives:** C. M. Suter and Arthur W. Weston
5. **Azlactones:** H. E. Carter
6. **Substitution and Addition Reactions of Thiocyanogen:** John L. Wood
7. **The Hofmann Reaction:** Everett L. Wallis and John F. Lane
8. **The Schmidt Reaction:** Hans Wolff
9. **The Curtius Reaction:** Peter A. S. Smith

Volume 4 (1948)

1. **The Diels-Alder Reaction with Maleic Anhydride:** Milton C. Kloetzel
2. **The Diels-Alder Reaction: Ethylenic and Acetylenic Dienophiles:** H. L. Holmes
3. **The Preparation of Amines by Reductive Alkylation:** William S. Emerson
4. **The Acyloins:** S. M. McElvain
5. **The Synthesis of Benzoins:** Walter S. Ide and Johannes S. Buck
6. **Synthesis of Benzoquinones by Oxidation:** James Cason
7. **The Rosenmund Reduction of Acid Chlorides to Aldehydes:** Erich Mosettig and Ralph Mozingo
8. **The Wolff-Kishner Reduction:** David Todd

Volume 5 (1949)

1. **The Synthesis of Acetylenes:** Thomas L. Jacobs
2. **Cyanoethylation:** Herman L. Bruson

3. **The Diels-Alder Reaction: Quinones and Other Cyclenones:** Lewis L. Butz and Anton W. Rytina
4. **Preparation of Aromatic Fluorine Compounds from Diazonium Fluroborates: The Schiemann Reaction:** Arthur Roe
5. **The Friedel and Crafts Reaction with Aliphatic Dibasic Acid Anhydrides:** Ernst Berliner
6. **The Gattermann-Koch Reaction:** Nathan N. Crounse
7. **The Leuckart Reaction:** Maurice L. Moore
8. **Selenium Dioxide Oxidation:** Norman Rabjohn
9. **The Hoesch Synthesis:** Paul E. Spoerri and Adrien S. DuBois
10. **The Darzens Glycidic Ester Condensation:** Melvin S. Newman and Barney J. Magerlein

Volume 6 (1951)

1. **The Stobbe Condensation:** William S. Johnson and Guido H. Daub
2. **The Preparation of 3,4-Dihydroisoquinolines and Related Compounds by the Bischler-Napieralski Reaction:** Wilson M. Whaley and Tutucorin R. Govindachari
3. **The Pictet-Spengler Synthesis of Tetrahydroisoquinolines and Related Compounds:** Wilson M. Whaley and Tutucorin R. Govindachari
4. **The Synthesis of Isoquinolines by the Pomeranz-Fritsch Reaction:** Walter J. Gensler
5. **The Oppenauer Oxidation:** Carl Djerassi
6. **The Synthesis of Phosphonic and Phosphinic Acids:** Gennady M. Kosolapoff
7. **The Halogen-Metal Interconversion Reaction with Organolithium Compounds:** Reuben G. Jones and Henry Gilman
8. **The Preparation of Thiazoles:** Richard H. Wiley, D. C. England, and Lyell C. Behr
9. **The Preparation of Thiophenes and Tetrahydrothiophenes:** Donald E. Wolf and Karl Folkers
10. **Reductions by Lithium Aluminum Hydride:** Weldon G. Brown

Volume 7 (1953)

1. **The Pechmann Reaction:** Suresh Sethna and Ragini Phadke
2. **The Skraup Synthesis of Quinolines:** R. H. F. Manske and Marshall Kulka
3. **Carbon–Carbon Alkylations with Amines and Ammonium Salts:** James H. Brewster and Ernest L. Eliel

2. **The Japp-Klingemann Reaction:** Robert R. Phillips

3. **The Michael Reaction:** Ernst D. Bergmann, David Ginsburg, and Raphael Pappo

Volume 11 (1960)

1. **The Beckmann Rearrangement:** L. Guy Donaruma and Walter Z. Heldt

2. **The Demjanov and Tiffeneau-Demjanov Ring Expansions:** Peter A. S. Smith and Donald R. Baer

3. **Arylation of Unsaturated Compounds by Diazonium Salts:** Christian S. Rondestvedt, Jr.

4. **The Favorskii Rearrangement of Haloketones:** Andrew S. Kende

5. **Olefins from Amines: The Hofmann Elimination Reaction and Amine Oxide Pyrolysis:** Arthur C. Cope and Elmer R. Trumbull

Volume 12 (1962)

1. **Cyclobutane Derivatives from Thermal Cycloaddition Reactions:** John D. Roberts and Clay M. Sharts

2. **The Preparation of Olefins by the Pyrolysis of Xanthates. The Chugaev Reaction:** Harold R. Nace

3. **The Synthesis of Aliphatic and Alicyclic Nitro Compounds:** Nathan Kornblum

4. **Synthesis of Peptides with Mixed Anhydrides:** Noel F. Albertson

5. **Desulfurization with Raney Nickel:** George R. Pettit and Eugene E. van Tamelen

Volume 13 (1963)

1. **Hydration of Olefins, Dienes, and Acetylenes via Hydroboration:** George Zweifel and Herbert C. Brown

2. **Halocyclopropanes from Halocarbenes:** William E. Parham and Edward E. Schweizer

3. **Free Radical Addition to Olefins to Form Carbon–Carbon Bonds:** Cheves Walling and Earl S. Huyser

4. **Formation of Carbon–Heteroatom Bonds by Free Radical Chain Additions to Carbon–Carbon Multiple Bonds:** F. W. Stacey and J. F. Harris, Jr.

Volume 14 (1965)

1. **The Chapman Rearrangement:** J. W. Schulenberg and S. Archer

2. **α-Amidoalkylations at Carbon:** Harold E. Zaugg and William B. Martin

3. **The Wittig Reaction:** Adalbert Maercker

Volume 15 (1967)

1. **The Dieckmann Condensation:** John P. Schaefer and Jordan J. Bloomfield
2. **The Knoevenagel Condensation:** G. Jones

Volume 16 (1968)

1. **The Aldol Condensation:** Arnold T. Nielsen and William J. Houlihan

Volume 17 (1969)

1. **The Synthesis of Substituted Ferrocenes and Other π-Cyclopentadienyl-Transition Metal Compounds:** Donald E. Bublitz and Kenneth L. Rinehart, Jr.
2. **The γ-Alkylation and γ-Arylation of Dianions of β-Dicarbonyl Compounds:** Thomas M. Harris and Constance M. Harris
3. **The Ritter Reaction:** L. I. Krimen and Donald J. Cota

Volume 18 (1970)

1. **Preparation of Ketones from the Reaction of Organolithium Reagents with Carboxylic Acids:** Margaret J. Jorgenson
2. **The Smiles and Related Rearrangements of Aromatic Systems:** W. E. Truce, Eunice M. Kreider, and William W. Brand
3. **The Reactions of Diazoacetic Esters with Alkenes, Alkynes, Heterocyclic, and Aromatic Compounds:** Vinod Dave and E. W. Warnhoff
4. **The Base-Promoted Rearrangements of Quaternary Ammonium Salts:** Stanley H. Pine

Volume 19 (1972)

1. **Conjugate Addition Reactions of Organocopper Reagents:** Gary H. Posner
2. **Formation of Carbon–Carbon Bonds via π-Allylnickel Compounds:** Martin F. Semmelhack
3. **The Thiele-Winter Acetoxylation of Quinones:** J. F. W. McOmie and J. M. Blatchly
4. **Oxidative Decarboxylation of Acids by Lead Tetraacetate:** Roger A. Sheldon and Jay K. Kochi

Volume 20 (1973)

1. **Cyclopropanes from Unsaturated Compounds, Methylene Iodide, and Zinc-Copper Couple:** H. E. Simmons, T. L. Cairns, Susan A. Vladuchick, and Connie M. Hoiness
2. **Sensitized Photooxygenation of Olefins:** R. W. Denny and A. Nickon
3. **The Synthesis of 5-Hydroxyindoles by the Nenitzescu Reaction:** George R. Allen, Jr.
4. **The Zinin Reaction of Nitroarenes:** H. K. Porter

Volume 21 (1974)

1. **Fluorination with Sulfur Tetrafluoride:** G. A. Boswell, Jr., W. C. Ripka, R. M. Scribner, and C. W. Tullock
2. **Modern Methods to Prepare Monofluoroaliphatic Compounds:** Clay M. Sharts and William A. Sheppard

Volume 22 (1975)

1. **The Claisen and Cope Rearrangements:** Sara Jane Rhoads and N. Rebecca Raulins
2. **Substitution Reactions Using Organocopper Reagents:** Gary H. Posner
3. **Clemmensen Reduction of Ketones in Anhydrous Organic Solvents:** E. Vedejs
4. **The Reformatsky Reaction:** Michael W. Rathke

Volume 23 (1976)

1. **Reduction and Related Reactions of α,β-Unsaturated Compounds with Metals in Liquid Ammonia:** Drury Caine
2. **The Acyloin Condensation:** Jordan J. Bloomfield, Dennis C. Owsley, and Janice M. Nelke
3. **Alkenes from Tosylhydrazones:** Robert H. Shapiro

Volume 24 (1976)

1. **Homogeneous Hydrogenation Catalysts in Organic Solvents:** Arthur J. Birch and David H. Williamson
2. **Ester Cleavages via S_N2-Type Dealkylation:** John E. McMurry
3. **Arylation of Unsaturated Compounds by Diazonium Salts (The Meerwein Arylation Reaction):** Christian S. Rondestvedt, Jr.
4. **Selenium Dioxide Oxidation:** Norman Rabjohn

Volume 25 (1977)

1. **The Ramberg-Bäcklund Rearrangement:** Leo A. Paquette
2. **Synthetic Applications of Phosphoryl-Stabilized Anions:** William S. Wadsworth, Jr.
3. **Hydrocyanation of Conjugated Carbonyl Compounds:** Wataru Nagata and Mitsuru Yoshioka

Volume 26 (1979)

1. **Heteroatom-Facilitated Lithiations:** Heinz W. Gschwend and Herman R. Rodriguez
2. **Intramolecular Reactions of Diazocarbonyl Compounds:** Steven D. Burke and Paul A. Grieco

Volume 27 (1982)

1. **Allylic and Benzylic Carbanions Substituted by Heteroatoms:** Jean-François Biellmann and Jean-Bernard Ducep

2. **Palladium-Catalyzed Vinylation of Organic Halides:** Richard F. Heck

Volume 28 (1982)

1. **The Reimer-Tiemann Reaction:** Hans Wynberg and Egbert W. Meijer

2. **The Friedländer Synthesis of Quinolines:** Chia-Chung Cheng and Shou-Jen Yan

3. **The Directed Aldol Reaction:** Teruaki Mukaiyama

Volume 29 (1983)

1. **Replacement of Alcoholic Hydroxy Groups by Halogens and Other Nucleophiles via Oxyphosphonium Intermediates:** Bertrand R. Castro

2. **Reductive Dehalogenation of Polyhalo Ketones with Low-Valent Metals and Related Reducing Agents:** Ryoji Noyori and Yoshihiro Hayakawa

3. **Base-Promoted Isomerizations of Epoxides:** Jack K. Crandall and Marcel Apparu

Volume 30 (1984)

1. **Photocyclization of Stilbenes and Related Molecules:** Frank B. Mallory and Clelia W. Mallory

2. **Olefin Synthesis via Deoxygenation of Vicinal Diols:** Eric Block

Volume 31 (1984)

1. **Addition and Substitution Reactions of Nitrile-Stabilized Carbanions:** Siméon Arseniyadis, Keith S. Kyler, and David S. Watt

Volume 32 (1984)

1. **The Intramolecular Diels-Alder Reaction:** Engelbert Ciganek

2. **Synthesis Using Alkyne-Derived Alkenyl- and Alkynylaluminum Compounds:** George Zweifel and Joseph A. Miller

Volume 33 (1985)

1. **Formation of Carbon–Carbon and Carbon–Heteroatom Bonds via Organoboranes and Organoborates:** Ei-Ichi Negishi and Michael J. Idacavage

2. **The Vinylcyclopropane-Cyclopentene Rearrangement:** Tomáš Hudlický, Toni M. Kutchan, and Saiyid M. Naqvi

Volume 34 (1985)

1. **Reductions by Metal Alkoxyaluminum Hydrides:** Jaroslav Málek
2. **Fluorination by Sulfur Tetrafluoride:** Chia-Lin J. Wang

Volume 35 (1988)

1. **The Beckmann Reactions: Rearrangements, Elimination-Additions, Fragmentations, and Rearrangement-Cyclizations:** Robert E. Gawley
2. **The Persulfate Oxidation of Phenols and Arylamines (The Elbs and the Boyland-Sims Oxidations):** E. J. Behrman
3. **Fluorination with Diethylaminosulfur Trifluoride and Related Aminofluorosulfuranes:** Miloš Hudlický

Volume 36 (1988)

1. **The [3 + 2] Nitrone-Olefin Cycloaddition Reaction:** Pat N. Confalone and Edward M. Huie
2. **Phosphorus Addition at sp^2 Carbon:** Robert Engel
3. **Reduction by Metal Alkoxyaluminum Hydrides. Part II. Carboxylic Acids and Derivatives, Nitrogen Compounds, and Sulfur Compounds:** Jaroslav Málek

Volume 37 (1989)

1. **Chiral Synthons by Ester Hydrolysis Catalyzed by Pig Liver Esterase:** Masaji Ohno and Masami Otsuka
2. **The Electrophilic Substitution of Allylsilanes and Vinylsilanes:** Ian Fleming, Jacques Dunoguès, and Roger Smithers

Volume 38 (1990)

1. **The Peterson Olefination Reaction:** David J. Ager
2. **Tandem Vicinal Difunctionalization: β-Addition to α,β-Unsaturated Carbonyl Substrates Followed by α-Functionalization:** Marc J. Chapdelaine and Martin Hulce
3. **The Nef Reaction:** Harold W. Pinnick

Volume 39 (1990)

1. **Lithioalkenes from Arenesulfonylhydrazones:** A. Richard Chamberlin and Steven H. Bloom
2. **The Polonovski Reaction:** David Grierson
3. **Oxidation of Alcohols to Carbonyl Compounds via Alkoxysulfonium Ylides: The Moffatt, Swern, and Related Oxidations:** Thomas T. Tidwell

Volume 40 (1991)

1. **The Pauson-Khand Cycloaddition Reaction for Synthesis of Cyclopentenones:** Neil E. Schore
2. **Reduction with Diimide:** Daniel J. Pasto and Richard T. Taylor
3. **The Pummerer Reaction of Sulfinyl Compounds:** Ottorino DeLucchi, Umberto Miotti, and Giorgio Modena
4. **The Catalyzed Nucleophilic Addition of Aldehydes to Electrophilic Double Bonds:** Hermann Stetter and Heinrich Kuhlmann

Volume 41 (1992)

1. **Divinylcyclopropane-Cycloheptadiene Rearrangement:** Tomáš Hudlický, Rulin Fan, Josephine W. Reed, and Kumar G. Gadamasetti
2. **Organocopper Reagents: Substitution, Conjugate Addition, Carbo/Metallo-cupration, and Other Reactions:** Bruce H. Lipshutz and Saumitra Sengupta

Volume 42 (1992)

1. **The Birch Reduction of Aromatic Compounds:** Peter W. Rabideau and Zbigniew Marcinow
2. **The Mitsunobu Reaction:** David L. Hughes

Volume 43 (1993)

1. **Carbonyl Methylenation and Alkylidenation Using Titanium-Based Reagents:** Stanley H. Pine
2. **Anion-Assisted Sigmatropic Rearrangements:** Stephen R. Wilson
3. **The Baeyer-Villiger Oxidation of Ketones and Aldehydes:** Grant R. Krow

Volume 44 (1993)

1. **Preparation of α,β-Unsaturated Carbonyl Compounds and Nitriles by Selenoxide Elimination:** Hans J. Reich and Susan Wollowitz
2. **Enone Olefin [2 + 2] Photochemical Cyclizations:** Michael T. Crimmins and Tracy L. Reinhold

Volume 45 (1994)

1. **The Nazarov Cyclization:** Karl L. Habermas, Scott E. Denmark, and Todd K. Jones
2. **Ketene Cycloadditions:** John Hyatt and Peter W. Raynolds

Volume 46 (1994)

1. **Tin(II) Enolates in the Aldol, Michael, and Related Reactions:** Teruaki Mukaiyama and Shū Kobayashi

2. **The [2,3]-Wittig Reaction:** Takeshi Nakai and Koichi Mikami

3. **Reductions with Samarium(II) Iodide:** Gary A. Molander

Volume 47 (1995)

1. **Lateral Lithiation Reactions Promoted by Heteroatomic Substituents:** Robin D. Clark and Alam Jahangir

2. **The Intramolecular Michael Reaction:** R. Daniel Little, Mohammad R. Masjedizadeh, Olof Wallquist (in part), and Jim I. McLoughlin (in part)

Volume 48 (1995)

1. **Asymmetric Epoxidation of Allylic Alcohols: The Katsuki-Sharpless Epoxidation Reaction:** Tsutomu Katsuki and Victor S. Martin

2. **Radical Cyclization Reactions:** B. Giese, B. Kopping, T. Göbel, J. Dickhaut, G. Thoma, K. J. Kulicke, and F. Trach

Volume 49 (1997)

1. **The Vilsmeier Reaction of Fully Conjugated Carbocycles and Heterocycles:** Gurnos Jones and Stephen P. Stanforth

2. **[6 + 4] Cycloaddition Reactions:** James H. Rigby

3. **Carbon–Carbon Bond-Forming Reactions Promoted by Trivalent Manganese:** Gagik G. Melikyan

Volume 50 (1997)

1. **The Stille Reaction:** Vittorio Farina, Venkat Krishnamurthy and William J. Scott

Volume 51 (1997)

1. **Asymmetric Aldol Reactions Using Boron Enolates:** Cameron J. Cowden and Ian Paterson

2. **The Catalyzed α-Hydroxyalkylation and α-Aminoalkylation of Activated Olefins (The Morita–Baylis–Hillman Reaction):** Engelbert Ciganek

3. **[4 + 3] Cycloaddition Reactions:** James H. Rigby and F. Christopher Pigge

Volume 52 (1998)

1. **The Retro–Diels–Alder Reaction. Part I. C–C Dienophiles:** Bruce Rickborn

2. **Enantioselective Reduction of Ketones:** Shinichi Itsuno

Volume 53 (1998)

1. **The Oxidation of Alcohols by Modified Oxochromium(VI)-Amine Complexes:** Frederick A. Luzzio

2. **The Retro-Diels-Alder Reaction. Part II. Dienophiles with One or More Heteroatoms:** Bruce Rickborn

Volume 54 (1999)

1. **Aromatic Substitution by the $S_{RN}1$ Reaction:** Roberto Rossi, Adriana B. Pierini, and Ana N. Santiago

2. **Oxidation of Carbonyl Compounds with Organohypervalent Iodine Reagents:** Robert M. Moriarty and Om Prakash

Volume 55 (1999)

1. **Synthesis of Nucleosides:** Helmut Vorbrüggen and Carmen Ruh-Pohlenz

Volume 56 (2000)

1. **The Hydroformylation Reaction:** Iwao Ojima, Chung-Ying Tsai, Maria Tzamarioudaki, and Dominique Bonafoux

2. **The Vilsmeier Reaction. 2. Reactions with Compounds Other Than Fully Conjugated Carbocycles and Heterocycles:** Gurnos Jones and Stephen P. Stanforth

Volume 57 (2001)

1. **Intermolecular Metal-Catalyzed Carbenoid Cyclopropanations:** Huw M. L. Davies and Evan G. Antoulinakis

2. **Oxidation of Phenolic Compounds with Organohypervalent Iodine Reagents:** Robert M. Moriarty and Om Prakash

3. **Synthetic Uses of Tosylmethyl Isocyanide (TosMIC):** Daan van Leusen and Albert M. van Leusen

Volume 58 (2001)

1. **Simmons-Smith Cyclopropanation Reaction:** André B. Charette and André Beauchemin

2. **Preparation and Applications of Functionalized Organozinc Compounds:** Paul Knochel, Nicolas Millot, Alain L. Rodriguez, and Charles E. Tucker

Volume 59 (2001)

1. **Reductive Aminations of Carbonyl Compounds with Borohydride and Borane Reducing Agents:** Ellen W. Baxter and Allen B. Reitz

Volume 60 (2002)

1. **Epoxide Migration (Payne Rearrangement) and Related Reactions:** Robert M. Hanson

2. **The Intramolecular Heck Reaction:** J. T. Link

Volume 61 (2002)

1. **[3 + 2] Cycloaddition of Trimethylenemethane and Its Synthetic Equivalents:** Shigeru Yamago and Eiichi Nakamura

2. **Dioxirane Epoxidation of Alkenes:** Waldemar Adam, Chantu R. Saha-Möller, and Cong-Gui Zhao

Volume 62 (2003)

1. **α-Hydroxylation of Enolates and Silyl Enol Ethers:** Bang-Chi Chen, Ping Zhou, Franklin A. Davis, and Engelbert Ciganek
2. **The Ramberg-Bäcklund Reaction:** Richard J. K. Taylor and Guy Casy
3. **The α-Hydroxy Ketone (α-Ketol) and Related Rearrangements:** Leo A. Paquette and John E. Hofferberth
4. **Transformation of Glycals into 2,3-Unsaturated Glycosyl Derivatives:** Robert J. Ferrier and Oleg A. Zubkov

Volume 63 (2004)

1. **The Biginelli Dihydropyrimidine Synthesis:** C. Oliver Kappe and Alexander Stadler
2. **Microbial Arene Oxidations:** Roy A. Johnson
3. **Cu, Ni, and Pd Mediated Homocoupling Reactions in Biaryl Syntheses: The Ullmann Reaction:** Todd D. Nelson and R. David Crouch

Volume 64 (2004)

1. **Additions of Allyl, Allenyl, and Propargylstannanes to Aldehydes and Imines:** Benjamin W. Gung
2. **Glycosylation with Sulfoxides and Sulfinates as Donors or Promoters:** David Crich and Linda B. L. Lim
3. **Addition of Organochromium Reagents to Carbonyl Compounds:** Kazuhiko Takai

Volume 65 (2005)

1. **The Passerini Reaction:** Luca Banfi and Renata Riva
2. **Diels-Alder Reactions of Imino Dienophiles:** Geoffrey R. Heintzelman, Ivona R. Meigh, Yogesh R. Mahajan, and Steven M. Weinreb

Volume 66 (2005)

1. **The Allylic Trihaloacetimidate Rearrangement:** Larry E. Overman and Nancy E. Carpenter
2. **Asymmetric Dihydroxylation of Alkenes:** Mark C. Noe, Michael A. Letavic, Sheri L. Snow, and Stuart McCombie

Volume 67 (2006)

1. **Catalytic Enantioselective Aldol Addition Reactions:** Erick M. Carreira, Alec Fettes, and Christiane Marti
2. **Benzylic Activation and Stereochemical Control in Reactions of Tricarbonyl(Arene)-Chromium Complexes:** Motokazu Uemura

Volume 68 (2007)

1. **Cotrimerizations of Acetylenic Compounds:** Nicolas Agenet, Olivier Buisine, Franck Slowinski, Vincent Gandon, Corinne Aubert, and Max Malacria

2. **Glycosylation on Polymer Supports:** Simone Bufali and Peter H. Seeberger

Volume 69 (2007)

1. **Dioxirane Oxidations of Compounds Other Than Alkenes:** Waldemar Adam, Cong-Gui Zhao, and Kavitha Jakka

2. **Electrophilic Fluorination with N–F Reagents:** Jérôme Baudoux and Dominique Cahard

Volume 70 (2008)

1. **The Catalytic Asymmetric Strecker Reaction:** Masakatsu Shibasaki, Motomu Kanai, and Tsuyoshi Mita

2. **The Synthesis of Phenols and Quinones via Fischer Carbene:** Marcey L. Waters and William D. Wulff

Volume 71 (2008)

1. **Ionic and Organometallic-Catalyzed Organosilane Reductions:** Gerald L. Larson and James L. Fry

Volume 72 (2008)

1. **Electrophilic Amination of Carbanions, Enolates, and Their Surrogates:** Engelbert Ciganek

2. **Desulfonylation Reactions:** Diego A. Alonso and Carmen Nájera

Volume 73 (2008)

1. **Allylboration of Carbonyl Compounds:** Hugo Lachance and Dennis G. Hall

Volume 74 (2009)

1. **Catalytic Asymmetric Hydrogenation of C=N Functions:** Hans-Ulrich Blaser and Felix Spindler

2. **Oxoammonium- and Nitroxide-Catalyzed Oxidations of Alcohols:** James M. Bobbitt, Christian Brückner, and Nabyl Merbouh

3. **Asymmetric Epoxidation of Electron-Deficient Alkenes:** Michael J. Porter and John Skidmore

Volume 75 (2011)

1. **Hydrocyanation of Alkenes and Alkynes:** T. V. RajanBabu

2. **Intermolecular C–H Insertions of Carbenoids:** Huw M. L. Davies and Phillip M. Pelphrey

3. **Cross-Coupling with Organosilicon Compounds:** Wen-Tau T. Chang, Russell C. Smith, Christopher S. Regens, Aaron D. Bailey, Nathan S. Werner, and Scott E. Denmark

4. **The Aza-Cope/Mannich Reaction:** Larry E. Overman, Philip G. Humphreys, and Gregory S. Welmaker

Volume 76 (2012)

1. **Hydrogen-Bonding-Mediated Directed Osmium Dihydroxylation:** Timothy J. Donohoe, Carole J. R. Bataille, and Paolo Innocenti

2. **Transition-Metal-Catalyzed α-Arylation of Enolates:** Damien Prim, Sylvain Marque, Anne Gaucher, and Jean-Marc Campagne

3. **Indoles via Palladium-Catalyzed Cyclization:** Sandro Cacchi, Giancarlo Fabrizi, and Antonella Goggiamani

Volume 77 (2012)

1. **The Kulinkovich Cyclopropanation of Carboxylic Acid Derivatives:** Jin Kun Cha and Oleg G. Kulinkovich

2. **The Barton-McCombie Reaction:** Stuart W. McCombie, William B. Motherwell, and Matthew Tozer

Volume 78 (2012)

1. **The Schmidt Reaction:** Aaron Wrobleski, Thomas C. Coombs, Chan Woo Huh, Sze-Wan Li, and Jeffrey Aubé

2. **The Neber Rearrangement:** William F. Berkowitz

3. **Twofold Extrusion Reactions:** Lynn James Guziec and Frank S. Guziec, Jr.

Volume 79 (2013)

1. **Cross-Coupling Reactions of Organotrifluoroborate Salts:** Gary A. Molander and Ludivine Jean-Gèrard

2. **Asymmetric Transformations by Deprotonation Using Chiral Lithium Amides:** Nigel S. Simpkins and Michael D. Weller

Volume 80 (2013)

1. **Catalytic, Asymmetric, Intramolecular Carbon-Hydrogen Insertion:** Michael P. Doyle, Yu Liu, and Maxim Ratnikov

2. **Cycloadditions of Carbonyl Ylides Derived from Diazocarbonyl Compounds:** David M. Hodgson, Agnès H. Labande, and Sengodagounder Muthusamy

Volume 81 (2013)

1. **The Krapcho Dealkoxycarbonylation Reaction of Esters with α-Electron-Withdrawing Substituents:** A. Paul Krapcho and Engelbert Ciganek

Volume 82 (2013)

1. **The McMurry Coupling and Related Reactions:** Takeshi Takeda and Akira Tsubouchi
2. **Catalytic Asymmetric Ketene [2 + 2] and [4 + 2] Cycloadditions:** Scott G. Nelson, Robert D. Dura, and Timothy J. Peelen

Volume 83 (2014)

1. **Iron-Catalyzed Cross-Coupling Reactions:** Eiichi Nakamura, Takuji Hatakeyama, Shingo Ito, Kentaro Ishizuka, Lauren Ilies, and Masaharu Nakamura
2. **The Boronic Acid Mannich Reaction:** Stephen G. Pyne and Minyan Tang

Volume 84 (2014)

1. **Molybdenum-Catalyzed Asymmetric Allylic Alkylations:** Christina Moberg
2. **The Wacker Oxidation:** Brian W. Michel, Laura D. Steffens, and Matthew S. Sigman

Volume 85 (2014)

1. **Copper-Catalyzed Amination of Aryl and Alkenyl Electrophiles:** Kevin H. Shaughnessy, Engelbert Ciganek, and Rebecca B. DeVasher

Volume 86 (2015)

1. **Ring-Expanding Carbonylation of Epoxides:** John W. Kramer, John M. Rowley, and Geoffrey W. Coates
2. **The Tishchenko Reaction:** Ari M. P. Koskinen and Antti O. Kataja

Volume 87 (2015)

1. **Oxidative Cleavage of Furans:** Pedro Merino
2. **Cycloaddition and Electrocyclic Reactions of Vinylketenes, Allenylketenes, and Alkynylketenes:** Nanyan Fu and Thomas Tidwell
3. **Carbozincation Reactions of Carbon-Carbon Multiple Bonds:** Genia Sklute and Hannah Cavender

Volume 88 (2015)

1. **Hydroamination of Alkenes:** Alexander L. Reznichenko and Kai C. Hultzsch

Volume 89 (2015)

1. **Olefin Ring-Closing Metathesis:** Larry Yet

Volume 90 (2016)

1. **The Catalytic, Enantioselective Michael Reaction:** Efraim Reyes, Uxue Uria, Jose L. Vicario, and Luisa Carrillo

Volume 91 (2016)

1. **Nucleophilic Additions of Perfluoroalkyl Groups:** Petr Beier, Mikhail Zibinsky, and G. K. Surya Prakash

Volume 92 (2016–2017)

1. **Gold-Catalyzed Cyclization of Alkynes with Alkenes and Arenes:** Antonio M. Echavarren, Michael E. Muratore, Verónica López-Carrillo, Ana Escribano-Cuesta, Núria Huguet, and Carla Obradors
2. **Cyclization of Vinyl and Aryl Azides into Pyrroles, Indoles, Carbazoles, and Related Fused Pyrroles:** William F. Berkowitz and Stuart W. McCombie

Volume 93 (2017)

1. **Enantioselective Rhodium-Catalyzed 1,4-Addition of Organoboron Reagents to Electron-Deficient Alkenes:** Alan R. Burns, Hon Wai Lam, and Iain D. Roy

Volume 94 (2017)

1. **[3 + 2] Dipolar Cycloadditions of Cyclic Nitrones with Alkenes:** Alberto Brandi, Francesca Cardona, Stefano Cicchi, Franca M. Cordero, and Andrea Goti

Volume 95 (2018)

1. **The Julia-Kocienski Olefination:** Paul R. Blakemore, Selena Milicevic Sephton, and Engelbert Ciganek
2. **Asymmetric Synthesis of β-Lactams by the Staudinger Reaction:** Aitor Landa, Antonia Mielgo, Mikel Oiarbide, and Claudio Palomo

Volume 96 (2018)

1. **Catalytic, Enantioselective Hydrogenation of Heteroaromatic Compounds:** Lei Shi and Yong-Gui Zhou
2. **Transition-Metal-Catalyzed Hydroacylation:** Vy M. Dong, Kevin G. M. Kou, and Diane N. Le

9. **Suzuki-Miyaura Cross-Coupling:** Alexander B. Pagett and Guy C. Lloyd-Jones

10. ***ortho*-Directed C-H Alkylation of Substituted Benzenes:** Yusuke Ano and Naoto Chatani

11. **Catalytic, Enantioselective, C-H Functionalization to Form Carbon-Carbon Bonds:** Qing-Feng Wu, Gang Chen, Jian He, and Jin-Quan Yu

12. **Catalytic, Enantioselective, Fluorination Reactions:** Richard T. Thornbury and F. Dean Toste

13. **Catalyst-Controlled Glycosylation:** Samuel M. Levi and Eric N. Jacobsen

14. **Palladium-Catalyzed Amination of Aryl Halides:** John F. Hartwig, Kevin H. Shaughnessy, Shashank Shekhar, and Rebecca A. Green

15. **Catalytic, Enantioselective, Copper-Boryl Additions to Alkenes:** Amir H. Hoveyda, Ming Joo Koh, Kyunga Lee, and Jaehee Lee

Volume 101 (2020)

1. **Dihydropyrans by Cycloadditions of Oxadienes:** Arnaud Martel, Robert Dhal, Catherine Gaulon, Mathieu Yves Laurent, and Gilles Dujardin

Volume 102 (2020)

1. **The Brook Rearrangement:** Lu Gao, Wenyu Yang, Ya Wu, and Zhenlei Song

2. **Alkyne Metathesis:** Daesung Lee, Ivan Volchkov, and Sang Young Yun

Volume 103 (2020)

1. **Transition-Metal-Catalyzed Alkyne Hydroarylation with Arylmetals and Aryl Halides:** Yoshihiko Yamamoto

2. **[3 + 2] Cycloadditions of Azomethine Imines:** Uroš Grošelj and Jurij Svete

3. **Propargylic Coupling Reactions via Bimetallic Alkyne Complexes: The Nicholas Reaction:** James R. Green and Kenneth M. Nicholas

Volume 104 (2020)

1. **Kinetic Resolution and Desymmetrization of Alcohols and Amines by Nonenzymatic, Enantioselective Acylation:** Aileen B. Frost, Elizabeth S. Munday, Stefania F. Musolino, Andrew D. Smith, Mark D. Greenhalgh and James E. Taylor

2. **The Piancatelli Reaction:** Lucile Marin, Emmanuelle Schulz, David Lebœuf and Vincent Gandon

3. **Transition-Metal-Mediated and Transition-Metal-Catalyzed Carbon–Fluorine Bond Formation:** Constanze N. Neumann and Tobias Ritter

Volume 105 (2021)

AUTHOR INDEX, VOLUMES 1–106

Volume number only is designated in this index

Organic Reactions, Vol. 106, Edited by P. Andrew Evans et al.

CHAPTER AND TOPIC INDEX, VOLUMES 1–106

Many chapters contain brief discussions of reactions and comparisons of alternative synthetic methods related to the reaction that is the subject of the chapter. These related reactions and alternative methods are not usually listed in this index. In this index, the volume number is in **boldface**, the chapter number is in ordinary type.

Organic Reactions, Vol. 106, Edited by P. Andrew Evans et al.